DRUG DESIGN OF ZINC-ENZYME INHIBITORS

Wiley Series in Drug Discovery and Development

Binghe Wang, Series Editor

Drug Delivery: Principles and Applications
Edited by Binghe Wang, Teruna Siahaan, and Richard A. Soltero

Computer Applications in Pharmaceutical Research and Development
Edited by Sean Ekins

Glycogen Synthase Kinase-3 (GSK-3) and Its Inhibitors: Drug Discovery and Development
Edited by Ana Martinez, Ana Castro, and Miguel Medina

Drug Transporters: Molecular Characterization and Role in Drug Disposition
Edited by Guofeng You and Marilyn E. Morris

Aminoglycoside Antibiotics: From Chemical Biology to Drug Discovery
Edited by Dev P. Arya

Drug-Drug Interactions in Pharmaceutical Development
Edited by Albert P. Li

Dopamine Transporters: Chemistry, Biology, and Pharmacology
Edited by Mark L. Trudell and Sari Izenwasser

Drug Design of Zinc-Enzyme Inhibitors: Functional, Structural, and Disease Applications
Edited by Claudiu T. Supuran and Jean-Yves Winum

DRUG DESIGN OF ZINC-ENZYME INHIBITORS

FUNCTIONAL, STRUCTURAL, AND DISEASE APPLICATIONS

Edited by

Claudiu T. Supuran
Jean-Yves Winum

WILEY

A JOHN WILEY & SONS, INC., PUBLICATION

Published by John Wiley & Sons, Inc., Hoboken, New Jersey
Published simultaneously in Canada

Library of Congress Cataloging-in-Publication Data:

Drug design of zinc-enzyme inhibitors : functional, structural, and
disease applications / [edited by] Claudiu T. Supuran, Jean-Yves Winum.
p. ; cm.
Includes bibliographical references and index.
ISBN 978-0-470-27500-9 (cloth)
1. Zinc enzymes–Inhibitors–Therapeutic use. 2. Drugs–Design. I.
Supuran, Claudiu T., 1962- II. Winum, Jean-Yves.
[DNLM: 1. Carbonic Anhydrase Inhibitors–pharmacokinetics. 2.
Carbonic Anhydrase Inhibitors–therapeutic use. 3. Drug Delivery
Systems. 4. Drug Design. 5. Zinc–chemistry. QU 143 D794 2009]
RM666.E548D475 2009
615′.19–dc22

2009005650

Printed in the United States of America

10 9 8 7 6 5 4 3 2 1

CONTENTS

PREFACE

In recent years, the life sciences research has considerably attracted scientists to investigate metalloenzymes and their modulators of activity (inhibitors and/or activators) to meet the challenges for improving human health by discovering new therapeutic targets.

This book mainly deals with the progress that has been made in the field of drug design and discovery of zinc metalloprotein inhibitors over the past years. Recent trends in zinc metalloenzymes are structured into five parts, comprising 40 chapters contributed by experts in the field from all over the world. Of these contributors, there are many who have contributed to this area for decades as scientists and have been recognized for the same.

The five parts of this book can be read as a whole or individually, independent of each other. In fact, the book not only caters to academic or industrial researchers in any of the areas related to pharmaceutical research and development but also to advanced undergraduates as well as graduates at the beginning of their research career, interested in specific topics of this field.

Part I (Chapter 1) outlines the importance of the zinc ion in biological systems and focuses on the importance of targeting zinc enzymes as a promising strategy for drug design and development.

Part II (Chapters 2–22) deals with one of the most studied zinc enzymes among all metalloproteins, the carbonic anhydrase (CA), and provides a comprehensive up-to-date review on the development of modulators of activity for both eukaryotic and prokaryotic CAs and their potential use in drug discovery. Part III (Chapters 23–27) brings to light the potential of matrix metalloproteinase/ADAM inhibitors as drug candidates. Part IV (Chapters 28–31) discusses the relevance of bacterial zinc protease as potential drug target and the use of inhibitors as anti-infective agents. Finally, Part V (Chapters 32–40) reviews the current and potential clinical applications of other zinc-containing enzymes in the treatment of cancer and viral and bacterial infections.

All the data given in this book provide a chemical, biological, and pharmacological framework for understanding the clinical utility of compounds targeting zinc metalloproteins for the treatment of various diseases.

All figures are available at ftp://ftp.wiley.com/public/sci_tech_med/zinc_enzyme.

We express our deepest gratitude to all our coworkers and colleagues who have contributed their highly informative manuscripts to this book on time and without whom this book would not have been possible.

We would also like to thank Prof. Binghe Wang who got this book included in the important collection of Wiley book series on drug discovery and development, which he is editing.

We also appreciate the editorial assistance and advice rendered with patience by Jonathan Rose and Lauren Hilger at John Wiley & Sons during the making of this book.

Montpellier, France JEAN-YVES WINUM

Florence, Italy CLAUDIU T. SUPURAN
February 2009

The editors, Dr. Jean-Yves Winum and Dr. Claudiu T. Supuran (Florence, Zaza restaurant)

CONTRIBUTORS

Mirko Altenkämper, Institut für Pharmazeutische Chemie, Philipps-Universität Marburg, Marbacher Weg 6, 35032 Marburg-Lahn, Germany

Vincenzo Alterio, Istituto di Biostrutture e Bioimmagini-CNR, Via Mezzocannone 16, 80134 Naples, Italy

Montserrat Andújar-Sánchez, Departmento Química Física, Bioquímica y Química Inorgánica, Universidad de Almería, Carretera Sacramento s/n Almería, 04120, Spain

Monika Barathova, Institute of Virology, Centre of Molecular Medicine, Slovak Academy of Sciences, Bratislava, Slovak Republic

Youngjoo Byun, Russell H. Morgan Department of Radiology and Radiological Sciences, Johns Hopkins Medical Institutions, Baltimore, MD 21212, USA

Ana Cámara-Artigas, Departmento Química Física, Bioquímica y Química Inorgánica, Universidad de Almería, Carretera Sacramento s/n Almería, 04120, Spain

Mauro Carcelli, Dipartimento di Chimica Generale ed Inorganica, Chimica Analitica, Chimica Fisica, Università di Parma, V.le Usberti 17/A, Campus Universitario, 43100 Parma, Italy

Joseph R. Casey, Membrane Protein Research Group, Departments of Physiology, University of Alberta, Edmonton, Canada T6G 2H7; Membrane Protein Research Group, Departments of Biochemistry, University of Alberta, Edmonton, Canada T6G 2H7

Alessandro Cecchi, Dipartimento di Chimica, Università degli Studi di Firenze, Via della Lastruccia 3, I-50019 Sesto Fiorentino (Firenze), Italy

Richard I. Christopherson, School of Molecular and Microbial Biosciences, University of Sydney, Sydney, NSW 2006, Australia

Laura Ciani, Dipartimento di Chimica, Università degli Studi di Firenze, Via della Lastruccia 3, I-50019 Sesto Fiorentino (Firenze), Italy

Katia D'Ambrosio, Istituto di Biostrutture e Bioimmagini-CNR, Via Mezzocannone 16, 80134 Naples, Italy

Giuseppina De Simone, Istituto di Biostrutture e Bioimmagini-CNR, Via Mezzocannone 16, 80134 Naples, Italy

Anna Di Fiore, Istituto di Biostrutture e Bioimmagini-CNR, Via Mezzocannone 16, 80134 Naples, Italy

Jean-Michel Dogné, Drug Design and Discovery Center, FUNDP, University of Namur, 61 rue de Bruxelles, 5000 Namur, Belgium

Johannes A. Eble, Center for Molecular Medicine, Department Vascular Matrix Biology, Excellence Cluster Cardio-Pulmonary System, Frankfurt University Hospital, Frankfurt am Main, Germany

James G. Ferry, Department of Biochemistry and Molecular Biology, Eberly College of Science, The Pennsylvania State University, University Park, PA 16802-4500, USA

Paul W. Finn, InhibOx Limited, Pembroke House, 36-37 Pembroke Street, Oxford OX1 1BP, UK

J. Mitchell Guss, Centenary Institute of Cancer Medicine and Cell Biology, Sydney, NSW 2042, Australia

Rebecca A. Hall, Department of Biosciences, University of Kent, Canterbury, Kent, CT2 7NJ, UK

Mika Hilvo, Institute of Medical Technology and School of Medicine, University of Tampere and Tampere University Hospital, Biokatu 6, 33014 University of Tampere, Finland

Marc A. Ilies, Department of Pharmaceutical Sciences, Temple University School of Pharmacy, 3307 North Broad Street, Philadelphia, PA 19140, USA

Alessio Innocenti, Laboratorio di Chimica Bioinorganica, Università degli Studi di Firenze, Room 188, Via della Lastruccia 3, I-50019 Sesto Fiorentino (Firenze), Italy

Vicente Jara-Pérez, Departmento Química Física, Bioquímica y Química Inorgánica, Universidad de Almería, Carretera Sacramento s/n Almería, 04120, Spain

Danielle E. Johnson, Membrane Protein Research Group, Departments of Physiology, University of Alberta, Edmonton, Canada T6G 2H7

Pascale Joseph, Centre d'Etudes d'Agents Pathogènes et Biotechnologies pour la Santé (CPBS) UMR 5236 CNRS-UM1-UM2, Université Montpellier II, cc100, Place E. Bataillon, 34095 Montpellier, France

Stephan Köhler, Centre d'Etudes d'Agents Pathogènes et Biotechnologies pour la Santé (CPBS) UMR 5236 CNRS-UM1-UM2, Université Montpellier II, cc100, Place E. Bataillon, 34095 Montpellier, France

Juraj Kopacek, Institute of Virology, Centre of Molecular Medicine, Slovak Academy of Sciences, Bratislava, Slovak Republic

Jerapan Krungkrai, Department of Biochemistry, Faculty of Medicine, Chulalongkorn University, 1873 Rama 4 Road, Pathumwan, Bangkok 10330, Thailand

Sudaratana R. Krungkrai, Unit of Biochemistry, Department of Medical Science, Faculty of Science, Rangsit University, Paholyothin Road, Patumthani 12000, Thailand

Mihwa Lee, School of Molecular and Microbial Biosciences, University of Sydney, Sydney, NSW 2006, Australia

Shawn E. Lupold, James Buchanan Brady Urological Institute, Johns Hopkins Medical Institutions, Baltimore, MD 21212, USA

Megan J. Maher, Centenary Institute of Cancer Medicine and Cell Biology, Sydney, NSW 2042, Australia

Alfonso Maresca, Laboratorio di Chimica Bioinorganica, Universita degli Studi di Firenze, Room 188, Via della Lastruccia 3, 50019 Sesto Fiorentino (Florence), Italy

Adriano Martinelli, Dipartimento di Scienze Farmaceutiche, Università di Pisa, Via Bonanno 6, 56126 Pisa, Italy

Bernard Masereel, Drug Design and Discovery Center, FUNDP, University of Namur, 61 rue de Bruxelles, 5000 Namur, Belgium

Antonio Mastrolorenzo, Dipartimento di Scienze Dermatologiche, Università degli Studi di Firenze, Villa Basilewsky, Via Lorenzo Magnifico 104, 50129 Firenze, Italy

Ronnie C. Mease, Russell H. Morgan Department of Radiology and Radiological Sciences, Johns Hopkins Medical Institutions, Baltimore, MD 21212, USA

Francesco Mincione, U.O. Oculistica Az. USL 3, Val di Nievole, Ospedale di Pescia, Pescia, Italy

Jean-Louis Montero, Institut des Biomolécules Max Mousseron (IBMM) UMR 5247 CNRS-UM1-UM2 Bãtiment de Recherche Max Mousseron, Ecole Nationale Supérieure de Chimie de Montpellier, 8 rue de l'Ecole Normale, 34296 Montpellier Cedex, France

Fritz. A. Mühlschlegel, Department of Biosciences, University of Kent, Canterbury, Kent, CT2 7NJ, UK; East Kent Hospitals NHS Trust Clinical Microbiology Service, Ashford, Kent, TN24 0LZ, UK

Hideaki Nagase, Department of Matrix Biology, The Kennedy Institute of Rheumatology Division, Imperial College London, London W6 8LH, UK

Nouri Neamati, Department of Pharmacology and Pharmaceutical Sciences, School of Pharmacy, University of Southern California, 1985 Zonal Avenue, PSC 304, Los Angeles, CA 90089, USA

Isao Nishimori, Department of Gastroenterology, Kochi Medical School, Nankoku, Kochi 783-8505, Japan

Elisa Nuti, Dipartimento di Scienze Farmaceutiche, Università di Pisa, Via Bonanno 6, 56126 Pisa, Italy

Uta-Maria Ohndorf, Department of Physiological Chemistry, Ruhr-University Bochum, MA2/141, 44801 Bochum, Germany

Regina Ortmann, Institut für Pharmazeutische Chemie, Philipps-Universität Marburg, Marbacher Weg 6, 35032 Marburg-Lahn, Germany

Seppo Parkkila, Institute of Medical Technology and School of Medicine, University of Tampere and Tampere University Hospital, Biokatu 6, 33014 University of Tampere, Finland

Jaromir Pastorek, Institute of Virology, Centre of Molecular Medicine, Slovak Academy of Sciences, Bratislava, Slovak Republic

Silvia Pastorekova, Institute of Virology, Centre of Molecular Medicine, Slovak Academy of Sciences, Bratislava, Slovak Republic

Martin G. Pomper, Russell H. Morgan Department of Radiology and Radiological Sciences, Johns Hopkins Medical Institutions, Baltimore, MD 21212, USA

Sandra Ristori, Dipartimento di Chimica, Università degli Studi di Firenze, Via della Lastruccia 3, I-50019 Sesto Fiorentino (Firenze), Italy

Dominga Rogolino, Dipartimento di Chimica Generale ed Inorganica, Chimica Analitica, Chimica Fisica, Università di Parma, V.le Usberti 17/A, Campus Universitario, 43100 Parma, Italy

Armando Rossello, Dipartimento di Scienze Farmaceutiche, Università di Pisa, Via Bonanno 6, 56126 Pisa, Italy

Eladio F. Sanchez, Research and Development Center, Ezequiel Dias Foundation, 30510-010 Belo Horizonte, MG, Brazil; Post-Graduate Program, Hospital Santa Casa, Belo Horizonte, Brazil

M. Amélia Santos, Centro de Química Estrutural, Instituto Superior Técnico-UTL, Av. Rovisco Pais 1, 1049-001 Lisboa, Portugal

Christine Schlicker, Department of Physiological Chemistry, Ruhr-University Bochum, MA2/141, 44801 Bochum, Germany

Martin Schlitzer, Institut für Pharmazeutische Chemie, Philipps-Universität Marburg, Marbacher Weg 6, 35032 Marburg-Lahn, Germany

Andrea Scozzafava, Laboratorio di Chimica Bioinorganica, Università degli Studi di Firenze, Polo Scientifico, Room 188, Via della Lastruccia 3, 50019 Sesto Fiorentino (Florence), Italy

Mario Sechi, Dipartimento Farmaco Chimico Tossicologico, Università di Sassari, Via Muroni 23/A, 07100 Sassari, Italy

Clemens Steegborn, Department of Physiological Chemistry, Ruhr-University Bochum, MA2/141, 44801 Bochum, Germany

Claudiu T. Supuran, Laboratorio di Chimica Bioinorganica, Università degli Studi di Firenze, Polo Scientifico, Room 188, Via della Lastruccia 3, 50019 Sesto Fiorentino (Florence), Italy

Hiroaki Takeuchi, Department of Laboratory Medicine, Kochi Medical School, Nankoku, Kochi 783-8505, Japan

Claudia Temperini, Dipartimento di Chimica, Università degli Studi di Firenze, Via della Lastruccia 3, I-50019 Sesto Fiorentino (Firenze), Italy

Anne Thiry, Drug Design and Discovery Center, FUNDP, University of Namur, 61 rue de Bruxelles, 5000 Namur, Belgium

Tiziano Tuccinardi, Dipartimento di Scienze Farmaceutiche, Università di Pisa, Via Bonanno 6, 56126 Pisa, Italy

François Turtaut, Institut des Biomolécules Max Mousseron (IBMM) UMR 5247 CNRS-UM1-UM2 Bâtiment de Recherche Max Mousseron, Ecole Nationale Supérieure de Chimie de Montpellier, 8 rue de l'Ecole Normale, 34296 Montpellier Cedex, France

Robert Visse, Department of Matrix Biology, The Kennedy Institute of Rheumatology Division, Imperial College London, London W6 8LH, UK

Daniela Vullo, Laboratorio di Chimica Bioinorganica, Università degli Studi di Firenze, Room 188, Via della Lastruccia 3, I-50019 Sesto Fiorentino (Firenze), Italy

Kunihiko Watanabe, Laboratory of Applied Microbiology, Division of Applied Life Sciences, Graduate School of Life and Environmental Sciences, Kyoto Prefectural University, Shimogamo, Sakyo, Kyoto 606-8522, Japan

Jean-Yves Winum, Institut des Biomolécules Max Mousseron (IBMM) UMR 5247 CNRS-UM1-UM2 Bâtiment de Recherche Max Mousseron, Ecole Nationale Supérieure de Chimie de Montpellier, 8 rue de l'Ecole Normale, 34296 Montpellier Cedex, France

PART I

INTRODUCTION

CHAPTER 1

Introduction to Zinc Enzymes as Drug Targets

CLAUDIU T. SUPURAN[1] and JEAN-YVES WINUM[2]

[1]Laboratorio di Chimica Bioinorganica, Università degli Studi di Firenze, Polo Scientifico, Room 188, Via della Lastruccia 3, 50019, Sesto Fiorentino (Florence), Italy
[2]Institut des Biomolécules Max Mousseron (IBMM) UMR 5247 CNRS-UM1-UM2 Bâtiment de Recherche Max Mousseron, Ecole Nationale Supérieure de Chimie de Montpellier, 8 rue de l'Ecole Normale, 34296, Montpellier Cedex, France

1.1 INTRODUCTION

Drug target is an old concept that was suggested at the end of the nineteenth and the beginning of the twentieth century by Ehrlich and Langley who developed the idea that compounds display biological activity by binding to cellular constituents.[1]

Commonly, most of the drug targets can be defined as human genome-derived proteins (or proteins belonging to pathogenic organisms) that undergo a selective interaction with compounds administered to treat or diagnose a disease.[2] Target identification and validation constitute the most important steps in the process of drug discovery. At present, there is an enormous interest in identifying and validating "druggable" targets in the human proteome and applying structure-based drug design to discover new therapies for important human diseases. The human genome is a huge reservoir of putative drug targets, and its sequencing has allowed identification of about 8000 targets of pharmacological interest. Nevertheless, for all classes of approved therapeutic drugs, around 300 targets have been disclosed with increasing frequency: 270 being encoded by the human genome and the remaining belonging to pathogenic organisms.[3,4] Several promising targets for drug intervention have been revealed in recent years, and their knowledge is helpful for molecular dissection of the mechanism of action of drugs and for predicting features that guide new drug design and the search for new targets.

According to Imming et al.,[4] drug targets can be divided into several categories: (i) enzymes, (ii) substrates, metabolites, and proteins, (iii) receptors, (iv) ion channels, (v) transport proteins, (vi) DNA/RNA and the ribosome, (vii) targets for monoclonal antibodies, (viii) various physicochemical mechanisms, and (ix) unknown mechanisms of action.

Among these different classes of drug targets, enzymes have long been considered valuable drug targets for the treatment of major human diseases, as several thousands of enzymes are encoded in the human genome, and they play a key role in virtually every physiological/pathological process. At present, at least 66 human enzymes and 20 bacterial, viral, fungal, or parasite enzymes are targets of approved drugs, for example, up to 40% of the known drug targets. Enzymes containing metals (metalloenzymes) are of increasing interest and importance, as the genetic consequences of metalloprotein regulation become better understood. The largest category of metalloproteins is constituted by zinc enzymes, with more than 300 representatives presently known, covering all major six enzyme classes established by the International Union of Biochemistry (i.e., oxidoreductases, transferases, hydrolases, lyases, isomerases, and ligases). Over the past few years, substantial evidence has been accumulated implicating the zinc enzymes in the pathophysiology and pathogenesis of a variety of human disorders ranging from infections to cancer. The relevance of zinc metalloproteins to biomedical sciences has increased much in the past few years, and modulation of their activity with small-molecule drugs, designed to interact with a clearly defined ligand binding site, constitutes a challenging area in drug design and discovery.[4–6] Furthermore, the availability of different high-resolution X-ray crystal structures of such enzymes and of their complexes with substrates and/or inhibitors has provided a wealth of information with a profound effect on the way we understand their biological functions.

In this chapter, we present an overview of the role of zinc in biological systems and explain why zinc proteins constitute promising targets for drug intervention.

1.2 IMPORTANCE OF ZINC IN BIOLOGICAL SYSTEMS: STRUCTURAL, REGULATORY, AND CATALYTIC ROLES

Among the transition and group II elements, zinc is the second most abundant metal, after iron, in all biological systems including microorganisms, plants, and animals. It is stable as dication (Zn^{2+}), has Lewis acid properties (it can accept a pair of electrons), and lacks redox activity, as it possesses a full d-shell d^{10} orbital. This ubiquitous element is considered an essential, nontoxic micronutrient, and its several biochemical roles regarding the structure and function of proteins, including enzymes, transcription factors, hormonal receptor sites, and biological membranes, have been recognized.[7,8] Zn(II) is highly regulated under normal physiological conditions, as this metal ion plays a key role in a wide variety of processes such as DNA and RNA synthesis, transmission of the genetic message, growth and development, signal transduction, apoptosis, brain and immune function, lipid metabolism, and so on.

In addition, the zinc ion is also closely involved in intracellular signaling and neuromodulation functions.[7,8]

Physiologically, approximately 98% of the total zinc in an organism is found within the cell (40% in the nucleus and 50% in cytoplasm, organelles, and specialized vesicles), while the remaining is found in the cell membrane.[7,9] The total zinc concentration in eukaryotic cells was reported to be in the high micromolar range, with a concentration around 200 μM.[10] Furthermore, zinc deficiency is detrimental in many aspects to the normal function of the organism, with notable effects on growth and immune functions.[7] The cytosolic concentration of free Zn^{2+} is very low and can be estimated in the subfemtomolar range, but it increases under oxidative stress conditions.[11]

At the molecular level, the intracellular Zn^{2+} is most often tightly bound to proteins considered an essential cofactor for hundreds of enzymes and thousands of metabolic and regulatory proteins, fulfilling both structural and catalytic roles.

1.2.1 The Structural Role of Zinc

Zinc plays an important role in the structure of proteins and cell membranes. In such structural site, it can be found either as a single metal ion or as part of a cluster of two or more ions, being coordinated only by amino acid residues with no bound solvent molecule(s). Thus, the metal ion ensures an essential role in the stabilization of the protein structure by creating or maintaining secondary/tertiary structural elements in the same manner as a disulfide bridge.[12] It can induce the correct folding of protein sequences as zinc fingers, zinc twists, or zinc clusters in numerous regulatory proteins and hormone receptors, contributing to the overall stability of these domains.[13] Zinc fingers are structurally diverse and are present in proteins that perform a broad range of functions in various cellular processes. The biological functions of zinc finger-containing proteins strongly depend on the zinc ion, which ensures integrity and stability and is critical for binding to DNA.[13] These structure-stabilizing motifs are as diverse as their functions and are associated with protein–nucleic acid recognition as well as protein–protein interactions.[14,15] The zinc ion may also be involved in the maintenance of the structure of chromatin and biomembranes, as it plays a crucial role in the regulation of their functions.[14]

The biological function of zinc is governed by the composition of its flexible coordination sphere. This can be a slightly distorted tetrahedral or a trigonal bipyramidal coordination polyhedron in most metalloproteins, with the metal ion coordinating three or four amino acid residues.[8,16–19] Structural sites are typically characterized by a zinc-centered tetrahedral coordination in which the metal ion is fully coordinated by four Cys residues via thiolate group, generally separated from a relatively short sequence in the protein (Fig. 1.1a). Other ligands may also compete with cysteines for binding Zn(II); the second most prevalent ligand is His, which is usually found in combination with Cys, forming structurally related "zinc finger" motifs (Fig. 1.1b).[16–19] Examples of non-Cys structural zinc site have also been reported, apart from the catalytic zinc site in matrix metalloproteinase (MMP) class

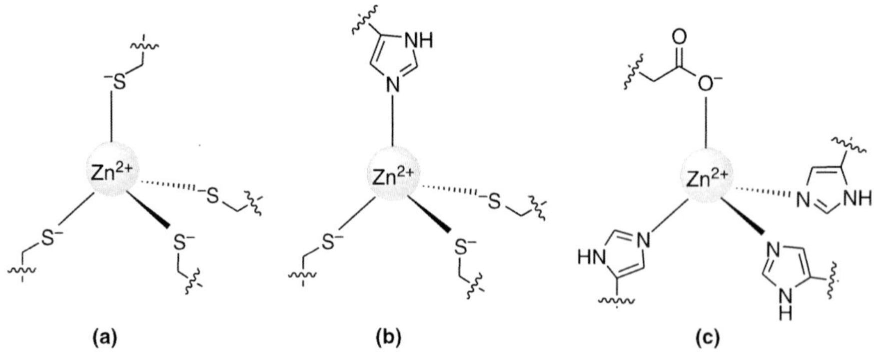

FIGURE 1.1 Structural zinc binding sites in metalloenzymes: (a) four Cys residues coordinated to the metal center; (b) 3 Cys + 1 His; (c) 3 His + 1 Asp/Glu residue coordinated to Zn(II).

of enzymes, with combinations of His and Asp residues coordinating the metal ion (Fig. 1.1c).

Structural zinc sites have important implications for the functioning of metalloproteins. By stabilizing and even inducing the local folding of protein subdomains in the immediate neighborhood of the metal site, one or more amino acid side chains can be orientated toward the active site, thus influencing the enzyme activity by affecting the chemical environment of the active center and/or by influencing the alignment of active site residues for catalysis.[16,17]

Besides structural zinc sites involved in protein functions, a number of protein interface zinc sites can be defined, where the zinc ion bridges proteins or their subunits, thus playing an important role in the organization of the quaternary structure and/or active site of the protein.[16,17,20,21] In these structural sites, zinc ions bridge the interfaces of proteins via ligands provided by different polypeptide chains and can cross-link the same protein leading to homodimers/trimers or tetramers. The link of two different proteins through such intermolecular ligands has also been observed. In such cases, zinc ligation is assumed by coordinating not only residues such as His, Glu, and Asp but also Cys, with two amino acid ligands supplied by both protein moieties or three amino acid ligands coming from one protein backbone and one ligand from another protein domain. The resulting protein interface zinc binding sites can function as catalytic, cocatalytic, or structural sites, playing a key role in transduction pathways that regulate a host of cellular functions.[16–21]

1.2.2 Catalytic and Cocatalytic Role of Zinc

In catalytic sites, zinc ions participate directly in the catalytic process and generally exhibit a distorted tetrahedral geometry with only three O/N/S ligands bound to the zinc ion, the fourth ligand being a water molecule that is an activated nucleophile for the catalytic process. The coordination number 5 can also be encountered for Zn(II) ions, with a trigonal bipyramidal geometry of the metal center. The zinc ion is essential

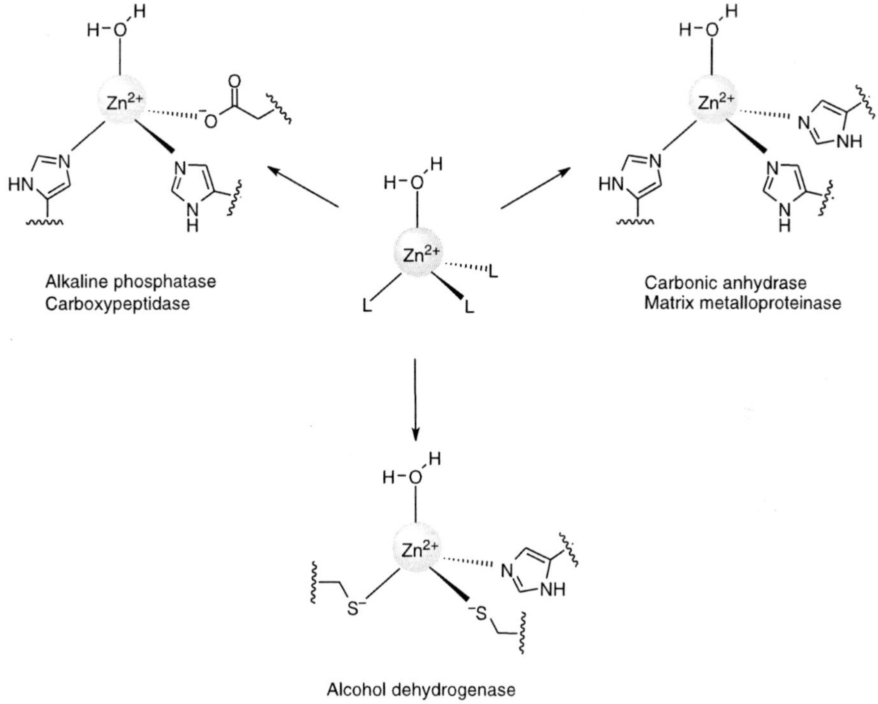

FIGURE 1.2 Catalytic zinc binding sites of some representative metalloenzymes.

for catalytic activity of more than 300 enzymes and is located at the core of the enzyme active site, participating directly in the catalytic mechanism through interaction with substrate molecules undergoing the transformation. The most common coordination feature of Zn(II) in catalytic sites is dominated by histidine side chains coordinating the metal ion, through interaction with the Nε atom of the imidazole ring. Other coordinated amino acid residues are Glu, Asp, and Cys. Due to the amphoteric properties of Zn^{2+}, a water molecule always participates in the coordination sphere as a fourth or fifth ligand. Catalytic zinc binding sites of some representative metalloenzymes are illustrated in Fig. 1.2, where the ligands (L) are three His residues for the lyases carbonic anhydrases (CAs) and two His residues for the hydrolase carboxypeptidase. The catalytic zinc site of alcohol dehydrogenase is the only one known so far where there is only one His residue bound to the metal ion, being also unique as two cysteine residues participate in the coordination.

This zinc-bound water molecule is crucial for the catalysis promoted by the metal center, as it can be either involved in the catalytic process as hydroxide ion or activated via polarizing effect of the neighboring amino acids of the active site, acting as a nucleophile per se.[16,17] Moreover, H_2O can be displaced by the substrate (S) or expanded upon interaction with the substrate (Fig. 1.3). The presence of water molecules in the coordination sphere is usually a distinguishing feature that allows to differentiate a catalytic zinc site from a structural one.[12]

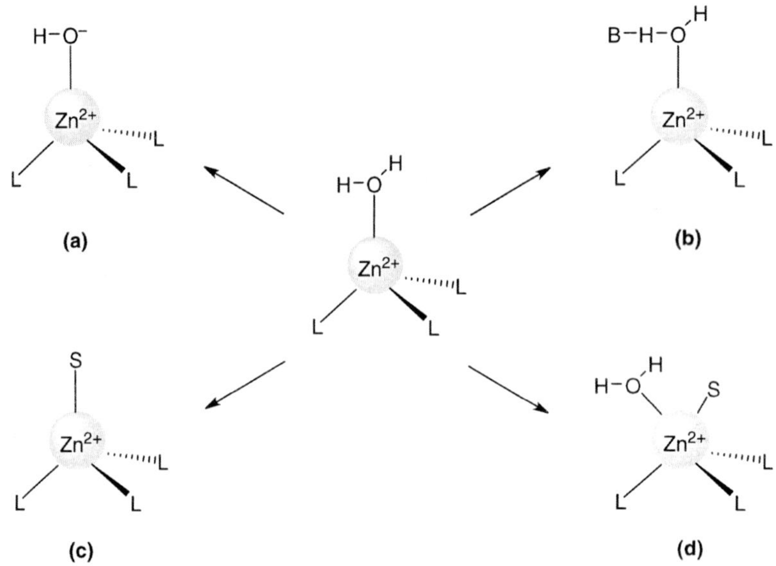

FIGURE 1.3 Role of the zinc-bound water molecule in catalysis: (a) ionization, (b) polarization, (c) displacement, and (d) expansion.[16,17]

Cocatalytic zinc sites can also be distinguished and are characteristic of multi-zinc-containing enzymes, with two or three metal ions in proximity, with two of them bridged by a side chain moiety of a single amino acid residue, such as Asp, Glu, or His, and sometimes a water molecule. Asp and His are preferred amino acids for these sites. No Cys ligands are found in such sites. Typically, the metal ions are separated by a short distance (around 3 Å) and bridged by at least one common ligand, frequently a water molecule or a carboxylate ligand of those mentioned above. The zinc ion can be bridged with another zinc ion or with another metal ion, such as Cu(II), for example, in Cu–Zn superoxide dismutases (SOD).[16–21]

In the past decade, there has been a great expansion in our knowledge of the role of metalloenzymes in the physiopathology of several diseases. Catalytic zinc sites provide convenient targets for drug intervention and the design and development of small-molecule drugs that can coordinate directly to the metal, displacing the zinc water in the active site and inhibiting the enzymes. This challenging research area has been extensively dealt with in this book.

1.3 ZINC METALLOPROTEINS AS DRUG TARGETS

1.3.1 Targeting Human Zinc Metalloenzymes

Since the identification of the first metalloenzyme, carbonic anhydrase, by Keilin and Mann in 1941,[22] more than 300 different enzymes requiring zinc as essential cofactor have been identified, showing their diverse and important physiological functions.

These enzymes are considered to be very attractive targets for drug therapy, and their inhibitors are included in the armamentarium of modern medicine against human diseases such as cardiovascular, neurological, infectious, and metabolic diseases, as well as cancer.[5,6]

Considering the importance and the diversity of zinc-containing enzymes, this book will focus on the zinc enzymes that are relevant for biomedical applications due to their well-known role in life-threatening diseases. For example, the two most investigated metalloproteins that will be considered in detail in this book are the carbonic anhydrases (dealt with in Part II) and the matrix metalloproteinases (dealt with in Part III).

Carbonic anhydrases (EC 4.2.1.1) that belong to the lyase family are ubiquitous zinc enzymes present in prokaryotes and eukaryotes, all over the phylogenetic tree. These are efficient catalysts for the hydration of carbon dioxide to bicarbonate and protons, playing crucial physiological/pathological roles in acid–base homeostasis, secretion of electrolytes, transport of ions, biosynthetic reactions, and tumorigenesis. These enzymes are of clinical relevance as some isoforms among the 15 known in humans are established drug targets, with many inhibitors that have been reported and developed as diuretics, antiglaucoma, anticancer, and antiobesity agents, or for the management of a variety of neurological disorders, including epilepsy and altitude disease.[23,24] Furthermore, a clear connection has recently been found between CA inhibition and lipogenesis (thus, CA inhibitors might be used as antiobesity agents) as well as tumorigenesis (antitumor drugs/diagnostic tools).[23,24] Thus, Chapters 2–22 will be dedicated not only to this class of enzymes and their inhibitors/activators from mammals (*Homo sapiens* being the most investigated one) but also to the various CAs recently characterized from many bacteria, archaea, protozoa, fungi, yeasts, and nematode species. Many such enzymes are now fully characterized kinetically, and their inhibition/activation studies on many classes of compounds reported, thus constituting an important starting point for the rational drug design of inhibitors with clinical applications.[23,24] Such research is very dynamic nowadays, and the near future may see the emergence of novel therapeutic agents targeting such enzymes.

Another essential class of zinc metalloproteins that will be taken into consideration is the superfamily of zinc endopeptidases, MMPs and ADAMs (a disintegrin and matrix metalloproteinase domain), which are dealt with in Chapters 23–27. MMPs are zinc endopeptidases that degrade both matrix and nonmatrix proteins. At least 23 MMPs are known in humans where they play an important role in morphogenesis and in a wide range of processes including tissue repair and remodeling. Their abnormal expression contributes to pathological processes, including arthritis, cancer, and cardiac and central nervous system diseases, and inhibition of MMPs has widely been sought as a strategy in the intervention of these disease processes.[5,6,25,26] A large number of MMP/ADAM inhibitors showing selectivity for the various members of this superfamily have been reported in the past few years holding considerable promises mainly in the anticancer and cardiovascular therapy.

Other zinc metalloenzymes of medical relevance, such as angiotensin-converting enzyme (ACE), histone deacetylase, prostate-specific membrane antigen (PSMA), and protein farnesyltransferase, among others, have already demonstrated a crucial therapeutic potential in various pathological, especially in cancer, neurodegenerative, and inflammatory diseases, and they are reviewed in Chapters 32–36. A special mention should be made of HIV integrase, which is a metalloenzyme containing zinc, (Chapter 37), but Zn(II) is not involved in the catalytic cycle (instead, it seems that Mg(II) is present at the active site). Considering the great importance that the treatment of HIV infection has nowadays gained and the fact that HIV integrase inhibitors were approved for clinical use in 2008, after a successful saga of research and development of more than 15 years, we decided to dedicate a chapter to this interesting zinc enzyme that is in fact not a real zinc enzyme. This is an exception, since, as the title mentions, the main focus of this book is the inhibition of zinc enzymes in which Zn(II) clearly has a catalytic role. Another very recent and quite promising antiviral target is constituted by the APOBEC3F/G family of enzymes that will be dealt with in Chapter 40.

1.3.2 Targeting Bacterial Zinc Metalloenzymes

Infectious diseases still remain the main cause of human deaths worldwide. The emergence and spread of pathogenic bacterial strains resistant to most classes of clinically used antibiotics have created the need for the development of such novel therapeutic agents as possessing a different mechanism of action. Development of new anti-infective agents with a novel mode of action and lacking cross-resistance to the existing drugs is a strong imperative of biomedical research of early twenty-first century, and a highly unaccomplished task until now. In the past 10 years, bacterial genome analysis allowed to define new essential bacterial genes and provided many details concerning the structure of bacterial proteins that play an important role in pathogenesis, with many such prokaryotic zinc metalloenzymes being identified. Metalloproteins that are essential for bacterial growth and are not required by mammalian cells constitute potential targets for antimicrobial drugs and form the basis for future therapies.[27,28] Several of these orphan (for the moment) targets, such as bacterial proteases, botulinum, tetanus and anthrax lethal factors (LAs), clostridial collagenases, and other bacterial proteases, will be dealt with in Chapters 28–31.

Characterization of many specific as well as ubiquitous proteases in both Gram-positive as well as Gram-negative pathogens has allowed the development of different classes of specific nanomolar-range inhibitors for bacterial proteases such as *Clostridium histolyticum* collagenase, *Botulinum* neurotoxin, and *Tetanus* neurotoxin.[29] Moreover, a number of approaches have been taken to identify inhibitors of the zinc-dependent metalloproteinase lethal factor, a critical component of anthrax toxin and an important potential target for small-molecule drugs.[30]

Identification of zinc metalloenzymes from bacterial genomes has allowed identification of new potential targets for the development of anti-infective agents. This strategy, which has already demonstrated promising results, constitutes a challenging area, considering all the possible targets available in the zinc metalloprotein family with potential therapeutic applications.

Several other chapters of the book deal wih zinc enzymes that are just beginning to be investigated in more detail, such as P-III metalloproteinase from a highly poisonous snake (Chapter 33), the histidinol dehydrogenases (which may constitute an interesting class of antibacterials, Chapter 38), or the dihydroorotase inhibitors (with potential for developing antimalarials, Chapter 39).

It is thus clear that the wealth of genomic, structural, biochemical, and synthetic data that has recently emerged in biomedical research of zinc enzymes and their inhibition enables us to dedicate this book to these fascinating fields. Although we clearly understand that due to the vastness of the field, it is not possible to deal with all important enzymes here, we have tried to make a comprehensive review of the literature data for the most relevant representatives, for their inhibitors, and for their biomedical applications.

REFERENCES

1. Silverman, R. B., Ed. *The Organic Chemistry of Drug Design and Drug Action*, 2nd ed.; Elsevier Academic Press, **2004**.
2. Landry, Y.; Gies, J.-P. Drugs and their molecular targets: an updated overview. *Fundam. Clin. Pharmacol.* **2008**, *22*, 1–18.
3. Overington, J. P.; Al-Lazikani, B.; Hopkins, A. L. How many drug targets are there? *Nat. Rev. Drug Discov.* **2006**, *5*, 993–996.
4. Imming, P.; Sinning, C.; Meyer, A. Drugs, their targets and the nature and number of drug targets. *Nat. Rev. Drug Discov.* **2006**, *5*, 821–834.
5. Anzellotti, A. I.; Farrell, N. P. Zinc metalloproteins as medicinal targets. *Chem. Soc. Rev.* **2008**, *37*, 1629–1651.
6. Jacobsen, F. E.; Lewis, J. A.; Cohen, S. E. The design of inhibitors for medicinally relevant metalloproteins. *ChemMedChem* **2007**, *2*, 152–171.
7. Vallee, B. L.; Falchuk, K. H. The biochemical basis of zinc physiology. *Physiol. Rev.* **1993**, *73*, 79–111.
8. McCall, K. A.; Huang, C. C.; Fierke, C. A. Function and mechanism of zinc metalloenzymes. *J. Nutr.* **2000**, *130*, 1437S–1446S.
9. Tapiero, H.; Tew, K. D. Trace elements in human physiology and pathology: zinc and metallothioneins. *Biomed. Pharmacother.* **2003**, *57*, 399–411.
10. Maret, W. Zinc biochemistry, physiology, and homeostasis: recent insights and current trends. *BioMetals* **2001**, *14*, 187–190.
11. Chang, C. J.; Jaworski, J.; Nolan, E. M.; Sheng, M.; Lippard, S. J. A tautomeric zinc sensor for ratiometric fluorescence imaging: application to nitric oxide-induced release of intracellular zinc. *Proc. Natl. Acad. Sci. USA* **2004**, *101*, 1129–1134.
12. Lee, Y.-M.; Lim, C. Physical basis of structural and catalytic Zn-binding sites in proteins. *J. Mol. Biol.* **2008**, *379*, 545–553.
13. Vallee, B. L.; Coleman, J. E.; Auld, D. S. Zinc fingers, zinc clusters, and zinc twists in DNA-binding protein domains. *Proc. Natl. Acad. Sci. USA* **1991**, *88*, 999–1003.
14. Cox, E. H.; McLendon, G. L. Zinc-dependent protein folding. *Curr. Opin. Chem. Biol.* **2000**, *4*, 162–165.
15. Gamsjaeger, R.; Liew, C. K.; Loughlin, F. E.; Crossley, M.; Mackay, J. P. Sticky fingers: zinc-fingers as protein-recognition motifs. *Trends Biochem. Sci.* **2007**, *32*, 63–70.

16. Auld, D. S.; Zinc enzymes. In *Encyclopedia of Inorganic Chemistry*, 2nd ed.; Bruce King, R., Ed.; John Wiley & Sons, Ltd: Chichester, **2005**; Vol. *IX*, pp 5885–5927.

17. Auld, D. S. Zinc coordination sphere in biochemical zinc sites. *BioMetals* **2001**, *14*, 271–313.

18. O'Dell B. L. Zinc plays both structural and catalytic roles in metalloproteins. *Nutr. Rev.* **1992**, *50*, 48–50.

19. Christianson, D. W. Structural biology of zinc. *Adv. Protein Chem.* **1991**, *42*, 281–355.

20. Maret, W. Protein interface zinc sites: the role of zinc in the supramolecular assembly of proteins and in transient protein–protein interactions. In *Handbook of Metalloproteins;* Messerschmidt, A.; Bode, W.; Cygler, M.,Eds.; John Wiley & Sons, Ltd: Chichester, **2004**; Vol. *3*, pp 432–441.

21. Maret, W. Exploring the zinc proteome. *J. Anal. At. Spectrom.* **2004**, *19*, 15–19.

22. Keilin, D.; Mann, T. Carbonic anhydrase. Purification and nature of the enzyme. *Biochem. J.* **1940**, *34*, 1163–1176.

23. Supuran, C. T. Carbonic anhydrases: novel therapeutic applications for inhibitors and activators. *Nat. Rev. Drug Discov.* **2008**, *7*, 168–181.

24. Supuran, C. T.; Scozzafava, A.; Conway, J., Eds.; *Carbonic Anhydrase: Its Inhibitors and Activators;* CRC Press: Boca Raton, FL, **2004**; pp 1–376, and references cited therein.

25. Tu, G.; Xu, W.; Huang, H.; Li, S. Progress in the development of matrix metalloproteinase inhibitors. *Curr. Med. Chem.* **2008**, *15*, 1388–1395.

26. Smith, H. J.; Simons, C., Eds.; *Proteinase and Peptidase Inhibition: Recent Potential Targets for Drug Development;* Taylor & Francis: London, **2002**.

27. White, R. J.; Margolis, P. S.; Trias, J.; Yuan, Z. Targeting metalloenzymes: a strategy that works. *Curr. Opin. Pharmacol.* **2003**, *3*, 502–507.

28. Winum, J.-Y.; Kohler, S.; Scozzafava, A.; Montero, J.-L.; Supuran, C. T. Targeting bacterial metalloenzymes: a new strategy for the development of anti-infective agents. *Anti-Infect. Agents Med. Chem.* **2008**, *7*, 169–179.

29. Supuran, C. T.; Scozzafava, A.; Clare, B. W. Bacterial protease inhibitors. *Med. Res. Rev.* **2002**, *22*, 329–372.

30. Turk, B. E. Discovery and development of anthrax lethal factor metalloproteinase inhibitors. *Curr. Pharm. Biotechnol.* **2008**, *9*, 24–33.

PART II

DRUG DESIGN OF CARBONIC ANHYDRASE INHIBITORS AND ACTIVATORS

CHAPTER 2

Carbonic Anhydrases as Drug Targets: General Presentation

CLAUDIU T. SUPURAN

Laboratorio di Chimica Bioinorganica, Università degli Studi di Firenze, Room 188, Via della Lastruccia 3, I-50019 Sesto Fiorentino (Firenze), Italy

2.1 INTRODUCTION

The carbonic anhydrases (CAs, EC 4.2.1.1) are ubiquitous zinc enzymes, present in prokaryotes and eukaryotes, encoded by five distinct, evolutionarily unrelated gene families: the α-CAs (present in vertebrates, bacteria, algae, and cytoplasm of green plants), the β-CAs (predominantly in bacteria, algae, and chloroplasts of both mono- and dicotyledons), the γ-CAs (mainly in archaea and some bacteria), and the δ- and ξ-CAs present in marine diatoms.[1–8] In mammals, 16 different α-CA isozymes or CA-related proteins (CARPs) were described (Table 2.1), with very different subcellular localization and tissue distribution.[1–8] Basically, there are several cytosolic forms (CA I–III, VII), four membrane-bound isozymes (CA IV, IX, XII, and XIV), one mitochondrial form (CA V), and a secreted CA isozyme, CA VI. These enzymes catalyze a very simple physiological reaction, the interconversion between carbon dioxide and the bicarbonate ion, and are thus involved in crucial physiological processes related to respiration and transport of CO_2/bicarbonate between metabolizing tissues and lungs, pH and CO_2 homeostasis, electrolyte secretion in a variety of tissues/organs, biosynthetic reactions (such as gluconeogenesis, lipogenesis, and ureagenesis), bone resorption, calcification, tumorigenicity, and many other physiologic or pathologic processes.[1–13] As will be discussed subsequently, many of these isozymes are important targets for the design of inhibitors or activators with clinical applications.

In addition to the physiological reaction, the reversible hydration of CO_2 to bicarbonate (reaction 2.1 in Fig. 2.1), α-CAs catalyze a variety of other reactions, such as the hydration of cyanate to carbamic acid or the hydration of cyanamide

TABLE 2.1 Kinetic Parameters for CO_2 Hydration Reaction Catalyzed by the 16 Vertebrate α-CA Isozymes, at 20°C and pH 7.5, and Their Subcellular Localization

Isozyme	k_{cat} (s^{-1})	K_m (mM)	k_{cat}/K_m ($M^{-1} s^{-1}$)	Subcellular Localization	Tissue/Organ Localization	References
hCA I	2.0×10^5	4.0	5.0×10^7	Cytosol	Erythrocytes, GI tract	1–3
hCA II	1.4×10^6	9.3	1.5×10^8	Cytosol	Erythrocytes, eye, GI tract, bone osteoclasts, kidney, lung, testis, brain	1–3
hCA III	1.0×10^4	33.3	3.0×10^5	Cytosol	Skeletal muscle, adipocytes	1–3
hCA IV	1.1×10^6	21.5	5.1×10^7	Membrane bound	Kidney, lung, pancreas, brain capillaries, colon, heart muscle	2, 3
hCA VA	2.9×10^5	10.0	2.9×10^7	Mitochondria	Liver	9
hCA VB	9.5×10^5	9.7	9.8×10^7	Mitochondria	Heart and skeletal muscle, pancreas, kidney, spinal cord, GI tract	9
hCA VI	3.4×10^5	6.9	4.9×10^7	Secreted (saliva/milk)	Salivary and mammary glands	10
hCA VII	9.5×10^5	11.4	8.3×10^7	Cytosol	CNS	11
hCA VIII	—	—	—	Cytosol	CNS	1–3
hCA IX	3.8×10^5	6.9	5.5×10^7	Transmembrane	Tumors; GI mucosa	1, 6
hCA X	—	—	—	Cytosol	CNS	1–3
hCA XI	—	—	—	Cytosol	CNS	1–3
hCA XII	4.2×10^5	12.0	3.5×10^7	Transmembrane	Renal, intestinal, reproductive epithelia, eye, tumors	12
hCA XIII	1.5×10^5	13.8	1.1×10^7	Cytosol	Kidney, brain, lung, gut, reproductive tract	13
hCA XIV	3.1×10^5	7.9	3.9×10^7	Transmembrane	Kidney, brain, liver	14
mCA XV	4.7×10^5	14.2	3.3×10^7	Membrane bound	Kidney	8, Unpublished research

h, human; m, mouse enzyme. hCA VIII, X, and XI are devoid of catalytic activity.

$O{=}C{=}O + H_2O \Leftrightarrow HCO_3^- + H^+$	(2.1)
$O{=}C{=}NH + H_2O \Leftrightarrow H_2NCOOH$	(2.2)
$HN{=}C{=}NH + H_2O \Leftrightarrow H_2NCONH_2$	(2.3)
$RCHO + H_2O \Leftrightarrow RCH(OH)_2$	(2.4)
$RCOOAr + H_2O \Leftrightarrow RCOOH + ArOH$	(2.5)
$RSO_3Ar + H_2O \Leftrightarrow RSO_3H + ArOH$	(2.6)
$ArF + H_2O \Leftrightarrow HF + ArOH$	(2.7)
(Ar = 2,4-dinitrophenyl)	
$PhCH_2OCOCl + H_2O \Leftrightarrow PhCH_2OH + CO_2 + HCl$	(2.8)
$RSO_2Cl + H_2O \Leftrightarrow RSO_3H + HCl$	(2.9)
(R = Me; Ph)	

FIGURE 2.1 Reactions catalyzed by α-CAs.

to urea (reactions 2.2 and 2.3); the aldehyde hydration to *gem*-diols (reaction 2.4); the hydrolysis of carboxylic or sulfonic (reactions 2.5 and 2.6), as well as other less investigated hydrolytic processes, such as those described by equations 2.7–2.9 in Fig. 2.1.[2,3] It is unclear at this moment whether α-CA-catalyzed reactions other than the CO_2 hydration have physiological significance. The X-ray crystal structure has been determined for six α-CAs (isozymes CA I–VA and XII–XIV)[1–6,12–14] and for representatives of the β- and γ-CA families (discussed subsequently).

2.2 CATALYTIC AND INHIBITION MECHANISMS OF CARBONIC ANHYDRASES

2.2.1 α-CAs

The metal ion (which is Zn(II) in all α-CAs investigated until now) is essential for catalysis.[1–8] X-ray crystallographic data showed that the metal ion is situated at the bottom of a 15 Å deep active site cleft (Fig. 2.2), being coordinated by three histidine residues (His94, His96, and His119) and a water molecule/hydroxide ion.[1–12,15,16] The zinc-bound water is also engaged in hydrogen bond interactions with the hydroxyl moiety of Thr199, which in turn is bridged to the carboxylate moiety of Glu106; these interactions enhance the nucleophilicity of the zinc-bound water molecule and orient the substrate (CO_2) in a favorable location for the nucleophilic attack (Fig. 2.3).[1–8]

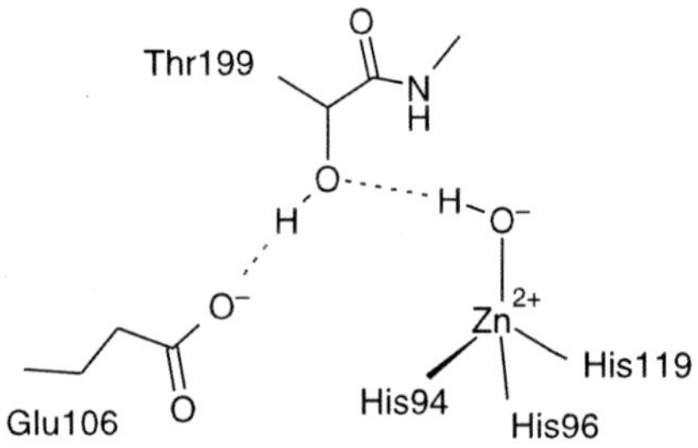

FIGURE 2.2 The Zn(II) ion coordination in the hCA II active site, with the three histidine ligands (His94, His96, and His119) and the gate-keeping residues (Thr199 and Glu106) shown.

The active form of the enzyme is the basic one, with hydroxide bound to Zn(II) (Fig 2.3a).[1–8] This strong nucleophile attacks the CO_2 molecule bound in a hydrophobic pocket in its neighborhood (the substrate binding site comprises residues Val121, Val143, and Leu198 in the case of the human isozyme CA II) (Fig. 2.3b), leading to the

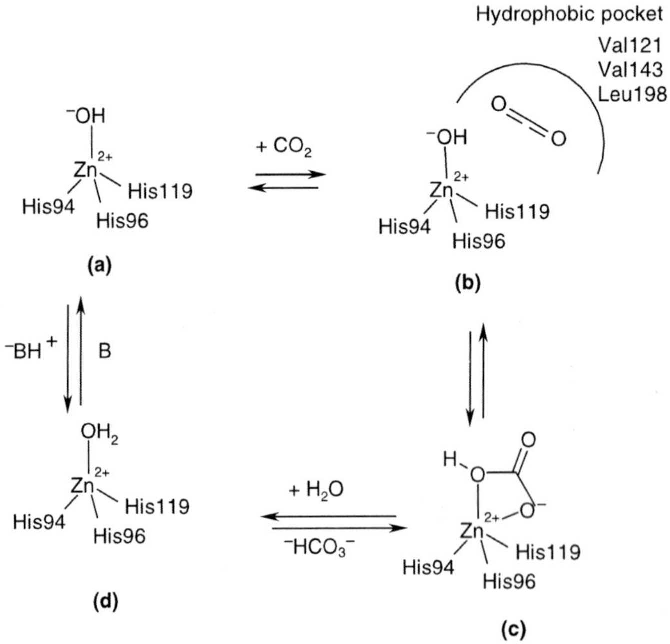

FIGURE 2.3 Schematic representation of the catalytic mechanism for the α-CAs catalyzed CO_2 hydration. The hydrophobic pocket for the binding of substrate(s) is shown schematically at step b.

formation of bicarbonate coordinated to Zn(II) (Fig. 2.3c). The bicarbonate ion is then displaced by a water molecule and liberated into solution, leading to the acid form of the enzyme, with water coordinated to Zn(II) (Fig. 2.3d), which is catalytically inactive.[1–3] To regenerate the basic form (a), a proton transfer reaction from the active site to the environment takes place, which may be assisted either by active site residues (such as His64—the proton shuttle in isozymes I, II, IV, VII, IX, and XII–XIV among others) or by buffers present in the medium. The process may be schematically represented by equations 2.10 and 2.11:

$$\text{E-Zn}^{2+}\text{-OH}^- + CO_2 \Leftrightarrow \text{E-Zn}^{2+}\text{-HCO}_3^- \overset{H_2O}{\Leftrightarrow} \text{E-Zn}^{2+}\text{-OH}_2 + HCO_3^- \qquad (2.10)$$

$$\text{E-Zn}^{2}\text{-OH}_2 \Leftrightarrow \text{E-Zn}^{2+}\text{-OH}^- + H^+ \qquad (2.11)$$

The rate-limiting step in catalysis is the second reaction, that is, the proton transfer that regenerates the zinc hydroxide species of the enzyme.[1–8,15,16] In the catalytically very active isozymes, such as CA II, IV, VII, and IX, the process is assisted by a histidine residue placed at the entrance of the active site (His64), as well as by a cluster of histidines (Fig. 2.2), which protrudes from the rim of the active site to the surface of the enzyme, thus ensuring a very efficient proton transfer process for the most efficient CA isozyme, CA II.[16] This also explains why CA II is one of the most active enzymes known (with a $k_{cat}/K_m = 1.5 \times 10^8\,M^{-1}\,s^{-1}$), approaching the limit of diffusion control, and has important consequences for the design of inhibitors with clinical applications.[1–3,15,16]

Two main classes of CA inhibitors (CAIs) are known: the metal complexing anions and the unsubstituted sulfonamides and their bioisosteres (sulfamates, sulfamides, etc.), which bind to the Zn(II) ion of the enzyme either by substituting the nonprotein zinc ligand (eq. 2.12 in Fig. 2.4) or by adding to the metal coordination sphere (eq. 2.13 in Fig. 2.4), generating trigonal bipyramidal species.[1–3,17–24] Sulfonamides bind in a tetrahedral geometry of the Zn(II) ion (Fig. 2.4), in deprotonated state, with the nitrogen atom of the sulfonamide moiety coordinated to Zn(II) and an extended network of hydrogen bonds, involving residues Thr199 and Glu106, also participating in the anchoring of the inhibitor molecule to the metal ion. The aromatic/heterocyclic part of the inhibitor (R) interacts with hydrophilic and hydrophobic residues of the cavity. Anions may bind either in tetrahedral geometry of the metal ion or in trigonal bipyramidal adducts, such as the thiocyanate adduct shown in Fig. 2.4b.[1–3,20]

There are at least 30 clinically used drugs/agents in clinical development reported to possess significant CA inhibitory properties (compounds **2.1–2.25**).[1] Many other such derivatives belonging to the sulfonamide, sulfamate, or sulfamide classes are constantly reported, being designed and synthesized by means of rational drug design processes.[25–34]

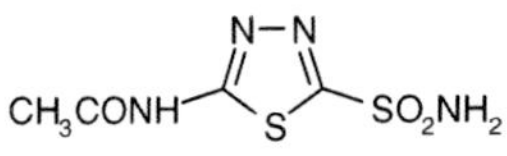

2.1

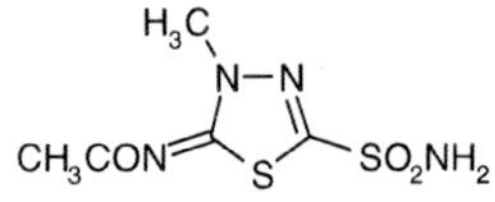

2.2

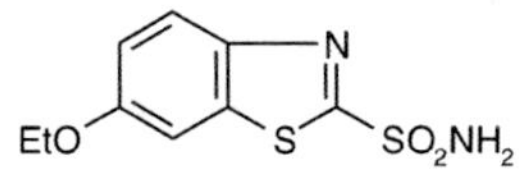

2.3

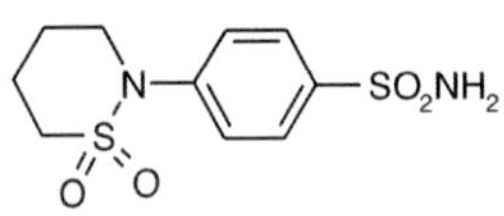

2.4

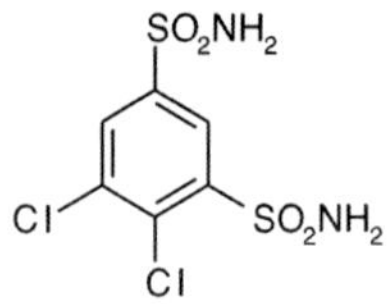

2.5

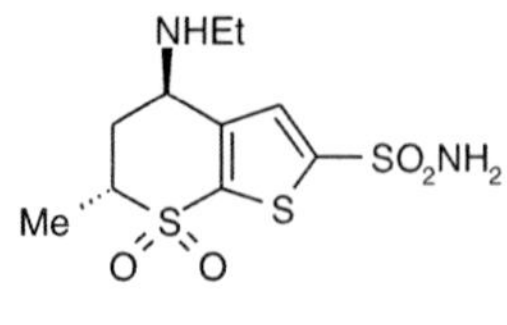

2.6

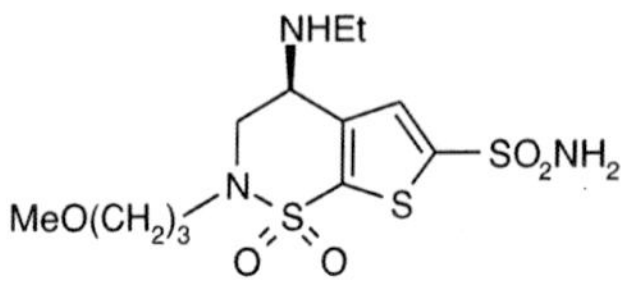

2.7

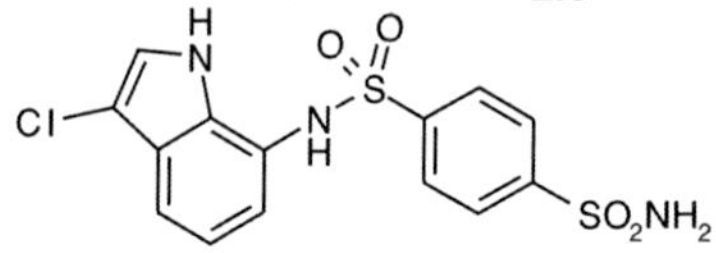

2.8

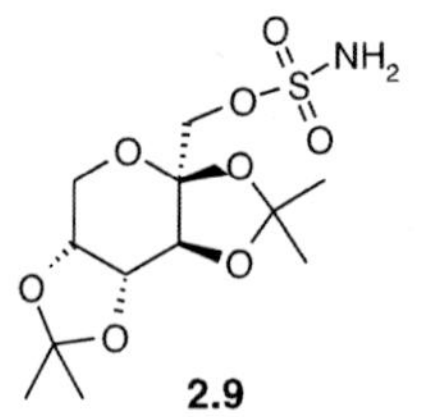

2.9

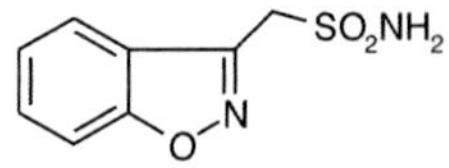

2.10

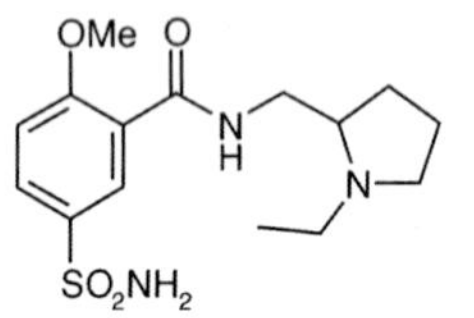

2.11

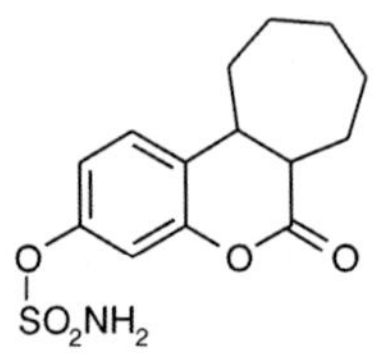

2.12

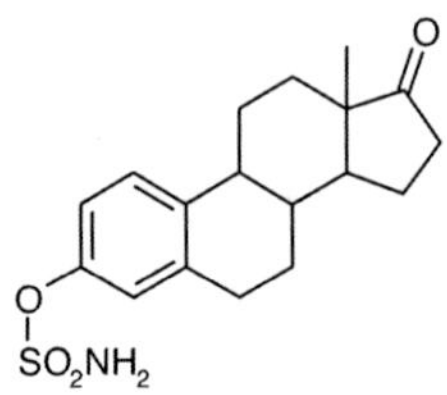

2.13

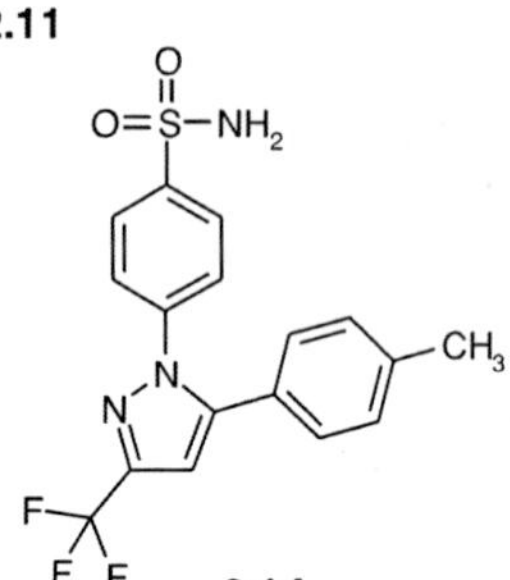

2.14

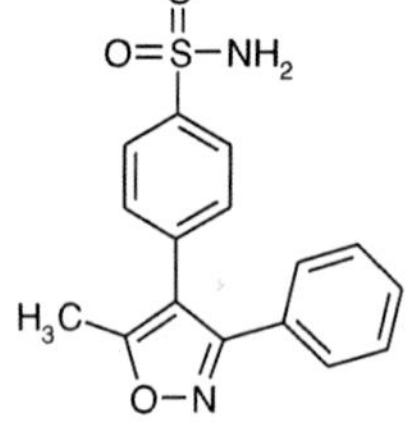

2.15

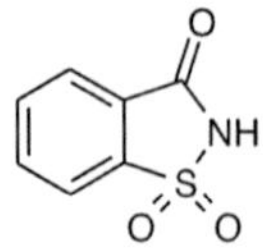

2.16

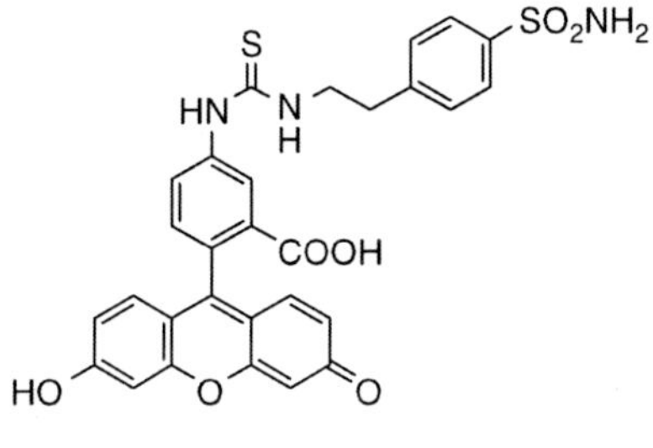

2.17

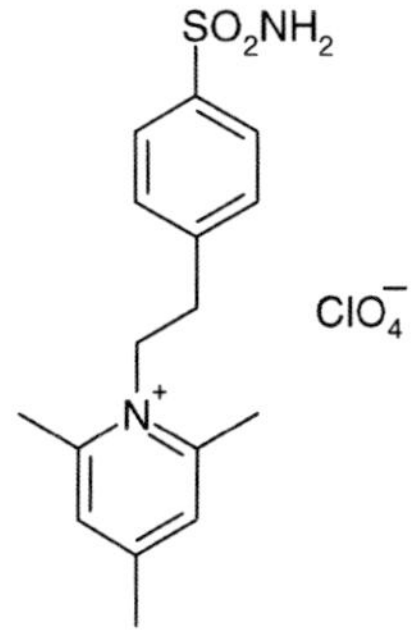

2.18

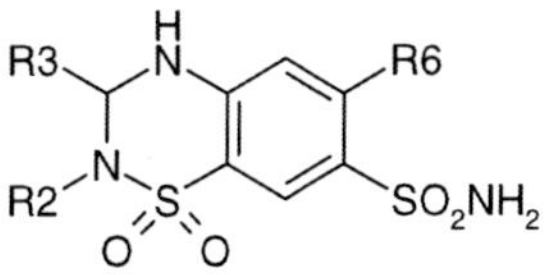

2.19

a: R2 = R3 = H, R6 = Cl, hydrochlorothiazide
b: R2 = R3 = H, R6 = CF_3, hydroflumethiazide
c: R2 = H, R3 = $PhCH_2$, R6 = CF_3, bendroflumethiazide
d: R2 = H, R3 = $CHCl_2$, R6 = Cl, trichloromethiazide
e: R2 = Me, R3 = $CH_2SCH_2CF_3$, R6 = Cl, polythiazide

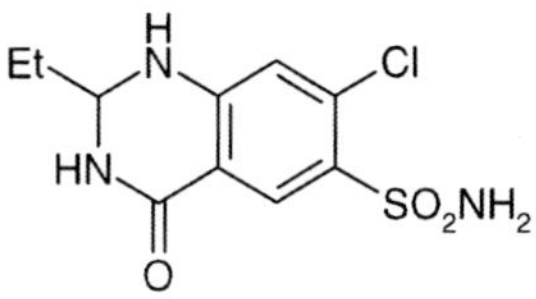

2.20

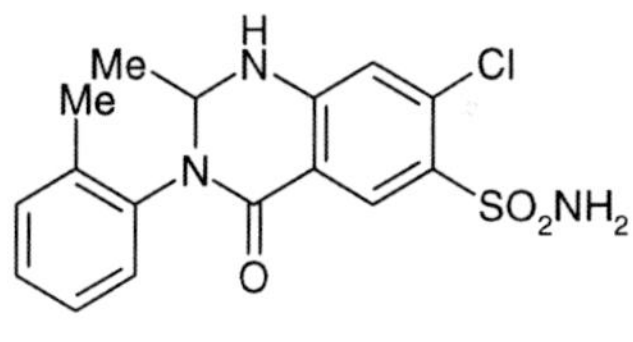

2.21

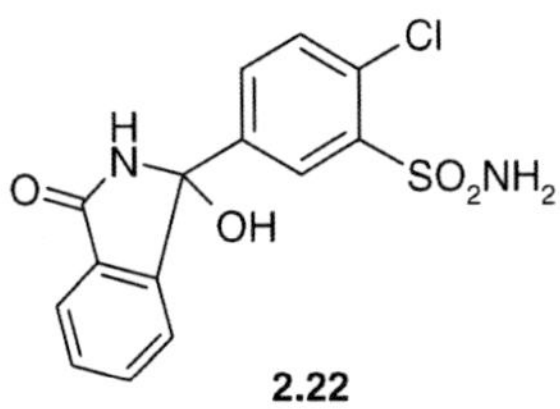

2.22

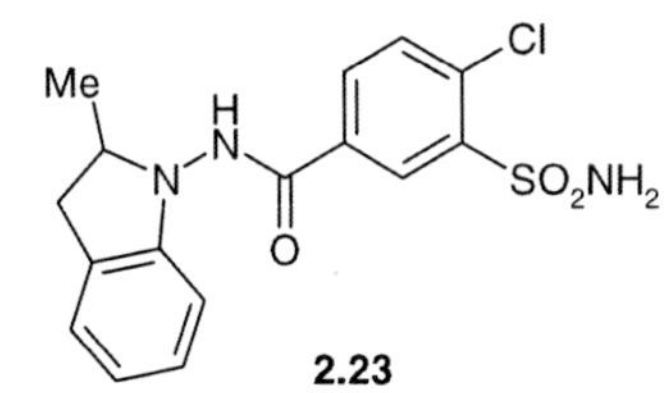

2.23

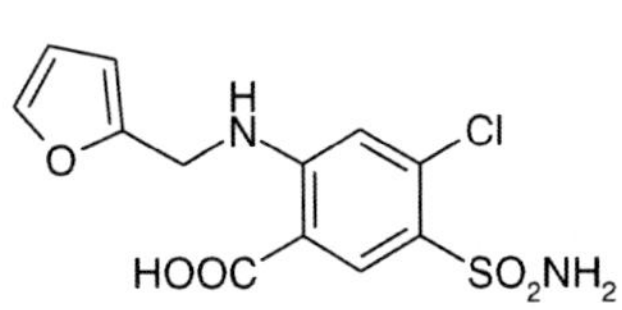

2.24

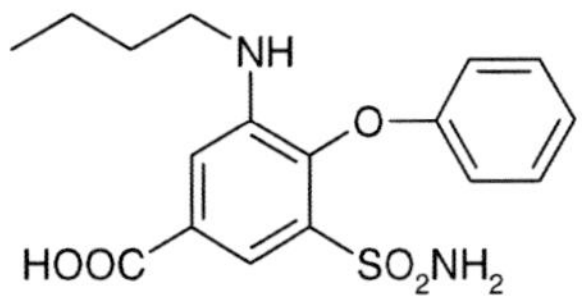

2.25

$$E\text{-}Zn^{2+}\text{-}OH_2 + I \Leftrightarrow E\text{-}Zn^{2+}\text{-}I + H_2O \qquad \text{(substitution)} \qquad (2.12)$$

Tetrahedral adduct

$$E\text{-}Zn^{2+}\text{-}OH_2 + I \Leftrightarrow E\text{-}Zn^{2+}\text{-}OH_2(I) \qquad \text{(addition)} \qquad (2.13)$$

Trigonal bipyramidal adduct

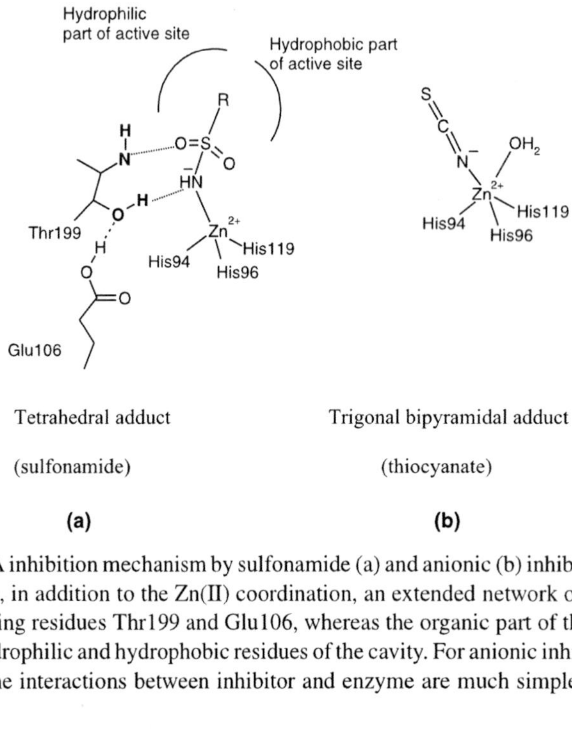

FIGURE 2.4 α-CA inhibition mechanism by sulfonamide (a) and anionic (b) inhibitors. In the case of sulfonamides, in addition to the Zn(II) coordination, an extended network of hydrogen bonds ensues, involving residues Thr199 and Glu106, whereas the organic part of the inhibitor (R) interacts with hydrophilic and hydrophobic residues of the cavity. For anionic inhibitors such as thiocyanate (b), the interactions between inhibitor and enzyme are much simpler.

Some of the clinically used compounds, such as acetazolamide (**2.1**), methazolamide (**2.2**), ethoxzolamide (**2.3**), sulthiame (**2.4**), and dichlorophenamide (**2.5**), are known for decades and were initially developed in the search of novel diuretics or antiepileptics in the 1950s and 1960s.[1,25–34] Although their diuretic use was not extensive,[1] it has been observed that such enzyme inhibitors may be employed for the systemic treatment of glaucoma.[1] Thus, many such drugs (e.g., acetazolamide, methazolamide, and dichlorophenamide) are still being used in ophthalmology (discussed subsequently), whereas two novel derivatives, dorzolamide (**2.6**) and brinzolamide (**2.7**), have been developed in the 1990s as topically acting antiglaucoma agents.[1] Inhibition data against all catalytically active human isozymes with some of the clinically used compounds are provided in Table 2.2.[1,25–34] These very diverse

TABLE 2.2 Inhibition Data with Some of the Clinically Used Sulfonamides/Sulfamates 2.1–2.25 Against Isozymes I–XIV (the Isoforms CA VIII, X, and XI are Devoid of Catalytic Activity and Probably Do Not Bind Sulfonamides as They Do Not Contain Zn(II) Ions)

	K_I (nM)																								
Isozyme[a]	2.1	2.2	2.3	2.4	2.5	2.6	2.7	2.8	2.9	2.10	2.11	2.12	2.13	2.14	2.15	2.16	2.17	2.18	2.19a	2.20	2.21	2.22	2.23	2.24	2.25
hCA I[b]	250	50	25	374	1200	50,000	45,000	31	250	56	12,000	3450	37	50,000	54,000	18,540	1300	4000	328	35,000	54,000	348	51,900	62	12,400
hCA II[b]	12	14	8	9	38	9	3	15	10	35	40	21	10	21	43	5950	45	21	290	1260	2000	138	2520	65	3050
hCA III[b]	2×10^5	7×10^5	1×10^6	6.3×10^5	6.8×10^5	7.7×10^5	1.1×10^5	10,400	7.8×10^5	2.2×10^6	10,600	7.0×10^4	6.5×10^5	7.4×10^4	7.8×10^4	1.0×10^6	1.3×10^6	3.1×10^5	7.9×10^5	nt[c]	6.1×10^5	1.1×10^4	2.3×10^5	3.2×10^6	nt
hCA IV[b]	74	6200	93	95	15,000	8500	3950	65	4900	8590	6.5×10^5	24	nt	880	1340	7920	650	60	427	nt	216	196	213	564	nt
hCA VA[b]	63	65	25	81	630	42	50	79	63	20	174	765	nt	794	912	10,060	134	88	4225	nt	750	917	890	499	nt
hCA VB[b]	54	62	19	91	21	33	30	23	30	6033	18	720	nt	93	88	7210	76	70	603	nt	312	9	274	322	nt
hCA VI[b]	11	10	43	134	79	10	0.9	47	45	89	0.8	653	nt	94	572	935	145	65	3655	nt	1714	1347	1606	245	nt
hCA VII[b]	2.5	2.1	0.8	6	26	3.5	2.8	122	0.9	117	3630	23	nt	2170	3900	10	18	15	5010	nt	2.1	2.8	0.23	513	nt
hCA IX[d]	25	27	34	43	50	52	37	24	58[e]	5.1	46	34	30	16	27	103	24	14	367	nt	320	23	36	420	nt
hCA XII[d]	5.7	3.4	22	56	50	3.5	3.0	3.4	3.8	11,000	3.9	12	7.5	18	13	633	5	7	355	nt	5.4	4.5	10	261	nt
mCA XIII[b]	17	19	50	1450	23	18	10	11	47	430	295	1050	nt	98	425	12,100	76	21	3885	nt	15	15	13	550	nt
hCA XIV[b]	41	43	25	1540	345	27	24	106	1460	5250	110	755	nt	689	107	773	33	13	4105	nt	5432	4130	4950	52	nt

[a] h: human; m: murine isozyme.
[b] Full-length enzyme.
[c] nt: not tested, data not available.
[d] Catalytic domain.
[e] The data against the full-length enzyme are of 1590 nM.

inhibition profiles of the various isozymes with derivatives **2.1–2.25** explain the very different clinical applications of the CAIs, ranging from diuretics and antiglaucoma agents to anticancer, antiobesity, and antiepileptic drugs.[1–8,25–34]

X-ray crystallographic structures are available for many adducts of sulfonamide/sulfamate/sulfamide inhibitors with isozymes CA I, II, IV, XII, XIII, and XIV.[16–34] In all these adducts, the deprotonated sulfonamide/sulfamate/sulfamide is coordinated to the Zn(II) ion of the enzyme, and its NH moiety participates in a hydrogen bond with the Oγ of Thr199, which in turn is engaged in another hydrogen bond to the carboxylate group of Glu106.[16–34] One of the oxygen atoms of the SO_2NH moiety also participates in a hydrogen bond with the backbone NH moiety of Thr199. Examples of various adducts of such inhibitors with CA II are provided for sulfonamide, sulfamate, and sulfamide inhibitors (Figs 2.5–2.7). It is obvious that all these zinc binding functions (ZBFs) are effective in designing potent CAIs.[16–34]

The different types of interactions by which a sulfonamide CAI achieves very high affinity (in the low nanomolar range) for the CA active site are illustrated in Figs 2.5–2.7 for sulthiame (**2.4**)[34] as sulfonamide inhibitor, a steroid sulfamate CA inhibitor, EMATE (**2.13**),[21] and the sulfamide analogue of the antiepileptic drug topiramate (**2.9**).[22] It may be observed that for these complexes, the ionized ZBF moiety (at the nitrogen NH_2 group) has replaced the hydroxyl ion coordinated to Zn(II) in the native enzyme (Zn–N distance of around 1.8–2.0 Å), with the metal ion remaining in its stable tetrahedral geometry, being coordinated in addition to the sulfonamidate nitrogen, by the imidazole nitrogens of His94, His96, and His119. The proton of the coordinated sulfonamidate nitrogen atom (NH^-) always makes a

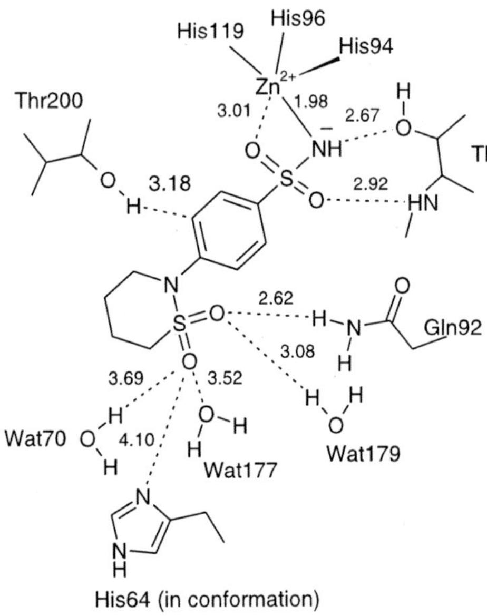

FIGURE 2.5 Schematic representation of the anticonvulsant sulfonamide sulthiame (**2.4**) bound within the hCA II active site (figures represent distances in angstroms).[34]

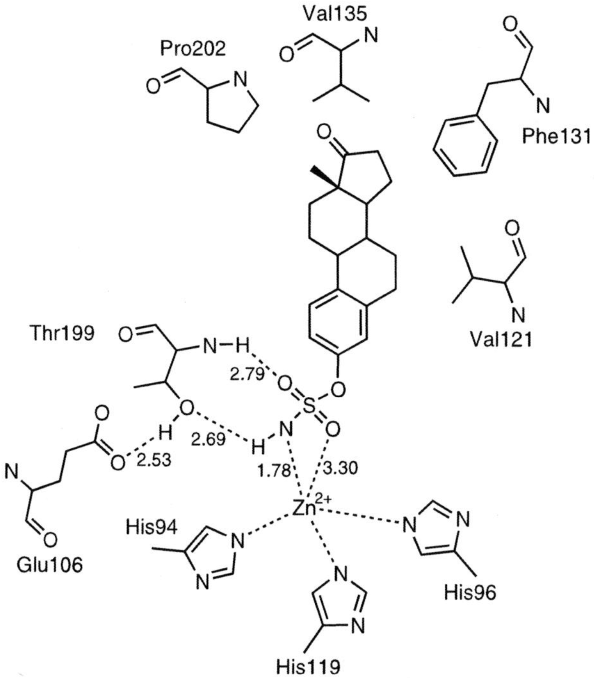

FIGURE 2.6 Binding of the sulfamate CA inhibitor EMATE (**2.13**) to hCA II (figures represent distances in angstroms).[21]

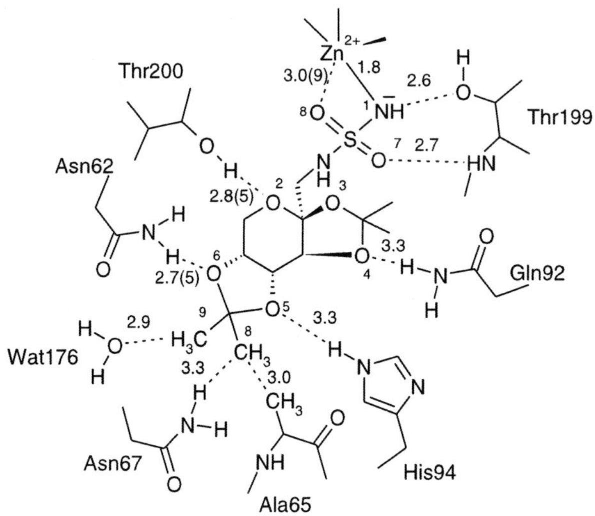

FIGURE 2.7 Binding of the sulfamide analogue of topiramate (**2.9**) to the hCA II-active site (figures represent distances in angstroms).[22]

hydrogen bond with the hydroxyl group of Thr199, which in turn accepts a hydrogen bond from the carboxylate of Glu106 (data not shown in Figs 2.5–2.7). One of the oxygen atoms of the ZBF moiety also makes a hydrogen bond with the backbone amide of Thr199, whereas the other one is semicoordinated to the catalytic Zn(II) ion (O–Zn distance of 3.0–3.3 Å). The scaffold of the inhibitors lies either in the hydrophilic (Fig. 2.5) or in the hydrophobic part (Fig. 2.6) of the active site cleft, or for bulkier derivatives may entirely fill it (Fig. 2.7), participating in extensive polar and van der Waals interactions with side chains of various amino acids lining these subsites (Figs 2.5–2.7).

As seen from the data in Table 2.2, most of the clinically used compounds **2.1–2.25** strongly inhibit many CA isozymes unselectively (such as CA I, II, IV, V, VII, IX, and XII–XIV), with affinities in the low nanomolar range for many of them, thus lacking selectivity toward the various isoforms with medicinal chemistry applications. The drug design of sulfonamide CAIs has recently been reviewed.[1,23] A major problem in the drug design of CAIs is obviously to obtain isozyme-selective inhibitors for the various clinical applications of these pharmacological agents.[1]

2.2.2 β-CAs

Many bacteria, archaea (such as *Methanobacterium thermoautotrophicum*), algae, and the chloroplasts of superior plants contain CAs belonging to the β-class.[4,5,35–42] The principal difference between these enzymes and the α-CAs already discussed consists in the fact that usually the β-CAs are oligomers, generally formed by 2–6 monomers of a molecular weight of 25–30 kDa. The X-ray structures of several such β-CAs are available at this moment: the enzyme isolated from the red alga *Porphyridium purpureum*, the enzyme from chloroplasts of *Pisum sativum*, another prokaryotic enzyme, this time isolated from *Escherichia coli*, and "cab," an enzyme isolated from the archaeon *M. thermoautotrophicum*.[35–39]

The *P. purpureum* CA monomer is composed of two internally repeating structures, folded as a pair of fundamentally equivalent motifs of an α/β-domain and three projecting α-helices. The motif is very distinct from that of either α- or γ-CAs. This homodimeric CA appeared like a tetramer with a pseudo 2-2-2 symmetry.[37] β-CAs are thus very different from the α-class enzymes. The Zn(II) ion is essential for catalysis in both families of enzymes, but its coordination is different and rather variable for the β-CAs; thus, in the prokaryotic β-CAs, the Zn(II) ion is coordinated by two cysteinate residues, an imidazole from a His residue, and a carboxylate from an Asp residue (Fig. 2.8a), whereas the chloroplast enzyme has the Zn(II) ion coordinated by the two cysteinates, the imidazole belonging to a His residue, and a water molecule (Fig. 2.8b).[35–38] The polypeptide chain folding and active site architecture are obviously very different from those of the CAs belonging to the α-class.

The structure of the β-CA from the dicotyledonous plant *P. sativum* at 1.93 Å resolution has also been reported.[36] The molecule assembles as an octamer with a novel dimer of dimers of dimers arrangement. The active site is located at the interface between two monomers, with Cys160, His220, and Cys223 binding the catalytic zinc ion and residues Asp162 (oriented by Arg164), Gly224, Gln151, Val184, Phe179, and

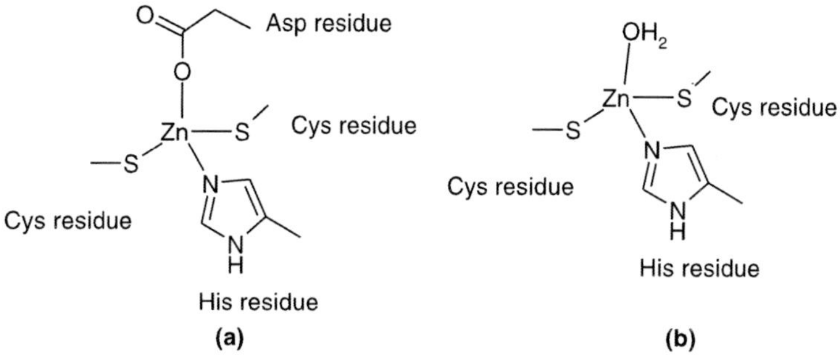

FIGURE 2.8 Schematic representation of the Zn(II) coordination sphere in β-CAs: (a) *P. purpureum*[37] and *E. coli*[35] enzymes; (b) *P. sativum* chloroplast and *M. thermoautotrophicum* enzyme,[36,38] as determined by X-ray crystallography.

Tyr205 interacting with acetic acid. The substrate binding groups have a one-to-one correspondence with the functional groups in the α-CA active site, with the corresponding residues being closely superimposable by a mirror plane. Therefore, despite differing folds, α- and β-CAs have converged upon a very similar active site design and are likely to share a common mechanism of action.[36]

Cab exists as a dimer with a subunit fold similar to that observed in plant-type β-CAs. The active site zinc ion was shown to be coordinated by the amino acid residues Cys32, His87, and Cys90, with the tetrahedral coordination completed by a water molecule.[39] The major difference between plant- and cab-type β-CAs is in the organization of the hydrophobic pocket (except for the zinc coordination already mentioned). The structure also revealed a HEPES buffer molecule bound 8 Å away from the active site zinc, which suggests a possible proton transfer pathway from the active site to the solvent.[39] No structural data are available at this moment regarding the binding of inhibitors to this type of CAs, except for the fact that acetate coordinates to the Zn(II) ion of the *P. sativum* enzyme.[36]

Since no water is directly coordinated to Zn(II) for some members of the β-CAs (Fig. 2.8a), the main problem is whether the zinc hydroxide mechanism presented in this chapter for the α-CAs is valid also for enzymes belonging to the β-family? A response to this question has been given recently by Nishimori et al., who proposed the catalytic mechanism shown in Fig. 2.9 for the enzyme isolated from *Helicobacter pylori* (hpβ-CA).[42]

Thus, it has been recently showed that the β-CA in the active site is "blocked" at pH 7.5 or lower, when the carboxylate of an aspartic acid coordinates as the fourth ligand to the Zn(II) ion (as in Fig. 2.9a).[42] However, at pH values over 8.3, an opening of the active site occurs, with the blocking aspartate forming a salt bridge with a conserved Arg residue in all β-CAs sequenced so far (Arg46 in the case of hpβ-CA), so that a water molecule/hydroxide ion has finally access to coordinate the metal ion for completion of its tetrahedral geometry. The catalytic mechanism of this β-CA possessing the "opened" active site is then rather identical to that of the α-class enzymes, with the substrate being probably bound in a hydrophobic pocket not far from the zinc-coordinated hydroxide (Fig. 2.9b), which attacks it with the formation of

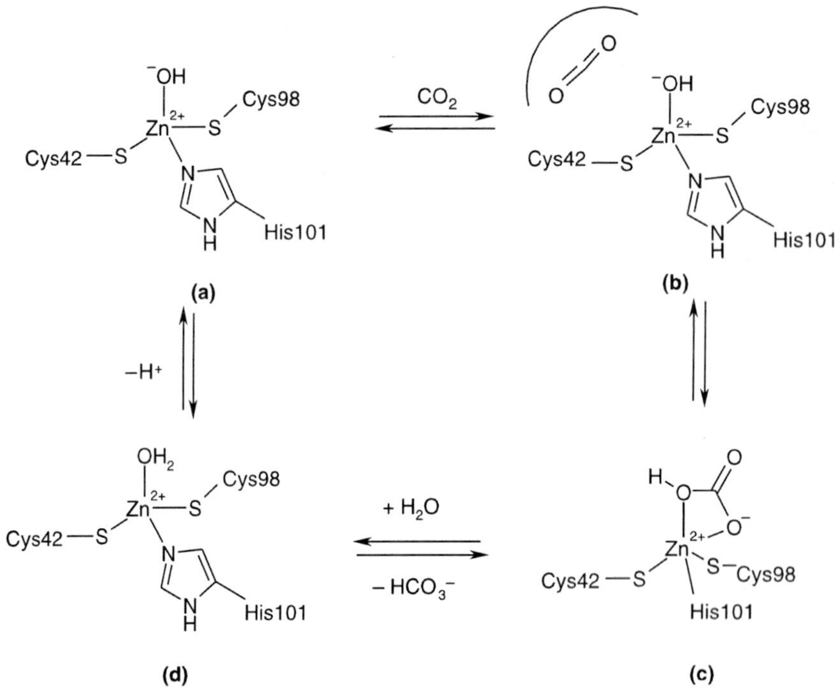

FIGURE 2.9 Proposed catalytic mechanism for prokaryotic β-CA from *H. pylori*.[42]

bidentately coordinated bicarbonate (Fig. 2.9c). This is then displaced by a water molecule and liberated in solution, with the formation of the acidic form of the enzyme, with water as the fourth zinc ligand (Fig. 2.9d). For generating the strong nucleophile with hydroxide coordinated to Zn(II), a proton transfer reaction must occur in the last step, with the formation of the catalytically active enzyme species (Fig. 2.9a). The proton transfer step is not well investigated until now in the β-class enzymes, this process being assisted by active site His residues in the α-CAs (e.g., His64 in hCA II and similar enzymes). This study was also the proof of concept showing that a β-class CA is a druggable target, since a very large number of sulfonamide and sulfamate inhibitors were tested as inhibitors, with some quite active derivatives being detected.[42]

2.2.3 γ-CAs

The prototype of the γ-class CAs, "Cam," has been isolated from the methanogenic archaeon *Methanosarcina thermophila*.[4,43–45] The crystal structures of zinc-containing and cobalt-substituted Cam were reported in the unbound form and cocrystallized with sulfate or bicarbonate.[43,44] Cam has several features that differentiate it from the α- and β-CAs. Thus, the protein fold is composed of a left-handed β-helix motif interrupted by three protruding loops and followed by short and long α-helices. The Cam monomer self-associates in a homotrimer with the approximate molecular weight of 70 kDa.[4,43,44] The Zn(II) ion within the active site is coordinated by three histidine residues, as in α-CAs, but is relative to the tetrahedral coordination geometry seen at the active site of α-CAs; the active site of this γ-CA contains additional metal-bound water ligands so that the overall coordination geometry is trigonal bipyramidal for the zinc-containing Cam and octahedral for the cobalt-substituted enzyme. Two of the His residues coordinating the metal ion belong to one monomer (monomer A) whereas the third one is from the adjacent monomer (monomer B). Thus, the three active sites are located at the interface between pairs of monomers.[4,43,44] The catalytic mechanism of γ-CAs was proposed to be similar with the one presented for the α-class enzymes. Still, the finding that Zn(II) is not tetracoordinated as originally reported but penta-coordinated,[43,44] with two water molecules bound to the metal ion, demonstrates that much is yet to be understood about these enzymes. At this moment, the zinc hydroxide mechanism is accepted as being valid for γ-CAs, as it is probable that an equilibrium exists between the trigonal bipyramidal and the tetrahedral species of the metal ion from the active site of the enzyme.

Ligands bound to the active site were shown to make contacts with the side chain of Glu62 in a manner that suggests this side chain to be probably protonated. In the uncomplexed zinc-containing Cam, the side chains of Glu62 and Glu84 appear to share a proton; in addition, Glu84 exhibits multiple conformations. This suggests that Glu84 may act as a proton shuttle, which is an important aspect of the reaction mechanism of α-CAs, for which a histidine active site residue generally plays this function, usually His64. Anions and sulfonamides were shown to bind to Cam, sometimes with high affinity, making the γ-CAs druggable targets for various biomedical or environmental applications.[40,45]

2.2.4 δ-CAs

X-ray absorption spectroscopy at the Zn *K*-edge indicates that the active site of the marine diatom *Thalassiosira weissflogii* CA (TWCA1) is strikingly similar to that of mammalian α-CAs. The zinc has three histidine ligands and a single water molecule, being quite different from the β-CAs of higher plants in which zinc is coordinated by two cysteine thiolates, one histidine, and a water molecule.[7,46] The diatom carbonic anhydrase shows no significant sequence similarity with other carbonic anhydrases and may represent an example of convergent evolution at the molecular level. In the same diatom, a rather perplexing discovery has been then made: the first cadmium-containing enzyme, which is a CA-type protein.[7] The marine diatom *T. weissflogii* growing under conditions of low zinc, typical of the marine environment, and in the presence of cadmium salts, led to increased levels of cellular CA activity, although the levels of TWCA1, the major intracellular Zn-requiring isoform of CA in *T. weissflogii*, remained low.[7] ^{109}Cd labeling comigrates with a protein band that showed this CA activity to be distinct from TWCA1 on native PAGE of radiolabeled *T. weissflogii* cell lysates. The levels of the Cd protein were modulated by CO_2 in a manner that was shown to be consistent with a role for this enzyme in carbon acquisition. Purification of the CA-active fraction leads to the isolation of a Cd-containing protein of 43 kDa, showing that *T. weissflogii* expresses a Cd-specific CA, which, particularly under conditions of Zn limitation, can replace the Zn enzyme TWCA1 in its carbon-concentrating mechanism.[7,46]

2.3 DISTRIBUTION OF CARBONIC ANHYDRASES

CAs were shown to be present in a multitude of prokaryotes, where these enzymes play important functions such as respiration, transport of carbon dioxide, and photosynthesis.[1–10,47–52] The possibility to develop CA inhibitor-based antibiotics/antifungals, by inhibiting bacterial/fungal CAs present in pathogenic species, raised much interest recently.[1,40–42,47–49] Thus, Muhlschlegel's group[41] investigated the ascomycete *Candida albicans*, the most common fungal pathogen in immunocompromised patients, for its ability to change morphology, from yeast to filamentous forms, in response to host environmental cues. Filamentation of this fungus is mediated by second messengers such as cyclic adenosine 3′,5′-monophosphate (cAMP) synthesized by adenylyl cyclase. The distantly related basidiomycete *Cryptococcus neoformans* is an encapsulated yeast that predominantly infects the central nervous system of the immunocompromised patients.[41] Similar to the morphological change in *C. albicans*, capsule biosynthesis in *C. neoformans* was shown to depend upon adenylyl cyclase activity. In a seminal paper, Muhlschlegel's group[41] demonstrated that physiological concentrations of CO_2/bicarbonate, formed as a result of the reaction catalyzed by CAs present in these organisms, induce filamentation in *C. albicans* by direct stimulation of adenylyl cyclase activity. Furthermore, the CO_2/bicarbonate equilibration by such CAs (belonging to the β-CA family) is essential for pathogenesis of *C. albicans* in niches where the available CO_2 is limited. The adenylyl cyclase from *C. neoformans* was also shown to be sensitive to physiological concentrations of CO_2/bicarbonate. Such data were demonstrated by using CA inhibitors, among others. Thus, the link between cAMP signaling and CO_2/bicarbonate sensing is conserved in fungi, and it was revealed that CO_2 sensing is an important mediator of fungal pathogenesis. Novel therapeutic agents, based on the inhibition of CAs present in these pathogens, could target such pathways at several levels to control fungal infections.[41]

H. pylori, a Gram-negative neutralophile discovered in the early 1980s, was shown to be associated with chronic gastritis, peptic ulcers, and, more recently, gastric cancer, the second most common tumor in humans.[42,47,48] Our group cloned and sequenced *H. pylori* α-class carbonic anhydrase (hpCA) from patients with different gastric mucosal lesions, including gastritis, ulcer, and cancer.[42,47,48] Although several polymorphisms such as ^{12}Ala, ^{13}Thr, ^{16}Ile, and ^{168}Phe were newly identified, there was no significant relevance of any polymorphism to gastric mucosal lesion types. A library of sulfonamides/sulfamates has been investigated for the inhibition of hpCA, whereas new derivatives have been obtained by attaching 4-*tert*-butyl-phenylcarboxamido/sulfonamido tails to benzenesulfonamide/1,3,4-thiadiazole-2-sulfonamide scaffolds. All types of activities for the inhibition of hpCA have been detected. Dorzolamide and simple 4-substituted benzenesulfonamides were weak inhibitors (K_I values of 873–4360 nM).

Sulfanilamide, orthanilamide, some of their derivatives, and indisulam showed better activity (K_I values of 413–640 nM), whereas most of the clinically used inhibitors such as methazolamide, ethoxzolamide, dichlorophenamide, brinzolamide, topiramate, zonisamide, and so on acted as medium potency inhibitors (K_I values of 105–378 nM). Some potent hpCA inhibitors were also detected (K_I values of 12–84 nM) among acetazolamide, 4-amino-6-chloro-1,3-benzenedisulfonamide, and some newly designed compounds incorporating lipophilic tails. Some of the newly prepared derivatives had selectivity ratios for inhibiting hpCA over hCA II in the range of 1.25–3.48, thus showing some selectivity for inhibiting the bacterial enzyme. Since hpCA is essential for the survival of the pathogen in acid, it might be used as a new pharmacologic tool in the management of drug-resistant *H. pylori*.[42,47,48]

The report of parasitic CAs by Krungkrai et al.[49] who discovered the presence of several different α-CAs in *Plasmodium falciparum*, one of the malaria-provoking protozoa, opens new vistas of the development of pharmacological agents based on CA inhibitors. Red cells infected by *P. falciparum* contained CA amounts approximately two times higher than those of normal red cells. The enzyme was then purified to homogeneity, showing a M_r of 32 kDa, being active in monomeric form. The parasite enzyme activity was sensitive to well-known sulfonamide-CA inhibitors such as sulfanilamide and acetazolamide.[49] A series of aromatic sulfonamides, most of which were Schiff's bases derived from sulfanilamide/homosulfanilamide/4-aminoethylbenzenesulfonamide and substituted aromatic aldehydes, or ureido-substituted sulfonamides, were investigated for *in vitro* inhibition of the malarial parasite enzyme (pfCA) and the growth of *P. falciparum*. Several inhibitors with affinity in the micromolar range (K_I values in the range of 0.080–1.230 μM) were detected, whereas the most potent of such derivatives were the clinically used sulfonamide CA inhibitor acetazolamide and 4-(3,4-dichlorophenylureidoethyl)-benzenesulfonamide, an inhibitor four times more effective than acetazolamide (K_I of 315 nM), which showed an inhibition constant of 80 nM against pfCA. The lipophilic 4-(3,4-dichlorophenylureidoethyl)-benzenesulfonamide was also an effective *in vitro* inhibitor for the growth of *P. falciparum* (IC_{50} of 2 μM), whereas acetazolamide achieved the same level of inhibition at 20 μM. This was the first study to prove that antimalarials possessing a novel mechanism of action can be obtained by inhibiting a critical enzyme for the life cycle of the parasite.[49] Indeed, by inhibiting pfCA, the synthesis of pyrimidines mediated by carbamoyl phosphate synthase is impaired in *P. falciparum* but not in the human host. Sulfonamide CA inhibitors have the potential for the development of novel antimalarial drugs.[49]

In higher plants, algae, and cyanobacteria, all members of the four CA families were found to be present.[5,50] For example, analysis of the *Arabidopsis* genome revealed that at least 14 different CAs are present in this plant, whereas in the unicellular green alga *Chlamydomonas reinhardtii*, 6 such enzymes are present.[5,50] In algae, CAs were found in mitochondria, chloroplast thylakoid, cytoplasm, and periplasmic space.[5] In C_3 dicotyledons, two types of CAs have been isolated, one in the chloroplast stroma and one in cytoplasm, whereas in C_4 plants, these enzymes are present in the mesophyll cells, where they provide bicarbonate to phosphoenolpyruvate (PEP) carboxylase, the first enzyme involved in CO_2 fixation into C_4 acids.[5] In CAM (crassulacean acid metabolism) plants, CAs probably present in the cytosol are quite abundant in chloroplasts, where they participate in CO_2 fixation, providing bicarbonate to PEP carboxylase[5]. These enzymes are quite abundant in the terrestrial vegetation and seem to be correlated with the content of atmospheric CO_2 and thus with the global warming processes.

In animals, and more specifically mammals, CAs are quite widespread (Table 2.1), as mentioned throughout this chapter. Since this field has recently been reviewed, the reader is advised to consult these particular papers for a detailed overview of the distribution and function of CAs in these organisms.[1,6,51,52]

CA activators (CAAs) have also been investigated in detail.[53–57] As the field has recently been reviewed, the reader is advised to consult these references.[58–60]

2.4 PHYSIOLOGICAL FUNCTIONS OF CARBONIC ANHYDRASES

It is not clear whether reactions catalyzed by CAs (Fig. 2.1), except for CO_2 hydration/bicarbonate dehydration, have physiological relevance.[2,3] Thus, at present, only reaction 2.1 is considered to be the physiological one in which these enzymes are involved.

In prokaryotes, as also shown in the preceding sections, CAs possess two general functions: (i) transport of CO_2/bicarbonate between different tissues of the organism and (ii) provision of CO_2/bicarbonate for enzymatic reactions. [4] In aquatic photosynthetic organisms, an additional role is that of a CO_2-concentrating mechanism, which helps overcome CO_2 limitation in the environment.[5,50] For example, in *C. reinhardtii*, this CO_2-concentrating mechanism is maintained by the pH gradient created across the chloroplast thylakoid membranes by photosystem II-mediated electron transport processes.[50] A large number of nonphotosynthetic prokaryotes catalyze reactions for which CA could be expected to provide CO_2/bicarbonate in the vicinity of the active site or to remove such compounds to improve the energetics of the reaction.[4] The large number of carboxylation/decarboxylation processes in which prokaryotic CAs may play such an important physiological function has recently been brilliantly reviewed by Smith and Ferry.[4]

In vertebrates, including *Homo sapiens*, the physiological functions of CAs have been thoroughly investigated over the past 70 years (Table 2.1).[1,6,51,52] Isozymes I, II, and IV have been found involved in respiration and regulation of the acid/base homeostasy.[1] These complex processes involve the transport of CO_2/bicarbonate between metabolizing tissues and excretion sites (lungs, kidneys), and these facilitate CO_2 elimination in capillaries and pulmonary microvasculature, elimination of H^+ ions in the renal tubules and collecting ducts, and reabsorption of bicarbonate in the brush border and thick ascending Henle loop in kidneys.[1] Usually, isozymes I, II, and IV are involved in these processes. By producing the bicarbonate-rich aqueous humor secretion (mediated by ciliary process isozymes CA II, IV, and XII) within the eye, CAs are involved in vision, and their misfunctioning leads to high intraocular pressure and glaucoma.[1] CA II is also involved in the bone development and function, such as the differentiation of osteoclasts or the provision of acid for bone resorption in osteoclasts. CAs are involved in the secretion of electrolytes in many other tissues/organs, such as CSF formation, by providing bicarbonate and by regulating the pH in the choroid plexus; saliva production in acinar and ductal cells; gastric acid production in the stomach parietal cells; bile production; pancreatic juice production; and intestinal ion transport.[1,51] CAs are also involved in gustation and olfaction, protection of gastrointestinal tract from extreme pH conditions (too acidic or too basic), regulation of pH and bicarbonate concentration in the seminal fluid, muscle functions, and adaptation to cellular stress. Some isozymes such as CA V are involved in molecular signaling processes, such as insulin secretion signaling in pancreas β-cells.[1,51] Isozymes II and VA are involved in important metabolic processes, as they provide bicarbonate for gluconeogenesis, fatty acids *de novo* biosynthesis, or pyrimidine base synthesis. Finally, some isozymes (such as CA IX and XII, CARP VIII) are quite abundant in tumors, as these are involved in oncogenesis and tumor progression.[6,51,52]

Although the physiological function of some mammal isozymes (CA I and III, the CARPs) is still unclear or is poorly understood, one may understand from the data already presented the importance of CAs for a host of physiological processes, in both normal and pathological states. This may explain why inhibitors of these enzymes found a place in clinical medicine as early as in 1954, with acetazolamide (**2.1**) being the first nonmercurial diuretic agent used clinically.[1] At present, inhibitors of these enzymes are widely used clinically as antiglaucoma agents, diuretics, and antiepileptics in the management of mountain sickness, gastric and duodenal ulcers, neurological disorders, or osteoporosis, among others.[1–3] The development of more specific agents is urgently needed because of the high number of isozymes present in the human body as well as the isolation of a large number of new representatives of the CAs in all kingdoms of living organisms. This is possible only by understanding in detail the catalytic and inhibition mechanisms of these enzymes. In fact, at present, much research is focused on at least five fronts in the drug design of pharmacological agents belonging to this class: (i) Antiglaucoma drugs with better profiles than dorzolamide and brinzolamide.[1] The target isozymes of such compounds are probably CA II and XII.[1] (ii) Anticancer drugs targeting primarily CA IX and XII, isozymes predominantly present in tumor cells.[6,32] (iii) Antiobesity agents, based on the reported effects of the strong CA inhibitors topiramate[17] and zonisamide,[18] which probably target the mitochondrial isoforms CA VA and/or CA VB. (iv) Anticonvulsants (probably targeting CA II, VII, XII, and XIV). (v) Antibacterials, antifungals, and other types of agents that target various CAs from pathogenic microorganisms such as the bacteria *H. pylori*, *Mycobacterium tuberculosis*, and so on; the protozoa *P. falciparum*; or the fungi *C. albicans* and *C. neoformans* among others.[40–42,47,48] The conclusion of this chapter is that these enzymes and their inhibitors are indeed remarkable: after many years of intense research in this field, these continue to offer interesting opportunities for the development of novel drugs and new diagnostic tools, or for understanding in greater depth the fundamental processes of the life sciences. The discovery of many such new CAs in various organisms all over the phylogenetic tree is also a clear signal that these ancient enzymes are involved in critical life processes and that perturbation of their activity may lead to novel ways of controlling widespread diseases.

ACKNOWLEDGMENTS

Research from the author's laboratory was supported in part by two EU grants of the sixth framework program (EUROXY and DeZnIT projects) and by a FP7 project (METOXIA).

REFERENCES

1. Supuran, C. T. Carbonic anhydrases: novel therapeutic applications for inhibitors and activators. *Nat. Rev. Drug. Discov.* **2008**, *7*, 168–181.
2. Supuran, C. T.; Scozzafava, A.; Conway, J., Eds. *Carbonic Anhydrase: Its Inhibitors and Activators;* CRC Press: Boca Raton, FL, **2004**; pp 1–363, and references cited therein.
3. (a) Supuran, C. T., Scozzafava, A.; Casini, A. Carbonic anhydrase inhibitors. *Med. Res. Rev.* **2003**, *23*, 146–189. (b) Krishnamurthy, V. M.; Kaufman, G. K.; Urbach, A. R.; Gitlin, I.; Gudiksen, K. L.; Weibel, D. B.; Whitesides, G. M. Carbonic anhydrase as a model for biophysical and physical-organic studies of proteins and protein–ligand binding. *Chem. Rev.* **2008**, *108*, 946–1051. (c) Stiti, M.; Cecchi, A.; Rami, M.; Abdaoui, M.; Barragan-Montero, V.; Scozzafava, A.; Guari, Y.; Winum, J. Y.; Supuran, C. T. Carbonic anhydrase inhibitor coated gold nanoparticles selectively inhibit the tumor-associated isoform IX over the cytosolic isozymes I and II. *J. Am. Chem. Soc.* **2008**, *130*, 16130–16131.
4. Smith, K. S.; Ferry, J. G. Prokaryotic carbonic anhydrases. *FEMS Microbiol. Rev.* **2000**, *24*, 335–366.
5. Badger, M. R.; Price, G. D. The role of carbonic anhydrase in photosynthesis. *Annu. Rev. Plant Physiol. Plant Mol. Biol.* **1994**, *45*, 369–392.
6. Thiry, A.; Dogné, J. M.; Masereel, B.; Supuran, C. T. Targeting tumor-associated carbonic anhydrase IX in cancer therapy. *Trends Pharmacol. Sci.* **2006**, *27*, 566–573.
7. (a) Xu, Y.; Feng, L.; Jeffrey, P. D.; Shi, Y.; Morel, F. M. Structure and metal exchange in the cadmium carbonic anhydrase of marine diatoms. *Nature* **2008**, *452*, 56–61. (b) Lane, T. W.; Morel, F. M. A biological function for cadmium in marine diatoms. *Proc. Natl. Acad. Sci. USA* **2000**, *97*, 4627–4631.
8. Hilvo, M.; Tolvanen, M.; Clark, A.; Shen, B.; Shah, G. N.; Waheed, A.; Halmi, P.; Hanninen, M.; Hamalainen, J. M.; Vihinen, M.; Sly, W. S.; Parkkila, S. Characterization of CA XV, a new GPI-anchored form of carbonic anhydrase. *Biochem. J.* **2005**, *392*, 83–92.
9. Nishimori, I.; Vullo, D.; Innocenti, A.; Scozzafava, A.; Mastrolorenzo, A.; Supuran, C. T. Carbonic anhydrase inhibitors. The mitochondrial isozyme VB as a new target for sulfonamide and sulfamate inhibitors. *J. Med. Chem.* **2005**, *48*, 7860–7866.
10. Nishimori, I.; Minakuchi, T.; Onishi, S.; Vullo, D.; Scozzafava, A.; Supuran, C. T. Carbonic anhydrase inhibitors. DNA cloning, characterization and inhibition studies of the human secretory isoform VI, a new target for sulfonamide and sulfamate inhibitors. *J. Med. Chem.* **2007**, *50*, 381–388.

11. Vullo, D.; Voipio, J.; Innocenti, A.; Rivera, C.; Ranki, H.; Scozzafava, A.; Kaila, K.; Supuran, C. T. Carbonic anhydrase inhibitors. Inhibition of the human cytosolic isozyme VII with aromatic and heterocyclic sulfonamides. *Bioorg. Med. Chem. Lett.* **2005**, *15*, 971–976.

12. Whittington, D. A.; Waheed, A.; Ulmasov, B.; Shah, G. N.; Grubb, J. H.; Sly, W. S.; Christianson, D. W. Crystal structure of the dimeric extracellular domain of human carbonic anhydrase XII, a bitopic membrane protein overexpressed in certain cancer tumor cells. *Proc. Natl. Acad. Sci. USA* **2001**, *98*, 9545–9550.

13. Lehtonen, J.; Shen, B.; Vihinen, M.; Casini, A.; Scozzafava, A.; Supuran, C. T.; Parkkila, A. K.; Saarnio, J.; Kivelä, A.; Waheed, A.; Sly, W. S.; Parkkila, S. Characterization of CA XIII, a novel member of the carbonic anhydrase isozyme family. *J. Biol. Chem.* **2004**, *279*, 2719–2727.

14. Di Fiore, A.; Monti, S. M.; Hilvo, M.; Parkkila, S.; Romano, V.; Scaloni, A.; Pedone, C.; Scozzafava, A.; Supuran, C. T.; De Simone, G. Crystal structure of human carbonic anhydrase XIII and its complex with the inhibitor acetazolamide. *Proteins* **2008**, *74*, 164–175.

15. Briganti, F.; Mangani, S.; Orioli, P.; Scozzafava, A.; Vernaglione, G.; Supuran, C. T. Carbonic anhydrase activators: X-ray crystallographic and spectroscopic investigations for the interaction of isozymes I and II with histamine. *Biochemistry* **1997**, *36*, 10384–10392.

16. Innocenti, A.; Scozzafava, A.; Parkkila, S.; Puccetti, L.; De Simone, G.; Supuran, C. T. Investigations of the esterase, phosphatase and sulfatase activities of the cytosolic mammalian carbonic anhydrase isoforms I, II and XIII with 4-nitrophenyl esters as substrates. *Bioorg. Med. Chem. Lett.* **2008**, *18*, 2267–2271.

17. Casini, A.; Antel, J.; Abbate, F.; Scozzafava, A.; David, S.; Waldeck, H.; Schäfer, S.; Supuran, C. T. Carbonic anhydrase inhibitors: SAR and X-ray crystallographic study for the interaction of sugar sulfamates/sulfamides with isozymes I, II and IV. *Bioorg. Med. Chem. Lett.* **2003**, *13*, 841–845.

18. De Simone, G.; Di Fiore A.; Menchise, V.; Pedone, C.; Antel, J.; Casini, A.; Scozzafava, A.; Wurl, M.; Supuran, C. T. Carbonic anhydrase inhibitors. Zonisamide is an effective inhibitor of the cytosolic isozyme II and mitochondrial isozyme V: solution and X-ray crystallographic studies. *Bioorg. Med. Chem. Lett.* **2005**, *15*, 2315–2320.

19. Menchise, V.; De Simone, G.; Alterio, V.; Di Fiore, A.; Pedone, C.; Scozzafava, A.; Supuran, C. T. Carbonic anhydrase inhibitors: stacking with Phe131 determines active site binding region of inhibitors as exemplified by the X-ray crystal structure of a membrane-impermeant antitumor sulfonamide complexed with isozyme II. *J. Med. Chem.* **2005**, *48*, 5721–5727.

20. Alterio, V.; Vitale, R. M.; Monti, S. M.; Pedone, C.; Scozzafava, A.; Cecchi, A.; De Simone, G.; Supuran, C. T. Carbonic anhydrase inhibitors: X-ray and molecular modeling study for the interaction of a fluorescent antitumor sulfonamide with isozyme II and IX. *J. Am. Chem. Soc.* **2006**, *128*, 8329–8335.

21. Abbate, F.; Winum, J. Y.; Potter, B. V.; Casini, A.; Montero, J. L.; Scozzafava, A.; Supuran, C. T. Carbonic anhydrase inhibitors: X-ray crystallographic structure of the adduct of human isozyme II with EMATE, a dual inhibitor of carbonic anhydrases and steroid sulfatase. *Bioorg. Med. Chem. Lett.* **2004**, *14*, 231–234.

22. Winum, J. Y.; Temperini, C.; El Cheikh, K.; Innocenti, A.; Vullo, D.; Ciattini, S.; Montero, J. L.; Scozzafava, A.; Supuran, C. T. Carbonic anhydrase inhibitors: clash with Ala65 as a

means for designing inhibitors with low affinity for the ubiquitous isozyme II, exemplified by the crystal structure of the topiramate sulfamide analogue. *J. Med. Chem.* **2006**, *49*, 7024–7031.

23. Supuran, C. T.; Casini, A.; Scozzafava, A. Development of sulfonamide carbonic anhydrase inhibitors. In *Carbonic Anhydrase: Its Inhibitors and Activators;* Supuran, C. T.; Scozzafava, A.; Conway, J., Eds.; CRC Press: Boca Raton, FL, **2004**; pp 67–148.
24. Pastorekova, S.; Parkkila, S.; Pastorek, J.; Supuran, C. T. Carbonic anhydrases: current state of the art, therapeutic applications and future prospects. *J. Enzyme Inhib. Med. Chem.* **2004**, *19*, 199–229.
25. Di Fiore, A.; Pedone, C.; D'Ambrosio, K.; Scozzafava, A.; De Simone, G.; Supuran, C. T. Carbonic anhydrase inhibitors: valdecoxib binds to a different active site region of the human isoform II as compared to the structurally related cyclooxygenase II "selective" inhibitor celecoxib. *Bioorg. Med. Chem. Lett.* **2006**, *16*, 437–442.
26. Scozzafava, A.; Briganti, F.; Ilies, M. A.; Supuran, C. T. Carbonic anhydrase inhibitors: synthesis of membrane-impermeant low molecular weight sulfonamides possessing *in vivo* selectivity for the membrane-bound versus cytosolic isozymes. *J. Med. Chem.* **2000**, *43*, 292–300.
27. Scozzafava, A.; Menabuoni, L.; Mincione, F.; Briganti, F.; Mincione, G.; Supuran, C. T. Carbonic anhydrase inhibitors: perfluoroalkyl/aryl-substituted derivatives of aromatic/heterocyclic sulfonamides as topical intraocular pressure-lowering agents with prolonged duration of action. *J. Med. Chem.* **2000**, *43*, 4542–4551.
28. Abbate, F.; Casini, A.; Owa, T.; Scozzafava, A.; Supuran, C. T. Carbonic anhydrase inhibitors: E7070, a sulfonamide anticancer agent, potently inhibits cytosolic isozymes I and II, and transmembrane, tumor-associated isozyme IX. *Bioorg. Med. Chem. Lett.* **2004**, *14*, 217–223.
29. Abbate, F.; Coetzee, A.; Casini, A.; Ciattini, S.; Scozzafava, A.; Supuran, C. T. Carbonic anhydrase inhibitors: X-ray crystallographic structure of the adduct of human isozyme II with the antipsychotic drug sulpiride. *Bioorg. Med. Chem. Lett.* **2004**, *14*, 337–341.
30. Weber, A.; Casini, A.; Heine, A.; Kuhn, D.; Supuran, C. T.; Scozzafava, A.; Klebe, G. Unexpected nanomolar inhibition of carbonic anhydrase by COX-2-selective celecoxib: new pharmacological opportunities due to related binding site recognition. *J. Med. Chem.* **2004**, *47*, 550–557.
31. Cecchi, A.; Hulikova, A.; Pastorek, J.; Pastorekova, S.; Scozzafava, A.; Winum, J. Y.; Montero, J. L.; Supuran, C. T. Carbonic anhydrase inhibitors. Design of fluorescent sulfonamides as probes of tumor-associated carbonic anhydrase IX that inhibit isozyme IX-mediated acidification of hypoxic tumors. *J. Med. Chem.* **2005**, *48*, 4834–4841.
32. Svastova, E.; Hulikova, A.; Rafajova, M.; Zat'ovicova, M.; Gibadulinova, A.; Casini, A.; Cecchi, A.; Scozzafava, A.; Supuran, C. T.; Pastorek, J.; Pastorekova, S. Hypoxia activates the capacity of tumor-associated carbonic anhydrase IX to acidify extracellular pH. *FEBS Lett.* **2004**, *577*, 439–445.
33. Köhler, K.; Hillebrecht, A.; Schulze Wischeler, J.; Innocenti, A.; Heine, A.; Supuran, C. T.; Klebe, G. Saccharin inhibits carbonic anhydrases: possible explanation for its unpleasant metallic aftertaste. *Angew. Chem. Int. Ed. Engl.* **2007**, *46*, 7697–7699.
34. Temperini, C.; Innocenti, A.; Mastrolorenzo, A.; Scozzafava, A.; Supuran, C. T. Carbonic anhydrase inhibitors. Interaction of the antiepileptic drug sulthiame with twelve mamma-

lian isoforms: kinetic and X-ray crystallographic studies. *Bioorg. Med. Chem. Lett.* **2007**, *17*, 4866–4872.

35. Cronk, J. D.; Endrizzi, J. A.; Cronk, M. R.; O'Neill, J. W.; Zhang, K. Y. Crystal structure of *E. coli* β-carbonic anhydrase, an enzyme with an unusual pH-dependent activity. *Protein Sci.* **2001**, *10*, 911–922.
36. Kimber, M. S.; Pai, E. F. The active site architecture of *Pisum sativum* β-carbonic anhydrase is a mirror image of that of α-carbonic anhydrases. *EMBO J.* **2000**, *19*, 2323–2330.
37. Mitsuhashi, S.; Mizushima, T.; Yamashita, E.; Yamamoto, M.; Kumasaka, T.; Moriyama, H.; Ueki, T.; Miyachi, S.; Tsukihara, T. X-ray structure of beta-carbonic anhydrase from the red alga, *Porphyridium purpureum*, reveals a novel catalytic site for CO_2 hydration. *J. Biol. Chem.* **2000**, *275*, 5521–5526.
38. Smith, K. S.; Ferry, J. G. A plant-type (β-class) carbonic anhydrase in the thermophilic methanoarchaeon *Methanobacterium thermoautotrophicum*. *J. Bacteriol.* **1999**, *181*, 6247–6253.
39. Strop, P.; Smith, K. S.; Iverson, T. M.; Ferry, J. G.; Rees, D. C. Crystal structure of the "cab"-type β-class carbonic anhydrase from the archaeon *Methanobacterium thermoautotrophicum*. *J. Biol. Chem.* **2001**, *276*, 10299–10305.
40. Zimmerman, S.; Innocenti, A.; Casini, A.; Ferry, J. G.; Scozzafava, A.; Supuran, C. T. Carbonic anhydrase inhibitors. Inhibition of the prokaryotic β and γ-class enzymes from Archaea with sulfonamides. *Bioorg. Med. Chem. Lett.* **2004**, *14*, 6001–6006.
41. Klengel, T.; Liang, W. J.; Chaloupka, J.; Ruoff, C.; Schroppel, K.; Naglik, J. R.; Eckert, S. E.; Mogensen, E. G.; Haynes, K.; Tuite, M. F.; Levin, L. R.; Buck, J.; Muhlschlegel, F. A. Fungal adenylyl cyclase integrates CO_2 sensing with cAMP signaling and virulence. *Curr. Biol.* **2005**, *15*, 2021–2026.
42. Nishimori, I.; Minakuchi, T.; Kohsaki, T.; Onishi, S.; Takeuchi, H.; Vullo, D.; Scozzafava, A.; Supuran, C. T. Carbonic anhydrase inhibitors: the beta-carbonic anhydrase from *Helicobacter pylori* is a new target for sulfonamide and sulfamate inhibitors. *Bioorg. Med. Chem. Lett.* **2007**, *17*, 3585–3594.
43. Kisker, C.; Schindelin, H.; Alber, B. E.; Ferry, J. G.; Rees, D. C. A left-handed beta-helix revealed by the crystal structure of a carbonic anhydrase from the archaeon *Methanosarcina thermophila*. *EMBO J.* **1996**, *15*, 2323–2330.
44. Iverson, T. M.; Alber, B. E.; Kisker, C.; Ferry, J. G.; Rees, D. C. A closer look at the active site of gamma-carbonic anhydrases: high resolution crystallographic studies of the carbonic anhydrase from *Methanosarcina thermophila*. *Biochemistry* **2000**, *39*, 9222–9231.
45. Innocenti, A.; Zimmerman, S.; Ferry, J. G.; Scozzafava, A.; Supuran, C. T. Carbonic anhydrase inhibitors. Inhibition of the zinc and cobalt γ-class enzyme from the archaeon *Methanosarcina thermophila* with anions. *Bioorg. Med. Chem. Lett.* **2004**, *14*, 3327–3331.
46. Cox, E. H.; McLendon, G. L.; Morel, F. M.; Lane, T. W.; Prince, R. C.; Pickering, I. J.; George, G. N. The active site structure of *Thalassiosira weissflogii* carbonic anhydrase 1. *Biochemistry* **2000**, *39*, 12128–12130.
47. Nishimori, I.; Minakuchi, T.; Morimoto, K.; Sano, S.; Onishi, S.; Takeuchi, H.; Vullo, D.; Scozzafava, A.; Supuran, C. T. Carbonic anhydrase inhibitors: DNA cloning and inhibition studies of the alpha-carbonic anhydrase from *Helicobacter pylori*, a new target for developing sulfonamide and sulfamate gastric drugs. *J. Med. Chem.* **2006**, *49*, 2117–2126.

48. Nishimori, I.; Vullo, D.; Minakuchi, T.; Morimoto, K.; Onishi, S.; Scozzafava, A.; Supuran, C. T. Carbonic anhydrase inhibitors: cloning and sulfonamide inhibition studies of a carboxyterminal truncated alpha-carbonic anhydrase from *Helicobacter pylori*. *Bioorg. Med. Chem. Lett.* **2006**, *16*, 2182–2188.

49. Krungkrai, J.; Scozzafava, A.; Reungprapavut, S.; Krungkrai, S. R.; Rattanajak, R.; Kamchonwongpaisan, S.; Supuran, C. T. Carbonic anhydrase inhibitors. Inhibition of *Plasmodium falciparum* carbonic anhydrase with aromatic sulfonamides: towards antimalarials with a novel mechanism of action? *Bioorg. Med. Chem.* **2005**, *13*, 483–489.

50. Park, Y.; Karlsson, J.; Rojdestvenski, I.; Pronina, N.; Klimov, V.; Oquist, G.; Samuelsson, G. Role of a novel photosystem II: associated carbonic anhydrase in photosynthetic carbon assimilation in *Chlamydomonas reinhardtii*. *FEBS Lett.* **1999**, *444*, 102–105.

51. Parkkila, S. An overview of the distribution and function of carbonic anhydrase in mammals. In *The Carbonic Anhydrases: New Horizons;* Chegwidden, W. R.; Edwards, Y.; Carter, N., Eds.; Birkhäuser Verlag: Basel, **2000**; pp 79–93.

52. Pastorek, J.; Pastorekova, S.; Callebaut, I.; Mornon, J. P.; Zelnik, V.; Opavsky, R.; Zatovicova, M.; Liao, S.; Portetelle, D.; Stanbridge, E. J.; Zavada, J.; Burny, A.; Kettmann, R. Cloning and characterization of MN, a human tumor-associated protein with a domain homologous to carbonic anhydrase and a putative helix-loop-helix DNA binding segment. *Oncogene* **1994**, *9*, 2877–2888.

53. Clare, B. W.; Supuran, C. T. Carbonic anhydrase activators. Part 3. Structure–activity correlations for a series of isozyme II activators. *J. Pharm. Sci.* **1994**, *83*, 768–779.

54. Temperini, C.; Innocenti, A.; Scozzafava, A.; Mastrolorenzo, A.; Supuran, C. T. Carbonic anhydrase activators: L-adrenaline plugs the active site entrance of isozyme II, activating better isoforms I, IV, VA, VII, and XIV. *Bioorg. Med. Chem. Lett.* **2007**, *17*, 628–635.

55. Temperini, C.; Scozzafava, A.; Vullo, D.; Supuran, C. T. Carbonic anhydrase activators. Activation of isozymes I, II, IV, VA, VII, and XIV with L- and D-histidine and crystallographic analysis of their adducts with isoform II: engineering proton-transfer processes within the active site of an enzyme. *Chemistry* **2006**, *12*, 7057–7066.

56. Temperini, C.; Scozzafava, A.; Vullo, D.; Supuran, C. T. Carbonic anhydrase activators. Activation of isoforms I, II, IV, VA, VII, and XIV with L- and D-phenylalanine and crystallographic analysis of their adducts with isozyme II: stereospecific recognition within the active site of an enzyme and its consequences for the drug design. *J. Med. Chem.* **2006**, *49*, 3019–3027.

57. Scozzafava, A.; Supuran, C. T. Carbonic anhydrase activators: human isozyme II is strongly activated by oligopeptides incorporating the carboxyterminal sequence of the bicarbonate anion exchanger AE1. *Bioorg. Med. Chem. Lett.* **2002**, *12*, 1177–1180.

58. Supuran, C. T.; Scozzafava, A. Activation of carbonic anhydrase isozymes. In *The Carbonic Anhydrases: New Horizons;* Chegwidden, W. R.; Carter, N.; Edwards, Y., Eds.; Birkhäuser Verlag: Basel, **2000**; pp 197–219.

59. Ilies, M; Scozzafava, A; Supuran, C. T. Carbonic anhydrase activators. In *Carbonic Anhydrase: Its Inhibitors and Activators;* Supuran, C. T.; Scozzafava, A.; Conway, J., Eds.; CRC Press: Boca Raton, FL, **2004**; pp 317–352.

60. Supuran, C. T.; Scozzafava, A. Carbonic anhydrase activators as potential anti-Alzheimer's disease agents. In *Protein Misfolding in Neurodegenerative Diseases: Mechanisms and Therapeutic Strategies;* Smith, H. J.; Simons, C.; Sewell, R. D. E.,Eds.; CRC Press: Boca Raton, FL, **2008**; pp 265–288.

CHAPTER 3

Zinc Binding Functions in the Design of Carbonic Anhydrase Inhibitors

JEAN-YVES WINUM[1], JEAN-LOUIS MONTERO[1], ANDREA SCOZZAFAVA[2], and CLAUDIU T. SUPURAN[2]

[1]Institut des Biomolécules Max Mousseron (IBMM) UMR 5247 CNRS-UM1-UM2 Bâtiment de Recherche Max Mousseron, Ecole Nationale Supérieure de Chimie de Montpellier, 8 rue de l'Ecole Normale, 34296 Montpellier Cedex, France

[2]Laboratorio di Chimica Bioinorganica, Università degli Studi di Firenze, Polo Scientifico, Room 188, Via della Lastruccia 3, 50019 Sesto Fiorentino (Florence), Italy

3.1 INTRODUCTION

Carbonic anhydrases (CAs, EC 4.2.1.1) constitute some of the most extensively studied zinc enzymes to date.[1–4] Belonging to a superfamily of ubiquitous zinc proteins present in prokaryotes and eukaryotes, CAs are encoded by five evolutionarily distinct, unrelated gene families: the α-class (present in vertebrates, bacteria, algae, and cytoplasm of green plants), the β-class (present predominantly in bacteria, algae, and chloroplasts), the γ-class (present mainly in archaea and some bacteria), and the δ- and ξ-classes (present in marine diatoms). The zinc ion is essential for the catalytic process of this very simple transformation, that is, the interconversion between carbon dioxide and bicarbonate with release of a proton. For the α-CA class, the metal ion is situated at the bottom of a 15 Å deep active cleft being coordinated by three histidine residues (His94, His96, and His119) and a water molecule/hydroxide ion in a tetrahedral geometry.[1–4]

Sixteen different isoforms have been described in mammalians, each differing in their relative hydrase activity, their subcellular localization, and their susceptibility to inhibition. Some of these catalytically active isozymes are critical in various fundamental physiological processes and have been clinically exploited for the treatment or prevention of various pathologies such as glaucoma, neurological disorders, and osteoporosis.[1] During the past few years, the discovery of CA isoforms implicated in pathological states has improved remarkably, and many carbonic anhydrase isoforms present in humans have been considered as targets for the drug design.[5] Several interesting studies have been made holding great promises for therapeutic use of carbonic anhydrase inhibitors, especially in oncology and in the treatment of obesity.[6]

The importance of developing isoform-specific CA inhibitors has recently emerged considering the undesired side effects caused by the first generation of inhibitors. Available clinical pharmacological agents are far from being perfect, as they possess many undesired side effects, mainly due to their lack of selectivity for the different isozymes. Development of isozyme-specific or at least organ-selective inhibitors would be highly beneficial both for obtaining novel types of drugs, devoid of major side effects, and for physiological studies in which specific/selective inhibitors may constitute valuable tools for understanding the physiology/physiopathology of these enzymes.[1,2] The catalytic and inhibition mechanisms of carbonic anhydrases are understood in great detail, and this was helpful for the design of potent inhibitors. From earlier studies, three different major basic structural elements have emerged as crucially important to the CA recognition pharmacophore. Modifications of the nature of the organic scaffold and tail moiety of the inhibitors to obtain tighter binding and eventually isozyme-specific CAIs have allowed to develop two general but complementary approaches:[2] (i) the "ring" approach, which consists of exploring a great variety of ring systems (aromatic or heteroaromatic) to which a sulfonamide ZBF is attached and is used for the discovery of the topical antiglaucoma agents dorzolamide and brinzolamide, and (ii) the "tail" approach, which consists of attaching different tails to the scaffolds of well-known aromatic/heterocyclic sulfonamides to modulate the physicochemical properties such as water solubility and enzyme binding capacity of these pharmacological agents.

Another type of structural variation was the nature of the anchor groups that can be incorporated into the inhibitors for coordinating the zinc ion. These zinc binding functions (ZBFs) or zinc binding groups tightly bind with the Zn^{2+} ion of the enzyme and the residues Thr199 and Glu106, two amino acids conserved in all CAs belonging to the α-class (Fig. 3.1). The ZBF is moreover directly linked to an aromatic or heterocyclic moiety that can interact inside the enzyme cavity with hydrophilic or hydrophobic residues. Inhibition of carbonic anhydrase is correlated with the coordination of the inhibitor molecule (in neutral or ionized state) to the catalytic metal ion with or without substitution of the metal-bound water molecule, which can also provide significant binding energy through supplementary interactions between such zinc binding functions and amino acid residues from the active site.

This key structural requirement was rarely taken into consideration, and until the beginning of the 1990s, sulfonamide moiety ($-SO_2NH_2$) was considered as zinc binding function above all else in the design of carbonic anhydrase inhibitors. Recently, several detailed studies regarding the possible modifications of the sulfonamide moiety compatible with the retention of strong binding to the enzyme have been reported.[7–9] In the past 10 years, much progress has been achieved in the conception of CAI, incorporating in their structure new anchor functions such as bioisosteric moieties of sulfonamide group, for example, sulfamate, sulfamide, and *N*-substituted

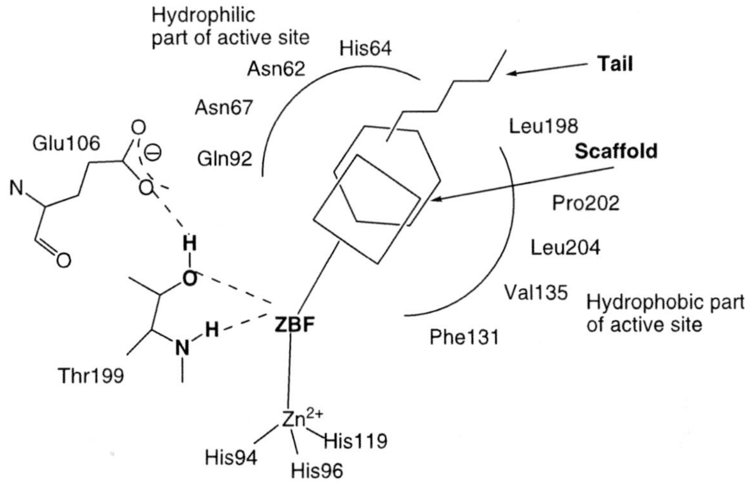

FIGURE 3.1 The general structure of an inhibitor complexed to the enzyme (α-CA) active site: ZBF, zinc binding function; the organic scaffold may be present or absent; the tail too. These structural elements interact with both the hydrophobic and hydrophilic halves of the active site (representative amino acid residues involved in binding are shown), whereas ZBF interacts with the Zn(II) ion as well as the neighboring residues Thr199 and Glu106.

sulfonamide/sulfamate/sulfamide directly linked to aromatic/heterocyclic or aliphatic moiety.[8,9] Other ZBGs have also emerged in this field as phosphate/phosphonate.[7]

Thus, exploration of new ZBGs is critical for the design of novel classes of CAIs to identify both new types of tight binding derivatives and compounds with a diverse inhibition profile as compared to the clinically used drugs, which generally indiscriminately inhibit many CA isoforms, thus leading to various side effects.

In this chapter, we will compile only the novel developments in the different zinc binding groups that were reported so far in the design of carbonic anhydrase inhibitors.

3.2 SULFONAMIDES AS CARBONIC ANHYDRASE INHIBITORS: LATEST DEVELOPMENTS

To date, the sulfonamide function ($R\text{-}SO_2NH_2$) is the most important and extensively used anchor group in the design of CAIs. First described in the case of sulfanilamide **3.1** by Mann and Keilin[10] in 1940, this ZBF was widely used in the conception of carbonic anhydrase inhibitors having important biomedical applications essentially as diuretic and antiglaucoma agents (acetazolamide **3.2**, methazolamide **3.3**, dichlorophenamide **3.4**, ethoxzolamide **3.5**, dorzolamide **3.6**, and brinzolamide **3.7**).[11] Some classical CAIs sulfonamides used in these pathologies are depicted in Fig. 3.2.

Researches in the field of sulfonamides as carbonic anhydrase inhibitors have been reviewed periodically, so the reader is directed to the previous literature.[1–4] Nevertheless, in this section, we will highlight the latest developments in this family of inhibitors since 2004.

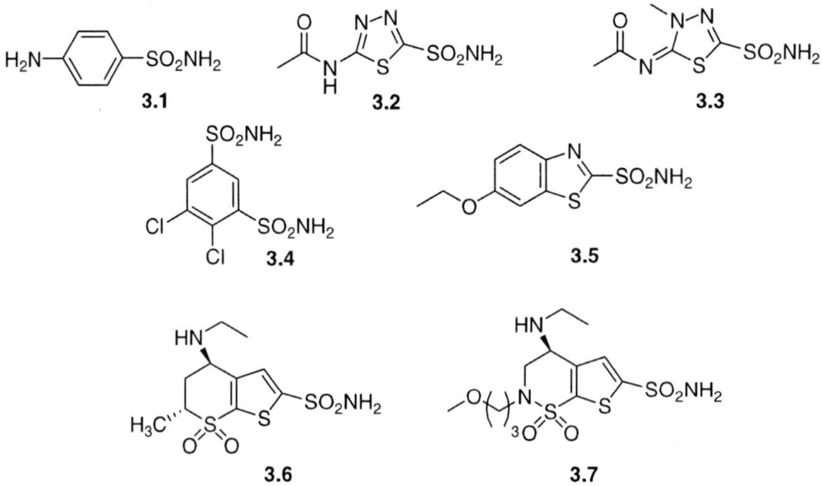

FIGURE 3.2 Some clinically used CAIs.

First of all, some sulfonamides, originally designed and described to act via another biochemical pathway, were revealed to be very good inhibitors of some CA isozymes. Indisulam **3.8** (Fig. 3.3), a sulfonamide antitumor drug discovered at Eisai Co., at present in clinical development for the treatment of solid tumors, is a cell cycle inhibitor and a potent CAI with inhibition constant against hCA II of 15 nM.[12] The antiepileptic zonisamide **3.9** was recently shown to possess significant inhibitory properties against many physiologically relevant CA isozymes (K_I(hCA II) = 35 nM).[13] The same was demonstrated for the antipsychotic sulpiride **3.10** (K_I(hCA II) = 40 nM).[14] The sulfonamide COX-2 inhibitors celecoxib **3.11**[15] and valdecoxib **3.12**[16] also act as potent inhibitors of many CA isozymes, and some of their clinical

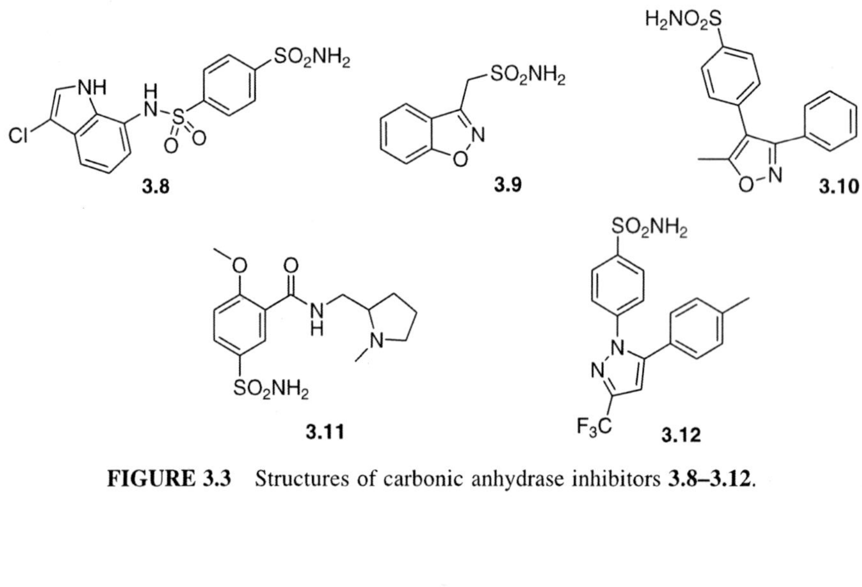

FIGURE 3.3 Structures of carbonic anhydrase inhibitors **3.8–3.12**.

applications (such as the prevention of some gastrointestinal tumors) are correlated with the strong inhibition of some CAs (**3.11**: K_I(hCA II) = 21 nM; **3.12**: K_I(hCA II) = 43 nM).

X-ray crystallographic structures are available for many adducts of sulfonamide inhibitors **3.8-3.12** with isozymes CA I, II, IV, and XIII.[12–16] In all these adducts, the classical sulfonamide ZBF binds in a tetrahedral geometry of the Zn(II) ion, in deprotonated state, with the nitrogen atom of the sulfonamide moiety coordinated to Zn(II). Moreover, this NH moiety also donates a hydrogen bond to the Oγ of Thr199, which in turn donates a hydrogen bond to the carboxylate group of Glu106. One of the oxygen atoms of the SO_2NH moiety also participates in a hydrogen bond with the backbone NH moiety of Thr199.[12–17] As seen from Fig. 3.1, extensive hydrophobic and van der Waals interactions between the heterocyclic/aromatic part of the inhibitor molecule and active site amino acid residues can assure a strong affinity of the inhibitor for the CA active site (which may arrive to the nanomolar range in some cases).[12–16] Even if numbers of inhibitors in sulfonamide series continue to be developed by different research groups in the past few years, in a more general way, the sulfonamide function has been used as zinc binding group model for the development of new concepts especially in the targeting of specific CA isoforms such as the transmembrane tumor-associated CA IX.[17] Thus, CA IX inhibitors activatable in hypoxic tumor have been the subject of a recent study[18] that reported sulfonamides incorporating 3,3′-dithiodipropionamide and 2,2′-dithiodibenzamido moieties obtained from aminosulfonamides and dithiodialiphatic/aromatic acyl halides. Most disulfide derivatives reported in this study showed weaker inhibitory activity, whereas the corresponding thiols acted as potent inhibitors of hCA I, II, and IX (with inhibition constants in the range 3.2–18 nM against the tumor-associated isoform).

The X-ray crystal structure of the most promising compound in its reduced form, 4-(2-mercaptophenylcarboxamido)benzenesulfonamide **3.13**, which as a disulfide **3.14** (Fig. 3.4) showed a K_I against hCA IX of 653 nM (in reduced form of 9.1 nM), in adduct with hCA II showed the inhibitor making a host of favorable interactions within the active site of the enzyme.[18] The same interactions were preserved in the adduct with hCA IX but, in addition, a hydrogen bond between the SH moiety of the inhibitor and the amide

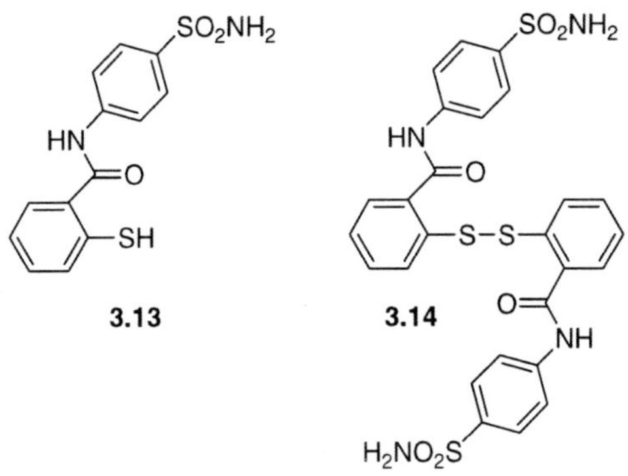

3.13 **3.14**

FIGURE 3.4 Structures of carbonic anhydrase inhibitors **3.13** and **3.14**.

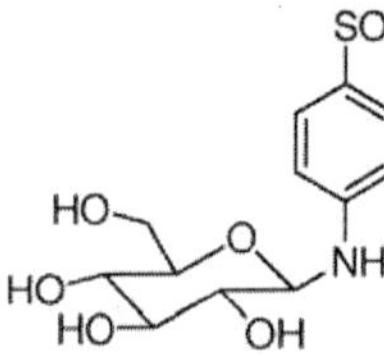

3.15

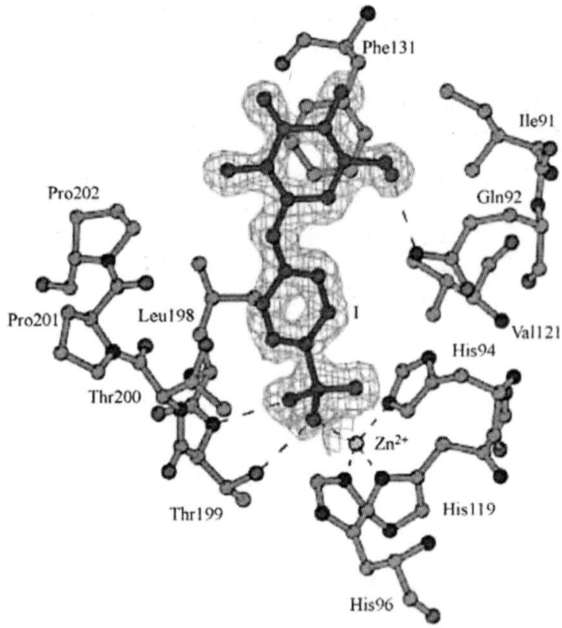

FIGURE 3.5 X-ray structure of hCA II– **3.15** complex.

nitrogen of Gln67 was evidenced, which may explain the almost two times more effective inhibition of the tumor-associated over the cytosolic isoform by this compound.

Another concept developed for designing selective CA inhibitors is the "sugar approach" that consists in the incorporation of glycosyl moieties in different aromatic/heterocyclic sulfonamide scaffolds.[19] This approach was first described in the development of effective antiglaucoma agents. A series of 4-sulfamoylphenylglycopyranosylamines have been prepared and tested on hCA II showing very effective inhibitory activity.[20] The high-resolution X-ray crystal structure of its *N*-(4-sulfamoylphenyl)-α-D-glucopyranosylamine **3.15** adduct with the target isoform involved in glaucoma, CA II, was also reported by this group (Fig. 3.5).[21] The crystal structure demonstrated that the sulfonamide zinc binding anchor and the phenyl ring of the inhibitor bound in the expected way, that is, being coordinated to the metal ion and filling the channel of the enzyme cavity, respectively. However, the glycosyl moiety responsible for the high water solubility of the compound was oriented toward the hydrophilic region of the active site, where other inhibitors were never observed to be bound until now.

3.3 THE SULFAMATE GROUP IN THE DESIGN OF CARBONIC ANHYDRASE INHIBITORS

The sulfamate group (-O-SO_2NH_2), a closely related variant of sulfonamide, has demonstrated very attractive possibilities for the design of various pharmacological

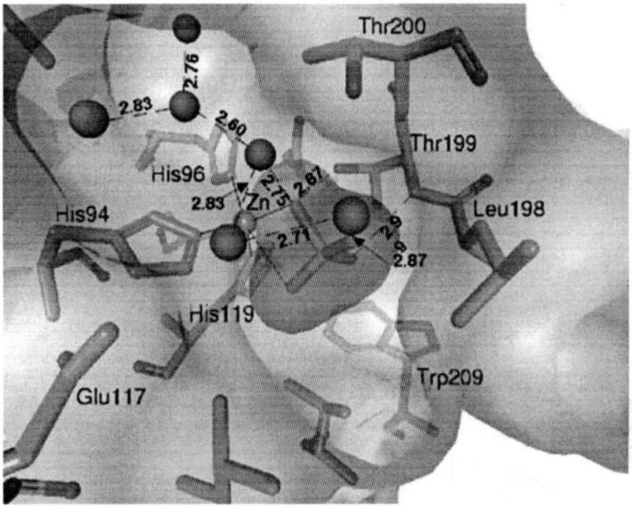

FIGURE 3.6 hCA II adduct with sulfamic acid **3.16** (as dianionic species) as determined by X-ray crystallography. The zinc coordination sphere and amino acid residues in the neighborhood of the bound inhibitor are shown (figures represent distances in angstroms).

agents, especially in the carbonic anhydrase field.[22,23] The investigation of the simplest sulfamate, that is, sulfamic acid (H_2NSO_3H, **3.16**), as CA I has been performed some years later by Briganti et al.[24] who showed that sulfamic acid is a moderately weak CA I, with K_I of 21 μM against the human isozyme hCA I and of 97 μM against hCA II for the esterase activity of these enzymes. The X-ray crystal structure of the adduct of this compound with the physiologically most important isozyme (hCA II) has been subsequently reported by the same group.[25] It was shown that sulfamic acid binds to the zinc ion of the enzyme as a dianion, via its $(NH)SO_3^{2-}$ sulfamate bianionic species (Zn–N distance of 2.07 Å), whereas the NH moiety of the inhibitor also participates in a hydrogen bond with Thr199 Oγ (Fig. 3.6). An additional third hydrogen bond is formed from this NH to an adjacent water molecule at a distance of 2.75 Å. The second nearest contact of the ligand to Zn, an oxygen atom, leading to extracoordination, is at a distance of about 3.07 Å from the metal ion, whereas the remaining two oxygens of the SO_3 moiety are involved in two other hydrogen bonds, one with the backbone NH group of Thr199 at a distance of 2.99 Å and the other with a water molecule at a distance of 2.87 Å. This extended extracoordination results in a distorted tetrahedral arrangement around the metal ion, the remaining three ligands of zinc being His94, His96, and His119 (as in the uninhibited enzyme).[25] In summary, this very simple inhibitor showed a large number of favorable contacts in the binding pocket of CA II and may be used as a lead molecule for the design of tighter binding CAIs.

The first sulfamate compounds that were investigated as CAIs, of type **3.17** (Fig. 3.7), were reported by Lo et al.[26] in the search of topically acting antiglaucoma agents. Indeed, several *m*- or *p*-imidazolyl-phenyloxyethyl/propyl sulfamates **3.17** were shown to possess CA inhibitory properties (IC_{50} in the range of 23–250 nM,

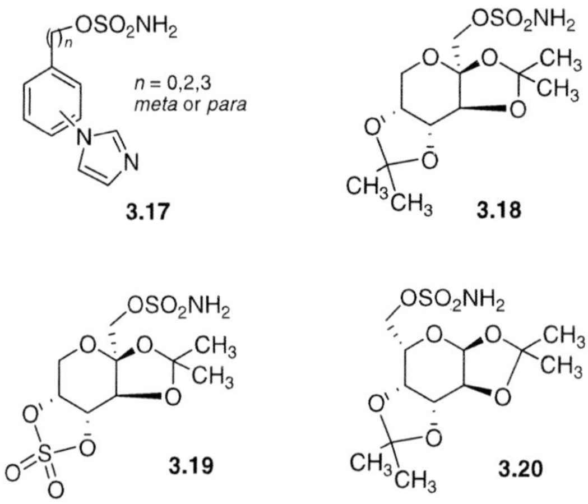

FIGURE 3.7 Structures of compounds **3.17–3.20.**

against bovine red cell CA) and to moderately decrease intraocular pressure in albino rabbits after topical administration directly into the eye.[26,27]

The two anticonvulsant sugar sulfamates topiramate **3.18** and RWJ-37497 **3.19**[28,29] were shown to behave as very potent CAIs and their X-ray crystal structures in complex with isozyme hCA II have been reported[30,31] (Fig. 3.7). The CA inhibitory effect may play an important role in the anticonvulsant activity as well as in the reported clinical side effects that might be encountered in patients treated with this drug (weight loss in human and animals and nephrolithiasis risk among others).

Both sugar derivatives **3.18** and **3.19** bind with their sulfamate moiety to zinc, resulting in a tetrahedral coordination of the metal ion (Fig. 3.8), which is coordinated to the three histidine residues as in the uninhibited enzyme (His94, His96, and His119) as well as the nitrogen atom of the sulfamate moiety, which is presumably in deprotonated state, as for the sulfonamide CAIs complexed within the active site of the enzyme.[1–4,11–16]

In addition, both compounds make two hydrogen bonds to the side chain oxygen atom of Thr199 and the backbone NH nitrogen atom of the same residue. The hydroxy group of Thr199 forms an additional hydrogen bond with Glu106, such that the Thr199 hydroxyl acts as a hydrogen bond acceptor for inhibitor binding. A total of eight hydrogen bonds between topiramate bound to the enzyme and amino acid residues from the cavity have been evidenced, which explain the very potent inhibitory activity of topiramate against hCA II (K_I of 5 nM).[31] Despite the similarity in anchoring to the Zn(II) ion, a surprising difference is observed in the binding mode of the two inhibitors with respect to the ring system. Topiramate also forms several hydrogen bonds to amino acid side chains in a hydrophilic binding pocket (Asn67 and Gln92) and to a water molecule that donates a hydrogen bond to Thr200 (Fig. 3.8a). In addition, this water interacts with the oxygen atom of the six-membered ring of **3.18**. RWJ-37497 **3.19**, on the other hand, shows a quite different binding mode in which the ring system is rotated by about 180° in comparison to that of topiramate. Therefore, surprisingly,

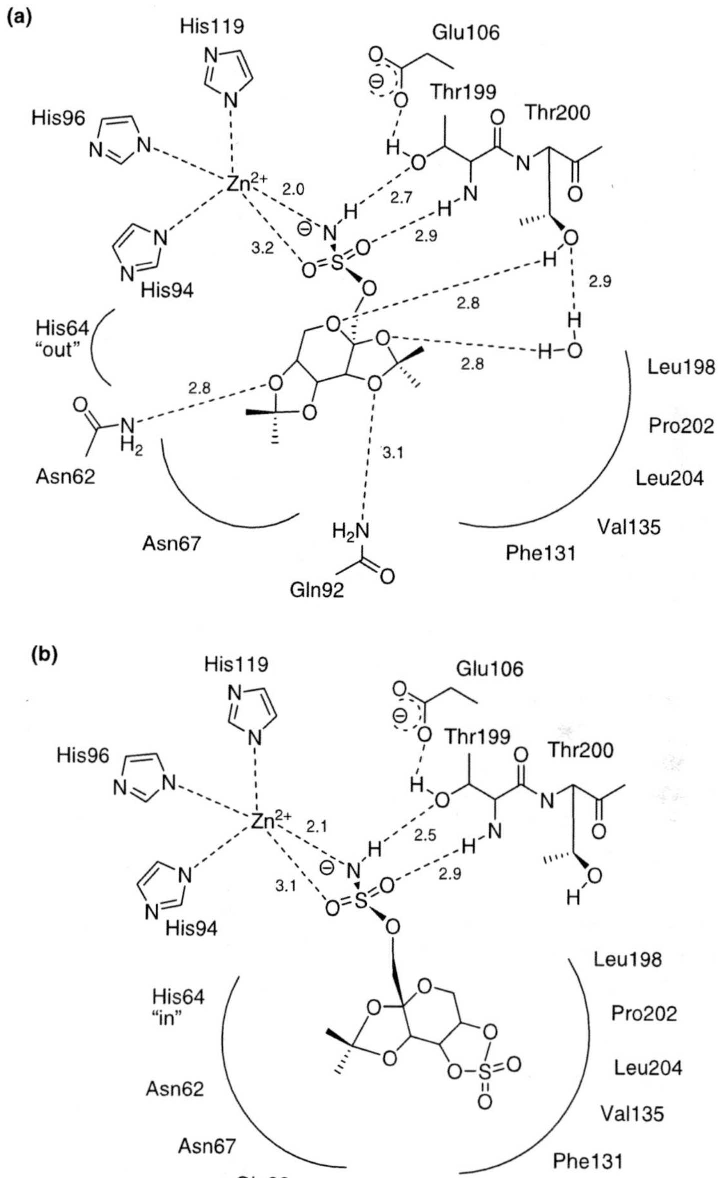

FIGURE 3.8 Schematic representation of the binding mode of topiramate **3.18** (a) and RWJ-37497 **3.19** to hCA II (b).[30,31] Hydrogen bonds between enzyme and inhibitor are shown as dotted lines. Distances are given in angstroms and are measured between the corresponding nonhydrogen atoms.[3]

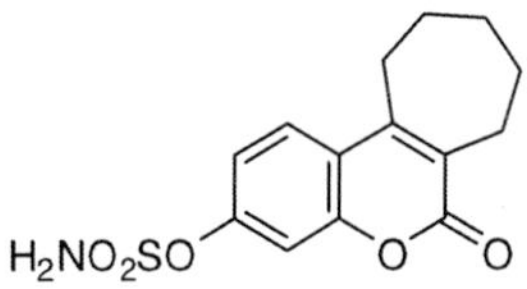

3.21

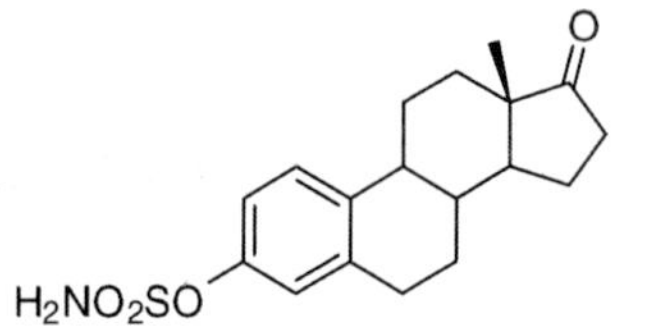

3.22

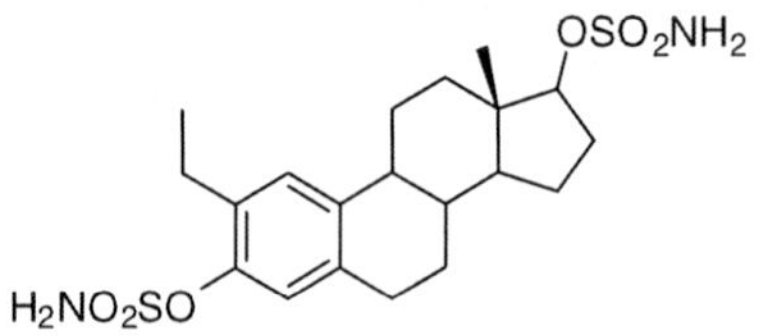

3.23

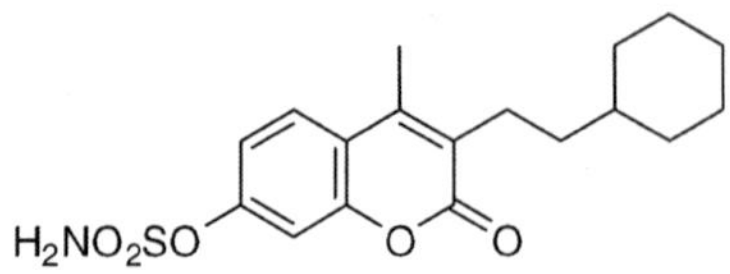

3.24

FIGURE 3.9 Structures of dual STS–CA inhibitors.

the cyclic sulfate group points toward a more hydrophobic pocket lined by residues Leu198, Pro202, and Phe131 (although this moiety is more hydrophilic than the corresponding diisopropylidene moiety of topiramate), and except for the sulfamate anchoring group, no further hydrogen bonds are observed (Fig. 3.8b). RWJ-37497 was reported to have an IC_{50} of 36 nM against hCA II.[22] It is interesting to note that the topiramate isomer **3.20**, possessing the sulfamate-methyl moiety in a different position as compared to **3.18**, is a much weaker CAI than the topiramate or RWJ-37497, having K_I values of 400 μM against hCA I, 16 μM against hCA II, and 27 μM against bCA IV (h, human; b, bovine; m, murine isozymes).[31]

Potter's and Reed's groups also investigated the CA inhibitory properties of several of the steroid sulfatase inhibitors (STSIs) developed during their research on novel therapies for hormone-dependent tumors.[32–34] Some of their best STSIs, such as 667COUMATE **3.21**, EMATE **3.22**, bis-sulfamate **3.23**, and the coumarin sulfamate **3.24**, were tested as hCA II inhibitors, showing to possess IC_{50} values of 17, 9, 290, and 15 nM, respectively (Fig. 3.9).[33,34]

It may be seen that except for the bis-sulfamate **3.23** possessing an ethyl group in *ortho* position of the sulfamate moiety (which presumably interferes with the coordination to the Zn(II) ion within the enzyme active site), the other sulfamates investigated showed very strong CA II inhibitory properties, which may be a beneficial feature for the antitumor effects of such pharmacological agents. The same group also reported docking studies of some of these inhibitors to hCA II[33] and hCA XII[34]. The predictions regarding EMATE **3.22** binding to hCA II[33] have been soon thereafter shown to be wrong, after the report of the X-ray crystal structure of **3.22** with hCA II, by this group[35] (Fig. 3.10).

The binding of EMATE **3.22** to the hCA II is similar to that of other sulfamates/ sulfonamides, considering the interactions of the zinc anchoring group (Fig. 3.10), but differs considerably when the steroidal scaffold of the inhibitor is analyzed. This part of the inhibitor interacts only within the hydrophobic half of the CA active site,

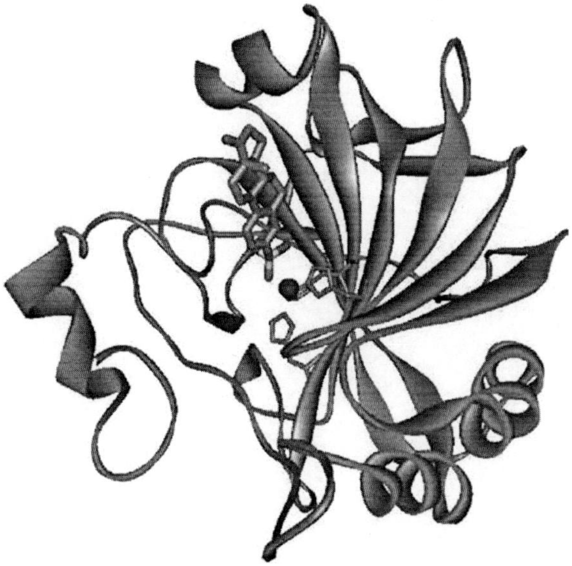

FIGURE 3.10 The hCA II–EMATE **3.22** adduct: the enzyme is shown as ribbon diagram, with the zinc ion (central sphere) and its protein ligands (His94, His96, and His119) shown. The inhibitor molecule lies toward the hydrophobic half of the enzyme cavity.[35]

interacting with residues Val121, Phe131, Val135, and Pro202 and leaving the hydrophilic half able to accommodate several water molecules not present in the uncomplexed enzyme. In addition, a very short bond of 1.78 Å between the zinc ion and the coordinated nitrogen atom of the sulfamate moiety is observed, which may explain the high affinity of this inhibitor for hCA II (K_I of 10 nM).[35]

But the most interesting studies regarding the design of CAIs of the sulfamate type with putative antitumor applications have been reported by Winum et al.[36,37] In these studies, detailed SAR data were obtained for two series of sulfamates, the first one preponderantly including aromatic and steroidal derivatives[36] (Table 3.1) whereas the second one including mostly aliphatic sulfamates[37] (Table 3.2), which were tested for their interaction not only with the red cell isozymes hCA I and II, but also for the first time with the tumor-associated isozyme CA IX, which is overexpressed in a large range of tumors.[3,17,38]

As seen from data of Table 3.1, very potent (nanomolar) inhibitors were detected against the three investigated CA isozymes.[36] Best hCA I inhibitors were phenyl sulfamate **3.25a**, and some of its 4-halogenoderivatives, as well as the aliphatic compound *n*-octyl sulfamate **3.25r** (K_I values in the range of 2.1–4.6 nM— these are among the most potent hCA I inhibitors ever reported!). Against hCA II, low nanomolar inhibitors (1.1–5 nM) were phenyl sulfamate **3.25a** and some of its 4-halogeno/nitro derivatives, *n*-octyl sulfamate **3.25r**, and estradiol-3,17β-disulfamate **3.25t** among others (Fig. 3.11). All the investigated sulfamates showed efficient CA IX inhibitory properties, with inhibition constants in the range of 18–63 nM. The best CA IX inhibitor detected so far was 4-chlorophenyl sulfamate

TABLE 3.1 Inhibition Data for Derivatives 3.25 Investigated by Winum et al.[36] Against Isozymes I, II, and IX

$RO\text{-}SO_2NH_2$

3.25a–x

		K_I^* (nM)		
Inhibitor	R	hCA I[a]	hCA II[a]	hCA IX[b]
Sulfamic acid (H_2NSO_3H) **3.16**		21,000	97,000	nt
Topiramate **3.18**		250	5	nt
EMATE **3.22**		37	10	30
3.25a	Ph	2.1	1.3	63
3.25b	4-Me-C_6H_4	3.8	1.9	59
3.25c	4-Ph-C_6H_4	113	95	50
3.25d	4-Cl-C_6H_4	4.6	1.1	18
3.25e	4-Br-C_6H_4	7.3	1.5	19
3.25f	4-I-C_6H_4	9.5	3.8	23
3.25g	4-MeO-C_6H_4	33	1.6	34
3.25h	4-PhO-C_6H_4	115	98	51
3.25i	4-AcNH-C_6H_4	37	18	45
3.25j	4-O_2N-C_6H_4	40	1.5	36
3.25k	4-NC-C_6H_4	480	149	41
3.25l	4-*t*-Bu-C_6H_4	43	2.9	33
3.25m	4-CF_3-C_6H_4	369	138	54
3.25n	C_6F_5	415	113	47
3.25o	C_6Cl_5	432	125	39
3.25p	2,4,6-$Cl_3C_6H_2$	454	138	37
3.25q	2-Naphthyl	103	63	40
3.25r	*n*-C_8H_{17}	3.5	2.7	25
3.25s	A	105	76	43
3.25t	A	6	5	58
3.25u	A	15	13	32
3.25v	A	31	27	44
3.25w	A	278	15	39
3.25x	A	28	23	26

nt: not tested; A: see structure in Fig. 3.11.
[a]Human (cloned) isozymes by the esterase method.
[b]Catalytic domain of human cloned isozyme by the CO_2 hydration method.

3.25d (K_I of 18 nM). These data are critical for the design of novel antitumor properties, mainly for hypoxic tumors that overexpress CA IX,[3,17,38] which are nonresponsive to radiation or chemotherapy. The antitumor properties of the STSIs[22,23,32] in clinical trials, on the other hand, may also be due to their potent inhibitory properties of CA isozymes involved in tumorigenicity, such as CA II and especially CA IX, and constitute an attractive dual mechanism of action for such antitumor agents.[22,23,32]

Considering the excellent (and unexpected) CA inhibitory properties of the only aliphatic sulfamate **3.21r** investigated in the first study,[36] the same group reported a

TABLE 3.2 Inhibition Data for Sulfamates 3.26 Investigated by Winum et al.[37] Against Isozymes hCA I, II, and IX

$RO\text{-}SO_2NH_2$

3.26a–ac

		K_i^*		
Inhibitor	R	hCA I^a(μM)	hCA II^a (nM)	hCA IX^b (nM)
3.26a	Me	40	6000	>1000
3.26b	Et	38	5500	>1000
3.26c	*n*-Pr	3.70	750	>1000
3.26d	*i*-Pr	690	>10,000	>1000
3.26e	*n*-Bu	3.10	70	>1000
3.26f	*n*-C_5H_{11}	0.71	58	126
3.26g	*n*-$C_{10}H_{21}$	0.53	0.7	23
3.26h	*n*-$C_{11}H_{23}$	0.43	4.7	17
3.26i	*n*-$C_{12}H_{25}$	0.27	10.0	9
3.26j	*n*-$C_{14}H_{29}$	0.15	87	15
3.26k	*n*-$C_{16}H_{33}$	58	97	22
3.26l	*n*-$C_{18}H_{37}$	65	129	120
3.26m	CF_3-CH_2	7.8	845	458
3.26n	*n*-$C_6F_{13}CH_2CH_2$	400	8000	335
3.26o	*n*-$C_8F_{17}CH_2CH_2$	>1000	9000	142
3.26p	$(CF_3)_2CH$	3.54	1580	279
3.26q	$CH_2{=}CHCH_2CH_2$	800	883	386
3.26r	$CH{\equiv}CCH_2CH_2$	990	5900	633
3.26s	$ClCH_2CH_2CH_2$	4.62	570	>1000
3.26t	*c*-C_5H_9		>1000	>10,000
3.26u	*c*-C_6H_{11}	59	60	>1000
3.26v	$PhCH_2$	0.76	3.4	14
3.26w	$PhCH_2CH_2$	0.41	1.1	12
3.26x	*p*-Me-$C_6H_4CH_2$	0.10	2.7	13
3.26y	*p*-Ph-$C_6H_4CH_2$	>1000	>10,000	>1000
3.26z	A (fluorenylmethyl)	38	4500	>1000
3.26aa	A (cholesteryl)	>1000	>10,000	>1000
3.26ab	A	0.40	13	65
3.26ac	A	0.41	23	76

A: see structure in Fig. 3.12.
[a]Human (cloned) isozymes by the CO_2 hydration method.
[b]Catalytic domain of human cloned isozyme by the CO_2 hydration method.

large series of aliphatic (and some polycyclic) sulfamates of type **3.26**, and their interaction with the same physiologically relevant three CA isozymes, that is, hCA I, II, and IX[31] (Table 3.2).

For this new series, the best hCA I inhibitor was *n*-tetradecyl sulfamate **3.26j** and some (substituted) benzyl/phenethyl sulfamates (inhibition constants in the low micromolar range). Against hCA II, low nanomolar inhibitors (0.7–3.4 nM) were *n*-decyl sulfamate **3.26g** and the (substituted)benzyl/phenethyl derivatives mentioned

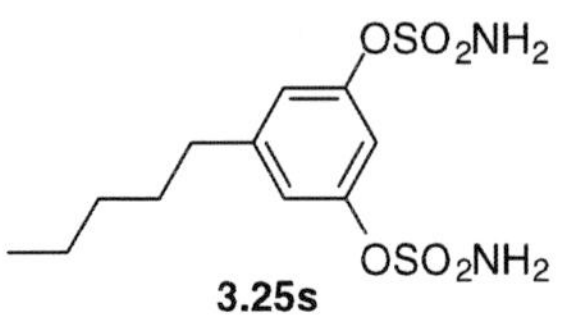

3.25s

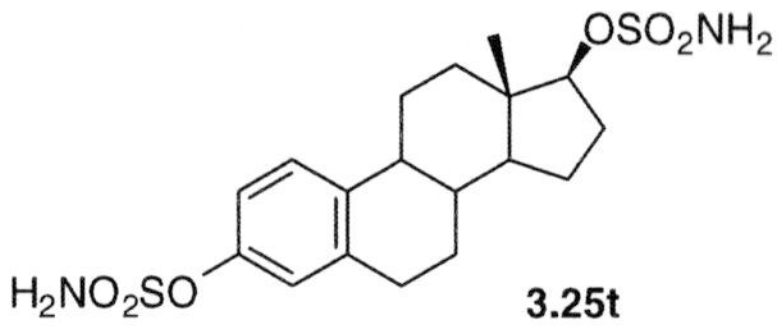

3.25t

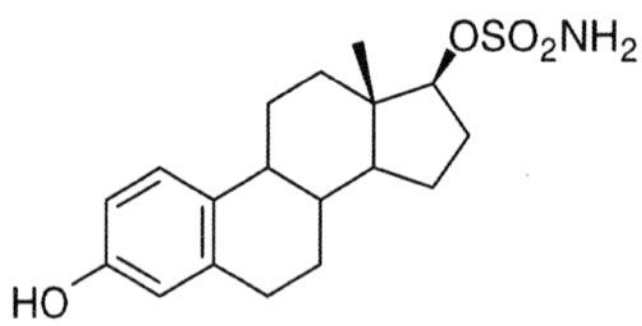

3.25u

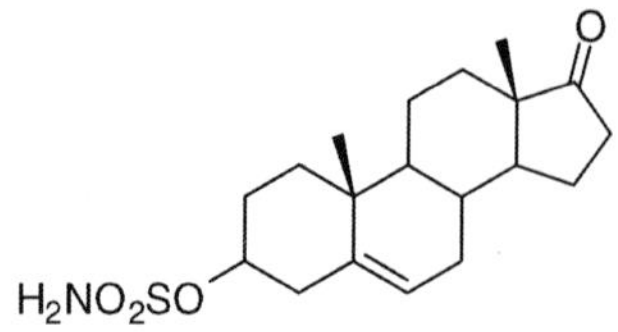

3.25v

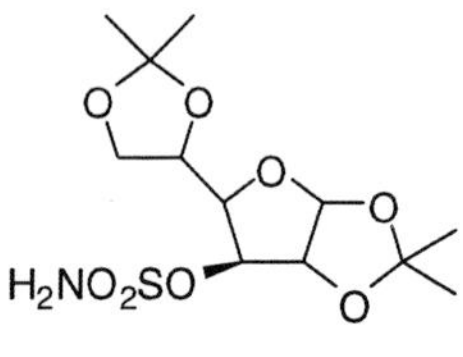

3.25w

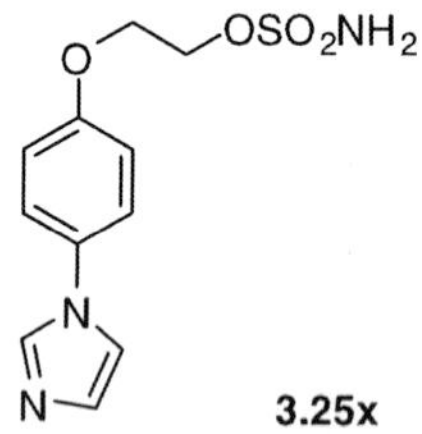

3.25x

FIGURE 3.11 Structures of inhibitors **3.25s–x**.

above, **3.26v–x**. Good CA II inhibition was also observed for the hydroxy/keto derivatives of dehydroepiandrosterone sulfamate **3.26ab** and **3.26ac** (Fig. 3.12). Efficient hCA IX inhibitory properties, with inhibition constants in the range of 9–23 nM, were observed for the aliphatic sulfamates C_{10}–C_{16} (with the best inhibitor

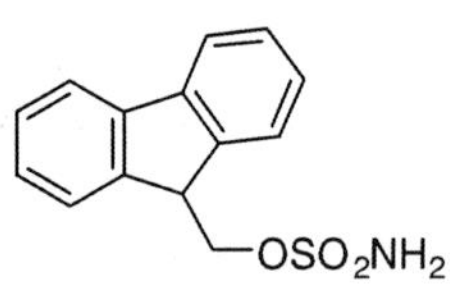

3.26z

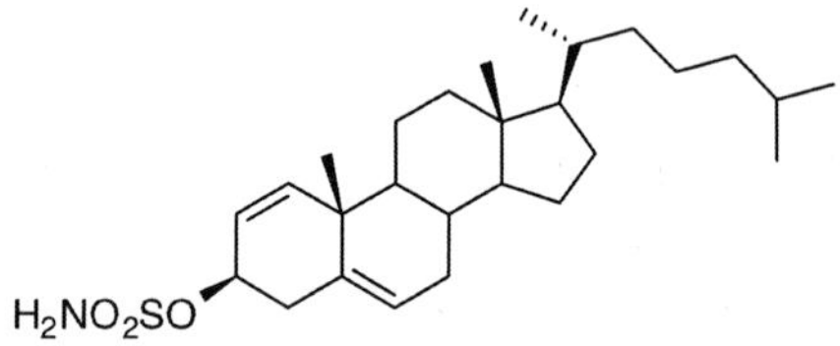

3.26aa

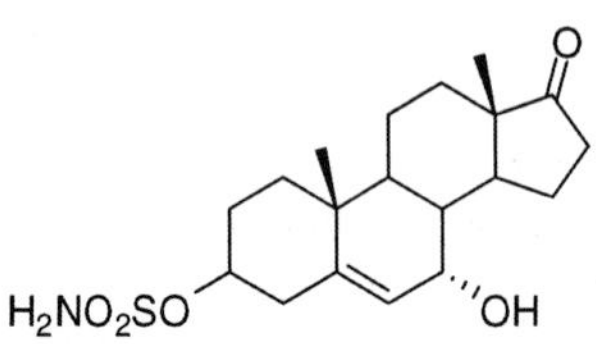

3.26ab

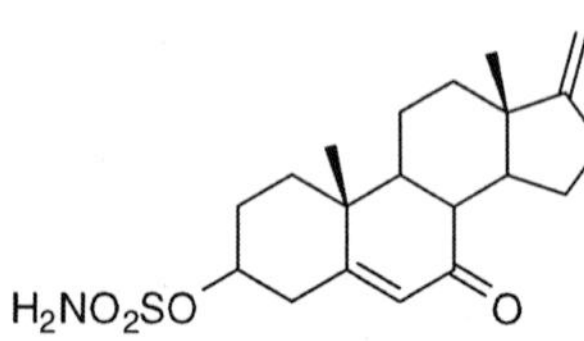

3.26ac

FIGURE 3.12 Structures of CA inhibitors **3.26z**, **3.26aa**, **3.26ab**, and **3.26ac**.

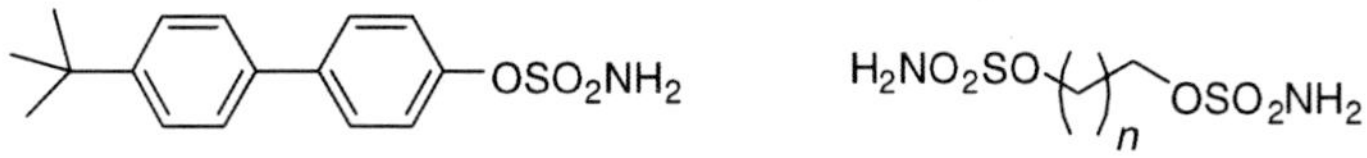

3.27 **3.28** $n = 1$–15

FIGURE 3.13 Structures of inhibitors **3.27** and **3.28**.

n-dodecyl derivative, K_I of 9 nM) and the (substituted)benzyl/phenethyl sulfamates **3.26v–x**.

Synthesis and inhibition studies of a series of aliphatic, aromatic, and polycyclic bis-sulfamates have been reported,[39] showing that each isozyme studied (hCA I, II, or IX) is best inhibited by a certain structural type of such compounds, with hCA I being best inhibited by resorcinyl-1,3-bis-sulfamate **3.25s** (K_I of 79 nM), hCA II by estradiol-3,17-bis-sulfamate **3.25t**, 4-*tert*-butyl-1,2-phenylene-bis-sulfamate **3.27**, and the aliphatic C_4–C_7 bis-sulfamates **3.28** (K_I values in the range of 5–9.7 nM), and CA IX by the C_7–C_{10} aliphatic bis-sulfamates **3.28** (K_I values in the range of 4–10 nM) (Fig. 3.13). Some of these derivatives are also more selective CA IX inhibitors compared to CA II inhibitors with selectivity ratios in the range of 2.63–10.43.

The inhibition profile of the three investigated isozymes with this type of compounds was rather different, allowing us to hope that the preparation of CA IX-selective inhibitors is possible. All these data demonstrate that in addition to the classical CAIs of the aromatic/heterocyclic sulfonamide type, sulfamate inhibitors can be designed with very strong affinity for the active site of different isozymes, a fact that might be relevant for obtaining diverse pharmacological agents that modulate the activity of these widespread enzymes.

3.4 THE SULFAMIDE MOTIF IN THE DESIGN OF CARBONIC ANHYDRASE INHIBITORS

Sulfamide motif attracts permanent interest as a lot of enzyme inhibitors containing sulfamide fragments exhibit a broad spectrum of activities.[40,41]

The first inhibition study of some CA isozymes with sulfamide ($H_2NSO_2NH_2$, **3.29**) has been reported in 1996 by Briganti et al.[24] It has been shown that this simple compound behaves as a weak inhibitor against the classical cytosolic isoforms CA I and II, with inhibition constants of 0.31 and 1.13 mM, respectively (for the physiological reaction catalyzed by these enzymes). Furthermore, working with the Co(II)-substituted CA II (in which the active site zinc ion has been replaced by the colored, paramagnetic Co(II) ion), it has been proved by means of electronic and ^{1}H-NMR spectroscopy (in paramagnetic systems) that the inhibitor directly coordinates to the metal ion within the enzyme active site, which presumably remains in its tetrahedral geometry, as in the wild-type, uninhibited enzyme.[24] This result was thereafter confirmed when the same group reported the high-resolution X-ray crystal structure of the adduct of sulfamide with human CA II[25] (Fig. 3.14).

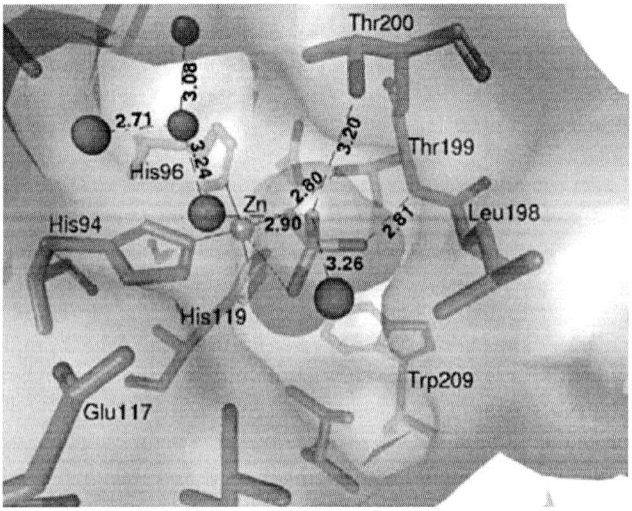

FIGURE 3.14 The hCA II– sulfamide **3.29** adduct: the inhibitor is shown in space fill and stick model, with the zinc ion (gray sphere), its ligands (His94, His96, and His119), water molecules (black balls), and other amino acid residues involved in the binding. Distances are in angstroms.[24]

As observed in Fig. 3.14, the inhibitor molecule, presumably as monoanion, is coordinated to the zinc ion by means of a nitrogen atom (Zn–N distance of 1.76 Å), similar to the sulfonamides for which such studies have been performed.[3] The same NH moiety coordinated to zinc participates in a hydrogen bond with the OH group of Thr199 (which in turn is hydrogen bonded to the carboxylate moiety of Glu106, these two amino acid residues being known as the "door-keepers" in the CA active site and are conserved in all α-CAs[3]). Another hydrogen bond then involves one of the oxygen atoms of sulfamide and the backbone NH of Thr200, which in turn participates in another hydrogen bond with this inhibitor: its OH group makes a 3.26 Å hydrogen bond with the second NH_2 moiety of sulfamide (the one noncoordinated to zinc). This second amino moiety also participates in two other hydrogen bonds with water molecules present in the active site. All these data showed for the first time that CAIs may presumably be designed from the sulfamide class of derivatives, as the lead compound (the simple inorganic sulfamide **3.29**), although being a weak inhibitor, makes a lot of favorable interaction with the amino acid residues in the CA II active site, at the same time possessing a derivatizable NH_2 group positioned in a favorable position for introducing bulkier moieties, in the search of more potent inhibitors. Indeed, in another study,[42] it has been then shown that this group can be derivatized by means of reactions with sulfonyl halides, arylsulfonyl isocyanates, or aromatic/heterocyclic aldehydes, leading to derivatives of types **3.30–3.32** (Fig. 3.15), some of which showed inhibition constants against isozymes CA I, II, and IV in the low nanomolar range.

The sulfonylated sulfamides **3.30**, the arylsulfonyl-carbonylsulfamides **3.31**, or the Schiff bases **3.32** incorporated a large variety of moieties R, belonging mainly to the aromatic/heterocyclic class. For derivatives **3.30**, some aliphatic derivatives have also

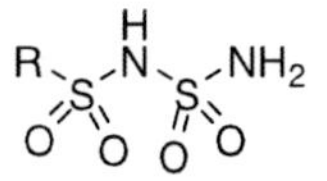

3.30

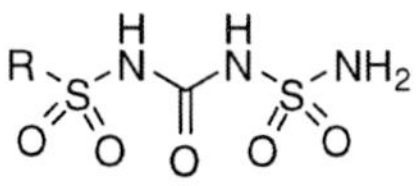

3.31

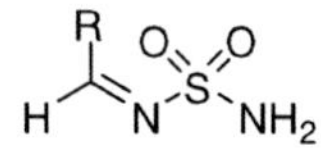

3.32

FIGURE 3.15 Structures of inhibitors **3.30–3.32**.

been prepared, but their activity was generally weaker than that of the aromatic/heterocyclic compounds.[42] Thus, the main conclusion of this work was that starting from a millimolar lead molecule, sulfamide **3.29**, low nanomolar CAIs could be obtained by means of very simple derivatization reactions.

In another study,[43] a series of *N,N*-disubstituted and *N*-substituted sulfamides of types **3.33** and **3.34** were prepared from the corresponding amines and *N*-(*tert*-butoxycarbonyl)-*N*-[4-(dimethylazaniumylidene)-1,4-dihydropyridin-1-ylsulfonyl] azanide or the unstable *N*-(*tert*-butoxycarbonyl)sulfamoyl chloride.[44] Being too bulky, the disubstituted compounds were ineffective as CAIs (Table 3.3), whereas monosubstituted derivatives (incorporating aliphatic, cyclic, and aromatic moieties) as well as a bis-sulfamide behaved as micro–nanomolar inhibitors of two cytosolic isozymes, hCA I and II, responsible for critical physiological processes in higher vertebrates. Aryl sulfamides were more effective than aliphatic derivatives. Low nanomolar inhibitors have been detected, which generally incorporated 4-substituted phenyl moieties in their molecule (Table 3.3). This was another interesting example of CAIs in which low nanomolar inhibitors were generated starting from an ineffective lead molecule.

A very interesting sulfamide derivative incorporating boron, of type **3.35** (Fig. 3.16), was designed and synthesized to obtain compounds with application in boron neutron capture therapy (in the management of tumors). This compound showed only moderate CA inhibitory properties, with inhibition constants in the range of 48–92 nM against isoforms hCA I, II, and IX.[45]

Our group reported[46] recently the X-ray crystal structure of the topiramate sulfamide analogue **3.36** (Fig. 3.16) in complex with hCA II. Compound **3.36** was revealed to be roughly 210 times a weaker hCA II inhibitor than the topiramate **3.18**. It is also a rather weak hCA I, IX, and XII inhibitor (affinity in the micromolar range) but shows nanomolar inhibitory activity against other CA isoforms such as CA VA, VB, VII, and XIII. The high-resolution X-ray structure of the hCA II–**3.36** adduct reveals tight binding within the active site, experienced by Zn(II) ion coordination through the deprotonated sulfamide moiety. Moreover, the organic scaffold participates in an extended network of hydrogen bonds with Thr199, Gln92, His94, Asn62, and Thr200. Its binding to hCA II is also similar to that of topiramate **3.18** except that an important clash between the 8-methyl moiety of **3.36** and the methyl group of Ala65 has been evidenced (Fig. 3.17).

This result supports our assay results[19,46] and provides rationale for the possibility to obtain CAIs with diminished affinity for hCA II while still maintaining tight binding for other isoforms. This could lead to compounds possessing fewer side effects owing to improved selectivity over this physiologically dominant isoform.

TABLE 3.3 Inhibition Data for Derivatives 3.33 and 3.34 Against the Cytosolic, Human Isozymes hCA I and hCA II by an Esterase Assay Method with 4-Nitrophenyl Acetate as Substrate[43]

$RR'N\text{-}SO_2NH_2$ **3.33** $NH_2SO_2\text{-}NH\text{-}RR'\text{-}NH\text{-}SO_2NH_2$ **3.34**

			K_I (nM)	
Inhibitor	R	R′	hCA I[a]	hCA II[a]
3.29 $H_2NSO_2NH_2$	–	–	35,000	82,000
3.33a	*n*-Bu	H	173	148
3.33b	Cyclohexyl	H	164	450
3.33c	2-Adamantyl	H	960	890
3.33d	$PhCH_2$	H	133	123
3.33d	*i*-Bu	*i*-Bu	>100,000	>100,000
3.33e	*i*-Pr	*i*-Pr	>100,000	>100,000
3.33f	Cyclohexyl	Cyclohexyl	>100,000	>100,000
3.33g	$PhCH_2$	$PhCH_2$	>100,000	647
3.33h	$\text{-}(CH_2)_5\text{-}$		155	148
3.33i	$\text{-}(CH_2)_6\text{-}$		163	131
3.33j	Ph	H	13	12
3.33k	$4\text{-Me-}C_6H_4$	H	15	13
3.33l	$4\text{-}CF_3\text{-}C_6H_4$	H	8	7
3.33m	$4\text{-Cl-}C_6H_4$	H	19	15
3.33n	$4\text{-Br-}C_6H_4$	H	23	21
3.33p	$4\text{-I-}C_6H_4$	H	18	17
3.33k	$4\text{-MeO-}C_6H_4$	H	14	11
3.33r	$4\text{-HO-}C_6H_4$	H	16	12
3.33s	$4\text{-}O_2N\text{-}C_6H_4$	H	18	13
3.33t	$4\text{-}EtO_2C\text{-}C_6H_4$	H	26	19
3.33u	$4\text{-NC-}C_6H_4$	H	20	16
3.33v	$4\text{-}Me_2N\text{-}C_6H_4$	H	17	21
3.33x	C_6F_5	H	34	32
3.33y	$3\text{-Benzoyl-}C_6H_4$	H	62	49
3.33z	2-Naphthyl	H	39	36
3.34	$(CH_2)_2\text{-SS-}(CH_2)_2$		149	27

[a]Human (cloned) isozymes by the esterase method.

3.5 *N*-SUBSTITUTED SULFONAMIDES/SULFAMATES/SULFAMIDES AS CARBONIC ANHYDRASE INHIBITORS

Krebs reported in 1948 that substitution of the sulfonamido moiety in compounds of type $ArSO_2NHR$ (i.e., R different of H) drastically reduces the CA inhibitory properties as compared to the corresponding derivatives possessing primary sulfonamido groups, $ArSO_2NH_2$.[47]

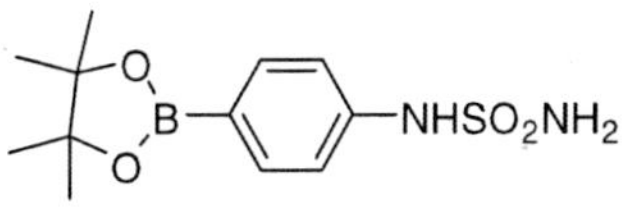

3.35

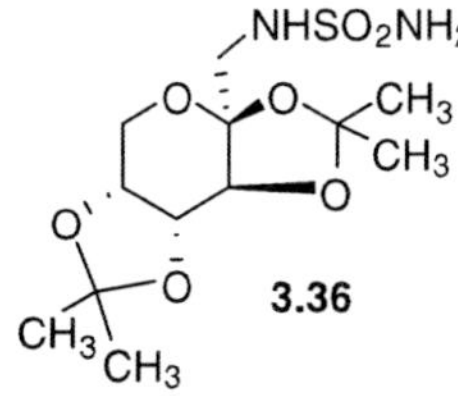

3.36

FIGURE 3.16 Structures of inhibitors **3.35** and **3.36**.

Recently, several detailed studies regarding the possible modifications of the sulfonamido moiety, compatible with the retention of strong binding to the enzyme, have been reported. Compounds of type **3.37**, **3.29**, and **3.16** were studied kinetically for inhibition of reactions catalyzed by CA I and II (CO_2 hydration and ester hydrolysis), but their binding to the enzyme has also been monitored spectroscopically by studying the electronic and ^{1}H-NMR spectra of adducts of such inhibitors with Co(II)-CA II[24] (Table 3.4).

Thus, for the series of derivatives with modified sulfonamido moieties **3.16**, **3.29**, and **3.37** (Table 3.4), it has been observed that the presence of bulky substituents at the sulfonamido moiety (such as phenylhydrazino, ureido, thioureido, guanidino, etc.) led to compounds with weak inhibitory properties, whereas moieties present in inorganic anion CAIs (such as NO, NCS, and N_3) or compact moieties substituting the sulfonamide nitrogen (such as OH, NH_2, CN, and halogeno) led to compounds with appreciable inhibitory properties. Thus, the *N*-hydroxy sulfonamide **3.37b**, the *N*-chloro-substituted derivatives **3.37j** and **3.37k**, and the nitroso and thiocyanato

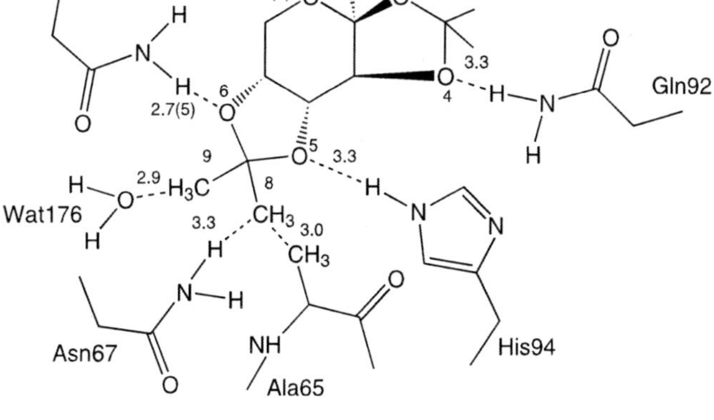

FIGURE 3.17 Binding of **3.36** to the hCA II active site. The figure represents distances in angstroms.

TABLE 3.4 Inhibition of hCA I and hCA II with Compounds Incorporating Modified Sulfonamide Moieties 3.37a–t, Sulfamide 3.29, and Sulfamic Acid 3.16 (Briganti et al.[24])

$4\text{-Me-}C_6H_4SO_2\text{-X}$	$H_2NSO_2NH_2$	$HOSO_2NH_2$
3.37a–t	**3.29**	**3.16**

		K_I (μM)[a]	
Inhibitor	X	hCA I	hCA II
3.16		21	390
3.29		310	1130
3.37a	NH_2	50	11
3.37b	NHOH	41	9
3.37c	NHOMe	220	173
3.37d	NO	35	24
3.37e	NCS	30	18
3.37f	N_3	27	45
3.37g	Imidazol-1-yl	160	34
3.37h	$NHNH_2$	70	53
3.37i	NHNHPh	>1000	120
3.37j	NHCl	19	2.1
3.37k	NCl_2	12	3.6
3.37m	NHCN	210	125
3.37n	$NHOCH_2COOH$	150	85
3.37p	OH	130	460
3.37q	SH	5	10
3.37r	$NHCONH_2$	>1000	460
3.37s	$NHCSNH_2$	>1000	410
3.37t	$NHC(NH)NH_2$	>1000	540

[a]Errors in the range of 5–10% of the shown data, from three different assays.

derivatives **3.37d** and **3.37e** possessed the same affinity for the two investigated isozymes as the unsubstituted sulfonamide **3.37a**. Interestingly, the thiosulfonic acid (as sodium salt) **3.37q** was one of the best inhibitors in this series, in contrast to the sulfonate (as sodium salt) **3.37p**, which behaved as a very weak inhibitor. Indeed, by using such compounds as leads, several series of much stronger inhibitors were then reported, possessing modified sulfonamido moieties as zinc binding functions, of the type SO_2NHOH, SO_2NHCN, $SO_2NHPO_3H_2$, $SO_2NHSO_2NH_2$, SO_2NHSO_3H, or $SO_2NHCH_2CONHOH$ among others.[42,48–50]

Thus, compounds such as **3.38–3.47** (Fig. 3.18), possessing *N*-cyano, *N*-hydroxy, or *N*-phosphoryl-sulfonamido moieties, or the related modified sulfamide/sulfamic acid zinc binding functions, and diverse alkyl, aryl, or heterocyclic moieties in their molecules, showed affinities in the low nanomolar range for hCA II (except for **3.44**), being equipotent or better inhibitors than the corresponding unsubstituted sulphonamides.[42,48–50] Compound **3.44** is a weak inhibitor of hCA II (affinity constant of

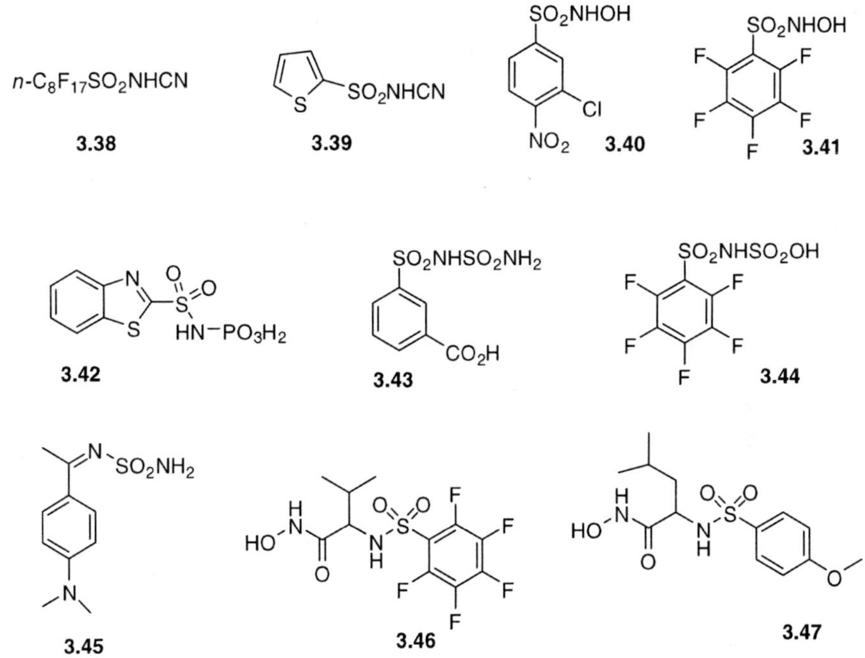

FIGURE 3.18 Structures of inhibitors **3.38–3.47**.

1.2 μM), but it has a much higher affinity (50 nM) for hCA I, thus being one of the most "selective" hCA I inhibitors reported until now.[42]

Sulfonylated amino acid hydroxamates were also shown to possess strong CA inhibitory properties.[49] Such hydroxamates generally act as potent inhibitors of metalloproteases containing catalytic zinc ions, such as the matrix metalloproteinases (MMPs) or the bacterial collagenases.[51] They bind to the Zn(II) ions present in these enzymes bidentately, coordinating through the hydroxamate (ionized) moiety.[51] Scolnik et al. showed that two simple hydroxamates, of the type RCONHOH (R=Me, CF_3), act as micromolar inhibitors of hCA II and bind to the Zn(II) ion of this enzyme, as demonstrated by X-ray crystallography.[52] By using these two derivatives as lead molecules, Scozzafava et al.[49] designed a series of sulfonylated amino acid hydroxamate derivatives possessing the general formula $RSO_2NHCH(R')CONHOH$ and showed that they bind to the Zn(II) ion of CA by means of electronic spectroscopic studies on the Co(II)-substituted CA. Some of these compounds, such as **3.46** and **3.47**, showed affinity in the low nanomolar range for the major CA isozymes (CA I, II, and IV), but substitution of the sulfonamide nitrogen by a benzyl or a substituted benzyl moiety led to a drastic reduction of the CA inhibitory properties and to an enhancement of the MMP inhibitory properties. Thus, between the two types of zinc enzymes, the zinc proteases and the CAs, there exists some cross-reactivity from the viewpoint of the hydroxamate inhibitors, but generally strong MMP inhibitors are weak CAIs, and vice versa.

All these data demonstrate that in addition to the classical CAIs of the aromatic/heterocyclic sulfonamide type, substitution of the $-NH_2$ with different motif can lead to inhibitors with potency greater than the corresponding unsubstituted sulfonamides, a fact that may be relevant for obtaining diverse pharmacological agents that modulate the activity of these widespread enzymes.

One of the best examples is the substitution of $-NH_2$ with $-NHOH$ that can give hydroxy sulfonamide (RSO_2NH-OH) or sulfamide ($RNHSO_2NH$-OH) derivatives, compounds closely related to hydroxamic acids R-CONHOH and hydroxyurea (R-NHCONHOH).

N-Hydroxyurea **3.48** and acetohydroxamic acid **3.49** were previously shown to act as weak CA inhibitors with K_I of 28 and 47 μM, respectively. Nevertheless, the X-ray crystal structure of the adduct of isozyme hCA II with **3.49** has been reported, showing that the hydroxamate moiety (deprotonated at the nitrogen atom) is coordinated in monodendate fashion to the Zn(II) ion, whereas its OH and CO groups participate in two hydrogen bonds with the OH moiety of Thr199 and with the backbone NH of the same residue.[52]

The first inhibition study of hydroxyurea **3.48** as CA inhibitor was published by Scozzafava et al. in 2003.[53] The X-ray crystal structure of the adduct hCA II with **3.48** was then recently reported by the same group who demonstrated that *N*-hydroxyurea **3.48** binds in a bidendate fashion to the Zn(II) ion within the hCA II active site by means of both nitrogen and oxygen atoms belonging to the NHOH group, with distances of these atoms from zinc in the range of 2.00–2.07 Å[54] (Fig. 3.19).

In addition, the oxygen of the carbonyl moiety of **3.48** makes a hydrogen bond with the backbone NH group of Thr199, which was also observed in the case of hCA II–**3.49** adduct. Moreover, a water molecule (not shown in Fig. 3.19) participates in a network of two hydrogen bonds involving the OH and NH of the NHOH moiety.

All these data underline that these two simple zinc binding functions bearing NH–OH moiety can be exploited for obtaining interesting and potent carbonic anhydrase inhibitors as well as other metalloenzymes inhibitors.

A same type of comparison can be made with sulfamide ($H_2NSO_2NH_2$, **3.29**) and hydroxysulfamide (H_2NSO_2NHOH, **3.50**). The first inhibition study of some CA

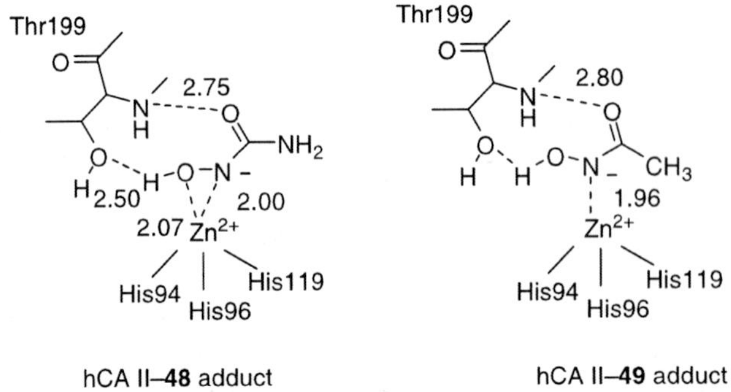

FIGURE 3.19 Schematic representation of the interaction between the inhibitors **3.48** and **3.49** with the hCA II active site. Distances are in angstroms.

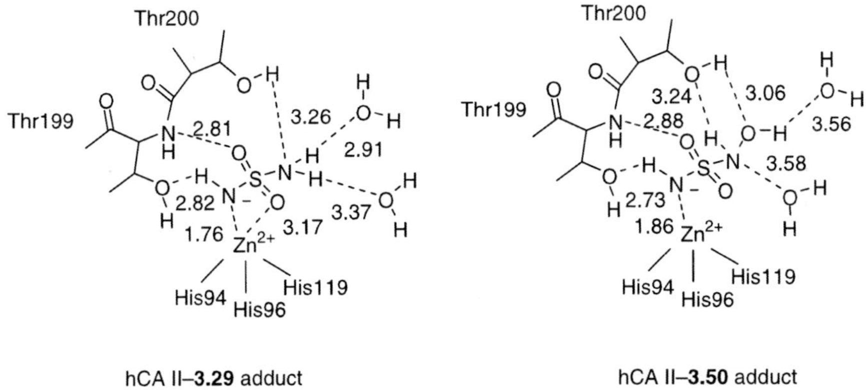

FIGURE 3.20 Schematic representation of the interaction between the inhibitors **3.29** and **3.50** with the hCA II active site. Distances are in angstroms.

isozymes with **3.20** has been reported by this group.[55] It has been shown that this simple compound behaves as a weak inhibitor against the classical cytosolic isoforms CA I and II, with inhibition constants of 0.31 and 1.13 mM, respectively (for the physiological reaction catalyzed by these enzymes). The same group reported the high-resolution X-ray crystal structure of the adduct of **3.20** with human CA II.[56]

As observed in Fig. 3.20, the sulfamide molecule **3.29** is coordinated to the zinc ion by means of a nitrogen atom (Zn–N distance of 1.76 Å), similar to the sulfonamides for which such studies have been performed.[3] The same NH moiety coordinated to zinc participates in a hydrogen bond with the OH group of Thr199. Another hydrogen bond then involves one of the oxygen atoms of sulfamide and the backbone NH of Thr200, which in turn participates in another hydrogen bond with this inhibitor: its OH group makes a 3.26 Å hydrogen bond with the second NH_2 moiety of sulfamide (the one noncoordinated to zinc). This second amino moiety also participates in two other hydrogen bonds with water molecules present in the active site.[56]

A study on the simple *N*-hydroxysulfamide **3.50** showed a highly enhanced CA inhibitory activity as compared to both sulfamide **3.29** and sulfamic acid **3.16**, compounds with which it is structurally related (Table 3.5).[57] Thus, *N*-hydroxysulfamide was approximately a 75-fold better hCA I inhibitor than the sulfamide and approximately 2000-fold better hCA II inhibitor than the sulfamide, whereas these factors were 11 for the inhibition of hCA IX and 9.8 for the inhibition of hCA XII.

Analysis of the three-dimensional structure[26] of the hCA II–**3.50** complex revealed that the inhibitor **3.50** is coordinated to the Zn(II) ion by the ionized, terminal H_2N moiety of the *N*-hydroxysulfamide molecule, although this is not the most acidic moiety present in *N*-hydroxysulfamide. The deprotonated primary amino group of the inhibitor is coordinated to the zinc ion, at a distance of 1.86 Å, intermediate between the very short distances evidenced for the sulfamide adduct (Zn–N of 1.76 Å). The NH moiety also coordinated to Zn(II) participating in a strong hydrogen bond with the OH moiety of Thr199 (of 2.73 Å), whereas one oxygen of the SO_2 moiety makes a second hydrogen bond (of 2.88 Å) with amide NH of the same amino acid residue (Fig. 3.20).

TABLE 3.5 Inhibition of Isozymes hCA I, II, IX, and XII with Sulfamide, Sulfamic Acid, N-Hydroxysulfamides 3.51, Sulfamides 3.52, and the Corresponding Sulfamates 3.53[58]

$RNHSO_2NHOH$	$RNHSO_2NH_2$	$ROSO_2NH_2$
3.51	**3.52**	**3.53**

		K_I (nM)^a			
Inhibitor	R	hCA I	hCA II	hCA IX	hCA XII
$H_2NSO_2NH_2$	–	0.31×10^6	1.13×10^6	9.6×10^3	13.2×10^3
$H_2NSO_3H^c$	–	0.21×10^5	0.39×10^6	9.2×10^3	10.7×10^3
3.50	H	4050	566	865	1340
3.51b	n-$C_{10}H_{23}$	5800	473	506	874
3.51c	n-$C_{12}H_{25}$	6000	89.6	485	539
3.51d	$PhCH_2$	8200	313	790	633
3.51e	4-$PhC_6H_4CH_2$	8100	50.5	353	372
3.52a	4-$CF_3C_6H_4$	8	7	26	48
3.53a	4-$CF_3C_6H_4$	369	138	54	103
3.52b	4-CNC_6H_4	20	16	30	45
3.53b	4-CNC_6H_4	480	149	41	76
3.52c	C_6F_5	34	32	40	19
3.53c	C_6F_5	415	113	47	34
3.52d	2-Naphthyl	39	36	38	30
3.53d	2-Naphthyl	103	63	40	62

^{a}Errors in the range of 5–10% of the shown data, from three different assays.

The other oxygen of the SO_2 moiety is at about 3.2 Å from the metal ion, as in many other sulfonamide–CA II adducts. The NH moiety of the NHOH functionality also participates in two hydrogen bonds, similar to the second H_2N moiety of sulfamide. Thus, a strong hydrogen bond with the OH moiety of Thr200 is observed (of 3.24 Å), and a second, weaker one with a water molecule of 3.58 Å. The OH group of the inhibitor molecule establishes again two hydrogen bonds, one with the same OH of Thr200 that participates in the hydrogen bond with the NH moiety of the inhibitor, this time of 3.06 Å, whereas the second one with another water molecule present within the active site of 3.56 Å. Thus, the presence of extra two hydrogen bonds in the hCA II–*N*-hydroxysulfamide adduct as compared to the hCA II–sulfamide adduct seems to be the main factor responsible for the enhanced affinity of the first inhibitor for the enzyme. In fact, these supplementary interactions are due to the presence of the additional OH moiety in the molecule of the tight binding inhibitor.

A study on substituted *N*-hydroxysulfamides **3.51** has also been described and their inhibitory activity has been investigated against the cytosolic (hCA I and II) and transmembrane tumor-associated (hCA IX and XII) CAs.[58] *N*-Hydroxysulfamide **3.50** was a more potent inhibitor as compared to sulfamide or sulfamic acid against all investigated isozymes, with inhibition constants in the range of 473 nm–4.05 μM. Its substituted *n*-decyl, *n*-dodecyl, benzyl, and biphenylmethyl derivatives were less inhibitory against hCA I (K_I values in the range of 5.8–8.2 μM) but more inhibitory against hCA II (K_I values in the range of 50.5–473 nM) (Table 3.5). The same situation was true for the tumor-associated isozymes, with K_I values in the range of 353–790 nM against hCA IX and 372–874 nM against hCA XII. Some sulfamides/sulfamates of types **3.52** and **3.53**, respectively, possessing similar substitution patterns have also been investigated for the inhibition of these isozymes, showing that in some particular cases sulfamides were more efficient inhibitors than the corresponding sulfamates. Potent CAIs targeting the cytosolic or tumor-associated CA isozymes can thus be designed from various classes of sulfonamides or sulfamides and their derivatives, considering the extensive interactions in which the inhibitor and the enzyme active site are engaged, based on the X-ray crystallographic data shown above and on this study that comparatively evaluated these compounds possessing different zinc binding groups and their relative efficiency in inhibiting various isoforms.[58]

3.6 PHOSPHATE–PHOSPHONATE-BASED CARBONIC ANHYDRASE INHIBITORS

Phosphate and phosphonate are important functions found in a large variety of molecule of biological interest, especially anticancer or antiviral drugs. The use of phosphonate motif in the design of carbonic anhydrase inhibitors was first reported by Fenesan et al. in 2000.[50] A small series of substituted aryl *N*-phosphoryl sulfonamides ($-SO_2NHPO(OH)_2$) **3.53–3.58** were prepared and studied kinetically for inhibition of reactions catalyzed by CA I and II (CO_2 hydration ester hydrolysis). Some derivatives (Fig. 3.21) exhibited strong inhibitory activity against CA II with inhibition constant below 20 nM for compounds **3.53e**, **3.53h**, **3.56**, and **3.57** (Table 3.6).

From this study, an important fact deserves to be mentioned. All the reported phosphoryl sulfonamides **3.53–3.58** are potent carbonic anhydrase inhibitors as the corresponding parent unsubstituted sulfonamides. By analogy, sulfamates possessing

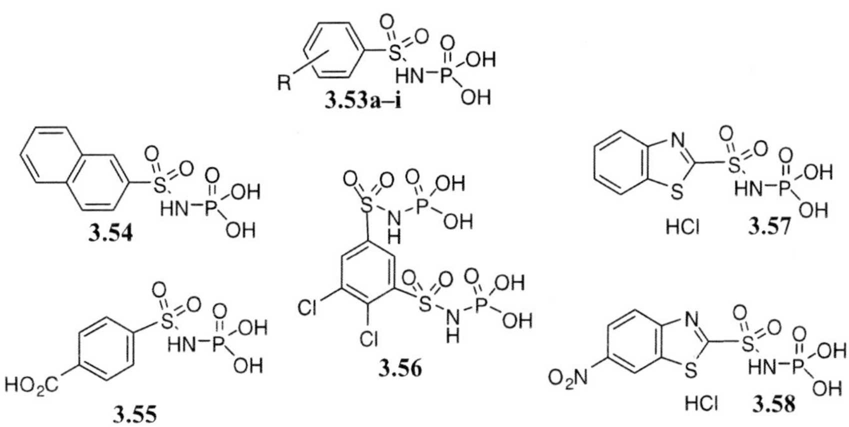

FIGURE 3.21 Structures of phosphoryl sulfonamides **3.53–3.58**.

TABLE 3.6 Inhibition of hCA I and hCA II with Phosphoryl Sulfonamides 3.53–3.58

		K_I (nM)[a]	
Inhibitor	R	hCA I[b]	hCA II[b]
3.53a	H	560	280
3.53b	4-F	485	190
3.53c	4-Cl	470	145
3.53d	4-Br	310	60
3.53e	4-I	42	8
3.53f	4-CH_3	215	54
3.53g	4-OCH_3	200	50
3.53h	4-NO_2	51	15
3.53i	2-NO_2	420	210
3.54		450	200
3.55		125	45
3.56		52	15
3.57		10	3
3.58		7	2

[a]Errors in the range of 5–10% of the reported value (from three different assays).
[b]Human (cloned) isozymes.

a phosphoryl moiety substituting the nitrogen atom (R-O-SO_2NH-PO_3H_2) have been reported by our group.[59] It was demonstrated that the phosphorylated sulfamate zinc binding group is very efficient for the design of low nanomolar CA inhibitors. Aliphatic compounds incorporating C_8–C_{16} chains lead to inhibitors with affinities of 8–16 nM against hCA I and 5–12 nM against hCA II (Table 3.7).[59] Compared to the parent sulfamate, the inhibitory activity was increased by addition of the phosphoryl motif.

A detailed inhibition study of the five carbonic anhydrase isoenzymes with inorganic phosphate (PO_4^{3-}, HPO_4^{2-}), carbamoyl phosphate ($H_2NCOOPO_3^{2-}$), and the antiviral phosphonate foscarnet (^{-}OOC-PO_3^{2-}) was described recently by

TABLE 3.7 CA Inhibition Data with Compounds (3.59–3.62) and Standard Inhibitors Against Human Isozymes hCA I and hCA II

		K_I (nM)[a]	
	Compound	hCA I[b]	hCA II[b]
3.59	n-C_8H_{17}-O-SO_2NH-PO_3H_2	8.2	5.3
3.60	n-$C_{12}H_{25}$-O-SO_2NH-PO_3H_2	10.5	9.9
3.61	n-$C_{14}H_{29}$-O-SO_2NH-PO_3H_2	14.6	11.9
3.62	n-$C_{16}H_{33}$-O-SO_2NH-PO_3H_2	16.1	11.2
Sulfanilamide **3.1**		28,000	300
Acetazolamide **3.2**		900	12
Topiramate **3.18**		250	5

[a]Errors in the range of 5–10% of the reported value (from three different assays).
[b]Human (cloned) isozymes.

Rusconi et al.,[60] demonstrating that the cytosolic ubiquitous isozyme hCA II and the transmembrane tumor-associated isozyme hCA IX were slightly inhibited whereas the cytosolic isozyme hCA I was activated by most of them. On the other hand, the membrane-associated isozyme hCA IV was the most sensitive to inhibition by these compounds, especially foscarnet **3.63** (K_I of 0.82 mM). hCA IV is very abundant in the kidneys, where it plays an important physiological function in the bicarbonate reabsorption and secretion of ammonium ion into urine among others. These results may explain some of the renal side effects of this antiviral drug.

These interesting preliminary results were completed in 2007 by the report of the interaction of the antiviral drug foscarnet **3.63** with 11 carbonic anhydrase isozymes and the X-ray crystal structure for the adduct of the drug with hCA I.[61] In this study, it was first demonstrated that foscarnet is a modulator of hCA I activity that can act either as an activator or as an inhibitor, depending on the duration of the incubation period. Inhibitory activity against CA I (K_I of 24.1 mM) was observed when foscarnet was incubated for a long period (3–5 days). The best inhibition profile was observed for the three following isozymes: the membrane-anchored CA V, and the cytosolic isoforms hCA VII and mCA XIII with K_I in the range of 0.56–0.87 mM. For the extracellular isoforms, for example, the secreted CA VI isoform found in milk or saliva, or the transmembrane isoforms CA IX, XII, and XIV that have the active site situated outside the cell, inhibition activity was observed in the range of 1.81–3.60 mM (see Table 3.8).

The hCA I–foscarnet **3.63** adduct was crystallized and showed that foscarnet coordinated to the zinc ion by means of the phosphonate moiety that possesses two negative charges, leading to additional electrostatic attraction between the positively

TABLE 3.8 Inhibition Constant of Foscarnet ($^-OOC\text{-}PO_3^{2-}$, 3.63) Trisodium Salt Against 11 CA Isozymes for the CO_2 Hydration Reaction at 20°C

Isoform	K_I (mM)[a]
hCA I	24.1
hCA II	14.2
hCA IV	0.82
hCA VA	41.7
hCA VB	11.8
hCA VI	1.81
hCA VII	0.56
hCA IX	2.21
hCA XII	1.29
hCA XIII	0.87
hCA XIV	3.60

h, human (cloned) isozymes; m, murine recombinant isoform. For all the isozymes, preincubation of enzyme with inhibitors for 15 min at room temperature, except for hCA I: preincubation for 5 days at 4°C.
[a]Errors in the range of 3–5% of the reported value (from three different assays).

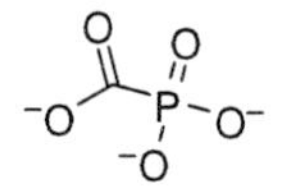

3.63

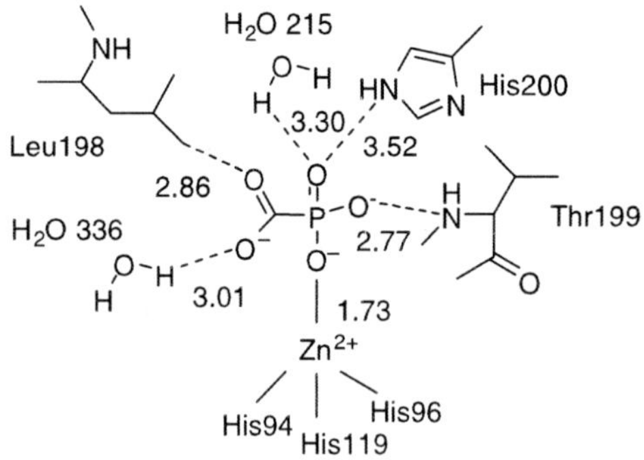

FIGURE 3.22 Schematic representation of the interaction between the inhibitor **3.63** and the hCA I active site. Distances are in angstroms.

charged dication (Zn^{2+}) and the negatively charged dianion, with the consequent reduction of the Zn–O distance (1.73 Å). This is to be compared with monoanion (RSO_2 NH−) inhibitors that coordinate to the zinc ion at a distance of 1.80–2.20 Å; (Fig. 3.22).

It is worth pointing that compared to all other CA inhibitor–adducts investigated until now by means of X-ray crystallography, no strong hydrogen bond was evidenced between the OH group of Thr199 and the inhibitor atom directly coordinated to the Zn(II) ion.

Moreover, a repulsive interaction between one methyl group of the side chain of Leu198 and the oxygen atom of the carboxylate moiety of foscarnet was observed. This type of clash was already observed by our group[46] for other CA inhibitors and may explain the relatively weak inhibitory activity of **3.63** against hCA I.

Another study realized by our group in 2005 presented the interactions of five hCA isozymes with organic phosphates and phosphonates, including methylphosphonic acid **3.64**, phenylphosphonic acid **3.65**, *N*-(phosphonoacetyl)-L-aspartic acid **3.66**, methylene diphosphonic acid **3.67**, the *O*-phosphates of serine **3.68**, and threonine **3.69**[62] (Fig. 3.23).

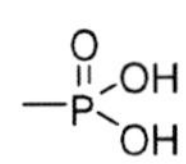

3.64

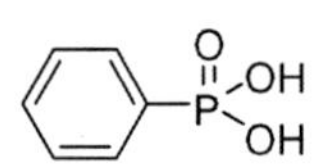

3.65

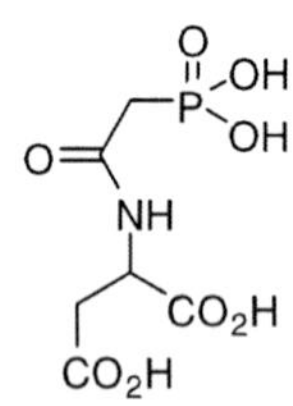

3.66

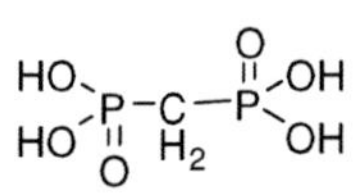

3.67

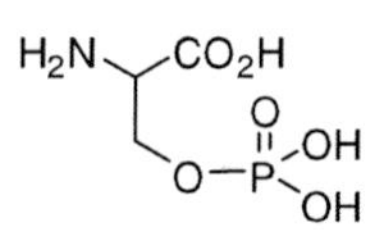

3.68

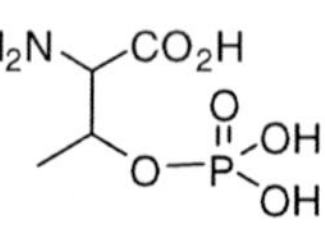

3.69

FIGURE 3.23 Structures of compounds **3.64–3.69**.

TABLE 3.9 Inhibition Constants of Organic Phosphonates/Phosphates Against hCA II, IV, V, and IX for the CO_2 Hydration Reaction at 20°C

	K_I (mM)[a]			
Inhibitor	hCA II	hCA IV	hCA V	hCA IX
3.64	98 nM	0.31	0.11	1.26
3.65	99 nM	5.4 μM	0.09	2.21
3.66	7.8 μM	79 nM	0.37	2.25
3.67	1.25	5.34	0.73	0.86
3.68	0.42	4.26	0.36	0.92
3.69	1.08	3.18	0.85	1.23

[a]Errors in the range of 3–5% of the reported value (from three different assays).

hCA I was activated by all these compounds with the best activators being **3.64** and **3.65** (K_A values of 0.10–1.20 μM). On the other hand, **3.64** and **3.65** were nanomolar inhibitors of hCA II (K_I values of 98–99 nM). Compound **3.66** demonstrated an inhibitory activity against hCA II of 7.8 mM whereas the other compounds were weak millimolar inhibitors of this isozyme (Table 3.9).

The membrane-associated isozyme hCA IV showed an interesting behavior toward this class of inhibitors. Compound **3.66** was the most efficient CA IV inhibitor with a K_I of 79 nM; however, **3.65** also showed efficient inhibition (K_I of 5.4 μM). The mitochondrial isozyme hCA V and the tumor-associated transmembrane isozyme CA IX were weakly inhibited by all these compounds (K_I in the range of 0.09–2.25 mM). It is important to note from this study the dramatic difference of activity of these derivatives against hCA IV as compared to hCA II. Thus, phosphonate motif constitutes an important zinc binding function that can be used for the design of isozyme-specific inhibitors.

All these data demonstrate that in addition to the classical sulfonamide motif, phosphonate function has strong affinity for the active site of different carbonic anhydrase isozymes and can be counted in the class of zinc binding function that can be used in the design of selective carbonic anhydrase inhibitors.

3.7 CONCLUSION

The design and discovery of carbonic anhydrase inhibitors continue to be the subject of intense research. Modulation of the zinc binding function in the structure of the inhibitors offers interesting opportunities for the development of novel drugs. Even if the classical sulfonamide function is always used as model ZBF, strong carbonic anhydrase inhibitors can be obtained varying the nature of the metal binding function. Bioisoteric sulfonamide moieties such as sulfamide, sulfamate, and their *N*-substituted analogues have proved to be ZBF of great interest for the design of compounds with selectivity and/or specificity for some of the medicinal chemistry targets belonging to this enzyme family. Other chemical classes of compounds such as phosphate and phosphonate have also been recently described. X-ray crystallography allowed a better understanding of the interactions between these new ZBFs and the enzyme active site that might be exploited for future investigation and discovery of new potent and selective carbonic anhydrase inhibitors.

REFERENCES

1. Supuran, C. T. Carbonic anhydrases: novel therapeutic applications for inhibitors and activators. *Nat. Rev. Drug Discov.* **2008**, *7*, 168–181.
2. Scozzafava, A.; Mastrolorenzo, A.; Supuran, C. T. Carbonic anhydrase inhibitors and activators and their use in therapy. *Expert Opin. Ther. Patents* **2006**, *16*, 1627–1664.
3. Supuran, C. T.; Scozzafava, A.; Conway, J., Eds. *Carbonic Anhydrase: Its Inhibitors and Activators;* CRC Press: Boca Raton, FL, **2004**; pp 1–376, and references cited therein.
4. Supuran, C. T.; Scozzafava, A.; Casini, A. Carbonic anhydrase inhibitors. *Med. Res. Rev.* **2003**, *23*, 146–189.
5. Supuran, C. T.; Scozzafava, A. Carbonic anhydrases as targets for medicinal chemistry. *Bioorg. Med. Chem.* **2007**, *15*, 4336–4350.
6. Supuran, C. T. Carbonic anhydrases: an overview. *Curr. Pharm. Des.* **2008**, *14*, 603–614.
7. Winum, J.-Y.; Scozzafava, A.; Montero, J.-L.; Supuran, C. T. Design of zinc binding functions for carbonic anhydrase inhibitors. *Curr. Pharm. Des.* **2008**, *14*, 615–621.
8. Winum, J. Y.; Scozzafava, A.; Montero, J. L.; Supuran, C. T. Metal binding functions in the design of carbonic anhydrase inhibitors. *Curr. Top. Med. Chem.* **2007**, *7*, 835–848.
9. Winum, J.-Y.; Scozzafava, A.; Montero, J.-L.; Supuran, C. T. New zinc binding motifs in the design of selective carbonic anhydrase inhibitors. *Mini Rev. Med. Chem.* **2006**, *6*, 921–936.
10. Mann, T.; Keilin, D. Sulphanilamide as a specific carbonic anhydrase inhibitor. *Nature* **1940**, *146*, 164–165.
11. Mincione, F.; Scozzafava, A.; Supuran, C. T. The development of topically acting carbonic anhydrase inhibitors as antiglaucoma agents. *Curr. Pharm. Des.* **2008**, *14*, 649–654.
12. Abbate, F.; Casini, A.; Owa, T.; Scozzafava, A.; Supuran, C. T. Carbonic anhydrase inhibitors: E7070, a sulfonamide anticancer agent, potently inhibits cytosolic isozymes I and II, and transmembrane, tumor-associated isozyme IX. *Bioorg. Med. Chem. Lett.* **2004**, *14*, 217–223.
13. De Simone, G.; Di Fiore, A.; Menchise, V.; Pedone, C.; Antel, J.; Casini, A.; Scozzafava, A.; Wurl, M.; Supuran, C. T. Carbonic anhydrase inhibitors. Zonisamide is an effective inhibitor of the cytosolic isozyme II and mitochondrial isozyme V: solution and X-ray crystallographic studies. *Bioorg. Med. Chem. Lett.* **2005**, *15*, 2315–2320.
14. Abbate, F.; Coetzee, A.; Casini, A.; Ciattini, S.; Scozzafava, A.; Supuran, C. T. Carbonic anhydrase inhibitors: X-ray crystallographic structure of the adduct of human isozyme II with the antipsychotic drug sulpiride. *Bioorg. Med. Chem. Lett.* **2004**, *14*, 337–341.
15. Weber, A.; Casini, A.; Heine, A.; Kuhn, D.; Supuran, C. T.; Scozzafava, A.; Klebe, G. Unexpected nanomolar inhibition of carbonic anhydrase by COX-2-selective celecoxib: new pharmacological opportunities due to related binding site recognition. *J. Med. Chem.* **2004**, *47*, 550–557.

16. Di Fiore, A.; Pedone, C.; D'Ambrosio, K.; Scozzafava, A.; De Simone, G.; Supuran, C. T. Carbonic anhydrase inhibitors: valdecoxib binds to a different active site region of the human isoform II as compared to the structurally related cyclooxygenase II "selective" inhibitor celecoxib. *Bioorg. Med. Chem. Lett.* **2006**, *16*, 437–442.

17. Winum, J.-Y.; Rami, M.; Scozzafava, A.; Montero, J.-L.; Supuran, C. Carbonic anhydrase IX: a new druggable target for the design of antitumor agents. *Med. Res. Rev.* **2008**, *28*, 445–463.

18. De Simone, G.; Vitale, R. M.; Di Fiore, A.; Pedone, C.; Scozzafava, A.; Montero, J.-L.; Winum, J.-Y.; Supuran, C.-T. Carbonic anhydrase inhibitors: hypoxia-activatable sulfonamides incorporating disulfide bonds that target the tumor-associated isoform IX. *J. Med. Chem.* **2006**, *49*, 5544–5551.

19. Winum, J.-Y.; Poulsen, S. A.; Supuran, C.-T. Therapeutic application of glycosidic carbonic anhydrase inhibitors. *Med. Res. Rev.* **2009**, *29*, 419–435.

20. Winum, J.-Y.; Casini, A.; Mincione, F.; Starnotti, M.; Montero, J.-L.; Scozzafava, A.; Supuran, C.-T. Carbonic anhydrase inhibitors: *N*-(*p*-sulfamoylphenyl)-alpha-D-glycopyranosylamines as topically acting antiglaucoma agents in hypertensive rabbits. *Bioorg. Med. Chem. Lett.* **2004**, *14*, 225–229.

21. Di Fiore, A.; Scozzafava, A.; Winum, J.-Y.; Montero, J.-L.; Pedone, C.; Supuran, C. T.; De Simone, G. Carbonic anhydrase inhibitors: binding of an antiglaucoma glycosylsulfanilamide derivative to human isoform II and its consequences for the drug design of enzyme inhibitors incorporating sugar moieties. *Bioorg. Med. Chem. Lett.* **2007**, *17*, 1726–1731.

22. Winum, J.-Y.; Scozzafava, A.; Montero, J.-L.; Supuran, C. T. Sulfamates and their therapeutic potential. *Med. Res. Rev.* **2005**, *25*, 186–228.

23. Winum, J.-Y.; Scozzafava, A.; Montero, J.-L.; Supuran, C. T. Therapeutic applications of sulfamates. *Expert Opin. Ther. Patents* **2004**, *14*, 1273–1308.

24. Briganti, F.; Pierattelli, A.; Scozzafava, A.; Supuran, C. T. Carbonic anhydrase inhibitors. Part 37. Novel classes of isozyme I and II inhibitors and their mechanism of action. Kinetic and spectroscopic investigations on native and cobalt-substituted enzymes. *Eur. J. Med. Chem.* **1996**, *31*, 1001–1010.

25. Abbate, F.; Supuran, C. T.; Scozzafava, A.; Orioli, P.; Stubbs, M. T.; Klebe, G. Nonaromatic sulfonamide group as an ideal anchor for potent human carbonic anhydrase inhibitors: role of hydrogen-bonding networks in ligand binding and drug design. *J. Med. Chem.* **2002**, *45*, 3583–3587.

26. Lo, Y. S.; Nolan, J. C.; Maren, T. H.; Welstead, W. J.; Gripshover, D. F.; Shamblee, D. A. Synthesis and physiochemical properties of sulfamate derivatives as topical antiglaucoma agents. *J. Med. Chem.* **1992**, *35*, 4790–4794.

27. Brechue, W. F.; Maren, T. H. Carbonic anhydrase inhibitory activity and ocular pharmacology of organic sulfamates. *J. Pharm. Exp. Ther.* **1992**, *264*, 670–675.

28. Maryanoff, B. E.; Nortey, S. O.; Gardocki, J. F.; Shank, R. P.; Dodgson, S. P. Anticonvulsant *O*-alkyl sulfamates. 2,3:4,5-Bis-*O*-(1-methyl ethylidene)-beta-D-fructopyranose sulfamate and related compounds. *J. Med. Chem.* **1987**, *30*, 880–887.

29. Maryanoff, B. E.; Costanzo, M. J.; Nortey, S. O.; Greco, M. N.; Shank, R. P.; Schupsky, J. J.; Ortegon, M. P.; Vaught, J. L. Structure–activity studies on anticonvulsant sugar sulfamates related to topiramate. Enhanced potency with cyclic sulfate derivatives. *J. Med. Chem.* **1998**, *41*, 1315–1343.

30. Recacha, R.; Costanzo, M. J.; Maryanoff, B. E.; Chattopadhyay, D. Crystal structure of human carbonic anhydrase II complexed with an anti-convulsant sugar sulphamate. *Biochem. J.* **2002**, *361*, 437–441.

31. Casini, A.; Antel, J.; Abbate, F.; Scozzafava, A.; David, S.; Waldeck, H.; Schafer, S.; Supuran, C. T. Carbonic anhydrase inhibitors: SAR and X-ray crystallographic study for the interaction of sugar sulfamates/sulfamides with isozymes I, II and IV. *Bioorg. Med. Chem. Lett.* **2003**, *13*, 841–845.

32. Nussbaumer, P.; Billich, A. Steroid sulfatase inhibitors. *Expert Opin. Ther. Patents* **2003**, *13*, 605–625.

33. Vicker, N.; Ho, Y.; Robinson, J.; Woo, L. L.; Purohit, A.; Reed, M. J.; Potter, B. V. Docking studies of sulphamate inhibitors of estrone sulfatase in human carbonic anhydrase II. *Bioorg. Med. Chem. Lett.* **2003**, *13*, 863–865.

34. Ho, Y. T.; Purohit, A.; Vicker, N.; Newman, S. P.; Robinson, J. J.; Leese, M. P.; Ganeshapillai, D.; Woo, L. W. L.; Potter, B. V.; Reed, M. J. Inhibition of carbonic anhydrase II by steroidal and non-steroidal sulfamates. *Biochem. Biophys. Res. Commun.* **2003**, *305*, 909–914.

35. Abbate, F.; Winum, J. Y.; Potter, B. V.; Casini, A.; Montero, J. L.; Scozzafava, A.; Supuran, C. T. Carbonic anhydrase inhibitors: X-ray crystallographic structure of the adduct of human isozyme II with EMATE, a dual inhibitor of carbonic anhydrases and steroid sulfatase. *Bioorg. Med. Chem. Lett.* **2004**, *14*, 231–234.

36. Winum, J. Y.; Vullo, D.; Casini, A.; Montero, J.-L.; Scozzafava, A.; Supuran, C. T. Carbonic anhydrase inhibitors. Inhibition of cytosolic isozymes I and II and transmembrane, tumor-associated isozyme IX with sulfamates including EMATE also acting as steroid sulfatase inhibitors. *J. Med. Chem.* **2003**, *46*, 2197–2204.

37. Winum, J. Y.; Vullo, D.; Casini, A.; Montero, J. L.; Scozzafava, A.; Supuran, C. T. Carbonic anhydrase inhibitors: inhibition of transmembrane, tumor-associated isozyme IX, and cytosolic isozymes I and II with aliphatic sulfamates. *J. Med. Chem.* **2003**, *46*, 5471–5477.

38. Thiry, A.; Dogné, J. M.; Masereel, B.; Supuran, C. T. Targeting tumor-associated carbonic anhydrase IX in cancer therapy. *Trends Pharmacol. Sci.* **2006**, *27*, 566–573.

39. Winum, J.-Y.; Pastorekova, S.; Jakubickova, L.; Montero, J.-L.; Scozzafava, A.; Pastorek, J.; Vullo, D.; Innocenti, A.; Supuran, C. T. Carbonic anhydrase inhibitors: synthesis and inhibition of cytosolic/tumor-associated carbonic anhydrase isozymes I, II, and IX with bis-sulfamates. *Bioorg. Med. Chem. Lett.* **2005**, *15*, 579–584.

40. Winum, J.-Y.; Scozzafava, A.; Montero, J.-L.; Supuran, C.-T. Therapeutic potential of sulfamides as enzyme inhibitors. *Med. Res. Rev.* **2006**, *26*, 767–792.

41. Winum, J.-Y.; Scozzafava, A.; Montero, J.-L.; Supuran, C.-T. The sulfamide motif in the design of enzyme inhibitors. *Expert Opin. Ther. Patents* **2006**, *16*, 27–47.

42. Scozzafava, A.; Banciu, M. D.; Popescu, A.; Supuran, C. T. Carbonic anhydrase inhibitors: inhibition of isozymes I, II, and IV by sulfamide and sulfamic acid derivatives. *J. Enzyme Inhib.* **2000**, *15*, 443–453.

43. Casini, A.; Winum, J.-Y.; Montero, J.-L.; Scozzafava, A.; Supuran, C. T. Carbonic anhydrase inhibitors: inhibition of cytosolic isozymes I and II with sulfamide derivatives. *Bioorg. Med. Chem. Lett.* **2003**, *13*, 837–840.

44. Winum, J.-Y.; Toupet, L.; Barragan, V.; Dewynter, G.; Montero, J.-L. *N*-(*tert*-Butoxycarbonyl)-*N*-[4-(dimethylazaniumylidene)-1,4-dihydropyridin-1-ylsulfonyl] azanide: a new

sulfamyolating agent. Structure and reactivity toward amines. *Org. Lett.* **2001**, *3*, 2241–2243.

45. Winum, J.-Y.; Cecchi, A.; Montero, J.-L.; Innocenti, A.; Scozzafava, A.; Supuran, C. T. Carbonic anhydrase inhibitors. Synthesis and inhibition of cytosolic/tumor-associated carbonic anhydrase isozymes I, II, and IX with boron-containing sulfonamides, sulfamides, and sulfamates: toward agents for boron neutron capture therapy of hypoxic tumors. *Bioorg. Med. Chem. Lett.* **2005**, *15*, 3302–3306.
46. Winum, J.-Y.; Temperini, C.; El Cheikh, K.; Innocenti, A.; Vullo, D.; Ciattini, S.; Montero, J.-L.; Scozzafava, A.; Supuran, C. T. Carbonic anhydrase inhibitors: clash with Ala65 as a means for designing inhibitors with low affinity for the ubiquitous isozyme II, exemplified by the crystal structure of the topiramate sulfamide analogue. *J. Med. Chem.* **2006**, *49*, 7024–7031.
47. Krebs, H. A. Inhibition of carbonic anhydrase by sulfonamides. *Biochem. J.* **1948**, *43*, 525–528.
48. Mincione, F.; Menabuoni, L.; Briganti, F.; Mincione, G.; Scozzafava, A.; Supuran, C. T. Carbonic anhydrase inhibitors. Inhibition of isozymes I, II and IV with *N*-hydroxysulfonamides. A novel class of intraocular pressure lowering agents. *J. Enzyme Inhib.* **1998**, *13*, 267–284.
49. Supuran, C. T.; Scozzafava, A.; Briganti, F. Carbonic anhydrase inhibitors: *N*-cyanosulfonamides: a new class of high affinity isozyme II and IV inhibitors. *J. Enzyme Inhib.* **1999**, *14*, 289–306.
50. Fenesan, I.; Popescu, R.; Scozzafava, A.; Crucin, V.; Mateiciuc, E.; Bauer, R.; Ilies, M. A.; Supuran, C. T. Carbonic anhydrase inhibitors. Phosphoryl sulfonamides. A new class of high affinity inhibitors of isozyme I. *J. Enzyme Inhib.* **2000**, *15*, 297–310.
51. Supuran, C. T.; Scozzafava, A. Applications of carbonic anhydrase inhibitors and activators in therapy. *Expert Opin. Ther. Patents* **2002**, *12*, 217–242.
52. Scolnik, L. R.; Clements, A. M.; Liao, J.; Crenshaw, L.; Hellberg, M.; May, J.; Dean, D. R.; Christianson, D. W. Novel binding mode of hydroxamate inhibitors to human carbonic anhydrase II. *J. Am. Chem. Soc.* **1997**, *119*, 850–851.
53. Scozzafava, A.; Supuran, C. T. Hydroxyurea is a carbonic anhydrase inhibitor. *Bioorg. Med. Chem.* **2003**, *11*, 2241–2246.
54. Temperini, C.; Innocenti, A.; Scozzafava, A.; Supuran, C. T. *N*-Hydroxyurea: a versatile zinc binding function in the design of metalloenzyme inhibitors. *Bioorg. Med. Chem. Lett.* **2006**, *16*, 4316–4320.
55. Briganti, F.; Pierattelli, A.; Scozzafava, A.; Supuran, C. T. Carbonic anhydrase inhibitors. Part 37. Novel classes of isozyme I and II inhibitors and their mechanism of action. Kinetic and spectroscopic investigations on native and cobalt-substituted enzymes. *Eur. J. Med. Chem.* **1996**, *31*, 1001–1010.
56. Abbate, F.; Supuran, C. T.; Scozzafava, A.; Orioli, P.; Stubbs, M. T.; Klebe, G. Nonaromatic sulfonamide group as an ideal anchor for potent human carbonic anhydrase inhibitors: role of hydrogen-bonding networks in ligand binding and drug design. *J. Med. Chem.* **2002**, *45*, 3583–3587.
57. Temperini, C.; Winum, J.-Y.; Montero, J.-L.; Scozzafava, A.; Supuran, C. T. Carbonic anhydrase inhibitors. The X-ray crystal structure of the adduct of *N*-hydroxysulfamide with the isozyme II explains why this new zinc binding function is effective in the design of potent inhibitors. *Bioorg. Med. Chem. Lett.* **2007**, *17*, 2797–2801.

58. Winum, J. Y.; Innocenti, A.; Nasr, J.; Montero, J.-L.; Scozzafava, A.; Vullo, D.; Supuran, C. T. Carbonic anhydrase inhibitors: synthesis and inhibition of cytosolic/tumor-associated carbonic anhydrase isozymes I, II, IX, and XII with *N*-hydroxysulfamides: a new zinc-binding function in the design of inhibitors. *Bioorg. Med. Chem. Lett.* **2005**, *15*, 2353–2358.

59. Bonnac, L.; Innocenti, A.; Winum, J.-Y.; Casini, A.; Montero, J.-L.; Scozzafava, A.; Barragan, V.; Supuran, C. T. Carbonic anhydrase inhibitors: aliphatic *N*-phosphorylated sulfamate. A novel zinc anchoring group leading to nanomolar inhibitors. *J. Enzyme Inhib. Med. Chem.* **2004**, *19*, 287–291.

60. Rusconi, S.; Innocenti, A.; Vullo, D.; Mastrolorenzo, A.; Scozzafava, A.; Supuran, C. T. Carbonic anhydrase inhibitors. Interaction of isozymes I, II, IV, V and IX with phosphates, carbamoyl phosphate, and the phosphonate antiviral drug foscarnet. *Bioorg. Med. Chem. Lett.* **2004**, *14*, 5763–5767.

61. Temperini, C.; Innocenti, A.; Guerri, A.; Scozzafava, A.; Rusconi, S.; Supuran, C. T. Phosph(on)ate as a zinc binding group in metalloenzyme inhibitors: X-ray crystal structure of the antiviral drug foscarnet complexed to human carbonic anhydrase I. *Bioorg. Med. Chem. Lett.* **2007**, *17*, 5763–5767.

62. Winum, J.-Y.; Innocenti, A.; Gagnard, V.; Montero, J.-L.; Scozzafava, A.; Vullo, D.; Supuran, C. T. Carbonic anhydrase inhibitors. Interaction of isozymes I, II, IV, V and IX with organic phosphates and phosphonates. *Bioorg. Med. Chem. Lett.* **2005**, *15*, 1683–1686.

CHAPTER 4

X-Ray Crystallography of Carbonic Anhydrase Inhibitors and Its Importance in Drug Design

VINCENZO ALTERIO[1], ANNA DI FIORE[1], KATIA D'AMBROSIO[1], CLAUDIU T. SUPURAN[2], and GIUSEPPINA DE SIMONE[1]

[1]Istituto di Biostrutture e Bioimmagini-CNR, Via Mezzocannone 16, 80134 Naples, Italy
[2]Laboratorio di Chimica Bioinorganica, Università degli Studi di Firenze, Room 188, Via della Lastruccia 3, I-50019 Sesto Fiorentino (Firenze), Italy

4.1 INTRODUCTION

Carbonic anhydrases (CAs, EC 4.2.1.1) constitute an ubiquitous family of metalloenzymes found in prokaryotes and eukaryotes that catalyze the reversible hydration of carbon dioxide to the bicarbonate ion ($CO_2 + H_2O \rightleftarrows HCO_3^- + H^+$). These proteins are encoded by four evolutionarily unrelated gene families: the α-CAs (in vertebrates, bacteria, algae, and cytoplasm of green plants), the β-CAs (predominantly in bacteria, algae, and chloroplasts), the γ-CAs (in archaea and some bacteria),[1–3] and the δ-CAs (in some marine diatoms).[3] There are no significant sequence homologies between representatives of the different CA families, but all members are zinc enzymes.[3]

Human CAs belong all to the α-class; to date, 15 isozymes have been identified, among which 12 are catalytically active (CA I–IV, VA–VB, VI–VII, IX, and XII–XIV), whereas the CA-related proteins (CARPs) VIII, X, and XI are devoid of any catalytic activity.[3] α-CA isozymes widely differ in their cellular localization. In particular, CA I, II, III, VII, and XIII reside in cytosol; CA IV, IX, XII, and XIV are associated with membranes; CA VA and VB occur in mitochondria; and CA VI is secreted.[2–12]

α-CA isozymes are widely distributed in many tissues and organs, where they play a crucial role in various physiological processes, such as CO_2/HCO_3^- transport between metabolizing tissues and lungs, pH and CO_2 homeostasis, electrolyte secretion, biosynthetic reactions (gluconeogenesis, lipogenesis, and ureagenesis), bone resorption, and tumorigenicity.[13–17] As a consequence, in the last years many of the CA isozymes have become important therapeutic targets for pharmaceutical research. However, most of the available CA-directed pharmacological agents are still far from being optimal drugs. In fact, they present various side effects, mainly because of their lack of selectivity for the different CA isozymes.[2,3] Thus, developing isozyme-specific CA inhibitors would be highly beneficial in obtaining novel classes of drugs. Prospects for achieving such a goal have not been very optimistic because of the high sequence similarity observed between various isozymes. However, recently a large number of X-ray crystallographic studies on different α-CA isozymes and CA–inhibitor complexes have provided a scientific basis for the rational drug design of more selective enzyme inhibitors.[3,7,16–28]

The aim of this chapter is to give a comprehensive update of the reported structural studies on CAs belonging to the α-class. In particular, in the first part of this review, we will examine the principal structural features of the α-CA isozymes for which the 3D structure has been solved so far. Next, we will summarize the current state of the art on complexes of the most thoroughly characterized CA isozyme, hCA II, with the principal classes of CA inhibitor, as determined by X-ray crystallography. Finally, some recent developments in the field of selective CA inhibitors will be also highlighted.

4.2 α-CARBONIC ANHYDRASE THREE-DIMENSIONAL STRUCTURES

To date, the three-dimensional structures of CA I, II, III, IV, VA, XII, XIII, and XIV from different mammals have been determined,[27–35] revealing that all these enzyme isoforms have a roughly ovoidal shape with approximate dimensions of $50 \times 40 \times 40$ Å^3 and a typical fold characterized by a central 10-stranded antiparallel β-sheet surrounded by several helices and additional β-strands (Fig. 4.1). As expected, on the base of high sequence identity, the three-dimensional structures of all these isozymes were similar and all the secondary structure elements were strictly conserved (Fig. 4.2). However, an accurate structural comparison revealed a number of small local structural differences that were mainly localized around residues 125–131 (amino acid numbering refers to hCA II), thus occurring both on the molecular surface (residues 125–130) and in the middle of the active site (residue 131).

The active site is located in a large, cone-shaped cavity that reaches almost to the center of the molecule. The catalytic zinc ion is located at the bottom of the cavity, coordinated by three conserved His residues and a water molecule/hydroxide ion.[36–39] The Zn^{2+}-bound solvent molecule is also engaged in H-bond interactions with another water molecule (called the deep water) and with the hydroxyl moiety of a conserved Thr residue (Thr199) that in turn is bridged to the carboxylate moiety of a conserved Glu residue (Glu106) (Fig. 4.3). These interactions enhance the nucleophilicity of the Zn^{2+}-bound water molecule and orient the CO_2 substrate in a location favorable for the nucleophilic attack (Fig. 4.4).[3,4,37,40] Finally, in all these enzyme isoforms, the active site cavity generally consists of two distinct portions made of hydrophobic or hydrophilic amino acids. In particular, residues in positions 121, 131, 141, 143, 198, and 207 delimit the hydrophobic region, while those in positions 62, 64, 67, and 92 identify the hydrophilic one (Fig. 4.2).[30]

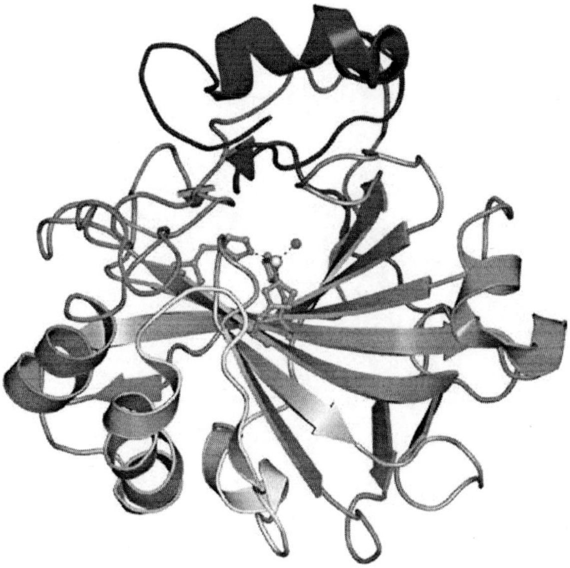

FIGURE 4.1 Ribbon diagram of hCA II, which has been chosen as representative CA isozyme. The active site Zn^{2+} ion coordination is also reported. (See the color version of this figure in Color Plates section.)

A careful comparison between the active sites of all these isozymes revealed a general conservation of the nature and conformation of most of the amino acids present within the active site cavity (Trp5, Tyr7, Gln92, Val121, Leu141, Val143, Leu198, Pro201, Pro202, Val207, Trp209) with few amino acid substitutions at positions 62, 65, 67, 69, 91, 131, 132, 135, 136, 200, and 204 (Fig. 4.2). However, since the latter residues have often been reported to have an important role in binding of inhibitors to CAs,[16,41–44] their variability could determine modifications of the specific isozyme interaction with inhibitor molecules.

4.3 THREE-DIMENSIONAL STRUCTURES OF hCA II/INHIBITOR COMPLEXES

Three main classes of CA inhibitors (CAIs) have been studied crystallographically: the ureates/hydroxamates, the metal complexing anions, and the unsubstituted sulfonamides and their bioisosteres such as sulfamates and sulfamides.[2,39,45,46] In the next section, the main features of the binding to hCA II active site of the most representative inhibitors belonging to these three classes will be schematically reviewed.

4.3.1 Binding of Ureates and Hydroxamates

In addition to the interconversion between carbon dioxide and bicarbonate, hCA II also catalyzes other nonphysiological reactions, among which the hydration of cyanamide

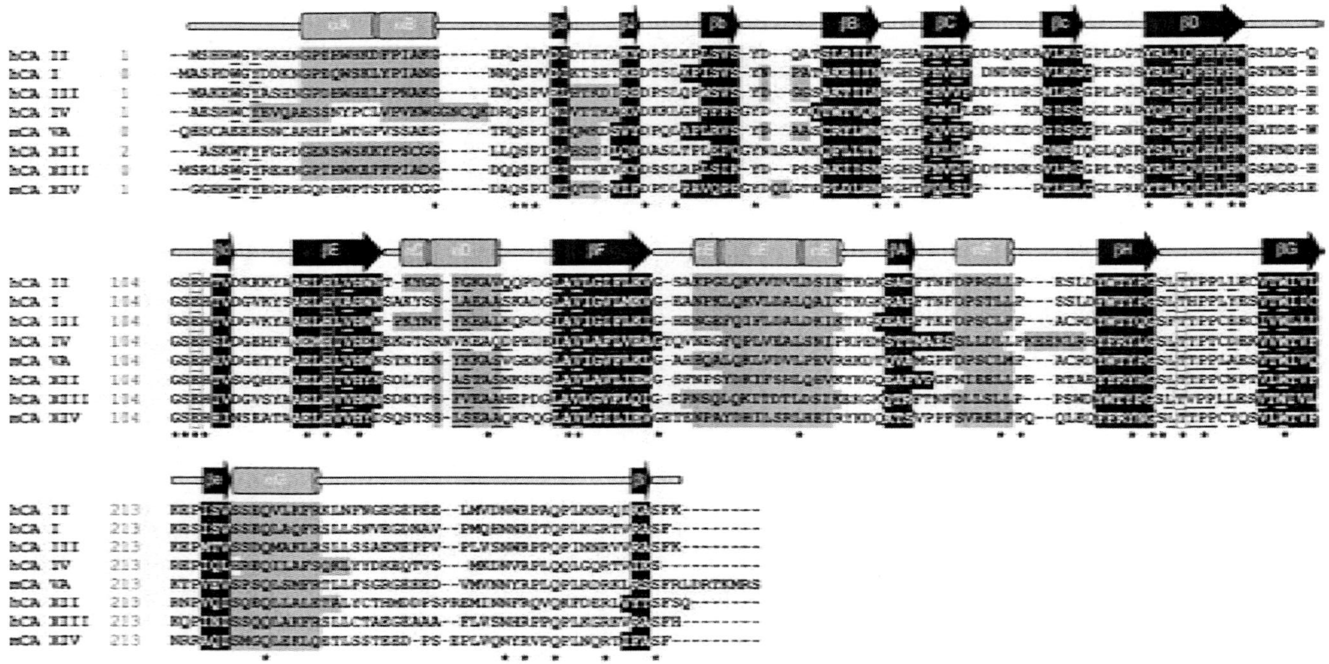

FIGURE 4.2 Sequence alignment of α-CAs with known three-dimensional structure. hCA II secondary structure elements are shown schematically; helices are represented by solid cylinders and β-strands as arrows. Helix and β-strand regions for the various isozymes are highlighted in grey and black, respectively. Conserved residues are starred, catalytic triad, Thr199 and Glu106, are boxed, while residues delimiting the active site cavity are underlined. β-strands and helix regions are named as reported by Eriksson et al.[30]

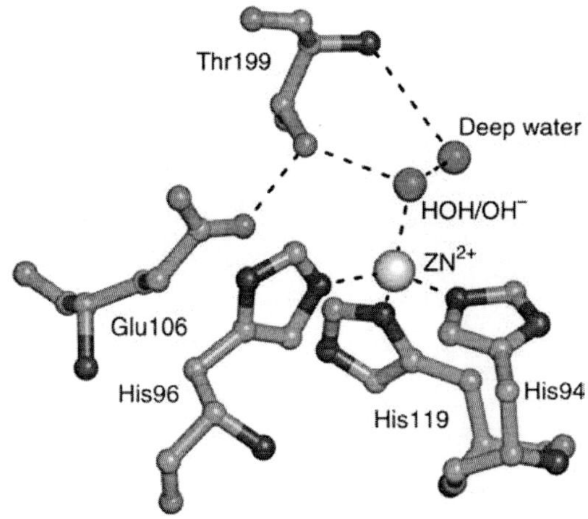

FIGURE 4.3 hCA II active site. The Zn^{2+} ion and its ligands are shown.

to urea ($HN{=}C{=}NH + H_2O \rightleftarrows H_2NCONH_2$). This reaction has been thoroughly characterized by spectroscopic, kinetic, and crystallographic techniques,[47,48] demonstrating that cyanamide interacts with the metal ion within the CA active site, adding to the coordination sphere and not substituting the metal-bound solvent molecule

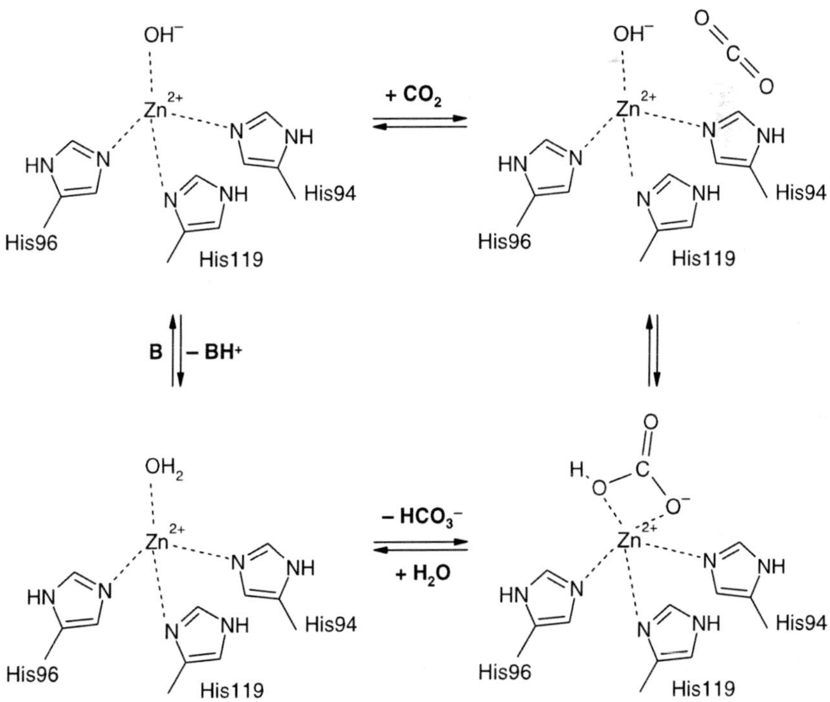

FIGURE 4.4 Schematic representation of the catalytic mechanism for the CA-catalyzed CO_2 hydration.

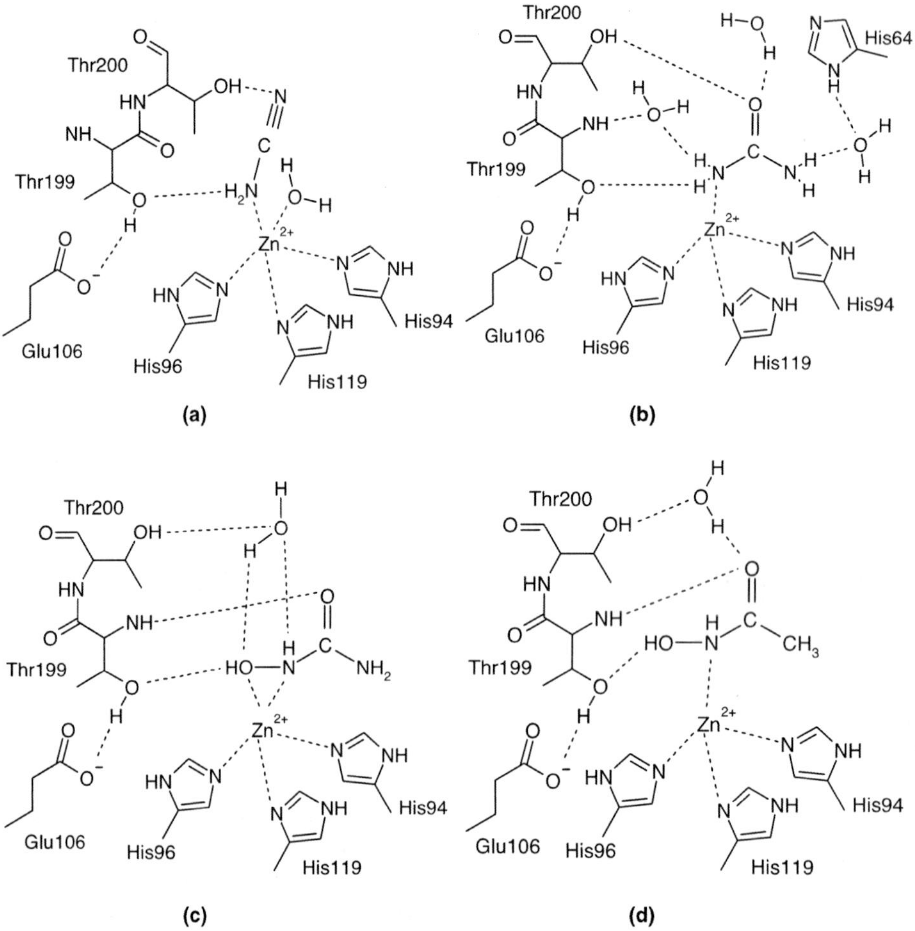

FIGURE 4.5 Schematic representation of the binding mode of (a) cyanamide, (b) urea, (c) *N*-hydroxyurea, and (d) acetohydroxamic acid.

(Fig. 4.5a).[47] It thereafter undergoes a nucleophilic attack from the metal-bound hydroxide ion, forming urea that remains bound to the metal, as observed in the X-ray crystal structure of hCA II soaked in cyanamide solutions for several hours.[47] The urea molecule is directly coordinated to the active site Zn^{2+} ion through a protonated nitrogen atom (Fig. 4.5b). Several hydrogen bonds involving active site residues Thr199 and Thr200, as well as three water molecules, further stabilize the hCA II–urea adduct.[47] Kinetic studies in solution further proved that urea acts as a tight binding inhibitor of hCA II, with very slow binding kinetics ($k_{on} = 2.5 \times 10^{-5}\ s^{-1}\ M^{-1}$).[47] These findings have been the starting point of new studies regarding the interaction of CAs with small molecules possessing appropriate structural features to act as new zinc binding functions. One such molecule is *N*-hydroxyurea, the simplest compound incorporating the hydroxamate functionality.[49] X-ray crystallography showed that *N*-hydroxyurea binds to the Zn^{2+} of hCA II active site in a bidentate mode by means of the oxygen and nitrogen atoms of the NHOH moiety, also participating in a network of hydrogen bonds with a water molecule and Thr199 (Fig. 4.5c).[50] It is worth noting that structural studies have demonstrated that the isosteric acetohydroxamic acid is coordinated in a monodentate fashion to the Zn^{2+} ion, whereas its OH and CO groups participate in two hydrogen bonds with Thr199OG and Thr199N atoms (Fig. 4.5d).[51] These findings suggest that the hydroxamate function can be usefully exploited for obtaining potent CA inhibitors due to its high versatility of binding to the metal ion present in the active site of these enzymes.

4.3.2 Binding of Inorganic Anions

Most monovalent anions act as CAs inhibitors with apparent dissociation constants that vary considerably from a few micromolar to about 1 M.[39] Crystallographic studies have revealed that four distinguishable binding modes are possible for these anions with respect to the Zn^{2+} ion in the active site: regular tetracoordination, distorted tetracoordination, pentacoordination, and inhibition without binding to the metal.[45] The diversity in behavior between different inhibitors can be ascribed to the structural features of the ligands. In particular, anions having a protonated ligand atom replace the zinc-bound solvent molecule with a regular tetrahedral coordination geometry and form a hydrogen bond with the Thr199OG atom. Examples of this kind of anions are hydrogen sulfite (HS^-)[52] and hydrogen bisulfite (HSO_3^-)[36] (Fig. 4.6a). When the anionic inhibitor lacks a protonated ligand atom, three different binding modes can be realized. Anions as bromide (Br^-)[53] and azide (N_3^-)[53,54] still coordinate the zinc ion with a distorted tetrahedral geometry, since they cannot form a hydrogen bond with Thr199OG atom (Fig. 4.6b). On the contrary, anions such as formate[36] and thiocyanate[55] bind to the enzyme active site by addition to the metal coordination sphere, generating a trigonal bipyramidal species (Fig. 4.6c). Finally, a last group of anions, including nitrate, cyanide, and cyanate,[52,56] does not seem to coordinate the zinc ion but to be located in a hydrophobic cavity near the Zn^{2+} and to be hydrogen bonded to the Thr199N atom (Fig. 4.6d), but these data need to be confirmed by more detailed studies.

4.3.3 Binding of Sulfonamide/Sulfamate/Sulfamide Inhibitors

Sulfonamides were early on shown to inhibit CAs in 1940.[57] Starting from this finding, a large number of kinetic, physiological, and pharmacological studies on these compounds have been performed.[3,46,58–60] A considerable number of crystal structures of CA–sulfonamide complexes have also been reported,[2,3] providing a detailed description of the binding mode of these compounds to the CA active site. In all the adducts, the sulfonamide group binds to the enzyme active site in its negatively charged deprotonated form,[61] with the NH^- moiety that replaces the zinc-bound water molecule and coordinates the metal ion with a regular tetrahedral geometry (Fig. 4.7). This NH^- moiety also participates in hydrogen bonding with the Thr199OG atom, which in turn is engaged in another hydrogen bonding with the carboxylate group of Glu106 residue.[4,33,37,62,63] One of the oxygen atoms of the sulfonamide group is hydrogen bonded to Thr199N atom (Fig. 4.7), while the other oxygen atom is about 3.2 Å away from the zinc ion.[2] These studies provided evidence that the sulfonamide

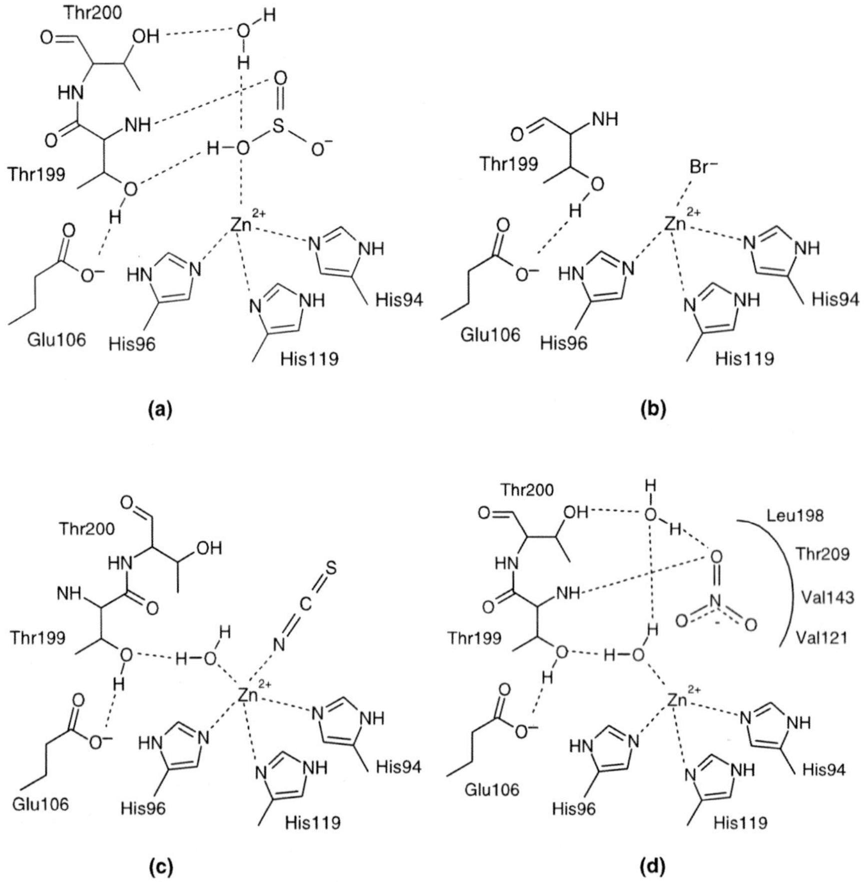

FIGURE 4.6 Schematic representation of the binding mode of some representative anions: (a) bisulfite, (b) bromide, (c) thiocyanate, and (d) nitrate.

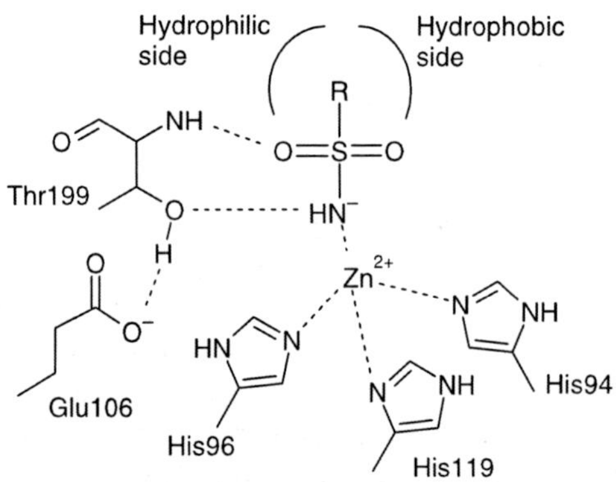

FIGURE 4.7 Schematic representation of the CA inhibition mechanism by a generic sulfonamide inhibitor.

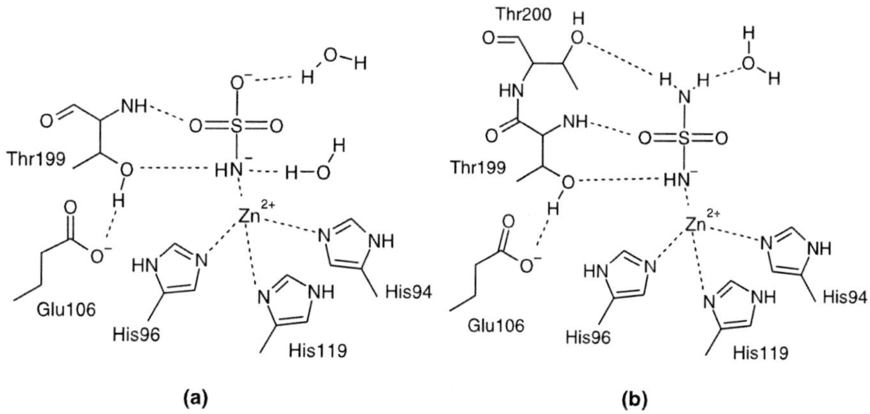

FIGURE 4.8 Binding mode of (a) sulfamate, supposedly binding as OSO_2NH^{2-}, and (b) sulfamide, supposedly present as $(NH_2)SO_2NH^-$, to the hCA II active site.

group represents an ideal ligand of the CA active site, since it combines the negative charge of the nitrogen with the positively charged zinc ion. Moreover, the presence of one proton on the coordinated nitrogen atom satisfies the hydrogen bond acceptor function of Thr199OG atom, which forms a strong bond with it.[2]

Recently, the crystal structures of hCA II adducts with the simplest parent compounds of all sulfonamides, that is, sulfamide ($H_2NSO_2NH_2$) and sulfamic acid (H_2NSO_3H), have been reported.[63] Both the sulfamic acid and the sulfamide bind to the CA II active site in their anionic form, coordinating to the Zn^{2+} with a regular tetrahedral geometry (Fig. 4.8). These studies highlighted that, despite structural similarities between these two inhibitors and the classical sulfonamides of the type R-SO_2NH_2, the substitution of the C-SO_2NH_2 with a N-SO_2NH_2/O-SO_2NH_2 bond in sulfamide/sulfamic acid leads to a different binding mode of these molecules to the enzyme active site. In particular, the latter two moieties are able to participate in a more intricate hydrogen bond network that differs from that formed by the classical sulfonamide inhibitors (Fig. 4.8).[63] These molecules have thus been suitably used as lead compounds for the design of new types of inhibitors, namely, the organic sulfamate and sulfamide derivatives.[2,3,58]

4.3.3.1 Benzenesulfonamides After the report of Mann and Keilin in 1940[57] that sulfanilamide (**4.1**) acts as a potent and specific CA inhibitor, many benzenesulfonamides were synthesized and investigated for their CA inhibitory action. These studies allowed to derive for this class of compounds the following structure–activity relationships: (i) 2-substituted and 2,4- and 3,4-disubstituted benzenesulfonamides are generally weaker inhibitors than 4-substituted derivatives; (ii) 4-substituents inducing good activities include halogens, acetamido and alkoxycarbonyl, and esters of 4-sulfamoylbenzoic acid; (iii) promising activities as well as desired physicochemical properties are seen for compounds possessing carboxy, hydrazido, ureido, thioureido, and methylamino moieties in position 4.[64,65] For a large number of these

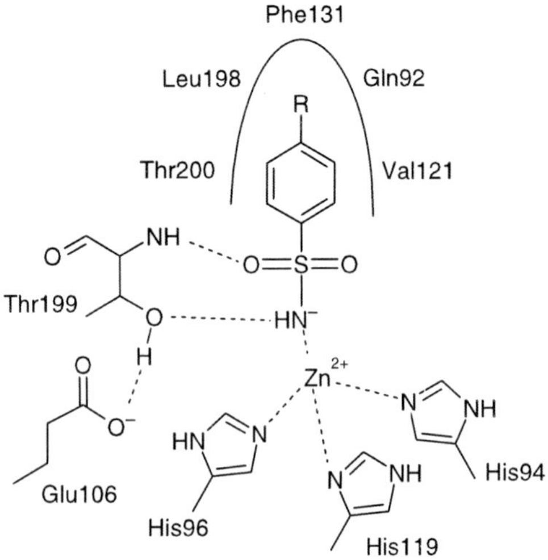

FIGURE 4.9 Schematic representation of the CA inhibition mechanism by a generic benzensulfonamide inhibitor.

compounds, the X-ray structures of the complexes with the physiologically most relevant isozyme, CA II, have been reported at a very good resolution providing important information on the binding modes of these molecules within the enzyme active site.[2,39,45,46] In particular, the phenyl sulfonamide moiety of all these inhibitors occupies the same position within the hCA II active site making several van der Waals interactions with the side chains of Gln92, Val121, Phe131, Leu198, and Thr200, in addition to the above-described Zn^{2+} coordination (Fig. 4.9). In contrast, different interactions of the moieties used to functionalize these molecules may occur. In the next section, the numerous benzenesulfonamide CAIs so far available have been grouped on the basis of either the functional groups used to derivatize the phenyl ring or the mechanism of action. Moreover, the main structural features of the adducts that some representative inhibitors form with hCA II have also been reviewed.

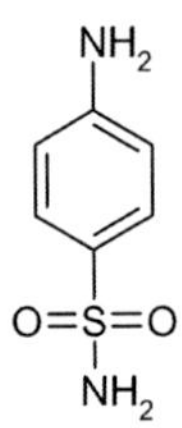

4.1

Sulfamoyl Benzenecarboxyamide Derivatives The reaction of 4-carboxybenzenesulfonamide (**4.2**) or 4-chloro-3-sulfamoyl benzoic acid (**4.3**) with different scaffolds possessing free amino groups leads to sulfamoyl-benzenecarboxyamide derivatives of type **4.4** and **4.5**, which constitute a particular class of CAIs.[28,66–72]

Several kinetic and structural studies on different types of these compounds are available at the moment in the literature showing that, depending from the nature of the R group, CAIs with varied affinity toward the different isozymes and diverse physicochemical properties can be obtained. As an example, the derivatization of **4.2** with carboxy-protected amino acids/dipeptides, or aromatic/heterocyclic sulfonamides, led to molecules that showed very strong *in vitro* affinity for the isozymes CA II and CA IV, which are involved in aqueous humor secretion within the eye. Moreover, these compounds presented an excellent penetrability through the cornea to reach ciliary processes CAs due to their good hydrophilicity/lipophilicity balance.[66] Thus, such compounds were easily formulated as eye drops either in solution or as suspensions at physiological pH values, and their *in vivo* efficacy in animal models of glaucoma was much higher than that of the clinically available topically acting antiglaucoma sulfonamides dorzolamide (**4.6**) and brinzolamide (**4.7**).[66]

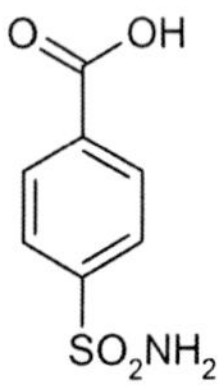

4.2

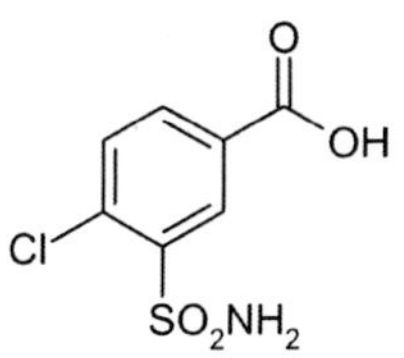

4.3

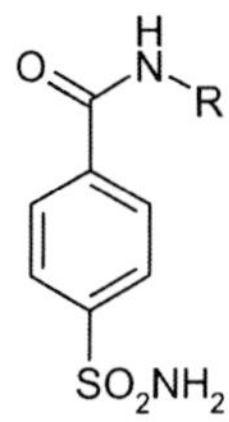

4.4

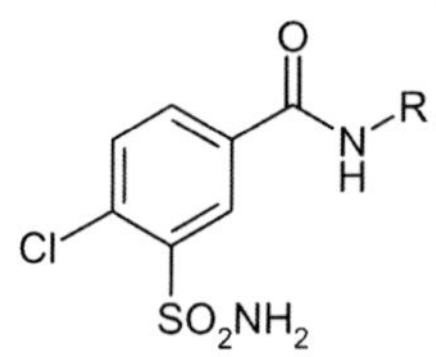

4.5

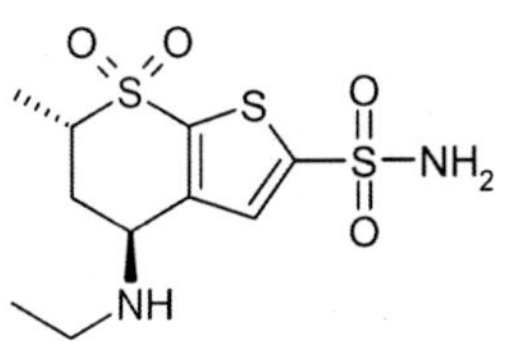

4.6

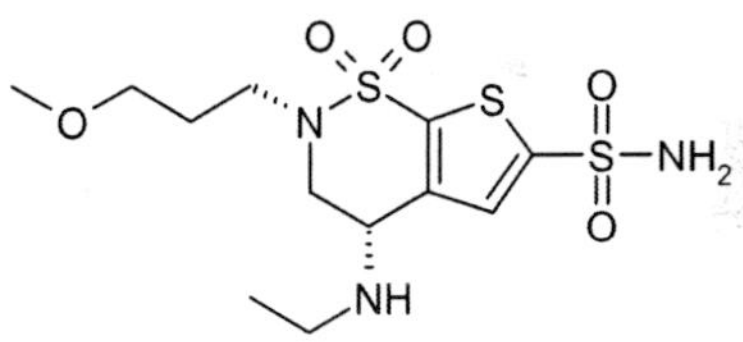

4.7

4-(Aminosulfonyl)-*N*-phenylmethylbenzamide (**4.8**) is another sulfamoyl-benzenecarboxyamide derivative that has been extensively used as lead compound to generate potent CAIs. This compound strongly inhibits hCA II with a dissociation constant of 2.1 nM.[67,73] The crystal structure of the hCA II–**4.8** adduct provided a structural explanation to the high affinity of the inhibitor toward the enzyme. In particular, besides the canonical interactions of the benzenesulfonamide moiety within the enzyme active site, an important edge-to-face interaction between the Phe131 and the benzyl ring of the inhibitor was observed.[74] The latter interaction suggested the presence of a weak but favorable electrostatic attraction between the partial positive charge on the ring hydrogen atoms of Phe131 and the partial negative charge above the ring plane of the inhibitor phenyl group. The effect of various fluorine substitution patterns of the inhibitor benzyl group (see compounds **4.9–4.12**) on the inhibitory properties of such compound has also been extensively studied from both kinetic and structural points of view,[24,73] showing that a tight binding inhibitor can be improved even further by simply substituting fluorine for hydrogen at specific locations to enhance weakly polar interactions with adjacent enzyme residues.[24]

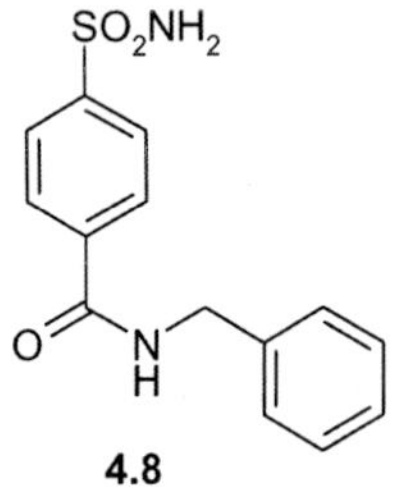

4.8

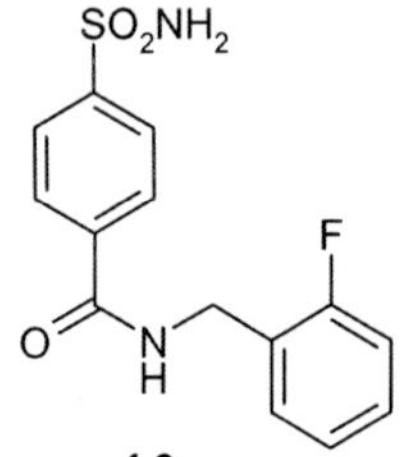

4.9

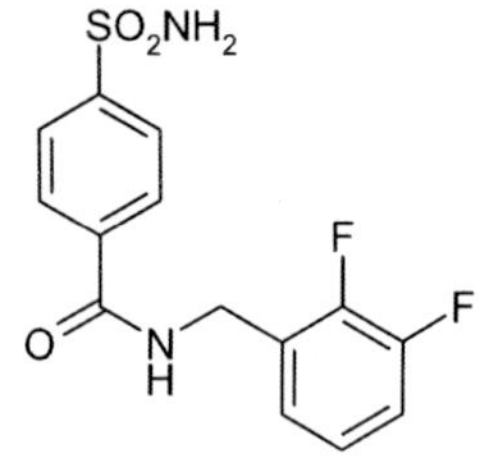

4.10

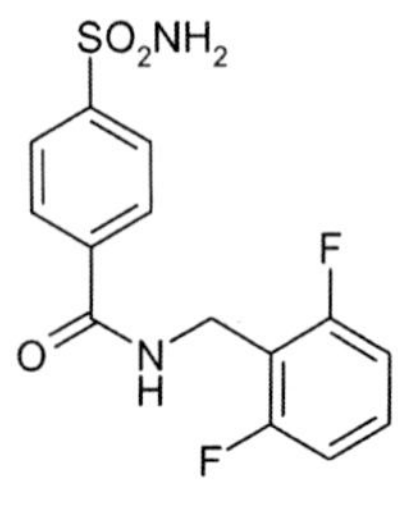

4.11

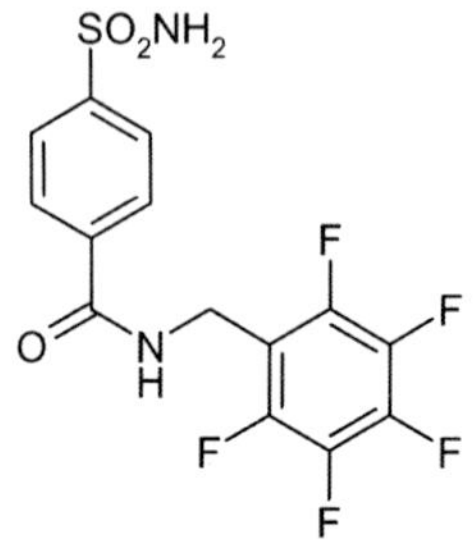

4.12

Several other carboxyamides structurally related to **4.4** were reported in the literature.[22,67–70] As an example, Whitesides' group reported the derivatization of 4-carboxy-benzenesulfonamide **4.2** with oligopeptidyl moieties generating compounds such as **4.13** and **4.14**.[22,67–69] In another series of such derivatives, oligoethylene glycol units were attached to **4.2** and the terminal hydroxyl moiety of the tail was derivatized by acyl-amino moieties, generating CAIs of type **4.15**.[70]

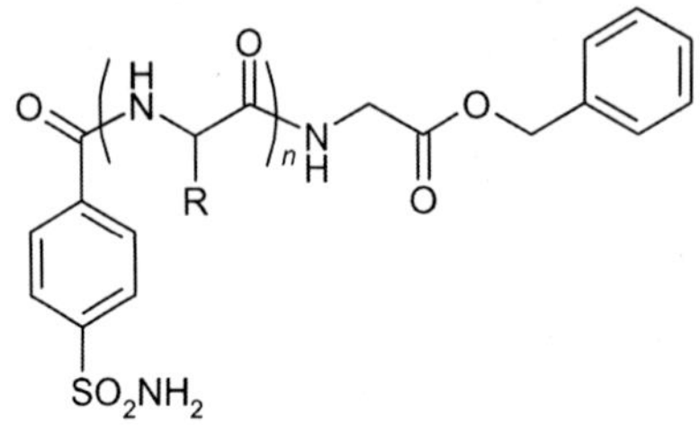

R = H; Ph; PhCH2; *n* = 2–4
4.13

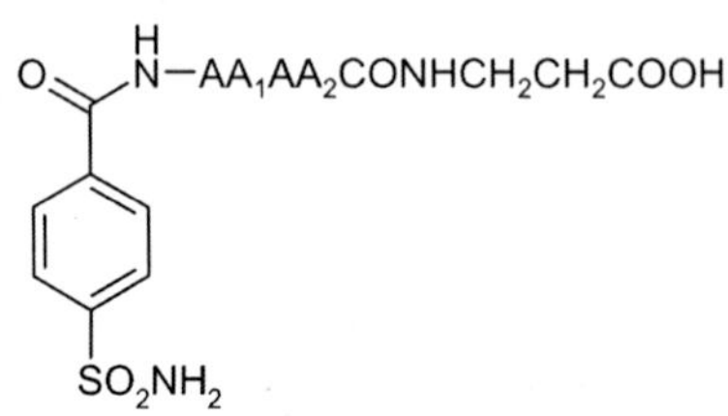

$AA_1 = AA_2$ = L-Leu; D-Leu; L-Thr; D-Thr; L-Ser; D-Ser; Gly
4.14

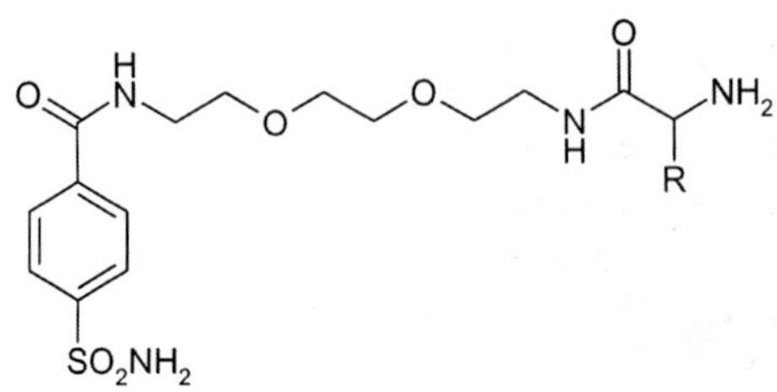

R = Gly; Leu; Lys; Glu; Phe
4.15

The effect of the length of repeating glycine and ethylene glycol (EG) units in compounds of type **4.13**, **4.15**, and **4.16** was investigated both structurally and kinetically.[22,74] These studies highlighted that the $(EG)_n$ group exhibits greater affinity for the CA II surface than does $(Gly)_n$ for inhibitors of comparable length. Moreover, the terminally unsubstitued inhibitors containing $(EG)_n$ or $(Gly)_n$ groups, such as **4.15** and **4.16**, bind 100-fold better than the unsubstituted arylsulfonamide.[22,75,76] Interestingly, as revealed by crystallographic studies, $(EG)_n$ and $(Gly)_n$ moieties show a common binding mode that may be due in part to the steering effect of the benzamide group, which apparently directs the linker chain primarily toward the hydrophobic side of the active site cleft.[22]

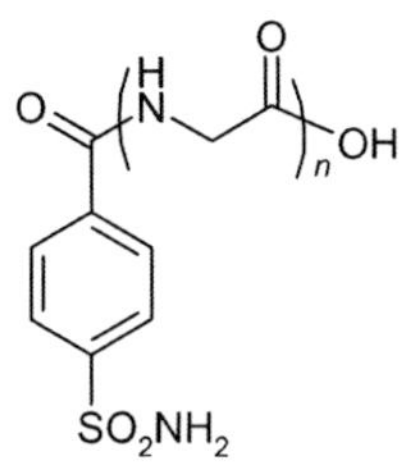

n = 1–6
4.16

The derivatization of **4.2** with a functional group that could independently bind to a second specific site of the CA enzymes has also been used by Srivastava and coworkers to obtain very strong CA inhibitors of type **4.17** and **4.18**.[71,72] In these molecules, the weak inhibitor benzenesulfonamide is conjugated to the iminodiacetate-Cu^{2+} (IDA-Cu^{2+}) moiety, which is supposed to interact with the surface-exposed His residues, via a spacer group, thus enhancing significantly the enzyme-inhibitor affinity. This "two-prong" approach is expected to serve as a general strategy for converting weak inhibitors of enzymes into tight binding inhibitors (Fig. 4.10). X-ray crystallographic

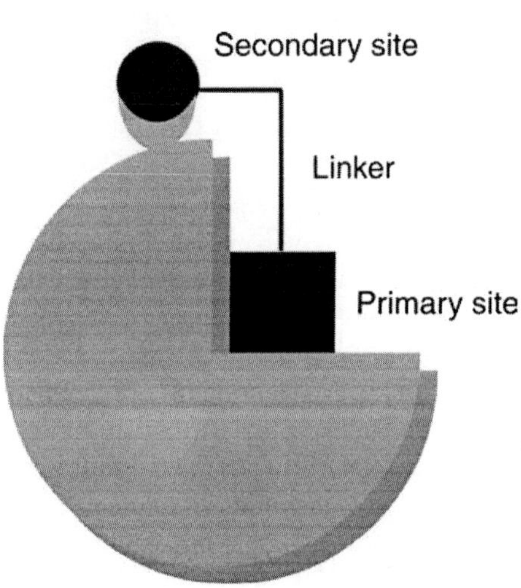

FIGURE 4.10 Interactions of a generic "two-prong" inhibitor with both primary and secondary binding sites of hCA II.

studies of the hCA II–**4.18** adduct reveal that, as expected, the benzenesulfonamide moiety is coordinated to the Zn^{2+} similarly to other benzenesulfonamide inhibitors, while the IDA-Cu^{2+} prong is bound to His64, thus suggesting that the high inhibition power of the inhibitor against CA II is related, in addition to the displacement of the catalytic nucleophile, to the inhibition of the catalytic proton shuttle His64.[71] However, due to the instability of copper–sulfonamide complexes (which liberate copper ions in solution) and the fact that only His64 complexation by the metal ions has been observed so far, the "two-prong" approach is not useful for obtaining isozyme-selective CAIs as originally stated by its discoverers.[71]

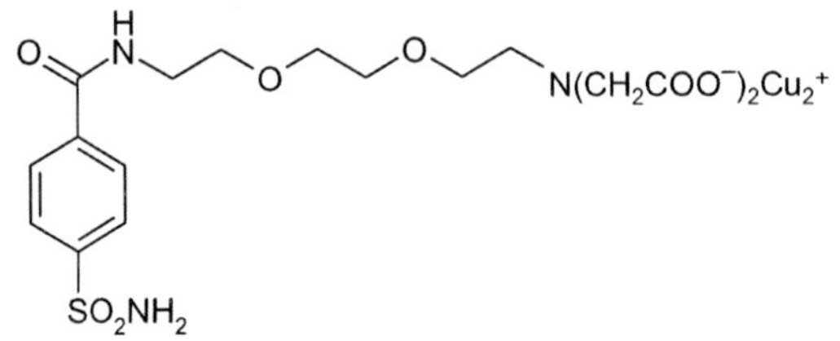

4.17

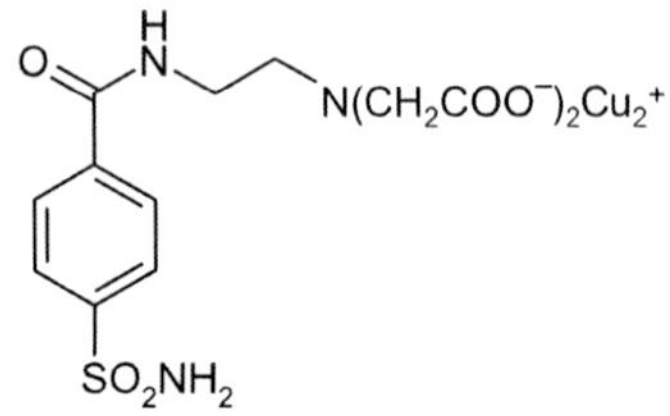

4.18

Benzenesulfonamides Containing a Thioureido Moiety The thioureas of type **4.19**, obtained from isothiocyanatosulfonamides (such as, for example, 4-isothiocyanatobenzenesulfonamide and its congeners) and amines, hydrazines, or amino acids,[77–80] represent an interesting class of derivatives that, among the various CAIs reported in the last few years, showed very promising applications.[77–80] Such compounds have generally an excellent water solubility and potent inhibitory activity against the cytosolic isozyme hCA II as well as the transmembrane, tumor-associated isozyme hCA IX, thus being interesting candidates for developing antiglaucoma/antitumor therapies based on them.[77–80] However, so far little structural information on the binding mode of such derivatives to CA active site is available. *N*-1-(4-Sulfamoylphenyl)-*N*-4-pentafluorophenyl-thiosemicarbazide (**4.20**) is an interesting member of this family, which has been demonstrated to be an inhibitor effective for hCA I, and very good for hCA II and hCA IX (inhibition constants in the range of 15–19 nM).[81] The X-ray crystal structure of the adduct with hCA II showed that the inhibitor binds within the hydrophobic half of the active site, making extensive and strong van der Waals contacts with amino acid residues Gln92, Val121, Phe131, Leu198, Thr200, and Pro202 in addition to the canonical coordination of the sulfonamide nitrogen to the Zn^{2+} ion. It is interesting to note that, although the organic scaffold of the inhibitor does not establish significant polar interactions with the enzyme (for example, no hydrogen bonds were evidenced between the C=S group or the nitrogen atoms belonging to the thiosemicarbazide moiety of the inhibitor and amino acid residues of the active site), the large number of hydrophobic contacts mentioned above can account for the good inhibitory properties of this molecule that are of the same magnitude as those of clinically used CAIs.[2,4,8]

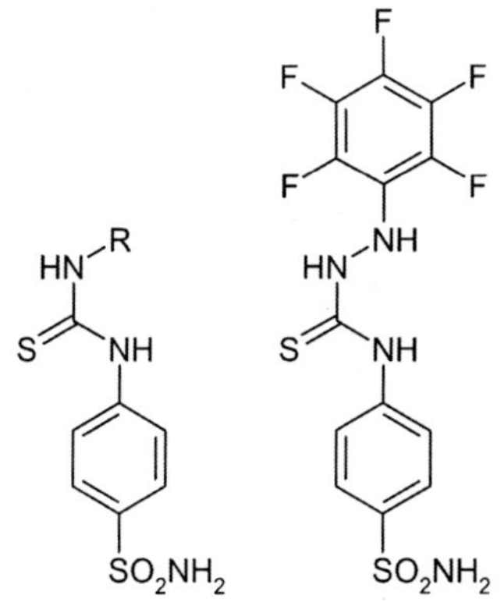

4.19 **4.20**

Fluorescent sulfonamides prepared either by the reaction of aminofluorescein derivative **4.21** with isothiocyanato-aromatic/heterocyclic sulfonamides **4.22**[77,80] or by the reaction of fluorescein isothiocyanate (**4.23**) with amino-substituted aromatic/heterocyclic sulfonamides **4.24**[82] constitute another interesting group of CAIs containing a thioureido moiety (Scheme 4.1). These molecules were reported to

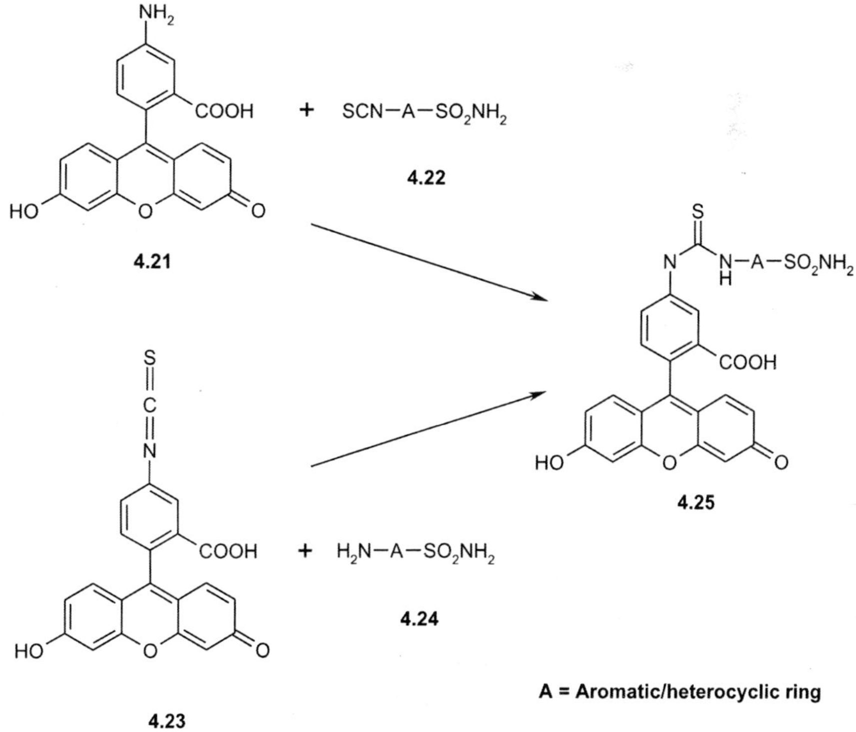

SCHEME 4.1

specifically target hypoxic tumors in which the isoform CA IX is overexpressed.[83,84] One of the most promising compounds in this series is derivative (4-sulfamoylphenylethylthioureido)fluorescein (**4.26**), which besides its strong capability to bind the cytosolic hCA II isozyme (K_d 0.30 nM) showed even higher affinity for the tumor-associated CA IX (K_d 0.64 nM). Thus, it is actually in clinical studies as an imaging tool for acute hypoxic tumors.[19] The X-ray crystal structure of **4.26** in complex with the cytosolic isoform hCA II has been reported together with a modeling study of the adduct that this inhibitor forms with the tumor-associated isoform hCA IX.[19] The crystallographic analysis showed that, beyond the canonical interactions of the benzenesulfonamide group of the inhibitor within the hCA II active site, the thioureido moiety was oriented toward the hydrophobic part of the active site cleft, establishing strong van der Waals interactions with residues Gln92, Val121, Phe131, Val135, Leu198, Thr199, Thr200, and Pro202, the 3-carboxy-amino-phenyl functionality was at van der Waals distance from Phe131, Gly132, and Val135, and the bulky tricyclic fluorescein moiety was located at the rim of the active site, on the protein surface, and strongly interacted with the α-helix formed by residues Asp130–Val135 (Fig. 4.11a). Molecular dynamic (MD) simulation studies revealed that all these interactions were preserved in the hCA IX–**4.26** adduct, even if an additional polar interaction was observed. In fact, in the latter case, the carbonyl moiety of the fluorescein tail of **4.26** participates in a strong hydrogen bond with the guanidine moiety of Arg130, an amino acid characteristic of the hCA IX active site (Fig. 4.11b). This interaction accounted for the roughly two times higher affinity of **4.26** for hCA IX over hCA II and explained why, *in vivo*, the compound specifically accumulated only in hypoxic tumors over-expressing CA IX and not in the normal tissues.[19] Thus, these studies constituted an interesting starting point for the further drug design of more isozyme-selective CA IX inhibitors with potential use as diagnostic tools or for the management of hypoxic tumors.

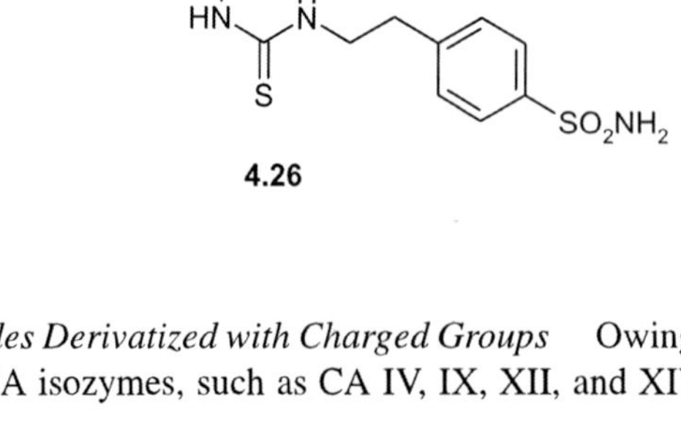

4.26

Benzenesulfonamides Derivatized with Charged Groups Owing to the extracellular location of some CA isozymes, such as CA IV, IX, XII, and XIV, it is theoretically

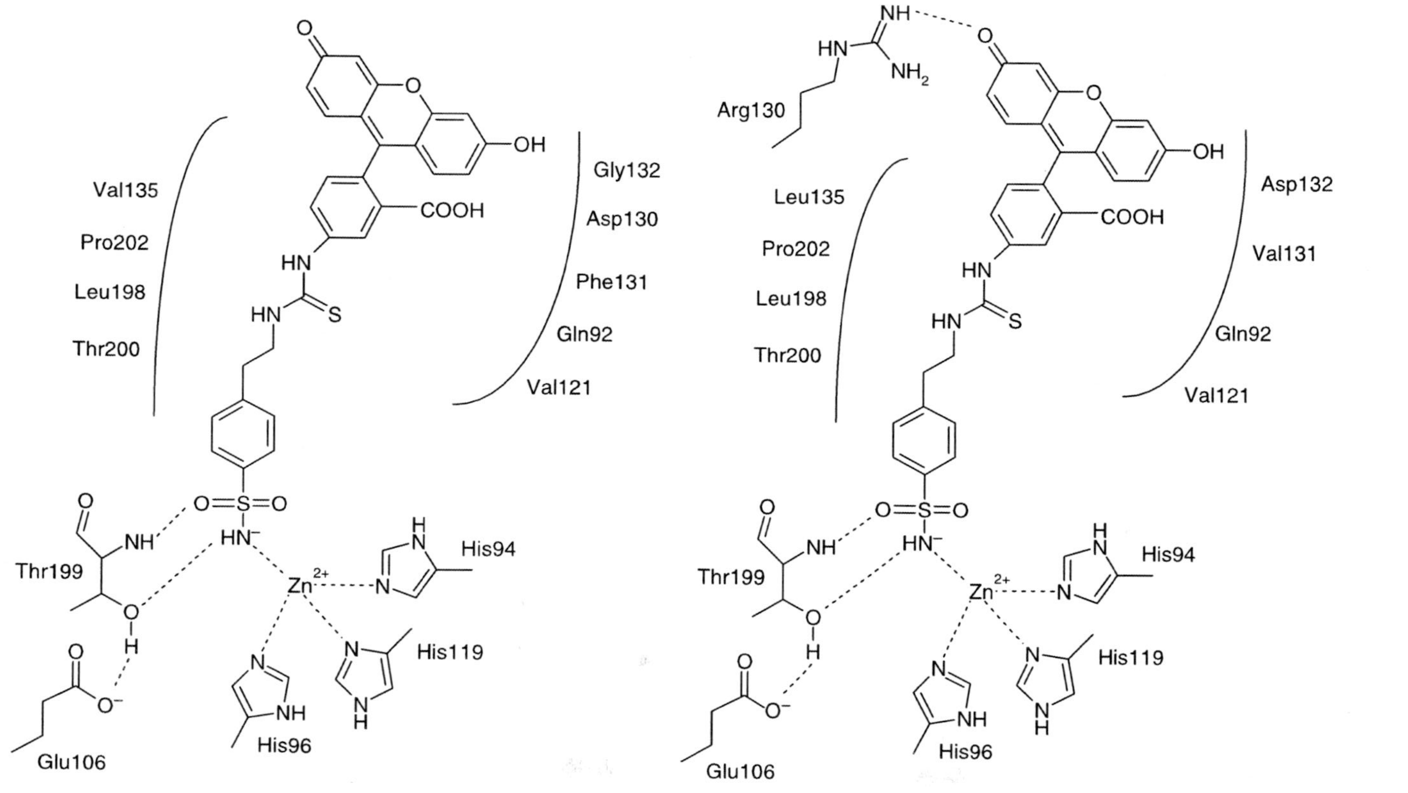

(a) (b)

FIGURE 4.11 (a) Active site region in the hCA II–**4.26** complex X-ray structure. (b) Model of the hCA IX–**4.26** adduct from MD simulations. Residues participating in recognition of the inhibitor molecule are reported.

possible to design membrane-impermeant CAIs that would selectively inhibit membrane-associated CAs without interacting with the cytosolic or mitochondrial isoforms. This possibility has been recently investigated through the design of positively charged sulfonamides[85–92] that generally incorporate pyridinium moieties.[85,89] Many representatives of such class of molecules have been reported in recent years, showing that these compounds may lead to effective CA inhibitors.[86–92] Interestingly, although isozyme-selective compounds have not been detected, it has been demonstrated that such derivatives were unable to cross the plasma membrane *in vivo*.[86,87] Such data were extremely important for specific *in vivo* inhibition of membrane-associated isozymes and consequently for the eventual development of novel anticancer therapies. Among these derivatives, one of the most promising compounds was the 1-*N*-(4-sulfamoylphenyl-ethyl)-2,4,6-trimethylpyridinium perchlorate (**4.27**).[16,85,90–92] In fact, kinetic studies demonstrated that this molecule was able to efficiently inhibit both CA II and IX (K_I values of 21 and 14 nM against CA II and IX, respectively).[16] However, *in vivo*, due to its membrane impermeability, it is likely that selective inhibition of the membrane-associated isozyme can be achieved without appreciable inhibition of the cytosolic isozyme CA II.

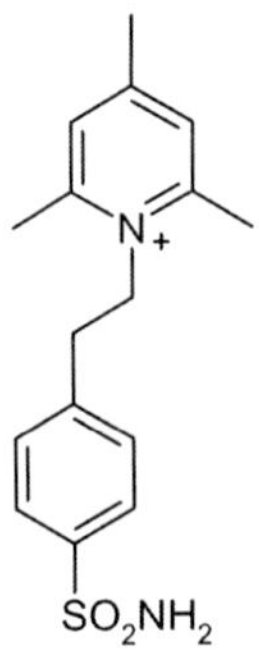

4.27

The X-ray crystal structure of **4.27** in complex with hCA II has been solved,[16] providing the molecular basis of the high affinity of this molecule toward CA active site. The crystallographic analysis showed that, beyond the canonical interactions of the benzenesulfonamide moiety within the active site channel, several other important stabilizing interactions were established by the 2,4,6-trimethylpyridinium functionality. In particular, this moiety was at van der Waals distance from the aliphatic chain of Ile91, pointing toward the hydrophilic half of the active site and making a face-to-face stacking with Phe131 aromatic ring. Its positively charged nitrogen was not involved in any direct binding to the protein, but it was stabilized by several dipolar interactions with water molecules present in the active site. A structural comparison of **4.27** with other benzenesulfonamide inhibitors containing an aromatic tail revealed that two main binding modes within the CA II active site can occur. Some inhibitors, such as **4.20** and **4.36**, bind with their tail within the

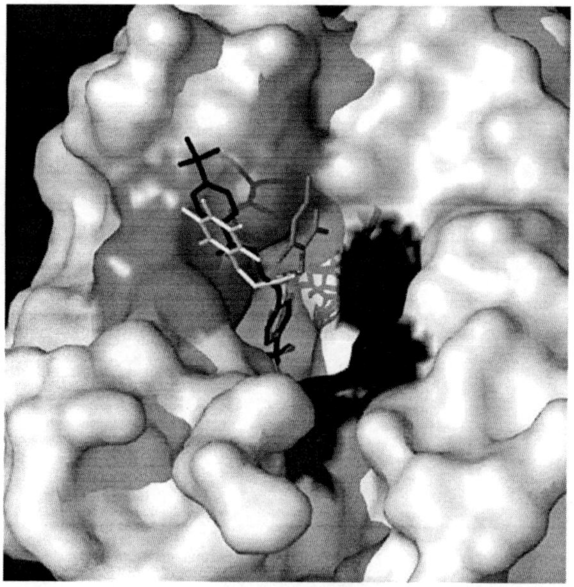

FIGURE 4.12 Solvent accessible surface of hCA II. The hydrophobic half of the active site cleft is shown in grey while the hydrophilic one in black. The superimposition of the inhibitors **4.27** (grey), **4.20** (white) and **4.36** (black), showing the two different binding mode of the aromatic tail, is also reported.

hydrophobic half of the active site, defined by residues Phe131, Val135, Leu198, Pro202 and Leu204. On the contrary, **4.27** and other derivatives bind with their tail pointing toward the hydrophilic half and making a strong face-to-face stacking interaction with Phe131. Thus, this interaction seems to steer the inhibitor toward this part of the active site, while the impossibility to participate in it leads to the binding within the hydrophobic region (Fig. 4.12).[16] Such findings evidenced a key role for residue Phe131 in the orientation of inhibitor molecules in the CA active site. Considering that different isozymes present diverse amino acids in position 131, this residue can represent a target to achieve inhibitors with isozyme selectivity.

The structural and functional consequences of positively charged, negatively charged, and neutral substituents on benzenesulfonamide CA inhibitors have also been investigated in detail in a systematic study of Srivastava and coworkers.[93] In particular, four simple benzenesulfonamide inhibitors, namely, compounds **4.28–4.31**, substituted at the *para* position with differently charged functional groups have been designed and synthesized. Moreover, the affinities and X-ray crystal structures of their complexes with hCA I and hCA II have been determined. These studies highlighted that a *para*-substituted positively charged amino group is more poorly tolerated in the active site of CA I compared to CA II, while a *para*-substituted negatively charged carboxylate substituent is tolerated equally well in the active sites of both CA isozymes. However, enzyme-inhibitor affinity increases in both cases upon neutralization of inhibitor charged groups by acylation or esterification.[93]

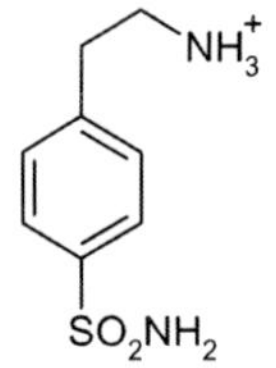

4.28

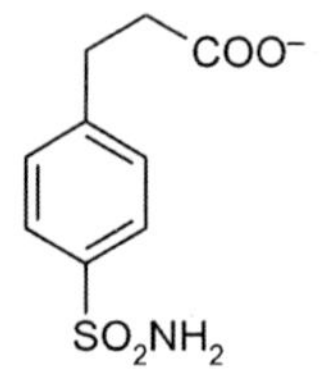

4.29

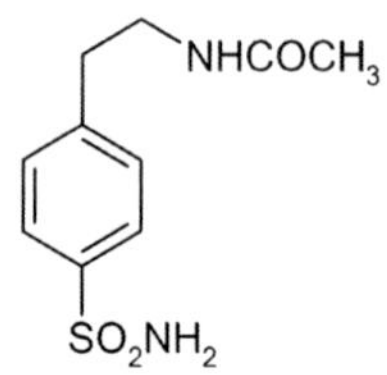

4.30

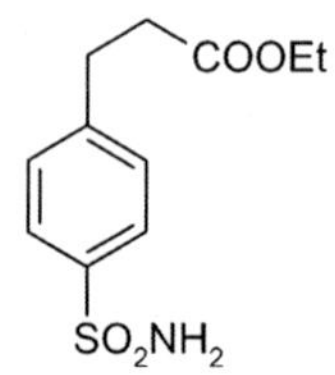

4.31

Benzenedisulfonamides Benzenesulfonamide derivatives that possess more than one sulfonamide group in their molecule were tested as CAIs either to search for more active inhibitors or to design compounds with a different biological activity, for example, saluretics and high ceiling diuretics (the CA inhibitory effect being secondary in the latter cases).[1,64] Two main classes of such compounds were extensively characterized so far: the 1,3-disulfamoylbenzenes of type **4.32–4.34**[20] and the bis-sulfonamides of type **4.35** derived from biphenyl, biphenylether, biphenylsulfide, biphenylmethane, or containing urea, guanidine, carboxyamido moieties, and so on as spacers.[18,44]

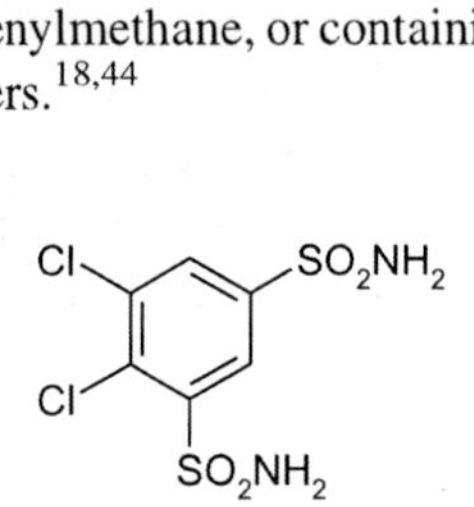

4.32

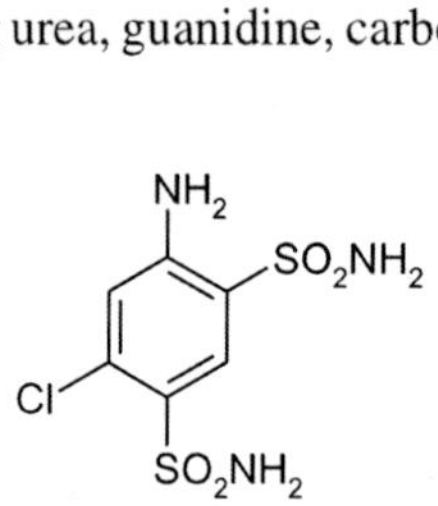

4.33

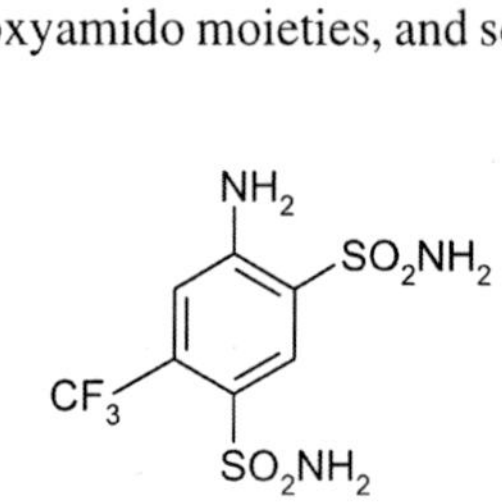

4.34

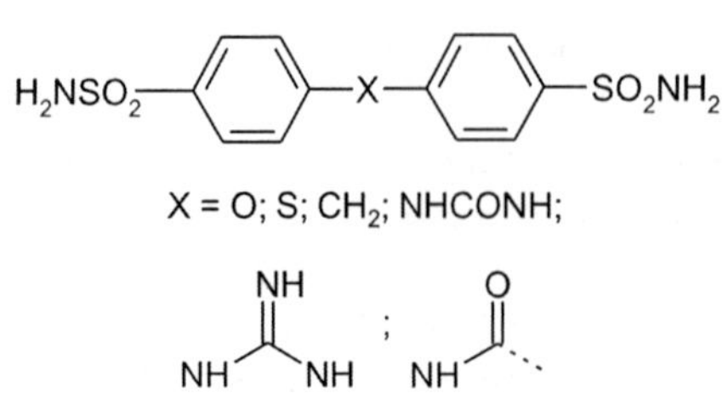

4.35

Dichlorophenamide (**4.32**) is a classical CAI belonging to the first generation of inhibitors that has been (and still is) largely clinically used for the management of glaucoma[94,95] and for the treatment of several neurological disorders.[94,96] This compound, together with its closely related analogues 6-chloro-4-amino-benzene-1,3-disulfonamide (**4.33**) and 6-trifluoromethyl-4-amino-benzene-1,3-disulfonamide (**4.34**), were investigated in detail in order to be used as starting compounds in drug design studies for obtaining anti-glaucoma agents. The inhibition properties of these three disulfonamides toward several mammalian and bacterial/archaeal isozymes were found to be quite peculiar and distinct from those of other clinically used derivatives. Particularly, **4.32** was found to be one of the best inhibitors of the mitochondrial isoform hCA VB, thus representing a good lead compound for antiobesity agents,[60,97] **4.33** was found to be a very good inhibitor for α-CA from *Helicobacter pylori,* and **4.34** was one of the best inhibitors of murine CA XIII.[20] The high-resolution structures of the three compounds **4.32–4.34** complexed to hCA II were also reported,[20] revealing a particular binding mode of these three inhibitors within the CA II active site cavity (Fig. 4.13). In particular, beyond the canonical coordination of the Zn^{2+} atom by means of the ionized sulfonamide NH^- group, the substituted phenyl moiety of each inhibitor is stabilized in the active site by various van der Waals and hydrogen bond interactions with residues delimiting the cavity (Fig. 4.13). The sulfonamide group in *meta*-position to the coordinated sulfamoyl moiety plays a key role in this stabilization, being oriented toward the hydrophilic side of the active site cleft in all the three adducts and establishing several hydrogen bonds with His64, Asn67, Gln92, and Thr200. Consequently, the plane of the phenyl moiety of the inhibitors appears to be rotated by 45° and tilted by 10° with respect to its most recurrent orientation in other CA II–benzenemonosulfonamide derivatives[8,96,98–105] (Fig. 4.14). Thus, these data suggest that the presence of an additional sulfonamide group in *meta*-position of hCA II benzenesulfonamide inhibitors can be favorably used to differently orient the phenyl moiety within the enzyme active site. As a consequence, a putative tail conjugated with the benzene-1,3-disulfonamide group can be opportunely oriented in the enzyme active site, leading to inhibitors with diverse inhibition profiles and selectivities for various mammalian, bacterial, or archaeal CAs.[20]

4-(4-Sulfamoylphenylcarboxamidoethyl)benzenesulfonamide (**4.36**) is an interesting compound, belonging to the class of bis-sulfonamides of type **4.35**, which has been shown to act as an efficient topical antiglaucoma sulfonamide with strong inhibitory properties toward hCA II (K_I of 5 nM).[66] The X-ray crystal structure of the adduct that this molecule forms with hCA II provided important information on the interaction mode of inhibitors containing two benzenesulfonamide moieties within the enzyme active site.[44] In particular, the structural analysis evidenced that, even if the interactions of the benzenesulfonamide zinc anchoring group of **4.36** are similar to that of other sulfonamides, the binding mode of the remaining organic scaffold of the inhibitor is considerably different (Fig. 4.15). In fact, this part of the inhibitor interacts only with the hydrophobic

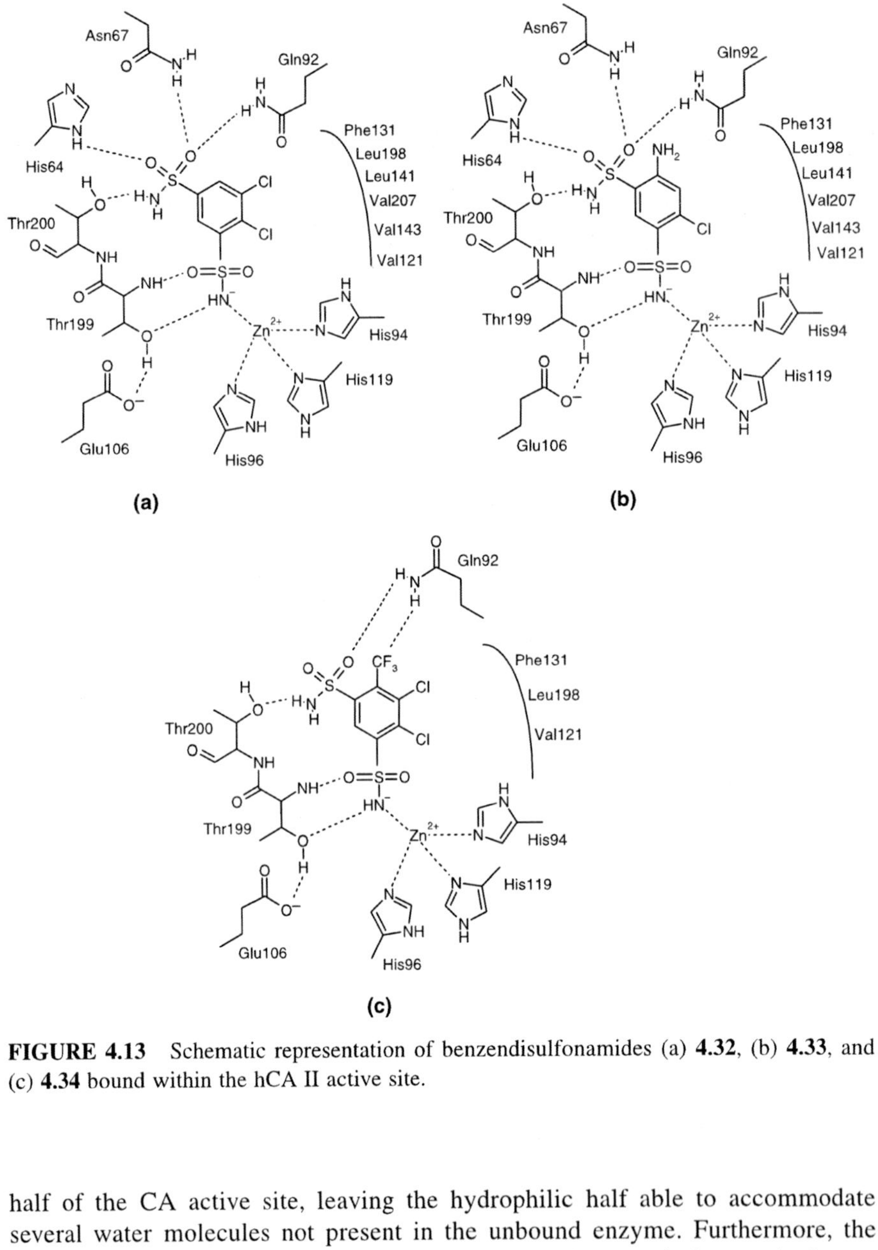

FIGURE 4.13 Schematic representation of benzendisulfonamides (a) **4.32**, (b) **4.33**, and (c) **4.34** bound within the hCA II active site.

half of the CA active site, leaving the hydrophilic half able to accommodate several water molecules not present in the unbound enzyme. Furthermore, the secondary sulfonamide moiety is well anchored to the protein by two hydrogen bonds involving residues Gly132 and Gln136, placed on the rim of the entrance to the active site cleft (Fig. 4.15). Finally, it is worth noting that only one sulfonamide moiety of the two "heads" that **4.36** possesses participates in the coordination of the metal ion when the inhibitor is bound within the enzyme active

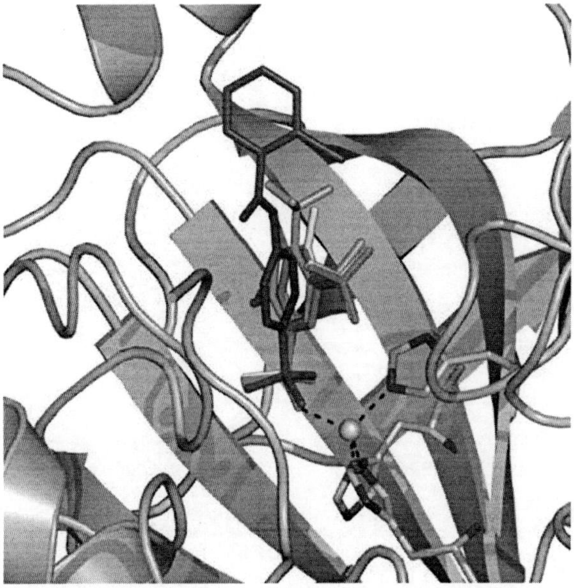

FIGURE 4.14 Superposition of hCA II–inhibitor adducts that highlights the different orientations of the benzene-1,3-disulfonamide inhibitors such as **4.32** (red), **4.33** (yellow), and **4.34** (orange) with respect to a classical monosulfonamide derivative such as **4.38** (blue). (See the color version of this figure in Color Plates section.)

site. Thus, the presence of two sulfamoyl moieties in the inhibitor molecule plays a role only in the proper orientation of the molecule within the active site, accounting for its high affinity toward the enzyme.[44]

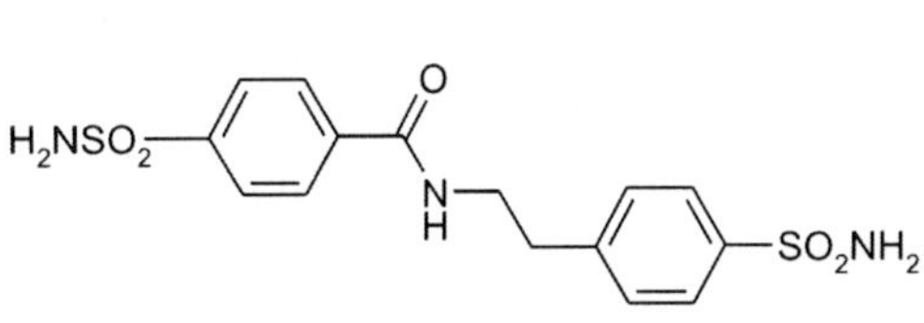

4.36

Hypoxia Activatable, Bioreductive Benzenesulfonamides The use of less toxic prodrugs that can be selectively activated in the tumor tissue, taking advantage of some exclusive aspects of tumor physiology, such as selective enzyme expression or hypoxia, represents an interesting approach for improving the selectivity of anticancer drugs.[106] Since CA IX is overexpressed in hypoxic tumors and is present only in low amounts in some parts of the normal gastrointestinal tract, this

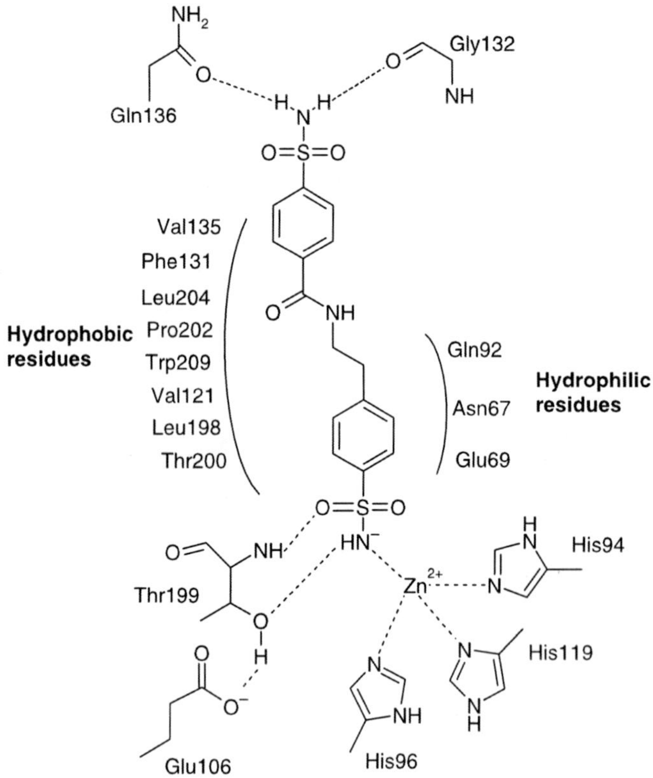

FIGURE 4.15 Schematic representation of inhibitor **4.36** binding within the hCA II active site.

enzyme may constitute an attractive target for the design of hypoxia activable prodrugs.[107–109] We have recently reported a new approach for designing bioreductive hypoxia activable CAIs[107] that utilizes disulfide derivatives of benzenesulfonamides. In principle, such disulfide-containing sulfonamides should be bulky and thus unable to bind within the restricted space of the CA active site, which normally can accommodate only one benzenesulfonamide moiety. In contrast, the corresponding thiol derivatives, obtained under the reducing conditions present in hypoxic tumors (see Scheme 4.2), are much less bulky than the sterically hindered disulfides, thus showing better CA inhibitory activity. Moreover, since this type of hypoxia activatable prodrug will be formed only in the cancer tissue, no monomeric sulfonamide derivative should be present outside the tumor cell. This should lead to a tumor-specific drug with potentially fewer side effects due to inhibition of other CA isoforms highly abundant in noncancer tissues, such as CA I, II, IV, or VA. One of the most promising compounds of this series is *N*,*N*″-(2,2′-dithio-dibenzoyl)bis-sulfonamide (**4.37**). This molecule shows very weak CA inhibitory activity with K_I's values in the range

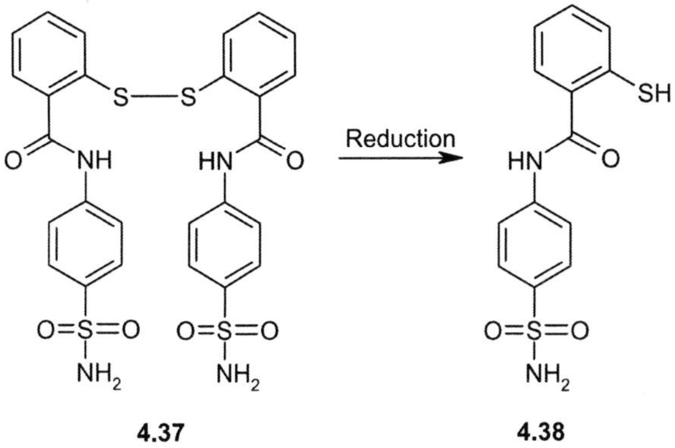

SCHEME 4.2

of 653–4975 nM against the CA I, II, and IX isoforms.[107] However, its monomeric derivative 4-(2-mercaptophenylcarboxamido)benzenesulfonamide (**4.38**) is a quite effective CA I inhibitor (K_I of 276 nM), and a very potent CA II and IX inhibitor, with inhibition constants of 16 and 9.1 nM, respectively.[107] The X-ray crystal structure of **4.38** in adduct with hCA II showed that the inhibitor makes a host of favorable interactions with the side chains of Gln92, Val121, Phe131, Leu198, Thr199, Thr200, Pro201, and Pro202, whereas the sulfamoyl moiety was coordinated to the Zn^{2+} ion (Fig. 4.16a). Modeling studies revealed that the same interactions were preserved in the

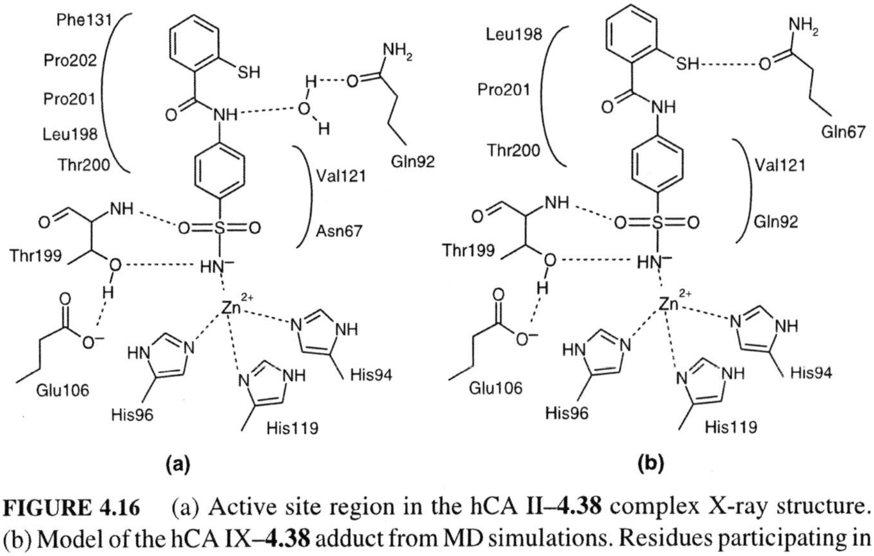

FIGURE 4.16 (a) Active site region in the hCA II–**4.38** complex X-ray structure. (b) Model of the hCA IX–**4.38** adduct from MD simulations. Residues participating in recognition of the inhibitor molecule are reported.

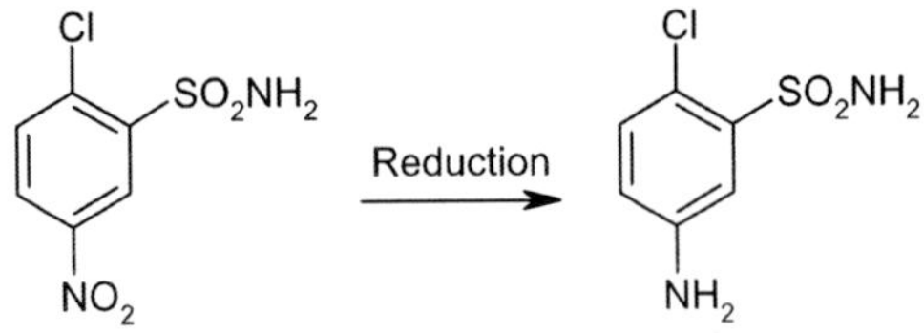

4.40 **4.41**

SCHEME 4.3

adduct of this compound with hCA IX, but in addition a strong polar interaction between Gln67 and the thiol group of the inhibitor was observed (Fig. 4.16b). This additional interaction may explain the almost two times more effective inhibition of the tumor-associated isozyme over the cytosolic isoform.[107] Further experiments are actually in progress to test *in vivo* the antitumor activity of these compounds.

2-Substituted-5-nitro-benzenesulfonamides[110] of type **4.39**, incorporating a large variety of secondary/tertiary amines, are another important class of bioreductive hypoxia activable CAIs useful for the treatment or imaging of hypoxic tumors. These compounds were generally ineffective inhibitors of the cytosolic isoform I, showed a better inhibition of the physiologically relevant CA II, and strongly inhibited the tumor-associated CA IX and XII.[108] Interestingly, some of these derivatives showed excellent selectivity ratios for the inhibition of the tumor-associated isozymes over the cytosolic ones.[108] 2-Chloro-5-nitrobenzenesulfonamide (**4.40**), which can be reduced (chemically or enzymatically) to the corresponding aminosulfonamide **4.41** (Scheme 4.3), represents the simplest member of this class of molecules. The X-ray crystal structure of the adduct hCA II–**4.40**, as well as the molecular modeling studies for the interaction of this inhibitor with hCA IX, allowed to identify a different pattern of hydrogen bonds responsible of the discrimination of the two isoforms for this type of bioreductive compound (Fig. 4.17).[108] In particular, Asn67 present in the hCA II active site is replaced in hCA IX by Gln67, which, as a consequence of its longer side chain, is able to form a stable hydrogen bond with the inhibitor NO_2 oxygen atoms (Fig. 4.17).

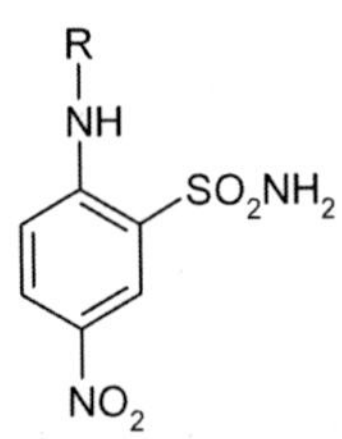

4.39

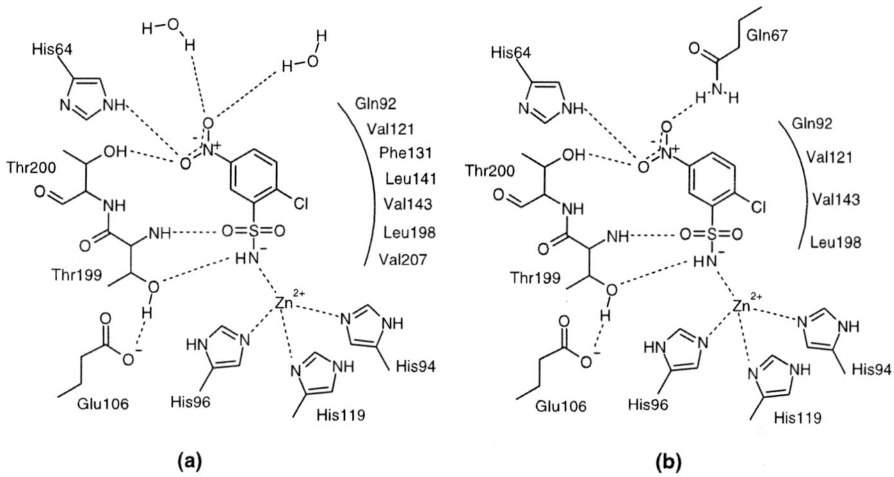

FIGURE 4.17 (a) Active site region in the hCA II–**4.40** complex X-ray structure. (b) Model of the hCA IX–**4.40** adduct from MD simulations. Residues participating in recognition of the inhibitor molecule are reported.

Benzenesulfonamides Acting as Dual CA and COX-2 Inhibitors Valdecoxib (**4.42**) and celecoxib (**4.43**) are two nonsteroidal anti-inflammatory drugs, acting as selective inhibitors of the inducible cyclooxygenase isoform 2 (COX-2).[111] Both compounds contain an unsubstituted arylsulfonamide group, which is a key feature of many CA inhibitors. Using kinetic studies, we recently demonstrated an unexpected nanomolar affinity of these compounds for several CA isozymes, such as CA I, II, IV, and IX.[4,17,43,112] Crystallographic studies on **4.42** and **4.43** in complex with CA II revealed that this inhibition is essentially mediated *via* binding of the sulfonamide group to the catalytic zinc ion. The structural analysis also revealed that the organic scaffolds of **4.42** and **4.43** were positioned in the CA active site in a completely different mode (Fig. 4.18). In particular, the phenylisoxazole moiety of **4.42** filled the active site channel of the enzyme and interacted with the side chains of Gln92, Val121, Leu198, Thr200, and Pro202, while the 3-phenyl group was located into a hydrophobic cavity, simultaneously establishing van der Waals interactions with the side chains of various hydrophobic residues (Ile91, Val121, Val135, Leu141, and Leu198) and a strong offset face-to-face stacking interaction with the aromatic ring of Phe131 (Fig. 4.18). In contrast, **4.43**, being more sterically hindered, completely filled the CA II active site, positioning the trifluoromethyl group in the hydrophobic part and the *p*-tolyl moiety in the hydrophilic region (Fig. 4.18). Thus, the capability of **4.43** bulky substituents to interact with both hydrophobic and hydrophilic parts of the active site may explain why it is approximately a two times stronger CA II inhibitor compared to **4.42**.[17,43] These studies highlight the cross-reactivity of these compounds with CAs and

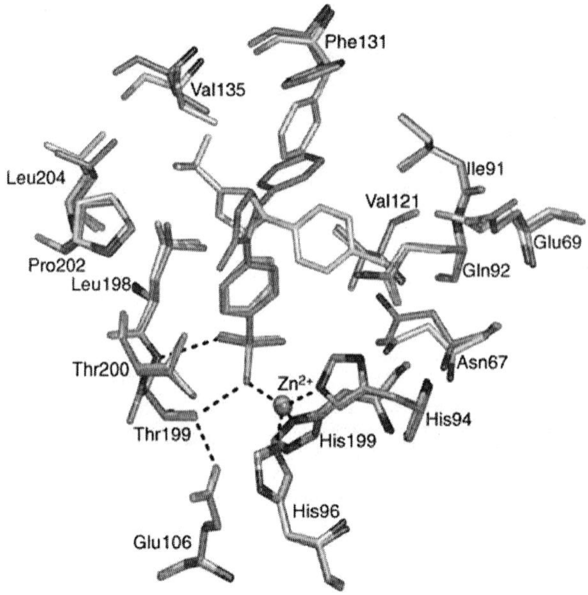

FIGURE 4.18 View of the hCA II active site complexed with valdecoxib **4.42** (magenta) and celecoxib **4.43** (cyan) brought to optimal structural overlay. (See the color version of this figure in Color Plates section.)

suggest new opportunities of these COX-2 selective nonsteroidal anti-inflammatory drugs in particular with respect to glaucoma and anticancer therapy.

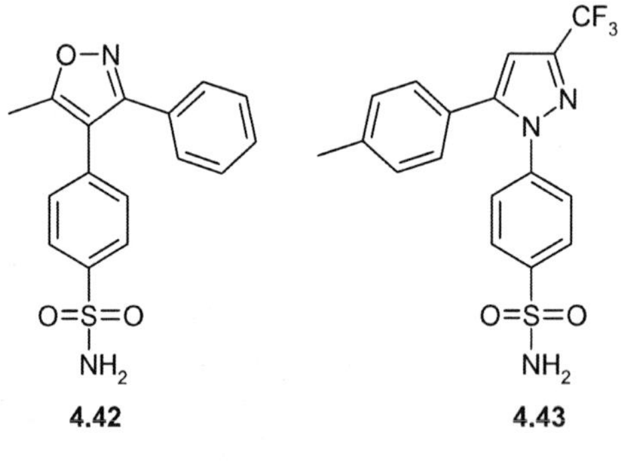

4.42 **4.43**

4.3.3.2 Heterocyclic Sulfonamides Investigation on heterocyclic sulfonamides as CAIs has been encouraged by the pioneeristic studies of Davenport in 1945[113] that reported that thiophene-2-sulfonamide was 40 times more active than sulfanilamide **4.1** as a CA inhibitor. Subsequently, the preparation of a very large series of heterocyclic sulfonamides derived from the most important ring systems (imidazole, alkyl and aryl imidazoles, benzimidazoles, benzothiazole, 1,2,4-triazole, thiazole, tetrazole and alkyl/aryl tetrazoles, 1,3,4 thiadiazole, pyrimidine, pyrazine, etc.) was also reported,[114,115] allowing to derive important structure/activity relationships for this kind of compounds. In particular, these studies highlighted that five-membered derivatives were more effective CAIs than six-membered ring compounds and that the presence of nitrogen and sulfur atoms within the ring led to the most potent inhibitors. Thus, extremely powerful inhibitors were found to be 5-substituted-1,3,4-thiadiazole-2-sulfonamide (**4.44**), 6-substituted-benzothiazole-2-sulfonamide (**4.45**), and 1,3,4-thiadiazoline-2-sulfonamides (**4.46**). Derivatives of thiophene-2-sulfonamide (**4.47**) with different substitution patterns were also shown to possess good inhibitory properties.[116] Finally, important results were also obtained from bicyclic derivatives of thieno-thiopyran-2-sulfonamides (**4.48**), which led to the first clinically used topically acting sulfonamide, namely, dorzolamide (**4.6**).[4,117] In the next section, the binding mode to the hCA II active site of the most representative inhibitors belonging to the heterocyclic sulfonamide group will be examined in detail.

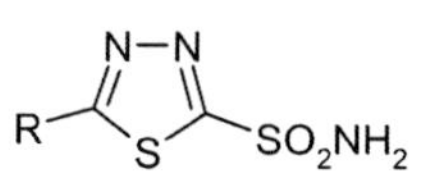

4.44

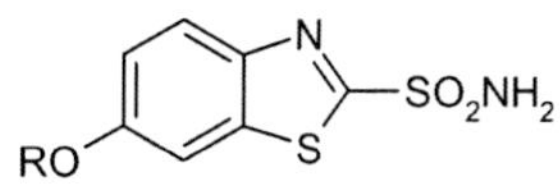

4.45

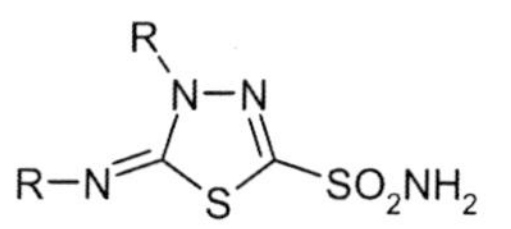

4.46

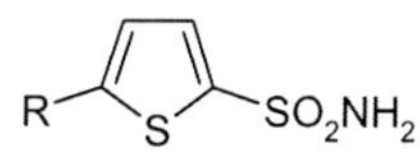

4.47

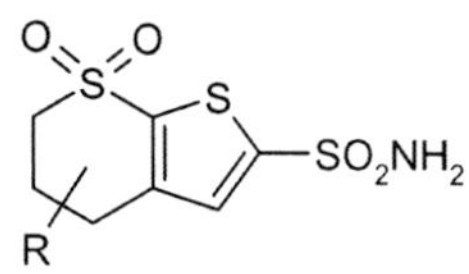

4.48

Thiophene and Thiadiazole Sulfonamide Derivatives The majority of structural studies on heterocyclic sulfonamides have been carried out on five-membered rings, such as thiophene, thiadiazole, and thiadiazoline derivatives (**4.47**, **4.44**, and **4.46**) in complex with isozyme II.[105,118–121] The structural analysis of these hCA II–inhibitor adducts revealed that the heterocyclic sulfonamide moiety of these inhibitors always occupies the same position within the hCA II active site, making several van der Waals interactions with the side chain of residues Gln92, His94, Val121, Leu198, Thr199, and Thr200 in addition to the canonical Zn^{2+} coordination (Fig. 4.19).[105,118–121] The thiadiazole ring, unlike the thiophene and the thiadiazoline ones, is able to establish additional hydrogen bond contacts with the hydroxyl group of Thr200 through its heterocyclic nitrogen atoms (Fig. 4.19b).[105,119,121]

Derivatization of the thiophene, 1,3,4-thiadiazole, and 1,3,4-thiadiazoline rings by introducing a functional group in position 5 results in CAIs with varied affinities and physicochemical properties (see compounds **4.49–4.56**).[114–116,120,122–128] As

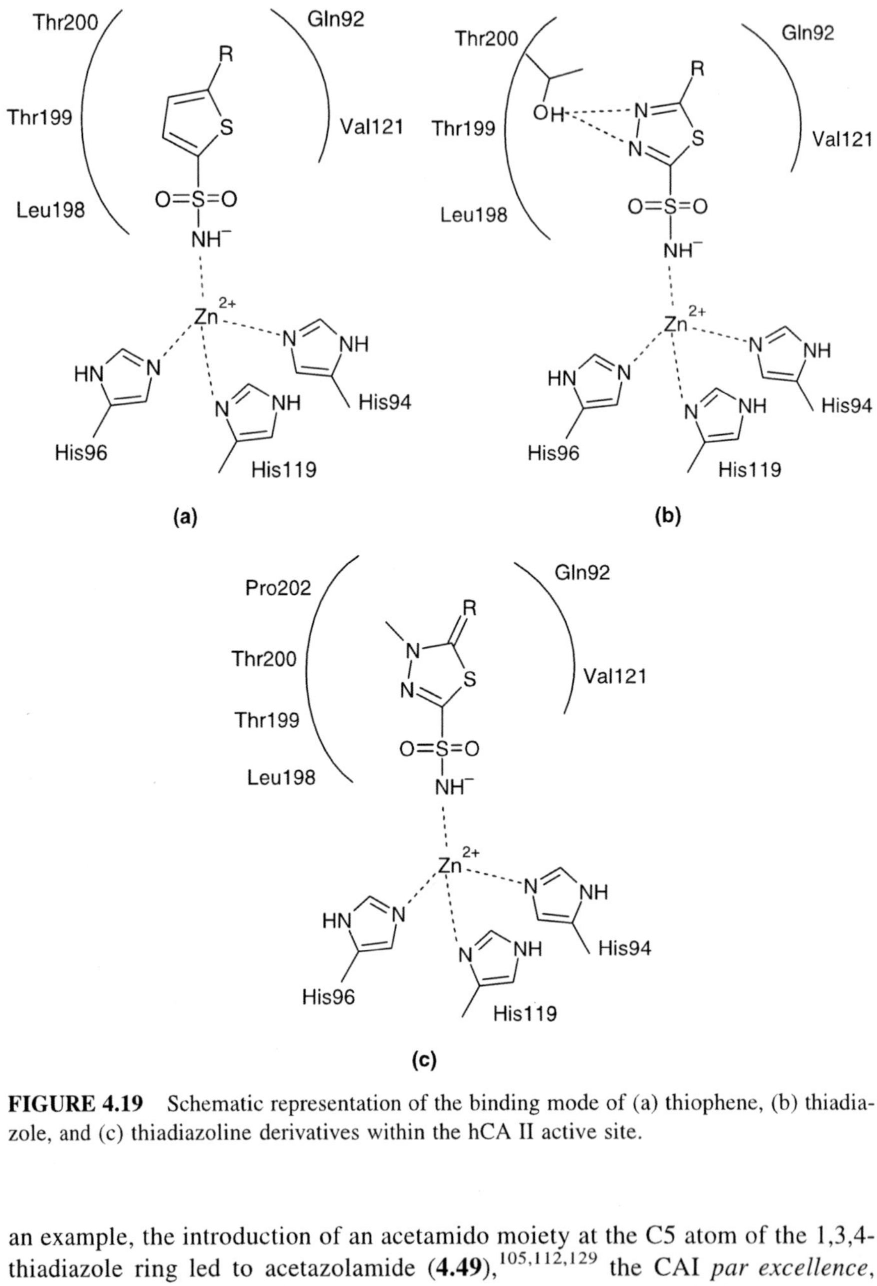

FIGURE 4.19 Schematic representation of the binding mode of (a) thiophene, (b) thiadiazole, and (c) thiadiazoline derivatives within the hCA II active site.

an example, the introduction of an acetamido moiety at the C5 atom of the 1,3,4-thiadiazole ring led to acetazolamide (**4.49**),[105,112,129] the CAI *par excellence*, which has been proved to possess very good pharmacological properties such as low toxicity and excellent bioavailability. This compound has been continuously used to manage glaucoma, gastroduodenal ulcers, and many other disorders since 1954.[112,129] The binding mode of **4.49** to the hCA II active site has been investigated in detail by means of X-ray crystallography revealing that, in addition to the already

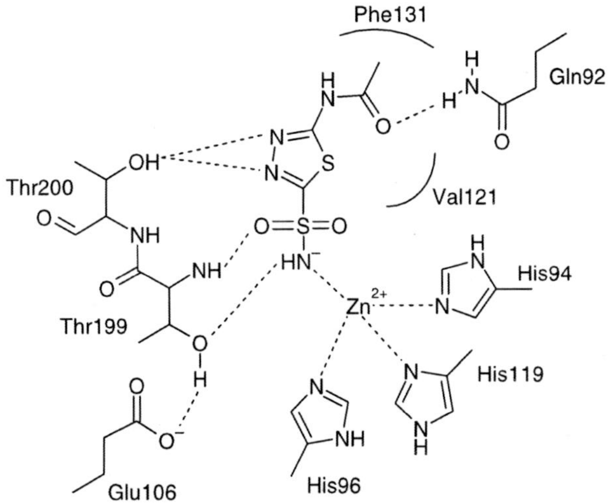

FIGURE 4.20 Schematic representation of inhibitor **4.49** binding within the hCA II active site.

described interactions of the 1,3,4 thiadiazole ring with the CA active site, the carbonyl oxygen of the acetamido moiety is hydrogen bonded to the side chain of Gln92, while the methyl group establishes van der Waals interactions with Phe131 and Gln92 (Fig. 4.20).[105]

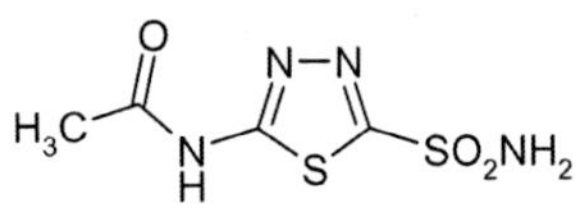

4.49

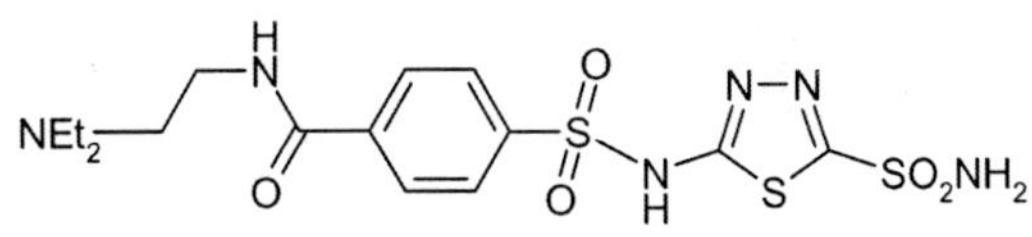

4.50

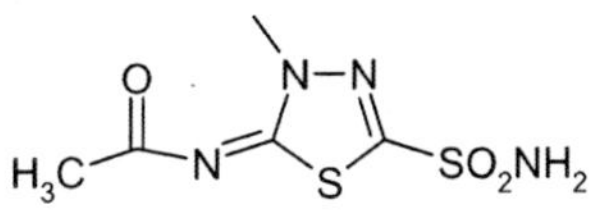

4.51

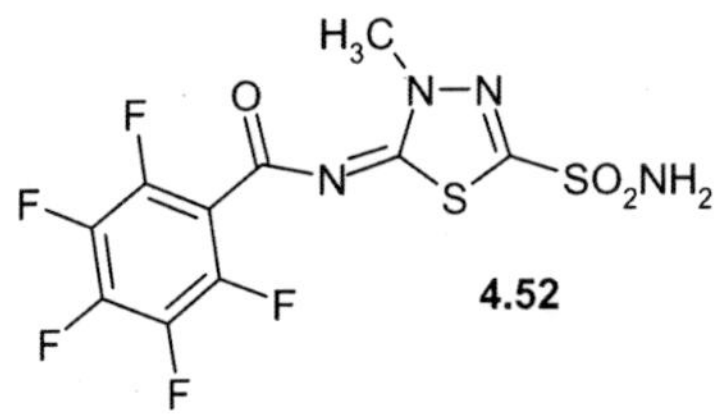

4.52

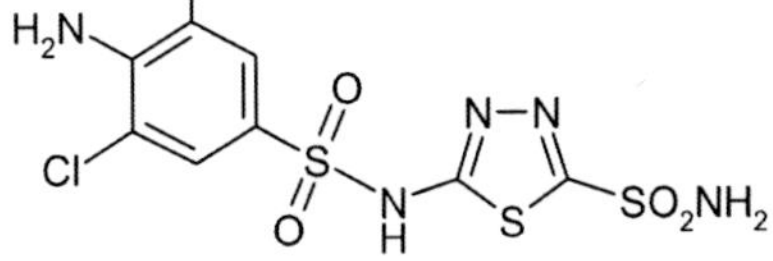

4.53

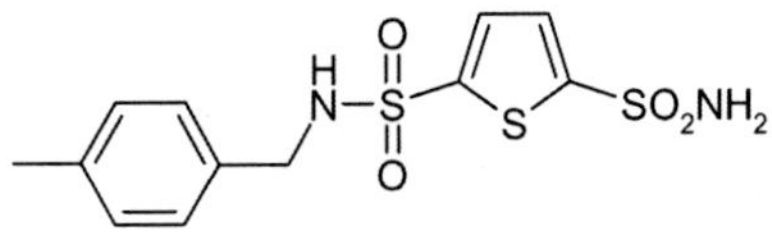

4.54

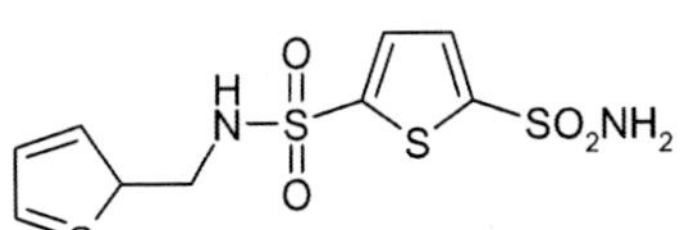

4.55

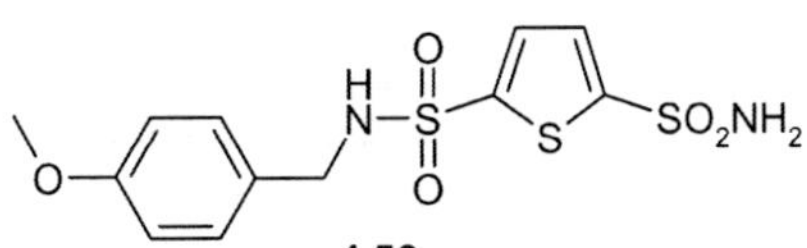

4.56

Methazolamide (**4.51**) is a thiadiazoline derivative, closely related to acetazolamide, that has been used in clinical medicine for more than 40 years.[1] Despite their very similar structure, methazolamide is more liposoluble than acetazolamide, and therefore it is characterized by a higher membrane permeability.[1] Due to its excellent pharmacological properties, it has been largely utilized as lead compound for the design of many other CAIs.[65,122,124,126] Among the compounds derived from methazolamide, 4-methyl-5-perfluorophenylcarboximido-δ^2-1,3,4-thiadiazoline-2-sulfonamide (**4.52**) is worth noting since it was recently shown to act as an efficient topical antiglaucoma drug.[126] Its binding to the hCA II active site is similar to that of other heterocyclic sulfonamide inhibitors considering the interactions of the sulfonamide and thiadiazoline ring but differs when the perfluorobenzoylimino fragment is analyzed. Indeed, a strong hydrogen bond between the carbonyl oxygen of the inhibitor and the Gln92NE2 atom is observed. Moreover, a strong stacking interaction with Phe131 was also present (Fig. 4.21).[118]

The introduction of a second sulfonamide group at the C5 atom of the thiophene- or 1,3,4-thiadiazole-2-sulfonamide core has also been largely used for designing important classes of pharmacological agents with improved physicochemical properties.[120,124] Important examples of compounds obtained in this way are constituted by molecules **4.50** and **4.53–4.56**.[119–121] The crystallographic analysis of the adducts of these compounds with hCA II revealed that the second sulfonamide group can establish several polar interactions with active site residues. In particular, in the case

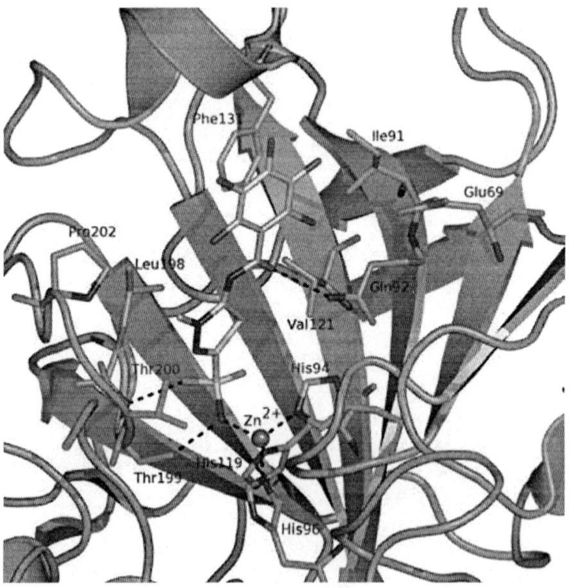

FIGURE 4.21 View of the active site region in the hCA II–**4.52** complex showing the residues participating in recognition of the inhibitor molecule. Hydrogen bonds and the active site Zn^{2+} ion coordination are also shown (dotted lines). (See the color version of this figure in Color Plates section.)

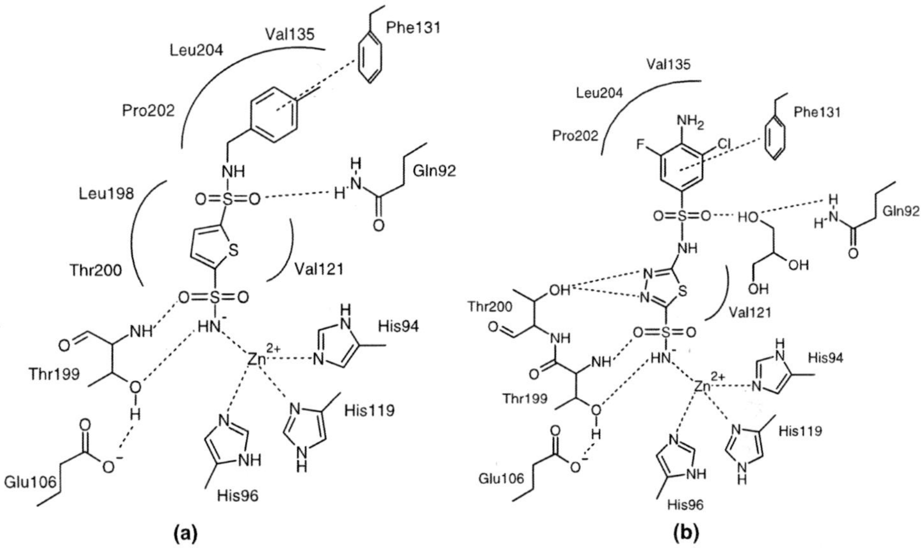

FIGURE 4.22 Schematic representation of inhibitors (a) **4.54** and (b) **4.53** within the hCA II active site.

of the thiophene disulfonamides, one of the oxygen atoms of the secondary 5-sulfonamide accepts a hydrogen bond from the Gln92NE2 atom[120] (Fig. 4.22a), while the secondary sulfonamide moiety in thiadiazole disulfonamides is not able to directly hydrogen bond the Gln92NE atom, but only through a bridging water and/or a glycerol molecule[119,121] (Fig. 4.22b).

Different aromatic tails have been introduced to the secondary sulfonamide group. Despite some sensible differences between these tails (ring type, ring substitutions, and tail length), a comparative structural analysis of all these compounds in complex with isoform II reveals that the aromatic tails are all superposable and point toward the hydrophobic part of the binding pocket, establishing an edge-to-face interaction with the phenyl ring of Phe131,[119–121] a residue known to play a critical role in the binding of inhibitors to hCA II.[16]

Bicyclic Sulfonamides Derived from Thiophene and Thiadiazole The derivatization of the previously reported thiophene disulfonamides by connecting the secondary 5-sulfonamide group with the C4 atom of the thiophene ring led to bicyclic compounds containing the thienothiazine-6-sulfonamide-1,1-dioxide skeleton (compounds **4.7** and **4.57–4.66**).[120] The CA inhibitory properties of these molecules were carefully investigated both kinetically and structurally by Boriack-Sjodin et al.,[120] highlighting that these bicyclic derivatives generally present a significant enhancement of the enzyme-inhibitor affinity with respect to the corresponding thiophene disulfonamides that present the same substitution of the secondary sulfonamide group.[120] Crystallographic studies suggested that this increase in binding affinity results from the preorganization of the secondary sulfonamide in an orientation that optimizes the interaction of the substituent with the hydrophobic patch defined by Phe131, Val135, Leu198, Pro202, and Leu204. As already observed for the corresponding thiophene disulfonamides, aromatic substituents on the *endo*-sulfonamide nitrogen establish a critical edge-to-face interaction with Phe131 side chain (Fig. 4.23).[120]

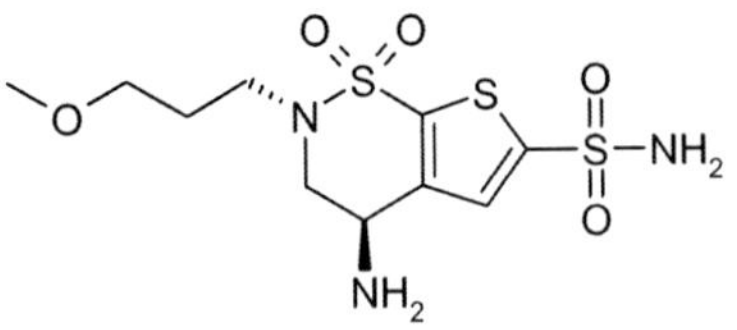

4.57

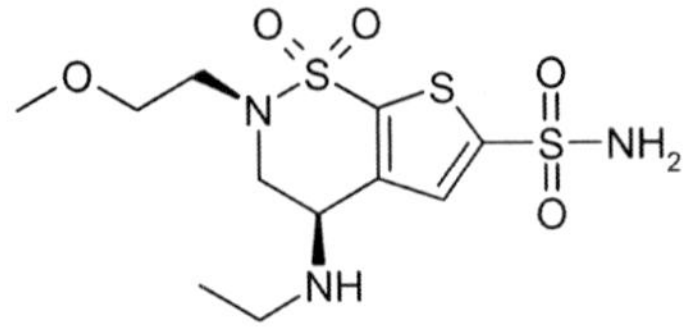

4.58

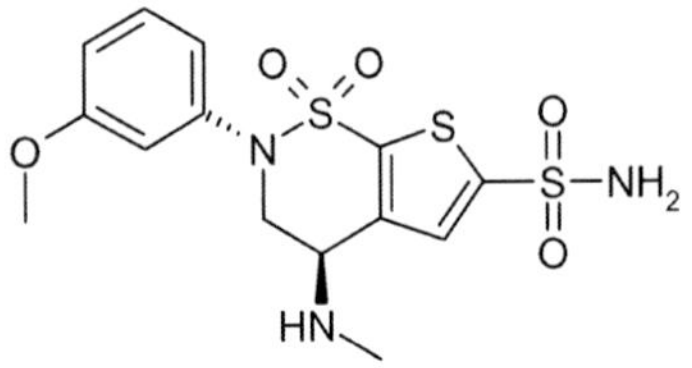

4.59

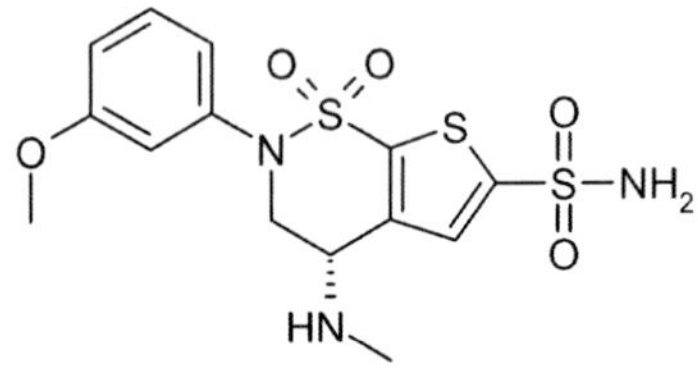

4.60

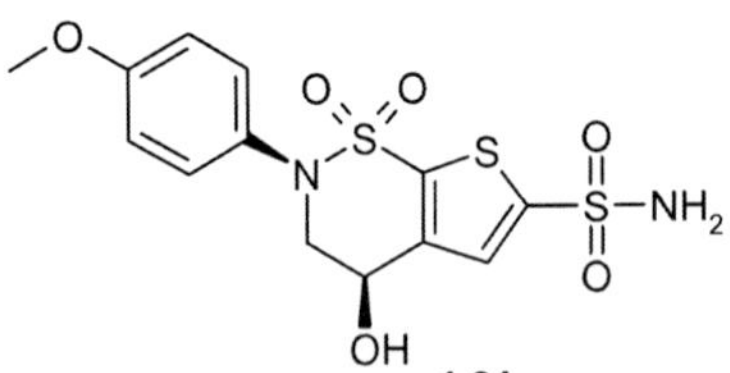

4.61

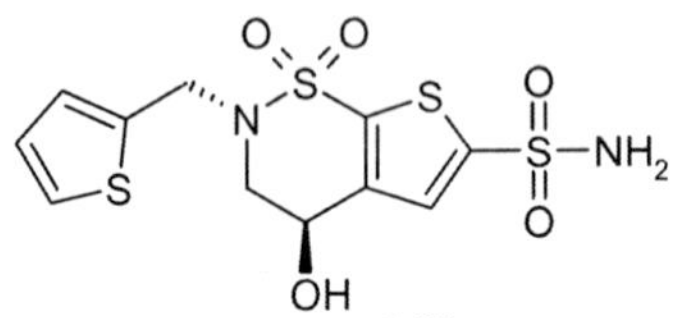

4.62

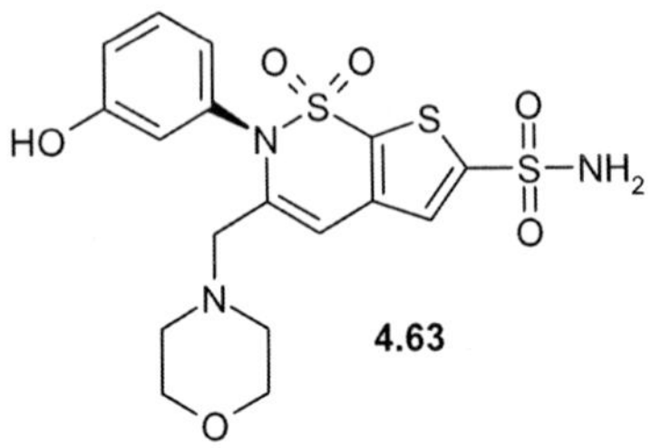

4.63

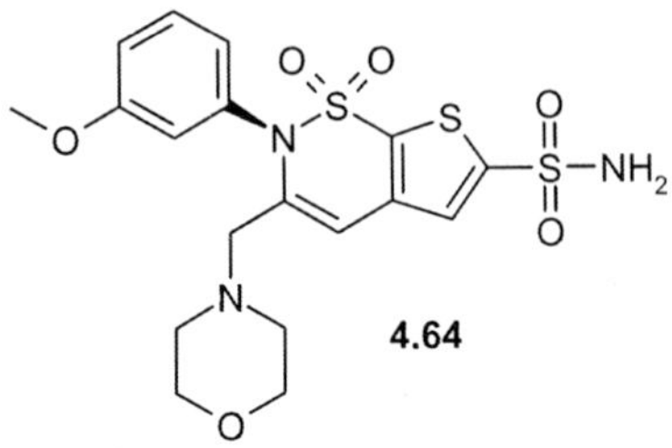

4.64

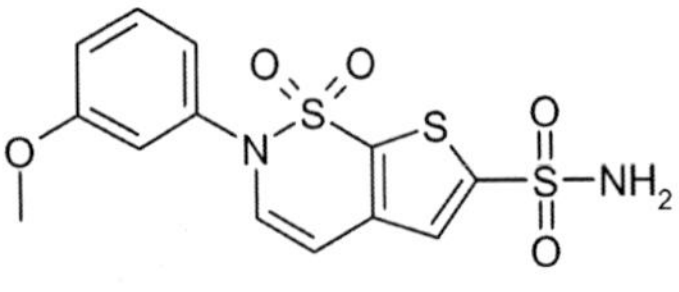

4.65

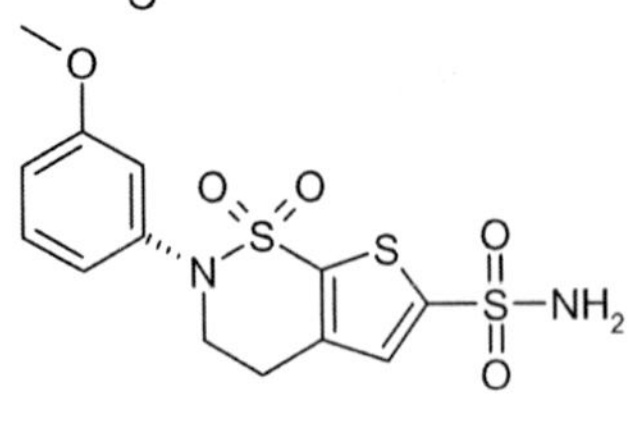

4.66

The effect of the substitution of the aromatic substituents on the *endo*-sulfonamide nitrogen with aliphatic ethers, such as in **4.58** and brinzolamide **4.7**, was also thoroughly investigated kinetically and structurally.[26,120] These studies revealed that the loss of the edge-to-face interaction between the aromatic substituent and Phe131

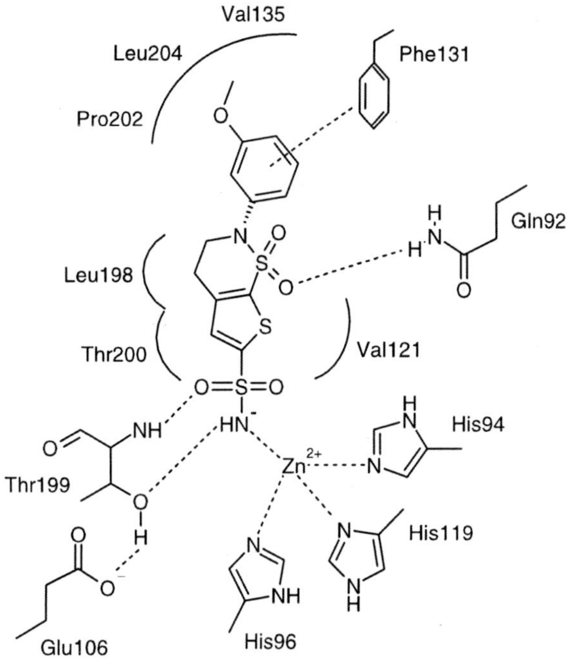

FIGURE 4.23 Schematic representation of inhibitor **4.66** binding within the hCA II active site.

has only a minor impact on enzyme-inhibitor affinity, as long as the substituted aliphatic group is sufficiently large to desolvate a correspondingly large hydrophobic patch in the enzyme active site.[120] Stereochemical and configurational variations at the C4 position of the thienothiazine-6-sulfonamide-1,1-dioxide skeleton also modulate enzyme-inhibitor affinity. In particular, the addition of a methylamino group with *S* stereochemistry changes the orientation of the ring compared to that in the absence of a C4 substituent. Consequently, the affinity of the inhibitor decreases dramatically, clearly indicating that bicyclic sulfonamides having substituents with *S* stereochemistry at C4 cannot be easily accommodated in the CA II active site.[120] In agreement with this observation, when a racemic mixture of the inhibitor is used, only the *R* enantiomers bind to the enzyme.[120]

The introduction of an additional degree of unsaturation between C3 and C4 in the six-membered thiazine ring, as observed in compound **4.65**, does not change the binding mode of the inhibitor. Therefore, neither a C4 substituent nor the conformational flexibility of the thiazine ring seems to be required to achieve enzyme-inhibitor affinity in the nanomolar or subnanomolar range.[120] On the contrary, the introduction of a substituent in position 3, as observed in compounds **4.63** and **4.64**, has a dramatic effect on the binding mode of the inhibitor.[130] In particular, the orientation of the aromatic substituents on the *endo*-sulfonamide nitrogen inside the active site changes dramatically. In fact, in the crystal structure of hCA II–**4.64** complex, the morpholino group is associated with the hydrophobic patch defined by Phe131, Val135, Pro202,

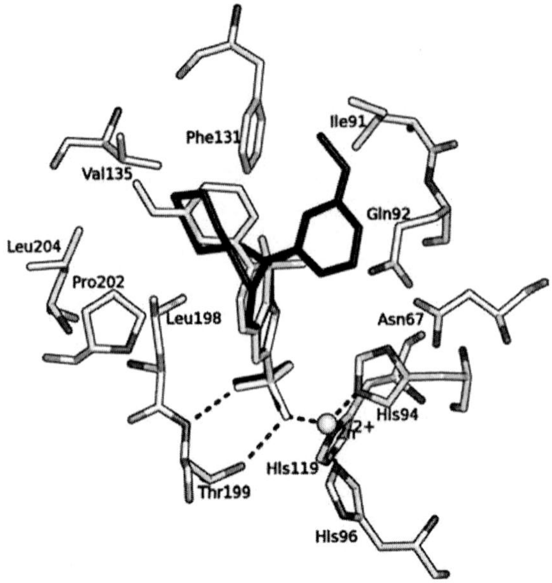

FIGURE 4.24 Superposition of hCA II–inhibitor adducts: **4.65** is reported in white and **4.64** in black.

and Leu204, generally occupied by the substituents on the *endo*-sulfonamide nitrogen in the absence of a C3 substituent (Fig. 4.24). Consequently, the *meta*-substituted phenyl group is orientated toward the opposite side of the active site, at van der Waals distance with Asn67, Ile91, Gln92, and Phe131 (Fig. 4.24).[130]

Ethoxzolamide (**4.67**) is another important bicyclic sulfonamide with CA inhibitory properties that has been largely used as drug (not so much nowadays) for the treatment of edema due to congestive heart failure, and for drug-induced edema, in addition to its application as antiglaucoma agent. It has also been used as lead molecule for the design of dorzolamide and brinzolamide. Although **4.67** presents a chemical structure quite similar to brinzolamide, it shows a rather different inhibition profile. In fact, while ethoxzolamide indiscriminately inhibits all CA isozymes except CA III,[131] in the nanomolar or subnanomolar range, brinzolamide presents selectivity for the inhibition of some isozymes over the others. Indeed, it does not inhibit appreciably CA I, CA III, and CA IV.[132] The structural analysis of the complex of ethoxzolamide with CA II and the comparison with the corresponding hCA II–brinzolamide adduct suggest that the different inhibition profiles of these two compounds can be ascribed to the presence of bulky substituents on the bicyclic ring system of brinzolamide.

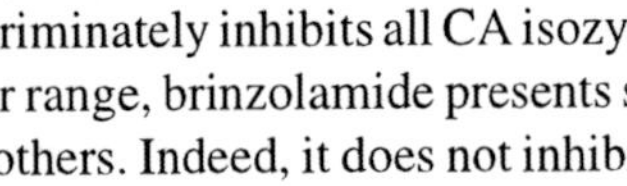

4.67

4.3.3.3 Sulfonamides Containing Other Ring Systems Many other annulated bicyclic systems, in which one of the rings incorporates a benzenesulfonamide moiety, were investigated for their CA inhibitory properties and many of these derivatives, such as indanesulfonamides of type **4.68–4.69**, thiazolinone derivatives of type **4.70**, and indolsulfonamides of type **4.71**, were also structurally characterized in their adduct with hCA II.[42,133,134] Among these compounds, indanesulfonamides, which were originally investigated to better understand the drug design of anticonvulsant CA inhibitors, have been thoroughly characterized. Indeed, Chazalette and coworkers in 2004[135] reported a large series of such derivatives, obtained by means of acylation reactions of some 1- and 2-amino derivatives of the parent unsubstituted compound **4.68** with alkyl- and/or arylcarboxylic acid chlorides. In particular, starting from the medium-potency CA II inhibitor **4.68** (K_I of 52 nM), much more effective CAIs were produced. Among these derivatives, the indane-2-valproylamido-5-sulfonamide (**4.69**) was one of the most interesting compounds, showing K_I values of 3.4 and 1.6 nM against CA II and I, respectively.[135] The pharmacological evaluation of this type of derivatives was initially tested only against CA I and II, and then extended to other isoforms with medicinal chemistry applications, showing these compounds being rather effective and selective against CA IX and VII.[136,137] The high-resolution crystal structures of both **4.68** and **4.69** in complex with the physiologically dominant human isoform II[133] provided important information on the molecular basis of the indanesulfonamides inhibitory properties. In particular, structural data showed that both the inhibitor molecules were located in the enzyme active site coordinating the Zn^{2+} ion by means of the deprotonated sulfonamide nitrogen and establishing a large number of hydrophobic and polar interactions (Fig. 4.25). However,

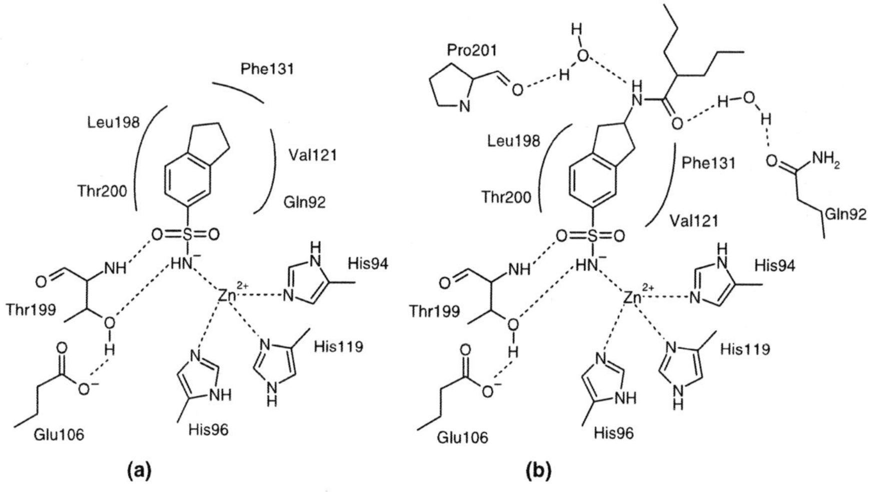

FIGURE 4.25 Schematic representation of inhibitors (a) **4.68** and (b) **4.69** within the hCA II active site.

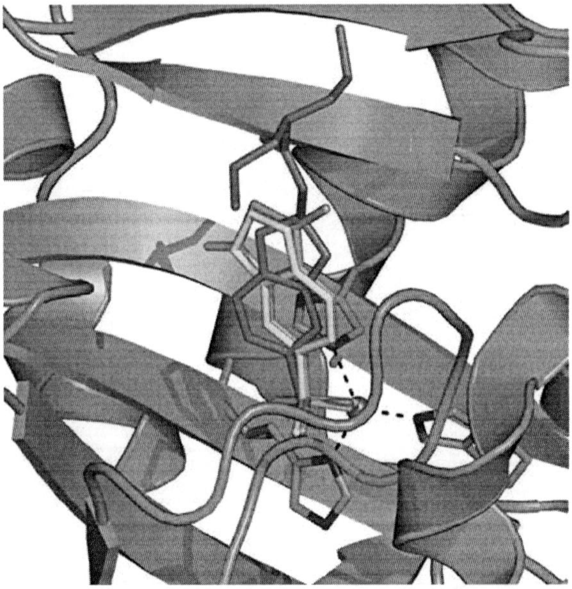

FIGURE 4.26 Superposition of hCA II–inhibitor adducts: **4.68** is reported in yellow, **4.69** in magenta, and **4.70** in cyan. (See the color version of this figure in Color Plates section.)

although these compounds have quite similar chemical structures, the arrangement of their indane moieties within the active site showed significant differences (Fig. 4.26). In fact, the indane ring plane within the hCA II–**4.69** complex was rotated about 180° with respect to that observed in hCA II–**4.68** counterpart. This structural rearrangement seemed to be necessary to avoid a steric clash between the bulky valproylamide functionality of **4.69** and the residues present in the hydrophobic region of the active site.[133] The hypothesis of a structural rearrangement of the inhibitor caused by the insertion of a bulky substituent on the indane ring found a confirmation in the crystal structure of the hCA II–**4.70** complex reported by Klebe's group (Fig. 4.26).[134] This molecule, developed using a virtual screening of some libraries of compounds, is a submicromolar hCA II inhibitor and is structurally related to **4.68** as it incorporates an unsubstituted thiazolinone ring. The structural overlay of **4.68**, **4.70**, and **4.69**, when bound to hCA II, showed a great level of similarity in the binding mode of the first two inhibitors, as compared to the latter one (Fig. 4.26). Taken together, these data suggested that the introduction of bulky substituents on bicyclic sulfonamides may strongly influence the binding mode of these molecules to the CA active site. Consequently, the introduction of bulky tails on bicyclic ring systems may represent a powerful strategy to induce the desired physicochemical properties (i.e., enhanced liposolubility) to an aromatic sulfonamide CA inhibitor or to obtain inhibitors with diverse inhibition profiles and selectivities for various mammalian CAs.

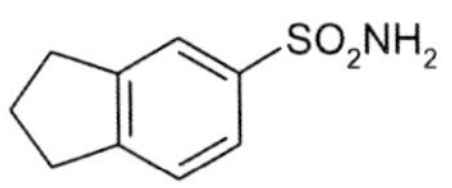

4.68

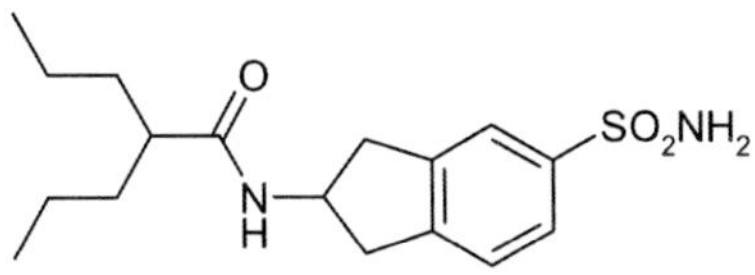

4.69

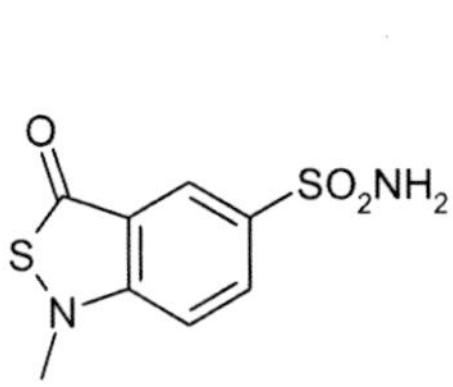

4.70

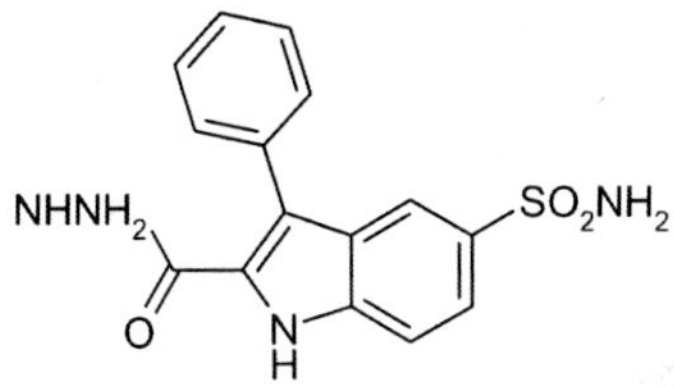

4.71

4.3.3.4 Aliphatic Sulfonamides Aliphatic sulfonamides of the type R-SO_2NH_2 (R = Me, $PhCH_2$) were originally investigated in the pioneeristic work of Maren in 1967[1] that showed that differently from aromatic/heterocyclic sulfonamides, these compounds were extremely weak CAIs. Successively, Maren and Conroy[138] reported that some types of aliphatic sulfonamides could act as very strong CAIs. In particular, these studies showed that an increasing number of halogen atoms in position 1 or 2, or both, from the sulfonamide moiety had the effect to enhance the acidity of the SO_2NH_2 moiety and consequently to improve the CA inhibitory properties. Thus, derivatives such as trifluoromethane sulfonamide (**4.72**) or nonafluorobutane-1-sulfonamide (**4.73**) were proved to act as nanomolar CAIs.[138]

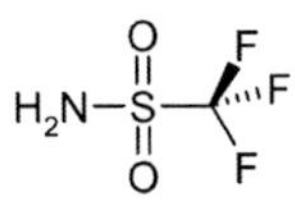

4.72

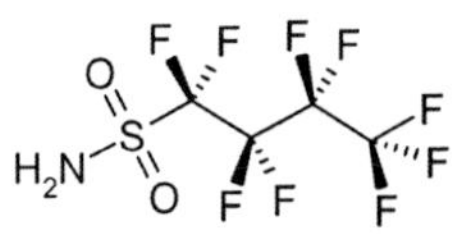

4.73

Starting from these observations, this type of relatively little investigated CAIs was object of great interest in the search of compounds with different and possibly improved selectivity/inhibition profiles, as well as enhanced solubility, compared to the classical sulfonamide CAIs. Unfortunately, until now only little structural information on the binding mode of these molecules to the CA active site is available. Trifluoromethane sulfonamide (**4.72**)[139] was the first aliphatic sulfonamide to be structurally characterized in its adduct with hCA II. This molecule is one of the most potent inhibitors of CA II, even if its clinical use has been prevented by its toxicity and limited bioavailability.[138] The significant CA inhibitory properties of this molecule were mainly ascribed to its very strong acidity due to the high electron withdrawing nature of the CF_3 moiety and to its close proximity to the sulfonamide group.[138] The

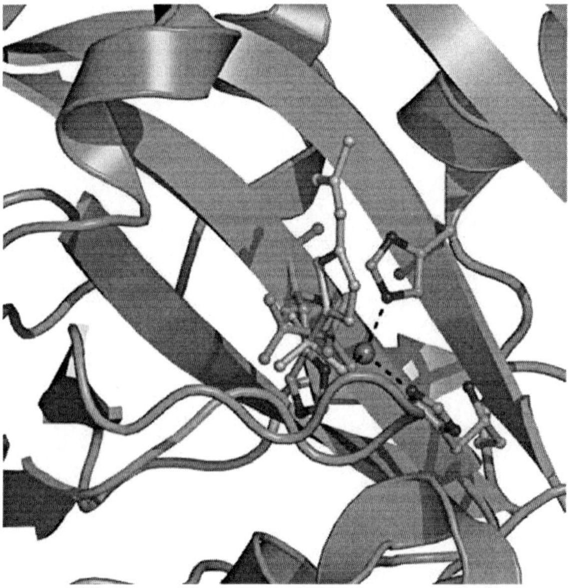

FIGURE 4.27 Superposition of hCA II–inhibitor adducts that highlights the different orientations of trifluoromethane sulfonamide **4.72** (green) with respect to a classical arylsulfonamide derivative such as acetazolamide **4.49** (cyan). (See the color version of this figure in Color Plates section.)

structural analysis revealed that the binding mode of this inhibitor to the CA II active site was considerably different from that of classical aromatic/heterocyclic sulfonamides. In fact, the small hydrophobic trifluoromethyl group was oriented inward toward the active site, making van der Waals contacts with residues Val121, Val143, Leu198, Thr199, and Trp209 (Fig. 4.27). This orientation is generally inaccessible for bulky aromatic groups that, instead, point outward from the active site cleft (Fig. 4.27).[139] Consequently, even if the position occupied by the sulfonamide nitrogen atom was similar to that observed for the other sulfonamide inhibitors, the positions of the two sulfonamide oxygens were different. In particular, the inhibitor was rotated by approximately 180° around the sulfur–nitrogen bond with respect to aromatic/heterocyclic sulfonamides, with both oxygens of the sulfonamide moiety pointing outward from the active site cleft (Fig. 4.27).[138]

The other structural information available for aliphatic sulfonamides regards different types of inhibitors, namely, compounds **4.74–4.77**, including the widely used antiepileptic drug zonisamide **4.74**.[21,140,141] These inhibitors are characterized by the presence of a heteroaromatic ring connected to the sulfonamide zinc binding group by an aliphatic spacer. X-ray crystallographic studies of the adducts of these inhibitors with hCA II evidenced that the tetrahedral geometry of the zinc ion and the hydrogen bond network between the sulfonamide moiety of all these inhibitors and the enzyme active site were all retained with respect to other hCA II–sulfonamide

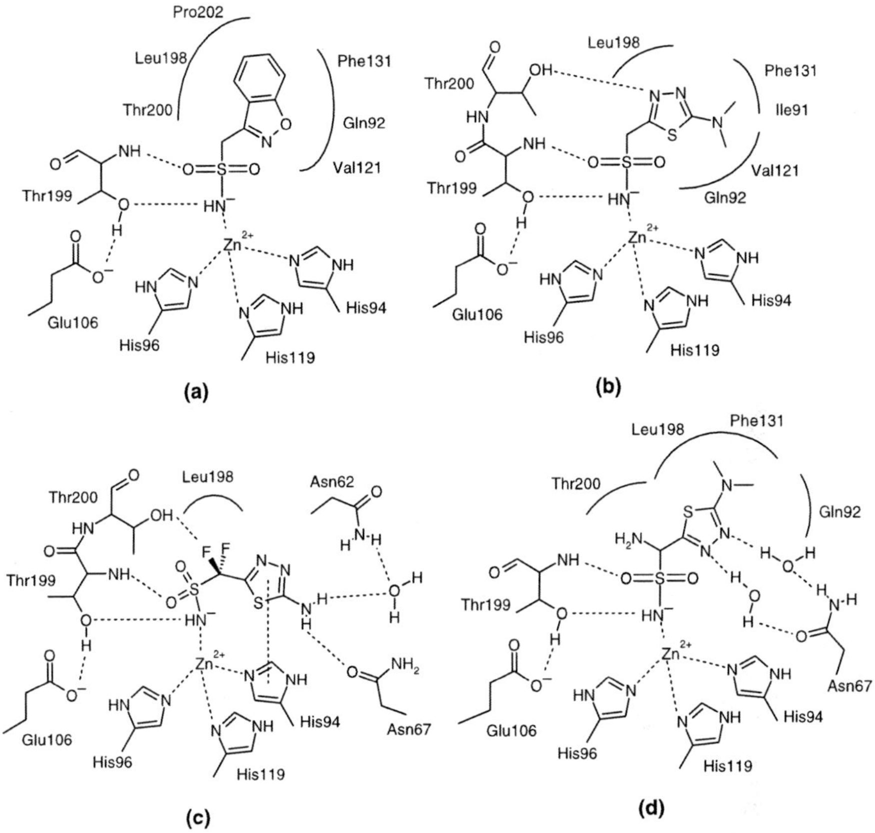

FIGURE 4.28 Schematic representation of aliphatic sulfonamides (a) **4.74**, (b) **4.75**, (c) **4.76**, and (d) **4.77** bound within the hCA II active site.

complexes structurally characterized so far (Fig. 4.28).[21,140,141] On the contrary, the orientation of the heterocyclic ring within the active site was completely different to that of structurally related compounds such as acetazolamide (Figs. 4.28 and 4.29). Indeed, the aliphatic spacer introduces an additional rotational freedom between the sulfonamide and the ring moiety allowing orientations inside the active site not accessible to the classical CAIs. The structural analysis also highlighted that, although compounds **4.74–4.77** have a similar organic scaffold, they present significant differences in the binding mode to the active site. Indeed, while the heterocyclic rings of inhibitors **4.74**, **4.75**, and **4.77** are quite well superimposable when bound to the enzyme, all of them pointing toward the hydrophobic part of the active site,[21,140,141] the inhibitor **4.76** adopts a completely different orientation, pointing toward the hydrophilic part of the active site and establishing a π-stacking interaction with the imidazole ring of His94 (Figs. 4.28c and 4.29).[141] This peculiar orientation was never observed in any other hCA II–inhibitor adducts and was mainly ascribed to

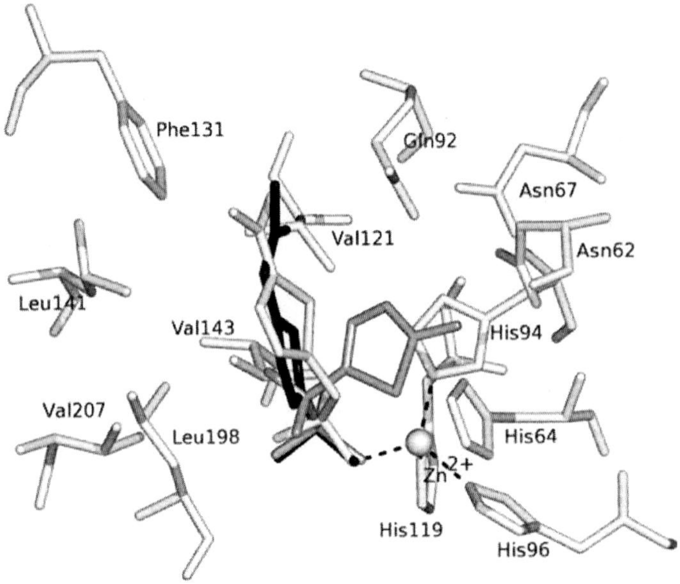

FIGURE 4.29 Superposition of hCA II–inhibitor adducts: **4.49** is reported in black, **4.75** in white, and **4.76** in grey.

the $-CF_2-$ spacer that formed a strong hydrogen bond with the Thr200OG atom through one of its two fluorine atoms (Fig. 4.28c). On the contrary, the spacer in the other three complexes[141] did not form any stabilizing interaction (Fig. 4.28a, b, and d). The unusual orientation of **4.76** was further stabilized by another hydrogen bond interaction between its NH_2 group and the Asn67OD1 atom (Fig. 4.28c).[141]

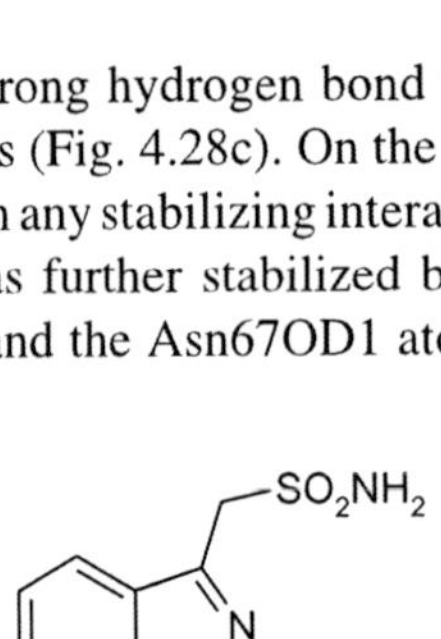

4.74

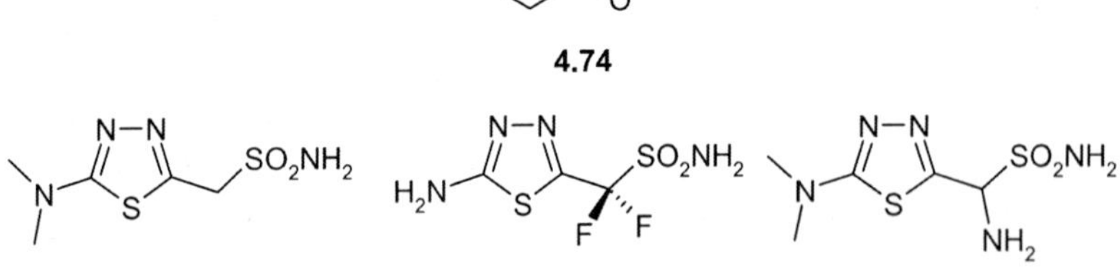

4.75 **4.76** **4.77**

These studies evidenced novel binding modes of aliphatic sulfonamides containing heterocyclic rings, suggesting that these compounds can represent an interesting alternative to the classical aromatic/heterocyclic sulfonamides in the development of CA isoform-specific inhibitors.

4.3.3.5 Sulfamate CAIs also Acting as Steroid Sulfatase Inhibitors A key therapeutic target for the treatment of estrogen-dependent tumors, that is, tumors with the hormones supporting their growth and development, is the steroid sulfatase (STS).[142,143] This enzyme catalyzes the hydrolysis of steroid sulfates, such as estrone-3-sulfate (E1S) to estrone (E1), which is the main source of estrogens in tumors,[142] and modulates the production of androstenediol (Adiol), which contributes to the estrogenic stimulation of hormone-dependent breast tumors. Owing to the role of STS in supporting tumor growth, several structurally diverse STS inhibitors have been developed in the past decade, with the irreversible types of inhibitors having a phenol sulfamate ester as their active pharmacophore.[142] One of the most important features of these sulfamates is the concomitant ability to interact with hCA II, since the sequestration by this enzyme in erythrocytes represents an important mechanism that optimizes their oral activity, pharmacokinetics, and represents a protection against first-pass metabolism.[144] As a consequence, in the last few years, to investigate in detail the interaction of many sulfamate-based STS inhibitors with hCA II, several X-ray structures of these compounds in complex with this enzyme have been reported.[145–150] Moreover, the finding that several sulfamates were also found to be potent inhibitors of CA IX,[151] whose expression is increased in many tumors, recently raised the intriguing possibility that the inhibition of CAs may contribute to the overall anticancer efficacy of this class of drug.

EMATE (estrone 3-*O*-sulfamate, **4.78**), was the first very potent STS inhibitor to be reported, even though its further development was not followed, since this compound was found to be highly estrogenic in rodents.[152] To overcome the side effects of EMATE, efforts were made in parallel by various research groups. The most notable approaches included the development of nonsteroidal mimics, such as STX64 (**4.79**), a STS inhibitor currently in Phase I clinical trials for the treatment of breast cancer, as well as the modification of the EMATE ring system or the introduction of substituent(s) at various positions of its steroidal scaffold to give derivatives such as **4.80**, **4.81**, and **4.82**[148–150,153,154] that were nonestrogenic but remained highly potent against STS[142] and hCA II. The X-ray crystal structures of their adducts with the latter enzyme[145,146,148–150] were also recently reported, showing that the sulfamate group of all these compounds binds to hCA II active site in a similar manner, with the deprotonated NH^- moiety coordinating the Zn^{2+} ion (Fig. 4.30). The steroid nucleus of **4.78**, **4.81**, and **4.82** and the tricyclic ring structure of **4.79** were rather well superposable (Fig. 4.31), all of them lying in the hydrophobic part of the active site pocket, establishing a large number of strong van der Waals interactions (see Fig. 4.30). Compound **4.82** established an additional hydrogen bond interaction with residue Gln136 (see Fig. 4.30e). Surprisingly, a different orientation was observed for compound **4.80**, bound to the Zn^{2+} ion in the active site with the 17-*O*-sulfamate group, rather than the 3-*O*-sulfamate one (Figs. 4.30c and 4.31). This finding was justified considering that, although the ionization of the alkyl sulfamate of 17-position is disfavored respect to the ionization of the aryl sulfamate of the 3-position, the 2-substituent could

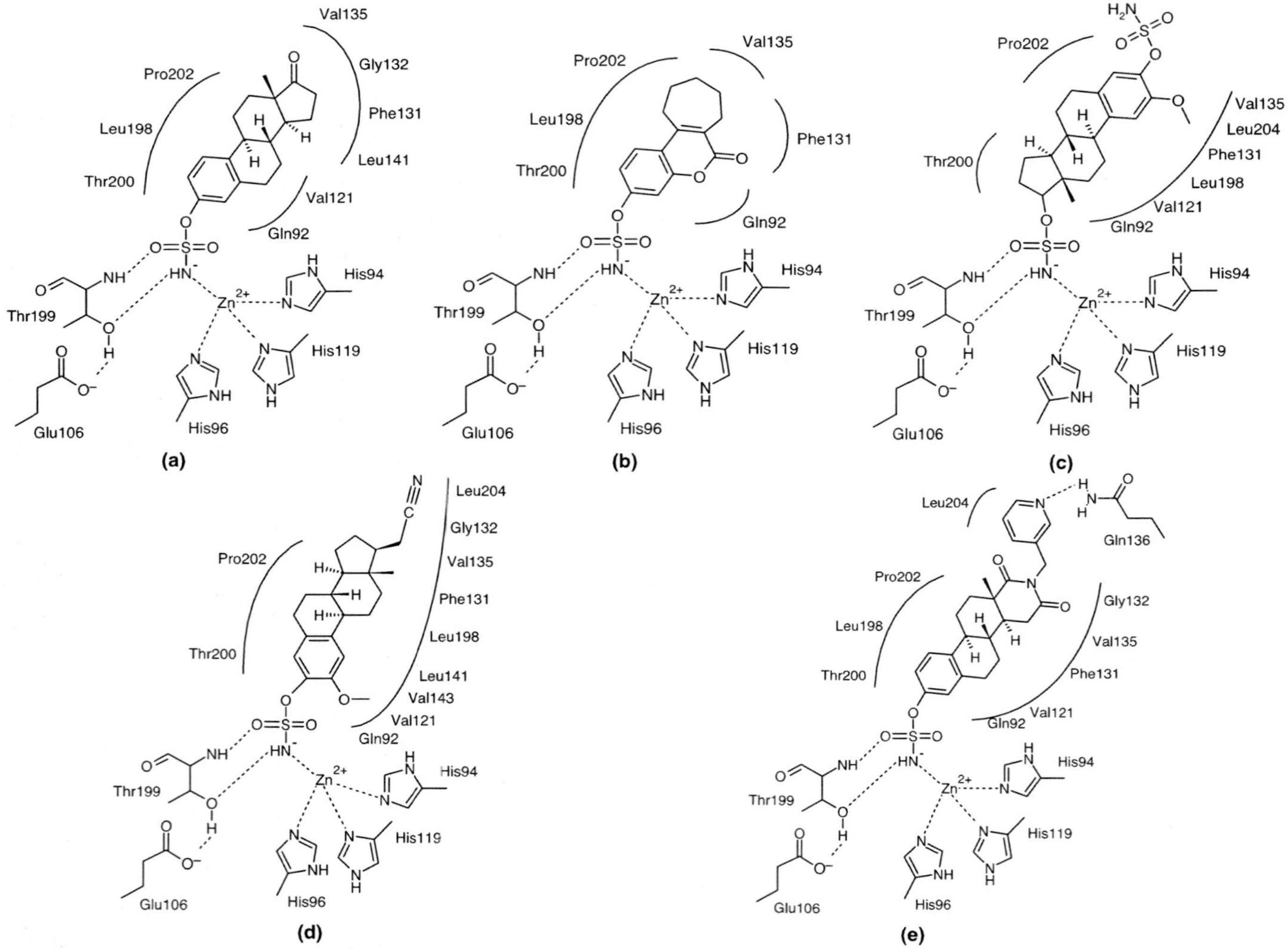

FIGURE 4.30 Schematic representation of sulfamate-based STS inhibitors (a) **4.78**, (b) **4.79**, (c) **4.80**, (d) **4.81**. and (e) **4.82** bound within the hCA II active site.

potentially render less favorable the interaction of the 3-*O*-sulfamate group with the zinc ion, causing a steric hindrance with proximal residues in the active site.[148]

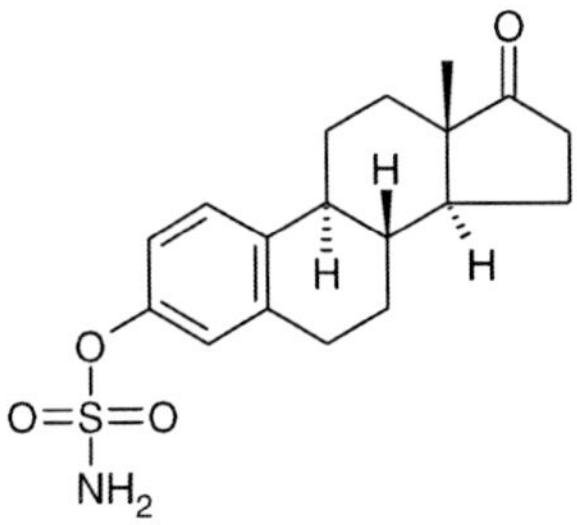

4.78

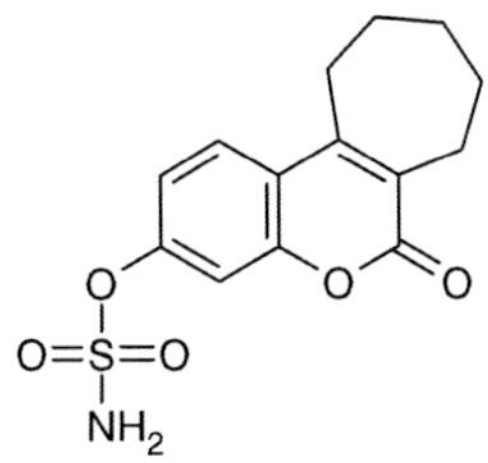

4.79

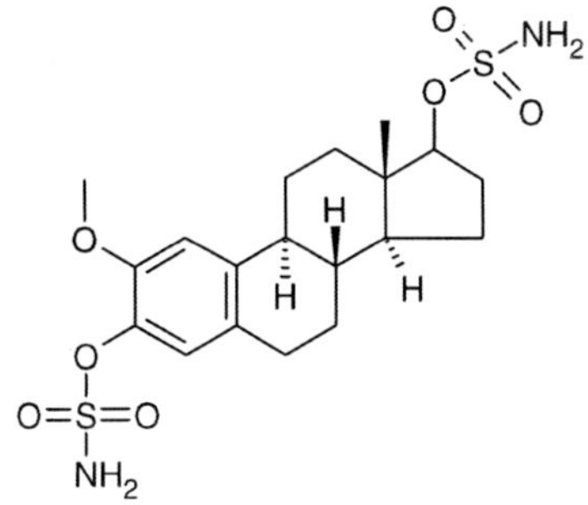

4.80

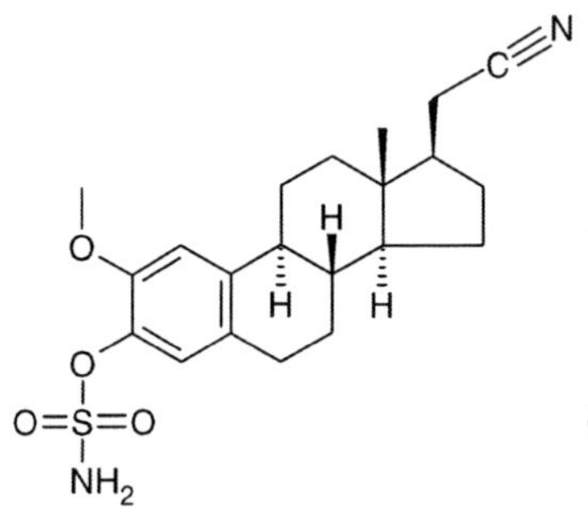

4.81

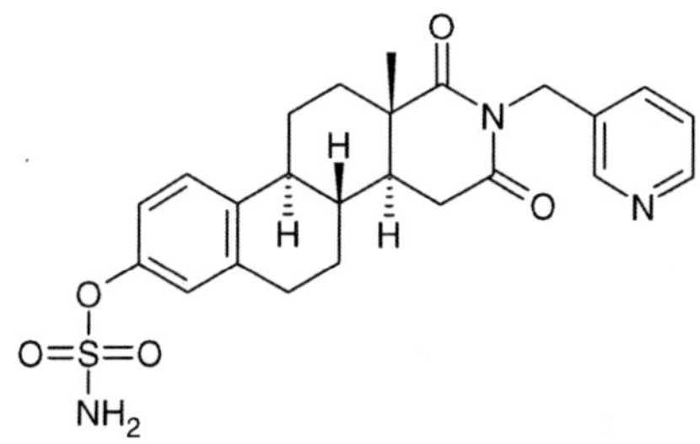

4.82

An alternative approach to the treatment of the hormone-dependent tumors consists in reducing levels of estrogens using aromatase inhibitors. As the inhibitors of aromatase given in conjunction with STS inhibitors would maximize estrogen depletion in treated patients, new drugs can be obtained by designing dual aromatase–steroid sulfatase inhibitors (DASIs) that inhibited both enzymes as single agents. Important example of DASIs included 4-[(4-*O*-sulfamoylbenzyl)(4-cyanophenyl)-amino]-4*H*-[1,2,4]triazole (**4.83**) and 4-[(3-bromo-4-*O*-sulfamoylbenzyl)-(4-cyano-phenyl)amino]-4*H*-[1,2,4]triazole (**4.84**).[155] These molecules, as observed for the aforementioned STS inhibitors, containing the sulfamate moiety were also able to bind the hCA II active site. In particular, compound **4.83** had IC_{50} value of 27 nM against CA II, while **4.84** was four to five times less potent with an IC_{50} value of 137 nM.[147] A detailed picture of the molecular interactions between these DASI compounds and hCA II was shown by their crystal structures.[147] Both molecules were located within the active site cavity and were stabilized by van der Waals interactions with several protein hydrophobic residues (Fig. 4.32). However, the binding of these compounds within the hCA II active site was similar, but not identical. Indeed, the introduction of the bromine in compound **4.84** appeared to rotate the sulfamate-bearing aromatic ring

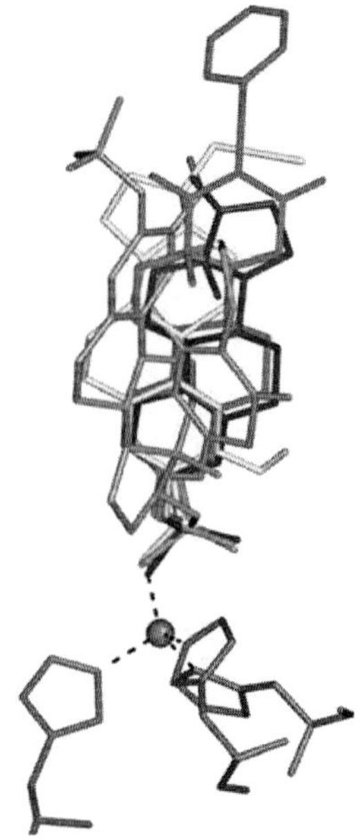

FIGURE 4.31 Superposition of hCA II–inhibitor adducts: **4.78** is reported in blue, **4.79** in green, **4.80** in orange, **4.81** in yellow, and **4.82** in red. (See the color version of this figure in Color Plates section.)

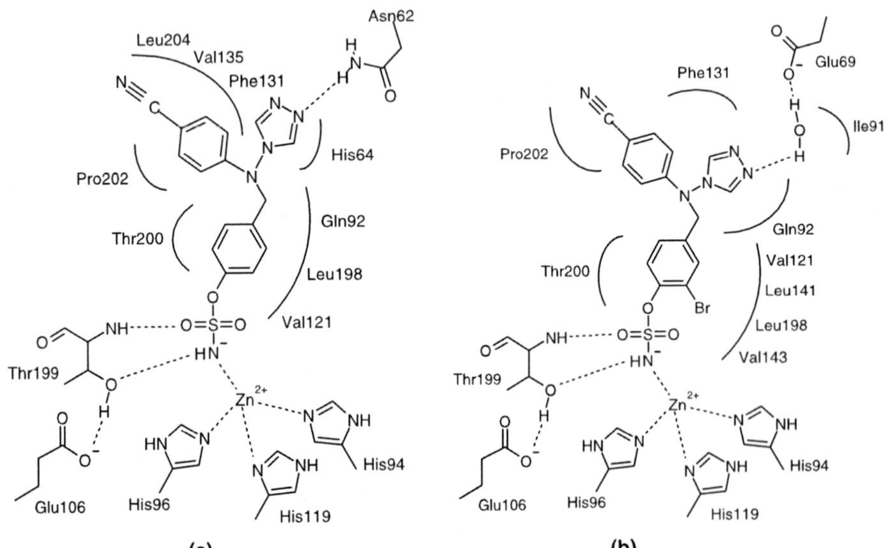

FIGURE 4.32 Schematic representation of dual aromatase-steroid sulfatase inhibitors (a) **4.83** and (b) **4.84** bound within the hCA II active site.

by about 30° away from Val121 and toward Pro202, and the ring was displaced by about 1.8 Å toward Val121. This was presumably due to the steric interactions between the protein and the large bromine atom. Thus, the positioning of the sulfamate containing aromatic ring and the accommodation of the large bromine atom appeared to be the major factors that resulted in a significantly lower affinity of **4.84** toward the CA II active site.

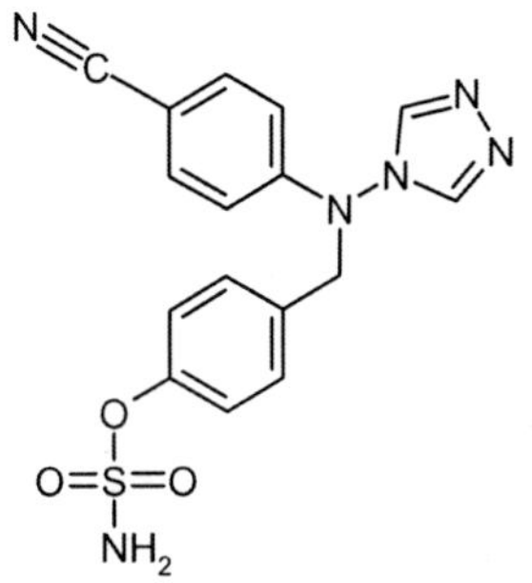

4.83

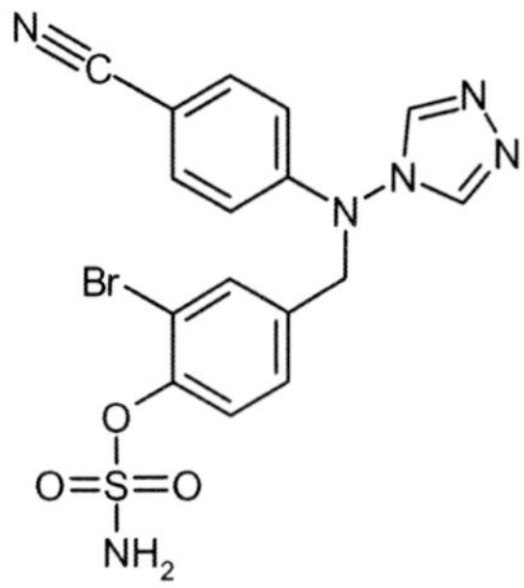

4.84

In summary, the analysis of structures of the hCA II–sulfamates adducts reviewed here, beyond to demonstrate that even small modifications to inhibitor structures can produce large modifications in binding mode and affinity, significantly expanded the knowledge of the interactions between sulfamate-based ligands and hCA II and supported more rational drug design strategies to test whether other classes of drugs can also be delivered by the inhibition of erythrocyte CA II upon sulfamoylation.

4.3.3.6 Sulfonamide/Sulfamate/Sulfamide Containing Sugar Moieties

Although the use of CAIs in the treatment of glaucoma represents a very useful tool in reducing elevated intraocular pressure, the systemic administration of such drugs often leads to unpleasant side effects as a consequence of the inhibition of CAs present in other tissues different from the eye. Thus, a great interest has been devoted in recent years to the development of topically effective antiglaucoma sulfonamides and important advances have indeed been achieved with dorzolamide and brinzolamide.[156,157] The attachment of sugar moieties to the scaffold of aromatic sulfonamides has also recently been proved to constitute an alternative efficient approach for the development of topically acting CAIs antiglaucoma agents.[158–161] In fact, the derivatization of 4-carboxy-benzenesulfonamide (**4.2**) or sulfanilamide (**4.1**) by means of tails incorporating various simple or more complex sugar moieties, such as in derivatives **4.85–4.93**, led to sulfonamides with an excellent CA inhibitory activity.[159,161–163] Moreover, due to the highly hydrophilic character of the sugar moieties, these sulfonamides presented a good water solubility, thus representing interesting drugs to be administered directly into the eye.

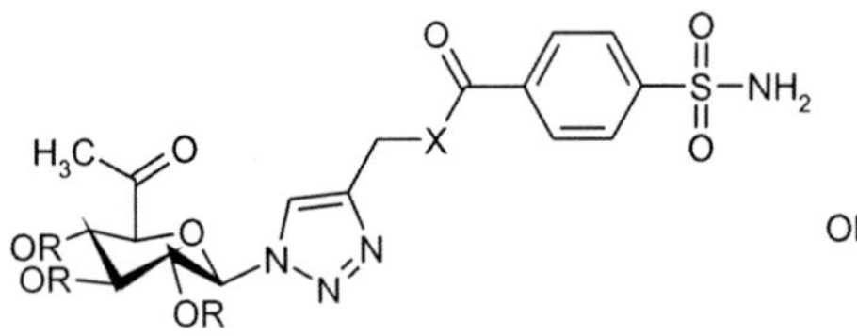

4.85: X = NH, R = Ac, **4.86**: X = NH, R = H
4.87: X = O, R = Ac, **4.88**: X = O, R = H

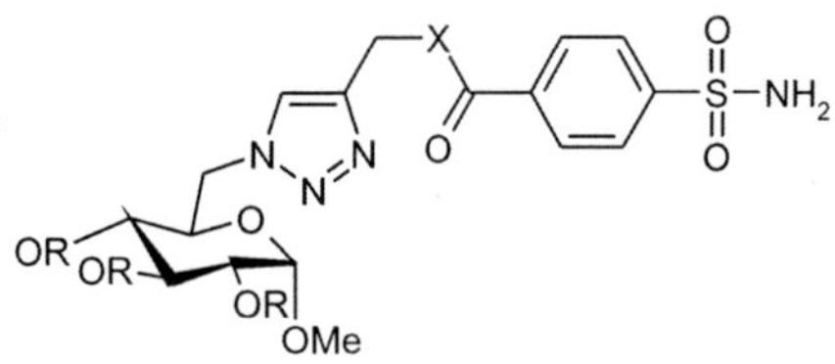

4.89: X = NH, R = Ac, **4.90**: X = NH, R = H
4.91: X = O, R = Ac, **4.92**: X = O, R = H

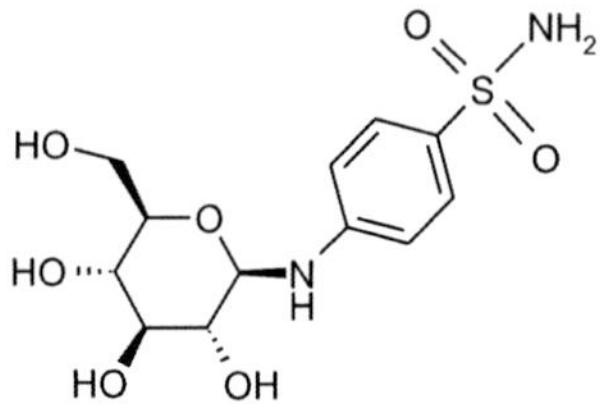

4.93

Until now, only one of these derivatives was structurally characterized for its interaction with hCA II, namely, *N*-(4-sulfamoylphenyl)-α-D-glucopyranosylamine (glucose derivative **4.93**), one of the most promising compounds of this series. In fact, this molecule was shown to be a potent inhibitor of several CA isozymes and a promising antiglaucoma agent with topical activity in an animal model of the disease.[161] The high-resolution X-ray crystal structure of the adduct hCA II–**4.93** was determined revealing the molecular basis of the strong enzyme inhibitory activity of this sugar derivative.[160] In particular, while the sulfonamide and the phenyl ring of the inhibitor bound in the canonical way to the enzyme, the glycosyl moiety, responsible for the high water solubility of the compound, was oriented toward the hydrophilic region of the active site, stabilized by a network of seven hydrogen bonds with five water molecules and the enzyme residues Pro201 and Gln92[160] (Fig. 4.33). It is interesting to observe that the types of interactions observed in the hCA II–**4.93** adduct were never reported previously in other hCA II–inhibitor complexes, and thus they could be suitably used for the rational drug design of other sugar-based CAIs.

Another class of CA inhibitors containing sugar moieties is represented by the widely used antiepileptic drug topiramate **4.94** and its analogues **4.95** and **4.96**. Topiramate presents a peculiar chemical structure derived from a monosaccharide bearing a sulfamate functional group. This drug possesses potent anticonvulsant effects as a consequence of a multifactorial mechanism of action: blockade of sodium channels and kainate/AMPA receptors, CO_2 retention secondary to inhibition of the red cell and brain CA isozymes, and enhancement of GABAergic transmission.[164–169] A side effect of this drug observed in obese patients was the loss of body weight,[170] although no certain pharmacological explanation of this phenomenon has been

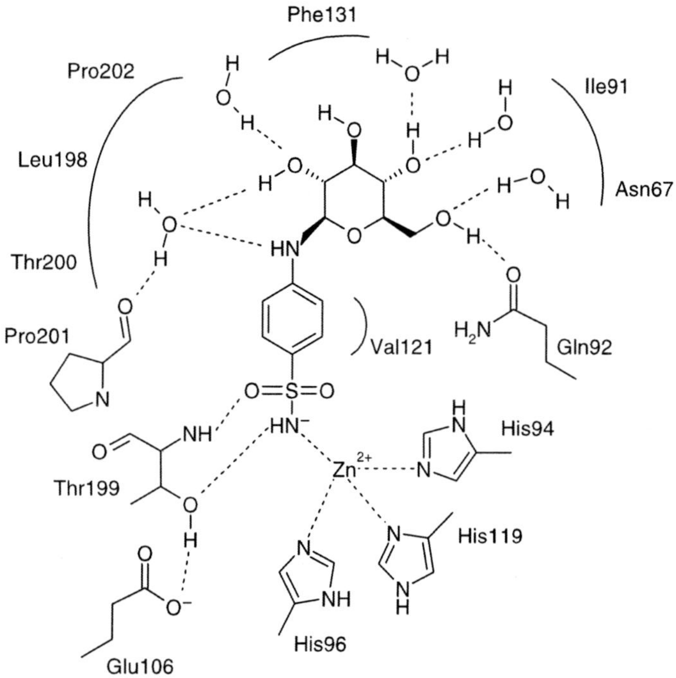

FIGURE 4.33 Schematic representation of inhibitor **4.93** binding within the hCA II active site.

provided so far. It was recently demonstrated that **4.94** is also a very potent inhibitor of several CA isozymes, among which are hCA II (K_I of 13.8 nM) and hCA VA (K_I of 25.4 nM).[23] To identify the basic molecular interactions that explain the high affinity of **4.94** for CAs, the X-ray crystal structure of the adduct hCA II–**4.94** was solved.[23] The main protein–inhibitor interactions observed in this structure are schematically depicted in Fig. 4.34a. According to this figure, apart from the common interactions of the sulfamate group within the hCA II active site, an extended network of hydrogen bonds between the inhibitor and some amino acid residues within the enzyme cavity is observed.

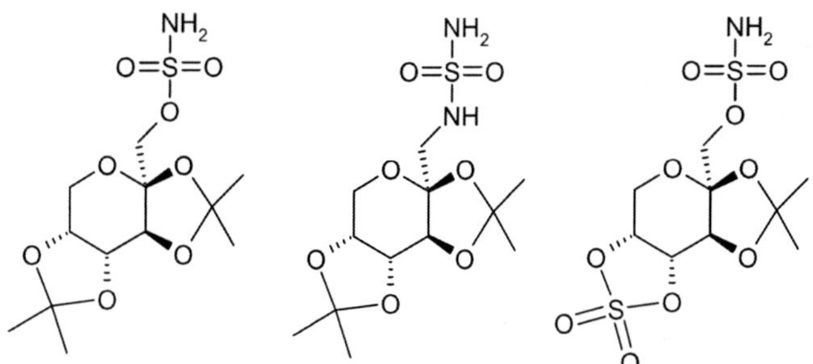

4.94 **4.95** **4.96**

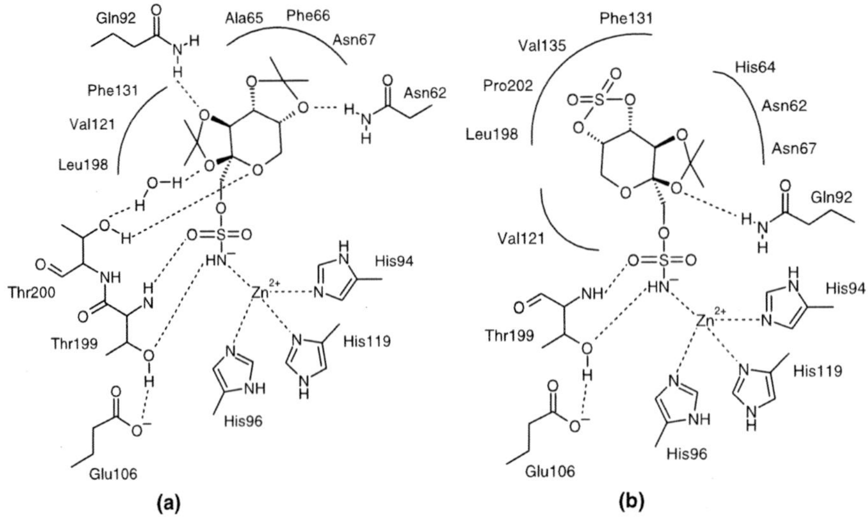

FIGURE 4.34 Schematic representation of inhibitors (a) **4.94** and (b) **4.96** bound within the hCA II active site.

The sulfamide analogue of **4.94** (compound **4.95**),[171] although differing only by the substitution of an oxygen atom with an NH moiety, is a much less potent hCA II inhibitor. The analysis of the crystal structure of the hCA II–**4.95** adduct explains well this finding. In fact, although **4.95** binds in nearly the same conformation and in the

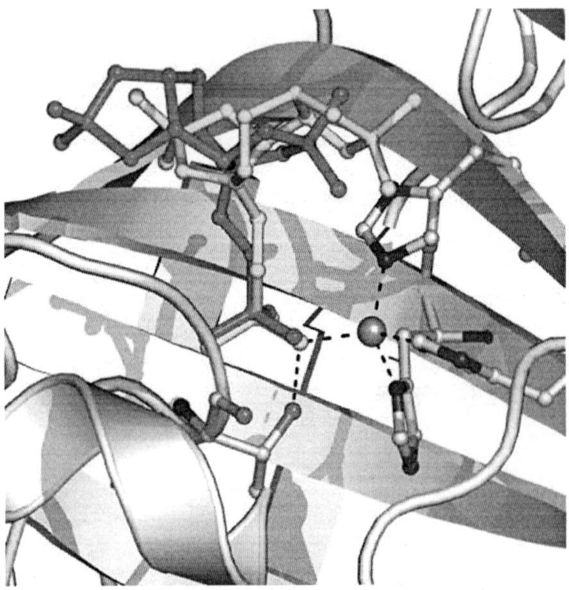

FIGURE 4.35 Superposition of hCA II–inhibitor adducts: **4.94** is reported in yellow and **4.96** in red. (See the color version of this figure in Color Plates section.)

same region of the active site compared to **4.94**, its C8 methyl group makes a clash with methyl side chain of Ala65,[41] thus explaining the weak binding of this compound to hCA II.

4.96[171] is another structural analogue of topiramate, where one of the di-isopropylidene moieties was substituted by a cyclic sulfate group. The X-ray structure of the hCA II–**4.96** complex has also been reported,[172] showing that the inhibitor binds to the enzyme active site in a completely different way as compared to **4.94** (Fig. 4.34b). In particular, while the sulfamate groups of these inhibitors are nearly superposable, their sugar-derivative moieties are rotated by about 180° within the enzyme active site (Fig. 4.35). This rotation causes in the hCA II–**4.96** complex a complete different pattern of hydrogen bond interactions and consequently a diverse inhibition capability toward hCA II.[172] These results suggest that even minor structural differences between two compounds, such as those observed between **4.94**, **4.95**, and **4.96**, may lead to a different arrangement in the CA active site and consequently to very different inhibitory effects.

4.4 CONCLUSIONS

A wealth of X-ray structural data have been accumulated in the past 15 years for CA–inhibitor complexes, including the two main classes of inhibitors: the pharmacologically relevant sulfonamides and their isosteres (sulfamates, sulfamides, ureates, and hydroxamates) and the simple inorganic anions. Although X-ray crystal structures are already available for the majority of the 12 catalytically active members of the α-CA family (i.e., isozymes I–VA, XII, XIII, and XIV), most of the reported complexes with inhibitors regard just isozyme II (and to a less extent isozyme I). These data are important for the drug design of isozyme-selective CAIs, a goal largely unattained for the moment, but for which important advances have been made in the last few years.

By considering the three main structural elements needed to be present in the molecule of a potent CAI, that is, a zinc binding group (ZBG), an organic scaffold, and one or more "tails" (side chains substituting the scaffold), important progress has been achieved in our understanding of the factors that govern both potency and selectivity against various CA isozymes for this class of pharmacological agents.

The ZBGs leading to potent CAIs may belong to various functionalities, with the classical sulfonamide one still constituting the main player. However, bioisosteres of the SO_2NH_2 moiety, such as the sulfamate and sulfamide groups, were shown to lead to equipotent inhibitors with the corresponding sulfonamides, whereas in some cases even stronger CAIs (than the sulfonamides) incorporating these functionalities have been evidenced. Furthermore, as shown extensively in this chapter, the sulfonamide, sulfamate, and sulfamide ZBGs bind in a very similar manner to the Zn^{2+} ion/amino acid residues within the CA active site in the enzyme–inhibitor complexes, analyzed by X-ray crystallography so far. In addition, in some cases, the sulfamides and sulfamates led to isozyme-selective CAIs (compared to the corresponding sulfonamides) due to structural conformational changes in the inhibitor molecule (not connected to the zinc binding of the anchoring group, but having to do with the organic scaffold/side chains incorporated into the inhibitor molecule and their interactions with the active site). This may be due to the different geometries of these molecules induced by diverse hybridizations/angles of the sulfamate/sulfamide ZBG compared to the sulfonamide one in which a $C\text{-}SO_2NH_2$ bond is present instead of the $X\text{-}SO_2NH_2$ one ($X = O$ or NH). Alternative ZBGs to the sulfonamide/sulfamate/sulfamide ones have also been explored in some detail by crystallographic methods. Thus, unsubstituted hydroxamates, *N*-hydroxy-urea, and urea can be considered as promising such moieties although few compounds were characterized thoroughly for the moment.

A multitude of various organic scaffolds were investigated in detail in CA–inhibitor complexes, with four main classes of compounds evidenced so far: aromatic, heterocyclic, aliphatic, and sugar derivatives. Aromatic (most of the time benzene or naphthalene derivatives) sulfonamides constitute by far the most extensively investigated class of CAIs, followed by aromatic sulfamates and sulfamides. Benzene-1,3-disulfonamides (of which dichlorophenamide is a clinically used drug) were also investigated in detail ultimately. The heterocyclic ring systems incorporating sulfonamide as ZBG are immediately second in importance to the benzenesulfonamides, considering the fact that they are contained in many of the clinically used, classical inhibitors (acetazolamide, methazolamide, ethoxzolamide), as well as in the second generation of such agents (dorzolamide and brinzolamide). Many of these CAIs incorporate five-membered sulphur-containing heterocycles (thiophene, 1,3,4-thiadiazole, 1,3,4-thiadiazoline) or annulated five-membered such rings (benzo-*b*-thiophene, benzothiazine, etc.). Recently, some indole-sulfonamide derivatives were also investigated in detail, this ring system conferring very interesting properties to the CAIs incorporating it (some degrees of isozyme selectivity, potent antitumor activity, etc). Several aliphatic derivatives were also finally investigated, such as the clinically used antiepileptic zonisamide and some 1,3,4-thiadiazole derivatives substituted with $CH_2SO_2NH_2$ or $CF_2SO_2NH_2$ moieties. The binding modes and the selectivity profiles for inhibition of mammalian CAs of these compounds are completely new and different from those of the aromatic/heterocyclic sulfonamides investigated earlier. Thus, more detailed investigations of the aliphatic derivatives such as CAIs (eventually incorporating other ZBGs than the sulfonamide one) are highly desirable and may lead to important advances. Several sugar derivatives (incorporating both sulfonamide and sulfamate and sulfamide ZBGs) were also characterized by means of X-ray crystallography, leading to very interesting results and some isozyme-selective inhibitors. For example, for the sulfamide analogue of topiramate, an important clash with Ala65 in the CA II complex has been evidenced, which explains why the compound is a roughly 200 times weaker hCA II inhibitor compared to topiramate. Since Ala65 is present only in hCA II and not in other mammalian CA isozymes, this finding affords a means for designing inhibitors that will bind hCA II with less efficacy, a feature highly desirable for many CAIs targeting other isoforms, such as CA VA/VB, VI, VII, IX, XII, XIII, or XIV. Indeed, the topiramate sulfamide retained a good inhibitory activity against most of these isozymes, while being a rather weak hCA II inhibitor. More studies on sugar derivatives such as CAIs are indeed needed to have more insights into the factors governing potency and selectivity toward various isozymes for this type of CAIs.

However, the main problem with the classical, clinically used compounds (also including the second-generation agents dorzolamide and brinzolamide) is related to the fact that they are promiscuous inhibitors of all (or most of the) CA isozymes found in mammals. Some low levels of isozyme selectivity are shown by dorzolamide and brinzolamide that have been designed in such a way to act as much weaker CA I than hCA II inhibitors, but similar to acetazolamide, methazolamide, and ethoxzolamide, these two second-generation inhibitors strongly inhibit the remaining 10 CA isozymes. Thus, considering only the ZBGs and the organic scaffold, it is practically impossible to design isoform-selective CAIs, as the interactions around the metal ion and the organic scaffold (normally positioned at the bottom and in the middle of the active site cavity, respectively) are basically the same between the inhibitors and most CA isozymes with medicinal chemistry applications. This also explains why the first- and second-generation CAIs are normally devoid of any isozyme selectivity. They are indeed rather small, compact molecules that bind deeply within the enzyme active site. However, around 10 years ago, we reported the "tail approach" that afforded the rather facile synthesis of a large number of CAIs starting from aromatic/heterocyclic scaffold, also containing derivatizable amino, imino, or hydroxy groups, to which various moieties (tails) were introduced by normal chemical modification reactions (acylation, alkylation, arylsulfonylation, condensation, etc.). In this way, it was possible to modulate both the physicochemical properties of the synthesized inhibitors (for example, by introducing tails that induce high water solubility, enhanced lipophilicity, or positive/negative charges leading to membrane impermeability, fluorescence, spin-labeled groups, etc.) and their affinity to the various isozymes, as the tail(s) usually interact with amino acid residues toward the exit of the active site or on its edge. In fact, these are the amino acids that are less conserved among the various mammalian CAs, and this explains why most of these novel generation inhibitors show much more interesting inhibition profiles compared to the classical ones. X-Ray crystal structures and homology modeling are available for some of these compounds, which proved that both favorable interactions and clashes with particular amino acids present only in some isozymes are critical for the inhibition profile and isozyme selectivity issues. However, this research is still in its infancy and much work must be done to better understand all factors governing selective inhibition and the design of isozyme-specific inhibitors.

The coming years will thus probably see the report of the X-ray crystal structures for the remaining mammalian CAs (CAVB, VI, VII, IX, and XV). It is also highly desirable to have such structures for the catalytically inactive members of the family (CAVIII, X, and XI) that are much less understood at this moment compared to the catalytically active enzymes. It is also desirable that interactions of the same inhibitor(s) with all CA isozymes should be characterized by X-ray crystallography, as in the last instance, this will indeed allow us to understand all factors governing selectivity in the drug design of CAIs. This is for the moment possible to some extent only for acetazolamide, as its crystal structures in complex with CA II, IV, VA, XII, XIII, and XIV are reported. However, as mentioned above, this is a too simple compound for allowing us to use these structural data in the drug design of isoform-selective compounds. Indeed, compounds with a more sophisticated chemical structure (such as those incorporating different ZBGs than the sulfonamide one, and probably various tails in their aromatic, heterocyclilc, aliphatic, and/or sugar scaffolds) may afford a deeper insight into phenomena governing these intricate but fascinating processes. Although the CAIs field is a small one, these findings may be relevant to the general drug design research, especially when enzyme families with a multitude of members and with similar active site features are targeted.

REFERENCES

1. Maren, T. H. Carbonic anhydrase: chemistry, physiology and inhibition. *Physiol. Rev.* **1967**, *47*, 595–781.
2. Supuran, C. T.; Scozzafava, A.; Conway, J. *Carbonic Anhydrase: Its Inhibitors and Activators*, CRC Press: Boca Raton, **2004**; pp 1–363.
3. Supuran, C. T. Carbonic anhydrases: novel therapeutic applications for inhibitors and activators. *Nat. Rev. Drug Discov.* **2008**, *7*, 168–181.
4. Supuran, C. T.; Scozzafava, A.; Casini, A. Carbonic anhydrase inhibitors. *Med. Res. Rev.* **2003**, *23*, 146–189.
5. Supuran, C. T. Carbonic anhydrases as drug targets: an overview. *Curr. Top. Med. Chem.* **2007**, *7*, 825–833.
6. Lehtonen, J. M.; Shen, B.; Vihinen, M.; Casini, A.; Scozzafava, A.; Supuran, C. T.; Parkkila, A.-K.; Saarnio, J.; Kivelä, A. J.; Waheed, A.; Sly, W. S.; Parkkila, S. Characterization of CA XIII, a novel member of the carbonic anhydrase isozyme family. *J. Biol. Chem.* **2004**, *279*, 2719–2727.
7. Köhler, K.; Hillebrecht, A.; Schulze Wischeler, J.; Innocenti, A.; Heine, A.; Supuran, C. T.; Klebe, G. Saccharin inhibits carbonic anhydrases: possible explanation for its unpleasant metallic aftertaste. *Angew. Chem. Int. Ed. Engl.* **2007**, *46*, 7697–7699.
8. Pastorekova, S.; Parkkila, S.; Pastorek, J.; Supuran, C. T. Carbonic anhydrases: current state of the art, therapeutic applications and future prospects. *J. Enzyme Inhib. Med. Chem.* **2004**, *19*, 199–229.
9. Scozzafava, A.; Mastrolorenzo, A.; Supuran, C. T. Modulation of carbonic anhydrase activity and its applications in therapy. *Expert Opin. Ther. Pat.* **2004**, *14*, 667–702.
10. Scozzafava, A.; Mastrolorenzo, A.; Supuran, C. T. Carbonic anhydrase inhibitors and activators and their use in therapy. *Expert. Opin. Ther. Pat.* **2006**, *16*, 1627–1664.
11. Supuran, C. T.; Scozzafava, A. Carbonic anhydrases as targets for medicinal chemistry. *Bioorg. Med. Chem.* **2007**, *15*, 4336–4350.
12. Winum, J.-Y.; Scozzafava, A.; Montero, J.-L.; Supuran, C. T. New zinc binding motifs in the design of selective carbonic anhydrase inhibitors. *Bioorg. Med. Chem.* **2006**, *6*, 921–936.
13. Hilvo, M.; Rafajová, M.; Pastoreková, S.; Pastorek, J.; Parkkila, S. Expression of carbonic anhydrase IX in mouse tissues. *J. Histochem. Cytochem.* **2004**, *52*, 1313–1322.

14. Kaunisto, K.; Parkkila, S.; Rajaniemi, H.; Waheed, A.; Grubb, J.; Sly, W. S. Carbonic anhydrase XIV: luminal expression suggests key role in renal acidification. *Kidney Int.* **2002**, *61*, 2111–2118.

15. Kyllönen, M. S.; Parkkila, S.; Rajaniemi, H.; Waheed, A.; Grubb, J. H.; Shah, G. N.; Sly, W. S.; Kaunisto, K. Localization of carbonic anhydrase XII to the basolateral membrane of H+-secreting cells of mouse and rat kidney. *J. Histochem. Cytochem.* **2003**, *51*, 1217–1224.

16. Menchise, V.; De Simone, G.; Alterio, V.; Di Fiore, A.; Pedone, C.; Scozzafava, A.; Supuran, C. T. Carbonic anhydrase inhibitors: stacking with Phe131 determines active site binding region of inhibitors as exemplified by the X-ray crystal structure of a membrane-impermeant antitumor sulfonamide complexed with isozyme II. *J. Med. Chem.* **2005**, *48*, 5721–5727.

17. Weber, A.; Casini, A.; Heine, A.; Kuhn, D.; Supuran, C. T.; Scozzafava, A.; Klebe, G. Unexpected nanomolar inhibition of carbonic anhydrase by COX-2-selective celecoxib: new pharmacological opportunities due to related binding site recognition. *J. Med. Chem.* **2004**, *47*, 550–557.

18. Abbate, F.; Casini, A.; Owa, T.; Scozzafava, A.; Supuran, C. T. Carbonic anhydrase inhibitors: E7070, a sulfonamide anticancer agent, potently inhibits cytosolic isozymes I and II, and transmembrane, tumor-associated isozyme IX. *Bioorg. Med. Chem. Lett.* **2004**, *14*, 217–223.

19. Alterio, V.; Vitale, R. M.; Monti, S. M.; Pedone, C.; Scozzafava, A.; Cecchi, A.; De Simone, G.; Supuran, C. T. Carbonic anhydrase inhibitors: X-ray and molecular modeling study for the interaction of a fluorescent antitumor sulfonamide with isozyme II and IX. *J. Am. Chem. Soc.* **2006**, *128*, 8329–8335.

20. Alterio, V.; De Simone, G.; Monti, S. M.; Scozzafava, A.; Supuran, C. T. Carbonic anhydrase inhibitors: inhibition of human, bacterial, and archaeal isozymes with benzene-1,3-disulfonamides—solution and crystallographic studies. *Bioorg. Med. Chem. Lett.* **2007**, *17*, 4201–4207.

21. De Simone, G.; Di Fiore, A.; Menchise, V.; Pedone, C.; Antel, J.; Casini, A.; Scozzafava, A.; Wurl, M.; Supuran, C. T. Carbonic anhydrase inhibitors. Zonisamide is an effective inhibitor of the cytosolic isozyme II and mitochondrial isozyme V: solution and X-ray crystallographic studies. *Bioorg. Med. Chem. Lett.* **2005**, *15*, 2315–2320.

22. Boriack, P. A.; Christianson, D. W.; Kingery-Wood, J.; Whitesides, G. M. Secondary interactions significantly removed from the sulfonamide binding pocket of carbonic anhydrase II influence inhibitor binding constants. *J. Med. Chem.* **1995**, *38*, 2286–2291.

23. Casini, A.; Antel, J.; Abbate, F.; Scozzafava, A.; David, S.; Waldeck, H.; Schafer, S.; Supuran, C. T. Carbonic anhydrase inhibitors: SAR and X-ray crystallographic study for the interaction of sugar sulfamates/sulfamides with isozymes I, II and IV. *Bioorg. Med. Chem. Lett.* **2003**, *13*, 841–845.

24. Kim, C.-Y.; Chang, J. S.; Doyon, J. B.; Baird, T. T.; Fierke, C. A.; Jain, A.; Christianson, D. W. Contribution of fluorine to protein–ligand affinity in the binding of fluoroaromatic inhibitors to carbonic anhydrase II. *J. Am. Chem. Soc.* **2000**, *122*, 12125–12134.

25. Smith, G. M.; Alexander, R. S.; Christianson, D. W.; McKeever, B. M.; Ponticello, G. S.; Springer, J. P.; Randall, W. C.; Baldwin, J. J.; Habecker, C. N. Positions of His-64 and a

bound water in human carbonic anhydrase II upon binding three structurally related inhibitors. *Protein Sci.* **1994**, *3*, 118–125.

26. Stams, T.; Chen, Y.; Boriack-Sjodin, P. A.; Hurt, J. D.; Liao, J.; May, J. A.; Dean, T.; Laipis, P.; Silverman, D. N.; Christianson, D. W. Structures of murine carbonic anhydrase IV and human carbonic anhydrase II complexed with brinzolamide: molecular basis of isozyme-drug discrimination. *Protein Sci.* **1998**, *7*, 556–563.
27. Whittington, D. A.; Waheed, A.; Ulmasov, B.; Shah, G. N.; Grubb, J. H.; Sly, W. S.; Christianson, D. W. Crystal structure of the dimeric extracellular domain of human carbonic anhydrase XII, a bitopic membrane protein overexpressed in certain cancer tumor cells. *Proc. Natl. Acad. Sci. USA* **2001**, *98*, 9545–9550.
28. Boriack-Sjodin, P. A.; Heck, R. W.; Laipis, P. J.; Silverman, D. N.; Christianson, D. W. Structure determination of murine mitochondrial carbonic anhydrase V at 2.45 Å resolution: implications for catalytic proton transfer and inhibitor design. *Proc. Natl. Acad. Sci. USA* **1995**, *92*, 10949–10953.
29. Kannan, K. K.; Ramanadham, M.; Jones, T. A. Structure, refinement, and function of carbonic anhydrase isozymes: refinement of human carbonic anhydrase I. *Ann. N.Y. Acad. Sci.* **1984**, *429*, 49–60.
30. Eriksson, A. E.; Jones, T. A.; Liljas, A. Refined structure of human carbonic anhydrase II at 2.0 Å resolution. *Proteins* **1988**, *4*, 274–282.
31. Eriksson, A. E.; Liljas, A. Refined structure of bovine carbonic anhydrase III at 2.0 Å resolution. *Proteins* **1993**, *16*, 29–42.
32. Duda, D. M.; Tu, C.; Fisher, S. Z.; An, H.; Yoshioka, C.; Govindasamy, L.; Laipis, P. J.; Agbandje-McKenna, M.; Silverman, D. N.; McKenna, R. Human carbonic anhydrase III: structural and kinetic study of catalysis and proton transfer. *Biochemistry* **2005**, *44*, 10046–10053.
33. Stams, T.; Nair, S. K.; Okuyama, T.; Waheed, A.; Sly, W. S.; Christianson, D. W. Crystal structure of the secretory form of membrane-associated human carbonic anhydrase IV at 2.8 Å resolution. *Proc. Natl. Acad. Sci. USA* **1996**, *93*, 13589–13594.
34. Whittington, D. A.; Grubb, J. H.; Waheed, A.; Shah, G. N.; Sly, W. S.; Christianson, D. W. Expression, assay, and structure of the extracellular domain of murine carbonic anhydrase XIV: implications for selective inhibition of membrane-associated isozymes. *J. Biol. Chem.* **2004**, *279*, 7223–7228.
35. Di Fiore, A.; Monti, S. M.; Hilvo, M.; Parkkila, S.; Romano, V.; Scaloni, A.; Pedone, C.; Scozzafava, A.; Supuran, C. T.; De Simone, G. Crystal structure of human carbonic anhydrase XIII and its complex with the inhibitor acetazolamide. *Proteins* **2008**, *74*, 164–175.
36. Håkansson, K.; Carlsson, M.; Svensson, L. A.; Liljas, A. Structure of native and apo carbonic anhydrase II and structure of some of its anion–ligand complexes. *J. Mol. Biol.* **1992**, *227*, 1192–1204.
37. Stams, T.; Christianson, D. W. X-ray crystallographic studies of mammalian carbonic anhydrase isozymes. In *The Carbonic Anhydrases: New Horizons;* Chegwidden, W. R.; Edwards, Y.; Carter, N.,Eds.; Birkhäuser Verlag: Basel, Switzerland, **2000**; pp 159–174.
38. Christianson, D. W.; Fierke, C. A. Carbonic anhydrase: evolution of the zinc binding site by nature and by design. *Acc. Chem. Res.* **1996**, *29*, 331–339.
39. Lindskog, S. Structure and mechanism of carbonic anhydrase. *Pharmacol. Ther.* **1997**, *74*, 1–20.

40. Lindskog, S.; Silverman, D. W. The catalytic mechanism of mammalian carbonic anhydrases. In *The Carbonic Anhydrases: New Horizons;* Chegwidden, W. R.; Edwards, Y. H.; Carter, N. D.,Eds.; Birkhäuser Verlag: Basel, Switzerland, **2000**; pp 175–195.

41. Winum, J. Y.; Temperini, C.; El Cheikh, K.; Innocenti, A.; Vullo, D.; Ciattini, S.; Montero, J. L.; Scozzafava, A.; Supuran, C. T. Carbonic anhydrase inhibitors: clash with Ala65 as a means for designing inhibitors with low affinity for the ubiquitous isozyme II, exemplified by the crystal structure of the topiramate sulfamide analogue. *J. Med. Chem.* **2006**, *49*, 7024–7031.

42. Güzel, O.; Temperini, C.; Innocenti, A.; Scozzafava, A.; Salman, A.; Supuran, C. T. Carbonic anhydrase inhibitors: interaction of 2-(hydrazinocarbonyl)-3-phenyl-1*H*-indole-5-sulfonamide with 12 mammalian isoforms: kinetic and X-ray crystallographic studies. *Bioorg. Med. Chem. Lett.* **2008**, *18*, 152–158.

43. Di Fiore, A.; Pedone, C.; D'Ambrosio, K.; Scozzafava, A.; De Simone, G.; Supuran, C. T. Carbonic anhydrase inhibitors: valdecoxib binds to a different active site region of the human isoform II as compared to the structurally related cyclooxygenase II "selective" inhibitor celecoxib. *Bioorg. Med. Chem. Lett.* **2006**, *16*, 437–442.

44. Casini, A.; Abbate, F.; Scozzafava, A.; Supuran,C. T. Carbonic anhydrase inhibitors: X-ray crystallographic structure of the adduct of human isozyme II with a bis-sulfonamide-two heads are better than one? *Bioorg. Med. Chem. Lett.* **2003**, *13*, 2759–2763.

45. Liljas, A.; Håkansson, K.; Jonsson, B. H.; Xue, Y. Inhibition and catalysis of carbonic anhydrase. Recent crystallographic analyses. *Eur. J. Biochem.* **1994**, *219*, 1–10.

46. Thiry, A.; Dogné, J. M.; Supuran, C. T.; Masereel, B. Anticonvulsant sulfonamides/sulfamates/sulfamides with carbonic anhydrase inhibitory activity: drug design and mechanism of action. *Curr. Pharm. Des.* **2008**, *14*, 661–671.

47. Briganti, F.; Mangani, S.; Scozzafava, A.; Vernaglione, G.; Supuran, C. T. Carbonic anhydrase catalyzes cyanamide hydration to urea: is it mimicking the physiological reaction? *J. Biol. Inorg. Chem.* **1999**, *4*, 528–536.

48. Guerri, A.; Briganti, F.; Scozzafava, A.; Supuran, C. T.; Mangani, S. Mechanism of cyanamide hydration catalyzed by carbonic anhydrase II suggested by cryogenic X-ray diffraction. *Biochemistry* **2000**, *39*, 12391–12397.

49. Scozzafava, A.; Supuran, C. T. Hydroxyurea is a carbonic anhydrase inhibitor. *Bioorg. Med. Chem.* **2003**, *11*, 2241–2246.

50. Temperini, C.; Innocenti, A.; Scozzafava, A.; Supuran, C. T. *N*-hydroxyurea: a versatile zinc binding function in the design of metalloenzyme inhibitors. *Bioorg. Med. Chem. Lett.* **2006**, *16*, 4316–4320.

51. Scolnick, L. R.; Clements, A. M.; Liao, J.; Crenshaw, L.; Hellberg, M.; May, J.; Dean, T. R.; Christianson, D. W. Novel binding mode of hydroxamate inhibitors to human carbonic anhydrase II. *J. Am. Chem. Soc.* **1997**, *119*, 850–851.

52. Mangani, S.; Håkansson, K. Crystallographic studies of the binding of protonated and unprotonated inhibitors to carbonic anhydrase using hydrogen sulfide and nitrate anions. *Eur. J. Biochem. 210;* **1992**, 867–871.

53. Jönsson, B. M.; Håkansson, K.; Liljas, A. The structure of human carbonic anhydrase II in complex with bromide and azide. *FEBS Lett.* **1992**, *322*, 186–190.

54. Nair, S. K.; Christianson, D. W. Crystallographic studies of azide binding to human carbonic anhydrase II. *Eur. J. Biochem.* **1993**, *213*, 507–515.

55. Eriksson, A. E.; Kylsten, P. M.; Jones, T. A.; Liljas, A. Crystallographic studies of inhibitor binding sites in human carbonic anhydrase II: a pentacoordinated binding of the SCN-ion to the zinc at high pH. *Proteins* **1983**, *4*, 283–293.

56. Lindahl, M.; Svensson, L. A.; Liljas, A. Metal poison inhibition of carbonic anhydrase. *Proteins* **1993**, *15*, 177–182.

57. Keilin, D.; Mann, T. Carbonic anhydrase. Purification and nature of the enzyme. *Biochem. J.* **1940**, *34*, 1163–1176.

58. Clare, B. W.; Supuran, C. T. A perspective on quantitative structure–activity relationships and carbonic anhydrase inhibitors. *Expert Opin. Drug Metab. Toxicol.* **2006**, *2*, 113–137.

59. Temperini, C.; Scozzafava, A.; Supuran, C. T. Carbonic anhydrase activation and the drug design. *Curr. Pharm. Des.* **2008**, *14*, 708–715.

60. De Simone, G.; Di Fiore, A.; Supuran, C. T. Are carbonic anhydrase inhibitors suitable for obtaining antiobesity drugs? *Curr. Pharm. Des.* **2008**, *14*, 655–660.

61. King, R. W; Burgen, A. S. Sulfonamide complexes of human carbonic anhydrases. Ultraviolet difference spectroscopy. *Biochim. Biophys. Acta* **1970**, *207*, 278–285.

62. Lindahl, M.; Vidgren, J.; Eriksson, E.; Habash, J.; Harrop, S.; Helliwell, J.; Liljas, A.; Lindeskog, M.; Walker, N.; Crystallographic studies of carbonic anhydrase inhibition. In *Carbonic Anhydrase: From Biochemistry and Genetics to Physiology and Clinical Medicine;* Botrè, F.; Gros, G.; Storey, B. T.,Eds.; VCH: Weinheim, Germany, **1991**; pp 111–118.

63. Abbate, F.; Supuran, C. T.; Scozzafava, A.; Orioli, P.; Stubbs, M. T.; Klebe, G. Nonaromatic sulfonamide group as an ideal anchor for potent human carbonic anhydrase inhibitors: role of hydrogen-bonding networks in ligand binding and drug design. *J. Med. Chem.* **2002**, *45*, 3583–3587.

64. Beasley, Y. M.; Overell, B. G.; Petrow, V.; Stephenson, O. Some *in vitro* inhibitors of carbonic anhydrase. *J. Pharm. Pharmacol.* **1958**, *10*, 696–705.

65. Supuran, C. T. Carbonic anhydrase inhibitors. In *Carbonic Anhydrase and Modulation of Physiological and Pathologic Processes in the Organism;* Puscas, I.,Ed.; Helicon Press: Timisoara, Romania, **1994**; pp 29–111.

66. Mincione, F.; Starnotti, M.; Menabuoni, L.; Scozzafava, A.; Casini, A.; Supuran, C. T. Carbonic anhydrase inhibitors: 4-sulfamoyl-benzenecarboxamides and 4-chloro-3-sulfamoyl-benzenecarboxamides with strong topical antiglaucoma properties. *Bioorg. Med. Chem. Lett.* **2001**, *11*, 1787–1791.

67. Jain, A.; Whitesides, G. M.; Alexander, R. S.; Christianson, D. W. Identification of two hydrophobic patches in the active-site cavity of human carbonic anhydrase II by solution-phase and solid-state studies and their use in the development of tight-binding inhibitors. *J. Med. Chem.* **1994**, *37*, 2100–2105.

68. Avila, L. Z.; Chu, Y. H.; Blossey, E. C.; Whitesides, G. M. Use of affinity capillary electrophoresis to determine kinetic and equilibrium constants for binding of arylsulfonamides to bovine carbonic anhydrase. *J. Med. Chem.* **1993**, *36*, 126–133.

69. Gao, J.; Qiao, S.; Whitesides, G. M. Increasing binding constants of ligands to carbonic anhydrase by using "greasy tails". *J. Med. Chem.* **1995**, *38*, 2292–2301.

70. Gao, J.; Cheng, X.; Chen, R.; Sigal, G. B.; Bruce, J. E.; Schwartz, B. L.; Hofstadler, S. A.; Anderson, G. A.; Smith, R. D.; Whiteside, G. M. Screening derivatized peptide libraries

for tight binding inhibitors to carbonic anhydrase II by electrospray ionization-mass spectrometry. *J. Med.Chem.* **1996**, *39*, 1949–1955.

71. Jude, K. M.; Banerjee, A. L.; Haldar, M. K.; Manokaran, S.; Roy, B.; Mallik, S.; Srivastava, D. K.; Christianson, D. W. Ultrahigh resolution crystal structures of human carbonic anhydrases I and II complexed with two-prong inhibitors reveal the molecular basis of high affinity. *J. Am. Chem. Soc.* **2006**, *128*, 3011–3018.

72. Roy, B. C.; Banerjee, A. L.; Swanson, M.; Jia, X. G.; Haldar, M. K.; Mallik, S.; Srivastava, D. K. Two-prong inhibitors for human carbonic anhydrase II. *J. Am. Chem. Soc.* **2004**, *126*, 13206–13207.

73. Doyon, J. B.; Jain, A. The pattern of fluorine substitution affects binding affinity in a small library of fluoroaromatic inhibitors for carbonic anhydrase. *Org. Lett.* **1999**, *1*, 183–185.

74. Cappalonga Bunn, A. M.; Alexander, R. S.; Christianson, D. W. Mapping protein–peptide affinity: binding of peptidylsulfonamide inhibitors to human carbonic anhydrase II. *J. Am. Chem. Soc.* **1994**, *116*, 5063–5068.

75. Jain, A.; Huang, S. G.; Whitesides, G. M. Lack of effect of the length of oligoglycine- and oligo(ethylene glycol)-derived *para*-substituents on the amnity of benzenesulfonamides for carbonic anhydrase II in solution. *J. Am. Chem. Soc.* **1994**, *116*, 5057–5062.

76. Taylor, P. W.; King, R. W.; Burgen, A. S. V. Kinetics of complex formation between human carbonic anhydrases and aromatic sulfonamides. *Biochemistry* **1970**, *9*, 2638–2645.

77. Casini, A.; Scozzafava, A.; Mincione, F.; Menabuoni, L.; Ilies, M. A.; Supuran, C. T. Carbonic anhydrase inhibitors: water-soluble 4-sulfamoylphenylthioureas as topical intraocular pressure-lowering agents with long-lasting effects. *J. Med. Chem.* **2000**, *43*, 4884–4892.

78. Casini, A.; Scozzafava, A.; Mincione, F.; Menabuoni, L.; Supuran, C. T. Carbonic anhydrase inhibitors: synthesis of water soluble sulfonamides incorporating a 4-sulfamoylphenylmethylthiourea scaffold, with potent intraocular pressure lowering properties. *J. Enzyme Inhib. Med. Chem.* **2002**, *17*, 333–343.

79. Cecchi, A.; Winum, J.-Y.; Innocenti, A.; Vullo, D.; Montero, J.-L.; Scozzafava, A.; Supuran, C. T. Carbonic anhydrase inhibitors: synthesis and inhibition of cytosolic/tumor-associated carbonic anhydrase isozymes I, II, and IX with sulfonamides derived from 4-isothiocyanato-benzolamide. *Bioorg. Med. Chem. Lett.* **2004**, *14*, 5775–5780.

80. Innocenti, A.; Casini, A.; Alcaro, M. C.; Papini, A. M.; Scozzafava, A.; Supuran, C. T. Carbonic anhydrase inhibitors: the first on-resin screening of a 4-sulfamoylphenylthiourea library. *J. Med. Chem.* **2004**, *47*, 5224–5229.

81. Di Fiore, A.; De Simone, G.; Menchise, V.; Pedone, C.; Casini, A.; Scozzafava, A.; Supuran, C. T. Carbonic anhydrase inhibitors: X-ray crystal structure of a benzenesulfonamide strong CA II and CA IX inhibitor bearing a pentafluorophenylaminothioureido tail in complex with isozyme II. *Bioorg. Med. Chem. Lett.* **2005**, *15*, 1937–1942.

82. Supuran, C. T.; Scozzafava, A.; Jurca, B. C.; Ilies, M. A. Carbonic anhydrase inhibitors. Part 49. Synthesis of substituted ureido and thioureido derivatives of aromatic/heterocyclic sulfonamides with increased affinities for isozyme I. *Eur. J. Med. Chem.* **1998**, *33*, 83–93.

83. Svastová, E.; Hulíková, A.; Rafajová, M.; Zat'ovicová, M.; Gibadulinová, A.; Casini, A.; Cecchi, A.; Scozzafava, A.; Supuran, C. T.; Pastorek, J.; Pastoreková, S. Hypoxia

activates the capacity of tumor-associated carbonic anhydrase IX to acidify extracellular pH. *FEBS Lett.* **2004**, *577*, 439–445.

84. Cecchi, A.; Hulikova, A.; Pastorek, J.; Pastoreková, S.; Scozzafava, A.; Winum, J. Y.; Montero, J. L.; Supuran, C. T. Carbonic anhydrase inhibitors. Design of fluorescent sulfonamides as probes of tumor-associated carbonic anhydrase IX that inhibit isozyme IX-mediated acidification of hypoxic tumors. *J. Med. Chem.* **2005**, *48*, 4834–4841.
85. Pastorekova, S.; Casini, A.; Scozzafava, A.; Vullo, D.; Pastorek, J.; Supuran, C. T. Carbonic anhydrase inhibitors: the first selective, membrane-impermeant inhibitors targeting the tumor-associated isozyme IX. *Bioorg. Med. Chem. Lett.* **2004**, *14*, 869–873.
86. Supuran, C. T.; Scozzafava, A.; Ilies, M. A.; Briganti, F. Carbonic anhydrase inhibitors: synthesis of sulfonamides incorporatine 2,4,6-trisubstituted-pyridinium-ethylcarboxamido moieties possessing membrane-impermeability and *in vivo* selectivity for the membrane-bound (CA IV) versus the cytosolic (CA I and CA II) isozymes. *J. Enzyme Inhib.* **2000**, *15*, 381–401.
87. Scozzafava, A.; Briganti, F.; Ilies, M. A.; Supuran, C. T. Carbonic anhydrase inhibitors: synthesis of membrane-impermeant low molecular weight sulfonamides possessing *in vivo* selectivity for the membrane-bound versus cytosolic isozymes. *J. Med. Chem.* **2000**, *43*, 292–300.
88. Supuran, C. T.; Scozzafava, A. Benzolamide is not a membrane-impermeant carbonic anhydrase inhibitor. *J. Enzyme Inhib. Med. Chem.* **2004**, *19*, 269–273.
89. Casey, J. R.; Morgan, P. E.; Vullo, D.; Scozzafava, A.; Mastrolorenzo, A.; Supuran, C. T. Carbonic anhydrase inhibitors. Design of selective, membrane-impermeant inhibitors targeting the human tumor-associated isozyme IX. *J. Med. Chem.* **2004**, *47*, 2337–2347.
90. Balaban, A. T.; Dinculescu, A.; Dorofeenko, G. N.; Fischer, G. W.; Koblik, A. V.; Mezheritskii, V. V.; Schroth, W. Pyrylium salts: syntheses, reactions and physical properties. In *Advances in Heterocyclic Chemistry;* Katritzky, A. R., Ed.; Academic Press: New York, **1982**; pp 8–360.
91. Supuran, C. T.; Manole, G.; Dinculescu, A.; Schiketanz, A.; Gheorghiu, M. D.; Puscas, I.; Balaban, A. T. Carbonic anhydrase inhibitors. V: Pyrylium salts in the synthesis of isozyme-specific inhibitors. *J. Pharm. Sci.* **1992**, *81*, 716–719.
92. Supuran, C. T.; Scozzafava, A.; Ilies, M. A.; Iorga, B.; Cristea, T.; Briganti, F.; Chiraleu, F.; Banciu, M. D. Carbonic anhydrase inhibitors. Part 53. Synthesis of substituted-pyridinium derivatives of aromatic sulfonamides: the first nonpolymeric membrane-impermeable inhibitors with selectivity for isozyme IV. *Eur. J. Med. Chem.* **1998**, *33*, 577–595.
93. Srivastava, D. K.; Jude, K. M.; Banerjee, A. L.; Haldar, M.; Manokaran, S.; Kooren, J.; Mallik, S.; Christianson, D. W. Structural analysis of charge discrimination in the binding of inhibitors to human carbonic anhydrases I and II. *J. Am. Chem. Soc.* **2007**, *129*, 5528–5537.
94. Supuran, C. T.; Scozzafava, A.; Casini, A. Development of sulfonamide carbonic anhydrase inhibitors. In *Carbonic Anhydrase: Its inhibitors and Activators;* Supuran, C. T.; Scozzafava, A.; Conway, J., Eds.; CRC Press: Boca Raton, FL, **2004**; pp 67–147.
95. Mincione, F.; Menabuoni, L.; Supuran, C. T. Clinical applications of the carbonic anhydrase inhibitors in ophthalmology. In *Carbonic Anhydrase, Its Inhibitors and Activators;* Supuran, C. T.; Scozzafava, A.; Conway, J., Eds.; CRC Press: Boca Raton, FL, **2004**; pp 243–254.

96. Tawil, R.; McDermott, M. P.; Brown, R., Jr.; Shapiro, B. C.; Ptacek, L. J.; McManis, P. G.; Dalakas, M. C.; Spector, S. A.; Mendell, J. R.; Hahn, A. F.; Griggs, R. C. Randomized trials of dichlorphenamide in the periodic paralyses. Working Group on Periodic Paralysis. *Ann. Neurol.* **2000**, *47*, 46–53.

97. De Simone, G.; Supuran, C. T. Antiobesity carbonic anhydrase inhibitors. *Curr. Top. Med. Chem.* **2007**, *7*, 879–884.

98. Kim, C. Y.; Chandra, P. P.; Jain, A.; Christianson, D. W. Fluoroaromatic–fluoroaromatic interactions between inhibitors bound in the crystal lattice of human carbonic anhydrase II. *J. Am. Chem. Soc.* **2001**, *123*, 9620–9627.

99. May, J. A.; Namil, A.; Chen, H. H.; Dantanarayana, A. P.; Dupré, B.; Liao, J. C. Quaternary ammonium substituted thieno[3,2-*e*]-1,2-thiazine-6-sulfonamide 1,1-dioxides: potential membrane-impermeable inhibitors of carbonic anhydrase. *Bioorg. Med. Chem.* **2006**, *14*, 2052–2059.

100. Nishimori, I.; Vullo, D.; Innocenti, A.; Scozzafava, A.; Mastrolorenzo, A.; Supuran, C. T. Carbonic anhydrase inhibitors. The mitochondrial isozyme VB as a new target for sulfonamide and sulfamate inhibitors. *J. Med. Chem.* **2005**, *48*, 7860.

101. Vullo, D.; Franchi, M.; Gallori, E.; Pastorek, J.; Scozzafava, A.; Pastorekova, S.; Supuran, C. T. Carbonic anhydrase inhibitors: inhibition of the tumor-associated isozyme IX with aromatic and heterocyclic sulfonamides. *Bioorg. Med. Chem. Lett.* **2003**, *13*, 1005–1009.

102. Winum, J. Y.; Dogné, J. M.; Casini, A.; de Leval, X.; Montero, J. L.; Scozzafava, A.; Vullo, D.; Innocenti, A.; Supuran, C. T. Carbonic anhydrase inhibitors: synthesis and inhibition of cytosolic/membrane-associated carbonic anhydrase isozymes I, II, and IX with sulfonamides incorporating hydrazino moieties. *J. Med. Chem.* **2005**, *48*, 2121–2125.

103. Abbate, F.; Coetzee, A.; Casini, A.; Ciattini, S.; Scozzafava, A.; Supuran, C. T. Carbonic anhydrase inhibitors: X-ray crystallographic structure of the adduct of human isozyme II with the antipsychotic drug sulpiride. *Bioorg. Med. Chem. Lett.* **2004**, *14*, 337–341.

104. Chakravarty, S.; Kannan, K. K. Drug–protein interactions. Refined structures of three sulfonamide drug complexes of human carbonic anhydrase I enzyme. *J. Mol. Biol.* **1994**, *243*, 298–309.

105. Vidgren, J.; Liljas, A.; Walker, N. P. C. Refined structure of the acetazolamide complex of human carbonic anhydrase II at 1.9 Å. *Int. J. Biol. Macromol.* **1990**, *12*, 342–344.

106. Denny, W. A. Tumor-activated prodrugs. A new approach to cancer therapy. *Cancer Invest.* **2004**, *22*, 604–619.

107. De Simone, G.; Vitale, R. M.; Di Fiore, A.; Pedone, C.; Scozzafava, A.; Montero, J.-L.; Winum, J.-Y.; Supuran, C. T. Carbonic anhydrase inhibitors: hypoxia-activatable sulfonamides incorporating disulfide bonds that target the tumor-associated isoform IX. *J. Med. Chem.* **2006**, *49*, 5544–5551.

108. D'Ambrosio, K.; Vitale, R. M.; Dogne, J. M.; Masereel, B.; Innocenti, A.; Scozzafava, A.; De Simone, G.; Supuran, C. T. Carbonic anhydrase inhibitors: bioreductive nitro-containing sulfonamides with selectivity for targeting the tumor associated isoforms IX and XII. *J. Med. Chem.* **2008**, *51*, 3230–3237.

109. Saczewski, F.; Sławinski, J.; Kornicka, A.; Brzozowski, Z.; Pomarnacka, E.; Innocenti, A.; Scozzafava, A.; Supuran, C. T. Carbonic anhydrase inhibitors. Inhibition of the cytosolic human isozymes I and II, and the transmembrane, tumor-associated isozymes IX and XII with substituted aromatic sulfonamides activatable in hypoxic tumors. *Bioorg. Med. Chem. Lett.* **2006**, *16*, 4846–4851.

110. Hanson, J.; Dogné, J. M.; Ghiotto, J.; Moray, A. L.; Kinsella, B. T.; Pirotte, B. Design, synthesis, and SAR study of a series of *N*-alkyl-*N'*-[2-(aryloxy)-5-nitrobenzenesulfonyl]ureas and -cyanoguanidine as selective antagonists of the TPalpha and TPbeta isoforms of the human thromboxane A2 receptor. *J. Med. Chem.* **2007**, *50*, 3928–3936.

111. FitzGerald, G. A.; Patrono, C. The coxibs, selective inhibitors of cyclooxygenase-2. *N. Engl. J. Med.* **2001**, *345*, 433–442.

112. Supuran, C. T.; Scozzafava, A. Carbonic anhydrase inhibitors and their therapeutic potential. *Expert Opin. Ther. Pat.* **2000**, *10*, 575–600.

113. Davemport, H. W. The inhibition of carbonic anhydrase by thiophene-2-sulfonamide and sulfanilamide. *J. Biol. Chem.* **1945**, *158*, 567–571.

114. Roblin, R. O., Jr.; Clapp, J. W. The preparation of heterocyclic sulfonamides. *J. Am. Chem. Soc.* **1950**, *72*, 4890–4892.

115. Miller, W. H.; Dessert, A. M.; Roblin, R. O., Jr. Heterocyclic sulfonamides as carbonic anhydrase inhibitors. *J. Am. Chem. Soc.* **1950**, *72*, 4893–4896.

116. Barnish, I. T.; Cross, P. E.; Dickinson, R. P.; Parry, M. J.; Randall, M. J. Cerebrovasodilatation through selective inhibition of the enzyme carbonic anhydrase. 3. 5-(Arylthio)-, 5-(arylsulfinyl)-, and 5-(arylsulfonyl)thiophene-2-sulfonamides. *J. Med. Chem.* **1981**, *24*, 959–964.

117. Maren, T. H. The development of topical carbonic anhydrase inhibitors. *J. Glaucoma* **1995**, *4*, 49–62.

118. Abbate, F.; Casini, A.; Scossafava, A.; Supuran, C. T. Carbonic anhydrase inhibitors: X-ray crystallographic structure of the adduct of human isozyme II with the perfluorobenzoyl analogue of methazolamide. Implications for the drug design of fluorinated inhibitors. *J. Enzyme Inhib. Med. Chem.* **2003**, *18*, 303–308.

119. Abbate, F.; Casini, A.; Scozzafava, A.; Supuran, C. T. Carbonic anhydrase inhibitors: X-ray crystallographic structure of the adduct of human isozyme II with a topically acting antiglaucoma sulfonamide. *Bioorg. Med. Chem. Lett.* **2004**, *14*, 2357–2361.

120. Boriack-Sjodin, P. A.; Zeitlin, S.; Chen, H. H.; Crenshaw, L.; Gross, S.; Dantanarayana, A.; Delgado, P.; May, J. A.; Dean, T.; Christianson, D. W. Structural analysis of inhibitor binding to human carbonic anhydrase II. *Protein Sci.* **1998**, *7*, 2483–2489.

121. Menchise, V.; De Simone, G.; Di Fiore, A.; Scozzafava, A.; Supuran, C. T. Carbonic anhydrase inhibitors: X-ray crystallographic studies for the binding of 5-amino-1,3,4-thiadiazole-2-sulfonamide and 5-(4-amino-3-chloro-5-fluorophenylsulfonamido)-1,3,4-thiadiazole-2-sulfonamide to human isoform II. *Bioorg. Med. Chem. Lett.* **2006**, *16*, 6204–6208.

122. Vaughan, J. R.; Eichler, J. A.; Anderson, G. W. Heterocyclic sulfonamides as carbonic anhydrase inhibitors, 2-acylamino and 2-sulfonamido-1,3,4-thiadiazole-5-sulfonamides. *J. Org. Chem.* **1956**, *21*, 700–771.

123. Young, R. W.; Wood, K. H.; Vaughan, J. R.; Anderson, G. W. 1,3,4-Thiadiazole and thiadiazolinesulfonamides as carbonic anhydrase inhibitors. Synthesis and structural studies. *J. Am. Chem. Soc.* **1956**, *78*, 4649–4654.

124. Supuran, C. T.; Ilies, M. A.; Scozzafava, A. Carbonic anhydrase inhibitors. Part 29. Interaction of isozymes I, II and IV with benzolamide-like derivatives. *Eur. J. Med. Chem.* **1998**, *33*, 739–752.

125. Casini, A.; Scozzafava, A.; Mincione, F.; Menabuoni, L.; Starnotti, M.; Supurana, C. T. Carbonic anhydrase inhibitors: topically acting antiglaucoma sulfonamides incorporating

esters and amides of 3- and 4-carboxybenzolamide. *Bioorg. Med. Chem. Lett.* **2003**, *13*, 2867–2873.

126. Scozzafava, A.; Menabuoni, L.; Mincione, F.; Briganti, F.; Mincione, G.; Supuran, C. T. Carbonic anhydrase inhibitors: perfluoroalkyl/aryl-substituted derivatives of aromatic/heterocyclic sulfonamides as topical intraocular pressure-lowering agents with prolonged duration of action. *J. Med. Chem.* **2000**, *43*, 4542–4551.

127. Ilies, M. A.; Vullo, D.; Pastorek, J.; Scozzafava, A.; Ilies, M.; Caproiu, M. T.; Pastorekova, S.; Supuran, C. T. Carbonic anhydrase inhibitors. Inhibition of tumor-associated isozyme IX by halogenosulfanilamide and halogenophenylaminobenzolamide derivatives. *J. Med. Chem.* **2003**, *46*, 2187–2196.

128. Chow, K.; Lai, R.; Holmes, J. M.; Wijono, M.; Wheeler, L. A.; Garst, M. E. 5-Substituted 3-thiopenesulfonamides as carbonic anhydrase inhibitors. *Eur. J. Med. Chem.* **1996**, *31*, 175–186.

129. Supuran, C. T.; Scozzafava, A. Applications of carbonic anhydrase inhibitors and activators in therapy. *Expert. Opin. Ther. Pat.* **2002**, *12*, 217–242.

130. Kim, C. Y.; Whittington, D. A.; Chang, J. S.; Liao, J.; May, J. A.; Christianson, D. W. Structural aspects of isozyme selectivity in the binding of inhibitors to carbonic anhydrases II and IV. *J. Med. Chem.* **2002**, *45*, 888–893.

131. Nishimori, I.; Minakuchi, T.; Onishi, S.; Vullo, D.; Cecchi, A.; Scozzafava, A.; Supuran, C. T. Carbonic anhydrase inhibitors: cloning, characterization, and inhibition studies of the cytosolic isozyme III with sulfonamides. *Bioorg. Med. Chem.* **2007**, *15*, 7229–7236.

132. Di Fiore, A.; Pedone, C.; Antel, J.; Waldeck, H.; Witte, A.; Wurl, M.; Scozzafava, A.; Supuran, C. T.; De Simone, G. Carbonic anhydrase inhibitors: the X-ray crystal structure of ethoxzolamide complexed to human isoform II reveals the importance of thr200 and gln92 for obtaining tight-binding inhibitors. *Bioorg. Med. Chem. Lett.* **2008**, *18*, 2669–2674.

133. D'Ambrosio, K.; Masereel, B.; Thiry, A.; Scozzafava, A.; Supuran, C. T.; De Simone, G. Carbonic anhydrase inhibitors: binding of indanesulfonamides to the human isoform II. *Chem. Med. Chem.* **2008**, *3*, 473–477.

134. Gruneberg, S.; Stubbs, M. T.; Klebe, G. Successful virtual screening for novel inhibitors of human carbonic anhydrase: strategy and experimental confirmation. *J. Med. Chem.* **2002**, *45*, 3588–3602.

135. Chazalette, C.; Masereel, B.; Rolin, S.; Thiry, A.; Scozzafava, A.; Innocenti, A.; Supuran, C. T. *Bioorg. Med. Chem. Lett.* **2004**, *14*, 5781–5786.

136. Thiry, A.; Ledecq, M.; Cecchi, A.; Supuran, C. T.; Wouters, J.; Dognè, J.-M.; Masereel, B. *J. Med. Chem.* **2006**, *49*, 2743–2749.

137. Thiry, A.; Masereel, B.; Dognè, J.-M.; Supuran, C. T.; Wouters, J.; Michaux, C.; *Chem. Med. Chem.* **2007**, *2*, 1273–1280.

138. Maren, T. H.; Conroy, C. W. A new class of carbonic anhydrase inhibitor. *J. Biol. Chem.* **1993**, *268*, 26233–26239.

139. Håkansson, K.; Liljas, A. The structure of a complex between carbonic anhydrase II and a new inhibitor, trifluoromethane sulfonamide. *FEBS Lett.* **1994**, *350*, 319–322.

140. Temperini, C.; Cecchi, A.; Boyle, N. A.; Scozzafava, A.; Cabeza, J. E.; Wentworth, P.; Blackburn, G. M.; Supuran, C. T. Carbonic anhydrase inhibitors. Interaction of 2-*N*, *N*-dimethylamino-1,3,4-thiadiazole-5-methanesulfonamide with 12 mammalian isoforms: kinetic and X-ray crystallographic studies. *Bioorg. Med. Chem. Lett.* **2008**, *18*, 999–1005.

141. Fisher, S. Z.; Govindasamy, L.; Boyle, N.; Agbandje-McKenna, M.; Silverman, D. N.; Blackburn, G. M.; McKenna, R. X-ray crystallographic studies reveal that the incorporation of spacer groups in carbonic anhydrase inhibitors causes alternate binding modes. *Acta Crystallogr. Sect. F. Struct. Biol. Cryst. Commun.* **2006**, *62*, 618–622.

142. Reed, M. J.; Purohit, A.; Newman, S. P.; Potter, B. V. L. Steroid sulfatase: molecular biology, regulation and inhibition. *Endocr. Rev.* **2005**, *26*, 171–202.

143. Stanway, S. J.; Purohit, A.; Woo, L. W. L.; Sufi, S.; Vigushin, D.; Ward, R.; Wilson, R.; Stanczyk, F. Z.; Dobbs, N.; Kulinkaya, E.; Elliott, M.; Potter, B. V. L.; Reed, M. J.; Coombs, R. C. Phase I study of STX64 (667 Coumate) in breast cancer patients: the first study of a steroid sulfatase inhibitor. *Clin. Cancer Res.* **2006**, *12*, 1585–1592.

144. Ireson, C. R.; Chander, S. K.; Purohit, A.; Parish, D. A.; Woo, L. W. L.; Potter, B. V. L.; Reed, M. J. Pharmacokinetics of the non-steroidal steroid sulfatase inhibitor 667-coumate and its sequestration into red blood cells. *Br. J. Cancer* **2004**, *91*, 1399–1404.

145. Abbate, F.; Winum, J. Y.; Potter, B. V. L.; Casini, A.; Montero, J. L.; Scozzafava, A.; Supuran, C. T. Carbonic anhydrase inhibitors: X-ray crystallographic structure of the adduct of human isozyme II with EMATE, a dual inhibitor of carbonic anhydrases and steroid sulfatase. *Bioorg. Med. Chem. Lett.* **2004**, *14*, 231–234.

146. Lloyd, M. D.; Pederick, R. L.; Natesh, R.; Woo, L. W. L.; Purohit, A.; Reed, M. J.; Acharya, K. R.; Potter, B. V. L. Crystal structure of human carbonic anhydrase II at 1.95 Å resolution in complex with 667-coumate, a novel anti-cancer agent. *Biochem. J.* **2005**, *385*, 715–720.

147. Lloyd, M. D.; Thiyagarajan, N.; Ho, Y. T.; Woo, L. W. L.; Sutcliffe, O. B.; Purohit, A.; Reed, M. J.; Acharya, K. R.; Potter, B. V. L. First crystal structures of human carbonic anhydrase II in complex with dual aromatase-steroid sulfatase inhibitors. *Biochemistry* **2005**, *44*, 6858–6866.

148. Leese, M. P.; Leblond, B.; Smith, A.; Newman, S. P.; Di Fiore, A.; De Simone, G.; Supuran, C. T.; Purohit, A.; Reed, M. J.; Potter, B. V. 2-Substituted estradiol bis-sulfamates, multitargeted antitumor agents: synthesis, *in vitro* SAR, protein crystallography, and *in vivo* activity. *J. Med. Chem.* **2006**, *49*, 7683–7696.

149. Leese, M. P.; Jourdan, F. L.; Gaukroger, K.; Mahon, M. F.; Newman, S. P.; Foster, P. A.; Stengel, C.; Regis-Lydi, S.; Ferrandis, E.; Di Fiore, A.; De Simone, G.; Supuran, C. T.; Purohit, A.; Reed, M. J.; Potter, B. V. Structure–activity relationships of C-17 cyano-substituted estratrienes as anticancer agents. *J. Med. Chem.* **2008**, *51*, 1295–1308.

150. Woo, L. W. L.; Fischer, D. S.; Sharland, C. M.; Trusselle, M.; Foster, P. A.; Chander, S. K.; Di Fiore, A.; Supuran, C. T.; De Simone, G.; Purohit, A.; Reed, M. J.; Potter, B. V. L. Anti-cancer steroid sulfatase inhibitors: synthesis of a potent fluorinated second generation agent, *in vitro* and *in vivo* activities, molecular modeling, and protein crystallography. *Mol. Cancer Ther.* **2008**, *7*, 2435–2444.

151. Winum, J.-Y.; Vullo, D.; Casini, A.; Montero, J.-L.; Scozzafava, A.; Supuran, C. T. Carbonic anhydrase inhibitors. Inhibition of cytosolic isozymes I and II and transmembrane, tumor-associated isozyme IX with sulfamates including EMATE also acting as steroid sulfatase inhibitors. *J. Med. Chem.* **2003**, *46*, 2197–2204.

152. Elger, W.; Schwarz, S.; Hedden, A. M.; Reddersen, G.; Schneider, B. Sulfamates of various estrogens are prodrugs with increased systemic and reduced hepatic estrogenicity at oral application. *J. Steroid Biochem. Mol. Biol.* **1995**, *55*, 395–403.

153. Fischer, D. S.; Woo, L. W.; Mahon, M. F.; Purohit, A.; Reed, M. J.; Potter, B. V. D-ring modified estrone derivatives as novel potent inhibitors of steroid sulfatase. *Bioorg. Med. Chem.* **2003**, *11*, 1685–1700.

154. Fischer, D. S.; Chander, S. K.; Woo, L. W.; Fenton, J. C.; Purohit, A.; Reed, M. J.; Potter, B. V. Novel D-ring modified steroid derivatives as potent, non-estrogenic, steroid sulfatase inhibitors with in vivo activity. *J. Steroid. Biochem. Mol. Biol.* **2003**, *84*, 343–349.

155. Woo, L. W. L.; Sutcliffe, O. B.; Bubert, C.; Grasso, A.; Chander, S. K.; Purohit, A.; Reed, M. J.; Potter, B. V. L. First dual aromatase-steroid sulfatase inhibitors. *J. Med. Chem.* **2003**, *46*, 3193–3196.

156. Sugrue, M. F. Pharmacological and ocular hypotensive properties of topical carbonic anhydrase inhibitors. *Prog. Retin. Eye Res.* **2000**, *19*, 87–112.

157. Silver, L. H.; Brinzolamide Dose–Response Study Group. Dose–response evaluation of the ocular hypotensive effect of brinzolamide ophthalmic suspension (Azopt). *Surv. Ophthalmol.* **2000**, *44*, (Suppl 2.), 147–153.

158. Colinas, P. A.; Bravo, R. D.; Vullo, D.; Scozzafava, A.; Supuran, C. T. Carbonic anhydrase inhibitors. Inhibition of cytosolic isoforms I and II, and extracellular isoforms IV, IX, and XII with sulfamides incorporating sugar moieties. *Bioorg. Med. Chem. Lett.* **2007**, *17*, 5086–5090.

159. Wilkinson, B. L.; Bornaghi, L. F.; Houston, T. A.; Innocenti, A.; Vullo, D.; Supuran, C. T.; Poulsen, S. A. Inhibition of membrane-associated carbonic anhydrase isozymes IX, XII and XIV with a library of glycoconjugate benzenesulfonamides. *Bioorg. Med. Chem. Lett.* **2007**, *17*, 987–992.

160. Di Fiore, A.; Scozzafava, A.; Winum, J. Y.; Montero, J. L.; Pedone, C.; Supuran, C. T.; De Simone, G. Carbonic anhydrase inhibitors: binding of an antiglaucoma glycosyl-sulfanilamide derivative to human isoform II and its consequences for the drug design of enzyme inhibitors incorporating sugar moieties. *Bioorg. Med. Chem. Lett.* **2007**, *17*, 1726–1731.

161. Winum, J. Y.; Casini, A.; Minzione, F.; Starnotti, M.; Montero, J. L.; Scozzafava, A.; Supuran, C. T. Carbonic anhydrase inhibitors: *N*-(*p*-sulfamoylphenyl)-alpha-D-glycopyranosylamines as topically acting antiglaucoma agents in hypertensive rabbits. *Bioorg. Med. Chem. Lett.* **2004**, *14*, 225–229.

162. Wilkinson, B. L.; Bornaghi, L. F.; Houston, T. A.; Innocenti, A.; Supuran, C. T.; Poulsen, S. A. A novel class of carbonic anhydrase inhibitors: glycoconjugate benzene sulfonamides prepared by "click-tailing". *J. Med. Chem.* **2006**, *49*, 6539–6548.

163. Wilkinson, B. L.; Bornaghi, L. F.; Houston, T. A.; Innocenti, A.; Vullo, D.; Supuran, C. T.; Poulsen, S. A. Carbonic anhydrase inhibitors: inhibition of isozymes I, II, and IX with triazole-linked *O*-glycosides of benzene sulfonamides. *J. Med. Chem.* **2007**, *50*, 1651–1657.

164. Shank, R. P.; Gardocki, J. F.; Vaught, J. L.; Davis, C. B.; Schupsky, J. J.; Raffa, R. B.; Dodgson, S. J.; Nortey, S. O.; Maryanoff, B. E. Topiramate: preclinical evaluation of structurally novel anticonvulsant. *Epilepsia* **1994**, *35*, 450–460.

165. Edmonds, H. L., Jr.; Jiang, Y. D.; Zhang, P. Y.; Shank, R. P. Anticonvulsant activity of topiramate and phenytoin in a rat model of ischemia-induced epilepsy. *Life Sci.* **1996**, *59*, PL127–PL131.

166. Stringer, J. L. A comparison of topiramate and acetazolamide on seizure duration and paired-pulse inhibition in the dentate gyrus of the rat. *Epilepsy Res.* **2000**, *40*, 147–153.

167. Sabers, A.; Gram, L. Newer anticonvulsants: comparative review of drug interactions and adverse effects. *Drugs* **2000**, *60*, 23–33.

168. Bourgeois, B. F. Pharmacokinetics and pharmacodynamics of topiramate. *J. Child. Neurol.* **2000**, *15*, S27–S30.

169. Bialer, M.; Johannessen, S. I.; Kupferberg, H. J.; Levy, R. H.; Loiseau, P.; Perucca, E. Progress report on new antiepileptic drugs: a summary of the Fifth Eilat Conference (EILAT V). *Epilepsy Res.* **2001**, *43*, 11–58.

170. Gordon, A.; Price, L. H. Mood stabilization and weight loss with topiramate. *Am. J. Psychiatry* **1999**, *156*, 968–969.

171. Maryanoff, B. E.; Costanzo, M. J.; Nortey, S. O.; Greco, M. N.; Shank, R. P.; Schupsky, J. J.; Ortegon, M. E.; Vaught, J. L. Structure–activity studies on anticonvulsant sugar sulfamates related to topiramate. Enhanced potency with cyclic sulfate derivatives. *J. Med. Chem.* **1998**, *41*, 1315–1343.

172. Recacha, R.; Costanzo, M. J.; Maryanoff, B. E.; Chattopadhyay, D. Crystal structure of human carbonic anhydrase II complexed with an anti-convulsant sugar sulfamate. *Biochem. J.* **2002**, *361*, 437–441.

CHAPTER 5

Antiglaucoma Carbonic Anhydrase Inhibitors as Ophthalmologic Drugs

FRANCESCO MINCIONE[1], ANDREA SCOZZAFAVA[2], and CLAUDIU T. SUPURAN[2]

[1]U.O. Oculistica Az. USL 3, Val di Nievole, Ospedale di Pescia, Pescia, Italy
[2]Laboratorio di Chimica Bioinorganica, Universita degli Studi di Firenze, Room 188, Via della Lastruccia 3, I-50019 Sesto Fiorentino (Firenze), Italy

5.1 INTRODUCTION

The pioneering studies of Friedenwald,[1] Kinsey,[2] and Kinsey and Barany[3] on the chemistry and dynamics of aqueous humor showed that the main constituent of this secretion is sodium bicarbonate. The next step was the identification of carbonic anhydrase (CA, EC 4.2.1.1) in the anterior uvea of the eye by Wistrand,[4] who demonstrated that this enzyme (present mainly in the ciliary processes) is responsible for the bicarbonate secretion, as a consequence of the hydration reaction of carbon dioxide with bicarbonate and protons. Becker[5] then showed that the sulfonamide CAI acetazolamide, **5.1**, produced a drop of the intraocular pressure (IOP) in experimental animals and humans, whereas Kinsey and Reddy[6] proved that this phenomenon is due to a reduced bicarbonate secretion as a consequence of CA inhibition by the sulfonamide drug. This was the beginning of a novel treatment for glaucoma, a condition affecting an increasing number of the population and is also the leading cause of blindness in the Western countries.[7]

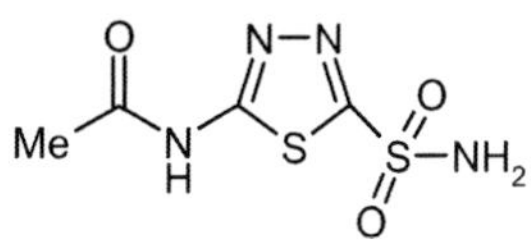

5.1

5.2 SULFONAMIDES IN THE TREATMENT OF GLAUCOMA

Glaucoma is a chronic, degenerative eye disease, characterized by high IOP that causes irreversible damage to the optic nerve head and as a result the progressive loss of visual function and eventually blindness.[8,9] Elevated IOP (ocular hypertension) is generally indicative of an early stage of the disease.[8,10,11] CAIs represent the most physiological treatment of glaucoma, since by inhibiting the ciliary process enzyme (the sulfonamide susceptible isozymes CA II and XII[12–15]), a reduced rate of bicarbonate and aqueous humor secretion is achieved, leading to a 25–30% decrease in IOP.[12–14] Indeed, acetazolamide **5.1**, methazolamide **5.2**, ethoxzolamide **5.3**, or dichlorophenamide **5.4** were and are still extensively used systemic drugs in the therapy of this disease,[12–17] as they all act as very efficient inhibitors of several CA isozymes, and principally of CA II, and CA XII, the isozymes thought to be involved in aqueous humor secretion at this moment.[12–15] It has previously considered[18] that CA IV is also involved in such processes, but we have recently proved that some clinically used antiglaucoma sulfonamides act as very weak human CA IV inhibitors, excluding thus the possibility that this isozyme plays a critical role in aqueous humor formation in *Homo sapiens*.[19]

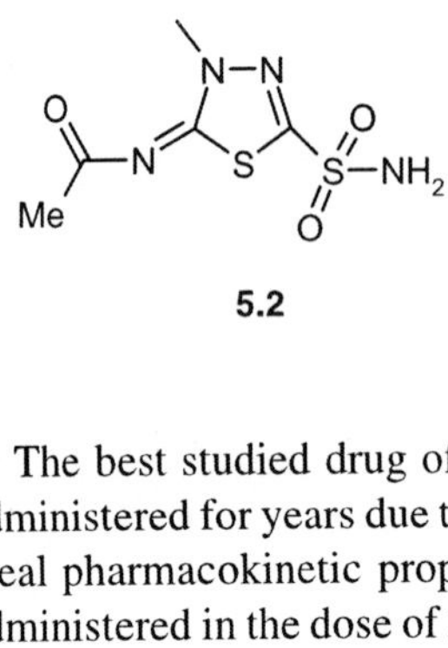

5.2

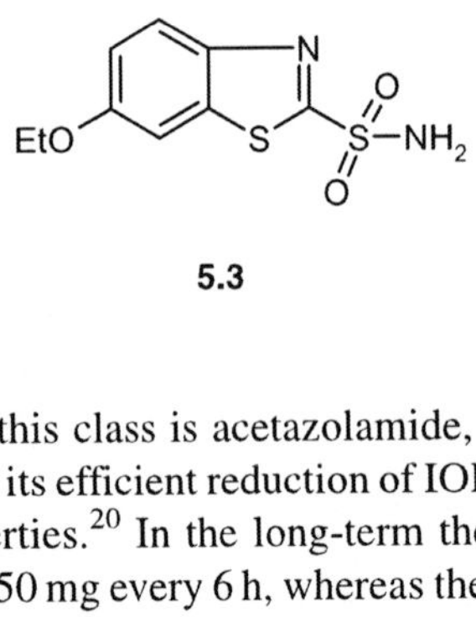

5.3

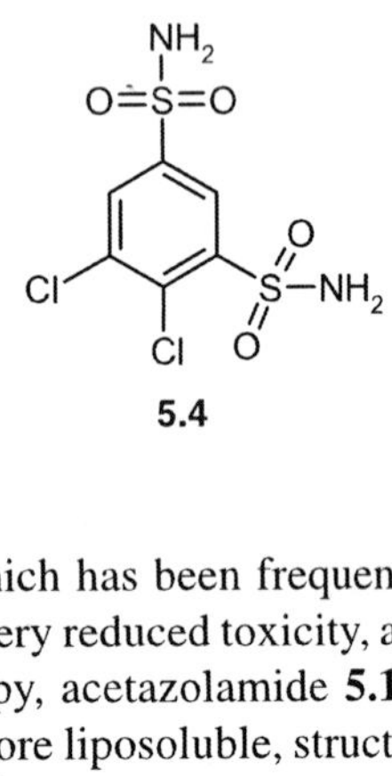

5.4

The best studied drug of this class is acetazolamide, which has been frequently administered for years due to its efficient reduction of IOP, very reduced toxicity, and ideal pharmacokinetic properties.[20] In the long-term therapy, acetazolamide **5.1** is administered in the dose of 250 mg every 6 h, whereas the more liposoluble, structurally related methazolamide **5.2** in dose of 25–100 mg three times daily; an equal dosage of dichlorophenamide **5.4** is also useful in reducing ocular hypertension.[8] But as CAs are ubiquitous enzymes in vertebrates, the administration of these systemic sulfonamides with such a high affinity for the enzyme leads to CA inhibition in other organs than the target one (i.e., the eye) and as a result, to undesired side effects of these drugs. The most frequent side effects are numbness and tingling of extremities, metallic taste, depression, fatigue, malaise, weight loss, decreased libido, gastrointestinal irritation, metabolic acidosis, renal calculi, and transient myopia.[8,18,21] Although producing all these unpleasant side effects, systemic sulfonamide CAIs of the type mentioned above (compounds **5.1–5.4**) are particularly useful in the management of glaucoma resistant to other antiglaucoma therapies and in the control of acute glaucoma attacks.[22] Even if the clinical introduction of the topically acting compounds **5.5** and **5.6** (see later in the text) initially seemed to have produced a marked loss of interest in acetazolamide and other systemic sulfonamides, there are several reasons why acetazolamide and the related drugs **5.2–5.4** still find a place in the clinical armamentarium of antiglaucoma drugs: (i) dorzolamide **5.5** reduces IOP by up to 23% as monotherapy and an extra 15% when combined with timolol, whereas acetazolamide **5.1** alone reduces IOP by 30%;[7,12,18] (ii) dorzolamide inhibits aqueous flow by 17%, whereas acetazolamide inhibits this parameter by 30%;[23] (iii) ocular burning and stinging was reported in 12–19% of patients undergoing topical dorzolamide treatment, whereas the other topically acting sulfonamide available, brinzolamide (**5.6**), is much less effective than acetazolamide and dorzolamide in reducing high IOP, for example, IOP reduction of 20% when brinzolamide was administered three times a day as 1% suspension.[24] Thus, systemically acting sulfonamide CAIs are many times preferred by ophthalmologists for the management of glaucoma, mainly due to their well-known pharmacological properties, low toxicity, and much lower cost compared to the newer drugs dorzolamide and brinzolamide.[22]

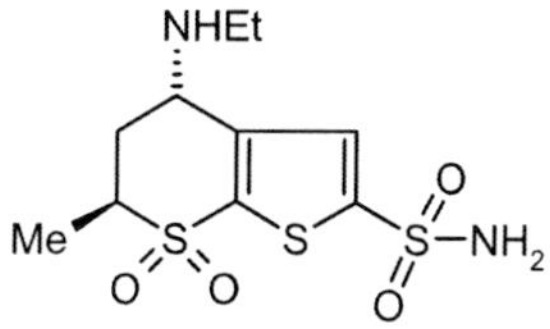

5.5

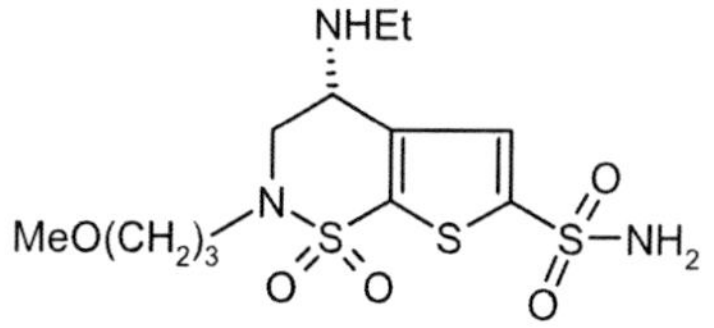

5.6

The idea to administer topically, directly into the eye, the sulfonamide CAI was already addressed by Becker[5]. This and other studies involving the clinically used compounds **5.1–5.4** only gave negative results, being concluded that sulfonamide CAIs are effective as antiglaucoma drugs only *via* the systemic route.[8,9] The lack of efficiency of sulfonamides **5.1–5.4** *via* the topical route was due to the fact that the drug was unable to arrive at the ciliary processes where CA is present.[9,21] The inadequate drug penetrability through the cornea was due to the fact that sulfonamides **5.1–5.4** possess inappropriate physicochemical properties for such a route of administration. Then, in 1983, in a seminal paper Maren et al.[21] postulated that a water-soluble sulfonamide, possessing a relatively balanced lipid solubility (to be able to penetrate through the cornea) as well as CA inhibitory properties strong enough, would be an effective IOP lowering drug *via* the topical route, but at that moment no inhibitors possessing such physicochemical properties existed, as the bio-organic chemistry of this class of compounds remained relatively unexplored.[25] Water-soluble sulfonamide CA inhibitors started to be developed in several laboratories soon thereafter, and the first such pharmacological agent dorzolamide **5.5** was launched by 1995.[18] A second compound, brinzolamide **5.6**, structurally very similar to dorzolamide, was approved for the topical treatment of glaucoma in 1999.[24]

Dorzolamide **5.5** and brinzolamide **5.6** are nanomolar CA II/XII inhibitors,[12,13–15] possess a good water solubility, are sufficiently liposoluble to penetrate through the cornea, and may be administered topically, directly into the eye, as a 2% water solution (of the dorzolamide hydrochloride salt) or as 1% suspension (as the brinzolamide hydrochloride salt) 2–3 times a day.[18,24] The two drugs are effective in reducing IOP and show fewer side effects compared to the systemically applied drugs. The observed side effects include stinging, burning, or reddening of the eye; blurred vision; pruritus; and bitter taste.[18,24] All but the last are probably due to the fact that dorzolamide (the best studied topical CAI) is the salt of a weak base with a very strong acid, so that the pH of the drug solution is rather acidic (generally around 5.5). The last side effect mentioned above is probably due to drug-laden lachrymal fluid draining into the oropharynx and inhibition of CA present in the saliva (CA VI) and the taste buds (CA II and VI), with the consequent accumulation of bicarbonate, and was seen with both systemic and topical CAIs.[18,24] Less is known for the moment regarding the side effects of brinzolamide, but it seems that this drug produces less stinging but more blurred vision compared to dorzolamide.[18,24] Unfortunately, dorzolamide showed some more serious side effects, such as contact allergy,[26] nephrolithiasis,[27] anorexia, depression, and dementia[28] and irreversible corneal decompensation in patients who already presented with corneal problems.[29] Thus, even if dorzolamide and brinzolamide indeed represent a major progress in the fight against glaucoma with therapies based on CAIs, novel types of topically effective inhibitors belonging to this class of pharmacological agents are stringently needed and much investigated.

One approach recently reported for obtaining novel types of such agents consisted in attaching water-solubilizing moieties to the molecules of aromatic/heterocyclic sulfonamides.[13–17,25] Such moieties included pyridine-carboximido; carboxypyridine-carboxamido, quinolinesulfonamido; picolinoyl, isonicotinoyl, perfluoroalkyl/arylsulfonyl, as well as amino acyl groups among others, whereas ring systems that have been derivatized by using the above-mentioned moieties included 2-, 3-, or 4-aminobenzenesulfonamides; 4-(ω-aminoalkyl)-benzenesulfonamides; 3-halogeno-substituted-sulfanilamides; 1,3-benzene-disulfonamides; 1,3,4-thiadiazole-2-sulfonamides; benzothiazole-2-sulfonamides; and thienothiopyran-2-sulfonamides among others, and they were chosen in such a way as to

TABLE 5.1 Fall of IOP of Normotensive Rabbits After Treatment with One Drop (50 μL) Solution of 2% CAIs of Types 5.7–5.12 Directly into the Eye, at 30, 60, and 90 min After Administration

		ΔIOP (mmHg)[b]			
Inhibitor	pH[a]	$t=0$	$t=30$ min	$t=60$ min	$t=90$ min
Dorzolamide	5.5	0	2.2 ± 0.1	4.1 ± 0.15	2.7 ± 0.1
5.5					
5.7	6.5	0	5.9 ± 0.2	11.2 ± 0.5	13.1 ± 0.3
5.8	6.5	0	5.4 ± 0.1	10.9 ± 0.4	12.5 ± 0.3
5.9	5.5	0	4.8 ± 0.1	8.2 ± 0.1	7.0 ± 0.2
5.10	7.0	0	4.0 ± 0.2	7.2 ± 0.1	9.0 ± 0.2
5.11	7.5	0	3.0 ± 0.1	7.2 ± 0.2	5.5 ± 0.3
5.12	5.5	0	4.9 ± 0.1	8.7 ± 0.3	6.9 ± 0.4

[a] The pH of the ophthalmic solution used in the experiments.

[b] $\Delta IOP = IOP_{control\ eye} - IOP_{treated\ eye}$; mean ± average spread ($n = 3$).

demonstrate that the proposed approach is a general one.[30–38] Compounds such as **5.7–5.12** showed 2–3 times more effective topical IOP lowering effects in rabbits compared to dorzolamide.[30–38] They possessed good water solubility (as hydrochlorides, triflates, or trifloroacetates), inhibition in the low nanomolar range against hCA II (the figures after the number characterizing the compound represent the inhibition constant against hCA II), good penetrability through the cornea, and very good IOP lowering properties in both normotensive and glaucomatous rabbits (Table 5.1). What is more important, this effect lasted for a prolonged period compared to the similar effect of dorzolamide.

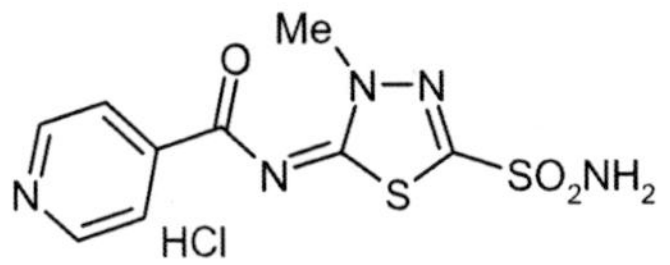

5.7 (3 nM, hCA II)

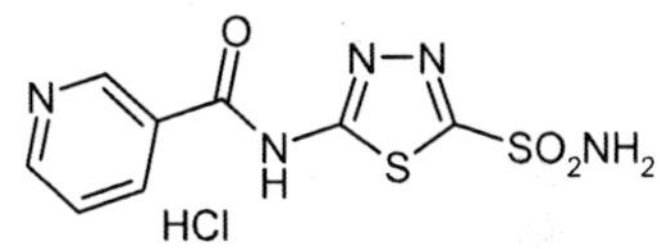

5.8 (2 nM, hCA II)

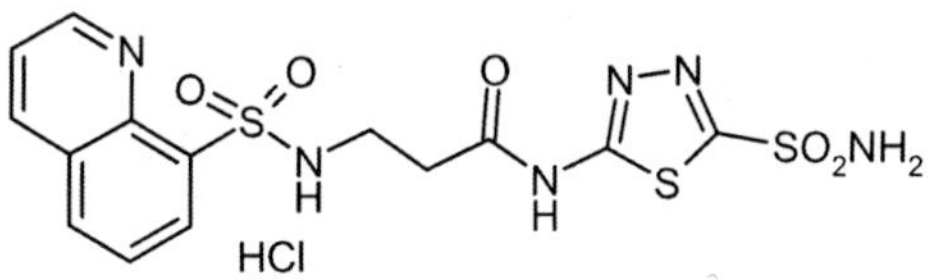

5.9 (3 nM, hCA II)

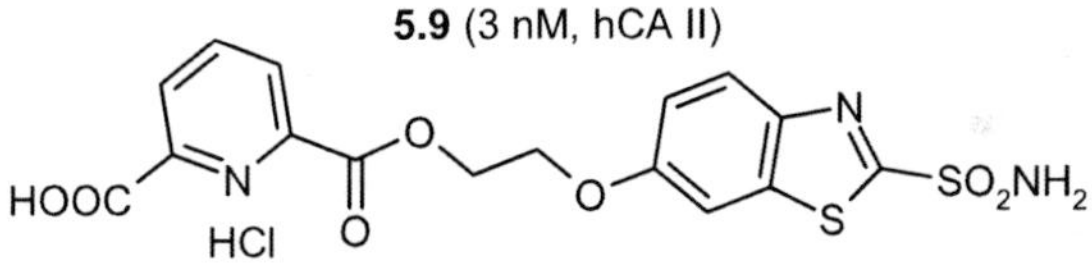

5.10 (3 nM, hCA II)

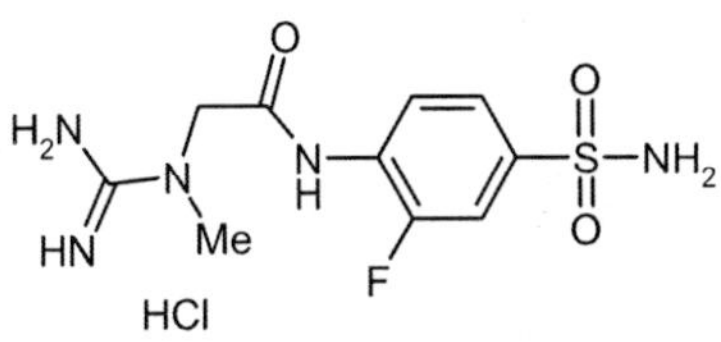

5.11 (10 nM, hCA II)

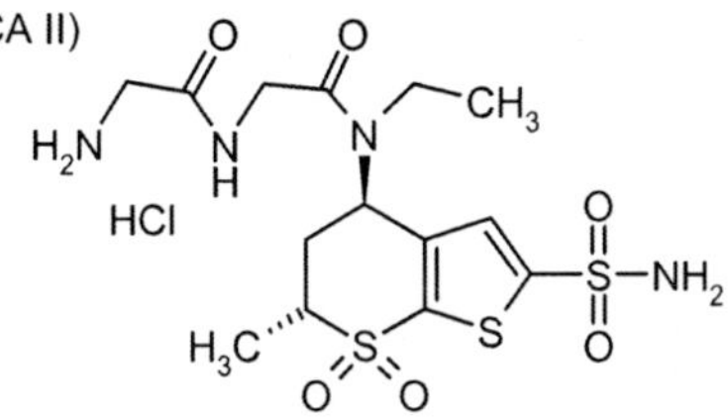

5.12 (6 nM, hCA II)

5.3 COMBINATION ANTIGLAUCOMA THERAPY OF CAIs WITH OTHER PHARMACOLOGICAL AGENTS

Another approach for improving topically administered CAIs, mainly dorzolamide or brinzolamide, consisted in formulating the eye drops containing the two sulfonamides in combination with other topically/systemically acting antiglaucoma drugs such as β-blockers, prostaglandins, or acetazolamide.[39–43] Xanthan gum (0.5%), which contains a high molecular weight polysaccharide possessing pseudoplastic properties and, in consequence, an unusually high ocular penetration, associated with dorzolamide hydrochloride (2% water solution), has also been used in association with dorzolamide, which is more efficient as an IOP lowering agent in this combination.[18] Dorzolamide hydrochloride 2% water solution (or brinzolamide hydrochloride, 1% suspension) was shown to effectively lower IOP when associated with the β-blocker timolol **5.13** (0.5%), with the prostaglandin derivatives latanoprost **5.14** (0.005–0.1%) or unoprostone **5.15** (0.005–0.1%), or with the systemically administered acetazolamide.[39–43] In all these cases, an additive effect of about 15% on IOP lowering of the two drugs has been observed, whereas side effects seemed to be reduced or of diminished intensity compared to the side effects of each drug alone.

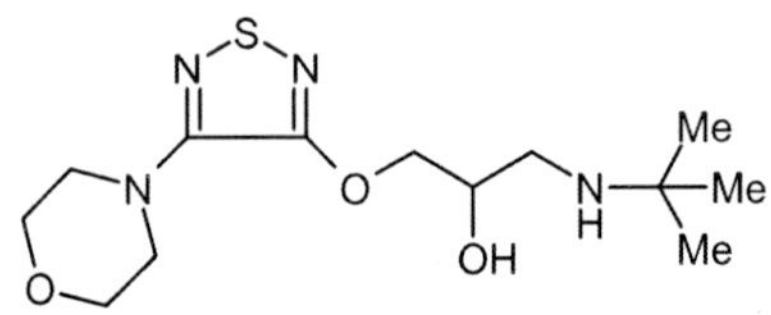

5.13

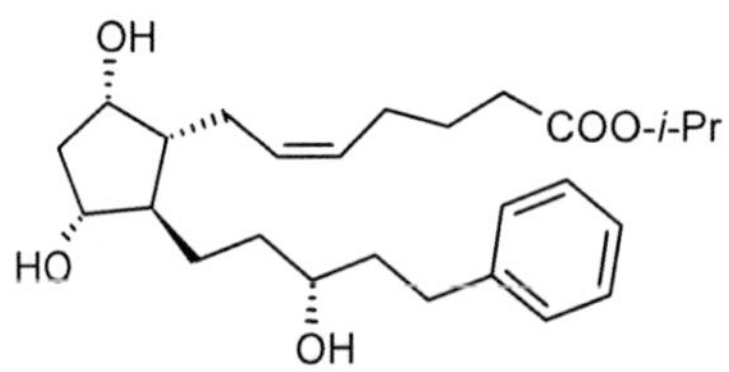

5.14

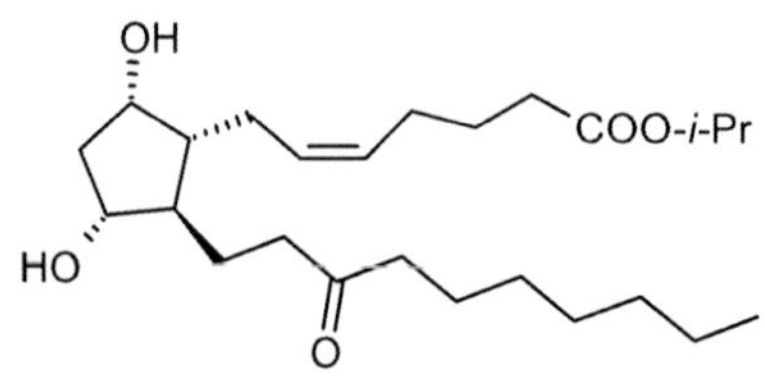

5.15

Complexes of several 1,3,4-thiadiazole-2-sulfonamide derivatives of type **5.16–5.19**, possessing strong CA inhibitory properties with hydroxypropyl-β-cyclodextrin (HPβCD), have also recently been investigated for IOP lowering properties, being known that the association of ophthalmologic drugs with cyclodextrins may lead to enhanced penetration of the drug within the eye.[44] Although the investigated CAIs **5.16–5.19** possessed very powerful inhibitory properties against the two CA isozymes involved in aqueous humor production within the eye, that is, CA II and IV (in the low nanomolar range), these compounds were topically ineffective as IOP lowering agents in normotensive/hypertensive rabbits due to their very low water solubility and impossibility to penetrate through the cornea to arrive at the ciliary process enzymes. On the contrary, the cyclodextrin–sulfonamide complexes proved to be effective and long-lasting IOP lowering agents in the two animal models of glaucoma mentioned above, leading to strong IOP lowering, as shown in Fig. 5.1, where data for the standard drug dorzolamide are also provided. It is obvious that the association of sulfonamide CAIs with cyclodextrins leads to a very efficient IOP lowering in this glaucoma animal model, constituting a thought-provoking approach for developing antiglaucoma medications for human use.

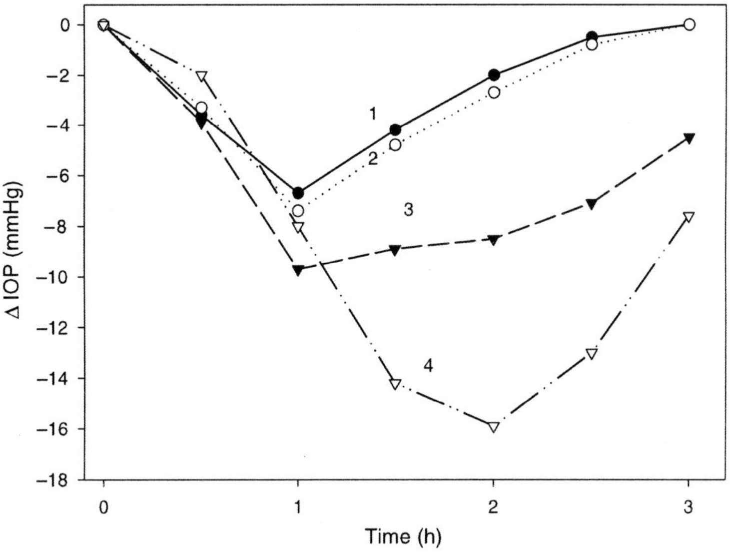

FIGURE 5.1 Effect of topically administered sulfonamide CA inhibitors (2% water solutions/suspensions) on the IOP of hypertensive albino rabbits (initial IOP of 33 ± 3 mmHg). Curve 1: dorzolamide **DZA** (hydrochloride salt, pH 5.5); curve 2: brinzolamide **BRZ** (2% suspension); curve 3: compound **5.24** (solution, pH 7.0); curve 4: compound **5.26** (solution, pH 7.0).

5.16–5.19

5.16: R = 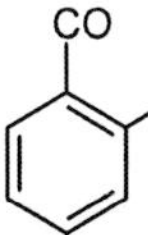

5.18: R = 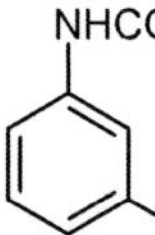

5.17: R = 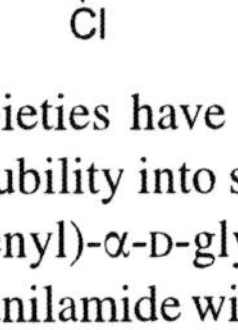

5.19: R = 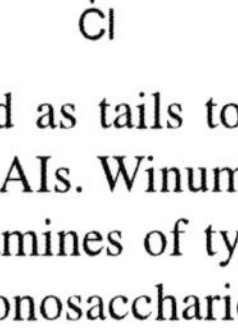

More recently, sugar moieties have been designed as tails to induce increased hydrophilicity and water solubility into sulfonamide CAIs. Winum et al.[45] reported a series of *N*-(*p*-sulfamoylphenyl)-α-D-glycopyranosylamines of type **5.20–5.27** prepared by the reaction of sulfanilamide with different monosaccharides in the presence of ammonium chloride. The new compounds were investigated for inhibition of CAs involved in aqueous humor secretion within the mammalian eye. Isozymes CA I and II were strongly inhibited by some of these compounds, with inhibition constants in the range of 510–1200 nM against CA I and 10–25 nM against CA II, similarly to clinically used sulfonamides, such as acetazolamide, methazolamide, dichlorophenamide, dorzolamide, and brinzolamide. The presence of sugar moieties in these molecules induced enhanced water solubility compared to other sulfonamides. In hypertensive rabbits (a widely used animal model of glaucoma), two of the new compounds, that is, **5.24** and **5.26**, showed strong and long-lasting IOP lowering, being more effective than dorzolamide and brinzolamide, the two clinically used, topically acting antiglaucoma sulfonamides with CA inhibitory properties (Fig. 5.1).

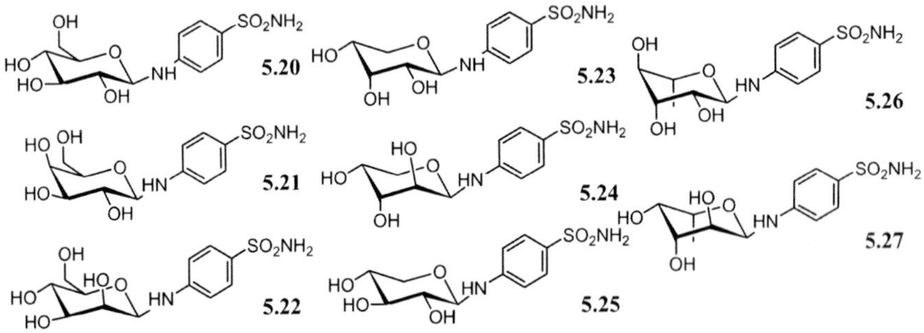

5.4 CAIs INCORPORATING NO-DONATING MOIETIES

The largest majority of the topically acting antiglaucoma drugs (the β-blockers, the prostaglandin analogues, the sympathomimetics, etc.), except for the sulfonamide CAIs, also act as vasoconstrictors.[14,39,40,46] Sulfonamide, such as acetazolamide, methazolamide, dorzolamide, or brinzolamide, in addition to efficiently lowering IOP after systemic (acetazolamide and methazolamide) or topical administration (dorzolamide and brinzolamide) show powerful vasodilating effects, connected with the concomitant raising of the carbon dioxide partial pressure in the blood.[46–50] The unusual property of this class of antiglaucoma drugs mentioned above prompted us to investigate the possible role of nitric oxide (NO) precursors/donors (such as arginine or sodium nitroprusside among others) in association with different topical antiglaucoma medications and also to design new compounds that incorporate both sulfonamide and NO-donating moieties in the same molecule.

It has been known since the early 1980s that the vascular relaxation caused by acetylcholine depends on the presence of the vascular endothelium, and this activity was ascribed to a labile humoral factor termed endothelium-derived relaxing factor, which was eventually attributed to nitric oxide.[46] The activity of NO as a vasodilator has been known for over 100 years. In addition, NO is released *in vivo* as a result of dosing a patient with amyl nitrite, nitroglycerin, or other nitrovasodilators. However, NO and its sources are not known to directly affect intraocular pressure.[46] We recently investigated whether the *in vivo* NO precursor arginine (Arg) as well as the inorganic salt with clinical applications sodium nitroprusside (SN, which easily liberated the

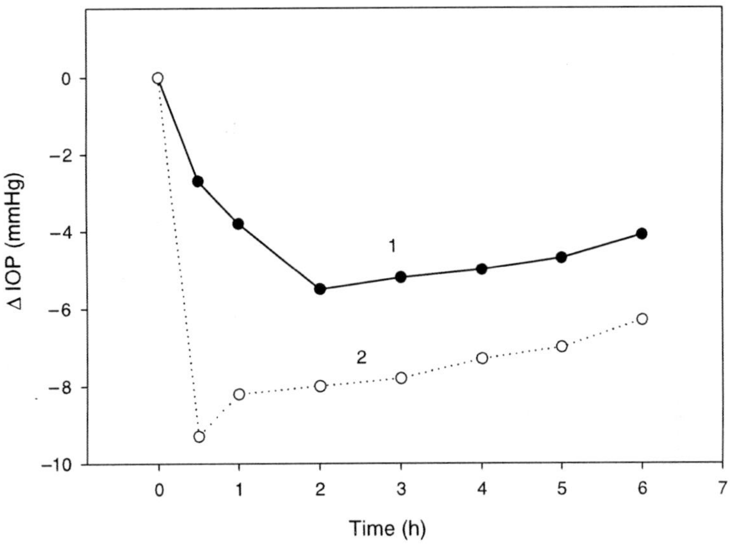

FIGURE 5.2 Effect of topically administered NO donors arginine and sodium nitroprusside on IOP lowering in normotensive albino rabbits (initial IOP = 18 ± 3 mmHg). Curve 1: Arg 2%; curve 2: Na nitroprusside 2%.

coordinated to Fe(III) NO molecule in solution) may have an effect on IOP in normotensive and hypertensive rabbits. As shown from data of Fig. 5.2, topically administered Arg or SN, in concentrations of around 2%, strongly lower IOP in this animal model of the disease.

Furthermore, we combined diverse IOP lowering drugs, such as dorzolamide (Figs 5.3 and 5.4) or timolol **5.13** (Fig. 5.5), with various concentrations of NO-donating agents (Arg or SN in concentrations from 1 to 3%) and again a very potent IOP lowering effect has been observed that is much more potent and long lasting than the individual drug alone.

In the next step, we reported an entire new class of sulfonamides incorporating nitrate esters in their molecule[46] that may also easily release NO in solution, of which compounds **5.28** and **5.29** are two well-studied representatives.[46] These compounds were shown to be low nanomolar CAIs for CA II, IV, IX, and XII and to strongly diminished elevated IOP in hypertensive albino rabbits when administered via the topical route (Fig. 5.4).

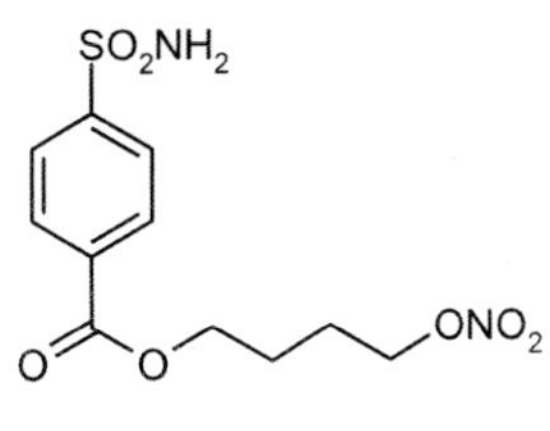

5.28

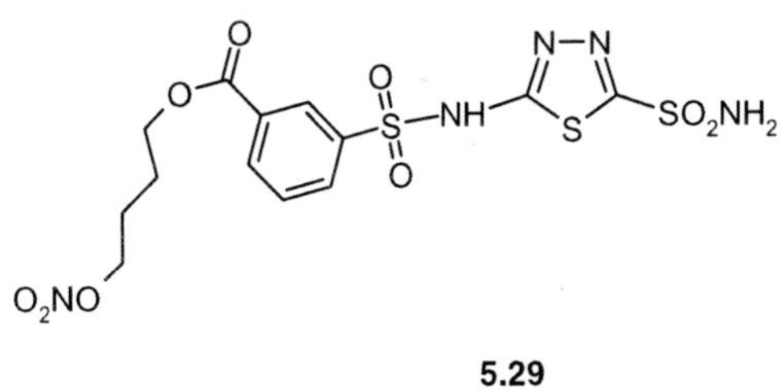

5.29

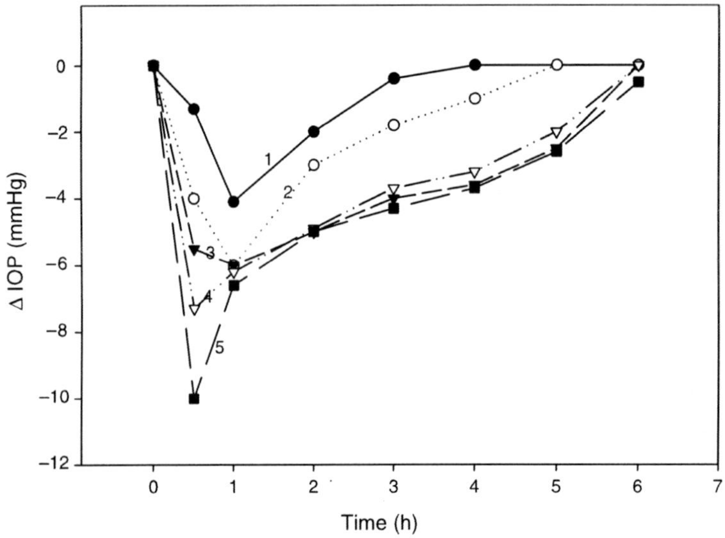

FIGURE 5.3 Effect of topically administered IOP lowering drugs (50 μL solutions) on the IOP of normotensive albino rabbits (initial IOP = 18 ± 3 mmHg). Curve 1: Dorzolamide (DZA) 2% (standard); curve 2: DZA 1% + Arg 1%; curve 3: DZA 1% + Arg 3%; curve 4: DZA 1% + Na nitroprusside 0.5%; curve 5: DZA 1% + Na nitroprusside 1%.

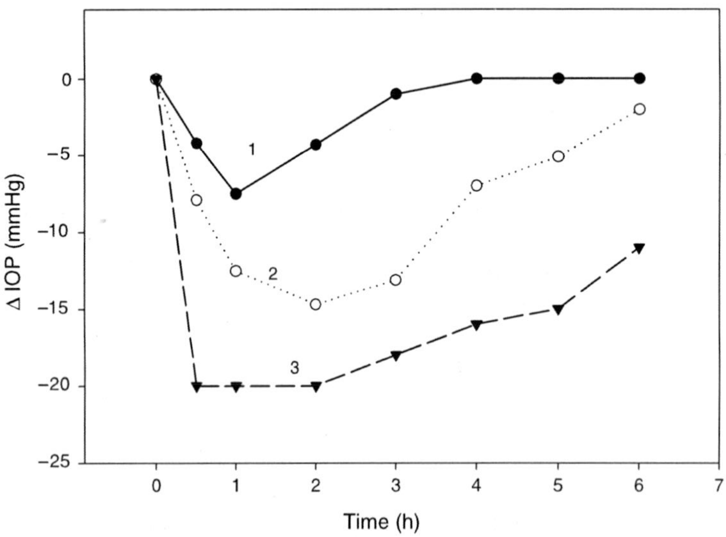

FIGURE 5.4 Effect of topically administered IOP lowering drugs (50 μL solutions) on the IOP of hypertensive albino rabbits (initial IOP = 30 ± 2 mmHg). Curve 1: Dorzolamide (DZA) 2% (standard); curve 2: Sulfonamide **5.28**[46] 1%; curve 3: Sulfonamide **28**[46] 2%.

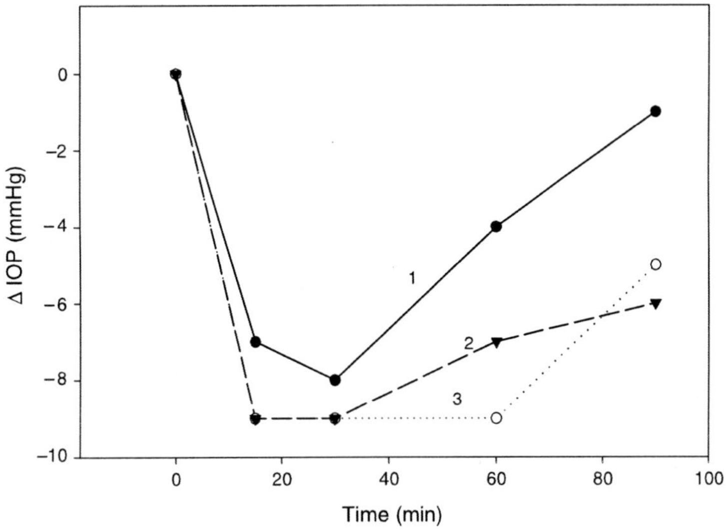

FIGURE 5.5 Effect of topically administered IOP lowering drugs (50 μL solutions) on the IOP of normotensive albino rabbits (initial IOP = 18 ± 3 mmHg). 1: Timolol **5.13** 2% (standard); 2: Timolol **5.13** 2% + Arg 2%; 3. Timolol **5.13** 1% + sodium nitroprusside 2%.

The mechanism of action of NO on IOP lowering is unknown for the moment, but it may be related to the direct inhibition of CA in addition to the well-known vasodilating effects of this molecule. Further studies are warranted to better understand these phenomena and to launch clinically this class of very promising pharmacological agents.

5.5 CAIs IN MACULAR EDEMA, MACULAR DEGENERATION, AND DIABETIC RETINOPATHY

Optic nerve blood flow is diminished in the eyes of primary open-angle glaucoma suspects and patients.[47] Since sulfonamides with CAI inhibitory properties act as vasodilators,[13,14] this may explain the use of such drugs in the treatment of retinal edema and age-related macular degeneration. In consequence, these pharmacological agents may represent a new approach for improving visual function. Thus, retinal edema (also referred to as cystoid macular edema) consists of a swelling process within the critically important central visual zone and may develop in association with a variety of ocular conditions such as diabetic retinopathy, ischemic retinopathies, intraocular surgery (such as cataract removal or implantation of plastic lenses), laser photocoagulation, and so on.[47–50] It is also common in patients affected by retinitis pigmentosa, a hereditary disorder leading to total blindness.[51] The precise mechanism by which the swelling process is triggered is uncertain, but natural metabolic toxins may play a role in this disease.[49] Macular degeneration on the other hand is characterized by fluid accumulation in the outer retina, accompanied by lipofuscin (a metabolic waste product) accumulation between photoreceptors and the villi of the retina pigment epithelium. These catabolic waste products (also denominated drusen) are generally cleared by the blood in the healthy retina, but with aging they tend to accumulate and coalesce, so that vast areas of the retinal photoreceptors become disengaged from their neighboring retinal pigment epithelial villi.[47–50] As a consequence of drusen confluencing in the foveal area, the affected sections of the retina become blind, which may trigger a macular degenerative disease, with a dramatic loss of the visual function. No satisfactory current therapy for this condition is known for the moment.[49,50]

The use of CAIs in the treatment of macular edema is based on the important observation[47] that acetazolamide **5.1** (as sodium salt) is effective in the treatment of this condition when administered systemically. A similar efficiency has also recently been reported for dorzolamide[48,49] after topical administration (obviously, without the side effects of the systemic inhibitor **5.1**). It is generally assumed that the disappearance of the edema and the improvement of visual function are independent of the hypotensive activity of the sulfonamide, being due to direct effects of the drug on the circulation in the retina.[49] Practically, acetazolamide, dorzolamide, or brinzolamide act as local vasodilators[13] and improve blood flow in this organ and, in consequence, clearing of metabolic waste products, drusen, and so on. The improvement of visual function after such a treatment (in early phases of the disease) seems to be very good.[49]

It has also been reported that acetazolamide (375 mg/day) is effective in the treatment of serous retinal detachment of various etiologies (this disease is also generally not amenable to treatment).[52] Additional studies should be done to establish whether the topically acting sulfonamide CAIs might be also used in this serious medical problem.

REFERENCES

1. Friedenwald, J. S. The formation of the intraocular fluid. *Am. J. Ophthalmol.* **1949**, *32*, 9–27.
2. Kinsey, V. E. Comparative chemistry of aqueous humor in posterior and anterior chambers of rabbit eye. *Arch. Ophthalmol.* **1953**, *50*, 401–417.
3. Kinsey, V. E.; Barany, E. The rate flow of aqueous humor. II. Derivation of rate of flow and its physiologic significance. *Am. J. Ophthalmol.* **1949**, *32*, 189–202.
4. Wistrand, P. J. Carbonic anhydrase in the anterior uvea of the rabbit. *Acta Physiol. Scand.* **1951**, *24*, 144–148.
5. Becker, B. The mechanism of the fall in intraocular pressure by the carbonic anhydrase inhibitor Diamox. *Am. J. Ophthalmol.* **1955**, *39*, 177–183.
6. Kinsey, V. E.; Reddy, D. V. N. Turnover of total carbon dioxide in aqueous humors and the effect thereon of acetazolamide. *Arch. Ophthalmol.* **1959**, *62*, 78–83.
7. Sugrue, M. F. The pharmacology of antiglaucoma drugs. *Pharmacol. Ther.* **1989**, *43*, 91–138.

8. Bartlett, J. D.; Jaanus, S. D. *Clinical Ocular Pharmacology;* Butterworth: Boston, **1989**; pp 254–263.

9. Maren, T. H. Carbonic anhydrase: chemistry, physiology and inhibition. *Physiol. Rev.* **1967**, *47*, 595–781.

10. Soltau, J. B.; Zimmerman, T. J. Changing paradigms in the medical treatment of glaucoma. *Surv. Ophthalmol.* **2002**, *47*, S2–S5.

11. Hoyng, P. F. J.; Kitazawa, Y. Medical treatment of normal tension glaucoma. *Surv. Ophthalmol.* **2002**, *47*, S116–S124.

12. Supuran, C. T. Carbonic anhydrases: novel therapeutic applications for inhibitors and activators. *Nat. Rev. Drug Discov.* **2008**, *7*, 168–181.

13. (a) Supuran, C. T.; Scozzafava, A. Carbonic anhydrase inhibitors and their therapeutic potential. *Expert Opin. Ther. Pat.* **2000**, *10*, 575–600. (b) Supuran, C. T.; Scozzafava, A. Carbonic anhydrase inhibitors. *Curr. Med. Chem. Immunol. Endo. Metab. Agents* **2001**, *1*, 61–97.

14. (a) Mincione, F.; Scozzafava, A.; Supuran, C. T. The development of topically acting carbonic anhydrase inhibitors as antiglaucoma agents. *Curr. Pharm. Des.* **2008**, *14*, 649–54. (b) Mincione, F.; Scozzafava, A.; Supuran, C. T. The development of topically acting carbonic anhydrase inhibitors as anti-glaucoma agents. *Curr. Top. Med. Chem.* **2007**, *7*, 849–854.

15. Mincione, F.; Starnotti, M.; Masini, E.; Bacciottini, L.; Scrivanti, C.; Casini, A.; Vullo, D.; Scozzafava, A.; Supuran, C. T. Carbonic anhydrase inhibitors: design of thioureido sulfonamides with potent isozyme II and XII inhibitory properties and intraocular pressure lowering activity in a rabbit model of glaucoma. *Bioorg. Med. Chem. Lett.* **2005**, *15*, 3821–3827.

16. Supuran, C. T.; Scozzafava, A. Applications of carbonic anhydrase inhibitors and activators in therapy. *Expert Opin. Ther. Pat.* **2002**, *12*, 217–242.

17. Supuran, C. T.; Scozzafava, A.; Casini, A. Carbonic anhydrase inhibitors. *Med. Res. Rev.* **2003**, *23*, 146–189.

18. Sugrue, M. F. Pharmacological and ocular hypotensive properties of topical carbonic anhydrase inhibitors. *Progr. Retinal Eye Res.* **2000**, *19*, 87–112.

19. Innocenti, A.; Firnges, M. A.; Antel, J.; Wurl, M.; Scozzafava, A.; Supuran, C. T. Carbonic anhydrase inhibitors. Inhibition of the membrane-bound human and bovine isozymes IV with sulfonamides. *Bioorg. Med. Chem. Lett.* **2005**, *15*, 1149–1154.

20. Wistrand, P. J.; Lindqvist, A. Design of carbonic anhydrase inhibitors and the relationship between the pharmacodynamics and pharmacokinetics of acetazolamide. In *Carbonic Anhydrase—From Biochemistry and Genetics to Physiology and Clinical Medicine;* Botrè, F.; Gros, G.; Storey, B. T., Eds.; VCH: Weinheim, **1991**; pp 352–378.

21. Maren, T. H.; Jankowska, L.; Sanyal, G.; Edelhauser, H. F. The transcorneal permeability of sulfonamide carbonic anhydrase innhibitors and their effect on aqueous humor secretion. *Exp. Eye Res.* **1983**, *36*, 457–480.

22. Kaur, I. P.; Smitha, R.; Aggarwal, D.; Kapil, M. Acetazolamide: future perspectives in topical glaucoma therapeutics. *Int. J. Pharm.* **2002**, *248*, 1–14.

23. Maus, T. L.; Larsson, L. I.; McLaren, J. W.; Brubaker, R. F. Comparison of dorzolamide and acetazolamide as suppressors of aqueous humor flow in humans. *Arch. Ophthalmol.* **1997**, *115*, 45–49.

24. Silver, L. H. Dose–response evaluation of the ocular hypotensive effect of brinzolamide ophthalmic suspension (Azopt). Brinzolamide dose–response study group. *Surv. Ophthalmol.* **2000**, *44* (Suppl. 2), 147–153.

25. Supuran, C. T.; Scozzafava, A.; Casini, A. Development of sulfonamide carbonic anhydrase inhibitors (CAIs). In *Carbonic Anhydrase—Its Inhibitors and Activators;* Supuran, C. T.; Scozzafava, A.; Conway, J.; CRC Press: Boca Raton, FL, **2004**; pp 67–147.

26. Aalto-Korte, K. Contact allergy to dorzolamide eyedrops. *Contact Dermatitis* **1998**, *39*, 206.

27. Carlsen, J.; Durcan, J.; Swartz, M.; Crandall, A. Nephrolithiasis with dorzolamide. *Arch. Ophthalmol.* **1999**, *117*, 1087–1088.

28. Thoe Schwartzenberg, G. W.; Trop, G. E. Anorexia, depression and dementia induced by dorzolamide eyedrops (Trusopt). *Can. J. Ophthalmol.* **1999**, *34*, 93–94.

29. Konowal, A.; Morrison, J. C.; Brown, S. V.; Cooke, D. L.; Maguire, L. J.; Verdier, D. V. Irreversible corneal decompensation in patients treated with topical dorzolamide. *Am. J. Ophthalmol.* **1999**, *127*, 403–406.

30. Borras, J.; Scozzafava, A.; Menabuoni, L.; Mincione, F.; Briganti, F.; Mincione, G.; Supuran, C. T. Carbonic anhydrase inhibitors: synthesis of water-soluble, topically effective intraocular pressure lowering aromatic/heterocyclic sulfonamides containing 8-quinoline-sulfonyl moieties: is the tail more important than the ring? *Bioorg. Med. Chem.* **1999**, *7*, 2397–2406.

31. Casini, A.; Scozzafava, A.; Mincione, F.; Menabuoni, L.; Ilies, M. A.; Supuran, C. T. Carbonic anhydrase inhibitors: water soluble 4-sulfamoylphenylthioureas as topical intraocular pressure lowering agents with long lasting effects. *J. Med. Chem.* **2000**, *43*, 4884–4892.

32. Menabuoni, L.; Scozzafava, A.; Mincione, F.; Briganti, F.; Mincione, G.; Supuran, C. T. Carbonic anhydrase inhibitors. Water-soluble, topically effective intraocular pressure lowering agents derived from isonicotinic acid and aromatic/heterocyclic sulfonamides: is the tail more important than the ring? *J. Enzyme Inhib.* **1999**, *14*, 457–474.

33. Mincione, G.; Menabuoni, L.; Briganti, F.; Scozzafava, A.; Mincione, F.; Supuran, C. T. Carbonic anhydrase inhibitors. Part 79. Synthesis and ocular pharmacology of topically acting sulfonamides incorporating GABA moieties in their molecule, with long-lasting intraocular pressure lowering properties. *Eur. J. Pharm. Sci.* **1999**, *9*, 185–199.

34. Scozzafava, A.; Menabuoni, L.; Mincione, F.; Briganti, F.; Mincione, G.; Supuran, C. T. Carbonic anhydrase inhibitors: synthesis of water-soluble, topically effective intraocular pressure lowering aromatic/heterocyclic sulfonamides containing cationic or anionic moieties: is the tail more important than the ring? *J. Med. Chem.* **1999**, *42*, 2641–2650.

35. Scozzafava, A.; Briganti, F.; Mincione, G.; Menabuoni, L.; Mincione, F.; Supuran, C. T. Sulfonamides possessing long-lasting intraocular pressure-lowering properties via the topical route. *J. Med. Chem.* **1999**, *42*, 3690–3700.

36. Scozzafava, A.; Menabuoni, L.; Mincione, F.; Briganti, F.; Mincione, G.; Supuran, C. T. Carbonic anhydrase inhibitors: perfluoroalkyl/aryl-substituted derivatives of aromatic/heterocyclic sulfonamides as topical intraocular pressure lowering agents with prolonged duration of action. *J. Med. Chem.* **2000**, *43*, 4542–4551.

37. Scozzafava, A.; Menabuoni, L.; Mincione, F.; Supuran, C. T. Carbonic anhydrase inhibitors. A general approach for the preparation of water soluble sulfonamides incorporating polyamino-polycarboxylate tails and of their metal complexes possessing long

lasting, topical intraocular pressure lowering properties. *J. Med. Chem.* **2002**, *45*, 1466–1476.

38. Supuran, C. T.; Scozzafava, A.; Menabuoni, L.; Mincione, F.; Briganti, F.; Mincione, G. Carbonic anhydrase inhibitors. Part 70. Synthesis and ocular pharmacology of a new class of water-soluble, topically effective intraocular pressure lowering agents derived from nicotinic acid and aromatic/heterocyclic sulfonamides. *Eur. J. Med. Chem.* **1999**, *34*, 799–808.
39. Wayman, L.; Larsson, L. I.; Maus, T.; Alm, A.; Brubaker, R. Comparison of dorzolamide and timolol as suppressors of aqueous humor flow in humans. *Arch. Ophthalmol.* **1997**, *115*, 1368–1371.
40. Alm, A. Prostaglandin derivates as ocular hypotensive agents. *Prog. Ret. Eye Res.* **1998**, *17*, 291–312.
41. Larsson, L. I.; Alm, A. Aqueous humor flow in human eyes treated with dorzolamide and different doses of acetazolamide. *Arch. Ophthalmol.* **1998**, *116*, 19–24.
42. Vanlandingham, B. D.; Brubaker, R. F. Combined effect of dorzolamide and latanoprost on the rate of aqueous humor flow. *Am. J. Ophthalmol.* **1998**, *126*, 191–196.
43. Higginbotham, E. J.; Diestelhorst, M.; Pfeiffer, N.; Rouland, J. F.; Alm, A. The efficacy and safety of unfixed and fixed combinations of latanoprost and other antiglaucoma medications. *Surv. Ophthalmol.* **2002**, *47*, S133–S140.
44. Maestrelli, F.; Mura, P.; Casini, A.; Mincione, F.; Scozzafava, A.; Supuran, C. T. Cyclodextrin complexes of sulfonamide carbonic anhydrase inhibitors as long lasting topically acting antiglaucoma agents. *J. Pharm. Sci.* **2002**, *91*, 2211–2219.
45. Winum, J. Y.; Casini, A.; Mincione, F.; Starnotti, M.; Montero, J. L.; Scozzafava, A.; Supuran, C. T. Carbonic anhydrase inhibitors: *N*-(*p*-sulfamoylphenyl)-α-D-glycopyranosylamines as topically acting antiglaucoma agents in hypertensive rabbits. *Bioorg. Med. Chem. Lett.* **2004**, *14*, 225–229.
46. Supuran, C.; Benedetti, F.; Biondi, S.; Ongini, E. Nitrate esters of carbonic anhydrase inhibitors. *WO Patent* 2008/071421, **2008**.
47. Cox, S. N.; Hay, E.; Bird, A. C. Treatment of chronic macular edema with acetazolamide. *Arch. Ophthalmol.* **1988**, *106*, 1190–1195.
48. Grover, S.; Fishman, G. A.; Fiscella, R. G.; Adelman, A. E. Efficacy of dorzolamide hydrochloride in the management of chronic cystoid macular edema in patients with retinitis pigmentosa. *Retina* **1997**, *17*, 222–231.
49. Sponsel, W. E.; Harrison, J.; Elliott, W. R.; Trigo, Y.; Kavanagh, J.; Harris, A. Dorzolamide hydrochloride and visual function in normal eyes. *Am. J. Ophthalmol.* **1997**, *123*, 759–766.
50. Barnes, G. E.; Li, B.; Dean, T.; Chandler, M. L. Increased optic nerve head blood flow after 1 week of twice daily brinzolamide treatment in Dutch-belted rabbits. *Surv Ophthalmol.* **2000**, *44* (Suppl. 2), 131–140.
51. Orzalesi, N.; Pierrottet, C.; Porta, A.; Aschero, M. Long-term treatment of retinitis pigmentosa with acetazolamide. A pilot study. *Graefes Arch. Clin. Exp. Ophthalmol.* **1993**, *231*, 254–256.
52. Gonzalez, C. Serous retinal detachment. Value of acetazolamide. *J. Fr. Ophthalmol.* **1992**, *15*, 529–536.

CHAPTER 6

Diuretics with Carbonic Anhydrase Inhibitory Activity: Toward Novel Applications for Sulfonamide Drugs

DANIELA VULLO, ALESSIO INNOCENTI, and CLAUDIU T. SUPURAN

Laboratorio di Chimica Bioinorganica, Università degli Studi di Firenze, Room 188, Via della Lastruccia 3, I-50019 Sesto Fiorentino (Firenze), Italy

6.1 INTRODUCTION

Thiazide diuretics, such as hydrochlorothiazide, chlorthalidone, or indapamide among others, were the first well-tolerated and efficient antihypertensive drugs that significantly reduced cardiovascular disease (CVD) morbidity and mortality in placebo-controlled clinical studies.[1–6] Being a fundamental therapeutic option not only for patients with CVD but also for those suffering from type II diabetes, obesity, and related metabolic complications, these drugs are highly prescribed.[1–6] As initially thought, their mechanism of action is more complicated than the diuretic/saluretic,[1–6] and it has recently been shown that many such sulfonamides exert a direct vasodilator effect by activating calcium-activated potassium channels (KCa),[4] which in turn depend on the pH control of the zinc enzyme carbonic anhydrase (CA, EC 4.2.1.1) that catalyzes the interconversion between carbon dioxide and bicarbonate, with a release of a proton.[7]

Acetazolamide (5-acetamido-1,3,4-thiadiazole-2-sulfonamide) **6.1** is a potent but promiscuous inhibitor of CAs and was the first nonmercurial diuretic to be used clinically in 1956.[7,8] Acetazolamide represents the prototype of a class of pharmacological agents with apparently limited therapeutic usefulness nowadays; however, they played a major role in the development of fundamental renal physiology and pharmacology, as well as in the design of many of the widely used diuretic agents such as the thiazide and high-ceiling diuretics.[1,7–12] In mammals, 16 different α-CA isozymes or CA-related proteins (CARP) have been described so far, with very different catalytic activity, subcellular localization, tissue distribution, and susceptibility to be inhibited by sulfonamides.[7–14] These enzymes catalyze a very simple physiological reaction, the interconversion between carbon dioxide and bicarbonate ion, and are thus involved in crucial physiological processes connected with respiration and transport of CO_2/bicarbonate between metabolizing tissues and lungs, pH and CO_2 homeostasis, electrolyte secretion in a variety of tissues/organs, biosynthetic reactions (such as gluconeogenesis, lipogenesis, and ureagenesis), bone resorption, calcification, tumorigenicity, and many other physiologic/pathologic processes.[7–14] CAs are highly abundant in the kidneys (a total concentration of about 8–10 μM has been estimated for this organ), and many isoforms have been shown to be present in various tissues of this organ.[15,16] CA isoforms, such as CA II, IV, VB, IX, XII and XIV, present in kidneys play a crucial role in at least three physiological processes: (i) the acid–base balance homeostasis (by secreting and excreting protons, due to the carbon dioxide hydration reaction catalyzed by these enzymes); (ii) the bicarbonate reabsorption processes, and (iii) and the renal NH_4^+ output. [16,17] These important functions are well localized in the different segments of the nephron: bicarbonate reabsorption occurs in the proximal tubule, whereas urinary acidification and NH_4^+ output occur in the distal tubule and collecting duct.[16,17] Following the administration of a CA inhibitor (CAI), such as acetazolamide **6.1**, the volume of urine promptly increases, and its normally acidic pH (of 6) becomes alkaline (around 8.2).[12,16,17] An increased amount of bicarbonate is eliminated into urine (120 times higher than the amount eliminated normally), together with Na^+ and K^+ as accompanying cations, whereas the amount of chloride excreted is diminished. The increased alkalinity of the urine is accompanied by a decrease in the excretion of titratable acid and ammonia, and in consequence by metabolic acidosis. This sequence of events is due to the inhibition of various CA isozymes in the proximal tubule, which leads to inhibition of H^+ secretion by this segment of the nephron. Inhibition of both cytosolic (CA II) and membrane-bound (CA IV and XIV) enzymes seems to be involved in the diuretic effects of the sulfonamides.[12,16–18] Inhibition of such CAs decreases the availability of protons for the Na^+–H^+ antiporter, which maintains a low proton concentration in the cell. The net effect of these processes is the transport of sodium bicarbonate from the tubular lumen to the interstitial space, followed by movement of the isotonically obligated water, and thereafter augmented diuresis. CAIs also increase phosphate excretion (by an unknown mechanism) but have little or no effect on the excretion of calcium or magnesium ions.[16,17]

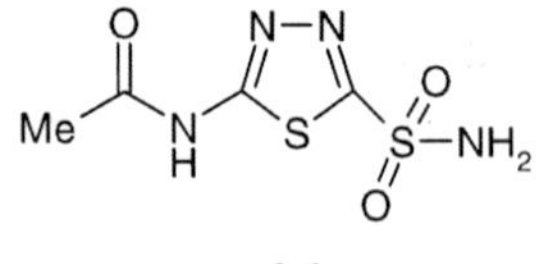

6.1

Acetazolamide **6.1** and structurally related sulfonamides, such as methazolamide **6.2**, ethoxzolamide **6.3**, and dichlorophenamide **6.4**, were and are still used for the treatment of edema due to congestive heart failure and for drug-induced edema, in addition to their applications as antiglaucoma agents.[7,8] However, these systemic CAIs generally produce many undesired side effects due to inhibition of CAs present in other organs than the kidneys (see Chapter 5). The structurally related compound to acetazolamide, benzolamide **6.5**, with an acidic pK_a of 3.2 for the secondary sulfonamide group is completely ionized at the physiological pH as sulfonamidate anion.[7,8] Its renal effect on bicarbonate excretion is around 10 times as potent as that of acetazolamide, the drug being maximally active at doses of 1 mg/kg and being actively and rapidly accumulated in the kidney, but its plasma half-life is only 20 min. All these facts make benzolamide a renal-specific CAI, but the compound has not been developed for wide clinical use due to its inappropriate pharmacokinetics, although some anecdotical reports indicate that it might be beneficial for patients suffering from chronic obstructive lung disease.[8]

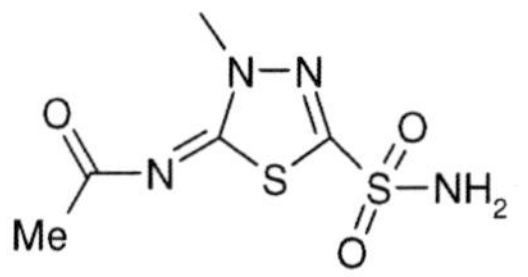

6.2

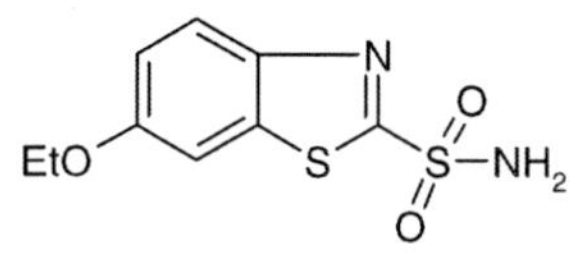

6.3

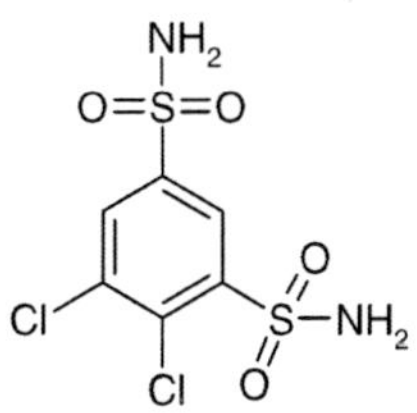

6.4

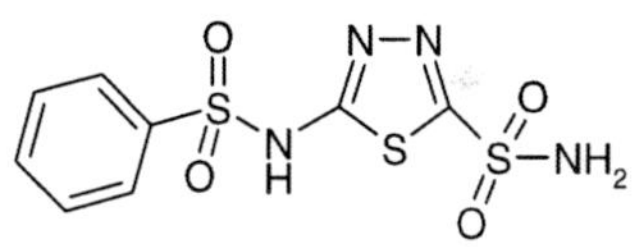

6.5

Using acetazolamide **6.1** as lead, a large number of other quite successful sulfonamide diuretics were developed in the 1960s and 1970s, such as the benzothiadiazines **6.6** (hydrochlorothiazide **6.6a**, hydroflumethiazide **6.6b**, and the like), quinethazone **6.7**, metolazone **6.8**, chlorthalidone **6.9**, indapamide **6.10**, furosemide **6.11**, and bumetanide **6.12**.[7,8,12,18] Some of them are among the most widely clinically used diuretics,[1–6] and as they all possess primary SO_2NH_2 moieties, acting as excellent zinc binding groups for the metal ion present within the CA active site,[7,8] it is to be expected that they should also have CA inhibitory properties. However, this issue has been investigated only in the 1960s or 1970s when these drugs were launched and when only one CA isozyme (i.e., CA II) was presumed to exist and be responsible for all the physiologic effects of the sulfonamide drugs. The interaction of some of these clinically used diuretics with all 12 catalytically active mammalian CA isoforms was reinvestigated recently.[7,12,18] The X-ray crystal structures of some of them (chlorthalidone **6.9** and indapamide **6.10**) in complex with CA II were also reported.[18] Such data are highly useful for the design of novel classes of more isozyme-selective CAIs and, most important, to understand some new pharmacological effects of these old drugs.

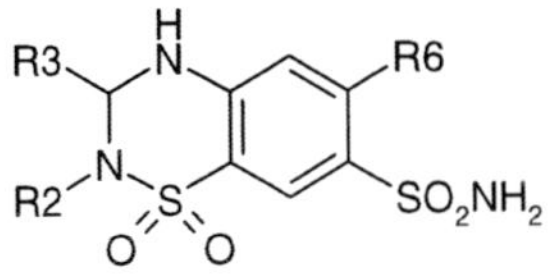

6.6

a: R2 = R3 = H, R6 = Cl, hydrochlorothiazide
b: R2 = R3 = H, R6 = CF_3, hydroflumethiazide
c: R2 = H, R3 = $PhCH_2$, R6 = CF_3, bendroflumethiazide
d: R2 = H, R3 = $CHCl_2$, R6 = Cl, trichloromethiazide
e: R2 = Me, R3 = $CH_2SCH_2CF_3$, R6 = Cl, polythiazide

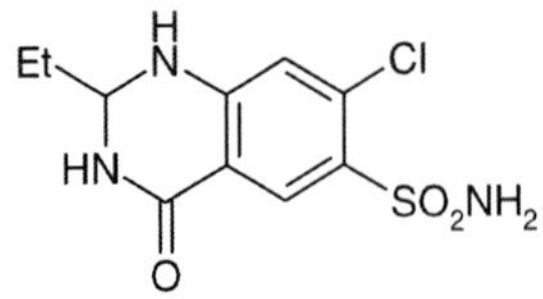

6.7

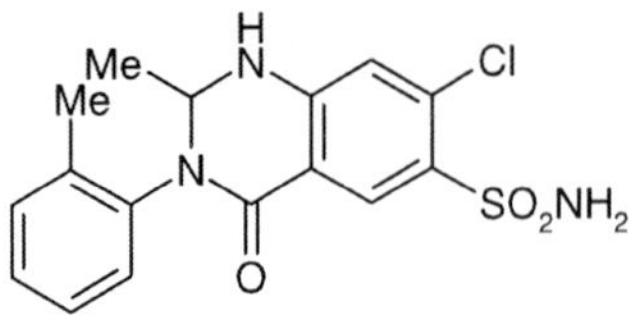

6.8

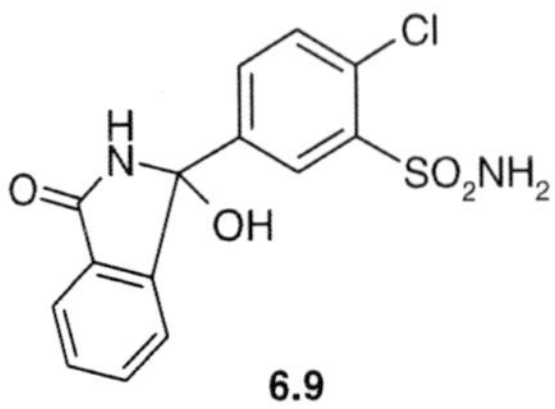

6.9

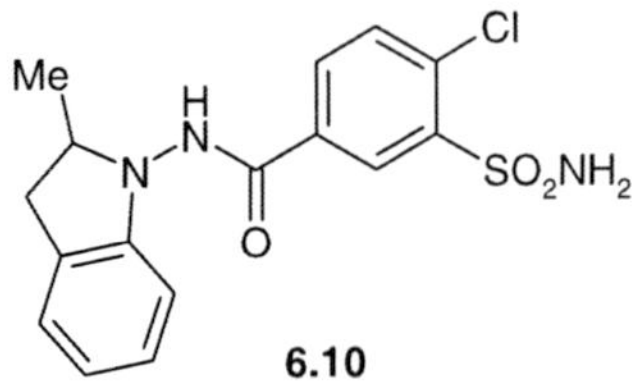

6.10

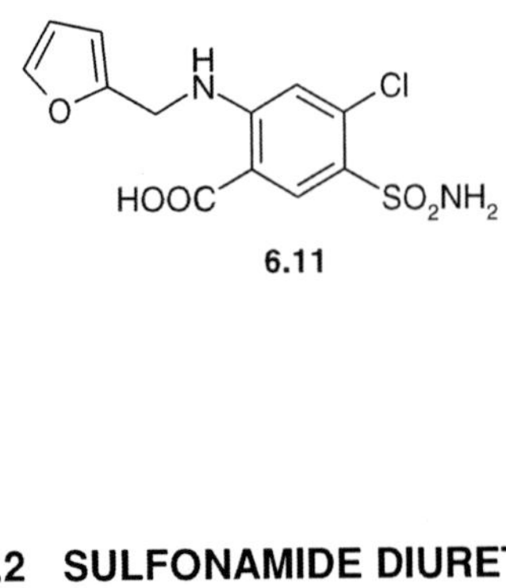

6.11

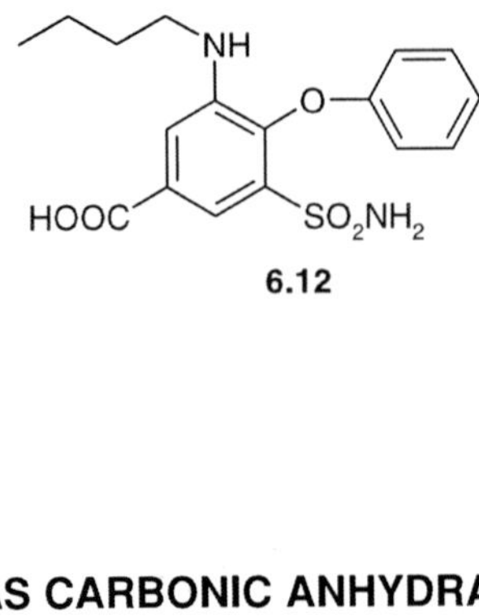

6.12

6.2 SULFONAMIDE DIURETICS AS CARBONIC ANHYDRASE INHIBITORS

In Table 6.1, the inhibition data of the 12 mammalian catalytically active CAs with drugs **6.1–6.12** are shown.[7] These data show that like the clinically used/orphan drug classical sulfonamide CAIs, that is, compounds **6.1–6.5**, the clinically used diuretics **6.6–6.12** also act as inhibitors of all 12 investigated CA isozymes, but with an inhibition profile quite different from that of inhibitors investigated earlier, particularly **6.1–6.5**. The following should be noted from these inhibition data. (i) Hydrochlorothiazide **6.6a** acts as a medium potency inhibitor of isoforms hCA I, II, VB, IX, and XII, with inhibition constants in the range of 290–603 nM, the compound being a weaker inhibitor of isoforms hCA VA, VI, VII, XIII, and XIV (K_I's in the range of 3.655–5.010 μM) and an exceedingly weak one against hCA III (K_I of 0.79 mM). (ii) Hydroflumethiazide **6.6b** shows an inhibition profile quite

TABLE 6.1 Inhibition Data with Some of the Clinically Used Sulfonamides 6.1–6.12 Against Isozymes I–XIV

	K_I (nM)												
Isozyme[a]	**6.1**	**6.2**	**6.3**	**6.4**	**6.5**	**6.6a**	**6.6b**	**6.7**	**6.8**	**6.9**	**6.10**	**6.11**	**6.12**
hCA I[b]	250	50	25	1200	15	328	2840	35,000	54,000	348	51,900	62	4930
hCA II[b]	12	14	8	38	9	290	435	1260	2000	138	2520	65	6980
hCA III[b]	2×10^5	7×10^5	1×10^6	6.8×10^5	1.4×10^5	7.9×10^5	8.7×10^5	nt	6.1×10^5	1.1×10^4	2.3×10^5	3.2×10^6	3.4×10^6
hCA IV[b]	74	6200	93	15,000	nt	427	4780	nt	216	196	213	564	303
hCA VA[b]	63	65	25	630	37	4225	10,200	nt	750	917	890	499	700
hCA VB[b]	54	62	19	21	34	603	429	nt	312	9	274	322	nt
hCA VI[b]	11	10	43	79	93	3655	8250	nt	1714	1347	1606	245	nt
hCA VII[b]	2.5	2.1	0.8	26	0.45	5010	433	nt	2.1	2.8	0.23	513	nt
hCA IX[c]	25	27	34	50	49	367	412	nt	320	23	36	420	25.8
hCA XII[c]	5.7	3.4	22	50	3.5	355	305	nt	5.4	4.5	10	261	21.1
mCA XIII[b]	17	19	50	23	nt	3885	15,400	nt	15	15	13	550	2570
hCA XIV[b]	41	43	25	345	33	4105	360	nt	5432	4130	4950	52	250

The isoforms CA VIII, X, and XI are devoid of catalytic activity and probably do not bind sulfonamides as they do not contain Zn(II) ions.

[a] h, human; m, murine isozyme; nt, not tested, data not available.

[b] Full-length enzyme.

[c] Catalytic domain.

distinct from that of the closely structurally related diuretic **6.6a**, being a rather efficient inhibitor of the isoforms hCA II, VB, VII, IX, XII, and XIV, with inhibition constants in the range of 305–435 nM. This sulfonamide is a weaker inhibitor of hCA I, IV, and VI (K_I's in the range of 2.84–8.25 μM) and shows very weak inhibition against isozymes hCA III, VA, and XIII (K_I's of 10.2–870 μM). It is already apparent that even small structural changes in the benzothiadiazine scaffold, such as the substitution of the chlorine atom in *ortho* to the sulfamoyl moiety by a trifluoromethyl group, as in the pair **6.6a/6.6b**, have dramatic consequences for the CA inhibitory properties of the two compounds, basically against all investigated CA iozymes (Table 6.1). (iii) Quinethazone **6.7** is the only diuretic among compounds **6.1–6.12** that is not approved for clinical use in Europe (but it is approved in the United States), and this derivative was not available to be investigated by us. Literature data[7] show it to be a very weak hCA I and a modest hCA II inhibitor, with inhibition constants in the range of 1.26–35 μM (Table 6.1). (iv) Metolazone **6.8** shows very weak hCA I and III inhibitory properties (K_I's in the range of 54–610 μM), being a low micromolar inhibitor (thus, not a very efficient one) of hCA II, VI, and XIV, with inhibition constants in the range of 1.714–5.432 μM. However, the drug is a medium potency inhibitor of isozymes hCA IV, VA, VB, and IX (K_I's in the range of 216–750 nM) and a very efficient one against hCA VII, hCA XII, and mCA XIII (K_I's in the range of 2.1–15 nM). (v) Chlorthalidone **6.9** also shows a very interesting inhibition profile, acting as a weak hCA III inhibitor (with a K_I of 11 μM, this compound is one of the most effective hCA III inhibitors ever detected among all known sulfonamides, except trifluoromethanesulfonamide that has a K_I of 0.9 μM)[9,19] and rather weak hCA VI and XIV inhibitor (K_I's in the range of 1.347–4.95 μM). Chlorthalidone is a moderate hCA VA inhibitor (K_I of 917 nM) and an effective or very effective inhibitor of the other mammalian CA isozymes. Thus, the ubiquitous hCA I and II, as well as hCA IV, show inhibition constants in the range of 138–348 nM, but isoforms VB, VII, IX, XII, and XIII are inhibited in the low nanomolar range (K_I's in the range of 2.8–23 nM). These results showing **6.9** to be such a strong CAI against many isoforms are quite unexpected, and considering the wide clinical use of the compound for the treatment of hypertension,[20] at least two issues can be raised: Is the inhibition of these CA isozymes responsible for some of the therapeutic effects of the drug (or for some side effects observed with it), and can these observations be useful for the design of CAIs with increased selectivity for some CA isoforms or for envisaging novel therapeutic applications of the drug (e.g., an adjuvant to antitumor therapies, considering its strong inhibitory effects against the tumor-associated CA isozymes IX and XII)?[21] We shall reply to some of these questions after considering in detail the inhibition profile of the remaining three drugs. (vi) Indapamide **6.10** acts as an inefficient CA I and III inhibitor (K_I's in the range of 51.9–230 μM) and is a weak inhibitor of isoforms CA II, VA, VI, and XIV (K_I's in the range of 890–4950 nM), but shows significant inhibitory activity against CA IV and VB (K_I's in the range of 213–274 nM) and excellent inhibition of CA VII, IX, XII, and XIII, with inhibition constants in the low nanomolar range (K_I's in the range of 0.23–36 nM). These data are indeed remarkable, considering the wide use of the drug as diuretic and its beneficial effects in patients with type 2 diabetes mellitus, as recently reported in several important clinical trials.[1–6,20] (vii) Furosemide **6.11** acts as a very weak hCA III inhibitor (K_I of 3200 μM), but it shows moderate inhibitory activity against many isoforms, such as CA IV, VA, VB, VI, VII, IX, XII, and XIII, with K_I's in the range of 261–564 nM. On the other hand, the compound is a much better inhibitor of CA I, II, and XIV, with K_I's in the range of 52–65 nM. (viii) Bumethanide is again an extremely weak hCA III inhibitor (K_I of 3400 μM), similarly to furosemide with which it is structurally related. However, bumethanide is also a weak inhibitor of hCA I, II, and XIII (K_I's in the range of 2570–6980 nM), probably due to the quite bulky phenoxy moiety in *ortho* to the sulfamoyl zinc binding group. The compound shows better inhibitory activity against isoforms CA IV, VA, and XIV (K_I's in the range of 250–700 nM) but excellent inhibition of the tumor-associated isoforms CA IX and XII (K_I's in the range of 21.1–25.8 nM, that is, the same order of magnitude as acetazolamide **6.1**, methazolamide **6.2**, or ethoxzolamide **6.3**). Thus, it may be observed that bumethanide is an effective inhibitor of only the tumor-associated isoforms CA IX and XII, thus discriminating between these drug targets and other CA isoforms that should not be inhibited by a cancer-specific CAI.

6.3 BINDING OF INDAPAMIDE TO HUMAN CARBONIC ANHYDRASE II

To understand at molecular level the interactions between indapamide **6.10** and the active site of a CA isozyme, the X-ray crystal structure of the indapamide adduct with the ubiquitous, highly investigated isoform hCA II was reported.[18] Interactions between the protein and Zn^{2+} ion were entirely preserved in the adduct, as in all other hCA II–sulfonamide/sulfamate/sulfamide complexes investigated so far.[22–24] A compact binding between the inhibitor and the enzyme active site was observed, similar to what was reported earlier for other such complexes, with the tetrahedral geometry of the Zn^{2+} binding site and the key hydrogen bonds between the SO_2NH_2 moiety of the inhibitor and enzyme active site all retained (Figs 6.1 and 6.2).[18,22–24] Thus, the ionized nitrogen atom of the sulfonamide group of **6.10** is coordinated to the zinc ion at a distance of 2.15 Å (Fig. 6.2). This nitrogen was also hydrogen bonded to the hydroxyl group of Thr199 (N–Thr199OG = 2.86 Å), which in turn interacted with the Glu106OE1 atom (2.5 Å, data not shown). One oxygen atom of the sulfonamide moiety was 3.06 Å away from the catalytic Zn^{2+} ion, being considered as weakly coordinated to the metal ion, whereas the second one participated in a hydrogen bond (of 3.19 Å) with the backbone amide group of Thr199.[18,22–24] His64 (in its *in* conformation) made strong van der Waals contacts (<4 Å) with the CONH moiety of the inhibitor, but these interactions cannot actually be considered as hydrogen bonds). A very strong interaction was, on the other hand, the offset face-to-face stacking between the annulated *ortho*-phenylene moiety of inhibitor **6.10** and the phenyl group of Phe131 (Figs 6.1 and 6.2), which has been observed previously for several other adducts of hCA II with sulfonamides such as the pyridinium derivative **6.13**[13] and sulpiride **6.14**.[23] Such a stacking interaction was in fact demonstrated to be

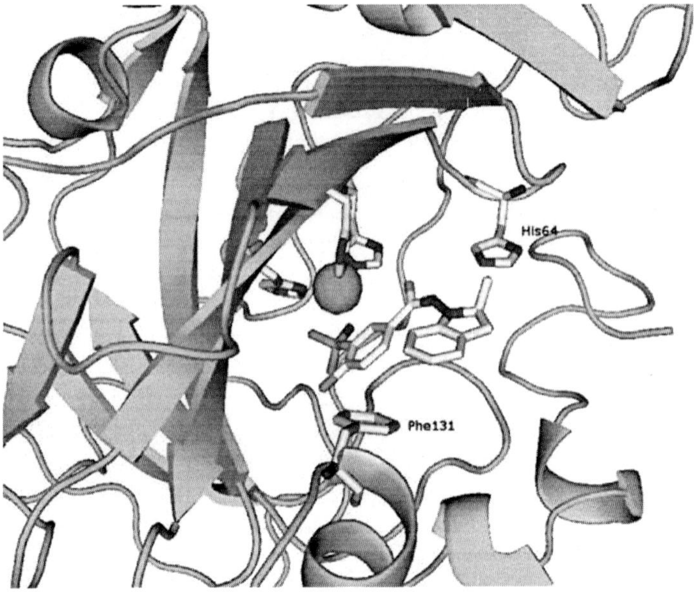

FIGURE 6.1 The hCA II–indapamide **6.10** complex. View of the zinc coordination sphere and neighboring amino acid residues (His64 and Phe131 among others) involved in the binding of the inhibitor **6.10**.

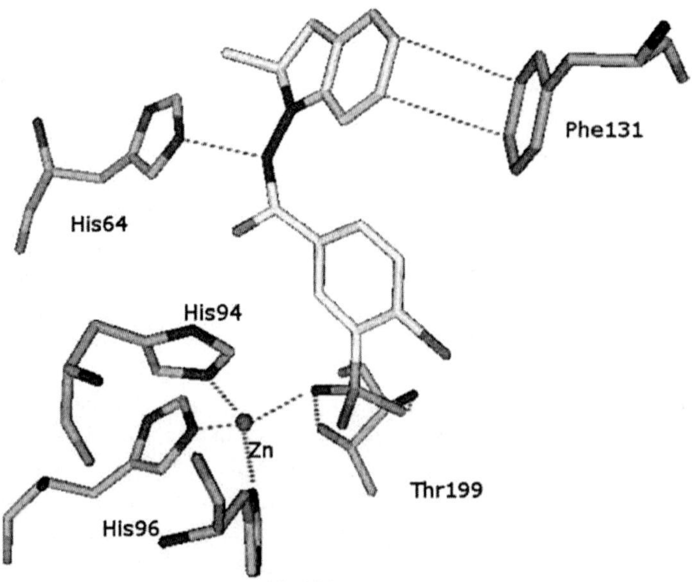

FIGURE 6.2 Detailed interactions in which indapamide **6.10** participates when bound within the hCA II active site. Active site residues coordinating the metal ion (His94, His96, and His119) as well as those involved in the binding of the inhibitor (His64, Phe131, and Thr199) are shown.

highly important for the orientation of the inhibitor within the active site and for the potency of a sulfonamide as CAI against this isoform.[13] Thus, to better understand the binding of **6.10** to hCA II, the superpositions of its hCA II adduct with those of dichlorophenamide **6.4**[24] as well as those of sulfonamides **6.13**[13] and **6.14**[23] are shown in Fig. 6.3. Two facts can be observed. First, a quite peculiar orientation for the phenyl ring of inhibitors **6.4** and **6.10** bound within the enzyme active site may be observed with respect to that of other benzene sulfonamides whose structures in complex with hCA II have been reported (Fig. 6.3a).[18,22–24] It may be seen that the plane of the phenyl moiety of the benzene-1,3-disulfonamide **6.4** and of indapamide **6.10** appears clearly rotated by almost 30° and tilted by approximately 10° with respect to its most recurrent orientation, as the derivatives **6.13** and **6.14** (Fig. 6.3b and c). This peculiar conformation may be ascribed to the presence of the secondary sulfonamide group in the *meta* position of **6.4**, and the corresponding bulky carbohydrazide substituent in the *meta* position to the sulfamoyl moiety of **6.10**, which probably orient the entire molecule in this unusual conformation when bound within the enzyme cavity. In addition, both these CAIs (i.e., **6.4** and **6.10**) possess a chlorine atom in *ortho* position to the sulfamoyl moiety coordinated to the Zn(II) present in the enzyme cavity, which could also influence the orientation of the organic scaffold due to steric impairment in the rather restricted environment near the metal ion.[24] In fact, Fig. 6.3a shows that the sulfamoyl-chloro-phenyl fragment present in the scaffold of inhibitors **6.4** and **6.10** is practically completely superposable in these two hCA II–sulfonamide complexes, whereas the second sulfamoyl group of dichlorophenamide **6.4** binds in the same active site region as the much bulkier carbohydrazide moiety of indapamide **6.10** (Fig. 6.3a). The second feature that is salient for the adduct of **6.10** with hCA II is with regard to the stacking interaction in which Phe131 participates with the phenylene moiety of the bicyclic ring present in indapamide. As seen from Figs 6.1–6.3, the two rings, that is, the *ortho*-phenylene moiety of the inhibitor **6.10** and the phenyl group of Phe131, are strictly parallel to each other, being at a distance of 3.41–3.46 Å. The same stacking has been previously seen between the phenyl moiety of Phe131 and the 2,4,6-trimethylpyridinium ring of the positively charged sulfonamide **6.13**,[13] or between the same phenyl moiety and the *N*-ethyl-pyrrolidine group of sulpiride **6.14**.[23] In both these complexes, this stacking interaction has been one of the main factors ensuring a strong affinity of the inhibitor for the hCA II active site. As seen from Fig. 6.3b and c, although the phenylene moiety of **6.10** binds in the same active site as the trimethylpyridinium ring of the positively charged sulfonamide **6.13** and the *N*-ethyl-pyrrolidine group of sulpiride **6.14**, these three cyclic moieties are not very well superposable with each other, which in part probably explain the net difference of activities of hCA II inhibitors between the three compounds. Thus, whereas **6.13** is a very strong hCA II inhibitor (K_I of 21 nM), sulpiride **6.14** is a slightly weaker one (K_I of 40 nM) and **6.10** is a moderate–weak inhibitor (K_I of 2.52 μM, Table 6.1). However, we consider this as a quite positive and interesting feature of indapamide as a CAI (as already discussed), which might be used to design compounds with reduced affinity for the ubiquitous, housekeeping enzyme hCA II, but with strong inhibitory properties against other isoforms such as hCA IV, VB, VII, IX, XII, or XIII, against which indapamide act as a much more efficient CAI (Table 6.1).

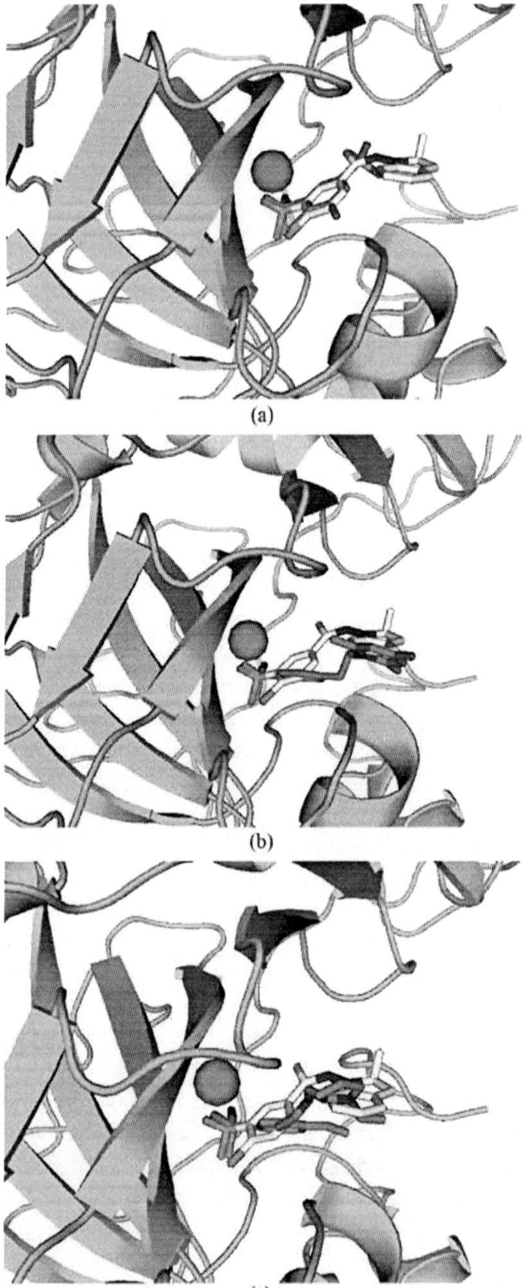

FIGURE 6.3 Superposition of (a) hCA II–indapamide **6.10** with hCA II–dichlorophenamide **6.4**; (b) hCA II–indapamide **6.10** with hCA II–pyridinium sulfonamide **6.13**; and (c) hCA II–indapamide **6.10** with hCA II–sulpiride **6.14** adducts.

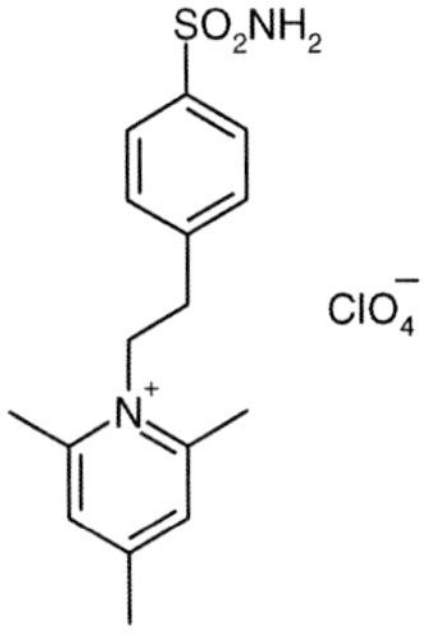

6.13

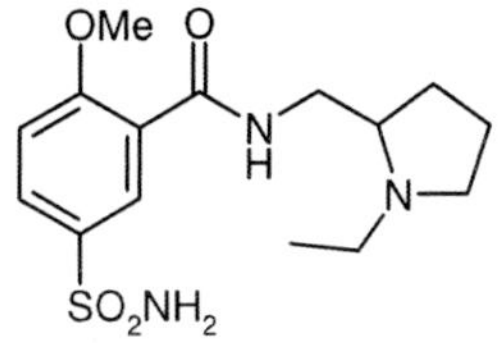

6.14

6.4 POTENTIAL NOVEL APPLICATIONS OF THE DIURETIC SULFONAMIDES IN THERAPY

The question arises as to what is the relevance of the above-mentioned findings for the drug design of CAIs with diverse pharmacological applications. We think that there are at least several aspects that need to be considered for answering this question. First, the classical sulfonamide diuretics **6.6–6.12** (widely used drugs to treat hypertension[1–6]) were considered until recently to be inactive as CAIs, due to the fact that they were launched in a period when only CA II was well known (and considered responsible for all physiologic effects of CAIs). It may indeed be observed that in contrast to the classical CAIs of type **6.1–6.5** (generally low nanomolar CA II inhibitors), all compounds **6.6–6.12** (except furosemide **6.11**) are much weaker inhibitors of this isozyme, usually in the micromolar range. Indeed, only furosemide **6.11** is a good CA II inhibitor among these diuretics, with a K_I of 65 nM, whereas all others show K_I's in the range of 138–6980 nM (Table 6.1). Again, with the exception of furosemide **6.11**, the diuretics **6.6–6.12** have low affinity for CA I, the other isoform known when these drugs have been discovered. However, data of Table 6.1 show that many of the drugs **6.6–6.12** appreciably inhibit CAs discovered after their introduction in clinical use, with some low nanomolar (or even subnanomolar) inhibitors against many of them. Examples of such situations are metholazone **6.8** against CA VII, XII, and XIII; chlorthalidone **6.9** against CA VB, VII, IX, XII, and XIII; indapamide **6.10** against CA VII, IX, XII, and XIII; furosemide **6.11** against CA I, II, and XIV; and bumethanide **6.12** against CA IX and XII among others (Table 6.1). As already mentioned, bumethanide **6.12** is a tumor-specific (targeting CA IX and XII) CAI, of equal potency to acetazolamide **6.1** but without the promiscuity of acetazolamide, which is a potent CAI against most mammalian isozymes. Indeed, bumethanide is a weak inhibitor of all other isoforms except CA IX and XII, which are overexpressed in tumors.[18] Indapamide **6.10** and chlorthalidone **6.9** are also strong inhibitors of the tumor-associated CAs, but they are also effective in inhibiting CA VII and XIII (Table 6.1). It is thus clear that these old drugs may indeed have newer applications in therapy as experimental agents in situations in which the selective inhibition of some CA isozymes is needed, and which cannot be obtained with the presently used compounds of types **6.1–6.5**.

A second important aspect of these findings is whether we can explain some recent observations of clinical trials in which such diuretic agents have been employed, and how this is reflected in the drug design process. Thus, a relevant question that arises is how these inhibitory activities of compounds of type **6.6–6.12** against CA isoforms known nowadays to play important physiological functions[7–14] are reflected by the pharmacological effects (or side effects profile) of these drugs? It is in fact known that many interesting classes of novel drugs have been discovered just by observing side effects of some clinically used agents.[7,8] Recently, it has been observed that indapamide **6.10** in combination with an ACE inhibitor (as diuretics) is highly beneficial for the treatment of patients with hypertension and type 2 diabetes.[1–6] Treatment with indapamide was also shown to lead to a significant decrease of plasma adiponectin concentration in patients with essential hypertension.[2,21] Adiponectin is considered to participate in the pathogenesis of carbohydrate metabolism disturbances often found in patients treated with other thiazide-type diuretics.[2] On the other hand, classical sulfonamide CAIs such as acetazolamide **6.1**, methazolamide **6.2** ethoxzolamide **6.3**, and other compounds possessing such properties are known to induce vasodilation in a variety of tissues and organs, including the kidneys, eye vasculature, brain vessels, and so on.[25–30] However, the exact mechanisms by which they produce this beneficial effect for many pathologies (e.g., hypertension, glaucoma, diabetic retinopathy, etc.), or the isoforms involved in it, are unknown for the moment.[25–30] In line with these studies,[25–30] a very recent report shows that indapamide **6.10** has a protective role against ischemia-induced injury and dysfunction of the blood–brain barrier, probably due to its vasodilating effects.[31,32] An organ-protective effect of indapamide in animal models of renal failure has also been reported, showing the drug to be beneficial in preventing damage to the capillary structures, the endothelium, and in reducing the hypertrophy of superficial glomeruli among others.[16,17,33] All these effects are probably mediated by inhibition of CAs present in blood vessels or in the kidneys, but no specific pharmacological or biochemical studies are available so far, except for these clinical observations mentioned here.[28–33] The lesson we learned from all these data is that probably many of the recently reported beneficial clinical properties of indapamide **6.10** are indeed due to its diuretic effects, but in conjunction with its strong inhibition of some CA isozymes (such as CA IV, VB, VII, IX, XII, and/or XIII) reported here for the first time, isoforms present in kidneys and blood vessels. This hypothesis may explain both the blood pressure lowering effects and organ-protective activity of the drug. For medicinal chemists, this means that probably it is possible to design sulfonamide CAIs possessing an inhibition profile similar to **6.10**, but with a stronger activity against the target isoform(s) involved in these pathologies, provided that it will be possible to understand which these isozymes are.

6.5 CONCLUSIONS

The widely used sulfonamide drugs, the benzothiadiazines and high-ceiling diuretics, among which hydrochlorothiazide, hydroflumethiazide, quinethazone, metolazone, chlorthalidone, indapamide, furosemide, and bumetanide, which contain primary sulfamoyl moieties acting as potential zinc binding functions, were recently investigated as inhibitors of 12 catalytically active mammalian CAs. These widely used drugs were launched in a period when only isoform CA II was known and considered physiologically/pharmacologically relevant, and thus no inhibition data against other CA isoforms were available in the literature until recently. Although acting as moderate–weak inhibitors of CA II, and CA I, all these drugs were shown to considerably inhibit other CA isozymes known nowadays to be involved in critical physiologic processes, among the many CAs present in vertebrates. Some low nanomolar (or even subnanomolar) inhibitors against such isoforms were detected, such as metholazone against CA VII, XII, and XIII; chlorthalidone against CA VB, VII, IX, XII, and XIII; indapamide against CA VII, IX, XII, and XIII; furosemide against CA I, II, and XIV; and bumethanide against CA IX and XII. The X-ray crystal structure of the CA II–indapamide adduct was also resolved at high resolution, showing features that may be useful for the drug design of novel classes of CAIs. We proposed that the recently observed beneficial effect of indapamide for the treatment of patients with hypertension and type 2 diabetes is due to the potent inhibition of CA isoforms present in kidneys and blood vessels, which explain both the blood pressure lowering effects and organ-protective activity of the drug. Thus, a class of old drugs may be useful to design CA inhibitors with completely novel applications in therapy.

REFERENCES

1. (a) Padilla, M. C.; Armas-Hernández, M. J.; Hernández, R. H.; Israili, Z. H.; Valasco, M. Update of diuretics in the treatment of hypertension. *Am. J. Ther.* **2007**, *14*, 154–160. (b) Smith, R. E.; Ashiya, M. Antihypertensive therapies. *Nat. Rev. Drug Discov.* **2007**, *6*, 597–598.
2. (a) Patel, A.; ADVANCE Collaborative Group; MacMahon, S.; Chalmers, J.; Neal, B.; Woodward, M.; Billot, L.; Harrap, S.; Poulter, N.; Marre, M.; Cooper, M.; Glasziou, P.; Grobbee, D. E.; Hamet, P.; Heller, S.; Liu, L. S.; Mancia, G.; Mogensen, C. E.; Pan, C. Y.; Rodgers, A.; Williams, B. Effects of a fixed combination of perindopril and indapamide on macrovascular and microvascular outcomes in patients with type 2 diabetes mellitus (the ADVANCE trial): a randomised controlled trial. *Lancet* **2007**, *370*, 829–840. (b) Piecha, G.; Adamczak, M.; Chudek, J.; Wiecek, A. Indapamide decreases plasma adiponectin concentration in patients with essential hypertension. *Kidney Blood Press. Res.* **2007**, *30*, 187–194. (c) Zannad, F. Benefits of first line combination of perindopril and indapamide in clinical practice for patients with hypertension and diabetes. *Am. J. Hypertens.* **2007**, *20*, 9S–14S. (d) Sica, D. A. Chlorthalidone: has it always been the best thiazide-type diuretic? *Hypertension* **2006**, *47*, 321–322.
3. (a) Nishioku, T.; Takata, F.; Yamauchi, A.; Sumi, N.; Yamamoto, I.; Fujino, A.; Naito, M.; Tsuruo, T.; Shuto, H.; Kataoka, Y. Protective action of indapamide, a thiazide-like

diuretic, on ischemia-induced injury and barrier dysfunction in mouse brain microvascular endothelial cells. *J. Pharmacol. Sci.* **2007**, *103*, 323–327. (b) Smith, D. H. Fixed-dose combination antihypertensives and reduction in target organ damage: are they all the same? *Am. J. Cardiovasc. Drugs* **2007**, *7*, 413–422. (c) Pickkers, P.; Garcha, R. S.; Schachter, M.; Smits, P.; Hughes, A. D. Inhibition of carbonic anhydrase accounts for the direct vascular effects of hydrochlorothiazide. *Hypertension* **1999**, *33*, 1043–1048.

4. (a) Dell'Omo, G.; Penno, G.; Del Prato, S.; Pedrinelli, R. Chlorthalidone improves endothelial-mediated vascular responses in hypertension complicated by nondiabetic metabolic syndrome. *J. Cardiovasc. Pharmacol. Ther.* **2005**, *10*, 265–272. (b) Struijker-Boudier, H. A. From macrocirculation to microcirculation: benefits of preterax. *Am. J. Hypertens.* **2007**, *20*, 15S–18S.
5. Patel, A. B.; Kostis, J. B.; Wilson, A. C.; Shea, M. L.; Pressel, S. L.; Davis, B. R. Long-term fatal outcomes in subjects with stroke or transient ischemic attack. Fourteen-year follow-up of the systolic hypertension in the elderly program. *Stroke* **2008**, *39*, 1084–1089.
6. (a) Black, H. R.; Davis, B.; Barzilay, J.; Nwachuku, C.; Baimbridge, C.; Marginean, H.; Wright, J. T., Jr.; Basile, J.; Wong, N. D.; Whelton, P.; Dart, R. A.; Thadani, U.; The Antihypertensive and Lipid-Lowering Treatment to Prevent Heart Attack Trial. Metabolic and clinical outcomes in nondiabetic individuals with the metabolic syndrome assigned to chlorthalidone, amlodipine, or lisinopril as initial treatment for hypertension: a report from the Antihypertensive and Lipid-Lowering Treatment to Prevent Heart Attack Trial (ALLHAT). *Diabetes Care* **2008**, *31*, 353–360. (b) Herraez-Fernandez, R.; Campins-Falco, P. Chromatographic separation of chlorthalidone enantiomers using β-cyclodextrins as chiral additives. *J. Chromatogr. B* **2000**, *740*, 169–177.
7. Supuran, C. T. Carbonic anhydrases: novel therapeutic applications for inhibitors and activators. *Nat. Rev. Drug Discov.* **2008**, *7*, 168–181.
8. Supuran, C. T.; Scozzafava, A.; Conway, J., Eds. *Carbonic Anhydrase: Its Inhibitors and Activators;* CRC Press: Boca Raton, FL, **2004**; pp 1–363, and references cited therein.
9. Supuran, C. T.; Casini, A.; Scozzafava, A. Development of sulfonamide carbonic anhydrase inhibitors. In *Carbonic Anhydrase: Its Inhibitors and Activators;* Supuran, C. T.; Scozzafava, A.; Conway, J., Eds.; CRC Press: Boca Raton, FL **2004**; pp 67–148.
10. Supuran, C. T.; Scozzafava, A.; Casini, A. Carbonic anhydrase inhibitors. *Med. Res. Rev.* **2003**, *23*, 146–189.
11. Supuran, C. T. Carbonic anhydrases as drug targets: an overview. *Curr. Top. Med. Chem.* **2007**, *7*, 825–833.
12. Supuran, C. T. Diuretics: from classical carbonic anhydrase inhibitors to novel applications of the sulfonamides. *Curr. Pharm. Des.* **2008**, *14*, 641–648.
13. Alterio, V.; Vitale, R. M.; Monti, S. M.; Pedone, C.; Scozzafava, A.; Cecchi, A.; De Simone, G.; Supuran, C. T. Carbonic anhydrase inhibitors: X-ray and molecular modeling study for the interaction of a fluorescent antitumor sulfonamide with isozyme II and IX. *J. Am. Chem. Soc.* **2006**, *128*, 8329–8335.
14. Menchise, V.; De Simone, G.; Alterio, V.; Di Fiore, A.; Pedone, C.; Scozzafava, A.; Supuran, C. T. Carbonic anhydrase inhibitors: stacking with Phe131 determines active site binding region of inhibitors as exemplified by the X-ray crystal structure of a membrane-impermeant antitumor sulfonamide complexed with isozyme II. *J. Med. Chem.* **2005**, *48*, 5721–5727.
15. Hilvo, M.; Rafajová, M.; Pastoreková, S.; Pastorek, J.; Parkkila, S. Expression of carbonic anhydrase IX in mouse tissues. *J. Histochem. Cytochem.* **2004**, *52*, 1313–1322.

16. Kaunisto, K.; Parkkila, S; Rajaniemi, H.; Waheed, A.; Grubb, J.; Sly, W. S. Carbonic anhydrase XIV: luminal expression suggests key role in renal acidification. *Kidney Int.* **2002**, *61*, 2111–2118.

17. Nawata, C. M.; Hung, C. C.; Tsui, T. K.; Wilson, J. M.; Wright, P. A.; Wood, C. M. Ammonia excretion in rainbow trout (*Oncorhynchus mykiss*): evidence for Rh glycoprotein and H^+-ATPase involvement. *Physiol. Genomics* **2007**, *31*, 463–474.

18. (a) Temperini, C.; Cecchi, A.; Scozzafava, A.; Supuran, C. T. Carbonic anhydrase inhibitors. Sulfonamide diuretics revisited: old leads for new applications? *Org. Biomol. Chem.* **2008**, *6*, 2499–2506. (b) Temperini, C.; Cecchi, A.; Scozzafava, A.; Supuran, C. T. Carbonic anhydrase inhibitors. Interaction of indapamide and related diuretics with twelve mammalian isozymes and X-ray crystallographic studies for the indapamide–isozyme II adduct. *Bioorg. Med. Chem. Lett.* **2008**, *18*, 2567–2573. (c) Temperini, C.; Cecchi, A.; Scozzafava, A.; Supuran, C. T. Carbonic anhydrase inhibitors. Comparison of chlorthalidone and indapamide X-ray crystal structures in adducts with isozyme II: when three water molecules and the keto–enol tautomerism make the difference. *J. Med. Chem.* **2009**, *52*, 322–328.

19. Nishimori, I.; Minakuchi, T.; Onishi, S.; Vullo, D.; Cecchi, A.; Scozzafava, A.; Supuran, C. T. Carbonic anhydrase inhibitors. Cloning, characterization and inhibition studies of the cytosolic isozyme III with sulfonamides. *Bioorg. Med. Chem.* **2007**, *15*, 7229–7236.

20. Elliott, W. J. Systemic hypertension. *Curr. Probl. Cardiol.* **2007**, *32*, 201–259.

21. Svastová, E.; Hulíková, A.; Rafajová, M.; Zat'ovicová, M.; Gibadulinová, A.; Casini, A.; Cecchi, A.; Scozzafava, A.; Supuran, C. T.; Pastorek, J.; Pastoreková, S. Hypoxia activates the capacity of tumor-associated carbonic anhydrase IX to acidify extracellular pH. *FEBS Lett.* **2004**, *577*, 439–445.

22. Abbate, F.; Winum, J. Y.; Potter, B. V.; Casini, A.; Montero, J. L.; Scozzafava, A.; Supuran, C. T. Carbonic anhydrase inhibitors: X-ray crystallographic structure of the adduct of human isozyme II with EMATE, a dual inhibitor of carbonic anhydrases and steroid sulfatase. *Bioorg. Med. Chem. Lett.* **2004**, *14*, 231–234.

23. Abbate, F.; Coetzee, A.; Casini, A.; Ciattini, S.; Scozzafava, A.; Supuran, C. T. Carbonic anhydrase inhibitors: X-ray crystallographic structure of the adduct of human isozyme II with the antipsychotic drug sulpiride. *Bioorg. Med. Chem. Lett.* **2004**, *14*, 337–341.

24. Alterio, V.; De Simone, G.; Monti, S. M.; Scozzafava, A.; Supuran, C. T. Carbonic anhydrase inhibitors: inhibition of human, bacterial, and archaeal isozymes with benzene-1,3-disulfonamides—solution and crystallographic studies. *Bioorg. Med. Chem. Lett.* **2007**, *17*, 4201–4207.

25. Taki, K.; Hirahara, K.; Tomita, S.; Totoki, T. Acetazolamide-induced increase in blood flow to rabbit organs is confirmed using colored microspheres. *Heart Vessels* **1998**, *13*, 63–67.

26. Taki, K.; Kato, H.; Endo, S.; Inada, K.; Totsuka, K. Cascade of acetazolamide-induced vasodilatation. *Res. Commun. Mol. Pathol. Pharmacol.* **1999**, *103*, 240–248.

27. Taki, K.; Oogushi, K.; Hirahara, K.; Gai, X.; Nagashima, F.; Tozuka, K. Preferential acetazolamide-induced vasodilation based on vessel size and organ: confirmation of peripheral vasodilation with use of colored microspheres. *Angiology* **2001**, *52*, 483–488.

28. Yamauchi, H.; Okazawa, H.; Kishibe, Y.; Sugimoto, K.; Takahashi, M. Effects of acetazolamide on cerebral blood flow, blood volume, and oxygen metabolism: a positron emission tomography study with healthy volunteers. *J. Cereb. Blood Flow Metab.* **2001**, *21*, 1472–1479.

29. Pedersen, D. B.; Koch Jensen, P.; la Cour, M.; Kiilgaard, J. F.; Eysteinsson, T.; Bang, K.; Wiencke, A. K.; Stefánsson, E. Carbonic anhydrase inhibition increases retinal oxygen tension and dilates retinal vessels. *Graefes Arch. Clin. Exp. Ophthalmol.* **2005**, *243*, 163–168.

30. Horita, Y.; Yakabe, K.; Tadokoro, M.; Suyama, N.; Hayashida, K.; Kawano, Y.; Miyazaki, M.; Kohno, S.; Taura, K. Renal circulatory effects of acetazolamide in patients with essential hypertension. *Am. J. Hypertens.* **2006**, *19*, 282–285.

31. Josefsson, A.; Sigurdsson, S. B.; Bang, K.; Eysteinsson, T. Dorzolamide induces vasodilatation in isolated pre-contracted bovine retinal arteries. *Exp. Eye Res.* **2004**, *78*, 215–221.

32. Nishioku, T.; Takata, F.; Yamauchi, A.; Sumi, N.; Yamamoto, I.; Fujino, A.; Naito, M.; Tsuruo, T.; Shuto, H.; Kataoka, Y. Protective action of indapamide, a thiazide-like diuretic, on ischemia-induced injury and barrier dysfunction in mouse brain microvascular endothelial cells. *J. Pharmacol. Sci.* **2007**, *103*, 323–327.

33. Struijker-Boudier, H. A. From macrocirculation to microcirculation: benefits of preterax. *Am. J. Hypertens.* **2007**, *20*, 15S–18S.

CHAPTER 7

Drug Design of Carbonic Anhydrase Inhibitors as Anticonvulsant Agents

ANNE THIRY[1], JEAN-MICHEL DOGNÉ[1], CLAUDIU T. SUPURAN[2], and BERNARD MASEREEL[1]

[1]Drug Design and Discovery Center, FUNDP, University of Namur, 61 rue de Bruxelles, 5000 Namur, Belgium
[2]Polo Scientifico, Laboratorio di Chimica Bioinorganica, Room 188, Università degli Studi di Firenze, Via della Lastruccia 3, 50019 Sesto Fiorentino, Florence, Italy

7.1 INTRODUCTION: FROM THE PHYSIOPATHOLOGY OF CONVULSIONS TO POTENTIAL TREATMENTS

Convulsions are among the most common neurological disorders in clinical medicine[1] and are a dysfunction in the brain gray matter arising because of an abnormal, excessive, paroxysmal, synchronous discharge in a population of neurons. The excessive excitability of neuron seizures is probably caused by ionic imbalance between the depolarized excitatory ions (sodium and calcium entry) and the hyperpolarized inhibitor ions (chloride influx and potassium outflux).[2–4] This instability comes from a membrane alteration of the K^+ conductance, a dysfunction of voltage-dependent sodium channels, or ionic ATP-dependent membrane transport. Moreover, a weak regulation of the excitatory and inhibitory neurotransmitters (i.e., glutamate and γ-aminobutyric acid, respectively) also contributes to neuronal excessive electric processes.[3]

Many causes of seizures such as metabolic trouble, infection, trauma, toxins, or epilepsy are described. Clinical and electroencephalogram manifestations of seizures are classified into two basic groups: partial and generalized. Partial seizures focally begin in a cortical site, and generalized seizures involve both hemispheres widely from the outset.[5] The behavioral manifestations of seizures are determined by the functions normally served by the cortical site where seizures are located. Simple partial seizures are associated with the preservation of the consciousness. There is usually impairment of consciousness during generalized seizures.[1,6,7] This classification contributes to establish a precise diagnostic of the type of seizure, thus determining the choice of the most appropriate treatment.

In the early 1920s, epilepsy or convulsions were treated with the bromides and phenobarbital.[8] Both drugs had major sedating, adverse effects and were frequently ineffective in completely controlling seizures. The ketogenic diet was then considered as a new treatment for epilepsy. Studies of the metabolic changes during this diet were undertaken and it was stated that ketone bodies were the immediate result of the oxidation of certain acids in the absence of sufficient glucose and postulated that they were anticonvulsant.[8,9] Furthermore, the discovery of phenytoin in 1938 had turned the attention of physicians to the development of new antiepileptic agents, keeping the ketogenic diet by the wayside. Several antiepileptic drugs were then discovered, and they are divided into two generations (Fig. 7.1): the first generation, before 1990, that

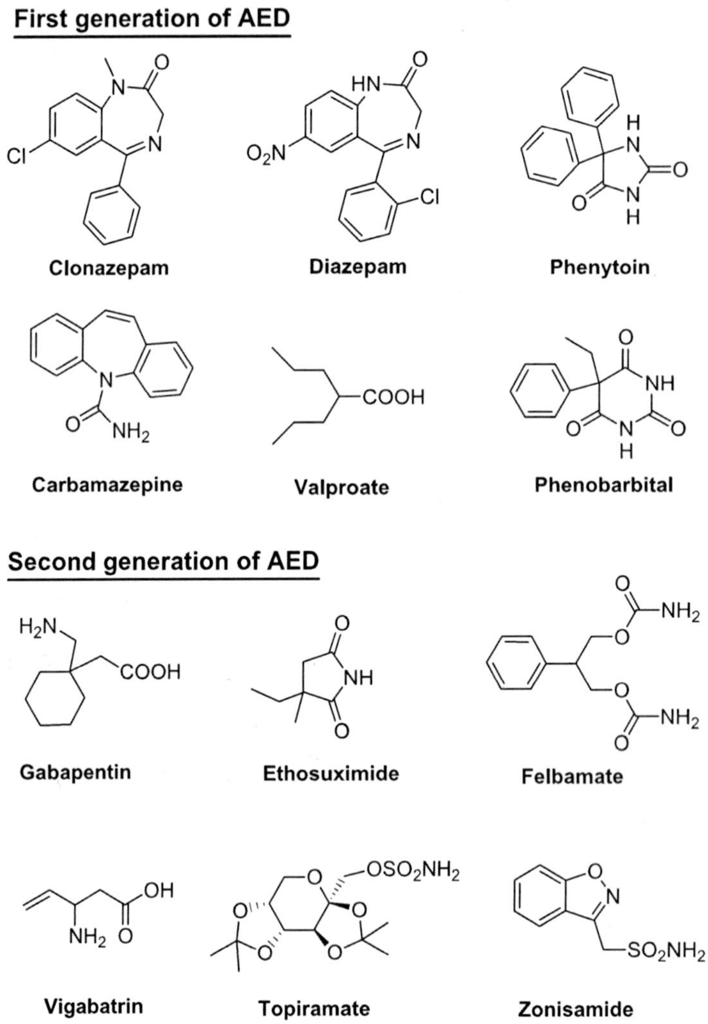

FIGURE 7.1 Chemical structures of antiepileptic drugs.

includes phenobarbital, phenytoin, benzodiazepine (diazepam, clonazepam), carbamazepine, and valproate and the second generation, from 1990, that includes among others gabapentin, ethosuximide, felbamate, vigabatrin, topiramate, and zonisamide. One of the last antiepileptic drugs (AED) developed is levetiracetam for which the mechanism of action is not completely understood. Most of the commercially available antiepileptic drugs target the potential-dependent ionic channels and the inhibitory or excitatory synaptic functions. Figure 7.2 illustrates the principal mechanism of action of the clinically used AED.

Efficacy of these AEDs was evaluated by Kwan and Brodie.[11] This study shows that 36% of the patients still suffered from ongoing seizures even with a combination of several AEDs. Moreover, numerous side effects and drug–drug interactions are also reported and they limit the clinical use of the currently available anticonvulsive drugs.[12] Other treatment options are thus considered, such as surgery, electrostimulation (vagus

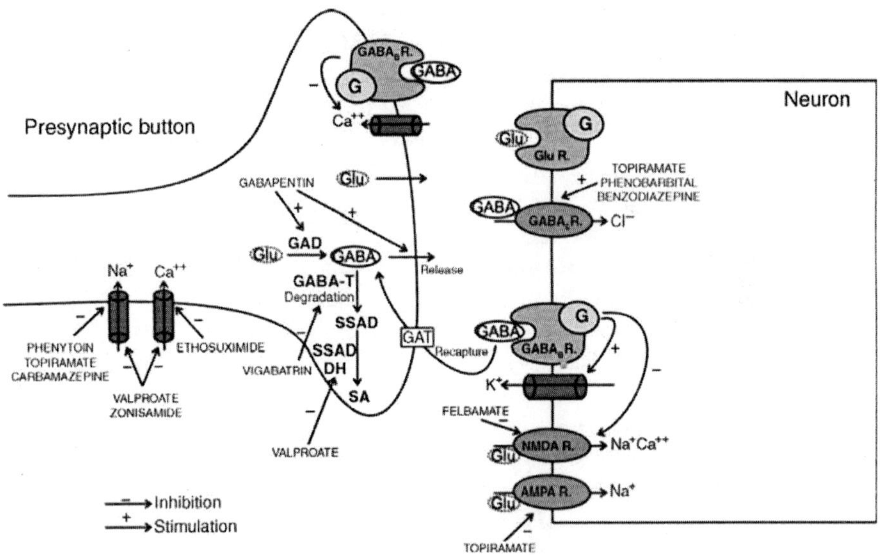

FIGURE 7.2 Overview of the mechanism of action of AEDs.[2,3,10] Phenytoin and carbamazepine block voltage-sensitive sodium channels. Phenobarbital and benzodiazepine enhance the GABA-mediated chloride flux that causes membrane hyperpolarization. Valproate closes the voltage-sensitive sodium channels and T-type calcium channels; it increases GABA synthesis and decreases GABA degradation. Vigabatrin is an irreversible inhibitor of GABA transaminase (GABA-T), the enzyme responsible for the degradation of GABA in the brain. Felbamate blocks the NMDA receptor, thus preventing the excitation of neurons. Gabapentin is a GABA analogue that activates the GABA release. Ethosuximide inhibits T-type calcium channels. Zonisamide blocks both sodium and calcium channels. On the one hand, topiramate blocks the voltage-sensitive sodium channel and the AMPA receptor and, on the other hand, it potentiates GABAergic transmission by the activation of $GABA_A$ receptors. Abbreviations: G = protein g; GABA = γ-aminobutyric acid; GABA-T = GABA transaminase; GAD = glutamic acid decarboxylase; GAT = GABA transporter; Glu = glutamate; R = receptor; SA = succinic acid; SSAD = succinic semialdehyde; SSADH = succinic semialdehyde dehydrogenase.

nerve or intracerebral), and radiosurgery.[12] Nevertheless, all these approaches together are not able to solve each clinical case of epilepsy. New ways for the development of antiepileptic drugs are required to better understand the epileptogenesis and the mechanisms of drug resistance. New anticonvulsant compounds structurally and pharmacologically different of the currently prescribed drugs might be designed.[12–14]

7.2 OVERVIEW OF THE CLINICAL USE OF CARBONIC ANHYDRASE INHIBITORS AS ANTICONVULSANT AGENTS

The use of carbonic anhydrase inhibitors as anticonvulsant drugs was first described in 1953 with acetazolamide and methazolamide (Fig. 7.3).[15] Modification of the environmental pH (alkalosis increases the neuronal excitability) is one potential factor that may contribute to seizure.[13,16–20] The pH buffering of the extracellular and intracellular spaces is mainly carried out by the CO_2/HCO_3^- buffer. The ratio of these two species is regulated by the activity of the carbonic anhydrase (CA, EC 4.2.1.1) that catalyses the reversible conversion of CO_2 and H_2O into H^+ and HCO_3^-.[21]

Acetazolamide and methazolamide are used either in combination therapy with other AED or in refractory epilepsy (treatment of partial, myoclonic, absence, and generalized tonic-clonic seizures uncontrolled by classical AED). Unfortunately, they do not provide effective long-term therapy and lead to a rapid tolerance.[22] Sulthiame (Fig. 7.3) is a clinical AED that has been used since 1964 and was found to exert remarkable anticonvulsant effects by CA inhibition.[23–28] It is largely used for symptomatic focal epilepsies in children or for benign epilepsy with centrotemporal spikes in adult patients.

The sulfamoyl group of zonisamide (Fig. 7.1), which belongs to the second generation of AED, was expected to suppress seizures in a similar way to acetazolamide through the inhibition of CA.[29] However, this does not appear to be the primary mechanism of action since zonisamide requires much higher doses than acetazolamide to achieve equivalent titration *in vivo*. Recently, it was described that topiramate (Fig. 7.1) was able to inhibit CA, contributing by this way to its anticonvulsant activity in addition to the other mechanisms of action described previously.[30–34]

The recent discovery of several carbonic anhydrase isozymes specifically expressed in the brain[13,35] stimulated the development of CA inhibitors with a high isozyme specificity. In fact, such drugs can be used as pharmacological tools to better understand the seizures involvement of CA. Furthermore, the design of novel specific CA inhibitor with fewer side effects than that of acetazolamide or methazolamide is of a great interest.

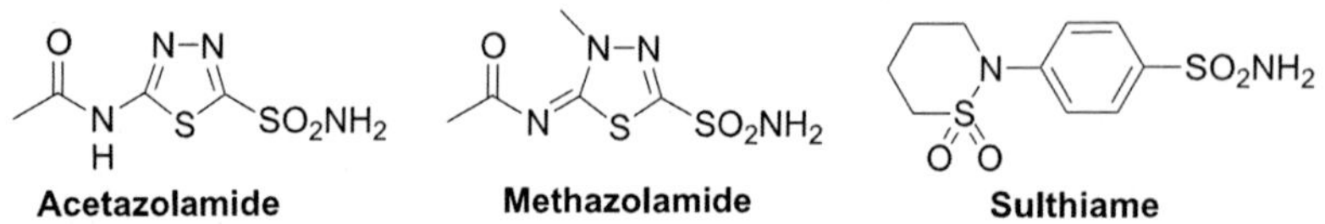

FIGURE 7.3 Structure of CA inhibitors clinically used as anticonvulsant agents.

7.3 EXPRESSION AND LOCALIZATION OF CAs IN CENTRAL NERVOUS SYSTEM

Several type of cells located in the central nervous system are found to express carbonic anhydrase (Fig. 7.4). CA II is expressed in choroid plexus, oligodendrocytes, myelinated tracts, and astrocytes in rodent brain.[13] CA III is present in rat and human choroid plexus. CA IV is located on the luminal membrane of cerebral capillaries and associated with the blood–brain barrier in rat and mouse.[36] It is also concentrated in layers III and VI in the cortex, hippocampus, and thalamus of rats. Before the discovery of a second CA V isoform,[37] isozyme V was found to be expressed in rodent nervous tissues, especially in astrocytes and neurons.[38] At present, two CA V were identified: CA VA and VB. Studies from mouse tissues showed that CA VA is limited to liver, skeletal muscle, and kidney, while CA VB is readily detected in most tissues. The brain expression of CA VII and CARP VIII is similar. Indeed, a relatively high expression is observed throughout the cortex, hippocampus, and thalamus.[35,39] Nevertheless, the CA VII expression is highly increased intrapyramidal at around postnatal day 12. This developmental expression promotes excitatory response evoked by intense GABAergic

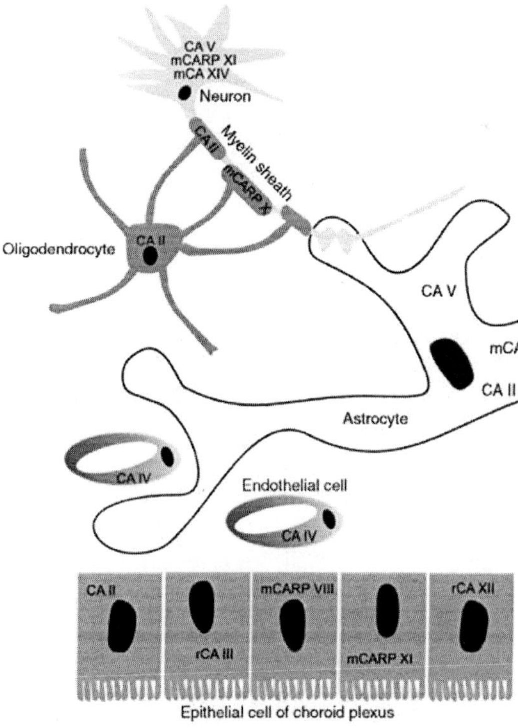

FIGURE 7.4 Schematic localization of CA isozymes in different elements of the CNS (mCA: isozymes expressed in mouse; rCA: isozymes expressed in rat; the others are expressed in both rat and mouse).[13,35,36,38,43–46]

activity.[39,40] The others CARPs (CARP X and XI) are constitutively expressed in CNS. Isozyme IX is situated in choroid plexus, but it is overexpressed in glioma, ependymoma, and other cerebral tumors. As CA IX, CA XII is also associated with cancer and is characterized by a expression pattern similar to that of CA IX.[41] Nevertheless, high levels of CA XII mRNA were seen in the dentate granule cells, media amygdala, choroid plexus, and hippocampus.[35] CA XIV is located on the plasma membrane of some neurons and on axons in the brain of both mouse and human. CA XV was only detected in the mice brain.[42]

Finally, Halmi and colleagues recently studied the expression of CA II, IV, VII, and XII in rat brain after kainic acid-induced status epilepticus.[35] In these conditions, the expression level of CA IV and VII was stable and not affected by kainic acid. In contrast, the expression of CA II and XII was induced by kainate.

7.4 INSIGHT INTO THE HYPOTHETICAL ROLES OF CA IN GENERATING CONVULSIONS

CAs catalyze the carbon dioxide hydration into bicarbonate anion and proton and are therefore characterized by various functions in the central nervous system (i.e., fluid and ion homeostasis,[21,34,47] formation of cerebrospinal fluid, regulation of its pH and ionic constituents, seizure activity,[48] regulation of GABAergic signaling,[39,49,50] respiratory response to carbon dioxide, generating bicarbonate for biosynthetic reactions,[51] and proliferation and differentiation[46,52]). The mechanism by which CAs can generate seizures is not completely understood. Nevertheless, several links between CAs and convulsions are discussed in the literature.[13]

1. CAs are responsible for the excitatory activity of glutamate receptors (NMDA type) by regulating the concentration of protons. Low extracellular concentration in protons is found to be related to high excitatory activity. In contrast, excessive amount of protons suppress the excitatory activity of the NMDA receptors and give a possible explanation for the anticonvulsant effects of CA deficiency or inhibition.[53]
2. CAs are involved in the regulation of GABAergic transmission by providing bicarbonate anion. The GABAergic transmission is mainly associated with inhibition/hyperpolarization through the entry of chloride ions in neurons. It provides a "stabilizing" influence on neuronal activity, and a decrease in this type of neurotransmission is often associated with epileptiform activity.[54] Nevertheless, the GABAergic transmission can also be exciting (depolarization) in certain cases, playing a crucial role in neuronal plasticity (responsible for memory and learning).[55] In this case, excessive depolarization is related to epileptiform activity. The ionic mechanism of this type of GABAergic excitation depends on the presence of bicarbonate anion, which is the only physiological ion in addition to chloride that is able to mediate a current through channels coupled to $GABA_A$ receptors. One CA isozyme, namely, CA VII in particular, has been pointed out for its involvement in this mechanism of GABAergic excitation.[13] Ethoxzolamide, a membrane-permeant CA, prevent these GABAergic excitations, while a CA membrane-impermeant

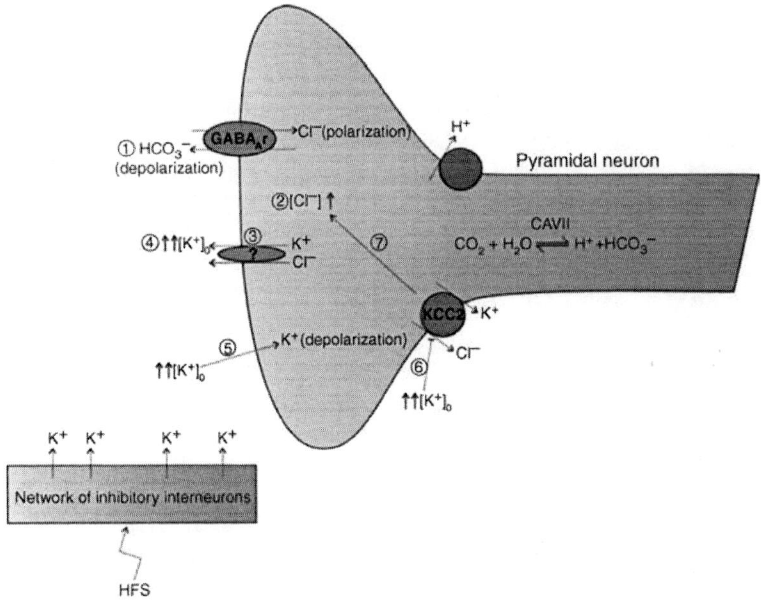

FIGURE 7.5 Mechanisms of $GABA_A$ receptor-dependent postsynaptic responses evoked in a pyramidal neuron. GABA binds to the $GABA_A$ receptor. It induces a massive chloride uptake and hyperpolarizes the neuron. The resting condition is reached by the action of the KCC2 transporter that can expel both K^+ and Cl^- to stabilize the intracellular concentration of chloride. A low intracellular concentration in chloride is necessary to allow a new hyperpolarization of the neuron. During a high-frequency stimulation (HFS), the neuronal depolarization is mediated by an efflux of bicarbonate through the $GABA_A$ receptor (1). In response to this depolarization, an accumulation of intraneuronal chloride is observed (2). Subsequently, the simultaneous efflux of Cl^- and K^+ occurs (through another cotransporter than the KCC2) in order to reach the initial chloride concentration (3). The extracellular K^+ concentration ($[K^+]_o$) is considerably increased (4) and can influence the membrane potential of pyramidal neurons (depolarization is mediated by plasmalemmal K^+ conductance) (5). The high extracellular concentration of K^+ also inhibits the KCC2 cotransporter leading to an accumulation of Cl^- inside the neuron and to the extruding of K^+ (3), maintaining therefore the neuronal excitability.[13,50,56,57]

CA inhibitor has no effect.[39] These results clearly show the implication of the cytosolic CA VII in the neuronal excitation as shown in Fig. 7.5.

3. Intracellular and extracellular pH changes have fundamental importance in regulating neuronal excitability.[13] Small changes in intracellular and extracellular pH alter the function of ligand-gated and voltage-gated channels. The pH regulation is principally done by CA IV and XIV.[42]
4. CA V present in astrocytes provides bicarbonate anion required for the normal functioning of pyruvate carboxylase, which is also expressed in these cells.[38] This isozyme can be related to seizures by the fact that its absence or its dysfunction might lead to a failure in pyruvate carboxylase function. Patients with pyruvate carboxylase deficiency develop severe neurological disorders including convulsions.[38]

In conclusion, the relation between CA and the physiopathological mechanisms of seizures remains unclear. Seizures cannot be attributed to one CA isozyme in particular. It seems that they are generated by complicated electrophysiological mechanisms including change in pH, neuronal transmission (GABAergic excitation), and action on excitatory receptors (NMDA) related to several CA isozymes (CA II, IV, V, VII, XII, and XIV).[13]

7.5 DEVELOPMENT OF NOVEL ANTICONVULSANT CA INHIBITORS

In humans and other mammals, sulfonamides (i.e., acetazolamide (**AZA**), methazolamide (**MZA**), zonisamide (**ZNS**), and sulthiame (**SUL**)) and sulfamate (i.e., topiramate (**TPM**)) inhibit CAs and act as anticonvulsant. Table 7.1 shows their CA inhibitory properties against CA isozymes that seem to be involved in convulsions. The analysis of the K_I against CA II shows that **AZA**, **MZA**, **TPM,** and **SUL** have a similar inhibitory potency, while **ZNS** is less active. **AZA** and **SUL** are the most active ones against CA IV. Each drug, except **ZNS**, exhibits a similar potency against CA VA and VB. **AZA**, **MZA**, **TPM,** and **SUL** show very potent hCA VII inhibitory properties with low nanomolar K_I's for this isozyme. **AZA**, **MZA**, and **TPM** provide a potent inhibition against CA XII. The best inhibitory potency against CA XIV is achieved by **AZA** and **MZA**.

In the course of the design of new carbonic anhydrase inhibitors as anticonvulsant drugs, acetazolamide, methazolamide, and topiramate were frequently used as lead compounds.

7.5.1 Acetazolamide and Methazolamide Analogues

Chufan and colleagues demonstrated that the lipophilic *tert*-butoxycarbonyl derivative of **AZA** (**7.1**) had a better anticonvulsant properties than the lead molecule **AZA** itself.[49] As lipophilicity is an important parameter to design drugs with a central action, lipophilic analogues of **AZA** were thus designed and synthesized.[62]

TABLE 7.1 Inhibitory Potency of Clinically Used Anticonvulsant Sulfonamides and Sulfamate Against the Human CA Isozymes with Hypothetical Convulsant Role

	K_I (nM)^a				
	AZA	MZA	ZNS	TPM	SUL
hCA II	12	14	35	10	7
hCA IV	74	6200	8590	4900	95
hCA VA	63	65	20.6	63	81
hCA VB	54	62	6033	30	91
hCA VII	2.5	2.1	117	0.9	6
hCA XII	5.7	3.4	11000	3.8	56
hCA XIV	41	43	5250	1460	1540

^{a} Stopped-flow method (CO_2 hydrase activity of the enzyme)[58]; Data are from Refs [27,59–61].

The CA inhibitory potency of these analogues was excellent and quite similar against the hCA II with K_I in the range of 1.2–10 nM, while some of them were less active against hCA I (K_I of 3–900 nM).[62] They possess different alkyl/arylcarboxamido/sulfonamido/ureido moieties in the position 5 of the thiadiazole ring and are characterized by a wide range of lipophilicity (compounds **7.2–7.11**; Clog *P* in the range 1.8 to −0.33). Those compounds were investigated for their anticonvulsant properties to establish structure–activity relationship that will help in the design of new anticonvulsant compounds. It was observed that the valproyl derivative of acetazolamide (5-valproylamido-1,3,4-thiadiazole-2-sulfonamide; **7.2**) and some structurally related derivatives such as 5-benzoylamido- (**7.4**), 5-toluenesulfonylamido- (**7.5**), 5-pivaloylamido- (**7.7**), and 5-adamantylcarboxamido-1,3,4-thiadiazole-2-sulfonamides (**7.9**) exhibited very strong anticonvulsant properties in the maximal electroshock seizure (MES) test in mice at 30 mg/kg.[62] The protection rate was 75–100% at 30 min and 25–100% at 3 h after intraperitoneal drug administration. All these compounds were quite lipophilic possessing Clog *P* in the range of 0.10–1.82. Other investigated sulfonamides (compounds **7.6**, **7.8**, **7.10**, and **7.11**) showed less effective protection (25–67% after 30 min and 3 h injection). These derivatives were generally less lipophilic than the first subseries except for compound **7.5**. The correlation of the lipophilicity (Clog *P*), the *in vitro* CA inhibition, and their *in vivo* anticonvulsant activity is not straightforward.[62] It seems that a strong CA inhibitor also possessing a good lipophilicity should intrinsically lead to powerful anticonvulsant properties. Nevertheless, parameters other than CA inhibition and lipophilicity (i.e., plasma protein binding, pK_a) may influence the *in vivo* anticonvulsant properties.[62]

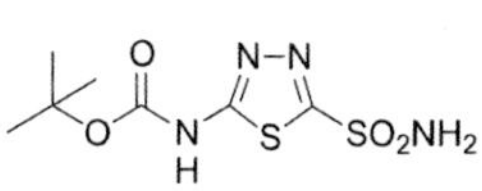

7.1

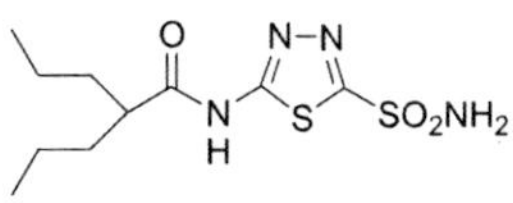

7.2

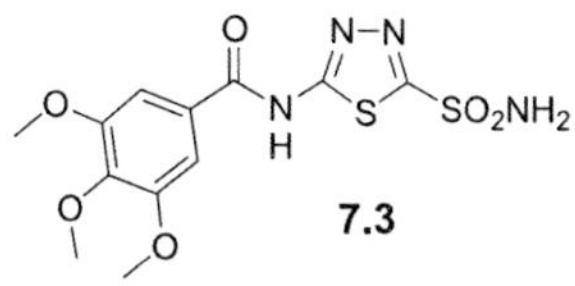

7.3

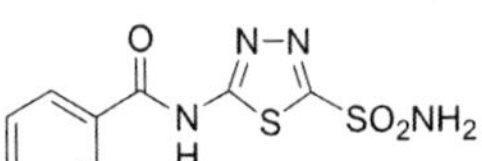

7.4

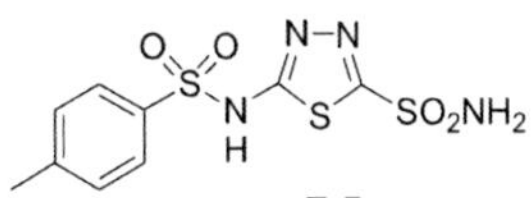

7.5

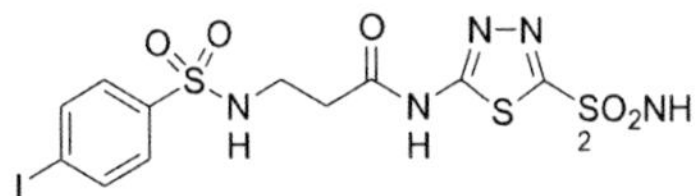

7.6

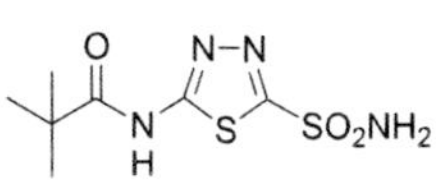

7.7

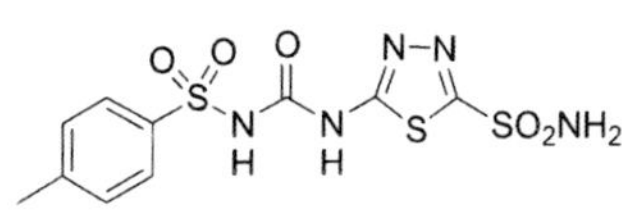

7.8

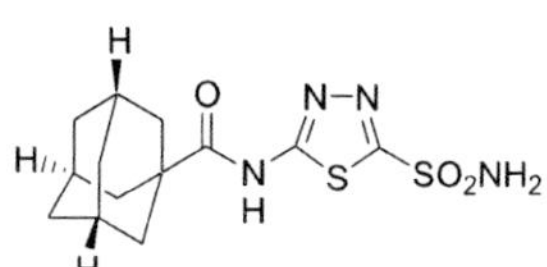

7.9

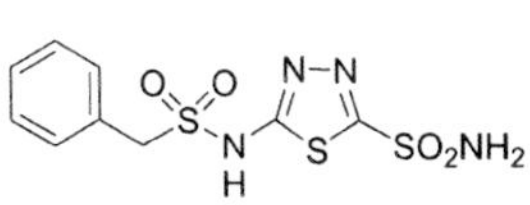

7.10

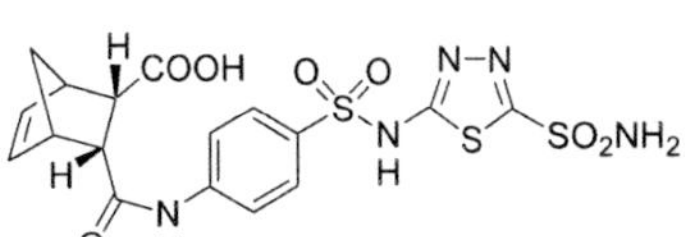

7.11

On the basis of the excellent result of 5-adamantylcarboxamido-1,3,4-thiadiazole-2-sulfonamide (**7.9**), which protect 92% of the mice in the MES test, two other adamantyl-containing moieties (5-adamantylureido- (**7.12**) and 5-adamantylmethyl-carboxamido-1,3,4-thiadiazole-2-sulfonamides (**7.13**)) were thus incorporated into the thiadiazole ring.[32] The evaluation of the CA inhibitory activity leads to the same range of K_I. Nevertheless, the anticonvulsant effect of these two new compounds was lower: 5-adamantylmethylcarboxamido-1,3,4-thiadiazole-2-sulfonamide (**7.13**) protected 75% of the mice, whereas it was 47% for the 5-adamantylureido-1,3,4-thiadiazole-2-sulfonamide (**7.12**).[32] The compound 5-adamantylcarboxamido-1,3,4-thiadiazole-2-sulfonamide (**7.9**) seems to be optimal for the protection of mice against induced electric convulsions. Dose–response curves were thus performed with this last compound and the ED_{50} was found to be 2.6 mg/kg.

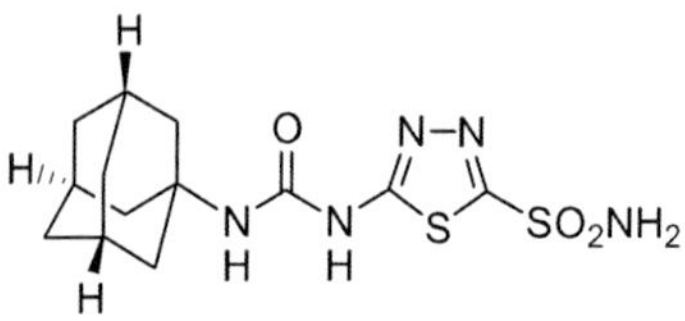

7.12

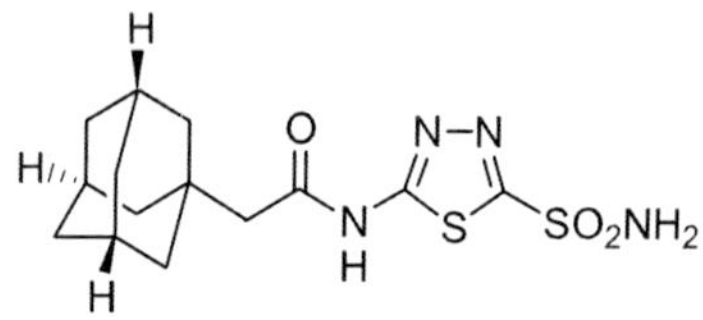

7.13

AZA analogues are thus found to inhibit CA and still retain their anticonvulsant properties. The adamantyl side chain was used as lead in the development of a new series of CA inhibitors as anticonvulsant drug. Other compounds should be further designed and developed on the basis of the leads **7.4**, **7.5**, and **7.7**.[62]

In comparison to the numerous **AZA** analogues that were developed, only a few methazolamide analogues were designed and synthesized. Four side chains (valproyl, adamantylcarboxamido, adamantylureido, or adamantylmethylcarboxamido) were incorporated into the **MZA** scaffold (**7.14–7.17**).[32,62] Their inhibitory potency against hCA II is excellent (K_I ranging from 7 to 14 nM) but is weaker against hCA I (K_I ranging from 70 to 800 nM). Anticonvulsant effect was performed with the adamantyl derivatives through the MES test: compound **7.15** protected 96% of the mice, compound **7.16** protected 73%, and compound **7.17** was less efficient with 42% of protected mice.[32] The dose–response curve performed with **7.15** led to an ED_{50} of 3.5 mg/kg.

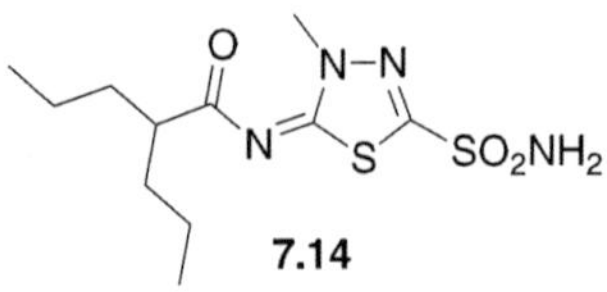

7.14

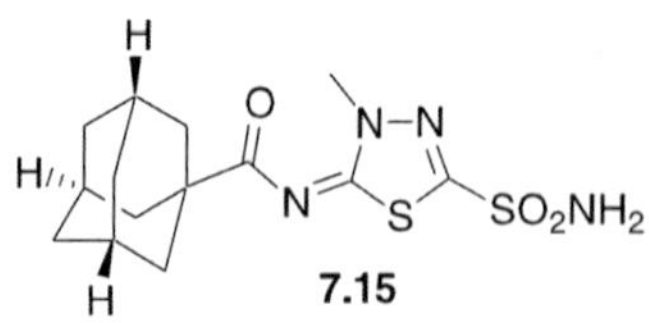

7.15

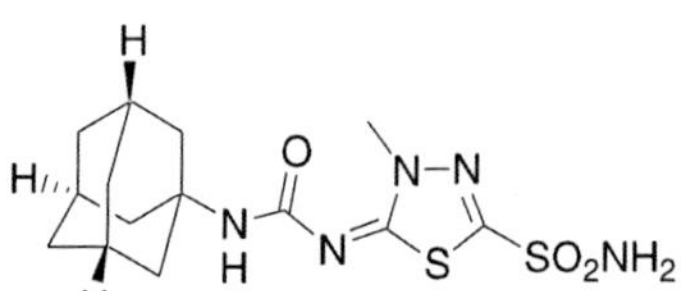

7.16

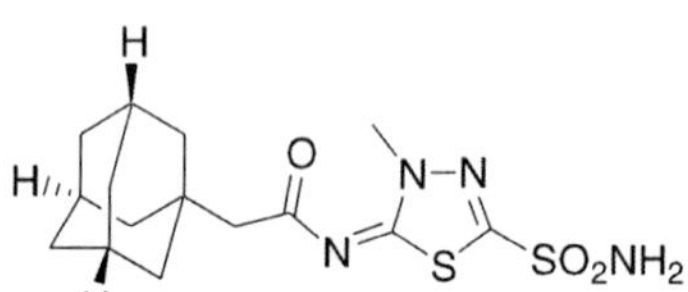

7.17

7.5.2 Aromatic/Heterocyclic Sulfonamides Incorporating Valproyl or Adamantyl Moieties

A large series of aromatic and heterocyclic sulfonamides were first designed as CA inhibitors with a topical antiglaucoma activity.[63–67] In recent work, the lipophilicity of this series was increased to design new anticonvulsant drug.[62] Valproic acid is a lipophilic drug widely used as AED. So, the valproyl moiety was selected and coupled to the aromatic or heterocyclic sulfonamide scaffolds previously described. The CA inhibitory activity (data available for hCA I and II) of these new derivatives is generally more potent when compared to their parent sulfonamide from which they were prepared. The enhanced inhibitory activity is due to the interaction of the valproyl side chain with the hydrophobic pocket of the CA active site. Among the synthesized derivatives, the heterocyclic sulfonamides were more active than the aromatic ones.[62] Unfortunately, no anticonvulsant properties are available for this series of compounds.

As the 5-adamantylcarboxamido-1,3,4-thiadiazole-2-sulfonamide (described as an **AZA** analogue) showed a promising *in vivo* anticonvulsant property,[62] the adamantyl side chain was incorporated into aromatic scaffolds to investigate the CA inhibitory activity and the anticonvulsant effect.[32] All those derivatives are also systematically more effective for the CA inhibition as compared to their parent sulfonamide from which they were prepared. Some of these molecules exhibited a weak inhibitory potency against the hCA I (K_I values ranging from 1 to 18.50 μM) and a good inhibitory potency against the hCA II (K_I values ranging from 28 to 265 nM).[32] Compounds characterized by a high CA II inhibitory potency and a lipophilicity propitious to brain penetration were evaluated for their anticonvulsant properties in the MES test at 30 mg/kg 3 h after intraperitoneal administration (compounds **7.18–7.21**).[32] Compounds **7.18b** and **7.21b** can be regarded as poor anticonvulsant drugs, as well as the aminomethylbenzensulfonamides (**7.18a–7.21a**). The molecules **7.19b** and **7.20b** protected mice against convulsions (67 and 75%, respectively).

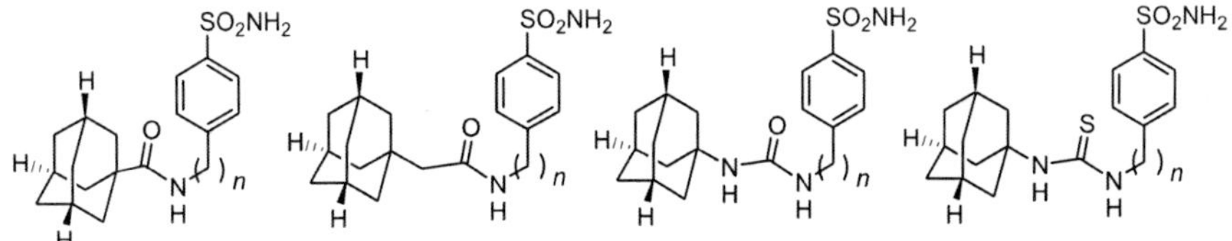

7.18a ($n = 1$) **7.18b** ($n = 2$); **7.19a** ($n = 1$) **7.19b** ($n = 2$); **7.20a** ($n = 1$) **7.20b** ($n = 2$); **7.21a** ($n = 1$) **7.21b** ($n = 2$)

7.5.3 Topiramate and Analogues

Topiramate is a sulfamate fructopyranose derivative currently used for treating partial onset. This "second-generation" AED has been shown to be clinically effective against simple and partial seizures and also against generalized tonic-clonic seizures.

Topiramate is characterized by several mechanisms of action including its ability (i) to block the voltage Na^+ channel and the kainite/AMPA receptor, (ii) to potentiate GABAergic transmission, and (iii) to inhibit CA. Nevertheless, the effect of **TPM** on the GABAergic transmission is supposed to be directly related to a decreased in the intracellular bicarbonate concentration, which may be caused by an inhibition of neuronal CA. **TPM** is therefore found to inhibit the cytosolic CA II and VII present in brain (Table 7.2).

Considering the interesting CA inhibitory properties of topiramate, Winum and colleagues designed and synthesized a series of derivatives incorporating the protected fructopyranose moiety of topiramate.[68] This tail was attached on aromatic or heterocyclic sulfonamide scaffold by a thiourea link. The new derivatives (compounds **7.22–7.28**) were active against hCA II (K_I ranging from 6 to 750 nM) and hCA VII (K_I ranging from 10 to 79 nM). Compounds **7.26** and **7.28** showed anticonvulsant properties in the MES test performed in mice. Intraperitoneal injected 2 h prior to induce convulsions, the fluorosulfanilamide derivative (**7.26**, 50 mg/kg) showed a more efficient anticonvulsant activity than topiramate (87.5and 69.7%, respectively). Compound **7.26** is also characterized by the best inhibitory potency against the hCA VII ($K_I = 10$ nM) but lower than topiramate ($K_I = 0.9$ nM) and acetazolamide ($K_I = 2.5$ nM). It is also an excellent CA II inhibitor ($K_I = 6$ nM).

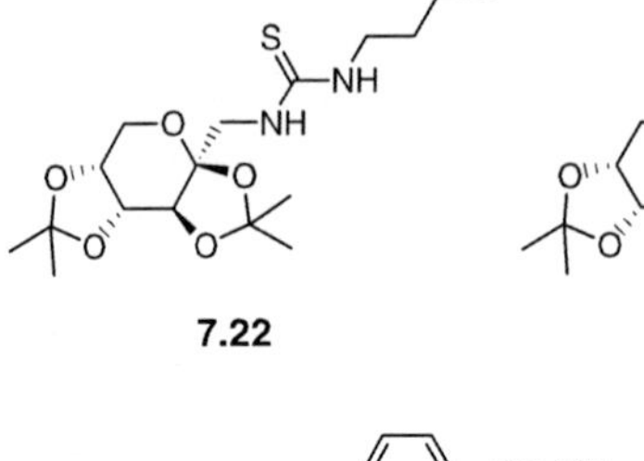

7.22

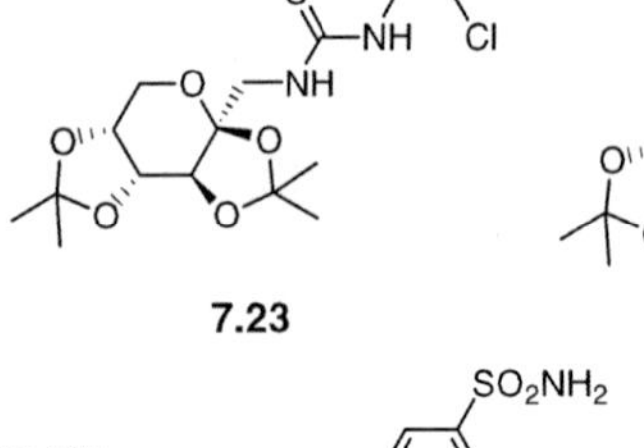

7.23

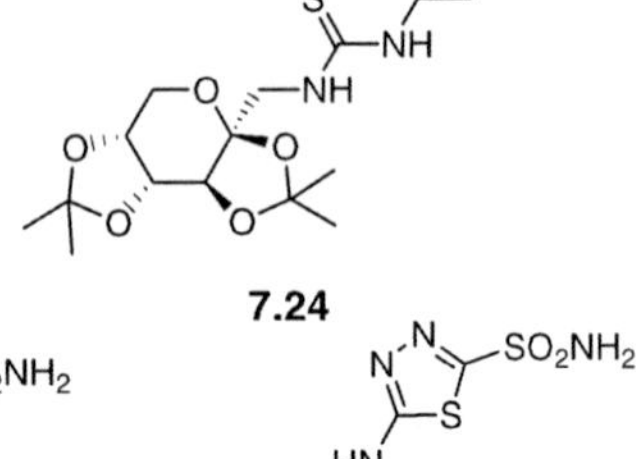

7.24

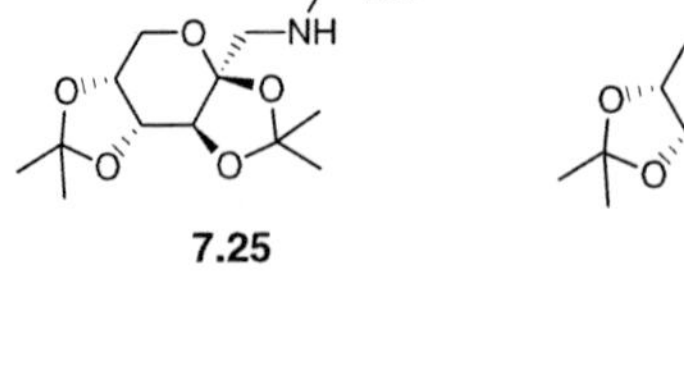

7.25

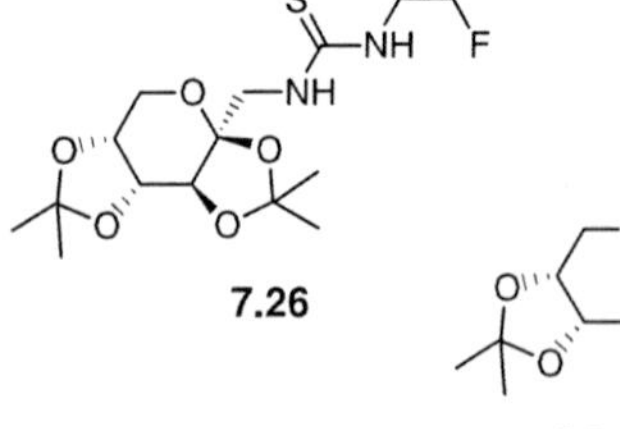

7.26

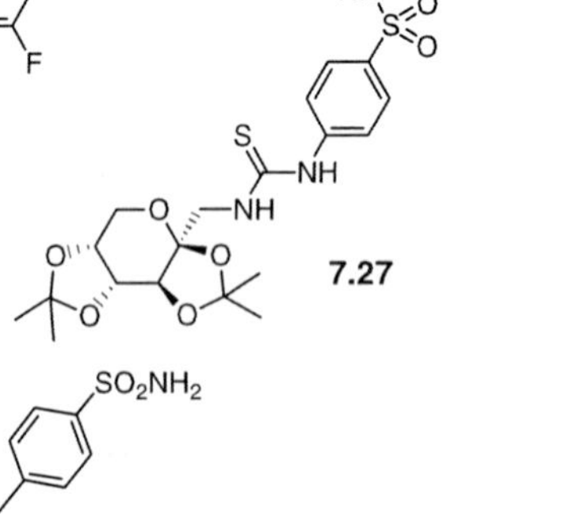

7.27

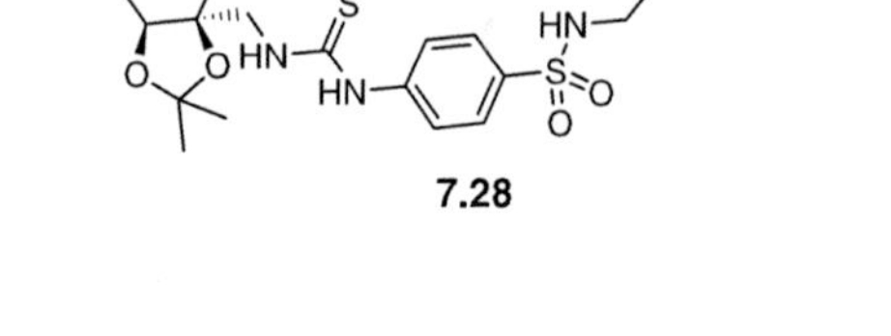

7.28

TABLE 7.2 Summary of the Inhibition Activities of Some Promising Indanesulfonamides

	K_I (nM)				MES test[c]
Compound	hCA II[a]	hCA VII[b]	hCA XII[b]	hCA XIV[b]	Percentage of Protected Mice
AZA	12	2.5	5.7	41	12.5
29	186	10	5.4	675	50
30	0.9	9.3	28	658	62.5
31	17	2.7	0.36	5.6	62.5
32	46	4.5	5.9	5.0	37.5

AZA, acetazolamide.
[a]Data are from Ref. 70.
[b]Data are from Ref. 69.
[c]The maximal electroshock seizure test was carried out 2 h after i.p. 50 mg/kg administration; $n = 8$ mice.

7.5.4 Indanesulfonamides

A library of indanesulfonamides was screened for the inhibition of the human carbonic anhydrase isoforms involved in neuronal excitation, that is, isoforms VII, XII, and XIV.[69] The inhibition pattern of these compounds was excellent with many nanomolar inhibitors detected (K_I's in the range of 0.78–10 nM against hCA VII, 0.32–56 nM against hCA XII, and 0.47–1030 nM against hCA XIV, respectively). The anticonvulsant activity of indanesulfonamides was examined by the maximal electroshock seizure test. At a dose of 50 mg/kg, three compounds **7.29**, **7.30**, and **7.31** protected at least 50% of the mice against the convulsions (Table 7.2). One compound **7.32** shows a moderate anticonvulsant activity (protection of 37.5%). Other derivatives were poor anticonvulsants but might be compared to acetazolamide (**AZA**), which was not able to protect mice against convulsions at a body weight dose of 50 mg/kg. These experiments showed that compounds that are powerful CA inhibitors do not constantly allow protection against convulsions through MES test.

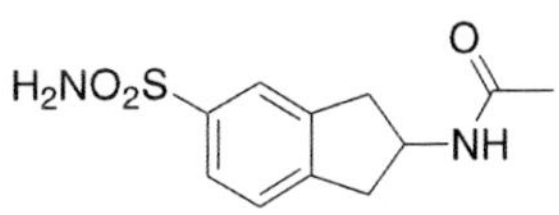

7.29

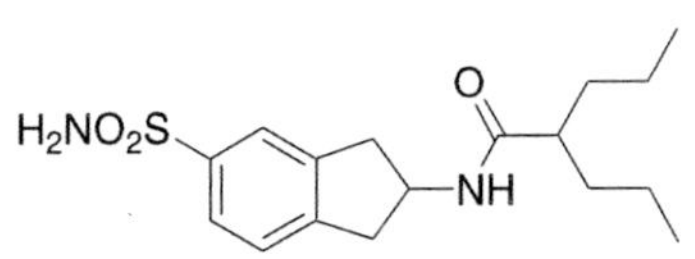

7.30

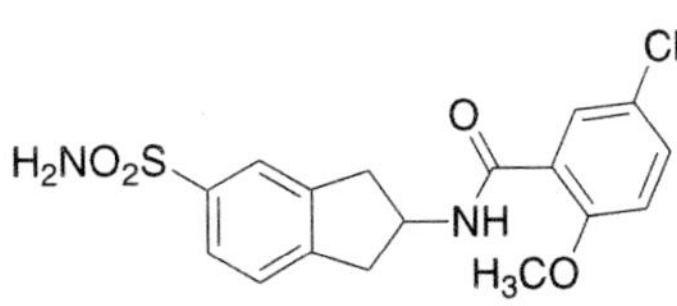

7.31

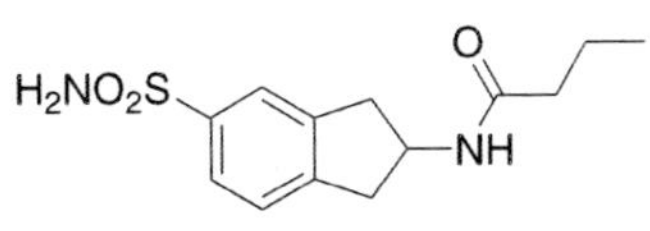

7.32

7.5.5 Toward Selective CA VII Inhibitors

The cytosolic CA VII is highly expressed in brain and contributes to the neuronal excitability by providing bicarbonate anion that can mediate current through channels coupled to $GABA_A$ receptor. This isozyme is thus a promising target in epilepsy seizures. Recently, hCA VII was cloned and several aromatic and heterocyclic sulfonamides were thus investigated for their CA VII inhibition.[61] The inhibitory potency was excellent with K_I values ranging from 4.3 to 210 nM. The study also showed that the inhibition profile of hCA VII is rather different from that of other cytosolic isozymes and thus provided a possibility for the design of more selective hCA VII inhibitors.[61] The design of more selective CA inhibitors is needed because such compounds may avoid side effects due to the inhibition of physiologically relevant CA isozymes. These novel selective inhibitors may be used as pharmacological tools to better understand the exact role of CA in epileptogenesis. At present, no selective CA VII inhibitor is described, but a molecular modeling study was conducted by our group that provided useful insight into the future design of selective CA VII inhibitors.[71]

As no three-dimensional structure is currently available, a model was built for hCA VII by homology with the crystal structure of hCA II.[71] The hCA VII active site was then characterized and compared to the active site of five other CA isozymes (I, II, IX, XII, and XIV). Sequence alignments were performed to identify the preserved and specific residues of hCA VII. Several changed residues were highlighted that are suspected to be important for the binding of hCA VII inhibitors. In this respect, Ser65 (located at the hydrophilic pocket), Lys91 and Thr130 (both located at the interface between hydrophilic and hydrophobic pockets), and Ser204 (located at the hydrophobic part of the active site) were found to be present only in hCA VII (Fig. 7.6) and can be specifically targeted to design more selective hCA VII inhibitors.[71]

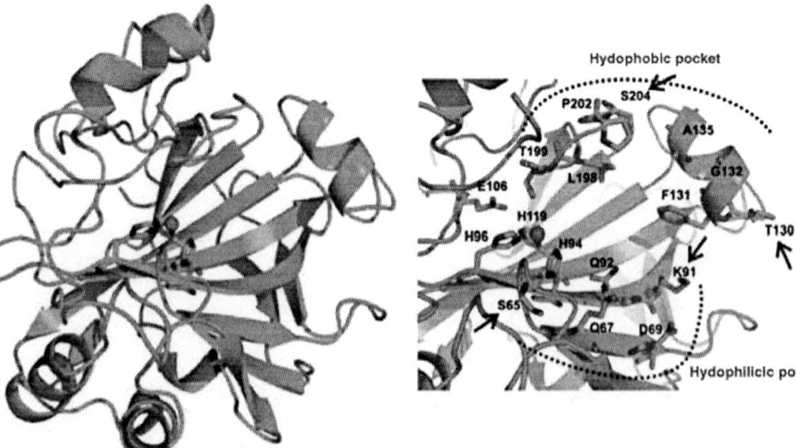

FIGURE 7.6 Model of CA VII (left) and zoom into the active site (right), where the specific amino acids residues Ser65, Lys91, Thr130, and Ser204 are indicated by black arrows.

Interesting findings have been discussed in the study of Winum to decrease the affinity of inhibitor for the ubiquitous cytosolic CA II isozyme.[72] The sulfamide analogue of the topiramate is 210 times less potent on CA II when compared to topiramate itself. To explain this feature, X-ray crystal structure of hCA II in adduct with the sulfamide analogue was solved.[72] This weak inhibition is due to a clash between a methyl group of the inhibitor and Ala65 present in CA II. This amino acid residue is only found in CA II, while it is changed by Ser in other CA isozymes.[21,71] The unfavorable interaction between a methyl and Ala65 should be thus exploited for the further design of more selective CA inhibitors.[72]

7.6 PERSPECTIVES IN DRUG DESIGN OF CA INHIBITORS AS ANTICONVULSANT

A large number of compounds that are presently marketed as AEDs exhibit a similar pharmacological profile, being active in the MES test and inactive in the subcutaneous pentylenetetrazol (PTZ) test, the other widely used seizure model.[73] This is the case for phenytoin, topiramate, zonisamide, and carbamazepine. Drugs, with the phenytoin-like profile of activity, have several mechanisms of action and some of them are characterized by a similar pharmacological action, which is the inhibition of the Na^+ channels.[73] Nevertheless, the knowledge of the mechanism of action of this kind of ligand is not straightforward.[74] To highlight the molecular requirements that are needed to provide an anticonvulsant effect, several pharmacophoric models were developed in the literature.[73–76] One of the proposed models is interesting for the design of new CA inhibitors with anticonvulsant property since acetazolamide was found (i) to exhibit a phenytoin-like profile and (ii) to superimpose the feature of this pharmacophore.[74] It is characterized by a hydrophobic chain coupled to a polar moiety placed in a well-defined conformation.[74]

7.7 CONCLUSIONS

Epilepsy is one of the most frequent neurological diseases and is characterized by recurrent seizures. Seizures are episodes of abnormal electrical activity in the brain and can result from other conditions (i.e., toxin, fever, trauma, . . .) than epilepsy disease. The mechanisms by which seizures occurred are not clearly established and the current available AEDs cannot solve entirely the problem of epilepsy treatment, particularly for patients suffering from refractory epilepsy. In this respect, new generation of AED should be thus designed.

Carbonic anhydrase inhibitors have been used to treat epilepsy since 1953, and despite their considerable side effects, they are still used clinically to treat refractory epilepsies. The recent discovery of several CAs specifically expressed in the brain has led to the emergence of new CA inhibitors as anticonvulsant agent. For this purpose, in addition to the analogues of acetazolamide and topiramate, several aromatic or heterocyclic sulfonamides incorporating valproyl or adamantyl moieties, indanesulfonamides were developed and evaluated through the MES test for their anticonvulsant effect. Several compounds were found to be interesting. However, further *in vivo* experiments (time course activity, other epileptic models, and therapeutic index by rotarod test) are warranted to confirm their interest in the treatment of epilepsy.

Underscoring the link between CA and seizures is complicated due to the widespread expression of CA isozymes in the brain and their multiple physiological roles in the CNS. CA II, IV, V, VII, XII, and XIV were pointed out for their contribution to generate neuronal excitation. The design and development of more selective CA inhibitors shall improve our knowledge of the functional relevance of CA isozyme in seizure. Nevertheless, as several molecular factors are involved in the etiology of seizures, new antiepileptic drug should not only target CA but also act on different pharmacological pathways.

ACKNOWLEDGMENTS

A.T. is very indebted to the Belgian F.R.S.-FNRS for the award of a research fellowship.

REFERENCES

1. Lothman, E. W.; Collins, R. C. Seizures. *Neurological Pathophysiology*, 3d ed.; Oxford University Press: New York, **1984**; 229–249.
2. Porter, R. J.; Meldrum, B. S. Antiépileptiques. *Pharmacologie fondamentale;* Padoue: **2006**; 379–400.
3. Thomas, P.; Arzimanoglou, A. *Epilepsies*, 3 ed.; Masson: Paris, **2003**.
4. Voilley, N. L'hyperexcitabilité neuronale. *Neurosciences*. **2002**, *5*, 174–188.
5. Commission on Classification and Terminology of the International League Against Epilepsy. Proposal for revised clinical and electroencephalographic classification of epileptic seizures. *Epilepsia* **1981**, *22*, 489–501.
6. Holmes, G. L. Classification of seizures and the epilepsies. *The Comprehensive Evaluation and Treatment of Epilepsy: A Practical Guide;* Academic Press, **1997**; pp 1–36.
7. McNamara, J. O. Drugs effective in the therapy of the epilepsies. *Goodman and Gilman's: The Pharmacological Basis of Therapeutics*, 9th ed; McGraw-Hill, **1995**; pp 461–486.
8. Freeman, J. M.; Kossoff, E. H.; Hartman, A. L. The ketogenic diet: one decade later. *Pediatrics* **2007**, *119*, 535–543.
9. Gamble, J. L.; Ross, G. S.; Tisdall, F. F. The metabolism of fixed base during fasting. *J. Biol. Chem.* **1923**, *57*, 633–695.
10. Brenner, G. M.; Stevens, C. W. Antiepileptic drugs. *Pharmacology*, 2nd ed.; W. B. Saunders Company, Philadelphia, **2006**; pp 209–220.
11. Kwan, P.; Brodie, M. J. Early identification of refractory epilepsy. *N. Engl. J. Med.* **2000**, *342*, 314–319.

12. Stefan, H.; Steinhoff, B. J. Emerging drugs for epilepsy and other treatment options. *Eur. J. Neurol.* **2007**, *14*, 1154–1161.

13. Thiry, A.; Dogné, J.-M.; Supuran, C. T.; Masereel, B. Carbonic anhydrase inhibitors as anticonvulsant agents. *Curr. Top. Med. Chem.* **2007**, *7*, 855–864.

14. Thiry, A.; Dogné, J.-M.; Supuran, C. T.; Masereel, B. Anticonvulsant sulfonamides/sulfamates/sulfamides with carbonic anhydrase inhibitory activity: drug design and mechanism of action. *Curr. Pharm. Des.* **2008**, *14*, 661–671.

15. Reiss, W. G.; Oles, K. S. Acetazolamide in the treatment of seizures. *Ann. Pharmacother.* **1996**, *30*, 514–519.

16. Aram, J. A.; Lodge, D. Epileptiform activity induced by alkalosis in rat neocortical slices: block by antagonists of *N*-methyl-D-Aspartate. *Neurosci. Lett.* **1987**, *83*, 345–350.

17. Aribi, A. M.; Stringer, J. L. Effects of antiepileptic drugs on extracellular pH regulation in the hippocampal CA1 region *in vivo*. *Epilepsy Res.* **2002**, *49*, 143–151.

18. Bonnet, U.; Wiemann, M.; Bingmann, D. CO2/HCO3(−)-withdrawal from the bath medium of hippocampal slices: biphasic effect on intracellular pH and bioelectric activity of CA3-neurons. *Brain Res.* **1998**, *796*, 161–170.

19. Velisek, L.; Dreier, J. P.; Stanton, P. K.; Heinemann, U.; Moshe, S. L. Lowering of extracellular pH suppresses low-Mg(2^+)-induces seizures in combined entorhinal cortex-hippocampal slices. *Exp. Brain Res.* **1994**, *101*, 44–52.

20. Xiong, Z. Q.; Saggau, P.; Stringer, J. L. Activity-dependent intracellular acidification correlates with the duration of seizure activity. *J. Neurosci.* **2000**, *20*, 1290–1296.

21. Supuran, C. T. *Carbonic Anhydrase. Its Inhibitors and Activators;* CRC Press: London, **2004**.

22. Katayama, F.; Miura, H.; Takanashi, S. Long-term effectiveness and side effects of acetazolamide as an adjunct to other anticonvulsants in the treatment of refractory epilepsies. *Brain Dev.* **2002**, *24*, 150–154.

23. Huppke, P.; Kohler, K.; Brockmann, K.; Stettner, G. M.; Gartner, J. Treatment of epilepsy in Rett syndrome. *Eur. J. Paediatr. Neurol.* **2007**, *11*, 10–16.

24. Koepp, M. J.; Patsalos, P. N.; Sander, J. W. Sulthiame in adults with refractory epilepsy and learning disability: an open trial. *Epilepsy Res.* **2002**, *50*, 277–282.

25. Leniger, T.; Wiemann, M.; Bingmann, D.; Widman, G.; Hufnagel, A.; Bonnet, U. Carbonic anhydrase inhibitor sulthiame reduces intracellular pH and epileptiform activity of hippocampal CA3 neurons. *Epilepsia* **2002**, *43*, 469–474.

26. Tanimukai, H.; Nishimura, T.; Sano, I. *N*-(4′Sulfamyphenyl)-butansultam(1–4) (Ospolot) as carbonic anhydrase inhibitor in the brain. *Klin. Wochenschr.* **1964**, *42*, 918–920.

27. Temperini, C.; Innocenti, A.; Mastrolorenzo, A.; Scozzafava, A.; Supuran, C. T. Carbonic anhydrase inhibitors. Interaction of the antiepileptic drug sulthiame with twelve mammalian isoforms: kinetic and X-ray crystallographic studies. *Bioorg. Med. Chem. Lett.* **2007**, *17*, 4866–4872.

28. Wirrell, E.; Ho, A. W.; Hamiwka, L. Sulthiame therapy for continuous spike and wave in slow-wave sleep. *Pediatr. Neurol.* **2006**, *35*, 204–208.

29. Leppik, I. E. Zonisamide: chemistry, mechanism of action, and pharmacokinetics. *Seizure* **2004**, *13* (Suppl. 1), S5–S9; discussion S10.

30. Dodgson, S. J.; Shank, R. P.; Maryanoff, B. E. Topiramate as an inhibitor of carbonic anhydrase isoenzymes. *Epilepsia* **2000**, *41* (Suppl. 1), S35–S39.

31. Herrero, A. I.; Del Olmo, N.; Gonzalez-Escalada, J. R.; Solis, J. M. Two new actions of topiramate: inhibition of depolarizing GABA(A)-mediated responses and activation of a potassium conductance. *Neuropharmacology* **2002**, *42*, 210–220.

32. Ilies, M. A.; Masereel, B.; Rolin, S.; Scozzafava, A.; Campeanu, G.; Cimpeanu, V.; Supuran, C. T. Carbonic anhydrase inhibitors: aromatic and heterocyclic sulfonamides incorporating adamantyl moieties with strong anticonvulsant activity. *Bioorg. Med. Chem.* **2004**, *12*, 2717–2726.

33. Shank, R. P.; Gardocki, J. F.; Streeter, A. J.; Maryanoff, B. E. An overview of the preclinical aspects of topiramate: pharmacology, pharmacokinetics, and mechanism of action. *Epilepsia* **2000**, *41* (Suppl. 1), S3–S9.

34. Supuran, C. T.; Scozzafava, A.; Casini, A. Carbonic anhydrase inhibitors. *Med. Res. Rev.* **2003**, *23*, 146–189.

35. Halmi, P.; Parkkila, S.; Honkaniemi, J. Expression of carbonic anhydrases II, IV, VII, VIII and XII in rat brain after kainic acid induced status epilepticus. *Neurochem. Int.* **2006**, *48*, 24–30.

36. Ghandour, M. S.; Langley, O. K.; Zhu, X. L.; Waheed, A.; Sly, W. S. Carbonic anhydrase IV on brain capillary endothelial cells: a marker associated with the blood–brain barrier. *Proc. Natl. Acad. Sci. USA* **1992**, *89*, 6823–6827.

37. Shah, G. N.; Hewett-Emmett, D.; Grubb, J. H.; Migas, M. C.; Fleming, R. E.; Waheed, A.; Sly, W. S. Mitochondrial carbonic anhydrase CA VB: differences in tissue distribution and pattern of evolution from those of CA VA suggest distinct physiological roles. *Proc. Natl. Acad. Sci. USA* **2000**, *97*, 1677–1682.

38. Ghandour, M. S.; Parkkila, A. K.; Parkkila, S.; Waheed, A.; Sly, W. S. Mitochondrial carbonic anhydrase in the nervous system: expression in neuronal and glial cells. *J. Neurochem.* **2000**, *75*, 2212–2220.

39. Ruusuvuori, E.; Li, H.; Huttu, K.; Palva, J. M.; Smirnov, S.; Rivera, C.; Kaila, K.; Voipio, J. Carbonic anhydrase isoform VII acts as a molecular switch in the development of synchronous gamma-frequency firing of hippocampal CA1 pyramidal cells. *J. Neurosci.* **2004**, *24*, 2699–2707.

40. Rivera, C.; Voipio, J.; Kaila, K. Two developmental switches in GABAergic signalling: the K^+-Cl^- cotransporter KCC2 and carbonic anhydrase CAVII. *J. Physiol.* **2005**, *562*, 27–36.

41. Ivanov, S.; Liao, S. Y.; Ivanova, A.; Danilkovitch-Miagkova, A.; Tarasova, N.; Weirich, G.; Merrill, M. J.; Proescholdt, M. A.; Oldfield, E. H.; Lee, J.; Zavada, J.; Waheed, A.; Sly, W.; Lerman, M. I.; Stanbridge, E. J. Expression of hypoxia-inducible cell-surface transmembrane carbonic anhydrases in human cancer. *Am. J. Pathol.* **2001**, *158*, 905–919.

42. Hilvo, M.; Tolvanen, M.; Clark, A.; Shen, B.; Shah, G. N.; Waheed, A.; Halmi, P.; Hanninen, M.; Hamalainen, J. M.; Vihinen, M.; Sly, W. S.; Parkkila, S. Characterization of CA XV, a new GPI-anchored form of carbonic anhydrase. *Biochem. J.* **2005**, *392*, 83–92.

43. Agnati, L. F.; Tinner, B.; Staines, W. A.; Vaananen, K.; Fuxe, K. On the cellular localization and distribution of carbonic anhydrase II immunoreactivity in the rat brain. *Brain Res.* **1995**, *676*, 10–24.

44. Nogradi, A.; Kelly, C.; Carter, N. D. Localization of acetazolamide-resistant carbonic anhydrase III in human and rat choroid plexus by immunocytochemistry and *in situ* hybridisation. *Neurosci. Lett.* **1993**, *151*, 162–165.

45. Parkkila, S.; Parkkila, A. K.; Rajaniemi, H.; Shah, G. N.; Grubb, J. H.; Waheed, A.; Sly, W. S. Expression of membrane-associated carbonic anhydrase XIVon neurons and axons in mouse and human brain. *Proc. Natl. Acad. Sci. USA* **2001**, *98*, 1918–1923.

46. Taniuchi, K.; Nishimori, I.; Takeuchi, T.; Ohtsuki, Y.; Onishi, S. cDNA cloning and developmental expression of murine carbonic anhydrase-related proteins VIII, X, and XI. *Mol. Brain Res.* **2002**, *109*, 207–215.

47. Maren, T. H. Carbonic anhydrase: chemistry, physiology, and inhibition. *Physiol. Rev.* **1967**, *47*, 595–781.

48. Anderson, R. E.; Engstrom, F. L.; Woodbury, D. M. Localization of carbonic anhydrase in the cerebrum and cerebellum of normal and audiogenic seizure mice. *Ann. N.Y. Acad. Sci.* **1984**, *429*, 502–504.

49. Chufan, E. E.; Pedregosa, J. C.; Baldini, O. N.; Bruno-Blanch, L. Anticonvulsant activity of analogues of acetazolamide. *Farmaco* **1999**, *54*, 838–841.

50. Rivera, C.; Voipio, J.; Kaila, K. Two developmental switches in GABAergic signalling: the K^+-Cl^- cotransporter KCC2 and carbonic anhydrase CAVII. *J. Physiol.* **2005**, *562*, 27–36.

51. Chegwidden, W. R.; Dodgson, S. J.; Spencer, I. M. The roles of carbonic anhydrase in metabolism, cell growth and cancer in animals. *The Carbonic Anhydrases: New Horizons;* Birkäuser Verlag: Basel, **2000**; 343–363.

52. Nishimori, I. Acatalytic CAs: carbonic anhydrase-related proteins. *Carbonic Anhydrase: Its Inhibitors and Activators;* CRC Press: Boca Raton, FL, **2004**; 25–43.

53. Tang, C. M.; Dichter, M.; Morad, M. Modulation of the *N*-methyl-D-aspartate channel by extracellular H^+. *Proc. Natl. Acad. Sci. USA* **1990**, *87*, 6445–6449.

54. Perez Velazquez, J. L. Bicarbonate-dependent depolarizing potentials in pyramidal cells and interneurons during epileptiform activity. *Eur. J. Neurosci.* **2003**, *18*, 1337–1342.

55. Marieb, E. N. *Anatomie et physiologie humaines;* De Boeck Université: Paris, Bruxelles, **1999**.

56. Ruusuvuori, E.; Li, H.; Huttu, K.; Palva, J. M.; Smirnov, S.; Rivera, C.; Kaila, K.; Voipio, J. Carbonic anhydrase isoform VII acts as a molecular switch in the development of synchronous gamma-frequency firing of hippocampal CA1 pyramidal cells. *J. Neurosci.* **2004**, *24*, 2699–2707.

57. Kaila, K.; Lamsa, K.; Smirnov, S.; Taira, T.; Voipio, J. Long-lasting GABA-mediated depolarization evoked by high-frequency stimulation in pyramidal neurons of rat hippocampal slice is attributable to a network-driven, bicarbonate-dependent K^+ transient. *J. Neurosci.* **1997**, *17*, 7662–7672.

58. Khalifah, R. G. The carbon dioxide hydration activity of carbonic anhydrase. I. Stop-flow kinetic studies on the native human isoenzymes B and C. *J. Biol. Chem.* **1971**, *246*, 2561–2573.

59. De Simone G.; Di Fiore A.; Menchise, V.; Pedone, C.; Antel, J.; Casini, A.; Scozzafava, A.; Wurl, M.; Supuran, C. T. Carbonic anhydrase inhibitors. Zonisamide is an effective inhibitor of the cytosolic isozyme II and mitochondrial isozyme V: solution and X-ray crystallographic studies. *Bioorg. Med. Chem. Lett.* **2005**, *15*, 2315–2320.

60. Nishimori, I.; Vullo, D.; Innocenti, A.; Scozzafava, A.; Mastrolorenzo, A.; Supuran, C. T. Carbonic anhydrase inhibitors: inhibition of the transmembrane isozyme XIV with sulfonamides. *Bioorg. Med. Chem. Lett.* **2005**, *15*, 3828–3833.

61. Vullo, D.; Voipio, J.; Innocenti, A.; Rivera, C.; Ranki, H.; Scozzafava, A.; Kaila, K.; Supuran, C. T. Carbonic anhydrase inhibitors. Inhibition of the human cytosolic isozyme VII with aromatic and heterocyclic sulfonamides. *Bioorg. Med. Chem. Lett.* **2005**, *15*, 971–976.

62. Masereel, B.; Rolin, S.; Abbate, F.; Scozzafava, A.; Supuran, C. T. Carbonic anhydrase inhibitors: anticonvulsant sulfonamides incorporating valproyl and other lipophilic moieties. *J. Med. Chem.* **2002**, *45*, 312–320.

63. Borras, J.; Scozzafava, A.; Menabuoni, L.; Mincione, F.; Briganti, F.; Mincione, G.; Supuran, C. T. Carbonic anhydrase inhibitors: synthesis of water-soluble, topically effective intraocular pressure lowering aromatic/heterocyclic sulfonamides containing 8-quinoline-sulfonyl moieties: is the tail more important than the ring? *Bioorg. Med. Chem.* **1999**, *7*, 2397–2406.

64. Casini, A.; Scozzafava, A.; Mincione, F.; Menabuoni, L.; Ilies, M. A.; Supuran, C. T. Carbonic anhydrase inhibitors: water-soluble 4-sulfamoylphenylthioureas as topical intraocular pressure-lowering agents with long-lasting effects. *J. Med. Chem.* **2000**, *43*, 4884–4892.

65. Scozzafava, A.; Briganti, F.; Mincione, G.; Menabuoni, L.; Mincione, F.; Supuran, C. T. Carbonic anhydrase inhibitors: synthesis of water-soluble, aminoacyl/dipeptidyl sulfonamides possessing long-lasting intraocular pressure-lowering properties via the topical route. *J. Med. Chem.* **1999**, *42*, 3690–3700.

66. Scozzafava, A.; Menabuoni, L.; Mincione, F.; Briganti, F.; Mincione, G.; Supuran, C. T. Carbonic anhydrase inhibitors. Synthesis of water-soluble, topically effective, intraocular pressure-lowering aromatic/heterocyclic sulfonamides containing cationic or anionic moieties: is the tail more important than the ring? *J. Med. Chem.* **1999**, *42*, 2641–2650.

67. Scozzafava, A.; Menabuoni, L.; Mincione, F.; Briganti, F.; Mincione, G.; Supuran, C. T. Carbonic anhydrase inhibitors: perfluoroalkyl/aryl-substituted derivatives of aromatic/heterocyclic sulfonamides as topical intraocular pressure-lowering agents with prolonged duration of action. *J. Med. Chem.* **2000**, *43*, 4542–4551.

68. Winum, J.-Y.; Thiry, A.; Cheikh, K. E.; Dogné, J.-M.; Montero, J.-L.; Vullo, D.; Scozzafava, A.; Masereel, B.; Supuran, C. T. Carbonic anhydrase inhibitors. Inhibition of isoforms I, II, IV, VA, VII, IX, and XIV with sulfonamides incorporating fructopyranose-thioureido tails. *Bioorg. Med. Chem. Lett.* **2007**, *17*, 2685–2691.

69. Thiry, A.; Rolin, S.; Vullo, D.; Frankart, A.; Scozzafava, A.; Dogné, J.-M.; Wouters, J.; Supuran, C. T.; Masereel, B. Indanesulfonamides as carbonic anhydrase inhibitors and anticonvulsant agents: structure–activity relationship and pharmacological evaluation. *Eur. J. Med. Chem.* **2008**, *43*, 2853–2860.

70. Chazalette, C.; Masereel, B.; Rolin, S.; Thiry, A.; Scozzafava, A.; Innocenti, A.; Supuran, C. T. Carbonic anhydrase inhibitors. Design of anticonvulsant sulfonamides incorporating indane moieties. *Bioorg. Med. Chem. Lett.* **2004**, *14*, 5781–5786.

71. Thiry, A.; Masereel, B.; Dogné, J.-M.; Supuran, C. T.; Wouters, J.; Michaux, C. Exploration of the binding mode of indanesulfonamides as selective inhibitors of human carbonic anhydrase type VII by targeting Lys91. *ChemMedChem* **2007**, *2*, 1273–1280.

72. Winum, J.-Y.; Temperini, C.; El Cheikh K. Innocenti, A.; Vullo, D.; Ciattini, S.; Montero, J.-L.; Scozzafava, A.; Supuran, C. T. Carbonic anhydrase inhibitors: clash with Ala65 as a means for designing inhibitors with low affinity for the ubiquitous isozyme II, exemplified by the crystal structure of the topiramate sulfamide analogue. *J. Med. Chem.* **2006**, *49*, 7024–7031.

73. Tasso, S. M.; Moon, S.; Bruno-Blanch, L. E.; Estiu, G. L. Characterization of the anticonvulsant profile of valpromide derivatives. *Bioorg. Med. Chem.* **2004**, *12*, 3857–3869.

74. Gavernet, L.; Barrios, I. A.; Cravero, M. S.; Bruno-Blanch, L. E. Design, synthesis, and anticonvulsant activity of some sulfamides. *Bioorg. Med. Chem.* **2007**, *15*, 5604–5614.

75. Dimmock, J. R.; Puthucode, R. N.; Smith, J. M.; Hetherington, M.; Quail, J. W.; Pugazhenthi, U.; Lechler, T.; Stables, J. P. (Aryloxy)aryl semicarbazones and related compounds: a novel class of anticonvulsant agents possessing high activity in the maximal electroshock screen. *J. Med. Chem.* **1996**, *39*, 3984–3997.

76. Pandeya, S. N.; Yogeeswari, P.; Stables, J. P. Synthesis and anticonvulsant activity of 4-bromophenyl substituted aryl semicarbazones. *Eur. J. Med. Chem.* **2000**, *35*, 879–886.

CHAPTER 8

Carbonic Anhydrase Inhibitors Targeting Cancer: Therapeutic, Immunologic, and Diagnostic Tools Targeting Isoforms IX and XII

SILVIA PASTOREKOVA, MONIKA BARATHOVA, JURAJ KOPACEK, and JAROMIR PASTOREK

Institute of Virology, Centre of Molecular Medicine, Slovak Academy of Sciences, Bratislava, Slovak Republic

8.1 INTRODUCTION

Carbonic anhydrases (CAs) are widespread metalloenzymes that use a zinc-activated hydroxide mechanism to catalyze the reversible conversion of carbon dioxide to carbonic acid in a simple reaction $CO_2 + H_2O \leftrightarrow HCO_3^- + H^+$.[1] CAs can either produce or consume bicarbonate and thereby participate in biosynthetic reactions, ion transport across the membranes, and acid–base balance via this activity. They are therefore essential for various physiological processes, and virtually every organ or tissue contains at least one CA isoform. In fact, adequate performance of metabolically active tissues, such as renal or gastric epithelium, depends on simultaneous action of several CA isoenzymes. Mammalian CAs exist in 16 isoforms, which can be divided according to subcellular localization into intracellular (CA I–III, VA, VB, VII, VIII, X, XI, XIII, XV) and extracellular (CA IV, VI, IX, XII, XIV), according to catalytic performance into active (CAI–VII, IX, XII–XV) and inactive (CA VIII, X, XI), and according to tissue distribution into widespread (CA II, IV, VB, XII, XIV) and confined to few tissues (CA I, III, VA, VI, VII).[2,3] Twelve active isoenzymes contain a conserved active site that involves three histidine residues participating in the coordination of a zinc ion and the fourth histidine residue functioning as a proton shuttle. Active CAs are mostly expressed in differentiated cells and are functionally involved in respiration, bone resorption, production of gastric acid and other body fluids, renal and testicular acidification, and so on. Loss or deregulated activity of certain isozymes has been associated with several diseases, including glaucoma, osteopetrosis, edema from heart and renal failure, neurological and neuromuscular disorders, and so on.[4]

Cancer represents a pathologic situation characterized by numerous aberrant features at molecular, cellular, and tissue levels. These include not only mutations and epigenetic changes that result in decreased cell death and increased proliferation (and hence increased requirement for energy and higher consumption of oxygen and nutrients), but also physiological abnormalities (such as lower delivery of oxygen and nutrients via insufficient vasculature) that lead to shift to anaerobic metabolism that generates acidic products and causes acid–base disbalance.[5] Such tumor metabolism clearly requires carbonic anhydrases as a part of regulatory machinery that allows the cancer cells to adapt to these abnormal conditions and survive. The mechanism of how carbonic anhydrases protect the cells in growing tumor tissue from physiological stresses has recently emerged from studies of two cell surface CA isoforms, namely, CA IX (almost exclusively associated with tumors) and CA XII (overexpressed in some tumor types). These two CA isoenzymes have attracted a lot of attention not only as functional constituents of tumor physiology, but also as cancer biomarkers and potential therapeutic targets.[6]

8.2 BIOCHEMICAL FEATURES OF CA IX AND CA XII

Carbonic anhydrase IX (CA IX), originally named MN protein, has been identified using a monoclonal antibody M75 as a cell density-regulated plasma membrane antigen in human HeLa cell line derived from carcinoma of the cervix.[7] Monomeric form of MN antigen was visualized by immunoblotting as a twin band of 58/54 kDa, which assembled into a 153 kDa trimer under nonreducing conditions. More important, expression of the MN antigen was found to correlate with the tumorigenic phenotype of somatic cell hybrids between HeLa and normal human fibroblasts *in vitro*.[8] Furthermore, the MN antigen was detected in various tumor cell lines and surgical tumor specimens, but not in the corresponding normal tissues.[8] These two pioneering studies uncovered the link of MN protein to cancer and suggested its possible usefulness as a tumor marker.

Sequence analysis of cDNA encoding MN antigen revealed its domain composition.[9] However, due to an error in the 5′ region of cDNA sequence, the N-terminal portion of MN protein was wrongly assumed to contain helix–loop–helix motif. This was corrected in the subsequent genomic analysis.[10] According to the corrected sequence, MN gene codes for 459-amino acid (aa) protein with 414-aa N-terminal extracellular part linked through the 20 aa hydrophobic transmembrane region (TM) with 25 aa C-terminal intracellular (IC) tail. Extracellular part is composed of 37 aa signal peptide, 59 aa region with similarity to keratan sulfate binding domain of a large proteoglycan aggrecan (PG) and a 257 aa carbonic anhydrase (CA) domain (see Fig. 8.1). While the N-terminal PG domain is unique for CA IX (it is not present in any other carbonic anhydrase isoenzyme), the CA domain is well conserved and shows

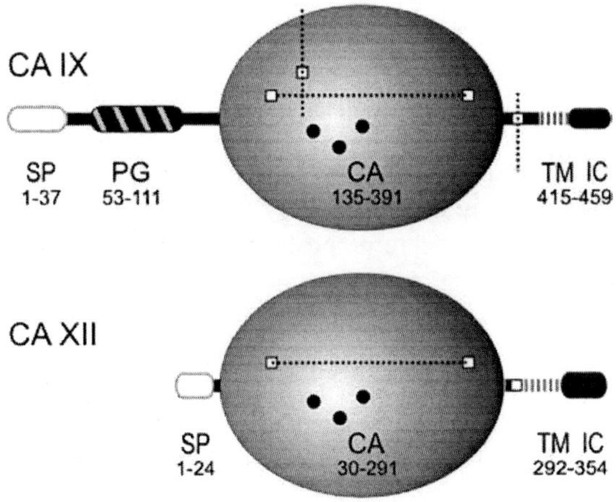

FIGURE 8.1 Domain composition of CA IX and XII isoenzymes. The proteins consist of signal peptides (SP, removed during the protein maturation), carbonic anhydrase domains, transmembrane regions, and intracellular tails. CA IX also contains a proteoglycan-like region (PG), which is absent in CA XII and other carbonic anhydrase isoforms. Conserved zinc binding histidines are indicated by black circles. Cysteines involved in folding or oligomerization are shown as white squares and their proposed contribution to disulfidic bonds is illustrated by dotted lines (horizontal for intramolecular bonds and vertical for intermolecular bonds).

a significant identity with CA VI, II, and other CA isoenzymes and contains all amino acids required for the catalytic activity. On this basis, MN antigen was renamed to carbonic anhydrase IX (as it was the ninth mammalian CA identified).[11]

In an independent line of research, monoclonal antibody G250 detected a renal cell carcinoma-associated antigen G250 that was subsequently recognized as a clinically relevant marker/and target of kidney cancer. Later on, it was shown to be identical with MN/CA IX, as reviewed elsewhere.[12]

Biochemical characteristics of CA IX have been recently studied in more detail using recombinant protein and its truncated variant produced in baculovirus system.[13] This study utilizing mass spectrometric approaches confirmed and extended earlier observations that CA IX is a glycoprotein containing a high mannose sugar chain linked to N^{346}. It also showed for the first time that the PG domain of CA IX is modified by O-linked keratan sulfate chain and proposed T^{115} as the attachment site. Biological importance of these modifications awaits further inspection. Furthermore, the study has demonstrated the presence of intramolecular S–S bond between C^{156} and C^{336} and intermolecular, trimer-stabilizing bond between C^{174} and C^{409}.

Cytoplasmic tail of CA IX has remained the least explored part of the protein although ongoing studies point to its high importance for proper functioning of CA IX (Hulikova et al., unpublished data). It contains three putative phosphorylation sites (T^{443}, S^{448}, Y^{449}), of which Y^{449} has been shown to mediate EGF-induced signal transduction to Akt kinase.[14]

CA IX belongs to the most active human CAs. Catalytic properties of the purified recombinant CA domain expressed in bacteria are comparable to CA II, with the proton transfer rate near 1 μs^{-1} and k_{cat}/K_m near 55 μM^{-1} s^{-1}.[15] Nevertheless, activity of the entire extracellular portion of CA IX (composed of PG and CA domains) expressed in baculovirus system is even higher than that of CA II.[13] Moreover, addition of $ZnCl_2$ increases the catalytic activity of the insect cell-derived human enzyme by approximately 10-fold, and it seems to possess the highest catalytic activity that has ever been measured for a human CA enzyme.

Homology model of CA IX based on the strong similarity of backbone conformations among the human CA isozymes revealed several differences between the residues in the active sites of CA IX and II, including Q^{67} instead of N^{67} and V^{131} instead of F^{131}.[15] These differences might modify the topology of the active site pocket and influence binding of inhibitors (and hence mediate selective action of some inhibitors toward CA IX as described below).

Although no crystal structure of CA IX has been obtained so far, there is a rapidly increasing amount of data that show the capacity of different classes of sulfonamide derivatives to inhibit the enzyme activity of CA IX.[1] Some compounds efficiently inhibit the activity of purified recombinant CA domain even at subnanomolar concentrations and work much better with CA IX than with other CA isozymes.[16,17]

As indicated by the numerical designation, carbonic anhydrase XII was identified a few years after CA IX in two independent studies. First, it was detected as an antigen overexpressed in about 10% of human renal cell carcinoma and cloned by serological expression screening with autologous antibodies.[18] Almost simultaneously, CA XII was cloned as a novel target of pVHL tumor suppressor protein using RNA differential display.[19]

cDNA sequencing revealed that CA XII is expressed as a 354 aa protein composed of 29 aa signal peptide, 261 aa CA domain, 9 aa juxtamembrane segment, 26 aa transmembrane domain, and a 29 aa cytoplasmic tail (Fig. 8.1). The extracellular part of CA XII contains all histidine residues required for the catalytic activity and two potential N-glycosylation sites (N^{52} and N^{136}). Molecular weight of the human CA XII expressed in transfected COS cells corresponded to 43–44 kDa and was reduced to 39 kDa by endoglycosidase treatment.[18]

Consistent with the conserved catalytic site, human CA XII is an active isoenzyme and its activity is similar to that of human CA IV.[20] In contrast to CA IX, the crystal structure of CA XII was reported and it was suggested that the active site cavity is very similar to CA II and IV. Single disulfidic bridge connects two cysteines (C^{23} and C^{203}). Short intracellular segment is placed in an opposite orientation to the extracellular enzyme domain. CA XII appears to form a dimer, in which the active site pockets face the extracellular space. The transmembrane domain of CA XII contains two putative amino acid motifs GXXXG and GXXXS that were proposed to contribute to dimerization. The cytoplasmic tail of CA XII is highly homologous to CA IX and also contains putative phosphorylation sites, although their positions and flanking sequences are not identical.

Despite many similarities, these two CA isoforms display many biochemical differences suggesting that they evolved to perform differential physiological roles.

8.3 TISSUE DISTRIBUTION OF CA IX AND CA XII

CA IX is naturally expressed in only few normal tissues, but its ectopic expression is induced in a wide spectrum of human tumors. In normal tissues, the most abundant expression of CA IX was found in the mucosa of the stomach and gallbladder.[21] Lower levels are expressed in the intestinal epithelium, where it is confined to the cryptal areas composed of proliferating cells.[22] Weak expression of CA IX was also observed in epithelia of pancreatic ducts, male reproductive organs, and lining cells of body cavity.[23,24,25]

On the other hand, CA IX is ectopically expressed at relatively high levels and with a high prevalence in different tumor tissues, whose normal counterparts do not contain this protein. These involve carcinomas of the cervix uteri, oesophagus, kidney, lung, breast, brain, and many other tumors.[26–32] For an unknown reason, CA IX is generally absent from the normal prostate tissues as well as from the prostate carcinomas. Tissues with high natural CA IX expression, such as stomach and gallbladder, usually lose or reduce CA IX upon conversion to carcinomas.[33,34] In the colonic epithelium, CA IX is present normally in the deep crypts and abnormally in the superficial adenomas and carcinomas, with the most intense staining seen in tumors with mucinous component.[35]

A high proportion of CA IX-positive specimens was found among the cervical, renal, and lung cancers. In the cervical cancer, CA IX is present in virtually all cervical carcinomas and the majority of cervical intraepithelial neoplasia.[26] Diffuse CA IX-positive staining signal in normal cervical tissues is found only in concurrent presence of dysplasia or carcinoma and therefore can be useful as an early diagnostic indicator of cervical neoplasia in Pap smears.[36,37] In the lung cancer, CA IX is not found in preneoplastic lesions, but is readily present in malignant tumors.[30] In the kidney cancer, CA IX protein expression is selectively linked with the most frequent carcinomas of renal clear cell (ccRCC) type. High levels of CA IX protein and/or mRNA are seen in primary, cystic, and metastatic ccRCCs, but not in benign lesions.[28,29] On the basis of the clear-cut division between the tissues with normal and ectopic expression of CA IX, as well as on the predominant association of CA IX with different types of tumors, CA IX was proposed as a promising tumor biomarker and further studies strongly supported this view.

CA XII is overexpressed in some tumor types, such as renal cell carcinomas (RCC), ovarian, breast, and cervical carcinomas, and in brain tumors,[25,33,38–42] but its relationship to cancer is generally not as tight as that of CA IX. In contrast to CA IX, expression of CA XII has been detected in many normal tissues.[25] Its mRNA is expressed in the normal adult kidney, pancreas, colon, prostate, ovary, testis, lung, and brain.[18] CA XII protein was also found in normal endometrium and other reproductive organs,[43,44] colon,[23] kidney,[45] eye,[46,47] and in developing embryo.[48] These data

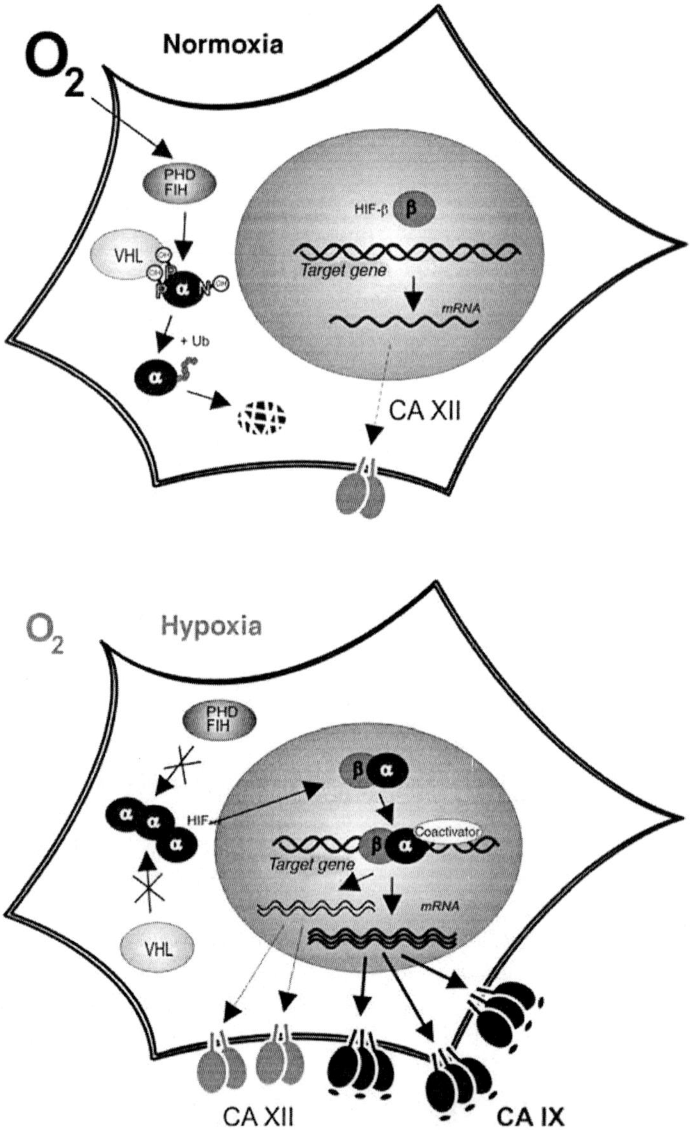

FIGURE 8.2 Mechanism of hypoxia-induced transcriptional activation of target genes, including the gene coding for CA IX. The upper part shows the situation in normoxia. In the presence of oxygen, prolyl hydroxylases (PHD), and factor inhibiting HIF modify α-subunit of hypoxia-inducible factor (HIF-α). Hydroxylation of conserved prolines in HIF-α molecule leads to recognition of HIF-α by VHL tumor suppressor protein, which mediates HIF-α ubiquitylation and degradation in proteasome. Thus, normoxic cells principally do not contain HIF-α and do not express CA IX, whereas CA XII expression can be activated by hypoxia-independent pathways as discussed in the chapter. The lower part shows the situation in hypoxia. The lack of oxygen maintains hydroxylases inactive so that they are unable to modify HIF-α. This allows HIF-α to escape recognition by VHL and prevents degradation. As a result,

suggest important physiological role for this enzyme in ion transport and fluid concentration in different normal organs.

8.4 REGULATION OF CA IX AND CA XII EXPRESSION

To understand regulation of CA IX expression, 5′ upstream region of the *CA9* gene was thoroughly investigated under conditions of high cell density known to increase the level of CA IX protein. CA9 promoter was localized to genomic sequence spanning 173 nucleotides upstream of the transcription initiation site and was shown to possess five regulatory regions containing several *cis*-acting elements.[49] AP1 and SP1 transcription factors bind proximally to transcription initiation site and their synergy is needed for the basic transcriptional activation of *CA9* gene.[50] The most important regulatory element of *CA9* promoter is localized between the SP1 binding site and the transcription start at position −10/−3 and consists of the nucleotide sequence 5′-TACGTGCA-3′ corresponding to a hypoxia response element (HRE).[51]

HRE element is recognized by a HIF transcription factor that assembles under hypoxic conditions from a constitutive β-subunit and an oxygen-regulated α-subunit. In normoxia, HIF-α is modified by prolyl hydroxylases at two conserved proline residues in the central oxygen-dependent degradation domain and subsequently bound and degraded via pVHL tumor suppressor protein.[52] Moreover, its transcriptional activity is blocked by factor inhibiting HIF (FIH)-mediated hydroxylation of arginine residue in the C-terminal transactivation domain.[53] When oxygen level is reduced and the cell is exposed to hypoxic or anoxic conditions, HIF-α avoids hydroxylation, accumulates, and following dimerization with HIF-β and interaction with cofactors becomes transcriptionally active[54] (Fig. 8.2). Presence of the functional HRE element juxtaposed to transcription start makes *CA9* a strong transcriptional target of HIF (namely, HIF-1) and places it among the genes regulated by hypoxia.[51] HRE element is also utilized in a density-induced transcription of *CA9*, but requires cooperation with neighboring SP1 binding site under these conditions. Activation of *CA9* transcription by increased cell density, which is linked with pericellular hypoxia, involves PI3 kinase pathway and subhypoxic levels of HIF-1α[55]. HRE-SP1 promoter motif is also essential for transcription of CA9 gene induced by activation of the MAP kinase pathway.[56]

Since the level of HIF-α is controlled by pVHL, it is not surprising that the expression of the wild-type *VHL* transgene can suppress the transcription of *CA9* mRNA in the normoxic cells and that *VHL* deletion or inactivating mutation leads to the release of *CA9* transcription.[19] Loss of functional pVHL is linked with a majority of

HIF-α accumulates in cytoplasm, enters the nucleus, interacts with coactivator, dimerizes with constitutive HIF-β-subunit, and forms an active transcription complex that induces transcription of genes containing hypoxia response element. CA IX-encoding gene is a direct target of HIF-1 and is strongly induced by hypoxia, whereas molecular mechanism of hypoxic induction of CA XII remains unknown.

clear cell renal cell carcinomas and provides explanation for the frequent presence of CA IX in ccRCC. Onset of CA IX expression is an early event occurring in morphologically normal cells within the renal tubules of patients with VHL disease and therefore provides a robust system for identification of early foci of VHL inactivation.[57] In addition, transcription of *CA9* gene can be modulated by methylation of CpG dinucleotides within the promoter region. In RCC cell lines and in tumors, expression of CA IX is also associated with hypomethylation of *CA9* promoter.[58,59] Methylation of −74 CpG site can also influence the expression of CA IX in the carcinoma cell lines of a different origin than RCC, where it seems to represent adverse factor modifying transcriptional response to cell density.[60]

Transcription of the *CA*12 gene has not been systematically investigated. Position and architecture of the promoter as well as identity of the critical regulatory factors remain to be elucidated. However, on the basis of the expression pattern in tissues, it could be recognized that expression of CA XII is also induced by hypoxia,[51,25,38] although it is still not known whether *CA*12 gene is a direct target of HIF and if yes, where is the relevant HRE element. Since induction of CA XII by hypoxia is weaker when compared to CA IX and the region proximal to transcription start of *CA*12 gene does not contain HRE consensus, it appears that hypoxia regulates *CA*12 expression via putative HRE elements localized in the distant 5′ upstream region. Similarly to CA IX, expression of CA XII is also inhibited by the wild-type VHL in RCC cell lines. However, suppression of CA12 mRNA requires the central pVHL domain involved in the HIF-1α binding as well as the C-terminal elongin binding domain.[19]

Moreover, it has been recently shown that expression of CA XII is highly correlated with estrogen receptor alpha (ER alpha) in human breast tumors and that *CA*12 gene is regulated by estrogen via ER alpha binding to a distal estrogen-responsive enhancer region in breast cancer cells. Upon addition of estradiol, ER alpha binds directly to this distal enhancer *in vivo*, resulting in the recruitment of RNA polymerase II and steroid receptor coactivators SRC-2 and SRC-3, and changes in histone acetylation.[61]

Interestingly, both CA9 and CA12 mRNA exist in two splicing forms. The splicing variant of CA9 lacks exons 8 and 9 and generates C-terminally truncated protein that is localized in the cytoplasm or secreted to extracellular space and behaves as dominant negative with respect to the wild-type form.[62] Its expression level is principally independent of hypoxia and tumor phenotype and can therefore mask or reduce the prognostic value of the wild-type *CA9* mRNA in transcriptional profiling studies based on primers specific for the 5′ part of the cDNA.

The splicing variant of *CA*12 mRNA lacks only a short segment coding for 11 amino acids and seems to be common in diffuse astrocytomas with poor prognosis.[42] The reasons for the association of CA12 splicing variant with aggressive brain tumors and its biological role are not clear and require further investigations.

CA IX has also been studied at the post-translational level. Although the protein is very stable (with a half-life corresponding to approximately 38 h) and persists in reoxygenated cells,[63] its abundance on the cell surface is modulated by both constitutive and inducible shedding to extracellular space, the latter being mediated by metalloproteinase TACE/ADAM17.[64] More important, the extracellular "soluble" form of CA IX has been proposed to indicate the presence of tumors and monitor the treatment outcome.[65] This proposal was further supported by a recent investigation of the clinical usefulness of serum CA IX that was performed to examine its correlation with tumor progression and track early recurrence after surgery for patients with localized renal cell cancer.[66] The mean serum CA IX levels were significantly higher in patients with metastatic disease compared to those with localized disease and in patients with localized disease compared to healthy controls. Higher CA IX levels were also significantly associated with postoperative recurrence.[66]

No data on stability and/or shedding of CA XII are available to date.

8.5 CLINICAL VALUE OF CA IX AND CA XII

Hypoxia often develops in solid tumors due to insufficient supply of oxygen by aberrant vasculature. To adapt to such physiological stress, hypoxic tumor cells acquire oncogenic alterations in metabolism, gain increased resistance to conventional anticancer treatment, and are more prone to mutations. Thus, tumor hypoxia is a clinically important phenomenon and detection of hypoxic areas within tumors is very important for selection of suitable treatment regimens and for prognosis of the disease development.[67] Current detection of hypoxia in tumors is based on different methods, such as polarographic oxygen Eppendorf histography using microelectrodes, or metabolic incorporation of chemical markers, for example, pimonidazole or nitroimidazole derivatives. However, these methods are either invasive or restrict measurements to easily accessible tumor sites and can be used only prospectively. In contrast, use of an intrinsic hypoxic marker is simple, reproducible, and can be performed on a routine basis, without the need for special equipment, using both prospective and retrospective samples.[68]

CA IX belongs to the most strongly induced proteins in response to hypoxia and was therefore suggested to serve as an intrinsic hypoxic marker with possible diagnostic, prognostic, and therapeutic value. CA IX expression was studied in different tumor types to examine its clinical relevance as a marker of hypoxia and was compared with different histopathological parameters, disease progression, treatment outcomes, and patient survival. Indeed, CA IX was found to correlate with other markers of hypoxia including pimonidazole, HIF-1α, VEGF, and GLUT-1, although their distributions do not fully overlap due to differential kinetics and magnitude of induction, as well as severity and duration of hypoxia. CA IX is present in a relatively wide perinecrotic area with median distance between a blood vessel and a beginning of CA IX expression of 80 μm (range 40–140) in head, neck, and bladder carcinomas[69,70] and of about 90 μm in non-small-lung carcinoma.[71] It has also been shown that that the cells expressing CA IX are viable, clonogenic, and resistant to killing by ionizing radiation.[72] Moreover, due to a long half-life, CA IX could be also present in tissues that have been hypoxic before.[63] It was recently shown that a comparison of the spatial distribution of CA IX and pimonidazole allows for the discrimination between hypoxic and reoxygenated areas.[73]

Hypoxia-related expression pattern of CA IX determines its link to clinical variables. CA IX significantly correlates with high tumor grade, necrosis, treatment outcome, and/or poor prognosis in patients with the breast and lung carcinomas,[74–81] with necrosis, advanced stage, and treatment outcome in head and neck cancer,[69,82] with necrosis, histological grade, and survival of patients with brain tumors,[32,83] with poor prognosis in oesophageal and gastric cancers,[84,85] and with metastases in primary cervical cancer.[40] In the cDNA microarray study of hypoxia transcriptome in human bladder cancer, CA IX was used as a surrogate marker of hypoxia to which array mRNA changes were correlated to define *in vivo* hypoxia profile and identify new hypoxia-regulated genes.[86] CA IX was also utilized as one of the 10 hypoxia-regulated genes to cluster a hypoxia metagene in head and neck squamous cell cancers, which was a significant prognostic factor in the published head and neck as well as in breast cancer data sets.[87]

CA IX expression is very high in clear cell renal cell carcinomas due to inactivating mutation in VHL gene.[88] Resulting activation of HIF pathway leads to constitutive upregulation of HIF targets (including CA IX) independently of physiological hypoxia.[89,57] In contrast to other tumor types, CA IX expression in RCC decreases with increasing tumor grade and stage most probably because it is a target of HIF-1α that is present very early in VHL disease but later disappears due to the onset of tumor growth-promoting isoform HIF-2α[57,90]. Accordingly, reduced CA IX levels are independently associated with poor prognosis in advanced RCC[91–95]. Nevertheless, levels of CA IX can be upregulated by interferons (IFN-α and IFN-γ), which are known to enhance the therapeutic responses in 5–25% of patients with metastatic RCC.[96] Furthermore, high CA IX expression was suggested as an important predictor of better outcome in RCC patients receiving interleukin-2-based therapy.[97,98] Noteworthy, it has been proposed that CA IX is the most significant molecular marker described in kidney cancer to date.[99,100]

Interestingly, in breast and lung cancer, CA IX expression correlates also with some other prognostic factors, including c-ErbB2,[31] epidermal growth factor receptor, c-ErbB2 and MUC-1,[76] EGFR and matrix metalloproteinase MMP-9,[101] osteopontin and CD44,[102] p53 and p300,[103] estrogen receptor, progesterone receptor, bcl-2, p53, c-ErbB2 and Ki-67,[77] cyclin E, cyclin A2, Ki-67,[78] lysyl oxidase, ephrin A1, galectin-1,[82] Slug,[104] and manganese superoxide dismutase.[105] Moreover, IL-6 induces Notch-3-dependent upregulation of carbonic anhydrase IX gene and thereby promotes a hypoxia-resistant/invasive phenotype in breast cells.[106]

CA XII isoform is less tightly regulated by hypoxia/pVHL pathway and less strongly linked to cancer as compared to CA IX.[74,25] In breast carcinoma in situ, expression of CA XII is associated with the absence of necrosis and low-grade lesions, indicating that it is primarily regulated by the factors related to differentiation.[74] In the colon, CA XII is normally expressed by the differentiated surface epithelium, and increased basal/deep mucosal expression is associated with increasing dysplasia and invasive tumor stage.[107] In the primary cervical cancer, CA12 mRNA expression was linked to a lower risk of metastasis, whereas CA9 showed the opposite correlation.[40] However, in diffuse astrocytomas, expression of CA XII correlated with poorer patient prognosis in both univariate and multivariate survival analyses.[42]

8.6 ROLES OF CA IX AND CA XII IN TUMOR CELLS

Because CA IX and XII are active extracellular enzymes involved in metabolism of CO_2, they have been implicated in acidification of extracellular microenvironment and at the same time in protection of cancer cells from the acidosis. Experimental evidence for this proposal comes from the studies of CA IX.[108–111] No experimental data from the studies of CA XII function in cancer have been published so far.

Hypoxia triggers the development of acidosis due to induction of metabolic shift from oxidative phosphorylation to anaerobic glycolysis that helps to sustain energy demands of tumor cells in low-oxygen conditions. HIF-1 upregulates expression of glucose transporters (GLUT-1, GLUT-3) and glycolytic enzymes, including LDH-A and LDH-5 (lactate dehydrogenases converting pyruvate to lactate) and PDK-1 (pyruvate dehydrogenase kinase preventing entry of pyruvate into the TCA cycle). As a result, the cells produce excess of lactate, protons, and CO_2.[112–114] To protect themselves from intracellular acidification that is incompatible with survival, tumor cells increase extrusion of these acidic catabolites that then accumulate in the pericellular space due to insufficient waste removal by defective vasculature. These extrusion mechanisms operating at the plasma membrane involve the H^+/monocarboxylate transporter (MCT), the Na^+/H^+ exchanger (NHE), and the vacuolar H^+/ATP pump. On the other hand, intracellular pH is neutralized by bicarbonate ions transported across the membrane to intracellular space via anion exchangers (AEs) and Na^+/bicarbonate cotransporters (NBCs).[5] Noteworthy, VHL/HIF pathways control expression of several components of the pH regulating machinery including AE2, NHE1, and MCT4.[115–117]

However, the bicarbonate import mechanism was difficult to explain before the identification of CA IX and XII, because the bicarbonate content in acidic pericellular space of tumor cells is generally very low. According to a recent concept, the cell surface carbonic anhydrases cooperate with bicarbonate transpoters and contribute to this mechanism by locally concentrated conversion of CO_2 to bicarbonate ions and protons.[118] While protons remain outside of cells and further acidify their microenvironment, bicarbonate ions are directly taken up by bicarbonate transporters (colocalizing and interacting with CAs) that bring them to intracellular space, where they buffer intracellular protons and thereby neutralize the intracellular space. Such CA-accelerated flux of bicarbonate across the membrane was initially described for noncancer carbonic anhydrase isoforms CA II and IV cooperating with bicarbonate transporters to form a "bicarbonate transport metabolon" in red blood cells and renal epithelial cells.[118]

As previously described, hypoxic tumor cells need pH regulation and ion movement to adapt to acidosis and thus induce expression of CA IX/XII to take part in these processes. Interestingly, the enzymatic activity of CA IX is insensitive to high lactate concentrations (in contrast to the other CA isoenzymes), thus allowing CA IX to work efficiently in the hypoxic tumor microenvironment, which is rich in lactate produced by glycolysis.[119] On the other hand, CA IX activity is inhibited by bicarbonate suggesting that it can preferentially catalyze the CO_2 hydration producing bicarbonate ions in pericellular tumor regions loaded with CO_2 and deprived of bicarbonate. Moreover,

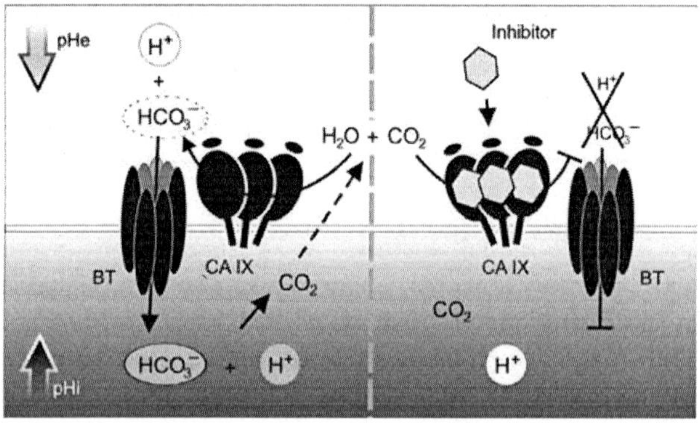

FIGURE 8.3 The proposed role of CA IX in the bicarbonate transport metabolon acting in hypoxic tumor cells. CA IX catalyzes hydration of pericellular CO_2 to bicarbonate and proton. Protons remain outside the cell and contribute to extracellular acidification. Bicarbonate transporters (BTs) transport the bicarbonate ions to cytoplasm, where they are converted back to CO_2 in a reaction that utilizes intracellular protons and leads to neutralization of intracellular pH. Based on the experimental data, the metabolon is activated in hypoxia. Sulfonamide inhibitors can inhibit production of bicarbonate and proton from pericellular carbon dioxide and block the activity of transport metabolon, thus perturbing intracellular pH neutralization and extracellular acidification.

CA IX was shown to interact with bicarbonate transporters via its extracellular catalytic domain.[120] These facts are well compatible with the concept of bicarbonate import for neutralization of intracellular pH of tumor cells (Fig. 8.3).

Functional involvement of CA IX in pH regulation is supported by three experimental evidences. First, CA IX contributes to acidification of the extracellular microenvironment of hypoxic cells,[108] second, CA IX minimizes the intracellular pH gradient and increases the extracellular pH gradient in the core of three-dimensional tumor spheroids,[110] and third, CA IX (and also CA XII) contribute to extracellular acidification and to more alkaline resting intracellular pH in response to a CO_2 load, and thereby support cell survival in acidosis.[111] Since these experiments were performed with constitutively expressed CA IX, the pH modulating effects are apparently related to CA IX catalytic activity, which is induced in hypoxia and/or acidosis.

This transcriptional and catalytic activation of CA IX by hypoxia/acidosis might have a strong impact on cancer progression, because maintenance of neutral intracellular pH is vital for cell proliferation and survival, whereas microenvironmental acidosis contributes to aggressive tumor phenotype by promoting invasion and metastasis.[112,121,122] In accord, transient RNA interference of *CA9* decreased the clonogenic survival of hypoxic tumor cells in vitro.[109] Moreover, silencing of *CA9* mRNA led to 40% reduction in tumor growth in vivo and invalidation of both CA IX and CA XII proteins gave an impressive 85% reduction.[111]

In addition to the involvement in pH control, CA IX has another important function related to cell adhesion and intercellular communication. CA IX mediates cell adhesion to solid support, and the cell adhesion site on N-terminal PG domain overlaps with the epitope for monoclonal antibody M75. Interestingly, acidic extracellular pH inhibits attachment of cells to CA IX, suggesting that CA IX adhesion capacity can be modulated by tumor microenvironment.[123,124] CA IX can also modulate E-cadherin-mediated cell-to-cell adhesion via the mechanism that involves direct binding of CA IX to β-catenin. This function of CA IX is consistent with a proposal that hypoxia can initiate tumor invasion via decreased E-cadherin-mediated cell–cell adhesion and offers a possibility that CA IX participates in this process.[125]

CA IX has been implicated also in signal transduction.[14] In renal carcinoma cells, tyrosine residue present in the intracellular tail of CA IX can be phosphorylated in EGF-dependent manner. Phosphorylated CA IX can interact with the regulatory subunit of PI3 kinase and contribute to Akt activation and thereby to cancer progression. Moreover, various mutations introduced into IC domain compromise acidification and/or adhesion capacity of CA IX (Hulikova et al., unpublished).

The role of CA XII in tumor biology remains less explored. In addition to CA XII contribution to pH control and tumor growth,[111] inhibition of CA XII has been associated with decreased *in vitro* invasion of renal carcinoma cells.[126]

8.7 MONOCLONAL ANTIBODIES AS TOOLS FOR IMAGING AND TARGETING OF CA IX

On the basis of the cell surface localization and the strong association with tumors, but rare expression in normal tissues, CA IX is an excellent target for detection and therapy of cancer using specific monoclonal antibodies. In contrast, clinical utilization of CA XII seems limited because this isoform is present in a broad range of normal tissues and its overexpression has been observed in a relatively low percentage of tumor tissues, particularly those with good prognosis (except brain tumors where CA XII expression correlates with aggressive phenotype[42]).

To date, there are several monoclonal antibodies useful for the CA IX detection purposes. Immunohistochemical detection of CA IX in human tissues has been predominantly performed with the M75 MAb that allowed for identification of CA IX and cloning of its cDNA.[7–9] M75 binds to a repetitive epitope in the N-terminal PG region, which is absent in the other carbonic anhydrase isoforms.[123] Due to recognition of a linear epitope, M75 can be used with different fixation and staining protocols, does not need demasking, and also works on old archival samples. It is therefore suitable for the routine survey of CA IX expression in tumor specimens for retrospective correlation studies and also for prediction of treatment outcome and stratification of patients. M75 is also a very valuable tool for the preclinical studies of the role of CA IX in cancer.

A series of additional monoclonal antibodies specific for the human CA IX protein has been generated using CA IX-deficient mice.[127] This approach has helped to avoid immunodominance of the PG region and led to production of several CA domain-specific MAbs. These new MAbs became useful for detection of the CA IX shed in body fluids.[65] The detection method is based on the combination of two noncompeting monoclonal antibodies, the PG domain-specific M75 MAb and CA domain-specific V/10 MAb. The same antibodies used in the same setting constitute a commercial

MN/CA IX ELISA offered by Siemens Medical solutions Diagnostics for detection of CA IX extracellular domain as a promising circulating cancer biomarker for patient selection, monitoring, and management.[128] As mentioned previously, this kit revealed a significant correlation between high levels of serum CA IX and tumor progression, as well as postoperative recurrence.[66]

The other well-known CA IX-specific monoclonal antibody G250 has been systematically investigated as a tool for clinical management of patients with renal carcinomas.[129] The G250 MAb and its humanized, chimeric, and bispecific variants have become the most important imaging and therapeutic tools targeting CA IX expressing tumors. Chimeric G250 (composed of variable regions of murine G250 and constant regions derived from human IgG) is recently produced by Wilex Co. under commercial name Rencarex®. Its effector therapeutic mechanism is ADCC (antibody-dependent cell cytotoxicity). Rencarex is currently in a pivotal Phase III trial as an adjuvant therapy of patients with nonmetastatic renal cell cancer. In Phase I and II studies with more than 100 RCC patients completed, Rencarex has shown good safety and tolerability and a promising efficacy profile.

In addition, the chimeric G250 MAb has recently entered a multicentric clinical trial as an *in vivo* imaging tool for noninvasive PET identification of RCC and potentially also for evaluation of early therapeutic response.[130]

Both G250 MAb and the recently generated VII/20 MAb bind to overlapping epitopes in CA domain of CA IX and are capable of internalizing in complex with CA IX.[131] Since receptor-mediated internalization is typical for signal-transduction molecules and is often associated with depletion of the cell surface receptor molecules, it is possible that these CA domain-specific antibodies (that do not cross-react with other CA isoforms) might function as direct modulators of CA IX function and might potentially have inherent anticancer potential independent of immune responses. This assumption is currently under investigation.

Monoclonal antibodies are not the only immunotherapeutic tools developed against CA IX expressing tumors. Other approaches include different types of vaccines (anti-idiotype, dendritic cell-based, oligopeptide, and chimeric protein vaccines) and genetically engineered cytotoxic cells that showed promising results when tested in preclinical settings as reviewed elsewhere.[132]

8.8 CARBONIC ANHYDRASE INHIBITORS FOR IMAGING AND TARGETING OF CA IX AND CA XII

Concept of carbonic anhydrase inhibitors as anticancer drugs has been first formulated by Supuran and coworkers on the basis of antiproliferative effects of some sulfonamide derivatives observed in cell lines derived from different types of human tumors.[133,134] However, sulfonamides generally show nonselective inhibition of different CA isoenzymes, and therefore their target could not be determined. According to our recent knowledge, at least in some cells lines, it could be CA IX and/or XII, which can be inhibited by many different classes of sulfonamides-derived or related compounds. In addition to classical sufonamides, such as acetazolamide, methazolamide, ethoxzolamide, and dichlorophenamide, good-to-excellent CA IX and/or XII inhibitory properties were proven for aromatic and heterocyclic sulfonamides,[135–137] halogenosulfanilamide and halogenophenylaminobenzolamide derivatives,[138] lipophilic sulfonamides,[139] sulfamates,[17,140] fluorine-containing sulfonamides,[16] sulfonamides incorporating 1,3,5-triazine and 1,2,4-triazine moieties,[141,142] sulfonamides derived from 4-isothiocyanato-benzolamide,[143] E7070 sulfonamide developed originally as anticancer agent blocking cell cycle,[144] sulfonamides incorporating thioureido-sulfanilyl scaffolds,[145] novel sulfanilamide/acetazolamide derivatives obtained by the tail approach,[146] sulfonamides incorporating hydrazine moieties,[147] 1,3,4-thiadiazole- and 1,2,4-triazole-thiols,[148] *N*-hydroxysulfamides,[149] polyfluorinated aromatic/heterocyclic sulfonamides,[150] benzo[*b*]thiophene 1,1-dioxide sulfonamides,[151] substituted di-fluoromethanesulfonamides,[152] indanesulfonamides,[153] and copper(II) complexes of polyamino-polycarboxylamido aromatic/heterocyclic sulfonamides.[154]

Interestingly, CA IX and XII can be also efficiently inhibited by nanomolar concentrations of celecoxib and valdecoxib sulfonamide inhibitors of cyclooxygense 2 (COX-2), which is a key enzyme of arachidonic acid metabolism involved in colorectal carcinoma.[155,156] In addition, CA IX can be inhibited by sulfamates, inhibitors of steroid sulfatase that play a role in the production of active steroids, and is implicated in breast cancer.[17] Also, sulthiame, a clinically used antiepileptic, is a potent inhibitor of CA IX and XII.[157]

Selectivity of inhibitors toward CA IX and/or XII can be achieved through modulation of their physical and chemical properties by various side chains and other modifications.[158,4,159] Certain alterations can introduce or improve the membrane impermeability so that the inhibitor binds only or predominantly to extracellular CAs.[160,161] The other modifications can affect the size or surface topology to fit better into active site cavity of CA IX/XII than into other isoforms. Some types of modifications improved inhibitors to work at subnanomolar concentrations[16] when analyzed against the recombinant catalytic domain of CA IX. These extensive studies revealed several compounds with a reasonable selectivity ratio favoring inhibition activities against CA IX/XII compared to other isoforms, in particular CA II. For example, the selectivity ratios for the inhibition of the tumor-associated CA IX and XII over the cytosolic isozymes CA I and II were in the range of 107–955 for glycosyl-thioureido-sulfonamides.[162] Another sophisticated strategy was used to generate hypoxia-activatable inhibitors. Different 2-mercapto-substituted-benzenesulfonamides and their disulfides/sulfones showed substaintially increased inhibition capacity than the corresponding oxidized (S-S type) derivatives.[163] The best representatives out of these differentially acting derivatives can serve as lead compounds for further development of CA IX/XII-specific inhibitors with anticancer potential.

The above inhibition data were obtained for the purified recombinant catalytic domains of CA IX and XII, and therefore it is difficult to predict their biological effects on cells in culture and in tumor tissue. Initial experimental data provide the proof of concept that inhibition of CA IX activity may affect pH regulation in hypoxic cells. Indeed, it has been demonstrated that the fluorescein-conjugated carbonic anhydrase inhibitor thioureido-homosulfanilamide (FITC-CAI) could bind only to hypoxic cells expressing CA IX, but neither to CA IX-positive normoxic cells nor to CA IX-negative controls.[108,164] Under reoxygenation, FITC-CAI is released from binding to CA IX.[165] Moreover, recent results show that this FITC-conjugated sulfonamide accumulates in the central hypoxic areas of tumor spheroids and that the zone of sulfonamide accumulation overlaps with the zone of CA IX expression (Hulikova et al., unpublished). Furthermore, the inhibition of CA IX activity slows down extracellular acidification by a greater extent at the spheroid core (which is usually affected by hypoxia).[111] This finding suggests that CA IX active site is accessible to sulfonamide only under hypoxic conditions. It is possible that CA IX isozyme undergoes structural changes triggered by hypoxia that opens the active site and allows inhibitor binding.

More important, exclusive binding of the FITC-CAI to hypoxic cells suggests that labeled CA IX-selective sulfonamides can be potentially used as tools for *in vivo* imaging of hypoxic tumors. In contrast to monoclonal antibodies that can visualize the CA IX-positive cells independently of their oxygenation, appropriately labeled sulfonamides should image only those CA IX-positive cells that are actually hypoxic.[166,167]

CA inhibitors were also proposed to serve as anticancer drugs targeting hypoxic tumor cells via inhibition of CA IX and XII enzyme function. Experimental data obtained with the transient knockdown of CA IX or deletion of its catalytic domain indicate that pH regulation facilitated by CA IX (and possibly also CA XII) is important for the survival of tumor cells in hypoxia. Therefore, it is quite plausible that inhibition of the CA catalytic activity could lead to perturbed pH regulation in hypoxic cells and reduction of their capability to adapt to hypoxic stress. Of course, even more striking effects could possibly be achieved by simultaneous treatment with inhibitors of bicarbonate transporters and lactate/proton extrusion molecules. Indeed, several sulfonamide derivatives were capable of reducing an extracellular acidification mediated by CA IX, suggesting that these compounds interfere with the activity of CA IX in cell culture exposed to hypoxia and deserve further investigation as anticancer drugs.[108,164,111]

Alternatively, CA inhibitors could be combined with conventional chemotherapeutic drugs whose uptake depends on pH gradient across the plasma membrane. For example, reduction of extracellular acidosis can increase the uptake and cytotoxic effects of weakly basic drugs, including doxorubicin.[168,169,122,112] Interestingly, chronic ingestion of sodium bicarbonate solution was shown to enhance the capacity of doxorubicin to decrease the tumor size.[122] Acetazolamide, a classical CA inhibitor, also reduced *in vivo* growth of tumor when given alone and produced additive tumor growth delays when administered in combination with various chemotherapeutic compounds.[170] These data need to be confirmed with the new generation of sulfonamides that are selective for CA IX and XII.

Sulfonamides that preferentially bind to CA IX or XII could be also potentially used for selective delivery of therapeutic moieties such as isotopes and cytotoxic agents to tumor cells.[159,167] To this end, boron-containing inhibitors with high affinity for CA IX have been designed and synthesized.[171] These inhibitors are potentially applicable in boron neutron capture therapy (BNTC) of tumors that are nonresponsive to classical therapeutic modalities.

8.9 CONCLUSION

Carbonic anhydrases CA IX and XII have become interesting molecules from the both research and clinical points of view. Their molecular and functional studies led to better understanding of pH control mechanisms operating in hypoxic cells and to elucidation of adaptive responses to hypoxia and acidosis leading to tumor progression.[5,172] Owing to tumor-associated and hypoxia-related expression pattern, these enzymes (particularly CA IX) can provide clinically useful diagnostic and/or prognostic information and can also serve for antibody-mediated immunotherapy of certain tumor types.[12] Functional involvement in cancer biology might allow utilization of CA IX and XII as molecular targets for function-perturbing or activity-inhibiting compounds, such as antibodies and inhibitors.[173] Numerous experimental and clinical studies do support diagnostic/prognostic value of CA IX and show that it is an excellent target for immunotherapy.[174] On the other hand, functional targeting of CA IX and XII as enzymes is only in an early stage of preclinical research and requires further efforts.

ACKNOWLEDGMENT

The research of the authors is supported by grants from the 7th Framework Program of the European Commission (Collaborative project METOXIA FP7-HEALTH-2007-222741) and from the Slovak Research and Development Support Agency (APVV-51-024805, APVV-51-024905).

SP is the member of the EORTC PathoBiology Group.

REFERENCES

1. Supuran, C. T. Carbonic anhydrases: novel therapeutic applications for inhibitors and activators. *Nat. Rev. Drug Discov.* **2008**, *7*, 168–181.
2. Parkkila, S. An overview of the distribution and function of carbonic anhydrase in mammals, In *The Carbonic Anhydrases: New Horizons;* Chegwidden, W. R.; Carter, N.; Edwards, Y., Eds.; Birkhauser Verlag: Basil, **2000**; pp 76–93.
3. Hilvo, M.; Supuran, C. T.; Parkkila, S. Characterization and inhibition of the recently discovered carbonic anhydrase isoforms CA XIII, XIV and XV. *Curr. Top. Med. Chem.* **2007**, *7*, 893–899.
4. Pastorekova, S.; Parkkila, S.; Pastorek, J.; Supuran, C. T. Carbonic anhydrases: current state of the art, therapeutic applications and future prospects. *J. Enzyme Inhib. Med. Chem.* **2004**, *19*, 199–229.
5. Brahimi-Horn, M. C.; Pouysségur, J. Hypoxia in cancer cell metabolism and pH regulation. *Essays Biochem.* **2007**, *43*, 165–178.
6. Pastorekova, S.; Parkkila, S.; Zavada, J. Tumor-associated carbonic anhydrases and their clinical significance. *Adv. Clin. Chem.* **2006**, *42*, 167–216.
7. Pastorekova, S.; Zavadova, Z.; Kostal, M.; Babusikova, O.; Zavada, J. A novel quasi-viral agent, MaTu, is a two-component system. *Virology* **1992**, *187*, 620–626.

8. Zavada, J.; Zavadova, Z.; Pastorekova, S.; Ciampor, F.; Pastorek, J.; Zelnik, V. Expression of MaTu-MN protein in human tumor cultures and in clinical specimens. *Int. J. Cancer.* **1993**, *54*, 268–274.

9. Pastorek, J.; Pastorekova, S.; Callebaut, I.; Mornon, J. P.; Zelnik, V.; Opavsky, R.; Zatovicova, M.; Liao, S.; portetelle, D.; Stanbridge, E. J.; Zavada, J.; Burny, A.; Kettmann, R.; et al. Cloning and characterization of MN, a human tumor-associated protein with a domain homologous to carbonic anhydrase and putative helix–loop–helix DNA binding segment. *Oncogene* **1994**, *9*, 2877–2888.

10. Opavsky, R.; Pastorekova, S.; Zelnik, V.; Gibadulinova, A.; Stanbridge, E. J.; Zavada, J.; Kettmann, R.; Pastorek, J. Human MN/CA9 gene, a novel member of the carbonic anhydrase family: structure and exon to protein domain relationships. *Genomics* **1996**, *33*, 480–487.

11. Hewett-Emmett, D.; Tashian, R. E. Functional diversity, conservation and convergence in the evolution of the alpha, beta and gamma carbonic anhydrase gene families. *Mol. Phylogenet. Evol.* **1996**, *5*, 50–77.

12. Oosterwijk, E. Carbonic anhydrase IX: historical and future perspectives. *BJU Int.* **2008**, *101* (Suppl. 4), 2–7.

13. Hilvo, M.; Baranauskiene, L.; Salzano, A. M.; Scaloni, A.; Matulis, D.; Innocenti, A.; Scozzafava, A.; Monti, S. M.; Di Fiore, A.; De Simone, G.; Lindfors, M.; Jänis, J.; Valjakka, J.; Pastoreková, S.; Pastorek, J.; Kulomaa, M. S.; Nordlund, H. R.; Supuran, C. T.; Parkkila, S. Biochemical characterization of CA IX: one of the most active carbonic anhydrase isozymes. *J. Biol. Chem.* **2008**, *283* 27799–27809. Aug 13. [Epub ahead of print].

14. Dorai, T.; Sawczuk, I. S.; Pastorek, J.; Wiernik, P. H.; Dutcher, J. P. Biological significance of overexpression of carbonic anhydrase IX in renal cell carcinoma: EGF-induced phosphorylation of carbonic anhydrase IX leads to activation of PI-3-kinase pathway in renal cell carcinoma. *Eur. J. Cancer* **2005**, *41*, 2935–2947.

15. Wingo, T.; Tu, C.; Laipis, P. J.; Silverman, D. N. The catalytic properties of human carbonic anhydrase IX. *Biochem. Biophys. Res. Commun.* **2001**, *288*, 666–669.

16. Vullo, D.; Scozzafava, A.; Pastorekova, S.; Pastorek, J.; Supuran, C. T. Carbonic anhydrase inhibitors: inhibition of the tumor-associated isozyme IX with fluorine-containing sulfonamides. The first subnanomolar CA IX inhibitor discovered. *Bioorg. Med. Chem. Lett.* **2004**, *14*, 2351–2356.

17. Winum, J. Y.; Vullo, D.; Casini, A.; Montero, J. L.; Scozzafava, A.; Supuran, C. T. Carbonic anhydrase inhibitors. Inhibition of cytosolic isozymes I and II and transmembrane, tumor-associated isozyme IX with sulfamates including EMATE also acting as steroid sulfatase inhibitors. *J. Med. Chem.* **2003**, *46*, 2197–2204.

18. Türeci, O.; Sahin, U.; Vollmar, E.; Siemer, S.; Göttert, E.; Seitz, G.; Parkkila, A. K.; Shah, G. N.; Grubb, J. H.; Pfreundschuh, M.; Sly, W. S. Human carbonic anhydrase XII: cDNA cloning, expression and chromosomal localization of a carbonic anhydrase gene that is overexpressed in some renal cancers. *Proc. Natl. Acad. Sci. USA* **1998**, *95*, 7608–7613.

19. Ivanov, S. V.; Kuzmin, I.; Wei, M. H.; Pack, S.; Geil, L.; Johnson, B. E.; Stanbridge, E. J.; Lerman, M. I. Down-regulation of transmembrane carbonic anhydrases in renal cell carcinoma cell lines by wild-type von Hippel-Lindau transgenes. *Proc. Natl. Acad. Sci. USA* **1998**, *95*, 12596–12601.

20. Ulmasov, B.; Waheed, A.; Shah, G. N.; Grubb, J. H.; Sly, W. S.; Tu, C.; Silverman, D. N. Purification and kinetic analysis of recombinant CA XII, a membrane carbonic

anhydrase overexpressed in certain cancers. *Proc. Natl. Acad. Sci. USA* **2000**, *97*, 14212–14217.

21. Pastorekova, S.; Parkkila, S.; Parkkila, A. K.; Opavsky, R.; Zelník, V.; Saarnio, J.; Pastorek, J. Carbonic anhydrase IX, MN/CA IX: analysis of stomach complementary DNA sequence and expression in human and rat alimentary tracts. *Gastroenterology* **1997**, *112*, 398–408.
22. Saarnio, J.; Parkkila, S.; Parkkila, A. K.; Waheed, A.; Casey, M. C.; Zhou, X. Y.; Pastorekova, S.; Pastorek, J.; Karttunen, T.; Haukipuro, K.; Kairaluoma, M. I.; Sly, W. S. Immunohistochemistry of carbonic anhydrase isozyme IX (MN/CA IX) in human gut reveals polarized expression in the epithelial cells with the highest proliferative capacity. *J. Histochem. Cytochem.* **1998**, *46*, 497–504.
23. Kivela, A.; Parkkila, S.; Saarnio, J.; Karttunen, T. J.; Kivela, J.; Parkkila, A. K.; Pastorekova, S.; Pastorek, J.; Waheed, A.; Sly, W. S.; Rajaniemi, H. Expression of transmembrane carbonic anhydrase isozymes IX and XII in normal human pancreas and pancreatic tumors. *Histochem. Cell. Biol.* **2000**, *114*, 197–204.
24. Karhumaa, P.; Kaunisto, K.; Parkkila, S.; Waheed, A.; Pastorekova, S.; Pastorek, J.; Sly, W. S.; Rajaniemi, H. Expression of the transmembrane carbonic anhydrases, CA IX and CA XII, in the human male excurrent ducts. *Mol. Hum. Reprod.* **2001**, *7*, 611–616.
25. Ivanov, S.; Liao, S. Y.; Ivanova, A.; Danilkovich-Miagkova, A.; Tarasova, N.; Weirich, G.; Merrill, M. J.; Proescholdt, M. A.; Oldfield, E. H.; Lee, J.; Zavada, J.; Waheed, A.; Sly, W.; Lerman, M. I.; Stanbridge, E. J. Expression of hypoxia-inducible cell-surface transmembrane carbonic anhydrases in human cancer. *Am. J. Pathol.* **2001**, *158*, 905–919.
26. Liao, S. Y.; Brewer, C.; Zavada, J.; Pastorek, J.; Pastorekova, S.; Manetta, A.; Berman, M. L.; DiSaia, P. J.; Stanbridge, E. J. Identification of the MN antigen as a diagnostic biomarker of cervical intraepithelial squamous and glandular neoplasia and cervical carcinomas. *Am. J. Pathol.* **1994**, *145*, 598–609.
27. Turner, J. R.; Odze, R. D.; Crum, C. P.; Resnick, M. B. MN antigen expression in normal, preneoplastic and neoplastic esophagus: a clinicopathological study of a new cancer-associated biomarker. *Hum. Pathol.* **1997**, *28*, 740–744.
28. Liao, S. Y.; Aurelio, O. N.; Jan, K.; Zavada, J.; Stanbridge, E. J. Identification of the MN/CA9 protein as a reliable diagnostic biomarker of clear cell carcinoma of the kidney. *Cancer Res.* **1997**, *57*, 2827–2831.
29. McKiernan, J. M.; Buttyan, R.; Bander, N. H.; Stifelman, M. D.; Katz, A. E.; Olsson, C. A.; Sawczuk, I. S. Expression of the tumour-associated gene MN: a potential biomarker for human renal cell carcinoma. *Cancer Res.* **1997**, *57*, 2362–2365.
30. Vermylen, P.; Roufosse, C.; Burny, A.; Verhest, A.; Bossehaerts, T.; Pastorekova, S.; Ninane, V.; Sculier, J. P. Carbonic anhydrase IX antigen differentiates between preneoplastic and malignant lesions in non-small cell lung carcinomas. *Eur. Respir. J.* **1999**, *14*, 806–811.
31. Bartosova, M.; Parkkila, S.; Pohlodek, K.; Karttunen, T. J.; Galbavy, S.; Mucha, V.; Harris, A. L.; Pastorek, J.; Pastorekova, S. Expression of carbonic anhydrase IX in breast is associated with malignant tissues and related to overexpression of c-erbB2. *J. Pathol.* **2002**, *197*, 1–8.
32. Haapasalo, J.; Nordfors, K.; Hilvo, M.; Rantala, I.; Soini, Y.; Parkkila, A. K.; Pastorekova, S.; Pastorek, J.; Parkkila, S.; Haapasalo, H. K. Expression of carbonic anhydrase IX in astrocytic tumors predicts poor prognosis. *Clin. Cancer Res.* **2006**, *12*, 473–477.

33. Leppilampi, M.; Saarnio, J.; Karttunen, T. J.; Kivelä, J.; Pastoreková, S.; Pastorek, J.; Waheed, A.; Sly, W. S.; Parkkila, S. Carbonic anhydrase isozymes IX and XII in gastric tumors. *World J. Gastroenterol.* **2003**, *9*, 1398–1403.

34. Saarnio, J.; Parkkila, S.; Parkkila, A. K.; Pastoreková, S.; Haukipuro, K.; Pastorek, J.; Juvonen, T.; Karttunen, T. J. Transmembrane carbonic anhydrase, MN/CA IX, is a potential biomarker for biliary tumours. *J. Hepatol.* **2001**, *35*, 643–649.

35. Saarnio, J.; Parkkila, S.; Parkkila, A. K.; Haukipuro, K.; Pastorekova, S.; Pastorek, J.; Kairaluoma, M. I.; Karttunen, T. J. Immunohistochemical study of colorectal tumors for expression of a novel transmembrane carbonic anhydrase, MN/CA IX, with potential value as a marker of cell proliferation. *Am. J. Pathol.* **1998**, *153*, 279–285.

36. Liao, S. Y.; Stanbridge, E. J. Expression of the MN antigen in cervical papanicolaou smears is an early diagnostic biomarker of cervical dysplasia. *Cancer Epidemiol. Biomarkers Prev.* **1996**, *5*, 549–557.

37. Liao, S. Y.; Stanbridge, E. J. Expression of MN/CA9 protein in Papanicolaou smears containing atypical glandular cells of undetermined significance is a diagnostic biomarker of cervical dysplasia and neoplasia. *Cancer* **2000**, *88*, 1108–1121.

38. Watson, P. H.; Chia, S. K.; Wykoff, C. C.; Han, C.; Leek, R. D.; Sly, W. S.; Gatter, K. C.; Ratcliffe, P. J.; Harris, A. L. Carbonic anhydrase XII is a marker of good in invasive breast carcinoma. *Br. J. Cancer* **2003**, *88*, 1065–1070.

39. Kivela, A. J.; Parkkila, S.; Saarnio, J.; Karttunen, T. J.; Kivela, J.; Parkkila, A. K.; Bartosova, M.; Mucha, V.; Novak, M.; Waheed, A.; Sly, W. S.; Rajaniemi, H.; Pastorekova, S.; Pastorek, J. Expression of von Hippel-Lindau tumor suppressor and tumor-associated carbonic anhydrases IX and XII in normal and neoplastic colorectal mucosa. *World J. Gastroenterol.* **2005**, *11*, 2616–2625.

40. Kim, J. Y.; Shin, H. J.; Kim, T. H.; Cho, K. H.; Shin, K. H.; Kim, B. K.; Roh, J. W.; Lee, S.; Park, S. Y.; Hwang, Y. J.; Han, I. O. Tumor-associated carbonic anhydrases are linked to metastases in primary cervical cancer. *J. Cancer Res. Clin. Oncol.* **2006**, *132*, 302–308.

41. Hynninen, P.; Vaskivuo, L.; Saarnio, J.; Haapasalo, H.; Kivelä, J.; Pastoreková, S.; Pastorek, J.; Waheed, A.; Sly, W. S.; Puistola, U.; Parkkila, S. Expression of transmembrane carbonic anhydrases IX and XII in ovarian tumours. *Histopathology* **2006**, *49*, 594–602.

42. Haapasalo, J.; Hilvo, M.; Nordfors, K.; Haapasalo, H.; Parkkila, S.; Hyrskyluoto, A.; Rantala, I.; Waheed, A.; Sly, W. S.; Pastorekova, S.; Pastorek, J.; Parkkila, A. K. Identification of an alternatively spliced isoform of carbonic anhydrase XII in diffusely infiltrating astrocytic gliomas. *Neuro Oncol.* **2008**, *10*, 131–138.

43. Hynninen, P.; Hamalainen, J. M.; Pastorekova, S.; Pastorek, J.; Waheed, A.; Sly, W. S.; Tomas, E.; Kirkinen, P.; Parkkila, S. Transmembrane carbonic anhydrase isozymes IX and XII in the female mouse reproductive tract. *Reprod. Biol. Endocrinol.* **2004**, *2*, 73.

44. Karhumaa, P.; Parkkila, S.; Tureci, O.; Waheed, A.; Grubb, J. H.; Shah, G.; Parkkila, S.; Kaunisto, K.; Tapanainen, J.; Sly, W. S.; Rajaniemi, H. Identification of carbonic anhydrase XII as the membrane isozyme expressed in the normal human endometrial epithelium. *Mol. Hum. Reprod.* **2000**, *6*, 68–74.

45. Parkkila, S.; Parkkila, A. K.; Saarnio, J.; Kivela, J.; Karttunen, T. J.; Kaunisto, K.; Waheed, A.; Sly, W. S.; Tureci, O.; Virtanen, I.; Rajaniemi, H. Expression of the membrane-associated carbonic anhydrase isozyme XII in the human kidney and renal tumors. *J. Histochem. Cytochem.* **2000**, *48*, 1601–1608.

46. Liao, S. Y.; Ivanov, S.; Ivanova, A.; Ghosh, S.; Cote, M. A.; Keefe, K.; Coca-Prados, M.; Stanbridge, E. J.; Lerman, M. I. Expression of cell surface transmembrane carbonic anhydrase genes CA9 and CA12 in the human eye: overexpression of CA12 (CAXII) in glaucoma. *J. Med. Genet.* **2003**, *40*, 257–261.

47. Gottsch, J. D.; Seitzman, G. D.; Margulies, E. H.; Bowers, A. L.; Michels, A. J.; Saha, S.; Jun, A. S.; Stark, W. J.; Liu, S. H. Gene expression in donor corneal endothelium. *Arch. Ophthalmol.* **2003**, *121*, 252–258.

48. Kallio, H.; Pastorekova, S.; Pastorek, J.; Waheed, A.; Sly, W. S.; Mannisto, S.; Heikinheimo, M.; Parkkila, S. Expression of carbonic anhydrases IX and XII during mouse embryonic development. *BMC Dev. Biol.* **2006**, *6*, 22.

49. Kaluz, S.; Kaluzova, M.; Opavsky, R.; Pastorekova, S.; Gibadulinova, A.; Dequiedt, F.; Kettmann, R.; Pastorek, J. Transcriptional regulation of the MN/CA9 gene coding for the tumor-associated carbonic anhydrase IX. Identification and characterization of a proximal silencer element. *J. Biol. Chem.* **1999**, *274*, 32588–32595.

50. Kaluzova, M.; Pastorekova, S.; Svastova, E.; Pastorek, J.; Stanbridge, E. J.; Kaluz, S. Characterization of the MN/CA9 proximal region: a role for SP and AP1 factors. *Biochem. J.* **2001**, *359*, 669–677.

51. Wykoff, C.; Beasley, N.; Watson, P.; Turner, K. J.; Pastorek, J.; Sibtain, A.; Wilson, G. D.; Turley, H.; Talks, K. L.; Maxwell, P. H.; Pugh, C. W.; Ratcliffe, P. J.; Harris, A. L. Hypoxia-inducible regulation of tumor-associated carbonic anhydrases. *Cancer Res.* **2000**, *60*, 7075–7083.

52. Kaelin, W. G., Jr.; Ratcliffe, P. J. Oxygen sensing by metazoans: the central role of the HIF hydroxylase pathway. *Mol. Cell.* **2008**, *30*, 393–402.

53. Coleman, M. L.; Ratcliffe, P. J. Oxygen sensing and hypoxia-induced responses. *Essays Biochem.* **2007**, *43*, 1–15.

54. Ruas, J. L.; Poellinger, L. Hypoxia-dependent activation of HIF into a transcriptional regulator. *Semin. Cell Dev. Biol.* **2005**, *16*, 514–522.

55. Kaluz, S.; Kaluzova, M.; Chrastina, A.; Olive, P. L.; Pastorekova, S.; Pastorek, J.; Lerman, M. I.; Stanbridge, E. J. Lowered oxygen tension induces expression of the hypoxia marker MN/carbonic anhydrase IX in the absence of hypoxia-inducible factor 1α stabilization: a role for phosphatidylinositol 3′-kinase. *Cancer Res.* **2002**, *62*, 4469–4477.

56. Kopacek, J.; Barathova, M.; Dequiedt, F.; Sepelakova, J.; Kettmann, R.; Pastorek, J.; Pastorekova, S. MAPK pathway contributes to density- and hypoxia-induced expression of the tumor-associated carbonic anhydrase IX. *Biochim. Biophys. Acta* **2005**, *1729*, 41–49.

57. Mandriota, S. J.; Turner, K. J.; Davies, D. R.; Murray, P. G.; Morgan, N. V.; Sowter, H. M.; Wykoff, C. C.; Maher, E. R.; Harris, A. L.; Ratcliffe, P. J.; Maxwell, P. H. HIF activation identifies early lesions in VHL kidneys: evidence for site-specific tumor suppressor function in the nephron. *Cancer Cell* **2002**, *1*, 459–468.

58. Cho, M.; Grabmaier, K.; Kitahori, Y.; Hiasa, Y.; Nakagawa, Y.; Uemura, H.; Hirao, Y.; Ohnishi, T.; Yoshikawa, K.; Ooesterwijk, E. Activation of the MN/CA9 gene is associated with hypomethylation in human renal cell carcinoma cell lines. *Mol. Carcinog.* **2000**, *27*, 184–189.

59. Cho, M.; Uemura, H.; Kim, S. C.; Kawada, Y.; Yoshida, K.; Hirao, Y.; Konishi, N.; Saga, S.; Oshikawa, K. Hypomethylation of the MN/CA9 promoter and upregulated MN/CA9 expression in human renal cell carcinoma. *Br. J. Cancer* **2001**, *85*, 563–567.

60. Jakubickova, L.; Biesova, Z.; Pastorekova, S.; Kettmann, R.; Pastorek, J. Methylation of the CA9 promoter can modulate expression of the tumor-associated carbonic anhydrase IX in dense carcinoma cell lines. *Int. J. Oncol.* **2005**, *26*, 1121–1127.

61. Barnett, D. H.; Sheng, S.; Charn, T. H.; Waheed, A.; Sly, W. S.; Lin, C. Y.; Liu, E. T.; Katzenellenbogen, B. S. Estrogen receptor regulation of carbonic anhydrase XII through a distal enhancer in breast cancer. *Cancer Res.* **2008**, *68*, 3505–3515.

62. Barathova, M.; Takacova, M.; Holotnakova, T.; Gibadulinova, A.; Ohradanova, A.; Zatovicova, M.; Hulikova, A.; Kopacek, J.; Parkkila, S.; Supuran, C. T.; Pastorekova, S.; Pastorek, J. Alternative splicing variant of the hypoxia marker carbonic anhydrase IX expressed independently of hypoxia and tumour phenotype. *Br. J. Cancer* **2008**, *98*, 129–136.

63. Rafajova, M.; Zatovicova, M.; Kettmann, R.; Pastorek, J.; Pastorekova, S. Induction by hypoxia combined with low glucose or low bicarbonate and high posttranslational stability upon reoxygenation contribute to carbonic anhydrase IX expression in cancer cells. *Int. J. Oncol.* **2004**, *24*, 995–1004.

64. Zatovicova, M.; Sedlakova, O.; Svastova, E.; Ohradanova, A.; Ciampor, F.; Arribas, J.; Pastorek, J.; Pastorekova, S. Ectodomain shedding of the hypoxia-induced carbonic anhydrase IX is a metalloprotease-dependent process regulated by TACE/ADAM17. *Br. J. Cancer* **2005**, *93*, 1267–1276.

65. Zavada, J.; Zavadova, Z.; Zatovicova, M.; Hyrsl, L.; Kawaciuk, I. Soluble form of carbonic anhydrase IX (CA IX) in the serum and urine of renal carcinoma patients. *Br. J. Cancer* **2003**, *89*, 1067–1071.

66. Li, G.; Feng, G.; Gentil-Perret, A.; Genin, C.; Tostain, J. Serum carbonic anhydrase 9 level is associated with postoperative recurrence of conventional renal cell cancer. *J. Urol.* **2008**, *180*, 510–513.

67. Vaupel, P.; Mayer, A. Hypoxia in cancer: significance and impact on clinical outcome. *Cancer Metastasis Rev.* **2007**, *26*, 225–239.

68. Moon, E. J.; Brizel, D. M.; Chi, J. T.; Dewhirst, M. W. The potential role of intrinsic hypoxia markers as prognostic variables in cancer. *Antioxid. Redox Signal.* **2007**, *9*, 1237–1294.

69. Beasley, N. J.; Wykoff, C. C.; Watson, P. H.; Leek, R.; Turley, H.; Gatter, K.; Pastorek, J.; Cox, G. J.; Ratcliffe, P.; Harris, A. L. Carbonic anhydrase IX, an endogenous hypoxia marker, expression in head and neck squamous cell carcinoma and its relationship to hypoxia, necrosis and microvessel density. *Cancer Res.* **2001**, *61*, 5262–5267.

70. Turner, K. J.; Crew, J. P.; Wykoff, C. C.; Watson, P. H.; Poulsom, R.; Pastorek, J.; Ratcliffe, P. J.; Cranston, D.; Harris, A. L. The hypoxia-inducible genes VEGF and CA9 are differentially regulated in superficial vs invasive bladder cancer. *Br. J. Cancer* **2002**, *86*, 1276–1282.

71. Swinson, D. E.; Jones, J. L.; Richardson, D.; Wykoff, C.; Turley, H.; Pastorek, J.; Taub, N.; Harris, A. L.; O'Byrne, K. J. Carbonic anhydrase IX expression, a novel surrogate marker of tumor hypoxia, is associated with a poor prognosis in non-small-cell lung cancer. *J. Clin. Oncol.* **2003**, *21*, 473–482.

72. Olive, P. L.; Aquino-Parsons, C.; MacPhail, S. H.; Laio, S.; Raleigh, J. A.; Lerman, M. I.; Stanbridge, E. J. Carbonic anhydrase 9 as an endogenous marker for hypoxic cells in cervical cancer. *Cancer Res.* **2001**, *61*, 8924–8929.

73. Shin, K. H.; Diaz-Gonzalez, J. A.; Russell, J.; Chen, Q.; Burgman, P.; Li, X. F.; Ling, C. C. Detecting changes in tumor hypoxia with carbonic anhydrase IX and pimonidazole. *Cancer Biol. Ther.* **2007**, *6*, 70–75.

74. Wykoff, C. C.; Beasley, N.; Watson, P. H.; Campo, L.; Chia, S. K.; English, R.; Pastorek, J.; Sly, W. S.; Ratcliffe, P.; Harris, A. L. Expression of the hypoxia-inducible and tumor-associated carbonic anhydrases in ductal carcinoma *in situ* of the breast. *Am. J. Pathol.* **2001**, *158*, 1011–1019.

75. Chia, S. K.; Wykoff, C. C.; Watson, P. H.; Han, C.; Leek, R. D.; Pastorek, J.; Gatter, K. C.; Ratcliffe, P.; Harris, A. L. Prognostic significance of a novel hypoxia-regulated marker, carbonic anhydrase IX, in invasive breast carcinoma. *J. Clin. Oncol.* **2001**, *19*, 3660–3668.

76. Giatromanolaki, A.; Koukourakis, M. I.; Sivridis, E.; Pastorek, J.; Wykoff, C. C.; Gatter, K. C.; Harris, A. L. Expression of hypoxia-inducible carbonic anhydrase-9 relates to angiogenic pathways and independently to poor outcome in non-small cell lung cancer. *Cancer Res.* **2001**, *61*, 7992–7998.

77. Generali, D.; Fox, S. B.; Berruti, A.; Brizzi, M. P.; Campo, L.; Bonardi, S.; Wigfield, S. M.; Bruzzi, P.; Bersiga, A.; Allevi, G.; Milani, M.; Aguggini, S.; Dogliotti, L.; Bottini, A.; Harris, A. L. Role of carbonic anhydrase IX expression in prediction of the efficacy and outcome of primary epirubicin/tamoxifen therapy for breast cancer. *Endocr. Relat. Cancer* **2006**, *13*, 921–930.

78. Brennan, D. J.; Jirstrom, K.; Kronblad, A.; Millikan, R. C.; Landberg, G.; Duffy, M. J.; Rydén, L.; Gallagher, W. M.; O'Brien, S. L. CA IX is an independent prognostic marker in premenopausal breast cancer patients with one to three positive lymph nodes and a putative marker of radiation resistance. *Clin. Cancer Res.* **2006**, *12*, 6421–6431.

79. Hussain, S. A.; Ganesan, R.; Reynolds, G.; Gross, L.; Stevens, A.; Pastorek, J.; Murray, P. G.; Perunovic, B.; Anwar, M. S.; Billingham, L.; James, N. D.; Spooner, D.; Poole, C. J.; Rea, D. W.; Palmer, D. H. Hypoxia-regulated carbonic anhydrase IX expression is associated with poor survival in patients with invasive breast cancer. *Br. J. Cancer* **2007**, *96*, 104–109.

80. Trastour, C.; Benizri, E.; Ettore, F.; Ramaioli, A.; Chamorey, E.; Pouysségur, J.; Berra, E. HIF-1alpha and CA IX staining in invasive breast carcinomas: prognosis and treatment outcome. *Int. J. Cancer* **2007**, *120*, 1451–1458.

81. Kon-no, H.; Ishii, G.; Nagai, K.; Yoshida, J.; Nishimura, M.; Nara, M.; Fujii, T.; Murata, Y.; Miyamoto, H.; Ochiai, A. Carbonic anhydrase IX expression is associated with tumor progression and a poor prognosis of lung adenocarcinoma. *Lung Cancer* **2006**, *54*, 409–418.

82. Le, Q. T.; Kong, C.; Lavori, P. W.; O'Byrne, K.; Erler, J. T.; Huang, X.; Chen, Y.; Cao, H.; Tibshirani, R.; Denko, N.; Giaccia, A. J.; Koong, A. C. Expression and prognostic significance of a panel of tissue hypoxia markers in head-and-neck squamous cell carcinomas. *Int. J. Radiat. Oncol. Biol. Phys.* **2007**, *69*, 167–175.

83. Korkolopoulou, P.; Perdiki, M.; Thymara, I.; Boviatsis, E.; Agrogiannis, G.; Kotsiakis, X.; Angelidakis, D.; Rologis, D.; Diamantopoulou, K.; Thomas-Tsagli, E.; Kaklamanis, L.; Gatter, K.; Patsouris, E. Expression of hypoxia-related tissue factors in astrocytic gliomas. A multivariate survival study with emphasis upon carbonic anhydrase IX. *Hum. Pathol.* **2007**, *38*, 629–638.

84. Chen, J.; Röcken, C.; Hoffmann, J.; Krüger, S.; Lendeckel, U.; Rocco, A.; Pastorekova, S.; Malfertheiner, P.; Ebert, M. P. Expression of carbonic anhydrase 9 at the invasion front of gastric cancers. *Gut* **2005**, *54*, 920–927.

85. Driessen, A.; Landuyt, W.; Pastorekova, S.; Moons, J.; Goethals, L.; Haustermans, K.; Nafteux, P.; Penninckx, F.; Geboes, K.; Lerut, T.; Ectors, N. Expression of carbonic

anhydrase IX (CA IX), a hypoxia-related protein, rather than vascular-endothelial growth factor (VEGF), a pro-angiogenic factor, correlates with an extremely poor prognosis in esophageal and gastric adenocarcinomas. *Ann. Surg.* **2006**, *243*, 334–340.

86. Ord, J. J.; Streeter, E. H.; Roberts, I. S.; Cranston, D.; Harris, A. L. Comparison of hypoxia transcriptome *in vitro* with *in vivo* gene expression in human bladder cancer. *Br. J. Cancer* **2005**, *93*, 346–354.

87. Winter, S. C.; Buffa, F. M.; Silva, P.; Miller, C.; Valentine, H. R.; Turley, H.; Shah, K. A.; Cox, G. J.; corbridge, R. J.; Homer, J. J.; Musgrove, B.; Slevin, N.; Sloan, P.; Price, P.; West, C. M.; Harris, A. L. Relation of a hypoxia metagene derived from head and neck cancer to prognosis of multiple cancers. *Cancer Res.* **2007**, *67*, 3441–3449.

88. Kaelin W. G., Jr. The von Hippel-Lindau tumor suppressor protein and clear cell renal carcinoma. *Clin. Cancer Res.* **2007**, *13*, 680s–684s.

89. Wiesener, M. S.; Munchenhagen, P. M.; Berger, I.; Morgan, N. V.; Roigas, J.; Schwiertz, A.; Jurgensen, J. S.; Gruber, G.; Maxwell, P. H.; Loning, S. A.; Frei, U.; Maher, E. R.; Grone, H. J.; Eckhardt, K. U. Constitutive activation of hypoxia-inducible genes related to overexpression of hypoxia-inducible factor-1α in clear cell renal carcinomas. *Cancer Res.* **2001**, *61*, 5215–5222.

90. Raval, R. R.; Lau, K. W.; Tran, M. G.; Sowter, H. M.; Mandriota, S. J.; Li, J. L.; Pugh, C. W.; Maxwell, P. H.; Harris, A. L.; Ratcliffe, P. J. Contrasting properties of hypoxia-inducible factor 1 (HIF-1) and HIF-2 in von Hippel-Lindau-associated renal cell carcinoma. *Mol. Cell Biol.* **2005**, *25*, 5675–5686.

91. Bui, M. H.; Seligson, D.; Han, K. R.; Pantuck, A. J.; Dorey, F. J.; Huang, Y.; Horvath, S.; Leibovich, B. C.; Chopra, S.; Liao, S. Y.; Stanbridge, E. J.; Lerman, M. I.; Palotie, A.; Figlin, R. A.; Belldegrun, A. S. Carbonic anhydrase IX is an independent predictor of survival in advanced renal clear cell carcinoma: implications for prognosis and therapy. *Clin. Cancer Res.* **2003**, *9*, 802–811.

92. Bui, M. H.; Visapaa, H.; Seligson, D.; Kim, H.; Han, K. R.; Huang, Y.; Horvath, S.; Stanbridge, E. J.; Palotie, A.; Figlin, R. A.; Belldegrun, A. S. Prognostic value of carbonic anhydrase IX and Ki67 as predictors of survival for renal clear cell carcinoma. *J. Urol.* **2004**, *171*, 2461–2466.

93. Sandlund, J.; Oosterwijk, E.; Grankvist, K.; Oosterwijk-Wakka, J.; Ljungberg, B.; Rasmuson, T. Prognostic impact of carbonic anhydrase IX expression in human renal cell carcinoma. *BJU Int.* **2007**, *100*, 556–560.

94. Patard, J. J.; Fergelot, P.; Karakiewicz, P. I.; Klatte, T.; Trinh, Q. D.; Rioux-Leclercq, N.; Said, J. W.; Belldegrun, A. S.; Pantuck, A. J. Low CAIX expression and absence of VHL gene mutation are associated with tumor aggressiveness and poor survival of clear cell renal cell carcinoma. *Int. J. Cancer* **2008**, *123*, 395–400.

95. Pantuck, A. J.; Klatte, T.; Seligson, D.; Atkins, M.; Belldegrun, A. Carbonic anhydrase IX as a predictive biomarker for clear cell renal cell carcinoma. *J. Clin. Oncol.* **2008**, *26*, 3105–3107.

96. Brouwers, A. H.; Frielink, C.; Oosterwijk, E.; Oyen, W. J.; Corstens, F. H.; Boerman, O. C. Interferons can upregulate the expression of the tumor associated antigen G250-MN/CA IX, a potential target for (radio)immunotherapy of renal cell carcinoma. *Cancer Biother. Radiopharm.* **2003**, *18*, 539–547.

97. Atkins, M.; Regan, M.; McDermott, D.; Mier, J.; Stanbridge, E.; Youmans, A.; Febbo, P.; Upton, M.; Lechpammer, M.; Signoretti, S. Carbonic anhydrase IX expression

predicts outcome of interleukin 2 therapy for renal cancer. *Clin. Cancer Res.* **2005**, *11*, 3714–3721.

98. Shuch, B.; Li, Z.; Belldegrun, A. S. Carbonic anhydrase IX and renal cell carcinoma: prognosis, response to systemic therapy, and future vaccine strategies. *BJU Int.* **2008**, *101* (Suppl. 4), 25–30.

99. Pantuck, A. J.; Zeng, G.; Belldegrun, A. S.; Figlin, R. A. Pathobiology, prognosis, and targeted therapy for renal cell carcinoma: exploiting the hypoxia-induced pathway. *Clin. Cancer Res.* **2003**, *9*, 4641–4652.

100. Tunuguntla, H. S.; Jorda, M. Diagnostic and prognostic molecular markers in renal cell carcinoma. *J. Urol.* **2008**, *179*, 2096–2102.

101. Swinson, D. E.; Jones, J. L.; Cox, G.; Richardson, D.; Harris, A. L.; O'Byrne, K. J. Hypoxia-inducible factor-1 alpha in non small cell lung cancer: relation to growth factor, protease and apoptosis pathways. *Int. J. Cancer* **2004**, *111*, 43–50.

102. Le, Q. T.; Chen, E.; Salim, A.; Cao, H.; Kong, C. S.; Whyte, R.; Donington, J.; Cannon, W.; Wakelee, H.; Tibshirani, R.; Mitchell, J. D.; Richardson, D.; Obyrne, K. J.; Koong, A. C.; Giaccia, A. J. An evaluation of tumor oxygenation and gene expression in patients with early stage non-small cell lung cancers. *Clin. Cancer Res.* **2006**, *12*, 1507–1514.

103. Vleugel, M. M.; Shvarts, D.; van der Wall, E.; van Diest, P. J. p300 and p53 levels determine activation of HIF-1 downstream targets in invasive breast cancer. *Hum. Pathol.* **2006**, *37*, 1085–1092.

104. Storci, G.; Sansone, P.; Trere, D.; Tavolari, S.; Taffurelli, M.; Ceccarelli, C.; Guarnieri, T.; Paterini, M.; Pariali, M.; Montanaro, L.; Santini, D.; Chieco, P.; Bonafe, M. The basal-like breast carcinoma phenotype is regulated by SLUG gene expression. *J. Pathol.* **2008**, *214*, 25–37.

105. Jarvela, S.; Parkkila, S.; Bragge, H.; Kahkonen, M.; Parkkila, A. K.; Soini, Y.; Pastorekova, S.; Pastorek, J.; Haapasalo, H. Carbonic anhydrase IX in oligodendroglial brain tumors. *BMC Cancer* **2008**, *8*, 1.

106. Sansone, P.; Storci, G.; Tavolari, S.; Guarnieri, T.; Giovannini, C.; Taffurelli, M.; Ceccarelli, C.; Santini, D.; Paterini, P.; Marcu, K. B.; Chieco, P.; Bonafe, M. IL-6 triggers malignant features in mammospheres from human ductal breast carcinoma and normal mammary gland. *J. Clin. Invest.* **2007**, *117*, 3988–4002.

107. Kivela, A.; Parkkila, S.; Saarnio, J.; Karttunen, T. J.; Kivela, J.; Parkkila, A. K.; Waheed, A.; Sly, W. S.; Grubb, J. H.; Shah, G.; Tureci, O.; Rajaniemi, H. Expression of a novel transmembrane carbonic anhydrase isozyme XII in normal human gut and colorectal tumors. *Am. J. Pathol.* **2000**, *156*, 577–584.

108. Svastova, E.; Hulikova, A.; Rafajova, M.; Zat'ovicova, M.; Gibadulinova, A.; Casini, A.; Scozzafava, A.; Supuran, C. T.; Pastorek, J.; Pastorekova, S. Hypoxia activates the capacity of tumor-associated carbonic anhydrase IX to acidify extracellular pH. *FEBS Lett.* **2004**, *577*, 439–495.

109. Robertson, N.; Potter, C.; Harris, A. L. Role of carbonic anhydrase IX in human tumor cell growth, survival, and invasion. *Cancer Res.* **2004**, *64*, 6160–6165.

110. Swietach, P.; Wigfield, S.; Cobden, P.; Supuran, C. T.; Harris, A. L.; Vaughan-Jones, R. D. Tumor-associated carbonic anhydrase 9 spatially coordinates intracellular pH in three-dimensional multicellular growths. *J. Biol. Chem.* **2008**, *283*, 20473–20483.

111. Chiche, J.; Ilc, K.; Laferrière, J.; Trottier, E.; Dayan, F.; Mazure, N. M.; Brahimi-Horn, M. C.; Pouysségur, J. Hypoxia-inducible carbonic anhydrase IX and XII promote tumor

cell growth by counteracting acidosis through the regulation of the intracellular pH. *Cancer Res.* **2009**, *69*, 358–368.

112. Stubbs, M.; McSheehy, P. M. J.; Griffiths, J. R.; Bashford, C. L. Causes and consequences of tumor acidity and implications for treatment. *Mol. Med. Today* **2000**, *6*, 15–19.

113. Helmlinger, G.; Sckell, A.; Dellian, M.; Forbes, N. S.; Jain, R. K. Acid production in glycolysis-impaired tumors provides new insights into tumor metabolism. *Clin. Cancer Res.* **2002**, *8*, 1284–1291.

114. Fukumura, D.; Jain, R. K. Tumor microenvironment abnormalities: causes, consequences, and strategies to normalize. *J. Cell. Biochem.* **2007**, *101*, 937–949.

115. Karumanchi, S. A.; Jiang, L.; Knebelmann, B.; Stuart-Tilley, A. K.; Alper, S. L.; Sukhatme, V. P. VHL tumor suppressor regulates Cl-/HCO3- exchange and Na^+/H^+ exchange activities in renal carcinoma cells. *Physiol. Genomics* **2001**, *5*, 119–128.

116. Shimoda, L. A.; Fallon, M.; Pisarcik, S.; Wang, J.; Semenza, G. L. HIF-1 regulates hypoxic induction of NHE1 expression and alkalinization of intracellular pH in pulmonary arterial myocytes. *Am. J. Physiol. Lung Cell. Mol. Physiol.* **2006**, *291*, L941–L949.

117. Ullah, M. S.; Davies, A. J.; Halestrap, A. P. The plasma membrane lactate transporter MCT4, but not MCT1, is up-regulated by hypoxia through a HIF-1alpha-dependent mechanism. *J. Biol. Chem.* **2006**, *281*, 9030–9037.

118. Casey, J. R. Why bicarbonate? *Biochem. Cell Biol.* **2006**, *84*, 930–939.

119. Innocenti, A.; Vullo, D.; Scozzafava, A.; Casey, J. R.; Supuran, C. T. Carbonic anhydrase inhibitors. Interaction of isozymes I, II, IV, V, and IX with carboxylates. *Bioorg. Med. Chem. Lett.* **2005**, *15*, 573–578.

120. Morgan, P. E.; Pastoreková, S.; Stuart-Tilley, A. K.; Alper, S. L.; Casey, J. R. Interactions of transmembrane carbonic anhydrase, CAIX, with bicarbonate transporters. *Am. J. Physiol. Cell Physiol.* **2007**, *293*, C738–C748.

121. Fukumura, D.; Jain, R. K. Tumor microvasculature and microenvironment: targets for anti-angiogenesis and normalization. *Microvasc. Res.* **2007**, *74*, 72–84.

122. Raghunand, N.; Gatenby, R. A.; Gillies, R. J. Microenvironmental and cellular consequences of altered blood flow in tumours. *Br. J. Radiol.* **2003**, *76*, S11–22.

123. Závada, J.; Závadová, Z.; Pastorek, J.; Biesová, Z.; Jezek, J.; Velek, J. Human tumour-associated cell adhesion protein MN/CA IX: identification of M75 epitope and of the region mediating cell adhesion. *Br. J. Cancer* **2000**, *82*, 1808–1813.

124. Zavadova, Z.; Zavada, J. Carbonic anhydrase IX (CA IX) mediates tumor cell interactions with microenvironment. *Oncol. Rep.* **2005**, *13*, 977–982.

125. Svastová, E.; Zilka, N.; Zat'ovicová, M.; Gibadulinová, A.; Ciampor, F.; Pastorek, J.; Pastoreková, S. Carbonic anhydrase IX reduces E-cadherin-mediated adhesion of MDCK cells via interaction with beta-catenin. *Exp. Cell Res.* **2003**, *290*, 332–345.

126. Parkkila, S.; Rajaniemi, H.; Parkkila, A. K.; Kivela, J.; Waheed, A.; Pastorekova, S.; Pastorek, J.; Sly, W. S. Carbonic anhydrase inhibitor suppresses invasion of renal cancer cells *in vitro*. *Proc. Natl. Acad. Sci. USA* **2000**, *97*, 2220–2224.

127. Zatovicova, M.; Tarabkova, K.; Svastova, E.; Gibadulinova, A.; Mucha, V.; Jakubickova, L.; Biesova, Z.; Rafajova, M.; Ortova Gut, M.; Parkkila, S.; Parkkila, A. K.; Waheed, A.; Sly, W. S.; Horak, I.; Pastorek, J.; Pastorekova, S. Monoclonal antibodies generated in carbonic anhydrase IX-deficient mice recognize different domains of tumour-associated hypoxia-induced carbonic anhydrase IX. *J. Immunol. Methods* **2003**, *282*, 117–134.

128. Carney, W. P. Circulating oncoproteins HER2/neu, EGFR and CAIX (MN) as novel cancer biomarkers. *Expert Rev. Mol. Diagn.* **2007**, *7*, 309–319.

129. Lam, J. S.; Pantuck, A. J.; Belldegrun, A. S.; Figlin, R. A. G250: a carbonic anhydrase IX monoclonal antibody. *Curr. Oncol. Rep.* **2005**, *7*, 109–115.

130. Divgi, C. The use of positron-emission tomography in the diagnosis of tumour phenotype. *BJU Int.* **2008**, *101*, 36–38.

131. Supalova, L.; Hulikova, A.; Barathova, M.; Gibadulinova, A.; Parkkila, S.; Supuran, C. T.; Pastorek, J.; Pastorekova, S.; Zatovicova, M. Characterization of G250 monoclonal antibody binding to human cancer-associated carbonic anhydrase IX. In Abstract booklet from *International CA IX Symposium: Functional and Clinical Aspects*, November 14, **2007**, Abstract No. 3.

132. Pastorekova, S.; Zavada, J. Carbonic anhydrase IX (CA IX) as a potential target for cancer therapy. *Cancer Ther.* **2004**, *2*, 245–262. Online: www.cancer-therapy.org.

133. Supuran, C.T.; Scozzafava, A. Carbonic anhydrase inhibitors and their therapeutic potential. *Expert Opin. Ther. Pat.* **2000**, *10*, 575–600.

134. Supuran, C. T.; Briganti, F.; Tilli, S.; Chegwidden, W. R.; Scozzafava, A. Carbonic anhydrase inhibitors: sulfonamides as antitumor agents? *Bioorg. Med. Chem. Lett.* **2001**, *9*, 703–714.

135. Vullo, D.; Franchi, M.; Gallori, E.; Pastorek, J.; Scozzafava, A.; Pastorekova, S.; Supuran, C. T. Carbonic anhydrase inhibitors. Inhibition of the tumor-associated isozyme IX with aromatic and heterocyclic sulfonamides. *Bioorg. Med. Chem. Lett.* **2003**, *13*, 1005–1009.

136. Jaiswal, M.; Khadikar, P. V.; Scozzafava, A.; Supuran, C. T. Carbonic anhydrase inhibitors: the first QSAR study on inhibition of tumor-associated isoenzyme IX with aromatic and heterocyclic sulfonamides. *Bioorg. Med. Chem. Lett.* **2004**, *14*, 3283–3290.

137. Ozensoy, O.; Pucetti, L.; Fasolis, G.; Arslan, O.; Scozzafava, A.; Supuran, C. T. Carbonic anhydrase inhibitors: inhibition of the tumor-associated isoenzymes IX and XII with a library of aromatic and heteroaromatic sulfonamides. *Bioorg. Med. Chem. Lett.* **2005**, *15*, 4862–4866.

138. Ilies, M. A.; Vullo, D.; Pastorek, J.; Scozzafava, A.; Ilies, M.; Caproiu, M. T.; Pastorekova, S.; Supuran, C. T. Carbonic anhydrase inhibitors. Inhibition of tumor-associated isozyme IX by halogenosulfanilamide and halogenophenylaminobenzolamide derivatives. *J. Med. Chem.* **2003**, *46*, 2187–2196.

139. Franchi, M.; Vullo, D.; Gallori, E.; Pastorek, J.; Russo, A.; Scozzafava, A.; Pastorekova, S.; Supuran, C. T. Carbonic anhydrase inhibitors. Inhibition of cytosolic isozymes I and II and transmembrane, cancer-associated isozyme IX with lipophilic sulfonamides. *J. Enzyme Inhib. Med. Chem.* **2003**, *18*, 333–338.

140. Winum, J. Y.; Pastorekova, S.; Jakubickova, L.; Montero, J. L.; Scozzafava, A.; Pastorek, J.; Vullo, D.; Innocenti, A.; Supuran, C. T. Carbonic anhydrase inhibitors: synthesis and inhibition of cytosolic/tumor-associated carbonic anhydrase isozymes I, II and IX with bis-sulfamates. *Bioorg. Med. Chem. Lett.* **2005**, *15*, 573–578.

141. Garaj, V.; Pucetti, L.; Fasolis, G.; Winum, J. Y.; Montero, J. L.; Scozzafava, A.; Vullo, D.; Innocenti, A.; Supuran, C. T. Carbonic anhydrase inhibitors: synthesis and inhibition of cytosolic/tumor-associated carbonic anhydrase isozymes I, II, and IX with sulfonamides incorporating 1,2,4-triazine moieties. *Bioorg. Med. Chem. Lett.* **2004**, *14*, 5427–5433.

142. Garaj, V.; Pucetti, L.; Fasolis, G.; Winum, J. Y.; Montero, J. L.; Scozzafava, A.; Supuran, C. T. Carbonic anhydrase inhibitors: novel sulfonamides incorporating 1,3,5-triazine

moieties as inhibitors of cytosolic and tumor-associated carbonic anhydrase isozymes I, II, and IX. *Bioorg. Med. Chem. Lett.* **2005**, *15*, 3102–3108.

143. Cecchi, A.; Winum, J. Y.; Innocenti, A.; Vullo, D.; Montero, J. L.; Scozzafava, A.; Supuran, C. T. Carbonic anhydrase inhibitors: synthesis and inhibition of cytosolic/tumor associated carbonic anhydrase isozymes I, II, and IX with sulfonamides derived from 4-isothiocyanato-benzolamide. *Bioorg. Med. Chem. Lett.* **2004**, *14*, 5775–5780.

144. Abbate, F.; Casini, A.; Owa, T.; Scozzafava, A.; Supuran, C. T. Carbonic anhydrase inhibitors: E7070, a sulfonamide anticancer agent, potently inhibits cytosolic isozymes I and II, and transmembrane, tumor-associated isozyme IX. *Bioorg. Med. Chem. Lett.* **2004**, *14*, 225–229.

145. Pucetti, L.; Fasolis, G.; Cecchi, A.; Winum, J. Y.; Gamberi, A.; Montero, J. L.; Scozzafava, A.; Supuran, C. T. Carbonic anhydrase inhibitors: synthesis and inhibition of cytosolic/tumor-associated carbonic anhydrase isozymes I, II and IX with sulfonamides incorporating thioureido-sulfanilyl scaffolds. *Bioorg. Med. Chem. Lett.* **2005**, *15*, 2359–2364.

146. Turkmen, H.; Dugrun, M.; Yilmaztekin, S.; Emul, M.; Innocenti, A.; Vullo, D.; Scozzafava, A.; Supuran, C. T. Carbonic anhydrase inhibitors: novel sulfanilamide/acetazolamide derivatives obtained by the tail approach and their interaction with the cytosolic isozymes I and II and the tumor-associated isozymes IX. *Bioorg. Med. Chem. Lett.* **2005**, *17*, 367–372.

147. Winum, J. Y.; Dogne, J. M.; Casini, A.; de Leval, X.; Montero, J. L.; Scozzafava, A.; Vullo, D.; Innocenti, A.; Supuran, C. T. Carbonic anhydrase inhibitors: synthesis and inhibition of cytosolic/tumor-associated carbonic anhydrase isozymes I, II and IX with sulfonamides incorporating hydrazine moieties. *J. Med. Chem.* **2005**, *48*, 2121–2125.

148. Almajan, G. L.; Innocenti, A.; Pucetti, L.; Manole, G.; Berbuceanu, S.; Saramet, I.; Scozzafava, A.; Supuran, C. T. Carbonic anhydrase inhibitors: inhibition of the cytosolic and tumor-associated carbonic anhydrase isozymes I, II and IX with a series of 1,3,4-thiadiazole- and 1,2,4-triazole-thiols. *Bioorg. Med. Chem. Lett.* **2005**, *15*, 2347–2352.

149. Winum, J. Y.; Innocenti, A.; Nasr, J.; Montero, J. L.; Scozzafava, A.; Vullo, D.; Supuran, C. T. Carbonic anhydrase inhibitors: synthesis and inhibition of cytosolic/tumor-associated carbonic anhydrase isozymes I, II and IX with *N*-hydroxysulfamides: a new zinc-binding function in the design of inhibitors. *Bioorg. Med. Chem. Lett.* **2005**, *15*, 2353–2358.

150. Pastorekova, S.; Vullo, D.; Casini, A.; Scozzafava, A.; Pastorek, J.; Nishimori, I.; Supuran, C. T. Carbonic anhydrase inhibitors: synthesis and inhibition of the tumor-associated isozymes IX and XII with polyfluorinated aromatic/heterocyclic sulfonamides. *J. Enzyme Inhib. Med. Chem.* **2005**, *20*, 211–217.

151. Innocenti, A.; Villar, R.; Martinez-Merino, V.; Gil, M. J.; Scozzafava, A.; Vullo, D.; Supuran, C. T. Carbonic anhydrase inhibitors: inhibition of cytosolic/tumor-associated carbonic anhydrase isozymes I, II and IX with benzo[*b*]thiophene 1,1 dioxide sulfonamides. *Bioorg. Med. Chem. Lett.* **2005**, *15*, 4872–4876.

152. Cecchi, A.; Taylor, S. D.; Liu, Y.; Hill, B.; Vullo, D.; Scozzafava, A.; Supuran, C. T. Carbonic anhydrase inhibitors: inhibition of the human isozymes I, II, VA and IX with a library of substituted difluoromethanesulfinamides. *Bioorg. Med. Chem. Lett.* **2005**, *15*, 5192–5196.

153. Thiry, A.; Ledecq, M.; Cecchi, A.; Dogne, J. M.; Wouters, J.; Supuran, C. T.; Masereel, B. Indanesulfonamides as carbonic anhydrase inhibitors. Toward structure-based design of selective inhibitors of the tumor-associated isozyme CA IX. *J. Med. Chem.* **2006**, *49*, 2743–2749.

154. Rami, M.; Winum, J. Y.; Innocenti, A.; Montero, J. L.; Scozzafava, A.; Supuran, C. T. Carbonic anhydrase inhibitors: copper(II) complexes of polyamino-polycarboxylamido aromatic/heterocyclic sulfonamides are very potent inhibitors of the tumor-associated isoforms IX and XII. *Bioorg. Med. Chem. Lett.* **2008**, *18*, 836–841.

155. Weber, A.; Casini, A.; Heine, A.; Kuhn, D.; Supuran, C. T.; Scozzafava, A.; Klebe, G. Unexpected nanomolar inhibition of carbonic anhydrase by COX-2-selective celecoxib: new pharmacological opportunities due to related binding site recognition. *J. Med. Chem.* **2004**, *47*, 550–557.

156. Dogné, J. M.; Thiry, A.; Pratico, D.; Masereel, B.; Supuran, C. T. Dual carbonic anhydrase—cyclooxygenase-2 inhibitors. *Curr. Top. Med. Chem.* **2007**, *7*, 885–891.

157. Temperini, C.; Innocenti, A.; Mastrolorenzo, A.; Scozzafava, A.; Supuran, C. T. Carbonic anhydrase inhibitors. Interaction of the antiepileptic drug sulthiame with twelve mammalian isoforms: kinetic and X-ray crystallographic studies. *Bioorg. Med. Chem. Lett.* **2007**, *17*, 4866–4872.

158. Scozzafava, A.; Mastrolorenzo, A.; Supuran, C.T. Modulation of carbonic anhydrase activity and its applications in therapy. *Expert Opin. Ther. Pat.* **2004**, *14*, 667–702.

159. Winum, J. Y.; Scozzafava, A.; Montero, J. L.; Supuran, C. T. Design of zinc binding functions for carbonic anhydrase inhibitors. *Curr. Pharm. Des.* **2008**, *14*, 615–621.

160. Casey, J. R.; Morgan, P. E.; Vullo, D.; Scozzafava, A.; Mastrolorenzo, A.; Supuran, C. T. Carbonic anhydrase inhibitors. Design of selective, membrane-impermeant inhibitors targeting the human tumor-associated isozyme IX. *J. Med. Chem.* **2004**, *47*, 2337–2347.

161. Pastorekova, S.; Casini, A.; Scozzafava, A.; Vullo, D.; Pastorek, J.; Supuran, C. T. Carbonic anhydrase inhibitors: the first selective, membrane-impermeant inhibitors targeting the tumor-associated isozyme IX. *Bioorg. Med. Chem. Lett.* **2004**, *14*, 869–873.

162. Smaine, F. Z.; Winum, J. Y.; Montero, J. L.; Regainia, Z.; Vullo, D.; Scozzafava, A.; Supuran, C. T. Carbonic anhydrase inhibitors: selective inhibition of the extracellular, tumor-associated isoforms IX and XII over isozymes I and II with glycosyl-thioureido-sulfonamides. *Bioorg. Med. Chem. Lett.* **2007**, *17*, 5096–5100.

163. Saczewski, F.; Sławiński, J.; Kornicka, A.; Brzozowski, Z.; Pomarnacka, E.; Innocenti, A.; Scozzafava, A.; Supuran, C. T. Carbonic anhydrase inhibitors. Inhibition of the cytosolic human isozymes I and II, and the transmembrane, tumor-associated isozymes IX and XII with substituted aromatic sulfonamides activatable in hypoxic tumors. *Bioorg. Med. Chem. Lett.* **2006**, *16*, 4846–4851.

164. Cecchi, A.; Hulikova, A.; Pastorek, J.; Pastorekova, S.; Scozzafava, A.; Winum, J. Y.; Montero, J. L.; Supuran, C. T. Carbonic anhydrase inhibitors. Design of fluorescent sulfonamides as probes of tumor-associated carbonic anhydrase IX that inhibit isozyme IX-mediated acidification of hypoxic tumors. *J. Med. Chem.* **2005**, *48*, 4834–4841.

165. Dubois, L.; Douma, K.; Supuran, C. T-; Chiu R. K.; van Zandvoort, M. A.; Pastoreková, S.; Scozzafava, A.; Wouters, B. G.; Lambin, P. Imaging the hypoxia surrogate marker CA IX requires expression and catalytic activity for binding fluorescent sulfonamide inhibitors. *Radiother. Oncol.* **2007**, *83*, 367–373.

166. Alterio, V.; Vitale, R. M.; Monti, S. M.; Pedone, C.; Scozzafava, A.; Cecchi, A.; De Simone, G.; Supuran, C. T. Carbonic anhydrase inhibitors: X-ray and molecular modeling study for the interaction of a fluorescent antitumor sulfonamide with isozyme II and IX. *J. Am. Chem. Soc.* **2006**, *128*, 8329–8335.

167. Cecchi, A.; Ciani, L.; Winum, J. Y.; Montero, J. L.; Scozzafava, A.; Ristori, S.; Supuran, C. T. Carbonic anhydrase inhibitors: design of spin-labeled sulfonamides incorporating TEMPO moieties as probes for cytosolic or transmembrane isozymes. *Bioorg. Med. Chem. Lett.* **2008**, *18*, 3475–3480.

168. Vukovic, V.; Tannock, I. F. Influence of the pH on cytotoxicity of paclitaxel, mitoxanthrone and topotecan. *Br. J. Cancer* **1997**, *75*, 1167–1172.

169. Kozin, S. V.; Shkarin, P.; Gerweck, L. E. The cell transmembrane pH gradient in tumors enhances cytotoxicity of specific weak acid chemotherapeutics. *Cancer Res.* **2001**, *61*, 4740–4743.

170. Teicher, B. A.; Liu, S. D.; Liu, J. T.; Holden, S. A.; Herman, T. S. A carbonic anhydrase inhibitor is potential modulator of cancer therapies. *Anticancer Res.* **1993**, *13*, 1549–1556.

171. Winum, J. Y.; Cecchi, A.; Montero, J. L.; Innocenti, A.; Scozzafava, A.; Supuran, C. T. Carbonic anhydrase inhibitors. Synthesis and inhibition of cytosolic/tumor-associated carbonic anhydrase isozymes I, II and IX with boron-containing sulfonamides, sulfamides and sulfamates: towards agents for boron neutron capture therapy of hypoxic tumors. *Bioorg. Med. Chem. Lett.* **2005**, *15*, 3302–3306.

172. Pastorekova, S.; Ratcliffe, P. J.; Pastorek, J. Molecular mechanisms of carbonic anhydrase IX-mediated pH regulation under hypoxia. *BJU Int.* **2008**, *101* (Suppl. 4), 8–15.

173. Thiry, A.; Supuran, C. T.; Masereel, B.; Dogné, J. M. Recent developments of carbonic anhydrase inhibitors as potential anticancer drugs. *J. Med. Chem.* **2008**, *51*, 3051–3056.

174. Potter, C. P.; Harris, A. L. Diagnostic, prognostic and therapeutic implications of carbonic anhydrases in cancer. *Br. J. Cancer* **2003**, *89*, 2–7.

CHAPTER 9

Fluorescent- and Spin-Labeled Sulfonamides as Probe for Carbonic Anhydrase IX

ALESSANDRO CECCHI, LAURA CIANI, SANDRA RISTORI, and CLAUDIU T. SUPURAN
Dipartimento di Chimica, Università degli Studi di Firenze, Via della Lastruccia 3, I-50019 Sesto Fiorentino (Firenze), Italy

9.1 INTRODUCTION

In the past years, a lot of research regarding the carbonic anhydrases (CAs) was focused on the involvement of these enzymes in many types of tumor. The first evidence was found by Pastorekova's group[1] who detected a new isoform (CA IX) in a human carcinoma cell line HeLa in 1992, followed 6 years later by two independent groups (Tureci[2] and Ivanov[3]) that published the cDNA sequence of another tumor-associated isozyme, CA XII. Until now, these are the only CAs that are associated and overexpressed, especially in tumor tissue that are characterized by hypoxic conditions.

Solid human tumors present several regions that are deficient of oxygen, principally due to an inadequate vasculature of the growing mass. This status is called "hypoxia," and in such condition the delivery of oxygen to the tumor cannot usually meet the demand of the tumor cells, so that large regions of the tumor exist in a chronic state of supply–demand mismatch.[4]

It should be noted that this phenomenon in tumor progression is accompanied by a dramatic change in the gene expression profile and several hypoxia-induced genes, including oncogenes, tumor suppressor genes, stress proteins, and cytokines,[5] have been described. Furthermore, hypoxic tumors are characterized by the strong shift from oxidative respiration to anaerobic glycolysis, the high interstitial fluid pressure (IFP), and acidic extracellular pH (pH_e).[6]

The low value of pH_e was traditionally attributed to the accumulation of lactic acid, excessively produced by glycolysis;[7] recently, Svastova et al.[8] demonstrate that the tumor-associated carbonic anhydrase IX is also directly involved in pH_e regulation and that this enzyme is the main contributor to such external acidification. Furthermore, it has been demonstrated that this acidity can be perturbed by deletion of the enzyme active site or using CA inhibitors (CAIs).[8] This finding is the starting point to study the inhibition of human hCA IX catalytic domain with selective CAIs to reduce the pH_e, and to use such compounds as potential antitumor agents.[8–12]

The hCA IX isoform was observed in many tumor tissues but only in a few normal ones, usually belonging to the alimentary tract.[13] Thus, hCA IX is overexpressed in tumors affecting kidney, liver, pancreas, esophagus, lung, colon, ovary, brain, skin, and breast.[14,15] In some tissues (such as pancreas, skin, and others), there is an increasing hCA IX expression from normal to tumor cells, whereas other tissues express hCA IX only in the malignant ones (such as kidney, breast, lung, and others). These features, with the fact that hCA IX is a HIF target, render the enzyme a significant biological element: hypoxia, in fact, is a clinically important tumor parameter[16] and this enzyme can play an important role as a potential marker of hypoxic tumor.[17] At the moment, hCA IX distribution is usually examined in relation to microvasculature density as a measure of angiogenesis, to the extent of necrosis as an indicator of severe hypoxia, to tumor stage and disease progression.[11]

In this chapter, the development of fluorescent sulfonamides, already used to demonstrate the involvement of CA IX in tumor acidification processes and its usefulness as diagnostic tools and/or therapeutic agents, will be reviewed.[8,12] The design, the inhibitory properties, and the X-ray diffraction data of such compounds, as well as their biological evaluation as possible markers of hypoxic tumor cell lines will also be analyzed. Furthermore, the mechanism of action and the possible use of such derivatives as potential antitumor agents will be explained. Finally, a new set of spin-labeled compounds will be reported[18] to explore their ability to target hypoxic tumors overexpressing hCA IX by EPR techniques, as well as for diagnostic/therapeutic applications.

9.2 FLUORESCENT SULFONAMIDE AS CA IX INHIBITORS

9.2.1 Design of Fluorescent Sulfonamides

The binding study of several fluorescent sulfonamides complexed with the tumor-associated isozyme hCA IX has been reported both in normoxic and hypoxic conditions.[12] The rational drug design of these compounds includes the fluorescein moiety as a tail because it presents a high quantum yield and an appropriate exciting and emission wavelength when present in various biologically active derivatives (i.e., 495 nm and 519 nm, respectively).[19] The "head" of these inhibitors, on the other hand, must guarantee good interaction with the enzyme active site. For this reason, benzenesulfonamides variously substituted in the phenyl ring, as well as the benzolamide system, were used, which are able to make many interactions that stabilize the enzyme–inhibitor complex.[20] Finally, a thioureido moiety was chosen as central linking part of the inhibitors to ensure better inhibition

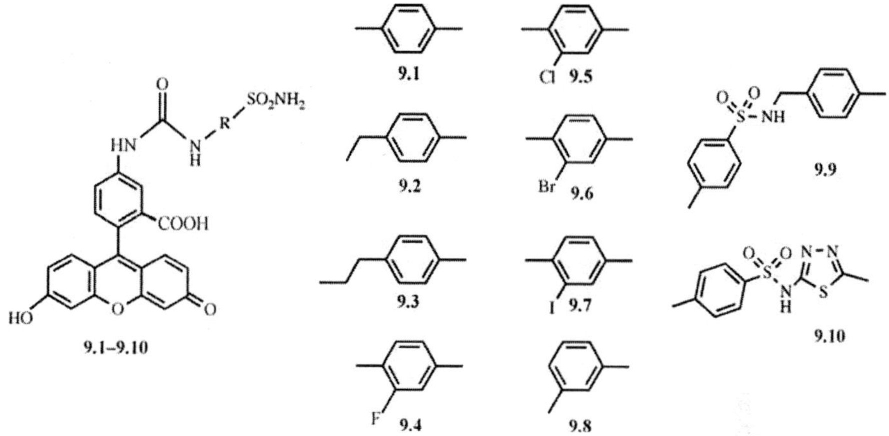

FIGURE 9.1 Chemical structures of fluorescent inhibitors **9.1–9.10**.

properties than the related ureido/carbonyl/carboxyl moieties.[21] The compounds **9.1–9.10** (Fig. 9.1) have been deeply investigated with inhibition and biological test[12] discussed in the following sections.

9.2.2 Inhibition Test and X-Ray Diffraction Studies

The inhibition properties of the sulfonamides **9.1–9.10** against the cytosolic isozymes hCA I and II and the transmembrane, tumor-associated isozyme hCA IX, together with those of standard, clinically used inhibitors (acetazolamide AZA, methazolamide MZA, ethoxzolamide EZA, dichlorophenamide DCP, and indisulam IND), are reported in Table 9.1.[12]

The data show the following structure–activity relationships:

(1) The fluorescent sulfonamides **9.1–9.10** act as moderate–weak inhibitors against the slow cytosolic isozyme hCA I, with inhibition constants in the range of 480–1500 nM.

(2) Compounds **9.1–9.10** behave as efficient inhibitors against the major cytosolic isozyme hCA II, with K_I values in a narrow range of 27–52 nM. Thus, the best hCA II inhibitors in this series were the aminobenzolamide derivative **9.10** and the sulfanilylhomosulfanilamide **9.9**, but the other compounds were only slightly less inhibitory than **9.9** and **9.10**. Despite this, all these compounds were less efficient hCA II inhibitors than the clinically used derivatives, which typically showed K_I values in the range of 8–15 nM.

(3) All these compounds were good inhibitors of hCA IX, with K_I values between 16 and 35 nM. The inhibition properties are similar to that of hCA II because the activity of these different derivatives does not change much. However, it is important to note how these compounds act as more efficient hCA IX than hCA II inhibitors. This is a remarkable finding, since possible drugs based on

TABLE 9.1 Inhibition Constant (K_I, nM) of Compounds 9.1–9.20

Compound	hCA I	hCA II	hCA IX	$K_{I hCA\ II}/K_{I hCA\ IX}$
AZA	900	12	25	0.48
MZA	780	14	27	0.52
EZA	25	8	34	0.23
DCP	1200	38	50	0.76
IND	31	15	24	0.62
9.1	1500	41	29	1.41
9.2	1450	44	26	1.69
9.3	1300	45	24	1.87
9.4	980	47	30	1.56
9.5	950	52	32	1.62
9.6	1100	43	35	1.23
9.7	1070	40	31	1.29
9.8	1400	52	34	1.53
9.9	630	34	20	1.70
9.10	480	27	16	1.68
9.11	179	41	41	1.00
9.12	204	42	35	1.20
9.13	182	28	39	0.72
9.14	2070	165	132	1.25
9.15	233	37	22	1.68
9.16	128	12	14	0.86
9.17	784	152	220	1.55
9.18	365	47	30	1.57
9.19	89	20	7	2.86
9.20	170	33	41	0.80

hCA IX inhibitors should bind as much as possible to the target cancer-associated isozymes (i.e., hCA IX and XII) but not to the other ubiquitous CA isozymes such as hCA II, IV, and V. It should also be noted that the hCA IX inhibitory properties of these fluorescent derivatives are in the same range as those of the clinically used sulfonamides, including indisulam, an antitumor sulfonamide in clinical trials.[10,11]

The inhibition mechanism of these fluorescent molecules was explained in detail by solving the X-ray structure of the complex between hCA II and compound **9.3**.[22] The inhibitor interaction does not generate relevant changes on the overall structure of the isoform II (Fig. 9.2). The zinc coordination and all the relevant amino acid residues lining the active site maintain the same conformation as in the native enzyme.[23] The inhibitor organic scaffold made a lot of hydrophobic contacts within the enzyme cavity, in contrast to the polar interactions observed for other hCA II–sulfonamide/sulfamate complexes.[24] It was noted that such compound oriented its phenethyl–thioureido moiety toward the hydrophobic part of the active site cleft, establishing many van der Waals interactions with residues Gln92, Val121, Phe131, Val135,

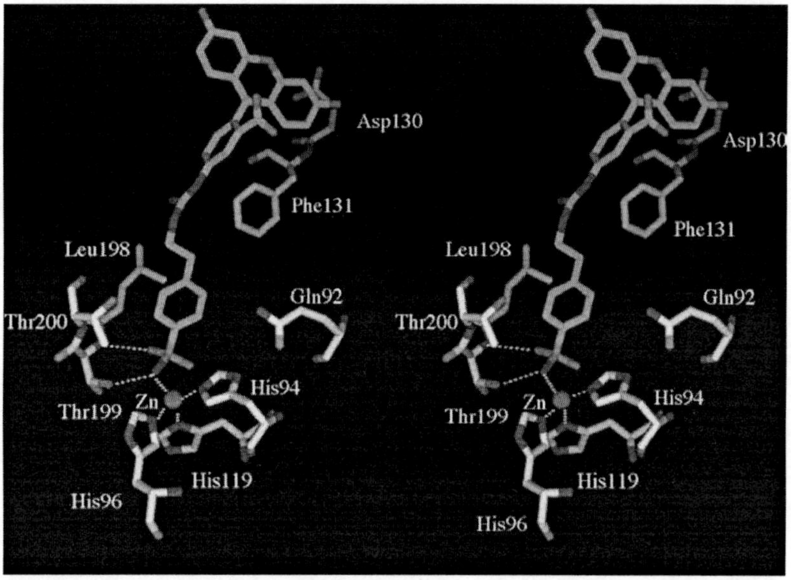

FIGURE 9.2 Stereoview of the Zn(II) coordination sphere and neighboring amino acid residues involved in the binding of compound **9.3** to the hCA II active site. (See the color version of this figure in Color Plates section.)

Leu198, Thr199, Thr200, and Pro202. Instead, the tricyclic fluorescein moiety was accommodated on the protein surface and stabilized strongly by interactions with the α-helix, formed by residues Asp130–Val135 and symmetry-related enzyme molecules.

For the complex hCA IX–compound **9.3**, a modeling study was performed.[22] The models of hCA IX catalytic domain were found using both hCA II and mCA XIV X-ray structures as templates either individually or in combination (sequence identity 34 and 44%, respectively). As expected, the hCA IX three-dimensional model was very similar to that of hCA II and on the basis of the high sequence/structural similarity between hCA IX and hCA II, it was not surprising that compound **9.3** showed rather small differences in the affinity toward both enzymes.

The model of the complex hCA IX–compound **9.3** and all kinds of interactions observed are given in Fig. 9.3. The analysis of this adduct reveals that all the active site residues were conserved with all the H-bonds ensuring a proper catalytic efficiency. Several polar and hydrophobic interactions stabilize the inhibitor within the hCA IX active site; in particular, it was noted that the sulfonamide group bound to the hCA IX active site in a way similar to that of hCA II–**9.3** complex. However, the mutation of Asp130–Arg and the different orientation of amino acid residues on hCA IX surface caused little changes in the bulky tricyclic fluorescence position; in fact, a new polar interaction was added to that the hCA II–**9.3** counterpart as shown in Fig. 9.3. Furthermore, independent of the Arg130 side chain starting orientation, all the 1 ns MD simulations carried out converged into a complex,

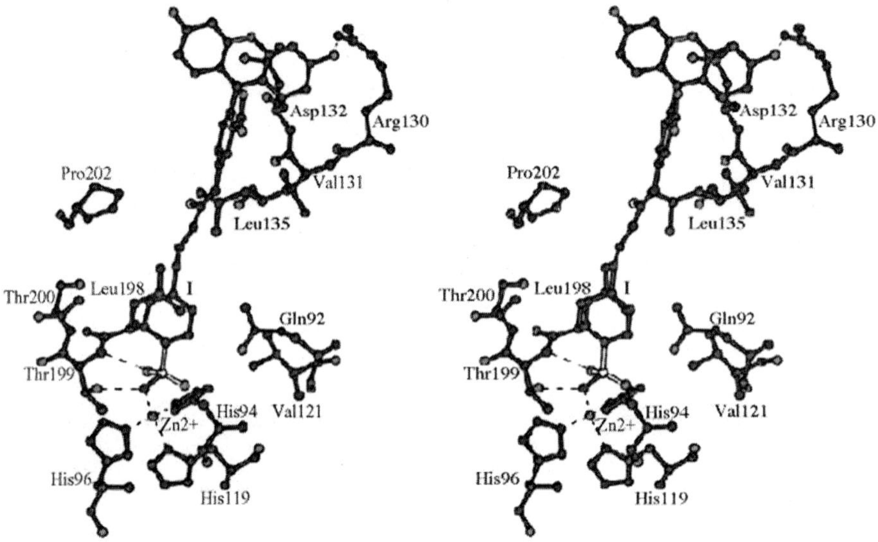

FIGURE 9.3 Stereoview of the active site region in the hCA IX–**9.3** complex.

which presented the stable hydrogen bond between Arg130 and the carbonyl group of compound **9.3**.

This polar interaction that is absent in hCA II may be considered as the unique structural feature accounting for the observed differences in binding affinity of ligand **9.3** toward hCA II and IX, but it is presumably quite important, since compound **9.3** is roughly two times more effective inhibitor of the tumor-associated isozyme (hCA IX) than the cytosolic one (hCA II).

9.2.3 Biological Evaluation

Several biological tests have been reported to prove the ability of these fluorescent derivatives to be employed as hCA IX inhibitors and as hypoxic tumor cell line markers. First, the permeability of such molecules through red blood cell membranes was measured;[12] in fact, since hCA IX is a transmembrane protein with the active site exposed out of the cell, derivatives that possess decreased permeability may lead to the selective inhibition of hCA IX and not of the cytosolic CA isozymes CA I or II. This is considered a very desirable property of a future drug belonging to this class of compounds. The derivatives **9.2** and **9.3** showed decreased membrane permeability at exposure times of 30–60 min but were more permeant after 48 h of exposure; this behavior is due to the carboxylic acid moiety present in the fluorescein tail that is deprotonated at physiological pH and leads to a decreased penetration through membranes.

Fluorescent sulfonamides **9.1–9.10** were then used by Svastova et al. to investigate the involvement of hCA IX in acidification of the external matrix in tumor cells.[8] Hypoxia induce the expression of hCA IX in tumor cells together with various

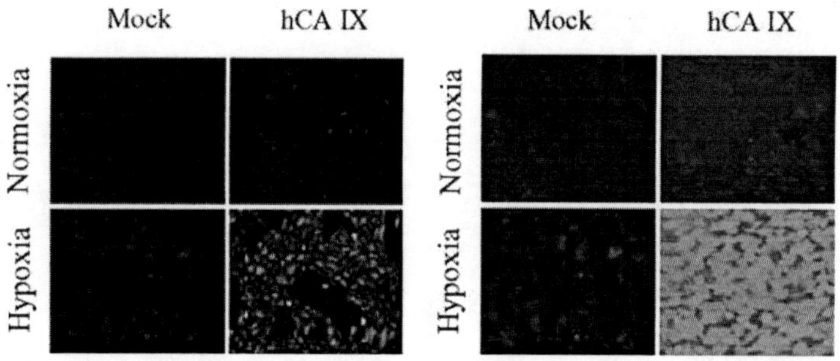

FIGURE 9.4 Immunofluorescence analysis of MDCK–CA IX cells treated with compounds **9.2** (left) and **9.3** (right).

components of anaerobic metabolism and acid extrusion pathways; this could complicate the determination of the contribution of hCA IX to the overall change in pH_e. Therefore, immortalized canine kidney epithelial cells (MDCK) were used that do not endogenously express CA IX, but were stably transfected to express the human CA IX protein in a constitutive manner. Immunofluorescence analysis of the transfected MDCK cells with compounds **9.12** and **9.13** show that these derivatives were perfect probes for such enzyme, as given in Fig. 9.4; in addition, it was noted that hCA IX was predominantly localized at the cell surface, although the membrane staining in hypoxic cells was less pronounced due to a hypoxia-induced perturbation of intercellular contacts. Moreover, the fluorescent derivatives were detected only in hypoxic MDCK–CA IX cells, and they were absent from their normoxic counterparts and from the mock transfected controls. In particular, the lack of fluorescence signal in the hCA IX–negative MDCK cells confirmed the selectivity of the inhibitors, which did not bind with other potentially present CA isoforms and indicated that only the hypoxic MDCK–CA IX cells contain the catalytically active hCA IX with the enzyme active site accessible to an inhibitor.

Finally, it was verified if these inhibitors were also able to reduce the acidic pH by inhibition of hCA IX. Four different hCA IX-selective inhibitors were tested, including compounds **9.2** and **9.3** that possess bigger selectivity ratio ($K_{I(hCA\ II)}/K_{I(hCA\ IX)}$) than the other fluorescent derivatives **9.1–9.10**.[12] All the sulfonamides tested were able to reduce the extracellular acidification of MDCK–CA IX cells in hypoxia and their effect on the normoxic pH_e was negligible, as showed in Fig. 9.5.

The extracellular pH of cervical carcinoma cells HeLa and SiHa was checked in presence of compound **9.2** to see whether the phenomenon of hCA IX-mediated acidification is applicable to tumor cells with endogenous hCA IX. Tumor cells, under hypoxic condition, coordinately express elevated levels of multiple HIF-1 targets,[25] including hCA IX; in addition, the activity of many components of the hypoxic pathway and related pH control mechanisms, such as ion transport across the plasma membrane, are abnormally increased to maintain the neutral intracellular pH. This explains the considerably decreased pH_e of the hypoxic versus normoxic HeLa and SiHa cells (Fig. 9.6). The acidosis was partially reduced by compound **9.2**, in support of the idea that activation of hCA IX is just one of the many consequences of hypoxia.

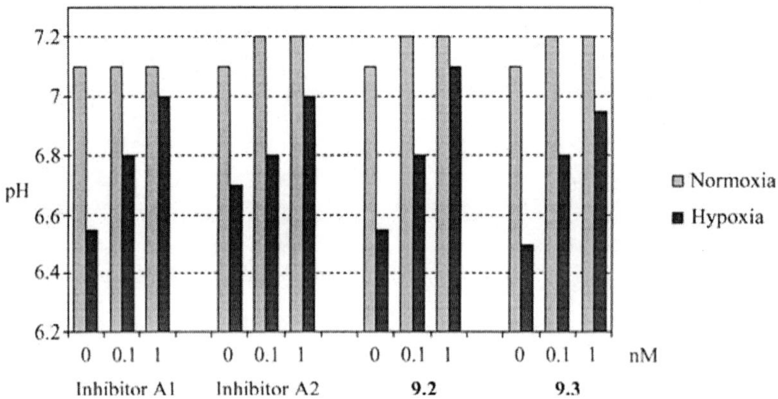

FIGURE 9.5 Values of pH_e in MDCK–CA IX cells treated with sulfonamides A1 (4-(2-aminoethyl)-benzenesulfonamide), A2 (4-(2,4,6-trimethyl-pyridinium-*N*-methylcarboxamido)-benzenesulfonamide), and compounds **9.2** and **9.3**.

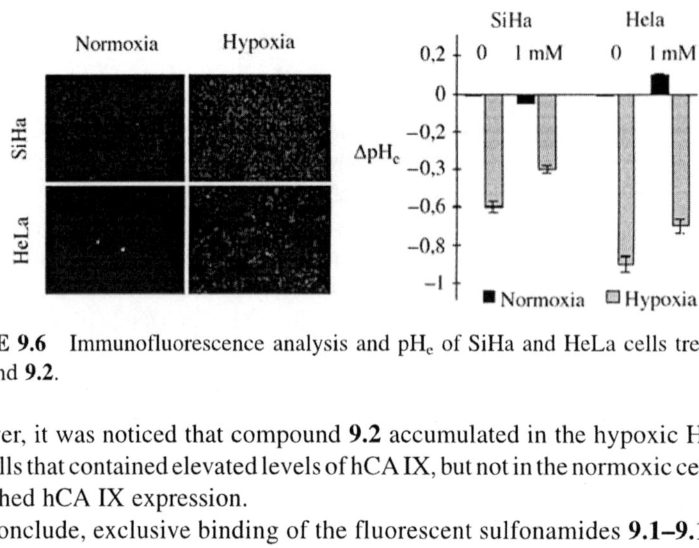

FIGURE 9.6 Immunofluorescence analysis and pH_e of SiHa and HeLa cells treated with compound **9.2**.

Moreover, it was noticed that compound **9.2** accumulated in the hypoxic HeLa and SiHa cells that contained elevated levels of hCA IX, but not in the normoxic cells with a diminished hCA IX expression.

To conclude, exclusive binding of the fluorescent sulfonamides **9.1–9.10** to the hypoxic cells that express activated hCA IX offers an attractive possibility for the use of similar sulfonamide-based compounds for the imaging purposes *in vivo*; moreover, hCA IX-selective sulfonamide derivatives may potentially serve as components of the therapeutic strategies designed to increase pH_e in the tumor microenvironment and thereby reduce the tumor aggressiveness and increase the drug uptake.

9.3 SPIN-LABELED SULFONAMIDE AS PROBES FOR CA IX

Electron paramagnetic resonance (EPR) technique has been largely used to study enzyme structures and their changes directly in solution. In fact, both continuous wave and pulsed EPR methods permit to unravel secondary structure elements, complex formation or domain arrangements.[26] The inhibitors possess different mobility, depending whether they are free in solution medium or they are bound to the enzyme active site; thus, CAIs with a radical moiety active in EPR measurements can produce diverse output signals, from which is possible to reveal the inhibitor rotational correlation time τ_e (defined as the time required for the label to rotate through one radian) or the maximal hyperfine splitting ($2T_{||}$).

Several spin-labeled sulfonamides with CAs have been studied to understand the mobility of the molecules in the enzyme–inhibitor complex and to clarify the topographic conformation of the enzyme active site.[27] This type of CAIs were in fact reported starting from 1970s, when not much was known of the binding of sulfonamides within the CA active site, because no X-ray crystallographic structures of any native isoform or in complex with inhibitors were available.

Chignell et al.[28] reported a series of spin-labeled sulfonamides (Fig. 9.7, structures **9.I–9.III**) in which the distance between the head of the inhibitors and the pyrrolidine/piperidine ring, which beams the radical nitroxide function, was gradually increased. In this way, it was possible to estimate the active site depth of hCA II that is in good agreement with crystallographic data, suggesting that the crystal and solution conformations of this enzyme are similar.[29] Furthermore, some of these inhibitors were more immobilized when bound to hCA I than to hCA II, showing that the first enzyme possess a narrower active site than the second one.[28,29]

Bovine CA B (bCA B) have been studied by Chesnut et al. with another spin-labeled CAI that possess an ester linkage between the radical piperidine *N*-oxide moiety and the benzenesulfonamide head (Fig. 9.7, structure **9.IV**).[30] In addition, Coleman et al. reported several studies of a radical *p*-hydrazido-benzenesulfonamide derivative (Fig. 9.7, structure **9.V**) against human and bovine CAs.[31,32]

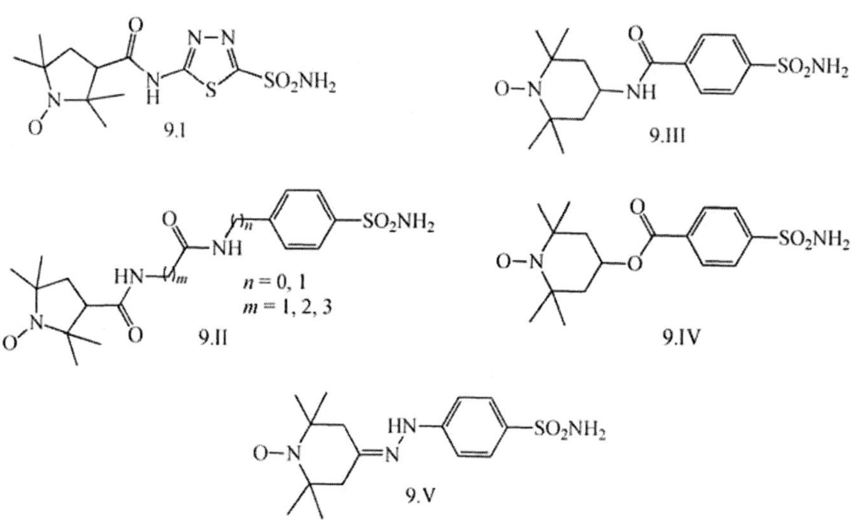

FIGURE 9.7 General structures of spin-labeled compounds **9.I–9.V**.

Recently, new series of spin-labeled thioureido-sulfonamides that behaved as strong hCA IX inhibitors with K_I in nanomolar range was described.[18] These compounds may be used as additional markers of tumor cell line, which overexpress hCA IX and could also give further information regarding the enzyme–inhibition complex in solution (i.e., inhibitor mobility, etc.), or developed as diagnostic tools.

9.3.1 Design of Spin-Labeled Sulfonamides Incorporating TEMPO Moieties

A novel and original set of radical TEMPO sulfonamides have been designed[18] using the classical tail strategy.[11,12,23] Thus, a thiourea linker, which was absent in any of the previous derivatives, was incorporated between a benzenesulfonamide head, present in the spin-labeled compounds **9.I–9.V** reported earlier, and a free radical TEMPO tail (Fig. 9.8).

The benzenesulfonamide moiety was chosen as "head" due to its ability to bind with the metal ion inside the enzyme active site, as shown by several X-ray crystallographic data.[11,33–35] The phenyl ring belonging to the benzenesulfonamide moiety was also substituted in positions 3 and 4 (with respect to the sulfamoyl group) with halogen atoms, as some of these derivatives were shown earlier to lead to potent CAIs.[20] As a central linker, a thiourea moiety was selected; in fact, several compounds incorporating this motif behave as potent hCA I, II, and IX inhibitors, as reported in literature.[36,37] A 2,2,6,6-tetramethyl-piperidine-1-oxyl moiety substituted in position 4 was chosen as tail; this type of tail is one of the most studied tails and it is considered a very suitable radical scaffold due to its properties (chemically stable, partially water soluble, nontoxic, simple EPR spectrum at ambient temperature).[38] Furthermore, the possibility to regulate the line width and intensity by tissue oxygen status or redox status renders

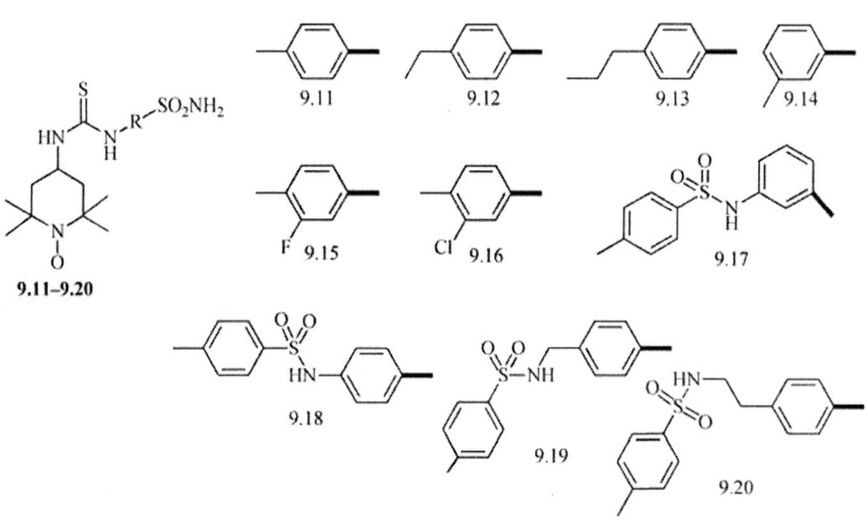

FIGURE 9.8 Structures of spin-labeled inhibitors **9.11–9.20**.

such molecules containing this radical scaffold to be incorporated in spin probes that interact with biomolecules such as enzymes.[39]

9.3.2 Inhibition Test and EPR Measurements

The CA inhibition studies of the radical compounds **9.11–9.20** against the cytosolic isozymes hCA I, II, and the tumor-associated isozyme hCA IX showed the following structure–activity relationships (Table 9.1):[18]

(1) The sulfonamides **9.11–9.20** inhibit the slow cytosolic isoform hCA I with inhibition constants in the range of 89–2070 nM. They exhibit medium inhibition properties (K_I 89–233 nM) for such enzymes, except for the derivatives **9.14, 9.17,** and **9.18**, which are less efficient. The reduced affinity for the molecules **9.14** and **9.17** is probably due to the fact that they possess a bulky moiety *meta* to the sulfamoyl zinc binding group, as proved in literature for other classes of derivatives bearing this substitution pattern.[40] Derivatives **9.11–9.13** show similar inhibition constants (179–204 nM) because the presence of a methyl or ethyl moiety between the aromatic scaffold and the thioureidic group does not lead to important differences in the inhibitory properties of such compounds.

(2) Against the ubiquitous, physiologically relevant isozyme hCA II, compounds **9.11–9.20** exhibit good K_I values, between 12 and 165 nM. Similar to hCA I values, the two compounds that have a *meta* substitution on the phenyl ring show bad inhibition results (165 and 152 nM for the molecules **9.14** and **9.17**, respectively). The derivatives **9.11–9.13** possess a compact behavior against this isoform, with K_I ranging from 28 to 42 nM. Furthermore, the substitution of the fluorine atom present in compound **9.15** with a chlorine (compound **9.16**) increases the affinity against the biological target from 37 to 12 nM. Sulfanilyl sulfonamides **9.18–9.20** were less efficient hCA II inhibitors than **9.16**, but they still possess good inhibition constants between 20 and 47 nM. Thus, we evidenced several types of substitution patterns of the TEMPO-containing CAIs that lead to efficient and low nanomolar inhibitors of the physiologically relevant isozyme hCA II.

(3) The spin-labeled compounds **9.11–9.20** inhibit the tumor-associated isozyme hCA IX (the main target for imaging or treatment purposes) with K_I's in the range of 7–220 nM. The derivatives **9.14** and **9.17** present the worst K_I against this isoform, similar to the hCA II inhibition already discussed, due to their *meta* aromatic substitution (132 and 220 nM, respectively). On the contrary, the other inhibitors possess good inhibition constant in the range of 7–41 nM. In addition, derivatives **9.16** and **9.19** have shown better K_I value for such enzyme than the clinically used drugs reported in Table 9.1.

(4) Inhibitors **9.12**, **9.15**, **9.18**, and **9.19** have also shown selectivity against the tumoral isoform hCA IX over the cytosolic hCA II, with a maximum for compound **9.19** (selectivity ratio = 2,9). These are interesting results,

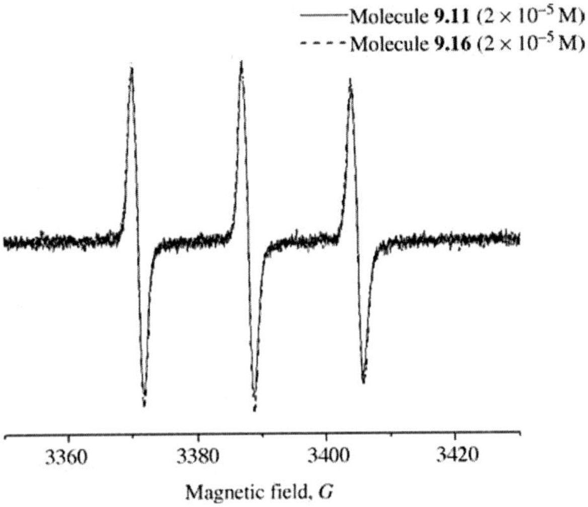

FIGURE 9.9 EPR signals of compounds **9.11** and **9.16** in solution.

considering that only few compounds reported in the literature selectively inhibit the tumor-associated isoform over the cytosolic CAs. This means that these derivatives will preferentially bind hCA IX (already overexpressed in hypoxic tumors)[1–3] and much less CA II, which is a housekeeping enzyme necessary for many physiological processes of the cell.[11]

To investigate the mobility of TEMPO-CAIs in solution and upon binding to the enzyme, the ESR spectra of the nitroxide were recorded both in the absence and in the presence of hCA II. The ESR spectra of compounds **9.11** and **9.16**, both free in solution and complexed to the enzyme, are shown in Figs 9.9 and 9.10. The signals of the spin-labeled CAIs dissolved in buffer were superposable as expected, considering that the only difference between compounds **9.11** and **9.16** is the chlorine atom that is placed quite far from the radical piperidine-*N*-oxide ring and it does not influence the output signal (Fig. 9.9). The molecules possess rapid movements in solution with hyperfine coupling typical of a polar medium ($\langle A_N \rangle = 17.1$). The ESR spectrum of compounds **9.11** and **9.16** significantly changes in the presence of hCA II to a slow motion signal, indicating that binding within the enzyme active site has efficiently occurred (Fig. 9.10). It should be noted that a very small fraction of labeled compounds did not interact with the enzyme; this fraction was found to be of the order of 0.1–3%, as measured from double integration of the different ESR peaks. However, in all cases, the ESR line shape of TEMPO-CAIs in the presence of hCA II was completely dominated by the spectrum of the bound molecules, which showed the anisotropic features characteristic to restricted motional conditions.[26–32]

These results prove that these free radical sulfonamides were able to strongly interact with the amino acid residues lining the cavity of the hCA II enzyme, making

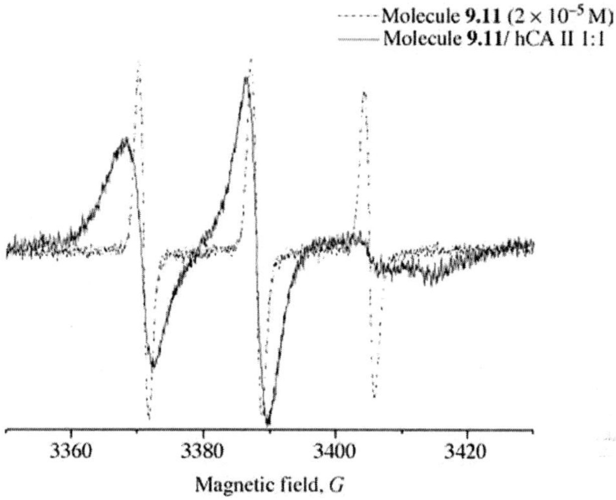

FIGURE 9.10 EPR signals of compounds **9.11** free in solution and complexed with hCA II.

them promising leads for the investigation of various CA isozymes by means of ESR techniques.

9.4 CONCLUSIONS

Fluorescent and spin-labeled sulfonamides were investigated as probes for the tumoral hCA IX. Fluorescein sulfonamides **9.1–9.10**, which act as potent CAIs, showed excellent properties as fluorescent markers for hypoxic tumor, due to their property to bind selectively with the membrane-associated hCA IX, without any interaction with the cytosolic hCA II. Thus, compounds **9.1–9.10** may be useful in the diagnosis of a large spectrum of tumors; particularly, compound **9.3** is in clinical studies as an imaging tool for acute hypoxic tumors.

Compounds **9.1–9.10** were also used to demonstrate the involvement of hCA IX in the acidification of external tumor matrix, together with the lactic acid accumulation during glycolysis. This is a very important discovery, considering that the low pH_e in the tumor environment is associated with bad tumor progression and it represents a barrier to drug delivery for many chemotherapeutic agents. These results also demonstrate that sulfonamides are able to enhance the pH_e by inhibition of hCA IX in tumor cells like HeLa and SiHa; they also represent the basis for a new approach in the treatment of cancer consisting in the inhibition of tumor-associated hCA IX.

Finally, the spin-labeled sulfonamides **9.11–9.20** have been reviewed. The data reported showed that EPR could be considered as an additional technique to reveal the complex enzyme–inhibitor formation in solution and to estimate the power of the binding between the sulfonamide and the carbonic anhydrases. Furthermore, with this type of analysis, it is possible to acquire information about the chemical environment and the mobility of the inhibitors bound to the enzyme active site.

REFERENCES

1. Pastorekova, S.; Zavadova, Z.; Kosal, L.; Babusikova, O.; Zavada, J. A novel quasi-viral agent, MaTu, is a two-component system. *Virology* **1992**, *187*, 620–626.
2. Turaci, O.; Sahin, U.; Vollmar, E.; Siemer, S.; Gottert, E.; Seitz, G.; Parkkila, A. K.; Shah, G. N.; Grubb, J. H.; Preundschuh, M.; Sly, W. S. Human carbonic anhydrase XII: cDNA cloning, expression, and chromosomal localization of a carbonic anhydrase gene that is overexpressed in some renal cell cancers. *Proc. Natl. Acad. Sci. USA* **1998**, *95*, 7608–7613.
3. Ivanov, S. V.; Kuzmin, I.; Wei, M. H.; Pack, S.; Geil, L.; Johnson, B. E.; Stanbridge, E. J.; Lerman, M. I. Down-regulation of transmembrane carbonic anhydrases in renal cell carcinoma cell lines by wild-type von Hippel–Lindau transgenes. *Proc. Natl. Acad. Sci. USA* **1998**, *95*, 12596–12601.
4. Cairns, R.; Papandreou, I.; Denko, N. Overcoming physiologic barriers to cancer treatment by molecularly targeting the tumor microenvironment. *Mol. Cancer Res.* **2006**, *4*, 61–70.
5. Vaupel, P. The role of hypoxia-induced factors in tumor progression. *Oncologist* **2004**, *5*, 10–17.
6. Helmlinger, G.; Yuan, F.; Dellian, M.; Jain, R. K. Interstitial pH and pO_2 gradients in solid tumors *in vivo*: high-resolution measurements reveal a lack of correlation. *Nat. Med.* **1997**, *3*, 177–182.
7. Warburg, O.; *The Metabolism of Tumours;* Arnold Constable: London, **1930**.
8. Svastova, E.; Hulýkova, A.; Rafajova, M.; Zatovicova, M.; Gibadulinova, A.; Casini, A.; Cecchi, A.; Scozzafava, A.; Supuran, C. T.; Pastorek, J.; Pastorekova, S. Hypoxia activates the capacity of tumor-associated carbonic anhydrase IX to acidify extracellular pH. *FEBS Lett.* **2004**, *577*, 439–445.
9. Chegwidden, W. R.; Spencer, I. M. Sulphonamide inhibitors of carbonic anhydrase inhibit the growth of human lymphoma cells in culture. *Inflammopharmacology* **1995**, *3*, 231–239.
10. Supuran, C. T.; Briganti, F.; Tilli, S.; Chegwidden, W. R.; Scozzafava, A. Carbonic anhydrase inhibitors: sulfonamides as antitumor agents? *Bioorg. Med. Chem.* **2001**, *9*, 703–714.
11. Supuran, C. T.; Scozzafava, A.; Conway, J.; *Carbonic Anhydrases: Its Inhibitors and Activators;* CRC Press (Taylor & Francis Group): Boca Raton, FL, **2004**.
12. Cecchi, A.; Hulikova, A.; Pastorek, J.; Pastorekova, S.; Scozzafava, A.; Winum, J. Y.; Montero, J. L.; Supuran, C. T. Carbonic anhydrase inhibitors. Design of fluorescent sulfonamides as probes of tumor-associated carbonic anhydrase IX that inhibit isozyme IX-mediated acidification of hypoxic tumors. *J. Med. Chem.* **2005**, *48*, 4834–4841.
13. Pastorekova, S.; Parkkila, S.; Parkkila, A. K.; Opavsky, R.; Zelnik, V.; Saarnio, J.; Pastorek, J. Carbonic anhydrase IX, MN/CA IX: analysis of stomach complementary DNA sequence and expression in human and rat alimentary tracts. *Gastroenterology* **1997**, *112*, 398–408.
14. Turner, J. R.; Odze, R. D.; Crum, C. P.; Resnick, M. B. MN antigen expression in normal, preneoplastic, and neoplastic esophagus: a clinicopathological study of a new cancer-associated biomarker. *Hum. Pathol.* **1997**, *28*, 740–744.
15. Ivanov, S.; Lia, S. Y.; Ivaniva, A.; Danilkovich-Miagkova, A.; Tarasova, N.; Weirich, G.; Merrill, M. J.; Proescholdt, M. A.; Oldfield, E. H.; Lee, J.; Zavada, J.; Waheed, A.; Sly, W.; Lerman, M. I.; Stanbridge, E. J. Expression of hypoxia-inducible cell-surface transmembrane carbonic anhydrases in human cancer. *Am. J. Pathol.* **2001**, *158*, 905–919.

16. Hockel, M.; Vaupel, P. Tumor hypoxia: definitions and current clinical, biologic, and molecular aspects. *J. Natl. Cancer Inst.* **2001**, *93*, 266–276.

17. Wykoff, C.; Beasley, N.; Watson, P.; Campo, L.; Chia, S. K.; English, R.; Pastorek, J.; Sly, W. S.; Ratcliffe, P.; Harris, A. Expression of the hypoxia-inducible and tumor-associated carbonic anhydrases in ductal carcinoma *in situ* of the breast. *Am. J. Pathol.* **2001**, *158*, 1011–1019.

18. Cecchi, A.; Ciani, L.; Winum, J. Y.; Montero, J. L.; Scozzafava, A.; Ristori, S.; Supuran, C. T. Carbonic anhydrase inhibitors: design of spin-labeled sulfonamides incorporating TEMPO moieties as probes for cytosolic or transmembrane isozymes. *Bioorg. Med. Chem. Lett.* **2008**, *18*, 3475–3480.

19. Kim, T. W.; Park, J. H.; Hong, J. I. Zn^{2+} fluorescent chemosensors and the influence of their spacer length on tuning Zn^{2+} selectivity. *J. Chem. Soc. Perkin Trans. 2* **2002**, 923–927.

20. Ilies, M.; Vullo, D.; Pastorek, J.; Scozzafava, A.; Ilies, M.; Caproiu, M. T.; Pastorekova, S.; Supuran, C. T. Carbonic anhydrase inhibitors. Inhibition of tumor-associated isozyme IX by halogenosulfanilamide and halogenophenylaminobenzolamide derivatives. *J. Med. Chem.* **2003**, *46*, 2187–2196.

21. Casini, A.; Scozzafava, A.; Mincione, F.; Menabuoni, L.; Ilies, M. A.; Supuran, C. T. Carbonic anhydrase inhibitors: water-soluble 4-sulfamoyl-phenylthioureas as topical intraocular pressure-lowering agents with long-lasting effects. *J. Med. Chem.* **2000**, *43*, 4884–4892.

22. Alterio, V.; Vitale, R. M.; Monti, S. M.; Pedone, C.; Scozzafava, A.; Cecchi, A.; De Simone, G.; Supuran, C. T. Carbonic anhydrase inhibitors: X-ray and molecular modeling study for the interaction of a fluorescent antitumor sulfonamide with isozyme II and IX. *J. Am. Chem. Soc.* **2006**, *128*, 8329–8335.

23. Supuran, C. T. Carbonic anhydrases: novel therapeutic applications for inhibitors and activators. *Nat. Rev. Drug Discov.* **2008**, *7*, 168–181.

24. Temperini, C.; Winum, J. Y.; Montero, J. L.; Scozzafava, A.; Supuran, C. T. Carbonic anhydrase inhibitors: the X-ray crystal structure of the adduct of *N*-hydroxysulfamide with isozyme II explains why this new zinc binding function is effective in the design of potent inhibitors. *Bioorg. Med. Chem. Lett.* **2007**, *17*, 2795–801.

25. Maxwell, P. H.; Wiesener, M. S.; Chang, G. W.; Clifford, S. C.; Vaux, E. C.; Cockman, M. E.; Wykof, C. C.; Pugh, C. W.; Maher, E. R.; Ratcliffe, P. J. The tumour suppressor protein VHL targets hypoxia-inducible factors for oxygen-dependent proteolysis. *Nature* **1999**, *399*, 271–275.

26. Schiemann, O.; Prisner, T. F. Long-range distance determinations in biomacromolecules by EPR spectroscopy. *Q. Rev. Biophys.* **2007**, *40*, 1–53.

27. Wee, V. T.; Feldmann, R. J.; Tanis, R. J.; Chignell, C. F. A comparative study of mammalian erythrocyte carbonic anhydrases employing spin-labeled analogues of inhibitory sulfonamides. *Mol. Pharmacol.* **1976**, *12*, 832–843.

28. Chignell, C. F.; Starkweather, D. K.; Erlich, R. H. Synthesis of some spin-labeled analogs of drug molecules. *J. Med. Chem.* **1972**, *15*, 876–878.

29. Erlich, R. H.; Starkweather, D. K.; Chignell, C. F. A spin-label study of human erythrocyte carbonic anhydrases B and C. *Mol. Pharmacol.* **1972**, *9*, 61–73.

30. Hower, J. F.; Henkens, R. W.; Chesnut, D. B. A spin-label investigation of the active site of an enzyme, bovine carbonic anhydrase. *J. Am. Chem. Soc.* **1971**, *93*, 6665–6671.

31. Taylor, J. S.; Mushak, P.; Coleman, J. E. Electron spin resonance studies of carbonic anhydrase: transition metal ions and spin-labeled sulfonamides. *Proc. Natl. Acad. Sci. USA* **1970**, *67*, 1410–1416.

32. Mushak, P.; Coleman, J. E. Electron spin resonance studies of spin-labeled carbonic anhydrase. *J. Biol. Chem.* **1972**, *247*, 373–380.

33. (a) Boriack, P. A.; Christianson, D. W.; Kingery-Wood, J.; Whitesides, G. M. J. Secondary interactions significantly removed from the sulfonamide binding pocket of carbonic anhydrase II influence inhibitor binding constants. *J. Med. Chem.* **1995**, *38*, 2286–2291. (b) Kim, C. Y.; Chang, J. S.; Doyon, J. B.; Baird, T. T.; Fierke, C. A.; Jain, A.; Christianson, D. W. Contribution of fluorine to protein–ligand affinity in the binding of fluoroaromatic inhibitors to carbonic anhydrase II. *J. Am. Chem. Soc.* **2000**, *122*, 12125–12134.

34. (a) Weber, A.; Casini, A.; Heine, A.; Kuhn, D.; Supuran, C. T.; Scozzafava, A.; Klebe, G. Unexpected nanomolar inhibition of carbonic anhydrase by COX-2-selective celecoxib: new pharmacological opportunities due to related binding site recognition. *J. Med. Chem.* **2004**, *47*, 550–557. (b) Menchise, V.; De Simone, G.; Alterio, V.; Di Fiore, A.; Pedone, C.; Scozzafava, A.; Supuran, C. T. Carbonic anhydrase inhibitors: stacking with Phe131 determines active site binding region of inhibitors as exemplified by the X-ray crystal structure of a membrane-impermeant antitumor sulfonamide complexed with isozyme II. *J. Med. Chem.* **2005**, *48*, 5721–5727.

35. (a) Temperini, C.; Cecchi, A.; Boyle, N. A.; Scozzafava, A.; Escribano Cabeza, J.; Wentworth, P., Jr.; Blackburn, G. M.; Supuran, C. T. Carbonic anhydrase inhibitors. Interaction of 2-*N,N*-dimethylamino-1,3,4-thiadiazole-5-methanesulfonamide with twelve mammalian isoforms: kinetic and X-ray crystallographic studies. *Bioorg. Med. Chem. Lett.* **2008**, *18*, 999–1005. (b) Temperini, C.; Cecchi, A.; Scozzafava, A.; Supuran, C. T. Carbonic anhydrase inhibitors. Interaction of indapamide and related diuretics with 12 mammalian isozymes and X-ray crystallographic studies for the indapamide–isozyme II adduct. *Bioorg. Med. Chem. Lett.* **2008**, *18*, 2567–2573.

36. (a) Scozzafava, A.; Supuran, C. T. Carbonic anhydrase inhibitors: ureido and thioureido derivatives of aromatic sulfonamides possessing increased affinities for isozyme I. A novel route to 2,5-disubstituted-1,3,4-thiadiazoles via thioureas, and their interaction with isozymes I, II and IV. *J. Enzyme Inhib.* **1998**, *13*, 103–123. (b) Supuran, C. T.; Scozzafava, A.; Jurca, B. C.; Ilies, M. Carbonic anhydrase inhibitors: Part 49: synthesis of substituted ureido and thioureido derivatives of aromatic/heterocyclic sulfonamides with increased affinities for isozyme I. *Eur. J. Med. Chem.* **1998**, *33*, 83–93.

37. (a) Cecchi, A.; Winum, J. Y.; Innocenti, A.; Vullo, D.; Montero, J. L.; Scozzafava, A.; Supuran, C. T. Carbonic anhydrase inhibitors: synthesis and inhibition of cytosolic/tumor-associated carbonic anhydrase isozymes I, II, and IX with sulfonamides derived from 4-isothiocyanato-benzolamide. *Bioorg. Med. Chem. Lett.* **2004**, *14*, 5775–5780. (b) Puccetti, L.; Fasolis, G.; Cecchi, A.; Winum, J. Y.; Gamberi, A.; Montero, J. L.; Scozzafava, A.; Supuran, C. T. Carbonic anhydrase inhibitors: synthesis and inhibition of cytosolic/tumor-associated carbonic anhydrase isozymes I, II and IX with sulfonamides incorporating thioureido–sulfanilyl scaffolds. *Biorg. Med. Chem. Lett.* **2005**, *15*, 2359–2364.

38. Wee, V. T.; Feldmann, R. J.; Tanis, R. J.; Chignell, C. F. A comparative study of mammalian erythrocyte carbonic anhydrases employing spin-labeled analogues of inhibitory sulfonamides. *Mol. Pharmacol.* **1976**, *12*, 832–843.

39. Sotgiu, A.; Mader, K.; Placidi, G.; Colacicchi, S.; Ursini, C. L.; Alecci, M. pH-sensitive imaging by low-frequency EPR: a model study for biological applications. *Phys. Med. Biol.* **1998**, *43*, 1921–1930.

40. Beasley, Y. M.; Overell, B. G.; Petrow, V.; Stephenson, O. Some *in vitro* inhibitors of carbonic anhydrase. *J. Pharm. Pharmacol.* **1958**, *10*, 696–705.

CHAPTER 10

Drug Design of Antiobesity Carbonic Anhydrase Inhibitors

GIUSEPPINA DE SIMONE[1] and CLAUDIU T. SUPURAN[2]

[1]Istituto di Biostrutture e Bioimmagini-CNR, Via Mezzocannone 16, 80134 Naples, Italy
[2]Laboratorio di Chimica Bioinorganica, Università degli Studi di Firenze, Room 188, Via della Lastruccia 3, I-50019 Sesto Fiorentino (Firenze), Italy

10.1 INTRODUCTION

Obesity is the most frequent metabolic disease in industrialized countries.[1–4] Nor are developing countries immune to this epidemic: according to the reports from China, for example, during 1989–1997, obesity doubled in women and almost tripled in men.[1–3] Obesity is also a risk factor for a variety of diseases such as type 2 diabetes, cardiovascular diseases, and various types of cancer,[5] and the prevalence of obesity-related diseases continues to increase. Obesity-associated morbidity and mortality have had an enormous impact on global health care and welfare systems and this economic burden is only destined to increase. Therefore, the development of strategies to reduce the worldwide obesity epidemic has recently become an important research goal.[5,6] Although diet and lifestyle changes should theoretically help control this condition, weight losses achieved with such approaches are quite modest and limited by high rates of recidivism and a compensatory slowing of the metabolism.[7,8] Thus, pharmacological interventions in the treatment of obesity are essential. Paradoxically, there are at present very few drugs available for the treatment of this disease, their mechanism of action is poorly understood, and their side effects are generally quite serious.[3]

Obesity is caused by an excessively positive energy balance, with the energy intake being greater than energy expenditure, even if the precise etiology of the disease is unknown.[3] Accordingly, depending on the mechanism of action on the energy balance, antiobesity drugs can be grouped into two large classes: those that reduce energy intake and those that stimulate energy expenditure. Drugs belonging to the first class can act

TABLE 10.1 Examples of Marketed Antiobesity Drugs

Compound	Company	Structure	Mechanism of Action
Phentermine	Medeva Pharmaceuticals Gate Pharmaceuticals	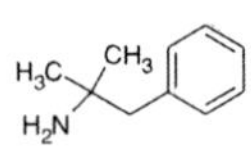	Increases levels of catecholamines, producing feeling of fullness
Sibutramine	Abbot	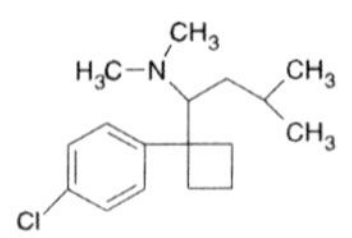	Monoamine-reuptake inhibitor (primarily norepinephrine and serotonin)
Orlistat	Roche GlaxoSmithKline	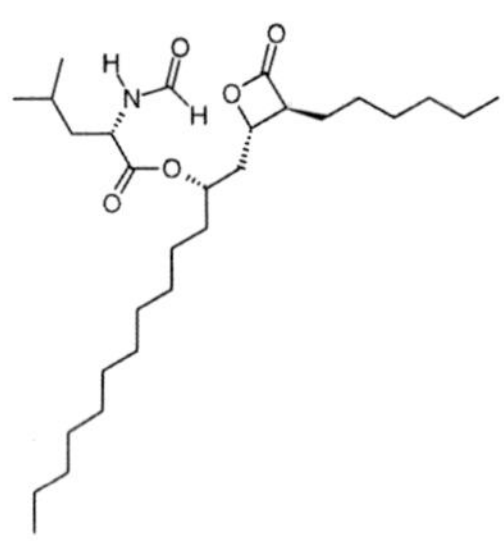	Inhibitor of gastric and pancreatic lipases
Rimonabant	Sanofi-Aventis	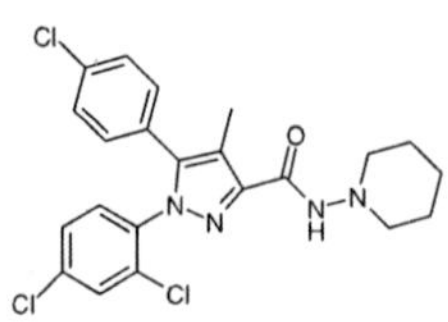	Endocannabinoid (CB1) receptor blocker

either by affecting appetite mechanisms to diminish food intake or by reducing fat absorption in the gastrointestinal tract, while drugs belonging to the second class operate on metabolic processes to facilitate energy expenditure.[9–11] The drug therapy for obesity belonging to the first class is dominated by four compounds (Table 10.1): Phentermine, which is indicated only for short-term treatment of obesity (a few weeks), and Sibutramine, Orlistat, and Rimonabant, which, in contrast, received approval for long-term use in treating the disease.[3,12,13]

Phentermine is a catecholaminergic drug with amphetamine-like properties, which acts as an appetite suppressant. It generally causes only modest body weight reduction and only for a few weeks. This drug is not recommended in patients with cardiovascular conditions such as advanced arteriosclerosis, symptomatic cardiovascular disease, and hypertension.[14] Until recently, two other drugs of this class were available, fenfluramine and dexfenfluramine, but they were both withdrawn from the market in 1997 due to their potential risk of valvular heart disease.[3]

Sibutramine is a tertiary amine that was originally developed as an antidepressant but subsequently approved by the Food and Drug Administration (FDA) for the long-term treatment of obesity. This drug is a monoamine-reuptake inhibitor, which mainly acts to increase satiety.[3,15] Sibutramine was also recently reported to stimulate thermogenesis; however, this secondary action plays a minor role in weight reduction.[16] Clinical studies demonstrated that sibutramine-associated weight loss is dose dependent,[17] takes place within the first 6 months of treatment, and may be maintained for at least 2 years.[18] The major side effects of this drug are dry mouth, nausea, paradoxically increased appetite, constipation, and trouble sleeping. Moreover, in some patients, it can generate small increases in blood pressure and pulse rate, thus suggesting potential cardiovascular toxic effects.

Orlistat is a semisynthetic derivative of an endogenous lipstatin produced by *Streptomyces toxytricini*.[3,19–23] It acts as a gastric and pancreatic lipase inhibitor, thus impairing the absorption of dietary fat.[3,20–23] Orlistat was shown to be effective in losing about 10% of body weight after 1 year of treatment, and this was associated with significant reductions in total cholesterol levels, LDL, but not triglyceride levels.[3,20–23] Commonly encountered side effects associated with the use of this drug are gastrointestinal and include oily spotting, fatty and oily stools, increased defecation, fecal urgency, flatulence, and abdominal cramping. However, these side effects are reduced when the patient follows a low-fat, low-calorie diet in concomitance with orlistat treatment.[3,20–23] Moreover, to avoid development of vitamin deficiency syndromes with the long-term use of the drug, supplementation of fat-soluble vitamins is recommended.[10,14]

Rimonabant is a selective endocannabinoid CB1 receptor antagonist, which mainly acts as an appetite suppressant, recently approved for the treatment of obesity.[12] Rimonabant clinical trials have provided very promising results:[12,13,24–26] Treatment with 20 mg/day of rimonabant plus a 2-year diet promoted modest but sustained reductions in weight and waist circumference and favorable changes in cardiometabolic risk factors.[12,13,24–26] The most important side effect of this drug is an increased occurrence of psychiatric disorders such as aggression, depression, anxiety, and irritability.[24–26]

It is clear from this brief overview that even if the medications currently approved for the treatment of obesity are reported to be successful in weight loss, all have undesirable side effects that require consideration. Thus, the development of antiobesity agents possessing different mechanisms of action is strongly needed. Fortunately, in recent years, pharmacological therapy for obesity has been in transition; expanding knowledge of the physiological mechanisms of body weight regulation has revealed new molecular targets, and hundreds of novel agents are under active development. Among these we should recall agents that affect the leptin/insulin/central nervous system pathways and the insulin metabolism/activity.[27] Potential leads for developing antiobesity drugs are also agents that increase glucagon-like peptide-1 activity (extendin 4, liraglutide, dipeptidyl peptidase IV inhibitors), increase protein YY3-36 activity, or decrease ghrelin activity, as well as amylin analogues (pramlintide).[27] In this chapter, we shall review another potential approach to treating obesity based on the inhibition of carbonic anhydrases (CAs, EC 4.2.1.1),[28] enzymes involved in several steps of *de novo* lipogenesis.[29,30]

10.2 CARBONIC ANHYDRASES AS NEW TARGETS IN THE TREATMENT OF OBESITY

CAs are ubiquitous metalloenzymes, which catalyze the reversible hydration of carbon dioxide according to the following reaction: $CO_2 + H_2O \leftrightarrow H^+ + HCO_3^-$.[28–31] At least 16 isozymes have been discovered until now in mammals, all belonging to the α-CA gene family. Thirteen of these isoforms are catalytically active (CA I–VA, VB, VI, VII, IX, and XII–XV), whereas the CA-related proteins (CARPs VIII, X, and XI) are devoid of catalytic activity.[28–34] The CA isozymes show various tissue expression patterns and different subcellular localizations: CA I, II, III, VII, and XIII reside in the cytosol, CA IV, IX, XII, XIV, and XV are associated with the cell membrane, CA VA and VB localize to mitochondria, and CA VI is secreted.[28–34]

CAs are also able to catalyze several other hydrolytic processes such as the hydration of cyanate to carbamic acid or of cyanamide to urea, aldehyde hydration to *gem*-diols, the hydrolysis of carboxylic or sulfonic acid esters, as well as other less investigated hydrolytic processes such as hydrolysis of halogeno derivatives and arylsulfonyl halides.[28–34] However, it is not yet known whether these other CA-catalyzed reactions have physiological importance in systems where these enzymes are present.[28–34]

Structural studies on isozymes I–V, XII, XIII, and XIV[35–42] isolated from different vertebrates have revealed that all these enzyme isoforms present a considerable degree of three-dimensional similarity. Moreover, all these enzymes contain an essential Zn^{2+} ion, localized at the bottom of a deep active site cleft, coordinated by three conserved His residues and a water molecule. The Zn^{2+}-bound water is also engaged in H-bond interactions with the hydroxyl moiety of a conserved Thr residue, which in turn is bridged to the carboxylate moiety of a conserved Glu residue. These interactions enhance the nucleophilicity of the Zn^{2+}-bound water molecule and orient CO_2 toward a favorable location for nucleophilic attack.[35–42]

CAs are present in all organisms throughout the phylogenetic tree, where their main function is to maintain an appropriate acid–base balance in organisms. Thus, they participate in several physiological processes such as CO_2 and HCO_3^- transport, bone resorption, production of body fluids, gluconeogenesis, ureagenesis, and lipogenesis.[28] Accordingly, in the past few years, CA isozymes have become an interesting target for the design of inhibitors or activators with biomedical applications.[28–34]

As mentioned above, CA VA and VB are the only CA isoforms found in mitochondria. These isozymes are involved in several biosynthetic processes, such as ureagenesis,[43] gluconeogenesis,[44] and lipogenesis, both in vertebrates and in invertebrates.[45–48] Indeed, the reaction catalyzed by these isozymes (probably assisted by the cytosolic isozyme CA II) ensures the presence of enough bicarbonate in several biosynthetic processes involving pyruvate carboxylase (PC), acetyl-CoA carboxylase (ACC), and carbamoyl phosphate synthetases I and II.[49–54] Figure 10.1 illustrates the transfer of acetyl groups from the mitochondrion to the cytosol for providing enough substrate for *de novo* lipogenesis.[46]

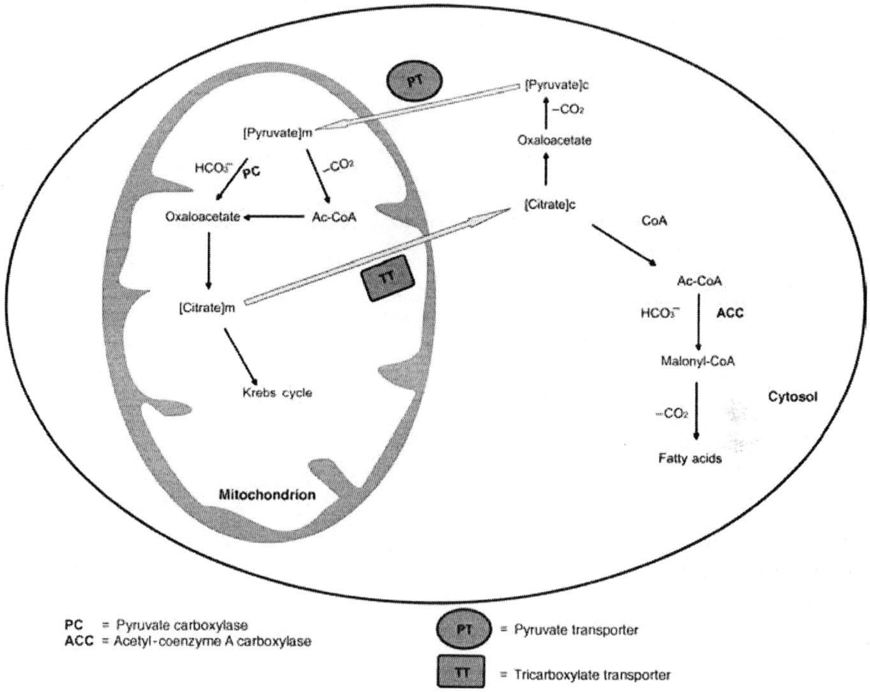

FIGURE 10.1 The role of carbonic anhydrase isozymes in fatty acid biosynthesis.

In practice, pyruvate is carboxylated to oxaloacetate in the presence of bicarbonate and PC. The bicarbonate needed for this process is generated under the catalytic influence of the mitochondrial isozymes CA VA and/or VB. The mitochondrial membrane is impermeant to acetyl-CoA that reacts with oxaloacetate, leading to citrate, which is then transported to the cytosol by means of the tricarboxylic acid transporter. Here, the citrate can be converted back to acetyl-CoA and oxaloacetate by ATP-citrate lyase. As oxaloacetate is unable to cross the mitochondrial membrane, its decarboxylation regenerates pyruvate, which can be then transported into the mitochondria by means of the pyruvate transporter.[33] The acetyl-CoA thus generated in the cytosol is used for the *de novo* lipogenesis, by carboxylation in the presence of ACC and bicarbonate, with formation of malonyl-CoA. The bicarbonate needed in this process is supplied by the CA II-catalyzed conversion of CO_2 to bicarbonate. Subsequent steps involving the sequential transfer of acetyl groups lead to longer chain fatty acids.[33] In conclusion, several CA isozymes are important for the entire process of fatty acid biosynthesis: CA VA or/and VB within the mitochondria (to provide enough substrate to PC), and CA II within the cytosol (for providing sufficient substrate to ACC). These observations suggest that CAs can represent a potential new target for the development of antiobesity drugs. Accordingly, several studies have provided evidence that inhibition of CAs by sulfonamides can reduce lipogenesis in adypocytes in cell culture.[48,49]

10.3 DRUG DESIGN OF ANTIOBESITY CARBONIC ANHYDRASE INHIBITORS

The main class of CA inhibitors (CAIs) is constituted by the unsubstituted sulfonamides and their bioisosteres (sulfamates, sulfamides, etc.), which bind to the Zn^{2+} ion of the enzyme, by substituting the nonprotein zinc ligand to generate a tetrahedral adduct.[28–34] Several such agents are clinically used drugs in the therapy of different pathologies such as glaucoma, acid–base disequilibria, epilepsy, and other minor neuromuscular disorders or as diuretics.[28–34]

Topiramate (**TPM**) (Fig. 10.2) is a sugar sulfamate derivative used as an antiepileptic. It possesses potent anticonvulsant effects due to a complex mechanism of action that involves blockage of sodium channels and kainate/AMPA receptors, enhancement of GABA (γ-aminobutyric acid)-ergic transmission and CO_2 retention secondary to inhibition of the red cell and brain CA isozymes.[55–60] It was recently demonstrated that this drug is also able to induce weight loss in epilepsy patients after pharmacological treatment.[61] Since the use of other CAIs was associated with weight loss,[32] it was presumed that this effect could be due to the inhibition of the CA isozymes involved in the biosynthesis of lipids. In support of this hypothesis, it was recently demonstrated that **TPM** is also a very potent inhibitor of several CA isozymes, including hCA II (K_I 13.8 nM) and VA (K_I 25.4 nM).[62] The atomic interactions that account for the high affinity of **TPM** for these two CA isoforms have been identified by means of a comparative structural study on the adducts that **TPM** forms with these two enzymes.[62,63] In particular, the X-ray crystal structure of the adduct hCA II/**TPM** was solved[62] and was used as template for a homology modeling and molecular dynamics (MD) simulation study on the adduct that **TPM** forms with hCA VA.[63] The main protein–inhibitor interactions observed in the hCA II/**TPM** structure are schematically depicted in Fig. 10.3a. In particular, as observed for other sulfamate/CA complexes,[28] the **TPM** sulfamate moiety is tetrahedrally coordinated to the zinc ion of the enzyme via its deprotonated nitrogen atom and hydrogen bonded to the Thr199 residue. An extended network of polar interactions between the sugar scaffold of the inhibitor and protein residues Asn62, Gln92, and Thr200 is also observed (see Fig. 10.3a). This network of hydrogen bonds stabilizes the complex, and explains the strong hCA II inhibitory properties of the drug.[62,63] Analysis of the hCA VA/**TPM** complex, as evidenced by the MD simulation study

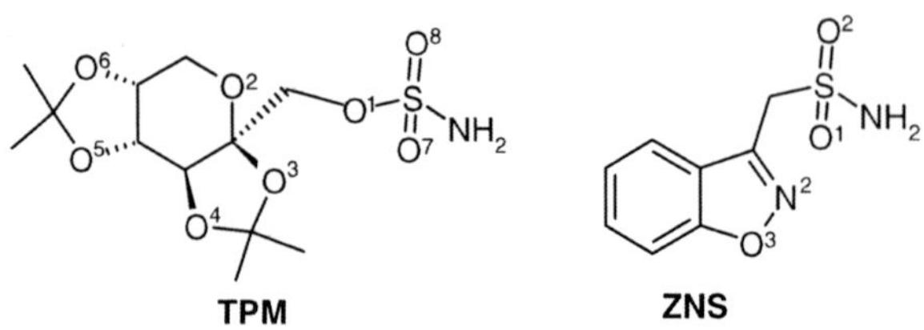

FIGURE 10.2 **TPM** and **ZNS** chemical structures.

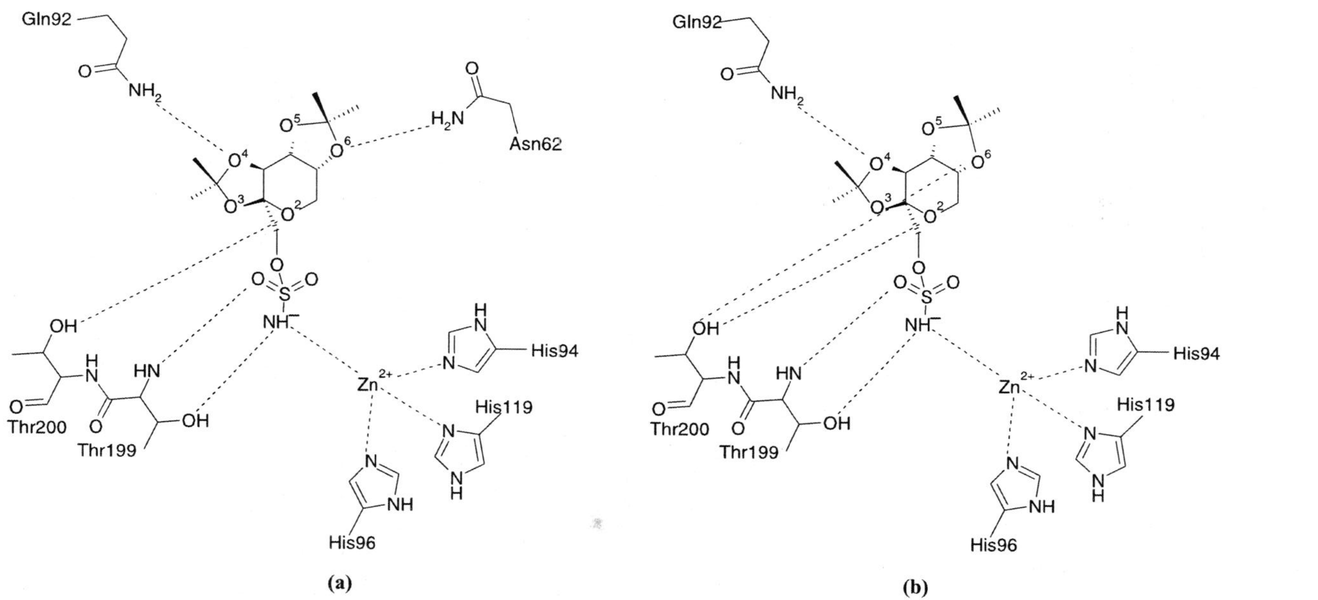

FIGURE 10.3 Active site region in the hCA II–TPM (a) and hCA VA–TPM (b) adducts. Only residues involved in hydrogen bonds and Zn^{2+} ion coordination are shown.

(Fig. 10.3b), reveals that while the sulfamate group of the inhibitor binds to the enzyme active site in a manner similar to that observed in the complex with isoform II, a different network of polar interactions of the sugar scaffold distinguishes the hCA VA/**TPM** adduct from the hCA II/**TPM** one (Fig. 10.3). These differences can justify the diverse binding affinity of **TPM** toward the two CA isozymes.[63]

Zonisamide (**ZNS**) (Fig. 10.2) is another antiepileptic drug used as adjunctive therapy for refractory partial seizures.[64] Due to the multiple mechanisms of action, **ZNS** shows a broad spectrum of anticonvulsant activities and has been effective in several types of seizures, including partial and generalized seizures, tonic–clonic seizures, and absence seizures in patients unresponsive to other anticonvulsants.[64] Recent studies demonstrated that **ZNS,** similar to **TPM,** is effective in reducing weight in obese patients and treating eating disorders, such as binge eating disorder, bulimia nervosa, or anorexia nervosa.[65,66] To investigate the molecular basis of **ZNS** antiobesity effects, the interaction of this sulfonamide drug with hCA II and hCA VA, the two CA isozymes involved in lipogenesis, was investigated by means of kinetic and structural studies.[63,67] In these studies, **ZNS** was demonstrated to strongly inhibit hCA II (K_I 47.6 nM) and hCA VA (K_I 20.6 nM).[63,67] The X-ray crystal structure of the adduct of hCA II/**ZNS** was solved,[67] providing a molecular explanation of the high affinity of this drug toward hCA II. In particular, these structural studies showed that **ZNS** was strongly bound within the hCA II active site, with the sulfonamide moiety participating in the classical interactions with the Zn^{2+} ion and the benzisoxazole ring establishing a large number of van der Waals interactions with residues located in the hydrophobic half of the active site cavity (Fig. 10.4a).[67] Starting from the hCA II/**ZNS** complex structure, homology modeling and MD simulation studies on the adduct hCA VA/**ZNS** were also performed,[63] revealing that the tetrahedral geometry of the Zn^{2+} binding site and the key hydrogen bonds between the sulfonamide moiety of the inhibitor and the enzyme active site were all maintained with respect to the adduct with isoform II. Moreover also in this case, the benzisoxazole ring of **ZNS** was also oriented toward the hydrophobic part of the active site cleft (Fig. 10.4b), establishing a large number of strong van der Waals interactions with enzyme residues. However, an additional polar interaction was observed in the hCA VA/**ZNS** adduct. In this case, the Gln92NE2 atom forms a bifurcated hydrogen bond with **ZNS**N2 and **ZNS**O3 atoms (see Fig. 10.4b). These structural data give a reasonable explanation for the experimentally observed affinities of **ZNS** toward hCA VA and hCA II.

In summary, similar to what was observed for **TPM,** these findings strongly suggest that the described clinical side effects of this drug, which caused weight loss in humans and animals,[65,66,68,69] may be due to the inhibition of the CA isozymes involved in lipogenesis. On the basis of these considerations, we believe that the inhibition of mitochondrial isoforms CA VA and VB, probably in conjunction with that of the ubiquitous cytosolic isoform CA II, may represent a potential target for novel antiobesity drugs.

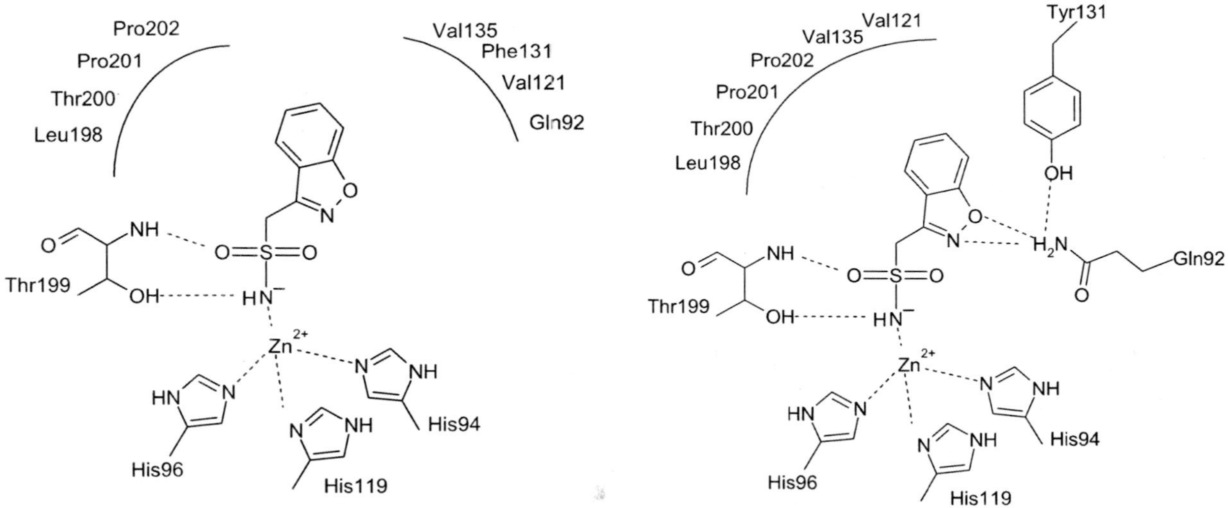

(a) (b)

FIGURE 10.4 Active site region in the hCA II–**ZNS** (a) and hCA VA–**ZNS** (b) adducts, showing residues participating in the recognition of the inhibitor molecules. Hydrogen bonds and the active site Zn^{2+} ion coordination are also shown.

ACKNOWLEDGMENTS

This work was financed by Italian National Research Council (RSTL_2007_id_0298).

REFERENCES

1. Friedman, J. M. A war on obesity, not the obese. *Science* **2003**, *299*, 856–859.
2. Hill, J. O.; Wyatt, H. R.; Reed, G. W.; Peters, J. C. Obesity and the environment: where do we go from here? *Science* **2003**, *299*, 853–855.
3. Ioannides-Demos, L. L.; Proietto, J.; Tonkin, A. M.; McNeil, J. J. Safety of drug therapies used for weight loss and treatment of obesity. *Drug Saf.* **2006**, *29*, 277–302.
4. Campfield, L. A.; Smith, F. J.; Burn, P. Strategies and potential molecular targets for obesity treatment. *Science* **1998**, *280*, 1383–1387.
5. Van Gaal, L.; Mertens, I. Effects of obesity on cardiovascular systems and blood pressure control, digestive disease and cancer. In *Clinical Obesity;* Kopelman, P.; Stock, M. J.,Eds.; Blackwell: Oxford, **1998**; pp 205–225.
6. Avenell, A.; Broom, J.; Brown, T. J.; Poobalan, A.; Aucott, L.; Stearns, S. C.; Smith, W. C.; Jung, R. T.; Campbell, M. K.; Grant, A. M. Systematic review of the long-term effects and economic consequences of treatments for obesity and implications for health improvement. *Health Technol. Assess.* **2004**, *8*, 1–182.
7. Lau, D. C. W. Synopsis of the 2006 Canadian clinical practice guidelines on the management and prevention of obesity in adults and children. *CMAJ* **2007**, *176*, 1103–1106.
8. Leibel, R. L.; Rosenbaum, M.; Hirsch, J. Changes in energy expenditure resulting from altered body weight. *N. Engl. J. Med.* **1995**, *332*, 621–628.
9. Halford, J. C. G. Pharmacotherapy for obesity. *Appetite* **2006**, *46*, 6–10.
10. Chaput, J. P.; Tremblay, A. Current and novel approaches to the drug therapy of obesity. *Eur. J. Clin. Pharmacol.* **2006**, *62*, 793–803.
11. Cheetham, S. C.; Jackson, H. C.; Vickers, S. P.; Dickinson, K.; Jones, R. B.; Heal, D. J. Novel targets for the treatment of obesity: a review of progress. *Drug Discov. Today Ther. Strateg.* **2004**, *1*, 227–235.
12. Gadde, K. M. Endocannabinoid receptor antagonists and other emerging pharmacological strategies for weight reduction. *Curr. Drug Targets Cardiovasc. Haematol. Disord.* **2005**, *5*, 549–556.
13. Pi-Sunyer, F. X.; Aronne, L. J.; Heshmati, H. M.; Devin, J.; Rosenstock, J. RIO-North America Study Group. Effect of rimonabant, a cannabinoid-1 receptor blocker, on weight and cardiometabolic risk factors in overweight or obese patients: RIO-North America: a randomized controlled trial. *JAMA* **2006**, *295*, 761–775.
14. Bays, H.; Dujovne, C. Pharmacotherapy of obesity: currently marketed and upcoming agents. *Am. J. Cardiovasc. Drugs* **2002**, *2*, 245–253.
15. Mcnelly, W.; Goa, K. L. Sibutramine. A review of its contribution to the management of obesity. *Drugs* **1998**, *56*, 1093–1124.
16. Lean, M. E. How does sibutramine work? *Int. J. Obes. Relat. Metab. Disord.* **2001**, *25* (Suppl. 4), S8–S11.

17. Bray, G. A.; Blackburn, G. L.; Ferguson, J. M.; Greenway, F. L.; Jain, A. K.; Mendel, C. M.; Mendels, J.; Ryan, D. H.; Schwartz, S. L.; Scheinbaum, M. L.; Seaton, T. B. Sibutramine produces dose-related weight loss. *Obes. Res.* **1999**, *7*, 189–198.

18. James, W. P.; Astrup, A.; Finer, N.; Hilsted, J.; Kopelman, P.; Rossner, S.; Saris, W. H.; Val Gaal, L. F. Effect of sibutramine on weight maintenance after weight loss: a randomised trial. STORM Study Group. Sibutramine Trial of Obesity Reduction and Maintenance. *Lancet* **2000**, *356*, 2119–2125.

19. Weibel, E. K.; Hadvary, P.; Hochuly, E.; Kupfer, E.; Lengsfeld, H. Lipstatin, an inhibitor of pancreatic lipase, produced by *Streptomyces toxytricini*, I: producing organism, fermentation, isolation and biological activity. *J. Antibiot. (Tokyo)* **1987**, *40*, 1081–1085.

20. Sidhaye, A.; Cheskin, L. J. Pharmacologic treatment of obesity. *Adv. Psychosom. Med.* **2006**, *27*, 42–52.

21. Davidson, M. H.; Hauptman, J.; DiGirolamo, M.; Foreyt, J. P.; Halsted, C. H.; Heber, D.; Heimburger, D. C.; Lucas, C. P.; Robbins, D. C.; Chung, J.; Heymsfield, S. B. Weight control and risk factor reduction in obese subjects treated for 2 years with orlistat. *JAMA* **1999**, *281*, 235–242.

22. Hollander, P. A.; Elbein, S. C.; Hirsch, I. B.; Kelley, D.; McGill, J.; Taylor, T.; Weiss, S. R.; Crockett, S. E.; Kaplan, R. A.; Comstock, J.; Lucas, C. P.; Lodewick, P. A.; Canovatchel, W.; Chung, J.; Hauptman, J. Role of orlistat in the treatment of obese patients with type 2 diabetes. *Diabetes Care* **1998**, *21*, 1288–1294.

23. Rössner, S.; Sjöström, L.; Noack, R.; Meinders, A. E.; Noseda, G. Weight loss, weight maintenance, and improved cardiovascular risk factors after 2 years treatment with orlistat for obesity. European Orlistat Obesity Study Group. *Obes. Res.* **2000**, *8*, 49–61.

24. Rucker, D.; Padwall, R.; Li, S. K.; Curioni, C.; Lau, D. C. Long term pharmacotherapy for obesity and overweight: updated meta-analysis. *BMJ* **2007**, *335*, 1194–1199.

25. Bray, G. A.; Ryan, D. H. Drug treatment of the overweight patient. *Gastroenterology* **2007**, *132*, 2239–2252.

26. Padwal, R. S.; Majumdar, S. R. Drug treatments for obesity: orlistat, sibutramine, and rimonabant. *Lancet* **2007**, *369*, 71–77.

27. Bays, H. E. Current and investigational antiobesity agents and obesity therapeutic treatment targets. *Obes. Res.* **2004**, *12*, 1197–1211.

28. Supuran, C. T. Carbonic anhydrases: novel therapeutic applications for inhibitors and activators. *Nat. Rev. Drug Discov.* **2008**, *7*, 168–181.

29. Supuran, C. T.; Scozzafava, A.; Casini, A. Carbonic anhydrase inhibitors. *Med. Res. Rev.* **2003**, *23*, 146–189.

30. Supuran, C. T. Carbonic anhydrase inhibitors in the treatment and prophylaxis of obesity. *Expert Opin. Ther. Pat.* **2003**, *13*, 1545–1550.

31. Tripp, B. C.; Smith, K.; Ferry, J. G. Carbonic anhydrase: new insights for an ancient enzyme. *J. Biol. Chem.* **2001**, *276*, 48615–48618.

32. Supuran, C. T.; Scozzafava, A. Carbonic anhydrase inhibitors and their therapeutic potential. *Expert Opin. Ther. Pat.* **2000**, *10*, 575–600.

33. Supuran, C. T.; Scozzafava, A. Applications of carbonic anhydrase inhibitors and activators in therapy. *Expert Opin. Ther. Pat.* **2002**, *12*, 217–242.

34. Supuran, C. T.; Scozzafava, A.; Conway, J., Eds.; *Carbonic Anhydrase: Its Inhibitors and Activators;* CRC Press: Boca Raton, FL, **2004**; 1–363.

35. Boriack-Sjodin, P. A.; Heck, R. W.; Laipis, P. J.; Silverman, D. N.; Christianson, D. W. Structure determination of murine mitochondrial carbonic anhydrase V at 2.45 Å resolution: implications for catalytic proton transfer and inhibitor design. *Proc. Natl. Acad. Sci. USA* **1995**, *92*, 10949–10953.

36. Whittington, D. A.; Waheed, A.; Ulmasov, B.; Shah, G. N.; Grubb, J. H.; Sly, W. S.; Christianson, D. W. Crystal structure of the dimeric extracellular domain of human carbonic anhydrase XII, a bitopic membrane protein overexpressed in certain cancer tumor cells. *Proc. Natl. Acad. Sci. USA* **2001**, *98*, 9545–9550.

37. Duda, D. M.; Tu, C.; Fisher, S. Z.; An, H.; Yoshioka, C.; Govindasamy, L.; Laipis, P. J.; Agbandje-McKenna, M.; Silverman, D. N.; McKenna, R. Human carbonic anhydrase III: structural and kinetic study of catalysis and proton transfer. *Biochemistry* **2005**, *44*, 10046–10053.

38. Eriksson, A. E., Jones, T. A., Liljas, A. Refined structure of human carbonic anhydrase II at 2.0 Å resolution. *Proteins* **1988**, *4*, 274–282.

39. Kannan, K. K.; Ramanadham, M.; Jones, T. A. Structure, refinement, and function of carbonic anhydrase isozymes: refinement of human carbonic anhydrase I. *Ann. N. Y. Acad. Sci.* **1984**, *429*, 49–60.

40. Stams, T.; Nair, S. K.; Okuyama, T.; Waheed, A.; Sly, W. S.; Christianson, D. W. Crystal structure of the secretory form of membrane-associated human carbonic anhydrase IV at 2.8 Å resolution. *Proc. Natl. Acad. Sci. USA* **1996**, *93*, 13589–13594.

41. Whittington, D. A.; Grubb, J. H.; Waheed, A.; Shah, G. N.; Sly, W. S.; Christianson, D. W. Expression, assay, and structure of the extracellular domain of murine carbonic anhydrase XIV: implications for selective inhibition of membrane-associated isozymes. *J. Biol. Chem.* **2004**, *279*, 7223–7228.

42. Di Fiore, A., Monti, S. M., Hilvo, M., Parkkila, S., Romano, V., Scaloni, A., Pedone, C., Scozzafava, A., Supuran, C. T., De Simone, G. Crystal structure of human carbonic anhydrase XIII and its complex with acetazolamide. *Proteins* **2009**, *74*, 164–175.

43. Dodgson, S. J. Inhibition of mitochondrial carbonic anhydrase: a discrepancy examined. *J. Appl. Physiol.* **1987**, *63*, 2134–2141.

44. Dodgson, S. J.; Cherian, K. Mitochondrial carbonic anhydrase is involved in rat renal glucose synthesis. *Am. J. Physiol.* **1989**, *257*, E791–E796.

45. Spencer, I. M.; Hargreaves, I.; Chegwidden, W. R. Effect of the carbonic anhydrase inhibitor acetazolamide on lipid synthesis in the locust. *Biochem. Soc. Trans.* **1988**, *16*, 973–974.

46. Chegwidden, W. R.; Spencer, I. M. Carbonic anhydrase provides bicarbonate for *de novo* lipogenesis in the locus. *Comp. Biochem. Physiol.* **1996**, *115B*, 247–254.

47. Chegwidden, W. R.; Dodgson, S. J.; Spencer, I. M. The roles of carbonic anhydrase in metabolism, cell growth and cancer in animals. In *The Carbonic Anhydrases: New Horizons;* Chegwidden, W. R.; Edwards, Y.; Carter, N., Eds.; Birkhäuser Verlag: Basel, **2000**; pp 343–363.

48. Lynch, C. J.; Fox, H.; Hazen, S. A.; Stanley, B. A.; Dodgson, S. J.; Lanoue, K. F. Role of hepatic carbonic anhydrase in *de novo* lipogenesis. *Biochem. J.* **1995**, *310*, 197–202.

49. Hazen, S. A.; Waheed, A.; Sly, W. S.; Lanoue, K. F.; Lynch, C. J. Differentiation-dependent expression of CAV and the role of carbonic anhydrase isozymes in pyruvate carboxylation in adipocytes. *FASEB J.* **1996**, *10*, 481–490.

50. Atwood, P. V. The structure and mechanism of action of pyruvate carboxylase. *Int. J. Biochem. Cell Biol.* **1995**, *27*, 231–249.

51. Alldred, J. B.; Reilly, K. E. Short-term regulation of acetyl CoA carboxylase in tissues of higher animals. *Prog. Lipid Res.* **1997**, *35*, 371–385.

52. Forster, R. E.; Dodgson, S. J. Membrane transport and provision of substrates for carbonic anhydrase in vertebrates. In *The Carbonic Anhydrases: New Horizons;* Chegwidden, W. R.; Edwards, Y.; Carter, N., Eds.; Birkhäuser Verlag: Basel, **2000**; pp 263–280.

53. Parkkila, S. An overview of the distribution and function of carbonic anhydrase isozymes in mammals. In *The Carbonic Anhydrases: New Horizons;* Chegwidden, W. R.; Edwards, Y.; Carter, N.,Eds.; Birkhäuser Verlag: Basel, **2000**; pp 79–94.

54. Parkkila, S.; Parkkila, A. K.; Kivelä, J. Role of carbonic anhydrase and its inhibitors in biological science related to gastroenterology, neurology and nephrology. In *Carbonic Anhydrase: Its Inhibitors and Activators;* Supuran, C. T.; Scozzafava, A.; Conway, J., Eds.; CRC Press: Boca Raton, FL, **2004**; pp 283–301.

55. Shank, R. P.; Gardocki, J. F.; Vaught, J. L.; Davis, C. B.; Schupsky, J. J.; Raffa, R. B.; Dodgson, S. J.; Nortey, S. O.; Maryanoff, B. E. Topiramate: preclinical evaluation of structurally novel anticonvulsant. *Epilepsia* **1994**, *35*, 450–460.

56. Edmonds, H. L.; Jiang, Y. D.; Zhang, P. Y.; Shank, R. P. Anticonvulsant activity of topiramate and phenytoin in a rat model of ischemia-induced epilepsy. *Life Sci.* **1996**, *59*, PL127–PL131.

57. Stringer, J. L. A comparison of topiramate and acetazolamide on seizure duration and paired-pulse inhibition in the dentate gyrus of the rat. *Epilepsy Res.* **2000**, *40*, 147–153.

58. Sabers, A.; Gram, L. Newer anticonvulsants: comparative review of drug interactions and adverse effects. *Drugs* **2000**, *60*, 23–33.

59. Bourgeois, B. F. D. Pharmacokinetics and pharmacodynamics of topiramate. *J. Child Neurol.* **2000**, *15*, S27–S30.

60. Bialer, M.; Johannessen, S. I.; Kupferberg, H. J.; Levy, R. H.; Loiseau, P.; Perucca, E. Progress report on new antiepileptic drugs: a summary of the Fifth Eilat Conference. *Epilepsy Res.* **2001**, *43*, 11–58.

61. Gordon, A.; Price, L. H. Mood stabilization and weight loss with topiramate. *Am. J. Psychiatry* **1999**, *156*, 968–969.

62. Casini, A.; Antel, J.; Abbate, F.; Scozzafava, A.; David, S.; Waldeck, H.; Schäfer, S.; Supuran, C. T. Carbonic anhydrase inhibitors: SAR and X-ray crystallographic study for the interaction of sugar sulfamates/sulfamides with isozymes I, II and IV. *Bioorg. Med. Chem. Lett.* **2003**, *13*, 841–845.

63. Vitale, R. M.; Pedone, C.; Amodeo, P.; Antel, J.; Wurl, M.; Scozzafava, A.; Supuran, C. T.; De Simone, G. Carbonic anhydrase inhibitors: molecular modeling study for the interaction of zonisamide and topiramate with isozyme VA. *Bioorg. Med. Chem.* **2007**, *15*, 4152–4158.

64. Zareba, G. Zonisamide: review of its use in epilepsy therapy. *Drug Today (Barc.)* **2005**, *41*, 589–597.

65. Kim, C. S. Zonisamide effective for weight loss in women. *J. Fam. Pract.* **2003**, *52*, 600–601.

66. Jennings, J. E. Zonisamide use in obesity and eating disorders. *PCT International Application* WO 2003092682, 2003, 18 pp.

67. De Simone, G.; Di Fiore, A.; Menchise, V.; Pedone, C.; Antel, J.; Casini, A.; Scozzafava, A.; Wurl, M.; Supuran, C. T. Carbonic anhydrase inhibitors. Zonisamide is an effective inhibitor of the cytosolic isozyme II and mitochondrial isozyme V: solution and X-ray crystallographic studies. *Bioorg. Med. Chem. Lett.* **2005**, *15*, 2315–2320.

68. Masuda, Y.; Karasawa, T. Inhibitory effect of zonisamide on human carbonic anhydrase *in vitro*. *Arzneimittelforschung* **1993**, *43*, 416–418.

69. Gadde, K. M.; Franciscy, D. M.; Wagner, H. R., 2nd; Krishnan, K. R. Zonisamide for weight loss in obese adults: a randomized controlled trial. *JAMA* **2003**, *289*, 1820–1825.

CHAPTER 11

Dual Carbonic Anhydrase and Cyclooxygenase-2 Inhibition

JEAN-MICHEL DOGNÉ[1], ANNE THIRY[1], BERNARD MASEREEL[1], and CLAUDIU T. SUPURAN[2]

[1]Drug Design and Discovery Center, University of Namur, 61 rue de Bruxelles, 5000 Namur, Belgium

[2]Laboratorio di Chimica Bioinorganica, Università degli Studi di Firenze, Polo Scientifico, Room 188, Via della Lastruccia 3, 50019 Sesto Fiorentino, Italy

11.1 INTRODUCTION

The sulfonamide moiety constitutes an important template for a series of classes of drugs, with several types of pharmacological agents among which are the antibacterial, anticarbonic anhydrase, diuretic, hypoglycemic, and anticancer agents. Rofecoxib, celecoxib, and valdecoxib are three nonsteroidal anti-inflammatory agents that were recently developed as COX-2 selective inhibitors. A key structural feature that characterizes celecoxib and valdecoxib is the presence of a primary sulfonamide moiety; rofecoxib possesses instead the isosteric and isoelectronic methylsulfone group in its molecule. It was recently shown that the sulfonamide COX-2 selective inhibitors (but not the methylsulfone ones) also act as nanomolar inhibitors of several isozymes of the metalloenzyme carbonic anhydrase (CA, EC 4.2.1.1), some of which (CA II, IX, and XII) are strongly involved in tumorigenesis. This was demonstrated in pharmacological studies and by the determination of the X-ray crystal structures for the adducts of celecoxib and valdecoxib with isozyme CA II. In this chapter, we will discuss how this dual activity may help to explain differences in clinical observation between sulfonamide and methylsulfone COX-2 inhibitors and to constitute an important mechanism of antitumor action.

11.2 NSAIDs AND CYCLOOXYGENASE INHIBITION

Nonsteroidal anti-inflammatory drugs (NSAIDs) act by interfering with the cyclooxygenase pathway. The cyclooxygenase enzyme (prostaglandin (PG) endoperoxide synthase or COX) catalyzes the transformation of arachidonic acid released from cellular membrane phospholipids by phospholipase A_2 into prostaglandin H_2 (PGH_2) and subsequently into prostaglandin G_2 (PGG_2). This reaction constitutes the first step in the biosynthesis of a series of derivatives collectively referred to as prostanoids and comprise the prostaglandins and thromboxane A_2 (TXA_2) (Fig. 11.1). Three primary COX isoenzymes are distinguished: COX-1 (constitutive), COX-2 (inducible), and COX-3, which is mainly detected in the central nervous system. A number of partial forms of COX-1 and COX-2 are also known, but their biological functions have not been well evaluated.[1] Different mechanisms stimulate the different types of cyclooxygenase. COX-1 is stimulated continuously in most tissues/cells and converts arachidonic acid to prostaglandins. These prostaglandins in turn stimulate body functions, such as stomach mucous production and kidney water excretion, as well as platelet formation (via thromboxane A_2). The location of the COX-1 enzyme dictates the functions of the prostaglandins it releases. Thus, COX-1 in the stomach wall produces prostaglandins that stimulate mucous production. This isozyme is also only expressed in platelets. Before 1991, COX-1 was the only isoform thought to be responsible for both physiological production of prostaglandins and their increased levels observed during inflammation processes. However, an inducible COX enzyme was identified as an isoform different from the constitutive enzyme and called COX-2 in opposition to the constitutive form COX-1. In contrast, the COX-2 enzyme is induced. It is not normally present in cells but its expression can be increased

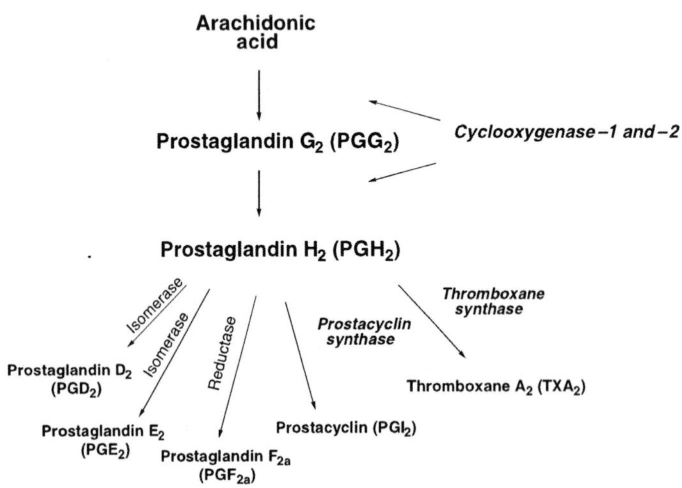

FIGURE 11.1 Cyclooxygenase pathway and chemical structures of arachidonic acid, prostaglandin endoperoxides (PGG_2 and PGH_2), prostaglandins (PGE_2, PGD_2, $PGF_{2\alpha}$), prostacyclin (PGI_2), and thromboxane A_2 (TXA_2).

dramatically by the action of macrophages.[2–6] COX-2 plays a key role in many inflammatory and proliferative reactions.[7] Experimental data have shown that prostaglandins have a central action in therapeutic targeting not only in the treatment of many inflammatory diseases but also in several types of human cancers.[8] Of importance, COX-2 expression is also observed in some tissues such as vascular endothelium, kidney, or brain under normal conditions, suggesting the involvement of COX-2 in the regulation of physiological processes.[4] Besides, the discovery of the COX isoforms led to establishing their importance in many nonarthritic or nonpain states where there is an inflammatory component of pathogenesis, including cancer,[8] Alzheimer's disease, and other neurodegenerative diseases.[9]

A splice variant of COX-1 mRNA, retaining intron 1, and given the names COX-3, COX-1b, or COX-1v, has also been identified in canine tissues as an acetaminophen-sensitive isoform but its functions are still unknown. COX-3 was suggested as the key to unlocking the mystery of the mechanism of action of acetaminophen. However, little is known regarding its physiological/pathological function or its inhibition in humans.[10,11] Moreover, attempts to explain the action of acetaminophen and derivatives by inhibition of a central COX-3 have been rejected by some authors. Thus, recently, Hinz et al. suggested that acetaminophen inhibited COX-2 by more than 80%, that is, to a degree comparable to NSAIDs and selective COX-2 inhibitors.[12,13]

Classical NSAIDs still represent the most commonly used medications for the treatment of pain and inflammation, but numerous well-described gastrointestinal and renal ADRs limit their use in daily practice. According to one meta-analysis,[14] approximately one-third of patients taking NSAIDs had gastric or duodenal ulcers, as diagnosed by endoscopy. Thus, for the identification and characterization of the role of COX-2 in inflammatory processes, selective inhibitors of COX-2 were developed to improve the safety of anti-inflammatory therapy in patients at elevated risk for gastrointestinal complications that are thought to be caused by depression of COX-1-derived mucosal prostanoids. They were not expected to be more efficacious analgesics than classical NSAIDs. Indeed, the rationale of this development was based on the hypothesis that this isoform mediates inflammation in several organs via the biosynthesis of prostaglandins E_2 and I_2 (or prostacyclin) and that COX-1 was the source of the same prostaglandins in the gastric epithelium, where they would act as cytoprotective mediators.[2,3,6,15]

The structural concept of COX-2 selective inhibition is based on the differences in amino acid sequences existing between COX-1 and COX-2. The differences in the amino acid sequences between COX isoforms are responsible for the differences in the enzyme structures and, especially, in the access to the COX catalytic site. Thus, the most significant difference between the isoenzymes, which allows selective inhibition, is the substitution of isoleucine at position 523 in COX-1 with valine in COX-2. The relatively smaller Val523 residue in COX-2 allows access to a hydrophobic side pocket in the enzyme coxibs that bind to this alternative site and are considered to be selective inhibitors of COX-2.[16]

Based on these structural differences, a large variety of COX-2 selective inhibitors have been developed in the past years and some of them have been marketed.[2,3] Celecoxib (Celebrex®) and rofecoxib (Vioxx) were the first two coxibs approved and

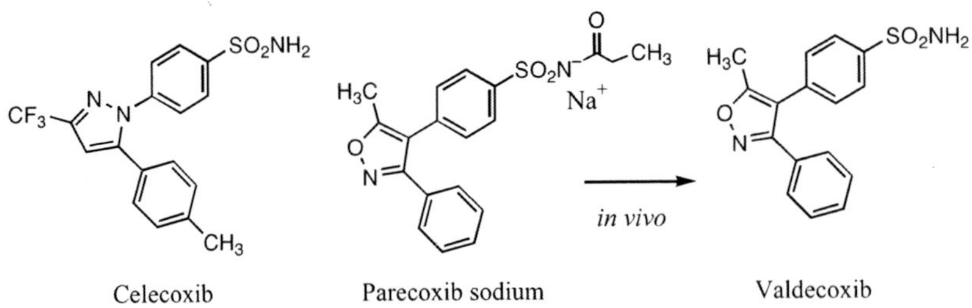

FIGURE 11.2 Chemical structures of "sulfonamide" COX-2 inhibitors.

launched in 1999. Valdecoxib (Bextra®) was approved by the FDA and launched in 2002. Three other coxibs were then approved in different countries: etoricoxib (Arcoxia®), its prodrug for parenteral use only, parecoxib sodium (Dynastat®), and lumiracoxib (Prexige®, Prexigem®) (Figs 11.2–11.4). Although found to have partially fulfilled their promises in terms of lowering gastrointestinal ADRs, an alarming turn of events took place at the end of September 2004 when Merck & Co. announced the voluntary withdrawal of rofecoxib (Vioxx) worldwide because of an increased risk of cardiovascular events, such as thrombotic events. Other coxibs were subsequently suspected to have this adverse reaction, although to a varying degree. This led the main international regulatory agencies to introduce specific cardiovascular warnings for the whole class of coxibs. However, the data on the potential for an increased risk of cardiovascular events with coxib use are still confusing and the risks associated with their use are still unclear. One of the critical questions still unanswered relates to the variation of the risk among the coxibs. The evidence base with regard to

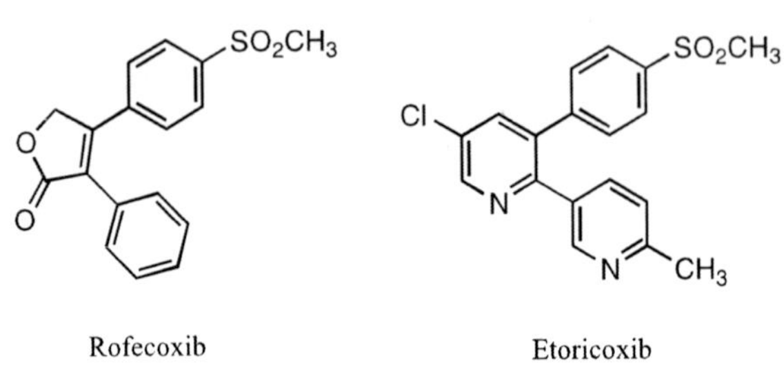

FIGURE 11.3 Chemical structures of "methylsulfone" COX-2 inhibitors.

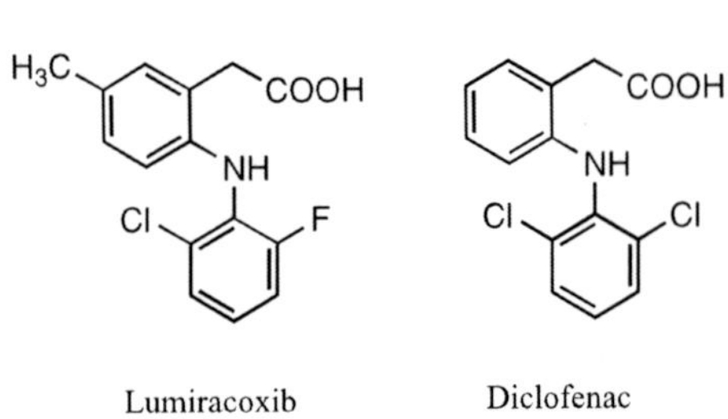

FIGURE 11.4 Chemical structures of two arylacetic acid NSAIDs: classical NSAID diclofenac and COX-2 inhibitor lumiracoxib.

the risk of cardiovascular events mainly consists of new data from clinical trials and epidemiological studies (studies of the causes of diseases in the population), information published in scientific journals, and postmarketing ADRs reporting. However, in addition to these data, an analysis of more fundamental studies is of great interest to support the class effect theory or to highlight differences between coxibs.

11.3 CYCLOOXYGENASE INHIBITION AND CARDIOVASCULAR RISKS

Is there a plausible mechanism that could explain the increased cardiovascular risks of rofecoxib? Is this risk associated with its selective COX-2 inhibition and, if so, is it applicable to other coxibs? These questions may be answered on the basis of the present knowledge of the physiological role of COX-2 and the mode of action and specificity of coxibs.

In terms of cardiovascular physiology, COX-1 and thromboxane synthase are constitutively expressed within platelets leading to TXA_2, a potent inducer of vasoconstriction and platelet aggregation. Studies performed in mice and humans revealed that COX-2 (and not COX-1 as initially sought) was the predominant source of prostacyclin by vascular endothelial cells in response to shear stress.[17] This is a key point since prostacyclin inhibits platelet aggregation, prevents vascular smooth muscle cell proliferation *in vitro*, and induces vascular smooth muscle relaxation. Low-dose aspirin selectively impairs COX-1-mediated TXA_2 production in platelets restoring the net antithrombotic balance. Nonspecific NSAIDs block both COX isoforms, and therefore they have a balanced effect of reducing the prothrombotic actions of TXA_2 and the antithrombotic properties of prostacyclin. In contrast, inhibition of COX-2 is associated with significant suppression of prostacyclin synthesis in human subjects,[18] potentially creating an alteration of such a delicate vascular homeostasis. Moreover, in the setting of atherosclerosis, COX-2 plays a greater role as a source of PGI_2 and more TXA_2 is produced; thus, inhibiting COX-2 has a more profound effect on prostanoid balance, favoring TXA_2 production and promoting platelet-dependent thrombosis.

The shift of the thromboxane/prostacyclin balance is a strong theoretical basis for an association between coxibs use and the occurrence of thrombotic phenomena. Other experimental studies further support the importance of the thromboxane/prostacyclin balance in the thrombotic risk. The interested reader is advised to read the review on prostaglandins and isoprostanes—therapeutic targets in atherogenesis—by the same author for more information on the topic.[3] Although the clinical relevance of these experimental results remained uncertain from a biochemical point of view, the suppression of prostacyclin formation might theoretically elevate blood pressure and increase prothrombotic risk associated with the rupture of an atherosclerotic plaque and facilitate myocardial infarction.

It is important to note that when all the coxibs have the same pharmacological profile, they differ in terms of COX-2/COX-1 selectivity ratios. Thus, Riendeau and collaborators compared the potency and selectivity of different COX inhibitors.[19] Selectivity ratios (COX-1/COX-2 IC_{50}) for the inhibition of COX-2 of 106, 35, 30, 7.6, and 7.3 were obtained for etoricoxib, rofecoxib, valdecoxib (and parecoxib), celecoxib, and nimesulide, respectively. By contrast, lower ratios were observed for classical NSAIDs diclofenac, etodolac, and meloxicam (two–threefold). Lumiracoxib, which was not evaluated in this study, has emerged as one of the most selective COX-2 inhibitors to date.[20] These *in vitro* data would suggest that celecoxib as nimesulide, a classical NSAID with preferential COX-2 inhibitory profile, can also partially inhibit COX-1. However, as for all coxibs, no effect on TXA_2 production or antiplatelet activity was reported with celecoxib in healthy volunteers at supratherapeutic doses while a suppression of urinary excretion of prostacyclin was observed.[18] In contrast, naproxen or ibuprofen produced statistically significant reduction in platelet aggregation, serum TXB_2 levels (TXA_2 metabolite), and increased bleeding time. Consequently, it clearly appears that at therapeutic dosage, all coxibs, including celecoxib that is the weakest coxib *in vitro*, are equal in depressing prostacyclin biosynthesis while having no significant impact on thromboxane production.[20] These data strongly support the hypothesis that the adverse cardiovascular events of these drugs are related to a class effect.

11.4 COX-2 INHIBITORS AND CA INHIBITION

Contradictory evidence exists about the increased risk of cardiovascular events with COX-2 inhibitors. While the cardiovascular risk of rofecoxib is clearly evidenced in many studies, the risk is more controversial with other coxibs such as celecoxib and lumiracoxib. Thus, in 2000, the results of the Vioxx Gastrointestinal Outcomes Research (VIGOR) trial revealed unexpected evidence of increased myocardial infarction and stroke.[21,22] When the Adenomatous Polyp Prevention on Vioxx trial (APPROVe),[23] comparing rofecoxib 25 mg/day with placebo in patients with a history of colorectal adenomas, reported an excess risk of thrombotic cardiovascular events in the rofecoxib group, Merck voluntarily withdrew the drug from the market in September 2004. While the VIGOR study revealed an increased risk of thrombotic events with rofecoxib, the CLASS (Celecoxib Long-Term Arthritis Safety)[24] and TARGET (Therapeutic Arthritis Research and Gastrointestinal Event Trial)[25] trials failed to demonstrate significant difference in cardiovascular side effects of NSAIDs and celecoxib (CLASS) and lumiracoxib (TARGET). However, there are many reports of increased risk of thrombotic cardiovascular events not only with rofecoxib but also with celecoxib and the third coxib available, valdecoxib. This led the FDA to issue a public health advisory concerning the use of all coxibs. In February 2005, the advisory panel of the FDA recommended that celecoxib and valdecoxib remain on the market, but advocated that "black box" warnings be added to the label.[5] At the same time, the European Medicines Agency (EMEA) imposed strong warnings on coxibs, recommending that they should not be prescribed to patients who have coronary heart disease or who have had a stroke and that they should be used with caution in patients at risk for vascular disease.

During the same period, different meta-analysis and clincal trials suggested clear differences between coxibs on renal events and hypertension. Indeed, although much attention has been focused on the putative prothrombotic effect of these agents, their cardiorenal and blood pressure elevating actions also contribute to cardiovascular risk. COX-2 is widely expressed throughout the kidney, and inhibition of this enzyme is responsible for blood pressure elevation via reduced glomerular filtration, salt, and water retention.[26] By studying the effect of a series of coxibs on carbonic anhydrase, we and others aimed at demonstrating that potential non-COX-2-dependent mechanism for these drugs may also provide mechanistic insights into the underlying blood pressure differences between COX-2 inhibitors.[27]

A key structural feature that characterizes celecoxib and valdecoxib is the presence of a primary sulfonamide moiety; rofecoxib possesses instead the isosteric and isoelectronic methylsulfone group in its molecule. Such unsubstituted sulfonamide moieties attached to an aromatic scaffold had the "theoretical" potential to inhibit the zinc enzyme carbonic anhydrase. Indeed, drugs containing sulfonamides are widely used in medicine. For example, unsubstituted sulfonamide constitutes the main class of inhibitors of the Zn^{2+}-containing enzyme carbonic anhydrase. Although COX-2 is not a Zn^{2+}-containing enzyme as is CA, the sulfonamide group of SC-558, a related *p*-bromo derivative of celecoxib, has been demonstrated as a major determinant of the protein binding interaction.

The carbonic anhydrases (EC 4.2.1.1) are ubiquitous metalloenzymes, present in prokaryotes and eukaryotes, being encoded by three distinct, evolutionarily unrelated gene families. In higher vertebrates, including humans, 16 different CA isozymes or CA-related proteins (CARP) were described, with very different subcellular localization and tissue distribution. Basically, there are several cytosolic forms (CA I–III, VII), four membrane-bound isozymes (CA IV, IX, XII, and XIV), two mitochondrial forms (CA VA and VB), and a secreted CA isozyme, CA VI, together with three acatalytic forms (isozymes CARP VIII, X, and XI).[28,29] These enzymes catalyze the interconversion between carbon dioxide and bicarbonate ion:

$$O = C = O + H_2O \Leftrightarrow HCO_3^- + H^+$$

Since CO_2 is generated in high amounts in all living organisms, CAs are involved in crucial physiological processes connected with respiration and transport of CO_2/bicarbonate between metabolizing tissues and lungs, pH and CO_2 homeostasis, electrolyte secretion in a variety of tissues/organs, biosynthetic reactions (such as gluconeogenesis, lipogenesis, and ureagenesis), bone resorption, calcification, tumorigenicity, and many other physiological or pathological processes. Sulfonamides, which are the most important carbonic anhydrase inhibitors (CAIs), bind in a tetrahedral geometry of the Zn^{2+} ion.[30] Several CAIs are clinically used, such as acetazolamide (AZA), methazolamide (MZA), ethoxzolamide (EZA), and dichlorophenamide (DCP), which have been marketed for many years as antiglaucoma agents, diuretics, antiepileptics, or for the management of other neuromuscular disorders. As antiglaucoma drugs, these agents act by reducing elevated intraocular pressure (IOP) characteristic to this disease. They represent the most efficient physiological treatment of glaucoma, since by inhibiting the ciliary process enzymes (the sulfonamide susceptible isozymes CA II and IV), a reduced rate of bicarbonate and aqueous humor secretion is achieved, which leads to a 25–30% decrease of IOP.

11.4.1 Pharmacological Evidences of CA Inhibition by Sulfonamide COX-2 Inhibitors

Both celecoxib and valdecoxib possess an unsubstituted arylsulfonamide group in their structure. Weber et al. demonstrated that the sulfonamide-type COX-2 inhibitors valdecoxib and celecoxib act as potent inhibitors of several CA II and IX isozymes, with affinity for some of them of the same order of magnitude as those of clinically used CAIs.[31] Thus, as seen from Table 11.1, valdecoxib and celecoxib exhibit efficient CA inhibitory potency against CA II and IX, with affinities comparable to those of dichlorophenamide, acetazolamide, methazolamide, and dorzolamide, whereas they inhibit with moderate potency the membrane-bound isozyme CA IV. On the other hand, they act as very weak CA I inhibitors, similar to dorzolamide. It is also important to note that the nonsulfonamide COX-2 inhibitor SC-560 and diclofenac, a classical NSAID, do not possess CA inhibitory activity.

Because this inhibition profile suggested that both COX-2 sulfonamide-containing inhibitors may possess antiglaucoma activities, they were administered systemically to hypertensive rabbits for one week. In this test, both sulfonamide COX-2 inhibitors demonstrate intraocular pressure- lowering properties, whereas the nonsulfonamides were inactive.

Finally, in another recent paper, our team demonstrated that valdecoxib and celecoxib also inhibited hCA XII with affinity values in the same range as those of the other sulfonamides used as reference CAI (dichlorophenamide, acetazolamide, methazolamide, and dorzolamide).

TABLE 11.1 CA Inhibition Data with Standard, Clinically Used Sulfonamide Inhibitors and COX-2 Selective Inhibitors (Table from Refs 28, 31, 33)

	K_I (nM)[a]			
Inhibitor	hCA I[b]	hCA II[b]	bCA IV[c]	hCA IX[d]
Celecoxib	50,000	21	290	16
Valdecoxib	54,000	43	340	27
Acetazolamide	250	11	70	25
Methazolamide	50	14	36	27
Dorzolamide	50,000	9	43	52
Dichlorophenamide	1100	38	380	50
SC-560	>100 μM	>100 μM	>100 μM	>100 μM

[a] Errors in the range of 5–10% of the reported value, from three determinations.
[b] Human cloned isozymes, esterase assay method.
[c] Isolated from bovine lung microsomes, esterase assay method.
[d] Human cloned isozyme, CO_2 hydrase assay method.[67]

The importance of sulfonamide moiety in inhibiting CA was recently confirmed in another pharmacological study performed in mature osteoclasts. The authors evidenced that CA II was expressed predominantly in mature osteoclasts, but not in the precursors. In agreement with the results already discussed, the activity of CA II expressed in osteoclasts was inhibited by sulfonamide-type COX-2 selective agents celecoxib, similar to CA II inhibitor acetazolamide, but not by a methylsulfone-type COX-2 inhibitor rofecoxib.[32]

11.4.2 Structural Evidences of CA Inhibition by Sulfonamide COX-2 Inhibitors

The X-ray crystal structures of the hCA II–celecoxib[31] and hCA II–valdecoxib[33] adducts have been resolved at high resolution, thus explaining at the molecular level why these COX-2 inhibitors also interact within the CA active site. It was demonstrated that both drugs were stabilized by polar and hydrophobic interactions within the hCA II active site. The hCAII–celecoxib adduct crystal structure revealed that this coxib adopts a similar binding mode compared to typical sulfonamide CAIs (Fig. 11.5).[31] It binds with its sulfonamide toward the zinc resulting in a tetrahedral coordination. The Zn^{2+} ion remained in its stable tetrahedral geometry, being

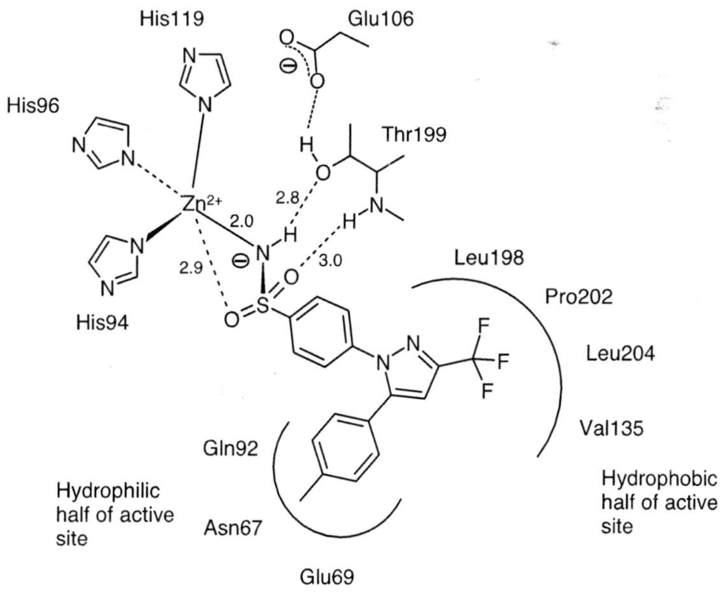

FIGURE 11.5 Schematic drawing of the interactions between hCA II and celecoxib-6. Hydrogen bonds formed between enzyme and the inhibitor are shown as dotted lines. Bond lengths are given in Å. Residues with a distance less than 4 Å around the inhibitor are schematically drawn. They form a hydrophobic subsite (Leu198, Pro202, Leu204, Val135, and Phe131) and a second more hydrophilic subsite (Gln92, Asn67, and Glu69).[28]

coordinated, in addition to the sulfonamidate nitrogen of celecoxib, by the imidazolic nitrogens of His94, His96, and His119. The proton attached to the sulfonamidate nitrogen atom of the inhibitor also makes a hydrogen bond with the hydroxyl group of Thr199, which in turn accepts a hydrogen bond from the carboxylic group of Glu106. One of the oxygen atoms of the coordinated sulfonamidate moiety makes a hydrogen bond with the backbone amide of Thr199, whereas the other one is semicoordinated to the catalytic Zn^{2+} ion. The benzenesulfonamide part of celecoxib lies in the hydrophobic part of the active site cleft, where it makes van der Waals contacts with the side chains of Leu198 and Pro202, whereas the *p*-toluyl group is oriented toward the hydrophilic part of it, making van der Waals contacts with residues Gln92, Asn67, and Glu69. The trifluoromethyl group of the inhibitor is also oriented toward the hydrophobic part of the active site, interacting with residues Leu198, Pro202, Leu204, and Val135, whereas the pyrazole heteroatoms do not make any hydrogen bonds or other types of interactions with amino acid residues of the active site.

Recently, the hCA II–valdecoxib adduct was reported.[33] Briefly, the sulfonamide moiety participates in the classical interactions with the Zn^{2+} ion as described for celecoxib and other sulfonamide-containing CA inhibitors. The phenyl–isoxazole moiety interacts with residues Gln92, Val111, Leu198, Thr200, and Pro202. Besides these interactions, the 3-phenyl group of the inhibitor is located in a hydrophobic pocket and it establishes van der Waals contacts with the aliphatic side chain of various hydrophobic residues and a strong offset face-to-face stacking interaction with the aromatic ring of Phe131.

Figure 11.6 shows a structural overlay of valdecoxib and celecoxib bound to hCA II, as determined by the superposition of hCA II active site residues.[33] As described above, in both cases, the sulfonamide moiety participates in the classical interactions with the Zn^{2+} and the organic scaffold of the inhibitor (i.e., the isoxazole ring of

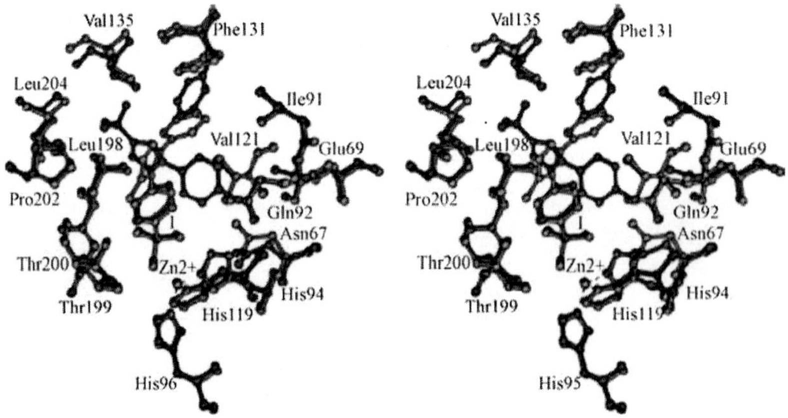

FIGURE 11.6 Stereoview of the hCA II active site complexed with valdecoxib 1 (red) and celecoxib-2 (blue) brought to optimal structural overlay.[33] (See the color version of this figure in Color Plates section.)

valdecoxib or the pyrazole ring of celecoxib) did not establish polar interactions with the enzyme active site, but participated in a large number of hydrophobic contacts. These similarities were reflected by a rather comparable value of K_I for the two inhibitors against hCA II. However, celecoxib binds in a completely different manner with hCA II as compared to valdecoxib, even though both drugs are structurally similar. Thus, celecoxib completely filled the entire CA II active site, with its trifluoromethyl group in the hydrophobic part of the active site and the *p*-toluyl moiety in the hydrophilic one (and this may also explain why it is approximately two times stronger hCA II inhibitor than valdecoxib). Consequently, the *p*-toluyl moiety of celecoxib did not establish any interaction with the side chain of Phe131. In contrast to this, valdecoxib was rotated by about 90° around the chemical bond connecting the benzenesulfonamide and the substituted isoxazole ring. This rotation placed the 3-phenyl substituent of the inhibitor in a different position and allowed, together with the aforementioned movement of Phe131, the strong stacking interaction with this aromatic residue. This is of importance since we recently demonstrated that just this interaction with Phe131 (or its absence) orients the active site binding region of inhibitors within the hCA II cavity.[34]

11.4.3 Potential Clinical Impact of CA Inhibition by Sulfonamide COX-2 Inhibitors

The demonstration of this nanomolar affinity of the sulfonamide COX-2 inhibitors celecoxib and valdecoxib for CA II led to this hypothesis that these two drugs could also be responsible for a diuretic effect that could counteract the renal hypertension induced by the COX-2 inhibition. This would explain why patients treated with rofecoxib demonstrate an increased incidence of hypertension and edema that is not seen in patients under celecoxib.[35–38] However, a recent clinical study did not confirm these results. Thus, Alper and collaborators investigated the inhibitory effects of celecoxib on renal carbonic anhydrase enzyme activity in human hypertensive subjects. Ten subjects with stable, treated hypertension were randomized to one of the three treatment sequences, which included, in differing order, 200 mg celecoxib twice a day, 250 mg acetazolamide twice a day, or placebo twice a day. Whereas acetazolamide caused a bicarbonate diuresis and a hyperchloremic metabolic acidosis, celecoxib appeared to have no detectable effect on renal carbonic anhydrase or acid–base homeostasis. Consequently, in this clinical trial, therapeutic doses of celecoxib did not appear to have a clinically significant inhibitory action on renal carbonic anhydrase. However, it should be noted that the main limitation of this study was its short-term duration.[39]

In the *in vitro* and *in vivo* studies already described, we mainly focused on the interaction of some coxibs with carbonic anhydrase, aiming at providing mechanistic insights into the underlying blood pressure differences between such coxibs. Other possible non-COX-2-dependent mechanisms may also contribute to such differences. These have been discussed elsewhere and include greater COX-2 selectivity of certain agents such as rofecoxib, the differing half-life, and the controversial aldosterone modulation by rofecoxib.[3,40]

11.5 CONCLUSIONS

The withdrawal of rofecoxib raised the question of cardiovascular safety of the entire class of COX-2 inhibitors. Major efforts are still being made to discover why cardiovascular reactions took place with rofecoxib and with all coxibs as a class. The significant reduction of prostacyclin production *in vivo* without affecting the COX-1-dependent thromboxane A_2 biosynthesis by platelets may theoretically support an association between all coxibs and the occurrence of cardiovascular side effects. Besides, their cardiorenal and blood pressure elevating actions also contribute to cardiovascular risk and, on this point, difference between coxibs have been reported. The demonstration of a nanomolar affinity of the sulfonamide COX-2 inhibitors celecoxib and valdecoxib for CA II suggested a possible additive diuretic effect that could counteract the renal hypertension induced by the COX-2 inhibition, although this was not confirmed in a short-term clinical trial. The study of potential non-COX-2-dependent mechanism for these drugs should continue to provide mechanistic insights into the underlying cardiovascular differences between COX-2 inhibitors.

It is also important to note that both the FDA and the EMEA issued cardiovascular warnings for all coxibs and classical NSAIDs. Thus, reports suggesting an apparent increase in cardiovascular events with some classical NSAIDs led the FDA to sensitize clinicians and patients to the emerging data on increased risk of thrombotic events not only with coxibs but also with NSAIDs in general. As part of its continuous monitoring of medicines, the EMEA also reviewed cardiovascular safety data on nonselective NSAIDs. In 2006, the Committee for Medicinal Products for Human Use (CHMP) has assessed information from previous reviews of safety data, new data from clinical trials and epidemiological studies, and data from the MEDAL program (comprising three long-term studies of the safety of a nonselective NSAID and a COX-2 inhibitor) and from previous assessments of non-selective NSAIDs and COX-2 inhibitors. The conclusions of the CHMP were that the possibility that nonselective NSAIDs may be associated with a small increase in the absolute risk for thrombotic events cannot be excluded, especially when used at high doses for long-term treatment.

In addition to the cardiovascular risk, coxibs also face specific concerns that are drug related. Thus, FDA concluded that the overall risk/benefit profile for valdecoxib was unfavorable and that valdecoxib lacked demonstrable advantage compared with other NSAIDs. In the context of the review of COX-2 inhibitors, the EMEA's CHMP has assessed the safety data of valdecoxib, in particular with regard to cardiovascular safety and serious skin reactions, and recommended that the marketing authorization be suspended. Indeed, in postmarketing experience, serious skin adverse reactions, most of which occurred within the first 2 weeks of treatment, have been reported in association with valdecoxib, and erythema multiforme has been reported in association with the use of parecoxib sodium. More recently, the EMEA and other countries have recommended the withdrawal of the marketing authorizations for all lumiracoxib-containing medicines because of the risk of serious side effects affecting the liver.

In addition to providing a possible explanation for why celecoxib and rofecoxib possess distinct clinical response profiles, the dual inhibition COX-2/hCA could support other therapeutic perspectives for these drugs. Thus, a number of *in vitro* and *in vivo* studies have suggested that COX-2 inhibitors possess anticancer properties. Indeed, because they affect mitogenesis, cellular adhesion, and apoptosis, prostaglandins appear to play a major role in the pathogenesis of several types of cancers such as head and neck, breast, lung, colon, pancreas, and prostate cancers. COX-2 expression has also been demonstrated to contribute to angiogenesis, which is a critical step in tumor development.[41] However, a recent study found that celecoxib, but not rofecoxib, inhibited growth of hematopoietic and epithelial cell lines that did not express COX-2.[42] The presence of a pharmacological effect in the absence of COX-2 suggests that the response to celecoxib is derived independent of prostanoid metabolism, at least in some cases. One possible explanation may be the dual activity on COX-2 and CA isozymes critical for the development and invasion of cancer cells, such as CA IX and XII. Thus, the tumor-associated isozymes CA IX and XII appear to regulate the extratumoral pH by their extracellular catalytic domain.[43] An intrinsic feature of the tumor phenotype cells is their lower extracellular pH (pH_e) than the normal cells.[44] Low pH_e benefits tumor cells by promoting invasiveness through multiple effects, including upregulation of growth factors and proteases, loss of intercellular adhesion, increased migration and metastasis, increased rate of mutation, and so on.[43,44] CA IX shows a restricted expression in normal tissues (gastrointestinal epithelial cells)[45] but is tightly associated with the different types of tumors derived from kidney, oesophagus, colon, lung, pancreas, liver, endometrium, ovary, brain, skin, and breast.[46–54] CA XII is present in many normal tissues and is overexpressed in some tumors.[43] The expression of the two proteins is strongly induced by hypoxia present in tumor.[43,55] Hypoxia is primarily a pathophysiological consequence of structurally and functionally disturbed microcirculation.[56] While the molecular mechanism of CA-XII-induced expression remains unclear, the expression of CA IX is regulated by the activation of the hypoxia-inducible transcription factor (HIF) that binds to hypoxia response element in the *CA9* promoter.[57] Therefore, CA IX has been proposed to serve as a marker of tumor hypoxia.[43,58] Both CA IX and XII are downregulated by the von Hippel–Lindau tumor suppressor protein (pVHL).[59] In some cancer cells (i.e., renal cell carcinomas), pVHL can be mutated leading to an overexpression of CA IX and XII. CA IX and XII should be considered as target for novel therapeutic application in the field of cancer. The involvement of some CAs and their sulfonamide inhibitors in cancer has been investigated. Many potent CA inhibitors derived from acetazolamide and ethoxzolamide were shown to inhibit the growth of several tumor cell lines *in vitro* and *in vivo*.[60–62] Most part, but not all, of the cell lines utilized are known to express CA IX and/or XII. Therefore, the observed antiproliferative effect of sulfonamides should be mediated by inhibition of these cancer-related isozymes. Moreover, the presence of an H^+ gradient across the membrane of tumor cells has implications for chemotherapy.[63] The environmental acidification in solid tumors may influence the uptake of some weakly basic anticancer drugs leading to a chemoresistance. Manipulation of the pH_e with specific CA IX and/or XII inhibitor will contribute to enhance the action of weakly basic drugs and will reduce the acquisition of metastasis phenotypes.[64–66]

In conclusion, the demonstration of a nanomolar activity of sulfonamide coxibs (celecoxib and valdecoxib) on different hCA may explain clinical differences with methylsulfone coxibs. This may help to explain differences in clinical observations between sulfonamide and methylsulfone COX-2 inhibitors. In addition, the dual COX-2 inhibitory activity with the inhibition of CA isozymes critical for the development and invasion of cancer cells, such as CA II, IX, and XII, may constitute an important mechanism of action of some potential antitumoral activity of coxibs such as celecoxib and valdecoxib.

REFERENCES

1. Roos, K. L.; Simmons, D. L. Cyclooxygenase variants: the role of alternative splicing. *Biochem. Biophys. Res. Commun.* **2005**, *338*, 62–69.
2. de Leval, X.; Julemont, F.; Benoit, V.; Frederich, M.; Pirotte, B.; Dogné, J.-M. First and second generations of COX-2 selective inhibitors. *Mini Rev. Med. Chem.* **2004**, *4*, 597–601.
3. Dogné, J.-M.; Supuran, C. T.; Pratico, D. Adverse cardiovascular effects of the coxibs. *J. Med. Chem.* **2005**, *48*, 2251–2257.
4. FitzGerald, G. A. COX-2 and beyond: approaches to prostaglandin inhibition in human disease. *Nat. Rev. Drug Discov.* **2003**, *2*, 879–890.
5. Antman, E. M.; DeMets, D.; Loscalzo, J. Cyclooxygenase inhibition and cardiovascular risk. *Circulation* **2005**, *111*, 759–770.
6. FitzGerald, G. A.; Patrono, C. The coxibs, selective inhibitors of cyclooxygenase-2. *N. Engl. J. Med.* **2001**, *345*, 433–442.
7. Cipollone, F.; Fazia, M. L. Cyclooxygenase-2 inhibition: vascular inflammation and cardiovascular risk. *Curr. Atheroscler. Rep.* **2006**, *8*, 245–251.
8. Harris, R. E.; Beebe-Donk, J.; Alshafie, G. A. Cancer chemoprevention by cyclooxygenase 2 (COX-2) blockade: results of case control studies. *Subcell. Biochem.* **2007**, *42*, 193–211.
9. Minghetti, L. Role of COX-2 in inflammatory and degenerative brain diseases. *Subcell. Biochem.* **2007**, *42*, 117–141.
10. Chandrasekharan, N. V.; Dai, H.; Roos, K. L.; Evanson, N. K.; Tomsik, J.; Elton, T. S.; Simmons, D. L. COX-3, a cyclooxygenase-1 variant inhibited by acetaminophen and other analgesic/antipyretic drugs: cloning, structure, and expression. *Proc. Natl. Acad. Sci. USA* **2002**, *99*, 13926–13931.
11. Qin, N.; Zhang, S. P.; Reitz, T. L.; Mei, J. M.; Flores, C. M. Cloning, expression, and functional characterization of human cyclooxygenase-1 splicing variants: evidence for intron 1 retention. *J. Pharmacol. Exp. Ther.* **2005**, *315*, 1198–1305.
12. Hinz, B.; Brune, K. Antipyretic analgesics: nonsteroidal antiinflammatory drugs, selective COX-2 inhibitors, paracetamol and pyrazolinones. *Handb. Exp. Pharmacol.* **2007**, *177*, 65–93.
13. Hinz, B.; Cheremina, O.; Brune, K. Acetaminophen (paracetamol) is a selective cyclooxygenase-2 inhibitor in man. *FASEB J.* **2008**, *22*, 383–390.
14. Huang, J. Q.; Sridhar, S.; Hunt, R. H. Role of *Helicobacter pylori* infection and nonsteroidal anti-inflammatory drugs in peptic-ulcer disease: a meta-analysis. *Lancet* **2002**, *359*, 14–22.
15. Pratico, D.; Dogné, J.-M. Selective cyclooxygenase-2 inhibitors development in cardiovascular medicine. *Circulation* **2005**, *111*, 1073–1079.

16. Garcia-Nieto, R.; Perez, C.; Checa, A.; Gago, F. Molecular model of the interaction between nimesulide and human cyclooxygenase-2. *Rheumatology (Oxford)* **1999**, *38* (Suppl. 1), 14–18.

17. Gimbrone, M. A., Jr.; Topper, J. N.; Nagel, T.; Anderson, K. R.; Garcia-Cardena, G. Endothelial dysfunction, hemodynamic forces, and atherogenesis. *Ann. N. Y. Acad. Sci.* **2000**, *902*, 230–239, discussion 239–240.

18. McAdam, B. F.; Catella-Lawson, F.; Mardini, I. A.; Kapoor, S.; Lawson, J. A.; FitzGerald, G. A. Systemic biosynthesis of prostacyclin by cyclooxygenase (COX)-2: the human pharmacology of a selective inhibitor of COX-2. *Proc. Natl. Acad. Sci. USA* **1999**, *96*, 272–277.

19. Riendeau, D.; Percival, M. D.; Brideau, C.; Charleson, S.; Dube, D.; Ethier, D.; Falgueyret, J. P.; Friesen, R. W.; Gordon, R.; Greig, G.; Guay, J.; Mancini, J.; Ouellet, M.; Wong, E.; Xu, L.; Boyce, S.; Visco, D.; Girard, Y.; Prasit, P.; Zamboni, R.; Rodger, I. W.; Gresser, M.; Ford-Hutchinson, A. W.; Young, R. N.; Chan, C. C. Etoricoxib (MK-0663): preclinical profile and comparison with other agents that selectively inhibit cyclooxygenase-2. *J. Pharmacol. Exp. Ther.* **2001**, *296*, 558–566.

20. Stichtenoth, D. O.; Frolich, J. C. The second generation of COX-2 inhibitors: what advantages do the newest offer? *Drugs* **2003**, *63*, 33–45.

21. Bombardier, C.; Laine, L.; Burgos-Vargas, R.; Davis, B.; Day, R.; Ferraz, M. B.; Hawkey, C. J.; Hochberg, M. C.; Kvien, T. K.; Schnitzer, T. J.; Weaver, A. Response to expression of concern regarding VIGOR study. *N. Engl. J. Med.* **2006**, *354*, 1196–1199.

22. Bombardier, C.; Laine, L.; Reicin, A.; Shapiro, D.; Burgos-Vargas, R.; Davis, B.; Day, R.; Ferraz, M. B.; Hawkey, C. J.; Hochberg, M. C.; Kvien, T. K.; Schnitzer, T. J. Comparison of upper gastrointestinal toxicity of rofecoxib and naproxen in patients with rheumatoid arthritis. VIGOR Study Group. *N. Engl. J. Med.* **2000**, *343*, 1520–1528.

23. Bresalier, R. S.; Sandler, R. S.; Quan, H.; Bolognese, J. A.; Oxenius, B.; Horgan, K.; Lines, C.; Riddell, R.; Morton, D.; Lanas, A.; Konstam, M. A.; Baron, J. A. Cardiovascular events associated with rofecoxib in a colorectal adenoma chemoprevention trial. *N. Engl. J. Med.* **2005**, *352*, 1092–1102.

24. Silverstein, F. E.; Faich, G.; Goldstein, J. L.; Simon, L. S.; Pincus, T.; Whelton, A.; Makuch, R.; Eisen, G.; Agrawal, N. M.; Stenson, W. F.; Burr, A. M.; Zhao, W. W.; Kent, J. D.; Lefkowith, J. B.; Verburg, K. M.; Geis, G. S. Gastrointestinal toxicity with celecoxib vs nonsteroidal anti-inflammatory drugs for osteoarthritis and rheumatoid arthritis: the CLASS study: a randomized controlled trial. Celecoxib Long-term Arthritis Safety Study. *JAMA* **2000**, *284*, 1147–1155.

25. Schnitzer, T. J.; Burmester, G. R.; Mysler, E.; Hochberg, M. C.; Doherty, M.; Ehrsam, E.; Gitton, X.; Krammer, G.; Mellein, B.; Matchaba, P.; Gimona, A.; Hawkey, C. J. Comparison of lumiracoxib with naproxen and ibuprofen in the Therapeutic Arthritis Research and Gastrointestinal Event Trial (TARGET), reduction in ulcer complications: randomised controlled trial. *Lancet* **2004**, *364*, 665–674.

26. Krum, H.; Aw, T. J.; Liew, D.; Haas, S. Blood pressure effects of COX-2 inhibitors. *J. Cardiovasc. Pharmacol.* **2006**, *47*, (Suppl. 1), S43–S48.

27. Dogné, J.-M.; Thiry, A.; Pratico, D.; Masereel, B.; Supuran, C. T. Dual carbonic anhydrase: cyclooxygenase-2 inhibitors. *Curr. Top. Med. Chem.* **2007**, *7*, 885–891.

28. Supuran, C. T.; Casini, A.; Mastrolorenzo, A.; Scozzafava, A. COX-2 selective inhibitors, carbonic anhydrase inhibition and anticancer properties of sulfonamides belonging to this class of pharmacological agents. *Mini Rev. Med. Chem.* **2004**, *4*, 625–632.

29. Supuran, C. T.; Scozzafava, A.; Casini, A. Carbonic anhydrase inhibitors. *Med. Res. Rev.* **2003**, *23*, 146–189.

30. Abbate, F.; Supuran, C. T.; Scozzafava, A.; Orioli, P.; Stubbs, M. T.; Klebe, G. Nonaromatic sulfonamide group as an ideal anchor for potent human carbonic anhydrase inhibitors: role of hydrogen-bonding networks in ligand binding and drug design. *J. Med. Chem.* **2002**, *45*, 3583–3587.

31. Weber, A.; Casini, A.; Heine, A.; Kuhn, D.; Supuran, C. T.; Scozzafava, A.; Klebe, G. Unexpected nanomolar inhibition of carbonic anhydrase by COX-2-selective celecoxib: new pharmacological opportunities due to related binding site recognition. *J. Med. Chem.* **2004**, *47*, 550–557.

32. Katagiri, M.; Ogasawara, T.; Hoshi, K.; Chikazu, D.; Kimoto, A.; Noguchi, M.; Sasamata, M.; Harada, S.; Akama, H.; Tazaki, H.; Chung, U. I.; Takato, T.; Nakamura, K.; Kawaguchi, H. Suppression of adjuvant-induced arthritic bone destruction by cyclooxygenase-2 selective agents with and without inhibitory potency against carbonic anhydrase II. *J. Bone Miner. Res.* **2006**, *21*, 219–227.

33. Di Fiore, A.; Pedone, C.; D'Ambrosio, K.; Scozzafava, A.; De Simone, G.; Supuran, C. T. Carbonic anhydrase inhibitors: valdecoxib binds to a different active site region of the human isoform II as compared to the structurally related cyclooxygenase II "selective" inhibitor celecoxib. *Bioorg. Med. Chem. Lett.* **2006**, *16*, 437–442.

34. Menchise, V.; De Simone, G.; Alterio, V.; Di Fiore, A.; Pedone, C.; Scozzafava, A.; Supuran, C. T. Carbonic anhydrase inhibitors: stacking with Phe131 determines active site binding region of inhibitors as exemplified by the X-ray crystal structure of a membrane-impermeant antitumor sulfonamide complexed with isozyme II. *J. Med. Chem.* **2005**, *48*, 5721–5727.

35. Whelton, A.; Fort, J. G.; Puma, J. A.; Normandin, D.; Bello, A. E.; Verburg, K. M. Cyclooxygenase-2: specific inhibitors and cardiorenal function: a randomized, controlled trial of celecoxib and rofecoxib in older hypertensive osteoarthritis patients. *Am. J. Ther.* **2001**, *8*, 85–95.

36. Whelton, A. Cyclooxygenase-2 inhibition and renal function. *Ann. Intern. Med.* **2001**, *134*, 1077–1078.

37. Whelton, A. COX-2-specific inhibitors and the kidney: effect on hypertension and oedema. *J. Hypertens. Suppl.* **2002**, *20*, S31–S35.

38. Whelton, A.; White, W. B.; Bello, A. E.; Puma, J. A.; Fort, J. G. Effects of celecoxib and rofecoxib on blood pressure and edema in patients > or =65 years of age with systemic hypertension and osteoarthritis. *Am. J. Cardiol.* **2002**, *90*, 959–963.

39. Alper, A. B., Jr.; Tomlin, H.; Sadhwani, U.; Whelton, A.; Puschett, J. Effects of the selective cyclooxygenase-2 inhibitor analgesic celecoxib on renal carbonic anhydrase enzyme activity: a randomized, controlled trial. *Am. J. Ther.* **2006**, *13*, 229–235.

40. Aw, T. J.; Liew, D.; Tofler, G. H.; Schneider, H. G.; Morel-Kopp, M. C.; Billah, B.; Krum, H. Can the blood pressure effects of COX-2 selective inhibitors be explained by changes in plasma aldosterone levels? *J. Hypertens.* **2006**, *24*, 1979–1984.

41. Julemont, F.; de Leval, X.; Michaux, C.; Renard, J. F.; Winum, J. Y.; Montero, J. L.; Damas, J.; Dogné, J.-M.; Pirotte, B. Design, synthesis, and pharmacological evaluation of pyridinic analogues of nimesulide as cyclooxygenase-2 selective inhibitors. *J. Med. Chem.* **2004**, *47*, 6749–6759.

42. Waskewich, C.; Blumenthal, R. D.; Li, H.; Stein, R.; Goldenberg, D. M.; Burton, J. Celecoxib exhibits the greatest potency amongst cyclooxygenase (COX) inhibitors for growth inhibition of COX-2-negative hematopoietic and epithelial cell lines. *Cancer Res.* **2002**, *62*, 2029–2033.

43. Pastoreková, S.; Pastorek, J. Cancer-related carbonic anhydrase isozymes and their inhibition. *Carbonic Anhydrase: Its Inhibitors and Activators;* CRC Press: London, **2004**; pp 255–281.

44. Stubbs, M.; McSheehy, P. M.; Griffiths, J. R.; Bashford, C. L. Causes and consequences of tumour acidity and implications for treatment. *Mol. Med. Today* **2000**, *6*, 15–19.

45. Gut, M. O.; Parkkila, S.; Vernerova, Z.; Rohde, E.; Zavada, J.; Hocker, M.; Pastorek, J.; Karttunen, T.; Gibadulinova, A.; Zavadova, Z.; Knobeloch, K. P.; Wiedenmann, B.; Svoboda, J.; Horak, I.; Pastorekova, S. Gastric hyperplasia in mice with targeted disruption of the carbonic anhydrase gene Car9. *Gastroenterology* **2002**, *113*, 1889–1903.

46. Bartosova, M.; Parkkila, S.; Pohlodek, K.; Karttunen, T. J.; Galbavy, S.; Mucha, V.; Harris, A. L.; Pastorek, J.; Pastorekova, S. Expression of carbonic anhydrase IX in breast is associated with malignant tissues and is related to overexpression of c-erbB2. *J. Pathol.* **2002**, *197*, 314–321.

47. Ivanov, S.; Liao, S. Y.; Ivanova, A.; Danilkovitch-Miagkova, A.; Tarasova, N.; Weirich, G.; Merrill, M. J.; Proescholdt, M. A.; Oldfield, E. H.; Lee, J.; Zavada, J.; Waheed, A.; Sly, W.; Lerman, M. I.; Stanbridge, E. J. Expression of hypoxia-inducible cell-surface transmembrane carbonic anhydrases in human cancer. *Am. J. Pathol.* **2001**, *158*, 905–919.

48. Saarnio, J.; Parkkila, S.; Parkkila, A. K.; Pastorekova, S.; Haukipuro, K.; Pastorek, J.; Juvonen, T.; Karttunen, T. J. Transmembrane carbonic anhydrase, MN/CA IX, is a potential biomarker for biliary tumours. *J. Hepatol.* **2001**, *35*, 643–649.

49. Kivela, A. J.; Parkkila, S.; Saarnio, J.; Karttunen, T. J.; Kivela, J.; Parkkila, A. K.; Pastorekova, S.; Pastorek, J.; Waheed, A.; Sly, W. S.; Rajaniemi, H. Expression of transmembrane carbonic anhydrase isoenzymes IX and XII in normal human pancreas and pancreatic tumours. *Histochem. Cell Biol.* **2000**, *114*, 197–204.

50. Vermylen, P.; Roufosse, C.; Burny, A.; Verhest, A.; Bosschaerts, T.; Pastorekova, S.; Ninane, V.; Sculier, J. P. Carbonic anhydrase IX antigen differentiates between preneoplastic malignant lesions in non-small cell lung carcinoma. *Eur. Respir. J.* **1999**, *14*, 806–811.

51. Saarnio, J.; Parkkila, S.; Parkkila, A. K.; Haukipuro, K.; Pastorekova, S.; Pastorek, J.; Kairaluoma, M. I.; Karttunen, T. J. Immunohistochemical study of colorectal tumors for expression of a novel transmembrane carbonic anhydrase, MN/CA IX, with potential value as a marker of cell proliferation. *Am. J. Pathol.* **1998**, *153*, 279–285.

52. McKiernan, J. M.; Buttyan, R.; Bander, N. H.; Stifelman, M. D.; Katz, A. E.; Chen, M. W.; Olsson, C. A.; Sawczuk, I. S. Expression of the tumor-associated gene MN: a potential biomarker for human renal cell carcinoma. *Cancer Res.* **1997**, *57*, 2362–2365.

53. Turner, J. R.; Odze, R. D.; Crum, C. P.; Resnick, M. B. MN antigen expression in normal, preneoplastic, and neoplastic esophagus: a clinicopathological study of a new cancer-associated biomarker. *Hum. Pathol.* **1997**, *28*, 740–744.

54. Liao, S. Y.; Aurelio, O. N.; Jan, K.; Zavada, J.; Stanbridge, E. J. Identification of the MN/CA9 protein as a reliable diagnostic biomarker of clear cell carcinoma of the kidney. *Cancer Res.* **1997**, *57*, 2827–2831.

55. Wykoff, C. C.; Beasley, N. J.; Watson, P. H.; Turner, K. J.; Pastorek, J.; Sibtain, A.; Wilson, G. D.; Turley, H.; Talks, K. L.; Maxwell, P. H.; Pugh, C. W.; Ratcliffe, P. J.; Harris, A. L. Hypoxia-inducible expression of tumor-associated carbonic anhydrases. *Cancer Res.* **2000**, *60*, 7075–7083.
56. Hockel, M.; Vaupel, P. Tumor hypoxia: definitions and current clinical, biologic, and molecular aspects. *J. Natl. Cancer Inst.* **2001**, *93*, 266–276.
57. Zhou, J.; Schmid, T.; Schnitzer, S.; Brune, B. Tumor hypoxia and cancer progression. *Cancer Lett.* **2005**, *237*, 10–21.
58. Loncaster, J. A.; Harris, A. L.; Davidson, S. E.; Logue, J. P.; Hunter, R. D.; Wycoff, C. C.; Pastorek, J.; Ratcliffe, P. J.; Stratford, I. J.; West, C. M. Carbonic anhydrase (CA IX) expression, a potential new intrinsic marker of hypoxia: correlations with tumor oxygen measurements and prognosis in locally advanced carcinoma of the cervix. *Cancer Res.* **2001**, *61*, 6394–6399.
59. Ivanov, S. V.; Kuzmin, I.; Wei, M. H.; Pack, S.; Geil, L.; Johnson, B. E.; Stanbridge, E. J.; Lerman, M. I. Down-regulation of transmembrane carbonic anhydrases in renal cell carcinoma cell lines by wild-type von Hippel–Lindau transgenes. *Proc. Natl. Acad. Sci. USA* **1998**, *95*, 11596–11601.
60. Supuran, C. T.; Briganti, F.; Tilli, S.; Chegwidden, W. R.; Scozzafava, A. Carbonic anhydrase inhibitors: sulfonamides as antitumor agents? *Bioorg. Med. Chem.* **2001**, *9*, 703–714.
61. Vullo, D.; Innocenti, A.; Nishimori, I.; Pastorek, J.; Scozzafava, A.; Pastorekova, S.; Supuran, C. T. Carbonic anhydrase inhibitors. Inhibition of the transmembrane isozyme XII with sulfonamides: a new target for the design of antitumor and antiglaucoma drugs? *Bioorg. Med. Chem. Lett.* **2005**, *15*, 963–969.
62. Xiang, Y.; Ma, B.; Li, T.; Gao, J. W.; Yu, H. M.; Li, X. J. Acetazolamide inhibits aquaporin-1 protein expression and angiogenesis. *Acta Pharmacol. Sin.* **2004**, *25*, 811–816.
63. Vukovic, V.; Tannock, I. F. Influence of low pH on cytotoxicity of paclitaxel, mitoxantrone and topotecan. *Br. J. Cancer* **1997**, *75*, 1167–1172.
64. Thiry, A.; Dogné, J.-M.; Masereel, B.; Supuran, C. T. Targeting tumor-associated carbonic anhydrase IX in cancer therapy. *Trends Pharmacol. Sci.* **2006**, *27*, 566–573.
65. Thiry, A.; Supuran, C. T.; Masereel, B.; Dogné, J.-M. Recent developments of carbonic anhydrase inhibitors as potential anticancer drugs. *J. Med. Chem.* **2008**, *51*, 3051–3056.
66. Winum, J.-Y.; Rami, M.; Scozzafava, A.; Montero, J. L.; Supuran, C. Carbonic anhydrase IX: a new druggable target for the design of antitumor agents. *Med. Res. Rev.* **2008**, *28*, 445–463.
67. Vullo, D.; Franchi, M.; Gallori, E.; Pastorek, J.; Scozzafava, A.; Pastorekova, S.; Supuran, C. T. Carbonic anhydrase inhibitors: inhibition of the tumor-associated isozyme IX with aromatic and heterocyclic sulfonamides. *Bioorg. Med. Chem. Lett.* **2003**, *13*, 1005–1009.

CHAPTER 12

Advances in the Inhibitory and Structural Investigations on Carbonic Anhydrase Isozymes XIII and XV

MIKA HILVO[1], GIUSEPPINA DE SIMONE[2], CLAUDIU T. SUPURAN[3], and SEPPO PARKKILA[1]

[1]Institute of Medical Technology and School of Medicine, University of Tampere and Tampere University Hospital, Biokatu 6, 33014 University of Tampere, Finland
[2]Istituto di Biostrutture e Bioimmagini-CNR, Via Mezzocannone 16, 80134 Naples, Italy
[3]Università degli Studi di Firenze, Polo Scientifico, Laboratorio di Chimica Bioinorganica, Room 188, Via della Lastruccia 3, 50019 Sesto Fiorentino (Florence), Italy

12.1 INTRODUCTION

Carbonic anhydrases (CAs) are zinc-containing metalloenzymes that catalyze the reversible hydration of carbon dioxide (CO_2) to bicarbonate (HCO_3) and a proton.[1] This reaction seems to be of great importance for the survival of cells and whole organisms because CAs are found ubiquitously in nature. All the mammalian CAs belong to the α-CA family, which consists of 12 active isozymes in primates and 13 active isozymes in other mammals. The mammalian isozymes are the cytosolic CA I, II, III, VII, and XIII,[2–8] the membrane-bound CA IV, IX, XII, XIV, and XV,[9–15] the mitochondrial CA VA and VB,[16,17] and the secretory CA VI.[18] The most recently identified isozymes are CA XIII and XV.[8,15] The latter is an exceptional CA isozyme because it is not expressed in humans or other primates, but is present in several other species. In this chapter, we will review the recently emerged inhibition and structural data on CA XIII and XV.

12.2 EXPRESSION AND CHARACTERIZATION OF CARBONIC ANHYDRASE XIII (CA XIII)

CA XIII is a cytosolic enzyme that was first characterized in 2004 and was shown to be phylogenetically closely related to the other cytosolic CA I, II, and III.[8] In this first study, the expression of this enzyme was investigated in humans and mice and it appeared that it has distinct differences in the two species.

In the human alimentary tract, CA XIII was expressed by the serous acinar cells and duct epithelial cells of the submandibular gland. In the gastric mucosa, it showed only very weak expression in the surface epithelial cells of the body and antrum segments and was found in the enterocytes in every segment of the human gut, especially in the jejunum and ileum. On the contrary, in the murine alimentary tract, the strongest expression was observed in the colon, whereas the more proximal segments of the gastrointestinal tract showed no, or only a faint, signal. Another study, which investigated the expression of mRNAs for mouse CA (mCA) isozymes, revealed that *Car13* mRNA was produced abundantly in the digestive system, with the highest expression found in the stomach and jejunum, followed by the esophagus, duodenum, ileum, colon, and pancreas.[19]

Interestingly, CA XIII was also shown to be expressed widely in the reproductive tissues.[8] In humans, different stages of developing sperm cells expressed CA XIII, but the enzyme was not present in the mouse testis. In the human female reproductive tract, CA XIII was abundantly present in the uterine cervix, and some endometrial glands were also found to be positive. CA XIII was also expressed in the mouse uterine epithelial cells. The expression of CA XIII in the reproductive tract suggests that this enzyme may be involved in the fertilization process. The bicarbonate present in ejaculate has been suggested to maintain the sperm motility until the cells enter the lumen of the uterus through the cervical canal.[20,21] In the female genital tract, the endometrial and oviductal epithelium may produce an alkaline environment to maintain sperm motility. CA XIII might contribute to the fertilization process by producing the appropriate bicarbonate concentration in the cervical and endometrial mucus.

In the human and mouse kidneys, CA XIII expression was found in the collecting ducts of the renal cortex and medulla; and weak CA XIII expression was also observed in the human glomerulus.[8] Finally, in mice, CA XIII was detected in the brain oligodendrocytes and lung alveolar cells, which most likely represent type II pneumocytes. Additionally, in humans, strong reverse transcription polymerase chain reaction RT-PCR reaction for CA13 mRNA was observed in the thymus and weaker signals were found in the spleen, prostate, and ovary. In mice, the RT-PCR results revealed high levels of *Car13* mRNA in the adult mouse spleen and 7- and 17-day-old embryos and weaker signals in the heart, skeletal muscle, and 11- and 15-day-old embryos.

12.3 CATALYTIC ACTIVITY AND INHIBITION OF CA XIII

The activities of both mCA and human CA (hCA) XIII enzymes have been determined to be quite low, with $k_{cat} = 8.3 \times 10^4\ s^{-1}$ and $k_{cat}/K_M = 4.3 \times 10^6\ M^{-1}\ s^{-1}$ for the mouse enzyme and $k_{cat} = 1.5 \times 10^5\ s^{-1}$ and $k_{cat}/K_M = 1.1 \times 10^7\ M^{-1}\ s^{-1}$ for the human enzyme (at pH 7.5 and 20°C).[22,23] Thus, CA XIII shows the second lowest catalytic activity preceding only CA III.

Currently, there are no inhibition data available for hCA XIII (apart from the K_I for acetazolamide (AZA), which is 16 nM[22]) whereas mCA XIII (mCA XIII) inhibition has been investigated with aromatic and heterocyclic sulfonamides in addition to anions.[24–26] mCA XIII has a higher affinity for both sulfonamide inhibitors (similar to CA II and IX) and anion inhibitors (similar to CA I) as compared to the very low activity enzyme, CA III. The inhibition profile of mCA XIII for most sulfonamides was compared only to human isozymes I, II, and IX[24]; for some other compounds, mCA XIII was studied in parallel with all the human isozymes I–XII and XIV.[25] The first study showed that all the investigated sulfonamides, except for the COX-2 selective inhibitor valdecoxib, were potent mCA XIII inhibitors, having inhibition constants in the range 1.3–56 nM.[24] Valdecoxib is a comparatively weak mCA XIII inhibitor, with an inhibition constant 0.425 μM, while it remains as a good potent inhibitor for isozymes II and IX. In another study, the clinically used antiepileptic drug sulthiame was a very low affinity inhibitor for mCA XIII, with a K_I value of 1.46 μM. Additionally, zonisamide showed quite weak inhibition, with an inhibition constant 0.43 μM.[25] The other clinically used sulfonamides, including AZA, methazolamide, dichlorophenamide, dorzolamide, and brinzolamide demonstrated potent mCA XIII inhibition with K_I's in the range 10–23 nM and the affinity of mCA XIII for these inhibitors was similar to that of CA II.[24,25] A group of simple aromatic sulfonamides, including sulfanilamide and halogenated sulfanilamides, homosulfanilamide, 4-aminoethyl-benzenesulfonamide, and orthanilamide as well as the clinical compounds ethoxzolamide and topiramate showed quite effective inhibition against mCA XIII, with K_I's in the range 32–56 nM. In general, these compounds behave as weak inhibitors of CA I, II, and IX.[24,25] The sulfanilyl-derived compounds obtained by reacting aminosulfonamides with 4-acetamido-benzenesulfonyl chloride (followed by deprotecting the amino group) showed good potent mCA XIII inhibitory properties, with K_I's in the range 1.3–2.4 nM. Because these compounds are generally much less effective as inhibitors for isozymes I, II, and IX, it seems likely that these inhibitors are quite selective toward mCA XIII.[24]

The inhibition of mCA XIII was also investigated with anions and the results were compared against human isozymes I, II, and V.[26] The results showed that the most potent mCA XIII inhibitors of this group included the metal poisons cyanate, cyanide, and thiocyanate, as well as sulfamide, which showed K_I's in the range 0.25 μM–0.74 mM. These inhibitors show quite similar effects against CA I and V. Another group of anions, including fluoride, iodide, azide, carbonate, and hydrogen sulfide, as well as phenylboronic and phenylarsonic acids, were shown to act as efficient mCA XIII inhibitors, having inhibition constants in the range 1.65–5.5 mM. Weaker inhibitory properties were shown by bromide, nitrate, nitrite, bisulfate, and sulfamic acid, which had K_I's in the range 12.6–75.5 mM. Chloride, bicarbonate (K_I's approximately 140 mM), and sulfate ($K_I = 267$ mM) were ineffective against mCA XIII. The most notable among these was bicarbonate, which was determined to be a potent inhibitor for

CA I ($K_I = 12$ mM) and less effective for isozymes II and V (K_I's = 80–85 mM). Surprisingly, bicarbonate seems to be an even more ineffective inhibitor for mCA XIII.[26]

The results of the anion inhibition studies raise interesting speculations on the function of CA XIII.[24] The inability of bicarbonate to inhibit CA XIII is particularly interesting with respect to the possible role of CA XIII in the reproductive tract. As it was discussed earlier, CA XIII may provide the appropriate bicarbonate concentration in the male and female reproductive tracts where it has to perform the catalysis in the presence of a high-concentration bicarbonate. Thus, the high inhibition constant is understandable and promotes the idea that CA XIII is an important enzyme in the regulation of acid–base balance in the male and female reproductive tracts.

The anion inhibition studies also revealed that CA XIII is resistant to chloride inhibition, similar to isozymes II and V.[26] It has been shown that CA II physically interacts with the Cl^-/HCO_3^- anion exchanger AE1,[27] and CA II and IV also interact with the Na^+/HCO_3^- cotransporters NBC1 and NBC3,[28,29] although the direct physical interaction of CA II with SLC4 bicarbonate transporters has been questioned recently.[30] However, a direct or indirect interaction of CA II with the anion exchanger proteins would be feasible because the resistance of this enzyme to the chloride and bicarbonate inhibition suggests that it is located in an environment containing high concentrations of these ions. Because CA XIII is resistant to these same anions, it raises the question of whether CA XIII could also interact with these or some other ion transport proteins.[26]

12.4 STRUCTURE OF HUMAN CA XIII

X-ray crystallographic characterization of hCA XIII in the unbound state and in complex with the inhibitor AZA was reported only recently.[23] As expected on the base of high-sequence identity, the hCA XIII 3-dimensional structure (Fig. 12.1) appeared to be very similar to that of CA II, which is the best-known isozyme of the CA family. The secondary structure elements are well conserved between these isozymes and the local structural differences occur at the surfaces of the protein molecules. As for the other CA isozymes, the most important residues in the active site are the three histidine residues (His94, His96, and His119), which coordinate the Zn^{2+} ion needed for the CA catalysis (Fig. 12.1).[23] The active site cavity consists of two distinct parts made up of either hydrophobic or hydrophilic amino acid residues. Comparison of CA XIII with the other isozymes, whose structures have been solved, revealed that most amino acids within the active site are conserved. However, CA XIII presented some peculiarities that make this isozyme unique within the CA isozyme family. The most striking difference between CA XIII and the other isozymes was the presence of a Val residue at position 200, where all the other isozymes possess a Thr residue, apart from CA I that has a His residue. Other residues in the active site, which were not unique to CA XIII but anyway present only in a subset of CA isozymes, were Ser62, Asn67, Asp69, Arg91, Val132, and Leu204.

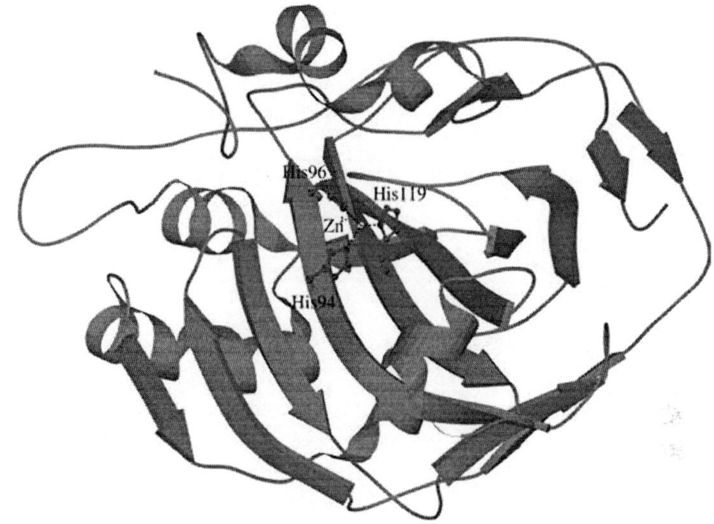

FIGURE 12.1 The structure of hCA XIII showing the three histidine residues and the Zn^{2+} ion essential for the catalytic activity.

It is interesting to note that although the structure of the active sites of CA II and XIII are very similar, these isozymes possess completely different level of activity.[23] CA II is an extremely fast enzyme, whereas CA XIII is the isozyme possessing the second lowest activity of the whole CA family. Therefore, it is likely that additional important structural features contributing to different activities may reside in other regions than within the active site itself. It has been previously proposed that the high activity of CA II is associated with a histidine cluster, which extends from the interior of the active site to the surface of the protein (His64, His3, His4, His10, His15, and His17).[31] This histidine cluster is probably involved in transfer of protons from the active site to the surrounding buffer. This proton transfer is extremely important for the CA catalysis because it is the rate-limiting step in the reaction. In fact, after CO_2 has been converted to HCO_3^-, the bicarbonate is replaced by a water molecule, generating the catalytically inactive form of the enzyme ($EZn^{2+}-OH_2$); thus, a proton has to be extruded from the active site to generate the catalytically active form of the enzyme ($EZn^{2+}-OH^-$). The lower catalytic efficiency measured for hCA XIII, with respect to CA II, is in good agreement with an isozyme structure where the histidine cluster is not present. In fact, apart from His64, no other histidine residue was present in the channel connecting the active site to the protein surface. Comparison of the structure and activity of CA XIII with the other cytosolic isozymes with known 3-dimensional structure, namely, CA I and III, further supports the hypothesis that a histidine cluster present at the entrance of the active site increases the CA catalytic activity.[23]

Although CA II and XIII have different catalytic activities, the binding to these isozymes of AZA, a classical CA inhibitor, is very similar.[22,23] In fact, most of the interactions stabilizing the inhibitor within CA active site were conserved in both

TABLE 12.1 Kinetics, Inhibition, and Subcellular Localization of Mammalian CA Isozymes

Isozyme	k_{cat} (s^{-1})	K_M (mM)	k_{cat}/K_M ($M^{-1} s^{-1}$)	K_I (AZA) (nM)	Subcellular Localization
hCA I[33,34]	2.0×10^5	4.0	5.0×10^7	250	Cytosol
hCA II[33,34]	1.4×10^6	9.3	1.5×10^8	12	Cytosol
hCA III[35,36]	1.3×10^4	52.0	$2.5–3.0 \times 10^5$	240,000	Cytosol
hCA IV[37–39]	1.1×10^6	21.5	5.1×10^7	74	Membrane bound
hCA VA[40]	2.9×10^5	10.0	2.9×10^7	63	Mitochondria
hCA VB[41]	9.5×10^5	9.7	9.8×10^7	54	Mitochondria
hCA VI[33]	3.4×10^5	6.9	4.9×10^7	11	Secreted
hCA VII[42]	9.5×10^5	11.4	8.3×10^7	2.5	Cytosol
hCA IX[a]	1.1×10^6	7.5	1.5×10^8	16	Transmembrane
hCA XII[43,44]	$4.0–4.2 \times 10^5$	12.0	$3.4–3.5 \times 10^7$	5.7	Transmembrane
hCA XIII[22]	1.5×10^5	13.8	1.1×10^7	16	Cytosol
hCA XIV[45]	3.1×10^5	7.9	3.9×10^7	41	Transmembrane
mCA XV[22]	4.7×10^5	14.2	3.3×10^7	72	Membrane bound

[a]PG + CA domains, Hilvo *et al.*, manuscript submitted.

isozymes. However, three additional hydrogen bond interactions were observed in the hCA II–AZA complex with respect to the hCA XIII–AZA complex. However, these differences did not have dramatic consequences on the affinity of the inhibitor for both isozymes, as proved by the comparable K_I's (Table 12.1). This finding may indicate that the combination of a higher number of interactions/repulsions involving several amino acids of the active site is necessary to influence the affinity of an inhibitor for a particular isozyme.

12.5 EXPRESSION AND CHARACTERIZATION OF CARBONIC ANHYDRASE XV (CA XV)

The most recently identified member of the mammalian CA family is CA XV that was characterized in 2005.[15] CA XV appeared to be a unique member of the CA family because the gene encoding this isozyme has become a nonprocessed pseudogene in humans and chimpanzees. In contrast, several other species appear to possess an active gene coding for CA XV. Sequence analysis revealed that this isozyme has been conserved throughout the evolution, from fish to mammals, but has become an inactive gene in terms of the evolutionary timescale quite recently.

Phylogenetic analysis indicated that CA XV is most closely related to CA IV.[15] Indeed, the biochemical properties of CA XV were shown to be very similar to those of CA IV. Both of these isozymes are *N*-glycosylated proteins that are attached to the cell membrane by a glycosylphosphatidylinositol (GPI) anchor. The expression of *Car15* mRNA in mouse tissues was investigated by RT-PCR and *in situ* hybridization techniques. Both methods clearly indicated that the expression of *Car15* occurs in few tissues. The only organ showing high expression for *Car15* was the kidney, where the highest signal was observed in the renal cortex, and lower expression was present in the medulla. The highest expression of CA XV in the renal cortex has been recently confirmed by immunohistochemistry (unpublished observations). In addition, the RT-PCR results indicated weak mRNA expression in the brain, testis, and 7- and 17-day-old embryos. Later, systematic study on CA mRNA expression in the mouse digestive system confirmed that the *Car15* expression in the digestive system is extremely low.[19]

The absence of CA XV in some species raises the questions of why it was lost during the evolution and why this protein is not needed for human physiology. It seems plausible that CA IV and CA XV are functionally redundant, and a highly active hCA IV may also cover the function of CA XV.

12.6 CATALYTIC ACTIVITY AND INHIBITION OF CA XV

Recombinant mCA XV has been produced recently in the baculovirus-insect cell expression system. The protein expressed from this eukaryotic system possessed $k_{cat} = 4.7 \times 10^5\,s^{-1}$ and $k_{cat}/K_M = 3.3 \times 10^7\,M^{-1}\,s^{-1}$ (at pH 7.5 and 20°C).[22] The activity of mCA XV is comparable to that of hCA XII and XIV, thus placing CA XV to the group of CAs showing moderate level of catalytic activity.

The first inhibition constant published for mCA XV was that of AZA, which was determined to be 72 nM.[22] Since this was nearly identical to the one obtained for hCA IV (74 nM), it was speculated that these enzymes might share a similar inhibition profile, because they are closely related to each other also in other aspects. However, recent results have shown that the inhibition profile of mCA XV is different from that of hCA IV. The clinically used CA inhibitors show comparatively weak inhibition for hCA IV although the K_I's for mCA XV are approximately in the range 60–75 nM (unpublished observations). The best mCA XV inhibitors found till now belong to the group of 2-(hydrazinocarbonyl)-3-phenyl-1*H*-indole-5-sulfonamides (Güzel et al., submitted). This lead molecule and its derivatives showed K_I's in the range 28–70 nM. The best molecule (28 nM) was a derivative that has a 4-tolyl group in the third position of the indole ring.

The inhibition of mCA XV has also been investigated with phenols, which are exceptional inhibitors in that they show competitive inhibition and bind to the active site in a different manner compared to the other CA inhibitors.[32] Mouse CA XV shows a different inhibition profile with phenols compared to the cytosolic human isozymes I and II. The inhibition results revealed that phenol and some of its 2-, 3-, and 4-substituted derivatives incorporating hydroxyl, fluoro, carboxy, and acetamido moieties were effective mCA XV inhibitors, with K_I's in the range 7.20–11.30 μM. Compounds incorporating 4-amino, 4-cyano, or 3-hydroxy groups were less effective, with inhibition constants of 335–434 μM. The best phenol inhibitor was clioquinol, whose K_I for mCA XV was 2.33 μM.

The crystal structure of CA XV has not been solved experimentally and at this moment the only structure of this enzyme is the one predicted by homology modeling using the previously-solved hCA IV structure as a template.[15] Detailed 3-dimensional structure of this enzyme could give insights to the function and inhibition of this enzyme and possibly would contribute to our understanding of the CA isozyme family as a whole.

12.7 CONCLUSION

During the 2000s, the cytosolic CA XIII and membrane-bound CA XV enzymes have been characterized. Based on the extensive bioinformatic analyses on current databases, these isozymes seem to represent the last members within the mammalian α-CA isozyme family. With the recent advances in the inhibition studies of mCA XV, the inhibition and activity profiles for all mammalian CA enzymes are available in the public domain. The catalytic activity and inhibition constants of hCA I–XIV and mCA XV are listed in Table 12.1.

With the addition of hCA XIII, the 3-dimensional structures have been solved for most of the CA isozymes. Future studies will hopefully reveal the high-resolution structures of the remaining isozymes. A comprehensive understanding of the details of the active sites of all CAs would provide a good basis for rational, isozyme-specific drug design targeting the different CA isozymes.

REFERENCES

1. Supuran, C. T. Carbonic anhydrases: novel therapeutic applications for inhibitors and activators. *Nat. Rev. Drug Discov.* **2008**, *7*, 168–181.
2. Andersson, B.; Nyman, P. O.; Strid, L. Amino acid sequence of human erythrocyte carbonic anhydrase B. *Biochem. Biophys. Res. Commun.* **1972**, *48*, 670–677.
3. Lin, K. T.; Deutsch, H. F. Human carbonic anhydrases. XI. The complete primary structure of carbonic anhydrase B. *J. Biol. Chem.* **1973**, *248*, 1885–1893.
4. Henderson, L. E.; Henriksson, D.; Nyman, P. O. Amino acid sequence of human erythrocyte carbonic anhydrase C. *Biochem. Biophys. Res. Commun.* **1973**, *52*, 1388–1394.
5. Lin, K. T.; Deutsch, H. F. Human carbonic anhydrases. XII. The complete primary structure of the C isozyme. *J. Biol. Chem.* **1974**, *249*, 2329–2337.
6. Koester, M. K.; Register, A. M.; Noltmann, E. A. Basic muscle protein, a third genetic locus isoenzyme of carbonic anhydrase? *Biochem. Biophys. Res. Commun.* **1977**, *76*, 196–204.
7. Montgomery, J. C.; Venta, P. J.; Eddy, R. L.; Fukushima, Y. S.; Shows, T. B.; et al. Characterization of the human gene for a newly discovered carbonic anhydrase, CA VII, and its localization to chromosome 16. *Genomics* **1991**, *11*, 835–848.
8. Lehtonen, J.; Shen, B.; Vihinen, M.; Casini, A.; Scozzafava, A.; et al. Characterization of CA XIII, a novel member of the carbonic anhydrase isozyme family. *J. Biol. Chem.* **2004**, *279*, 2719–2727.
9. Wistrand, P. J.; Knuuttila, K. G. Renal membrane-bound carbonic anhydrase. Purification and properties. *Kidney Int.* **1989**, *35*, 851–859.

10. Pastoreková, S.; Závadová, Z.; Kostál, M.; Babusiková, O.; Závada, J. A novel quasi-viral agent, MaTu, is a two-component system. *Virology* **1992**, *187*, 620–626.

11. Opavský, R.; Pastoreková, S.; Zelnik, V.; Gibadulinová, A.; Stanbridge, E. J.; et al. Human MN/CA9 gene, a novel member of the carbonic anhydrase family: structure and exon to protein domain relationships. *Genomics* **1996**, *33*, 480–487.

12. Türeci, Ö.; Sahin, U.; Vollmar, E.; Siemer, S.; Göttert, E.; et al. Human carbonic anhydrase XII: cDNA cloning, expression, and chromosomal localization of a carbonic anhydrase gene that is overexpressed in some renal cell cancers. *Proc. Natl. Acad. Sci. USA.* **1998**, *95*, 7608–7613.

13. Mori, K.; Ogawa, Y.; Ebihara, K.; Tamura, N.; Tashiro, K.; et al. Isolation and characterization of CA XIV, a novel membrane-bound carbonic anhydrase from mouse kidney. *J. Biol. Chem.* **1999**, *274*, 15701–15705.

14. Fujikawa-Adachi, K.; Nishimori, I.; Taguchi, T.; Onishi, S. Human carbonic anhydrase XIV (CA14): cDNA cloning, mRNA expression, and mapping to chromosome 1. *Genomics* **1999**, *61*, 74–81.

15. Hilvo, M.; Tolvanen, M.; Clark, A.; Shen, B.; Shah, G. N.; et al. Characterization of CA XV, a new GPI-anchored form of carbonic anhydrase. *Biochem. J.* **2005**, *392*, 83–92.

16. Dodgson, S. J.; Forster, R. E. II; Storey B. T.; Mela, L. Mitochondrial carbonic anhydrase. *Proc. Natl. Acad. Sci. USA.* **1980**, *77*, 5562–5566.

17. Fujikawa-Adachi, K.; Nishimori, I.; Taguchi, T.; Onishi, S. Human mitochondrial carbonic anhydrase VB. cDNA cloning, mRNA expression, subcellular localization, and mapping to chromosome x. *J. Biol. Chem.* **1999**, *274*, 21228–21233.

18. Fernley, R. T.; Wright, R. D.; Coghlan, J. P. A novel carbonic anhydrase from the ovine parotid gland. *FEBS Lett.* **1979**, *105*, 299–302.

19. Pan, P. W.; Rodriguez, A.; Parkkila, S. A systematic quantification of carbonic anhydrase transcripts in the mouse digestive system. *BMC Mol. Biol.* **2007**, *8*, 22.

20. Kaunisto, K.; Parkkila, S.; Tammela, T.; Rönnberg, L.; Rajaniemi, H. Immunohistochemical localization of carbonic anhydrase isoenzymes in the human male reproductive tract. *Histochemistry* **1990**, *94*, 381–386.

21. Okamura, N.; Tajima, Y.; Soejima, A.; Masuda, H.; Sugita, Y. Sodium bicarbonate in seminal plasma stimulates the motility of mammalian spermatozoa through direct activation of adenylate cyclase. *J. Biol. Chem.* **1985**, *260*, 9699–9705.

22. Hilvo, M.; Innocenti, A.; Monti, S. M.; De Simone, G.; Supuran, C. T.; et al. Recent advances in research on the most novel carbonic anhydrases, CA XIII and XV. *Curr. Pharm. Des.* **2008**, *14*, 672–678.

23. Di Fiore, A.; Monti, S. M.; Hilvo, M.; Parkkila, S.; Romano, V.; et al. Crystal structure of human carbonic anhydrase XIII and its complex with the inhibitor acetazolamide. *Proteins* **2009**, *74*, 164–175.

24. Lehtonen, J. M.; Parkkila, S.; Vullo, D.; Casini, A.; Scozzafava, A.; et al. Carbonic anhydrase inhibitors. Inhibition of cytosolic isozyme XIII with aromatic and heterocyclic sulfonamides: a novel target for the drug design. *Bioorg. Med. Chem. Lett.* **2004**, *14*, 3757–3762.

25. Temperini, C.; Innocenti, A.; Mastrolorenzo, A.; Scozzafava, A.; Supuran, C. T. Carbonic anhydrase inhibitors. Interaction of the antiepileptic drug sulthiame with twelve

mammalian isoforms: kinetic and X-ray crystallographic studies. *Bioorg. Med. Chem. Lett.* **2007**, *17*, 4866–4872.

26. Innocenti, A.; Lehtonen, J. M.; Parkkila, S.; Scozzafava, A.; Supuran, C. T. Carbonic anhydrase inhibitors. Inhibition of the newly isolated murine isozyme XIII with anions. *Bioorg. Med. Chem. Lett.* **2004**, *14*, 5435–5439.
27. Sterling, D.; Reithmeier, R. A.; Casey, J. R. A transport metabolon. Functional interaction of carbonic anhydrase II and chloride/bicarbonate exchangers. *J. Biol. Chem.* **2001**, *276*, 47886–47894.
28. Alvarez, B. V.; Loiselle, F. B.; Supuran, C. T.; Schwartz, G. J.; Casey, J. R. Direct extracellular interaction between carbonic anhydrase IV and the human NBC1 sodium/bicarbonate co-transporter. *Biochemistry* **2003**, *42*, 12321–12329.
29. Loiselle, F. B.; Morgan, P. E.; Alvarez, B. V.; Casey, J. R. Regulation of the human NBC3 Na+/HCO3- cotransporter by carbonic anhydrase II and PKA. *Am. J. Physiol. Cell. Physiol.* **2004**, *286*, C1423–1433.
30. Piermarini, P. M.; Kim, E. Y.; Boron, W. F. Evidence against a direct interaction between intracellular carbonic anhydrase II and pure C-terminal domains of SLC4 bicarbonate transporters. *J. Biol. Chem.* **2007**, *282*, 1409–1421.
31. Briganti, F.; Mangani, S.; Orioli, P.; Scozzafava, A.; Vernaglione, G.; et al. Carbonic anhydrase activators: X-ray crystallographic and spectroscopic investigations for the interaction of isozymes I and II with histamine. *Biochemistry* **1997**, *36*, 10384–10392.
32. Innocenti, A.; Hilvo, M.; Scozzafava, A.; Parkkila, S.; Supuran, C. T. Carbonic anhydrase inhibitors: inhibition of the new membrane-associated isoform XV with phenols. *Bioorg. Med. Chem. Lett.* **2008**.
33. Nishimori, I.; Innocenti, A.; Vullo, D.; Scozzafava, A.; Supuran, C. T. Carbonic anhydrase inhibitors. Inhibition studies of the human secretory isoform VI with anions. *Bioorg. Med. Chem. Lett.* **2007**, *17*, 1037–1042.
34. Khalifah, R. G. The carbon dioxide hydration activity of carbonic anhydrase. I. Stop-flow kinetic studies on the native human isoenzymes B and C. *J. Biol. Chem.* **1971**, *246*, 2561–2573.
35. Nishimori, I.; Minakuchi, T.; Onishi, S.; Vullo, D.; Cecchi, A.; et al. Carbonic anhydrase inhibitors: cloning, characterization, and inhibition studies of the cytosolic isozyme III with sulfonamides. *Bioorg. Med. Chem.* **2007**, *15*, 7229–7236.
36. Tu, C.; Chen, X.; Ren, X.; LoGrasso, P. V.; Jewell, D. A.; et al. Interactions of active-site residues and catalytic activity of human carbonic anhydrase III. *J. Biol. Chem.* **1994**, *269*, 23002–23006.
37. Innocenti, A.; Firnges, M. A.; Antel, J.; Wurl, M.; Scozzafava, A.; et al. Carbonic anhydrase inhibitors. Inhibition of the membrane-bound human and bovine isozymes IV with sulfonamides. *Bioorg. Med. Chem. Lett.* **2005**, *15*, 1149–1154.
38. Stams, T.; Christianson, D.W. X-ray crystallographic studies of mammalian carbonic anhydrase isozymes. In *The Carbonic Anhydrases: New Horizons;* Chegwidden, W.R.; Carter, N. D., Edwards, Y. H., Eds.; Birkhäuser Verlag: Basel, **2000**; pp 159–174.
39. Baird, T. T., Jr.; Waheed, A.; Okuyama, T.; Sly, W. S.; Fierke, C. A. Catalysis and inhibition of human carbonic anhydrase IV. *Biochemistry* **1997**, *36*, 2669–2678.
40. Franchi, M.; Vullo, D.; Gallori, E.; Antel, J.; Wurl, M.; et al. Carbonic anhydrase inhibitors: inhibition of human and murine mitochondrial isozymes V with anions. *Bioorg. Med. Chem. Lett.* **2003**, *13*, 2857–2861.

41. Nishimori, I.; Vullo, D.; Innocenti, A.; Scozzafava, A.; Mastrolorenzo, A.; et al. Carbonic anhydrase inhibitors. The mitochondrial isozyme VB as a new target for sulfonamide and sulfamate inhibitors. *J. Med. Chem.* **2005**, *48*, 7860–7866.

42. Vullo, D.; Voipio, J.; Innocenti, A.; Rivera, C.; Ranki, H.; et al. Carbonic anhydrase inhibitors. Inhibition of the human cytosolic isozyme VII with aromatic and heterocyclic sulfonamides. *Bioorg. Med. Chem. Lett.* **2005**, *15*, 971–976.

43. Vullo, D.; Innocenti, A.; Nishimori, I.; Pastorek, J.; Scozzafava, A.; et al. Carbonic anhydrase inhibitors. Inhibition of the transmembrane isozyme XII with sulfonamides—a new target for the design of antitumor and antiglaucoma drugs? *Bioorg. Med. Chem. Lett.* **2005**, *15*, 963–969.

44. Ulmasov, B.; Waheed, A.; Shah, G. N.; Grubb, J. H.; Sly, W. S.; et al. Purification and kinetic analysis of recombinant CA XII, a membrane carbonic anhydrase overexpressed in certain cancers. *Proc. Natl. Acad. Sci. USA.* **2000**, *97*, 14212–14217.

45. Nishimori, I.; Vullo, D.; Innocenti, A.; Scozzafava, A.; Mastrolorenzo, A.; et al. Carbonic anhydrase inhibitors: inhibition of the transmembrane isozyme XIV with sulfonamides. *Bioorg. Med. Chem. Lett.* **2005**, *15*, 3828–3833.

CHAPTER 13

Mechanism and Inhibition of the β-Class and γ-Class Carbonic Anhydrases

JAMES G. FERRY[1] and CLAUDIU T. SUPURAN[2]

[1]Department of Biochemistry and Molecular Biology, Eberly College of Science, The Pennsylvania State University, University Park, PA 16802-4500, USA
[2]Laboratorio di Chimica Bioinorganica, Università degli Studi di Firenze, Room 188, Via della Lastruccia 3, I-50019 Sesto Fiorentino (Firenze), Italy

13.1 INTRODUCTION

There are five independently evolved classes (α, β, γ, δ, and ζ) of carbonic anhydrases (CAs) with no known significant structural or sequence identity among them; however, all have an active site metal and catalyze the reversible hydration of carbon dioxide ($CO_2 + H_2O = HCO_3^- + H^+$). Zinc functions as the active site metal for all characterized CAs, although some can substitute other metals that may be more physiologically significant. Iron is thought to function *in vivo* for the archetype of the γ-class (Cam)[1] and the archetype of the ζ-class is proposed to utilize either cadmium or zinc *in vivo.*[2]

A representative CA from at least one class is found in all cell types spanning all the three domains of life (eukarya, bacteria, and archaea), reflecting a fundamental importance of this enzyme to life on Earth. Indeed, it is likely the γ-class played a role in the physiology of ancient life-forms.[3] The α-class is primarily found in mammals, although a few are present in pathogenic prokaryotes.[4] Enzymes from the β-class have been isolated, and putative γ-class CAs have been identified, from all three domains of life.[3] CAs have a diversity of functions that underscore the importance of this enzyme across the spectrum of Nature. Mammalian CAs are found in every tissue for which there are 16 isozymes that have multiple functions in transport and metabolism.[5] The β-class CAs from aerobic prokaryotes are implicated in maintaining internal pH and CO_2/bicarbonate balances required for biosynthetic reactions.[4] In anaerobic prokaryotes, β-class CAs are implicated in the transport of

CO_2 and bicarbonate across the cytoplasmic membrane that regulates pH and facilitates acquisition of substrates and product removal required for growth. Plants and phototrophic prokaryotes employ β-class CAs for transport and concentration of CO_2 for photosynthesis. The δ- and ζ-classes in marine diatoms also function to concentrate CO_2 for photosynthesis.[6] Although widely distributed in Nature, physiological roles for the γ-class CAs are not as well documented.

The overwhelming majority of research has focused on α-class isozymes from mammals with relatively less attention to the remaining classes. Of particular interest is the inhibition patterns of CAs that are naturally relevant for understanding substrate binding and the catalytic mechanism.[7] Here we review the β- and γ-class CAs focusing on the catalytic mechanisms and inhibition patterns.

13.2 ACTIVE SITES AND CATALYTIC MECHANISM

13.2.1 The γ-Class

Characterization of the γ-class archetype (Cam) from the strictly anaerobic *Methanosarcina thermophila* has been performed with the enzyme overproduced in *Escherichia coli* and purified aerobically that contains zinc in the active site.[8] However, when the enzyme is purified anaerobically from *E. coli*, iron is present in the active site that suggests it may be the physiologically relevant metal *in vivo*.[1] The active site can also be reconstituted with cobalt and the enzyme has robust activity with all three metals, although the iron-containing enzyme is most active. Crystal structures of Co- and Zn-substituted Cam identify a homotrimer with an overall left-handed β-helical fold.[9,10] The active site metal is ligated with three histidine residues with one monomer contributing one histidine ligand and the adjacent monomer contributing the remaining two histidine ligands. The cobalt is further coordinated with three water molecules whereas zinc is coordinated with two water molecules.[10] A crystal structure is unavailable for Fe-Cam, although it is proposed to have a metal coordination sphere similar to Co-substituted Cam.[1] The orientation of the metal-bound catalytic water in Co-substituted Cam is slightly different from that of the catalytic water in the Zn-substituted enzyme, although both are within hydrogen bonding distance to glutamine-75 and asparagine-202 that are essential for catalysis.[10,11]

Kinetic analyses of Cam are consistent with the two-step ping-pong mechanism for the reversible hydration of carbon dioxide to bicarbonate proposed for the α-class enzymes,[11,12] as shown in the following equations where E represents enzyme residues, M is metal, and B is buffer.

$$\text{E-M}^{2+}\text{-OH}^- + CO_2 = \text{E-M}^{2+}\text{-HCO}_3^- \quad (13.1a)$$

$$\text{E-M}^{2+}\text{-HCO}_3^- + H_2O = \text{E-M}^{2+}\text{-H}_2\text{O} + HCO_3^- \quad (13.1b)$$

$$\text{E-M}^{2+}\text{-H}_2\text{O} = \text{H}^+\text{-E - M}^{2+}\text{-OH}^- \quad (13.2a)$$

$$\text{H}^+\text{-E-M}^{2+}\text{-OH}^- + \text{B} = \text{E-M}^{2+}\text{-OH}^- + \text{BH}^+ \quad (13.2b)$$

In the first step (13.1a), the lone pair electrons of the metal-bound hydroxide attack CO_2producing metal-bound HCO_3^- that is displaced by water (13.1b). In the second step (13.2b), a proton is extracted from the metal-bound water and transferred to bulk solvent or buffer 13.2b. As for several α-class isozymes, Cam has a $k_{cat} > 10^4\ s^{-1}$ that is faster than the fastest rate at which protons can transfer from the zinc-bound water with a pK_a of 7 to water. CAs with slower CO_2 hydration rates transfer the proton directly to buffer or water. However, CAs with $k_{cat} > 10^4\ s^{-1}$ must transfer the proton from the zinc-bound water to an intermediate proton shuttle residue (H^+-E in (13.2) a and b) and then to an external buffer molecule. Glutamate-84, residing on an acidic loop exposed to solvent (Fig. 13.1), has been shown as a key proton shuttle residue in Cam.[13] The histidine proton shuttle residue in the α-class human CA II assumes two conformations that facilitate proton transfer. Thus, that glutamate-84 in Cam can assume two conformations (Fig. 13.1) is a further support for a proton transfer function for this residue.

In the α-class, residues glutamate-106 and threonine-199 hydrogen bond with the zinc-bound hydroxide via the threonine-199 Oγ atom. This hydrogen bond network facilitates lowering the pK_a of the metal-bound hydroxide to near neutrality and optimally positions the lone pair of electrons for attack on CO_2.[14–16] Further, the backbone amide of threonine-199 hydrogen bonds and stabilizes transition states of HCO_3^-.[14,15] Catalytically essential residues have been identified in the γ-class Cam by the replacement of active site residues via site-directed mutagenesis and kinetic analyses of the variant enzymes.[11,16] Glutamine-75 and asparagine-73 are proposed to

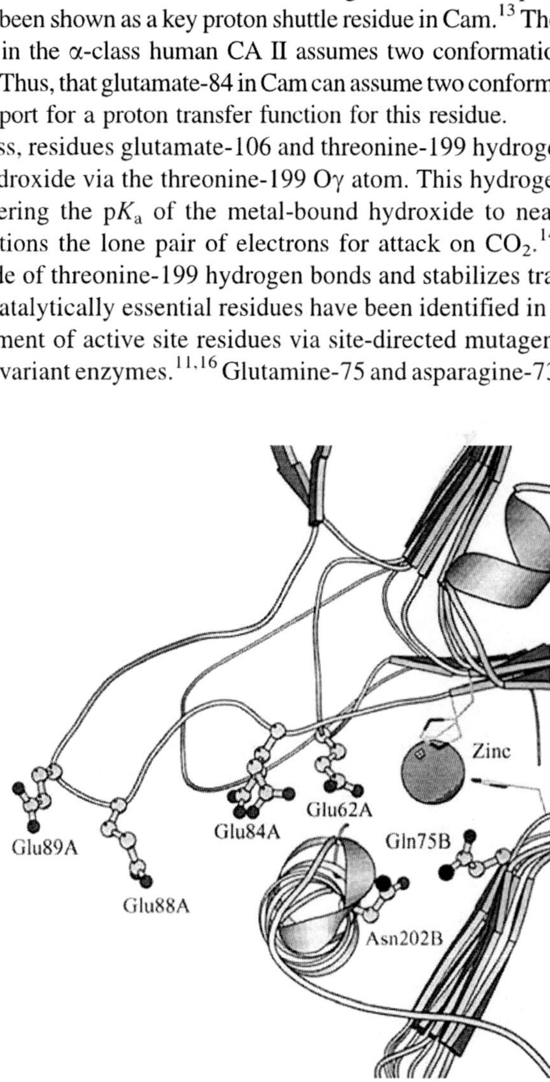

FIGURE 13.1 The active site of the γ-class Cam from *M. thermophila*. Two conformations are shown for glutamine-84. Reproduced by permission.[13]

participate in a hydrogen bond network to orient and increase the nucleophilicity of the metal-bound hydroxide.[11] Asparagine-202 is proposed to stabilize the transition state analogous to threonine-199 of the α-class.[11] Glutamate-62 was shown to be essential for CO_2 hydration (step 1).[13] Based on these results, a catalytic mechanism has been proposed, as shown in Fig. 13.2.[11] The mechanism is based on the zinc enzyme for which the metal is coordinated by two water molecules (Fig. 13.2a). Glutamine-62

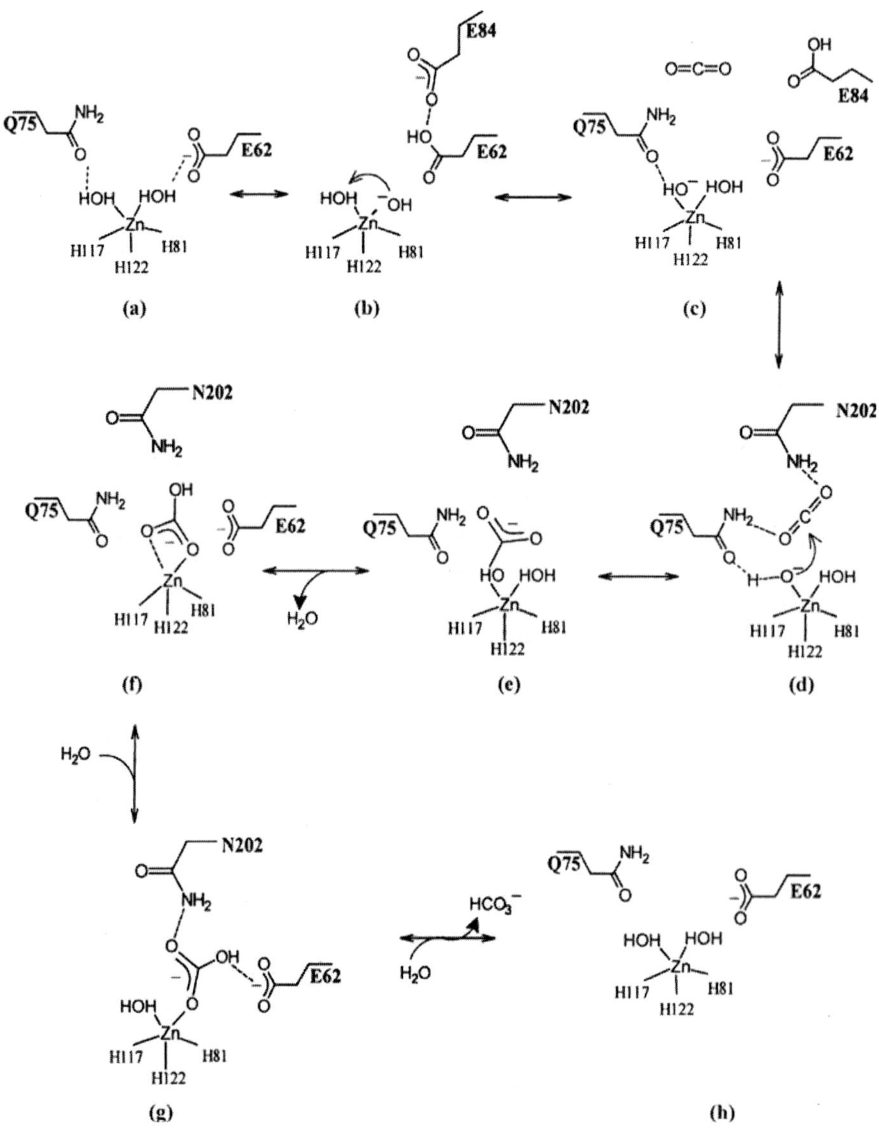

FIGURE 13.2 Catalytic mechanism proposed for the γ-class Cam from *M. thermophila*. Reproduced by permission.[11]

extracts a proton from one of the metal-bound water molecules that then extracts a proton from the adjacent water producing Zn-OH^- (Fig. 13.2b). The Zn-OH^- attacks CO_2 resulting in a bound HCO_3^- that displaces a water molecule (Fig. 13.2c–f). The HCO_3^- undergoes a bidentate transition state where the proton either rotates or transfers to the nonmetal-bound oxygen of the HCO_3^-. Glutamine-62 hydrogen bonds with the hydroxyl of HCO_3^-, thereby destabilizing it (Fig. 13.2g). An incoming water molecule further destabilizes the HCO_3^- by replacing one of the bound oxygens. A second incoming water molecule completely displaces the HCO_3^- resulting in product removal and regeneration of the active site (Fig. 13.2h).

Activation of Cam with several natural and nonnatural amino acids and aromatic/heterocyclic amines show a profile with natural L- and D-amino acids robustly different from those of the α-class enzymes.[17] Most compounds showed medium activating efficacy toward Cam. 2-Pyridylmethylamine and 1-(2-aminoethyl)-piperazine were effective toward Zn-substituted Cam (K_A's of 10.1–11.4 μM), whereas serotonin, L-adrenaline, and 2-pyridylmethylamine were most effective toward the Co-substituted enzyme (K_A's of 0.97–8.9 μM). Thus, the activation mechanisms of the α- and β-class CAs appear similar, although the profiles with various compounds differ dramatically between the classes reflecting diversity in the active sites and catalytic mechanism.

13.2.2 The β-Class

Crystal structures for the β-class reveal two subclasses based on the zinc coordination sphere that are defined here as the "canonical" and "noncanonical" subclasses. Both subclasses have a conserved aspartate in the zinc coordination sphere with zinc coordinated by one histidine, two cysteines, and a fourth ligand. Water or a substrate analogue is the fourth ligand in the canonical subclass, whereas the conserved aspartate is the fourth ligand in the noncanonical subclass. Thus, in the noncanonical subclass, a water molecule is not coordinated to zinc, which is a major departure from the mechanism of the well-studied α- and γ-classes for which the zinc-bound water when deprotonated yields a hydroxyl group for nucleophilic attack on CO_2 producing HCO_3^-.

13.2.2.1 The Canonical Subclass This subclass is defined here based on similarities with the well studied α-class zinc hydroxide mechanism for catalysis. Crystal structures for the canonical subclass are available for enzymes from the common pea plant (*Pisum sativum*),[18] a methane-producing thermophilic species from the archaea domain (*Methanobacterium thermoautotrophicum*),[19] a human pathogen from the bacteria domain (*Mycobacterium tuberculosis*, homologue Rv1284),[20] and a photosynthetic prokaryote from the bacteria domain (*Halothiobacillus neapolitans*).[21]

The structure of the enzyme from *P. sativum* reveals an octamer in a dimer of dimers arrangement.[18] The active site is located at the interface between two monomers with a very narrow hydrophobic channel leading to the bulk solvent that would require substantial rearrangement for a molecule larger than water molecule to enter the active site. Indeed, an essential rearrangement for catalysis is consistent with inhibition of the *P. sativum* enzyme by sulfonamides that would otherwise be too bulky to enter the active site.[22] The crystal structure[18] shows acetate, a component of the crystallization buffer, as the fourth ligand to zinc that acts as a substrate analogue of HCO_3^-. The active site functional groups correspond to that of the α-class to an extent that the corresponding residues approximate a mirror image. Thus, despite independent evolution, the canonical β-class and α-class appear to have converged on a nearly identical active site design with the exception of the conserved aspartate in the pea enzyme that is hydrogen bonded to an arginine. Kinetic analyses show that the pea enzyme is nearly as fast as the α-class HCA II, is activated by increasing pH, and displays a hydrogen isotope effect indicating a rate-limiting proton transfer step consistent with a zinc hydroxide mechanism.[22,23] The structural and kinetic similarities between the *P. sativum* and the α-class has led to a proposed mechanism for the canonical subclass shown in Fig. 13.3 for the HCO_3^- dehydration reaction based in part on the crystal structure.[18] In the mechanism, HCO_3^- is bound in the active site (Fig. 13.3a) similar to that shown for the acetic acid analogue in the crystal structure. The acetic acid forms hydrogen bonds with glycine-224, glutamate-151, and the conserved aspartate-162. Loss of water from the bound HCO_3^- yields CO_2 (Fig. 13.3b) that diffuses out of the active site leaving a zinc-bound hydroxide (Fig. 13.3c). Finally, the bound hydroxide accepts a proton from a buffer molecule or bulk solvent leaving water bound to the zinc (Fig. 13.3d). However, the route of

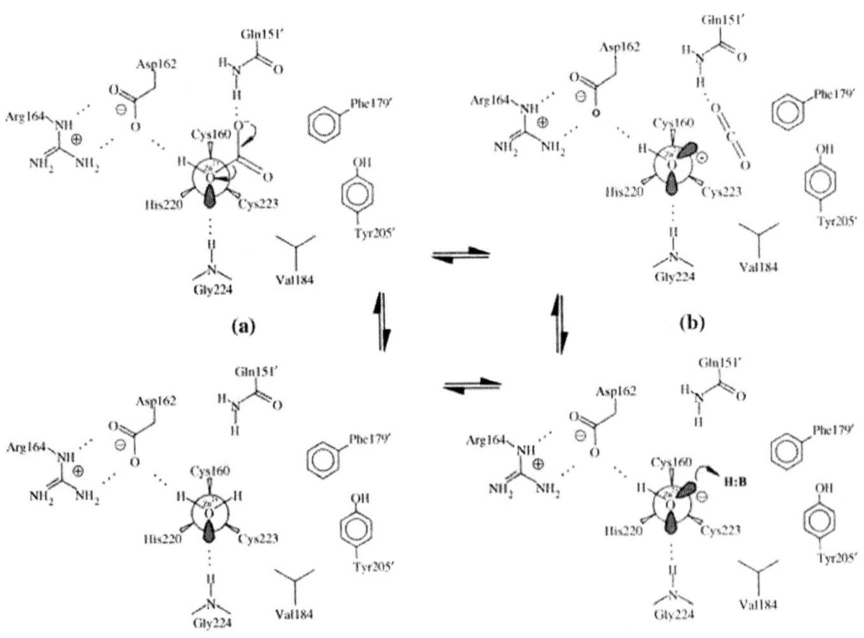

FIGURE 13.3 Mechanism proposed for the *P. sativum* CA representing the canonical subclass of the β-class. Reproduced by permission.[18]

proton transfer from the active site to bulk solvent is not clear from the crystal structure. Thus, the conserved aspartate is proposed to hydrogen bond to HCO_3^- analogous to the threonine-199 "gate keeper" of the α-class rather than to proton transfer as proposed for the noncanonical subclass. In the pea enzyme, the aspartate is sequestered from coordinating zinc by two hydrogen bonds to the guanidinium group of arginine (Fig. 13.3), opposed to the noncanonical *Porphyridium purpureum* enzyme for which the conserved aspartate is postulated to coordinate zinc in a regulatory mechanism.

The crystal structure of the *M. tuberculosis* Rv1284 homologue (Fig. 13.4a) shows water as the fourth ligand and the conserved aspartate hydrogen bonded to an arginine characteristic of the canonical subclass.[20] The active site cavity is nearly fully closed. The enzyme is reported to contain only 0.3 equivalents of zinc and 0.18 of nickel, which places the type and role of metals in question. No CA activity could be detected for the enzyme leading to the designation of a "β-CA-like" protein. Interestingly, the enzyme was reported to be essential based on mutagenesis in strain H37Rv[24] and upregulation of the encoding gene under the starvation conditions used to model persistent bacteria.[25]

The canonical subclass CA (Cab) from *M. thermoautotrophicum* is the first β-class CA characterized from the archaea domain and the first from a thermophilic methane-producing species.[26] Indeed, the zinc-containing dimeric enzyme is stable up to 75°C. The crystal structure[19] shows that the active site cavity is open to the bulk solvent to an extent that a HEPES buffer molecule is present 8 Å from the active site zinc (Fig. 13.5), which is a departure from the *P. sativum* and *M. tuberculosis* enzymes. Cab has a CO_2 hydration activity with a k_{cat} of 1.7×10^4 s^{-1} and kinetic analyses are consistent with a zinc hydroxide mechanism.[26,27] The crystal structure shows water as the fourth ligand to zinc, also consistent with a zinc hydroxide mechanism.[19] Characteristic of the canonical subclass, the conserved aspartate-34 is tethered to the adjacent arginine-36

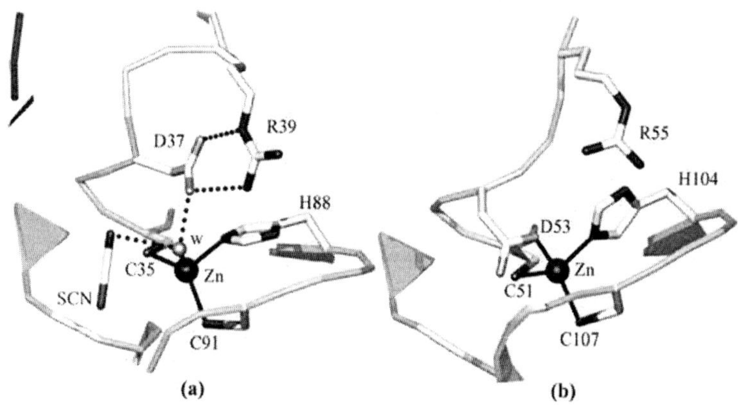

FIGURE 13.4 Active site structures of the *M. tuberculosis* Rv1284 (a) and Rv3588c (b) β-class CAs showing residues involved in metal chelation. The structures are shown in ball-and-stick models where the hydrogen bonds are indicated by black dotted lines. Reproduced by permission.[20]

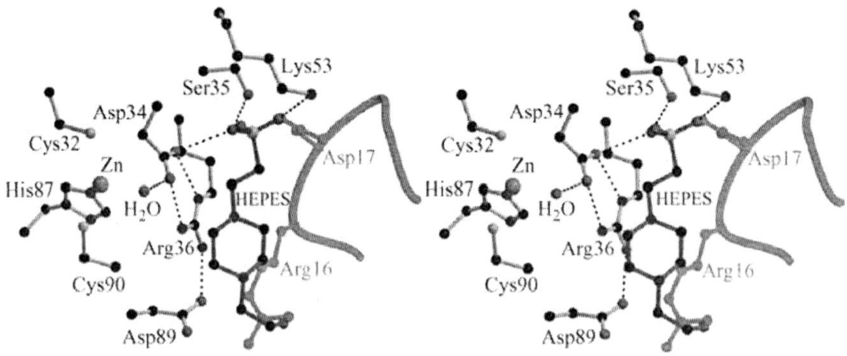

FIGURE 13.5 Stereoview of the active site of Cab from *M. thermoautotrophicum* representing the canonical β-class subclass. Reproduced by permission.[19]

(Fig. 13.5). Roles for the conserved aspartate and adjacent arginine of the *M. thermoautotrophicum* enzyme were probed by the replacement with alanine via site-directed mutagenesis and kinetic analyses of the variant enzymes.[28] The aspartate→alanine replacement reduced the k_{cat}/K_m value, although not to an extent supporting an essential role in the reaction mechanism. However, a substantially larger reduction in k_{cat}/K_m was observed for the arginine→alanine variant suggesting that the conserved aspartate-34 once free of the salt bridge with arginine-36 is able to coordinate zinc. Consistent with this interpretation, the double variant had greater activity compared with the arginine→alanine variant. Imidazole rescued activity of both the aspartate→alanine and double variants, a result suggesting a role for these residues in the proton transfer step.

As for Cam, activation of Cab with natural and nonnatural amino acids and aromatic/heterocyclic amines show a profile with natural, L- and D-amino acids different from the α-class enzymes.[17] Most compounds showed medium efficacy toward Cab except for D-Phe and L-Tyr (K_A's of 10.3–10.5 μM). Thus, the activation mechanisms of the α- and β-classes appear similar, although the profiles with various compounds between these classes are different reflecting the diversity in active sites and mechanisms.

The crystal structure[21] of the canonical subclass CA from *H. neapolitans* (CsoSCA) shows an active site architecture most similar to the *P. sativum* enzyme with the conserved aspartate hydrogen bonded to an arginine (Fig. 13.6). Thus, a catalytic mechanism was proposed for CsoSCA similar to that proposed for the *P. sativum* enzyme in which the conserved aspartate functions in a "gate keeper" role. Although the zinc coordination sphere identifies it as a canonical subclass as defined in this review, the authors suggest that CsoSCA is the archetype of a novel subclass based primarily on the unusual domain structure relative to all other β-class enzymes for which active sites are organized in pairs.[21] The active sites of CAs from *E. coli*, *P. sativum*, *M. thermoautotrophicum*, *M. tuberculosis*, and *Haemophilus influenzae* are organized in pairs by homodimerization. For the *P. purpureum* enzyme, the two active site domains are contained in one polypeptide that presumably evolved by

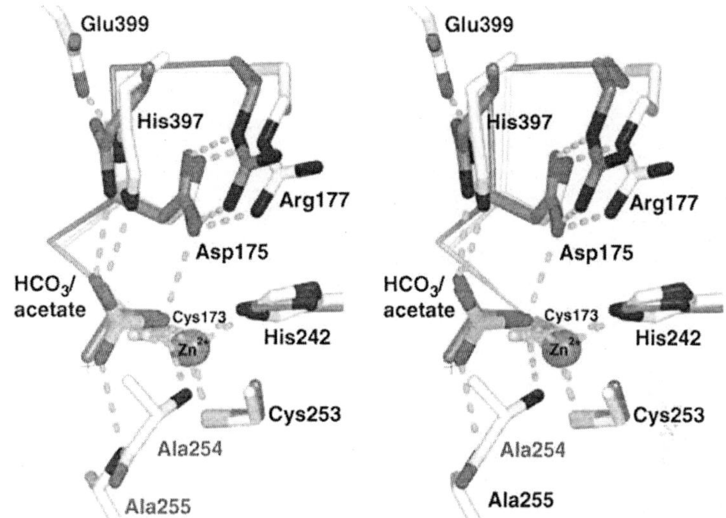

FIGURE 13.6 Stereoview of the active site of CsoSCA from *Halothiobacillus neapolitanus* (shown in light gray) superimposed on the active site of *P. sativum* (shown in dark gray). Bicarbonate was modeled into the active site, based on the position of acetate in the structure from *P. sativum*. Reproduced by permission.[21] (See the color version of this figure in Color Plates section.)

duplication through gene fusion. The CsoSCA enzyme also appears to have evolved two active site domains by a gene duplication, although only one of the domains contains a functional active site whereas the other has diverged to the extent that one of the ligands to zinc is missing and without metal is catalytically incompetent.[21]

13.2.2.2 The Noncanonical Subclass

The noncanonical subclass includes enzymes from the red alga *P. purpureum* and the pathogenic prokaryotes *E. coli*, *M. tuberculosis* (homologue Rv3588c), and *H. influenzae*. Unlike the other subclasses, the enzymes are characterized by very low CO_2 hydration activity when assayed below pH 7.0. These fundamental differences have led to the proposal for an alternative catalytic mechanism by the noncanonical subclass based on the crystal structure of the enzyme from *P. purpureum*.[29] In the postulated mechanism (Fig. 13.7), the zinc-bound aspartate extracts a proton from a water molecule hydrogen bonded to the residue yielding a nucleophilic hydroxide (Fig. 13.7, step 1). The protonated aspartate is released from the zinc and the nucleophilic hydroxide migrates and binds to zinc (Fig. 13.7, step 2). The zinc-bound hydroxide then attacks CO_2 generating zinc-bound HCO_3^-. The proton of the protonated aspartate is transferred to bulk solvent or buffer through an undetermined pathway (Fig. 13.7, step 3). The zinc-bound HCO_3^- is displaced by the deprotonated aspartate releasing HCO_3^- (Fig. 13.7, step 4). In the final step (Fig. 13.7, step 5), a water molecule binds to the carboxyl oxygen of the zinc-bound aspartate ready for another round of catalysis. The low activity at neutral pH for this subclass is consistent with this hypothesis since coordination to zinc is essential

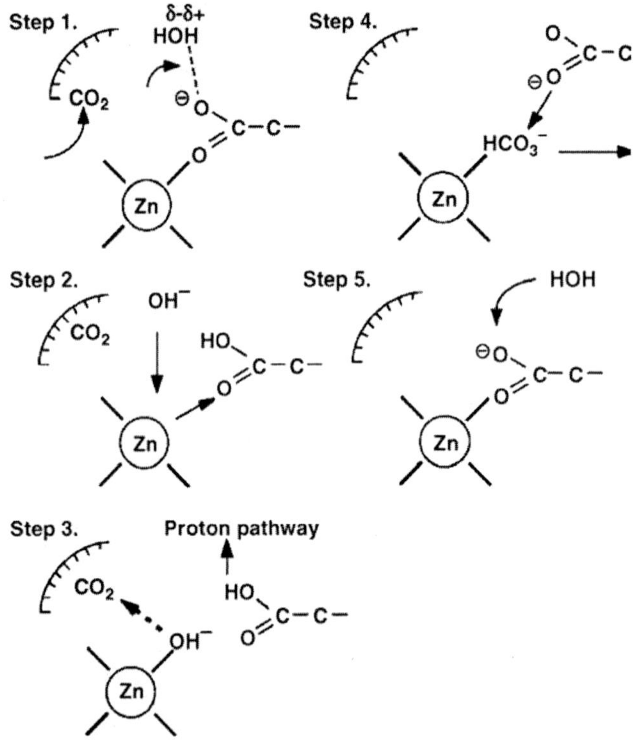

FIGURE 13.7 Proposed mechanism for the noncanonical subclass of the β-class of CAs based on the crystal structure of the enzyme from *P. purpureum*. Reproduced by permission.[29]

for the metal to act as a Lewis acid lowering the pK_a of water to near neutrality. Analysis of the crystal structure of the enzyme from *E. coli* has led to the prediction that conformational changes of the aspartate and adjacent residues allow interconversion between active and inactive forms serving a physiological role in regulating activity.[30]

The crystal structure of the *H. influenzae* enzyme[31] is identical in overall fold and zinc coordination geometry of the *E. coli* enzyme.[30] The active site structures of both enzymes (Fig. 13.8a) reveal a noncatalytic HCO_3^- binding site that is postulated to regulate activity by initiating a pH-dependent and cooperative switch between active and inactive conformations.[31] Thus, HCO_3^- acts as both a regulator and substrate. A mechanism is postulated for this regulation, assuming a zinc hydroxide mechanism for catalysis, wherein aspartate displaces zinc-bound water when HCO_3^- occupies the noncatalytic site and the enzyme is inactivated.[31] Active site conformations have been postulated for the inactive (T) and active (R) states based on structures of the *H. influenzae* and *E. coli* enzymes (Fig. 13.8a) and the enzyme from *P. sativum* (Fig. 13.8b), representing enzymes of the canonical subclass that lack a noncatalytic HCO_3^- binding site.[31] The inactive "T" state is stabilized by HCO_3^- through hydrogen bonding to residues allowing aspartate ligation to zinc that displaces zinc-bound water. In the *P. sativum* enzyme, the aspartate is engaged in a salt bridge

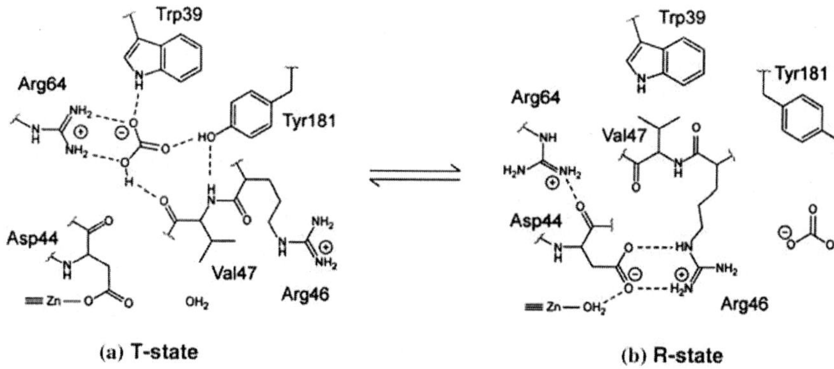

FIGURE 13.8 Active site structures of the noncanonical subclass β-class enzymes from *H. influenzae* and *E. coli* representing the inactive "T" state model stabilized by HCO_3^- and the canonical subclass β-class enzyme from *P. sativum* representing the active "R" state" model. Reproduced by permission.[31]

with arginine-46, although the aspartate is also hydrogen bonded to the zinc-bound water yielding the active "R" state. That HCO_3^- serves a regulatory role is supported by kinetic data suggesting a role for HCO_3^- in allosteric control of activity.[31] The kinetic analyses show an abrupt pH-dependent transition from active to inactive enzyme near pH 8.0 occurs in both CO_2 hydration and HCO_3^- dehydration directions that indicate that the applicable ionizations are not essential for catalysis. Furthermore, the enzyme is most active at high pH in both directions. Finally, the structural features identified for the noncatalytic HCO_3^- binding site are also present in the *P. purpureum* and *M. tuberculosis* enzymes, although the physiological role for the proposed HCO_3^- regulation for the noncanonical subclass has yet to be determined.

The crystal structure of the homodimeric *M. tuberculosis* Rv3588c CA (Fig. 13.4b) shows the conserved aspartate ligated to the active site zinc preventing a potential salt bridge to an adjacent arginine.[20] Thus, the structure conforms with the noncanonical subclass. The active site is open and directly exposed to the bulk solvent as opposed to the canonical subclass Rv1284 enzyme for which the active site is nearly completely shielded from solvent.

13.3 INHIBITORS OF THE β- AND γ-CLASS CAs

Inhibitor studies have been reported for several isozymes of the mammalian α-class with medical applications[7]; however, relatively little is understood regarding inhibitors of the remaining classes. Studies on a β-class CA from *Helicobacter pylori*[32] showed strong inhibition by many sulfonamides/sulfamates with K_I's in the range of 24–45 nM, including acetazolamide, ethoxzolamide, topiramate, and sulpiride that are clinically used drugs. Recently, inhibition with anions (halogenides, pseudohalogenides, bicarbonate, carbonate, nitrate, nitrite, hydrogen sulfide, bisulfite, perchlorate, and sulfate) was reported for the β-class CAs from the pathogenic fungi *Candida albicans* (Nce103) and *Cryptococcus neoformans* (Can2).[33] The *C. neoformans* enzyme was weakly inhibited by cyanide and sulfamic acid with K_I's of 8.22–13.56 mM, whereas all other anions were more robust inhibitors. The enzyme from *C. albicans* was strongly inhibited by cyanide and carbonate with K_I's of 10–11 μM, although weakly inhibited by sulfate, phenylboronic, and phenyl arsonic acid with K_I's of 14.15–30.85 mM. The results show that pathogenic, fungal β-CAs may be targets for the development of antifungals with a novel mechanism of action.

13.3.1 Inhibitors of Cab and Cam Representing the β- and γ-Classes

13.3.1.1 Sulfonamides The archaeal β-class Cab and γ-class Cam CAs respond differently to sulfonamides than the α-class CAs.[34] Crystal structures of the α-class CAs show that the inhibitor binds zinc via the nitrogen of the NH_2SO_2 moiety, which displaces the metal-bound catalytic hydroxide.[35–38] The inhibitor also hydrogen bonds with threonine-199 via the oxygen atoms on the R-NH_2SO_2 group which disrupts catalysis. Sulfonamides are stronger inhibitors of Zn- and Co-substituted Cam than for either the α-class CAs or the canonical β-class Cab. It was suggested that the increased affinity of the Zn- and Co-substituted Cam for sulfonamides may be a consequence of different active site architectures, such that the active site of Cam may allow entry of bulkier compounds without significant rearrangement. Furthermore, catalytically important glutamine-75 and asparagine-202 may be better positioned to interact strongly with negative atoms of the inhibitor such as the oxygen atoms of R-NH_4SO_2 that inhibits catalysis more effectively. Different inhibition patterns between Zn- and Co-substituted Cam have been observed for some sulfonamides[34]; for example, sulfamate ($HOSO_2NH_2$) is a more effective inhibitor of the Zn-substituted Cam compared to Co-substituted Cam.[34] However, the least effective inhibitor of the Zn-substituted Cam is sulfamide ($H_2NSO_2NH_2$) with a K_I of ~70 μM, which is unexpected since it is structurally similar to sulfamate.

In general, the β-class Cab is less inhibited by sulfonamides compared to the γ-class Cam.[34] Least effective are sulfamic acid ($HOSO_2NH_2$) and sulfamide ($H_2NSO_2NH_2$) with K_I's in the mM range comparable with the α-class human CA II. Ethoxzolamide, acetazolamide, and topiramate are the most effective sulfonamide inhibitors with K_I's equal to or less than 23.5 μM. The more effective inhibitors are the heterocyclic sulfonamides and simple substituted sulfonamide derivatives. Halogenated and bulky sulfonamide derivatives are less effective with K_I values in the range of 50–127 μM. The bulkier sulfonamides may be weaker inhibitors because of the physical active site limitations that hinder entry or binding. Data reported for a few sulfonamides indicate comparable inhibition of the canonical subclass β-class CA from *P. sativum*,[22] suggesting that the most effective β-class sulfonamide inhibitor is ethoxzolamide.

13.3.1.2 Anions The α-class CAs are inhibited with anions binding to the 4- or 5-coordinate zinc.[7,39] Similar to the α-class CAs, the archaeal CAs Cab (β-class) and Cam (γ-class) are less inhibited by anions than sulfonamides,[34,40,41] although several anions inhibit Cab to a lesser extent than either Cam or the α-class CAs.[40] Thiocyanate and hydrogen sulfide are the most effective anion inhibitors of Cab, with K_I values of 0.52 and 0.70 mM, respectively. However, Cab has low affinity for cyanide and azide, which is a surprising result since these anions are known to be effective toward four-coordinate metalloenzymes.[40] The result could be explained if the active site architecture of Cab hinders anion binding.

With the exception of the α-class human CAI isozyme, anions are stronger inhibitors of Cam than the α-class and Cab representing the canonical β-class subclass.[40,41] Halides are weak inhibitors of Zn- and Co-substituted Cam with the exception of iodide, which is a moderate inhibitor of Co–Cam compared to other anions. The anions CNO^-, SCN^-, and CN^- that bind in a monodentate fashion to metal are most effective inhibitors of Zn-substituted Cam and less effective inhibitors of the Co-substituted enzyme.[41] On the other hand, anionic inhibitors that bind in a bidentate fashion (HCO_3^-, CO_3^-, NO_3^-, and HSO_3^-) are the most effective for Co-substituted Cam.[41] The different coordination geometries of the five-coordinate Zn-substituted and six-coordinate Co-substituted Cam could explain the different inhibition patterns. In tetrahedrally coordinated zinc ions, monodentate anions replace the coordinating water, add a ligand to the tetrahedral zinc to form the preferred trigonal-bipyramidal geometry, or replace one of the two coordinating water molecules in the trigonal bipyramidally coordinated zinc. Cobalt is most stable with an octahedral geometry as for Co-substituted Cam,[10] explaining why it has a higher affinity for bidentate anions that are able to displace either the catalytic water or one of the two remaining coordinating water molecules.

Cab and Zn-Cam are strongly inhibited by phenylboronic acid and phenylarsonic acid, anions known to inhibit metalloenzymes.[40–43] Conversely, phenylboronic acid and phenylarsonic acid are much weaker anionic inhibitors of α-class CAs suggesting that these anions may be specific inhibitors of prokaryotic CAs. Clearly, further investigation is warranted to test this hypothesis.

REFERENCES

1. Tripp, B. C.; Bell, C. B.; Cruz, F.; Krebs, C.; Ferry, J. G. A role for iron in an ancient carbonic anhydrase. *J. Biol. Chem.* **2004**, *279*, 6683–6687.
2. Xu, Y.; Feng, L.; Jeffrey, P. D.; Shi, Y.; Morel, F. M. Structure and metal exchange in the cadmium carbonic anhydrase of marine diatoms. *Nature* **2008**, *452*(7183), 56–61.
3. Tripp, B. C.; Smith, K.; Ferry, J. G. Carbonic anhydrase: new insights for an ancient enzyme. *J. Biol. Chem.* **2001**, *276*(52), 48615–48618.
4. Smith, K. S.; Ferry, J. G. Procaryotic carbonic anhydrases. *FEMS Microbiol. Rev.* **2000**, *24*, 335–366.
5. Henry, R. P. Multiple roles of carbonic anhydrase in cellular transport and metabolism. *Annu. Rev. Physiol.* **1996**, *58*, 523–538.
6. Park, H.; Song, B.; Morel, F. M. Diversity of the cadmium-containing carbonic anhydrase in marine diatoms and natural waters. *Environ. Microbiol.* **2007**, *9*(2), 403–413.
7. Supuran, C. T.; Scozzafava, A.; Casini, A. Carbonic anhydrase inhibitors. *Med. Res. Rev.* **2003**, *23*(2), 146–189.

8. Alber, B. E.; Ferry, J. G. Characterization of heterologously produced carbonic anhydrase from *Methanosarcina thermophila*. *J. Bacteriol.* **1996**, *178*, 3270–3274.

9. Kisker, C.; Schindelin, H.; Alber, B. E.; Ferry, J. G.; Rees, D. C. A left-handed beta-helix revealed by the crystal structure of a carbonic anhydrase from the archaeon *Methanosarcina thermophila*. *EMBO J.* **1996**, *15*, 2323–2330.

10. Iverson, T. M.; Alber, B. E.; Kisker, C.; Ferry, J. G.; Rees, D. C. A closer look at the active site of gamma-carbonic anhydrases: high resolution crystallographic studies of the carbonic anhydrase from *Methanosarcina thermophila*. *Biochemistry* **2000**, *39*, 9222–9231.

11. Zimmerman, S. A.; Ferry, J. G. Proposal for a hydrogen bond network in the active site of the prototypic gamma-class carbonic anhydrase. *Biochemistry* **2006**, *45*(16), 5149–5157.

12. Alber, B. E.; Colangelo, C. M.; Dong, J.; Stålhandske, C. M. V.; Baird, T. T.; Tu, C.; Fierke, C. A.; Silverman, D. N.; Scott, R. A.; Ferry, J. G. Kinetic and spectroscopic characterization of the gamma carbonic anhydrase from the methanoarchaeon *Methanosarcina thermophila*. *Biochemistry* **1999**, *38*, 13119–13128.

13. Tripp, B. C.; Ferry, J. G. A structure–function study of a proton transport pathway in a novel γ-class carbonic anhydrase from *Methanosarcina thermophila*. *Biochemistry* **2000**, *39*, 9232–9240.

14. Liang, Z. W.; Xue, Y. F.; Behravan, G.; Jonsson, B. H.; Lindskog, S. Importance of the conserved active-site residues Tyr7, Glu106 and Thr199 for the catalytic function of human carbonic anhydrase-II. *Eur. J. Biochem.* **1993**, *211*, 821–827.

15. Krebs, J. F.; Ippolito, J. A.; Christianson, D. W.; Fierke, C. A. Structural and functional importance of a conserved hydrogen bond network in human carbonic anhydrase-II. *J. Biol. Chem.* **1993**, *268*, 27458–27466.

16. Merz, K. M. Insights into the function of the zinc hydroxide–Thr199–Glu106 hydrogen bonding network in carbonic anhydrases. *J. Mol. Biol.* **1990**, *214*, 799–802.

17. Innocenti, A.; Zimmerman, S.A.; Scozzafava, A.; Ferry, J.G.; Supuran, C.T. Carbonic anhydrase activators. Activation of the archaeal β-class (Cab) and γ-class (Cam) carbonic anhydrases with amino acids and amines. *Bioorg. Med. Chem. Lett.* **2008**, *18*, 6194–6198.

18. Kimber, M. S.; Pai, E. F. The active site architecture of *Pisum sativum* β-carbonic anhydrase is a mirror image of that of α-carbonic anhydrases. *EMBO J.* **2000**, *19*(7), 1407–1418.

19. Strop, P.; Smith, K. S.; Iverson, T. M.; Ferry, J. G.; Rees, D. C. Crystal structure of the "cab"-type β-class carbonic anhydrase from the archaeon *Methanobacterium thermoautotrophicum*. *J. Biol. Chem.* **2001**, *276*(13), 10299–10305.

20. Suarez Covarrubias, A.; Larsson, A. M.; Hogbom, M.; Lindberg, J.; Bergfors, T.; Bjorkelid, C.; Mowbray, S. L.; Unge, T.; Jones, T. A. Structure and function of carbonic anhydrases from *Mycobacterium tuberculosis*. *J. Biol. Chem.* **2005**, *280*(19), 18782–18789.

21. Sawaya, M. R.; Cannon, G. C.; Heinhorst, S.; Tanaka, S.; Williams, E. B.; Yeates, T. O.; Kerfeld, C. A. The structure of β-carbonic anhydrase from the carboxysomal shell reveals a distinct subclass with one active site for the price of two. *J. Biol. Chem.* **2006**, *281*(11), 7546–7555.

22. Johansson, I.-M.; Forsman, C. Kinetic studies of pea carbonic anhydrase. *Eur. J. Biochem.* **1993**, *218*, 439–446.

23. Johansson, I. M.; Forsman, C. Solvent hydrogen isotope effects and anion inhibition of CO_2 hydration catalysed by carbonic anhydrase from *Pisum sativum*. *Eur. J. Biochem.* **1994**, *224*, 901–907.

24. Sassetti, C. M.; Boyd, D. H.; Rubin, E. J. Genes required for mycobacterial growth defined by high density mutagenesis. *Mol. Microbiol.* **2003**, *48*(1), 77–84.

25. Betts, J. C.; Lukey, P. T.; Robb, L. C.; McAdam, R. A.; Duncan, K. Evaluation of a nutrient starvation model of *Mycobacterium tuberculosis* persistence by gene and protein expression profiling. *Mol. Microbiol.* **2002**, *43*(3), 717–731.

26. Smith, K. S.; Ferry, J. G. A plant-type (beta-class) carbonic anhydrase in the thermophilic methanoarchaeon *Methanobacterium thermoautotrophicum*. *J. Bacteriol.* **1999**, *181*(20), 6247–6253.

27. Smith, K. S.; Cosper, N. J.; Stalhandske, C.; Scott, R. A.; Ferry, J. G. Structural and kinetic characterization of an archaeal β-class carbonic anhydrase. *J. Bacteriol.* **2000**, *182*(23), 6605–6613.

28. Smith, K. S.; Ingram-Smith, C.; Ferry, J. G. Roles of the conserved aspartate and arginine in the catalytic mechanism of an archaeal β-class carbonic anhydrase. *J. Bacteriol.* **2002**, *184*, 4240–4245.

29. Mitsuhashi, S.; Mizushima, T.; Yamashita, E.; Yamamoto, M.; Kumasaka, T.; Moriyama, H.; Ueki, T.; Miyachi, S.; Tsukihara, T. X-ray structure of beta-carbonic anhydrase from the red alga, *Porphyridium purpureum*, reveals a novel catalytic site for CO_2 hydration. *J. Biol. Chem.* **2000**, *275*(8), 5521–5526.

30. Cronk, J. D.; Endrizzi, J. A.; Cronk, M. R.; O'Neill, J. W.; Zhang, K. Y. Crystal structure of *E. coli* β-carbonic anhydrase, an enzyme with an unusual pH-dependent activity. *Protein Sci.* **2001**, *10*(5), 911–922.

31. Cronk, J. D.; Rowlett, R. S.; Zhang, K. Y.; Tu, C.; Endrizzi, J. A.; Lee, J.; Gareiss, P. C.; Preiss, J. R. Identification of a novel noncatalytic bicarbonate binding site in eubacterial β-carbonic anhydrase. *Biochemistry* **2006**, *45*(14), 4351–4361.

32. Nishimori, I.; Minakuchi, T.; Kohsaki, T.; Onishi, S.; Takeuchi, H.; Vullo, D.; Scozzafava, A.; Supuran, C. T. Carbonic anhydrase inhibitors: the beta-carbonic anhydrase from *Helicobacter pylori* is a new target for sulfonamide and sulfamate inhibitors. *Bioorg. Med. Chem. Lett.* **2007**, *17*(13), 3585–3594.

33. Innocenti, A.; Muhlschlegel, F. A.; Hall, R. A.; Steegborn, C.; Scozzafava, A.; Supuran, C. T. Carbonic anhydrase inhibitors: inhibition of the β-class enzymes from the fungal pathogens *Candida albicans* and *Cryptococcus neoformans* with simple anions. *Bioorg. Med. Chem. Lett.* **2008**, *18*(18), 5066–5070.

34. Zimmerman, S.; Innocenti, A.; Casini, A.; Ferry, J. G.; Scozzafava, A.; Supuran, C. T. Carbonic anhydrase inhibitors. Inhibition of the prokaryotic β and γ-class enzymes from archaea with sulfonamides. *Bioorg. Med. Chem. Lett.* **2004**, *14*, 6001–6006.

35. Huang, S.; Xue, Y.; Sauer-Eriksson, E.; Chirica, L.; Lindskog, S.; Jonsson, B. H. Crystal structure of carbonic anhydrase from *Neisseria gonorrhoeae* and its complex with the inhibitor acetazolamide. *J. Mol. Biol.* **1998**, *283*(1), 301–310.

36. Abbate, F.; Supuran, C. T.; Scozzafava, A.; Orioli, P.; Stubbs, M. T.; Klebe, G. Nonaromatic sulfonamide group as an ideal anchor for potent human carbonic anhydrase inhibitors: role of hydrogen-bonding networks in ligand binding and drug design. *J. Med. Chem.* **2002**, *45*(17), 3583–3587.

37. Boriack-Sjodin, P. A.; Zeitlin, S.; Chen, H. H.; Crenshaw, L.; Gross, S.; Dantanarayana, A.; Delgado, P.; May, J. A.; Dean, T.; Christianson, D. W. Structural analysis of inhibitor binding to human carbonic anhydrase II. *Protein Sci.* **1998**, *7*(12), 2483–2489.

38. Nair, S. K.; Krebs, J. F.; Christianson, D. W.; Fierke, C. A. Structural basis of inhibitor affinity to variants of human carbonic anhydrase II. *Biochemistry* **1995**, *34*, 3981–3989.

39. Supuran, C. T. Carbonic anhydrases as drug targets: an overview. *Curr. Top. Med. Chem.* **2007**, *7*(9), 825–833.

40. Innocenti, A.; Zimmerman, S.; Ferry, J. G.; Scozzafava, A.; Supuran, C. T. Carbonic anhydrase inhibitors. Inhibition of the beta-class enzyme from the methanoarchaeon *Methanobacterium thermoautotrophicum* (Cab) with anions. *Bioorg. Med. Chem. Lett.* **2004**, *14*(17), 4563–4567.

41. Innocenti, A.; Zimmerman, S.; Ferry, J. G.; Scozzafava, A.; Supuran, C. T. Carbonic anhydrase inhibitors. Inhibition of the zinc and cobalt γ-class enzyme from the archaeon *Methanosarcina thermophila* with anions. *Bioorg. Med. Chem. Lett.* **2004**, *14*, 3327–3331.

42. Christianson, D. W.; Cox, J. D. Catalysis by metal-activated hydroxide in zinc and manganese metalloenzymes. *Annu. Rev. Biochem.* **1999**, *68*, 33–57.

43. Beesley, T.; Gascoyne, N.; Knott-Hunziker, V.; Petursson, S.; Waley, S. G.; Jaurin, B.; Grundstrom, T. The inhibition of class C beta-lactamases by boronic acids. *Biochem. J.* **1983**, *209*(1), 229–233.

CHAPTER 14

Fungal and Nematode Carbonic Anhydrases: Their Inhibition in Drug Design

REBECCA A. HALL[1] and FRITZ. A. MÜHLSCHLEGEL[1,2]

[1]Department of Biosciences, University of Kent, Canterbury, Kent, CT2 7NJ, UK
[2]East Kent Hospitals NHS Trust Clinical Microbiology Service, Ashford, Kent, TN24 0LZ, UK

14.1 INTRODUCTION

This chapter aims to provide a review of recent research on fungal and nematode carbonic anhydrases (CAs). A brief introduction to each species covered within the review will be provided to outline why the species is a good model system, that is, to highlight the organism's medical significance. Information on transcriptional regulation, protein activity, and inhibitor studies is included where known. This information will lead to discussions regarding carbonic anhydrases, in these systems, as potential therapeutic targets. In addition, new research characterizing one of the *Caenorhabditis elegans* carbonic anhydrases, CAH-4, will be discussed as an opening into new methods for the treatment of pathogenic nematode species.

In the section on fungal carbonic anhydrases, genes and proteins are designated by *NCE103* and Nce103p, respectively. Mutations are denoted by *nce103*. In the section on nematode carbonic anhydrases, *cah-4* represents genes while CAH-4 represents proteins. Spliced isoforms are represented by CAH-4a.

14.2 FUNGAL CARBONIC ANHYDRASES

14.2.1 *Saccharomyces cerevisiae NCE103*

14.2.1.1 *S. cerevisiae* *S. cerevisiae* is one of the most studied budding yeasts and is routinely used as a eukaryotic model. The genome of *S. cerevisiae* was completely sequenced in 1996 and was the first eukaryotic genome sequenced.[1] The genome is comprised of 6275 genes condensed into 16 chromosomes. Although not a pathogen, *S. cerevisiae* has many industrial purposes, including fermentation processes. In addition, the availability of the genomic knockout library (contains all viable null mutations) permits high-throughput screening of therapeutic compounds, and research done in this model organism can be directly transplanted into pathogenic fungi where research techniques may be limited.

14.2.1.2 Discovery of NCE103 (β-Carbonic Anhydrase) The gene YNL036w located on chromosome 14, more commonly known as *NCE103* (*n*on-*c*lassical *e*xport), was initially identified by Cleaves et al. in 1996 as a protein involved in a novel protein secretion pathway.[2] Disruption of *NCE103* alleles resulted in growth retardation when grown under standard laboratory conditions on glucose-rich media.[3,4] Initially, this phenotype was attributed to oxygen stress, as growth was restored in hypoxic atmospheres and environments enriched with carbon dioxide.[3,5] In addition, growth of the *S. cerevisiae* null mutant was complemented by heterologous expression of carbonic anhydrases from a variety of different species, including human CA II.[4] This suggested that *NCE103* could encode a CA. Initial reports failed to identify CA activity in *S. cerevisiae* cell lysates.[3] However, ScNce103p was able to complement the growth of an *Escherichia coli* CA-deficient strain, and subsequent cell lysates confirmed that the Nce103p protein displayed CA activity.[4] Comparisons of ScNce103p with other known CA sequences identified a CA catalytic site predicting Nce103p to be a member of the β-CA family.

14.2.1.3 Functionality of Nce103p Other CA-deficient microorganisms that also show impaired growth under standard laboratory conditions have been reported.[6–8] In some cases, it was established that supplementation of CO_2 was required to provide an initial carbon source to initiate growth.[8] After this point there was no further requirement for elevated CO_2 levels. However, in *S. cerevisiae*, this is not the case. Aguilera et al. reported that the replacement of CO_2 with N_2 or air after an initial period results in the termination of growth.[9] Therefore, in the case of *S. cerevisiae*, constant elevated CO_2 levels are required to promote growth in the *nce103* mutant strain.

Bicarbonate is an essential substrate for carboxylation reactions. Yeast metabolism contains four essential carboxylation reactions catalyzed by pyruvate carboxylase,[10,11] acetyl-CoA carboxylase,[12]carbamoyl phosphate synthase (CPSase), and phosphoribosylaminoimidazole (AIR) carboxylase. Supplementation of the end products into media for three of these pathways involving pyruvate carboxylase, acetyl-CoA carboxylase, and CPSase complemented the growth of the *nce103* null in air.[9] Therefore, Nce103p is required to provide sufficient bicarbonate for essential metabolic carboxylation reactions.

14.2.1.4 Transcriptional Analysis of NCE103 The *S. cerevisiae NCE103* is differentially regulated in response to many environmental signals. *NCE103* is reportedly upregulated upon encountering heat shock, nitrogen starvation, high pH, and DNA damaging agents.[13–15] These global transcriptional approaches would tend to suggest that *NCE103* is upregulated as a general stress response. In addition, available carbon sources also affect the transcriptional level of Sc*NCE103*. Environments with elevated CO_2 transcriptionally repress *NCE103*.[5,16] This is hypothesized to result from diffusion of CO_2 across the membrane increasing local CO_2 concentrations within the cytoplasm. At high concentrations, CO_2 undergoes spontaneous conversion to bicarbonate, which is then available for carboxylation reactions. Therefore, in high CO_2 environments, the function of Nce103p is redundant, resulting in its suppression. These transcriptional data agree with the mutant phenotype, where Nce103p is only required for growth in normal atmospheric conditions when bicarbonate could be limited.[5] Furthermore, from global transcriptional analysis approaches, *NCE103* is induced after phagocytosis along with several other genes involved in the glyoxylate cycle, suggesting that metabolism is a key component of fungal pathogenesis.[17]

14.2.2 *Cryptococcus neoformans CAN2*

14.2.2.1 C. neoformans *C. neoformans* is a basidiomycete naturally found in soil contaminated with pigeon guano. Healthy individuals do not normally suffer from *C. neoformans* infections. However, upon inhalation into the alveoli of immunosuppressed individuals, for example AIDS patients and people undergoing chemotherapy, *C. neoformans* is able to colonize deep tissues and transverse into the nervous system where it can cause cryptococcal meningitis. One important factor that accounts for this organism's pathogenicity is the production of a polysaccharide capsule that surrounds it. Capsule biosynthesis is regulated by a number of environmental factors that are normally found in the host. For example, serum,[18] pH, and carbon dioxide levels[18,19] are known to promote capsule biosynthesis to protect the organism from host defenses. It is believed that increasing the understanding of the mechanisms that contribute to the organism's pathogenicity may lead to the identification of novel therapeutic targets.

14.2.2.2 Identification of CAN2 Bahn et al. and Mogensen et al. identified two *C. neoformans* genes that shared identity with the *S. cerevisiae NCE103* β-CA, which were subsequently designated *CAN1* and *CAN2* for carbonic anhydrase 1 and 2.[20,21] *CAN2* can easily be detected and is expressed at high levels, while *CAN1* could only be detected by RT-PCR with low expression levels, suggesting that *CAN2* is the major CA isoform in *C. neoformans*.[20] Can2p contains one histidine and two cystine residues in the active site for metal ion binding, suggesting that it is a member of the β-CA family.

14.2.2.3 Functionality of Can1 and Can2 *C. neoformans* strains without a functional *CAN1* gene are able to maintain growth in both normal atmospheres and atmospheres supplemented with carbon dioxide (5%), while *CAN2* mutants are only able to grow when supplemented with CO_2.[20,21] This phenotype was annotated as high CO_2 requiring (HCR), but this phenotype has also been documented for many other organisms that lack functional CA isoforms. Therefore, *CAN2*, but not *CAN1*, is essential for promoting growth in environments where CO_2 is limiting.

Mogenesen et al. characterized *CAN2* from *C. neoformans*. They confirmed that expression of *CAN2* alone in a CA deficient *E. coli* strain was sufficient to complement the growth defect of the *E. coli* in low CO_2 conditions.[21] This result was also confirmed by Bahn et al. where *CAN2* was successfully used to complement the growth defect of the *nce103* mutant of *S. cerevisiae*.[20] Therefore, *CAN2* encodes a functional CA. Biochemical characterization of Can2p confirmed that it has significant CA activity and is inhibited by commercially available carbonic anhydrase inhibitors, like ethoxyzolamide and acetazolamide,[21] as assayed by the electrometric method originally described by Wilbur and Anderson.[22] The reported IC_{50} for hCA II with ethoxyzolamide is 50 nM, while the IC_{50} for Can2p was reported to be 100 μM.[21] This may not be surprising, considering that ethoxyzolamide was designed as a specific α-class inhibitor;[23] however, it does suggest that more specific β-class CA inhibitors will be required if CA is to be a therapeutic target. Ethoxyzolamide has been tested for its ability to affect the growth of *C. neoformans*. A concentration of 3 mM ethoxyzolamide is sufficient to reduce the growth rate of *C. neoformans* in a wild-type background in low CO_2 environments, but not in high CO_2 conditions.[21] However, for therapeutic purposes, these concentrations are extremely high. This may be caused by two attributing factors: (1) The inhibitor is not specific for β-CA and, therefore, a higher dose is required, as is suggested by the biochemical data. (2) The capsule and fungal cell wall surrounding the organism provides it with some protection from the inhibitor. Therefore, future inhibitors should be screened for the ability to transverse the capsule and cell wall to have maximal bioactivity.

The growth of *S. cerevisiae nce103* mutants could be complemented in low CO_2 environments by the addition of certain fatty acids. Therefore, Bahn et al. tested whether intermediates of fatty acid biosynthesis or purine and pyrimidine biosynthesis were able to complement the growth of the *can2* mutant in low CO_2 conditions. Palmitate, and to a certain degree myristate, rescue the growth of the *can2* mutant in a dose-dependent manner, although not as efficiently as supplementing with CO_2.[20] Therefore, Bahn et al. concluded that the growth defect of the *can2* mutant is in part due to insufficient bicarbonate being provided to fulfill requirements of fatty acid biosynthesis, as predicated for *S. cerevisiae*.

One of the major virulent determinants of *C. neoformans* is its ability to synthesize a capsule, protecting itself from the host's immune system. As capsule biosynthesis is upregulated upon exposure to environments with high CO_2 concentrations, Bahn et al. investigated the role of Can1 and Can2 in capsule biosynthesis. Mutations in either *can1* or *can2* failed to reduce capsule biosynthesis upon exposure to CO_2.[20] Bahn et al. suggested two potential hypotheses for their result: (i) Can1 and Can2 are not required for capsule biosynthesis and are, therefore, not required for virulence. This would favor the argument that it is actually the CO_2 and not the bicarbonate that has impact on capsule biosynthesis. (ii) In elevated CO_2 environments, the spontaneous conversion of CO_2 to bicarbonate is sufficient to promote capsule biosynthesis. This would suggest that it is specifically the bicarbonate that induces capsule biosynthesis. Many favor the second hypothesis. Fungal species contain a soluble adenylyl cyclase (sAC) that is reportedly activated by bicarbonate ions.[21,24] Therefore, Mogensen et al. hypothesize that Can2p is required to produce bicarbonate, which then directly

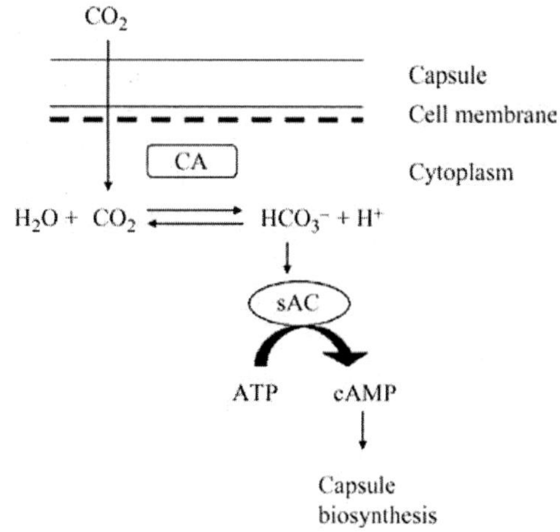

FIGURE 14.1 Carbon dioxide sensing in *C. neoformans*. Under aerobic conditions, carbon dioxide diffuses into the cytoplasm where it is converted into bicarbonate by the carbonic anhydrase, Can2. The local intracellular concentrations of bicarbonate directly activate the fungal soluble adenylyl cyclase, increasing concentrations of the secondary messenger, cAMP. Downstream effectors in the camp-signaling cascade then upregulate expression of specific genes required for capsule biosynthesis. Scheme modified from Ref. 21.

activates sAC to increase cAMP output and promote capsule biosynthesis.[21] Mogensen et al. also conclude that in elevated CO_2, the function of Can2 is dispensable due to the spontaneous hydration of CO_2[21] (Fig. 14.1).

Observations that *can2* mutants showed an increase in mating filaments compared to wild-type strains prompted the exploration into carbonic anhydrase function in *C. neoformans* mating. Indeed, incubation of wild-type *C. neoformans* in atmospheres of high CO_2 inhibits mating and is assumed to result from increased intracellular bicarbonate levels.[20] Bahn et al. hypothesize that under these conditions, in the *can2* strain, the spontaneous hydration of CO_2 to HCO_3^- is sufficient to promote growth, but is not sufficient to interfere with mating.[20] Although *can2* mutants produce mating filaments, they are reportedly deficient in sporulation due to inhibition of cell fusion.[20] As increased cAMP signaling is required for sporulation in *S. cerevisiae,* it is probable that cAMP levels mediate sporulation in *C. neoformans*. Therefore, Can2 could be required to produce bicarbonate to increase intracellular cAMP concentrations, through direct interaction with adenylyl cyclase, which is essential for spore formation.[20]

14.2.2.4 Regulation of Can2 In contrast to *NCE103* from *S. cerevisiae*, *CAN2* is not regulated at the transcriptional level in *C. neoformans*.[20] Therefore, although *CAN2* is not required for growth in environments of elevated CO_2, there appears to be no regulation at the transcriptional level to prevent *CAN2* expression. This may be due to the fact that *CAN2* function is still required under these conditions for alternative functions, or that regulation is predominantly at the posttranscriptional level, which has so far not been investigated.

14.2.3 *Candida albicans NCE103*

14.2.3.1 C. albicans *C. albicans* is an opportunistic fungal pathogen of humans. In healthy individuals, *C. albicans* forms part of the gastrointestinal flora and rarely causes infections. However, when individuals become immunosuppressed due to chemotherapy, AIDS, or even as a consequence of age, *C. albicans* is able to disseminate and cause chronic and systemic disease. One important factor that contributes to *C. albicans*' success as a pathogen is its polymorphic nature. *C. albicans* grows as yeast, pseudohyphal, true hyphal, and chlamydospore forms. The ability to switch between the yeast and hyphal forms is regulated by many host environmental factors, including serum, pH, temperature, CO_2, and available carbon source.[24–27]

14.2.3.2 Identification of NCE103 Carbon dioxide is a strong promoter of filamentation in *C. albicans* (Fig. 14.2). In systemic infections, *C. albicans* disseminates into the blood stream. Here, in addition to serum, *C. albicans* is exposed to high concentrations of CO_2 (5.5%). This phenomenon has been dubbed "CO_2 sensing" and has opened up a wide and expansive new field in *C. albicans* research. This field has so far opened new doors for the diagnosis of *Candida* species in hospital environments.[28] *C. albicans* is the only *Candida* species that displays a filamentous phenotype in the presence of CO_2, allowing diagnosis and treatment within 48 h of the sample being received. Attempts to characterize the CO_2 sensing pathway have so far highlighted two proteins that play a significant role. The first is adenylyl cyclase. Adenylyl cyclase produces the secondary messenger, cAMP, which is required for filamentation in *C. albicans*.[29] Adenylyl cyclase in *C. albicans* is termed a soluble adenylyl cyclase as it does not posses the typical 12-membrane spanning domains. The sAC is directly activated by CO_2/HCO_3^-.[24] The second is the β-carbonic anhydrase, Nce103p. Nce103p was identified through its similarity to *S. cerevisiae* Nce103p.

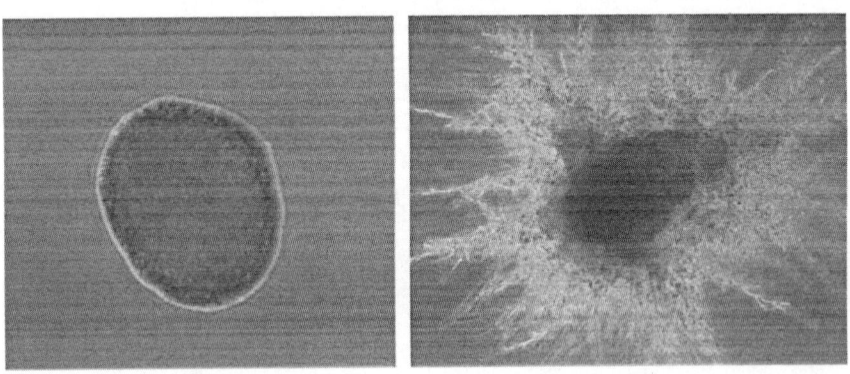

(a) (b)

FIGURE 14.2 Effect of carbon dioxide on *C. albicans* morphology. Under aerobic conditions, *C. albicans* grows in yeast form (a). However, supplementation of 5% carbon dioxide into the atmosphere induces extensive filamentation (b), which is required for virulence of the pathogen. Image taken by Kara J. Turner on a Leica DMR microscope at 100× magnification using a DC300F digital camera.

14.2.3.3 Functionality of Nce103p To identify the function Nce103p plays in the CO_2 sensing pathway, Klengel et al. constructed homozygote knockouts of *nce103* in two backgrounds (CAI4 and BWP17).[24] In parallel with *S. cerevisiae*, the *nce103* mutant strains were unable to grow in conditions where the CO_2 concentration was low (0.033%), but proliferated in environments enriched with CO_2 (5.5%).[24]

Klengel et al. biochemically characterized the Nce103p protein as an N-terminal GST fusion.[24] Nce103p is a functional carbonic anhydrase and is inhibited by ethoxyzolamide (0.5 mM).[24] Klengel et al. also showed that the *nce103* mutant is avirulent when tested in epithelial models of infection where the CO_2 concentration is low, but the Nce103p does not play a role in establishing infection in host sights where the CO_2 concentration is elevated.[24]

14.2.3.4 Regulation of NCE103 Regulation of *NCE103* in *C. albicans* in response to CO_2 has not been studied so far. In *S. cerevisiae*, *NCE103* is downregulated at the transcriptional level in conditions when carbonic anhydrase is not required for survival (i.e., in environments with elevated CO_2).[5] Therefore, as *nce103* mutants have the same phenotype, it would be tempting to speculate that the *NCE103* in *C. albicans* would be regulated in a similar manner. However, carbonic anhydrase mutants in other fungal species also behave in a similar manner as in *S. cerevisiae*, but so far these have been shown not to be regulated in response to elevated CO_2, at least at the transcriptional level.[20] Therefore, although functionally similar, different species appear to have developed different regulatory mechanisms. However, from global transcriptional approaches, *NCE103* is significantly upregulated upon immediate adhesion to surfaces and during the formation of biofilms.[30] The exact function of *NCE103* in biofilm formation in *C. albicans* has not been fully investigated, but it may be required to provide bicarbonate ions for intermediate metabolism. Identifying and comparing how each of the carbonic anhydrases in different fungal species is regulated could be an important aim for future research.

14.2.4 Other Fungal Species

14.2.4.1 Candida glabrata *C. glabrata* is smaller than *C. albicans* (1–4 μm compared to 4–6 μm), has a haploid genome, and is nondimorphic.[31] In fact, *C. glabrata* appears to be more closely related to *S. cerevisiae* than to other *Candida* species, and it was initially classified as *Torulopsis glabrata.* Like *C. albicans*, *C. glabrata* is a commensal organism of the skin, vagina, oral, and gastrointestinal tracts. Although *C. albicans* is the most frequently isolated *Candida* species from infected patients, *C. glabrata*, which was originally thought to be nonpathogenic, is rapidly becoming a medical concern due to its increased prevalence in immunosuppressed individuals and AIDS patients.[32] This is of clinical importance as *C. glabrata* shows reduced susceptibility to common azoles, which are routinely used to treat fungal infections.[33] In addition, virulence factors associated with *C. glabrata* are not well understood when compared to *C. albicans*, making *C. glabrata* an important fungal pathogen for future research. The increase of *C. glabrata* in superficial infections, as well as in mucosal and systemic infection, suggests that *C. glabrata* could be a potential

```
C.albicans          -------------------------MGRENILKYQLEH-DHESDLVTEKDQSLLLDNNNN 34
C.tropicalis        -------------------------MGRENILKYQLEH-DHETDLSISPPSGTTTPENKP 34
C.parapsilosis      -------------------------MGRENILHYLQEHNETDKELEFDAKNGSQQSKNEI 35
C.guilliermondii    MPPQFSHHTLVLFRSYRQLGTLSRIMGRENILRYQLEH-DNETDLALADYLKTSARAS-- 57
C.lipolytica        -------------------------MAQKPLFRFGATD---------------------- 13
C.glabrata          -------------------------MTKDTIFTLTGKC---------------------- 13
                                             * :. ::

C.albicans          LNGMNNTIKTHPVRVSSGNHNNFPFTLSSESTLQDFLNNNKFFVDSIKHNHGNQIFDLNG 94
C.tropicalis        IR---------PVRVS-GTHTNFPFTLSPDSTLQDYLHNNKYFVDSIDHNHGNDIFYLNG 84
C.parapsilosis      EIP--------QFKRKVTSHSNFPFTLSPDSTITDYLNNNKFYVDSIKHNHGDQIFELNG 87
C.guilliermondii    ------------------SHSHTPFLLLKESTKKDFLDNNRYFVDSIKHNHSSEVFELNG 99
C.lipolytica        -------------------------------SLATLLERNEKWANRVSSVR-PSLFPTNA 41
C.glabrata          -------------------------------ALDNILDANRQWAQAMHRSQ-PQLFPTNG 41
                                                   :    *. *. :.: :   :  .:*  *.

C.albicans          QGQSPHTLWIGCSDSRAGDQCLATLPGEIFVHRNIANIVNANDISSQGVIQFAIDVLKVK 154
C.tropicalis        KGQTPHTLWIGCSDSRAGEQCLATLPGEIFVHRNIANIVNANDISSQGVIQFAVDVLKVR 144
C.parapsilosis      KGQSPHTLWIGCSDSRAGEQCLATLPGEIFVHRNIANIVNSNDFSSQGVIQFAIDVLKVK 147
C.guilliermondii    RGQSPHTLWIGCSDSRAGESCLATLPGEIFTHRNIANIITASDISSQGIIQFAIDVLKVK 159
C.lipolytica        QGQAPKILWIGCSDSRAGEGCLDLLPGEVFVHRNIANLLPDSDFSSLSVIQFAVQVLKVR 101
C.glabrata          QGQDPHTLFIGCSDSRYNEDCLGVLPGEIFTLKTVANICHTDDHSLLATLEFAILNLKVN 101
                    :** *: *:******* .: **  ****:*. :.:**:   .* *  . ::**:  ***.

C.albicans          KIIVCGHTDCGGIWASLSK----KKIGGVLDLWLNPVRHIRAANLKLLEEYNQDPKLKAK 210
C.tropicalis        KIIVCGHTDCGGIWASLSK----KKIGGVLDLWLNPVRHTRAANLKLLNELNDKPREKAK 200
C.parapsilosis      KIIVCGHTDCGGIWASLSS----KKIGGVLDLWLNPVRHIRAQNLKLLEQYNHEPKLKAR 203
C.guilliermondii    KIIVCGHTDCGGVWASLSS----KKIGGVLDLWLNPIRHIRAANLKLLNEYNDDPKMKAR 215
C.lipolytica        HIIVCGHYDCGGVWSSLTS----KKLG-IIDHWLRPIRDTKVRHKAMLDAIE-DPKDKCA 155
C.glabrata          RIILCGHTDCGGIKTCLLGRESIKESCPHLYEHLDDIEDLVESHESELNQLD-NICSKSK 160
                    :**:*** ****: :.*      *:    :   *  :..    :   *:  :  .   *.

C.albicans          KLAELNVISSVTALKRHPSASVALKKN----EIEVWGMLYDVATGYLSQVEIPQDEFEDL 266
C.tropicalis        KLAELNVIASVTALKRHPSASMALKKG----EIEVWGMMYDVATGYLSQVEIPDDEFEDL 256
C.parapsilosis      KLAELNVIASVIALKRHPSASTALKQG----KIEVWGMIYDVASGYLSELEIPQDEFHEL 259
C.guilliermondii    KLAELNVVLSVMALKRHPSASMALKNG----EIEVWGMMYDVSTGYLNELEIPDDEFEDL 271
C.lipolytica        RLVELNVCAQVNNLKRNTVIIEAQGER----DLQIHGVVYDPGSGLLKEIMVPEDEYAED 211
C.glabrata          LMSHRNVERQLQRLLQIPVVQDALRNSNQDHEFNIFGLVYNVDSGLVDVVREVYGNQQK- 219
                     : . **  .:  * : .    *  :     .::: *::*:  :* :. :     .: .

C.albicans          FHVHDEHDEEEYNPH 281
C.tropicalis        FHVNDE-DDEEFNPH 270
C.parapsilosis      FHVSDEEDASAHNAH 274
C.guilliermondii    FHVHDE--SDDINPH 284
C.lipolytica        YFVSDA--GELVH-- 222
C.glabrata          ---------------
```

FIGURE 14.3 Sequence alignment of putative carbonic anhydrases from *Candida* species using ClustalW (www.ebi.ac.uk/clustalw/). (*) indicates 100% conservation of amino acids, (:) represents conserved amino acids according to amino acid properties, while (.) designates substitutions that are only partially conserved in accordance to amino acid properties.

target for CA inhibition. The genome of *C. glabrata* is predicted to encode one β-CA isoform from the open reading frame designated *CAGL0G01540g* (Fig. 14.3). So far, there is little information regarding the characterization of this putative CA. However, the discovery of the importance of CA isoforms in aerobic growth in other fungal species will surely prompt investigation into the CA of *C. glabrata*.

14.2.4.2 *Aspergillus*

The *Aspergillus* genus contains approximately 200 species of filamentous fungi, which have commercial as well as medical importance. The common route of *Aspergillus* infection is through inhalation of spores from the atmosphere. In healthy individuals, these spores are eradicated by the immune system, but in immunocompromised patients, spores accumulate in the lung. *Aspergillus* species are capable of causing allergic disease as well as systemic infections. *Aspergillus* species are also agriculturally important, as they cause fungal infections in crops, especially grain crops. Environmental pH sensing in *Aspergillus fumigatus* has been extensively studied,[26] but carbon dioxide sensing and the involvement of CA isoforms in aerobic growth are not well understood. The genomes of *A. fumigatus, Aspergillus niger*, and *Aspergillus terreus* are predicted to contain four CA isoforms, while *Aspergillus clavatus, Aspergillus flavus*, and *Aspergillus nidulans* are predicted to contain three CA isoforms. The presence of so many isoforms is surprising, considering that other fungal species studied to date only have one or potentially two isoforms. In addition, the CA isoform in *Aspergillus oryzae* is predicted to be a member of the α-class, which is intriguing as so far all CA isoforms investigated in fungal species have been of the β-class. At present, documentation on the precise function, cellular localization, and biochemical characterization of the different isoforms is not known, but it would be interesting to identify which isoforms are essential for growth in aerobic environments, and if there is any functional compensation between the different isoforms.

14.3 CARBONIC ANHYDRASE INHIBITORS AS ANTIFUNGALS

Due to the increase of immunocompromised patients and the growth of antifungal resistance, new methods for the treatment of fungal infections are rapidly needed. The discovery that removal or inhibition of carbonic anhydrases in fungal species prevents proliferation in normal aerobic environments has promoted the consideration of CA inhibitors as antifungals. So far, screening of compound libraries has proved a successful way to identify potential CA inhibitors.[34] From these, additional compounds can be synthesized to increase CA isoform specificity and bioactivity. Furthermore, investments into the structural characterization of fungal CA isoforms and modeling approaches have the potential to design optimal CA inhibitors through *in silico* methods.

The use of CA inhibitors as antifungals evolved from the fact that, in epithelial virulence models, the *C. albicans* carbonic anhydrase mutant is avirulent.[24] Therefore, carbonic anhydrase inhibitors have the potential to affect against skin infections. *Dermatophytes* are a collection of 40 fungal species from the genera *Microsporum, Trichophyton*, and *Epidermophyton*. These fungal parasites reside within the nonliving epidermis layers of the skin and nails obtaining nutrients from the keratinized tissue. These species cannot penetrate into the skin of immunocompetent individuals with inflammatory responses resulting from metabolic by-products secreted by the parasite. Therefore, during infection, *dermatophytes* are constantly in contact with the atmosphere where CO_2 levels would be minimal, environments where carbonic anhydrase inhibitors have been shown to be effective. In addition, the fungus *Malazessia furfur* is part of the skin natural flora, but has implications in dandruff and atopic dermatitis (AD).[35] In both these conditions, *M. furfur* is exposed to the atmosphere and, therefore, has the potential to be treated by CA inhibitors. Carbon dioxide sensing in *dermatophytes* or *M. furfur* has so far not been investigated, but with the number of microorganisms where CAs are shown to be required for growth under normal atmospheric conditions rapidly expanding, this theory is certainly plausible. Although CAIs have been shown to mimic the effects of a CA mutant strain in aerobic environments under standard laboratory conditions[21] (reduced growth rates), results from virulence models using wild-type strains in the presence of CAIs are yet to be published.

14.4 NEMATODE CARBONIC ANHYDRASES

14.4.1 *C. elegans* Carbonic Anhydrases

14.4.1.1 C. elegans *C. elegans* is a free living, nonpathogenic nematode. This organism gained model status through several key attributes, including the ease in which the nematode can be cultured within a laboratory environment, the consistence of progeny to contain 959 somatic cells, and its relatively short lifespan (2–3 weeks). Therefore, it is not surprising that *C. elegans* was the first multicellular organism to have its genome completely sequenced. The *C. elegans* genome contains approximately 20,000 predicted open reading frames. Although annotations are known for thousands of genes, many genes still encode proteins of unknown function. This list is gradually shortening through the development of global posttranscriptional gene silencing screens, commonly known as RNA-mediated interference (RNAi), which was first identified in *C. elegans.*[36] Subsequently, A.Z. Fire and C.C. Mello were awarded the Nobel Prize in 2006 for the discovery of RNAi.

Even though *C. elegans* has established itself as a model organism, many aspects of the nematode's basic biology have yet to be explored. For example, in the *C. elegans* genome, seven genes are predicted to encode carbonic anhydrases. Due to alternative splicing of several transcripts, as confirmed by cDNA sequencing, it is probable that the nematode has 10 carbonic anhydrases in total (www.wormbase.org). However, in comparison to human carbonic anhydrases where tissue-specific expression and cellular localization have been established for all 15 isoforms, the *C. elegans* carbonic anhydrases remain predominantly uncharacterized. Global RNAi screening programs have so far not identified associated phenotypes with the loss of individual carbonic anhydrases, even though the human isoforms are known to be associated with tumor formation,[37–41] pH regulation, and oxidative stress,[42] and in many microorganisms carbonic anhydrases are involved in carbon dioxide sensing.[20,21,24,43]

14.4.1.2 CAH-4: A pH-Regulated Carbonic Anhydrase The increased knowledge of environmental sensing in microorganisms has sparked interest into how many organisms respond to changes in environmental cues. To this end, transcriptional profiling has enabled us to answer these important questions. Investigations into the effects of environmental pH on *C. elegans* identified that one (*cah-4*) of the seven carbonic anhydrases was differentially regulated in response to pH. Transcript levels of *cah-4* were significantly upregulated (five fold) in alkaline conditions (pH 9) when compared to neutral or acidic (pH 4) environments. The carbonic anhydrase *cah-4* is alternatively spliced, producing two isoforms *cah-4a* and *cah-4b*, but it was specifically the *cah-4b* isoform that was differentially regulated in alkaline environments.[44]

Sequence alignments of CAH-4 with other known carbonic anhydrases suggest that this carbonic anhydrase is at most 37% identical to other known carbonic anhydrases. The histidine residues involved in zinc metal binding are conserved, suggesting that both CAH-4a and CAH-4b are members of the α-class (Fig. 14.4).

```
Human               ------------------------SHHWGYGKHN-GPEHWHKDFPIAKGERQSPVDIDTH 35
Mouse               ------------------------SHHWGYSKHN-GPENWHKDFPIANGDRQSPVDIDTA 35
Bovine              ------------------------SHHWGYGKHN-GPEHWHKDFPIANGERQSPVDIDTK 35
Chick               ------------------------SHHWGYDSHN-GPAHWHEHFPIANGERQSPIAISTK 35
CeCAH-4             MAPPQVRRSARLSKRCQEEKVKLQKKNVGFKAKSKSAKKSNKKFKKAAAQRQSPIDIVPQ 60
                                                *              *  *   ****  *

Human               TAKYDP---SLKPLSVSYDQATSLRILNNGHAFNVEFDDSQDKAVLKGGPLDGTYRLIQF 92
Mouse               TAHHDP---ALQPLLISYDKAASKSIVNNGHSFNVEFDDSQDNAVLKGGPLSDSYRLIQF 92
Bovine              AVVQDP---ALKPLALVYGEATSRRMVNNGHSFNVEYDDSQDKAVLKDGPLTGTYRLVQF 92
Chick               AARYDP---ALKPLSFSYDAGTAKAIVNNGHSFNVEFDDSSDKSVLQGGALDGVYRLVQF 92
CeCAH-4             HVCCDTDVCKADALNIDYKSGDCCDVLVSEGGFLVNVKRNCGTFLTANHLPSSKFALAQF 120
                        *        *   *              *                       * **

Human               HFHWGSLDGQGSEHTVDKKKYAAELHLVHWNTKYGDFGKAVQQPDGLAVLGIFLKVGSAK 152
Mouse               HFHWGSSDGQGSEHTVNKKKYAAELHLVHWNTKYGDFGKAVQQPDGLAVLGYFLKIGPAS 152
Bovine              HFHWGSSDDQGSEHTVDRKKYAAELHLVHWNTKYGDFGTAAQQPDGLAVVGVFLKVGDAN 152
Chick               HIHWGSCEGQGSEHTVDGVKYDAELHIVHWNVKYGKFAEALKHPDGLAVVGIFMKVGNAK 152
CeCAH-4             HAHWGSNSKEGSEHFLDGKQLSGEVHFVFWNTSYESFNVALSKPDGLAVVGVFLKEGKYN 180
                    * ****    ****          * * * **  *  *  *   ****** * * * *

Human               PGLQKVVDVLDSIKTKGKSADFTN-FDPRGLLP--ESLDYWTYPGSLTTPPLLECVTWIV 209
Mouse               QGLQKVLEALHSIKTKGKRAAFAN-FDPCSLLP--GNLDYWTYPGSLTTPPLLECVTWIV 209
Bovine              PALQKVLDALDSIKTKGKSTDFPN-FDPGSLLP--NVLDYWTYPGSLTTPPLLESVTWIV 209
Chick               PEIQKVVDALNSIQTKGKQASFTN-FDPTGLLP--PCRDYWTYPGSLTTPPLHECVIWHV 209
CeCAH-4             DNYHGLIDTVRKATGNATPIAMPKDFHIEHLLPSPDKREFVTYLGSLTTPPYNECVIWTL 240
                                              *    ***       ** *******  * * *

Human               LKEPISVSSEQVLKFRKLNFNGEGEPEELMVDNWRPAQPLKNRQIKASFK- 259
Mouse               LREPITVSSEQMSHFRTLNFNEEGDAEEAMVDNWRPAQPLKNRKIKASFK- 259
Bovine              LKEPISVSSQQMLKFRTLNFNAEGEPELLMLANWRPAQPLKNRQVRGFPK- 259
Chick               LKEPITVSSEQMCKLRGLCFSAENEPVCRMVDNWRPCQPLKSREVRASFQ- 259
CeCAH-4             FTEPVEVSFGQLNVLR----------NIIPANHRACQDRCDREIRSSFNF 280
                      **  **  *    *                * *  *    *
```

FIGURE 14.4 Multiple sequence alignment of carbonic anhydrases from different species using ClustalW (www.ebi.ac.uk/clustalw/). The *C. elegans* CAH-4a was aligned with type 2 CA from various species. (*) indicates 100% conservation of amino acids. Histidine residues 93, 95, and 118 in the human isoform coordinate binding of the catalytic zinc ion and are present in all the five isoforms (highlighted residues).

In addition to being regulated by pH, *cah-4* is reportedly regulated by hypoxic conditions.[45] Hypoxia inducible factor (HIF-1) is a transcription factor that differentially regulates genes in response to hypoxia (<21% oxygen). HIF-1 is tightly controlled through posttranslational modifications.[46] Under normal oxygen conditions, the α-subunit of HIF-1 is hydroxylated in the oxygen-dependent degradation domain.[47,48] This hydroxylation results in rapid degradation of the α-subunit via the ubiquitin-dependent proteasome pathway involving the von Hippel–Lindau (VHL-1) tumor suppressor.[49] During exposure to hypoxic conditions, hydroxylation is suppressed, thus allowing the accumulation of the HIF-1 α-subunit within the nucleus. Inspection of the promoter region of *cah-4* reveals an HIF-1 binding site 245 bp downstream of the ATG start codon. Investigations into the role of HIF-1 and VHL-1 in the transcriptional regulation of *cah-4* under different environmental pH conditions were investigated on a small scale. In HIF-1 mutants, the levels of *cah-4* were suppressed under different pH conditions, suggesting a role for HIF-1, but in VHL-1 mutants, where HIF-1-regulated genes are reportedly constitutively expressed, the expression of *cah-4* was still pH dependent. Therefore, it is plausible that the transcription factor, HIF-1, could play a role in *cah-4* regulation, but this is likely to be a complex interaction between multiple signaling pathways, which requires further investigation. The discovery of hypoxic regulated genes in nematodes is not surprising, considering that nematodes can penetrate deep into human tissue where oxygen levels are depleted. In fact, a CA isoform from a pathogenic nematode species has also been identified that has the potential to be regulated by many environmental factors encountered within the host environment, including pH, carbon dioxide levels, and hypoxia.[50,51]

14.4.1.3 Characterization of CAH-4 Functional analysis of both CAH-4a and CAH-4b, using a CA-deficient strain of *E. coli* that can only grow in atmospheres supplemented with a high concentration of carbon dioxide, confirmed that both isoforms are active carbonic anhydrases as *E. coli* growth was restored in air conditions upon expression of either *C. elegans* CA isoforms.

Biochemical analysis of recombinant, purified protein identified CAH-4b as the most active isoform, while CAH-4a only displayed minimal CA activity (Fig. 14.5 and Table 14.1). Extensive inhibitor screening on CAH-4b recombinant protein confirmed that its activity was inhibited by commercially available benzenesulfonamide CA inhibitors, including acetazolamide and ethoxyzolamide, while the less complex benzenesulfonamides were relatively weak CAH-4b inhibitors.[44] In addition, K_I values for CAH-4b were considerably lower than that for the human carbonic anhydrase II for some of the compounds tested (Table 14.2 and Fig. 14.6), suggesting that nematode-specific carbonic anhydrase inhibitors are not a far stretch. This thought is further promoted by the discovery that carbonic anhydrase inhibitors can be used to treat parasites that cause malaria.[52]

RNAi targeting both carbonic anhydrase isoforms failed to identify phenotypes linked with reduced transcript levels under standard laboratory conditions.[53–55] RNAi targeting *cah-4* does not result in pH sensitivity at pH 4, 7, or 9.[5,44] However, compensation activities of the other carbonic anhydrase isoforms remain unknown.

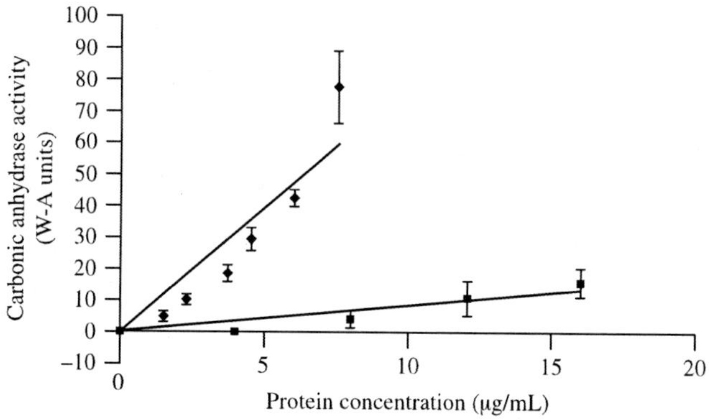

FIGURE 14.5 Carbonic anhydrase activities of CAH-4b (●) and CAH-4a (■) hexa-His tagged proteins in 10 mM HEPES pH 7.5. Error bars represent the standard deviation where $n = 3$.

14.4.1.4 Carbonic Anhydrase Inhibitor Design To identify whether carbonic anhydrase inhibitors have any effect on *C. elegans*, survival compounds that were shown *in vitro* to have high K_I constants were tested *in vivo*. However, survival rates were not significantly effective when using concentrations 30,000× the K_I value (K_I value for acetazolamide against CAH-4 is 35 nM, and 1 mM was used for *in vivo* screening). Therefore, FITC-labeled inhibitors were tested to identify permeability of the inhibitors. Preliminary results identified that the inhibitor was present in the intestinal lumen of the nematode, but it did not spread throughout the organism (Fig. 14.7). Therefore, it is probable that tested compounds have a low biological activity due to inaccessibility of the target, as previously described for *Plasmodium falciparum*.[52] As the inhibitor was identified in the intestine, it suggests that uptake of drugs is predominately through the mouth. Therefore, the thick collagen cuticle that surrounds nematodes may not pose a problem when designing therapeutic compounds. However, the polarity of the compound and the uptake mechanism from the intestine should be considered to ensure maximal biological activity is reached without affecting the potency of the drug.

TABLE 14.1 Comparison of Enzymatic Activities of the Cytoplasmic Human Carbonic Anhydrase Isoenzymes I–III (hCA I, II, and III), the Mitochondrial Carbonic Anhydrase Isoenzyme (hCA VA), and the *C. elegans* CAH-4b protein. Activity was Determined with Stopped-Flow Methods as Previously Described[60] (Data Modified from Hall et al.[44])

Isoenzyme	K_{cat} (s^{-1})	K_I (nM) (Acetazolamide)
hCA I	2.0×10^5	250
hCA II	1.4×10^6	12
hCA III	1.0×10^4	300,000
hCA VA	2.9×10^5	63
CAH-4	7.2×10^5	35

TABLE 14.2 Inhibitory Constants for Human Carbonic Anhydrase Isoenzymes I (hCA I) and the *C. elegans* CAH-4b as Determined with Stopped-Flow Methods. Claudiu T. Supuran Synthesized Compounds A–D, While Compound E is Commercially Available (Data Modified from Hall et al.[44])

Compound	K_I hCA I (nM)	K_I CAH-4b (nM)	Selectivity Ratio hCA II/CAH-4b
A 16	164	39	1.17
B 17	185	40	1.25
C 21	55	39	2.05
D 22	21,000	38	3.28
E 23	23,000	38	3.5

14.4.2 Parasitic Nematodes

For a pathogenic nematode to establish a successful infection, it must first penetrate into the host's defenses. This can be achieved by several methods, depending on the type of nematode. For example, nematodes that develop within arthropods gain direct entry into the tissues and bloodstream through the wound made by the insect, other species are ingested and gain entry into the gastrointestinal track, while the free-living nematodes actively penetrate into the skin. Once inside the host, environmental cues like pH, carbon dioxide levels, and temperature signal exsheathment[56] (shedding of the outer L_2 cuticle), which is critical for the pathogenicity of the nematode (Fig. 14.8). Incubation of *Haemonchus contortus* nematodes in media supplemented with ethoxyzolamide prevents the exsheathment process, suggesting a role for CA.[57] DeRosa et al. investigated a CA isoform homologous to *C. elegans* CAH-6 (78% homology) and hCA III (55% homology) for its involvement in the exsheathment of *Ostertagia ostertagi* nematodes that infect intestines of cattle.[51] Although DeRosa et al. did not

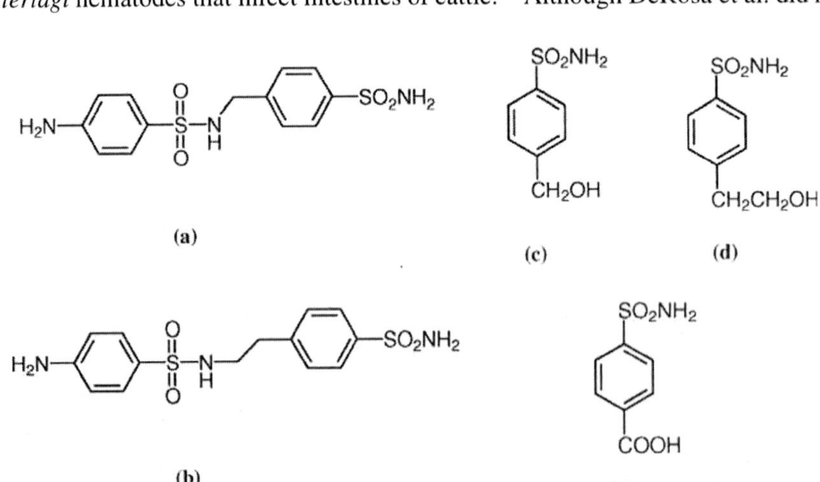

FIGURE 14.6 Structures of compounds shown to inhibit CAH-4b with an increased specificity compared to hCA II (modified from Hall et al.[44]).

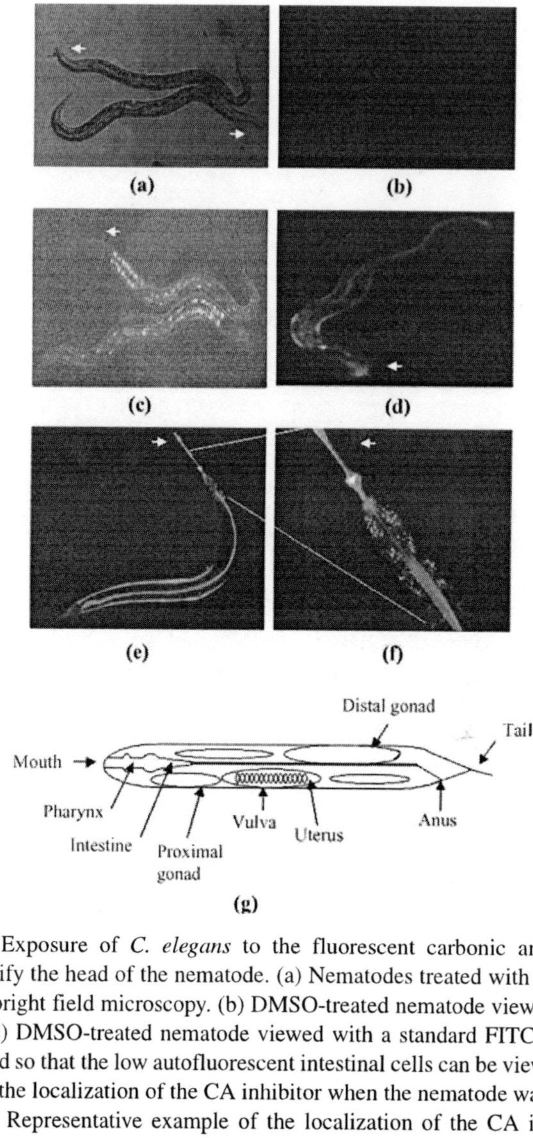

FIGURE 14.7 Exposure of *C. elegans* to the fluorescent carbonic anhydrase inhibitor. Arrowheads identify the head of the nematode. (a) Nematodes treated with equivalent volume of DMSO under bright field microscopy. (b) DMSO-treated nematode viewed with a standard FITC filter set. (c) DMSO-treated nematode viewed with a standard FITC filter set with the brightness adjusted so that the low autofluorescent intestinal cells can be viewed. (d) Representative example of the localization of the CA inhibitor when the nematode was exposed to pH 9 environments. (e) Representative example of the localization of the CA inhibitor when the nematode was exposed to pH 7 environments. (f) Enlargement of the anterior intestine section to show the localization of the CA inhibitor in a small proportion of the intestinal cells. All images were taken at 10× magnification. (g) Schematic diagram of the *C. elegans* anatomy (not to scale).

confirm that this CA isoform was involved in the process of exsheathment, the transcriptional regulation of the CA suggests that it may function in developmental processes directly after exsheathment.[51] However, with nematodes predicted to contain multiple CA isoforms, it is likely that another isoform is involved in initiating the process. Therefore, further investigations into other isoforms in *H. contortus* and other pathogenic species are clearly required.

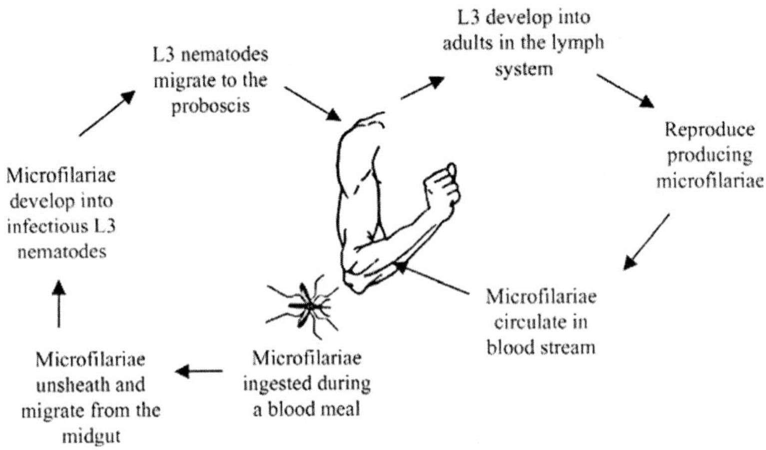

FIGURE 14.8 Life cycle of the filarial pathogen *B. malayi*. (modified from www.cdc.gov).

14.5 CARBONIC ANHYDRASE INHIBITORS AS A NEW TREATMENT OPTION AGAINST NEMATODE PARASITIC DISEASE

At present, treatment options against parasitic nematode species like *Onchocerca volvulus*, *Wuchereria bancrofti*, and *Brugia malayi* are limited. Ivermectin (Mectizan®) is the most common treatment available, especially in developing countries. Treatment consists in giving individuals doses of 150 μg/Kg twice annually. The side effects of Ivermectin include fever, loss of sight, severe pain, and swelling of limbs, all of which depend on the microfilarial (mf) load of the individual being treated[58,59] (the higher the mf load, the most severe the side effects). In addition, Ivermectin only reduces microfilarial loads, although a long-term treatment with Ivermectin has been shown to reduce fertility in adult females.[58] Therefore, treatment options that are cost-effective, as effective as Ivermectin, and have considerably less side effects are greatly sought after.

Interfering with the parasite's life cycle seems an obvious way to prevent the spread of infection, especially as the infective process is governed by many environmental signals. The discovery that carbon dioxide and pH play a role in the organism's pathogenicity and that CA inhibitors can prevent the exsheathment process suggests that CA inhibitors could be one effective solution. Before designing CA inhibitors for the treatment of pathogenic nematodes, one would have to identify the medical significance of the inhibitors in the environments where infection occurs.

14.6 CONCLUDING REMARKS

Carbonic anhydrases are widespread. Recent advances in cancer biology and environmental sensing has greatly increased the scientific interest in these important enzymes and their inhibition by small compounds. In fact, a recent review by Supuran highlights that CA inhibitors have the potential to be used as antiobesity, antimalarial, and anticancer therapies in addition to being anti-infectives (bacterial as well as fungal).[34] So far, 15 different isoforms have been identified in humans and the number of carbonic anhydrases identified in microorganisms is constantly increasing. Since low CO_2 environments are encountered by virtually all organisms, it is highly likely that CA isoforms will be identified in the majority of species. But is targeting an enzyme that is so widely distributed for therapeutic purposes a good idea? As more and more chemical libraries are being screened for compounds with the potential to inhibit CA activity and the synthesis of new compounds is also increasing, it is likely that isoform-specific CA inhibitors will be identified; in fact, some are already commercially available. For example, ethoxyzolamide was designed to be a β-CA inhibitor. But is isoform specificity enough? As discussed previously, the therapeutic action of CA inhibitors lies only in niches where the CO_2 concentration is limited, such as the skin, the eye, or where there is constant exposure to the atmosphere. These environments not only provide refuge for the pathogen that is to be treated, but also support microbes that form part of the natural flora. Therefore, although it may be possible to prevent inhibition of the human isoforms through isoform-specific inhibitors, these will not be sufficient to prevent inhibition of CA in other microbes that are likely to have the same isoform as the pathogen. This factor must be considered in the design of any therapeutic compound. For example, penicillin will be effective against different gram-positive bacteria, not just the one causing the infection. Protection of the natural flora is achieved through controlled administration of these compounds, avoiding prolonged treatments. Therefore, compounds that target a multiplicity of organisms can be used as a therapeutic as long as the course of treatment is suitably controlled.

However, despite taking these measures into consideration, CA inhibitors have been shown to successfully treat malaria, which suggests that for some organisms, CAs are potential therapeutic targets. The problem with targeting CA in fungal or bacterial species is that for some organisms, deletion of the gene does not kill the organism in low CO_2, but merely inhibits growth. Therefore, treatment with CA inhibitors alone could potentially reduce the spread of infection, but it will not kill the pathogen. In this case, CA inhibitors would be classed as having "fungal-static" rather than "fungicidal" properties. There are many bacterial static compounds, including tetracycline, chloramphenicol, and macrolides, that are routinely used to treat infections. Therefore, carbonic anhydrase inhibitors would suppress the growth of the fungus, thus allowing the immune system to combat the infection.

In conclusion, carbonic anhydrase inhibitors have the potential to be developed as therapeutic compounds against fungal infections, where the infection is contained within a niche continually exposed to the atmosphere. In the case of parasitic nematodes, carbonic anhydrases and their inhibitors are receiving renewed interest due to the recent expansion of carbonic anhydrases as therapeutic targets in fungal species. Although a relatively new field of research, there is definitely scope and potential for future studies concerning carbonic anhydrases, especially as one of the key steps in the nematodes life cycle, exsheathment, which is essential for pathogenicity, is regulated by many environmental factors that have already been shown to regulate CA activity.

REFERENCES

1. Goffeau, A.; Barrell, B. G.; Bussey, H.; Davis, R. W.; Dujon, B.; Feldmann, H.; Galibert, F.; Hoheisel, J. D.; Jacq, C.; Johnston, M.; Louis, E. J.; Mewes, H. W.; Murakami, Y.; Philippsen, P.; Tettelin, H.; Oliver, S. G. Life with 6000 genes. *Science* **1996**, *274*(5287), 546–567.
2. Cleves, A. E.; Cooper, D. N.; Barondes, S. H.; Kelly, R. B. A new pathway for protein export in *Saccharomyces cerevisiae*. *J. Cell Biol.* **1996**, *133*(5), 1017–1026.
3. Götz, R.; Gnann, A.; Zimmermann, F. K. Deletion of the carbonic anhydrase-like gene *NCE103* of the yeast *Saccharomyces cerevisiae* causes an oxygen-sensitive growth defect. *Yeast* **1999**, *15*(10A), 855–864.
4. Clark, D.; Rowlett, R. S.; Coleman, J. R.; Klessig, D. F. Complementation of the yeast deletion mutant Delta *NCE103* by members of the beta class of carbonic anhydrases is dependent on carbonic anhydrase activity rather than on antioxidant activity. *Biochem. J.* **2004**, *379*(3), 609–615.
5. Amoroso, G.; Morell-Avrahov, L.; Muller, D.; Klug, K.; Sultemeyer, D. The gene *NCE103* (YNL036w) from *Saccharomyces cerevisiae* encodes a functional carbonic anhydrase and its transcription is regulated by the concentration of inorganic carbon in the medium. *Mol. Microbiol.* **2005**, *56*(2), 549–558.
6. Merlin, C.; Masters, M.; McAteer, S.; Coulson, A. Why is carbonic anhydrase essential to *Escherichia coli*? *J. Bacteriol.* **2003**, *185*(21), 6415–6424.
7. Mitsuhashi, S.; Ohnishi, J.; Hayashi, M. M. I. A gene homologous to beta-type carbonic anhydrase is essential for the growth of *Corynebacterium glutamicum* under atmospheric conditions. *Appl. Microbiol. Biotechnol.* **2004**, *63*(5), 592–601.
8. Kusian, B.; Sultemeyer, D.; Bowien, B. Carbonic anhydrase is essential for growth of *Ralstonia eutropha* at ambient CO_2 concentrations. *J. Bacteriol.* **2002**, *184*(18), 5018–5026.
9. Aguilera, J.; Van Dijken, J. P.; De Winde, J. H.; Pronk, J. T. Carbonic anhydrase (Nce103p): an essential biosynthetic enzyme for growth of *Saccharomyces cerevisiae* at atmospheric carbon dioxide pressure. *Biochem. J.* **2005**, *391*(2), 311–316.
10. Stucka, R.; Dequin, S.; Salmon, J.; Gancedo, C. DNA sequences in chromosomes II and VII code for pyruvate carboxylase isoenzymes in *Saccharomyces cerevisiae*: analysis of pyruvate carboxylase-deficient strains. *Mol. Gen. Genet.* **1991**, *229*(2), 307–315.
11. Brewster, N. K.; Val, D. L.; Walker, M. E.; Wallace, J. C. Regulation of pyruvate carboxylase isozyme (PYC1, PYC2) gene expression in *Saccharomyces cerevisiae* during fermentative and nonfermentative growth. *Arch. Biochem. Biophys.* **1994**, *311* (1), 62–71.
12. Roggenkamp, R.; Numa, S.; Schweizer, E. Fatty acid-requiring mutant of *Saccharomyces cerevisiae* defective in acetyl-CoA carboxylase. *Proc. Natl. Acad. Sci. USA* **1980**, *77*(4), 1814–1817.
13. Gasch, A. P.; Spellman, P. T.; Kao, C. M.; Carmel-Harel, O.; Eisen, M. B.; Storz, G.; Botstein, D.; Brown, P. O. Genomic expression programs in the response of yeast cells to environmental changes. *Mol. Biol. Cell* **2000**, *11*(12), 4241–4257.
14. Gasch, A. P.; Huang, M.; Metzner, S.; Botstein, D.; Elledge, S. J.; Brown, P. O. Genomic expression responses to DNA-damaging agents and the regulatory role of the yeast ATR homolog Mec1p. *Mol. Biol. Cell* **2001**, *12*(10), 2987–3003.

15. Causton, H. C.; Ren, B.; Koh, S. S.; Harbison, C. T.; Kanin, E.; Jennings, E. G.; Lee, T. I.; True, H. L.; Lander, E. S.; Young, R. A. Remodeling of yeast genome expression in response to environmental changes. *Mol. Biol. Cell* **2001**, *12*(2), 323–337.

16. Aguilera, J.; Petit, T.; Winde, J. H.; Pronk, J. T. Physiological and genome-wide transcriptional responses of *Saccharomyces cerevisiae* to high carbon dioxide concentrations. *FEMS Yeast Res.* **2005**, *5*(6–7), 579–593.

17. Lorenz, M. C.; Fink, G. R. The glyoxylate cycle is required for fungal virulence. *Nature* **2001**, *412*, 83–86.

18. Zaragoza, O.; Fries, B. C.; Casadevall, A. Induction of capsule growth in *Cryptococcus neoformans* by mammalian serum and CO_2. *Infect. Immun.* **2003**, *71*(11), 6155–6164.

19. Granger, D. L.; Perfect, J. R.; Durack, D. T. Virulence of *Cryptococcus neoformans*. Regulation of capsule synthesis by carbon dioxide. *J. Clin. Invest.* **1985**, *76*, 508–516.

20. Bahn, Y.-S.; Cox, G. M.; Perfect, J. R.; Heitman, J. Carbonic anhydrase and CO_2 sensing during *Cryptococcus neoformans* growth, differentiation, and virulence. *Curr. Biol.* **2005**, *15*(22), 2013–2020.

21. Mogensen, E. G.; Janbon, G.; Chaloupka, J.; Steegborn, C.; Fu, M. S.; Moyrand, F.; Klengel, T.; Pearson, D. S.; Geeves, M. A.; Buck, J.; Levin, L. R.; Mühlschlegel, F. A. *Cryptococcus neoformans* senses CO_2 through the carbonic anhydrase Can2 and the adenylyl cyclase Cac1. *Eukaryot. Cell* **2006**, *5*(1), 103–111.

22. Wilbur, K. M.; Anderson, N. G. Electrometric and colorimetric determination of carbonic anhydrase. *J. Biol. Chem.* **1948**, *176*(1), 147–154.

23. Zimmerman, S.; Innocenti, A.; Casini, A.; Ferry, J. G.; Scozzafava, A.; Supuran, C. T. Carbonic anhydrase inhibitors. Inhibition of the prokaryotic beta- and gamma-class enzymes from archaea with sulfonamides. *Bio. Med. Chem. Lett.* **2004**, *14*(24), 6001–6006.

24. Klengel, T.; Liang, W.-J.; Chaloupka, J.; Ruoff, C.; Schroppel, K.; Naglik, J. R.; Eckert, S. E.; Mogensen, E. G.; Haynes, K.; Tuite, M. F.; Levin, L. R.; Buck, J.; Muhlschlegel, F. A. Fungal adenylyl cyclase integrates CO_2 sensing with cAMP signaling and virulence. *Curr. Biol.* **2005**, *15*(22), 2021–2026.

25. Odds, F. C.; *Candida and Candidosis*, 2nd ed.; Bailliere Tindall: London, **1988**.

26. Penalva, M. A.; Arst, H. N., Jr. Regulation of gene expression by ambient pH in filamentous fungi and yeasts. *Microbiol. Mol. Biol. Rev.* **2002**, *66*(3), 426–446.

27. Barnesa, R. A.; Vale, L. 'Spiking' as a rapid method for differentiation of *Candida albicans* from other yeast species. *J. Hosp. Infect.* **2005**, *60*(1), 78–80.

28. Sheth, C. C.; Johnson, E.; Baker, M. E.; Haynes, K.; Mühlschlegel, F. A. Phenotypic identification of *Candida albicans* by growth on chocolate agar. *Med. Mycol.* **2005**, *43*(8), 735–738.

29. Rocha, C. R. C.; Schroppel, K.; Harcus, D.; Marcil, A.; Dignard, D.; Taylor, B. N.; Thomas, D. Y.; Whiteway, M.; Leberer, E. Signaling through adenylyl cyclase is essential for hyphal growth and virulence in the pathogenic fungus *Candida albicans*. *Mol. Biol. Cell* **2001**, *12* (11), 3631–3643.

30. Murillo, L. A.; Newport, G.; Lan, C.-Y.; Habelitz, S.; Dungan, J.; Agabian, N. M. Genome-wide transcription profiling of the early phase of biofilm formation by *Candida albicans*. *Eukaryot. Cell* **2005**, *4*(9), 1562–1573.

31. Fidel, P. L., Jr.; Vazquez, J. A.; Sobel, J. D. *Candida glabrata:* review of epidemiology, pathogenesis, and clinical disease with comparison to *C. albicans*. *Clin. Microbiol. Rev.* **1999**, *12*(1), 80–96.

32. Wingard, J. R.; Merz, W. G.; Rinaldi, M. G.; Miller, C. B.; Karp, J. E.; Saral, R. Association of *Torulopsis glabrata* infections with fluconazole prophylaxis in neutropenic bone marrow transplant patients. *Antimicrob. Agents Chemother.* **1993**, *37*(9), 1847–1849.

33. Hitchcock, C. A.; Pye, G. W.; Troke, P. F.; Johnson, E. M.; Warnock, D. W. Fluconazole resistance in *Candida glabrata. Antimicrob. Agents Chemother.* **1993**, *37*(9), 1962–1965.

34. Supuran, C. T. Carbonic anhydrases: novel therapeutic applications for inhibitors and activators. *Nat. Rev. Drug Discov.* **2008**, *7*, 168–181.

35. Baker, B. S. The role of microorganisms in atopic dermatitis. British Society for Immunology. *Clin. Exp. Immunol.* **2006**, *144*, 1–9.

36. Fire, A.; Xu, S.; Montgomery, M. K.; Kostas, S. A.; Driver, S. E.; Mello, C. C. Potent and specific genetic interference by double-stranded RNA in *Caenorhabditis elegans. Nature* **1998**, *391*(6669), 806–811.

37. Tureci, O.; Sahin, U.; Vollmar, E.; Siemer, S.; Gottert, E.; Seitz, G.; Parkkila, A.-K.; Shah, G. N.; Grubb, J. H.; Pfreundschuh, M.; Sly, W. S. Human carbonic anhydrase XII: cDNA cloning, expression, and chromosomal localization of a carbonic anhydrase gene that is overexpressed in some renal cell cancers. *Proc. Natl. Acad. Sci. USA* **1998**, *95*(13), 7608–7613.

38. Liao, S. Y.; Aurelio, O. N.; Jan, K.; Zavada, J.; Stanbridge, E. J. Identification of the MN/CA9 protein as a reliable diagnostic biomarker of clear cell carcinoma of the kidney. *Cancer Res.* **1997**, *57*(14), 2827–2831.

39. Bui, M. H. T.; Seligson, D.; Han, K.-R.; Pantuck, A. J.; Dorey, F. J.; Huang, Y.; Horvath, S.; Leibovich, B. C.; Chopra, S.; Liao, S.-Y.; Stanbridge, E.; Lerman, M. I.; Palotie, A.; Figlin, R. A.; Belldegrun, A. S. Carbonic anhydrase IX is an independent predictor of survival in advanced renal clear cell carcinoma: implications for prognosis and therapy. *Clin. Cancer Res.* **2003**, *9*(2), 802–811.

40. Loncaster, J. A.; Harris, A. L.; Davidson, S. E.; Logue, J. P.; Hunter, R. D.; Wycoff, C. C.; Pastorek, J.; Ratcliffe, P. J.; Stratford, I. J.; West, C. M. L. Carbonic anhydrase (CA IX) expression, a potential new intrinsic marker of hypoxia: correlations with tumor oxygen measurements and prognosis in locally advanced carcinoma of the cervix. *Cancer Res.* **2001**, *61*(17), 6394–6399.

41. Thiry, A.; Dogne, J.-M.; Masereel, B.; Supuran, C. T. Targeting tumor-associated carbonic anhydrase IX in cancer therapy. *Trends Pharmacol. Sci.* **2006**, *27*, 566–573.

42. Raisanen, S. R.; Lehenkari, P.; Tasanen, M.; Rahkila, P.; Harkonen, P. L.; Vaananen, H. K. Carbonic anhydrase III protects cells from hydrogen peroxide-induced apoptosis. *FASEB J.* **1999**, *13*(3), 513–522.

43. Bahn, Y.-S.; Muhlschlegel, F. A. CO_2 sensing in fungi and beyond. *Curr. Opin. Microbiol.* **2006**, *9*(6), 572–578.

44. Hall, R. A.; Vullo, D.; Innocenti, A. A.; Scozzafava, A.; Supuran, C. T.; Klappa, P.; Mühlschlegel, F. A. External pH influences the transcriptional profile of the carbonic anhydrase, CAH-4b in *Caenorhabditis elegans. Mol. Biochem. Parasitol.* **2008**, *161*(2), 140–149.

45. Bishop, T.; Lau, K. W.; Epstein, A. C. R.; Kim, S. K.; Jiang, M.; Rourke, D.; Pugh, C. W.; Gleadle, J. M.; Taylor, M. S.; Hodgkin, J.; Ratcliffe, P. J. Genetic analysis of pathways regulated by the von Hippel–Lindau tumor suppressor in *Caenorhabditis elegans. PLoS Biol.* **2004**, *2*(10), e289.

46. Wang, G. L.; Semenza, G. L. Characterization of hypoxia-inducible factor 1 and regulation of DNA binding activity by hypoxia. *J. Biol. Chem.* **1993**, *268*(29), 21513–21518.

47. Ivan, M.; Kondo, K.; Yang, H.; Kim, W.; Valiando, J.; Ohh, M.; Salic, A.; Asara, J. M.; Lane, W. S.; Kaelin, W. G., Jr. HIF-alpha targeted for VHL-mediated destruction by proline hydroxylation: implications for O_2 sensing. *Science* **2001**, *292*(5516), 464–468.

48. Jaakkola, P.; Mole, D. R.; Tian, Y.-M.; Wilson, M. I.; Gielbert, J.; Gaskell, S. J.; Kriegsheim, A. v.; Hebestreit, H. F.; Mukherji, M.; Schofield, C. J.; Maxwell, P. H.; Pugh, C. W.; Ratcliffe, P. J. Targeting of HIF-alpha to the von Hippel–Lindau ubiquitylation complex by O_2-regulated prolyl hydroxylation. *Science* **2001**, *292*(5516), 468–472.

49. Kamura, T.; Sato, S.; Iwai, K.; Czyzyk-Krzeska, M.; Conaway, R. C.; Conaway, J. W. Activation of HIF1alpha ubiquitination by a reconstituted von Hippel–Lindau (VHL) tumor suppressor complex. *Proc. Natl. Acad. Sci. USA* **2000**, *97*(19), 10430–10435.

50. DeRosa, A. A. Isolation and characterization of carbonic anhydrase from *Ostertagia ostertagi.* PhD thesis, Louisiana State University and Agricultural Mechanical College, Louisiana, **2004**.

51. DeRosa, A. A.; Chirgwin, S. R.; Williams, J. C.; Klei, T. R. Isolation and characterization of a gene encoding carbonic anhydrase from *Ostertagia ostertagi* and quantitative measurement of expression during *in vivo* exsheathment. *Vet. Parasitol.* **2008**, *154*(1–2), 58–66.

52. Krungkrai, J.; Scozzafava, A.; Reungprapavut, S.; Krungkrai, S. R.; Rattanajak, R.; Kamchonwongpaisan, S.; Supuran, C. T. Carbonic anhydrase inhibitors. Inhibition of *Plasmodium falciparum* carbonic anhydrase with aromatic sulfonamides: towards antimalarials with a novel mechanism of action? *Bio. Med. Chem.* **2005**, *13*(2), 483–489.

53. Simmer, F.; Moorman, C.; van der Linden, A. M.; Kuijk, E.; van den Berghe, P. V. E.; Kamath, R. S.; Fraser, A. G.; Ahringer, J.; Plasterk, R. H. A. Genome-wide RNAi of *C. elegans* using the hypersensitive rrf-3 strain reveals novel gene functions. *PLoS Biol.* **2003**, *1*(1), 77–84.

54. Gonczy, P.; Echeverri, C.; Oegema, K.; Coulson, A.; Jones, S. J. M.; Copley, R. R.; Duperon, J.; Oegema, J.; Brehm, M.; Cassin, E.; Hannak, E.; Kirkham, M.; Pichler, S.; Flohrs, K.; Goessen, A.; Leidel, S.; Alleaume, A.-M.; Martin, C.; Ozlu, N.; Bork, P.; Hyman, A. A. Functional genomic analysis of cell division in *C. elegans* using RNAi of genes on chromosome III. *Nature* **2000**, *408*(6810), 331–336.

55. Kamath, R. S.; Fraser, A. G.; Dong, Y.; Poulin, G.; Durbin, R.; Gotta, M.; Kanapin, A.; Le Bot, N.; Moreno, S.; Sohrmann, M.; Welchman, D. P.; Zipperlen, P.; Ahringer, J. Systematic functional analysis of the *Caenorhabditis elegans* genome using RNAi. *Nature* **2003**, *421*(6920), 231–237.

56. Rogers, W. P. The physiology of infective processes of nematode parasites; the stimulus from the animal host. *Proc. R. Soc. Lond. B Biol. Sci.* **1960**, *152*(948), 367–386.

57. Davey, K. G.; Sommerville, R. I.; Rogers, W. P. The effect of ethoxyzolamide, an analogue of insect juvenile hormone, nor-adrenaline and iodine on changes in the optical path difference in the excretory cells and oesophagus during exsheathment in *Haemonchus contortus*. *Int. J. Parasitol.* **1982**, *12*(6), 509–513.

58. Cupp, E. W.; Duke, B. O.; Mackenzie, C. D.; Guzman, J. R.; Vieira, J. C.; Mendez-Galvan, J.; Castro, J.; Richards, F.; Sauerbrey, M.; Dominguez, A.; Eversole, R. R.; Cupp, M. S. The effects of long-term community level treatment with ivermectin (Mectizan®) on adult *Onchocerca volvulus* in Latin America. *Am. J. Trop. Med. Hyg.* **2004**, *71*(5), 602–607.

59. Kipp, W.; Bamhuhiiga, J.; Rubaale, T.; Buttner, D. W. Adverse reaction to ivermectin treatment in *Simulium neavei*-transmitted onchocerciasis. *Am. J. Trop. Med. Hyg.* **2003**, *69* (6), 621–623.

60. Khalifah, R. G. The carbon dioxide hydration activity of carbonic anhydrase. *J. Biol. Chem.* **1971**, *246*(8), 2561–2573.

CHAPTER 15

Crystallographic Studies on Carbonic Anhydrases from Fungal Pathogens for Structure-Assisted Drug Development

UTA-MARIA OHNDORF, CHRISTINE SCHLICKER, and CLEMENS STEEGBORN

Department of Physiological Chemistry, Ruhr-University Bochum, MA2/141, 44801 Bochum, Germany

15.1 CARBONIC ANHYDRASE FAMILIES AND STRUCTURES

Carbonic anhydrases (CAs, EC 4.2.1.1) are ubiquitous metalloenzymes that catalyze the reversible hydration of carbon dioxide to bicarbonate and a proton (Fig. 15.1a).[1] They contribute to a wide variety of important physiological functions such as electrolyte secretion or carbohydrate anabolism.[2] CAs are found in all domains of life, with many organisms harboring several CA isoforms.[3] The CA family is divided into five evolutionarily independent classes.[1] The α-class comprises enzymes from bacteria, algae, plant cytosols, and vertebrates. In humans, 15 α-CA isoforms (I–XV) contribute to cellular functions, such as respiration and pH regulation.[2,4] β-CAs are predominantly found in bacteria, fungi, algae, and plant chloroplasts, and γ-CAs are mainly found in archaea and in some bacteria. δ-CAs and the recently described ζ-CAs are only known in marine diatoms so far.[1,5]

Although the different CA families are unrelated in overall structure (Fig. 15.1b), the catalytic centers of CAs generally contain a tetrahedrally coordinated Zn(II) (Fig. 15.1a), with the exception of the δ-CA CDCA1, which contains a Cd(II) ion instead,[5] and γ-CAs, which have also been found with Fe(II) as active site ion.[6] The catalytic ion is coordinated by three conserved residues: Three histidines in α- and γ-class CAs, and two cysteines and one histidine in the β-class (Fig. 15.1a).[1] The fourth Zn(II) ligand binding site is occupied by a water molecule. In the catalytic cycle, the water molecule is activated through proton abstraction and formation of a metal hydroxide used for

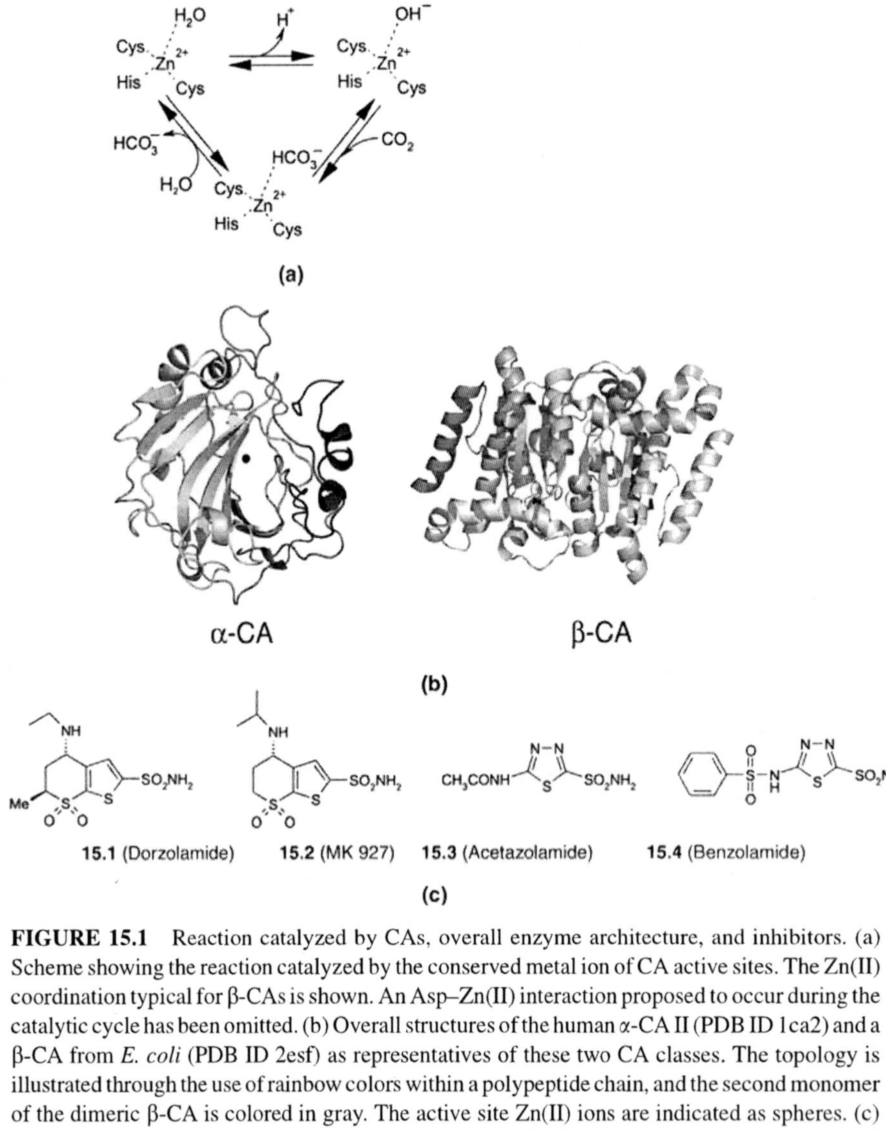

FIGURE 15.1 Reaction catalyzed by CAs, overall enzyme architecture, and inhibitors. (a) Scheme showing the reaction catalyzed by the conserved metal ion of CA active sites. The Zn(II) coordination typical for β-CAs is shown. An Asp–Zn(II) interaction proposed to occur during the catalytic cycle has been omitted. (b) Overall structures of the human α-CA II (PDB ID 1ca2) and a β-CA from *E. coli* (PDB ID 2esf) as representatives of these two CA classes. The topology is illustrated through the use of rainbow colors within a polypeptide chain, and the second monomer of the dimeric β-CA is colored in gray. The active site Zn(II) ions are indicated as spheres. (c) Known inhibitors for CA enzymes. Dorzolamide **15.1**, developed using a structure-based approach starting from MK927 **15.2**; acetazolamide **15.3**; and benzolamide **15.4**.

hydration of CO_2. At least the α-CAs, however, can catalyze additional hydrolysis reactions, for example, of esters or phosphates.[3]

The α-class is the best studied CA family.[1,3] Several crystal structures of different α-CA isoenzymes and their complexes with ligands and a wealth of biochemical studies have revealed their general architecture (Fig. 15.1b) as well as mechanistic details.[1,3,7] They form globular monomers, with a conserved histidine serving as proton shuttle during water activation. Our understanding of β-class CAs has lagged behind that for β-class enzymes. The first crystal structure for a β-class CA, from the red alga *Porphyridium pupureum*, was only reported in 2000.[8] The β-class enzymes were found not only sharing some features with α-CAs, such as the active site zinc ion, but also displaying pronounced differences. The overall structure displays two subdomains with a gap in between. By putting the smaller subdomain into this gap of a partner monomer, the β-class enzymes form dimeric building blocks (Fig. 15.1b), which then assemble further to form tetramers, hexamers, or even octamers as physiologically active forms.[1] The γ-CA "Cam" from *Methanosarcina thermophila* has yet another architecture and forms a homotrimer. Three histidine residues coordinate the active site Zn(II) ion, similar to the monomeric α-class, but these histidines are situated in two different monomers.[9]

15.2 CARBONIC ANHYDRASES AS DRUG TARGETS

15.2.1 Carbonic Anhydrase Inhibitors

CAs are well known for their "drug-ability."[4,10] Several mammalian CA isozymes are implicated in various diseases, and treatment often involves the application of sulfonamides that inhibit carbonic anhydrase activity.[4] A well-known example is dorzolamide (**15.1**; Fig. 15.1c), an α-CA inhibitor developed for the treatment of glaucomas through inhibition of CA II.[11] This compound is also the first example of drug development predominantly based on structural information. In the mid-1980s, researchers at Merck & Co. used the crystal structure of a CA complex with the thienothiopyran sulfonamide MK927 (**15.2**; Fig. 15.1c) to design the slightly modified compound **15.1**, which showed improved drug properties.[12]

Similar to many metalloenzymes, CAs are inhibited by inorganic, metal-complexing anions,[13] but the larger organic compounds with higher inhibition specificity are also known, such as dorzolamide (**15.1**), MK927 (**15.2**), acetazolamide (**15.3**), and benzolamide (**15.4**) (Fig. 15.1c).[4,10] These CA inhibitors can be described as constructs of two building blocks. A chemical moiety for interaction with the active site Zn^{2+} ion, called zinc binding group (ZBG), is combined with various chemical groups for interactions with isoenzyme-specific active site pockets, such as carbon and heterocyclic ring systems. The dominant ZBG for CA inhibitors are the sulfonamides and their bioisosteres (sulfamates, sulfamides, etc.). Examples for therapeutically used sulfonamide-based compounds are **15.1** and **15.3**.

15.2.2 Targeting β-CAs for the Development of Antifungal Drugs

The β-class enzymes seem to predominantly occur in bacterial and fungal species and have no counterparts in mammals, which exclusively display α-CAs.[1,4] The β-class architecture highly differs from that of the α-class, and phylogenetic analyses indicate a significant diversity even within the β-class.[14] Thus, targeting these enzymes with specific inhibitors should result in specific effects on the targeted microbe with little side effects on the mammalian host and its microflora. A function of β-class CAs in fungal virulence and thus the suitability of targeting these enzymes to prevent virulent growth of pathogens have recently been demonstrated for *Cryptococcus neoformans* and *Candida albicans*.[15–17] *C. neoformans* is able to colonize and to transverse into the nervous system of immunosuppressed individuals, where it can cause life-threatening meningoencephalitis.[18] The fungus experiences a dramatic raise in CO_2 concentration during transitions from the natural environment (~0.03% CO_2) to its mammalian host (~5% CO_2). This increase in CO_2 promotes biosynthesis of a polysaccharide capsule, an important *C. neoformans* virulence factor.[19] The CO_2-sensing system of *C. neoformans* includes two prominent enzymes, the carbonic anhydrase Can2 and the fungal adenylyl cyclase Cac1. Can2 catalyzes the formation of bicarbonate and a proton from CO_2 and water. The bicarbonate activates the adenylyl cyclase Cac1, which promotes capsule biosynthesis.[15,17] In an analogous manner, the CA NCE103 from *C. albicans* is part of a CO_2-sensing system, triggering filamentous growth, a major virulence attribute of this pathogen.[16] Fungal carbonic anhydrases, including Can2 and NCE103, are essential for survival of *C. neoformans* serotype A in its natural environment and for *C. albicans* infections in epithelial virulence models.[16] Consistently, cultivation of *C. neoformans* on medium supplemented with CA inhibitors prevented growth of *C. neoformans* in low CO_2 conditions typical for natural environments (given subsequently).[20] Thus, these β-CAs constitute attractive targets for development of antifungals that target horizontal transmission and treatment of superficial skin infections. These novel compounds for treatment of fungal infections are urgently needed due to an increase in the number of immunocompromised patients and a growth of antifungal resistances.[21,22]

15.3 STRUCTURAL CHARACTERIZATION OF β-CAs FROM PATHOGENIC FUNGI

15.3.1 Overall Structure of Can2

The recently solved crystal structure of Can2 led to the first fungal β-CA structure.[20] It revealed molecular details that contribute to the specific function of this group of enzymes and that could help in the development of specific inhibitors. The Can2 structure was solved at 1.34 Å by using Patterson search methods ("molecular replacement") with an *Escherichia coli* β-CA structure as a search model (Fig. 15.2). In the crystal structure, two Can2 monomers form a tightly packed dimer (Fig. 15.2b), consistent with the observation that Can2 is dimeric in solution.[17]

The overall fold of Can2 (Fig. 15.2) with N- and C-terminal subdomains is similar to other β-CAs, except for an N-terminal extension, which sticks out of the bulk protein structure. The N-terminal subdomain of the Can2 core is formed by four antiparallel α-helices (α1–4; Fig. 15.2a). The C-terminal subdomain contains a five-stranded β-sheet, with four parallel β-strands (β1–4) and β5, formed by the Can2 C-terminus, attached in antiparallel orientation. The C-terminal subdomain is packed between the N- and C-terminal domains of a partner monomer, resulting in a tight dimer interaction

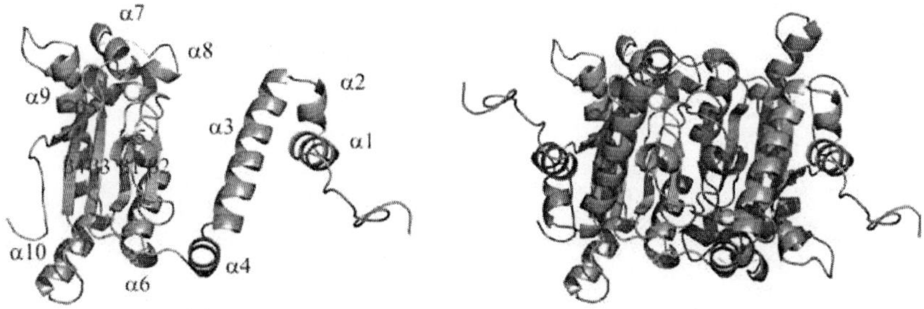

(a) (b)

FIGURE 15.2 Crystal structure of the β-class CA Can2 from *C. neoformans*. (a) Overall structure of the Can2 monomer, with labeled secondary structure elements and the active site Zn (II) as a sphere. (b) Overall structure of the active dimer of Can2, with the two monomers colored in different gray shades, respectively. The Zn(II) ions of the two symmetric active sites are shown as spheres.

(Fig. 15.2b). This overall structure resembles the known structures of the two known β-class subtypes, the "cab type" and "plant type" (given subsequently).

The known "plant type" structures comprise a CA from *Pisum sativum* (PDB entry 1ekj[23]), *Porphyridium purpureum* (PDB 1ddz[8]), *E. coli* (PDB 1i6p[24] and 1t75), *Haemophilus influenzae* (PDB 2a8c[25]), and *Mycobacterium tuberculosis* Rv3588c (PDB 1ym3[26] and 2a5v[27]). The "cab type" structures comprise a CA from *Methanobacterium thermoautotrophicum* (PDB entry 1g5c[28]) and *M. tuberculosis* Rv1284 (PDB 1ylk[26]). Can2 appears more closely related to the "plant type" β-CAs based on an overlay of their overall structures,[20] which was confirmed by analysis of its active site (given subsequently). Major differences occur, however, in the N-terminal region (given subsequently) and in the C-terminus, which is shorter than those of other "plant type" β-CAs.

15.3.2 The Can2 Active Site and Its Interaction with an N-Terminal Extension

The catalytic zinc ion of the Can2 active site, located between β1/2 and α7 (Fig. 15.2a), is coordinated by Cys68, His124, and Cys127 (Fig. 15.3a). The fourth coordination site is occupied by a water or hydroxyl molecule. The position of the fourth ligand is stabilized through a hydrogen bond to Asp70, which in turn interacts with Arg72. The three zinc ligands plus the Asp/Arg pair, which has previously been proposed to be involved in proton shuffling[1] and catalysis,[23] are the only five residues completely conserved in β-CAs. However, the "plant type" subgroup of β-CAs further contains three additional conserved active site residues corresponding to Can2 Gln59, Phe87, and Tyr109.[1,23] Their presence in Can2 confirms its classification as "plant type". The β-CAs with variable residues at these positions are called the "Cab type."[1] The two subfamilies show different susceptibility for inhibition by known CA inhibitors,[29] which can be rationalized from analysis of their active sites (given subsequently), indicating that highly specific inhibitors for individual β-CAs can be developed.

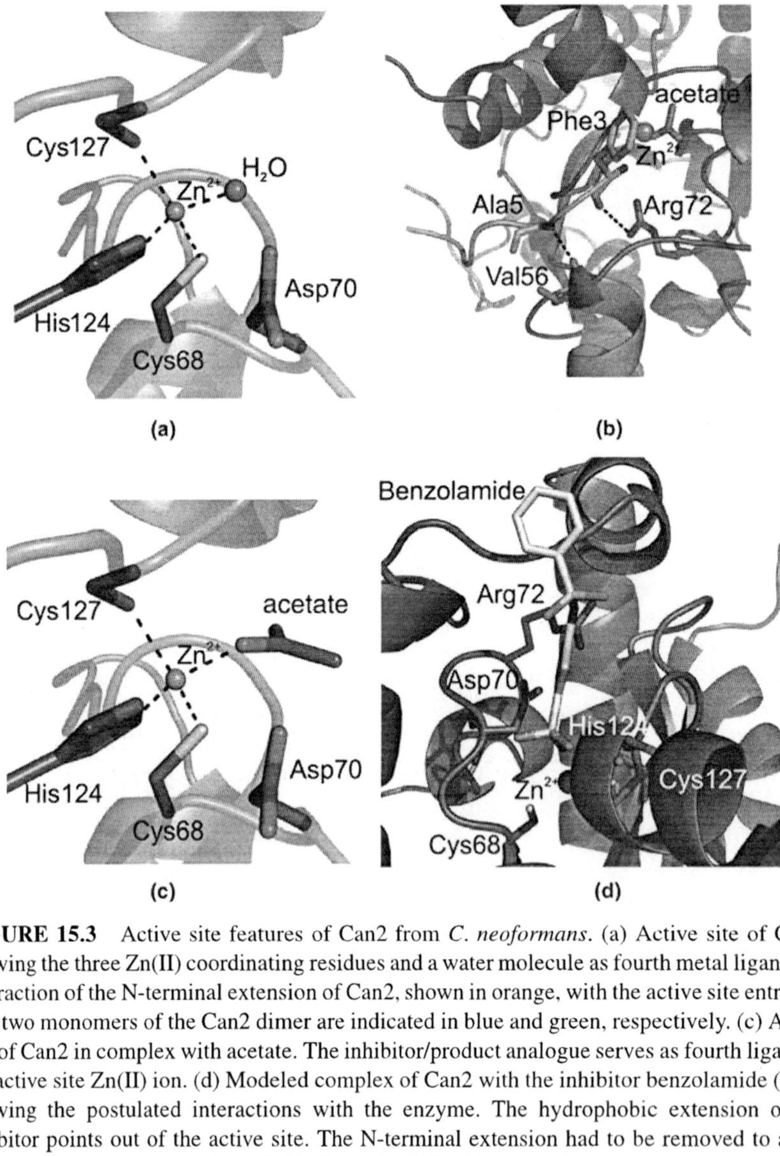

FIGURE 15.3 Active site features of Can2 from *C. neoformans*. (a) Active site of Can2, showing the three Zn(II) coordinating residues and a water molecule as fourth metal ligand. (b) Interaction of the N-terminal extension of Can2, shown in orange, with the active site entrance. The two monomers of the Can2 dimer are indicated in blue and green, respectively. (c) Active site of Can2 in complex with acetate. The inhibitor/product analogue serves as fourth ligand of the active site Zn(II) ion. (d) Modeled complex of Can2 with the inhibitor benzolamide (**15.4**) showing the postulated interactions with the enzyme. The hydrophobic extension of the inhibitor points out of the active site. The N-terminal extension had to be removed to avoid a clash. (See the color version of this figure in Color Plates section.)

The catalytic cores of Can2 and NCE103 of the fungal CO_2-sensing systems have N-terminal sequence extensions of about 26 and 63 residues, respectively, compared to most other β-CA sequences.[20] In Can2, these residues form two α-helices and an extended structure sticking out from the body of the dimer (Fig. 15.2b). The extension is oriented toward a symmetry-related dimer or folded back onto the dimer itself (in the acetate complex structure described subsequently), where it interacts with a surface groove on top of the active site (Fig. 15.3b). Interestingly, the side chain of Arg72, the conserved residue orienting the active site residue Asp70, forms a hydrogen bond to the carboxyl oxygen of Phe3 within this extension. Thus, the N-terminal extension of Can2 not only covers the active site entrance but also directly interacts with a conserved active site residue. In other β-CAs, there are various higher oligomerization states,[1] but none exploits an interaction resembling the one observed in the Can2 crystals. Considering that Can2 behaves as a dimer under all *in vitro* conditions tested,[17] it can be assumed that this interaction mode of the N-terminus within the dimer could be of physiological relevance. The interaction with active site residues, especially Arg72, could influence the enzymes activity, with the more extended conformation of the N-terminus constituting a different activation state (activated or inhibited). Interestingly, binding of potent inhibitors such as benzolamide requires a dramatic reorientation of the N-terminus (given subsequently), leading to the speculation that the N-terminus mediates a regulation mechanism or protein/protein interaction specific for this family of fungal enzymes.[20]

15.4 INHIBITION OF FUNGAL β-CAs: A STRUCTURE-ASSISTED APPROACH

15.4.1 Can2 Inhibition by Sulfonamides and Sulfamates

Catalytic activity of Can2 and its general sensitivity to inhibition by compounds of the sulfonamide/sulfamate class is similar to other β-class CAs and α-class CA enzymes.[30] Can2 inhibition tests were reported for a panel of drugs in therapeutic use as α-CA inhibitors, such as **15.1** and **15.3**, and a set of 22 simple sulfonamides widely used for the design of more potent CA inhibitors.[20] The sulfonamides/sulfamates inhibit Can2 appreciably, but their potency varies several orders of magnitude (K_I values 10.5 nM–32 μM). The highest potency was found for **15.3** ($K_I = 10.5$ nM), followed by **15.4** ($K_I = 23$ nM). Inhibitors also showing K_I values below 100 nM were aminobenzolamide, methazolamide, ethoxyzolamide, and brinzolamide, which belong to rather different classes of sulfonamides with various scaffolds.[20] Compared to the two major cytosolic isoforms in mammals hCA I and II, most compounds inhibited Can2 more effectively than hCA I but more weakly than hCA II.[10] A detectable correlation between the effects on hCA II and Can2 indicates that further efforts will be needed to obtain more specific compounds, either through further variations on sulfonamides or by introducing completely new compound classes (given subsequently).

Growth of *C. neoformans* could be inhibited *in vitro* by supplementing the medium with one of the two most potent Can2 inhibitors identified, **15.3** and **15.4**.[20] The concentrations of **15.3** or **15.4** used for growth inhibition *in vitro* (3 mM) would be too high for therapeutic purposes, while the concentrations needed for potent inhibition of the purified enzymes are several orders of magnitude lower. This discrepancy is likely to be attributed to the fungal cell wall, which constitutes a barrier for many substances. Future drug development efforts, thus, should aim not only at increasing specificity but also at identifying compounds able to transverse fungal capsules and cell walls more efficiently.

15.4.2 Can2 Complex with the Inhibitor and Product Analogue Acetate

The differences in active sites between mammalian α-CAs and Can2 should enable to refine the current inhibitors into more specific Can2-targeting drugs. To rationalize the binding mode of sulfonamides, such as **15.4**, cocrystallization with Can2 was attempted. In the structure, however, an acetate molecule from the crystallization solution bound to the Zn^{2+} ion instead of the inhibitor.[20] The acetate molecule, which is an analogue of the product bicarbonate, acts as fourth ligand of the catalytic Zn^{2+} ion and is located close to OD2 of Asp70 and NE2 of His124, and there might be a weak interaction with the conserved Gln59. The overall structure is very similar to that of the Can2 structure without ligand, except that the N-terminal extension in one dimer interacts with the dimer itself instead of a neighboring dimer (already mentioned). The conformations of most active site residues stay nearly unchanged (Fig. 15.3c), with some of the interacting residues showing shifts of about 0.5–0.7 Å toward the acetate molecule. Only the conserved Tyr109 slightly moves away from the ligand to provide space in a position similar to an acetate complex of the β-CA from *P. sativum*.[23]

The fact that acetate present in the crystallization solution competed successfully with the tight binding inhibitor **15.4** indicates that acetate, and possibly other organic carboxylates, could serve as good ligands for the Zn(II) in Can2 and possibly in the related β-CA NCE103. In fact, testing the effect of acetate on Can2 activity confirmed that it inhibits Can2 with a K_I of 10 μM.[20] In β-CAs, it has been suggested that the conserved Asp70 coordinates to the Zn(II) ion upon release of the product bicarbonate.[1,8] Existence of such an interaction could explain the surprisingly high affinity of the active site Zn(II) for acetate.

To rationalize the interactions of bulkier inhibitors of the sulfonamide family with Can2, a model of a Can2/**15.4** complex was generated.[20] The sulfonamide group was positioned based on the acetate orientation, with the NH_2 group binding to the zinc ion (Fig. 15.3d). The remaining part of the inhibitor then occupies a channel from the active site toward the solvent. This orientation would result in a clash with the N-terminus that interacts with the surface area at the entrance of this channel, indicating that this "lid" can be reoriented to allow open access to the active site (already mentioned). After removing the N-terminal "lid," the bottleneck for inhibitor binding is formed by the side chain of Tyr109 and by the backbone of Gly128 and Gly129, which have to shift slightly away to accommodate the inhibitor. The phenyl group of the inhibitor is located between Asn113 and Val114 of α6′ and the α4/β1 loop (Gly58 and Gln59). The finding that the "plant type" characteristic residues Tyr109 and Gln59 appear to contribute to inhibitor accommodation likely explains the previously observed differences in inhibitor sensitivity between the two β-CA subtypes.[29] The observations from the Can2/acetate complex and from the modeled complex with **15.4** could serve as direction for future drug development efforts, for example, in identifying residues contributing to specificity and whose dynamics have to be considered.

15.5 OUTLOOK: STRUCTURE-ASSISTED DRUG DEVELOPMENT AGAINST CAN2 AND NCE103

The role of the β-class CA enzymes Can2 and NCE103 in regulating virulence mechanisms in fungal pathogens, together with the known "drugability" of CAs and the pronounced differences of β-CAs to the human α-CA, makes these enzymes attractive targets for the development of novel antifungals. First inhibition studies have identified submicromolar inhibitors, but these compounds do not yet show sufficient specificity for these β-CAs, and they show effects against the fungal organisms only at excessive concentrations. Future drug development efforts should, therefore, not only aim at increasing the ability of compounds to discriminate between α-class and β-class CAs but also at identifying the compound classes able to efficiently transverse the fungal capsules and cell walls.

The finding from cocrystallization experiments that acetate successfully competes with **15.4** indicates that carboxylate-based inhibitors might be useful for the development of highly potent Can2 inhibitors, which is further substantiated by the high affinity to Can2 measured for this anion. Testing other simple anions[30] showed that Can2 is not very sensitive to inhibition by cyanide. Carbonate, in contrast, is a submillimolar inhibitor for Can2 and NCE103, clearly different to its effects on human CA I and CA II. These findings indicate again that even the ZBG can contribute to the discriminatory power of CA inhibitors.

The structural analysis of Can2 indicates residues that contribute to binding of the available inhibitors or are in positions to form contacts with other compounds. Furthermore, the structures indicate that some residues, such as Gln59 and Tyr108, have to shift upon ligand binding, giving hints about active site features and conformations that will be helpful for drug development efforts. Based on models of enzyme/inhibitor complexes, such as Can2/benzolamide, modified compounds can now be designed and subsequently be tested *in vitro* and *in vivo*. Furthermore, virtual ligand screening techniques (docking calculations) can be used for searching novel compounds, possibly from compound classes completely unrelated to present CA inhibitors. This approach might lead to the identification of drugs better able to transverse fungal cell walls. Structural analysis of complexes with improved compounds could then be employed for further compound refinement. Finally, future structural and biochemical studies should be directed toward better understanding the role of the Can2 N-terminus with the aim to develop more specific inhibitors by exploiting their overlapping sites of action.

Inhibition of the fungal enzyme Can2 is thus a wonderful example of a drug development effort that can clearly benefit from structural studies. The crystal structure has led to a better understanding of active site details and identification of a novel ZBG. It will help to refine ligands further and to identify novel ones, and it serves as a starting point toward a closer examination of unique Can2 features such as the extended N-terminus.

ACKNOWLEDGMENT

Work on fungal CO_2 sensing in our lab is supported by Deutsche Forschungsgemeinschaft within priority program SPP1160 (grant STE1701/2).

REFERENCES

1. Tripp, B. C.; Smith, K.; Ferry, J. G. Carbonic anhydrase: new insights for an ancient enzyme. *J. Biol. Chem.* **2001**, *276*, 48615–48618.
2. Breton, S. The cellular physiology of carbonic anhydrases. *JOP* **2001**, *2*, 159–164.
3. Supuran, C. T. Carbonic anhydrases: an overview. *Curr. Pharm. Des.* **2008**, *14*, 603–614.
4. Supuran, C. T. Carbonic anhydrases: novel therapeutic applications for inhibitors and activators. *Nat. Rev. Drug Discov.* **2008**, *7*, 168–181.
5. Xu, Y.; Feng, L.; Jeffrey, P. D.; Shi, Y.; Morel, F. M. Structure and metal exchange in the cadmium carbonic anhydrase of marine diatoms. *Nature* **2008**, *452*, 56–61.
6. Zimmerman, S. A.; Ferry, J. G.; Supuran, C. T. Inhibition of the archaeal beta-class (Cab) and gamma-class (Cam) carbonic anhydrases. *Curr. Top. Med. Chem.* **2007**, *7*, 901–908.
7. Christianson, D. W.; Cox, J. D. Catalysis by metal-activated hydroxide in zinc and manganese metalloenzymes. *Annu. Rev. Biochem.* **1999**, *68*, 33–57.
8. Mitsuhashi, S.; Mizushima, T.; Yamashita, E.; Yamamoto, M.; Kumasaka, T.; et al. X-ray structure of beta-carbonic anhydrase from the red alga, *Porphyridium purpureum*, reveals a novel catalytic site for CO_2 hydration. *J. Biol. Chem.* **2000**, *275*, 5521–5526.
9. Kisker, C.; Schindelin, H.; Alber, B. E.; Ferry, J. G.; Rees, D. C. A left-hand beta-helix revealed by the crystal structure of a carbonic anhydrase from the archaeon *Methanosarcina thermophila. EMBO J.* **1996**, *15*, 2323–2330.
10. Supuran, C. T.; Scozzafava, A.; Casini, A. Carbonic anhydrase inhibitors. *Med. Res. Rev.* **2003**, *23*, 146–189.
11. Mincione, F.; Scozzafava, A.; Supuran, C. T. The development of topically acting carbonic anhydrase inhibitors as antiglaucoma agents. *Curr. Pharm. Des.* **2008**, *14*, 649–654.
12. Greer, J.; Erickson, J. W.; Baldwin, J. J.; Varney, M. D. Application of the three-dimensional structures of protein target molecules in structure-based drug design. *J. Med. Chem.* **1994**, *37*, 1035–1054.
13. Ilies, M.; Banciu, M. Nonsulfonamide carbonic anhydrase inhibitors. *Carbonic Anhydrase: Its Inhibitors and Activators;* CRC: Boca Raton, FL, **2004**; pp 209–242.
14. Smith, K. S.; Jakubzick, C.; Whittam, T. S.; Ferry, J. G. Carbonic anhydrase is an ancient enzyme widespread in prokaryotes. *Proc. Natl. Acad. Sci. USA* **1999**, *96*, 15184–15189.
15. Bahn, Y. S.; Cox, G. M.; Perfect, J. R.; Heitman, J. Carbonic anhydrase and CO_2 sensing during *Cryptococcus neoformans* growth, differentiation, and virulence. *Curr. Biol.* **2005**, *15*, 2013–2020.

16. Klengel, T.; Liang, W. J.; Chaloupka, J.; Ruoff, C.; Schroppel, K.; et al. Fungal adenylyl cyclase integrates CO_2 sensing with cAMP signaling and virulence. *Curr. Biol.* **2005**, *15*, 2021–2026.

17. Mogensen, E. G.; Janbon, G.; Chaloupka, J.; Steegborn, C.; Fu, M. S.; et al. *Cryptococcus neoformans* senses CO_2 through the carbonic anhydrase Can2 and the adenylyl cyclase Cac1. *Eukaryot. Cell* **2006**, *5*, 103–111.

18. Lin, X.; Heitman, J. The biology of the *Cryptococcus neoformans* species complex. *Annu. Rev. Microbiol.* **2006**, *60*, 69–105.

19. Granger, D. L.; Perfect, J. R.; Durack, D. T. Virulence of *Cryptococcus neoformans*. Regulation of capsule synthesis by carbon dioxide. *J. Clin. Invest.* **1985**, *76*, 508–516.

20. Schlicker, C.; Hall, R. A.; Vullo, D.; Middelhaufe, S.; Gertz, M.; et al. Structure and inhibition of the CO_2-sensing carbonic anhydrase Can2 from the pathogenic fungus *Cryptococcus neoformans*. *J. Mol. Biol.* **2009**, *385*, 1207–1220.

21. Cannon, R. D.; Lamping, E.; Holmes, A. R.; Niimi, K.; Tanabe, K.; et al. *Candida albicans* drug resistance another way to cope with stress. *Microbiology* **2007**, *153*, 3211–3217.

22. Cowen, L. E.; Steinbach, W. J. Stress, drugs, and evolution: the role of cellular signaling in fungal drug resistance. *Eukaryot. Cell* **2008**, *7*, 747–764.

23. Kimber, M. S.; Pai, E. F. The active site architecture of *Pisum sativum* beta-carbonic anhydrase is a mirror image of that of alpha-carbonic anhydrases. *EMBO J.* **2000**, *19*, 1407–1418.

24. Cronk, J. D.; Endrizzi, J. A.; Cronk, M. R.; O'Neill, J. W.; Zhang, K. Y. Crystal structure of *E. coli* beta-carbonic anhydrase, an enzyme with an unusual pH-dependent activity. *Protein Sci.* **2001**, *10*, 911–922.

25. Cronk, J. D.; Rowlett, R. S.; Zhang, K. Y.; Tu, C.; Endrizzi, J. A.; et al. Identification of a novel noncatalytic bicarbonate binding site in eubacterial beta-carbonic anhydrase. *Biochemistry* **2006**, *45*, 4351–4361.

26. Suarez Covarrubias, A.; Larsson, A. M.; Hogbom, M.; Lindberg, J.; Bergfors, T.; et al. Structure and function of carbonic anhydrases from *Mycobacterium tuberculosis*. *J. Biol. Chem.* **2005**, *280*, 18782–18789.

27. Covarrubias, A. S.; Bergfors, T.; Jones, T. A.; Hogbom, M. Structural mechanics of the pH-dependent activity of beta-carbonic anhydrase from *Mycobacterium tuberculosis*. *J. Biol. Chem.* **2006**, *281*, 4993–4999.

28. Strop, P.; Smith, K. S.; Iverson, T. M.; Ferry, J. G.; Rees, D. C. Crystal structure of the "cab"-type beta-class carbonic anhydrase from the archaeon *Methanobacterium thermoautotrophicum*. *J. Biol. Chem.* **2001**, *276*, 10299–10305.

29. Smith, K. S.; Ferry, J. G. Prokaryotic carbonic anhydrases. *FEMS Microbiol. Rev.* **2000**, *24*, 335–366.

30. Innocenti, A.; Vullo, D.; Scozzafava, A.; Casey, J. R.; Supuran, C. Carbonic anhydrase inhibitors. Interaction of isozymes I, II, IV, V, and IX with carboxylates. *Bioorg. Med. Chem. Lett.* **2005**, *15*, 573–578.

CHAPTER 16

Malaria Parasite Carbonic Anhydrase and Its Inhibition in the Development of Novel Therapies of Malaria

JERAPAN KRUNGKRAI[1], SUDARATANA R. KRUNGKRAI[2], and CLAUDIU T. SUPURAN[3]

[1]Department of Biochemistry, Faculty of Medicine, Chulalongkorn University, 1873 Rama 4 Road, Pathumwan, Bangkok 10330, Thailand
[2]Unit of Biochemistry, Department of Medical Science, Faculty of Science, Rangsit University, Paholyothin Road, Patumthani 12000, Thailand
[3]Polo Scientifico, Laboratorio di Chimica Bioinorganica, Università degli Studi di Firenze, Room 188, Via della Lastruccia 3, 50019 Sesto Fiorentino (Florence), Italy

16.1 INTRODUCTION

Malaria, an important parasitic disease in humans, is caused by the genus *Plasmodium*, classified in phylum Apicomplexa of subkingdom protozoa.[1] The disease afflicts 515 million people and kills 1.5–2.7 million of them each year, mainly children in sub-Saharan Africa.[2–6] *P. falciparum* is responsible for the majority of deaths.[2,6] In addition to the lack of effective vector control and vaccine, the limitation and toxicity of antimalarial drugs in current use, and the spread of drug-resistant malaria parasites accompanied by a worldwide resurgence of malaria highlight the need to develop quickly more effective and less toxic new antimalarial drugs with different mechanism of action.[5,7–10] In general, drug screening procedures have rarely been applied to this disease, and there is a paucity of information on a number of metabolic pathways that can be exploited for chemotherapy.[7–9] A better understanding of biochemical differences between the parasite and the human may provide new targets for intervention in the disease.[10]

This chapter will describe the state of the art of the enzyme carbonic anhydrase (CA) in protozoa and helminthes with special emphasis on the CA of the human malaria parasite *P. falciparum*. Putative metabolic roles of the parasite CA are proposed. Aromatic sulfonamides behave as potent inhibitors, with the 4-(3,4-dichlorophenylureidoethyl)-benzenesulfonamide as the most effective inhibitor for *in vitro P. falciparum* growth. Thus, sulfonamide CA inhibitors may have the potential for the development of novel antimalarial drugs. This information provides further support that the CA has an essential role in parasite metabolism.

16.2 CARBONIC ANHYDRASE OF PROTOZOA AND HELMINTHES

The enzyme carbonic anhydrase has been identified in all organisms so far examined: animals, plants, yeast, archaea, and bacteria.[11–17] The α-CA isozymes of mammals, in particular the human and bovine, have been thoroughly investigated.[18–22] Using the bioinformatics approach[23,24], with the amino acid sequence of all available CA as query, only 10 putative α-CAs were identified in the protozoan and helminthes parasites surveyed in the database. The amino acid sequences of these parasites are not similar to those of the mosquito (*Aedes aegypti*) and human isozymes CA I, II, III, and VI. *In silico* analyses show the limited information of the putative α-CA genes in all parasite genomes, due in part to the low percentage of amino acid identity among the parasite CAs. Table 16.1 shows the 10 putative α-CAs with accession numbers deposited in the genome database, compared to the amino acid sequences of mosquito and some human CA isozymes.

In helminthes, the CA genes are identified in *Ascaris* (parasitic nematode) and *Caenorhabditis* (free-living nematode), also having very low identity between themselves. The *Ascaris* CA has the highest identity to the mosquito and human CA III (31–32%), whereas the *Caenorhabditis* CA has only 25% identity to the human CA II and III. In protozoa, CA genes are identified in *Plasmodium, Theileria, Trypanosoma, Leishmania,* and *Entamoeba*. The *Plasmodium* CA has very low identity (<25%) to the mosquito and the human CA I, II, III, and VI isozymes,[25,26] is closely related to the *Theileria* enzyme, and is distinct from the other protozoan enzymes. The predicted protein sequences of the malaria parasites, however, exhibit high amino acid identity among themselves.

16.3 CATALYSIS AND STRUCTURE OF CARBONIC ANHYDRASE

The CAs, also known as carbonate anhydrases (EC 4.2.1.1), are Zn^{2+}-metalloenzymes that catalyze the reversible hydration of CO_2 in forming the bicarbonate ion (HCO_3^-) and protons.[11]

$$CO_2 + H_2O \leftrightarrow HCO_3^- + H^+ \qquad (16.1)$$

CA was first purified from bovine red cells in 1933,[12] followed by the identification of several isozymes ubiquitously distributed in all organisms so far examined.[13–15] The CAs in protozoa and helminthes parasites are found sparsely.[25–28] Recent studies on the biochemistry and the crystal structure of CAs from various organisms reveal that they

TABLE 16.1 α-Carbonic Anhydrases in Protozoa and Helminthes Parasites (16.1–16.10), Mosquito (16.11), and Human (16.12–16.15). The Numbers are Percentage of Amino Acids Identity Between Pairs of Organisms

Organism[a]	1	2	3	4	5	6	7	8	9	10	11	12	13	14	15
P. falciparum	100	42	40	47	7	3	7	20	2	8	12	8	3	8	10
P. chabaudi	42	100	65	84	13	4	8	17	6	5	12	10	9	6	7
P. yoelii	40	65	100	85	10	10	6	26	8	8	10	19	17	16	13
P. berghei	47	84	85	100	9	5	8	24	5	2	1	7	5	6	12
Ascaris	7	13	10	9	100	15	6	24	17	4	32	27	24	31	24
Caenorhabditis	3	4	10	5	15	100	9	10	10	5	18	23	25	25	19
Entamoeba	7	8	6	8	6	9	100	6	4	20	12	3	2	3	5
Theileria	20	17	26	24	24	10	6	100	11	7	18	14	17	19	18
Trypanosoma	2	6	8	5	17	10	4	11	100	2	21	18	20	15	18
Leishmania	8	5	8	2	4	5	20	7	2	100	5	4	2	1	6
Mosquito	12	12	16	10	32	18	12	18	21	5	100	29	28	28	22
Human CAI	8	10	19	7	27	23	3	14	18	4	29	100	60	53	31
Human CA II	3	9	17	5	24	25	2	17	20	2	28	60	100	58	33
Human CA III	8	6	16	6	31	25	3	19	15	1	28	53	58	100	32
Human CA VI	10	7	13	12	24	19	5	18	18	6	22	31	33	32	100

[a]Name of organisms	Accession numbers	Number of amino acids
1 = *P. falciparum*	AAN35993	418
2 = *P. chabaudi*	CAH77957	486
3 = *P. yoelii*	XP_726574	728
4 = *P. berghei*	XP_676575	363
5 = *Ascaris suum*	BU606588	202
6 = *Caenorhabditis elegans*	AAB53054	319
7 = *Entamoeba histolytica*	EAL51300	188
8 = *Theileria annulata*	CAI74572	616
9 = *Trypanosoma brucei*	XP_828900	422
10 = *Leishmania major*	CAJ02106	236
11 = *Aedes aegypti*	AAL72625	298
12 = *H. sapiens* (CA I)	P00915	261
13 = *H. sapiens* (CA II)	P00918	260
14 = *H. sapiens* (CA III)	P07451	260
15 = *H. Sapiens* (CA VI)	P23280	308

evolved independently and are divided into four unrelated families.[29–32] These are the α-CAs (mainly in mammals, vertebrates, bacteria, algae, protozoa, and cytoplasm of plant), the β-CAs (ubiquitously in bacteria, algae, and plant chloroplast), the γ-CAs (predominantly in archaea and some bacteria), and the δ-CAs (present in marine diatoms).[18,33] In mammalian tissues, 16 α-CA isozymes and CA-related proteins have been identified where they function in diverse essential processes.[19,33–37]

The α, β, and γ families bear no significant amino acid sequence identity and exhibit structural differences, except that their active sites function with a single zinc ion (Zn^{2+}) essential for catalysis. The catalytic and inhibition mechanism of α-CAs is well established.[11,18,19] X-ray crystallographic data of the human CA II showed that Zn^{2+} is

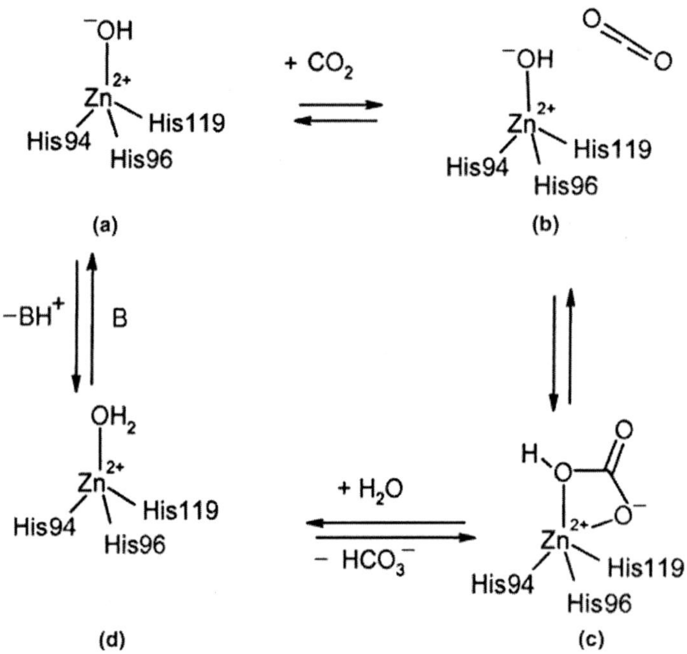

FIGURE 16.1 A catalytic mechanism for a carbonic anhydrase-catalyzed reaction of H_2O and CO_2.

located at the bottom of a 15 Å deep active site cleft, being coordinated by three histidine residues (His94, His96, and His119) and a water molecule/hydroxide ion. The histidine cluster (His4, His10, His15, His17, and His64) is also of critical importance in the catalytic cycle of the enzyme. The overall enzyme-catalyzed reaction of CA is illustrated (Fig. 16.1). The active form of the enzyme is maintained with hydroxide bound to Zn^{2+} (Fig. 16.1a). This strong nucleophile attacks the CO_2 molecule bound in a hydrophobic pocket in its neighborhood (Fig. 16.1b), leading to the formation of HCO_3^- coordinated to Zn^{2+} (Figure 16.1c). The HCO_3^- is then displaced by a water molecule and liberated into solution, leading to the acid form of the enzyme that is catalytically inactive with water coordinated to Zn^{2+} (Fig. 16.1d). The basic form **A** is then regenerated by a proton transfer reaction from the active site to its environments, for example, the active site His64.

The type of α-CA kinetic mechanism reveals a zinc hydroxide catalysis that also extends to the β- and γ-CA families.[31,38] This can be summarized as

$$E{-}Zn^{2+}{-}OH^-{+}CO_2 \leftrightarrow E\text{-}Zn^{2+}{-}HCO_3^- \tag{16.2}$$

$$E\text{-}Zn^{2+} - HCO_3^- + H_2O \leftrightarrow E\text{-}Zn^{2+} - H_2O + HCO_3^- \tag{16.3}$$

$$E\text{-}Zn^{2+} - H_2O \leftrightarrow {}^{+}H - E\text{-}Zn^{2+} - OH^- \tag{16.4}$$

$${}^{+}H\text{-}E\text{-}Zn^{2+} - OH^- \leftrightarrow E - Zn^{2+} - OH^- + H^+ \tag{16.5}$$

where E is the CA enzyme, and reaction 16.5 is the rate-limiting step in catalysis; that is, the proton transfer that regenerates the zinc hydroxide species of the enzyme.

The X-ray crystallographic structures of many adducts of the CAs with aliphatic, aromatic, or heterocyclic sulfonamides have been reported,[11,30,39,40] illustrating that the sulfonamide inhibitor is directly bound to Zn^{2+} of the enzyme through the sulfonamide moiety. The interactions between the bound inhibitor and the enzyme active site are critical for the affinity of this class of inhibitors to the different isozymes and, obviously, for the design of novel drugs exploiting the sulfonamide-based structure. Furthermore, at least 25 clinically used drugs have been reported to possess significant CA inhibitory properties.[19] Thus, CA inhibition and activation have therapeutic applications the for treatment of many human diseases, for instance, Alzheimer, glaucoma, edema, obesity, cancer, epilepsy, and osteoporosis.[18–20,33,34,38,40]

16.4 CELL BIOLOGY AND BIOCHEMISTRY OF MALARIA PARASITE

The term "malaria" disease is derived from the Italian word "mal'aria," which means bad air. It is estimated that more than 100 species of malaria parasites are known to infect numerous animal species such as reptiles, birds, and various mammals. Only four species can infect humans in over 100 countries: *P. falciparum, P. vivax, P. malariae*, and *P. ovale*.[41] The most widespread malaria parasite is *P. vivax*, but its infections are rarely fatal. *P. malariae* is found worldwide with relatively low frequency, while the least common parasite is *P. ovale*. The disease is transmitted through the bite of an infected female *Anopheles* mosquito. The life cycle of the parasite is relatively complex, having several stages of its development, in both the mosquito and its human host. In *P. falciparum*, the disease symptoms manifested during asexual development include fever, chills, prostration, severe anemia, delirium, metabolic acidosis, cerebral malaria, multiorgan system failure, coma, and death.[5,42]

By comparison to the well-characterized genomics of human host (size = 2900 Mb) and *Anopheles* vector (size = 278 Mb) (Table 16.2), the complete genome sequences of several *Plasmodium* species, including *P. vivax* and *P. falciparum*, are currently being completed and understood. The *P. falciparum* nuclear genome is composed of 22.8 Mb distributed over 14 chromosomes ranging in size from 0.64 to 3.29 Mb. The parasite nuclear genome is the most (A + T) rich sequenced to date, with an average (A + T) composition of 80.6% (reaching up to ~90% in introns and intergenic regions). The nuclear genome contains a full set of tRNA to bind all codons and several single 5S–5.8S–18S–28S rRNA genes. *P. falciparum* also contains two extrachromosomal genomes: The mitochondrial genome is a 6 kb tandemly repeated linear element with an (A + T) composition of 69%, encoding only three proteins for the subunit of its electron transport system.[42] The circular 35 kb plastid-like genome in the apicoplast organelle has an (A + T) composition of 86% and encodes for 30 known proteins (Table 16.3).

TABLE 16.2 Comparative Genomics of *P. falciparum*, *P. vivax*, Mosquito (*Anopheles*), and Human

Characteristic	*P. Falciparum*	*P. Vivax*	Mosquito	Human
Size (Mb)	22.8	24	278	2,900
(A +T) content (%)	80.6	60	65	59
Number of genes	5268	5126	14,000	31,000
Hypothetical protein (%)	60	—	—	—
Gene density (kb per gene)	4.3	4.4	—	—
Genes with introns (%)	54	—	—	>90
Percent gene coding	53	—	7	<5
Number of exons per gene	2.4	—	3.6	—
Microsattelites frequency	++	+	−	++++
SNPs (site)[a]	10,000	—	444,000	1,420,000

[a] SNPs (single nucleotide polymorphisms) of *P. falciparum* is determined only in chromosome 3 (~403 sites), and the value is based on a calculation for all chromosomes (assuming 1 SNP site per 2.3 kb).

Comparative genomics of *P. falciparum* and *Arabidopsis thaliana* shows the most similarity.[10]

Biochemical studies in *P. falciparum* have been restricted primarily to the intraerythrocytic stage of the life cycle because of the difficulty of obtaining suitable quantities of parasite material from the other life cycle stages. From the genome

TABLE 16.3 Malaria Parasites have Three Genomes: One Nuclear and Two Organellar Genomes

Genome	Size	Genes and Proteins	tRNA Genes	Protein Targeting
Nuclear DNAs	22.8 Mb	5268	43	4471 (to cytosol/ membrane)
Chromosome 1	0.64 Mb	143	0	—
2	0.95 Mb	223	1	—
3	1.06 Mb	239	2	—
4	1.20 Mb	237	5	—
5	1.34 Mb	312	5	—
6	1.38 Mb	312	3	—
7	1.35 Mb	277	7	—
8	1.32 Mb	295	0	—
9	1.54 Mb	365	0	—
10	1.69 Mb	403	0	—
11	2.04 Mb	492	2	—
12	2.27 Mb	526	3	—
13	2.75 Mb	672	5	—
14	3.29 Mb	769	2	—
Mitochondrial DNA	6 kb	3	Import	246 (to organelle)
Apicoplast DNA	35 kb	30	import	551 (to organelle)

sequence analysis, identifying essential metabolic pathways for the parasite survival, 5268 protein-coding genes are predicted. Of the predicted proteins, 733 (~14%) have been identified as enzymes involved in parasite metabolisms. During the erythrocytic stage, the parasite degrades up to 80% of the hemoglobin in the host cell for their amino acid requirement. The degradation occurs in a lysosomal food vacuole.[9] The parasite relies on anaerobic glycolysis for ATP production by obtaining glucose from the host. The essential metabolisms, as well as biochemical pathways that are being elucidated, include the apicoplast fatty acid synthesis, synthesis of heme, isoprenoid, and ubiquinone; synthesis of amino acids, folate, and other coenzymes biosynthesis; glycerolipid and glycosylphosphatidylinositol metabolism; pentose phosphate pathway; and limited mitochondrial oxygen uptake and electron transport system.[9,10,42]

Moreover, the parasite utilizes purines and pyrimidines for DNA/RNA synthesis during its exponential growth and replication. The parasite, known as a purine auxotroph, is incapable of *de novo* purine synthesis owing to the missing enzymes in the biosynthetic pathway. It salvages the preformed purine bases/nucleosides (e.g., hypoxanthine and adenosine) from the human host and converts them to their mono-, di- and triphosphates. The parasite can synthesize pyrimidines *de novo* from HCO_3^-, adenosine 5′-triphosphate (ATP), glutamine (Gln), aspartate (Asp), and 5-phosphoribosyl-1-pyrophosphate (PRPP).[43–54] These unique properties on both purine and pyrimidine requirements of the parasite are key differences from the human host, in which both functional *de novo* and salvage pathways of the purine and pyrimidine synthesis exist.[46,55–58]

16.5 MALARIA PARASITE α-CARBONIC ANHYDRASE

In 1998, Sein and Aikawa demonstrated the *in situ* CA activity for the first time in the *P. falciparum*-infected red cells by using electron microscopy and CA-specific Hanssen's stain.[27] Recently, we have demonstrated the existence of the CA enzymatic activities in *P. falciparum* and in the mouse parasite *P. berghei*.[28,59]

The *P. falciparum*-infected red cells contain CA activity about two times higher than those of normal and uninfected red cells. The three asexual developmental stages, ring (young), trophozoite (growing), and schizont (mature), show stage-dependent activity, with the specific activity of the CA enzyme increasing as the parasite matures (Table 16.4). At least three CA isozymes (PfCA1, PfCA2, and PfCA3) have been identified to date in *P. falciparum*.[28] These malarial enzymes belong to the α-CA family. The major *P. falciparum* α-CA isozyme, namely, PfCA1, which is completely inhibited by acetazolamide (AZA), has been purified and extensively characterized. In *P. berghei*, a fivefold increase in total activity of the CA enzyme in the infected, compared to the uninfected and normal, red cells has been observed. At least four isozymes (PbCA1, PbCA2, PbCa3, and PbCA4) were demonstrated in *P. berghei*.[59] The α isozymes PbCA2 and PbCA3 are major forms. All four *P. berghei* CA activities are completely inhibited by AZA. The purified and native PfCA1 enzyme has a K_i value of AZA higher than those of the human CA II and bacterial CAs. Notably, the yeast CA, the plant CA, and the mammalian CA III are peculiarly insensitive to AZA.[60–67]

TABLE 16.4 Stage-Dependent α-Carbonic Anhydrase Activity in Three Developmental Forms of *P. falciparum*, Comparing to Normal and Infected Red Cells

	Activity[a]	
Cell	Unit/10^9 Cells	Unit/mg Protein
Normal red cell	420 ± 35[b]	0.07 ± 0.01
Infected red cell[c]	750 ± 60	0.20 ± 0.03
Parasites		
Ring	85 ± 6	0.13 ± 0.01
Trophozoite	640 ± 80	0.28 ± 0.03
Schizont	1540 ± 110	0.42 ± 0.04

[a] Unit of enzyme activity is expressed as nmol/min at 37°C.
[b] Values are mean ± S.D., taken from 8–10 sample preparations.
[c] Red cells are infected with more than 90% parasitemia of mixed stages of ring (46%), trophozoite (44%), and schizont (10%).

With the completion of genome sequencing for *P. falciparum*'[10] a search for nucleotide sequences that encode CA isozymes in the malaria parasites was performed. By applying the bioinformatics approach, TBLASTN searching of the genome database using protein CA sequences obtained from other organisms yields an open reading frame similar to the α-CAs from various organisms, including human. The single copy CA gene was found in *P. falciparum* and in three species of mouse malaria parasites *P. chabaudi*, *P. yoelii*, and *P. berghei* (Fig. 16.2). The primary amino acid sequence of the *P. falciparum* gene has 47, 42, and 40% identity with *P. berghei*, *P. chabaudi*, and *P. yoelii*, respectively. High identity (>80%) was observed among the three rodent malarial parasites, *P. chabaudi*, *P. yoelii*, and *P. berghei*. Low identity (<25%) of the malarial sequences is found when compared to the mosquito *Aedes* and human CA I, II, III, and VI sequences (Table 16.1). The active site residues responsible for binding of substrate and catalysis are, nevertheless, highly conserved among the four *Plasmodium* species.[25,26]

At present, only the full-length *PfCA1* gene encoding the major form of *P. falciparum* CA has been cloned and functionally expressed in *Escherichia coli* by using a pET-15b vector.[59] The recombinant PfCA1 protein shows authenticity to the native enzyme purified from *in vitro P. falciparum* culture.[28] The kinetic parameters including K_m, k_{cat}, and K_i of the inhibitor AZA are also found to be similar between the native and recombinant enzymes (Table 16.5). The recombinant protein obtained is used for drug screening test for a mechanism-based drug design, especially for aromatic sulfonamide CA inhibitors.

FIGURE 16.2 Multiple alignments of the amino acid sequences of *P. falciparum*, *P. chabaudi*, *P. berghei*, and *P. yoelii* CAs, deduced from the continuously single open reading frame of the parasite genes. The identical amino acids and conservative replacements are shown by star and dot symbols, respectively.

```
P. yoelii           MKHIIFLSIVLCFCDNVMYNNYVERMLFELPNNITDDLNSDPIVEYKIKEKKNDNIDINK 60
P. berghei          ------------------------------------------------------------
P. chabaudi         ------------------------------------------------------------
P. falciparum       ------------------------------------------------------------

P. yoelii           DVRHWDIEINEHKDDPNIQRNIEWHDNNDGNGNNSGNNSGNNNGNNSGNNNDDNNDNDYG 120
P. berghei          ------------------------------------------------------------
P. chabaudi         ----------------NIQR---------------------------SNEGNDDNNDN--- 15
P. falciparum       ------------------------------------------------------------

P. yoelii           NDKNWEYNSNYNDEEFERQNENERDEFSLKNEVEKNSEERKERAFDESNEYADFENMNDL 180
P. berghei          -------------------NENERNGFSLKNEVEKNPEERKDTPFDEYNEYANFENMN-- 39
P. chabaudi         ----WQYHSNYNDKQSESQNENERNEFSLKNEMEKKPEEIKDTQFDKYNEYDDFENVD-- 69
P. falciparum       ---MLEMIDKYN--THFVQTTKPYYEFNVTNLT--NSKKKKKKKRENHLIGSGENMQ-- 51
                                       . :    *.:.*  ..:.:: *.    : :   . **::

P. yoelii           ENMNNIEKEKKKYFEDMQS-KYVEDNTSDGNKEYMGEMKNQQNE---------YEQNEHQ 230
P. berghei          ---NNFEKNKKKYFEDMQS-TYMEDKKNVDNKEYMDEVKNKKIE---------YQKN--- 83
P. chabaudi         ---NNFEENKRKHFEDMQS-EDMEDKKRADNKEYADWIEDKKRSDNRGYTGGMEDKNSAG 125
P. falciparum       -------KKDEKNIKDFHINDYEIDGKTIHNKENKDSFKMNKNK-----------LNDNE 93
                           ::..* ::*::      * .   ***  . .: :: .          . *.

P. yoelii           KNEYEQNEHQQNEYEQNEHQQNEYEQNEHQQNEYE-----QNEHQQNEYEQNEHQQYEQN 285
P. berghei          ------------------------------------------------------------
P. chabaudi         NRSYIDGVEDKNSASNKEYISWMDDKNSASNKGYAGWTDDKNGASNKEYAGWRDDKQIED 185
P. falciparum       ELFYMDNILSYKPNKKKLFTYSFSENEGNSEKE----------------ETLYNFKNMKN 137
                    .    ..  . .  ...       ...  ...        ..  ....       .  ..

P. yoelii           EQNEEGNIKNGMIQNNENLSFNYAKHGMDWNVGICKNGKYQSPVDLHMHTLKERELKNLS 345
P. berghei          ---EENNIKNGIIQYNDNLSFDYSKHGMDWNVGICKNGKYQSPVDLHMHTLKERELKNLS 140
P. chabaudi         HRNEENNTKSDTTQDNDNLSFDYSKQGVNWDVGVCKNGKYQSPVDLHMHTLKERELKNLS 245
P. falciparum       INSVQNNIN-KTFLYNKLKNVDYYEHGYNWDIGQCKTGKYQSPVDLPMKDLKERELKNIS 196
                      . :.* :.     *.  ..:* ::* :*::* **.********* *: ********:*

P. yoelii           DFYLNAFYDNDEYSWNNFNRPWFKGDIFYYYENLINKIIINRQNNMFKIKASNNEIIPFG 405
P. berghei          DFYLNAFYDNDEYSWNNYNKPWFKGDIFYYYENLINKIIINRQNNMFKIKASNNEIIPFG 200
P. chabaudi         DFYLNAFYDNDEYSWNNYNRPWFKGDIFYYYENLINKIIINRQNNMFKIKASNNEIIPFG 305
P. falciparum       DVYLN-LFDDDNYAWNNYNKPWMKGDFFYYYEYFIKKIVINRQNNIFQIKAARDGIIPFG 255
                    *.***:::*:*:*:***:*:**:***:***** :*:**:******:*:***:.: *****

P. yoelii           VLFTTDEPTIFYSHHINFHSPSEHTFEGSGNRRHIEMQIYHSTNEIYDYDENKWN----- 460
P. berghei          VLFTTDEPAIFYSHHINFHSPSEHTFEGSGNRRHIEMQIYHSTNEIYDYDENKWN----- 255
P. chabaudi         VLFTTDEPAIFYSHHINFHSPSEHTFEGSGNRRHIEMQIYHSTNEIYDYDESKWN----- 360
P. falciparum       VLFTTEQPAMFYADQIHFHAPSEHTFQGSGNRREIEMQIFHSTNYFYDIQDDKSKYKKKY 315
                    *****::*::**:.:*:**:******:******.*****:**** :** ::.* :

P. yoelii           GVFEKKNYKKKNNETNIQ-----HSYILTFLMNSLSNPHLGQQYTKNKKRNKRSKSLYNS 515
P. berghei          GVFGKKTYKKKNNETNIQ-----HSYILTFLMNSLSNPHLSQQYTKNKKRNKR------- 303
P. chabaudi         GILGKKKNQKKNNETNIQ-----HSYILTFLRNSLSNPHLGHQNTKNKKRNKRSKS-YNN 414
P. falciparum       GLHIYNNLKKNSKETSKKDSSRYHSYLMSFLMNSLSNEQLQNKYNK-KKRIKKMK----- 369
                    *:   :. :*:.:**. :     ***:::** ***** :* :: .*.*** *: .:  ..

P. yoelii           IRLDENGKNTKRENQYQVISITFSSAEIDKSTINNFKKLPSEKFLKTILEASQNVPVGSD 575
P. berghei          ------------KSQYQVISITFSSAEIDKSTINNFKKLPSEKFLKTILEGSQNVPVGSG 351
P. chabaudi         IQLGRNGKNTKRINQYQVISITFSSAEIDNSTINNFKKLPSEKFLKTILEGTQNIPVGSG 474
P. falciparum       -------------NQYEVISITFTSAEINASTINAFKKLPSEKFLRTIINVSSAVHVGSG 416
                                 .**:******:****: **** *********:**::  :. : ***.

P. yoelii           PKLVNLKKPLNLNSLLMMLNMKSMEFFAYHGSSTSPDCNENVHWKVAKKSLPISTETMLK 635
P. berghei          ---------------------EKMSSFIFHKKV--------------------------- 363
P. chabaudi         ---------------------KKIYSLIFHKNV--------------------------- 486
P. falciparum       ---------------------NK------------------------------------- 418

P. yoelii           FYNMLKKTTPDYNASDNDNFRALQNVQGNIHNYGRVYLIQGFPVQLLISSALTTSEDKNV 695
P. berghei          ------------------------------------------------------------
P. chabaudi         ------------------------------------------------------------
P. falciparum       ------------------------------------------------------------

P. yoelii           IENIKLAYSKSSGNYIYFNLIFLLLIFMFLQNY 728
P. berghei          ---------------------------------
P. chabaudi         ---------------------------------
P. falciparum       ---------------------------------
```

TABLE 16.5 Comparison of Kinetic and Inhibitory Constants of Human Red Cell CA II, Native and Recombinant *P. falciparum* Carbonic Anhydrases (PfCA)

Enzyme	K_m (mM)	k_{cat} (min^{-1})	K_i^{AZA} (nM)	$K_i^{SFA^a}$ (μM)
Human CA II	10.1 ± 0.8^b	74.1 ± 5.7	99 ± 6	145 ± 2
Native PfCA	3.7 ± 0.2	10.4 ± 1.2	247 ± 14	56 ± 4
Recombinant PfCA	5.6 ± 0.3	8.2 ± 1.6	315 ± 26	84 ± 10

[a] SFA = sulfanilamide (4-aminobenzenesulfonamide).
[b] The values are mean ± S.D., taken from 3–4 separate experiments of the enzyme preparations.

16.6 MALARIA PARASITE α-CARBONIC ANHYDRASE INHIBITORS

Since the pyrimidine biosynthetic pathway represents a key difference between the parasite and its human host, the pathway constitutes an important feature for the possible targeting of PfCA in the design of novel antimalarials. The PfCA catalyzes the formation of HCO_3^- as a substrate for the first enzyme of the pyrimidine pathway, carbamoylphosphate synthetase II (CPS II). Such compounds should possess a different mechanism of action compared to the presently known drugs, most of which are rather toxic and lead to the emergence of drug resistance.[5,7–9]

It is established that α-CAs are strongly inhibited by aromatic/heterocyclic sulfonamides, which bind in deprotonated state to Zn^{2+} within the enzyme active site.[11,15,19,30,40,68–70] Of note, some compounds belonging to this class, such as AZA, methazolamide, dichlorophenamide, or indisulam among others, are widely used pharmacological agents, mainly as diuretics, antiglaucoma, antiepileptics, or anticancer agents.[11,18,19,30,33,34] Indisulam is in advanced clinical trials for the treatment of solid tumors.[11,68]

A series of aromatic sulfonamides, derivatives of sulfamide, sulfanilamide, homosulfanilamide, and 4-aminoethylbenzenesulfonamide, of compounds **16.1–16.18**, has been investigated for recombinant PfCA inhibition *in vitro*. Details of the sulfonamide synthesis are beyond the scope of this chapter. Sulfonamides **16.1–16.18** have been prepared as previously reported.[71–77] Their structures are shown in Fig. 16.3.

Mainly in the search of isozyme-specific CA inhibitors (CAIs) or for the design of novel topically acting antiglaucoma agents or CAIs with potential applications as antitumor agents, a library of sulfonamides has been previously reported.[71–77] As glimpse from the structures of these derivatives, aromatic sulfonamides are investigated for the differences in the spacers between the benzenesulfonamide and the derivatized amino moieties. Compound **16.8** is one exception, which is a sulfamide Schiff's base possessing a completely different zinc binding function, and for which we have recently shown by X-ray crystallography that additional stabilization of the E-I adduct is achieved owing to the presence of the heteroatom.[72,78] Most of the investigated derivatives are Schiff's bases obtained from sulfanilamide, homosulfanilamide, or 4-aminoethylbenzene-sulfonamide and different substituted aromatic aldehydes (compounds **16.1–16.7** and **16.9–16.15**). To see whether the derivatization of the amino moiety of the starting aminosulfonamides is a critical factor for PfCA

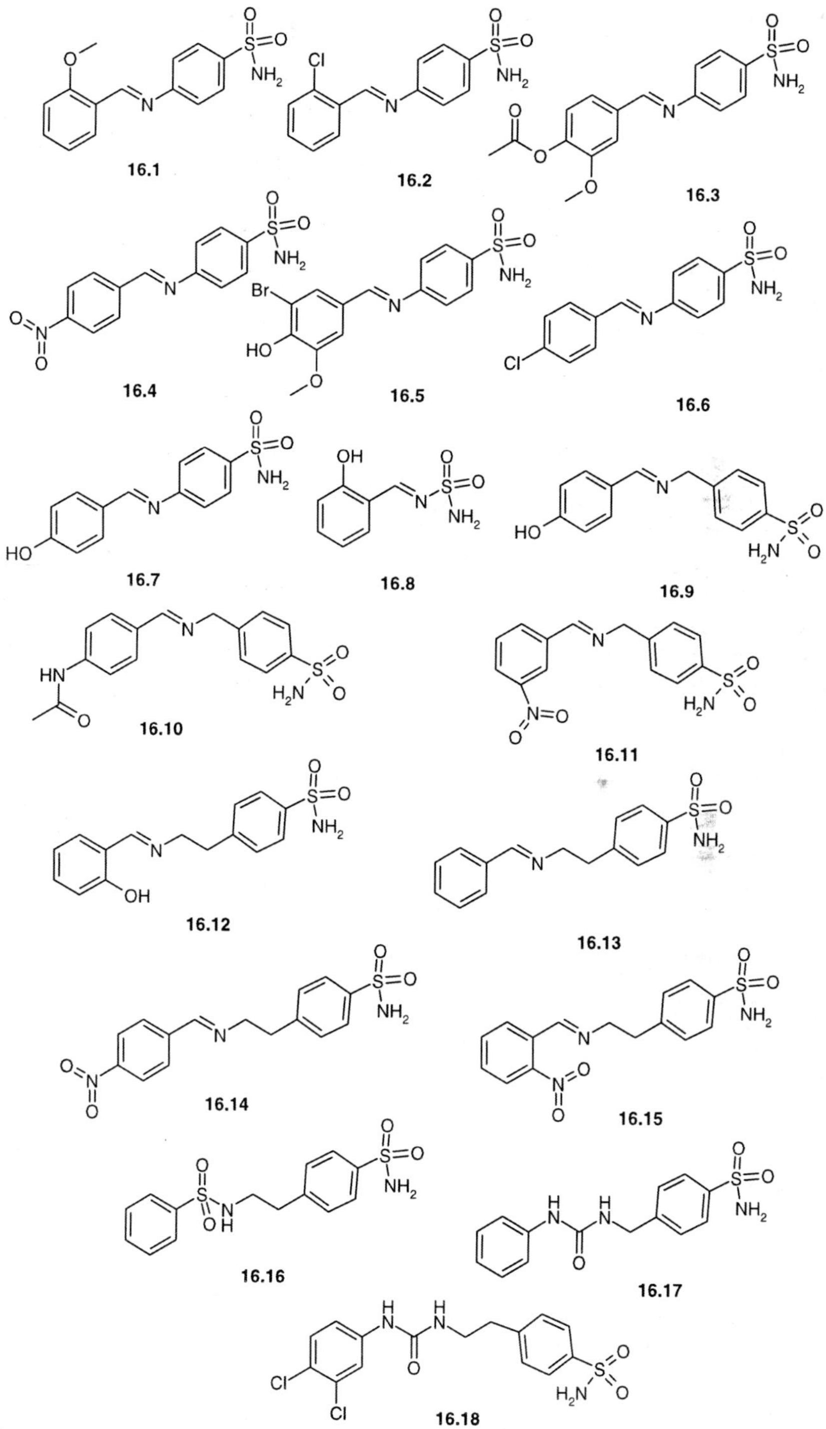

FIGURE 16.3 Structures of aromatic/heterocyclic sulfonamide carbonic anhydrase inhibitors **16.1–16.18**.

TABLE 16.6 α-Carbonic Anhydrase Inhibition Data Against Human Isozymes I and II and *P. falciparum* Enzyme PfCA with Sulfonamides 16.1–16.18 and Acetazolamide AZA

	K_i (νM)		
Inhibitor	PfCA	Human CA I	Human CA II
16.1	0.670	42	0.17
16.2	0.535	13	0.29
16.3	4.100	3	0.10
16.4	6.980	13	0.05
16.5	1.230	12	0.04
16.6	0.735	25	0.28
16.7	7.470	14	0.19
16.8	>10	13	0.04
16.9	0.620	8	0.07
16.10	>10	10	0.11
16.11	3.260	14	0.18
16.12	0.465	11	0.13
16.13	0.560	12	0.14
16.14	0.500	10	0.02
16.15	0.770	1	0.25
16.16	0.824	0.69	0.28
16.17	0.335	8	0.105
16.18	0.080	0.12	0.13
AZA	0.315	0.2	0.07

inhibition, comparisons of derivatives possessing diverse moieties than the azomethine one of the sulfonamido (compound **16.16**) or ureido type (compounds **16.17** and **16.18**) are performed.

Inhibition data against two human red cell isozymes, human CA I and CA II, as well as PfCA with sulfonamides **16.1–16.18** and AZA as standard inhibitor, are summarized in Table 16.6.[79] The following structure–activity relationship (SAR) can be summarized from the data of Table 16.6:

1. The first group of derivatives, such as compounds **16.3, 16.4, 16.7, 16.8, 16.10**, and **16.11**, acts as low potency or ineffective CA inhibitors of the malarial enzyme, with K_i against PfCA in the range of 3.2–7.4 μM for compounds **16.3, 16.4, 16.7**, and **16.11**, or even higher than 10 μM in the case of compounds **16.8** and **16.10**. With the exception of compound **16.8**, which is a sulfamide Schiff's base, the other ineffective PfCA inhibitors from this subgroup are all Schiff's bases derived from sulfanilamide/homosulfanilamide and aromatic aldehydes possessing various substituents at the aromatic moiety. Thus, the first SAR conclusion is that the nature of the group(s) substituting the aromatic ring of the aldehyde from which the Schiff's base was obtained is an important parameter for the PfCA inhibitory activity of these derivatives. It should also be stressed

that these compounds are much more potent inhibitors of the major human isozymes, that is, human CA II (K_is in the range of 0.04–0.19 μM), whereas they behave as more ineffective human CA I inhibitors (K_is in the range of 3–14 μM).

2. The second group of derivatives, such as compounds **16.1, 16.2, 16.5, 16.6, 16.9, 16.13, 16.15**, and **16.16**, act as medium potency PfCA inhibitors, with K_is in the range of 0.54–1.23 μM. Except for compound **16.16**, which is a sulfonylated aminosulfonamide, all other derivatives in this subgroup are the Schiff's bases derived from sulfanilamide/homosulfanilamide/4-aminoethyl-benzenesulfonamide, whereas the nature of the aldehyde from which they are obtained is the same as for compounds described above. Thus, the first SAR conclusion mentioned above is reinforced, being also possible to hypothesize that increasing the length of the parent sulfonamide, that is, from sulfanilamide to 4-aminoethylbenzenesulfonamide, from which the Schiff's base was obtained, seems also to be beneficial for enhancing affinity for the malarial enzyme, a situation generally also true for the other two α-CAs, that is, human CA I and II. Moieties substituting the aldehyde part of the molecule leading to enhanced PfCA inhibitory properties are 2-methoxyphenyl; 2- or 4-chlorophenyl; 2- or 4-hydroxyphenyl, and 3-methoxy-4-hydroxy-5-bromophenyl among others. It should also be mentioned that the unsubstituted, benzaldehyde derived Schiff's base **16.13** shows a good inhibitory activity. Also, comparing derivatives **16.13** and **16.16**, which are quite similar except for the chemical functionality by which the tail is attached to the sulfonamide part (i.e., Schiff's base for compound **16.13** and secondary sulfonamide for compound **16.16**), it is clear that the compound **16.13** is better for the PfCA inhibitory properties compared to the compound **16.16**.[11,15,30,40] The inhibition profile of these derivatives against the human isozymes CA I and II is somewhat similar with that of the derivatives discussed above, being more efficient human CA II inhibitors compared to their inhibition of the malarial enzyme and less effective human CA I inhibitors.
3. The third group of several derivatives, such as compounds **16.12, 16.14, 16.17, 16.18** and the clinically used drug AZA, shows much more effective PfCA inhibitory properties, with K_is in the range of 80 nM–0.50 μM. Two of these derivatives are 4-aminoethylbenzenesulfonamide derived Schiff's bases, two are ureido derivatives of homosulfanilamide/4-aminoethylbenzenesulfonamide, whereas AZA is the only heterocyclic sulfonamide. It is interesting to note that the most effective PfCA inhibitor is the urea derivative **16.18**, which is almost four times more effective than the clinically used drug AZA, being at the same time a rather efficient human CA I and II inhibitor, with K_is in the range of 120–130 nM. AZA, on the other hand, is a stronger human CA II inhibitor ($K_i = 70$ nM) and a less effective human CA I inhibitor ($K_i = 200$ nM) compared to the compound **16.18**. Thus, from the small library of derivatives, it is apparent that Schiff's bases and urea-based aromatic sulfonamides, mainly derived from homosulfanilamide/4-aminoethyl-benzenesulfonamide, lead to potent PfCA inhibitors, and they may also appreciably inhibit the human red cell isozymes

CA I and II. The nature of the groups substituting the aromatic-ureido or aromatic-azomethine fragment of the molecule is a critical parameter for the CA inhibitory activities of these aromatic sulfonamide derivatives, both against the malaria and against the human enzymes.

16.7 ANTIMALARIAL PROPERTIES OF α-CARBONIC ANHYDRASE INHIBITORS

By using [^{3}H]hypoxanthine incorporation for monitoring the growth of *P. falciparum* in *in vitro* culture[45,80] (with a starting parasitemia of 0.5%, mixed stages), the 50% inhibitory concentration (IC_{50}) for AZA is 20 μM. Based on morphology examination, the effect of AZA was more pronounced in the ring/trophozoite forms than the schizont stage of *P. falciparum*, as shown by clumping of nucleus and collapsing cytoplasm (Fig. 16.4). This is consistent with the stage-dependent activity of the enzyme in that more maturing parasites contain higher activity. Pretreatment of the human red cell with AZA, which totally abolished the host enzyme activity, showed no pronounced

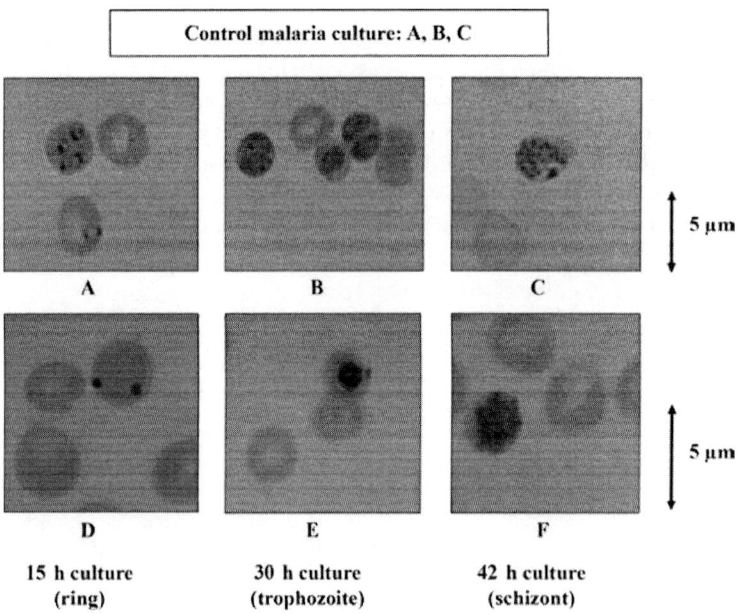

FIGURE 16.4 Effect of acetazolamide on *P. falciparum* morphology during an intraerythrocytic cycle (ring, trophozoite, and schizont stages). The morphological changes were examined in the absence (panels A, B, and C; control) or in the presence of 100 μM AZA (panels D, E, and F; AZA-treated culture) at various times of the parasite culture starting with ring stage.

effect on the parasite invasion. It is then concluded that the CAI directly affects the parasite carbonic anhydrase and eventually leads to death of the parasite in the host red cell.

With the three groups of sulfonamides, the compounds **16.1–16.17** are ineffective for *P. falciparum* growth inhibition ($IC_{50} > 50$ μM). The only promising compound appears to be ureido-sulfonamide derivative **16.18**. While AZA has medium potency efficiency for the inhibition of *P. falciparum,* the compound **16.18** on the other hand is 10 times as effective an inhibitor, with an IC_{50} of 2 μM. The enhanced efficacy of this compound compared to AZA may be explained by the fact that the compound **16.18** is a much more liposoluble derivative than AZA, and thus its penetration through membranes is facilitated. AZA, which is a very polar molecule, is expected to have some difficulty in crossing biological membranes, and this may explain the 10 times lower activity, although the difference in inhibition constants between the two derivatives is only fourfold. The antimalarial effect of both compounds has also been examined based on the morphological differences in the different life stages (Fig. 16.5). Significant inhibition of the parasite growth, as evidenced by decreasing % parasitemia, is observed in the culture when the compound **16.18** is presented at 5 μM.

These recent studies on the CA inhibitor affecting the *P. falciparum* growth indicates the potential use of sulfonamide CA inhibitors for the treatment of malaria. This also provides that antimalarial drugs possessing a novel mechanism of action can be obtained by inhibiting a critical enzyme in a metabolism for the life cycle of the malaria parasite.

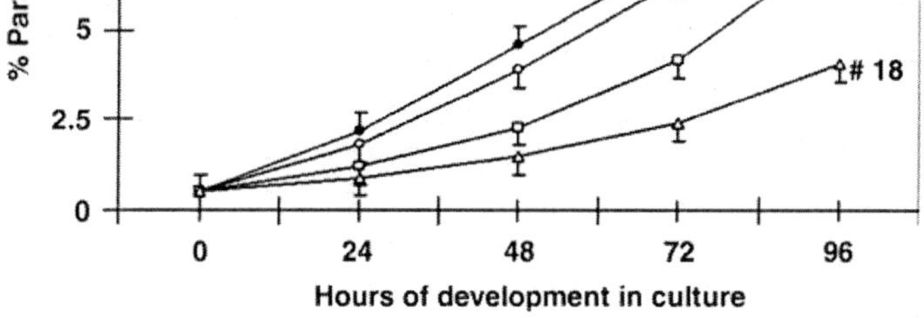

FIGURE 16.5 Antimalarial activity of *P. falciparum* carbonic anhydrase inhibitors, 4-aminobenzenesulfonamide (SFA), acetazolamide, and 4-(3,4-dichlorophenyl-ureidoethyl) benzenesulfonamide (#18). *P. falciparum* culture was started with 0.5% parasitemia at 2.5% red cell suspension and monitored every 24 h for 96 h at 37°C. The parasite was grown in the absence of inhibitor as control (●). 50 μM SFA (○), 20 μM AZA (□), or 5 μM (Δ) #18 was presented during the 96 h growth.

16.8 METABOLIC ROLES OF MALARIA PARASITE α-CARBONIC ANHYDRASE

Human red cells infected with *P. falciparum* have more CA enzymatic activity than do the uninfected red cells, with CA activity increasing parallel with other metabolic activities as the parasite matures from ring to schizont stages.[28] This is expected as *P. falciparum* CA is involved in the first biosynthetic step leading to pyrimidines, that is, the synthesis of carbamoylphosphate from Gln, ATP, and HCO_3^- catalyzed by enzyme CPS II (Fig. 16.6).[25,26] It is well-known that in this reaction HCO_3^-, and not CO_2, is the true substrate, and thus the interconversion between these two species, catalyzed by the parasite CA, is critical for the entire metabolic pathway leading to pyrimidine *de novo* biosynthesis, and they are responsible for the synthesis of DNA/RNA and lipid.

The lines of evidence have suggested that the putative function of CA provides HCO_3^- for CPS II in the pyrimidine biosynthesis, phosphoenolpyruvate carboxykinase,

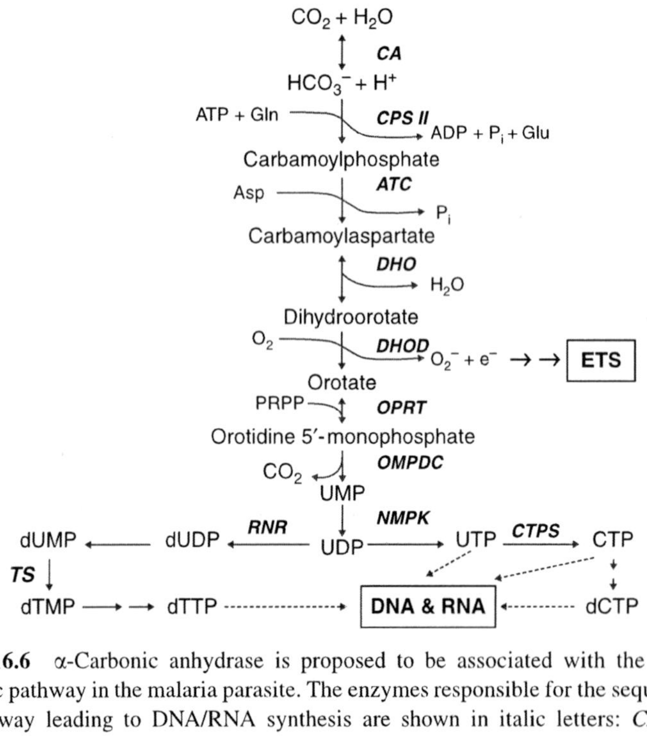

FIGURE 16.6 α-Carbonic anhydrase is proposed to be associated with the pyrimidine biosynthetic pathway in the malaria parasite. The enzymes responsible for the sequential steps in the pathway leading to DNA/RNA synthesis are shown in italic letters: *CA*, carbonic anhydrase; *CPS II*, carbamoylphosphate synthetase II; *ATC*, aspartate transcarbamoylase; *DHO*, dihydroorotase; *DHOD*, dihydroorotate dehydrogenase; *OPRT*, orotate phosphoribosyltransferase; *OMPDC*, orotidine 5′-monophosphate decarboxylase; *NMPK*, nucleoside monophosphate kinase; *RNR*, ribonucleotide reductase; *TS*, thymidylate synthetase; *CTPS*, cytidine 5′-triphosphate synthetase. The mitochondrial electron transport system (ETS) is linked to the enzyme DHOD of the pyrimidine pathway, functioning as electron disposal.

phosphoenolpyruvate carboxylase, and acetyl CoA carboxylase for fatty acid synthesis.[16,17,19,81,82] In addition, the enzyme may play an additional role in HCO_3^- transport across cell membrane [16,35] and maintain intracellular pH in the malaria parasite.[56,57,83] Intracellular pH regulation is necessary for the balance of electrolytes and transport of Na^+, K^+, and H^+ in the malaria parasite during its development in the host red cells. Furthermore, *P. falciparum* has been shown to be capable of CO_2 fixing activity.[84] It may be possible that the enzyme in the malaria parasite may be responsible for this function. At least it plays a role in CO_2/HCO_3^- equilibration of all involved metabolisms.

16.9 CONCLUSION AND PERSPECTIVE

In both protozoa and helminthes, it is expected that each should have their own CA activity, as demonstrated in all species of mammals, plants, archaea, and bacteria examined so far. In malaria parasites, there are at least four species known to contain putative genes encoding the α-CA family. One *P. falciparum* gene has been cloned and functionally expressed in *E. coli*. The recombinant enzyme is catalytically active and has authentic properties similar to the wild-type enzyme purified directly from *P. falciparum* cultures. A series of aromatic sulfonamides were tested for inhibition to the malarial enzyme, with the most potent such derivatives identified to be the clinically used sulfonamide CA inhibitor acetazolamide and the lipophilic 4-(3,4-dichlorophenylureidoethyl)benzenesulfonamide. This is the first evidence proving that antimalarials possessing a novel mechanism of action can be obtained by inhibiting a critical enzyme in the life cycle of the parasite. Indeed, by inhibiting malarial CA, the synthesis of pyrimidines mediated by carbamoylphosphate synthetase II is impaired in *P. falciparum* but not in the host, since the human host is able to obtain pyrimidines by the salvage pathway. Sulfonamide CA inhibitors may have the potential for the development of novel antimalarial drugs. Compounds with potent CA inhibitory properties should be further investigated. In addition, such compounds must possess a balanced lipo/hydrosolubility to achieve a good bioavailability. Work is still in progress in our laboratories for evaluating even more potent malarial CA inhibitors and testing antimalarial activities both *in vitro* and *in vivo*. Furthermore, the evolutionary relationship of the parasite enzymes to other organisms is still unclear. Finally, the functional roles of the enzyme in parasite metabolism need to be further investigated. Our ultimate goal is the elucidation of the 3D structure of the parasite CA for rationale drug design, lending further insights into its differences from the equivalent enzyme in human.

ACKNOWLEDGMENTS

This study was supported by the UNDP/World Bank/WHO Special Programme for Research and Training in Tropical Diseases, the National Science and Technology Development Agency of Thailand, and the Thailand Research Fund.

REFERENCES

1. Cox, F. E. G. *Modern Parasitology;* Blackwell Scientific Publications: Oxford, **1982**; pp. 1–346.
2. Marsh, K. Malaria disaster in Africa. *Lancet* **1998**, *352*, 924–925.
3. Guerin, P. J.; Olliaro, P.; Nosten, F.; Druilhe, P.; Laxminarayan, R.; Blinka, F.; Kilama, W. L.; Ford, N.; White, N. J. Malaria: current status of control, diagnosis, treatment, and a proposed agenda for research and development. *Lancet Infect. Dis.* **2002**, *2*, 564–573.
4. Attaran, A.; Barnes, K. I.; Curtis, C.; d'Alessandro, U.; Fanello, C. I.; Galinski, M. R.; Kokwaro, G.; Loareesuwan, S.; Makanga, M.; Mutabingwa, T. K.; Talisuna, A.; Trape, J. F.; Watkins, W. M. WHO, the Global Fund, and medical malpractice in malaria treatment. *Lancet* **2004**, *36*, 237–240.
5. White, N. J. Antimalarial drug resistance. *J. Clin. Invest.* **2004**, *113*, 1084–1092.
6. Snow, R. W.; Guerra, C. A.; Noor, A. M.; Myint, H. Y.; Hay, S. I. The global distribution of clinical episodes of *Plasmodium falciparum* malaria. *Nature* **2005**, *434*, 214–217.
7. Pink, R.; Hudson, A.; Mouries, M.-A.; Bendig, M. Opportunities and challenges in antiparasitic drug discovery. *Nat. Rev. Drug Discov.* **2005**, *4*, 727–740.
8. Hopkins, A. L.; Witty, M. J.; Nwaka, S. Mission possible. *Nature* **2007**, *449*, 166–169.
9. Ridley, R. G. Medical need, scientific opportunity and the drive for antimalarial drugs. *Nature* **2002**, *415*, 686–693.
10. Gardner, M. J.; Hall, N.; Fung, E.; White, O.; Berriman, M.; Hyman, R. W.; Carlton, J. M.; Pain, A.; Nelson, K. E.; Bowman, S; Paulsen, I. T.; James, K; Eisen, J. A.; Rutherford, K.; Salzberg, S. L.; Craig, A.; Kyes, S.; Chan, M.; Nene, V.; Shallom, S. J.; Suh, B.; Peterson, J.; Angiuoli, S.; Pertea, M.; Allen, J.; Selengut, J.; Haft, D.; Mather, M. W.; Vaidya, A. B.; Martin, D. M. A.; Fairlamb, A. H.; Fraunholz, M. J.; Roos, D. S.; Ralph, S. A.; McFadden, G.; Cummings, L. M.; Subramanian, G. M.; Mungall, C.; Venter, J. C.; Carucci, D. J.; Hoffman, S. L.; Newbold, C.; Davis, R. W.; Fraser, C. M.; Barrell, B. Genome sequence of the human malaria parasite *Plasmodium falciparum. Nature* **2002**, *419*, 498–511.
11. Supuran, C. T.; Scozzafava, A.; Conway, J. *Carbonic Anhydrase: Its Inhibitors and Activators;* CRC Press: Boca Raton, FL, **2004**; pp. 1–363.
12. Meldrum, N. U.; Roughton, F. J. W. Carbonic anhydrase: its preparation and properties. *J. Physiol.* **1933**, *80*, 113–142.
13. Tashian, R. E. Genetic variation and evolution of the carboxylic esterases and carbonic anhydrases of primate erythrocytes. *Am. J. Hum. Genet.* **1965**, *17*, 257–272.
14. Smith, K. S.; Jakubzick, C.; Whittam, T. S.; Ferry, J. G. Carbonic anhydrase is an ancient enzyme widespread in prokaryotes. *Proc. Natl. Acad. Sci. USA* **1999**, *96*, 15184–15189.
15. Winum, J. Y.; Scozzafava, A.; Montero, J. L.; Supuran, C. T. Sulfamates and their therapeutic potential. *Med. Res. Rev.* **2005**, *25*, 186–228.
16. Fujikawa, A. K.; Nishimori, I.; Taguchi, T.; Onishi, S. Human carbonic anhydrase XIV (CA14): cDNA cloning, mRNA expression, and mapping to chromosome 1. *Genomics* **1999**, *61*, 74–81.
17. Lindskog, S. Carbonic anhydrase. *Adv. Inorg. Biochem.* **1982**, *4*, 115–170.
18. Supuran, C. T.; Scozzafava, A. Carbonic anhydrases as targets for medicinal chemistry. *Bioorg. Med. Chem.* **2007**, *15*, 4336–4350.

19. Supuran, C. T. Carbonic anhydrases: novel therapeutic applications for inhibitors and activators. *Nat. Rev. Drug Discov.* **2008**, *7*, 168–181.

20. Boriack-Sjodin, P. A.; Zeitlin, S.; Chen, H. H.; Crenshaw, L.; Gross, S.; Dantanarayana, A.; Delgado, P.; May, J. A.; Dean, T.; Christianson, D. W. Structural analysis of inhibitor binding to human carbonic anhydrase II. *Protein Sci.* **1998**, *7*, 2483–2489.

21. Hunt, J. A.; Ahmed, M.; Fierke, C. A. Metal binding specificity in carbonic anhydrase is influenced by conserved hydrophobic core residues. *Biochemistry* **1999**, *38*, 9054–9062.

22. Fisher, Z.; Prada, J. A. H.; Tu, C.; Duda, D.; Yoshioka, C.; An, H.; Govindasamy, L.; Silverman, D. N.; McKenna, R. Structural and kinetic characterization of active-site histidine as a proton shuttle in catalysis by human carbonic anhydrase II. *Biochemistry* **2005**, *44*, 1097–1105.

23. Altschul, S. F.; Madden, T. L.; Schaffer, A. A.; Zang, J.; Zang, Z.; Miller, W.; Lipman, D. J. Gapped BLAST and PSI-BLAST: a new generation of protein database search programs. *Nucleic Acids Res.* **1997**, *25*, 3389–3402.

24. Thompson, J. D.; Higgins, D. G.; Gibson, T. J. CLUSTALW: improving the sensitivity of progressive multiple sequence alignment through sequence weighting, position-specific gap penalties and weight matrix choice. *Nucleic Acids Res.* **1994**, *22*, 4673–4680.

25. Krungkrai, J.; Krungkrai, S. R.; Supuran, C. T. Malarial parasite carbonic anhydrase and its inhibitors. *Curr. Top. Med. Chem.* **2007**, *7*, 909–917.

26. Krungkrai, J.; Supuran, C. T. The alpha-carbonic anhydrase from the malaria parasite and its inhibition. *Curr. Pharm. Des.* **2008**, *14*, 631–640.

27. Sein, K. K.; Aikawa, M. The pivotal role of carbonic anhydrase in malaria infection. *Med. Hypotheses* **1998**, *5*, 9–23.

28. Krungkrai, S. R.; Suraveratum, N.; Rochanakij, S.; Krungkrai, J. Characterisation of carbonic anhydrase in *Plasmodium falciparum*. *Int. J. Parasitol.* **2001**, *31*, 661–668.

29. Tripp, B. C.; Smith, K.; Ferry, J. G. Carbonic anhydrase: new insights for an ancient enzyme. *J. Biol. Chem.* **2001**, *276*, 48615–48618.

30. Scozzafava, A.; Mastrolorenzo, A.; Supuran, C. T. Modulation of carbonic anhydrase activity and its applications in therapy. *Expert Opin. Ther. Pat.* **2004**, *14*, 667–702.

31. Tripp, B. C.; Bell, C. B., III; Cruz, F.; Krebs, C.; Ferry, J. G. A role of iron in an ancient carbonic anhydrase. *J. Biol. Chem.* **2004**, *279*, 6883–6887.

32. Liljas, A.; Laurberg, M. A wheel invented three times: the molecular structures of the three carbonic anhydrases. *EMBO Rep.* **2000**, *1*, 16–17.

33. Supuran, C. T. Carbonic anhydrase as drug targets: an overview. *Curr. Top. Med. Chem.* **2007**, *7*, 825–833.

34. Cupuran, C. T.; Mastrolorenzo, A.; Scozzafava, A. Carbonic anhydrase inhibitors and activators and their use in therapy. *Expert Opin. Ther. Pat.* **2006**, *16*, 1627–1664.

35. Mizumori, M.; Meyerowitz, J.; Takeuchi, T.; Lim, S.; Lee, P.; Supuran, C. T.; Guth, P. H.; Engel, E.; Kaunitz, J. D.; Akiba, Y. Epithelial carbonic anhydrases facilitate pCO_2 and pH regulation in rat duodenal mucosa. *J. Physiol.* **2006**, *573*, 827–842.

36. Shah, G. N.; Ulmasov, B.; Waheed, A.; Becker, T.; Makani, S.; Svichar, N.; Chesler, M.; Sly, W. S. Carbonic anhydrase IV and XIV knockout mice: roles of the respective carbonic anhydrases in buffering the extracellular space in brain. *Proc. Natl. Acad. Sci. USA* **2005**, *102*, 16771–16776.

37. Kivela, A.; Kivela, J.; Saarnio, J.; Parkkila, S. Carbonic anhydrases in normal gastrointestinal tract and gastrointestinal tumors. *World J. Gastroenterol.* **2005**, *11*, 155–163.
38. Lindskog, S. Structure and mechanism of carbonic anhydrase. *Pharmacol. Ther.* **1997**, *74*, 1–20.
39. Chakravarty, S.; Kannan, K. K. Drug–protein interactions: refined structures of three sulfonamide drug complexes of human carbonic anhydrase I enzyme. *J. Mol. Biol.* **1994**, *243*, 298–309.
40. Supuran, C. T.; Scozzafava, A.; Casini, A. Carbonic anhydrase inhibitors. *Med. Res. Rev.* **2003**, *23*, 146–189.
41. Hyde, J. E. Drug-resistant malaria. *FEBS J.* **2007**, *274*, 4688–4698.
42. Krungkrai, J.; Kanchanaphum, P.; Pongsabut, S.; Krungkrai, S. R. Putative metabolic roles of the mitochondria in asexual blood stages and gametocytes of *Plasmodium falciparum*. *Southeast Asian J. Trop. Med. Public Health* **2008**, *1*, 31–49.
43. Krungkrai, J.; Cerami, A.; Henderson, G. B. Pyrimidine biosynthesis in parasitic protozoa: purification of a monofunctional dihydroorotase from *Plasmodium berghei* and *Crithidia fasciculata*. *Biochemistry* **1990**, *29*, 6270–6275.
44. Krungkrai, J.; Cerami, A.; Henderson, G. B. Purification and characterization of dihydroorotate dehydrogenase from the rodent malaria parasite *Plasmodium berghei*. *Biochemistry* **1991**, *30*, 1934–1939.
45. Krungkrai, J.; Krungkrai, S. R.; Phakanont, K. Antimalarial activity of orotate analogs that inhibit dihydroorotase and dihydroorotate dehydrogenase. *Biochem. Pharmacol.* **1992**, *43*, 1295–1301.
46. Krungkrai, J. Dihydroorotase and dihydroorotate dehydrogenase as a target for antimalarial drugs. *Drugs Future* **1993**, *18*, 441–450.
47. Krungkrai, J. Purification, characterization and localization of mitochondrial dihydroorotate dehydrogenase in *Plasmodium falciparum*, human malaria parasite. *Biochim. Biophys. Acta* **1995**, *1243*, 351–360.
48. Krungkrai, J.; Prapunwattana, P.; Wichitkul, C.; Reungprapavut, S.; Krungkrai, S. R.; Horii, T. Molecular biology and biochemistry of malarial parasite pyrimidine biosynthetic pathway. *Southeast Asian J. Trop. Med. Public Health* **2003**, *34* (Suppl. 2), 32–43.
49. Krungkrai, S. R.; Aoki, S.; Palacpac, N. M. Q.; Sato, D.; Mitamura, T.; Krungkrai, J.; Horii, T. Human malaria parasite orotate phosphoribosyltransferase: functional expression, characterization of kinetic reaction mechanism and inhibition profile. *Mol. Biochem. Parasitol.* **2004**, *134*, 245–255.
50. Krungkrai, S. R.; Prapunwattana, P.; Horii, T.; Krungkrai, J. Orotate phosphoribosyltransferase and orotidine 5′-monophosphate decarboxylase exist as multienzyme complex in human malaria parasite *Plasmodium falciparum*. *Biochem. Biophys. Res. Commun.* **2004**, *318*, 1012–1018.
51. Krungkrai, S. R.; DelFraino, B. J.; Smiley, J. A.; Prapunwattana, P.; Mitamura, M.; Horii, T.; Krungkrai, J. A novel enzyme complex of orotate phosphoribosyltransferase and orotidine 5′-monophosphate decarboxylase in human malaria parasite *Plasmodium falciparum*: physical association, kinetics and inhibition characterization. *Biochemistry* **2005**, *44*, 1643–1652.
52. Ginsburg, H. Progress in *in silico* functional genomics: the malaria parasite metabolic pathways database. *Trends Parasitol.* **2006**, *22*, 238–240.

53. Krungkrai, S. R.; Kusakari, Y.; Tokuoka, K.; Inoue, T.; Adachi, H.; Matsumura, H.; Takano, K.; Murakami, S.; Mori, Y.; Kai, Y.; Krungkrai, J.; Horii, T. Crystallization and preliminary crystallographic analysis of orotidine 5′-monophosphate decarboxylase from human malaria parasite *Plasmodium falciparum*. *Acta Crystallogr. Sect. F Struct. Biol. Cryst. Commun.* **2006**, *F62*, 542–545.

54. Tokuoka, K.; Kusakari, Y.; Krungkrai S. R.; Matsumura, H.; Krungkrai, J.; Horii, T.; Inoue, T. Structural basis for the decarboxylation of orotidine 5′-monophosphate (OMP) by *Plasmodium falciparum* OMP decarboxylase. *J. Biochem.* **2008**, *143*, 69–78.

55. Krungkrai, S. R.; Wutipraditkul, N.; Krungkrai, J. Dihydroorotase of human malarial parasite *Plasmodium falciparum* differs from host enzyme. *Biochem. Biophys. Res. Commun.* **2008**, *366*, 821–826.

56. Sherman, I. W. Biochemistry of *Plasmodium* (malaria parasites). *Microbiol. Rev.* **1979**, *43*, 453–495.

57. Scheibel, L. W. Plasmodial metabolism: carbohydrate. In *Malaria;* Wernsdorfer, W. H.; McGregor, I. Eds.; Churchill Livingstone: New York, **1988**; Vol. 1, pp. 171–217.

58. Krungkrai, J.; Krungkrai, S. Malaria parasite: genomics, biochemistry and drug target for antimalarial development. *Chula. Med. J.* **2006**, *50*, 127–142.

59. Reungprapavut, S.; Krungkrai, S. R.; Krungkrai, J. *Plasmodium falciparum* carbonic anhydrase is a possible target for malaria chemotherapy. *J. Enzyme Inhib. Med. Chem.* **2004**, *19*, 249–256.

60. Adler, L.; Brundell, J.; Falkbring, S. O.; Nyman, P. O. Carbonic anhydrase from *Neisseria sicca*, strain 6021 I. Bacterial growth and purification of the enzyme. *Biochim. Biophys. Acta* **1972**, *284*, 298–310.

61. Brundell, J.; Falkbring, S. O.; Nyman, P. O. Carbonic anhydrase from *Neisseria sicca*, strain 6021 II. Properties of the purified enzyme. *Biochim. Biophys. Acta* **1972**, *284*, 311–323.

62. Garg, L. C. Catalytic activity and inhibition of carbonic anhydrase of rat tissues. *Biochem. Pharmacol.* **1974**, *23*, 3153–3161.

63. Graham, D.; Reed, M. L.; Patterson, B. D.; Hockley, D. G.; Dwyer, M. R. Chemical properties, distribution, and physiology of plant and algal carbonic anhydrases. *Ann. N.Y. Acad. Sci.* **1984**, *429*, 222–237.

64. King, R. W.; Garg, L. C.; Huckson, J.; Maren, T. H. The isolation and partial characterization of sulfonamide-resistant carbonic anhydrases from the liver of the male rat. *Mol. Pharmacol.* **1974**, *10*, 335–343.

65. Shoaf, W. T.; Jones, M. E. Carbonic anhydrase of microorganisms I: an enzyme from baker's yeast which catalyzes the formation of carbamate from ammonium bicarbonate solutions. *Arch. Biochem. Biophys.* **1970**, *139*, 130–142.

66. Khalifah, R. G. The carbon dioxide hydration activity of carbonic anhydrase. I. Slow-flow kinetic studies on the native human isoenzymes B and C. *J. Biol. Chem.* **1971**, *246*, 2561–2573.

67. Soltes, R. E.; Mulligan, M. E.; Coleman, J. R. Identification and characterization of a gene encoding a vertebrate-type carbonic anhydrase in cyanobacteria. *J. Bacteriol.* **1997**, *179*, 769–774.

68. Pastorekova, S.; Parkkila, S.; Pastorek, J.; Supuran, C. T. Carbonic anhydrases: current state of the art, therapeutic applications and future prospects. *J. Enzyme Inhib. Med. Chem.* **2004**, *19*, 199–229.

69. Lehtonen, J.; Shen, B.; Vihinen, M.; Casini, A.; Scozzafava, A.; Supuran, C. T.; Parkkila, A. K.; Saarnio, J.; Kivela, A. J.; Waheed, A.; Sly, W. S.; Parkkila, S. Characterization of CA XIII, a novel member of the carbonic anhydrase isozyme family. *J. Biol. Chem.* **2004**, *279*, 2719–2727.

70. Wistrand, P. J.; Lindqvist, A. Design of carbonic anhydrase inhibitors and the relationship between the pharmacodymanics and pharmacokinetics of acetazolamide. In *Carbonic Anhydrase. From Biochemistry and Genetics to Physiology and Clinical Medicine;* Botrè, F.; Gros, G.; Storey, B. T. Eds.; VCH: Weinheim, **1991**; pp. 352–378.

71. Supuran, C. T.; Scozzafava, A.; Popescu, A.; Bobes-Tureac, R.; Banciu, A.; Creanga, A.; Bobes-Tureac, G.; Banciu, M. D. Carbonic anhydrase inhibitors. Part 43. Schiff bases derived from aromatic sulfonamides: towards more specific inhibitors for membrane-bound versus cytosolic isozymes. *Eur. J. Med. Chem.* **1997**, *32*, 445–452.

72. Supuran, C. T.; Nicolae, A.; Popescu, A. Carbonic anhydrase inhibitors. Part 35. Synthesis of Schiff bases derived from sulfanilamide and aromatic aldehyde: the first inhibitors with equally high affinity towards cytosolic and membrane-bound isozymes. *Eur. J. Med. Chem.* **1996**, *31*, 431–438.

73. Scozzafava, A.; Banciu, M. D.; Popescu, A.; Supuran, C. T. Carbonic anhydrase inhibitors: inhibition of isoenzymes I, II and IV by sulfamide and sulfamic acid derivatives. *J. Enzyme Inhib.* **2000**, *15*, 443–453.

74. Scozzafava, A.; Banciu, M. D.; Popescu, A.; Supuran, C. T. Carbonic anhydrase inhibitors: synthesis of Schiff bases of hydroxybenzaldehydes with aromatic sulfonamides and their reactions with arylsulfonyl isocyanates. *J. Enzyme Inhib.* **2000**, *15*, 533–546.

75. Popescu, A.; Simion, A.; Scozzafava, A.; Briganti, F.; Supuran, C. T. Carbonic anhydrase inhibitors: Schiff bases of some aromatic sulfonamides and their metal complexes: toward more selective inhibitors of carbonic anhydrase isozyme IV. *J. Enzyme Inhib.* **1999**, *14*, 407–423.

76. Clare, B. W.; Supuran, C. T. Carbonic anhydrase inhibitors. Part 61. Quantum chemical QSAR of a group of benzenedisulfonamides. *Eur. J. Med. Chem.* **1999**, *34*, 463–474.

77. Supuran, C. T.; Scozzafava, A.; Jurca, B. C.; Ilies, M. A. Carbonic anhydrase inhibitors. Part 49. Synthesis of substituted ureido and thioureido derivatives of aromatic/heterocyclic sulfonamides with increased affinities for isozyme I. *Eur. J. Med. Chem.* **1998**, *33*, 83–93.

78. Abbate, F.; Supuran, C. T.; Scozzafava, A.; Orioli, P.; Stubbs, M.; Klebe, G. Non-aromatic sulfonamide group as an ideal anchor for potent human carbonic anhydrase inhibitors: role of hydrogen-bonding networks in ligand binding and drug design. *J. Med. Chem.* **2002**, *45*, 3583–3587.

79. Krungkrai, J.; Scozzafava, A.; Reungprapavut, S.; Krungkrai, S. R.; Rattanajak, R.; Kamchonwongpaisan, S.; Supuran, C. T. Carbonic anhydrase inhibitors. Inhibition of *Plasmodium falciparum* carbonic anhydrase with aromatic sulfonamides: towards antimalarials with a novel mechanism of action? *Bioorg. Med.Chem.* **2005**, *13*, 483–489.

80. Trager, W.; Jensen, J. B. Human malaria parasites in continuous culture. *Science* **1976**, *193*, 673–675.

81. Sigler, K.; Hofer, M. Mechanism of acid extrusion in yeast. *Biochim. Biophys. Acta* **1071**, *1991*, 375–391.

82. Dodgson, S. J. The carbonic anhydrases: overview of their importance in cellular physiology and in molecular genetics. In *The Carbonic Anhydrases: Cellular Physiology*

and Molecular Genetics; Dodgson, S. J.; Tashian, R. E.; Gros, G.; Carter, N. D., Eds.; Plenum Press: New York, **1991**; pp. 3–14.

83. Ginsburg, H.; Divo, A. A.; Geary, T. G.; Boland, M. T.; Jensen, J. B. Effect of mitochondrial inhibitors on intraerythrocytic *Plasmodium falciparum* in *in vitro* cultures. *J. Protozool.* **1986**, *33*, 121–125.

84. Blum, J. J.; Ginsburg, H. Absence of alpha-ketoglutarate dehydrogenase activity and presence of CO_2-fixing activity in *Plasmodium falciparum* grown *in vitro* in human erythrocytes. *J. Protozool.* **1984**, *31*, 167–169.

CHAPTER 17

Inhibitors of *Helicobacter pylori* α- and β-Carbonic Anhydrases as Novel Drugs for Gastroduodenal Diseases

ISAO NISHIMORI[1], HIROAKI TAKEUCHI[2], and CLAUDIU T. SUPURAN[3]

[1]Department of Gastroenterology, Kochi Medical School, Nankoku, Kochi 783-8505, Japan
[2]Department of Laboratory Medicine, Kochi Medical School, Nankoku, Kochi 783-8505, Japan
[3]Laboratorio di Chimica Bioinorganica, Università degli Studi di Firenze, Room 188, Via della Lastruccia 3, I-50019 Sesto Fiorentino (Firenze), Italy

17.1 INTRODUCTION

Helicobacter pylori, a gram-negative spiral bacterium, was discovered to associate with peptic ulcers in 1983 by Warren and Marshall.[1] It is one of the most common organisms causing chronic infections, which has been found in approximately one-third of individuals in the Western world and in most of the population in underdeveloped countries.[2] This organism attacks the gastric epithelial surface, resulting in chronic gastritis, and can cause more severe diseases, including gastric carcinomas and lymphomas, peptic ulcers, and severe diarrhea.[3]

In patients with such gastroduodenal diseases, eradication of *H. pylori* has become the main therapeutic goal.[3] The recommended therapy consists of a proton pump inhibitor (PPI) and two antibiotics, mainly amoxicillin and clarithromycin, as the first-line eradication triple therapy.[3–5] Although this treatment has been shown to be effective in a number of clinical trials, several meta-analyses have revealed that the success rates of eradication varied widely ranging 70–95%. At present, the efficacy (eradication rate) has been decreasing mainly due to increased resistance of the bacteria to the antibiotics.[5–7] Following failure of eradication by the first-line treatment, consensus publications, such as the Maastricht 2–2000 report, have recommended a second-line quadruple therapy using PPI, bismuth salts, metronidazole, and tetracycline.[4] A recent meta-analysis showed that this treatment is effective,[8] but it has also been frequently associated with eradication failure in more than 20% of cases in certain countries or regions.[7–9] Reasons for eradication failure include *H. pylori* resistance to metronidazole[10,11] and considerable side-effect rates of metronidazole and tetracycline resulting in reduced patients compliance.[12] Accordingly, there is a real need for the development of alternative therapies, especially exploiting novel targets that are free of the problems arising with currently available drugs.

17.2 MOLECULAR NATURE OF α- AND β-CAs FROM *H. pylori*

Among bacteria, *H. pylori* has the unique ability to grow in the stomach presenting highly acidic conditions at pH values as low as 1.4.[13] Therefore, the pathogen has evolved specialized processes for survival in acid, which maintain its cytoplasmic pH at around 6.4.[13] At least two kinds of enzymes are involved in these processes: urease[13] and carbonic anhydrase (CA),[13–15] which play an important role in the urea and bicarbonate metabolism, as well as the acid resistance of the bacterium described subsequently.[13,14]

To data, three genome projects of *H. pylori* using three independent strains had been completed: the project with strain 26695 by TIGR sequencing center (GenBank: AE000511),[16] strain J99 by ASTRA research center (AE001439),[17] and strain HPAG1 by Washington University (CP000241).[18] The GenBank database search on these *H. pylori* genome sequences of approximately 1.6 Mbp long identified two completely different DNA clones of CA. One CA clone belongs to the α-class CA (hpαCA) (locus tag in each strain; HP1186 in strain 26695; jhp1112 in strain J99, and HPAG1_1126 in strain HPAG1) and another belongs to the β-class of CA (hpβCA) (locus tag in each strain; HP0004 in strain 26695; jhp0004 in strain J99, and HPAG1_0004 in strain HPAG1).

hpαCA composes 247 amino acid residues, which shows 27–36% similarity with other α-class bacterial CAs, including *Klebsiella pneumoniae*, *Neisseria gonorrhoeae*, *Enterococcus faecalis*, *Anabaena PCC7120*, *Erwinia carotovora*, and *Synechococcus PCC7942*.[19] Surprisingly, hpαCA amino acid sequence shows relatively high similarity (23–36%) with 15 kinds of human CAs (hCA), although there is a long evolutionary distance between two species, indicating that CAs posses a fundamental biological function. An alignment of hpαCA with hCA II and hCA I shows that 19 and 17 out of 36 active site residues are conserved, respectively (Fig. 17.1). Especially, 11 out of 19 residues known to form a network of hydrogen bonds are well conserved between hpαCA and hCA I and II, including the three zinc-liganded His residues (at position 94, 96, and 119) and several important residues for the binding of the substrates, inhibitors, and activators, such as Tyr7, His64, and Leu198.[20,21] However, Asn67 in hCA II is replaced with Ser67 in hCA I and Lys67 in hpαCA, which is one of the reasons to explain the lower catalytic activity of hCA I and hpαCA as compared to hCA II (Fig. 17.2). Asn67 in hCA II plays an important role in catalytic activity, as this amino acid is near the proton shuttle of the enzyme (His64) and is also in contact with many inhibitors or activators, which bind within the active site.[22,23]

hpβCA is composed of 221 amino acid residues, and shows 25–34% similarity with other β-class bacterial CAs, including the two enzymes from *Escherichia coli*, the ones

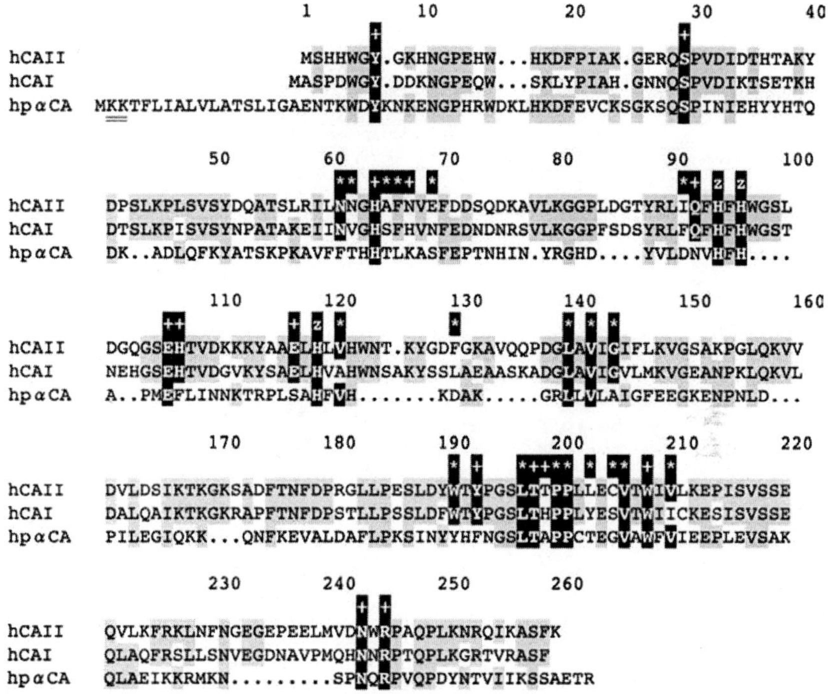

FIGURE 17.1 Alignment of α-class CA of *H. pylori* (hpαCA) with human CA II (hCA II). Thirty-six active site residues previously defined as forming the active site are indicated by a mixture of asterisk, "plus," and "z" signs above the hCA II sequence.[20,21] Seventeen residues known to participate in a network of hydrogen bonds and being involved in the binding of inhibitors/activators[22,23] are indicated by "plus" and "z" above the sequence; the latter sign indicates the three zinc-liganded histidine residues (His94, 96, and 119). Conserved amino acids in the three isoforms are indicated by a closed box. A double underline at the N-terminal indicates putative cytoplasmic retention signal (Lys–Lys).

from *Synechococcus elongatus*, *Brucella suis*, and *Haemophilus influenzae* (Fig. 17.3).[19,24–29] It has been reported that most of the bacterial β-CAs are composed of three sequential components: (i) an N-terminal arm including two α-helices (H1 and H2), (ii) a zinc binding core including three β-sheets (S1–S3) and two α-helices (H3 and H4), and (iii) a C-terminal subdomain including two β-sheets (S4 and S5) and five α-helices (H5–H9) (Fig. 17.3).[24,25] In agreement with previous reports,[24–27] the amino acid sequence of the zinc binding core of hpβCA is highly conserved among other bacterial β-CAs sequenced to date. These residues include the zinc(II)-coordinating amino acids Cys42, Asp44, His98, and Cys101 (indicated by triangles in Fig. 17.3; residue numbers are based on the *E. coli* CynT2 numbering system[24]). The principal difference between α- and β-CAs consists in the fact that generally β-CAs are considered to be oligomers, usually formed of 2–6 monomers of molecular weight of 25–30 kDa.[24–29] However, oligomerization of hpβCA has not been evidenced so far.

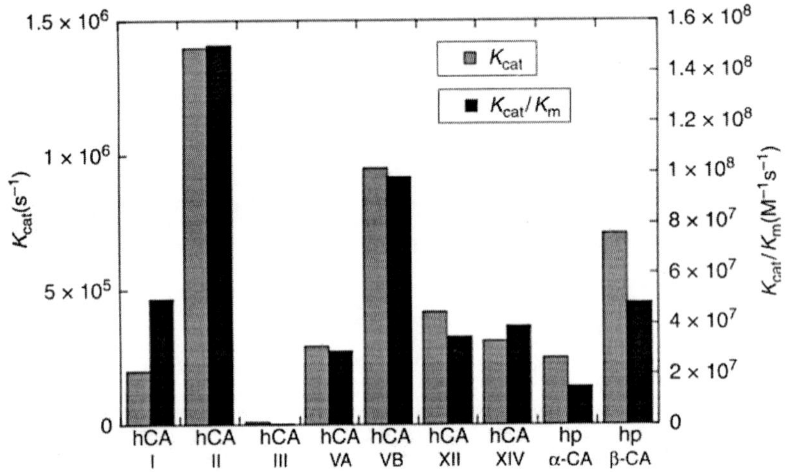

FIGURE 17.2 Kinetic parameters for CO_2 hydration reaction catalyzed by some human (h) α-CA isozymes (at 20°C and pH 7.5) and *H. pylori* (hp) α-CA (at pH 8.9 and 25°C) and β-CA (at pH 8.3 and 20 °C).[34]

17.3 KINETIC AND INHIBITION PROPERTIES OF α- AND β-CAs FROM *H. pylori*

In 2001, Chirica et al.[30] produced a recombinant protein of hpαCA using an *E. coli* expression system and showed that the enzyme had a catalytic activity similar to that of the human slow isoform hCA I (highly abundant in red blood cells and the gastrointestinal tracts[20,21]). They also reported that hpCA was susceptible to inhibition by sulfonamides (and thiocyanate).[30] We also obtained DNA clones of hpαCA and hpβCA and successfully showed that the recombinant protein had significant catalytic CA activity (Fig. 17.2) and was inhibited by a panel of sulfonamides and sulfamates (Table 17.1).[31–33]

The catalytic activity of recombinant, purified hpαCA and hpβCA for the physiologic reaction (CO_2 hydration), in comparison with that of several α-CAs of human origin, such as hCA I–III (cytosolic isozymes), hCA VA and VB (mitochondrial isoforms), and hCA XII and XIV (transmembrane isozymes), are shown in Table 17.1. It may be observed that hpαCA and hpβCA are catalytically efficient CAs, possessing a high enzymatic activity (the β-class enzyme is 3.2 times more active than the α-CA from this bacterium).[33] Furthermore, this activity is almost identical (as k_{cat}/K_m value) to that of hCA I, whereas the K_m value of the bacterial enzyme is closer to that of hCA II than to that of hCA I. In fact, hpβCA is a medium efficient CA, possessing a catalytic activity higher than that of hCA III, VA, XII, and XIV among others. Only hCA VB and especially hCA II, one of the best catalysts known in nature, show a better activity than hpβCA.[20,21]

Inhibition data against hpαCA, hpβCA, and hCA I and II (the host isozymes for the bacteria) *in vitro* are provided in Table 17.1.[31–34] Data of Table 17.1 show that hpβCA is inhibited by all 47 derivatives (sulfonamides and one sulfamate), with an inhibition profile completely distinct of those of the α-class enzymes of human (hCA I and II) or

```
                                 H1              H2      S1         H3
H. pylori, cons.           VKAFLGALEFQENEYEELKELYESLKTKQKPHTLFISCVDSRVVPNLITGTK
E. coli, CynT             MKEIIDGFLKFQREAFPKREALFKQLATQQSPRTLFISCSDSRLVPELVTQRG
S. elongatus              MRKLIEGLRHFRTSYYPSHRDLFEQFAKGQHPRVLFITCSDSRIDPNLITQSG
B. suis          PMKNDHSPDQRTLSELFEHNRQWAAEKQEKDPEYFSRLLSSQRPEFLWIGCSDSRVPANVVTGLQ
E. coli, CynT2          1 MKDIDTLISNNALWSKMLVEEDPGFFEKLAQAQKPRFLWIGCSDSRVPAERLTGLE 56
H. influenzae             MDKIKQLFANNYSWAQRMKEENSTYFKELADHQTPHYLWIGCSDSRVPAEKLTNLE

                   S2                   H4             S3     H5
H. pylori, cons. PGELYVIRNMGNVIPPKTSYKESLSTMASIEYAIVHVGVQNLIICGHSDCGACGSIHLINDGITKAKT
E. coli, CynT    MGELFVIRNAGNLIPPYGPEP--GGVSASVEYAVAALRVSDIVICGHSNCGAMTAIASCQCM---DHM
S. elongatus     MGELFVIRNAGNLIPPFGAAN--GGEGASIEYAIAALNIEHVVVCGHSHCGAMKGLLKLNQLQ--EDM
B. suis          PGEVFVHRNVANLVHRAD-----LNLLSVLEFAVGVLEIKHIIVCGHYGCGGVRAAMDGY------GH
E. coli, CynT2   PGELFVHRNVANLVIHTD-----LNCLSVVQYAVDVLEVEHIIICGHYGCGGVQAAVENP------EL 122
H. influenzae    PGELFVHRNVANQVIHTD-----FNCLSVVQYAVDVLKIEHIIICGHTNCGGIHAAMADK------DL

                     H6         H7          H8             H9          S4
H. pylori, cons. PYIADWIQFLEPIKEELKNHPQFSNHFAKRSWLTERLNVRLQLNNLLSYDFIQERVMNNE-LKIFGWH
E. coli, CynT    PAVSHWLRYADSARVVNEARPHSDLPSKAAAM-VRE-NVIAQLANLQTHPSVRLALEEGR-IALHGWV
S. elongatus     PLVYDWLQHAQATRRLVLDNYSGYETDDLVEILVAE-NVLTQIENLKTYPIVRSRLFQGK-LQIFGWI
B. suis          GIIDNWLQPIRDIAQANQAELDTIENTQDRLDRLCELSVSSQVESLSRTPVLQSAWKDGKDIIVHGWM
E. coli, CynT2   GLINNWLLHIRDIWFKHSS-LLGEMPQERRLDTLCELNVMEQVYNLGHSTIMQSAWKRGQKVTIHGWA 198
H. influenzae    GLINNWLLHIRDIWFKHGH-LLGKLSPEKRADMLTKINVAEQVYNLGRTSIVKSAWERGQKLSLHGWV

                       S5
H. pylori, cons. YIIETGRIYNYNFESHFFEPIEETIKQRKSHENF
E. coli, CynT    YDIESGSIAAFDGATRQFVPLAANPRVCAIPLRQPTAA
S. elongatus     YEVESGEVLQISRTSSDDTGIDECPVRLPGSQEKAILGRCVVPLTEEV(+40)
B. suis          YNLKDGLLRDIGCDCTRNALQFACQPAE
E. coli, CynT2   TXIHDGLLRDLDVTATNRETLEQRYRHGISNLKLKHANHK 237
H. influenzae    YDVNDGFLVDQGVMATSRETLEISYRNAIARLSILDEENILKKDHLENT
```

FIGURE 17.3 Consensus amino acid sequence of hpβCA (*H. pylori* cons.), which consists of the most commonly used amino acid residues among the 15 *H. pylori* strains from patients with gastritis, as compared to those of the β-CAs from *E. coli* (two genes products, CynT and cynT2), *S. elongatus, B. suis,* and *H. influenzae*. The secondary structural features are indicated above the alignment (helices as bars and strands as arrows) according to the sequential subdomains. Arrowheads indicate the zinc(II)-coordinating amino acid residues: Cys42, Asp44, His98, and Cys101 (residue numbers are based on the *E. coli* CynT2 numbering system[24]).

bacterial (hpαCA) origin. Thus, a number of aromatic/heterocyclic simple sulfonamides, such as compounds **17.1–17.10**, **17.13**, **17.24**, **17.25**, and **17.31** showed inefficient hpβCA inhibitory activity, with inhibition constants in the range of 1.1–24.8 μM. Weak inhibitory activity, with K_I's in the range of 128–973 nM was then showed by compounds **17.10**, **17.11**, **17.14**, **17.21–17.23**, **MZA**, **BRZ**, **ZNS**, **IND**, **17.26**, **17.27**, **17.32**, and **17.33**. It may be observed that these compounds belong to heterogeneous classes of sulfonamides, possessing various substitution patterns. Potent hpβCA inhibitory action was then detected for many derivatives, among which are compounds **17.15–17.17**, **DCP**, **DZA**, **BZA**, **17.28**, and **17.34**. These compounds showed K_I's in the range of 54–105 nM. The best hpβCA inhibitors (K_I's in the range of 24–45 nM) were the following derivatives: **17.18–17.20**, **AAZ**, **EZA**, **TPM**, **SLP**, **17.29**, **17.30**, **17.35**, and **17.36**. Some of them are clinically used drugs (the CA inhibitor par excellence, **AAZ**, as well as the structurally related **EZA**; the antiepileptic topiramate **TPM**, the antipsychotic sulpiride **SLP**, and the recently reported derivatives possessing lipophilic *tert*-butyl tails **17.29**, **17.30**, **17.35**, and **17.36**). As

TABLE 17.1 Human (h) hCA I, II, and *H. pylori* (hp) α/βCA Inhibition Data with Compounds 17.1–17.24 and the Clinically Used Derivatives AAZ–IND, and Newly Designed Inhibitors 17.25–17.36[34]

	$K_I{}^a$ (nM)			
Inhibitor	hCA I[b]	hCA II[b]	hpαCA[b]	hpβCA[b]
1	45,400	295	426	16,400
2	25,000	240	454	1845
3	28,000	300	316	8650
4	78,500	320	430	2470
5	25,000	170	873	2360
6	21,000	160	1150	3500
7	8300	60	1230	1359
8	9800	110	378	1463
9	6500	40	452	1235
10	6000	70	510	1146
11	5800	63	412	973
12	8400	75	49	640
13	8600	60	323	2590
14	9300	19	549	768
15	6	2	268	64
16	164	46	131	87
17	185	50	114	71
18	109	33	84	38
19	95	30	207	39
20	690	12	105	37
21	55	80	876	236
22	21,000	125	1134	218
23	23,000	133	1052	450
24	24,000	125	541	15,250
AAZ	250	12	21	40
MZA	50	14	225	176
EZA	25	8	193	33
DCP	1200	38	378	105
DZA	50,000	9	4360	73
BRZ	45,000	3	210	128
BZA	15	9	315	54
TPM	250	10	172	32
ZNS	56	35	231	254
SLP	1200	40	204	35
IND	31	15	413	143
25	12,300	241	539	23,500
26	10,750	210	316	241
27	14,250	133	79	158
28	12,300	241	539	101
29	13,270	127	62	44
30	541	18	13	28
31	14,700	354	640	24,800
32	9620	203	318	213
33	13,000	119	60	150

TABLE 17.1 (*Continued*)

	$K_I{}^b$ (nM)			
Inhibitor	hCA I[c]	hCA II[c]	hpαCA[c]	hpβCA[c]
34	12,150	104	31	96
35	12,045	94	27	45
36	338	15	12	24

[a] Errors in the range of 5–10% of the shown data, from three different assays.
[b] Human/hp recombinant isozymes, stopped-flow CO_2 hydrase assay method.[20,21]

many of these derivatives were also quite effective hpαCA inhibitors (but also hCA I and II inhibitors, Table 17.1), dual inhibition of α- and/or β-class CAs of *H. pylori* could represent a useful and novel means for the management of gastritis/gastric ulcers, as well as gastric cancer.

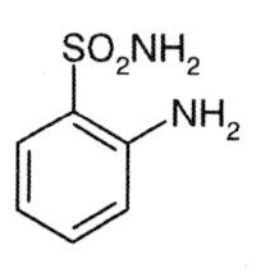

17.1

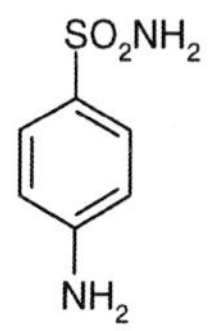

17.2

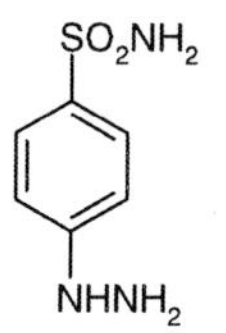

17.3

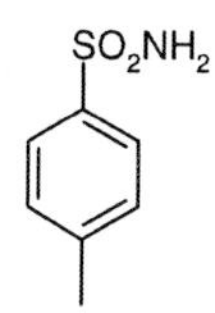

17.4

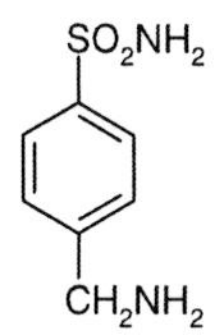

17.5

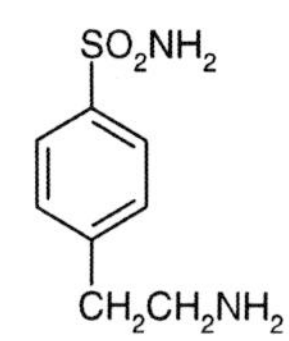

17.6

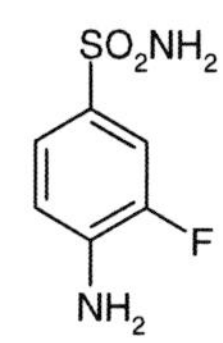

17.7

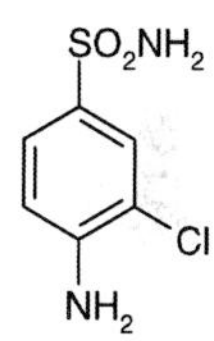

17.8

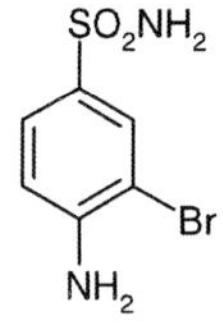

17.9

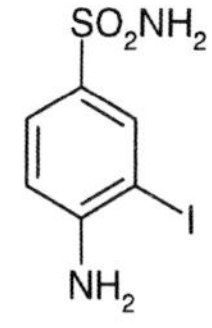

17.10

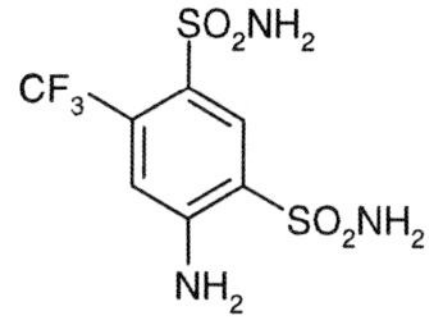

17.11

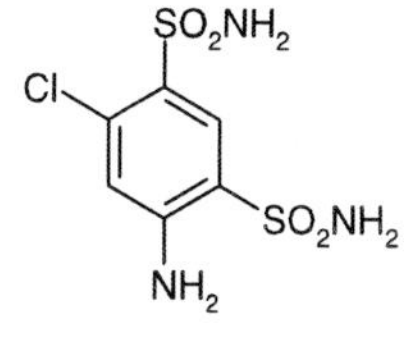

17.12

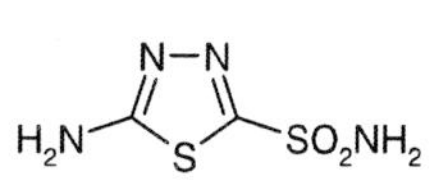

17.13

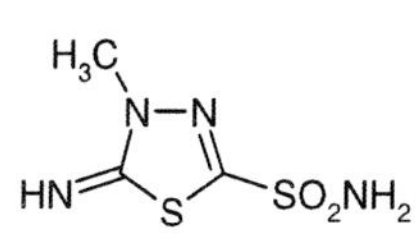

17.14

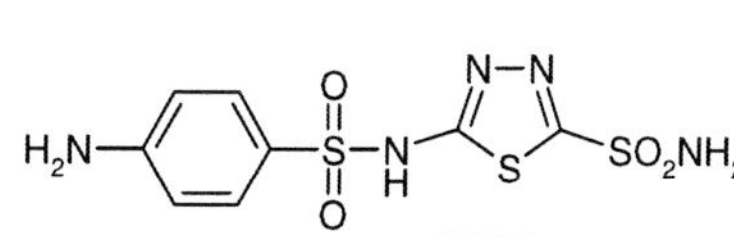

17.15

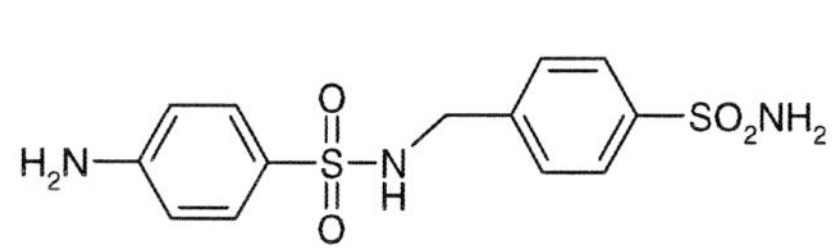

17.16

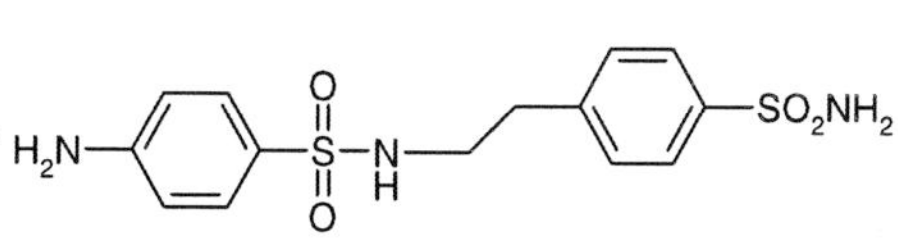

17.17

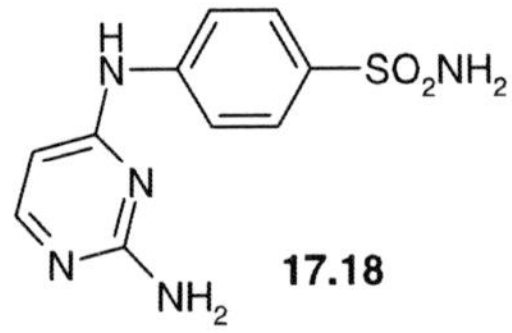

17.18

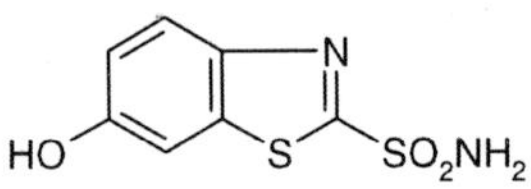

17.19

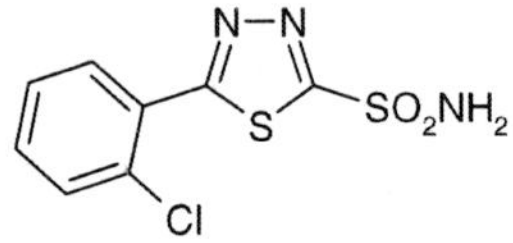

17.20

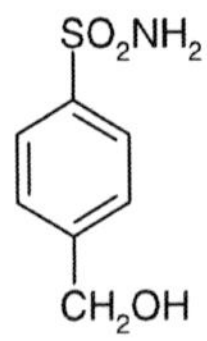

17.21

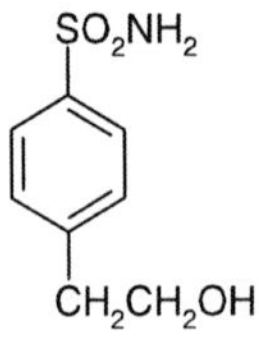

17.22

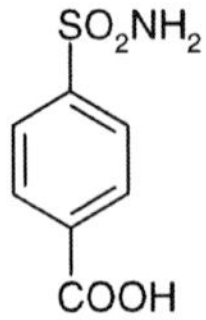

17.23

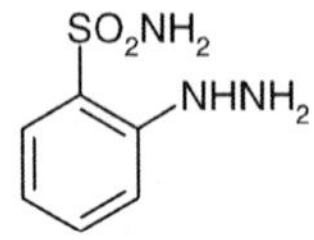

17.24

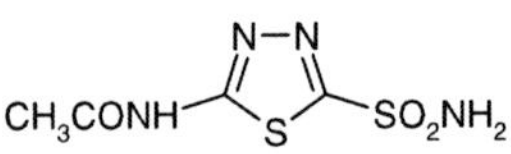

AAZ

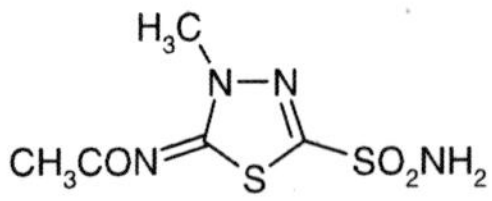

MZA

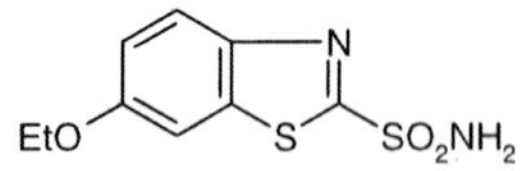

EZA

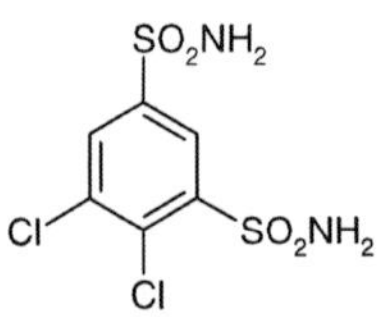

DCP

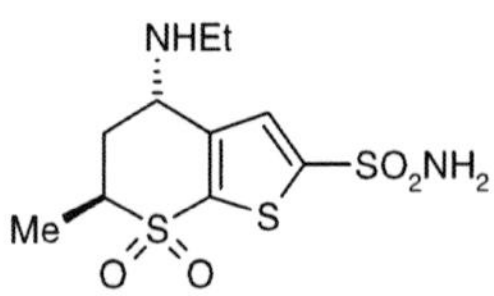

DZA

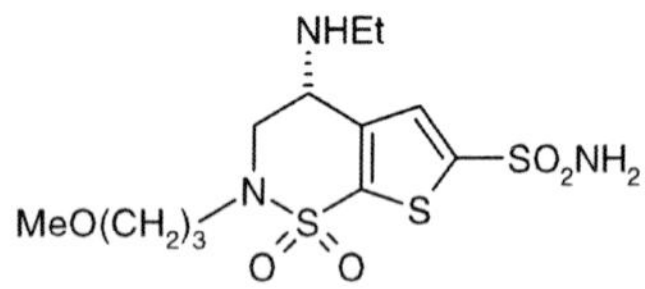

BRZ

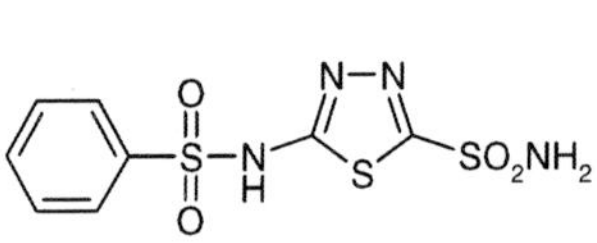

BZA

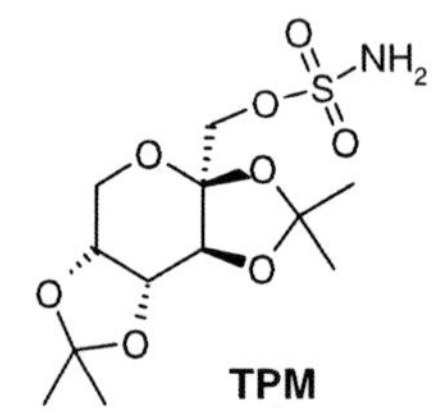

TPM

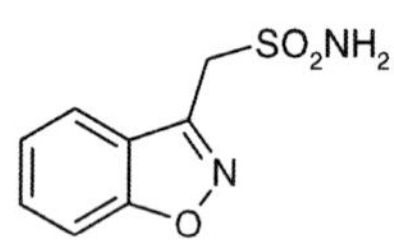

ZNS

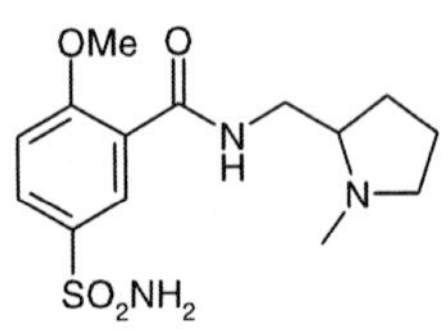

SLP

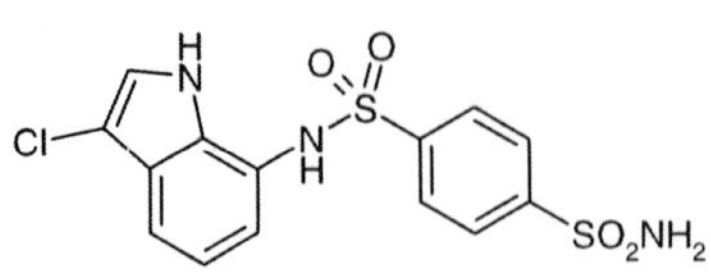

IND

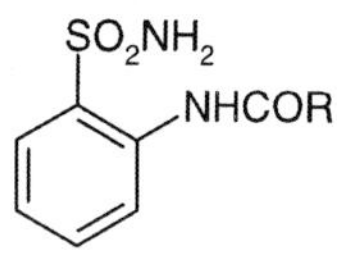

17.25

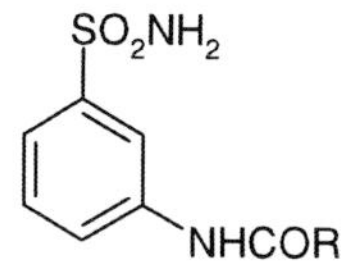

17.26

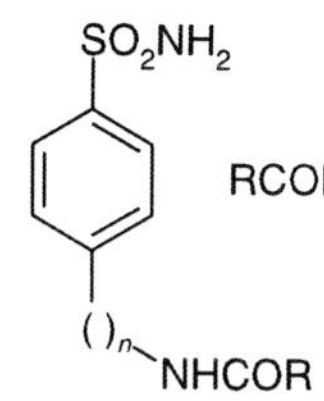

17.27: $n = 0$
17.28: $n = 1$
17.29: $n = 2$

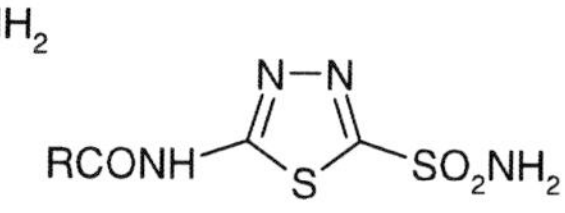

17.30

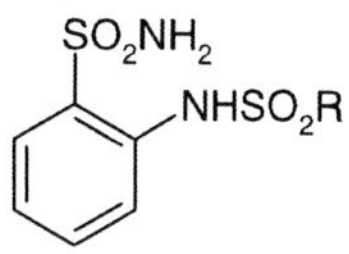

17.31

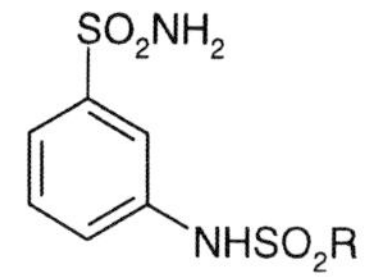

17.32

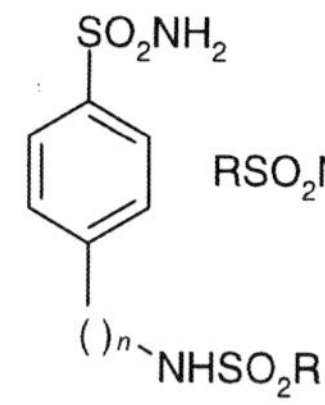

17.33: $n = 0$
17.34: $n = 1$
17.35: $n = 2$

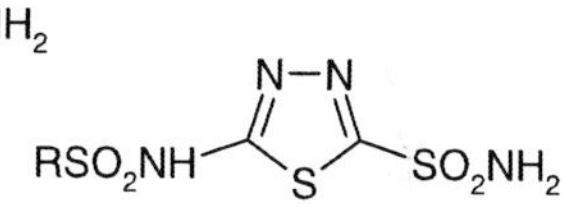

17.36

R = 4-*t*-Bu-C_6H_4

17.4 SUBCELLULAR LOCALIZATION AND BIOLOGICAL FUNCTION OF α- AND β-CAs IN *H. Pylori*

Chirica et al. studied subcellular localization of hpαCA and hpβCA by using electron microscopy with immunonegative staining and SDS-digested freeze-fracture immunogold labeling and then showed that hpαCA was attached to the outer membrane and that hpβCA was localized in the cytosol, on the cytosolic side of the inner membrane and on the outer membrane facing the periplasmic space.[35] Since hpαCA contains an N-terminal cytoplasmic retention signal (Lys–Lys) preceding a membrane insertion signal anchor sequence (see Fig. 17.1), it probably is localized on the outer membrane and expose its active site in the periplasmic space. This was supported by Western blot analysis, which showed hpαCA was present in the membrane fraction of *H. pylori* homogenate.[14] The presence of hpβCA in the cytosol and on the cytosolic side of the inner membrane is agreeable with the cytosolic localization of other bacterial β-CA.[36] The location of hpβCA on the outer membrane is somewhat puzzling in possibly artifact due to some leakage of enzyme during the freeze-fracture process.[35]

What are biological roles of periplasmic hpαCA and cytosolic hpβCA in *H. pylori*? Stähler et al. reported that hpαCA and hpβCA deletion mutants as well as the double mutant displayed a significant decrease in urease activity, indicating a metabolic link of these two enzyme types, urease and CA.[15] They also showed, to somewhat strange, that only the hpβCA deletion mutant showed a clearly reduced growth at pH 6.0–6.25

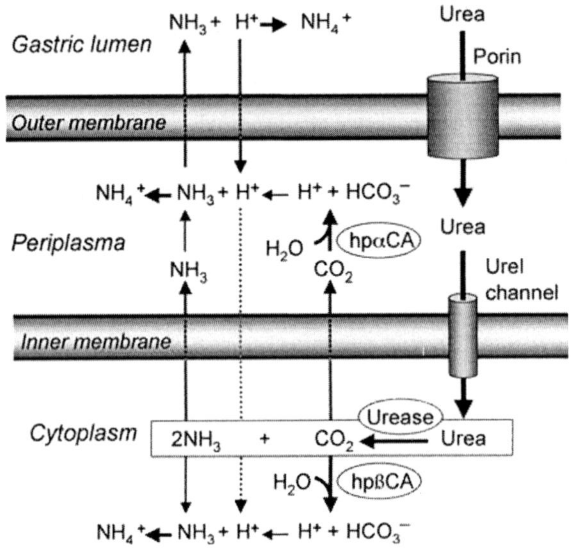

FIGURE 17.4 A model for the role of urease and α- and β-CA in pH regulation in *H. pylori*.[34] Under acidic condition, urea moves into the cytoplasm through the UreI channel. In the cytoplasm, two HN_3 and one CO_2 molecules are produced from urea in the presence of urease activity. CO_2 in the periplasm and cytoplasm is hydrated by α- and β-CA, respectively, resulting in the production of H^+ and bicarbonate. The proton is consumed by NH_3 to form HN_4^+ ions in the periplasm and cytoplasm.

for 48 h culture compared to the parental strain, hpαCA deletion mutant and the double mutant, suggesting the cytosolic hpβCA is only important for acid resistance when the periplasmic hpαCA is functional.[15] In the other study by Marcus et al., hpαCA deletion mutant showed an approximately 3 $\log_{10}$ decrease in survival after 30 min of exposure to pH 2.0 as compared to the parental strain.[14] These discrepant findings are probably due to the different culture conditions employed in the studies. In the presence of a CA inhibitor against both hpαCA ($pK_I = 225$ nM) and hpβCA ($pK_I = 176$ nM) (Table 17.1), acetazolamide (AAZ), a certain strain of *H. pylori* (ATCC 43504), also showed an approximately 3 $\log_{10}$ decrease in survival.[14] These findings indicate that both hpαCA and hpβCA might be essential for the bacterial survival.

Marcus et al. has proved that hpαCA is essential for the acid acclimation and survival of the pathogen and then proposed a model for the role of urease and hpαCA in the maintenance of periplasmic pH in acid.[13,14] We added the role of hpβCA and extended their model (Fig. 17.4, modified from the figure reported by Marcus et al.[14]). They employed CA deletion mutants of *H. pylori* and showed that the generation of NH_3 by urease and bicarbonate by hpαCA has a major role in the regulation of the periplasmic pH 6.1 and inner membrane potential −101 mV under acidic conditions.[14] It has also been proved that the absence of hpαCA activity decreased membrane integrity in acid medium, as observed by using membrane-permeant and membrane-impermeant fluorescent DNA dyes.[14] The increase in membrane potential and cytoplasmic buffering following urea addition was observed in the wild-type organisms in acid, but was absent in the hpαCA deletion mutant, although the urease remained fully functional.[14] Thus, buffering of the periplasm to a pH consistent with viability depends not only on the ammonia efflux from the cytoplasm but also on the conversion of CO_2 produced by urease to bicarbonate by the periplasmic hpαCA (Fig. 17.4). In fact, the pK_a of the carbonic acid/bicarbonate couple of around 6.1 is very appropriate for such a task, unlike the ammonium/ammonia buffer, which having a pK_a of 9.2 is less useful for buffering the periplasm to pH values close to neutrality.

The similar reactions to the one mentioned above for the periplasm will occur in the cytoplasm, that is, hydration of CO_2 by cytoplasmic hpβCA producing bicarbonate to buffer the cytoplasmic pH at around 8.0 (Fig. 17.4).[13,37] Another product by hpβCA, protons are neutralized by NH_3, the product by urease. NH_3, can also neutralize entering protons, which could occur in the highly acidic condition such as in the stomach. Accordingly, in cooperation with cytoplasmic urease, hpαCA and hpβCA work for acid acclimation of *H. pylori* in its periplasm and cytoplasm, respectively, suggesting that these CAs may be attractive drug targets for developing anti-*H. pylori* agents.

17.5 THERAPEUTIC USAGE OF CA INHIBITORS IN GASTRODUODENAL DISEASES

Before the development of eradication therapy against *H. pylori*, patients with peptic ulcer diseases have been treated with H2-receptor antagonist (H2 antagonist) or PPI. Apart from these antiacid secretion drugs, in 1980s, effectiveness of a CA inhibitor, acetazolamide (**AAZ**), for peptic ulcer diseases was reported; treatment with 25–35 mg/kg body weight/day of **AAZ** for 30 days achieved 96–97% of gastric and duodenal ulcer healing.[38,39] At this time, pharmaceutical mechanism expected for **AAZ** was inhibition of acid secretion, possibly due to inhibition of α-CA activity functioning as cytoplasmic pH buffering system in the parietal cells of the patients.[40] It was reported, indeed, that treatment with **AAZ** for 10 days suppressed basal secretion of gastric acid by 92% and histamine-stimulated secretion by 83% in patients with gastric and duodenal ulcer.[39] At that time, however, inhibitory effect of **AAZ** on *H. pylori* CA had not been considered at all. Interestingly, **AAZ** was shown to be sufficiently effective not only in ulcer healing but also in prevention of ulcer recurrence. Two years after antiulcer therapy was discontinued, the recurrence rate in patients treated with **AAZ** (6.2%) was significantly lower than that with classical antiacid drugs, such as 6–9 g/day of aluminum hydroxide and magnesium carbonate (34%).[38] It is well known to date that even in patients treated with H2 antagonist or PPI, the recurrence rate of peptic ulcer is extremely high of 60%.[41] Surprisingly, the recurrence rate of ulcer diseases following **AAZ** treatment is similar to that achieved by the triple eradication therapy.[41] Taken together, these clinical evidences suggest that, apart from antiacid secretion, **AAZ** might have additional pharmaceutical function, probably certain direct effect on *H. pylori*, a causative pathogen for peptide ulcer disease.

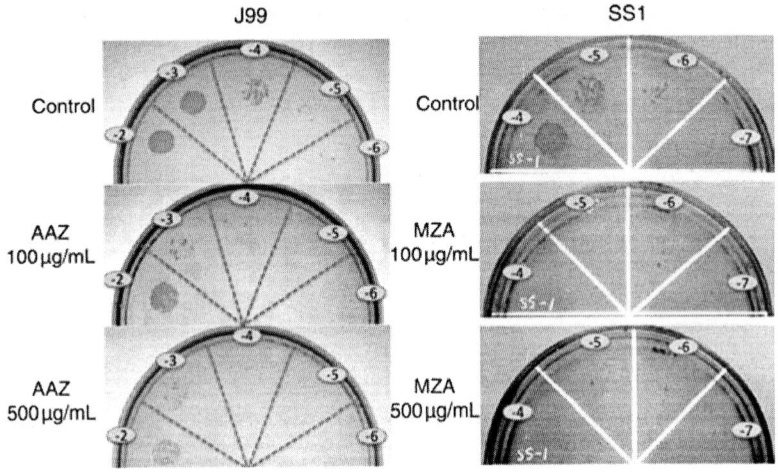

FIGURE 17.5 Inhibition effect of acetazolamide (AAZ) and methazolamide (MZA) on growth of *H. pylori* strains, J99 and SS1.[34] Bacteria are cultivated in *Brucella* broth supplemented with 10% horse serum at 37°C under microaerobic conditions as previously reported.[46] Bacteria in the early stationary phase (OD_{590}: 0.9–1.0) are diluted by 2–7 $\log_{10}$ (numbers indicated in the figure) and then subjected to colony forming assay in the absence and presence of AAZ and MZA at 100 and 500 μg/mL of final concentration.

Early *in vitro* study of the potent and clinically used CA inhibitor, **AAZ** ($pK_I = 21$ nM for hpαCA and 40 nM for hpβCA) (Table 17.1), show that it did not inhibit bacterial growth at the concentration of 5 μM.[35] However, in the presence of 1 mM of **AAZ**, there was an approximately 3 $\log_{10}$ decrease in acid survival of *H. pylori*.[14] We also have had a preliminary data that CA inhibitors, **AAZ** and methazolamide (**MZA**), displayed growth inhibition of certain strains of *H. pylori* in colony formation assay (Fig. 17.5).[34] **MZA** possesses weaker K_I than **AAZ** but strong inhibition activity against hpαCA and hpβCA (Table 17.1). Interestingly, the growth of strain SS1 was totally inhibited at 100 μg/mL of **MZA**, but there was no inhibitory effect of the drug at this concentration on growth of other strain, 11637.[34] Recently, we studied susceptibility of *H. pylori* to sulpiride (SLP), an antiulcer and antipsychotic drug clinically used, by the killing assay.[42] **SLP** is a strong inhibitor against hpβCA as well as **AAZ** and possesses comparable inhibition activity on hpα CA as **MZA** (Table 17.1). **SLP** at concentration greater than 200 μg/mL successfully showed killing effects in a dose-dependent manner on five different strains of *H. pylori*, including strains resistant to clarithromycin, metronidazole, and/or ampicillin.[42] Interestingly, *H. pylori* (strain SS1) treated with 500 μg/mL of **SLP** showed single or aggregated coccoid form; whereas the untreated bacteria were mostly short rods.[42] These findings indicate that certain CA inhibitors against both α- and β-CAs of *H. pylori*, such as **AAZ**, **MZA**, and **SLP**, could be applied for the development of alternative therapies, which would be free of the problems arising with currently available drugs.

To date, there is only one preliminary study of clinical evaluation of CA inhibitor for eradication of *H. pylori*. Graham's group tried to eradicate *H. pylori* by treating with

1 g/day of **AAZ** for 4 days in eight volunteers with active *H. pylori* infection.[43] However, no one turned off the positive urea breath test after the treatment. Several reasons for their negative result are possible. First, dosage of **AAZ** was possibly insufficient to eradicate *H. pylori*. As already described, previous clinical studies, which successfully showed 96–97% of ulcer healing, used 25–35 mg/kg body weight/day of **AAZ**.[38] Second, duration of drug administration was too short. CA inhibitors, including **AAZ** and **MZA**, were shown to have inhibitory effect on the bacterial growth, but they were not found to have killing effect against *H. pylori*. Even if they have the killing effect, 4 days treatment is still too short, since the triple therapy including two potent antibiotics also requires at least 7 days administration.[3–5] As described above, hpαCA and/or hpβCA are necessary to obtain urease activity for *H. pyori*[15] and urease is essential for gastric colonalization of the bacteria,[44,45] suggesting CA inhibitors possibly work by inhibition of efficient colonalization of the bacteria. In such case, it takes much longer period to obtain the effect of CA inhibitors.

17.6 FUTURE DIRECTION

As describe herein, together with urease, two CAs of *H. pylori* play central role in acid acclimation to survive in the highly acid condition of the stomach. Several CA inhibitors had been shown to have inhibitory effect on the growth of *H. pylori in vitro*, as well as on the inhibition of acid secretion from the parietal cells.[38] These findings strongly recommend clinical use of CA inhibitor as a novel agent for eradication therapy of *H. pylori*. This can include, of course, combination therapy of a CA inhibitor with certain kinds of antibiotics. Fortunately, any serious complication of **AAZ** was not observed in previous clinical studies; only appearance of mild paresthesias in the limb and around the mouth was reported.[38,43] For the development of alternative eradication therapies that possess different pharmacological mechanism from previously used drugs, clinical trials with carefully designed protocol using a panel of CA inhibitors (see Table 17.1) should be warranted.

REFERENCES

1. Marshall, B. J.; Warren, J. R. Unidentified curved bacilli in the stomach of patients with gastritis and peptic ulceration. *Lancet* **1984**, *1*, 1311–13115.
2. Penston, J. G.; McColl, K. E. Eradication of *Helicobacter pylori*: an objective assessment of current therapies. *Br. J. Clin. Pharmacol.* **1997**, *43*, 223–243.
3. Suerbaum, S.; Michetti, P. *Helicobacter pylori* infection. *N. Engl. J. Med.* **2002**, *347*, 1175–1186.
4. Bytzer, P.; O'Morain, C. Treatment of *Helicobacter pylori*. *Helicobacter* **2005**, *10* (Suppl. 1), 40–46.
5. Lam, S. K.; Talley, N. J. Report of the 1997 Asia Pacific Consensus Conference on the management of *Helicobacter pylori* infection. *J. Gastroenterol. Hepatol.* **1998**, *13*, 1–12.

6. Huang, J. Q.; Hunt, R. H. Treatment after failure: the problem of "non-responders". *Gut* **1999**, *45* (Suppl. 1), I40–I44.

7. Laheij, R. J.; Rossum, L. G.; Jansen, J. B.; Straatman, H.; Verbeek, A. L. Evaluation of treatment regimens to cure *Helicobacter pylori* infection: a meta-analysis. *Aliment. Pharmacol. Ther.* **1999**, *13*, 857–864.

8. Fischbach, L. A.; van Zanten, S.; Dickason, J. Meta-analysis: the efficacy, adverse events, and adherence related to first-line anti-*Helicobacter pylori* quadruple therapies. *Aliment. Pharmacol. Ther.* **2004**, *20*, 1071–1082.

9. Gisbert, J. P.; Pajares, J. M. *Helicobacter pylori* therapy: first-line options and rescue regimen. *Dig. Dis.* **2001**, *19*, 134–143.

10. Chi, C. H.; Lin, C. Y.; Sheu, B. S.; Yang, H. B.; Huang, A. H.; Wu, J. J. Quadruple therapy containing amoxicillin and tetracycline is an effective regimen to rescue failed triple therapy by overcoming the antimicrobial resistance of *Helicobacter pylori. Aliment. Pharmacol. Ther.* **2003**, *18*, 347–353.

11. Lamouliatte, H.; Mégraud, F.; Delchier, J. C.; Bretagne, J. F.; Courillon-Mallet, A.; De Korwin, J. D.; Fauchère, J. L.; Labigne, A.; Fléjou, J. F.; Barthelemy, P.; Multicentre Study Group. Second-line treatment for failure to eradicate *Helicobacter pylori*: a randomized trial comparing four treatment strategies. *Aliment. Pharmacol. Ther.* **2003**, *18*, 791–797.

12. Mantzaris, G. J.; Petraki, K.; Archavlis, E.; Amberiadis, P.; Christoforidis, P.; Kourtessas, D.; Chiotakakou, E.; Triantafyllou,G. Omeprazole triple therapy versus omeprazole quadruple therapy for healing duodenal ulcer and eradication of *Helicobacter pylori* infection: a 24-month follow-up study. *Eur. J. Gastroenterol. Hepatol.* **2002**, *14*, 1237–1243.

13. Sachs, G.; Weeks, D. L.; Wen, Y.; : Marcus, E. A.; Scott,D. R. Acid acclimation by *Helicobacter pylori. Physiology (Bethesda)* **2005**, *20*, 429–438.

14. Marcus,E. A.; Moshfegh, A. P.; Sachs, G.; Scott,D. R. The periplasmic α-carbonic anhydrase activity of *Helicobacter pylori* is essential for acid acclimation. *J. Bacteriol.* **2005**, *187*, 729–738.

15. Stähler, F. N.; Ganter, L.; Lederer, K.; Kist, M.; Bereswill, S. Mutational analysis of the *Helicobacter pylori* carbonic anhydrases. *FEMS Immunol. Med. Microbiol.* **2005**, *44*, 183–189.

16. Tomb, J. F.; White, O.; Kerlavage, A. R.; Clayton, R. A.; Sutton, G. G.; Fleischmann, R. D.; Ketchum, K. A.; Klenk, H. P.; Gill, S.; Dougherty, B. A.; Nelson, K.; Quackenbush, J.; Zhou, L.; Kirkness, E. F.; Peterson, S.; Loftus, B.; Richardson, D.; Dodson, R.; Khalak, H. G.; Glodek, A.; McKenney, K.; Fitzegerald, L. M.; Lee, N.; Adams, M. D.; Hickey, E. K.; Berg, D. E.; Gocayne, J. D.; Utterback, T. R.; Peterso, L.; Wallin, E.; Hayes, W. S.; Borodovsky, M.; Karp, P. D.; Smith, H. O.; Fraser, C. M.; Venter, J. C. The complete genome sequence of the gastric pathogen *Helicobacter pylori. Nature* **1997**, *388*, 539–547.

17. Alm, R. A.; Ling, L. S.; Moir, D. T.; King, B. L.; Brown, E. D.; Doig, P. C.; Smith, D. R.; Noonan, B.; Guild, B. C.; De Jonge, B. L.; Carmel, G.; Tummino, P. J.; Caruso, A.; Uria-Nickelsen, M.; Mills, D. M.; Ives, C.; Gibson, R.; Merberg, D.; Mills, S. D.; Jiang, Q.; Taylor, D. E.; Vovis, G. F.; Trust, T. J. Genomic-sequence comparison of two unrelated isolates of the human gastric pathogen *Helicobacter pylori. Nature* **1999**, *397*, 176–180.

18. Oh, J. D.; Kling-Bäckhed, H.; Giannakis, M.; Xu, J.; Fulton, R. S.; Fulton, L. A.; Cordum, H. S.; Wang, C.; Elliott, G.; Edwards, J.; Mardis, E. R.; Engstrand, L. G.; Gordon, J. I. The

complete genome sequence of a chronic atrophic gastritis *Helicobacter pylori* strain: evolution during disease progression. *Proc. Natl. Acad. Sci. USA* **2006**, *103*, 9999–10004.

19. Smith, K. S.; Jakubzick, C.; Whittam, T. S.; Ferry, J. G. Carbonic anhydrase is an ancient enzyme widespread in prokaryotes. *Proc. Natl. Acad. Sci. USA* **1999**, *96*, 15184–15189.
20. Supuran, C. T.; Scozzafava, A.; Casini, A. Development of sulfonamide carbonic anhydrase inhibitors. In *Carbonic Anhydrases: Its Inhibitors and Activators;* Supuran, C. T.; Scozzafava,A.; Conway, J., Eds.; CRC Press: Boca Raton, FL, **2004**; pp 67–147.
21. Supuran, C. T.; Scozzafava, A. Carbonic anhydrases as targets for medicinal chemistry. *Bioorg. Med. Chem.* **2007**, *15*, 4336–4350.
22. Alterio, V.; De Simone, G.; Monti, S. M.; Scozzafava, A.; Supuran, C. T. Carbonic anhydrase inhibitors: inhibition of human, bacterial, and archaeal isozymes with benzene-1,3-disulfonamides: solution and crystallographic studies. *Bioorg. Med. Chem. Lett.* **2007**, *17*, 4201–4207.
23. Temperini, C.; Scozzafava, A.; Vullo, D.; Supuran, C. T. Carbonic anhydrase activators. Activation of isozymes I, II, IV, VA, VII, and XIV with L- and D-histidine and crystallographic analysis of their adducts with isoform II: engineering proton-transfer processes within the active site of an enzyme. *Chemistry* **2006**, *12*, 7057–7066.
24. Cronk, J. D.; Endrizzi, J. A.; Cronk, M. R.; O'neill, J. W.; Zhang, K. Y. Crystal structure of *E. coli* beta-carbonic anhydrase, an enzyme with an unusual pH-dependent activity. *Protein Sci.* **2001**, *10*, 911–922.
25. Mitsuhashi, S.; Mizushima, T.; Yamashita, E.; Yamamoto, M.; Kumasaka, T.; Moriyama, H.; Ueki, T.; Miyachi, S.; Tsukihara, T. X-ray structure of beta-carbonic anhydrase from the red alga, *Porphyridium purpureum*, reveals a novel catalytic site for CO_2 hydration. *J. Biol. Chem.* **2000**, *275*, 5521–5526.
26. Kimber, M. S.; Pai, E. F. The active site architecture of *Pisum sativum* beta-carbonic anhydrase is a mirror image of that of alpha-carbonic anhydrases. *EMBO J.* **2000**, *19*, 1407–1418.
27. Hiltonen, T.; Björkbacka, H.; Forsman, C.; Clarke, A. K.; Samuelsson, G. Intracellular beta-carbonic anhydrase of the unicellular green alga *Coccomyxa*. Cloning of the cDNA and characterization of the functional enzyme overexpressed in *Escherichia coli*. *Plant Physiol.* **1998**, *117*, 1341–1349.
28. Strop, P.; Smith, K. S.; Iverson, T. M.; Ferry, J. G.; Rees,D. C. Crystal structure of the "cab"-type beta class carbonic anhydrase from the archaeon *Methanobacterium thermoautotrophicum*. *J. Biol. Chem.* **2001**, *276*, 10299–10305.
29. Smith, K. S.; Ferry, J. G. Prokaryotic carbonic anhydrases. *FEMS Microbiol. Rev.* **2000**, *24*, 335–366.
30. Chirica, L. C.; Elleby, B.; Lindskog, S.; Cloning, expression and some properties of alpha-carbonic anhydrase from *Helicobacter pylori*. *Biochim. Biophys. Acta.* **2001**, *1544*, 55–63.
31. Nishimori, I.; Minakuchi, T.; Morimoto, K.; Sano, S.; Onishi, S.; Takeuchi, H.; Vullo, D.; Scozzafava, A.; Supuran, C. T. Carbonic anhydrase inhibitors: DNA cloning and inhibition studies of the alpha-carbonic anhydrase from *Helicobacter pylori*, a new target for developing sulfonamide and sulfamate gastric drugs. *J. Med. Chem.* **2006**, *49*, 2117–2126.
32. Nishimori, I.; Vullo, D.; Minakuchi, T.; Morimoto, K.; Onishi, S.; Scozzafava, A.; Supuran,C. T. Carbonic anhydrase inhibitors: cloning and sulfonamide inhibition studies of a carboxyterminal truncated alpha-carbonic anhydrase from *Helicobacter pylori*. *Bioorg. Med. Chem. Lett.* **2006**, *16*, 2182–2188.

33. Nishimori, I.; Minakuchi, T.; Kohsaki, T.; Onishi, S.; Takeuchi, H.; Vullo, D.; Scozzafava, A.; Supuran, C. T. Carbonic anhydrase inhibitors: the beta-carbonic anhydrase from *Helicobacter pylori* is a new target for sulfonamide and sulfamate inhibitors. *Bioorg. Med. Chem. Lett.* **2007**, *17*, 3585–3594.

34. Nishimori, I.; Onishi, S.; Takeuchi, H.; Supuran, C. T. The alpha and beta classes carbonic anhydrases from *Helicobacter pylori* as novel drug targets. *Curr. Pharm. Des.* **2008**, *14*, 622–630.

35. Chirica, L. C.; Petersson, C.; Hurtig, M.; Jonsson, B. H.; Borén, T.; Lindskog, S. Expression and localization of alpha- and beta-carbonic anhydrase in *Helicobacter pylori*. *Biochim. Biophys. Acta.* **2002**, *1601*, 192–199.

36. Guilloton, M. B.; Korte, J. J.; Lamblin, A. F.; Fuchs, J. A.; Anderson, P. M. Carbonic anhydrase in *Escherichia coli*. A product of the cyn operon. *J. Biol. Chem.* **1992**, *267*, 3731–3734.

37. Scott, D. R.; Weeks, D.; Hong, D. C.; Postius, S.; Melchers, K.; Sachs, G. The role of internal urease in acid resistance of *Helicobacter pylori*. *Gastroenterology* **1998**, *114*, 58–70.

38. Puscas, I. Treatment of gastroduodenal ulcers with carbonic anhydrase inhibitors. *Ann. N.Y. Acad. Sci.* **1984**, *429*, 587–591.

39. Valean, S.; Vlaicu, R.; Ionescu, I. Treatment of gastric ulcer with carbonic anhydrase inhibitors. *Ann. N.Y. Acad. Sci.* **1984**, *429*, 597–600.

40. Davenport, H. The mechanisms of acid secretion. *Gastroenterology* **1968**, *54* (Suppl. 4), 702–705.

41. Penton, J. G.; McColl, K. E. Eradication of *Helicobacter pylori*: an objective assessment of current therapies. *Br. J. Clin. Pharmacol.* **1997**, *43*, 223–243.

42. Morishita, S.; Nishimori, I.; Minakuchi, T.; Onishi, S.; Takeuchi, H.; Sugiura, T.; Vullo, D.; Scozzafava, A.; Supuran,C. T. Cloning, polymorphism and inhibition of β-carbonic anhydrase of *Helicobacter pylori*. *J. Gastroenterol.* **2008**, *43*, 849–857.

43. Shahidzadeh, R.; Opekun, A.; Shiotani, A.; Graham, D. Y. Effect of the carbonic anhydrase inhibitor, acetazolamide, on *Helicobacter pylori* infection *in vivo*: a pilot study. *Helicobacter* **2005**, *10*, 136–138.

44. Andrutis, K. A.; Fox, J. G.; Schauer, D. B.; Marini, R. P.; Murphy, J. C.; Yan, L.; Solnick, J. V. Inability of an isogenic urease-negative mutant strain of *Helicobacter mustelae* to colonize the ferret stomach. *Infect. Immun.* **1995**, *63*, 3722–3725.

45. Eaton, K. A.; Krakowka, S. Effect of gastric pH on urease-dependent colonization of gnotobiotic piglets by *Helicobacter pylori*. *Infect. Immun.* **1994**, *62*, 3604–3607.

46. Takeuchi, H.; Nakazawa, T.; Okamoto, T.; Shirai, M.; Kimoto, M.; Nishioka, M.; Con, S. A.; Morimoto, N.; Sugiura, T. Cell elongation and cell death of *Helicobacter pylori* is modulated by the disruption of *cdrA* (cell division-related gene A). *Microbiol. Immunol.* **2006**, *50*, 487–497.

CHAPTER 18

QSAR of Carbonic Anhydrase Inhibitors and Their Impact on Drug Design

ADRIANO MARTINELLI and TIZIANO TUCCINARDI
Dipartimento di Scienze Farmaceutiche, Università di Pisa, Via Bonanno 6, 56126 Pisa, Italy

18.1 INTRODUCTION*

Carbonic anhydrases (CAs) are ubiquitous in all kingdoms, such as archaea, bacteria, algae, and green plants as well as higher animals including vertebrates. They are encoded by five distinct gene families: the α-CAs (present in vertebrates), the β-CAs (mainly present in bacteria and plants), the γ-CAs (mainly present in archaea), and the δ- and ε-CAs (present in marine diatoms and chemolithoautotrophic bacteria, respectively).

In vertebrates, CAs are widely distributed zinc metalloenzymes and are implicated in a variety of physiological functions. These enzymes catalyze the reversible hydration of CO_2 to form HCO_3^-, which is involved in many biosynthetic reactions, among which are gluconeogenesis, lipogenesis, the synthesis of certain amino acids, and pyrimidine nucleotide biosynthesis. Moreover, these enzymes are involved in pH homeostasis, bone resorption, ion transport, electrolyte secretion in a variety of tissues, calcification, and tumorigenesis.[1] Thus, it is not surprising that many of their isozymes have been discovered as important targets for inhibitors with clinical applications.

Quantitative structure–activity relationships (QSARs) are widely used in modeling a variety of physicochemical parameters as well as the biological activity of chemically active compounds. Molecular descriptors, which are numerical representations of molecular structures, are used to perform QSAR analysis. Up to now, a large number of substituted derivatives of aromatic/heterocyclic, and more recently aliphatic, sulfonamides have been synthesized and tested for their CA inhibitory activity, and subsequently many QSAR studies have been reported.

*A brief introduction to QSAR and 3D-QSAR is reported in Chapter 27.

18.2 QSAR STUDIES

Figure 18.1 shows the general scaffolds of the CA ligands that have been investigated by means of QSAR techniques. All the compounds except those characterized by scaffold G possess a sulfonamide substituent as a zinc binding group (ZBG).

Table 18.1 reports all the main QSAR models for CA ligands. A large number of models have been reported for CA I, II, and IV; one QSAR model is reported also for CA VI, IX, and XIII. In some cases, two (CA I and CA II) or three (CA I, II, and IV) isoforms together were taken into consideration.

Over the past 10 years, QSAR studies on CA inhibitors have not made use of classical descriptors like Hammet's σ constant or molecular refractivity (MR), and the octanol/water partition coefficient (log *P*) has only been considered in some cases, according to either a linear[2,3,8,23,24] or a quadratic[11,16] relationship.

Two kinds of descriptors have typically been used for developing these models: topological indexes[5–7,9,11,12,15,17–21,23,25,28–30,32] and quantum mechanically calculated parameters.[2–4,8,10,14,22,24,26,31] One of the most widely used topological indexes is the Kier's first-order valence molecular connectivity index ${}^1\chi^v$, which describes the degree of branching, connectivity of atoms, and unsaturation in the molecule.[5,6,11,12,16,20] Other commonly used descriptors are the electrotopological state (E-state) index of atoms (*S*, which takes into account the electronic effects of atoms/substituents), the Randic connectivity index ${}^m\chi$, and the Balaban (*J*), Wiener (*W*), and

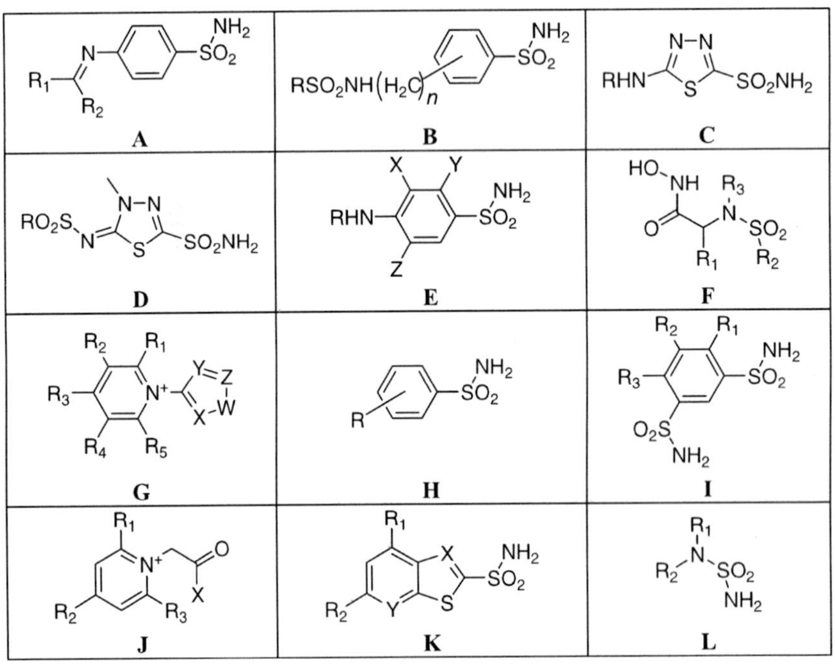

FIGURE 18.1 General scaffold of the CA inhibitors used for the development of QSAR models.

TABLE 18.1 Main QSAR Models of CA Inhibitors

CA	Compound	QSAR Equation	Statistics	Reference
CA I	A	$\log K_I = 0.1003(\pm 0.0189)D_X \pm 0.0949(\pm 0.0203)D_Z +3.133(\pm 0.635)D_I -736.7(\pm 95.9)Q_H +0.3893 (\pm 0.0473)\log P +80.81(\pm 16.52)S_M^E -907.7 (\pm 148.0)(\pm S_P^E -0.5S_0^E)+472.5(\pm 60.1)$	$n=35$, $r_{cv}^2 = 0.683$, $s=0.190$, $F_{3.66}=19.0$	2
CA I	B	$\log IC_{50} = 5.79(\pm 2.12)Q_{S7}+4.26(\pm 1.01)Q_{H17} -0.0326(\pm 0.011)\mu_X+0.299(\pm 0.108)E_H +0.600(\pm 0.145)E_L-0.00212(\pm 0.00114)\Pi_{yy} -0.0209(\pm 0.0088)\Delta H_S+3.13(\pm 0.96)Q_{C1} -10.30(\pm 5.58)$	$n=72$, $r_{cv}^2 = 0.512$, $s=0.22$, $F_{3.10}=12.8$	3
CA I	C, D	$\log IC_{50} = 0.00929(\pm 0.00138)\Pi_{xx}-0.00572(\pm 0.00231)\Pi_{zz} -13.04(\pm 2.87)Q_{Nr2}+17.07(\pm 4.59)Q_{S1}+1.560(\pm 0.790)Q_{S2} +0.0690(\pm 0.0182)\mu_x-50.83(\pm 12.29)$	$n=40$, $r_{cv}^2 = 0.628$, $s=0.289$, $F_{2.87}=16.78$	4
CA I	F	$\log(\pm 1/K_I) = -0.177(\pm 0.060)^1\chi^v -0.335(\pm 0.103)S_S+0.513(\pm 0.161)S_N +0.427(\pm 0.121)I+5.755(\pm 1.181)$	$n=23$, $r=0.969$, $s=0.13$, $F_{4.19}=68.89$	5
CA I	C, D, H, K	$K_I = -0.2435(\pm 0.0334)^1\chi^v -0.8008(\pm 0.1736)IP_2 -2.0818(\pm 0.1703)IP_3+5.3835$	$n=41$, $r=0.9251$, $SE=0.4546$, $F=73.224$	6
CA I	G	$K_A(\pm\mu M) = 0.0002(\pm\int^{Ac})^2+0.013(\pm\int^{Ac})+0.21$	$n=42$	7
CA I	A	$\log C_I = -20.9(\pm 2.1)Q_S+98.7(\pm 8.1)Q_H+0.354(\pm 0.026)\log P -0.736(\pm 0.096)E_{SH}+0.221(\pm 0.057)E_H-0.226(\pm 0.032)\cos 2\Phi_H -0.042(\pm 0.036)\sin 2\Phi_H+0.021(\pm 0.023)\cos 4\Phi_L -0.717(\pm 0.068)\sin 4\Phi_L-23.7(\pm 6.0)$	$n=27$, $r_{cv}^2 = 0.876$, $s=0.08$, $F_{4.01}=43.8$	8
CA I	C, D, H, K	$\log K_i = 4.0181-0.3617(\pm 0.0323)^1\chi + 1.0254(\pm 0.2407)J -0.8013(\pm 0.1678)IP_1-1.3744(\pm 0.1498)IP_2 +0.5054(\pm 0.2647)IP_3$	$n=49$, $r=0.9499$, $SE=0.3974$, $F=79.447$	9

(continued)

TABLE 18.1 *(Continued)*

CA	Compound	QSAR Equation	Statistics	Reference
CA I	C, D, H, K	$\log K_I = 0.2208(\pm 2.7)\cos 2\Phi_H - 0.4274(\pm 5.6)\sin 2\Phi_H - 1.2627(\pm 5.3)\cos 4\Phi_L + 1.3262(\pm 12.7)\sin 4\Phi_L + 0.0210(\pm 4.2)\Delta H_S + 0.5637(\pm 7.7)\pi_{tail} + 0.5422(\pm 3.2)E_L + 7.9811(\pm 10.8)Q_M - 95.57(\pm 10.9)Q_N - 0.0040(\pm 3.6)\Pi_{zz} + 26.67(\pm 2.9)Q_H - 0.421(\pm 5.7)E_{SH} - 106.10$	$n = 144$, $r^2_{cv} = 0.802$, $s = 0.41$	10
CA I	C, D, H, K	$\log(\pm 1/K_I) = 0.699(\pm 0.346)\log P + 1.122(\pm 0.436)(\pm \log P)^2 - 3.032(\pm 1.754)^1\chi^v + 0.284(\pm 0.133)(\pm^1\chi^v)^2 + 12.538(\pm 5 : 399)$	$n = 25$, $r = 0.940$, $r^2_{cv} = 0.80$, $s = 0.50$, $F_{4.20} = 38.06$	11
CA I	J	$\log(\pm 1/K_I) = 0.671(\pm 0.530)^1\chi^v - 0.022(\pm 0.022)(\pm^1\chi^v)^2 + 0.870(\pm 0.389)I + 0.503(\pm 0.265)I_1 + 1.486(\pm 0.242)I_A + 0.972(\pm 3.165)$	$n = 94$, $r = 0.881$, $r^2_{cv} = 0.80$, $s = 0.48$, $F_{5.88} = 61.11$	11
CA I	C, D, H, K	$\log K_I = -0.4869(\pm 0.0523)^1\chi^v + 0.2686(\pm 0.1912)IP_1 - 1.1730(\pm 0.1705)IP_3$	$n = 50$, $r = 0.9031$, $SE = 0.5444$, $F = 67.824$	12
CA I	C, D, H, K	$\log K_I = -1.359 + 0.641(\pm 0.321)\delta(\pm -SO_2NH_2) - 1.619(\pm 0.363)I_1 - 2.834(\pm 0.436)I_2$	$n = 20$, $r = 0.899$, $r^2_{cv} = 0.807$, $F = 22.279$	13
CA I	L	$pK_I = [2.6265(\pm 1.2956)] + E_{HOMO}[0.4548(\pm 0.1182)] + E_{LUMO}[-0.7881(\pm 0.3071)] + \text{PMI-}X[-0.0014(\pm 0.0020)]$	$n_{tr} = 16$, $r = 0.9625$, $r^2_{cv} = 0.8952$, $s = 0.1393$, $F = 50.4040$, $n_{ts} = 6$	14
CA I	X	$\log K_A = -4.2817 + 4.1976(\pm 1.2697)^2\chi^{shape} - 3.7718(\pm 0.9688)^3\chi^{shape} + 0.0091(\pm 0.0056)\text{PNSA-1} - 0.0853(\pm 0.0372)\text{PPSA-2}$	$n = 17$, $SE = 2.5655$, $F = 4.729$	15
CA II	A	$\log K_I = 3.441(\pm 0.850)D_1 - 312.12(\pm 104.52)Q_H + 1.5494 \pm 0.2580)I_P - 1541.8(\pm 196.4)(\pm S^E_P - 0.5S^E_0) + 196.4(\pm 62.2)$	$n = 35$, $r^2_{cv} = 0.715$, $s = 0.318$, $F_{2.98} = 28.2$	2
CA II	B	$\log IC_{50} = 0.00263(\pm 0.00134)\Pi_{yy} + 0.00367(\pm 0.00123)\Pi_{zz} - 1.12(\pm 0.244)Q_{N13} + 0.0239(\pm 0.0117)\mu + 0.644(\pm 0.173)E_L - 0.115(\pm 0.046)\log P - 4.16(\pm 1.14)Q_{O_2} - 0.592(\pm 0.095)I_3 - 7.50(\pm 2.26)$	$n = 72$, $r^2_{cv} = 0.594$, $s = 0.24$, $F_{1.75} = 16.9$	3

CA II	X	$\log(\pm 1/IC_{50}) = 0.71\,{}^1\chi^v - 0.35\,{}^0\chi^v + 0.66\,{}^1\kappa - 0.23\,{}^2\kappa - 0.53\text{apol} + 0.67I + 0.18\log P - 0.37(\pm\log P)^2 + 0.46$	$n_{tr} = 280$, $r^2_{cv} = 0.82$, $s = 0.98$, $n_{ts} = 57$, $r^2_{pred} = 0.81$	16
CA II	C, D	$\log IC_{50} = 0.00892(\pm 0.00120)\Pi_{xx} - 6.68(\pm 1.56)Q_{Cr1} + 18.97(\pm 4.94)Q_{S1} - 0.736(\pm 0.266)E_H + 0.0667(\pm 0.0211)\mu_x - 0.0417(\pm 0.0160)\mu_z + 0.0275(\pm 0.0081)\Delta H_S - 64.15(\pm 14.45)$	$n = 40$, $r^2_{cv} = 0.475$, $s = 0.304$, $F_{2.47} = 11.70$	4
CA II	B, C, E	$pK_I = -0.0054(\pm 0.0012)W + 0.0037(\pm 8.9737\times 10^{-4})S_z + 1.2519(\pm 0.1542)J + 0.8344(\pm 0.2559)IP_1 + 1.2944(\pm 0.1577)IP_2 - 0.3717$	$n = 39$, $r = 0.9401$, $SE = 0.3235$, $F = 40.559$	17
CA II	A	$\log K_I = -2.3666(\pm 0.7414)J - 0.8239(\pm 0.1520)IP_1 + 4.8475$	$n = 35$, $r = 0.8418$, $s = 0.3664$, $F = 38.917$	18
CA II	F	$\log(\pm 1/K_I) = -0.089(\pm 0.087)\,{}^1\chi^v - 0.353(\pm 0.128)S_S + 0.597(\pm 0.209)S_N + 0.305(\pm 0.159)I + 5.065(\pm 1.561)$	$n = 28$, $r = 0.920$, $s = 0.19$, $F_{4.23} = 31.55$	5
CA II	C, D, H, K	$K_I = -0.0661(\pm 0.0164)\,{}^1\chi^v - 0.5308(\pm 0.0820)IP_2 - 1.4832(\pm 0.0810)IP_3 + 2.5489$	$n = 41$, $r = 0.9551$, $SE = 0.2148$, $F = 128.057$	6
CA II	H	$\log K_c = 304.2(\pm 64.80)\text{Max } 1{-}eR_O - 3.096(\pm 0.189)J + 2.916(\pm 0.205)\text{Min } n\text{-}n\ R_{C\text{-}H} + 12.20(\pm 0.533) \times \text{Avg } B_O - 67.01(\pm 5.261)$	$n = 29$, $r = 0.991$, $r^2_{cv} = 0.971$, $s = 0.204$, $F = 323.5$	16
CA II	A	$\log K_I = -2.815\times 10^{-3}(\pm 3.833\times 10^{-4})\text{WNSA} + 0.331(\pm 0.058)\text{RPCS} + 0.269(\pm 0.185)$	$n = 35$, $r = 0.869$, $r^2_{cv} = 0.702$, $s = 0.325$, $F = 49.45$	19
CA II	H	$\log K_I = 4.5778 + 0.1593(\pm 0.0812)\,{}^1\chi^v + 0.5928(\pm 0.1246)\sigma + 0.3306(\pm 0.0828)\pi$	$n = 16$, $r = 0.9655$, $SE = 0.1725$, $F = 89.3702$	20
CA II	A	$\log C_I = 62.2(\pm 9.4)Q_H - 0.160(\pm 0.033)\log P - 0.482(\pm 0.108)E_{SH} + 0.990(\pm 0.355)E_{SL} - 0.263(\pm 0.059)\cos 2\Phi_H + 0.222(\pm 0.080)\sin 2\Phi_H + 0.066(\pm 0.061)\cos 4\Phi_L + 0.270(\pm 0.056)\sin 4\Phi_L - 46.7(\pm 8.1)$	$n = 27$, $r^2_{cv} = 0.780$, $s = 0.18$, $F_{1.68} = 22.9$	8
CA II	H	$\log K_c = 18.3126 - 6.538(\pm 0.6004)J + 0.7504(\pm 0.1999)\,{}^1\chi - 0.0025(\pm 8.6322\times 10^{-4})W + 1.2837(\pm 0.1651)I$	$n = 29$, $r = 0.9873$, $SE = 0.2412$, $F = 231.859$	21

(continued)

TABLE 18.1 *(Continued)*

CA	Compound	QSAR Equation	Statistics	Reference
CA II	C, D, H, K	$\log K_I = 2.2447-0.1593(\pm0.0219)^1\chi+0.3728(\pm0.1656)J -0.5423(\pm0.1140)IP_1-0.9305(\pm0.1018)IP_2+0.1013(\pm0.1799)IP_3$	$n=49$, $r=0.9166$, $SE=0.2700$, $F=45.172$	9
CA II	C, D, H, K	$\log K_I = 7.53(\pm10.2)Q_M-10.29(\pm4.4)Q_C-1.32(\pm9.5)E_H -0.709(\pm4.8)E_{5L}+0.8105(\pm6.8)\pi_{tail} +0.0354(\pm8.6)\Delta H_S-38.65(\pm18.1)Q_O-0.1660(\pm2.1)\cos2\Phi_H +0.173(\pm11.7)\sin2\Phi_H-0.8583(\pm3.8)\cos4\Phi_L -0.1503(\pm2.1)\sin4\Phi_L+53.41$	$n=144$, $r^2_{cv}=0.854$, $s=0.39$, $F=85.06$	22
CA II	C, D, H, K	$\log K_I = -0.166(\pm2.1)\cos2\Phi_H+0.810(\pm11.6)\sin2\Phi_H -0.858(\pm3.9)\cos4\Phi_L-0.150(\pm2.0)\sin4\Phi_L-10.29(\pm4.4)Q_C +0.0354(\pm8.6)\Delta H_S-1.324(\pm9.5)E_H+0.467(\pm6.8)\pi_{tail} +7.53(\pm10.2)Q_M-38.65(\pm18.0)Q_O-0.709(\pm4.8)E_{SL}-53.41$	$n=144$, $r^2_{cv}=0.854$, $s=0.39$	10
CA II	C, D, H, K	$\log(\pm1/K_I) = 0.530(\pm0.311)\log P+0.911(\pm0.576)(\pm\log P)^2 -1.193(\pm1.187)^1\chi^v+0.120(\pm0.090)(\pm^1\chi^v)^2+9.663(\pm3:645)$	$n=23$, $r=0.921$, $r^2_{cv}=0.71$, $s=0:32$, $F_{4.18}=24.98$	11
CA II	J	$\log(\pm1/K_I) = 0.402(\pm0.247)^1\chi^v-0.014(\pm0.010)(\pm^1\chi^v)^2 +0.435(\pm0.181)I+0.489(\pm0.124)I_1+0.885(\pm0.112)I_A +4.546(\pm1.473)$	$n=94$, $r=0.903$, $r^2_{cv}=0.78$, $s=0.22$, $F_{5.88}=77.87$	11
CA II	H	$\log K_I = 1.6434+0.4033(\pm0:0546)\log P+0.4350(\pm0.0585)^2\chi +3.0451(\pm0.1104)IP_1+1.0549(\pm0.1383)IP_2$	$n=29$, $s=0.2180$, $r=0.9897$, $r^2_{cv}=0.9790$, $F=286.500$	23
CA II	C, D, H, K	$\log K_I = -0.1593(\pm0.0188)^1\chi^v-0.4052(\pm0.0940)IP_2 -0.8649(\pm0.0907)IP_3+3.0080$	$n=48$, $r=0.9226$, $SE=0.2542$, $F=83.853$	12
CA II	H	$\log K_I = 341.593-362.447(\pm126.664)MNC-0.629(\pm0.122)DMy -0.176(\pm0.057)\chi-0.259(\pm0.129)+0.217(\pm0.049)DMz +0.043(\pm0.013)H_f+2.132(\pm0.380)\log P$	$n=29$, $r=0.993$, $SE=0.187$, $F=223.468$	24
CA II	H	$\log K_I = 0.71(\pm0.67)^0\chi_{inf}-0.30(\pm1.37)^1\chi_{inf} -0.90(\pm0.41)^0\chi^v_{inf}-1.01(\pm0.65)^1\chi_{inf} -0.87(\pm0.21)N\text{-rings}+5.49(\pm2.77)$	$n=47$, $r=08541$, $r^2_{cv}=0.7296$, $F=22.12$	25

CA II	C, D, I, K	$\log K_I = 5.869 + 0.0017T_e - 0.225\mu + 0.0091S + 0.403\chi$	$n = 18$, $r^2_{cv} = 0.789$, $F = 19.4$	26
CA II	L	$pK_I = [2.4909(\pm 1.4516)] + E_{HOMO}[0.4819(\pm 0.1325)] + E_{LUMO}[-1.0320(\pm 0.3441)] + \text{PMI-}X[-0.0022(\pm 0.0022)]$	$n_{tr} = 16$, $r = 0.9644$, $r^2_{cv} = 0.8699$, $s = 0.1561$, $F = 53.3471$, $n_{ts} = 7$	14
CA II	X	$\log K_A = 2.1997 - 4.8511(\pm 0.1545)^2\chi^0 - 2.0137(\pm 0.6515)^1\chi^{shape} - 5.1231(\pm 0.9448)^3\chi^{shape} + 0.0061(\pm 0.0029)\text{PNSA-1}$	$n = 17$, $SE = 1.0391$, $F = 25.236$	15
CA II	H	$\log K_I = -2.205 + 1.739\text{NBR} - 0.279\text{NCA} + 0.182\text{NNA} + 0.282\text{ALP} + 0.977^2\text{AIC}$	$n = 36$, $r^2_{cv} = 0.777$, $F = 31.54$	27
CA II	E, H, I	$\log K_I = 534095.2166 + 0.5421(\pm 0.1363)^2\chi - 0.6371(\pm 0.1267)^3\chi - 0.3663(\pm 0.0783)^2\chi^v - 0.1761(\pm 0.0777)^3\chi^{shape} - 0.0060(\pm 0.0025)\text{PNSA-2} - 0.0535(\pm 0.0185)\text{DPSA}-3 + 534099.9358(\pm 239377.0497)\text{FPSA}-1 + 534099.9094(\pm 239377.0614)\text{FNSA-1}$	$n = 47$, $r^2_{cv} = 0.8014$, $SE = 0.1307$, $F = 23.920$	28
CA IV	B	$\log IC_{50} = -0.891(\pm 0.317)Q_{N13} + 0.322(\pm 0.134)E_H + 0.377(\pm 0.166)E_L - 0.921(\pm 0.367)D_1 - 0.184(\pm 0.060)\log P - 2.885(\pm 1.108)Q_{O_2} - 0.350(\pm 0.101)I_3 + 0.45(\pm 2.16)$	$n = 72$, $r^2_{cv} = 0.591$, $s = 0.25$, $F_{1.75} = 17.9$	3
CA IV	C, D	$\log IC_{50} = 0.00731(\pm 0.000965)\Pi_{xx} - 5.570(\pm 1.431)Q_{Cr1} + 11.46(\pm 4.90)Q_{S1} + 0.0602(\pm 0.0201)\mu_x - 37.16(\pm 13.68)$	$n = 40$, $r^2_{cv} = 0.444$, $s = 0.34$, $F_{1.48} = 15.07$	4
CA IV	F	$\log(\pm 1/K_I) = -0.158(\pm 0.038)^1\chi^v - 0.231(\pm 0.062)S_S + 0.338(\pm 0.101)S_N + 0.364(\pm 0.076)I + 6.580(\pm 0.742)$	$n = 28$, $r = 0.978$, $s = 0.09$, $F_{4.23} = 124.24$	5
CA IV	C, D, H, K	$K_I = -0.1608(\pm 0.0216)^1\chi^v - 0.7079(\pm 0.1093)IP_2 - 1.3426(\pm 0.1070)IP_3 + 3.5781$	$n = 41$, $r = 0.9307$, $SE = 0.2862$, $F = 79.787$	6
CA IV	C, D, H, K	$\log K_I = 2.7452 - 0.1811(\pm 0.0250)^1\chi + 0.5123(\pm 0.1866)J - 0.8039(\pm 0.1300)IP_1 - 0.9189(\pm 0.1161)IP_2 + 0.2796(\pm 0.2052)IP_3$	$n = 49$, $r = 0.9077$, $SE = 0.3080$, $F = 40.223$.	9
CA IV	C, D, H, K	$\log K_I = 0.414(\pm 4.3)\cos 2\Phi_H - 1.085(\pm 11.7)\sin 2\Phi_H - 2.053(\pm 7.5)\cos 4\Phi_L + 0.289(\pm 3.3)\sin 4\Phi_L + 0.0381(\pm 6.7)\Delta H_S - 1.221(\pm 6.8)E_H + 0.364(\pm 4.1)\pi_{tail} + 0.430(\pm 2.7)E_L + 5.049(\pm 5.6)Q_M - 34.1(\pm 5.9)Q_N - 0.0025(\pm 2.2)\Pi_{xx} - 35.50$	$n = 144$, $r^2_{cv} = 0.831$, $s = 0.45$	10

(continued)

TABLE 18.1 *(Continued)*

CA	Compound	QSAR Equation	Statistics	Reference
CA IV	C, D, H, K	$\log(\pm 1/K_I) = 0.603(\pm 0.301)\log P + 0.590(\pm 0.378)(\pm \log P)^2 - 1.591(\pm 1.589)^1\chi^v + 0.164(\pm 0.123)(\pm {}^1\chi^v)^2 + 9.946(\pm 4:859)$	$n = 24$, $r = 0.907$, $r^2_{cv} = 0.70$, $s = 0.43$, $F_{4.19} = 22.11$	11
CA IV	J	$\log(\pm 1/K_I) = 0.265(\pm 0.058)\mathrm{Pol} - 0.200(\pm 0.078)\log P - 0.020(\pm 0.014)(\pm \log P)^2 + 0.252(\pm 0.082)I_1 + 0.664(\pm 0.078)I_A + 5.369(\pm 0.352)$	$n = 96$, $r = 0.903$, $r^2_{cv} = 0.79$, $s = 0.16$, $F_{5.90} = 79.21$	11
CA IV	C, D, H, K	$\log K_I = -0.2309(\pm 0.0281)^1\chi^v - 0.4644(\pm 0.1303)\mathrm{IP}_2 - 0.8581(\pm 0.1218)\mathrm{IP}_3$	$n = 50$, $r = 0.8730$, $SE = 0.3629$, $F = 49.134$	12
CA VI	X	$\log K_A = 2.0758 - 0.4411(\pm 0.1912)^3\chi^{shape} - 0.0044(\pm 0.0044)\mathrm{TMSA} + 0.0252(\pm 0.0119)\mathrm{PPSA\text{-}2} + 0.0027(\pm 0.0025)\mathrm{PNSA\text{-}1}$	$n = 17$, $SE = 0.3110$, $F = 1.897$	15
CA IX	C, D, H, K	$\log K_I = 3.9171 - 0.5583(\pm 0.3274)^1\chi - 0.1447(\pm 0.212)^2\chi + 0.4171(\pm 0.2403)^2\chi^v + 0.0053(\pm 0.0038)\mathrm{PI} + 0.3771(\pm 0.1517)I_2 + 0.3480(\pm 0.1142)I_3$	$n = 32$, $r = 0.8158$, $SE = 0.2748$, $F = 8.2903$	29
CA XIII	C, D, H, K	$\log K_I = 0.013 + 1.341(\pm 0.446)J_{hetv} - 1.336 \times 10_{-3}(\pm 1.943 \times 10^{-4})W + 0.181(\pm 0.057)^1\chi^v - 0.798(\pm 0.324)J_{hetp}$	$n = 14$, $r = 0.988$, $SE = 0.091$, $F = 89.813$	30
CAs	X	$\log K_I = -3.51(\pm 0.62)Q_c + 41.7(\pm 17.2)S^E_N - 1.17(\pm 0.25)D_1 + 0.0827(\pm 0.0120)\mu_{S-N} - 0.0157(\pm 0.0032)\Delta H_S - 0.0205(\pm 0.00315)\Pi + 0.968(\pm 0.161)E_{L-H} + 7.03(\pm 5.43)$	$n = 36$, $r^2_{cv} = 0.658$, $s = 0.25$, $F_{2.84} = 17.92$	31
CAs	I	$\log \mathrm{IC}_{50} = 11.4514(\pm 1:7713)^1\chi - 0.1874(\pm 0.0279)\mathrm{PI} + 0.2918(\pm 0.1635)I_2 - 44.5688$	$n = 21$, $r = 0.8814$, $SE = 0.2917$, $F = 19.7259$	32

X, compounds characterized by different scaffolds; n, number of compounds;

n_{tr}, number of compounds in the training set; r, correlation coefficient; SE, standard error of estimation; F, variance ratio at the specified degree of freedom; s, standard deviation; r^2_{cv}, square of cross-validated correlation coefficient; n_{ts}, number of compounds in the test set; r^2_{pred}, square of correlation coefficient of the test set.

Szeged (Sz) indexes. In most cases, also indicator parameters (I) were considered to account for structural differences in the training set.

Quantum mechanical parameters have generally been calculated at the semi-empirical level, but more sophisticated methods were also used until the DFT level. Among these types of parameters, it was possible to find the Mulliken charge on atoms (Q), the dipole moment (μ), the polarizability components (Π), the energy of the orbitals (E), and the angles between the node in frontier occupied π orbitals and the ZBG group (Φ_H, Φ_L).

More than 50 CA QSAR models have been published, and more than 1500 CA ligands have been investigated. However, with only three exceptions, all the compounds were used for constructing the models; that is, they belonged to training sets. As already reported by Golbraikh and Tropsha,[33] the training set statistics are insufficient for assessing the predictivity of QSAR models, and it is fundamental to evaluate the correlation between the predicted and observed activities of compounds from external test sets. Hence, as most of the reported QSAR studies did not use external test sets, the predictive reliability of these models cannot be estimated and only their interpolative ability can be evaluated. In other words, the results can furnish important information for understanding the role of the molecular features in determining activity and selectivity, but their utility for driving the design of new ligands should be considered with caution.

Recently, Kumar and coworkers reported human CA I and II QSAR models for a series of sulfamide derivatives with known inhibitory activity.[14] The data set was divided into a training set of 16 compounds and a test set of 6 compounds for CA I, and a training set of 16 compounds and a test set of 7 compounds for CA II, using the random selection method. A multiple linear regression analysis (MLRA) was performed using molecular descriptors as independent parameters and inhibitory activity as dependent parameters. For the molecular descriptors, the highest occupied molecular orbital energy (E_{HOMO}), lowest unoccupied molecular orbital energy (E_{LUMO}), and principal moment of inertia at the X-axis (PMI-X) were used. For the CA I, the best model showed an r of 0.9625, a standard deviation of 0.1393, and a cross-validated r^2_{cv} of 0.8952. The QSAR analysis indicated that the inhibitory activity was improved by the electron withdrawing ability of the substituents together with their moderate hindrance. The validity of the model was also checked using the test set of six compounds that were not included in the model development. The model had a good predictive squared correlation coefficient ($r^2_{pred} = 0.6016$). For the CA II, the best model showed an r of 0.9644, a standard deviation of 0.1561, and an r^2_{cv} of 0.8699, obtained with leave-one-out (LOO) cross-validation. The validity of the model was also externally checked using a test set of seven compounds that were not included in the model development, and the model showed a good external predictivity ($r^2_{pred} = 0.7662$).

In 1999, Gao and Bajorath reported the comparison between binary and conventional 2D-QSAR models for a set of 280 CA II inhibitors characterized by a high degree of diversity.[16] For the development of the binary QSAR model, 55 compounds were defined as inactive and 225 compounds as active. The best model was obtained with a combination of six molecular descriptors: two connectivity indexes (${}^1\chi^v$ and ${}^2\chi^v$), one Kier shape index (${}^1\kappa$), one hydrogen bond acceptor parameter, the log P, and one descriptor for the number of sulfonamide fragments (fn). The LOO cross-validated accuracy was 96% for the active compounds, 82% for the inactive compounds, and 93% for all the compounds.

The QSAR equations for the 2D-QSAR models are listed in Table 18.1. They were obtained by using seven molecular descriptors: the zero- and first-order atomic valence connectivity indexes (${}^0\chi^v$ and ${}^1\chi^v$), the Kier first and second shape indexes (${}^1\kappa$ and ${}^2\kappa$), the sum of atomic polarizability (apol), and the presence of unsubstituted sulfonamido group and the log *P*. The statistical results highlighted a LOO cross-validated correlation coefficient of 0.82 and a standard error of estimation of 0.98.

The molecular descriptors used were similar to that employed for the development of the binary QSAR model, suggesting that both binary and 2D-QSAR studies captured similar structural features of the inhibitors.

To evaluate the predictive value of both QSAR models, 57 known CA II inhibitors that were not included in the training set were analyzed. For the binary QSAR model, 80% of the inactive compounds and 94% of the active compounds were correctly predicted. For the 2D-QSAR model, 81% ($r^2_{pred} = 0.81$) of the variance in the test set was accounted for. On the basis of the results, both QSAR models have significant predictive power, suggesting the possibility of using binary QSAR to efficiently screen a large compound database for compounds with likely activity and applying conventional 2D-QSAR to aid in the identification of the most active compounds.

Tarko and Supuran recently reported the first QSAR model for human CA XIV inhibitors.[34] For each of the 55 sulfonamides, approximately 1800 descriptors were calculated. After statistical calculations, the best QSAR model was characterized by 285 significant descriptors and 46 compounds (nine outliers were eliminated). The model showed a correlation coefficient (r^2) of 0.8911, a Pearson cross-validated square correlation (r^2_{cv}) of 0.8567, and a Fischer value (*F*) of 67.07.

The analysis was also extended to the ability of these compounds to inhibit CA XII, but the qualities of the QSAR equations were weak ($r^2 = 0.1771$, $r^2_{cv} = 0.1088$, and $F = 5.70$), suggesting that some compounds may have slightly different inhibition mechanisms against the two investigated CA isoforms.

Using a database of 340 aromatic/heterocyclic sulfonamides, Hemmateenejad and coworkers reported the development of QSAR models for CA isoenzyme types I, II, IV, and IX obtained by applying MLRA with stepwise selection of variables.[35] The sulfonamide derivatives used in this study belonged to a wide variety of molecular families, and to account for the structural effects on the CA inhibitory activity of the compounds, a variety of molecular descriptors were used. Specifically, the log *P* and hydration energy (HE) were considered as descriptors for hydrophobic effects; the steric effects were considered by means of surface area, volume, and topological indices; and the electronic descriptors such as E_{HOMO}, E_{LUMO}, and local charges were derived from AM1 calculations. Indices of electronegativity, electrophilicity, hardness, and softness were calculated from the HOMO and LUMO energies. Constitutional descriptors were used to describe the effect of different fragments of the molecules on their inhibitory activity. Among the investigated sulfonamide derivatives, the inhibitory activity toward the CA I isoenzyme was available for 312 molecules. Among them, 250 molecules were used as a calibration set and the rest as prediction set molecules. The best model was developed removing nine outliers and showed a correlation coefficient (r^2) of 0.904, a LOO cross-validated correlation coefficient (r^2_{cv}) of 0.881, and a Fischer (F) value of 173.2. Furthermore, the prediction of the 62 compounds that were not included in the training set showed a standard deviation of errors of prediction (SDEP) of 0.57. For the CA II subtype, 312 molecules were available, and 250 were used as the training set and 62 as the external test set. Removing 10 outliers, the obtained statistical results for the best model were similar to those obtained for the CA I activity. The model showed an r^2 of 0.893, an r^2_{cv} of 0.882, and a Fischer value of 138.997. The prediction of the activity for the external test set showed an SDEP of 0.46. The model for CA IV inhibition was developed using a training set of 95 compounds; considering one compound as an outlier, the best QSAR model showed an r^2 of 0.889, an r^2_{cv} of 0.823, and a Fischer value of 78.186, and the prediction of the external test made up of 20 compounds showed an SDEP of 0.46. Finally, the best QSAR model for CA IX inhibition was obtained by using 104 compounds for the calibration, and was validated using 30 prediction set molecules. This model had higher statistical quality relative to the models obtained for the other types of CA isoenzymes. It showed an r^2 of 0.935, an r^2_{cv} of 0.909, a Fischer value of 104.29, and an external test set SDEP of 0.57.

18.3 NONCLASSICAL QSAR STUDIES

In 2001, Gao tested the application of 3D H-suppressed BCUT metrics (BCUTs) in binary QSAR analysis, using the same 337 ligands investigated in 1999.[36] BCUTs are an extension of Burden's parameters, which are based on a combination of the atomic number for each atom and a description of the nominal bond type for adjacent and nonadjacent atoms, and incorporate both connectivity information and atomic properties related to intermolecular interactions.[37]

Using a training set of 287 CA II inhibitors, the best binary QSAR model was obtained with a combination of 23 molecular descriptors including four connectivity indexes ($^2\chi$, $^0\chi^v$, $^1\chi^v$, and $^2\chi^v$), three shape indexes ($^2\kappa$, $^1\kappa_\alpha$, $^3\kappa_\alpha$), log P, and 15 BCUTs. The LOO cross-validated accuracy was 91% for the active compounds, 92% for the inactive compounds, and 91% for all the compounds. The prediction of the activity of the 50 compounds that were not included in the training set resulted in a correct estimation for 90% of the inactive compounds and 98% of the active compounds. The comparison of these statistical results with those reported for the previous published binary QSAR reveals only a slight improvement of the BCUT binary QSAR model (see Table 18.2); however, the two binary QSAR models were not obtained with the same training and test sets.

In 2002, Gao and coworkers explored the possibility of applying a genetic algorithm (GA) as a variable selection method for developing binary QSAR models.[38] The QSAR models were developed using the 280 CA II inhibitors already analyzed. The GA parameters used an initial population of 200, a reproductive population size of 10 chromosomes, and a mutation rate of 5%. A total of 1400 combinations of molecular descriptors were explored, and within 60 GA generations, a binary QSAR model was obtained with a LOO cross-validated accuracy of 90% for actives, 94% for

TABLE 18.2 Comparison of Binary (1–3) and SVM (4) QSAR Models

			Training Set Errors, LOO Cross-Validation (%)			Test Set Errors (%)			
Model	Training Set	Test Set	Active	Inactive	Overall	Active	Inactive	Overall	Reference
1	280	57	96	82	93	94	80	92	16
2	287	50	91	92	91	98	90	96	36
3	280	57	90	94	91	91	100	93	38
4	280	57	98	80	95	96	80	95	40

inactives, and 91% for all the compounds. Thus, the obtained QSAR model was also tested for its ability to predict the activity of 57 CA II inhibitors not included in the training set. As a result, 91% of the active compounds were correctly predicted, and all inactive compounds (100%) were correctly predicted.

Comparing this model to the binary QSAR models described above, the last approach assures a better predictive ability (see Table 18.2), also improving the speed of the calculations.

In 2002, Mattioni and Jurs reported the development of QSAR and classification models for a set of 142 inhibitors of CA I, II and IV.[39] The models were developed using topological, geometric, and electronic features, and MLRA, and were accompanied by a simulated annealing optimization algorithm employed to survey the descriptor pool to find models with low training set rms errors. Descriptors derived from this analysis were then passed to a three-layer computational neural network (CNN) for analysis. The CNNs consisted of three layers, input, hidden, and output, and were able to generate nonlinear models with the descriptors to produce predicted values comparable to the experimental values. Finally, a nonlinear feature selection was performed on the obtained results by using simulated annealing and GA routines accompanied by a nonlinear CNN.

For CA I, an 8-5-1 CNN committee produced the best model, obtaining rms errors of 0.105 log K_I ($r^2=0.994$), 0.133 log K_I ($r^2=0.990$), and 0.208 log K_I ($r^2=0.980$) for the training, cross-validation, and prediction sets, respectively. For CA II, the fully nonlinear CNN model obtained with a 9-5-1 CNN committee produced the best model, resulting in rms errors of 0.140 log K_I ($r^2=0.992$), 0.163 log K_I ($r^2=0.990$), and 0.231 log K_I ($r^2=0.971$) for the training, cross-validation, and prediction sets, respectively. Finally, a committee of five 8-5-1 CNNs produced the best nonlinear CA IV model. Rms errors for the training, cross-validation, and prediction sets of 0.147 log K_I ($r^2=0.992$), 0.170 log K_I ($r^2=0.995$), and 0.211 log K_I ($r^2=0.991$), respectively, were obtained from the model. In addition, classification models were developed using *k*-nearest neighbor (*k*NN) analysis. As a result, a three-descriptor model proved to be able to label compounds as active or inactive inhibitors.

In 2003, Zernov and coworkers tested the possibility of using support vector machines (SVM) for estimating the activity of CA II inhibitors.[40] To compare the SVM results with those reported by applying the binary QSAR methodology, the authors used the series of 337 CA II inhibitors that were already used by Gao and coworkers[16,36,38] and used the same six descriptors employed by Gao and Bajorat (two connectivity indexes, one Keir shape index, one hydrogen bond acceptor parameter, the calculated octanol/water partition coefficient, and one descriptor for the number of sulfonamide fragments) to develop the binary QSAR model. The statistical results for the best SVM model showed a LOO cross-validated accuracy of 98% for actives, 80% for inactives, and 95% for all the compounds. Regarding the prediction of the activity of the 57 CA II inhibitors not included in the training set, 96% of the active compounds were correctly predicted, and 80% of the inactive compounds were correctly predicted.

Table 18.2 summarizes the statistical results of the three reported binary QSAR models and the best SVM model. From these data, the SVM model shows the best statistical results in terms of training set errors and close to the best in terms of external test set predictivity.

In 2006, Popelier and Smith developed a QSAR model for 22 CA II inhibitors using the quantum topological molecular similarity (QTMS) approach.[41] This method uses quantum chemical topology (QCT) to define electronic descriptors drawn from *ab initio* wave functions of geometry-optimized molecules. Essentially this method proposes a chemometric analysis using QCT descriptors calculated from quantum mechanical calculations. Molecular electron densities contain special points, named critical points, where the gradient of the electron density vanishes. Among the various types of critical points, the bond critical point (BCP) can be investigated for developing QCT descriptors. The BCP appears between two nuclei that are bonded, and certain functions, if evaluated at a given BCP, are able to characterize the corresponding bond and produce the QCT descriptors.

The best model showed good statistical results, in particular a correlation coefficient (r^2) of 0.9483 and a cross-validated correlation coefficient (q^2) of 0.914 obtained with two latent variables (LVs).

Very recently, Eroglu and Türmen reported QSAR studies of 30 CA II inhibitors using molecular descriptors calculated by means of DFT/B3LYP quantum mechanics studies.[42] The QSAR models were obtained by using quantum mechanical descriptors: dipole moment, average polarizability, ionization potential, electron affinity, LUMO energy, HOMO energy, total energy at 0 K, entropy at 298 K, electronegativity, hardness, electrophilicity, and differences between HOMO and LUMO energies. The best model showed a correlation coefficient (r^2) of 0.847, a LOO cross-validated coefficient (r^2_{cv}) of 0.775, and a Fischer value (F) of 34.67, and was developed using the dipole moment, the electronegativity, the entropy at 298 K, and the total energy at 0 K as molecular descriptors.

The quantum mechanical descriptors used to build the model were also calculated using the semiempirical AM1 method, and the statistical results obtained from the QSAR obtained with these descriptors clearly demonstrates that the DFT-based quantum mechanical descriptors led to better correlation relationships than the corresponding descriptors based on the AM1 method.

Using the same approach, the authors published a second QSAR analysis for a set of 18 CA II inhibitors.[26] The best model was developed using the dipole moment, the electronegativity, the entropy at 298 K, and the total energy at 0 K as molecular descriptors, and the statistical results showed a correlation coefficient (r^2) of 0.857, a LOO cross-validated coefficient (r^2_{cv}) of 0.789, and a Fischer value (F) of 19.4.

Finally, in 2007, Jalali-Heravi and Kyani reported the application of the genetic algorithm-kernel partial least square (GA-KPLS) as a nonlinear feature selection method, and this approach was combined with artificial neural network (ANN)[43] to develop a nonlinear QSAR model for predicting activities of a series of 114 substituted aromatic sulfonamides as CA II inhibitors.[44] The KPLS has two main advantages: simplicity and the possibility of application in a wide range of nonlinearities because one can choose different kinds of suitable kernels. After the development of the GA-KPLS nonlinear feature selection method, the eight chosen descriptors were considered as input for the ANN calculations. The results of the best QSAR model obtained with the GA-KPLS approach were compared to those of the QSAR model obtained with GA-PLS and the linear feature selection technique. The statistical parameters obtained by LOO cross-validation for GA-PLS-ANN, GA-KPLS-ANN, and the linear QSAR models highlighted that the GA-KPLS-ANN model was the best, with an r^2 of 0.899 and an r^2_{cv} of 0.800, suggesting that the nonlinear feature selection method of GA-KPLS in choosing variables may be a promising approach.

18.4 3D-QSAR STUDIES

To date, 21 3D-QSAR models of CAs have been published, 5 models have been proposed for CA I and IV, 10 models for CA II, and 1 model has been proposed for CA IX inhibition (Table 18.3, Fig. 18.2).

Most of the reported models were developed using ligands characterized by different central scaffolds, and some models were also reported using ligands with different ZBG.

Hillebrecht and coworkers reported the development of 3D-QSAR models for CA I, II, and IV using the CoMFA and CoMSIA methods, and also testing the possibility of using the AFMoC approach (adaptation of fields for molecular comparison).[45] The last approach is based on the knowledge-based scoring function DrugScore,[51] which is used for calculating interaction fields, and is obtained by multiplying the DrugScore potentials with weights derived from Gaussian functions (with $\sigma = 0.85$) which were centered at the ligand atom positions.

From the literature, an initial data set of 1748 compounds was retrieved and was reduced to a structurally diverse set of 173 ligands; 96 compounds were used for developing the models (training set) and 48 for testing the predictive ability (external test set). The ligands alignment was carried out using a receptor-guided method. The molecules were manually placed into the binding pocket of X-ray structures of CA I, II, and IV, matching the ligands onto corresponding atoms in the crystal structure references. Once the anchor fragment was superimposed onto the crystal coordinates, the remaining torsions of the ligand were adjusted to achieve maximal overlap to an appropriate crystal structure template, and the system was then minimized, keeping the protein and ZBG of the ligands fixed. Using this method, three different alignments were generated, using CA I, II, and IV as X-ray structures. The CoMFA, CoMSIA, and

TABLE 18.3 Main 3D-QSAR Models of CA Inhibitors

CA	Compound	Alignment	Statistics	Software	Reference
CA I	A–I	Receptor	$n_{tr} = 96$, $q^2 = 0.799$, $F = 325.628$, $NC = 5$, $n_{ts} = 48$, $r^2_{pred} = 0.746$	CoMFA	45
CA I	A–I	Receptor	$n_{tr} = 96$, $q^2 = 0.840$, $F = 387.312$, $NC = 4$, $n_{ts} = 48$, $r^2_{pred} = 0.816$	CoMSIA	45
CA I	A–I	Receptor	$n_{tr} = 96$, $q^2 = 0.820$, $F = 278.600$, $NC = 6$, $n_{ts} = 48$, $r^2_{pred} = 0.781$	AFMoC	45
CA I	A–D, G, F	Receptor	$n_{tr} = 60$, $q^2 = 0.650$, $F = 211.6$, $NC = 5$, $n_{ts} = 27$, $r^2_{pred} = 0.60$	CoMFA	46
CA I	A–D, G, F	Receptor	$n_{tr} = 60$, $q^2 = 0.658$, $F = 161.4$, $NC = 4$, $n_{ts} = 27$, $r^2_{pred} = 0.68$	CoMSIA	46
CA II	A–I	Receptor	$n_{tr} = 96$, $q^2 = 0.853$, $F = 423.840$, $NC = 6$, $n_{ts} = 48$, $r^2_{pred} = 0.767$	CoMFA	45
CA II	A-I	Receptor	$n_{tr} = 96$, $q^2 = 0.860$, $F = 453.478$, $NC = 5$, $n_{ts} = 48$, $r^2_{pred} = 0.825$	CoMSIA	45
CA II	A–I	Receptor	$n_{tr} = 96$, $q^2 = 0.791$, $F = 244.000$, $NC = 4$, $n_{ts} = 48$, $r^2_{pred} = 0.658$	AFMoC	45
CA II	A–D, G, F	Receptor	$n_{tr} = 60$, $q^2 = 0.568$, $F = 156.2$, $NC = 4$, $n_{ts} = 27$, $r^2_{pred} = 0.58$	CoMFA	46
CA II	A–D, G, F	Receptor	$n_{tr} = 60$, $q^2 = 0.645$, $F = 166.0$, $NC = 5$, $n_{ts} = 27$, $r^2_{pred} = 0.66$	CoMSIA	46
CA II	J	Docking	$n_{tr} = 51$, $q^2 = 0.623$, $F = 498.1$, $NC = 6$, $n_{ts} = 9$, $r^2_{pred} = 0.74$	CoMFA	47
CA II	J	Docking	$n_{tr} = 51$, $q^2 = 0.562$, $F = 392.0$, $NC = 8$, $n_{ts} = 10$, $r^2_{pred} = 0.62$	CoMSIA	47
CA II	A–D, K–N	Docking	$n_{tr} = 220$, $q^2 = 0.72$, $NC = 4$, $n_{ts} = 15$, $SDEP_{ts} = 0.44$, $n_{ts} = 28$, $SDEP_{ts} = 0.63$, $n_{ts} = 31$, $SDEP_{ts} = 0.41$	Golpe	48
CA II	A–I	Ligand	$n_{tr} = 138$, $q^2 = 0.798$, $F = 243.854$, $NC = 4$, $n_{ts} = 663$, $r^2_{pred} = 0.454$	CoMFA	49

(continued)

TABLE 18.3 *(Continued)*

CA	Compound	Alignment	Statistics	Software	Reference
CA II	A–I	Ligand	$n_{tr}=138$, $q^2=0.822$, $F=441.585$, $NC=2$, $n_{ts}=663$, $r^2_{pred}=0.482$	CoMSIA	49
CA IV	A–I	Receptor	$n_{tr}=96$, $q^2=0.822$, $F=398.011$, $NC=5$, $n_{ts}=48$, $r^2_{pred}=0.838$	CoMFA	45
CA IV	A–I	Receptor	$n_{tr}=96$, $q^2=0.851$, $F=466.890$, $NC=4$, $n_{ts}=48$, $r^2_{pred}=0.833$	CoMSIA	45
CA IV	A–I	Receptor	$n_{tr}=96$, $q^2=0.753$, $F=191.100$, $NC=4$, $n_{ts}=48$, $r^2_{pred}=0.590$	AFMoC	45
CA IV	A–D, G, F	Receptor	$n_{tr}=60$, $q^2=0.599$, $F=293.6$, $NC=7$, $n_{ts}=27$, $r^2_{pred}=0.63$	CoMFA	46
CA IV	A–D, G, F	Receptor	$n_{tr}=60$, $q^2=0.674$, $F=201.7$, $NC=6$, $n_{ts}=27$, $r^2_{pred}=0.62$	CoMSIA	46
CA IX	A–D, K	Docking	$n_{tr}=87$, $q^2=0.74$, $NC=3$, $n_{ts}=20$, $SDEP_{ts}=0.58$, $n_{ts}=17$, $SDEP_{ts}=0.52$	Golpe	50

n_{tr}, number of compounds in the training set; F, variance ratio at the specified degree of freedom; q^2, square of cross-validated correlation coefficient; n_{ts}, number of compounds in the test set; NC, principal components; r^2_{pred}, square of the correlation coefficient of the prediction of the external test set; $SDEP_{pr}$, standard deviation of error of the prediction of the external test set.

AFMoC analyses revealed that the alignment obtained by using the CA II X-ray structure showed the best results, and among the three methods, CoMSIA showed the best performance. As shown in Table 18.3, for CA I, the CoMSIA model for five components showed a correlation coefficient (q^2) of 0.799, a Fisher value (F) of 325.628, and a predictive r^2 for the external test set of 0.746; for CA II the model with four component showed a q^2 of 0.791, an F of 244.000, and a predictive r^2 for the external test set of 0.658. Finally, for CA IV the CoMSIA model highlighted, with four components, a q^2 of 0.822, an F of 398.011, and a predictive r^2 for the external test set of 0.838. To analyze the selectivity in a quantitative manner, the authors developed 3D-QSAR models using two different approaches, using differences of pK_I values as independent variables and subtracting the pK_I values from the 3D-QSAR models developed for each CA subtype.

Table 18.4 shows the best results using both methods. They were obtained by using the CoMSIA approach and the alignment developed using the CA II X-ray crystal structure.

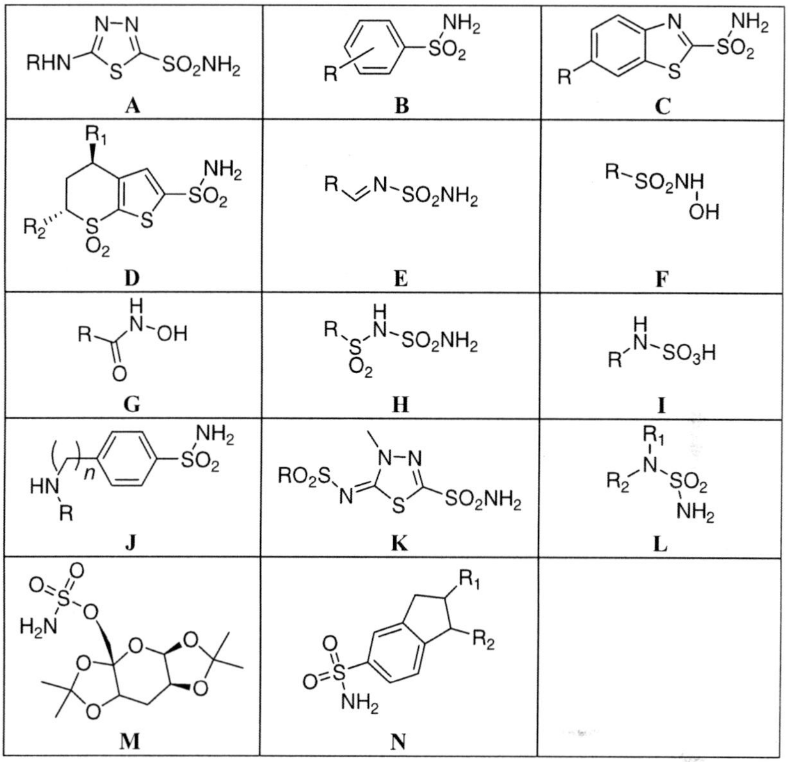

FIGURE 18.2 General scaffold of the CA inhibitors used for the development of 3D-QSAR models.

The results clearly suggest that the two approaches gave good results in terms of predictive ability; moreover, these results were very similar and it is not possible to identify which is the best method.

In 2007, Tuccinardi and coworkers reported a docking-based 3D-QSAR model for CA II inhibition.[48] Applying a cross-docking study on 40 CA II inhibitor X-ray structures, a reliable method for docking CA II inhibitors using Gold software[52] was

TABLE 18.4 Statistical Results of CoMSIA Analysis of the Selectivity Data Using ΔpK_I as the Dependent Variable[45]

		NC	q^2	F	r^2_{pred}
CA I/II	pK_I I–pK_I II	4	0.786	330.962	0.791
	pK_I I–pK_I II estrapol	5	0.775	340.394	0.771
CA I/IV	pK_I I–IV	5	0.737	283.130	0.741
	pK_I I–pK_I IV estrapol	5	0.738	277.463	0.651
CA II/IV	pK_I II–IV	4	0.667	179.450	0.565
	pK_I II–pK_IIV estrapol	5	0.641	179.265	0.480

tuned. Using the best approach, a training set of 220 was docked into the CA II binding site, and for each ligand, the best docked structure was chosen for developing the docking-based alignment. The best 3D-QSAR model, developed using Golpe software,[53] showed a q^2 of 0.72 and an r^2 of 0.88 for four components; furthermore, to test the predictive reliability of the model, three different external test sets were used, and their activity prediction showed an SDEP value between 0.41 and 0.63. The authors also developed another 3D-QSAR model using the same training and test sets and aligning the molecules using a shape similarity approach; however, the docking-based method showed the best results.

The same authors also reported the development of the first 3D-QSAR model for CA IX. To date, there are no reported X-ray structures for the CA IX, and therefore a homology model was built using CA XIV as a template. Then, using the docking procedure used for the development of the docking-based CA II 3D-QSAR model described above, 124 compounds were docked inside the optimized CA IX homology model. The best docking poses of 87 of these inhibitors were used as a training set for developing the 3D-QSAR models. The best model showed a q^2 of 0.74 and an r^2 of 0.89 for three components; furthermore, to test the predictive ability of the model, two external test sets of 20 and 17 compounds were used, and their activity prediction showed SDEPs of 0.58 and 0.52, respectively.

Finally, a very interesting study was recently published by Hillebrecht and Klebe.[49] A training set of 144 ligands characterized by nine scaffolds was subjected to the automated FlexS[54] alignment procedure, which required for each of the nine scaffolds an anchoring fragment used to start the incremental FlexS construction algorithm and a reference ligand for spatial alignment. This training set was used for developing QSAR models for CA II inhibition and CA I/II selectivity (using the difference of the pK_I values toward hCA I and II as independent variables). The best CoMFA and CoMSIA models obtained with this alignment procedure were compared to those obtained by using a manual alignment followed by minimization of the binding pocket and also to 2D-QSAR models derived from fragments (MACCS keys[55]) and properties (VSA descriptors[56]). For the CA II 3D-QSAR models, all the different approaches resulted in very similar q^2 values between 0.790 and 0.860. All the CA I/II selectivity models showed a significant decrease of the q^2 value (with respect to the CA II models), which was more pronounced for the 2D-QSAR models.

To test the reliability of the FlexS and 2D-QSAR models, a large external test set of 663 sulfonamide type inhibitors was used. The prediction of the activity of these compounds highlighted that the CoMSIA model exhibited the best results (r^2_{pred} $pK_I = 0.482$), with CoMFA exhibiting slightly worse values (r^2_{pred} $pK_I = 0.454$). The MACCS approach (r^2_{pred} $pK_I = 0.302$) performed significantly better than VSA (r^2_{pred} $pK_I = -0.710$); however, its predictive power was clearly worse compared to the 3D techniques. None of the approaches was able to give decent numerical predictions of the CA I/II selectivity; however, the results of categorical predictivity suggested that the 3D models can still give crude estimates about selectivity. These results clearly confirm that to assess the predictivity of QSAR models, the training set statistics are insufficient and it is fundamental to evaluate the correlation between the predicted and observed activities of compounds from an external test set.

18.5 CONCLUSIONS

In the field of CA computational research, the past decade has been highly dynamic and productive. Many QSAR models have been published using a wide number of compounds, characterized by a good level of structural diversity. These models were mainly built by using several nonclassical descriptors; only in a few cases, the log *P* descriptor was used, whereas various kinds of topological indexes and quantum mechanically calculated descriptors were widely used, together with indicator parameters able to take into account structural differences.

All the reported 3D-QSAR models were validated internally and on external test sets. They were based on various alignment methods that were able to take into account the ligand and receptor characteristics. The large amount and diversity of compounds used for building some of these models also suggested the possibility they could be used for virtual screening studies.

Most of the reported QSAR models have been developed for the inhibition of the CA I, II, and IV, and models concerning CA VI, IX, and XIII have been reported only in the past 2 years. The most recent research strongly suggests that the CA subtypes such as CA VA, VB, IX, and XIV are druggable targets;[57] therefore, following the development of new compounds able to act on these subtypes, we expect that in the next few years the QSAR studies will be focused on the analysis of these new targets.

Another important aspect is related to the fact that only a few compounds have been identified that exhibit selectivity for any CA isoform. Even if important results have been obtained in the design of compounds with high selectivity for CA VA, IX, and XIII isoforms over CA II, the development of highly specific CA ligands that are able to discriminate among the different members of CA isoforms remains a challenge. For this goal, QSAR approaches seem to be a very promising tool.

REFERENCES

1. Chegwidden, W. R.; Dodgson, S. J.; Spencer, I. M. The roles of carbonic anhydrase in metabolism, cell growth and cancer in animals. In *The Carbonic Anhydrases: New Horizons*, Chegwidden, W. R.; Carter, N. D.; Edwards, Y. H., Eds.; Birkhauser Verlag: Basel, Switzerland, **2000**; pp 343–363.
2. Supuran, C. T.; Clare, B. W. Carbonic anhydrase inhibitors. Part 47: Quantum chemical quantitative structure–activity relationships for a group of sulfanilamide Schiff base inhibitors of carbonic anhydrase. *Eur. J. Med. Chem.* **1998**, *33*, 489–500.
3. Clare, B. W.; Supuran, C. T. Carbonic anhydrase inhibitors. Part 61: Quantum chemical QSAR of a group of benzenedisulfonamides. *Eur. J. Med. Chem.* **1999**, *34*, 463–474.
4. Supuran, C. T.; Clare, B. W. Carbonic anhydrase inhibitors. Part 57: Quantum chemical QSAR of a group of 1,3,4-thiadiazole- and 1,3,4-thiadiazoline disulfonamides with carbonic anhydrase inhibitory properties. *Eur. J. Med. Chem.* **1999**, *34*, 41–50.
5. Gupta, S. P.; Maheswaran, V.; Pande, V.; Kumar, D. A comparative QSAR study on carbonic anhydrase and matrix metalloproteinase inhibition by sulfonylated amino acid hydroxamates *J. Enzyme Inhib. Med. Chem.* **2003**, *18*, 7–13.

6. Agrawal, V. K.; Bano, S.; Supuran, C. T.; Khadikar, P. V. QSAR study on carbonic anhydrase inhibitors: aromatic/heterocyclic sulfonamides containing 8-quinoline-sulfonyl moieties, with topical activity as antiglaucoma agents. *Eur. J. Med. Chem.* **2004**, *39*, 593–600.

7. Bajaja, S.; Sambia, S. S.; Madan, A. K. Prediction of carbonic anhydrase activation by tri-/tetrasubstituted-pyridinium-azole compounds: a computational approach using novel topochemical descriptor. *QSAR Comb. Sci.* **2004**, *23*, 506–514.

8. Supuran, C. T.; Clare, B. W. Quantum theoretic QSAR of benzene derivatives: some enzyme inhibitors. *J. Enzyme Inhib. Med. Chem.* **2004**, *19*, 237–248.

9. Agrawal, V. K.; Banerji, M.; Gupta, M.; Singh, J.; Khadikar, P. V.; Supuran, C. T. QSAR study on carbonic anhydrase inhibitors: water-soluble sulfonamides incorporating beta-alanyl moieties, possessing long lasting-intra ocular pressure lowering properties—a molecular connectivity approach. *Eur. J. Med. Chem.* **2005**, *40*, 1002–1012.

10. Clare, B. W.; Supuran, C. T. A physically interpretable quantum-theoretic QSAR for some carbonic anhydrase inhibitors with diverse aromatic rings, obtained by a new QSAR procedure. *Bioorg. Med. Chem.* **2005**, *13*, 2197–2211.

11. Gupta, S. P.; Kumaran, S. A quantitative structure–activity relationship study on some aromatic/heterocyclic sulfonamides and their charged derivatives acting as carbonic anhydrase inhibitors. *J. Enzyme Inhib. Med. Chem.* **2005**, *20*, 251–259.

12. Agrawal, V. K.; Singh, J.; Khadikar, P. V.; Supuran, C. T. QSAR study on topically acting sulfonamides incorporating GABA moieties: a molecular connectivity approach. *Bioorg. Med. Chem. Lett.* **2006**, *16*, 2044–2051.

13. Singha, S.; Singhb, J.; Inglec, M.; Mishra, R.; Khadikar, P. V. A QSAR study on carbonic anhydrase inhibition: predicting log K_I(hCAI) by using (SO_2NH_2) NMR chemical shift as a molecular descriptor. *ARKIVOC* **2006**, *16*, 1–15.

14. Kumar, S.; Singh, V.; Tiwari, M. Quantitative structure–activity relationship studies of sulfamide derivatives as carbonic anhydrase inhibitor: as antiglaucoma agents. *Med. Chem.* **2007**, *3*, 379–386.

15. Singh, J.; Shaik, B.; Singh, S.; Sikhima, S.; Agrawal, V. K.; Khadikar, P. V.; Supuran, C. T. QSAR studies on the activation of the human carbonic anhydrase cytosolic isoforms I and II and secretory isozyme VI with amino acids and amines. *Bioorg. Med. Chem.* **2007**, *15*, 6501–6509.

16. Gao, H.; Bajorath, J. Comparison of binary and 2D QSAR analyses using inhibitors of human carbonic anhydrase II as a test case. *Mol. Divers.* **1999**, *4*, 115–130.

17. Agrawal, V. K.; Sharma, R.; Khadikar, P. V. QSAR studies on carbonic anhydrase inhibitors: a case of ureido and thioureido derivatives of aromatic/heterocyclic sulfonamides. *Bioorg. Med. Chem.* **2002**, *10*, 2993–2999.

18. Agrawal, V. K.; Srivastava, S.; Khadikar, P. V.; Supuran, C. T. Quantitative structure–activity relationship study on sulfanilamide Schiff's bases: carbonic anhydrase (CA) inhibitors. *Bioorg. Med. Chem.* **2003**, *11*, 5353–5362.

19. Balaban, A. T.; Basak, S. C.; Beteringhe, A.; Mills, D.; Supuran, C. T. QSAR study using topological indices for inhibition of carbonic anhydrase II by sulfanilamides and Schiff bases. *Mol. Divers.* **2004**, *8*, 401–412.

20. Jaiswal, M.; Khadikar, P. V.; Supuran, C. T. Topological modeling of lipophilicity, diuretic activity, and carbonic inhibition activity of benzene sulfonamides: a molecular connectivity approach. *Bioorg. Med. Chem. Lett.* **2004**, *14*, 5661–5666.

21. Thakur, A.; Thakur, M.; Khadikar, P. V.; Supuran, C. T.; Sudele, P. QSAR study on benzenesulphonamide carbonic anhydrase inhibitors: topological approach using Balaban index. *Bioorg. Med. Chem.* **2004**, *12*, 789–793.

22. Clare, B. W.; Supuran, C. T. Predictive flip regression: a technique for QSAR of derivatives of symmetric molecules. *J. Chem. Inf. Model.* **2005**, *45*, 1385–1391.

23. Khadikar, P. V.; Sharma, V.; Karmarkar, S.; Supuran, C. T. QSAR studies on benzene sulfonamide carbonic anhydrase inhibitors: need of hydrophobic parameter for topological modeling of binding constants of sulfonamides to human CA-II. *Bioorg. Med. Chem. Lett.* **2005**, *15*, 923–930.

24. Khadikar, P. V.; Deeb, O.; Jaber, A.; Singh, J.; Agrawal, V. K.; Singh, S.; Lakhwani, M. Development of quantitative structure–activity relationship for a set of carbonic anhydrase inhibitors: use of quantum and chemical descriptors. *Lett. Drug Des. Discov.* **2006**, *3*, 622–635.

25. Melagraki, G.; Afantitis, A.; Sarimveis, H.; Igglessi-Markopoulou, O.; Supuran, C. T. QSAR study on *para*-substituted aromatic sulfonamides as carbonic anhydrase II inhibitors using topological information indices. *Bioorg. Med. Chem.* **2006**, *14*, 1108–1114.

26. Eroğlu, E.; Türkmen, H.; Güler, S.; Palaz, S.; Oltulu, O. A DFT-based QSARs study of acetazolamide/sulfanilamide derivatives with carbonic anhydrase (CA-II) isozyme inhibitory activity. *Int. J. Mol. Sci.* **2007**, *8*, 145–155.

27. Eroglu, E. Some QSAR studies for a group of sulfonamide Schiff base as carbonic anhydrase CA II inhibitors. *Int. J. Mol. Sci.* **2008**, *9*, 181–197.

28. Singh, J.; Shaik, B.; Singh, S.; Agrawal, V. K.; Khadikar, P. V.; Deeb, O.; Supuran, C. T. Comparative QSAR study on *para*-substituted aromatic sulphonamides as CAII inhibitors: information versus topological (distance-based and connectivity) indices. *Chem. Biol. Drug Des.* **2008**, *71*, 244–259.

29. Jaiswal, M.; Khadikar, P. V.; Scozzafava, A.; Supuran, C. T. Carbonic anhydrase inhibitors: the first QSAR study on inhibition of tumor-associated isoenzyme IX with aromatic and heterocyclic sulfonamides. *Bioorg. Med. Chem. Lett.* **2004**, *14*, 3283–3290.

30. Singh, J.; Singh, S.; Thakur, S.; Lakhwani, M.; Khadikar, P. V.; Agrawal, V. K.; Supuran, C. T. QSAR study on murine recombinant isozyme mCAXIII: topological vs structural descriptors. *ARKIVOC* **2006**, *14*, 103–118.

31. Clare, B. W.; Supuran, C. T. Carbonic anhydrase inhibitors. Part 86: A QSAR study on some sulfonamide drugs which lower intra-ocular pressure, using the ACE non-linear statistical method. *Eur. J. Med. Chem.* **2000**, *35*, 859–865.

32. Jaiswal, M.; Khadikar, P. V.; Supuran, C. T. QSAR study on CA inhibitory activity of disulfonamides: effect of halogen substitution. *Bioorg. Med. Chem.* **2004**, *12*, 2477–2482.

33. Golbraikh, A.; Tropsha, A. Beware of q2! *J. Mol. Graph. Model.* **2002**, *20*, 269–276.

34. Tarko, L.; Supuran, C. T. QSAR studies for the inhibition of the transmembrane isozymes XII and XIV of human carbonic anhydrase with a series of sulfonamides. *Bioorg. Med. Chem.* **2007**, *15*, 5666–5671.

35. Hemmateenejad, B.; Miri, R.; Jafarpourb, M.; Tabarzadb, M.; Shamsipurd, M. Exploring QSAR for the inhibitory activity of a large set of aromatic/heterocyclic sulfonamides toward four different isoenzymes of carbonic anhydrase. *QSAR Comb. Sci.* **2007**, *26*, 1065–1075.

36. Gao, H. Application of BCUT metrics and genetic algorithm in binary QSAR analysis. *J. Chem. Inf. Comput. Sci.* **2001**, *41*, 402–407.

37. Burden, F. R. Molecular identification number for substructure searches. *J. Chem. Inf. Comput. Sci.* **1989**, *29*, 225–227.

38. Gao, H.; Lajiness, M. S.; Van Drie, J. Enhancement of binary QSAR analysis by a GA-based variable selection method. *J. Mol. Graph. Model.* **2002**, *20*, 259–268.

39. Mattioni, B. E.; Jurs, P. C. Development of quantitative structure–activity relationship and classification models for a set of carbonic anhydrase inhibitors. *J. Chem. Inf. Comput. Sci.* **2002**, *42*, 94–102.

40. Zernov, V. V.; Balakin, K. V.; Ivaschenko, A. A.; Savchuk, N. P.; Pletnev, I. V. Drug discovery using support vector machines. The case studies of drug-likeness, agrochemical-likeness, and enzyme inhibition predictions. *J. Chem. Inf. Comput. Sci.* **2003**, *43*, 2048–2056.

41. Popelier, P. L.; Smith, P. J. QSAR models based on quantum topological molecular similarity. *Eur. J. Med. Chem.* **2006**, *41*, 862–873.

42. Eroğlu, E.; Türkmen, H. A DFT-based quantum theoretic QSAR study of aromatic and heterocyclic sulfonamides as carbonic anhydrase inhibitors against isozyme, CA-II. *J. Mol. Graph. Model.* **2007**, *26*, 701–708.

43. Baskin, I. I.; Palyulin, V. A.; Zefirov, N. S. Neural networks in building QSAR models. *Methods Mol. Biol.* **2008**, *458*, 137–158.

44. Jalali-Heravi, M.; Kyani, A. Application of genetic algorithm-kernel partial least square as a novel nonlinear feature selection method: activity of carbonic anhydrase II inhibitors. *Eur. J. Med. Chem.* **2007**, *42*, 649–659.

45. Hillebrecht, A.; Supuran, C. T.; Klebe, G. Integrated approach using protein and ligand information to analyze selectivity- and affinity-determining features of carbonic anhydrase isozymes. *ChemMedChem* **2006**, *1*, 839–853.

46. Weber, A.; Böhm, M.; Supuran, C. T.; Scozzafava, A.; Sotriffer, C. A.; Klebe, G. 3D QSAR selectivity analyses of carbonic anhydrase inhibitors: insights for the design of isozyme selective inhibitors. *J. Chem. Inf. Model.* **2006**, *46*, 2737–2760.

47. Huang, H.; Pan, X.; Tan, N.; Zeng, G.; Ji, C. 3D-QSAR study of sulfonamide inhibitors of human carbonic anhydrase II. *Eur. J. Med. Chem.* **2007**, *42*, 365–372.

48. Tuccinardi, T.; Nuti, E.; Ortore, G.; Supuran, C. T.; Rossello, A.; Martinelli, A. Analysis of human carbonic anhydrase II: docking reliability and receptor-based 3D-QSAR study. *J. Chem. Inf. Model.* **2007**, *47*, 515–525.

49. Hillebrecht, A.; Klebe, G. Use of 3D QSAR models for database screening: a feasibility study. *J. Chem. Inf. Model.* **2008**, *48*, 384–396.

50. Tuccinardi, T.; Ortore, G.; Rossello, A.; Supuran, C. T.; Martinelli, A. Homology modeling and receptor-based 3D-QSAR study of carbonic anhydrase IX. *J. Chem. Inf. Model.* **2007**, *47*, 2253–2262.

51. Gohlke, H.; Hendlich, M.; Klebe, G. Knowledge-based scoring function to predict protein–ligand interactions. *J. Mol. Biol.* **2000**, *295*, 337–356.

52. Jones, G.; Willett, P.; Glen, R. C.; Leach, A. R.; Taylor, R. Development and validation of a genetic algorithm for flexible docking. *J. Mol. Biol.* **1997**, *267*, 727–748.

53. *GOLPE 4.5*; Multivariate Infometric Analysis Srl.: Perugia, Italy, **1999**.

54. Lemmen, C.; Lengauer, T.; Klebe, G. FLEXS: a method for fast flexible ligand superposition. *J. Med. Chem.* **1998**, *41*, 4502–20.

55. MDL Information Systems, Inc., 14600 Catalina Street, San Leandro, CA 94577.

56. Labute, P. A widely applicable set of descriptors. *J. Mol. Graph. Model.* **2000**, *18*, 464–477.

57. Supuran, C. T. Carbonic anhydrases: novel therapeutic applications for inhibitors and activators. *Nat. Rev. Drug Discov.* **2008**, *7*, 168–181.

CHAPTER 19

Selectivity Issues in the Design of CA Inhibitors

CLAUDIU T. SUPURAN[1] and JEAN-YVES WINUM[2]

[1]Laboratorio di Chimica Bioinorganica, Polo Scientifico, Università degli Studi di Firenze, Room. 188, Via della Lastruccia 3, 50019 Sesto Fiorentino, Florence, Italy
[2]Institut des Biomolécules Max Mousseron (IBMM) UMR 5247 CNRS-UM1-UM2 Bâtiment de Recherche Max Mousseron, Ecole Nationale Supérieure de Chimie de Montpellier, 8 rue de l'Ecole Normale, 34296 Montpellier Cedex, France

19.1 INTRODUCTION

Carbonic anhydrases (CAs) are old and universal zinc metalloproteins, which are having fundamental functions in many life processes. These metalloenzymes are encoded by five distinct class of gene families evolutionarily unrelated and are present in all kingdoms: the α-CAs described in vertebrates, bacteria, algae, and cytoplasm of green plants; the β-CAs found predominantly in bacteria, algae, and chloroplasts of both mono- and dicotyledons; the γ-CAs present mainly in archaea and some bacteria; and the ε-CAs that were evidenced in some marine diatoms.[1,2]

By catalyzing one of the most important physiological reactions, for example, the reversible hydration of carbon dioxide to bicarbonate and proton, CAs play important physiological and physiopathological functions. In higher vertebrates, 16 different α-CA isozymes or CA-related proteins (CARP) have been identified, and they clearly differ in their cell localization, tissue distributions, and functions (Table 19.1).

Among the catalytic active forms, we can distinguish five cytoplasmic isozymes: CA I, II, III, VII, and XIII; five membrane-associated isoforms having active site outside the cell (CA IV, IX, XII, XIV, and XV); two mitochondrial (CA VA and VB); and one secreted (CA VI) (Fig. 19.1). Three carbonic anhydrase-related proteins (CARPs) are known (CARP VIII, X, and XI) that present no catalytic activities.[1–3]

The 13 active isoforms are involved in many essential biological processes including respiration and transport of CO_2/bicarbonate between metabolizing tissues and lungs, pH and CO_2 homeostasis, electrolyte secretion, biosynthetic reaction (such

TABLE 19.1 Higher Vertebrate α-CA Isozymes, Their Relative CO_2 Hydrase Activity, Affinity for Sulfonamide Inhibitors, and Subcellular Localization

Isozyme Localization	Catalytic Activity (CO_2 Hydration)	Affinity for Sulfonamides	Subcellular
CA I	Moderate	Medium	Cytosol
CA II	High	Very high	Cytosol
CA III	Very low	Very low	Cytosol
CA IV	High	High	Membrane-bound
CA VA	Low–moderate[a]	High	Mitochondria
CA VB	High	High	Mitochondria
CA VI saliva/milk	Moderate	High	Secreted into
CA VII	High	Very high	Cytosol
CARP VIII	Acatalytic	[b]	Cytosol
CA IX	Moderate–high	High	Transmembrane
CARP X	Acatalytic	[b]	Secreted
CARP XI	Acatalytic	[b]	Secreted
CA XII	Low	Very high	Transmembrane
CA XIII	Moderate	Medium–high	Cytosol
CA XIV	Moderate	High	Transmembrane
CA XV	Low	Unknown	Membrane-bound

[a] Low at pH = 7.4, moderate at pH = 8.2 or higher.
[b] The native CARP isozymes do not contain Zn(II), so that their affinity for the sulfonamide inhibitors has not been measured. With site-directed mutagenesis, it is possible to modify these proteins and transform them in enzymes with CA-like activity that probably are inhibited by sulfonamides, but no detailed studies on this subject are available presently.

as gluconeogenesis, lipogenesis, and ureagenesis), bone resorption, calcification, tumorigenicity, and also many other physiological processes in humans (except CA XV, which is not expressed in human tissues).[1–3] Thus, involvement of CAs in various physiological processes implies that deregulated expression and/or abnormal performance of some isoforms may have important pathological consequences. Considering the wide distribution of the 12 active CA isozymes in many cells, tissues, and organs, the modulation of CA activity, either by inhibition or activation, offers important biomedical options in the design of therapeutic agents useful in the management or prevention of many diseases where CA expression and/or activity are dysregulated.[3–5]

The actual available pharmacological agents (Fig. 19.2) are far from being perfect, as they possess many undesired side effects, mainly due to their lack of selectivity for the different CAs isoforms (Table 19.2).

One of the most active isoforms (CA II) is an extremely abundant in most human cells and is often indiscriminately inhibited irrespective of whether it plays a key role in a disease or is just coexpressed in the same tissue and elsewhere in the body leading to deleterious effects.[3–5] Most clinically used CA inhibitors such as sulfonamide/sulfamate inhibitors have very high affinities against the ubiquitous CA II. So the main challenge is finding the derivatives that should not act as very strong inhibitors of this isozyme and at the same time, preserving good affinity for another target CA such as CA VA and VB that are involved in adipogenesis or CA IX, and XII that are involved

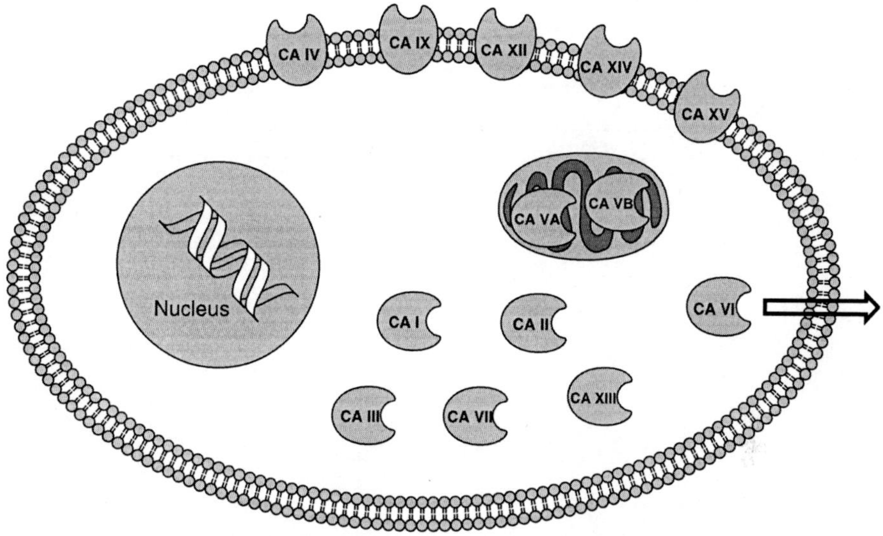

FIGURE 19.1 Locations of active CA isoenzymes in a schematic cell model. CA I, II, III, VII, and XIII are cytosolic isoforms; VA and VB are mitochondrial; CA VI is secretory; and CA IV, IX, XII, XIV, and XV are membrane-bound.

in tumorigenesis. Moreover, finding inhibitors with higher affinity to CA I, as compared to the sulfonamide-avide isozyme CA II constitutes an important goal, mainly because the physiological function of CA I, is still a mystery although this protein is very abundant in many mammals including humans.

So, the development of isozyme-specific or at least organ-selective inhibitors would be highly beneficial both for obtaining novel types of drugs, devoid of major side effects, as well as for physiological studies in which the specific/selective inhibitors may constitute valuable tools for understanding the physiology/physiopathology of these enzymes. Some progresses have been registered recently in the design of compounds with some selectivity toward various CAs and these data will be reviewed in this chapter.

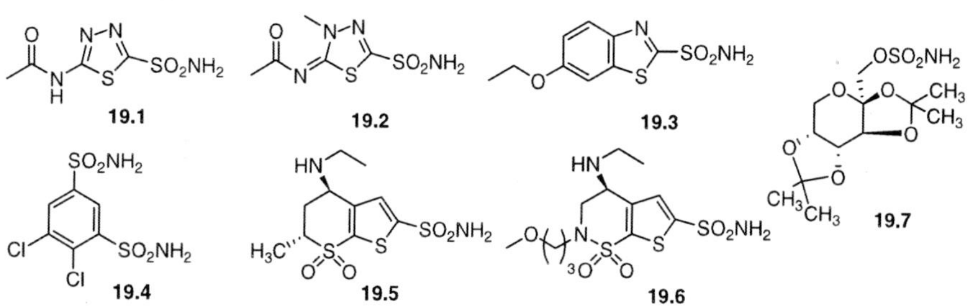

FIGURE 19.2 Structure of clinically used carbonic anhydrase inhibitors.

TABLE 19.2 Inhibition Data with the Clinically Used Sulfonamides 19.1–19.6 and the Clinically Used Sulfamate 19.7 (Topiramate)

	K_I (nM)						
Isozyme	**19.1**	**19.2**	**19.3**	**19.4**	**19.5**	**19.6**	**19.7**
hCA I[a]	250	50	25	1200	50,000	45,000	250
hCA II[a]	12	14	8	38	9	3	10
hCA III[a]	3.10^5	1.10^5	5000	nt	8000	nt	nt
hCA IV[a]	74	6200	93	15,000	8500	nt	4900
hCA VA[a]	63	65	25	630	42	50	63
hCA VB[a]	54	62	19	21	33	30	30
hCA VI[a]	11	10	43	79	10	0.9	45
hCA VII[a]	2.5	2.1	0.8	26	3.5	2.8	0.9
hCA IX[b]	25	27	34	50	52	37	58[c]
hCA XII[b]	5.7	3.4	22	50	3.5	3.0	3.8
mCA XIII[a]	17	19	nt	23	18	nt	47
hCA XIV[a]	41	43	25	345	27	24	1460

h, human; m, murine isozyme; nt, not tested (no data available).
[a] Full length enzyme.
[b] Catalytic domain.
[c] The data against the full length enzyme are of 1590 nM.

19.2 SELECTIVE INHIBITION OF CYTOSOLIC ISOZYMES VERSUS CA II

19.2.1 Isozyme CA I

CA I is a well-known cytoplasmic carbonic anhydrase isoform (molecular weight of approximately 30 kDa) expressed in a variety of cells and tissues such as erythrocytes, capillary, and corneal endothelium, lens of the eye, islets of Langerhans, fetal membranes, and placenta. It is present also in the gastrointestinal tract especially in the epithelium of the esophagus, jejunum, ileum, and colon.[1,2]

The active site architecture of CA I is characterized by a higher number of histidine residues compared to CA II isozyme. Thus, in addition to the zinc ligands (His94, His96, and His119), and the residue His64 that plays an important role in catalysis CA I presents three additional residues in the active site: His67, His200, and His243.[6] Another important difference between the two isozymes is that CA II contains a histidine cluster, consisting of residues: His64 and His4 (these two residues possess a flexible conformation in the crystal structure), His3, His10, His15, and His17 (prolonging from the middle of the active site to the rim of the cavity, and protruding on the surface of the protein),[6] which is absent in CA I. These two isozymes also possess a different affinity for the two main classes of inhibitors: CA I has larger affinity than CA II for anions (such as cyanide, thiocyanate, cyanate, halides, etc.), whereas CA II has generally a higher affinity for sulfonamides as compared to CA I (see also Table 15.2).[1,7] As a consequence, it is relatively difficult to obtain sulfonamide inhibitors with higher affinity for CA I than for CA II, although the two

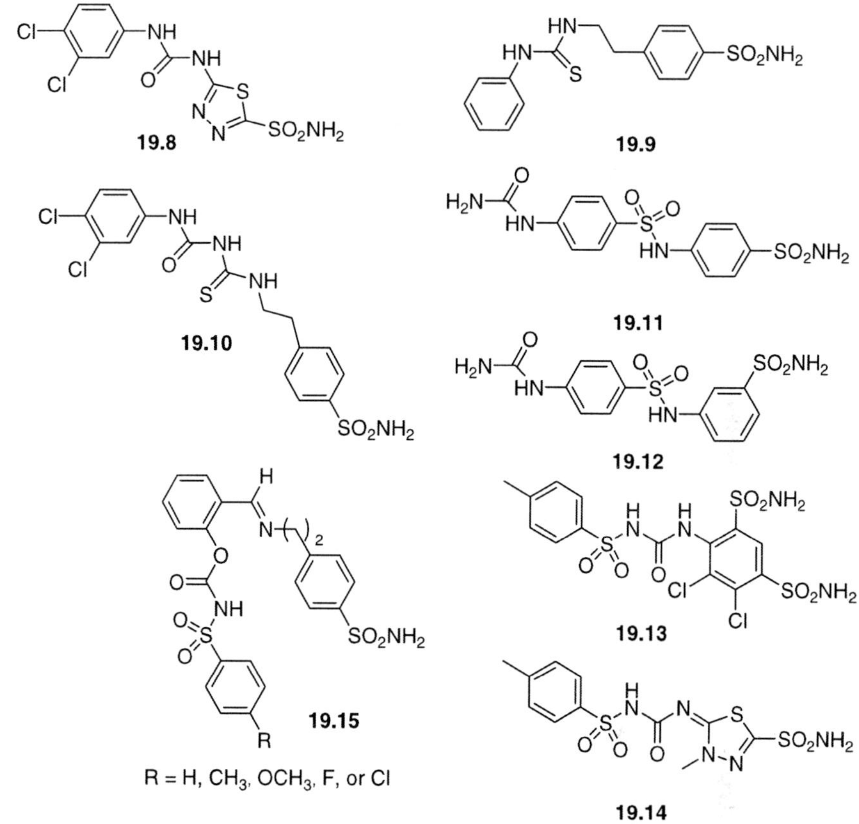

FIGURE 19.3 Structure of CA inhibitors **19.8–19.15**.

isozymes possess significant differences in the active site architecture. The first such compounds were only recently reported by this group,[8–11] and were discovered serendipitously, by screening of a large number of sulfonamides possessing different structural motifs in their molecules. Remarkably, all the compounds possessing higher affinity for CA I as compared to CA II (and IV), of types **19.8–19.15** (Fig. 19.3), contain ureido or thioureido moieties in their molecules. Their inhibition data against the three isozymes mentioned above are listed in Table 19.3.

Such isozyme I avid inhibitors belong both to the aromatic sulfonamide class, as well as to the heterocyclic sulfonamide class, whereas the ureido/thioureido moieties present in their molecules may be unsubstituted or substituted with bulkier groups (3,4-dichlorophenyl; phenyl; substituted-phenylsulfonyl, etc.). It must also be mentioned that compounds **19.15** containing arylsulfonylcarbamate moieties instead of the arylsulfonylureido ones were investigated in more details.[8–11] These compounds also inhibit significantly isozymes II and IV, and thus, they are not really isozyme-I specific, but represent anyhow an important step toward the generation of isozyme specific CAIs. It must also be noted that dorzolamide **19.5** has a very low affinity for hCA I, but its deethylated metabolite is a very potent inhibitor of this isozyme.[1,7]

TABLE 19.3 Inhibition of Isozymes I, II, and IV with Compounds 19.8–19.15 Showing Selectivity Toward One of These Isoforms

	K_I (nM)		
Inhibitor	hCA I	hCA II	bCA IV
19.8	3	6	8
19.9	50	53	70
19.10	7	10	24
19.11	3	8	20
19.12	4	10	25
19.13	8	12	14
19.14	4	5	11
19.15 (R = Me)	40	110	120
19.15 (R = Cl)	60	100	160

19.2.2 Isozyme CA III

CA III has the lowest activity of all the isozymes and is relatively resistant to acetazolamide. Abundantly present in type I fiber red skeletal muscle, this isozyme has also been detected in liver, epithelium of the salivary gland ducts, colon, bronchi, male genital tract and adipocytes.

Although the structure of this isozyme is relatively similar to that of hCA II, CA III has a CO_2 hydration activity of approximately 0.3% that of hCA II, as it does not possess a His residue in position 64, but a Lys residue, which is much less effective as a proton shuttle.[1,12] Furthermore, position 198 in CA III is occupied by a Phe, possessing a very bulky side chain, whereas the water bound to Zn(II) has a pK_a around 5.5.[1,12] All these particularities may explain the low catalytic activity of CA III, as well as its insensitivity to sulfonamide inhibitors, which do not have space enough to bind in the neighborhood of the Zn(II) ion, principally due to the steric impairment of Phe198. In fact, only the very small sulfonamide $CF_3SO_2NH_2$ acts as an efficient CA III inhibitor, with an inhibition constant of 0.9 μM (on the other hand, this compound is a nanomolar inhibitor of CA I, II, and IV).[1,12] Other sulfonamides (such as acetazolamide, methazolamide, etc.) inhibit CA III, with inhibition constants in the millimolar range (Table 19.2).[1]

19.2.3 Clash with Ala65 in CA Active Site for Designing Compounds with Low Affinity to CA II

The sulfamide analogue **19.16** of the antiepileptic drug topiramate **19.7** is a 210 times less potent inhibitor of isozyme CA II as compared to topiramate, but effectively inhibits isozymes CA VA, VB, VII, XIII, and XIV (for K_I's in the range 21–35 nM, see Fig. 15.4). Its weak binding to CA II was recently shown to be due to a clash between one methyl group of the inhibitor and Ala65, an amino acid residue unique only to the ubiquitous isozyme CA II, and may be exploited for the drug design of compounds with lower affinity for this ubiquitous isozyme.[7]

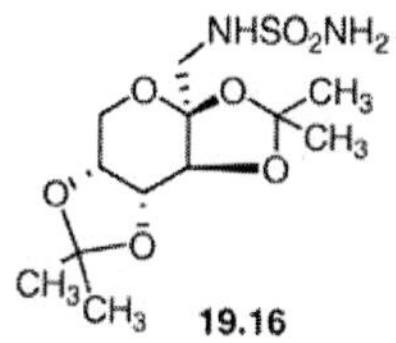

19.16

K_I (nM)	
hCA I	3450
hCA II	2135
hCA IV	941
hCA VA	32
hCA VB	21
hCA VII	35
hCA IX	4580
hCA XII	1875
mCA XIII	30
hCA XIV	25

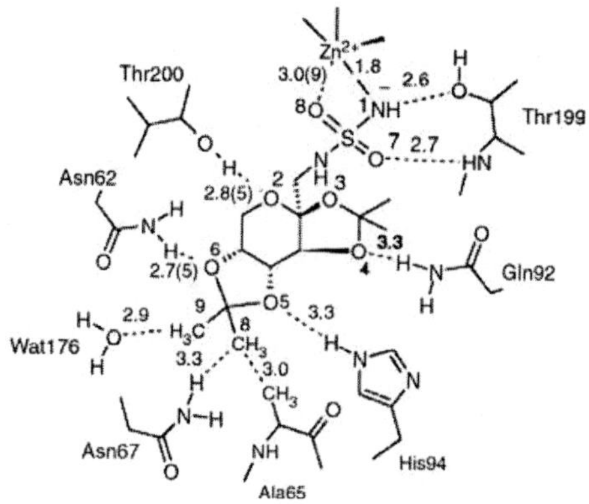

FIGURE 19.4 Structure, inhibitory activities, and schemactic representation of the interactions of the topiramate sulfamide analogue **19.16** with hCA II active site residues.

As shown by X-ray crystallography (Fig. 19.4),[7] the sulfamide analogue binds to CA II with the deprotonated sulfamide moiety coordinated to Zn(II), and the organic scaffold making an extended network of hydrogen bonds with Thr199, Gln92, His94, Asn62, and Thr200. In particular, the ionized nitrogen atom of the sulfamide group of **19.16** is coordinated to the zinc ion at a distance of 1.80 Å, much shorter than the corresponding distance in the topiramate **19.7** adduct (1.97 Å). The N1 nitrogen is also hydrogen bonded to the hydroxyl group of Thr199 (N $\cdots$ Thr199OG = 2.7 Å), which in turn interacts with the Glu106OE1 atom (2.5 Å). The inhibitor O7 atom is hydrogen bonded to the backbone amide of Thr199 (ThrN $\cdots$ O2 = 2.7 Å), whereas the O8 atom is 3.09 Å away from the catalytic Zn^{2+} ion, being considered as weakly coordinated to the metal ion.[7] All these interactions have also been observed in the adduct of hCA II with topiramate **19.7**, but the corresponding distances are slightly different.[13] The second NH group of **19.16**, similarly with the corresponding oxygen atom of topiramate,[13] does not participate in any interaction with the protein or the metal ion. This is another important difference between the sulfamide/sulfamate inhibitors discussed here. The endocyclic sugar oxygen of **19.16** (O2) makes a hydrogen bond of 2.85 Å with the hydroxyl moiety of Thr200 (the same interaction is present in the topiramate adduct, where the distance is of 2.84 Å).[13] Three oxygen atoms of the dioxolane rings of **19.16**, that is, O4, O5, and O6, participate in three hydrogen bonds with Gln92, His94, and Asn62, respectively (Fig. 19.4). One of them (involving O5) was not observed in the topiramate adduct,[14] whereas the other two are present in both adducts,[7,13] although the distances between the corresponding pairs of atoms are rather different. Thus, the distance between O6 and the amide nitrogen of Asn62 is in the range 2.7–3.0 Å for the two adducts of hCA II with **19.16** and **19.7**, respectively, whereas the interaction with Gln92 is a strong one in the case of topiramate (distance of 2.8 Å) and much weaker in the topiramate sulfamide adduct (distance of 3.3 Å).

However, the most important differences in the structures of the adducts of topiramate **19.7** and its sulfamide analogue **19.16** in complex with hCA II; regard the C8 and C9 methyl groups of the inhibitors. In the case of the topiramate adduct, the

C8 group is in van der Waals contacts (distances of 3.5 Å[13]) with the methyl group of Ala65 and the carboxamide moiety of Asn67. For the topiramate sulfamide **19.16** adduct, the corresponding distances are much shorter (in the range 3.0–3.3 Å) leading to an important clash between the C8 methyl and the methyl side chain of Ala65. In fact this was the main reason why we experienced difficulties in fitting the electron density of the inhibitor **19.16** in this region of the active site, which is sterically hindered by the presence of these amino acid side chains (Ala65 and Asn67), and also by the presence of the second methyl group of the inhibitor (C9) and a water molecule with which it is in close contact (Wat176 ··· C9 = 2.9 Å).[7] These repulsive interactions were not observed in the hCA II–topiramate **19.7** adduct,[13] and they constitute the only possible explanation regarding the important differences of activity between the two compounds, with topiramate being approximately 210 times a better hCA II inhibitor as compared to its sulfamide analogue.

But how relevant are such findings for the design of better drugs based on CAIs? Obviously, clash interactions between an inhibitor and an amino acid residue within the active site, which may prevent the strong binding of the inhibitor, may be useful to design compounds with selectivity for some isoforms, provided that the interacting residue is present only in isoforms that should be not inhibited. A close look at the amino acid sequences of all known human CAs[1] shows that only CA II has Ala in position 65, whereas this amino acid is Ser in CA I, IV, VB, VII, IX, XII, and XIII, Thr in CA III, VI, and XIV, and Leu in CA VA. We do not want to imply that just one amino acid substitution may change the binding affinity of an inhibitor for the active site, as it is clear that isozymes possessing Ser65, such as hCA I, IV, IX, or XII are only slightly inhibited by sulfamide **19.16**, whereas other isoforms possessing the same amino acid, such as CA VB, VII, or XIII, are quite well inhibited by this compound. What we wish to stress is that by resolving the X-ray crystal structure of hCA II in adduct with **19.16**, we evidenced an amino acid residue which is unique to the hCA II active site and that its clashing with some moieties of an inhibitor may constitute a powerful means for designing more selective CAIs, with weaker binding to this ubiquitous isozyme, but preserving a strong affinity for other isoforms. As far as we know this was the first example in which CA II-sparing inhibitors were evidenced and this fact was explained at molecular level.[7] Such findings may be thus quite useful in designing compounds with less CA II inhibitory activity, but which maintain strong inhibition of clinically relevant isoforms such as CA VA, VB, IX, XII, XIII, or XIV among others.

19.3 SELECTIVE INHIBITION OF MEMBRANE-BOUND CAs (CA IV, CA IX, CA XII, AND CA XIV)

At least four CA isozymes (CA IV, IX, XII, and XIV) are associated with cell membranes, with the enzyme active site always oriented extracellularly.[1,15] Some of these isozymes were shown to play pivotal physiological roles (such as CA IV in the eyes, lungs, and kidneys; CA IX in the gastric mucosa and many tumor cells; and CA XII in tumors and also in some normal epithelia)[1,15] whereas the function of the other such isozymes (CA XIV) is not well understood for the moment.

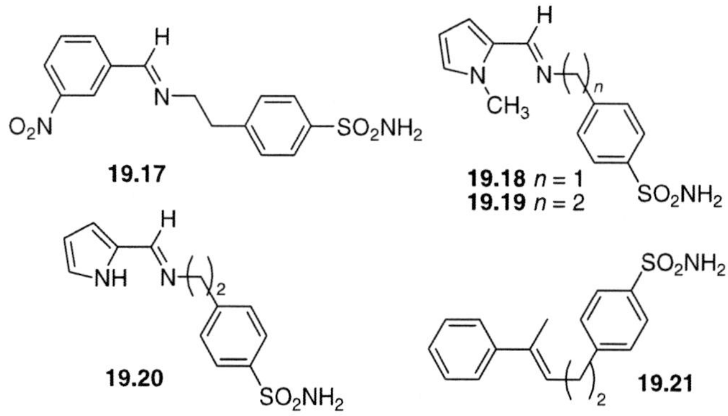

FIGURE 19.5 Structure of CA inhibitors **19.17–19.21**.

19.3.1 Isozyme CA IV

First membrane-associated CA isozyme to be described, isozyme CA IV contains only one histidine residue within its active site, His64, which plays a critical role in catalysis as in hCA II, as proton shuttle residue between the active site and the environment.[1,15] The most characteristic feature of the active site of this isozyme is related to the presence of four cysteine residues, which form two disulfide bonds, situated at the entrance within the cavity (Cys6–Cys11G and Cys23–Cys203, respectively).[15] These residues occupy practically the same region of the active site as the histidine cluster in hCA II[6] and it was hypothesized that this might be the most relevant aspect explaining the difference in affinity for sulfonamide inhibitors of these two isozymes.[16] Even so, similar to CA I, the first compounds with some specificity for CA IV, of type **19.17–19.21** (Fig. 19.5), were again discovered serendipitously, and they all belong to the same class of Schiff bases of aromatic/heterocyclic sulfonamides.[17–20]

Only Schiff bases of aromatic sulfonamides were investigated in some detail and it was shown that the best CA IV inhibition patterns are connected with the presence of heterocyclic moieties (in the original aldehyde used for the preparation of the Schiff base) or aromatic moieties substituted with electron attracting groups, such as the nitro one.[17–20] Such compounds also appreciably inhibited CA II, and to a smaller extent CA I (Table 19.4).[17–20]

TABLE 19.4 Inhibition of Isozymes I, II, and IV with Compounds 19.17–19.21

	K_I (nM)		
Inhibitor	hCA I	hCA II	bCA IV
19.17	1100	150	140
19.18	200	20	10
19.19	200	10	8
19.20	620	12	10
19.21	180	15	12

Due to the extracellular location of these isozymes, it is possible to design membrane-impermeant CAIs, which in this way become specific inhibitors for the membrane-associated CAs. This possibility has been fully explored in this laboratory, by designing positively-charged sulfonamides.[16,21–25]

A program of developing cationic sulfonamides has been initiated in our laboratory, using QAS **19.22**(Fig. 19.6) as lead molecule (which is also a relatively weak CAI, with micromolar affinity for CA II).[1] The first such compounds, of types **19.23–19.26**, were prepared by reaction of aromatic/heterocyclic sulfonamides containing free NH_2 groups with pyrylium salts, affording pyridinium derivatives.[16,21,22] These compounds were moderately active CA II and IV inhibitors, with affinities in the 10^{-6}–10^{-7} M range. By using QSAR data of this laboratory,[22] it has been shown that the increased CA II and IV inhibitory properties of aromatic/heterocyclic sulfonamides are connected with the presence of elongated inhibitor molecules (on the axis passing through the Zn(II) of the enzyme, the sulfonamide nitrogen atom, and the long axis of the inhibitor molecule itself). In consequence, such "elongated" molecules have been designed[23–25] by reacting pyrylium salts with amino acids (such as glycine or β-alanine) and coupling of the pyridinium derivatives with the aromatic/heterocyclic sulfonamides possessing free amino, hydroxy, imino, or hydroxyl moieties. The inhibitors obtained in this way, such as **19.27–19.30**, showed nanomolar

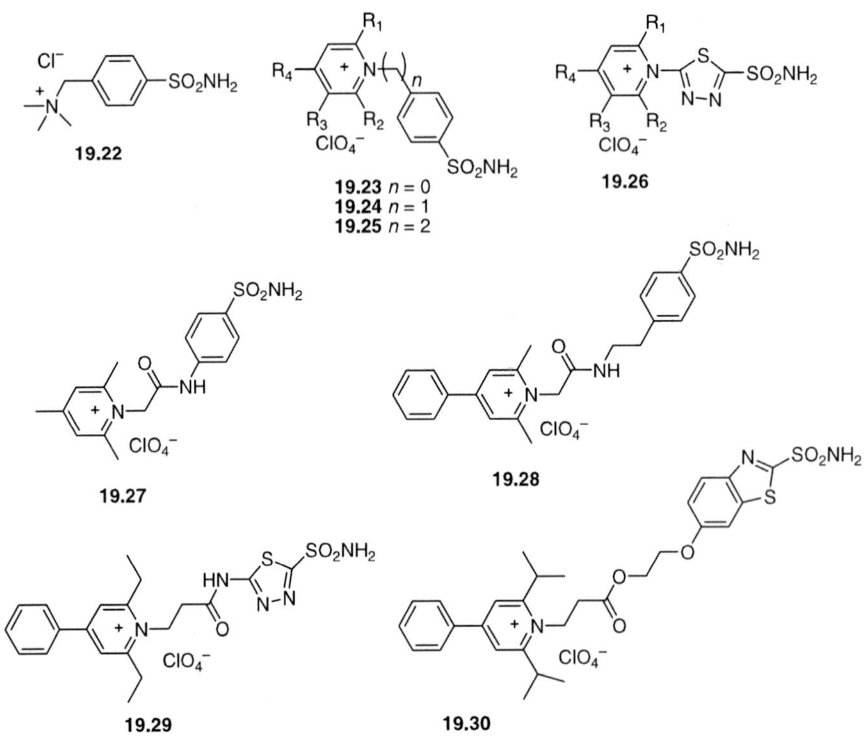

FIGURE 19.6 Structure of cationic sulfonamides **19.22–19.30**.

affinities both for CA II, as well as CA IV and IX, and more importantly, they were unable to cross the plasma membranes *in vivo*.[16,21–25] In the two model systems (human red cells and perfusion experiments in rats, respectively), this new class of potent, positively charged CAIs, was able to discriminate for the membrane-bound versus the cytosolic isozymes, selectively inhibiting only CA IV.[16,21–25] Such data are important not only for the specific *in vivo* inhibition of membrane-associated isozymes but also for the development of some novel anticancer therapies, as it has been shown that many hypoxic tumor cells predominantly express only some membrane-associated CA isozymes, such as CA IX and XII.[14]

19.3.2 Isozyme CA IX

The membrane-associated isoform CA IX, whose expression is normally restricted to the mucosa of alimentary tract, is highly overexpressed in a broad spectrum of hypoxic human tumors.[14,26] Upregulated expression of CA IX was reported in carcinoma cells in various organ including oesophagus, lung, kidney, colon and rectum, breast, cervix, head and neck, and bladder. This protein constitutes an endogenous marker of cellular hypoxia, a natural phenotype of solid tumor[27], and itsprognostic potential has been demonstrated in various clinical studies.[28]

By using positively-charged derivatives of type **19.25**, Svastova et al.[29] proved that the tumor-associated isozyme CA IX, and not the lactic acid, is the main contributor to acidification of tumors because its extracellular enzyme domain is highly active, its expression is induced by hypoxia and correlates with poor prognosis. CA IX acidifies pH of the culture medium in hypoxia but not in normoxia, independently of the lactic acid production. Sulfonamide CA IX-selective inhibitors belonging to various classes (among which the positively-charged derivatives **19.24** and **19.25**) were observed to bind only to hypoxic cells containing CA IX and to reverse the tumor acidification processes mediated by the enzyme.

Using fluorescent sulfonamide that preferentially inhibit the activity of CA IX expressed only under hypoxic conditions,[30,31] Dubois and colleagues demonstrated that to enable this inhibitors to bind to CA IX: not only the expression of CA IX is necessary but also the presence of active targeted protein is necessary, which is found only during hypoxia.[32]

As it was previously shown that many sulfonamides possess appreciable tumor cell growth inhibitory properties *in vitro* and *in vivo*,[1] such findings constituted the proof-of-concept that the anticancer therapies based on tumor-associated CA isozyme inhibition can be developed and also offer interesting tools for investigating hypoxic tumors as well as for their imaging.

Many CAIs with low nanomolar K_I have been developed in the past 5 years.[33] Among them, only one library of inhibitors showed selectivity against CA IX versus the cytosolic hCA II. Aromatic benzene sulfonamides incorporating triazinyl moieties prepared by our group,[34,35] inhibited hCA IX with K_I's in the range of nanomolar. Among the most potent and selective hCA IX inhibitors obtained up to now, three compounds were found with K_I in the subnanomolar range 0.12–0.34 nM. The chlorotriazinyl-sulfanilamide **19.31** as well as the bis-ethoxytriazinyl derivatives of

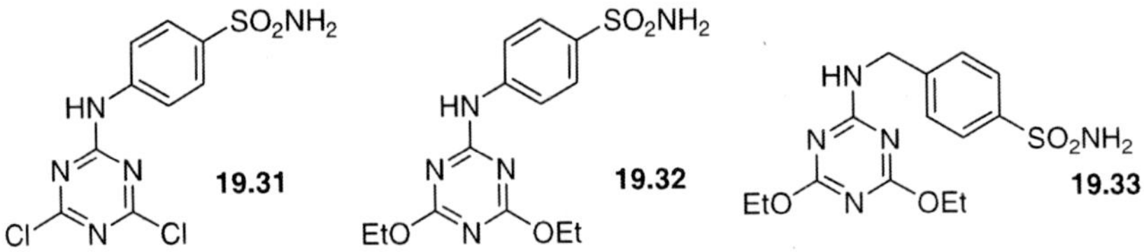

19.31 *K*I(hCA IX) = 0.15 nM; *K*I(hCA II)/*K*I(hCA IX) = 706

19.32 *K*I(hCA IX) = 0.12 nM; *K*I(hCA II)/*K*I(hCA IX) =166

19.33 *K*I(hCA IX) = 0.34 nM; *K*I(hCA II)/*K*I(hCA IX) = 247

FIGURE 19.7 Structure of triazinyl CAIs **19.31–19.33**.

sulfanilamide **19.32** and homosulfanilamide **19.33** showed selectivity ratios for CA IX over CA II inhibition in the range 166–706, thus having a much higher affinity for the tumor-associated isozyme. (Fig. 19.7). All the data currently available on the selective inhibition of this isoform confirm the possible use of inhibition of the tumor associated CA isozyme IX in the management of hypoxic tumors, which do not respond to the classical chemo- and radiotherapy.

19.4 CONCLUSION

Carbonic anhydrase belongs to a family of enzymes with many isoforms (in humans 15 CAs are known, 12 of which are catalytically active),[1–3] in which one isozyme (CA II) is ubiquitous in most cells in humans and its inhibition may be deleterious. Many biochemical, physiological, and pharmacological novel data demonstrate that the selective inhibition of other isoforms other than CA II, which are involved in specific physiologic/pathologic processes (such as CA IX and XII involved in tumorigenesis or CA VA and VB involved in adipogenesis) or show a restricted localization only in some tissues/organs (such as CA VA, VII, or XIII)[1] may lead to drugs with less severe side effects.

REFERENCES

1. Supuran, C. T.; Scozzafava, A.; Conway J., Eds.; *Carbonic anhydrase—Its inhibitors and activators;* CRC Press: Boca Raton, FL, **2004**; pp 1–376, and references cited therein.
2. Pastorekova, S.; Parkkila, S.; Pastorek, J.; Supuran, C. T. Carbonic anhydrases: current state of the art, therapeutic applications and future prospects. *J. Enzyme Inhib. Med. Chem.* **2004**, *19*, 199–229.
3. Supuran, C. T. Carbonic anhydrases: novel therapeutic applications for inhibitors and activators. *Nat. Rev. Drug. Discov.* **2008**, *7*, 168–181.
4. Scozzafava, A.; Mastrolorenzo, A.; Supuran, C. T. Carbonic anhydrase inhibitors and activators and their use in therapy. *Expert Opin. Ther. Patents* **2006**, *16*, 1627–1664.
5. Supuran, C. T.; Scozzafava, A.; Casini, A. Carbonic anhydrase inhibitors. *Med. Res. Rev.* **2003**, *23*, 146–189.

6. Briganti, F.; Mangani, S.; Orioli, P.; Scozzafava, A.; Vernaglione, G.; Supuran, C. T. Carbonic anhydrase activators: X-ray crystallographic and spectroscopic investigations for the interaction of isozymes I and II with histamine. *Biochemistry* **1997**, *36*, 10384–10392.
7. Winum, J. Y.; Temperini, C.; El Cheikh, K.; Innocenti, A.; Vullo, D.; Ciattini, S.; Montero, J. L.; Scozzafava, A.; Supuran, C. T. Carbonic anhydrase inhibitors: clash with Ala65 as a means for designing inhibitors with low affinity for the ubiquitous isozyme II, exemplified by the crystal structure of the topiramate sulfamide analogue. *J. Med. Chem.* **2006**, *49*, 7024–7031.
8. Supuran, C. T.; Scozzafava, A.; Jurca, B. C.; Ilies, M. A. Carbonic anhydrase inhibitors. Part 49: Synthesis of substituted ureido and thioureido derivatives of aromatic/heterocyclic sulfonamides with increased affinities for isozyme I. *Eur J. Med. Chem.* **1998**, *33*, 83–93.
9. Scozzafava, A.; Supuran, C. T. Carbonic anhydrase inhibitors: ureido and thioureido derivatives of aromatic sulfonamides possessing increased affinities for isozyme I. A novel route to 2,5-disubstituted-1,3,4-thiadiazoles via thioureas, and their interaction with isozymes I, II and IV. *J. Enzyme Inhib.* **1998**, *13*, 103–123.
10. Scozzafava, A.; Supuran, C. T. Carbonic anhydrase inhibitors Arylsulfonylureido- and arylureido-substituted aromatic and heterocyclic sulfonamides: towards selective inhibitors of carbonic anhydrase isozyme I. *J. Enzyme Inhib.* **1999**, *14*, 343–363.
11. Scozzafava, A.; Banciu, M. D.; Popescu, A.; Supuran, C. T. Carbonic anhydrase inhibitors: synthesis of Schiff bases of hydroxybenzaldehydes with aromatic sulfonamides and their reactions with arylsulfonyl isocyanates. *J. Enzyme Inhib.* **2000**, *15*, 533–546.
12. Eriksson, A. E.; Liljas, A. Refined structure of bovine carbonic anhydrase III at 2.0A resolution. *Proteins: Struct. Fund. Genet.* **1993**, *16*, 29–42.
13. Casini, A.; Antel, J.; Abbate, F.; Scozzafava, A.; David, S.; Waldeck, H.; Schäfer, S.; Supuran, C. T. Carbonic anhydrase inhibitors: SAR and X-ray crystallographic study for the interaction of sugar sulfamates/sulfamides with isozymes I, II and IV. *Bioorg. Med. Chem. Lett.* **2003**, *13*, 841–845.
14. Thiry, A.; Dogné, J.-M.; Masereel, B.; Supuran, C. T. Targeting tumor-associated carbonic anhydrase IX in cancer therapy. *Trends Pharmacol. Sci.* **2006**, *27*, 566–573.
15. Whittington, D. A.; Waheed, A.; Ulmasov, B.; Shah, G. N.; Grubb, J. H.; Sly, W. S.; Christianson, D. W. Crystal structure of the dimeric extracellular domain of human carbonic anhydrase XII, a bitopic membrane protein overexpressed in certain cancer tumor cells. *Proc. Natl. Acad. Sci. USA.* **2001**, *98*, 9545–9550.
16. Supuran, C. T.; Scozzafava, A.; Ilies, M. A.; Iorga, B.; Cristea, T.; Briganti, F.; Chiraleu, F.; Banciu, M. D. Carbonic anhydrase inhibitors. Part 53: Synthesis of substituted-pyridinium derivatives of aromatic sulfonamides: The first non-polymeric membrane-impermeable inhibitors with selectivity for isozyme IV. *Eur. J. Med. Chem.* **1998**, *33*, 577–594.
17. Supuran, C. T.; Nicolae, A.; Popescu, A. Carbonic anhydrase inhibitors. Part 35: Synthesis of Schiff bases derived from sulfanilamide and aromatic aldehydes: the first inhibitors with equally high affinity towards cytosolic and membrane-bound isozymes. *Eur. J. Med. Chem.* **1996**, *31*, 431–438.
18. Supuran, C. T.; Popescu, A.; Ilisiu, M.; Costandache, A.; Banciu, M. D. Carbonic anhydrase inhibitors. Part 36: Inhibition of isozymes I and II with Schiff bases derived from chalkones and aromatic/heterocyclic sulfonamides. *Eur. J. Med. Chem.* **1996**, *31*, 439–447.

19. Supuran, C. T.; Scozzafava, A.; Popescu, A.; Bobes-Tureac, R.; Banciu, A.; Creanga, A.; Bobes-Tureac, G.; Banciu, M. D. Carbonic anhydrase inhibitors. Part 43: Schiff bases derived from aromatic sulfonamides: towards more specific inhibitors for membrane-bound versus cytosolic isozymes. *Eur. J. Med. Chem.* **1997**, *32*, 445–452.

20. Popescu, A.; Simion, A.; Scozzafava, A.; Briganti, F.; Supuran, C. T. Carbonic anhydrase inhibitors Schiff bases of some aromatic sulfonamides and their metal complexes: towards more selective inhibitors of carbonic anhydrase isozyme IV. *J. Enzyme Inhib.* **1999**, *14*, 407–423.

21. Supuran, C. T.; Manole, G.; Dinculescu, A.; Schiketanz, A.; Gheorghiu, M. D.; Puscas, I.; Balaban, A. T. Carbonic anhydrase inhibitors V: pyrylium salts in the synthesis of isozyme-specific inhibitors. *J. Pharm. Sci.* **1992**, *81*, 716–719.

22. Supuran, C. T.; Clare, B. W. Carbonic anhydrase inhibitors. Part 24: A quantitative structure-activity relationship study of positively charged sulfonamide inhibitors. *Eur. J. Med. Chem.* **1995**, *30*, 687–696.

23. Supuran, C. T.; Ilies, M. A.; Scozzafava, A. Carbonic anhydrase inhibitors. Part 29 1: Interaction of isozymes I, II and IV with benzolamide-like derivatives. *Eur. J. Med. Chem.* **1998**, *33*, 739–751.

24. Supuran, C. T.; Scozzafava, A.; Ilies, M. A.; Briganti, F. Carbonic anhydrase inhibitors: synthesis of sulfonamides incorporating 2,4 6-trisubstituted-pyridinium-ethylcarboxamido moieties possessing membrane-impermeability and in vivo selectivity for the membrane-bound (CA IV) versus the cytosolic (CA I and CA II) isozymes. *J. Enzyme Inhib.* **2000**, *15*, 381–401.

25. Scozzafava, A.; Briganti, F.; Ilies, M. A.; Supuran, C. T. Carbonic anhydrase inhibitors: synthesis of membrane-impermeant low molecular weight sulfonamides possessing in vivo selectivity for the membrane-bound versus cytosolic isozymes. *J. Med. Chem.* **2000**, *43*, 292–300.

26. Robertson, N.; Potter, C.; Harris, A. L. Role of carbonic anhydrase IX in human tumor cell growth, survival, and invasion. *Cancer Res.* **2004**, *64*, 6160–6165.

27. Harris, A. L. Hypoxia—a key regulatory factor in tumor growth. *Nature Rev. Cancer* **2002**, 2, 38–47.

28. Potter, C. P. S.; Harris, A. L. Diagnostic, prognostic and therapeutic implications of carbonic anhydrases in cancer. *Brit. J. Cancer* **2003**, *89*, 2–7.

29. Svastova, E.; Hulikova, A.; Rafajova, M.; Zatovicova, M.; Gibadulinova, A.; Casini, A.; Cecchi, A.; Scozzafava, A.; Supuran, C. T.; Pastorek, J.; Pastorekova, S. Hypoxia activates the capacity of tumor-associated carbonic anhydrase IX to acidify extracellular pH. *FEBS Lett.* **2004**, *577*, 439–445.

30. Cecchi, A.; Hulikova, A.; Pastorek, J.; Pastorekova, S.; Scozzafava, A.; Winum, J. -Y.; Montero, J. -L.; Supuran, C. T. Carbonic anhydrase inhibitors design of fluorescent sulfonamides as probes of tumor-associated carbonic anhydrase IX that inhibit isozyme IX-mediated acidification of hypoxic tumors. *J. Med. Chem.* **2005**, *48*, 4834–4841.

31. Alterio, V.; Vitale, R. M.; Monti, S. M.; Pedone, C.; Scozzafava, A.; Cecchi, A.; De Simone, G.; Supuran, C. T. Carbonic anhydrase inhibitors: X-ray and molecular modeling study for the interaction of a fluorescent antitumor sulfonamide with isozyme II and IX. *J. Am. Chem. Soc.* **2006**, *128*, 8329–8335.

32. Dubois, L.; Douma, K.; Supuran, C. T.; Chiu, R. K.; van Zandvoort, M.; Pastorekova, S.; Scozzafava, A.; Wouters, B. G.; Lambin, P. Imaging the hypoxia surrogate marker CA IX

requires expression and catalytic activity for binding fluorescent sulfonamide inhibitors. *Radiother. Oncol.* **2007**, *83*, 367–373.

33. Winum, J. -Y.; Rami, M.; Scozzafava, A.; Montero, J. -L.; Supuran, C. Carbonic anhydrase IX: a new druggable target for the design of antitumor agents. *Med. Res. Rev.* **2008**, *28*, 445–463.

34. Garaj, V.; Pucetti, L.; Fasolis, G.; Winum, J. -Y.; Montero, J. -L.; Scozzafava, A.; Vullo, D.; Innocenti, A.; Supuran, C. T. Carbonic anhydrase inhibitors: synthesis and inhibition of cytosolic/tumor-associated carbonic anhydrase isozymes I, II, and IX with sulfonamides incorporating 1,2 4-triazine moieties. *Bioorg. Med. Chem. Lett.* **2004**, *14*, 5427–5433.

35. Garaj, V.; Pucetti, L.; Fasolis, G.; Winum, J. -Y.; Montero, J. -L.; Scozzafava, A.; Vullo, D.; Innocenti, A.; Supuran, C. T. Carbonic anhydrase inhibitors: Novel sulfonamides incorporating 1,3,5-triazine moieties as inhibitors of the cytosolic and tumour-associated carbonic anhydrase isozymes I, II and IX. *Bioorg. Med. Chem. Lett.* **2005**, *15*, 3102–3108.

CHAPTER 20

Bicarbonate Transport Metabolons

DANIELLE E. JOHNSON[1] and JOSEPH R. CASEY[1,2]

[1]Membrane Protein Research Group, Department of Physiology, University of Alberta, Edmonton, Canada T6G 2H7

[2]Membrane Protein Research Group, Department of Biochemistry, University of Alberta, Edmonton, Canada T6G 2H7

20.1 BIOCHEMISTRY OF BICARBONATE

Bicarbonate (HCO_3^-) is an interesting and important molecule. It is a labile substrate, as its form changes subject to the coupled reactions: CO_2 (gas) $\Leftrightarrow$ CO_2 (dissolved) + $H_2O \Leftrightarrow H_2CO_3 \Leftrightarrow HCO_3^- + H^+ \Leftrightarrow CO_3^{2-} + 2H^+$.[1] The reaction $CO_2 + H_2O \Leftrightarrow H_2CO_3 \Leftrightarrow HCO_3^- + H^+$ occurs spontaneously but is also catalyzed by carbonic anhydrase enzymes. The equilibrium between gaseous and dissolved CO_2 varies with partial pressure of CO_2, temperature, and pH, thus adding further complexity to the system. As bicarbonate changes form, pH does too, so understanding the chemistry behind this molecule is necessary to understand pH and pH regulation.

HCO_3^- is involved in pH and cell volume regulation, and carbon dioxide (CO_2) metabolism. Precise regulation of pH is essential to whole-body homeostasis, as biochemical processes occur within a narrow optimal pH window.[1] The CO_2/HCO_3^- system serves as the cell's main pH buffer to limit changes in intra- and extracellular pH. Under physiological conditions, cellular HCO_3^- levels are around 25 mmol/L and metabolic acid will be consumed by the conversion of HCO_3^- to CO_2.[1] CO_2 is membrane permeant and can cross the plasma membrane of mammalian cells by diffusion. Bicarbonate, on the other hand, is membrane impermeant and requires bicarbonate transporters (BTs) to move across the membrane. Transport of the base, bicarbonate, into a cell will alkalinize it, while movement of bicarbonate out of a cell will acidify it. Mammalian kidneys are involved in pH regulation, as they reabsorb nearly all the HCO_3^- that is filtered through them, using a series of bicarbonate transporters.[1] The secretion of HCO_3^- must be minimized to prevent cellular and whole-body acidosis.

Bicarbonate transporters also contribute to cellular volume regulation, which is an important homeostatic mechanism necessary for cell survival.[2] For example, the coordinated activation of a Cl^-/HCO_3^- exchanger and a Na^+/H^+ exchanger is able to load cells with NaCl to deal with a hyperosmotic challenge without a change in pH.[1,3]

Carbon dioxide, the waste product of mitochondrial respiratory oxidation, is continuously produced by mammalian cells and must be removed.[1] In peripheral tissues, membrane-permeant CO_2 diffuses into the erythrocyte, where its hydration is catalyzed by cytosolic carbonic anhydrase II. The resulting membrane-impermeant HCO_3^- is transported into the plasma by the Cl^-/HCO_3^- exchanger AE1, thus increasing CO_2 carrying capacity of blood. In the lungs, the process is reversed and HCO_3^- is transported back into the erythrocyte, dehydrated, and the resulting CO_2 diffuses across the erythrocyte and alveolar membranes and is exhaled.[4,5] Erythrocytes contain $\sim 10^6$ copies of CA II and 1.2×10^6 copies of AE1,[4] and maximal Cl^-/HCO_3^- exchange in erythrocytes depends on CA II.[4,6]

Since HCO_3^- has key biological functions, the translocation of HCO_3^- across cell membranes is carefully controlled by HCO_3^- transport proteins. The focus of this chapter is to examine the interactions of HCO_3^- transport proteins with carbonic anhydrases to form bicarbonate transport metabolons.

20.2 BICARBONATE TRANSPORT PROTEINS

Bicarbonate transport proteins facilitate the movement of membrane-impermeant bicarbonate across biological membranes.[1] Human bicarbonate transport proteins cluster phylogenetically into three classes: electroneutral Cl^-/HCO_3^- exchangers of the SLC4A family (solute carrier 4A), Na^+-coupled HCO_3^- cotransporters (SLC4A family), and anion transporters of the SLC26 family (Fig. 20.1).[7] These bicarbonate transport proteins differ in their tissue localization and mechanism of action (Table 20.1).

The anion exchanger (AE) family is composed of three isoforms (AE1, AE2, and AE3) that share 65% amino acid sequence identity in their transmembrane regions.[7–9] AEs facilitate the electroneutral exchange of Cl^- for HCO_3^- across the plasma membrane of mammalian cells, with a turnover rate of about $5 \times 10^4\ s^{-1}$.[9] AEs can also transport other small inorganic and organic anions, including bromide, fluoride, iodide, phosphate, and sulphate, with varying transport rates.

Human AE1 is comprised of a 43 kDa amino terminal cytoplasmic domain that interacts with cytoskeletal proteins and glycolytic enzymes,[10] a 55 kDa membrane spanning domain responsible for Cl^-/HCO_3^- exchange activity,[4,5] and a short 33-amino acid cytoplasmic carboxyl terminal domain that contains the binding site for CA II.[5,11] AE1 is expressed in the erythrocyte plasma membrane (eAE1) and in the basolateral membrane of α-intercalated cells in the kidney (kAE1). kAE1 lacks the N-terminal 65 amino acids of eAE1.[7] Mutations in the human AE1 gene cause autosomal dominant spherocytic anemia, Southeast Asian ovalocytosis, and distal renal tubular acidosis (dRTA).[12]

AE2 is the most widely expressed isoform and is found in the basolateral membrane of many epithelial cells.[7] AE2 activity is strongly inhibited by protons, but activated by hypertonicity and low concentrations of ammonium ions, whereas AE1 activity is

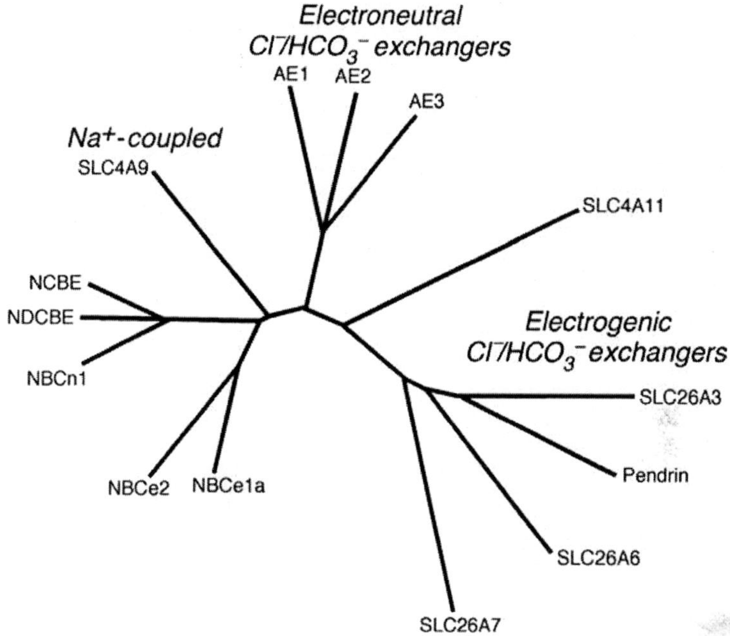

FIGURE 20.1 Phylogenetic tree of human bicarbonate transporter genes. Amino acid sequences corresponding to human bicarbonate transporter genes were aligned using Clustal W software (http://align.genome.jp/). Phylogenetic relationships were established using Clustal W software. The length of each line in the plot denotes relative evolutionary distance. The three families of bicarbonate transporters are labeled.

unaffected by all these parameters.[7] AE1/AE2 chimera studies demonstrated that the pH response of AE2 activity is due to the transmembrane domain ("sensor domain") and that the cytoplasmic domain ("modifier domain") serves to increase sensitivity to protons.

AE3 is expressed predominantly in excitable tissues, including brain, heart, and retina, and throughout the gastrointestinal tract.[7] Two transcripts that differ in their N-termini[9] are encoded by the AE3 gene, AE3 full length (AE3fl) and AE3 cardiac (AE3c). Both are expressed in heart and retina.[13]

The second family of bicarbonate transporters are the Na^+/HCO_3^- cotransporters (NBCs) that facilitate the cotransport of Na^+ and HCO_3^- across the plasma membrane with an either electroneutral (NBC3/NBCn1) or electrogenic ($2/3HCO_3^-$:$1Na^+$) (NBCe1, NBCe2/NBC4) mechanism.[8] The Na^+-coupled HCO_3^- transporter family is composed of NBCe1, NBCe2, NBCn1, NDCBE, and NCBE.[14] NBCs have a widespread tissue distribution, including pancreas, kidney, and heart.[15]

Four members of the SLC26a family have established Cl^-/HCO_3^- exchange activity. Individual paralogs differ significantly in their transported anion specificity.[16] SLC26A4 (Pendrin) is insensitive to intra- and extracellular pH, while SLC26A3 (DRA) is inhibited by intracellular acidification and activated by intracellular alkalinization, ammonium, and hypertonicity.[7] Genetic diseases associated with SLC26 proteins include chloride-losing diarrhea (SLC26A3, DRA) and goiter/

TABLE 20.1 Bicarbonate Transport Proteins and Carbonic Anhydrase Interactions

Transport Protein	Other Names	Mechanism/Charge Movement	Identified CA Interactions	Citations
AE1	SLC4A1, band 3, kAE1, eAE1	Cl^-/HCO_3^- exchange/ electroneutral	CA II, IV, IX	4,5,11, 24–26,34,39
AE2	SLC4A2	Cl^-/HCO_3^- exchange/ electroneutral	CA II, IV, IX	4,5,25,34,39
AE3	SLC4A3 AE3c, AE3fl	Cl^-/HCO_3^- exchange/ electroneutral	CA II, IV, IX, XIV	4,25,34,38,39
NBCe1	SLC4A4, NBC1, hhNBC, pNBC, kNBC1	Na^+/HCO_3^- cotransport/−2 or −1	CA II, IV	14,15,49,50
NBCe2	NBC4, SLC4A5	Na^+/HCO_3^- cotransport/−1 or 0		
NBCn1	SLC4A7, NBC3 (NBC2, SLC4A6)	Na^+/HCO_3^- cotransport/ electroneutral	CA II	48
SLC4A9	AE4	Na^+/HCO_3^- cotransport electroneutral		
NDCBE	SLC4A8, NDAE1 (kNBC-3)	Na^+-dependent Cl^-/HCO_3^- exchange and (electroneutral NBC: splice variant)/ electroneutral		
NCBE	SLC4A10 NBCn2	Na^+-dependent Cl^-/HCO_3^- exchange or Na^+/HCO_3^- cotransport/ electroneutral		
SLC4A11	BTR1, NaBC1	$Na^+/H_2BO_4^-$ cotransport? HCO_3^- transport?/ electrogenic		
SLC26A3	DRA, CLD	Cl^-/HCO_3^- exchange/ electroneutral	CA II (functional only)	39
Pendrin	SLC26A4, PDS	Cl^-/HCO_3^- exchange; also I^-/electroneutral		
SLC26A6	PAT-1, CFEX	Cl^-/HCO_3^- exchange; also oxalate and formate/ electroneutral	CA II	46
SLC26A7		Cl^-/HCO_3^- exchange	CA II (functional only)	39

Bracketed names are not generally accepted, but are found in the literature. Mechanism/charge movement refers to the mode of bicarbonate transport. In the column on identified CA interactions, "(functional only)" means that no physical interaction was required but an effect of CA activity on the rate of HCO_3^- transport is established.

deafness syndrome (SLC26A4, Pendrin).[7,16] SLC26A6 carries out both Cl^-/HCO_3^- and Cl^-/OH^- exchange and is the predominant anion exchanger in the heart.[17] SLC26A7 is a Cl^-/HCO_3^- exchanger localized predominantly in the basolateral membrane of gastric parietal cells.[18]

20.3 THE BICARBONATE TRANSPORT METABOLON

20.3.1 Overview

A metabolon is a physical complex of enzymes in a linked metabolic pathway that functions to maximize flux of substrate through the pathway by the direct transfer of an intermediate between the active site of two enzymes that catalyze sequential reactions, thereby limiting the loss of substrate through diffusion.[4,19–21] This is also known as substrate channeling.[22] Metabolons include coupled enzymes in glycolysis, citric acid cycle, and urea cycle, as well as DNA, RNA, and protein biosynthesis.[19,20]

The linked physiological function (catalysis and transport of HCO_3^-) and the widespread tissue distribution of both bicarbonate transporters and carbonic anhydrases suggest that they could form a complex; CAs both supply the HCO_3^- substrate for transport and remove HCO_3^- following transport.[23] A bicarbonate transport metabolon, composed of a bicarbonate transporter and a CA protein, has thus been proposed on the basis of physical and functional interactions between bicarbonate transporters and carbonic anhydrases (Fig. 20.2).[4,5,11,19,24,25]

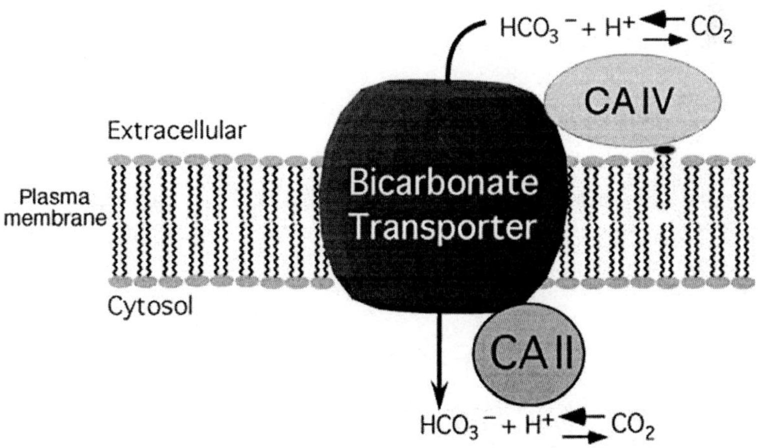

FIGURE 20.2 The bicarbonate transport metabolon. Carbonic anhydrases (CAs) catalyze the reversible conversion of CO_2 to HCO_3^- and H^+. Some bicarbonate transporters directly bind both the soluble cytosolic enzyme CA II and the extracellular enzyme CA IV, anchored to the cell surface via a glycosylphosphatidyl inositol linkage. These direct interactions facilitate the rate of HCO_3^- transport by preventing depletion of HCO_3^- at the *cis* side of the transporter and minimizing it on the *trans* side, thereby maintaining the magnitude of the transmembrane HCO_3^- gradient. This gradient is the primary driving force for HCO_3^- transport.

20.3.2 Initial Characterization

An interaction between the Cl^-/HCO_3^- exchanger, AE1, and CA II in erythrocytes was initially found by examining the effect of the extracellular stilbene-disulfonate inhibitor (DIDS) of AE1 on the fluorescence of a labeled bovine CA II.[26] In this experiment, the kinetics of CA II binding to the CA inhibitor, dansylsulfonamide (DNSA), were measured fluorometrically. Binding of DIDS to AE1 altered the CA II–DNSA binding kinetics.[26] This provided evidence for a CA II/AE1 physical complex.

A direct, physical interaction between CA II and the C-terminus of AE1 was later shown in erythrocytes or erythrocyte ghost membranes by immunolocalization, coimmunoprecipitation, and microtiter plate assays.[5,11] Colocalization between AE1 and CA II was suggested by immunofluorescence experiments performed on tomato lectin-clustered AE1 ghost membranes (tomato lectin binds the extracellular sugar moieties present on the extracellular surface of AE1); immunofluorescence revealed parallel clustering of CA II at the cytosolic membrane surface.[11] This indicates that CA II is physically tethered to AE1 in the membrane. Moreover, CA II and AE1 were coimmunoprecipitated from solubilized erythrocyte membranes.[11] To investigate this physical interaction further, CA II was immobilized on microtiter plates in a solid-phase binding assay and the binding of full-length AE1 or AE1 membrane domain was investigated.[11] Both polypeptides bound saturably to the immobilized CA II, and the binding was decreased when AE1 or AE1 membrane domain was preincubated with an antibody directed against the C-terminus of AE1. Together, these data suggest that the C-terminus of AE1 is involved in the interaction with CA II and that this interaction is specific. In further investigation of the AE1–CA II interaction, a glutathione S-transferase (GST)-fusion protein encoding the C terminal region of AE1 (residues 879–911) (GST-AE1Ct) was used as a probe in an affinity blotting assay, solid-phase assay, and affinity chromatography.[11] These experiments all showed that the GST-Ct bound CA II with high affinity.[5,11]

A series of truncation mutants of the AE1Ct sequence were characterized to narrow the region of AE1Ct involved in association with CA II[5]; amino acids 879–890 of human AE1 were sufficient to bind CA II.[5] This carbonic anhydrase binding (CAB) site is immediately adjacent to the lipid bilayer, as the last transmembrane segment of AE1 is predicted to end at position 877, and thus CA II binding would bring CA II within close proximity to the bicarbonate translocation mechanism of AE1. CA II would then be ideally located to hydrate incoming CO_2 and supply AE1 with HCO_3^- for transport.[5]

An electrostatic interaction between CA II and the AE1 C-terminus was suggested by the pH and ionic strength dependence of their interaction.[5,24] Tantalizingly, the C-terminus of AE1 contains three short acidic regions. Among these, the most membrane proximal was identified as responsible for CA II binding. Further point mutations showed that a precise acidic cluster of residues ($D^{887}ADD$) was required for CA II binding.[5] Within this sequence, at least two out of three acidic residues are necessary for CA II binding, as DADA, DAAD, and AADD mutants still bound CA II with similar levels to DADD. A conserved leucine residue (L886) in AE1 was also found to be essential for the AE1–CA II interaction and indicates that both ionic and nonpolar interactions are involved.[24] The consensus CA II binding motif on AE1 was thus refined to consist of a hydrophobic amino acid followed by four residues, at least two of which are acidic.

What part of CA II binds AE1? The electrostatic nature of their interaction and the requirement for an acid region on AE1 suggested a complementary basic region on CA II that would form the binding site. The CA II crystal structure reveals that the amino terminal region of CA II contains multiple histidine residues along with lysine and arginine residues that together form a basic patch extending from the surface of CA II that could interact with the highly acidic C-terminus of AE1.[5,11,27] A series of truncation and point mutants of CA II showed that five histidine residues in CA II (His3, His4, His10, His15, and His17) were required for binding to GST-AE1Ct in a microtiter plate binding assay.[24] These data indicate that the AE1 binding site in CA II is located within the first 17 amino acids and that the association is mediated by electrostatic interactions involving histidine residues.[24] The histidine-rich region may act as an "electrostatic highway" to funnel HCO_3^- or Cl^- to and from the active sites of CA II and AE1.[22]

AE2 has 60% C-terminal sequence identity with AE1, but the corresponding acidic cluster in AE2 is LDANE.[5] Mutation of AE1's LDADD sequence to LDANE did not affect AE1 binding to CA II, and a GST fusion protein containing the C-terminus of AE2 was able to bind CA II, demonstrating that CA II physically interacts with both AE1 and AE2.

Erythrocytes contain both CA II and CAI, but CAI does not bind AE1 despite 60% sequence identity to CA II.[24] This is likely due to the absence of the basic histidine patch found in CA II.[24]

20.3.3 Physiological Significance

The physiological role of the CA II/AE1 interaction was investigated by measurement of changes in intracellular pH associated with AE-mediated Cl^-/HCO_3^- exchange in transfected HEK293 cells.[4] First, AE1-mediated Cl^-/HCO_3^- exchange was inhibited by 50–60% in the presence of the sulfonamide CA inhibitor, acetazolamide, which indicates that CA II is required for full HCO_3^- transport activity.[4] Acetazolamide inhibits CA enzymatic activity without direct effects on AE1 anion exchange.[6,28] Second, transport activity of AE1 mutants that were previously shown to be unable to bind CA II (LNANN, LAAAA)[5] was assessed. Wild-type AE1 had about 10 times the activity as the mutants, yet the mutants had similar total amounts of protein expression and equal or greater plasma membrane expression; the decrease in Cl^-/HCO_3^- exchange was thus not due to defects in processing the mutants to the cell surface but due to failure to associate with CA II.[4] A 20-fold overexpression of CA II had no effect on wild-type or mutant AE1-mediated Cl^-/HCO_3^- exchange, indicating that endogenous levels of CA II are not rate limiting and that overexpression was not able to rescue the activity of the mutants.[4]

To examine the nature and importance of the CA II/AE1 interaction, a catalytically inactive CA II mutant (CA II V143Y)[29] was overexpressed in HEK293 cells in conjunction with AE1.[4] Wild-type CA II and CA II V143Y bound GST-AE1Ct with similar affinities in a microtiter plate assay and expression of AE1 was not altered with varying levels of wild-type or mutant CA II. Cells cotransfected with AE1 and increasing amounts of CA II V143Y cDNA decreased AE1-mediated Cl^-/HCO_3^- rates by up to 60%.[4] Alternatively, increasing amounts of wild-type CA II cDNA had no effect.[4] This suggests that overexpressed CA II V143Y displaces endogenous wild-type CA II from the binding site on AE1.[8] AE2 and AE3c contain acidic C-terminal sequences similar to AE1, and their exchange activity was also decreased by coexpression with CA II V143Y.[4] Taken together, these results indicate that the CA II/AE physical interaction is required for full activity of a variety of bicarbonate transport proteins.

Two possible mechanisms may explain the increased Cl^-/HCO_3^- exchange activity of these SLC4 proteins when bound to CA II.[12] CA II may contribute to substrate channeling to and/or from the anion translocation pathway, as proposed,[4] or it could allosterically upregulate AE1 activity at the cell surface by changes of the conformational state of the bicarbonate transporter.[12] The latter possibility is inconsistent with the finding that catalytically inactive CA II, able to bind AE1, *decreased* AE1 transport activity. The allosteric mechanism was further tested by the effect that kAE1 C-terminal mutations had on Cl^-/HCO_3^- and Cl^-/Cl^- exchange in *Xenopus* oocytes.[12] The rationale behind these experiments was that a change in the Cl^-/Cl^- exchange rate upon CA II activation would be evidence of a CA II-induced conformational change. The dRTA truncation mutant AE1 901X exhibited normal Cl^-/Cl^- and Cl^-/HCO_3^- exchange, while further truncations (896X, 891X) showed no Cl^-/HCO_3^- exchange, but preserved Cl^-/Cl^- exchange. The retention of Cl^-/Cl^- exchange by the mutants incapable of Cl^-/HCO_3^- exchange does not support the hypothesis that CA II binding allosterically alters the conformation of AE1 such that it could upregulate transport of another nonbicarbonate ion, but it remains consistent with the channeling hypothesis.[12]

Coexpression studies of wild-type and mutant AE1 proteins further probed metabolon behavior; AE1 forms dimers in which each AE1 subunit is sufficient for HCO_3^- transport function.[30,31] The heterozygous hAE1 mutant (AE1 Δ400–408, SAO), associated with Southeast Asian ovalocytosis and dRTA, is expressed at the cell surface, but lacks Cl^-/HCO_3^- exchange activity.[12] mAE1 E699Q is also expressed at the cell surface and exhibits severely impaired Cl^-/HCO_3^- exchange. Both of these mutants retain the CA II binding site. When the mutants were coexpressed in *Xenopus* oocytes with AE1 LDAAA, which cannot bind CA II, they were able to rescue the Cl^-/HCO_3^- exchange activity of AE1 LDAAA by 50%.[12] The ability of the heterodimers (homodimers presumably remain inactive) of SAO/LDAAA and E699Q/LDAAA to rescue Cl^-/HCO_3^- exchange is attributable to the functional proximity of the C-terminal tail of one protomer (SAO, E699Q) to the anion translocation pathway in the adjacent one (LDAAA).[12] In other words, the CA II molecule does not need to be bound to the same protomer that catalyzes transport; CA II binding at the adjacent monomer within the dimer unit brings the enzyme sufficiently close to the transport site to activate Cl^-/HCO_3^- exchange.

Another mechanism by which metabolon interaction facilitates coupled flux is that association of CA II with AE proteins activates CA II's catalytic rate. Di-, tri-, tetrapeptides containing all or part of the LDADD motif increased CA II activity over fivefold.[32] These data strengthen the CA/AE metabolon hypothesis.

20.3.4 Interactions with Membrane-Anchored CAs

Some members of the CA family have their catalytic sites anchored to the extracellular surface. Cell surface glycosylphosphatidyl inositol (GPI)-anchored carbonic anhydrase IV (CA IV) has a high degree of similarity to the cytosolic CAs and shares 36% sequence identity with CA II.[33] In addition, CA IV has comparable catalytic activity to CA II.[33] Studies to isolate the contribution of CA IV to the rate of AE-mediated Cl^-/HCO_3^- exchange were hindered by the inability to use pharmacological inhibitors of CA II, as membrane-permeant compounds would also block extracellular CA IV activity. To circumvent this problem, the inactive CA II V143Y mutant was used to suppress endogenous CA II activity.[25] CA IV was found to be necessary for maximal AE1-mediated Cl^-/HCO_3^- exchange when assayed in cotransfected HEK293 cells.[25,34] Overexpression of CA IV was able to rescue AE1, AE2, and AE3 activity in the presence of the inactive CA II V143Y mutant, thus indicating that there is a functional interaction between the anion exchangers and CA IV.[25] This enhancement was shown to be specific to CA activity, as treatment with acetazolamide abolished the increase in transport rate.

A physical association between CA IV and AE1 was demonstrated by comigration of AE1 and CA IV on sucrose gradients, blot overlay assays, and GST pull-down assays.[25] CA IV binds specifically to the fourth extracellular loop of AE1 (EC4), and this interaction forms the extracellular component of a bicarbonate transport metabolon.[25] EC4 is the largest extracellular loop in AE1. In addition, it has been proposed that a region in EC4 (Arg656–Ile661) forms the outer vestibule that funnels anions to and from the transport site.[25,35] Thus, CA IV is very close to the active site of AE1, where it can be most effective in HCO_3^- dehydration and maximizing the transmembrane HCO_3^- gradient local to the transporter (Fig. 20.2). The structures of AE2 and AE3 differ somewhat from AE1 in that EC3 is larger than EC4, but a similar physical interaction probably exists.[25,36,37]

AE3, CA II, and CA XIV (another extracellular-linked CA isoform) are involved in pH_i, [Cl^-], [HCO_3^-], and volume regulation in the Müller and horizontal cells of the human retina.[38] The retina is a highly metabolically active tissue that produces large amounts of CO_2 and H^+, which must be eliminated in order to maintain intracellular and extracellular pH. Studies of AE3 knockout mice demonstrate that the gene product is essential in maintaining normal retinal function.[38] Furthermore, $AE3^{-/-}$ mice have compensatory elevated expression of NBCe1, CA II, and CA XIV.[38] However, pH regulation and cell function are still compromised leading to apoptotic cell death. A CA XIV-deficient mouse has similar retinal deficiencies, further suggesting a link between AE3 and CA XIV.[38] A physical link between the two proteins was also found (Alvarez et al., submitted for publication). AE3-mediated HCO_3^- flux may be maximized by AE3's interaction with CA XIV and CA II, thus maintaining the transmembrane HCO_3^- gradient local to the transporter to maximize HCO_3^- flux. This association may represent a mechanism for the disposal of CO_2 and H^+. This provides further evidence for a physical interaction between CAs and AEs in health and disease.

CA IX's catalytic domain is anchored to the extracellular surface of cells by a single transmembrane segment.[39] In normal tissues, CA IX is expressed in gastric, intestinal, and gallbladder mucosa, where it may play a role in acid secretion.[40] In addition, CA IX is expressed in many types of carcinomas where tumor growth usually results in hypoxia in poorly perfused regions of tissue, leading to the upregulation of glycolysis and the production of excess lactic acid.[41] Tumor growth is promoted by an acidic extracellular milieu that is amplified by CA IX. In addition, CA IX has been shown to induce near-uniform intracellular pH throughout model solid tumors (cultured spheroids of RT112 cells derived from bladder carcinoma).[42] CA IX may facilitate CO_2 diffusion in the spheroid's unstirred extracellular space to increase tumor growth and survival.

AE2 is also highly expressed in the stomach and involved in basolateral parietal cell uptake of Cl^- destined for HCl secretion and extrusion of HCO_3^- generated intracellularly during acid secretion.[43] Another basolateral membrane anion transporter, SLC26A7, has also been proposed to contribute to HCl secretion in the parietal cell.[18] Since CA IX localizes to the basolateral membrane, and basolateral HCO_3^- efflux is required for apical acid secretion, could CA IX interact with AE2 or SLC26A7?[39,40] In HEK293 cells, CA IX increased AE-mediated Cl^-/HCO_3^- exchange when coexpressed with AE1, AE2, and AE3, but not SLC26A7.[39] Treatment of cells with acetazolamide did decrease SLC26A7 Cl^-/HCO_3^- exchange activity, indicating that CA activity is necessary for maximal transport. Further support is provided by the finding that SLC26A7 mRNA and protein expression levels are severely reduced in CA II-deficient mice.[44] Coexpression with CA II V143Y did not, however, alter SLC26A7 transport, which suggests that cytosolic CA II binding is not required for full SLC26A7 activity. This is similar to the phenomenon seen with SLC26A3, which will be discussed later.[45] AE2-mediated Cl^-/HCO_3^- exchange was decreased by CA II V143Y overexpression, and this decrease was rescued by coexpression with CA IX, which supports a functional interaction between CA II and IX, and AE2.

Immunoprecipitation experiments demonstrated that CA IX physically interacts with AE1, AE2, and AE3, but not with SLC26A7.[39] This remains consistent with the functional data. As mentioned previously, CA IX is a transmembrane protein with four domains: an N-terminal proteoglycan-like domain (amino acids 53–111), an extracellular catalytic domain (amino acids 135–391), a single transmembrane domain (amino acids 415–436), and a short intracellular C-terminal tail (amino acids 436–459).[39,40] Constructs that lack the catalytic domain (ΔC-CA IX) and the proteoglycan domain (ΔPG-CA IX) were used to determine which portion(s) of CA IX bind AE2.[39] Coimmunoprecipitation and GST pull-down approaches demonstrated that AE2 binds the catalytic domain of CA IX.

20.3.5 SLC26A/CA Interactions

Amino acid sequence alignment reveals that the C-terminal tails of all known bicarbonate transport proteins (Fig. 20.1), except for SLC26A3 and SLC26A7, contain at least one CA II binding motif (a hydrophobic residue followed by a cluster of acidic resides).[45] SLC26A3 was able to bind immobilized CA II with a much lower affinity and capacity than AE1, and its Cl^-/HCO_3^- exchange activity was not altered by overexpression of CA II V143Y. Thus, it appears that there is not a physical interaction between SLC26A3 and CA II, which makes it (and SLC26A7) unique among bicarbonate transport proteins.[45] SLC26A3 does, however, require cytosolic CA II activity, as inhibition of CA II with acetazolamide decreased SLC26A3 activity.[45] Conversely, extracellular CA inhibitors had no effect. CA II and SLC26A3 are functionally coupled despite the lack of physical interaction.

SLC26A6 transports several anions, including sulfate, formate, oxalate, nitrate, and iodide, but the most physiologically relevant transport modes are Cl^-/HCO_3^- exchange and Cl^-/OH^- exchange.[46] SLC26A6 has important functions in renal proximal tubule cells, where it is involved in NaCl reabsorption, and in pancreatic duct cells, where it is involved in secretion of basic pancreatic fluid.[46]

Possible interactions between SLC26A6 and CA were investigated, as other SLC26 anion exchangers have been shown to interact functionally, but not physically with CA II and/or CA IV.[39,45] Glutathione S-transferase fusion to various regions of the C-terminal cytoplasmic domain of SLC26A6 revealed that SLC26A6 D546–F549 region bound CA II.[46] Cl^-/HCO_3^- exchange activity assays in the absence or presence of acetazolamide tested the physiological significance of this binding. As with other anion exchange proteins, CA II inhibition decreased exchange activity by approximately 50%, which reveals that functional CA II is required for full SLC26A6 transport activity.[46] Overexpression of CA II V143Y also decreased transport rates, consistent with a required physical interaction between SLC26A6 and CA II. Further to this, mutation of the SLC26A6 CA II binding site decreased anion transport, but not cell surface processing.[46]

Angiotensin II (AngII) activates protein kinase C (PKC) and inhibits pancreatic bicarbonate secretion through effects on an anion exchanger on the luminal membrane of pancreatic ducts, where SLC26A6 also localizes.[46] To test the role that PKC might have on SLC26A6 activity, Cl^-/HCO_3^- exchange activity was assessed in HEK293 cells transfected with SLC26A6 cDNA and AT_{1a} AngII receptor cDNA.[46] Upon AngII treatment, Cl^-/HCO_3^- exchange activity was decreased by ~40%, and this decrease was eliminated by pretreatment with the broad-spectrum PKC inhibitor chelerythrine (CHE), consistent with effects mediated by PKC.[46] There are five consensus PKC phosphorylation sites in the cytoplasmic C-terminus of SLC26A6:SQK (553), SPK (582–584), TLK (636–638), SLK (667–669), and TKK (706–708).[46] S553 and S582 are close to the SLC26A6 CA II binding site, which raised the possibility that phosphorylation might alter CA II binding. To investigate the possible role of SLC26A6 phosphorylation by PKC, HEK293 cells were transfected with wild-type SLC26A6, or two PKC consensus site mutants. Both mutants were fully functional. SLC26A6-S582A transport activity was inhibited by PKC to an extent similar to wild type. SLC26A6-S553A was, however, unaffected by PKC activation with PMA (phorbol 12-myristate 13-acetate ester), thus indicating that S553 is important for the regulation of SLC26A6 activity by PKC.[46] Furthermore, CA II binding to wild-type SLC26A6 was drastically reduced in the presence of PMA, suggesting that PKC phosphorylation disrupts the binding of CA II to SLC26A6 and that this is the cause of decreased anion exchange activity.

Both the CA II binding site and the identified PKC phosphorylation site localize to a conserved surface loop in the STAS domain[47] as revealed by homology with a high-resolution structure of a bacterial STAS domain.

PKC thus inhibits SLC26A6 Cl^-/HCO_3^- exchange through disruption of the bicarbonate transport metabolon by displacement of CA II from the surface of SLC26A6.[46] This is an example of bicarbonate transport metabolon disruption, where an enzyme that catalyzes the production of a transport substrate is displaced from the binding site on the surface of the transporter.[46] This represents a new mechanism for regulation of membrane transporter function.

20.3.6 CA Interactions with Na^+-Coupled HCO_3^- Transporters

The sodium bicarbonate cotransporters comprise the third group of bicarbonate transport proteins. As with other members of the BT family, they are involved in regulation of intracellular pH and HCO_3^- metabolism. NBC3 is an electroneutral Na^+/HCO_3^- transporter that is involved in intracellular pH regulation in heart, skeletal muscle, and kidney.[48] As with other bicarbonate transporters, a region in the C-terminus of NBC3 (D1135–D1136) bound CA II, and this interaction was essential for maximal HCO_3^- transport.[48] NBCe1 is an electrogenic Na^+/HCO_3^- cotransporter that operates with either a 3:1 or 2:1 $HCO_3^-:Na^+$ stoichiometry.[48] Splicing variant NBCe1a (kNBCe1) plays the central role in HCO_3^- reabsorption in the basolateral membranes of the proximal tubule in conjunction with the luminal Na^+/H^+ exchanger, NHE3, and a H^+-ATPase, which secrete acid.[15] NBCe1a thus mediates HCO_3^- efflux from the cell to the blood and normally works with a stoichiometry of $3HCO_3^-$ per Na^+. NBCe1b (pNBCe1) is expressed in the basolateral membranes of pancreatic duct cells where it transports HCO_3^- into the cell and functions with a $2HCO_3^-$ per Na^+ stoichiometry.[15] CA IV and NBCe1 colocalize in mammalian kidney, pancreas, and heart.[15]

CA IVoverexpression increases the rate of recovery from acid load in HEK293 cells cotransfected with NBCe1 and CA IV cDNA.[15] Sequence conservation across the SLC4A family of bicarbonate transporters (AEs–NBCs) (Fig. 20.1) suggested that there could be a conserved mode of CA IV binding. Mutation of NBCe1 in the fourth extracellular loop (G767T-NBCe1) abolished the CA IV-induced increase in transport and abrogated physical binding in GST pull-down assays, consistent with a role of this loop in the interaction with CA IV.[15] CA IV and NBCe1 could also be coimmunoprecipitated from lysates of mouse kidney.[15] Membrane-permeant acetazolamide and a membrane-impermeant CA inhibitor both reduced NBCe1 activity.

Intracellular CA II also has a role in maximizing NBCe1 activity. Two acidic motifs in NBCe1 (L^{958}DDVand D^{986}NDD) are involved in binding CA II, as demonstrated by GST pull-down assays.[49] The NBCe1-mediated recovery after acid load was inhibited by ~40% when overexpressed with CA II V143Y, consistent with a dominant negative effect.[49] Similarly, electrophysiological recordings of NBCe1 expressed in a mouse proximal convoluted tubule cell line (mPCT) showed that CA II increased NBCe1 short-circuit current through the transporter.[14,50] Taken together, data indicate that

NBCe1 physically and functionally interacts with both CA II and IV to form a bicarbonate transport metabolon.

Phosphorylation also modulates NBCe1 transport activity, possibly through a bicarbonate transport metabolon. The transport stoichiometry of NBCe1a is shifted from 3:1 HCO_3^-:Na^+ to 2:1 by cAMP-dependent PKA phosphorylation of Ser982 (NBCe1a).[50] Three aspartate residues (D^{986}NDD) involved in binding CA II are close to the PKA phosphorylation site, and acetazolamide decreases NBCe1a activity only when the transporter is unphosphorylated (3:1 stoichiometry).[50] This suggests that PKA has an effect on CA II's interaction with NBCe1a and thus activity.[50] PKA-dependent phosphorylation of NBCe1a in the proximal tubule also reduces transport activity of NBCe1a.[51]

The increased negative charge on NBCe1a-Ser982-phosphate may shift the stoichiometry by interference with the binding of HCO_3^- to NBCe1a by modification of the electric field around the binding site.[50] The nearby D^{986}NDD acidic cluster might also play a role in the shift in stoichiometry. Mutation of all the three aspartate residues in the D^{986}NDD motif to asparagines (N^{986}NNN) decreased CA II binding by ~90%, confirming the physical interaction between CA II and NBCe1a.[49] Furthermore, 8-Br-cAMP (a PKA activator) had no effect on the stoichiometry of the N^{986}NNN mutant, suggesting that the D^{986}NDD motif is indeed required for the shift in stoichiometry. Moreover, the PKA-induced shift in stoichiometry is dependent on D986 and D988, but not D989.[49,50] The PKA-induced shift in stoichiometry from 3:1 HCO_3^-:Na^+ to 2:1 is thus dependent on the phosphorylation of Ser982 and requires the presence of D986 and D988 of the CA II binding motif. This shift in stoichiometry decreases NBCe1a transport and renders the transporter's activity insensitive to acetazolamide, thus suggesting that phosphorylation interferes with CA II binding. Gross et al. proposed a model where CA II binds to NBCe1a when the transporter functions in the 3:1 mode (NBCe1a-Ser982 unphosphorylated).[50] Upon phosphorylation of NBCe1a-Ser982, CA II dissociates from the transporter. This explains the decreased effect of CA inhibition.

20.3.7 Bicarbonate Transport Metabolon Controversies

Some controversy surrounds the bicarbonate transport metabolon. Lu et al. were unable to show a difference in NBCe1-associated current (I_{NBC}) in NBCe1 expressing *Xenopus* oocytes coinjected with either CA II protein or Tris.[52] They have also not been able to see a difference in I_{NBC} when NBC was fused to the N-terminus of CA II. Thus, they conclude that CA II does not enhance the current carried by NBC but instead only accelerates the intracellular hydration of CO_2.[52] Note that in renal proximal tubule cells the HCO_3^-:Na^+ stoichiometry is 3:1, while in oocytes the stoichiometry is 2:1,[52] which is significant in the context of the work of Gross et al. discussed previously. Furthermore, in other cell types when unphosphorylated NBCe1 works with a 3:1 HCO_3^-:Na^+ stoichiometry, its transport is decreased with acetazolamide treatment. When phosphorylated, NBCe1 works with a 2:1 HCO_3^-:Na^+ stoichiometry and is not sensitive to acetazolamide treatment. CA II may then only interact physically and functionally with NBCe1, when active in the 3:1 stoichiometry; the lack of CA II-enhanced activity of NBCe1 in these experiments with *Xenopus* oocytes may be due to the transporter's function with 2:1 stoichiometry. Becker et al., however, showed that injection or coexpression of CA II did not change the reversal potential in NBCe1 expressing oocytes, in line with NBCe1 operation in 2:1 mode when interacting with CA II.[14]

The physical interaction between BTs and CAs was further investigated by two solid-phase binding assays: an ELISA (enzyme-linked immunosorbent assay) and SPR (surface plasmon resonance).[53] Three SLC4 members were investigated: AE1 (SLC4A1), NBCe1 (SLC4A4) and NDCBE (SLC4A8). GST-SLC4-Ct fusion proteins and pure SLC4-Ct peptides (similar constructs with GST cleaved off) were used. Consistent with previous results, the GST-SLC4-Ct fusion proteins bound immobilized CA II better than GST alone, but no binding between CA II and the pure SLC4-Ct peptides was observed. When the orientation was reversed and SLC4-Ct pure peptides GST or GST-fusion proteins were immobilized, CA II bound GST to a greater extent than the fusion proteins.[53] In addition, SPR was also unable to detect an interaction between CA II and BTs. The SPR experiment is, however, potentially confounded by the way that CA II was immobilized to the SPR chip. CA II was coupled using amine-directed chemistry, which would have linked CA II to the SPR chip via the primary amine both at the CA II N-terminus and at lysines 9 and 24. All three of the amines reside in the N-terminal region required for AE1 binding, thus presenting significant steric constraint to AE1 binding. The authors conclude that a bicarbonate transport metabolon may exist but that CA II may not bind directly to the bicarbonate transporters.[53] These results disagree with the findings of other groups.[4,5,11,24]

Recently, the interaction between NBCe1 and CA II was further investigated by expressing NBCe1 in *Xenopus* oocytes, with and without injected or coexpressed CA II.[14] Two CA II mutants were used, CA II V143Y and an N-terminal mutant with six exchanged amino acids in the N-terminal tail (CA II-HEX), which prevents binding to AE1. Transport activity of NBCe1 was determined by simultaneous measurement of membrane current $[Na^+]_i$ and $[H^+]_i$ (calculated from pH) of voltage-clamped oocytes during a switch from HEPES buffer to CO_2/HCO_3^- buffer.[14] Membrane conductance was also determined. CA II (either injected or coexpressed) increased membrane current, membrane conductance, and the rate of cytosolic $[Na^+]$ and $[H^+]$ change in NBCe1 expressing oocytes following a switch to CO_2/HCO_3^- buffer to a greater extent than NBCe1 alone. Incubation with CA inhibitor 6-ethoxy-2-benzothiazole-sulfonamide (EZA) abolished these increases. Coexpression with CA II V143Y also abolished these increases, consistent with an essential physical interaction. CA II-HEX expressed in oocytes showed similar responses to CA II wild type, suggesting that this mutant can still interact with NBCe1.

These results differ from those reported by Lu et al., who analyzed membrane current only in terms of its slope conductance that was not changed by injection of CA II.[52] There were several differences between the Becker and Lu studies related to the amount of CA II injected, the amplitude of slope conductance, and the level of NBC expression.[14] Becker et al. detected a small current of 2.2 μS in CA II/NBCe1 expressing oocytes, which could have been missed in the Lu et al. study due to their higher slope conductance and lack of multiple current measurement approaches.[14] In summary, the results of Becker et al. demonstrate that CA II enhances the activity of NBCe1 and supports the conclusion that NBCe1 forms a bicarbonate transport metabolon with CA II.[14,49,50]

20.3.8 Summary

Intracellular and extracellular carbonic anhydrases interact physically and/or functionally with many bicarbonate transport proteins (Table 20.1). Cytosolic CA II is able to physically interact with AE1, AE2, AE3, NBCe1, NBC3, and SLC26A6, but not SLC26A3 or SLC26A7. Extracellular GPI-linked CA IV physically interacts with AE1, AE2, AE3, and NBCe1. Extracellular CA IX interacts with AE1, AE2, and AE3, but not SLC26A7. CA XIV interacts with AE3. The physical interaction of all of the above CA II/BT pairs is able to enhance activity. CA II is also required to maximize SLC26A3 and SLC26A7 activity.

As defined previously, a metabolon is a physical complex of enzymes in a linked metabolic pathway that functions to maximize flux of substrate through the pathway. There is extensive evidence supporting a bicarbonate transport metabolon. This metabolon would maximize HCO_3^- flux through the bicarbonate transporter by maximizing the transmembrane HCO_3^- gradient local to the bicarbonate transporter. In HCO_3^- efflux mode, this would be accomplished by localization of cytosolic CA II within close proximity to the membrane, where CA II would be ideally located to hydrate incoming CO_2 and supply the transporter with HCO_3^- for transport.[5] CA II would ensure that there is a continued pool of substrate for the BT to transport. For example, AE1 has a high turnover rate of $5 \times 10^4\ s^{-1}$ and would rapidly deplete substrate without proximal CA II, which has an even greater turnover rate of $10^6\ s^{-1}$.[4] Extracellular membrane-linked carbonic anhydrases (CA IV, CA IX) will rapidly deplete extruded HCO_3^- by conversion to CO_2, thus maximizing the transmembrane HCO_3^- flux to maximize transport rate.[25] Modulation of the interaction between CA II with bicarbonate transport proteins would provide a mechanism to regulate bicarbonate transport.[4,19] In conclusion, the interaction of bicarbonate transporters with intracellular and extracellular carbonic anhydrases is a universal component of bicarbonate transport physiology and that such interactions with linked metabolic enzymes suggest a general mechanism to modulate membrane transport activity that extends beyond bicarbonate transporters.[15]

20.4 OTHER ACID/BASE TRANSPORTER CARBONIC ANHYDRASE METABOLONS

20.4.1 Na^+/H^+ Exchanger and CA II

Na^+/H^+ exchangers (NHEs) mediate the electroneutral exchange of one intracellular H^+ for one extracellular Na^+ to mediate essential roles in the regulation of intracellular pH, cell volume, and sodium flux.[54] NHEs are also involved in cytoskeletal organization, heart disease (including ischemic and reperfusion injury and cardiac hypertrophy), and cancer.[55,56] Among the nine isoforms of mammalian NHEs, NHE1 is the most thoroughly characterized and is ubiquitously expressed. Briefly, NHE1 is composed of an N-terminal membrane domain that mediates Na^+/H^+ exchange and a large C-terminal cytoplasmic domain that is involved in the regulation of exchanger activity and cytoskeletal interactions.[54,55] NHE1 is activated by angiotensin, endothelin, and α_1-adrenergic stimulation, all of which increase phosphorylation of the C-terminal domain by protein kinases.[54]

Since carbonic anhydrases are involved in proton production, it is reasonable to hypothesize that an association between NHEs and CAs could facilitate proton removal.[57] Indeed, early studies demonstrated that the CA inhibitor acetazolamide caused a decrease in NHE activity.[58] NHE1 and CA II also colocalize and create pH microdomains in oligodendrocytes.[59] A solid-phase binding assay revealed that there was a physical interaction between the C-terminus of NHE1 (tagged with an N-terminal GST or a C-terminal polyhistidine tag) and CA II and that this interaction was increased in acidic medium, presumably by increased electrostatic interactions.[57] This interaction was shown to be specific, as an antibody against the C-terminal 178 amino acids of NHE1 blocked the association. Results were confirmed with an affinity blotting technique and immunoprecipitation of stably transfected cell lysates. The physiological relevance of the NHE1/CA II interaction was examined by the measurement of the rate of pH recovery after acid load in NHE1-transfected cells with or without CA II.[57] Cotransfection with CA II increased NHE1 activity. This increase was abolished by treatment with acetazolamide and by cotransfection with the dominant negative CA II V143Y. These results show that there is a physical interaction between NHE1 and CA II and that this interaction is required for maximal Na^+/H^+ exchange.

Phosphorylation of NHE1 increases its activity, so the effect of phosphorylation on CA II binding to NHE1 was investigated via CA II binding assays with phosphorylated and nonphosphorylated NHE1 C-terminal protein.[57] Phosphorylated NHE1 C-terminal protein bound significantly higher amounts of CA II than the nonphosphorylated protein, consistent with a model in which phosphorylation stimulates NHE1 activity by enhanced CA II binding. This is another example of regulation of transport function by modulating the formation of a bicarbonate transport metabolon.

Recent studies using fusion proteins of various C-terminal regions of NHE1 established that CA II binds the penultimate 13 amino acids on the cytoplasmic tail of NHE1 (^{790}RIQRCLSDPGPHP^{802}).[56] Results were confirmed by gel overlay assays, GST pull-down assays, and solid-phase binding assays. It came as a surprise that this region did not contain the acidic stretch of amino acids (^{735}EEDEDDD^{759}) that were predicted to bind CA II based on their similarity to the acidic stretch of amino acids in AE1. In addition, mutation of the acidic residues to neutral residues had no effect on CA II binding. Of the penultimate 13 amino acids, S796 and D797 were shown to be essential for CA II binding.[56] Phosphorylation of the last 26 amino acids of NHE1 did not affect CA II binding although, as reported previously, phosphorylation of a larger portion of the C-terminal tail increased CA II binding.[56,57] These results indicate that although S796 can be phosphorylated, phosphorylation at a more proximal site is responsible for increased CA II binding. There may be a region in the C-terminus of

NHE1 involved in the reduction of CA II binding, possibly by blockade of the CA II binding site.[56] Phosphorylation of NHE1 appears to prevent this blockage. NHE1 and CA II physically and functionally interact to form a transport metabolon, and this metabolon is regulated by phosphorylation of NHE1.

20.4.2 Monocarboxylate Transporter 1 (MCT1) and CA II

Monocarboxylate transporters (MCTs) transport lactate or pyruvate in an electroneutral $1H^+$:1 organic anion transport mode.[60] The 14 MCT isoforms in the SLC16 gene family are expressed in most tissues, with special importance in highly active tissues such as muscle and brain.[60] In the brain, MCT1 and MCT2 work in concert to shuttle lactate from glial cells to neurons; MCT1 facilitates the export of lactate, which is then taken up by MCT2.[61] In skeletal muscle, MCT1, found in oxidative slow twitch fibers, mediates the import of lactate that is released from glycolytic fast twitch muscle cells via MCT3 and MCT4.[60,62]

MCT1, when expressed in *Xenopus* oocytes, mediates the import of lactate and H^+ in both HEPES and CO_2/HCO_3^- buffer, and transport (measured by changes in pH_i upon application of lactate) is enhanced by the injection of CA protein.[61] The increase in H^+ flux, however, was insensitive to the CA inhibitor, ethoxzolamide (EZA). This suggests that CA II enhances the transport activity of MCT, but does not require CA II's enzymatic activity. Further, coexpression with the catalytically inactive CA II V143Y also caused an increase in transport activity.[60] Both the MCT1 protein and the MCT1D56 C-terminal mutant (lacking the C-terminal 56 amino acids) bind immobilized CA in pull-down assays.[61] This suggests that these 56 amino acids in the C-terminus of MCT1 (which include two acidic amino acid clusters, 456–458 and 489–491) are not involved in binding CA II, unlike many bicarbonate transport proteins. The C-terminus is, however, necessary for the enhanced transport in the presence of CA II, as no increase in transport is seen with the C-terminal mutant.[61] In another set of experiments, coexpression with CA II-HEX (an N-terminal CA II mutant shown previously as unable to bind AE1) was unable to increase the rate of MCT1 transport, suggesting that the interaction between MCT1 and CA II is mediated by the same motif as that found with AE1.

Injection of CAI into MCT1 expressing oocytes had no effect on MCT1 activity, so that increased amounts of CA catalytic activity are not sufficient to increase MCT1 activity.[60] It also suggests that a unique aspect of CA II that is not found in CAI is required for the enhanced MCT1 activity.

Together, these results indicate that CA II accelerates the removal of protons away from the MCT1 domain at the inner face of the cell membrane and prevents the formation of an acidic microdomain, which would slow transport.[60] Rapid removal of H^+ would prevent the dissipation of the transmembrane H^+ gradient and thus enhance H^+/lactate cotransport. Immunofluorescence reveals that CA II accumulates at the inner face of the plasma membrane, perhaps due to an accumulation of negative charges imparted by anionic phospholipids in the inner leaflet and by physical interaction with acid/base transporters.[60,63] The fate of imported H^+ was examined with the pH-sensitive dye, BCECF-AM, to monitor the distribution of H^+ along the inner face of the oocyte membrane upon focal application of lactate to MCT1 expressing oocytes in the presence or absence of injected CA II.[60] In CA II-injected oocytes, the rate of fluorescent signal decreased to 50% at a distance of 323 μm away from the area of lactate application, while in H_2O-injected oocytes the rate decreased to 50% at 108 μm.[60] This supports the hypothesis that CA II is involved in the removal of H^+ from the inner face of MCT1 to prevent the formation of an acidic pH microdomain.

20.4.3 N-Type Sodium-Dependent Neutral Amino Acid Transporter Isoform 3 (SNAT3) and CA II

SNAT3 catalyzes the electroneutral cotransport of $1Na^+$ and glutamine in exchange for $1H^+$.[64] Transport of glutamine (non-neuroexcitatory amino acid) is essential to the glutamate–glutamine cycle in the brain, which is important to replenish neuronal glutamate (neurotransmitter) stores.[65] Transport of glutamine in *Xenopus* oocytes elicits currents presumably associated with uncoupled H^+ conductance.[64] CA II suppressed this cation conductance in the presence of CO_2/HCO_3^- buffer, but no enhancement of substrate transport was observed with CA II either injected or coexpressed.[64] No CA II binding motifs were found in SNAT3, so if there is a direct interaction between SNAT3 and CA II, it does not occur through the canonical CAB motif. CA II's catalytic activity was required for the suppression of membrane conductance associated with SNAT3, as CA II V143Y did not affect membrane conductance, and ethoxyzolamide restored membrane conductance in the presence of wild-type CA II.[64] In conclusion, the catalytic activity of CA II affects SNAT3-associated cation conductance and demonstrates yet another example wherein CA II activity is important for acid/base membrane transport.

20.5 NONCATALYTIC ROLE OF THE CA METABOLON

Studies of the MCT1 monocarboxylate transporter showed that CA II associates with the cytosolic surface of the plasma membrane and that MCT1 transport rate was enhanced by catalytically *inactive* CA II.[60] This led to the proposal that CA II might function to form a network to funnel H^+ away from the cytosolic mouth of MCT1 (Fig. 20.3). In such a model, CA II binds to the surface of a H^+ accumulating transporter, such as MCT1. In addition, CA II is able to bind the inner surface of the plasma membrane, as supported by immunofluorescent observations. Histidine residues on the surface of CA II could form dissociable sites for H^+ binding. An interaction between CA II and the transporter would therefore form the initial component of a network of CA II molecules to move H^+ outward from the transporter. This CA II network would facilitate H^+ movement away form the mouth of the transporter, eliminating the localized H^+ accumulation, which would otherwise reduce the rate of substrate movement across the plasma membrane. H^+ movement through the cytosol is normally extraordinarily slow, as indicated by studies of H^+ diffusion through cardiomyocyte cytosol, which revealed that more

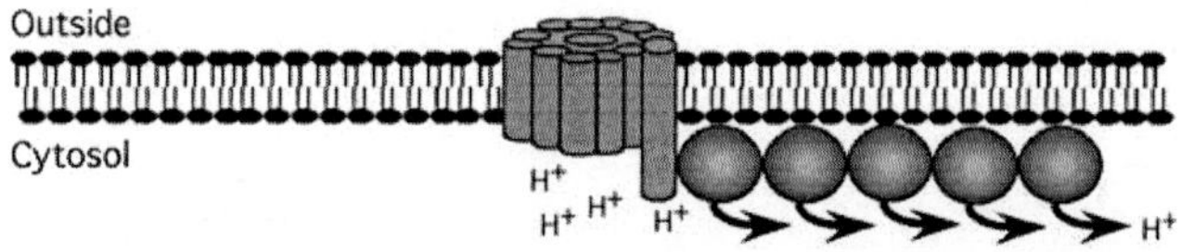

FIGURE 20.3 Noncatalytic role of membrane-associated CA II. CA II (gray filled ovals) associates with both the cytosolic surface of the plasma membrane, via electrostatic interaction with negative phospholipid head groups, and with the cytoplasmic regions of plasma membrane transporters. Histidine residues on the surface of CA II form ionizable groups to form an exchangeable H^+ binding network. CA II thus facilitates the movement of H^+ away from the transporter to prevent localized H^+ accumulation that could slow down transmembrane substrate flux (as proposed from work on MCT1).[60]

than 30 s are required for a H^+ front to move the length of a cardiomyocyte.[66] Countering the development of H^+ gradients catalytically and noncatalytically is the central function of CA metabolons.

20.5.1 Acknowledgments

DEJ and JRC are supported, respectively, by graduate studentship and Scientist awards from the Alberta Heritage Foundation for Medical Research. Research in the laboratory of JRC is supported by Canadian Institutes of Health Research and the Heart and Stroke Foundation. We thank Dr. Holger Becker for discussions of his model concerning catalysis-independent roles of CA II. We thank Dr. Gonzalo Vilas and Ms. Pamela Bonar for comments this manuscript.

REFERENCES

1. Casey, J. R. Why bicarbonate? *Biochem. Cell Biol.* **2006**, *84*, 930–939.
2. Perlman, D. F.; Goldstein, L. The anion exchanger as an osmolyte channel in the skate erythrocyte. *Neurochem. Res.* **2004**, *29*, 9–15.
3. Mason, M. J.; Smith, J. D.; Garcia-Soto, J. J.; Grinstein, S. Internal pH-sensitive site couples Cl^-/HCO_3^- exchange to Na^+/H^+ antiport in lymphocytes. *Am. J. Physiol.* **1989**, *256*, C428–C433.
4. Sterling, D.; Reithmeier, R. A.; Casey, J. R. A transport metabolon functional interaction of carbonic anhydrase II and chloride/bicarbonate exchangers. *J. Biol. Chem.* **2001**, *276*, 47886–47894.
5. Vince, J. W.; Reithmeier, R. A. Identification of the carbonic anhydrase II binding site in the Cl^-/HCO_3^- anion exchanger AE1. *Biochemistry* **2000**, *39*, 5527–5533.
6. Cousin, J. L.; Motais, R.; Sola, F. Transmembrane exchange of chloride with bicarbonate ion in mammalian red blood cells: evidence for a sulphonamide-sensitive "carrier". *J. Physiol.* **1975**, *253*, 385–399.
7. Alper, S. L.; Chernova, M. N.; Stewart, A. K. Regulation of Na^+-independent Cl^-/HCO_3^- exchangers by pH. *JOP* **2001**, *2*, 171–175.

8. McMurtrie, H. L.; Cleary, H. J.; Alvarez, B. V.; Loiselle, F. B.; Sterling, D.; Morgan, P. E.; Johnson, D. E.; Casey, J. R. The bicarbonate transport metabolon. *J. Enzyme Inhib. Med. Chem.* **2004**, *19*, 231–236.
9. Sterling, D.; Casey, J. R. Bicarbonate transport proteins. *Biochem. Cell Biol.* **2002**, *80*, 483–497.
10. Low, P. S. Structure and function of the cytoplasmic domain of band 3: center of erythrocyte membrane–peripheral protein interactions. *Biochim. Biophys. Acta* **1986**, *864*, 145–167.
11. Vince, J. W.; Reithmeier, R. A. Carbonic anhydrase II binds to the carboxyl terminus of human band 3, the erythrocyte Cl^-/HCO_3^- exchanger. *J. Biol. Chem.* **1998**, *273*, 28430–28437.
12. Dahl, N. K.; Jiang, L.; Chernova, M. N.; Stuart-Tilley, A. K.; Shmukler, B. E.; Alper, S. L. Deficient HCO_3^- transport in an AE1 mutant with normal Cl^- transport can be rescued by carbonic anhydrase II presented on an adjacent AE1 protomer. *J. Biol. Chem.* **2003**, *278*, 44949–44958.
13. Linn, S. C.; Askew, G. R.; Menon, A. G.; Shull, G. E. Conservation of an AE3 Cl^-/HCO_3^- exchanger cardiac-specific exon and promoter region and AE3 mRNA expression patterns in murine and human hearts. *Circ. Res.* **1995**, *76*, 584–591.
14. Becker, H. M.; Deitmer, J. W. Carbonic anhydrase II increases the activity of the human electrogenic Na^+/HCO_3^- cotransporter. *J. Biol. Chem.* **2007**, *282*, 13508–13521.
15. Alvarez, B. V.; Loiselle, F. B.; Supuran, C. T.; Schwartz, G. J.; Casey, J. R. Direct extracellular interaction between carbonic anhydrase IV and the human NBC1 sodium/bicarbonate co-transporter. *Biochemistry* **2003**, *42*, 12321–12329.
16. Mount, D. B.; Romero, M. F. The SLC26 gene family of multifunctional anion exchangers. *Pflugers Arch.* **2004**, *447*, 710–721.
17. Alvarez, B. V.; Kieller, D. M.; Quon, A. L.; Markovich, D.; Casey, J. R. Slc26a6: a cardiac chloride-hydroxyl exchanger and predominant chloride-bicarbonate exchanger of the mouse heart. *J. Physiol.* **2004**, *561*, 721–734.
18. Petrovic, S.; Ju, X.; Barone, S.; Seidler, U.; Alper, S. L.; Lohi, H.; Kere, J.; Soleimani, M. Identification of a basolateral Cl^-/HCO_3^- exchanger specific to gastric parietal cells. *Am. J. Physiol. Gastrointest. Liver Physiol.* **2003**, *284*, G1093–G1103.
19. Reithmeier, R. A. A membrane metabolon linking carbonic anhydrase with chloride/bicarbonate anion exchangers. *Blood Cells Mol. Dis.* **2001**, *27*, 85–89.
20. Srere, P. A. Complexes of sequential metabolic enzymes. *Annu. Rev. Biochem.* **1987**, *56*, 89–124.
21. Srere, P. A. The metabolon. *Trends Biochem. Sci.* **1985**, *10*, 109–110.
22. Miles, E. W.; Rhee, S.; Davies, D. R. The molecular basis of substrate channelling. *J. Biol. Chem.* **1999**, *274*, 12193–12196.
23. Sterling, D.; Reithmeier, R. A.; Casey, J. R. Carbonic anhydrase: in the driver's seat for bicarbonate transport. *JOP* **2001**, *2*, 165–170.
24. Vince, J. W.; Carlsson, U.; Reithmeier, R. A. Localization of the Cl^-/HCO_3^- anion exchanger binding site to the amino-terminal region of carbonic anhydrase II. *Biochemistry* **2000**, *39*, 13344–13349.
25. Sterling, D.; Alvarez, B. V.; Casey, J. R. The extracellular component of a transport metabolon. Extracellular loop 4 of the human AE1 Cl^-/HCO_3^- exchanger binds carbonic anhydrase IV. *J. Biol. Chem.* **2002**, *277*, 25239–25246.

26. Kifor, G.; Toon, M. R.; Janoshazi, A.; Solomon, A. K. Interaction between red cell membrane band 3 and cytosolic carbonic anhydrase. *J. Membr. Biol.* **1993**, *134*, 169–179.

27. Eriksson, A. E.; Jones, T. A. Liljas, A. Refined structure of human carbonic anhydrase II at 2 0 Å resolution. *Proteins* **1988**, *4*, 274–282.

28. Cousin, J. L.; Motais, R. The role of carbonic anhydrase inhibitors on anion permeability into ox red blood cells. *J. Physiol.* **1976**, *256*, 61–80.

29. Fierke, C. A.; Calderone, T. L.; Krebs, J. F. Functional consequences of engineering the hydrophobic pocket of carbonic anhydrase II. *Biochemistry* **1991**, *30*, 11054–11063.

30. Jennings, M. L. Oligomeric structure and the anion transport function of human erythrocyte band 3 protein. *J. Membr. Biol.* **1984**, *80*, 105–117.

31. Wieth, J. O. Bicarbonate exchange through the human red cell membrane determined with [^{14}C] bicarbonate. *J. Physiol.* **1979**, *294*, 521–539.

32. Scozzafava, A.; Supuran, C. T. Carbonic anhydrase activators: human isozyme II is strongly activated by oligopeptides incorporating the carboxyterminal sequence of the bicarbonate anion exchanger AE1. *Bioorg. Med. Chem. Lett.* **2002**, *12*, 1177–1180.

33. Baird, T. T., Jr.; Waheed, A.; Okuyama, T.; Sly, W. S.; Fierke, C. A. Catalysis and inhibition of human carbonic anhydrase IV. *Biochemistry* **1997**, *36*, 2669–2678.

34. Waheed, A.; Zhu, X. L.; Sly, W. S.; Wetzel, P.; Gros, G. Rat skeletal muscle membrane associated carbonic anhydrase is 39-kDa, glycosylated GPI-anchored CA IV. *Arch. Biochem. Biophys.* **1992**, *294*, 550–556.

35. Tang, X. B.; Fujinaga, J.; Kopito, R.; Casey, J. R. Topology of the region surrounding Glu681 of human AE1 protein, the erythrocyte anion exchanger. *J. Biol. Chem.* **1998**, *273*, 22545–22553.

36. Zolotarev, A. S.; Chernova, M. N.; Yannoukakos, D.; Alper, S. L. Proteolytic cleavage sites of native AE2 anion exchanger in gastric mucosal membranes. *Biochemistry* **1996**, *35*, 10367–10376.

37. Kudrycki, K. E.; Newman, P. R.; Shull, G. E. cDNA cloning and tissue distribution of mRNAs for two proteins that are related to the band 3 Cl^-/HCO_3^- exchanger. *J. Biol. Chem.* **1990**, *265*, 462–471.

38. Alvarez, B. V.; Gilmour, G. S.; Mema, S. C.; Martin, B. T.; Shull, G. E.; Casey, J. R.; Sauve, Y. Blindness caused by deficiency in AE3 chloride/bicarbonate exchanger. *PLoS ONE* **2007**, *2*, e839.

39. Morgan, P. E.; Pastorekova, S.; Stuart-Tilley, A. K.; Alper, S. L.; Casey, J. R. Interactions of transmembrane carbonic anhydrase, CAIX, with bicarbonate transporters. *Am. J. Physiol. Cell Physiol.* **2007**, *293*, C738–C748.

40. Pastorekova, S.; Parkkila, S.; Parkkila, A. K.; Opavský, R.; Zelnik, V.; Saarnio, J.; Pastorek, J. Carbonic anhydrase IX MN/CA IX: analysis of stomach complementary DNA sequence and expression in human and rat alimentary tracts. *Gastroenterology* **1997**, *112*, 398–408.

41. Swietach, P.; Vaughan-Jones, R. D.; Harris, A. L. Regulation of tumor pH and the role of carbonic anhydrase 9. *Cancer Metastasis Rev.* **2007**, *26*, 299–310.

42. Swietach, P.; Wigfield, S.; Cobden, P.; Supuran, C. T.; Harris, A. L.; Vaughan-Jones, R. D. Tumor-associated carbonic anhydrase 9 spatially co-ordinates intracellular pH in three-dimensional multicellular growths. *J. Biol. Chem.* **2008**.

43. Gawenis, L. R.; Ledoussal, C.; Judd, L. M.; Prasad, V.; Alper, S. L.; Stuart-Tilley, A.; Woo, A. L.; Grisham, C.; Sanford, L. P.; Doetschman, T.; Miller, M. L.; Shull, G. E. Mice with a targeted disruption of the AE2 Cl^-/HCO_3^- exchanger are achlorhydric. *J. Biol. Chem.* **2004**, *279*, 30531–30539.

44. Sun, X.; Soleimani, M.; Petrovic,S. Decreased expression of Slc26a4 (Pendrin) and Slc26a7 in the kidneys of carbonic anhydrase II-deficient mice. *Cell. Physiol. Biochem.* **2008**, *21*, 95–108.

45. Sterling, D.; Brown, N. J.; Supuran, C. T.; Casey, J. R. The functional and physical relationship between the DRA bicarbonate transporter and carbonic anhydrase II. *Am. J. Physiol. Cell Physiol.* **2002**, *283*, C1522–C1529.

46. Alvarez, B. V.; Vilas, G. L.; Casey, J. R. Metabolon disruption: a mechanism that regulates bicarbonate transport. *EMBO J.* **2005**, *24*, 2499–2511.

47. Aravind, L.; Koonin, E. V. The STAS domain: a link between anion transporters and antisigma-factor antagonists. *Curr. Biol.* **2000**, *10*, R53–R55.

48. Loiselle, F. B.; Morgan, P. E.; Alvarez, B. V.; Casey, J. R. Regulation of the human NBC3 Na^+/HCO_3^- cotransporter by carbonic anhydrase II and PKA. *Am. J. Physiol. Cell Physiol.* **2004**, *286*, C1423–C1433.

49. Pushkin, A.; Abuladze, N.; Gross, E.; Newman, D.; Tatishchev, S.; Lee, I.; Fedotoff, O.; Bondar, G.; Azimov, R.; Ngyuen, M.; Kurtz, I. Molecular mechanism of kNBC1–carbonic anhydrase II interaction in proximal tubule cells. *J. Physiol.* **2004**, *559*, 55–65.

50. Gross, E.; Pushkin, A.; Abuladze, N.; Fedotoff, O.; Kurtz, I. Regulation of the sodium bicarbonate cotransporter kNBC1 function: role of Asp(986), Asp(988) and kNBC1-carbonic anhydrase II binding. *J. Physiol.* **2002**, *544*, 679–685.

51. Soleimani, M.; Burnham, C. E. Physiologic and molecular aspects of the $Na^+:HCO_3^-$ cotransporter in health and disease processes. *Kidney Int.* **2000**, *57*, 371–384.

52. Lu, J.; Daly, C. M.; Parker, M. D.; Gill, H. S.; Piermarini, P. M.; Pelletier, M. F.; Boron, W. F. Effect of human carbonic anhydrase II on the activity of the human electrogenic Na^+/HCO_3^- cotransporter NBCe1-A in *Xenopus* oocytes. *J. Biol. Chem.* **2006**, *281*, 19241–19250.

53. Piermarini, P. M.; Kim, E. Y.; Boron, W. F. Evidence against a direct interaction between intracellular carbonic anhydrase II and pure C-terminal domains of SLC4 bicarbonate transporters. *J. Biol. Chem.* **2007**, *282*, 1409–1421.

54. Fliegel, L. Regulation of myocardial Na^+/H^+ exchanger activity. *Basic Res. Cardiol.* **2001**, *96*, 301–305.

55. Slepkov, E. R.; Rainey, J. K.; Sykes, B. D.; Fliegel, L. Structural functional analysis of the Na^+/H^+ exchanger. *Biochem. J.* **2007**, *401*, 623–633.

56. Li, X.; Liu, Y.; Alvarez, B. V.; Casey, J. R.; Fliegel, L. A novel carbonic anhydrase II binding site regulates NHE1 activity. *Biochemistry* **2006**, *45*, 2414–2424.

57. Li, X.; Alvarez, B.; Casey, J. R.; Reithmeier, R. A.; Fliegel, L. Carbonic anhydrase II binds to and enhances activity of the Na^+/H^+ exchanger. *J. Biol. Chem.* **2002**, *277*, 36085–36091.

58. Wu, Q.; Pierce, W. M., Jr.; Delamere, N. A. Cytoplasmic pH responses to carbonic anhydrase inhibitors in cultured rabbit nonpigmented ciliary epithelium. *J. Membr. Biol.* **1998**, *162*, 31–38.

59. Ro, H. A.; Carson, J. H. pH microdomains in oligodendrocytes. *J. Biol. Chem.* **2004**, *279*, 37115–37123.

60. Becker, H. M.; Deitmer, J. W. Non-enzymatic proton handling by carbonic anhydrase II during H^+-lactate cotransport via monocarboxylate transporter 1. *J. Biol. Chem.* **2008.**

61. Becker, H. M.; Hirnet, D.; Fecher-Trost, C.; Sultemeyer, D.; Deitmer, J. W. Transport activity of MCT1 expressed in *Xenopus* oocytes is increased by interaction with carbonic anhydrase. *J. Biol. Chem.* **2005**, *280*, 39882–39889.

62. Bergersen, L. H. Is lactate food for neurons? Comparison of monocarboxylate transporter subtypes in brain and muscle. *Neuroscience* **2007**, *145*, 11–19.

63. Yeung, T.; Grinstein, S. Lipid signaling and the modulation of surface charge during phagocytosis. *Immunol. Rev.* **2007**, *219*, 17–36.

64. Weise, A.; Becker, H. M.; Deitmer, J. W. Enzymatic suppression of the membrane conductance associated with the glutamine transporter SNAT3 expressed in *Xenopus* oocytes by carbonic anhydrase II. *J. Gen. Physiol.* **2007**, *130*, 203–215.

65. Broer, S.; Brookes, N. Transfer of glutamine between astrocytes and neurons. *J. Neurochem.* **2001**, *77*, 705–719.

66. Zaniboni, M.; Swietach, P.; Rossini, A.; Yamamoto, T.; Spitzer, K. W.; Vaughan-Jones, R. D. Intracellular proton mobility and buffering power in cardiac ventricular myocytes from rat, rabbit, and guinea pig. *Am. J. Physiol. Heart Circ. Physiol.* **2003**, *285*, H1236–H1246.

CHAPTER 21

Metal Complexes of Sulfonamides as Dual Carbonic Anhydrase Inhibitors

MARC A. ILIES

Department of Pharmaceutical Sciences, Temple University School of Pharmacy, 3307 North Broad Street, Philadelphia, PA 19140, USA

21.1 INTRODUCTION

The interaction of proteins with metal ions is of utmost importance in nature, being encoded in the structures of natural amino acids that constitute the building blocks of life. Metal ions are among the most important cofactors in proteins, where they can play structural and/or catalytic roles. Various metalloprotein structures provide insights into the protein–metal partnership, revealing metal ion sites that control or are controlled by protein conformational changes.[1–3]

The carbonic anhydrase (CA) represents itself the ultimate example of a perfectly evolved metalloprotein, being able to catalyze its physiological reaction—the hydration of carbon dioxide to yield a bicarbonate ion and a proton—with a rate approaching diffusion control ($K_{cat}/K_m = 1.5 \times 10^8\ M^{-1}\ s^{-1}$ for isozyme CA II).[4–7] In the fastest isozyme (CA II), a zinc ion is coordinated by three histidine residues (His94, His96, and His119), the fourth ligand being a water molecule. The metal is essential to catalysis and is used to decrease the pK_a of bound water, which can be readily deprotonated and transformed into a hydroxide ion in what constitutes the active form of the enzyme. This powerful nucleophile is properly oriented by adjacent residues Thr199 and Glu106 to attack the substrate (CO_2), generating the bicarbonate ion. Displacement of the bound HCO_3^- ion by another water molecule closes the catalytic cycle and generates the acidic form of the enzyme, catalytically inactive.[3,8,9] The deprotonation of the zinc-bound water molecule is the rate-determining step of the catalytic cycle. In CA II, the proton transfer is internally assisted by His64, which shuttles it from the active site to the periphery of the protein.[10,11] It is believed that a histidine cluster (His4, His3, His17, His15, and His10) ensures an active proton transfer interface between His64 and the buffer systems present in the environment.[12]

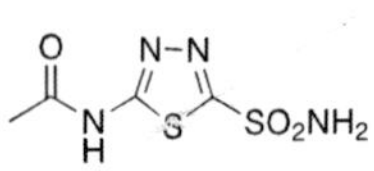

Acetazolamide, H_2AAZ (**21.1**)

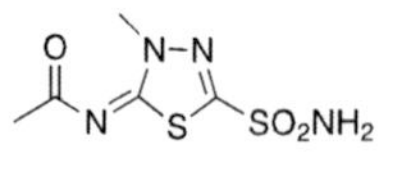

Methazolamide, HMZA (**21.2**)

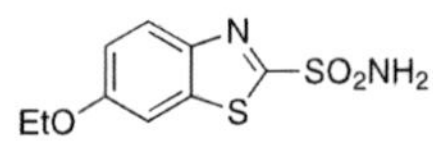

Ethoxzolamide, HEZA (**21.3**)

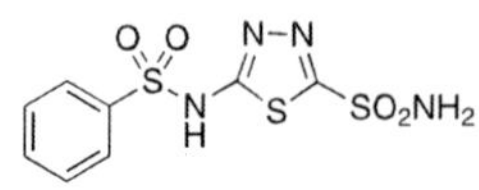

Benzolamide, H_2BZA (**21.4**)

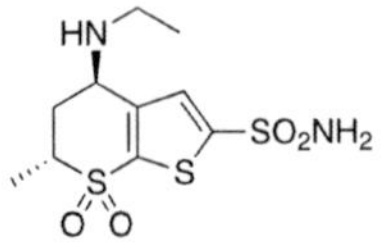

Dorzolamide, HDZA (**21.5**)

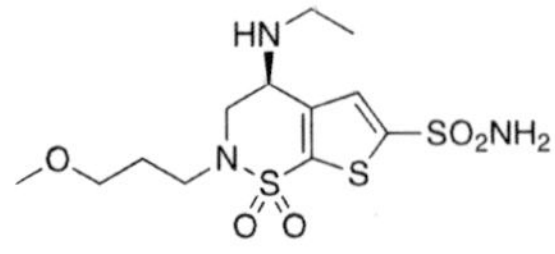

Brinzolamide, HBRA (**21.6**)

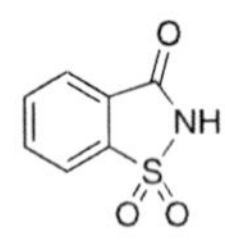

Saccharin, HSAC (**21.7**)

CHART 21.1

The zinc ion, the proton shuttle (His64), and the histidine cluster are essential structural elements for the interaction of various classes of inhibitors with the enzyme. The inhibition of carbonic anhydrase and its isozymes was investigated almost immediately after its discovery [13,14] and played an important role in understanding its function and catalytic mechanism.[15,16] Carbonic anhydrase inhibitors (CAIs) represent an active field of research for over seven decades and it was periodically reviewed.[4–6,16–23] Two main classes of CAI are known: inorganic anions (e.g., CNO^-, CN^-, HS^-, SCN^-, and N_3^-) [24,25] and unsubstituted sulfonamides. The latter class is responsible for the main impact of CAI in therapy, aromatic and heterocyclic sulfonamides such as acetazolamide (H_2AAZ, **21.1**), methazolamide (HMZA, **21.2**), ethoxzolamide (HEZA, **21.3**), benzolamide (H_2BZA, **21.4**), and the topically active dorzolamide (HDZA, **21.5**) and brinzolamide (HBRA, **21.6**) being clinically used as diuretics, and in the treatment of glaucoma, epilepsy, gastric ulcers, and so on.[5,6,19,20,26] (Chart 21.1) Recently, other nonsulfonamide CAIs (sulfamates, sulfamides, and hydroxamates) have been reported to act as potent CAI.[24,27] These classes of inhibitors are examined in detail in the previous chapters.

21.2 SULFONAMIDES AS COMPLEXING AGENTS

All the above-mentioned inhibitors bind to the Zn^{2+} ion, either by displacing the H_2O/HO^- ligand and generating a tetracoordinated species, or by inserting themselves into the coordination sphere of the metal and yielding trigonal bipyramidal pentacoordinated entities.[5,24] Unsubstituted aromatic/heterocyclic sulfonamides belong to the first class of ligands, binding to the Zn^{2+} ion in the ionized form (RSO_2NH^-) and generating a symmetrical all-N tetrahedral geometry for the metal center, further stabilized by hydrogen bonding of the sulfonamidate ion with Thr199.[28,29] However, these structural details have been available only in recent years due to the advancement in X-ray crystallography techniques.[4,10,29] For many years, the exact nature of sulfonamide interaction with the zinc ion in the active site of CA was unclear since these compounds were generally considered weak complexing agents (ligands) in aqueous medium due to the internal conjugation of the sulfonamidate anion. This reason may justify why the coordination chemistry of these widely used clinical agents has been rarely investigated until recently. Historically, the first sulfonamide metal complex—the silver salt of sulfanilamide—was obtained in 1941,[30] but the ability of the sulfonamidate anion to act as σ-donor ligand for cations began to be revealed in the early 1980s by Bult,[31] who studied various complexes of substituted sulfanilamides, and by Malecki et al.,[32] who explored several copper complexes of aromatic and heterocyclic sulfonamides.

A thorough investigation of the metal complexes as a new class of CAI was largely undertaken by Supuran's and Boras's groups starting in late 1980s (*vide infra*) and the field was periodically reviewed.[20,33–35] Thus, it has been proved that clinically used heterocyclic sulfonamides **21.1–21.6** are actually versatile ligands that can generate interesting coordination species with a wide range of transitional and main group mono-, di-, tri-, and tetravalent ions (*vide infra*), but the striking discovery was that in many cases the complexes were 10–100 times more powerful CA inhibitors than the sulfonamides used to generate them. This unexpected behavior was later explained by Supuran[36] to be the consequence of a *dual-action mechanism*: in dilute solutions (such as the ones used in enzymatic assays), the complexes partially dissociate into sulfonamide anions and metal ions, species that can associate with different areas of the active site of the enzyme. Sulfonamide anion will bind to the Zn^{2+} ion displacing the H_2O/HO^- ligand (critical for catalysis), whereas metal ions will associate with the active site residues displaying good ligand properties. The proton shuttle system of the enzyme (His64 and the histidine cluster) was thought to be involved [12,34] due to excellent coordination properties of the histidine's imidazole ring and taking into consideration the proved ability of Hg^{2+} and Cu^{2+} to act as proton shuttle inhibitors.[9,37–39]

This synergetic inhibition effect will be explored in what follows from the perspective of sulfonamide ligands used, emphasizing on the structural characteristics of the metal complexes and their impact on the CAI potency and pharmacological properties of the coordination compounds.

21.3 ACETAZOLAMIDE COMPLEXES

Acetazolamide (H_2AAZ, **21.1**) was the most extensively used sulfonamide CAI in the preparation of metal complexes, owing to its clinical importance and to its versatility as a ligand.[20,34,35] In the heterocyclic sulfonamides (as well as in the aromatic ones), the electron-withdrawing effect of the heterocyclic/aromatic ring considerably increases the acidity of the sulfonamidic protons, which can be easily dissociated generating the sulfonamidate anion—an effective σ donor ligand for cations. Besides the sulfonamide group, acetazolamide can act as a ligand through the acetamido moiety, which contains a second ionizable proton. Ferrer et al.[40] determined potentiometrically the pK_as of the two ionizable groups in water (and in aqueous/ethanolic medium) and found a value of 7.19 (7.52) for the sulfonamide group and 8.65 (9.41) for the acetamido one, respectively. At neutral pH, the nonionized species A coexists with both monoionized species B and C (and their resonance structures such as E) and the di-ionized species D (Scheme 21.1), thus conferring H_2AAZ great versatility as ligand.

Consequently, the structures of its various metal complexes will depend on (i) the intrinsic donor capability and metal affinity of each A–D species, (ii) the nature of the

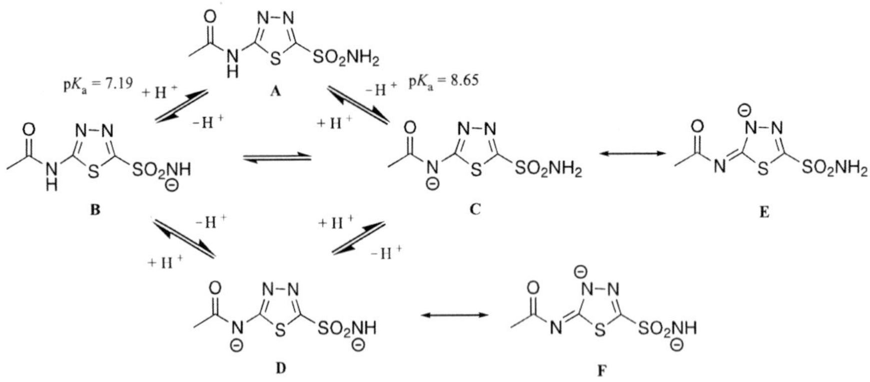

SCHEME 21.1 Ionization equilibrium of acetazolamide in water; all species can coordinate metal ions, inducing great versatility to this ligand.

metal ion and its own ligand affinity and stereoelectronic demands, (iii) the steric factors related to the crystal packing of the metal and ligands, and (iv) stabilizing factors such as hydrogen bonding between the ligands in the complex. Other essential factors are the method used to prepare the complex, the nature of solvent and base used to ionize the ligand, the pH, coordination/chelation properties of the base and of the other chemical entities present in solution, and so on.[20,34,35,41] Under the influence of these factors, acetazolamide can act as monodentate (through the ionized sulfonamidic group or the N3/N4 thiadiazol atoms), bidentate, or bridging bidentate ligand (through both the previously mentioned groups simultaneously), and even as a tridentate coordinating entity in some cases—as revealed by spectroscopic and X-ray diffraction data (*vide infra*).

Ferrer et al.[42] reported the first acetazolamide complexes, prepared by mixing H_2AAZ, MCl_2, (M=Zn, Co) in alcoholic solution in the presence of excess ammonia. The resulting complexes were characterized by elemental analysis and spectroscopic methods (IR, UV, ^1H- and ^{13}C-NMR) and thermogravimetric analyses, yielding the compositions $[Co(HAAZ)_2(NH_3)_2]$ and $[Zn(HAAZ)_2(NH_3)_2]$; it was also established that in both complexes the metal ions have a pseudotetrahedral surrounding, with acetazolamide as a monodentate ligand through the sulfonamidic group. Later on, Hartmann and Vahrenkamp[43] were able to confirm the structure for the Zn complex by X-ray crystallography (Fig. 21.1a). Borras's group subsequently reported the Ni(II) complex of acetazolamide,[44] prepared in the same conditions as the Co and Zn congeners (alcoholic solution, excess NH_3), for which the formula $[Ni(HAAZ)_2(NH_3)_4]$ was found. Its exact structure was determined by X-ray crystallography when it was revealed that the Ni(II) has a distorted octahedral geometry, being coordinated by the endocyclic N4 atom of acetazolamide, which is deprotonated at the acetamido moiety (Fig. 21.1b). The involvement of limit structures of types B and E was proved by means of IR spectroscopy (the SO_2 vibrations are not shifted to lower wavenumbers) and by comparing the bond distances of the ligand in the complex versus the normal distances determined for the free compound (structure A, Scheme 21.1).[45] The same group subsequently reported the structure of the dinuclear

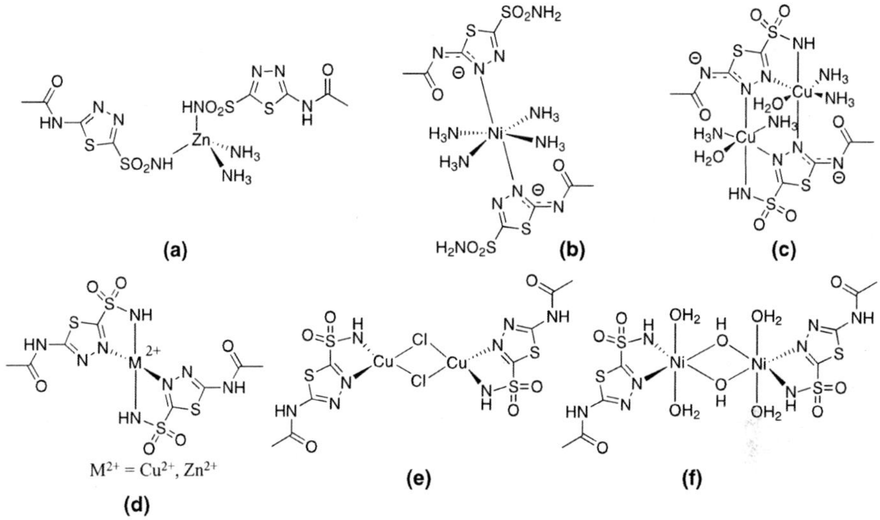

FIGURE 21.1 Representative coordination compounds of acetazolamide.

Cu(II) complex, prepared also using ammonia as base, and having formula $[Cu(HAAZ)(NH_3)_2(OH_2)]_2{\cdot}2H_2O$. X-ray crystallography revealed a polynuclear derivative in which acetazolamide acts in a tridentate fashion (structures D and F in Scheme 21.1), interacting with the copper ions through the ionized sulfonamidic nitrogen and the N3 and N4 atoms of the heterocyclic ring. Each metal ion is also coordinated by two ammonia molecules and a water molecule in a tetragonally elongated octahedral geometry (Fig. 21.1c).[46] Mention must be made that this complex is not very stable at room temperature, gradually losing ammonia in time.

A breakthrough in the field was done by Supuran's report in early 1990[47] that showed that coordination compounds of acetazolamide are actually very strong CAIs. He refluxed the acetazolamide sodium salt (NaHAAZ) and copper salts, in methanol, at different molar ratios of reagents, yielding two new compounds: $[Cu(HAAZ)_2]$ and $[Cu(HAAZ)Cl]_2$. Their structures, together with those of $[Zn(HAAZ)_2]$ and $[Ni(HAAZ)(OH)(OH_2)_2]_2{\cdot}6H_2O$, involved a bidentate coordination of acetazolamide through the ionized sulfonamide nitrogen and the endocyclic N3 atom (Fig. 21.1d–f).[47]

The true novelty of this study was the finding that the metal complexes are in fact stronger CA inhibitors than the parent sulfonamide, later explained via a double-action inhibition mechanism. Thus, due to relatively weak complexing ability of the sulfonamides, their metal complexes can easily dissociate into sulfonamide anions (RSO_2NH^-, represented as L in the following equation) and metal ions, both of which are efficient in inhibiting the enzyme:[36]

$$[ML_n] \rightleftharpoons [ML_{n-1}]+L \rightleftharpoons \ldots \rightleftharpoons M^{n+}+nL$$

Following these seminal communications, both groups reported a large number of metal complexes with main group and transition metal ions. Their main structural and inhibitory properties are summarized in Table 21.1.[20,34,35,41]

TABLE 21.1 Acetazolamide (H_2AAZ) Complexes and Their Structural And Biological Characteristics

Metal Ion	Acetazolamide Complex	Base	Acetazolamide as Ligand	Metal Coordination	CA I, CA II Inhibition IC_{50} (nM)		Reference
					CA I	CA II	
	H_2AAZ				580	10	
Be(II)	$[Be(HAAZ)_2]$	NaOH	Bidentate, B	Tetrahedral		8	48
Mg(II)	$[Mg(HAAZ)_2(H_2O)_2]$	NaOH	Bidentate, B	Octahedral		5	48
M(III)	$[La(HAAZ)_3]\cdot 8H_2O$	NaOH	Bidentate, B		180	0.1	49
	$[Ce(HAAZ)_3]\cdot 8H_2O$	NaOH	Bidentate, B		200	0.1	49
	$[Pr(HAAZ)_3]\cdot 8H_2O$	NaOH	Bidentate, B		80	0.06	49
	$[Nd(HAAZ)_3]\cdot 8H_2O$	NaOH	Bidentate, B		80	0.04	49
	$[Sm(HAAZ)_3]\cdot 8H_2O$	NaOH	Bidentate, B		70	0.02	49
	$[Eu(HAAZ)_3]\cdot 8H_2O$	NaOH	Bidentate, B		50	0.01	49
	$[Gd(HAAZ)_3]\cdot 8H_2O$	NaOH	Bidentate, B		60	0.02	49
	$[Tb(HAAZ)_3]\cdot 8H_2O$	NaOH	Bidentate, B		50	0.01	49
	$[Dy(HAAZ)_3]\cdot 8H_2O$	NaOH	Bidentate, B		50	0.02	49
	$[Ho(HAAZ)_3]\cdot 8H_2O$	NaOH	Bidentate, B		50	0.01	49
	$[Er(HAAZ)_3]\cdot 8H_2O$	NaOH	Bidentate, B		40	0.01	49
	$[Tm(HAAZ)_3]\cdot 8H_2O$	NaOH	Bidentate, B		30	0.01	49
	$[Yb(HAAZ)_3]\cdot 8H_2O$	NaOH	Bidentate, B		30	0.01	49
Ce(IV)	$[Ce(HAAZ)_4]\cdot 6H_2O$	NaOH	Bidentate, B			0.15	50
Th(IV)	$[Th(HAAZ)_4]\cdot 4H_2O$	NaOH	Bidentate, B			0.13	50
U(VI)	$[UO_2(HAAZ)_2(OH_2)]\cdot 2H_2O$	NaOH	Bidentate, B	Pentagonal bipyramidal		0.09	50
V(IV)	$[VO(HAAZ)_2]$	NaOH	Bidentate, B	Square pyramidal		6	50
Fe(III)	$[Fe(HAAZ)_3]\cdot 8H_2O$	NaOH	Bidentate, B	Octahedral		3	51
Ru(III)	$[Ru(HAAZ)_3]$	NaOH	Bidentate, B	Octahedral	800	1.9	52
Co(II)	$[Co(HAAZ)_2(NH_3)_2]$	NH_3	Monodentate, B	Tetrahedral		40	33,42,53,54
Rh(III)	$[Rh(HAAZ)_3]$	NaOH	Bidentate, B	Octahedral	1600	0.2	52

Ni(II)	$[Ni(HAAZ)_2(NH_3)_4]$		NH_3	Monodentate, C/E	Octahedral	50	33,44,53,54
	$K_2Ni_2[(HAAZ)_3]\cdot 5H_2O$	KOH		Octahedral		7	33,54,55
	$[Ni(HAAZ)(OH)(OH_2)_2]_2\cdot 6H_2O$	NaOH	Bidentate, B	Pseudo-octahedral		30	47,54
Cu(II)	$[Cu(AAZ)(NH_3)_2(OH_2)]_2\cdot 2H_2O$	NH_3	Tridentate, D/F	Pseudo-octahedral		70	33,46,53,54
	$[Cu(HAAZ)_2]$	NaOH	Bidentate, B	Tetrahedral		10	47,54
	$[Cu(HAAZ)Cl]_2$	NaOH	Bidentate, B	Tetrahedral		40	47,54
Ag(I)	$[Ag_2(HAAZ)]$	—				3	31,33,54,55
	$[Ag_2(HAAZ)(NH_3)]$	NH_3				4	33,55
Au(III)	$[Au(HAAZ)_2]Cl$	NaOH	Bidentate, B	Square planar	200	0.3	52
Zn(II)	$[Zn(HAAZ)_2(NH_3)_2]$	NH_3	Monodentate, B	Tetrahedral		20	33,42,43, 53,54
	$[Zn(HAAZ)_2]$	NaOH	Bidentate, B	Tetrahedral		90	47,54
Cd(II)	$[Cd(AAZ)]\cdot 3/2H_2O$	Bu_4NOH				7	33,54,55
Hg(II)	$[Hg(AAZ)]\cdot 1/2H_2O$	KOH				2	33,54,55
Al(III)	$[Al(HAAZ)_3]\cdot 8H_2O$	NaOH	Bidentate, B	Octahedral		0.8	56
Ga(III)	$[Ga(HAAZ)_3]\cdot 8H_2O$	NaOH	Bidentate, B	Octahedral		0.6	56
In(III)	$[In(HAAZ)_3]\cdot 8H_2O$	NaOH	Bidentate, B	Octahedral		0.7	56
Tl(III)	$[Tl(HAAZ)_3]\cdot 8H_2O$	NaOH	Bidentate, B	Octahedral		0.3	56
Pb(II)	$[Pb(HAAZ)_2(OH_2)_2]$	NaOH	Bidentate, B	Octahedral		1	48

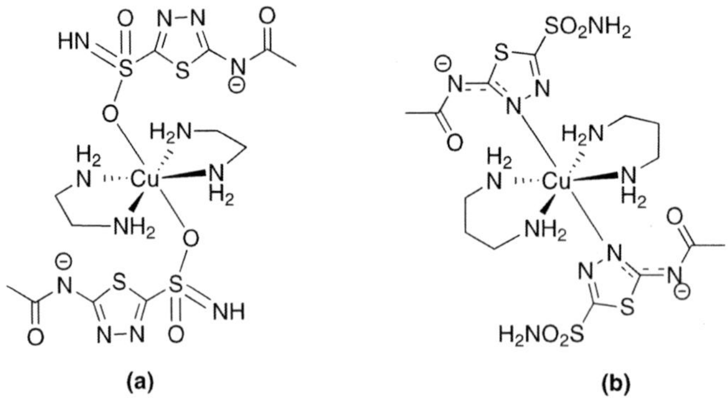

FIGURE 21.2 Cu(II) complexes of acetazolamide generated in the presence of chelating organic bases (a) ethylenediamine (en) and (b) 1,3-propanediamine (tn).

One can observe very good CA inhibitory properties of all metal complexes, generally better than the parent ligand, against both CA isozymes. Cations generating very powerful inhibitors were Ag(I), Hg(II), Pb(II), and all trivalent cations (especially lanthanides), Ce(IV), Th(IV), and U(VI).

Mention must be made that the Spanish group also reported acetazolamide complexes of Co(II), Ni(II), and Cu(II) obtained in the presence of organic bases such as ethylamine, diethylamine, triethylamine, ethylenediamine (en), and 1,3-propanediamine (tn).[53,57] Although the CA inhibitory properties of these compounds were never investigated, they exhibited very interesting structures, as revealed by X-ray crystallography (Fig. 21.2).[53] In both cases, the acetazolamide molecule is deprotonated at the amidic N, while the copper ions are lying on symmetry centers, displaying an elongated tetragonally distorted octahedral environment. In the ethylenediamine complex [Cu(AAZ)$_2$(en)$_2$] (Fig. 21.2a), the coordination of the metal ion is achieved through an oxygen atom of the sulfonamide group, which is probably in its tautomeric S=NH form hydrogen bonded intermolecularly with the ionized amide group from the acetazolamide ligand of the neighboring complex. The 1,3-propanediamine complex [Cu(HAAZ)$_2$(tn)$_2$] structure (Fig. 21.2b) closely resembles the structure of the [Ni(HAAZ)$_2$(NH$_3$)$_4$] complex (Fig. 21.1b), with Cu(II) being coordinated by the endocyclic N4 atom of acetazolamide, which acts again as a monodentate ligand.[53]

21.4 ACETAZOLAMIDE-RELATED COMPOUNDS AND THEIR COMPLEXES

The success of acetazolamide in the clinical management of glaucoma, epilepsy, mountain, and altitude sickness, and the possibility to further enhance its potency and modify its pharmacokinetics profile via various hydrophilic and lipophilic moieties, prompted the synthesis and investigation of the biological activity of a large number of acetazolamide-related compounds such as structures **21.8–21.17** and their metal complexes (Fig. 21.3).[54,58–66]

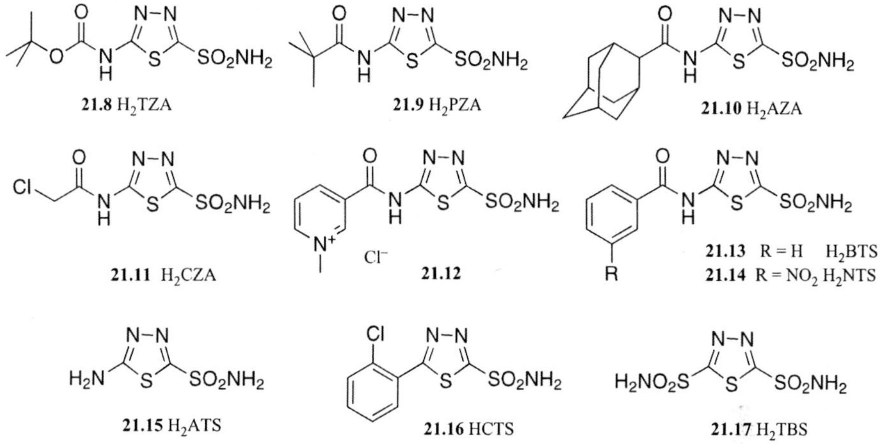

FIGURE 21.3 Acetazolamide-like derivatives.

Intrigued by the versatility of acetazolamide as ligand, in particular toward the Cu(II) ions that yielded complexes with interesting biological activity such as the anticonvulsant $K_6[Cu_2(AAZ)_5(OH_2)_2]$,[67] Pedregosa et al.[60] decided to increase the steric bulk around the N4 of the thiadiazole ring of H_2AAZ with a *tert*-butoxy group, while keeping the rest of the backbone intact. The new *t*-butyloxycarbonylamido ligand **21.8** (H_2TZA) generated, as expected, a slightly different coordination complex than acetazolamide when treated with Cu(II) and ammonia under the same experimental conditions[46] as used for H_2AAZ (Fig. 21.4). In the complex $[Cu(TZA)(NH_3)_2]_2$, the metal ions are coordinated through the ionized nitrogen of the sulfonamide moiety and the N3 of the thiadiazole ring in a bidentate fashion, similarly to acetazolamide complex. The N4 atom of the thiadiazole ring coordinates the other copper ion generating the dimeric structure, mirroring the case

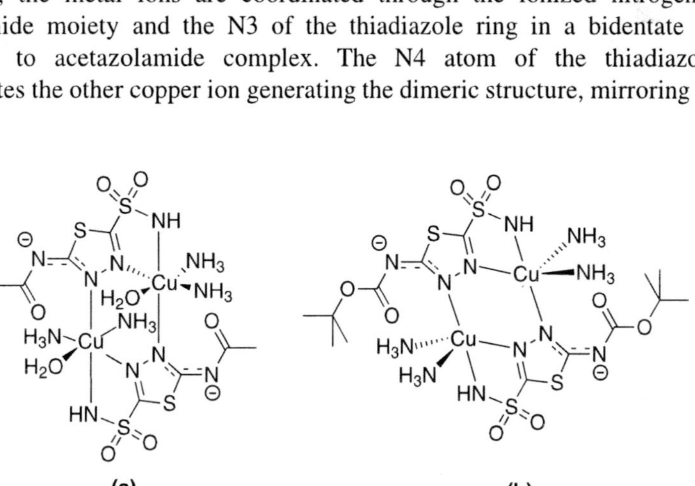

FIGURE 21.4 Comparison between the binuclear Cu(II) complexes of (a) acetazolamide **21.1** and (b) 5-*t*-butyloxycarbonylamido-1,3,4-thiadiazole-2-sulfonamide **21.8**, revealing the change in coordination sphere and geometry of the metal induced by the presence of lipophilic *t*-butoxy group.

of $[Cu(AAZ)(NH_3)_2(OH_2)]_2 \cdot 2H_2O$. Mention must be made that the amide group of ligand **21.8** is more acidic than the one in acetazolamide, therefore being ionized at the working pH to a much greater extent. However, in contrast to the acetazolamide complex, in the case of $[Cu(TZA)(NH_3)_2]_2$ the increased steric bulk and the hydrophobic microenvironment induced by the presence of *t*-butoxy group modify the hydrogen bond network in the crystal, change the Cu(II) coordination from octahedral to distorted square pyramidal, and prevent the coordination of water molecules at the metal center.[60,61]

Supuran et al. reported the synthesis of structurally related pivaloyl derivative **21.9** (H_2PZA)[58,59] and the coordination properties of this ligand toward various metal ions, obtained using NaOH as base. The resulted coordinative compounds were generally stronger CA inhibitors than the parent sulfonamide, with Zn(II), Cd(II), Hg(II), Pb(II), and lanthanide ions generating the most potent inhibitors, similarly to the case of acetazolamide. The coordination behavior of this ligand was also similar to the behavior of H_2AAZ.[58,59] Structurally related with H_2PZA is the adamantyl derivative H_2AZA **21.10**, introduced by the same group shortly after,[68] as a more lipophilic version of the pivaloyl derivative **21.9**. The compound proved to be a very potent CAI itself ($K_I = 10$ nM against CA II, more potent than acetazolamide), but its coordination compounds with Fe(III), Co(II), Ni(II), Cu(II), Zn(II), and Al(III) were even stronger inhibitors, with K_Is between 0.5 and 7 nM against CA II, displaying also a good inhibition profile against CA I and IV. Moreover, the study revealed that the tetrahedral Zn(II) complex $[Zn(HAZA)_2]$ and the octahedral Cu(II) complex $[Cu(HAZA)_2(OH_2)_2]$ were able to act topically, penetrating the cornea of experimental animals (normotensive and glaucomatous rabbits) and inhibiting the CA isozymes in the ciliary processes of the eyes. The decrease of the intraocular pressure (ΔIOP) was similar (for $[Cu(HAZA)_2(OH_2)_2]$) or higher (in the case of $[Zn(HAZA)_2]$) than the action of the clinically used antiglaucoma drug dorzolamide **21.5**. More important, the duration of action of the IOP reduction with $[Zn(HAZA)_2]$ was much longer than the effect of dorzolamide at the same dose.[68] The inhibitor itself is not topically active, revealing that physicochemical properties of the drug are dramatically changed by the interaction with the metal ion. The metal complex possesses the optimal balance between hydrophilicity and lipophilicity, shown by Maren to be essential for topical action.[69]

These observations fueled the exploration of other acetazolamide-like, lipophilic inhibitors. Thus, Supuran's group reintroduced 5-chloroacetamido-1,3,4-thiadiazole-2-sulfonamide **21.11** (H_2CZA)[63,70] and explored the structural and biological properties of its coordinative compounds with a large number of metal ions (Table 21.2).[63,64]

Efficient inhibitors were obtained with Zn(II), Cd(II), Hg(II), and Pb(II) ions, with a good inhibition profile against all three isozymes. The Zn(II) coordination compound $[Zn(HCZA)_2]$ also showed topical antiglaucoma action, being able to reduce the intraocular pressure to a higher extent than dorzolamide, while displaying an extended duration of action.[63]

Continuing the quest for acetazolamide-like ligands with special pharmacological properties, Jitianu et al. explored (substituted)benzoylamido derivatives **21.13**

TABLE 21.2 5-Chloroacetamido-1,3,4-thiadiazole-2-sulfonamide (H_2CZA) Complexes and Their Structural and Biological Characteristics

Metal Ion	H_2CZA Complex	Base	H_2CZA as Ligand	Metal Coordination	CA I, CA II, CA IV Inhibition K_I (nM)			ΔIOP[a] (mmHg)	Reference
	H_2CZA				640	5	24	0(30′); 0(1 h)	63
Be(II)	$[Be(HCZA)_2]$	NaOH	Bidentate	Tetrahedral	120	5	12		63
V(IV)	$K_2[VO(HCZA)_4]$	KOH	Bidentate	Square pyramidal	230	4.1	19		64
Fe(II)	$[Fe(HCZA)_2Cl_2]$	KOH	Bidentate	Octahedral	255	4.4	16		64
Co(II)	$[Co(HCZA)_2(NH_3)_2]$	NH_3	Bidentate	Octahedral	180	3.6	12		64
	$[Co_2(HCZA)_2(OH_2)_2](OH)_2$	KOH	Bidentate	Tetrahedral	360	4.2	21		64
	$[Co(HCZA)_2Cl_2(OH_2)_2]$	KOH	Bidentate	Tetrahedral	200	3.9	15		64
	$[Co(HCZA)_4(OH_2)_2]$	KOH	Bidentate	Tetrahedral	175	3.5	13		64
Ni(II)	$[Ni(HCZA)_2(NH_3)_2]$	NH_3	Bidentate	Octahedral	220	3.7	14		64
	$[Ni(HCZA)_2(OH)_2]$	KOH	Bidentate	Octahedral	205	3.5	11		64
Cu(II)	$[Cu_2(H_2CZA)_2(Cl)_4]$		Monodentate		190	3.2	9		64
	$[Cu_2(HCZA)(OCOCH_3)_2]$		Bidentate		85	2.5	8		64
	$[Cu_2(HCZA)(OCOCH_3)_2]_n$		Bidentate		155	3.9	12		64
Zn(II)	$[Zn(HCZA)_2]$	NaOH	Bidentate	Tetrahedral	40	3	9	8.0(30′); 8.1(1 h)	63
Cd(II)	$[Cd(HCZA)_2]$	NaOH	Bidentate	Tetrahedral	40	3	10		63
Hg(II)	$[Hg(HCZA)_2]$	NaOH	Bidentate	Tetrahedral	9	2	5		63
Al(III)	$[Al(HCZA)_3]$	NaOH	Bidentate	Pseudo-octahedral	80	4	16		63
Pb(II)	$[Pb(HCZA)_2(OH_2)_2]$	NaOH	Bidentate	Pseudo-octahedral	15	5	10		63
	Dorzolamide (HDZA, **21.5**)							2.2(30′); 4.1(1 h)	63

[a] The drop in intraocular pressure was measured at 30′ and 1 h after administration of a 2% solution of inhibitor directly into the rabbit's eye; ΔIOP = IOP control eye − IOP treated eye.

(H_2BTS) and **21.14** (H_2NTS).[65,66] Taking into account the good biological properties obtained previously with the metal complexes of positively charged sulfonamide **21.12**,[54] the authors explored a large set of metal ions in conjunction with the two sulfonamides. Again, the best CA inhibition profiles were obtained when Zn(II), Hg(II), and Pb(II) ions were used, similarly to acetazolamide and its related lipophilic ligands. Therefore, it became of interest to explore the coordination behavior of the more hydrophilic 5-amino-1,3,4-thiadiazol-2-sulfonamide H_2ATS, **21.15**—the synthetic precursor of acetazolamide.[71] Both Supuran's and Borras's groups had investigated this compound, showing the similarities and differences between this ligand and acetazolamide. It was revealed that in the presence of strong bases the aminosulfonamide **21.15** generally coordinates in a bidentate fashion, through the N atom of the ionized sulfonamido moiety and the N3 of the thiadiazole ring, similarly to the structure B of acetazolamide.[63,64] The carbonic anhydrase inhibition properties of the coordination compounds of H_2ATS are generally one order of magnitude better than the parent ligand, with Cd(II), Hg(II), and Pb(II) generating the most potent inhibitors (Table 21.3).

The amine group is not involved in complexation, probably due to its conjugation with the heterocyclic ring and the sulfonamido moiety. This conjugation increases the electronic density on the aromatic ring and enhances the ability of this ligand to involve the heterocyclic N3 in metal complexation. The complex of H_2ATS with Zn(II) generated using ammonia as base has a different structure than the acetazolamide congener: in $[Zn(HATS)_2(NH_3)]\cdot H_2O$,[73] the HATS ionized ligand bridges two Zn(II) ions forming infinite chains Zn-ligand, thus acting in a bidentate fashion although acetazolamide was monodentate in the same conditions (*vide supra*). However, the same conjugation decreases the acidity of the sulfonamide group, and therefore its coordination ability will depend directly on the base strength and the nature of the metal ion. In the presence of dipropylenetriamine (dipn), a polyamine with medium basicity and good coordination properties, the HATS is able to coordinate Cu(II) ions in the usual manner; if dipropylenetriamine is replaced with the less basic diethylenetriamine (dien) and the Cu(II) ion is substituted for Zn(II), then the HATS is displaced from the coordination sphere and acts just as counter-ion.[72] These particularities in electronic structure and complexation behavior confer again special physicochemical properties for the Zn coordination compound of 5-amino-1,3,4-thiadiazole-2-sulfonamide; this coordination compound was also found to be a topically active CAI, decreasing the intraocular pressure similarly to dorzolamide.[63] Mention must be made that the parent ligand H_2ATS was totally devoid of these pharmacological properties. Similar topical intraocular pressure lowering properties were observed for Zn(II) and Cu(II) complexes of structurally related ligand 5-(2-chlorophenyl)-1,3,4-thiadiazole-2-sulfonamide **21.16** (HCTS). Again, the ligand itself was devoid of such ability.[74] Its complexation behavior was similar to H_2ATS, acting in a bidentate fashion through the ionized sulfonamide nitrogen and the N3 atom of the heterocyclic ring. The 1,3,4-thiadiazole-2, 5-sulfonamide **21.17** (H_2TBS)[75,76] can be considered an extreme case of this subclass of ligands, its double bidentate binding yielding polynuclear complexes such as the zinc one depicted in Fig. 21.5.[77,78]

TABLE 21.3 5-Amino-1,3,4-Thiadiazole-2-Sulfonamide (H_2ATS) Complexes and Their Structural and Biological Characteristics

Metal Ion	5-Amino-1,3,4-thiadiazole-2-sulfonamide Complex	Base	Aminosulfonamide as Ligand	Metal Coordination	CA I, CA II, CA IV Inhibition K_I (nM)			Reference
	H_2ATS				1550	230	780	63,71
Be(II)	$[Be(HATS)_2]$	NaOH	Bidentate	Tetrahedral	1050	190	540	63
Mg(II)	$[Mg(HATS)_2]\cdot 3H_2O$	NaOH	Bidentate	Tetrahedral	350	110	220	63
Cr(III)	$[Cr_2(H_2ATS)_2Cl_4(OH)_2]$	NaOH	Bidentate	Octahedral	740	87	540	64
	$[Cr(H_2ATS)_2Cl_2]Cl\cdot 8H_2O$	NaOH	Bidentate	Octahedral	680	79	490	64
Ni(II)	$[Ni(dien)_2]$ $(HATS)Cl\cdot H_2O$	dien	—	Octahedral				72
Cu(II)	$[Cu(HATS)_2(dipn)]$	dipn	Bidentate	Pseudo-octahedral				72
Zn(II)	$[Zn(HATS)_2(NH_3)]\cdot H_2O$	NH_3	Bidentate	Tetrahedral				73
	$[Zn(HATS)_2]$	NaOH	Bidentate	Tetrahedral	50	15	25	63
Cd(II)	$[Cd(HATS)_2]$	NaOH	Bidentate	Tetrahedral	40	14	19	63
Hg(II)	$[Hg(HATS)_2]$	NaOH	Bidentate	Tetrahedral	12	7	10	63
Al(III)	$[Al(HATS)_3]$	NaOH	Bidentate	Pseudo-octahedral	240	76	110	63
Pb(II)	$[Pb(HATS)_2(OH_2)_2]$	NaOH	Bidentate	Pseudo-octahedral	80	10	26	63

dien = diethylenetriamine, dipn = dipropylenetriamine.

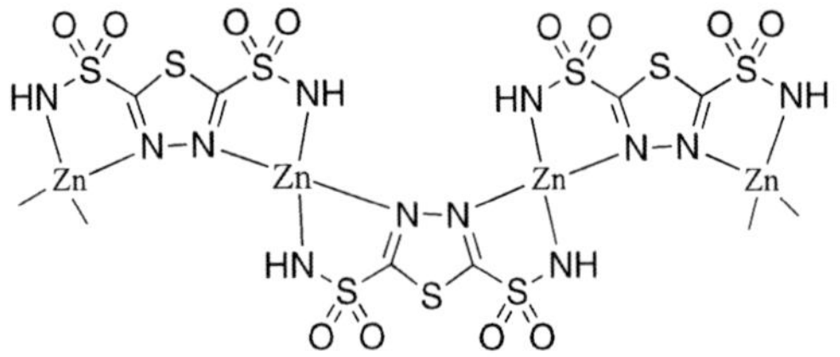

FIGURE 21.5 Schematic representation of the structure of polynuclear Zn complex of 1,3,4-thiadiazole-2,5-bissulfonamide.

21.5 METHAZOLAMIDE COMPLEXES

Methazolamide (HMZA, **21.2**) is a potent carbonic anhydrase inhibitor, used mainly for the management of different forms of glaucoma. Structurally, it is closely related to acetazolamide, albeit having the N4 of the thiadiazole ring methylated and an exocyclic C4=N bond. The involvement of the nitrogen atom of the amido group in this double bond reduces drastically the ability of this atom/group to participate in interactions with the metal ions. Therefore, the coordination abilities of methazolamide are limited to the sulfonamide group ($pK_a = 7.4$) and to the N3 atom of the thiadiazole ring, similarly to the "classical" complexation pattern of acetazolamide (Scheme 21.2).[33]

Supuran's and Borras's groups reported almost simultaneously metal complexes of methazolamide with a large variety of ions (Table 21.4).[20,33–35,41] Supuran used in all cases the sodium salt of methazolamide, obtained *in situ* from the parent compound and sodium hydroxide, while the Spanish group explored various bases with good complexing properties, such as ammonia and pyridine. Under these conditions, Alzuet et al. revealed via X-ray crystallography a monodentate binding of methazolamide (through the ionized sulfonamide group) in all complexes.[79–82] Representative structures are the distorted octahedral Ni(II) complexes $[Ni(MZA)_2(NH_3)_4]$,[80] $[Ni(MZA)_2(Py)_2(OH_2)_2]$,[79] and the tetrahedral Zn(II) complex $[Zn(MZA)_2(NH_3)_2]$[82] shown in Fig. 21.6.

On the basis of spectroscopic, conductometric, and thermogravimetric data, Supuran's group was able to assign a bidentate binding of methazolamide to all complexes obtained from the sodium salt of the ligand, as well as the geometries of metal ions Al(III), Ru(III), Rh(III) (all octahedral), and Au(III) (square planar) in their corresponding coordinative compounds. Biological testing of these dual

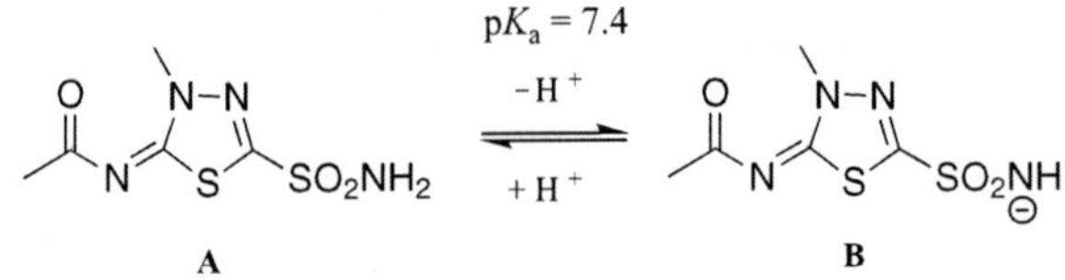

SCHEME 21.2 Ionization equilibrium of methazolamide in water.

TABLE 21.4 Methazolamide (HMZA) Complexes and Their Structural and Biological Characteristics

Metal Ion	Methazolamide Complex	Base	Methazolamide as Ligand	Metal Coordination	CA I, CA II Inhibition IC_{50} (nM)		Reference
	HMZA				940	20	
M(III)	$[La(MZA)_3]\cdot 8H_2O$	NaOH	Bidentate		480	7.5	83
	$[Ce(MZA)_3]\cdot 8H_2O$	NaOH	Bidentate		490	5.7	83
	$[Pr(MZA)_3]\cdot 8H_2O$	NaOH	Bidentate		410	6.4	83
	$[Nd(MZA)_3]\cdot 8H_2O$	NaOH	Bidentate		530	0.9	83
	$[Sm(MZA)_3]\cdot 8H_2O$	NaOH	Bidentate		470	1.5	83
	$[Eu(MZA)_3]\cdot 8H_2O$	NaOH	Bidentate		400	0.8	83
	$[Gd(MZA)_3]\cdot 8H_2O$	NaOH	Bidentate		370	1.3	83
	$[Tb(MZA)_3]\cdot 8H_2O$	NaOH	Bidentate		330	1.0	83
	$[Dy(MZA)_3]\cdot 8H_2O$	NaOH	Bidentate		250	0.8	83
	$[Ho(MZA)_3]\cdot 8H_2O$	NaOH	Bidentate		350	0.8	83
	$[Er(MZA)_3]\cdot 8H_2O$	NaOH	Bidentate		200	0.9	83
	$[Tm(MZA)_3]\cdot 8H_2O$	NaOH	Bidentate		220	0.7	83
	$[Yb(MZA)_3]\cdot 8H_2O$	NaOH	Bidentate		200	0.7	83
Ru(III)	$[Ru(MZA)_3]$	NaOH	Bidentate	Octahedral	170	5.4	52
Co(II)	$[Co(MZA)_2(Py)_2(OH_2)_2]$	Pyridine	Monodentate	Octahedral (distorted)			79
	$[Co(MZA)_2(NH_3)_2]$	NH_3	Monodentate	Octahedral (distorted)			82
Rh(III)	$[Rh(MZA)_3]$	NaOH	Bidentate	Octahedral	530	1.8	52
Ni(II)	$[Ni(MZA)_2(Py)_2(OH_2)_2]$	Pyridine	Monodentate	Octahedral (distorted)			79
	$[Ni(MZA)_2(NH_3)_4]$	NH_3	Monodentate	Octahedral (distorted)			80
Cu(II)	$[Cu(MZA)_2(Py)_2(OH_2)_2]$	Pyridine	Monodentate	Octahedral (distorted)			79
	$[Cu(MZA)_2(NH_3)_3(OH_2)]$	NH_3	Monodentate	Octahedral (distorted)			82
Au(III)	$[Au(MZA)_2]Cl$	NaOH	Bidentate	Square planar	40	1.8	52
Zn(II)	$[Zn(MZA)_2(NH_3)_2]$	NH_3	Monodentate	Tetrahedral (distorted)			82
Al(III)	$[Al(MZA)_3]\cdot 8H_2O$	NaOH	Bidentate	Octahedral		9	54

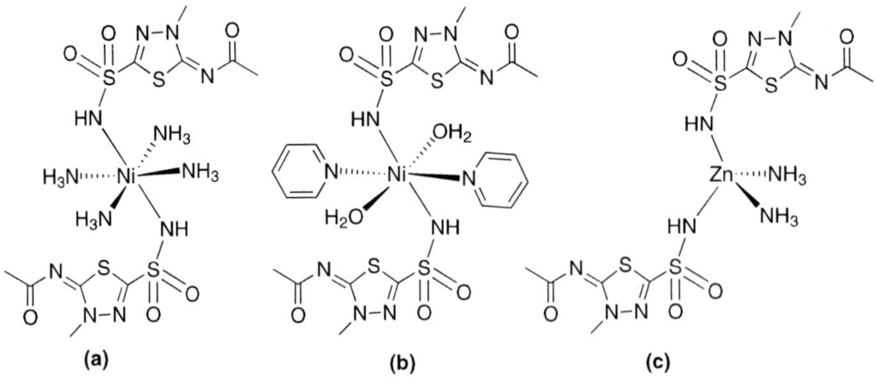

FIGURE 21.6 Representative Ni(II) (a, b) and Zn(II) (c) complexes of methazolamide, generated in the presence of bases with good complexing properties. Note the monodentate binding of the ligand under these conditions.

inhibitors against the two carbonic anhydrase isozymes CA I and II revealed that similarly to the case of acetazolamide, the metal complexes were more potent inhibitors than the parent ligand, lanthanides generating again the most potent representatives from all the tested coordinative compounds (Table 21.4). The activity increases with the increase of the atomic weight of the metal, probably correlated with a decrease in the stability constant of the complex, while moving to higher atomic weights.[54]

Mention must be made that the Cu(II) complex of methazolamide [$Cu(MZA)_2(NH_3)_3(OH_2)$] is an effective anticonvulsant, more powerful than the parent sulfonamide, as proved by Alzuet et al.[82] in a maximal electroshock (MES) seizure model.

21.6 ETHOXZOLAMIDE COMPLEXES

Coordination compounds of ethoxzolamide (**21.3**, HEZA) and their CA inhibitory properties were reported only by Supuran's group.[20,33–35,41] The authors showed that coordination chemistry of this compound is highly similar to that of methazolamide, the ligand binding the metal ions in a bidentate fashion through the nitrogen of the ionized sulfonamido group and the heterocyclic N3 (Table 21.5).

The geometries of the metal ions in HEZA coordination compounds were found to be generally identical to those displayed by the same ions in the complexes with acetazolamide and/or methazolamide, although exceptions were noticed too; for example, the Zn(II) and Ni(II) complexes of ethoxzolamide are very similar to the corresponding acetazolamide derivatives in both the donor system and the geometry of metal ions (see, for example, Fig. 21.1 structures D and F).[85] However, the Cu(II) complex of ethoxzolamide is octahedral, containing two coordinated water molecules, while the acetazolamide derivative was found to be tetrahedral.

Exceptions with the classical *N,N*-bidentate coordination behavior of ethoxzolamide were revealed in its Pd(II) and Pt(II) derivatives.[85] Based on physicochemical

TABLE 21.5 Ethoxzolamide (HEZA) Complexes and Their Structural and Biological Characteristics

Metal Ion	Ethoxzolamide Complex	Base	Ethoxzolamide as Ligand	Metal Coordination	CA I, CA II Inhibition IC_{50} (nM)		Reference
	HEZA				300	10	
Be(II)	$[Be(EZA)_2]$	NaOH	Bidentate	Tetrahedral		0.9	48
Mg(II)	$[Mg(EZA)_2(OH_2)_2]$	NaOH	Bidentate	Octahedral		0.7	48
M(III)	$[La(EZA)_3]\cdot 12\ H_2O$	NaOH	Bidentate		300	0.1	83,84
	$[Ce(EZA)_3]\cdot 8\ H_2O$	NaOH	Bidentate		380	0.6	83,84
	$[Pr(EZA)_3]\cdot 12\ H_2O$	NaOH	Bidentate		500	0.08	83
	$[Nd(EZA)_3]\cdot 8\ H_2O$	NaOH	Bidentate		50	0.1	83
	$[Sm(EZA)_3]\cdot 8\ H_2O$	NaOH	Bidentate		50	0.07	83
	$[Eu(EZA)_3]\cdot 12\ H_2O$	NaOH	Bidentate		40	0.08	83
	$[Gd(EZA)_3]\cdot 12\ H_2O$	NaOH	Bidentate		50	0.07	83
	$[Tb(EZA)_3]\cdot 8\ H_2O$	NaOH	Bidentate		30	0.06	83
	$[Dy(EZA)_3]\cdot 8\ H_2O$	NaOH	Bidentate		40	0.04	83
	$[Ho(EZA)_3]\cdot 8\ H_2O$	NaOH	Bidentate		10	0.04	83
	$[Er(EZA)_3]\cdot 8\ H_2O$	NaOH	Bidentate		0.7	0.04	83,84
	$[Tm(EZA)_3]\cdot 12\ H_2O$	NaOH	Bidentate		20	0.05	83
	$[Yb(EZA)_3]\cdot 12\ H_2O$	NaOH	Bidentate		20	0.05	83
Ce(IV)	$[Ce(EZA)_4]\cdot 6\ H_2O$	NaOH	Bidentate			0.18	50
Th(IV)	$[Th(EZA)_4]\cdot 4\ H_2O$	NaOH	Bidentate			0.01	50
U(VI)	$[UO_2(EZA)_2(OH_2)]\cdot 2\ H_2O$	NaOH	Bidentate	Pentagonal bipyramidal		0.01	50
V(IV)	$[VO(EZA)_2]$	NaOH	Bidentate	Square pyramidal		0.95	50
Fe(III)	$[Fe(EZA)_3]\cdot 6\ H_2O$	NaOH	Bidentate	Octahedral		0.8	51
Ru(III)	$[Ru(EZA)_3]$	NaOH	Bidentate	Octahedral	30	0.04	52
Rh(III)	$[Rh(EZA)_3]$	NaOH	Bidentate	Octahedral	10	0.02	52
Ni(II)	$[Ni(EZA)(OH_2)_2(OH)]\cdot 6\ H_2O$	NaOH	Bidentate	Pseudo-octahedral		75	85

(*continued*)

TABLE 21.5 *(Continued)*

Metal Ion	Ethoxzolamide Complex	Base	Ethoxzolamide as Ligand	Metal Coordination	CA I, CA II Inhibition IC_{50} (nM)		Reference
					CA I	CA II	
Pt(II)	$[Pt(EZA)Cl]_2$	NaOH	Bidentate	Square planar		28	85
Cu(II)	$[Cu(EZA)_2(OH_2)_2]$	NaOH	Bidentate	Octahedral		9	47
Au(III)	$[Au(EZA)_2]Cl$	NaOH	Bidentate	Square planar	10	0.01	52
Zn(II)	$[Zn(EZA)_2]$	NaOH	Bidentate	Tetrahedral		9	47
Al(III)	$[Al(EZA)_3]\cdot 8\ H_2O$	NaOH	Bidentate	Octahedral		0.3	56
Ga(III)	$[Ga(EZA)_3]\cdot 8\ H_2O$	NaOH	Bidentate	Octahedral		0.2	56
In(III)	$[In(EZA)_3]\cdot 8\ H_2O$	NaOH	Bidentate	Octahedral		0.2	56
Tl(III)	$[Tl(EZA)_3]\cdot 8\ H_2O$	NaOH	bidentate	Octahedral		0.1	56
Pb(II)	$[Pb(EZA)_2(OH_2)_2]$	NaOH	Bidentate	Octahedral		0.2	48

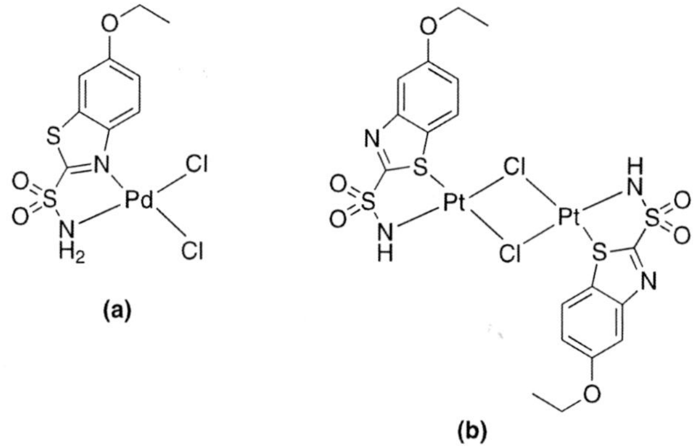

FIGURE 21.7 Unusual coordination behavior of ethoxzolamide in its complexes with Pd(II) and Pt(II).[85]

measurements, it was determined that in the $[Pd(EZA)Cl_2]$ derivative ethoxzolamide coordinates the palladium ion through the *unionized* sulfonamidic nitrogen and the endocyclic N3, together with two chloride ions, in a square planar geometry (Fig. 21.7a). The platinum complex is also square planar, but binuclear, coordination of the two metal ions, being achieved in this case through the ionized sulfonamide nitrogen and the heterocyclic *sulfur* atom S1, with two chloride ions acting as bridging ligands (Fig. 21.7b). These two derivatives are the only reported examples in which ethoxzolamide shows this exotic coordination preference.

Biological data from Table 21.5 are show that, with the exception of binuclear compounds of Ni(II) and Pd(II), the metal complexes of ethoxzolamide are extremely powerful CA inhibitors, more efficient that the parent ligand (which is quite a potent inhibitor itself). The most potent inhibitors were the coordinative compounds of Au(III), Th(IV), and U(VI), also found very efficient when acetazolamide or methazolamide was functioning as ligands.

21.7 BENZOLAMIDE COMPLEXES

Benzolamide (5-phenylsulfonamido-1,3,4-thiadiazole-2-sulfonamide, **21.4**, H_2BZA), an orphan drug,[86] occupies a special place between the nanomolar inhibitors of CA due to its highly acidic secondary sulfonamido group ($pK_a = 3.2$). This group is ionized at physiologic pH, reducing the membrane penetrability of the drug and conferring partial selectivity against membrane isozymes.[21,86,87] Therefore, benzolamide was a valuable tool for the studies of renal physiology, being used for elucidating the role of various CA isozymes in excretion and kidney function.[86] Its ligand properties are also influenced by the acidic secondary sulfonamido group, which can act in tandem with the primary sulfonamidic moiety, thus making benzolamide a versatile ligand, similarly to acetazolamide (Scheme 21.3).

The coordination compounds of H_2BZA were studies by both Borras's and Supuran's groups and are summarized in Table 21.6.

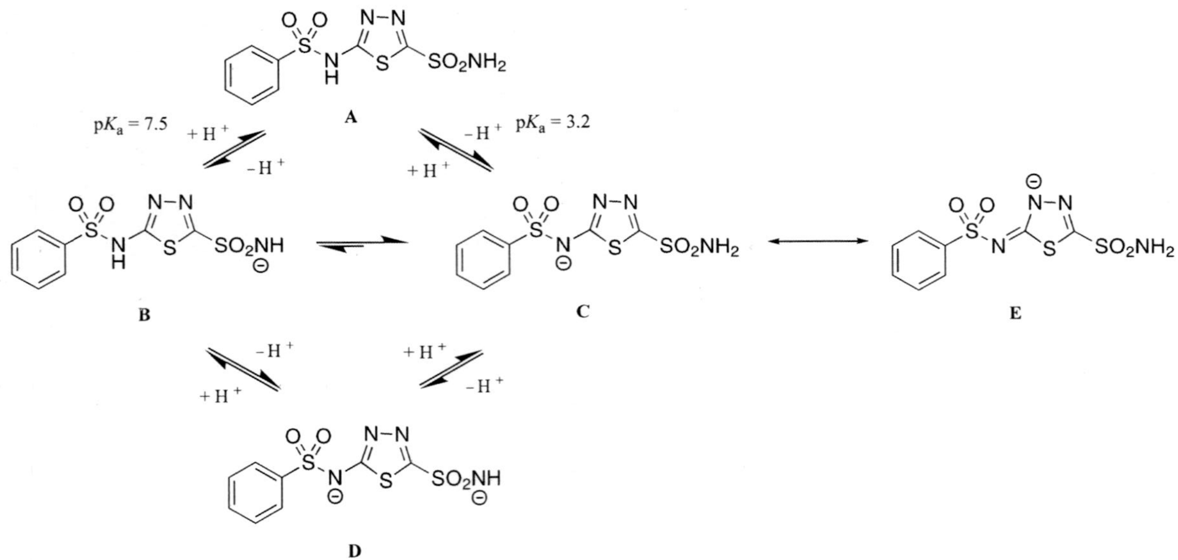

SCHEME 21.3 Ionization equilibrium of benzolamide in water.

TABLE 21.6 Benzolamide (H_2BZA) Complexes and Their Structural and Biological Characteristics

Metal Ion	Benzolamide Complex	Base*	Benzolamide as Ligand	Metal Coordination	CA I, CA II, CA IV Inhibition: IC_{50} (nM)			Reference
	H_2BZA				25	5	13	
Cu(II)	$[Cu(BZA)(NH_3)_4]$	NH_3	Monodentate, D	Square pyramidal				88
	$[Cu(BZA)(dien)(OH_2)]$	dien	Monodentate, D	Square pyramidal				88
	$[Cu(BZA)(dipn)(OH_2)]$	dipn	Monodentate, D	Square pyramidal				88
	$[Cu(HBZA)_2(dien)]$	dien	Monodentate, E	Square pyramidal				89
Zn(II)	$[[Zn_2(BZA)_2(NH_3)_4]\cdot 2H_2O]\infty$	NH_3	Bidentate, D	Tetrahedral				90
	$[Zn(HBZA)_2(dien)]$	dien	Monodentate, E	Square pyramidal				89
Cd(II)	$[Cd(BZA)(NH_3)_4]$	NH_3			2	1.8	3	91
Al(III)	$[Al(HBZA)_3(OH_2)_3]$	NaOH	Octahedral		6	3	8	91
	$[Al(BZA)_3(NH_3)_3]$	NH_3	Octahedral		5	2	9	91
	$[Al(BZA)_3(Py)_3]$	Py	Octahedral		7	1.7	3	91
Pb(II)	$[Pb(BZA)_2(NH_3)_4]$	NH_3	Octahedral		1	1.1	4	91
	$[Pb(BZA)_2(Py)_4]$	Py	Octahedral		2	1.2	5	91

*dien = diethylentriamine; dipn = dipropylenetriamine; Py = pyridine.

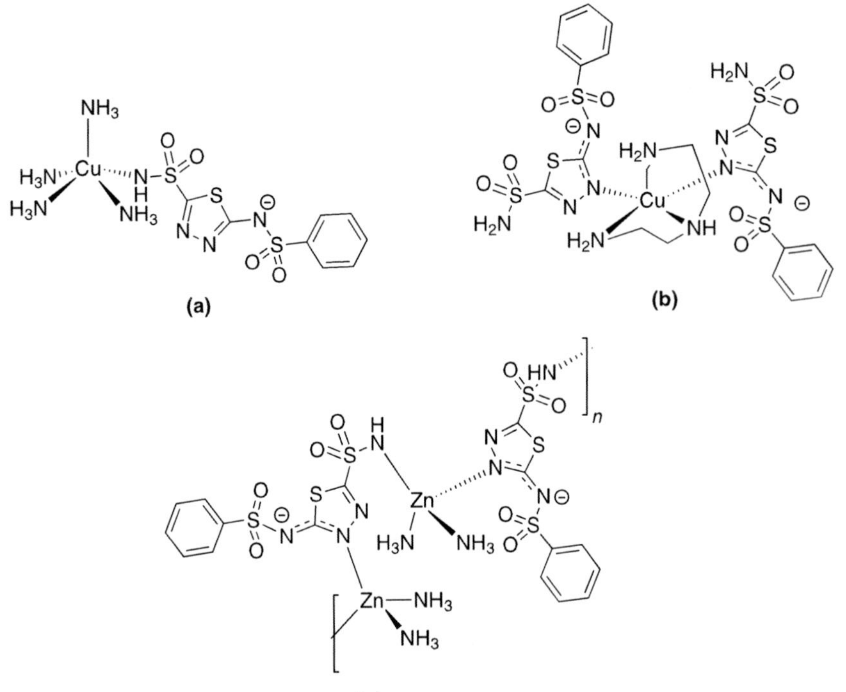

FIGURE 21.8 Representative metal complexes of benzolamide, showing various coordination behaviors of the ligand.

An unexpected coordination behavior of benzolamide was found in the square pyramidal complex [Cu(BZA)(NH_3)$_4$], reported by Alzuet et al.,[88] in which the compound, although it is dideprotonated, is acting as a monodentate ligand through the N of the primary sulfonamide group (Fig. 21.8). This finding is in contrast with the structure observed for the similar complex with acetazolamide [Cu(AAZ)(NH_3)$_2$(OH_2)$_2$]·$2H_2O$ (Fig. 21.1), in which the doubly ionized ligand coordinates the Cu(II) ion through the N of primary sulfonamide and the two N atoms of the thiadiazole ring (*vide supra*). Based on spectroscopic data, a similar coordination behavior was proposed for the coordinative compounds [Cu(BZA)(dien)(OH_2)] and [Cu(BZA)(dipn)(OH_2)].[89]

However, when the amount of benzolamide was doubled (molar ratio of Cu(II)/H_2BZA/dien = 1/2/1), a different complex was obtained, namely, [Cu(HBZA)$_2$(dien)].[89] In this new complex, the benzolamide is monoionized and coordinates the Cu(II) ion through the N4 of the thiadiazole ring. The geometry of the metal is square pyramidal, being coordinated by two benzolamide ligands and one (tridentate) diethylenetriamine molecule. The structure of the corresponding Zn complex, [Zn(HBZA)$_2$(dien)], is identical, with the same metal coordination (Fig. 21.8b).[89]

The same group[90] also reported the polymeric Zn(II) complex [[Zn$_2$(BZA)$_2$(NH_3)$_2$]·H_2O]$_\infty$ obtained from Zn(II), benzolamide, and ammonia. Benzolamide is dideprotonated at both sulfonamidic groups, but in this case it acts as a bidentate ligand, bridging two metal centers through the N of the primary sulfonamide group and the N4 of the thiadiazole ring (Fig. 21.8c). The structure consists of infinite chains connected via hydrogen bonds.[90]

The differences in the complexing ability of benzolamide versus acetazolamide were explained by assuming that in the case of acetazolamide the negative charge on the amide N generated after ionization is significantly delocalized on the N4 of the thiadiazole ring (Scheme 21.1 structures D and E), enhancing the donation properties of this atom, whereas in the case of benzolamide the anion C is conjugated mainly internally (with the oxygen atoms), the thiadiazole ring being partially "decoupled" from the secondary sulfonamide anion. As a direct consequence, benzolamide has sometimes a reduced ability to bind metal ions through the heterocyclic N4 or through the secondary sulfonamide moiety (even though it is ionized in all metal complexes). The above-mentioned crystal structures show the versatility of this ligand, whose coordination behavior can change under the influence of base strength and pH, molar ratio of the reagents, also depending on the nature of the metal partner.[35,88–90]

21.8 THIENOTHIOPYRAN SULFONAMIDE COMPLEXES

As mentioned previously, the classical CA inhibitors such as acetazolamide **21.1** are very efficient in reducing the elevated intraocular pressure (IOP) associated with glaucoma, but their use is associated with unpleasant side effects due to inhibition of various CA isozymes in red blood cells, lungs, kidneys, and other tissues. Various attempts to administer these drugs topically (directly into the eye) failed and the topical delivery route was believed to be not feasible. This dogma was changed in 1983 when Maren et al.[69] showed that the trifluorinated analogue of methazolamide **21.18** (a very potent albeit unstable CAI) was able to penetrate cornea and to inhibit the CAs from the ciliary processes of the eye, thus reducing intraocular pressure and acting as anti-glaucoma drug. An extensive synthetic effort from Merck[92,93] followed this initial discovery and yielded topically active thienothiopyran sulfonamides such as sezolamide **21.19** (HSZA) and its congener dorzolamide **21.5** (HDZA) (Fig. 21.9). Dorzolamide was approved in 1995 in the United States for the management of glaucoma, followed by brinzolamide **21.6** (made by Alcon) in 1998.

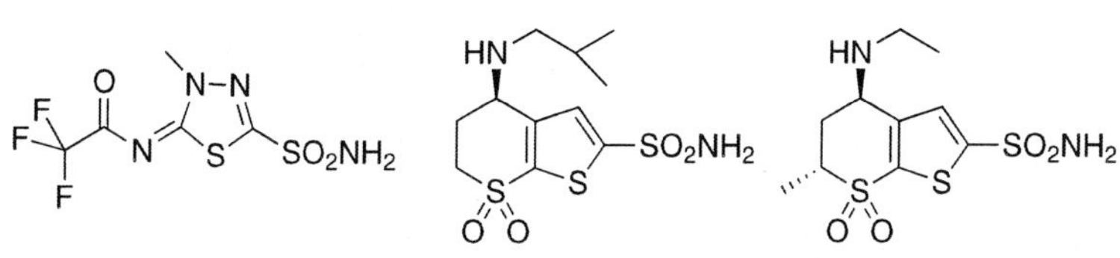

Trifluoromethazolamide (**21.18**) Sezolamide, HSZA (**21.19**) Dorzolamide, HDZA (**21.5**)

FIGURE 21.9 Topically active carbonic anhydrase inhibitors.

TABLE 21.7 Sezolamide and Dorzolamide Complexes and Their Structural and Biological Characteristics

Metal Ion	HSZA/HDZA Complex	Base	HSZA/ HDZA as Ligand	Metal Coordination	CA II Inhibition IC_{50} (nM)	References
	HSZA				0.6	94,95
Co(II)	$[Co(SZA)_2]$	NaOH	Bidentate	Pseudotetrahedral	0.15	94,95
Cu(II)	$[Cu(SZA)_2(OH_2)_2]$	NaOH	Bidentate	Octahedral	0.09	94,95
Zn(II)	$[Zn(SZA)_2]$ HDZA	NaOH	Bidentate	Pseudotetrahedral	0.15	94,95
					0.6	94,95
Co(II)	$[Co(DZA)_2]$	NaOH	Bidentate	Pseudotetrahedral	0.10	94,95
Cu(II)	$[Cu(DZA)_2(OH_2)_2]$	NaOH	Bidentate	Octahedral	0.12	94,95
Zn(II)	$[Zn(DZA)_2]$	NaOH	Bidentate	Pseudotetrahedral	0.07	94,95

Supuran investigated the coordination behavior of sezolamide and dorzolamide,[94,95] motivated by the structural differences exhibited by these compounds when compared with classical CAIs. Both compounds have the primary sulfonamide group attached to a thiophene ring and no other ionizable group. He found that the coordination behavior of HSZA and HDZA is similar as expected. Both compounds act against Co(II), Cu(II), and Zn(II) ions as bidentate ligands through the ionized sulfonamidic nitrogen and the endocyclic sulphur atom of the thiophene ring, similarly to ethoxzolamide in its Pt(II) complex (Fig. 21.7b). All metal complexes were stronger CA inhibitors than parent sulfonamides (Table 21.7)

21.9 SACCHARIN COMPLEXES

Metal complexes of saccharin **21.7**, a weak CA inhibitor,[96] showed probably the most dramatic potency increase as a consequence of the dual inhibition mechanism of action of this class of CAI. In all its complexes, it acts as a monodentate ligand, interacting with the metal ions through the ionized nitrogen atom of the isothiazole ring. Saccharin is also the strongest acid among all these ligands ($pK_a = 1.30$), and therefore in all syntheses its sodium salt (NaSAC) was used to generate the coordinative compounds (Table 21.8).[33,34]

Only one crystal structure is available for this class of complexes, namely, the Cu(II) complex of saccharin.[97] The geometry of the Cu(II) ion is octahedral, with four water molecules occupying the equatorial positions, while two saccharin ligands are placed (*trans*) in the axial ones, coordinating the metal through the nitrogen atom of the isothiazole ring.

As mentioned previously, these coordination compounds of saccharin are, without exception, far more potent than the parent inhibitor. The most efficient ions were proved to be Hg(II), Pb(II), and Zn(II), inducing a three orders of magnitude increase in inhibition efficiency.

TABLE 21.8 Saccharin (HSAC) Complexes and Their Structural and Biological Characteristics

Metal Ion	Saccharin Complex	Metal Coordination	CA II Inhibition IC_{50} (μM)	Reference
	NaSAC		90	
Be(II)	$[Be(SAC)_2(OH_2)_2]$	Tetrahedral	8.4	48
Mg(II)	$[Mg(SAC)_2(H_2O)_4]\cdot 2H_2O$	Octahedral	5.7	48
Ce(IV)	$[Ce(SAC)_4]\cdot 8H_2O$		0.86	50
Th(IV)	$[Th(SAC)_4]\cdot 6H_2O$		0.92	50
U(VI)	$[UO_2(SAC)_2]\cdot 3H_2O$	Pentagonal bipyramidal	0.8	50
V(IV)	$[VO(SAC)_2]\cdot 2H_2O$	Square pyramidal	1.3	50
Fe(III)	$[Fe(SAC)_3(OH_2)_3]\cdot 2H_2O$	Octahedral	0.6	51
Co(II)	$[Mg(SAC)_2(OH_2)_4]\cdot 2H_2O$	Octahedral	4.9	48
Ni(II)	$[Ni(SAC)_2(OH_2)_2]\cdot 4H_2O$	Pseudo-octahedral	7.6	48,97
Cu(II)	$[Cu(SAC)_2(OH_2)_4]\cdot 2H_2O$	Octahedral	0.6	54,97
Zn(II)	$[Zn(SAC)_2(OH_2)_2]$	Tetrahedral	0.2	54,97
Cd(II)	$[Cd(SAC)_2(OH_2)_2]$	Tetrahedral	0.9	48
Hg(II)	$[Hg(SAC)_2(OH_2)_2]$	Tetrahedral	0.1	48
	$[Hg(SAC)_2]_n$	Octahedral	3.8	48
Pb(II)	$[Pb(SAC)_2(OH_2)_2]\cdot 6H_2O$	Tetrahedral	0.2	48

21.10 RECENT DEVELOPMENTS, TRENDS, AND PERSPECTIVES

The field of metal complexes as carbonic anhydrase inhibitors remains a dynamic research topic, with many compounds being synthesized, characterized, and biologically tested, aiming toward improvements in "traditional" applications, as well as toward identifying and validating new biologically relevant applications for this class of carbonic anhydrase inhibitors.

Thus, Scozzafava et al.[91] and Ilies et al.[98] revealed the ability of some Zn(II), Mg(II), Cu(II), and especially Al(III) complexes of acetazolamide **21.1**, methazolamide **21.2**, ethoxzolamide **21.3**, and benzolamide **21.4** to act as antacids, as treatment alternatives (alone or in combination with other drugs) for gastric secretion imbalances.

Supuran's group, building on the structure–antiglaucoma activity correlation studies mentioned previously, introduced new ligands with good water solubility balanced with a medium lipophilicity such as 5-(3,4-dichlorophenylureido)-1,3,4-thiadiazole-2-sulfonamide **21.20**, *p*-fluorobenzolamide **21.21**, diethylenetriamino pentaacetic (dtpa) bis(5-amino-1,3,4-thiadiazole-2-sulfonamide) **21.22** (DTPA-M2), or its ethylenebis(oxyethylenenitrilo) tetraacetic acid congener **21.23** (EGTA-M) and related compounds.[99–102] In conjunction with Zn(II) or Cu(II) ions, these ligands generated coordinative compounds with very potent CA inhibitory

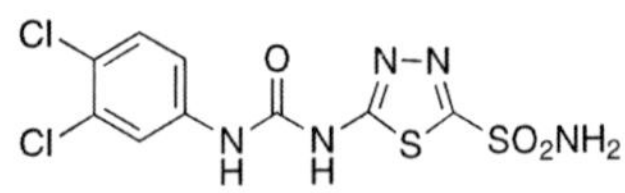

H_2CUS **21.20**

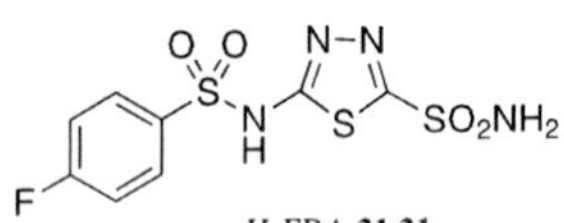

H_2FBA **21.21**

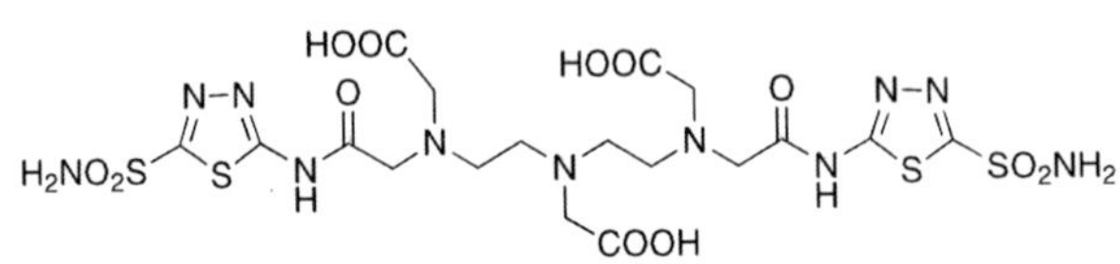

DTPA-M2 **21.22**

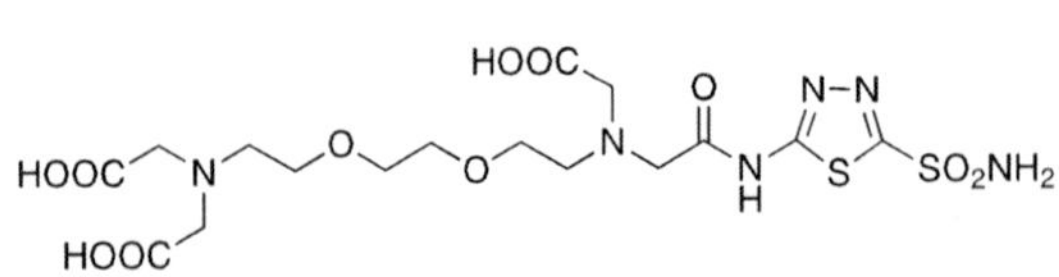

EGTA-M **21.23**

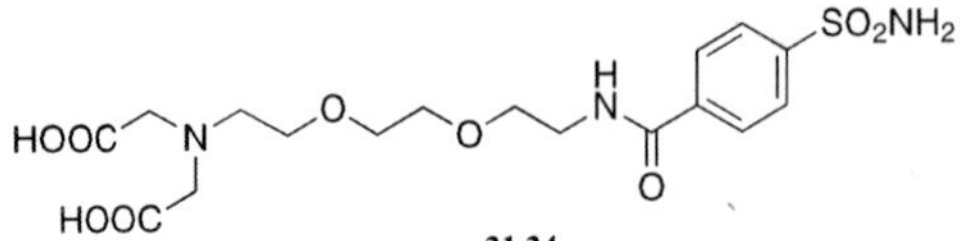

21.24

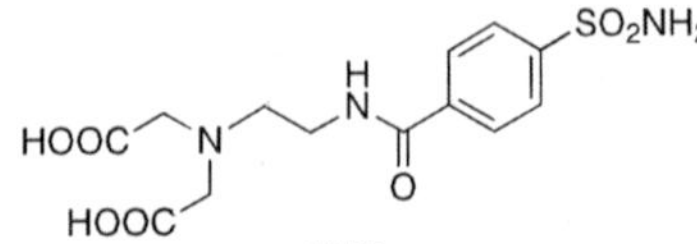

21.25

CHART 21.2

properties doubled with excellent corneal penetrability. The reduction in intraocular pressure generated by these metal complexes was stronger and lasted longer than clinically used dorzolamide **21.5**, making them good candidates for new antiglaucoma drugs (Chart 21.2).[99–102]

In this context, an interesting development was proposed by Srivastava's[103,104] and Christianson's groups[104] that generated isozyme-specific inhibitors by targeting the histidine cluster in CA II with Cu(II) complexes of ligands **21.24**, **21.25** and congeners. Their design exploits the coordination ability of imidazole rings from histidine residues to enhance the selectivity of heterocyclic sulfonamides via a classical two-prong approach[105] in which one prong of the inhibitor binds to the zinc ion, while the second arm of the molecule interacts with the histidine cluster via the copper ion, forming an internal metal complex where some of the ligands are specific moieties from the isozyme's backbone. Various designs were synthesized and tested, proving the feasibility of the idea. Mention must be made that in a very recent development from Supuran's group, it was revealed[106] that Cu(II) complexes of dtpa derivative **21.22** and structurally related congeners are very potent inhibitors of the tumor-associated CA isoforms IX and XII, thus opening the use of metal complexes of sulfonamides as novel anticancer agents, another exciting potential application of this versatile class of carbonic anhydrase inbitors.

REFERENCES

1. Licini, G.; Scrimin, P. Metal-ion-binding peptides: from catalysis to protein tagging. *Angew. Chem. Int. Ed. Engl.* **2003**, *42*(38), 4572–4575.
2. Barondeau, D. P.; Getzoff, E. D. Structural insights into protein–metal ion partnerships. *Curr. Opin. Struct. Biol.* **2004**, *14*(6), 765–774.
3. Christianson, D. W.; Cox, J. D. Catalysis by metal-activated hydroxide in zinc and manganese metalloenzymes. *Annu. Rev. Biochem.* **1999**, *68*, 33–57.
4. Supuran, C. T.; Scozzafava, A.; Conway, J., Eds. *Carbonic Anhydrase: Its Inhibitors and Activators;* CRC Press: Boca Raton, FL, **2004**; pp 1–363.
5. Supuran, C. T. Carbonic anhydrases: novel therapeutic applications for inhibitors and activators. *Nat. Rev. Drug Discov.* **2008**, *7*, 168–181.
6. Supuran, C. T.; Scozzafava, A. Carbonic anhydrases as targets for medicinal chemistry. *Bioorg. Med. Chem.* **2007**, *15*(13), 4336–4350.
7. Supuran, C. T. Carbonic anhydrases: catalytic and inhibition mechanisms, distribution and physiological roles. In *Carbonic Anhydrase: Its Inhibitors and Activators;* Supuran, C. T.; Scozzafava, A.; Conway, J., Eds.; CRC Press: Boca Raton, FL, **2004**; pp 1–23.
8. Lindskog, S.; Silverman, D. N. The catalytic mechanism of mammalian carbonic anhydrases. *EXS* **2000**, *90*, 175–195.
9. Silverman, D. N.; Lindskog, S. The catalytic mechanism of carbonic anhydrase: implications of a rate-limiting protolysis of water. *Acc. Chem. Res.* **1988**, *21*(1), 30–36.
10. Stams, T.; Christianson, D. W. X-ray crystallographic studies of mammalian carbonic anhydrase isozymes. *EXS* **2000** (90), 159–174.

11. Christianson, D. W.; Fierke, C. A. Carbonic anhydrase: evolution of the zinc binding site by nature and by design. *Acc. Chem. Res.* **1996**, *29*(7), 331–339.
12. Briganti, F.; Mangani, S.; Orioli, P.; Scozzafava, A.; Vernaglione, G.; Supuran, C. T. Carbonic anhydrase activators: X-ray crystallographic and spectroscopic investigations for the interaction of isozymes I and II with histamine. *Biochemistry* **1997**, *36*(34), 10384–10392.
13. Meldrum, N. U.; Roughton, F. J. W. Carbonic anhydrase. Its preparation and properties. *J. Physiol.* **1933**, *80*, 113–142.
14. Forster, R. E. Remarks on the discovery of carbonic anhydrase. *EXS* **2000**, *90*, 1–11.
15. Maren, T. H. Carbonic anhydrase: the middle years, 1945–1960, and introduction to pharmacology of sulfonamides. *Ann. N.Y. Acad. Sci.* **1984**, *429*, 10–17.
16. Maren, T. H.; Sanyal, G. The activity of sulfonamides and anions against the carbonic anhydrases of animals, plants, and bacteria. *Annu. Rev. Pharmacol. Toxicol.* **1983**, *23*, 439–459.
17. Supuran, C. T. Therapeutic applications of the carbonic anhydrase inhibitors. *Therapy* **2007**, *4*(3), 355–378.
18. Scozzafava, A.; Mastrolorenzo, A.; Supuran, C. T. Carbonic anhydrase inhibitors and activators and their use in therapy. *Expert Opin. Ther. Pat.* **2006**, *16*(12), 1627–1664.
19. Supuran, C. T.; Casini, A.; Scozzafava, A. Development of sulfonamide carbonic anhydrase inhibitors. In *Carbonic Anhydrase: Its Inhibitors and Activators;* Supuran, C. T.; Scozzafava, A.; Conway, J., Eds.; CRC Press: Boca Raton, FL, **2004**; pp 67–147.
20. Supuran, C. T.; Scozzafava, A.; Casini, A. Carbonic anhydrase inhibitors. *Med. Res. Rev.* **2003**, *23*(2), 146–189.
21. Maren, T. H. Carbonic anhydrase: chemistry, physiology, and inhibition. *Physiol. Rev.* **1967**, *47*(4), 595–781.
22. Supuran, C. T.; Scozzafava, A. Carbonic anhydrase inhibitors and their therapeutic potential. *Expert Opin. Ther. Pat.* **2000**, *10*(5), 575–600.
23. Maren, T. H.; Clare, B. W.; Supuran, C. T. Structure–activity studies of sulfonamide carbonic anhydrase inhibitors. *Roum. Chem. Quart. Rev.* **1994**, *2*(4), 259–282.
24. Ilies, M. A.; Banciu, M. D. Nonsulfonamide carbonic anhydrase inhibitors. In *Carbonic Anhydrase: Its Inhibitors and Activators;* Supuran, C. T.; Scozzafava, A.; Conway, J., Eds.; CRC Press: Boca Raton, FL, **2004**; pp 209–241.
25. Bertini, I.; Luchinat, C.; Scozzafava, A. Carbonic anhydrase: an insight into the zinc binding site and into the active cavity through metal substitution. *Struct. Bonding* **1981**, *48*, 45–92.
26. Krishnamurthy, V. M.; Kaufman, G. K.; Urbach, A. R.; Gitlin, I.; Gudiksen, K. L.; Weibel, D. B.; Whitesides, G. M. Carbonic anhydrase as a model for biophysical and physical-organic studies of proteins and protein–ligand binding. *Chem. Rev.* **2008**, *108*(3), 946–1051.
27. Winum, J. Y.; Scozzafava, A.; Montero, J. L.; Supuran, C. T. Metal binding functions in the design of carbonic anhydrase inhibitors. *Curr. Top. Med. Chem.* **2007**, *7*(9), 835–848.
28. Xue, Y.; Liljas, A.; Jonsson, B. H.; Lindskog, S. Structural analysis of the zinc hydroxide-Thr-199-Glu-106 hydrogen-bond network in human carbonic anhydrase II. *Proteins* **1993**, *17*(1), 93–106.

29. Lindahl, M.; Vidgren, J.; Eriksson, E.; Habash, J.; Harrop, S.; Helliwell, J.; Liljas, A.; Lindeskog, M.; Walker, N. Crystallographic studies of carbonic anhydrase inhibition. *Carbonic Anhydrase, Proceedings of International Workshop*, **1991**, 111–118.

30. Braun, C. E.; Towle, J. L. N1-Silver derivatives of sulfanilamide and some related compounds. *J. Am. Chem. Soc.* **1941**, *63*, 3523.

31. Bult, A. Metal complexes of sulfanilamides in pharmaceutical analysis and therapy. *Met. Ions Biol. Systs.* **1983**, *16*, 261–278.

32. Malecki, F.; Staroscik, R.; Weiss-Gradzinska, W. Complex formation in copper(II)-sulfonamide systems. *Pharmazie* **1984**, *39*(3), 158–160.

33. Supuran, C. T. Carbonic anhydrase inhibitors. In *Carbonic anhydrase and modulation of physiologic and pathologic processes in the organism. Enzyme activators and inhibitors;* Puscas, I., Ed.; Helicon: Timisoara, **1994**; pp 30–111.

34. Alzuet, G.; Ferrer, S.; Borras, J.; Supuran, C. T. Complexes of heterocyclic sulfonamides: a class of potent, dual carbonic anhydrase inhibitors. *Roum. Chem. Quart. Rev.* **1994**, *2*(4), 283–300.

35. Borras, J.; Alzuet, G.; Ferrer, S.; Supuran, C. T. Metal complexes of heterocyclic sulfonamides as carbonic anhydrase inhibitors. In *Carbonic Anhydrase: Its Inhibitors and Activators;* Supuran, C. T.; Scozzafava, A;. Conway, J., Eds.; CRC Press: Boca Raton, FL, **2004**; pp 183–207.

36. Luca, C.; Barboiu, M.; Supuran, C. T. Carbonic anhydrase inhibitors. 7. Stability constants of complex inhibitors and their mechanism of action. *Rev. Roum. Chim.* **1991**, *36*(9–10), 1169–1173.

37. Magid, E. The activity of carbonic anhydrases B and C from human erythrocytes and the inhibition of the enzymes by copper. *Scand. J. Haematol.* **1967**, *4*(4), 257–270.

38. Tu, C.; Wynns, G. C.; Silverman, D. N. Inhibition by cupric ions of 18O exchange catalyzed by human carbonic anhydrase II. Relation to the interaction between carbonic anhydrase and hemoglobin. *J. Biol. Chem.* **1981**, *256*(18), 9466–9470.

39. Eriksson, A. E.; Kylsten, P. M.; Jones, T. A.; Liljas, A. Crystallographic studies of inhibitor binding sites in human carbonic anhydrase II: a pentacoordinated binding of the SCN- ion to the zinc at high pH. *Proteins* **1988**, *4*(4), 283–293.

40. Ferrer, S.; Borras, J.; Garcia-Espana, E. Complex formation equilibria between acetazolamide (5-acetamido-1,3,4-thiadiazole-2-sulfonamide), a potent inhibitor of carbonic anhydrase, and Zn(II), Co(II), Ni(II) and Cu(II)in aqueous and ethanol-aqueous solutions. *J. Inorg. Biochem.* **1990**, *39*, 297–306.

41. Raper, E. S. Copper complexes of heterocyclic thioamides and related ligands. *Coord. Chem. Rev.* **1994**, *129*, 91–156.

42. Ferrer, S.; Jimenez, A.; Borras, J.; Synthesis and Characterization of Acetazolamide Complexes of Co(II) and Zn(II). *Inorg. Chim. Acta* **1987**, *129*, 103–106.

43. Hartmann, U.; Vahrenkamp, H. A zinc complex of the carbonic anhydrase inhibitor acetazolamide (aaaH): crystal structure of $(aaa)_2Zn(NH_3)_2$. *Inorg. Chem.* **1991**, *30*, 4676–4677.

44. Ferrer, S.; Borras, J.; Miratvilles, C.; Fuertes, A. Coordination behavior of acetazolamide (5-acetamido-1,3,4-thiadiazole-2-sulfonamide): synthesis, crystal structure, and properties of bis(acetazolamidato)tetraamminenickel (II). *Inorg. Chem.* **1989**, *28*, 160–163.

45. Mathew, M.; Palenik, G. J. Crystal and molecular structure of acetazolamide (5-acetamido-1,3,4-thiadiazole-2-sulphonamide), a potent inhibitor of carbonic anhydrase. *J. Chem. Soc. Perkin Trans.* **1974**, *2*, 532–536.

46. Ferrer, S.; Borras, J.; Miratvilles, C.; Fuertes, A. Synthesis and characterization of copper(II)-acetazolamide (5-acetamido-1,3,4-thiadiazole-2-sulfonamide) complexes. Crystal Structure of dimeric $[Cu(Acm)(NH_3)_2(OH_2)]_2{\cdot}2H_2O$. *Inorg. Chem.* **1990**, *29*, 206–210.

47. Supuran, C. T.; Andruh, M.; Puscas, I. Metal complexes of sulfonamides: a novel class of carbonic anhydrase inhibitors. *Rev. Roum. Chim.* **1990**, *35*(3), 393–399.

48. Supuran, C. T.; Loloiu, G.; Manole, G. Carbonic anhydrase inhibitors. Part 15. Complex inhibitors containing main-group and transitional divalent cations. *Rev. Roum. Chim.* **1993**, *38*(1), 115–122.

49. Supuran, C. T.; Manole, G.; Manzatu, I. Carbonic anhydrase inhibitors. Part 12. Lanthanide complexes with acetazolamide as dual inhibitors. *Rev. Roum. Chim.* **1992**, *37*(6), 739–744.

50. Supuran, C. T. Carbonic anhydrase inhibitors. Part 16. Complex inhibitors containing metal ions in high oxidation states V(IV); Ce(IV); Th(IV); U(VI). *Rev. Roum. Chim.* **1993**, *38*(2), 229–236.

51. Supuran, C. T.; Loloiu, G.; Manole, G. Carbonic anhydrase inhibitors. Part 14. Iron (III) complexes of heterocyclic sulfonamides and mercaptans as dual inhibitors. *Rev. Roum. Chim.* **1992**, *37*(10), 1181–1189.

52. Manole, G.; Maior, O.; Supuran, C. T. Carbonic anhydrase inhibitors. Part 17. Complexes of heterocyclic sulfonamides with Ru(III), Rh(III) and Au(III) are very strong dual inhibitors. *Rev. Roum. Chim.* **1993**, *38*(4), 475–479.

53. Ferrer, S.; Haasnoot, J. G.; de Graaff, R. A. G.; Reedijk, J.; Borras, J. Synthesis, crystal structure and properties of two acetazolamide (5-acetamido-1,3,4-thiadiazole-2-sulfonamide) complexes: bis(5-acetamidato-1,3,4-thiadiazole-2-sulfonamide-O)bis(1,2-ethanediamine)copper(II) and bis(5-acetamidato-1,3,4-thiadiazole-2-sulfonamide-N)bis (1,3-propanediamine)copper(II); an unusually weak ambidentate anionic ligand. *Inorg. Chim. Acta* **1992**, *192*, 129–138.

54. Supuran, C. T. Carbonic anhydrase inhibitors. Part 13. Complex-type mechanism-based inhibitors. *Rev. Roum. Chim.* **1992**, *37*(7), 849–855.

55. Ferrer, S.; Alzuet, G.; Borras, J. Synthesis and characterization of acetazolamide complexes of Ni(II), Cd(II), Hg(II) and Ag(I). *J. Inorg. Biochem.* **1989**, *37*, 163–174.

56. Supuran, C. T.; Andruh, M. Carbonic anhydrase inhibitors. Part 18. Coordination compounds of heterocyclic sulfonamides with main group trivalent cations are potent isozyme II inhibitors. *Rev. Roum. Chim.* **1994**, *39*(10), 1229–1234.

57. Alzuet, G.; Ferrer, S.; Borras, J. Acetazolamide-M(II) [M(II) = Co(II), Ni(II), Cu(II)] complexes with ethylamine, diethylamine, triethylamine, and potassium hydroxide. *J. Inorg. Biochem.* **1991**, *42*, 79–86.

58. Almajan, L. G.; Supuran, C. T. Carbonic anhydrase inhibitors. Part 30. Complexes of 5-pivaloylamido-1,3,4-thiadiazole-2-sulfonamide with trivalent metal ions. *Rev. Roum. Chim.* **1997**, *42*(7), 593–597.

59. Supuran, C. T.; Almajan, L. G. Carbonic anhydrase inhibitors. Part 34. Synthesis and biological activity of coordination compounds of 5-pivaloylamido-1,3,

4-thiadiazole-2-sulfonamide with divalent metal ions. *Main Group Met. Chem.* **1996**, *8*, 347–351.

60. Pedregosa, J. C.; Casanova, J.; Alzuet, G.; Borras, J.; Garcia Granda, S.; Diaz, M. R.; Gutierrez-Rodriguez, A. Metal complexes of 5-tertbutyloxycarbonylamido-1,3,4-thiadiazole-2-sulfonamide (B-H_2ats), a carbonic anhydrase inhibitor. Synthesis and characterization of the copper(II) complex. Crystal structures of B-H_2ats and the [Cu (B-ats)$(NH_3)_2]_2$ dimer complex. *Inorg. Chim. Acta* **1995**, *232*, 117–124.

61. Chufan, E. E.; Pedregosa, J. C.; Ferrer, S.; Borras, J. Spectroscopic behavior of metal-drug complexes. Infrared spectra of Cu(II) dimer complexes with acetazolamide (H_2acm) and an analogue sulfonamide (B-H_2ats). *Vibr. Spectrosc.* **1999**, *20*, 35–45.

62. Chufan, E. E.; Pedregosa, J. C.; Baldini, O. N.; Bruno-Blanch, L. Anticonvulsant activity of analogues of acetazolamide. *Farmaco* **1999**, *54*, 838–841.

63. Supuran, C. T.; Scozzafava, A.; Jitianu, A. Carbonic anhydrase inhibitors. Part 54. Metal complexes of heterocyclic sulfonamides: a new class of antiglaucoma agents. *Met. Based Drugs* **1997**, *4*, 307–315.

64. Supuran, C. T.; Scozzafava, A.; Briganti, F.; Ilies, M. A.; Jitianu, A. Carbonic anhydrase inhibitors. Part 55. Metal complexes of 1,3,4-thiadiazole-2-sulfonamide derivatives: *in vitro* inhibition studies with carbonic anhydrase isozymes I, II and IV. *Met. Based Drugs* **1998**, *5*(2), 103–114.

65. Jitianu, A.; Ilies, M. A.; Scozzafava, A.; Supuran, C. T. Complexes with biologically active ligands. Part 8. Synthesis and carbonic anhydrase inhibitory activity of 5-benzoylamido- and 5-(3-nitro-benzoylamido)-1,3,4-thiadiazole-2-sulfonamide and their metal complexes. *Main Group Met. Chem.* **1997**, *20*, 151–156.

66. Jitianu, A.; Ilies, M. A.; Briganti, F.; Scozzafava, A.; Supuran, C. T. Complexes with biologically active ligands. Part 9. Metal complexes containing of 5-benzoylamino- and 5-(3-nitrobenzoylamino)-1,3,4-thiadiazole-2-sulfonamide as carbonic anhydrase inhibitors. *Met. Based Drugs* **1997**, *4*, 1–8.

67. Alzuet, G.; Ferrer, S.; Borras, J.; Sorenson, J. R. J. Anticonvulsant properties of copper acetazolamide complexes. *J. Inorg. Biochem.* **1994**, *55*, 147–151.

68. Supuran, C. T.; Mincione, F.; Scozzafava, A.; Briganti, F.; Mincione, G.; Ilies, M. A. Carbonic anhydrase inhibitors. Part 52. Metal complexes of heterocyclic sulfonamides: a new class of strong topical intraocular pressure-lowering agents with potential use as antiglaucoma drugs. *Eur. J. Med. Chem.* **1998**, *33*, 247–254.

69. Maren, T. H.; Jankowska, L.; Sanyal, G.; Edelhauser, H. F. The transcorneal permeability of sulfonamide carbonic anhydrase inhibitors and their effect on aqueous humor secretion. *Exp. Eye Res.* **1983**, *36*(4), 457–479.

70. Vaughan, J. R., Jr.; Eichler, J. A.; Anderson, G. W. Heterocyclic sulfonamides as carbonic anhydrase inhibitors. 2-Acylamido- and 2-sulfonamido-1,3,4-thiadiazole-5-sulfonamides. *J. Org. Chem.* **1956**, *21*, 700–701.

71. Young, R. W.; Wood, K. H.; Eichler, J. A.; Vaughan, J. R., Jr.; Anderson, G. W. 1,3,4-Thiadiazole- and thiadiazolinesulfonamides as carbonic anhydrase inhibitors. Synthesis and structural studies. *J. Am. Chem. Soc.* **1956**, *78*, 4649–4654.

72. Chufan, E. E.; Garcia-Granda, S.; Diaz, M. R.; Borras, J.; Pedregosa, J. C. Several coordination modes of 5-amino-1,3,4-thiadiazole-2-sulfonamide (Hats) with Cu(II), Ni(II) and Zn(II): mimetic ternary complexes of carbonic anhydrase-inhibitor. *J. Coord. Chem.* **2001**, *54*(3–4), 303–312.

73. Borja, P.; Alzuet, G.; Casanova, J.; Server-Carrio, J.; Borras, J.; Martinez-Ripoll, M.; Supuran, C. T. Zn complexes of carbonic anhydrase inhibitors: Crystal structure of [Zn (5-amino-1,3,4-thiadiazole-2-sulfonamidate)$_2$(NH$_3$)]·H$_2$O. Carbonic anhydrase inhibitory activity. *Main Group Met. Chem.* **1998**, *21*, 279–292.

74. Briganti, F.; Tilli, S.; Mincione, G.; Mincione, F.; Menabuoni, L.; Supuran, C. T. Carbonic anhydrase inhibitors. Metal complexes of 5-(2-chlorophenyl)-1,3,4-thiadiazole-2-sulfonamide with topical intraocular pressure lowering properties: the influence of metal ions upon the pharmacological activity. *J. Enzyme Inhib.* **2000**, *15*(2), 185–200.

75. Roblin, R. O.; Clapp, J. W. The preparation of heterocyclic sulfonamides. *J. Am. Chem. Soc.* **1950**, *72*, 4890–4892.

76. Supuran, C. T.; Conroy, C. W.; Maren, T. H. Carbonic anhydrase inhibitors. Synthesis and inhibitory properties of 1,3,4-thiadiazole-2,5-bissulfonamide. *Eur. J. Med. Chem.* **1996**, *31*, 843–846.

77. Supuran, C. T. Metal complexes of 1,3,4-thiadiazole-2,5-disulfonamide are strong dual carbonic anhydrase inhibitors, although the ligand possesses very weak such properties. *Met. Based Drugs* **1995**, *2*, 331–336.

78. Supuran, C. T. Carbonic anhydrase inhibitors. Part 26. Preparation and inhibitory properties of coordination compounds of 1,3,4-thiadiazole-2,5-disulfonamide. *Rev. Roum. Chim.* **1996**, *41*(7–8), 653–659.

79. Alzuet, G.; Ferrer, S.; Borras, J.; Castineiras, A.; Solans, X.; Font-Bardia, M. Coordination compounds of methazolamide: synthesis, spectroscopic studies and crystal structures of [M(macm)$_2$(py)$_2$(OH$_2$)$_2$] [M=Co(II), Ni(II) and Cu(II)]. *Polyhedron* **1992**, *22*, 2849–2856.

80. Alzuet, G.; Ferrer, S.; Borras, J.; Solans, X.; Font-Bardia, M. Coordination behaviour of methazolamide [*N*-(-4-methyl-2-sulfamoyl-Δ^2-1,3,4-thiadiazolin-5-ylidene)]acetamide, an inhibitor of carbonic anhydrase enzyme. Synthesis, crystal structure and properties of bis(methazolamidate)tetrammine nickel (II). *Inorg. Chim. Acta* **1993**, *203*(2), 257–261.

81. Alzuet, G.; Ferrer, S.; Casanova, J.; Borras, J.; Castineiras, A. A Co(III) complex of carbonic anhydrase inhibitor methazolamide and the amino-imino "aib" ligand formed by reaction of acetone and ammonia. *Inorg. Chim. Acta* **1993**, *205*, 79–84.

82. Alzuet, G.; Casanova, J.; Ramirez, J. A.; Borras, J.; Carugo, O. Metal complexes of the carbonic anhydrase inhibitor methazolamide: crystal structure of the Zn(macm)$_2$(NH$_3$)$_2$. Anticonvulsant properties of Cu(macm)$_2$(NH$_3$)$_2$(H$_2$O). *J. Inorg. Biochem.* **1995**, *57*, 219–234.

83. Supuran, C. T.; Manole, G.; Andruh, M. Carbonic anhydrase inhibitors. Part 11. Coordination compounds of heterocyclic sulfonamides with lanthanides are potent inhibitors of isozymes I and II. *J. Inorg. Biochem.* **1993**, *49*(2), 97–103.

84. Supuran, C. T.; Stefan, R.; Manole, G.; Puscas, I.; Andruh, M. Carbonic anhydrase inhibitors. 8. Complexes of ethoxzolamide with lanthanides are powerful inhibitors of isozymes I and II. *Rev. Roum. Chim.* **1991**, *36*(9–10), 1175–1179.

85. Andruh, M.; Cristurean, E.; Stefan, R.; Supuran, C. T. Carbonic anhydrase inhibitors. 6. Novel coordination compounds of Pd(II), Pt(II) and Ni(II) with 6-ethoxy-benzothiazole-2-sulfonamide. *Rev. Roum. Chim.* **1991**, *36*, 727–732.

86. Maren, T. H. Benzolamide: a renal carbonic anhydrase inhibitor. In *Orphan Drugs;* Karch, F. E., Ed.; Dekker: New York, **1982**; pp 89–115.

87. Supuran, C. T.; Scozzafava, A. Benzolamide is not a membrane-impermeant carbonic anhydrase inhibitor. *J Enzyme Inhib. Med. Chem.* **2004**, *19*(3), 269–273.

88. Alzuet, G.; Casanova, J.; Borras, J.; Garcia Granda, S.; Gutierrez-Rodriguez, A.; Supuran, C. T. Copper complexes modelling the interaction between benzolamide and Cu-substituted carbonic anhydrase: crystal structure of $Cu(bz)(NH_3)_4$ complex. *Inorg. Chim. Acta* **1998**, *273*, 334–338.

89. Alzuet, G.; Ferrer-Llusar, S.; Borras, J.; Martinez-Manez, R. New Cu(II) and Zn(II) complexes of benzolamide with diethylentriamine: synthesis, spectroscopy and X-ray structures. *Polyhedron* **2000**, *19*, 725–730.

90. Alzuet, G.; Ferrer-Llusar, S.; Borras, J.; Server-Carrio, J.; Martinez-Manez, R. Coordinative versatility of the carbonic anhydrase inhibitor benzolamide in zinc and copper model compounds. *J. Inorg. Biochem.* **1999**, *75*, 189–198.

91. Scozzafava, A.; Ilies, M. A.; Supuran, C. T. Carbonic anhydrase inhibitors. Part 89. Metal complexes of benzolamide with strong enzyme inhibitory and putative antiulcer properties. *Rev. Roum. Chim.* **2000**, *45*(7–8), 771–778.

92. Ponticello, G. S.; Freedman, M. B.; Habecker, C. N.; Lyle, P. A.; Schwam, H.; Varga, S. L.; Christy, M. E.; Randall, W. C.; Baldwin, J. J. Thienothiopyran-2-sulfonamides: a novel class of water-soluble carbonic anhydrase inhibitors. *J. Med. Chem.* **1987**, *30*(4), 591–597.

93. Baldwin, J. J.; Ponticello, G. S.; Anderson, P. S.; Christy, M. E.; Murcko, M. A.; Randall, W. C.; Schwam, H.; Sugrue, M. F.; Springer, J. P., Gautheron, P., et al. Thienothiopyran-2-sulfonamides: novel topically active carbonic anhydrase inhibitors for the treatment of glaucoma. *J. Med. Chem.* **1989**, *32*(12), 2510–2513.

94. Supuran, C. T. Thienothiopyransulfonamides as complexing agents for the preparation of dual carbonic anhydrase inhibitors. *Met. Based Drugs* **1995**, *2*, 327–330.

95. Supuran, C. T. Carbonic anhydrase inhibitors. Part 25. Thienothiopyran sulfonamides: a novel class of complexing agents for the preparation of dual enzyme inhibitors. *Rev. Roum. Chim.* **1996**, *41*(5–6), 495–499.

96. Supuran, C. T.; Banciu, M. D. Carbonic anhydrase inhibitors. Inhibitors with modified sulfonamido groups and their interaction with the zinc enzyme. *Rev. Roum. Chim.* **1991**, *36*(11–12), 1345–1353.

97. Ahmed, K. J.; Habib, A.; Haider, S. Z.; Malik, K. M. A.; Hursthousse, M. B. The preparation and X-ray crystal structure of a saccharin complex of copper(II). *Inorg. Chim. Acta* **1981**, *56*, L37.

98. Ilies, M. A.; Supuran, C. T.; Scozzafava, A. Carbonic anhydrase inhibitors. Part 91. Metal complexes of heterocyclic sulfonamides as potential pharmacological agents in the treatment of gastric acid secretion imbalances. *Met. Based Drugs* **2000**, *7*(2), 57–62.

99. Supuran, C. T.; Scozzafava, A.; Mincione, F.; Menabuoni, L.; Briganti, F.; Mincione, G.; Jitianu, M. Carbonic anhydrase inhibitors. Part 60. The topical intraocular pressure-lowering properties of metal complexes of a heterocyclic sulfonamide: influence of the metal ion upon biological activity. *Eur. J. Med. Chem.* **1999**, *34*(7–8), 585–595.

100. Supuran, C. T.; Scozzafava, A.; Menabuoni, L.; Mincione, F.; Briganti, F.; Mincione, G. Carbonic anhydrase inhibitors. Part 72. Synthesis and antiglaucoma properties of metal complexes of p-fluorobenzolamide. *Met. Based Drugs* **1999**, *6*, 67–74.

101. Scozzafava, A.; Menabuoni, L.; Mincione, F.; Mincione, G.; Supuran, C. T. Carbonic anhydrase inhibitors: synthesis of sulfonamides incorporating dtpa tails and of their zinc complexes with powerful topical antiglaucoma properties. *Bioorg. Med. Chem. Lett.* **2001**, *11*(4), 575–582.

102. Scozzafava, A.; Menabuoni, L.; Mincione, F.; Supuran, C. T. Carbonic anhydrase inhibitors. A general approach for the preparation of water-soluble sulfonamides incorporating polyamino-polycarboxylate tails and of their metal complexes possessing long-lasting, topical intraocular pressure-lowering properties. *J. Med. Chem.* **2002**, *45*(7), 1466–1476.

103. Banerjee, A. L.; Swanson, M.; Roy, B. C.; Jia, X.; Haldar, M. K.; Mallik, S.; Srivastava, D. K. Protein surface-assisted enhancement in the binding affinity of an inhibitor for recombinant human carbonic anhydrase II. *J. Am. Chem. Soc.* **2004**, *126*(35), 10875–10883.

104. Jude, K. M.; Banerjee, A. L.; Haldar, M. K.; Manokaran, S.; Roy, B.; Mallik, S.; Srivastava, D. K.; Christianson, D. W. Ultrahigh resolution crystal structures of human carbonic anhydrases I and II complexed with "two-prong" inhibitors reveal the molecular basis of high affinity. *J. Am. Chem. Soc.* **2006**, *128*(9), 3011–3018.

105. Mammen, M.; Choi, S. K.; Whitesides, G. M. Polyvalent interactions in biological systems: implications for design and use of multivalent ligands and inhibitors. *Angew. Chem. Int. Ed. Eng.* **1998**, *37*, 2754–2794.

106. Rami, M.; Winum, J. Y.; Innocenti, A.; Montero, J. L.; Scozzafava, A.; Supuran, C. T. Carbonic anhydrase inhibitors: copper(II) complexes of polyamino-polycarboxylamido aromatic/heterocyclic sulfonamides are very potent inhibitors of the tumor-associated isoforms IX and XII. *Bioorg. Med. Chem. Lett.* **2008**, *18*(2), 836–841.

CHAPTER 22

Drug Design Studies of Carbonic Anhydrase Activators

CLAUDIA TEMPERINI, ANDREA SCOZZAFAVA, and CLAUDIU T. SUPURAN

Dipartimento di Chimica, Università degli Studi di Firenze, Via della Lastruccia 3, I-50019 Sesto Fiorentino (Firenze), Italy

22.1 INTRODUCTION

Carbonic anhydrases (CA, EC 4.2.1.1) carry out a number of biosynthetic reactions, including the reversible hydration of CO_2 to form bicarbonate. Possessing a (His)3-coordinated active site Zn^{2+}, these enzymes are potentially inhibited by sulfonamides,[1] numerous CA–inhibitors complexes have been extensively characterized, leading to a detailed understanding of the catalytic and inhibition mechanisms and also to several valuable drugs.[2]

Another CA pathway, less "evident"[3] but no less important, is the activation mechanism that constituted a controversial issue immediately after it was first described.[4] Thus, activation of crude human red cell enzyme (a mixture of isozymes CA I and II) by different compounds, such as histamine, amino acids, and some purine derivatives, has been reported[5] without arriving at a clear-cut answer regarding the mere existence of such class of CA activity modulators. This topic, then, received little attention from the scientific community for at least two reasons: (i) the statement by Clark and Perrin that activators of CA do not exist[6] and (ii) the idea that the reported activation is not a phenomenon per se but an artifact generally due to restoration of CA activity possibly lost in the presence of adventitious metal ions or other impurities (or due to enzyme adsorption at interfaces, or even due to enzyme denaturation followed by renaturation in the presence of activators).[7] Leiner,[3] the researcher who played an important role in discovering this class of modulators of CA activity, observed that the activation was readily detected when working with highly purified enzyme preparations, and this may explain the large discrepancies between the different early studies describing this phenomenon. Recently, the X-ray structures of the complexes between the human isozymes hCA II with different activators corroborated the existence of this class of modulators of enzyme activity as well as elucidated their mechanism of action at the molecular level.[8,9]

22.2 CARBONIC ANHYDRASES ACTIVATION MECHANISM

The binding of CA activators (CAAs) to various isozymes, such as CA I, II, IV, VA, VII, XIII, and XIV, was studied by electronic spectroscopy, kinetic, and X-ray crystallographic techniques. CA activation mechanism can be described as follows:

$$\begin{aligned} EZn^{2+} - OH_2 + A \leftrightarrows [EZn^{2+} - OH_2 \ldots A \leftrightarrows EZn^{2+} - OH^- \ldots AH^+] \\ \leftrightarrows EZn^{2+} - OH^- + AH^+ \text{ enzyme - activator complexes} \end{aligned}$$

In the equation, the activator interferes in the rate-determining step of the catalytic cycle, that is, the shuttling of protons between the active site and the reaction medium, a process that in most CA isoforms is assisted by a histidine residue (His64, CA I numbering) placed in the middle of the active site cavity.[10,11] In the presence of CAAs, there is possibility of alternative proton transfer pathways, involving a protonatable moiety of the activator bound within the enzyme active site, which explains the enhanced overall catalytic efficiency.

CA activation phenomenon at the molecular level was better understood by solving the X-ray structure of adducts of hCA II with histamine **22.1**, L- and D-histidine **22.2**, L- and D-phenylalanine **22.3**, and L-adrenaline **22.4** as well as the adduct of hCA I with L-histidine (Fig. 22.1).[12–16]

As previously reported, the effects of several carbonic anhydrase activators, with different chemical structures, on synaptic plasticity and spatial memory indicate that

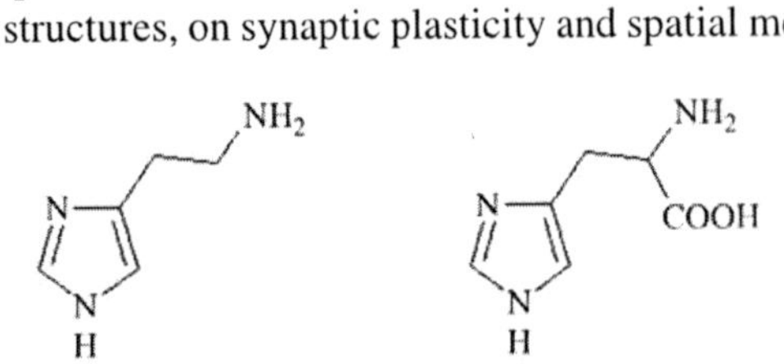

22.1 **22.2**

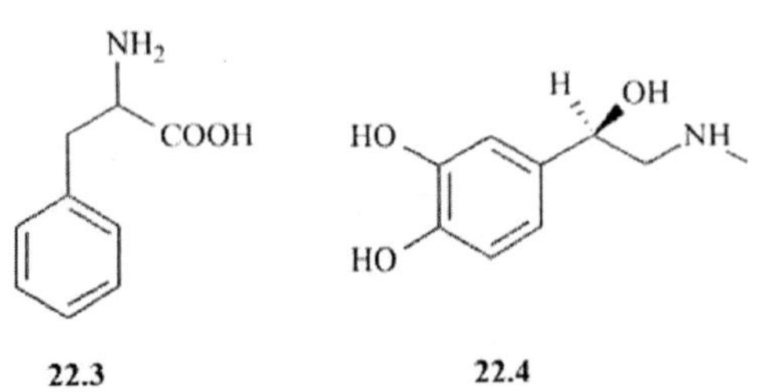

22.3 **22.4**

FIGURE 22.1 Chemical structure of CAA of the amine and amino acid type **22.1–22.4**.

their common action on CA activity provides the underlying mechanism responsible for attention-gated center of learning. In fact, some CAAs (such as phenylalanine and imidazole) administered to experimental animals may produce an important pharmacological enhancement of synaptic efficacy, spatial learning, and memory. The study proves that this class of relatively unexplored enzyme modulators may have pharmacological applications in conditions in which learning and memory are impaired, for example, Alzheimer's disease or aging. It needs to be remembered that the levels of CA are significantly diminished in the brain of patients affected by Alzheimer's disease and these facts strongly support the involvement of different CA isozymes in cognitive functions.[10,11]

22.3 DRUG DESIGN AND MOLECULAR STRUCTURE

Activation studies of CA isoforms I and II with amino acids and related compounds (including natural/synthetic amino acids, their esters, *N*-alkyl, *N*-acyl, or pyridinium derivatives) have been done.[2] It was observed as follows: (i) The most powerful CAAs in this series of compounds were L-histidine, L-proline, and L-homoproline, together with aromatic amino acids, structurally related to L-phenylalanine (e.g., compounds incorporating substituted phenyl rings, such as 4-halogeno-, 4-hydroxy-, or 4-amino-phenyl moieties, among others). In fact, X-ray crystallographic data (for the hCA II–histamine adduct, and the hCA II–azide-phenylalanine ternary complex)[12,13] clearly demonstrated that the aryl/hetaryl moieties present in the activator molecule increase the stability of the enzyme–activator complexes. (ii) Derivatization of the amino or carboxylic groups generally diminished the activator efficiency, mainly due to decrease in the charge on the most electronegative atom of the molecule (through induced electronic effects), as it was later rationalized by means of QSAR calculations.[17,18] However, some *N*-derivatized or carboxy-derivatized amino acid derivatives still showed good CA-activation properties against isozymes hCA I and II. (iii) A strong correlation has been observed between the pK_a value of the activator molecule and its activity, with compounds possessing a pK_a in the range of 6.5–8.0 for at least one deprotonatable moiety, leading to the best CA I and II activating properties. All these data clearly showed that a CAA must possess specific steric and electronic requirements for good activity. First, it must fit within the restricted active site cavity of the enzyme, but should also interact favorably with amino acid residues present in the activator binding pocket. Second, an ideal CAA should possess a moiety able to participate in proton transfer processes, best suited with a pK_a in the range of 6.5–8.0 units.

The first X-ray crystallographic structure of adduct of the main human isoform, hCA II, with histamine, showed that the activator molecule is bound at the entrance of the active site cavity, anchored by hydrogen bonds to three amino acid side chains and to a water molecule. These hydrogen bonds involve only the nitrogen atoms of the imidazole moiety of histamine (the Nδ1 and Nε2 are engaged in hydrogen bonds with the side chains of Asn62, His64, and Gln92 and with Wat152), whereas the aliphatic amino group is not experiencing any contact with the enzyme but is extending away from the cavity into the solvent.[8] Positioned in such a favorable way, histamine facilitates the rate-limiting step of CA catalysis, that is, the proton transfer processes between the active site and the environment,[5] and also allows its easy derivatization (at the aliphatic amino group) to obtain stronger activators.[12–14]

The same activator binding site as the one in the hCA II–histamine adduct was then shown to be occupied by phenylalanine, the second activator for which the X-ray structure has been reported (as a ternary complex, hCA II–Phe–azide).[9] In similar way to phenylalanine, several other aromatic amino acids show good CA activating properties, for example, histidine. This last amino acid (His) was used as a lead molecule for obtaining new types of tight binding CAAs.

22.4 ACTIVATION OF CARBONIC ANHYDRASE I

22.4.1 L-Histidine Complex

The physiological function of hCA I is mainly unknown at this moment, although this enzyme is present in very high amounts in the red blood cells (there is 10 μM of hCA I in this tissue) and in the gastrointestinal tract.[26]

The binding affinity of L-histidine for hCA I is high, with a constant of 30 nM, being on the other hand a weaker hCA II activator, with an affinity constant of around 10 μM.[13,26]

Although the sequence similarity between isozymes hCA I and II is quite high,[6] at least two critical amino acid residues from the active site are different, that is, those at position 200 (His in hCA I and Thr in hCA II), and 67 (His in hCA I again and Asn in hCA II).[12] Indeed, the two bulky histidines (His200 and His67) lead to a highly restricted active site for hCA I as compared to hCA II. This may explain the lower catalytic activity of isoform I, as well as the fact that it is usually 100 times less prone than hCA II to be inhibited by sulfonamides. As His200 is close to the Zn(II) ion, this residue may probably influence the binding of inhibitors, substrates, or activators to the enzyme cavity. In fact, it is well established that the corresponding residue in hCA II (i.e., Thr200) interacts by hydrogen bonds with inhibitors of the sulfamate/sulfamide type, as demonstrated by means of X-ray crystallography earlier. On the other hand, His67 is situated in the middle of the active site cavity, near His64 (conserved in both isoforms hCA I and II). Another amino acid residue maintained in the two isozymes is Thr199, involved in hydrogen bond with the zinc-bound water in all CAs investigated until now. Thus, we expect a quite diverse pattern of interaction between the hCA I active site cavity and the activators, as compared to the corresponding adducts with isozyme hCA II (Fig. 22.2).

The activator binding site of hCA I is rather different from that of hCA II case in which the same activator, L-His, was shown to bind at the entrance of the cavity, being anchored by several strong hydrogen bonds to His64, Asn67, and Gln92. In the case of the hCA I/L-His adduct, it may be observed that the activator is bound much deeper, positioned in a inner task just under the enzyme active site. It bridges two hydrogen bonds, the zinc-bound hydroxide ion and the Nε imidazole atom of His200 (an amino acid residue characteristic for this isozyme, as in hCA II there is a Thr in position

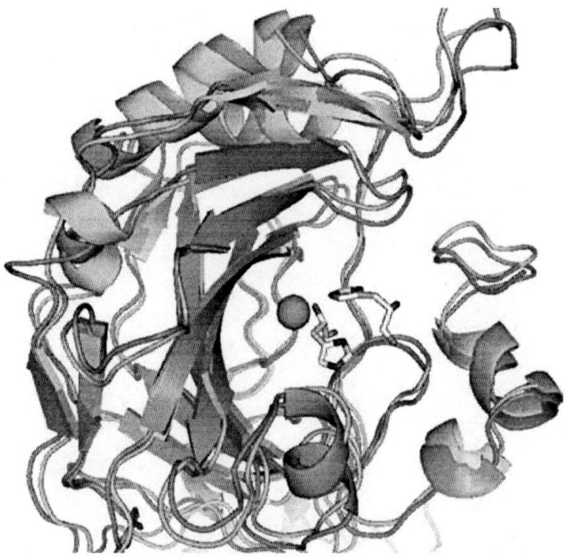

FIGURE 22.2 Superposition of the complexes hCA II/L- His with the hCA I/L-His.

200).[13] The orientation of the activator molecule is quite different in the two complexes, as it is the carboxylate moiety of L-His participating in the main interactions with the hCA I active site (in fact, the COO^- points toward the zinc ion, whereas the imidazole moiety of L-His points toward the Zn(II) ion in the hCA II adduct) and also participating in a network of four hydrogen bonds with three amino acid residues and a water molecule bound within the cavity. In addition, the zinc-bound hydroxide is directly hydrogen bonded to the activator molecule in the hCA I adduct, but is bridged by three different water molecules in the hCA II complex (Fig. 22.3).

This study clearly illustrates that very minor differences in the active site architecture of similar isozymes lead to a completely different pattern of interactions with small molecules acting as activators/inhibitors. It may also lead to design better hCA I activators, useful in patients affected by a genetic syndrome of hCA II deficiency (who possess normal levels of hCA I) or in increasing synaptic efficiency. It is well established[21] that levels of various CA isozymes are diminished in patients affected by Alzheimer's disease or in the aging population.

22.5 ACTIVATION OF CARBONIC ANHYDRASE II

22.5.1 L/D Histidine Complexes

Activation of six CA isoforms, hCA I, II, IV, VA, VII, and XIV, with L- and D-histidine **22.2** has been investigated by kinetic and X-ray crystallographic methods. L-His behaved as a potent activator of isozymes I, VA, VII, and XIV (activation constants in

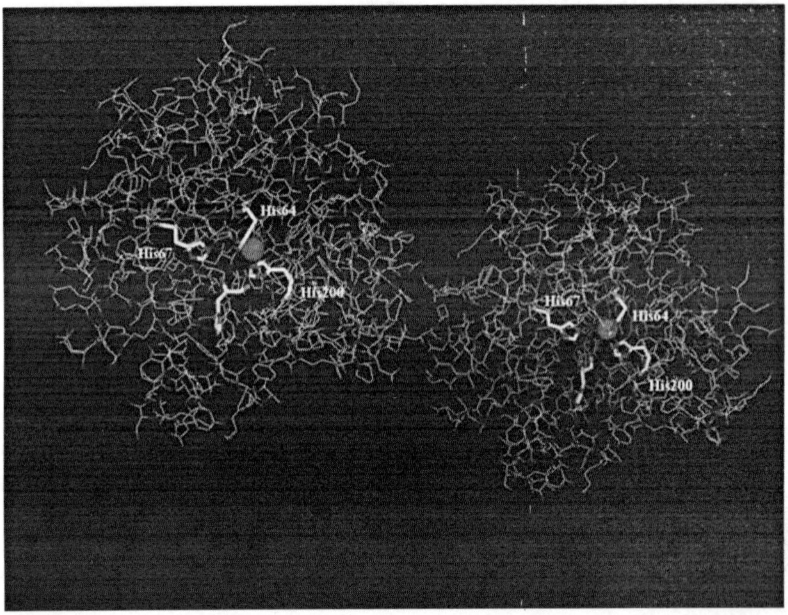

FIGURE 22.3 Overall view of the hCA I/L-His complex. The Zn(II) ion (pink sphere) and the activator molecule (yellow) are shown. (See the color version of this figure in Color Plates section.)

the range of 0.03–1.34 μM) and a weaker activator of isoforms II and IV (K_A in the range of 7.3–10.9 μM). D-His showed good hCA I, VA, and VII activatory properties (K_A in the range of 0.09–0.71 μM), being a moderate hCA XIV (K_A (2.37 μM) and a weak hCA II and IV activator (K_A in the range of 12.3–43.5 μM) (Table 22.1). The

TABLE 22.1 Activation of hCA Isozymes I, II, IV, VA, VII, and XIV with L- and D-Phenylalanine, L- and D-Histidine, and L-Adrenaline at 25°C for the CO_2 Hydration Reaction

Isozyme	k_{cat} [a]	$(k_{cat})_{L\text{-Phe}}$ [b] (s^{-1})	$(k_{cat})_{D\text{-Phe}}$ [b] (s^{-1})	$(k_{cat})_{L\text{-His}}$ [b] (s^{-1})	$(k_{cat})_{D\text{-His}}$ [b] (s^{-1})	$(k_{cat})_{L\text{-Adre}}$ [b] (s^{-1})
hCA I [c]	2.0×10^5	19.8×10^5	2.3×10^5	13.4×10^5	9.1×10^5	10.4×10^5
hCA II [c]	1.4×10^6	5.7×10^6	5.2×10^6	4.3×10^6	2.7×10^6	2.0×10^6
hCA IV [d]	1.2×10^6	1.6×10^6	1.4×10^6	4.3×10^6	3.8×10^6	3.5×10^6
hCA VA [e]	2.9×10^5	4.3×10^5	9.7×10^5	9.8×10^5	12.0×10^5	8.6×10^5
hCA VII [a]	9.5×10^5	14.6×10^5	15.8×10^5	16.7×10^5	15.4×10^5	15.2×10^5
hCA XIV [e]	3.1×10^5	12.5×10^5	6.1×10^5	11.4×10^5	8.5×10^5	10.0×10^5

Standard errors were in the range of 5–10% of the reported values.

[a] Observed catalytic rate without activator.

[b] Observed catalytic rate in the presence of 10 μM activator.

[c] *Human* recombinant isozymes.

[d] Truncated *human* recombinant isozyme lacking the first 20 amino acid residues.

[e] Full-length, *human* recombinant isoforms.

X-ray structure of the hCA II–L-His/D-His complexes showed the activators to be anchored at the entrance of the active site cavity, participating in an extended network of hydrogen bonds with different amino acid residues and water molecules that may explain their different potency as well as interaction patterns with various CA isozymes. The His64, Asn67, and Gln92 amino acid residues interact with L-histidine imidazole moiety, whereas three water molecules connected the activator to the zinc-bound water. For the D-His adduct, the amino acid residues involved in recognition of the activator molecule were Trp5, His64, and Pro201, and two water molecules connected the zinc-bound water to the activator in an extended network of hydrogen bonds (Fig. 22.4). This is the first study showing the very different binding mode for stereoisomeric activators within the CA II active site, with consequences for the overall proton transfer processes that are rate determining for the catalytic cycle. It also points out the differences of activation efficiency between various isozymes with structurally related activators, which may be exploited for designing alternative proton transfer pathways, useful both for a better understanding of the catalytic mechanism as well as for obtaining pharmacologically useful derivatives, mainly for the management of Alzheimer's disease, a condition in which the dysfunction of CA II activity has recently been evidenced.

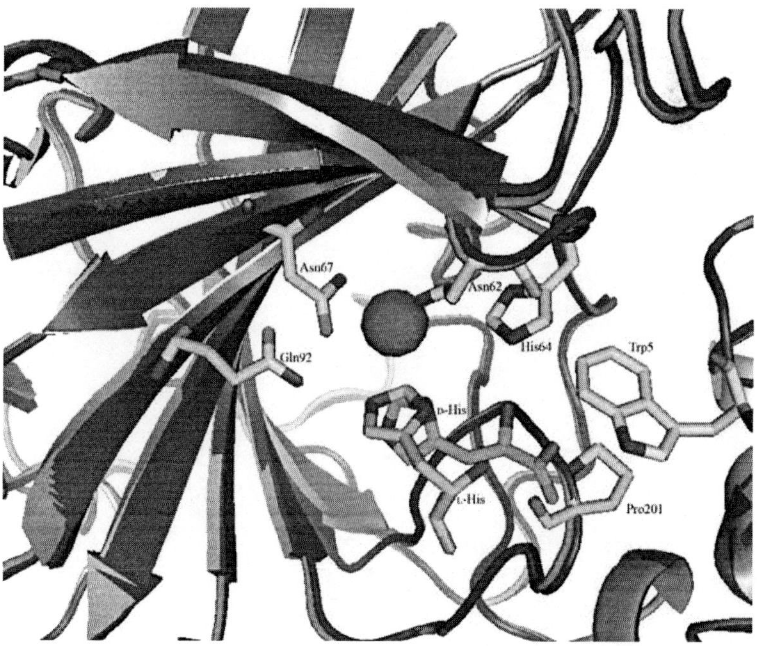

FIGURE 22.4 Superposition of the two hCA II adducts with L/D-His, with the zinc ion and amino acids residues present in the activator binding site. (See the color version of this figure in Color Plates section.)

22.5.2 L/D Phenylalanine Complexes

Recently, redox proteomics studies revealed a decreased catalytic activity of CA I and II in the brain of patients affected by Alzheimer's and Parkinson's diseases. These disorders are generated by a fall of reactions that occur in frontal cortex and hyppocampus.[21–25] Thus, in the brain of these patients, the dysfunctions of CA I and/or CA II activity leads to imbalances of the extra- and intracellular pH, which may trigger aggregation of proteins, thus contributing to progression of the disease. Probably, the agents that may restore to a certain degree the catalytic activity of these particular isozymes (CA I and II) or increase that of other CAs present in the brain (such as isoforms CA IV, VA, VII, and XIV) might lead to conceptually novel approaches for the management of cognitive disorders.[25]

To rationalize the CA activating properties of L-Phe and D-Phe, X-ray crystal structures of their adducts with the physiologically most relevant isoform, that is, hCA II, have been solved, in addition to kinetic studies with all isoforms found in the human brain, hCA I, II, IV, VII, and XIV.

It was observed that the two stereoisomers bind differently to the enzyme. Unlike the two previously investigated activators histamine[12] or histidine[14], in the hCA II/L-Phe and hCA II/D-Phe **22.3** complexes, the pattern of hydrogen bonds and hydrophobic contacts with amino acid residues, present within the activator binding site, are completely different (Fig. 22.5). Thus, in the hCA II/L-Phe adduct, the activator molecule is anchored by its amino group by means of two hydrogen bonds with the indole nitrogen of Trp5, and with one of the imidazolic nitrogen atoms of His64. In fact, some of these amino acid residues belong to the activator binding pocket, and they

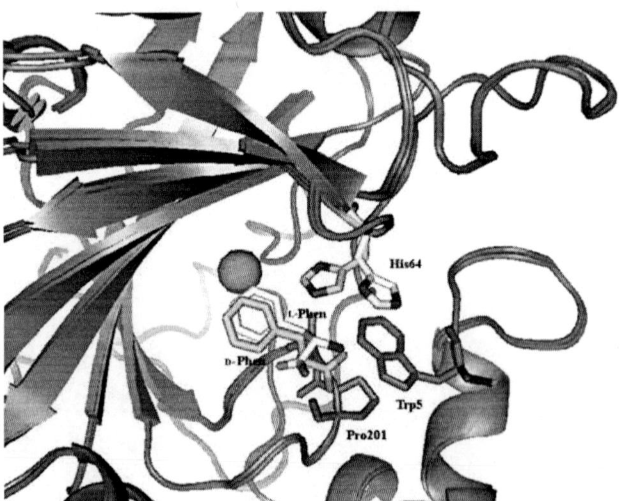

FIGURE 22.5 Superposition of the two hCA II adducts with L/D-Phe, with the zinc ion and amino acids residues present in the activator binding site. His64 is in the "out" conformation in the L-Phe adduct, and in the "in" conformation in the D-Phe adduct.

were shown to be involved in the binding of the other activators investigated earlier. For the hCA II/D-Phe adduct, the same four amino acid residues interacted with the activator but in a completely different manner as compared to the hCA II/L-Phe adduct. The amino group of the activator participates in two hydrogen bonds with the carbonyl oxygen of Pro201 and the OH moiety of Thr200, whereas the carboxylate moiety of D-Phe makes a hydrogen bond with the indole NH of Trp5 and a good contact with a nitrogen atom of His64.

It may be observed that although binding in the same region of the active site, the orientation of the two aromatic rings and of the amino and carboxy moieties of the activators is rather different. Thus, the phenyl rings point in both cases toward the inner part of the cavity, but they cannot be superposed. The same may be noticed regarding the carboxylate and amino moieties, which point toward different parts of the active site. These data clearly show that the stereospecific recognition within the active site of hCA II is responsible for the activation efficacy of different compounds and may also lead to novel directions in the drug design of more efficient and also isozyme-specific activators.

22.5.3 L-Adrenaline Complex

L-Adrenaline (epinephrine) **22.4**, one of the neurotransmitter catecholamines released by the sympathetic nervous system and adrenal medulla, is involved in the regulation of blood pressure, vasoconstriction, cardiac stimulation, relaxation of the smooth muscles (such as the bronchial ones), and several metabolic processes.[28] As a consequence, **22.4** has a variety of clinical uses, such as among others, for relieving respiratory distress in asthma, in treating hypersensitivity reactions due to various allergens, cardiac arrest, or as a topical hemostatic agent. The study of adrenaline activating effects on CA II (of bovine origin, bCA II)[29] showed that the compound is a weaker CAA as compared to histamine, aromatic/heterocyclic amino acids, or other structurally related amines investigated in the same study. However, since adrenaline is such an important endogenous compound, and its concentrations in blood or other tissues seem to be rather high, in the range of 2–5 μM, we decided to investigate in more detail its interaction with various physiologically relevant CA isozymes,[2] such as CA I, II, IV, VA, VII, and XIV (all of them present among others in the brain),[30,31] mentioned subsequently.

X-ray studies show that the activator molecule binds at the entrance of the cavity, interacting with amino acid residues and water molecules that stabilize its binding to the enzyme. It should be stressed that the side chain of His64, an amino acid residue extremely important in the CA catalytic cycle, was observed with both its two characteristic conformations, the 'in' and 'out' ones, although in other CA–activator adducts investigated earlier, His64 adopted only the out conformation.[32,33] Thus, L-adrenaline participates in an extended network of hydrogen bonds involving five water molecules and several amino acid residues, when bound to the hCA II active site.

In contrast to other activators, L-adrenaline adopts an extended conformation, perpendicular to that of histamine, so that the activator molecule plugs the

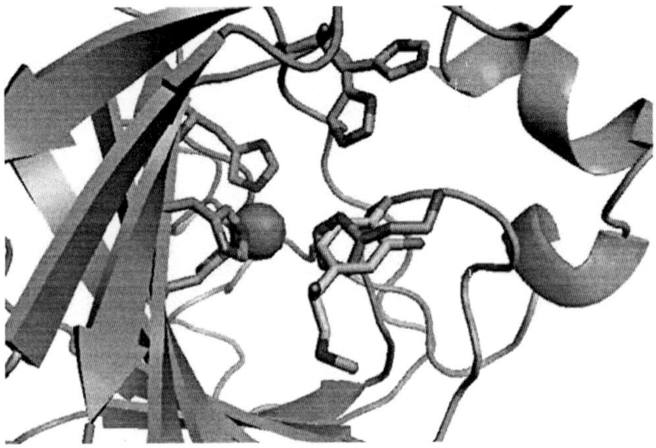

FIGURE 22.6 Superposition of the hCA II–histamine and the hCA II–L-adrenaline complexes. The proton shuttle, His64, is present in both the "in" and the "out" conformations. (See the color version of this figure in Color Plates section.)

entrance of the active site cavity, obstructing it almost completely (Fig. 22.6). In this conformation, it is unable to facilitate the shuttling of protons between the active site and the environment, because the pK_a's of its protonatable moieties are in the range of 8.6–11.34. On the contrary, histamine bound to the enzyme active site adopts a conformation that allows its imidazolic moiety (with a pK_a around 7) to easily participate in proton shuttling, similar to residue His64, the natural proton shuttle amino acid in the CA II active site. These findings thus explain that both the steric requirements (orientation in which the activator binds within the active site) and electronic factors (pK_a of the proton shuttle moiety) are important for a compound to act as an effective CA activator, and may shed light on the recognition processes by metalloenzymes of ligands that do not directly interact with the metal ion.

22.6 ACTIVATION OF HUMAN BRAIN CARBONIC ANHYDRASE ISOFORMS

Among the 16 CA isoforms characterized so far in mammals, only 6 isozymes are present in human brain: the cytosolic forms, hCA I and II, which are ubiquitous in the human body and quite abundant in the brain, and isoform VII, mainly restricted to the brain.[6] hCA IV and hCA XIV have an extracellular active site, being membrane-associated (CA IV) or transmembrane (CA XIV) isoforms, quite abundant in the brain (and other tissues too)[6,7] where they seem to play important physiological roles in pH homeostasis.[19–21] Finally, CA VA is a mitochondrial isozyme involved in several biosynthetic processes such as gluconeogenesis, ureagenesis, and lipogenesis among others.[22,23]

The interaction of the activator molecules with the physiologically relevant human brain CA isozymes, that is, hCA I, II, IV, VA, VII, and XIV, was investigated by kinetic technique. Data of Table 22.1 show the activation constants of all isoforms investigated with L'D-histidine, L'D-phenilalanine, and L-adrenaline activators. It should also be noted that there are two types of CA isoforms from catalytic point of view: the low activity ones (CA I-like), including hCA I, VA, and XIV (k_{cat} values in the range of 2.0–3.1 $\times$ 10^5 s^{-1}), and the high activity ones (CA II-like), among which are hCA II, IV, and VII (k_{cat} values in the range of 0.95–1.4 $\times$ 10^6 s^{-1}). It must be stressed that the activators had no influence on the K_M values, as the Michaelis–Menten constants were identical with or without activator, but a very strong influence has been observed on k_{cat}. This parameter is generally greatly enhanced in the presence of activators, proving that it is just the rate-determining step, that is, the proton transfer between the active site and the reaction medium, which is favored by the activators, as already proved in the first CA activation study on histamine.[16] These data clearly show at least two important facts: (i) There are net differences of activating efficacy of the two stereoisomers against various brain CAs. Some isoforms, such as CA I, II, IV, and XIV, are better activated by the L-amino acid, whereas others (such as CA VA and VII) show a better activation profile with D-amino acid. This fact is quite important as it points out that on one hand diverse CA isozymes probably possess different activator binding sites, and on the other hand, such structural differences may lead to the discovery (or the development) of isozyme-selective CAAs. (ii) By selecting different CA isozymes and diverse CAAs possessing different stereochemistry among the many classes of such derivatives already reported,[1,2,5,6,22] it is possible to engineer proton transfer processes between the active site of the enzyme and the reaction medium, enhancing the catalytic efficiency of a particular isoform. Such phenomena are critical for better understanding the catalytic mechanisms of enzymes for which the rate-determining step is a proton transfer reaction (as the CAs) as well as, in this particular case, for the drug design of more efficient and possibly isozyme-specific CAAs, some of which may show biomedical applications for the management of CA deficiencies, Alzheimer's disease, or aging, among others.

22.7 CONCLUSIONS

Carbonic anhydrase has been a therapeutic target for many years, as gauged by the considerable effort focused on developing high affinity inhibitors of CA and the more recent activators of CA.

The X-ray structure of the complexes between CA and activators show a different binding modes of stereoisomeric molecules within hCA II/hCA I active site, with consequences for overall proton transfer processes.

The kinetic studies point out the differences of activation efficiency between various isozymes with structurally related activators, which may be exploited for designing alternative proton transfer pathways, useful both for a better understanding of the catalytic mechanism as well as for obtaining pharmacologically useful derivatives.

REFERENCES

1. Mann, T.; Keilin, D. Sulphanilamide as a specific inhibitor of carbonic anhydrase. *Nature* **1940**, *146*, 164–165.
2. (a) Supuran, C. T.; Scozzafava, A. Carbonic anhydrase inhibitors. *Curr. Med. Chem. Imm. Endoc. Metab. Agents* **2001**, *1*, 61–97. (b) Supuran, C. T.; Scozzafava, A. Carbonic anhydrase inhibitors and their therapeutic potential. *Expert Opin. Ther. Pat.* **2000**, *10*, 575–600.
3. (a) Leiner, M. Das Ferment Kohlensäureanhydrase im Tierkörper. *Naturwissenschaften* **1940**, *28*, 316–317. (b) Leiner, M.; Leiner, G. Die Aktivatoren der Kohlensäureanhydrase. *Naturwissenschaften* **1941**, *29*, 195–197. (c) Leiner, M.; Leiner, G. Die Messmethoden zur Untersuchung der katalytischen Wirksamkeit der Kohlensäureanhydrase. *Biochem. Z.* **1941**, *311*, 119–145.
4. (a) Kiese, M. Die Aktivierung der Kohlensäureanhydrase. *Naturwissenschaften* **1941**, *29*, 116–117. (b) van Goor, H. Carbonic anhydrase: its properties, distribution and significance for carbon dioxide transport. *Enzymologia* **1948**, *13*, 73–164.
5. Supuran, C. T.; Scozzafava, A. Activation of carbonic anhydrase isozymes. In *The Carbonic Anhydrases: New Horizons;* Chegwidden, W. R.; Carter, N.; Edwards, Y., Eds.; Birkhauser Verlag: Basel, Switzerland, **2000**; pp 197–219.
6. Clark, A. M.; Perrin, D. D. A reinvestigation of the question of activators of carbonic anhydrase. *Biochem. J.* **1951**, *48*, 495–503.
7. (a) Roughton, F. J. W.; Booth, H. The effect of substrate concentration, pH and other factors upon the activity of carbonic anhydrase. *Biochem. J.* **1946**, *40*, 319–329. (b) Maren, T. H. Carbonic anhydrase: chemistry, physiology and inhibition. *Physiol. Rev.* **1967**, *47*, 595–781.
8. Lindskog, S. Structure and mechanism of carbonic anhydrase. *Pharmacol. Ther.* **1997**, *74*, 1–20.
9. Sly, W. S.; Hu, P. Y. Human carbonic anhydrases and carbonic anhydrase deficiencies. *Annu. Rev. Biochem.* **1995**, *64*, 375–401.
10. Stams, T.; Christianson, D. W. X-ray crystallographic studies of mammalian carbonic anhydrase isozymes. In *The Carbonic Anhydrases: New Horizons;* Chegwidden, W. R.; Edwards, Y.; Carter, N., Eds.; Birkhäuser Verlag: Basel, Switzerland, **2000**; pp 159–174.
11. Meier-Ruge, W.; Iwangoff, P.; Reichlmeier, K. Neurochemical enzyme changes in Alzheimer's and Pick's disease. *Arch. Gerontol. Geriatr.* **1984**, *3*, 161–165.
12. Briganti, F.; Mangani, S.; Orioli, P.; Scozzafava, A.; Vernaglione, G.; Supuran, C. T. Carbonic anhydrase activators: X-ray crystallographic and spectroscopic investigations for the interaction of isozymes I and II with histamine. *Biochemistry* **1997**, *36*, 10384–10392.
13. Briganti, F.; Iaconi, V.; Mangani, S.; Orioli, P.; Scozzafava, A.; Vernaglione, G.; Supuran, C. T. A ternary complex of carbonic anhydrase: X-ray crystallographic structure of the adduct of human carbonic anhydrase II with the activator phenylalanine and the inhibitor azide. *Inorg. Chim. Acta* **1998**, *275–276*, 295–300.
14. Temperini, C.; Scozzafava, A.; Puccetti, L.; Supuran, C. T. Carbonic anhydrase activators: X-ray crystal structure of the adduct of human isozyme II with L-histidine as a platform for the design of stronger activators. *Bioorg. Med. Chem. Lett.* **2005**, *15*, 5136–5141.
15. Temperini, C.; Scozzafava, A.; Vullo, D.; Supuran, C. T. Carbonic anhydrase activators. Activation of isoforms I, II, IV, VA, VII, and XIV with L- and D-phenylalanine and

crystallographic analysis of their adducts with isozyme II: stereospecific recognition within the active site of an enzyme and its consequences for the drug design. *J. Med. Chem.* **2006**, *49*, 3019–3027.

16. Temperini, C.; Scozzafava, A.; Vullo, D.; Supuran, C. T. Carbonic anhydrase activators. Activation of isozymes I, II, IV, VA, VII, and XIV with L- and D-histidine and crystallographic analysis of their adducts with isoform II: engineering proton-transfer processes within the active site of an enzyme. *Chemistry* **2006**, *12*, 7057–7066.
17. Temperini, C.; Innocenti, A.; Scozzafava, A.; Mastrolorenzo, A.; Supuran, C. T. Carbonic anhydrase activators: L-adrenaline plugs the active site entrance of isozyme II, activating better isoforms I, IV, VA, VII, and XIV. *Bioorg. Med. Chem. Lett.* **2007**, *17*, 628–635.
18. Clare, B. W.; Supuran, C. T. Carbonic anhydrase activators. Part 3. Structure–activity correlations for a series of isozyme II activators. *J. Pharm. Sci.* **1994**, *83*, 768–779.
19. Sun, M. K.; Alkon, D. L. Pharmacological enhancement of synaptic efficacy, spatial learning and memory through carbonic anhydrase activation in rats. *J. Pharmacol. Exp. Ther.* **2001**, *29*, 961–967.
20. Sun, M. K.; Alkon, D. L. Carbonic anhydrase gating of attention: memory therapy and enhancement. *Trends Pharmacol. Sci.* **2002**, *23*, 83–89.
21. Korolainen, M. A.; Goldsteins, G.; Nyman, T. A.; Alafuzoff, I.; Koistinaho, J.; Pirttila, T. Oxidative modification of proteins in the frontal cortex of Alzheimer's disease brain. *Neurobiol. Aging* **2006**, *27*, 42–53.
22. Sultana, R.; Poon, H. F.; Cai, J.; Pierce, W. M.; Merchant, M.; Klein, J. B.; Markesbery, W. R.; Butterfield, D. A. Identification of nitrated proteins in Alzheimer's disease brain using a redox proteomics approach. *Neurobiol. Dis.* **2006**, *22*, 76–87.
23. Poon, H. F.; Frasier, M.; Shreve, N.; Calabrese, V.; Wolozin, B.; Butterfield, D. A. Mitochondrial associated metabolic proteins are selectively oxidized in A30P α-synuclein transgenic mice: a model of familial Parkinson's disease. *Neurobiol. Dis.* **2005**, *18*, 492–498.
24. Izumi, M.; Yamazaki, H.; Nakabayashi, H.; Owada, M. Magnetic resonance imaging of the brain in phenylketonuria. *No To Hattatsu* **2006**, *38*, 27–31.
25. Parkkila, S.; Parkkila, A. K.; Kivela, J. Role of carbonic anhydrase and its inhibitors in biological science related to gastroenterology, neurology and nephrology. In *Carbonic Anhydrase: Its Inhibitors and Activators;* Supuran, C. T.; Scozzafava, A.; Conway, J., Eds.; CRC Press: Boca Raton, FL, **2004**; pp 283–302.
26. Khalifah, R. G. The carbon dioxide hydration activity of carbonic anhydrase. I. Stop-flow kinetic studies on the native human isoenzymes B and C. *J. Biol. Chem.* **1971**, *246*, 2561–2573.
27. Hoffman, B. B.; Lefkowitz, R. J. Catecholamines, sympathomimetic drugs, and adrenergic receptor antagonists. In *Goodman & Gilman's The Pharmacological Basis of Therapeutics*, Hardman, J. G.; Limbird, L. E.; Molinoff, P. B.; Ruddon, R. W.; Goodman Gilman, A., 9th ed.; McGraw-Hill: New York, **1996**; p 199.
28. Scozzafava, A.; Mastrolorenzo, A.; Supuran, C. T. Modulation of carbonic anhydrase activity and its applications in therapy. *Expert Opin. Ther. Pat.* **2004**, *14*, 667–702.
29. Supuran, C. T.; Scozzafava, A. Carbonic anhydrase activators as potential anti-Alzheimer's disease agents. In *Protein Misfolding in Neurodegenerative Diseases: Mechanisms and Therapeutic Strategies;* Smith, H. J; Simons, C.; Sewell, R. D. E., Eds.; CRC Press: Boca Raton, FL, **2008**; pp 265–288.

30. Innocenti, A.; Firnges, M. A.; Antel, J.; Wurl, M.; Scozzafava, A.; Supuran, C. T. Carbonic anhydrase inhibitors. Inhibition of the membrane-bound human and bovine isozymes IV with sulfonamides. *Bioorg. Med. Chem. Lett.* **2005**, *15*, 1149–1154.

31. Nair, S. K.; Christianson, D. W. Unexpected pH-dependent conformation of His-64, the proton shuttle of carbonic anhydrase II. *J. Am. Chem. Soc.* **1991**, *113*, 9455–9458.

32. Christianson, D. W.; Fierke, C. A. Carbonic anhydrase: evolution of the zinc binding site by nature and by design. *Acc. Chem. Res.* **1996**, *29*, 331–339.

33. Igbo, I. N.; Reigel, C. E., Jr.; Greene, I. M.; Kenny, A. D. Effect of reserpine pretreatment on avian erythrocyte carbonic anhydrase activation by isoproterenol. *Pharmacology* **1994**, *49*, 112–120.

PART III

DRUG DESIGN OF MATRIX METALLOPROTEINASE INHIBITORS

CHAPTER 23

Matrix Metalloproteinases: An Overview

HIDEAKI NAGASE and ROBERT VISSE

Department of Matrix Biology, The Kennedy Institute of Rheumatology Division, Imperial College London, London W6 8LH, UK

23.1 INTRODUCTION

Matrix metalloproteinases (MMPs), also called matrixins, are a group of structurally related zinc metalloendopeptidases. They are either secreted from the cell or bound to plasma membranes. Matrixins are found in vertebrates, fruit fly (*Drosophila melanogaster*), nematodes (*Caenorhabditis elegans*), hydra (*Hydra vulgaris*), and plants (*Arabidopsis*). The human genome has 24 MMP genes, one of which is duplicate, and therefore 23 different MMP proteins are encoded. There are 2 genes in *Drosophila*, 6 genes in *C. elegans*, 7 genes in *Ciona*, 26 genes in zebrafish (*Danio revio*). They are classified as the matrixin subfamily of metalloproteinase family M10 in the MEROPS database (http://www.merops.sanger.ac.uk).

Most MMPs digest extracellular matrix (ECM) molecules at neutral pH. Therefore, they are considered to play key roles in many biological processes, such as embryonic development, organ morphogenesis, ovulation, embryo implantation, nerve growth, bone remodeling, wound healing, angiogenesis, and apoptosis.[1–3] Under such physiological conditions, the activities of MMPs are regulated in accordance with matrix synthesis and requirement of cellular environments.[2,3] Cleavage of specific ECM molecules may allow expressing of their cryptic functions. For example, collagenase-cleaved collagen I fragments promote keratinocyte migration in skin[4] and osteocleast activation in bone.[5] More recent studies have also shown that MMPs have broader activities by acting on many non-ECM molecules; for example, they can activate or inactivate cytokines and chemokines, release growth factors from the matrix or binding proteins, and shed cell surface molecules.[6,7] Many cells have the ability to synthesize MMPs, but their production is tightly regulated by growth factors, cytokines, physical stress to the cell, oncogenic transformation, hormones, cell–cell and cell–ECM interactions, and such regulation is cell type dependent.[3] Aberrant activities of MMPs are often associated with the progression of diseases such as cancer,[3,8] arthritis,[9–11] cardiovascular disease,[12] neurodegenerative diseases,[13] periodontal disease, skin ulceration, gastric ulcer, corneal ulceration, liver fibrosis, emphysema, fibrotic lung, and many others.[1]

In this chapter, we describe structural features and functions of human MMPs and their endogenous inhibitors TIMPs. We also discuss newer approaches to regulate activities of MMPs.

23.2 THE MATRIXIN FAMILY MEMBERS

23.2.1 Earlier Members and Nomenclature

Collagens are major components of skin, tendon, bone, cartilage, blood vessels, and basement membranes. They consist of three α-chains that intertwine with each other to form triple helices. This triple helical structure, particularly those of interstitial collagen types I, II, and III, makes them resistant to the action of most proteinases. Vertebrate collagenase was first discovered by Gross and Lapiere in tadpole tails, skins, and gills undergoing metamorphosis.[14] This discovery stimulated many investigators to seek for collagenolytic activity in mammalian tissues, since collagen breakdown is considered to be essential in normal tissue repair and remodeling and in diseases such as skin ulceration, arthritis, cancer, and periodontal disease. The first human collagenase purified was from rheumatoid synovium.[15] As tadpole collagenase, it cleaved native type I collagen into 3/4 and 1/4 fragments. It also exhibited weak activity on gelatin (heat-denatured collagen), but it was distinguished from gelatinase, a metalloproteinase that readily degrades gelatin but not native collagen. A third metalloproteinase was also isolated from human rheumatoid synovial cells in culture.[16] The enzyme had a broader proteolytic activity than collagenase and it degraded a number of ECM molecules such as proteoglycans, fibronectin, laminin, and type IV collagen, but it failed to cleave fibrillar type I collagen. It was called "matrix metalloproteinase 3" and abbreviated as "MMP-3," referring to collagenase as "MMP-1" and gelatinase as "MMP-2" produced by the same cell system. The enzymatic properties of human MMP-3 were similar to those of metalloproteinases called "proteoglycanase"[17] from rabbit bone, and "procollagenase activator"[18] and "stromelysin"[19] from rabbit synovial fibroblasts. cDNA cloning of these metalloproteinases proved that they are the orthologues.[20]

Following the MMP numbering system introduced by Okada et al.,[16] metallotelopeptidase,[21] a metalloproteinase that degraded the 3/4 fragment of type I and II collagens generated by collagenase,[22] and acid metalloproteinase[23] were designated as MMP-4, MMP-5, and MMP-6, respectively. However, further biochemical studies indicated that acid metalloproteinase was MMP-3,[24] 3/4 collagen fragment degrading metalloproteinase was MMP-2, and telopeptidase was most likely MMP-3. Another metalloproteinase was discovered by Woessner and Taplin[25] in postpartum rat uterus.

This enzyme had proteolytic activity on ECM substrates similar to MMP-3 but was distinguished from it because its zymogen was smaller in molecular mass than the zymogen of MMP-3 (proMMP-3). The enzyme was designated as MMP-7 and later called "matrilysin 1."

Definitive identification of MMPs relied on cDNA cloning. When Goldberg et al.[26] elucidated the first primary structure of MMP-1 (fibroblast collagenase), it was found to be homologous to an epidermal growth factor-induced rat gene product called transin whose function was not known at that time.[27] Soon after transin was shown to have proteolytic activity[28] and cDNA cloning of human MMP-3 indicated that it is a rat orthologue of MMP-3.[29] Biochemical characterization and cDNA cloning indicated that collagenase and gelatinase from neutrophils are genetically different from those secreted from fibroblasts. Neutrophil collagenase was designated as MMP-8 and neutrophil gelatinase as MMP-9.

The primary structures of several MMPs deduced by cDNA cloning revealed a number of common features. They encode a signal peptide, a propeptide domain, a catalytic metalloproteinase domain, a hinge (linker) region, and a hemopexin-like (Hpx) domain. The propeptides have the so-called "cysteine switch" sequence PRCGXPD, and the catalytic domains have the predicted zinc ion binding motif HEXXHXXGXXH, where three histidine coordinate to Zn^{2+}. Thus, MMPs are synthesized as preproproteins and secreted from the cell as a zymogen form (proMMPs). ProMMPs are activated by the proteolytic removal of the propeptide, and one of the unique properties is that many proMMPs are activated by nonproteolytic agents such as mercurial compounds (e.g., $HgCl_2$, 4-aminophenylmercuric acetate), chaotropic agents, and sodium dodecyl sulphate. These common features are used to distinguish MMPs from other metalloendopeptidases.

In addition to the zinc binding motif HEXGHXXGXXH in the catalytic domain, there is a methionine conserved in all MMPs; eight residues downstream it forms a "Met-turn" and supports the active site structure around the catalytic Zn^{2+}. The zinc binding motif and the conserved Met-turn are also found in ADAM (a disintegrin and metalloproteinase), ADAMTS (ADAM with thrombospondin motifs), astacin, pappalysin, and bacterial serralysin metalloproteinase families. These zinc metalloproteinases are collectively called "Metzincins,"[30] but what distinguishes MMPs from other metzincins is the primary structure of the catalytic domain and the composition of noncatalytic ancillary domains.

There were two naming systems for the MMP family members: one is to use trivial names such as collagenase 1, gelatinase A, and stromelysin 1 and the other is the MMP numbering system.[31] The MMPs are grouped into collagenases, gelatinases, stromelysins, matrilysins, membrane-type MMPs, and others based on domain organization and substrate preference. This grouping is somewhat historical since more recently discovered members are grouped into others even though they have structural similarity to the members of the above subgroups.

Huxley-Jones et al.[32] grouped MMPs into six subgroups based on phylogenetic relationship: subgroup A, MMP-19, MMP-26, and MMP-28; subgroup B, MMP-11, MMP-21, and MMP-23; subgroup C, MMP-17 and MMP-25, which are glycosylphosphatidylinositol (GPI)-anchored MMPs; subgroup D, MMP-1, MMP-3, MMP-8,

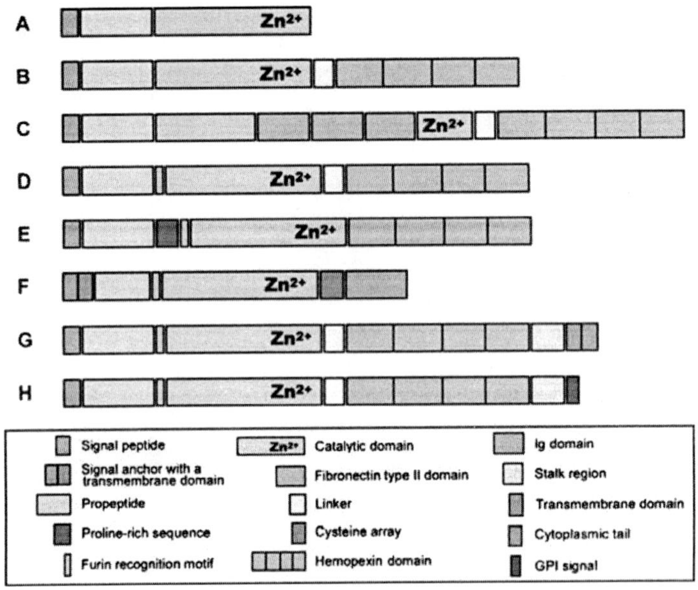

FIGURE 23.1 Domain structure of the matrix metalloproteinase family. See Table 23.1 for the domain arrangements of specific MMPs.

MMP-10, MMP-12, MMP-13, and MMP-27 whose genes are all clustered in chromosome 11q21–24; subgroup E, MMP-14, MMP-15, MMP-16, and MMP-24, which are type I membrane-type MMPs (MT-MMPs); and subgroup F, MMP-2, MMP-7, MMP-9, and MMP-20.

In his chapter, we use the conventional grouping and discuss their biochemical properties (Fig. 23.1 and Table 23.1).

23.2.2 Collagenases

MMP-1 (collagenases 1, fibroblast collagenase), MMP-8 (collagenase 2, neutrophil collagenase), and MMP-13 (collagenase 3) have a common activity and cleave interstitial fibrillar collagens into the N-terminal 3/4 and C-terminal 1/4 fragments. They also have activities against other ECM molecules and soluble proteins such as aggrecan core protein, fibronectin and α1-proteinase inhibitor. The catalytic domain of collagenases alone are capable of cleaving noncollagenous proteins, but not native triple helical collagens. The latter activity requires the linker and Hpx domains. Structural alteration of the linker region significantly reduces collagenolytic activity,[33,34] suggesting that the cooperation between the catalytic domain and the Hpx domain is important for collagenolytic activity.

MMP-2 (gelatinase A)[35] and MMP-14 (MT1-MMP)[36] also digest collagen in a similar manner, but they have different domain organizations, and therefore they are grouped into gelatinase and MT-MMP subgroups, respectively.

TABLE 23.1 Mammalian MMPs

Enzyme	MMP	Domain Composition	Human Chromosome
Soluble-type			
Collagenases			
Interstitial collagenase (collagenase 1)	MMP-1	B	11q22–23
Neutrophil collagenase (collagenase 2)	MMP-8	B	11q21–22
Collagenase 3	MMP-13	B	11q22.3
Collagenase 4 (*Xenopus*)	MMP-18	B	
Gelatinases			
Gelatinase A	MMP-2	C	16q13
Gelatinase B	MMP-9	C	20q12.2–13.1
Stromelysins			
Stromelysin 1	MMP-3	B	11q23
Stromelysin 2	MMP-10	B	11q22.3–23
Stromelysin 3	MMP-11	D	11q11.2
Matrilysins			
Matrilysin 1	MMP-7	A	11q21–22
Matrilysin 2	MMP-26	A	11p15
Others			
Metalloelastase	MMP-12	B	11q22.2–22.3
(No name)	MMP-19	B	12q14
Enamelysin	MMP-20	B	11q22.3
(No name)	MMP-21	E	10
CA-MMP	MMP-23[a]	F	1p36.3
(No name)	MMP-27	B	11q24
Epilysin	MMP-28	D	17q21.1
Membrane-types			
Transmembrane			
MT1-MMP	MMP-14	G	14q11–12
MT2-MMP	MMP-15	G	15q13-q21
MT3-MMP	MMP-16	G	15q13q21
MT5-MMP	MMP-24	G	20q11.2
GPI anchored			
MT4-MMP	MMP-17	H	12q24.3
MT6-MMP	MMP-25	H	16p13.3

[a] Two identical genes are found in head-to-head arrangement in chromosome 1.

23.2.3 Gelatinases

MMP-2 (gelatinase A) and MMP-9 (gelatinase B) are in this subgroup. The unique structural feature of the two gelatinases is that they have three repeats of fibronectin type II motif attached to the catalytic metalloproteinase domain. These repeats interact with gelatins, collagens, and laminins. They degrade denatured collagens (gelatins) most effectively among MMPs. They also digest other ECM components including collagen IV, V, and XI and aggrecan core protein. MMP-2 digests native collagens I, II, and III into 3/4 and 1/4 fragments like collagenases, but MMP-9 does not. Intracellular

MMP-2 was found in cardiomyocytes, and it digests troponin I,[37] myosin light chain,[38] and poly(ADP-ribose) polymerase,[39] which may contribute to cardiac dysfunction. While proMMP-9 is activated by a number of proteinases including MMP-3, the activators of proMMP-2 are MT-MMPs, especially MT1-MMP, which is considered to be a major activator on the cell surface.

23.2.4 Stromelysins

MMP-3 (stromelysin 1), MMP-10 (stromelysin 2), and MMP-11 (stromelysin 3) are assigned to this subgroup. MMP-3 and MMP-10 have a domain arrangement similar to that of collagenases, but they do not cleave fibrillar collagens. Both enzymes have similar substrate specificity and digest a number of ECM molecules and participate in proMMP activation. However, MMP-11 is distantly related, as seen from evolutionary classification.[32] MMP-11 has only weak activity toward ECM molecules.[40] While MMP-3 and MMP-10 are secreted from the cells as inactive zymogens, MMP-11 is activated intracellularly by furin and secreted from the cell as an active enzyme.[41] An alternatively spliced intracellular form of MMP-11 that lacks the signal peptide and the prodomain has been reported, but its function is not known.[42]

23.2.5 Matrilysins

MMP-7 (matrilysins 1) and MMP-26 (matrilysins 2) belong to this subgroup. The main feature of the two enzymes is that they consist of only a prodomain and a catalytic domain. MMP-7 is produced in epithelial cells and secreted apically. *In vitro* studies have shown that it digests ECM molecules, but more physiological activity is to activate pro-α-defensin and to release cell surface molecules such as FAS ligand, tumor necrosis factor-α, syndecan 1, and E-cadherin.[43] MMP-26 is expressed in the endometrium and in some carcinomas, and it has an ability to digest several ECM molecules.[44]

23.2.6 MT-MMPs

There are six members that belong to this subgroup. Four of them (MMP-14, MMP-15, MMP-16, and MMP-24) are type I transmembrane proteins with a C-terminal cytoplasmic domain. They are phylogenetically closely related and grouped together.[32] The other two are GPI-anchored proteins (MMP-17 and MMP-25). All six MT-MMPs have a furin recognition sequence RX[R/K]R at the junction of the prodomain and the catalytic domain. Thus, they are likely to be activated intracellularly and active on the cell surface. All MT-MMPs, except MT4-MMP (MMP-17), can activate proMMP-2.[45] Among them, MT1-MMP was the most extensively studied, and it plays a key role in cell migration and matrix invasion and angiogenesis.[46]

MT1-MMP (MMP-14) has collagenolytic activity on collagens I, II, and III,[36,47] and it also activates proMMP-2 and proMMP-13.[48] MT1-MMP null mice exhibit skeletal abnormalities during postnatal development, which is attributed to decreased collagenolysis.[49]

23.2.7 Other MMPs

Seven MMPs are not grouped into the above subgroup categories partially because different trivial names are given to them or characterization of enzymatic activity has not been sufficient. Nevertheless, MMP-12 (metalloelastase), MMP-20 (enamelysin), and MMP-27 genes are located on chromosome 11q22–24, where collagenases, stromelysins, and matrilysins are found (Table 23.1) and have a domain arrangement similar to collagenases and stromelysins, but they do not cleave fibrillar collagens. Thus, they may be assigned to the stromelysin subgroup.

MMP-12 was originally discovered in macrophages, but it is also found in hypertrophic chrondrocytes[50] and osteoclasts.[51] It degrades elastin and a number of ECM proteins. It is essential for macrophage migration.[52]

MMP-19 has a typical MMP domain organization including a hemopexin domain (see Table 23.1 and Fig. 23.1), but it has an unusual latency motif in the propeptide, an additional cysteine in the catalytic domain, and a C-terminal extension after hemopexin domain.[53,54] It is widely expressed in normal human tissues[53] and considered to play a role in tissue remodeling, wound healing, and keratinocyte proliferation and migration by cleaving insulin-like growth factor binding protein 3 (IGFBP-3) and the laminin-5 γ2 chain.[55,56] The expression of MMP-19 is downregulated in invasive carcinomas.[57] It is a negative regulator of early steps of tumor angiogenesis and invasion[58], and is essential for T-cell development and T-cell-mediated immune responses.[59]

MMP-20 (enamelysin) is a tooth-specific MMP expressed in newly formed tooth enamel. It processes amelogenin.[60] MMP-20 null mice have a defective enamel organ morphogenesis owing to the lack of amelogenin processing.[61] Amelogenin imperfecta, a genetic disorder with defective enamel formation, is due to mutations in amelogenin at MMP-20 cleavage sites.[62]

MMP-21 has a domain organization similar to most MMPs with a furin activation sequence, but it lacks the linker region.[63] It is expressed in various fetal and adult tissues, and it is also expressed in basal and squamous cell carcinoma[64], macrophages of granulomatous skin lesions, and fibroblasts in dermatofibromas.[65] It has gelatinolytic activity, but physiological substrates of the enzymes are not known.

MMP-23 has a unique domain organization. It lacks the cysteine switch motif in the propeptide and the Hpx domain. The latter is substituted by a cysteine-rich immunoglobulin-like domain.[66] The propeptide harbors a furin recognition motif and a transmembrane domain at the N-terminus (type II transmembrane protein). Therefore, the enzyme detached from the membrane anchored propeptide by a proprotein convertase.[67] Two identical copies of the MMP-23 gene are located on human chromosome 1p36.3 loci in a head-to-head arrangement. It is expressed predominantly in ovary, testis, and prostate, suggesting a specialized role in the reproductive process,[66] but physiological substrates of the enzyme are not known.

MMP-27 was first cloned from a chicken embryo fibroblast cDNA library[68] and originally designated as MMP-22. The chicken enzyme digests gelatin and casein, and it autodigests, but physiological substrates of both chicken and mammalian MMP-27 are not known. MMP-27 is expressed in B lymphocytes, and the level of its expression increases when treated with anti-IgG/IgM in culture.[69]

MMP-28 (epilysin) has a domain organization similar to MMP-1, but like MMP-11 the propeptide harbors a furin recognition sequence. It is expressed in many tissues such as lung, placenta, heart gastrointestinal tract, and testis.[70] The enzyme expressed in basal keratinocytes of skin is considered to function in wound repair.[71] Overexpression of recombinant MMP-28 in lung adenocarcinoma cells cleaves E-cadherin and activates latent TGF-β complex, which causes irreversible epithelial–mesenchymal transition and upregulates MT1-MMP and MMP-9, and results in collagen invasion.[72] MMP-28 is also elevated in cartilage from patients with osteoarthritis[73] and rheumatoid arthritis.[74]

23.3 THREE-DIMENSIONAL STRUCTURES

In 1994, 3D structures of the catalytic domain of MMP-1 and MMP-8 were solved by X-ray crystallography by several groups.[75–78] Since then numerous structures of the catalytic domains with synthetic inhibitors including MMP-2, MMP-3, MMP-7, MMP-9, MMP-10, MMP-11, MMP-12, MMP-13, MMP-14, and MMP-16 have been solved by both X-ray crystallography and NMR spectroscopy (see Maskos[79] for review). Other structures include proMMP-3 without the linker and the Hpx domain[80], proMMP-9 without the linker and the Hpx domain[81], proMMP-1[82], proMMP-2,[83] the proMMP-2–TIMP-2 complex[84], MMP-3 catalytic domain (cat)–TIMP-1 complex,[85] MMP-14 cat–TIMP-2 complex,[86] and the full-length MMP-1.[87,88] Those studies were useful in understanding the mechanism of catalysis and in designing selective inhibitors. Figure 23.2 shows the crystal structure of proMMP-2–TIMP-2 complex and individual domains found in MMPs. The overall polypeptide folds of prodomains, catalytic domains, and Hpx domains from other MMPs are very similar.

23.3.1 Prodomain

The structures of four prodomains were solved in proMMP-1[88], proMMP-2,[83] proMMP-3,[80] and proMMP-9.[81] They consist of three α-chains and connective loops (Fig. 23.2). The "cysteine switch" PRCGXPD lies in the substrate binding groove of the catalytic domain and forms hydrogen bonds similar to the substrate that were deduced from the peptidic inhibitor binding[89], but the direction to which the cysteine switch segment sits in the groove is opposite to that of a peptide substrate (Fig. 23.3a and b). The SH group of the cysteine interacts with the catalytic Zn^{2+}. Upon activation of proMMP, the Cys–Zn^{2+} interaction is disrupted. This enables Zn^{2+} to bind to H_2O necessary for peptide hydrolysis. Disruption of the Cys–Zn^{2+} interaction can be triggered by the proteolysis of the "bait region" located between helix 1 and helix 2, or structural perturbation of the prodomain by chemicals such as chaotropic agents, SDS, mercurial compounds, or oxidation of the SH group.[90] The activation by reactive oxygen species such as hypochlorous acid (HOCl) generated by myeloperoxidase from phagocytes is important physiologically, as shown for proMMP-7[91] NO for S-nitrosylation of proMMP-9.[92] On the other hand, HOCl may inactivate MMP activity by modifying tryptophan–glycine residue of the catalytic domain through the

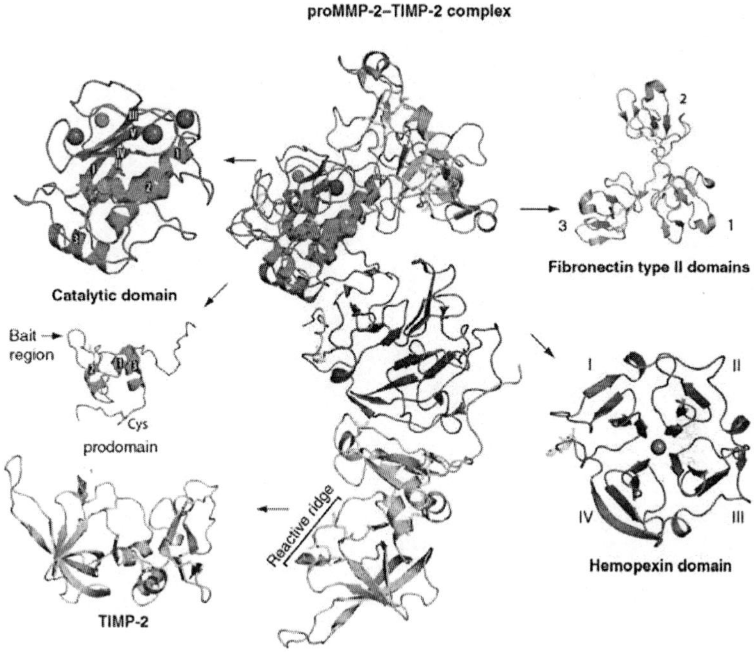

FIGURE 23.2 Three-dimensional structure of the human proMMP-2–TIMP-2 complex. In the center a ribbon presentation of the complex of proMMP-2 with TIMP-2 binding through its C-terminus to the Hpx domain. The structures of TIMP-2, and the individual proMMP-2 domains, the prodomain, the catalytic domain, fibronectin type II domains, and the Hpx domain, are also represented as ribbon structures separately. The figure was prepared with Pymol using the PDB ID: 1GXD, and 1FBL (for ribbon structure of separate cat and Hpx domains). (See color version of this figure in Color Plates section.)

formation of a cyclic indole-amide species, in which tryptophan cross-links to the main chain nitrogen of the adjacent glycine to form an aromatic six-membered ring.[93] The structures of the proteinase susceptible bait region of proMMP-1, proMMP-3, and proMMP-9 were not resolved due to the flexible nature of this region, but in the case of proMMP-2, it was resolved because it is stabilized by a disulfide bond (Fig. 23.2). The propeptide of MMPs are considered to play an important role in folding of the enzyme during synthesis.

23.3.2 Catalytic Domain

The 3D structures of 11 different MMP catalytic domains have indicated that the overall polypeptide folds superimpose well and that all have a five-stranded β-sheet and three α-helices, two zinc ions, and one to three calcium ions (Fig. 23.2). Of the two zinc ions, one is catalytic and the other is structural. The catalytic zinc is coordinated with three histidines in the HEXGHXXGXXH sequence. The structural zinc is coordinated by His149Nε2, Asp151, His164Nε2, and His177Nδ1

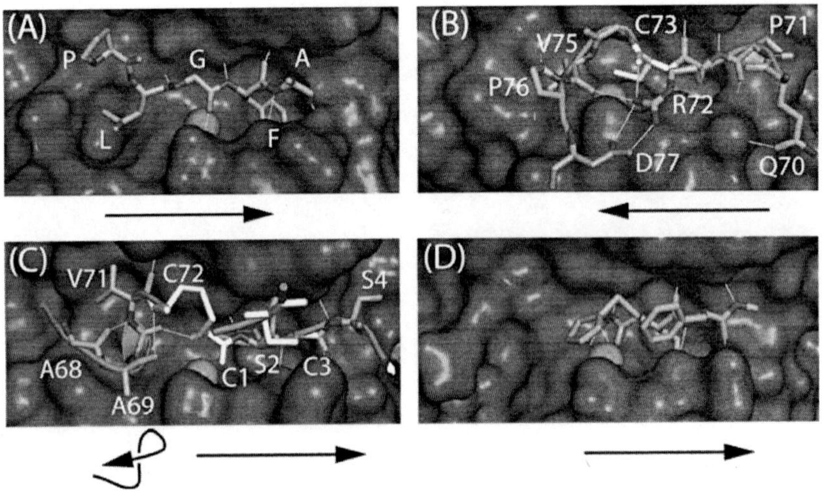

FIGURE 23.3 The active site cleft of matrix metalloproteinases. Interaction of the active site cleft (solid surface with the Zn^{2+} shown as a sphere) with a substrate peptide, the prodomain cysteine switch, a tissue inhibitor of metalloproteinases (TIMP) and a small molecule inhibitor are shown as stick models in A-D respectively. A) The peptide PLGFA docked in the active site of MMP-1 (based on PDB ID: 1FBL). P binds in the P3 pocket, L in the P2 pocket, and F in the S1′ pocket respectively. Hydrogen bonds are shown as thin dotted lines. B) Human proMMP-1 (PDB ID: 1SU3), the prodomain sequence QPRCGVPDVA (70–79) containing the cysteine switch is shown. Note C73 coordinating the active site Zn^{2+}, and the salt bridge between R72 and D77, which stabilises the cysteine switch loop. C) TIMP-2 interacting with MMP-13 (PDB ID: 2E2D). The N-terminal residues C1TCS4 and the CD loop residues SSAVC (68–72) are shown (S68 and S69 are present as A in the structure). D) The small molecule inhibitor BB94 bound in the active site of MMP-12 (PDB ID: 1JK3). Figure prepared with Pymol.

(in MMP-1), and these residues are conserved in MMPs. The substrate binding site cleft is created by β-strand IV, helix 2, and a stretch of random coil adjacent to the COOH-terminus of helix 2. In the orientation presented in Fig. 23.2, a peptide substrate binds to the binding site from left (N-terminal of the substrate) to right (C-terminal of the substrate) of the catalytic domain and the binding is dictated by the structure of the substrate binding site, including a pocket called the "S1′" located to the right of the zinc ion. This pocket is hydrophobic in nature, but variable in depth and shape among MMPs. It is therefore one of the determining factors for substrate specificity of MMPs. Nevertheless, for protein substrate recognition, particularly large extended ECM molecules, substrate binding subsites may extend not only in the catalytic domain but also to sites in the noncatalytic domains, called exosites (see next sections).

23.3.3 Fibronectin Type II Domain

Three repeats of fibronectin type II (Fn II) domain are found in the two gelatinases and these domains are important for their substrate specificity for gelatin, collagen IV, and laminin. They are inserted into the loop between β-strand V and helix 2 of the catalytic domains (Figs 23.1 and 23.2). The module has highly conserved hydrophobic residues and two cystines. The Fn II domains in MMP-2 and MMP-9 have similar conformation. Each domain consists of two antiparallel β-sheets connected with a short α-helix that are stabilized by two disulfide bonds. However, the placements of the three Fn II domains in the two gelatinases are different. After superimposing the catalytic domains of proMMP-2 and proMMP-9, Fn II domain 1 and domain 3 fall roughly in the same places, but the position of domain 2 differs. Domain 2 of proMMP-2 has an area that interacts with the catalytic domain, but the corresponding domain of proMMP-9 is rotated and twisted away from the catalytic domain without making these contacts.[81] Domain 3 in the two progelatinases makes contact with the propeptide and with the catalytic domain.[81,83]

23.3.4 Linker Region

The linker region, also called hinge region, connects the catalytic domain and the Hpx domain but it is variable in length. These regions are considered to be flexible as shown for MMP-9,[94] but there are a number of prolines.[82,83,87,88] Therefore, the linker may have some structural constraint suited for specific functions. Mutation of this region in MMP-1[34] and MMP-8[33] significantly decreased the collagenolytic activity, supporting this notion. It may be postulated that correct movement and rearrangement between the catalytic and Hpx domains is critical for collagenases to express collagenolytic activity. The linker regions of MMP-9 and MMP-14 are N- and O-glycosylated and these sugar chains have significant effects on the cellular biochemistry of these MMPs.[95,96]

23.3.5 Hpx Domain

The Hpx domain is found in most matrixins. It has an ellipsoidal disk shape, and the overall folds of Hpxs from MMP-1, MMP-2, and MMP-13 are very similar, being arranged as a four-bladed β-propeller structure with a single disulfide bond between the first and the fourth blade (Fig. 23.2).[82,87,88,97,98] Each blade is made up of four antiparallel β-strands, and the four β-sheets have similar scaffolds and are arranged almost symmetrically around a central core axis. The center of the propeller generally contains one calcium ion and a chloride. The Hpx domains are essential for collagenases to degrade triple helical collagens and for MT1-MMP to dimerize on the cell surface. The latter event is essential for MT1-MMP to activate proMMP-2[99] and to cleave fibrillar collagens.[47]

23.4 SUBSTRATE SPECIFICITY

The substrate specificities of matrixins have been studied using either proteins or a series of synthetic peptide substrates with systematic replacement of their residue,[100] phage display systems, and proteomic screening.[101] The studies have indicated that most MMPs cleave a peptide bond before a residue (referred as the P1′ residue) with a hydrophobic side chain, such as Leu, Ile, Met, Phe, or Tyr, but the P1 residue located at the left of the scissile bond is not defined.[100] The hydrophobic P1′ residues fit into the S1′ specificity pocket of the enzyme, whose size and shape differ considerably among MMPs.[102] In addition to the S1′ pocket, other substrate binding sites are also important for the substrate specificity.[100] In particular, Pro in the P3 site (the third residue before the scissile bond) is generally favored by most MMPs. It is, however, notable that not only the primary structure but also the local topology of the substrate influences the enzymic activity. For example, heat-denatured collagen I (gelatine I) is a poor substrate of MMP-1 at 37°C compared to native collagen I, but both substrates are equally susceptible to MMP-1 at 25°C. These observations together with that the collagenase locally unwinds triple helical collagens before it cleaves the peptide bonds[103] indicate that structural changes that occur in triple helical collagen upon interacting with collagenase is more favored as a substrate than more extensively denatured gelatin.

23.4.1 Noncatalytic Domains of MMPs in Substrate Specificity

The matrixins are multidomain metalloproteinases, and noncatalytic ancillary domains play important roles for some MMPs to recognize natural ECM substrates. The Fn II domains of gelatinases and the Hpx domains of collagenases and MT1-MMP are prime examples. Deletion of Fn II from gelatinases significantly reduces their activities on type IV collagen, elastin, and gelatin, but it does not affect hydrolysis of small peptides.[104,105] Their interaction with specific ECM molecules probably allows the catalytic site to be oriented to the scissile peptide bond. For collagenases to cleave triple helical collagens, the presence of the Hpx domain is essential. Chung et al.[103] have demonstrated that collagenases locally unwind the triple helical collagen chains before they cleave the peptide bonds. Based on the molecular modeling, it is postulated that the collagen binding groove consists of the catalytic domain and the Hpx domain, and effective unwinding may require cooperative molecular interactions between the two domains and the triple helical substrate. The observation that proMMP-1 does not bind to collagen I, but the activated collagenase does,[106,107] may be explained from the structural changes that take place in MMP-1 upon activation of proMMP-1. In proMMP-1, the prodomain interacts with both the catalytic and hemopexin domains, forming a "closed" configuration[82] compared to an active form of MMP-1[87,88] and the prodomain partially blocks the putative collagen binding site. The removal of the prodomain opens the groove ("open" form). It is thus postulated that the open form of collagenase binds to triple helical collagen. These possibilities, however, need to be structurally investigated.

Another important feature of the Hpx domain is that those of progelatinases tightly bind to endogenous inhibitors, tissue inhibitors of metalloproteinases (TIMPs). TIMP-2 was first discovered as a complex with proMMP-2.[108] Similarly, proMMP-9 was found in the medium as a complex with TIMP-1.[109] These interactions are through the Hpx domain of each zymogen and the C-terminal noninhibitory domain of the TIMP molecule. Therefore, these complexes have MMP inhibitory activity. The complex formation of proMMP-2 and with TIMP-2 (see Fig. 23.2) is particularly important for the activation of proMMP-2 by MT1-MMP.[46] Since the proMMP-2–TIMP-2 complex can inhibit active MMPs, the complex binds to an active MT1-MMP on the cell surface. This interaction presumably orients the propeptide of proMMP-2 to an adjacent active MT1-MMP, and the specific arrangement of the two MT1-MMP molecules is driven by dimerization of the hemopexin domains of MT1-MMP. TIMP-3 and TIMP-4 also form a complex with proMMP-2 in a manner similar to TIMP-2, and the proMMP-2–TIMP-4 complex can bind to MT1-MMP, but it is nonproductive in terms of MMP-2 activation.[110] The biological significance of the proMMP-2–TIMP-3 complex is not known.

The Hpx domain of ProMMP-9 binds to TIMP-1 and TIMP-3, but the biological significance of these complexes is not clear, except that proMMP-9–TIMP complexes inhibit metalloproteinases and activation of proMMP-9 of the complex by MMPs is restricted.

23.5 ENDOGENOUS INHIBITORS OF MMPs

Early studies have indicated that human serum contains anticollagenase activity.[111] It was identified as α_2-macroglobulin.[112,113] Collagenase inhibitor activity was also found in the medium of cultured skin fibroblasts in 1975.[114] In the same year, a small molecular weight serum protein called "β1-anticollagenase" was reported.[115] The extract of bovine cartilage showed the presence of a collagenase inhibitor.[116] In 1979, a collagenase inhibitor of 25–31 kDa was purified from the conditioned medium of cultured human skin fibroblasts[117] and human tendon.[118] It was later called "tissue inhibitor of metalloproteinases," abbreviated as "TIMP," as it inhibited not only collagenase, but also gelatinases and proteoglycanase (MMP-3).[119] It is now designated as TIMP-1 among four TIMPs in humans. TIMPs form a complex with MMPs in a 1 : 1 stoichiometry, and TIMPs are the major endogenous inhibitors of MMPs in the tissue.

Several other proteins have been reported to inhibit selected members of MMPs. A secreted form of β-amyloid precursor protein was reported to inhibit MMP-2.[120] A C-terminal fragment of procollagen C-proteinase enhancer protein inhibits MMP-2,[121] and RECK, a GPI-anchored glycoprotein that suppresses angiogenesis, inhibits MMP-2, MMP-9, and MMP-14.[122] However, the mechanisms of action of these proteins are not known. Tissue factor pathway inhibitor-2, a serine proteinase inhibitor, was reported to inhibit MMP-1 and MMP-2,[123] but this effect could not be confirmed.[124]

23.5.1 α_2-Macroglobin

Human α_2-macroglobin is a glycoprotein of 725 kDa consisting of four identical subunits of 180 kDa. It is a general proteinase inhibitor and inhibits most endopepidases, regardless of their catalytic mechanism. The interaction between a proteinase and α_2-macroglobin is triggered by proteolytic action of the enzyme on the macroglobulin and subsequent entrapment of the proteinase within the macroglobulin.[125] MMP activities in the fluid phase are mainly regulated by α_2-macroglobulin and related proteins. Although interstial collagens are the main substrates of MMP-1 (collagenase 1), α_2-macroglobin is about 150 times better substrate (more rapidly cleaved) than type I collagen for MMP-1.[126]

23.5.2 Inhibition Profiles of TIMPs

TIMPs regulate MMP activities in the tissue. They inhibit all MMPs tested so far with a nanomolar to subnanomolar K_i, although TIMP-1 is a poor inhibitor of MT1-MMP, MT3-MMP, MT5-MMP, and MMP-19.[127] Recent studies have shown that TIMPs can inhibit a broader spectrum of metalloproteinases. For example, TIMP-1 inhibits ADAM-10,[128] while TIMP-2 inhibits ADAM12.[129] TIMP-3 inhibits ADAM10,[128] ADAM12,[129] and ADAM17,[130] as well as ADAMTSs (ADAMTS-1, ADAMTS-2, ADAMTS-4, and ADAMTS-5).[131–133] Possibly because of this broad inhibitory spectrum, TIMP-3 ablation in mice causes a number of disorders including lung emphysema-like alveolar damage[134] and faster apoptosis of mammary epithelial cells after weaning,[135] whereas TIMP-1 null mice and TIMP-2 null mice do not exhibit obvious abnormalities unless those animals are challenged. This indicates that TIMP-3 is a major regulator of metalloproteinase activities *in vivo*.

23.5.3 Inhibition Mechanism of TIMPs and TIMP Variants

The primary structure of TIMP-1 was deduced from cDNA sequencing in 1985.[136] The four TIMPs are homologous to each other and consist of 184–194 amino acids with 12 conserved cysteines. The six disulfide bond arrangements were determined for TIMP-1,[137] which showed that it consists of an N-terminal domain and a C-terminal domain, each containing three disulfide bonds. The N-terminal domain alone is fully functional for MMP inhibition.[138]

The first 3D structure solved for the N-terminal domain of TIMP-2 by NMR spectroscopy was reported in 1994.[139] It revealed of a β-barrel structure with oligosaccharide–oligonucleotide binding (OB) fold and two small α-helices at the C-terminal end, but the mechanism to inhibit MMPs was still an enigma. A possible MMP binding site of TIMP-1 was first proposed to be around Val69–Cys70, based on the biochemical studies that the cleavage of the Val69–Cys70 bond of TIMP-1 inactivated the inhibitor, but this cleavage was protected when TIMP-1 formed a complex with MMP-3.[140] The NMR structure of N-TIMP-2 showed that Val69 is a part of exposed ridge-like structure formed together with the N-terminal segment of the inhibitor where Cys1 and Cys70 are linked by disulfide bond. Mutagenesis studies around the ridge region, which altered the reactivity of TIMP-1 with MMPs,[141] supported this notion. The exact mechanism by which TIMP inhibits MMPs was elucidated by the crystal structure of the TIMP-1–MMP-3 catalytic domain complexes.[85] It showed that the N-terminal four residues Cys1–Thr–Cys–Val4 and the residues Glu67–Ser–Val–Cys70 that are linked by a disulfide bond form a reactive ridge, which slots into the active site of the MMP. The catalytic zinc ion of MMP-3 is bidentately chelated by the N-terminal amino group and the carbonyl group of Cys1, which expels the water molecule bound to the zinc. This reactive ridge region occupies about 75% of the protein–protein interaction in the case of the complex of the catalytic domain of MMP-3 and TIMP-1 and the side chain of Thr2 enters into the S1′ pocket of the enzyme. The crystal structure of TIMP-2–MMP-14 catalytic domain showed a complex formation similar to TIMP-1 and MMP-3 (Fig. 23.3c), but TIMP-2 has a longer AB loop (see Fig. 23.4) that makes additional contact with MMP-14.[86]

Numerous mutagenesis studies were carried out to alter the specificity of TIMPs using N-terminal inhibitory domains (N-TIMPs). The residues in TIMP-1, TIMP-2, and TIMP-3 that showed major changes in inhibition specificity are highlighted in Fig. 23.3. Mutation of Thr2 of N-TIMP-1 greatly altered the affinity for MMPs.[142] This residue is likely to interact with S1′ pocket of the enzyme, but there was a very poor correlation with residues affecting substrate specificity. Substitution of Thr2 to glycine essentially inactivated TIMP-1 for MMPs.[142] Recent studies of Hamze et al.[143] however showed that this mutant still maintained MMP-9 inhibition with K_i value of 2.1 nM, while it is essentially inactive for MMP-2 ($K_i > 40$ μM). Additional mutations at positions Val4 and Ser68 and their combination generated inhibitors discriminatory between MMP-1, MMP-2, and MMP-3,[144] indicating that the MMP reactive ridge of TIMPs may be modified to make them more selective. TIMP-2 has a longer AB loop than TIMP-1 or TIMP-3. Ile35, Tyr36, and Asn38 of TIMP-2 in the AB loop fit into a special cavity on the surface of the MMP-14 molecule, and mutation of Tyr36 indicated that it is a key residue for this interaction (Fig. 23.4).[145]

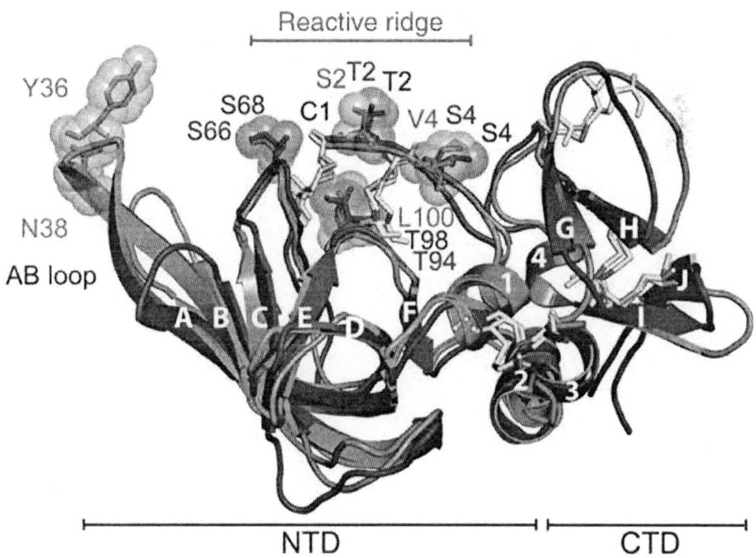

FIGURE 23.4 Superimposition of TIMP-1, TIMP-2, and the N-terminal domain of TIMP-3 shown as ribbon structures. Beta-strands are labelled A to J and alpha helices 1–4. Individual residues discussed in the text are shown as sticks with transparent VanderWaals spheres. Disulfide bonds are shown as sticks. Based on PDB IDs: 1UEA for TIMP-1, 1BQQ for TIMP-2 and 3CKI for TIMP-3, superimpositions and figure prepared with Pymol. (See color version of this figure in Color Plates section.)

TIMP-1 is a poor inhibitor of MT1-MMP, MMP-19, and tumor necrosis factor-α converting enzyme (TACE/ADAM17), but it gains reactivity for all three metalloproteinases by replacing Thr98 with Leu found in TIMP-2 and TIMP-3.[146,147] A further improvement was achieved by replacing the AB loop with that of TIMP-3.[147] TIMP-2 does not inhibit TACE, but it was also converted to inhibit TACE by replacing the AB loop with that of TIMP-3 in combination with S2T/A70S/V71L mutation.[148] Full-length TIMP-4 is a weak inhibitor of TACE, but truncation of the C-terminal domain increased the reactivity, and replacement with the TIMP-3 AB loop further improved the inhibitory activity with a subnanomolar K_I.[149]

Carbamylation of the N-terminus[150] or addition of an extra Ala at the N-terminus[151] inactivated TIMPs, suggesting that free N-terminal amino group of TIMPs is important to chelate the catalytic zinc ion of the MMPs. However, Wei *et al.*[152] have reported that the addition of Ala to the N-terminus of TIMP-3 did not significantly alter the ability to inhibit ADAM17, although it lost its inhibitory activity for MMPs. Mutation of Thr2 of TIMP-3 to Gly also inactivated TIMP-3 for most MMPs but retained good inhibitory activity for ADAM17. These studies suggest that ADAMs may have more flexibility to interact with these mutants. What is also not clear is why the glycine mutation at position 2 does not inhibit most MMPs.

23.6 SYNTHETIC INHIBITORS OF MATRIXINS

Because elevated activities of matrixins are closely associated with many diseases, numerous MMP inhibitors have been designed and synthesized during past 25 years or so. The development of synthetic inhibitors of MMPs was initially based on the use of a peptide sequence recognized by the target metalloproteinase to which a chelating moiety is introduced to interact with the zinc ion of the active site. The most commonly used chelating group was the hydroxamate group. This strategy generated several potent nonpeptidic inhibitors, and those that had shown efficacy in animal models were brought to clinical trials, mostly for cancer and some for rheumatoid arthritis and osteoarthritis. However, they failed to show efficacy in human diseases.[9] In addition, some inhibitors caused side effects including musculoskeletal pain, tendonitis, and in some cases, mild anemia with elevated levels of liver enzymes.[9,153] These adverse effects must come from poor selectivity displayed by this class of inhibitors. The human genome encodes 62 metalloproteinases belonging to the metzincin superfamily that include 23 MMPs, 13 ADAMs, 19 ADAMTSs, 6 astacins, and 1 pappalysin. These enzymes have similar basic active site structure, which may make them susceptible to some extent to the same inhibitor. Currently, we have only limited knowledge of the precise function of these enzymes *in vivo*, but some studies in animal models have suggested that their importance in biology and the consequence of blocking their activities may cause adverse effects. For example, the deletion of the MMP-8 gene increased skin tumor susceptibility to male mice,[154] and apoE/MMP double knockout mice indicated that MMP-3 and MMP-9 had a protective role against atherosclerotic plaque growth and instability.[155] Deletion of the MMP-14 or ADAM15 gene in the mouse caused arthritis-like lesions, and MMP-2 null mice developed more severe arthritis than the wild type when challenged with an antibody-induced arthritis model.[156] These observations emphasize that high degree of specificity of metalloproteinase inhibition is essential to avoid toxicity.

New generation variants of the hydroxamate-based inhibitors, which are relatively more specific, are now being evaluated.[157] Newer trends in inhibitor design are also in evidence. Thiol-, hydroxypyrone-, and barbiturate-based inhibitors have been described, as well as phosphinates that act as transition state analogues. More recently, new developments have also made for highly selective inhibitors of MMP-13; those compounds bind deep into the S1′ pocket of the enzyme but lacks a chelating group.[158,159] High selectivity was attained with those compounds probably because they lack the binding energy derived from zinc binding groups whose properties are broadly shared by other metalloproteinases.

23.7 CONCLUSIONS AND PROSPECTS

Structural and functional studies have provided clues as to how we can manipulate elevated MMP activities found under pathological conditions, and newer approaches are being developed to design and synthesize MMP inhibitors for potential therapeutic intervention. While the development of highly specific synthetic active site directed inhibitors of metalloproteinases remains a challenge due to the nature of their active sites, sophisticated theoretical and experimental approaches could provide further insights into this challenge by exploiting specific structural features of each individual metalloproteinase.

Alternative strategies for the development of metalloproteinase inhibitors may be to target noncatalytic domains of certain MMPs. For instance, the Hpx domain of collagenase is essential for the specificity to cleave collagen and that of MT1-MMP (MMP-14) is essential for cell surface clustering as part of its collagenolytic activity and proMMP 2 activation. Fn II domains of MMP-2 and MMP-9 that interact with specific extracellular collagenous sites can be the targets. The production of specific antibody fragments for these domains may be worthy of screening. Such proteins could inhibit the enzyme in a more specific way than active site directed inhibitors. An example may be seen in autoantibodies found in patients with thrombotic thrombocytopenic purpura, which causes microangiopathic hemolytic anemia, renal failure, neurologic dysfunction, and fever. Those autoantibodies react with the noncatalytic ancillary domain of ADAMTS-13 and inhibit the enzyme activity on von Willebrand factor.[160] This supports the concept that the ancillary domains could be good targets for developing exosite or allosteric inhibitors with high specificity. Recent approaches to discover small molecules as allosteric inhibitors might provide useful paradigms for this type of approach.[161,162]

We have learned a great deal about the importance of the matixins in both biological and pathological processes, but it is also acknowledged that their involvement in those processes is complex and we need to continue to investigate the exact function of each MMP under both physiological and pathological conditions. For successful development of MMP inhibitor therapy, methods for effectively monitoring the clinical efficacy of MMP inhibitors during clinical trials in the form of appropriate biomarkers are essential. Imaging MMP activities *in vivo* is an attractive approach. We remain optimistic that these barriers can be overcome as our knowledge of this area is rapidly increasing.

ACKNOWLEDGMENTS

This work was supported by grants from the Wellcome Trust, Arthritis Research Campaign, National Institutes of Health (AR40994), and European Union Framework 5 Program.

REFERENCES

1. Woessner, J. F., Jr. The matrix metalloproteinase family. In *Matrix Metalloproteinases;* Parks, W. C.; Mecham, R. P., Eds.; Academic Press: San Diego, **1998**; pp1–14.
2. Nagase, H.; Woessner, J. F. Matrix metalloproteinases. *J. Biol. Chem.* **1999**, *274*, 21491–21494.
3. Sternlicht, M. D.; Werb, Z. How matrix metalloproteinases regulate cell behavior. *Annu. Rev. Cell Dev. Biol.* **2001**, *17*, 463–516.
4. Pilcher, B. K.; Dumin, J. A.; Sudbeck, B. D.; Krane, S. M.; Welgus, H. G.; Parks, W. C. The activity of collagenase-1 is required for keratinocyte migration on a type I collagen matrix. *J. Cell Biol.* **1997**, *137*, 1445–1457.
5. Zhao, W. G.; Byrne, M. H.; Boyce, B. F.; Krane, S. M. Bone resorption induced by parathyroid hormone is strikingly diminished in collagenase-resistant mutant mice. *J. Clin. Invest.* **1999**, *103*, 517–524.
6. Nagase, H.; Visse, R.; Murphy, G. Structure and function of matrix metalloproteinases and TIMPs. *Cardiovasc. Res.* **2006**, *69*, 562–573.
7. Cauwe, B.; Van den Steen, P. E.; Opdenakker, G. The biochemical, biological, and pathological kaleidoscope of cell surface substrates processed by matrix metalloproteinases. *Crit. Rev. Biochem. Mol. Biol.* **2007**, *42*, 113–185.
8. Egeblad, M.; Werb, Z. New functions for the matrix metalloproteinases in cancer progression. *Nat. Rev. Cancer* **2002**, *2*, 161–174.
9. Milner, J. M.; Cawston, T. E. Matrix metalloproteinase knockout studies and the potential use of matrix metalloproteinase inhibitors in the rheumatic diseases. *Curr. Drug Targets Inflamm. Allergy* **2005**, *4*, 363–375.
10. Burrage, P. S.; Brinckerhoff, C. E. Molecular targets in osteoarthritis: metalloproteinases and their inhibitors. *Curr. Drug Targets* **2007**, *8*, 293–303.
11. Murphy, G.; Nagase, H. Reappraising metalloproteinases in rheumatoid arthritis and osteoarthritis: destruction or repair? *Nat. Clin. Pract. Rheumatol.* **2008**, *4*, 128–135.
12. Newby, A. C. Dual role of matrix metalloproteinases (matrixins) in intimal thickening and atherosclerotic plaque rupture. *Physiol. Rev.* **2005**, *85*, 1–31.
13. Yong, V. W.; Power, C.; Forsyth, P.; Edwards, D. R. Metalloproteinases in biology and pathology of the nervous system. *Nat. Rev. Neurosci.* **2001**, *2*, 502–511.

14. Gross, J.; Lapiere, C. M. Collagenolytic activity in amphibian tissues: a tissue culture assay. *Proc. Natl. Acad. Sci. USA* **1962**, *48*, 1014–1022.

15. Woolley, D. E.; Glanville, R. W.; Crossley, M. J.; Evanson, J. M. Purification of rheumatoid synovial collagenase and its action on soluble and insoluble collagen. *Eur. J. Biochem.* **1975**, *54*, 611–622.

16. Okada, Y.; Nagase, H.; Harris, E. D., Jr. A metalloproteinase from human rheumatoid synovial fibroblasts that digests connective tissue matrix components. Purification and characterization. *J. Biol. Chem.* **1986**, *261*, 14245–14255.

17. Galloway, W. A.; Murphy, G.; Sandy, J. D.; Gavrilovic, J.; Cawston, T. E.; Reynolds, J. J. Purification and characterization of a rabbit bone metalloproteinase that degrades proteoglycan and other connective-tissue components. *Biochem. J.* **1983**, *209*, 741–752.

18. Vater, C. A.; Nagase, H.; Harris, E. D., Jr. Purification of an endogenous activator of procollagenase from rabbit synovial fibroblast culture medium. *J. Biol. Chem.* **1983**, *258*, 9374–9382.

19. Chin, J. R.; Murphy, G.; Werb, Z. Stromelysin, a connective tissue-degrading metalloendopeptidase secreted by stimulated rabbit synovial fibroblasts in parallel with collagenase. Biosynthesis, isolation, characterization, and substrates. *J. Biol. Chem.* **1985**, *260*, 12367–12376.

20. Murphy, G.; Nagase, H.; Brinckerhoff, C. E. Relationship of procollagenase activator, stromelysin and matrix metalloproteinase 3. *Coll. Relat. Res.* **1988**, *8*, 389–391.

21. Goldberg, H. A.; Scott, P. G. Isolation from cultured porcine gingival explants of a neutral proteinase with collagen telopeptidase activity. *Connect. Tissue Res.* **1986**, *15*, 209–219.

22. Overall, C. M.; Sodek, J. Initial characterization of a neutral metalloproteinase, active on native 3/4-collagen fragments, synthesized by ROS 17/2.8 osteoblastic cells, periodontal fibroblasts, and identified in gingival crevicular fluid. *J. Dent. Res.* **1987**, *66*, 1271–1282.

23. Azzo, W.; Woessner, J. F., Jr. Purification and characterization of an acid metalloproteinase from human articular cartilage. *J. Biol. Chem.* **1986**, *261*, 5434–5441.

24. Gunja-Smith, Z.; Nagase, H.; Woessner, J. F., Jr. Purification of the neutral proteoglycan-degrading metalloproteinase from human articular cartilage tissue and its identification as stromelysin matrix metalloproteinase-3. *Biochem. J.* **1989**, *258*, 115–119.

25. Woessner, J. F., Jr.; Taplin, C. J. Purification and properties of a small latent matrix metalloproteinase of the rat uterus. *J. Biol. Chem.* **1988**, *263*, 16918–16925.

26. Goldberg, G. I.; Wilhelm, S. M.; Kronberger, A.; Bauer, E. A.; Grant, G. A.; Eisen, A. Z. Human fibroblast collagenase. Complete primary structure and homology to an oncogene transformation-induced rat protein. *J. Biol. Chem.* **1986**, *261*, 6600–6605.

27. Matrisian, L. M.; Glaichenhaus, N.; Gesnel, M.-C.; Breathnach, R. Epidermal growth factor and oncogenes induce transcription of the same cellular mRNA in rat fibroblasts. *EMBO J.* **1985**, *4*, 1435–1440.

28. Matrisian, L. M.; Bowden, G. T.; Krieg, P.; Furstenberger, G.; Briand, J.-P.; Leroy, P.; Breathnach, R. The mRNA coding for the secreted protease transin is expressed more abundantly in malignant than in benign tumors. *Proc. Natl. Acad. Sci. USA* **1986**, *83*, 9413–9417.

29. Whitham, S. E.; Murphy, G.; Angel, P.; Rahmsdorf, H. J.; Smith, B. J.; Lyons, A.; Harris, T. J.; Reynolds, J. J.; Herrlich, P.; Docherty, A. J. P. Comparison of human stromelysin and collagenase by cloning and sequence analysis. *Biochem. J.* **1986**, *240*, 913–916.

30. Bode, W.; Gomis-Rüth, F. X.; Stöcker, W. Astacins, serralysins, snake venom and matrix metalloproteinases exhibit identical zinc-binding environments (HEXXHXXGXXH and Met-turn) and topologies and should be grouped into a common family, the 'metzincins'. *FEBS Lett.* **1993**, *331*, 134–140.

31. Nagase, H.; Barrett, A. J.; Woessner, J. F., Jr. Nomenclature and glossary of the matrix metalloproteinases. *Matrix Suppl.* **1992**, *1*, 421–424.

32. Huxley-Jones, J.; Clarke, T. K.; Beck, C.; Toubaris, G.; Robertson, D. L.; Boot-Handford, R. P. The evolution of the vertebrate metzincins; insights from Ciona intestinalis and Danio rerio. *BMC Evol. Biol.* **2007**, *7*, 63.

33. Knäuper, V.; Docherty, A. J. P.; Smith, B.; Tschesche, H.; Murphy, G. Analysis of the contribution of the hinge region of human neutrophil collagenase (HNC, MMP-8) to stability and collagenolytic activity by alanine scanning mutagenesis. *FEBS Lett.* **1997**, *405*, 60–64.

34. Tsukada, H.; Pourmotabbed, T. Unexpected crucial role of residue 272 in substrate specificity of fibroblast collagenase. *J. Biol. Chem.* **2002**, *277*, 27378–27384.

35. Aimes, R. T.; Quigley, J. P. Matrix metalloproteinase-2 is an interstitial collagenase. Inhibitor-free enzyme catalyzes the cleavage of collagen fibrils and soluble native type I collagen generating the specific 3/4- and 1/4-length fragments. *J. Biol. Chem.* **1995**, *270*, 5872–5876.

36. Ohuchi, E.; Imai, K.; Fujii, Y.; Sato, H.; Seiki, M.; Okada, Y. Membrane type 1 matrix metalloproteinase digests interstitial collagens and other extracellular matrix macromolecules. *J. Biol. Chem.* **1997**, *272*, 2446–2451.

37. Wang, W.; Schulze, C. J.; Suarez-Pinzon, W. L.; Dyck, J. R.; Sawicki, G.; Schulz, R. Intracellular action of matrix metalloproteinase-2 accounts for acute myocardial ischemia and reperfusion injury. *Circulation* **2002**, *106*, 1543–1549.

38. Sawicki, G.; Leon, H.; Sawicka, J.; Sariahmetoglu, M.; Schulze, C. J.; Scott, P. G.; Szczesna-Cordary, D.; Schulz, R. Degradation of myosin light chain in isolated rat hearts subjected to ischemia-reperfusion injury: a new intracellular target for matrix metalloproteinase-2. *Circulation* **2005**, *112*, 544–552.

39. Kwan, J. A.; Schulze, C. J.; Wang, W.; Leon, H.; Sariahmetoglu, M.; Sung, M.; Sawicka, J.; Sims, D. E.; Sawicki, G.; Schulz, R. Matrix metalloproteinase-2 (MMP-2) is present in the nucleus of cardiac myocytes and is capable of cleaving poly(ADP-ribose) polymerase (PARP) *in vitro*. *FASEB J.* **2004**, *18*, 690–692.

40. Murphy, G.; Segain, J. P.; O'Shea, M.; Cockett, M.; Ioannou, C.; Lefebvre, O.; Chambon, P.; Basset, P. The 28-kDa N-terminal domain of mouse stromelysin-3 has the general properties of a weak metalloproteinase. *J. Biol. Chem.* **1993**, *268*, 15435–15441.

41. Pei, D.; Weiss, S. J. Furin-dependent intracellular activation of the human stromelysin-3 zymogen. *Nature* **1995**, *375*, 244–247.

42. Luo, D.; Mari, B.; Stoll, I.; Anglard, P. Alternative splicing and promoter usage generates an intracellular stromelysin 3 isoform directly translated as an active matrix metalloproteinase. *J. Biol. Chem.* **2002**, *277*, 25527–25536.

43. Parks, W. C.; Wilson, C. L.; Lopez-Boado, Y. S. Matrix metalloproteinases as modulators of inflammation and innate immunity. *Nat. Rev. Immunol.* **2004**, *4*, 617–629.

44. Marchenko, N. D.; Marchenko, G. N.; Weinreb, R. N.; Lindsey, J. D.; Kyshtoobayeva, A.; Crawford, H. C.; Strongin, A. Y. Beta-catenin regulates the gene of MMP-26, a novel

metalloproteinase expressed both in carcinomas and normal epithelial cells. *Int. J. Biochem. Cell Biol.* **2004**, *36*, 942–956.

45. English, W. R.; Puente, X. S.; Freije, J. M. P.; Knäuper, V.; Amour, A.; Merryweather, A.; López-Otín, C.; Murphy, G. Membrane type 4 matrix metalloproteinase (MMP17) has tumor necrosis factor-alpha convertase activity but does not activate pro-MMP2. *J. Biol. Chem.* **2000**, *275*, 14046–14055.
46. Itoh, Y.; Seiki, M. MT1-MMP: a potent modifier of pericellular microenvironment. *J. Cell Physiol.* **2006**, *206*, 1–8.
47. Itoh, Y.; Ito, N.; Nagase, H.; Evans, R. D.; Bird, S. A.; Seiki, M. Cell surface collagenolysis requires homodimerization of the membrane-bound collagenase MT1-MMP. *Mol. Biol. Cell* **2006**, *17*, 5390–5399.
48. Knäuper, V.; Will, H.; López-Otín, C.; Smith, B.; Atkinson, S. J.; Stanton, H.; Hembry, R. M.; Murphy, G. Cellular mechanisms for human procollagenase-3 (MMP-13) activation. Evidence that MT1-MMP (MMP-14) and gelatinase A (MMP-2) are able to generate active enzyme. *J. Biol. Chem.* **1996**, *271*, 17124–17131.
49. Holmbeck, K.; Bianco, P.; Caterina, J.; Yamada, S.; Kromer, M.; Kuznetsov, S. A.; Mankani, M.; Robey, P. G.; Poole, A. R.; Pidoux, I.; Ward, J. M.; Birkedal-Hansen, H. MT1-MMP-deficient mice develop dwarfism, osteopenia, arthritis, and connective tissue disease due to inadequate collagen turnover. *Cell* **1999**, *99*, 81–92.
50. Kerkela, E.; Bohling, T.; Herva, R.; Uria, J. A.; Saarialho-Kere, U. Human macrophage metalloelastase (MMP-12) expression is induced in chondrocytes during fetal development and malignant transformation. *Bone* **2001**, *29*, 487–493.
51. Hou, P.; Troen, T.; Ovejero, M. C.; Kirkegaard, T.; Andersen, T. L.; Byrjalsen, I.; Ferreras, M.; Sato, T.; Shapiro, S. D.; Foged, N. T.; Delaisse, J. M. Matrix metalloproteinase-12 (MMP-12) in osteoclasts: new lesson on the involvement of MMPs in bone resorption. *Bone* **2004**, *34*, 37–47.
52. Shipley, J. M.; Wesselschmidt, R. L.; Kobayashi, D. K.; Ley, T. J.; Shapiro, S. D. Metalloelastase is required for macrophage-mediated proteolysis and matrix invasion in mice. *Proc. Natl. Acad. Sci. USA* **1996**, *93*, 3942–3946.
53. Pendas, A. M.; Knäuper, V.; Puente, X. S.; Llano, E.; Mattei, M. G.; Apte, S.; Murphy, G.; López-Otín, C. Identification and characterization of a novel human matrix metalloproteinase with unique structural characteristics, chromosomal location, and tissue distribution. *J. Biol. Chem.* **1997**, *272*, 4281–4286.
54. Sedlacek, R.; Mauch, S.; Kolb, B.; Schatzlein, C.; Eibel, H.; Peter, H. H.; Schmitt, J.; Krawinkel, U. Matrix metalloproteinase MMP-19 (RASI 1) is expressed on the surface of activated peripheral blood mononuclear cells and is detected as an autoantigen in rheumatoid arthritis. *Immunobiology* **1998**, *198*, 408–423.
55. Sadowski, T.; Dietrich, S.; Koschinsky, F.; Sedlacek, R. Matrix metalloproteinase 19 regulates insulin-like growth factor-mediated proliferation, migration, and adhesion in human keratinocytes through proteolysis of insulin-like growth factor binding protein-3. *Mol. Biol. Cell* **2003**, *14*, 4569–4580.
56. Sadowski, T.; Dietrich, S.; Koschinsky, F.; Ludwig, A.; Proksch, E.; Titz, B.; Sedlacek, R. Matrix metalloproteinase 19 processes the laminin 5 gamma 2 chain and induces epithelial cell migration. *Cell Mol. Life Sci.* **2005**, *62*, 870–880.
57. Impola, U.; Jeskanen, L.; Ravanti, L.; Syrjanen, S.; Baldursson, B.; Kähäri, V. M.; Saarialho-Kere, U. Expression of matrix metalloproteinase (MMP)-7 and MMP-13 and

loss of MMP-19 and p16 are associated with malignant progression in chronic wounds. *Br. J. Dermatol.* **2005**, *152*, 720–726.

58. Jost, M.; Folgueras, A. R.; Frerart, F.; Pendas, A. M.; Blacher, S.; Houard, X.; Berndt, S.; Munaut, C.; Cataldo, D.; Alvarez, J.; Melen-Lamalle, L.; Foidart, J. M.; López-Otín, C.; Noel, A. Earlier onset of tumoral angiogenesis in matrix metalloproteinase-19-deficient mice. *Cancer Res.* **2006**, *66*, 5234–5241.

59. Beck, I. M.; Ruckert, R.; Brandt, K.; Mueller, M. S.; Sadowski, T.; Brauer, R.; Schirmacher, P.; Mentlein, R.; Sedlacek, R. MMP19 is essential for T cell development and T cell-mediated cutaneous immune responses. *PLoS ONE* **2008**, *3*, e2343.

60. Ryu, O. H.; Fincham, A. G.; Hu, C. C.; Zhang, C.; Qian, Q.; Bartlett, J. D.; Simmer, J. P. Characterization of recombinant pig enamelysin activity and cleavage of recombinant pig and mouse amelogenins. *J. Dent. Res.* **1999**, *78*, 743–750.

61. Bartlett, J. D.; Beniash, E.; Lee, D. H.; Smith, C. E. Decreased mineral content in MMP-20 null mouse enamel is prominent during the maturation stage. *J. Dent. Res.* **2004**, *83*, 909–913.

62. Li, W.; Gibson, C. W.; Abrams, W. R.; Andrews, D. W.; DenBesten, P. K. Reduced hydrolysis of amelogenin may result in X-linked amelogenesis imperfecta. *Matrix Biol.* **2001**, *19*, 755–760.

63. Ahokas, K.; Lohi, J.; Lohi, H.; Elomaa, O.; Karjalainen-Lindsberg, M. L.; Kere, J.; Saarialho-Kere, U. Matrix metalloproteinase-21, the human orthologue for XMMP, is expressed during fetal development and in cancer. *Gene* **2002**, *301*, 31–41.

64. Ahokas, K.; Lohi, J.; Illman, S. A.; Llano, E.; Elomaa, O.; Impola, U.; Karjalainen-Lindsberg, M. L.; Saarialho-Kere, U. Matrix metalloproteinase-21 is expressed epithelially during development and in cancer and is up-regulated by transforming growth factor-beta1 in keratinocytes. *Lab. Invest.* **2003**, *83*, 1887–1899.

65. Skoog, T.; Ahokas, K.; Orsmark, C.; Jeskanen, L.; Isaka, K.; Saarialho-Kere, U. MMP-21 is expressed by macrophages and fibroblasts *in vivo* and in culture. *Exp. Dermatol.* **2006**, *15*, 775–783.

66. Velasco, G.; Pendas, A. M.; Fueyo, A.; Knauper, V.; Murphy, G.; Lopez-Otin, C. Cloning and characterization of human MMP-23, a new matrix metalloproteinase predominantly expressed in reproductive tissues and lacking conserved domains in other family members. *J. Biol. Chem.* **1999**, *274*, 4570–4576.

67. Pei, D. Q.; Kang, T. B.; Qi, H. X. Cysteine array matrix metalloproteinase (CA-MMP)/MMP-23 is a type II transmembrane matrix metalloproteinase regulated by a single cleavage for both secretion and activation. *J. Biol. Chem.* **2000**, *275*, 33988–33997.

68. Yang, M. Z.; Kurkinen, M. Cloning and characterization of a novel matrix metalloproteinase (MMP), CMMP, from chicken embryo fibroblasts: CMMP, *Xenopus* XMMP, and human MMP-19 have a conserved unique cysteine in the catalytic domain. *J. Biol. Chem.* **1998**, *273*, 17893–17900.

69. Bar-Or, A.; Nuttall, R. K.; Duddy, M.; Alter, A.; Kim, H. J.; Ifergan, I.; Pennington, C. J.; Bourgoin, P.; Edwards, D. R.; Yong, V. W. Analyses of all matrix metalloproteinase members in leukocytes emphasize monocytes as major inflammatory mediators in multiple sclerosis. *Brain* **2003**, *126*, 2738–2749.

70. Lohi, J.; Wilson, C. L.; Roby, J. D.; Parks, W. C. Epilysin, a novel human matrix metalloproteinase (MMP-28) expressed in testis and keratinocytes and in response to injury. *J. Biol. Chem.* **2001**, *276*, 10134–10144.

71. Saarialho-Kere, U.; Kerkela, E.; Jahkola, T.; Suomela, S.; Keski-Oja, J.; Lohi, J. Epilysin (MMP-28) expression is associated with cell proliferation during epithelial repair. *J. Invest. Dermatol.* **2002**, *119*, 14–21.

72. Illman, S. A.; Lehti, K.; Keski-Oja, J.; Lohi, J. Epilysin (MMP-28) induces TGF-beta mediated epithelial to mesenchymal transition in lung carcinoma cells. *J. Cell Sci.* **2006**, *119*, 3856–3865.

73. Kevorkian, L.; Young, D. A.; Darrah, C.; Donell, S. T.; Shepstone, L.; Porter, S.; Brockbank, S. M.; Edwards, D. R.; Parker, A. E.; Clark, I. M. Expression profiling of metalloproteinases and their inhibitors in cartilage. *Arthritis Rheum.* **2004**, *50*, 131–141.

74. Momohara, S.; Okamoto, H.; Komiya, K.; Ikari, K.; Takeuchi, M.; Tomatsu, T.; Kamatani, N. Matrix metalloproteinase 28/epilysin expression in cartilage from patients with rheumatoid arthritis and osteoarthritis: comment on the article by Kevorkian et al. *Arthritis Rheum.* **2004**, *50*, 4074–4075.

75. Lovejoy, B.; Cleasby, A.; Hassell, A. M.; Longley, K.; Luther, M. A.; Weigl, D.; McGeehan, G.; McElroy, A. B.; Drewry, D.; Lambert, M. H.; et al. Structure of the catalytic domain of fibroblast collagenase complexed with an inhibitor. *Science* **1994**, *263*, 375–377.

76. Bode, W.; Reinemer, P.; Huber, R.; Kleine, T.; Schnierer, S.; Tschesche, H. The X-ray crystal structure of the catalytic domain of human neutrophil collagenase inhibited by a substrate analogue reveals the essentials for catalysis and specificity. *EMBO J.* **1994**, *13*, 1263–1269.

77. Borkakoti, N.; Winkler, F. K.; Williams, D. H.; D'Arcy, A.; Broadhurst, M. J.; Brown, P. A.; Johnson, W. H.; Murray, E. J. Structure of the catalytic domain of human fibroblast collagenase complexed with an inhibitor. *Nat. Struct. Biol.* **1994**, *1*, 106–110.

78. Stams, T.; Spurlino, J. C.; Smith, D. L.; Wahl, R. C.; Ho, T. F.; Qoronfleh, M. W.; Banks, T. M.; Rubin, B. Structure of human neutrophil collagenase reveals large S1′ specificity pocket. *Nat. Struct. Biol.* **1994**, *1*, 119–123.

79. Maskos, K. Crystal structures of MMPs in complex with physiological and pharmacological inhibitors. *Biochimie* **2005**, *87*, 249–263.

80. Becker, J. W.; Marcy, A. I.; Rokosz, L. L.; Axel, M. G.; Burbaum, J. J.; Fitzgerald, P. M.; Cameron, P. M.; Esser, C. K.; Hagmann, W. K.; Hermes, J. D.; et al. Stromelysin-1: three-dimensional structure of the inhibited catalytic domain and of the C-truncated proenzyme. *Protein Sci.* **1995**, *4*, 1966–1976.

81. Elkins, P. A.; Ho, Y. S.; Smith, W. W.; Janson, C. A.; D'Alessio, K. J.; McQueney, M. S.; Cummings, M. D.; Romanic, A. M. Structure of the C-terminally truncated human ProMMP9, a gelatin-binding matrix metalloproteinase. *Acta Crystallogr. D Biol. Crystallogr.* **2002**, *58*, 1182–1192.

82. Jozic, D.; Bourenkov, G.; Lim, N. H.; Visse, R.; Nagase, H.; Bode, W.; Maskos, K. X-ray structure of human proMMP-1: new insights into procollagenase activation and collagen binding. *J. Biol. Chem.* **2005**, *280*, 9578–9585.

83. Morgunova, E.; Tuuttila, A.; Bergmann, U.; Isupov, M.; Lindqvist, Y.; Schneider, G.; Tryggvason, K. Structure of human pro-matrix metalloproteinase-2: activation mechanism revealed. *Science* **1999**, *284*, 1667–1670.

84. Morgunova, E.; Tuuttila, A.; Bergmann, U.; Tryggvason, K. Structural insight into the complex formation of latent matrix metalloproteinase 2 with tissue inhibitor of metalloproteinase 2. *Proc. Natl. Acad. Sci. USA* **2002**, *99*, 7414–7419.

85. Gomis-Rüth, F. X.; Maskos, K.; Betz, M.; Bergner, A.; Huber, R.; Suzuki, K.; Yoshida, N.; Nagase, H.; Brew, K.; Bourenkov, G. P.; Bartunik, H.; Bode, W. Mechanism of inhibition of the human matrix metalloproteinase stromelysin-1 by TIMP-1. *Nature* **1997**, *389*, 77–81.

86. Fernandez-Catalan, C.; Bode, W.; Huber, R.; Turk, D.; Calvete, J. J.; Lichte, A.; Tschesche, H.; Maskos, K. Crystal structure of the complex formed by the membrane type 1-matrix metalloproteinase with the tissue inhibitor of metalloproteinases-2, the soluble progelatinase a receptor. *EMBO J.* **1998**, *17*, 5238–5248.

87. Li, J.; Brick, P.; O'Hare, M. C.; Skarzynski, T.; Lloyd, L. F.; Curry, V. A.; Clark, I. M.; Bigg, H. F.; Hazleman, B. L.; Cawston, T. E.; et al. Structure of full-length porcine synovial collagenase reveals a C-terminal domain containing a calcium-linked, four-bladed beta-propeller. *Structure* **1995**, *3*, 541–549.

88. Iyer, S.; Visse, R.; Nagase, H.; Acharya, K. R. Crystal structure of an active form of human MMP-1. *J. Mol. Biol.* **2006**, *362*, 78–88.

89. Grams, F.; Reinemer, P.; Powers, J. C.; Kleine, T.; Pieper, M.; Tschesche, H.; Huber, R.; Bode, W. X-ray structures of human neutrophil collagenase complexed with peptide hydroxamate and peptide thiol inhibitors. Implications for substrate binding and rational drug design. *Eur. J. Biochem.* **1995**, *228*, 830–841.

90. Nagase, H. Activation mechanisms of matrix metalloproteinases. *Biol. Chem.* **1997**, *378*, 151–160.

91. Fu, X.; Kassim, S. Y.; Parks, W. C.; Heinecke, J. W. Hypochlorous acid oxygenates the cysteine switch domain of pro-matrilysin (MMP-7). A mechanism for matrix metalloproteinase activation and atherosclerotic plaque rupture by myeloperoxidase. *J. Biol. Chem.* **2001**, *276*, 41279–41287.

92. Gu, Z.; Kaul, M.; Yan, B.; Kridel, S. J.; Cui, J.; Strongin, A.; Smith, J. W.; Liddington, R. C.; Lipton, S. A. S-nitrosylation of matrix metalloproteinases: signaling pathway to neuronal cell death. *Science* **2002**, *297*, 1186–1190.

93. Fu, X.; Kao, J. L.; Bergt, C.; Kassim, S. Y.; Huq, N. P.; d'Avignon, A.; Parks, W. C.; Mecham, R. P.; Heinecke, J. W. Oxidative cross-linking of tryptophan to glycine restrains matrix metalloproteinase activity: specific structural motifs control protein oxidation. *J. Biol. Chem.* **2004**, *279*, 6209–6212.

94. Rosenblum, G.; Van den Steen, P. E.; Cohen, S. R.; Grossmann, J. G.; Frenkel, J.; Sertchook, R.; Slack, N.; Strange, R. W.; Opdenakker, G.; Sagi, I. Insights into the structure and domain flexibility of full-length pro-matrix metalloproteinase-9/gelatinase B. *Structure* **2007**, *15*, 1227–1236.

95. Van den Steen, P. E.; Van, A. I.; Hvidberg, V.; Piccard, H.; Fiten, P.; Jacobsen, C.; Moestrup, S. K.; Fry, S.; Royle, L.; Wormald, M. R.; Wallis, R.; Rudd, P. M.; Dwek, R. A.; Opdenakker, G. The hemopexin and O-glycosylated domains tune gelatinase B/MMP-9 bioavailability via inhibition and binding to cargo receptors. *J. Biol. Chem.* **2006**, *281*, 18626–18637.

96. Wu, Y. I.; Munshi, H. G.; Sen, R.; Snipas, S. J.; Salvesen, G. S.; Fridman, R.; Stack, M. S. Glycosylation broadens the substrate profile of membrane type 1 matrix metalloproteinase. *J. Biol. Chem.* **2004**, *279*, 8278–8289.

97. Libson, A. M.; Gittis, A. G.; Collier, I. E.; Marmer, B. L.; Goldberg, G. I.; Lattman, E. E. Crystal structure of the haemopexin-like C-terminal domain of gelatinase A. *Nat. Struct. Biol.* **1995**, 2, 938–942.

98. Gomis-Rüth, F. X.; Gohlke, U.; Betz, M.; Knäuper, V.; Murphy, G.; López-Otín, C.; Bode, W. The helping hand of collagenase-3 (MMP-13): 2.7 Å crystal structure of its C-terminal haemopexin-like domain. *J. Mol. Biol.* **1996**, *264*, 556–566.

99. Itoh, Y.; Takamura, A.; Ito, N.; Maru, Y.; Sato, H.; Suenaga, N.; Aoki, T.; Seiki, M. Homophilic complex formation of MT1-MMP facilitates proMMP-2 activation on the cell surface and promotes tumor cell invasion. *EMBO J.* **2001**, *20*, 4782–4793.

100. Nagase, H. Substrate specificity of MMPs. In *Matrix Metalloproteinase Inhibitors in Cancer Therapy;* Clendeninn, N. J.; Appelt, K., Eds.; Humana Press: Totowa, NJ, **2001**; pp39–66.

101. Overall, C. M.; Blobel, C. P. In search of partners: linking extracellular proteases to substrates. *Nat. Rev. Mol. Cell Biol.* **2007**, *8*, 245–257.

102. Bode, W.; Fernandez-Catalan, C.; Tschesche, H.; Grams, F.; Nagase, H.; Maskos, K. Structural properties of matrix metalloproteinases. *Cell. Mol. Life Sci.* **1999**, *55*, 639–652.

103. Chung, L.; Dinakarpandian, D.; Yoshida, N.; Lauer-Fields, J. L.; Fields, G. B.; Visse, R.; Nagase, H. Collagenase unwinds triple-helical collagen prior to peptide bond hydrolysis. *EMBO J.* **2004**, *23*, 3020–3030.

104. Murphy, G.; Nguyen, Q.; Cockett, M. I.; Atkinson, S. J.; Allan, J. A.; Knight, C. G.; Willenbrock, F.; Docherty, A. J. P. Assessment of the role of the fibronectin-like domain of gelatinase A by analysis of a deletion mutant. *J. Biol. Chem.* **1994**, *269*, 6632–6636.

105. Shipley, J. M.; Doyle, G. A.; Fliszar, C. J.; Ye, Q. Z.; Johnson, L. L.; Shapiro, S. D.; Welgus, H. G.; Senior, R. M. The structural basis for the elastolytic activity of the 92-kDa and 72-kDa gelatinases. Role of the fibronectin type II-like repeats. *J. Biol. Chem.* **1996**, *271*, 4335–4341.

106. Welgus, H. G.; Jeffrey, J. J.; Eisen, A. Z.; Roswit, W. T.; Stricklin, G. P. Human skin fibroblast collagenase: interaction with substrate and inhibitor. *Coll. Relat. Res.* **1985**, *5*, 167–179.

107. Allan, J. A.; Hembry, R. M.; Angal, S.; Reynolds, J. J.; Murphy, G. Binding of latent and high Mr active forms of stromelysin to collagen is mediated by the C-terminal domain. *J. Cell Sci.* **1991**, *99*, 789–795.

108. Stetler-Stevenson, W. G.; Krutzsch, H. C.; Liotta, L. A. Tissue inhibitor of metalloproteinase (TIMP-2). A new member of the metalloproteinase inhibitor family. *J. Biol. Chem.* **1989**, *264*, 17374–17378.

109. Wilhelm, S. M.; Collier, I. E.; Marmer, B. L.; Eisen, A. Z.; Grant, G. A.; Goldberg, G. I.; SV40-transformed human lung fibroblasts secrete a 92-kDa type IV collagenase which is identical to that secreted by normal human macrophages [published erratum appears in J Biol Chem 1990 Dec 25;265(36):22570]. *J. Biol. Chem.* **1989**, *264*, 17213–17221.

110. Bigg, H. F.; Morrison, C. J.; Butler, G. S.; Bogoyevitch, M. A.; Wang, Z. P.; Soloway, P. D.; Overall, C. M. Tissue inhibitor of metalloproteinases-4 inhibits but does not support the activation of gelatinase A via efficient inhibition of membrane type I-matrix metalloproteinase. *Cancer Res.* **2001**, *61*, 3610–3618.

111. Eisen, A. Z.; Bloch, K. J.; Sakai, T. Inhibition of human skin collagenase by human serum. *J. Lab. Clin. Med.* **1970**, *75*, 258–263.

112. Abe, S.; Nagai, Y. Interaction between tadpole collagenase and human α_2-macroglobulin. *Biochim. Biophys. Acta* **1972**, *278*, 125–132.

113. Sakamoto, S.; Goldhaber, P.; Glimcher, M. J. Further studies on the nature of the components in serum which inhibit mouse bone collagenase. *Calcif. Tissue Res.* **1972**, *10*, 280–288.

114. Bauer, E. A.; Stricklin, G. P.; Jeffrey, J. J.; Eisen, A. Z. Collagenase production by human skin fibroblasts. *Biochem. Biophys. Res. Commun.* **1975**, *64*, 232–240.

115. Woolley, D. E.; Roberts, D. R.; Evanson, J. M. Inhibition of human collagenase activity by a small molecular weight serum protein. *Biochem. Biophys. Res. Commun.* **1975**, *66*, 747–754.

116. Kuettner, K. E.; Hiti, J.; Eisenstein, R.; Harper, E. Collagenase inhibition by cationic proteins derived from cartilage and aorta. *Biochem. Biophys. Res. Commun.* **1976**, *72*, 40–46.

117. Welgus, H. G.; Stricklin, G. P.; Eisen, A. Z.; Bauer, E. A.; Cooney, R. V.; Jeffrey, J. J. A specific inhibitor of vertebrate collagenase produced by human skin fibroblasts. *J. Biol. Chem.* **1979**, *254*, 1938–1943.

118. Vater, C. A.; Mainardi, C. L.; Harris, E. D., Jr. Inhibitor of human collagenase from cultures of human tendon. *J. Biol. Chem.* **1979**, *254*, 3045–3053.

119. Cawston, T. E.; Galloway, W. A.; Mercer, E.; Murphy, G.; Reynolds, J. J. Purification of rabbit bone inhibitor of collagenase. *Biochem. J.* **1981**, *195*, 159–165.

120. Higashi, S.; Miyazaki, K. Novel processing of beta-amyloid precursor protein catalyzed by membrane type 1 matrix metalloproteinase releases a fragment lacking the inhibitor domain against gelatinase A. *Biochemistry* **2003**, *42*, 6514–6526.

121. Mott, J. D.; Thomas, C. L.; Rosenbach, M. T.; Takahara, K.; Greenspan, D. S.; Banda, M. J. Post-translational proteolytic processing of procollagen C-terminal proteinase enhancer releases a metalloproteinase inhibitor. *J. Biol. Chem.* **2000**, *275*, 1384–1390.

122. Oh, J.; Takahashi, R.; Kondo, S.; Mizoguchi, A.; Adachi, E.; Sasahara, R. M.; Nishimura, S.; Imamura, Y.; Kitayama, H.; Alexander, D. B.; Ide, C.; Horan, T. P.; Arakawa, T.; Yoshida, H.; Nishikawa, S. I.; Itoh, Y.; Seiki, M.; Itohara, S.; Takahashi, C.; Noda, M. The membrane-anchored MMP inhibitor RECK is a key regulator of extracellular matrix integrity and angiogenesis. *Cell* **2001**, *107*, 789–800.

123. Herman, M. P.; Sukhova, G. K.; Kisiel, W.; Foster, D.; Kehry, M. R.; Libby, P.; Schonbeck, U. Tissue factor pathway inhibitor-2 is a novel inhibitor of matrix metalloproteinases with implications for atherosclerosis. *J. Clin. Invest.* **2001**, *107*, 1117–1126.

124. Du, X.; Chand, H. S.; Kisiel, W. Human tissue factor pathway inhibitor-2 does not bind or inhibit activated matrix metalloproteinase-1. *Biochim. Biophys. Acta* **2003**, *1621*, 242–245.

125. Barrett, A. J.; Starkey, P. M. The interaction of alpha 2-macroglobulin with proteinases. Characteristics and specificity of the reaction, and a hypothesis concerning its molecular mechanism. *Biochem. J.* **1973**, *133*, 709–724.

126. Enghild, J. J.; Salvesen, G.; Brew, K.; Nagase, H. Interaction of human rheumatoid synovial collagenase (matrix metalloproteinase 1) and stromelysin (matrix metalloproteinase 3) with human α_2-macroglobulin and chicken ovostatin. Binding kinetics and identification of matrix metalloproteinase cleavage sites. *J. Biol. Chem.* **1989**, *264*, 8779–8785.

127. Baker, A. H.; Edwards, D. R.; Murphy, G. Metalloproteinase inhibitors: biological actions and therapeutic opportunities. *J. Cell Sci.* **2002**, *115*, 3719–3727.

128. Amour, A.; Knight, C. G.; Webster, A.; Slocombe, P. M.; Stephens, P. E.; Knauper, V.; Docherty, A. J. P.; Murphy, G. The *in vitro* activity of ADAM-10 is inhibited by TIMP-1 and TIMP-3. *FEBS Lett.* **2000**, *473*, 275–279.

129. Jacobsen, J.; Visse, R.; Sorensen, H. P.; Enghild, J. J.; Brew, K.; Wewer, U. M.; Nagase, H. Catalytic properties of ADAM12 and its domain deletion mutants. *Biochemistry* **2008**, *47*, 537–547.

130. Amour, A.; Hutton, M.; Knauper, V.; Slocombe, P. M.; Webster, A.; Butler, M.; Knight, C. G.; Smith, B. J.; Docherty, A. J.; Murphy, G. Inhibition of the metalloproteinase domain of mouse TACE. *Ann. N. Y. Acad. Sci.* **1999**, *878*, 728–731.

131. Rodriguez-Manzaneque, J. C.; Westling, J.; Thai, S. N.; Luque, A.; Knäuper, V.; Murphy, G.; Sandy, J. D.; Iruela-Arispe, M. L. ADAMTS1 cleaves aggrecan at multiple sites and is differentially inhibited by metalloproteinase inhibitors. *Biochem. Biophys. Res. Commun.* **2002**, *293*, 501–508.

132. Wang, W. M.; Ge, G.; Lim, N. H.; Nagase, H.; Greenspan, D. S. TIMP-3 inhibits the procollagen N-proteinase ADAMTS-2. *Biochem. J.* **2006**, *398*, 515–519.

133. Kashiwagi, M.; Tortorella, M.; Nagase, H.; Brew, K. TIMP-3 is a potent inhibitor of aggrecanase 1 (ADAM-TS4) and aggrecanase 2 (ADAM-TS5). *J. Biol. Chem.* **2001**, *276*, 12501–12504.

134. Leco, K. J.; Waterhouse, P.; Sanchez, O. H.; Gowing, K. L. M.; Poole, A. R.; Wakeham, A.; Mak, T. W.; Khokha, R. Spontaneous air space enlargement in the lungs of mice lacking tissue inhibitor of metalloproteinases-3 (TIMP-3). *J. Clin. Invest.* **2001**, *108*, 817–829.

135. Fata, J. E.; Leco, K. J.; Voura, E. B.; Yu, H. Y.; Waterhouse, P.; Murphy, G.; Moorehead, R. A.; Khokha, R. Accelerated apoptosis in the Timp-3-deficient mammary gland. *J. Clin. Invest* **2001**, *108*, 831–841.

136. Docherty, A. J. P.; Lyons, A.; Smith, B. J.; Wright, E. M.; Stephens, P. E.; Harris, T. J. R.; Murphy, G.; Reynolds, J. J. Sequence of human tissue inhibitor of metalloproteinases and its identity to erythroid-potentiating activity. *Nature* **1985**, *318*, 66–69.

137. Williamson, R. A.; Marston, F. A. O.; Angal, S.; Koklitis, P.; Panico, M.; Morris, H. R.; Carne, A. F.; Smith, B. J.; Harris, T. J. R.; Freedman, R. B. Disulphide bond assignment in human tissue inhibitor of metalloproteinases (TIMP). *Biochem. J.* **1990**, *268*, 267–274.

138. Murphy, G.; Houbrechts, A.; Cockett, M. I.; Williamson, R. A.; O'Shea, M.; Docherty, A. J. P. The N-terminal domain of tissue inhibitor of metalloproteinases retains metalloproteinase inhibitory activity [published erratum appears in Biochemistry 1991 Oct 22;30 (42):10362]. *Biochemistry* **1991**, *30*, 8097–8102.

139. Williamson, R. A.; Martorell, G.; Carr, M. D.; Murphy, G.; Docherty, A. J. P.; Freedman, R. B.; Feeney, J. Solution structure of the active domain of tissue inhibitor of metalloproteinases-2. A new member of the OB fold protein family. *Biochemistry* **1994**, *33*, 11745–11759.

140. Nagase, H.; Suzuki, K.; Cawston, T. E.; Brew, K. Involvement of a region near valine-69 of tissue inhibitor of metalloproteinases (TIMP)-1 in the interaction with matrix metalloproteinase 3 (stromelysin 1). *Biochem. J.* **1997**, *325*, 163–167.

141. Huang, W.; Meng, Q.; Suzuki, K.; Nagase, H.; Brew, K. Mutational study of the amino-terminal domain of human tissue inhibitor of metalloproteinases I (TIMP-1) locates an inhibitory region for matrix metalloproteinases. *J. Biol. Chem.* **1997**, *272*, 22086–22091.

142. Meng, Q.; Malinovskii, V.; Huang, W.; Hu, Y. J.; Chung, L.; Nagase, H.; Bode, W.; Maskos, K.; Brew, K. Residue 2 of TIMP-1 is a major determinant of affinity and specificity for matrix metalloproteinases but effects of substitutions do not correlate with those of the corresponding P1′ residue of substrate. *J. Biol. Chem.* **1999**, *274*, 10184–10189.

143. Hamze, A. B.; Wei, S.; Bahudhanapati, H.; Kota, S.; Acharya, K. R.; Brew, K. Constraining specificity in the N-domain of tissue inhibitor of metalloproteinases-1; gelatinase-selective inhibitors. *Protein Sci.* **2007**, *16*, 1905–1913.

144. Wei, S.; Chen, Y.; Chung, L.; Nagase, H.; Brew, K. Protein engineering of the tissue inhibitor of metalloproteinase 1 (TIMP-1) inhibitory domain. In search of selective matrix metalloproteinase inhibitors. *J. Biol. Chem.* **2003**, *278*, 9831–9834.

145. Williamson, R. A.; Hutton, M.; Vogt, G.; Rapti, M.; Knäuper, V.; Carr, M. D.; Murphy, G. Tyrosine 36 plays a critical role in the interaction of the AB loop of tissue inhibitor of metalloproteinases-2 with matrix metalloproteinase-14. *J. Biol. Chem.* **2001**, *276*, 32966–32970.

146. Lee, M. H.; Rapti, M.; Murphy, G. Unveiling the surface epitopes that render tissue inhibitor of metalloproteinase-1 inactive against membrane type 1-matrix metalloproteinase. *J. Biol. Chem.* **2003**, *278*, 40224–40230.

147. Lee, M. H.; Rapti, M.; Knäuper, V.; Murphy, G. Threonine 98, the pivotal residue of tissue inhibitor of metalloproteinases (TIMP)-1 in metalloproteinase recognition. *J. Biol. Chem.* **2004**, *279*, 17562–17569.

148. Lee, M. H.; Rapti, M.; Murphy, G. Delineating the molecular basis of the inactivity of tissue inhibitor of metalloproteinase-2 against tumor necrosis factor-alpha-converting enzyme. *J. Biol. Chem.* **2004**, *279*, 45121–45129.

149. Lee, M. H.; Rapti, M.; Murphy, G. Total conversion of tissue inhibitor of metalloproteinase (TIMP) for specific metalloproteinase targeting: fine-tuning TIMP-4 for optimal inhibition of tumor necrosis factor-{alpha}-converting enzyme. *J. Biol. Chem.* **2005**, *280*, 15967–15975.

150. Higashi, S.; Miyazaki, K. Reactive site-modified tissue inhibitor of metalloproteinases-2 inhibits the cell-mediated activation of progelatinase A. *J. Biol. Chem.* **1999**, *274*, 10497–10504.

151. Wingfield, P. T.; Sax, J. K.; Stahl, S. J.; Kaufman, J.; Palmer, I.; Chung, V.; Corcoran, M. L.; Kleiner, D. E.; Stetler-Stevenson, W. G. Biophysical and functional characterization of full-length, recombinant human tissue inhibitor of metalloproteinases-2 (TIMP-2) produced in *Escherichia coli*: comparison of wild type and amino-terminal alanine appended variant with implications for the mechanism of TIMP functions. *J. Biol. Chem.* **1999**, *274*, 21362–21368.

152. Wei, S.; Kashiwagi, M.; Kota, S.; Xie, Z.; Nagase, H.; Brew, K. Reactive site mutations in tissue inhibitor of metalloproteinase-3 disrupt inhibition of matrix metalloproteinases but not tumor necrosis factor-alpha-converting enzyme. *J. Biol. Chem.* **2005**, *280*, 32877–32882.

153. Zucker, S.; Cao, J.; Chen, W. T. Critical appraisal of the use of matrix metalloproteinase inhibitors in cancer treatment. *Oncogene* **2000**, *19*, 6642–6650.

154. Balbin, M.; Fueyo, A.; Tester, A. M.; Pendas, A. M.; Pitiot, A. S.; Astudillo, A.; Overall, C. M.; Shapiro, S. D.; Lopez-Otin, C. Loss of collagenase-2 confers increased skin tumor susceptibility to male mice. *Nat. Genet.* **2003**, *35*, 252–257.

155. Johnson, J. L.; George, S. J.; Newby, A. C.; Jackson, C. L. Divergent effects of matrix metalloproteinases 3, 7, 9, and 12 on atherosclerotic plaque stability in mouse brachiocephalic arteries. *Proc. Natl. Acad. Sci. USA* **2005**, *102*, 15575–15580.

156. Itoh, T.; Matsuda, H.; Tanioka, M.; Kuwabara, K.; Itohara, S.; Suzuki, R. The role of matrix metalloproteinase-2 and matrix metalloproteinase-9 in antibody-induced arthritis. *J. Immunol.* **2002**, *169*, 2643–2647.

157. Fisher, J. F.; Mobashery, S. Recent advances in MMP inhibitor design. *Cancer Metastasis Rev.* **2006**, *25*, 115–136.

158. Engel, C. K.; Pirard, B.; Schimanski, S.; Kirsch, R.; Habermann, J.; Klingler, O.; Schlotte, V.; Weithmann, K. U.; Wendt, K. U. Structural basis for the highly selective inhibition of MMP-13. *Chem. Biol.* **2005**, *12*, 181–189.

159. Johnson, A. R.; Pavlovsky, A. G.; Ortwine, D. F.; Prior, F.; Man, C. F.; Bornemeier, D. A.; Banotai, C. A.; Mueller, W. T.; McConnell, P.; Yan, C.; Baragi, V.; Lesch, C.; Roark, W. H.; Wilson, M.; Datta, K.; Guzman, R.; Han, H. K.; Dyer, R. D. Discovery and characterization of a novel inhibitor of matrix metalloprotease-13 (MMP13) that reduces cartilage damage *in vivo* without joint fibroplasia side effects. *J. Biol. Chem.* **2007**, *282*, 27781–27791.

160. Soejima, K.; Matsumoto, M.; Kokame, K.; Yagi, H.; Ishizashi, H.; Maeda, H.; Nozaki, C.; Miyata, T.; Fujimura, Y.; Nakagaki, T. ADAMTS-13 cysteine-rich/spacer domains are functionally essential for von Willebrand factor cleavage. *Blood* **2003**, *102*, 3232–3237.

161. Hardy, J. A.; Wells, J. A. Searching for new allosteric sites in enzymes. *Curr. Opin. Struct. Biol.* **2004**, *14*, 706–715.

162. Hardy, J. A.; Lam, J.; Nguyen, J. T.; O'Brien, T.; Wells, J. A. Discovery of an allosteric site in the caspases. *Proc. Natl. Acad. Sci. USA* **2004**, *101*, 12461–12466.

163. Wisniewska, M.; Goettig, P.; Maskos, K.; Belouski, E.; Winters, D.; Hecht, R.; Black, R.; Bode, W. Structural determinants of the ADAM inhibition by TIMP-3: crystal structure of the TACE-N-TIMP-3 complex. *J. Mol. Biol.* **2008**, *381*, 1307–1319.

CHAPTER 24

MMP Inhibitors Based on Earlier Succinimide Strategies: From Early to New Approaches

M. AMÉLIA SANTOS

Centro de Química Estrutural, Instituto Superior Técnico-UTL, Av. Rovisco Pais 1, 1049-001 Lisboa, Portugal

24.1 INTRODUCTION

Matrix metalloproteinases (MMPs, matrixins) are a family of the metzincin group of zinc-dependent endopeptidases that promote the hydrolysis and cleavage of peptide bonds with subsequent turnover of basement membranes and most extracellular matrix (ECM) components. These proteinases play an important role in many physiological processes such as embryogenesis, normal tissue remodeling, wound healing, angiogenesis, but they are deregulated in diverse diseases such as arthritis, tumor metastasis, periodontal diseases, neurological, and cardiovascular diseases.[1–3] In physiological conditions, MMP activity is controlled by their gene transcriptional regulation, zymogen activation and by their endogenous inhibitors, the tissue inhibitors of metalloproteinases (TIMPs), to prevent the uncontrolled destruction of body tissues.[4–6] However, under pathological conditions, this regulation is modified or disrupted, and leads to overexpression of MMPs followed by accelerated matrix degradation. The later has been associated with several pathologies, including cancer, osteoarthritis (OA), rheumatoid arthritis (RA), cardiovascular diseases, and periodontitis.[7–9] As a consequence, the MMP family has emerged as an attractive pharmaceutical target, with several possible approaches for therapeutic intervention, namely, blocking MMP gene transcription, proenzyme activation, or active site-directed MMP inhibition. Currently, 27 MMPs, sharing a number of analogies, have been identified in humans with different substrate and disease specificity, as summarized in Table 24.1.[10,11] For the past three decades, the inhibition of MMPs has been widely

TABLE 24.1 The Matrix Metalloproteinase Family*

MMP Number	Enzyme Name(s)	Therapeutic Areas
MMP-1	Collagenase-1, fibroblast collagenase	Arthritis, cancer, periodontal disease
MMP-2	Gelatinase A; type IV collagenase	Cancer, MS, stroke, angiogenesis
MMP-3	Stromelysin-1	Cancer, arthritis
MMP-7	Matrilysin	Cancer
MMP-8	Collagenase-2, neutrophil collagenase	
MMP-9	Gelatinase-B, type V collagenase	Cancer, MS, stroke
MMP-10	Stromelysin-2	
MMP-11	Stromelysin-3	Cancer
MMP-12	Metalloelastase, macrophase elastase	Emphysema
MMP-13	Collagenase-3	Arthritis
MMP-14 and -17	Membrane-type	Cancer

*A selection, from Ref. 10.

sought as a main strategy in intervention of these disease processes.[7,12–15] A large number of synthetic MMP inhibitors (MMPIs) have been disclosed and entered clinical trials as potential treatments for cancer and arthritis.[10–12,16,17] However, the advanced clinical trials failed for the major part of the compounds, apparently due to lack of efficacy and musculoskeletal side effects. Several critical analyses performed on this failure have concluded that it can be mainly due to the cross-reactivity of the tested compounds and incorrect design of clinical trials.[15,18] However, further studies have been performed recently and they indicate that the inhibition of MMPs is still regarded as a valid strategy for several pathologies, including cancer.[7,19]

The earliest generation of MMPIs usually combines a zinc-binding moiety (ZBG, mostly of hydroxamates type) with a wide variety of structural molecular segments, which mimics the binding mode of the peptidic substrate. The second generations of MMPI have been associated with the knowledge of 3-dimensional (3D) structure of MMPs and also to the use of alternative ZBG.[20,21] The latest generation of MMPI has been orientated toward a even higher selectivity for target and antitarget MMPs.[22] Several reviews on MMPI are available, some including a general overview of the most recent results of this field.[7,23–25] This chapter will focus mostly not only on the development of inhibitors with peptide scaffold based on succinic and malonic hydroxamate but also on the evolutions occurred in this class of compounds, including truncated analogues, leading to new inhibitors with increased specificity.

24.2 THE MATRIX METALLOPROTEINASE FAMILY

The MMPs are a family of structurally related zinc-containing endopeptidases able to degrade all the protein components of ECM. They can degrade and deliver ECM from the tissues, activating cell-matrix and cell–cell interactions, cell migration, growth, differentiation, and apoptosis, as well as inflammation and angiogenesis. Therefore, they are involved in a number of not only normal but also abnormal cellular processes such as tumor growth, invasion, metastasis, and arthritis, and so they have been an important target for therapeutics in many inflammatory, malignant and degenerative diseases.[7,15] The activity of MMPs can be regulated by multiple mechanisms, namely, through gene expression and secretion in an inactive pro-enzyme form that must be cleaved to become active, and often it is the result of a complex protein cascade. Normally, the homeostasis of MMPs is maintained by their tissue endogenous inhibitors (TIMPS), but imbalances between MMPs and TIMPs that are involved in pathological processes, in which there is an uncontrolled overexpression of MMPs. Currently, at least 27 members of the mammalian MMP family have been identified and studied. Most of them fall into the following major subclasses, based on their substrate target: collagenases (MMP-1, -8, and -13), which degrade fibrillary collagen; gelatinases (MMP-2 and-9), which degrade denaturated and basement membrane collagens; stromelysins (MMP-3, -10, and -11), which degrade proteoglycans and glycoproteins; matrilysin (MMP-7) which degrade nonfibrillar collagen, fibronectinn and laminin; and macrophage elastase (MMP-12), which degrade many substrates, including elastin.[11,16] Table 24.1 lists the MMPs of current therapeutic interest, showing that they are mostly targeted to treat cancer or autoimmune diseases such as arthritis.

Human MMP isoenzymes possess three common domain structures (see Fig. 24.1): the prepeptide, propeptide, and the catalytic domains (e.g., MMP-7), but most of them contain an additional C-terminal hemopexin-like domain (collagenases, stromelysins, and metalloelastases) and further ones, like the transmembrane domain (e.g., MMP-14 and -15). The catalytic domain has a conservative motif of three histidine residues binding the Zn^{2+} ion. The prepeptide domain is a signal peptide that targets the MMPs for secretion. The propetide contains a conservative cysteine residue, which thiol group coordinates to the zinc atom, hiding the catalytic site from the substrate and so maintaining the pro-MMP in its inactive form (zymogen). Therefore, MMP activation involves cleavage of the prodomain and disruption of that Cys-Zn interaction (Cys switch). The C-terminal hemopexin-like domain is involved on the binding to the substrates and TIMPs. There is also a praline-rich hinge region, a flexible linker peptide that links the N-terminal catalytic and C-terminal hemopexin domain.

FIGURE 24.1 Example of structure of human matrix metalloproteinase domains, namely, simple Hemopexin domain-containing MMPs (MMP-1, -3, -8, -10, -12, -13, -18, -19, -20, -22, and -27. Pre, prepeptide, as the signal sequence; Pro, propeptide with a free zinc-ligating thiol (SH) group; catalytic, catalytic domain contains the highly conserved zinc-binding site with three coordinating histidines; H, hinge region; Hpx, hemopexin-like domain.

24.3 STRUCTURAL BASIS FOR THE MMP INHIBITION

The insight into the structural determinants of the MMP isozymes and their interactions with specific substrates or inhibitors, namely, with the catalytic domain, have been recognized of particular relevance for selective inhibitor design. The MMP structures and their interactions with inhibitors have been investigated by X-ray crystallography, nuclear magnetic resonance techniques, and some homology modeling.[11,12,26]

All the inhibitors of MMPs have their inhibitory capacity correlated with their binding to the catalytic metal ion, which is surrounded by a conserved active-site sequence motif, HEXXHXXGXXH, with three histidine residues coordinating the zinc, and also a glutamic acid residue, which facilitates catalysis.

Thus, any MMP inhibitor must contain a zinc-binding group (ZBG) attached to a framework, which should also interact with the catalytic site residues. The hydroxamate was found to be the preferred chelator for the design of MMP inhibitors. In fact, the hydoxamate (HA) moiety possesses two oxygen donor atoms, which bidentately bind to the catalytic Zn(II) ion of the enzyme, acquiring a distorted trigonal bipyramidal coordination geometry. Also, the hydroxamate anion forms a short and strong hydrogen bond with the carboxylate moiety of the Glu219, while the NH hydroxamate participates in a hydrogen bond with the carbonyl oxygen of Ala182 (see below).[11,12,27] Thus the existence of several strong interactions may be responsible for the higher affinity of HA-based MMPI, as compared with the MMPI inhibitors belonging to other chemical classes, such as carboxylates, thiols, and so on.

On the other hand, as stated above, the various MMPs exhibit different selectivity for the various matrix protein substrates. Therefore, the understanding of the substrate selectivity appeared as an important aid to design selective MMP inhibitors.

24.3.1 MMP Active Site and Binding to Inhibitors

Initial studies on the design of specific inhibitors were focused on determining the sequence about the cleavage site, and then many inhibitors were derived by substituting the scissile peptide bond by a ZBG.

The MMP active site is located in a large cleft having the catalytic zinc situated almost in the middle and partially exposed, facilitating the access and binding to the substrate or inhibitor. It has been proposed that, for the MMP proteolytic activity, the substrate should contain at least six amino acids (three on each side of the scissile bond). The non-prime (S_{3}-S_1 subsites) and prime (S_3'-S_1' subsites) sides of the inhibitor or substrate binds to, respectively, the left- and right-hand side of the catalytic Zn (see Fig. 24.2).

However, the non-prime side is much more explored to the solvent than the prime side (right-hand side), thus making weaker interactions with the inhibitors. The prime-side parts of the active site are much better defined, with a deep S_1' pocket and two shallow S_2' and S_3' pockets, so that the inhibitor which bind this side of the active site have higher affinity and selectivity. Crystal structures and NMR of known MMP-inhibitor complexes show that the inhibitors mostly bind to the non-prime

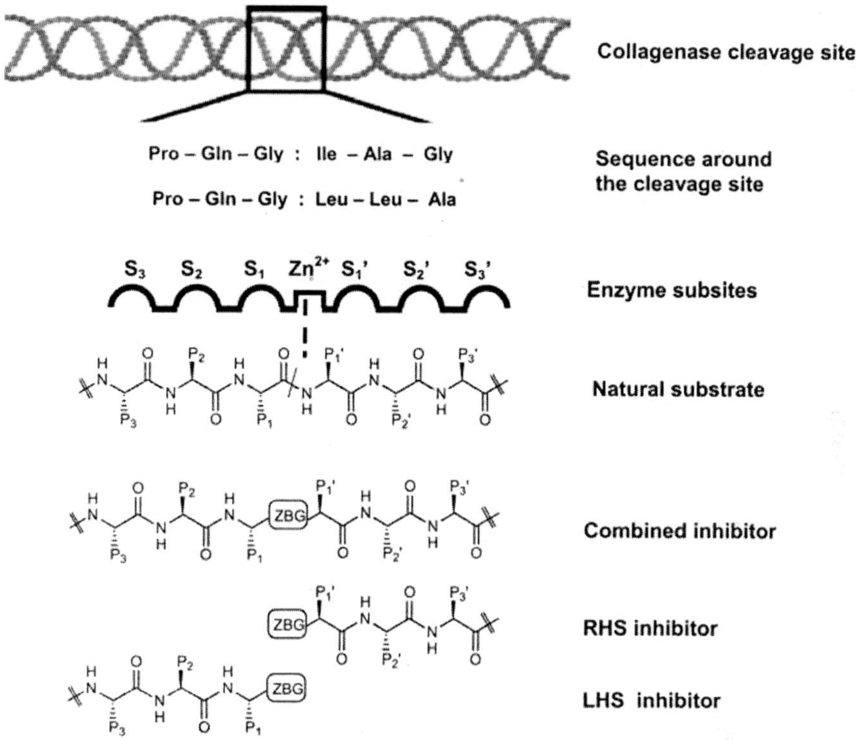

FIGURE 24.2 Representation of the design of matrix metalloproteinase inhibitors based on the peptide sequence around the cleavage site of collagen. ZBG, zinc-binding group; RHS, right-hand side; LHS, left-hand side. (Adapted from Ref. 12)

side of the enzyme.[10–12] Therefore, much of the efforts for the design of potent MMPIs have been concentrated on studying the S_1'-S_3' subsites and corresponding on the inhibitor P_i' residues, while the non-prime side has been much less successfully utilized for this proposal.[10,28,29]

The hydrophobic S'_1 subpocket is localized in close proximity to the zinc ion and forms a kind of tunnel that penetrates the surface of the enzyme. Studies indicate that, although this S_1' subsite is very flexible, allowing conformational changes, it has been considered the most determinant for the selectivity and so the most common target for MMPI selective design.[30] The selectivity seems to be mainly due to differences on the dept of the S_1' pocket. Based on X-ray crystallography, NMR analysis and homology modeling, the MMPs have been classified in two broad structural classes, those with a comparatively deep S_1' pocket (MMP-2, -3, -8, -9, and -13) and those with a shallow S_1' pocket (MMP-1, -7, and -11). Thus, inhibitors with long P_1' groups are more selective than those with smaller P_1' groups.[23] Concerning the S_2' subsite, as this is a solvent-exposed cleft, there is a general preference for hydrophobic P_2' residues. The S_3' subset is a comparatively ill-defined solvent exposed region, located at the edge of the

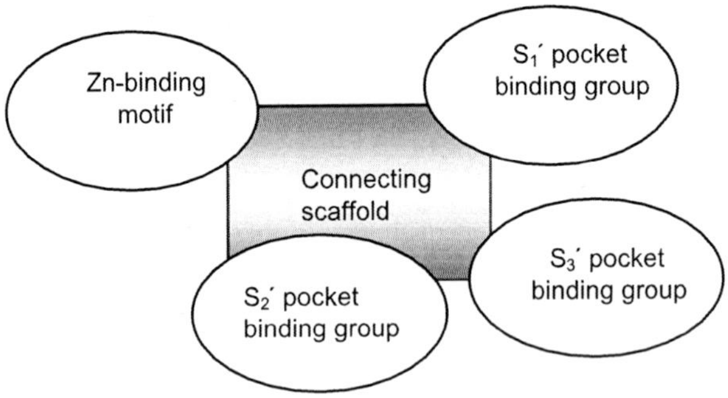

FIGURE 24.3 A simplified model of MMPI-binding site.

catalytic domain and so the P_3' residue has a very modest effect on the inhibitor selectivity.[2,11]

Thus, on the base of 3D similarity of the binding sites of MMP inhibitors, a simplified binding site architecture has been proposed for a huge number of MMP (see Fig. 24.3).[31]

24.4 INHIBITORS OF MATRIX METALLOPROTEINASES

Inhibition of MMPs has been widely sought as a strategy in intervention of these disease processes. The design and clinical use of synthetic MMP inhibitors (MMPI) to mimic the *in vivo* effects of TIMPs in diseases has attracted the attention from pharmaceutical industry for more than two decades.

The early approach to the MMPI development followed a substrate–base design, generally combining a zinc-binding group (usually a hydroxamate) with other structural residues, namely, a peptide-like structure which mimics the *in vivo* effects of TIMPs in diseases, specially the binding of the peptide substrate at the site of cleavage.[12]

As stated above, the MMPI that mimics substrates can bind to one or both the prime side and the non-prime side, but compounds of the first type revealed as much stronger inhibitors and so they were much more investigated. However, these early compounds were mostly nonspecific, acting as potent but broad-spectrum inhibitors. Whereas, a large number of broad-spectrum MMP inhibitors have been developed over the past decade, these inhibitors have not met the promise and expectations in clinical trials. As they present broad-spectrum inhibition, besides the MMPs, they often target other metalloproteinases, a feature that has been considered one of the potential problems responsible for the therapeutic inefficacy of MMPs inhibitors.

With the availability of X-ray crystallography and NMR methods for studying the structure of MMPs, as well as the simulating computational methods, the field of MMPI has evolved toward a second generation of compounds with structure-based design, which includes different scaffolds (nonpeptide), different structural motifs to improve selectivity,[10,16,32] different zinc-binding groups [17,21] and a few compounds even without any zinc-binding group (allosteric inhibitors).[33] Several generations of MMP inhibitors, showing improved selectivity for various MMPs, have been reported in the past few years and they can be promising drugs for effective targeting of the important selected enzymes.

24.5 ZINC-BINDING GROUPS. HYDROXAMATE INHIBITORS

The hydroxamate moiety, CONHOH, has been the ZBG mostly used in the development of MMP inhibitors although many other nonhydroxamate ZBGs, such as carboxylate, sulphydryl, and phosphonate, have been identified in the past,[11,34] and in the present, mostly due to metabolic lability attributed to this ZBG.[35] Specifically, there are recent reports on MMP inhibitors based on carboxylate,[36] thiol,[37] phosphinic,[38] and a variety of heteroccycles, namely, pyrone[39] and pyrimidinone compounds.[40,41]

However, since long time, the hydroxamate has been recognized to be the strongest ZBG for the inhibition of MMPs.[42] In fact, as stated above, the hydroxamate acts as a bidentate ligand forming a strong chelate with Zn(II) ion of the active site; on the other hand, the position of the hydroxamate anion and the amine proton seems suitable to form a hydrogen bond with a carbonyl oxygen of the enzyme backbones (Glu219 and Ala182), thus enhancing the inhibitor-enzyme interaction at the zinc site[27] (see Fig. 24.4). The remaining residues (R_1, R_2, etc.) also play a significant role on filling specific pockets of the enzyme, thus being relevant for the selectivity of the inhibitors.

In the next two sections, we will be focused on hydroxamate peptidomimetic and nonpeptidomimetic MMP inhibitors, with emphasis on the recent trends in this area of medicinal chemistry. Several main classes of peptidomimetics will be reported:

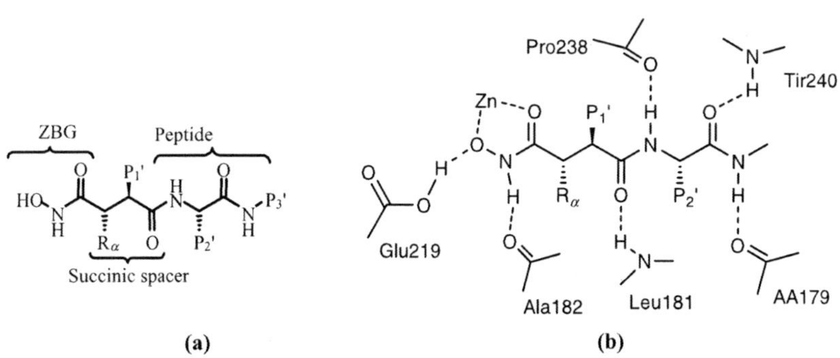

FIGURE 24.4 (a) Generic structure for peptidic succinic hydroxamate inhibitors showing the R_α P_1', P_2 and P_3'substituent groups interacting, respectively, with the S_1, S_1', S_2 and S_3' enzyme subsites and (b) some binding interactions on the MPP8- inhibitor complex. (from Ref. 12)

the succinyl hydroxamates and their derivatives, including truncated analogues; the malonic acid hydroxamate. Among the truncated analogues, special reference will be given to iminodiacetyl hydroxamates and a brief reference will be given to the sulphonamide hydroxamate based inhibitors.

24.6 PEPTIDOMIMETIC HYDROXAMATE INHIBITORS

24.6.1 Succinyl Hydroxamate Inhibitors

The succinyl hydroxamate-based MMPI belong to the first generation of inhibitors, with peptide-like backbone to mimic the substrate site of cleavage by the enzyme. The succinyl hydroxamate MMPIs have been proved to be much stronger inhibitors than the analogues derived from malonic acid,[43] and so most of the early research in this area was focused on the series of succinyl hydroxamates,[11] namely, on modification of their side chains.

24.6.1.1 Earlier Succinyl Inhibitors. Batimastat and Marimastat The investigation on earlier MMP inhibitors started from peptidic succinyl scaffolds and a lot of search was dedicated to their derivation, to find the best P'_i substituents that could maximize the interaction with a specific enzyme, and improve the selectivity and the capacity of the inhibitory activity. Among the group of the succinyl hydroxamate MMPIs developed, Batimastat (BB-94) (**24.1**) and Marimastat (BB-2516) (**24.2**) (see Fig. 24.5) are the most important ones and they were tested

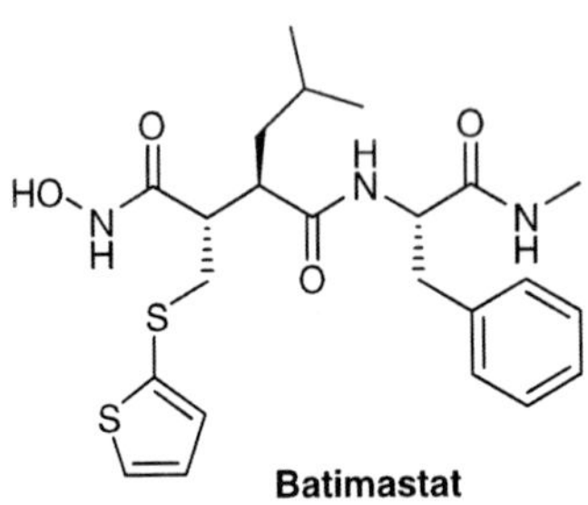

Batimastat

BB-94 (**1**)

IC_{50}		
	MMP-1	10 nM
	MMP-2	4 nM
	MMP-3	20 nM
	MMP-8	10 nM
	MMP-9	1 nM
	MMP-14	3 nM

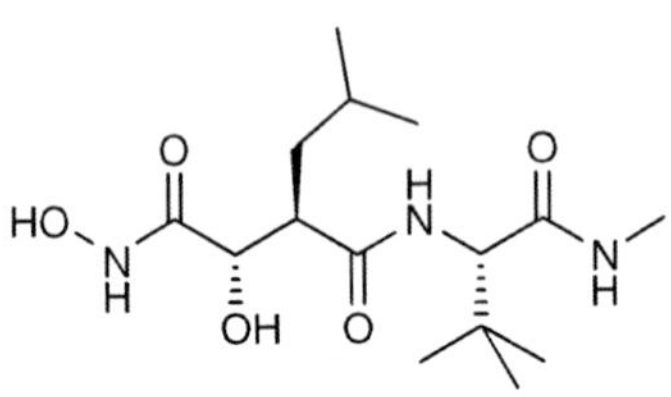

Marimastat

BB-2516 (**2**)

IC_{50}		
	MMP-1	5 nM
	MMP-2	8 nM
	MMP-3	200 nM
	MMP-7	20 nM
	MMP-8	2 nM
	MMP-9	3 nM
	MMP-14	1.8 nM

FIGURE 24.5 Structure of the two earlier MMP succinyl inhibitors: Batimastat and Marimastat.

in humans. They both are collagen-based peptidomimetic MMPIs and were discovered by British Biotech Pharmaceutical.[11] They are generally broad-spectrum MMPIs and showed good *in vivo* activity in several diseases models, and later they entered for clinical studies.

Batimastat (**24.1**) showed high inhibitory activity against several enzymes (MMP-1, -2, -3, -7, and -9) and was the first MMPI to enter clinical testing (phase I, 1994).[44] Batimastat and Marimastat are closed related inhibitors, but with different α-substituent groups (P_1), namely, a thienylthiomethylene and a hydroxyl group, respectively. Because of the poor solubility, Batimastat could not be administered orally neither intravenously. However, the direct injectable administration (parenterally: intraperitoneal or intrapleural) caused serious problems so that, although phase III trial was initiated, it was closed soon. The major side effects were found along efficacy improvements, namely, the higher dosage administration of Batimastat results, after 3–5 months, in a muskulosketal problem, which manifests itself as syndrome (MSS syndrome) with musculoskeletal pain and inflammation.[45]

Unlike Batimastat, Marimastat is orally active with long plasma half-life of 8–10 h.[46] Marimastat has undergone several phase III clinical trials for several types of cancers and demonstrates to reduce the rate of tumor development in a variety of malignancies.[16] It was evidenced as a high potent and reversible MMPI with IC_{50} values in the nanomolar range against MMP-1, -2, -3, -7, -9, and -12.[47] However, in spite of being well absorbed and tolerated, either administered alone or in combination with cytotoxic drugs (e.g., paclitaxel or carboplatin) for several cancer types, with positive responses, Marimastat also present some reversible musculoskeletal problems, which occurred in a dose- and time-dependent manner.[48] Despite this limitation, Marimastat continued being considered as effective as some conventional therapies, namely, in treatment of pancreatic carcinoma patients, and showed to improve survival in advanced gastric cancer.[49]

In summary, these peptide-mimetic early inhibitors were potent (nanomolar range) but, apparently because of their broad-spectrum activity, they present some pharmacokinetic limitations and some other problems. Of special concern are the musculoskeletal side effects, which may result from impairment of normal tissue remodeling governed by some MMPs (e.g., MMP-1) and/or by shedases such as TNF-α convertase, with the concomitant excessive matrix deposition leading to fibrosis.[8,50] Therefore, the lack of activity of inhibitors against MMP-1 has been considered as an important factor to reduce the side effects of broad-spectrum MMPIs.[51]

24.6.1.2 Variations in Succinyl Hydroxamate Inhibitors The properties of Batimastat and Marimastat prompted several researchers to perform multiple structure–activity relationship (SAR) studies on right-hand side MMP inhibitors with succinyl scaffold. The main conclusions about the most determinant positional groups and their characteristics, in relation with specificity and activity, are schematically represented in Fig. 24.6, and they can be summarized as follows:

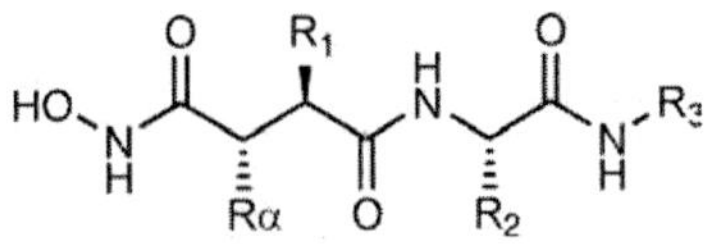

FIGURE 24.6 Summary of structure–activity relationships for right-hand side succinic hydroxamate MMP inhibitors. (Adapted from Ref. 11)

(1) The substituent group at the α-carbon atom (P_1) can be modified to improve the oral bioavailability and the selectivity toward MMP-1 and -3; it can also be cyclized, together with the P_2' substituent group.
(2) P_1' is the major determinant group for the selectivity and activity of the inhibitor; although the size and shape depend on the enzyme target S_i'pocket, it should be a lipophilic group.
(3) -P_2' and P_3' can include a wide range of different type of substituent groups but, usually, large groups can improve the oral bioavailability.

Due to the above reported limitations in the field of the strong peptide-like MMPI (e.g., succinyl hydroxamate inhibitors), further developments were carried out by several researchers and pharmaceutical laboratories[11,23] that were aimed at improving the most determinant properties such as selectivity and bioavailability. They were mostly based on the knowledge of structural data (X-ray crystallography and NMR studies) of an inhibitor complexed within the active site of an enzyme that was then used to model the complex and build other refined models. In this case, the design strategy was a combination of the designed substrate- and structural-based approaches.

This involves a number of variations to improve the selectivity and the bioavailability, namely, changes of the substituent groups of the α-carbon atom, variations of the P_1'-P_3'moieties and cyclized analogues and also retro-hydroxamate analogues.

Modifications on P_1 and Cyclized Analogues The X-ray crystal structure of the complex Batimastat–human neutrophil collagenase[52] showed that the α-thienylthiomethylene group, P_1 substituent (α-positioned to the hydroxamic acid moiety), and the P_2'phenylalanine side chain pointed away from the active site and in close proximity to each other. This observation suggested that a macrocyclic structure connecting these two groups would not disturb the binding mode and could favorably alter the pharmacokinetic profile for these peptide-mimetic inhibitors. This strategy was followed by two independent research groups, leading to the identification of the same optimized compound, SE-205 (**24.3**) (macrocycle 3), which presented equipotent to the uncyclized Batimastat (broad-spectrum and nanomolar order) but with a substantial increase in the aqueous solubility. Also, variations on the macrocycle size or on the terminal amide group lead to little effects on the *in vivo* MMP inhibitory activity.[53,54]

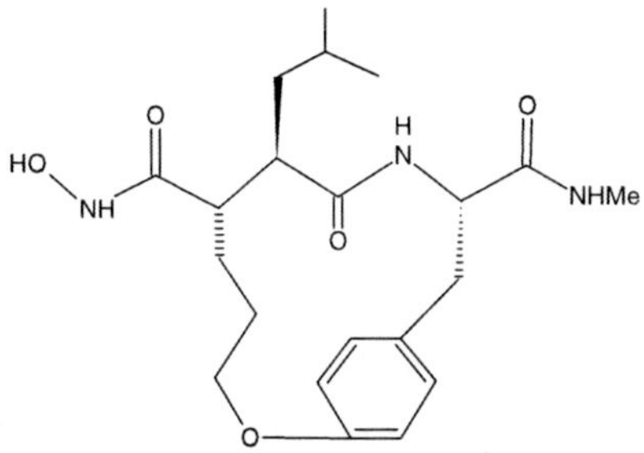

SE-205 (24.3)

Noteworthy is the fact that a variety of residues were also included in the macrocyclic ring (13–16-membered), namely, amide, carbamate, alkyl, sufonamido, amino, as P_1-P_2' linkers, and some of them showed high inhibitory activity against the TNF-α (tumor necrosis factor α) converting enzyme (TACE).[55]

Besides the above referred variations on the P_1 substituent (thienylthiomethylene and hydroxyl group) which led, respectively, to Batimastat (BB-94) (**24.1**)[56] and Marimastat (BB-2516) (**24.2**),[57] the introduction of the allyl and hydroxymethyl groups led to the compounds BB-1101 (**24.4**)[58] and TNF-484 (**24.5**),[59] respectively, all of them were equipotent broad-spectrum inhibitors.

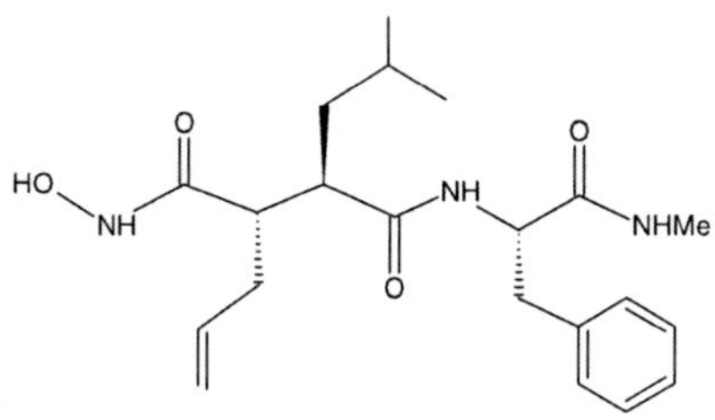

BB-1101 (24.4)

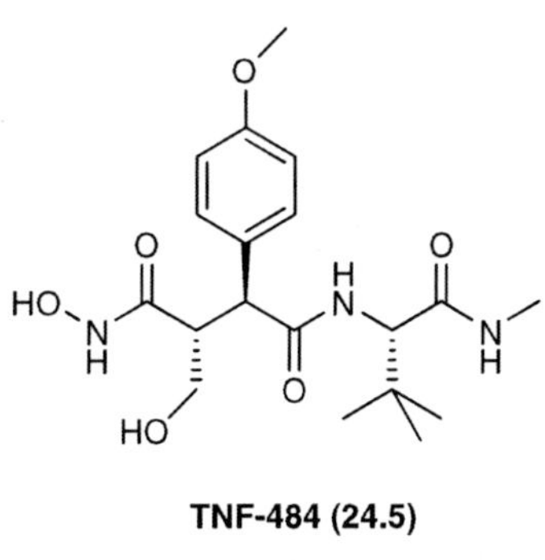

TNF-484 (24.5)

Modifications on P_1 groups have a major influence on the aqueous solubility and oral availability, due to their interaction with the solvent molecule (mainly achieved for Marimastat). Although these variations produce minor MMP inhibition selectivity, BB-1101 (**24.4**)[11] and mostly TNF-484 (**24.5**), (which was resulted from modification at the P_1'group by introduction of a β-aryl (*p*-metoxyphenyl) group) presented also potent inhibitory activity against TACE. This is an important feature because, since a long time, this dual MMP/TACE inhibitory activity has been considered of high relevance for some pathologic processes that involve both inflammation and matrix remodeling (e.g., arthritis[61] and multiple sclerosis).[62] Further modification on P_1 group, namely, the replacement of the hydroxymethyl (a potentially metabolic weak point due oxidation/glucorinidation) in TNF-484 by other groups (e.g., hydantoin residue) did not lead to any improvement on inhibitory activity, but weakened the pharmacokinetic profile due to the higher clearance.[63]

P_1' *Group Modifications* As previously referred, the selectivity of MMPI can be potentially achieved by taking advantage of differences in the size and depth of the S_1' pocket of the various MMPs. X-ray crystallographic analysis evidenced that the S_1' pocket is longer for a major part of the enzymes (MMP-2, -3, -8, and -13) but it is short for a few MMPs (e.g., MMP-1 and -7).[11] Consequently, the incorporation of small P_1' groups generally leads to broad-spectrum activity, whereas the extended P_1' groups tend to present a selective inhibition for the deep pocket enzymes over the short pocket enzymes. For example, on the macrocyclic derivative **24.3** with 700-fold selectivity of MMP-2 over MMP-1 was obtained by changing R_1 from $-(CH_2)_3OCH_2Ph$ to a $-(CH_2)_3$(4-tolyl) group.[23]

Variations at the P_2' *and* P_3' *Groups* The subsite S_2' is partially solvent-exposed, while S_3' is even much more solvent exposed and, thus, the corresponding P_2' and P_3' groups on the inhibitors can be comparatively polar. However, differences on the residues surrounding these subsites have been used for selectivity purposes. For example, P_3' group is usually larger in MMP-1 as in MMP-3 inhibitors.[23]

On the other side, although both MMP-2 and -3 have been described to contain deep S_1' pockets, for the development of MMP inhibitors as anticancer agents, MMP-2 represents a preferential target. Therefore, new succinyl hydroxamate have been developed with selective inhibition of MMP-2 over MMP-3, namely, Ilomastat (GM-6001) (**24.6**), having an indole as the P_2'group. Also, some analogues of Ilomastat with an isobutylidene group at S_1' (providing approximately a 100-fold greater selectivity for MMP-2 inhibition over MMP-3) and modifications for the S_2' group (e.g., indole to modulate the selectivity of MMP-2 over MMP-3) and S_3'subsites (**24.7**) to increase their overall hydrophobicity.[64]

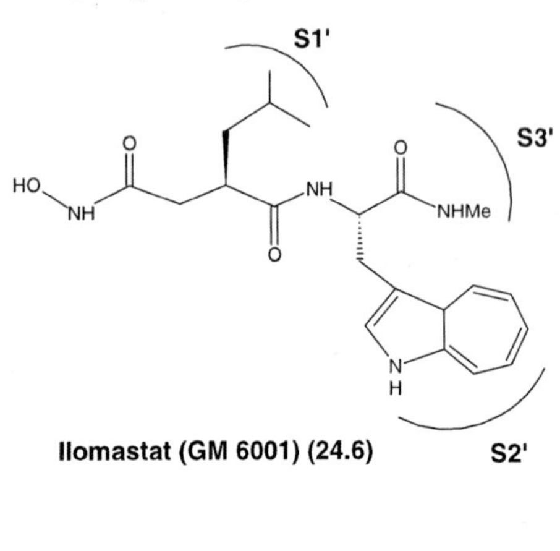

Ilomastat (GM 6001) (24.6)

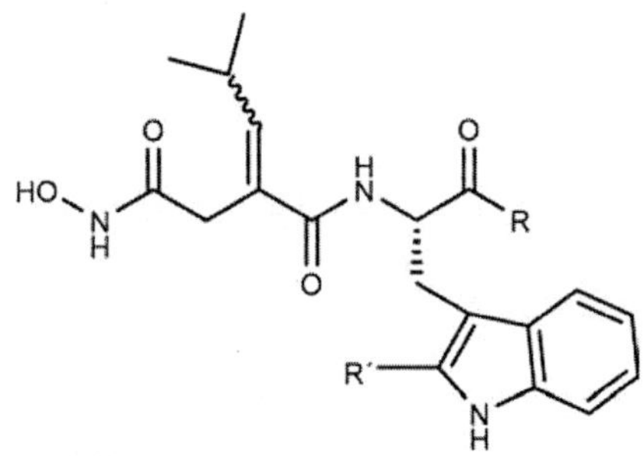

1 R = NHMe, R' = H

2 R = OEt, R' = H

3 R = NHMe, R' = $(CH_2)_2NMe_2$

24.7

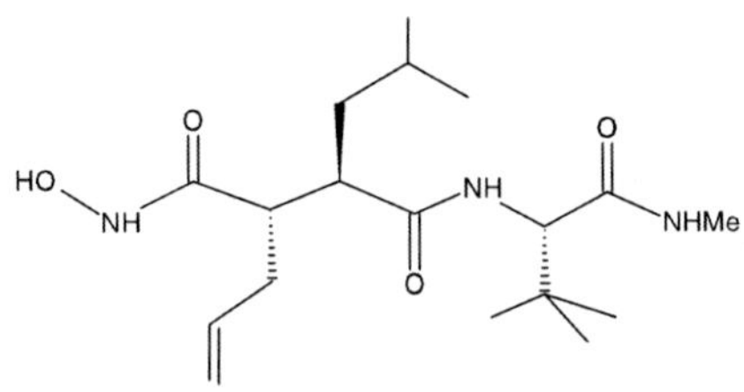

Ro 31-9790 (24.8)

IC_{50}	MMP-1	10 nM
	MMP-2	8 nM
	MMP-3	700 nM
	MMP-14	1.9 nM

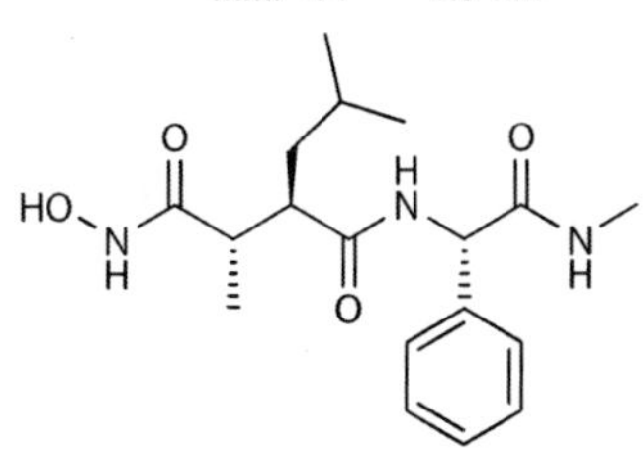

KB-R7785 (24.9)

Interestingly, the introduction of a *tert*-butyl as P_2' (Ro 31-9790) (**24.8**) group also induces the same type of selectivity (700:8). Furthermore, both *tert*-butyl (KB-R7785) (**24.9**) or phenyl group (**24.10**) at the P_2' position can promote the oral activity and improve the pharmacokinetic profile, a feature that may be attributed to the role of the amide as a shielding moiety (as it was proposed for the *tert*-butyl as α-substituent to the hydroxamate group).[65]

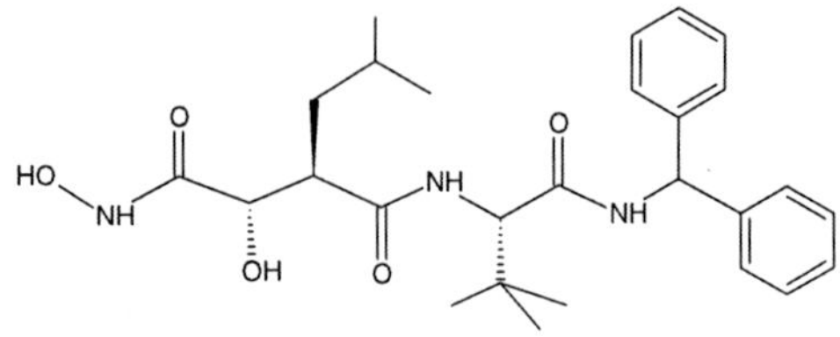

24.10

IC_{50}	MMP-1	600 nM
	MMP-2	3000 nM
	MMP-3	50 nM
	MMP-7	4 nM

Other modifications performed on P_2' and P_3' substituents, included their cyclization to form a lactam, but for the optimized ring size only a 10-fold increase was achieved compared with the acyclic analogues.

Modifications at the C-Terminal Amide Group (P_3') Several modification o the P_3' position have been successfully performed, namely, by replacing the labile terminal methyl amide of Marimastat (**24.2**) by an aryl amide moiety (e.g., benzhydril) (**24.10**), which led to the improvement of the selectivity of MMP-3 and -7 over MMP-1 and -2.[11,66]

Variations on the size of the P_3' group showed a, comparatively, little impact on the potency of MMP-3 inhibition. This is according to the X-ray crystallography which showed that the S_3' site is open and solvent exposed, whereas the inhibition of MMP-2 was found to be the most dependent on P_3'.[67]

24.6.1.3 *Truncations on the P_2'-P_3' Group of Succinyl Hydroxamate Inhibitors* Some authors have further replaced the P_2'-P_3' amide bond by different functionalities, in some cases leading to truncations of the initial scaffolds with three amide bonds. For example, in Ilomastat (**24.6**), the substitution of that bond by the imidazole group (**24.11**) results the decrease of the inhibitory potency against all MMPs tested, but induced a selectivity enhancement for MMP-1 and MMP-7 over MMP-3.[68]

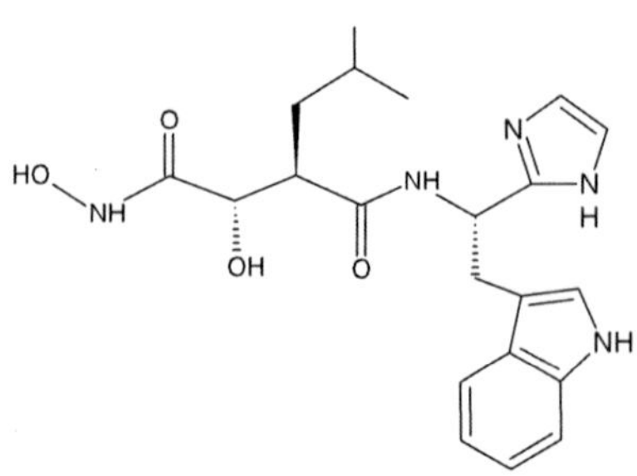

24.11

IC_{50}	MMP-1	3 nM
	MMP-3	280 nM
	MMP-7	184 nM

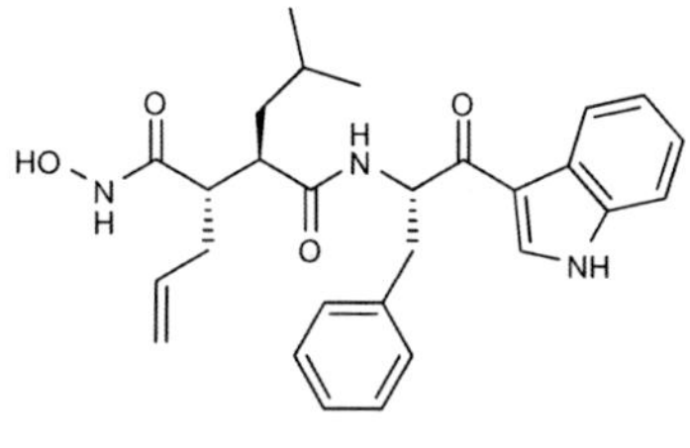

24.12

IC_{50}	MMP-1	1.1 nM
	MMP-2	1.1 nM
	MMP-3	2.3 nM
	MMP-7	2.2 nM

The P_2'-P_3' amide bond was also replaced of by aryl or heteroaryl ketones.[22] In particular, an analogue of BB-1101 was investigated with that amide bond replaced by 3-acyl indole (**24.12**).[69,70]

The incorporation of a C-terminal ketone was also investigated in the macrocyclic series (P_1-P_2' cyclized), as analogues of SE-205, with P_3' group as phenyl ketones.[53]

It was found that the inhibitory activity of these ketone derivatives is comparable to that of amide analogous, but they displayed some improvement on the oral availability with, namely, arylamino ketones.

24.6.1.4 Other Mutations. Isosteric Analogues Many other modifications on earlier peptidic succinyl hydroxamate have also been reported, but only the most representative ones have been included herein. Noteworthy is a very recent work on analogues of this class of peptidomimetic inhibitors, which studied the effect of replacing the CH_2–CO–NH–OH chain of the succinyl hydroxamate by the NH–CO–NH–OH moiety, inserted in the same peptidic scaffold.[71] Thus, according to that design approach, peptidyl 3-substituted 1-hydroxyureas could be admitted as isosteric analogues of the potent earlier peptidic succinyl hydroxamate MMP inhibitors (e.g., Batimastat, see Fig. 24.7), and so several parallel models have been studied with the same type of substituent groups.

Based on the quite amount of similarities between these two classes of isosteric analogues, they could be expected to present parallel behaviors, in terms of fitting into

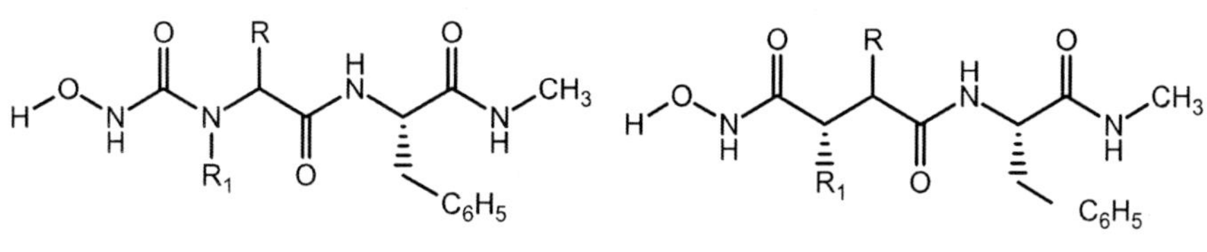

FIGURE 24.7 (a) Peptidyl 3-substituted 1-hydroxy-urea and (b) the analogous succinyl hydroxamic acid. (Adapted from Ref. 71)

the subpockets of the enzyme active site. However, the results indicated a drastic decrease in the potency of the *N*-hydroxy-urea derivatives with respect to the hydroxamate analogues, generally, changing from nano- to micromolar range. Therefore, in spite of being isosteric analogues, that extra nitrogen accounts for considerable differences on zinc-binding efficacy. This can be due, at least partially, to the electron-withdrawing effect of the nitrogen atom, which may lead to a drastic change from a strong bidentate to a weaker monodentate binding mode, as found with other hydroxyureia nonpeptidic MMP inhibitors.[71]

24.6.2 Malonyl Hydroxamate Inhibitors

The interest for introducing structural variations on the succinyl hydroxamate inhibitors may also be on the basis of the development of malonic acid hydroxamate derivatives.[2,72] They can be admitted as the result of truncation of the earlier compounds, succinyl hydroxamate-based MMPIs, namely, through the fusion of two methylene groups of the succinyl spacer (between ZBG and the peptide backbone). They appeared as a new lead structure for MMPI. However, inhibition studies demonstrated that the malonyl hydroxamic acid derivatives (**24–13, 24–14**) were much less effective inhibitors than the corresponding succinyl analogues (one methylene group as spacer). X-ray studies of the malonyl hydroxamate inhibitor–enzyme complex, aided the rationalization of these results, showed that these inhibitors bind to the enzyme through a nonsubstrate like mode, in opposition to the succinyl derivatives (see Fig. 24.8).[72,73]

Based on the structure-based knowledge, a large variety of malonic hydroxamate based derivatives have been described, most of them containing the peptide-like moiety (**24.13**). Some improvements on the peptidic tail of the malonyl hydroxamates to fit the S_2'pocket, namely, changes into another peptidic or to a nonpeptidic tail, modifications on the P_1'group to increase the interaction with the S_1'pocket, and on the

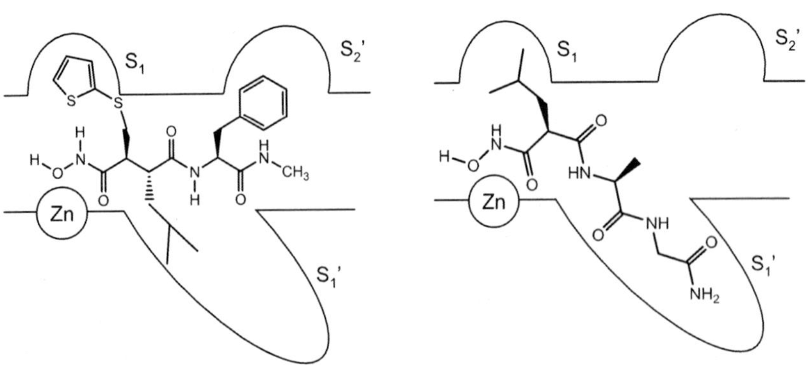

FIGURE 24.8 Schematic representation of the substrate-like binding mode of Batimastat (left) and the nonsubstrate-like binding mode of a malonic acid base hydroxamate inhibitor (right) to MMP-8. (From Ref. 72)

P_1 group (e.g., from isobutyl to a benzyl group), led to the discovery of low molecular weight nonpeptidic inhibitors (**24.14**) with 500–1000-fold improved inhibitory potency.[72,74,75]

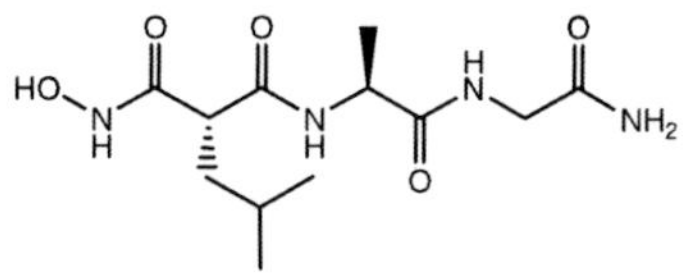

24.13

IC_{50} = 121 μM, MMP-8 ....

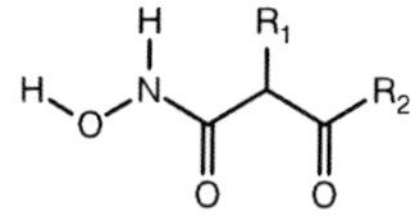

24.14

IC_{50} = 0.24 μM, MMP-8

(R_1 = isobutyl; R_2 = NH-*n*-octyl)

However, generally, their activity was lower than that of the corresponding succinyl hydroxamates, and so further research for the optimization of malonic acid based derivatives as MMP inhibitors was not pursuit as happened with the succinyl analogues.

24.7 EVOLUTION FROM PEPTIDIC TO NONPEPTIDIC INHIBITORS

In spite of their complexity and the inherently high effort involved on the development of the early peptide-mimetic inhibitors, such as Batimastat and Marimastat, they are now admitted as pioneering MMPI,[76] and a number of variations has been reported.

24.7.1 Succinyl Hydroxamate Inhibitors

The most recently reported trends involve the truncation of the peptide-like backbone and even its substitution by nonpeptide ones. Following the same tendency of modifying the structure of pseudopeptide succinyl hydroxamate, namely at the P_2'-P_3' molecular fragment, another important compound was identified (Trocade or Ro 32-3555) as a very strong collagenase inhibitor with potential application as an antiarthritis agent.[77,78]

In this case, there is a truncation of P_2'-P_3' group and, thus leading to the transformation of peptidic into a nonpeptidic succinyl hydroxamate inhibitor. The introduction of a cyclic imide group at the P_1 (α-substituent) seems to provide a favorable balance between the active-site interaction and the solvation, in spite of the removal of the H-bonds associated with succinyl hydroxamates with the P_2' group. The introduction of the cyclopentylmethyl at the P_1', instead of the usual *tert*-butyl, improve efficacy, thus suggesting a better complementarity with the S_1' subsite. The fact that Trocade presents selectivity for the three members of collagenase family (MMP-1, -8, and -13) over MMP-2, -3, and -9, together its good oral bioavailability and cartilage protecting roles evidenced in animal models, make it a promising drug candidate for the therapy of both rheumatoid arthritis and osteoarthritis.[77]

Among many other variations associated with the truncation of P_2'-P_3' group, it is noteworthy that its replacement by a hydrazide or a piperidine moiety leading to Ro 32-7315 (**24.15**) and Ro 32-3555 (**24.16**), respectively. The first compound presented moderate MMP inhibition, but potent TACE inhibitory activity. These two compounds were selected for clinical tests.[79] It is also interesting that the modification made on TNF-385 (**24.5**), namely, the replacement of the P_2'-P_3' group by a pipecolic amide, which also leads to a dual MMP/TACE inhibitors that entered in preclinical trials (**24.17**).[63]

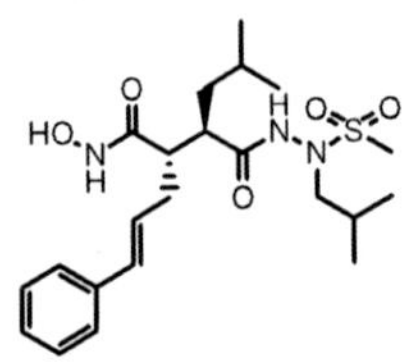

24.15

IC_{50}	MMP-1	500 nM
	MMP-2	250 nM
	MMP-3	210 nM
	MMP-7	310 nM
	MMP-9	100 nM
	MMP-12	11 nM
	MMP-13	110 nM
	TACE	5 nM

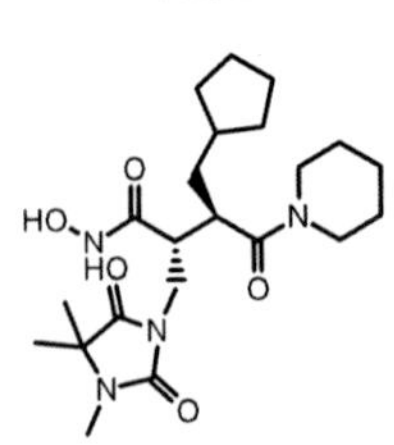

24.16

IC_{50}	MMP-1	3 nM
	MMP-2	154 nM
	MMP-3	527 nM
	MMP-8	4 nM
	MMP-9	59 nM
	MMP-13	3 nM

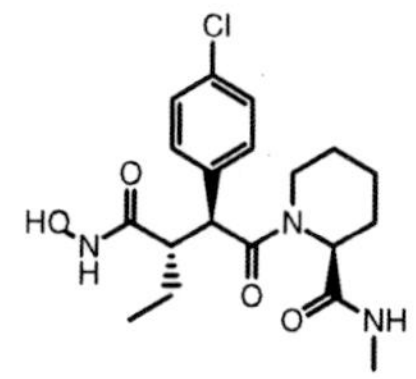

24.17

IC_{50}	MMP-1	0.07 nM
	MMP-3	5 nM
	MMP-13	2 nM
	TACE	5 nM

24.7.2 Iminodiacetyl Hydroxamate Inhibitors

Following the general interest of improving the bioavailability of the matrix metalloproteinase inhibitors by substituting the earlier peptidic-based succinyl hydroxamates by smaller size nonpeptidic analogues, a new class of nonpeptidic hydroxamate inhibitors, based on iminodiacetic (IDA) scaffolds, was developed in our group.[80] These iminodiacetic-hydroxamate (IDA-HA) inhibitors can also be admitted as truncated mimics of potent peptidic succinyl-hydroxamate MMPIs, such as Marimastat or Batimastat. In this case, the truncations occur at the succinic spacer (two atoms on the *beta* and *gamma* positions compared with the hydroxamate moiety) and at the peptidic part to eliminate the amide bonds (see Fig. 24.9). Three positions of the IDA scaffold can be easily provided with a variety of substituent groups, for improving the interaction with different pockets at the active site, and concomitant enhancement on the selectivity and activity on the inhibition of a specific MMP target. The main variations on the IDA-hydroxamate scaffold for the development of MMPIs include the following substituents: R_1, with a wide variety of lipophilic groups, mostly aromatic, eventually containing some heteroatom; R_2, as hydroxyl or alkylaryl groups eventually containing heteroatom; R_α, as a H atom or an *i*-propyl group. The hydroxamate moiety (amine nitrogen) can also be cyclized with R_2, leading to formation of a cyclic compound (IDACYHA).[81] Noteworthy is also the fact that, generally, these nonpeptidic IDA-hydroxamate based compounds are more easily made than the earlier peptidic inhibitors.

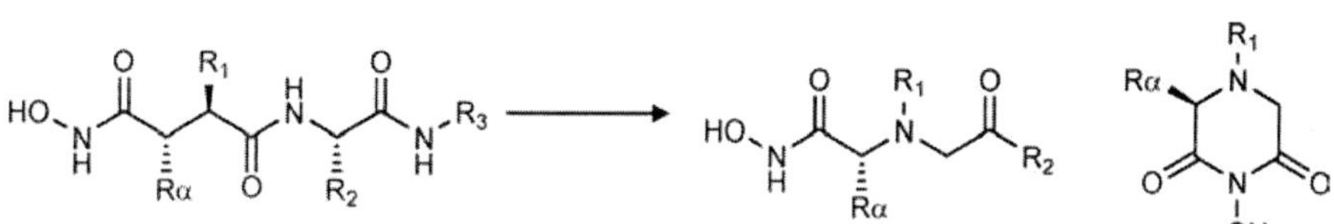

FIGURE 24.9 Schematic representation of the scaffold truncation from succinyl-HA to IDA-HA based MMPIs.

The design approach for IDA-HA inhibitors followed three main stages. Firstly, quite simple compounds were designed and studied, by introducing only modifications on R_1 of the IDA-HA skeleton, while the other groups were kept unsubstituted ($R_2 = OH$; $R_\alpha = H$).[82] Thus, a set of *N*-substituted IDA-HA derivatives, namely, *N*-alkylaryl-, *N*-alkyl/cycloalkyl-, and *N*-sulfonylaryl-iminodiacetic hydroxamate acids were developed, aimed at improving the potency and selectivity versus MMP-2 and -9 (two usual MMP targets in cancer therapy) and sparing MMP-1 and-14 antitargets. The screening of MMP inhibitory activity indicated that the IDA-HA derivatives with more polar and hydrophilic *N*-substituent groups (R_1) led to lower inhibitory activity than those with the more hydrophobic ones (e.g., *N*-alkylaryl and *N*-sulfonylaryl). This is in agreement with the work of other authors who have shown that the interaction with the S1′ pocket is favored by hydrofobic groups.[11,83,84] A series of compounds with different hydrophobic groups, namely, *N*-alkylaryl and the corresponding *N*-sulfonylaryl analogues (aryl as *p*-metoxi-phenyl, *p*-phenoxy-phenyl, and *p*-biphenyl) was screened against a set of MMPs. The results clearly indicated that for these two types of compounds, the substitution of a methylene by a sulfonyl group leads to a considerable increase on inhibitory potency, with IC_{50} changing from the micro- to the nanomolar ranges.[85] Among the *N*-sulfonylaryl-based inhibitors, the use of bulkier/lengthier but also more flexible/adaptable bis-aryl as *N*-substituent groups (*p*-phenoxybenzene group) (**24.18**) showed a considerable improvement on their inhibitory potency against the MMPs characterized by a deep S_1' pocket, such as MMP-2, -8, -9, -13, -14 and -16, with IC_{50} ranging from 1 to 30 nM, according to the following order MMP-2 > MMP-13 > MMP-9 > MMP-8 > MMP-16 > MMP-14, and -27). This is according to others.[24,86]

Furthermore, the *N*-arylsulfonamide-based inhibitor **24.18** showed also a very low activity toward MMP-1, an antitarget MMP which inhibition has been associated with the musculoskeletal side effects.[87] Modeling studies with an MMP-2 model corroborated the bioassay results, showing that the *N*-sulfonyl-(*p*-phenoxy-phenyl) is the best fitted group in the S1′pocket. This is because the sulfonyl group enables extra H-bond interactions with oxygen atoms of amino acid residues, but mostly due to its bent spatial disposition, allowing a lipophilic interaction of the second phenyl ring with lipophilic residues of this pocket. Furthermore, the hydroxamate group interacts with zinc and forms two H bonds with Ala165 and Glu202 residues, while the carboxylate group is orientated toward the S_2' region.

The high activity and MMP-2/-1 selectivity of **24.18** lead to further investigation on the effect of the R_2 group (P_2'substituent). Some improvements on MMP inhibitory activity were associated with the change of OH by 4-phenyl-piperazine-propyl or 4-sulfamoyl-phenyl-ethyl (SPE) group (**24.19**). Furthermore, for compounds containing this type of substituents, high inhibitory activities (nanomolar range) and selectivities, against tumor-associated carbonic anhydrase isoforms (e.g., CA IX), were also detected and thus, compound **24.19** evidenced potential as a double targeting drug, which may be revealed as an important feature for some types of malignant tumor processes.[88]

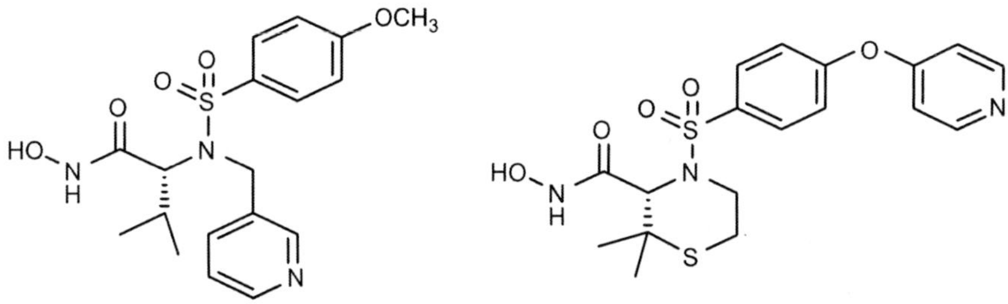

CGS-27023A (24.22) **Prinomastat (24.23)**

FIGURE 24.10 Examples of structures of second generation of MMP inhibitors.

Modifications at R_α group by insertion of the (R)-isopropyl group (**24.20**) led also to improvement of the inhibitory activity against all tested MMPs, as compared with the nonsubstituted analogue, although with some lost on the selectivity of MMP-2 over MMP-1.

The novel cyclic hydroxamate compounds (IDACYHA) (**24.21**) are analogues of iminodiacetic-hydroxamate inhibitors (**24.20**), but with the hydroxamate moiety cyclized with R_2 group. Compound **24.21** is equipotent of the noncyclic analogue (**24.20**) but with improved selectivity for MMP-2 over MMP-1. Although these new cyclic ZBG-containing derivatives are still under investigation, their high stability in aqueous solutions associated with the evidenced high potency and selectivity, make them potential candidates to pursuit in preclinical trials.

24.7.3 *N*-Sulfonyl Aminoacid Hydroxamate Inhibitors

A new generation of MMP inhibitors based on sulfonylated amino acid hydroxamates have also recently emerged, mostly based on the structural protein information and molecular modeling. These nonpeptidic compounds can ultimately be admitted as the result of a further extreme truncation of the peptide-like succinyl hydroxamate inhibitors. The first compound to be developed for clinical trials was an orally active broad spectrum (CGS-27023A) (**24.22**)[89] and then Prinomastat (AG-3340) (**24.23**)[51] (Fig. 24.10), which present excellent potency against MMP-2 (sub-nM), and with some MMP-1 selectivity (10 nM). In recent years, this new generation of MMPIs has been extensively explored to improve MMPI selectivity and bioavailability, as described in next chapter.[22,90]

24.8 SYNOPSIS. FROM EARLIER PEPTIDIC SUCCINYL HYDROXAMATE MMP INHIBITORS: MODIFICATIONS AND TRUNCATIONS

Along this chapter, besides the description and discussion of the earlier MMP inhibitors based on peptidic succinic hydroxamate inhibitors, some emphasis was

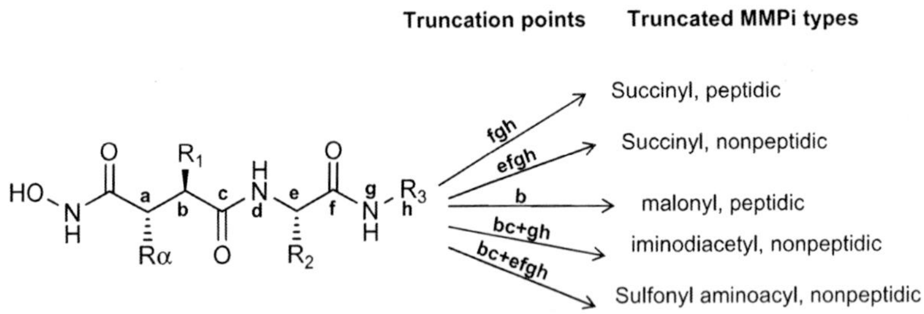

FIGURE 24.11 Graphical summary of evolution of hydroxamate-based MMP inhibitors.

also put on the structure–activity relationship associated with substituent modification on the starting peptide succinyl hydroxamate scaffold and on their evolution in/from peptide mimetics toward the nonpeptide derivatives. In some cases those evolutions appeared as truncations of the early succinyl skeletons, in small steps toward new generation of compounds, and they were mostly aimed at improving selectivity, potency, bioavailability, and feasibility. A graphical summary of the above-described evolution on hydroxamate inhibitors is presented in Fig. 24.11 aimed at facilitating a general view and understanding of those main changes.

This picture illustrates how the earlier substrate-based succinyl peptidic hydroxamate inhibitors appear as pioneers of the recently developed "truncated" structural-based peptidic and nonpeptidic derivatives, some of them belonging to the new generation of MMP inhibitors. As substrate-based earlier compounds have been modified aided by structural knowledge, distinction between these two types of compounds is becoming more tenure.

24.9 CONCLUSIONS AND PERSPECTIVES

The first generation of MMP inhibitors, as succinyl hydroxamate compounds with peptide like scaffolds, was developed on substrate-based knowledge and some of them were tested in phase III trials in humans for different inflammatory and malignant diseases. They lead to potent broad-spectrum MMP inhibitors, but their benefit thus far has not been encouraging in terms of therapeutic efficacy. Therefore, starting from the earlier succinyl hydroxamate compounds, mostly based on SAR studies, a variety of modifications and truncations have been accomplished toward new generations of inhibitors and lead to compounds which still continue to be modified, aided by structural knowledge, to improve the oral bioavailability and the suitability for chronic administration, as well as the specificity to avoid cross-reactivity. Therefore, by modifying skeletons, substituent motifs and also the ZBG, it is hopped that the problems such as side effects, (e.g., musculoskeletal syndrome) and delivery limitations to the disease site will be resolved. In the future, besides the development of more suitable inhibitors, studies have also be directed toward a better understanding of the up-regulation of these enzymes as well as the influence and importance of specific MMPs in certain types of cancer to the MMPIs as effective anticancer drugs.

REFERENCES

1. Woessner, J. F.; Nagase, H. *Matrix Metalloproteinases and TIMPs*. Oxford University Press: Oxford, **2000**; pp 1–223.
2. Supuran, C. T.; Scozzafava, A. Matrix metalloproteinases (MMPs). In *Proteinase and Peptidase Inhibition: Recent Potential Targets for Drug Development*, Smith, H.J.; Simons, C., Eds.; Taylors & Francis: London, New York, **2002**; pp 35–61.
3. Nelson, A.R.; Fingleton, B.; Matrisian, L. M. Matrix metalloproteinases: clinical implications. *J. Clin. Oncol.* **2000**, *18*, 1135–1149.
4. Nagase, H.; Suzuki, K.; Itoh, Y.; Kan, C. C.; Gehring, M. R.; Huang, W.; Brew, K. Involvement of tissue inhibitors of metalloproteinases (TIMPs) during matrix metalloproteinase activation. *Adv. Exp. Med. Biol.* **1996**, *389*, 23–31.
5. Shapiro, S. H. Matrix metalloproteinase degradation of extracellular matrix: biological consequences. *Curr. Opin. Cell Biol.* **1998**, *10*, 602–608.
6. Sternlicht, M.; Werb, Z. How matrix metalloproteinases regulate cell behaviour. *Annu. Rev. Cell Dev. Biol.* **2001**, *17*, 463–516.
7. Fingleton, B.; Matrix metalloproteinase as valid clinical targets. *Curr. Pharm. Des.* **2007**, *13*, 333–346.
8. Bigg, H. F.; Rowan, A. D. The inhibition of metalloproteinases as a therapeutic target in rheumatoid arthritis and osteoarthritis. *Curr. Opin. Pharmacol.* **2001**, *1*, 314–320.
9. Ashley, R. A. Clinical trials of a matrix metalloproteinase inhibitor in human periodontal disease. *Ann. NY Acad. Sci.* **1999**, *878*, 335–346.
10. Rao, B. G. Recent developments in the design of specific matrix metalloproteinase inhibitors aided by structural and computational studies. *Curr. Pharmaceut. Des.* **2005**, *11*, 295–322.
11. Whittaker, M.; Floyd, C. D.; Brown, P.; Gearing, A. J. Design and therapeutic application of matrix metalloproteinase inhibitors. *Chem. Rev.* **1999**, *99*, 2735–2776.
12. Babine, R. E.; Bender, S. L. Molecular recognition of protein–ligand complexes: applications to drug design. *Chem. Rev.* **1997**, *97*, 1359–1472.
13. Woessner, J. F., Jr. Matrix metalloproteinases inhibition. From the Jurassic to the third millennium. *Ann. NY. Acad. Sci.* **1999**, *878*, 388–403.
14. Chambers, A. F.; Matrisian, L. M. Changing views of the role of matrix metalloproteinases in metastasis. *J. Nac. Cancer Inst.* **1997**, *89*, 1260–1270.
15. Coussens, L. M.; Fingleton, B.; Matrisian, L. M. Matrix metalloproteinase inhibitors and cancer: trials and tribulations. *Science* **2002**, *295*, 2387–2392.
16. Skiles, J. W.; Gonnella, N. C.; Jeng, A. The design, structure and clinical update of small molecular weight matrix metalloproteinase inhibitors. *Curr. Med. Chem.* **2004**, *11*, 2911–2977.

17. Breuer, E.; Frant, J.; Reich, R. Recent non-hydroxamate matrix metalloproteinase inhibitor. *Exp. Opin. Ther. Pat.* **2005**, *15*, 253–269.

18. Overall, C. M.; Lopez-Otin, C. Strategies for MMP inhibition in cancer innovations for the post-trial era. *Nat. Rev. Cancer* **2002**, *2*, 657–652.

19. Mannello, F.; Tonti, G.; Papa, S. Matrix metalloproteinase inhibitors as targets of anticancer therapeutics. *Curr. Cancer Drug Targets* **2005**, *5*, 285–298.

20. Matter, H.; Schudok, M. Recent advances in the design of matrix metalloproteinase inhibitors. *Curr. Opin. Drug Discov. Devel.* **2004**, *7*, 513–535.

21. Jacobsen, F. E.; Lewis, J. A.; Cohen, S. M. The design of inhibitors for medicinally relevant metalloproteins. *Chem. Med. Chem.* **2007**, *2*, 152–171.

22. Nuti, E.; Tuccinardi, T.; Rossello, A. Matrix metalloproteinase inhibitors: new challenges in the era of post broad-spectrum inhibitors. *Curr. Pharm. Des.* **2007**, *13*, 2087–2100.

23. Wada, C. K. The evolution of matrix metalloproteinase inhibitors. drug discovery program at Abbott laboratories. *Curr. Top. Med. Chem.* **2004**, *4*, 1255–1267.

24. Fisher, J. F.; Mobashery, S. Recent advances in MMP inhibitor design. *Cancer Metast. Rev.* **2006**, *25*, 115–136.

25. Pirard, B. Insight into the structural determinants for selective inhibition of matrix metalloproteinases. *Drug Discov. Today* **2007**, *12*, 640–646.

26. Visse, R.; Nagase, H. Matrix metallopronases and tissue inhibitor of metalloproteases: structure, function and biochemistry. *Ciac. Res.* **2003**, *92*, 827–839.

27. Supuran, C. T.; Casini, A.; Scozzafava, A. Protease inhibitors of sulphonamide type: anticancer, antiinflammatory and antiviral agents. *Med. Res. Rev.* **2003**, *23*, 535–558. 1–33.

28. Borkakoti, N. Matrix metalloproteases: variations on a theme. *Progr. Biophys. Mol. Biol.* **1998**, *70*, 73–94.

29. Terp, G. E.; Cruciani, G.; Christensen, I. T.; Jorgensen, F. S. Structural differences of matrix metalloproteinases—by the GRID/CPCA approach. *J. Med. Chem.* **2002**, *45*, 2675–2684.

30. Pirard, B.; Matter, H. Matrix metalloproteinase target family landscape: a chemomatrix approach to ligand selectivity based on protein binding site analysis. *J. Med. Chem.* **2006**, *49*, 51–69.

31. Dorman, G.; Kocsis-Sommer, K.; Spadonic, C.; Peter, F. MMP inhibitors in cardiac diseases: an update. *Recent Pat. Cardiovasc. Drug Discov.* **2007**, *2*, 186–194.

32. Brown, S.; Meroueh, S. O.; Fridman, R.; Mobashery, S. Quest for selectivity in inhibition of matrix metalloproteinase inhibitors. *Curr. Top. Med. Chem.* **2004**, *4*, 1227–1238.

33. Morales, R.; Perrier, S.; Florent, J. -M.; Beltra, J.; Dufour, S.; De Mendez, I.; Manceau, P.; Tertre, A.; Moreau, F.; Compere, D.; Dublanchet, A. -C.; O'Gara, M. Crystal structures of novel non-peptidic, non-zinc chelating inhibitors bound to MMP-12. *J. Mol. Biol.* **2004**, *341*, 1063–1076.

34. (a) Broadhurst, M. J.; Johnson, W. H.; Lawton, G.; Handa, B. K.; Machin, P. J. Phosphinic acid derivative. *European Patent* EPO276436, **1998**. (b) Brown, P. D. Clinical trials of a low molecular weight matrix metalloproteinase inhibitor in cancer. *Ann NY. Acad. Sci.* **1994**, *732*, 217–221; (c) Darlak, K.; Miller, R. B.; Stack, M. S.; Spatola, A. F.; Gray, R. D.

Thiol-based inhibitors of mammalian collagenase. Substituted amide and peptide derivatives of the leucine analogue, 2-[R,S)-mercaptomethyl]-4-methylpentanoic acid. *J. Biol. Chem.* **1990**, *265*, 5199–5205. (d) Markwell, R. E.; Hunter, D. J.; Ward, R. W. Peptides with collagenase inhibiting activity. *European Patent* EP0320118, **1989**. (e) Markwell, R. E.; Hunter, D. J.; Ward, R. W. Peptides with collagenase inhibiting activity. *World Patent* WO9309136, **1989**. (f) Odake, S.; Morita, Y.; Morikawa, T.; Yoshida, N.; Hori, H.; Nagai, Y. Inhibition of matrix metalloproteinases by peptidyl hydroxamic acids. *Biochem. Biophys. Res. Comm.* **1994**, *199*, 1442–1446.

35. Matter, H.; Schudok, M. Recent advances in the design of matrix metalloproteinase inhibitors. *Curr. Opin. Drug Discov. Devel.* **2004**, *7*, 513–535.
36. Wu, J.; Rush, T. S., III; Hotchandani, R.; Du, X.; Geck, M.; Collins, E.; Xu, Z.-B.; Skotnicki, J.; Levin, J. I.; Lovering, F. E. Identification of potent and selective MMP-13 inhibitors. *Bioorg. Med. Chem. Lett.* **2005**, *15*, 4105–4109.
37. Bernardo, M. M.; Brown, S.; Li, Z. H.; Fridman, R.; Mobashery, S. Design, synthesis and characterization of potent. Slow-binding inhibitors that are selective for gelatinases. *J. Biol. Chem.* **2002**, *277*, 11201–11207.
38. Matziari, M.; Beau, F.; Cuniasse, P.; Dive, V.; Yiotakis, A. Evaluation of P_1'-diversified phosphinic peptides leads to the development of highly selective inhibitors of MMP-11. *J. Med. Chem.* **2004**, *47*, 325–336.
39. Puerta, D. T.; Mongan, J.; Tran, B. L.; Mc Cammon, J. A.; Cohen, S. M. Potent, selective pyrone-based inhibitors of stromelysin-1. *J. Amer. Chem. Soc.* **2005**, *127*, 14148–14149.
40. Blagg, J. A.; Noe, M. C.; Wolf-Gouveia, L. A.; Reiter, L. A.; Laird, E. R.; Chang, S. P. Potent pyrimidinetrione-based inhibitors of MMP-13 with enhanced selectivity over MMP-14. *Bioorg. Med. Chem. Lett.* **2005**, *15*, 1807–1810.
41. Esteves, M. A.; Cachudo, A.; Ribeiro, C.; Chaves, S.; Santos, M. A., New hydroxypyrimidinone chelators as potential matrix metalloproteinase inhibitors. In *Metal Ions in Biology and Medicine*, Alpoim M. C.; Morais P. V.; Santos, M. A.; Cristovão, A. J.; Centeno, J.; Collery, Ph., Eds; John Libbey Eurotext: Paris, **2006**; Vol. *9*, pp 35–39.
42. Castelhano, A. L.; Billedeau, R.; Dewdney, N.; Donnelly, S.; Horne, S.; Kurz, L. J.; Liak, T. J.; Martin, R.; Uppington, R.; Yuan, Z.; Krantz, A. Novel indolactam-based inhibitors of matrix metalloproteinases. *Bioorg. Med. Chem. Lett.* **1995**, *5*, 1415–1420.
43. Johnson, W. H.; Roberts, N. A.; Borkakoti, N. Collagenase inhibitors: their design and potential therapeutic use. *J. Enzyme Inhib.* **1987**, *2*, 1–22.
44. Rothenberg, M. L.; Nelson, A. R.; Hande, K. R. New drugs on the horizon: matrix metalloproteinase inhibitors. *Stem Cells* **1999**, *1*, 237–240.
45. Hutchinson, J. W.; Tierney, G. M.; Parsons, S. L.; Davis, T. R. C. Dupuytren's disease and frosen shoulder induced by treatment with a matrix metalloproteinase inhibitor. *J. Bone Joint Surg. Br.* **1998**, *80-B*, 907–908.
46. Rasmussen, H. S.; McCann, P. P. Matrix metalloproteinase inhibition as a novel anticancer strategy: a review with special focus on Batimastat and Marimastat. *Pharmacol. Ther.* **1997**, *75*, 69–75.
47. Hoekstra, R.; Skens, F. A. L. M.; Verwej, J. Matrix metalloproteinase inhibitors: current developments and future perspectives. *Oncologist* **2001**, *6*, 415–427.
48. Rothemberg, M. L.; Nelson, A. R.; Hande, K. N. New drugs on the horizon: matrix metalloproteinase inhibitors. *Oncologist* **1998**, *3*, 271–274.

49. Folgueras, A. R.; Pendas, A. M.; Sanchez, L. M.; Lopez-Otin, C. Matrix metalloproteinases in cancer: from new functions to improved inhibition strategies. *Int. J. Dev. Biol.* **2004**, *48*, 411–424.

50. Dahlberg, L.; Billinghhurs, R. C.; Manner, P.; Nelson, S.; Webb, G.; Ionescu, M.; Reiner, A.; Tanzer, M.; Zukor, D.; Chen, J.; Van Wart, H. E.; Poole, A. R. Selective enhancement of collagenase-mediated cleavage of resident type II collagen in cultured osteoarthritic cartilage and arrest with a synthetic inhibitor that spares collagenase 1 (matrix metalloproteinase 1). *Arthritis Rheum.* **2000**, *43*, 673–682.

51. Scatena, R. Prinomastat, a hydroxamate-based matrix metalloproteinase inhibitor. a novel pharmaceutical approach for tissue remodelling-related diseases. *Exp. Opin. Invest. Drugs* **2000**, *9*, 2159.

52. Grams, F.; Crimmin, M.; Hinnes, L.; Huxley, P.; Pieper, M.; Teschesche, H.; Bode, W. Structure determination and analysis of human neutrophil collagenase complexed with a hydroxamate inhibitor. *Biochemistry* **1995**, *34*, 14012–14120.

53. Xue, C. -B.; He, X.; Roderick, J.; DeGrado, W. F.; Cherney, R. J.; Hardman, K. D.; Nelson, D. J.; Copeland, R.; Jafee, A.; Decicco, B. D.; Design, C. P. Synthesis of cyclic inhibitors of matrix metalloproteinases and TNF-α production. *J. Med. Chem.* **1998**, *41*, 1745–1748.

54. Steinman, D. H.; Curtin, M. L.; Garland, R. B.; Davidsen, S. K.; Heyman, H. R.; Holms, J. H.; Albert, D. H.; Magoc, T. J.; Nagy, I. B.; Marcotte, P. A.; Li, J.; Morgan, D. W.; Hutchins, C.; Summers, J. B. The design synthesis and structure–activity relationships of a series of macrocyclic MMP inhibitors. *Bioorg. Med. Chem. Lett.* **1998**, *8*, 2087–2092.

55. Xue, C. -B.; Voss, M. E.; Nelson, D. J.; Duan, J. J. -W.; Cherney, R. J.; Jacobson, I. C.; He, X.; Roderick, J.; Chen, L.; Corbett, R. L.; Wang, L.; Meyer, D. T.; Kennedy, K.; DeGrado, W. F.; Hardman, K. D.; Teleha, C. A.; Jaffee, B. D.; Liu, R. -Q.; Copeland, R. A.; Covington, M. B.; Christ, D. D.; Trzaskos, J. M.; Newton, R. C.; Magolda, R. L.; Wexler, R. R.; Decicco, C. P. Design, synthesis, and structure–activity relationships of macrocyclic hydroxamic acids that inhibit tumor necrosis factor release *in vitro* and *in vivo*. *J. Med. Chem.* **2001**, *44*, 2636–2660.

56. Campion, C.; Davidson, A. H.; Dickens, J. P.; Crimmin, M. J. *PCT Patent Application* WO9005719, **1990**; *Chem. Abstr.* **1990**, *113*, 212677.

57. Crimmin, M. J.; Beckett, P. R.; Davis, M. H. *PCT Patent Application* WO9421625, **1994**; *Chem. Abstr.* **1995**, *122*, 188173.

58. Crimmin, M. J.; Ayscough, A. P.; Beckett, R. P. *PCT Patent Application* WO9424140, **1994**; *Chem. Abstr.* **1995**, *123*, 144644.

59. Kottirsch, G.; Koch, G.; Feifel, R.; Neumann, U. β-Aryl-succinic acid hydroxamates as dual inhibitors of matrix metalloproteinases and tumor necrosis factor alpha converting enzyme. *J. Med. Chem. 45*, **2002**, 2289–2293.

60. Koch, G.; Kottirsch, G.; Wietfeld, B.; Kuster, E. Process development of a dual MMP/TNF inhibitor (SDZ 242-484). *Org. Process Res. Dev. 6*, **2002**, 652–659.

61. Di Martino, M.; Wolff, C.; High, W.; Stoup, G.; Hoffman, S.; Laydon, J.; Lee, J. C.; Bertolini, D.; Galloway, W. A.; Crimmin, M. J.; Davis, M.; Davies, S. Anti-arthritic activity of hydroxamic acid-based pseudopeptide inhibitors of matrix metalloproteinases and TNF-α processing. *Inflamm. Res. 46*, **1997**, 211–215.

62. Clements, J. M.; Cossins, J. A.; Wells, G. M. A.; Corkill, D. J.; Helfrich, K.; Wood, L. M.; Pigott, R.; Stabler, G.; Ward, G.A.; Gearing, A. J. H.; Miller, K. M. Matrix metalloproteinase expression during experimental autoimmune encephalomyelitis and effects of a combined matrix metalloproteinase and tumour necrosis factor-α inhibitor. *J. Neuroimmunol. 74*, **1997**, 85–94.

63. Janser, P.; Neumann, U.; Miltz, W.; Feifel, R.; Buhl, T. A cassette-dosing approach for improvement of oral bioavailability of dual TACE/MMP inhibitors. *Bioorg. Med. Chem. Lett. 16*, **2006**, 2632–2636.

64. Marcq, V.; Mirand, C.; Decarme, M.; Emonard, H.; Hornebeck, W. MMPs inhibitors: new succinyl hydroxamates with selective inhibition of MMP-2 over MMP-3. *Bioorg. Med. Chem. 13*, **2003**, 2843–2846.

65. Hirayama, R.; Yamamoto, M.; Tsukida, T.; Matsuo, K.; Obata, Y.; Sakamoto, F.; Ikeda, S. Synthesis and biological evaluation of orally active matrix metalloproteinase inhibitors. *Bioorg. Med. Chem. 5*, **1997**, 765–778.

66. Fray, M. J.; Dickinson, R. P. Discovery of potent and selective succinyl hydroxamate inhibitors of matrix metalloprotease-3 (stromelysin-1). *Bioorg. Med. Chem. Lett. 11*, **2001**, 571–574.

67. Fray, M. J.; Burslem, M. F.; Dickinson, R. P. Selectivity of inhibition of matrix metalloproteases MMP-3 and MMP-2 by succinyl hydroxamates and their carboxylic acid analogues is dependent on P_3' group chirality. *Bioorg. Med. Chem. Lett. 11*, **2001**, 567–570.

68. Chen, J. J.; Zhang, Y.; Hammond, S.; Dewdney, N.; Ho, T.; Lin, X.; Browner, M. F.; Castelhano, A. L. *Bioorg. Med. Chem. Lett.* **1996**, *6*, 1601.

69. Sheppard, G. S.; Florjancic, A. S.; Giesler, J. R.; Xu, L.; Guo, Y.; Davidsen, S. K.; Marcotte, P. A.; Elmore, I.; Albert, D. H.; Magoc, T. J.; Bouska, J. J.; Goodfellow, C. L.; Morgan, D. W.; Summers, J. B. Aryl ketones as novel replacements for the C-terminal amide bond of succinyl hydroxamate MMP inhibitors. *Bioorg. Med. Chem. Lett.* **1998**, *8*, 3251–3256.

70. Sheppard, G. S.; Pireh, D.; Carrera, G. M., Jr.; Bures, M. G.; Heyman, H. R.; Steinman, D. H.; Davidsen, S. K.; Phillips, J. G.; Guinn, D. E.; May, P. D.; Conway, R. G.; Rhein, D. A.; Calhoun, W. C.; Albert, D. H.; Magoc, T. J.; Carter, G. W.; Summers, J. B., Jr. 3-(2-3-Pyridinyl)thiazolidin-4-oyl)indoles, a novel series of platelet activating factor antagonists. *J. Med. Chem.* **1994**, *37*, 2011–2032.

71. Campestre, C.; Tortorella, P.; Agamennone, M.; Preziuso, S.; Biasone, A.; Nuti, E.; Rossello, A.; Gallina, C. Peptidy 3-substituted 1-hydroxyureas as isosteric analogues of succinyl hydroxamate MMP inhibitors. *Eur. J. Med. Chem.* **2008**, *43*, 1008–1014.

72. Von Roedern, E. G.; Brandstetter, H.; Engh, R. A.; Bode, W.; Grams, F.; Moroder, L. Bis-substituted malonic acid hydroxamate derivatives as inhibitors of human neutrophil collagenase (MMP8). *J. Med. Chem.* **1998**, *41*, 3041–3047.

73. Grams, F.; Reinemer, P.; Powers, J.C.; Kleine, T.; Pieper, M.; Tschesche, H.; Huber, R.; Bode, W. X-ray structures of human neutrophil collagenase complexed with peptide hydroxamate and peptide thiol inhibitors. *Eur. J. Biochem.* **1995**, *228*, 830–841.

74. Von Roedern, E. G.; Grams, F.; Brandstetter, H.; Moroder, L. Design and synthesis of malonic acid-based inhibitors of human neutrophil collagenase (MMP8). *J. Med. Chem.* **1998**, *41*, 339–345.

75. Brandstetter, H.; Engh, R. A.; Von Roedern, E. G.; Moroder, L.; Huber, R.; Bode, W.; Grams, F. Structure of malonic acid-based inhibitors bound to human neutrophil collagenase. a new binding mode explains apparently anomalous data. *Protein Sci.* **1998**, *7*, 1303–1309.
76. Fisher, J. F.; Mobashery, S. Recent advances in MMP inhibitor Design. *Cancer Metastasis Rev.* **2006**, *25*, 115–136.
77. Bottomley, K. M.; Johnson, W. H.; Walter, D. S. Matrix metalloproteinase inhibitors in arthritis. *J. Enzyme Inhib.* **1998**, *13*, 79–192.
78. Lewis, E. J.; Bishop, J.; Bottomley, K. M. K.; Bradshaw, D.; Brewster, M.; Broadhurst, M. J.; Brown, P. A.; Budd, J. M.; Elliott, L.; Greenham, A. K.; Johnson, W. H.; Nixon, J. S.; Rose, F.; Sutton, B.; Wilson, K. Ro 32-3555, an orally active collagenase inhibitor, prevents cartilage breakdown *in vitro* and *in vivo*. *Br. J. Pharmacol.* **1997**, *121*, 540–546.
79. Broadhurst, M. J.; Johnson, W. H.; Walter, D. S. *GB Patent Application* GB 2326881A, **1999**; *Chem. Abstr.* **1999**, *130*, 110060.
80. Santos, M. A.; Marques, S. M.; Gil, M.; Tegoni, M.; Chaves, S. New metalloenzyme inhibitors with potential clinic application. *PT Patente* 103003; **2003**.
81. Santos, M. A. Iminodiacetyl-monohydroxamate derivatives as potent and selective MMP inhibitors. In *Metal Ions in Biology and Medicine,* Alpoim, M. C.; Morais, P. V.; Santos M. A., Cristovão A. J., Centeno J., Collery Ph., Eds; John Libbey Eurotext: Paris, **2006**; Vol. *9*, pp 117–121.
82. Marques, S. M.; Rossello, A.; Santos, M. A. Metalloenzyme inhibitors with potential medical application: 1-hydroxy-piperazine-2,6-dione as a new zinc binding group. *PT Patente* 103897, **2007**.
83. Bode, W.; Fernandez-Catalan, C.; Grams, F.; Gomis-Rüth, F. -X.; Nagase, H.; Tschesche, H.; Maskos, K. Insights into MMP-TIMP interactions. *Ann. NY Acad. Sci.* **1999**, *878*, 73–91.
84. Kontogiorgis, C. A.; Papaioannou, P.; Hadjipavlou-Litina, D. J. Matrix metalloproteinase inhibitors: a review on pharmacophore mapping and (Q)SARs results. *Curr. Med. Chem.* **2005**, *12*, 339–355.
85. Santos, M. A.; Marques, S. M.; Tuccinardi, T.; Carelli, P.; Panelli, L.; Rossello, A. Design, synthesis and molecular modelling study of iminodiacetyl monohydroxamic acid derivatives as MMP inhibitors. *Bioorg. Med. Chem.* **2006**, *14*, 7539–7550.
86. Hanessian, S.; Moitessier, N. Structure-based design, synthesis, and hemopsin 2 (BACE) inhibitory activity of carbocyclic and heterocyclic peptidomimetics. *Curr. Top. Med. Chem.* **2004**, *4*, 1269–1287.
87. Overall, C. M.; Kleifeld, O. Validating matrix metalloproteinases as drug targets and anti-targets for cancer therapy. *Nat. Rev. Cancer* **2006**, *6*, 227–239.
88. Marques, S. M.; Nuti, E.; Rossello, A.; Supuran, C. T.; Tuccinardi, T.; Adriano Martinelli, A.; and Santos, M. A. Dual Inhibitors of Matrix Metalloproteinases and Carbonic Anhydrases: Iminodiacetyl-Based Hydroxamate Benzenesulfonamide Conjugates. *J. Med. Chem.* **2008**, *51*, 7968–7979.
89. MacPherson, L. J.; Bayburt, E. K.; Capparelli, M. P.; Carroll, B. J.; Goldstein, R.; Justice, M. R.; Zhu, L.; Hu, S.; Melton, R. A.; Fryer, L.; Goldberg, R. L.; Doughty, J. R.; Spirito, S.; Blancuzzi, V.; Wilson, D.; O'Byrne, E. M.; Ganu, V.; Parker, D. T. Discovery of CGS 27023A, a non-peptidic, potent and orally active stromelysin that blocks cartilage degradation in rabbits. *J. Med. Chem.* **1997**, *40*, 2525.

90. Reiter, L. A.; Robinson, R. P.; McClure, K. F.; Jones, C. S.; Reese, M. R.; Mitchell, P. G.; Otterness, I. G.; Bliven, M. L.; Liras, J.; Cortina, S. R.; Donahue, K. M.; Eskra, J. D.; Griffiths, R. J.; Lame, M. E.; Lopez-Anaya, A.; Martinelli, G. J.; McGahee, S. M.; Yocum, S. A.; Lopresti-Morrow L. L.; Tobiassen L. M.; Vaughn-Bowser, M. L. Pyran-containing sulfonamide hydroxamic acids: potent MMP inhibitors that spare MMP-1. *Bioorg. Med. Chem. Lett.* **2004**, *14*, 3389–3395.

CHAPTER 25

Drug Design of Sulfonylated MMP Inhibitors

ARMANDO ROSSELLO and ELISA NUTI

Dipartimento di Scienze Farmaceutiche, Università di Pisa, Via Bonanno 6, 56126 Pisa, Italy

25.1 INTRODUCTION

A deficiency in one of the mechanisms of control that regulates the physiological levels of matrix metalloproteinases (MMPs) can be linked to many serious pathologies. These diseases are characterized by a strong breakdown of structural tissues by overexpressed MMPs from resident tissue cells and/or by invasive tumor and/or inflammatory cells. Specific degenerative tissue remodeling by MMPs has been pointed out in tumors and pathologies of the central and peripheral nervous systems, the cardiovascular system, the gastrointestinal system, the musculoskeletal system, and also in degenerative diseases of the kidney, lung, skin, eye, and mucous membranes. Hence, the use of inhibitors of MMPs could be an important way to control some of the serious pathologies that are characterized by uncontrolled overexpression of proteolytic activity in degenerate tissues. This is the reason why many of the principal pharmaceutical companies have developed very expensive studies in this field over the past 20 years.

Two generations of synthetic matrix metalloproteinase inhibitors (MMPI) have been developed in this long period, but at present there are only two drugs launched on the market: Periostat® (doxycycline, **25.1**), a tetracycline used for periodontal disease, and glucosamine sulfate **25.2**, used for osteoarthritis (Fig. 25.1).[1,2]

The use of these two old drugs for new therapeutic targets seems to be a very limited result for so much effort. What happened during these years?

Figure 25.2 represents a recently reported analysis of the literature over a period of 10 years, from 1995 to 2005; these years were the more productive years in the development of new classes of MMP synthetic inhibitors. This study collected the literature on MMPI of those years according to the type of document (i.e., articles in

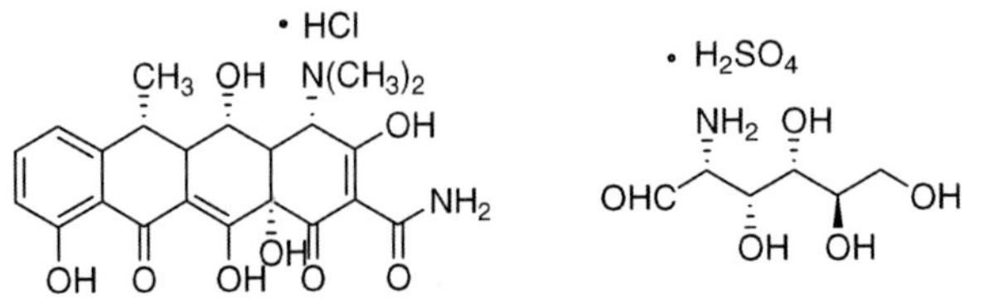

25.1, Doxycycline hydrochloride **25.2**, Glucosamine sulfate

FIGURE 25.1 MMP inhibitors on the market.

journals, patents, trials, and MMPI and imaging literature (use of MMPI as diagnostics)).[3]

Looking at the graphical trends in those years for the complete literature, articles, patents, and trials, it was possible to identify two historical periods, ranging between 1995 and 1999 for the first period and between 1999 and 2003 for the second period, with two peaks of maximum scientific production on the topic of MMP synthetic inhibitors. An analysis of the literature around these two peaks, based on the characteristics of selectivity and of the trials conducted, led us to identify two generations of inhibitors. In the first period (1995–1999), peptidomimetic hydroxamate inhibitors, like batimastat **25.3** from British Biotech, which is highly potent but not selective, were mainly reported. Clinical trials with these first-generation inhibitors gave very discouraging results, especially in tumor pathologies. Considering

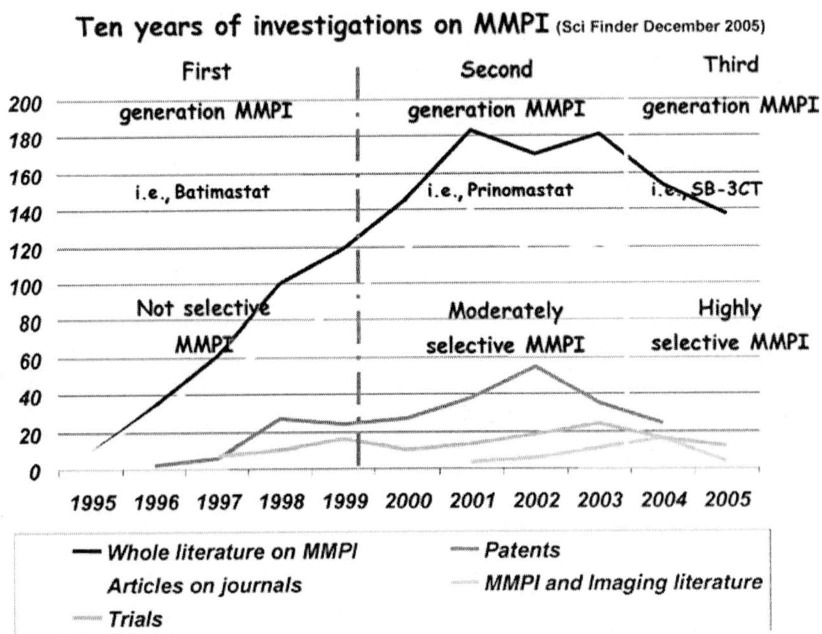

FIGURE 25.2 An analysis of the literature on MMPI from 1995 to 2005.

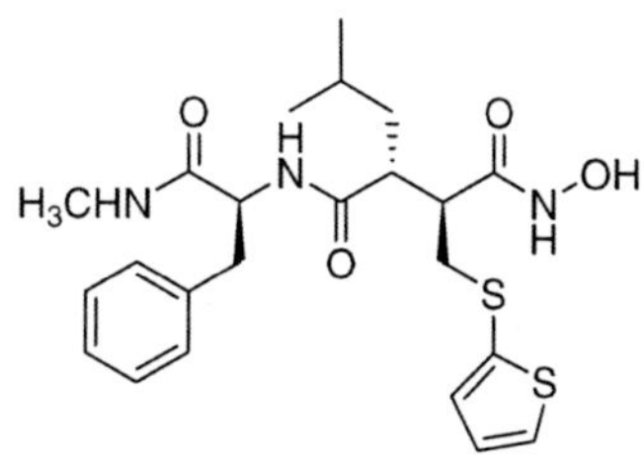

25.3, Batimastat

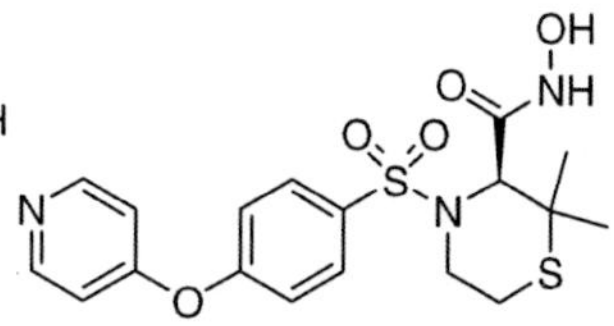

25.4, Prinomastat

FIGURE 25.3 Compounds representative of the first and the second generation of MMPI.

these data, in the period between 1999 and 2003, many efforts were directed toward developing a more selective second generation of inhibitors against the specific MMPs believed to be involved in the different pathologies. Many new classes of moderately selective MMPIs were developed, the most representative of which being the sulfonamido-based inhibitors, like prinomastat **25.4** (Fig. 25.3). This period was the most productive one for the companies that brought a large number of new classes of inhibitors to enter the clinical trials.[4]

During and after 2003, there was a decrease in the total number of works produced, but there were strong differences in the MMPI profile of the published new inhibitors (especially in the patent literature). New compounds were targeted versus specific MMPs, which were demonstrated in these recent years to be a real target for the relevant pathologies. In fact, only in the past few years gene-directed studies with transgenic animal models were developed to demonstrate which MMPs are involved in each disease. These experimental trials defined the currently accepted classification of MMPs based on their relevant functions in physiopathological processes. Today, some MMPs are classified as antitarget MMPs and others as target MMPs. The inhibition of MMPs conducted without highly selective drugs could cause serious problems during therapy, due to the imbalance of the physiological tissue remodeling linked to the inhibition of the antitarget MMPs.[5] Hence, at present, we have a third generation of MMPI, designed to inhibit only the target MMPs involved in each pathology. These new compounds show K_I value ratios of inhibition between antitarget/target MMPs that are higher than a factor of 1000 on isolate enzymes. A representative compound of this new class could be the highly selective mechanism-based MMP-2 inhibitor SB-3CT **25.5**, developed by Mobashery et al. (Fig. 25.4).[6]

At present, new challenges are represented by the development of new MMPIs bearing hydroxamate and more effective zinc binding groups (ZBGs) and by the development of new allosteric nonzinc binding inhibitors, devoid of the ZBGs (new

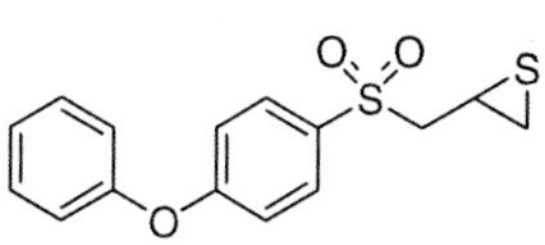

25.5, SB-3CT

FIGURE 25.4 Third generation of MMPI: the mechanism-based MMP-2 inhibitor SB-3CT.

emerging fourth generation of MMPIs, able to target with high potency specific MMP only).

In this chapter, we shall provide an overview of the whole literature (journals and patents) regarding the most representative sulfonylated MMPI belonging to these sequential generations of inhibitors, with particular emphasis on those belonging to the more recent generations.

25.2 MECHANISM OF ACTION OF SYNTHETIC MMP INHIBITORS

Today, from crystallographic data, two zinc atoms are known to be present in MMPs: a stable structural tetracoordinated atom, which is linked to three histidines and an aspartate, and another key zinc atom with catalytic activity.[7] It has been demonstrated that the catalytic zinc ion in the active site of MMPs is directly involved in the degradation of extracellular matrix components. For example, in the latent form of gelatinase A (proMMP-2), the active site presents the zinc bound to three histidine residues (His374, His378, and His384) and blocked by the sulfhydrylic group of a cysteine (Cys73) present in the prodomaïn.[8] During activation, the disjunction of the propeptide (cysteine–switch) generates a proactive–proform of MMP-2 that in a second phase involves the dissociation of this residue. The prodomain shedding, autocatalytically removed or mediated by other proteases, activates the catalytic site, making it accessible to the substrate.[9,10]

On the basis of structural information, a mechanism has been proposed for the catalytic activity of collagenase I (MMP-1) that can be representative for all this large family of zinc endoproteases. In the first step, the catalytic zinc ion is tetrahedrally coordinated to three histidines and a water molecule. During proteolysis, before the peptide binding to the catalytic site, the water donates a proton to Glu219 (1HFC sequence number),[11] which transfers it to the nitrogen of the scissile amide bond that is stabilized by Ala182; this is followed by the generation of a salt bridge between Glu219 and the free amine of the cleaved substrate.

Figure 25.5 summarizes the passages from the proenzyme activation to the substrate hydrolysis in a typical known MMP.[12]

Synthetic MMPIs can be classified into two large classes of compounds, considering their binding mode to the catalytic binding site (see Fig. 25.6). The first class represents the more abundant known family of synthetic inhibitors, developed in about two decades. They are all zinc binder inhibitors (ZBIs), where **A** are ZBGs.[13]

More recently, a second class of no binder inhibitors (NBIs), devoid of any zinc-chelating group, has been developed.[14]

In both classes of compounds, the potency and selectivity depend on their accessory peptide–mimic chains (P1–P*n* and/or P1′–P′*n* groups) able to interact with the recognition sites on the enzyme catalytic region (S1–S*n* and/or S1′–S′*n*) or/and on the exocites.[15]

ZBIs, the most studied family of MMPIs, have been developed as mimics of the transition-state intermediates (Fig. 25.5). Until now, a large number of different ZBGs able to compete with the substrate in zinc coordination have been studied in the inhibitors development.[13]

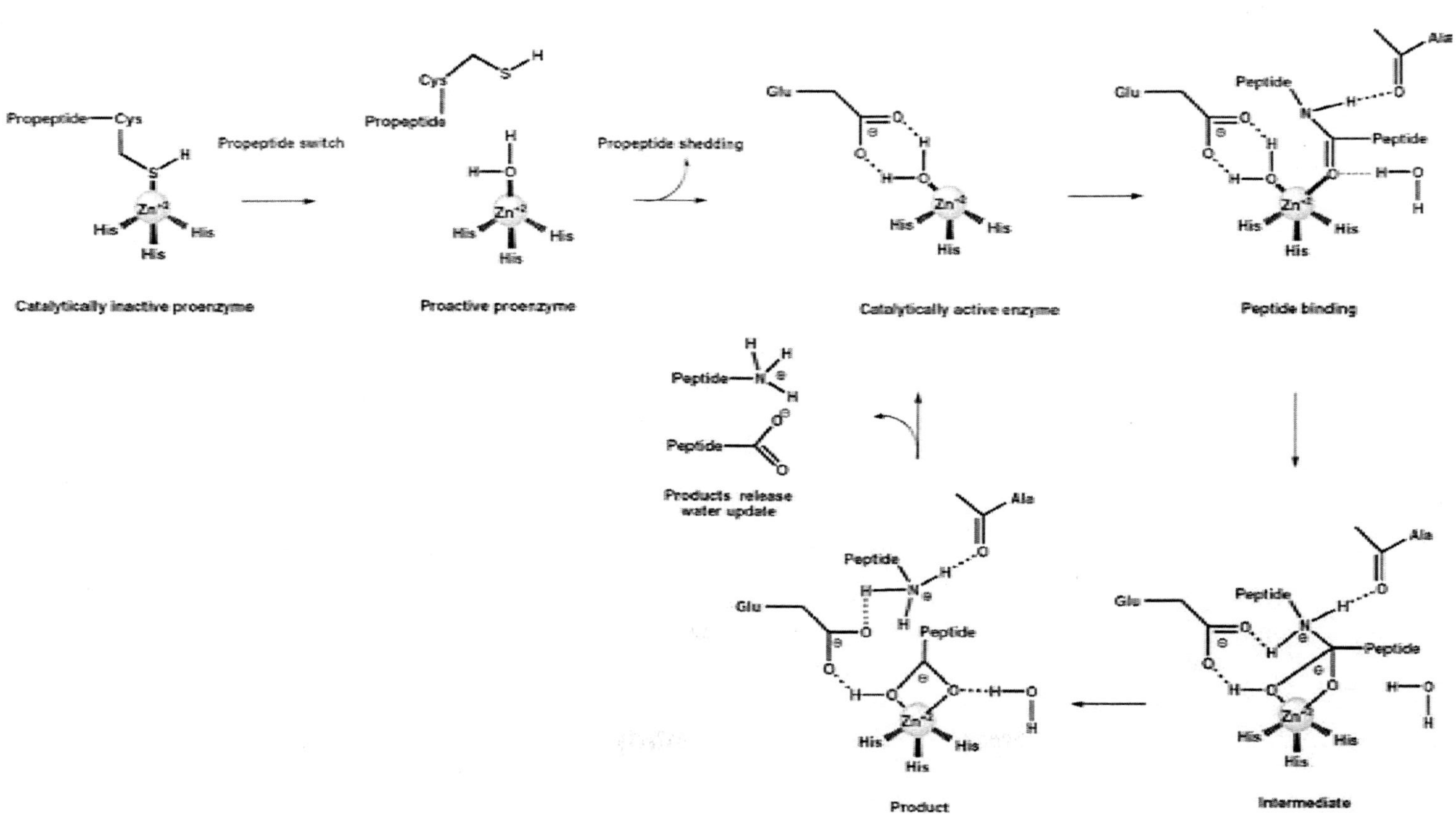

FIGURE 25.5 Pro-MMPs activation and catalytic mechanism of substrate hydrolysis.

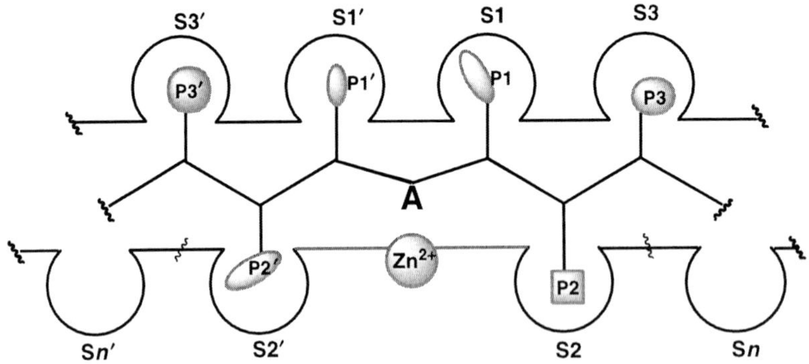

FIGURE 25.6 MMPI binding regions.

Common ZBGs of MMPI include hydroxamates **25.6**, retrohydroxamates **25.7**, carboxylates **25.8**, thiolates **25.9**, phosphonates **25.10**, phosphinates **25.11**, and, more recently, pyrimidinetriones **25.12** (Fig. 25.7).[15–17]

Among these, the hydroxamate group has proven to be one the most potent (100–2000 times compared with their carboxylate analogues).[16]

Crystallographic data have revealed that the zinc catalytic atom, in the presence of hydroxamate inhibitors, is linked to three histidines and to the ligand with a pentacoordination in a bipyramidal-trigonal geometry, considering that the hydroxamate group behaves like a bidentate ligand (see, for example, the complex 1OV0 of a sulfonylated hydroxamate with the catalytic domain of MMP-2 reported in Fig. 25.8).[18–20]

In this chapter, we will describe more accurately MMP inhibitors belonging to the sulfonylated hydroxamate/carboxylate series, which are the most studied ones. The final section will be concerned with sulfonyl-based inhibitors bearing new ZBGs.

25.3 DESIGN OF SULFONYL-BASED MMP INHIBITORS

Sulfonyl-based compounds represent the most abundant group of ligands for MMPs after peptidomimetic inhibitors. Sulfonyl-based MMPIs include sulfonamides, tertiary and secondary, and sulfones. The studies on these scaffolds began in the mid-1990s, when CGS 27023A (known also as MMI-270, compound **25.6**), an orally active tertiary sulfonamide able to potently inhibit many MMPs, was discovered by Ciba-Geigy researchers. This potent sulfonamide, together with two other secondary sulfonamides discovered in the same period (1998–1999) by the Shionogi (hydroxamate **25.13**) and Warner-Lambert (carboxylate **25.14**) companies, can be considered as the progenitors of all sulfonyl-based MMPIs. A schematic image of sulfonylated compounds is represented by structure **B** reported in Fig. 25.9.

In this chapter, **B** will be used to describe the evolution of the principal and well-characterized structural classes of sulfonyl-based MMPIs discovered in the past two decades.

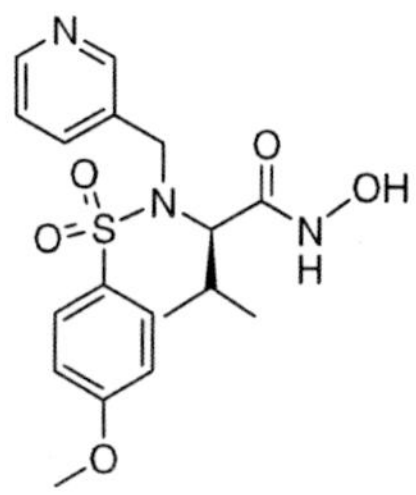

25.6, Hydroxamic acid, that is, MMI270

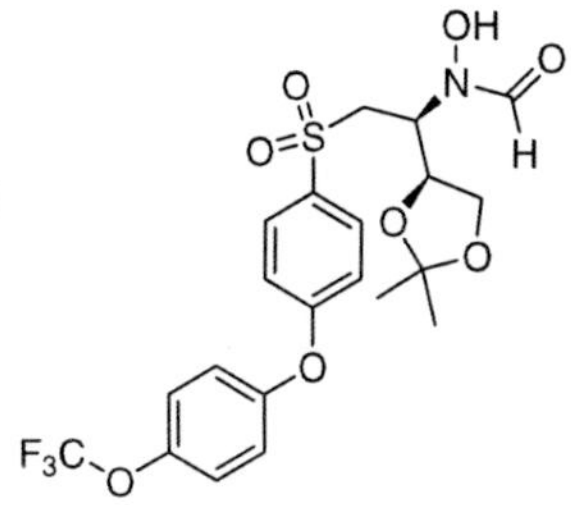

25.7, Retrohydroxamate, that is, ABT-518

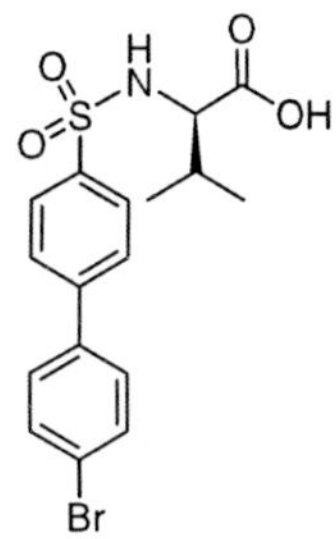

25.8, Carboxylic acid, that is, PD166793

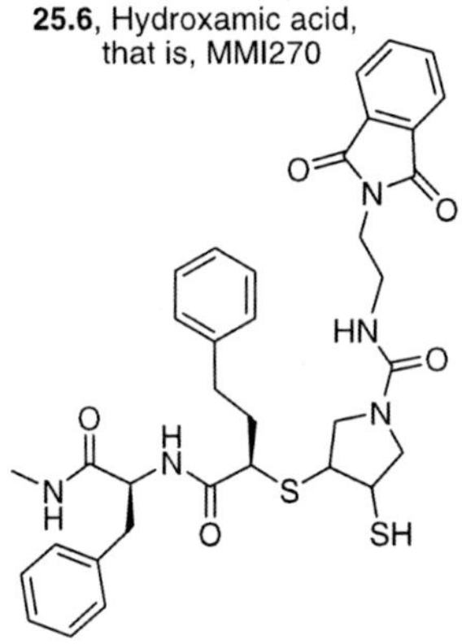

25.9, Thiolates, that is, YHJ- 223

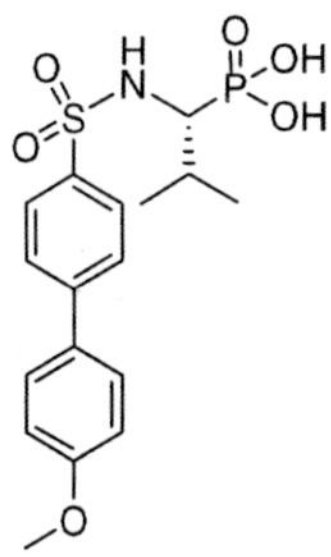

25.10, Phosphonate acid, that is, CAS RN: 873586-62-0

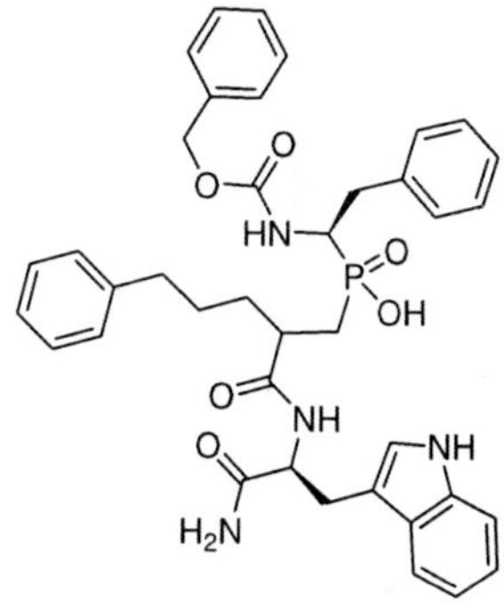

25.11, Phosphinate acid, that is, RXP-03

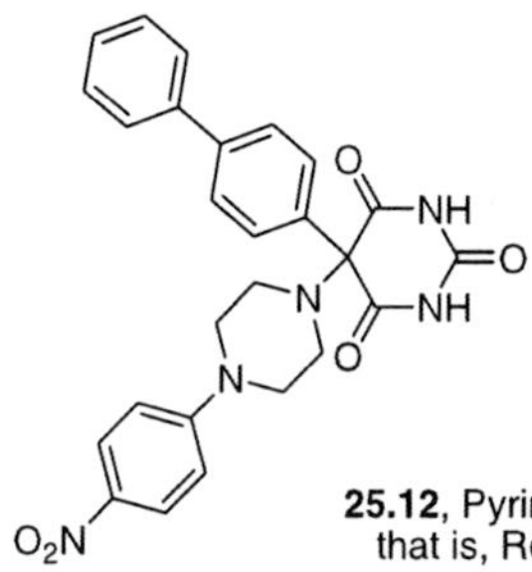

25.12, Pyrimidinetrione, that is, Ro 28-2653

FIGURE 25.7 Examples of common ZBGs used in known MMPI.

25.3.1 Sulfonamido-Based MMPI

25.3.1.1 Tertiary Sulfonamido-Based MMPI Tertiary Sulfonamido-Based MMPIs represent an evolutionary family directly derived from CGS 27023A, **25.6** (MMI270); they include four principal classes of inhibitors: CGS 27023A analogues, conformationally restrained sulfonamides, reversed sulfonamides, and the more recent *N-O* alkyl sulfonamides.

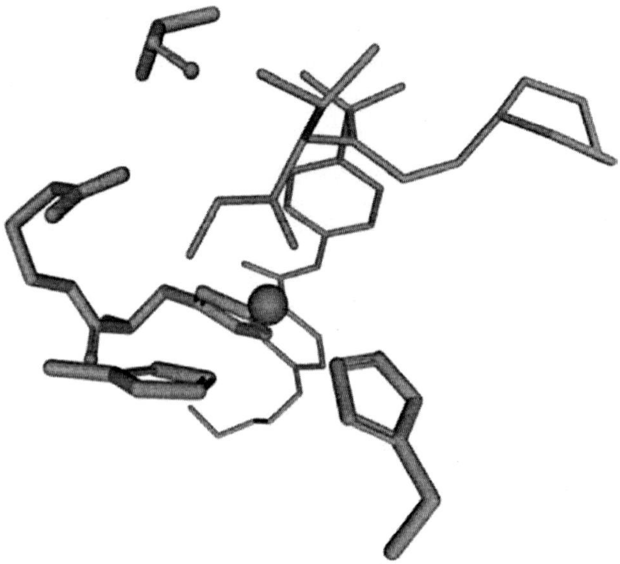

FIGURE 25.8 1OV0 Complex (MMP-2CD with a sulfonylated MMPI).

CGS 27023A Analogues At the Ciba-Geigy Corporation in the beginning of the 1990s, many efforts were made to find a lead compound among MMP-3 inhibitors to be used in the treatment of arthritis.[21] First, structure–activity relationship (SAR) studies were developed on 4-substituted benzenesulfonamides of the glycine hydroxamate, varying the P2′ substituent and the R-substituent on the 4′-phenyl ring (compounds **25.15–25.35**).

Table 25.1 shows the results obtained by *in vitro* tests on MMP-3 for these first compounds. They indicated that in glycinylsulfonamides, the combination of an alkyl

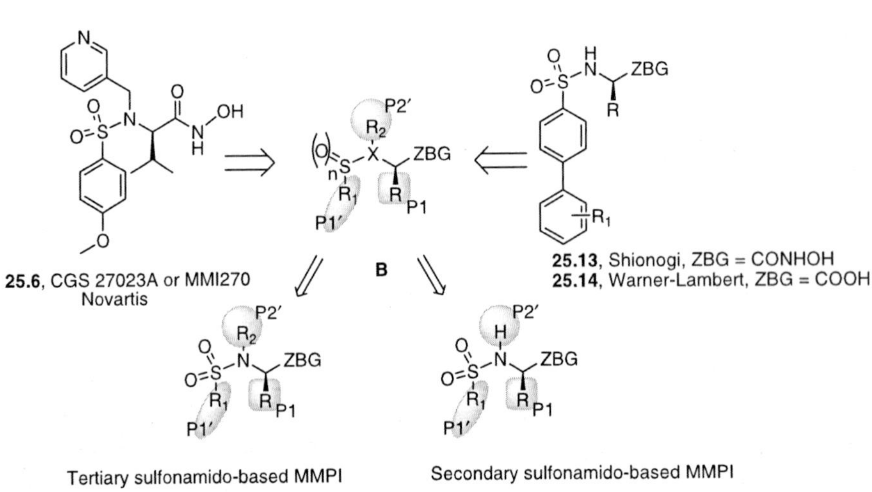

FIGURE 25.9 Examples of sulfonylated MMPI: tertiary and secondary sulfonamides.

TABLE 25.1 SAR Studies on Benzenesulfonamido-glycinylhydroxamates (Ciba-Geigy). Effects of P1–P1′–P2′ Substitution on MMP-3 Inhibition

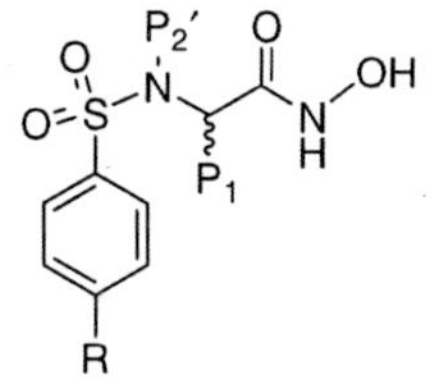

Compound	P1	P2′	R	K_I (μM)
25.15	H	$CH_2CH(CH_3)_2$	H	1.76
25.16	H	$C(CH_3)_3$	OCH_3	2.5
25.17	H	$CH_2CH(CH_3)_2$	OCH_3	0.13
25.18	H	$CH_2CH(CH_3)_2$	Cl	0.94
25.19	H	$CH_2CH(CH_3)_2$	CH_3	3.27
25.20	H	$CH_2CH(CH_3)_2$	F	3.27
25.21	H	$CH_2CH(CH_3)_2$	$N(CH_3)_2$	4.40
25.22	H	$CH_2CH(CH_3)_2$	NH_2	5.00
25.23	H	$CH_2CH(CH_3)_2$	CF_3	5.35
25.24	H	$CH_2CH(CH_3)_2$	NO_2	Inactive
25.25	H	$CH_2CH(CH_3)_2$	SO_2CH_3	Inactive
25.26	H	$CH_2CH(CH_3)_2$	$O\text{-}(CH_2)_3CH_3$	0.06
25.27	H	$CH_2CH(CH_3)_2$	$O\text{-}(CH_2)_2CH(CH_3)_2$	0.08
25.28	H	$CH_2CH(CH_3)_2$	$O\text{-}(CH_2)_5CH_3$	0.06
25.29	H	$CH_2CH(CH_3)_2$	$O\text{-}CH_2(C_6H_{11})$	0.06
25.30	H	$CH_2CH(CH_3)_2$	$O\text{-}CH(CH_3)_2$	0.34
25.31	H	$CH_2CH(CH_3)_2$	$O\text{-}(CH_2)_2\text{-}O\text{-}C_2H_5$	1.13
25.32	H	$CH(CH_3)\ CH_2\ CH_3$	OCH_3	1.82
25.33	H	$CH_2C_6H_5$	OCH_3	0.07
25.34	H	$CH_2CH_2C_6H_5$	OCH_3	0.06
25.35	H	$CH_2(3\text{-}C_5H_4N)$	OCH_3	0.11
25.36	$R\text{-}CH_3$	$CH_2C_6H_5$	OCH_3	0.04
25.37	$S\text{-}CH_3$	$CH_2C_6H_5$	OCH_3	0.14
25.38	$R\text{-}CH(CH_3)_2$	$CH_2C_6H_5$	OCH_3	0.03
25.39	$R\text{-}CH_2C_6H_5$	$CH_2C_6H_5$	OCH_3	0.03
25.40	$R\text{-}(CH_2)_4NHCbz$	$CH_2C_6H_5$	OCH_3	0.07
25.41	$R\text{-}(CH_2)_4N(CH_3)_2$	$CH_2C_6H_5$	OCH_3	0.07
25.42	$R\text{-}(CH_2)_4N(CH_2)_2O$	$CH_2C_6H_5$	OCH_3	0.05
25.43	$R\text{-}(CH_2)_4NCON(CH_2)_2O$	$CH_2C_6H_5$	OCH_3	0.02
25.44	$R\text{-}(CH_2)_4NCOCH_2N(CH_3)_2$	$CH_2C_6H_5$	OCH_3	0.03
25.6	$R\text{-}CH(CH_3)_2$	$CH_2(3\text{-}C_5H_4N)$	OCH_3	0.04

substituent in P2′ with ethereal substituents (as a methoxy one) in the *para* position of the phenyl ring provided good inhibitors for this enzyme, with K_I values in the submicromolar range (see compounds **25.17** and **25.26–25.35**). On the contrary, the use of different R-substituents on the phenyl ring, such as methyl, fluorine, amino, nitro, or sulfonylmethyl groups, seemed to reduce the affinity for this enzyme (see compounds **25.19–25.25**).

The further substitution with alkyl groups of the H-atom in the P1 position generated an asymmetry center on these molecules. The enzyme interacted better with the *R*-enantiomer, as is evident by comparing the two methyl derivatives *N*-benzyl-substituted **25.36** and **25.37**. The substitution of this methyl group with a phenyl, as in compound **25.39**, or with more bulky alkylic substituents carrying lipophilic or polar moieties was able to maintain a good activity on MMP-3 (see compounds **25.38–25.44**).

The introduction of polar groups in the P1 position, as in compounds **25.40–25.44**, instead of hydrophobic residues, as in their analogues **25.36–25.39**, improved their biopharmacological properties, maintaining the *in vitro* activity and increasing their *in vivo* potency; these improved properties were also observed after administration *per os* on a rabbit model of cartilage degradation. Compound **25.6** (CGS 27023A) showed the best biopharmacological profile, and considering its good oral absorption and bioavailability, it became the more studied nonsuccinyl-based hydroxamate. Therefore, a very large number of synthetic analogues of this compound have been developed in many laboratories around the world.[22] The resolution of the complex 1BM6 of the catalytic domain of MMP-3 with CGS 27023A[23] contributed to clarify the binding mode of this type of sulfonamides in the active site of these zinc endopeptidases and opened the way to an intensive design of sulfonamido-based MMPIs.

Figure 25.10 shows how these inhibitors can bind to the catalytic domain of MMPs. They are hydroxamate binders of the catalytic zinc atom. The sulfonamide moiety was fundamental for the binding to MMPs, since it was able to give a hydrogen bond between an oxygen atom of the SO_2 group and the backbone of the protein, near the S1′ pocket (the so-called selectivity pocket). The bidentate chelation of the catalytic zinc atom and this peculiar hydrogen bond with the protein backbone was determinant to

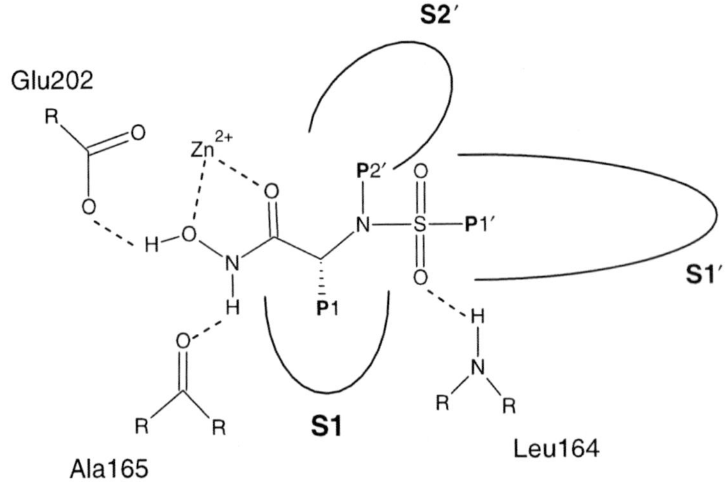

FIGURE 25.10 Binding mode of sulfonamido-based MMPI in MMP-3.

better orient the P1, P1′, and P2′ substituents of the inhibitor in the respective S1, S1′, and S2′ pockets. It was also very useful to orient the selectivity of action among the various MMPs that are quite similar in that region. The major differences among the various MMPs, which can be exploited to drive the design of selective inhibitors, were in the size and geometry of the S1′ pocket and, to a minor extent, of the S1 and S2′ regions sited on the surface of these proteins oriented to the solvent.[24]

Further developments in this field were concerned with the changes in the groups substituting the α-carbon atom, the length of the chain between the hydroxamate and the amino group (from one to two carbon atoms), and the usual variations of the P1′–P3′ moieties.[15]

The insertion of a thioethereal group in P1 on the scaffold of CGS 27023A (in place of the *i*-propyl alkyl chain) led to an increase in the inhibition potency toward the deep pocket MMPs, despite the presence of a relatively small substituent in P1′, such as the 4-methoxy benzene group. In fact, compound **25.45** (Fig. 25.11), developed by Hanessian's group,[25] showed a potent inhibitory activity for the deep pocket MMPs (MMP-2, MMP-3, MMP-9, and MMP-13) and a selectivity of about 150 times for MMP-2 over MMP-1. This fact was also established by molecular modeling studies from the same research group.[26]

More recently, a large number of other analogues of **25.6**, arylsulfonyl hydroxamates derived from glycine, L-alanine, L-valine, and L-leucine, possessing *N*-benzyl or *N*-benzyl-substituted moieties, were reported by Supuran et al. with IC_{50} values in the nanomolar range against MMP-1, MMP-2, MMP-8, and MMP-9 (compounds **25.46a–g**, **25.47–25.49**, Table 25.2).[27,28] The most important parameters influencing the activity in these classes of MMPIs were as follows: (1) a bulky group substituting the amino acid moiety in the P2′ position (with nitrobenzyl derivatives more active than chlorobenzyl derivatives, which in turn were more active than the unsubstituted benzyl derivatives) and (2) the nature of the alkyl/arylsulfonyl moieties in P1′. Aromatic derivatives were generally much more active than aliphatic ones, except for the perfluorobutyl and perfluorooctylsulfonyl compounds, which showed very good inhibitory effects, similar to those of simple aromatic compounds incorporating substituted phenyl moieties (such as *p*-methoxyphenyl, *p*-aminophenyl,

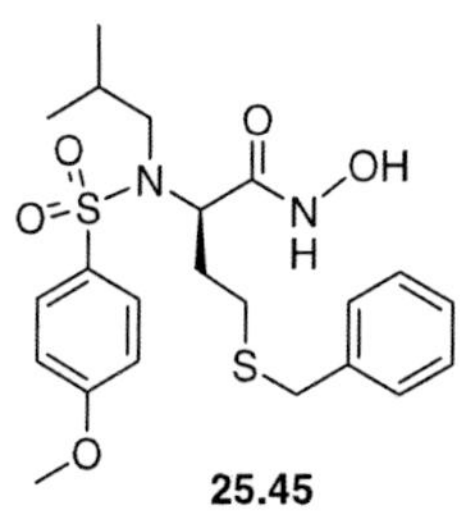

25.45

(IC_{50}, nM)

MMP-1 : 104; MMP-2 : 0.7;
MMP-3 : 0.7; MMP-9 : 2.5;
MMP-13 : 12

FIGURE 25.11 Structure of MMP inhibitor **25.45**.

TABLE 25.2 Arylsulfonyl Hydroxamates Derived from Glycine, L-Alanine, L-Valine, and L-Leucine, Possessing *N*-Benzyl or *N*-Benzyl-substituted Moieties

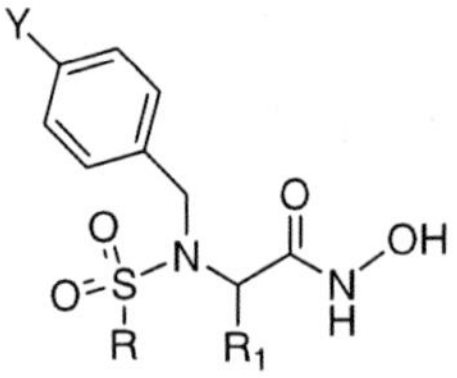

Compound	Y	R	R_1	K_I (nM) MMP-1	MMP-2	MMP-8	MMP-9
25.46a	H	n-C_4F_9	H	75	12	120	8.1
25.46b	H	n-C_8F_{17}	H	98	2.7	8.6	5.1
25.46c	H	C_6F_5	H	8.5	1.6	5.4	3.2
25.46d	NO_2	n-C_4F_9	H	62	1.5	2.4	2.0
25.46e	NO_2	n-C_8F_{17}	H	79	0.9	1.3	1.3
25.46f	NO_2	C_6F_5	H	3.0	0.7	0.1	0.6
25.46g	NO_2	3-$CF_3C_6H_4$	H	5.2	1.1	0.7	0.8
25.47	NO_2	C_6F_5	Me	3.1	0.6	0.1	0.7
25.48	NO_2	C_6F_5	i-Pr	3.0	0.5	0.2	0.6
25.49	NO_2	C_6F_5	i-Bu	2.5	0.5	0.1	0.5

p-halophenylsulfonyl, etc.). The most promising aromatic substitutions were those including perfluorophenylsulfonyl or 3-trifluoromethylbenzenesulfonyl, among others.

More recently, Santos et al.[29] improved studies on this type of sulfonamides evaluating the effects on potency and selectivity of MMP inhibition of other types of substituents in P1′ and P2′. A more exhaustive discussion of these results will be developed in the next chapter.

All these findings led to the development of other new classes of tertiary sulfonamido-based hydroxamates with increased potency and selectivity against the targeted MMPs, such as conformationally restrained tertiary sulfonamides, reversed tertiary sulfonamides, and *N-O* alkyl tertiary sulfonamides.

Conformationally Restrained Tertiary Sulfonamido-Based Hydroxamates Considering the above cited differences between the P1 and P2′ positions, the introduction of a conformational restriction was considered to be able to increase the potency and selectivity of these inhibitors on some MMPs that are overexpressed in tumors and in rheumatoid arthritis, such as the two gelatinases (MMP-2 and MMP-9), stromelysin-1 (MMP-3), and collagenase-3 (MMP-13). For this reason, some analogues containing morpholino, piperidino, piperazino, thiomorpholino, azepino, diazepino, and thiazepino cycles in their P1–P2′ positions were rapidly developed (Fig. 25.12).[30–35]

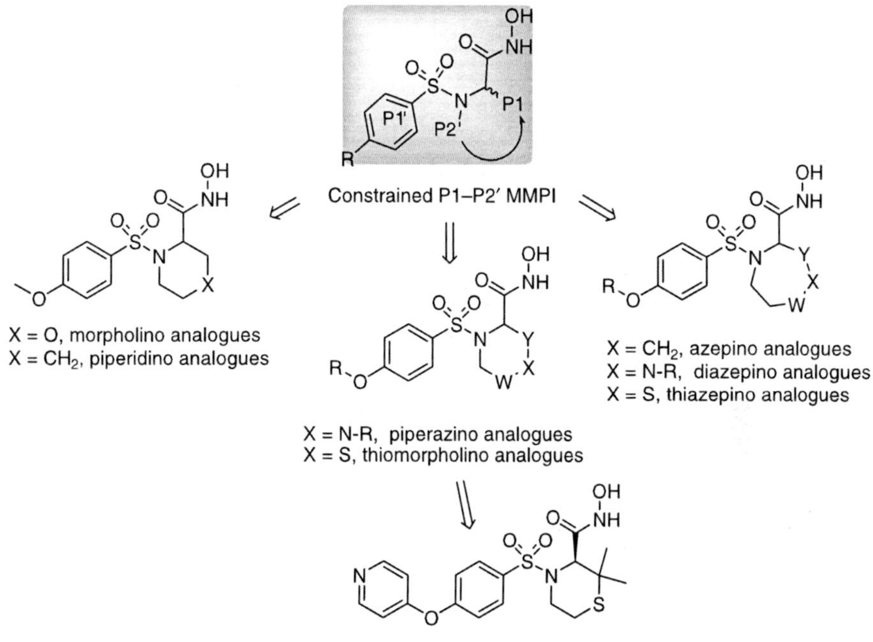

FIGURE 25.12 P1–P2′ conformationally restrained tertiary sulfonamido-based hydroxamates.

In a first study of SAR in this field, some cyclic 4-methoxy benzenesulfonamides (the same group present in P1′ position of CGS 27023A) were synthesized to evaluate the effect of both ring size and presence of a ring heteroatom on enzyme inhibition (Table 25.3).[35]

The inhibitors based on either the six- or seven-membered rings were quite potent against collagenase-1 (MMP-1) (compounds **25.52–25.54**, **25.57**, and **25.58**). The corresponding five-membered proline- and cysteine-based inhibitors (**25.50** and **25.51**) were found to be almost one order of magnitude less potent than the larger rings. The presence of a ring heteroatom in both the six- and seven-membered series did not significantly influence the potency or selectivity of the inhibitor (compare compounds **25.52–25.54**). In the six-membered ring series, substitution of the ring carbon (piperidine ring) either with an oxygen (morpholine ring) or with a sulfur atom (thiomorpholine ring) had only slight effects on enzyme inhibition. The piperazine-based inhibitor (compound **25.56**, substitution with –NH) seemed to be somewhat less potent. On the contrary, the further substitution of the ring nitrogen in the piperazine series led to a significant increase in potency (compare also compounds in Table 25.4). Oxidation of the sulfur atom in compound **25.54** to the corresponding sulfone **25.55** led to a significant decrease in potency.

In this SAR on conformationally restrained P1–P2′ sulfonamides, some new P2′ C-2 *gem*-dimethyl analogues were synthesized and analyzed for their MMPs

TABLE 25.3 Effect of Ring Size and Ring Heteroatom on MMP Inhibition

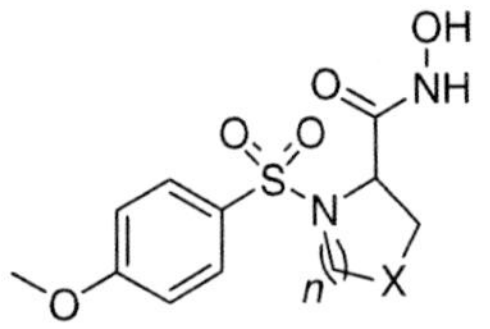

Compound	*n*	X	IC$_{50}$ (nM) MMP-1	MMP-3	MMP-7	MMP-13
25.50	1	CH_2	297	258	nd	nd
25.51	1	S	638	175	nd	nd
25.52	2	CH_2	23	14	313	0.5
25.53	2	O	49.5	5.8	nd	1.0
25.54	2	S	62	15	2070	1.0
25.55	2	SO_2	688	737	12,848	18.4
25.56	2	NH	175	48	2390	1.7
25.57	3	CH_2	49	7.0	895	1.8
25.58	3	S	64	4.3	1102	1.9

inhibition properties. In fact, it was hypothesized that a steric shielding of the hydroxamic group, for example, using a *gem*-dimethyl group introduced adjacent to this functionality, could prevent or hinder the degradation of this important functionality. As a result, it could avoid the metabolic toxicity linked to the potential

TABLE 25.4 Effect of *gem*-Dimethyl Group and Ring Size on MMP Inhibition

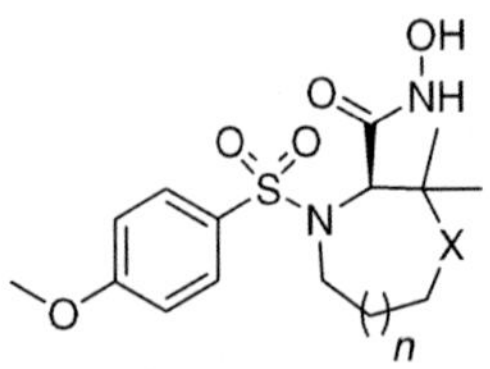

Compound	*n*	X	IC$_{50}$ (nM) MMP-1	MMP-2	MMP-3	MMP-7	MMP-8	MMP-9	MMP-13
25.59	0	S	0.4	1.4	0.7	23	0.7	0.9	1.0
25.60	0	SO_2	13	14	18.5	nd	nd	nd	4.3
25.61	1	S	0.8	2.7	0.7	30	1.4	1.9	0.9
***R*-25.61**	1	S	7780	2905	>10,000	>10,000	881	1390	2330
25.62	1	SO_2	1.9	7.8	6.9	41	2.4	3.6	2.1
25.63	1	CH_2	2.4	1.4	3.3	18	1.9	0.5	nd
25.64	3	S	3.0	2.4	7.2	30	0.6	3.7	1.0
25.6			49.5	9.1	17	106	4.4	4.3	4.3

release of free hydroxylamine and to the reduction of the zinc binding properties, due to the hydrolysis of the hydroxamic group to its corresponding carboxylic acid. The introduction of this *gem*-dimethyl group on these heterocyclic inhibitors gave a dramatic increase in potency in both the thiomorpholine (six-membered ring) and thiazepine (seven-membered ring) series (compare values in Table 25.4 with those in Table 25.3). The thiomorpholino-based inhibitors, containing the *gem*-dimethyl group, resulted in extremely potent inhibitors of both MMP-1 and MMP-3.

An increase in potency of almost two orders of magnitude was observed against MMP-1 with the addition of the *gem*-dimethyl group (compare compound **25.54** in Table 25.3 with compound **25.59** in Table 25.4). A significant increase (at least one order of magnitude) in potency was also observed against MMP-3 and MMP-7 when the *gem*-dimethyl group was introduced. Interestingly, the influence of the *gem*-dimethyl group on the inhibition of MMP-13 was much less significant. The effect on *in vitro* inhibition was also observed in the sulfone series (compare compound **25.55** in Table 25.3 with compound **25.60** in Table 25.4). The sulfone **25.60** was a potent inhibitor of MMP-1 and MMP-3, although it was two orders of magnitude less potent than the parent thiazine **25.59**. On these bases, it was confirmed that the *gem*-dimethyl group had a dramatic effect on the potency of this series of compounds. The thiazepines containing the C-2 *gem*-dimethyl group were determined to be as potent inhibitors of both MMP-1 and MMP-3 as the corresponding thiazine-based compounds. The major disparity between the thiomorpholine and thiazepine-based inhibitors was observed upon oxidation of the ring sulfur to the corresponding sulfone. Oxidation of the ring sulfur in the thiazepine series did not result in any significant decrease in potency (compare thiazines **25.59** and **25.60** with thiazepines **25.61** and **25.62**). This was a useful result, as the metabolic oxidation of a sulfide to the corresponding sulfoxide and sulfone is a well-known process.[36] The corresponding azepine analogue **25.63** was almost identical with respect to the profile observed with the thiazepine **25.61**. Apparently, the sulfur atom did not significantly influence the potency of the molecule, even in the *gem*-dimethyl series. The nine-membered thiazonane **25.64** was prepared to determine if the enzyme active site could accommodate such large rings. This compound was found to be as potent as the corresponding thiomorpholine analogue **25.61**. To improve the knowledge of these thiomorpholine derivatives, the influence of the amino acid configuration on their C-1 stereogenic center on inhibitory properties was also examined. The distomer (*R*)-**25.61**, prepared and tested for *in vitro* activity, was a very poor inhibitor of the MMPs, with potencies three to four orders of magnitude lower than that of the corresponding eutomer **25.61** (see below).[35]

Considering the good results obtained with the thiazepine ring in the conformational restriction on P1–P2′, the sulfonamide portion of these molecules and its influence on enzyme inhibition were also examined (Table 25.5).

Replacement of the 4-methoxy group of **25.65** (already reported in Table 25.4 as **25.61**) with a 4-bromo substituent led to a significant shift in enzyme selectivity for MMP-1 over MMP-3 (compare **25.65** and **25.66** with **25.67** and **25.68**). Moreover, the 2-methyl-4-bromophenylsulfonamide **25.69** was found to be even more selective for MMP-1. On the contrary, the results obtained with the 4-butoxy substituent were

TABLE 25.5 ***In Vitro*** **Inhibitory Profile of Thiazepine MMPI**

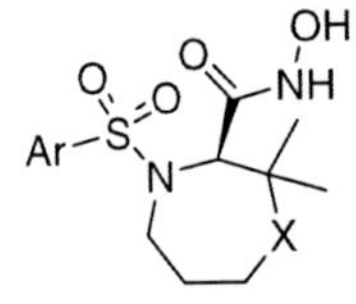

			IC_{50} (nM)				
Compound	X	Ar	MMP-1	MMP-2	MMP-3	MMP-7	MMP-13
25.65	S	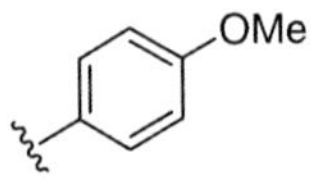	0.8	2.7	0.7	30	0.9
25.66	SO_2	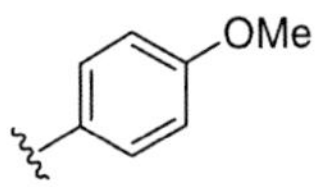	1.9	7.8	6.9	41	2.1
25.67	S	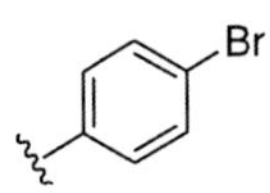	0.7	6.8	10	17	1.3
25.68	SO_2	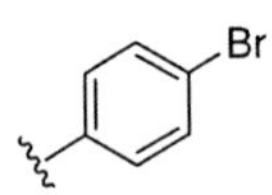	1.2	20	90	79	2.7
25.69	S	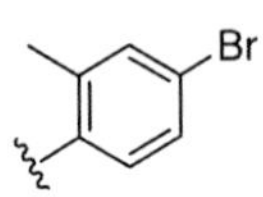	0.5	24	88	354	9.2
25.70	S	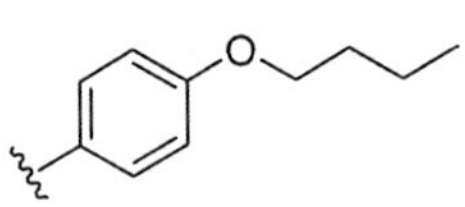	18	1.0	6.6	nd	2.3
25.71	SO_2	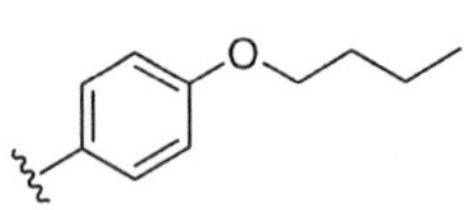	19	1.2	2.7	27	0.4
25.72	S	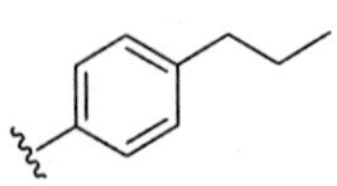	22	35	46	637	7.5
25.73	S	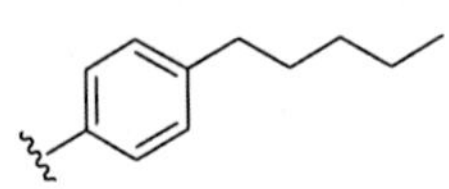	61	nd	25	330	2.1
25.74	S	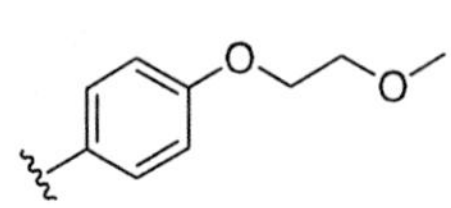	42	11.4	17	509	5.6

TABLE 25.5 (*Continued*)

25.75	S	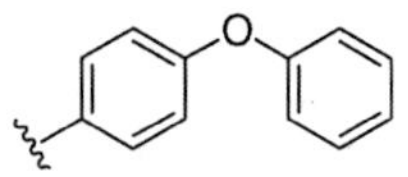	2.3	nd	11	16	<0.4
25.76	S	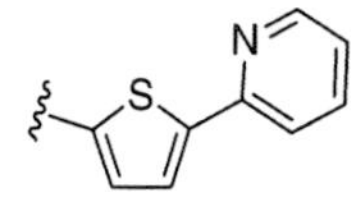	1.0	1.0	1.1	170	<0.2
25.77	S	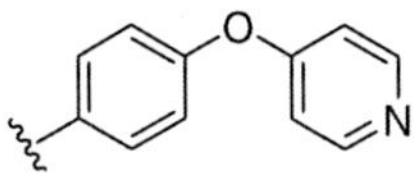	5.7	<0.4	6.3	105	<0.4
25.78	S	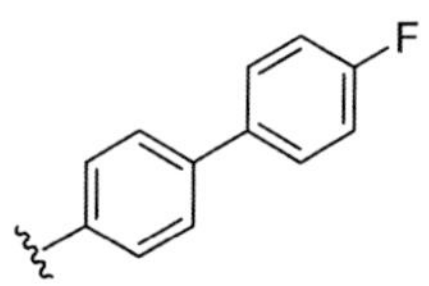	2.9	1.3	6.5	<100	<1.0

unexpected. A slight decrease in potency was observed for both MMP-1 and MMP-3 with compounds **25.70** and **25.71**. This fact seemed to be inconsistent with previous observations on compounds that possessed a long P1′ substituent because they were expected to be more potent on MMP-3 than on MMP-1.[16] The same considerations could be made for the 4-alkylated phenylsulfonamides **25.72** and **25.73** and the 4-alkyldiether **25.74**.

The aryl ethers **25.75** and **25.77**, the heterocyclic sulfonamide **25.76**, and the biphenylsulfonamide **25.78** were all found to be broad-spectrum inhibitors, with increased activity on MMP-13. The same unexpectedly high potency of inhibitors with large P1′ groups toward MMP-1 has also been observed by Agouron Pharmaceuticals with the thiomorpholino-based inhibitor AG3340, **25.4** (given subsequently).[37]

In MMP-1, the S1′ pocket is small and closed by the nonconserved residue Arg214 (leucine in MMP-2). Although this residue shows conformational flexibility, it is known that the low inhibition potency of many inhibitors against MMP-1 is often due to the presence of large P1′ groups that would require large conformational changes to interact with the S1′ pocket.[38] Recent simulations have shown that the presence of the 4-methoxyphenyl group (as in CGS 27023A, **25.6**) in the S1′ pocket was able to determine the movement of the Arg214. In the initial structure, this residue was stabilized by four H bonds, three of them with Leu235 and Thr241 (two H bonds) and a fourth intramolecular H bond between the carboxy and amino group of the same Arg214 (see Fig. 25.13a). During the simulation, the rotation of the δ-torsional of the arginine with the disappearance of the intramolecular H bond and of the H bond with Leu235 was observed, thus determining an increase in the pocket depth (see Fig. 25.13b). This fact was also confirmed for the 4-(4′-pyridine)-oxyphenyl group

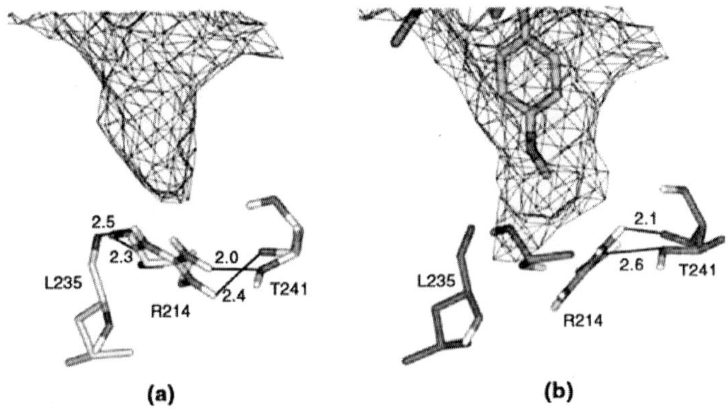

FIGURE 25.13 Analysis of the S1′ cavity in the X-ray structure (a) and after modeling with CGS 27023A (b). The presence of the *p*-methoxyphenyl group determines the movement of Arg214 and an increase in cavity depth.

of prinomastat (**25.4**), due to the same type of movement for Arg214, previously observed for CGS 27023A **25.6**.

The only differences described were in a larger shift of this residue, with enlargement and stretching of the S1′ pocket, and the formation of a new H bond between the final pyridine nitrogen of this inhibitor and Thr241.[39]

In summary, the results obtained with the thiazepines showed that the introduction of a *gem*-dimethyl group on the ring was able to give potent and broad-spectrum inhibitors, which were not significantly influenced by the P1′ sulfonamide substituent. Further substitutions on the thiazepine ring or a change in the ring size seemed to be unable to modify their enzyme inhibition profile (Table 25.6). For example, the parent inhibitors **25.79** and **25.80**, containing either a methoxymethyl ether or a hydroxyl group at C-6, remained potent and broad-spectrum inhibitors, similar to compound **25.65**.

This fact was also reiterated in the nine-membered oxathiazonines **25.81** and **25.82** that remained potent MMPIs. The ring size and the incorporation of another heteroatom on the ring had only a slight effect on the inhibition profile. On the contrary, the addition of a C-5 phenyl substituent caused a significant decrease of potency on MMP-1 and MMP-3. Finally, interestingly enough, by inserting a C-3 methyl group on the 1,4-thiazepine ring, as in derivative **25.85**, this inhibitor was able to maintain a good activity on MMP-2 and MMP-13, sparing especially MMP-7 and, to a minor extent, MMP-1.

To complete these studies, at the Procter & Gamble Laboratories, the importance of the hydroxamic acid moiety for *in vitro* activity was also evaluated (Fig. 25.14). In this case, some changes were made in the ZBG to determine if either a carboxylic acid or an *N*-alkyl-substituted hydroxamic acid derivative would retain any *in vitro* activity. The changing of the hydroxamic ZBG of **25.65** and **25.75** with a carboxylic group had a dramatic effect. In fact, the resulting carboxylic acid analogues **25.86** and **25.87** were inactive against all of the MMPs considered (with IC_{50}'s > 10 μM). On the contrary, the *N*-methylhydroxamic acids **25.88** and **25.89** were found to be only three orders of magnitude less potent than the parent hydroxamic acids **25.65** and **25.75**. Finally, as

TABLE 25.6 ***In Vitro*** **Inhibitory Profile of Thiazepine MMP Inhibitors**

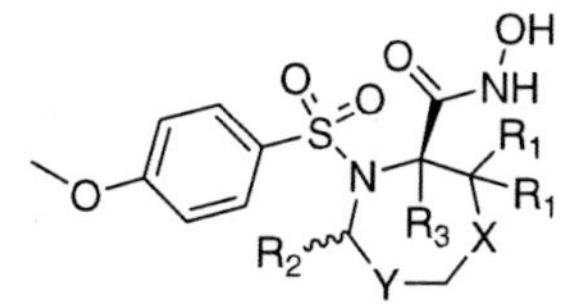

Compound	R_1	X	Y	R_2	R_3	IC$_{50}$ (nM) MMP-1	MMP-2	MMP-3	MMP-7	MMP-13
25.65	Me	S	$-CH_2-$	H	H	0.8	2.7	0.7	30	0.9
25.79	Me	S	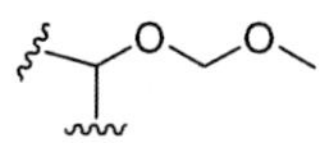	H	H	1.0	1.8	1.0	38	1.4
25.80	Me	SO_2	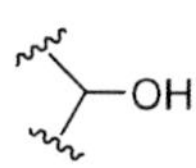	H	H	1.9	5.2	7.3	nd	1.3
25.81	Me	S	$-CH_2OCH_2-$	H	H	<1.0	3.5	0.7	<100	0.2
25.82	Me	SO_2	$-CH_2OCH_2-$	H	H	1.9	nd	45	183	3.3
25.83	Me	S	$-CH_2-$	α-Ph	H	236	nd	26	54	137
25.84	Me	S	$-CH_2-$	β-Ph	H	271	nd	33	110	2.5
25.85	H	S	$-CH_2-$	H	Me	217	4.2	31	17,651	4.0

seen above for the carboxylic acids **25.86** and **25.87**, the *N*-methoxyhydroxamic acid **25.90** was found to be completely inactive against the MMPs tested. These results clearly demonstrate the importance of the hydroxamic acid group in the thiazepine series for *in vitro* inhibition.

Table 25.7 reports the MMPs inhibition profiles of the other conformationally restrained analogues **25.91–25.109**, developed in the same period by Pfizer, Agouron, Wyeth, and Procter & Gamble Pharmaceuticals, with some heterocyclic rings, preferentially six-membered rings, substituted by bulky aromatic sulfonamides on

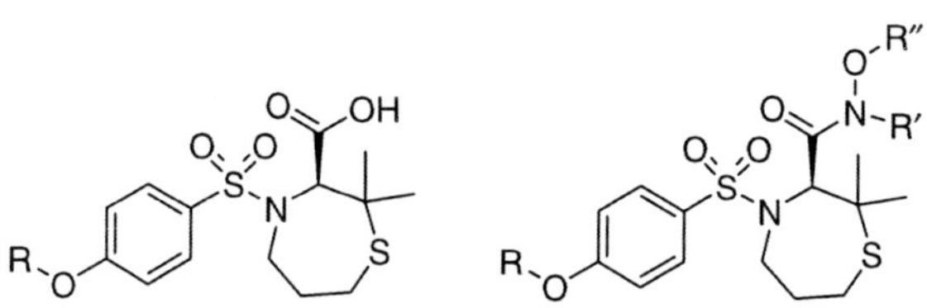

25.86, R = CH_3
25.87, R = Ph

25.88, R = CH_3; R′ = CH_3; R″ = H
25.89, R = Ph; R′ = CH_3; R″ = H
25.90, R = Ph; R′ = H; R″= CH_3

FIGURE 25.14 Modification of the ZBG in thiazepine MMPI.

TABLE 25.7 ***In Vitro* Inhibitory Profile of Other Conformationally Restrained Sulfonamido-based Analogues**

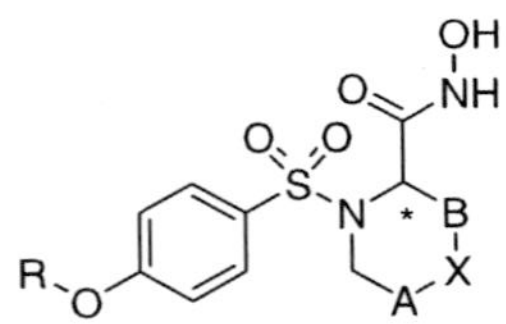

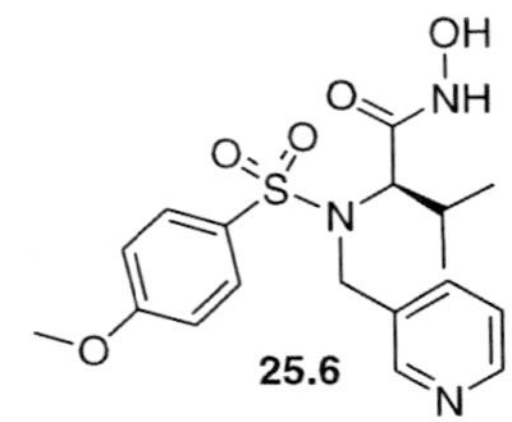

25.6

Compound	*	X	A	B	R	K_I (nM)[a]					
						MMP-1	MMP-2	MMP-3	MMP-7	MMP-9	MMP-13
25.91	*R/S*	NH	CH_2	CH_2	4-Br-Ph	68	0.77	1.8	263	—	1.1
25.92	*R*	NH	CH_2	CH_2	4-Cl-Ph	34	0.41	1.0	171	—	0.36
25.93	*R*	NH	CH_2	CH_2	4-F-Ph	176	0.41	1.9	2060	—	0.013
25.94	*R*	C=O	NH	CH_2	4-F-Ph	187	0.15	5.1	1840	—	—
25.95	*R*	O	CH_2	CH_2	4-Cl-Ph	8.9	0.031	0.093	77	—	0.021
25.96	*R/S*	S	CH_2	CH_2	4-Br-Ph	—	0.04	0.33	169	—	—
25.97	*R/S*	NCH_3	CH_2	CH_2	Ph	—	0.86	6.2	560	—	—
25.98	*R/S*	NCH_3	CH_2	CH_2	4-Br-Ph	257	0.035	1.4	1860	—	0.022
25.99	*R/S*	$NCOCH_3$	CH_2	CH_2	Ph	—	0.11	0.64	113	—	0.050
25.100	*R/S*	$NCOC(CH_3)_3$	CH_2	CH_2	4-Br-Ph	60	0.025	0.73	378	—	0.070
25.101	*R*	$NCOC(CH_3)_3$	CH_2	CH_2	4-Cl-Ph	—	0.007	0.31	142	—	0.06
25.102	*R/S*	$NCOC(CH_3)_3$	CH_2	CH_2	4-CN-Ph	580	0.70	14	3570	—	1.97
25.103	*R/S*	$NCOC(CH_3)_3$	CH_2	CH_2	2-Py	550	0.72	17	2850	—	1.64
25.104	*R/S*	NH	C=O	$C(CH_3)_2$	4-F-Ph	5.9	0.07	0.84	—	—	0.07
25.105	*S*	S	CH_2	$C(CH_3)_2$	4-Cl-Ph	1.3	0.017	0.06	—	—	0.001
25.106	*R/S*	NCH_2Ph	CH_2CH_2	CH_2	CH_3	445	—	—	—	31	65
25.107	*R/S*	$NC(O)CH_2NHBoc$	CH_2CH_2	CH_2	CH_3	690	—	—	—	23	16
25.108	*R/S*	$NC(O)CH_2NH_2$	CH_2CH_2	CH_2	CH_3	703	—	—	—	157	46
25.109	*R/S*	NCOPh	CH_2CH_2	CH_2	Ph	22	—	—	—	1.2	1.3
25.4[b]	*S*	S	CH_2	$C(CH_3)_2$	4-Py	8.2	0.083	0.27	54	<0.4	0.038
25.6[c]			33	20	43	106	8	4.3			

[a] Fluorescence based assay on isolate and activated enzymes.
[b] AG3340 (Prinomastat).
[c] CGS27023A (MMI270).

their P1′ position. Also in this case, the new inhibitors were compared with the noncyclic tertiary sulfonamide CGS 27023A **25.6**.

As shown in this table, in many of the reported examples, there was an improvement of inhibition potency on some of the MMPs, such as MMP-2, MMP-3, and MMP-13, with respect to MMP-1 and especially MMP-7, passing from **25.6** to the analogues, conformationally restrained in their P1–P2′ positions. This is particularly evident, for example, in the piperazino (**25.91–25.93** and **25.97–25.104**), morpholino (**25.95**), and thiomorpholino (**25.96**, **25.105**, **25.4**) cyclic sulfonamides. Also, these examples confirmed the chiral requirements for the C(1) stereogenic center necessary to have a good fitting with these enzymes (compare chiral compounds **25.92–25.95**, **25.101**, **25.105**, and **25.4**). The good effect of the C-2 *gem*-dimethyl substitution was also reconfirmed in the thiomorpholine analogues described here.

The best results were obtained with the chiral (*S*)-C(1), C-2 *gem*-dimethyl, thiomorpholine **25.4**, AG3340 (prinomastat, from Agouron Pharmaceuticals), the first example of a restrained inhibitor entered in clinical cancer trials.[30,34,40]

As in the case of the first generation of MMPIs, the major efforts were mainly dedicated to the development of inhibitors for cancer in the second generation.

The first clinical results obtained with prinomastat, belonging to the second generation of MMPIs, were disappointing and quite similar to those of the first generation of MMPIs. In fact, phase III trials with **25.4** for advanced prostate and lung cancer were stopped in 2000 because no therapeutic effects were observed using it in combination with chemotherapy, as compared with chemotherapy alone.[41] The lack of selectivity contributed to the failure of the trials with the first generation of MMPIs; another contribution to its failure was the combined and undervalued conduction of the trials in patients with advanced metastatic disease.[17]

With regard to the second generation, **25.4** was designed to be a selective inhibitor of some specific MMPs. Unfortunately, the selectivity of **25.4** was still not sufficient to cause inhibition of cancer growth at the maximum tolerated doses. Further trials using higher doses of **25.4** proved to be toxic, causing the musculoskeletal syndrome (MSS).

One of the probable causes of MSS could be the inhibition of MMP-1, which is expressed in both normal and diseased joints, differently from MMP-13, which seems to be particularly upregulated in arthritic cartilage.[42] For this reason, specific inhibitors devoid of MMP-1 activity still belonging to the conformationally restrained sulfonamido-based hydroxamates class were developed to reduce these side effects. A series of arylhydroxamate sulfonamides active against MMP-2 and MMP-13 was described by Barta and coworkers.[43] They planned to insert a long substituent into P1′ to achieve selectivity over MMP-1, and they synthesized compound **25.110** (Fig. 25.15), which showed nanomolar activity on MMP-2 and MMP-13 and spared MMP-1. X-ray crystallography revealed that the sulfonyl group of the inhibitor made a single hydrogen bond with Leu160 and that the piperidine-*O*-phenyl moiety extended into the S1′ subsite, making van der Waals contact with the protein.

Selective inhibitors of MMP-9, MMP-13, and TACE have been identified by Wyeth researchers[44] based on an anthranilic acid scaffold. Among these sulfonamide derivatives, the 4-pyridyl ether **25.111** (Fig. 25.15), a potent MMP-9 and MMP-13

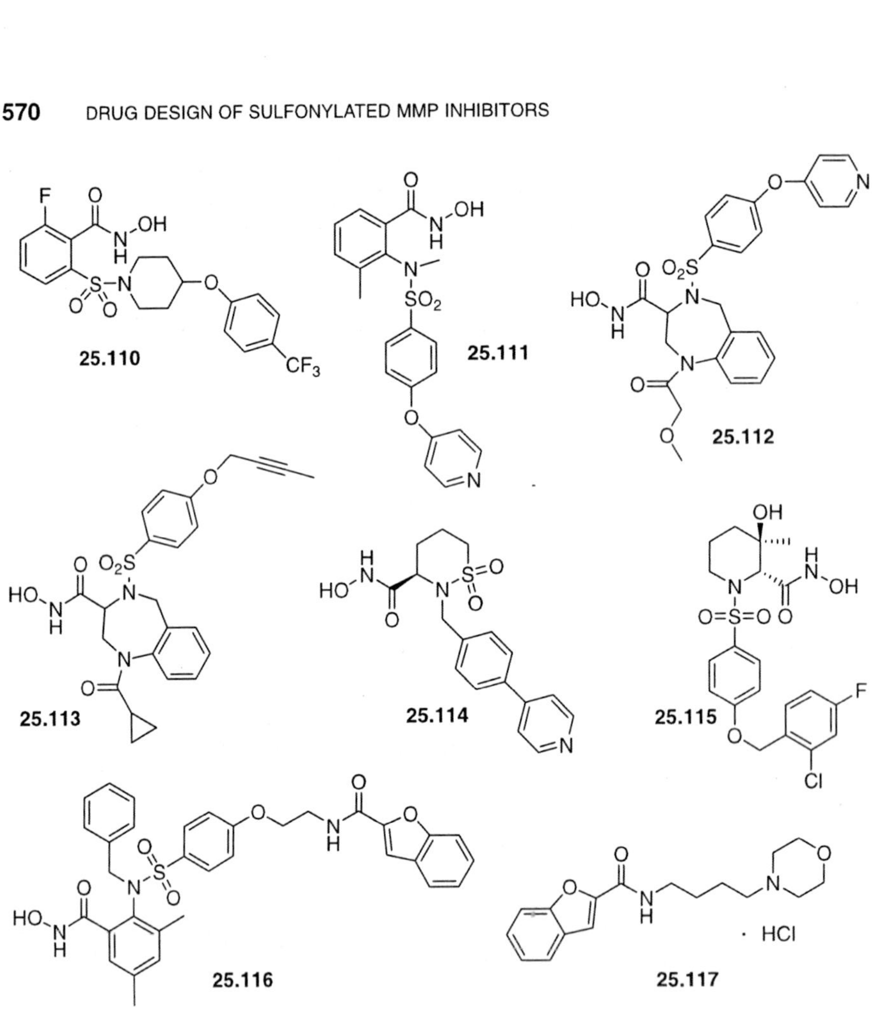

FIGURE 25.15 Structures of MMP inhibitors **25.110–25.117**.

inhibitor, displayed the highest selectivity over MMP-1, due to the length of the P1′ alkoxy moiety. This compound also demonstrated oral activity in a rat sponge-wrapped cartilage model.

The same Wyeth researchers recently published an interesting study on a series of cyclic sulfonamidic hydroxamates, designed using a benzodiazepine template as a framework.[45] To provide compounds with improved selectivity for MMP-13, they introduced bulky substituents in the aryl sulfonic portion. Compound **25.112**, again with the 4-pyridyl ether in P1′, displayed a 381-fold selectivity for MMP-13 over MMP-1 and, meanwhile, **25.113**, with a 2-ynyloxy group in P1′, showed an 84-fold preference for TACE over MMP-1 (Fig. 25.15).

Looking for new templates that could offer different selectivity profiles, sultam hydroxamates were proposed as novel MMPIs by Cherney and coworkers.[46] Compound **25.114** (Fig. 25.15) was found to be potent and selective for MMP-2, MMP-9, and MMP-13 over MMP-1 and TACE. The crystal structure of **25.114** complexed with MMP-13 indicated that the pro-(*S*) sultam sulfonyl was able to form a hydrogen bond with Leu185 and that the *N*-methylene of the sultam gave the bisaryl the critical turn necessary to align the bisaryls into the S1′ pocket. Initial pharmacokinetic assays suggested that the sultam hydroxamates were orally bioavailable.

Recently, Pfizer researchers focused their attention on finding compounds able to inhibit both MMP-13 and aggrecanase so as to protect both the aggrecan and type II collagen components of the cartilage from degradation. Such compounds could be useful for the treatment of osteoarthritis. This kind of selectivity profile was achieved with a series of 3,3-dimethyl-5-hydroxy pipecolic hydroxamic acids discovered by Noe et al.,[47] which possessed potent inhibitory activity versus aggrecanase and MMP-13 with selectivity versus MMP-1 and TACE. Successively, with the aim of reducing the metabolic liability of this first series of compounds, they explored the effects of introducing a polar functionality at the 3-position of the piperidine ring.[47] This strategy produced compound **25.115** (Fig. 25.15), which showed not only an excellent activity on MMP-13 and aggrecanase but also an improved bioavailability and lower metabolic clearance compared to the analogous compounds of the previous series.

With the aim of discovering a very selective MMP-13 inhibitor, Chen et al.[48] described a compound, WAY-170523 **25.116**, derived from the combination of a known potent broad-spectrum MMP inhibitor (bearing a hydroxamate moiety) with a weak but selective MMP-13 inhibitor (compound **25.117**) that occupied the S1′ pocket of MMP-13 without interacting with the catalytic zinc. The hybrid compound showed an IC_{50} of 17 nM for MMP-13 and a strong selectivity against the other MMPs tested (Table 25.8).

Reversed Tertiary Sulfonamido-Based Hydroxamates Considering the relevant interest for MMP-3, an important enzyme involved in arthritis and chronic nonhealing wounds, in a recent development of SAR studies for highly selective MMP-3 inhibitors at Pfizer Pharmaceuticals, Whitlock et al. discovered a new class of reversed tertiary sulfonamido-based hydroxamates that showed a high index of selectivity among MMPs. These results seem to be linked to a reversed 4-biaryl piperidine sulfonamide core, which binds specifically at the S1′ subsite of MMP-3. Modeling of **25.118** into the published MMP-2 catalytic domain crystal structure suggested that a substitution in the *meta*-position of ring B could induce unfavorable interactions with the loop forming the S1′ pocket and, therefore, reduce MMP-2

TABLE 25.8 ***In Vitro*** **Inhibitory Profile of MMP-13 Inhibitors 25.110–25.117 Able to Spare MMP-1**

	IC_{50} (nM)				
Compound	MMP-1	MMP-2	MMP-9	MMP-13	TACE
25.110	> 10,000	3.3	—	12.2	—
25.111	3245	—	7.0	4.0	—
25.112	763	—	2.0	2.0	157
25.113	841	—	33.0	29.0	10.0
25.114	1085	1.0	10.0	3.0	—
25.115	310	—	—	0.9	—
25.116	> 100,000	—	945	17	> 1000
25.117	Inactive	—	Inactive	90% (10 μM)	Inactive

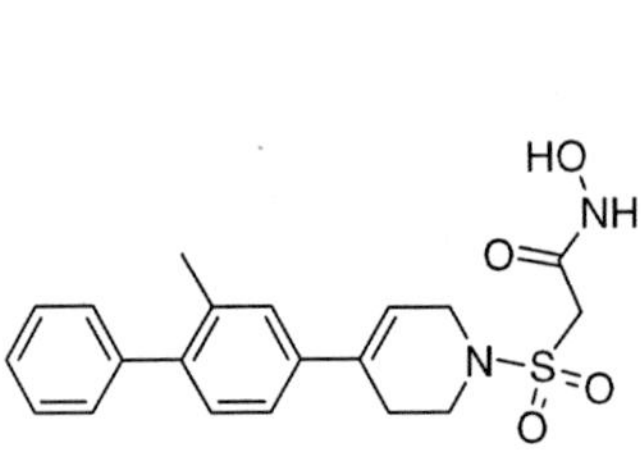

25.118

(IC_{50}, nM)
MMP-3: 16; MMP-2: 320

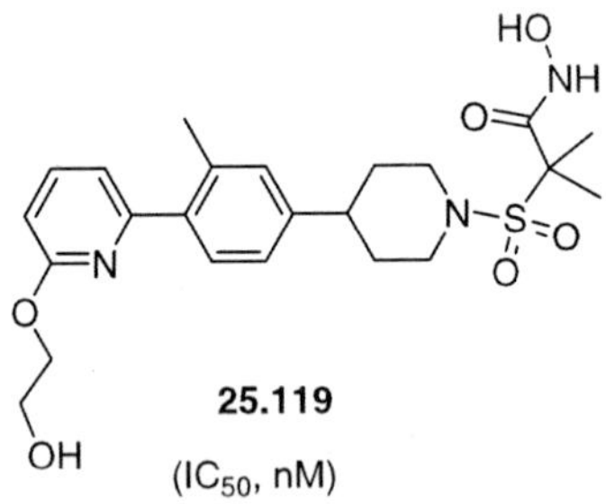

25.119

(IC_{50}, nM)
MMP-3: 1; MMP-1: 14,000; MMP-2: 529;
MMP-9: 2400; MMP-14: 20,000

FIGURE 25.16 Reversed tertiary sulfonamido-based inhibitors **25.118** and **25.119**.

enzyme inhibition. This loop region is three residues shorter in MMP-2 compared to MMP-3. This observation led to the synthesis of compound **25.119**, which is an example of a selective MMP-3 inhibitor that is able to spare MMP-1, MMP-2, MMP-9, and MMP-14 with high efficiency (Fig. 25.16)[49]

N-O-Alkyl Tertiary Sulfonamido-Based MMPI Recently, a new family of tertiary sulfonamides was developed at Pisa University starting from the original structure of CGS 27023A, **25.6.** Some *N*-Oxa-alkyl analogues of **25.6** were reported, such as compound **25.121** (Fig. 25.17), which can be considered an analogue of the already known *N-i*-butyl-biphenylsulfonamide **25.120**.[50] In this new type of sulfonamide, the

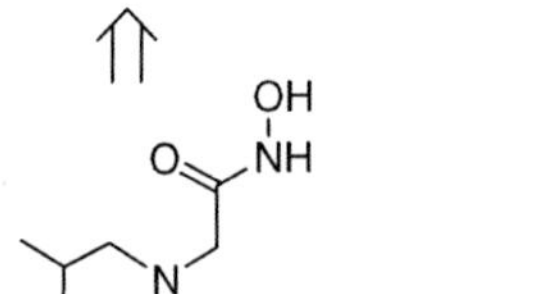

Tertiary sulfonamido-based MMPI

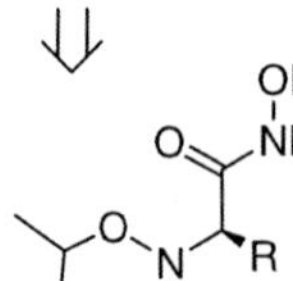

Tertiary *N*-*O*-alkylsulfonamido-based MMPI

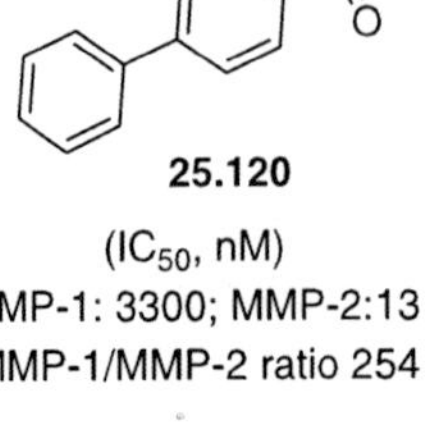

25.120

(IC_{50}, nM)
MMP-1: 3300; MMP-2:13
MMP-1/MMP-2 ratio 254

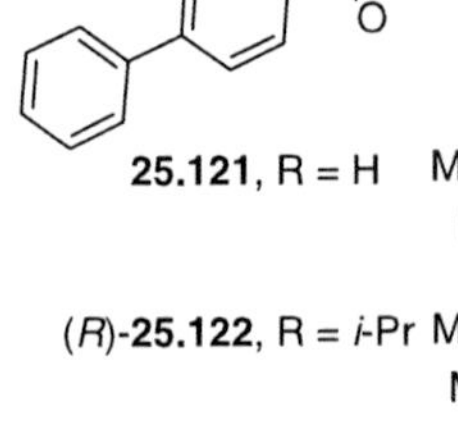

(IC_{50}, nM)
25.121, R = H MMP-1: 12,500; MMP-2: 12.4
MMP-1/MMP-2 ratio 1004

(*R*)-**25.122**, R = *i*-Pr MMP-1: 500; MMP-2: 0.80
MMP-1/MMP-2 ratio 625

FIGURE 25.17 Development of *N-O* alkyl tertiary sulfonamido-based MMPI.

methylenic carbon atom of the isobutyl group, in the P2′ position, was replaced by an oxygen atom. This new functionality (the oxyaminic oxygen) produced a structural modification of the molecular backbone able to improve the selectivity of these inhibitors for MMP-2.

In fact, the *N-O-i*-propyl biphenylsulfonamide **25.121** was equipotent to its *N*-methylene analogue **25.120** on MMP-2 but four times more selective on MMP-1. Finally, **25.121** also proved to be effective in a model of invasion on matrigel carried out on cellular lines of fibrosarcoma HT1080 (tumoral cells overexpressing MMP-2).[51] The introduction of an alkyl substituent in position α to the hydroxamic acid, such as in compound **25.122**, was able to improve the binding with a hydrophobic region in the S1 site, thus increasing potency and maintaining selectivity against the targeted MMP-2. Inhibitor **25.122** was proven to be effective in an *in vitro* model of angiogenesis.[52] More exhaustive kinetic studies carried out with **25.121**, **25.122**, and other N–O sulfonamides of this new family of MMPIs seemed to confirm the existence of a different mechanism of binding for these inhibitors with the various MMPs studied.[53]

25.3.1.2 Secondary Sulfonamido-Based MMPI In parallel to the development of tertiary sulfonamides, some gelatinase inhibitors belonging to the secondary sulfonamido-based family were developed initially by the Shionogi Company (Fig. 25.18).

Tamura et al.[54] reported that several 4′-biphenylsulfonamides of unnatural D-amino acids were potent inhibitors of several S1′ deep pocket MMPs, such as the two gelatinases (MMP-2 and MMP-9). The hydroxamic acid (*R*)-**25.123**, a biphenylsulfonamide of D-phenylalanine, was initially a very potent and almost selective MMPI (see Table 25.9). However, these hydroxamic acid derivatives belonging to the

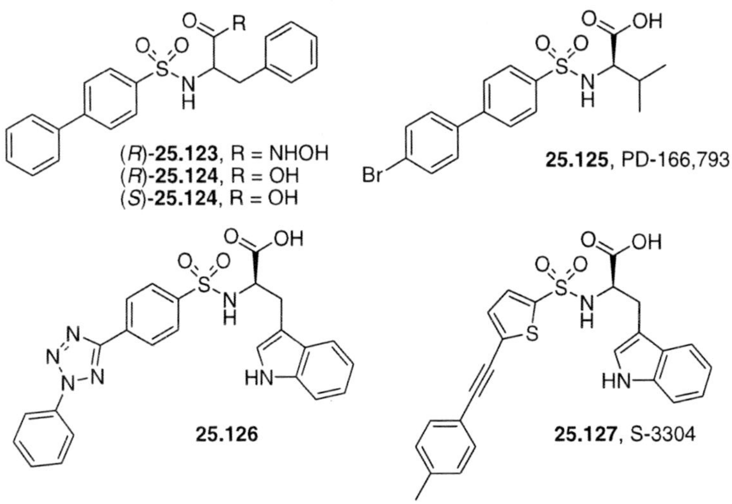

(*R*)-**25.123**, R = NHOH
(*R*)-**25.124**, R = OH
(*S*)-**25.124**, R = OH

25.125, PD-166,793

25.126

25.127, S-3304

FIGURE 25.18 Examples of secondary sulfonamido-based MMPI.

TABLE 25.9 ***In Vitro* Inhibitory Profile of Shionogi Secondary Sulfonamido-based MMPI**

	IC_{50} (nM)					
Compound	MMP-1	MMP-2	MMP-3	MMP-7	MMP-9	MMP-14
25.123	970	12	> 1000	800	16	17
25.124	—	310	—	—	240	—
(*S*)-**25.124**	—	—	—	—	800	—
25.125	3200	5	12	—	8300	—
25.126	> 1000	19	> 1000	> 1000	32	—
25.127	> 1000	20	> 1000	> 1000	23	—

secondary sulfonamido-based MMPI were found to be chemically and pharmacokinetically unstable, releasing *in vivo* hydroxylamine, a known carcinogenic compound.[55] Moreover, these hydroxamic acids were poorly absorbed by oral administration. On the other hand, carboxylic acid inhibitors were less potent *in vitro* on isolate enzymes but showed better pharmacokinetic properties than the corresponding hydroxamic acids. In fact, the hydroxamic acid derivative (*R*)-**25.123** showed a maximum plasma concentration of 2.5 μM after *per os* administration of 200 mg/kg in mice, and its carboxylic acid analogue (*R*)-**25.124**, about 20 times less potent on gelatinases *in vitro*, showed a maximum plasma concentration of 291 μM under the same experimental conditions. Therefore, the carboxylic acid derivatives of many unnatural D-amino acids were developed, which resulted in highly selective and orally active MMP-2 and MMP-9 inhibitors. Generally, valine and phenylalanine derivatives were less potent than D-tryptophan inhibitors. Moreover, the insertion of a suitable spacer, such as azo, a tetrazole, a triple bond, an amide, or an oxygen atom, between the two phenyl rings of the biphenylsulfonamide moiety of these compounds resulted in the inhibitory activity being retained and enhanced (see, for example, compound **25.126**); meanwhile, insertion of methylene, carbonyl, and urea groups led to poor inhibitory activities. The isosteric replacement of the phenyl ring linked to the sulfonamide derivatives, with a thiophene ring directly attached to the sulfonamide group, not only enhanced the inhibitory activity in this class but also improved the pharmacokinetic properties. Some of these carboxylic derivatives were able to significantly suppress the lung colonization of Lewis lung carcinoma cells of *in vivo* animal cancer models. In addition, the antitumor activity was also demonstrated in a human lung cancer model. Daily oral administration of compound **25.126** resulted in prolonged survival of Ma44-bearing mice.

A highly selective and potent MMP-2/MMP-9 inhibitor **25.127**, known as S-3304, was tested in a phase I clinical trial. It was very well tolerated and produced inhibition of gelatinase activity in tumor biopsies at the lowest dose tested, a dose that produced very little toxicity. In this trial, 50% of the patients with renal cell carcinoma had stable disease, and further testing of this disease is ongoing.[56]

Highly selective MMPIs were also developed at Agouron Phamaceuticals in the same period. The carboxylic acid **25.128** and its hydroxamic acid analogue **25.129** are examples of very good MMPIs that are selective for MMP-2 (Fig. 25.19). In these two

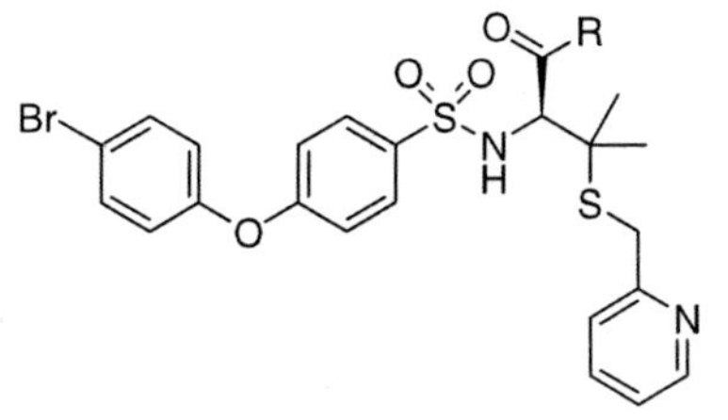

(K_I, nM)

25.128, R = OH   MMP-1: 1500; MMP-2: 7; MMP-3: 34,000
MMP-7: 75,000; MMP-13: 178

25.129, R = NHOH   MMP-1:0.26; MMP-2: 0.009; MMP-3: 4.1
MMP-7:7.9; MMP-13: 0.009

FIGURE 25.19 Structures of MMP inhibitors **25.128** and **25.129**.

examples, some important findings emerged from the previous SAR studies, such as the use of the C-α of the appropriate configuration bearing a proper substituent, the *gem*-dimethyl substitution to increase metabolic stability, a pyridyl substituent to improve bioavailability, and a large P1′ sulfonamido substituent to better interact with the deep S1′ pocket of some MMPs, such as MMP-2 and MMP-13. In this case, the hydroxamate, **25.129**, was more potent (K_I in the pM range) but less selective than the corresponding carboxylic acid, **25.128**.[57]

Recent SAR studies on this type of MMPI have disclosed that the introduction of groups, such as acylamides or ethers bearing hindered heterocyclic rings like polysubstituted benzoxazoles, on the 4′-position of the biphenylsulfonamide (P1′ position) was able to increase selectivity and potency against some MMPs and aggreganases (ADAMTS) involved in osteoarticular diseases.[58–60] The MMPI **25.130** (Fig. 25.20) represented a significant example in this class. It showed very high selectivity ratios MMP1/MMP13 (>200,000 times) and an interesting selectivity ratio for MMP-2/MMP-13 (75 times). A similar or major trend of selectivity was observed in many of

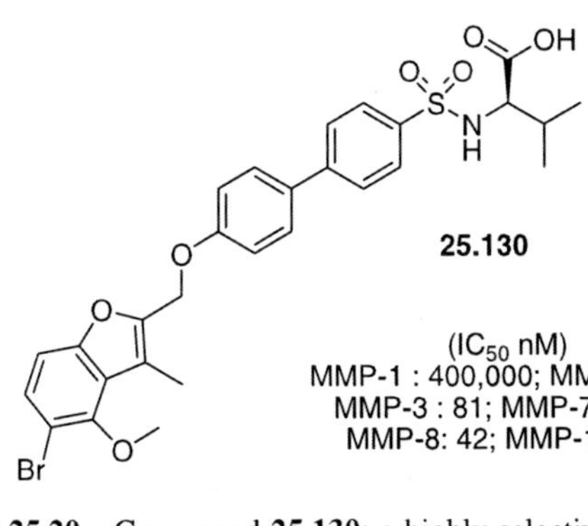

25.130

(IC_{50} nM)
MMP-1 : 400,000; MMP-2 :135;
MMP-3 : 81; MMP-7: 1100;
MMP-8: 42; MMP-13: 1.8

FIGURE 25.20 Compound **25.130**: a highly selective MMP-13 inhibitor.

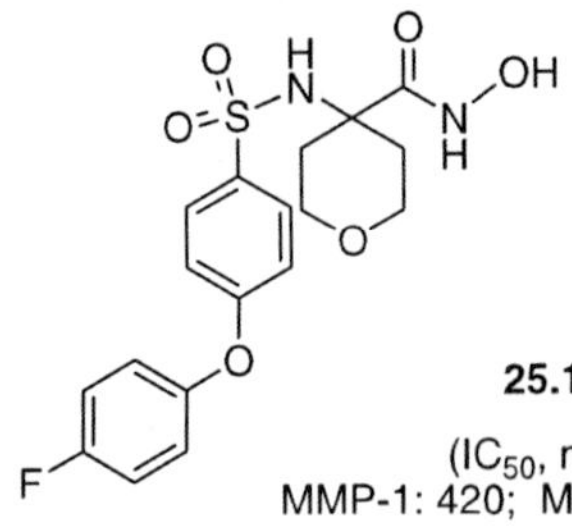

25.131

(IC_{50}, nM)
MMP-1: 420; MMP-13: 0.75

FIGURE 25.21 Structure of MMP-13 inhibitor **25.131**.

the reported inhibitors. Moreover, these orally active compounds showed good pharmacokinetic properties.

In other SAR studies by Pfizer on MMP-13 inhibitors, some new secondary sulfonamido-based inhibitors were described, characterized by the presence of a spirocycle adjacent to the ZBG. It was confirmed that the insertion of a sterically hindered system, such as the pyran one, was able to block the metabolic degradation while increasing MMPs selectivity. Compound **25.131** (Fig. 25.21) was a representative example of this family of inhibitors that was relatively selective for MMP-13. Extensive biopharmacological studies conducted with this inhibitor confirmed its good bioavailability and reduced liver metabolism.[61]

25.3.2 Sulfone-Based MMPI

Replacing the arylsulfonamido moiety of the tertiary sulfonamido-based hydroxamate analogues of **25.6** with an arylsulfone one, such as in the case of the derivative β-(arylsulfonyl)hydroxamic acid **25.132** (Fig. 25.22), it was possible to obtain a new family of sulfonylated MMPIs.[62]

These first results, obtained at Rhône-Poulenc, were very interesting for the activity obtained against the S1′ deep pocket MMPs tested (MMP-2 and MMP-3) and against MMP-1. Also, in this case, as previously seen for Hanessian's inhibitors, the results seem to be independent from the small 4-methoxy sulfonamide substituent in P1′.

These exciting results led to new and improved studies on SAR on this new class of MMPI.

Compound **25.132** proved to be a moderate inhibitor of phosphodiesterase type 4 (PDE4) ($IC_{50} = 9\ \mu M$); therefore, with the aim of increasing selectivity for MMP over PDE4 inhibition, a spirocyclic group was incorporated adjacent to the sulfonyl moiety, similar to compound **25.133** (RS-113,456 Fig. 25.23).[63,64] Oral bioavailability and half-life were improved in this type of sulfones, simply by shifting the spirocyclic substituent to the α-position with respect to the hydroxamate group, as in **25.134** (RS-130,830).[65] Comparative X-ray crystallographic analyses of both α- and β-sulfone-based MMPIs, **25.133** and **25.134**, bound to the catalytic domain of collagenase 3 (MMP-13) showed that the two compounds adopted quite similar conformations in the binding mode to the catalytic site.[38]

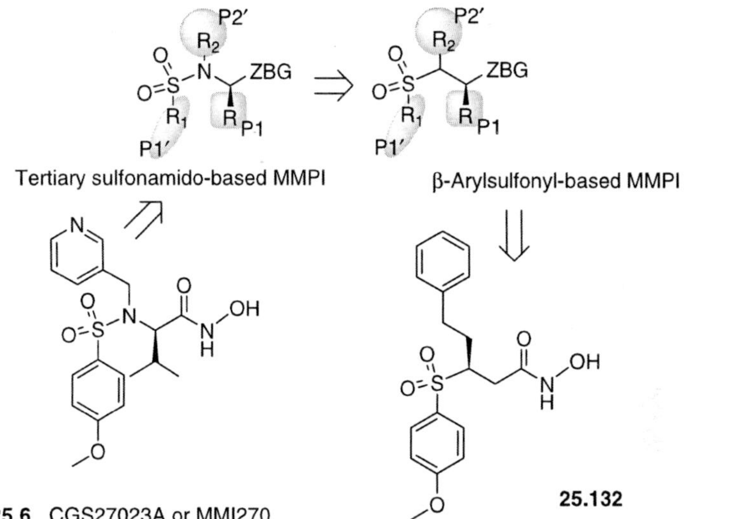

25.6, CGS27023A or MMI270
(K_I, nM)
MMP-1: 33; MMP-2: 20; MMP-3: 43

25.132
(K_I, nM)
MMP-1: 4000; MMP-2: 30; MMP-3: 200

FIGURE 25.22 Development of sulfone-based MMP inhibitors.

An X-ray crystal structure of compound **25.135** (RS-104,966, an analogue of **25.133** devoid of the 4′-chlorine atom on the biphenylsulfonamide) demonstrated the particular binding mode of this class of inhibitors to the catalytic domain of MMP-1. This analysis showed that an induced fitting of MMP inhibitors with large P1′ substituents in MMP-1 could occur by Arg214 adopting a new position, thereby creating a larger open S1′ pocket, as previously mentioned in this chapter.

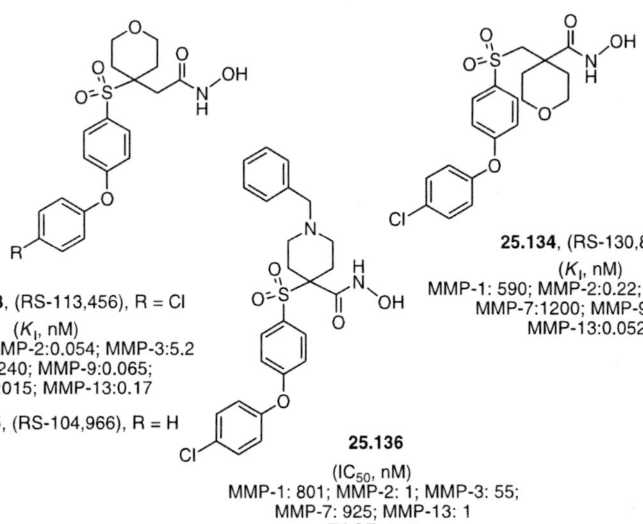

25.133, (RS-113,456), R = Cl
(K_I, nM)
MMP-1: 70; MMP-2:0.054; MMP-3:5.2
MMP-7:240; MMP-9:0.065;
MMP-12:015; MMP-13:0.17

25.135, (RS-104,966), R = H

25.136
(IC_{50}, nM)
MMP-1: 801; MMP-2: 1; MMP-3: 55;
MMP-7: 925; MMP-13: 1
TACE: 300

25.134, (RS-130,830)
(K_I, nM)
MMP-1: 590; MMP-2:0.22; MMP-3:9.3;
MMP-7:1200; MMP-9:0.58;
MMP-13:0.052

FIGURE 25.23 Structures of MMP inhibitors **25.133–25.136**.

MMPI **25.133** was orally active and able to reduce flow-mediated arterial enlargement in a rat arteriovenous fistula model.[65] Its analogue, β-sulfone-based **25.134**, underwent clinical trials for the treatment of osteoarthritis. The presence of a pyran spirocycle adjacent to the hydroxamate introduced a sterically hindered system able to block the metabolic degradation while increasing the MMPs selectivity (Fig. 25.23).

In a development of sulfone-based MMPI at Wyeth, it was discovered that shortening the distance between the sulfone and the hydroxamic acid led to improved biopharmacological properties of the inhibitors. Compound **25.136** (Fig. 25.23), an *N*-hydroxy-α-sulfonyl acetamide with a piperidine spirocycle adjacent to the ZBG, instead of the pyran of the above-cited MMPI **25.133–25.135**, was effective for the treatment of osteoarthritis.[66] SAR studies showed that in this class of compounds, the extended aromatic sulfonyl groups in P1′ were necessary for excellent activity on deep pocket S1′ MMPs. Moreover, the aliphatic basic amino group of the cycle was found to play a key role in activity *in vivo*. Compound **25.136** represented a good MMP-9/MMP-13 inhibitor with an appreciable activity over the tumor necrosis factor α-convertase (TACE or ADAM-17). This is a member of the parent family of the MMPs known as disintegrin and metalloproteinases or ADAMs, which are involved in the activation of the tumor necrosis factor α on the cell membrane surfaces and are responsible for the shedding of many other membrane-bound proteins. Overexpression of MMP-13 and TACE seems to be related to osteoarthritis.[67] In a further development of these studies, the same authors indicated that modifications of the sulfur oxidation state were able to guide selectivity and potency between these two types of metalloproteinases (MMPs-TACE).[68]

At the Searle company, new *N*-hydroxy-α-sulfonyl acetamides with a piperidine spirocycle adjacent to the hydroxamic acid were studied, possessing a 4′-thiophenoxyphenyl sulfone substituent in their P1′ position like compound **25.137**, SC-276 (Fig. 25.24).[69] This 4′-thiophenoxyphenyl sulfone, despite its similar MMP-2/MMP-13 subnanomolar K_I values with respect to the parent 4′-phenoxyphenyl sulfone, showed an improved MMP-1 sparing. SC-276 was a potent orally active antiangiogenic (in a mouse model of bFGF-stimulated corneal neovascularization) and anticancer agent in an implanted mouse model MX-1 (in association with paclitaxel).

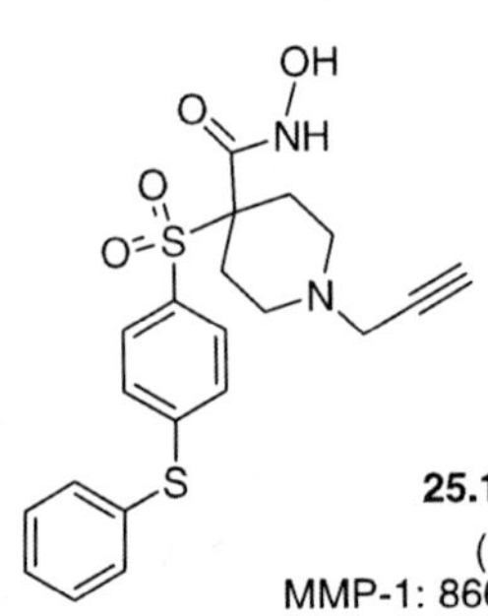

25.137, SC-276

(K_I, nM)
MMP-1: 8660; MMP-2: 0.33 ;
MMP-13: 0.40

FIGURE 25.24 Structure of MMP inhibitor **25.137**.

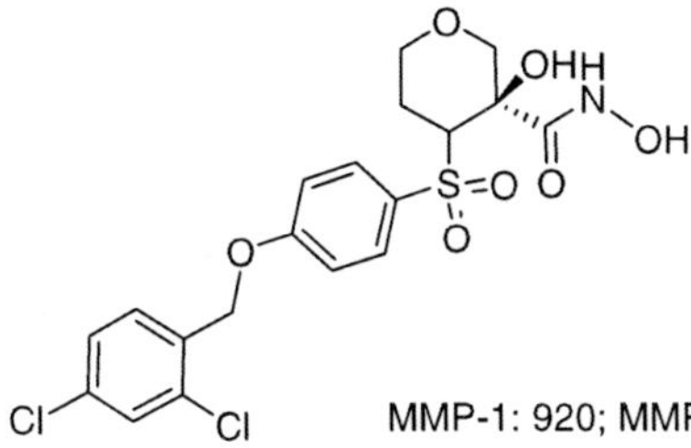

25.138

(IC_{50}, nM)
MMP-1: 920; MMP-13: 0.95; Aggrecanase: 8.1

FIGURE 25.25 Compound **25.138**: an MMP-13 and aggrecanase inhibitor.

These findings could be explained with the recent results on cancer degradome disclosed by Overall and Lopez-Otin, where some MMPs have been classified as target or antitarget in cancer therapy. MMP-2 was a potent target MMP in cancer therapy.[5] Moreover, a selectivity ratio MMP-1/MMP-2 of more than 26,000 times could delineate a safe therapeutic profile to avoid the putative role of the MMP-1 in the development of MSS, as hypothesized for some old MMPIs belonging to the first and second generations.[17]

More recently, Pfizer researchers[70] also synthesized a new class of nonsulfonamidic hydroxamates as potential MMP-13 and aggrecanase inhibitors. Starting from pipecolic acid-based derivatives reported as TACE inhibitors,[71] the same group developed 3-hydroxy-4-arylsulfonyltetrahydropyranyl-3-hydroxamic acids to improve the metabolic stability of the previous compounds. In fact, as already reported, the hydroxamic acid group is particularly subject to hydrolysis, reduction, and glucuronidation. These sulfones, presenting an increased polarity and steric hindrance of the hydroxamate functionality, had better clearance. Moreover, compound **25.138** (Fig. 25.25) was shown to potently inhibit both MMP-13 and aggrecanase, as a result of the combination of an *ortho*- and a *para*-chloro substitution on the benzylic ring.

Another strategy followed to achieve selective MMP inhibitors was the development of mechanism-based thiirane inhibitors. Bernardo et al.[72] reported the synthesis and kinetic characterization of slow binding inhibitors that covalently bind to the MMP active site. These inhibitors were selective for gelatinases (MMP-2 and MMP-9) and were successively improved to obtain compound **25.139**, which proved to be selective only for MMP-2.[73] In these thiirane-containing inhibitors, the sulfur first coordinates with the catalytic zinc ion, and this interaction then activates the thiirane for nucleophilic attack by the active site glutamate (Glu404 in MMP-2), a process that covalently modifies the enzyme, causing a loss of activity (Fig. 25.26). These inhibitors were the first example of suicide inhibitors of MMPs.

Recently, Zhang et al.[74] at Johnson & Johnson Pharmaceuticals developed a new SAR study on the P1 (C-α to the ZBG) and P1′ positions of the sulfone-based carboxylic acid **25.140** (Fig. 25.27). In summary, these inhibitors were potent against MMP-2/MMP-9 and selective over MMP-1. The SAR study on both P1′ and P1/P2′ showed that a straight, hydrophobic P1′ group with the optimal length was preferred. On the contrary, an amido or imido group was critical for MMP-9 inhibition. Some of the described carboxylic acids exhibited better pharmacokinetic properties in rats compared to their hydroxamate analogues.

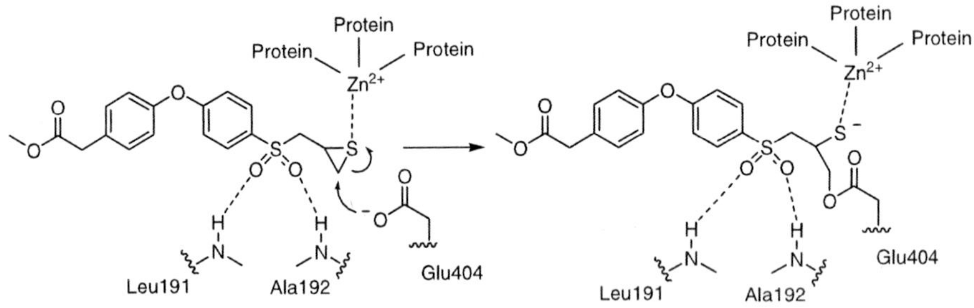

25.139

(K_I, nM)
MMP-1: 11,000; MMP-2: 50; MMP-14: 590

FIGURE 25.26 Mechanism of action of thiirane-containing inhibitors.

More recently, Rossello et al.,[75] on the basis of their previous works on MMPIs, disclosed a new highly selective class of MMPIs that are able to discriminate between MMP-2 and MMP-9, sparing MMP-1 and MMP-14; this was done using a 2-thioaryl-phenyl **A** or a 2-thioaryl-benzyl **B** template (Fig. 25.28) and bearing different ZBGs, such as hydroxamic acids, carboxylic acids, phosphonic acids, and pyrimidintriones.

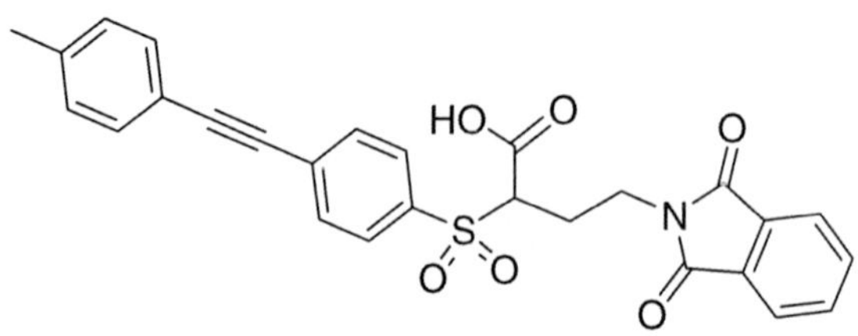

25.140

(IC_{50}, nM)
MMP-1: >1000; MMP-2: 0.7; MMP-3: 28;
MMP-9: 0.5; MMP-13: 1.4

FIGURE 25.27 Structure of carboxylic acid inhibitor **25.140**.

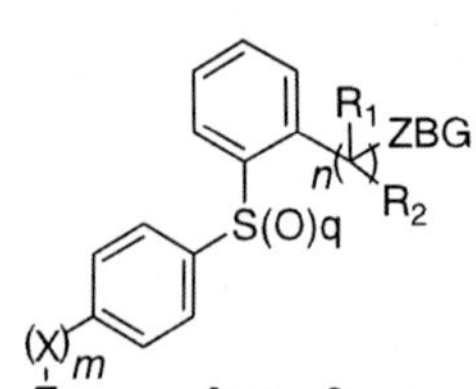

A: $n = 0$
B: $n = 1$

FIGURE 25.28 General structure of new MMP inhibitors.

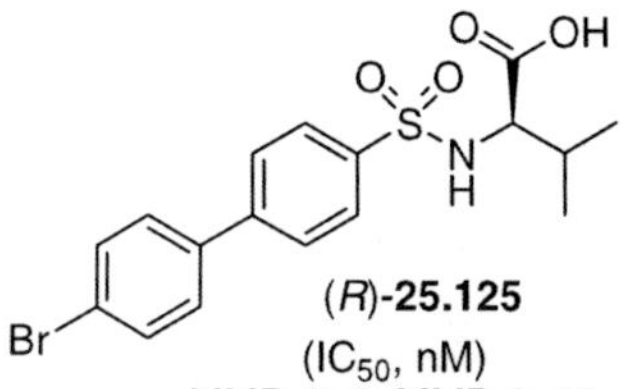

(*R*)-**25.125**
(IC_{50}, nM)
MMP-2: 5; MMP-3: 55

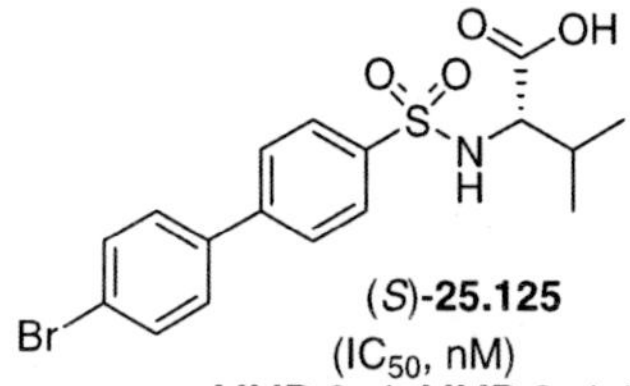

(*S*)-**25.125**
(IC_{50}, nM)
MMP-2: 4; MMP-3: 4.4

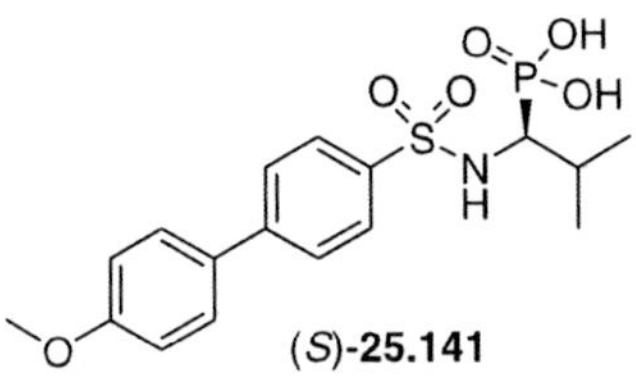

(*S*)-**25.141**
(K_I, nM)
MMP-2: 1200; MMP-3:<15% (0.1 mM);
MMP-8: 700

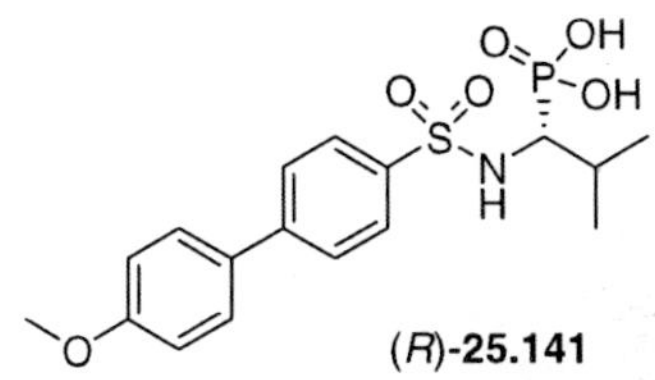

(*R*)-**25.141**
(K_I, nM)
MMP-2: 5; MMP-3: 40; MMP-8: 0.6

FIGURE 25.29 Enantiomers of phosphonate inhibitor **25.141** and their carboxylate analogues (*R*)-**25.125** and (*S*)-**25.125**.

25.3.3 Sulfonamido-Based Inhibitors with New ZBGs

In another study, Mazza et al.[76] clarified the binding mode of [1-(4′-methoxybiphenyl-4-sulfonylamino)-2-methylpropyl]phosphonate **25.141** (Fig. 25.29) using crystal structures of the complexes of its *R*- and *S*-enantiomers with MMP-8. These two enantiomers were compared to the carboxylate analogue inhibitors (*R*)-**25.125** and (*S*)-**25.125** (enantiomers of PD-166,793, previously described). The most potent enantiomer was the *S* for carboxylates (about 12-fold against MMP-3) and the *R* for phosphonates (1000-fold against MMP-8 and MMP-2). These values established this compound as the most active phosphonate MMP inhibitor made to date.

The crystal structures of each enantiomer with MMP-8 showed nearly identical phosphonate coordination to the zinc and to Glu198, similar biphenyl insertion into S1′ deep pocket, and similar arrangement for the α-isopropyl substituent in the S1 pocket. The structural arrangement that enabled both enantiomers to bind was possible by a rotation of the sulfonamide chain and also by means of a different torsional angle for the biphenyl in S1′. Four differences between the MMP-8-bound *R*- and *S*-aminophosphonates accounted for the 1000-fold difference in the MMP-8 K_I values. The *R*-enantiomer had better hydrophobic interactions with the MMP-8 for both the isopropyl and the biphenyl and an additional hydrogen bond between the sulfonamide nitrogen and the Ala160 carbonyl. Moreover, it had a hydrogen bond between the sulfonamide oxygen and the Ala160 amide nitrogen. The same authors showed the importance of the stereochemical features of inhibitor recognition by the MMPs. They showed that the absolute configuration of (*S*)-**25.125** was the same as for these aminophosphonates. In the same paper, in a comparison with the analogue hydroxamates of these two couples of chiral MMPIs, it was established that the better bound

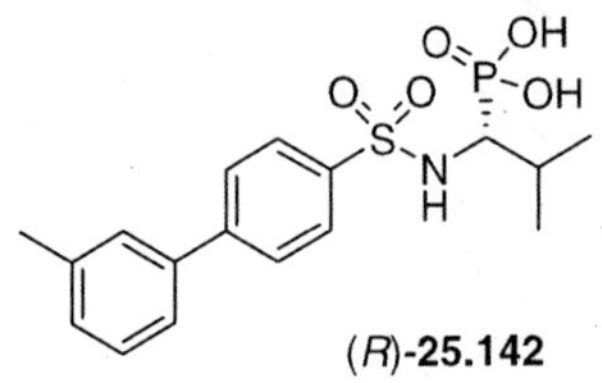

(*R*)-**25.142**

(IC_{50}, nM)
MMP-1: 320; MMP-2: 24; MMP-3: 230;
MMP-8: 0.4; MMP-9: 64
ADAM-10: >10,000; ADAM-17: >10,000

FIGURE 25.30 Structure of phosphonate **25.142**: a potent MMP-8 inhibitor.

hydroxamate cognate inhibitor had the opposite absolute configuration (*R*) with respect to the acid (*S*)-**25.125**. The mutual ability of both inhibitor and MMP to fit, so as to attain optimal zinc binding group interactions, showed the importance of the stereochemical requisites in SAR studies on MMPIs during the development with new ZBGs.

The development of SAR studies on the P1′ position for this type of *N*-sulfonylaminophosphonate inhibitors has discovered the potent subnanomolar inhibitor of MMP-8, the phosphonate (*R*)-**25.142** (Fig. 25.30).[77]

Folgueras et al.[78] successively demonstrated that the administration of the MMP-8 selective inhibitor **25.142** to mice with autoimmune encephalomyelitis was able to reduce the severity of the disease. Based on these findings, it could be concluded that MMP-8 plays an important role in EAE development and that this enzyme may be a novel therapeutic target in human neuroinflammatory diseases, such as multiple sclerosis.

In another study, Zhang et al.[79] disclosed some new tertiary and secondary arylsulfonamides bearing a 1-hydroxy-2-pyridinone as the ZBG[80–82], which were potent MMP-9 inhibitors for use in the treatment of ischemic stroke. The hydroxypyridinone tertiary sulfonamide **25.143** (Fig. 25.31), bearing a 4′-Cl-substituted-4-phenoxyphenyl

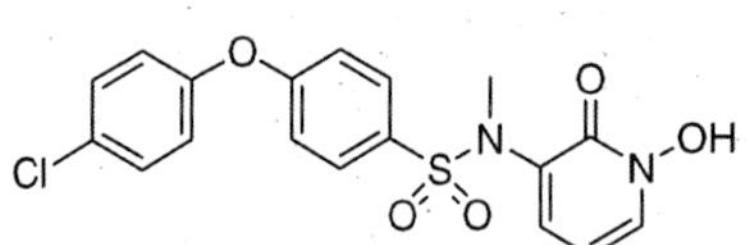

25.143

(IC_{50}, nM)
MMP-1: >1000; MMP-2: 5.0; MMP-3: 57;
MMP-9: 2.4; MMP13: 2.5

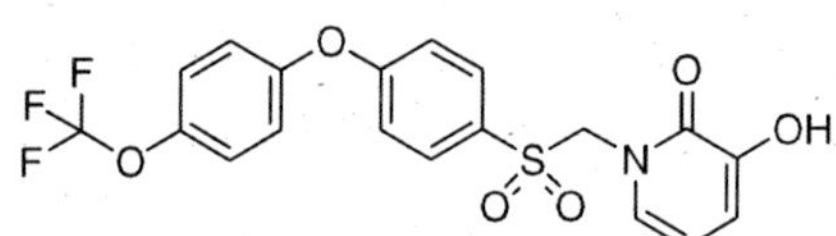

25.144

(IC_{50}, nM)
MMP-1: >1000; MMP-2: 13; MMP-3: 203;
MMP-9: 4; MMP13: 7.3

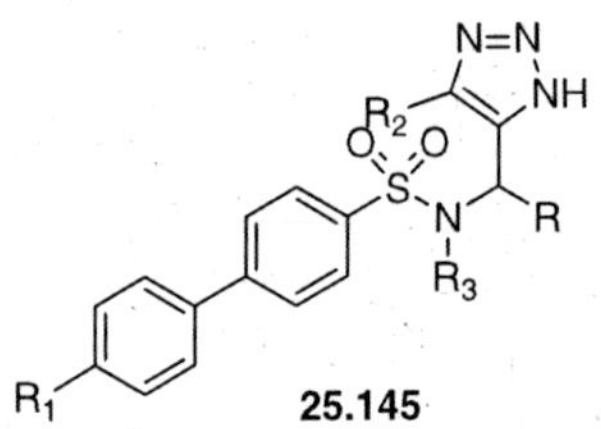

25.145

FIGURE 25.31 Examples of published MMPI with new ZBGs.

sulfonamide group in the P1′ position, was a good MMP-9 and MMP-2 inhibitor that was able to spare MMP-1 activity. This inhibitor showed a good pharmacokinetic profile and a good efficacy in a mouse model of transient cerebral ischemia of the midcerebral artery occlusion.[79]

Continuing their studies on new MMPIs bearing different ZBGs, Zhang et al. discovered a new class of sulfones that were somewhat structurally related to the above family of 2-pyridinone-based MMP-9/MMP-2 selective inhibitors. Compound **25.144** (Fig. 25.31) was a representative of this new class. In this family of inhibitors, the 3-hydroxy-2-pyridinone ZBG was used, linking a methylenesulfonyl moiety bearing some aromatic hindered phenolic ethers on the N(1).[83]

As can be seen, the 3-hydroxy-2-pyridinone inhibitor **25.144** was equipotent and more selective for MMP-2/MMP-9 than the previously described hydroxypyridinone tertiary sulfonamide **25.143**.

Finally, Fryer et al. patented several new sulfonamido-based MMPIs with general formula **25.145** (Fig. 25.31) bearing a triazolyl group as ZBG and showing IC_{50} values lower than 10 μM on MMP-13. These compounds were substituted biphenylsulfonamides of generally secondary type, $R_3 = H$, and tertiary type, $R_3 =$ alkyl. They presented one or two alkyl groups (R) in the P1 position, and moreover, alkyl groups (R_2) could be substituted on the triazolyl ZBG.[84]

25.4 CONCLUSIONS

This chapter summarizes the main studies on drug design of sulfonylated MMP inhibitors published in the past 15 years. A large number of compounds have been reported, the vast majority of which presented a hydroxamate or a carboxylate as a ZBG. The latest developments in this field have led to the discovery of new ZBG-bearing inhibitors, in an attempt to overcome toxicity and pharmacokinetic issues related to the use of hydroxamate derivatives. Moreover, many positive functions of MMPs have recently been recognized, and therefore, new generations of MMPIs are emerging, especially ones designed to block only the target MMPs and spare the antitarget ones. Results from clinical trials conducted on selective MMPIs as anticancer and antinflammatory agents will point out the right way to be followed by researchers in the coming years.

REFERENCES

1. Hanemaaijer, R.; Visser, H.; Koolwijk, P.; Sorsa, T.; Salo, T.; Golub, L. M.; van Hinsbergh, V. W. Inhibition of MMP synthesis by doxycycline and chemically modified tetracyclines (CMTs) in human endothelial cells. *Adv. Dent. Res.* **1998**, *12*, 114–118.
2. Rajapakse, N.; Mendis, E.; Kim, M. M.; Kim, S. K. Sulfated glucosamine inhibits MMP-2 and MMP-9 expressions in human fibrosarcoma cells. *Bioorg. Med. Chem.* **2007**, *15*, 4891–4896.

3. Nuti, E.; Tuccinardi, T.; Rossello, A. matrix metalloproteinase inhibitors: new challenges in the era of post broad-spectrum inhibitors. *Curr. Pharm. Des.* **2007**, *13*, 2087–2100.

4. Hidalgo, M. Development of matrix metalloproteinase inhibitors in cancer therapy. *J. Nat. Cancer Inst.* **2001**, *93*, 178–193.

5. Overall, C. M.; Kleifeld, O. Validating matrix metalloproteinases as drug targets and anti-targets for cancer therapy. *Nat. Rev. Cancer* **2006**, *6*, 227–239.

6. Brown, S.; Bernardo, M. M.; Li, Z. H.; Kotra, L. P.; Tanaka, Y.; Fridman, R.; Mobashery, S. Potent and selective mechanism-based inhibition of gelatinases. *J. Am. Chem. Soc.* **2000**, *122*, 6799–6800.

7. Nagase, H.; Woessner, J. F. Matrix metalloproteinases. *J. Biol. Chem.* **1999**, *274*, 21491–21494.

8. Morgunova, E.; Tuuttila, A.; Bergmann, U.; Tryggvason, K. Structural insight into the complex formation of latent matrix metalloproteinase 2 with tissue inhibitor of metalloproteinase 2. *Proc. Natl. Acad. Sci. USA* **2002**, *99*, 7414–7419.

9. Morgunova, E.; Tuuttila, A.; Bergmann, U.; Isupov, M.; Lindqvist, Y.; Schneider, G.; Tryggvason, K. Structure of human pro-matrix metalloproteinase-2: activation mechanism revealed. *Science* **1999**, *284*, 1667–1670.

10. Ra, H. J.; Parks, W. C. Control of matrix metalloproteinase catalytic activity. *Matrix Biol.* **2007**, *26*, 587–596.

11. Spurlino, J. C.; Smallwood, A. M.; Carlton, D. D.; Banks, T. M.; Vavra, K. J.; Johnšon, J. S.; Cook, E. R.; Falvo, J.; Wahl, R. C.; Pulvino, T. A.; Wendoloski, J. J.; Smith, D. L. Å structure of mature truncated human fibroblast collagenase. *Proteins* **1994**, *19*, 98–109.

12. Leung, D.; Abbenante, G.; Fairlie, D. P. Protease inhibitors: current status and future prospects. *J. Med. Chem.* **2000**, *43*, 305–341.

13. Jacobsen, F. E.; Lewis, J. A.; Cohen, S. M. The design of inhibitors of medicinally relevant metalloproteins. *Chem. Med. Chem.* **2007**, 2, 152–171.

14. Breuer, E.; Frant, J.; Reich, R. Recent non-hydroxamate matrix metalloproteinase inhibitors. *Expert Opin. Ther. Pat.* **2005**, *15*, 253–269.

15. Whittaker, M.; Floyd, C. D.; Brown, P.; Gearing, A. J. H. Design and therapeutic application of matrix metalloproteinase inhibitors. *Chem. Rev.* **1999**, *99*, 2735–2776.

16. Babine, R. E.; Bender, S. L. Molecular recognition of protein–ligand complexes: applications to drug design. *Chem. Rev.* **1997**, *97*, 1359–1472.

17. Coussens, L. M.; Fingleton, B.; Matrisian, L. M. Matrix metalloproteinase inhibitors and cancer: trials and tribulations. *Science* **2002**, *295*, 2387–2390.

18. Oreola, A.; Donini, T.; Kollman, P. A. Calculation and prediction of binding free energies for the matrix metalloproteinases. *J. Med. Chem.* **2000**, *43*, 4180–4188.

19. Browner, M. F.; Smith, W. W.; Castelhano, A. L. Crystal structures of matrilysin–inhibitor complexes. *Biochemistry* **1995**, *34*, 6602–6610.

20. Rowsell, S.; Hawatin, P.; Minshull, C. A.; Jepson, H.; Brockbank, S. M. V.; Barratt, D. G.; Slater, A. M.; McPheat, W. L.; Waterson, D.; Henney, A. M.; Pauptit, R. A. Crystal structure of human MMP9 in complex with a reverse hydroxamate inhibitor. *J. Mol. Biol.* **2002**, *319*, 173–181.

21. MacPherson, L. J.; Parker, D. T. Arylsulfonamido-substituted hydroxamic acids. EP 606046, **1994**.

22. MacPherson, L. J.; Bayburt, E. K.; Capparelli, M. P.; Carroll, B. J.; Goldstein, R.; Justice, M. R.; Zhu, L. J.; Hu, S. I.; Melton, R. A.; Fryer, L.; Goldberg, R. L.; Doughty, J. R.; Spirito, S.; Blancuzzi, V.; Wilson, D. O.; Byrne, E. M.; Ganu, V.; Parker, D. T. Discovery of CGS 27023A, a non-peptidic, potent, and orally active stromelysin inhibitor that blocks cartilage degradation in rabbits. *J. Med. Chem.* **1997**, *40*, 2525–2532.

23. Li, Y. C.; Zhang, X.; Melton, R.; Ganu, V.; Gonnella, N. C. Solution structure of the catalytic domain of human stromelysin-1 complexed to a potent, nonpeptidic inhibitor. *Biochemistry* **1998**, *37*, 14048–14056.

24. Terp, G. E.; Cruciani, G.; Christensen, I. T.; Jorgensen, F. S. Structural differences of matrix metalloproteinases with potential implications for inhibitor selectivity examined by the GRID/CPCA approach. *J. Med. Chem.* **2002**, *45*, 2675–2684.

25. Hanessian, S.; Bouzbouz, S.; Boudon, A.; Tucker, G. C.; Peyroulan, D. Picking the S1, S1′ and S2′ pockets of matrix metalloproteinases. A niche for potent acyclic sulfonamide inhibitors. *Bioorg. Med. Chem. Lett.* **1999**, *9*, 1691–1696.

26. Hanessian, S.; Moitessier, N. Sulfonamide-based acyclic and conformationally constrained MMP inhibitors: from computer-assisted design to nanomolar compounds. *Curr. Top. Med. Chem.* **2004**, *4*, 1269–1287.

27. Scozzafava, A.; Supuran, C. T. Protease inhibitors. Synthesis of potent matrix metalloproteinase and bacterial collagenase inhibitors incorporating *N*-4-nitrobenzylsulfonyl glycine hydroxamate moieties. *J. Med. Chem.* **2000**, *43*, 1858–1865.

28. Clare, B. W.; Scozzafava, A.; Supuran, C. T. Protease inhibitors: synthesis of a series of bacterial collagenase inhibitors of the sulfonyl amino acyl hydroxamate type. *J. Med. Chem.* **2001**, *44*, 2253–2258.

29. Santos, M. A.; Marques, S. M.; Tuccinardi, T.; Carelli, P.; Panelli, L.; Rossello, A. Design, synthesis and molecular modeling study of iminodiacetyl monohydroxamic acid derivatives as MMP inhibitors. *Bioorg. Med. Chem.* **2006**, *14*, 7539–7550.

30. Zook, S. E.; Dagnino, R., Jr.; Deason, M. E.; Bender, S. L.; Melnick, M. J. Metalloproteinase inhibitors, pharmaceutical composition containing them and their pharmaceutical uses, and methods and intermediates useful for their preparation. WO 9720824, **1997**.

31. Piscopio, A. D.; Rizzi, J. P. Arylsulfonyl hydroxamic acid derivatives as MMP and TNF inhibitors. WO 9633172, **1996**.

32. De, B.; Natchus, M. G.; Pikul, S.; Almstead, N. G.; Matthews, R. S.; Taiwo, Y. O.; Cheng, M. 1,4-Heterocyclic metalloprotease inhibitors. WO 9808825, **1998**.

33. De, B.; Natchus, M. G.; Pikul, S.; Almstead, N. G.; Matthews, R. S.; Taiwo, Y. O.; Cheng, M. Heterocyclic metalloprotease inhibitors. WO 9808827, **1998**.

34. Levin, J. I.; DiJoseph, J. F.; Killar, L. M.; Sung, A.; Walter, T.; Sharr, M. A.; Roth, C. E.; Skotnicki, J. S.; Albright, J. D. The synthesis and biological activity of a novel series of diazepine MMP inhibitors. *Bioorg. Med. Chem. Lett.* **1998**, *8*, 2657–2662.

35. Almstead, N. G.; Bradley, R. S.; Pikul, S.; De, B.; Natchus, M. G.; Taiwo, Y. O.; Gu, F.; Williams, L. E.; Hynd, B. A.; Janusz, M. J.; Dunaway, C. M.; Mieling, G. E. Design, synthesis, and biological evaluation of potent thiazine- and thiazepine-based matrix metalloproteinase inhibitors. *J. Med. Chem.* **1999**, *42*, 4547–4562.

36. Damani, L. A., Ed. Metabolism of sulphur functional groups. *Sulphur-Containing Drugs and Related Organic Compounds: Chemistry, Biochemistry and Toxicology*, Vol. 1, Parts A and B; Ellis Horwood Limited: London, **1989**.

37. Shalinsky, D. R.; Brekken, J.; Zou, H.; McDermott, C. D.; Forsyth, P.; Edwards, D.; Margosiak, S.; Bender, S.; Truitt, G.; Wood, A.; Varki, N. M.; Appelt, K. Broad antitumour and antiangiogenic activities of AG3340, a potent and selective MMP inhibitor undergoing advanced oncology clinical trials. *Ann. N.Y. Acad. Sci.* **1999**, *878*, 236–270.

38. Lovejoy, B.; Welch, A. R.; Carr, S.; Luong, C.; Broka, C.; Hendricks, R. T.; Campbell, J. A.; Walker, K. A.; Martin, R.; Van Wart, H.; Browner, M. F. *Nat. Struct. Biol.* **1999**, *6*, 217–221.

39. Tuccinardi, T.; Martinelli, A.; Nuti, E.; Carelli, P.; Balzano, F.; Uccello-Barretta, G.; Murphy, G.; Rossello, A. Amber force field implementation, molecular modelling study, synthesis and MMP-1/MMP-2 inhibition profile of (*R*)- and (*S*)-*N*-hydroxy-2-(*N*-isopropoxybiphenyl-4-ylsulfonamido)-3-methylbutanamides. *Bioorg. Med. Chem.* **2006**, *14*, 4260–4276.

40. Santos, O.; McDermott, C. D.; Daniels, R. G.; Appelt, K. Rodent pharmacokinetic and anti-tumour efficacy studies with a series of synthetic inhibitors of matrix metalloproteinases. *Clin. Exp. Metastasis.* **1997**, *15*, 499–508.

41. Peterson, J. T. The importance of estimating the therapeutic index in the development of matrix metalloproteinase inhibitors. *Cardiovas. Res.* **2006**, *69*, 677–687.

42. Borden, P.; Solymar, D.; Sucharczuk, A.; Lindman, B.; Cannon, P.; Heller, R. A. *J. Biol. Chem.* **1996**, *271*, 23577–23581.

43. Barta, T. E.; Becker, D. P.; Bedell, L. J.; De Crescenzo, G. A.; McDonald, J. J.; Munie, G. E.; Rao, S.; Shieh, H. S.; Stegeman, R.; Stevens, A. M.; Villamil, C. I. *Bioorg. Med. Chem. Lett.* **2000**, *10*, 2815–2817.

44. Levin, J. I.; Chen, J.; Du, M.; Hogan, M.; Kincaid, S.; Nelson, F. C.; Venkatesan, A. M.; Wehr, T.; Zask, A.; DiJoseph, J.; Killar, L. M.; Skala, S.; Sung, A.; Sharr, M.; Roth, C.; Jin, G.; Cowling, R.; Mohler, K. M.; Black, R. A.; March, C. J.; Skotnicki, J. S. *Bioorg. Med. Chem. Lett.* **2001**, *11*, 2189–2192.

45. Nelson, F. C.; Delos Santos, E.; Levin, J. I.; Chen, J. M.; Skotnicki, J. S.; Di Joseph, J.; Sharr, M.; Sung, A.; Killar, L. M.; Cowling, R.; Jin, G.; Roth, C. E.; Albright, J. D. *Bioorg. Med. Chem. Lett.* **2002**, *12*, 2867–2870.

46. Cherney, R. J.; Mo, R.; Meyer, D. T.; Hardman, K. D.; Liu, R. Q.; Covington, M. B.; Qian, M.; Wasserman, Z. R.; Christ, D. D.; Trzaskos, J. M.; Newton, R. C.; Decicco, C. P. *J. Med. Chem.* **2004**, *47*, 2981–2983.

47. Noe, M. C.; Natarajan, V.; Snow, S. L.; Mitchell, P. G.; Lopresti-Morrow, L.; Reeves, L. M.; Yocum, S. A.; Carty, T. J.; Barberia, J. A.; Sweeney, F. J.; Liras, J. L.; Vaughn, M.; Hardink, J. R.; Hawkins, J. M.; Tokar, C. *Bioorg. Med. Chem. Lett.* **2005**, *15*, 2808–2811.

48. Chen, J. M.; Nelson, F. C.; Levin, J. I.; Mobilio, D.; Moy, F. J.; NilaKantan, R.; Zask, A.; Powers, R. *J. Am. Chem. Soc.* **2000**, *122*, 9648–9654.

49. Whitlock, G. A.; Dack, K. N.; Dickinson, R. P.; Lewis, M. L. A novel series of highly selective inhibitors of MMP-3. *Bioorg. Med. Chem. Lett.* **2007**, *17*, 6750–6753.

50. Zook, S. E.; Dagnino, R., Jr.; Deason, M. E.; Bender, S. L.; Melnick, M. J. Metalloproteinase inhibitors, pharmaceutical composition containing them and their pharmaceutical uses, and methods and intermediates useful for their preparation. WO 9720824, **1997**.

51. Rossello, A.; Nuti, E.; Orlandini, E.; Carelli, P.; Rapposelli, S.; Macchia, M.; Minutolo, F.; Carbonaro, L.; Albini, A.; Benelli, R.; Cercignani, G.; Murphy, G.; Balsamo, A. New *N*-arylsulfonyl-*N*-alkoxyaminoacetohydroxamic acids as selective inhibitors of gelatinase A (MMP-2). *Bioorg. Med. Chem.* **2004**, *12*, 2441–2450.

52. Rossello, A.; Nuti, E.; Carelli, P.; Orlandini, E.; Macchia, M.; Nencetti, S.; Zandomeneghi, M.; Balzano, F.; Uccello Barretta, G.; Albini, A.; Benelli, R.; Cercignani, G.; Murphy, G.; Balsamo, A. *N-i*-propoxy-*N*-biphenylsulfonylaminobutylhydroxamic acids as potent and selective inhibitors of MMP-2 and MT1-MMP. *Bioorg. Med. Chem. Lett.* **2005**, *15*, 1321–1326.

53. Nuti, E.; Conti, C.; Avramova, S. I.; Orlandini, E.; Nencetti, S.; Cercignani, G.; Murphy, G.; Rossello, A. Synthetic MMP inhibitors: slow-binding behavior of enantioselective N–O-sulfonamido-based inhibitors directed to specific matrixins. *Frontiers in CNS and Oncology Medicinal Chemistry, ACS-EFMC, Siena, Italy*, October **2007**.

54. Tamura, Y.; Watanabe, F.; Nakatani, T.; Yasui, K.; Fuji, M.; Komurasaki, T.; Tsuzuki, H.; Maekawa, R.; Yoshioka, T.; Kawada, K.; Sugita, K.; Ohtani, M. Highly selective and orally active inhibitors of type IV collagenase (MMP-9 and MMP-2): *N*-sulfonylamino acid derivatives. *J. Med. Chem.* **1998**, *41*, 640–649.

55. Hodgson, J. Remodeling MMPIs. Matrix metalloproteinase inhibitors will be approved as drugs, probably this year, but questions remain concerning their specificity, bioavailability, and potential long-term toxicity. *Biotechnology* **1995**, *13*, 554–557.

56. Chiappori, A. A.; Eckhardt, S. G.; Bukowski, R.; Sullivan, D. M.; Ikeda, M.; Yano, Y.; Yamada-Sawada, T.; Kambayashi, Y.; Tanaka, K.; Javle, M. M.; Mekhail, T.; O'Bryant, C. L.; Creaven, P. J. A phase I pharmacokinetic and pharmacodynamic study of S-3304, a novel matrix metalloproteinase inhibitor, in patients with advanced and refractory solid tumours. *Clin. Cancer Res.* **2007**, *13*, 2091–2099.

57. Bender, S. L.; Melwin, A. A. Metalloproteinase inhibitors, pharmaceutical composition containing them and their pharmaceutical compositions. US 5985900, **1999**.

58. Hu, Y.; Xiang, J. S.; DiGrandi, M. J.; Du, X.; Ipek, M.; Laakso, L. M.; Li, J.; Li, W.; Rush, T. S.; Schmid, J.; Skotnicki, J. S.; Tam, S.; Thomason, J. R.; Wang, Q.; Levin, J. I. Potent, selective, and orally bioavailable matrix metalloproteinase-13 inhibitors for the treatment of osteoarthritis. *Bioorg. Med. Chem.* **2005**, *13*, 6629–6644.

59. Xiang, J. S.; Hu, Y.; Rush, T. S.; Thomason, J. R.; Ipek, M.; Sum, P. E.; Abrous, L.; Sabatini, J. J.; Georgiadis, K.; Reifenberg, E.; Majumdar, M.; Morris, E. A.; Tama, S. Synthesis and biological evaluation of biphenylsulfonamide carboxylate aggrecanase-1 inhibitors. *Bioorg. Med. Chem. Lett.* **2006**, *16*, 311–316.

60. Li, J.; Rush, T. S.; Li, W.; DeVincentis, D.; Du, X.; Hu, Y.; Thomason, J. R.; Xiang, J. S.; Skotnicki, J. S.; Tam, S.; Cunningham, K. M.; Chockalingam, P. S.; Morris, E. A.; Levin, J. I. Synthesis and SAR of highly selective MMP-13 inhibitors. *Bioorg. Med. Chem. Lett.* **2005**, *15*, 4961–4966.

61. Reiter, L. A.; Robinson, R. P.; McClure, K. F.; Jones, C. S.; Reese, M. R.; Mitchell, P. G.; Otterness, I. G.; Bliven, M. L.; Liras, J.; Cortina, S. R.; Donahue, K. M.; Eskra, J. D.; Griffiths, R. J.; Lame, M. E.; Lopez-Anaya, A.; Martinelli, G. J.; McGahee, S. M.; Yocum, S. A.; Lopresti-Morrow, L. L.; Tobiassen, L. M.; Vaughn-Bowser, M. L. Pyran-containing sulfonamide hydroxamic acids: potent MMP inhibitors that spare MMP-1. *Bioorg. Med. Chem. Lett.* **2004**, *14*, 3389–3395.

62. Groneberg, R. D.; Burns, C. J.; Morrissette, M. M.; Ullrich, J. W.; Morris, R. L.; Darnbrough, S.; Djuric, S. W.; Condon, S. M.; McGeehan, G. M.; Labaudiniere, R.; Neuenschwander, K.; Scotese, A. C.; Kline, J. A. Dual inhibition of phosphodiesterase 4 and matrix metalloproteinases by an (arylsulfonyl)hydroxamic acid template. *J. Med. Chem.* **1999**, *42*, 541–544.

63. Bender, S. L.; Broka, C. A.; Campbell, J. A.; Castelhano, A. L.; Fisher, L. E.; Hendricks, R. T.; Sarma, K. Preparation of arylthioalkanoates and analogs as matrix metalloprotease inhibitors. EP 780386, **1997**.

64. Burns, C. J.; Groneberg, R. D.; Salvino, J. M.; McGeehan, G.; Condon, S. M.; Morris, R.; Morrissette, M. M.; Matthew, R.; Darnbrough, S.; Neuenschwander, K.; Scotese, A.; Djuric, S. W.; Ullrich, J. W.; Labaudiniere, R. Nanomolar inhibitors for two distinct biological target families from a single synthetic sequence: a next step in combinatorial library design? *Angew. Chem. Int. Ed.* **1998**, *37*, 2848–2850.

65. Abbruzzese, T. A.; Guzman, R. J.; Martin, R. L.; Yee, C.; Zarins, C. K.; Dalman, R. L. Matrix metalloproteinase inhibition limits arterial enlargement in a rodent arteriovenous fistula model. *Surgery* **1998**, *124*, 328–335.

66. Aranapakam, V.; Davis, J. M.; Grosu, G. T.; Baker, J.; Ellingboe, J.; Zask, A.; Levin, J. I.; Sandanayaka, V. P.; Du, M.; Skotnicki, J. S.; DiJoseph, J. F.; Sung, A.; Sharr, M. A.; Killar, L. M.; Walter, T.; Jin, G.; Cowling, R.; Tillett, J.; Zhao, W.; McDevitt, J.; Xu, Z. B. Synthesis and structure–activity relationship of *N*-substituted 4-arylsulfonylpiperidine-4-hydroxamic acids as novel, orally active matrix metalloproteinase inhibitors for the treatment of osteoarthritis. *J. Med. Chem.* **2003**, *46*, 2376–2396.

67. Aranapakam, V.; Grosu, G. T.; Davis, J. M.; Hu, B.; Ellingboe, J.; Baker, J.; Skotnicki, J. S.; Zask, A.; DiJoseph, J. F.; Sung, A.; Sharr, M. A.; Killar, L. M.; Walter, T.; Jin, G.; Cowling, R. Synthesis and structure–activity relationship of α-sulfonylhydroxamic acids as novel, orally active matrix metalloproteinase inhibitors for the treatment of osteoarthritis. *J. Med. Chem.* **2003**, *46*, 2361–2375.

68. Aranapakam, V.; Davis, J. M.; Grosu, G. T.; Baker, J.; Zask, A.; Levin, J. I.; Ellingboe, J.; Skotnicki, J. S.; DiJoseph, J. F.; Sung, A.; Jin, G.; Xu, W.; McCarthy, D. J.; Barone, D. Synthesis and structure–activity relationships of 4-alkynyloxy phenyl sulfanyl, sulfinyl, and sulfonyl alkyl hydroxamates as tumour necrosis factor-converting enzyme and matrix metalloproteinase inhibitors. *J. Med. Chem.* **2004**, *47*, 6255–6269.

69. Becker, D. P.; Villamil, C. I.; Barta, T. E.; Bedell, L. J.; Boehm, T. L.; DeCrescenzo, G. A.; Freskos, J. N.; Getman, D. P.; Hockerman, S.; Heintz, R.; Howard, S. C.; Li, M. H.; McDonald, J. J.; Carron, C. P.; Funckes-Shippy, C. L.; Mehta, P. P.; Munie, G. E.; Swearingen, C. A. Synthesis and structure–activity relationships of α- and β-piperidine sulfone hydroxamic acid matrix metalloproteinase inhibitors with oral antitumour efficacy. *J. Med. Chem.* **2005**, *48*, 6713–6730.

70. Noe, M. C.; Snow, S. L.; Wolf-Gouveia, L. A.; Mitchell, P. G.; Lopresti-Morrow, L.; Reeves, L. M.; Yocum, S. A.; Liras, J. L.; Vaughn, M. 3-Hydroxy-4-arylsulfonyltetrahydropyranyl-3-hydroxamic acids are novel inhibitors of MMP-13 and aggrecanase. *Bioorg. Med. Chem. Lett.* **2004**, *14*, 4727–4730.

71. Letavic, M. A.; Axt, M. Z.; Barberia, J. A.; Barberia, J. T.; Carty, T. J.; Danley, D. E.; Geoghegan, K. F.; Halim, N. S.; Hoth, L. R.; Kamath, A. V.; Laird, E. R.; Lopresti-Morrow, L. L.; McClure, K. F.; Mitchell, P. G.; Natarajan, V.; Noe, M. C.; Pandit, J.; Reeves, L. M.; Schulte, G. K.; Snow, S. L.; Sweeney, F. J.; Tan, D. H.; Yu, C. H. Synthesis and biological activity of selective pipecolic acid-based TNF-converting enzyme (TACE) inhibitors. *Bioorg. Med. Chem. Lett.* **2002**, *12*, 1387–1390.

72. Bernardo, M. M.; Brown, S.; Li, Z. H.; Fridman, R.; Mobashery, S. Design, synthesis, and characterization of potent, slow-binding inhibitors that are selective for gelatinases. *J. Biol. Chem.* **2002**, *277*, 11201–11207.

73. Ikejiri, M.; Bernardo, M. M.; Merouch, S. O.; Brown, S.; Chang, M.; Fridman, R.; Mobashery, S. Design, synthesis, and evaluation of a mechanism-based inhibitor for gelatinase A. *J. Org. Chem.* **2005**, *70*, 5709–5712.

74. Zhang, Y. M.; Fan, X.; Xiang, B.; Chakravarty, D.; Scannevin, R.; Burke, S.; Karnachi, P.; Rhodes, K.; Jackson, P. Synthesis and SAR of α-sulfonylcarboxylic acids as potent matrix metalloproteinase inhibitors. *Bioorg. Med. Chem. Lett.* **2006**, *16*, 3096–3100.

75. Rossello, A.; Nuti, E.; Orlandini, E.; Balsamo, A.; Panelli, L. Inhibitors of zinc proteases thioaryl substituted and their use. WO 2008015139, **2008**.

76. Pochetti, G.; Gavuzzo, E.; Campestre, C.; Agamennone, M.; Tortorella, P.; Consalvi, V.; Gallina, C.; Hiller, O.; Tschesche, H.; Tucker, P. A.; Mazza, F. Structural insight into the stereoselective inhibition of MMP-8 by enantiomeric sulfonamide phosphonates. *J. Med. Chem.* **2006**, *49*, 923–931.

77. Biasone, A.; Tortorella, P.; Campestre, C.; Agamennone, M.; Preziuso, S.; Chiappini, M.; Nuti, E.; Carelli, P.; Rossello, A.; Mazza, F.; Gallina, C. α-Biphenylsulfonylamino 2-methylpropyl phosphonates: enantioselective synthesis and selective inhibition of MMPs. *Bioorg. Med. Chem.* **2007**, *15*, 791–799.

78. Folgueras, A. R.; Fueyo, A.; Garcia-Suarez, O.; Cox, J.; Astudillo, A.; Tortorella, P.; Campestre, C.; Gutierrez-Fernandez, A.; Fanjul-Fernandez, M.; Pennington, C. J.; Edwards, D. R.; Overall, C. M.; Lopez-Otin, C. Collagenase-2 deficiency or inhibition impairs experimental autoimmune encephalomyelitis in mice. *J. Biol. Chem.* **2008**, *283*, 9465–9474.

79. Zhang, Y. M.; Fan, X.; Chakaravarty, D.; Xiang, B.; Scannevin, R. H.; Huang, Z.; Ma, J.; Burke, S. L.; Karnachi, P.; Rhodes, K. J.; Jackson, P. F. 1-Hydroxy-2-pyridinone-based MMP inhibitors: synthesis and biological evaluation for the treatment of ischemic stroke. *Bioorg. Med. Chem. Lett.* **2008**, *18*, 409–413.

80. Puerta, D. T.; Cohen, S. M. Examination of novel zinc-binding groups for use in matrix metalloproteinase inhibitors. *Inorg. Chem.* **2003**, *42*, 3423–3430.

81. Puerta, D. T.; Lewis, J. A.; Cohen, S.M. New beginnings for matrix metalloproteinase inhibitors: identification of high-affinity zinc-binding groups. *J. Am. Chem. Soc.* **2004**, *126*, 8388–8389.

82. Puerta, D. T.; Cohen, S. M.; Lewis, J. A. Metalloprotein inhibitors. US 20070117848, **2007**.

83. Zhang, Y. M.; Fan, X.; Yang, S. M.; Scannevin, R. H.; Burke, S. L.; Rhodes, K. J.; Jackson, P. F. Syntheses and *in vitro* evaluation of arylsulfone-based MMP inhibitors with heterocycle-derived zinc-binding groups (ZBGs). *Bioorg. Med. Chem. Lett.* **2008**, *18*, 405–408.

84. Fryer, A.; Frizzi, J. Alkylsulfonamido substituted triazoles as matrix metalloproteinase inhibitors. WO 20080244784 A1, **2008**.

CHAPTER 26

ADAMs and ADAMTs Selective Synthetic Inhibitors

ARMANDO ROSSELLO[1], ELISA NUTI[1], and ALFONSO MARESCA[2]

[1]Dipartimento di Scienze Farmaceutiche, Università di Pisa, Via Bonanno 6, 56126 Pisa, Italy
[2]Laboratorio di Chimica Bioinorganica, Universita degli Studi di Firenze, Room. 188, Via della Lastruccia 3, 50019 Sesto Fiorentino (Florence), Italy

26.1 INTRODUCTION

26.1.1 ADAM Family Members: Membrane-Anchored ADAMs and Secreted ADAMTSs

Nowadays, there are innumerable evidences of the key roles of proteolytic enzymes, belonging to the large superfamily of metzincin,[1] in many serious degenerative diseases involving every structural and functional tissue. Any imbalance between the physiological levels and/or activities of these enzymes is central to the pathology, and any tissue of the body could be damaged. Cancer, arthritis, diabetes, and many other diseases involving functions and tissues of the central and peripheral nervous systems, cardiocirculatory, muscle–skeletal, respiratory, gastrointestinal, lymphatic, reproductive, urinary, glandular systems, eye, and skin could be involved.[2–7]

The superfamily of these zinc endopeptidases includes four families: the astacins (e.g., crayfish collagenolytic enzyme and bone morphogenetic protein-1), the reprolysins or adamalysins (snake venom proteinases and a disintegrin and metalloproteinase (ADAMs)), the serralysins (bacterial proteinases), and the matrix metalloproteinases (MMPs), also denoted as matrixins.[8] In addition to the traditional four families within the metzincin superfamily, a fifth family of metalloproteinases consisting of two members has been cloned and characterized.[9] Both demonstrate several features, suggesting that they belong to the metzincin superfamily, and have been termed as "pappalysins." This metalloproteinase family currently consists of pregnancy-associated plasma protein-A1 (PAPP-A1) and

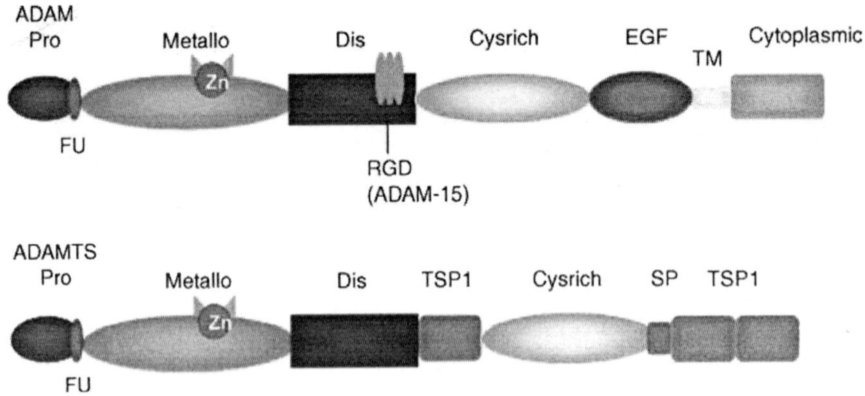

FIGURE 26.1 Structure of ADAM and ADAMTS proteinases. ADAM members are composed of common domains including propeptide (pro), metalloproteinase (metallo), disintegrin (dis) with a conserved RGD domain for ADAM-15, cystein-rich (cysrich), EGF-like (EGF), transmembrane (TM), and cytoplasmic domains. ADAMTS contain thrombospondin motifs (TSP1) and spacer domain (SP) but lack EGF-like transmembrane and cytoplasmic domains. Some proteinases contain in addition a sequence recognized by furin-like enzymes (Fu).[2]

pregnancy-associated plasma protein-A2. On the basis of the homology around the "Met-turn," pappalysins are most similar to the serralysin family.

ADAM family members, as cited above, display sequence similarities with the reprolysin family of snake venomases, and two subgroups are distinguished in the family: the membrane-anchored ADAMs and the secreted ADAMTSs.

Figure 26.1 reports the different domains of the two subfamilies of ADAMs that show complementary functions with features of proteinases and adhesion molecules.[10,11] Also, in these metzincins, as seen in the previous chapters on MMPs, the prodomain maintains the metalloproteinase domain inactive and has the ability to unveil catalytic site through a cysteine switch mechanism upon activation by various processes. The furin recognition site (RXXR sequence), placed between pro- and metalloproteinase domain, is believed to participate in intracellular activation of many ADAMs (ADAM-9, ADAM-12, ADAM-15, and ADAM-17) by the action of furin-like proprotein convertases in the trans-Golgi network.[12]

The metalloproteinase domain is characterized by a conserved HEXGH sequence shared with MMPs and confers the catalytic activity. In every case, there are some ADAMs that display alterations in this sequence and this fact causes them to loss of proteolytic activity.[13] Regarding the disintegrin domain, widely described as able to interact with integrins, mediating cell–cell and cell–matrix interactions,[11,14,15] a caution should be taken because Takeda et al. have shown that the disintegrin domain is not available for protein binding due to protein folding.[16] Finally, the carboxy terminal end is composed of a cystein-rich domain involved in cell–cell fusion,[17] an EGF-like domain, a transmembrane domain, and a cytoplasmic tail containing phosphorylation sites and SH3 binding domains.[18] At present, of about 40 known

ADAMs, 19 are expressed in *Homo sapiens*. Among those, only 12 display a proteolytic activity.[19] The human ADAMTS family comprises 19 ADAMTS genes.[20–22] These are characterized by the presence of an additional thrombospondin type I (TSP-I) motifs in their C-terminal part. On the contrary, with respect to the ADAMs, ADAMTS are devoid of the EGF-like transmembrane and cytoplasmic domains.[23,24] Some of them have one or two additional specific C-terminal modules, such as a mucin domain (ADAMTS-7 and ADAMTS-12), a GON domain (ADAMTS-20 and ADAMTS-9), two CUB domains (ADAMTS-13), and/or a PLAC domain (ADAMTS-2, ADAMTS-3, ADAMTS-10, ADAMTS-12, ADAMTS-14, ADAMTS-17, and ADAMTS-19). Although ADAMTSs are soluble proteins, many of them bind to the extracellular matrix through their thrombospondin motifs or their spacer region.[24] ADAMTS are regulated through a proteolytic process occurring at the furin-like recognition site located between the pro- and catalytic domains. ADAMTS-10 and ADAMTS-12 are an exception to this regulatory mechanism.[21,22] Their natural inhibitor, the tissue inhibitor of metalloproteinases (TIMPs), demonstrates selectivity of inhibition between the ADAMs and the ADAMTSs. This relevant fact, which is in contrast to that observed for the parent family of MMP enzymes, could be useful in the oriented drug target design.[25–28] This fact has been demonstrated, for example, for ADAM-17, exclusively inhibited by TIMP-3, for ADAM-10, inhibited by TIMP-1 and TIMP-3 but not by TIMP-2 and TIMP-4, or again, for ADAM-8 and ADAM-9 not controlled by TIMPs.[29,30]

26.1.2 Tumor Necrosis Factor-α Converting Enzyme (TACE or ADAM-17)

Tumor necrosis factor-α (TNF-α) converting enzyme (TACE) and other ADAM proteases (those containing a disintegrin and a metalloprotease domain) have emerged as potential targets in some therapeutic areas such as arthritis, cancer, diabetes, cardiovascular diseases, HIV, and cachexia.[2–7] TACE is the first ADAM protease to process the known physiological substrate and inflammatory cytokine, membrane-bound precursor TNF-α, to its mature soluble form. Later on, TACE was shown to be required for several different processing events such as tumor growth factor-α precursor and amyloid precursor protein (APP) cleavage.[31] Moreover, this sheddase is able to process many other membrane proteins and receptors as the epidermal growth factor receptor (EGFR),[32] the ErbB family of receptor tyrosine kinases (e.g., the ER3),[33,34] and adhesion molecules such as ALCAM.[35]

There is strong evidence suggesting that TACE is the first physiological TNF-α converting enzyme. TACE was purified for its ability to process precursor TNF-α and a peptide that spans the cleavage sequence of the proform of the cytokine at the correct cleavage sequence, and the activity was completely inhibited by a metalloproteinase inhibitor of TNF-α release.[36,37] T cells derived from knockout mice that have a disruption in the metalloproteinase consensus sequence (HEXXH) found in the exon that encodes TACE ($TACE^{\Delta Zn/\Delta Zn}$) lose 90% of their ability to process precursor TNF-α.[36] This is especially important as two groups also identified ADAM-10 (sometimes known as MADM) as a TNF-α converting enzyme.[38,39] However, the

k_{cat}/K_m for the processing of a peptide substrate that spans the cleavage sequence of precursor TNF-α is 10-fold lower for ADAM-10 than for ADAM-17.[40] Moreover, a potent (<10 nM) inhibitor of ADAM-10, but not of TACE, does not inhibit TNF release from peripheral blood mononuclear cells stimulated with lipopolysaccharide, thus providing more evidence of the involvement of TACE as the main TNF-α converting enzyme.

A role for TACE in the processing of other substrates comes from the analysis of the phenotype of TACE$^{\Delta Zn/\Delta Zn}$ mice. The TACE$^{\Delta Zn/\Delta Zn}$ mice die between the embryonic day 17.5 and the day of birth and exhibit a failure to fuse their eyelids, have thinned corneas, lack a conjunctival sac, and have a wavy hair phenotype along with several epidermal defects.[41] This phenotype is characteristic of mice that have a disruption in the TGF-α gene.[42,43] Additional defects identified are epithelial maturation of multiple organs and defects in the spongiotrophoblast layer of the placenta. These defects are reminiscent of mice that have a deletion of the EGFR.[44–46] The findings point to a general defect in the processing of multiple growth factors of the EGF family of ligands, such as epidermal growth factor (EGF) (HB-EGF), amphiregulin, and TGF-α. Such is the case with TGF-α because the growth factor is deficient in fibroblasts taken from the TACE$^{\Delta Zn/\Delta Zn}$ mice that have been immortalized by *ras* transformation.[41] Other putative substrates for TACE that are inhibited by hydroxamic acid inhibitors and that have been identified using the TACE$^{\Delta Zn/\Delta Zn}$ mice are L-selectin[41], TNF receptor I (TNF-RI[47]) and II (TNF-RII[41]), APP,[48] interleukin-1 (IL-1) receptor II (IL-1RII[47]), interleukin-6 receptor,[49] and Erb-B4 receptor.[50] A substrate that is inhibited by metalloproteinase inhibitors, but the secretion of which is not impaired in the TACE$^{\Delta Zn/\Delta Zn}$ mice, is angiotensin converting enzyme.[51] Finally, TNF-related activation-induced cytokine (TRANCE), an activator of osteoclasts, and the Notch receptor were also recently described as substrates for TACE.[52,53] There is some preliminary *in vitro* biochemical evidence suggesting that TACE directly processes some of these substrates. For example, peptides spanning the cleavage sequences of the α-secretase site of APP, L-selectin, and the N-terminal processing site of TGF-α are all processed at the correct positions.

Recently, the relationship between TACE and activated leukocyte cell adhesion molecule (ALCAM) has been discovered.[35] ALCAM mediates cell–cell clustering through homophilic (ALCAM–ALCAM) and heterophilic (ALCAM–CD6) interactions. ALCAM is expressed at the cell surface of ovarian epithelial cancer (EOC) cells by an ALCAM-specific recombinant antibody, and ALCAM can be internalized through a clathrin-dependent pathway.[35] The cytoplasmic localization of ALCAM is a marker of poor outcome in advanced-stage ovarian carcinoma patients, compared to its membrane expression. ALCAM can be trimmed from the cell membrane by proteolysis, as reported for other adhesion molecules, leading to the generation of a soluble ALCAM form (sALCAM) that contains the great part of the extracellular domain. ALCAM shedding from EOC is enhanced by stimuli such as EGF, which activates ADAM-17, and can be blocked by inhibitors of ADAMs and by ADAM-17 silencing. A recombinant antibody blocking ALCAM adhesive functions significantly increased EOC cell motility in wound-healing assays, whereas inhibitors of ADAM-17 inhibited EOC cell motility and invasiveness triggered by EGF. Altogether, these data suggest that TACE-mediated release of ALCAM adhesive functions may play a role in EOC cell motility and invasiveness.[35]

An explanation for the results described with the TACE$^{\Delta Zn/\Delta Zn}$ mice is that the mutant TACE is acting as a dominant negative by interfering with protein folding and/or some other mechanism in which active TACE is required indirectly for substrate processing. An example of the latter would be a mechanism in which perhaps TACE dimerizes with other disintegrin metalloproteinases and indirectly affects their catalytic efficiency. In fact, a dominant negative form of TACE has been described that retains the disintegrin domain and cytoplasmic tail but lacks the pro- and catalytic domains.[54] However, a more full-length TACE$^{\Delta Zn}$ mutant does not prevent the processing of TNF, L-selectin, or TNF-RII when cotransfected into COS cells, which indicates that the mutation leads to a loss of function rather than a dominant negative effect.[55] Experiments with full-length substrates and more comprehensive inhibitor studies might aid in the determination of whether TACE is the physiological convertase for many of the putative substrates identified by the gene knockout studies.

Preliminary biochemical data suggest that different constructs of TACE are necessary for turnover of substrate to product. For example, using cells from the TACE$^{\Delta Zn/\Delta Zn}$ mice that are deficient in processing, a chimeric protein encoding the pro- and catalytic domains of TACE and downstream domains from ADAM-10 (but lacking an ADAM cytoplasmic tail) was found to be sufficient for cleavage of TNF and TNF-RII, whereas the downstream domains of TACE (e.g., disintegrin and cysteine rich) were also needed to specify IL-1RII shedding.[47] These results suggest that TACE might regulate which substrates it utilizes by existing in different enzyme forms (isolated catalytic versus catalytic–disintegrin–cysteine-rich versus full length). An alternative mechanism could be that processing of certain substrates depends on where the TACE is localized. Although most of the TACE exists in a perinuclear compartment, it is also present on the cell membrane.[56,57] PMA stimulation downregulates TACE on the cell surface,[58] a finding that is consistent with TACE relocalizing to the trans-Golgi, where it is needed to process substrates such as precursor TNF-α.

The structure of the catalytic domain of TACE, bound to a hydroxamate inhibitor, has been determined using X-ray crystallography.[59] The overall structures of TACE, adamalysin II, and atrolysin C are similar, with α-helices packed above and below a central β-sheet. This fold is generally similar to that of the MMP family although many of the peripheral loops have different conformations. The structures are especially similar around the catalytic site, in which TACE, adamalysin, and MMPs all bind zinc with the same geometry. In all cases, the active site cleft is deeper on the "primed" (right-hand) side of the zinc, corresponding to the P1′, P2′, and P3′ residues of the substrate, than on the "unprimed" (left-hand) side, corresponding to the P1, P2, and P3 residues. The S1′ pocket is the deepest and it appears to be especially large and deep in adamalysin and atrolysin.[60] The S1′ pocket is somewhat shallower in TACE and has a different shape that connects below Leu384 and Ala439 with the S3′ pocket.[59] The TACE structure has been solved with various different inhibitors. In most cases, the inhibitors contain a "head group," such as hydroxamate, that ligates the zinc, a peptide or peptide-mimetic backbone that lies in the "primed" (right-hand) side of the active site cleft, and a lipophilic substituent directed into the S1′ pocket.[61–63]

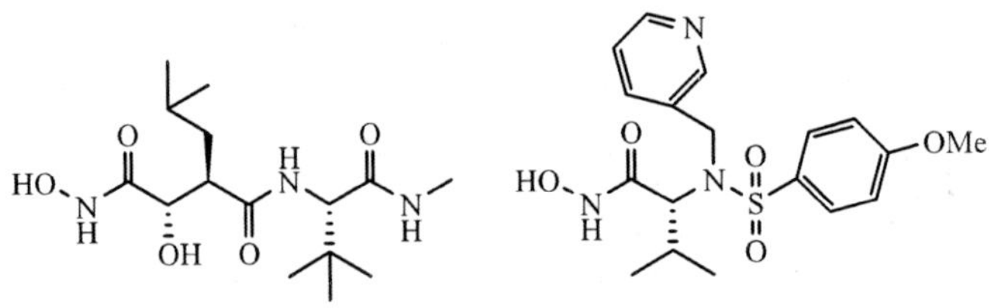

26.1 26.2

FIGURE 26.2 Prototypical matrix metalloproteases and tumor necrosis factor-α converting enzyme (TACE) inhibitors, marimastat **26.1**, and CGS27023 **26.2**.[7]

26.2 TACE SYNTHETIC INHIBITORS

Historically, similarities in the catalytic domain between MMPs and TACE have driven the starting design of inhibitors. To date, TACE inhibitors reported in the literature show almost exclusively a hydroxamate moiety as ZBG, whereas their features of selectivity and potency are modulated by the scaffold characteristics that can be of many different kinds. The two first examples of TACE inhibitors are the succinyl hydroxamate marimastat (**26.1**) and the peptidomimetic/sulfonamide hydroxamate CGS27023A (**26.2**), shown in Fig. 26.2. They are two archetypal well-known MMPIs that have been widely used for the development of MMPI and TACE inhibitors.

The hydroxamate BB2516 (marimastat, **26.1**) was potent against TACE ($K_I = 22$ nM), ADAM-9 ($K_I = 274$ nM), and MMPs (MMP-1, $K_I = 1$ nM; MMP-3, $K_I = 68$ nM; MMP-9, $K_I = 1$ nM; and MMP-13, $K_I = 0.1$ nM).[64] Marimastat and the sulfonamide CGS27023A (**26.2**) (TACE ($K_I = 54$ nM), ADAM-9 ($K_I = 1$ nM), and MMPs (MMP-1, $K_I = 11$ nM; MMP-3, $K_I = 16$ nM; MMP-9, $K_I = 3$ nM; and MMP-13, $K_I = 5$ nM))[64,65], also an inhibitor of cell-free TACE, cellular TNF secretion[66], and ADAM-9, continue to serve as basic templates for the design of TACE inhibitors disclosed in the literature. Marimastat docking into TACE pocket is shown in Fig. 26.3. The P1′ isobutyl group is directed into the S1′ pocket, but does not fill it, which suggests that stronger affinity could be obtained with larger P1′ substituents. Larger groups could also be accommodated at P3′ giving an increase in potency for the inhibition of TNF release.[67]

Many other molecules have been disclosed in the literature of TACE inhibitors (Fig. 26.4). Derivatization of the position alpha to the hydroxamate in the succinate inhibitor template led to potent inhibitors of TACE with submicromolar IC_{50} values for inhibition of TNF release in human whole blood assay (WBA).[68] The best compound, **26.3**, possessed a K_I value of 0.57 nM against TACE and an IC_{50} value of 280 nM against TNF release in human whole blood assay.

The *N*-hydroxyformamide, or "reverse hydroxamate" functional group, can serve to bind to zinc in MMPs and thus has been explored as a zinc chelator in TACE and MMP inhibitors.

The synthesis and biological activity of the *N*-hydroxyformamide GW4459, **26.4,** has been described.[69] This molecule was a potent inhibitor of TACE ($K_I = 4$ nM) and

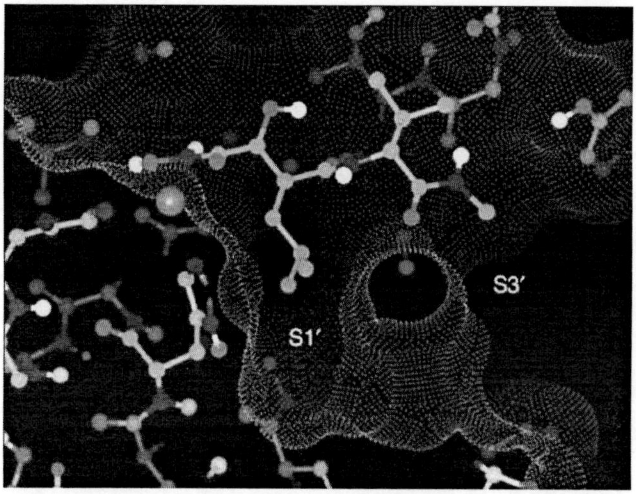

FIGURE 26.3 Model of marimastat **26.1** bound to human TACE. Hydrogen, oxygen and nitrogen are represented in white, gray and dark gray, respectively. Zinc atom is the bright gray sphere, see left side. Carbon atoms are clear gray in the protein and the inhibitor, respectively. The surface of the protein is depicted with a Connolly dot surface.[7]

MMPs, and inhibited TNF release from cultured MonoMac-6 cells ($IC_{50} = 34$ nM). Furthermore, GW4459 administered subcutaneously reduced TNF levels in the pleural fluid of zymosan-treated rats. Related compounds were synthesized possessing a methyl substitution at the P2′ arginine side chain vicinal to the peptide backbone, with the aim of endowing the compound with stability and oral bioavailability.

An example of branched methyl substitution at P2′ in the *N*-hydroxyformamide inhibitor class was GW3333 **26.5**, an inhibitor of TACE and MMPs. Furthermore, the compound was orally active with a long half-life in rat and dog. The compound effectively inhibited TNF release in mice subjected to LPS administration.[67]

TACE inhibitors have been designed that possess P2′ amino amide mimetics or truncated P2′-P3′ substituents. The cyclic hydrazide **26.6** was a potent inhibitor of TNF release from THP-1 cells ($IC_{50} = 268$ nM).[70] Although no data on direct TACE enzyme inhibition were available, **26.6** was similar to Ro327315 **26.7**, which inhibited TACE ($K_I = 3$ nM) and TNF release from THP-1 cells ($IC_{50} = 375$ nM)[71] and has entered phase I clinical trials. The morpholine derivative **26.8** was a P1-P2′ constrained analogue of **26.2** and was reported to inhibit TACE with a K_I value of 10 nM.[72] Compound **26.8** was selective against collagenase-1 ($K_I = 3500$ nM) and inhibited collagenase-3 with a K_I value of 3 nM. Despite the presence of three asymmetric centers, the morpholine ring of **26.8** and its congeners were easily synthesized from threonine. The elaboration of anthranilic acid and related bicyclic aromatic skeletons into TACE and MMP inhibitors was an example of substantial innovation in TACE/MMP inhibitor design.[73] The anthranilic hydroxamate **26.9** inhibited TACE with a K_I value of 40 nM. Remarkably, adjustment of the arylsulfonyl substituent from methoxy to 2-butynyloxy produced compound **26.10** with a K_I value of 15 nM against TACE and

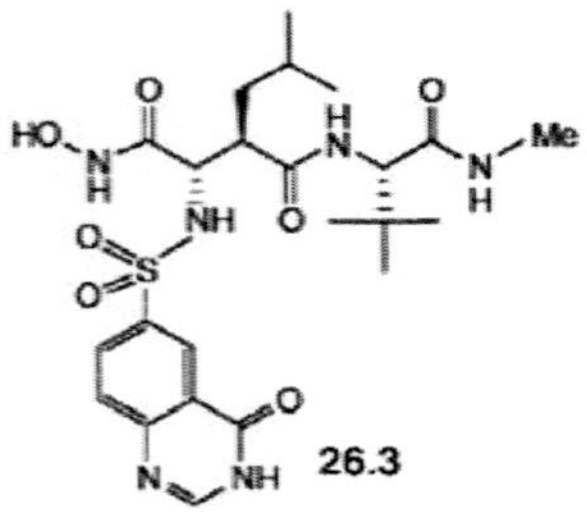

26.3

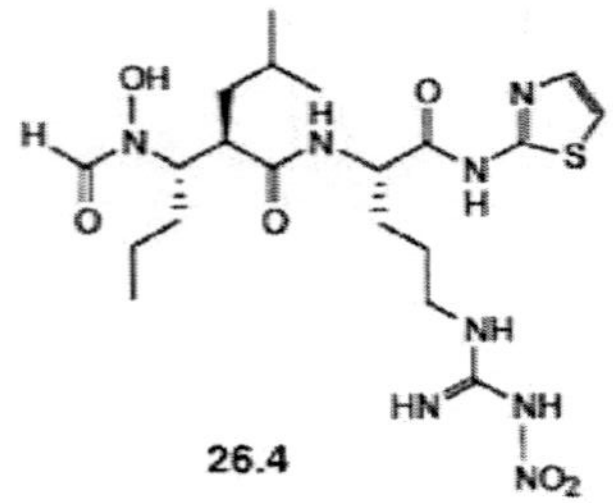

26.4

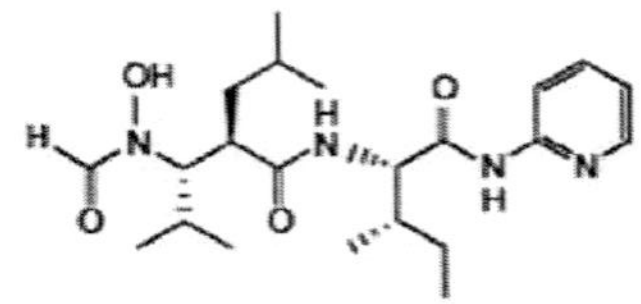

26.5

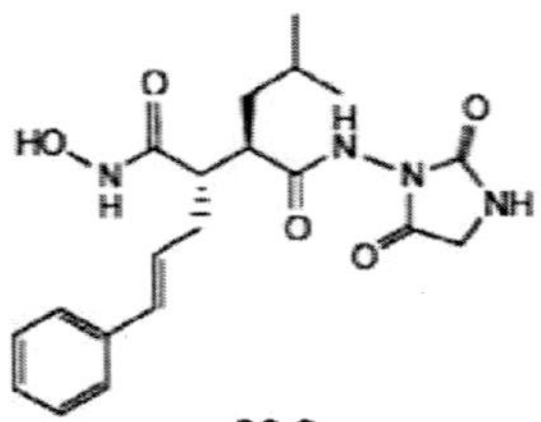

26.6

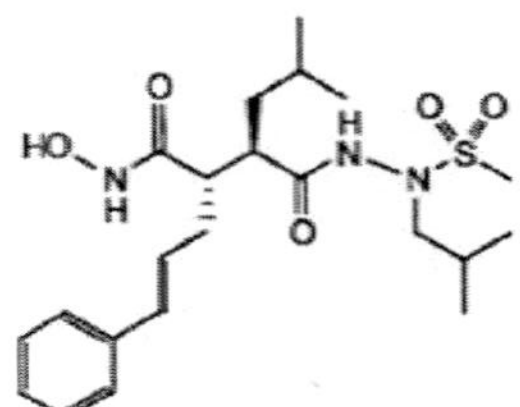

26.7

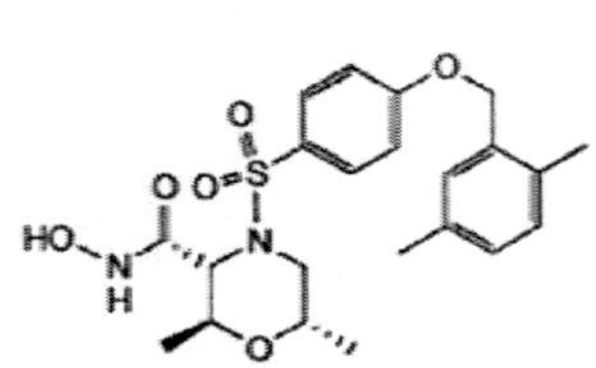

26.8

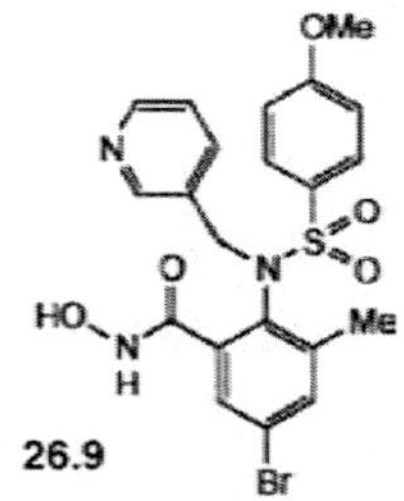

26.9

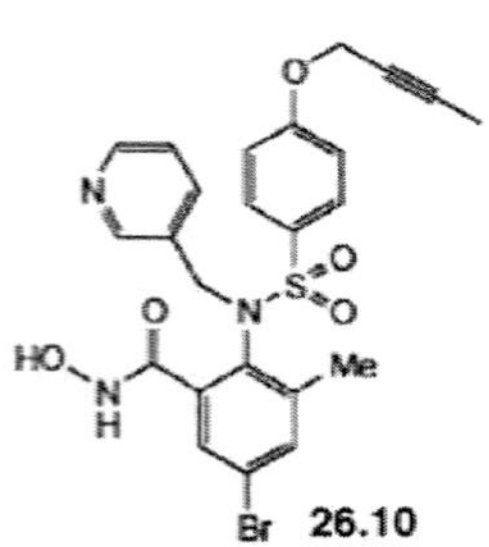

26.10

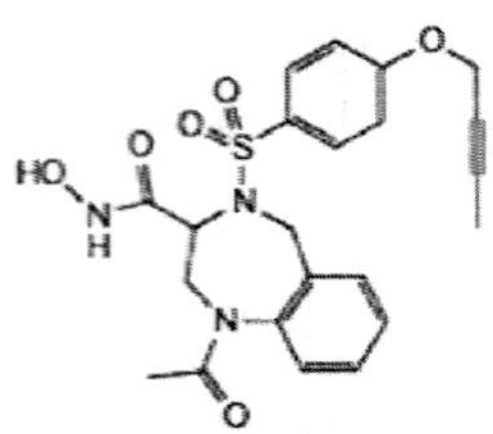

26.11

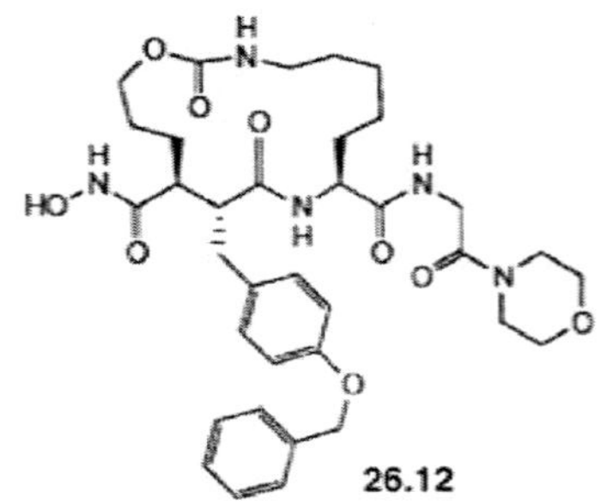

26.12

FIGURE 26.4 Tumor necrosis factor-α converting enzyme inhibitors **26.3–26.12**.

concurrent selectivity against collagenase-1 and collagenase-3. The butynyloxy substituent fits well in the angled S1′ subsite of TACE.[59] Such a modification on the benzazepine **26.11** has the same effect, producing a molecule with a K_I value of 16 nM against TACE and selectivity against collagenase-1. Certain succinyl hydroxamates, when bridged between P1 and P2′, afforded potent MMP/TACE inhbitors.[74] The macrocyclic carbamate **26.12** possessed K_I values of 18 and 5000 nM against TACE and collagenase-1, respectively. Members of this inhibitor class were orally bioavailable and reduced TNF titer *in vivo*.

26.2.1 Semirigid Succinamides and Cyclic Amides as Selective TACE Inhibitors

With the aim of developing selective and potent TACE inhibitors with improved biopharmacological properties such as good oral bioavailability and good inhibition of TNF production in *in vivo* models, many compounds were designed by inserting the structure of the succinyl hydroxamates in semirigid cyclic systems, as shown in Fig. 26.5. This arrangement provided an appropriate orientation between the ZBG and the P1′ specific substituent. On this basis, some classes of inhibitors were developed, in particular some piperidinecarboxamides such as **26.13** and analogues.[75] The inhibitor **26.13** (IM491), developed by Xue and colleague, was potent against TACE ($IC_{50} = 6.2$ nM), particularly selective over some MMPs, such as MMP-1, MMP-2, and MMP-9 (IC_{50} in the μM range), and more relevantly, highly potent in the *in vivo* inhibition of the TNF release with an $IC_{50} = 20$ nM (WBA analysis, human whole blood assay).[75]

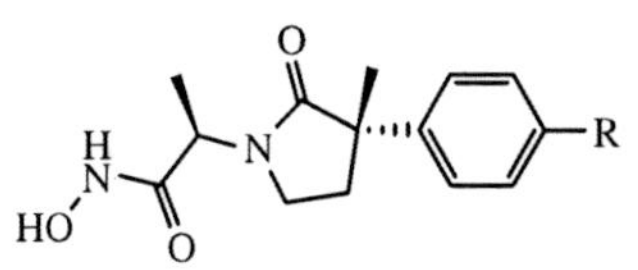

26.13

26.13a: R = H

26.13b: R =

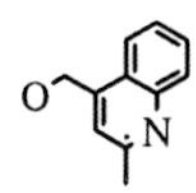

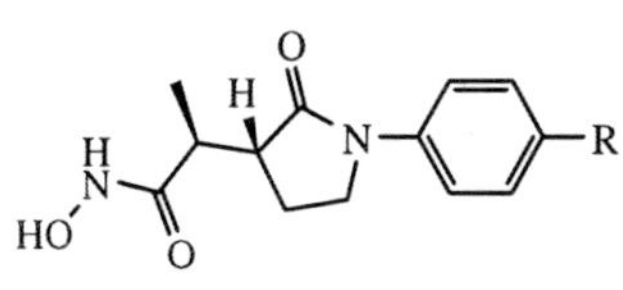

26.14

26.14a: R = H

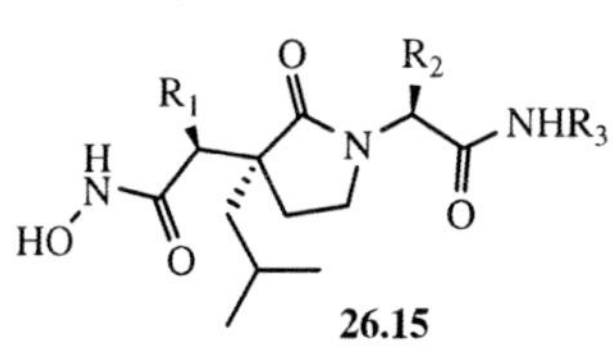

26.15

FIGURE 26.5 **IM491** (a succinyl hydroxamate scaffold inserted into a piperidinecarboxamide semirigid system) and representative γ-lactam inhibitors of TACE (**26.14** and **26.15**) and MMP (**26.16**).[75,76]

Recently, new TACE inhibitors were discovered by Bristol-Myers Squibb researchers using a *N*-hydroxy-2-(2-oxo-3-pyrrolidinyl)acetamide scaffold (**26.15**).[76] Incorporation of the TACE selective (2-methyl-4-quinolinyl)methoxy P1′ group produced a series of highly potent inhibitors of TACE that were selective over MMP-1, MMP-2, and MMP-9. To understand the development of such molecules, the previously reported lactam **26.14** (Fig. 26.5)[77] should be considered.

The lactam scaffold was designed by taking advantage of the structure of MMP-3 and the apparent similarity between the active site regions of TACE and MMP-3, revealed from early TACE inhibitors. A docking study of a lactam analogue **26.14b** in MMP-3 suggested that the hydroxamate group binds to the catalytic zinc ion in a bidentate fashion and the carbonyl group of the γ-lactam formed two important hydrogen bonds with the NH groups of Val163 and Leu164 (Fig. 26.6). A structure–activity relationship (SAR) study indicated that the lactam scaffold binds to TACE in a similar way.[77] Optimization of **26.14a** led to the discovery of [(2-methyl-4-quinolinyl)methoxy]phenyl P1′ group as the critical determinant of selectivity for TACE relative to MMPs.

As shown in Fig. 26.6, an analogue of the new lactam, **26.15a**, is nicely supposed with **26.14a**. A related lactam, **26.16**, has been reported as an MMP inhibitor. However, lactam **26.15** was distinct from **26.16** in that the [(2-methyl-4-quinolinyl)methoxy] phenyl P1′ group in **26.15** was attached to the nitrogen of the lactam ring, whereas lactam **26.16** was designed to mimic the more traditional binding mode of succinates

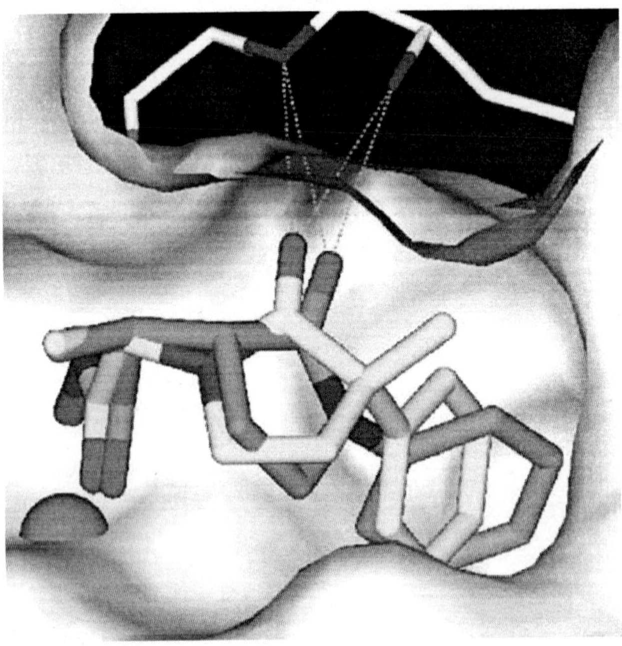

FIGURE 26.6 Model of lactams **26.14a** and **26.15a** in MMP-3. Nitrogen atoms are dark gray, oxygen atoms gray, carbon atoms of **26.14a** white, and those of **26.15a** pale gray. Residues of the MMP-3 in the backbone of the strand just above the active site (Val163–Leu164) are white. Other amino acids are omitted for clarity.[76]

and required the opposite configuration at α-position of the lactam to project the isobutyl group to the S1′ site of MMPs.

A large number of analogues (compounds **15b–t**, **26.17–26.20**, Table 26.1) were synthesized and tested on semipurified porcine TACE (pTACE) from porcine spleen. Selectivity profile was evaluated using MMP-2 and MMP-9 as representative members with deep S1′ pocket and MMP-1 as a representative with shallow S1′. Cellular activity was assessed using LPS-stimulated human WBA where inhibition of TNF-α production was measured.[76]

Because the (2-methyl-4-quinolinyl)methoxy P1′ group has been optimized for potency and selectivity in the *N*-hydroxy-2-(2-oxo-1-pyrrolidinyl)acetamide series (lactam **26.14**, Fig. 26.5), this group was kept constant in most of the SAR studies discussed for these compounds. It was found that compound **26.15b** ($R_1 = Me$, $R_2 = R_3 = H$, Table 26.1) was extremely potent in the pTACE (porcine TACE) assay with an IC_{50} of 1 nM. This compound also exhibited an excellent selectivity (at least 1200-fold) relative to MMP-1, MMP-2, and MMP-9. Inversion of the stereocenter at α-position of the hydroxamate group resulted in 50-fold loss of pTACE potency (**26.15c**, $R_2 = Me$, $R_1 = R_3 = H$). These data were consistent with the proposed binding conformation and lent validation to the prediction that lactam **26.15** could mimic the binding of **26.14** (Fig. 26.6). In the proposed conformation, the R_2 group, when other than a hydrogen, clashes with the backbone of the protein, whereas the R_1 group points toward solvent and hence is expected to tolerate variations.

Despite its affinity for pTACE, lactam **26.15b** was only moderately effective in inhibiting TNF-α production in WBA ($IC_{50} = 1.57\ \mu M$, Table 26.1). Activity in WBA is complicated by multiple factors and the foremost of which are protein binding and cell membrane permeability (most of pro-TNF-α processing occurs intracellularly). Because the R_1 group was predicted to be exposed to solvent and to have minimal contact with the protein (Fig. 26.6), they tried to modulate the WBA potency using different R_1 groups. The hydroxy and amino groups (**26.15e,f**) had essentially no effect on the IC_{50} value of pTACE or WBA (Table 26.1) compared to **26.15b**. The dimethylamino compound **26.15g** was sevenfold less active for pTACE compared to **26.15b**. However, **26.15g** remained as potent as **26.15b** in the WBA, which could be in part attributed to the relatively high free fraction in human serum (31% unbound). Pyrroly substitution (**26.15h**) attenuated pTACE activity. Acetylamino compound **26.15i** gave comparable potency in pTACE and WBA to **26.15b** but had decreased selectivity over MMP-2. Benzamide and *n*-butylcarbamate analogues **26.15j** and **26.15k** were less potent in the cell assay. Compounds **26.15l–n** were highly potent in pTACE assay ($IC_{50} < 1$ nM). Notably, the morpholinyl carbamate **26.15m** resulted in the first submicromolar inhibitor in the WBA (0.84 mM) for the series, a twofold improvement over **26.15b**. The WBA activity of **26.15l** and **26.15n** did not match with their pTACE affinity. The Boc-protected amino analogue **26.15o** was potent in the WBA with an IC_{50} of 0.42 μM. Even though it picked up moderate MMP activity, **26.15o** was still 700-fold selective over MMP-1 and approximately 300-fold selective over MMP-2 and MMP-9. Lactam **26.15o** also had good permeability in Caco-2 assay ($P_{app} = 11 \times 10^{-6}$ cm/s), indicating that it might have good oral absorption *in vivo*.

TABLE 26.1 ***In Vitro*** **Potency of Lactams 26.15b–t and 26.17–26.20 in pTACE, MMP-1, MMP-2, and MMP-9, and WBA** [76]

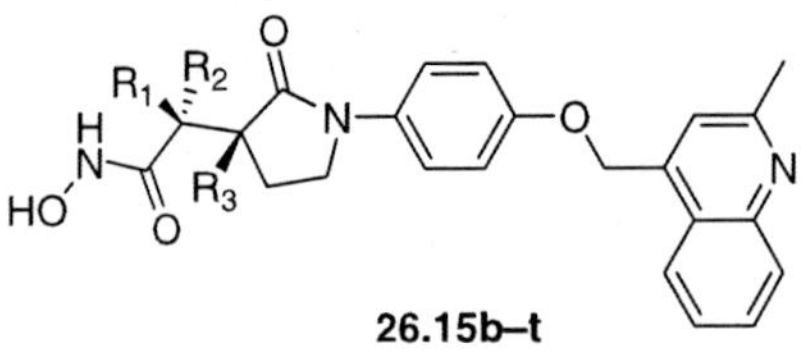

26.15b–t

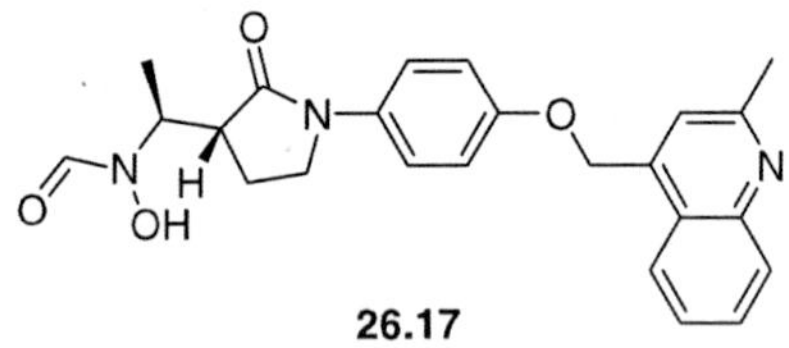

26.17

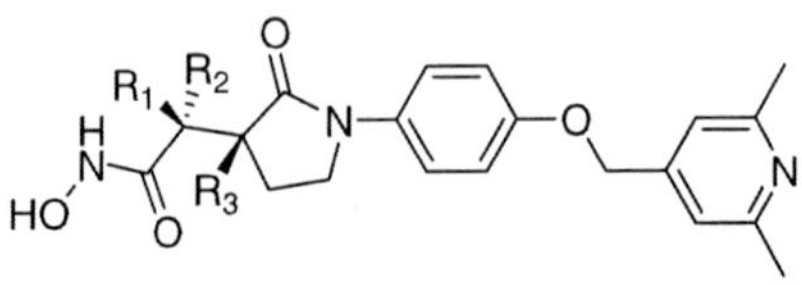

26.18, 26.19

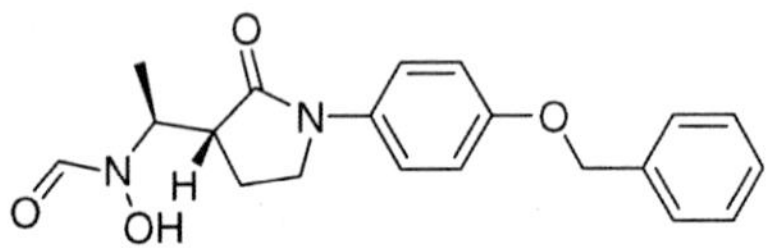

26.20

Compound	R_1	R_2	R_3	pTACE IC_{50}, nM[a]	MMP-1 K_I, nM[a]	MMP-2 K_I, nM[a]	MMP-9 K_I, nM[a]	WBA IC_{50}, μM[b]
26.15b	Me	H	H	1	>5000	1267	>2000	1.57
26.15c	H	Me	H	52	>5000	>3000	>2000	>3
26.15d	H	*n*-Pr	H	140	>5000	>3000	>2000	—[c]
26.15e	OH	H	H	3	>5000	>3000	>2000	2.4
26.15f	NH_2	H	H	2	>5000	446	1964	1.7
26.15g	NMe_2	H	H	7	>5000	>3000	>2000	1.5
26.15h	1-pyrrolyl	H	H	26	>5000	>3000	>2000	>10
26.15i	NHAc	H	H	1	>5000	194	>2000	1.56
26.15j	NHC(O)Ph	H	H	1	>5000	435	>2000	9.1
26.15k	$NHCO_2$-*n*-Bu	H	H	2	1627	305	>2000	6.83
26.15l	NHPiv	H	H	<1	>5000	1244	>2000	4.95
26.15m	NHC(O)-4-morpholinyl	H	H	<1	>5000	805	>2000	0.84
26.15n	$NHCO_2Me$	H	H	<1	>5000	482	>2000	2.65
26.15o	NHBoc	H	H	1	788	324	291	0.42
26.15p[d]	H	H	NH_2	5	>5000	>3000	>2000	4.8
26.15q[d]	Me	H	NH_2	2	>5000	>3000	>2000	1.1
26.15r[d]	*n*-Pr	H	NH_2	4	>5000	>3000	>2000	>3
26.15s[d]	Me	H	NMe_2	12	>5000	433	1299	>3
26.15t[d]	H	*n*-Pr	NH_2	405	>5000	>3000	>2000	>10
26.17	—	—	—	2	>5000	>3000	>2000	8.7
26.18	NH_2	H	H	50	>5000	>3000	>2000	>10
26.19	NHBoc	H	H	6	2238	>3000	—[c]	>3
26.20	—	—	—	111	>5000	567	>2000	>10

[a]pTACE IC_{50} and MMP K_I values are from a single determination.

[b]Inhibition of TNF-α release in WBA was determined with three donors.

[c]Not tested.

[d]Compounds **26.15p–t** were tested as a racemic mixture.

In an attempt to further improve cellular potency, an amino group was introduced to the 3-position of the pyrrolidinone (R_3). Unfunctionalized amino group yielded potent and selective TACE inhibitors (**26.15p–r**, racemic mixture) regardless of whether R_1 was present (methyl and *n*-propyl) or not. None of them improved WBA potency, with **26.15q** being the most promising at 1.1 μM. Dimethylamino R_3 analogue (**26.15s**) was sixfold less potent for pTACE than the amino compound **26.15q**. Similar to the SAR trend observed with R_3 being hydrogen, inversion of stereocenter at the α-position of the hydroxamate resulted in a 100-fold loss of potency (**26.15t** versus **26.15r**).

N-Hydroxyformamide (retrohydroxamate) group has been reported to be an effective zinc binding group to replace the commonly used hydroxamate group in MMP and TACE inhibitors. Hence, *N*-hydroxyformamide analogue **26.17** was prepared and tested. Indeed, it was highly potent for pTACE (2 nM) and exhibited excellent selectivity relative to the three MMPs (at least 1000-fold). However, it did not offer any advantages over the conventional hydroxamate in the WBA and was approximately fivefold less potent than **26.15b**.

Two analogues (**26.18** and **26.19**) with (2,6-dimethylpyridinyl)methoxy group in place of the (2-methyl-4-quinolinyl)methoxy P1′ group were evaluated. In both cases, a significant loss of enzyme and cell activity was observed (**26.18** versus **26.15f**, **26.19** versus **26.15o**). For the *N*-hydroxyformamide-derived inhibitor, an analogue with benzyloxy P1′ group (**26.20**) also had a diminished TACE affinity (50-fold) compared to **26.17**. These data demonstrated that the (2-methyl-4-quinolinyl)methoxy group remained a superior P1′ group for this new series of lactam TACE inhibitors. The discovery of this series added diversity to the TACE inhibitors portfolio, thus offering new opportunity to develop TACE inhibitors for the treatment of rheumatoid arthritis.

In several efforts to discover compounds that had both pTACE (porcine TACE) and WBA activities, several series of cyclic-β-aminohydroxamic acids (Fig. 26.7) were designed and synthesized (**26.21–26.24**).[78]

The first series was formed by cyclizing α,β to the hydroxamic acid, giving type **26.22** compounds (Fig. 26.8).

Two promising new scaffolds belonging to this class were the tetrahydropyran-β-aminohydroxamic acid **26.22a** and pyrrolidine-β-aminohydroxamic acid **26.22b**.[79,80] Compound **26.22a** showed high potency against pTACE ($K_I = 0.15$ nM), good WBA-LPS inhibition of TACE release ($IC_{50} = 130$ nM), and very high TACE inhibition sparing many MMPs (MMP-1, MMP-2, MMP-8, MMP-9, MMP-13, MMP14, MMP-15, and MMP-16), some other ADAMs (ADAM-9 and ADAM-10), and ADAMTS (ADAMTS4 and ADAMTS5) with K_I values ranging in four to six orders of magnitude more higher. The pyrrolidinyl analogue **26.22b** was a potent TACE inhibitor, in the subnanomolar range, able to inhibit effectively the release of TACE at 109 nM in the WBA test.[80] This inhibitor was highly selective over many of the MMPs considered, sparing MMP-1, MMP-2, MMP-8, MMP-9, MMP-14, and MMP-15 with K_I values ranging between four and five orders of magnitude more higher. More recently, a new family of tetrahydropyranyl-β-aminohydroxamic acids, containing on their P1′ position a new promising (2-substituted-1*H*-benzo[*d*]imidazol-1-yl)methyl)benzamide substituent, was discovered. The tetrahydropiranyl derivative **26.22c** was potent against TACE inhibition ($IC_{50} = 1.4$ nM) and exceptionally selective over a wide panel

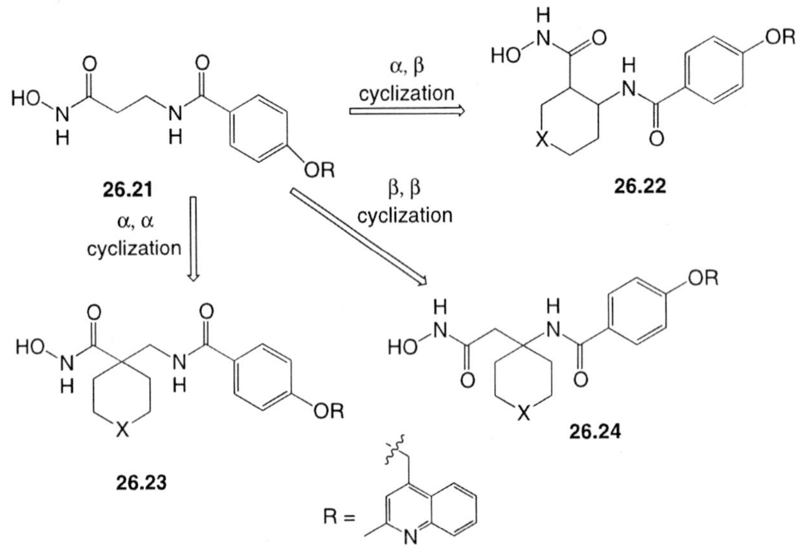

FIGURE 26.7 Strategy in the design of cyclic TACE inhibitors derived from β-amino-hydroxamic acid type **26.21**, **26.22–26.24** series.[78]

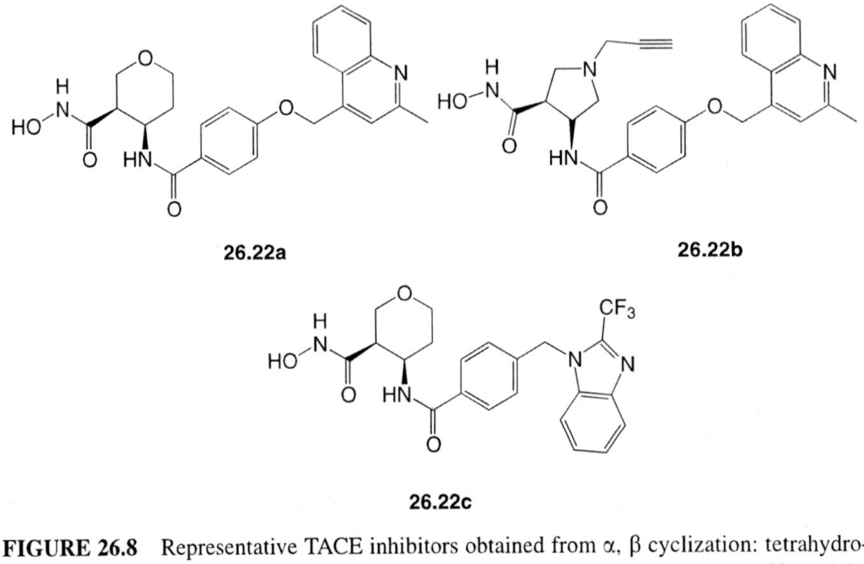

FIGURE 26.8 Representative TACE inhibitors obtained from α, β cyclization: tetrahydro-pyran-β-aminohydroxamic acid **26.22a**, pyrrolidine-β-aminohydroxamic acid **26.22b**, and the highly potent and selective tetrahydropyranyl derivative **26.22c** containing in P1′ a new (2-substituted-1*H*-benzo[*d*]imidazol-1-yl)methyl)benzamide substituent.[79,80]

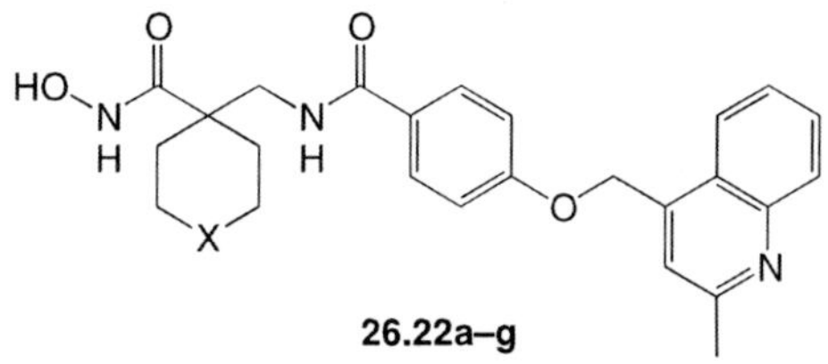

26.23a, X = CH_2; **26.23b**, X = NBoc; **26.23c**, X = NH; **26.23d**, X = N-Propyl; **26.23e**, X = N(CO)-*t*Bu; **26.23f**, X = N-SO_2Me; **26.23g**, X = O

FIGURE 26.9 α,α-Cyclic-β-aminohydroxamic acids **26.23a–g**.[78]

of secreted MMPs such as MMP-1, MMP-2, MMP-3, MMP-7, MMP-8, MMP-9, MMP-10, and MMP-13 (IC_{50} or K_I values >8–10 μM), some membrane-type MMPs such as MMP-14, MMP-15, and MMP-16 ($IC_{50} > 10$ μM), and also against some ADAMs such as ADAM-10 ($IC_{50} > 10$ μM) and ADAMTS such as ADAM-TS1 ($IC_{50} > 1$ μM), ADAMTS-4 ($IC_{50} = 6.7$ μM), and ADAM-TS5 ($IC_{50} = 1$ μM). Moreover, it was able to suppress LPS-induced TNF-α in human whole blood assay ($IC_{50} = 232$ nM) with an improved oral bioavailability on rat and dog models and, importantly, following oral dosing, **26.22c** displayed an ED_{50} of 1.9 mg/kg in an acute model of inflammation.[80]

As already mentioned, other two series of inhibitors were obtained by cyclizing α,α and β,β to the hydroxamic acid (Fig. 26.7). The α,α-cyclization (type **26.23** compounds) was designed to help *in vivo* stability of the hydroxamic moiety through steric hindrance since this was known to be a problem in other series of inhibitors. Some of the α,α-cyclic-β-aminohydroxamic acids (**26.23a–g**) (Fig. 26.9) showed to be potent TACE inhibitors (IC_{50} in the nanometer range) but poor inhibitors in the assay of WBA-LPS inhibition of TACE release ($IC_{50} > 3$ μM).[78]

The β,β-cyclization (type **26.24** inhibitors) was also examined. In this case, the **26.24a–z** derivatives included nonbasic derivatives such as amides, sulfonamides, ureas of piperidines, and a tetrahydropyran. Also, in this group of compounds the activities were good on pTACE with IC_{50} ranging between 0.35 and 3 nM, and moreover the WBA inhibitions were good for many of them. Compound **26.24r**, N-acetylated on the piperidine nitrogen, was very good in the WBA test with an IC_{50} value of 23 nM. On the contrary, the Caco-2 cell permeability was not improved significantly in this type of inhibitors. Many of these compounds were also evaluated in an LPS mouse model and were found to have good anti-TNF-α activity. Compounds **26.24c,l,r,y** and **z** had ED_{50}'s less than 8 mg/kg after oral administration in the mouse model (Table 26.2). Despite the low to marginal Caco-2 values, **26.24c** and **26.24z** were also examined in an expanded panel of other MMPs (MMP-3, MMP-7, MMP-14, MMP-15, and MMP-16) and some ADAMTS (ADAMTS-1, ADAMTS-4, and ADAMTS-5). Both compounds have shown good selectivity (>1000) for pTACE relative to MMP-1, MMP-2, MMP-9, MMP-13, MMP-14, MMP-15, and MMP-16. Compound **26.24c** also showed over 50-fold selectivity for pTACE over MMP-3, MMP-8, ADAMTS-1, and ADAMTS-5 and 10-fold selectivity over MMP-7.

TABLE 26.2 ***In Vitro* and *In Vivo* Profile of Some β,β- Cyclic-β-aminohydroxamic acids (26.24 Type)**[78]

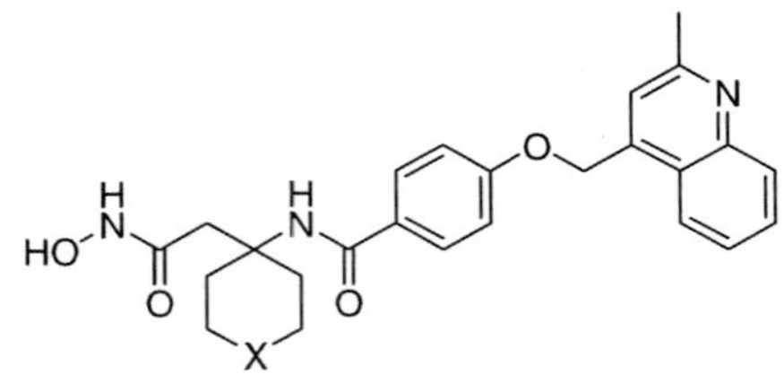

26.24a–z

Compound	X	pTACE[a]	WBA[a]	MMP-1[a]	MMP-2[a]	MMP-9[a]	MMP-13[a]	Caco-2 P_{app} ($\times 10^{-6}$ cm/s)	LPS mouse[b] (10 mg/kg/p.o.)
26.24a	CH_2	1.4	737	>4949	>3333	>2128	>5025	3.6	—
26.24b	NBoc	7.7	370	>4949	>3333	>2128	>5025	8.2	—
26.24c	NH	1.9	36	>4949	>3333	>2128	>5025	0.0	$ED_{50} < 3.0$ mg/kg
26.24d	*N*-Methyl	3.1	109	>4949	>3333	>2128	>5025	0.1	17%
26.24e	*N*-Ethyl	2.2	26	>4949	>3333	>2128	>5025	0.1	—
26.24f	*N*-Propyl	2.0	96	>4949	>3333	>2128	>5025	0.1	—
26.24g	*N*-^{t}Bu	3.7	56	>4949	>3333	>2128	>5025	0.1	29%
26.24h	*N*-$CH_2{}^t$Bu	2.6	143	>4949	>3333	>2128	>5025	2.4	38%
26.24i	*N*-3,3-Dimethyl-propargyl	2.3	490	>4949	>3333	>2128	>5025	3.2	—
26.24j	*N*-Phenyl	1.9	>1500	>4949	>3333	>2128	>5025	0.5	—
26.24k	*N*-Benzyl	1.8	290	>4949	>3333	>2128	>5025	0.1	—
26.24l	*N*-$(CH_3)_2COO^t$Bu	1.9	272	>4949	>3333	>2128	2462	5.3	$ED_{50} = 7.8$ mg/kg

26.24m	*N*-$(CH_3)_2COOH$	3.2	50	>4949	>3333	>2128	>5025	0.1	24%
26.24n	*N*-2-(4,5-Dihydrothiazole)	1.2	30	>4949	>3333	>2128	>5025	0.1	100%[c]
26.24o	*N*-$CH_2CH_2SO_2Me$	1.2	45	>4949	>3333	>2128	>5025	0.81	—
26.24p	*N*-$CH_2CH_2SO_2Et$	1.4	119	>4949	>3333	>2128	>5025	0.55	—
26.24q	*N*-CH_2CH_2F	2.6	90	>4949	>3333	>2128	3136	0.55	—
26.24r	*N*-C(O)Me	1.9	23	>4949	>3333	>2128	>5025	0.1	$ED_{50} = 1.4$ mg/kg
26.24s	*N*-C(O)Et	2.7	39	>4949	>3333	>2128	>5025	0.1	5%
26.24t	*N*-C(O)^{i}Pr	1.2	89	>4949	>3333	>2128	>5025	0.2	17%
26.24u	*N*-C(O)^{t}Bu	2.1	102	>4949	>3333	>2128	>5025	0.88	$ED_{50} = 25.5$ mg/kg
26.24v	*N*-C(O)$CH_2{}^t$Bu	1.9	212	>4949	>3333	>2128	>5025	0.79	—
26.24w	*N*-C(O)^{i}Bu	2.2	105	>4949	>3333	>2128	>5025	0.56	48%
26.24x	*N*-SO_2Me	1.0	54	>4949	>3333	>2128	>5025	—	—
26.24y	*N*-C(O)NMe_2	2.8	152	>4949	>3333	>2128	>5025	—	$ED_{50} < .3.0$ mg/kg
26.24z	O	0.35[d]	150	>4949	>3333	>2128	>5025	0.8	$ED_{50} = 4.7$ mg/kg

[a]pTACE IC50, WBA IC50 (inhibition of TNF-α release in WBA), and MMP K_I values are nM.
[b]LPS mouse model studies.
[c]LPS mouse for **26.24n** i.p. administration.
[d]k_i value.

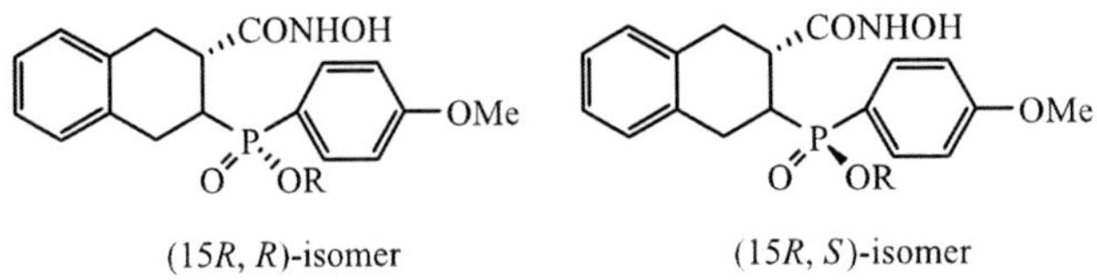

26.25a: R = Et
26.26a: R = $CH_2CH_2CF_3$

26.25b: R = Et
26.26b: R = $CH_2CH_2CF_3$

FIGURE 26.10 Structures of phosphonamide-based MMP inhibitors.[82]

Compound **26.24z** was found to be the more selective of the two with more than 1000-fold selectivity over MMP-8, ADAMTS-1, ADAMTS-4, and ADAMTS-5, as well as more than 80-fold selectivity over MMP-3 and MMP-7. These two inhibitors showed good metabolic stability in an *in vitro* study using liver S9 cells. Despite having low to moderate Caco-2 permeability, **26.24z** showed good oral anti-TNF-α activity in the LPS model. Pharmacokinetic studies demonstrated that it was orally bioavailable in dog and rat animal models.[78]

26.2.2 Phosphonamido-Based TACE Inhibitors

A novel series of phosphonamido-based TACE inhibitors has been developed by Sawa et al.,[81] represented in Fig. 26.10.

They discovered that the (*R*)-isomer at the phosphorus atom, **26.25a**, showed potent inhibitory activity toward TACE, collagenase-1 (MMP-1), stromelysin-1 (MMP-3), and gelatinase B (MMP-9), while the (*S*)-isomer, **26.25b**, was almost inactive for those enzymes (Table 26.3).[82]

The introduction of fluorine atoms into the ester moiety of the (*S*)-isomers led to highly potent and selective inhibition against MMP-1 (**26.26b**, K_I: 6.23 nM for MMP-1, and selectivities for MMP-3 and MMP-9 are >104- and 66-fold, respectively).[83] This unexpected appearance of the inhibitory activity of the (*S*)-isomer could be explained by the switching of the binding mode in the enzyme, as shown in Fig. 26.11.

Therefore, this observation was extended to cover other types of metalloproteases, and TACE as a prototypical enzyme since its active site was found to be very similar to those of MMPs. Initially, a series of various ester derivatives of **26.25b** were tested, but none of compounds showed inhibitory activity against TACE in the nanomolar range. Based on modeling studies using X-ray structure of TACE, it was anticipated that the ester groups could not be oriented correctly for the occupancy of the S1′ pocket due to the restriction of the bicyclic tetrahydroisoquinoline structure. Therefore, it was considered that acyclic compounds such as **26.27** could allow greater flexibility of the ester groups in the compounds (Fig. 26.12).[82]

The D-leucine derivatives **26.28a** and **26.28b** were tested *in vitro* for their ability to inhibit TACE, collagenase-1 (MMP-1), stromelysin-1 (MMP-3), and gelatinase B (MMP-9) (Table 26.3). (*R*)-Isomer **26.28a** was a potent inhibitor of TACE ($K_I = 5.06$ nM) and also a moderate inhibitor of MMPs ($K_I = 196$, 69.8, and 165 nM for MMP-1, MMP-3, and MMP-9, respectively). It is worth noting that although the (*S*)-isomer of

TABLE 26.3 ***In Vitro*** **Profile of Phosphonamide Derivatives**[82]

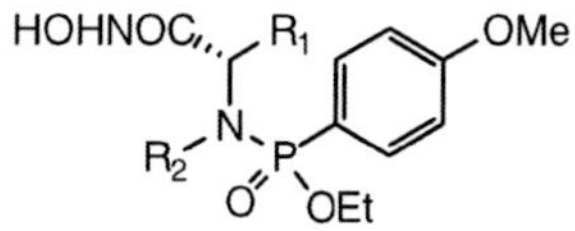

Compound	R_1	R_2	TACE K_I	MMP-1	MMP-3	MMP-9
26.25a (*R*)	Tetrahydro- isoquinoline ring	H	7.15	4.59	5.20	5.05
26.25b (*S*)	Tetrahydro- isoquinoline ring	H	(41% @1000)[a]	(26% @1000)[a]	(9% @1000)[a]	(18% @1000)[a]
26.28a (*R*)	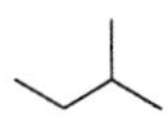	H	5.06	196	69.8	165
26.28b (*S*)	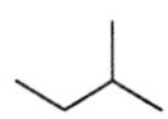	H	76.4	(10% @1000)[a]	(17% @1000)[a]	(17% @1000)[a]
26.29a (*R*)	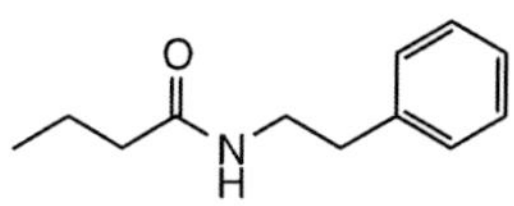	H	27.6	40.4	7.29	23.7
26.29b (*S*)	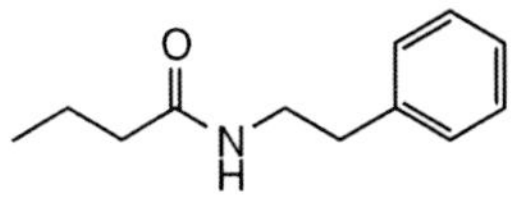	H	100	(27% @1000)[a]	(27% @1000)[a]	(27% @1000)[a]

[a]% inhibition at the concentration (nM).

the tetrahydroisoquinoline derivative **26.25b** showed no inhibition for TACE, the (*S*)-isomer of the D-leucine derivative **26.28b** exhibited inhibitory activity against TACE ($K_I = 76.4$ nM). An additional substituent capable of interacting with S1/S2 site could provide an opportunity to modify not only potency but also the selectivity. Thus, D-glutamate derivatives **26.29a** and **26.29b** were prepared and tested for their ability to inhibit these enzymes. (*R*)-Isomer **26.29a** was also potent inhibitor of TACE with K_I value of 27.6 nM; inhibition of TACE was decreased, the inhibition of MMPs was increased significantly compared to the leucine derivative **26.28a**, whereas the inhibitory activity of compound **26.29b** was slightly decreased against TACE ($K_I = 100$ nM).

Compound **26.29b** was still selective against TACE, while a significant enhancement of potency for MMPs was observed in **26.29a**. Modeling study of **26.28b** with the active site of TACE suggested that the flexible structure of this compound could adopt the switched binding mode as shown in Fig. 26.13.[82]

These studies revealed the potential of the phosphonamide derivatives as a new type of MMP inhibitors and provided an alternative concept in the design of selective inhibitors.

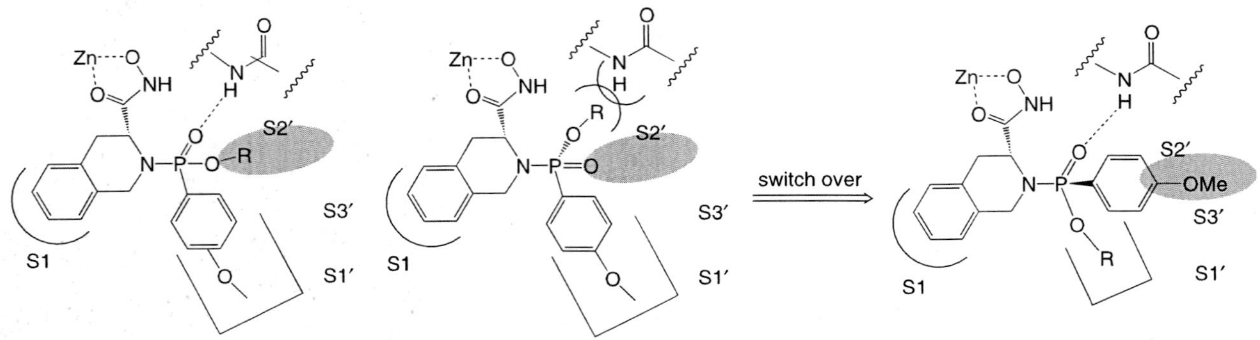

(*R*)-form (*S*)-form (inactive binding mode) (*S*)-form (active binding mode)

FIGURE 26.11 Expected binding mode of the phosphonamide inhibitors in MMPs.[82]

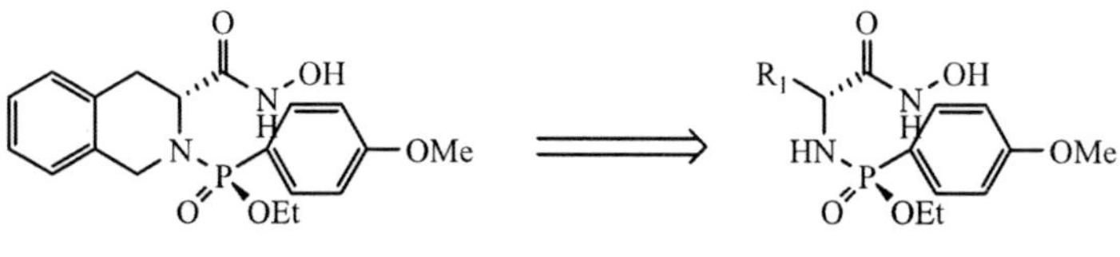

26.25b 26.27

FIGURE 26.12 Design of new phosphonamide derivatives.[82]

26.2.3 Sulfonamido- and Sulfone-Based TACE Inhibitors

In the last few years, in addition to the amide-based and phosphonamide-based TACE inhibitors, other sulfonamido-based and sulfone-based families were developed, derived from the structure of CGS27023A, as discussed previously. On this family of sulfonamides, a wide SAR study was carried out at Wyeth by Levin and colleagues obtaining TMI-2, **26.30** (Fig. 26.14), a potent TACE inhibitor ($IC_{50} = 2$ nM), slightly selective *in vitro* (more than 250-fold selective over MMP-1, MMP-7, MMP-9, MMP-14, and ADAM-10) that was able to reduce, in a cell-based assay, TNF-α production in WBA ($IC_{50} = 1$ μM). Remarkable was the high efficiency in the inhibition of the spontaneous release of TNF-α in human synovium tissue explants of rheumatoid arthritis patients ($IC_{50} = 0.8$ μM). The *in vitro* efficacy of TMI-2 in the inhibition of TNF-α production (WBA) was replayed *in vivo* in mice ($ED_{50} = 3$ mg/kg) and in a rat model of arthritis, where p.o. treatment with TMI-2 was highly effective in reducing joint arthritis scores. Also, in a semitherapeutic collagen-induced arthritis model in mice, TMI-2 was highly effective in reducing disease severity scores after oral administration (100 mg/kg twice per day). Results obtained with the potent TACE inhibitor TMI-2 were important to ensure that the TACE inhibitors are effective and

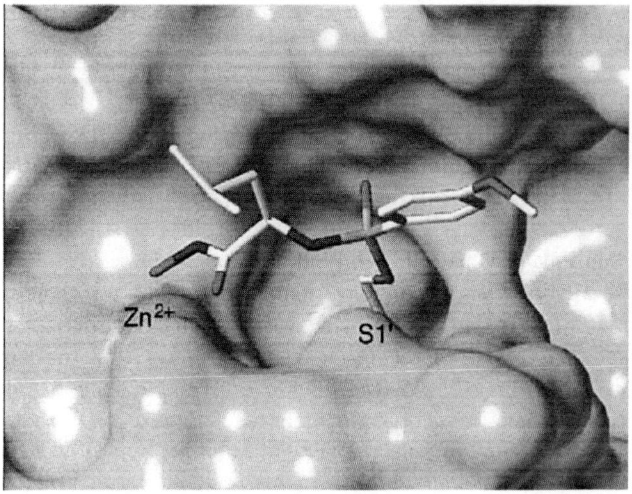

FIGURE 26.13 Docking model of **26.28b** in TACE. Oxygen, nitrogen, phosphorous and carbon are represented in dark gray, black, gray and white, respectively.[82]

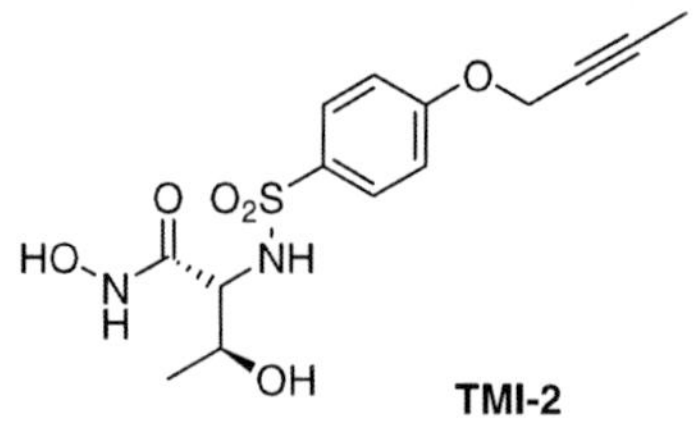

FIGURE 26.14 Structure of **TMI-2** (**26.30**).[84,85]

beneficial in the treatment of rheumatoid arthritis, as well as other TNF-mediated inflammatory autoimmune diseases.[84,85]

In parallel to TACE inhibitors derived from the structure of CGS27023A, **26.2**, cyclic sulfonamides, originating from the thiomorpholine scaffold of prinomastat (a potent MMPI, described in Chapter 19), were also obtained targeting TACE. In particular, two such analogues, thiomorpholine sulfonamides **26.31** and **26.32** (Fig. 26.15), were orally bioavailable dual TACE/MMP inhibitors with excellent potency in human whole blood assay.[86] Moreover, **26.32** entered clinical trials for the treatment of RA.[87] While these thiomorpholines showed little selectivity for TACE versus MMPs, a design of new semirigid scaffolds on the P1–P3′ positions, considering the differences in geometry between the S1–S3′ pockets of the MMPs and TACE, led to the α-sulfone piperidines. The piperidinesulfone **26.33** exhibited good *in vitro* potency against TACE and some selectivity for TACE over MMP-1, MMP-9, and MMP-13.[88] Similarly, 4,4-piperidine β-sulfone hydroxamic acid **26.34** had excellent TACE activity ($IC_{50} = 1.5$ nM) and greater than 150-fold selectivity over both MMP-2 and MMP-13.[89]

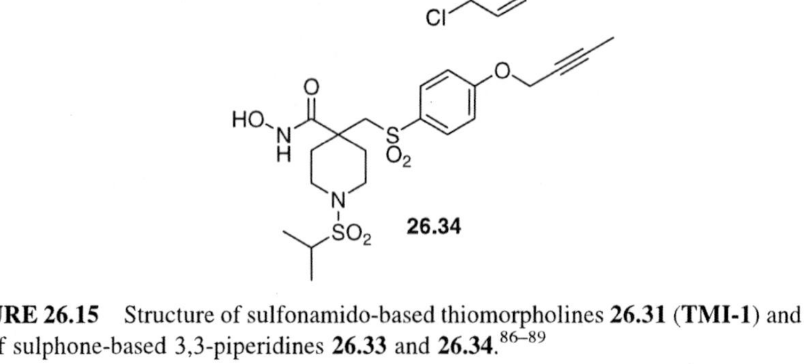

FIGURE 26.15 Structure of sulfonamido-based thiomorpholines **26.31** (**TMI-1**) and **26.32**, and of sulphone-based 3,3-piperidines **26.33** and **26.34**.[86–89]

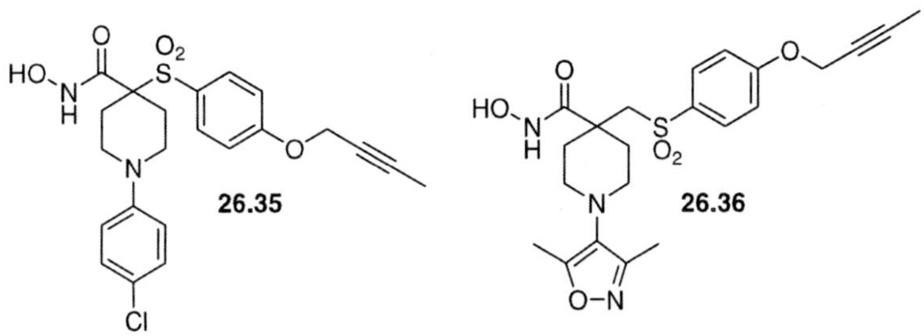

FIGURE 26.16 Structure of α-sulfone piperidine **26.35** and β-sulfone hydroxamic acid **26.36**.[89,90]

Further studies on these two subclasses of hydroxamic acids led to some highly selective and potent TACE inhibitors such as the α-sulfone piperidine **26.35**, which has shown an IC_{50} value of 50 nM on TACE, sparing MMP-1, MMP-9, and MMP-13 over 10 μM, and the β-sulfone hydroxamic acid **26.36** (Fig. 26.16), which was more potent against TACE ($IC_{50} = 2.2$ nM) and selective sparing some MMPs such as MMP-1, MMP-2, MMP-9, MMP-13, and MMP-14.[88,90] Some of the inhibitors developed in these two subfamilies resulted in potent *in vivo* inhibitors of TNF-α release in mouse model at 100 mg/kg p.o.[89]

26.2.4 Emerging New Scaffolds as Selective TACE Inhibitors: Cyclopropyl Hydroxamic Acids, an Example

In 2008, Zhu et al. discovered a new pharmachoforic model able to interact selectively with TACE sparing the MMPs. In these studies, the families of benzylcyclopropyl hydroxamic acids, such as **26.37a**, and of phenylcyclopropyl hydroxamic acids, such as **26.37b**, have been disclosed.[91]

26.37a and **26.37b** differed in the stereochemistry of cyclopropyl ring, displayed good to excellent selectivity over five MMPs and BMP-1 (PCP), but only **26.37a** was inactive toward ADAM-10 (Fig. 26.17). Surprisingly, X-ray crystal structures of complexes of **26.37a** and **26.37b** with TACE revealed an unexpected, reversed binding mode between these two inhibitors.

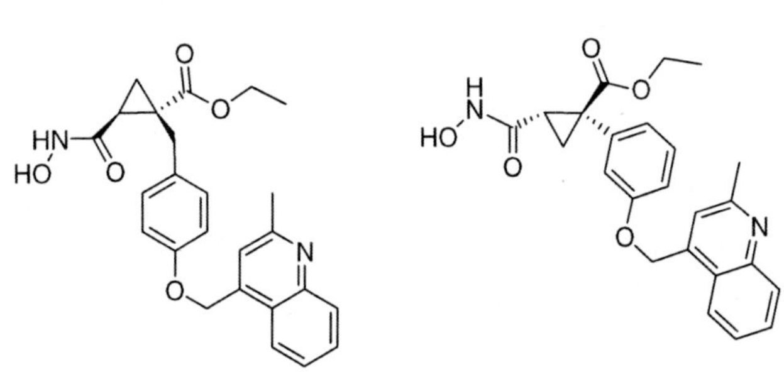

26.37a **26.37b**

FIGURE 26.17 New cyclopropyl hydroxamic acids **26.37a** and **26.37b**.[91]

Hydroxamate **26.37a** projected the quinoline group toward S3′ subsite through the S1′ pocket, and hydroxamate **26.37b** was extended in the opposite direction with the carbomethoxy and quinoline groups pointing toward S1 and S3 subsites, respectively. This different binding mode was attributed to the differences in the selectivity profile of two inhibitors for TACE and provided a solid basis to the previously described hypothesis that TACE inhibitors might have already taken advantage of both primed and unprimed site binding pockets to obtain selectivity. Moreover, the unprimed site binding mode appeared more efficient in terms of inhibitor selectivity over MMPs (IC_{50} > 13–1600 times on MMP-1, MMP-2, MMP-3, MMP-7, and MMP-14) and over procollagen C-proteinase (PCP or bone morphogenetic protein-1, IC_{50} > 167 times). This finding can be linked to the shape of S3 pocket of TACE, which is different in TACE with respect to the MMPs. In fact, it has been recently recognized as very useful to develop selective TACE inhibitors.[92] However, this binding mode was not useful to discriminate TACE selectively with respect to other ADAMs such as ADAM-10 (Fig. 26.18).

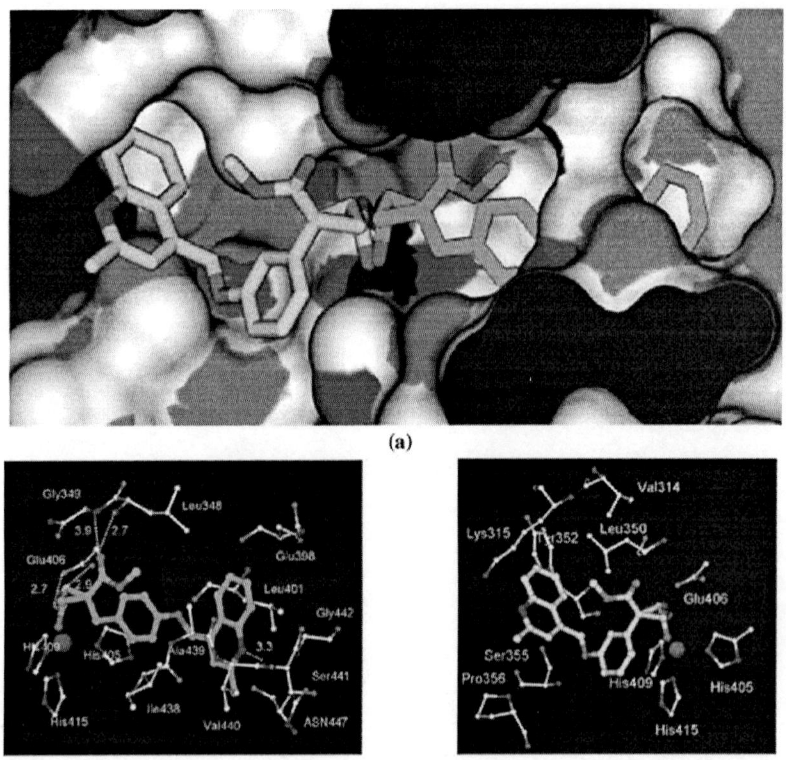

FIGURE 26.18 (a) Superimposition of X-ray crystal structures of compounds **26.37a** and **26.37b** in the TACE enzyme. (b) Key interaction of **26.37a** (dark gray) with TACE enzyme. (c) Key interactions of **26.37b** (light gray) with TACE enzyme.[91]

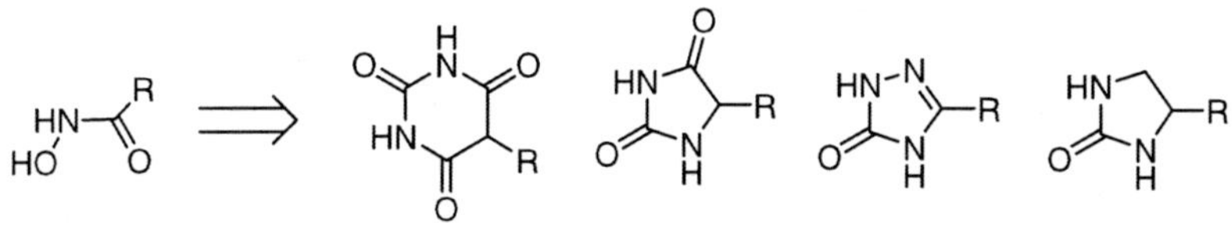

26.38 26.39 26.40 26.41

FIGURE 26.19 New ZBGs developed in the field of TACE inhibitors: pyrimidinetriones, **26.38**, hydantoins, **26.39**, triazolons, **26.40**, and imidazolons **26.41**.

26.2.5 Novel Nonhydroxamate Inhibitors of TACE

Similar to what happened in MMPI research field, in the last few years in the development field of TACE inhibitors, different alternatives to the hydroxamic acid were sought due to the toxicity and metabolic instability of this chelating group. Therefore, some highly potent and selective new TACE inhibitors devoid of this ZBG were developed.

In particular, some cyclic urea derivatives such as pyrimidinetriones, **26.38**, hydantoins, **26.39**, triazolones, **26.40**, and imidazolones, **26.41**, have been developed with good results (Fig. 26.19).

Using a pyrimidine-2,4,6-trione motif as a zinc binding group, a series of selective inhibitors of TACE was discovered by Duan and colleagues[93] for the treatment of rheumatoid arthritis and other inflammatory diseases. They developed a SAR study to find an appropriate geometry between the efficient ZBG (trione) and the proper quinolyl P1′ substituent (Table 26.4).

TABLE 26.4 ***In Vitro*** **Potency in pTACE of Pyrimidinetrione-Based Quinolyl P1′ Substituted**[93]

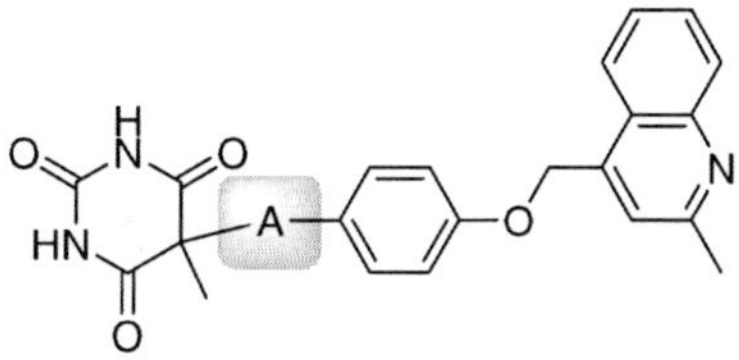

Compound	A	IC_{50} (μM)
26.42	—	1.03
26.43	–O–	1.30
26.44	$-CH_2-$	2.20
26.45	$-CH_2CH_2-$	49.0
26.46	$-CH_2CH_2CH_2-$	>100
26.47	$-CH_2CH{=}CH-$ (trans)	11.0
26.48	$-CH_2CH_2C(O)-$	0.80
26.49	$-CH_2NHC(O)-$	0.026
26.50	–NHC(O)–	>100
26.51	$-CH_2NHC(O)CH_2-$	>100
26.52	$-CH_2C(O)NH-$	12.0
26.53	–CH(4-pyridinyl)NHC(O)–	2.36

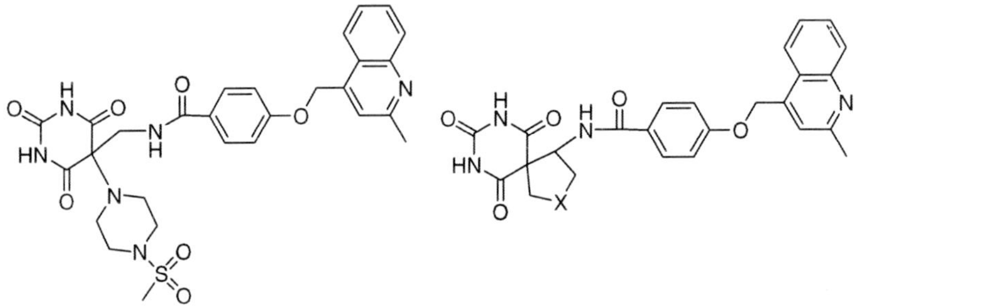

26.54 **26.55**, X = CH_2, $(CH_2)_n$, O, NH, NR, NCOR, NSO_2R

FIGURE 26.20 Selective pyrimidinetrione **26.54** (IC_{50} (nM): 2 (TACE); 2170 (MMP-2); >4500 (MMP-3); >6370 (MMP-7); >1020 (MMP-12); >>1000 (aggrecanase)) and scaffold **26.55** of the spyrocyclic analogues.[93]

The inhibitory activity, evaluated using pTACE, gave some important results. In particular, the elongation with an oxygen atom (**26.43**) or an isosteric methylene group (**26.44**) between the pyrimidinetrione and the phenyl group did not significantly alter the activity on pTACE with respect to the lead **26.42**. On the contrary, a further elongation with ethylenic or propylenic chains (**26.45** or **26.46**) resulted in a dramatic decrease of potency. The best results were obtained with an amide side chain as in **26.49** ($IC_{50} = 0.026\ \mu M$), which was able to restore or increase significantly the potency of inhibition.

Inhibition potency on TACE of amide **26.49** (nanomolar range) was very interesting because this inhibitor was selective with respect to many MMPs with almost two orders of magnitude. Moreover, a substitution with piperidine and piperazine substituents at 5-position of the pyrimidinetrione was able to improve affinity for TACE (**26.54**, $IC_{50} = 2$ nM) (Fig. 26.20). Unfortunately, this and other substitutions on this position, such as the insertion of this C(5) carbon atom in a spirocyclic function involving the carbon atom of the CH_2 of the amide side chain on P1′ (scaffold **26.55**), despite the ability of the new inhibitors to maintain a good potency against and selectivity for TACE, were unable to give a good TNF-α inhibition in WBA.[93]

More recently, some other cyclic urea derivatives have been studied as monodentate chelators of the TACE catalytic zinc atom.[94,95] In fact, a molecular modeling study revealed that this type of nonhydroxamate ZBGs participate in monodentate metal interactions as compared to the bidentate hydroxamates. Additional hydrogen bonds in the vicinity of the active site were able to increase the binding of these inhibitors and consequently restore the potency of inhibition.[94] This fact appeared to be different from that observed for the pyrimidinetrione complexes with MMPs, as seen in a crystal structure of MMP-8 complexed with this type of chelator.[96] On this basis, some selective hydantoins, such as the potent derivative **26.56**, were developed for TACE inhibition (Fig. 26.21). Compound **26.56** resulted in a good TACE inhibitor ($IC_{50} = 9$ nM) able to spare MMPs, such as MMP-2, MMP-3, MMP-7; MMP-12, and MMP-13, more than 1 μM.

Finally, thiol-based derivatives have also been used in the case of nonhydroxamate TACE inhibitors, as previously seen in the MMPI field. Starting from the structure of a simple hydroxamic sulphonamido-based TACE inhibitor **26.57** and passing through

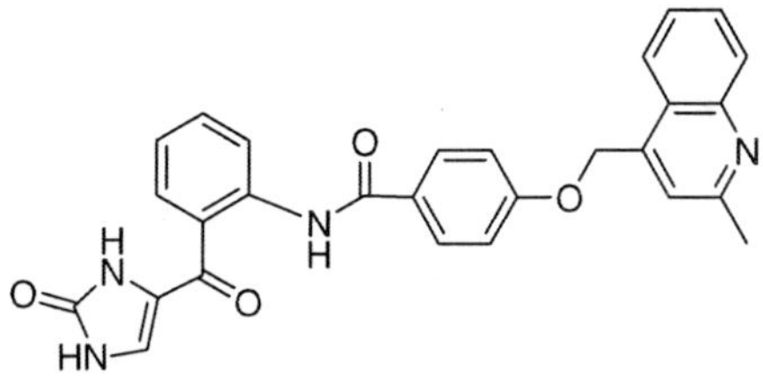

26.56

FIGURE 26.21 Potent and selective TACE inhibitor **26.56**.[95]

the structure of the cyclic sulphonamido-based thiol **26.58**, the potent sulfone-based thiol **26.59** (Fig. 26.22) was developed obtaining good potency ($IC_{50} = 10$ nM) and high selectivity over MMPs, sparing them at more that two to three orders of magnitude (MMP-2, MMP-7, MMP-8, MMP-9, and MMP-13).[97]

26.2.6 New Trends in ADAMs Inhibition

Other important studies on TACE prodomain have been conducted recently.[98]

It has been shown previously that TACE prodomain (TACE-Pro) seemed to act as an inhibitor of this enzyme because the activity of TACE was recovered only upon its removal.[99] The inhibitory potency of TACE-Pro was much lower against the complete TACE ectodomain (catalytic plus disintegrin/cysteine-rich domains). This indicates that the disintegrin/cysteine-rich domain may play a role in displacing the prodomain of TACE from the catalytic domain upon processing by furin or a furin-like enzyme. It has been found that the disintegrin/cysteine-rich domain can remove the prodomain from the procatalytic complex in *trans*. It is possible that part of the disintegrin/cysteine-rich domain of TACE may be positioned next to the substrate binding cleft. Therefore, it may sterically hinder the interaction of the prodomain with this surface of the catalytic domain. TACE-Pro includes a cysteine switch box (PKVCGY186), a feature present in most metzincins, including matrix metalloproteinases and ADAMs.

The cysteine switch has been shown to be a key element in the procatalytic interaction, even if, within the ADAM family, the role of this motif has been studied in ADAM-9, ADAM-10, and ADAM-12. For ADAM-9 and ADAM-10, a Cys to Ala

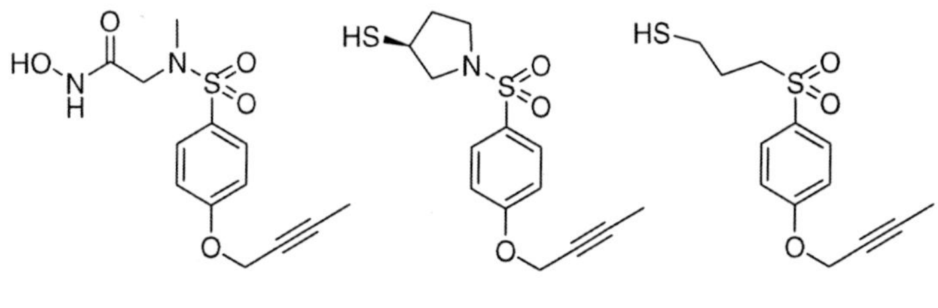

26.57 **26.58** **26.59**

FIGURE 26.22 Thiol sulfone-based selective TACE inhibitors derived from known hydroxamate sulfonamido-based **26.57**.[97]

mutation in the switch box prevented the production of functional enzyme, probably because of protein misfolding. In the case of ADAM-12, the cysteine residue does not seem to be required for secretion, but a mutation to alanine renders the prodomain incapable of inhibiting the activity of the catalytic domain.

It has been proposed that the prodomains of metzincins act as inhibitors of their catalytic domains through a mechanism that involves ligation of the cysteinyl thiol within the cysteine switch box to the zinc ion in the active site. The cysteine switch present in TACE-Pro appeared to be important for the inhibition of this enzyme because thiol-modifying reagents such as 4-aminophenylmercuric acetate and octylthioglucoside promoted prodomain release from the catalytic domain and, therefore, enzyme activation.[99]

It has been reported that an intact cysteine switch is not required for the inhibition of TACE by its prodomain. A TACE-Pro variant carrying a Cys to Ala mutation at position 184 proved to have similar inhibitory potency to that of its wild-type counterpart. Other residues adjacent to this central cysteine residue in the switch box do not seem to be required for the procatalytic domain interaction either.

Thus, the biophysical properties of the prodomain of TACE suggest a novel pathway in the inhibition of this enzyme, mainly regarding the cysteine switch mechanism for the maintenance of the zymogen state. However, other studies to know the possible importance of other regions outside the cysteine switch box of TACE, in maintaining the latency of the enzyme, are actually underway.

More recently, the recombinant mouse ADAM-10 prodomain, purified from *Escherichia coli*, has proved to be a potent competitive inhibitor of the human ADAM-10 catalytic/disintegrin domain, with a K_I value of 48 nM. Moreover, the mouse ADAM-10 prodomain is a selective inhibitor as it only weakly inhibits other ADAM family proteinases in the micromolar range and does not inhibit members of the matrix metalloproteinase family under similar conditions. Mouse prodomains of ADAM-17 and ADAM-8 do not inhibit their respective enzymes, indicating that ADAM-10 inhibition by its prodomain is unique. In cell-based assays, it has been shown that the ADAM-10 prodomain inhibits betacellulin shedding, demonstrating that it could be of potential use as a therapeutic agent to treat cancer.[100]

26.3 SELECTIVE INHIBITORS OF OTHER ADAMs

26.3.1 ADAM-10 Inhibitors

In the last few years, with the discovery and characterization of the physiopathologic roles of the new entry in the ADAM family, some other specific metzincins have been targeted.

In this field, a particular interest has been directed to ADAM-10, a sheddase able to process the epidermal growth factor receptor-2 (HER-2 or ErbB-2).[101–103] HER-2 is a key tyrosine kinase receptor activable by homo- or heterodimerization with other HER family members or by proteolytic cleavage (shedding) of the extracellular domain. These activations are able to give key intracellular signal transduction pathways in cell proliferation, differentiation, motility, adhesion, and survival.[104] Overexpression of the oncogene HER-2/neu has been associated with aggressive pathogenesis and poor prognosis in nonsmall-cell lung cancer, ovarian cancer, and breast cancer patients.[101] In addition, elevated plasma levels of HER-2 extracellular domain have been associated with increased metastatic potential and overall survival in patients with breast cancer.[102,105] Therefore, inhibition of ADAM-10, responsible for HER-2 extracellular domain shedding, may be therapeutically desirable for treating cancer patients who overexpress HER-2, such as breast cancer patients.

The first example of a study on ADAM-10 inhibitors dates back to 2003. Banner et al.[106] at the Exelixis, Inc. patented some new sulfonamido-based hydroxamates belonging to the well-known class of the MMPI containing a 1-phenylsufonylpiperazine nucleus (**26.60**) (Fig. 26.23). Compounds **26.61a–c** are examples reported in the patent that have shown interesting activities on ADAM-10 with respect to ADAM-17 and some other enzymes belonging to the family of the MMPs. Tests were reported as a fluorescent peptide technology on isolated and activated enzymes. Compounds **26.61a–c** were good inhibitors against ADAM-10 with IC_{50} values below 50 nM. **26.61a** and **26.61b** were slightly selective, nearly 20 times, for ADAM-17 and MMP-3 and more selective, about 400 times, for MMP-1. No selectivity was found for MMP-2 and MMP-13. The best inhibitor was **26.61c**, able to spare MMP-1 (>400 times), MMP-3 (~400 times), and MMP-13 (~20 times).[106]

Recently, a group at the Incyte Corporation, starting from the structure of the above-described scaffold of the piperidinecarboxamide **26.13** (IM491), a potent TACE

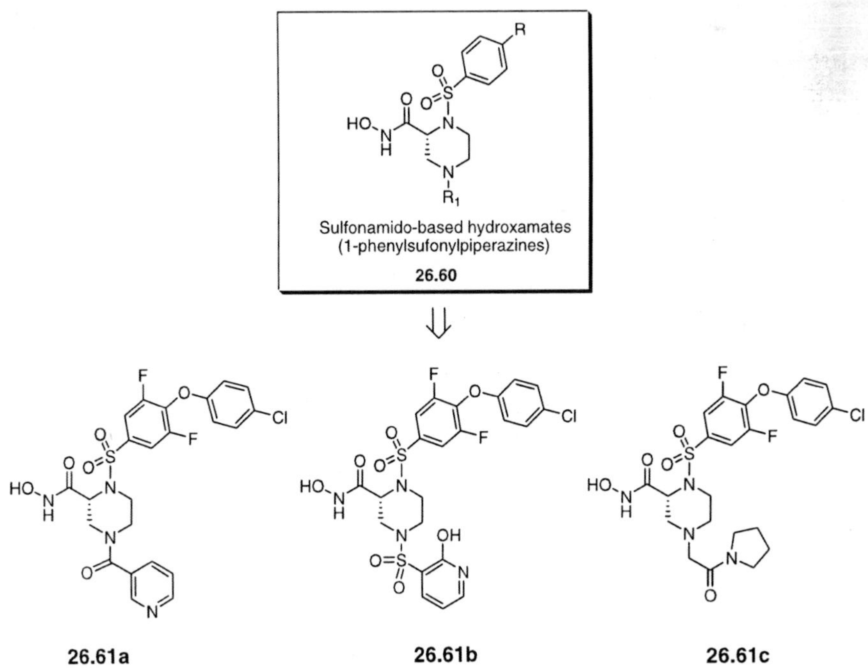

FIGURE 26.23 1-Phenylsufonylpiperazine scaffold of MMPI **26.60** as template for ADAM-10 inhibitors.[106]

(ADAM-17) inhibitor belonging to the class of the semirigid succinamides, has developed a lead optimization targeting ADAM-10. The group's attention was directed first to the succinate scaffold and then to the P1′ substituent. Two points were particularly investigated: the substitution and stiffening of the piperidine scaffold to obtain an appropriate geometry between the hydroxamic ZBG and the P1′ substituent and the peculiarity of the P1′ bicyclic substituent. The insertion of a spirocyclopropyl group into the C(4) of the piperidine scaffold seemed to be good to better define the spatial arrangement of P1′ side chain and proper selection of P1′ bicyclic system to achieve a noncoplanar orientation between the two rings that proved to be the major determinant of inhibitor selectivity. Furthermore, the selectivity profile depended on the type of substitution of piperidine nitrogen (P2′ position), which seemed to enhance P1′–S1′ specific interactions by affecting favorably the overall conformation of the inhibitor.[107] The structure of **26.62** is representative of these new inhibitors. This compound was highly specific for ADAM-10 (IC_{50} = 97 nM) over ADAM-17 (IC_{50} = 2045 nM) and ADAM-33 (IC_{50} = 4125 nM), five secreted MMPs (such as MMP-1, MMP-2 MMP-3, MMP-7, and MMP-9, with their IC50 values ranging 5000 nM), and two membrane-type MMPs (such as MMP-14 and MMP-15 with IC_{50} values >5000 nM). In this compound, only one of the MMPs tested resulted less selective, MMP-12 with an IC_{50} = 564 nM (Fig. 26.24).[108,109] Another spirocyclopropylpiperidine analogue, possessing a little carbamoyl moiety on the piperidine nitrogen in P2′ position, compound **26.63** (INCB3619) was an ADAM-10/ADAM-17-specific inhibitor (IC_{50} values of 22 and 14 nM, respectively) possessing good selectivity over other ADAMs (such as ADAM-8, ADAM-9, and ADAM-33 with IC_{50} values ranging between two and three orders of magnitude) and over other secreted MMPs such as MMP-1, MMP-3, and MMP-7.[107,109,110] Another analogue, compound INCB7839, well tolerated in refractory cancer patients, of this class of dual ADAM-10/TACE inhibitors possessing improved selectivity over MMP-2, MMP-9, and MMP-14 is in clinical trials, although its structure has not been published.[111–113]

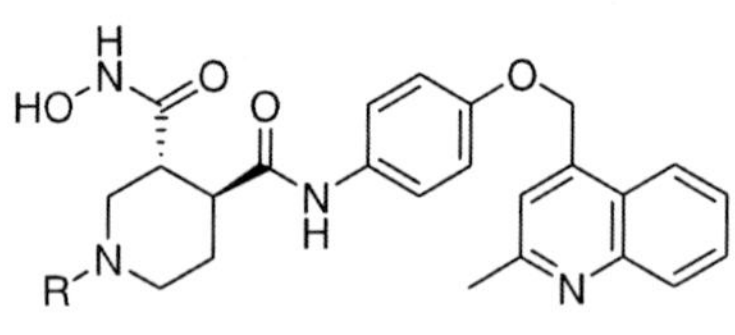

26.13: IM491

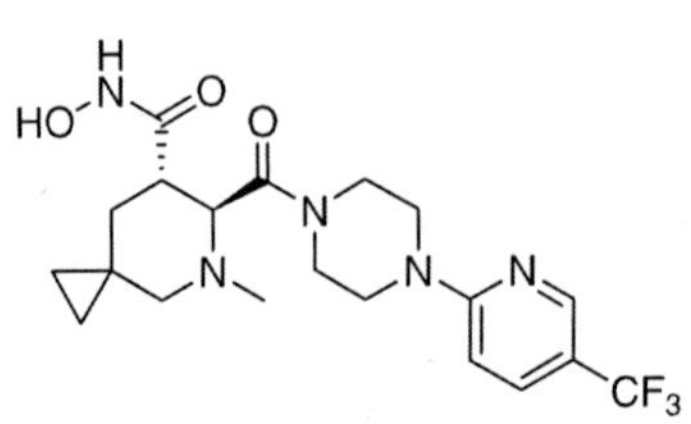

26.62

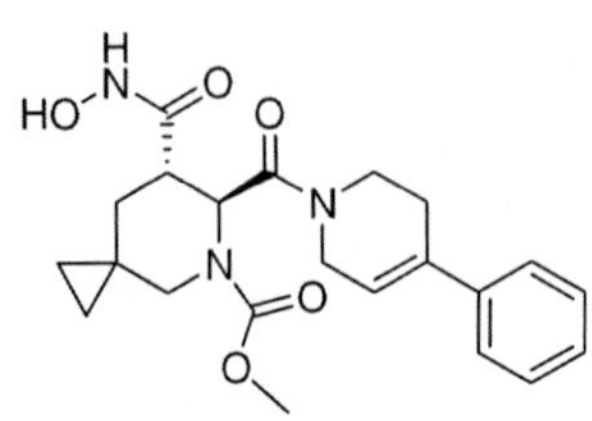

26.63

FIGURE 26.24 Incyte Corporation ADAM-10 inhibitors.

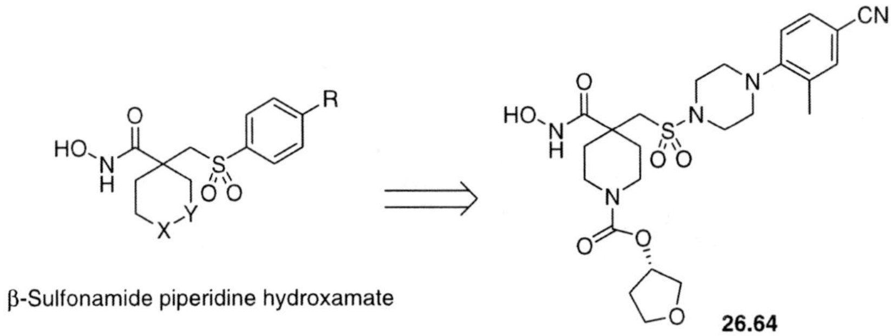

FIGURE 26.25 Development of ADAM-10 inhibitor 26.64 from beta-sulfonamide piperidine hydroxamate scaffold of MMPI and ADAM-17 inhibitors.

In another project, Burns et al.,[114] working on the β-sulfonamide piperidine hydroxamate scaffold of MMPI and ADAM-17 inhibitors, described previously, were able to improve selectivity for ADAM-10 over secreted MMPs such as MMP-1, MMP-2, MMP-3, and MMP-9. The β-sulfonamide piperidine hydroxamate **26.64** (Fig. 26.25) was a potent ADAM-10 inhibitor ($IC_{50} = 26$ nM) based on a different core scaffold (with respect to similar inhibitors of MMPs) but retained similar structural characteristics in P1′ position to INCB3619 (**26.63**) that seem to be necessary for ADAM-10 selectivity.

Finally, Zhou et al.[115] have patented another family of ADAM-10 inhibitors of (hetero)cyclic *N*-hydroxy-carboxamide structure (derived from a parallel study on analogues of piperidinecarboxamide **26.13**, IM491), useful as inhibitors of matrix metalloproteinases, sheddases, and ADAMs. The described compounds were proposed to be useful in treating diseases such as rheumatoid arthritis, psoriasis, neoplastic diseases, allergies, and all those diseases wherein inhibition of MMPs is desirable. The hydroxamate derivative **26.65** of piperidinecarboxamide structure can be described as representative of this family of inhibitors. These compounds were semirigid succinamides that showed a classical P1′ substituent for ADAM-10 S1′ pocket and a different side chain in their P1 position. This side chain was different, with respect to the spyrocyclopropyl one of type **26.62** compounds, and was a (*R*)-pyrrolidin-3-ol. Finally, these compounds were devoid of the *N*-alkyl or of the *N*-acyl substituent on their P2′ position preferentially. The new compounds showed IC_{50} in the range of about 10 nM to about 10 μM for Her2 sheddase inhibition, TACE inhibition, and MMP2, MMP12, and MMP3 inhibition (Fig. 26.26).

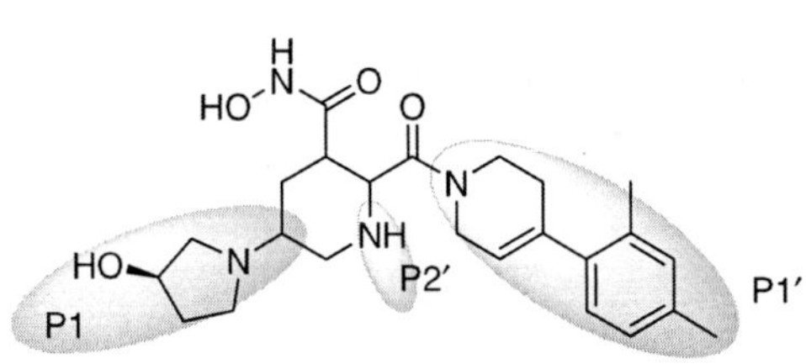

26.65

FIGURE 26.26 (Hetero)cyclic *N*-hydroxy-carboxamide inhibitors of ADAM-10.

26.3.2 ADAM-12 Inhibitors

Of the other ADAMs, ADAM-12 is the only one for which data of designing inhibitors have been published.

ADAM-12 plays a key role in the development of cardiac hypertrophy by shedding heparin binding epidermal growth factor.[116] Therefore, inhibition of ADAM-12 could be a potent therapeutic strategy for cardiac hypertrophy and congestive heart failure.[116–118] On this basis, Oh et al.[120] have developed a series of potent and selective inhibitors of ADAM-12 that were discovered using computational screening of a virtual library. The initial structure-based virtual screening selected 64 compounds from a 3D database of 67,062 molecules, containing a ZBG moiety such as carboxyl, hydroxyl, sulfur, or hydroxamate. Virtual screening of this library for the ADAM-12 pharmacophore resulted in 1217 compounds as initial hits that were further refined by consensus scoring functions.[119] Applying a criterion of top 30% of each score, 64 compounds were finally selected for biological testing. Figure 26.27 shows one of the selected compounds, **26.66**, bound to the pharmacophore of ADAM-12.

Compound **26.66** and its analogues, **26.67**, **26.68**, and **26.69** (Fig. 26.28), tested in a cell-based ADAM-12 activity assay, showed to be selective and potent to this enzyme. The inhibitor **26.66** was the most potent on ADAM-12 (IC_{50} = 17 nM) followed by its analogues **26.69** (IC_{50} = 25 nM), **26.67** (IC_{50} = 41 nM) and **26.68** (IC_{50} = 43 nM). Moreover, some of these compounds, **26.66**, **26.67** and **26.68** spared ADAM-10 with

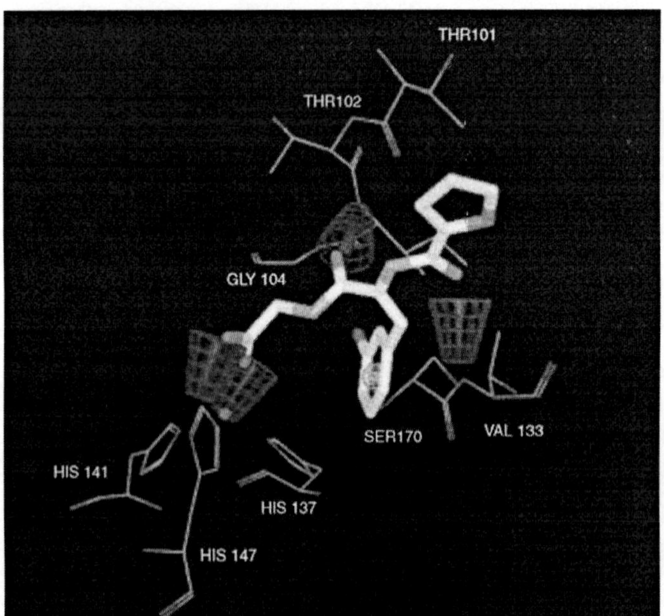

FIGURE 26.27 Interaction between the pharmacophore of the Zn^{2+} binding site of ADAM-12 and a selected compound **26.66**. Dark gray conical shape and light gray circle represent hydrogen bond donor and hydrophobic core, respectively.

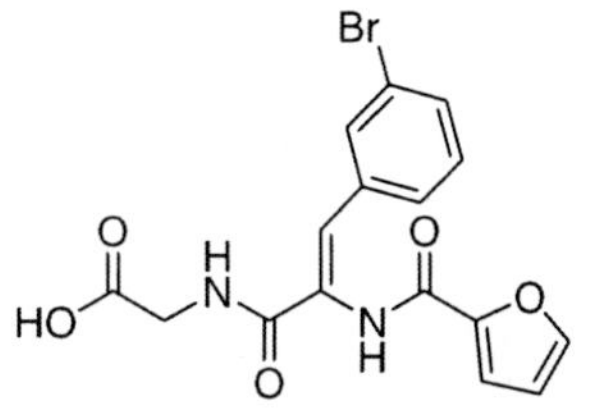

26.66

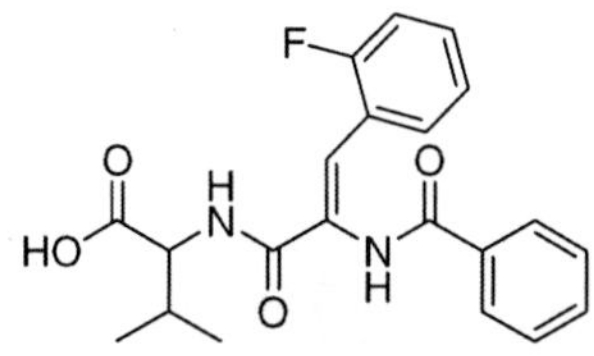

26.67

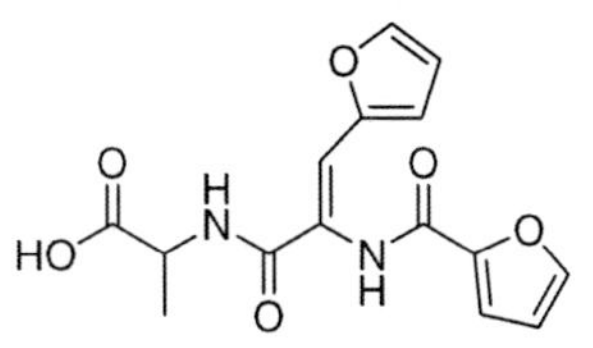

26.68

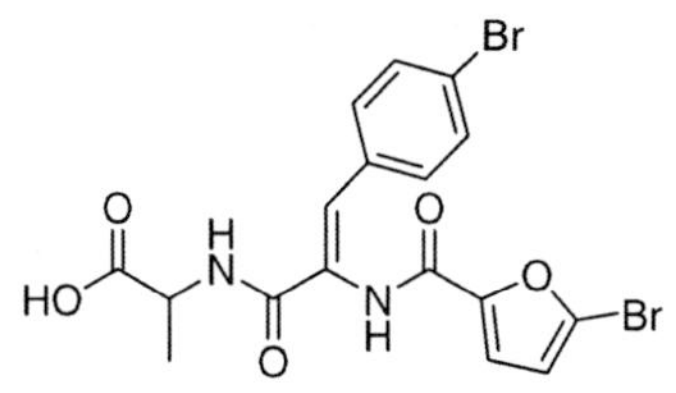

26.69

FIGURE 26.28 Selective inhibitors of ADAM-12.

about three orders of magnitude, being active on this other enzyme only in the high micromolar range ($IC_{50} \sim 10\,\mu M$).[120]

26.4 ADAM-TS INHIBITORS

Recently, key roles of the ADAM-TS (a disintegrin and metalloproteinase with thrombospondin motifs) proteinases have been discovered in some connective tissue pathologies such as Ehlers–Danlos syndrome type VII C, Weill-Marchesani syndrome, encephalomyelitis, and arthritis. On this basis, these metzincins are considered as potential therapeutic targets for the treatment of such conditions. As described above, the synthesis and activity of ADAM-TS proteinases are regulated at multiple levels: transcription, RNA splicing, translation, proteolytic processing, cofactor stimulation, and inhibition, each of which represents a possible point of therapeutic intervention. Recent research suggests that, in addition to the direct inhibition of ADAM-TS proteinases with low molecular weight nonpeptidic inhibitors, targeting the transcription and protein processing of these enzymes could be effective therapeutic approaches.[121]

At present, two ADAM-TS family components are under intensive evaluation for drug target and therapy design: ADAM-TS4 (or aggrecanase-1) and ADAM-TS5 (or aggrecanase-2). Both are able to cleave aggrecan at the physiologically relevant Glu373-Ala374 peptide bond (four other ADAM-TS such as ADAM-TS1, ADAM-TS8, ADAM-TS9, and ADAM-TS15 are also able to cleave the same peptide bond).[122–124]

Aggrecan is an important structural component of the proteoglycan that provides the elasticity and compressive resistance to the articular cartilage. Also, many matrix metalloproteinases (MMP-1, MMP-2, MMP-3, MMP-7, MMP-8, MMP-9, and

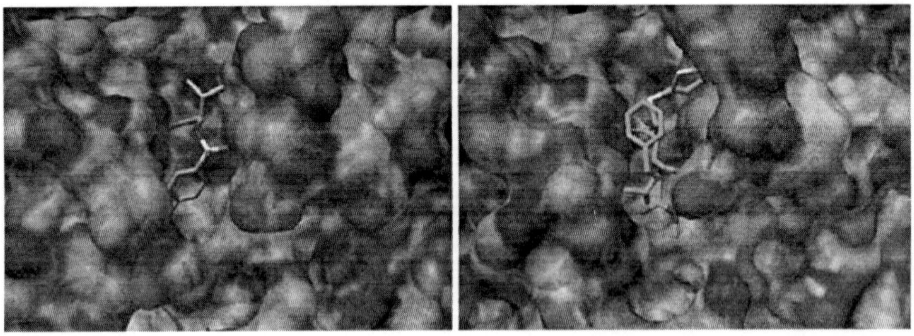

FIGURE 26.29 X-Ray structures of ADAM-TS4 complexed with *N*-({4′-[(4-isobutyrylphenoxy)methyl]biphenyl- 4-yl}sulfonyl)-D-valine, light gray structure, left image and ADAM-TS5 complexed with Batimastat (**26.70**), opaque gray structure, right image.[138]

MMP-13) have been shown to cleave aggrecan *in vitro* but their proteolytic activity is at the Asn341–Phe342 site.[125–128] G-1 fragments resulting from this cleavage site have been identified within articular cartilage bound to hyaluronic acid.[126] Recently, aggrecanase-1 was isolated, cloned, and expressed by Arner and coworkers.[129] This enzyme effectively cleaves the core protein within the interglobular domain (IGD) at the Glu373–Ala374 bond. Aggrecan fragments resulting from this cleavage have been identified in the synovial fluid of patients with osteoarthritis, inflammatory joint disease, and joint injury, suggesting that aggrecanase plays a key role in the catabolism of aggrecan in human arthritic disease.[130,131]

More recent studies with knockout mice have indicated that ADAM-TS5 is responsible for cartilage destruction by aggrecan degradation in mice.[132,133] In human chondrocytes, however, ADAM-TS4 is the aggrecanase induced by treatment with cytokines such as interleukin-1, whereas the ADAM-TS5 expression is constitutive.[134–136]

On this basis, considering the peculiar physiopathological roles of these two ADAM-TS on the cartilage destruction in osteoarthritis (OA), some synthetic inhibitors have been designed and tested targeting them.

X-Ray structures of ADAM-TS4 and ADAM-TS5 have been disclosed very recently and this fact could be useful for the design of new inhibitors as seen above for the new classes of MMPs and ADAM-17 inhibitors.[137–139]Figure 26.29 reports the structures of complexes of these two aggrecanases with two inhibitors of MMPs: ADAM-TS4 is complexed with *N*-({4′-[(4-isobutyrylphenoxy)methyl]biphenyl-4-yl}sulfonyl)-D-valine and ADAM-TS5 is complexed with Batimastat (**26.70**).[138]

26.4.1 ADAM-TS4 (Aggrecanase-1) Inhibitors

The first studies with inhibitors on this enzyme were developed with hydroxamic acids of succinylamides belonging to the old class of a wide range of MMPIs, discussed previously in the chapters on metzincins. Yao et al. at the DuPont Pharmaceuticals developed a pharmacophore model of the P1′ site, specific for aggrecanase, that was defined using the specificity studies of the matrix metalloproteinases and the similar biological activity of aggrecanase and MMP-8. Incorporation of the side chain of a tyrosine residue into compound **26.71** as the P1′ group provided compound **26.72** with a modest selectivity for aggrecanase over MMP-1, MMP-2, and MMP-9 (Fig. 26.30). A *cis*-(1*S*)-(2*R*)-amino-2-indanol scaffold was incorporated as a tyrosine mimic (P2′) to conformationally constrain **26.72**, and further optimization resulted in compound **26.73**, a potent, selective, and orally bioavailable inhibitor of aggrecanase.[140]

Further SAR studies on the P1 position of these inhibitors, confirmed the importance of P1 substitution in the selectivity profile. In fact, inhibition potency as well as selectivity over MMP-9 was significantly enhanced by rational P1 optimization leading to analogues such as **26.74,** which also displayed a more than 100 times increase in selectivity on MMP-1 and MMP-2 (Fig. 26.30).[140]

In a parallel way to the studies on MMPI and TACE inhibitors, some sulfones and sulfonamides were developed at Pfizer by Noe et al., targeting high-selectivity MMP-13 and ADAM-TS4, to obtain effective dual-target inhibitors for the OA disease. In fact, together with the two aggrecanases, high levels of MMP-13 (collagenase-3) were

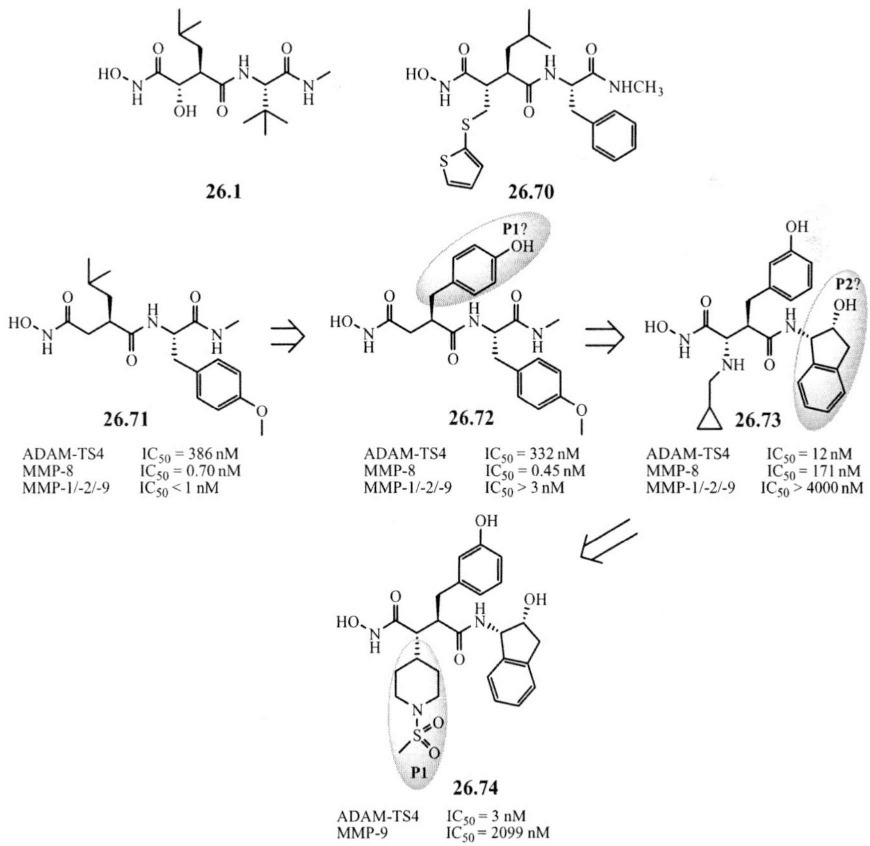

FIGURE 26.30 ADAM-TS4 inhibitors.

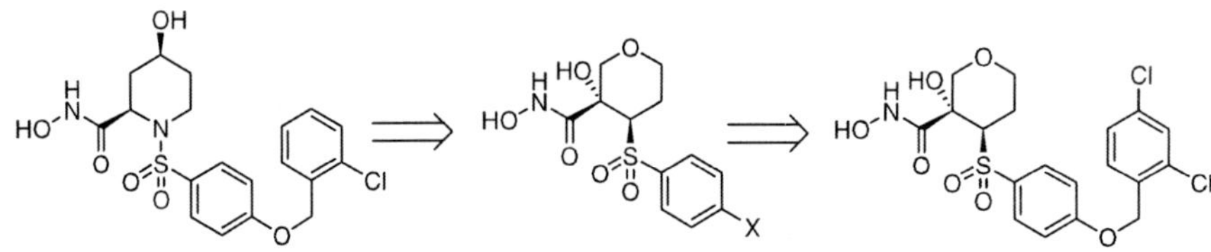

26.75 **26.76** **26.76a**

TACE IC_{50} = 36 nM
MMP-13 IC_{50} = 12 nM
MMP-1 IC_{50} = 29000 nM

ADAM-TS4 IC_{50} = 8.1 nM
MMP-13 IC_{50} = 0.95 nM
MMP-1 IC_{50} = 920 nM
MMP-2 IC_{50} = 12 nM
MMP-3 IC_{50} = 55 nM
MMP-9 IC_{50} = 5.3 nM
MMP-12 IC_{50} = 0.95 nM
MMP-14 IC_{50} = 150 nM

FIGURE 26.31 ADAM-TS4/MMP-13 inhibitors developed by Pfizer.

found in the osteoarthritic joints.[141] Starting from the potent and selective dual-target (MMP-13/TACE) inhibitor **26.75**, a pipecolic sulfonamido-based inhibitor devoid of inhibition of aggrecanase and suffering from metabolic lability of the hydroxamic acid moiety, some new morpholino sulfones of type **26.76** were developed. Compounds **26.76** were designed to obtain new inhibitors more polar and more stable at the hydroxamic acid moiety for the insertion of an hydroxy substituent on the C(α) into the hydroxamate ZBG on the tetrahydropyrane cycle (Fig. 26.31).

One of them, the inhibitor **26.76a**, was potent against aggreganase and MMP-13, sparing a little MMP-1 (see Fig. 26.31). The pharmacokinetic parameters of compound **26.76a** in rat show low clearance (Clp = 9.0 mL/min/kg) and a long $t_{1/2}$ (5 h). Unfortunately, oral bioavailability was poor. Finally, these compounds, even if similar to other TACE inhibitors, were found totally inactive on TNF-α release in WBA, showing their inability to inhibit TACE at 100 μM.[142]

In another study on dual-target aggrecanase-1/MMP-13 inhibitors, the same authors developed some hydroxypipecolic sulfonamido-based hydroxamates to obtain potency against the two targeted enzymes, selectivity for the other MMPs and ADAMs, and better biopharmacological profiles. Therefore, starting from previous studies on MMPI and TACE inhibitors, where pipecolic acid-based inhibitors such as **26.77** were discovered, new series of *ortho*-methyl or halo-substituted compounds, such as **26.78**, showed to be potent ($IC_{50} < 500$ nM) against both MMP-13 and aggrecanase. Unfortunately, these compounds exhibited poor pharmacokinetic properties. On this basis, they incorporated additional substituents at the piperidine 3-position with the intent of improving the metabolic stability of the hydroxamate group by increasing steric hindrance, thereby limiting access to metabolic enzymes. Compound **26.79**, a 3,3-dimethyl-5-hydroxypipecolic hydroxamate-based inhibitor belonging to this class, had good dual inhibition potency against ADAM-TS4 and MMP-13 (IC_{50} = 2.1 and 3.0 nM, respectively), a moderate selectivity for MMP-1, and a good activity for TACE, sparing the release of TNF-α in the WBA. An improvement of pharmacokinetic properties was found in these new types of inhibitors compared to the previous type **26.76** inhibitors.[143]

Considering the risk of metabolic lability of the hydroxamic acid (ZBG), potentially able to undergo hydrolysis, reduction, and glucuronidation *in vivo*, the introduction of a polar functionality at the 3-position of the piperidine ring as in type **26.80** compounds was analyzed.[144] In an SAR analysis on these compounds were observed the influences of the substitution and stereochemistry on the C(3) of the pipecolic acid cycle and the effects of the substitution on the phenyl ring of the benzyloxy terminal group on P1′. Regarding the C(3) substitution, it was seen that both aggrecanase and MMP-1 potency was slightly influenced by the pendant alkyl group and stereochemistry of the 3-position. Better results on aggrecanase inhibition were obtained using a 3 (*R*)-methyl substituent. Regarding the above-mentioned phenyl ring on P1′ position, the use of a p-fluorine atom was found important for the potency against aggrecanase, while the introduction of an *ortho*-substituent, especially a chlorine, was able to increase a little selectivity to MMP-1 (see Fig. 26.32). Compound **26.80a** was found potent against ADAM-TS4 and MMP-13 and poorly able to activate the release of TNF-α in the WBA ($IC_{50} = 10{,}400$). In spite of its poor selectivity for the other MMPs, this compound showed a potent inhibition of IL-1-induced aggrecan ($IC_{50} = 10.3$ nM) and collagen ($IC_{50} = 100$ nM) degradation in bovine nasal cartilage explants. Moreover in human osteoarthritic cartilage, **26.80a** was able to inhibit IL-1-induced aggrecan degradation with an IC_{50} of 6.0 nM. The pharmacokinetic properties and *in vivo* activity observed with this compound suggested that it could provide a useful starting point for the discovery of agents for the treatment of osteoarthritis.[144]

In another effort in this field, Xiang et al.[145] at Wyeth developed a SAR in the class of the biphenyl sulfonamido-based carboxylic acid inhibitors of MMPs to differentiate inhibition properties of ADAM-TS4 from those of some MMPs, maintaining the dual-target activity with MMP-13. Starting from observations on the two enantiomers of CGS 27023A, where the (*S*)-isomer was found totally inactive on ADAM-TS4, some acids belonging to the biphenylsulfonamide family of MMPI were developed by studying the effects of different groups on their P1, P1′, and P2′ positions. Relating the chiral carbon atom on the C(α) to the carboxylic acid (ZBG), the necessity to have a (*R*)-configuration also in the biphenyl series was confirmed, to maintain the activity on ADAM-TS4. Moreover, only a hydrogen or a small alkyl group, such as methyl, could be sustained on the P2′ position. The introduction of more hindered groups led to a decrease in potency. The authors investigated more in deep the substitution on the P1′ biphenyl group, observing the necessity to have a 4′ substituent on the terminal phenyl ring because other substitutions were not accepted. The best results were obtained when a 2-oxymethyl-3-methylbenzofuranyl substituent was introduced. Compound **26.81** showed to be a potent, dual-target inhibitor of ADAM-TS4/MMP-13 with IC_{50} values of 0.7 μM and 4.4 nM, respectively. Moreover, it was highly selective over MMP-1 and MMP-14 ($IC_{50} > 100{,}000$ nM for MMP-1 and $IC_{50} = 3000$ nM for MMP-14). This inhibitor demonstrated to possess an excellent cartilage penetration, a good inhibition of proteoglycan degradation in a cell-based assay (IL-1 stimulated bovine cartilage explant assay, 89% inhibition of proteoglycan at 10 μg/mL), and a good oral bioavailability in animal models (Fig. 26.33).

In a recent patent, Inaba T. et al.[146] at Japan Tobacco, Inc. developed a new class of dual-target ADAM-TS4/MMP-13 inhibitors belonging to the above-mentioned

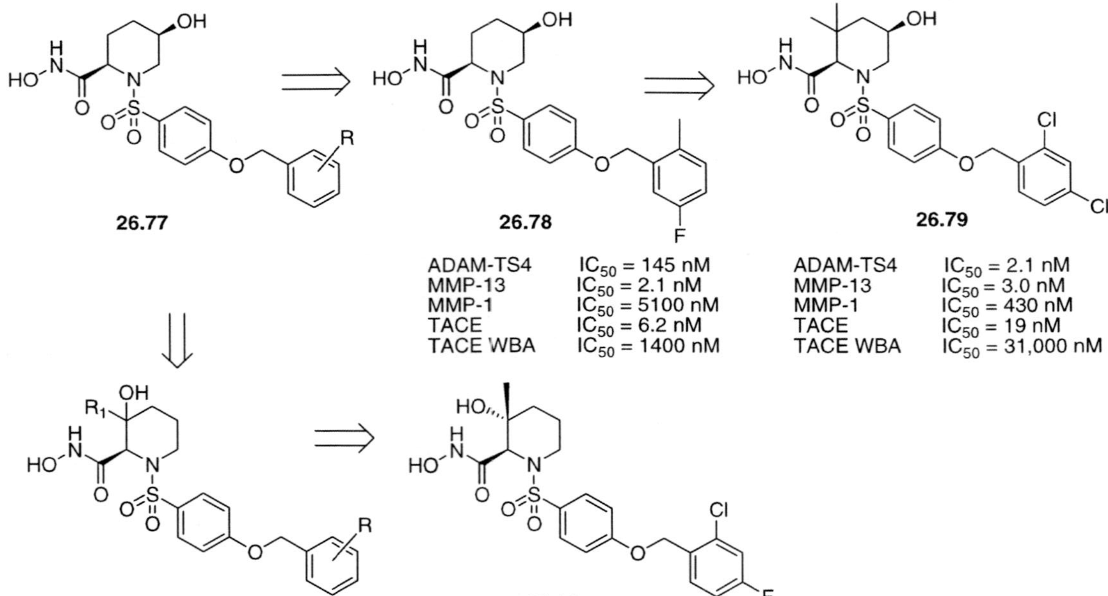

26.77

26.78

ADAM-TS4	IC_{50} = 145 nM
MMP-13	IC_{50} = 2.1 nM
MMP-1	IC_{50} = 5100 nM
TACE	IC_{50} = 6.2 nM
TACE WBA	IC_{50} = 1400 nM

26.79

ADAM-TS4	IC_{50} = 2.1 nM
MMP-13	IC_{50} = 3.0 nM
MMP-1	IC_{50} = 430 nM
TACE	IC_{50} = 19 nM
TACE WBA	IC_{50} = 31,000 nM

26.80

26.80a

ADAM-TS4	IC_{50} = 1.5 nM
MMP-13	IC_{50} = 0.9 nM
MMP-1	IC_{50} = 310 nM
MMP-2	IC_{50} = 4 nM
MMP-3	IC_{50} = 5 nM
MMP-8	IC_{50} = 1 nM
MMP-9	IC_{50} = 11 nM
MMP-12	IC_{50} = 0.9 nM
MMP-14	IC_{50} = 83 nM
TACE WBA	IC_{50} = 10400 nM

FIGURE 26.32 ADAM-TS4/MMP-13 inhibitors developed by Pfizer with improved pharmacokinetic properties.

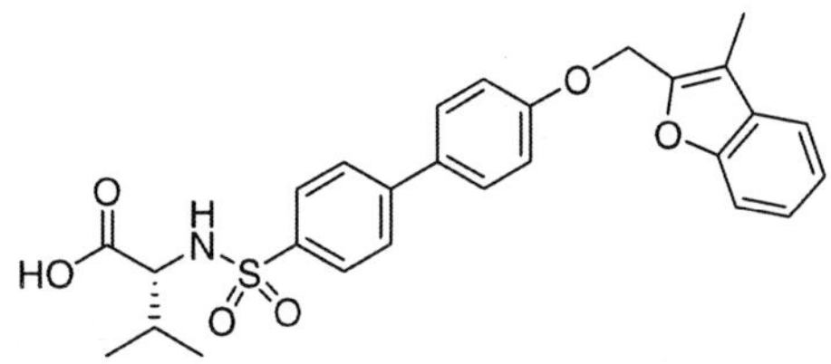

26.81

FIGURE 26.33 ADAM-TS4/MMP-13 inhibitor developed by Wyeth.

biphenylsulfonamido-based hydroxamates, inserting a spyro-cyclopropyl aryl-substituted chain into their P1 position to investigate the binding regions in S1 on the two enzymes (see type **26.82** compounds, Fig. 26.34). Moreover, a large library of P1′ substituted byphenyls or etherocyclic extended analogues was studied. Compound **26.82a**, a molecule described in the patent taken as example, was equipotent against aggrecanase-1 and MMP-13 with IC_{50} values in the nanomolar range, sparing MMP-1 ($IC_{50} > 10000$ nM). No *in vivo* results were reported.[146]

In another recent patent, Gallagher et al., at Atlantos, Inc.,[147] developed a new class of heterobicylic metalloproteinase inhibitors of type **26.84a–c**, where a central heterobicyclic system was linked at two symmetrical or unsymmetrical lateral chains. These compounds, in some cases, were similar in structure to the known class of pyridin-4,6-carboxamide-based selective inhibitors of MMP-13 (nonzinc binding ligands for S1′-S1″, developed from Engel et al.[148] at the Sanofi-Aventis) such as compound **26.83**, a potent inhibitor of MMP-13 ($IC_{50} = 8$ nM) that was inactive on the other MMPs ($IC_{50} > 10{,}000$ nM). For this structural similarity, even if not characterized as regard to the mechanism of inhibition in the patent, compounds **26.84a–c** could be pointed as nonzinc binding inhibitors, probably interacting with the S1′-S1″ sites of MMP-13 and ADAM-TS4. As an example of the activity of these inhibitors, compound **26.85** showed IC_{50} values ranging between 300–1000 nM on these two target enzymes (Fig. 26.35).[147]

Finally, Takizawa et al.[149] demonstrated the direct role of calcium pentosan polysulfate **26.86** (a calcium salt of a natural product, isolated from *Fagus silvatica*,

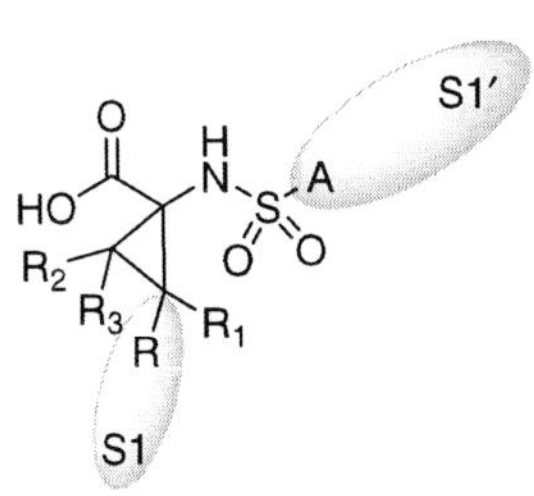

26.82

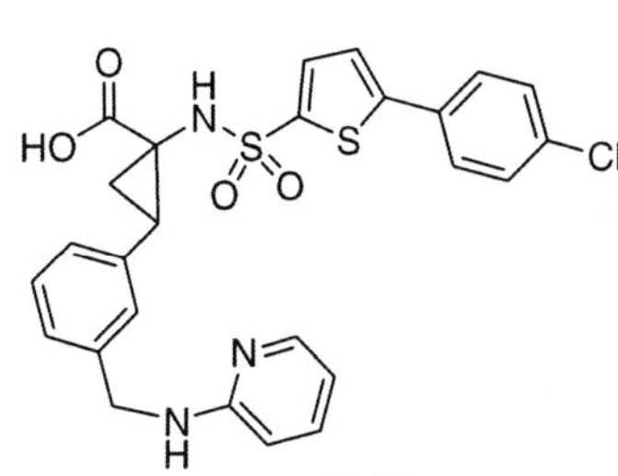

26.82a

FIGURE 26.34 Biphenylsulfonamido-based hydroxamates as ADAM-TS4/MMP-13 inhibitors.

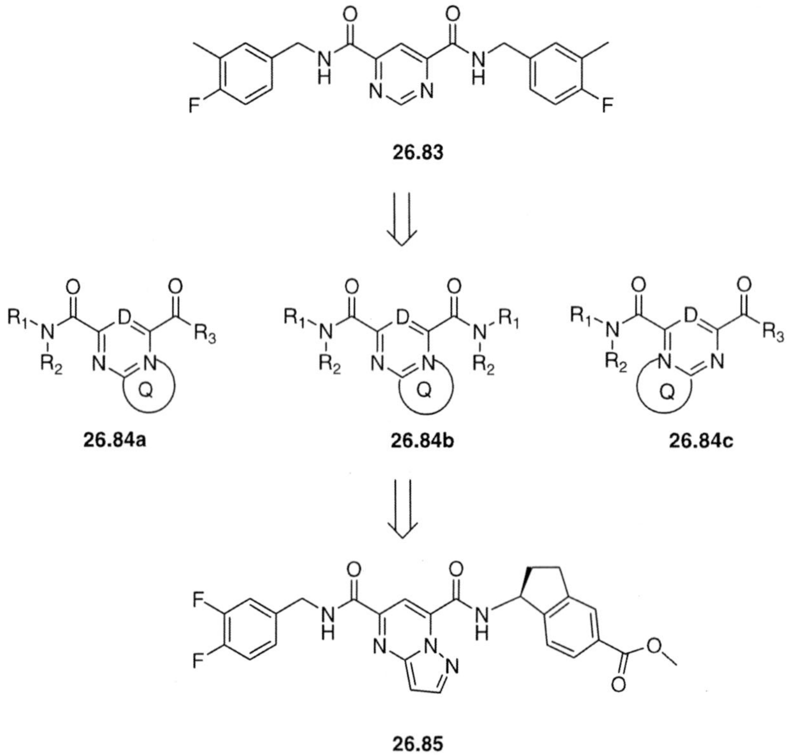

FIGURE 26.35 A new class of heterobicylic ADAM-TS4/MMP-13 inhibitors structurally related to non-zinc binding MMP-13 inhibitor 26.83.

that shows a linear xylan backbone (pentosan) containing on average one 4-*O*-methyl-glucuronate side chain linked to the 2-position on every 10th xylose ring) in the inhibition of enzymatic activity of ADAM-TS4 in osteoarthritic chondrocytes (Fig. 26.36). This work could support rationally the protective effects of this salt and other polysulfated polysaccharide salts on the osteoporosis disease, as observed by Cullis-Hill at Arthropharm Pty Ltd.[150] These observations on the ability of this agent to inhibit ADAM-TS4 and increase TIMP-3 production lend further support to the assignment of this agent to the class of disease modifying drugs for OA.

26.4.2 ADAM-TS5 (Aggrecanase-2) Inhibitors

As described in previous section, a recent work has demonstrated significantly reduced OA severity for ADAM-TS5 knockout mice in a surgically induced instability model.[132] ADAM-TS5 has also been shown to be the major ADAMTS in a mouse model of inflammatory arthritis.[132,133] Thus, the inhibition of ADAM-TS5 may therefore protect cartilage from damage and provide the first potential therapy to halt and/or reverse the progression of OA. A large library of compounds, possessing various ZBGs, was tested in an ADAM-TS5 HTS (high-throughput screening) by

Calcium pentosan polysulfate

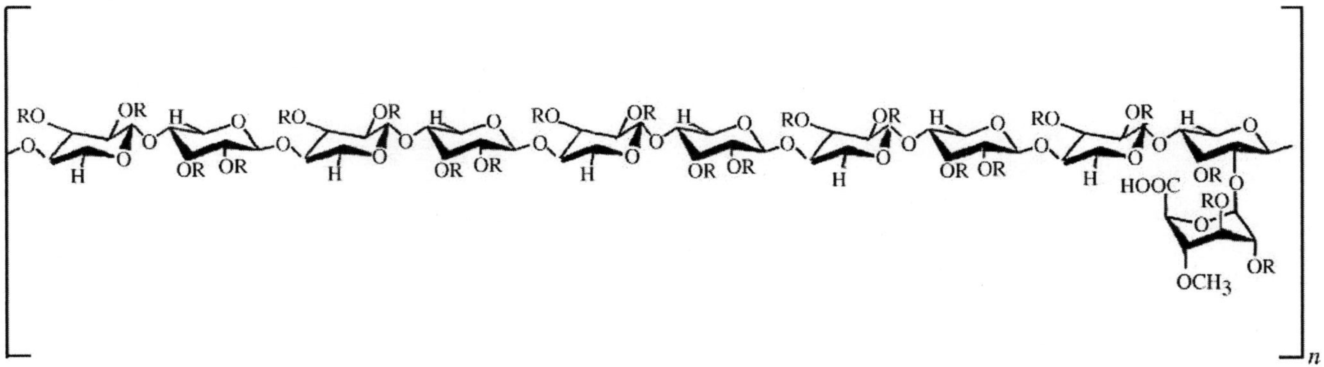

86 R = SO_3^{-} $Ca^{1/2+}$

FIGURE 26.36 Structure of Calcium pentosan polysulfate.

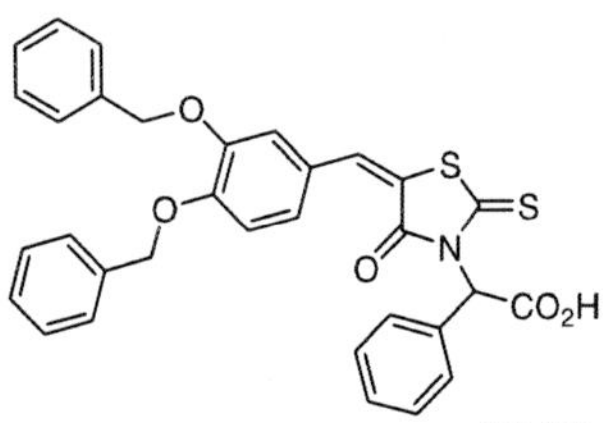

26.87

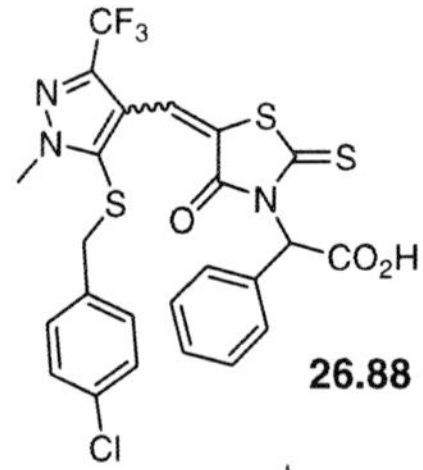

26.88

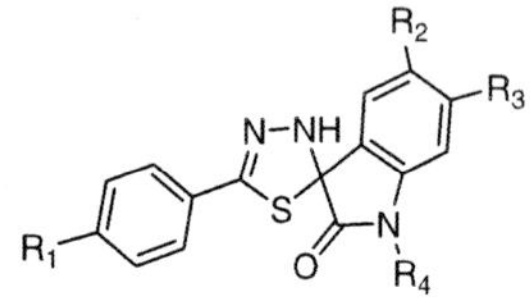

26.89

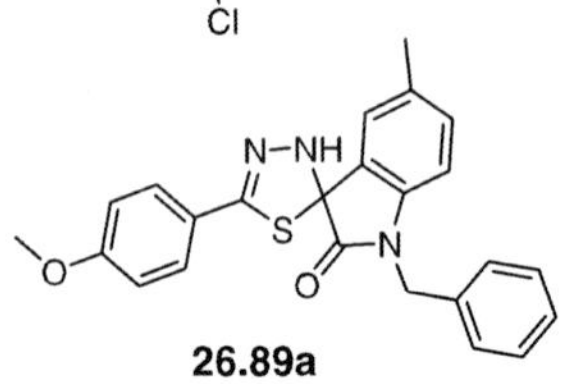

26.89a

FIGURE 26.37 ADAM-TS5 inhibitors.

Bursavich et al.[151] at Wyeth to discover new inhibitors. So, some thioxothiazolidin-4-ones were found to be active and a SAR was developed to optimize their activities (Fig. 26.37). Compound **26.87,** belonging to this class of inhibitors, resulted the first to be active lower in the micromolar range (IC_{50} = 900 nM). Proceeding with this project, Gilbert et al.[152] discovered another group of promising thioxothiazolidin-4-ones, ADAM-TS5 inhibitors. In fact, the introduction of a pyrazolyl nucleus instead of the phenyl ring linked to the side exomethylenic chain was able to give more potent compounds. The inhibitor **26.88** (Fig. 26.37) resulted was sufficiently potent against ADAM-TS5 (IC_{50} = 1.1 μM) slightly sparing ADAM-TS4 (IC_{50} = 44 μM).

In the process of these studies, another class of more selective aggrecanase-2 inhibitors has been discovered at Wyeth[153]. In fact, the 5′-phenyl-3′*H*-spiro[indoline-3,2′-[1,3,4]thiadiazol]-2-ones of type **26.89** resulted were more selective and potent than the previously described thioxothiazolidin-4-ones. This new scaffold seemed to represent an important pharmacophoric structure for the selective inhibition of ADAM-TS5. Compound **26.89a** was potent against this target enzyme, lower than the micromolar range (IC_{50} = 640 nM on aggrecanase-2), sparing ADAM-TS4 (IC_{50} > 22,000 nM) and some MMPs such as MMP-12 (IC_{50} > 22,000 nM) and MMP-13 (IC_{50} > 100,000 nM). Unfortunately, no mechanism of inhibition for cell- or animal-based experiments was reported to better characterize these new families of inhibitors (Fig. 26.37).

Finally, in the recent patent literature, some new glutamate-based inhibitors have been reported by Wyeth[154] as modulators of metalloproteinase activity that were able to modulate activity between the two aggrecanases (ADAM-TS4 and ADAM-TS5) sparing the MMPs activity.

The inhibitor **26.90**, cited as an example of this new family of inhibitors, showed high affinity to aggecanase-1 (IC_{50} = 19 nM) sparing around two orders of magnitude aggrecanase-2 (IC_{50} = 970 nM) and more relevantly sparing the MMPs such as MMP-1 (IC_{50} > 200,000 nM), MMP-2 (IC_{50} = 7600 nM), MMP-13 (IC_{50} > 50,000 nM), and MMP-14 (IC_{50} > 200,000 nM) (Fig. 26.38).

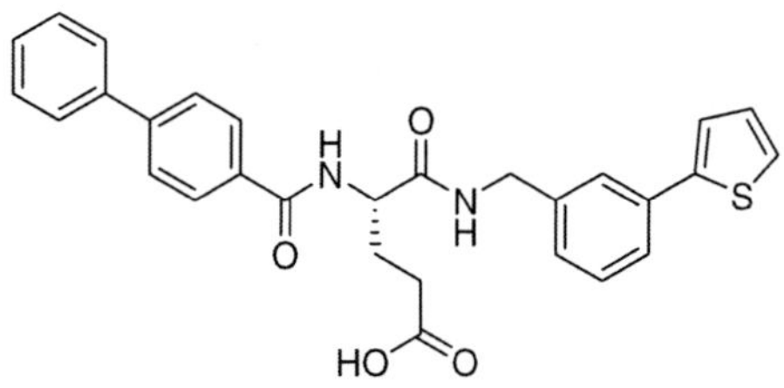

26.90

FIGURE 26.38 Aggrecanase-1 selective inhibitor 26.90.

26.5 CONCLUSIONS

This chapter reports the more recent studies on drug design of ADAMs and ADAM-TS inhibitors published in the last few years. A large number of compounds have been discovered following a parallel development with respect to MMPIs. These new compounds, which target ADAMs and ADAM-TSs, are structurally related to the MMPIs in many cases. This fact is linked to the similarities in the regions adjacent to the catalytic zinc cofactor between these classes of metzincins. A strong effort in this field has been made to develop new selective classes of inhibitors devoid of activity on MMPs or active only on those MMPs validated as target for each pathology.

Overall, particular care should be taken in the development of new compounds for clinical trials. In fact, an improved proteomic and genomic profile of the new clinical candidates should be made before starting trials to avoid the errors that occurred in the past decade with the first generation of MMP inhibitors. In fact, these families of proteases contain structurally and/or functionally still uncharacterized enzymes that may be involved in many regulatory cell life and death cascades in physiology and pathology in the human body.

REFERENCES

1. Fowlkes, J. L.; Winkler, M. K. Survey exploring the interface between metallo-proteinase activity and growth factor and cytokine bioavailability. *Cytokine Growth Factor Rev.* **2002**, *13*, 277–287.
2. Rocks, N.; Paulissen, G.; El Hour, M.; Quesada, F.; Crahay, C.; Gueders, M.; Foidart, J. M.; Noel, A.; Cataldo, D. Emerging roles of ADAM and ADAMTS metalloproteinases in cancer. *Biochimie* **2008**, *90*, 369–379.
3. Mochizuki, S.; Okada, Y. ADAMs in cancer cell proliferation and progression. *Cancer Sci.* **2007**, *98*, 621–628.
4. Rowan, A. D.; Litherland, G. J.; Hui, W.; Milner, J. M. Metalloproteinases as potential therapeutic targets in arthritis treatment. *Expert Opin. Ther. Targets* **2008**, *12*, 1–18.

5. Yang, P. K.; Baker, A.; Hagg, T. The ADAMs family: coordinators of nervous system development, plasticity and repair. *Prog. Neurobiol.* **2006**, *79*, 73–94.
6. Lovering, F.; Zhang, Y. Therapeutic potential of TACE inhibitors in stroke. *Curr. Drug Targets* **2005**, *4*, 161–168.
7. Moss, M. L.; White, J. M.; Andrews, L.; Andrews, R. C. TACE and other ADAM proteases as targets for drug discovery. *Drug Discov. Today* **2001**, *6*, 417–426.
8. Stocker, W.; Bode, W. Structural features of a superfamily of zinc-endopeptidases: the metzincins. *Curr. Opin. Struct. Biol.* **1995**, *5*, 383–390.
9. Boldt, H. B.; Overgaard, M. T.; Laurensen, L. S.; Weyer, K.; Sottrup-Jensen, L.; Oxvig, C. Mutational analysis of the proteolytic domain of pregnancy-associated plasma protein-A (PAPP-A): classification as a metzincin. *Biochem. J.* **2001**, *358*, 359–367.
10. Blobel, C. P. Metalloprotease-disintegrins: links to cell adhesion and cleavage of TNF alpha and Notch. *Cell* **1997**, *90*, 589–592.
11. Zhang, X. P.; Kamata, T.; Yokoyama, K.; Puzon-McLaughlin, W.; Takada, Y. Specific interaction of the recombinant disintegrin-like domain of MDC-15 (metargidin, ADAM-15) with integrin alpha v beta 3. *J. Biol. Chem.* **1998**, *273*, 7345–7350.
12. Loechel, F.; Gilpin, B. J.; Engvall, E.; Albrechtsen, R.; Wewer, U. M. Human ADAM 12 (meltrin alpha) is an active metalloprotease. *J. Biol. Chem.* **1998**, *273*, 16993–16997.
13. Black, R. A.; White, J. M. ADAMs: focus on the protease domain. *Curr. Opin. Cell Biol.* **1998**, *10*, 654–659.
14. Eto, K.; Huet, C.; Tarui, T.; Kupriyanov, S.; Liu, H. Z.; Puzon-McLaughlin, W.; Zhang, X. P.; Sheppard, D.; Engvall, E.; Takada, Y. Functional classification of ADAMs based on a conserved motif for binding to integrin alpha 9 beta 1: implications for sperm-egg binding and other cell interactions. *J. Biol. Chem.* **2002**, *277*, 17804–17810.
15. Reiss, K.; Ludwig, A.; Saftig, P. Breaking up the tie: disintegrin-like metalloproteinases as regulators of cell migration in inflammation and invasion. *Pharmacol. Ther.* **2006**, *111*, 985–1006.
16. Takeda, S.; Igarashi, T.; Mori, H.; Araki, S. Crystal structures of VAP1 reveal ADAMs' MDC domain architecture and its unique C-shaped scaffold. *EMBO J.* **2006**, *25*, 2388–2396.
17. Huovila, A. P.; Almeida, E. A.; White, J. M. ADAMs and cell fusion. *Curr. Opin. Cell Biol.* **1996**, *8*, 692–699.
18. Stone, A. L.; Kroeger, M.; Sang, Q. X. Structure-function analysis of the ADAM family of disintegrin-like and metalloproteinase-containing proteins. *J. Protein Chem.* **1999**, *18*, 447–465.
19. Seals, D. F.; Courtneidge, S. A. The ADAMs family of metalloproteases: multidomain proteins with multiple functions. *Genes Dev.* **2003**, *17*, 7–30.
20. Apte, S. S. A disintegrin-like and metalloprotease (reprolysin type) with thrombospondin type 1 motifs: the ADAMTS family. *Int. J. Biochem. Cell Biol.* **2004**, *36*, 981–985.
21. Cal, S.; Obaya, A. J.; Llamazares, M.; Garabaya, C.; Quesada, V.; Lopez-Otin, C. Cloning, expression analysis, and structural characterization of seven novel human ADAMTSs, a family of metalloproteinases with disintegrin and thrombospondin-1 domains. *Gene* **2002**, *83*, 49–62.

22. Porter, S.; Clark, I. M.; Kevorkian, L.; Edwards, D. R. The ADAMTS metalloproteinases. *Biochem. J.* **2005**, *386*, 15–27.

23. Kaushal, G. P.; Shah, S. V. The new kids on the block: ADAMTSs, potentially multifunctional metalloproteinases of the ADAM family. *J. Clin. Invest.* **2000**, *105*, 1335–1337.

24. Tang, B. L.; Hong, W. ADAMTS: a novel family of proteases with an ADAM protease domain and thrombospondin 1 repeats. *FEBS Lett.* **1999**, *445*, 223–225.

25. Amour, A.; Knight, C. G.; Webster, A.; Slocombe, P. M.; Stephens, P. E.; Knäuper, V.; Docherty, A. J.; Murphy, G. The *in vitro* activity of ADAM-10 is inhibited by TIMP-1 and TIMP-3. *FEBS Lett.* **2000**, *473*, 275–279.

26. Kashiwagi, M.; Tortorella, M.; Nagase, H.; Brew, K. TIMP-3 is a potent inhibitor of aggrecanase 1 (ADAMTS4) and aggrecanase 2 (ADAMTS5). *J. Biol. Chem.* **2001**, *276*, 12501–12504.

27. Murphy, G.; Knäuper, V.; Lee, M. H.; Amour, A.; Worley, J. R.; Hutton, M.; Atkinson, S.; Rapti, M.; Williamson, R. Role of TIMPs (tissue inhibitors of metalloproteinases) in pericellular proteolysis: the specificity is in the detail. *Biochem. Soc. Symp.* **2003**, *70*, 65–80.

28. Nagase, H.; Brew, K. Designing TIMP (tissue inhibitor of metalloproteinases) variants that are selective metalloproteinase inhibitors. *Biochem. Soc. Symp.* **2003**, *70*, 201–212.

29. Amour, A.; Knight, C. G.; English, W. R.; Webster, A.; Slocombe, P. M.; Knäuper, V.; Docherty, A. J.; Becherer, J. D.; Blobel, C. P.; Murphy, G. The enzymatic activity of ADAM8 and ADAM9 is not regulated by TIMPs. *FEBS Lett.* **2002**, *524*, 154–158.

30. Wayne, G. J.; Deng, S. J.; Amour, A.; Borman, S.; Matico, R.; Carter, H. L.; Murphy, G. TIMP-3 inhibition of ADAMTS-4 (Aggrecanase-1) is regulated by interactions between aggrecan and the C-terminal domain of ADAMTS-4. *J. Biol. Chem.* **2007**, *282*, 20991–20998.

31. (a) Küper, C.;Bartels, H.; Fraek, M. L.; Beck, F. X.; Neuhofer, W. Ectodomain shedding of pro-TGF-α is required for COX-2 induction and cell survival in renal medullary cells exposed to osmotic stress. *Am. J. Physiol. Cell Physiol.* **2007**, *293*, C1971–C1982. (b) Auerbach, I. D.; Vinters, H. V. Effects of anoxia and hypoxia on amyloid precursor protein processing in cerebral microvascular smooth muscle cells. *J. Neuropathol. Exp. Neurol.* **2006**, *65*, 610–620.

32. Kenny, P. A. TACE: a new target in epidermal growth factor receptor dependent tumors. *Differentiation* **2007**, *75*, 800–808.

33. Zhou, B. B.; Peyton, M.; He, B.; Liu, C.; Girard, L.; Caudler, E.; Lo, Y.; Baribaud, F.; Mikami, I.; Reguart, N.; Yang, G.; Li, Y.; Yao, W.; Vaddi, K.; Gazdar, A. F.; Friedman, S. M.; Jablons, D. M.; Newton, R. C.; Fridman, J. S.; Minna, J. D.; Schrle, P. A. Targeting ADAM-mediated ligand cleavage to inhibit HER3 and EGFR pathways in non-small cell lung cancer. *Cancer Cell* **2006**, *10*, 39–50.

34. Fridman, J. S.; Caulder, E.; Hansbury, M.; Liu, X.; Yang, G.; Wang, Q.; Lo, Y.; Zhou, B. B.; Pan, M.; Thomas, S. M.; Grandis, J. R.; Zhuo, J.; Yao, W.; Newton, R. C.; Friedman, S. M.; Scherle, P. A.; Vaddi, K. Selective inhibition of ADAM metalloproteases as a novel approach for modulating ErbB pathways in cancer. *Clin. Cancer Res.* **2007**, *13*, 1892–1902.

35. Rosso, O.; Piazza, T.; Bongarzone, I.; Rossello, A.; Mezzanzanica, D.; Canevari, S.; Orengo, A. M.; Puppo, A.; Ferrini, S.; Fabbi, M. The ALCAM shedding by the

metalloprotease ADAM17/TACE is involved in motility of ovarian carcinoma cells. *Mol. Cancer Res.* **2007**, *5*, 1246–1253.

36. Black, R. A.; Rauch, C. T.; Kozlosky, C. J.; Peschon, J. J.; Slack, J. L.; Wolfson, M. F.; Castner, B. J.; Stocking, K. L.; Reddy, P.; Srinivasan, S.; Nelson, N.; Boiani, N.; Schooley, K. A.; Gerhart, M.; Davis, R.; Fitzner, J. N.; Johnson, R. S.; Paxton, R. J.; March, C. J.; Cerretti, D. P. A metalloproteinase disintegrin that releases tumour-necrosis factor-α from cells. *Nature* **1997**, *385*, 729–732.
37. Moss, M. L.; Jin, S. L.; Milla, M. E.; Bickett, D. M.; Burkhart, W.; Carter, H. L.; Chen, W. J.; Clay, W. C.; Didsbury, J. R.; Hassler, D.; Hoffman, C. R.; Kost, T. A.; Lambert, M. H.; Leesnitzer, M. A.; McCauley, P.; McGeehan, G.; Mitchell, J.; Moyer, M.; Pahel, G.; Rocque, W.; Overton, L. K.; Schoenen, F.; Seaton, T.; Su, J. L.; Becherer, J. D. et al. Cloning of a disintegrin metalloproteinase that processes precursor tumor-necrosis factor-α. *Nature* **1997**, *385*, 733–736.
38. Rosendahl, M. S.; Ko, S. C.; Long, D. L.; Brewer, M. T.; Rosenzweig, B.; Hedl, E.; Anderson, L.; Pyle, S. M.; Moreland, J.; Meyers, M. A.; Kohno, T.; Lyons, D.; Lichenstein, H. S. Identification and characterization of a pro-tumor necrosis factor-α-processing enzyme from the ADAM family of zinc metalloproteases. *J. Biol. Chem.* **1997**, *272*, 24588–24593.
39. Lunn, C. A.; Fan, X.; Dalie, B.; Miller, K.; Zavodny, P. J.; Narula, S. K.; Lundell, D. Purification of ADAM 10 from bovine spleen as a TNF-α convertase. *FEBS Lett.* **1997**, *400*, 333–335.
40. Moss, M.; Becherer, J. D.; Milla, M.; Pahel, G.; Lambert, M.; Andrews, R.; Frye, S.; Haffner, C.; Cowan, D.; Maloney, P. TNF-α convertase. In *Metalloproteinases as Targets for Anti-Inflammatory Drugs;* Bradshaw, D.; Nixon, J. S.; Bottomley, K., Eds.; Birkhauser: Basel, **1999**; pp 87–204.
41. Peschon, J. J.; Slack, J. L.; Reddy, P.; Stocking, K. L.; Sunnarborg, S. W.; Lee, D. C.; Russell, W. E.; Castner, B. J.; Johnson, R. S.; Fitzner, J. N.; Boyce, R. W.; Nelson, N.; Kozlosky, C. J.; Wolfson, M. F.; Rauch, C. T.; Cerretti, D. P.; Paxton, R. J.; March, C. J.; Black, R. A. An essential role for ectodomain shedding in mammalian development. *Science* **1998**, *282*, 1281–1284.
42. Luetteke, N. C.; Qiu, T. H.; Peiffer, R. L.; Oliver, P.; Smithies, O.; Lee, D. C. TGF-α deficiency results in hair follicle and eye abnormalities in targeted and waved-1 mice. *Cell* **1993**, *73*, 263–278.
43. Mann, G. B.; Fowler, K. J.; Gabriel, A.; Nice, E. C.; Williams, R. L.; Dunn, A. R. Mice with a null mutation of the TGF-α gene have abnormal skin architecture, wavy hair, and curly whiskers and often develop corneal inflammation. *Cell* **1993**, *73*, 249–261.
44. Miettinen, P. J.; Berger, J. E.; Meneses, J.; Phung, Y.; Pedersen, R. A.; Werb, Z.; Derynck, R. Epithelial immaturity and multiorgan failure in mice lacking epidermal growth factor receptor. *Nature* **1995**, *376*, 337–341.
45. Threadgill, D. W.; Dlugosz, A. A.; Hansen, L. A.; Tennenbaum, T.; Lichti, U.; Yee, D.; LaMantia, C.; Mourton, T.; Herrup, K.; Harris, R. C. et al. Targeted disruption of mouse EGF receptor: effect of genetic background on mutant phenotype. *Science* **1995**, *269*, 230–234.
46. Sibilia, M.; Wagner, E. F. Strain-dependent epithelial defects in mice lacking the EGF receptor. *Science* **1995**, *269*, 234–238.

47. Reddy, P.; Slack, J. L.; Davis, R.; Cerretti, D. P.; Kozlosky, C. J.; Blanton, R. A.; Shows, D.; Peschon, J. J.; Black, R. A. Functional analysis of the domain structure of tumor necrosis factor-α converting enzyme. *J. Biol. Chem.* **2000**, *275*, 14608–14614.

48. Buxbaum, J. D.; Liu, K. N.; Luo, Y.; Slack, J. L.; Stocking, K. L.; Peschon, J. J.; Johnson, R. S.; Castner, B. J.; Cerretti, D. P.; Black, R. A. Evidence that tumor necrosis factor-α converting enzyme is involved in regulated α-secretase cleavage of the Alzheimer amyloid protein precursor. *J. Biol. Chem.* **1998**, *273*, 27765–27767.

49. Althoff, K.; Reddy, P.; Voltz, N.; Rose-John, S.; Müllberg, J. Shedding of interleukin-6 receptor and tumor necrosis factor α. Contribution of the stalk sequence to the cleavage pattern of transmembrane proteins. *Eur. J. Biochem.* **2000**, *267*, 2624–2631.

50. Rio, C.; Buxbaum, J. D.; Peschon, J. J.; Corfas, G. Tumor necrosis factor-α-converting enzyme is required for cleavage of erbB4/HER4. *J. Biol. Chem.* **2000**, *275*, 10379–10387.

51. Sadhukhan, R.; Santhamma, K. R.; Reddy, P.; Peschon, J. J.; Black, R. A.; Sen, I. Unaltered cleavage and secretion of angiotensin-converting enzyme in tumor necrosis factor-α-converting enzyme-deficient mice. *J. Biol. Chem.* **1999**, *274*, 10511–10516.

52. Lum, L.; Wong, B. R.; Josien, R.; Becherer, J. D.; Erdjument-Bromage, H.; Schlöndorff, J.; Tempst, P.; Choi, Y.; Blobel, C. P. Evidence for the role of tumour necrosis factor-α (TNF-α)-converting enzyme-like protease in shedding of TRANCE, a TNF family member involved in osteoclastogenesis and dendritic cell survival. *J. Biol. Chem.* **1999**, *274*, 13613–13618.

53. Brou, C.; Logeat, F.; Gupta, N.; Bessia, C.; LeBail, O.; Doedens, J. R.; Cumano, A.; Roux, P.; Black, R. A.; Israël, A. A novel proteolytic cleavage involved in Notch signaling: the role of the disintegrin-metalloprotease TACE. *Mol. Cell* **2000**, *5*, 207–216.

54. Solomon, K. A.; Pesti, N.; Wu, G.; Newton, R. C. Cutting edge: a dominant-negative form of TNF-α converting enzyme inhibits proTNF and TNFRII secretion. *J. Immunol.* **1999**, *163*, 4105–4108.

55. Schlondorff, J.; Blobel, C. P. Metalloprotease-disintegrins: modular proteins capable of promoting cell–cell interactions and triggering signals by protein-ectodomain shedding. *J. Cell Sci.* **1999**, *112*, 3603–3617.

56. Black, R. A.; Rauch, C. T.; Kozlosky, C. J.; Peschon, J. J.; Slack, J. L.; Wolfson, M. F.; Castner, B. J.; Stocking, K. L.; Reddy, P.; Srinivasan, S.; Nelson, N.; Boiani, N.; Schooley, K. A.; Gerhart, M.; Davis, R.; Fitzner, J. N.; Johnson, R. S.; Paxton, R. J.; March, C. J.; Cerretti, D. P. A metalloproteinase disintegrin that releases tumour-necrosis factor-α from cells. *Nature* **1997**, *385*, 729–732.

57. Schlöndorff, J.; Becherer, J. D.; Blobel, C. P. Intracellular maturation and localization of the tumour necrosis factor-α convertase (TACE). *Biochem. J.* **2000**, *347*, 131–138.

58. Doedens, J. R.; Black, R. A. Stimulation-induced downregulation of tumor necrosis factor-α converting enzyme. *J. Biol. Chem.* **2000**, *275*, 14598–14607.

59. Maskos, K.; Fernandez-Catalan, C.; Huber, R.; Bourenkov, G. P.; Bartunik, H.; Ellestad, G. A.; Reddy, P.; Wolfson, M. F.; Rauch, C. T.; Castner, B. J.; Davis, R.; Clarke, H. R.; Petersen, M.; Fitzner, J. N.; Cerretti, D. P.; March, C. J.; Paxton, R. J.; Black, R. A.; Bode, W. Crystal structure of the catalytic domain of human tumor necrosis factor-α-converting enzyme. *Proc. Natl. Acad. Sci. USA* **1998**, *95*, 3408–3412.

60. Botos, I.; Scapozza, L.; Zhang, D.; Liotta, L. A.; Meyer, E. F.; Batimastat, a potent matrix metalloproteinase inhibitor, exhibits an unexpected mode of binding. *Proc. Natl. Acad. Sci. USA* **1996**, *93*, 2749–2754.

61. Cirilli, M.; Gallina, C.; Gavuzzo, E.; Giordano, C.; Gomis-Rüth, F. X.; Gorini, B.; Kress, L. F.; Mazza, F.; Paradisi, M. P.; Pochetti, G.; Politi, V. 2 Å X-ray structure of adamalysin II complexed with a peptide phosphonate inhibitor adopting a retro-binding mode. *FEBS Lett.* **1997**, *418*, 319–322.

62. Gomis-Rüth, F. X.; Meyer, E. F.; Kress, L. F.; Politi, V. Structures of adamalysin II with peptidic inhibitors. Implications for the design of tumor necrosis factor-α convertase inhibitors. *Protein Sci.* **1998**, *7*, 283–292.

63. Gonnella, N. C.; Li, Y. C.; Zhang, X.; Paris, C. G. Bioactive conformation of a potent stromelysin inhibitor determined by X-nucleus filtered and multidimensional NMR spectroscopy. *Bioorg. Med. Chem.* **1997**, *5*, 2193–2201.

64. Roghani, M.; Becherer, J. D.; Moss, M. L.; Atherton, R. E.; Erdjument-Bromage, H.; Arribas, J.; Blackburn, R. K.; Weskamp, G.; Tempst, P.; Blobel, C. P. Metalloprotease-disintegrin MDC9: intracellular maturation and catalytic activity. *J. Biol. Chem.* **1999**, *274*, 3531–3540.

65. Parker, D. T.; O'Byrne, E.; Goldemberg, R.; Ganu, V.; Melton, R.; Blancuzzi, V.; Wilson, D.; Hu, S. -I.; Roberts, E. D.; Singh, H.; Ludewing, R.; MacPerson, L. J. The development of 27023A: a novel, potent, and orally active matrix metalloprotease inhibitor. *Fourth International Conference, Inflammation Research Association, White Haven, PA, USA*, **1994**, p 73 (abstracts).

66. Barberia, J. T.; Carty, T. J.; Georghean, K. F. Development of a cell-based assay for TNF-α release which avoids transcription and translation steps. *Fifth International Conference, Inflammation Research Association, Hershey, PA, USA*, **1996**, p 1 (abstracts).

67. Moss, M. L.; White, J. M.; Lambert, M. H.; Andrews, R. C. TACE and ADAM proteases as target for drug discovery. *Drug Discov. Today* **2001**, *6*, 417–426.

68. Barlaam, B.; Bird, T. G.; Lambert-Van Der Brempt, C.; Campbell, D.; Foster, S. J.; Maciewicz, R. New α-substituted succinate-based hydroxamic acids as TNF-α convertase inhibitors. *J. Med. Chem.* **1999**, *42*, 4890–4908.

69. (a) Andrews, R. C.; Anderson, M. W.; Stanford, J. B.; Bubacz, D. G.; Chan, J. H.; Cowan, D. J.; Gaul, M. D.; McDougald, D. L.; Musso, D. L.; Rabinowitz, M. H.; Wiethe, R. W. *Brit. UK Pat. Appl.* GB 2348198, **2000.** (b) Rabinowitz, M. Synthesis and evaluation of novel *N*-hydroxy formamide inhibitors of tumor necrosis factor-α converting enzyme. *220th National Meeting of the American* Chemical Society, Washington, DC; American Chemical Society, Washington, DC, **2000.**

70. Broadhurst, M. J.; Johnson, W. H.; Walter, D. S. Cyclic hydrazine derivatives as TNF-α inhibitors.*PCT Int. Appl.* WO 0035885 A1, **2000.**

71. Nixon, J. S.; Bottomley, K. M. Metalloproteinase inhibitors: new opportunities for the treatment of rheumatoid arthritis and osteoarthritis. *Exp. Opin. Invest. Drugs* **2000**, *9*, 1469–1478.

72. McClure, K. F.; Noe, M. C.; Letavic, M. A.; Chupak, L. S. Preparation of 4-phenylsulfonyl-3- morpholinehydroxamic acids and analogs as tumor necrosis factor-α- convertase inhibitors. *PCT Int. Appl.* WO 0009492 A1, **2000.**

73. (a) Levin, J. I.; Du Mila, T.; Venkatesan, A. M.; Nelson, F. C.; Zask, A.; Gu, Y. The preparation and use of ortho-sulfonamido aryl hydroxamic acids as matrix metalloproteinase and TACE inhibitors. *PCT Int. Appl.* WO 9816503, **1998.** (b) Levin, J. I.; Du, M. T.; DiJoseph, J. F.; Killar, L. M.; Sung, A.; Walter, T.; Sharr, M. A.; Roth, C. E.; Moy, F. J.; Powers, R.; Jin, G.; Cowling, R.; Skotnicki, J. S. The discovery of anthranilic acid-based

MMP inhibitors. Part 1: SAR of the 3-position. *Bioorg. Med. Chem. Lett.* **2001**, *11*, 235–238.

74. Xue, C.-B.; Chemey, R. J.; Decicco, C. P.; DeGrado, W. F.; He, X.; Hodge, C. N.; Jacobson, I. C.; Magolda, R. L.; Arner, E. C.; Duma, J.; Nelson, D. J. *Patent Appl.* WO9718207, **1997**.

75. (a) Xue, C. B.; He, X.; Roderick, J.; Corbett, R. L.; Duan, J. J.; Liu, R. Q.; Covington, M. B.; Newton, R. C.; Trzaskos, J. M.; Magolda, R. L.; Wexler, R. R.; Decicco, C. P. Rational design, synthesis and structure–activity relationships of a cyclic succinate series of TNF-α converting enzyme inhibitors. Part 1: lead Identification. *Bioorg. Med Chem. Lett.* **2003**, *13*, 4293–4297. (b) Xue, C. B.; He, X.; Roderick, J.; Corbett, R. L.; Duan, J. J.; Liu, R. Q.; Covington, M. B.; Qian, M.; Ribadeneira, M. D.; Vaddi, K.; Christ, D. D.; Newton, R. C.; Trzaskos, J. M.; Magolda, R. L.; Wexler, R. R.; Decicco, C. P. Rational design, synthesis and structure–activity relationships of a cyclic succinate series of TNF-α converting enzyme inhibitors. Part 2: lead Optimization. *Bioorg. Med. Chem. Lett.* **2003**, *13*, 4299–4304.

76. Duan, J. J.; Lu, Z.; Xue, C. B.; He, X.; Seng, J. L.; Roderick, J. J.; Wasserman, Z. R.; Liu, R. Q.; Covington, M. B.; Magolda, R. L.; Newton, R. C.; Trzaskos, J. M.; Decicco, C. P. Discovery of *N*-Hydroxy-2-(2-oxo-3-pyrrolidinyl)acetamides as potent and selective inhibitors of tumor necrosis factor-α converting enzyme (TACE). *Bioorg. Med. Chem. Lett.* **2003**, *13*, 2035–2040.

77. Duan, J. J.; Chen, L.; Wasserman, Z. R.; Lu, Z.; Liu, R. Q.; Covington, M. B.; Qian, M.; Hardman, K. D.; Magolda, R. L.; Newton, R. C.; Christ, D. D.; Wexler, R. R.; Decicco, C. P. Discovery of γ-lactam hydroxamic acids as selective inhibitors of tumor necrosis factor α converting enzyme: design, synthesis, and structure–activity relationships. *J. Med. Chem.* **2002**, *45*, 4954–4957.

78. Gilmore, J. L.; King, B. W.; Harris, C.; Maduskuie, T.; Mercer, S. E.; Liu, R. Q.; Covington, M. B.; Qian, M.; Ribadeneria, M. D.; Vaddi, K.; Trzaskos, J. M.; Newton, R. C.; Decicco, C. P.; Duan, J. J. Synthesis and structure–activity relationship of a novel, achiral series of TNF-α converting enzyme inhibitors. *Bioorg. Med. Chem. Lett.* **2006**, *16*, 2699–2704.

79. Duan, J. J.; Chen, L.; Lu, Z.; Xue, C. B.; Liu, R. Q.; Covington, M. B.; Qian, M.; Wasserman, Z. R.; Vaddi, K.; Christ, D. D.; Trzaskos, J. M.; Newton, R. C.; Decicco, C. P. Discovery of β-benzamido hydroxamic acids as potent, selective, and orally bioavailable TACE inhibitors. *Bioorg. Med. Chem. Lett.* **2008**, *18*, 241–247.

80. (a) Ott, G. R.; Asakawa, N.; Lu, Z.; Liu, R. Q.; Covington, M. B.; Vaddi, K.; Qian, M.; Newton, R. C.; Christ, D. D.; Traskos, J. M.; Decicco, C. P.; Duan, J. J. α,β-Cyclic-β-benzamido hydroxamic acids: novel templates for the design, synthesis, and evaluation of selective inhibitors of TNF-alpha converting enzyme (TACE). *Bioorg. Med. Chem. Lett.* **2008**, *18*, 694–699. (b) Ott, G. R.; Asakawa, N.; Lu, Z.; Anand, R.; Liu, R. Q.; Covington, M. B.; Vaddi, K.; Qian, M.; Newton, R. C.; Christ, D. D.; Trzaskos, J. M.; Duan, J. J. Potent, exceptionally selective, orally bioavailable inhibitors of TNF-α converting enzyme (TACE): novel 2-substituted-1*H*-benzo[*d*]imidazol-1-yl)methyl)benzamide P1′ substituents. *Bioorg. Med. Chem. Lett.* **2008**, *18*, 1577–1582.

81. Sawa, M.; Kiyoi, T.; Kurokawa, K.; Kumihara, H.; Yamamoto, M.; Miyasaka, T.; Ito, Y.; Hirayama, R.; Inoue, T.; Kirii, Y.; Nishiwaki, E.; Ohmoto, H.; Maeda, Y.; Ishibushi, E.; Inoue, Y.; Yoshino, K.; Kondo, H. New type of metalloproteinase inhibitor: design and

synthesis of new phosphonamide-based hydroxamic acids. *J. Med. Chem.* **2002**, *45*, 919–929.

82. Sawa, M.; Kiyoi, T.; Kurokawa, K.; Kumihara, H.; Yamamoto, M.; Miyasaka, T.; Ito, Y.; Hirayama, R.; Inoue, T.; Kirii, Y.; Nishiwaki, E.; Ohmoto, H.; Maeda, Y.; Ishibushi, E.; Inoue, Y.; Yoshino, K.; Kondo, H. Discovery of selective phosphonamide-based inhibitors of tumor necrosis factor-α converting enzyme (TACE). *Bioorg. Med. Chem. Lett.* **2003**, *13*, 2021–2024.
83. Sawa, M.; Kondo, H.; Nishimura, S. Encounter with unexpected collagenase-1 selective inhibitor: switchover of inhibitor binding pocket induced by fluorine atom. *Bioorg. Med. Chem. Lett.* **2002**, *12*, 581–584.
84. Levin, J. I.; Chen, J. M.; Cheung, K.; Cole, D.; Crago, C.; Santos, E. D.; Du, X.; Khafizova, G.; MacEwan, G.; Niu, C.; Salaski, E. J.; Zask, A.; Cummons, T.; Sung, A.; Xu, J.; Zhang, Y.; Xu, W.; Ayral-Kaloustian, S.; Jin, G.; Cowling, R.; Barone, D.; Mohler, K. M.; Black, R. A.; Skotnicki, J. S. Acetylenic TACE inhibitors. Part 1: SAR of the acyclic sulfonamide hydroxamates. *Bioorg. Med Chem. Lett.* **2003**, *13*, 2799–2803.
85. Zhang, Y.; Hegen, M.; Xu, J.; Keith, J. C., Jr.; Jin, G.; Du, X.; Cummons, T.; Sheppard, B. J.; Sun, L.; Zhu, Y.; Rao, V. R.; Wang, Q.; Xu, W.; Cowling, R.; Nickerson-Nutter, C. L.; Gibbons, J.; Skotnicki, J.; Lin, L. L.; Levin, J. Characterization of (2*R*, 3*S*)-2-({[4-(2-butynyloxy)phenyl]sulfonyl}amino)-*N*,3-dihydroxybutanamide, a potent and selective inhibitor of TNF-α converting enzyme. *Int. Immunopharmacol.* **2004**, *4*, 1845–1857.
86. (a) Zhang, Y.; Xu, J.; Levin, J.; Hegen, M.; Li, G.; Robertshaw, H.; Brennan, F.; Cummons, T.; Clarke, D.; Vansell, N.; Nickerson-Nutter, C.; Barone, D.; Mohler, K.; Black, R.; Skotnicki, J.; Gibbons, J.; Feldmann, M.; Frost, P.; Larsen, G.; Lin, L. L. Identification and characterization of 4-[[4-(2-butynyloxy)phenyl]sulfonyl]-*N*-hydroxy-2,2-dimethyl-(3*S*)thiomorpholinecarboxamide (TMI-1), a novel dual tumor necrosis factor-alpha-converting enzyme/matrix metalloprotease inhibitor for the treatment of rheumatoid arthritis. *J. Pharmcol. Exp. Ther.* **2004**, *309*, 348–355. (b) Levin, J. I.; Chen, J. M.; Laakso, L. M.; Du, M.; Du, X.; Venkatesan, A. M.; Sandanayaka, V.; Zask, A.; Xu, J.; Xu, W.; Zhang, Y.; Skotnicki, J. S. Acetylenic TACE inhibitors. Part 2: SAR of six-membered cyclic sulfonamide hydroxamates. *Bioorg. Med. Chem. Lett.* **2005**, *15*, 4345–4349.
87. Levin, J. I.; Chen, J. M.; Laakso, L. M.; Du, M.; Schmid, J.; Xu, W.; Cummons, T.; Xu, J.; Jin, G.; Barone, D.; Skotnicki, J. S. Acetylenic TACE inhibitors. Part 3: thiomorpholine sulfonamide hydroxamates. *Bioorg. Med. Chem. Lett.* **2006**, *16*, 1605–1609.
88. Venkatesan, A. M.; Davis, J. M.; Grosu, G. T.; Baker, J.; Zask, A.; Levin, J. I.; Ellingboe, J.; Skotnicki, J. S.; Dijoseph, J. F.; Sung, A.; Jin, G.; Xu, W.; McCarthy, D. J.; Barone, D. Synthesis and structure–activity relationships of 4-alkynyloxy phenyl sulfanyl, sulfinyl, and sulfonyl alkyl hydroxamates as tumor necrosis factor-alpha converting enzyme and matrix metalloproteinase inhibitors. *J. Med. Chem.* **2004**, *47*, 6255–6269.
89. Kaapjoo Park, K.; Aplasca, A.; Du, M. T.; Sun, L.; Zhu, Y.; Zhang, Y.; Levin, J. I. Design and synthesis of butynyloxyphenyl b-sulfone piperidine hydroxamates as TACE inhibitors. *Bioorg. Med. Chem. Lett.* **2006**, *16*, 3927–3931.
90. Condon, J. S.; Joseph-McCarthy, D.; Levin, J. I.; Lombart, H. G.; Lovering, F. E.; Sun, L.; Wang, W.; Xu, W.; Zhang, Y. Identification of potent and selective TACE inhibitors via the S1 pocket. *Bioorg. Med. Chem. Lett.* **2007**, *17*, 34–39.

91. Zhu, Z.; Mazzola, R.; Sinning, L.; McKittrick, B.; Niu, X.; Lundell, D.; Sun, J.; Orth, P.; Guo, Z.; Madison, V.; Ingram, R.; Beyer, B. M. Discovery of novel hydroxamates as highly potent tumor necrosis factor-α converting enzyme inhibitors. Part I: discovery of two binding modes. *J. Med. Chem.* **2008**, *51*, 725–736.

92. Lukacova, V.; Zhang, Y.; Kroll, D. M.; Raha, S.; Comez, D.; Balaz, S. A comparison of the binding sites of matrix metalloproteinases and tumor necrosis factor-alpha converting enzyme: implications for selectivity. *J. Med. Chem.* **2005**, *48*, 2361–2370.

93. Duan, J. J.; Chen, L.; Lu, Z.; Jiang, B.; Asakawa, N.; Sheppeck, J. E., 2nd; Liu R. Q.; Covington, M. B.; Pitts, W.; Kim, S. H.; Decicco, C. P. Discovery of low nanomolar non-hydroxamate inhibitors of tumor necrosis factor-α converting enzyme (TACE). *Bioorg. Med. Chem. Lett.* **2007**, *17*, 266–271.

94. Sheppeck, J. E., 2nd; Tebben A.; Gilmore, J. L.; Yang, A.; Wasserman, Z. R.; Decicco, C. P.; Duan, J. J. A molecular modeling analysis of novel non-hydroxamate inhibitors of TACE. *Bioorg. Med. Chem. Lett.* **2007**, *17*, 1408–1412.

95. Sheppeck, J. E., 2nd; Gilmore J. L.; Tebben, A.; Xue, C. B.; Liu, R. Q.; Decicco, C. P.; Duan, J. J. Hydantoins, triazolones, and imidazolones as selective non-hydroxamate inhibitors of tumor necrosis factor-α converting enzyme (TACE). *Bioorg. Med. Chem. Lett.* **2007**, *17*, 2769–2774.

96. Brandstetter, H.; Grams, F.; Glitz, D.; Lang, A.; Huber, R.; Bode, W.; Krell, H. W.; Engh, R. A. The 1.8-Å crystal structure of a matrix metalloproteinase 8-barbiturate inhibitor complex reveals a previously unobserved mechanism for collagenase substrate recognition. *J. Biol. Chem.* **2001**, *276*, 17405–17412.

97. Bandarage, U. K.; Wang, T.; Come, J. H.; Perola, E.; Wei, Y.; Rao, B. G. Novel thiol-based TACE inhibitors. Part 2: rational design, synthesis, and SAR of thiol-containing aryl sulfones. *Bioorg. Med. Chem. Lett.* **2008**, *18*, 44–48.

98. Gonzales, P. E.; Solomon, A.; Miller, A. B.; Leesnitzer, M. A.; Sagi, I.; Milla, M. E. Inhibition of the tumor necrosis factor-α-converting enzyme by its pro domain. *J. Biol. Chem.* **2004**, *279*, 31638–31645.

99. Milla, M. E.; Leesnitzer, M. A.; Moss, M. L.; Clay, W. C.; Carter, H. L.; Miller, A. B.; Su, J. L.; Lambert, M. H.; Willard, D. H.; Sheeley, D. M.; Kost, T. A.; Burkhar, W.; Moyer, M.; Blackburn, R. K.; Pahel, G. L.; Mitchell, J. L.; Hoffman, C. R.; Becherer, J. D. Specific sequence elements are required for the expression of functional tumor necrosis factor-α-converting enzyme (TACE). *J. Biol. Chem.* **1999**, *274*, 30563–31645.

100. Moss, M. L.; Bomar, M.; Liu, Q.; Sage, H.; Dempsey, P.; Lenhart, P. M.; Gillispie, P. A.; Stoeck, A.; Wildeboer, D.; Bartsch, J. W.; Palmisano, R.; Zhou, P. The ADAM10 prodomain is a specific inhibitor of ADAM-10 proteolytic activity and inhibits cellular shedding events. *J. Biol. Chem.* **2007**, *282*, 35712–35721.

101. Mass, R. D. The HER receptor family: a rich target for therapeutic development. *Int. J. Radiat. Oncol. Biol. Phys.* **2004**, *58*, 932–940.

102. Carney, W. P.; Neumann, R.; Lipton, A.; Leitzel, K.; Ali, S.; Price, C. P. Potential clinical utility of serum HER-2/neu oncoprotein concentrations in patients with breast cancer. *Clin. Chem.* **2003**, *49*, 1579–1598.

103. Arribas, J.; Borroto, A. Protein ectodomain shedding. *Chem. Rev.* **2002**, *102*, 4627–4637.

104. Marmor, M. D.; Skaria, K. B.; Yarden, Y. Signal transduction and oncogenesis by ErbB/HER receptors. *Int. J. Radiat. Oncol. Biol. Phys.* **2004**, *58*, 903–913.

105. Molina, M. A.; Saez, R.; Ramsey, E. E.; Garcia-Barchino, M.-J.; Rojo, F.; Evans, A. J.; Albanell, J.; Keenan, E. J.; Lluch, A.; Garcia-Conde, J.; Baselga, J.; Clinton, G. M. NH2-terminal truncated HER-2 protein but not full-length receptor is associated with nodal metastasis in human breast cancer. *Clin. Cancer Res.* **2002**, *8*, 347–353.

106. Banner, L. C.; Co, E. W.; Jammalamadaka, V.; Nuss, J. M.; Kim, M. H.; Le, D. T.; Lew, A.; Mac Morrison, B.; Mamo, S.; Wen, Z.; Xu, W. Human ADAM-10 Inhibitors. WO 03106381, **2003**.

107. Yao, W.; Zhuo, J.; Burns, D. M.; Li, Y. L.; Qian, D. Q.; Zhang, C.; He, C.; Xu, M.; Shi, E.; Li, Y.; Marando, C. A.; Covington, M. B.; Yang, G.; Liu, X.; Pan, M.; Fridman, J. S.; Scherle, P.; Wasserman, Z. R.; Hollis, G.; Vaddi, K.; Yeleswaram, S.; Newton, R.; Friedman, S.; Metcalf, B. Design and identification of selective HER-2 sheddase inhibitors via P1′ manipulation and unconventional P2′ perturbations to induce a molecular metamorphosis. *Bioorg. Med. Chem. Lett.* **2008**, *18*, 159–163.

108. Liu, X.; Fridman, J. S.; Wang, Q.; Caulder, E.; Yang, G.; Covington, M.; Liu, C.; Marando, C.; Zhuo, J.; Li, Y.; Yao, W.; Vaddi, K.; Newton, R. C.; Scherle, P. A.; Friedman, S. M. Selective inhibition of ADAM metalloproteases blocks HER-2 extracellular domain (ECD) cleavage and potentiates the anti-tumor effects of trastuzumab. *Cancer Biol. Ther.* **2006**, *5*, 648–656.

109. Zhou, B. B.; Peyton, M.; He, B.; Liu, C.; Girard, L.; Caudler, E.; Lo, Y.; Baribaud, F.; Mikami, I.; Reguart, N.; Yang, G.; Li, Y.; Yao, W.; Vaddi, K.; Gazdar, A. F.; Friedman, S. M.; Jablons, D. M.; Newton, R. C.; Fridman, J. S.; Minna, J. D.; Scherle, D. A. Targeting ADAM-mediated ligand cleavage to inhibit HER3 and EGFR pathways in non-small cell lung cancer. *Cancer Cell* **2006**, *10*, 39–50.

110. Yao, W.; Zhuo, J.; Burns, D. M.; Xu, M.; Zhang, C.; Li, Y. L.; Qian, D. Q.; He, C.; Weng, L.; Shi, E.; Lin, Q.; Agrios, C.; Burn, T. C.; Caulder, E.; Covington, M. B.; Fridman, J. S.; Friedman, S.; Katiyar, K.; Hollis, G.; Li, Y.; Liu, C.; LiuX.; Marando, C. A.; Newton, R.; Pan, M.; Scherle, P.; Taylor, N.; Vaddi, K.; Wasserman, Z. R.; Wynn, R.; Yeleswaram, S.; Jalluri, R.; Bower, M.; Zhou, B. B.; Metcalf, B. Discovery of a potent, selective, and orally active human epidermal growth factor receptor-2 sheddase inhibitor for the treatment of cancer. *J. Med. Chem.* **2007**, *50*, 603–606.

111. Infante, J.; Burris, H. A.; Lewis, N.; Donehower, R.; Redman, J.; Friedman, S.; Scherle, P.; Fridman, J.; Li, J.; Emm, T.; Troy, S.; Eckhardt, S. G. *30th San Antonio Breast Cancer Symposium; Poster No. 6064; 2007 December 13–16, San Antonio, TX, USA*, www.incyte.com.

112. Witters, L.; Scherle, P.; Friedman, S.; Fridman, J.; Caulder, E.; Newton, R.; Lipton, A. Synergistic inhibition with a dual epidermal growth factor receptor/HER-2/neu tyrosine kinase inhibitor and a disintegrin and metalloprotease inhibitor. *Cancer Res.* **2008**, *68*, 7083–7089.

113. Wenquing, Y.; Zhou, J.; Xu, M.; Zhang, F.; Metcalf, B. Aza spiro alcane derivatives as inhibitors of metalloproteinases. *PCT* WO 04096139, **2004**.

114. Burns, D. M.; He, C.; Li, Y.; Scherle, P.; Liu, X.; Marando, C. A.; Covington, M. B.; Yang, G.; Pan, M.; Turner, S.; Fridman, J. S.; Hollis, G.; Vaddi, K.; Yeleswaram, S.; Newton, R.; Friedman, S.; Metcalf, B.; Yao, W. Conversion of an MMP-potent scaffold to an MMP-selective HER-2 sheddase inhibitor via scaffold hybridization and subtle P01 permutations. *Bioorg. Med. Chem. Lett.* **2008**, *18*, 560–564.

115. Li, Y.-L.; Zhou, J.; Burns, D.; Yao, W. A preparation of (hetero)cyclic *N*-hydroxy-carboxamide derivatives, useful as inhibitors of matrix metalloproteinases (MMPs). *PCT* WO 05037826, **2005**.

116. Schaub, M. C.; Hefti, M. A.; Zuellig, R.A.; Morano, I. Modulation of contractility in human cardiac hypertrophy by myosin essential light chain isoforms. *Cardiovasc. Res.* **1998**, *37*, 381–404.

117. Elliott, P.; McKenna, W. J. Hypertrophic cardiomyopathy. *Lancet* **2004**, *363*, 1881–1891.

118. Asakura, M.; Kitakaze, M.; Takashima, S.; Liao, Y.; Ishikura, F.; Yoshinaka, T.; Ohmoto, H.; Node, K.; Yoshino, K.; Ishiguro, H.; Asanuma, H.; Sanada, S.; Matsumura, Y.; Takeda, H.; Beppu, S.; Tada, M.; Hori, M.; Higashiyama, S. Cardiac hypertrophy is inhibited by antagonism of ADAM12 processing of HB-EGF: metalloproteinase inhibitors as a new therapy. *Nat. Med.* **2002**, *8*, 35–40.

119. Grüneberg, S.; Stubbs, M. T.; Klebe, G. Successful virtual screening for novel inhibitors of human carbonic anhydrase: strategy and experimental confirmation. *J. Med. Chem.* **2002**, *45*, 3588–3602.

120. Oh, M.; Im, I.; Lee, Y. J.; Kim, Y. H.; Yoon, J. H.; Park, H. G.; Higashiyama, S.; Kim, Y. C.; Park, W. J. Structure-based virtual screening and biological evaluation of potent and selective ADAM12 inhibitors. *Bioorg. Med. Chem. Lett.* **2004**, *14*, 6071–6074.

121. Jones, G. C. ADAMTS Proteinases: potential therapeutic targets? *Curr. Pharm. Biotechnol.* **2006**, *7*, 25–31.

122. Takizawa, M.; Yatabe, T.; Okada, A.; Chijiiwa, M.; Ghosh, P.; Okada, Y. Calcium pentosan polysulfate directly inhibits enzymatic activity of ADAMTS4 (aggrecanase-1) in osteoarthritic chondrocytes. *FEBS Lett.* **2008**, *582*, 2945–2949.

123. Okada, Y. Proteinases and matrix degradation. In *Kelley's Textbook of Rheumatology*, Firestein, G. S.; Budd, R. C.; Harris, E. D., Jr.; McInnes, I. B. Ruddy, S., Sergent, J. S. 8th ed.; Elsevier Saunders: Philadelphia.

124. Porter, S.; Clark, I. M.; Kevorkian, L.; Edwards, D. R. The ADAMTS metalloproteinases. *Biochem. J.* **2005**, *386*, 15–27.

125. Fosang, A. J.; Neame, P. J.; Last, K.; Hardhingham, T. E.; Murphy, G.; Hamilton, J. A. The interglobular domain of cartilage aggrecan is cleaved by PUMP, gelatinases, and cathepsin B. *J. Biol. Chem.* **1992**, *267*, 19470–19474.

126. Flannery, C. R.; Lark, M. W.; Sandy, J. D. Identification of a stromelysin cleavage site within the interglobular domain of human aggrecan. Evidence for proteolysis at this site *in vivo* in human articular cartilage. *J. Biol. Chem.* **1992**, *267*, 1008–1014.

127. Fosang, A. J.; Last, K.; Knauper, V.; Neame, P. J.; Murphy, G.; Hardingham, T. E.; Tschesche, H.; Hamilton, J. A. Fibroblast and neutrophil collagenases cleave at two sites in the cartilage aggrecan interglobular domain. *Biochem. J.* **1993**, *295*, 273–276.

128. Fosang, A. J.; Last, K.; Knauper, V.; Murphy, G.; Neame, P. J. Degradation of cartilage aggrecan by collagenase-3 (MMP-13). *FEBS Lett.* **1996**, *380*, 17–20.

129. Tortorella, M. D.; Burn, T. C.; Pratta, M. A.; Abbaszade, I.; Hollis, J. M.; Liu, R.; Rosenfeld, S. A.; Copeland, R. A.; Decicco, C. P.; Wynn, R.; Rockwell, A.; Yang, F.; Duke, J. L.; Solomon, K.; George, H.; Bruckner, R.; Nagase, H.; Itoh, Y.; Ellis, D. M.; Ross, H.; Wiswall, B. H.; Murphy, K.; Hillman, M. C., Jr.; Hollis, G. F.; Newton, R. C.; Magolda, R. L.; Trzaskos, J. M.; Arner, E. C. Purification and cloning of aggrecanase-1: a member of the ADAMTS family of proteins. *Science* **1999**, *284*, 1664–1666.

130. Arner, E. C.; Hughes, C. E.; Decicco, C. P.; Caterson, B.; Tortorella, M. D. Cytokine-induced cartilage proteoglycan degradation is mediated by aggrecanase. *Osteoarthritis Cartilage* **1998**, *6*, 214–228.

131. Arner, E. C.; Pratta, M. A.; Trzaskos, J. M.; Decicco, C. P.; Tortorella, M. D. Generation and characterization of aggrecanase. A soluble, cartilage-derived aggrecan-degrading activity. *J. Biol. Chem.* **1999**, *274*, 6594–6601.

132. Glasson, S. S.; Askew, R.; Sheppard, B.; Carito, B.; Blanchet, T.; Ma, H. L.; Flannery, C. R.; Peluso, D.; Kanki, K.; Yang, Z.; Majumdar, M. K.; Morris, E. A. Deletion of active ADAMTS5 prevents cartilage degradation in a murine model of osteoarthritis. *Nature* **2005**, *434*, 644–648.

133. Stanton, H.; Rogerson, F. M.; East, C. J.; Golub, S. B.; Lawlor, K. E.; Meeker, C. T.; Little, C. B.; Last, K.; Farmer, P. J.; Campbell, I. K.; Fourie, A. M.; Fosang, A. J. ADAMTS5 is the major aggrecanase in mouse cartilage *in vivo* and *in vitro*. *Nature* **2005**, *434*, 648–652.

134. Bau, B.; Gebhard, P. M.; Haag, J.; Knorr, T.; Bartnik, E.; Aigner, T. Relative messenger RNA expression profiling of collagenases and aggrecanases in human articular chondrocytes *in vivo* and *in vitro*. *Arthritis Rheum.* **2002**, *46*, 2648–2657.

135. Moulharat, N.; Lesur, C.; Thomas, M.; Rolland-Valognes, G.; Pastoureau, P.; Anract, P.; De Ceuninck, F.; Sabatini, M. Effects of transforming growth factor-beta on aggrecanase production and proteoglycan degradation by human chondrocytes *in vitro*. *Osteoarthritis Cartilage* **2004**, *12*, 296–305.

136. Hui, W.; Barksby, H. E.; Young, D. A.; Cawston, T. E.; McKie, N.; Rowan, A. D. Oncostatin M in combination with tumour necrosis factor alpha induces a chondrocyte membrane associated aggrecanase that is distinct from ADAMTS aggrecanase-1 or -2. *Ann. Rheum. Dis.* **2005**, *64*, 1624–1632.

137. Gerhardt, S.; Hassall, G.; Hawtin, P.; McCall, E.; Flavell, L.; Minshull, C.; Hargreaves, D.; Ting, A.; Pauptit, R. A.; Parker, A. E.; Abbott, W. M. Crystal structures of human ADAMTS-1 reveal a conserved catalytic domain and a disintegrin-like domain with a fold homologous to cysteine-rich domains. *J. Mol. Biol.* **2007**, *373*, 891–902.

138. Mosyak, L.; Georgiadis, K.; Shane, T.; Svenson, K.; Hebert, T.; McDonagh, T.; Mackie, S.; Olland, S.; Lin, L.; Zhong, X.; Kriz, R.; Reifenberg, E. L.; Collins-Racie, L. A.; Corcoran, C.; Freeman, B.; Zollner, R.; Marvell, T.; Vera, M.; Sum, P. E.; Lavallie, E. R.; Stahl, M.; Somers, W. Crystal structures of the two major aggrecan degrading enzymes, ADAMTS4 and ADAMTS5. *Protein Sci.* **2008**, *17*, 16–21.

139. Shieh, H. S.; Mathis, K. J.; Williams, J. M.; Hills, R. L.; Wiese, J. F.; Benson, T. E.; Kiefer, J. R.; Marino, M. H.; Carroll, J. N.; Leone, J. W.; Malfait, A. M.; Arner, E. C.; Tortorella, M. D.; Tomasselli, A. High resolution crystal structure of the catalytic domain of ADAMTS-5 (aggrecanase-2). *J. Biol. Chem.* **2008**, *283*, 1501–1507.

140. (a) Yao, W.; Wasserman, Z. R.; Chao, M.; Reddy, G.; Shi, E.; Liu, R. Q.; Covington, M. B.; Arner, E. C.; Pratta, M. A.; Tortorella, M.; Magolda, R. L.; Newton, R.; Qian, M. X.; Ribadeneira, M. D.; Christ, D.; Wexler, R. R.; Decicco, C. P. Design and synthesis of a series of (2*R*)-N4-Hydroxy-2-(3-hydroxybenzyl)-N1-[(1*S*,2*R*)-2-hydroxy-2,3-dihydro-1*H*-inden-1-yl]butanediamide derivatives as potent, selective, and orally bioavailable aggrecanase inhibitors. *J. Med. Chem.* **2001**, *44*, 3347–3350. (b) Cherney, R. J.; Mo, R.; Meyer, D. T.; Wang, L.; Yao, W.; Wasserman, Z. R.; Liu, R. Q.; Covington, M. B.; Tortorella, M. D.; Arner, E. C.; Qian, M.; Christ, D. D.; Trzaskos, J. M.; Newton, R. C.; Magolda, R. L.; Decicco, C. P. *Bioorg. Med. Chem. Lett.* **2003**, *13*, 1297.

141. Mitchell, P. G.; Magna, H. A.; Reeves, L. M.; Lopresti-Morrow, L. L.; Yocum, S. A.; Rosner, P. J.; Geoghegan, K. F.; Hambor, J. E. Cloning, expression, and type II

collagenolytic activity of matrix metalloproteinase-13 from human osteoarthritic cartilage. *J. Clin. Invest.* **1996**, *97*, 761–768.

142. Noe, M. C.; Snow, S. L.; Wolf-Gouveia, L. A.; Mitchell, P. G.; Lopresti-Morrow, L.; Reeves, L. M.; Yocum, S. A.; Liras, J. L.; Vaughn, M. 3-Hydroxy-4-arylsulfonyltetrahydropyranyl-3-hydroxamic acids are novel inhibitors of MMP-13 and aggrecanase. *Bioorg. Med. Chem. Lett.* **2004**, *14*, 4727–4730.

143. Noe, M. C.; Natarajan, V.; Snow, S. L.; Mitchell, P. G.; Lopresti-Morrow, L.; Reeves, L. M.; Yocum, S. A.; Carty, T. J.; Barberia, J. A.; Sweeney, F. J.; Liras, J. L.; Vaughn, M.; Hardink, J. R.; Hawkins, J. M.; Tokar, C. Discovery of 3,3-dimethyl-5-hydroxypipecolic hydroxamate-based inhibitors of aggrecanase and MMP-13. *Bioorg. Med. Chem. Lett.* **2005**, *15*, 2808–2811.

144. Noe, M. C.; Natarajan, V.; Snow, S. L.; Wolf-Gouveia, L. A.; Mitchell, P. G.; Lopresti-Morrow, L.; Reeves, L. M.; Yocum, S. A.; Otterness, I.; Bliven, M. A.; Carty, T. J.; Barberia, J. T.; Sweeney, F. J.; Liras, J. L.; Vaughn, M. Discovery of 3-OH-3-methylpipecolic hydroxamates: Potent orally active inhibitors of aggrecanase and MMP-13. *Bioorg. Med. Chem. Lett.* **2005**, *15*, 3385–3388.

145. Xiang, J. S.; Hu, Y.; Rush, T. S.; Thomason, J. R.; Ipek, M.; Sum, P.-E.; Abrous, L.; Sabatini, J. J.; Georgiadis, K.; Reifenberg, E.; Majumdar, M.; Morrisc, E. A.; Tam, S. Synthesis and biological evaluation of biphenylsulfonamide carboxylate aggrecanase-1 inhibitors. *Bioorg. Med. Chem. Lett.* **2006**, *16*, 311–316.

146. Inaba, T.; Haas, J.; Shiozaki, M.; Litman, N. M.; Yasue, K.; Andrews, S. W.; Sakai, A.; Fryer, A. M.; Matsuo, T.; Laird, E. R.; Suma, A.; Shinozaki, Y.; Hori, Y.; Imai, H.; Negoro, T. Cyclopropane compounds and pharmaceutical use thereof. *PCT* WO 058884, **2005**.

147. Gallagher, B. M.; Van Veldhuizen, J.; Deng, H.; Sucholeiki, I.; Hochguertel, M.; Bluhm, H.; Taveras, A. G. Heterobicyclic metalloproteinase inhibitors. PCT/US 2007/0155737.

148. Engel, C. K.; Pirard, B.; Schimanski, S.; Kirsch, R.; Habermann, J.; Klinger, O.; Schlotte, V.; Weithmann, K. U.; Wendt, U. Structural basis for the highly selective inhibition of MMP-13. *Chem. Biol.* **2005**, *12*, 181–189.

149. Takizawaa, M.; Yatabea, T.; Okadaa, A.; Chijiiwaa, M.; Mochizukia, S.; Ghoshb, P.; Okada, Y. Calcium pentosan polysulfate directly inhibits enzymatic activity of ADAMTS4 (aggrecanase-1) in osteoarthritic chondrocytes. *FEBS Lett.* **2008**, *582*, 2945–2949.

150. Cullis-Hill, D. Treatment of osteoporosis. *PCT* WO 02/41901, **2002**.

151. Bursavich, M. G.; Gilbert, A. M.; Lombardi, S.; Georgiadis, K. E.; Reifenberg, E.; Flannery, C. R.; Morris, E. A. Synthesis and evaluation of aryl thioxothiazolidinone inhibitors of ADAMTS-5 (Aggrecanase-2). *Bioorg. Med. Chem. Lett.* **2007**, *17*, 1185–1188.

152. Gilbert, A. M.; Bursavich, M. G.; Lombardi, S.; Georgiadis, K. E.; Reifenberg, E.; Flannery, C. R.; Morris, E. A. 5-((1*H*-Pyrazol-4-yl)methylene)-2-thioxothiazolidin-4-one inhibitors of ADAMTS-5. *Bioorg. Med. Chem. Lett.* **2007**, *17*, 1189–1192.

153. (a) Gilbert, A. M.; Bursavich, M. G.; Lombardi, S.; Georgiadis, K. E.; Reifenberg, E.; Flannery, C. R.; Morris, E. A. 5′-Phenyl-3′*H*-spiro[indoline-3,2′-[1,3,4]thiadiazol]-2-one inhibitors of ADAMTS-5 (Aggrecanase-2). *Bioorg. Med. Chem. Lett.* **2007**, *17*, 5630–5633.(b) Bursavich, M. G.; Lombardi, S.; Gilbert, A. M.PCT/US2007/013530.

154. Sum, P.-E.; How, D. B.; Sabatini, J. J.; Xiang, J. S.; Ipek, M.; Feyfant, E.PCT/US2006/027066.

CHAPTER 27

QSAR Studies of MMP Inhibitors

TIZIANO TUCCINARDI and ADRIANO MARTINELLI

Dipartimento di Scienze Farmaceutiche, Università di Pisa, Via Bonanno 6, 56126 Pisa, Italy

27.1 INTRODUCTION

Quantitative structure–activity relationship (QSAR) is the most direct approach for trying to correlate structural or property descriptors of compounds with activities. This methodology, since its advent more than 40 years ago,[1] has become more and more useful for understanding ligand–receptor interaction mechanisms.[2] It has also been utilized for the evaluation of absorption, distribution, metabolism, and excretion (ADME).[3]

The descriptors can be obtained through experimental measurements, but today they are generally calculated through computational methods and can include parameters to account for structural features or physicochemical properties. The activities used in QSAR are generally obtained by biochemical or pharmacological assays.

QSAR uses extra-thermodynamically derived and computationally calculated descriptors to correlate biological activity in isolated systems and *in vivo*. Three molecular descriptor types are mainly used in QSAR analysis: electronic, hydrophobic, and steric. In many cases, they are able to describe critical receptor–ligand interactions.[2]

Generally, QSAR models are built on the basis of a linear SAR hypothesis, and therefore multiple linear regression (MLR) techniques are used; however, nonlinear models have also been built through neural networks (NN) or artificial intelligence approaches.[4,5]

The quality of a QSAR model mainly depends on the type and quality of the data, and is applicable only to compounds possessing molecular structures similar to those used to build the model. In all case, a model's predictive ability needs to be carefully validated.[6]

More recently, the availability of more sophisticated and potent hardware and software resources has allowed the development of 3D-QSAR methods in which conformation dependent descriptors obtained from molecular interaction fields (MIFs) are used, and models are calculated by using partial least squares (PLS) algorithms.[7,8]

QSAR and 3D-QSAR models are used not only for elucidating the interaction mechanisms of ligands but also for optimizing their biological profile. Furthermore, it is also possible to use these models for screening chemical databases or virtual libraries to rationalize the drug design process.

27.2 QSAR STUDIES

Many MMP inhibitors (MMPIs) have been studied in terms of QSAR analysis, and Fig. 27.1 shows the main scaffolds.

Table 27.1 reports all the main published QSAR models with their equations and the main statistical data.

At least 22 scaffolds have been investigated by means of QSAR techniques. All the derivatives were characterized by the presence of the hydroxamic acid as the zinc binding group (ZBG) and a substituted sulfonyl group. The only exception concerns the compounds studied by Verma, Jamloki, Gupta, and coworkers[13,19,21] (scaffolds E, F, and S).

To the best of our knowledge, 85 QSAR studies have been published in the past 8 years. As shown in Fig. 27.2, these studies investigated MMP-1, -2, -3, -7, -8, -9, and -13. In particular, MMP-1 and -13 were the most studied, whereas only two studies were reported for MMP-7.

In most cases, some indicator variables (*I*) were used together with physicochemical descriptors to take into account the structural differences among the compounds in the data set.

The main physicochemical parameters used for investigating the molecules are related to the hydrophobicity. In fact, more than one-third of the published QSAR models were developed using the correlation between the hydrophobicity of the substituent/molecule and the inhibitory activity. Many QSAR models were developed using a linear correlation between the octanol/water partition coefficient ($\log P$ or $C\log P$) and π (that takes into account the hydrophobicity of the substituents); in a few cases, a parabolic correlation of $C\log P$ was used.

Other parameters used in the QSAR studies are topological indices like Kier's first-order valence molecular connectivity index of the substituents/molecules (${}^1\chi^v$) and Kier's third alpha modified shape index (${}^3K_\alpha$), the electrotopological state (E-state) indices (*S*) of nitrogen and sulfur atoms, which take into account electronic effects of atoms/substituents, the polarizability of the molecules (Pol), the number of valence electrons (NVE), the molar refractivity (MR), the Hammett electronic parameters (σ), Broto–Moreau's autocorrelation coefficients (ATS), Moran's indices (MATS), and Geary's coefficients (GATS).

Hundreds of MMPIs were analyzed by means of the QSAR studies reported in Table 27.1; however, with only a few exceptions, all the compounds were used for constructing the models, that is, they belonged to training sets. As already reported by

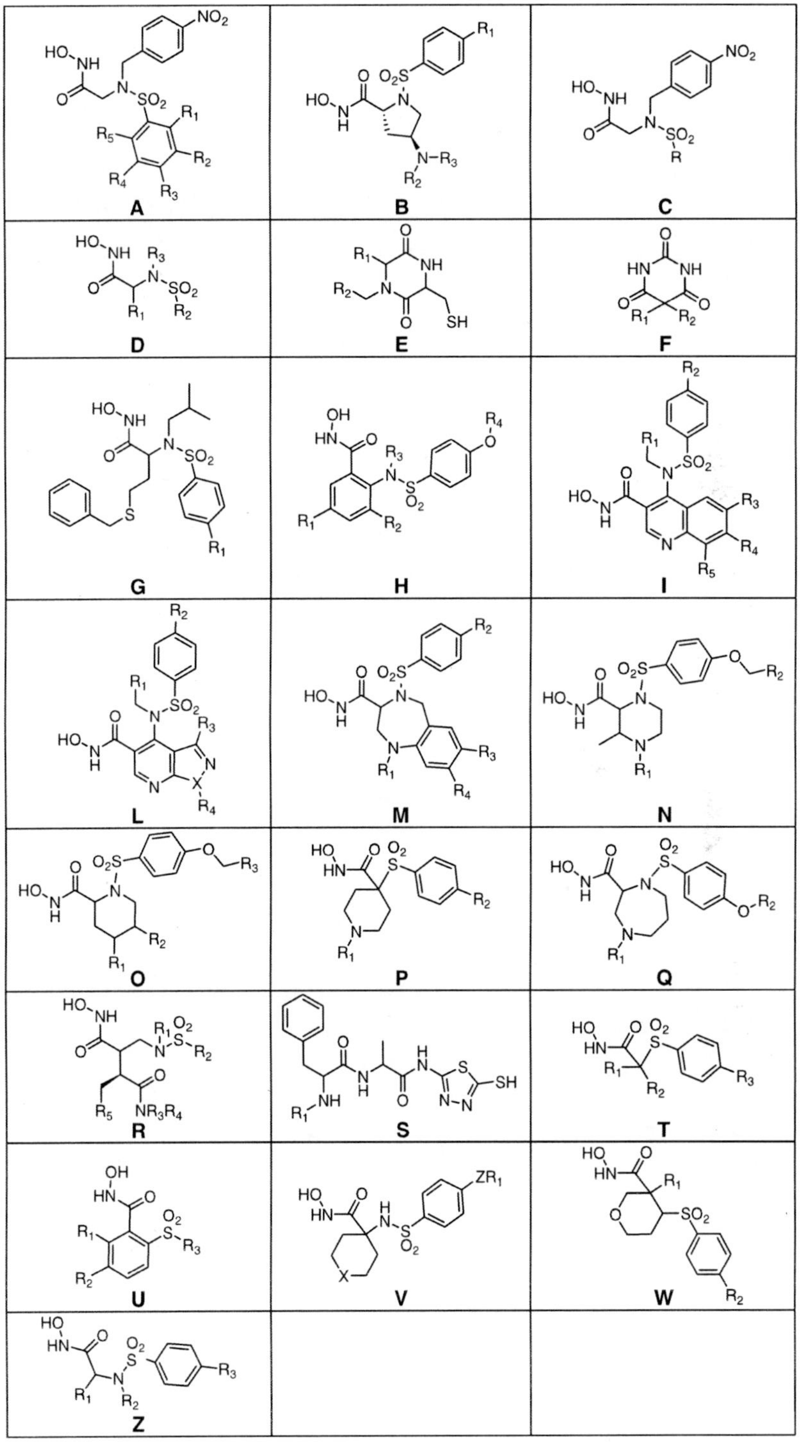

FIGURE 27.1 General scaffold of the MMP inhibitors used for the development of QSAR models.

TABLE 27.1 Main QSAR Models on MMP Inhibitors

MMP	Compound	QSAR Equation	Statistics	Reference
MMP-1	A	$\log[1/K_I(M)] = 0.550(\pm0.264)\pi_{0+m}-0.576(\pm0.332)I_{vic}+0.587(\pm0.367)\sigma_m-0.634(\pm0.383)MR_0+1.752(\pm0.482)$	$n=19$, $r=0.878$, $SEE=0.172$, $F_{4,14}=11.8$	9
MMP-1	B	$\log(1/IC_{50}) = 6.775(\pm0.268)-0.286(\pm0.166)^1\chi_R^v+0.748(\pm0.309)I_w$	$n=19$, $r=0.901$, $r^2_{cv}=0.63$, $s=0.23$, $F_{2,16}=34.71(6.23)$	10
MMP-1	B	$\log(1/IC_{50}) = -4.721(\pm1.093)^1\chi_R^v+1.858(\pm0.387)(^1\chi_R^v)^2+5.678(\pm2.511)S_S+1.444(\pm0.712)I_Y+42.213(\pm14.811)$	$n=26$, $r=0.933$, $r^2_{cv}=0.78$, $s=0.27$, $F_{4,21}=35.36(4.37)$	10
MMP-1	C	$\log(1/K_I) = 7.441(\pm0.114)+0.198(\pm0.122)S_N+0.893(\pm0.259)D$	$n=28$, $r=0.830$, $s=0.17$, $F_{2,25}=27.67(5.57)$	11
MMP-1	D	$\log(1/K_I) = 0.194(\pm0.082)^1\chi^v+0.423(\pm0.144)S_S-0.862(\pm0.242)S_N+0.75(\pm0.142)I+8.859(\pm1.614)$	$n=31$, $r=0.945$, $s=0.17$, $F_{4,26}=53.78(4.14)$	12
MMP-1	E	$\log(1/C) = 0.038(\pm0.011)NVE+1.85(\pm1.22)$	$n=6$, $q^2=0.887$, $s=0.076$	
MMP-1	H	$\log(1/IC_{50}) = 7.286(\pm0.331)-2.743(\pm1.279)\log P+1.098(\pm0.683)(\log P)^2$	$n=7$, $r=0.960$, $r^2_{cv}=0.80$, $s=0.18$, $F_{1,4}=23.53(21.20)$	14
MMP-1	H	$\log(1/IC_{50}) = 1.020(\pm0.396)I_2+0.596(\pm0.487)I_3-0.192(\pm0.118)\log P+5.979(\pm0.432)$	$n=16$, $r=0.919$, $r^2_{cv}=0.67$, $s=0.28$, $F_{3,12}=21.86(5.95)$	14
MMP-1	H	$\log(1/IC_{50}) = 0.234(\pm0.147)I_{4,Br}+0.317(\pm0.186)I_3+0.629(\pm0.173)I_2+0.534(\pm0.251)I_1-0.062(\pm0.060)\log P+6.801(\pm0.150)$	$n=19$, $r=0.935$, $r^2_{cv}=0.74$, $s=0.13$, $F_{5,13}=18.15(4.86)$	14

MMP-1	H	$\log(1/IC_{50}) = 0.410(\pm0.158)Pol + 0.773(\pm0.325)I_1 + 5.960(\pm0.824)$	$n = 19$, $r = 0.914$, $r^2_{cv} = 0.79$, $s = 0.19$, $F_{2.14} = 40.71(6.51)$	14
MMP-1	I	$\log(1/IC_{50}) = 0.840(\pm0.292)I_{2,M} - 0.176(\pm0.136)C\log P + 6.576(\pm0.597)$	$n = 19$, $r = 0.914$, $r^2_{cv} = 0.73$, $s = 0.21$, $F_{2.16} = 40.75(6.23)$	15
MMP-1	L	$\log(1/IC_{50}) = 6.895(\pm0.588) - 2.116(\pm0.882)C\log P + 0.928(\pm0.372)(C\log P)^2 + 0.462(\pm0.408)I_{1,pyr}$	$n = 10$, $r = 0.954$, $r^2_{cv} = 0.71$, $s = 0.24$, $F_{3.6} = 20.12(9.78)$	15
MMP-1	M	$\log(1/IC_{50}) = 1.123(\pm0.347)I_{R1} - 1.929(\pm0.972)C\log P + 0.259(\pm0.170)(C\log P)^2 + 9.635(\pm1.276)$	$n = 16$, $r = 0.911$, $r^2_{cv} = 0.72$, $s = 0.27$, $F_{3.12} = 19.55(5.95)$	16
MMP-1	M	$\log(1/IC_{50}) = 0.444(\pm0.108)I_{R1-A} - 2.740(\pm0.968)C\log P + 0.672(\pm0.235)(C\log P)^2 + 8.734(\pm0.950)$	$n = 12$, $r = 0.981$, $r^2_{cv} = 0.91$, $s = 0.06$, $F_{3.8} = 69.69(7.59)$	16
MMP-1	N	$\log(1/IC_{50}) = 0.965(\pm0.473)C\log P - 0.173(\pm0.073)(C\log P)^2 - 1.008(\pm0.288)I_{1,H}$	$n = 13$, $r = 0.947$, $r^2_{cv} = 0.67$, $s = 0.16$, $F_{3.9} = 26.16(6.99)$	17
MMP-1	O	$\log(1/IC_{50}) = 0.527(\pm0.234)\pi_{X,R} + 0.289(\pm0.212)I_{R1} + 4.846(\pm0.128)$	$n = 15$, $r = 0.844$, $r^2_{cv} = 0.58$, $s = 0.17$, $F_{2.12} = 14.90(6.93)$	17
MMP-1	P	$\log(1/IC_{50}) = 0.523(\pm0.180)C\log P - 0.102(\pm0.044)(C\log P)^2 + 0.592(\pm0.170)I_{1,OMe} + 4.984(\pm0.249)$	$n = 38$, $r = 0.850$, $r^2_{cv} = 0.64$, $s = 0.21$, $F_{3.34} = 28.95(4.42)$	17
MMP-1	R	$\log(1/IC_{50}) = 7.600(\pm0.334) - 0.404(\pm0.263)C\log P + 0.085(\pm0.051)(C\log P)^2 + 0.935(\pm0.212)I_R + 0.388(\pm0.201)I_{R2}$	$n = 18$, $r = 0.954$, $r^2_{cv} = 0.75$, $s = 0.15$, $F_{4.13} = 32.80(5.20)$	18
MMP-1	D	$\log(1/IC_{50}) = 6.869(\pm0.975) - 1.367(\pm1.068)C\log P + 0.551(\pm0.283)(C\log P)^2 - 0.673(\pm0.464)I_{1,NH} - 0.785(\pm0.315)I_{2,H}$	$n = 15$, $r = 0.961$, $r^2_{cv} = 0.86$, $s = 0.27$, $F_{4.10} = 30.28(5.99)$	18

(continued)

TABLE 27.1 (*Continued*)

MMP	Compound	QSAR Equation	Statistics	Reference
MMP-1	S	$\log(1/K_I) = 4.39725(\pm 1.03418) + {}^1\chi_c^v - [0.207152(\pm 0.0851363)] + 3K_\alpha[0.201545(\pm 0.0927689)] + a\_nF[0.0832763(\pm 0.0471238)]$	$n = 27$, $r = 0.913$, $q^2 = 0.734$, SEE $= 0.122$, $F_{3.23} = 38.645$ (4.765)	19
MMP-1	P, T	$\log(10^6/IC_{50}) = -17.557MATS4m - 5.396MATS3v + 17.908MATS6v - 4.396MATS5e - 4.375MATS6e + 10.359GATS6v - 5.118GATS7v + 15.274$	$n = 63$, $r^2_{cv} = 0.559$, $s_{cv} = 0.312$, $n_{ts} = 10$, $r^2_{EP} = 0.664$, $s_{EP} = 0.282$	20
MMP-1	S	$\log(1/K_I) = 1.522(\pm 0.398)S_S - 3.690(\pm 0.779)S_N - 1.624(\pm 0.420)I + 22.598(\pm 4.183)$	$n = 24$, $r = 0.934$, $r^2_{cv} = 0.81$, $s = 0.11$, $F_{3.20} = 45.83(4.94)$	21
MMP-1	V	$\log(1/IC_{50}) = 0.935(\pm 0.219)I_{1,Ph} + 0.961(\pm 0.429)\pi_{3,4}(R^1) + 5.128(\pm 0.162)$	$n = 20$, $r = 0.924$, $r^2_{cv} = 0.79$, $s = 0.23$, $F_{2.17} = 49.42(6.11)$	22
MMP-1	W	$\log(1/IC_{50}) = 4.843(\pm 0.202) - 4.207(\pm 3.357)\pi_4(R^2) + 8.113(\pm 5.153)[\pi_4(R^2)]2 + 1.806(\pm 0.413)I_{2,OPh}$	$n = 12$, $r = 0.971$, $r^2_{cv} = 0.87$, $s = 0.22$, $F_{3.8} = 43.21(7.59)$	22
MMP-1	Z	$\log(10^6/IC_{50}) = -166.804MATS4m - 51.519MATS8m - 13.020MATS3v + 4.817GATS1e - 6.913GATS2e + 219.442$	$n = 26$, $q^2 = 0.745$, $s = 0.383$	23
MMP-2	A	$\log[1/K_I(\mu M)] = 5.695(\pm 1.852)\sigma_m - 5.499(\pm 2.404)\sigma_{m^2} - 1.515(\pm 0.622)MR_m - 0.362(\pm 0.333)\pi_o + 2.158(\pm 0.550)$	$n = 18$, $r = 0.911$, SEE $= 0.202$, $F_{4.13} = 15.9$	9
MMP-2	B	$\log(1/IC_{50}) = 8.369(\pm 0.414) + 0.202(\pm 0.106)\,{}^1\chi_N^v$	$n = 9$, $r = 0.863$, $r^2_{cv} = 0.56$, $s = 0.23$, $F_{1.7} = 20.46(12.25)$	10

MMP-2	B	$\log(1/IC_{50}) = -1.965(\pm0.823)S_S - 0.217(\pm0.195)\,^1\chi^v_N - 2.195(\pm4.679)$	$n=9$, $r=0.926$, $r^2_{cv}=0.76$, $s=0.15$, $F_{2.6}=18.18(10.92)$	10
MMP-2	C	$\log(1/K_I) = 9.447(\pm1.052) - 1.172(\pm0.489)S_N + 0.432(\pm0.312)S_S + 0.149(\pm0.082)\,^1\chi^v + 0.898(\pm0.207)D$	$n=33$, $r=0.928$, $s=0.19$, $F_{4.28}=43.47(4.07)$	11
MMP-2	D	$\log(1/K_I) = 0.204(\pm0.080)\,^1\chi^v + 0.198(\pm0.127)S_S - 0.682(\pm0.211)S_N + 0.616(\pm0.144)I + 7.938(\pm1.563)I$ $= 1$ if the R_2 group is C_6F_5, otherwise, $= 0$	$n=31$, $r=0.964$, $s=0.21$, $F_{4.34}=111(3.93)$	12
MMP-2	R	$\log(1/IC_{50}) = 0.648(\pm0.133)Pol + 0.496(\pm0.234)I_R + 1.088(\pm0.236)I_{R2} + 3.260(\pm0.633)$	$n=19$, $r=0.971$, $r^2_{cv}=0.90$, $s=0.19$, $F_{3.15}=81.88(5.42)$	18
MMP-2	S	$\log(1/K_I) = 2.92822(\pm0.860004) + {}^3K_\alpha[0.255415(\pm0.122169)] + a\_nF[0.110207(\pm0.0556175)] + I[0.357692(\pm0.191531)]$	$n=27$, $r=0.893$, $q^2=0.682$, SEE $=0.152$, $F_{3.23}=30.175(4.765)$	19
MMP-2	S	$\log(1/K_I) = 0.853(\pm0.309)S_S - 2.331(\pm0.598)S_N - 0.683(\pm0.325)I + 15.530(\pm3.227)$	$n=24$, $r=0.952$, $r^2_{cv}=0.87$, $s=0.09$, $F_{3.20}=64.35(4.94)$	21
MMP-2	T	$\log(1/IC_{50}) = 3.528(\pm2.400)C\log P - 0.894(\pm0.592)(C\log P)^2 + 1.975(\pm0.458)I_1 - 1.352(\pm0.711)I_{1,N} + 3.860(\pm2.162)$	$n=22$, $r=0.916$, $r^2_{cv}=0.82$, $s=0.42$, $F_{4.17}=46.66(4.67)$	24
MMP-2	Z	$\log(10^6/IC_{50}) = -0.042ATS2v - 110.925MATS4m - 55.421MATS8m - 8.431MATS3v + 2.379GATS6e + 5.227GATS2p + 162.799$	$n=32$, $q^2=0.721$, $s=0.420$	23
MMP-3	B	$\log(1/IC_{50}) = -0.501(\pm0.206)\,^1\chi^v_N + 0.050(\pm0.029)(^1\chi^v_N)^2 + 0.656(\pm0.218)I_w - 2.946(\pm0.945)S_S - 7.850(\pm5.066)$	$n=20$, $r=0.916$, $r^2_{cv}=0.65$, $s=0.18$, $F_{4.15}=19.58(4.89)$	10

(continued)

TABLE 27.1 *(Continued)*

MMP	Compound	QSAR Equation	Statistics	Reference
MMP-3	B	$\log(1/\mathrm{IC}_{50}) = 7.864(\pm0.278) - 0.137(\pm0.080)\,^1\chi^v_N + 0.561(\pm0.177)\mathrm{I_R}$	$n=26$, $r=0.843$, $r^2_{cv}=0.63$, $s=0.14$, $F_{2.23}=28.27(5.66)$	10
MMP-3	R	$\log(1/\mathrm{IC}_{50}) = 0.818(\pm0.195)\mathrm{Pol} + 0.692(\pm0.324)I_R + 1.363(\pm0.328)I_{R2} + 2.398(\pm0.944)$	$n=19$, $r=0.963$, $r^2_{cv}=0.89$, $s=0.26$, $F_{3.15}=64.04(5.42)$	18
MMP-3	Z	$\log(10^6/\mathrm{IC}_{50}) = -112.340\mathrm{MATS4m} - 15.666\mathrm{MATS3v} + 8.716\mathrm{MATS6e} + 2.379\mathrm{GATS6e} - 1.457\mathrm{GATS8e} + 7.200\mathrm{GATS2p} + 104.088$	$n=30$, $q^2=0.581$, $s=0.431$	23
MMP-7	B	$\log(1/\mathrm{IC}_{50}) = 0.175(\pm0.084)\,^1\chi^v_N + 1.863(\pm0.982)\mathrm{S_S} + 0.405(\pm0.234)I_w + 15.319(\pm5.224)$	$n=12$, $r=0.866$, $r^2_{cv}=0.47$, $s=0.10$, $F_{3.8}=8.01(7.59)$	10
MMP-7	B	$\log(1/\mathrm{IC}_{50}) = 1.843(\pm1.541)\,^1\chi^v_N - 0.274(\pm0.247)(^1\chi^v_N)^2 - 1.709(\pm1.025)\,^1\chi^v_R + 0.743(\pm0.348)(^1\chi^v_R)^2 - 1.811(\pm0.931)\mathrm{S_S} + 0.928(\pm0.346)I_R - 7.950(\pm5.200)$	$n=27$, $r=0.903$, $r^2_{cv}=0.68$, $s=0.25$, $F_{6.20}=14.71(3.87)$	10
MMP-8	A	$\log[1/K_I(\mu\mathrm{M})] = 7.676(\pm2.222)\sigma_m - 6.572(\pm2.884){\sigma_m}^2 - 2.573(\pm0.747)\mathrm{MR_m} - 0.461(\pm0.397)\pi_o + 2.342(\pm0.659)$	$n=18$, $r=0.939$, $\mathrm{SEE}=0.242$, $F_{4.13}=24.4$	9
MMP-8	C	$\log(1/\mathrm{K_I}) = 8.205(\pm0.163) - 0.466(\pm0.172)\mathrm{S_N} + 0.909(\pm0.300)D$	$n=31$, $r=0.844$, $s=0.24$, $F_{2.28}=34.64(5.45)$	11
MMP-8	D	$\log(1/K_I) = 0.244(\pm0.118)\,^1\chi^v + 0.264(\pm0.194)\mathrm{S_S} - 0.757(\pm0.320)\mathrm{S_N} + 0.667(\pm0.219)\mathrm{I} + 7.933(\pm2.366)$ $I = 1$ if the R_2 group is C_6F_5; otherwise, $= 0$	$n=37$, $r=0.940$, $s=0.31$, $F_{4.32}=60.20(3.97)$	12
MMP-8	R	$\log(1/\mathrm{IC}_{50}) = 0.425(\pm0.158)\mathrm{Pol} + 0.783(\pm0.257)I_R + 1.045(\pm0.262)I_{R2} + 5.459(\pm0.762)$	$n=18$, $r=0.956$, $r^2_{cv}=0.85$, $s=0.20$, $F_{3.14}=49.69(5.56)$	18

MMP-8	S	$\log(1/K_I) = [2.96998(\pm0.987275) + {}^3K_\alpha[0.242856(\pm0.140256)] + a\_nF[0.272465(\pm0.063659)] + I[0.413534(\pm0.245314)]$	$n = 25$, $r = 0.916$, $q^2 = 0.772$, SEE $= 0.170$, $F_{3.21} = 36.846(4.874)$	19
MMP-8	S	$\log(1/K_I) = 0.977(\pm0.377)S_S - 3.579(\pm0.552)S_N + 19.495(\pm3.296)$	$n = 24$, $r = 0.957$, $r^2_{cv} = 0.90$, $s = 0.21$, $F_{2.21} = 113.03(5.78)$	21
MMP-9	A	$\log[1/K_I(M)] = 6.581(\pm2.173)\sigma_m - 6.302(\pm2.820)\sigma_m{}^2 - 1.796(\pm0.730)MR_m - 0.526(\pm0.390)\pi_o + 2.161(\pm0.645)$	$n = 18$, $r = 0.910$, SEE $= 0.237$, $F_{4.13} = 15.7$	9
MMP-9	C	$\log(1/K_I) = 9.581(\pm1.332) - 1.249(\pm0.618)S_N + 0.512(\pm0.394)S_S + 0.166(\pm0.103){}^1\chi^v + 1.057(\pm0.262)D$	$n = 33$, $r = 0.902$, $s = 0.23$, $F_{4.28} = 30.50(4.07)$	11
MMP-9	D	$\log(1/K_I) = 0.283(\pm0.106){}^1\chi^v + 0.433(\pm0.143)S_N + 0.683(\pm0.240)I + 5.510(\pm0.951)$ $I = 1$ if the R_2 group is C_6F_5, otherwise, $= 0$	$n = 37$, $r = 0.920$, $s = 0.34$, $F_{3.33} = 60.70(4.44)$	12
MMP-9	F	$\log(1/C) = 0.037(\pm0.013)NVE + 2.04(\pm1.70)$	$n = 7$, $q^2 = 0.843$, $s = 0.318$	13
MMP-9	H	$\log(1/IC_{50}) = 8.525(\pm0.193) - 0.576(\pm0.147)\log P$	$n = 10$, $r = 0.954$, $r^2_{cv} = 0.86$, $s = 0.15$, $F_{1.8} = 81.56(11.26)$	14
MMP-9	H	$\log(1/IC_{50}) = 8.336(\pm0.492) - 0.265(\pm0.183)\log P - 1.241(\pm0.725)I_1 + 1.183(\pm0.691)I_4$	$n = 19$, $r = 0.882$, $r^2_{cv} = 0.64$, $s = 0.50$, $F_{3.15} = 17.47(5.42)$	14
MMP-9	H	$\log(1/IC_{50}) = 0.695(\pm0.172)I_3 - 0.154(\pm0.133)\log P + 0.064(\pm0.039)(\log P)^2 + 8.143(\pm0.122)$	$n = 19$, $r = 0.931$, $r^2_{cv} = 0.74$, $s = 0.16$, $F_{2.16} = 32.65(6.23)$	14
MMP-9	H	$\log(1/IC_{50}) = 0.503(\pm0.477)Pol - 1.806(\pm0.567)I_{1,CC} - 0.807(\pm0.559)I_{1,N} + 5.916(\pm2.137)$	$n = 16$, $r = 0.920$, $r^2_{cv} = 0.64$, $s = 0.37$, $F_{3.12} = 21.92(5.95)$	14

(*continued*)

TABLE 27.1 *(Continued)*

MMP	Compound	QSAR Equation	Statistics	Reference
MMP-9	I	$\log(1/\mathrm{IC}_{50}) = 1.391(\pm0.440)I_{2,\mathrm{M}} - 0.461(\pm0.414)I_6 - 0.619(\pm0.312)I_7 - 0.161(\pm0.153)C\log P + 7.680(\pm0.747)$	$n = 18$, $r = 0.945$, $r^2_{cv} = 0.60$, $s = 0.25$, $F_{4.13} = 27.28(5.20)$	15
MMP-9	L	$\log(1/\mathrm{IC}_{50}) = 2.192(\pm0.340)\mathrm{Pol} - 0.866(\pm0.256)I_{2,\mathrm{CC}} - 2.077(\pm1.603)$	$n = 10$, $r = 0.991$, $r^2_{cv} = 0.95$, $s = 0.15$, $F_{2.7} = 199.82(9.55)$	15
MMP-9	M	$\log(1/\mathrm{IC}_{50}) = 0.119(\pm0.074)\mathrm{S_N} - 1.149(\pm0.355)I_{\mathrm{R2-CC}} - 0.798(\pm0.440)I_{\mathrm{AC}} + 11.169(\pm1.535)$	$n = 17$, $r = 0.932$, $r^2_{cv} = 0.65$, $s = 0.25$, $F_{3.14} = 28.47(5.56)$	16
MMP-9	P	$\log(1/\mathrm{IC}_{50}) = 0.909(\pm0.263)\mathrm{C}\log P - 0.128(\pm0.061)(C\log P)^2 + 0.770(\pm0.285)I_{1,\mathrm{PhCl}} + 6.686(\pm0.284)$	$n = 42$, $r = 0.891$, $r^2_{cv} = 0.74$, $s = 0.31$, $\mathrm{F}_{3.38} = 48.79(4.34)$	17
MMP-9	Q	$\log(1/\mathrm{IC}_{50}) = 0.339(\pm0.148)C\log P + 0.853(\pm0.479)I_{1,\mathrm{COPh}} + 7.217(\pm0.329)$	$n = 10$, $r = 0.942$, $r^2_{cv} = 0.70$, $s = 0.25$, $F_{2.7} = 27.51(9.55)$	17
MMP-9	D	$\log(1/\mathrm{IC}_{50}) = 0.612(\pm0.222)\mathrm{C}\log P + 1.363(\pm0.499)I_{\mathrm{CC}} - 0.940(\pm0.337)I_{2,\mathrm{H}} + 4.946(\pm0.682)$	$n = 17$, $r = 0.944$, $r^2_{cv} = 0.80$, $s = 0.29$, $\mathrm{F}_{3.13} = 35.32(5.74)$	18
MMP-9	S	$\log(1/K_\mathrm{I}) = 2.59183(\pm1.567515) + {}^3K_\alpha[0.293886(\pm0.222674)] + \mathrm{a\_}nF[0.237027(\pm0.101373)] + I[0.789148(\pm0.349099)]$	$n = 27$, $r = 0.885$, $q^2 = 0.676$, $\mathrm{SEE} = 0.276$, $F_{3.23} = 27.834(4.765)$	19
MMP-9	P, T	$\log(10^6/\mathrm{IC}_{50}) = -0.005\mathrm{ATS6m} + 0.018\mathrm{ATS3e} + 8.881\mathrm{MATS2e} - 7.718\mathrm{MATS4e} - 4.655\mathrm{GATS1v} + 14.788\mathrm{GATS1e} + 2.379\mathrm{GATS6p} - 4.571$	$n = 66$, $r^2_{cv} = 0.605$, $s_{cv} = 0.416$, $n_{ts} = 12$, $r^2_{EP} = 0.713$, $s_{EP} = 0.415$	20

MMP-9	S	$\log(1/K_I) = 0.990(\pm0.259)S_S - 3.132(\pm0.380)S_N + 18.383(\pm2.267)$	$n = 24$, $r = 0.971$, $r^2_{cv} = 0.93$, $s = 0.15$, $F_{2.21} = 170.65(5.78)$	21
MMP-9	Z	$\log(10^6/IC_{50}) = 109.844MATS2m - 51.167MATS4m + 8.380MATS2v + 13.240MATS6e + 7.864GATS6e + 13.015GATS2p - 72.742$	$n = 32$, $q^2 = 0.644$, $s = 0.478$	23
MMP-13	B	$\log(1/IC_{50}) = 8.098(\pm0.269) + 0.065(\pm0.054)^1\chi^v_N + 0.379(\pm0.124)^1\chi^v_R + 0.535(\pm0.211)I_W$	$n = 18$, $r = 0.900$, $r^2_{cv} = 0.72$, $s = 0.15$, $F_{3.14} = 19.87(5.56)$	10
MMP-13	B	$\log(1/IC_{50}) = 1.158(\pm0.964)^1\chi^v_N - 0.187(\pm0.151)(^1\chi^v_N)2 - 0.799(\pm0.487)S_S + 0.871(\pm0.213)I_R + 2.102(\pm3.042)$	$n = 20$, $r = 0.915$, $r^2_{cv} = 0.73$, $s = 0.14$, $F_{4.15} = 19.38(4.89)$	10
MMP-13	G	$\log(1/C) = 0.025(\pm0.011)NVE + 3.73(\pm1.95)$	$n = 4$, $q^2 = 0.925$, $s = 0.058$	13
MMP-13	H	$\log(1/IC_{50}) = 8.699(\pm0.302) - 1.240(\pm0.247)\log P$	$n = 9$, $r = 0.976$, $r^2_{cv} = 0.92$, $s = 0.23$, $F_{1.7} = 140.93(12.25)$	14
MMP-13	H	$\log(1/IC_{50}) = 7.902(\pm0.366) - 0.184(\pm0.138)\log P - 1.051(\pm0.552)I_1 + 1.079(\pm0.626)I_3 + 1.341(\pm0.527)I_4$	$n = 22$, $r = 0.919$, $r^2_{cv} = 0.75$, $s = 0.38$, $F_{4.17} = 23.06(4.67)$	14
MMP-13	H	$\log(1/IC_{50}) = 0.410(\pm0.158)Pol + 0.773(\pm0.325)I_1 + 5.960(\pm0.824)$	$n = 19$, $r = 0.914$, $r^2_{cv} = 0.79$, $s = 0.19$, $F_{2.14} = 40.71(6.51)$	14
MMP-13	H	$\log(1/IC_{50}) = 8.220(\pm0.467) - 0.320(\pm0.192)\log P - 0.891(\pm0.422)I_{1,CC} - 0.531(\pm0.403)I_{1,N}$	$n = 17$, $r = 0.910$, $r^2_{cv} = 0.70$, $s = 0.29$, $F_{3.13} = 20.78(5.74)$	14
MMP-13	I	$\log(1/IC_{50}) = 9.691(\pm0.382) - 0.396(\pm0.114)C\log P$	$n = 18$, $r = 0.879$, $r^2_{cv} = 0.73$, $s = 0.21$, $F_{1.16} = 54.63(8.53)$	15

(continued)

TABLE 27.1 (*Continued*)

MMP	Compound	QSAR Equation	Statistics	Reference
MMP-13	L	$\log(1/IC_{50}) = 1.730(\pm0.596)\mathrm{Pol} - 1.106(\pm0.449)I_{2,CC} + 0.233(\pm2.810)$	$n = 10$, $r = 0.971$, $r^2_{cv} = 0.88$, $s = 0.26$, $F_{2,7} = 56.77(9.55)$	15
MMP-13	M	$\log(1/IC_{50}) = 0.115(\pm0.065)\mathrm{SN} - 0.885(\pm0.314)I_{R2-CC} - 0.834(\pm0.390)\mathrm{IAC} + 11.205(\pm1.361)$	$n = 17$, $r = 0.927$, $r^2_{cv} = 0.73$, $s = 0.23$, $F_{3,14} = 26.34(5.56)$	16
MMP-13	M	$\log(1/IC_{50}) = 0.079(\pm0.055)S_N - 1.519(\pm0.747)\ C\log P + 0.234(\pm0.125)(C\log P)2 + 11.190(\pm1.861)$	$n = 13$, $r = 0.853$, $r^2_{cv} = 0.20$, $s = 0.16$, $F_{3,9} = 8.01(6.99)$	16
MMP-13	P	$\log(1/IC_{50}) = 0.726(\pm0.192)\ C\log P - 0.141(\pm0.045)(C\log P)^2 - 0.265(\pm0.177)I_{1,OMe} + 8.000(\pm0.265)$	$n = 41$, $r = 0.850$, $r^2_{cv} = 0.63$, $s = 0.23$, $F_{3,37} = 31.90(4.36)$	17
MMP-13	Q	$\log(1/IC_{50}) = 0.168(\pm0.135)C\log P + 0.834(\pm0.415)I_{1,COPh} + 7.574(\pm0.304)$	$n = 10$, $r = 0.914$, $r^2_{cv} = 0.71$, $s = 0.22$, $F_{2,7} = 17.75(9.55)$	17
MMP-13	R	$\log(1/IC_{50}) = 0.501(\pm0.212)\mathrm{Pol} + 0.993(\pm0.424)I_R + 0.802(\pm0.319)I_{R2} + 4.296(\pm0.944)$	$n = 16$, $r = 0.947$, $r^2_{cv} = 0.72$, $s = 0.24$, $F_{3,12} = 34.55(5.95)$	18
MMP-13	D	$\log(1/IC_{50}) = 0.644(\pm0.251)\ C\log P + 1.270(\pm0.556)I_{CC} + 5.085(\pm0.777)$	$n = 15$, $r = 0.890$, $r^2_{cv} = 0.68$, $s = 0.33$, $F_{2,12} = 22.90(6.93)$	18
MMP-13	P, T	$\log(10^6/IC_{50}) = 0.017\mathrm{ATS3m} + 11.363\mathrm{MATS6v} - 1.118\mathrm{MATS6e} - 1.826\mathrm{GATS1v} + 11.911\mathrm{GATS6v} + 6.097\mathrm{GATS1e} + 1.297\mathrm{GATS4e} - 11.139$	$n_{tr} = 68$, $r^2_{cv} = 0.598$, $s_{cv} = 0.376$, $n_{ts} = 12$, $r^2_{EP} = 0.727$, $s_{EP} = 0.429$	20

MMP-13	T	$\log(1/IC_{50}) = 5.485(\pm 2.130)\,C\log P - 1.334(\pm 0.519)(C\log P)^2 + 1.866(\pm 0.406)I_1 + 1.280(\pm 1.945)$	$n = 22$, $r = 0.910$, $r^2_{cv} = 0.86$, $s = 0.38$, $F_{4.17} = 60.61(4.67)$	24
MMP-13	V	$\log(1/IC_{50}) = 0.962(\pm 0.265)I_{1,Ph} - 1.163(\pm 0.840)\pi_2(R^1) + 7.948(\pm 0.193)$	$n = 20$, $r = 0.884$, $r^2_{cv} = 0.69$, $s = 0.27$, $F_{2.17} = 30.30(6.11)$	22
MMP-13	W	$\log(1/IC_{50}) = 0.772(\pm 0.625)\pi_4(R^2) - 12.768(\pm 4.248)\pi_2(R^2) + 16.953(\pm 6.207)[\pi_2(R^2)]^2 - 1.088(\pm 0.692)I_1 + 8.715(\pm 0.326)$	$n = 18$, $r = 0.926$, $r^2_{cv} = 0.68$, $s = 0.38$, $F_{4.13} = 19.60(5.20)$	22
MMP-13	Z	$\log(10^6/IC_{50}) = -86.944MATS4m - 33.782MATS8m - 5.375MATS3v + 8.188MATS1p + 1.103GATS5e + 122.338$	$n = 29$, $q^2 = 0.703$, $s = 0.397$	23

n, number of compounds; n_{tr}, number of compounds in the training set; r, correlation coefficient; SEE, standard error of estimate; F, variance ratio at the specified degree of freedom; s, standard deviation; r^2_{cv} and q^2, square of cross-validated correlation coefficient; n_{ts}, number of compounds in the test set; s_{EP}, standard deviation of the test set; r^2_{EP}, square of correlation coefficient of the test set.

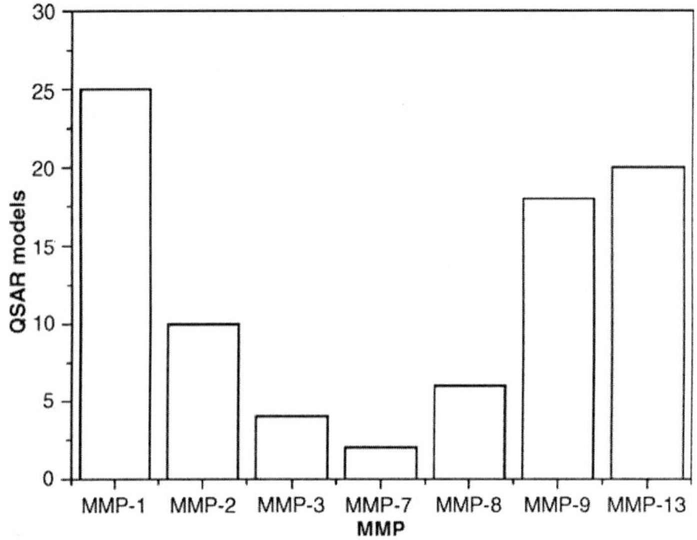

FIGURE 27.2 Evaluation of the number of MMP QSAR studies reported in the past 8 years.

Golbraikh and Tropsha,[25] to assess the predictivity of QSAR models, the training set statistics are insufficient and it is necessary to evaluate the correlation between the predicted and observed activities of compounds from external test sets. Since most of the reported QSAR studies did not use an external test set, the predictive reliability of these models cannot be estimated and only their interpolative ability can be evaluated. In other words, the obtained results can furnish important information for understanding the role of the molecular features in determining the activity and selectivity, but their use for driving the design of new ligands should be considered with caution.

As already reported in the chapter reporting the QSAR of CAs, judgment of the reliability of QSAR models should be mainly developed by evaluating the correlation between the predicted and observed activities of compounds from an external test set. Among all the QSAR studies reported in Table 27.1, only a few models were validated using an external test set. In particular, among all the QSAR studies reported in Table 27.1, it is noteworthy that only the models reported by Fernández and Caballero were validated using an external test set.[20] In this study, the authors developed linear and nonlinear predictive models for identifying the main molecular features that determine the selectivity of a set of 80 *N*-hydroxy-α-phenylsulfonylacetamide derivatives in the inhibition of MMP-1, MMP-9, and MMP-13. The considered molecular data set included the selective compounds that are potent inhibitors of MMP-9 and MMP-13 and moderate inhibitors of MMP-1. About 60 compounds were used for the training set, and 10–12 compounds were used for the external test set. The structure–activity relationships were established using both MLR and Bayesian-regularized genetic neural network (BRGNN) approaches, obtaining six models.

The three MLR models were able to explain the data variance and were quite stable to the inclusion–exclusion of compounds as measured by the square of the cross-

TABLE 27.2 QSAR Results for MMP-1, -9, and -13 Obtained by Using the BRGNN Approach[20]

		Training Set			Test Set		
MMP	Descriptors	n	r^2_{cv}	s_{cv}	n	r^2_{EP}	s_{EP}
MMP-1	ATS3e, MATS3m, MATS3e, MATS5e, MATS6e, GATS1v, GATS7p	63	0.601	0.377	10	0.781	0.362
MMP-9	ATS6m, MATS2m, MATS5v, MATS1e, GATS4v, GATS5e, GATS4p	66	0.692	0.421	12	0.814	0.332
MMP-13	ATS3m, ATS6m, MATS1v, GATS7v, GATS3e, GATS4e, GATS6p	68	0.647	0.384	12	0.785	0.429

n, number of compounds; r^2_{cv}: square of cross-validated correlation coefficient; s_{cv}, standard deviation of cross-validation; n_{ts}, number of compounds in the test set; r^2_{EP}, square of correlation coefficient of the test set; s_{EP}, standard deviation of the test set.

validated correlation coefficients (see Table 27.1). Moreover, the MLR models were able to describe the test set variances with squares of the correlation coefficients better than 0.6. Despite the agreeable results obtained using the MLR method, the authors carried out an additional search using the BRGNN approach[26] and the statistical results reveal that the NN approaches surpass the results achieved by MLR in regard to the fitness and predictive capacity ($r^2_{cv} > 0.6$ and $r^2_{EP} > 0.7$ for all the MMP models; see Table 27.2).

Analysis of the relevance of the physicochemical properties in the BRGNNR models showed that the atomic polarizability terms have a poor contribution in all MMPs; furthermore, MMP-9 and -13 contribution profiles are very similar, and the model describing the inhibitory activity of MMP-1 is mainly influenced by the atomic Sanderson electronegativity terms.

27.3 NONCLASSICAL QSAR METHODS

Two new QSAR methods have been developed and tested using MMPs. In 2005, Matter and coworkers[27] reported a "QSAR-by-NMR" study in which they correlated the biological activity of 20 MMP-3 inhibitors with the ^{15}N and ^{1}H amide chemical shift differences of the amino acids surrounding the inhibitor-binding site. To detect ligand binding by ^{15}N and ^{1}H amide chemical shift differences in two-dimensional ^{15}N-heteronuclear single-quantum correlation spectra (^{15}N-HSQC), partial-least-squares regressions (PLS) were used to correlate the biochemical binding affinities with the ^{15}N and ^{1}H amide chemical shift differences in congeneric series relative to uncomplexed protein spectra. The chemical shift of backbone amide $^{1}H/^{15}P$ resonances in ^{15}N-MMP-3 HSQC spectra were extracted for 81 amino acids surrounding the inhibitor-binding site in the MMP-3 catalytic domain, and referenced to the 2D-HSQC spectra of uncomplexed MMP-3 under identical conditions. This resulted in a data table of 162 chemical shift differences as descriptors in the X-block and 20 biological activities as dependent variables, and the subsequent PLS analysis and cross-validation provided a model with three PLS components and a predictivity (r^2_{cv}) value of 0.657, calculated through the leave-one-out (LOO) cross-validation method.

Beyond the quantitative results of this work, the most important aspect concerns the possibility of highlighting the chemical shift, and thus the residues, that are principally related to the activity of the different ligands. A more positive chemical shift difference for a particular residue is related to an increase of binding affinity, whereas negative PLS coefficients indicate that either positive chemical shift differences decrease the affinity or negative chemical shift differences increase the affinity.

Since the binding of different ligands to a protein determines different changes in the binding site, this kind of analysis could be very useful in investigating the residues principally involved in the binding of different ligands, extracting information about ligand binding modes and thus linking the statistical model to known structure–activity relationships. For the MMP-3 model, the authors found a positive ^{15}N shift contribution for Ala120 that can be due to bulky substituents in S_1' that influence the Phe196 and Leu197 side chain conformations next to Ala120. The influence for Ala165, involved in a hydrogen bond with an inhibitor sulfonamide oxygen, is due to changes in hydrogen-bond geometry, or in neighboring residues due to conformation caused by ligands able to have an optimal fit into S_1'. Other important positive ^{15}N shift contributions have been reported for Gly159, His201, and Arg233, whereas negative ^{15}N shift contributions were observed for Asp158 and Tyr168. Finally, the positive ^{1}H shift coefficient for His179 could be explained by its hydrogen bond to the carbonyl oxygen of His166, as the latter residue is directly involved in ligand recognition, and the negative ^{1}H shift contributions are observed for Gly204, which is also close to His201/205 and thus directly affected by possible conformational changes upon ligand binding.

This approach, in combination with an interpretation in structural terms, may be able to provide an understanding of protein–ligand interaction features in structure-driven drug discovery.

Finally, in 2006 Prathipati and Saxena reported the development of binary QSAR models of MMP-3 using LUDI and MOE scoring functions.[28] The binary QSAR is a relatively new methodology in which biological activity is expressed in a "binary" format (1 = active and 0 = inactive) and is correlated with molecular descriptors. In this method, the set of computed molecular descriptors is transformed into a set of decorrelated and normalized variables, and the probability distribution is estimated based on Bayes' theorem.[29] Using a training set of 60 MMP-3 inhibitors, a test set of 20 compounds and 999 nonbinders, the authors developed a binary QSAR model, derived using the MOE scoring function, which showed a good quality. The performance of the model was assessed by measuring four parameters: (a) accuracy on active compounds, *c0/m0*; (b) accuracy on inactive compounds, *c1/m1*; (c) overall accuracy on all of the compounds, $(c0 + c1)/(m0 + m1)$, where $m0$ represents the number of active compounds, $m1$ is the number of inactive compounds, $c0$ is the number of active compounds correctly labeled by the QSAR model, and $c1$ is the number of inactive compounds correctly predicted by the QSAR model. The fourth parameter measured was the *enrichment factor*, which indicates the relative enrichment of active compounds in the set of instances predicted to be active in relation to the fraction of active compounds in the original data set. The results obtained suggested that this kind of approach may be useful as a preliminary screening layer in a multilayered virtual screening paradigm.

27.4 3D-QSAR STUDIES

With respect to the published QSAR models, the number of 3D-QSAR models is smaller. As shown in Fig. 27.3, to date only 8 different scaffolds have been analyzed and only 17 models concerning 4 MMP subtypes (MMP-1, -3, -8 and -13) have been reported (see Table 27.3).

Among the 17 3D-QSAR models, 4 were developed by using compounds possessing the hydroxamate group as ZBG, 4 were developed by using compounds possessing the carboxylate group as ZBG, whereas the remaining 9 models were developed by using training sets in which both hydroxamate and carboxylate were considered.

In 1999, Matter and coworkers reported one of the first 3D-QSAR concerning MMPs.[35] A series of 90 2-(arylsulfonyl)-1,2,3,4-tetrahydroisoquinoline-3-carboxylates and hydroxamates (scaffold D and E in Fig. 27.3) as inhibitors of MMP-8 were synthesized and investigated by 3D-QSAR techniques (CoMFA and CoMSIA).

After the analysis of key protein–ligand interactions, the candidate molecules were manually docked into the active site. Subsequently, the protein–ligand complex for a reference compound was minimized, treating all the ligand atoms and protein residues within a sphere of 4 Å as flexible, while the remaining receptor was used to compute nonbonded interactions. All the other compounds were built accordingly, docked into MMP-8, and minimized using a rigid receptor from the first minimization. This alignment was used to develop the CoMFA and CoMSIA models. Using a 2 Å grid spacing, a CoMFA model with an r^2_{cv} value of 0.569 for five components and a conventional r^2 of 0.905 was obtained. To further validate the model, the effects of the alignment and of the probe atom choice were analyzed, suggesting only a slight dependence of the model on the absolute orientation and on the chosen probe. Similar results were obtained using the CoMSIA approach, the best model with an r^2_{cv} value of 0.478 for seven components and a conventional r^2 of 0.924 was obtained. Finally, the CoMFA and CoMSIA correlation coefficients were found to be in agreement with the MMP-8-ligand interactions in terms of steric, electrostatic, and hydrophobic complementarity.

After 7 months, Matter and Schwab reported a further 3D-QSAR study for the same 2-(arylsulfonyl)-1,2,3,4-tetrahydroisoquinoline-3-carboxylates and hydroxamates derivatives.[34] Using the same alignment method and the same computational procedures, they reported the development of CoMFA and CoMSIA models for the activity

TABLE 27.3 Main 3D-QSAR Models on MMP Inhibitors

MMP	Compound	Alignment	Statistics	Software	Reference
MMP-1	A[a]	Ligand	$n_{tr}=60$, $r^2_{cv}=0.649$, $F=224.17$, $NC=5$, $n_{ts}=24$	CoMFA	30
MMP-1	A[a]	Ligand	$n_{tr}=60$, $r^2_{cv}=0.730$, $F=175.64$, $NC=6$, $n_{ts}=24$	CoMSIA	30
MMP-3	B	Ligand	$n_{tr}=41$, $r^2_{cv}=0.70$, $NC=1$, $n_{ts}=10$, $r^2_{cv}=0.82$, $SD_{pr}=0.32$	CoMFA	31
MMP-3	B	Docking	$n_{Tr}=41$, $r^2_{cv}=0.42$, $NC=1$, $n_{ts}=10$, $r^2{}_{pr}=0.57$, $SD_{pr}=0.53$	CoMFA	31
MMP-3	C	Docking	$n_{tr}=22$, $r^2_{cv}=0.592$, $F=297.36$, $NC=2$, $n_{ts}=4$	CoMFA	32
MMP-3	C	Ligand	$n_{tr}=12$, $r^2_{cv}=-0.216$, $F=168.42$, $NC=6$, $n_{ts}=4$	CoMFA	32
MMP-3	D, E	Ligand	$n_{tr}=27$, $r^2_{cv}=0.738$, $F=202.52$, $n_{ts}=4$	CoMFA	33
MMP-3	D, E	Ligand	$n_{tr}=33$, $r^2_{cv}=0.608$, $F=21.99$, $n_{ts}=4$	CoMSIA	33
MMP-3	F, G	Ligand	$n_{tr}=31$, $r^2_{cv}=0.522$, $F=256.04$, $n_{ts}=4$	CoMFA	33
MMP-3	D, E	Docking	$n_{tr}=90$, $r^2_{cv}=0.563$, $SD_{cv}=0.629$, $NC=6$	CoMFA	34
MMP-3	D, E	Docking	$n_{tr}=90$, $r^2_{cv}=0.413$, $SD_{cv}=0.738$, $NC=8$	CoMSIA	34
MMP-3	D, E	Docking	$n_{tr}=90$, $\mathrm{r}^2_{cv}=0.795$, $SD_{cv}=0.413$, $NC=5$	Golpe	34
MMP-8	D, E	Docking	$n_{tr}=90$, $r^2_{cv}=0.729$, $SD_{cv}=0.512$, $NC=4$	Golpe	34
MMP-8	D, E	Docking	$n_{tr}=90$, $r^2_{cv}=0.569$, $SD_{cv}=0.685$, $NC=5$	CoMFA	35
MMP-8	D, E	Docking	$n_{tr}=90$, $r^2_{cv}=0.478$, $SD_{cv}=0.763$, $NC=7$	CoMSIA	35
MMP-13	H	Ligand	$n_{tr}=33$, $r^2_{cv}=0.616$, $F=19.98$, $n_{ts}=6$	MFA	36
MMP-13	H	Ligand	$n_{tr}=33$, $r^2_{cv}=0.681$, $F=19.80$, $n_{ts}=6$	RSA	36

n_{tr}, number of compounds in the training set; F, variance ratio at the specified degree of freedom; r^2_{cv}, square of cross-validated correlation coefficient; n_{ts}, number of compounds in the test set; SD_{cv}, standard deviation of error in the cross-validation; NC, principal components; r^2_{pr}, square of correlation coefficient of the prediction of the external test set; SD_{pr}, standard deviation of error of the prediction of the external test set.

[a]Beyond compounds of generic scaffold A, this study also used 12 structurally more diversified compounds.

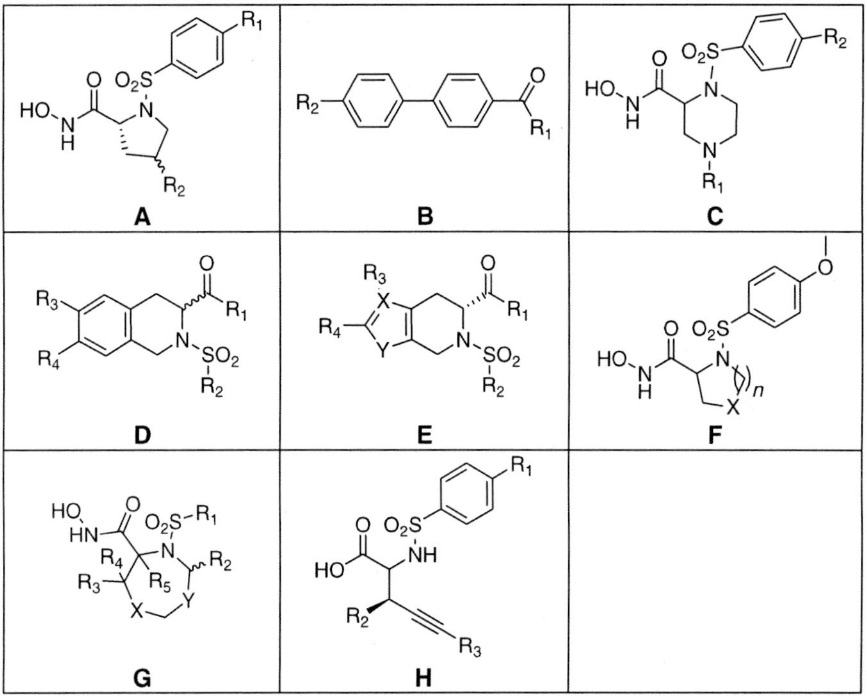

FIGURE 27.3 General scaffold of the MMP inhibitors used for the development of 3D-QSAR models.

of the compounds against MMP-3; moreover, they also developed 3D-QSAR models for MMP-3 and MMP-8 using the Golpe software. The CoMFA and CoMSIA results for MMP-3 were very similar to those obtained for MMP-8. The CoMFA model showed an r^2_{cv} value of 0.563 for six components and a conventional r^2 of 0.944, and the CoMSIA model showed an r^2_{cv} value of 0.413 for eight components and a conventional r^2 of 0.957. The 3D-QSAR models obtained with the Golpe software were developed using the phenolic OH probe and possessed better statistical results. Using D-optimal preselection of variables and fractional factorial design (FFD), the MMP-3 3D-QSAR model showed an r^2_{cv} value of 0.795 with five PLS components and an r^2 of 0.967, and the MMP-8 model showed an r^2_{cv} value of 0.729 with four PLS components and an r^2 of 0.934.

Then, to investigate the selectivity between MMP-3 and MMP-8, a 3D-QSAR model was developed by means of Golpe using the ratio between the logarithm of the IC_{50} of MMP-8 and MMP-3 as a quantitative selectivity measure for the dependent variable in PLS. Also for this model, the GRID interaction energies of the phenolic OH probe were used as descriptors. The PLS model after FFD-based variable selection resulted in an r^2_{cv} value of 0.532 for three components and an r^2 of 0.831. From a statistical point of view, the model has a certain degree of significance, but compared with the 3D-QSAR developed by using the MMP-3 and MMP-8 affinity, it showed worse statistical values, suggesting that the models of higher consistency are obtained using a single biological observable rather than by introducing a dependent variable as a function of different IC_{50}.

Using a subset of the compounds investigated by Matter and coworkers, Amin and Welsh developed other MMP-3 CoMFA and CoMSIA models.[33] The crystal structure containing one of the analyzed compounds complexed with MMP-8 provided the alignment template for the remaining 38 molecules. The best CoMFA model was obtained by removing seven compounds that exhibited a conformationally restrained ring structure unlike that found in the remainder of the data set (subset A), and removing three compounds for which a large difference existed between the experimental and the predicted biological activity (subset B). The developed model showed an r^2_{cv} value of 0.608, an r^2 of 0.835, and an F of 21.986. Different from the CoMFA analysis, the best CoMSIA model ($r^2_{cv} = 0.608$, $r^2 = 0.835$, and $F = 21.986$) was obtained by omitting subset B and another compound for which a large difference existed between the experimental determined and the predicted biological activity and using Gasteiger–Hückel charges. The CoMFA and CoMSIA models were applied to derive the biological activity values of four MMPIs originally omitted from the training set and ranging from inactive ($pIC_{50} = 5.000$) to moderately active ($pIC_{50} = 7.699$). The CoMSIA-predicted activities of the test set compounds were, except for one compound, much closer to the experimental values than those predicted by CoMFA.

In the same paper, the authors also reported the development of CoMFA and CoMSIA models for 35 thiazine- and thiazepine-based compounds. The best CoMFA model was obtained by omitting the four compounds with the large residual values, and showed an r^2_{cv} value of 0.522, an r^2 of 0.985, and an F of 256.04. The model was again applied to derive the biological activity values of five MMPIs originally omitted from the training set, ranging from fairly active ($pIC_{50} = 6.757$) to very active ($pIC_{50} = 8.201$). With the exception of one compound, the predicted values were close to the experimental biological activity data, with very low residuals.

Finally, the CoMSIA model for this training set was not predictive; the best model yielded an r^2_{cv} value of 0.103 and an r^2 of 0.157 with an extremely high standard error of estimate.

In 2001, Amin and Welsh reported the 3D-QSAR study for piperazine-based MMP-3 inhibitors.[32] Based on the preliminary CoMFA results, the initial data set of 39 compounds was divided in two subsets (I and II). The compounds of the first subset were built directly from the crystal structure of a reference compound of the subset I, whereas the compounds of the second subset were built directly from the crystal structure of a reference compound of subset II. For the subset II, the CoMFA results were not good ($r^2_{cv} = -0.216$, $r^2 = 0.994$, and $F = 168.42$), whereas for the subset I the best CoMFA results were obtained by docking each inhibitor into the MMP-3 binding cleft, after which ligand conformation was manually altered to achieve the best fit to the S_1' and S_1'-S_2' binding pockets. This alignment produced a CoMFA model with a high self-consistency ($r^2 = 0.989$) and acceptable internal predictive ability ($r^2_{cv} = 0.592$). This model was also used for predicting the activity of four compounds originally omitted from the training set, and the CoMFA-predicted activities of these compounds were extremely close to the experimentally determined values.

In 2001, Muegge and Podlogar investigated the possibility of developing a CoMFA model for MMP-3 inhibitors using a docking-based alignment,[31] and compared the results with those obtained using an atom-based alignment. Using Dock 4.0 software[37] combined with the PMF[38] scoring protocol, a training set of 51 biphenyl carboxylic acid MMP-3 inhibitors was docked into the crystal structure of the MMP-3 binding site. All structures were optimized against the electrostatic and steric fields using appropriate conformation of a reference inhibitor as the template structure. This alignment was used for developing the CoMFA model, which showed good statistical results ($n = 41$, $r^2_{cv} = 0.42$, and NC = 1). Furthermore, 10 compounds were omitted from the training set and the activity prediction of this test set showed an r^2 value of 0.57 and a SD of 0.53. The second CoMFA model was developed using an atom-based alignment. All the compounds were aligned using a reference compound as a template for a manual alignment of the inhibitors. The statistical results suggested that the atom-based alignment showed the best results since the obtained CoMFA model showed an r^2_{cv} of 0.70, and the prediction of the external test set showed an $r^2_{predict}$ of 0.82 and an $SD_{predict}$ of 0.32. However, analyzing the protein, the statistically best CoMFA model created by the atom-based alignment was inconsistent with the MMP-3 crystal structure, different from the docking-based CoMFA model that was consistent with the ligand-binding site of the target protein.

With regards to MMP-1, Tsai and Lin reported a CoMFA and CoMSIA model of 72 proline-based plus 12 structurally more diversified nonproline MMPIs.[30] For the alignment, some atoms on the structure of the most active inhibitor in the series were treated as the correspondence points for performing the point alignment. Using the different alignment points, many alignments were analyzed by the CoMFA, CoMSIA, and PLS programs and the best alignment showed a CoMFA model with five principal components with an r^2_{cv} value of 0.649, an r^2 of 0.954, and an F of 224.17. The corresponding best CoMSIA model showed an r^2_{cv} value of 0.730, an r^2 of 0.952, and an F of 175.64, with six principal components. The corresponding features of this CoMSIA model were used for the construction of a pharmacophore hypothesis with the Catalyst 4.9 program; the training set was extended to include 11 structurally more diversified and nonproline inhibitors, and the authors found that the activities predicted by the first hypothesis were statistically as good as those predicted by the CoMSIA model from which the hypothesis was derived.

Finally, Sarma and coworkers reported a 3D-QSAR for MMP-13 by using molecular field analysis (MFA) and receptor surface analysis (RSA) methodologies on a series of 39 inhibitors.[36]

All the molecules were manually aligned to the most active molecule by considering the significant common substructure, then further refinement was developed by using a molecular shape analysis.

The MFA model was developed by analyzing all the aligned molecules by using CH_3 and H^+ probes for steric and electrostatic interactions, respectively, and showed an r^2_{cv} value of 0.616, an r^2 of 0.822, and an F of 19.98. For the development of the RSA model, the aligned compounds were reconsidered for the generation of receptor surface, which in principle represents a virtual active site of the target. The interaction energies of all the molecules were evaluated within this receptor surface by using 3D-descriptors derived from the van der Waals and electrostatic interaction energies between the receptor surface and CH_3 and H^+ groups as probes. The final model showed an r^2_{cv} value of 0.847, an r^2 of 0.681, and an F of 19.80. Furthermore, the predictive ability of both the MFA and RSA models were evaluated by predicting the biological activities of a test set of six molecules originally omitted from the training set, obtaining good results, in particular for the RSA model. From these results, it seems that the RSA model better predicts when compared with the MFA model, and agrees well with the experimental results in the test set as well as in the training set.

27.5 CONCLUSIONS

This chapter summarizes the main MMP QSAR studies. The number of MMPIs is large; however, a careful examination of the structures shows that there is much redundancy of information in these structures, and most of the analyzed compounds have hydroxamate as a ZBG.

More than 80 MMP QSAR models have been published, but almost all models have been internally validated by using a LOO cross-validated analysis and, with only few exceptions, none of them used external test sets for testing the predictive ability.

In comparisons with analogous studies carried on for carbonic anhydrase (CA) inhibitors, these studies are generally developed by using training sets characterized by a lower number of compounds with greater structural similarity among them, and a larger use of classical descriptors such as log *P*.

The reported 3D-QSAR models were validated internally and also by means of the prediction of external test sets. Many of these models were developed by using the docking approach as a tool for aligning the MMPIs, whereas the remaining models were developed by using an alignment based on the characteristics of the ligands. The first alignment approach seems to give slightly more reliable results; however, the increase of structural information provided by X-ray structures and the publication of new data will help the tuning of boot alignment methods.

Despite the known 3D structure of several catalytic domains of MMPs, the development of highly specific MMPIs that have ability for discriminating among the different members of this protease family remains the strongest challenge. In our opinion, to accomplish this task, the future trend of QSAR studies, having a larger amount of biological data concerning MMPIs tested on a wide spectrum of MMPs at their disposal, should be able to target the analysis of the selectivity.

REFERENCES

1. Hansch, C.; Maloney, P. P.; Fujita, T.; Muir, R. M. Correlation of biological activity of phenoxyacetic acids with Hammett substituent constants and partition coefficients. *Nature* **1962**, *194*, 178–180.
2. Selassie, C. D.; Garg, R.; Kapur, S.; Kurup, A.; Verma, R. P.; Mekapati, S. B.; Hansch, C. Comparative QSAR and the radical toxicity of various functional groups. *Chem. Rev.* **2002**, *102*, 2585–2605.
3. Hansch, C.; Leo, A.; Mekapati, S. B.; Kurup, A. QSAR and ADME. *Bioorg. Med. Chem.* **2004**, *12*, 3391–3400.

4. Doucet, J. P.; Barbault, F.; Xia, H.; Panaye, A.; Fan, B. Nonlinear SVM approaches to QSPR/QSAR studies and drug design. *Curr. Computer-Aided Drug Des.* **2007**, *3*, 263–289.

5. Arodz, T.; Dudek, A. Z. Multivariate modeling and analysis in drug discovery. *Curr. Computer-Aided Drug Des.* **2007**, *3*, 240–247.

6. Selassie, C. D.; Mekapati, S. B.; Verma, R. P. QSAR: then and now. *Curr. Top. Med. Chem.* **2002**, *2*, 1357–1379.

7. Wade, R. C. Calculation and application of molecular interaction fields. *Meth. Princ. Med. Chem.* **2006**, *27*, 27–42.

8. Cruciani, G.; Carosati, E.; Wade, R. C.; Baroni, M. Characterization of protein-binding sites and ligands using molecular interaction fields. *Compr. Med. Chem. II* **2006**, *4*, 237–253.

9. Roy, K.; Pal, D. K.; De, A. U.; Sengupta, C. QSAR of matrix metalloproteinase inhibitor *N*-[(substituted phenyl)sulfonyl]-*N*-4-nitrobenzylglycine hydroxamates using LFER model. *Drug Des. Discov.* **2001**, *17*, 315–323.

10. Gupta, S. P.; Kumar, D.; Kumaran, S. A quantitative structure–activity relationship study of hydroxamate matrix metalloproteinase inhibitors derived from functionalized 4-aminoprolines. *Bioorg. Med. Chem.* **2003**, *11*, 1975–1981.

11. Kumar, D.; Gupta, S. P. A quantitative structure–activity relationship study on some matrix metalloproteinase and collagenase inhibitors. *Bioorg. Med. Chem.* **2003**, *11*, 421–426.

12. Gupta, S. P.; Maheswaran, V.; Pande, V.; Kumar, D. A comparative QSAR study on carbonic anhydrase and matrix metalloproteinase inhibition by sulfonylated amino acid hydroxamates. *Enzyme Inhib. Med. Chem.* **2003**, *18*, 7–13.

13. Verma, R. P.; Kurup, A.; Hansch, C. On the role of polarizability in QSAR. *Bioorg. Med. Chem.* **2005**, *13*, 237–255.

14. Gupta, S. P.; Kumaran, S. A quantitative structure–activity relationship study on some series of anthranilic acid-based matrix metalloproteinase inhibitors. *Bioorg. Med. Chem.* **2005**, *13*, 5454–5462.

15. Gupta, S. P.; Kumaran, S. Quantitative structure–activity relationship studies on matrix metalloproteinase inhibitors: bicyclic heteroaryl hydroxamic acid analogs. *Lett. Drug Des. Disc.* **2005**, *2*, 522–528.

16. Gupta, S. P.; Kumaran, S. Quantitative structure–activity relationship studies on benzodiazepine hydroxamic acid inhibitors of matrix metalloproteinases and tumor necrosis factor-α converting enzyme. *Asian J Biochem.* **2006**, *1*, 47–56.

17. Gupta, S. P.; Kumaran, S. quantitative structure–activity relationship studies on matrix metalloproteinases inhibitors: piperazine piperidine and diazepine hydroxamic acid analogs. *Asian J Biochem.* **2006**, *1*, 211–223.

18. Gupta, S. P.; Kumaran, S. Quantitative structure–activity relationship studies on matrix metalloproteinase inhibitors: hydroxamic acid analogs. *Med. Chem.* **2006**, *2*, 243–250.

19. Jamloki, A.; Karthikeyan, C.; Hari Narayana Moorthy, N. S.; Trivedi, P. QSAR analysis of some 5-amino-2-mercapto-1,3,4-thiadiazole based inhibitors of matrix metalloproteinases and bacterial collagenase. *Bioorg. Med. Chem. Lett.* **2006**, *16*, 3847–3854.

20. Fernández, M.; Caballero, J. QSAR modeling of matrix metalloproteinase inhibition by *N*-hydroxy-alpha-phenylsulfonylacetamide derivatives. *Bioorg. Med. Chem.* **2007**, *15*, 6298–6310.

21. Gupta, S. P.; Bagaria, P.; Satuluri, A. K.; Venkata, S. A quantitative structure–activity relationship study on some aryl sulfonyl amido and ureido derivatives acting as matrix

metalloproteinase and clostridium histolyticum collagenase inhibitors. *Lett. Drug Des. Disc.* **2007**, *4*, 496–501.

22. Kumaran, S.; Gupta, S. P. A quantitative structure–activity relationship study on some novel series of hydroxamic acid analogs acting as matrix metalloproteinase inhibitors. *Med. Chem.* **2007**, *3*, 167–173.
23. Fernández, M.; Caballero, J.; Tundidor-Camba, A. Linear and nonlinear QSAR study of *N*-hydroxy-2-[(phenylsulfonyl)amino]acetamide derivatives as matrix metalloproteinase inhibitors. *Bioorg. Med. Chem.* **2006**, *14*, 4137–4150.
24. Kumaran, S.; Gupta, S. P. A quantitative structure–activity relationship study on matrix metalloproteinase inhibitors: piperidine sulfonamide aryl hydroxamic acid analogs. *J. Enzyme Inhib. Med. Chem.* **2007**, *22*, 23–27.
25. Golbraikh, A.; Tropsha, A. Beware of q2! *J. Mol. Graph. Model.* **2002**, *20*, 269–276.
26. Caballero, J.; Fernández, M. Linear and nonlinear modeling of antifungal activity of some heterocyclic ring derivatives using multiple linear regression and Bayesian-regularized neural networks. *J. Mol. Model.* **2006**, *12*, 168–181.
27. Matter, H.; Schudok, M.; Elshorst, B.; Jacobs, D. M.; Saxena, K.; Kogler, H. QSAR-by-NMR: quantitative insights into structural determinants for binding affinity by analysis of 1H/15N chemical shift differences in MMP-3 ligands. *Bioorg. Med. Chem. Lett.* **2005**, *15*, 1779–1783.
28. Prathipati, P.; Saxena, A. K. Evaluation of binary QSAR models derived from LUDI and MOE scoring functions for structure based virtual screening. *J. Chem. Inf. Model.* **2006**, *46*, 39–51.
29. Labute, P.; Nilar, S.; Williams, C. A probabilistic approach to high throughput drug discovery. *Comb. Chem. High Throughput Screen.* **2002**, *5*, 135–145.
30. Tsai, K. C.; Lin, T. H. A ligand-based molecular modeling study on some matrix metalloproteinase-1 inhibitors using several 3D QSAR techniques. *J. Chem. Inf. Comput. Sci.* **2004**, *44*, 1857–1871.
31. Muegge, I.; Podlogar, B. L. 3D-Quantitative structure–activity relationship of biphenyl carboxylic acid MMP-3 inhibitors: exploring automated docking as alignment method. *QSAR* **2001**, *20*, 215–222.
32. Amin, E. A.; Welsh, W. J. Three-dimensional quantitative structure–activity relationship (3D-QSAR) models for a novel class of piperazine-based stromelysin-1 (MMP-3) inhibitors: applying a "divide and conquer" strategy. *J. Med. Chem.* **2001**, *44*, 3849–3855.
33. Amin, E. A.; Welsh, W. J. Highly predictive CoMFA and CoMSIA models for two series of stromelysin-1 (MMP-3) inhibitors elucidate S1′ and S1-S2′ binding modes. *J. Chem. Inf. Model.* **2006**, *46*, 1775–1783.
34. Matter, H.; Schwab, W. Affinity and selectivity of matrix metalloproteinase inhibitors: a chemometrical study from the perspective of ligands and proteins. *J. Med. Chem.* **1999**, *42*, 4506–4523.
35. Matter, H.; Schwab, W.; Barbier, D.; Billen, G.; Haase, B.; Neises, B.; Schudok, M.; Thorwart, W.; Schreuder, H.; Brachvogel, V.; Lönze, P.; Weithmann, K. U. Quantitative structure–activity relationship of human neutrophil collagenase (MMP-8) inhibitors using comparative molecular field analysis and X-ray structure analysis. *J. Med. Chem.* **1999**, *42*, 1908–1920.

36. Sarma, J. A.; Rambabu, G.; Srikanth, K.; Raveendra, D.; Vithal, M. Analogue based design of MMP-13 (collagenase-3) inhibitors. *Bioorg. Med. Chem. Lett.* **2002**, *12*, 2689–2693.
37. Kuntz, I. D.; Blaney, J. M.; Oatley, S. J.; Langridge, R. L. A geometric approach to macromolecule–ligand interactions. *J. Mol. Biol.* **1982**, *161*, 269–288.
38. Muegge, I.; Martin, Y. C. A general and fast scoring function for protein–ligand interactions: a simplified potential approach. *J. Med. Chem.* **1999**, *42*, 791–804.

PART IV

DRUG DESIGN OF BACTERIAL ZINC PROTEASE INHIBITORS

CHAPTER 28

Bacterial Zinc Proteases as Orphan Targets

CLAUDIU T. SUPURAN

Dipartimento di Chimica, Università degli Studi di Firenze, Laboratorio di Chimica Inorganica e Bioinorganica, Via della Lastruccia 3, 50019 Sesto Fiorentino (Firenze), Italy

28.1 INTRODUCTION

Proteases (PRs), also denominated proteinases or peptidases, constitute one of the largest functional groups of proteins, with more than 560 members actually described.[1] By hydrolyzing one of the most important chemical bonds present in biomolecules, that is, the peptide bond, PRs play crucial functions in organisms all over the phylogenetic tree, starting from viruses, bacteria, protozoa, metazoa, or fungi, and ending with plants and animals. Numerous practical applications in biotechnology of such enzymes, and the understanding that PRs are important targets for the drug design, ultimately fueled much research in this field.[1] Although much progress has been registered in the design and clinical applications of viral PR inhibitors (mainly those targeted against the aspartic protease of HIV),[2–4] not the same situation is true for the bacterial proteases.[5,6] PRs are widespread in all types of bacteria, where they are involved in critical processes such as colonization and evasion of host immune defenses, acquisition of nutrients for growth and proliferation, facilitation of dissemination, or tissue damage during infection.[5,6] Even more subtle roles for bacterial PRs have recently been evidenced in the interaction between host and the invading microorganisms: interruption of cascade activation pathways, disruption of cytokine network, excision of cell surface receptors, and inactivation of host protease inhibitors (PIs).[5–8] Regulation of proteolysis is a critical element of the host immune system and plays an important role in the induction of pro- and anti-inflammatory reactions in response to infection. Some bacterial species take advantage of these processes and recruit host proteinases to their surface to counteract the host attack.[15] Nevertheless, it is surprising that there is little or no regulation of bacterial PRs by plasma-derived PIs, for instance, the serpins (**ser**ine **p**rotease **in**hibitors) that are present in relatively high concentrations in plasma.[5] Even worse for the host is that bacteria developed strategies for neutralizing such plasma-derived PIs, ensuring in this way an efficient attack of the invaded organism.[5–8] Taking into account all these facts, it is obvious that bacterial PRs may represent very attractive targets for the development of novel types of antibiotics, since inhibition of such critical enzymes would presumably lead to the death of the invading pathogen.[5] Until now all the antibiotics used in clinical practice share a common mechanism of action, acting as inhibitors of the bacterial cell wall biosynthesis or affecting protein synthesis on ribosomes and not intervening in more fundamental metabolic processes of the pathogen. Considering the specific role that bacterial PRs play in such critical steps for the successful invasion of the host[5,7] and the constant emergence of antibiotic resistance,[9] it is crucial to develop bacterial PIs as a novel antibiotic class. Mention should be made that no drugs belonging to this class of pharmacological agents are available at present for clinical use, although some progress has nonetheless been registered. Thus, many possible targets for the drug design will be discussed, together with the recent progress achieved in understanding the PR types present in pathogenic bacteria, as well as the inhibitors for such enzymes. In this chapter, the discussion will be restricted to metallo-PRs present in *Eubacteriae*, since the *Archaeobacteriae* are generally nonpathogenic. Also, PRs of viral, fungal, protozoan, or other parasitic origin will not be considered herein.

Five catalytic types of PRs have been recognized so far, in which serine, threonine, cysteine, or aspartic groups as well as metal ions play a primary role in catalysis. All these types of enzymes are present in bacteria.[1] The first three types of PRs are catalytically very different from the aspartic and metallo-PRs, mainly because the nucleophile of the catalytic site is part of an amino acid in the first case, whereas it is an activated water molecule for the second group of such enzymes. Thus, acyl enzyme intermediates are formed only in the reactions of the Ser/Thr/Cys PRs, and only these peptidases can readily act as transferases.[1]

The classification of PRs used in this book is based on the classical one of Rawlings and Barrett[1,10] in which the catalytic type of protein represents the top level in the hierarchical classification. According to this rule, the PRs can be divided into clans based on three-dimensional protein folding and into families based on evolutionary relationships of the primary sequence. The terminology used in describing the specificity of PRs depends on a model in which the catalytic site is considered to be flanked on one or both sides by specificity subsites, each able to accommodate the side chain of a single amino acid residue, as originally proposed by Berger and Schechter,[11] and adopted thereafter by many researchers.[1] These sites are numbered from the catalytic site, S1, S2, ... S*n* toward the N-terminus of the substrate, and S1′, S2′, ... S*n*′ toward the C-terminus. The residues they accommodate are numbered P1, P2, ... P*n*, and P1′, P2′, ... P*n*′, respectively, as follows (the catalytic site of the enzyme is marked "*", and the scissile peptide bond of the substrate as "#"). The same formalism is also valid for enzyme inhibitors that bind to the catalytic site:

$$\text{Substrate/Inhibitor: -P3-P2-P1\# P1'-P2'-P3'-}$$

$$\text{Enzyme:} - \text{S3} - \text{S2} - \text{S1* S1'} - \text{S2'} - \text{S3'} -$$

For each enzyme with potential therapeutic use per se or for the development of inhibitors useful as possible antibiotics, the main characteristics will be mentioned, such as the EC classification (when available), the catalytic and family type (in the classification of Barrett et al.),[1] the organisms in which it is present, the preferred scissile bond(s), and eventually the type of *in vivo* attacked substrate, relevant for the pathogenic effects of the considered bacterial species. Furthermore, the latest developments in the design of inhibitors for such PRs will be discussed, as well as possibilities to consider such inhibitors as leads for the drug design of novel classes of antibiotics.

Metalloproteases (MPRs) are the hydrolases in which the nucleophilic attack on the scissile peptide bond is mediated by a water molecule coordinated to a divalent metal ion (usually Zn(II) but sometimes cobalt or manganese may also activate the water molecule) or bridged to a dimetallic center (two Zn(II) ions, or one Zn(II) and one Co(II)/Mn(II) ions, etc.).[1] The catalytical metal ion is coordinated by amino acid moieties present within the active site, usually three in number (the most frequent being His, Glu, and Asp), the fourth ligand being, as already mentioned, a water molecule/hydroxide ion.[12] Thus, MPRs are divided into two main groups depending on the number of metal ions required for catalysis: in the majority of the described MPRs, only one metal ion is required, but in some families there are two metal ions that act cocatalytically. All the metallopeptidases in which cobalt or manganese is essential for activity require two metal ions, but there are also families of zinc-dependent MPR in which two zinc ions are cocatalytic.[1,12] The best studied dinuclear zinc PRs are exopeptidases such as the *Aeromonas proteolytica* aminopeptidase (AMP), the *Streptomyces griseus* aminopeptidase (SGAP), carboxypeptidase G2 (CPG2) of pseudomonads, as well as methionyl aminopeptidase of *Escherichia coli*.[1,12] In such MPRs with binuclear metal centers, five amino acids residues act as ligands (predominantly by means of carboxylate moieties), and one of them ligates both metal ions, acting as bridging ligand. All MPRs with two catalytic metal ions so far described are exopeptidases, whereas MPRs with one catalytic metal ion may be exo- or endopeptidases.[1,12] Only the zinc-containing MPRs will be discussed here.

MPRs are widely spread in all types of bacteria (Table 28.1), being critical virulence factors, and play various pathogenic roles in infection.[6,13–15] Thus, in local bacterial infections, such as keratitis, dermatitis, and pneumonia, MPRs function as decisive virulence determinants, being generated at the site of infection and causing necrotic or hemorrhagic tissue damage through digestion of structural tissular components.[6,13] Furthermore, such *in situ* generated PRs also enhance vascular permeability, leading to the formation of oedematous lesions through generation of inflammation mediators such as histamine, bradykinins, kallikreins, or thrombin, allowing further dissemination of the infection through the systemic circulation.[6,13,14] In the case of systemic infections, such as septicemia, MPRs act as a synergistic virulence factor, provoking a disordered proteolysis of various plasma proteins, causing imbalances of the proteinase–proteinase inhibitor equilibrium, thus disturbing the physiological homeostasis and eliciting an immunocompromised state of the host.[6,14] MPRs also trigger the disintegration of the iron-carrier

TABLE 28.1 Bacterial Metalloproteases, with their Preferred Scissile Bond and Organisms from Which They were Isolated

EC	Protease	Family	Organism(s)	Preferred Scissile Bond (P1#P1′)
NC	Listeria metalloprotease (Mpl)	M4	*L. monocytogenes*	Unspecific PR (thermolysin-like)
3.4.24.30	Coccolysin	M4	*E. faecalis*	Phe#Phe; Gly#Phe; Gly#Leu; Pro#Phe
NC	Hemagglutinin/ protease	M4	*V. cholerae*; *Helicobacter pylori*	Gly#Phe; Gly#Leu; Ser#Met; #Ser
3.4.24.26	Pseudolysin	M4	*P. aeruginosa*	Phe#Xaa; Gly#Leu
NC	Legionella metalloendopeptidase	M4	*L. pneumophila*	Unknown
NC	Camelysin	M4	*B. cereus*	Unknown
NC	rSVP	M4	*Salinivibrio* spp.	Unknown
NC	ZmpA, ZmpB	M4	*Burkholderia* spp.	Unknown
3.4.24.3	Vibrio collagenase	M9	*V. alginolyticus*; *V. parahaemolyticus*	Xaa#Gly
NC	*V. vulnificus* MPR	M9	*V. vulnificus*	Unknown
3.4.24.3	Clostridium collagenases	M9	*C. histolyticum*; *C. perfringens*	Xaa#Gly
NC	Nonhaemolytic enterotoxin	M9?	*B. cereus*	Xaa#Gly
3.4.24.40	Serralysin	M10	*Serratia* spp.; *P. aeruginosa; E. chrysanthemi*	Xaa#Gly; Xaa#Ala
NC	Arazyme	M10	*S. proteamaculans*	Unknown
3.4.24.40	Aeruginolysin	M10	*P. aeruginosa*	Leu#Gly; Gly#Gly
NC	Mirabilysin	M10	*P. mirabilis*	Leu#Gly (in IgA)
3.4.24.74	Fragilysin	M10	*B. fragilis*	Leu#Gly; Gly#Leu
NC	Flavastacin	M12	*F. meningosepticum*	Xaa#Asp
3.4.17.18	Carboxypeptidase T	M14	*Streptomyces* spp.	Xaa#Xaa-COOH
3.4.11.10	Leucyl aminopeptidase	M17	*Chlamydia trachomatis*; *E. coli*; *Haemophilus influenzae*; *M. tuberculosis*; *Mycoplasma genitalium*; *Rickettsia prowazekii*; *S. typhimurium*	H_2N-Leu#Xaa

TABLE 28.1 ***(Continued)***

EC	Protease	Family	Organism(s)	Preferred Scissile Bond (P1#P1′)
3.4.17.11	Glutamate carboxypeptidase	M20	*Pseudomonas* spp.; *Flavobacterium* spp.; *Acinetobacter* spp.	Xaa#GluCOOH
NC	VanX D,D-dipeptidase	M19	*Enterococcus* spp.; *Synechocystis* spp.	D-Ala#D-Ala[a]
3.4.24.57	*O*-sialoglycoprotein endopeptidase	M22	*H. influenzae*; *M. leprae*; *Mycoplasma genitalium*; *P. haemolytica*	Arg#Asp[b]
3.4.24.32	β-Lytic metalloendopeptidase	M23	*L. enzymogenes*	*N*-acetylmuramoyl#Ala; Gly#(ε-amino)Lys
NC	Staphylolysin	M23	*Aeromonas hydrophila*; *P. aeruginosa*	Gly#Gly[c]
NC	LytM	M23	*S. aureus*	Gly#Gly
3.4.24.13	IgA-specific metalloendopeptidase	M26	*Streptococcus* spp.; *Neisseria* spp.; *Haemophilus* spp.; *Ureaplasma* spp.; *Clostridium* spp.; *Capnocytophaga* spp.; *Prevotella* spp.	Pro#Thr; Pro#Ser
3.4.24.68	Tentoxilysin (tetanus neurotoxin)	M27	*Clostridium tetani*	Gln#Phe
3.4.24.69	Bontoxilysin (botulinum neurotoxin)	M27	*Clostridium botulinum*; *C. barati*; *C. butyricum*	Gln#Phe; Gln#Arg; Lys#Ala, Arg#Ile
NC	Anthrax toxin lethal factor	M34	*Bacillus anthracis*	Unknown
3.4.24.75	Lysostaphin	M37	*Staphylococcus staphylolyticus*; *S. simulans*	Gly#Gly[c]
3.4.24.29	Aureolysin	M40	*S. aureus*	Xaa#Leu
NC	AAA proteases (FtsH)	M41	Ubiquitous	Unknown

NC = not classified; M = metallo-PR family.

[a]The dipeptide is the enzyme substrate.

[b]From *O*-sialoglycoproteins.

[c]From pentaglycine cross-linking peptides of *S. aureus* peptidoglycan.

proteins of the host, which has, as a consequence, an enhanced uptake of iron, an essential element for the growth of bacteria. Some bacterial toxins, such as some enterotoxins (cholera toxin) are activated by MPRs, whereas other toxins, such as the botulinum (BoNT) and tetanus neurotoxins (TeNT) are metal proteases themselves.[16,17] It is thus clear that such MPRs constitute very important potential targets for the drug design of novel types of antibiotics.

28.2 METALLOPROTEASES OF THE THERMOLYSIN FAMILY (M4)

Thermolysin, an unspecific protease secreted by the gram-positive thermophilic bacterium *Bacillus thermoproteolyticus* is an extracellular 34.6 kDa metalloendopeptidase, without pharmacological applications, but which has been much investigated (it was the first MPR for which the X-ray structure was available[12]) and used to generate active site models for the design of inhibitors for mammalian zinc PRs (such as neprilysin).[18] This and related M4 family proteases contain a Zn(II) ion coordinated by two histidine residues belonging to a His-Glu-Xaa-Xaa-His (HEXXH) motif, whereas the third zinc ligand is a glutamate residue, at least 14 residues C-terminal to the last His of this motif. The zinc-coordinated water molecule also establishes a hydrogen bond with the carboxylate moiety of the glutamate belonging to the HEXXH motif, which strongly enhances the nucleophilicity of this water molecule.[1,12,18]

Several MPRs belonging to the M4 family were isolated in different bacteria, which might have important applications for the drug design of new antibiotics (Table 28.1). Thus, *Listeria monocytogenes* is a gram-positive nonspore-forming, facultative intracellular rod-shaped bacterium that is capable of causing sepsis and CNS infections in humans and animals.[19] The *L. monocytogenes* Mpl protease contains 510 amino acids and has a predicted molecular mass of 57.4 kDa. Following cleavage of the amino-terminal 24-amino acid signal sequence, the 55 kDa inactive zymogen is secreted to the external medium, whereas mature active protease possessing a molecular mass of 36 kDa is formed after further processing of the N-terminal 180 amino acids.[19] The enzyme is inhibited by the metal chelators of the polyamino-polycarboxylic acid type (EDTA, EGTA), by 1,10-phenanthroline, and by the thermolysin-type protease inhibitor phosphoramidon. Its activity is stimulated by low concentrations (0.1 mM) of zinc chloride, but inhibited by high concentrations (0.5 mM). The specificity of cleavage by the Mpl protease has not been defined, and no specific inhibitors were designed, although a model of its active site has recently been reported on the basis of its homology with thermolysin.[19]

Enterococcus faecalis is frequently identified as the etiologic agent of several opportunistic infections (such as soft tissue and urinary tract infections, intra-abdominal abscesses, and root canal infections), and as the causative agent in several cases of endocarditis, secondary bacteremia, and food poisoning. *Enterococci* account for nearly 10% of all nosocomial infections and constitute a significant treatment challenge due to their multidrug resistance properties. One of the well-studied virulence factors of

E. faecalis is a secreted bacterial protease, termed coccolysin or gelatinase, which has been shown to contribute to the process of biofilm formation.[20,21] Coccolysin is synthesized as a preproenzyme consisting of a signal sequence, a putative propeptide, and the mature enzyme. *E. faecalis* coccolysin/gelatinase requires C-terminal processing for full activation of protease activity, making it a unique enzyme among the members of the M4 family of proteases of gram-positive bacteria. It has been proposed, based on the specificity profile of the MPR isolated from this pathogen, that the extracellular production of this protease is associated with the above-mentioned clinical conditions.[20] Coccolysin inactivates human endothelin-1 by hydrolyzing this peptide primarily at the Ser5#Leu6 and the His16#Leu17 bonds and the human big endothelin at several bonds involving hydrophobic amino acid residues (similar to thermolysis or Mpl, this is also a relatively unspecific protease).[20] The degradation of endothelin by coccolysin resembles the peptidolytic processing of endothelin by thermolysin. Because *E. faecalis* is associated with a large number of infectious diseases, it is probable that the manifestation of inflammatory conditions in the presence of this organism is related to the coccolysin-catalyzed inactivation of endothelin, but no inhibitors directed against this protease have been reported for the moment.[20]

Cholera is an infectious disease caused by the bacterium *Vibrio cholerae* and characterized by severe vomiting and watery diarrhea.[22] These bacteria produce a cytolytic toxin named cytolysin/hemolysin, which is encoded by the hlyA gene. The cytolysin is produced as a 79 kDa precursor form (pro-HlyA) after cleavage of the signal peptide of prepro-HlyA. The pro-HlyA is then processed to a 65 kDa mature cytolysin (mature HlyA) after cleavage of the 15 kDa amino-terminal peptide (pro region) of the 79 kDa precursor, usually at the bond between Ala157 and Asn158. The hemagglutinin/protease, a major thermolysin-like protease of *V. cholerae*, processes the pro-HlyA to the 65 kDa mature form of the protein. Along with this, the protease-processed HlyA drastically increases hemolytic activity (the N-terminal amino acid of the mature form of cytolysin generated by HA/protease was Phe151).[22] Both hemagglutinating and protease functions of this protease are inhibited by chelating agents, including 2-(*N*-hydroxycarboxamido)-4-methyl pentanoyl-L-Ala-Gly-NH_2 (Zincov), a hydroxamic acid derivative specifically designed to inhibit zinc metalloproteases,[1] but no other specific or more potent inhibitors of this enzyme have been reported. Since the proteolytic activity of hemagglutinin/protease plays a key role in activation of toxins of *V. cholerae*,[23] finding such inhibitors would constitute an interesting alternative to the treatment of this disease. On the other hand, a *V. cholerae* strain defective in hemagglutinin/protease constitutes an anticholera vaccine candidate, which has been examined for safety and immunogenicity in healthy adult volunteers, showing promising activity.[23] A homology modeling study of hemagglutinin/protease from *V. cholerae* in the presence of inhibitor HPI [*N*-(1-carboxy-3-phenylpropyl)-phenylalanyl-alpha-aspargine] has recently been reported.[24] The 3D structure was predicted based on the sequence homology with *Pseudomonas aeruginosa* elastase (pseudolysin, mentioned subsequently). Comparison of the 3D structures of the two MPRs revealed a similarity for the binding of this inhibitor, highlighting the key catalytic residues as well as the residues at the S1 and S1′ binding subsites to be the same.[24]

Pseudolysin is the present name of the most abundant extracellular endopeptidase of *P. aeruginosa*, an opportunistic pathogen that may cause life-threatening infections in compromised patients with underlying respiratory disease like bronchiectasis, cystic fibrosis, and diffuse panbronchiolitis (this protease is commonly called *P. aeruginosa* elastase). Most strains of *P. aeruginosa* produce some kind of protease with broad substrate specificities during the infectious state in the host. *P. aeruginosa* elastase has a tissue-damaging proteolytic activity and is capable of degrading plasma proteins such as immunoglobulins, complement factors, and cytokines.[25] Destruction of the arterial elastic laminae in human systemic *P. aeruginosa* infections was the first evidence that *P. aeruginosa* secretes an elastinolytic protease. Pseudolysin is the major extracellular virulence factor of *P. aeruginosa*.[25,26] The contribution of pseudolysin to disease may be direct, causing tissue destruction, damaging some cell functions, or indirect, promoting virulence by interfering with host defense mechanisms.

Metal chelators, including EDTA, EGTA, 1,10-phenanthroline, and tetraethylene pentamine, inhibit the activity of pseudolysin. Phosphoramidon, phosphoryl-dipeptides such as phosphoryl-Leu-Phe and phosphoryl-Leu-Trp, and peptides containing thiol or hydroxamate groups such as $HSCH_2CONH$-Phe-Leu, $HSCH_2CH(CH_2C_6H_5)CO$-Ala-Gly-NH_2, or $HONHCOCH(CH_2C_6H_5)$-CO-Ala-Gly-NH_2, are potent reversible inhibitors.[27] Inhibition by these compounds is not specific to pseudolysin since other M4 family PRs are also inhibited.[27]

Pseudolysin may contribute in many ways to different diseases, so that effective/specific inhibitors would be important as potential antibiotics. For example, pseudolysin is probably responsible for the destruction of arterial elastic laminae in the vasculitis observed in cases of *Pseudomonas* septicemia.[26,27] This PR may also induce septic shock through activation of the Hageman factor-dependent kinin system; activation of the host kinin cascade may also be involved in the pathogenesis of skin burns infected by this pathogen, in which case pseudolysin appears to support growth and invasiveness of the organisms.[27] By rapidly degrading corneal proteoglycans, pseudolysin causes severe corneal destruction during *Pseudomonas* keratitis but it may also affect corneal damage indirectly, by activating endogenous corneal proteinases.[27] In *Pseudomonas* pneumonia, pseudolysin may cause lung damage with hemorrhages and necrosis of alveolar septal cells, destroying alveolar epithelial cell junctions, which increases epithelial permeability to macromolecules.[27] Pseudolysin also seems to be involved in the chronic disease caused by *P. aeruginosa* in cystic fibrosis patients.[27]

Molecular dynamics (MD) simulations and theoretical affinity predictions were used to gain molecular insight into pseudolysin inhibition.[28] Four low molecular weight inhibitors (H2N-Gly-Ala-COCH2CONHOH, Z-L-Leu-OH (Z = benzyloxycarbonyl), Z-L-PheOH, and phosphoramidon, $K_I = 0.2–340\ \mu M$) were docked at their putative binding sites, MD simulations were performed for 5.0 ns, and the free energy of binding was calculated by the linear interaction energy method. The number and the contact surface area of stabilizing hydrophobic, aromatic, and hydrogen bonding interactions was shown to reflect the affinity differences between the inhibitors. The calculations indicated that inhibitors interact with pseudolysin via the rigid active site loop, but that also contact sites outside this loop contributing significantly to the free energy of association.[28]

An epidemic of pneumonia that broke out among attendees at a convention of the American Legion in Philadelphia in 1976 was ultimately traced to a novel organism, *Legionella pneumophila*, isolated in the water-cooling units of the air-conditioning system of the convention hotel.[29] The organism is a facultative intracellular parasite in alveolar macrophages, and it secretes a thermolysin-like metalloprotease, which is one of the principal pathogenic factors in Legionnaire's disease due to its cytotoxic, tissue-destructive, and phagocyte-inhibitory properties.[29,30] The protease is cytotoxic to both neutrophils and monocyte/macrophages and interferes with the binding of natural killer cells to their target cells. Activity of this enzyme is inhibited by various metal chelators such as EDTA, EGTA, and 1,10-phenanthroline, but not by DTT (10 mM). Potent inhibition is obtained with phosphoramidon and the phosphoramidate analogue Z-GlyP(=O)Leu-Ala.[29] No specific inhibitors for the *Legionella* metalloendopeptidase have been designed until now.

Bacillus cereus frequently causes food poisoning or nosocomial diseases. Vegetative cells express the novel surface metalloproteinase camelysin (casein-cleaving metalloproteinase) during exponential growth on complex, peptide-rich media.[31] Camelysin spontaneously migrates from the surface of intact bacterial cells to preformed liposomes. The complete sequence of the camelysin-encoding gene, calY, was determined by reverse PCR on the basis of the N-terminal sequence and some internal tryptic cleavage peptides. The calY gene codes for a polypeptide of 21.569 kDa with a putative signal peptide of 27 amino acids preceding the mature protein (19.056 kDa). Although the predicted amino acid sequence of CalY does not exhibit a typical metalloprotease consensus sequence, high-pressure liquid chromatography-purified camelysin contains one zinc ion per protein molecule. Matrix-assisted laser desorption ionization time-of-flight mass spectrometry and tryptic peptide mass fingerprinting confirmed the identity of this zinc binding protein as CalY. Disruption of the calY gene results in a strong decrease in the cell-bound proteolytic activity on various substrates.[31]

A novel zinc metalloprotease from the moderately halophilic bacterium *Salinivibrio* spp. (strain AF-2004) has been cloned, sequenced, and reported to the GenBank recently.[32] Nucleotide sequence analysis of the selected clone revealed a single open reading frame (ORF) of 1833 bp encoding 611 amino acids. The deduced amino acid sequence encodes a zinc metalloprotease HEXXH-E consensus motif that is highly conserved in the M4 family of proteases. The primary amino acid sequence alignment search in the database revealed a moderate homology between the deduced amino acid sequence and the known zinc metalloproteases, including vibriolysin from *Vibrio vulnificus* and *P. aeruginosa* elastase (pseudolysin). Active recombinant enzyme was also obtained. Mass spectrometric fingerprinting of trypsin-digested analysis identified the processed mature protease that starts at Ala200 of the full-length protein. Although this result suggested a mature protein of 412 amino acids (44.8 kDa), electrospray ionization mass spectrometry revealed that the molecular mass of the purified protein (rSVP) was only 34.2 kDa, which indicates a further cleavage site at the carboxy terminal part of the protein.[32]

The genus *Burkholderia* contains the primary pathogens *Burkholderia pseudomallei* and *Burkholderia mallei* and several other species that emerged as opportunistic pathogens in persons suffering from cystic fibrosis or chronic granulomatous disease and in immunocompromised individuals.[33] *Burkholderia* species utilize quorum-sensing (QS) systems that rely on *N*-acyl-homoserine lactone (AHL) signal molecules to express virulence factors and other functions in a population-density-dependent manner. Most *Burkholderia* species employ the CepIR QS system, which relies on *N*-octanoyl-homoserine lactone.[33] However, some strains harbor multiple QS systems and produce numerous AHLs. *Burkholderia cenocepacia* protease ZmpA is expressed as a preproenzyme typical of thermolysin-like proteases such as *P. aeruginosa* LasB and *B. thermoproteolyticus* thermolysin.[34] Recombinant His6–prepro-ZmpA (of 62 kDa) was obtained, being observed that the fusion protein was autoproteolytically cleaved into 36 kDa (mature ZmpA) and 27 kDa peptides. The activity of the recombinant ZmpA was inhibited by EDTA and 1,10-phenanthroline, indicating that it is indeed a zinc MPR. ZmpA, however, was not inhibited by phosphoramidon, a classical inhibitor of the thermolysin-like proteases. The refolded mature ZmpA enzyme was proteolytically active against various substrates, including hide powder azure, type IV collagen, fibronectin, neutrophil α-1 proteinase inhibitor, α-2-macroglobulin, and γ-interferon, suggesting that *B. cenocepacia* ZmpA may cause either direct tissue damage to the host or damage to host tissues through a modulation of the host's immune system.[34] *B. cenocepacia* has an additional MPR, which was designated ZmpB.[35] The zmpB gene is present in the same species as zmpA and was detected in *B. cepacia*, *B. cenocepacia*, *Burkholderia stabilis*, *Burkholderia ambifaria*, and *Burkholderia pyrrocinia* but was absent from *Burkholderia multivorans*, *Burkholderia vietnamiensis*, *Burkholderia dolosa*, and *Burkholderia anthina*.[35] ZmpB has a predicted preproenzyme structure typical of thermolysin-like proteases and is distantly related to *B. cereus* bacillolysin. ZmpB was expressed as a 63 kDa preproenzyme precursor that was autocatalytically cleaved into mature ZmpB (35 kDa) and a 27 kDa prepropeptide. EDTA, 1,10-phenanthroline, and Zn(II) ions inhibited the enzyme activity, indicating that it is a metalloprotease. ZmpB had proteolytic activity against α-1 proteinase inhibitor, α-2-macrogobulin, type IV collagen, fibronectin, lactoferrin, transferrin, and immunoglobulins. *B. cenocepacia* zmpB and zmpA zmpB mutants had no proteolytic activity against casein and were less virulent in a rat agar bead chronic infection model, indicating that zmpB is involved in *B. cenocepacia* virulence. Expression of zmpB was regulated by both the CepIR and CciIR quorum sensing systems.[34,35]

28.3 METALLOPROTEASES OF THE M9 FAMILY (*Vibrio* AND *Clostridium* COLLAGENASES)

Family M9 contains bacterial collagenases from *Vibrio* and *Clostridium*, as well as a number of other collagenolytic bacterial endopeptidases less investigated for the moment.[1] The *Clostridium* collagenase (ChC, the best studied enzyme in this family) is only distantly related to the *Vibrio* collagenase, and had previously been considered to be the sole member of the now defunct M31 family.[1] Both the *Vibrio* and *Clostridium* collagenases are secreted enzymes, and are synthesized as precursors. The metal ion ligands have only recently been determined for ChC[36]: two of the zinc ligands occur in the His^{415} ExxH motif, while the third one is Glu447. Similar to the vertebrate matrix metalloproteinases (MMPs, that also degrade collagen), ChC is a multiunit protein, consisting of four segments, S1, S2a, S2b, and S3, with S1 incorporating the catalytic domain.[36,37]

Vibrio alginolyticus chemovar *iophagus* is a nonpathogenic marine bacterium isolated from cured hides, which rapidly lyses collagen under aerobic conditions.[38] The 82 kDa collagenase responsible for this activity (containing one Zn(II) ion/polypeptide chain) cleaves native collagen much more rapidly than does vertebrate interstitial collagenase (MMP-1).[38] In the first step, this enzyme acts on collagen in a manner similar to the MMPs, by attacking at a point three-quarters of the way from the amino-terminus, although the bond preferentially cleaved is different: Xaa#Gly for *Vibrio* collagenase, instead of Gly#Leu or Gly#Ile for the MMPs.[38] The use of *Vibrio* collagenase in tissue cell dispersions, elastin purification, and selective cleavages of gene products in biotechnology is similar to the same use of ChC, which will be discussed subsequently. However, more important applications in human therapy are in the removal of necrotic tissues from burns, ulcers, and decubitus ulcers because of its strong and specific activity against native collagen, a characteristic it also shares with the *Clostridium* collagenase.[38]

V. vulnificus is on the other hand ubiquitous in aquatic environments, causing only occasionally serious or fatal infections in humans, such as invasive septicemia contracted through consumption of raw seafood, as well as wound infections acquired through contact with brackish or marine waters.[39] *V. vulnificus* produces various virulence factors, including polysaccharide capsule, type IV pili, hemolysin, and proteolytic enzymes. A 45 kDa metalloprotease is thought to be the causative factor of the skin lesions, because the purified protease enhances vascular permeability through generation of chemical mediators and also induces serious hemorrhagic damage through digestion of the vascular basement membrane.[39] The other bacteria, *V. vulnificus*, also regulates the protease production through the quorum-sensing system depending on bacterial cell density. The N-terminal propeptide of *V. vulnificus* extracellular metalloprotease was recently shown to be both an inhibitor and a substrate for the enzyme.[40]

In another related species, *Vibrio anguillarum*, a secreted processing protease for the EmpA metalloprotease, denominated Epp., has been identified and characterized. It is not clear whether it belongs to the M9 family, and what its inhibition profile is.[41]

A mechanism for inhibiting the activity of transcription factors is their sequestration to the membrane from which they are released by proteolysis when needed. Acting in contrast to this inhibition mechanism are the virulence regulators ToxR and TcpP proteins of *V. cholerae*, which are localized to the inner membrane of the cell, where they bind promoter DNA and activate gene expression.[42] TcpP is rapidly degraded in the absence of another protein, TcpH. Regulators of TcpP stability were identified, one of which was YaeL, a membrane-localized zinc MPR as responsible for degrading

TcpP in the absence of TcpH.[42] The YaeL-dependent degradation pathway is active in TcpH(+) cells under conditions that are not favorable for virulence gene activation. This work expands the knowledge of YaeL-dependent processing in the bacterial cell and reveals an unexpected layer of virulence gene regulation in *V. cholerae*. It is unknown to which MPR family YaeL belongs or what are the classes of inhibitors acting on it.

Clostridium histolyticum is a pathogenic anaerobe that causes gas gangrene and other infections such as bacterial corneal keratitis.[43–45] All strains of this bacterium elaborate a collagenase (ChC), possibly used as a means to invade the host and to degrade its protein for nutritional purposes.[43] There are at least seven collagenase isozymes, with molecular masses ranging from 68 to 130 kDa, that have been purified to homogeneity.[44] They are designated as class I (α, β, γ, and η) and class II (δ, ε, and ξ) enzymes, based on a variety of criteria, including their relative activities toward collagen as well as synthetic substrates.[44] In fact, the crude homogenate of *C. histolyticum* is one of the most efficient systems known for the degradation of connective tissue, being able to hydrolyze triple helical regions of collagen under physiological conditions, as well as a multitude of synthetic peptide substrates.[43–45] The initial proteolytic events in this collagenase-mediated hydrolysis of type I, II, and III collagens have been delineated: the enzymes initially attack all three collagens at distinct hyperreactive sites, cleaving between Yaa#Gly bonds in the repeating Gly–Xaa–Yaa collagen sequence.[43] The hyperreactivity of these cleavage sites appears to be more related to local conformational features of the collagen fold than to the surrounding sequence. Thus, in contrast to the vertebrate collagenases, ChC degrades collagen into small peptides.[43,46]

ChC could not be crystallized for the moment, and its three-dimensional structure is thus not available. Only recently some electronic spectroscopic studies of Co(II)-substituted ChC have been reported,[47] thus offering interesting information regarding the binding of inhibitors within the active site of this bacterial protease. The Co(II)-substituted ChC retains catalytic activity, similar to the native zinc enzyme,[43] and possesses a pH-dependent electronic absorption spectrum, with a maximum centered at 585 nm and a shoulder at 530 nm, this spectrum being relatively similar to that of Co(II)-substituted carboxypeptidase A or thermolysin,[48] two enzymes in which the Zn(II) ion is coordinated – such as in ChC – by two histidines and a glutamate.[49] The inhibition of ChC is treated in detail in Chapter 30.

Recently, a collagenolytic and gelatinolytic MPR has been isolated in *B. cereus*, this protein having some sequence homology with the *C. histolyticum* and *Clostridium perfringens* collagenases. This 105 kDa protease presumably belongs to the M9 family, but little is known for the moment on its role in infection, inhibition, and so on.[50]

28.4 SERRALYSIN AND RELATED M10 PROTEASES; PROTEASES OF THE M12 FAMILY

MPRs belonging to the clan MB have two of the three zinc ligand His residues in an HEXXH motif, but unlike clan MA discussed above, their third zinc ligand is again a

His (or an Asp in rarer cases) in the extended motif **H**EX X**H**XXGXX(**H/D**) (the zinc ligands are evidenced in bold characters).[1] The Glu next to the first histidine is predicted to have the same role in catalysis as Glu142 of thermolysin and other MA metalloproteases already discussed (see Section 28.4.1), and the conserved Gly allows the formation of a β-turn that brings the zinc ligands together.[1] The endopeptidases from clan MB are also known as metzincins, because there is a conserved Met (Met145 in astacin) in a turn that underlies the active site. This clan includes very important enzymes, such as the MMPs as well as several bacterial MPRs that will be discussed subsequently.

Serralysin, a single-chain MPR with a molecular mass of about 55 kDa isolated from *Serratia* as well as some other bacteria (*Pseudomonas or Erwinia chrysanthemi*), shows a broad specificity with a preference for small- to medium-sized and hydrophobic residues in P1′ position (notably Gly and Ala).[51] Serralysin is considered as one of the virulence factors produced during *Serratia* or *Pseudomonas* infection, though its importance seems to be less than that of other toxins, such as aeruginolysin from *P. aeruginosa*.[52]

Activity of this MPR is inhibited by EDTA, tetramethylenepentamine, and 1,10-phenanthroline, but can be regained by addition of divalent metal ions. The serralysins are not inhibited by the classical phosphoramidons that are thermolysin inhibitors.[52] The physiological function of serralysins is not clear, but presumably, they play a role in nutrient digestion/uptake by the bacteria. Thus, some potent and specific inhibitors of this enzyme would be beneficial both for better understanding the contribution of the protease as virulence factor as well as for the development of antibiotics against these pathogens. A novel serralysin MPR was recently reported in the radiation resistant bacterium *Deinococcus radiodurans*.[53]

Serratia proteamaculans, bacteria isolated from the digestive tract of a spider, produce an extracellular protease named arazyme, with an estimated molecular mass of 51.5 kDa.[54] The purified enzyme was characterized as having high activities at wide pH and temperature ranges. This protease efficiently hydrolyzed a broad range of protein substrates, including albumin, keratin, and collagen. The dependence of enzymatic activities on the presence of metal ions such as calcium and zinc indicated that the enzyme is a MPR, being strongly inhibited by 1,10-phenanthroline and EDTA. The nucleotide sequence revealed that the deduced amino acid sequences shared extensive similarity with those of the serralysin family of MPRs from other enteric bacteria. A gene (inh) encoding a putative protease inhibitor was also identified immediately adjacent to the araA structural gene.[54]

Another protease that has been isolated from various strains of *P. aeruginosa* such as IFO 3080, IFO 3455, T 30, or from 18 strains isolated from infected patients were at first called *P. aeruginosa* alkaline proteinases (because of their pH optimum in the alkaline domain) and later as aeruginolysin.[55] Aeruginolysin can cause a wide range of pathogenic effects in hosts infected with *P. aeruginosa*, including tissue degradation, spreading of infection and septicemia, inactivation of defense-oriented proteins, including immunoglobulins, lysozyme, and transferrin, being thus an important virulence factor.[55] Aeruginolysin is inhibited by metal ion chelators such as EDTA and 1,10-phenanthroline as well as by peptidyl-mercaptoanilides such as Bz-Phe-Arg-SH (K_I = 7.5 mM), but is insensitive to phosphoramidon and zincov that inhibit many metalloproteinases of clan MA.[55] No potent inhibitors of this protease were reported for the moment.

Among the virulence components known to be expressed by the urinary tract pathogen, *Proteus mirabilis* is a MPR of 55 kDA, belonging to the M10 family, referred to as mirabilysin. This is an extracellular protease that can be isolated when *P. mirabilis* is grown on media containing a suitable substrate such as skim milk agar.[56] Thus, the protease produced during *Proteus* infections degrades among other proteins urinary tract IgA. Recombinant mirabilysin also degrades human IgA1 and IgG in a time-dependent manner resulting in complete digestion of the IgA1 substrate into numerous smaller fragments.[56]

Fragilysin, another metalloproteinase belonging to the metzincin clan, has been isolated as the *Bacteroides fragilis* enterotoxin.[57] *B. fragilis* is an anaerobic bacterium that is a part of the normal flora found in the large intestines of humans and most other mammals at about the same concentration as *E. coli*, but similar to *E. coli*, some strains of *B. fragilis* can produce toxin. Such enterotoxigenic strains were shown to produce a 20 kDa zinc metalloprotease toxin, which has been associated with diarrheal disease in animals and young children.[57] Despite more than a decade of research by several laboratories, the mechanism of action of the enterotoxin remained unknown until it was showed that the enterotoxin gene codes for a signature zinc binding motif belonging to the metzincin clan.[58] Fragilysin is cytopathic to intestinal epithelial cells and induces fluid secretion and tissue damage in ligated intestinal loops. This is the first reported example of a protease causing diarrhea. Fragilysin causes proteolytic degradation of the tight junctions and basement membranes (the extracellular matrix) of tissue-cultured epithelial cells, and this is probably the major mechanism by which the intestinal epithelium is altered *in vivo*.[58] Fragilysin also may activate host proteases and inactivate protease inhibitors, leading to exacerbation of the destruction.[58] Fragilysin hydrolyzes a number of proteins in the extracellular matrix of epithelial cells, which is probably how this enzyme causes its pathological effects on the intestine.[57–59] It has been shown in tissue culture experiments that the enzyme contributes to the invasion of intracellular bacterial species into epithelial cells by the proteolytic degradation of the extracellular matrix. Since the intestinal tract has to be resistant to the action of proteases in general, because it is constantly bathed in proteases, it is not at all clear how this protease can have such an unusual effect on this tissue. Enzymatic activity of fragilysin was inhibited by metal chelators (EDTA, DTPA) but not by inhibitors of other classes of proteases. In addition, cytotoxic activity of the enterotoxin on human carcinoma HT-29 cells was inhibited by acetoxymethyl ester EDTA.[60] No potent/selective fragilysin inhibitors were reported, although such compounds would be very desirable as potential antibiotics. Not very much is known regarding the inhibition of the M12 family protease flavastacin, isolated from the pathogen *Flavobacterium meningosepticum*, which cleaves peptide bonds of the type Xaa#Asp.[1]

28.5 BACTERIAL METALLOEXOPEPTIDASES (M14 AND M17 FAMILY PROTEASES)

Metallocarboxypeptidases were purified from *S. griseus* and *Thermoactinomyces vulgaris*, a bacterium that shares a number of biochemical traits with bacilli.[61] Although the physiological role of carboxypeptidase T (CPT)—the enzyme isolated from these bacteria—and related bacterial enzymes remains to be elucidated, it seems plausible that these exopeptidases are involved in degradation of protein substrates, which is consistent with their unusually broad specificity.[61] The crystal structure of CPT from *T. vulgaris* has been reported, being shown that this enzyme is a remote homologue of mammalian zinc carboxypeptidases.[61] In spite of the low degree of amino acid sequence identity, the three-dimensional structure of CPT is very similar to that of pancreatic carboxypeptidases A and B, with the active site being located at the C-edge of the central part of the beta sheet. Amino acid residues directly involved in catalysis and binding of the C-terminal carboxyl of the substrate are strictly conserved in CPT, CPA, and CPB, suggesting that the catalytic mechanism of the three carboxypeptidases is similar. Little is known regarding inhibition of CPT at this moment.[61]

Enzymes similar with the mammalian leucyl aminopeptidase have been isolated in many bacteria (Table 28.1) and they are referred to as bacterial leucyl aminopeptidases (LAPs) or PepA aminopeptidases. The LAPs of *E. coli* and *Salmonella typhimurium* are the best characterized of such bacterial enzymes and exhibit peptidase activities similar to their mammalian counterparts.[62] These are enzymes with broad substrate specificity that catalyze the release of N-terminal amino acids from peptides, especially leucyl and methionyl substrates.[62] Initially, unlike the zinc-dependent mammalian enzymes, the bacterial enzymes were considered to contain two Mn^{2+} ions, but it was later shown that they contain two Zn^{2+} ions, similar to the mammalian LAP.[62,63] The active sites of PepA from *E. coli* and LAP from bovine lens are isostructural, in both structures, a bicarbonate anion being bound to an arginine side chain (Arg356 in PepA and Arg336 in bovine lens LAP) very near to the two catalytic zinc ions.[63] Peptidase B (PepB) of *Salmonella enterica* serovar *Typhimurium* is one of three broad-specificity aminopeptidases found in this organism, being a 427-amino acid (46.36 kDa) protein, which has been assigned to the LAP structural family.[64] The *A. proteolytica* aminopeptidase, *Pseudomonas* spp. (RS-16), carboxypeptidase G2, and *S. griseus* aminopeptidase are other zinc-dependent exopeptidases with cocatalytic zinc ion centers and a conserved aminopeptidase fold, probably possessing a three-dimensional structure similar to that of the prostate-specific membrane antigen (PSMA) and the transferrin receptor (TfR).[65]

In *E. coli* and *S. typhimurium*, PepA functions as an aminopeptidase to hydrolyze exogenous peptides as a source of amino acids, to promote protein turnover during starvation, and probably to degrade abnormal proteins.[62,63] Thus, inhibitors of such enzymes might be important for the design of new antibiotics. Indeed, Huntington et al.[66] reported that peptide-derived thiols of the general structure **28.1** are potent, slow binding inhibitors of the aminopeptidase from *A. proteolytica*, with K_I values in the range of 2.5–57 nM. To investigate the nature of the interaction of these thiol-based inhibitors with the dinuclear active site of AAP, the electronic absorption and EPR spectra of Co(II)Co(II)-, Co(II)Zn(II)-, and Zn(II)Co(II)-AAP in the presence of the strongest binding inhibitor have been recorded, being shown that both [CoZn(AAP)] and [ZnCo(AAP)] in the presence of such inhibitors exhibited an absorption band centered at 320 nm characteristic of an Sz→Co(II) ligand–metal charge-transfer band. It was shown that the inhibitor is interacting with each active site metal ion: practically, the inhibitor interacts weakly with one of the metal ions in the dinuclear site, and the crystallographically identified micro-OH(H) bridge, which has been shown to mediate electronic interaction of the Co(II) ions, is likely broken upon inhibitor binding. These data suggested that the thiolate moiety of the inhibitor may bind to either of the metal ions in the dinuclear active site of AAP but does not bridge the dinuclear cluster. These compounds constitute interesting lead molecules for developing more potent LAP inhibitors with potential antibiotic properties.

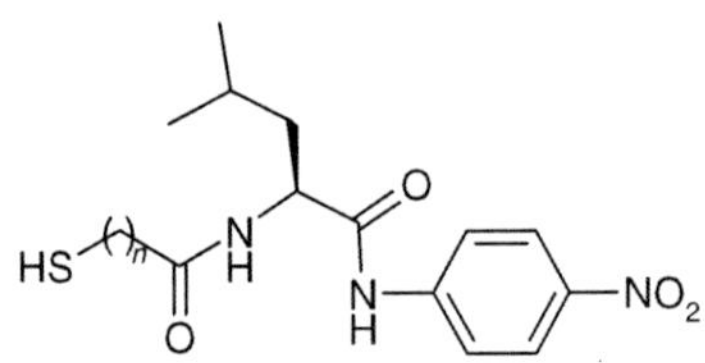

28.1 n = 1–3

28.6 PROTEASES OF THE M19, M20, M22, M23, M26, AND M27 FAMILIES

Proteases able to release the C-terminal glutamate residues from a wide range of *N*-acylating moieties, including peptidyl, aminoacyl, benzoyl, benzyloxycarbonyl, folyl, and pteroyl derivatives, are denominated glutamate carboxypeptidases (GCPs), and such enzymes were isolated only in prokaryotes (Table 28.1).[67] On the other hand, enzymes with similar activity, which hydrolyze the γ-glutamyl tail of antifolate and folate polyglutamates (named glutamyl hydrolases), were isolated in higher vertebrates,[68] but both the prokaryotic as well as the eukaryotic such enzymes have important pharmacological applications in the treatment of cancer. The prime interest in such enzymes was due to their ability to cleave the glutamate residue from folate polyglutamates **28.2** and, more significantly, folate analogues, such as the chemotherapeutic agent methotrexate **28.3**, an inhibitor of dihydrofolate reductase.[67,68] This property provided the opportunity to assess these enzymes as antitumor agents through depletion of reduced folates, essential cofactors in DNA synthesis, and as rescue agents against methotrexate toxicity.[67] Progress was limited through availability of the enzyme until isolation of carboxypeptidase G from *Pseudomonas* spp. strain RS-16.[67] Bacterial glutamate carboxypeptidases are characterized as Zn(II)-requiring exopeptidases with specificity for glutamate, whereas γ-glutamyl hydrolase (EC 3.4.19.9) is a cysteine protease and glutamate carboxypeptidase II, a metalloprotease belonging to the M28 family (but since these are mammalian enzymes, they will not be dealt with here).[67,68]

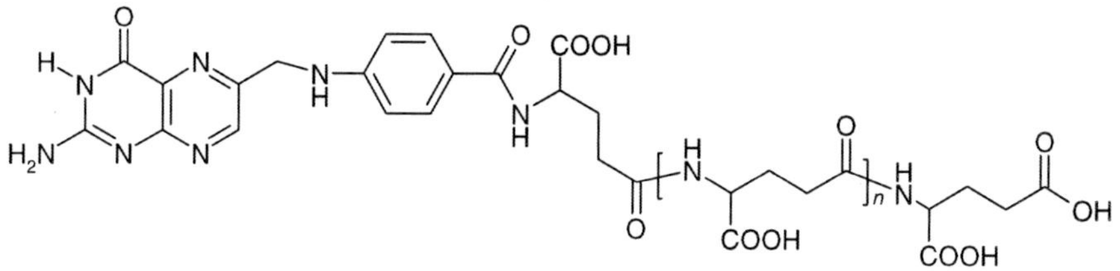

28.2 *n* = 0–7

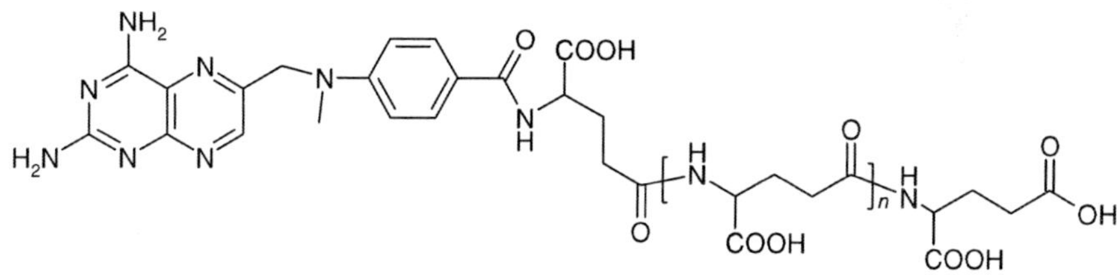

28.3 *n* = 3 (Methotrexate)

Carboxypeptidase G2 is a dimeric protein of 83.6 kDa containing two Zn(II) ions per subunit, and there are no disulfide bonds in its molecule.[67] In both native and recombinant forms, the enzyme is located in the periplasmic space, being targeted by a 22-amino acid signal peptide. Carboxypeptidase G2 has been crystallized and its structure determined at 2.5 Å resolution.[69] Each subunit of the molecular dimer consists of a large catalytic domain containing two Zn(II) ions at the active site and a separate smaller domain that forms the dimer interface. The two active sites in the dimer are more than 60 Å apart and are presumably independent; each contains a symmetric distribution of carboxylate and histidine ligands around the dimetallic center, the two zinc ions being at a distance of 3.2 Å.[69]

The enzyme has been developed for clinical use in two settings: (i) As a rescue agent during high dose (3–30 $g\ m^{-2}$) methotrexate therapy, against a range of cancers.[67] The principle of "rescue" is well established using leucovorin (5-formyltetrahydrofolate) to rapidly restore cellular levels of reduced folates depleted by the inhibitory action of methotrexate on dihydrofolate reductase. Removal of methotrexate then relies on patient hydration and renal function. If the latter is impaired, then high circulating levels of methotrexate can result in bone marrow and other toxicities. GCPs can thus be used to rapidly eliminate methotrexate from the circulation. Rescue protocols have been extended to include intrathecal, as opposed to systemic, methotrexate therapy. (ii) By using targeted carboxypeptidase G2 to remove the glutamate residue from prodrugs, releasing a highly cytotoxic agent at tumor sites.[67] This method has been given the acronym ADEPT (antibody-directed enzyme prodrug therapy). Targeting is achieved by covalent linkage of carboxypeptidase G2 to tumor-associated antibodies and monoclonal antibody fragments, including antihuman chorionic gonadotrophin, antihuman carcinoembryonic antigen, or antihuman c-erbB2 protooncogene product among others.[67,70] Once delivered to the tumor site, the enzyme is used to activate prodrugs, predominantly based on glutamyl derivatives of nitrogen mustard compounds such as 4-[(2-mesyloxyethyl)(2-chloroethyl)amino] benzoyl glutamic acid and 4-[*N*,*N*,-bis(2-iodoethyl)amino]phenol linked to glutamic acid.[67] Clinical studies emerging by using such approaches are very promising.[67]

There are at least 17 families of MPRs that could not yet been assigned to clans.[1] Of these, 10 possess the HEXXH motif that includes two zinc ligands and the catalytic Glu that is found in clan MA and clan MB, whereas for 7 families, the HEXXH motif has not been determined, nor any motifs similar to those in any clans of metallopeptidase.[1] The bacterial enzymes discussed subsequently all belong to these 17 families of less characterized proteases.

The zinc-containing D-alanyl-D-alanine (D-Ala-D-Ala) dipeptidase VanX has been detected in both gram-positive and gram-negative bacteria, being associated with resistance to the glycopeptide antibiotic vancomycin.[71–73] It appears that this enzyme has at least three distinct physiological roles: (i) In pathogenic vancomycin-resistant *enterococci* (such as *Enterococcus faecium*), vanX is part of a five-gene cluster that is switched on to reprogram cell wall biosynthesis to produce peptidoglycan chain precursors terminating in the depsipeptide D-Ala-D-lactate rather than in the dipeptide D-Ala-D-Ala. This modified peptidoglycan exhibits a 1000-fold decrease in affinity for vancomycin (accounting for the observed phenotypic resistance) due to the loss of one hydrogen bond between the antibiotic molecule and the amide NH of the dipeptide (now replaced by the ester bond of the depsipeptide).[72] (ii) In the glycopeptide antibiotic producers, such as *Streptomyces toyocaensis* (which secretes a vancomycin-like antibiotic) and *Amylocatopsis orientalis*, a vanHAX operon, may have coevolved with antibiotic biosynthesis genes to provide immunity by reprogramming cell wall termini to D-Ala-D-lactate as antibiotic biosynthesis proceeds and the toxic compound accumulates. (iii) In the gram-negative bacterium *E. coli*, which is never challenged by the glycopeptide antibiotics because such molecules are unable to penetrate into the outer membrane permeability barrier, a vanX homologue (named ddpX) is cotranscribed with a putative dipeptide transport system (ddpABCDF) in stationary phase by the transcription factor RpoS (σ^s).[72] The combined action of DdpX and the permease would permit hydrolysis of D-Ala-D-Ala transported back into the cytoplasm from the periplasm as cell wall cross-links are refashioned, and D-Ala resulting in hydrolysis could then be oxidized as an energy source for cell survival under starvation conditions.[72]

VanX is a 202-amino acid polypeptide (23.5 kDa) existing in solution as a homodimer, and it contains 1.0 mol of bound Zn(II) per mol of polypeptide. It is a D-,D-dipeptidase, catalyzing the hydrolysis of the bacterial cell wall biosynthesis intermediate D-Ala-D-Ala into two molar equivalents of D-alanine; as mentioned above, VanX does not catalyze the hydrolysis of the depsipeptide D-Ala-D-lactate. The exact mechanism by which VanX preferentially hydrolyzes the amide substrate versus its kinetically and thermodynamically more favorable ester analogue has not yet been determined and is the subject of current research efforts. VanX is resistant to the action of β-lactam antibiotics but is inhibited by transition-state substrate analogues or by chelating agents. Dithiols such as DTT and dithioerythritol are also potent time-dependent micromolar inhibitors whose mechanism of inactivation is postulated to occur by the formation of an enzyme–metal–dithiol ternary complex.[71–73]

The crystal structure of the *Enterococcus faecius* enzyme has recently been determined: the Zn(II) ligands are His116, Asp123, and His184, whereas Glu181 hydrogen bonds the zinc-coordinated water molecule.[74] The most important finding of this study was that the active site is small and constricted (of around 150 $\AA^3$), which may make rational design of VanX inhibitors a significant medicinal chemistry challenge.[73,74] Unfortunately, potent, nanomolar inhibitors of VanX have not been reported for the moment, although such compounds might have an important clinical impact on the treatment of infections resistant to vancomycin.

Another enzyme potentially important for the development of nonclassical antibiotics is *O*-sialoglycoprotein endopeptidase, abbreviated as glycoprotease, which derives its name from its unique specificity for the proteolytic cleavage of glycoproteins. To date, all known substrates of this enzyme are glycoproteins rich in sialoglycans O-linked to threonine or serine residues.[75] The enzyme was discovered in *Pasteurella haemolytica* A1 (an organism associated with fibrinous pneumonia in cattle) due to its ability to hydrolyze human erythrocyte glycophorin A.[75] The high specificity of the *P. haemolytica* glycoprotease for cell surface *O*-sialoglycoproteins has made the enzyme a useful tool in cell surface antigen studies, since the lack of action of the glycoprotease against protein substrates that do not bear O-linked glycans means that cell surface molecules and their roles can be delineated by its use. However, there is as yet no clear indication of the role of the enzyme in the pathogenesis of fibrinous pneumonia in cattle, although animals vaccinated with *P. haemolytica* culture proteins supplemented with the recombinant glycoprotease fusion protein rGcp-F show enhanced protection against experimental challenge with live pathogen.[75]

The soil bacterium *Lysobacter enzymogenes* has a remarkable ability to lyse other bacteria and some soil nematodes, a property that allowed the isolation of a protease named β-lytic protease, since this enzyme has the ability to cause lysis of bacterial cell walls and differentiate it from α-lytic protease. This metalloendopeptidase lyses *Arthrobacter globiformis*, *Micrococcus luteus*, and *Staphylococcus aureus* cells, thus inhibiting the growth of sensitive organisms and potentially serving as an antimicrobial agent.[76] This 19.1 kDa protein (containing 178-amino acid residues) incorporates two disulfide bonds between residues 65–111 and 155–168 and one zinc ion per mole, which is essential for activity. It does not contain the consensus zinc binding site HEXXH but has a conserved HXH sequence that is probably implicated in zinc binding.[76]

Staphylolysin (lysostaphin) is the new name suggested for *P. aeruginosa* LasA endopeptidase that causes lysis of *S. aureus* cells. Recent studies show that staphylolysin is a secreted endopeptidase that can slowly degrade insoluble elastin, rendering it a better substrate for pseudolysin and other endopeptidases.[77] Staphylolysin cleaves peptide bonds within the pentaglycine cross-linking peptides of *S. aureus* peptidoglycan, thus leading to cell lysis. It also lyses cell walls of *Micrococcus radiodurans* and *Gaffkia tetragena*, which contain di- and triglycine sequences in their interpeptides, respectively.[77] Staphylolysin is a 19.965 kDa protein, consisting of 182 residues; its amino acid sequence is ~40% identical to those of *Achromobacter lyticus* and *L. enzymogenes* β-lytic endopeptidases. Staphylolysin contains four conserved Cys residues (which by analogy to the *Lysobacter* enzyme are likely to be engaged in disulfide bonds) and one zinc ion. Although it does not possess the classical zinc binding motif (HEXXH), a conserved HXH sequence has been evidenced to be the potential zinc binding site (where X = Leu121).[77]

Staphylolysin appears to play a role in pathogenesis of several infections, including corneal infections: in a mouse model of *P. aeruginosa* corneal infection, a defined lasA mutant causes mild to no disease following infection; application of 5 mg of purified staphylolysin to scarified mouse corneas produced an acute toxic reaction.[77] In experimental models of lung infections, bacterial production of staphylolysin increases the virulence of *P. aeruginosa*. Staphylolysin inhibits the growth of *S. aureus* and thus may play a role in acquiring certain niches, such as the cystic fibrosis lung environment. Its ability to stimulate elastin degradation by other endopeptidases, as demonstrated *in vitro*, suggests that it may contribute to the tissue destruction associated with *P. aeruginosa* infections. No inhibitors of this enzyme were reported, but the measurement of antibodies against *S. aureus* staphylolysin was used to discriminate between complicated and uncomplicated *S. aureus* septicaemia.[78]

LytM, an autolysin from *S. aureus*, is a zinc-dependent glycyl-glycine endopeptidase with a characteristic HxH motif that belongs to the lysostaphin family (M23/37) of MPRs. The 1.3A crystal structure of LytM, the first structure of a lysostaphin-type peptidase, has recently been solved.[79] In the LytM structure, the zinc ion is tetrahedrally coordinated by the side chains of N117, H210, D214, and H293, the second histidine of the HxH motif. Although close to the active site, H291, the first histidine of the HxH motif, is not directly involved in zinc coordination and there is no water molecule in the coordination sphere of the Zn^{2+}, suggesting that the crystal structure shows a latent form of the enzyme. The "asparagine switch"[79] in LytM was also proposed to be analogous to the "cysteine switch" in promatrix metalloproteases (see Chapter 23).

Lysostaphin-type peptidases and D-Ala-D-Ala metallopeptidases have also been shown to possess similar active sites and share a core folding motif in otherwise highly divergent folds.[80] The central Zn^{2+} is tetrahedrally coordinated by two histidines, an aspartate and a water molecule in both classes of enzymes. The Zn^{2+} chelating residues occur in the order histidine, aspartate, histidine in all sequences and contact the metal via the N-epsilon, the O-delta, and the N-delta, respectively. The identity of the other active site residues varies, but in all enzymes of known structure except for VanX, a conserved histidine is present two residues upstream of the second histidine ligand to the Zn^{2+}. As the same arrangement of active site residues is also found in the N-terminal, cryptic peptidase domain of sonic hedgehog, it was proposed that this arrangement of active site residues be called the "LAS" arrangement, because it is present in lysostaphin-type enzymes, D-Ala-D-Ala metallopeptidases, and in the cryptic peptidase in the N-domain of sonic hedgehog.[80]

IgA proteinases were recognized by identification of unusual fragments of immunoglobulin A (IgA) in cell-free fluids of the human digestive tract and since this cleaving activity was also found in normal human saliva, it appeared to be of bacterial origin, not from the pancreas or intestine. This was confirmed by finding an IgA cleaving activity in culture filtrates of many bacteria colonizing or infecting human beings.[81] IgA proteinases are a group of endopeptidases produced by pathogenically important bacteria belonging to the genera *Streptococcus*, *Neisseria*, *Haemophilus*, *Ureaplasma*, *Clostridium*, *Capnocytophaga*, and *Prevotella*.[81] All these proteinases cleave the glycosylated hinge region of human IgA, a flexible peptide stretch that lies within the heavy chain, between the F_{ab} and F_c domains.[81] An analysis of 13 immunoglobulin A1 (IgA1) protease genes (iga) of strains of *Streptococcus pneumoniae*, *Streptococcus oralis*, *Streptococcus mitis*, and *Streptococcus sanguis* showed all of them to encode proteins with molecular masses of approximately 200 kDa, containing the sequence motif HEMTH and an aspartate residue 20-amino acids downstream, which are characteristic of Zn MPRs. In addition, all had a typical gram-positive cell wall anchor motif, LPNTG, which, in contrast to such motifs in other known streptococcal and staphylococcal proteins, was located in their N-terminal parts.[82]

The independent evolution of several distinct classes of enzymes with a similar biological function points to the importance of inactivation of IgA1 for colonization by bacterial mucosal pathogens. Yet no specific role for any IgA proteinase in infection has been proven for the moment. IgA proteinase activity is, in principle, capable of interfering with most functions of IgA because cleavage separates the antigen binding F_{ab} from the effector F_c domains of the molecule, and this not only facilitates bacterial evasion of immunity by establishing an immunodeficiency toward the microorganism but may also contribute to immune evasion by the secretory IgA cleavage products themselves.[81] No inhibitors active in the submillimolar range have been successfully made for these enzymes. Attempts to synthesize small, substrate-based inhibitors for the metallo-type IgA proteinases have been unsuccessful, while millimolar inhibitors have been made for the serine-type enzymes.

Bacterial toxins belonging to the zinc MPR class such as tentoxilysin (tetanus neurotoxin), botulinum neurotoxin (both belonging to the M27 class of MPRs), and anthrax toxin lethal factor (class M34) are treated in a separate chapter.

28.7 AAA PROTEASES AND OTHER UNCLASSIFIED PROTEASES

AAA proteases (ATP-ases associated with a variety of cellular activities) represent a conserved class of ATP-dependent zinc proteases that mediate the degradation of membrane proteins in bacteria, mitochondria, and chloroplasts.[83,84] These proteins combine proteolytic and chaperone-like activities, forming a membrane-integrated "quality control" system, and their inactivation/inhibition provokes severe defects in various organisms.[83,84] They are probably present in all bacteria, being some of the few proteases essential for viability of the microorganisms.[83,84] Such proteins were first detected in *E. coli*, being denominated FtsH. FtsH is an ATP- and Zn^{2+}-dependent metalloprotease with a molecular mass of about 70 kDa, being anchored to the cytoplasmic membrane via two transmembrane regions in such a way that the very short amino and the long carboxy termini are exposed into the cytoplasm.[83]

A 200-amino acid residue long module contains the ATP binding site, whereas the second transmembrane segment, which neighbors the C-terminal cytoplasmic region of FtsH, participates not only in its membrane anchoring but also in its protease activity (ascribed to an HEXXH zinc binding motif) and homo-oligomerization.[85] The specificity of peptide bond cleavage reaction by this protease has not been elucidated, but it was shown that FtsH cleaves the C-terminal side of hydrophobic residues and produces a characteristic set of small peptides (<30 kDa) without releasing a large intermediate.[86] Thus, FtsH recognizes the unfolded structure of proteins and progressively digests them at the expense of ATP.[83–86] In the absence of substrate proteins, FtsH hydrolyzes ATP with an apparent K_m value for ATP of about 80 mM.[83–86] Proteolytic activity of FtsH is inhibited by chelators such as 1,10-phenanthroline and EDTA, as well as by vanadate, whereas it is insensitive to *N*-ethylmaleimide, azide, KNO_3, and PMSF.[83–86]

In *E. coli* (but presumably in many other bacteria), FtsH (or its homologues) forms a complex with a two periplasmically exposed membrane proteins, HflK and HflC. FtsH is required for proteolytic degradation of some unstable proteins that include both soluble regulatory proteins such as sigma-32 (heat-shock sigma factor) and phage lambda CII (transcriptional activator), as well as membrane proteins, including uncomplexed forms of SecY (that forms the translocon together with SecE and SecG) and a subunit of the FO complex of the H^+–ATPase.[84] Its activity can be modulated by the HflKC proteins, by an another membrane protein designated YccA that can transiently associate with both the FtsH and the HflKC proteins, or by the small peptides such as CIII encoded by phage lambda (involved in lysogenization) or SpoVM (needed for sporulation) encoded by *Bacillus subtilis*. Besides being a protease, there is circumstantial evidence that FtsH also acts as a molecular chaperone, being designated as a "charonin."[84]

Since AAA proteases are essential for the viability of the bacterial cells, developing inhibitors targeted against them would lead to a new class of efficient antibiotics. Still, such studies have not been performed for the moment as this class of proteases has only recently been studied in some detail; however, they are very promising targets for the drug design of antibiotics of the future.

It was recently shown that FtsH, the membrane-bound ATP-dependent zinc metalloprotease that proteolytically regulates the levels of specific membrane and cytoplasmic proteins, is also crucial to the human pathogen *Mycobacterium tuberculosis*.[87] As the substrates of MtFtsH in mycobacteria are not known, it has been examined whether recombinant MtFtsH could complement the lethality of a Delta-ftsH3::kan mutation in *E. coli* and elicit proteolytic activity against the known substrates of *E. coli* FtsH, namely, heat shock transcription factor sigma32 protein, protein translocation subunit SecY, and bacteriophage lambdaCII repressor protein. The MtFtsH protein could not only efficiently complement lethality of Delta-ftsH3::kan mutation in *E. coli* but could also degrade all three heterologous substrates with specificity when expressed in ftsH-null cells of *E. coli*. These observations revealed the degree of conservation in the mechanisms of substrate recognition and cellular processes involving FtsH protease of *M. tuberculosis* and *E. coli*.[87]

Thus, FtsH has N-terminally located transmembrane segments and a main cytosolic region consisting of AAA-ATPase and Zn^{2+}-metalloprotease domains. It was recently shown to form a homohexamer, which is further complexed with an oligomer of the membrane-bound modulating factor HflKC.[88] FtsH degrades a set of short-lived proteins, enabling cellular regulation at the level of protein stability. FtsH also degrades some misassembled membrane proteins, contributing to their quality maintenance. It is an energy-utilizing and processive endopeptidase with a special ability to dislocate membrane protein substrates out of the membrane, for which its own membrane-embedded nature is essential.[88]

E. coli HtpX is a putative membrane-bound zinc MPR that has been suggested to participate in the proteolytic quality control of membrane proteins in conjunction with FtsH.[89] HtpX was biochemically characterized and its proteolytic activities against membrane and soluble proteins were confirmed. HtpX underwent self-degradation upon cell disruption or membrane solubilization. In the presence of zinc, HtpX also degraded casein and cleaved a solubilized membrane protein SecY. Such results showed that HtpX is a zinc-dependent endoprotease member of the membrane-localized proteolytic system in *E. coli*.[89]

28.8 CONCLUSIONS

As seen throughout this chapter, there is a wealth of data on the presence of zinc proteases in many bacterial species, some of which causing serious disease. Furthermore, many of these pathogens developed resistance to the clinically used antibiotics, which normally target bacterial cell wall biosynthesis or affect protein synthesis on ribosomes. In these organisms, MPRs play critical functions related to colonization and evasion of host immune defenses, acquisition of nutrients for growth and proliferation, facilitation of dissemination, or tissue damage during infection. However, only in the last decade, such targets started to be investigated in detail. Thus, there emerged a lot of interesting data on the cloning, purification, and biochemical characterization of many such zinc proteases, from both gram-positive as well as gram-negative pathogens. However, for the moment, a few potent, specific inhibitors for such bacterial proteases have been reported except for *C. histolyticum* collagenase, botulinum, and tetanus neurotoxin and anthrax lethal factor. No inhibitors of the critically important and ubiquitous IgA-specific metallopeptidases or AAA proteases have been reported to date, although such compounds would presumably constitute a new class of highly effective antibiotics.

REFERENCES

1. Barrett, A. J.; Rawlings, N. D.; Woessner, J. F., Jr., Eds. *Handbook of Proteolytic Enzymes (CD-ROM);* Academic Press: London, **1998**; Chapters 1–569.
2. (a) Supuran, C. T.; Scozzafava, A.; Mastrolorenzo, A. Bacterial proteases: current therapeutic use and future prospects for the development of new antibiotics. *Expert Opin.*

Ther. Pat. **2001**, *11*, 221–259. (b) Supuran, C. T.; Scozzafava, A.; Clare, B. W. Bacterial protease inhibitors. *Med. Res. Rev.* **2002**, *22*, 329–372.

3. (a) Mastrolorenzo, A.; Rusconi, S.; Scozzafava, A.; Barbaro, G.; Supuran, C. T. Inhibitors of HIV-1 protease: current state of the art 10 years after their introduction. From antiretroviral drugs to antifungal, antibacterial and antitumor agents based on aspartic protease inhibitors. *Curr. Med. Chem.* **2007**, *14*, 2734–2748. (b) Barbaro, G.; Scozzafava, A.; Mastrolorenzo, A.; Supuran, C. T. Highly active antiretroviral therapy: current state of the art, new agents and their pharmacological interactions useful for improving therapeutic outcome. *Curr. Pharm. Des.* **2005**, *11*, 1805–1843.
4. Wlodawer, A.; Gustchina, A. Structural and biochemical studies of retroviral proteases. *Biochim. Biophys. Acta* **2000**, *1477*, 16–34.
5. Travis, J.; Potempa, J. Bacterial proteinases as targets for the development of second generation antibiotics. *Biochim. Biophys. Acta* **2000**, *1477*, 35–50.
6. Miyoshi, S. I.; Shinoda, S. Bacterial metalloproteases as the toxic factor in infection. *J. Toxicol. Toxin. Rev.* **1997**, *16*, 177–194.
7. Maeda, H. Role of microbial proteases in pathogenesis. *Microbiol. Immunol.* **1996**, *40*, 685–699.
8. Rice, S. A.; Givskov, M.; Steinberg, P.; Kjelleberg, S. Bacterial signals and antagonists: the interaction between bacteria and higher organisms. *J. Mol. Microbiol. Biotechnol.* **1999**, *1*, 23–31.
9. Wright, G. D. Resisting resistance: new strategies for battling superbugs. *Chem. Biol.* **2000**, *7*, R127–R132.
10. Rawlings, N. D.; Barrett, A. J. Evolutionary families of peptidases. *Biochem. J.* **1993**, *290*, 205–218.
11. Berger, A.; Schechter, I. Mapping the active site of papain with the aid of peptide substrates and inhibitors. *Philos. Trans. R. Soc. Lond.* **1970**, *257*, 249–264.
12. Coleman, J. E. Zinc enzymes. *Curr. Opin. Chem. Biol.* **1998**, *2*, 222–234.
13. Shinoda, S.; Miyoshi, S. I.; Wakae, H.; Rahman, M.; Tomochika, K. I. Bacterial proteases as pathogenic factors, with special emphasis on *vibrio* proteases. *J. Toxicol. Toxin. Rev.* **1996**, *15*, 327–339.
14. Vollmer, P.; Walev, I.; Rose-John, S.; Bhakdi, S. Novel pathogenic mechanism of microbial metalloproteinase: liberation of membrane-anchored molecules in biologically active form exemplified by studies with the human interleukin-6 receptor. *Infect. Immun.* **1996**, *64*, 3646–3651.
15. Påhlman, L. I.; Marx, P. F.; Mörgelin, M.; Lukomski, S.; Meijers, J. C.; Herwald, H. Thrombin-activatable fibrinolysis inhibitor binds to *Streptococcus pyogenes* by interacting with collagen-like proteins A and B. *J. Biol. Chem.* **2007**, *282*, 24873–24881.
16. Schiavo, G.; Matteoli, M.; Montecucco, C. Neurotoxins affecting neuroexocytosis. *Physiol. Rev.* **2000**, *80*, 717–766.
17. Humeau, Y.; Doussau, F.; Grant, N. J.; Poulain, B. How botulinum and tetanus neurotoxins block neurotransmitter release. *Biochimie* **2000**, *82*, 427–446.
18. Beynon, R. J.; Beaumont, A. Thermolysin. In *Handbook of Proteolytic Enzymes (CD-ROM)*; Barrett, A. J.; Rawlings, N. D.; Woessner, J. F., Jr., Eds.; Academic Press: London, **1998**; Chapter 351.

19. Coffey, A.; van den Burg, B.; Veltman, R.; Abee, T. Characteristics of the biologically active 35-kDa metalloprotease virulence factor from *Listeria monocytogenes*. *J. Appl. Microbiol.* **2000**, *88*, 132–141.

20. Makinen, P. L.; Makinen, K. K. The *Enterococcus faecalis* extracellular metalloendopeptidase (EC 3.4.24.30; coccolysin) inactivates human endothelin at bonds involving hydrophobic amino acid residues. *Biochem. Biophys. Res. Commun.* **1994**, *200*, 981–985.

21. Del Papa, M. F.; Hancock, L. E.; Thomas, V. C.; Perego, M. Full activation of *Enterococcus faecalis* gelatinase by a C-terminal proteolytic cleavage. *J. Bacteriol.* **2007**, *189*, 8835–8843.

22. Nagamune, K.; Yamamoto, K.; Naka, A.; Matsuyama, J.; Miwatani, T.; Honda, T. *In vitro* proteolytic processing and activation of the recombinant precursor of El Tor cytolysin/hemolysin (pro-HlyA) of *Vibrio cholerae* by soluble hemagglutinin/protease of *V. cholerae*, trypsin, and other proteases. *Infect. Immun.* **1996**, *64*, 4655–4658.

23. Benitez, J. A.; Garcia, L.; Silva, A.; Garcia, H.; Fando, R.; Cedre, B. Preliminary assessment of the safety and immunogenicity of a new CTXPhi-negative, hemagglutinin/protease-defective El Tor strain as a cholera vaccine candidate. *Infect. Immun.* **1999**, *67*, 539–545.

24. Lutfullah, G.; Amin, F.; Khan, Z.; Azhar, N.; Azim, M. K.; Noor, S.; Shoukat, K. Homology modeling of hemagglutinin/protease [HA/P (vibriolysin)] from *Vibrio cholerae*: sequence comparison, residue interactions and molecular mechanism. *Protein J.* **2008**, *27*, 105–114.

25. Kon, Y.; Tsukada, H.; Hasegawa, T.; Igarashi, K.; Wada, K.; Suzuki, E.; Arakawa, M.; Gejyo, F. The role of *Pseudomonas aeruginosa* elastase as a potent inflammatory factor in a rat air pouch inflammation model. *FEMS Immunol. Med. Microbiol.* **1999**, *25*, 313–321.

26. Pesci, E. C.; Milbank, J. B.; Pearson, J. P.; McKnight, S.; Kende, A. S.; Greenberg, E. P.; Iglewski, B. H. Quinolone signaling in the cell-to-cell communication system of *Pseudomonas aeruginosa*. *Proc. Natl. Acad. Sci. USA* **1999**, *96*, 11229–11234.

27. Kessler, E.; Ohman, D. E. Pseudolysin. In *Handbook of Proteolytic Enzymes (CD-ROM);* Barrett, A. J.; Rawlings, N. D.; Woessner, J. F., Jr., Eds.; Academic Press: London, **1998**; Chapter 357.

28. Adekoya, O. A.; Willassen, N. P.; Sylte, I. Molecular insight into pseudolysin inhibition using the MM-PBSA and LIE methods. *J. Struct. Biol.* **2006**, *153*, 129–144.

29. Woessner, J. F. *Legionella* metalloendopeptidase. In *Handbook of Proteolytic Enzymes (CD-ROM);* Barrett, A. J.; Rawlings, N. D.; Woessner, J. F., Jr., Eds.; Academic Press: London, **1998**; Chapter 358.

30. Moffat, J. F.; Edelstein, P. H.; Regula, D. P.; Cirillo, J. D.; Tompkins, L. S. Effects of an isogenic Zn-metalloprotease-deficient mutant of *Legionella pneumophila* in a guinea-pig pneumonia model. *Mol. Microbiol.* **1994**, *12*, 693–705.

31. Grass, G.; Schierhorn, A.; Sorkau, E.; Müller, H.; Rücknagel, P.; Nies, D. H.; Fricke, B. Camelysin is a novel surface metalloproteinase from *Bacillus cereus*. *Infect. Immun.* **2004**, *72*, 219–228.

32. Karbalaei-Heidari, H. R.; Ziaee, A. A.; Amoozegar, M. A.; Cheburkin, Y.; Budisa, N. Molecular cloning and sequence analysis of a novel zinc-metalloprotease gene from the *Salinivibrio* sp. strain AF-2004 and its extracellular expression in *E. coli*. *Gene* **2008**, *408*, 196–203.

33. Sokol, P. A.; Malott, R. J.; Riedel, K.; Eberl, L. Communication systems in the genus *Burkholderia*: global regulators and targets for novel antipathogenic drugs. *Future Microbiol.* **2007**, *2*, 555–563.
34. Kooi, C.; Corbett, C. R.; Sokol, P. A. Functional analysis of the *Burkholderia cenocepacia* ZmpA metalloprotease. *J. Bacteriol.* **2005**, *187*, 4421–4429.
35. Kooi, C.; Subsin, B.; Chen, R.; Pohorelic, B.; Sokol, P. A. *Burkholderia cenocepacia* ZmpB is a broad-specificity zinc metalloprotease involved in virulence. *Infect. Immun.* **2006**, *74*, 4083–4093.
36. Jung, C. M.; Matsushita, O.; Katayama, S.; Minami, J.; Sakurai, J.; Okabe, A. Identification of metal ligands in the *Clostridium histolyticum* ColH collagenase. *J. Bacteriol.* **1999**, *181*, 2816–2822.
37. Matsushita, O.; Jung, C. M.; Minami, J.; Katayama, S.; Nishi, N.; Okabe, A. A study of the collagen-binding domain of a 116-kDa *Clostridium histolyticum* collagenase. *J. Biol. Chem.* **1998**, *273*, 3643–3648.
38. Fukushima, J.; Okuda, K. *Vibrio* collagenase. In *Handbook of Proteolytic Enzymes (CD-ROM);* Barrett, A. J.; Rawlings, N. D.; Woessner, J. F., Jr., Eds.; Academic Press: London, **1998**; Chapter 367.
39. Miyoshi, S. *Vibrio vulnificus* infection and metalloprotease. *J. Dermatol.* **2006**, *33*, 589–595.
40. Chang, A. K.; Park, J. W.; Lee, E. H.; Lee, J. S. The N-terminal propeptide of *Vibrio vulnificus* extracellular metalloprotease is both an inhibitor of and a substrate for the enzyme. *J. Bacteriol.* **2007**, *189*, 6832–6838.
41. Varina, M.; Denkin, S. M.; Staroscik, A. M.; Nelson, D. R. Identification and characterization of Epp, the secreted processing protease for the *Vibrio anguillarum* EmpA metalloprotease. *J. Bacteriol.* **2008**, *190*, 6589–6597.
42. Matson, J. S.; DiRita, V. J. Degradation of the membrane-localized virulence activator TcpP by the YaeL protease in *Vibrio cholerae*. *Proc. Natl. Acad. Sci. USA* **2005**, *102*, 16403–16408.
43. Van Wart, H. E. *Clostridium* collagenase. In *Handbook of Proteolytic Enzymes (CD-ROM);* Barrett, A. J.; Rawlings, N. D.; Woessner, J. F., Jr., Eds.; Academic Press: London, **1998**; Chapter 368.
44. Bond, M. D.; Van Wart, H. E. Purification and separation of individual collagenases of *Clostridium histolyticum* using red dye ligand chromatography. *Biochemistry* **1984**, *23*, 3077–3085.
45. Bond, M. D.; Van Wart, H. E. Characterization of the individual collagenases from *Clostridium histolyticum*. *Biochemistry* **1984**, *23*, 3085–3091.
46. Jung, W.; Winter, H. Considerations for the use of clostridial collagenase in clinical practice. *Clin. Drug Invest.* **1998**, *15*, 245–252.
47. Scozzafava, A.; Supuran, C. T. Protease inhibitors: synthesis of potent bacterial collagenase and matrix metalloproteinase inhibitors incorporating *N*-4-nitrobenzylsulfonylglycine hydroxamate moieties. *J. Med. Chem.* **2000**, *43*, 1858–1865.
48. Valee, B. L.; Auld, D. S. Zinc coordination, function and structure of zinc enzymes and other proteins. *Biochemistry* **1990**, *29*, 5647–5659.
49. Larsen, K. S.; Zhang, K.; Auld, D. S. D-Phe complexes of zinc and cobalt carboxypeptidase A. *J. Inorg. Biochem.* **1996**, *64*, 149–162.
50. Lund, T.; Granum, P. E. The 105-kDa protein component of *Bacillus cereus* non-hemolytic enterotoxin is a metalloprotease with gelatinolytic and collagenolytic activity. *FEMS Microbiol. Lett.* **1999**, *178*, 355–361.

51. Louis, D.; Bernillon, J.; Paisse, J. O.; Wallach, J. M. Use of liquid chromatography–mass spectrometry coupling for monitoring the serralysin-catalyzed hydrolysis of a peptide library. *J. Chromatogr. B. Biomed. Sci. Appl.* **1999**, *732*, 271–276.

52. Baumann, U. Serralysin. In *Handbook of Proteolytic Enzymes (CD-ROM);* Barrett, A. J.; Rawlings, N. D.; Woessner, J. F., Jr., Eds.; Academic Press: London, **1998**; Chapter 386.

53. Basu, B.; Apte, S. K. A novel serralysin metalloprotease from *Deinococcus radiodurans. Biochim. Biophys. Acta* **2008**, *1784*, 1256–1264.

54. Kwak, J.; Lee, K.; Shin, D. H.; Maeng, J. S.; Park, D. S.; Oh, H. W.; Son, K. H.; Bae, K. S.; Park, H. Y. Biochemical and genetic characterization of arazyme, an extracellular metalloprotease produced from *Serratia proteamaculans* HY-3. *J. Microbiol. Biotechnol.* **2007**, *17*, 761–768.

55. Morihara, K. Aeruginolysin. In *Handbook of Proteolytic Enzymes (CD-ROM);* Barrett, A. J.; Rawlings, N. D.; Woessner, J. F., Jr., Eds.; Academic Press: London, **1998**; Chapter 387.

56. Belas, R.Mirabilysin. In *Handbook of Proteolytic Enzymes (CD-ROM);* Barrett, A. J.; Rawlings, N. D.; Woessner, J. F., Jr., Eds.; Academic Press: London, **1998**; Chapter 388.

57. Franco, A. A.; Cheng, R. K.; Chung, G. T.; Wu, S.; Oh, H. B.; Sears, C. L. Molecular evolution of the pathogenicity island of enterotoxigenic *Bacteroides fragilis* strains. *J. Bacteriol.* **1999**, *181*, 6623–6633.

58. Obiso, R. J.; Bevan, D. R.; Wilkins, T. D. Molecular modeling and analysis of the *Bacteroides fragilis* toxin. *Clin. Infect. Dis.* **1997**, *25*, S153–S155.

59. Moncrief, J. S.; Duncan, A. J.; Wright, R. L.; Barroso, L. A.; Wilkins, T. D. Molecular characterization of the fragilysin pathogenicity islet of enterotoxigenic *Bacteroides fragilis. Infect. Immun.* **1998**, *66*, 1735–1739.

60. Moncrief, J. S.; Obiso, R. J.; Barroso, L. A.; Kling, J. J.; Wright, R. L.; Van Tassell, R. L.; Lyerly, D. M.; Wilkins, T. D. The enterotoxin of *Bacteroides fragilis* is a metalloprotease. *Infect. Immun.* **1995**, *63*, 175–181.

61. Teplyakov, A.; Polyakov, K.; Obmolova, G. Crystal structure of carboxypeptidase T from *Thermoactinomyces vulgaris. Eur. J. Biochem.* **1992**, *208*, 281–288.

62. Strater, N.; Sherratt, D. J.; Colloms, S. D. X-ray structure of aminopeptidase A from *Escherichia coli* and a model for the nucleoprotein complex in Xer site-specific recombination. *EMBO J.* **1999**, *18*, 4513–4522.

63. Strater, N.; Sun, L.; Kantrowitz, E. R.; Lipscomb, W. N. A bicarbonate ion as a general base in the mechanism of peptide hydrolysis by dizinc leucine aminopeptidase. *Proc. Natl. Acad. Sci. USA* **1999**, *96*, 11151–11155.

64. Mathew, Z.; Knox, T. M.; Miller, C. G. *Salmonella enterica* serovar *typhimurium* peptidase B is a leucyl aminopeptidase with specificity for acidic amino acids. *J. Bacteriol.* **2000**, *182*, 3383–3393.

65. Mahadevan, D.; Saldanha, J. W. The extracellular regions of PSMA and the transferrin receptor contain an aminopeptidase domain: implications for drug design. *Protein Sci.* **1999**, *8*, 2546–2549.

66. Huntington, K. M.; Bienvenue, D. L.; Wei, Y.; Bennett, B.; Holz, R. C.; Pei, D. Slow-binding inhibition of the aminopeptidase from *Aeromonas proteolytica* by peptide thiols: synthesis and spectroscopic characterization. *Biochemistry* **1999**, *38*, 15587–15596.

67. Sherwood, R. F.; Melton, R. G. Glutamate carboxypeptidase. In *Handbook of Proteolytic Enzymes (CD-ROM);* Barrett, A. J.; Rawlings, N. D.; Woessner, J. F., Jr., Eds.; Academic Press: London, **1998**; Chapter 483.

68. Galivan, J.; Ryan, T. J.; Chave, K.; Rhee, M.; Yao, R.; Yin, D. Glutamyl hydrolase: pharmacological role and enzymatic characterization. *Pharmacol. Ther.* **2000**, *85*, 207–215.

69. Rowsell, S.; Pauptit, R. A.; Tucker, A. D.; Melton, R. G.; Blow, D. M.; Brick, P. Crystal structure of carboxypeptidase G2, a bacterial enzyme with applications in cancer therapy. *Structure* **1997**, *5*, 337–347.

70. Melton, R. G.; Sherwood, R. F. Antibody-enzyme conjugates for cancer therapy. *J. Natl. Cancer Inst.* **1996**, *88*, 153–165.

71. Boger, D. L. Vancomycin, teicoplanin and ramoplanin: synthetic and mechanistic studies. *Med. Res. Rev.* **2001**, *21*, 356–381.

72. Lessard, I. A.; Walsh, C. T. VanX, a bacterial D-alanyl-D-alanine dipeptidase: resistance, immunity, or survival function? *Proc. Natl. Acad. Sci. USA* **1999**, *96*, 11028–11032.

73. Wu, Z.; Walsh, C. T. Dithiol compounds: potent, time-dependent inhibitors of VanX, a zinc-dependent D,D-dipeptidase required for vancomycin resistance in *Enterococcus faecium*. *J. Am. Chem. Soc.* **1996**, *118*, 1785–1786.

74. Bussiere, D. E.; Pratt, S. D.; Katz, L.; Severin, J. M.; Holzman, T.; Park, C. The structure of VanX reveals a novel amino-dipeptidase involved in mediating transposon-based vancomycin resistance. *Mol. Cell* **1998**, *2*, 75–84.

75. Mellors, A. *O*-sialoglycoprotein endopeptidase. In *Handbook of Proteolytic Enzymes (CD-ROM);* Barrett, A. J.; Rawlings, N. D.; Woessner, J. F., Jr., Eds.; Academic Press: London, **1998**; Chapter 505.

76. Kessler, E. β-Lytic metalloendopeptidase. In *Handbook of Proteolytic Enzymes (CD-ROM);* Barrett, A. J.; Rawlings, N. D.; Woessner, J. F., Jr., Eds.; Academic Press: London, **1998**; Chapter 506.

77. Kessler, E.; Ohman, D. E. Staphylolysin. In *Handbook of Proteolytic Enzymes (CD-ROM);* Barrett, A. J.; Rawlings, N. D.; Woessner, J. F., Jr., Eds.; Academic Press: London, **1998**; Chapter k507.

78. Ryding, U.; Renneberg, J.; Rollof, J.; Christensson, B. Antibody response to *Staphylococcus aureus* whole cell, lipase and staphylolysin in patients with *S. aureus* infections. *FEMS Microbiol. Immunol.* **1992**, *4*, 105–110.

79. Odintsov, S. G.; Sabala, I.; Marcyjaniak, M.; Bochtler, M. Latent LytM at 1.3 Å resolution. *J. Mol. Biol.* **2004**, *335*, 775–785.

80. Bochtler, M.; Odintsov, S. G.; Marcyjaniak, M.; Sabala, I. Similar active sites in lysostaphins and D-Ala-D-Ala metallopeptidases. *Protein Sci.* **2004**, *13*, 854–861.

81. Plaut, A. G. IgA-specific metalloendopeptidase. In *Handbook of Proteolytic Enzymes (CD-ROM);* Barrett, A. J.; Rawlings, N. D.; Woessner, J. F., Jr., Eds.; Academic Press: London, **1998**; Chapter 508.

82. Poulsen, K.; Reinholdt, J.; Jespersgaard, C.; Boye, K.; Brown, T. A.; Hauge, M.; Kilian, M. A comprehensive genetic study of streptococcal immunoglobulin A1 proteases: evidence for recombination within and between species. *Infect. Immun.* **1998**, *66*, 181–190.

83. Langer, T. AAA proteases: cellular machines for degrading membrane proteins. *Trends Biol. Sci.* **2000**, *25*, 247–251.

84. Schumann, W. FtsH: a single chain chaperonin? *FEMS Microbiol. Rev.* **1999**, *23*, 1–11.

85. Makino, S.; Makino, T.; Abe, K.; Hashimoto, J.; Tatsuta, T.; Kitagawa, M.; Mori, H.; Ogura, T.; Fujii, T.; Fushinobu, S.; Wakagi, T.; Matsuzawa, H. Second transmembrane segment of FtsH plays a role in its proteolytic activity and homo-oligomerization. *FEBS Lett.* **1999**, *460*, 554–558.

86. Asahara, Y.; Atsuta, K.; Motohashi, K.; Taguchi, H.; Yohda, M.; Yoshida, M. FtsH recognizes proteins with unfolded structure and hydrolyzes the carboxyl side of hydrophobic residues. *J. Biochem. (Tokyo)* **2000**, *127*, 931–937.

87. Srinivasan, R.; Anilkumar, G.; Rajeswari, H.; Ajitkumar, P. *FEMS Microbiol. Lett.* **2006**, *259*, 97–105.

88. Ito, K.; Akiyama, Y. Cellular functions, mechanism of action, and regulation of FtsH protease. *Annu. Rev. Microbiol.* **2005**, *59*, 211–231.

89. Sakoh, M.; Ito, K.; Akiyama, Y. Proteolytic activity of HtpX, a membrane-bound and stress-controlled protease from *Escherichia coli*. *J. Biol. Chem.* **2005**, *280*, 33305–33310.

CHAPTER 29

Botulinus Toxin, Tetanus Toxin, and Anthrax Lethal Factor Inhibitors

ANTONIO MASTROLORENZO[1] and CLAUDIU T. SUPURAN[2]

[1]Dipartimento di Scienze Dermatologiche, Università degli Studi di Firenze, Villa Basilewsky, Via Lorenzo Magnifico 104, 50129 Firenze, Italy
[2]Laboratorio di Chimica Inorganica e Bioinorganica, Dipartimento di Chimica, Università degli Studi di Firenze, Via della Lastruccia 3, 50019 Sesto Fiorentino, Firenze, Italy

29.1 THE TETANUS AND BOTULINUM NEUROTOXINS

Clostridium tetani and *Clostridium botulinum* (but also *Clostridium barati* and *Clostridium butirycum*) are strictly anaerobic pathogens that provoke tetanus and botulism, respectively.[1,2] The first disease has already been described by Hippocrates of Kos in the fourth century BC, being characterized by an often fatal spastic paralysis with contraction of skeletal muscles that work one against the other; botulism, on the other hand, is a neurologic syndrome of vertebrates, characterized by the loss of function of peripheral cholinergic synapses.[1,2] This functional loss may be highly variable, from minimal unnoticed effects up to a generalized flaccid paralysis. Such a manifestation accounts for the fact that botulism was first described only at the beginning of the nineteenth century and infant botulism was identified only 20 years ago.[1,2] Both diseases are caused entirely by clostridial neurotoxins, the tetanus neurotoxin (TeNT) and botulinum neurotoxin (BoNT, of which seven distinct serotypes are known, BoNT/A-G), respectively. These toxins constitute a group of bacterial metalloproteases (MPRs) belonging to the M27 family (see also Chapter 28) with unique properties activity.[1–5]

BoNT and TeNT are translated as single chain protoxins that are subsequently cleaved within a disulphide loop situated about one-third from the amino-terminus, generating the fully active di-chain 150 kDa neurotoxin.[1,2] This is composed of a light (L) chain of approximately 50 kDa and a heavy (H) chain of 100 kDa linked by an interchain disulphide bridge and by noncovalent interactions.[1,2,5] The protease activity is associated to the L chain, whereas the heavy chain is important for the membrane translocation of the L chain into the neuronal cytosol, playing a critical role in pore formation.[1,2] The carboxyterminal part of the H chain may also be subdivided in two subdomains: a lectin-like domain and a β-trefoil fold and is probably involved in the toxin binding to presynaptic membranes via a double interaction, most likely with two different molecules of the nerve terminal.[1,2] The zinc endopeptidase activity of the L chain is inhibited when the toxin is intact, being expressed only after reduction of the disulphide bond bridging the L and H chains.[1,2] One Zn(II) ion is present bound to the L chain, which is critical for the protease activity of these neurotoxins.[1,3] The Zn(II) of these MPRs is coordinated by two histidine residues (TeNT numbering: His222 and His226, belonging to the consensus zinc-binding site HEXXH), by the carboxylate group of Glu261, in addition to the water molecule critical for catalysis.[1,2] As for many other MPRs discussed here, the glutamate residue (Glu223 in this case) of the consensus zinc-binding sequence mentioned above, is also critical in catalysis, as it participates in hydrogen bonding and activation of the zinc-bound water molecule.[1,2] The catalytic domain of the BoNT/A contains both α-helix and β-strand secondary structures, having little similarity with other MPRs, except for the zinc binding motif.[5] The protease active site of these enzymes is quite deep, accommodating at least 16 amino acid residues, which makes them among the most "bulky" metalloproteases studied so far.[1,5] This may have very important consequences for the successful design of active-site directed inhibitors of the protease activity of BoNT/TeNT.

The only known proteolytic substrate of TeNT is a 120-residue protein anchored to the membrane of cell vesicles, which has been termed not only VAMP (vesicle-associated membrane protein) but also synaptobrevin and cellubrevin.[1,2] Similarly, BoNTs/B, D, F, and G cleave specifically VAMP at different single peptide bonds (see also Chapter 28).[1,2,4] The three VAMP isoforms, VAMP-1, VAMP-2, and cellubrevin, of human and mice are cleaved at the same site by BoNT/B and TeNT, this site being Gln76-Phe77.[1–4] VAMPs of other species may carry mutations at the cleavage/recognition site(s) that render them neurotoxin insensitive, as is the case of the rat and chicken VAMP-1, which are not cleaved by BoNT/B.[3] BoNTs also possess two other intracellular targets in addition to VAMP: SNAP-25 (25 kDa synaptosomal-associated protein) and syntaxin, both of which are cleaved between Gln-Arg (by BoNT/A), Lys-Ala (by BoNT/C), Arg-Ala (by BoNT/C), or Gln-Lys (different neurotoxins) residues.[1,2] One must mention that syntaxin is a 35 kDa type II membrane protein located in the neuronal plasmalemma, being important for neuronal development and survival, possibly modulating calcium entry within the neurons.[1] SNAP-25 is a palmitoylated protein of the CNS, highly conserved in the evolutionarily tree (from yeasts to humans), being also involved in the calcium-dependent phase of neurotransmitter release at exocytotic sites.[1,2]

Recently several more X-ray crystal structures of various fragments of these proteases. Alone or complexed with substrates/inhibitors were reported. Thus, the crystal structures of intact neurotoxin type B (BoNT/B) and its complex with sialyllactose, was determined at 1.8 and 2.6 Å resolution, respectively.[6] These structural data provided insight into its catalytic and binding sites. The position of the belt region in BoNT/B is different from that in BoNT/A; this observation presenting interesting possibilities for designing specific inhibitors that could be used to block the activity of this neurotoxin. The structures of BoNT/B and its complex with sialyllactose provide a detailed description of the active site and a model for interactions between the toxin and its cell surface receptor. The latter may provide valuable information for recombinant vaccine development.[6] Another work reported the crystal structure of the catalytic light chain (LC) of BoNT type G (BoNT/G-LC) at 2.35 Å resolution.[7] The structure revealed a C-terminal beta-sheet that is critical for LC oligomerization and is unlike that seen in the other LC structures reported earòier.[5,6] Its structural comparison with thermolysin and the available pool of LC structures revealed important serotype differences that are likely to be involved in substrate recognition of the P1′ residue. In addition, structural and sequence analyses have identified a potential exosite of BoNT/G-LC that recognizes a SNARE recognition motif of VAMP.[7] As the seven serotypes (A–G) of BoNT function through their proteolytic cleavage of one of the three proteins (SNAP-25, Syntaxin, and VAMP) that form the SNARE complex required for synaptic vesicle fusion, it is crucial to understand the structural determinants of these differences between the various serotypes. As mentioned above, the different BoNTs have very specific protease recognition requirements, between 15 and 50 amino acids in length depending on the serotype. However, the structural details involved in substrate recognition remain largely unknown. Recently it was reported the 1.65 Å resolution crystal structure of the catalytic domain of BoNT serotype D (BoNT/D-LC), which provided insight into the protein–protein binding interaction and final proteolysis of VAMP-2.[8] Structural analysis has identified a hydrophobic pocket potentially involved in substrate recognition of the P1′ VAMP residue (Leu60) and a second remote site for recognition of the V1 SNARE motif that is critical for activity.[8] A structural comparison of BoNT/D-LC with BoNT/F-LC that also recognizes VAMP-2 one residue away from the BoNT/D-LC site provided additional molecular details about the unique serotype specific activities. In particular, BoNT/D prefers a hydrophobic interaction for the V1 motif of VAMP-2, while BoNT/F adopts a more hydrophilic strategy for recognition of the same V1 motif.[8] The 3-dimensional crystal structure of the apo botulinum neurotoxin serotype B catalytic domain (BoNT/B-LC) has also been determined to 2.2 Å resolution, and the complex of cleaved Sb2 with the catalytic domain (Sb2-BoNT/B-LC) has been determined to 2.0 Å resolution.[9] A comparison of the holotoxin catalytic domain and the isolated BoNT/B-LC structure showed a rearrangement of three active site loops. This rearrangement exposed the BoNT/B active site. The Sb2-BoNT/B-LC structure illustrated two distinct binding regions, which explains the specificity of each botulinum neurotoxin for its synaptic vesicle protein. Such observations provided an explanation for the proposed cooperativity between binding of full-length substrate and catalysis and suggested a mechanism of synaptobrevin proteolysis employed by the clostridial neurotoxins.[9] As mentioned earlier, the substrate specificity and the peptide bond cleavage selectivity of their catalytic domains are different for the seven BoNT serotypes. The reason for this unique specificity of botulinum neurotoxins is still not well understood.[10] If an inhibitor leading to a therapeutic drug common to all serotypes is to be developed, it is essential to understand the differences in their 3-dimensional structures that empower them with unique characteristics. Accordingly, high-resolution structures of all serotypes are required and, toward achieving this goal, the crystal structure of the catalytic domain of BoNT type E has been determined to 2.1 Å resolution recently.[10] The crystal structure of the inactive mutant Glu212→Gln of this protein has also been reported.[10] While the overall conformation is unaltered in the active site, the position of the nucleophilic water changes in the mutant, thereby causing it to lose its ability to activate the catalytic reaction. These structures explain the importance of the nucleophilic water activated by the zinc ion and the role of the negative charge on Glu212. The structural differences responsible for the loss of activity of the mutant provide a common model for the catalytic pathway of *Clostridium* neurotoxins because Glu212 is conserved and has a similar role in all serotypes, as illustrated above. This or a more nonconservative mutant (e.g., Glu212→Ala) could provide thus a novel, genetically modified protein vaccine for botulinum.[10]

Although BoNT/B and TeNT cleave VAMP-2 at the same scissile bond, their mechanism(s) of VAMP-2 recognition is not clear.[11] Mapping experiments showed that the residues 60–87 of VAMP-2 were sufficient for efficient cleavage by BoNT/B and that residues 40–87 of VAMP-2 were sufficient for efficient TeNT cleavage. Alanine-scanning mutagenesis and kinetic analysis identified three regions within VAMP-2 that were recognized by BoNT/B and TeNT: residues adjacent to the site of scissile bond cleavage (cleavage region) and residues located within N-terminal and C-terminal regions relative to the cleavage region. Analysis of residues within the cleavage region showed that the mutations at the P7, P4, P2, and P1′ residues of VAMP-2 had the greatest inhibition of LC/B cleavage (≥32-fold), whereas mutations at P7, P4, P1′, and P2′ residues of VAMP-2 had the greatest inhibition of LC/TeNT cleavage (≥64-fold). Residues within the cleavage region influenced catalysis, whereas residues N-terminal and C-terminal to the cleavage region influenced binding affinity. Thus, BoNT/B and TeNT possess similar organization but have the unique residues to recognize and cleave VAMP-2. These studies provide new, important insights into how the clostridial neurotoxins recognize their substrates.[11]

Tetanus and botulinum type B neurotoxins enzymatically cleave the same substrate, vesicle-associated membrane proteins, at the same peptide bond although the optimum length of substrate peptide required for cleavage by them is different.[12] Recently, the first experimentally determined 3-dimensional structure of the catalytic domain of tetanus neurotoxin has been presented.[12] The structure provided the insight into the active site of TeNT proteolytic activity and the importance of the nucleophilic water and the role of the zinc ion, which are similar to those discussed above for BoNT. The probable reason for different modes of binding of vesicle-associated membrane protein to BoNT type B and the TeNT was also discussed. The structure provides a basis for designing a novel recombinant vaccine or structure-based drugs for tetanus (see later in the text).[12]

Tetanus is acquired by contamination of wounds with spores of toxigenic strains of *Clostridium tetani*, which are ubiquitous, but highly enriched in feces from many animals including farmyard animals. Under anaerobic conditions, spores germinate, and tetanus neurotoxin is produced and released upon autolysis. The toxin spreads in the body and binds, to an as yet unidentified receptor of the presynaptic terminal of the neuromuscular junction and after endocytosis, the toxin migrates retroaxonally inside the motor neuron and reaches the spinal cord. The toxin is then released in the intersynaptic space and it enters small synaptic vesicles when they expose their lumen to the outside following fusion with the presynaptic membrane and release of neurotransmitter.[1,2] Synaptic vesicles are then endocytosed and their lumen is acidified by a vacuolar ATPase proton pump to drive the reuptake of neurotransmitter. Such lumenal acidification is essential for intoxication to occur because tetanus neurotoxin at low pH changes conformation and becomes hydrophobic.[1] In such a form, the toxin inserts in the lipid bilayer of the vesicle membrane and somehow the H domain manages to translocate the L chain into the cytosol, with the consequent cleavage of VAMP.[1,2] Such specific proteolysis is sufficient to cause a prolonged blockade of neuroexocytosis, and this demonstrates the essential role of VAMP in neuroexocytosis.[1,2]

Clostridial spores harboring the genes encoding botulinum neurotoxins are widespread in the environment and may germinate in anaerobic foods, and produce and release the neurotoxins. Eating poisoned food causes botulism, following transcytosis of botulinum neurotoxin across the intestinal epithelial layer and its spread through the body. Botulinum neurotoxins bind the unmyelinated presynaptic membrane of motor neurons and of other cholinergic synapses. This results in a flaccid paralysis as well as in alterations of the autonomic nervous system.[1,2]

29.2 INHIBITORS OF BoNT AND TeNT ZINC PROTEASES

Till recently, very few BoNT/TeNT protease inhibitors were described in the literature. However, the events that followed the terroristic attacks of September 2001 and the possible use of these toxins for terroristic purposes prompted much research in this field. Considering the intracellular location of the toxins, membrane permeant inhibitors of the protease activity of these enzymes would be helpful therapeutic agents for the treatment of tetanus and botulism, as at the present time there is no effective drug therapy to prevent the progressive evolution of tetanus or botulism following intoxication or infection.[13] The first compounds that potently inhibit these zinc-proteases have been reported by Roques's group.[13,14] One must mention that similarly to many other metalloproteases, BoNT and TeNT are inhibited by chelating agents such as EDTA, 1,10-phenantroline or captopril, generally in millimolar concentrations, and thus, such compounds do not have practical applications for the treatment of diseases provoked by these neurotoxins.[1,2]

Thus, Martin et al.[13,14] recently showed that the various β-aminothiols of type **29.1–29.7** act as micromolar inhibitors (affinity in the 35–250 μM range) against the protease activity of the TeNT light chain. The compounds incorporating the aromatic ring, and mainly those containing the free sulfonamide group were the best inhibitors. This is an extremely interesting result, as carbonic anhydrase (CA, EC 4.2.1.1), another zinc enzyme possessing CO_2 hydrase and esterase (but not protease) activities is strongly inhibited by aliphatic/aromatic sulfonamides, with the inhibitor coordinated to the metal ion by means of the ionized sulfonamide nitrogen (see Chapters 1–20).[15] The fact that in the entire series of derivatives reported by the French group,[13,14] the most active were those containing the SO_2NH_2 moiety, might signify that similarly to the case of inhibiting CA, such sulfonamides may bind to the Zn(II) ion within the metalloprotease active site (alone, or in chelate form, together with the thiolate sulphur atom of the aminothiol molecule). The same group reported that the dipeptidyl-based aminothiols incorporating aliphatic/aromatic sulfonamide moieties, of types **29.8–29.11** show good inhibitory properties against both TeNT as well as BoNT/B.[14] This is the first example of potent synthetic inhibitors for these metalloproteases and open very interesting potential applications of such protease inhibitors for the treatment of botulism/tetanus.

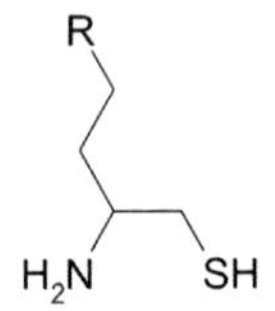

29.1 R = $CONH_2$ (250)
29.2 R = SO_3Na (300)
29.3 R = SO_2NH_2 (100)

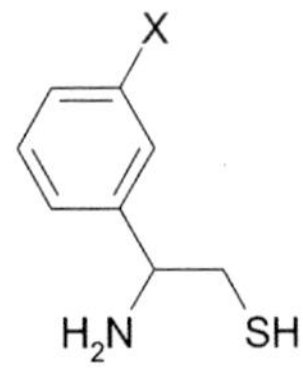

29.4 X = COOH (250)
29.5 X = $CONH_2$ (125)
29.6 X = SO_3Na (40)
29.7 X = SO_2NH_2 (35)

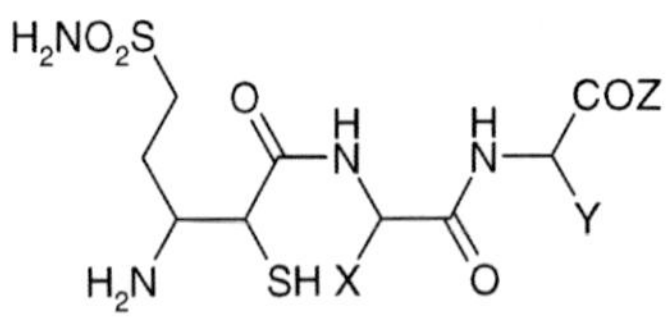

No	X	Y	Z	K_I (μM) TeNT	K_I (μM) BoNT
29.8	*i*-Pr	*i*-Bu	OH	10	15
29.9	*i*-Pr	*i*-Bu	$NHCH_2Ph$	8	10
29.10	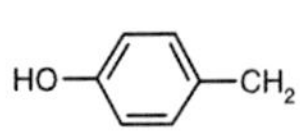	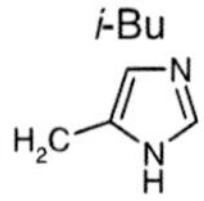	$NHCH_2Ph$	3.5	5

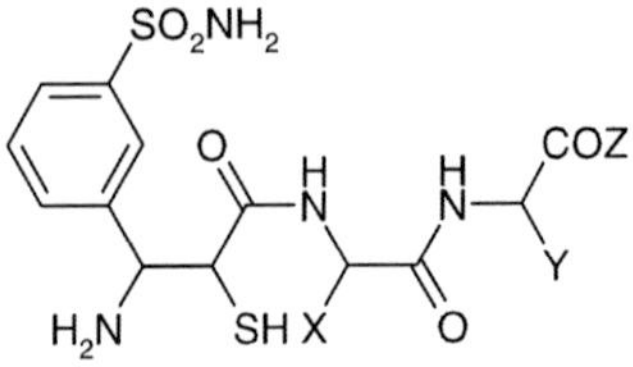

No	X	Y	Z	K_I (μM) TeNT	K_I (μM) BoNT
29.11	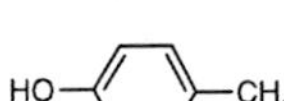	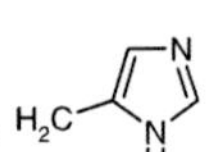	$NHCH_2Ph$	3.5	2

Burnett et al.[16] used a high-throughput assay to identify the small molecules that inhibit the metalloprotease activity of BoNT serotype A light chain (BoNT/A LC). Conformational analyses of these compounds, in conjunction with the molecular docking studies, were used to predict structural features that contribute to inhibitor binding and potency. Based on these results, a common pharmacophore for BoNT/A LC inhibitors was proposed. Among them, the active structures (in the micromolar range) were silver sulfadiazine **29.12**, michellamine B **29.13**, NSC 357756 **29.14**, NSC 119889 **29.15**, and the dinitroderivative **29.16**.

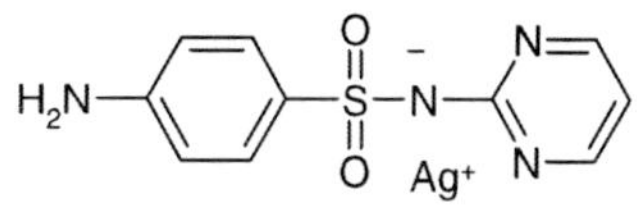

29.12

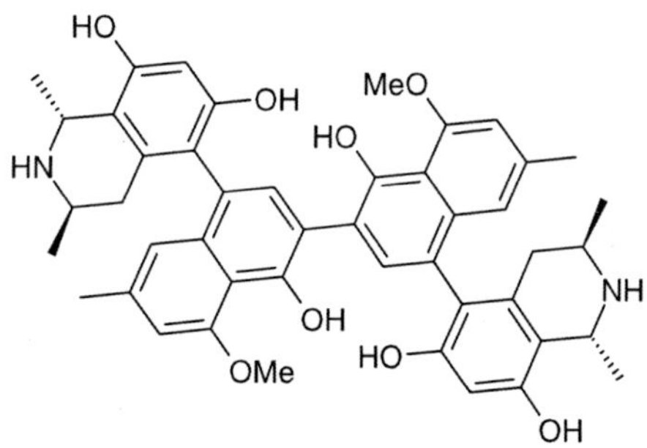

29.13

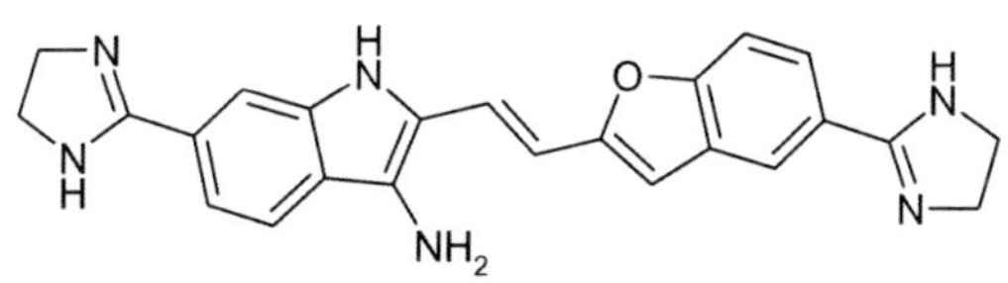

29.14

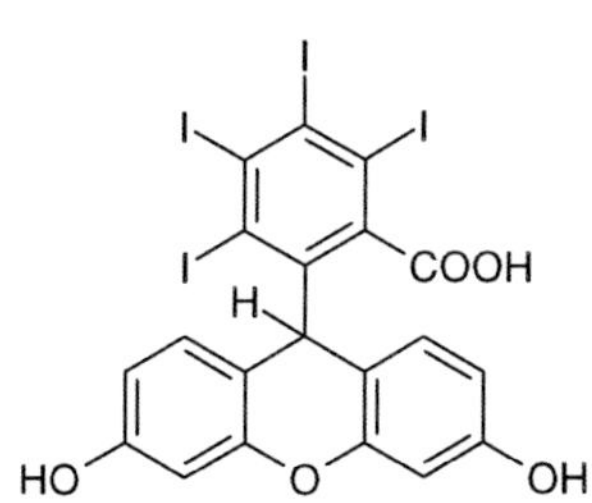

29.15

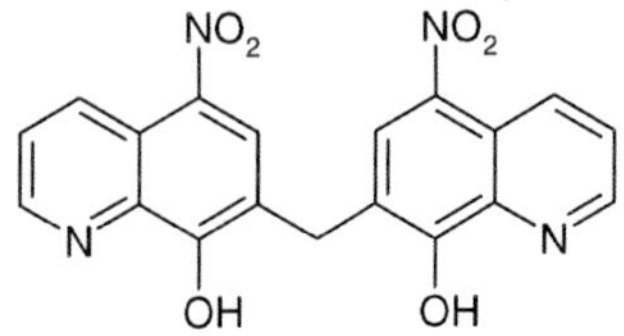

29.16

More recently, the same group identified other BoNT serotype A (BoNT/A) light chain (LC) inhibitors.[17] Of these, several (including antimalarial drugs) contained a 4-amino-7-chloroquinoline substructures and a separate positively charged, ionizable amine component. The same antimalarials have also been found to interfere with BoNT/A translocation into neurons, via pH elevation of the toxin-mediated endosome. Thus, this structural class of small molecules may serve as dual-function BoNT/A inhibitors (IC_{50}'s ranged from 3.2 to 17 μM). Molecular docking indicated that the binding modes for the new compounds are consistent when compared with those of the other inhibitors that were identified earlier.[16]

29.3 ANTHRAX TOXIN LETHAL FACTOR

Bacillus anthracis secretes three proteins, which associate in binary combinations to form toxic complexes at the surface of mammalian cells: receptor-bound protective antigen (PA) is proteolytically activated, yielding a 63 kDa fragment which oligomerizes into heptamers, which bind edema factor (EF) or lethal factor (LF) to form the toxic complexes.[18] Anthrax toxin lethal factor is one of the three components that are collectively termed anthrax toxin and it is a zinc MPR belonging to the M34 family (see also Chapter 28 of this book). The combination of PA (82.7 kDa) and LF (90.2 kDa) is designated lethal toxin because it kills certain animals and rapidly lyses macrophages. The combination of PA and EF (88.8 kDa), termed edema toxin, inhibits phagocytes by increasing cAMP concentrations to unphysiologic levels.[18,19] Both LF and EF enter cells by binding to furin-activated, receptor-bound protective antigen.[18,19] The EF is a calmodulin-dependent adenylate cyclase. LF is a metalloprotease, which cleaves specifically isoforms 1, 2, and 3 of MAPK kinases within their N-terminal tail thus interfering in a major pathway of signaling from the plasma membrane to the cell nucleus.[20–22] Indeed, macrophages, which are killed by PA + LF, are protected by certain hydrophobic peptides such as Leu-NH_2 (EC_{50} 1 mM), Phe-NH_2 (EC_{50} 0.2 mM), Leu-CH_2Cl (EC_{50} 0.1 mM), and bestatin (EC_{50} 0.2 mM).[23]

The crystal structure of LF and its complex with the N terminus of MAPKK-2 were determined recently.[24] LF comprises four domains: domain I binds the membrane-translocating component of anthrax toxin, the protective antigen (PA); domains II, III, and IV together create a long deep groove that holds the 16-residue N-terminal tail of MAPKK-2 before cleavage. Domain II resembles the ADP-ribosylating toxin from *Bacillus cereus*, but the active site has been mutated and recruited to augment substrate recognition. Domain III is inserted into domain II, and seems to have arisen from a repeated duplication of a structural element of domain II. Domain IV is distantly related to the zinc MPR family, and contains the catalytic center, resembling somehow domain I. The structure thus reveals a protein that has evolved through a process of gene duplication, mutation and fusion, into an enzyme with high and unusual specificity.[24] The active site contains the HEXXH motif typical of zinc MPRs that is located on an alpha helix at the bottom of it. The two His residues within this motif, His686 and His690 along with Glu735 (which is located on a separate alpha helix) coordinate the Zn(II) ion (as for other MPRs discussed here, see Chapters 28 and 30) whereas the fourth ligand is a water molecule/hydroxide ion which is the nucleophile acting on the scissile peptide bond.[24] The LF active site has an residue in close proximity (3.3 Å) to the zinc ion, which is essential for catalysis, Tyr728.[25] Tyr728 may serve to stabilize the negative charge on the scissile bond carbonyl oxygen in the transition state, through its OH moiety (analogously to the oxyanion hole for serine proteases mechanisms). [25,26]

29.4 ANTHRAX TOXIN LETHAL FACTOR INHIBITORS

The limited success of antibiotic therapy for anthrax infection has motivated investigation of complementary therapeutic strategies that target the bacteria's secreted toxin.[27] The zinc-dependent MPR LF is a critical component of anthrax toxin and an important potential target for small molecule drugs, as emerged ultimately although synthetic and assay efforts from many groups. In the past few years, a number of approaches have been taken to identify LF inhibitors, from generating conventional metal chelating substrate analogs (e.g., hydroxamates) to random screening of diverse compound libraries. These efforts have produced several different classes of specific nanomolar range inhibitors. Some compounds have fared well in animal models for anthrax toxemia and infection, and these inhibitors and their derivatives may form the basis for future therapies to treat the disease in humans.[27]

The first compounds that were used in the design of LF inhibitors were peptidomimetics similar to the protease substrate, incorporating a hydroxamate as a zinc-binding group. One of the first interesting leads which led to a low micromolar ionhibitor was **29.17**.[28] Using this peptide hydroxamate as lead, a group from Merck prepared and assayed as in vitro LF inhibitors a very large series of hydroxamates, with a reduced peptidomimetic character, which led to the low nanomolar inhibitor (2R)-2-[(4-fluoro-3-methylphenyl)sulfonylamino]-*N*-hydroxy-2-(tetrahydro-2H-pyran-4-yl) acetamide **29.18** (K_I of 24 nM) which is also not a strong inhibitor of the host (MMP) proteinases.[29,30]

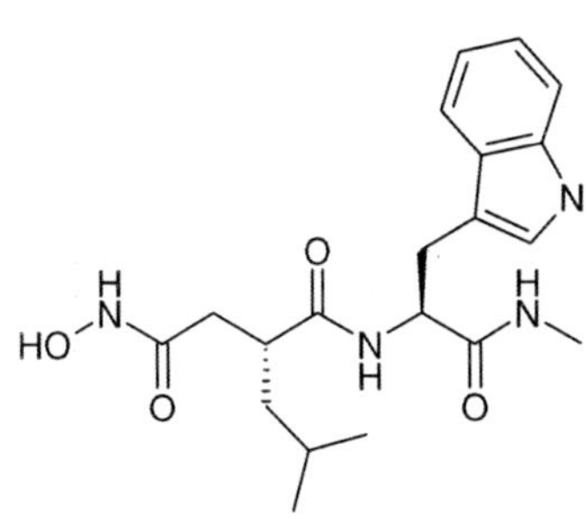

29.17 (K_I = 2.1 μM)

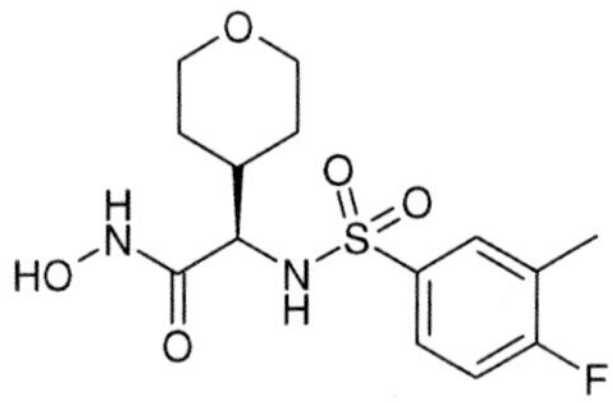

29.18 (K_I = 24 nM)

As hydroxamates have different drawbacks (toxicity, indiscriminate inhibition of most metalloenzymes, etc.), various efforts were taken to detect nonhydroxamate, competitive LF inhibitors. Some of these compounds, such as **29.14** (also active as a BoNT inhibitor, see above) and derivatives **29.19–29.23** belong to very heterogeneous chemical classes and show activity in the nanomolar–micromolar range.[27,31]

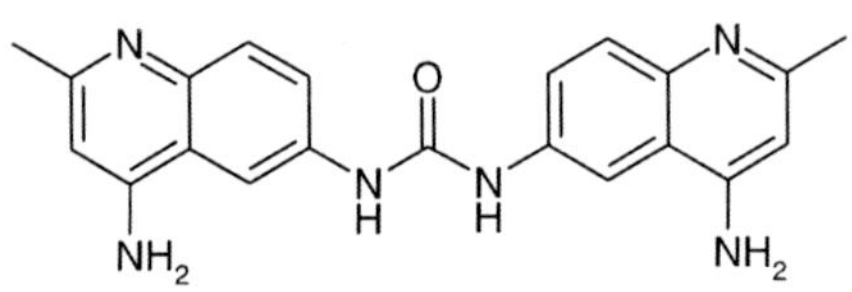

29.19 (K_I = 0.5 μM)

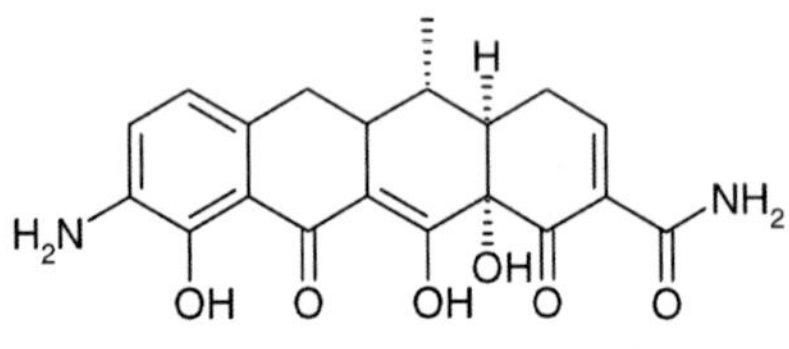

29.20 (K_I = 5 μM)

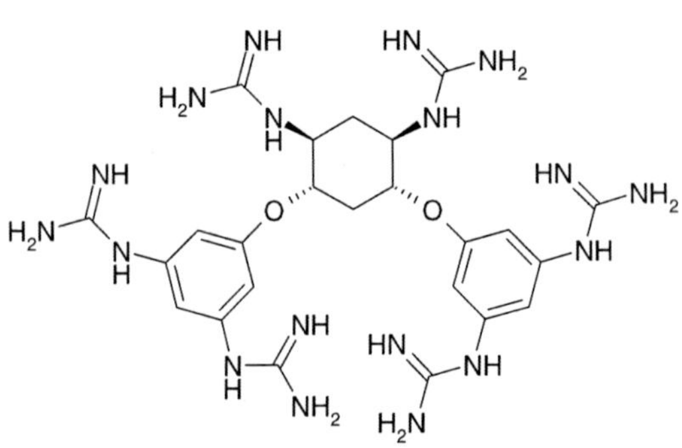

29.21 (K_I = 65 nM)

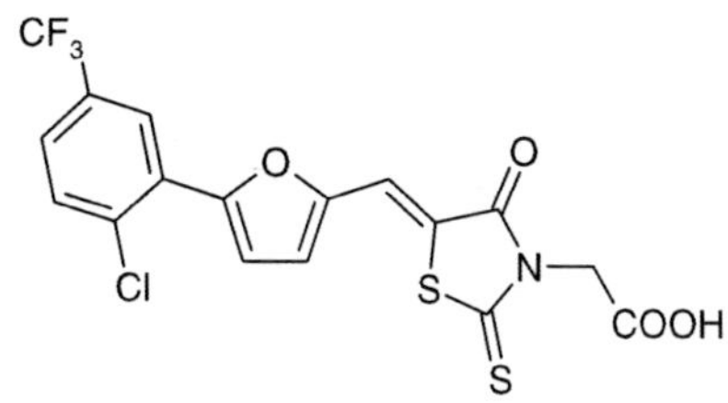

29.22 (K_I = 32 nM)

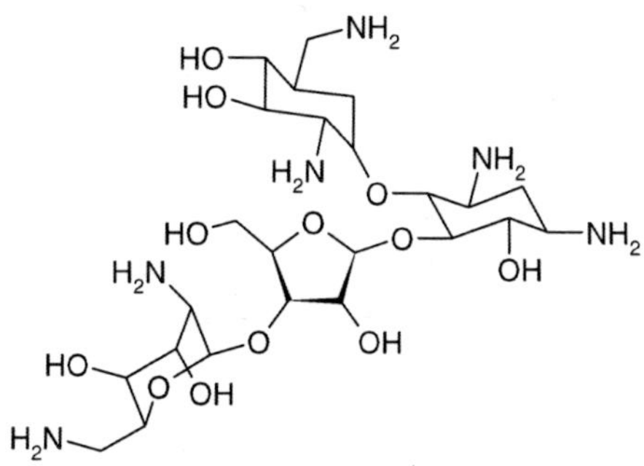

29.23 (K_I = 37 nM)

A completely different assay method, developed by Mrksich's group led to noncompetitive LF inhibitors of types **29.24–29.27**, again belonging to very different chemical classes, and with potencies in the 100 nM–4.3 μM range.[32,33]

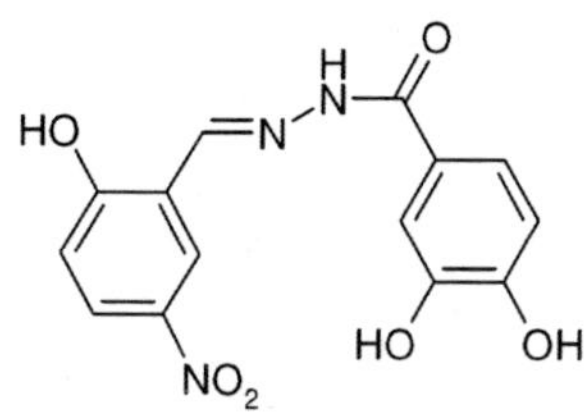

29.24 (K_I = 1.1 μM)

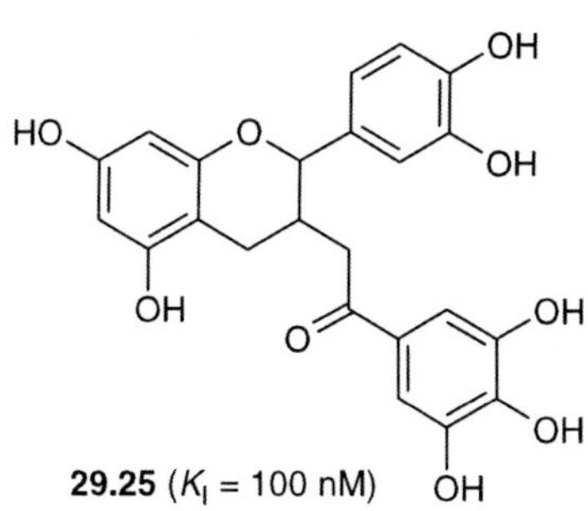

29.25 (K_I = 100 nM)

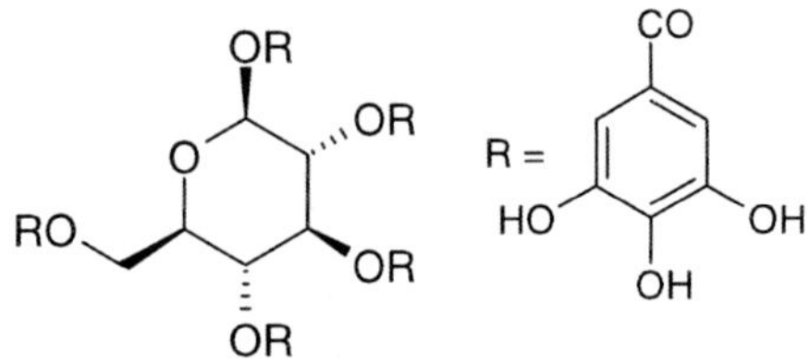

29.26 (K_I = 1.8 μM)

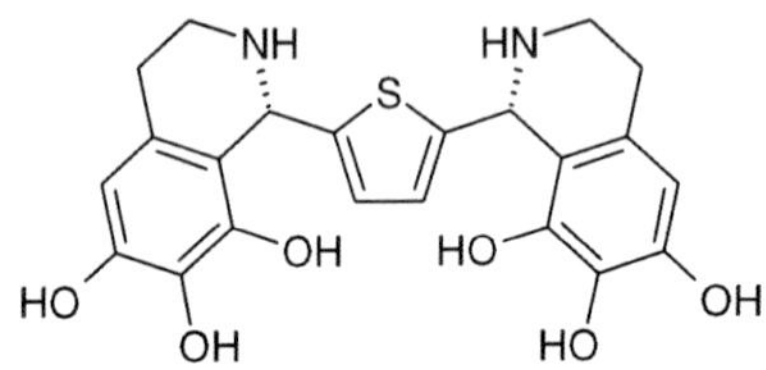

29.27 (K_I = 4.3 μM)

Thus, it is clear that a wealth of structures of novel lead compounds are available now, together with several low nanomolar LF inhibitors that are also effective *in vivo*. It is hoped that soon bacterial protease inhibitors belonging to this class will be clinically available.

29.5 CONCLUSION

The high specificity of the action of BoNT provided the basis for its clinical use the therapy of a variety of human diseases caused by hyperfunction of cholinergic terminals.[1,34–36] Thus, injection of minute amounts of neurotoxin into the muscle(s) to be paralyzed leads to a depression of the symptoms for a period of several months, and this is the best available treatment for dystonias, strabismus,[34] facial wrinkling, brow position,[35] as well as palmar and axillary hyperhidrosis.[36] Mainly BoNT/A is used for such treatments, but some other types of modified toxins are also investigated clinically.[2] The cosmetic use of BoNT serotype A involves low doses of toxin administered for facial wrinkles and hyperhidrosis. The structural and functional properties of BoNT/A can affect the degree and duration of effect. Actively using the injected muscle is favorable as it exposes more receptors to BoNT/A. Divided doses of BoNT/A at an interval of more than 3 days may be longer lasting than single dose by blocking nascent neuronal sprouts. Antibodies are unlikely to be effective in BoNT/A neutralization because of the large area of receptor interaction. Several commonly used drugs including zinc and chloroquine can interact with BoNT/A, necessitating dosage adjustment for optimum effect. Serotype E (BoNT/E) can emerge as an antidote for BoNT/A for cosmetic use.[37]

More recently, it emerged that the efficacy of BoNT/A is evident in the symptomatic therapy of disorders in which muscular hyperactivity plays a prominent role, such as focal dystonias and hemifacial spasm.[38] In these disorders, BoNT/A is considered as the first-line therapy. BoNT/A is also beneficial in the treatment of both adults and children with spasticity of various causes. The pain that frequently accompanies these conditions is effectively reduced by BoNT/A. A genuine analgesic effect for BoNT/A unrelated to skeletal muscle spasmolysis has been suggested on the basis of *in vitro* and *in vivo* (animal) data.[38] However, studies in humans designed to detect such an effect were negative, as were controlled studies of BoNT/A in patients with primary headache disorders. BoNT/A also acts on cholinergic synapses of the autonomic nervous system, and injection of BoNT/A into salivary glands significantly decreases the production of saliva. This may be beneficial for patients with Parkinson's disease, in whom the excessive production of saliva may be problematic. Overall, BoNT/A has been confirmed as an efficacious, predictable and well tolerated drug in an ever-increasing number of neurological disorders.[38] In fact, the use of BoNT to treat disorders of the salivary glands is increasing in popularity in recent years.[39] Recent reports of the use of BoNT in glandular hypersecretion suggest overall favorable results with minimal side-effects. However, with few randomized clinical trials at the moment, it means that data are limited with respect to candidate suitability, treatment dosages, frequency and duration of treatment. A report of such cases managed with botulinum toxin and a review on the current data on use of BoNT to treat salivary gland disorders such as Frey's syndrome, excessive salivation (sialorrhoea), focal and general hyperhidrosis, excessive lacrimation and chronic rhinitis has recently been published.[39]

The proper management of pain is a critical issue in the practice of medicine. Despite the availability of a large number of analgesic medications, management of pain that is refractory to conventional treatments remains a challenge for both clinicians and surgeons.[40] BoNT has recently emerged as a potential novel approach to control pain. Animal studies have revealed a number of mechanisms by which BoNTs can influence and alleviate chronic pain, including inhibition of pain peptide release from nerve terminals and sensory ganglia, anti-inflammatory and antiglutaminergic effects, reduction of sympathetic neural discharge, and inhibition of muscle spindle discharge.[40] In humans, prospective, placebo-controlled, double-blind studies have also provided evidence for effectiveness of BoNT therapy in a number of painful disorders. These include cervical dystonia, pelvic pain, low back pain, plantar fasciitis, postsurgical painful spasms, myofascial pain syndromes, migraine, and chronic daily headaches.[40] Long-term studies on cervical dystonia and low back pain have demonstrated safety and sustained efficacy after repeated injections.[40]

The last applications are those in urology.[41,42] BoNT has been shown to be effective when used in the management of lower urinary tract dysfunction, prostatic disorders and more recently vaginismus in women suffering with pelvic floor muscle tension.[41] The proposed mechanisms of action of BoNT-A in the bladder provide new insights that might help to understand better the complex machinery of the bladder wall and the importance of the afferent signaling mechanisms. BoNT-A might thus become to be considered a small revolution in the understanding and treatment of bladder overactivity. However, substantially more clinical and basic scientific data are needed before we can truly claim to comprehend how it influences lower urinary tract function, or its role in managing urological problems.[42]

REFERENCES

1. Schiavo, G.; Matteoli, M.; Montecucco, C. Neurotoxins affecting neuroexocytosis. *Physiol. Rev.* **2000**, *80*, 717–766.
2. Humeau, Y.; Doussau, F.; Grant, N. J.; Poulain, B. How botulinum and tetanus neurotoxins block neurotransmitter release. *Biochimie* **2000**, *82*, 427–446.
3. Schiavo, G.; Benfenati, F.; Poulain, B.; Rossetto, O.; Polverino de Laureto, P.; DasGupta, B. R.; Montecucco, C. Tetanus and botulinum-B neurotoxins block neurotransmitter release by proteolytic cleavage of synaptobrevin. *Nature* **1992**, *359*, 832–834.
4. Schiavo, G.; Rossetto, O.; Catsicas, S.; Polverino de Laureto, P.; DasGupta, B. R.; Benfenati, F.; Montecucco, C. Identification of the nerve terminal targets of botulinum neurotoxins serotypes A, D and E. *J. Biol. Chem.* **1993**, *268*, 23784–23787.
5. Lacy, D. B.; Tepp, W.; Cohen, A. C.; DasGupta, B. R.; Stevens, R. C. Crystal structure of botulinum neurotoxin type A and implications for toxicity. *Nat. Struct. Biol.* **1998**, *5*, 898–902.
6. Swaminathan, S.; Eswaramoorthy, S. Structural analysis of the catalytic and binding sites of *Clostridium botulinum* neurotoxin B. *Nat. Struct. Biol.* **2000**, *7*, 617–619.
7. Arndt, J. W.; Yu, W.; Bi, F.; Stevens, R. C.; Crystal structure of botulinum neurotoxin type G light chain: serotype divergence in substrate, recognition. *Biochemistry* **2005**, *44*, 9574–9580.
8. Arndt, J. W.; Chai, Q.; Christian, T.; Stevens, R. C. Structure of botulinum neurotoxin type D light chain at 1.65 Å resolution: repercussions for VAMP-2 substrate specificity. *Biochemistry* **2006**, *45*, 3255–3262.
9. Hanson, M. A.; Stevens, R. C. Cocrystal structure of synaptobrevin-II bound to botulinum neurotoxin type B at 2.0 Å resolution. *Nat. Struct. Biol.* **2000**, *7*, 687–692.
10. Agarwal, R.; Eswaramoorthy, S.; Kumaran, D.; Binz, T.; Swaminathan, S. Structural analysis of botulinum neurotoxin type E catalytic domain and its mutant Glu212→Gln reveals the pivotal role of the Glu212 carboxylate in the catalytic pathway. *Biochemistry* **2004**, *43*, 6637–6644.
11. Chen, S.; Hall, C.; Barbieri, J. T. Substrate recognition of VAMP-2 by botulinum neurotoxin B and tetanus neurotoxin. *J. Biol. Chem.* **2008**, *283*, 21153–21159.
12. Rao, K. N.; Kumaran, D.; Binz, T.; Swaminathan, S. Structural analysis of the catalytic domain of tetanus neurotoxin. *Toxicon* **2005**, *45*, 929–939.
13. Martin, L.; Cornille, F.; Coric, P.; Roques, B. P.; Fournié-Zaluski, M. C. β-Amino-thiols inhibit the zinc metallopeptidase activity of tetanus toxin light chain. *J. Med. Chem.* **1998**, *41*, 3450–3460.
14. Martin, L.; Cornille, F.; Turcaud, S.; Meudal, H.; Roques, B. P.; Fournié-Zaluski, M. C. Metallopeptidase inhibitors of tetanus toxin: a combinatorial approach. *J. Med. Chem.* **1999**, *49*, 515–525.
15. Supuran, C. T. Carbonic anhydrases: novel therapeutic applications for inhibitors and activators. *Nat. Rev. Drug Discov.* **2008**, *7*, 168–181.
16. Burnett, J. C.; Schmidt, J. J.; Stafford, R. G.; Panchal, R. G.; Nguyen, T. L.; Hermone, A. R.; Vennerstrom, J. L.; McGrath, C. F.; Lane, D. J.; Sausville, E. A.; Zaharevitz, D. W.; Gussio, R.; Bavari, S. Novel small molecule inhibitors of botulinum neurotoxin A metalloprotease activity. *Biochem. Biophys. Res. Commun.* **2003**, *310*, 84–93.

17. Burnett, J. C.; Opsenica, D.; Sriraghavan, K.; Panchal, R. G.; Ruthel, G.; Hermone, A. R.; Nguyen, T. L.; Kenny, T. A.; Lane, D. J.; McGrath, C. F.; Schmidt, J. J.; Vennerstrom, J. L.; Gussio, R.; Solaja, B. A.; Bavari, S. A refined pharmacophore identifies potent 4-amino-7-chloroquinoline-based inhibitors of the botulinum neurotoxin serotype A metalloprotease. *J. Med. Chem.* **2007**, *50*, 2127–2136.

18. Elliott, J. L.; Mogridge, J.; Collier, R. J. A quantitative study of the interactions of *Bacillus anthracis* edema factor and lethal factor with activated protective antigen. *Biochemistry* **2000**, *39*, 6706–6713.

19. Leppla, S. H. Anthrax toxin lethal factor. In *Handbook of Proteolytic Enzymes (CD-ROM);* Barrett, A. J.; Rawlings, N. D.; Woessner, J. F. Jr.,Eds.; Academic Press: London, **1998**; Chapter 511.

20. Duesbery, N. S.; Webb, C. P.; Leppla, S. H.; Gordon, V. M.; Klimpel, K. R.; Copeland, T. D.; Ahn, N. G.; Oskarsson, M. K.; Fukasawa, K.; Paull, K. D.; Vande Woude, G. F. Proteolytic inactivation of MAP-kinase-kinase by anthrax lethal factor. *Science* **1998**, *280*, 734–737.

21. Vitale, G.; Pellizzari, R.; Recchi, C.; Napolitani, G.; Mock, M.; Montecucco, C. Anthrax lethal factor cleaves the N-terminus of MAPKKs and induces tyrosine/threonine phosphorylation of MAPKs in cultured macrophages. *Biochem. Biophys. Res. Commun.* **1998**, *248*, 706–711.

22. Pellizzari, R.; Guidi-Rontani, C.; Vitale, G.; Mock, M.; Montecucco, C. Anthrax lethal factor cleaves MKK3 in macrophages and inhibits the LPS/IFNgamma-induced release of NO and TNF-alpha. *FEBS Lett.* **1999**, *462*, 199–204.

23. Menard, A.; Papini, E.; Mock, M.; Montecucco, C. The cytotoxic activity of *Bacillus anthracis* lethal factor is inhibited by leukotriene A4 hydrolase and metallopeptidase inhibitors. *Biochem. J.* **1996**, *320*, 687–691.

24. Pannifer, A. D.; Wong, T. Y.; Schwarzenbacher, R.; Renatus, M.; Petosa, C.; Bienkowska, J.; Lacy, D. B.; Collier, R. J.; Park, S.; Leppla, S. H.; Hanna, P.; Liddington, R. C. Crystal structure of the anthrax lethal factor. *Nature* **2001**, *414*, 160–161.

25. Tonello, F.; Naletto, L.; Romanello, V.; Dal Molin, F.; Montecucco, C. Tyrosine-728 and glutamic acid-735 are essential for the metalloproteolytic activity of the lethal factor of *Bacillus anthracis*. *Biochem. Biophys. Res. Commun.* **2004**, *313*, 496–502.

26. De Simone, G.; Menchise, V.; Omaggio, S.; Pedone, C.; Scozzafava, A.; Supuran, C. T. Design of weakly basic thrombin inhibitors incorporating novel P1 binding functions: Molecular and X-ray crystallographic studies. *Biochemistry* **2003**, *42*, 9013–9021.

27. Turk, B. E. Discovery and development of anthrax lethal factor metalloproteinase inhibitors. *Curr. Pharm. Biotechnol.* **2008**, *9*, 24–33.

28. Gowravaram, M. R.; Tomczuk, B. E.; Johnson, J. S.; Delecki, D.; Cook, E. R.; Ghose, A. K.; Mathiowetz, A. M.; Spurlino, J. C.; Rubin, B.; Smith, D. L.; Pulvino, T.; Wahl, R. C. Inhibition of matrix metalloproteinases by hydroxamates containing heteroatom-based modifications of the P1′ group. *J. Med. Chem.* **1995**, *38*, 2570–2581.

29. Shoop, W. L.; Xiong, Y.; Wiltsie, J.; Woods, A.; Guo, J.; Pivnichny, J. V.; Felcetto, T.; Michael, B. F.; Bansal, A.; Cummings, R. T.; Cunningham, B. R.; Friedlander, A. M.; Douglas, C. M.; Patel, S. B.; Wisniewski, D.; Scapin, G.; Salowe, S. P.; Zaller, D. M.; Chapman, K. T.; Scolnick, E. M.; Schmatz, D. M.; Bartizal, K.; MacCoss, M.; Hermes, J. D. Anthrax lethal factor inhibition. *Proc. Natl. Acad. Sci. USA.* **2005**, *102*, 7958–7963.

30. Xiong, Y.; Wiltsie, J.; Woods, A.; Guo, J.; Pivnichny, J. V.; Tang, W.; Bansal, A.; Cummings, R. T.; Cunningham, B. R.; Friedlander, A. M.; Douglas, C. M.; Salowe, S. P.; Zaller, D. M.; Scolnick, E. M.; Schmatz, D. M.; Bartizal, K.; Hermes, J. D.; MacCoss, M.; Chapman, K. T. The discovery of a potent and selective lethal factor inhibitor for adjunct therapy of anthrax infection. *Bioorg. Med. Chem. Lett.* **2006**, *16*, 964–968.

31. Johnson, S. L.; Jung, D.; Forino, M.; Chen, Y.; Satterthwait, A.; Rozanov, D. V.; Strongin, A. Y.; Pellecchia, M. Anthrax lethal factor protease inhibitors: synthesis, SAR, and structure-based 3D QSAR studies. *J. Med. Chem.* **2006**, *49*, 27–30.

32. Min, D. H.; Tang, W. J.; Mrksich, M. Chemical screening by mass spectrometry to identify inhibitors of anthrax lethal factor. *Nat. Biotechnol.* **2004**, *22*, 717–723.

33. Soelaiman, S.; Wei, B. Q.; Bergson, P.; Lee, Y. S.; Shen, Y.; Mrksich, M.; Shoichet, B. K.; Tang, W. J. Structure-based inhibitor discovery against adenylyl cyclase toxins from pathogenic bacteria that cause anthrax and whooping cough. *J. Biol. Chem.* **2003**, *278*, 25990–25997.

34. Scott, A. B. Botulinum toxin injection into extraocular muscles as an alternative to strabismus surgery. *Ophthalmology* **1980**, *87*, 1044–1049.

35. Huang, W.; Foster, J. A.; Rogachefsky, A. S. Pharmacology of botulinum toxin. *J. Am. Acad. Dermatol.* **2000**, *43*, 249–259.

36. Solomon, B. A.; Hayman, R. Botulinum toxin type A for palmar and digital hyperhidrosis. *J. Am. Acad. Dermatol.* **2000**, *42*, 1026–1029.

37. Eapen, B. R. Molecular biology of botulinum neurotoxin serotype A: a cosmetic perspective. *J. Cosmet. Dermatol.* **2008**, *7*, 221–225.

38. Schulte-Mattler, W. J. Use of botulinum toxin A in adult neurological disorders: efficacy, tolerability and safety. *CNS Drugs* **2008**, *22*, 725–738.

39. Laing, T. A.; Laing, M. E.; O'Sullivan, S. T. Botulinum toxin for treatment of glandular hypersecretory disorders. *J. Plast. Reconstr. Aesthet. Surg.* **2008**, *61*, 1024–1028.

40. Jabbari, B. Botulinum neurotoxins in the treatment of refractory pain. *Nat. Clin. Pract. Neurol.* **2008**, *4*, 676–685.

41. Eccleston, K. J.; Woolley, P. D. Botulinum toxin for urogenital conditions. *Int. J. STD AIDS* **2008**, *19*, 797–799.

42. De Ridder, D. Botulinum toxin: future developments. *BJU Int.* **2008**, *102* (Suppl. 1), 20–22.

CHAPTER 30

Clostridium histolyticum Collagenase Inhibitors in the Drug Design

CLAUDIU T. SUPURAN

Dipartimento di Chimica, Laboratorio di Chimica Inorganica e Bioinorganica, Università degli Studi di Firenze, Via della Lastruccia 3, 50019, Sesto Fiorentino (Firenze), Italy

30.1 INTRODUCTION

The M9 family of metalloproteases (MPRs) contains among others the bacterial collagenases from *Vibrio* and *Clostridium*, as well as a number of other collagenolytic bacterial endopeptidases less investigated for the moment (see also Chapter 28).[1] The *Clostridium* collagenase (ChC, E.C. 3.4.24.3), the best studied enzyme in this family, is only distantly related to the *Vibrio* collagenase and had previously been considered to be the sole member of the now defunct M31 family.[1] Both the *Vibrio* and *Clostridium* collagenases are secreted enzymes and are synthesized as precursors. The metal ion ligands have only recently been determined for ChC[2]: two of the zinc ligands occur in the His415ExxH motif, while the third one is Glu447. Similarly to the vertebrate matrix metalloproteinases (MMPs that also degrade collagen), ChC is a multiunit protein, consisting of four segments, S1, S2a, S2b and S3, with S1 incorporating the catalytic domain.[2–5]

Clostridium histolyticum is a pathogenic anaerobe that causes gas gangrene and other infections such as bacterial corneal keratitis.[5–7] All strains of this bacterium elaborate a collagenase, possibly used as a means to invade the host and to degrade its protein for nutritional purposes.[5] There are at least seven collagenase isozymes, with molecular masses ranging from 68 to 130 kDa, that have been purified to homogeneity.[4,6] They are designated as class I (α, β, γ, and η) and class II (δ, ε, and ξ) enzymes, based on a variety of criteria including their relative activities toward collagen as well as synthetic substrates.[6] In fact, the crude homogenate of *Clostridium histolyticum* is one of the most efficient systems known for the degradation of connective tissue, being able to hydrolyze triple helical regions of collagen under physiological conditions, as well as a multitude of synthetic peptide substrates.[4–7] The initial proteolytic events in this collagenase-mediated hydrolysis of type I, II, and III collagens have been delineated: the enzymes initially attack all three collagens at distinct hyperreactive sites, cleaving between Yaa#Gly bonds in the repeating Gly-Xaa-Yaa collagen sequence.[5] The hyperreactivity of these cleavage sites appears to be more related to local conformational features of the collagen fold than to the surrounding sequence. Thus, in contrast to the vertebrate collagenases, ChC degrades collagen into small peptides.[6,8]

ChC has been crystallized very recently, but only a preliminary three-dimensional structure at low resolution has been reported by Brandstetter's group for the moment.[9] Thus, the catalytic domain of ChC has been cloned, recombinantly expressed in *Escherichia coli* and purified by affinity and size exclusion column chromatographic methods.[9] Crystals of the catalytic domain were obtained from 0.12 M sodium citrate and 23% (v/v) PEG 3350 at 293 K. The crystals diffracted to 2.75 Å resolution and belonged to an orthorhombic space group, with unit cell parameters $a = 57$, $b = 109$, and $c = 181$ Å. This unit cell is consistent with the presence of one enzyme molecule per asymmetric unit and a solvent content of approximately 53%.[9] Hopefully, the high-resolution structure will be available soon that will allow to understand better the active site topology of this fascinating enzyme.

Some electronic spectroscopic studies of Co(II)-substituted ChC have also been reported,[10] offering thus interesting information regarding the binding of inhibitors within the active site of this bacterial protease. The Co(II)-substituted ChC retains catalytic activity, similarly to the native zinc enzyme,[5] and possesses a pH-dependent electronic absorption spectrum, with a maximum centered at 585 nm and a shoulder at 530 nm, and this spectrum being relatively similar to that of Co(II)-substituted carboxypeptidase A or thermolysin,[11] two enzymes in which the Zn(II) ion is coordinated—such as in ChC—by two histidines (His415 and His419) and the carboxylate of a glutamate residue (Glu447).[12] In the presence of the bidentate sulfonylated hydroxamate inhibitors (described later), major changes of this spectrum were evidenced: three absorption maxima instead of the two mentioned above appeared, at 501–505, 562–563, and 597–598 nm, respectively. The spectra were of low intensity (molar absorbances around 80–120 $M^{-1}\,cm^{-1}$ for the first two maxima

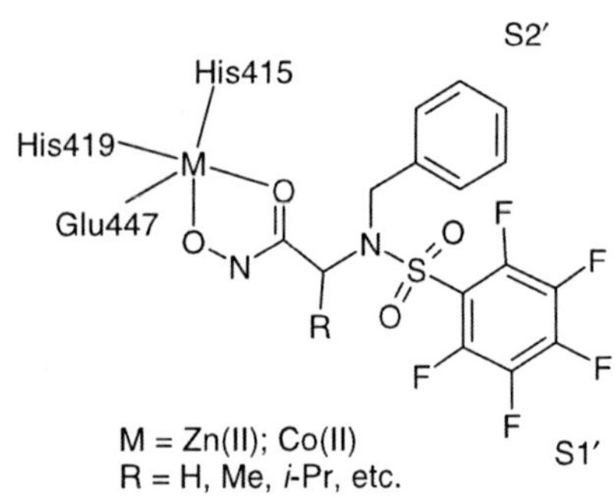

FIGURE 30.1 Proposed schematic binding of a sulfonylated amino acid inhibitor within the active site of ChC.[10]

and of around 11–15 M^{-1} cm^{-1} for the last one), being characteristic of Co(II) in pentacoordinated geometry.[10] It was thus concluded that the sulfonylated amino acid hydroxamates, the most potent class of ChC inhibitors, bind to the metal ion within the enzyme active site, leading to pentacoordinated Co(II) ions,[10] as shown schematically in Fig. 30.1.

30.2 HYDROXAMATES AS ChC INHIBITORS

Considering the above binding mode of hydroxamates to the metal ion within the ChC active site (Fig. 30.1) and the fact that this class of compounds generally has a good affinity for many zinc enzymes, an extensive program of obtaining ChC inhibitors based on the hydroxamate zinc binding function was initiated by this group.[10,13–21] A large series of sulfonylated amino acid hydroxamate ChC inhibitors of types **30.1–30.4** were reported, considering the MMP inhibitors[13] of the same type as lead molecules for obtaining high-affinity ligands for this bacterial protease.[10,13–21] These compounds were shown to inhibit both bacterial and vertebrate collagenases.

These studies showed that the sulfonylated amino acid hydroxamates represent a very potent class of ChC inhibitors (Table 30.1), but some structurally related arylsulfonylureas, arylureas, or sulfenamido-4-nitrobenzyl-Gly derivatives were also proved to inhibit this enzyme.[10,13–22] It was observed that the S1 binding moiety of the

TABLE 30.1 Inhibition of ChC and MMPs with Sulfonylated Hydroxamates 30.1–30.4

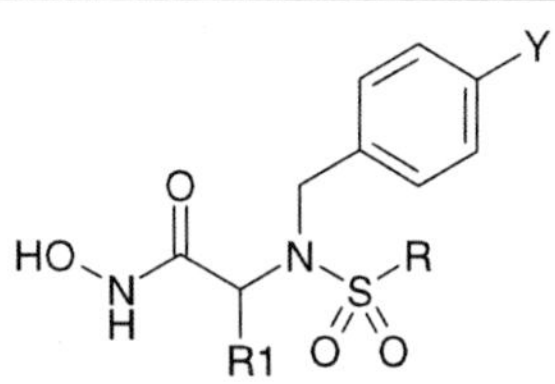

30.1 – 30.4

				K_I (nM)				
Compound	Y	R	R1	MMP-1	MMP-2	MMP-8	MMP-9	ChC
30.1a	H	*n*-C_4F_9	H	75	12	120	8.1	13
30.1b	H	*n*-C_8F_{17}	H	98	2.7	8.6	5.1	9
30.1c	H	C_6F_5	H	8.5	1.6	5.4	3.2	6
30.1d	NO_2	*n*-C_4F_9	H	62	1.5	2.4	2.0	12
30.1e	NO_2	*n*-C_8F_{17}	H	79	0.9	1.3	1.3	8
30.1f	NO_2	C_6F_5	H	3.0	0.7	0.1	0.6	5
30.1g	NO_2	3-$CF_3C_6H_4$	H	5.2	1.1	0.7	0.8	5
30.2a	NO_2	C_6F_5	Me	3.1	0.6	0.1	0.7	5
30.3a	NO_2	C_6F_5	*i*-Pr	3.0	0.5	0.2	0.6	4
30.4a	NO_2	C_6F_5	*i*-Bu	2.5	0.5	0.1	0.5	3

arylsulfonamide type, previously investigated for obtaining nonpeptide MMP inhibitors,[13] can be efficiently substituted by related moieties such as alkylsulfonyl-; arylsulfenyl-; arylsulfonylureido-; arylureido- or benzoyl-thioureido, without loss of the MMP/ChC inhibitory properties. In the large series of alkyl/arylsulfonamido derivatives investigated as MPR inhibitors, the best ChC inhibitory properties were correlated with the presence of perfluoroalkylsulfonyl, perfluorophenylsulfonyl, 3-trifluoromethylphenylsulfonyl; 3-chloro-4-nitro-phenylsulfonyl; 3-/4-protected-amino-phenylsulfonyl; 3-/4-carboxy-phenylsulfonyl moieties as S1′ anchoring group. Such derivatives possessed inhibition constants in the range of 5–10 nM against ChC. These data indicated that ChC is similar to a short-pocket MMP, eventually possessing a slightly wider neck than MMP-1.[10,13–22] A quantum theoretic QSAR study for the sulfonylated hydroxamate ChC inhibitors has also been obtained, explaining theoretically some of these experimental findings.[23]

30.3 1,3,4-THIADIAZOLE-2-THIONES AS ChC INHIBITORS

ChC inhibitors incorporating 5-amino-2-mercapto-1,3,4-thiadiazole zinc binding functions were also reported.[24] A series of compounds of types **30.5** and **30.6** was prepared by reaction of arylsulfonyl isocyanates or arylsulfonyl halides with phenylalanyl-alanine, followed by coupling with 5-amino-2-mercapto-1,3,4-thiadiazole in the presence of carbodiimides.[24] These new compounds were assayed as inhibitors of human MMP-1, MMP-2, MMP-8, and MMP-9 and of the collagenase isolated from the anaerobe *Clostridium histolyticum* (ChC, Table 30.2).

The new derivatives were proved to be powerful inhibitors of these metalloproteases, with activities in the low micromolar range for some of the target enzymes, depending on the substitution pattern at the arylsulfonyl(ureido) moieties. The following SAR was observed: (i) most of the new thiadiazoles **30.5** and **30.6** were micromolar inhibitors of MMP-1, MMP-2, MMP-8, MMP-9, and ChC; (ii) substitution pattern at the sulfonamido/sulfonylureido moiety of **30.5** and **30.6** was the primary factor influencing MMP/ChC inhibitory properties, with the sulfonylureido derivatives **30.6** generally more active than the sulfonamides **30.5**. Thus, for the first subgroup of compounds (**30.6**), low micromolar inhibitory activity was observed against MMP-2, MMP-8, MMP-9, and ChC (with inhibition constants in the range 0.1–8 μM), whereas these derivatives were slightly less active against MMP-1 (inhibition constants in the range 10–19 μM). All the substitution patterns of the aryl moieties in the sulfonylureido function of these derivatives were inducing good metalloprotease inhibitory activities, with the 4-fluorophenyl/4-chlorophenyl ones being the most effective; (iii) among the sulfonamides **30.5**, only few derivatives were as active as the sulfonylureido compounds **30.6**. Thus, the polyhalogeno-substituted compounds **30.5m–30.5o** as well as the 4-acetamidophenylsulfonyl derivative **30.5k** were among the most effective inhibitors in this subseries, with K_I's in the range of 1–17 μM against MMP-2, MMP -8, MMP-9, and ChC and 9–19 μM against MMP-1, respectively. Other substitution patterns that led to effective MMP/BP inhibitors were those incorporating monohalogeno-phenylsulfonyl, nitrophenylsulfonyl,

TABLE 30.2 Inhibition of ChC MMPs and with Thiadiazoles 30.5a–x and 30.6a–e

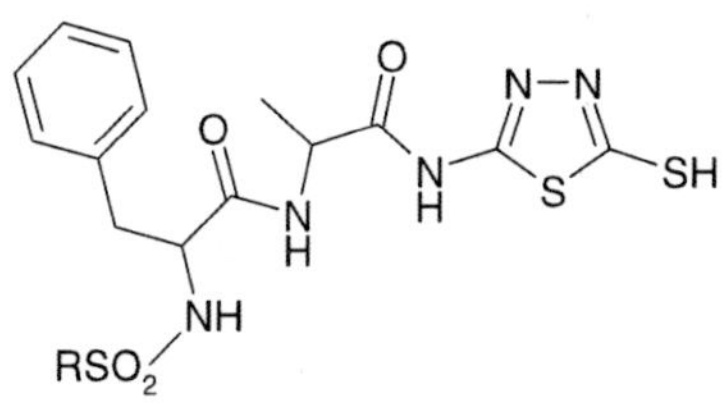

30.5

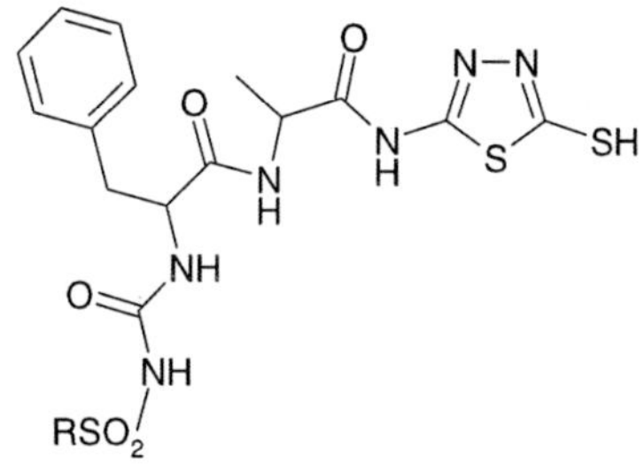

30.6

		K_I (μM)				
Compound	R	MMP-1[a]	MMP-2[a]	MMP-8[a]	MMP-9[a]	ChC[b]
30.5a	C_6H_5-	22	18	24	15	15
30.5b	$PhCH_2-$	24	16	25	16	19
30.5c	4-F-C_6H_4-	19	17	18	14	21
30.5d	4-Cl-C_6H_4-	20	16	18	15	20
30.5e	4-Br-C_6H_4-	24	15	20	21	16
30.5f	4-I-C_6H_4-	21	17	19	18	14
30.5g	4-CH_3-C_6H_4-	30	20	23	20	15
30.5h	4-O_2N-C_6H_4-	15	12	10	18	13
30.5i	3-O_2N-C_6H_4-	18	12	13	18	12
30.5j	2-O_2N-C_6H_4-	29	15	17	14	15
30.5k	4-AcNH-C_6H_4-	14	8	9	10	16
30.5m	C_6F_5-	9	6	1	2	5
30.5n	3-CF_3-C_6H_4	13	11	3	5	9
30.5o	2,5-$Cl_2C_6H_3$	19	10	9	17	15
30.5p	4-MeO-C_6H_4-	28	18	20	21	14
30.5q	2,4,6-$Me_3C_6H_2$-	33	24	25	29	17
30.5r	1-Naphthyl	80	36	40	44	35
30.5s	2-Naphthyl	72	33	45	53	40
30.5t	5-Me_2N-1-naphthyl–	87	40	48	50	43
30.5u	2-Thienyl	14	13	12	10	9
30.5v	Quinoline-8-yl	74	41	39	45	36
30.5x	Camphor-10-yl	69	35	34	40	25
30.6a	C_6H_5-	19	5	5	4	3
30.6b	4-F-C_6H_4-	12	2	0.1	0.3	0.2
30.6c	4-Cl-C_6H_4-	10	4	0.2	0.4	0.3
30.6d	4-Me-C_6H_4-	15	7	5	5	4
30.6e	2-Me-C_6H_4-	19	8	6	7	7

[a] With the thioester substrate Ac-ProLeuGly-S-LeuLeuGlyOEt, spectrophotometrically.[24]

[b] With FALGPA (furanacryloyl-leucyl-glycyl-prolyl-alanine) as substrate, spectrophotometrically.[24]

2-thienylsulfonyl, benzylsulfonyl of phenysulfonyl, among others. These compounds were slightly less effective than the previously mentioned ones, with K_I's in the range of 10–33 μM against MMP-2, MMP -8, MMP-9, and ChC and 14–30 μM against MMP-1, respectively. The most ineffective inhibitors in this subgroup were those incorporating bicyclic aromatic moieties of the naphthyl or quinoline type (**30.5r–30.5t** and **30.5v**), as well as the 10-camphor-sulfonyl derivative **30.5x**. These compounds showed K_I's in the range of 25–53 μM against MMP-2, MMP-8, MMP-9, and ChC and 69–80 μM against MMP-1; (iv) the affinity of the different metalloproteases investigated here for this class of inhibitors generally varied in the following order: MMP- 8 > ChC > MMP-9 > MMP-2 > MMP-1 for the sulfonylureido derivatives **30.6** and in the order MMP-9 > MMP-2 ≅ ChC > MMP-8 > MMP-1 for most of the derivatives **30.5** (although for some derivatives, such as **30.5m–30.5o**, the best inhibition was seen against MMP-8 as for compounds **30.6** discussed previously).

30.4 FATTY ACIDS AS ChC INHIBITORS

As the carboxylate zinc binding group is similar to the hydroxamate one, the ability of some fatty acids to inhibit the activity of ChC has also been invstigated.[25] It has been found that some fatty acids inhibited collagenase at concentrations between 50 and 500 mM. The best ChC inhibitors were the saturated fatty acids with C16–C19 unbranched chains. Unfortunately, this type of ChC inhibitors has not been studied extensively except for this important first report.[25]

30.5 APPLICATIONS OF THE ChC INHIBITORS

ChC inhibitors of the above-mentioned type might be very useful for the treatment of bacterial keratitis.[26–29] Thus, it was reported that collagen shields applied to the corneas of patients affected by bacterial keratitis degrade rapidly, often within a few hours.[26] Once treatment brings the infection under control, subsequently applied collagen shields degrade more slowly, being shown that the rate of collagen shield degradation may be a clinically useful index of collagenase activity on the ocular surface. Ultrastructural studies of collagen shields from patients with acute bacterial keratitis revealed irregular degradation of shield matrix with no evidence of adherence of microorganisms or inflammatory cells.[26] Coincubation of de-epithelialized rabbit corneas and collagen shields resulted in the inhibition of the digestion of the rabbit corneas. Collagen shields may inhibit corneal collagen degradation in infectious ulceration and melting disorders by effectively competing for collagenase on the ocular surface.[26] Furthermore, combining this inhibitory effect with the one of an endogenous collagenase inhibitor would highly facilitate the healing in this very serious eye disease.[26–29] Corneal collagen shields as a drug delivery device were also investigated for the treatment of bacterial keratitis.[30] Thus, the effectiveness of topical antibiotic treatment, with and without the use of corneal collagen shields, in a rabbit model of *Pseudomonas* keratitis, showed that collagen shields hydrated in tobramycin and supplemented with topical tobramycin were highly effective in the sustained treatment of experimental *Pseudomonas* keratitis.[30] Treatment of keratitis with antibiotic-impregnated collagen shields may reduce the need for very frequent application of topical drops but may be more effective with topical drop supplementation to increase the amount of drug available over the course of therapy.[26–29]

Another important application of *Clostridium* (and *Vibrio*) collagenases is in dermatology, in the debridement of dermal ulcers, in patients with chronic nonhealing wounds, such as pressure sores, venous leg ulcers, and diabetic ulcers, among others.[29,30] In such cases, treatment of the wound with bacterial proteases has a debriding effect due to proteolysis of the connective tissue by the MPR, whereas some proteolysis products may promote healing themselves by a mechanism little understood for the moment, probably involving the migration and stimulation of activity of important cells such as wound macrophages, fibroblasts, and keratinocytes.[29,30] The collagen cleavage products after treatment with bacterial collagenases also show chemoattractant properties for diverse cells involved in wound healing processes. These enzymes may also be used in the treatment of burn wounds.[29,30] Chronic wounds are characterized by failure in wound healing response and a delay in healing or nonclosure of the wounds.[30] A major difference between acute wounds and chronic wounds is the imbalance of proteinase inhibitors and proteinase activity that regulates the degradation and regeneration of the extracellular matrix proteins. Collagen and collagen/oxidized regenerated cellulose dressings act as a competitive substrate for MMP-2, MMP-9, and bacterial collagenase and influence this imbalance positively. Both wound dressings, approved for chronic wound treatment, the bovine collagen type I sponge and the oxidized regenerated cellulose collagen sponge, did not differ significantly in their sorption profiles for all enzymes. In general, binding was enhanced with a longer incubation time. The density of the device and the accessible surface, which can be controlled by the manufacturing process, are the crucial factors for the efficiency of the wound dressing.[30]

30.6 CONCLUSIONS

The bacterial collagenases from *Clostridium histolyticum* (ChC, EC 3.4.24.3) is a 116 kDa protein belonging to the M31 metalloproteinase family, being able to efficiently hydrolyze triple helical regions of collagens under physiological conditions, as well as an entire range of synthetic peptide substrates. ChC is involved in the pathogenicity of this and related clostridia, such as, among others, *Clostridium perfringens* that causes human gas gangrene and food poisoning. The best ChC inhibitors reported until now belong to the sulfonyl amino acyl hydroxamates incorporating alkyl/arylsulfonyl amino acid moieties. Many such hydroxamates proved to be effective bacterial collagenase inhibitors, the main contributor to activity being the substitution pattern at the sulfonamido moiety. Best ChC inhibitors were those containing pentafluorophenylsulfonyl, 3- and 4-protected-aminophenylsulfonyl P1 groups, among others, with affinities in the low nanomolar range. Other inhibitors incorporate 1,3,4-thiadiazole-2-thiol as zinc binding group, and they were low micromolar ChC inhibitors. Some fatty acids (carboxylate as zinc binding group) were also investigated for the inhibition of ChC. These inhibitors may be useful for the treatment of keratitis with antibiotic-impregnated collagen shields, which may reduce the need for very frequent application of topical drops and may be more effective than classical antibiotics. However, as for other bacterial protease inhibitors, no clinically approved drugs are available at this moment.

REFERENCES

1. Barrett, A.J.; Rawlings, N.D.; Woessner, J.F., Jr., Eds. *Handbook of Proteolytic Enzymes (CD-ROM);* Academic Press: London, **1998**; Chapters 1–569.
2. Jung, C. M.; Matsushita, O.; Katayama, S.; Minami, J.; Sakurai, J.; Okabe, A. Identification of metal ligands in the *Clostridium histolyticum* ColH collagenase. *J. Bacteriol.* **1999**, *181*, 2816–2822.
3. Matsushita, O.; Jung, C. M.; Minami, J.; Katayama, S.; Nishi, N.; Okabe, A. A study of the collagen-binding domain of a 116-kDa *Clostridium histolyticum* collagenase. *J. Biol. Chem.* **1998**, *273*, 3643–3648.
4. Eckhard, U.; Schönauer, E.; Ducka, P.; Briza, P.; Nüss, D.; Brandstetter, H. Biochemical characterization of the catalytic domains of three different clostridial collagenases. *Biol. Chem.* **2009**, *390*, 11–18.
5. Van Wart, H.E. *Clostridium* collagenase. *Handbook of Proteolytic Enzymes (CD-ROM);* Barrett, A.J.; Rawlings, N.D.; Woessner, J.F., Jr., Eds.; Academic Press: London, **1998**; Chapter 368.
6. Bond, M. D.; Van Wart, H. E. Purification and separation of individual collagenases of *Clostridium histolyticum* using red dye ligand chromatography. *Biochemistry* **1984**, *23*, 3077–3085.
7. Bond, M. D.; Van Wart, H. E. Characterization of the individual collagenases from *Clostridium histolyticum*. *Biochemistry* **1984**, *23*, 3085–3091.
8. Jung, W.; Winter, H. Considerations for the use of clostridial collagenase in clinical practice. *Clin. Drug Invest.* **1998**, *15*, 245–252.
9. Eckhard, U.; Nüss, D.; Ducka, P.; Schönauer, E.; Brandstetter, H. Crystallization and preliminary X-ray characterization of the catalytic domain of collagenase G from *Clostridium histolyticum*. *Acta Crystallogr. Sect. F Struct. Biol. Cryst. Commun.* **2008**, *64*, 419–421.
10. Scozzafava, A.; Supuran, C. T. Protease inhibitors: synthesis of potent bacterial collagenase and matrix metalloproteinase inhibitors incorporating *N*-4-nitrobenzylsulfonylglycine hydroxamate moieties. *J. Med. Chem.* **2000**, *43*, 1858–1865.
11. Valee, B. L.; Auld, D. S. Zinc coordination, function and structure of zinc enzymes and other proteins. *Biochemistry* **1990**, *29*, 5647–5659.
12. Larsen, K. S.; Zhang, K.; Auld, D. S. D-Phe complexes of zinc and cobalt carboxypeptidase A. *J. Inorg. Biochem.* **1996**, *64*, 149–162.
13. Scozzafava, A.; Supuran, C. T. Carbonic anhydrase and matrix metalloproteinase inhibitors. Sulfonylated amino acid hydroxamates with MMP inhibitory properties act as efficient inhibitors of carbonic anhydrase isozymes I, II and IV, and *N*-hydroxysulfonamides inhibit both these zinc enzymes. *J. Med. Chem.* **2000**, *43*, 3677–3687.
14. Scozzafava, A.; Supuran, C. T. Protease inhibitors. Part 5. Alkyl/arylsulfonyl- and arylsulfonylureido-/arylureido-glycine hydroxamate inhibitors of *Clostridium histolyticum* collagenase. *Eur. J. Med. Chem.* **2000**, *35*, 299–307.

15. Supuran, C. T.; Briganti, F.; Mincione, G.; Scozzafava, A. Protease inhibitors: synthesis of L-alanine hydroxamate sulfonylated derivatives as inhibitors of *Clostridium histolyticum* collagenase. *J. Enzyme Inhib.* **2000**, *15*, 111–128.

16. Supuran, C. T.; Scozzafava, A. Protease inhibitors. Part 7. Inhibition of *Clostridium histolyticum* collagenase with sulfonylated derivatives of L-valine hydroxamate. *Eur. J. Pharm. Sci.* **2000**, *10*, 67–76.

17. Scozzafava, A.; Supuran, C. T. Protease inhibitors. Part 8. Synthesis of potent *Clostridium histolyticum* collagenase inhibitors incorporating sulfonylated L-alanine hydroxamate moieties. *Bioorg. Med. Chem.* **2000**, *8*, 637–645.

18. Scozzafava, A.; Supuran, C. T. Protease inhibitors. Part 9. Synthesis of *Clostridium histolyticum* collagenase inhibitors incorporating sulfonyl-L-alanine hydroxamate moieties. *Bioorg. Med. Chem. Lett.* **2000**, *10*, 499–502.

19. Scozzafava, A.; Ilies, M. A.; Manole, G.; Supuran, C. T. Protease inhibitors. Part 12. Synthesis of potent matrix metalloproteinase and bacterial collagenase inhibitors incorporating sulfonylated *N*-4-nitrobenzyl-β-alanine hydroxamate moieties. *Eur. J. Pharm. Sci.* **2000**, *11*, 69–79.

20. Clare, B. W.; Scozzafava, A.; Supuran, C. T. Protease inhibitors. Synthesis of a series of bacterial collagenase inhibitors of the sulfonyl amino acyl hydroxamate type. *J. Med. Chem.* **2001**, *44*, 2253–2258.

21. Ilies, M.; Banciu, M. D.; Scozzafava, A.; Ilies, M. A.; Caproiu, M. T.; Supuran, C. T. Protease inhibitors: synthesis of bacterial collagenase and matrix metalloproteinase inhibitors incorporating arylsulfonylureido and 5-dibenzo-suberenyl/suberyl moieties. *Bioorg. Med. Chem.* **2003**, *11*, 2227–2239.

22. Santos, M. A.; Marques, S.; Gil, M.; Tegoni, M.; Scozzafava, A.; Supuran, C. T. Protease inhibitors: synthesis of bacterial collagenase and matrix metalloproteinase inhibitors incorporating succinyl hydroxamate and iminodiacetic acid hydroxamate moieties. *J. Enzyme Inhib. Med. Chem.* **2003**, *18*, 233–242.

23. Supuran, C. T.; Clare, B. W. Quantum theoretic QSAR of benzene derivatives: some enzyme inhibitors. *J. Enzyme Inhib. Med. Chem.* **2004**, *19*, 237–248.

24. Scozzafava, A.; Supuran, C. T. Protease inhibitors: synthesis of matrix metalloproteinase and bacterial collagenase inhibitors incorporating 5-amino-2-mercapto-1,3,4-thiadiazole zinc binding functions. *Bioorg. Med. Chem. Lett.* **2002**, *12*, 2667–2672.

25. Rennert, B.; Melzig, M. F. Free fatty acids inhibit the activity of *Clostridium histolyticum* collagenase and human neutrophil elastase. *Planta Med.* **2002**, *68*, 767–769.

26. Clinch, T. E.; Hobden, J. A.; Hill, J. M.; O'Callaghan, R. J.; Engel, L. S.; Kaufmann, H. E. Collagen shields containing tobramycin for sustained therapy (24 hours) of experimental *Pseudomonas* keratitis. *CLAO J.* **1992**, *18*, 245–247.

27. Kuwano, M.; Horibe, Y.; Kawashima, Y. Effect of collagen cross-linking in collagen corneal shields on ocular drug delivery. *J. Ocul. Pharmacol. Ther.* **1997**, *13*, 31–40.

28. Harding, K. G.; Bale, S.; Llewellyn, M.; Baggot, J.; Robbins, K. A pilot study of *Clostridium* collagenase (Collagenase ABC$^{(TM)}$) ointment in the debridement of dermal ulcers. *Clin. Drug Invest.* **1996**, *11*, 139–144.

29. Jung, W.; Winter, H. Considerations for the use of Clostridial collagenase in clinical practice. *Clin. Drug Invest.* **1998**, *15*, 245–252.

30. Metzmacher, I.; Ruth, P.; Abel, M.; Friess, W. *In vitro* binding of matrix metalloproteinase-2 (MMP-2), MMP-9, and bacterial collagenase on collagenous wound dressings. *Wound Repair Regen.* **2007**, *15*, 549–555.

CHAPTER 31

Other Bacterial Zinc Peptidases as Potential Drug Targets

KUNIHIKO WATANABE

Laboratory of Applied Microbiology, Division of Applied Life Sciences, Graduate School of Life and Environmental Sciences, Kyoto Prefectural University, Shimogamo, Sakyo, Kyoto 606-8522, Japan

31.1 METALLO-β-LACTAMASE

β-Lactam antibiotics are clinically important drugs because of a lethal action to a wide spectrum of bacteria. However, some bacteria produce β-lactamases that catalyze the hydrolysis of the four-membered ring β-lactam and consequently this abolish its antibiotic activity.[1] Therefore, the enzymes play a critical role in the nosocomial infection by pathogenic bacteria that are resistant to these essential antibiotics.[2] β-Lactamases are grouped into four classes A–D according to sequence homology.[3,4] Class A, C, and D enzymes commonly have a serine residue as a nucleophile at the active site and have been extensively studied.[1] On the other hand, class B β-lactamases use one or possibly two zinc (II) ions as cofactors and are capable of hydrolyzing a large variety of β-lactam antibiotics including third-generation cephalosporins and carbapenems. Thus, the group of enzymes is termed as metallo-β-lactamase. However, structural alignments confirm that class B β-lactamases are no more closely related to each other than are the three classes of serine β-lactamases, classes A, C, and D. As a result, metallo-β-lactamases show a striking diversity in the active site structures, catalytic features, and metal ion requirements and, therefore, distinguished further into the B1, B2, and B3 subclasses.[5,6] While the B1 (*Bacillus cereus* BcII,[7,8] *Bacteroides fragilis* CcrA[9]) and B3 enzymes (*Stenotrophomonas maltophilia* L1,[10] *Legionella gormanii* FEZ-1[11]) exhibit broad substrate specificities, the activity of B2 enzymes (*Aeromonas hydrophila* CphA[12]) is restricted to carbapenems. In contrast to B1 and B3 enzymes, which exhibit maximum activity in the presence of two zinc atoms, the B2 enzymes are inhibited probably when a second Zn^{2+} ion is bound to a low-affinity

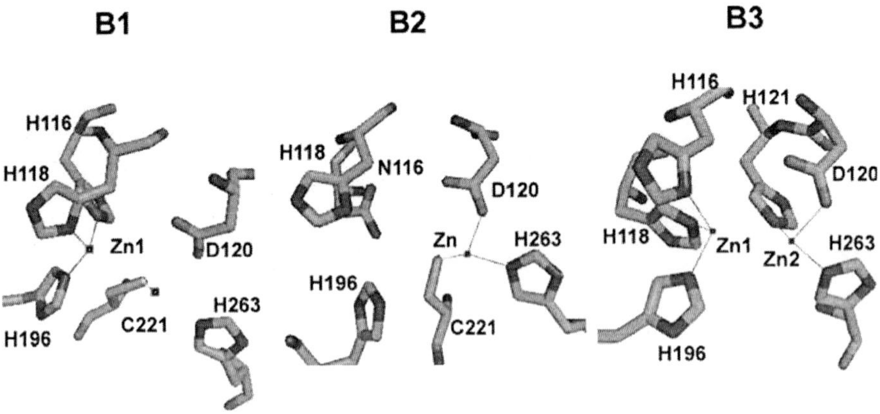

FIGURE 31.1 Metallo-β-lactamase zinc binding sites. Three classes (B1, B2, and B3) of metallo-β-lactamases are depicted with zinc atoms and coordinating residues. The structure of B1 is derived from *B. cereus* BcII (protein data bank code, 1BC2), that of B2 is from *A. hydrophila* CphA (1X8G), and that of B3 is from *S. maltophilia* L1 (1SML). (See the color version of this figure in Color Plates section.)

site.[13] Moreover, based on the molecular structures of metallo-β-lactamases from the three subclasses solved by X-ray crystallography, it is found that there are major differences in the residues that coordinate two Zn atoms (Fig. 31.1).[7–12] Three subclasses are equipped with the fundamental residues that act as Zn ligands (Table 31.1). In B1 and B3 enzymes, three histidine ligands, His116, His118, and His196, in the so called "3-H site" coordinate one Zn ion (Zn1) with a water/OH^- molecule.[8,9] In addition, in B1 enzymes, the coordination polyhedron of another Zn ion (Zn2) is provided by Asp120, Cys221, His263, and one or two water molecules at "DCH site." Similarly, in B3 enzymes, the zinc ion (Zn2) is bound to Asp120, His121, His263, and one or two water molecules at "DHH site," where Cys221 is replaced with Ser and is no longer a metal ligand.[10] In contrast, B2 enzymes have an asparagine at the equivalent position to His116 in the 3-H site; therefore, this site may not function for zinc ion (Zn1) binding, and there is no evidence based on the structural analysis that the inhibitory zinc binds as a second Zn^{2+} ion in addition to a first one in "DHC site."[12]

For metallo-β-lactamases, there are no clinically available inhibitors. Metallo-β-lactamases are resistant to inhibitors of serine β-lactamase, such as clavulanic acid, and a clinically useful inhibitor of metallo-β-lactamases has been sought.

TABLE 31.1 Zn^{2+} Ligands in Three Classes of Metallo-β-lactamases

	3-H Site			DCH/DHH Site			
Subclass	116	118	196	120	121	221	263
B1	His	His	His	Asp	—	Cys	His
B2	(Asn)	(His)	(His)	Asp	—	Cys	His
B3	His/Gln	His	His	Asp	His	—	His

A number of compounds have been synthesized and investigated as inhibitors of metallo-β-lactamases, such as trifluoromethyl alcohols and ketones,[14] hydroxamates,[15] and thiols,[16,17] containing cysteinyl peptides,[18,19] thioester derivatives,[20–24] biphenyl tetrazoles,[25] succinic acid derivatives,[26] tricyclic natural products,[27] and sulfonyl hydrazones.[28] Generally, the side chains of inhibitors bind in a predominantly hydrophobic pocket, while their functional groups interact with both zinc ions. In the case of thiol inhibitors, the sulfur displaces the hydroxide ion that bridges the two zinc ions.[29] With respect to B2 enzymes, the reported inhibitors have exhibited a lower efficiency than that shown toward the other subgroups B1 and B3. A B2 enzyme, *A. hydrophila* CphA is competitively inhibited by 2-picolinic acid and one of its derivatives, pyridine-2,4-dicarboxylic acid.[30]

31.2 *Legionella pneumophila* METALLOPEPTIDASE

Legionella pneumophila is the causative agent of Legionaires' disease pneumonia, a severe systemic disease characterized by acute pneumonia. Its infection occurs by inhalation of contaminated water droplets from aerosol generating devices. Simultaneously, this bacterium is an intracellular parasite of aquatic amoebae and human macrophages. Although the relation with the mechanism of infection is not fully understood yet, *L. pneumophila* is recognized to secrete many extracellular enzymes including a metalloprotease from the periplasm to the exterior by a type II general secretion pathway.[31] The major secretory metalloprotease, variously known as ProA or MspA, is a 38 kDa zinc metalloprotease, which also exhibits hemolytic and cytotoxic properties.[32–37] The metalloprotease is functionally and structurally homologous to the *Pseudomonas aeruginosa* elastase, now known as pseudolysin.[38] The gene for the metalloprotease was cloned, sequenced, and classified to family M4. The metal ion chelators EDTA, EGTA, and *o*-phenanthroline are all effective in inhibiting the protease activity. Significant reduction of activity is observed at chelator concentration as low as 0.1 mM.[39] The protease is not affected by incubation with dithiothreitol, PMSF, or chymostatin and soybean trypsin inhibitors. Inhibition of the protease activity by EDTA can be restored by the addition of various metal ions such as Zn^{2+}, Fe^{2+}, Mn^{2+}, and Cu^{2+}.

31.3 *Vibrio cholerae* METALLOPROTEINASE

Cholera is an acute diarrheal disease caused by the infection of *Vibrio cholerae* strains producing a variety of virulence determinants, such as the toxin-coregulated pilus, cholera toxin, and other factors required to multiply and survive in the mammalian host. *V. cholerae* produces a soluble Zn-metalloprotease, hemagglutinin/protease (HapA), encoded by *hapA*.[40] HapA hydrolyzes physiologically important proteins of intestinal epithelial cells including mucin, fibronectin, and lactoferin[41] and causes morphological changes in cultured epithelial cells.[42–44] Therefore, HapA is most likely thought to be an essential virulent factor by showing other results including microarray studies.[42,45,46] However, any study cannot provide a conclusive proof for the pathogenesis of the disease.[47] Interestingly, HapA has been shown to contribute to reactogenicity of cholera vaccine strain in human.[48]

HapA protease undergoes several steps of processing, including cleavage of the signal peptide of the proprotein (69.3 kDa) to generate the mature N-terminal form (45 kDa) and a further proteolytic processing of its C-terminal region, generating the 32 kDa form.[40] The metalloprotease is not inhibited by PMSF, but it is drastically inhibited by EDTA, Zincov (a zinc chelator), and α_2-macroglobulin.[49] The homology modeling of HapA is presented in the presence of inhibitor HPI [*N*-(1-carboxy-3-phenylpropyl)-phenylalanyl-α-asparagine].[50] Comparison of the 3D structures of HapA with *Pseudomonas aeruginosa* elastase reveals a remarkable similarity having a conserved $\alpha + \beta$ domain. The inhibitor HPI shows similar binding features as seen in other metalloproteases of M4 peptidase family.

31.4 BACTERIAL Pz PEPTIDASES

Pz peptidases are called after the synthetic substrate 4-phenylazobenzyloxycarbonyl-Pro-Leu-Gly-Pro-D-Arg (Pz-PLGPR), which contains the collagen-specific tripeptide sequence. The polypeptide chains of collagen molecules are composed of numerous repeats of a tripeptide unit, Gly-Pro-X, where X is often proline and is post-translationally modified to hydroxyproline. Pz-PLGPR has been used as a substrate in the simple assay method for bacterial collagenases.[51] Bacterial collagenases hydrolyze collagens and cleave the synthetic substrate Pz-PLGPR at the Leu–Gly bond. However, there are some distinct peptidases that hydrolyzed Pz-PLGPR but do not act on protein substrates containing collagens.[52] Pz peptidases activity is considered to be essential for collagen biodegradation in collagen-degrading microorganisms, in collaboration with their collagenolytic proteases. A collagen-degrading thermophile, *Geobacillus collagenovorans* MO-1, was reported to produce two different Pz peptidases intracellularly.[53] The Pz peptidases named A and B are purified and characterized extensively. There are many similarities between them in their catalytic properties and localization; however, they have different molecular masses and shared no antigenic groups against the respective antibodies. Their primary structures clarified from the cloned genes showed that they belong to the M3 family, in particular, the M3B subfamily, but that there is lower identity (22%). Therefore, the pair of Pz peptidases A and B occurs in *G. collagenovorans* MO-1 as isozymes. Pz peptidase A shows the highest identities to *Bacillus cereus* ATCC10987 putative peptidase (protein ID: AAS40381), *Bacillus anthracis* strain Ames putative peptidase (protein ID: AAP25299), and *Clostridium perfringens* strain 13 putative peptidase (protein ID: BAB82029), while Pz peptidase B shows that to *B. anthracis* strain Ames oligopeptidase F-like peptidase (protein ID: AAP25167), *B. cereus* ATCC10987 oligopeptidase F (protein ID: AAS40243), and *Bacillus subtilis* strain 168 (protein ID: CAB13011). The alignments are shown in Fig. 31.2. Interestingly, such a couple of two Pz peptidases are also present in other bacteria, for example, *B. anthracis* strain Ames and *B. cereus* ATCC10987. It is more intriguing that the two

```
Pz-peptidase A        YKAPFIFSNFTGTSGDIDVLTHEAGHAFQVYES-RHYEIPEYNWPTLEAC 383
B.cer/unassigned      YKAPFIFSNFNGTSGDIDVLTHEAGHAFQVYES-RKFEIPEYNWPTYEAC 383
B.ant/unassigned      YKAPFIFSNFNGTSGDIDVLTHEAGHAFQVYES-RKFEIPEYNWPTYEAC 383
C.per/unassigned      YKAPFIFSNFNGTAGDVDVLTHEAGHAFQNFRS-SWIEMQECQWPTMESC 383
Pz-peptidase B        -THPYILLNWQDNVNNLFTLVHEFGHSVHSYYT-RKTQPYPYANYSIFVA 422
B.ant/oligoF          -TNPYILMNWHDNVNNLFTLAHEFGHSVHSYYT-RKTQPHVYGDYSIFVA 422
B.cer/oligoF          -TNPYILMNWHDNVNNLFTLAHEFGHSVHSYYT-RKTQPHVYGDYSIFVA 422
B.sub/oligoF          -TNPYILMNWHNNVNNLFTLVHEFGHSVHSYYT-RKHQPYPYGNYSIFVA 422
TOP                   FTKPTADAPSLLQHDEVETYFHEFGHVMHQLCSQAEFAMFSGTHVERDFV 501
                       . *          .:: .  ** ** .:   :

Pz-peptidase A        EIHSMSMEFFTWPWMKLFFKEDAEKYQFYHLSDALLFLPYG--VAVDEFQ 431
B.cer/unassigned      EIHSMSMEFFTWPWMKLFFEEDADKYYFSHLSSALLFLPYG--VSVDEYQ 431
B.ant/unassigned      EIHSMSMEFFTWPWMKLFFEEDADKYYFSHLSSALLFLPYG--VSVDEYQ 431
C.per/unassigned      EIHSMSMEFFTWPWMYLFFKEDTDKYKFYHLGDAIKFIPYG--VTVDEFQ 431
Pz-pepeptidase B      EVASTCNEALLNDYLLKTMDEEKKRLYLLNHYLEGFRGTVFRQTMFAEFE 472
B.ant/oligoF          EVASTCNEALLNDYLLKTTEDKKERLYLLNHYLEGFRGTVFRQTMFAEFE 472
B.cer/oligoF          EVASTCNEALLNDYLLKTTEDKKERLYLLNHYLEGFRGTVFRQTMFAEFE 472
B.sub/oligoF          EVASTTNEALLGEYLLNNLEDEKQRLYILNHMLEGFRGTVFRQTMFAEFE 472
TOP                   EAPSQMLENWVWEQEPLLRMSRHYRTGSAVPRELLEKLIESRQANTGLFN 551
                      *: *   *            ..  :                       ::

Pz-peptidaseA         HFVYENPNATPAERKQAWRAIERKYMPTKDYDGNDYLERGGFWQRQSHIY 481
B.cer/unassigned      HYVYENPEASPEERKTAWRNIEKKYLPHRDYEDNDYLERGGFWQRQGHIY 481
B.ant/unassigned      HYVYENPEASPEERKTAWRNIEKKYLPHRDYEDNDYLERGGFWQRQGHIY 481
C.per/unassigned      HFVYENPECTPKERKDAWRRIEKKYLPHKNYDECDFLERGGWWFQQNHIF 481
Pz-peptidaseB         HMIHIKAQEGEALTADSLTSLYYELNKKYFGDDIVVDQEIGLEWARIPHF 522
B.ant/oligoF          HIIHKKVQEGHAVTPDMLTEIYYDLNKKYFGDALVIDEEIGLEWSRIPHF 522
B.cer/oligoF          HIIHKKVQEGHAVTPDMLTEIYYDLNKKYFGDALVIDEEIGLEWSRIPHF 522
B.sub/oligoF          HLIHTKAQEGEPLTPELLTNVYYDLNKKYFGDGMVIDKEIGLEWSRIPHF 522
TOP                   LRQIVLAKVDQVLHTQTDVDPAEEYARLCQEILGVPATPGTNMPATFGHL 601
                                             .

Pz-peptidaseA         ---TTAFYYIDYTLAQICAFQFWKRSRENYKEAWNDYLTLCRQGGSKPFT 528
B.cer/unassigned      ---SSPFYYIDYTLAQICALQFWKRARDNRQEAWEDYVNLCQQGGSKSFL 528
B.ant/unassigned      ---SSPFYYIDYTLAQICALQFWKRARDNRQEAWEDYVNLCQQGGSKSFL 528
C.per/unassigned      ---LSPFYYIDYTLAQICALQFWKKDRENHEKAWEDYLNLCKIGGTKTFL 528
Pz-peptidaseB         ----YYNYYV-YQYATGFSAATALSKQILEEGEPAVKRYI-EFLKAGSSD 567
B.ant/oligoF          ----YYNYYV-YQYATGFSAATALSKQILEEGQPAVERYINEFLKAGSSD 567
B.cer/oligoF          ----YYNYYV-YQYATGFSAATALSKQILEEGQPAVERYINEFLKAGSSD 567
B.sub/oligoF          ----YYNYYV-YQYATGYSAAQALSSQILKEGKPAVDRYI-DFLKAGSSQ 567
TOP                   AGGYDAQYYG-YLWSEVYSMDMFHTRFKQEGVLSPKVGMDYRTSILRPGG 650
                         :   **  *   .  :                       :
```

FIGURE 31.2 Amino acid sequence alignments for the critical regions of Pz peptidases A and B and oligopeptidases showing the highest homology. The abbreviations are *Bacillus cereus* ATCC 10987 unassigned peptidase (B.cer/unassigned: protein ID AAS40381), *Bacillus anthracis* strain Ames unassigned peptidase (B.ant/unassigned: protein ID AAP25299); *Clostridium perfringens* strain 13 unassigned peptidase (C.per/unassigned: protein ID BAB82029); *B. anthracis* strain Ames oligopeptidase F-like peptidase (B.ant/OligoF: protein ID AAP25167.1); *B. cereus* ATCC 10987 oligopeptidase F-like peptidase (B.cer/OligoF: protein ID AAS40243.1), and *B. subtilis* 168 oligopeptidase F-like peptidase (B.sub/OligoF: protein ID CAB13011.1). In addition, thimet oligopeptidase (TOP; protein ID NP_003240) is shown. Strictly conserved amino acid residues are indicated with (*); semiconservative ones are with (:), and less conservative ones, with (.).

strains equipped with two Pz peptidase homologues are both known to be pathogenic and to infect animals. Those enzymes are plausibly integrated into a mechanism in which pathogenic microorganisms attach to the skin of animals, decompose collagens, and finally help in the display of pathogenity. Therefore, the pair of enzymes may be potential target to inhibit the pathogenic bacteria.

Pz peptidases have been studied from various angles in all forms, including those in eukaryotes, from humans to yeast.[54] Although early studies on eukaryotic Pz peptidases supported the idea that the enzyme might physiologically participate in the collagen turnover, they were recognized as different from collagenases in spite of having the same hydrolytic site and are then renamed as thimet oligopeptidases (TOPs).[55] Bacterial Pz peptidases share lower identities with TOPs in the overall primary sequence, but a significant resemblance in the vicinity of the catalytic site (Fig. 31.2).[53] X-ray crystallographic analysis was completed for human TOP,[56] but a preliminary X-ray data for *G. collagenovorans* Pz peptidase A was only reported.[57]

G. collagenovorans Pz peptidases A and B showed no aminopeptidase activity on X-*p*-nitroanilide (X: Pro, Gly, Leu, Arg), Gly-Pro-*p*-nitroanilide, CBZ-Gly-Pro-*p*-nitroanilide, and no hydrolytic activity toward protein substrates, collagens, gelatin, elastin, keratin, and casein.[53] The Pz peptidases hydrolyze 4–13-mer oligopeptides (synthetic peptides and hormones) containing Gly-Pro and equivalent sequences, but the cleavage sites are different for longer hormone peptides, such as substance P and neurotensin.[53] *G. collagenovorans* Pz peptidases A and B are 100% inhibited by EDTA and 1,10-phenanthroline, while thiol reagents such as dithiothreitol and *N*-ethylmaleimide inhibit the enzyme activities partly. Three phosphine peptides, benzyloxycarbonyl (=Z) –(L,D)-Phe-(PO_2CH_2)–(L,D)-Ala-Lys-Tyr, Z–(L,D)-Phe-(PO_2CH_2)–(L,D)-Ala-Lys-Ser, and Gly-Pro-Phe(PO_2CH_2)Gly-Pro-Nle show potential inhibition of peptidases A and B in a range from 10 to 100 nM for K_I.[58,59]

31.5 IMMUNOGLOBULIN A1 PROTEINASE FROM *Streptococcus pneumoniae*

Streptococcus pneumoniae causes inflammation as a prominent feature of bacterial infection in both human and animals. Although pneumococcal cell wall and the toxin pneumolysis are well known to induce inflammation, other bacterial factors influencing the inflammation responses are under investigation. There are two to four surface located zinc metalloproteases in *S. pneumoniae*, probably having a role in inducing inflammation. In particular, one of the metalloproteases specifically cleaves human immunoglobulin A11 (IgA1) within the hinge region that links the Fab_α antigen binding portion of the antibody to the Fc portion.[60,61] This pneumococcal IgA1 protease belongs to M26 family and is a postproline endopeptidase, hydrolyzing the peptide bonds between Pro-Thr or Pro-Ser in the hinge region of IgA1; however, because these bonds are absent in the hinge regions of IgA2 and other IgAs from species other than primates, different types of IgAs are resistant to this protease.[62] Thus, it becomes impossible to employ experimental animals for the research of the relation of IgA1 protease with inflammation.[63]

The IgA1 proteases are synthesized as a single-chain protein of about 200 kDa (about 1700 residues). The typical Gram-positive cell wall anchor motif "LPXTG" is located in the N-terminal part, in contrast to such motifs in other known streptococcal and staphylococcal proteins at C-terminal.[64] The protease could be cleaved by the sortase and anchored in the membrane via the two potential N-terminal transmembrane domains, whereas the propeptide located before to the LPXTG motif would remain attached to the cell wall peptidoglycan by an amide bond. The C-terminal part following the cell wall anchor motif contains about 10 tandem repeats of a sequence of 20 residues (VXPXQVXXXPEYXGXXXGAX). There is no clear evidence of post-translational processing of the protein, but the IgA peptidases of family M26 are tightly associated with the bacterial cell surface.

Although the IgA1 protease was overexpressed in *Escherichia coli* and purified to homogeneity,[65] no crystal structure is available for the IgA1 protease, and the functional structure of the peptidase unit cannot be defined with experimental data. Active site residues for members of this enzyme occur in HEXXH motif that functions as coordinating residues of one zinc atom. Therefore, the common inhibitors are zinc chelators such as EDTA and 1,10-phenanthroline. No specific inhibitors of this enzyme had been found that act in the submillimolar range and are safe to be administered to humans.[60] However, doxycycline and a chemically modified tetracycline inhibit the IgA1 protease *in vitro* at low micromolar concentrations. The IC_{50} (the concentration of an agent required to inhibit the enzymatic activity by 50%) for 6-deoxy-6-demethyl-4-dedimethylaminotetracycline (the so-called CMT-3, Fig. 31.3) is between 10 and 50 μM, while the estimated IC_{50} for doxycycline is approximately 15 μM.[66] The compound is a modified form of tetracycline in which the C4 dimethylamino group has been removed and thus lacks the functional groups that interact with the bacterial ribosome.[67] Thus, the compound does not show the antibacterial activity *in vivo*, and it becomes unnecessary to select for the development of antibiotic-resistant strains. On the other hand, tetracycline pyrozole, where the metal binding functional groups are absent, is a poor inhibitor of the protease activity even in the presence of an activator $CaCl_2$. The function of the specific inhibitor is to remove zinc from the IgA1 protease. It is noted that the inhibitors were shown to reduce tissue destruction cascade including extracellular matrix degradation and thus severity and progression of disease.[68]

In addition to the IgA1 protease, ZmpB and ZmpC are identified as homologous zinc metalloproteases belonging to the same family M26.[69,70] The genes for both metalloproteases are putative. ZmpB is suggested to be a novel virulence factor capable of inducing inflammation in the lower respiratory tract, while ZmpC may

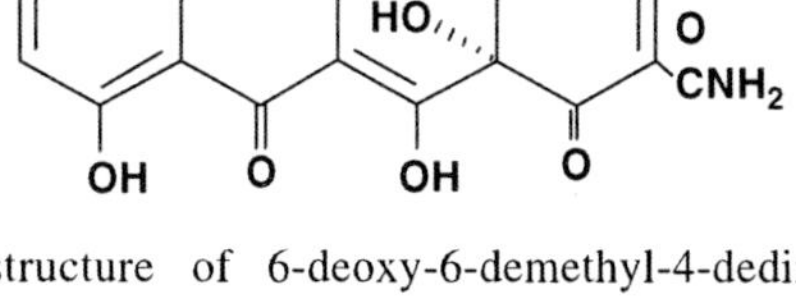

FIGURE 31.3 The structure of 6-deoxy-6-demethyl-4-dedimethylaminotetracycline (CMT-3).

specifically cleave human matrix metalloproteinase MMP-9 and play a role in pneumococcal virulence and pathogenicity in the lung.

31.6 *E. coli* StcE PROTEASE

StcE protease is a protein produced by an *E. coli* O157:H7 strain that causes diarrhea, hemorrhagic colitis, and the hemolytic uremic syndrome. The gene for StcE protease is included in the virulence plasmid pO157 carried by the strain. The name of StcE was derived from secreted protease of C1 esterase inhibitor from enterohemorrhagic *E. coli.*[71] This protease specifically cleaves C1 esterase inhibitor (C1-INH), a member of the serine protease inhibitor family (serpins), but StcE-cleaved C1-INH holds functional serpin activity to inhibit the classical complement cascade.[72] Serpins inhibit the protease activity by changing the conformation of the proteases via insertion of the serpin reactive center loop; the insertion causes displacement of active serine residue from its catalytic counterparts.[73] StcE protease binds and localizes functional C1-INH to cell membranes, effectively enhancing the ability of C1-INH to regulate complement effectors at sites of potential lytic complex formation. In addition, the protease cleaves highly glycosylated proteins, such as Muc7 and gp340, found at high levels in saliva, and thus reduces the viscosity of human saliva.[74] Therefore, it is suggested that StcE protease contributes to intimate adherence of enterohemorrhagic *E. coli* O157:H7 to host human cell.

StcE protease contains a N-terminal signal peptide in the open reading frame composed of 898 amino acids and is secreted by the closely linked *etp* type II secretion system. The molecular weight is deduced as 98 kDa and the p*I* is estimated as 6.27.[71] This protease belongs to a M66 family that does not have a conserved methionine C-terminal to the active site motif. The active site residues of StcE peptidase are located in an HEXXH motif. The third metal ligand is located six residues C-terminal to the motif. No crystal structure is obtained for the protease.

StcE protease is a robust enzyme that is resistant to proteolysis by other proteases and antibody interference, active in buffers at pH 6.1–9.0, in the presence of NaCl ranging from 0 to 600 mM, at a temperatures ranging from 4 to 55°C, and most characteristically in the presence of low concentrations of detergents.[74] These stable properties are biologically relevant to physiological conditions in human intestine and colon. Until now, no specific inhibitor is reported for StcE protease. However, as these systems by enterohemorrhagic *E. coli* may cause life threatening disease, StcE could be a great target for the treatment of these infections.

31.7 D-ALANYL-D-ALANINE DIPEPTIDASE FROM VANCOMYCIN-RESISITANT ENTEROCOCCI

D-Alanyl-D-alanine dipeptidase has been highlighted since vancomycin-resistant enterococci appeared in 1988.[75,76] It is because this metallopeptidase was found to be significantly related to the resistant mechanism gained by the acquisition of five

TABLE 31.2 Inhibition Constants of Dithiols Against VanX[86]

Thiol Inhibitors	$K_I{}^a$ (μM)
1,2-ethanedithiol	1.8
2,3-dimercapto-1-propanol	0.32
2,3-dimercapto-1-propanesulfonic acid	0.19
1,3-propanedithiol	17
dithiothreitol	7.3

[a] The final inhibition constant was estimated by Dixon plot of steady-state kinetics.

essential genes containing a transposon and clustered in an operon.[77,78] The antibiotic vancomycin shows the antimicrobial activity by binding to the usual D-alanyl-D-alanine (D-Ala-D-Ala) termini of the peptidoglycan precursor side chain, thereby inhibiting peptideglycan synthesis in collaboration with penicillin binding proteins.[79] However, vancomycin resistance uses a strategy of reprogramming the termini of peptideglycan intermediates in cell wall cross-linking steps from D-Ala-D-Ala termini to D-alanyl-D-lactate (D-Ala-D-Lac) termini.[80] This mechanism is directed by newly acquired genes including the gene *vanX* for D-alanyl-D-alanine dipeptidase.[80] Vancomycin binds to D-Ala-D-Lac only with a 1000-fold lower affinity than to D-Ala-D-Ala, and as a result, the resistant bacteria can synthesize peptidoglycans without any disturbance by vancomycin.[81]

D-Alanyl-D-alanine dipeptidase VanX is a zinc-dependent dipeptidase that strictly hydrolyzes D-Ala-D-Ala to produce D-alanine. This cytosolic metallopeptidase belongs to a M15 family and is composed of two identical subunits with the molecular mass of 23 kDa. However, subsequent studies have revealed that VanX aggregates at higher concentration.[82] X-ray crystallographic analysis revealed that a zinc ion occupies a central position inside the cavity and is coordinated with three residues and a water molecule.[83] From the proposed reaction mechanism, competitive ligands binding to zinc ion replace the water molecule and/or the binding residues in the active site, resulting in VanX inhibition.[83–85] Three classes of inhibitors are introduced by two different groups. As one class of inhibitors, the incorporation of a dithiol moiety into D-Ala-D-Ala substrate scaffolding was indicated as a more potent and specific inhibitor (Table 31.2).[86] Particularly, 2,3-dimercapto-1-propanesulfonic acid and 2,3-dimercapto-1-propanol are the most potent ones with K_I values up to 104-fold tighter than substrate K_m. Since the oxidative degradation by the dithiol compounds does not affect VanX activity, and the metal content analysis showed that zinc content in VanX remains at the same stoichiometric level after treatment of VanX with dithiothreitol, it is proposed that the two thiol groups, as in the case of mercaptan–thermolysin complex, act as two ligands to the zinc ion in the active site with a five-membered ring transition state analogue (Fig. 31.4).

As another class of inhibitors, several phosphinate analogues of D-Ala-D-Ala are shown as potent slow binding inhibitors for VanX.[87,88] Among D-3-[(1-aminoethyl)-phosphinyl]-D-2-methylpropionic acid (Fig. 31.5) showed a time-dependent onset of

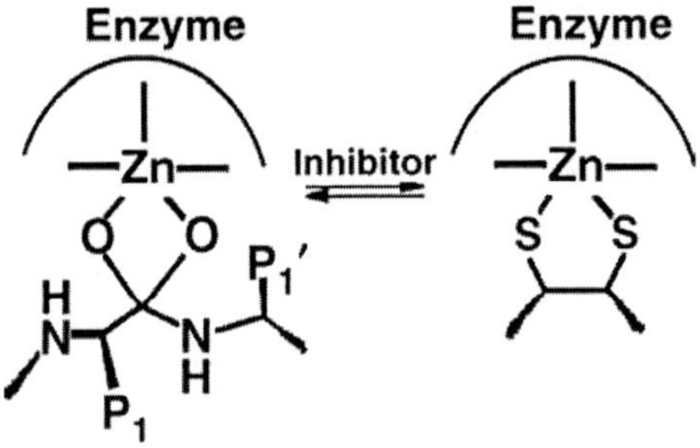

FIGURE 31.4 Inhibition mechanism of D-alanyl-D-alanine dipeptidase VanX by dithol.[86]

inhibition with a final K_I of 0.47 μM, followed by a time-dependent return to uninhibited steady-state rates upon dilution of the enzyme–inhibitor mixture.[87] Racemic mixtures 3-[(1-aminoethyl)phosphinyl]-2-ethylpropionic acid and 3-[(1-aminoethyl)phosphinyl]-2-propylpropionic acid (Fig. 31.5) inhibited with estimated final K_I of 0.35 and 1.7 μM, respectively. But those for the D-, D-diastereomer should be one-quarter of the values (90 nM and 0.44 μM) listed for the two analogues. This result indicates the possibility to further optimize phosphine peptide inhibitor design by changing the side chain.

As the third class of inhibitors, a dipeptide-like inhibitor D-Ala-D-Gly(SΦ*p*-CHF2)-OH was reported.[89] Interestingly, VanX hydrolyzes this inhibitor and generates a highly reactive 4-thioqiunone fluoromethide (Fig. 31.6) that covalently reacts with nucleophilic residues of the enzyme, resulting in irreversible inhibition of VanX (Fig. 31.6). Therefore, the inhibitor induces a suicide inactivation to VanX. Inactivation

(1)
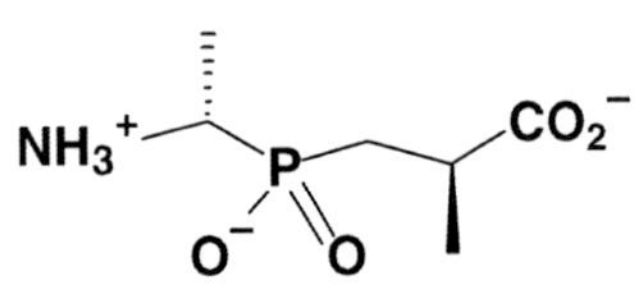

(2)
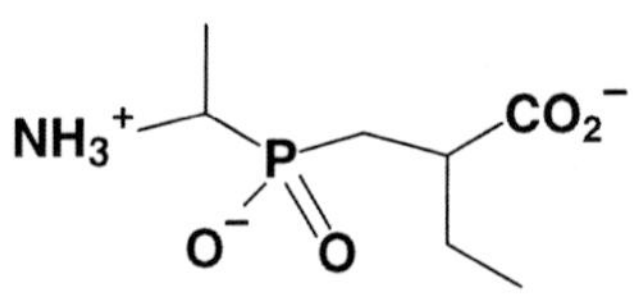

(3)
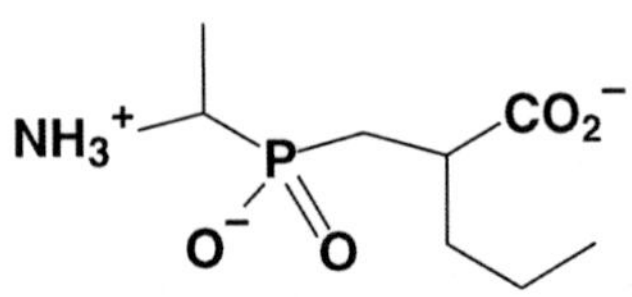

FIGURE 31.5 The structure of phosphinate inhibitor (1) D-3-[(1-aminoethyl)phosphinyl]-D-2-methylpropionic acid, (2) 3-[(1-aminoethyl)phosphinyl]-2-ethylpropionic acid, and (3) 3-[(1-aminoethyl)phosphinyl]-2-propylpropionic acid.

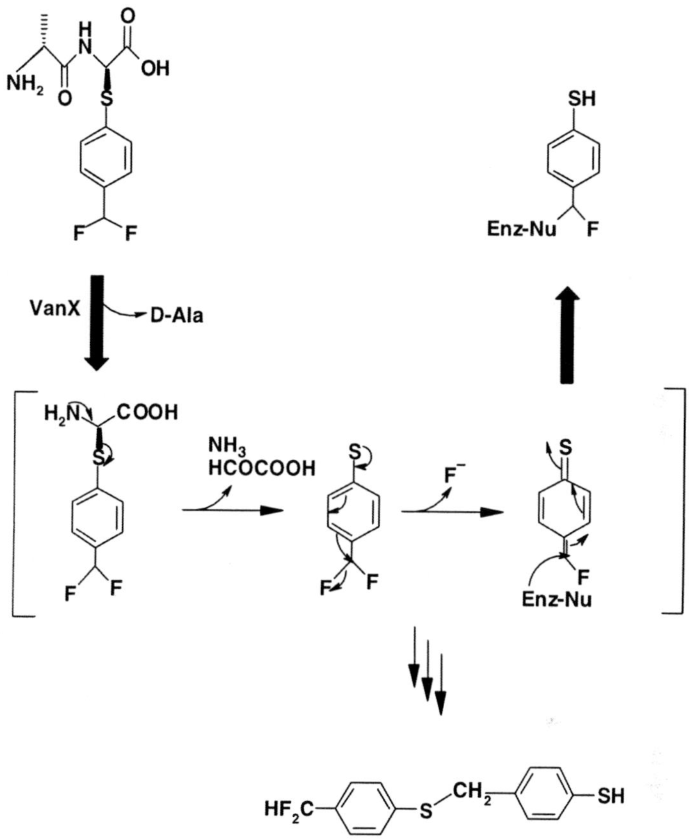

FIGURE 31.6 The inhibition mechanism of VanX by a dipeptide-like inhibitor D-Ala-D-Gly (SΦ*p*-CHF_2)–OH.[89]

was associated with the elimination of fluoride ion as deduced from ^{19}F NMR spectroscopy analysis and with the production of fluorinated thiophenol dimmer (Fig. 31.6). Inhibition of VanX by D-Ala-D-Gly(SΦ*p*-CHF2)-OH was time dependent ($K_{irr} = 30\,\mu M$; $k_{inact} = 7.3\,min^{-1}$).

REFERENCES

1. Frère, J. M. β-Lactamases and bacterial resistance to antibiotics. *Mol. Microbiol.* **1995**, *16*, 385–395.
2. Livermore, D. M.; Woodford, N. The β-lactamase threat in *Enterobacteriaceae*, *Pseudomonas* and *Acinetobacter*. *Trends Microbiol.* **2006**, *14*, 413–420.
3. Frère, J. M.; Galleni, M.; Bush, K.; Dideberg, O. Is it necessary to change the classification of β-lactamases? *J. Antimicrob. Chemother.* **2005**, *55*, 1051–1053.

4. Hall, B. G.; Salipante, S. J.; Barlow, M. The metallo-β-lactamases fall into two distinct phylogenetic groups. *J. Mol. Evol.* **2003**, *57*, 249–254.

5. Walsh, T. R.; Toleman, M. A.; Poirel, L.; Nordmann, P. Metallo-β-lactamases: the quiet before the storm? *Clin. Microbiol. Rev.* **2005**, *18*, 306–325.

6. Garau, G.; Di Guilmi, A. M.; Hall, B. G. Structure-based phylogeny of the metallo-β-lactamases. *Antimicrob. Agents Chemother.* **2005**, *49*, 2778–2784.

7. Carfi, A.; Pares, S.; Duée, E.; Galleni, M.; Duez, C.; Frère, J. M.; Dideberg, O. The 3-D structure of a zinc metallo-β-lactamase from *Bacillus cereus* reveals a new type of protein fold. *EMBO J.* **1995**, *14*, 4914–4921.

8. Fabiane, S. M.; Sohi, M. K.; Wan, T.; Payne, D. J.; Bateson, J. H.; Mitchell, T.; Sutton, B. J. Crystal structure of the zinc-dependent β-lactamase from *Bacillus cereus* at 1.9 Å resolution: binuclear active site with features of a mononuclear enzyme. *Biochemistry* **1998**, *37*, 12404–12411.

9. Concha, N. O.; Rasmussen, B. A.; Bush, K.; Herzberg, O. Crystal structure of the wide-spectrum binuclear zinc β-lactamase from *Bacteroides fragilis*. *Structure* **1996**, *4*, 823–836.

10. Ullah, J. H.; Walsh, T. R.; Taylor, I. A.; Emery, D. C.; Verma, C. S.; Gamblin, S. J.; Spencer, J. The crystal structure of the L1 metallo-β-lactamase from *Stenotrophomonas maltophilia* at 1.7 Å resolution. *J. Mol. Biol.* **1998**, *284*, 125–136.

11. García-Sáez, I.; Mercuri, P. S.; Papamicael, C.; Kahn, R.; Frère, J. M.; Galleni, M.; Rossolini, G. M.; Dideberg, O. Three-dimensional structure of FEZ-1, a monomeric subclass B3 metallo-β-lactamase from *Fluoribacter gormanii*, in native form and in complex with D-captopril. *J. Mol. Biol.* **2003**, *325*, 651–660.

12. Garau, G.; Bebrone, C.; Anne, C.; Galleni, M.; Frère, J. M.; Dideberg, O. A metallo-β-lactamase enzyme in action: crystal structures of the monozinc carbapenemase CphA and its complex with biapenem. *J. Mol. Biol.* **2005**, *345*, 785–795.

13. Hernandez-Valladares, M.; Felici, A.; Weber, G.; Adolph, H. W.; Zeppezauer, M.; Rossolini, G. M.; Amicosante, G.; Frère, J. M.; Galleni, M. Zn(II) dependence of the *Aeromonas hydrophila* AE036 metallo-β-lactamase activity and stability. *Biochemistry* **1997**, *36*, 11534–11541.

14. Walter, M. W.; Felici, A.; Galleni, M.; Soto, R. P.; Adlington, R. M.; Baldwin, J. E.; Frère, J.-M.; Gololobov, M.; Schofield, C. J. Trifluoromethyl alcohol and ketone inhibitors of metallo-β-lactamases. *Bioorg. Med. Chem. Lett.* **1996**, *6*, 2455–2458.

15. Walter, M. W.; Hernandez-Valladares, M.; Adlington, R. M.; Amicosante, G.; Baldwin, J. E.; Frère, J. M.; Galleni, M.; Rossolini, G. M.; Schofield, C. J. Hydroxamate inhibitors of *Aeromonas hydrophila* AE036 metallo-β-lactamase. *Bioorg. Chem.* **1999**, *27*, 35–40.

16. Arawaka, Y.; Shibata, N.; Shibayama, K.; Kurokawa, H.; Yagi, T.; Fujiwara, H.; Goto, M. Convenient test for screening metallo-β-lactamase-producing Gram-negative bacteria by using thiol compounds. *J. Clin. Microbiol.* **2000**, *38*, 40–43.

17. Goto, M.; Takahashi, T.; Yamashita, F.; Koreeda, A.; Mori, H.; Ohta, M.; Arakawa, Y. Inhibition of the metallo-β-lactamase produced from *Serratia marcescens* by thiol compounds. *Biol. Pharm. Bull.* **1997**, *20*, 1136–1140.

18. Bounaga, S.; Laws, A. P.; Galleni, M.; Page, M. I. The mechanism of catalysis and the inhibition of the *Bacillus cereus* zinc-dependent β-lactamase. *Biochem. J.* **1998**, *331*, 703–711.

19. Bounaga, S.; Galleni, M.; Laws, A. P.; Page, M. I. Cysteinyl peptide inhibitors of *Bacillus cereus* zinc β-lactamase. *Bioorg. Med. Chem.* **2001**, *9*, 503–510.

20. Greenlee, M. L.; Laub, J. B.; Balkovec, J. M.; Hammond, M. L.; Hammond, G. G.; Pompliano, D. L.; Epstein-Toney, J. H. Synthesis and SAR of thioester and thiol inhibitors of IMP-1 metallo-β-lactamase. *Bioorg. Med. Chem. Lett.* **1999**, *9*, 2549–2554.

21. Payne, D. J.; Bateson, J. H.; Gasson, B. C.; Proctor, D.; Khushi, T.; Farmer, T. H.; Toldon, D. A.; Bell, D.; Skett, P. W.; Marshall, A. C.; Reid, R.; Ghosez, L.; Combret, Y.; Marchand-Brynaert, J. Inhibition of metallo-β-lactamases by a series of mercaptoacetic acid thiol ester derivatives. *Antimicrob. Agents Chemother.* **1997**, *41*, 135–140.

22. Payne, D. J.; Bateson, J. H.; Gasson, B. C.; Khushi, T.; Proctor, D.; Pearson, S. C.; Reid, R. Inhibition of metallo-β-lactamases by a series of thiol ester derivatives of mercaptophenylacetic acid. *FEMS Microbiol. Lett.* **1997**, *157*, 171–175.

23. Hammond, G. G.; Huber, J. L.; Greenlee, M. L.; Laub, J. B.; Young, K.; Silver, L. L.; Balkovec, J. M.; Pryor, K. D.; Wu, J. K.; Leiting, B.; Pompliano, D. L.; Toney, J. H. Inhibition of IMP-1 metallo-β-lactamase and sensitisation of IMP-1-producing bacteria by thioester derivatives. *FEMS Microbiol. Lett.* **1999**, *179*, 289–296.

24. Mollard, C.; Moali, C.; Papamicael, C.; Damblon, C.; Vessilier, S.; Amicosante, G.; Schofield, C. J.; Galleni, M.; Frère, J. M.; Roberts, G. C. K. Thiomandelic acid, a broad-spectrum inhibitor of zinc β-lactamases: kinetic and spectroscopic studies. *Biol. Chem.* **2001**, *276*, 45015–45023.

25. Toney, J. H.; Cleary, K. A.; Hammond, G. G.; Yuan, X.; May, W. J.; Hutchins, S. M.; Ashton, W. T.; Vanderwall, D. E. Structure–activity relationships of biphenyl tetrazoles as metallo-β-lactamase inhibitors. *Bioorg. Med. Chem. Lett.* **1999**, *9*, 2741–2746.

26. Toney, J. H.; Hammond, G. G.; Fitzgerald, P. M.; Sharma, N.; Balkovec, J. M.; Rouen, G. P.; Olson, S. H.; Hammond, M. L.; Greenlee, M. L.; Gao, Y. D. Succinic acids as potent inhibitors of plasmid-borne IMP-1 metallo-β-lactamase. *J. Biol. Chem.* **2001**, *276*, 31913–31918.

27. Payne, D. J.; Hueso-Rodríguez, J. A.; Boyd, H.; Concha, N. O.; Janson, C. A.; Gilpin, M.; Bateson, J. H.; Cheever, C.; Niconovich, N. L.; Pearson, S.; Rittenhouse, S.; Tew, D.; Díez, E.; Pérez, P.; De La Fuente, J.; Rees, M.; Rivera-Sagredo, A. Identification of a series of tricyclic natural products as potent broad-spectrum inhibitors of metallo-β-lactamases. *Antimicrob. Agents Chemother.* **2002**, *46*, 1880–1886.

28. Siemann, S.; Evanoff, D. P.; Marrone, L.; Clarke, A. J.; Viswantha, T.; Dimitrenko, G. I. N-arylsulfonyl hydrazones as inhibitors of IMP-1 metallo-β-lactamase. *Antimicrob. Agents Chemother.* **2002**, *46*, 2450–2457.

29. Heinz, U.; Bauer, R.; Wommer, S.; Meyer-Klaucke, W.; Papamichaels, C.; Bateson, J.; Adolph, H. W. Coordination geometries of metal ions in D- or L-captopril-inhibited metallo-β-lactamases. *J. Biol. Chem.* **2003**, *278*, 20659–20666.

30. Horsfall, L. E.; Garau, G.; Liénard, B. M.; Dideberg, O.; Schofield, C. J.; Frère, J. M.; Galleni, M. Competitive inhibitors of the CphA metallo-β-lactamase from *Aeromonas hydrophila*. *Antimicrob. Agents Chemother.* **2007**, *51*, 2136–2142.

31. Rossier, O.; Dao, J.; Cianciotto, N. P. The type II secretion system of *Legionella pneumophila* elaborates two aminopeptidases, as well as a metalloprotease that contributes to differential infection among protozoan hosts. *Appl. Environ. Microbiol.* **2008**, *74*, 753–761.

32. Hales, L. M.; Shuman, H. A. *Legionella pneumophila* contains a type II general secretion pathway required for growth in amoebae as well as for secretion of the Msp protease. *Infect. Immun.* **1999**, *67*, 3662–3666.

33. Liles, M. R.; Edelstein, P. H.; Cianciotto, N. P. The prepilin peptidase is required for protein secretion by and the virulence of the intracellular pathogen *Legionella pneumophila*. *Mol. Microbiol.* **1999**, *31*, 959–970.

34. Moffat, J. F.; Black, W. J.; Tompkins, L. S. Further molecular characterization of the cloned *Legionella pneumophila* zinc metalloprotease. *Infect. Immun.* **1994**, *62*, 751–753.

35. Quinn, F. D.; Tompkins, L. S. Analysis of a cloned sequence of *Legionella pneumophila* encoding a 38 kD metalloprotease possessing haemolytic and cytotoxic activities. *Mol. Microbiol.* **1989**, *3*, 797–805.

36. Szeto, L.; Shuman, H. A. The *Legionella pneumophila* major secretory protein, a protease, is not required for intracellular growth or cell killing. *Infect. Immun.* **1990**, *58*, 2585–2592.

37. Keen, M. G.; Hoffman, P. S. Characterization of a *Legionella pneumophila* extracellular protease exhibiting haemolytic and cytotoxic activities. *Infect. Immun.* **1989**, *57*, 732–738.

38. Black, W. G.; Quinn, F. D.; Tompkins, L. S. *Legionella pneumophila* zinc metalloprotease is structurally and functionally homologous to *Pseudomonas aeruginosa* elastase. *J. Bacteriol.* **1990**, *172*, 2608–2613.

39. Dreyfus, L. A.; Iglewski, B. H. Purification and characterization of an extracellular protease of *Legionella pneumophila*. *Infect. Immun.* **1986**, *51*, 736–743.

40. Häse, C. C.; Finkelstein, R. A. Cloning and nucleotide sequence of the *Vibrio cholerae* hemagglutinin/protease (HA/protease) gene and construction of an HA/protease-negative strain. *J. Bacteriol.* **1991**, *173*, 3311–3317.

41. Finkelstein, R. A.; Boesman-Finkelstein, M.; Holt, P. *Vibrio cholerae* hemagglutinin/lectin/protease hydrolyzes fibronectin and ovomucin: F.M. Burnet revisited. *Proc. Natl. Acad. Sci. USA* **1983**, *80*, 1092–1095.

42. Mel, S. F.; Fullner, K. J.; Wimer-Mackin, S.; Lencer, W. I.; Mekalanos, J. J. Association of protease activity in *Vibrio cholerae* vaccine strains with decreases in transcellular epithelial resistance of polarized T84 intestinal epithelial cells. *Infect. Immun.* **2000**, *68*, 6487–6492.

43. Wu, Z.; Milton, D.; Nybom, P.; Sjö, A.; Magnusson, K. E. *Vibrio cholerae* hemagglutinin/protease (HA/protease) causes morphological changes in cultured epithelial cells and perturbs their paracellular barrier function. *Microb. Pathog.* **1996**, *21*, 111–123.

44. Wu, Z.; Nybom, P.; Magnusson, K. E. Distinct effects of *Vibrio cholerae* haemagglutinin/protease on the structure and localization of the tight junction-associated proteins occludin and ZO-1. *Cell. Microbiol.* **2000**, *2*, 11–17.

45. Silva, A. J.; Pham, K.; Benitez, J. A. Haemagglutinin/protease expression and mucin gel penetration in El Tor biotype *Vibrio cholerae*. *Microbiology* **2003**, *149*, 1883–1891.

46. Xu, Q.; Dziejman, M.; Mekalanos, J. J. Determination of the transcriptome of *Vibrio cholerae* during intraintestinal growth and midexponential phase *in vitro*. *Proc. Natl Acad. Sci. USA* **2003**, *100*, 1286–1291.

47. Ghosh, A.; Saha, D. R.; Hoque, K. M.; Asakuna, M.; Yamasaki, S.; Koley, H.; Das, S. S.; Chakrabarti, M. K.; Pal, A. Enterotoxigenicity of mature 45-kilodalton and processed 35-kilodalton forms of hemagglutinin protease purified from a cholera toxin gene-negative *Vibrio cholerae* non-O1, non-O139 strain. *Infect. Immun.* **2006**, *74*, 2937–2946.

48. Benítez, J. A.; García, L.; Silva, A.; García, H.; Fando, R.; Cedré, B.; Pérez, A.; Campos, J.; Rodríguez, B. L.; Pérez, J. L.; Valmaseda, T.; Pérez, O.; Pérez, A.; Ramírez, M.; Ledón, T.;

Jidy M. D.; Lastre, M.; Bravo, L.; Sierra, G. Preliminary assessment of the safety and immunogenicity of a new CTXPhi-negative, hemagglutinin/protease-defective El Tor strain as a cholera vaccine candidate. *Infect. Immun.* **1999**, *67*, 539–545.

49. Crowther, R. S.; Roomi, N. W.; Fahim, R. E.; Forstner, J. F. *Vibrio cholerae* metalloproteinase degrades intestinal mucin and facilitates enterotoxin-induced secretion from rat intestine. *Biochim. Biophys. Acta* **1987**, *924*, 393–402.

50. Lutfullah, G.; Amin, F.; Khan, Z.; Azhar, N.; Azim, M. K.; Noor, S.; Shoukat, K. Homology modeling of hemagglutinin/protease [HA/P (vibriolysin)] from *Vibrio Cholerae:* sequence comparison, residue interactions and molecular mechanism. *Protein J.* **2008**, *27*, 105–114.

51. Wünsch, E.; Heidrich, H. G. Zur quantitiativen Bestimmung der Kollagenase. *Hoppe-Seyler's Z. Physiol. Chem.* **1963**, *333*, 149–151.

52. Watanabe, K. Collagenolytic proteases from bacteria. *Appl. Microbiol. Biotechnol.* **2004**, *63*, 520–526.

53. Miyake, R.; Shigeri, Y.; Tatsu, Y.; Yumoto, N.; Umekawa, M.; Tsujimoto, Y.; Matsui, H.; Watanabe, K. Two thimet oligopeptidase-like Pz peptidases produced by a collagen-degrading thermophile, *Geobacillus collagenovorans* MO-1. *J. Bacteriol.* **2005**, *187*, 4140–4148.

54. Ibrahim-Granet, O.; D'Enfert, C. The *Aspergillus fumigatus mepB* gene encodes an 82 kDa intracellular metalloproteinase structurally related to mammalian thimet oligopeptidases. *Microbiology* **1997**, *143*, 2247–2253.

55. Barrett, A. J.; Chen, J. M. TOP. In *Handbook of Proteolytic Enzymes*, 2nd ed; Barrett, A. J.; Rawlings, N. D.; Woessner J. F., Eds.; Elsevier: **2004**; pp. 352–356.

56. Ray, K.; Hines, C. S.; Coll-Rodriguez, J.; Rodgers, D. W. Crystal structure of human thimet oligopeptidase provides insight into substrate recognition, regulation, and localization. *J. Biol. Chem.* **2004**, *279*, 20480–20489.

57. Kawasaki, A.; Nakano, H.; Tsujimoto, Y.; Matsui, H.; Shimizu, T.; Nakatsu, T.; Kato, H.; Watanabe, K. Crystallization and preliminary X-ray crystallographic studies of Pz peptidase A from *Geobacillus collagenovorans* MO-1. *Acta. Crystallogr. Sect. F Struct. Biol. Cryst. Commun.* **2007**, *63*, 142–144.

58. Jiracek, J.; Yiotakis, A.; Vincent, B.; Checler, F.; Dive, V. Development of the first potent and selective inhibitor of the zinc endopeptidase neurolysin using a systematic approach based on combinatorial chemistry of phosphinic peptides. *J. Biol. Chem.* **1996**, *271*, 19606–19611.

59. Sugihara, Y.; Kawasaki, A.; Tsujimoto, Y.; Matsui, H.; Watanabe, K. Potencies of phosphine peptide inhibitors of mammalian thimet oligopeptidase and neurolysin on two bacterial Pz peptidases. *Biosci. Biotechnol. Biochem.* **2007**, *71*, 594–597.

60. Plaut, A. G.; Wright, A. Immunoglobulin A-metallo-type specific prolyl endopeptidases. *Methods Enzymol.* **1995**, *248*, 634–642.

61. Plaut, A. G. IgA1-specific metalloendopeptidase. In *Handbook of Proteolytic Enzymes;* 2 edn; Barrett, A. J.; Rawlings, N. D.; Woessner, J. F., Eds.; Elsevier: London, **2004**; pp 772–776.

62. Senior, B. W.; Dunlop, J. I.; Batten, M. R.; Kilian, M.; Woof, J. M. Cleavage of a recombinant human immunoglobulin A2 (IgA2)-IgA1 hybrid antibody by certain bacterial IgA1 proteases. *Infect. Immun.* **2000**, *68*, 463–469.

63. Qiu, J.; Brackee, G. P.; Plaut, A. G. Analysis of the specificity of bacterial immunoglobulin A (IgA) proteases by a comparative study of ape serum IgAs as substrates. *Infect. Immun.* **1996**, *64*, 933–937.

64. Schneewind, O.; Mihaylova-Petkov, D.; Model, P. Cell wall sorting signals in surface proteins of gram-positive bacteria. *EMBO J.* **1993**, *12*, 4803–4811.

65. Romanello, V.; Marcacci, M.; Dal Molin, F.; Moschioni, M.; Censini, S.; Covacci, A.; Baritussio, A. G.; Montecucco, C.; Tonello, F. Cloning, expression, purification, and characterization of *Streptococcus pneumoniae* IgA1 protease. *Protein Expr. Purif.* **2006**, *45*, 142–149.

66. Walker, S. G.; Carnu, O. I.; Tüter, G.; Ryan, M. E. The immunoglobulin A1 proteinase from *Streptococcus pneumoniae* is inhibited by tetracycline compounds. *FEMS Immunol. Med. Microbiol.* **2006**, *48*, 218–222.

67. Nelson, M. L. Chemical and biological dynamics of tetracyclines. *Adv. Dent. Res.* **1998**, *12*, 5–11.

68. Grenier, D.; Plamondon, P.; Sorsa, T.; Lee, H. M.; McNamara, T.; Ramamurthy, N. S.; Golub, L. M.; Teronen, O.; Mayrand, D. Inhibition of proteolytic, serpinolytic, and progelatinase-b activation activities of periodontopathogens by doxycycline and the non-antimicrobial chemically modified tetracycline derivatives. *J. Periodontol.* **2002**, *73*, 79–85.

69. Blue, C. E.; Paterson, G. K.; Kerr, A. R.; Bergé, M.; Claverys, J. P.; Mitchell, T. J. ZmpB, a novel virulence factor of *Streptococcus pneumoniae* that induces tumor necrosis factor α production in the respiratory tract. *Infect. Immun.* **2003**, *71*, 4925–4935.

70. Oggioni, M. R.; Memmi, G.; Maggi, T.; Chiavolini, D.; Iannelli, F.; Pozzi, G. Pneumococcal zinc metalloproteinase ZmpC cleaves human matrix metalloproteinase 9 and is a virulence factor in experimental pneumonia. *Mol. Microbiol.* **2003**, *49*, 795–805.

71. Lathem, W. W.; Grys, T. E.; Witowski, S. E.; Torres, A. G.; Kaper, J. B.; Tarr, P. I.; Welch, R. A. StcE, a metalloprotease secreted by *Escherichia coli* O157:H7, specifically cleaves C1 esterase inhibitor. *Mol. Microbiol.* **2002**, *45*, 277–288.

72. Lathem, W. W.; Bergsbaken, T.; Welch, R. A. Potentiation of C1 esterase inhibitor by StcE, a metalloprotease secreted by *Escherichia coli* O157:H7. *J. Exp. Med.* **2004**, *199*, 1077–1087.

73. Huntington, J. A.; Read, R. J.; Carrell, R. W. Structure of a serpin–protease complex shows inhibition by deformation. *Nature* **2000**, *407*, 923–926.

74. Grys, T. E.; Siegel, M. B.; Lathem, W. W.; Welch, R. A. The StcE protease contributes to intimate adherence of enterohemorrhagic *Escherichia coli* O157:H7 to host cells. *Infect. Immun.* **2005**, *73*, 1295–1303.

75. Uttley, A. H.; Collins, C. H.; Naidoo, J.; George, R. C. Vancomycin-resistant enterococci. *Lancet* **1988**, *1*, 57–58.

76. Leclercq, R.; Derlot, E.; Duval, J.; Courvalin, P. Plasmid-mediated resistance to vancomycin and teicoplanin in *Enterococcus faecium*. *N. Engl. J. Med.* **1988**, *319*, 157–161.

77. Arthur, M.; Depardieu, F.; Cabanié, L.; Reynolds, P.; Courvalin, P. Requirement of the VanY and VanX D,D-peptidases for glycopeptide resistance in enterococci. *Mol. Microbiol.* **1998**, *30*, 819–830.

78. Arthur, M.; Molinas, C.; Courvalin, P. The VanS-VanR two-component regulatory system controls synthesis of depsipeptide peptidoglycan precursors in *Enterococcus faecium* BM4147. *J. Bacteriol.* **1992**, *174*, 2582–2591.

79. Walsh, C. T.; Fisher, S. L.; Park, I. S.; Prahalad, M.; Wu, Z. Bacterial resistance to vancomycin: five genes and one missing hydrogen bond tell the story. *Chem. Biol.* **1996**, *3*, 21–28.

80. Lessard, I. A.; Walsh, C. T. VanX, a bacterial D-alanyl-D-alanine dipeptidase: resistance, immunity, or survival function? *Proc. Natl. Acad. Sci. USA* **1999**, *96*, 11028–11032.

81. Bugg, T. D.; Wright, G. D.; Dutka-Malen, S.; Arthur, M.; Courvalin, P.; Walsh, C. T. Molecular basis for vancomycin resistance in *Enterococcus faecium* BM4147: biosynthesis of a depsipeptide peptidoglycan precursor by vancomycin resistance proteins VanH and VanA. *Biochemistry* **1991**, *30*, 10408–10415.

82. Breece, R. M.; Costello, A.; Bennett, B.; Sigdel, T. K.; Matthews, M. L.; Tierney, D. L.; Crowder, M. W. A five-coordinate metal center in Co(II)-substituted VanX. *J. Biol. Chem.* **2005**, *280*, 11074–11081.

83. Bussiere, D. E.; Pratt, S. D.; Katz, L.; Severin, J. M.; Holzman, T.; Park, C. H. The structure of VanX reveals a novel amino-dipeptidase involved in mediating transposon-based vancomycin resistance. *Mol. Cell.* **1998**, *2*, 75–84.

84. Gao, Y. Glycopeptide antibiotics and development of inhibitors to overcome vancomycin resistance. *Nat. Prod. Rep.* **2002**, *19*, 100–107.

85. Lessard, I. A.; Walsh, C. T. Mutational analysis of active-site residues of the enterococcal D-ala-D-Ala dipeptidase VanX and comparison with *Escherichia coli*D-ala-D-Ala ligase and D-ala-D-Ala carboxypeptidase VanY. *Chem. Biol.* **1999**, *6*, 177–187.

86. Wu, Z.; Walsh, C. T. Dithiol compounds: potent, time-dependent inhibitors of VanX, a zinc-dependent D, D-dipeptidase required for vancomycin resistance in *Enterococcus faecium*. *J. Am. Chem. Soc.* **1996**, *118*, 1785–1786.

87. Wu, Z.; Walsh, C. T. Phosphinate analogs of D-, D-dipeptides: slow-binding inhibition and proteolysis protection of VanX, a D-, D-dipeptidase required for vancomycin resistance in *Enterococcus faecium*. *Proc. Natl. Acad. Sci. USA* **1995**, *92*, 11603–11607.

88. Ellsworth, B. A.; Tom, N. J.; Bartlett, P. A. Synthesis and evaluation of inhibitors of bacterial D-alanine:D-alanine ligases. *Chem. Biol.* **1996**, *3*, 37–44.

89. Aráoz, R.; Anhalt, E.; René, L.; Badet-Denisot, M. A.; Courvalin, P.; Badet, B. Mechanism-based inactivation of VanX, a D-alanyl-D-alanine dipeptidase necessary for vancomycin resistance. *Biochemistry* **2000**, *39*, 15971–15979.

PART V

DRUG DESIGN STUDIES OF OTHER ZINC-CONTAINING ENZYMES

CHAPTER 32

Angiotensin Converting Enzyme (ACE) Inhibitors

ANA CÁMARA-ARTIGAS, VICENTE JARA-PÉREZ, and MONTSERRAT ANDÚJAR-SÁNCHEZ

Departmento Química Física, Bioquímica y Química Inorgánica, Universidad de Almería, Carretera Sacramento s/n Almería, 04120, Spain

32.1 INTRODUCTION: BRIEF HISTORY OF ACE INHIBITORS

Hypertension is an important disease that affects more than a fourth of the adult population in the world[1] and it is a major risk factor for myocardial infarction, stroke, and renal failure. The control of blood pressure is crucial in the prevention of these adverse outcomes and to regulate high blood pressure, drugs from a number of different classes are used, either alone or in combination.[2] Finding new effective drugs for hypertension treatment is a difficult task and in some cases these drugs were discovered by chance observations. It is the case of the accidental discovery of the secondary effect in the blood pressure of the β-blockers designed to block the effects of adrenaline on the β-receptors in heart for the treatment of angina.[3]

In the mid-1970s, Cushman and Ondetti developed the first orally active antihypertensive inhibitor using "rational drug design." It was captopril[4] and the target protein was the angiotensin I converting enzyme (ACE, E.C. 3.4.15.1). This enzyme, also known as peptidyl-dipeptidase A or kininase II, was first isolated in 1956 and it is a chloride-dependent metalloenzyme that cleaves a dipeptide from the carboxyl terminus of the decapeptide angiotensin I (AGI), Asp-Arg-Val-Tyr-Ile-His-Pro-Phe-His-Leu, to form the potent vasopressor angiotensin II (AGII),[5] Asp-Arg-Val-Tyr-Ile-His-Pro-Phe. Enzyme kinetic experiments with isolated ACE indicated that this enzyme was a zinc metallopeptidase with a catalytic center similar to that of carboxypeptidase A. At that moment, the availability of the three-dimensional structure of the carboxypeptidase A[6] allowed to tackle the development of ACE inhibitors based on the structure of the active site of this enzyme. At the same time, researchers at the John Vane lab from the

Squibb Institute for Medical Research had evidence that the venom peptides isolated from a Brazilian viper inhibited ACE.[7] The best results were obtained with teprotide (Pyr-Tpr-Pro-Arg-Pro-Gln-Ile-Pro-Pro) and this nonapeptide was the first ACE inhibitor to be studied in hypertensive patients with promising results in lowering blood pressure. The major problem with this inhibitor was its lack of oral activity. The design of new synthetic venom peptide analogues were based on structure–activity studies and these studies yield a succinyl amino acid derivative, D-2-methylsuccinyl-L-proline, where the replacement of the carboxyl group by a sulfhydryl group gave the potent ACE inhibitor captopril. The assumption of the sulfhydryl group binding to the active site zinc atom was supported by the structural studies with other metalloprotease, the thermolysin.[8]

Captopril becomes the first available commercial ACE inhibitor compound, and since this drug was approved for use in 1981, many highly potent ACE inhibitors have subsequently been developed.[4,9] All of these inhibitors interact with the active site zinc atom, but the zinc-coordinating group is different. In this way, a chemical structure classification of ACE inhibitors can be made based on the zinc-interacting group in sulfhydryl-containing ACE inhibitors, dicarboxyl-containing ACE inhibitors, and phosphorus-containing ACE inhibitors.[10] Besides, the therapeutic effect of ACE inhibitors in lowering the high blood pressure, some of these drugs have also been proved to be successful in the treatment of symptomatic or asymptomatic left ventricular systolic dysfunction, postmyocardial infarction, renal failure, and diabetic nephropathy. Because of its effectiveness, ACE inhibitors have been the favorite therapy for the treatment of the high blood pressure in the past 20 years,[11] but they have important side effects such as cough and angioneurotic edema[12,13] and important drug–drug interactions.[10]

There are many reviews on ACE that compile most of the information of the past three decades. A compilation of all the biochemical information about ACE has been done recently by Corvol et al. in the *Handbook of Proteolytic Enzymes*.[14] This review does not include the three-dimensional structural information recently available, however, it is present in the reviews by Riordan[15] and Acharya et al.[16] Owing to the relevant use of ACE as target drugs for the high blood pressure control, there are a large number of reviews focused on ACE inhibitors and its pharmaceutical use, for example, Sleight,[17] Menard and Patchett,[18] Jackson et al.,[19] Matsusaka and Ichikawa,[20] Antonaccio,[21] Erdos,[22] and Baudin.[23]

The present work is a review of the biochemical information on ACE inhibitors and the enzyme, focusing especially on the achievements of ACE knowledge highly increased in the past 5 years by the availability of new structural information.

32.2 ACE DISTRIBUTION IN TISSUES AND ORGANISM

Two distinct forms of ACE are known in humans, the somatic form (sACE) that is found in many tissues, and the germinal form (tACE) that is found exclusively in testes and plays a role in male fertility.[24,25] The physiological role of ACE is to cleave a single C-terminal dipeptide from AGI to produce the potent vasopressor AGII, but the enzyme also hydrolyze other peptides, such as the vasodilatory peptide bradykinin (Arg-Pro-Pro-Gly-Phe-Ser-Pro-Phe-Arg) that is inactivated by the sequential removal of two C-terminal dipeptides[26]. In fact, bradykinin is the most favorable substrate for ACE with a k_{cat}/K_m value of 3900–5000 $mM^{-1} s^{-1}$, while AGI has only a k_{cat}/K_m value of 147–189 $mM^{-1} s^{-1}$.[14] Thus, ACE is a nonspecific dipeptidase that hydrolyzes the C-terminal dipeptide from peptides where neither an aspartate nor a glutamate is the terminal residue and proline is not in the next position.

In the last 1980s, the first cDNA sequence of the somatic form of the human ACE was determined by Soubrier et al.[27] and the presence of a high degree of internal homology was revealed. The sequence of human sACE has 1306 residues in a single polypeptide chain with a SDS-PAGE estimated molecular weight of about 140 kDa.[28] The polypeptide chain is composed of two domains, the N-domain and the C-domain, and each domain contains an active site with the typical zinc binding motif HEXXH that is found in many zinc peptidases. The two histidines and the downstream glutamate are ligands for the zinc cofactor, required for the peptidase catalytic activity. The high internal homology found in this single polypeptide chain allows to suggest that sACE was the result of a gene duplication event during evolution.[27] Supporting this point of view, tACE has approximately the half of molecular weight of sACE with only 732 residues, which contains a single catalytic site, and the amino acid sequence shows that residues 68–732 of tACE are identical to residues 642–1306 of sACE. Besides, the somatic and germinal forms of ACE mRNA are transcribed from the same gene using alternative promoters,[29] with sACE being transcribed from a promoter region upstream of the duplication and tACE from a promoter within intron 12.[30] Both forms of the protein are type I C-terminal membrane-anchored ectoenzymes, with the hydrophobic transmembrane anchor present in the common C-terminal domain.[31] It consists of a 17-amino acid hydrophobic anchor located 30 amino acid residues from the cytosolic carboxyl terminus.[27] The 17-amino acid residue of the hydrophobic anchor must be structured in a transmembrane-spanning α-helix and its role in the membrane anchoring of ACE has been demonstrated by studying a carboxyl terminal-truncated cDNA ($ACE_{\Delta COOH}$), which is secreted faster in the medium of Chinese hamster ovary (CHO) cells used to synthesize the recombinant protein.[32] The membrane-bound ACE is solubilized by cleavage within its juxtamembrane stalk region through an endogenous proteolytic enzyme, termed ACE secretase or sheddase, that is a zinc metalloprotease and also an integral membrane protein.[33,34] Despite the matching sequence of the COOH-terminal of sACE and tACE, the shedding of sACE from the cytoplasmatic membrane is different from that of tACE. Woodman et al.[35] have demonstrated that the cleavages sites in sACE and tACE, Arg1203-Ser1204 and Arg627-Ser628, respectively, are identical and differences in the shedding must be attributed to another cause. Several studies support that the proteolysis of the juxtamembrane stalk region of the C-domain of ACE to release the enzyme from the cytoplasmatic membrane seems to be regulated by the N-domain of ACE.[36,37] Beldent et al.[36] proposed two alternative mechanism for the N-domain regulation of shedding: one possibility is that the N-domain establish some interactions with the juxtamembrane stalk region of the C-domain, and another is that the amino-terminal domain of ACE affects the conformation of the membrane-bound secretase itself, via protein–protein interactions. Recent studies on ACE phosphorylation show a new possibility of regulation of the shedding process by kinases. Santhamma et al.[38] demonstrated that phosphorylation of a tyrosine located in the distal ectodomain of tACE facilitates the cleavage of the enzyme. However, other authors indicated that in human sACE, the phosphorylated residue is Ser1207,[39,40] which is located in a highly conserved 13-amino acid sequence at the C-terminal end of the enzyme.

A soluble form of ACE is detected in plasma and other biological fluids,[41] but this soluble form has been proved to be the result of the proteolytic cleavage of the membrane-anchoring COOH-terminal of the enzyme using antibody raised against the C-terminal cytoplasmatic domain of s-ACE.[42] The role of this soluble form of ACE is not clear, but abnormal levels of soluble ACE in plasma have been related to several diseases, such as sarcoidosis, diabetes mellitus, Gaucher's disease, leprosy, and hyperthyroidism.[43,44] Moreover, an increase in plasma ACE activity determined soon after the onset of myocardial infarction has been suggested to be an important predictor of the development of left ventricular dilation 1 year after infarction and may identify patients at risk.[45]

A high level of sequence homology among the different somatic mammalian ACEs has been found[14] and this high homology is kept even in ACE-like enzymes found in other nonmammalian species such as the *Torpedo california*[46] and *Torpedo marmorata*[47,48] electric organ membranes, the goldfish *Carassius auratus*[49] and the chicken.[50] Besides, single domain homologous ACE-like enzymes have also been found in insects such as the *Musca domestica*[51] and *Drosophila melanogaster.*[52] Some of the invertebrate ACEs are soluble enzymes as they lack the hydrophobic sequence that anchor the enzyme to the membrane.[53] *D. melanogaster* enzyme has been named AnCE and shares the 42% of the amino acid sequence when compared with the C-domain of human sACE, and this percentage even markedly increase to 70% when compared with the regions surrounding the active site.[54] A second ACE-like protein, termed ACEr, has also been identified in *D. melanogaster.* Both ACE-like proteins appear to be alternatively expressed during *D. melanogaster* pupal development,[55] suggesting different roles for the two enzymes.

In the past decade, new homologues of ACE have been found and of them, the most interesting is ACE2 that was simultaneously discovered by two groups of researchers in 2000. This new ACE homologue was found by screening from expressed sequence tag (EST) database[56] and also by 5′ sequencing of human heart failure ventricular cDNA library.[57] The gene encoding ACE2 is located on the chromosome X and in contrast to the rather ubiquitous ACE, mammalian ACE2 was initially described in the heart, kidney, and testis.[58] Besides the presence of ACE2 in mammalian species, ACE2-like enzymes have also been described recently to be present in nonmammalian vertebrates.[59] ACE2 is a single domain ACE homologue containing only 805 residues and, like ACE, is a type I integral membrane protein. Both proteins share 61% of their sequences in the catalytic site that contains a single HEXXH zinc binding motif, but ACE2 functions exclusively as carboxypeptidase, removing a single residue from the C-terminus of its substrate and some significant differences with ACE can be found in its substrate specificity. Although ACE2 proteolyses AGI and AGII,[56] bradykinin is not a substrate and the enzyme is not inhibited by classical ACE inhibitors such as captopril, lisinopril, or enalaprilat. In the past years, there has been increasing interest to develop drugs that target ACE2 to control the blood pressure using activators of this enzyme.[60,61]

Recently, the Turner's group has found another single domain ACE-like protein that has been called ACE3[62] and it is present in several mammalian genomes, but there is no evidence that ACE3 gene is expressed in humans. Moreover, the presence of deletions and insertions in the sequence indicate that in humans, ACE3 is a pseudogene contained within the ACE gene. In mouse, rat, cow, and dog ACE3, the catalytic Glu is replaced by Gln (HQXXH) in the putative zinc binding motif, indicating that in these species, ACE3 would lack catalytic activity as a zinc metalloprotease.

The availability of the sequence genomes of many bacteria in the past years allows the search of the putative ACE encoding genomic DNA sequences in bacterial species, and this search reveals their presence in a broad range of bacteria, suggesting that ACE is an ancestral enzyme.[63] To study the functionality of this putative bacterial ACEs, Corvol et al. expressed the 2 kb open reading frame encoding a 672-amino acid soluble protein containing a single active site from the phytopathogenic bacterium *Xanthomonas axonopodis pv. Citri* (XcACE). *In vitro* expression and biochemical characterization demonstrate that XcACE is a functional 72 kDa dipeptidyl-carboxypeptidase that can hydrolyze AGI into AGII and is sensitive to ACE inhibitors and chloride ions concentration.

The new genetic techniques also provide a powerful tool to study ACE differential expression and functionality in the diverse organs. Using a promoter-swapping strategy, Shen et al.[64] studied several mouse lines with unique ACE expression patterns. These researchers have discovered many *in vivo* functions for ACE apart from its well-known role in cardiovascular system. For example, ACE expression can influence the immune response to some type of tumors in mice.[65] In addition, these genetically modified mice allow to study whether there are true physiologic differences in the function of the ACE in N- and C-domains (discussed subsequently).[66]

32.3 ACE THREE-DIMENSIONAL STRUCTURE

The first ACE crystal structure was determined in 2003.[67] Due to the large size of the protein, the only technique applicable for structural determination at atomic level is the X-ray diffraction, but the bottleneck of this technique is the difficulty in obtaining single crystals of good quality for the diffraction pattern to emerge, and ACE has several characteristics that make its crystallization very difficult: besides the large size (180 kDa) and the membrane-anchored segment, ACE is heavily glycosylated.[27] To crystallize the ACE, the protein's and carbohydrates must be removed or modified. Acharya's group manages to overcome the oligosaccharides microheterogeneity on the surface of the protein using the expression of ACE in the presence of a glucosidase I inhibitor, *N*-butyldeoxynojirimycin, that inhibits formation of complex oligosaccharide[68] and the mutagenesis of the N-linked asparagine residues to glutamine.[69] At the same time, Kim et al. attempted to crystallize an ACE homologue protein, the *D. melanogaster* ACE (AnCE) isolated from high-expression Hi5 insect cells. This protein lacks the carboxyl terminal membrane-anchoring hydrophobic sequence and has only three N-linked glycosylation sites that are important for its stability;[70] these authors also achieved the X-ray structure determination in 2003.[54]

tACE structure has an ellipsoid-shaped form, and is mainly constituted by α-helices. The most outstanding characteristic of tACE structure is the presence of a central cavity or channel that extends to about 30 Å into the molecule dividing it into two "subdomains": subdomain I and II.[67] Subdomain I is the zinc-containing domain and is composed of residues 40–120, 299–400, and 425–440. Subdomain II is composed of residues 122–298, 408–425, and 439–617. The boundaries of the cavity are provided by helices α13 (375–393), α14 (407–422), α15 (440–472), α17 (521–540), and strand β4 (355–358). The three N-terminal helices that cover the cavity contain several charged residues and restrict the access of large polypeptides to the active site cleft, and this feature of the structure likely accounts for the enzyme's inability to hydrolyze large folded substrates.[71]

It is interesting to remark that although the drugs inhibiting ACE have been developed based on the previous known three-dimensional structures of the carboxypeptidase A and thermolysin, tACE does not show much structural homology with these enzymes. Structural comparison of tACE using the DALI server[72] shows high structural homology with several proteins. Besides the structural similarities of neurolysin and the *Pyrococcus furiosus* carboxypeptidase reported by Sturrock et al.,[71] the structure of several peptidases have been recently shown to have even higher structural homology, such as the *Escherichia Coli* dipeptidylcarboxypeptidase[73] or the thimet oligopeptidase.[74] All the homologous structures found until now correspond to the single domain ACE structures.

The first structure obtained for the N-domain of sACE was determined in 2006.[75] The structures of both domains of ACE have been determined in the presence of the inhibitor lisinopril and, in addition, tACE has also been determined in the presence of enalaprilat, captopril[76] and the selective inhibitor RXPA380.[77] In spite of the advances achieved in the crystal determination of ACE, so far only the crystal structures of single domain ACE have been determined. The details of ACE crystal structures have been given in Table 32.1. The resolution of ACE structures range from 3 to 1.8 Å and the quality of the model structure highly depend on the maximum resolution of the crystal diffraction; hence, the ligand and solvent molecules are modeled more accurately in the higher resolution structures. The known structures show that both N- and C-domains are potential binding sites for the substrates and inhibitors. Analysis of the inhibitor-bound structures show that in all the cases, the inhibitors directly interact with the catalytic Zn^{2+} ion in the active site of the enzyme through the thiol group of captopril, the carboxylate groups of enalaprilat and lisinopril, and the phosphate group in RXPA380.

Comparison of the C-domain and N-domain structures complexed with lisinopril allow the identification of the peptidase subsites S′2, S′1, and S1. A superimposition of the active site residues and lisinopril molecules yields an rmsd of 0.45 Å and most of the residues in contact with the inhibitor molecule are conserved, but some significant differences can be noted that can account for the higher selectivity shown for the lisinopril by the C-domain. The subsite S′2 is formed by the residues that interact with

TABLE 32.1 ACE and ACE Homologue Structures

Structure	PDB Code	Resolution (Å)	Ligand	References
Drosophila AnCE	1J36	2.4	Lisinopril	54
Drosophila AnCE	1J37	2.4	Captopril	54
Drosophila AnCE	1J38	2.6	—	54
tACE	1O86	2.0	Lisinopril	67
tACE	1O8A	2.0	—	67
ACE2	1R42	2.2	—	80
ACE2	1R4L	3	MLN-4760	80
tACE	1UZE	1.8	Enalapril	76
tACE	1UZF	2	Captopril	76
N-domain ACE	2C6F	3.0	—	75
N-domain ACE	2C6N	3.0	Lisinopril	75
tACE G13 mutant	2IUL	2.0	ZN	82
tACE G13 mutant	2IUL	2.0	CL	82
tACE G13 mutant	2IUX	2.8	NXA	82
tACE	2OC2	2.25	RXPA380	77

the prolyl carboxyl terminal moiety of lisinopril. Several aromatic residues are present in this pocket of the enzyme and are conserved in both domains: Tyr520, Tyr523, Phe527, and Phe457 in C-domain and Tyr498, Tyr501, Phe505, and Phe435 in N-domain. A positive charge in the active site has been proposed to stabilize the carboxyl terminal moiety of the inhibitor and Lys511 at the C-domain or Lys489 at the N-domain, placed at 2.93 Å and 2.68 Å, respectively, can account for this role (Fig. 32.1a). Naqvi et al.[78]

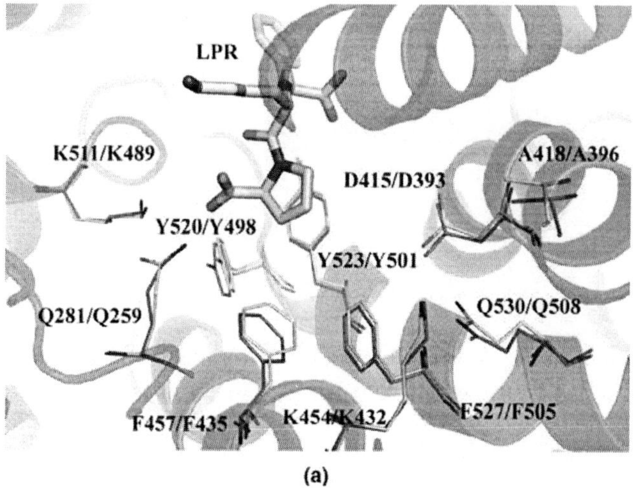

(a)

FIGURE 32.1 Comparison of the active site of C-domain (clear grey) and N-domain (dark grey) ACE lisinopril-bound structures. (a) Subsite S′2 composed of the prolyl carboxyl terminal moiety of lisinopril molecule. (b) Subsite S′1 composed of the lysyl moiety of the lisinopril molecule. (c) Subsite S1 composed of the phenyl moiety of lisinopril molecule.

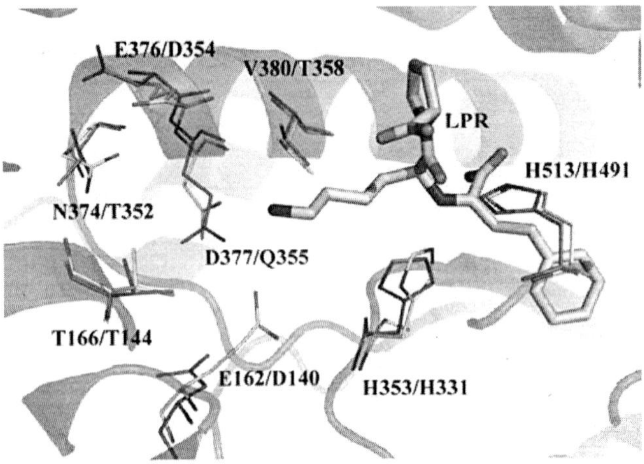

(b)

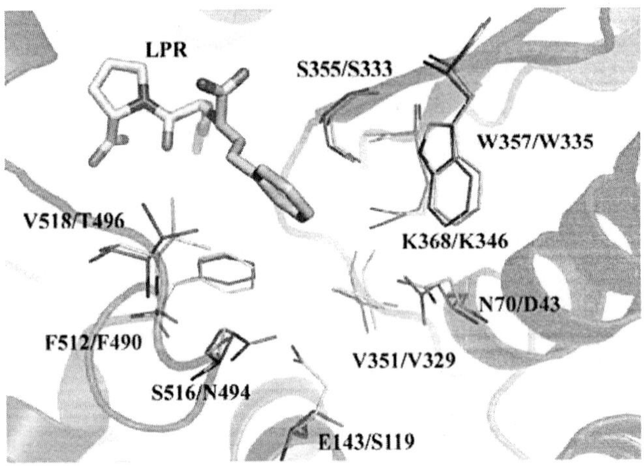

(c)

FIGURE 32.1 (*Continued*)

have studied by direct mutagenesis the role played by Lys511 and Tyr520 in the C-terminal carboxylate docking of the peptide to be hydrolyzed. This C-terminal carboxylate docking is believed to contribute to the cleavage of the C-terminal dipeptide by the stabilization of the transition state and this viewpoint is supported by the experiment conducted with the Lys511Ala/Tyr520Phe double mutant where a >10,000-fold decrease in the loss of hydrolytic activity is observed.

The lysyl moiety of the lisinopril molecule binds to the S′1 subsite, where the Glu162 in the C-domain is forming a salt bridge with the lysyl side chain of lisinopril at 3.45 Å distance (Fig. 32.1b). Glu162 is replaced by Asp140 in the N-domain and the distance to the lysyl chain increases to 6.46 Å. It can be expected that the shorter side chain of Asp140 makes this interaction weaker; in addition, aspartic side chain is oriented in the opposite direction, facing the carbonyl of Cys348 in subdomain I at a distance of 3.6 Å. In addition, in the C-domain, the positive charge of lysyl chain is compensated by the presence of additional acid residues such as Asp377 (4.23 Å) and Glu376 (5.45 Å). In fact, Asp377 is forming a water-mediated interaction with the lysyl group. In addition, some potential hydrogen bonds between the amino acids in subdomain I and amino acids in the subdomain II can be established, such as Glu376 and Asn285 (3.21 Å). In the N-domain, Glu376 is replaced by Asp354 and this aspartic acid is at hydrogen bond distance from Glu262 (3.18 Å) in the subdomain II. This hydrogen bond feature between subdomain I and II can also be found in the structure of the homologous ACE, ACE2, and its structure has been determined in two different conformations: a closet conformation, equivalent to those determined for the N- and C-domain ACE, and an unligated open conformation. In the ACE2 closet conformation, the equivalent residues Thr276 and Asp367 are at hydrogen bond distance (3.5 Å) and in the open conformation of the enzyme, the distance between both residues increase to 11.32 Å. These hydrogen bonds, together with some salt bridges (Glu145–Lys363 in ACE2), are important as they are present in the closet but not in the open conformation and can play some role in conformational changes and stabilization of the closet conformation.

At the active site, there are two histidine residues that can provide the proposed hydrogen bond that link the carbonyl moiety of the inhibitor to the protein placed between the subsite S′1 and S′2, His513 (3.11 Å) and His353 (2.76 Å) in the C-domain and His491 (2.98 Å) and His331 (3.28 Å) in the N-domain. Actually, site-directed mutagenesis studies show that His513 play an important role in the ACE catalysis.[79]

The last subsite in the lisinopril-bound structure is the S1 subsite, formed by the protein residues that interact with the phenyl moiety of lisinopril. This subsite is not clearly defined in tACE as it opens up into a broad cavity where Ser355 and Val518 form the entry residues. The phenyl moiety of lisinopril is placed between both residues that are separated at 9 Å distance (Fig. 32.1c).

When the coordinates of lisinopril:ACE bound structure is compared with those of enalaprilat, captopril, and the highly C-domain-specific phosphinic inhibitor RXPA380,[77] despite the differences in the zinc coordination moieties, the inhibitor coordinating atoms is similarly positioned. Several hydrogen bonds are established between the protein and the inhibitors that can play some role in the specificity of the inhibitor binding. In effect, the binding of RXPA380 to the C-domain involves more direct interactions than the previously determined ACE inhibitor structures, but the specificity of RXPA380 for the C-domain does not appear to depend on hydrogen bonding to the protein, as all of these residues implied are conserved in the N-domain. Corradi et al.[77] proposed that the specificity of RXPA380 by the C-domain rely on the shape and polarity of the active site. In this way, when the active sites of the C- and N-domains are compared, there are three hydrophobic residues Phe391, Val379, and Val380 that are changed by the polar residues Tyr369, Ser357, and Thr358 in the N-domain.

The structure of sACE is still unsolved and the functionality of the two sites present in sACE is yet to be known. However, the structure of the ACE homologue ACE2[80] might give us new insights into the enzymatic mechanism of ACE. Similar to ACE, the single domain ACE2 can be further divided into two subdomains (I and II), which form the two sides of a long and deep cleft in the active site. When ACE2 inhibitor-bound structure is compared with the ACE structures in the N- and C-domains, the largest difference is found in the insertion of a loop between α-helices 7 and 8 in the subdomain II of ACE2. If the insertion loop is not taken into account, an rms deviation of 0.53 Å is obtained for the superposition of tACE and ACE2. The most interesting aspect of this protein structure is the big conformational movement observed upon inhibitor binding. The two subdomains undergo a large inhibitor-dependent hinge-bending movement of one catalytic subdomain relative to the other (~16°) that causes the deep open cleft in the unbound form of the enzyme to close around the inhibitor. In the active site, 13 tACE residues are conserved in the ACE2, and the differences in substrate specificity and inhibitor binding must be attributed to the 8 active site tACE residues that are substituted in ACE2. In fact, when the active site of tACE and ACE2 are superposed, the active site residue Gln281 substitutions by Arg273 in ACE2 appear to eliminate the S′2 substrate binding subsite, and this can explain why ACE acts as peptidyl dipeptidase and why ACE2 has only carboxypeptidase activity.

The open and closet conformations observed in ACE2 have not been observed in the C-domain or N-domain ACE structures, and both domain structures do not reveal any striking difference in the conformation of the inhibitor binding site before and after inhibitor binding. In fact, the unbound form of the C- and N-domains have been solved only in a closet conformation, and this closet conformation can be explained by the evidence that both structures are not a real unbound structure as the fourth coordination position of the zinc active site atom is occupied by an acetate molecule. Indeed, the early structures of thermolysin in the absence of ligands were proven to be bound structures, but the low resolution of the X-ray data precluded the identification of the ligand bound to the active site.[81] Besides, hinge-bending motions have also been proposed for thermolysin and related neutral proteases before and after the binding of the ligand. Watermeyer et al.[82] identified the hinge regions in ACE2 and compared the residues present in these regions with those in tACE and conducted a normal mode analysis with both structures; based on the sequence composition and conservation of the hinge regions, they proposed the conservation of the hinge mechanism among the homologous structures of ACE. Moreover, the studies conducted with tACE reveal a similar hinge motion in both domains of sACE, as tACE is essentially identical to the C-domain of sACE.

Another important structural feature of ACE structures is the presence of chloride sites. The requirement of chloride was first proposed by Skeggs et al.[5] and Bunning et al.[83] demonstrated that AGI cannot be cleaved in the absence of chloride ions. This feature is almost unique among metalloproteases, but the mechanism of the chloride enzymatic activity regulation is still unclear. C- and N-domains of sACE exhibit similar catalytic activities toward AGI, bradykinin, and substance P, but the activity of the C-domain highly depends on chloride ion concentration while N-domain is active even in the absence of chloride and is fully activated to a relatively low concentration of chloride.[32] Also, the inhibitions of the N- and C-domains depend on chloride concentration. The crystallographic structures of tACE (C-domain) and N-domain of sACE show the presence of chloride ions bound to the enzyme. In the case of

C-domain, two sites chloride I and II have been reported, which are bound to Arg186-Arg489-Trp485 and Arg522-Tyr224, respectively. The distances between these chloride ions and the catalytic zinc ion are ~20 and ~10 Å, respectively. Although chloride II site is placed near the active site, no direct interactions between the chloride atom and the inhibitors are observed. The chloride II site is also observed in the N-domain structure, where the chloride ion is ligated to Arg500 and Tyr157, but not to the chloride I site where Arg186 has been replaced by His164. The dependence on chloride concentration has also been reported for the ACE2 enzymatic activity,[58] but in this case, only the chloride I site is occupied and the α-helices surrounding this chloride site show a significant displacement upon inhibitor binding. In the N-domain of sACE, Arg500 is bound to both the chloride ion and Glu389, which is coordinating the zinc atom in the active site, and upon inhibitor binding, the distance between Arg500 and Glu389 is reduced from 4.4 to 3.5 Å.[84] Therefore, upon inhibitor binding, ACE structures show some changes at the chloride sites that might play an important role in the ACE activation. Moreover, mutagenesis data[78] and kinetic studies[85] supported that the chloride linked to Arg1098 (Arg522 at the C-domain) at the subsite S′2 serves to stabilize the substrate P′2 arginine side chain in bradykinin, and an interaction between the anion and the substrate in addition to the anion–enzyme interaction has been suggested.

It is important to note that the differences between the separated N- and C-domains in related reaction rates for AGI hydrolysis and the chloride activation constants for this reaction are not identical to those observed between the domains in full-length sACE and could reflect the contribution of interdomain interactions to the catalytic reaction.[85,86] To clarify this domain interaction in sACE, Corradi et al.[75] have conducted some modeling studies on sACE structure. These authors determined the first structure of the N-domain and then proposed two different models of sACE based on the available crystallographic results. In sACE, N- and C-domains are linked by an interdomain sequence of about 17 residues (residues 612–628) that is susceptible of proteolysis by endoproteinase Asp-N.[87] This linker was included in the N-domain construct to visualize this region in the three-dimensional structure and allow the modeling of the whole sACE. The linker is placed next to the β-hairpin formed by residues 126–138 that shows a high flexibility. Furthermore, there is an N-glycosylation sequon in this flexible loop of the N-domain and in the top part of the C-domain lid; Corradi et al.[75] proposed that interdomain interactions and movements may be mediated or aided by sugars.

It is well known that oligosaccharides on glycoproteins play a variety of roles in processes such as protein secretion, specific recognition of other glycoproteins, protection of glycoproteins against proteolytic degradation, and the insertion or proper orientation of glycoproteins.[88] As we noted before, ACE is extensively glycosylated and the number and position of sites of glycosylation and the structure of the oligosaccharide chains depend on the ACE source. sACE sequence contains 17 potential N-linked glycosylation sites but only 7 of these sites were found to be glycosylated in human sACE,[89] the asparagine residues 9, 82, 117, 480, 913, 1196. Although no Ser/Thr-rich region indicative of O-glycosylation has been found in sACE,[27] the unique N-terminal sequence of 36 amino acid residues is heavily

O-glycosylated in tACE.[90] In addition, six N-linked glycosylation sites have been identified in human tACE by matrix-assisted laser desorption ionization/time-of-flight/mass spectrometry of peptides generated by proteolytic and cyanogen's bromide digestion.[68] These glycosylation sites were found at the asparagine residues 72, 90, 109, 155, 337, and 586.[68] Deglycosylation of mature sACE and tACE has been shown that does not abolish its enzymatic activity[90,91] and the specific role of these carbohydrates in ACE is not known. Studies conducted with unglycosylated or partially glycosylated ACEs suggest that glycosylation plays an important role in the membrane targeting and release of ACE, possibly by affecting the folding of the polypeptide and its recognition by a variety of enzymes in the folding and transport machinaries.[92] Moreover, Sadhukhan and Sen[93] reported that mutations at individual N-linked glycosylation sites in rabbit tACE resulted in different efficiencies in enzyme release, which suggests that N-linked glycans at each site may make different contributions to ACE transport and release.

32.4 KINETIC AND MOLECULAR BASIS OF ACE INHIBITION

As noted earlier, ACE catalyzes the last step in the synthesis of the vasoconstrictor AGII from AGI and the metabolic inactivation of the vasodilator bradykinin. In both cases, ACE acts as a peptidyl dipeptidase removing the C-terminal dipeptide from the substrate, but ACE can also act as an endopeptidase on some substrates, such as substance P,[94] cholecystokinin,[95] enkephalins,[31] and luliberin (LHRH, luteinizing hormone-releasing hormone)[96], that are amidated at the C-terminus. Several researches[94,96–98] show that for substrates with amidated C-termini, ACE not only displays exopeptidase activity but also acts as an endopeptidase. Given that many metalloproteinases closely related to ACE, such as neprilysin and endothelin-converting enzyme, are endopeptidases, it is likely that ACE evolved from an endopeptidase ancestor. In fact, the most ancestral ACE enzyme described to date is the bacterial XcACE that, as human ACE, hydrolyzes a broad range of substrates and, in addition to the dicarboxypeptidase activity, also acts as an endopeptidase.[63] Therefore, ACE might have a more general role in the metabolism of biologically active peptides.

The two domains of sACE are highly homologous in sequence and both bear a functional catalytic site,[32] but each domain displays some unlike enzymatic properties. For example, the rate of hydrolysis of AGI and substance P is faster in the C-domain as compared to the N-domain, but the N-terminal cleavage of luliberin is performed faster by the N-domain active site.[99] Besides, the hemoregulatory peptide *N*-acetyl-Ser-Asp-Lys-Pro (AcSDKP) is hydrolyzed more efficiently by the N-domain.[100] The selectivity can be also extended to the inhibitors. A detailed study of Wei et al.[101] with the whole sACE and mutants containing only one domain (N- or C-domain) and the broadly used ACE inhibitors captopril, enalaprilat, lisinopril, and trandolaprilat show that the relative potency of these ACE inhibitors is trandolaprilat > lisinopril > enalaprilat > captopril for the C-domain and trandolaprilat > captopril > enalaprilat > lisinopril for the N-domain. Recently, high

TABLE 32.2 Binding Constant of ACE Inhibitors

Domain	Substrate	[Cl$^-$] (mM)	K_I (M^{-1}) Lisinopril	K_I (M^{-1}) Captopril	K_I (M^{-1}) Enalaprilat	
sACE	Hip-His-Leu	300	2.7×10^9	7.7×10^8	1.5×10^9	101
N-domain			2.3×10^8	1.1×10^9	3.8×10^8	
C-domain			4.2×10^9	7.1×10^8	1.6×10^9	
N-domain		20	2.4×10^8	1.1×10^9	3.2×10^8	
C-domain			3.7×10^8	9.0×10^7	1.3×10^8	
sACE	Hip-Lys-Pro	20	2.6×10^8	2.4×10^8		158
N-domain			2.6×10^8	5.8×10^7		
C-domain			1.4×10^8	6.0×10^7		
sACE		100	5.8×10^8	9.3×10^8		
N-domain			5.8×10^8	2.3×10^8		
C-domain			2.1×10^9	2.8×10^8		
sACE	Hip-His-Leu	300	2.6×10^9	2.9×10^8		
N-domain			1.5×10^8	2.4×10^9		
C-domain			3.4×10^9	4.8×10^8		
SACE	AcsDKP	50	4.2×10^9	4.2×10^9		
N-domain			1.9×10^9	1.2×10^{10}		
C-domain			3.7×10^9	8.8×10^8		
sACE	AngI	50	2.8×10^9	2.5×10^8		
N-domain			4.9×10^8	7.6×10^7		
C-domain			5.1×10^8	7.0×10^7		
SACE	Cbz-Phe-His-Leu	150	5.6×10^9	1.7×10^9		105
N-domain			5.0×10^9	2.0×10^9		
C-domain			8.3×10^9	1.1×10^8		
sACE	ITC	300	1.9×10^8	7.5×10^7	6.2×10^8	84

Enzymatic assays were performed at 37°C and ITC binding constant at 25°C.

specificity N- and C-domain inhibitors have been synthesized: the RXP407 that is an N-domain-specific inhibitor[102] and the RXPA380 that selectively inhibits the C-domain.[103] Table 32.2 shows the values obtained for the binding constant to sACE of several inhibitors broadly studied. Most of the inhibitor constants reported in the bibliography for ACE have been determined by means of kinetic measurements that have been developed in the presence of different sodium chloride concentrations and substrates, but recently the measurement of ACE inhibitor constants by a thermodynamic method allows a more accurate determination of the value of the binding constant and the characterization of the thermodynamic driven force of the binding process.[84] The comparison of the binding constant determined by means of calorimetric techniques with the values obtained by means of kinetic methods shows that the values obtained by means of enzymatic measurements are one order of magnitude higher than those values obtained by means of the displacement method using isothermal titration calorimetry (ITC), but this discrepancy can be attributed to different experimental conditions.[84] The binding of captopril, enalaprilat, and lisinopril to sACE[84,104] was characterized by a positive enthalpy and entropy, which indicates that the binding of these inhibitors was entropically driven. The three inhibitors studied show almost the same positive enthalpy change upon binding, and enthalpically favorable interactions, as hydrogen bond formation, could play a role in the specificity of the binding process. However, the contribution to the energy of binding of the hydrogen bonds established between the inhibitors and the residues at the active site of the enzyme is overcome by the entropic contribution. The large contribution of entropy is indicative of hydrophobic effect, where the increase of water disorder after ligand binding can play an important role. In addition, ITC is an accurate technique to determine the stoichiometry of binding in tight binding process and the experiments conducted with three strong ACE inhibitors (lisinopril, enalaprilat, captopril) yields a stoichiometry of only 1 mol of inhibitor per mol of monomer of sACE.[104,105] Moreover, the large negative heat capacity obtained in the ITC measurements[84] indicates a decreased exposure of hydrophobic surface, as occurring in the burial of surface within a protein interface. In addition, studies performed by Skirgello et al.[86] reveal that the kinetic analysis of the simultaneous hydrolysis of two substrates by sACE demonstrated a competition between these substrates for binding to sACE and the binding of a substrate at a unique active site makes the other one unavailable for either the same or a different substrate.

To understand the *in vivo* role of the two ACE domains, Fuchs et al.[66] used gene targeting to create mice with point mutations in the ACE C-domain zinc binding motif and the results of this study shows that the C-domain of ACE was the predominant site of AGI cleavage *in vivo*. The molecular bases of this finding and the stoichiometry found in the titration experiment would be explained by the postulated hinge motion between the closet–open conformations and a strong negative cooperativity between the two sites.

In addition, the N- and C-domains also show differences in the dependence of the enzymatic activity on chloride ion concentration,[32,101] but its physiological significance is not well understood yet. Primarily, chloride ions produce an enhancement of substrate binding and studies with several ACE inhibitors indicate that while the N-domain inhibition seems to be unaffected by the sodium chloride concentration, the C-domain is highly affected and the K_I of the inhibitors can show changes of one order of magnitude.[101]

In spite of the large amount of kinetic data collected for many years, the lack of ACE structures has limited the studies about the mechanism of the enzyme and the catalytic mechanism of ACE is yet to be elucidated. However, the catalytic mechanism proposed for ACE is based upon other zinc metalloenzymes with the same HEXXH metal binding motif. In fact, ACE belongs to the gluzincin family (clan MA) of metalloproteases, of which thermolysin is the prototypical member.[106] In brief, the mechanism proposed for thermolysin catalysis involves the displacement of a zinc-bound water molecule by the approaching substrate followed by the attack of this water molecule on the scissile carbonyl bond to form an oxyanion. The attack of the water molecule is facilitated by an active site glutamic acid, Glu143, and the resulting tetrahedral intermediate subsequently decomposes to yield the products. The negative charge on the scissile bond carbonyl oxygen that develops during transition state binding is stabilized by hydrogen bonding interactions with a protonated histidine, His231, and a tyrosine in the active site, Tyr157. The protonated histidine keep its state

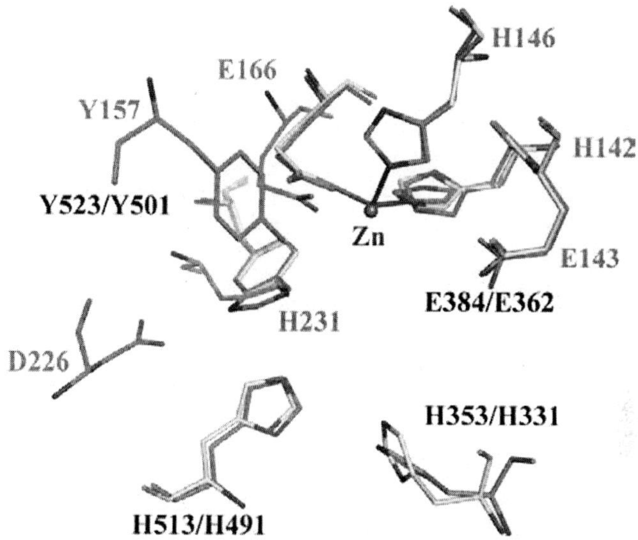

FIGURE 32.2 Active site comparison of C-domain (yellow) and N-domain (cyan) ACE structures with the active site of the thermolysin (green). (See the color version of this figure in Color Plates section.)

through a hydrogen bonding interaction with an aspartic residue, Asp 226. The crucial stabilization of the oxyanion by His231 occurs after the formation of the Michaelis enzyme–substrate complex and greatly influences catalytic rate.[107,108]

On the basis of the role proposed for active site residues in the thermolysin, a superposition of the active site of thermolysin and C- and N-domain ACE structures will allow us to identify the potential equivalent residues in the active site of ACE and propose a hypothetical catalytic mechanism (Fig. 32.2). The HEXXH metal binding motif is formed by HEMGH, residues 383–387/361–365 (residues in the C-domain/residues in the N-domain), and the glutamic acid coordinating the zinc atom is Glu411/Glu389. The active site zinc ion shows a tetrahedral coordination, where the fourth ligand is an oxygen atom of an acetate molecule in the unbound form, or the carboxyl coordinating group of the inhibitor lisinopril in the lisinopril-bound structure. The inhibitor lisinopril closely mimics the transition state configuration of the scissile bond. Glu384/Glu362 is placed at the same position of Glu143 that acts as a general base during catalysis abstracting a proton from the metal-bound water and shuttling this proton to the peptide amide nitrogen of the substrate. Protonation of this amide nitrogen makes it a better leaving group, thereby facilitating cleavage of the amide bond. At the position occupied by His231 in the thermolysin structure, no histidine residue is found in C- or N-domain ACE structures. In thermolysin, the NE2 atom of His231 is positioned at a distance of 4.23 Å from the Zn atom and has been proposed to play a role in the transition state stabilization as hydrogen bond acceptor that would stabilize the tetrahedral transition state. Instead, a histidine residue Tyr523/Tyr501 has been found at the same position at a distance of 4.4 Å from the Zn atom and it can play the role of thermolysin Tyr157, stabilizing the tetrahedral intermediate. Besides,

Tzakos et al.[109] proposed that Tyr523/Tyr501, in the presence of chloride ions, induces the stabilization of the transition state by locking Arg522 away from Asp465. The presence of a tyrosine residue instead of histidine in the position occupied by His231 in thermolysin is a common feature in all members of ACE family. Moreover, a conserved histidine residue is found at the active site, as well as in ACE structural homologues such as neurolysin and *P. furiosus* carboxypeptidase. This histidine residue corresponds to His513/His491, with His513 (His1089) being the same histidine residue as proposed by Fernandez et al.,[79] to be equivalent to His231 as shown by direct mutagenesis studies. The mutation of His513 to Ala or Leu in the human ACE C-domain causes a small increase in the Michaelis constant for the substrate AGI at the same time that produces a large decrease in the catalytic constant. The loss in transition state binding energy in both mutants can be correlated with the loss of a strong hydrogen bond in the tetrahedral intermediate and equivalent values have been found for the same mutation in a thermolysin-like enzyme.[110] Studies on the

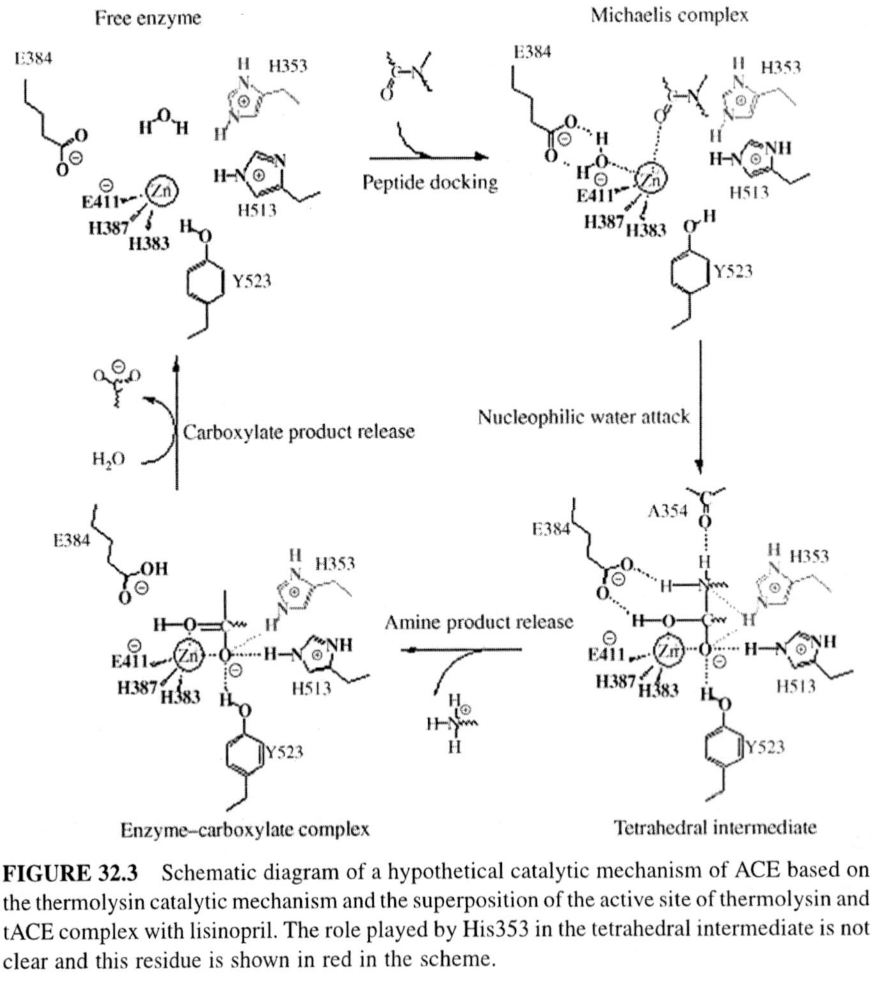

FIGURE 32.3 Schematic diagram of a hypothetical catalytic mechanism of ACE based on the thermolysin catalytic mechanism and the superposition of the active site of thermolysin and tACE complex with lisinopril. The role played by His353 in the tetrahedral intermediate is not clear and this residue is shown in red in the scheme.

pH dependence of the enzymatic activity of the His513 Ala mutant of C-domain of human ACE demonstrated that His513 is the major determinant of the alkaline pH optimum of the enzyme.[79] His513 is a part of the subsite S1′ and it is hydrogen bonded to the carbonyl of lisinopril, which is also linked through the strong hydrogen bond to His353 (His331 at the N-domain) at the same subsite of the enzyme. The distance between His513 and Tyr523 is hardly the same as that between His231 and Tyr157 in thermolysin, so it is probable that they exert the same role in the active site but with a different orientation. A schematic diagram of a hypothetical catalytic mechanism of ACE based on the thermolysin catalytic mechanism is shown in Fig. 32.3. The role played by His353 in the tetrahedral intermediate is not clear, but the structure of the tACE complex with lisinopril shows that the residues Tyr523, His513, and His353 are hydrogen bond distance of lisinopril O1 atom and can stabilize the tetrahedral transition state. Besides, lisinopril N1 atom is hydrogen bond distance of carbonyl oxygen atom of Ala 354 (2.92 Å) and His353 (3.24 Å) and both residues can play some role in the stabilization of sp^3 nitrogen in the transition state, but this hydrogen bond feature is different in the N-domain, where the equivalent His331 cannot form this interaction. All the residues implied in this mechanism are conserved in the N-domain of ACE and AnCE structures. In ACE2, the only exception is that hydrogen bond to the nitrogen of the scissile bond is provided by the carbonyl oxygen atom of Pro346.

It is unquestionable that the three-dimensional structure of ACE now paves the way for the solution of the catalytic mechanism of the enzyme and this knowledge will help the rational design of a new generation of ACE inhibitors.

32.5 THE RENIN-ANGIOTENSIN SYSTEM

To better understand the pharmacological use of ACE inhibition, we must take into account the renin-angiotensin-aldosterone neurohormonal system (RAS). Many discoveries have helped in the understanding of the RAS since several researchers between 1930s and 1940s described that the relationship between kidney and blood pressure was the result of a complicated enzymatic hormonal system, where the renal enzyme renin and the vasopressor substance angiotensin play a key role. RAS participates in the homeostatic control of arterial pressure, tissue perfusion, and extracellular volume through the formation of the active hormone AGII by sequential proteolytic cleavage of its precursors. In this way, a decrease in the formation of the AGII and its blockage in the RAS by angiotensin receptor blockers (ARBs) are central to most strategies to high blood pressure treatment.

In brief, the mechanism of this complex system begins with the initial RAS activation that results in renin synthesized from both renal and extra-renal tissues. The system is activated in response to stimuli such as glomerular hypoperfusion, sympathetic stimulation, and decreased delivery of chloride anion to the juxtaglomerular cells of the renal macula densa. In the first synthetic step, renin catalyzes the rate-limiting formation of AGII, in which hepatically synthesized α2-globulin angiotensinogen is converted into decapeptide AGI. In the second step, ACE catalyzes the conversion of AGI to AGII. The octapeptide AGII is recognized by two G-protein-coupled receptors

(GPCRs), AT_1 and AT_2. Although AT_1 is responsible for most of the known actions of AGII, such as vasoconstriction, facilitation of sympathetic transmission, stimulation of aldosterone release, and promotion of cellular growth, much less is known about the function of the AT_2 receptor, but recent studies suggest that it may play a role in mediating antiproliferation, cellular differentiation, apoptosis, and vasodilatation.

The significance of the study of RAS is that several genetic studies have shown gene polymorphism–drug interactions. A better knowledge of this system can help in the most satisfactory drug prescription for high blood pressure treatment.[111] In the past years, significant progress in knowledge regarding the RAS has been made, including the discovery of novel bioactive peptides, additional specific receptors, alternative pathways of AGII generation, and additional roles for precursor components other than AGII synthesis.[112–114] ACE2 discovery in 2000 introduces a new level of complexity in the RAS. ACE2 differs from ACE in their substrate affinity and although ACE activity leads to the formation of the vasoconstrictor AGII, ACE2 seems to be involved in the formation of vasodilator mediators.

32.6 ACE INHIBITORS AND PHARMACEUTICAL USE

ACE inhibitors have a broader use than blood pressure control. These inhibitors not only play a crucial role in the treatment of hypertension by decreasing the AGII production but can also be used in the therapy of patients with left ventricular dysfunction, postmyocardial infarction, diabetes mellitus, and renal disease.[10] Since Cushman and Ondetti[115] proved in the early 1970s that ACE activity was inhibited by several natural peptides, many antihypertensive drugs have been developed. The chronic condition of the disease may be preferably treated using drugs that can be taken orally by the patient. To avoid the degradation by proteases, Ondetti and Cushman developed a new approach, the synthesis of ACE inhibitors designed as synthetic pseudopeptides. The studies conducted with the inhibitors from the venom of *Bothrops jararaca* led to an active site model to design new and strong ACE inhibitors and this design was so successful that the first pseudopeptidic inhibitor, captopril, is still one of the most used antihypertensive drugs.

The most important interaction of ACE inhibitors with the active site of the enzyme relies on the contact between a strong chelating group and the zinc atom at the active site. ACE inhibitors classification is done on the basis of the type of zinc-chelating group, and those used as drugs belong to the following three classes of reactive compounds: sulfhydryl-containing inhibitors, dicarboxyl-containing inhibitors, and phosphorus-containing inhibitors. Tables 32.3–32.5 show the three different classes of ACE inhibitors compiled in the PUBMED/Medline database. Most of these inhibitors show nanomolar binding constants and are processed after oral intake to obtain the active form that inhibits ACE. By analogy with the carboxypeptidase A and from the results of the study of ACE activity in the presence of different substrates and peptide-based inhibitors, Cushman and Ondetti developed a simplified sketch of the active siteof ACE using the Schechter and Berger nomenclature for the active site of proteases[116] and the subsites, as indicated in the tables.

TABLE 32.3 Sulfhydryl-Containing ACE Inhibitors

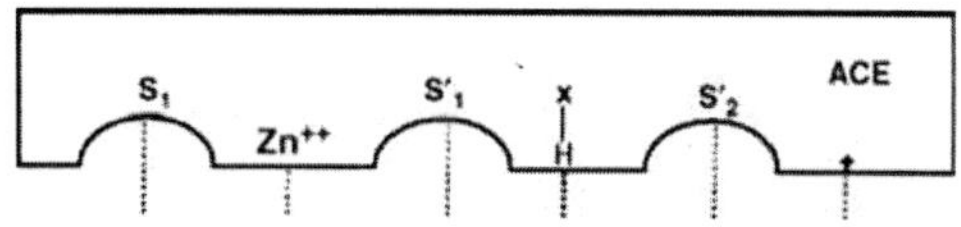

Alacepril	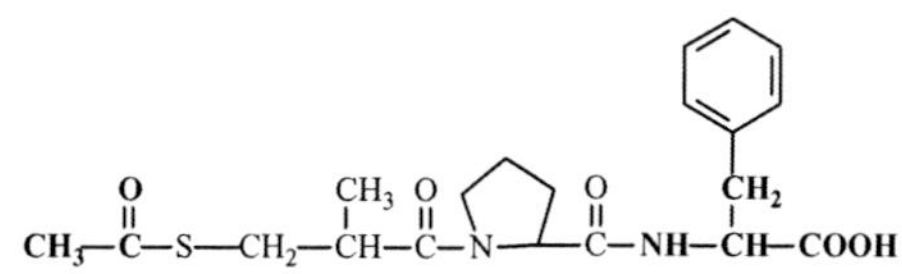
Altiopril	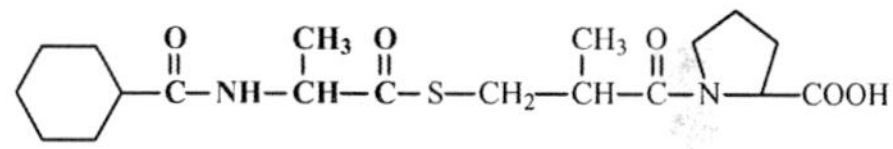
Captopril	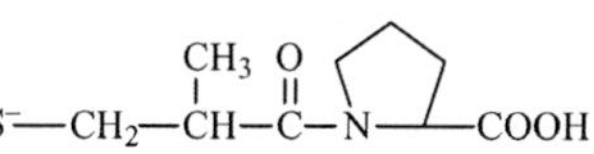
Pivalopril	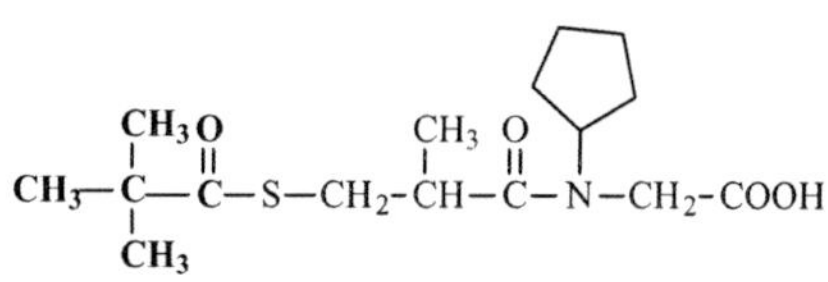
Rentiapril	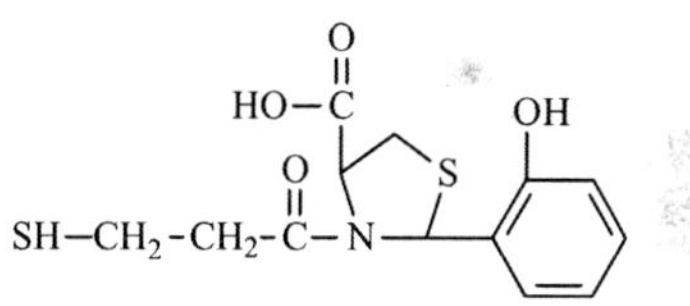
Zofenopril	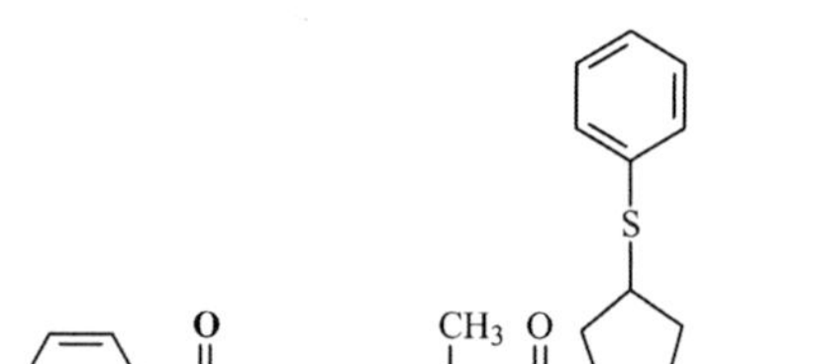

Most of the ACE inhibitors have been reported to be able to bond with both domains; however, in the past years, there has been increasing interest in a new class of ACE inhibitors called domain-selective ACE inhibitors. They do not have a clinical application yet, but we have included them because of their significance in the study of the bases of the domain selectivity.

ACE inhibitors have different pharmacokinetic profiles and the physician must take into account the physical condition of the patient to select the most appropriate drug for the hypertension control.

TABLE 32.4 Dicarboxyl-Containing ACE Inhibitors

	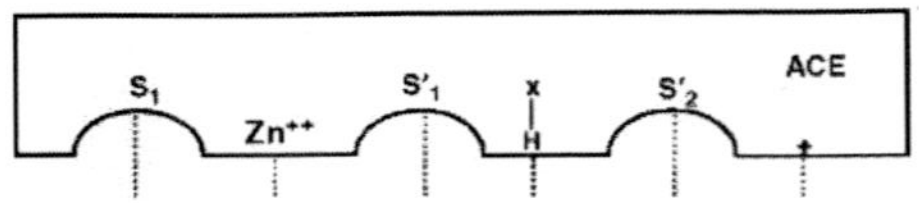		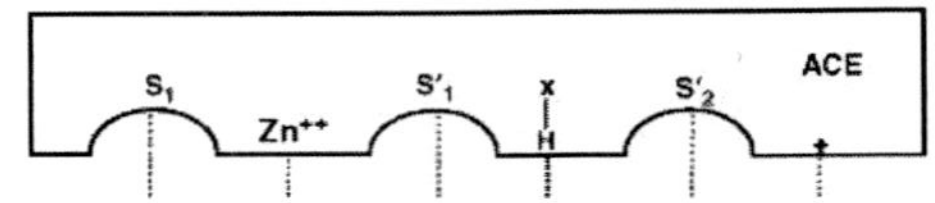
Benazepril	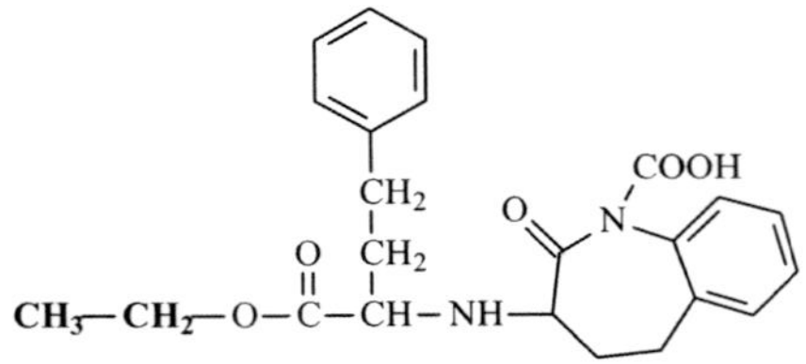	Pentopril	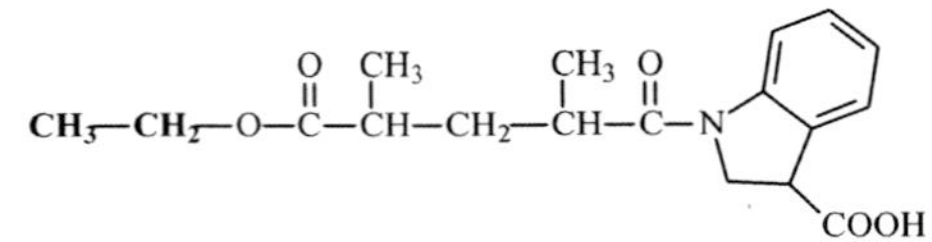
Cilazapril	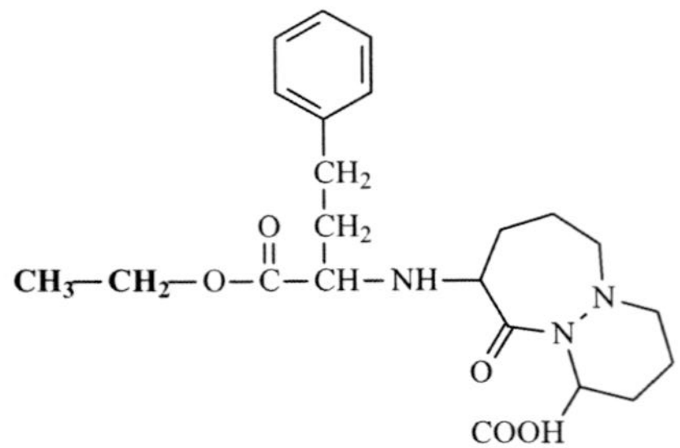	Perindopril	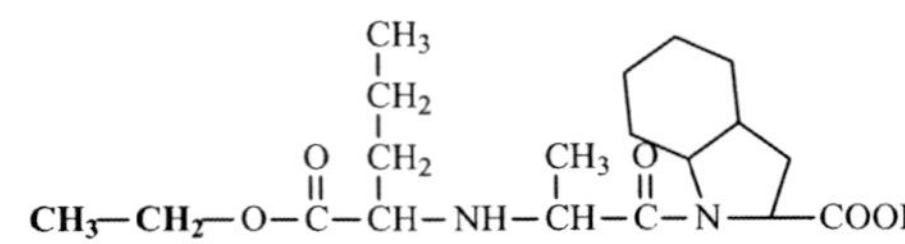
Delapril	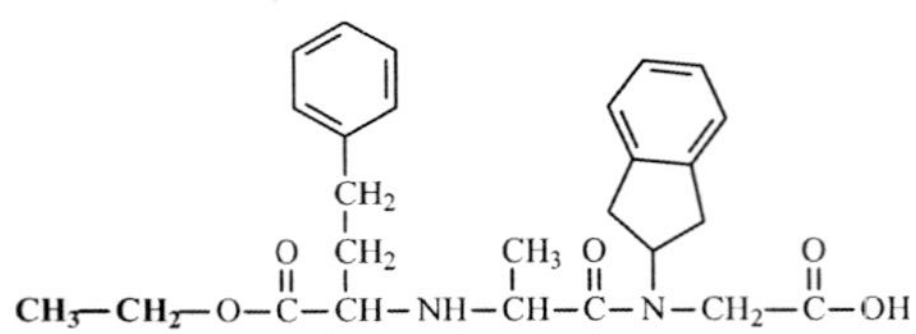	Quinapril	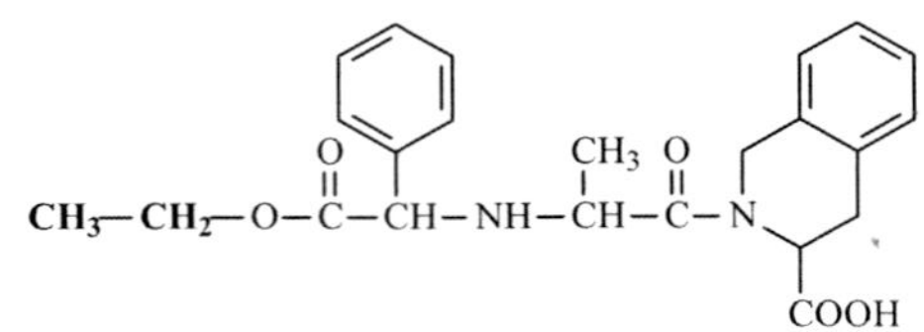
Enalapril	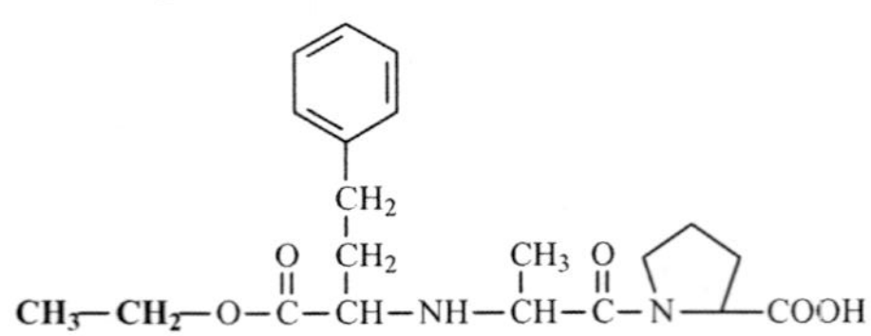	Ramipril	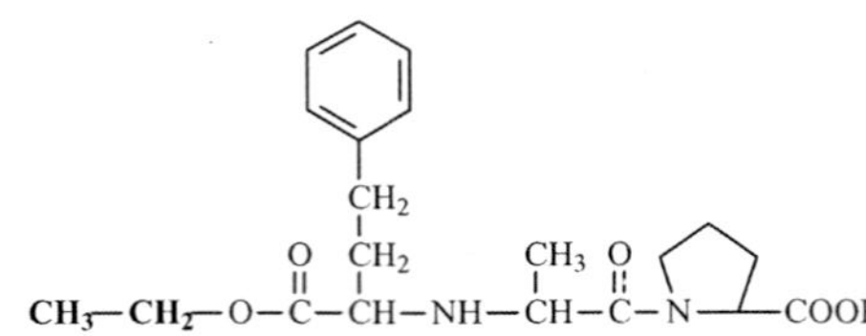

Imidapril	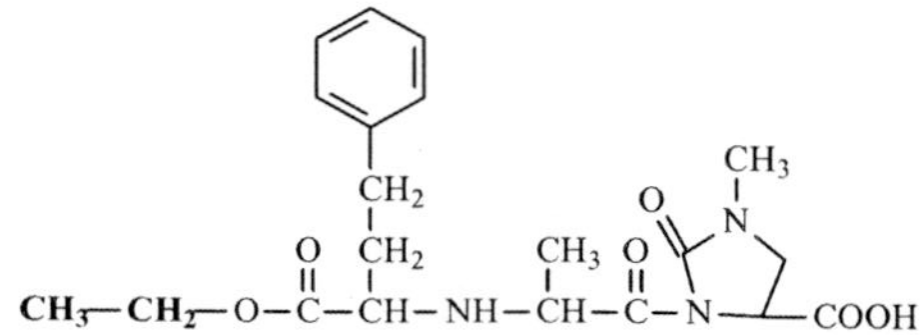
Libenzapril	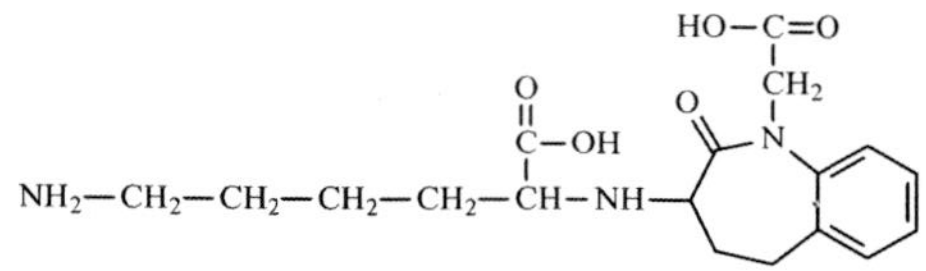
Lisinopril	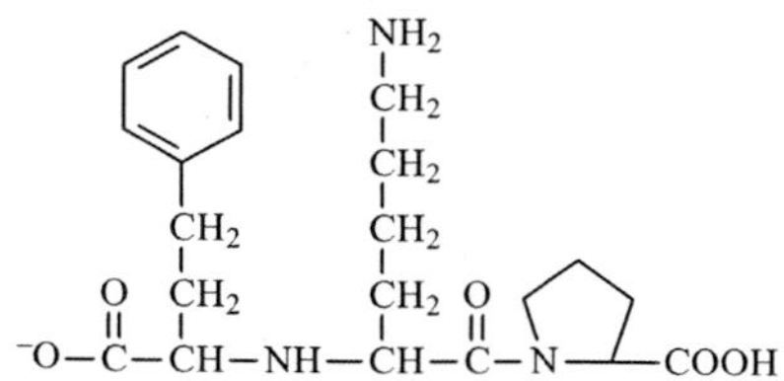
Moexipril	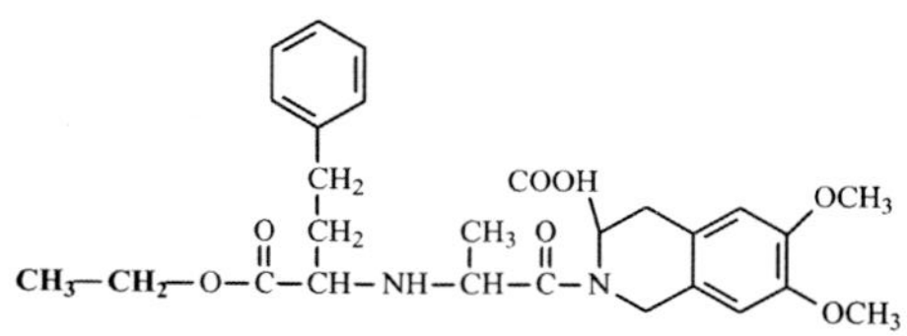
Omapatrilat	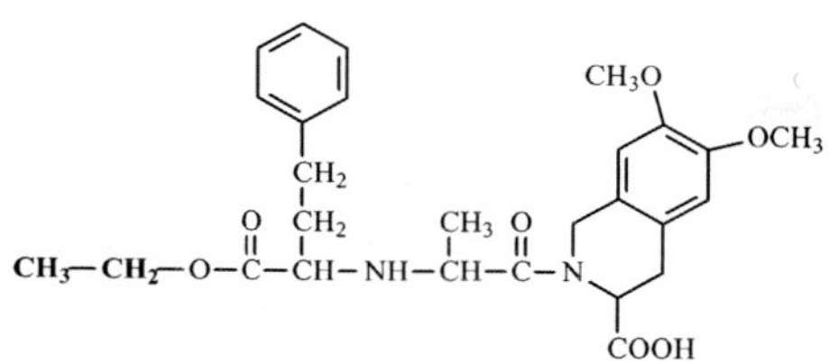
Spirapril	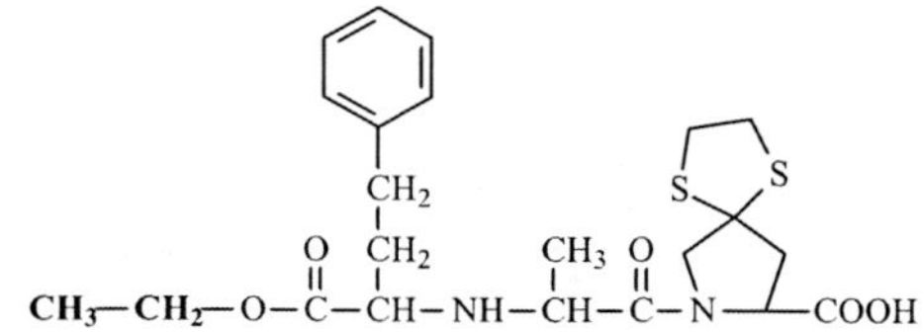
Telmisartan	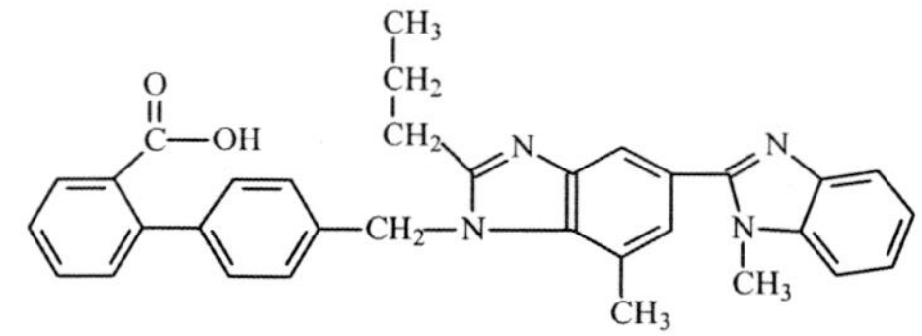
Temocapril	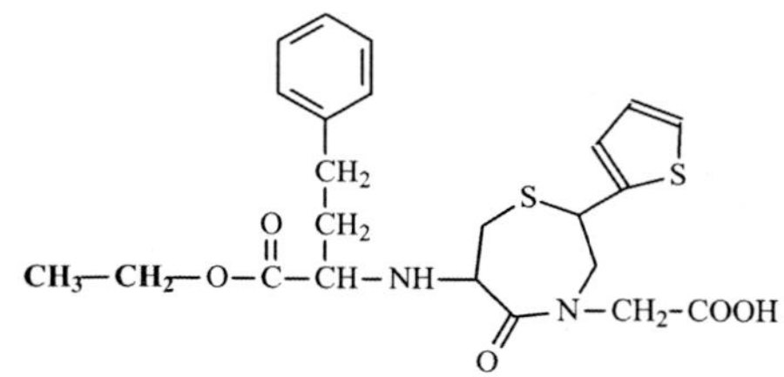
Trandolapril	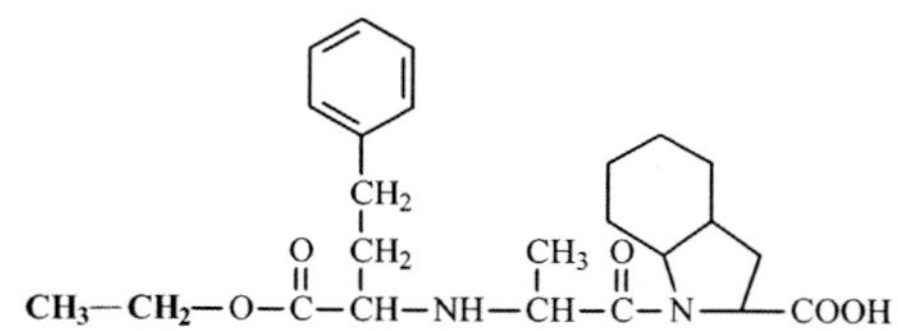

TABLE 32.5 Phosphorous-Containing ACE Inhibitors

	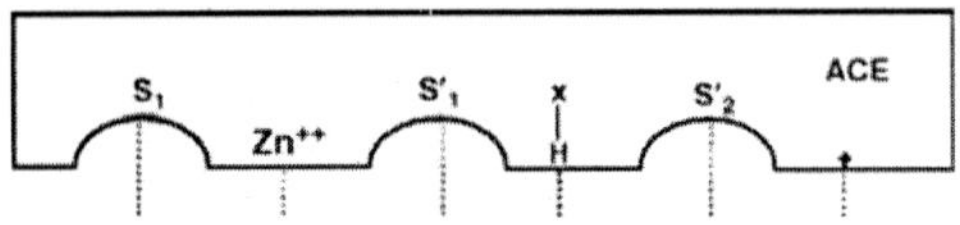
Ceranapril	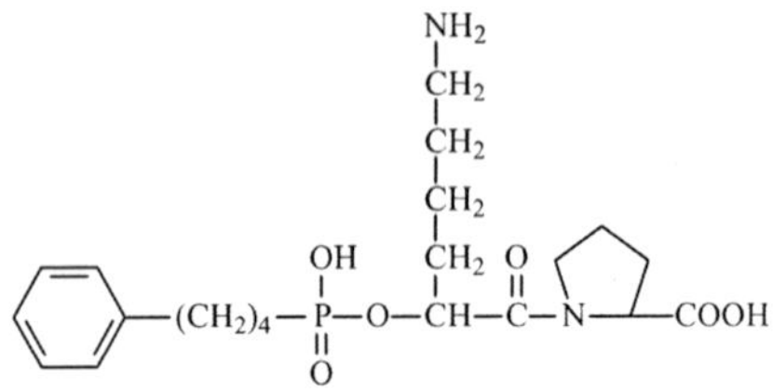
Fosinopril	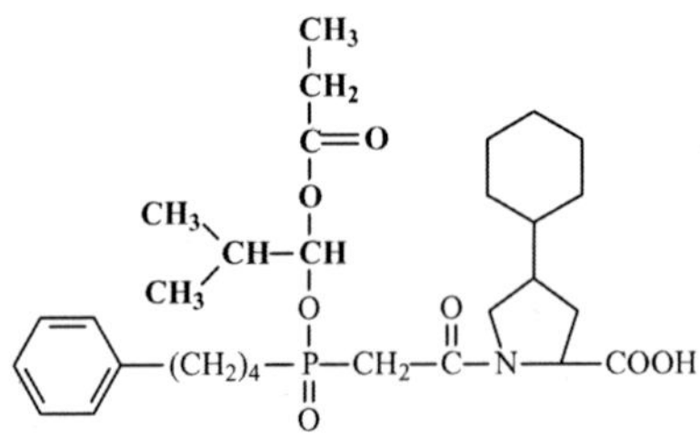
RXPA380	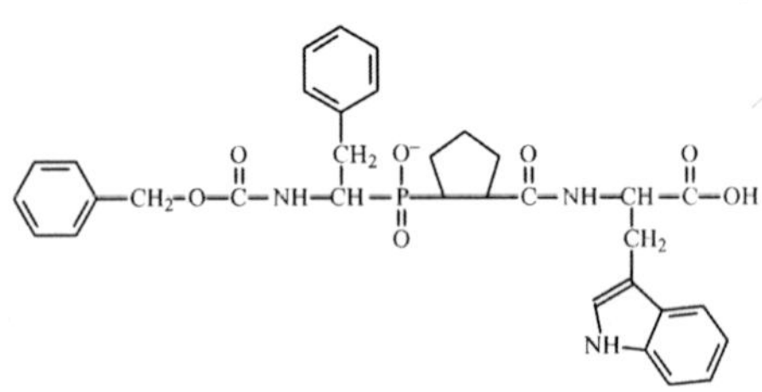
RXP407	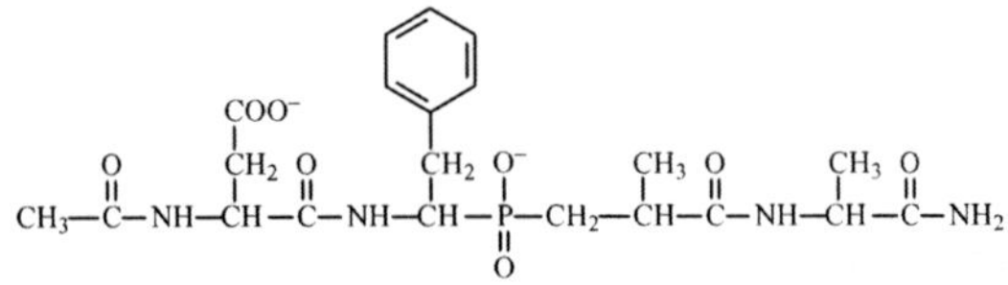

32.6.1 Sulfhydryl-Containing ACE Inhibitors

Captopril was the first ACE inhibitor designed for commercial use and also the simplest because it is only comprised of the P1′ and P2′ residues and a thiol zinc binding group. Other commercial thiol-based inhibitors are alacepril, altiopril, rentiapril, pivalopril, and zofenopril. All the newer sulfhydryl-containing ACE inhibitors differ from captopril since they are prodrugs, and among them alacepril and probably altiopril are converted to captopril *in vivo*.[117] In fact, alacepril forms captopril as final product *in vivo* through an active intermediate metabolite, desacetylalacepril, which has sympathoinhibitory action and potentiates the antihypertensive effect of alacepril.[118]

In altiopril, the hydrolysis-resistant masking of the mercapto group is obtained by sterification[119] and it is approximately 23 times less potent than captopril.

Rentiapril is characterized by higher lipid solubility, and in contrast to captopril, it has also been demonstrated to be taken up by brain tissue following oral administration.[120]

Zofenopril shows a high inhibitory effect on ACE together with significant tissue selectivity and a long duration of action. Orally administered zofenopril is rapidly converted into the active zofenoprilat. Other properties shown by this drug, such as antioxidant activity and cardiovascular protection, make zofenopril potentially suitable for the treatment, and possibly prevention, of several cardiovascular diseases.[121] The use of such sulfhydryl-containing ACE inhibitor is limited because of its adverse effects, for example neutropenia, nephrotic syndrome, and skin rash,[122–124] that appear to be related to the presence of the sulfhydryl group in these ACE inhibitors.[125]

The sulfhydryl-containing inhibitors may exert cardiac effects due to the presence of the thiol moiety. It has been suggested that the sulfhydryl group acts as scavenger of free radicals and/or as antioxidant agent.[126] These actions may play an important role after myocardial infarction because of their beneficial effects on reperfusion-induced depression of contractility and reperfusion arrhythmias.[127,128] Another effect of sulfhydryl group is vasodilatation, caused by a direct effect or potentiation of nitrovasodilatators.[129]

32.6.2 Dicarboxyl-Containing ACE Inhibitors

In general, carboxyl-containing inhibitors are more potent than captopril, but may have a limited bioavailability.[130] This problem has been largely overcome by developing prodrugs, which have to be converted into the active form. The bioactivation takes place principally at hepatic level by means of enzymatic hydrolysis and deesterification that will liberate the metabolite responsible for the therapeutic action. The active metabolites have a higher liposolubility, which implies a better facility to reach the organs of interest.[131]

Enalapril was the first dicarboxyl-containing ACE inhibitors designed and it was introduced in 1986 for hypertension treatment. Enalapril, also known as MK-421, is an esterified prodrug, which is converted by the liver to the bioactive potent ACE inhibitor enalaprilat and its excretion is primarily renal.[132] In case of extensive liver damage, enalaprilat is available for intravenous administration. Closely related to enalaprilat, lisonopril bears a lysil moiety, but in this case, it is the active form drug and does not undergo metabolism and is excreted entirely unchanged in the urine. It is one of the few dicarboxyl-containing ACE inhibitors that act as such. Most of the dicarboxyl-containing ACE inhibitors have the same mechanism as that of enalapril. In fact, ramipril,[133] cilazapril,[134] trandolapril,[135] imidapril,[136] spirapril[137] benazepril,[138] quinapril,[139] delapril,[140] and perindopril[141] are converted into ramiprilat, cilazaprilat, trandolaprilat, imidaprilat, spiraprilat, benazeprilat, quinaprilat, delaprilat, and perindoprilat, respectively, by hydrolysis of the ester group to form the biologically active metabolite. Most of the diacid metabolites are potent inhibitors of ACE activity, whereas prodrug has almost no effect, showing that conversion of the parent molecule is essential to obtain the pharmacological activity in these drugs. In addition, some of these drugs are unsuitable for oral administration because they are poorly absorbed, for example, benazepril.[142]

Because ACE and neural endopeptidase (NEP) share some common catalytic mechanisms, it has been possible to design dual NEP/ACE inhibitors, also called vasopeptidase inhibitors.[143] Omapatrilat is a potent conformationally constrained peptidomimetic vasopeptidase inhibitor with a similar nanomolar inhibitory constant for both NEP and ACE.[144] Combined NEP and ACE inhibition has been proposed as a new therapeutic strategy to hypertension and congestive heart failure.

Other dicarboxylic ACE inhibitors are included in Table 32.4.

32.6.3 Phosphorus-Containing ACE Inhibitors

The potencies of phosphorus-containing ACE inhibitors are comparable with captopril but they have a longer duration of action.[126] Fosinopril is one of the third class of ACE inhibitors that is commercially available. It contains a phosphinate group capable of specific binding to the active site of ACE. Fosinopril is administered orally as an inactive sodium salt and the conversion of fosinopril to its active metabolite, the diacid fosinoprilat, occurs in the gastrointestinal mucosa and liver.[145] Fosinopril has balanced elimination, 50% renal and 50% hepatic, in individuals with normal renal function.[145]

The most interesting ACE inhibitors of this group are the domain selective inhibitors. These inhibitors have been developed to avoid some ACE inhibitors side effects that have been attributed to the modified levels in nonangiotensin peptides, especially an increase in bradikinin concentration, but they also are an invaluable help to unveil the differential functional roles of two active sites of ACE. The first N-domain-specific and potent inhibitor of ACE, the RXP407, was described by Dive et al.,[146] and it was found by screening phosphinic peptide libraries. As compared to most conventional ACE inhibitors, RXP407 possesses an unusual structural feature, an aspartate in the P2 position. Acharya et al.[16] indicates the presence of an arginine near the aspartate of RXP407 in the N-domain S2 subsite, which can establish a salt bridge with the inhibitor and can explain its selectivity for the N-domain ACE inhibition. In the C-domain, this arginine is replaced by a glutamate in the corresponding S2 subsite, resulting in much lower inhibition potency for the C-domain, which binds the inhibitor with a constant three orders of magnitude lower.

Another selective phosphinic inhibitor is RXPA380,[147] but in this case, it is C-domain specific and interacts with the S1–S2′ subsites of the enzyme active cleft. The analysis of contribution of each RXPA380 residue revealed that both pseudoproline and tryptophan residues in the P1 and P2 positions of RXPA380 play a critical role in the selectivity of this inhibitor for the C-domain.[103] This selectivity is not due to a preference of the C-domain for inhibitors bearing pseudoproline and tryptophan residues, but it rather reflects the poor accommodation of these inhibitor residues by the N-domain. Of the 12 residues that surround the tryptophan side chain of RXPA380 in the C-domain, 5 are different in the N-domain. These differences in the S2 composition between the N- and C-domains are suggested to contribute to RXPA380 selectivity.

32.6.4 Others

Some few ACE inhibitors interact with the zinc atom of different groups already mentioned. Among those, the hydroxamic acid inhibition is based on the interaction of the hydroxamic group with the zinc atom at the active site and shows a noteworthy specificity and potency.[148,149] A comparative study of a series of monoamidic derivatives of *cis*- and *trans*-1,2-cyclohexanedicarboxylic and 1,2-cyclopentanedicarboxylic acids bearing a carboxylic, sulfhydrylic, or hydroxamic group as zinc–ligand group shows that the inhibitory activity against sACE is higher for the hydroxamates than their corresponding carboxylic or sulphydryl analogues.[149] The stereochemistry of carboxylate and hydroxamic acid zinc binding functionalities also play a role in the potency of the inhibitor, with the carboxylates and hydroxamates bearing the *S*-stereochemistry more potent than those with the *R*-stereochemistry. Idrapril is the prototype of this class of ACE inhibitors, and is characterized by the substitution of the proline-like moiety with an alicyclic dicarboxylic acid.[148]

Ketomethylene derivatives have been broadly used as protease inhibitors and 5-*S*-5-benzamido-4-*oxo*-6-phenylhexanoyl-L-proline (keto-ACE)[150] was the first ketomethylene compound to be described as ACE inhibitor with an appreciable activity and domain selectivity. This compound shows a moderate selectivity by the C-domain sACE and inhibits the hydrolysis of AGI and bradykinin via the C-domain at concentration one order of magnitude lower than via the N-domain of sACE. In the ketomethylene derivatives, the ketone group interacts with the zinc atom at the catalytic site. Based on keto-ACE and the known ACE structures, several new compounds have been designed and synthesized. Some of them have been found to inhibit ACE in the nanomolar range and in some cases show differential affinity by the N- or C-domain of sACE.[151,152] These compounds contain a ketomethylene isostere replacement at the scissile bond that is believed to mimic the tetrahedral transition state of the proteolytic reaction.[153] Natesh et al.[67] has attributed the C-domain selectivity to the presence of a bulky P1 group and P2 benzyl ring that may interact in the deep S′2 subsite. So, the selectivity of keto-ACE and analogues has been proposed to be consequence of the replacement of the Phe391 at the active site of C-domain by Tyr369 in the N-domain that have a less favorable staking interaction with the benzyl ring of keto-ACE.[152]

32.7 NEW PERSPECTIVES IN DRUG DESIGN

The availability of the three-dimensional structures of ACE has enabled to design new drugs based on the structures of the N- and C-domains. These structures are the results of advances in molecular biology as well as in protein crystallography, as the structures of the single domain ACEs were obtained in an expression system with modification in the membrane-anchored part of the structures and in the glycosides content. Moreover, these expression systems also allow regular modification of key residues in the active site to investigate their particular role. This experimental information is an invaluable help for *in silico* design of new drugs. Acharya et al.[16] published a review on ACE structure-based drug design just after the first structure of tACE appeared.[67] In this review, the authors indicated that as in other important enzyme target of drugs, the high-resolution structures provided the detail knowledge of ACE active site and enabled new strategies to re-design drugs to develop a "second generation" of domain selective inhibitors. In fact, there are several new studies that take the advantage of this structural information, for example, the work of Tzakos et al.,[109,154,155] and it is being used for the design of new drugs.[151,156]

There has been keen interest in the development of domain selective inhibitors. This is based on the hypothesis that although the C-domain has been proven to be the one that mainly hydrolyzes AGI *in vivo*,[66] bradikinin is hydrolyzed by both domains. Most of the adverse effects of the use of ACE inhibitors come from the increase of bradikinin levels and a selective inhibition of the C-domain will allow the degradation of bradikinin by the N-domain, avoiding ACE inhibitor side effects. In the same way, N-domain selective inhibitors must have some beneficial effects too, as some important physiological peptides are hydrolyzed preferentially by this domain. Nevertheless, the study of ACE inhibitors differential effect *in vitro* and *in vivo* indicates domain interaction, so any attempt to develop domain selective inhibitors as effective antihypertensive drugs with additional beneficial side effects might be hampered by the little information available until now about the way both ACE domains interact *in vitro*. Thus, it is essential to fully understand the mechanism of both the domains as they work together *in vivo*.

32.8 CONCLUSIONS

Nowadays the major cause of death in the developed world is attributed to cardiovascular diseases. Among these diseases, we find coronary heart disease, cerebrovascular disease, peripheral vascular disease, and hypertension. Of these, hypertension is the most widespread cardiovascular disease and an important contributor to coronary heart disease and cerebrovascular disease morbidity and mortality. Hypertension has been treated using different drugs such as thiazide drugs, antagonists at β-adrenoreceptors, antagonists at α-adrenoreceptors, Ca^{2+} influx inhibitors, antagonists at the AGII receptor subtype 1, and ACE inhibitors.[2] Of these, AGII receptor blockers and ACE inhibitors are the most widely used as they have been proven to be very effective for the hypertension treatment. Both drugs target at the AGII: one inhibits its production and the other inhibits its action. The recent development of a vaccine that targets AGII[157] is a new strategy to high blood pressure treatment. Although this vaccine is tested and proved to be effective and free of collateral effects, the classical therapy with drugs is the only treatment for this disease. Besides, in some patients, the use of ACE inhibitors is highly recommended for its beneficial side effects in heart failure and other pathological conditions.[10]

Since the development of the first effective drug for hypertension that targets ACE in the 1970s, this enzyme has been central to the design of inhibitors to hamper the AGII production. For many years, the design of ACE inhibitors was based on the knowledge of the active site of Zn proteases of known structures, such as carboxypeptidase A and thermolysin. This "first generation" of ACE inhibitors was a successful example of "rational drug design" and one of the best examples of serendipity in science. The single domain structures of ACE available until now show that ACE is unrelated to the carboxypeptidases and the active site of ACE is alike to that of thermolysin and the proposed enzyme mechanism of ACE is based on that anticipated for this enzyme. However, significant differences are present among the residues that participated in the proteolytic mechanism and the availability of the crystal structures of ACE complexed with different inhibitors enables to establish its catalytic mechanism and the design of "second generation inhibitors" based on the functionality of the residues in the active site. However, sACE structure is still to be determined and the possibility of a conformational change associated with the binding of substrates/inhibitors in the active site of ACE makes the enzyme mechanism's scenario more complex and an interesting subject of study. To increase this interest, the recent discovery that the C-domain seems to be the only active domain *in vivo* arises the new question about the role of the N-domain. ACE inhibitors, besides its pharmacological importance, may be the key to clarify the catalytic mechanism of this complex enzyme.

ACKNOWLEDGMENTS

This research was funded by grant BIO2006-15517-C02-02 from the Spanish Ministry of Education and Sciences and grant CVI-292 to the research team from the Andalussian Government. We thank to Prof. Jose Carlos Redondo for revising the text.

REFERENCES

1. Staessen, J. A.; Wang, J.; Bianchi, G.; Birkenhager, W. H. Essential hypertension. *Lancet* **2003**, *361*, 1629–1641.
2. Small, R. Antihypertensive drugs. *Anaesth Intensive Care Med.* **2006**, *7*, 298–302.
3. Black, J. W.; Stephenson, J. S. Pharmacology of a new adrenergic beta-receptor-blocking compound (Nethalide). *Lancet* **1962**, *2*, 311–314.
4. Cushman, D. W.; Ondetti, M. A. Design of angiotensin converting enzyme inhibitors. *Nat. Med.* **1999**, *5*, 1110–1113.
5. Skeggs, L. T.; Kahn, J. R.; Shumway, N. P. The preparation and function of the hypertensin-converting enzyme. *J. Exp. Med.* **1956**, *103*, 295–299.
6. Lipscomb, W. N.; Reeke, G. N., Jr.; Hartsuck, J. A.; Quiocho, F. A.; Bethge, P. H. The structure of carboxypeptidase A. 8. Atomic interpretation at 0.2 nm resolution, a new study of the complex of glycyl-L-tyrosine with CPA, and mechanistic deductions. *Philos. Trans. R. Soc. Lond. B Biol. Sci.* **1970**, *257*, 177–214.

7. Greene, L. J.; Camargo, A. C.; Krieger, E. M.; Stewart, J. M.; Ferreira, S. H. Inhibition of the conversion of angiotensin I to II and potentiation of bradykinin by small peptides present in *Bothrops jararaca* venom. *Circ. Res.* **1972**, *31* (Suppl. 2), 62–71.
8. Monzingo, A. F.; Matthews, B. W. Structure of a mercaptan–thermolysin complex illustrates mode of inhibition of zinc proteases by substrate-analogue mercaptans. *Biochemistry* **1982**, *21*, 3390–3394.
9. Patchett, A. A.; Cordes, E. H. The design and properties of *N*-carboxyalkyldipeptide inhibitors of angiotensin-converting enzyme. *Adv. Enzymol. Relat. Areas Mol. Biol.* **1985**, *57*, 1–84.
10. Wong, J.; Patel, R. A.; Kowey, P. R. The clinical use of angiotensin-converting enzyme inhibitors. *Prog. Cardiovasc. Dis.* **2004**, *47*, 116–130.
11. Chobanian, A. V.; Bakris, G. L.; Black, H. R.; Cushman, W. C.; Green, L. A.;et al. The Seventh Report of the Joint National Committee on Prevention, Detection, Evaluation, and Treatment of High Blood Pressure: the JNC 7 report. *JAMA* **2003**, *289*, 2560–2572.
12. Adam, A.; Cugno, M.; Molinaro, G.; Perez, M.; Lepage, Y.;et al. Aminopeptidase P in individuals with a history of angio-oedema on ACE inhibitors. *Lancet* **2002**, *359*, 2088–2089.
13. Bicket, D. P. Using ACE inhibitors appropriately. *Am. Fam. Physician* **2002**, *66*, 461–468.
14. Corvol, P.; Eyries, M.; Soubrier, F. Peptidyl-dipeptidase A/angiotensin I-converting enzyme. *Handbook of Proteolytic Enzymes*, Barrett, A. J.; Rawlings, N. D.; Woessner, J. F., eds. 2nd ed.; Academic Press: San Diego, **2004**; pp 332–346.
15. Riordan, J. F. Angiotensin-I-converting enzyme and its relatives. *Genome Biol.* **2003**, *4*, 225.
16. Acharya, K. R.; Sturrock, E. D.; Riordan, J. F.; Ehlers, M. R. Ace revisited: a new target for structure-based drug design. *Nat. Rev. Drug Discov.* **2003**, *2*, 891–902.
17. Sleight, P. The role of angiotensin-converting enzyme inhibitors in the treatment of hypertension. *Curr. Cardiol. Rep.* **2001**, *3*, 511–518.
18. Menard, J.; Patchett, A. A. Angiotensin-converting enzyme inhibitors. *Adv. Protein Chem.* **2001**, *56*, 13–75.
19. Jackson, B.; Mendelsohn, F. A.; Johnston, C. I. Angiotensin-converting enzyme inhibition: prospects for the future. *J. Cardiovasc. Pharmacol.* **1991**, *18* (Suppl. 7), S4–S8.
20. Matsusaka, T.; Ichikawa, I. Biological functions of angiotensin and its receptors. *Annu. Rev. Physiol.* **1997**, *59*, 395–412.
21. Antonaccio, M. J. Angiotensin converting enzyme (ACE) inhibitors. *Annu. Rev. Pharmacol. Toxicol.* **1982**, *22*, 57–87.
22. Erdos, E. G. Angiotensin I converting enzyme and the changes in our concepts through the years. Lewis K. Dahl memorial lecture. *Hypertension* **1990**, *16*, 363–370.
23. Baudin, B. New aspects on angiotensin-converting enzyme: from gene to disease. *Clin. Chem. Lab. Med.* **2002**, *40*, 256–265.
24. Corvol, P. ACE sets up fertilization. *Nat. Med.* **2005**, *11*, 118–119.
25. Fuchs, S.; Frenzel, K.; Hubert, C.; Lyng, R.; Muller, L.; et al. Male fertility is dependent on dipeptidase activity of testis ACE. *Nat. Med.* **2005**, *11*, 1140–1142, author reply 1142–1143.
26. Erdos, E. G.; Yang, H. Y. An enzyme in microsomal fraction of kidney that inactivates bradykinin. *Life Sci.* **1967**, *6*, 569–574.

27. Soubrier, F.; Alhenc-Gelas, F.; Hubert, C.; Allegrini, J.; John, M.; et al. Two putative active centers in human angiotensin I-converting enzyme revealed by molecular cloning. *Proc. Natl. Acad. Sci. USA* **1988**, *85*, 9386–9390.

28. Lanzillo, J. J.; Stevens, J.; Dasarathy, Y.; Yotsumoto, H.; Fanburg, B. L. Angiotensin-converting enzyme from human tissues. Physicochemical, catalytic, and immunological properties. *J. Biol. Chem.* **1985**, *260*, 14938–14944.

29. Hubert, C.; Houot, A. M.; Corvol, P.; Soubrier, F. Structure of the angiotensin I-converting enzyme gene. Two alternate promoters correspond to evolutionary steps of a duplicated gene. *J. Biol. Chem.* **1991**, *266*, 15377–15383.

30. Langford, K. G.; Shai, S. Y.; Howard, T. E.; Kovac, M. J.; Overbeek, P. A.; et al. Transgenic mice demonstrate a testis-specific promoter for angiotensin-converting enzyme. *J. Biol. Chem.* **1991**, *266*, 15559–15562.

31. Hooper, N. M.; Turner, A. J.; Isolation of two differentially glycosylated forms of peptidyl-dipeptidase A (angiotensin converting enzyme) from pig brain: a re-evaluation of their role in neuropeptide metabolism. *Biochem. J.* **1987**, *241*, 625–633.

32. Wei, L.; Alhenc-Gelas, F.; Corvol, P.; Clauser, E. The two homologous domains of human angiotensin I-converting enzyme are both catalytically active. *J. Biol. Chem.* **1991**, *266*, 9002–9008.

33. Parvathy, S.; Oppong, S. Y.; Karran, E. H.; Buckle, D. R.; Turner, A. J.; et al. Angiotensin-converting enzyme secretase is inhibited by zinc metalloprotease inhibitors and requires its substrate to be inserted in a lipid bilayer. *Biochem. J.* **1997**, *327* (Part 1), 37–43.

34. Oppong, S. Y.; Hooper, N. M. Characterization of a secretase activity which releases angiotensin-converting enzyme from the membrane. *Biochem. J.* **1993**, *292* (Part 2), 597–603.

35. Woodman, Z. L.; Oppong, S. Y.; Cook, S.; Hooper, N. M.; Schwager, S. L.; et al. Shedding of somatic angiotensin-converting enzyme (ACE) is inefficient compared with testis ACE despite cleavage at identical stalk sites. *Biochem. J.* **2000**, *347* (Part 3), 711–718.

36. Beldent, V.; Michaud, A.; Bonnefoy, C.; Chauvet, M. T.; Corvol, P. Cell surface localization of proteolysis of human endothelial angiotensin I-converting enzyme. Effect of the amino-terminal domain in the solubilization process. *J. Biol. Chem.* **1995**, *270*, 28962–28969.

37. Woodman, Z. L.; Schwager, S. L.; Redelinghuys, P.; Carmona, A. K.; Ehlers, M. R.; et al. The N-domain of somatic angiotensin-converting enzyme negatively regulates ectodomain shedding and catalytic activity. *Biochem. J.* **2005**, *389*, 739–744.

38. Santhamma, K. R.; Sadhukhan, R.; Kinter, M.; Chattopadhyay, S.; McCue, B.; et al. Role of tyrosine phosphorylation in the regulation of cleavage secretion of angiotensin-converting enzyme. *J. Biol. Chem.* **2004**, *279*, 40227–40236.

39. Kohlstedt, K.; Shoghi, F.; Muller-Esterl, W.; Busse, R.; Fleming, I. CK2 phosphorylates the angiotensin-converting enzyme and regulates its retention in the endothelial cell plasma membrane. *Circ. Res.* **2002**, *91*, 749–756.

40. Fleming, I.; Kohlstedt, K.; Busse, R. The tissue renin-angiotensin system and intracellular signalling. *Curr. Opin. Nephrol. Hypertens.* **2006**, *15*, 8–13.

41. Hooper, N. M. Angiotensin converting enzyme: implications from molecular biology for its physiological functions. *Int. J. Biochem.* **1991**, *23*, 641–647.

42. Wei, L.; Alhenc-Gelas, F.; Soubrier, F.; Michaud, A.; Corvol, P.; et al. Expression and characterization of recombinant human angiotensin I-converting enzyme. Evidence for a

C-terminal transmembrane anchor and for a proteolytic processing of the secreted recombinant and plasma enzymes. *J. Biol. Chem.* **1991**, *266*, 5540–5546.

43. Studdy, P. R.; Lapworth, R.; Bird, R. Angiotensin-converting enzyme and its clinical significance: a review. *J. Clin. Pathol.* **1983**, *36*, 938–947.
44. Erdos, E. G.; Skidgel, R. A. The angiotensin I-converting enzyme. *Lab. Invest.* **1987**, *56*, 345–348.
45. Oosterga, M.; Voors, A. A.; de Kam, P. J.; Schunkert, H.; Pinto, Y. M.; et al. Plasma angiotensin-converting enzyme activity and left ventricular dilation after myocardial infarction. *Circulation* **1997**, *95*, 2607–2609.
46. Altstein, M.; Dudai, Y.; Vogel, Z. Angiotensin-converting enzyme associated with *Torpedo california* electric organ membranes. *J. Neurosci. Res.* **1987**, *18*, 333–340.
47. Turner, A. J.; Hryszko, J.; Hooper, N. M.; Dowdall, M. J. Purification and characterization of a peptidyl dipeptidase resembling angiotensin converting enzyme from the electric organ of *Torpedo marmorata*. *J. Neurochem.* **1987**, *48*, 910–916.
48. Gunes, H. V.; Ata, N.; Degirmenci, I.; Basaran, A.; Timuralp, B.; et al. Frequency of angiotensin-converting enzyme gene polymorphism in Turkish hypertensive patients. *Int. J. Clin. Pract.* **2004**, *58*, 838–843.
49. Okawara, Y.; Kobayashi, H. Enhancement of water intake by captopril (SQ14225), an angiotensin I-converting enzyme inhibitor, in the goldfish, *Carassius auratus*. *Gen. Comp. Endocrinol.* **1988**, *69*, 114–118.
50. Esther, C. R., Jr.; Thomas, K. E.; Bernstein, K. E. Chicken lacks the testis specific isozyme of angiotensin converting enzyme found in mammals. *Biochem. Biophys. Res. Commun.* **1994**, *205*, 1916–1921.
51. Lamango, N.; Isaac, R. E. Identification of an ACE-like peptidyl dipeptidase activity in the housefly, *Musca domestica*. *Biochem. Soc. Trans.* **1993**, *21* (Part 3), 245S.
52. Cornell, M. J.; Williams, T. A.; Lamango, N. S.; Coates, D.; Corvol, P.; et al. Cloning and expression of an evolutionary conserved single-domain angiotensin converting enzyme from *Drosophila melanogaster*. *J. Biol. Chem.* **1995**, *270*, 13613–13619.
53. Isaac, R. E. Peptidyl-dipeptidase A (invertebrate). *Handbook of Proteolytic Enzymes*, 2nd ed.; Academic Press: London, **2004**; pp 347–349.
54. Kim, H. M.; Shin, D. R.; Yoo, O. J.; Lee, H.; Lee, J. O. Crystal structure of *Drosophila* angiotensin I-converting enzyme bound to captopril and lisinopril. *FEBS Lett.* **2003**, *538*, 65–70.
55. Houard, X.; Williams, T. A.; Michaud, A.; Dani, P.; Isaac, R. E.; et al. The *Drosophila melanogaster*-related angiotensin-I-converting enzymes Acer and Ance: distinct enzymic characteristics and alternative expression during pupal development. *Eur. J. Biochem.* **1998**, *257*, 599–606.
56. Tipnis, S. R.; Hooper, N. M.; Hyde, R.; Karran, E.; Christie, G.; et al. A human homolog of angiotensin-converting enzyme. Cloning and functional expression as a captopril-insensitive carboxypeptidase. *J. Biol. Chem.* **2000**, *275*, 33238–33243.
57. Donoghue, M.; Hsieh, F.; Baronas, E.; Godbout, K.; Gosselin, M.; et al. A novel angiotensin-converting enzyme-related carboxypeptidase (ACE2) converts angiotensin I to angiotensin 1–9. *Circ. Res.* **2000**, *87*, E1–E9.
58. Guy, J. L.; Lambert, D. W.; Warner, F. J.; Hooper, N. M.; Turner, A. J. Membrane-associated zinc peptidase families: comparing ACE and ACE2. *Biochim. Biophys. Acta* **2005**, *1751*, 2–8.

59. Chou, C. F.; Loh, C. B.; Foo, Y. K.; Shen, S.; Fielding, B. C.; et al. ACE2 orthologues in non-mammalian vertebrates (*Danio*, *Gallus*, *Fugu*, *Tetraodon* and *Xenopus*). *Gene* **2006**, *377*, 46–55.

60. Ferreira, A. J.; Raizada, M. K. Are we poised to target ACE2 for the next generation of antihypertensives? *J. Mol. Med.* **2008**, *86*, 685–690.

61. Turner, A. J.; Hooper, N. M. The angiotensin-converting enzyme gene family: genomics and pharmacology. *Trends Pharmacol. Sci.* **2002**, *23*, 177–183.

62. Rella, M.; Elliot, J. L.; Revett, T. J.; Lanfear, J.; Phelan, A.; et al. Identification and characterisation of the angiotensin converting enzyme-3 (ACE3) gene: a novel mammalian homologue of ACE. *BMC Genomics* **2007**, *8*, 194.

63. Riviere, G.; Michaud, A.; Corradi, H. R.; Sturrock, E. D.; Ravi Acharya, K.; et al. Characterization of the first angiotensin-converting like enzyme in bacteria: ancestor ACE is already active. *Gene* **2007**, *399*, 81–90.

64. Shen, X. Z.; Xiao, H. D.; Li, P.; Lin, C. X.; Billet, S.; et al. New insights into the role of angiotensin-converting enzyme obtained from the analysis of genetically modified mice. *J. Mol. Med.* **2008**, *86*, 679–684.

65. Shen, X. Z.; Xiao, H. D.; Li, P.; Lin, C. X.; Fuchs, S.; et al. Tissue specific expression of angiotensin converting enzyme: a new way to study an old friend. *Int. Immunopharmacol.* **2008**, *8*, 171–176.

66. Fuchs, S.; Xiao, H. D.; Hubert, C.; Michaud, A.; Campbell, D. J.; et al. Angiotensin-converting enzyme C-terminal catalytic domain is the main site of angiotensin I cleavage *in vivo*. *Hypertension* **2008**, *51*, 267–274.

67. Natesh, R.; Schwager, S. L.; Sturrock, E. D.; Acharya, K. R. Crystal structure of the human angiotensin-converting enzyme–lisinopril complex. *Nature* **2003**, *421*, 551–554.

68. Yu, X. C.; Sturrock, E. D.; Wu, Z.; Biemann, K.; Ehlers, M. R.; et al. Identification of N-linked glycosylation sites in human testis angiotensin-converting enzyme and expression of an active deglycosylated form. *J. Biol. Chem.* **1997**, *272*, 3511–3519.

69. Gordon, K.; Redelinghuys, P.; Schwager, S. L.; Ehlers, M. R.; Papageorgiou, A. C.; et al. Deglycosylation, processing and crystallization of human testis angiotensin-converting enzyme. *Biochem. J.* **2003**, *371*, 437–442.

70. Williams, T. A.; Michaud, A.; Houard, X.; Chauvet, M. T.; Soubrier, F.; et al. *Drosophila melanogaster* angiotensin I-converting enzyme expressed in *Pichia pastoris* resembles the C-domain of the mammalian homologue and does not require glycosylation for secretion and enzymic activity. *Biochem. J.* **1996**, *318* (Part 1), 125–131.

71. Sturrock, E. D.; Natesh, R.; van Rooyen, J. M.; Acharya, K. R. Structure of angiotensin I-converting enzyme. *Cell Mol. Life Sci.* **2004**, *61*, 2677–2686.

72. Holm, L.; Sander, C. Protein folds and families: sequence and structure alignments. *Nucleic Acids Res.* **1999**, *27*, 244–247.

73. Comellas-Bigler, M.; Lang, R.; Bode, W.; Maskos, K. Crystal structure of the *E. coli* dipeptidyl carboxypeptidase Dcp: further indication of a ligand-dependent hinge movement mechanism. *J. Mol. Biol.* **2005**, *349*, 99–112.

74. Ray, K.; Hines, C. S.; Coll-Rodriguez, J.; Rodgers, D. W. Crystal structure of human thimet oligopeptidase provides insight into substrate recognition, regulation, and localization. *J. Biol. Chem.* **2004**, *279*, 20480–20489.

75. Corradi, H. R.; Schwager, S. L.; Nchinda, A. T.; Sturrock, E. D.; Acharya, K. R. Crystal structure of the N-domain of human somatic angiotensin I-converting enzyme provides

a structural basis for domain-specific inhibitor design. *J. Mol. Biol.* **2006**, *357*, 964–974.

76. Natesh, R.; Schwager, S. L.; Evans, H. R.; Sturrock, E. D.; Acharya, K. R. Structural details on the binding of antihypertensive drugs captopril and enalaprilat to human testicular angiotensin I-converting enzyme. *Biochemistry* **2004**, *43*, 8718–8724.
77. Corradi, H. R.; Chitapi, I.; Sewell, B. T.; Georgiadis, D.; Dive, V.; et al. The structure of testis angiotensin-converting enzyme in complex with the C-domain-specific inhibitor RXPA380. *Biochemistry* **2007**, *46*, 5473–5478.
78. Naqvi, N.; Liu, K.; Graham, R. M.; Husain, A. Molecular basis of exopeptidase activity in the C-terminal domain of human angiotensin I-converting enzyme: insights into the origins of its exopeptidase activity. *J. Biol. Chem.* **2005**, *280*, 6669–6675.
79. Fernandez, M.; Liu, X.; Wouters, M. A.; Heyberger, S.; Husain, A. Angiotensin I-converting enzyme transition state stabilization by HIS1089: evidence for a catalytic mechanism distinct from other gluzincin metalloproteinases. *J. Biol. Chem.* **2001**, *276*, 4998–5004.
80. Towler, P.; Staker, B.; Prasad, S. G.; Menon, S.; Tang, J.; et al. ACE2 X-ray structures reveal a large hinge-bending motion important for inhibitor binding and catalysis. *J. Biol. Chem.* **2004**, *279*, 17996–18007.
81. Holland, D. R.; Tronrud, D. E.; Pley, H. W.; Flaherty, K. M.; Stark, W.; et al. Structural comparison suggests that thermolysin and related neutral proteases undergo hinge-bending motion during catalysis. *Biochemistry* **1992**, *31*, 11310–11316.
82. Watermeyer, J. M.; Sewell, B. T.; Schwager, S. L.; Natesh, R.; Corradi, H. R.; et al. Structure of testis ACE glycosylation mutants and evidence for conserved domain movement. *Biochemistry* **2006**, *45*, 12654–12663.
83. Bunning, P.; Holmquist, B.; Riordan, J. F. Substrate specificity and kinetic characteristics of angiotensin converting enzyme. *Biochemistry* **1983**, *22*, 103–110.
84. Andujar-Sanchez, M.; Jara-Perez, V.; Camara-Artigas, A. Thermodynamic determination of the binding constants of angiotensin-converting enzyme inhibitors by a displacement method. *FEBS Lett.* **2007**, *581*, 3449–3454.
85. Liu, X.; Fernandez, M.; Wouters, M. A.; Heyberger, S.; Husain, A. Arg(1098) is critical for the chloride dependence of human angiotensin I-converting enzyme C-domain catalytic activity. *J. Biol. Chem.* **2001**, *276*, 33518–33525.
86. Skirgello, O. E.; Binevski, P. V.; Pozdnev, V. F.; Kost, O. A. Kinetic probes for inter-domain co-operation in human somatic angiotensin-converting enzyme. *Biochem. J.* **2005**, *391*, 641–647.
87. Sturrock, E. D.; Danilov, S. M.; Riordan, J. F. Limited proteolysis of human kidney angiotensin-converting enzyme and generation of catalytically active N- and C-terminal domains. *Biochem. Biophys. Res. Commun.* **1997**, *236*, 16–19.
88. Olden, K.; Pratt, R. M.; Jaworski, C.; Yamada, K. M. Evidence for role of glycoprotein carbohydrates in membrane transport: specific inhibition by tunicamycin. *Proc. Natl. Acad. Sci. USA* **1979**, *76*, 791–795.
89. Ripka, J. E.; Ryan, J. W.; Valido, F. A.; Chung, A. Y.; Peterson, C. M.; et al. N-glycosylation of forms of angiotensin converting enzyme from four mammalian species. *Biochem. Biophys. Res. Commun.* **1993**, *196*, 503–508.
90. Ehlers, M. R.; Chen, Y. N.; Riordan, J. F. The unique N-terminal sequence of testis angiotensin-converting enzyme is heavily O-glycosylated and unessential for activity or stability. *Biochem. Biophys. Res. Commun.* **1992**, *183*, 199–205.

91. Baudin, B.; Alves, N.; Pilon, A.; Beneteau-Burnat, B.; Giboudeau, J. Structural and biological roles of glycosylations in pulmonary angiotensin I-converting enzyme. *Glycobiology* **1997**, *7*, 565–570.

92. Kasturi, S.; Jabbar, M. A.; Sen, G. C.; Sen, I. Role of glycosylation in the biosynthesis and activity of rabbit testicular angiotensin-converting enzyme. *Biochemistry* **1994**, *33*, 6228–6234.

93. Sadhukhan, R.; Sen, I. Different glycosylation requirements for the synthesis of enzymatically active angiotensin-converting enzyme in mammalian cells and yeast. *J. Biol. Chem.* **1996**, *271*, 6429–6434.

94. Cascieri, M. A.; Bull, H. G.; Mumford, R. A.; Patchett, A. A.; Thornberry, N. A.; et al. Carboxyl-terminal tripeptidyl hydrolysis of substance P by purified rabbit lung angiotensin-converting enzyme and the potentiation of substance P activity *in vivo* by captopril and MK-422. *Mol. Pharmacol.* **1984**, *25*, 287–293.

95. Dubreuil, P.; Fulcrand, P.; Rodriguez, M.; Fulcrand, H.; Laur, J.; et al. Novel activity of angiotensin-converting enzyme. Hydrolysis of cholecystokinin and gastrin analogues with release of the amidated C-terminal dipeptide. *Biochem. J.* **1989**, *262*, 125–130.

96. Skidgel, R. A.; Erdos, E. G. Novel activity of human angiotensin I converting enzyme: release of the NH2- and COOH-terminal tripeptides from the luteinizing hormone-releasing hormone. *Proc. Natl. Acad. Sci. USA* **1985**, *82*, 1025–1029.

97. Skidgel, R. A.; Engelbrecht, S.; Johnson, A. R.; Erdos, E. G. Hydrolysis of substance P and neurotensin by converting enzyme and neutral endopeptidase. *Peptides* **1984**, *5*, 769–776.

98. Yokosawa, H.; Endo, S.; Ohgaki, Y.; Maeyama, J.; Ishii, S. Hydrolysis of substance P and its analogs by angiotensin-converting enzyme from rat lung. Characterization of endopeptidase activity of the enzyme. *J. Biochem.* **1985**, *98*, 1293–1299.

99. Jaspard, E.; Wei, L.; Alhenc-Gelas, F. Differences in the properties and enzymatic specificities of the two active sites of angiotensin I-converting enzyme (kininase II). Studies with bradykinin and other natural peptides. *J. Biol. Chem.* **1993**, *268*, 9496–9503.

100. Rousseau, A.; Michaud, A.; Chauvet, M. T.; Lenfant, M.; Corvol, P. The hemoregulatory peptide *N*-acetyl-Ser-Asp-Lys-Pro is a natural and specific substrate of the N-terminal active site of human angiotensin-converting enzyme. *J. Biol. Chem.* **1995**, *270*, 3656–3661.

101. Wei, L.; Clauser, E.; Alhenc-Gelas, F.; Corvol, P. The two homologous domains of human angiotensin I-converting enzyme interact differently with competitive inhibitors. *J. Biol. Chem.* **1992**, *267*, 13398–13405.

102. Vazeux, G.; Cotton, J.; Cuniasse, P.; Dive, V. Potency and selectivity of RXP407 on human, rat, and mouse angiotensin-converting enzyme. *Biochem. Pharmacol.* **2001**, *61*, 835–841.

103. Georgiadis, D.; Cuniasse, P.; Cotton, J.; Yiotakis, A.; Dive, V. Structural determinants of RXPA380, a potent and highly selective inhibitor of the angiotensin-converting enzyme C-domain. *Biochemistry* **2004**, *43*, 8048–8054.

104. Andujar-Sanchez, M.; Camara-Artigas, A.; Jara-Perez, V. A calorimetric study of the binding of lisinopril, enalaprilat and captopril to angiotensin-converting enzyme. *Biophys. Chem.* **2004**, *111*, 183–189.

105. Binevski, P. V.; Sizova, E. A.; Pozdnev, V. F.; Kost, O. A. Evidence for the negative cooperativity of the two active sites within bovine somatic angiotensin-converting enzyme. *FEBS Lett.* **2003**, *550*, 84–88.

106. Rawlings, N. D.; Morton, F. R.; Kok, C. Y.; Kong, J.; Barrett, A. J. MEROPS: the peptidase database. *Nucleic Acids Res.* **2008**, *36*, D320–D325.

107. Holden, H. M.; Matthews, B. W. The binding of L-valyl-L-tryptophan to crystalline thermolysin illustrates the mode of interaction of a product of peptide hydrolysis. *J. Biol. Chem.* **1988**, *263*, 3256–3260.

108. Pelmenschikov, V.; Blomberg, M. R.; Siegbahn, P. E. A theoretical study of the mechanism for peptide hydrolysis by thermolysin. *J. Biol. Inorg. Chem.* **2002**, *7*, 284–298.

109. Tzakos, A. G.; Galanis, A. S.; Spyroulias, G. A.; Cordopatis, P.; Manessi-Zoupa, E.; et al. Structure–function discrimination of the N- and C-catalytic domains of human angiotensin-converting enzyme: implications for Cl-activation and peptide hydrolysis mechanisms. *Protein Eng.* **2003**, *16*, 993–1003.

110. Dion, N.; Le Moual, H.; Crine, P.; Boileau, G. Kinetic evidence that His-711 of neutral endopeptidase 24.11 is involved in stabilization of the transition state. *FEBS Lett.* **1993**, *318*, 301–304.

111. Arnett, D. K.; Claas, S. A.; Glasser, S. P. Pharmacogenetics of antihypertensive treatment. *Vascul. Pharmacol.* **2006**, *44*, 107–118.

112. Chappell, M. C.; Modrall, J. G.; Diz, D. I.; Ferrario, C. M. Novel aspects of the renal renin-angiotensin system: angiotensin-(1–7), ACE2 and blood pressure regulation. *Contrib. Nephrol.* **2004**, *143*, 77–89.

113. Kurdi, M.; De Mello, W. C.; Booz, G. W. Working outside the system: an update on the unconventional behavior of the renin-angiotensin system components. *Int. J. Biochem. Cell Biol.* **2005**, *37*, 1357–1367.

114. Carey, R. M.; Siragy, H. M. Newly recognized components of the renin-angiotensin system: potential roles in cardiovascular and renal regulation. *Endocr. Rev.* **2003**, *24*, 261–271.

115. Cushman, D. W. C.; Hong, S.; Sabo, E. F.; Ondetti, M. A. Angiotensin Converting Enzyme Inhibitors: Evolution of a New Class of Antihypertensive Drugs; Urban & Schwarzenberg: Baltimore, **1981**; pp 3–25.

116. Schechter, I.; Berger, A. On the size of the active site in proteases. I. Papain. *Biochem. Biophys. Res. Commun.* **1967**, *27*, 157–162.

117. Salvetti, A. Newer ACE inhibitors. A look at the future. *Drugs* **1990**, *40*, 800–828.

118. Utsunomiya, K.; Ikeda, Y.; Alacepril Study Group. Beneficial effect of alacepril, a new angiotensin-converting enzyme inhibitor on albuminuria and glycemic state: an open multicenter trial. *J. Diabet. Complications* **1991**, *5*, 165–166.

119. Sakai, K.; Aono, J.; Shiraki, Y.; Hinohara, Y.; Akamatsu, A.; et al. MC-838 (altiopril calcium): a novel orally active angiotensin converting enzyme inhibitor. *Tohoku J. Exp. Med.* **1987**, *152*, 363–374.

120. Unger, T.; Yukimura, T.; Marin-Grez, M.; Lang, R. E.; Rascher, W.; et al. SA446, a new orally active converting enzyme inhibitor: antihypertensive action and comparison with captopril in stroke-prone spontaneously hypersensitive rats. *Eur. J. Pharmacol.* **1982**, *78*, 411–420.

121. Borghi, C.; Bacchelli, S.; Degli Esposti, D.; Ambrosioni, E. A review of the angiotensin-converting enzyme inhibitor, zofenopril, in the treatment of cardiovascular diseases. *Expert Opin. Pharmacother.* **2004**, *5*, 1965–1977.

122. DiBianco, R. Adverse reactions with angiotensin converting enzyme (ACE) inhibitors. *Med. Toxicol.* **1986**, *1*, 122–141.

123. Furness, P. N.; Goodfield, M. J.; MacLennan, K. A.; Stevens, A.; Millard, L. G. Severe cutaneous reactions to captopril and enalapril; histological study and comparison with early mycosis fungoides. *J. Clin. Pathol.* **1986**, *39*, 902–907.

124. Frohlich, E. D.; Cooper, R. A.; Lewis, E. J. Review of the overall experience of captopril in hypertension. *Arch. Intern. Med.* **1984**, *144*, 1441–1444.

125. Brown, N. J.; Vaughan, D. E. Angiotensin-converting enzyme inhibitors. *Circulation* **1998**, *97*, 1411–1420.

126. Voors, A.; Herre Kingma, J.; van Gilst, W. Drug differences between ACE-inhibitors in experimental settings and clinical practice. *J. Cardiovasc. Risk* **1995**, *2*, 413–422.

127. Pi, X. J.; Chen, X. Captopril and ramiprilat protect against free radical injury in isolated working rat hearts. *J. Mol. Cell Cardiol.* **1989**, *21*, 1261–1271.

128. Przyklenk, K.; Kloner, R. A. Angiotensin converting enzyme inhibitors improve contractile function of stunned myocardium by different mechanisms of action. *Am. Heart J.* **1991**, *121*, 1319–1330.

129. van Gilst, W. H.; de Graeff, P. A.; de Leeuw, M. J.; Scholtens, E.; Wesseling, H. Converting enzyme inhibitors and the role of the sulfhydryl group in the potentiation of exo- and endogenous nitrovasodilators. *J. Cardiovasc. Pharmacol.* **1991**, *18*, 429–436.

130. Ulm, E. H. Enalapril maleate (MK-421), a potent, nonsulfhydryl angiotensin-converting enzyme inhibitor: absorption, disposition, and metabolism in man. *Drug Metab. Rev.* **1983**, *14*, 99–110.

131. Kaplan, H. R.; Cohen, D. M.; Essenburg, A. D.; Major, T. C.; Mertz, T. E.; et al. CI-906 and CI-907: new orally active nonsulfhydryl angiotensin-converting enzyme inhibitors. *Fed. Proc.* **1984**, *43*, 1326–1329.

132. Ulm, E. H.; Hichens, M.; Gomez, H. J.; Till, A. E.; Hand, E.; et al. Enalapril maleate and a lysine analogue (MK-521): disposition in man. *Br. J. Clin. Pharmacol.* **1982**, *14*, 357–362.

133. Ball, S. G.; Robertson, J. I. Clinical pharmacology of ramipril. *Am. J. Cardiol.* **1987**, *59*, 23D–27D.

134. Williams, P. E.; Brown, A. N.; Rajaguru, S.; Francis, R. J.; Walters, G. E.; et al. The pharmacokinetics and bioavailability of cilazapril in normal man. *Br. J. Clin. Pharmacol.* **1989**, *27* (Suppl. 2), 181S–188S.

135. Guay, D. R. Trandolapril: a newer angiotensin-converting enzyme inhibitor. *Clin. Ther.* **2003**, *25*, 713–775.

136. Kubo, M.; Kato, J.; Ochiai, T.; Ishida, R. Pharmacological studies on (4*S*)-1-methyl-3-[(2*S*)-2-[*N*-((1*S*)-1-ethoxycarbonyl-3-phenylpropyl)amino] propionyl]-2-*oxo*-imidazolidine-4-carboxylic acid hydrochloride (TA-6366), a new ACE inhibitor: I. ACE inhibitory and anti-hypertensive activities. *Jpn. J. Pharmacol.* **1990**, *53*, 201–210.

137. Sybertz, E. J.; Watkins, R. W.; Ahn, H. S.; Baum, T.; La Rocca, P.; et al. Pharmacologic, metabolic, and toxicologic profile of spirapril (SCH 33844), a new angiotensin converting inhibitor. *J. Cardiovasc. Pharmacol.* **1987**, *10* (Suppl. 7), S105–S108.

138. Kaiser, G.; Ackermann, R.; Sioufi, A. Pharmacokinetics of a new angiotensin-converting enzyme inhibitor, benazepril hydrochloride, in special populations. *Am. Heart J.* **1989**, *117*, 746–751.

139. Klutchko, S.; Blankley, C. J.; Fleming, R. W.; Hinkley, J. M.; Werner, A. E.; et al. Synthesis of novel angiotensin converting enzyme inhibitor quinapril and related compounds. A divergence of structure–activity relationships for non-sulfhydryl and sulfhydryl types. *J. Med. Chem.* **1986**, *29*, 1953–1961.

140. Razzetti, R.; Acerbi, D. Pharmacokinetic and pharmacologic properties of delapril, a lipophilic nonsulfhydryl angiotensin-converting enzyme inhibitor. *Am. J. Cardiol.* **1995**, *75*, 7F–12F.

141. Parker, E.; Aarons, L.; Rowland, M.; Resplandy, G. The pharmacokinetics of perindoprilat in normal volunteers and patients: influence of age and disease state. *Eur. J. Pharm. Sci.* **2005**, *26*, 104–113.

142. Graf, P.; Frueh, F.; Schmid, K. Determination of the angiotensin converting enzyme inhibitor benazeprilat in plasma and urine by an enzymic method. *J. Chromatogr.* **1988**, *425*, 353–361.

143. Azizi, M.; Massien, C.; Michaud, A.; Corvol, P. *In vitro* and *in vivo* inhibition of the 2 active sites of ACE by omapatrilat, a vasopeptidase inhibitor. *Hypertension* **2000**, *35*, 1226–1231.

144. Robl, J. A.; Sun, C. Q.; Stevenson, J.; Ryono, D. E.; Simpkins, L. M.; et al. Dual metalloprotease inhibitors: mercaptoacetyl-based fused heterocyclic dipeptide mimetics as inhibitors of angiotensin-converting enzyme and neutral endopeptidase. *J. Med. Chem.* **1997**, *40*, 1570–1577.

145. Singhvi, S. M.; Duchin, K. L.; Morrison, R. A.; Willard, D. A.; Everett, D. W.; et al. Disposition of fosinopril sodium in healthy subjects. *Br. J. Clin. Pharmacol.* **1988**, *25*, 9–15.

146. Dive, V.; Cotton, J.; Yiotakis, A.; Michaud, A.; Vassiliou, S.; et al. RXP 407, a phosphinic peptide is a potent inhibitor of angiotensin I converting enzyme able to differentiate between its two active sites. *Proc. Natl. Acad. Sci. USA* **1999**, *96*, 4330–4335.

147. Georgiadis, D.; Beau, F.; Czarny, B.; Cotton, J.; Yiotakis, A.; et al. Roles of the two active sites of somatic angiotensin-converting enzyme in the cleavage of angiotensin I and bradykinin: insights from selective inhibitors. *Circ. Res.* **2003**, *93*, 148–154.

148. Subissi, A.; Criscuoli, M.; Sardelli, G.; Guelfi, M.; Giachetti, A. Pharmacology of idrapril: a new class of angiotensin converting enzyme inhibitors. *J. Cardiovasc. Pharmacol.* **1992**, *20*, 139–146.

149. Turbanti, L.; Cerbai, G.; Di Bugno, C.; Giorgi, R.; Garzelli, G.; et al. 1,2-Cyclomethylenecarboxylic monoamide hydroxamic derivatives. A novel class of non-amino acid angiotensin converting enzyme inhibitors. *J. Med. Chem.* **1993**, *36*, 699–707.

150. Deddish, P. A.; Marcic, B.; Jackman, H. L.; Wang, H. Z.; Skidgel, R. A.; et al. N-domain-specific substrate and C-domain inhibitors of angiotensin-converting enzyme: angiotensin-(1–7) and keto-ACE. *Hypertension* **1998**, *31*, 912–917.

151. Nchinda, A. T.; Chibale, K.; Redelinghuys, P.; Sturrock, E. D. Synthesis of novel keto-ACE analogues as domain-selective angiotensin I-converting enzyme inhibitors. *Bioorg. Med. Chem. Lett.* **2006**, *16*, 4612–4615.

152. Redelinghuys, P.; Nchinda, A. T.; Chibale, K.; Sturrock, E. D. Novel ketomethylene inhibitors of angiotensin I-converting enzyme (ACE): inhibition and molecular modelling. *Biol. Chem.* **2006**, *387*, 461–466.

153. Meyer, R. F.; Nicolaides, E. D.; Tinney, F. J.; Lunney, E. A.; Holmes, A.; et al. Novel synthesis of (*S*)-1-[5-(benzoylamino)-1,4-dioxo-6-phenylhexyl]-L-proline and analogues: potent angiotensin converting enzyme inhibitors. *J. Med. Chem.* **1981**, *24*, 964–969.

154. Galanis, A. S.; Spyroulias, G. A.; Pierattelli, R.; Tzakos, A.; Troganis, A.; et al. Zinc binding in peptide models of angiotensin-I converting enzyme active sites studied through 1H-NMR and chemical shift perturbation mapping. *Biopolymers* **2003**, *69*, 244–252.

155. Tzakos, A. G.; Naqvi, N.; Comporozos, K.; Pierattelli, R.; Theodorou, V.; et al. The molecular basis for the selection of captopril *cis* and *trans* conformations by angiotensin I converting enzyme. *Bioorg. Med. Chem. Lett.* **2006**, *16*, 5084–5087.

156. Nchinda, A. T.; Chibale, K.; Redelinghuys, P.; Sturrock, E. D. Synthesis and molecular modeling of a lisinopril-tryptophan analogue inhibitor of angiotensin I-converting enzyme. *Bioorg. Med. Chem. Lett.* **2006**, *16*, 4616–4619.

157. Tissot, A. C.; Maurer, P.; Nussberger, J.; Sabat, R.; Pfister, T.; et al. Effect of immunisation against angiotensin II with CYT006-AngQb on ambulatory blood pressure: a double-blind, randomised, placebo-controlled phase IIa study. *Lancet* **2008**, *371*, 821–827.

158. Michaud, A.; Williams, T. A.; Chauvet, M. T.; Corvol, P. Substrate dependence of angiotensin I-converting enzyme inhibition: captopril displays a partial selectivity for inhibition of *N*-acetyl-seryl-aspartyl-lysyl-proline hydrolysis compared with that of angiotensin I. *Mol. Pharmacol.* **1997**, *51*, 1070–1076.

CHAPTER 33

P-III Metalloproteinase (Leucurolysin-B) from *Bothrops leucurus* Venom: Isolation and Possible Inhibition

ELADIO F. SANCHEZ[1,3] and JOHANNES A. EBLE[2]

[1]Research and Development Center, Ezequiel Dias Foundation, 30510-010 Belo Horizonte, MG, Brazil

[2]Center for Molecular Medicine, Department Vascular Matrix Biology, Excellence Cluster Cardio-Pulmonary System, Frankfurt University Hospital, Frankfurt am Main, Germany

[3]Post-Graduate Program, Hospital Santa Casa, Belo Horizonte, Brazil

33.1 INTRODUCTION

Snake venoms are very complex mixtures of proteins and peptides with diverse pharmacological effects, which are able to affect several physiological systems. Among them, we found enzyme-generating factors or proteinases capable of mediating, directly or indirectly, the early events of hemorrhage and inflammatory processes. It is estimated that about two-thirds of the living snake species produce toxic secretions in specialized, exocrine dental glands located in the upper jaw.[1] Local tissue damage including hemorrhage, necrosis, and edema is a common occurrence resulting from snakebite poisoning by the majority of viperid snakes (pit vipers and true vipers). Furthermore, systemic alterations such as hemorrhage in many internal organs, coagulopathy, circulatory shock, and renal failure may also occur.[2,3] These effects are induced by several components including metalloproteinases[4–6] and phospholipases A_2,[7] among others. Local bleeding at the bite site as well as systemic bleeding involves the direct damage of microvessels, provoked by hemorrhagic proteinases, combined with a variety of effects that viperid venoms exert in the hemostatic system.[4,8,9]

The SVMPs (adamalysins) are members of the reprolysin family (Clan M12) of metalloproteinases, which are also called ADAMs (*a* *d*isintegrin-like *a*nd *m*etallo-proteinase)/MDC (*m*etalloproteinase, *d*isintegrin, *c*ysteine rich). They are secreted as zymogen precursors, with a cysteine-switch mechanism that inhibits catalytic activity.[4] Together with the matrixins (matrix metalloproteinases, MMPs), astacins, leishmanolysin, and serralysins (large bacterial zinc endopeptidases), the SVMPs are grouped in the superfamily of "metzincins" because there is a conserved methionine (e.g., Met169 in leuc-B) in a turn that underlies the active site.[10–13] Like the other metzincins, the SVMPs comprise an elongated zinc binding consensus sequence **HEXXHXXGXXH** and a so-called "Met-turn," forming the hydrophobic base of the catalytic zinc binding site. On the basis of the protein structure/molecular size, they are classified into four major groups P-I (protein class I) to P-IV,[4,13,14] with different group members often found in the same venom. P-I refers to the short metalloproteinases consisting of the protease domain only. Yet, within P-I class these proteinases can have either weak or no hemorrhagic activity. However, the exact mechanism of venom-induced hemorrhage is not fully explained. P-II SVMPs are synthesized with metalloprotease and disintegrin domains, and there is increasing evidence that most of them are readily cleaved to yield a disintegrin as well as the P-I class metalloproteases. The high molecular mass hemorrhagic toxins (P-III) have a third domain, the cysteine-rich domain following the metalloprotease domain. As a general rule, the P-III class contains more potent hemorrhagic toxins than the single-domain proteinases. It has been shown that the noncatalytic disintegrin-like and cysteine-rich domains are inhibitors of platelet aggregation[9,15] and trigger the release of cytokines,[16] which explains, at least in part, their hemorrhagic potentiating effect. The P-IV class has a similar domain structure to the P-III, but with an additional disulfide-linked C-type lectin domain (e.g., RVV-X and VLFXA from *Vipera* genus). The cDNA analysis of factor X activator from *Vipera lebetina* (VLFXA) revealed that this protein is synthesized from three independent RNAs that code for a P-III metalloprotease (heavy chain) and two type-C lectins (light chains).[17] In addition, recent submissions about RVV-X sequences (GenBank accession nos. Q7LZ61, Q4PRD1, and Q4PRD2) reinforce this observation. Thus, these data suggest that P-IV proteins are a modification of P-III SVMPs. Furthermore, SVMPs are collectively able *in vitro* and *in vivo* to degrade all kinds of extracellular matrix (ECM) protein components such as interstitial and basement membrane collagens, fibronectin, proteoglycans, nidogen/enactin (EN), laminin (LM), and aggrecan and thus could be associated with disruption of connective tissue and hemostasis as well as with apoptosis and inflammation.[5,14]

The venom of the Brazilian pit viper *Bothrops leucurus* contains P-I and P-III class metallopeptidases. Two of them have recently been reported as leucurolysin-a and leucurolysin-B (leuc-B), respectively.[18,19] Leucurolysins belong to the metzincin superfamily of metalloproteinases and are the founding members of the reprolysin family. Leuc-B is the most potent hemorrhagic toxin isolated to date from *B. leucurus* venom and is considered to be responsible for the main complex alterations in the normal hemostatic system, which potentially contribute to the local and systemic bleeding in envenomed victims.[19]

33.1.1 Occurrence and Purification

In restricted areas of the northeast of Brazil, including the States of Ceara and Bahia to Espirito Santo in the southeast, where snakebite envenomations represent a relevant public health problem, white-tailed jararaca snake (*B. leucurus*) is the leading cause of human accidents. Leuc-B is produced and stored in the venom glands. The enzyme has been isolated to homogeneity by a three-step purification procedure described here in detail.[19] *First step*: The crude venom (2 g) of *B. leucurus* was dissolved in 12 mL of 50 mM ammonium acetate buffer (pH 7.3) containing 0.3 M NaCl and centrifuged at 6000 × *g* to remove the insoluble materials. The solution (1680 mg) was then applied to two (100 cm × 2.5 cm each) columns in series packed with Sephacryl S-200 as described.[19] Using this column the venom was separated into seven fractions (1–7) as shown in Fig. 33.1 (first step). Proteolytic activity on dimethylcasein (DMC) and fibrin was found mainly in fractions 1 and 4. Fraction 1 (tubes 133–137) that contained the high molecular mass proteins showing hemorrhagic (~55%) and proteolytic (~22%) activities was concentrated for further purification. Fraction 4 (tubes 185–191) contained proteins with Mr values of approximately 24 kDa, exhibiting proteolytic (29%) and hemorrhagic (5%) activities. A metalloendopeptidase of 23 kDa named leuc-a was first purified from this fraction. Leuc-a has been characterized as a P-I class showing direct fibrinolytic activity without hemorrhagic effect.[18,20] *Second step*: A concentrated fraction 1 containing 29.7 mg was then subjected to gel filtration chromatography on Sephacryl S-300 (100 cm × 1.5 cm) (Fig. 33.1, second step). The major fraction II (tubes 24–29) containing the bulk of the hemorrhagic and proteolytic activities was desalted and lyophilized. SDS-PAGE analysis (12% gel) of fraction II showed the presence of other protein bands. *Third step*: The collected material from the second step (26 mg) was further purified by anion-exchange chromatography on a column of DEAE Sepharose CL 6B (20 cm × 1.6 cm), at pH 7.5, using a gradient of NaCl as eluent. Chromatography of fraction II on the DEAE column resulted in four protein peaks (a–d) as shown in Fig. 33.1 (third step). Proteolytic activity on DMC was found in fractions a, b, and d, however, the hemorrhagic effect was concentrated in fraction d. Active fraction d that eluted at a concentration of 0.1–0.2 M NaCl corresponds to leuc-B. The name leucurolysin-B is based on the consideration of the snake species (*leucur-*) and because it is the second metalloendopeptidase to be isolated from *B. leucurus* venom. The extent of purification and the yield at each step are summarized in Table 33.1. A 20.6-fold purification was achieved with a yield of 24.5% with respect to hemorrhagic activity. Leuc-B represented 1.2% (w/w) of the total venom protein.

33.1.2 Structural Chemistry

N-terminal sequence analysis of leuc-B by automated Edman degradation failed, strongly suggesting that the N-terminal amino acid of the proteinase is blocked, a common finding in this class of enzymes.[19,21,22] However, the tryptic peptide fragments corresponding to residues 86–422 of other SVMPs were obtained in high yield and confirmed the sequence from 86 to 422 (Fig. 33.2). The amino acid sequence analysis of leuc-B establishes that it is a member of the P-III class of SVMPs, comprising an N-terminal metalloproteinase, a disintegrin-like domain, and a C-terminal cysteine-rich domain. The multidomain structure is in agreement with the common precursor model of snake venom metalloproteinase/disintegrin.[23] By

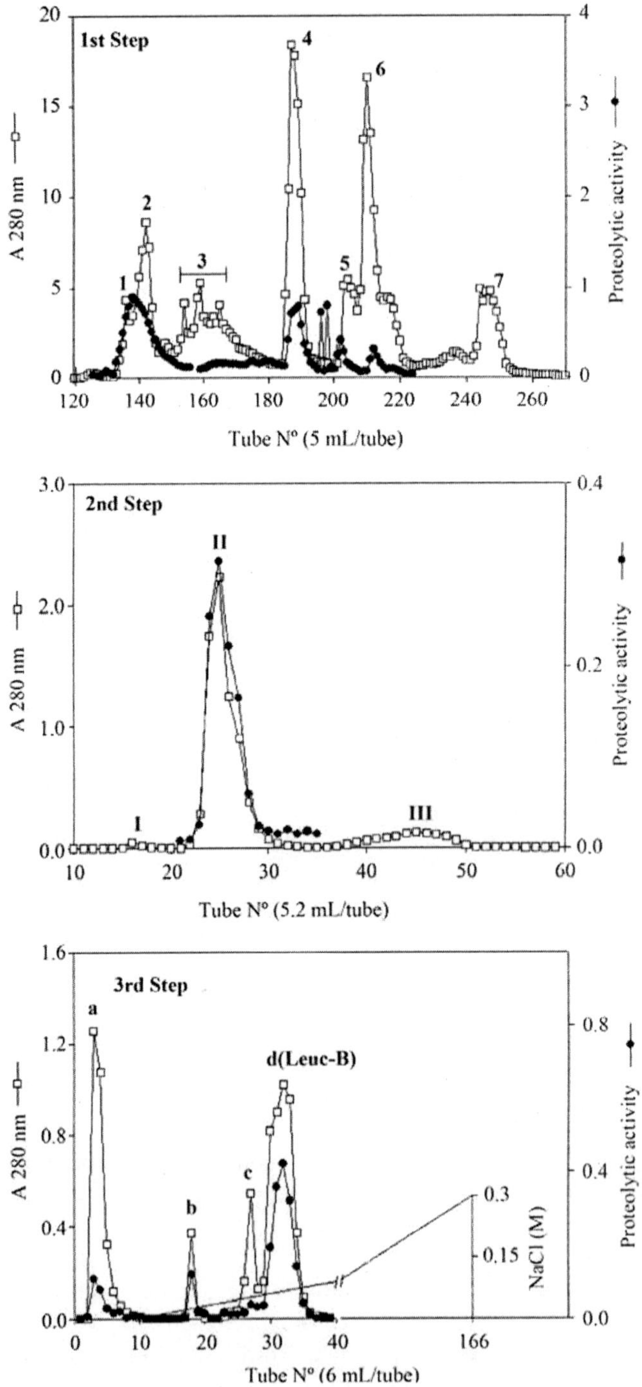
1st Step
A 280 nm
Proteolytic activity
Tube N° (5 mL/tube)
2nd Step
II
I
III
Tube N° (5.2 mL/tube)
3rd Step
a
b
c
d(Leuc-B)
NaCl (M)
Tube N° (6 mL/tube)

TABLE 33.1 Purification of Leucurolysin-B from *B. leucurus* Snake Venom

				Hemorrhagic Activity				Proteolytic Activity			
	Protein		MDH	Sp. Activity (U/mg[a] × 10^3)	Total	Yield		Sp. Activity (U/mg[c] × 10^3)	Total	Yield	
Step	mg	%	(μg)		(U)	(%)	PF[b]		(U)	(%)	PF[b]
Crude venom	1680	100	0.24	1.6	2688	100	1	8.5	14,280	100	1
Sephacryl S-200	29.7	1.8	0.90	4.5	134	4.9	3	35.4	1051	7.4	4
Sephacryl S-300	26.0	1.6	0.10	10	260	9.7	6	25.2	655	4.6	3
DEAE Sepharose CL-6B	20.0	1.2	0.03	33	660	24.5	20.6	34.7	694	4.8	4

[a] One unit of hemorrhagic activity is one minimum hemorrhagic dose (MHD) per milligram.
[b] PF: purification factor.
[c] One unit of DMC-hydrolyzing activity was defined as ΔA_{340}/min. The enzyme activity was expressed as units/mg.

SDS-PAGE, the native proteinase has a calculated Mr of 55 kDa (Fig. 33.3a, lane N) with three potential glycosylation sites at Asn_{183}, Asn_{385}, and Asn_{400}. Glycoprotein bands observed on SDS-PAGE are often broad due to the heterogeneous glycosylation pattern; thus, we have examined the Asn-linked sugar chains of leuc-B. The enzyme was digested with N-glycosidase F (PNGase F) that specifically cleaves off Asn-linked chains from glycoproteins. When the carbohydrate structures were removed from the native protein (55 kDa) by PNGase F treatment, the deglycosylated protein migrated as a smaller molecular size corresponding to 40 kDa (Fig. 33.3a, lane D: deglycosylated; lane N: native). These results suggest that glycosylation at one, two, or three of these sites gives rise to the differences in the calculated Mr of these proteins. In addition, glycan chains can significantly alter protein conformation and may

FIGURE 33.1 Purification of leuc-B from *B. leucurus* venom. *First step*: Lyophilized *B. leucurus* venom (2 g, 1680 mg protein) was loaded onto two (100 cm × 2.5 cm each) columns in series of Sephacryl S-200. The columns were equilibrated and eluted with 50 mM ammonium acetate buffer (pH 7.3) containing 0.3 M NaCl, at a flow rate of 7 mL/h. The tubes 133–137 (active fraction 1) containing proteolytic and hemorrhagic activities were pooled and concentrated for the next step. *Second step*: Fraction 1 was subsequently applied to a Sephacryl S-300 (100 cm × 1.5 cm) column, which was equilibrated and eluted with 25 mM Hepes buffer, pH 7.5, containing 2 mM $CaCl_2$ at a flow rate of 7 mL/h. The active fraction II (tubes 24–29) was dialyzed against distilled water and lyophilized. *Third step*: Fraction II was applied to a DEAE ion-exchange column (20 cm × 1.6 cm). The column was equilibrated with 50 mM Hepes buffer, pH 7.5, containing 2 mM $CaCl_2$. Elution was performed at a flow rate of 12 mL/h with a buffer gradient as indicated. The active fractions (tubes 30–34) containing leuc-B were pooled.

```
           1        10        20        30        40        50
Leuc-B  -----------------------------------------------------
Jarar   EQQRYDPYKYIEFFVVVDQGTVTKNNGDLDKIKARMYELANIVNEIPRYLYMH
Bothr   EQQKYNPFRYVELFIVVDQGMVTKNNGDLDKIKARMYELANIVNEILRYLYMH
HR1B    EQQRF-PRRYIKLAIVVDHGIVTKHHGNLKKIRKWIYQLVNTINNIYRSLNIL
Leuc-a  ------SPRYIELVVVADHGMFKKYNSNLNTIRKWVHEMLNTVNGFYRSMNVD

               60        70        80        90       100
Leuc-B  ---------------------------------DTVLLNRISHDNAQLLTAIVF
Jarar   VALVGLEIWSNGDKITVKPDVDYTLNSFAEWRKTDLLTRKKHDNAQLLTAIDF
Bothr   AALVGLEIWSNGDKITVKPDVDYTLNSFAEWRKTDLLTRKKHDNAQLLTAIDF
HR1B    VALVYLEIWSKQNKITVQSASNVTLDLFGDWRESVLLKQRSHDCAQLLTTIDF
Leuc-a  ASLVNLEVWSKKDLIKVEKDSSKTLTSFGEWRERDLLPIRSHDHAQLLTVIFL
                                        :  **    **  ***** * :
           110       120       130       140       150
Leuc-B  NENVIGKAYTGGMCDPRYSVGVVMDHSPINRLVADTMAHEMGHNLGIHHDTGS
Jarar   NGPTIGYAYIGSMCHPKRSVGIVQDYSPINLVVAVIMAHEMGHNLGIHHDTGS
Bothr   NGPTIGYAYIGSMCHPKRSVAIVEDYSPINLVVAVIMAHEMGHNLGIHHDTDF
HR1B    DGPTIGKAYTASMCDPKRSVGIVQDYSPINLVVAVIMTHEMGHNLGIPHDGNS
Leuc-a  DEETIGIAYTAGMCDLSQMVAWGQDY------VAVSMAHELGHNLGMRHDGNQ
        :   **  ** : **      :*   *:****  :**   *  **  **** *  **
        160       170       180       190       200       210
Leuc-B  CSCGGHSCIMSRVISHQPLQYFSNCSYIEYWDFITKLNPQCILNEPLRTDIVS
Jarar   CSCGDYPCIMGPTISNEPSKFFSNCSYIQCWDFIMNHNPECIINEPLGTDIIS
Bothr   CSCGDYPCIMGPTISNEPSKFFSNCSYIQCWDFIMKENPQCILNEPLGTDIVS
HR1B    CTCGGFPCIMSPMISDPPSELFSNCSKAYYQTFLTDHKPQCILNAPSKTDIVS
Leuc-a  CHCNAPSCIMADTLSKGLSFEFSDCSQNQYQTYLTKHNPQCILNKP
        * *     ***    :*         **:**:        :      :*:**:* *   ***:*
              220       230       240       250       260
Leuc-B  PPVCGNELLEMGEECDCGSPRNCRDLCCDAATCKLHSWVECESGECCDQCRFI
Jarar   PPVCGNELLEVGEECDCGTPENCQNECCDAATCKLKSGSQCGHGDCCEQCKFS
Bothr   PPVCGNELLEVGEECDCGTPENCQNECCDAATCKLKSGSQCGHGDCCEQCKFS
HR1B    PPVCGNELLEAGEECDCGSPENCQYQCCDAASCKLHSWVKCESGECCDQCRFR
        ***********  *******:*  **:    *****:***:*     *   *:**:**:*
            270       280       290       300       310
Leuc-B  KAGNVCRPPRKECDVAEACTGQSAQCPTDDFKRNGQPCLNNYAYCYQGNCPIM
Jarar   KSGTECRASMSECDPAEHCTGQSSECPADVFHKNGQPCLDNYGYCYNGNCPIM
Bothr   KSGTECRASMSECDPAEHCTGQSSECPADVFHKNGQPCLDNYGYCYNGNCPIM
HR1B    TAGTECRAAESECDIPESCTGQSADCPTDRFHRNGQFCLYNHGYCYNGKCPIM
           *  **     ***   *  **** * :** *  *:   ***  **  *  :***:*:****
        320       330       340       350       360       370
Leuc-B  YHQCYALFGSDATMAQDSCFQVNKKGNEYFYCRLENGINIPCAQEDVKCGRLF
Jarar   YHQCYALFGADVYEAEDSCFKDNQKGNYYGYCRKENGKKIPCAPEDVKCGRLY
Bothr   YHQCYALFGADVYEAEDSCFKDNQKGNYYGYCRKENGKKIPCAPEDVKCGRLY
HR1B    FYQCYFLFGSNATVAEDDCFNNNKKGDKYFYCRKENEKYIPCAQEDVKCGRLF
         :*** ***  :     *:*  **: *:**: * ***  **    ****  ********
              380       390       400       410       420
Leuc-B  CHNMKYEQD--CNYSDR------GMVDNGTKCAEGKVCNSNR-----QAYQR
Jarar   CKDNSPGQNNPCKMFYSNDDEHKGMVLPGTKCADGKVCSNGHCVDVATAY
Bothr   CKDNSPGQNNPCKMFYSNDDEHKGMVLPGTKCADGKVCSNGHCVDVATAY
HR1B    CDNK----KYPCHYNYSEDLDF-GMVDHGTKCADGKVCSNRQCVDVNEAYKS
        *  :     *:   *                ***  *****:****                **:
```

consequently modulate the functional activity of a protein as well as protein–protein interactions. Furthermore, gelatin zymography analysis of leuc-B was performed to visualize its proteolytic activity. The zymogram showed a cleared zone of the gel associated with a 55-kDa molecule (Fig. 33.3b).

The sequence of the amino terminal half (residues 86–205) of leuc-B contains a metalloproteinase structure similar to that of the P-I class SVMPs such as leuc-a, isolated from the same *B. leucurus* venom[18] (Fig. 33.2) as well as the metalloproteinase domains of high Mr hemorrhagins from snake venoms (Fig. 33.2). On the basis of the sequence homology data, the motif **HEXXHXXGXXH** from residues 145–155, as well as the methionine-turn, CIM (residues 167–169), is also involved in the zinc binding of the metzincin superfamily.[24,25] As shown in Fig. 33.2, leuc-B shows extensive similarity to other P-III SVMPs, including jararhagin and bothropasin from *B. jararaca* and HR1B from *Trimeresurus flavoviridis*. In its disintegrin-like domain, leuc-B contains a conserved putative integrin-interacting motif ECD (residues 278–280 in place of the RGD motif of disintegrins), structurally related to the family of cell surface and secreted integral membrane glycoproteins that are generally called ADAMs/MDC.[13,26,27] Both SVMPs and ADAMs belong to the reprolysin family of metalloproteinases. The disintegrin-like domain that has high homology to the disintegrins containing an RGD loop of the disintegrins is thought to be involved in platelet aggregation inhibitory activity, as reported for some P-III peptidases.[15,28]

Comparison of the primary structures of class P-III and P-IV of SVMPs has indicated that the number and location of cysteinyl residues in the metalloproteinase domain might be the major differentiating factor associated with their biological activities.[13] Commonly, there are seven cysteinyl residues in the proteinase domain, six of which are highly conserved and the seventh is variable. On the basis of this observation, researchers have divided these proteins into subclasses of P-IIIa and P-IIIb. For instance, the toxins jararhagin and bothropasin (*B. jararaca*), catrocollastatin (*Crotalus atrox*), and acurhagin (*Agkistrodon acutus*) have a cysteinyl residue at the position 189[19] (leuc-B numbering, Fig. 33.2) in the proteinase domain.[13,28] It is possible that the seventh cysteinyl residue of leuc-B may be at position 61 as in the case of bilitoxin-1 (*Agkistrodon bilineatus*).[29] In the former case, the Cys residue at position

FIGURE 33.2 Comparison of the amino acid sequence of leucurolysin-B with the sequences of other SVMPs. The amino acid sequence of leuc-B (this work) was aligned with those of other P-III SVMPs, jararhagin (Jarar) from *B. jararaca* (GenBank accession no. P30431), bothropasin (Bothr) from *B. jararaca* (AAC61986), HR1B (trimerelysin I) from *T. flavoviridis*,[36] and leuc-a, a P-I SVMP from the same *B. leucurus*.[20] Zinc binding motif (HEMGHNLGIH) and the methionine 169 of the basement (Met-turn) are invariant and are expressed in boldface. Disintegrin-like (ECD) sequences are expressed in boldface. Gaps were inserted to obtain maximum degrees of similarity. Putative N-linked glycosylation sites are in gray box. PyroGlu residue is conventionally assigned as residue 1.[57] Numbers on the top indicate the residue number of leuc-B. The spacer domains are in gray. Positions fully conserved (*); strongly similar residues (:)

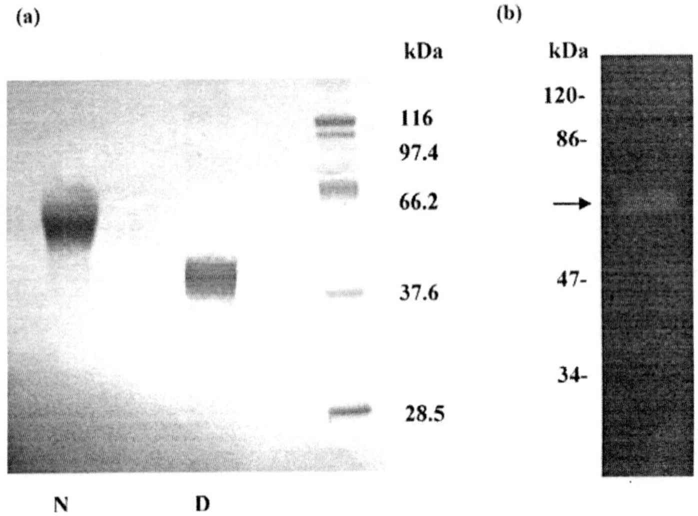

FIGURE 33.3 SDS-PAGE and zymogram analyses of native and deglycosylated leuc-B. (a) Purified leuc-B before (lane N) and after treatment with N-glycosidase F (two units) at 37°C for 22 h (lane D). Samples (5 μg per lane) were then subjected to SDS-PAGE (12% gel) under reducing conditions and stained with Coomassie blue. Molecular mass markers are indicated at the right. (b) Leuc-B (10 μg) was subjected to a 10% gelatin zymography, subsequently to renaturing and incubation overnight. Proteolytic activity of leuc-B can be seen as the clear band (55 kDa) in the darkly stained gel.

61 may be either unpaired or involved in disulfide bond exchange with one of the conserved site cysteins. As stressed earlier, these are speculations and require experimental data for proof. It has been suggested that these proteinases (subclass P-IIIa) can undergo proteolysis/autolysis during secretion or in the venom glands to produce an active tandem domain product comprising the disintegrin-like and cysteine-rich domains (DC domains). On the other hand, members of the subclass P-IIIb, including leuc-B, HR1b, and others, have a tyrosine residue at this position[13,19] (Fig. 33.2). Moreover, it has been suggested that a new subclass P-IIIc, including the two SVMPs stejnihaging-A (accession no. DQ195154) and stejnihaging-B (accession no. DQ195153) from *T. stejnegeri*, which together with HR1b shared the common cysteinyl residue at the position 100 in the proteinase domain, exists. Data in the literature indicate that most of the metalloproteinase domain in P-III share six conserved cysteinyl residues at positions 120, 160, 162, 167, 184, and 200 (leuc-B numbering).[13,19,30] Moreover, all the P-III SVMPs from elapid venoms, such as kaouthiagin (*Naja kaouthia*) and ohagin (*Ophiophagus hanah*), have six conserved cysteinyl residues in the proteinase domain, but the seventh cysteinyl residue was not found.[30]. Although in the cysteine-rich domain, the number and position of cysteinyl residues of the elapid Asian snakes ohagin and kaouthiagin are identical with those of viperidae SVMPs, two additional cysteinyl residues were found in this domain of mocarhagin (GenPep AA51550), BmPM (*Bungarus multicinctus*) and BfMP (*B. fasciatus*).[30] In some of the P-III class (e.g., atrolysin A and jararhagin), it has

been shown that this domain is involved in collagen binding.[15,32–34] In contrast, in the disintegrin-like domain, all the cysteinyl residues are conserved in both viperid and elapid SVMPs. It is known that all the proteins are proteolytically active having functional metalloproteinase domains. In addition to this proteolytic activity, there are a variety of functions associated with P-III SVMPs, which suggests that the disintegrin-like and cysteine-rich domains may be important in their biological activities. As documented elsewhere, the disintegrin-like domains of these proteases are considered capable of binding receptors and thus could serve a double function: targeting the enzyme to a specific cell surface integrin for subsequent cleavage and inactivation as well as for inhibiting integrin binding to the matrix ligand. [13,15,27,31] However, the exact role(s) that the disintegrin-like domain plays in the context of P-III enzymes' biological effect has not been totally elucidated. Further investigations are needed to elucidate the structure–function relationship of this domain in the overall function of these proteins.

33.1.3 Stability Properties

The purified preparation was stable for several weeks either frozen or at 4°C; however, prolonged storage accompanied by freezing and thawing results in some autolysis, which becomes manifest by the appearance of one or two bands on SDS-PAGE, even when no obvious loss in activity was evident. In addition, the stability of leuc-B was tested by exposing it to various pH values and temperatures. The general proteolytic activity of the enzyme was tested by the sensitive DMC assay and on other globular protein substrates. The enzymatic activity was more stable in the alkaline region. Thus, the enzyme shows maximal activity at pH 7.5–9.0 and at 30°C but irreversibly loses its activity above 50°C.

33.1.4 Activity and Specificity

Leuc-B exhibits strong proteolytic activity of 35 units/mg when DMC is used as a substrate. To characterize the peptide bonds that are cleaved by the enzyme and the manner in which leuc-B acts on potential ECM and plasma proteins substrates we used type IV and I collagens and their gelatins, laminin-111, enactin/nidogen, vitronectin, fibronectin (FN), and fibrinogen (Fg). It is known that the multidomain SVMPs (P-III class) are significantly more potent in inducing hemorrhage (about 30-fold) than the single-domain P-I class, which suggests that the disintegrin-like and cysteine-rich domains may function in the recognition and binding of substrates, and in hemorrhagic potency.[4–6] We have demonstrated that leuc-B (minimum hemorrhagic dose, MHD 30 ng, in rabbit) and mutalysin-I (mut-I, MHD = 0.5 μg, in rabbit) showed restricted specificity and potent hemorrhagic effect, while the P-I enzymes such as leuc-a are nonhemorrhagic[18] and mutalysin-II (mut-II, *L.m. muta*) is weakly hemorrhagic.[35,36] Thus, when oxidized insulin B-chain was used as a model substrate, leuc-B cleaved only the Ala_{14}–Leu_{15} bond at a slow rate (Table 33.2). Under similar experimental conditions, the hemorrhagic toxin mut-I from bushmaster (*L.m. muta*) venom showed identical specificity. Thus, mut-I cleaves only the Ala_{14}–Leu_{15} bond of insulin B-chain

TABLE 33.2 Proteolytic Specificity of Some SVMPs on Oxidized Insulin B-Chain

SO_3H (Cys 7), SO_3H (Cys 19)

1 ... 10 ... 15 ... 20 ... 25

F-V-N-Q-H-L-C-G-S-H-L-V-E-A-L-Y-L-V-C-G-E-R-G-F-F-Y-T-P-K-A

	N3–Q4	H5–L6	H10–L11	A14–L15	Y16–L17	F24–F25
P-III						
Leuc-B				↑		
Mut-I[35]				↑		
Atrolysin a[4]	↑	↑	↑	↑	↑	
HP-IV[37]				↑		
Hf-b[38]				↑		
P-I						
Leuc-a[18]				↑	↑	
Neuwiedase[64]				↑	↑	
Fibrolase[8]				↑		
Mut-II[35]		↑	↑	↑		↑

at a slow rate (specific activity 1/1000 that of mut-II). In agreement with these results, hydrolysis of the collagenase substrate Abz–Pro–Leu–Gly–Leu–Leu–Gly–Arg–EDDnp (Abz, *o*-aminobenzoic acid; EDDnp, *N*-(2,4-dinitrophenyl)ethylenediamine) by mut-I at the Leu–Leu bond was 7060-fold lower than that by mut-II.[35] Other P-III toxins, for example, the hemorrhagic protease IV from timber rattlesnake[37] and the hemorrhagic factor b from the Chinese habu snake,[38] were also characterized by their restricted substrate specificity and high hemorrhagic effects. The substrate specificity of several P-I and P-III SVMPs, whether hemorrhagic or nonhemorrhagic, against oxidized insulin-B chain is summarized in Table 33.2. These results indicate that the substrate specificity of each SVMP is distinct and characteristic. In purified leuc-B, none of the following enzyme activities contained in the original venom was detectable: neither coagulant activity, arginine (TAME)-hydrolyzing, and BAPNA-amydolitic activity, nor L-amino acid oxidase and phospholipase A_2 activities.

Most SVMPs reported to date induce hemorrhage by degrading components of the ECM and by hydrolyzing various blood coagulation factors. Leuc-B hydrolyzes fibrinogen and fibrin *in vitro*. The fibrinogenolytic activity of leuc-B was determined by incubating it with human Fg (0–60 min) that is also required for platelet aggregation. As shown in Fig. 33.4a, the enzyme can degrade Fg to unclottable products with the Aα-chains being rapidly degraded. Similarly, leuc-B was able to completely degrade the α-chains of fibrin in a time-dependent (Fig. 33.4b) and dose-dependent (not shown) manner without significant enzymatic cleavage of β-chains. Furthermore, its fibrinolytic effect is direct and does not rely on the plasminogen activation. Consequently, leuc-B is characterized as an anticoagulant α-fibrinogenase.[19] Like other P-III SVMPs, leuc-B can cause severe bleeding in envenomed animals by interfering with the blood coagulation cascade and hemostatic plug formation as well

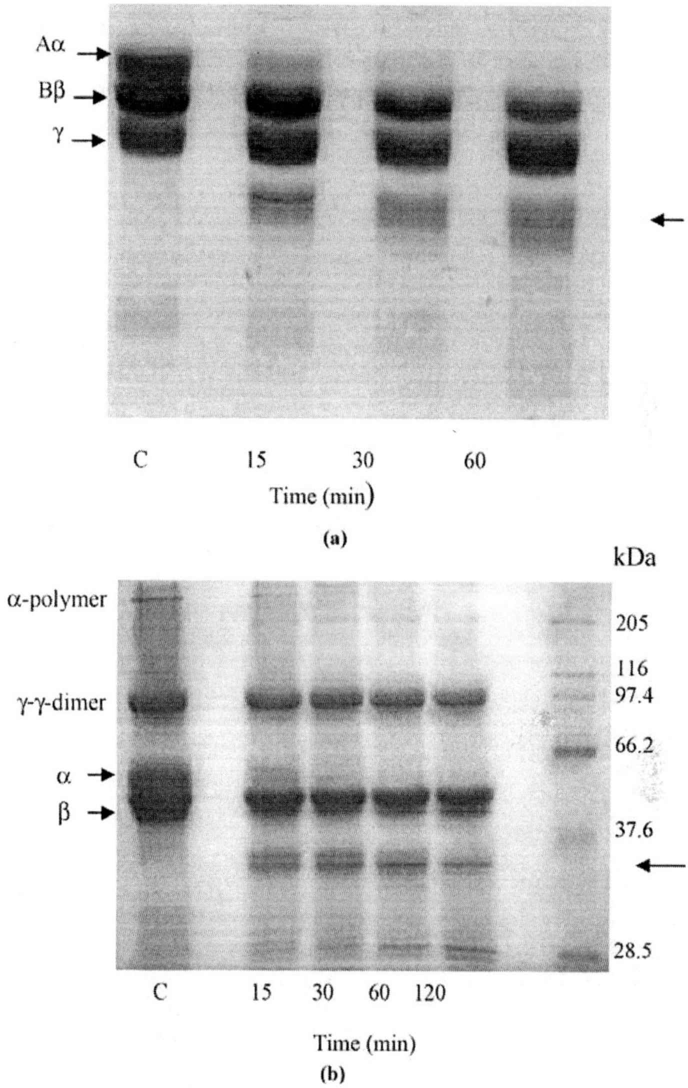

FIGURE 33.4 Digestion of fibrinogen and fibrin by leuc-B. (a) Samples of 500-μL human fibrinogen (2.5 mg/mL) were incubated with leuc-B (3 μg) for the indicated time intervals at 37°C. At each interval, aliquots of 50 μL were withdrawn and mixed with an equal volume of denaturing solution (10 M urea, 4% β-mercaptoethanol, and 4% SDS). Samples were reduced and denatured overnight at room temperature and then subjected to SDS-PAGE (12%). The positions of three polypeptide chains of fibrinogen control (lane C), Aα, Bβ, and γ are shown as indicated. (b) Human fibrin was incubated with leuc-B for the indicated time intervals and aliquots (5 μg) were subjected to SDS-PAGE (12%) under reducing conditions. Fibrin control (no leuc-B) after 120 min incubation is shown (lane C). Molecular mass markers are shown at the right. Arrows on the right-hand side of the gels indicate the main degradation products.

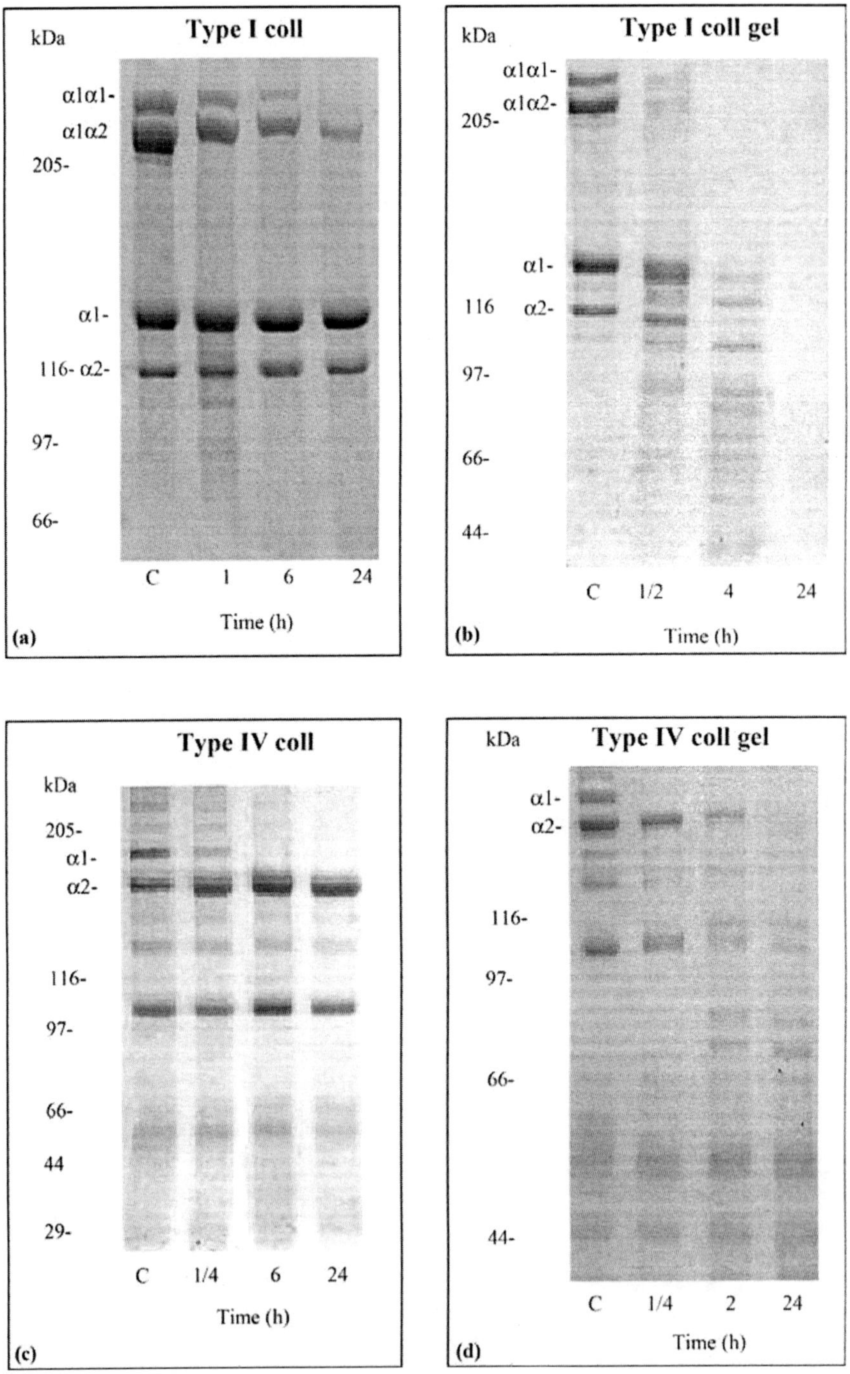
kDa
Type I coll
α1α1-
α1α2
205-
α1-
116- α2-
97-
66-
C 1 6 24
Time (h)
(a)
kDa
Type I coll gel
α1α1-
α1α2-
205-
α1-
116 α2-
97-
66-
44-
C 1/2 4 24
Time (h)
(b)
Type IV coll
kDa
205-
α1-
α2-
116-
97-
66-
44
29-
C 1/4 6 24
Time (h)
(c)
kDa
Type IV coll gel
α1-
α2-
116-
97-
66-
44-
C 1/4 2 24
Time (h)
(d)

as by degrading the ECM components surrounding blood vessels of envenomed animals. In *in vivo* models, disruption of microvessels becomes evident within a few minutes of an intradermal injection of leuc-B as well as by other SVMPs, while *in vitro* the proteolytic attack of these enzymes on ECM proteins is slow.

33.1.5 Degradation of ECM and Basement Membrane Components

The action of SVMPs on proteins of the ECM has been studied and the mechanism of hemorrhage induction investigated by several authors.[5,39,40] For instance, the larger P-III atrolysins (a and e, *C. atrox*) degraded ECM proteins more potently than smaller P-I atrolysins (c, b, and d) found in the same venom. This observation is positively correlated with the potency of these enzymes to induce hemorrhage.[39,40] It is well known that the proteolytic disruption of capillary BM is an important element in the sequence of events leading to the appearance of hemorrhage by hemorrhagic toxins. The proteolytic activity on ECM components, for example, type IV collagen, EN, and vitronectin, as well as on some plasma proteins (FN, Fg) reflects the fact that leuc-B is the most potent hemorrhagic toxin in *B. leucurus* venom. Under our experimental conditions, the interstitial type I collagen was readily degraded by leuc-B. As shown in Fig. 33.5, the first targets to be degraded are the naturally cross-linked α1α1-chain homodimers followed by the cross-linked α1α2-chain heterodimers. Type IV collagen is considered a major structural component in the BM, which serves as the foundation to which endothelial, epithelial, smooth muscle, fat, and nerve cells are anchored. Despite its stability against other proteases, this collagen was rapidly cleaved in 1 h at the α1-chain (Mr ~ 200 kDa). Its α2-chain was more resistant to hydrolysis, but its molecular mass also changed slightly (Fig. 33.5). In this context, selective degradation of type IV collagen has been associated with the ability of hemorrhagic toxins from snake venoms to induce hemorrhage,[38–41] considering the important scaffolding role that type IV collagen plays in the architecture of the BM.[42] After denaturation into gelatins, type I and IV collagen substrates were degraded completely, emphasizing the proteolytic stability of the collagen triple helix (Fig. 33.5).

Similar to hemorrhagic toxins from other sources such as a metalloproteinase from *Vibrio vulnificus*, leuc-B degrades type IV collagen but not LM.[43] Laminins are the most abundant glycoproteins in BMs and bind to collagen type IV, heparan sulfate proteoglycan, nidogen/enactin, and themselves to create an integrated structure within the BM. As substrate for leuc-B we have used the laminin-111/nidogen (LM–EN)

FIGURE 33.5 SDS-PAGE analysis of type I and IV collagens and their gelatins after digestion with leuc-B. Purified type I and IV collagens (left) and their gelatin (right) samples were incubated (37°C) for different periods, as indicated at the bottom of each panel with leuc-B at molar ratio of 1:50 (enzyme/substrate). Digestion products obtained were analyzed by 5% SDS-PAGE. Typical type I collagen chains (cross-linked α1α1 and α1α2, chain dimers, and monomeric α1 and α2) are indicated at left (top). Typical type IV collagen chains (α1, α2) are indicated at left (bottom). Samples of denatured type I (left top) and IV collagen (left bottom) gelatin samples were incubated at 25°C for different periods (bottom of figures) with leuc-B at ratio of 1:50 (enzyme/substrate). Products obtained were analyzed by 7.5% SDS-PAGE.

complex, purified from Engelbreth–Holm–Swarm (EHS) tumors as previously described.[44,45] Laminin-111 appears as three major bands, α1 of approximately 400 kDa and the comigrating bands of β1 and γ1 of approximately 220 and 210 kDa, respectively. In the SDS-PAGE analyses, laminin-111 appears to be resistant to proteolysis by leuc-B; no evidence of degradation products was observed even after 24 h incubation.[19] Nidogen/enactin was first described as a Mr 158,000 sulfated glycoprotein present in BM.[46] Leuc-B totally cleaves EN substrate giving rise to two major degradation products of 100 and 50 kDa, respectively, after 1 h,[19] in accordance with the high protease susceptibility of nidogen. As nidogen/enactin cross-links the collagen IV and laminin networks within the BM, the fast cleavage of nidogen by leuc-B may substantially contribute to the effective destruction of the subendothelial BM, its leakage, and thus to the formation of hemorrhages.

FN that plays a central role in cell adhesion exists in a soluble protomeric isoform in micromolar concentration in blood plasma and in an insoluble multimeric isoform in the ECM based on different splice variants.[47] FNs are composed of structurally similar polypeptides varying in size between 210 and 250 kDa. Plasma FN appears as a band of approximately 210 kDa. From SDS-PAGE analysis, the digestion pattern of FN by leuc-B showed a major degradation product of ~80 kDa after 1 h incubation, followed by other fragments that appeared below the FN control after 6 h.[19] Plasma vitronectin is known to be involved in the processes that occur subsequent to platelet stimulation and blood clot formation, such as the binding of heparin, angiogenic sprouting of endothelial cells, as well as fibrinolysis. There are two physiologically occurring forms of vitronectin: a 75 kDa one-chain form and the 65 kDa nicked two-chain form. The band of the 75 kDa form appears to be digested more rapidly by the proteinase, leading to the 65 kDa band. The major degradation product has an apparent Mr of ~42 kDa after 1 h incubation.[19] From these experiments and data in the literature, proteolysis of the ECM proteins, especially type IV collagen and nidogen/enactin, may account for the disruption of the subendothelial BM and thus for the hemorrhagic activity of SVMPs. However, the *in vitro* proteolysis of ECM components is slow compared with hemorrhagic effect *in vivo*. Perhaps the cleavage of certain key matrix molecules, such as nidogen/enactin, which due to their cross-linking function have a prominent role in BM architecture and stability, may explain the detrimental effect of leuc-B on the blood vessel wall. However, the exact mechanism of venom-induced hemorrhage is not fully understood yet.

33.1.6 Effect of Leuc-B on Platelet Function and Binding to α1β1 and α2β1 Integrins

Proteins from snake venoms that modulate platelet adhesive interactions are mainly from either of two major structural families: the C-type lectin-like family or the metalloproteinase/disintegrins. It has been widely documented that the majority of SVMPs reported in the literature have inhibitory effect on platelet aggregation induced by several agonists.[9,48,49] The inhibitory effects of SVMPs on platelet aggregation were attributed to their multiple functions targeting platelet ligands and/or membrane receptors. Leuc-B has a disintegrin-like domain containing the ECD sequence instead of the typical RGD/KGD of disintegrins, which is the binding motif of several cell surface integrins and is highly similar to that of other P-III SVMPs. Thus, the proteinase was able to suppresses the collagen-induced platelet aggregation in a dose-dependent manner, when tested with platelet-rich plasma.[19] As both native and EDTA-inactivated proteinases exhibited distinct potency to block platelet aggregation, it was evident that their capability to block platelet aggregation does not depend on their enzymatic activity. However, the molecular mechanism of this effect remains to be evaluated in the near future. In connection with this, the integrins $\alpha1\beta1$ and $\alpha2\beta1$ are important cellular receptors for collagens. Whereas $\alpha2\beta1$ integrin is the only collagen binding integrin on platelets, endothelial cells mediate their attachment to collagens by both integrin receptors $\alpha1\beta1$ and $\alpha2\beta1$. A hallmark of the integrins is the ability of individual family members to recognize several ligands. Previous studies have shown that jararhagin (P-III class) inhibits collagen-induced platelet aggregation by binding to $\alpha2\beta1$ integrin, thereby blocking collagen binding and its subsequent cell surface receptor-mediated signaling.[50,51]

In addition, it has been reported that a cyclic RKKH peptide derived from the metalloproteinase domain of jararhagin recognizes the collagen binding I domain within the integrin $\alpha2$ subunit and thus prevents collagen binding to $\alpha2\beta1$ integrin.[52] Other studies have indicated that peptides derived from all three domains can inhibit $\alpha2\beta1$,[15,33] and therefore, SVMPs might contain several integrin recognition sites. A primary goal of many structure–function analyses in the integrin research field has been the reduction of macromolecular ligands to minimal recognition sequences.[53] This endeavor has been highly successful, and many amino acid motifs of large ECM proteins have been tested for their potential to interfere with integrin–ligand interaction.[54] The prototype example is the RGD sequence. A number of P-III toxins, for example, atrolysin A (*C. atrox*)[28] and more recently, halysase (*Glodius halys*)[55] and AAVI (*A. acutus*)[56], also bind collagen and inhibit platelet aggregation. As shown in Fig. 33.2, the sequence RKKH located in the metalloproteinase domain is distant from the active site of jararhagin, which has been showed to be involved in blocking collagen binding to the $\alpha2$ I-domain.[52] This sequence is replaced by the sequence RISH at the corresponding position in leuc-B. ELISA binding assays indicated that leuc-B does not bind to the collagen binding integrins $\alpha1\beta1$ and $\alpha2\beta1$, whereas an interaction of $\alpha2\beta1$ integrin with jararhagin in a cation-independent manner was evidenced.[19,51] These observations are in agreement with Ref. 51. Why the P-III SVMP-derived collagen antagonists have different working mechanisms is unknown. Further investigations on the structure–function relationship of leuc-B would be valuable to understand the mechanism and biological significance of disintegrin-like and cysteine-rich domains of SVMPs and ADAM proteins.

33.1.7 Inhibition and Cross-Reactivity Studies

The effects of some protein inhibitors and cations on proteolytic activity of leuc-B are summarized in Table 33.3. It is well known that the active site of all metzincins is based on the elongated zinc binding motif (**HEXXHXXGXXH**). Further, a structurally conserved methionine residue, located in the so-called Met-turn, is an invariable

TABLE 33.3 Effect of Some Reagents on Proteolytic Activity of Leucurolysin-B

Compound Added	Concentration (mM)	Residual Activity (%)
Enzyme	—	100
EDTA	3	3.3 ± 1.2
PMSF	3	85 ± 4.5
SBTI	200 μg/mL	65 ± 2.9
DDT	2	0
Ca^{+2}	2	114 ± 4.3
Zn^{+2}	2	0
Mg^{+2}	2	125 ± 3.2
$Zn^{+2} + Ca^{+2}$	2 (Each)	0
$Zn^{+2} + Mg^{+2}$	2 (Each)	0
Ca^{+2} + EDTA	2 (Each)	0
Mg^{+2} + EDTA	2 (Each)	0

The purified leuc-B (1 g) was treated with each reagent at the indicated concentrations in 50 mM Hepes buffer, pH 8.0 (1 mL final volume), at 37°C for 30 min. Enzymatic activity was measured with DMC as substrate. These values represent the mean ± SD (n = 4).

feature of all known proteinases of this superfamily.[10] Therefore, SVMPs are dependent on the presence of Zn^{2+} for enzymatic activity and Ca^{2+} for structural stability.[57] In accordance with the data, leuc-B loses its DMC hydrolyzing and hemorrhagic activities after treatment with the metal ion chelator EDTA. Although, Ca^{2+} or Zn^{2+} are required for catalysis, Mg^{2+} and Ca^{2+} enhanced leuc-B activity. In contrast, 2 mM of Zn^{2+} and Cu^{2+} inhibited its proteolytic activity. It has been established that excess of zinc (up to 1 mM) inhibits thermolysin's catalytic activity[58] as well as other metalloproteinases, for example, leuc-a.[18] Furthermore, the addition of Mg^{2+} or Ca^{2+} to solution of leuc-B inactivated with EDTA or Zn^{2+} failed to restore its enzymatic activity. Recent advances in enzyme inhibition have shown that the chelating agents including EDTA and BAPTA (bis(2-aminophenoxy)ethane-*N*,*N*,*N*′,*N*′-tetraacetic acid) were effective in blocking the hemorrhagic effect of *Echis ocellatus* venom as well as the P-III hemorrhagic toxin EoVMP2, while the developed synthetic matrix metalloproteinase inhibitors (MMPIs, marimastat, AG-3340, CGS-270, and Bay-129566) exhibited more variation in their efficiencies.[59] It is known that all protein inhibitors prevent access of substrates to the proteins' catalytic centers through steric hindrance. In addition, the high degree of structural and functional homology between SVMPs and their mammalian relatives MMPs suggests that substrate–inhibitor interactions between these subfamilies of metzincins are likely to be analogous.

The main plasma proteinase inhibitor is α2-macroglobulin (α2-M) that has the unique property of binding and inhibiting the great majority of proteinases regardless of their catalytic mechanism.[60] The venom enzyme was incubated with increasing amounts of the human plasma proteinase inhibitor (5 min at 37°C), and proteolytic activity was assayed on DMC. The results indicated that the P-III leuc-B was not significantly inhibited by α2-M, while the activity of the nonhemorrhagic leuc-a (P-I class) was completely abolished at a 1:1 enzyme/inhibitor ratio (Fig. 33.6) (Sanchez

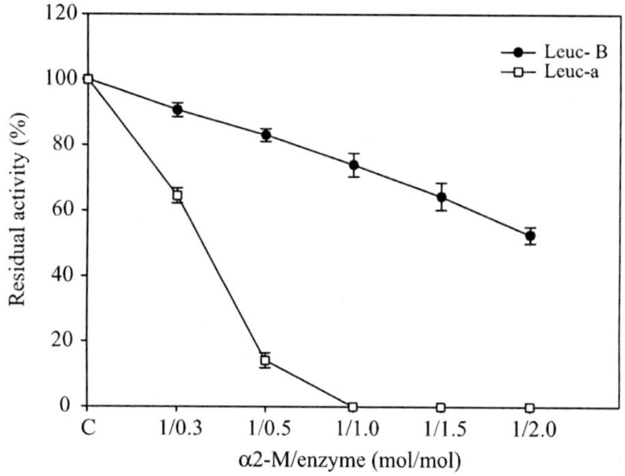

FIGURE 33.6 Stoichiometry of inhibition of leucurolysin-B and -a by α2-macroglobulin. Varying amounts of α2-M were incubated with leucurolysins at molar ratios indicated on the abscissa (inhibitor/enzyme = 0.3–2.0). The residual protease activity is expressed as percentage of the original activity measured on DMC as substrate.

et al., manuscript in preparation). The stoichiometry of 1:1 in the α2-M/enzyme complex was also determined for the homologous P-I SVMPs such as mutalysin-II[9] and atrolysin C/D from *C. atrox.*[61] Effects of α2-M against some larger multidomain SVMPs were also studied. Atrolysin a (Ht-a) from *C. atrox,*[61] mutalysin I from bushmaster venom,[9] and jararhagin (*B. jararaca*)[62] were not inhibited by α2-M, and this has been attributed to the low affinity of these enzymes for the bait region of α2-M. The resistance of leuc-B to inhibition by the principal plasma proteinase inhibitor could allow the active enzyme to persist in the circulation. Therefore, the data offer a possible explanation why leuc-B contributes significantly not only to local but also to systemic bleeding associated with *B. leucurus* envenomation.

There is evidence that P-III class metalloproteases are widely distributed in *Bothrops* and *L.m. muta* venoms.[6,13,19] Thus, we have evaluated the immunological reactivity of rabbit anti-leuc-B antibodies against the Brazilian pit viper (*B. leucurus, B. alternatus, B jararaca, L.m. muta*) venoms and against leuc-a, a nonhemorrhagic P-I class (23 kDa) by ELISAand Western blotting. In titration experiments, the antibodies raised against leuc-B displayed high level of immunoreactivity with its respective antigen and with close phylogenetic links to *B. leucurus*. However, anti-leuc-B antibodies revealed moderate or weak reactivity with either *L.m. muta* venom or leuc-a, P-I class (Fig. 33.7). These differences were also evident when rabbit anti-mut-I antibodies were used. The high level of cross-reactivity against *Bothrops* venoms suggests that SVMPs comprise a high percentage of the protein classes found in these venoms. The data also suggest the existence of several proteins with varying Mr values (20–100 kDa) that show common or similar epitopes in different classes of SVMPs such as those in P-III class. The low reactivity of anti-leuc-B antibodies with a small nonhemorrhagic leuc-a may indicate that additional domains or motifs characteristic of a P-III/P-IV SVMPs have higher levels of antigenicity than the metalloproteinase

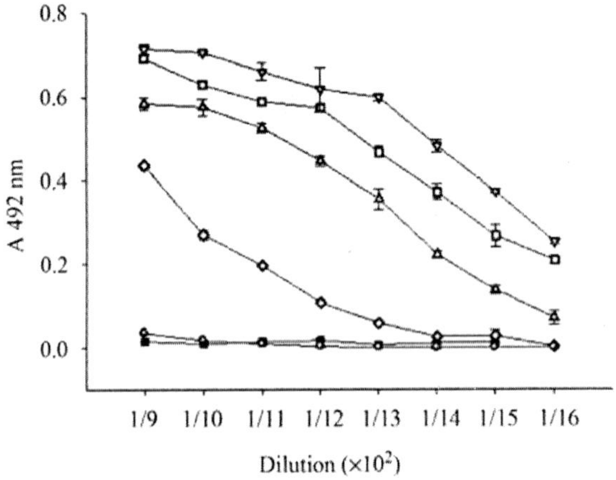

FIGURE 33.7 Cross-reactivity of rabbit anti-leuc-B antibodies against crotalinae venoms and SVMPs by ELISA. Standard concentrations (0.5 μg/well) of leuc-B or venoms were used to challenge the ELISA system with serial dilutions of anti-leuc-B antibodies. Leuc-B (▽) was used as a positive control and normal rabbit serum (◎) was used as a negative control.⊙, *B. leucurus*; Δ, *B. alternatus*; ◇, *L.m. muta*; ◆, leuc-a. The binding was visualized by incubation with peroxidase-coupled anti-rabbit IgG (diluted 1:1000). Each point represents the mean ± SD ($n = 4$).

domain, as suggested for SVMPs from *Echis ocellatus* venom.[63] The antigenic cross-reactivity was also evident by Western blotting. The antibodies raised against the 55-kDa leuc-B detected antigens of molecular size of approximately 55, 33, and 22 kDa in *B. leucurus* venom (data not shown). This may indicate the conservation of key residues or motifs in the exposed domain. The bands at ~55 and 22 kDa revealed by leuc-B antibodies have similar Mr values to leuc-B and leuc-a, respectively.

Furthermore, neutralization of leuc-B′ proteolytic activity by four commercial antivenoms, (i) polyvalent bothropic antivenom (α-B, FUNED, Brazil), (ii) anti-lachetic/bothropic (α-L/B, FUNED, Brazil), (iii) monovalent anti-lachetic (α-L, Butantan Institute, Brazil), and (iv) monospecific anti-crotalic antivenom (α-C, FUNED, Brazil), and by the rabbit anti-leuc-B antibodies was measured. The antigenic pool used to produce the antiserum against *Bothrops* venom (aB) contained venoms from *B. jararaca* (50%), *B. alternatus* (12.5%), *B. neuwiedi* (12.5%), *B. jararacussu* (12.5%), and *B. moojeni* (12.5%) (*Source:* FUNED). Anti-lachetic/bothropic antivenom was raised against Brazilian *L. muta* and *Bothrops* venoms (produced at FUNED). Monospecific anti-lachetic (produced by Butantan Institute, Brazil) and anti-crotalic (*Source:* FUNED) antivenoms were raised against Brazilian *L.m. muta* and *C. durissus terrificus* venoms, respectively. Neutralization of DMC hydrolyzing activity of leuc-B by these therapeutic antivenoms (all at 1:20 dilution), which are currently used clinically, is shown in Fig. 33.8. Surprisingly, the *in vitro* assay shows that anti-lachetic (α-L) and anti-lachetic/bothropic antivenoms (α-L/B) were slightly more effective (66% and 62% inhibition, respectively) than anti-bothropic antivenom (α-B) in neutralizing the proteolytic activity (55%) of leuc-B.

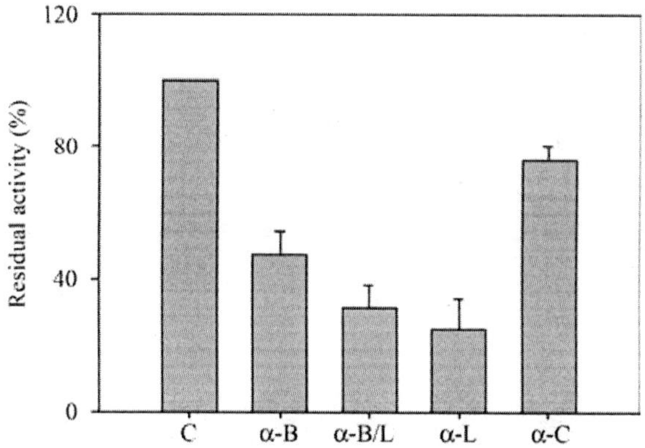

FIGURE 33.8 Inhibition of proteolytic activity of leuc-B by Brazilian commercial antivenoms. One microgram of leuc-B was preincubated with α-B, α-B/L, α-L, or α-C antivenoms (all at 1:20 dilution) for 10 min at 37°C. Proteinase activity in the mixture was assayed with DMC as substrate. As control marked with C, leuc-B activity was measured without any antibody. Values represent percentage of control and are presented as mean ± SD ($n = 4$).

On the other hand, 22% inhibition of leuc-B′ proteolytic activity was observed by anticrotalic antivenom (α-C). The main reason for the moderate efficacy of α-B antivenom to neutralize the proteolytic effect of leuc-B may be its lack of specificity and also because the venoms of species of *Bothrops* included in the antigenic pool do not contain *B. leucurus* venom. In an earlier report, we have demonstrated that the α-B antivenom gave low titers against *B. leucurus*.[65] The fact that structural differences between *L.m. muta* and *Bothrops* toxins determine the antigenic properties could also explain the moderate efficacy of α-B antivenom in neutralizing the proteolytic activity of the hemorrhagic enzyme leuc-B. A specific antibody (JD9) against the C-terminal (disintegrin-like/cysteine-rich) domain of jararhagin possesses extensive immunological reactivity to venom components in snake venoms of distinct species and genera.[66] On the other hand, different epitopes present on SVMPs or on other toxins, which are not or partially neutralized by α-B antivenom, are probably involved in the full expression of the toxicity of the enzymes. In connection with this, the inefficiency of the bothropic antivenom (α-B) on the neutralization of the coagulant effect of *L.m. muta* venom has been reported.[67]

In conclusion, hemorrhagic metalloproteinase, leuc-B, has been purified from *B. leucurus* (white-tailed jararaca) snake venom. Based on the amino acid sequence analysis of leuc-B, we have determined the proteinase to be a P-III class SVMP. The sequence is highly similar to those of P-III SVMPs comprising metalloproteinase, disintegrin-like and Cys-rich domains. Like members of the related ADAM protein family, leuc-B has an ECD sequence in the disintegrin-like domain, which might have a synergistic disturbing activity for platelet aggregation induced by collagen. However, the proteinase does not directly bind to $\alpha 1\beta 1$ and $\alpha 2\beta 1$ integrins, the cellular receptors for collagen. Thus, it is interesting to speculate whether snake venom toxins continue to provide new and novel tools to study collagen receptors for collagen, as well as other platelet targets. Proteolytic degradation of one or more ECM and plasma proteins may lead to the destruction of structural integrity of ECM, resulting in hemorrhage as well as drastic alterations of the hemostatic system. Antibodies raised against leuc-B showed higher cross-reactivity with other P-III proteins and with several *Bothrops* venoms, but low reactivity with the P-I leuc-a. Moderate neutralizing effect of leuc-B and whole venom proteolytic activities by therapeutic α-B serum suggest that this antivenom is lacking antibodies to all the potential toxins within *B. leucurus* venom.

ACKNOWLEDGMENTS

This work was supported by the Brazilan agency FAPEMIG (grants CBB 359/06 and CBB APQ 2673-4.01/07 to E.F.S) and DFG (grant Eb177/4-2 to J.A.E). The German–Brazilian cooperation and exchange was made possible by the DAAD and CAPES (PROBAL D/05/30352).

REFERENCES

1. Kochva, E. The origin of snakes and evolution of the venom apparatus. *Toxicon* **1987**, *25*, 65–106.
2. Warrell, D. A. Snakebites in Central and South America: epidemiology, clinical features, and clinical management. In *The Venomous Reptiles of the Western Hemisphere;* Campbell, J. A.; Lamar, W. W., Cornell University Press: Ithaca, NY, **2004**; pp 709–761.
3. Ownby, C. L. Locally acting agents: myotoxins, hemorrhagic toxins and dermonecrotic factors. In *Handbook of Toxinology;* Shier, W. T.; Mebs, D., Eds.; Marcel Dekker: New York, **1990**; pp 602–654.
4. Bjarnason, J. B.; Fox, J. W. Hemorrhagic metalloproteinases from snake venoms. *Pharmacol. Ther.* **1994**, *62*, 325–372.
5. Gutierrez, J. M.; Rucavado, A.; Escalante, T.; Diaz, C. Hemorrhage induced by snake venom metalloproteinases: biochemical and biophysical mechanisms involved in micro-vessel damage. *Toxicon* **2005**, *45*, 997–1011.
6. Sanchez, E. F. Mutalysins. In *Handbook of Proteolytic Enzymes*, 2nd ed.; Barrett, A. J.; Rawlings, N.; Woessner, F. J., Eds.; Elsevier Academic Press, **2004**; pp 692–694.
7. Gutierrez, J. M.; Lamonte, B. Phospholipase A_2 myotoxins from *Bothrops* snake venoms. In *Venom Phospholipase A_2 Enzymes. Structure, Function and Mechanism;* Kini, R. M., Ed.; John Wiley & Sons, Ltd., **1997**; pp 321–352.
8. Markland, F. S. Snake venoms and the hemostatic system. *Toxicon* **1998**, *36*, 1749–1800.
9. Estevao-Costa, M. I.; Diniz, C. R.; Magalhaes, A.; Markland, F. S.; Sanchez, E. F. Action of metalloproteinases mutalysin I and II on several components of the hemostatic and fibrinolytic systems. *Thromb. Res.* **2000**, *99*, 363–376.
10. Stöcker, W.; Grams, F.; Baumann, U.; Reinemer, P.; Gomis-Rüth, F. X.; McKay, D. B.; Bode, W. The metzincins—topological and sequential relations between the astacins,

adamalysins, serralysins, and matrixins (collagenases) define a superfamily of zinc-peptidases. *Protein Sci.* **1995**, *4*, 823–840.

11. Gomis-Rüth, F. X.; Meyer, E. F.; Kress, L. F.; Politi, V. Structures of adamalysin II with peptidic inhibitors. Implications for the design of tumor necrosis factor α convertase inhibitor. *Protein Sci.* **1998**, *7*, 283–292.
12. Wolfsberg, T. G.; Primakoff, P.; Myles, D. G.; White, J. M. ADAM, a novel family of membrane proteins containing a disintegrin and metaloprotease domain-multipotential functions in cell–cell and cell–matrix interactions. *J. Cell Biol.* **1995**, *131*, 275–278.
13. Fox, J. W.; Serrano, S. M. T. Structural considerations of the snake venom metalloproteinases, key members of the M12 reprolysin family of metalloproteinases. *Toxicon* **2005**, *45*, 969–985.
14. Wan, S. G.; Jin, Y.; Lee, W. H.; Zhang, Y. A snake venom metalloproteinase that inhibited cell proliferation and induced morphological changes of ECV304 cells. *Toxicon* **2006**, *47*, 480–489.
15. Jia, L. G.; Wang, X. M.; Shannon, G. D.; Bjarnason, J. B.; Fox, J. W. Function of disintegrin-like/cysteine-rich domains of atroxlysin A. *J. Biol. Chem.* **1997**, *272*, 13094–13102.
16. Schattner, M.; Fritzen, M.; Ventura, J.; de Alburquerque, M. J. C.; Pozner, R. G.; Moura da Silva, A. M.; Chudzinski-Tavassi, A. M. The snake venom metalloproteases berythractivase and jararhagin activate endothelial cells. *Biol. Chem.* **2005**, *386*, 369–374.
17. Siigur, E.; Aaspollu, A.; Trummal, K.; Tonismagi, K.; Tammiste, I.; Kallinen, N.; Siigur, J. Factor X activator from *Vipera lebetina* venom is synthesized from different genes. *Biochim. Biophys. Acta* **2004**, *1702*, 41–41.
18. Bello, C. A.; Hermogenes, A. L.; Magalhaes, A.; Veiga, S. S.; Gremski, L. H.; Richardson, M.; Sanchez, E. F. Isolation and biochemical characterization of a fibrinolytic proteinase from *Bothrops leucurus* (white-tailed-jararaca) snake venom. *Biochimie* **2006**, *88*, 189–200.
19. Sanchez, E. F.; Gabriel, L. M.; Gontijo, S.; Gremski, H. L.; Veiga, S. S.; Evangelista, K. S.; Eble, J. A.; Richardson, M. Structural and functional characterization of a P-III metalloproteinase, leucurolysin-B, from *Bothrops leucurus* venom. *Arch. Biochem. Biophys.* **2007**, *468*, 193–204.
20. Gremski, L. H.; Chaim, O. M.; Paludo, K. S.; Sade, Y. B.; Otuki, M. F.; Richardson, M.; Gremski, W.; Sanchez, E. F.; Veiga, S. S. Cytotoxic, thrombolytic, and edematogenic activities of leucurolysin-a, a metalloproteinase from *Bothrops leucurus* snake venom. *Toxicon* **2007**, *50*, 120–134.
21. Wang, W. J.; Shih, Ch.; Huang, T. F. Primary structure and antiplatelet mechanism of a snake venom metalloproteinase, acurhagin, from *Agkistrodon acutus* venom. *Biochimie* **2005**, *87*, 1065–1077.
22. Takeya, H.; Oda, K.; Miyata, T.; Omori-Satoh, T.; Iwanaga, S. Snake venom hemorrhagic protein HR-1b. *J. Biol. Chem.* **1990**, *265*, 16068–16073.
23. Kini, R. M.; Evans, H. J. Structural domains in venom proteins: evidences that metalloproteinases and nonenzymatic platelet aggregation inhibitors (disintegrins) from snake venoms are derived by proteolysis from common precursor. *Toxicon* **1992**, *30*, 265–293.
24. Jongeneel, C. V.; Bouvier, J.; Bairoch, A. A unique signature identifies a family of zinc-dependent metalloproteinases. *FEBS Lett.* **1989**, *242*, 211–214.

25. Stöcker, W.; Grams, F.; Baumann, U.; Reinemer, P.; Gomis-Rüth, F. -X.; McKey, D. B.; Bode, W. The metzincins—topological and sequential relations between the astacins, adamalysins, serralysins, and matrixins (collagenases) define a superfamily of of zinc-peptidases. *Protein Sci.* **1995**, *4*, 823–840.
26. Roberts, C. M.; Tani, P. H.; Bridges, L. C.; Laszik, Z. MDC-L, a novel metalloprotease disintegrin cysteine-rich protein family member expressed by human lymphocytes. *J. Biol. Chem.* **1999**, *274*, 29251–29259.
27. Lu, X.; Lu, M. F.; Kakkar, V. V. Snake venom metalloproteinase containing a disintegrin-like domain, its structure–activity relationships at interacting with integrins. *Curr. Med. Chem.-Cardiovasc. Hematol. Agents* **2005**, *3*, 249–260.
28. Shimokawa, K.; Shannon, J. D.; Jia, L. -G.; Fox, J. W. Sequence and biological activity of catrocollastatin-C: a disintegrin-like/cysteine-rich two domain protein from *Crotalus atrox* venom. *Arch. Biochem. Biophys.* **1997**, *343*, 35–43.
29. Nikai, T.; Taniguchi, K.; Komori, Y.; Masuda, K.; Fox, J. W.; Sugihara, H. Primary structure and functional characterization of bilitoxin-1, a novel dimeric P-II snake venom metalloproteinase from *Agkistrodon bilineatus* venom. *Arch. Biochem. Biophys.* **2000**, *378*, 6–15.
30. Guo, X. -X.; Zeng, L.; Lee, W. H.; Zhang, Y.; Jin, Y. Isolation and cloning of a metalloproteinase from king cobra snake venom. *Toxicon* **2007**, *48*, 954–965.
31. Ito, M.; Hamako, J.; Sakurai, Y.; Matsumoto, M.; Fujimura, Y.; Suzuki, M.; Hashimoto, K.; Titani, K.; Matsui, T. Complete amino acid sequence of kaouthiagin, a novel cobra venom metalloproteinase with two disintegrin-like sequences. *Biochemistry* **2001**, *40*, 4503–4511.
32. Jia, L. G.; Bjarnason, J. B.; Fox, J. W. Function of the cysteine-rich domain of the hemorrhagic metalloproteinase atrolysin A: collagen targeting and inhibition of platelet aggregation. *Arch. Biochem. Biophys.* **2000**, *373*, 281–286.
33. Kamiguti, A. S.; Gallagher, P.; Marcinkiewicz, C.; Theakston, R. D. G.; Zuzel, M.; Fox, J. W. Identification of sites in the cysteine-rich domain of the class P-III snake venom metalloproteinases responsible for inhibition of platelet function. *FEBS Lett.* **2003**, *549*, 129–134.
34. Moura da Silva, A. M.; Ramos, O. H.; Baldo, C.; Niland, S.; Hansen, U.; Ventura, J. S.; Furlan, S.; Butera, D.; Della-Casa, M. S.; Tanjoni, I.; Clissa, P. B.; Fernandes, I.; Chudzinski-Tavassi, A. M.; Eble, J. A. Collagen binding is a key factor for the hemorrhagic activity of snake venom metalloproteinases. *Biochimie* **2008**, *90*, 484–492.
35. Sanchez, E. F.; Cordeiro, M. N.; Oliveira, E. B.; Juliano, L.; Prado, E. S.; Diniz, C. R. Proteolytic specificity of two hemorrhagic factors, LHF-I and LHF-II, isolated from the venom of the bushmaster snake (*Lachesis muta muta*). *Toxicon* **1995**, *33*, 1061–1069.
36. Agero, U.; Arantes, R. M. E.; Lacerda-Queiroz, N.; Mesquita, O. N.; Magalhães, A.; Sanchez, E. F.; Carvalho-Tavares, J. Effect of mutalysin-II on vascular recanalization after thrombosis induction in the ear of the hairless mice model. *Toxicon* **2007**, *50*, 698–706.
37. Civello, D. J.; Duong, H. L.; Geren, C. R. Isolation and characterization of a hemorrhagic proteinase from timber rattlesnake venom. *Biochemistry* **1983**, *22*, 749–755.
38. Nikai, T.; Mori, N.; Kishida, Y.; Kato, Ch.; Takenata, T.; Murakami, S.; Shigezane, H.; Sugihara, H. Isolation and characterization of hemorrhagic factors a and b from the venom of the Chinese habu snake (*Trimeresurus mucrosquamatus*). *Biochim. Biophys. Acta.* **1985**, *838*, 122–131.

39. Bjarnason, J. B.; Hamilton, D.; Fox, J. W. Studies on the mechanism of hemorrhage production by five proteolytic hemorrhagic toxins from *Crotalus atrox* venom. *Biol. Chem. Hoppe-Seyler* **1988**, *369*, 121–129.

40. Baramova, E. N.; Shannon, J. D.; Bjarnason, J. B.; Fox, J. W. Degradation of extracellular matrix proteins by hemorrhagic metalloproteinases. *Arch. Biochem. Biophys.* **1989**, *275*, 63–71.

41. Mashiko, H.; Takahashi, H. Hemorrhagic factors from snake venoms. II, structures of hemorrhagic factors and types and mechanisms of hemorrhage. *J. Toxicol. Toxin Rev.* **1998**, *17*, 493–512.

42. Kalluri, R. Basement membranes: structure, assembly and role in tumor angiogenesis. *Nat. Rev. Cancer* **2003**, *3*, 422–433.

43. Miyoshi, S. I.; Nakasawa, H.; Kawata, K.; Tomochika, K. I.; Tobe, K.; Shinoda, S. Characterization of the hemorrhagic reaction caused by *Vibrio vulnificus* metalloproteinase, a member of the thermolysin family. *Infect. Immun.* **1998**, *66*, 4851–4855.

44. Timpl, R.; Rohde, H.; Robey, P. G.; Rennard, S. I.; Foidartc, J. M.; Martin, G. R. Laminin—a glycoprotein from basement membranes. *J. Biol. Chem.* **1979**, *254*, 9933–9937.

45. Paulsson, M.; Aumaillay, M.; Deutzmann, R.; Timpl, R.; Beck, K. Laminin-nidogen complex: extraction with chelating agents and structural characterization. *Eur. J. Biochem.* **1987**, *166*, 11–19.

46. Paulson, M. Basement membrane proteins: structure, assembly, and cellular interactions. *Crit. Rev. Biochem. Mol. Biol.* **1992**, *27*, 93–127.

47. Magnus, K.; Mosher, D. F. Fibronectin. Structure, assembly, and cardiovascular implications. *Arterioscler. Thromb. Vasc. Biol.* **1998**, *18*, 1363–1370.

48. Kamiguti, S. A. Platelets as targets of snake venom metalloproteinases. *Toxicon* **2005**, *45*, 1041–1049.

49. Wijeyewckrema, L. C.; Berndt, M. C.; Andrews, R. K. Snake venom probes of platelet adhesion receptors and their ligands. *Toxicon* **2005**, *45*, 1051–1061.

50. Kamiguti, A. S.; Hay, C. M. H.; Zuzel, M. Inhibition of collagen-induced platelet aggregation as the result of cleavage of $\alpha 2\beta 1$-integrin by the snake venom metalloproteinase jararhagin. *Biochem. J.* **1996**, *320*, 635–641.

51. Zigrino, P.; Kamiguti, A. S.; Eble, J.; Drescher, C.; Nischt, R.; Fox, J. W.; Mauch, C. The reprolysin jararhagin, a snake venom metalloproteinase, functions as a fibrilar collagen agonist involved in fibroblast cell adhesion and signaling. *J. Biol. Chem.* **2002**, *277*, 40528–40535.

52. Nylmalm, Y.; Puranen, J. S.; Nyholm, T. K. M.; Käpylä, J.; Kidron, H.; Pentikäinen, O. T.; Airenne, T. T.; Heino, J.; Slotte, J. P.; Johonson, M. S.; Salminen, T. A. Jararhagin-derived RKKH peptides induce structural changes in $\alpha 1$I domain of human integrin $\alpha 1\beta 1$. *J. Biol. Chem.* **2004**, *279*, 7962–7970.

53. Ruoslahti, E. RGD and other recognition sequences for integrins. *Annu. Rev. Cell. Biol.* **1996**, *12*, 679–715.

54. Eble, J. A. The molecular basis of integrin-extracellular matrix interactions. *Osteoar. Cartilage* **2001**, *9A*, S131–S140.

55. You, W. -K.; Jang, Y. -J.; Chung, K. -H.; Kim, D. -S. Functional roles of the two distinct domains of halysase, a snake venom metalloprotease, to inhibit human platelet aggregation. *Biochem. Biophys. Res. Commun.* **2006**, *339*, 964–970.

56. Wang, W. -J. Purification and functional characterization of AAV1, a novel P-III metalloproteinase, from Formosan *Agkistrodon acutus* venom. *Biochimie* **2007**, *89*, 105–115.

57. Gong, W.; Zhu, X.; Liu, S.; Teng, M.; Niu, L. Crystal structures of acutolysin A, a three disulfide hemorrhagic zinc metalloproteinase from *Agkistrodon acutus*. *J. Mol. Biol.* **1998**, *283*, 657–668.

58. Holland, D. R.; Hausrath, A. C.; Juers, D.; Matthews, B. W. Structural analysis of zinc substitutions in the active site of thermolysin. *Protein Sci.* **1995**, *4*, 1955–1965.

59. Howes, J. M.; Theakston, R. D. G.; Laing, G. D. Neutralization of the haemorrhagic activities of viperine snake venoms and venom metalloproteinases using synthetic peptide inhibitors and chelators. *Toxicon* **2007**, *49*, 734–739.

60. Starkey, P. M.; Barrett, A. J. Evolution of α2-macroglobulin. *Biochem. J.* **1982**, *205*, 91–95.

61. Baramova, E. N.; Shannon, J. D.; Bjarnason, J. B.; Gonias, S. L.; Fox, J. W. Interaction of hemorrhagic metalloproteinases with human α2-macroglobulin. *Biochemistry* **1990**, *29*, 1069–1074.

62. Kamiguti, A. S.; Desmond, H. P.; Theakston, R. D. G.; Hay, C. R. M.; Zuzel, M. Ineffectiveness of the inhibition of the main hemorrhagic metalloproteinase from *Bothrops jararaca* venom by its only plasma inhibitor α2-macroglobulin. *Biochim. Biophys. Acta.* **1994**, *1200*, 307–314.

63. Howes, J. M.; Theakston, R. D. G.; Laing, G. D. Antigenic relationships and relative immunogenicities of isolated metalloproteinases from *Echis ocellatus* venom. *Toxicon* **2005**, *45*, 677–680.

64. Rodrigues, V. M.; Soares, A. M.; Guerra-Sá, R.; Rodrigues, V.; Fontes, M. R. M.; Giglio, J. R. Structural and functional characterization of neuwiedase, a nonhemorrhagic fibrinogenolytic metaloprotease from *Bothrops neuwiedi* snake venom. *Arch. Biochem. Biophys.* **2000**, *381*, 213–224.

65. Camey, K. U.; Velarde, D. T.; Sanchez, E. F. Pharmacological characterization and neutralization of the venoms used in the production of *Bothropic antivenom* in Brazil. *Toxicon* **2002**, *40*, 501–509.

66. Harrison, R. A.; Wüster, W.; Theakston, R. D. G. The conserved structure of snake venom toxins confers extensive immunological cross-reactivity to toxin-specific antibody. *Toxicon* **2003**, *41*, 441–449.

67. Bard, R.; de Lima, J. C.; Sa Neto, R. P.; de Oliveira, S. G.; dos Santos, M. C. Ineficácia do antiveneno botropico na neutralização da atividade coagulante do veneno de *Lachesis muta muta*. Relato de caso e comprovação experimental. *Rev. Inst. Méd. Trop. S. Paulo* **1994**, *36*, 77–81.

CHAPTER 34

CaaX-PROTEIN PRENYLTRANSFERASE INHIBITORS

MARTIN SCHLITZER, REGINA ORTMANN, and MIRKO ALTENKÄMPER

Institut für Pharmazeutische Chemie, Philipps-Universität Marburg, Marbacher Weg 6, 35032 Marburg-Lahn, Germany

34.1 INTRODUCTION

Numerous proteins are synthesized in biologically inactive forms and have to undergo posttranslational modifications to fulfill their roles in cellular signal transduction. A series of posttranslational events is initiated by the transfer of either a farnesyl or one or two geranylgeranyl residues from the respective prenyl diphosphates to the cysteine thiol near the C-terminus of a protein. There are three distinct protein prenyl transferases: farnesyltransferase (FTase) and geranylgeranyltransferases-I and -II (GGTase-I and GGTase-II). Substrates of GGTase-II are proteins of the Rab family of small GTPases typically containing two geranylgeranylated cysteine residues adjacent to their C-termini (XXCC, XCXC, or CCXX). Farnesyltransferase and geranylgeranyltransferase-I are also called CaaX-prenyltransferases since their substrate proteins comprise the so-called CaaX (Ca_1a_2X or CAAX) consensus sequence at their C-termini, where C stands for the invariable prenyl acceptor cysteine, a for amino acids that normally but not necessarily carry aliphatic side chains and X for a variable amino acid. This terminal amino acid X is the major determinant whether this particular protein will be farnesylated or geranylgeranylated. Proteins terminating in serine, methionine, glutamine, or alanine that are recognized by FTase, proteins with a C-terminal leucine residue that are substrates of GGTase-I. Interest in protein prenylation and, especially in protein farnesylation, arose in the early 1990s from the observation that Ras proteins, which mutated forms are found in a variety of human cancers, need to be farnesylated to be active. Meanwhile, up to 250 proteins terminating in a CaaX sequence have been identified. Table 34.1 gives an overview over the proteins most likely being involved in the antiproliferative activity of CaaX-protein prenylation inhibitors.[1–4]

TABLE 34.1 Farnesylated Proteins Possibly Involved in the Antiproliferative Activity of Farnesyltransferase Inhibitors

Protein	CaaX	Alternatively Prenylated	Function
H-Ras	CVLS	No	Signal transduction
N-Ras	CVVM	Yes	Signal transduction
K-Ras	CVIM	Yes	Signal transduction
RhoB	CKVL	Yes	Signal transduction
Rheb	CSVM	No	Signal transduction
CENP-E	CKTQ	No	Kinesin motor protein
CENP-F	CKVQ	No	Chromosome passenger protein
PRL1	CCIQ	Yes	Tyrosine phosphatase
PRL2	CCVQ	Yes	Tyrosine phosphatase
PRL3	CCVM	Yes	Tyrosine phosphatase
HDJ2	CQTS	No	Chaperone
Prelamin A	CSIM	No	Nuclear envelope protein

In addition to protein prenylation, the CaaX motif serves as a substrate for additional posttranslational modifications following on the attachment of the prenyl residue to the cysteine side chain. Prenylation targets the protein to the membrane of the endoplasmatic reticulum, where the terminal aaX-tripeptide is removed by an endoprotease called Ras converting enzyme 1 (Rce1). Subsequently, the carboxyl group of the now C-terminal *S*-prenylated cysteine is converted to the corresponding methyl ester by the enzyme isoprenylcysteine carboxymethyltransferase (ICMT) with S-adenosylmethionine serving as methyl group donor. Cleavage of the aaX-tripeptide enhances membrane association 2-fold, aaX cleavage and subsequent carboxy methylation 4-fold. Some proteins undergo additional modification through the *S*-acylation (mainly by palmitoyl residues) at downstream cysteine residues. These posttranslational events altogether create hydrophobic domains, which facilitate critical protein–protein interactions or membrane attachment of the so modified proteins (Fig. 34.1).

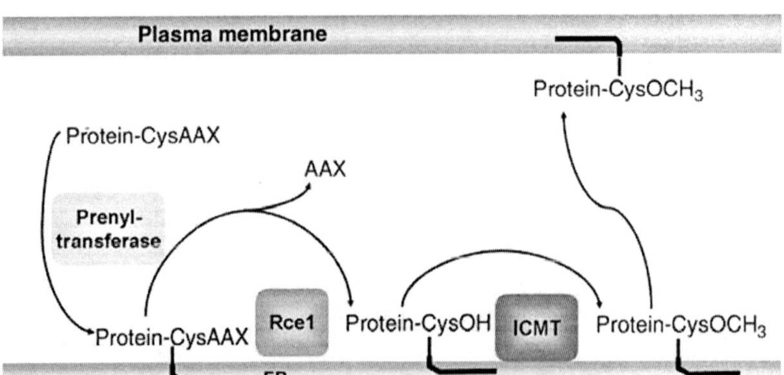

FIGURE 34.1 Posttranslational modification of proteins terminating with the CaaX sequence.

34.2 CaaX-PRENYLTRANSFERASES STRUCTURE AND MECHANISM

FTase and GGTase-I are heterodimers sharing a common α-subunit (48 kDa), but being different in the β-subunit (46 kDa in FTase and 42 kDa in GGTase-I). Although the β-subunits share only 25% sequence identity, they are very similar in structure. A funnel shaped cavity 14 Å deep and approximately 15 Å wide opens up at the interface of the two subunits forming the enzyme's active site. A zinc ion is essential for the activity of both enzymes. It is coordinated by three strictly conserved residues Asp297β, Cys299β, and His362β in FTase (Fig. 34.2) and Asp269β, Cys271β, and His321β in GGTase-I. In the empty enzyme or the binary enzyme FPP complex, the fourth ligand position is either occupied by a water molecule or by the second carboxylate oxygen from Asp279β making this residue a bidentate ligand. The nature of the fourth ligand cannot be unambiguously assigned on the basis of X-ray crystal structures. Calculations demonstrated both alternatives are lying in close energetic proximity, even though the bidentate hypothesis has a somewhat lower energy. Further calculations indicate that the mono-bidentate conversion is reversible and very fast and that probably both states exist in equilibrium. Therefore, a carboxylate-shift mechanism might assist the coordination/displacement of the substrate/product sulfur ligand in course of the farnesylation reaction.[5]

Briefly, the farnesyltransferase reaction starts with the formation of the binary enzyme FPP complex. Then, the CaaX tetrapeptide substrate is bound followed by the transfer of the farnesyl residue to the thiol of the cysteine side chain. Upon binding of

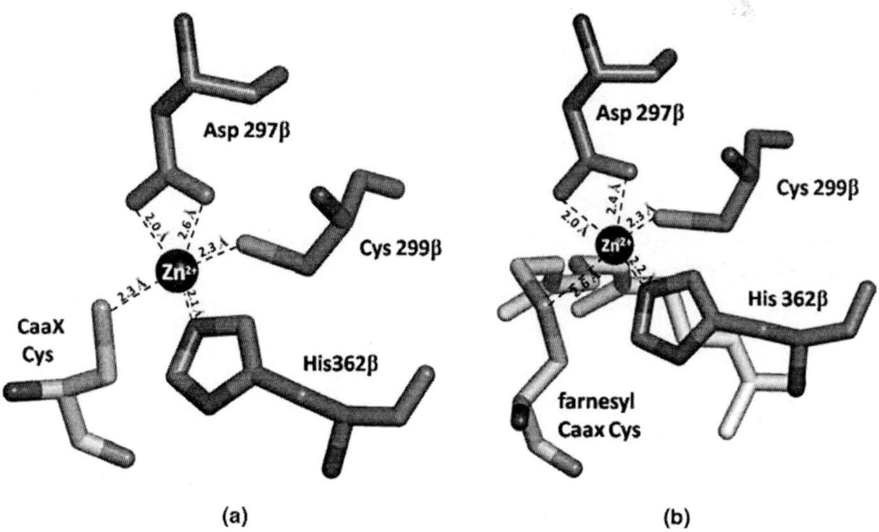

FIGURE 34.2 Zinc (sphere) is essential for prenyltransferase's activity. It is coordinated by three active site amino acid side chains and (a) the thiolate of the CaaX peptide substrate and (b) the thioether of the prenylated peptide (shown for FTase; pictures based on pdb entries: 1TN6 and 1KZP).

an additional molecule FPP, the farnesylated tetrapeptide changes its conformation from an extended one to a β-turn. Concurrently, the attached farnesyl residue moves from the farnesyl substrate-binding site to a shallow solvent exposed groove, called exit groove. Subsequent dissociation of the farnesylated peptide is the rate-limiting step of the farnesyltransferase reaction. The GGTase-I reaction proceeds analogously (Figs 34.3 and 34.4).

The diphosphate moiety of both FPP and GGPP binds in a positively charged cleft, formed by Lys164α, His248β, Arg291β, and Tyr300β in FTase and Lys 164α, His219β, Arg263β, Lys266β, and Tyr 272β in GGTase-I. Binding sites of the prenyl portions are very similar in FTase and GGTase-I being a cavity lined with conserved aromatic residues. The primary differences are at the bottom of this cavity, where the fourth isoprene unit of GGPP binds in GGTase-I. Here, FTase has the bulky side chains of Typ102β and Tyr365β, whereas the corresponding positions in GGTase-I are occupied by the smaller side chains of Thr49β and Phe324β to create space for the fourth isoprene unit of GGPP, which cannot fit into FTase (Fig. 34.5). For this situation, the allegory of both enzymes functioning as simple molecular rulers has been created. Both enzymes are able to bind both isoprene diphosphates FPP and GGPP. In case of GGPP binding by FTase the isoprene diphosphate acts as a competitive antagonist, whereas FPP is a weak substrate of GGTase-I. In both prenyltransferases, the isoprenoid forms a large part of the Ca_1a_2X-tetrapeptide binding surface with its second and third isoprene (and fourth in case of GGTase-I) units in direct contact with the peptide, especially with the terminal a_2X-portion.

The Ca_1a_2X tetrapeptide of the substrate protein is bound in an extended conformation with the cysteine's thiol directly coordinated to the zinc ion (Fig. 34.6; Table 34.2). Approximately 12 Å away from this cysteine, the carboxyl group of the C-terminal X amino acid forms hydrogen bonds to Gln167α (in both FTase and GGTase-I) and a buried water molecule coordinated by His149β, Glu198β, and Arg202β (His121β, Glu169β, and Arg173β in GGTase-I). Apart from an additional hydrogen bond between the backbone carbonyl of the amino acid in the a_2 position and the side chain of Arg202β (Arg173β in GGTase-I) these are the only directional interactions between the peptide substrate and the prenyltransferases. The presence of two fixed anchor points, namely, the cysteine's thiol and the C-terminal carboxyl group discriminates against all other peptide sequences either too long or too short or lacking the cysteine at the correct position. The a_1 amino acid side chain is exposed to the solvent and consequently, any amino acid is acceptable in this position.

Charged amino acid side chains as, for example, Asn in the Rap2a protein, are able to form a hydrogen bond to FTase. In GGTase-I, but not in FTase, a water molecule mediates an additional hydrogen bond between the carbonyl oxygen of the a_1 amino acid and His201α. The terminal a_2X dipeptide is buried in the active site and therefore, these two amino acids are the major determinants of peptide selectivity through steric and electrostatic complementarity between the side chain of amino acid a_2 and the a_2-pocket and the so-called specificity pocket and the terminal X amino acid. The a_2-binding pocket in FTase is smaller and has more aromatic character than the corresponding binding site in GGTase-I. Generally, small amino acids Leu, Val,

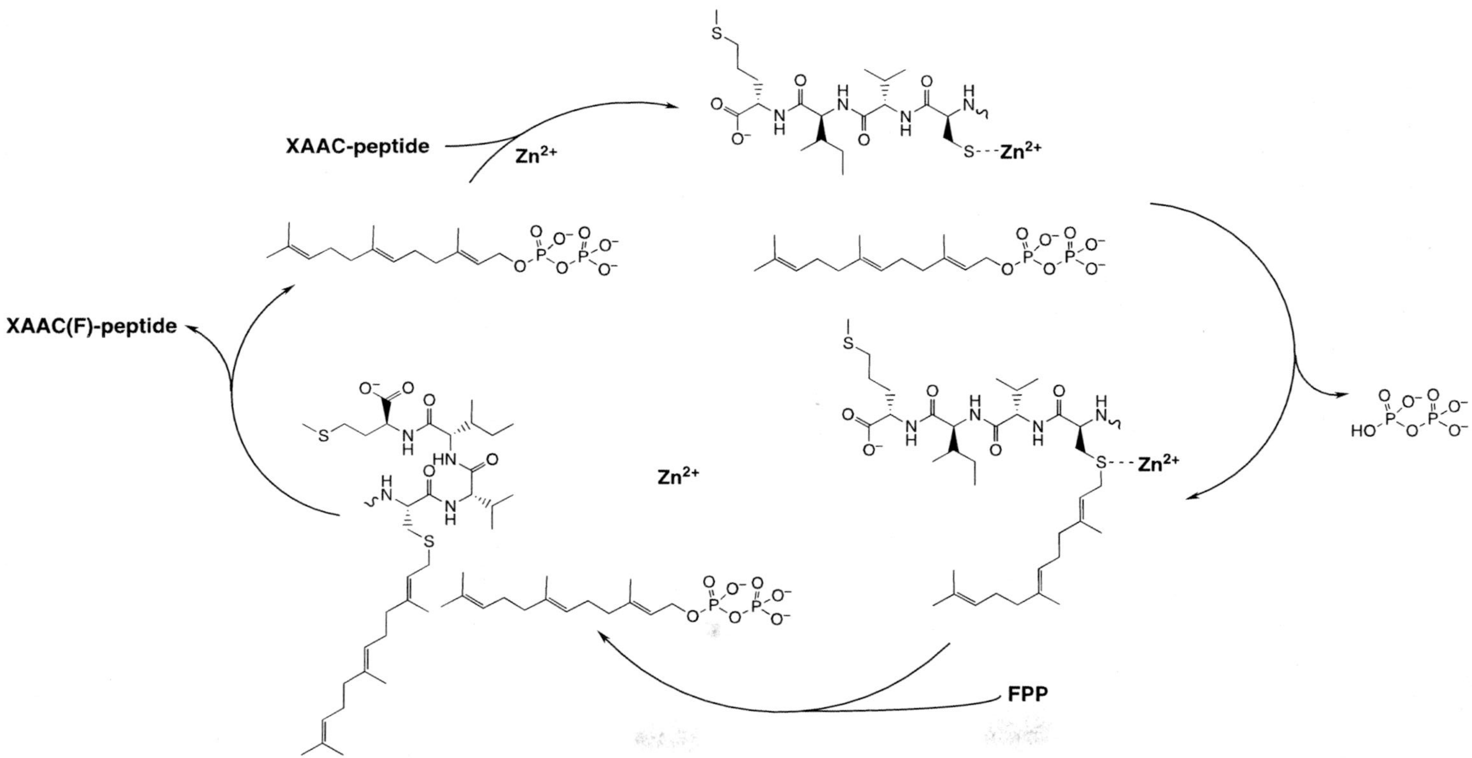

FIGURE 34.3 Schematic representation of the reaction path of FTase. The GGTase-I reaction proceeds analogously.

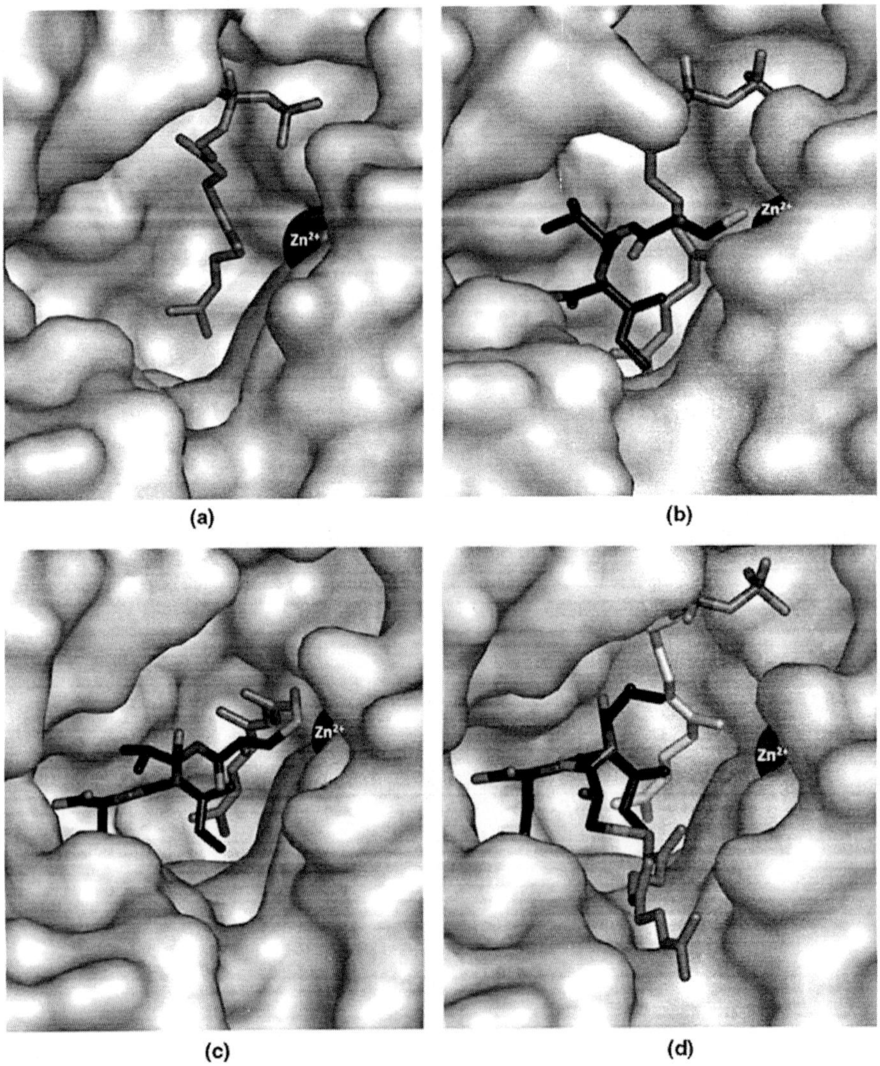

FIGURE 34.4 Reaction path of FTase: (a) FTase FPP complex, (b) FTase FPP CVIM complex, (c) farnesylated product in extended conformation, and (d) farnesylated product in β-turn conformation after the subsequent FPP has bound. The catalytic zinc ion is represented as a dark sphere. (pdb Entries: 1FT2, 1D8D, 1JCR, 1KZP, and 1KZO.)

and Ile predominate in the substrates of both prenyltransferases, however, there are subtle preferences: Val is more often found in substrates of FTase, whereas Leu is more often present in GGTase-I substrates. This is consistent with the different sizes of the a_2-binding sites in FTase and GGTase-I Fig. 34.7, suggesting that the a_2 residue may influence peptide substrate preference. The specificity pockets of both FTase and GGTase-I are too small to bind bulky amino acids like Tyr, Trp, or Arg. The specificity

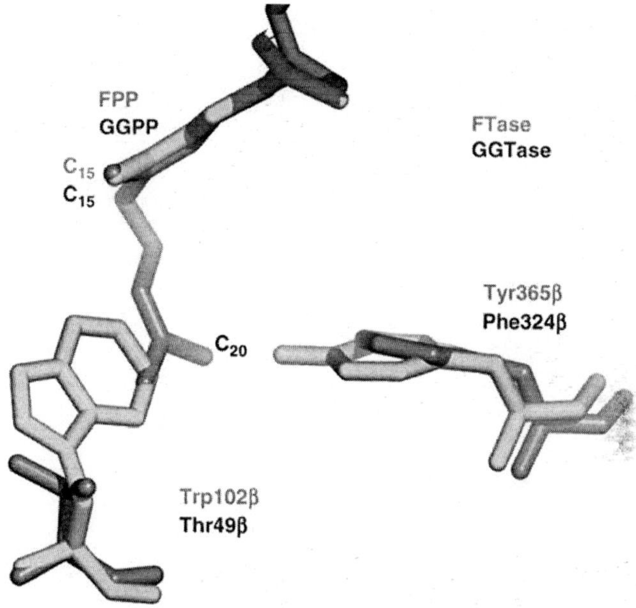

FIGURE 34.5 Differences in the prenyl-binding sites of FTase and GGTase-I. (Picture based on pdb entries: 1KZO and 1N4P.)

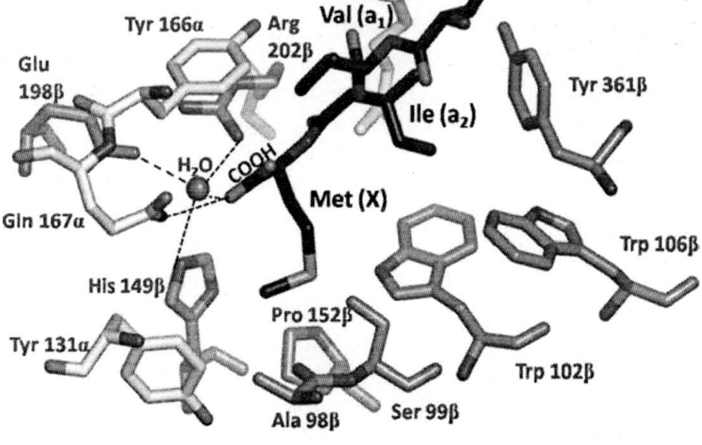

FIGURE 34.6 Binding of a CVIM substrate peptide. Polar interactions are indicated as dotted lines. (Picture based on pdb entry: 1TN6.) (See the color version of this figure in Color Plates section.)

TABLE 34.2 Binding Sites of FTase and GGTase-I

	FTase		GGTase-I	
Residue	Binding Site	Accepts	Binding Site	Accepts
C	Bound Zn^{2+}	Cys only	Bound Zn^{2+}	Cys only
a_1	Solvent accessible	No restrictions	Solvent accessible	No restrictions
a_2	Trp102β, Trp106β, Tyr 361β, FPP-isoprene 3	Val, Ile, Phe, Thr, Met	Thr49β, Phe53β, Leu320β, GGPP-isoprenes 3 and 4, X residue	Ile, Leu, Val, Phe, Tyr, Thr, Met
X	Specificity pocket: Tyr131α, Ala98β, Ser99β, Trp102β, His149β, Ala151β, Pro152β	Met, Gln; Ser, Ala, Thr, Cys with H_2O	Thr49β, His121β, Ala123β, Phe174β, GGPP-isoprene 4, A2-residue	Leu, Ile, Val, Phe
	Alternative X-binding site: Leu96β, Ser99β, Trp102β, Trp106β, Ala151β, FPP-isoprene 3, A_2-residue	Phe		

pocket of FTase either accommodate the similar shaped hydrophobic or polar side chains of methionine and glutamine, respectively, or smaller residues like cysteine, serine, threonine, or alanine. In the latter case, the remaining space is filled by a buried water and stabilized by a network of hydrogen bonds. Charged amino acids such as lysine, aspartate or glutamate cannot be accommodated in the FTase's specificity pocket. This pocket is also too small to accept phenylalanine's side chain, however, it can bind into an adjacent hydrophobic cavity. This alternative X-binding site can also accommodate leucine, asparagines, or histidine. GGTase's specificity pocket is much more hydrophobic in nature and therefore discriminates against all polar or charged amino acid side chains of similar shape as leucine and against all smaller amino acids binding in conjunction with a buried water molecule. Thus, the GGTase-I specificity pocket can only accept leucine, phenylalanine and methionine. In GGTase-I, there is no alternative X-binding site.

Some proteins as K-Ras4B and RhoB for instance are substrates of both prenyltransferases. This cross reactivity appears to be a function of their unique aaX sequences rather than any special conformation, particularly of the X amino acid. This may be exemplified with the proteins RhoB and Rab2b. The CaaX boxes of RhoB (CKVL) and Rab2b (CVIL) differ only in the central aa dipepeptide. The C-terminal amino acid Leu makes both proteins substrates of GGTase-I, while only RhoB can also

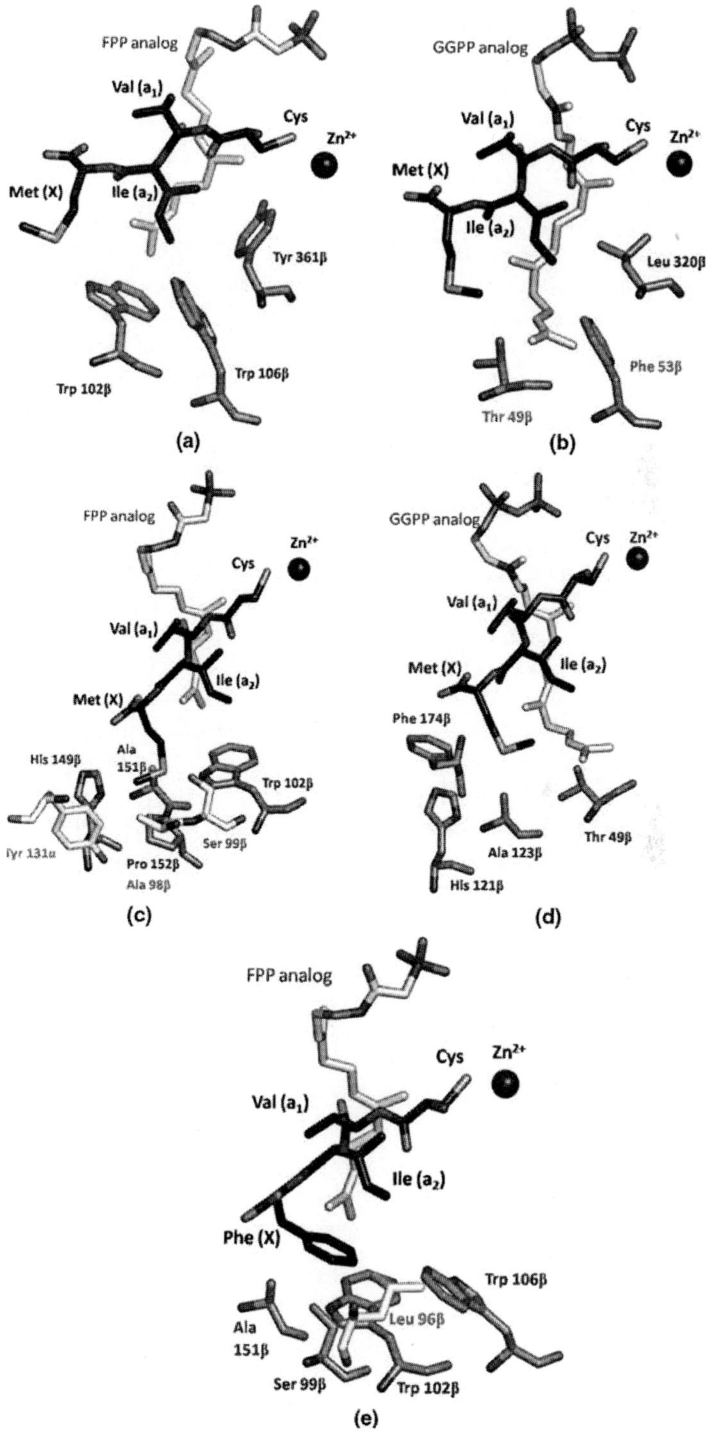

FIGURE 34.7 a_2-Binding site of (a) FTase, (b) GGTase-I; specificity pocket of (c) FTase, (d) GGTase-I, and (e) alternative X-binding site of FTase. (Pictures based on pdb entries: 1D8D, 1TNO, and 1TN7).

be farnesylated. The CKVL sequence of RhoB has been modeled into FTase's active site. This modeling placed the C-terminal Leu side chain into the alternative X-binding site. Occupation of the a_2-binding site by the smaller Val side chain makes clashes with the adjoining Leu residue less likely as with the larger residues as, for example, Ile. The a_1 amino acid Lys can adopt numerous conformations several of which permit direct- or water-mediated hydrogen bonds to the enzymes providing additional stabilization. Therefore, it is the special combination of residues in the a_1-, a_2-, and X-positions that allows RhoB to function as a substrate for FTase in addition to GGTase-I.

In FTase as well as in GGTase-I, there is no conformational change in the CaaX peptide during the prenylation reaction. In the ternary substrate complexes the prenyl-C-1 is approximately 7 Å away from the metal bound thiolate ion, a distance that has to be bridged during the prenylation reaction. It has been postulated that this achieved through a rotation involving the first two isoprene units. By combining the molecular mechanics and quantum chemical calculations, it has been shown that this reaction path is energetically feasible and the most probable way the prenylation reaction proceeds.[6] The binding conformation of the β-phosphate is maintained during this movement. Inversion of conformation is observed at the prenyl-C-1 atom. The developing negative charge is mainly localized on the α-phosphate and stabilized through interactions with Lys164α and Tyr300β (Tyr272β in GGTase-I) (Fig. 34.8). FTase, but not GGTase-I needs millimolar levels of Mg^{2+} to be active. It is believed that this magnesium ion (itself complexed by Asp352β) helps to stabilize the developing charge on the α-phosphate, while in GGTase-I this is done by the protonated side chain of Lys311β.

The rate-limiting step of the prenylation reaction is not the prenyl group transfer described above, but the release of the reaction product, the prenylated peptide. This is initiated by the binding of a fresh isoprenoide diphosphate molecule. This causes the

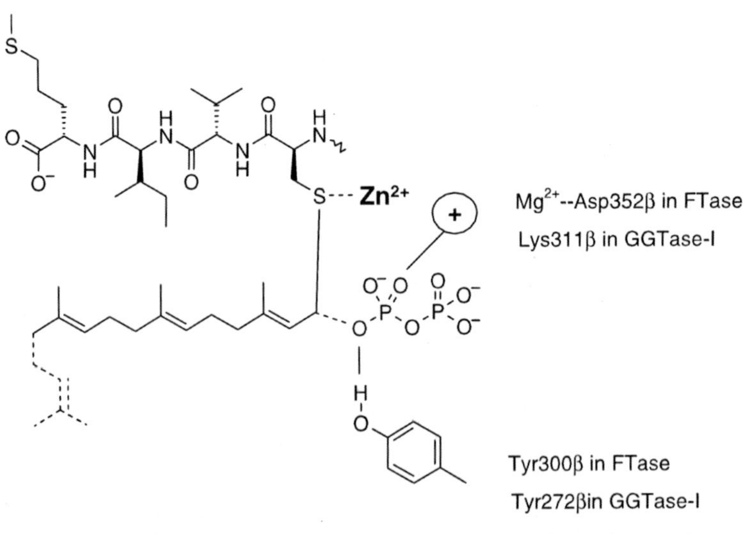

FIGURE 34.8 Schematic representation of the prenyl transfer reaction.

prenyl group of the prenylated peptide to move to a new binding site called exit groove. This movement is accompanied by a conformational change in the CaaX peptide from the extended conformation into a type-I β-turn (Fig. 34.4d). In this conformation, the three terminal AAX amino acids make extensive van der Waals contacts with the new isoprenoid diphosphate substrate, suggesting that the AAX sequence may modulate subsequent product release. It has been proposed that this substrate mediated product release provides a mechanism for a regulated handover of the prenylated protein to the next processing enzyme (this paragraph is based on the literature provided in Refs[7–9]).

34.3 CaaX-PRENYLTRANSFERASE INHIBITORS

Prenyltransferase inhibitors may be divided into three categories: Compounds competitive with the prenyl diphosphate substrate, compounds competitive with the CaaX tetrapeptide substrate, and inhibitors which are designed to occupy both the peptide and the prenyl binding site, the so-called bisubstrate analogue inhibitors. Among these classes, the compounds that occupy the peptide substrate-binding site constitute by far the most important class of farnesyltransferase inhibitors. The development of farnesyltransferase inhibitors has been described in a number of excellent reviews.[10–23] For the original papers, the reader is referred to these reviews. Original references will be cited only for inhibitor development not covered by these reviews.

34.3.1 CaaX-competitive Inhibitors

CaaX tetrapeptides like CVIM (**34.1**) (Fig. 34.9) are alternative substrates inhibiting peptide farnesylation. Through the replacement of the a_2 amino acid isoleucine by phenylalanine a nonsubstrate competitive inhibitor **34.2** with an IC_{50} of 25 nM has been obtained. This tetrapeptide with an aromatic side chain in the a_2 position adopts a nonsubstrate conformation in which the cysteine Cα is displaced by 3 Å in comparison to the substrate conformation so that the inhibitor interferes with the movement of the prenyl residue necessary for thioether formation. The aromatic side chain has been incorporated into a rigid tetrahydroisoquinoline carboxylic acid moiety as in **34.3**. In an attempt to make these peptidic inhibitors more stable, the two *N*-terminal amid moieties have been reduced to the corresponding amines (B581 (**34.4**)). In L-739,750 (**34.5**) one "reduced amid" has been replaced by an ether linkage to prevent intramolecular cyclization to an inactive diketopiperazine. Additionally, the methionine's thioether has been oxidized to the corresponding sulfone to enhance stability. The possibility of shortening the tetrapeptide CaaX was explored with Ca_1a_2 tripeptide analogues like **34.6** having a_2 side chain with a large hydrophobic group.[24]

To reduce the peptidic nature of the inhibitors, the central aa dipeptide has been replaced by a variety of nonpeptidic spacers as, for example, benzodiazepinone (BZA-2b (**34.7**)), 4-amino-2-phenylbenzoic acid (FTI-276 (**34.9**)), 6-aminonaphthoic acid (RPR 113829 (**34.10**)), and 4-aminopiperidine (**34.11**)[25] (Fig. 34.10). Omission of the methionine portion in the benzodiazepine-based inhibitor **34.8** led to a considerable reduction in enzyme inhibitory activity. It is interesting to note that both the

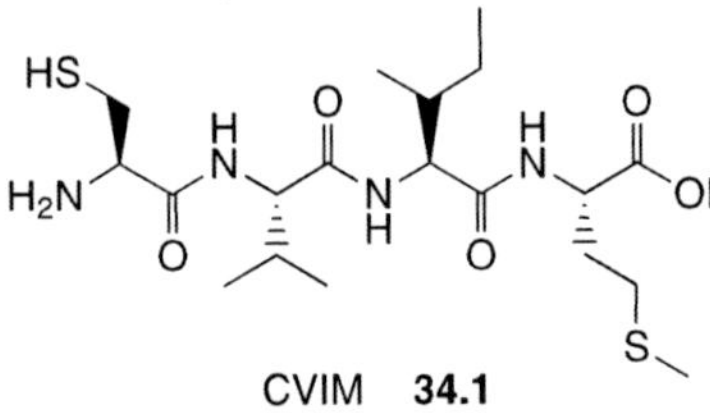

CVIM **34.1**

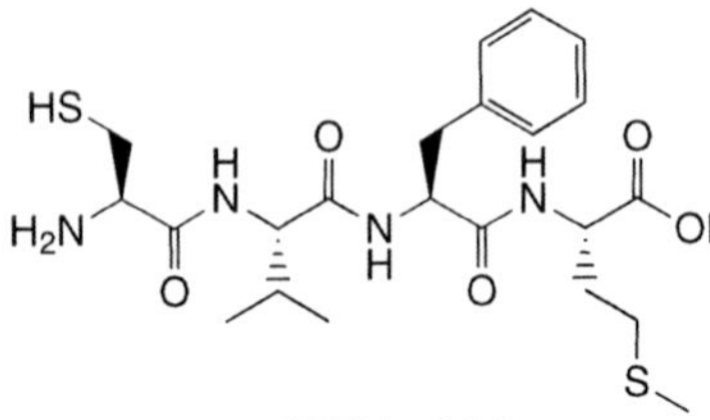

CVFM **34.2**

IC_{50} = 25 nM

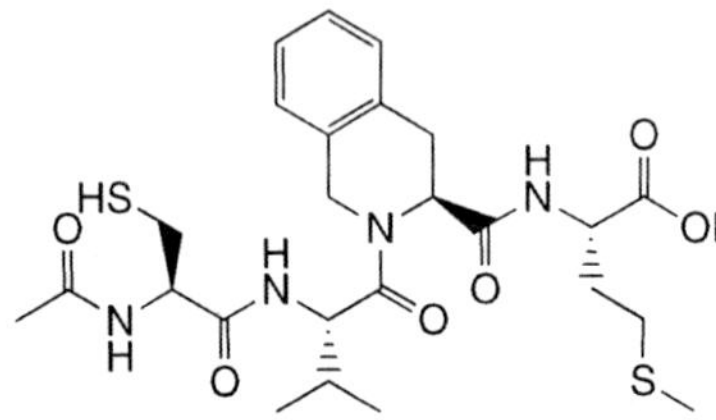

34.3

IC_{50} = 1 nM

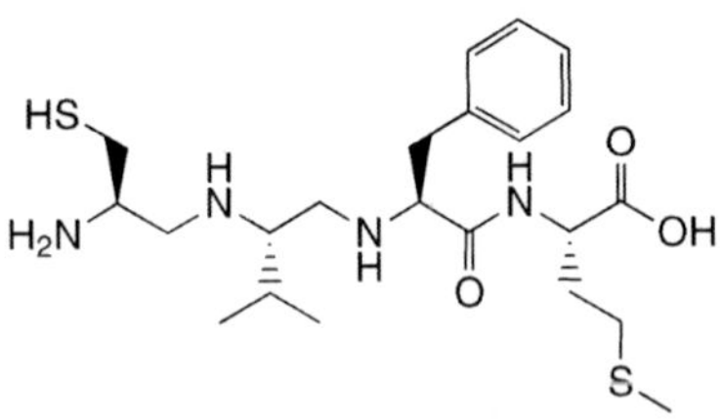

B581 **34.4**

IC_{50} = 21 nM

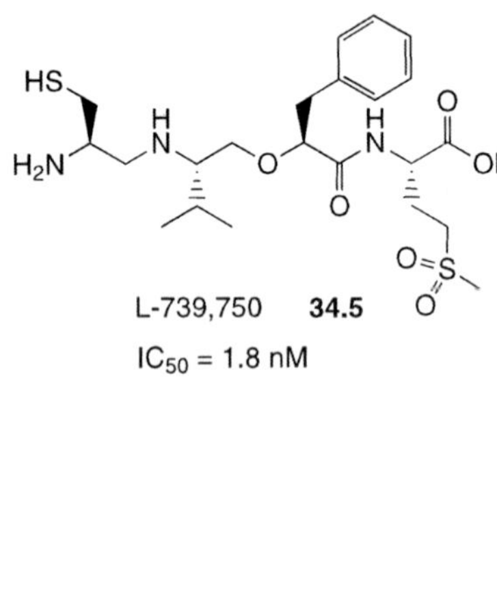

L-739,750 **34.5**

IC_{50} = 1.8 nM

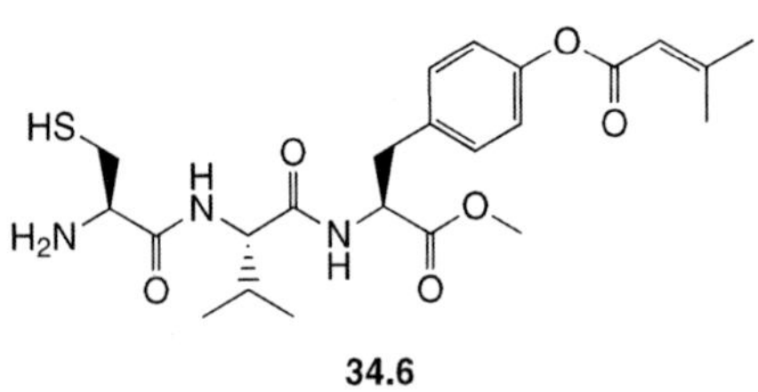

34.6

IC_{50} = 500 nM

FIGURE 34.9 Peptide and peptidomimetic inhibitors.

benzodiazepine derivative BDZ-2b (**34.7**) and the *p*-aminobenzoic acid derivative FTI-276 (**34.9**) are highly active farnesyltransferase inhibitors although they are based on different design assumptions. Both inhibitors have been designed before the crystal structures of FTase in complex with its substrates have been available. NMR experiments suggested that the CaaX peptide substrate will adopt a β-turn conformation,

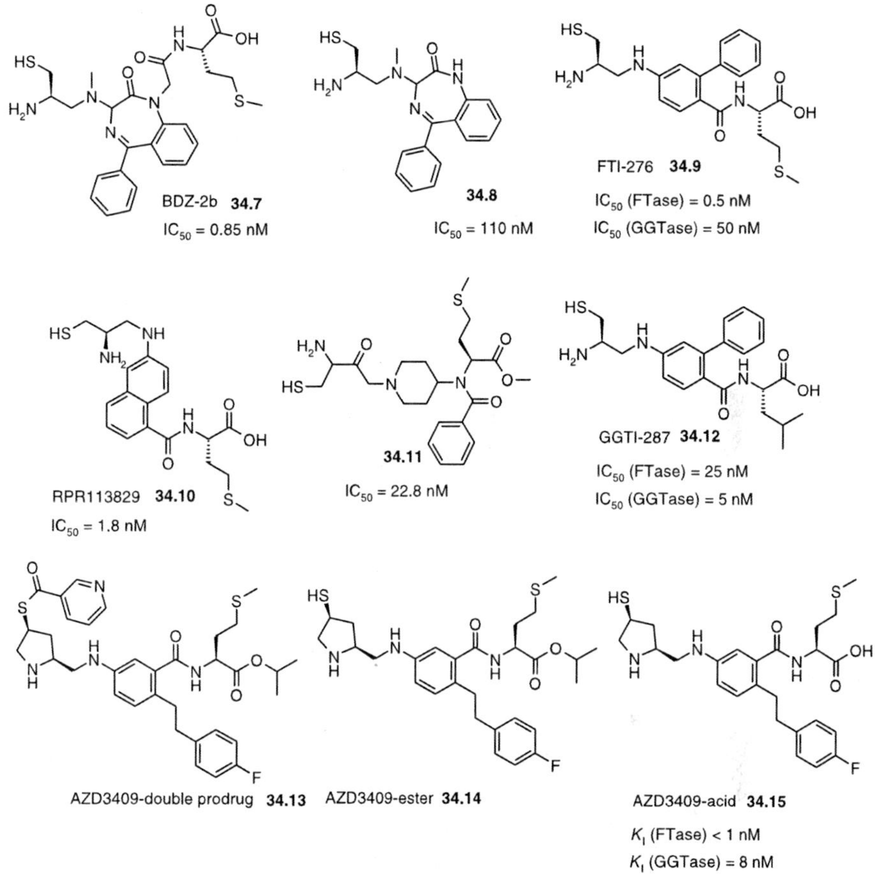

FIGURE 34.10 Peptidomimetic inhibitors.

which is mimicked by the benzodiazepine while the design of the FTI series has been based on the assumption of an extended conformation. Later, crystal structures demonstrated that the peptide substrate indeed adopts an extended conformation but rearranges into a β-turn like conformation after the prenyl transfer and the binding of an additional FPP has taken place. Replacement of the terminal methionine by leucine transformed the farnesyltransferase inhibitor FTI-276 (**34.9**) into the geranylgeranyl-transferase-I inhibitor GGTI-287 (**34.12**). With all these peptide derivatives or peptidomimetics bearing a carboxyl group, an ester prodrug strategy is mandatory to facilitate the membrane penetration and to attain cellular activity.

AstraZeneca has recently brought a drug called AZD3409 (**34.13**) to phase I of clinical evaluation.[26] AZD3409 is a double prodrug, which is converted into the active metabolite in two steps. In plasma, the thioester **34.13** is rapidly cleaved to yield the major circulating metabolite called AZD3409-ester (**34.14**). After penetration into the cell, the ester **34.14** is converted into the active drug called AZD-acid (**34.15**).

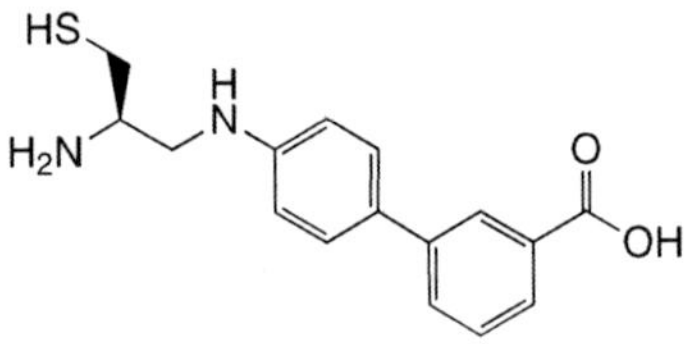

FTI-265 **34.16**

IC_{50} = 150 nM

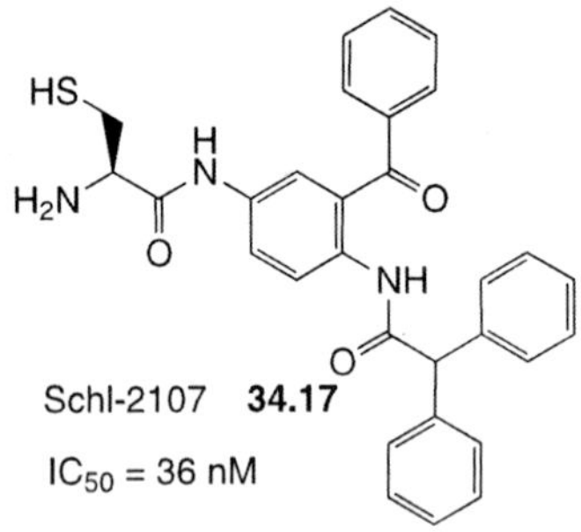

Schl-2107 **34.17**

IC_{50} = 36 nM

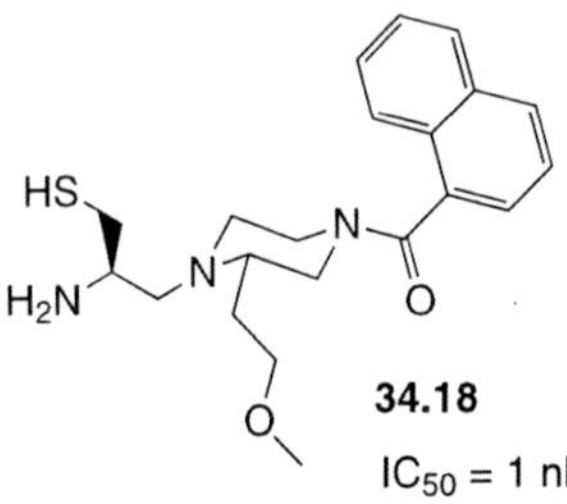

34.18

IC_{50} = 1 nM

FIGURE 34.11 Farnesyltransferase inhibitors lacking the X amino acid.

Structurally, there are some resemblances to the Hamilton's FTI series. AZD3409 (**34.13**) is a representative of the so-called dual prenylation inhibitors, compounds that display considerable activity against both CaaX prenyltransferases.

The design approach to reduce the peptidic nature of the CaaX-analogues has been carried further by the replacement of the C-terminal X amino acid by a carboxyphenyl residue as in FTI-265 (**34.16**). Furthermore, it has been demonstrated that even the carboxyl group is not necessary to obtain submicromolar inhibitors. Earlier examples are series of piperazine- (**34.18**) or benzophenone- (**34.17**) based inhibitors as Schl-2107[27] (Fig. 34.11).

All these inhibitors described so far contain an unmodified cysteinyl residue or an aminopropanthiol as zinc coordinating moiety. However, due to its sensitivity toward oxidation and to the potential to cause nonmechanism based toxicity, free thiols are normally undesired in potential drugs. Therefore, considerable efforts have been made to replace the thiol by alternative zinc coordinating substructures. Early attempts with phenols, lactams, or carboxylic acids have been met with limited success. More successful was the use of different nitrogen-containing heterocycles. In the early times of the development of nonthiol farnesyltransferase inhibitors numerous different derivatives of nitrogen heterocycles have been evaluated. From these studies, the imidazolyl residue emerged as the most appropriate zinc coordinating substructure and became something like a standard moiety of the so-called nonthiol or second generation farnesyltransferase inhibitors. The (substituted) imidazolyl residue has been connected to aaX mimetics used already in conjunction with the cysteinyl or aminopropanthiol residue and numerous other molecules designed to occupy variable portions of the farnesyltransferase's peptide-binding site. Often, a lipophilic residue

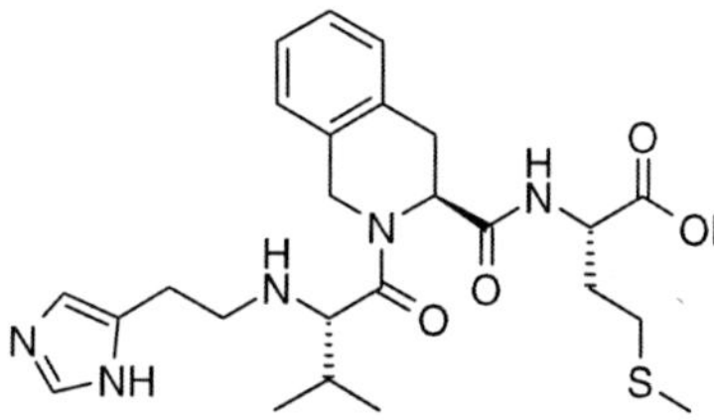

34.19

IC_{50} = 0.79 nM

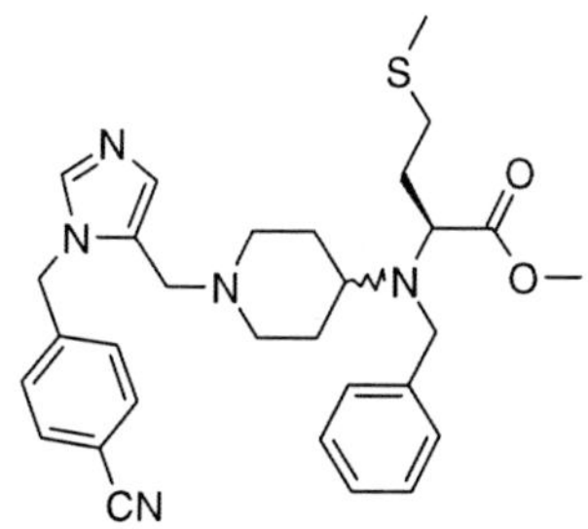

34.20

IC_{50}(FTase) = 2.35 nM

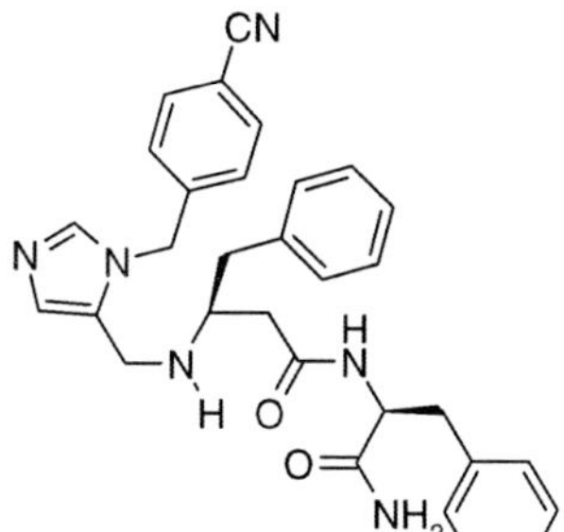

34.21

IC_{50}(FTase) = 28.5 nM

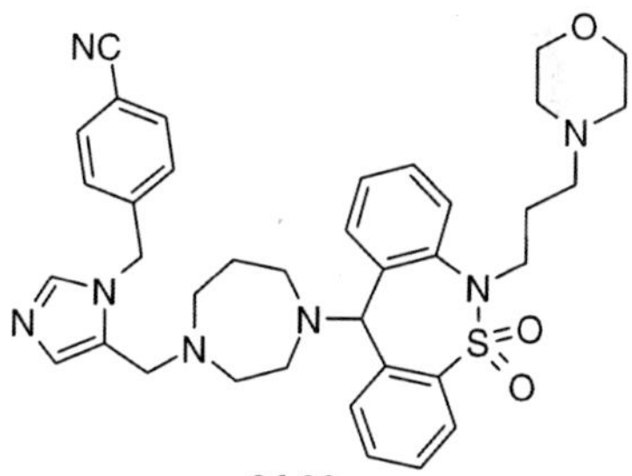

34.22

IC_{50}(FTase) = 17.3 nM

FIGURE 34.12 Nonthiol farnesyltransferase inhibitors.

like 4-cyanobenzyl is added to imidazolyl. In case of L-778,123 (vide infra) a crystal structure revealed that this residue binds into the peptide-binding site and makes contacts with parts of the farnesyl residue of the FPP substrate and the side chain of Tyr166α.[28] Possibly, there is also a directed interaction between the cyano group and the side chain of Arg202β.

The tetrahydroisoquinoline carboxylic acid and the 4-aminopiperidine scaffolds have also been used for the preparation of nonthiol farnesyltransferase inhibitors (**34.19**, **34.20**)[29,30] (Fig. 34.12). In an attempt to freeze the presumed active conformations of such 4-aminopiperidine based inhibitors and structurally related homophenylalanines (e.g., **34.21**)[31] 1,4-diazepinyl substituted dibenzothiazepines, as inhibitor **34.22**, for example, have been made.[32]

Merck researchers also used their piperazine scaffold of **34.18** for the design of nonthiol FTIs through the combination with the cyanobenzylimidazole cysteine replacement (Fig. 34.13). One of the earlier examples is inhibitor **34.23**. Further development has let to a compound named L-778,123 (**34.24**), which is one of the farnesyltransferase inhibitors that are, or have been, in clinical evaluation. In contrast to **34.18**, L-778,123 (**34.24**) also inhibits GGTase-I (IC_{50} = 100 nM) what explains its activity against K-Ras dependent tumors since K-Ras undergoes alternative prenylation through GGTase-I in case FTase activity is inhibited by appropriate drugs.

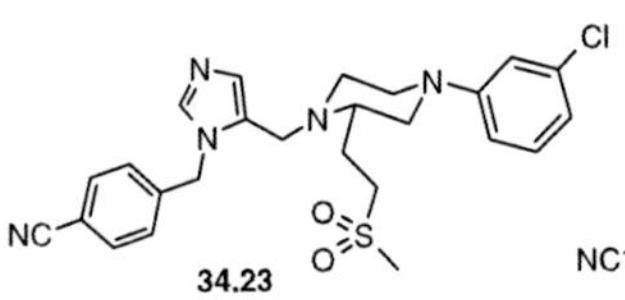

34.23

IC_{50} = 0.18 nM

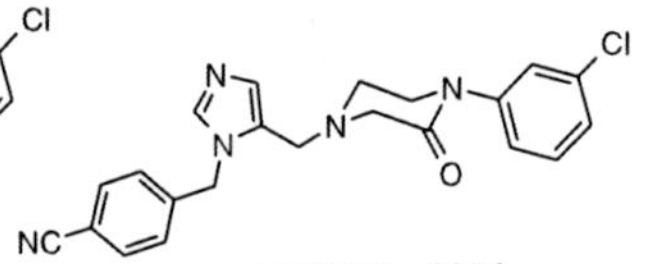

L-778,123 **34.24**

IC_{50} = 2 nM

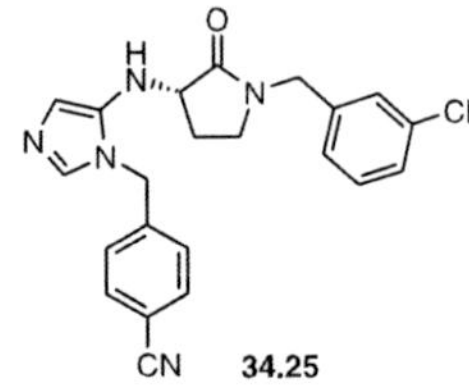

34.25

IC_{50}(FTase) = 1.9 nM

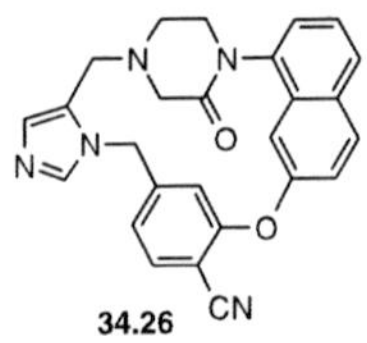

34.26

IC_{50}(FTase) = 0.1 nM

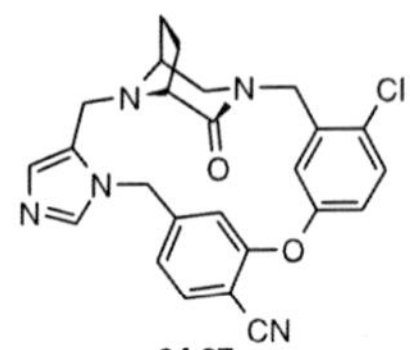

34.27

IC_{50}(FTase) = 0.13 nM

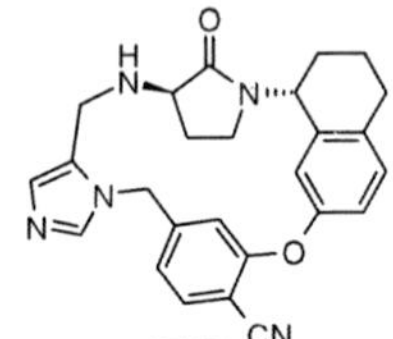

34.28

IC_{50}(FTase) = 20 pM

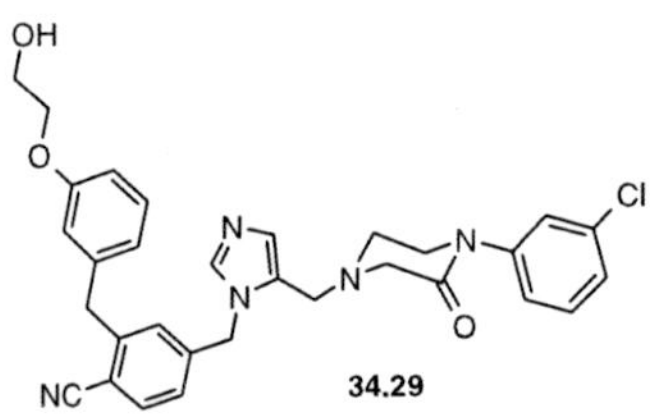

34.29

IC_{50}(FTase) = 2.5 nM

IC_{50}(GGTase) = 1.3 nM

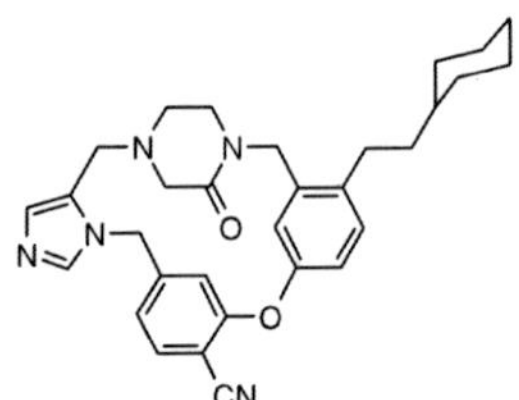

34.30

IC_{50}(FTase) = 1.7 nM

IC_{50}(GGTase) = 1.6 nM

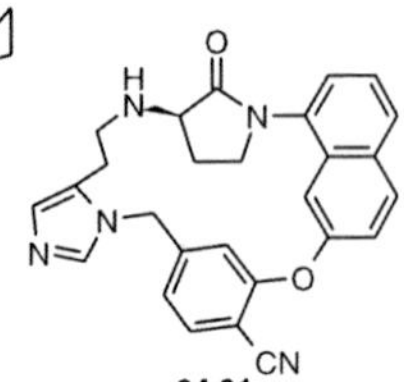

34.31

IC_{50}(FTase) = 1.1 nM

IC_{50}(GGTase) = 5.5 nM

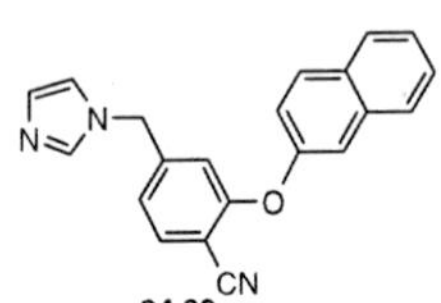

34.32

IC_{50}(FTase) = 1.1 nM

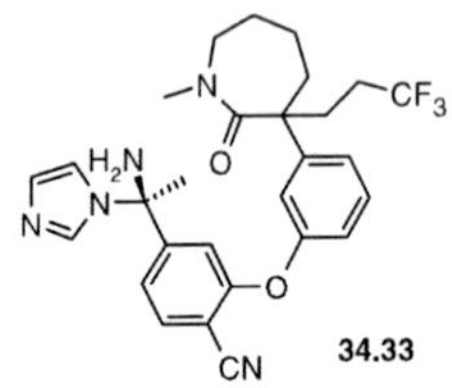

34.33

IC_{50}(FTase) = 0.15 nM

IC_{50}(GGTase) = 1.3 nM

FIGURE 34.13 Merck inhibitors based on the piperidine and 3-aminopyrrolidinone scaffold and inhibitors derived from these compounds.

Therefore, L-778,123 (**34.24**) (Fig. 34.14) is one earlier example of the so-called dual prenylation inhibitors, compounds that display significant activity against GGTase-I as well as against FTase. In an additional series, the central piperidine has been replaced by a 3-aminopyrrolidinone leading to the inhibitors (e.g., **34.25**) of comparable activity.

On the basis of the piperazinone and the 3-aminopyrrolidone scaffold, macrocyclic inhibitors **34.26–34.28** have been prepared with the intention to freeze the presumed active conformation of these inhibitors. Especially the pyrrolidione based inhibitor **34.28** displays exceptional *in vitro* activity.[33] Addition of a spacious lipophilic residue improved GGTase-I inhibitory activity in case of the conformationally unrestricted

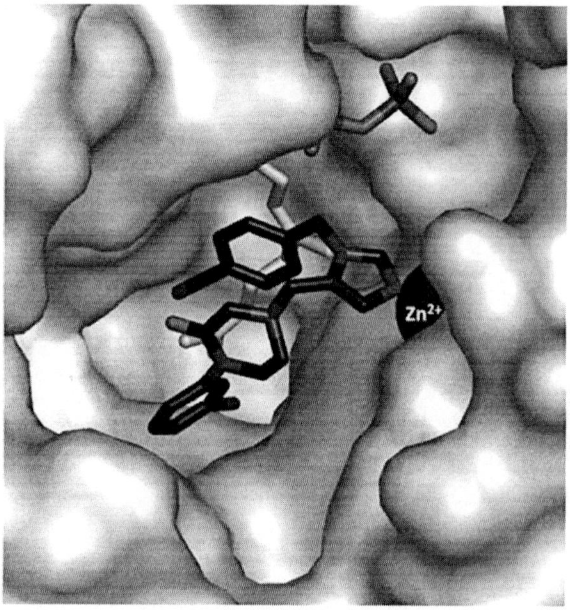

FIGURE 34.14 Crystal structure of L-778,123 (**34.24**) bound to FTase (pdb code: 1S63).[28] (See the color version of this figure in Color Plates section.)

(**34.29**) as well as of the macrocyclic piperidone (**34.30**) derivatives yielding true dual prenylation inhibitors. Dual prenylation inhibitors (e.g., **34.31**) were also obtained through methylene insertion into the 3-aminopyrrolidinone based macrocycles.[34] Modeling studies indicated that the spatial arrangement of the cyanophenyl and the chlorophenyl residues of L-778,123 (**34.24**) may also be achieved by a simple biarylether arrangement as in **34.32**. Further development ultimatively let to ε-caprolactame-substituted aryl-cyanophenylethers (e.g., **34.33**) with good dual prenylation inhibitory activities.

When drug developers of Bristol-Myers Squibb became aware of the piperazine-based FTIs of Merck described above, they found that these compounds did fit to their own pharmacophore model. Subsequently, BMS researchers expanded the piperazine to homopiperazine and fused one of the two aromatic residues demanded by their model to the homopiperazine to obtain a tetrahydrobenzodiazepine scaffold. Compound **34.34** is an example for the results of these early design efforts.[35] Extensive SAR studies finally let to BMS-214662 (**34.35**), which is in clinical development[36] (Fig. 34.15). The crystal structure of a BMS-214662-FPP-FTase complex (Fig. 34.16) showed the inhibitor in the peptide-binding site with the imidazole nitrogen coordinated to the catalytic zinc ion. Beside this polar interaction, the inhibitor forms only van der Waals contacts with the protein and the farnesyl residue. The thienylsulfonyl substituent is mostly exposed to the solvent. The more than 1000-fold selectivity of BMS-214662 (**34.35**) toward FTase is due to selective aromatic stacking interactions

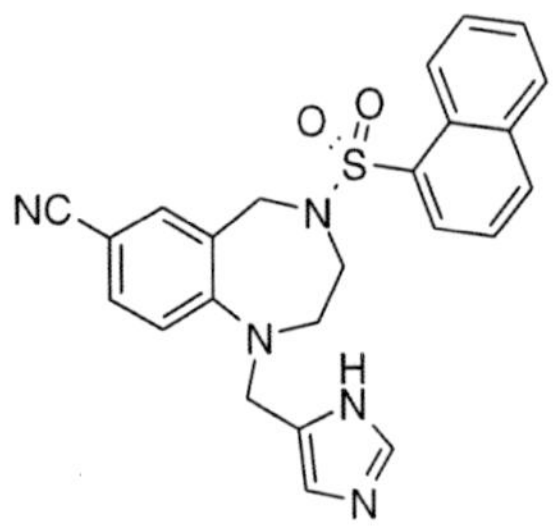

34.34

IC_{50} = 480 nM

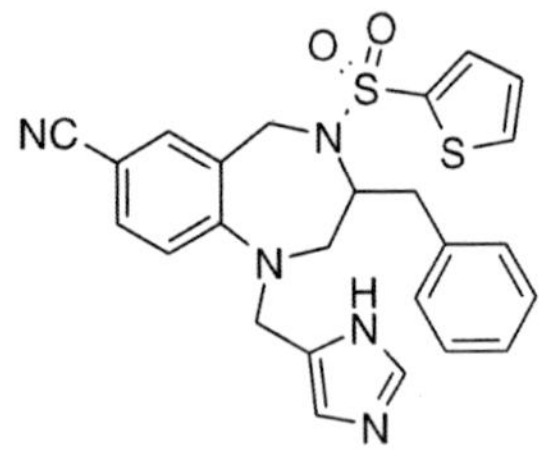

BMS-214662 **34.35**

IC_{50} = 1.35 nM

FIGURE 34.15 Tetrahydrobenzodiazepine-based inhibitors from Bristol-Myers Squibb.

with the FTase's a_2-binding site. Such interactions are not possible with the less aromatic a_2-binding site of GGTase-I.[37]

Using their benzoylmethione and benzoylleucine substructure for the connection with imidazolyl residues, the Hamilton group obtained nonthiol prenylation inhibitors as exemplified by FTI-2148 (**34.36**), GGTI-2151 (**34.37**), and GGTI-2154 (**34.38**)[38,39] (Fig. 34.17).

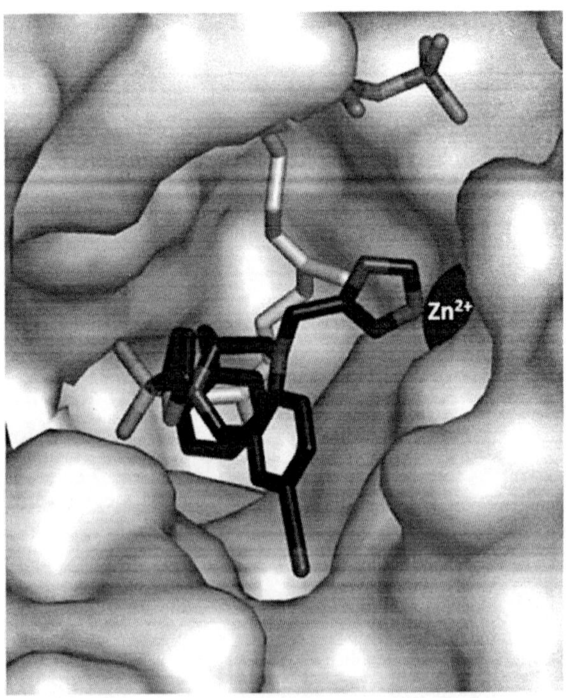

FIGURE 34.16 Crystal structure of tetrahydrobenzodiazepine BMS-214662 (**34.35**).

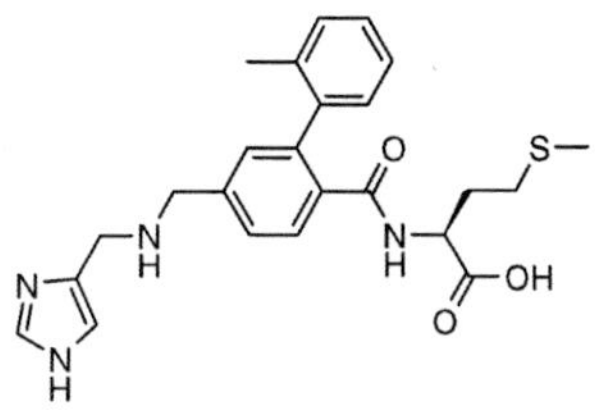

FTI-2148 **34.36**

IC_{50} (FTase) = 0.82 nM

IC_{50} (GGTase-I) = 1700 nM

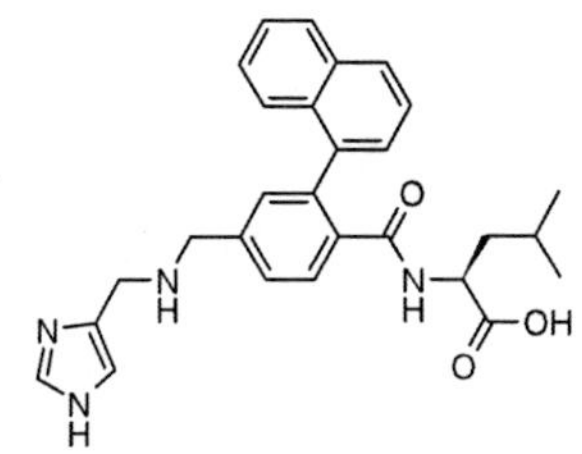

GGTI-2151 **34.37**

IC_{50} (FTase) > 10,000 nM

IC_{50} (GGTase-I) = 44 nM

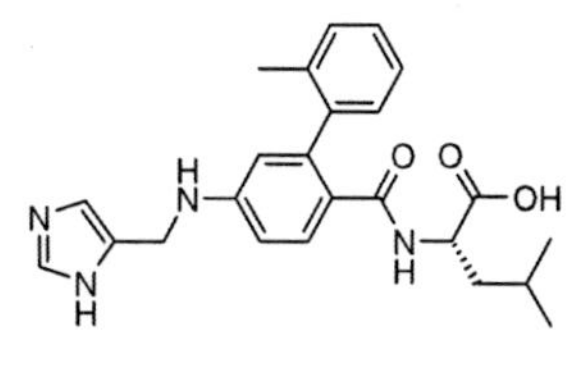

GGTI-2154 **34.38**

IC_{50}(FTase) = 5600 nM

IC_{50}(GGTase-I) = 21 nM

FIGURE 34.17 Benzoylmethione- and benzoylleucine-based farnesyl- and geranylgeranyl-transferase inhibitors.

Development of another farnesyltransferase inhibitor in clinical evaluation began with a systematic search of Schering-Plough compound libraries. Among the compounds identified were three tricyclic benzocycloheptapyridine heterocycles initially developed as potential antihistaminics. The compound with the highest activity against farnesyltransferase, SCH 44342 (**34.39**) (Fig. 34.18), became the

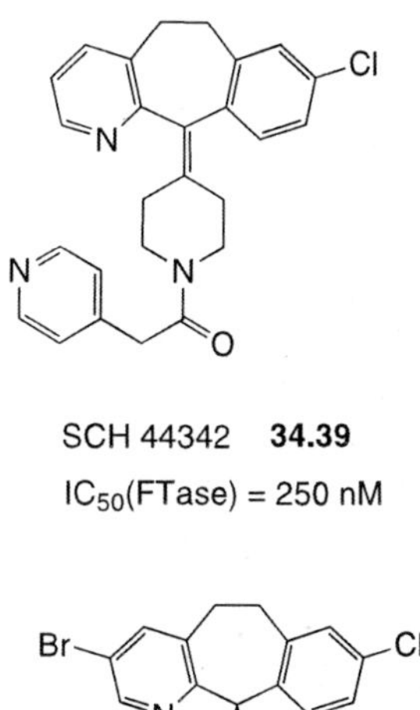

SCH 44342 **34.39**

IC_{50}(FTase) = 250 nM

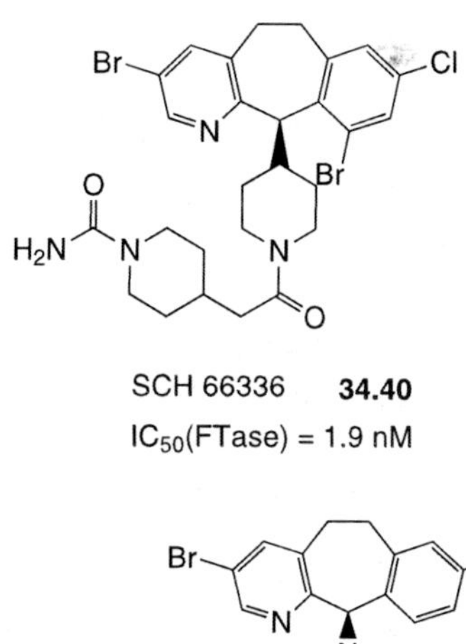

SCH 66336 **34.40**

IC_{50}(FTase) = 1.9 nM

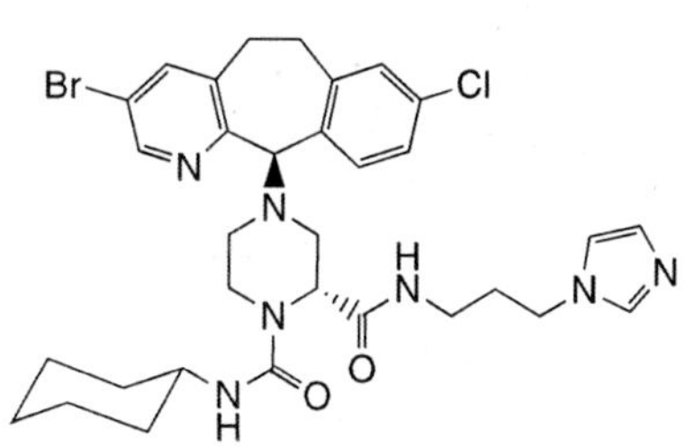

SCH 211618 **34.41**

IC_{50}(FTase) = 0.04 nM

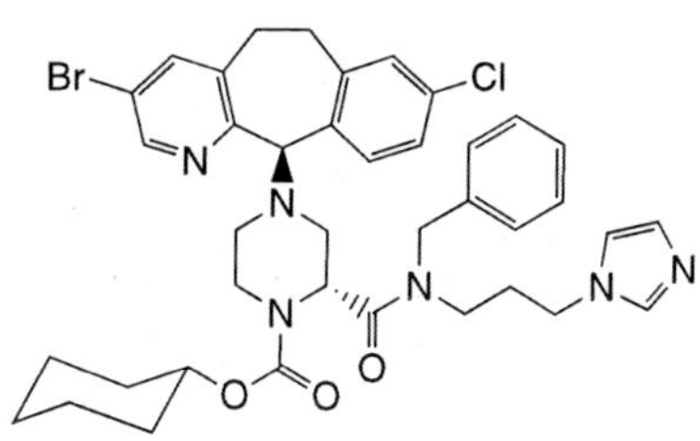

SCH 226374 **34.42**

IC_{50}(FTase) = 0.036 nM

FIGURE 34.18 Tricyclic benzocycloheptapyridine inhibitors.

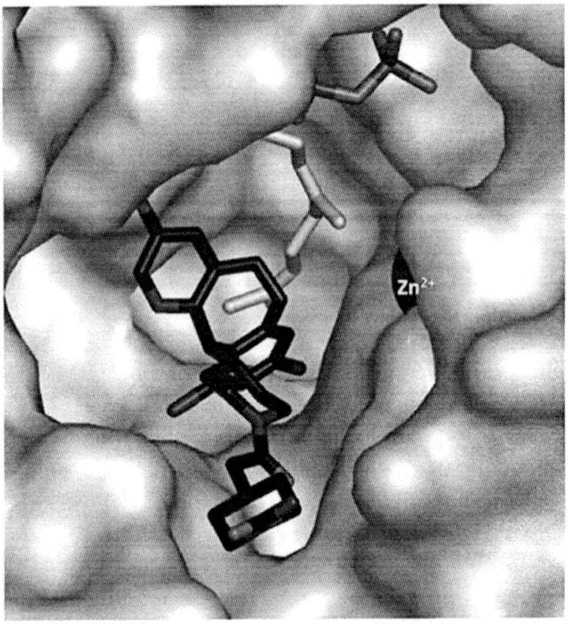

FIGURE 34.19 Crystal structure of SCH 66336 (**34.40**) (pdb entry: 1O5M).

lead structure for further development which led to SCH 66336 (**34.40**), which is now in clinical development as lonafarnib (Sarasar®). In contrast to all other inhibitors described so far, SCH 66336 (**34.40**) lacks any zinc coordinating moiety. The benzocycloheptapyridyl tricycle of SCH 66336 (**34.40**) occupies large portions of the peptide-binding site making hydrophobic interactions with several aromatic amino acid side chains (Fig. 34.19). The appended piperidyl urea moiety extends into the exit groove, which is normally occupied by the farnesyl residue after farnesyl transfer and rearrangement of the CaaX tetrapeptide from an extended into a β-turn conformation has taken place.[40–42] Therefore, SCH 66336 (**34.40**) may also be regarded as a product analogue. To gain the additional affinity through coordination of the catalytic zinc, an imidazole-containing side chain has been added leading to compounds as SCH 211618 (**34.41**), for example, with picomolar activity. Metabolic stability has been improved by addition of a benzyl residue to carboxamid nitrogen (SCH 226374 (**34.42**)).[41]

The discovery that the antifungal 14-α-demethylase inhibitor ketoconazole had also some potential antitumor activity spurred Janssen's entry into Ras-targeted research. Screening of a library of antifungal compounds yielded 36 hits of which 33 contained an imidazole known to be able to coordinate zinc ions as they are essential for farnesyltransferase's activity. Among the hits, **34.43** was the most active one (Fig. 34.20). The following development resulted in the clinical candidate R115777 (tipifarnib; Zarnestra® (**34.44**)), structurally not too different from the initial lead **34.43**.[43] R115777 (**34.44**) binds into the CaaX peptide substrate-binding site mainly by aromatic stacking interactions. The drug adopts a U shape stabilized by stacking between the two chloro-substituted phenyl residues (Fig. 34.21). The imidazole

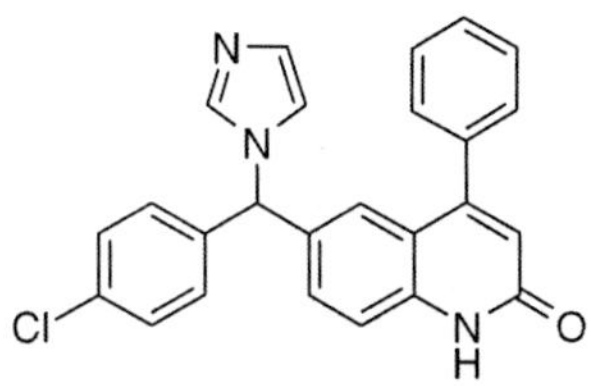

34.43

IC_{50} = 180 nM

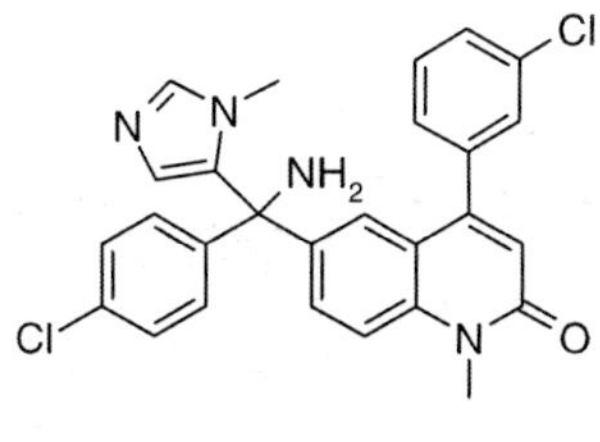

R115777 **34.44**

IC_{50} = 0.8 nM

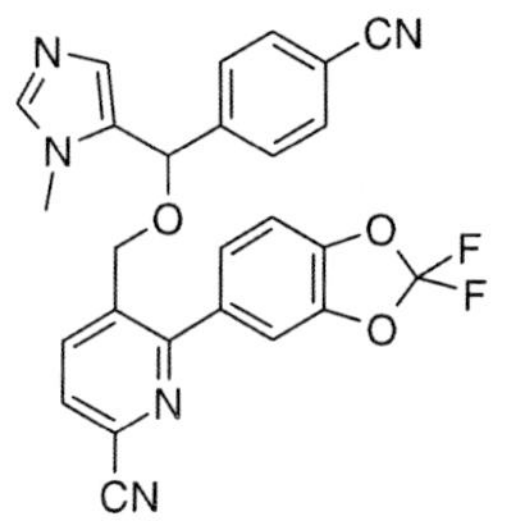

34.45

IC_{50}(FTase) = 0.18 nM

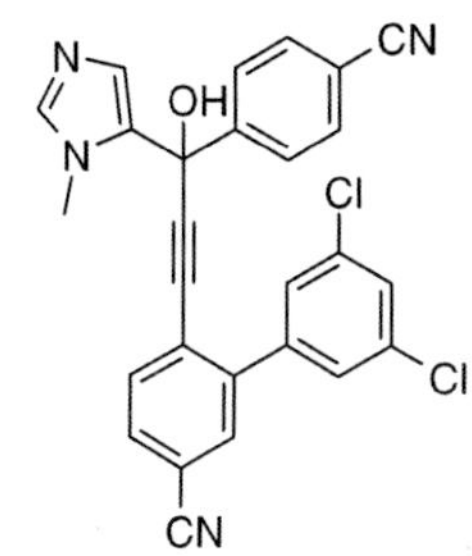

34.46

IC_{50}(FTase) = 0.079 nM

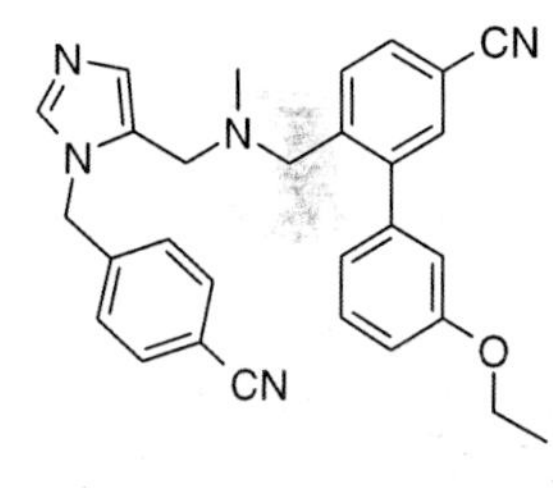

34.47

IC_{50}(FTase) = 0.034 nM

FIGURE 34.20 R115777-like inhibitors.

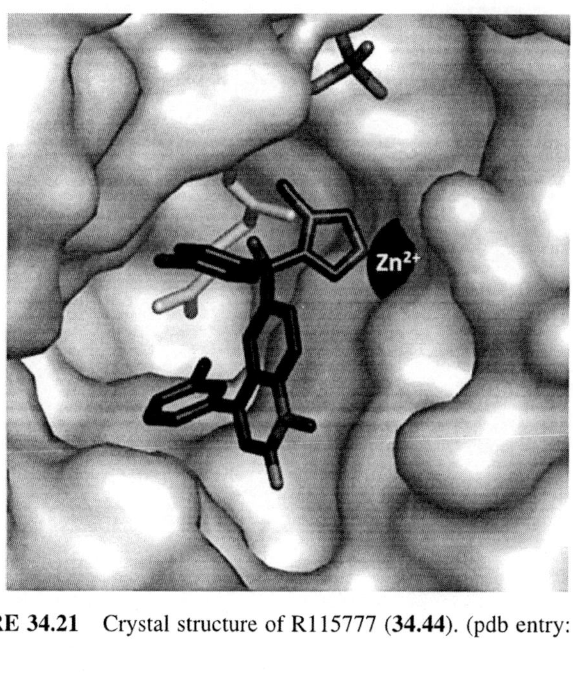

FIGURE 34.21 Crystal structure of R115777 (**34.44**). (pdb entry: 1SA4).

nitrogen juts out from the apex of the turn and coordinates the catalytic zinc. As with BMS-214662 (**34.35**) selectivity toward FTase is due to selective aromatic stacking interactions with the FTase's a_2-binding site, which cannot occur in the less aromatic active site of GGTase-I.[37] In 14 papers, the researchers from Abbott laboratories published a vast number of inhibitors derived from the tipifarnib structure. Few examples (**34.45–34.47**) are shown in Fig. 34.20.[44–46]

Except for the benzocycloheptapyridyl compounds, all inhibitors possess a zinc-binding moiety, either a thiol or a nitrogen-containing heterocycle. Using molecular modeling, our group has identified an additional lipophilic-binding site adjacent to the active site that we called the "far aryl-binding site." Connecting biaryl moieties via an acryloyl residue with our benzophenone-based aaX-mimetic scaffold, we obtained farnesyltransferase inhibitors (as **34.48** (Fig. 34.22), for example) with nanomolar activity devoid of a zinc binding substructures. Subsequently, we developed these compounds mainly as antiprotozoic agents (vide infra).[47–49]

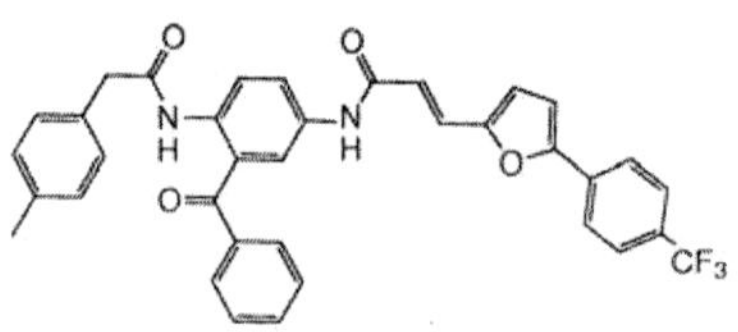

Schl-5199 **34.48**

IC_{50} = 4 nM

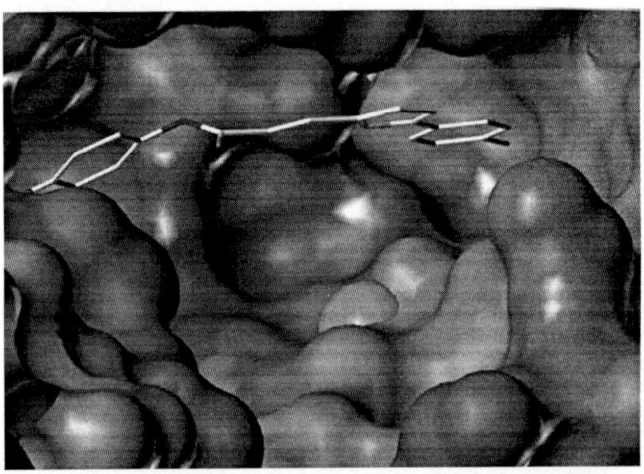

FIGURE 34.22 Benzophenone inhibitor Schl-5199 (**34.48**) that uses the so-called far aryl-binding site adjacent to the active site. Details of farnesyltransferase's active site and the far aryl-binding site. Lipophilic (brown) and hydrophilic (green to blue) properties are displayed on the Connolly surface. The structural zinc is shown as a magenta sphere. Biarylacryloyl-substituted benzophenone inhibitor deeply buries into a lipophilic area in the far aryl-binding site. (See the color version of this figure in Color Plates section.)

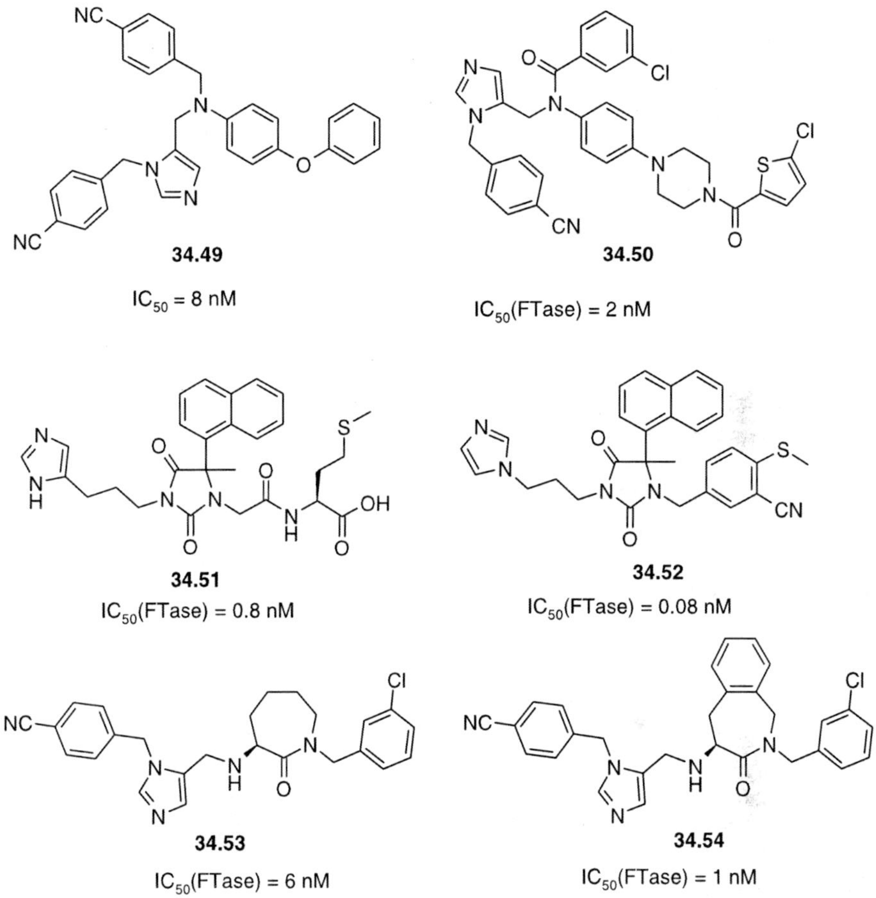

34.49 IC_{50} = 8 nM

34.50 IC_{50}(FTase) = 2 nM

34.51 IC_{50}(FTase) = 0.8 nM

34.52 IC_{50}(FTase) = 0.08 nM

34.53 IC_{50}(FTase) = 6 nM

34.54 IC_{50}(FTase) = 1 nM

FIGURE 34.23 Farnesyl transferase inhibitors with novel structural elements.

Figures 34.23 and 34.24 show representative examples of farnesyltransferase inhibitors with new structural elements published in the last years. Inhibitor **34.49** is based on a simple diphenyl ether, designed to occupy parts of the peptide-binding site.[50] An even more extended structure has been used in the phenylpiperidine derivative **34.50**. According to dockings, this molecule is able to fill almost the whole peptide-binding site of FTase.[51] Subnanomolar activity has been obtained with an aa-mimetic scaffold based on imidazolidinedione **34.51**. Replacement of the terminal methionine by a substituted benzyl resulted in even higher activity **34.52**.[52] Compounds **34.53** and **34.54** are two examples of farnesyltransferase inhibitors based on β-amino-ε-caprolactame.[53,54]

In inhibitor **34.55**, the standard imidazole has been replaced by a triazole heterocyclus.[55] Through compound library screening the piperidone **34.56** has been discovered as a novel lead structure.[56] Structural modification led to inhibitors (e.g., **34.57**) with improved activity and metabolic stability.[57]

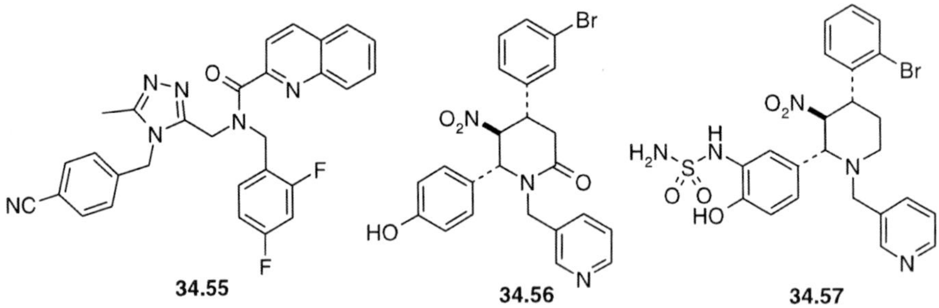

FIGURE 34.24 Piperidin-based farnesyltransferase inhibitors.

34.3.2 GGTase-I Inhibitors

While selectivity against farnesyltransferase has been the design objective for longtime, the observation that many proteins undergo alternative prenylation let appear the inhibition of GGTase-I or both CaaX prenyltransferases attractive. As described in the preceding paragraphs, the structural modifications of the molecules once designed as farnesyltransferase inhibitors turned them into dual prenylation inhibitors with more or less equal activity against both prenyltransferases or inhibitors with some selectivity toward GGTase-I. In addition some groups re-embarked into the design of prenylation inhibitors now with the intention to obtain GGTase-I inhibitors.

Screening of a library of 138 heterocyclic compounds yielded the dihydropyrrol and the tetrahydropyridine carboxylic acid **34.58** and **34.59** (Fig. 34.25) as novel leads. Through minor structural modifications, GGTase-I inhibitory activity could significantly be improved (**34.60** and **34.61**).[58,59] Another library screening yielded a pyrazole-based inhibitor with an IC_{50} against GGTase of approximately 10 μM. Structure optimization led to a highly selective compound named GGTI-DU40 (**34.62**).[60] The Hamilton group replaced their 2-arylbenzoic acid aa-mimetic by a 3-arylpiperazone scaffold to yield a new class of GGTase-I inhibitors as GGTI-2418 (**34.63**).[61] Combination of structural elements of GGTI-2154 (**34.38**) (vide supra) and GGTI-2418 (**34.63**) yield another new class of GGTase-I inhibitors (e.g., **34.64**).[62]

34.3.3 FPP Competitive Inhibitors

Work on FPP competitive inhibitors has been far less intense than on inhibitors interfering with peptide substrate binding. There may several reasons be named for this:

- FTase binds FPP very tightly ($K_D \approx 2$ nM), and therefore any FPP competitive inhibitor will face stiff competition from the cells FPP pool.

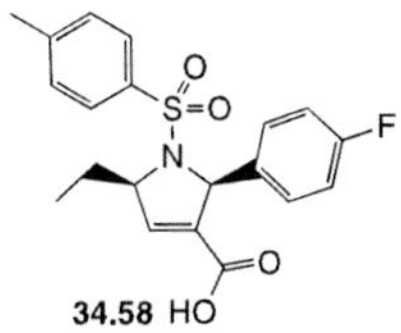

34.58

IC_{50}(GGTase-I) = 200 μM

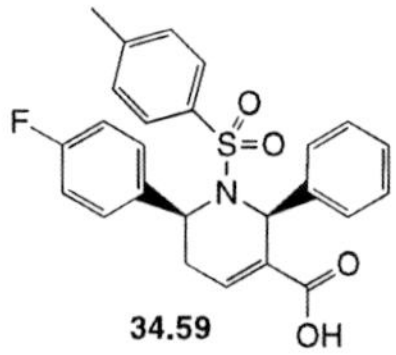

34.59

IC_{50}(GGTase-I) = 120 μM

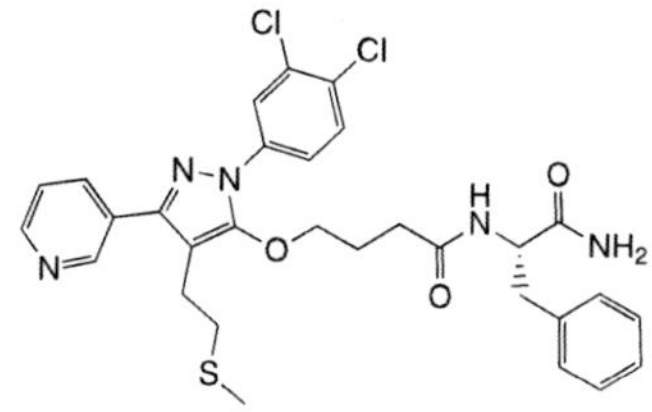

GGTI-DU40 **34.62**

IC_{50}(GGTase-I) = 8.24 nM

IC_{50}(FTase) > 2 mM

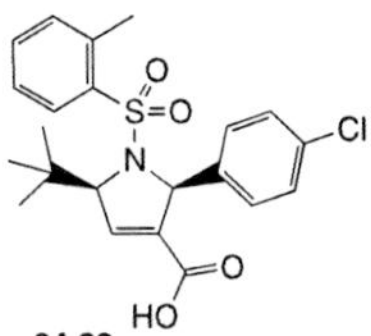

34.60

IC_{50}(GGTase-I) = 466 nM

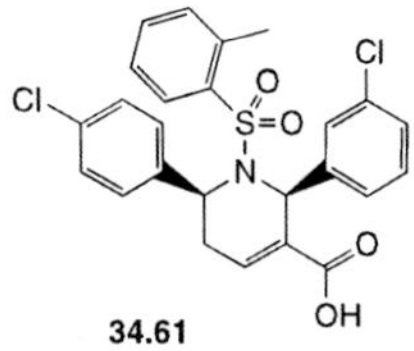

34.61

IC_{50}(GGTase-I) = 313 nM

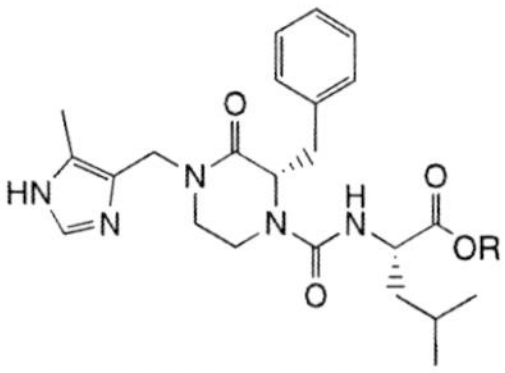

GGTI-2418 **34.63**

IC_{50}(GGTase-I) = 9.5 nM

IC_{50}(FTase) = 53 mM

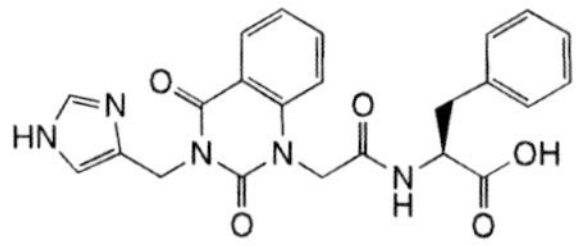

34.64

IC_{50}(GGTase-I) = 66 nM

IC_{50}(FTase) = 3500 nM

FIGURE 34.25 GGTase inhibitors.

- FPP is used by several different enzymes, so that a more unspecific action could be anticipated.
- The pyrophosphate moiety is highly susceptible to hydrolytic and enzymatic cleavage and even inhibitors possessing pyrophosphate mimetics will still bear a considerable negative charge making a prodrug strategy necessary for successful cell delivery.

Structural variations of the farnesyl residue surprisingly often led not to FPP competitive inhibitors but to alternative substrates of the enzyme. Some examples are shown in Fig. 34.26a. Figure 34.26b shows the most active example of a successful variation of the farnesyl residue leading to a competitive inhibitor.

In another approach, labile pyrophosphate moiety has been replaced by more stable mimetics. Representative examples are shown in Fig. 34.26c. Compound **34.70** is one of the rare examples where it has been attempted to replace the farnesyl as well as the pyrophosphate by nonprenyl and nonpyrophosphate substructures (Fig. 34.27).

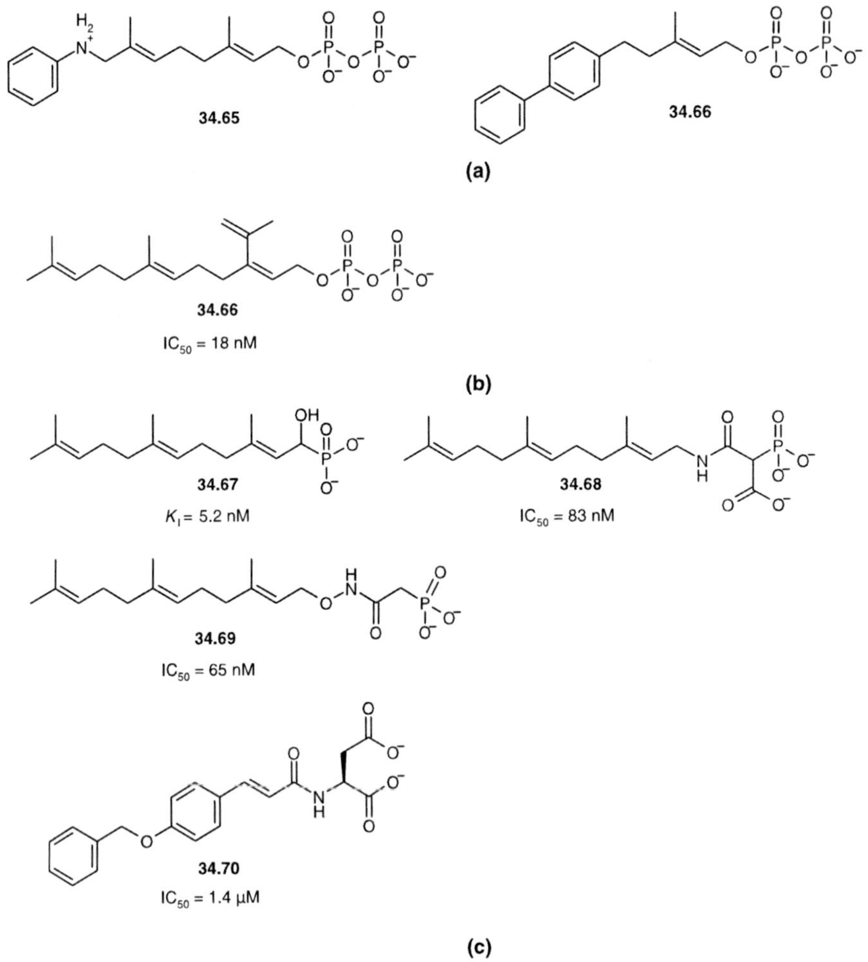

FIGURE 34.26 FPP-competitive farnesyltransferase inhibitors.

Through elongation of the prenyl residue by one isoprene unit, GGTase-I inhibitors have been obtained.

Library screening yielded several structurally diverse FPP-competitive inhibitors, which have been served as starting points for further structural modifications. The results of these approaches are shown in Fig. 34.28.

34.3.4 Bisubstrate Analogue Inhibitors

Bisubstrate analogue farnesyltransferase inhibitors are compounds, which are designed to occupy both the farnesyl and the peptide-binding site. Early bisubstrate analogues are composed of the terminal aaX tripeptide motive, a linker that may

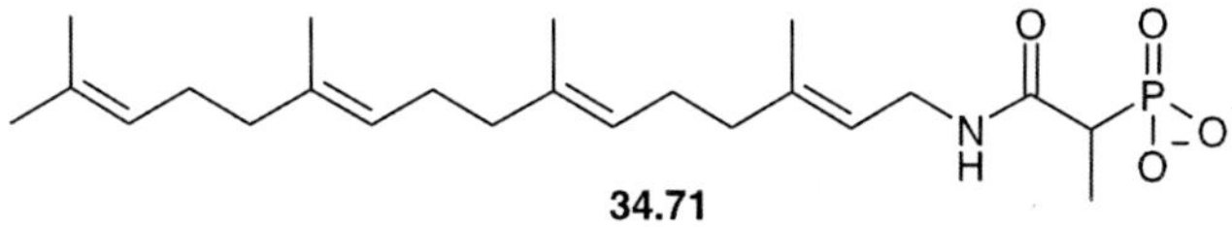

34.71

IC_{50} = 111 nM

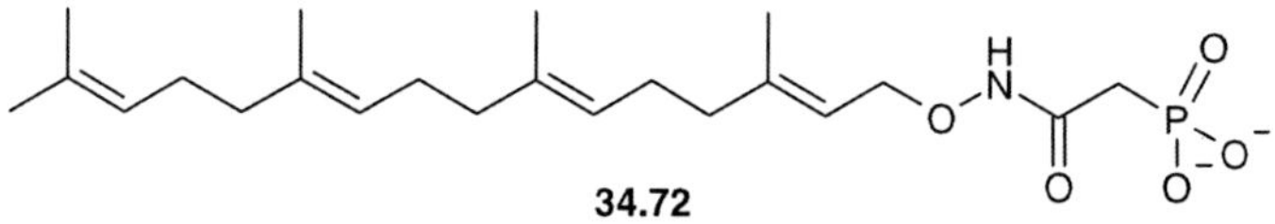

34.72

IC_{50} = 66 nM

FIGURE 34.27 GGPP-competitive inhibitors.

contain a moiety able to complex the enzyme bound zinc ion, and a farnesyl residue. Two representative examples (**34.77** and **34.78**) are shown in Fig. 34.29. Our group set out to design compounds we called nonprenylic, nonpeptidic bisubstrate analogue farnesyltransferase inhibitors, which should be in contrast to the other bisubstrate inhibitors, be devoid of any prenylic and peptidic substructures and be composed of a farnesyl and an aaX-peptide mimetic connected by an appropriate linker. The palmitoyl-β-alanyl substituted benzophenone Schl-872 (**34.79**) represents the final result of our design efforts.[48]

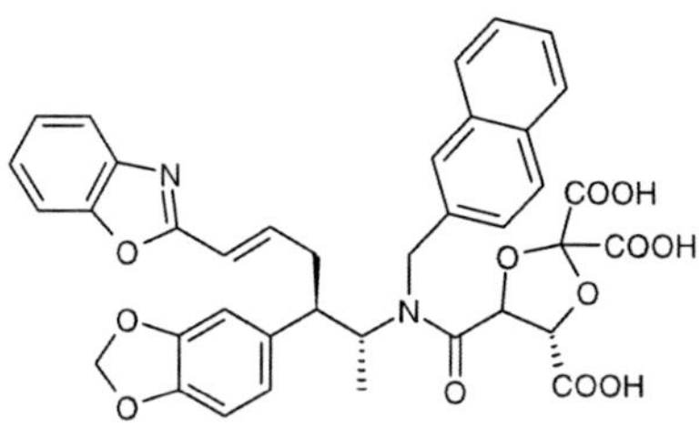

J104,871 **34.73**

IC_{50} = 3.9 nM

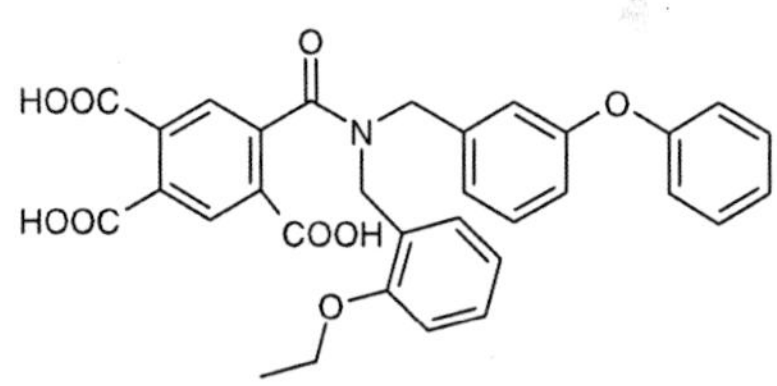

A-176120 **34.74**

IC_{50} = 1.2 nM

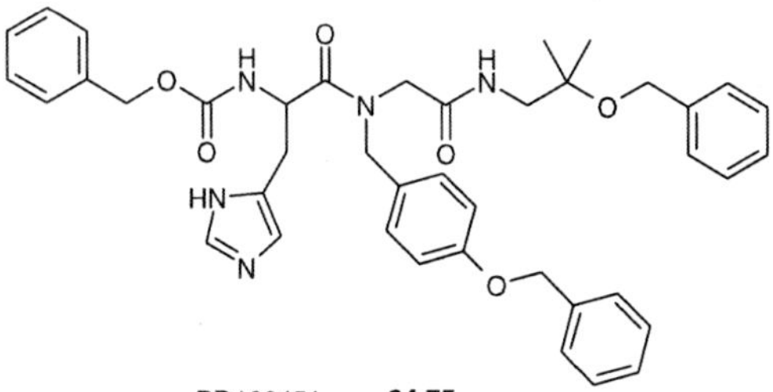

PD169451 **34.75**

IC_{50} = 4 nM

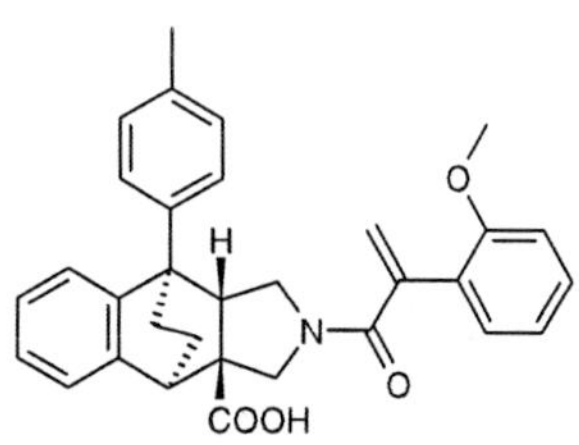

RPR130401 **34.76**

IC_{50} = 3.9 nM

FIGURE 34.28 FPP-competitive FTIs obtained from library screening.

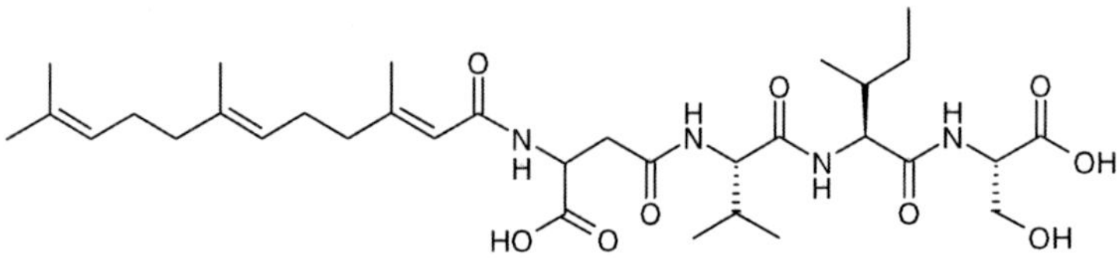

34.77 IC_{50} = 33 nM

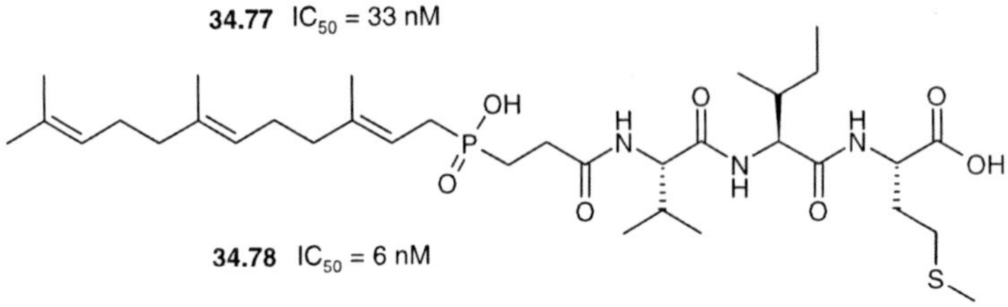

34.78 IC_{50} = 6 nM

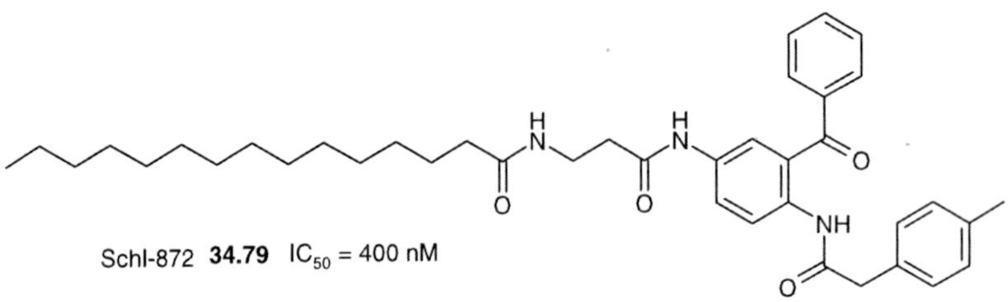

Schl-872 **34.79** IC_{50} = 400 nM

FIGURE 34.29 Bisubstrate-analogue farnesyltransferase inhibitors.

34.3.5 Natural Products

Screening of natural products has furnished a number of compounds active against prenyltransferases. Often, activity is weak in comparison with the synthetic compounds described above. Several reviews cover this issue. Selected natural compounds are shown in Fig. 34.30. Pepticinnamin is a bisubstrate analogue inhibitor.

Two newer examples are shown in Fig. 34.31. Citrafungin **34.86** displays remarkable selectivity to GGTase-I.[63,64]

34.4 PRECLINICAL ACTIVITY

Farnesyltransferase inhibitors block the growth of a large number of human tumor cell lines. FTIs induce G2/M accumulation in a number of tumor cell lines and G1 arrest in cells harboring mutant H-Ras. Under special circumstances, when cells are deprived of serum or substratum attachment, FTIs can induce apoptosis. In contrast, GGTase-I selective inhibitors can cause both cell cycle arrest in G1 and apoptosis. Tumor cells that express wild-type p53 are especially sensitive to FTI-induced growth inhibition. However, also the growth of tumor xenografts that lack functional p53 has been inhibited by FTIs. The development of FTIs has been initiated with the intention to block Ras-driven tumor development. Indeed, H-Ras driven tumors proved especially sensitive to FTIs, while N- or K-Ras-driven tumors, which constitute the majority of

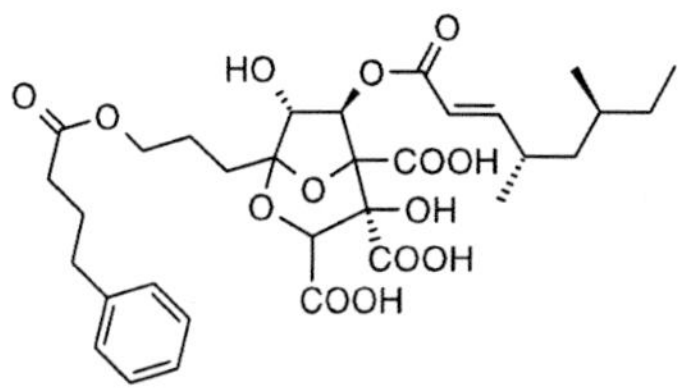

zaragonic acid analogue **34.80**
IC_{50} = 50 nM

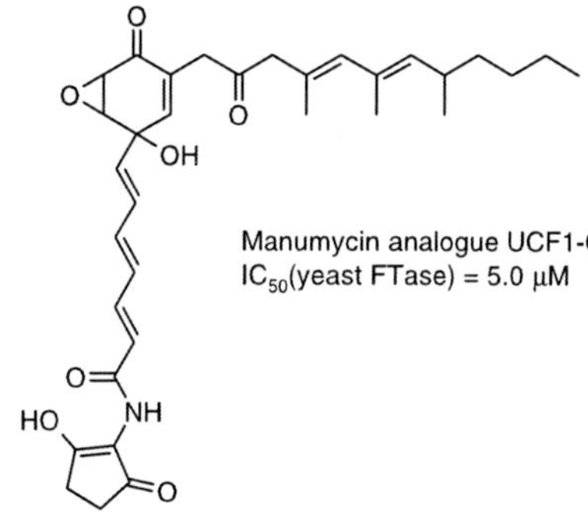

Manumycin analogue UCF1-C **34.81**
IC_{50}(yeast FTase) = 5.0 μM

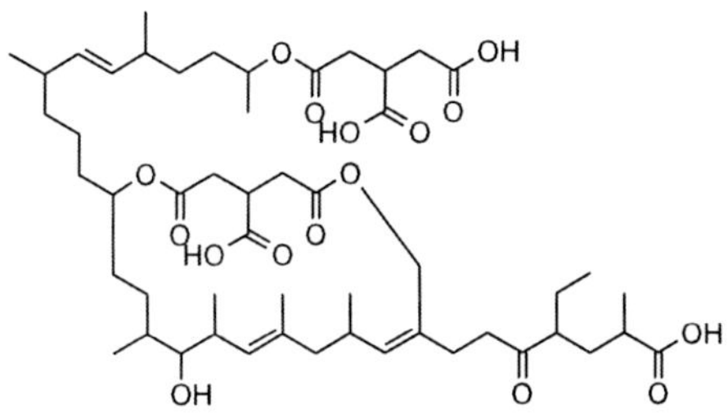

actinoplanic acid A **34.82**
IC_{50}(human FTase) = 230 nM

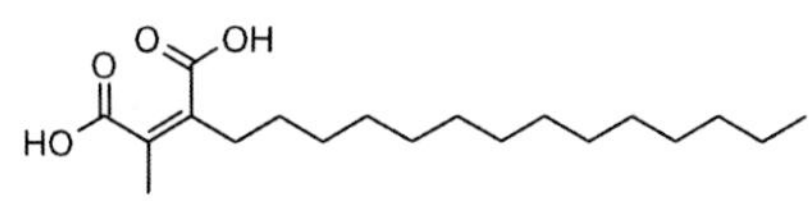

chaetomellic acid A **34.83**
IC_{50}(human FTase) = 60 nM

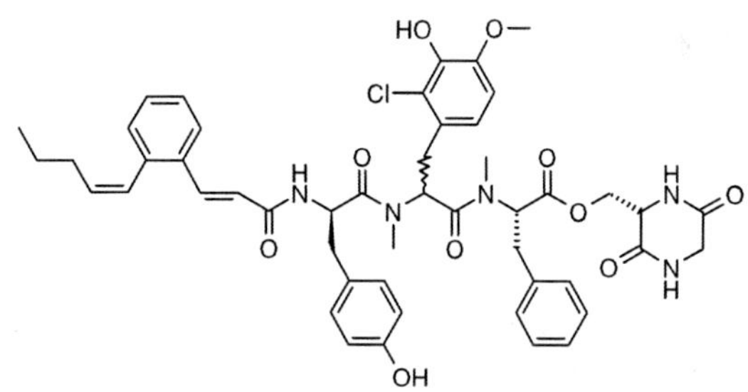

pepticinnamin **34.84**
IC_{50}(human FTase) = 100 nM

FIGURE 34.30 Farnesyltransferase inhibitors from natural sources.

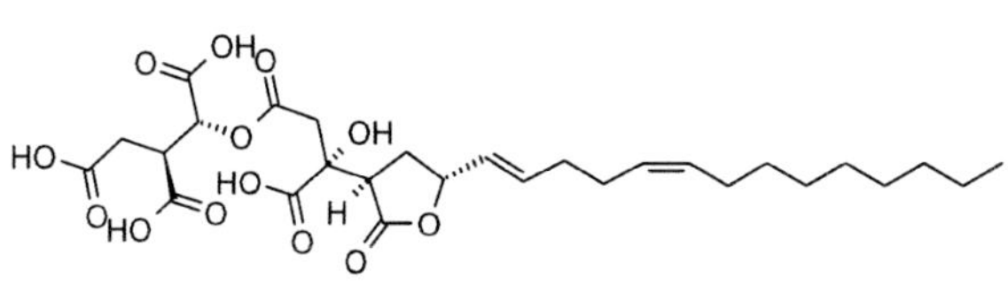

citrafungin B **34.86**
IC_{50}(ScFTase) = 174 μM
IC_{50}(humanGGTase-I) = 900 nM

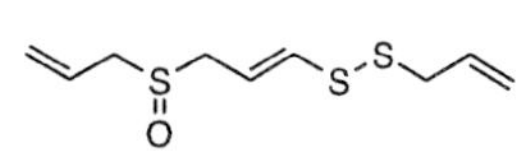

ajoene **34.87**
IC_{50}(FTase) = 13.0 μM
IC_{50}(GGTase-I) = 6.9 μM

FIGURE 34.31 Natural products.

solid tumors, often showed increased survival. This has been attributed to the so-called alternative prenylation; which means that in case farnesyltransferase activity is blocked by specific inhibitors, these proteins become substrates of geranylgeranyltransferase-I that leads to fully functional geranylgeranylated N- and K-Ras proteins. This is in part the reason for the recently renewed interest in the design of dual prenylation inhibitors or specific GGTIs. Furthermore, there may be other geranylgeranylated proteins involved in tumorgenesis.

To make a long story of the mechanism of prenyltransferase inhibitors short, one has to state that in spite of nearly two decades of intense research we are far from understanding the molecular pharmacology of prenyltransferase inhibitors. The real targets have still to be determinded.

Nevertheless, FTIs and GGTIs do inhibit the growth of tumor cells *in vitro* and in a variety of xenograft models (for example, CAPAN-2 pancreatic, LoVo colon, C32 melanoma, DU-human prostate, NCI-H460 lung, A549 lung, MiaPaCa pancreatic, HTC116 colon, MDA-435 breast and ES2, IGROV, and TOV112D ovarian) as well as in transgenic mice harboring mutant H-, K-, or N-Ras genes.

In addition, FTIs have shown enhanced *in vitro* activity in combination with taxanes, cisplatin, MEK kinase inhibitor, tamoxifen, cyclin-dependent kinase inhibitors, 5-fluorouracil, and the Bcr-Abl kinase inhibitor STI-517 (imatinib, Gleevec®). In xenografts, enhanced activity has been reported for the combination with paclitaxel, cyclophosphamid, 5-fluorouracil, and vincristine. FTIs also sensitize tumor cells to radiation. The molecular mechanism of the synergism between FTIs and taxanes may be that FTIs arrest cells in the G2/M state what makes them more sensitive to taxanes that enhance microtubule stability in the M phase. Furthermore, lonafarnib has been demonstrated to inhibit MDR1-mediated drug transport with an IC_{50} of 3 μM. MDR1-mediated drug efflux is a major mechanism by which cancer cells become resistant toward cytotoxic drugs.[1,3,65,66]

34.5 CLINICAL ACTIVITY

The clinical activity of farnesyltransferase inhibitors can also be summarized in one sentence: As single agents they displayed some activity against hematological cancers while they were disappointing against solid tumors.

Allegedly six farnesyltransferase inhibitors are or have been in clinical investigation. For L-744,832, the isopropyl ester of L-739,750 (**34.5**) (Fig. 34.9) and FTI-277, the methyl ester of FTI-276 (9) (Fig. 34.10) no results are available in public domain. For the other four FTIs, namely, tipifarnib (R115777, Zarnestra (**34.44**) Fig. 34.20), lonafarnib (SCH 66336 (**34.40**), Sarasar Fig. 34.18, BMS-214662 (**34.35**) (Fig. 34.15), and L-778,123 (**34.24**) (Fig. 34.13), the maximum tolerated dose has been determined in the first trials as usual for anticancer drugs. In addition, the results of a phase 1 study with the dual prenylation inhibitor AZD3409 (**34.13**), Fig. 34.10) are available. Table 34.3 gives an overview over side effects and dose limiting toxicities. Due to QTc time prolongation that may result in dangerous heart arrhythmias, further development of L-778,123 (**34.24**) has been discontinued.

TABLE 34.3 Dose Limiting Toxicities and Other Side Effects of Clinical Evaluated FTIs

Drug	Dose Limiting Toxicity	Other Side Effects
Tipifarnib	Myelosuppression, neurotoxicity	Nausea, vomiting, fatigue
Lonafarnib	Diarrhea, nausea, vomiting	Fatigue
BMS-214662	Diarrhea, nausea, vomiting, creatinine elevation, acute pancreatitis, renal failure, hypokalemia, cardiovascular toxicity	
AZD3409	Diarrhea	Orthostatic hypotension, paresthesia, nausea, abdominal pain, dizziness

As summarized in the first sentence, only in trial involving hematological cancer significant response rates have been seen. In contrast, with solid cancers, only few, if any, responses have been seen together with some cases of stable disease for some time. FTIs may prove more effective in the ongoing trials in which they are evaluated in combination with other anticancer agents. In anticancer therapy, the belief exists that one can expect the highest antitumor activity when the drug is given in the highest dose a patient can tolerate. Therefore, the maximum tolerated dose is often the end point of those studies. While this is appropriate for cytotoxic drugs, it may not be the best end point for the signal transduction modulators as farnesyltransferase inhibitors. Here, the biochemical effect would possibly more appropriate. Unfortunately, since the targets of prenylation inhibitors are yet to be identified, there is no good marker for the extent of prenylation inhibition necessary to achieve the desired clinical antitumor effect.[1,3,4,66,67]

So, more biochemical as well as clinical research is necessary to determine the value of prenylation inhibitors in cancer therapy.

34.6 FTIs AS ANTI-INFECTIVE AGENTS

In addition to mammals, farnesyltransferases have been detected in other eukaryotic organisms including the pathogenic protozoa *Giardia lamblia*'[68] *Schistosoma mansoni*,[69] *Plasmodium falciparum*,[70,71] *Trypanosoma brucei*,[72,73] *Trypanosoma cruzi*,[74] *Leishmania* spp.,[74] *Toxoplasma gondii*,[75] and *Entamoeba histolytica*.[76] GGTase-I has been found in *Trypanosoma cruzi*[77] and *Entamoeba histolytica*.[78] Several unspecified CaaX-peptidomimetic GGTase-I inhibitors displayed very low potency against *T. cruzi* GGTase.[77] The high activity of farnesyltransferase inhibitors against *P. falciparum* and *T. brucei* has been attributed to the apparent absence of GGTase-I and the lack of alternative prenylation of critical farnesylated proteins in these organisms.[77] No evaluations of farnesyltransferase inhibitors against *G. lamblia* and *S. mansoni* have been reported. Farnesyltransferase inhibitors have also been evaluated against fungal and viral infections. Most work has been done in the development of antiparasitic farnesyltransferase inhibitors. In protozoa, the target proteins are largely unknown.[71,79]

34.6.1 Malaria

In the early development of farnesyltransferase inhibitors against malaria,[80] most of the inhibitors had the drawback of relative selectivity against the human enzyme and insufficient activity against cultured parasites. For example, FTI-2148 (**34.36**) inhibits the human FTase with an IC_{50} of 0.82 nM while it has an IC_{50} of 15 nM against *Pf*FTase. Against cultured parasites, this compound is virtually inactive (IC_{50} > 66 μM). The benzylester (**34.87**) (Fig. 34.32) showed improved *in vitro* activity (IC_{50} = 150 nM) and activity in a mouse model (ED_{50} ≈ 95 μmol (50 mg)/kg).[81]

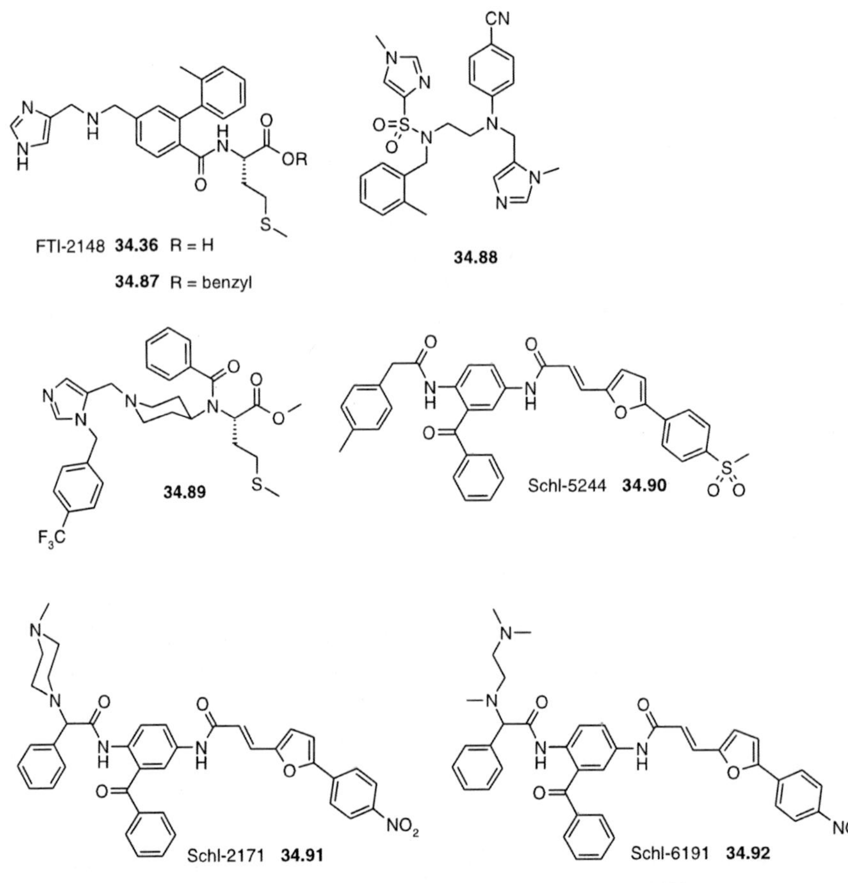

FIGURE 34.32 Anti-malarial farnesyltransferase inhibitors.

Apparently, work on this class of compounds has been discontinued since the same group came up with a novel structural simple lead (**34.88**), which inhibits *Pf*FTase with an IC_{50} of 0.6 nM and the growth of cultured parasites with IC_{50} values of 93 nM (3D7 strain) and 150 nM (K1 strain), respectively.[82] Other antimalarial farnesyltransferase inhibitors are based on the piperidinylbenzamide scaffold. The most active compound (**34.89**) of this series inhibits the growth of a chloroquine-resistant strain with an IC_{50} of 850 nM.[30] Our group has developed a class of benzophenone-based farnesyltransferase inhibitors.[48] Extensive structural variation yielded inhibitor Schl-5244 (**34.90**), which displayed an IC_{50} of 37 nM against the Dd2 strain but failed to kill the parasites in a mouse model.[83] Through introduction of a methylpiperazinyl residue (Schl-2171 (**34.91**)), *in vitro* activity was reduced ($IC_{50} = 270$ nM); however, this compound was active in *P. vinkei* infected mice with an ED_{50} value of 45 µmol (30 mg)/kg ip.[84] Enhancement of conformational flexibility through the replacement of the piperazinyl substituent by an *N,N,N'*-trimethylethylenediamine in Schl-6191 (**34.92**) considerably improved *in vitro* ($IC_{50} = 32$ nM) and *in vivo* ($ED_{50} = 24$ µmol (16 mg)/kg) activity [84a].

While most of the above-mentioned farnesyltransferase inhibitors were designed with their antimalarial properties in mind, researchers of the University of Washington followed a different approach, which they called "piggy backing". They tried to make use of the considerable effort invested by the pharmaceutical industry in the development of farnesyltransferase inhibitors as potential cancer therapeutics. Arguing that toxicity levels of farnesyltransferase inhibitors shown in clinical trials are also acceptable for antimalarial therapy such drugs could possibly also used as antimalarials what would save a lot of development costs.[85,86] Unfortunately, even BMS-214662 (**34.35**), which was the most active antimalarial compound among the farnesyltransferase inhibitors in clinical development, displayed insufficient parasite growth suppression activity ($IC_{50} = 180$ nM). Fortunately, the developer of BMS-214682 (**34.35**) possessed a library of tetrahydroquinolines structurally related to the benzodiazepine **34.35**. One of these tetrahydroquinolines, BMS-386914 (**34.93**) (Fig. 34.33) became the new lead. BMS-386914 (**34.93**) inhibits *Pf*FTase with an IC_{50} of 0.9 nM and the growth of 4 *P. falciparum* laboratory strains with IC_{50}'s between 5 and 20 nM. Due to the lack of oral absorption, BMS-386914 (**34.93**) has been administered through an implanted osmotic pump that liberated 380 µmol (200 mg)/kg and day. With this dosing 60% of *P. berghei* infected mice were cured.[87] Extensive structural variation of the tetrahydroquinolines[88] led to PB 93 (**34.94**), which inhibits *Pf*FTase with an IC_{50} of 0.58 nM and the growth of 2 *P. falciparum* laboratory strains with IC_{50}'s of 15 and 16 nM, respectively. Oral administration of 89 µmol (50 mg)/kg 3 times a day for 3 days suppressed parasitemia in *P. berghei* infected rats beyond detection level. However, recrudescence was seen in 20% of the animals.[89] To prevent oxidative desalkylation of the imidazolylmethyl group, a carbonyl has been introduced into the 2-position of the tetrahydroquinoline core yielding 2-oxotetrahydroquinoline with higher metabolic stability. The IC_{50} value of the most active compound **34.95** against *Pf*FTase is <0.5 nM, the growth of the 3D7 and the K1 strain is inhibited with IC_{50}'s of 35 and 30 nM, respectively.[90]

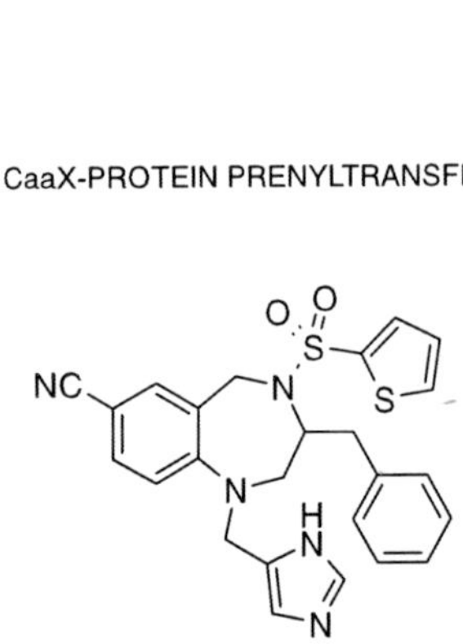

BMS-214662 **34.35**

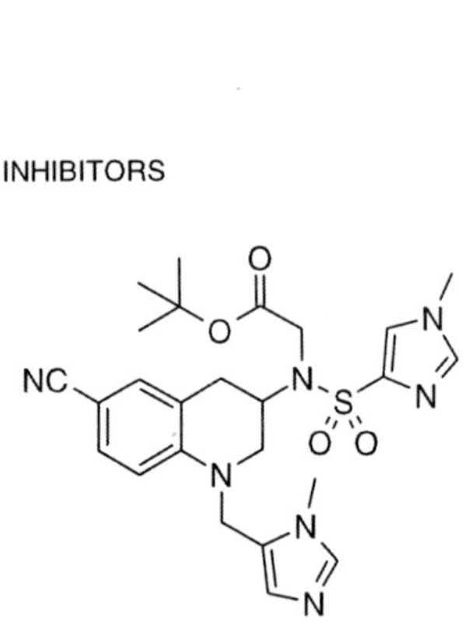

BMS-386914 **34.93**
IC_{50}(*Pf*FTase) = 0.9 nM

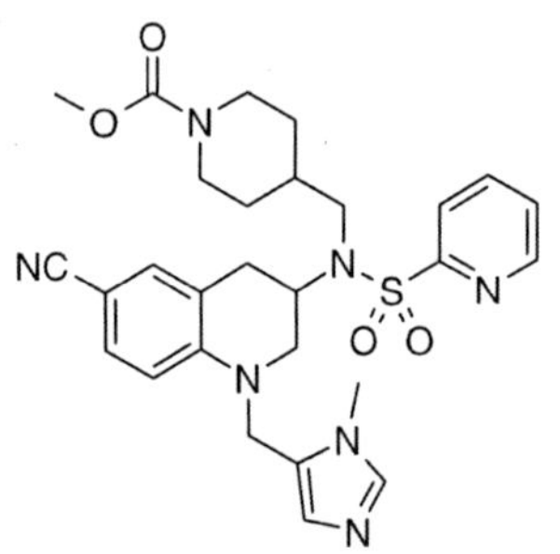

PB93 **34.94**
IC_{50}(*Pf*FTase) = 0.58 nM

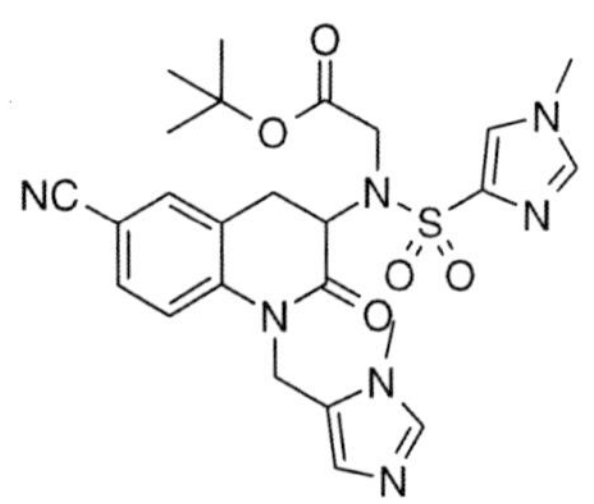

34.95
IC_{50}(*Pf*FTase) < 0.5 nM

FIGURE 34.33 Anti-malarial farnesyltransferase inhibitors based on benzodiazepine and tetrahydroquinoline-scaffold.

34.6.2 African Sleeping Sickness

The benzodiazepine farnesyltransferase inhibitor BMS-214662 (**34.35**) inhibits *Trypanonsoma brucei* farnesyltransferase (*Tb*FTase) with an IC_{50} of 1.9 nM and parasite growth with an IC_{50} of 200 nM. The tetrahydroquinoline BMS-386914 (**34.8**) is slightly less active with the IC_{50} values of 50 nM against *Tb*FTase and 500 nM against cultured parasites. Tipifarnip (**34.44**) displays also high activity against isolated enzyme (IC_{50} = 3.3 nM), but is less active against the *T. brucei* parasites (IC_{50} = 2.5 μM).[85] FTI-276 (**34.9**) inhibits *Tb*FTase with an IC_{50} of 1.7 nM and its methyl ester prodrug FTI-277 (**34.96**) inhibits parasite growth with an IC_{50} of 700 nM. Further development of this class replaced the terminal deoxycysteine by different imidazole-containing moieties yielding compounds with comparable activity against *Tb*FTase (FTI-2148 (**34.36**): IC_{50} = 1.8 nM and **34.98**: IC_{50} = 5.9 nM, respectively), but considerably improved activity against cultured parasites (FTI-2153 (**34.97**): IC_{50} against *T. brucei brucei* 25 nM, against *T. brucei rhodesiense* 2.6 nM; **34.99**: IC_{50} against *T. brucei brucei* 100 nM, against *T. brucei rhodesiense* 1.5 nM)).[91] Out of our series of benzophenone-based farnesyltransferase inhibitors the *N*-boc-phenylalanine derivative Schl-180022 (**34.100**) turned out to be the most active one with an IC_{50} of 34 nM against *T. brucei brucei* parasites (unpublished results).

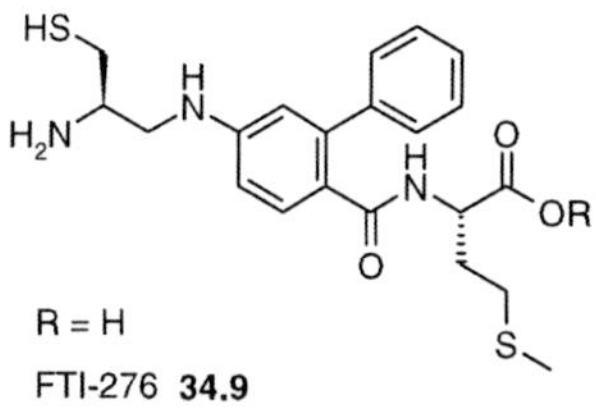

R = H
FTI-276 **34.9**
IC_{50}(*Tb*FTase) = 1.7 nM

R = CH_3
FTI-277 **34.96**

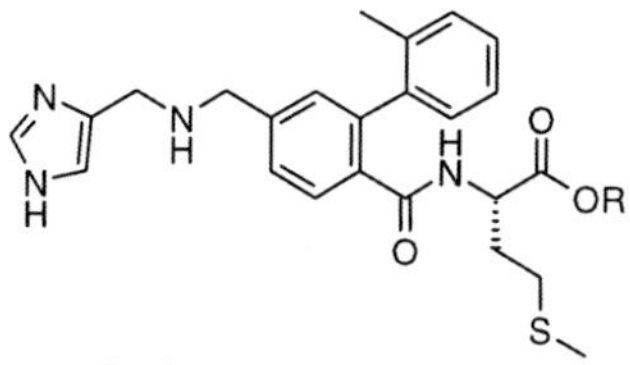

R = H
FTI-2148 **34.36**
IC_{50}(*Tb*FTase) = 1.8 nM

R = CH_3
FTI-2153 **34.97**

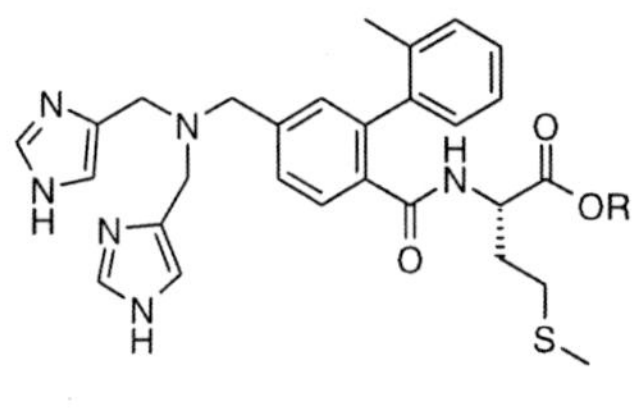

R = H **34.98**
IC_{50}(*Tb*FTase) = 5.9 nM

R = CH_3 **34.99**

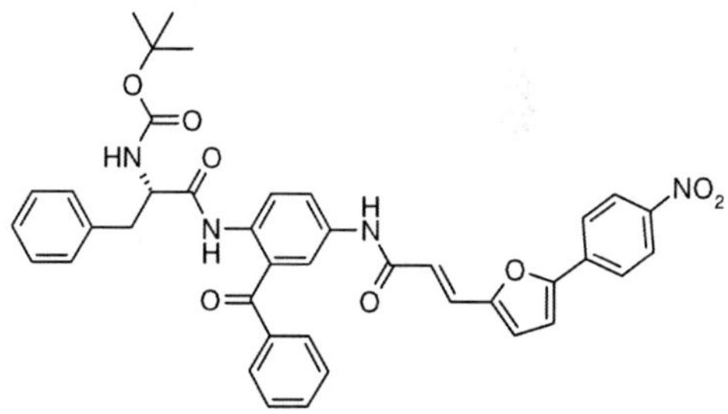

Schl-18022 **34.100**

FIGURE 34.34 Farnesyltransferase inhibitors active against *Trypanosoma brucei.*

34.6.3 American Trypanosomiasis

FTI-277 (**34.96**) (Fig. 34.34) the methylester prodrug of FTI-276 (**34.9**) displayed an IC_{50} value of 8 μM against *Trypanosoma cruzi* amastigotes. Tipifarnib (**34.44**) inhibits farnesyltransferase from *T. cruzi* with an IC_{50} of 75 nM, which is less than its activity against cultured parasites (IC_{50} = 4 nM). Later, it was shown that the tipifarnibs activity against *T. cruzi* parasites is due to the inhibition of a 14-sterol-demethylase.[92] From our benzophenone-based farnesyltransferase inhibitors, the *R*-phenylalanine derivative Schl-6071 (**34.101**) (Fig. 34.35) displayed an IC_{50} of 1 nM against *T. cruzi*

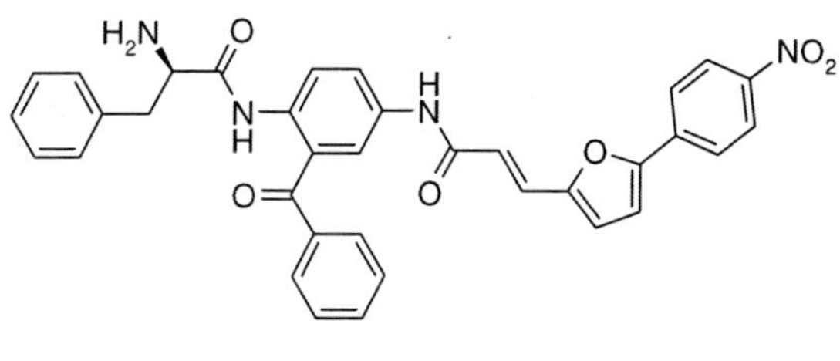

Schl-6071 **34.101**

FIGURE 34.35 Benzophenone farnesyltransferase inhibitors with activity against *Trypanosoma cruci.*

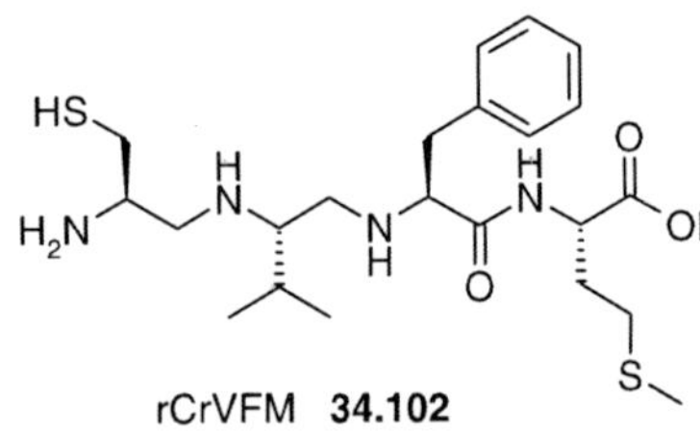

rCrVFM **34.102**

FIGURE 34.36 Peptidomimetic farnesyltransferase inhibitors inhibiting *Leishmania*.

parasites. Oral dosing of 10 μmol (6.4 mg)/kg and led to a 80% survival rate of infected mice in comparison to 16% of the untreated control group.[93]

34.6.4 Leishmaniasis

Only few farnesyltransferase inhibitors have been tested against *Leishmania* spp. The reduced peptide rCrVFM (**34.102**) (Fig. 34.36) inhibited the growth of *L. mexicana* insect forms with an IC_{50} of 30 μM.[94] Some of our benzophenone-based FTIs showed growth inhibition of *L. brasiliensis* at concentrations between 1 and 10 μM (unpublished results).

34.6.5 Toxoplasmosis

Apparently, only few farnesyltransferase inhibitors have been evaluated against *Tg*FTase. Among these compounds, the FPP mimetic **34.103** (Fig. 34.37) showed excellent activity ($IC_{50} = 950$ pm).[75]

34.6.6 Amebiasis

Only five farnesyltransferase inhibitors have been tested against *Entamoeba histolytica*. Among them, FTI-276 (**34.9**) was one of the most active inhibitors with an IC_{50} of 2.4 μM.[76]

34.6.7 FTIs As Antiviral Agents

Hepatitis C virus (HCV) RNA replication and assembly of the viral replication complex requires one or more host proteins that are geranylgeranylated. Consequently,

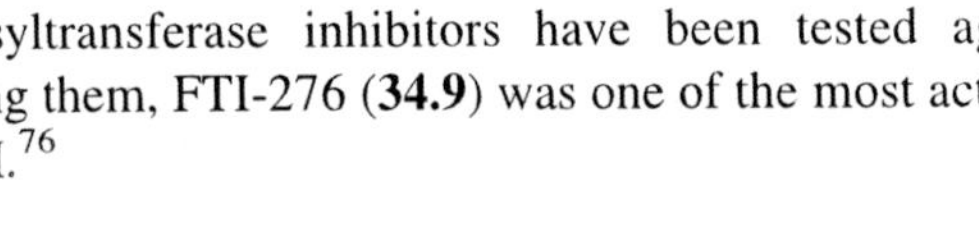

34.103

FIGURE 34.37 FPP-mimetic farnesyltransferase inhibitors inhibiting *Toxoplasma gondii*.

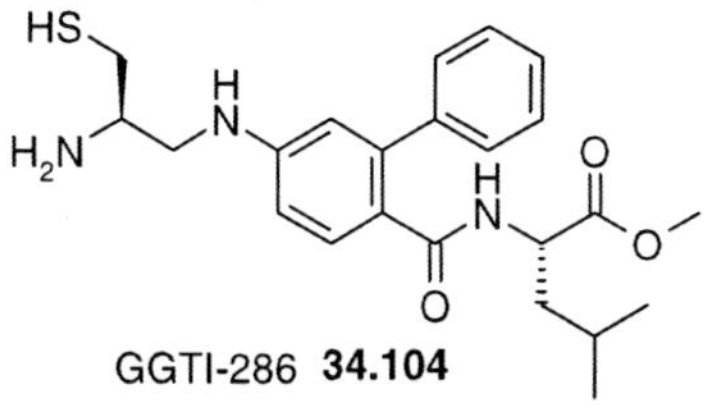

GGTI-286 **34.104**

FIGURE 34.38 Structure of the GGT-ase inhibitor GGTI-286.

GGTase inhibitor GGTI-286 (**34.104**) (Fig. 34.38) inhibited *in vitro* HCV replication, while a farnesyltransferase inhibitor was ineffective.[95] Hepatitis D virus (HDV) can worsen the liver disease in patients coinfected with hepatitis B virus (HBV). HDV requires HBV surface proteins for its envelope, but can replicate on its own once inside the cell. The HDV genome codes only two proteins, two isoforms of the delta antigen. HDV viron assembly is critically dependent on the farnesylation of the large delta antigen isoform. Anti-HDV activity has first been shown for the benzodiazepine farnesyltransferase inhibitor BDZ-5A.[96] Later prevention of complete, infectious HDV virons was also achieved with the structurally unrelated FTI-277 (**34.96**, see Fig. 34.34).[97] This compound and the related nonthiol FTI-2153 (the methylester prodrug of FTI-2148 (**34.36**), see Fig. 34.34) cleared viremia in a mouse model.[98] These findings present a perspective for the development of a treatment for an otherwise hardly to treat viral infection.

34.6.8 FTIs As Antifungal Agents

In several pathogenic fungi protein farnesylation has been demonstrated. In most cases, farnesyl or geranylgeranyltransferase inhibitors showed visible effects on growth or differentiation only at concentrations of several hundred micromoles per liter.[99–102] This has been attributed to alternative prenylation of cellular target proteins[103] or insufficient cell permeation of the inhibitors. The last effect could be demonstrated for the GGTase inhibitor **34.105** (Fig. 34.39) (IC_{50} (*Ca*GGTase) <5 nM) which MIC is above 300 μM. Also the methyl ester did not perform better in the cell-based assay. In contrast, the alanyl derivative **34.106**, designed to be incorporated into the cell using an amino acid transporter, had an MIC of 2.3 μM.[104]

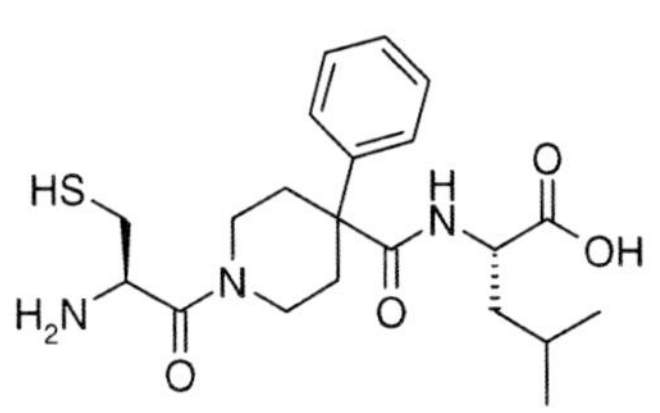

34.105
IC_{50}(Candida GGTase-I) < 5 nM

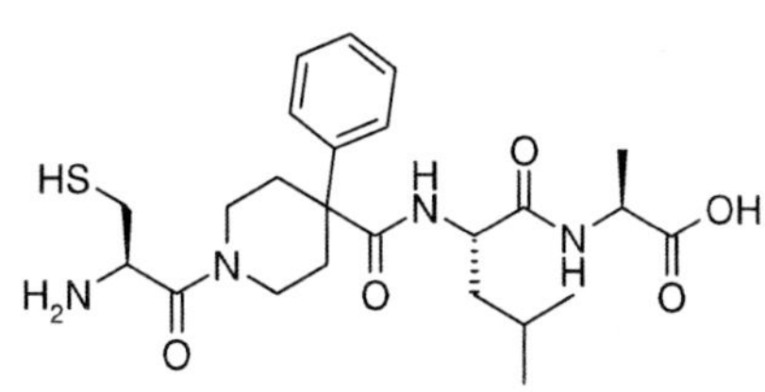

34.106
IC_{50}(Candida GGTase-I) = 10 nM

FIGURE 34.39 Antifungal farnesyltransferase inhibitor and its alanine prodrug.

34.7 CONCLUSION

Through structural biology, the mechanism of CaaX prenyltransferases is understood on molecular level as is the binding of a number of inhibitors to this enzyme. Drug development has furnished a large number of structural different, highly active inhibitors, either specific to one of the two prenyltransferases (FTase or GGTase-I) or equally active against both (the dual prenylation inhibitors). From preclinical investigations farnesyltransferase inhibitors arose as a highly promising class of novel anticancer agents although the exact mechanism by which they exert their antiproliferative effect has not been understood and still is not. In clinical evaluation, the farnesyltransferase inhibitors could not hold the promises from the preclinical studies showing some clinical effect against hematologic malignancies but being disappointing against solid tumors. Maybe, the combination of farnesyl- and geranylgeranyltransferase inhibitors with established anticancer therapeutics will provide better results. In any case, more research is necessary to understand the underlying biology. Prenylation inhibitors showed some promising activity against preclinical models of protozoic infections. Here also, more development is necessary to provide inhibitors that may be suited for clinical evaluation.

REFERENCES

1. Appels, N. M.; Beijnen, J. H.; Schellens, J. H. Development of farnesyl transferase inhibitors: a review. *Oncologist* **2005**, *10*, 565–578.
2. Margaritora, S.; Cesario, A.; Porziella, V.; Granone, P.; Catassi, A.; Russo, P. Farnesyltransferase Inhibitors: overview of their action and role in solid malignancy therapy. *Lett. Drug Des. Discov.* **2005**, *2*, 26–35.
3. Basso, A. D.; Kirschmeier, P.; Bishop, W. R. Lipid posttranslational modifications. Farnesyl transferase inhibitors. *J. Lipid Res.* **2006**, *47*, 15–31.
4. Konstantinopoulos, P. A.; Karamouzis, M. V.; Papavassiliou, A. G. Post-translational modifications and regulation of the RAS superfamily of GTPases as anticancer targets. *Nat. Rev. Drug Discov.* **2007**, *6*, 541–555.
5. Sousa, S. F.; Fernandes, P. A.; Ramos, M. J. Farnesyltransferase—new insights into the zinc-coordination sphere paradigm: evidence for a carboxylate-shift mechanism. *Biophys. J.* **2005**, *88*, 483–494.
6. Sousa, S. F.; Fernandes, P. A.; Ramos, M. J. Theoretical studies on farnesyltransferase: the distances paradox explained. *Proteins* **2007**, *66*, 205–218.
7. Lane, K. T.; Beese, L. S. Thematic review series: lipid posttranslational modifications. Structural biology of protein farnesyltransferase and geranylgeranyltransferase type I. *J. Lipid Res.* **2006**, *47*, 681–699.
8. Reid, T. S.; Terry, K. L.; Casey, P. J.; Beese, L. S. Crystallographic analysis of CaaX prenyltransferases complexed with substrates defines rules of protein substrate selectivity. *J Mol Biol* **2004**, *343*, 417–433.
9. Long, S. B.; Casey, P. J.; Beese, L. S. Reaction path of protein farnesyltransferase at atomic resolution. *Nature* **2002**, *419*, 645–650.

10. Qian, Y.; Sebti, S. M.; Hamilton, A. D. Farnesyltransferase as a target for anticancer drug design. *Biopolymers* **1997**, *43*, 25–41.

11. Leonard, D. M. Ras farnesyltransferase: a new therapeutic target. *J. Med. Chem.* **1997**, *40*, 2971–2990.

12. Ayral-Kaloustian, S.; Salaski, E. J. Protein farnesyltransferase inhibitors. *Curr. Med. Chem.* **2002**, *9*, 1003–1032.

13. Graham, S. L. Review Oncologic, Endocrine & Metabolic: Inhibitors of protein farnesylation: a new approach to cancer chemotherapy. *Expert Opin. Ther. Pat.* **1995**, *5*, 1269–1285.

14. Ohkanda, J.; Blaskovich, M. A.; Sebti, S. M.; Hamilton, A. D. The development of protein farnesyltransferase inhibitors as signaling-based anticancer agents. *Prog. Cell Cycle Res.* **2003**, *5*, 211–217.

15. Ohkanda, J.; Knowles, D. B.; Blaskovich, M. A.; Sebti, S. M.; Hamilton, A. D. Inhibitors of protein farnesyltransferase as novel anticancer agents. *Curr. Top. Med. Chem.* **2002**, *2*, 303–323.

16. Huang, C.; Rokosz, L.; Farnesyltransferase inhibitors: recent advances. *Expert Opin. Ther. Pat.* **2004**, *14*, 175–186.

17. Bell, I. M. Inhibitors of farnesyltransferase: a rational approach to cancer chemotherapy? *J. Med. Chem.* **2004**, *47*, 1869–1878.

18. Puntambekar, D. S.; Giridhar, R.; Yadav, M. R. Inhibition of farnesyltransferase: a rational approach to treat cancer? *J. Enzyme Inhib. Med. Chem.* **2007**, *22*, 127–140.

19. Gibbs, R. A.; Zahn, T. J.; Sebolt-Leopold, J. S. Non-peptidic prenyltransferase inhibitors: diverse structural classes and surprising anti-cancer mechanisms. *Curr. Med. Chem.* **2001**, *8*, 1437–1465.

20. End, D. W. Farnesyl protein transferase inhibitors and other therapies targeting the Ras signal transduction pathway. *Invest. New Drugs* **1999**, *17*, 241–258.

21. Halazy, S.; Gotteland, J. -P.; Lamothe, M.; Perrin, D.; Hill, B. T. Rationally designed FPTase inhibitors. *Drugs Future* **1997**, *22*,1133–1146.

22. Sebti, S. M.; Adjei, A. A. Farnesyltransferase inhibitors. *Semin. Oncol.* **2004**, *31*, 28–39.

23. Sebti, S. M.; Hamilton, A. D. New approaches to anticancer drug design based on the inhibition of farnesyltransferase. *Drug Discov. Today* **1998**, *3*, 26–33.

24. Lee, H. Y.; Sohn, J. H.; Kwon, B. M. Development of tripeptidyl farnesyltransferase inhibitors. *Bioorg. Med. Chem. Lett.* **2002**, *12*, 1599–1602.

25. Houssin, R.; Pommery, J.; Salaun, M. C.; Deweer, S.; Goossens, J. F.; Chavatte, P.; Henichart, J. P. Design, synthesis, and pharmacological evaluation of new farnesyl protein transferase inhibitors. *J. Med. Chem.* **2002**, *45*, 533–536.

26. Appels, N. M.; Bolijn, M. J.; Chan, K.; Stephens, T. C.; Hoctin-Boes, G.; Middleton, M.; Beijnen, J. H.; de Bono, J. S.; Harris, A. L.; Schellens, J. H. Phase I pharmacokinetic and pharmacodynamic study of the prenyl transferase inhibitor AZD3409 in patients with advanced cancer. *Br. J. Cancer* **2008**, *98*, 1951–1958.

27. Sakowski, J.; Bohm, M.; Sattler, I.; Dahse, H. M.; Schlitzer, M. Synthesis, molecular modeling, and structure–activity relationship of benzophenone-based CAAX-peptidomimetic farnesyltransferase inhibitors. *J. Med. Chem.* **2001**, *44*, 2886–2899.

28. Reid, T. S.; Long, S. B.; Beese, L. S. Crystallographic analysis reveals that anticancer clinical candidate L-778,123 inhibits protein farnesyltransferase and geranylgeranyltransferase-I by different binding modes. *Biochemistry* **2004**, *43*, 9000–9008.

29. Millet, R.; Domarkas, J.; Houssin, R.; Gilleron, P.; Goossens, J. F.; Chavatte, P.; Loge, C.; Pommery, N.; Pommery, J.; Henichart, J. P. Potent and selective farnesyl transferase inhibitors. *J. Med. Chem.* **2004**, *47*, 6812–6820.

30. Ryckebusch, A.; Gilleron, P.; Millet, R.; Houssin, R.; Lemoine, A.; Pommery, N.; Grellier, P.; Sergheraert, C.; Henichart, J. P. Novel *N*-(4-piperidinyl)benzamide antimalarials with mammalian protein farnesyltransferase inhibitory activity. *Chem. Pharm. Bull. (Tokyo)* **2005**, *53*, 1324–1326.

31. Gilleron, P.; Millet, R.; Houssin, R.; Wlodarczyk, N.; Farce, A.; Lemoine, A.; Goossens, J. F.; Chavatte, P.; Pommery, N.; Henichart, J. P. Solid-phase synthesis and pharmacological evaluation of a library of peptidomimetics as potential farnesyltransferase inhibitors: an approach to new lead compounds. *Eur. J. Med. Chem.* **2006**, *41*, 745–755.

32. Gilleron, P.; Wlodarczyk, N.; Houssin, R.; Farce, A.; Laconde, G.; Goossens, J. F.; Lemoine, A.; Pommery, N.; Henichart, J. P.; Millet, R. Design, synthesis and biological evaluation of substituted dioxodibenzothiazepines and dibenzocycloheptanes as farnesyltransferase inhibitors. *Bioorg. Med. Chem. Lett.* **2007**, *17*, 5465–5471.

33. Dinsmore, C. J.; Bell, I. M. Inhibitors of farnesyltransferase and geranylgeranyltransferase-I for antitumor therapy: substrate-based design, conformational constraint and biological activity. *Curr. Top. Med. Chem.* **2003**, *3*, 1075–1093.

34. El Oualid, F.; Cohen, L. H.; van der Marel, G. A.; Overhand, M. Inhibitors of protein: geranylgeranyl transferases. *Curr. Med. Chem.* **2006**, *13*, 2385–2427.

35. Ding, C. Z.; Batorsky, R.; Bhide, R.; Chao, H. J.; Cho, Y.; Chong, S.; Gullo-Brown, J.; Guo, P.; Kim, S. H.; Lee, F.; Leftheris, K.; Miller, A.; Mitt, T.; Patel, M.; Penhallow, B. A.; Ricca, C.; Rose, W. C.; Schmidt, R.; Slusarchyk, W. A.; Vite, G.; Yan, N.; Manne, V.; Hunt, J. T. Discovery and structure–activity relationships of imidazole-containing tetrahydrobenzodiazepine inhibitors of farnesyltransferase. *J. Med. Chem.* **1999**, *42*, 5241–5253.

36. Hunt, J. T.; Ding, C. Z.; Batorsky, R.; Bednarz, M.; Bhide, R.; Cho, Y.; Chong, S.; Chao, S.; Gullo-Brown, J.; Guo, P.; Kim, S. H.; Lee, F. Y.; Leftheris, K.; Miller, A.; Mitt, T.; Patel, M.; Penhallow, B. A.; Ricca, C.; Rose, W. C.; Schmidt, R.; Slusarchyk, W. A.; Vite, G.; Manne, V. Discovery of (R)-7-cyano-2,3,4, 5-tetrahydro-1-(1H-imidazol-4-ylmethyl)-3- (phenylmethyl)-4-(2-thienylsulfonyl)-1H-1,4-benzodiazepine (BMS-214662), a farnesyltransferase inhibitor with potent preclinical antitumor activity. *J. Med. Chem.* **2000**, *43*, 3587–3595.

37. Reid, T. S.; Beese, L. S. Crystal structures of the anticancer clinical candidates R115777 (Tipifarnib) and BMS-214662 complexed with protein farnesyltransferase suggest a mechanism of FTI selectivity. *Biochemistry* **2004**, *43*, 6877–6884.

38. Ohkanda, J.; Strickland, C. L.; Blaskovich, M. A.; Carrico, D.; Lockman, J. W.; Vogt, A.; Bucher, C. J.; Sun, J.; Qian, Y.; Knowles, D.; Pusateri, E. E.; Sebti, S. M.; Hamilton, A. D. Structure-based design of imidazole-containing peptidomimetic inhibitors of protein farnesyltransferase. *Org. Biomol. Chem.* **2006**, *4*, 482–492.

39. Vasudevan, A.; Qian, Y.; Vogt, A.; Blaskovich, M. A.; Ohkanda, J.; Sebti, S. M.; Hamilton, A. D. Potent, highly selective, and non-thiol inhibitors of protein geranylgeranyltransferase-I. *J. Med. Chem.* **1999**, *42*, 1333–1340.

40. Ganguly, A. K.; Doll, R. J.; Girijavallabhan, V. M. Farnesyl protein transferase inhibition: a novel approach to anti-tumor therapy. The discovery and development of SCH 66336. *Curr. Med. Chem.* **2001**, *8*, 1419–1436.

41. Taveras, A. G.; Kirschmeier, P.; Baum, C. M. Sch-66336 (Sarasar) and other benzocycloheptapyridyl farnesyl protein transferase inhibitors: discovery, biology and clinical observations. *Curr. Top. Med. Chem.* **2003**, *3*, 1103–1114.

42. Strickland, C. L.; Weber, P. C.; Windsor, W. T.; Wu, Z.; Le, H. V.; Albanese, M. M.; Alvarez, C. S.; Cesarz, D.; del Rosario, J.; Deskus, J.; Mallams, A. K.; Njoroge, F. G.; Piwinski, J. J.; Remiszewski, S.; Rossman, R. R.; Taveras, A. G.; Vibulbhan, B.; Doll, R. J.; Girijavallabhan, V. M.; Ganguly, A. K. Tricyclic farnesyl protein transferase inhibitors: crystallographic and calorimetric studies of structure–activity relationships. *J. Med. Chem.* **1999**, *42*, 2125–2135.

43. Venet, M.; End, D.; Angibaud, P. Farnesyl protein transferase inhibitor ZARNESTRA R115777—history of a discovery. *Curr. Top. Med. Chem.* **2003**, *3*, 1095–1102.

44. Tong, Y.; Lin, N. H.; Wang, L.; Hasvold, L.; Wang, W.; Leonard, N.; Li, T.; Li, Q.; Cohen, J.; Gu, W. Z.; Zhang, H.; Stoll, V.; Bauch, J.; Marsh, K.; Rosenberg, S. H.; Sham, H. L. Discovery of potent imidazole and cyanophenyl containing farnesyltransferase inhibitors with improved oral bioavailability. *Bioorg. Med. Chem. Lett.* **2003**, *13*, 1571–1574.

45. Lin, N. H.; Wang, L.; Cohen, J.; Gu, W. Z.; Frost, D.; Zhang, H.; Rosenberg, S.; Sham, H. Synthesis and biological evaluation of 4-[3-biphenyl-2-yl-1-hydroxy-1-(3-methyl-3H-imidazol-4-yl)-prop-2-ynyl]-1-yl-benzonitrile as novel farnesyltransferase inhibitor. *Bioorg. Med. Chem. Lett.* **2003**, *13*, 1293–1296.

46. Lin, N. H.; Wang, L.; Wang, X.; Wang, G. T.; Cohen, J.; Gu, W. Z.; Zhang, H.; Rosenberg, S. H.; Sham, H. L. Synthesis and biological evaluation of 1-benzyl-5-(3-biphenyl-2-yl-propyl)-1H-imidazole as novel farnesyltransferase inhibitor. *Bioorg. Med. Chem. Lett.* **2004**, *14*, 5057–5062.

47. Bohm, M.; Mitsch, A.; Wissner, P.; Sattler, I.; Schlitzer, M. Exploration of novel aryl binding sites of farnesyltransferase using molecular modeling and benzophenone-based farnesyltransferase inhibitors. *J. Med. Chem.* **2001**, *44*, 3117–3124.

48. Schlitzer, M. Structure based design of benzophenone-based non-thiol farnesyltransferase inhibitors. *Curr. Pharm. Des.* **2002**, *8*, 1713–22.

49. Mitsch, A.; Wissner, P.; Silber, K.; Haebel, P.; Sattler, I.; Klebe, G.; Schlitzer, M. Non-thiol farnesyltransferase inhibitors: *N*-(4-tolylacetylamino-3-benzoylphenyl)-3-arylfurylacrylic acid amides. *Bioorg. Med. Chem.* **2004**, *12*, 4585–4600.

50. Dinsmore, C. J.; Williams, T. M.; O'Neill, T. J.; Liu, D.; Rands, E.; Culberson, J. C.; Lobell, R. B.; Koblan, K. S.; Kohl, N. E.; Gibbs, J. B.; Oliff, A. I.; Graham, S. L.; Hartman, G. D. Imidazole-containing diarylether and diarylsulfone inhibitors of farnesyl-protein transferase. *Bioorg. Med. Chem. Lett.* **1999**, *9*, 3301–3306.

51. Perez, M.; Maraval, C.; Dumond, S.; Lamothe, M.; Schambel, P.; Etievant, C.; Hill, B. Synthesis and evaluation of a novel series of farnesyl protein transferase inhibitors as non-peptidic CAAX tetrapeptide analogues. *Bioorg. Med. Chem. Lett.* **2003**, *13*, 1455–1458.

52. Lee, J.; Kim, J.; Koh, J. S.; Chung, H. H.; Kim, K. H. Hydantoin derivatives as non-peptidic inhibitors of Ras farnesyl transferase. *Bioorg. Med. Chem. Lett.* **2006**, *16*, 1954–1956.

53. Le Diguarher, T.; Ortuno, J. C.; Dorey, G.; Shanks, D.; Guilbaud, N.; Pierre, A.; Fauchere, J. L.; Hickman, J. A.; Tucker, G. C.; Casara, P. J. Parallel liquid synthesis of *N,N′*-Disubstituted 3-amino azepin-2-ones as potent and specific farnesyl transferase inhibitors. *Bioorg. Med. Chem.* **2003**, *11*, 3193–3204.

54. Le Diguarher, T.; Ortuno, J. C.; Shanks, D.; Guilbaud, N.; Pierre, A.; Raimbaud, E.; Fauchere, J. L.; Hickman, J. A.; Tucker, G. C.; Casara, P. J. Synthesis of *N,N′*-disubstituted 3-aminobenzo[c] and [d]azepin-2-ones as potent and specific farnesyl transferase inhibitors. *Bioorg. Med. Chem. Lett.* **2004**, *14*, 767–771.

55. Saha, A. K.; Liu, L.; Simoneaux, R.; DeCorte, B.; Meyer, C.; Skrzat, S.; Breslin, H. J.; Kukla, M. J.; End, D. W. Novel triazole based inhibitors of Ras farnesyl transferase. *Bioorg. Med. Chem. Lett.* **2005**, *15*, 5407–5411.

56. Nara, S.; Tanaka, R.; Eishima, J.; Hara, M.; Takahashi, Y.; Otaki, S.; Foglesong, R. J.; Hughes, P. F.; Turkington, S.; Kanda, Y. Discovery and structure–activity relationships of novel piperidine inhibitors of farnesyltransferase. *J. Med. Chem.* **2003**, *46*, 2467–2473.

57. Tanaka, R.; Rubio, A.; Harn, N. K.; Gernert, D.; Grese, T. A.; Eishima, J.; Hara, M.; Yoda, N.; Ohashi, R.; Kuwabara, T.; Soga, S.; Akinaga, S.; Nara, S.; Kanda, Y. Design and synthesis of piperidine farnesyltransferase inhibitors with reduced glucuronidation potential. *Bioorg. Med. Chem.* **2007**, *15*, 1363–1382.

58. Castellano, S.; Fiji, H. D.; Kinderman, S. S.; Watanabe, M.; Leon, P.; Tamanoi, F.; Kwon, O. Small-molecule inhibitors of protein geranylgeranyltransferase type I. *J. Am. Chem. Soc.* **2007**, *129*, 5843–5845.

59. Watanabe, M.; Fiji, H. D.; Guo, L.; Chan, L.; Kinderman, S. S.; Slamon, D. J.; Kwon, O.; Tamanoi, F. Inhibitors of protein geranylgeranyltransferase I and Rab geranylgeranyltransferase identified from a library of allenoate-derived compounds. *J. Biol. Chem.* **2008**, *283*, 9571–9579.

60. Peterson, Y. K.; Kelly, P.; Weinbaum, C. A.; Casey, P. J. A novel protein geranylgeranyltransferase-I inhibitor with high potency, selectivity, and cellular activity. *J. Biol. Chem.* **2006**, *281*, 12445–12450.

61. Peng, H.; Carrico, D.; Thai, V.; Blaskovich, M.; Bucher, C.; Pusateri, E. E.; Sebti, S. M.; Hamilton, A. D. Synthesis and evaluation of potent, highly-selective, 3-aryl-piperazinone inhibitors of protein geranylgeranyltransferase-I. *Org. Biomol. Chem.* **2006**, *4*, 1768–1784.

62. Carrico, D.; Blaskovich, M. A.; Bucher, C. J.; Sebti, S. M.; Hamilton, A. D. Design, synthesis, and evaluation of potent and selective benzoyleneurea-based inhibitors of protein geranylgeranyltransferase-I. *Bioorg. Med. Chem.* **2005**, *13*, 677–688.

63. Singh, S. B.; Zink, D. L.; Doss, G. A.; Polishook, J. D.; Ruby, C.; Register, E.; Kelly, T. M.; Bonfiglio, C.; Williamson, J. M.; Kelly, R. Citrafungins A and B, two new fungal metabolite inhibitors of GGTase I with antifungal activity. *Org. Lett.* **2004**, *6*, 337–340.

64. Ferri, N.; Yokoyama, K.; Sadilek, M.; Paoletti, R.; Apitz-Castro, R.; Gelb, M. H.; Corsini, A. Ajoene, a garlic compound, inhibits protein prenylation and arterial smooth muscle cell proliferation. *Br. J. Pharmacol.* **2003**, *138*, 811–818.

65. Philips, M. R.; Cox, A. D. Geranylgeranyltransferase I as a target for anti-cancer drugs. *J. Clin. Invest.* **2007**, *117*, 1223–1225.

66. Blum, R.; Cox, A. D.; Kloog, Y. Inhibitors of chronically active Ras: potential for treatment of human malignancies. *Recent Pat. Anticancer Drug Discov.* **2008**, *3*, 31–47.

67. Caraglia, M.; Budillon, A.; Tagliaferri, P.; Marra, M.; Abbruzzese, A.; Caponigro, F. Isoprenylation of intracellular proteins as a new target for the therapy of human neoplasms: preclinical and clinical implications. *Curr. Drug Targets* **2005**, *6*, 301–323.

68. Lujan, H. D.; Mowatt, M. R.; Chen, G. Z.; Nash, T. E. Isoprenylation of proteins in the protozoan *Giardia lamblia. Mol. Biochem. Parasitol.* **1995**, *72*, 121–127.

69. Chen, G. Z.; Bennett, J. L. Characterization of mevalonate-labeled lipids isolated from parasite proteins in *Schistosoma mansoni. Mol. Biochem. Parasitol.* **1993**, *59*, 287–292.

70. Chakrabarti, D.; Azam, T.; DelVecchio, C.; Qiu, L.; Park, Y. I.; Allen, C. M. Protein prenyl transferase activities of *Plasmodium falciparum. Mol. Biochem. Parasitol.* **1998**, *94*, 175–184.

71. Chakrabarti, D.; Da Silva, T.; Barger, J.; Paquette, S.; Patel, H.; Patterson, S.; Allen, C. M. Protein farnesyltransferase and protein prenylation in *Plasmodium falciparum. J. Biol. Chem.* **2002**, *277*, 42066–42073.

72. Yokoyama, K.; Trobridge, P.; Buckner, F. S.; Van Voorhis, W. C.; Stuart, K. D.; Gelb, M. H. Protein farnesyltransferase from *Trypanosoma brucei.* A heterodimer of 61- and 65-kDa subunits as a new target for antiparasite therapeutics. *J. Biol. Chem.* **1998**, *273*, 26497–26505.

73. Buckner, F. S.; Yokoyama, K.; Nguyen, L.; Grewal, A.; Erdjument-Bromage, H.; Tempst, P.; Strickland, C. L.; Xiao, L.; Van Voorhis, W. C.; Gelb, M. H. Cloning, heterologous expression, and distinct substrate specificity of protein farnesyltransferase from *Trypanosoma brucei. J. Biol. Chem.* **2000**, *275*, 21870–21876.

74. Buckner, F. S.; Eastman, R. T.; Nepomuceno-Silva, J. L.; Speelmon, E. C.; Myler, P. J.; Van Voorhis, W. C.; Yokoyama, K. Cloning, heterologous expression, and substrate specificities of protein farnesyltransferases from *Trypanosoma cruzi* and *Leishmania major. Mol. Biochem. Parasitol.* **2002**, *122*, 181–188.

75. Ibrahim, M.; Azzouz, N.; Gerold, P.; Schwarz, R. T. Identification and characterisation of *Toxoplasma gondii* protein farnesyltransferase. *Int. J. Parasitol.* **2001**, *31*, 1489–1497.

76. Kumagai, M.; Makioka, A.; Takeuchi, T.; Nozaki, T. Molecular cloning and characterization of a protein farnesyltransferase from the enteric protozoan parasite *Entamoeba histolytica. J. Biol. Chem.* **2004**, *279*, 2316–2323.

77. Yokoyama, K.; Gillespie, J. R.; Van Voorhis, W. C.; Buckner, F. S.; Gelb, M. H. Protein geranylgeranyltransferase-I of *Trypanosoma cruzi. Mol. Biochem. Parasitol.* **2008**, *157*, 32–43.

78. Makioka, A.; Kumagai, M.; Takeuchi, T.; Nozaki, T. Characterization of protein geranylgeranyltransferase I from the enteric protist *Entamoeba histolytica. Mol. Biochem. Parasitol.* **2006**, *145*, 216–225.

79. Pendyala, P. R.; Ayong, L.; Eatrides, J.; Schreiber, M.; Pham, C.; Chakrabarti, R.; Fidock, D. A.; Allen, C. M.; Chakrabarti, D. Characterization of a PRL protein tyrosine phosphatase from *Plasmodium falciparum. Mol. Biochem. Parasitol.* **2008**, *158*, 1–10.

80. Schlitzer, M. Inhibition of farnesyltransferase as a strategy for the development of novel anti-malarials. *Curr. Med. Chem. Antiinfect Agents* **2005**, *4*, 277–286.

81. Carrico, D.; Ohkanda, J.; Kendrick, H.; Yokoyama, K.; Blaskovich, M. A.; Bucher, C. J.; Buckner, F. S.; Van Voorhis, W. C.; Chakrabarti, D.; Croft, S. L.; Gelb, M. H.; Sebti, S. M.; Hamilton, A. D. *In vitro* and *in vivo* antimalarial activity of peptidomimetic protein farnesyltransferase inhibitors with improved membrane permeability. *Bioorg. Med. Chem.* **2004**, *12*, 6517–6526.

82. Glenn, M. P.; Chang, S. Y.; Hucke, O.; Verlinde, C. L.; Rivas, K.; Horney, C.; Yokoyama, K.; Buckner, F. S.; Pendyala, P. R.; Chakrabarti, D.; Gelb, M.; Van Voorhis, W. C.; Sebti, S. M.; Hamilton, A. D. Structurally simple farnesyltransferase inhibitors arrest the growth of malaria parasites. *Angew. Chem. Int. Ed. Engl.* **2005**, *44*, 4903–4906.

83. Wiesner, J.; Mitsch, A.; Jomaa, H.; Schlitzer, M. Structure–activity relationships of novel anti-malarial agents. Part 7: *N*-(3-benzoyl-4-tolylacetylaminophenyl)-3-(5-aryl-2-furyl) acrylic acid amides with polar moieties. *Bioorg. Med. Chem. Lett.* **2003**, *13*, 2159–2161.

84. Wiesner, J.; Kettler, K.; Sakowski, J.; Ortmann, R.; Katzin, A. M.; Kimura, E. A.; Silber, K.; Klebe, G.; Jomaa, H.; Schlitzer, M. Farnesyltransferase inhibitors inhibit the growth of malaria parasites *in vitro* and *in vivo*. *Angew. Chem. Int. Ed. Engl.* **2004**, *43*, 251–254.

84a. Kohring, K.; Wiesner, J.; Altenkämper, M.; Sakowski, J.; Silber, K.; Hillebrecht, A.; Haebel, P.; Dahse, H. M.; Ortmann, R.; Jomaa, H.; Klebe, G.; Schlitzer, M. Development of benzophenone-based farnesyltransferase inhibitors as novel antimalarials. *ChemMedChem* **2008**, *3*(8),1217–1231.

85. Buckner, F. S.; Eastman, R. T.; Yokoyama, K.; Gelb, M. H.; Van Voorhis, W. C. Protein farnesyl transferase inhibitors for the treatment of malaria and African trypanosomiasis. *Curr. Opin. Investig. Drugs* **2005**, *6*, 791–797.

86. Eastman, R. T.; Buckner, F. S.; Yokoyama, K.; Gelb, M. H.; Van Voorhis, W. C. Thematic review series: lipid posttranslational modifications. Fighting parasitic disease by blocking protein farnesylation. *J. Lipid Res.* **2006**, *47*, 233–240.

87. Nallan, L.; Bauer, K. D.; Bendale, P.; Rivas, K.; Yokoyama, K.; Horney, C. P.; Pendyala, P. R.; Floyd, D.; Lombardo, L. J.; Williams, D. K.; Hamilton, A.; Sebti, S.; Windsor, W. T.; Weber, P. C.; Buckner, F. S.; Chakrabarti, D.; Gelb, M. H.; Van Voorhis, W. C. Protein farnesyltransferase inhibitors exhibit potent antimalarial activity. *J. Med. Chem.* **2005**, *48*, 3704–3713.

88. Bendale, P.; Olepu, S.; Suryadevara, P. K.; Bulbule, V.; Rivas, K.; Nallan, L.; Smart, B.; Yokoyama, K.; Ankala, S.; Pendyala, P. R.; Floyd, D.; Lombardo, L. J.; Williams, D. K.; Buckner, F.S.; Chakrabarti, D.; Verlinde, C. L.; Van Voorhis, W. C.; Gelb, M.H. Second generation tetrahydroquinoline-based protein farnesyltransferase inhibitors as antimalarials. *J. Med. Chem.* **2007**, *50*, 4585–4605.

89. Van Voorhis, W. C.; Rivas, K. L.; Bendale, P.; Nallan, L.; Horney, C.; Barrett, L. K.; Bauer, K. D.; Smart, B. P.; Ankala, S.; Hucke, O.; Verlinde, C. L.; Chakrabarti, D.; Strickland, C.; Yokoyama, K.; Buckner, F. S.; Hamilton, A. D.; Williams, D. K.; Lombardo, L. J.; Floyd, D.; Gelb, M. H. Efficacy, pharmacokinetics, and metabolism of tetrahydroquinoline inhibitors of *Plasmodium falciparum* protein farnesyltransferase. *Antimicrob. Agents Chemother.* **2007**, *51*, 3659–3671.

90. Bulbule, V. J.; Rivas, K.; Verlinde, C. L.; Van Voorhis, W. C.; Gelb, M. H. 2-Oxotetrahydroquinoline-based antimalarials with high potency and metabolic stability. *J. Med. Chem.* **2008**, *51*, 384–387.

91. Ohkanda, J.; Buckner, F. S.; Lockman, J. W.; Yokoyama, K.; Carrico, D.; Eastman, R.; de Luca-Fradley, K.; Davies, W.; Croft, S. L.; Van Voorhis, W. C.; Gelb, M. H.; Sebti, S. M.; Hamilton, A. D. Design and synthesis of peptidomimetic protein farnesyltransferase inhibitors as anti-*Trypanosoma brucei* agents. *J. Med. Chem.* **2004**, *47*, 432–445.

92. Hucke, O.; Gelb, M. H.; Verlinde, C. L.; Buckner,F. S. The protein farnesyltransferase inhibitor Tipifarnib as a new lead for the development of drugs against Chagas disease. *J. Med. Chem.* **2005**, *48*, 5415–5418.

93. Esteva, M. I.; Kettler, K.; Maidana, C.; Fichera, L.; Ruiz, A.M.; Bontempi, E. J.; Andersson, B.; Dahse, H. M.; Haebel, P.; Ortmann, R.; Klebe, G.; Schlitzer, M. Benzophenone-based farnesyltransferase inhibitors with high activity against *Trypanosoma cruzi*. *J. Med. Chem.* **2005**, *48*, 7186–7191.

94. Yokoyama, K.; Trobridge, P.; Buckner, F. S.; Scholten, J.; Stuart, K. D.; Van Voorhis, W. C.; Gelb, M. H. The effects of protein farnesyltransferase inhibitors on trypanosomatids: inhibition of protein farnesylation and cell growth. *Mol. Biochem. Parasitol.* **1998**, *94*, 87–97.

95. Ye, J.; Wang, C.; Sumpter, R., Jr.; Brown, M. S.; Goldstein, J. L.; Gale, M., Jr. Disruption of hepatitis C virus RNA replication through inhibition of host protein geranylgeranylation. *Proc. Natl. Acad. Sci. USA.* **2003**, *100*, 15865–15870.

96. Glenn, J. S.; Marsters, J. C., Jr.; Greenberg, H. B. Use of a prenylation inhibitor as a novel antiviral agent. *J. Virol.* **1998**, *72*, 9303–9306.

97. Bordier, B. B.; Marion, P. L.; Ohashi, K.; Kay, M. A.; Greenberg, H. B.; Casey, J. L.; Glenn, J. S. A prenylation inhibitor prevents production of infectious hepatitis delta virus particles. *J. Virol.* **2002**, *76*, 10465–10472.

98. Bordier, B. B.; Ohkanda, J.; Liu, P.; Lee, S. Y.; Salazar, F. H.; Marion, P. L.; Ohashi, K.; Meuse, L.; Kay, M. A.; Casey, J. L.; Sebti, S. M.; Hamilton, A. D.; Glenn, J. S. *In vivo* antiviral efficacy of prenylation inhibitors against hepatitis delta virus. *J. Clin. Invest.* **2003**, *112*, 407–414.

99. Song, J. L.; White, T. C. RAM2: an essential gene in the prenylation pathway of *Candida albicans*. *Microbiology* **2003**, *149*, 249–259.

100. McGeady, P.; Logan, D. A.; Wansley, D. L. A protein-farnesyl transferase inhibitor interferes with the serum-induced conversion of *Candida albicans* from a cellular yeast form to a filamentous form. *FEMS Microbiol. Lett.* **2002**, *213*, 41–44.

101. Vallim, M. A.; Fernandes, L.; Alspaugh, J. A. The RAM1 gene encoding a protein-farnesyltransferase beta-subunit homologue is essential in *Cryptococcus neoformans*. *Microbiology* **2004**, *150*, 1925–1935.

102. Fernandes, L.; Paes, H. C.; Tavares, A. H.; Silva, S. S.; Dantas, A.; Soares, C. M.; Torres, F. A.; Felipe, M. S. Transcriptional profile of ras1 and ras2 and the potential role of farnesylation in the dimorphism of the human pathogen *Paracoccidioides brasiliensis*. *FEMS Yeast Res.* **2008**, *8*, 300–310.

103. Kelly, R.; Card, D.; Register, E.; Mazur, P.; Kelly, T.; Tanaka, K. I.; Onishi, J.; Williamson, J. M.; Fan, H.; Satoh, T.; Kurtz, M. Geranylgeranyltransferase I of *Candida albicans*: null mutants or enzyme inhibitors produce unexpected phenotypes. *J. Bacteriol.* **2000**, *182*, 704–713.

104. Murthi, K. K.; Smith, S. E.; Kluge, A. F.; Bergnes, G.; Bureau, P.; Berlin, V. Antifungal activity of a *Candida albicans* GGTase I inhibitor-alanine conjugate. Inhibition of Rho1p prenylation in *C. albicans*. *Bioorg. Med. Chem. Lett.* **2003**, *13*, 1935–1937.

CHAPTER 35

Histone Deacetylase Inhibitors

PAUL W. FINN
InhibOx Limited, Pembroke House, 36-37 Pembroke Street, Oxford OX1 1BP, UK

35.1 INTRODUCTION

35.1.1 Histones and Chromatin Structure

DNA in eukaryotic cells occurs as chromatin, wherein the DNA is complexed with histone proteins. There are five main classes of histones, H1, H2A, H2B, H3, and H4. The amino acid sequences of histones H2A, H2B, H3, and H4 show remarkable conservation between species, whereas H1 varies somewhat, and in some cases is replaced by another histone, for example, H5. Four pairs of each of H2A, H2B, H3, and H4 together form a disk-shaped octomeric protein core, around which DNA (about 150 bp) is wound to form a nucleosome. Individual nucleosomes are connected by short stretches of linker DNA associated with another histone molecule (H1, or in certain cases, H5) to form a structure resembling a beaded string, which is itself arranged in a helical stack, known as a solenoid. The four nucleosomal histones consist of two domains: a C-terminal domain located in the nucleosome core and an N-terminal "tail" extending out of the nucleosome. These tails are rich in the basic amino acids lysine and arginine and contact the phosphate groups of DNA.

The majority of histones are synthesized during the S phase of the cell cycle, and newly synthesized histones quickly enter the nucleus to become associated with DNA. Likewise, within minutes of its synthesis, new DNA becomes associated with histones in nucleosomal structures.

Histones are subject to a variety of posttranslational modifications, including acetylation, methylation, phosphorylation, and ubiquitination, particularly occurring in the histone tails but also at certain positions in the globular domains. The functional consequences of these complex and coordinated set of posttranslational modifications are still not fully understood but they are under intense study, a part of the burgeoning field of epigenetics. However, it is already clear that there is a complex network of site-specific and interdependent histone modifications, associated with binding of transcription factors and the regulation of gene expression, which has been referred to as the "histone code," and that chromatin is a highly dynamic structure.

The link between the acetylation status of histones and the transcription of genes first appeared in the literature in 1964.[1] Since then, a growing body of evidence supports the idea that acetylation of histones, in particular on the tails of H3 and H4, is linked with an "open" state of chromatin that is accessible to the cellular transcriptional machinery and in which gene transcription can occur. Conversely, deacetylated histones are linked with a "closed" and transcriptionally silent DNA. This is consistent with a simple physical model of chromatin in which the electrostatic stabilization arising from interaction of positively charged lysine residues in histone with negatively charged phosphates on the DNA backbone is weakened, because acetylation removes the positive charge from the ω-amino group of lysine. The enzymes that catalyze and reversibly regulate the acetylation status of histones are the histone acetyltransferases[2] that add acetyl groups and the histone deacetylases (HDACs), which remove them. HDACs function as part of large multiprotein complexes, which are tethered to the promoter and repress transcription. Well-characterized transcriptional repressors such as Mad,[3] pRb,[4] nuclear receptors,[5] and YY1[6] associate with HDAC complexes to exert their repressor function.

35.1.2 Histone Deacetylases

HDACs are members of an ancient enzyme family found in animals, plants, fungi, and bacteria. Eleven zinc-dependent HDACs have been identified in humans. These have been divided into three classes, based on a phylogenetic analysis of all HDAC-related proteins in all fully sequenced free-living organisms.[7] HDACs 1–3 and 8 form the class I enzymes and are related to the yeast HDAC Rpd3, HDACs 4–7 and 9–10 form class II (related to Hda1), and HDAC11 is the only member of class IV. HDAC6 and HDAC10 represent a subclass of class II enzymes in having two catalytic domains. Class III (the sirtuins) consists of seven members in humans. These enzymes are structurally distinct from classes I, II, and IV, being NAD-dependent and related to the yeast protein Sir2. They are less studied than the other enzyme classes, but are of growing interest because their overexpression leads to increased life span in yeast and because of their potential involvement in metabolic diseases. As the sirtuins are not zinc dependent, they and their inhibitors will not be discussed further here.

All three non-sirtuin HDAC classes (including class IV) exist in eubacteria. Phylogenetic analysis of bacterial HDAC relatives suggests that all three HDAC classes precede the evolution of histone proteins and raises the possibility that the primary activity of some "histone" deacetylase enzymes is directed against nonhistone substrates. Indeed, a growing number of nonhistone proteins are known to be acetylated, and although our knowledge of the "acetylome"[8] is far from complete, it is becoming clear that there are important nonhistone substrates for HDACs. For example, the transcription factor and tumor suppressor p53 and NF-κB, which is involved in antiapoptotic responses, are regulated by acetylation.[9,10] In addition, it is not only nuclear proteins that are regulated by acetylation, but both the microtubule component tubulin[11] and the chaperone protein HSP90[12] are also deacetylated by cytoplasmic HDAC6. Other acetylated proteins include the protooncogene c-Myb, estrogen receptor alpha, and Ku70, a suppressor of apoptosis. These are only representative examples of a long and growing list of proteins and cellular functions in which acetylation is involved. Further support for this idea comes from the cellular and tissue distribution of HDAC enzymes. The class I enzymes are nuclear and are universally expressed. On the other hand, the class II enzymes shuttle between nucleus and cytoplasm and have a more restricted tissue distribution; HDACs 4 and 5, for example, are highly expressed in cardiomyocytes and involved in cardiac hypertrophy.[13] However, we are far from a complete understanding of the regulatory role of acetylation and of the individual HDAC isoforms. For example, mice lacking histone deacetylase 6 have hyperacetylated tubulin and HSP90 but are viable and develop normally.[14] Another indication of the complexity of the field comes from a comparison of the expression profiles of wild-type and HDAC1-deficient embryonic stem cells that showed deregulation of only 7% of genes, whereas a view based on the simple electrostatic mechanism described above would have expected much more widespread effects. Although perhaps smaller in number than might naively be expected, the list of genes whose expression is affected by HDAC inhibitors includes raf-1, cyclin A, VEGF, Bcl-2, and many others involved in cell cycle progression and proliferation and DNA synthesis. Therefore, HDAC inhibitors have received great attention as potential treatments for cancer and other diseases.

35.2 HDAC INHIBITORS

35.2.1 Natural Products

Several natural product inhibitors of HDACs have been identified and are shown in Fig. 35.1. Trichostatin A (TSA) was initially identified as a fungistatic antibiotic from *Streptomyces*.[15] TSA causes cell cycle arrest at both G1 and G2 phases, reverts the transformed phenotype of different cell lines, and induces differentiation of Friend leukemia cells.[16,17] TSA is a potent (nanomolar) and selective inhibitor of HDACs, and the observed cell cycle arrest induced by TSA correlates with an increased expression of gelsolin,[18] an actin regulatory protein that is downregulated in malignant breast cancer.[19] The fungal metabolite, depudecin, induces hyperacetylation of histones in a dose-dependent manner, although it is a relatively weak (micromolar) inhibitor of HDAC.[20] It can revert the transformed phenotype of NIH3T3 cells caused by *v-ras*,[21] although its structure is not the one to excite the interest of medicinal chemists. Very different structurally are the trapoxin cyclic tetrapeptides, of which there are several members and which are potent irreversible inhibitors of HDAC, presumably via their epoxide functionality.[22] Intriguingly, another fungal cyclic tetrapeptide, apicidin, lacks the epoxide functionality of the trapoxins, instead possessing an ethyl ketone.[23] Another naturally occurring polypeptide, FK228, contains an internal disulfide that acts as a prodrug of the

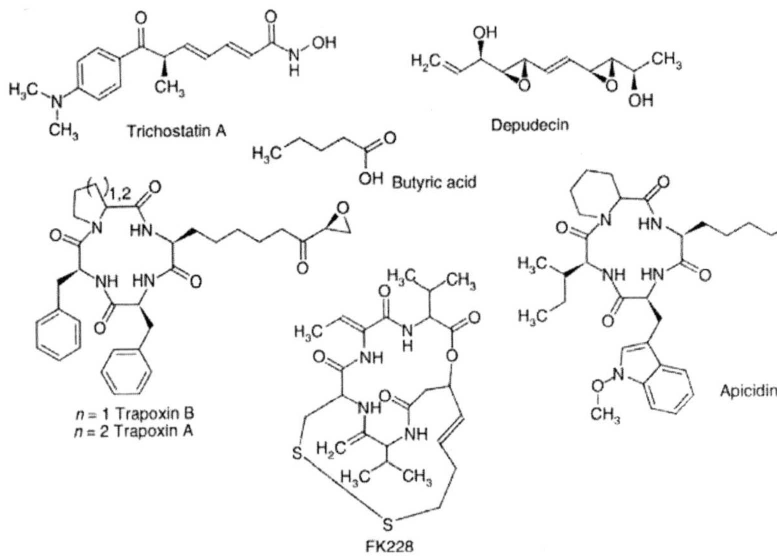

FIGURE 35.1 Structures of natural product HDAC inhibitors

active form in which the thiol can be generated under the reducing conditions inside the cell and covalently interact with the HDAC enzyme.[24]

Also to be mentioned in this section are butyrate and related short-chain fatty acids such as penylbutyrate and the structurally related anticonvulsant valproic acid (VPA). These have been demonstrated to inhibit HDACs, albeit at very high (millimolar) concentrations.[25] At the high concentrations required to achieve significant HDAC inhibition, the short-chain fatty acids are likely to have pleiotropic effects, but they have been observed to affect histone acetylation levels in a dose-dependent manner, as would be expected of an HDAC inhibitor.

35.3 HDAC STRUCTURAL INFORMATION

Structural data on HDACs and related proteins provides a rationalization of the inhibitory mechanism of many of these structurally diverse natural products. In 1999, the structures of a HDAC enzyme homologue HDLP (histone deacetylase-like protein) from the hyperthermophilic bacterium *Aquifex aeolicus* and its complex with TSA and the synthetic inhibitor SAHA (Fig. 35.2) were published.[26] This structure revealed an active site consisting of a predominantly hydrophobic cylindrical tunnel, at the bottom of which is located the catalytic zinc and two Asp–His charge-relay systems, establishing the mechanism of HDAC inhibition. The tubular pocket is suitable for accommodating the acetyl-lysine side chain of substrates with the catalytic zinc atom being suitably positioned at the bottom of the cavity such that a coordinated water/hydroxide ion can attack the amide carbonyl. In the *Aquifex* enzyme complex with TSA, the zinc is pentacoordinated, from interactions with two aspartate residues and one histidine and two interactions from the oxygen atoms of the hydroxamic acid.

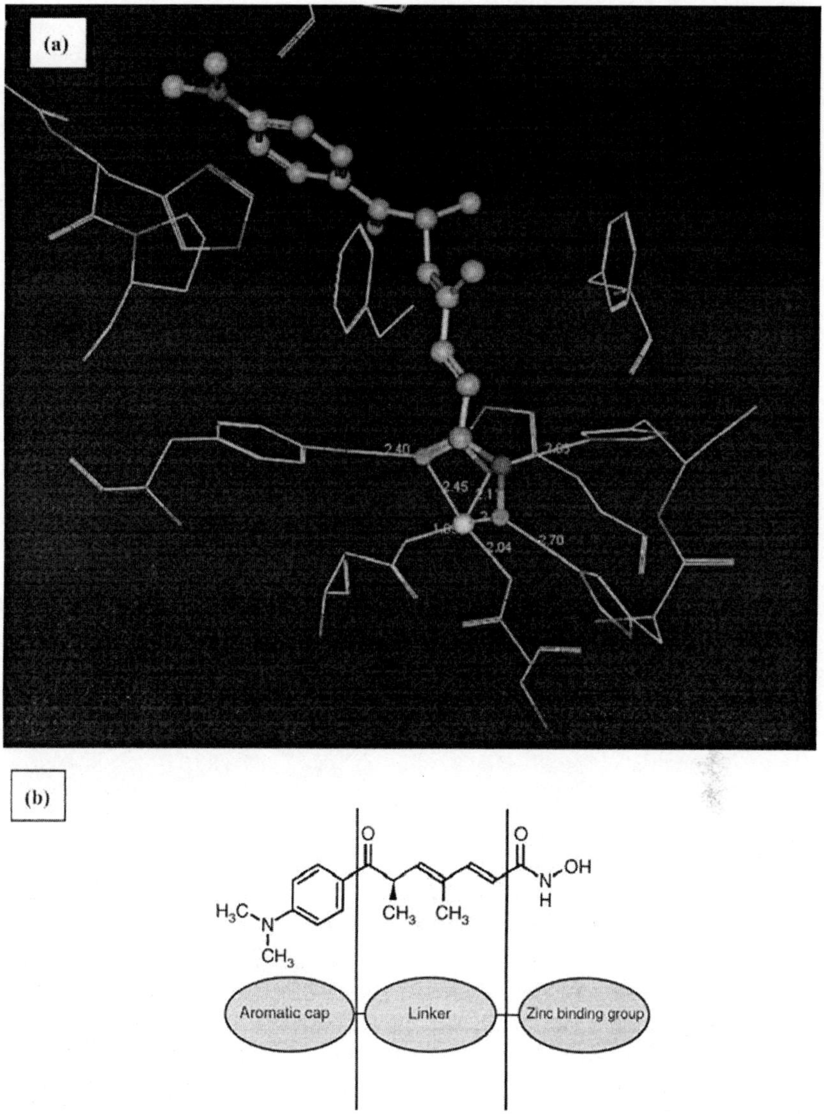

FIGURE 35.2 (a) TSA in the binding site of HDLP. The pentacoordination of the zinc and hydrogen bonding interactions to the hydroxamic acid moiety are shown with green lines; the interactions between the central portion and phenylalanine residues can also be seen. (b) A simple three-component pharmacophore, illustrated for TSA, consisting of a zinc binding group, a linker, and a capping group. (See the color version of this figure in Color Plates section.)

These two oxygen atoms also make hydrogen bonds, in one case to the side chain of His142 and in the other with the hydroxyl of Tyr306. The nitrogen of the hydroxamic acid donates a hydrogen bond to His143. Taken together, these interactions rationalize the strength of the hydroxamic acid group as a zinc ligand, a group that has been much exploited in synthetic inhibitor design, as will be discussed below, and also explain the much weaker interactions of carboxylic acids. The alkyl portion of TSA makes hydrophobic contacts with the tunnel, in particular with the aromatic side chains of two phenylalanine residues. The tunnel is quite narrow, precluding the presence of bulky groups and concordant with the straight chain alkyl moieties found in the natural products. At the surface of the enzyme, the tunnel opens out where the aromatic portion of TSA can make additional interactions with surface features. Thus, a model for HDAC inhibitors consisting of three elements is established in which a zinc binding group is connected to a hydrophobic linker (occupying the cylindrically shaped pocket) that in turn is connected to a "capping" group that can interact with the residues at the surface of the protein.

Comparison of the HDAC sequences reveals a high level of sequence conservation across the catalytic domains of the class I and II HDACs, and the catalytic residues and those that line the hydrophobic tunnel are particularly highly conserved. Thus, there was an expectation that the three-dimensional structures of the enzymes would be similarly conserved. The determination of the structure of the class I enzyme human HDAC8 confirmed the overall architecture,[27] and the HDAC8 structure in complex with several hydroxamic acid inhibitors is now in the public domain. As expected, the most significant differences were observed in the length and structure of the loops surrounding the active site, including the presence of two potassium ions in the HDAC8 structure. One of these interacts with key catalytic residues and CD data suggest a direct role of potassium in the fold stabilization of the enzyme. The high-resolution crystal structure of a catalytically inactive HDAC8 active-site mutant, Tyr306Phe, bound to an acetylated peptidic substrate helps clarify the role of active-site residues in the deacetylation reaction and substrate recognition. The structure of a catalytically inactive mutant of HDAC8 with a substrate showed an unexpected role for a conserved residue at the active-site rim, Asp101, in positioning the substrate by directly interacting with the peptidic backbone and imposing a constrained *cis*-conformation. A similar interaction was also seen in a hydroxamate inhibitor–HDAC8 structure. The crucial role of Asp101 in substrate and inhibitor recognition was confirmed by activity and binding assays of wild-type HDAC8 and Asp101Ala, Tyr306Phe, and Asp101Ala/Tyr306Phe mutants.[28] Very recently, the structure of HDAC7 has been deposited in the protein databank (PDB codes 3C0Z and 3C10) and analysis of these structures may shed light on differences between class I and class II enzymes that may be exploited in inhibitor design.

35.4 SYNTHETIC HYDROXAMIC ACIDS

These represent the largest class of HDAC inhibitors with many new examples being described each year. The best known compound is suberoylanilide hydroxamic acid (SAHA), originally identified from a screen for inducers of differentiation of murine erythroleukemia cells. It was only subsequently that SAHA was identified as a potent inhibitor of HDAC.[29] The crystal structure of SAHA in HDLP shows, as expected, a very similar mode of interaction to TSA. A very large number of SAHA

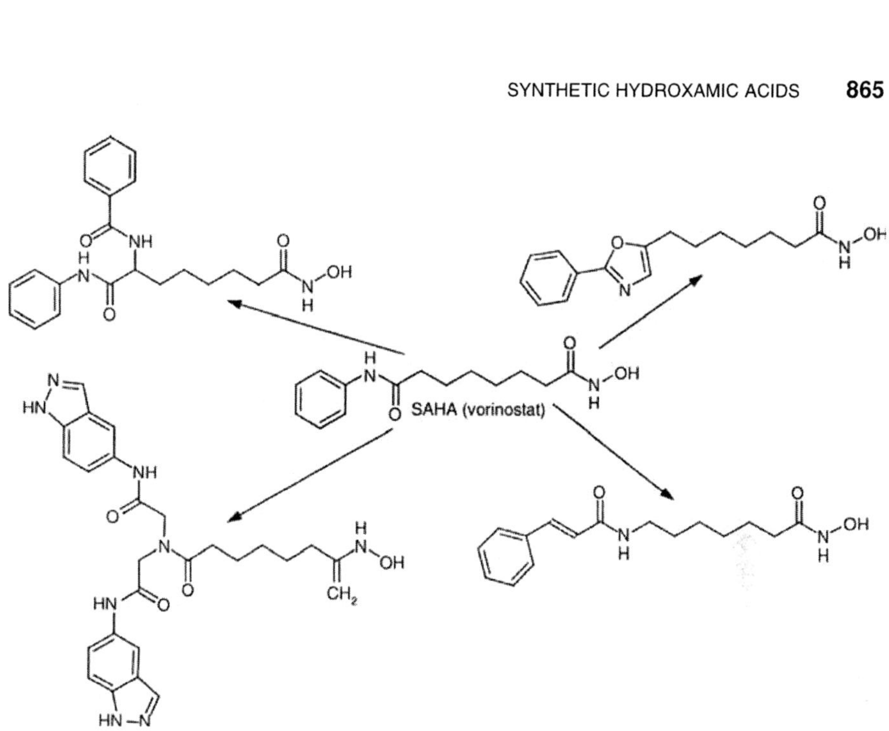

FIGURE 35.3 Structures of SAHA and related compounds.

analogues have been reported by the Aton/Merck groups and many others. The introduction of branching alpha to the SAHA amide can generate compounds more potent than SAHA in both enzyme and cellular assays.[30] Further elaboration of this branching strategy, still within the framework of SAHA analogues, has been reported in the patent literature.[31] The replacement of the amide unit of SAHA with other moieties can also yield potent compounds: examples reported in the literature include urea, sulfur, and various heterocycles. Some representative compounds are shown in Fig. 35.3.

The saturated alkyl linker of SAHA and analogues is highly flexible. One would expect that the restriction of flexibility occurring on binding would require the payment of an energy penalty because of loss of entropy and therefore more rigid linkers may have an advantage in this regard. Several groups have reported HDAC inhibitors that possess the more conformationally constrained cinnamoyl linker. TopoTarget (Prolifix) have reported a series using sulfonamide and cinnamic acid as the linking group to produce a series of highly potent HDAC inhibitors[32] exemplified by belinostat. The best compounds have a meta-linkage at the aromatic ring and with the sulfonamide attached with the sulfur connected to the aromatic ring. Novartis have also developed cinnamoyl-containing inhibitors such as LBH-589, in this case with a *para*-linkage to an amine containing chain, some with branching at the amine. As was the case with the alkyl linker compounds, many groups have reported variations on this theme, including the replacement of the cinnamoyl aromatic ring with heterocycles or bicyclic systems and a wide variety of capping groups (Fig. 35.4).

Direct connection of an aromatic ring to hydroxamic acid has also been shown to deliver active compounds. The *para*-substituted phenyl hydroxamic acids can provide

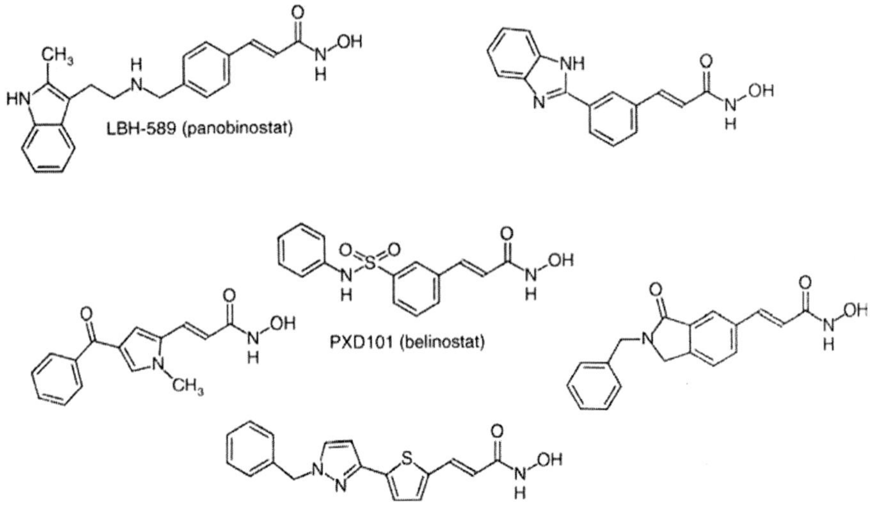

FIGURE 35.4 Structures of PXD101 and related cinnamic acid-containing inhibitors.

nanomolar HDAC inhibition and there are several reported series where a heteroaromatic group has been used, in particular thiophene, but also benzothiophene, benzofuran, thiazole, benzimidazole, and pyrimidine (Fig. 35.5).

As will be discussed further below, several of the hydroxamic acid-containing molecules described above have entered the clinic. However, the hydroxamic acid group is regarded with suspicion by some medicinal chemists because of concerns over its frequently observed short half-life and concerns over selectivity and toxicity. Therefore, there has been a considerable body of research directed toward identifying

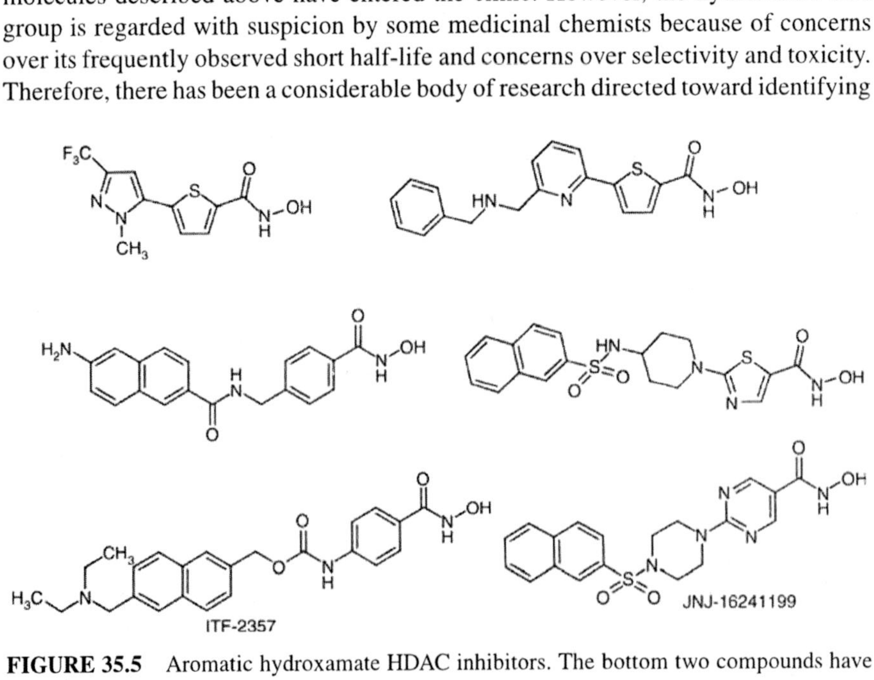

FIGURE 35.5 Aromatic hydroxamate HDAC inhibitors. The bottom two compounds have entered clinical trials.

HDAC inhibitors containing alternatives to hydroxamic acid as the zinc binding functionality.

The 2-aminophenylamide now known as MS-275 was reported in 1999.[33] This and related compounds are significantly weaker inhibitors of HDACs than the hydroxamic acids, with IC_{50}'s typically in the micromolar rather than nanomolar range. The 2-aminophenyl amide is discussed in the literature as the zinc binding group, although direct experimental evidence for this is lacking. Nevertheless, despite their poorer *in vitro* profiles, compounds containing this group have shown activity in animal models and acceptable pharmacokinetic profiles, and three molecules from this class are in clinical trials. Recently, novel 2-aminophenyl-amides were obtained using an iterative design procedure that led to the identification of compounds that are significantly more potent than those from this class described previously, with the best possessing an IC_{50} of 41 nM together with good cellular and pharmacokinetic profiles.[34]

Unlike the 2-aminophenylamides, thiols are very well precedented zinc binding groups and a number of thiols have been reported as inhibitors of HDAC. The natural product FK228, which possesses a masked thiol, has already been mentioned above. Among the first synthetic thiols to be reported were thiol-based SAHA analogues. These are potent inhibitors,[35] with decreased activity of a thioacetate analogue and complete loss of activity of the corresponding methyl sulfide compound, indicating that a thiol is necessary for zinc binding. Compounds of this type have poor cellular activity, but a combination of a thioester "prodrug" group together with optimization of the aromatic capping group has provided a compound with a similar enzyme and antiproliferative activity to SAHA.[36] The introduction of an amide group into the alkyl linker has also produced active compounds where the amide is spaced one methylene group from the thiol. The best such compounds show activity equivalent to or better than SAHA.[37] It is possible that the carbonyl of the amide is able to interact with zinc in addition to the thiol, thus giving these compounds bidentate rather than monodentate zinc binding groups.

Electrophilic ketones have also been investigated as HDAC inhibitors. Representative examples are the α-keto esters, α-keto amides,[38] and α-keto oxazoles.[39] These compounds displayed excellent enzyme inhibition, but were poorly active in cellular and/or *in vivo* models because of poor stability and rapid conversion to inactive alcohols. In a very interesting recent development, potent alkyl ketone inhibitors of HDAC have been described. The unusual zinc binding group of the natural product apicidin has already been remarked upon above. Taking apicidin as a starting point, workers at Merck identified via directed screening of compounds containing the unusual L-2-amino-8-oxodecanoic residue a lower molecular weight nonmacrocyclic compound that was subsequently optimized. The optimized compounds show levels of cellular activity equivalent to current clinical candidates and cause tumor growth inhibition in a xenograft model.[40] The presence of bulky aromatic capping groups in these compounds may indicate that the intrinsic contribution to binding from the alkyl ketone is lower than that from hydroxamic acid (as might be expected on physical grounds) and therefore more binding energy needs to be obtained from these interactions. The same group has also recently

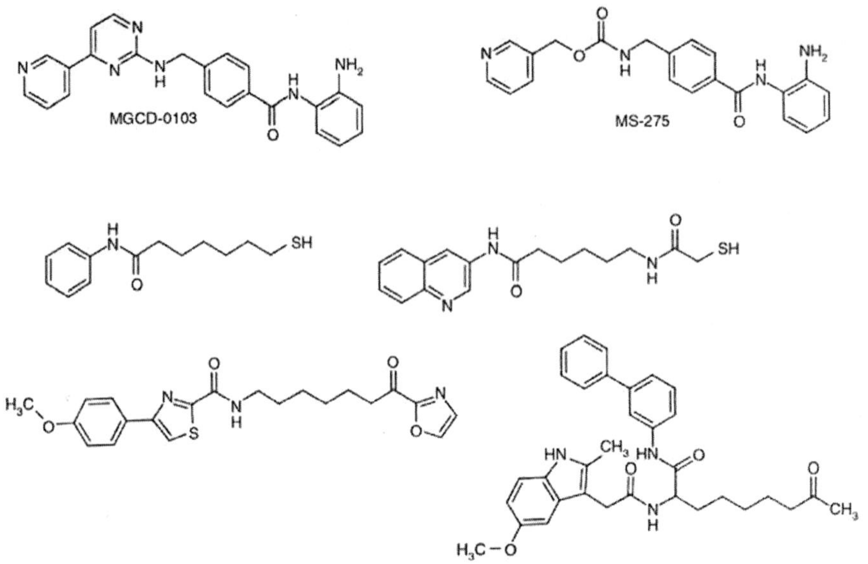

FIGURE 35.6 2-Aminophenylamide, thiol, and ketone HDAC inhibitors. The two 2-aminophenylamide inhibitors have entered clinical trials.

reported a novel series of trifluoroacetylthiophenes together with crystal structure information indicating that the compounds bind in their hydrated form (Fig. 35.6).[41]

35.5 THERAPEUTIC APPLICATIONS OF HDAC INHIBITORS

35.5.1 Cancer

The therapeutic area that has received by far the most attention for HDAC inhibitors is cancer. Overexpression of HDACs can mediate tumor cell proliferation,[42] and deregulation of HDAC recruitment to promoters appears to be one of the mechanisms by which these enzymes contribute to tumorigenesis. *In vitro*, HDAC inhibitors inhibit the growth of fibroblast cells by causing cell cycle arrest in the G1 and G2 phases and can lead to the terminal differentiation and loss of transforming potential of a variety of transformed cell lines.[43,44] TSA and SAHA have been reported to inhibit cell growth, induce terminal differentiation, and prevent the formation of tumors in mice, and similar effects on cell cycle and differentiation have been observed with a number of deacetylase inhibitors.

The clear involvement of HDACs in the control of cell proliferation and differentiation suggested that aberrant HDAC activity may play a role in cancer. The most direct demonstration that deacetylases contribute to cancer development comes from the analysis of different acute promyelocytic leukemias (APL). In most APL patients, a translocation of chromosomes 15 and 17 (t(15;17)) results in the expression of a fusion protein containing the N-terminal portion of PML gene product linked to most of RARα (retinoic acid receptor). In some cases, a different translocation (t(11;17)) causes the fusion between the zinc finger protein PLZF and RARα. In the absence of ligand, the wild-type RARα represses target genes by tethering HDAC repressor complexes to the promoter DNA. During normal hematopoiesis, retinoic acid (RA) binds RARα and displaces the repressor complex, allowing expression of genes implicated in myeloid differentiation. The RARα fusion proteins occurring in APL patients are no longer responsive to physiological levels of RA and they interfere with the expression of the RA inducible genes that promote myeloid differentiation. This results in a clonal expansion of promyelocytic cells and development of leukemia. *In vitro* experiments have shown that TSA is capable of restoring RA responsiveness to the fusion RARα proteins and of allowing myeloid differentiation. These results establish a link between HDACs and oncogenesis and suggest that HDACs are potential targets for pharmaceutical intervention in APL patients.[45–47]

Furthermore, different lines of evidence suggest that HDACs may be important therapeutic targets in other types of cancer. Cell lines derived from many different cancers (prostate, colorectal, breast, neuronal, hepatic) are induced to differentiate by HDAC inhibitors.[48] A number of HDAC inhibitors have been studied in animal models of cancer. They reduce tumor growth and prolong the life span of mice bearing different types of transplanted tumors, including melanoma, leukemia, colon, lung and gastric carcinomas, and so on.[49] SAHA has also been reported to be effective in preventing the formation of mammary tumors in rats and lung tumors in mice.[50]

One concern that was expressed early in the development of HDAC inhibitors was potential toxicity. A mechanism of action that putatively acts relatively nondiscriminately at the level of gene expression might be expected to be toxic. However, gene profiling studies show that a relatively small percentage of genes are up- or down-regulated in response to HDAC inhibitor treatment and the empirical evidence from preclinical (and clinical) studies is that the toxicity profile of HDAC inhibitors is rather benign.

Based on this encouraging preclinical profile, a number of HDAC inhibitors have been taken into clinical trials. One of the first and most clinically advanced compounds is SAHA (also known as vorinostat). SAHA has demonstrated activity in a number of clinical studies including T-cell and B-cell lymphomas, AML, and mesothelioma. In a phase IIb study on T-cell lymphoma, approximately 30% of patients responded to an oral dose of 400 mg per day.[51] Vorinostat has recently been approved by the FDA for the treatment of the cutaneous manifestations of advanced, refractory cutaneous T-cell lymphoma under the name Zolinza. Another hydroxamic acid HDAC inhibitor in advanced clinical trail is belinostat. The phase I data have been recently published, showing the compound to be well tolerated, with dose-dependent pharmacodynamic effects and promising antitumor activity.[52] Further studies have indicated activity in peripheral T-cell lymphoma where encouraging data have been reported from a phase II trial and a fast track designation has been granted by the FDA for the development program for belinostat for relapsed or refractory PTCL after at least one prior systemic therapy. Nonhydroxamate HDAC inhibitors that have entered the clinic include MS-275, where elevated levels of H3 acetylation have been observed. FK228 has also shown clinical activity in lymphoma[53] and AML. However, unlike other HDAC inhibitors, cardiotoxicity, including tachycardia and prolonged QTc interval, has been observed[54] with this HDAC inhibitor. Despite their weak inhibition of HDAC in enzyme assays, short-chain fatty acids have also been taken into the clinic. VPA has been tested in acute myelogenous leukemia and myodysplastic syndromes. High serum concentrations were achieved, 44% of patients had tumor responses, and side effects were mild. Thus, HDAC inhibitors appear to show particularly good activity as single agents in the treatment of hematological disease. At present, it is unclear whether this stems from a specifically vulnerable pathway or mechanism of action in these tumor types or just that this is an area that has been more investigated. Many studies in both hematological and solid tumors are ongoing that will shed light on this question.

As noted above, single agent activity has been observed with several HDAC inhibitors. However, given their mechanism of action, the greatest therapeutic potential for HDAC inhibitors may lie in their use in combination with other agents. A large number of preclinical combination studies have been published where HDAC inhibitors have been combined with chemotherapeutic agents, nuclear receptor ligands, or signal transduction inhibitors. For example, FK228 and TSA acted synergistically with the DNA methyltransferase inhibitor 2-deoxy-5′-azacytidine in causing apoptotic cell death in human lung cancer cells.[55] Resistance to cisplatin *in vitro* and *in vivo* can arise from the loss of the DNA mismatch repair pathway due to methylation of the hMLH1 (mutL homologue 1) gene promoter. The cisplatin-resistant cell line A2780/cp70 is eightfold more resistant to cisplatin than the nonresistant cell line and has the hMLH1 gene methylated. Treatment with an inhibitor of DNA methyltransferase, 2-deoxy-5′-azacytidine, results in a partial reversal of DNA methylation, reexpression of hMLH1, and sensitization to cisplatin both *in vitro* and *in vivo*. Treatment of A2780/cp70 tumor-bearing mice with 2-deoxy-5′-azacytidine followed by belinostat results in a marked increase in the number of cells that reexpress MLH1. Since the clinical use of DNA methyltransferase inhibitors may be limited by toxicity and eventual remethylation of genes, the combination of 2-deoxy-5′-azacytidine and belinostat could have a role in increasing the efficacy of chemotherapy in patients with tumors that lack MLH1 expression due to hMLH1 gene promoter methylation.[56]

In another example, synergistic inhibition of proliferation and clonogenicity was obtained when HCT116 cells were incubated with belinostat and 5-fluorouracil. 5-fluorouracil combined with belinostat also increased DNA fragmentation and PARP cleavage in HCT116 cells. Incubation with belinostat downregulated thymidylate synthase expression in HCT116 cells, which may provide a mechanistic explanation for the observed synergy. *In vivo* studies, using mouse HT29 and HCT116 xenograft models, showed improved reductions in tumor volume compared to single compound, when belinostat and 5-fluorouracil were combined.[57] In keeping with this preclinical evidence, many combination trials are currently ongoing in a wide range of tumor types. Encouragingly, the preclinical findings of synergistic or additive effects of combinations of HDAC inhibitors with other anticancer agents appear to be translating into the clinical setting. For example, clinical activity has also been reported for belinostat in a phase II study in ovarian cancer where the treatment regime is in combination with two cytotoxic drugs, carboplatin and paclitaxel. Substantial antitumor activity was seen with an overall response rate of 43% in patients with platinum-sensitive and platinum-resistant tumors, including patients with a platinum-free interval of less than 3 months. Encouragingly, this treatment regimen is well tolerated with a safety profile consistent with that observed with chemotherapy alone. Given the large number of ongoing clinical studies, more examples are likely to emerge in the next few years.

35.5.2 Other Therapeutic Areas

Cancer is by far the most studied therapeutic area where HDAC inhibitors may be beneficial, but there is a large and growing literature pointing to their potential in others.

In recent years, HDAC inhibitors have emerged as potential anti-inflammatory drugs because of the involvement in inflammation of a number of proinflammatory genes and transcription factors such as NF-κB; many of the underlying processes are common between cancer and inflammation. SAHA has been shown to reduce the levels of TNF-α and other cytokines in a LPS-stimulated inflammatory animal model.[58] TSA has been shown to be effective in an animal model of rheumatoid arthritis.[59] Psoriasis is a common chronic disfiguring skin disease that is characterized by well-demarcated, red, hardened scaly plaques: these may be limited or widespread. The prevalence rate of psoriasis is approximately 2%, and while the disease is rarely fatal, it clearly has serious detrimental effects upon the quality of life of the patient, further compounded by the lack of effective therapies. There is therefore a large unmet clinical need for effective and safe drugs for this condition. Psoriasis is a disease of complex etiology. While there is clearly a genetic component, with a number of gene loci being involved, there are also undefined environmental triggers. Whatever the ultimate cause of psoriasis, at the cellular level, it is characterized by local T-cell mediated inflammation, by keratinocyte hyperproliferation, and by localized angiogenesis. These are all processes in which histone deacetylases have been implicated,[60–62] suggesting that psoriasis may be another disease with a strong inflammatory component that may be amenable to therapy with HDAC inhibitors. Thus, the data from animal models of a variety of inflammatory diseases suggest that HDAC inhibitors may have utility in inflammatory diseases. However, concerns have been expressed regarding the long-term safety of these agents compared to currently used anti-inflammatory therapies such as glucocorticoids.[63] Perhaps in part because of these concerns, despite the promising preclinical data, no clinical investigations of HDAC inhibitors in inflammatory disease have yet been initiated.

Neurodegenerative diseases such as Huntington's disease and spinal muscular atrophies are devastating diseases with few therapeutic options. HDAC inhibitors offer a new therapeutic modality because of their ability to activate disease-modifying genes and correct histone acetylation homeostasis. Histones associated with down-regulated genes have been shown to be hypo-acetylated in Huntington's disease models.[64] Activity of SAHA in a mouse model of Huntington's disease has been observed,[65] and a phase II study to evaluate the safety, tolerability, and clinical impact of 15 g daily of the short-chain fatty acid sodium phenylbutyrate in Huntington's disease is ongoing.

Apicidin was originally identified as an inhibitor of the histone deacetylase of apicomplexan parasites and to be orally and parenterally active *in vivo* against *Plasmodium berghei* malaria in mice.[23] Activity in hydroxamic acid HDAC inhibitors has also been identified.[66] A key concern regarding the development of inhibitors for use in parasitic diseases is the possible side effects that could arise from the inhibition of human HDACs. In this regard, the recent development of a model of *Plasmodium falciparum* HDAC-1 may be of help in designing selective inhibitors.[67]

35.5.3 Subtype-Selective HDAC Inhibitors

With 11 human zinc-containing HDACs, an obvious question concerns the isoform selectivity profile of inhibitors and whether a particular selectivity profile offers advantages for particular therapeutic areas or from the point of view of toxicity profile. Work in this area is still at an early stage. In part, this has been because of the difficulty of expressing some of the HDAC family members in pure and active form, as they are frequently involved in multiprotein complexes. However, the expression of several isoforms in a baculovirus expression system has been achieved. In a recent comparative study, the activity of a diverse set of 10 HDAC inhibitors was determined against four class I (HDACs 1, 2, 3, and 8) and four class II enzymes (HDACs 4, 6, 7, and 9). Concordant with most of the literature data, the small-molecule hydroxamic acid inhibitors, TSA, and those in clinical studies, SAHA, belinostat, and panobinostat, are broad-spectrum inhibitors across the classes I and II. All the compounds were poorer inhibitors of HDAC8 than the other class I enzymes. There is an indication that the metal ion at the active site of HDAC8 may be Fe(II) rather than Zn(II),[68] which may contribute to the observed spectrum of activity. The two molecules of the benzamide inhibitor class studied, MS-275 and a closely related molecule MGCD0103, had selectivity for class I over class II, with MS-275 showing particular selectivity for HDAC1 and MGCD0103 for HDAC1/2. Apicidin was also class I selective, although it appears to be inactive against HDAC1. As expected, compounds that inhibited class I enzymes both raised histone acetylation levels and were antiproliferative toward HeLa cells. Only compounds with class II activity were able to increase acetylation of tubulin, which is in agreement with this activity being catalyzed by HDAC6.

Thus, the molecules that have entered clinical trials to date have been pan-inhibitors of the class I and II HDACs or show a modest selectivity for class I. There is therefore a real need for compounds with genuine isoform selectivity both as tool compounds to help better understand the biological roles of HDACs and potentially as therapeutic compounds.

The development of a novel series of alkyl ketone inhibitors based on apicidin, as recently reported by workers at Merck, was discussed above. The isoform selectivity of these compounds has also been studied and they show reasonable selectivity for class I over class II enzymes. Interestingly, this includes potent activity against HDAC1, something that apicidin lacks. Small changes to the alkyl group could modulate isoform selectivity; for example, the ethyl ketone analogue retains activity against HDACs 1–3, but loses activity against HDAC6. In contrast to the class I selectivity identified for these alkyl ketones, the 5-(trifluoroacetyl)thiophene-2-carboxamides[41] show modest (10-fold) selectivity for HDACs 4 and 6 over class I enzymes. This study also throws interesting light on the subtle interplay between the nature of the zinc binding group and the linker region of HDAC inhibitors. With thiophene-2-carboxamide as the scaffold, small changes to the trifluoromethyl ZBG proved detrimental (alcohol and ketal inactive, large losses of activity for difluoromethyl and pentafluoroethyl ketones). Unlike the alkyl ketones, when a methyl ketone is placed on this thiophene scaffold, it is inactive. Concerning more radical replacement with other known ZBGs, hydroxamic acid (but not carboxylic acid) produced potent compounds, as would be expected, but with the selectivity switched modestly in favor of class I. Boronic acid, methylsulfone, and phosphonic acids were all inactive. Replacement of the thiophene with the isomeric 3-(trifluoromethyl) thiophene or with thiazole gave active compounds, pyrrole analogues showed some activity only against HDAC6, while phenyl and pyridine analogues were inactive. Finally, activity was less sensitive to changes in the capping group. Thus, very modest bioisosteric changes such as thiophenyl to phenyl can have large effects on activity, and the ability of groups such as methyl ketone to function as ZBGs has been shown to be highly dependent on the linker to which they are attached, which thus plays a more active role in determining the overall HDAC inhibitory effect than its name perhaps implies.

By screening a library of about 7000 1,3-dioxane-containing molecules based on the SAHA framework, prepared by solid-phase synthesis, a compound was identified, tubacin, that is a selective inhibitor of HDAC6 over class I HDACs.[69] Tubacin inhibits α-tubulin acetylation while having no effect on the acetylation status of histones or cellular proliferation. Site-directed mutants indicated that only one of the two active sites within HDAC6 was the tubulin deacetylase. Using molecular dynamics simulations, a computational study has compared the binding of SAHA and tubacin to structural models of HDACs 1, 6, and 8.[70] For these compounds, selectivity is postulated to arise from specific interactions between four residues on the protein surface at the opening of the active site (an aspartic acid in one loop, and two phenylalanines and a methionine in a nonpolar loop) and the inhibitors. The narrower opening to the active site of HDAC6, together with the greater bulk of tubacin, allows it to make more efficient interactions with this isoform.

35.6 CONCLUSIONS

The large and growing number of scientific articles and patent publications in the field of HDAC inhibitors demonstrates that it is currently one of the "hottest topics" within the already hot topic of epigenetic modulators. Impressive activity has been observed in preclinical cancer models. Despite initial concerns, potent inhibition of HDAC does not appear to be associated with significant toxicities, at least within the context of the doses and schedules that have been developed for cancer therapy. Several molecules are in clinical development and one, Zolinza (SAHA), has already reached the market, validating HDAC as a clinical target for cancer. Many studies are ongoing and one may expect other HDAC inhibitors to enter the market in the coming years. The toxicity profile noted above provides encouragement that they may find a role in combination with existing cytotoxic regimens, bringing improved efficacy, reduced side effects, or a combination of both. The development of HDAC inhibitors in other therapeutic areas is much less well advanced. For many of these areas, toxicity may be a more important issue because of both the seriousness of the diseases and the likely necessity in many cases for long-term treatment. However, this is also an area of very active current research.

Much remains to be learned about the specific mechanisms by which HDAC inhibitors achieve their biological effects. The compounds that have entered clinical evaluation either are pan-inhibitors of the class I and II enzymes or have some modest class I selectivity. These molecules alter the levels of histone acetylation and this has been used as a biomarker of activity, but whether this is directly related to the effects of inhibitors, or simply correlative, is currently unknown. The many nonhistone proteins that are regulated via acetylation, including a number of important cytoplasmic proteins such as tubulin and HSP90, indicate that there may be at least a nonhistone component to the activity of HDAC inhibitors. It is also not clear to what extent the observed anticancer activity arises from the broad spectrum of inhibition of the current compounds or whether a specific inhibitor of, say, HDAC1 would be equally effective but with an improved safety profile. Studies of the correlation between histone acetylation, gene expression, and tumor response will help to shed light on these important issues. The identification of isoform-specific HDAC inhibitors has proven difficult because of the high sequence homology between the isoforms and the difficulty in obtaining assays of the individual recombinant enzymes because of their tendency to be associated with other proteins (including themselves) in multiprotein complexes. Such inhibitors would be invaluable in elucidating the biological role of HDACs and the pharmacological implications and would be highly complementary to genetic studies such as gene knockouts and those with siRNA, particularly because of the involvement of HDACs in multiprotein complexes.

ACKNOWLEDGMENT

I thank Sarah Finn for proofreading of the manuscript.

REFERENCES

1. Allfrey, V. G.; Faulkner, R.; Mirsky, A. E. Acetylation and methylation of histones and their possible role in the regulation of RNA synthesis. *Proc. Natl. Acad. Sci. USA* **1964**, *51*, 786–794.
2. Roth, S. Y.; Denu, J. M.; Allis, C. D. Histone acetyltransferases. *Annu. Rev. Biochem.* **2001**, *70*, 81–120.

3. Laherty, C. D.; Yang, W. M.; Sun, J. M.; Davie, J. R.; Seto, E.; Eisenman, R. N. Histone deacetylases associated with the mSin3 corepressor mediate mad transcriptional repression. *Cell* **1997**, *89*, 349–356.

4. Brehm, A.; Miska, E. A.; McCance, D. J.; Reid, J. L.; Bannister, A. J.; Kouzarides, T. Retinoblastoma protein recruits histone deacetylase to repress transcription. *Nature* **1998**, *391*, 597–601.

5. Wong, J.; Patterton, D.; Imhof, A.; Guschin, D.; Shi, Y. B.; Wolffe, A. P. Distinct requirements for chromatin assembly in transcriptional repression by thyroid hormone receptor and histone deacetylase. *EMBO J.* **1998**, *17*, 520–534.

6. Yang, W. M.; Inouye, C.; Zeng, Y.; Bearss, D.; Seto, E. Transcriptional repression of YY1 is mediated by interaction with a mammalian homolog of the yeast global regulator RPD3. *Proc. Natl. Acad. Sci. USA* **1996**, *93*, 12845–12850.

7. Gregoretti, I. V.; Lee, Y. M.; Goodson, H. V. Molecular evolution of the histone deacetylase family: functional implications of phylogenetic analysis. *J. Mol. Biol.* **2004**, *338*, 17–31.

8. Minucci, S.; Pelicci, P. G. Histone deacetylase inhibitors and the promise of epigenetic (and more) treatments for cancer. *Nat. Rev. Cancer* **2006**, *6*, 38–51.

9. Bode, A. M.; Dong, Z. Post-translational modification of p52 in tumorigenesis. *Nat. Rev. Cancer* **2004**, *4*, 793–805.

10. Chen, L. F.; Greene, W. C. Regulation of distinct biological activities of the NF-kappaB transcription factor complex by acetylation. *J. Mol. Med.* **2003**, *81*, 549–557.

11. Zhang, Y.; Li, N.; Caron, C.; Matthias, G.; Hess, D.; Khochbin, S.; Matthias, P. HDAC-6 interacts with and deacetylates tubulin and microtubules *in vivo*. *EMBO J.* **2003**, *22*, 1168–1179.

12. Kovacs, J. J.; Murphy, P. J.; Gaillard, S.; Zhao, X.; Wu, J. T.; Nicchitta, C. V.; Yoshida, M.; Toft, D. O.; Pratt, W. B.; Yao, T. P. HDAC6 regulates Hsp90 acetylation and chaperone-dependent activation of glucocorticoid receptor. *Mol. Cell* **2005**, *18*, 601–607.

13. Zhang, C. L.; McKinsey, T. A.; Chang, S.; Antos, C. L.; Hill, J. A.; Olsen, E. N. Class II histone deacetylases act as signal-responsive repressors of cardiac hypertrophy. *Cell* **2002**, *110*, 479–488.

14. Zhang, Y.; Kwon, S.; Yamaguchi, T.; Cubizolles, F.; Rousseaux, S.; Kneissel, M.; Cao, C.; Li, N.; Cheng, H. L.; Chua, K.; Lombard, D.; Mizeracki, A.; Matthias, G.; Alt, F. W.; Khochbin, S.; Matthias, P. Mice lacking histone deacetylase 6 have hyperacetylated tubulin but are viable and develop normally. *Mol. Cell. Biol.* **2008**, *28*, 1688–1701.

15. Tsuji, N.; Kobayashi, M.; Nagashima, K.; Wakisaki, Y.; Koizumi, K. A new antifungal antibiotic, trichostatin. *J. Antibiot.* **1976**, *29*, 1–6.

16. Yoshida, M.; Beppu, T. Reversible arrest of proliferation of rat 3Y1 fibroblasts in both G1 and G2 phases by trichostatin A. *Exp. Cell Res.* **1988**, *177*, 122–131.

17. Yoshida, M.; Kijima, M.; Akita, M.; Beppu, T. Potent and specific inhibition of mammalian histone deacetylase both *in vivo* and *in vitro* by trichostatin A. *J. Biol. Chem.* **1990**, *265*, 17174–17179.

18. Hoshikawa, Y.; Kwon, H. J.; Yoshida, M.; Horinouchi, S.; Beppu, T. Trichostatin A induces morphological changes and gelsolin expression by inhibiting histone deacetylase in human carcinoma cell lines. *Exp. Cell Res.* **1994**, *214*, 189–197.

19. Mielnicki, L. M.; Ying, A. M.; Head, K. L.; Asch, H. L.; Asch, B. B. Epigenetic regulation of gelsolin expression in human breast cancer cells. *Exp. Cell Res.* **1999**, *249*, 161–176.

20. Kwon, H. J.; Owa, T.; Hassig, C. H.; Shimada, J.; Schreiber, S. L. Depudecin induces morphological reversion of transformed fibroblasts via the inhibition of histone deacetylase. *Proc. Natl. Acad. Sci. USA* **1998**, *95*, 3356–3361.
21. Matsumoto, M.; Matsutani, S.; Sugita, K.; Yoshida, H.; Hayashi, F.; Terui, Y.; Nakai, H.; Uotani, N.; Kawamura, Y.; Matsumoto, K. Depudecin: a novel compound inducing the flat phenotype of NIH3T3 cells doubly transformed by ras- and src-oncogene, produced by *Alternaria brassicicola*. *J. Antibiot.* **1992**, *45*, 879–885.
22. Kijima, M.; Yosjida, M.; Sugita, K.; Horinouchi, S.; Beppu, T. Trapoxin, an antitumor cyclic tetrapeptide, is an irreversible inhibitor of mammalian histone deacetylase. *J. Biol. Chem.* **1993**, *268*, 22429–22435.
23. Darkin-Rattray, S. J.; Gurnett, A. M.; Myers, R. W.; Dulski, P. M.; Crumley, T. M.; Allocco, J. J.; Cannova, C.; Meinke, P. T.; Colletti, S. L.; Bednarek, M. A.; Singh, S. B.; Goetz, M. A.; Dombrowski, A. W.; Polishook, J. D.; Schmatz, D. M. Apicidin: a novel antiprotozoal agent that inhibits parasite histone deacetylase. *Proc. Natl. Acad. Sci. USA* **1996**, *93*, 13143–13147.
24. Ueda, H.; Nakajima, H.; Hori, Y.; Fujita, T.; Nishimura, M.; Goto, T.; Okuhara, M. FR901228, a novel antitumor bicyclic depsipeptide produced by *Chromobacterium violaceum* No. 968. I. Taxonomy, fermentation, isolation, physico-chemical and biological properties, and antitumor activity. *J. Antibiot.* **1994**, *47*, 30–310.
25. Newmark, H. L.; Lupton, J. R.; Young, C. W. Butyrate as a differentiating agent: pharmacokinetics, analogues and current status. *Cancer Lett.* **2001**, *78*, 1–5.
26. Finnin, M. S.; Donigian, J. R.; Cohen, A.; Richon, V. M.; Rifkind, R. A.: Marks, P. A.; Breslow, R.; Pavletich, N. P. Structures of a histone deacetylase homologue bound to the TSA and SAHA inhibitors. *Nature* **1999**, *401*, 188–193 .
27. Vannini, A.; Volpari, C.; Filocamo, G.; Casavola, E. C.; Brunetti, M.; Renzoni, D.; Chakravarty, P.; Paolini, C.; De Francesco, R.; Gallinari, P.; Steinkuhler, C.; Di Marco, S. Crystal structure of a eukaryotic zinc-dependent histone deacetylase, human HDAC8, complexed with a hydroxamic acid inhibitor. *Proc. Natl. Acad. Sci. USA 101* **2004**, 15064–15069.
28. Vannini, A.; Volpari, C.; Gallinari, P.; Jones, P.; Mattu, M.; Carfi, A.; Defrancesco, R.; Steinkuhler, C.; Di Marco, S. Substrate binding to histone deacetylases as revealed by crystal structure of HDAC8–substrate complex. *EMBO Rep.* **2008**, *8*, 879–884.
29. Richon, V. M.; Emiliani, S.; Verdin, E.; Webb, Y.; Breslow, R.; Rifkind, R. A.; Marks, P. A. A class of hybrid polar inducers of transformed cell differentiation inhibits histone deacetylases. *Proc. Natl. Acad. Sci. USA* **1998**, *95*, 3003–3007.
30. Belvedere, S.; Witter, D. J.; Yan, Y.; Secrist, P.; Richon, V.; Miller, T. A. Aminosuberoyl hydroxamic acids (ASHAs): a potent new class of HDAC inhibitors. *Bioorg. Med. Chem. Lett.* **2007**, *17*, 3969–3971.
31. Miller, T. A.; Witter, D. J.; Belvedere, S. Diamine and iminodiacetic acid hydroxamic acid derivatives. WO2005053610, **2005**.
32. Finn, P. W.; Bandara, M.; Butcher, C.; Finn, A.; Hollinshead, R.; Khan, N.; Law, N.; Murthy, S.; Romero, R.; Watkins, C.; Andrianov, V.; Bokaldere, R. M.; Dikovska, K.; Gailite, V.; Loza, E.; Piskunova, I.; Starchenkov, I.; Vorona, M.; Kalvinsh, I. Novel sulphonamide derivatives as inhibitors of histone deacetylase. *Helv. Chim. Acta* **2005**, *88*, 1630–1657.
33. Suzuki, T.; Ando, T.; Tsuchiya, K.; Fukazawa, N. Synthesis and histone deacetylase inhibitory activity of new benzamide derivatives. *J. Med. Chem.* **1999**, *42*, 3001–3003.

34. Siliphaivanh, P.; Harrington, P.; Witter, D. J.; Otte, K.; Tempest, P.; Kattar, S.; Kral, A. M.; Fleming, J. C.; Deshmukh, S. V.; Harsch, A.; Secrist, P. J.; Miller, T. A. Design of novel histone deacetylase inhibitors. *Bioorg. Med. Chem. Lett.* **2007**, *17*, 4619–4624.

35. Suzuki, T.; Kouketsu, A.; Matsuura, A.; Kohara, A.; Ninomoya, S.-I.; Kohda, K.; Miyata, N. Thiol-based SAHA analogues as potent histone deacetylase inhibitors. *Bioorg. Med. Chem. Lett.* **2004**, *14*, 3313–3317.

36. Suzuki, T.; Nagano, Y.; Kouketsu, A.; Matsuura, A.; Maruyama, S.; Kurotaki, M.; Nakagawa, H.; Miyata, N. Novel inhibitors of histone deacetylases: design, synthesis, enzyme inhibition and cancer cell growth inhibition of SAHA-based non-hydroxamates. *J. Med. Chem.* **2005**, *48*, 1019–1032.

37. Gu, W.; Nusinzon, I.; Smith, R. D.; Horvath, C. M.; Silverman, R. B. Carbonyl- and sulphur-containing analogs of suberoylamide hydroxamic acid: potent inhibition of histone deacetylases. *Bioorg. Med. Chem. Lett.* **2006**, *14*, 3320–3329.

38. Wada, K. C.; Frey, R. R.; Ji, Z.; Curtin, M. L.; Garland, R. B.; Holms, J. H.; Li, J.; Pease, L. J.; Guo, J.; Glaser, K. B.; Marcotte, P. A.; Richardson, P. L.; Murphy, S. S.; Bouska, J. J.; Tapang, P.; Magoc, T. J.; Albert, D. H.; Davidsen, S. K.; Michaelides, M. R. α-Keto amides as inhibitors of histone deacetylase. *Bioorg. Med. Chem. Lett.* **2003**, *13*, 3331–3335.

39. Vasudevan, A.; Ji, Z.; Frey, R. R.; Wada, K. C.; Steinman, D.; Heyman, H. R.; Guo, Y.; Curtin, M. L.; GGuo, J.; Li, J.; Glaser, K. B.; Marcotte, P. A.; Bouska, J. J.; Albert, D. H.; Davidsen, S. K.; Michaelides, M. R. Heterocyclic ketones as inhibitors of histone deacetylase. *Bioorg. Med. Chem. Lett.* **2003**, *13*, 3909–3913.

40. Jones, P.; Altamura, S.; De Francesco, R.; Paz, O. G.; Kinzel, O.; Mesiti, G.; Monteagudo, E.; Pescatore, G.; Rowley, M.; Verdirame, M.; Steinkuehler, C. A novel series of potent and selective ketone histone deacetylase inhibitors with antitumor activity *in vivo*. *J. Med. Chem.* **2008**, *51*, 2350–2353.

41. Jones, P.; Bottomley, M. J.; Carfi, A.; Cecchetti, O.; Ferrigno, F.; Lo Surdo, P.; Ontoria, J. M.; Rowley, M.; Scarpelli, R.; Schultz-Fademrecht, C.; Steinkuehler, C. 2-Trifluoroacetylthiophenes, a novel series of potent and selective class II histone deacetylase inhibitors. *Bioorg. Med. Chem. Lett.* **2008**, DOI:10.1016/j.bmcl.2008.02.026.

42. Kouzarides, T. Histone acetylases and deacetylases in cell proliferation. *Curr. Opin. Genet. Dev.* **1999**, *9*, 40–48.

43. Richon, V. M.; Webb, Y.; Merger, R.; Sheppard, T.; Jursic, B.; Ngo, L.; Civoli, F.; Breslow, R.; Rifkind, R. A.; Marks, P. A. Second generation hybrid polar compounds are potent inducers of transformed cell differentiation. *Proc. Natl. Acad. Sci. USA* **1996**, *93*, 5705–5708.

44. Kim, Y. B.; Lee, K. H.; Sugita, K.; Yoshida, M.; Horinouchi, S. Oxamflatin is a novel antitumor compound that inhibits mammalian histone deacetylase. *Oncogene* **1999**, *18*, 2461–2470.

45. Kitamura, K.; Hoshi, S.; Koike, M.; Kiyoi, H.; Saito, H.; Naoe, T. Histone deacetylase inhibitor but not arsenic trioxide differentiates acute promyelocytic leukaemia cells with t (11;17) in combination with all-*trans* retinoic acid. *Br. J. Haematol.* **2000**, *108*, 696–702.

46. David, G.; Alland, L.; Hong, S. H.; Wong, C. W.; DePinho, R. A.; Dejean, A. Histone deacetylase associated with mSin3A mediates repression by the acute promyelocytic leukemia-associated PLZF protein. *Oncogene* **1998**, *16*, 2549–2556.

47. Lin, R. J.; Nagy, L.; Inoue, S.; Shao, W.; Miller, W. H., Jr.; Evans, R. M. Role of the histone deacetylase complex in acute promyelocytic leukaemia. *Nature* **1998**, *391*, 811–814.

48. Yoshida, M.; Horinouchi, S. Trichostatin and leptomycin. Inhibition of histone deacetylation and signal-dependent nuclear export. *Ann. N.Y. Acad. Sci.* **1999**, *886*, 23–36.

49. Ueda, H.; Manda, T.; Matsumoto, S.; Mukumoto, S.; Nishigaki, F.; Kawamura, I.; Shimomura, K. FR901228, a novel antitumor bicyclic depsipeptide produced by *Chromobacterium violaceum* No. 968. III. Antitumor activities on experimental tumors in mice. *J. Antibiot.* **1994**, *47*, 315–323.

50. Desai, D.; Das, A.; Cohen, L.; el-Bayoumy, K.; Amin, S. Chemopreventive efficacy of suberoylanilide hydroxamic acid (SAHA) against 4-(methylnitrosamino)-1-(3-pyridyl)-1-butanone (NNK)-induced lung tumorigenesis in female A/J mice. *Anticancer Res.* **2003**, *23*, 499–503.

51. Olsen, E.; Kim, Y. H.; Kuzel, T.,et al. Vorinostat (suberoylanilide hydroxamic acid, SAHA) is clinically active in advanced cutaneous T-cell lymphoma (CTCL): results of a phase IIb trial. *J. Clin. Oncol.* **2006**, *24*(18S), ASCO Annual Meeting Proceedings Part I, Abstract 7500.

52. Steele, N. L.; Plumb, J. A.; Vidal, L.; Tjornelund, J.; Knoblauch, P.; Rasmussen, A.; Ooi, C. E.; Buhl-Jensen, P.; Brown, R.; Evans, T. R. J.; De Bono, J. S. A phase 1 pharmacokinetic and pharmacodynamic study of the histone deacetylase inhibitor belinostat in patients with advanced solid tumors. *Clin. Cancer Res.* **2008**, *14*, 804–810.

53. Piekarz, R. L.; Robey, R.; Sandor, V.; Bakke, S.; Wilson, W. H.; Dahmoush, L.; Kingma, D. M.; Turner, M. L.; Altemus, R.; Bates, S. E. Inhibitor of histone deacetylase, depsipeptide (FR901228), in the treatment of peripheral and cutaneous T-cell lymphoma: a case report. *Blood* **2001**, *98*, 2865–2868.

54. Shah, M. H.; Binkley, P.; Chan, K.; Xiao, J.; Arbogast, D.; Collamore, M.; Farra, Y.: Young, D.; Grever, M. Cardiotoxicity of histone deacetylase inhibitor depsipeptide in patients with metastatic neuroendocrine tumors. *Clin. Cancer Res.* **2006**, *12*, 3997–4003.

55. Zhu, W.-G.; Lakshmanan, R. R.; Beal, M. D.; Otterson, G. A. DNA methyltransferase inhibition enhances apoptosis induced by histone deacetylase inhibitors. *Cancer Res.* **2001**, *61*, 1327–1333.

56. Plumb, J. A.; Steele, N.; Finn, P. W.; Brown, R. Epigenetic approaches to cancer therapy. *Biochem. Soc. Trans.* **2004**, *32*, 1095–1097.

57. Tumber, A.; Collins, L. S.; Petersen, K. D.; Thougaard, A.; Christiansen, S. J.; Dejligbjerg, M.; Buhl Jensen, P.; Sehested, M.; Ritchie, J. W. A. The histone deacetylase inhibitor PXD101 synergises with 5-fluorouracil to inhibit colon cancer cell growth *in vitro* and *in vivo*. *Cancer Chemother. Pharmacol.* **2007**, *60*, 275–283.

58. Leoni, F.; Zaliani, A.; Bertolini, G.; Porro, G.; Pagani, P.; Pozzi, P.; Donà, G.; Fossati, G.; Sozzani, S.; Azam, T.; Bufler, P.; Fantuzzi, G.; Goncharo, I.; Kim, S. H.; Pomerantz, B. J.; Reznikov, L. L.; Siegmund, B.; Dinarello, C. A.; Mascagni, P. The antitumor histone deacetylase inhibitor suberoylanilide hydroxamic acid exhibits anti-inflammatory properties via suppression of cytokines. *Proc. Natl. Acad. Sci. USA* **2002**, *99*, 2995–3000.

59. Chung, Y.-L.; Lee, M.-Y.; wang, A.-J.; Yao, L.-F. A therapeutic strategy uses histone deacetylase inhibitors to modulate the expression of genes involved in the pathogenesis of rheumatoid arthritis. *Mol. Ther.* **2003**, *8*, 707–717.

60. Saunders, N.; Dicker, A.; Popa, C.; Jones, S.; Dahler, A. Histone deacetylase inhibitors as potential anti-skin cancer agents. *Cancer Res.* **1999**, *59*, 399–404.

61. Takahashi, I.; Miyaji, H.; Yoshida, T.; Sato, S.; Mizukami, T. Selective inhibition of IL-2 gene expression by trichostatin A, a potent inhibitor of mammalian histone deacetylase. *J. Antibiot.* **1996**, *49*, 453–457.

62. Kim, M. S.; Kwon, H. J.; Lee, Y. M.; Baek, J. H.; Jang, J. E.; Lee, S. W.; Moon, E. J.; Kim, H. S.; Lee, S. K.; Chung, H. Y.; Kim, C. W.; Kim, K. W. Histone deacetylases induce angiogenesis by negative regulation of tumor suppressor genes. *Nat. Med.* **2001**, *7*, 437–443.

63. Adcock, I. M. HDAC inhibitors as anti-inflammatory agents. *Br. J. Pharmacol.* **2007**, *150*, 829–831.

64. Sadri-Vakili, G.; Bouzou, B.; Benn, C. L.; Kim, M. O.; Chawla, P.; Overland, R. P.; Glajch, K. E.; Xia, E.; Qiu, Z.; Hersch, S. M.; Clark, T. W.; Yohrling, G. J.; Cha, J. H. Histones associated with downregulated genes are hypo-acetylated in Huntington's disease models. *Hum. Mol. Genet.* **2007**, *16*, 1293–1306.

65. Hockly, E.; Richon, V. M.; Woodman, B.; Smith, D. L.; Zhou, X.; Rosa, E.; Sathasivam, K.; Ghazi-Noori, S.; Mahal, A.; Lowden, P. A.; Steffan, J. S.; Marsh, J. L.; Thompson, L. M.; Lewis, C. M.; Marks, P. A.; Bates, G. P. Suberoylanilide hydroxamic acid, a histone deacetylase inhibitor, ameliorates motor deficits in a mouse model of Huntington's disease. *Proc. Natl. Acad. Sci. USA* **2003**, *100*, 2041–2046.

66. Mai, A.; Cerbara, I.; Valente, S.; Massa, S.; Walker, L. A.; Tekwani, B. L. Antimalarial and antileishmanial activities of aroyl-pyrrolyl-hydroxyamides, a new class of histone deacetylase inhibitors. *Antimicrob. Agents Chemother.* **2004**, *48*, 1435–1436.

67. Mukherjee, P.; Pradhan, A.; Shah, F.; Tekwani, B. L.; Avery, M. A. Structural insights into the *Plasmodium falciparum* histone deacetylase 1 (PfHDAC-1): a novel target for the development of antimalarial therapy. *Bioorg. Med. Chem.* **2008**, *16*, 5254–5365.

68. Grant, S. L.; Gattis, S. G.; Fierke, C. A. Catalytic activity and inhibition of human histone deacetylase 8 is dependent on the identity of the active site metal ion. *Biochemistry* **2006**, *45*, 6170–6178.

69. Haggarty, S. J.; Koeller, K. M.; Wong, J. C.; Grozinger, C. M.; Schreiber,S. L. Domain selective small-molecule inhibitor of histone deacetylase 6 (HDAC6)-mediated tubulin deacetylation. *Proc. Natl. Acad. Sci. USA* **2003**, *100*, 4389–4394.

70. Estiu, G.; Greenberg, E.; Harrison, C. B.; Kwiatkowski, N. P.; Mazitschek, R.; Bradner, J. E.; Wiest, O. Structural origin of selectivity in class II-selective histone deacetylase inhibitors. *J. Med. Chem.* **2008**, *51*, 2898–906.

CHAPTER 36

Recent Development of Diagnostic and Therapeutic Agents Targeting Glutamate Carboxypeptidase II (GCPII)

YOUNGJOO BYUN[1], RONNIE C. MEASE[1], SHAWN E. LUPOLD[2], and MARTIN G. POMPER[1]

[1]Russell H. Morgan Department of Radiology and Radiological Sciences, Johns Hopkins Medical Institutions, Baltimore, MD 21212, USA
[2]James Buchanan Brady Urological Institute, Johns Hopkins Medical Institutions, Baltimore, MD 21212, USA

36.1 INTRODUCTION

Glutamate carboxypeptidase II (GCPII) is a type II transmembrane metalloenzyme. GCPII was first reported to exhibit NAALADase (*N*-acetyl-α-linked acidic dipeptidase) activity, whereby it cleaves *N*-acetylaspartylglutamate (NAAG) to release *N*-acetylaspartate (NAA) and glutamate,[1] the most abundant excitatory neurotransmitter in the brain. Excessive production of glutamate in the synaptic cleft is associated with a number of neurological diseases including ischemia, traumatic brain injury, neuropathic pain, amyotrophic lateral sclerosis, diabetic polyneuropathy, and schizophrenia.[2–6] Inhibition of GCPII can lead to the decrease of glutamate concentration and the increase of NAAG. NAAG itself acts as an agonist at group III metabotropic glutamate receptors (mGlu3). Activation of mGlu3 receptors inhibits the release of glutamate and increases transforming growth factor β (TGF-β) levels, thus playing a protective role in the brain.[7] GCPII has also been identified in non-neuronal tissues including kidney, prostate, and small intestine, where it can have different functions. Accordingly, GCPII has also been referred to as the prostate-specific membrane antigen (PSMA) and folate hydrolase 1 (FOLH1). The highest expression of PSMA is found on the surface of prostate cancer cells, in particular, in androgen-independent, advanced disease. Because of that expression profile, PSMA has been considered as an attractive biomarker for the diagnosis and prognosis of prostate cancer.[8,9] The FOLH1 activity of GCPII hydrolyzes γ-linked glutamates from poly-γ-glutamyl tetrahydrofolate to produce folic acid, which can then be processed by folic acid receptors.[10] This activity is known to be involved in the uptake of dietary folate in the small intestine and may provide a growth advantage to prostate tumors.[11,12] GCPII has also been identified in the neovasculature of most solid tumors.[13] That vascular expression adds to the value of GCPII as a cancer imaging and therapeutic target. In addition to the carboxypeptidase activity of GCPII, the protein has the ability to internalize certain bound ligands.[14–16] The ability to internalize ligands makes PSMA an attractive target for intracellular drug delivery. Herein we will focus on the recent development of GCPII ligands (inhibitors and substrates), which can potentially be used as targeting, therapeutic, or diagnostic agents.

36.2 BACKGROUND

In the brain GCPII acts as a zinc metalloenzyme located on the surface of glial cells. GCPII is composed of 750 amino acids with a short cytosolic 19-amino acid segment, a single 22-amino acid membrane-spanning segment, and a large extracellular 709-amino acid segment. Five N-terminal amino acids of the cytosolic domain are required for the internalization of bound ligands, such as antibodies, which takes place through a clathrin-dependent endocytic mechanism.[17] The bulky extracellular segment consists of three distinct domains: the protease-like domain, the apical domain, and the C-terminal domain. The extracellular portion of GCPII exhibits enzymatic activity and cleaves two natural substrates, NAAG and poly-γ-glutamyl folates (Fig. 36.1). Recent GCPII crystal structures with potent or weak inhibitors have revealed detailed information on the binding modes of a number of GCPII inhibitors within the active site.[18–23] The first X-ray crystal structure of GCPII without a ligand (PDB ID: 1Z8L, 3.5 Å) was resolved by Bjorkman and coworkers.[22] The homodimer of GCPII has an overall similarity to transferrin receptor 1 (TfR1),[24] which is a type II membrane glycoprotein associated with iron transport. An early homology model of GCPII was built using the crystal structure of TfR1 (PDB ID: 1DE4) as a template, which has only 28% sequence identity and 46% sequence similarity to GCPII.[25] The homology model exhibited a precise prediction that two zinc ions are located at the interface of the S1 and S1′ sites and are coordinated by His377, Asp387, Glu425, Asp453, and His553. Docking studies of 2-(phosphonomethyl)pentanedioic acid (2-PMPA, **36.1**, Fig. 36.4) with the GCPII homology model predicted that the phosphonate group coordinates with Zn^{2+} ions and the pentanedioic acid groups of 2-PMPA have hydrogen bonding interactions with Arg534 and Arg536 in the S1 site.[25] However, the recent GCPII crystal structures with 2-PMPA showed the strong interaction of the pentanedioic acid moiety with Arg210 and Lys699 in the S1′ site rather than with the predicted Arg534 and Arg536 residues.[20,23]

The GCPII active site is composed of two distinct subpockets that form a "glutamate-sensor" S1′ site and an amphiphilic S1 site (Fig. 36.2). The cylinder-like ~20 Å deep tunnel region exists adjacent to the S1 site and extends to the surface of the enzyme (Fig. 36.2).[22] Two zinc ions are found at the interface between the S1′ and

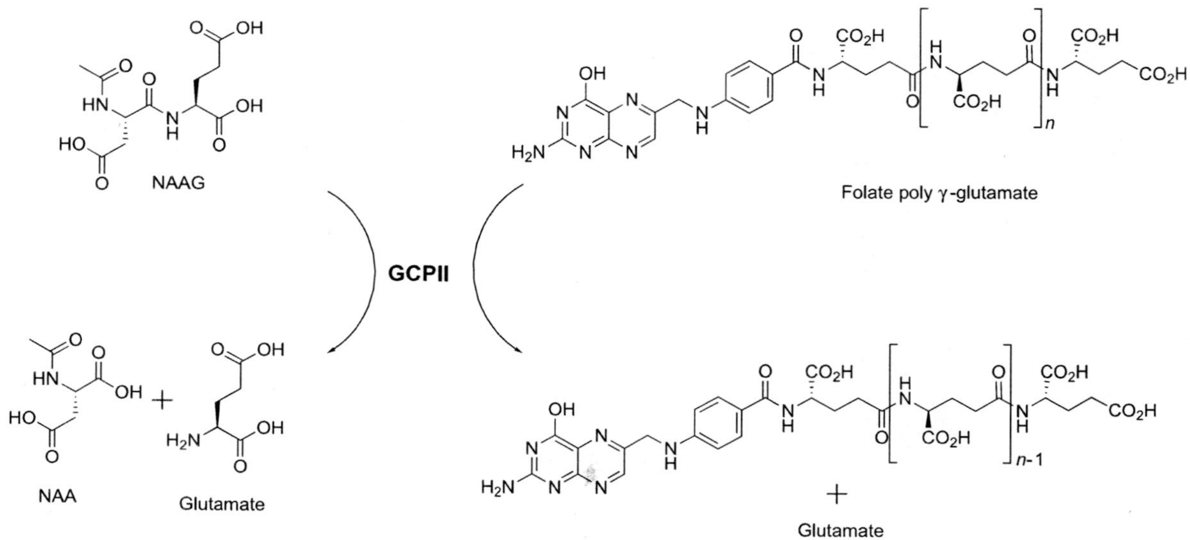

FIGURE 36.1 Catabolism of two natural substrates (NAAG and folate poly γ-glutamate) by GCPII.

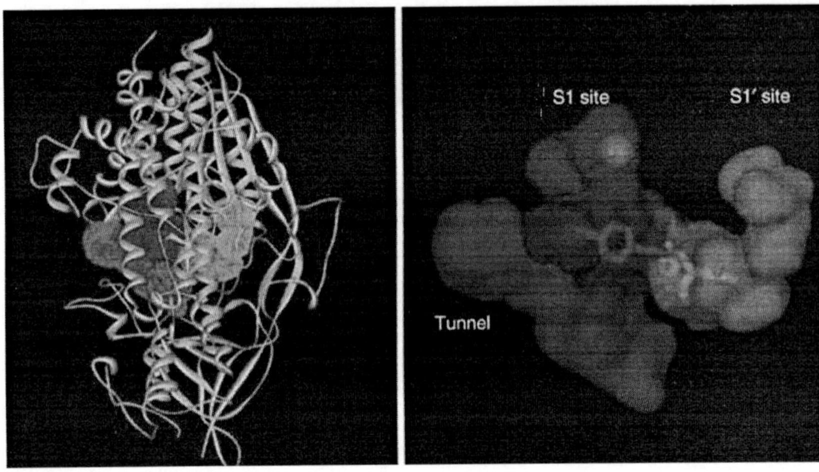

FIGURE 36.2 (a) 3D ectodomain GCPII structure (b) The S1 and S1′ binding pockets and the tunnel region in the active site of GCPII. (See the color version of this figure in Color Plates section.)

S1 sites and are coordinated with side chains of the five amino acids predicted in the homology modeling studies.[22] The zinc ions activate the carbonyl oxygen of the amide bond of NAAG and facilitate the hydrolysis of NAAG by Glu424.[22] A significant binding interaction between GCPII substrates/inhibitors and the S1′ site has been demonstrated.[20] The S1′ site is highly sensitive to modification, thus, minimal changes within the P1′ site significantly affect the binding affinities of GCPII substrates and inhibitors. The glutamate moiety found in NAAG and poly-γ-glutamyl folate is the most preferred functional group to reside within the S1′ site. The α- and the γ-carboxylates of glutamate in the P1′ site form two strong salt bridges with Arg210 and Lys699, respectively. In addition, the α- and the γ-carboxylates provide hydrogen bonding interactions with Tyr552, Tyr700, and Asn257. Site-directed mutagenesis studies of five key amino acids (Arg210, Lys699, Asn257, Tyr552, and Tyr700) in the S1′ site showed a substantial increase in K_m values and a decrease in k_{cat} values.[26] In particular, the R210K mutant exhibited a 700-fold increase in K_m compared to wild-type GCPII.[26] Therefore, productive substitution of glutamate with a bioisostere in the P1′ site is extremely limited, and only the 3-carboxybenzyl group in a series of thiol-based GCPII inhibitors retained GCPII inhibitory activity[27] (see Section 36.3.2). Those key amino acids in the S1′ site are completely conserved among the mammalian GCI orthologs and GCPII homologs.[28,29] Although the S1′ site has an exclusive preference for glutamate, it was postulated to accommodate the glutamate bioisostere through an induced-fit mechanism of β-hairpin residues (Lys699 and Tyr700).[21] On the other hand, the S1 site has been known to tolerate significant modification of the natural substrate and steric variations in the P1 region of inhibitors and substrates are abundant. The S1 site consists of a hydrophilic, "arginine patch" region (Arg463, Arg534, and Arg536) and a hydrophobic region (Gly548, Tyr549, and Tyr552), which render the S1 site amphiphilic. A chloride ion is located in the S1 site and is coordinated with Arg534 and Arg536. The arginine patch residues undergo a conformational change when GCPII binds to bulky inhibitors such as the urea-based inhibitors and phosphopeptide analogs.[18] The stacking conformation of the side chains of Arg463, Arg534, and Arg536 shifts when the enzyme assumes the binding conformation, generating a subpocket in the S1 site. The new site can accommodate and stabilize an aromatic ring via cation–π interactions.[30]

There are two assay methods to evaluate GCPII inhibitory activity *in vitro*: radioactivity-based and fluorescence-based. The radioactivity-based assay uses [^{3}H]NAAG (30 nM) as a substrate and measures [^{3}H]glutamate released upon hydrolysis. The liberated [^{3}H]glutamate is partitioned by anion exchange chromatography and quantified using scintillation counting. The fluorescence-based assay uses nonradiolabeled NAAG (4 μM) and measures the fluorescence of resorufin, which is oxidized from AmplexRed by the coupled actions of glutamate oxidase and horseradish peroxidase.[1,31] Table 36.1 summarizes the two *in vitro* assay methods that have been widely applied to determine *in vitro* IC_{50} values of GCPII inhibitors. Recently, urea-based GCPII inhibitors were evaluated by both the methods.[32] The results showed that both methods produced similar K_I values and that there were no differences between the two assays after converting IC_{50} values into K_I values using the Cheng–Prusoff equation.[32,33] Throughout this chapter, however, we refer to IC_{50} values for the majority of the presented *in vitro* results.

The genetic sequence of human GCPII was discovered through the study of a monoclonal antibody, 7E11-C5, directed toward prostate cancer. This antibody was developed against the membrane fraction of the prostate cancer cell line, LNCaP, and strongly stained tumor cells in radical prostatectomy specimens.[34] The protein target of 7E11-C5, PSMA, was then purified by immunoprecipitation, and the resulting peptide fragments were sequenced. Degenerate PCR primers, based on those peptide sequences, successfully amplified a cDNA product that was cloned and sequenced.[9] Fragments of the cDNA sequence demonstrated homology with the human transferrin receptor, however, the full sequence was later found to be highly homologous to rat brain NAALADase.[35] Indeed, expression of PSMA cDNA in cultured cells bestowed glutamate carboxypeptidase activity upon human cells. Further, 7E11-C5 immunoprecipitates from human brain were active for glutamate carboxypeptidase, and PSMA cDNA sequences could be cloned from brain mRNA. Thus, it was concluded that PSMA and NAALADase arise from the same gene, *GCPII*. In 1998, O'Keefe and colleagues cloned the gene encoding GCPII from the short arm of chromosome 11 (11p11–p12).[36] The convergence of work from several different fields has resulted in several names for products transcribed from the same gene, including GCPII, PSMA,

TABLE 36.1 Comparison of *In Vitro* Assay Methods

Method	Incubation Time	NAAG Concentration	Detection Tool
Radioactivity-based assay	10–20 min	Radiolabeled NAA[^{3}H]G (30 nM)	Scintillation counter
Fluorescence-based assay	120 min	NAAG (4 μM)	Fluorescence measurement

FOLH1, and NAALADase I. A second, similar gene, the PSMA-like gene, exists on the long arm of chromosome 11 (11q14).[37] Genetic studies must be carefully designed to distinguish between these two genes.

In prostate cancer, PSMA mRNA levels are significantly higher than in any other tissue.[38] In addition to the elevated expression in prostate cancer, the PSMA transcript is alternatively spliced. Specifically, splicing in normal prostate cells predominantly produces a cytoplasmic protein, PSM′, whereas in cancer, the full-length transmembrane protein, PSMA, is produced.[39] This may further explain the elevated PSMA protein levels detected with 7E11-C5, which recognizes a cytoplasmic epitope present only in full-length PSMA.[40] Thus, two separate factors lead to elevated PSMA protein levels on prostate cancer cells, elevated expression and alternative splicing. In addition to the PSM′ transcript, several other alternate splice forms and alternate transcription start sites of PSMA have been reported.

The mechanisms behind elevated PSMA expression and alternative splicing in prostate cancer have not been fully elucidated. The upstream PSMA promoter region has been characterized and it maintains prostate-selective gene expression.[36] However, the transcription factors driving the high levels seen in cancer are not known. Many prostate-related transcripts, such as prostate-specific antigen (PSA), are driven by steroid-activated androgen receptor (AR). PSMA expression, on the other hand, is significantly repressed by the activated AR.[38] The gene region responsible for androgen repression is not located in the promoter region, but rather in an enhancer region within the third intron of PSMA.[41] Androgen regulation of PSMA may partially account for the elevated expression in advanced prostate tumors, where patients have been treated with antiandrogen therapy or castration. Collectively, the unique properties of PSMA, including elevated expression in prostate cancer, alternative splicing for cell surface display, and androgen repression, make PSMA one of the best targets for advanced, metastatic prostate cancer lesions.

36.3 DEVELOPMENT OF GCPII INHIBITORS

Since 2-PMPA (**36.1**, Fig. 36.4) was reported as a potent GCPII inhibitor in 1996,[42] extensive efforts have been made to discover and develop additional inhibitors of this enzyme. Initially, the main strategy was to find zinc binding groups linked to a glutamate moiety. Three zinc-affinity functional groups, including phosphonate (phosphite and phosphoramidate), thiol, and urea, have been identified and have shown high affinities for zinc ions. A great number of low molecular weight (MW) GCPII inhibitors in those three classes have been synthesized and evaluated biologically. Interested readers can find more information on clinical implications of these compounds in recent reviews.[4,43–45]

36.3.1 Phosphonate-, Phosphite-, and Phosphoramidate-Based GCPII Inhibitors

Jackson et al. reported a series of phosphonate analogs, including 2-PMPA (**36.1**, $IC_{50} = 0.3$ nM) that has been considered the most potent GCPII inhibitor to date.[42]

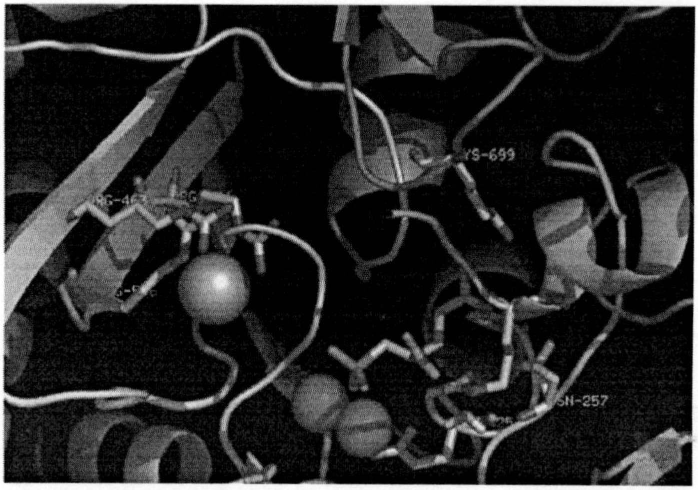

FIGURE 36.3 Active site of GCPII with 2-PMPA (PDB ID: 2PVW). Zinc ions (pink spheres) and chloride ion (white sphere). The carbon atoms backbone of 2-PMPA is shown in yellow. The picture was generated using Pymol program. (See the color version of this figure in Color Plates section.)

2-PMPA exhibited neuroprotective activity in an animal stroke model as well as in *in vitro* and *in vivo* studies using an ischemia model.[6,46,47] However, the mechanism regarding how the highly hydrophilic 2-PMPA penetrates the blood–brain barrier (BBB) is still unknown. According to the GCPII crystal structure in complex with 2-PMPA, the glutamate portion of 2-PMPA binds to the S1′ site and oxygen atoms from the phosphonate strongly interact with two zinc ions (Fig. 36.3).[23] Extensive structure–activity relationship (SAR) studies of 2-PMPA analogs have been undertaken by Guilford Pharmaceuticals (MGI Pharma/Eisai). Compound **36.2** (700 nM, Fig. 36.4), in which the glutamate portion of **36.1** is substituted with aspartate, demonstrated a significant decrease in GCPII inhibitory activity, indicating the importance of spacer length between the α- and the γ-carboxylates at the P1′ site.[42] Alteration of the spacer length between the phosphate group and the α-carboyxlate of glutamate also dramatically affected GCPII potency. Compounds **36.3** and **36.4** (Fig. 36.4) with methyl or propyl spacers exhibited an approximately two-fold decrease in potency.[48] Compounds **36.5** ($IC_{50} = 1.9$ nM, Fig. 36.4) and **36.6** ($IC_{50} = 9$ nM, Fig. 36.4), which contain an additional carboxylic acid at the P1 site, were found to be slightly less active than 2-PMPA.[18] However, introduction of hydroxamate acid (**36.7**, 90 nM, Fig. 36.4) instead of carboxylic acid at the P1 site decreased GCPII inhibition by more than 10-fold.[49] Replacement of two hydroxyl groups of phosphonic acid in 2-PMPA with methoxy groups (**36.8** and **36.9**, Fig. 36.4) also decreased GCPII inhibitory activity, indicating that the free OH groups are crucial for coordinating with zinc ions.[49] A recent GCPII crystal structure with 2-PMPA confirmed the strong chelation of the two phosphonate oxygens with the active site zinc ions.[20] More recently, Miller's group reported the synthesis of enantiomerically pure

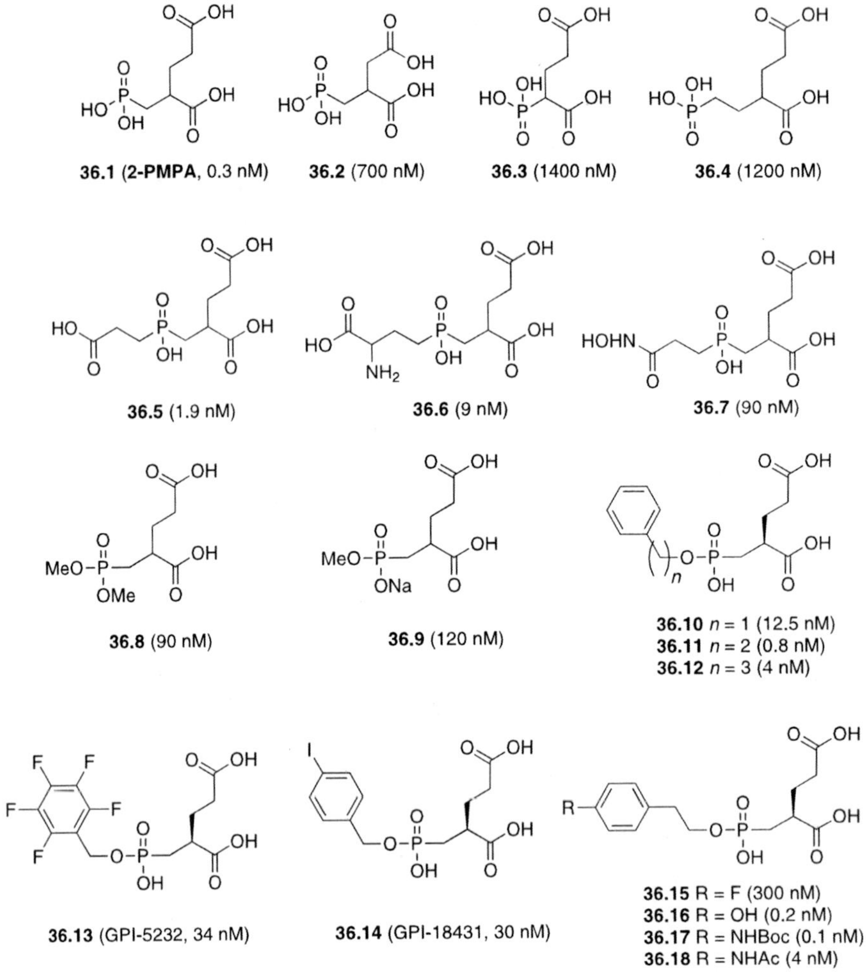

FIGURE 36.4 Phosphonate- or phosphite-based GCPII inhibitors.

phenylalkylphosphonic acid analogs.[50] *In vitro* activity of compound **36.11** (IC_{50} = 0.8 nM, Fig. 36.4), which contains an ethyl linker between the phenyl ring and the phosphonate moiety, is comparable to 2-PMPA and is more potent than the corresponding methyl- (**36.10**, IC_{50} = 12.5 nM, Fig. 36.4) and propyl-linked analogs (**36.12**, IC_{50} = 4 nM, Fig. 36.4).[50] Among methylphosphite analogs, **36.13** (GPI-5232, IC_{50} = 34 nM, Fig. 36.4) and **36.14** (GPI-18431, IC_{50} = 30 nM, Fig. 36.4) were extensively studied to investigate *in vivo* efficacy and to elucidate the binding mode of phosphonate-based GCPII inhibitors. Compound **36.13** showed an improved neuroprotective effect in the rat middle cerebral artery occlusion (MCAO) stroke model with extended therapeutic kinetics compared to 2-PMPA (120 min for **36.13** versus 90 min for 2-PMPA).[51] According to the X-ray crystal complex of GCPII with GPI-18431 (**36.14**), the glutamate portion of **36.14** occupies

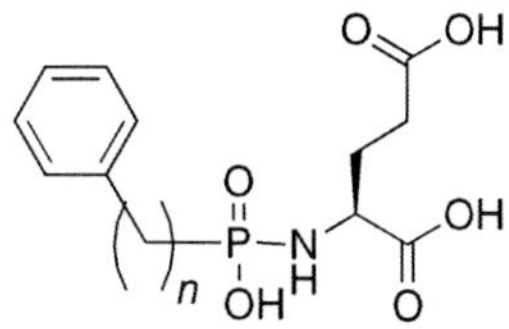

36.19 $n = 0$ (603 nM)
36.20 $n = 1$ (459 nM)
36.21 $n = 2$ (159 nM)
36.22 $n = 3$ (98 nM)
36.23 $n = 4$ (107 nM)
36.24 $n = 5$ (70 nM)
36.25 $n = 6$ (77 nM)

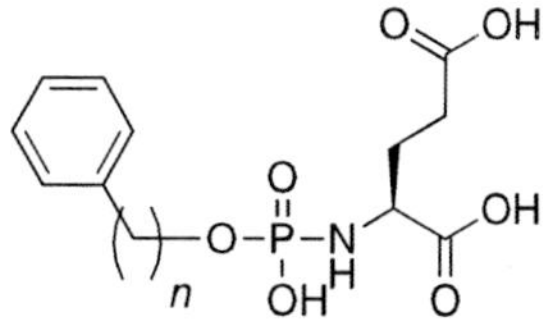

36.26 $n = 1$ (12 nM)
36.27 $n = 3$ (20 nM)

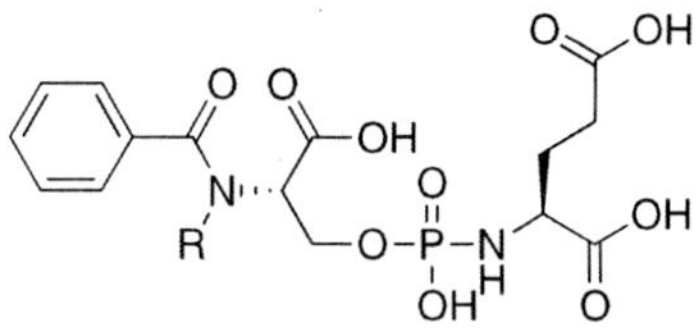

36.28 R = H (35 nM)
36.29 R = CH_3 (55 nM)

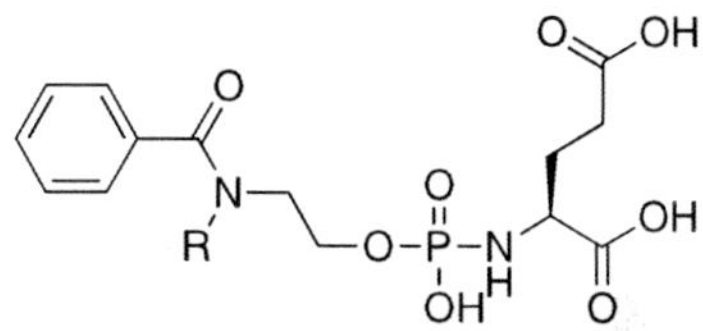

36.30 R = H (1800 nM)
36.31 R = CH_3 (280 nM)

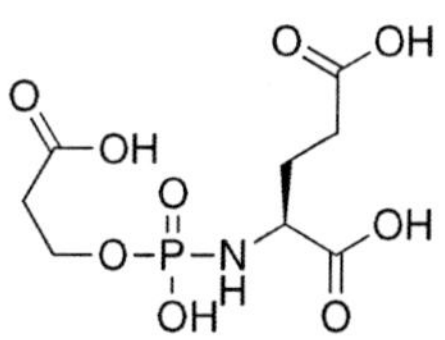

36.32 (4100 nM)

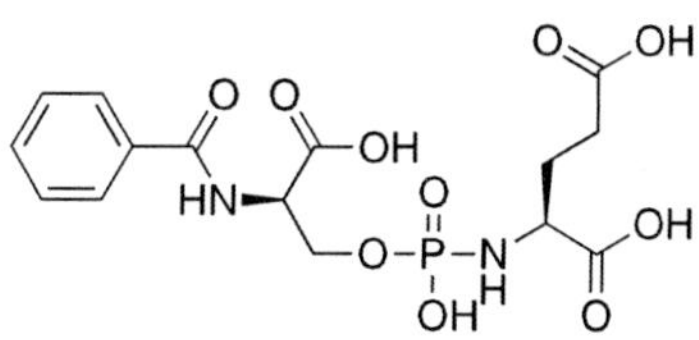

36.33 (700 nM)

FIGURE 36.5 Phosphonamidate- and phosphoramidate-based GCPII inhibitors.

the S1′ site and makes strong salt bridges with Arg210 and Lys699 in a fashion similar to that of 2-PMPA.[21] The *p*-iodobenzyl moiety of **36.14** partially occupies the S1 pocket, hydrogen-bonded with Arg536 and Ser454, and interacts with Tyr549 and Tyr552 through hydrophobic interaction.[21] Ethyl-linked phosphonic acid analogs **36.16–36.18** (Fig. 36.4), which have a variety of functional groups at the *para*-position of the phenyl ring, demonstrate potent *in vitro* GCPII inhibitory activities in the range of 0.1–4 nM.[50]

Phosphoramidate analogs **36.19–36.27** (Fig. 36.5) with linkers of different lengths between the phenyl ring and the phosphoramidate moiety were synthesized and their *in vitro* activities were evaluated.[52,53] Compared to phosphonate-based GCPII inhibitors, compounds **36.19–36.25** are about 10–500-fold less active despite structural similarity to the phosphonates. However, phosphoramidates **36.26** (12 nM) and **36.27** (20 nM) showed comparable GCPII inhibition potency to the corresponding phosphonates. Recently, the Berkman group reported the inhibition mechanism of phosphoramidates **36.28–36.33** (Fig. 36.5).[54] Interestingly, compounds **36.28** and **36.29** showed pseudoirreversible binding patterns in enzymatic studies while phosphoramidates lacking either a carboxylic acid (**36.30** and **36.31**) or a hydro-phobic group (**36.32**) in the P1 site exhibited moderately reversible inhibition.[54] The carboxylic acid group and the hydrophobic group at the P1 site might stabilize the enzyme–inhibitor complex by providing electrostatic interactions with the arginine patch (Arg534 and Arg536) and by hydrophobic interaction within the S1

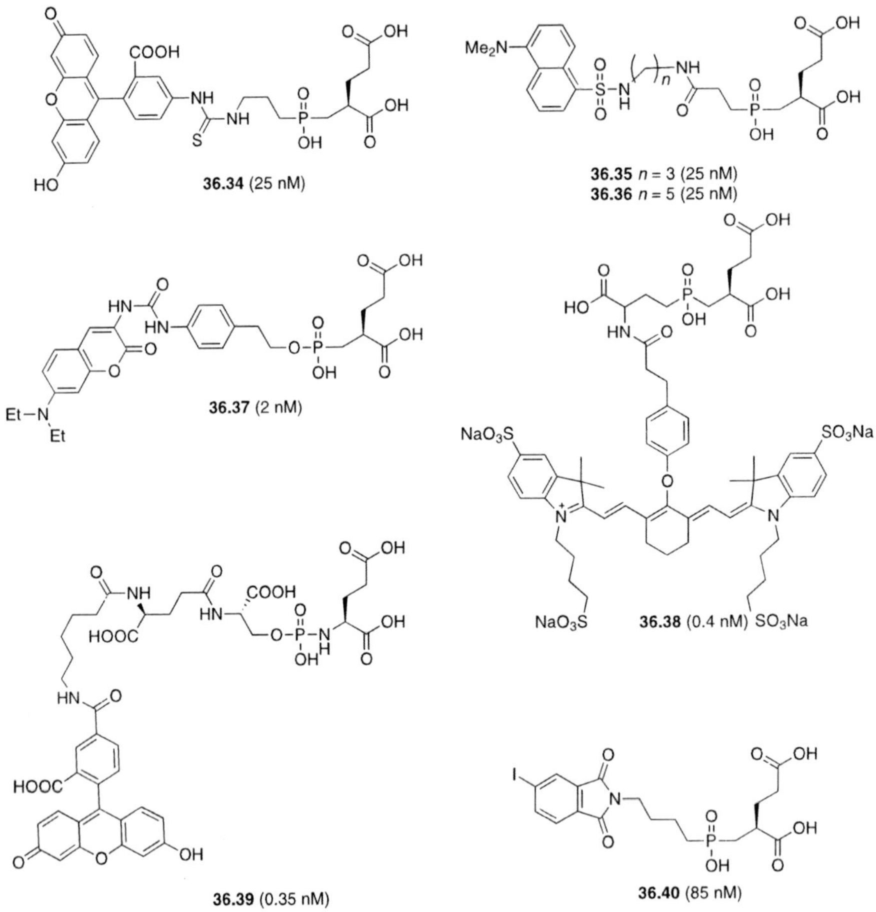

FIGURE 36.6 Phosphonate-based optical imaging agents.

site, respectively. These kinetic data are supported by cellular studies where incubation of the pseudoirreversible inhibitor **36.28** with PSMA-positive LNCaP cells induced greater internalization of PSMA than the reversible inhibitor 2-PMPA. However, compound **36.33**, a diastereoisomer of **36.29**, showed time-dependent reversible binding, implying the importance of orientation of the inhibitor within the P1 active site.

Compound **36.34** (Fig. 36.6) conjugated with fluorescein demonstrated an IC_{50} of 25 nM, implying that introduction of a bulky moiety at the P1 site is tolerated well.[49] Dansyl-linked phosphates **36.35** and **36.36** (Fig. 36.6) also showed similar *in vitro* GCPII inhibitory potencies.[49] Compounds **36.34** and **36.35** exhibited strong binding to LNCaP (GCPII+) cells in *in vitro* fluorescence microscopy studies. Coumarin-linked phosphonate **36.37** (2 nM, Fig. 36.6) was reported as an optical imaging agent to detect prostate cancer cells.[50] Compound **36.38** (Fig. 36.6), which contains a near-infrared fluorescent indocyanine dye, was synthesized for the sensitive and real-time detection

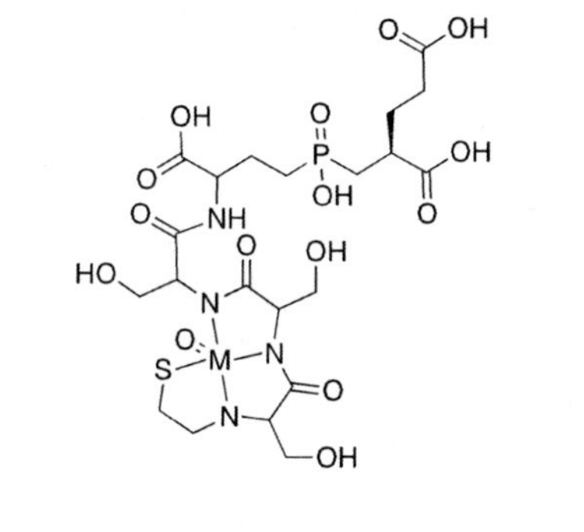

36.41 M = Re (34.6 nM)
36.42 M = ^{99m}Tc

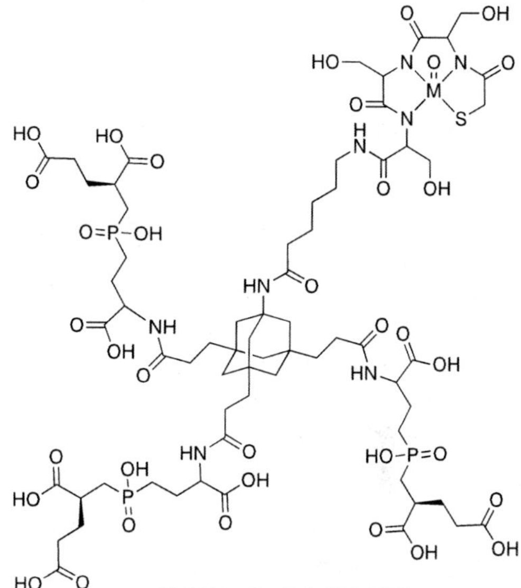

36.43 M = Re (2.9 nM in PBS)
36.44 M = ^{99m}Tc

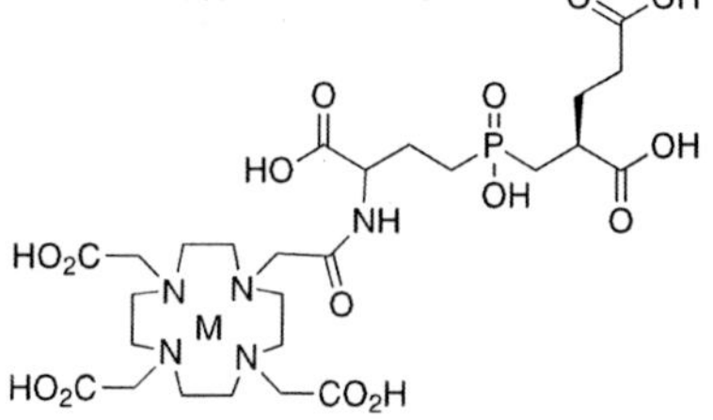

36.45 M = Gd^{3+} (29 nM)
36.46 M = In^{3+} (22.3 nM)
36.47 M = Y^{3+} (12.7 nM)

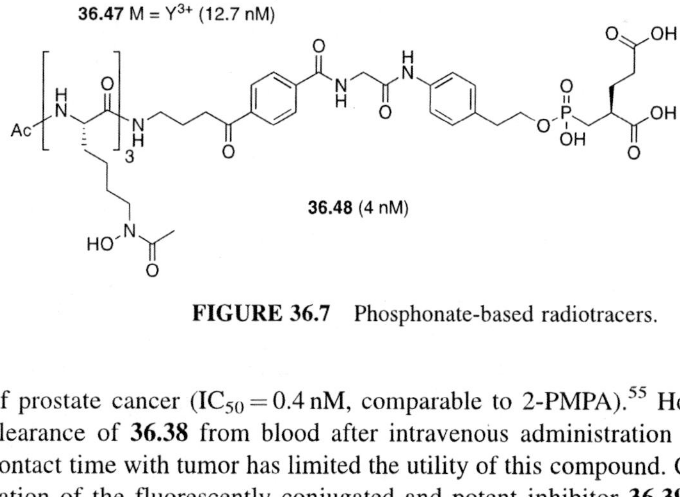

36.48 (4 nM)

FIGURE 36.7 Phosphonate-based radiotracers.

of prostate cancer ($IC_{50} = 0.4$ nM, comparable to 2-PMPA).[55] However, the rapid clearance of **36.38** from blood after intravenous administration and the minimal contact time with tumor has limited the utility of this compound. Cellular internalization of the fluorescently conjugated and potent inhibitor **36.39** (Fig. 36.6) was observed in perinuclear endosomes after 30 min of incubation at 37°C.[56] Compound **36.40** (85 nM, Fig. 36.6), generated in a nonradioactive form, was reported where the nonradioactive iodine atom could potentially be exchanged with radioactive iodine atoms for imaging (^{124}I and ^{123}I) or therapy (^{131}I).[49]

Technetium-99m-labeled **36.42** and **36.44** (Fig. 36.7) were obtained using a simple, cartridge-based, solid-phase prelabeled *S*-acetylmercaptoacetyltriserine (MAS_3) chelator.[57] However, no *in vivo* imaging studies have been reported. The corresponding rhenium analogs **36.41** and **36.43** (Fig. 36.7) exhibited strong GCPII inhibitory activities (35 and 3 nM, respectively). Multivalent, adamantane-trimerized **36.43** showed a 100-fold stronger binding affinity than its monovalent adamantine-monomerized counterpart (structure not shown) in an *in vitro* cell assay.[57] Recently, 1,4,7,10-tetraazacyclododecane-*N*,*N*,*N*,*N*-tetraacetic acid (DOTA) chelators complexed with gadolinium (**36.45**), indium (**36.46**), and yttrium (**36.47**) were reported by Frangioni et al. (Fig. 36.7).[58] Gd-DOTA complex **36.45** could potentially be used for magnetic resonance imaging, **36.46** for single photon emission computed tomography (SPECT) imaging, and **36.47** for radiotherapy through coordination with ^{90}Y. More recently, the siderophore-containing phosphate **36.48** (4 nM, Fig. 36.7) was reported. That compound contains a tri(*N*-hydroxyacetamidyllysine) moiety, which can coordinate iron or gadolinium for MRI applications.[59]

36.3.2 Thiol-Based GCPII Inhibitors

Due to low penetration across the BBB of hydrophilic phosphate or phosphite-based GCPII inhibitors, their therapeutic or diagnostic use in neurological diseases has been limited. The rationale behind the design of thiol-based GCPII inhibitors includes the high affinity of the thiol group for zinc ions observed during the development of metalloprotease inhibitors.[48] Captopril, a potent inhibitor of angiotensin-converting enzyme, was used as a template.[60] Although this class of GCPII inhibitors is overall less potent than the phosphonate-based inhibitors, it has been expected to provide better penetration of the BBB due to higher lipophilicity. The first report of a series of thiol-containing GCPII inhibitors and their *in vitro* and *in vivo* biological evaluation was published by Majer et al.[48] Extensive SAR studies showed that the optimal alkyl length between the carboxylate and thiol groups involves a propyl group (**36.52**, 2-(3-mercaptopropyl)pentanedioic acid [2-MPPA]), which is at least 10-fold more potent than the next longer or shorter alkyl spacers (Fig. 36.8). Originally, the thiol group was assumed to be coordinated with zinc ions in the active site. Indeed, thiol-masked compounds **36.55** and **36.56** (Fig. 36.8) were inactive even at high μM concentration. *In vivo* treatment studies of diabetic rats with 2-MPPA (30 mg/kg/day) via oral administration daily prevented hyperalgesia and myelinated fiber degeneration and promoted nerve fiber regeneration compared to control groups.[61] In that experiment, diabetic rats treated with 2-MPPA did not exhibit increased blood glucose levels and/or a decreased body weight. 2-MPPA is at present in clinical trials for the treatment of diabetic neuropathy.

Further SAR studies of thiol-based GCPII inhibitors focused on the modification of the glutamate portion of 2-MPPA using a carboxybenzyl group as a surrogate of the γ-carboxylate of glutamate (Fig. 36.8).[27] The optimal linker length between the thiol and carboxylic acid groups in this series is also propyl (**36.58**, 15 nM, Fig. 36.8), which is the same spacer length in the 2-MPPA series. *meta*- and *para*-Substituted carboxybenzyl analogs (**36.58** and **36.59**, Fig. 36.8) exhibited improved GCPII inhibitory activity compared to the *ortho*-substituted compound (**36.57**, 1700 nM, Fig. 36.8). Removal or replacement of the carboxylic acid group from the phenyl ring with other functional groups, including methyl ester (**36.65**,

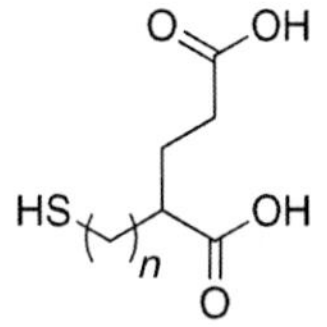

36.49 *n* = 0 (2600 nM)
36.50 *n* = 1 (6000 nM)
36.51 *n* = 2 (580 nM)
36.52 *n* = 3 (**2-MPPA**, 90 nM)
36.53 *n* = 4 (1100 nM)
36.54 *n* = 5 (2600 nM)

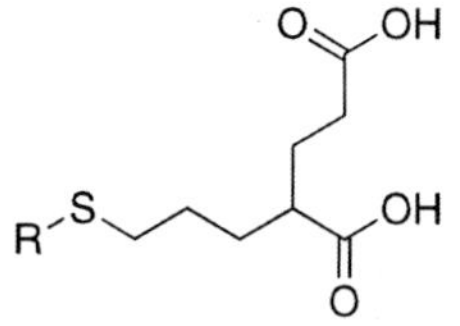

36.55 R = CH_3 (6500 nM)
36.56 R = CPh_3 (>10,000 nM)

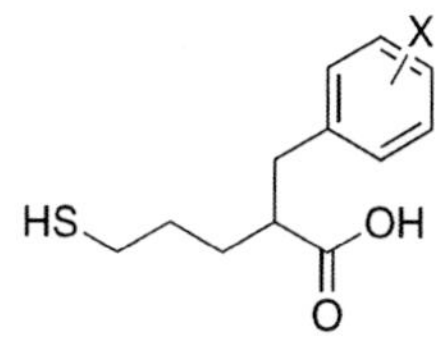

36.57 X = *o*-COOH (1700 nM)
36.58 X = *m*-COOH (15 nM)
36.59 X = *p*-COOH (63 nM)

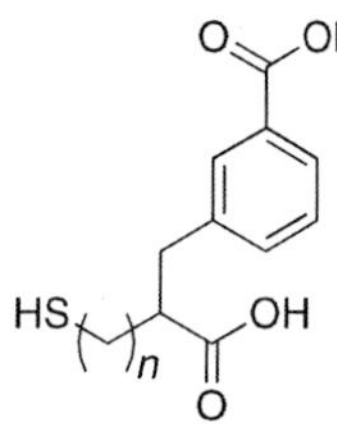

36.60 *n* = 1 (440 nM)
36.61 *n* = 2 (1100 nM)
36.58 *n* = 3 (15 nM)
36.62 *n* = 4 (390 nM)
36.63 *n* = 5 (190 nM)

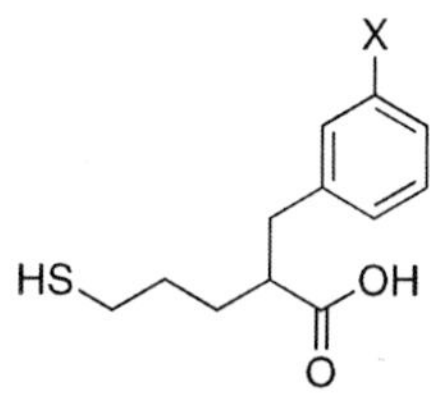

36.64 X = H (1400 nM)
36.65 X = CO_2CH_3 (2700 nM)
36.66 X = $CONH_2$ (2200 nM)
36.67 X = CN (1800 nM)

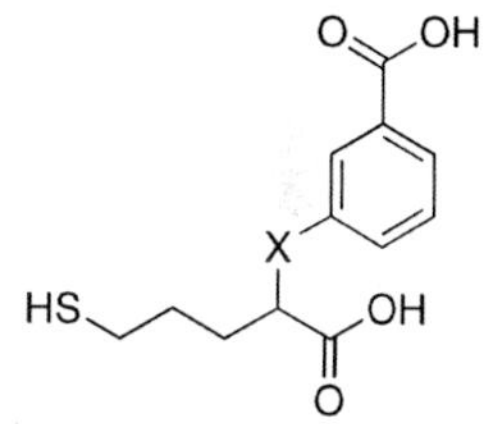

36.68 X = O (14 nM)
36.69 X = S (32 nM)

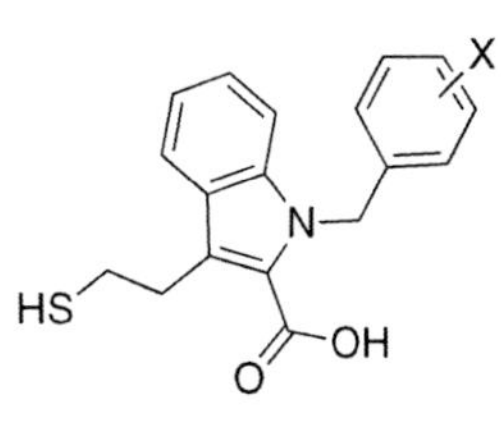

36.70 X = *o*-COOH (18.5 nM)
36.71 X = *m*-COOH (55.3 nM)
36.72 X = *p*-COOH (66.5 nM)
36.73 X = H (12100 nM)

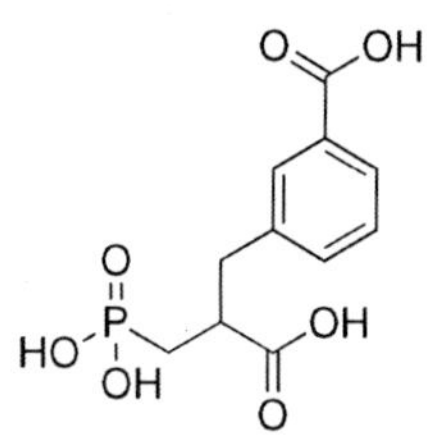

36.74 (120 nM)

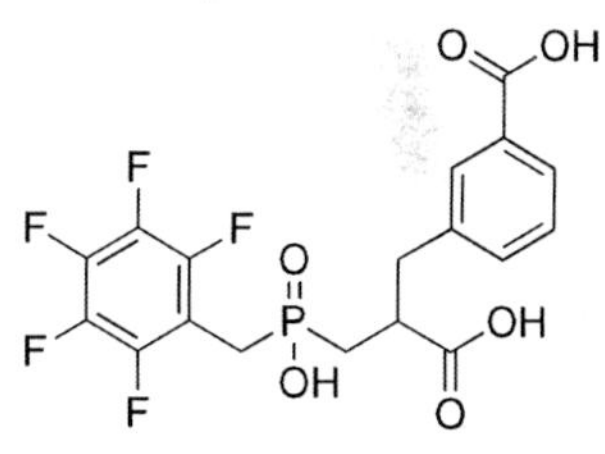

36.75 (2400 nM)

FIGURE 36.8 Thiol-based GCPII inhibitors.

2700 nM, Fig. 36.8), carboxyamide (**36.66**, 2200 nM, Fig. 36.8), and nitrile (**36.67**, 1800 nM, Fig. 36.8) resulted in decreased or lost activity. However, substitution of the benzylic carbon with oxygen (**36.68**, 14 nM, Fig. 36.8) or sulfur (**36.69**, 32 nM, Fig. 36.8) maintained *in vitro* GCPII inhibitory potency. In the *in vivo* rat chronic constriction injury model of neuropathic pain, two potent compounds, **36.58** and **36.68**, showed reduced hyperalgesia compared to the control group on days 8 and 12, respectively. Another class of thiol-based GCPII inhibitors (**36.70–36.72**, Fig. 36.8), which contain an indole ring, is worth noting although there have been no reported *in vivo* studies.[62] The position of the carboxylic acid group in the phenyl

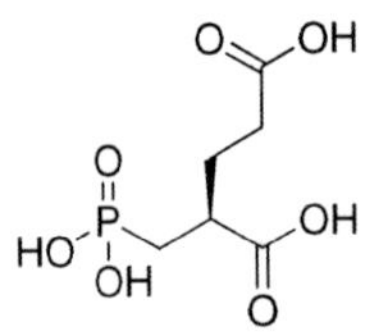

36.76 (*S*, 0.1 nM)

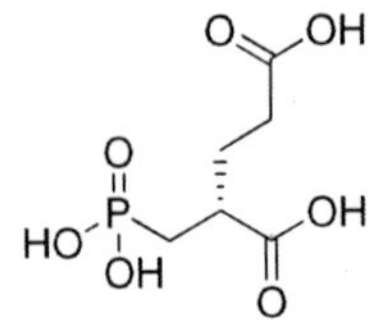

36.77 (*R*, 30 nM)

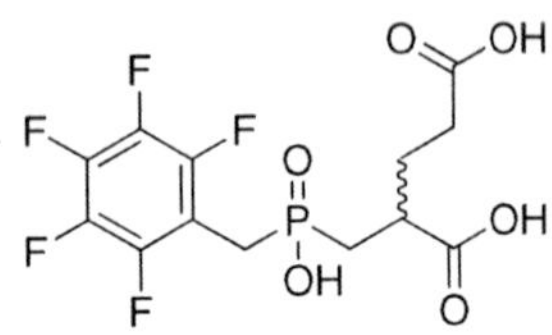

36.78 (82 nM)

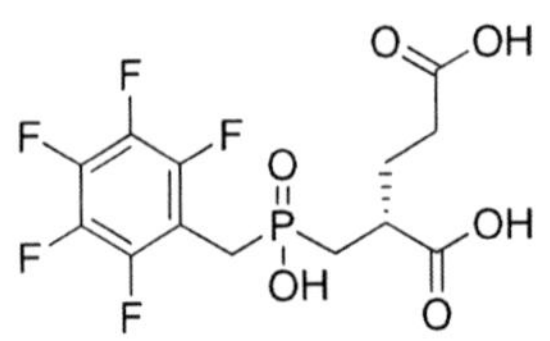

36.79 (*R*, 1400 nM)

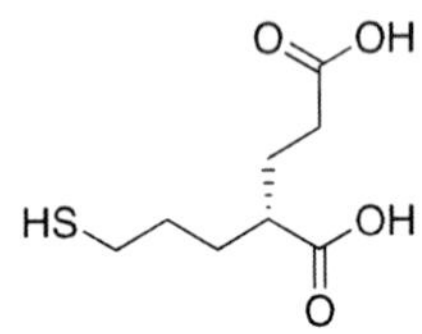

36.80 (*S*, 67 nM)

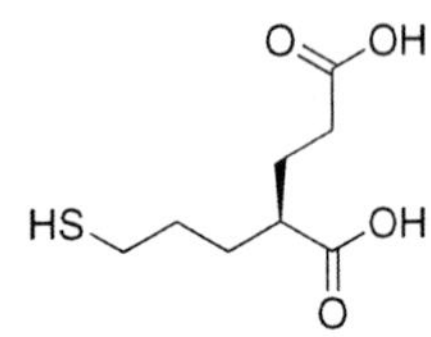

36.81(*R*, 85 nM)

FIGURE 36.9 Effect of stereochemistry on GCPII activity.

ring did not show any significant difference regarding *in vitro* potency. However, *o*-carboxybenzyl indole **36.70** (18.5 nM) was more potent than the *m*- (**36.71**, 55.3 nM) and *p*-carboxybenzyl (**36.72**, 66.5 nM) counterparts.[62] However, removal of the carboxylic acid moiety from the phenyl ring (**36.73**) abolished GCPII inhibitory activity. Introduction of a carboxybenzyl group at the P1′ site in the phosphonate-based series (**36.74** and **36.75**) significantly reduced GCPII inhibition, indicating that the carboxybenzyl group is a unique bioisostere of glutamate in the thiol-based GCPII inhibitor series and that its binding mode at the GCPII active site might differ from that of the phosphite-based compounds.

Studies of an enantiospecific effect of potent thiol- and phosphite-based inhibitors on GCPII inhibition also supported the hypothesis of a unique binding mode for thiol-based analogs. (*S*)-2-PMPA (**36.76**, 0.1 nM, Fig. 36.9), which has the absolute stereochemistry found in natural L-(*S*)-glutamate, is 300-fold more potent than (*R*)-2-PMPA (**36.71**, 30 nM, Fig. 36.9).[63] Only (*S*)-2-PMPA was found in the GCPII/2-PMPA crystal structure despite the use of racemic 2-PMPA, confirming that (*S*)-isomer is more active than the (*R*)-isomer.[20] Another phosphite analog, **36.78** (Fig. 36.9), also showed the same trend (34 nM for (*S*)-isomer versus 1400 nM for (*R*)-isomer **36.79**). However, the thiol-based GCPII inhibitor 2-MPPA (**36.52**, Fig. 36.8) did not show any preference for absolute stereochemistry between two isomers (**36.80** and **36.81**, Fig. 36.9).[63]

36.3.3 Urea-Based GCPII Inhibitors

The first urea-based GCPII inhibitor **36.83** (45 nM, Fig. 36.10) was derived from **36.82** (21 nM, Figure 36.10), which is a mGlu3 agonist as well as a GCPII inhibitor. Inhibitor **36.83** was synthesized by Kozikowski and coworkers where the $CH_2P(O)(OH)CH_2$ unit was replaced with a urea moiety.[64] Although **36.83** was twofold less active than

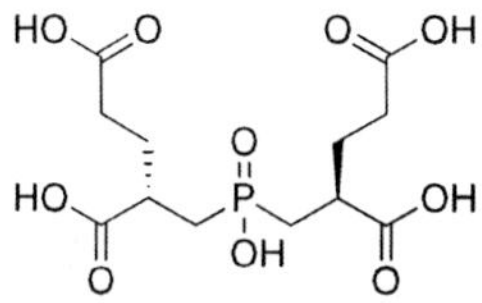

36.82 (21 nM)

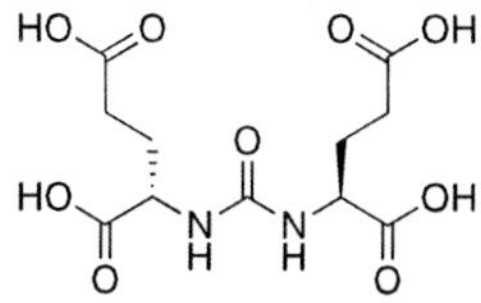

36.83 (47 nM)

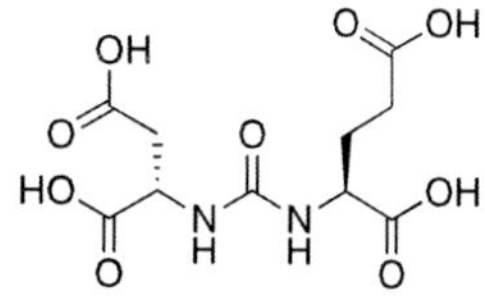

36.84 (46 nM)

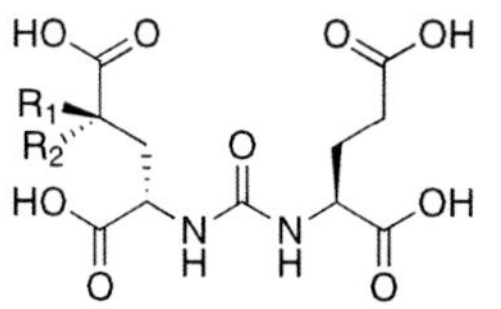

36.85 R_1 = Bn, R_2 = H (95 nM)
36.86 R_1 = H, R_2 = CH_3 (14 nM)

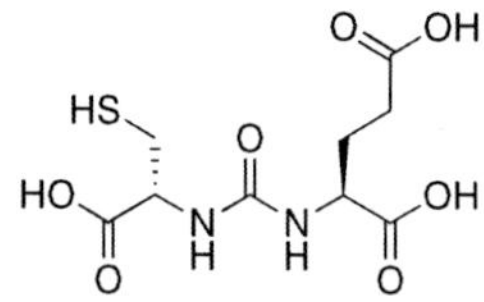

36.87 (6.9 nM)

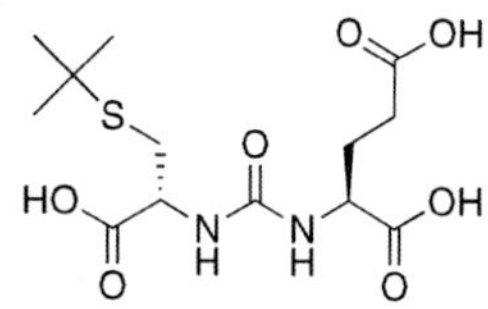

36.88 (29 nM)

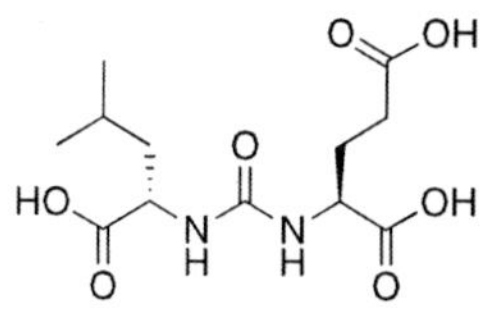

36.89 (ZJ-43, 0.8 nM)

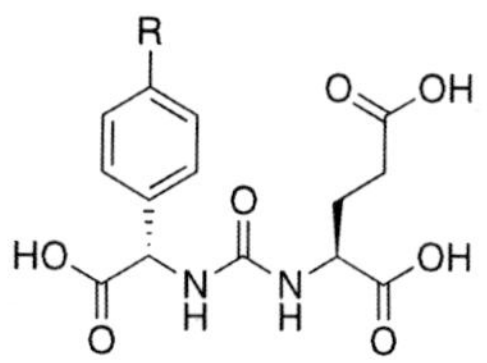

36.90 R = H (2.1 nM)
36.91 R = OH (12 nM)

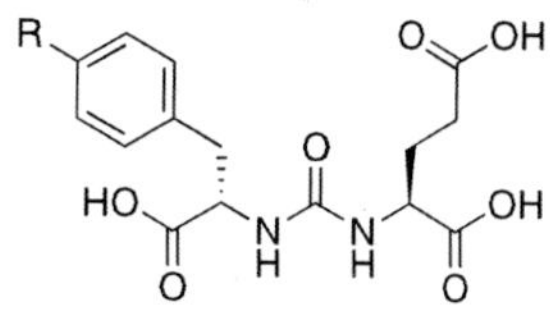

36.92 R = H (5.9 nM)
36.93 R = OH (3 nM)

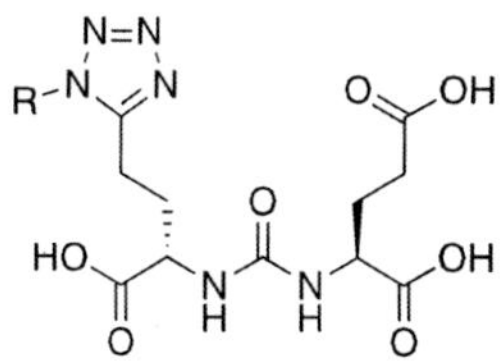

36.94 R = H (0.9 nM)
36.95 R = CH_2CH_2CN (5.3 nM)

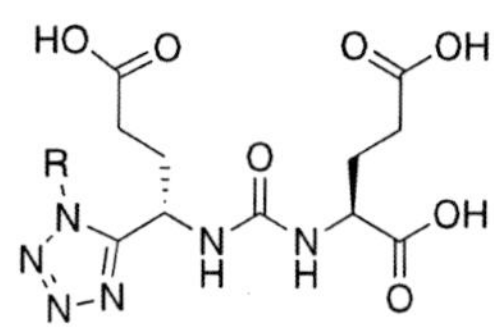

36.96 R = CH_2CH_2CN (2711 nM)
36.97 R = H (335 nM)

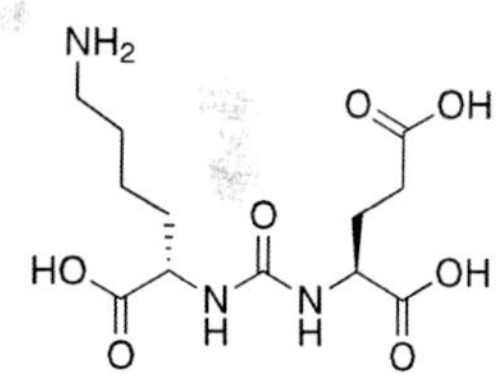

36.98 R = H (450 nM)

FIGURE 36.10 Urea-based GCPII inhibitors.

36.82, it was potent enough to be used as a lead compound of the urea-based GCPII inhibitors. Modifications of the P1 site were well tolerated and did not decrease the GCPII potency. Compound **36.84** (Fig. 36.10), which contains (*S*)-aspartate instead of (*S*)-glutamate, was comparable in inhibitory capacity to **36.83**. However, the stereochemistry of glutamate in the P1 site proved critical and the corresponding (*R*)-Glu-C(O)-(*S*)-Glu showed only 25% inhibition at 1 μM concentration.[64] Methyl-branched glutamate (**36.86**, 14 nM, Fig. 36.10) is more potent than benzyl-branched analog (**36.85**, 95 nM, Fig. 36.10).[31] Introduction of amino acids that are more lipophilic than glutamate, including cysteine (**36.87** and **36.88**), leucine (**36.89**), phenylglycine (**36.90** and **36.91**), and tyrosine (**36.92** and **36.93**), increased GCPII potency to low nM IC_{50} concentrations (Fig. 36.10).[31] Replacement of the γ-carboxylate in the P1 site with tetrazole (**36.94** and **36.95**), a carboxylate bioisostere, increased GCPII inhibition

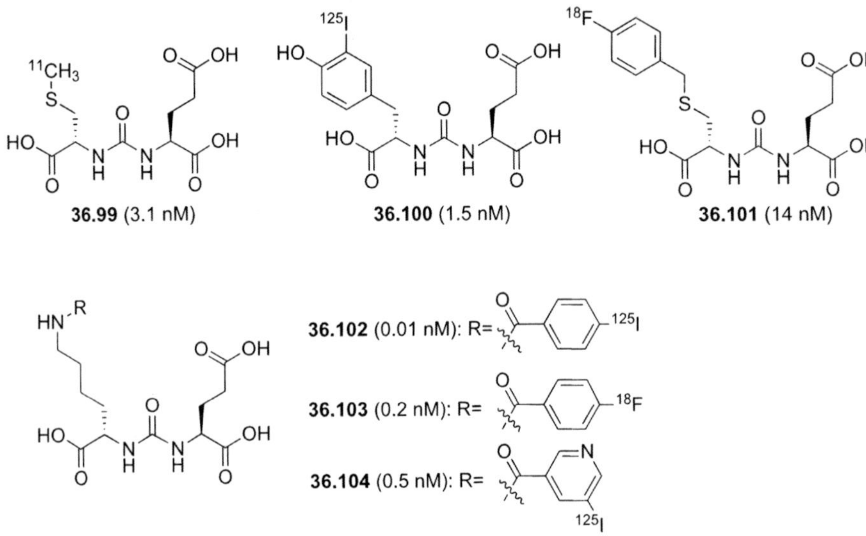

FIGURE 36.11 Urea-based PET and SPECT imaging agents for prostate cancer.

activities while replacement of the α-carboxylate (**36.96** and **36.97**) decreased activity by 75-fold (Fig. 36.10).[31] This result indicated that the α-carboxylate binding site is sensitive to steric variation. Recent GCPII crystal structures with urea analogs confirmed that the α-carboxylate at the P1 site is projected into the narrow space of the S1 site and that it makes strong electrostatic and hydrogen bonding interactions with Arg534, Arg536, and Asn519.[30]

[^{11}C]-labeled **36.99** (3.1 nM, Fig. 36.11) and ^{125}I-labeled **36.100** (1.5 nM, Fig. 36.11) were prepared from **36.87** and **36.93** in one step as imaging agents for prostate cancer. *In vivo* imaging and biodistribution studies showed preferential uptake of **36.99** and **36.100** in PSMA-positive LNCaP-derived tumors compared to PSMA-negative MCF-7- and PC-3-derived tumors.[65] Thirty minutes after intravenous injection, **36.99** and **36.100** showed high tumor to muscle ratios of 10.8 and 4.7, respectively. Recently, ^{18}F-labeled urea-based GCPII inhibitor **36.101** (14 nM, Fig. 36.11) was reported as an imaging probe for prostate cancer using positron emission tomography (PET) by Pomper and coworkers.[66] Biodistribution studies of **36.101** with PC3-PSMA expressing/PC3-control tumor-bearing mice showed high uptake of **36.101** in the PSMA+ tumors and achieved a tumor/muscle ratio of 20:1 at 2 h after intravenous injection.[66] The other urea-based GCPII inhibitors (**36.102–36.104**, Fig. 36.11), which mask the terminal ε-amine of Lys at the P1 site with substituted benzoyl or pyridyl groups, exhibited over 1000-fold stronger GCPII inhibition than the free lysine analog (**36.98**, 450 nM) and also demonstrated high PSMA-specific uptake in PSMA+ tumors.[32] The X-ray crystal structure of GCPII in complex with **36.102** (Fig. 36.12) explained its high GCPII binding affinity. The iodophenyl ring of **36.102** is fully inserted into the subpocket, which is generated by three arginine residues (Arg463, Arg534, and Arg536), thus enhancing the tight

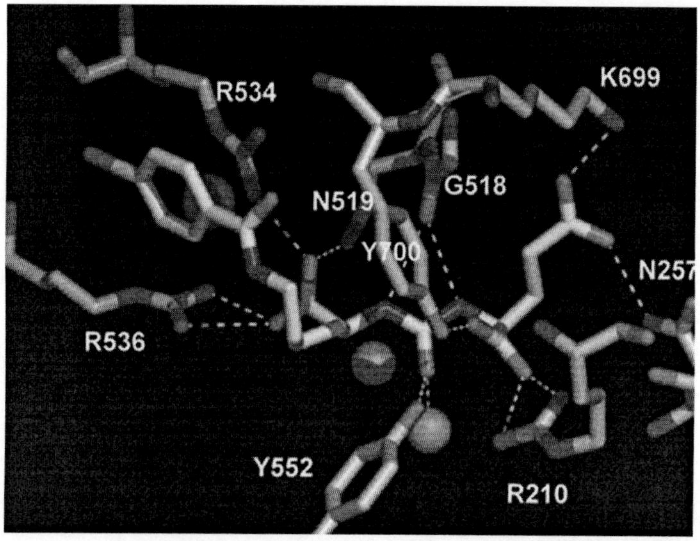

FIGURE 36.12 Crystal structure of GCPII with **36.102**. (See the color version of this figure in Color Plates section.)

binding of **36.102** with GCPII.[30] Compound **36.103** (0.2 nM), also a positron-emitter, provided site-selective tumor uptake at 1 h after injection of the radiopharmaceutical (Fig. 36.13). Notably, compounds of this class labeled with ^{123}I are being used in an initial human imaging trial, sponsored by Molecular Insight Pharmaceuticals, Inc. (John Babich, Cambridge, MA, personal communication).

Kozikowski's group reported a GCPII inhibitor–doxorubicin conjugate (**36.105**, Fig. 36.14) for potential prostate cancer therapy.[67] On the basis of the SAR studies of urea-based GCPII inhibitors, they introduced the bulky doxorubicin at the P1 site using a linker, which occupies the S1 site and tunnel region of GCPII. Although **36.105** demonstrated high GCPII binding affinity (41 nM), it showed poor antitumor activity. Recently, Pomper's group reported urea-based SPECT imaging agents (**36.106–36.108**, Fig. 36.14). In those compounds the bulky $^{99m}Tc/Re$ chelating groups are connected to the lysine-glutamate urea with a linking group.[68] Technetium-99m-labeled **36.106** (10.8 nM for Re) showed 7.9% injected dose per gram in PSMA+ tumors at 30 min after injection and a PSMA-selective ratio of 44:1 at 2 h postinjection when comparing PSMA+ PC-3 to PSMA− PC-3 control tumors in the same animal.

36.3.4 GCPII Inhibitors with Glutamate Bioisostere

Among glutamate bioisosteres, quisqualic acid (**36.109**, 9.5 μM) and willaridine (**36.110**, 67 μM) showed weak inhibitory activity of GCPII (Fig. 36.15).[20] However, they were stronger than glutamate (428 μM) itself.[20] On the other hand, α-amino-3-hydroxyl-5-methyl-4-isoxazole-propionate (**36.111**, >10 mM, Fig. 36.15) did not

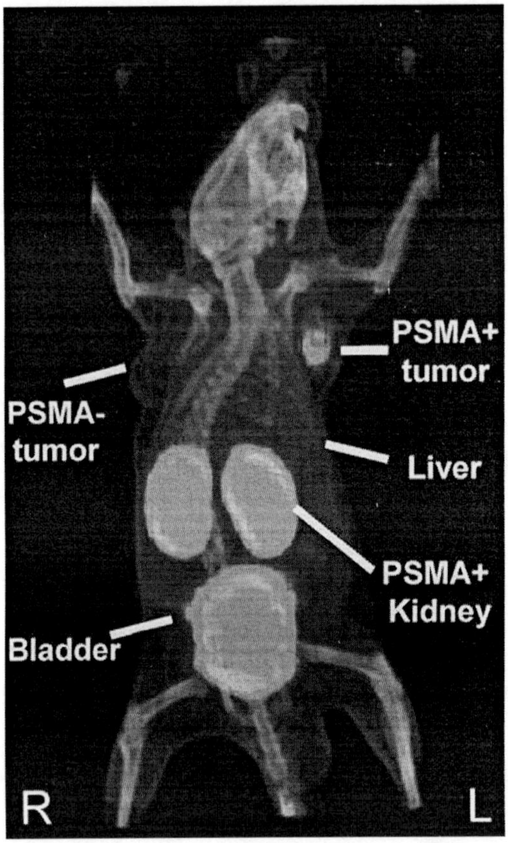

FIGURE 36.13 Sixty minute image of **36.103** demonstrating PSMA+ tumor target specificity.

show any GCPII activity up to mM concentrations. (L)-Serine-*O*-sulfate (**36.112**, 438 μM) and (L)-serine-*O*-phosphate (**36.113**, 523 μM) were comparable to glutamate in inhibitory capacity (Fig. 36.15). Because their inhibitory activities were better than glutamate itself, quisqualic acid and willaridine could be used as bioisosteres of the glutamate moiety of potential GCPII inhibitors. The GCPII crystal structure with **36.109** showed that the oxazolidine ring occupies the S1′ site in a pattern similar to the γ-carboxylate of glutamate.[20] The pK_a value of the NH in the oxazolidine ring (pK_a = 4.2) is very close to the γ-carboxylic acid (pK_a = 4.4). Conformationally restricted glutamate analogs **36.114** and **36.115** (Fig. 36.15) were more potent than unrestricted NAAG. *Cis*-isomer **36.115** (100 nM) proved nine-fold more active than the *trans*-isomer **36.114** (900 nM).[69] The tetrazole (**36.116**, 14.9 nM, Fig. 36.15), which contains an acidic hydrogen, can act as a surrogate for the γ-carboxylate of glutamate in a series of urea-based inhibitors, while NH-masked tetrazole **36.117** (Fig. 36.15) abolished GCPII inhibitory activity.[31] Although the hydrophobic channel generated by Phe209 and Leu428 in the S1′ site indicated that this region may be

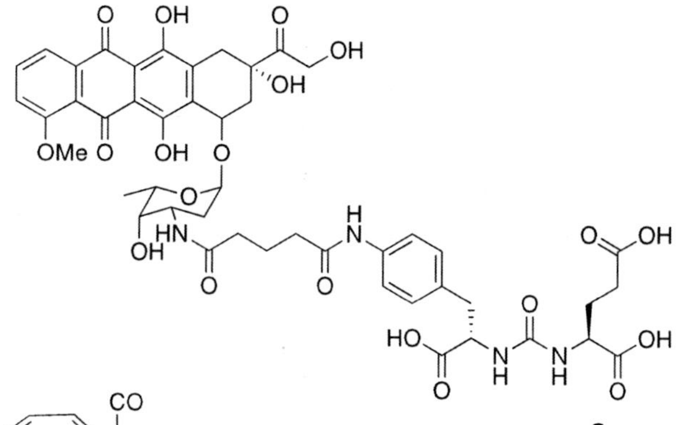

36.105 (41 nM)

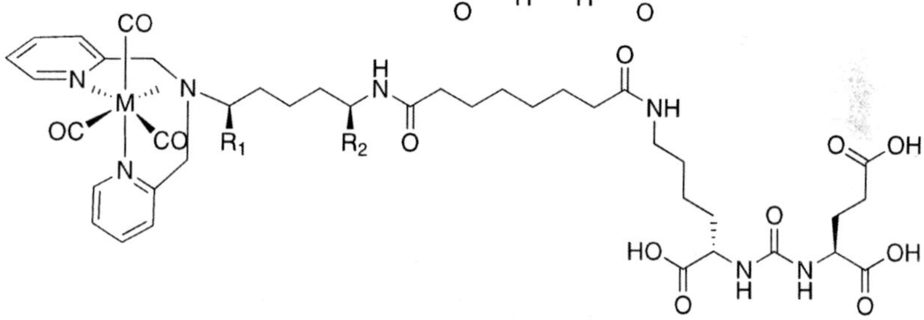

Compound	Chelates	R_1	R_2	Metal	K_I (nM)	Labeling
36.106	2-Pyridine	H	CO_2H	-/Re	15.3/10.8	^{99m}Tc
36.107	2-Pyridine	CO_2H	H	-/Re	1.1/10.3	^{99m}Tc
36.108	2-Quinoline	H	CO_2H	-/Re	0.2/0.5	^{99m}Tc

FIGURE 36.14 Urea-based GCPII inhibitors with linkers to interact with tunnel region.

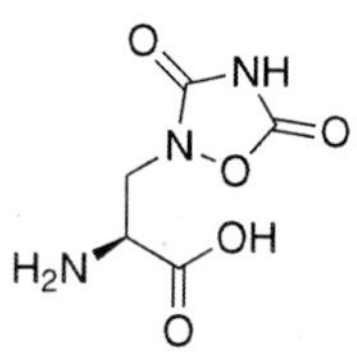

36.109 (9.5 μM)

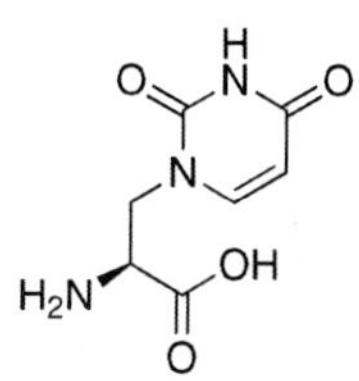

36.110 (67 μM)

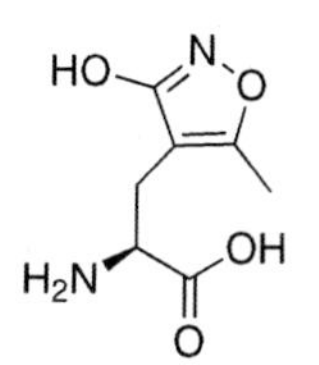

36.111 (>10 mM)

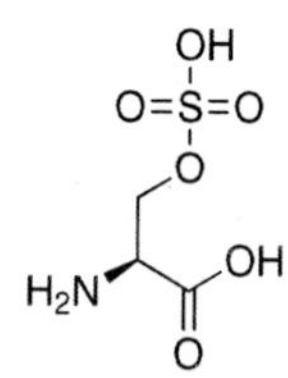

36.112 (438 μM)

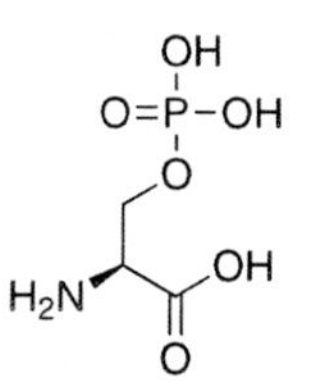

36.113 (523 μM)

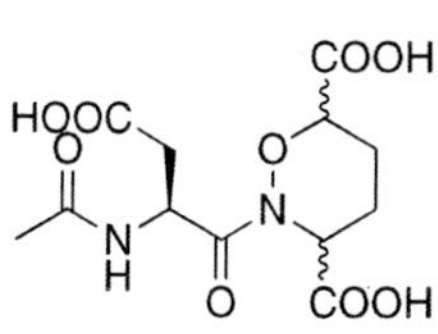

36.114 (trans, 900 nM)
36.115 (cis, 100 nM)

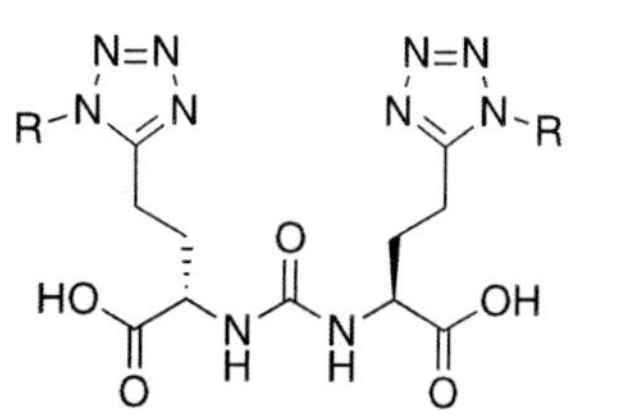

36.116 R = H (14.9 nM)
36.117 R = CH_2CH_2CN (4434 nM)

FIGURE 36.15 Glutamate bioisostere analogs.

amenable toward structural modification of the glutamate moiety, to date there has been limited success in modification of the P1′ site of GCPII inhibitors.

36.4 GCPII-BASED SUBSTRATES

Little progress has been made in developing therapeutic or diagnostic GCPII-based substrates when compared to GCPII-based inhibitors. Dipeptides **36.118–36.122** (Fig. 36.16) have been kinetically characterized using an assay based on high-performance liquid chromatography and recombinant human GCPII.[18] The natural GCPII substrate, NAAG, showed the best k_{cat}/K_m ($10^3\ s^{-1}\ M^{-1}$) among the tested dipeptides. (D)-Enantiomers of α-linked dipeptides **36.118** and **36.120** in the P1 site were more than 10-fold less effective than the L-enantiomers, NAAG itself, and **36.119** (Fig. 36.16). However, γ-linked Glu–Glu (**36.121** and **36.122**) did not show any difference between two diastereoisomers. Berkman and coworkers reported chromophore-containing analogs **36.123–36.126** (Fig. 36.16), which contain γ-linked (L)-Glu–(L)-Glu at the C-terminus. 4-Phenylazobenzyl (4-Pab)-linked dipeptide **36.123** ($K_m = 415$ nM, $V_{max} = 237$ nM/min) was cleaved by GCPII more efficiently than the

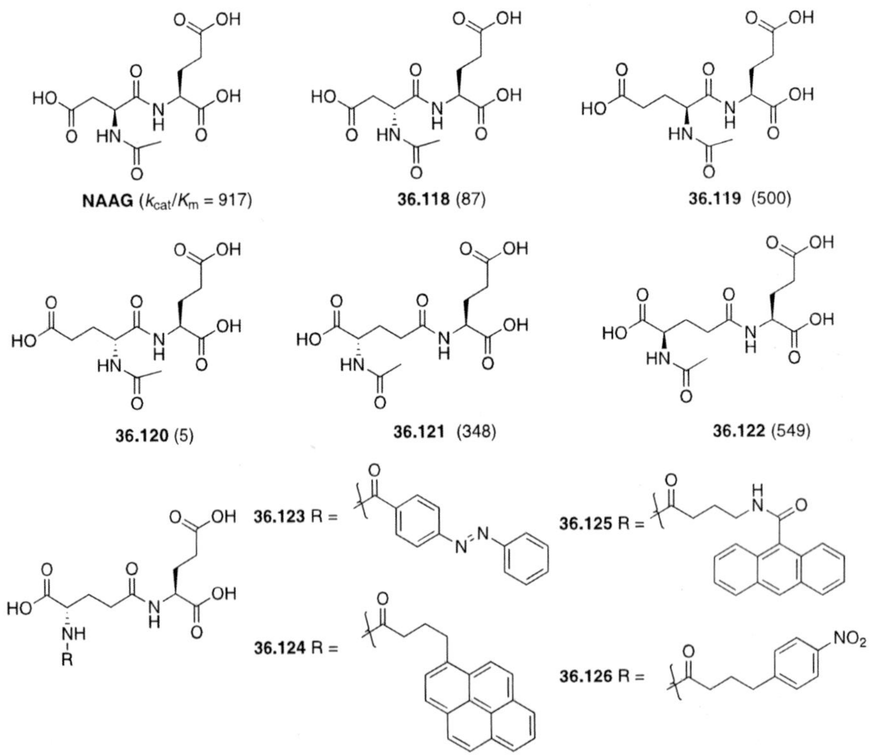

FIGURE 36.16 GCPII-based substrates.

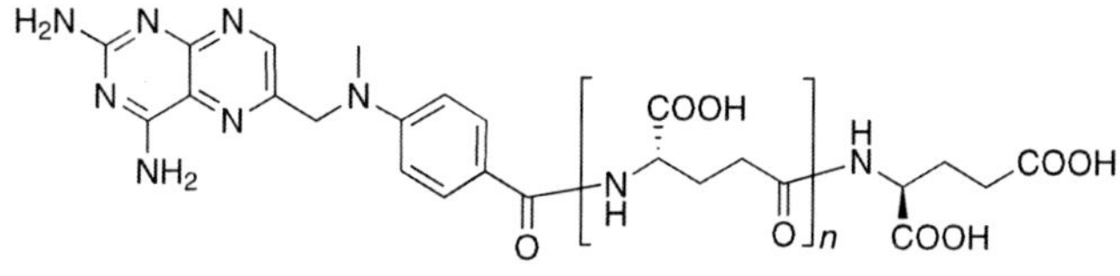

36.127

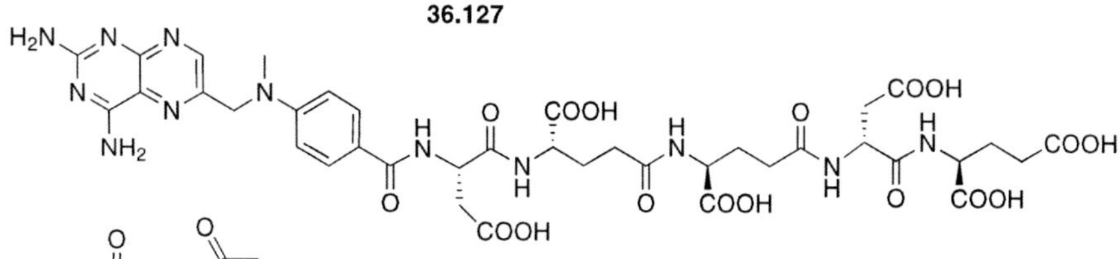

36.128

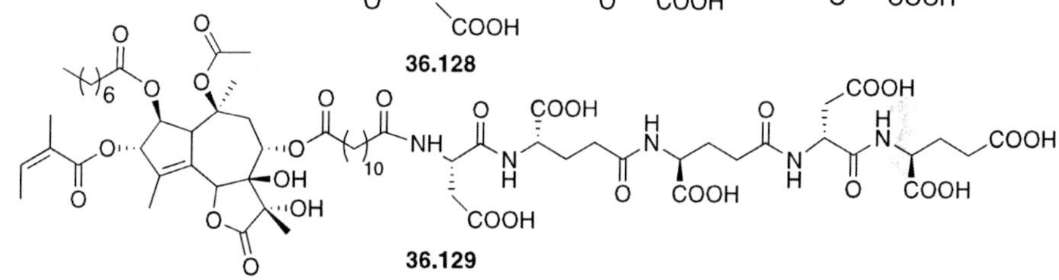

36.129

FIGURE 36.17 GCPII substrates conjugated with anticancer agents.

pyrene-linked (**36.124**, $V_{max} = 115$ nM/min), anthracene-linked (**36.125**, $V_{max} = 171$ nM/min), or 4-nitrophenyl-linked dipeptides (**36.126**, $V_{max} = 90$ nM/min).[70] Stereochemical inversion of (L)-Glu into (D)-Glu ($K_m = 643$ nM, $V_{max} = 80$ nM/min) at the P1 site of **36.123** maintained kinetic properties expected of a viable substrate, while **36.123** conjugated with the α-linked Glu–Glu and α- or β-linked Asp–Glu showed significantly reduced GCPII substrate characteristics.

Denmeade and coworkers screened a number of polypeptide conjugates composed of Glu- and Asp-linked peptides conjugated to the APA (4-*N*[*N*-2,4-diamino-6-pteridinyl-methyl]-*N*-methylaminobenzoate) portion of methotrexate to identify GCPII-selective substrates that might serve as therapeutic agents against prostate cancer.[71] Among them, **36.127** (Fig. 36.17) was most efficiently hydrolyzed by GCPII, but it was unstable in plasma. Another analog containing α-linked or γ-linked Asp and Glu (**36.128**, Fig. 36.17) was stable in plasma and showed more toxicity against TSU prostate tumor cells in the presence of GCPII than TSU cells without GCPII. They also reported that a thapsigargin-based pentapeptide (**36.129**, Fig. 36.17) was stable to mouse and human serum protease and was cleaved by GCPII to produce the shorter peptide thapsigargin conjugates.[72]

36.5 MACROMOLECULAR PSMA LIGANDS

The initial studies characterizing PSMA as a target for prostate tumor cells stemmed from macromolecules rather than small molecule ligands or substrates. The antibody that identified PSMA, 7E11-C5, was radiolabeled and clinically translated into the prostate imaging agent ^{111}In-capromab pendetide. Capromab pendetide, also known as ProstaScint™ and CYT-356, was approved by the U.S. Food and Drug Administration in 1996 for imaging of prostate cancer in patients at high risk for local lymph node metastasis. Studies found that ProstaScint™ could identify soft tissue metastasis in about two-thirds of presurgical patients,[73] however, it was not as successful in imaging bone metastasis. The intracellular epitope of 7E11-C5 may limit the ability of this antibody to identify these and other lesions. Because antibodies are generally too large to diffuse passively through the cell membrane, ProstaScint™ most likely binds to nonviable prostate cells or cell fragments, although it may also be actively internalized.

With the initial, moderate success of ProstaScint™, it was hypothesized that ligands targeted toward the extracellular portion of PSMA would have even greater success in identifying tumors *in vivo*. In 1997, Liu and colleagues generated several antibodies to the extracellular portion of PSMA and proved that they, unlike 7E11-C5, could indeed bind viable cells.[13] The most successful antibody from those studies, J591, has been translated into several clinical trials as a therapeutic targeting agent.[74] In radiotherapy trials, [^{177}Lu]J591 also showed promise as an imaging agent where it was capable of identifying some bony metastasis that were not visible by standard bone scans.[75] Preclinical and clinical studies are still ongoing with this humanized antibody.

Antibodies have the advantage of high affinity and avidity, however, their large size and long serum half-life can limit their efficacy for imaging. Therefore, several groups developed smaller macromolecules with superior pharmacokinetic properties to target PSMA. An affinity selection technique, known as biopanning, was applied to identify PSMA binding molecules from high-diversity random-sequence peptide libraries. Two separate groups biopanned phage display peptide libraries against purified extracellular PSMA to identify prostate cancer targeting peptides.[76,77] Interestingly, the peptides identified by Aggarwal and colleagues were competitive inhibitors of GCPII where the peptides identified by Lupold and Rodriquez enhanced enzymatic activity in a dose-dependent fashion. It is, therefore, likely that the R5-XC1 peptide, from the latter study, binds and stabilizes either the PSMA active site or PSMA homodimers. Nevertheless, follow-up studies from an independent group at Harvard confirmed the ability of R5-XC1 to target PSMA expressing cells selectively and revealed that the peptide could be used to deliver a lytic cationic amphipathic peptide.[78]

In 2002, a similar biopanning strategy was applied to identify nuclease-stabilized RNA aptamers that bind to the cell surface PSMA.[79] RNA aptamers are short linear RNAs that fold into unique three-dimensional structures capable of binding target sites in a manner comparable to antibodies.[80] Further, aptamers can be stabilized against nuclease degradation by replacing the 2′-hydroxyl of pyrimidine residues with 2′-fluoro or 2′-amino groups.[81] RNA aptamers are becoming an attractive new class of therapeutics due to their thermostability, large-scale chemical production, ability for chemical modification, short serum half-lives, lack of immunogenicity, and high affinities. The anti-PSMA aptamers, xPSM-A9 and xPSM-A10, bind two separate extracellular epitopes with nanomolar affinity and are the first reported aptamers targeted to a cell surface cancer antigen.[79] The xPSM-A10 aptamer was truncated to the minimal functional unit of 56 nucleotides, or approximately 18.5 kDa, and termed xPSM-A10-3.[79] Since the original publication, several groups have used the reported sequences to synthesize and apply these aptamers as targeting agents. The broadest application of anti-PSMA aptamers has been in the targeting of nanoparticles.[14,82–92] The aptamers have two unique ends (5′ or 3′), which can be modified with functional groups for nanoparticle conjugation. The restriction to a single functional group overcomes many of the difficulties associated with peptide or antibody targeting strategies, where multiple reactive moieties can cross-link to generate complex, unsolicited products. In addition to nanoparticle targeting, anti-PSMA aptamers have been used to target small interfering RNAs (siRNAs) specifically to PSMA expressing cells and tumors.[93–95] At present, this is the only reported means to knock down gene expression selectively in prostate cancer cells through RNA interference.

36.6 CONCLUSIONS

Three classes of GCPII ligands have been successfully identified in the past decade—low molecular weight inhibitors, substrates, and macromolecular agents. Phosphonate- and urea-based low MW inhibitors have been crystallized with the ectodomain of GCPII. These inhibitors have the potential to be used as (1) therapeutic agents for the treatment of glutamate-associated neurological diseases, (2) diagnostic agents for prostate cancer, and (3) radiotherapeutic agents for solid tumors (due to PSMA in the neovasculature) when labeled with α- or β-emitting radionuclides. Little progress has been made in the area of GCPII-based substrates, requiring further investigation. Larger MW targeting ligands, including peptides, antibodies, and aptamers, have been applied to target a variety of radionuclides, drugs, small interfering RNAs, and nanoparticles to prostate cancer cells and tumors. Although several compounds targeting GCPII have successfully entered human imaging trials, further optimization of current lead compounds and the discovery of new leads are warranted to provide agents with greater specificity and improved pharmacokinetics for use *in vivo*.

REFERENCES

1. Robinson, M. B.; Blakely, R. D.; Couto, R.; Coyle, J. T. Hydrolysis of the brain dipeptide *N*-acetyl-L-aspartyl-L-glutamate. Identification and characterization of a novel *N*-acetylated alpha-linked acidic dipeptidase activity from rat brain. *J. Biol. Chem.* **1987**, *262*, 14498–14506.
2. Olszewski, R. T.; Bukhari, N.; Zhou, J.; Kozikowski, A. P.; Wroblewski, J. T.; Shamimi-Noori, S.; Wroblewska, B.; Bzdega, T.; Vicini, S.; Barton, F. B.; Neale, J. H. NAAG peptidase inhibition reduces locomotor activity and some stereotypes in the PCP model of schizophrenia via group II mGluR. *J. Neurochem.* **2004**, *89*, 876–885.
3. Bacich, D. J.; Wozniak, K. M.; Lu, X. C. M.; O'Keefe, D. S.; Callizot, N.; Heston, W. D. W.; Slusher, B. S. Mice lacking glutamate carboxypeptidase II are protected from peripheral neuropathy and ischemic brain injury. *J. Neurochem.* **2005**, *95*, 314–323.

4. Neale, J. H.; Olszewski, R. T.; Gehl, L. M.; Wroblewska, B.; Bzdega, T. The neurotransmitter *N*-acetylaspartylglutamate in models of pain, ALS, diabetic neuropathy, CNS injury and schizophrenia. *Trends Pharmacol. Sci.* **2005**, *26*, 477–484.

5. Ghadge, G. D.; Slusher, B. S.; Bodner, A.; Dal Canto, M.; Wozniak, K.; Thomas, A. G.; Rojas, C.; Tsukamoto, T.; Majer, P.; Miller, R. J.; Monti, A. L.; Roos, R. P. Glutamate carboxypeptidase II inhibition protects motor neurons from death in familial amyotrophic lateral sclerosis models. *Proc. Natl. Acad. Sci. USA* **2003**, *100*, 9554–9559.

6. Slusher, B. S.; Vornov, J.; Thomas, A. G.; Hurn, P. D.; Harukuni, I.; Bhardwaj, A.; Traystman, R. J.; Robinson, M. B.; Britton, P.; Lu, X. C. M.; Tortella, F. C.; Wozniak, S. K. M.; Yudkoff, M.; Potier, B. M.; Jackson, P. F. Selective inhibition of NAALADase, which converts NAAG to glutamate, reduces ischemic brain injury. *Nat. Med.* **1999**, *5*, 1396–1402.

7. Thomas, A. G.; Liu, W.; Olkowski, J. L.; Tang, Z.; Lin, Q.; Lu, X.-C. M.; Slusher, B. S. Neuroprotection mediated by glutamate carboxypeptidase II (NAALADase) inhibition requires TGF-beta. *Eur. J. Pharmacol.* **2001**, *430*, 33–40.

8. Bostwick, D. G.; Pacelli, A.; Blute, M.; Roche, P.; Murphy, G. P. Prostate specific membrane antigen expression in prostatic intraepithelial neoplasia and adenocarcinoma: a study of 184 cases. *Cancer* **1998**, *82*, 2256–2261.

9. Israeli, R. S.; Powell, C. T.; Fair, W. R.; Heston, W. D. Molecular cloning of a complementary DNA encoding a prostate-specific membrane antigen. *Cancer Res.* **1993**, *53*, 227–230.

10. Pinto, J. T.; Suffoletto, B. P.; Berzin, T. M.; Qiao, C. H.; Lin, S.; Tong, W. P.; May, F.; Mukherjee, B.; Heston, W. D. Prostate-specific membrane antigen: a novel folate hydrolase in human prostatic carcinoma cells. *Clin. Cancer Res.* **1996**, *2*, 1445–1451.

11. Yao, V.; Bacich, D. J. Prostate specific membrane antigen (PSMA) expression gives prostate cancer cells a growth advantage in a physiologically relevant folate environment *in vitro*. *Prostate* **2006**, *66*, 867–875.

12. Yao, V.; Parwani, A.; Maier, C.; Heston, W. D.; Bacich, D. J. Moderate expression of prostate-specific membrane antigen, a tissue differentiation antigen and folate hydrolase, facilitates prostate carcinogenesis. *Cancer Res.* **2008**, *68*, 9070–9077.

13. Liu, H.; Moy, P.; Kim, S.; Xia, Y.; Rajasekaran, A.; Navarro, V.; Knudsen, B.; Bander, N. H. Monoclonal antibodies to the extracellular domain of prostate-specific membrane antigen also react with tumor vascular endothelium. *Cancer Res.* **1997**, *57*, 3629–3634.

14. Farokhzad, O. C.; Cheng, J.; Teply, B. A.; Sherifi, I.; Jon, S.; Kantoff, P. W.; Richie, J. P.; Langer, R. Targeted nanoparticle-aptamer bioconjugates for cancer chemotherapy *in vivo*. *Proc. Natl. Acad. Sci. USA* **2006**, *103*, 6315–6320.

15. Liu, H.; Rajasekaran, A. K.; Moy, P.; Xia, Y.; Kim, S.; Navarro, V.; Rahmati, R.; Bander, N. H. Constitutive and antibody-induced internalization of prostate-specific membrane antigen. *Cancer Res.* **1998**, *58*, 4055–4060.

16. Liu, T.; Wu, L. Y.; Kazak, M.; Berkman, C. E. Cell-surface labeling and internalization by a fluorescent inhibitor of prostate-specific membrane antigen. *Prostate* **2008**, *68*, 955–964.

17. Rajasekaran, S. A.; Anilkumar, G.; Oshima, E.; Bowie, J. U.; Liu, H.; Heston, W.; Bander, N. H.; Rajasekaran, A. K. A novel cytoplasmic tail MXXXL motif mediates the internalization of prostate-specific membrane antigen. *Mol. Biol. Cell.* **2003**, *14*, 4835–4845.

18. Barinka, C.; Hlouchova, K.; Rovenska, M.; Majer, P.; Dauter, M.; Hin, N.; Ko, Y.-S.; Tsukamoto, T.; Slusher, B. S.; Konvalinka, J.; Lubkowski, J. Structural basis of interactions

between human glutamate carboxypeptidase II and its substrate analogs. *J. Mol. Biol.* **2008**, *376*, 1438–1450.

19. Barinka, C.; Starkova, J.; Konvalinka, J.; Lubkowski, J. A high-resolution structure of ligand-free human glutamate carboxypeptidase II. *Acta Crystallogr. Sect. F: Struct. Biol. Cryst. Commun.* **2007**, *F63*, 150–153.
20. Barinka, C.; Rovenska, M.; Mlcochova, P.; Hlouchova, K.; Plechanovova, A.; Majer, P.; Tsukamoto, T.; Slusher, B. S.; Konvalinka, J.; Lubkowski, J. Structural insight into the pharmacophore pocket of human glutamate carboxypeptidase II. *J. Med. Chem.* **2007**, *50*, 3267–3273.
21. Mesters, J. R.; Barinka, C.; Li, W.; Tsukamoto, T.; Majer, P.; Slusher, B. S.; Konvalinka, J.; Hilgenfeld, R. Structure of glutamate carboxypeptidase II, a drug target in neuronal damage and prostate cancer. *EMBO J.* **2006**, *25*, 1375–1384.
22. Davis, M. I.; Bennett, M. J.; Thomas, L. M.; Bjorkman, P. J. Crystal structure of prostate-specific membrane antigen, a tumor marker and peptidase. *Proc. Natl. Acad. Sci. USA* **2005**, *102*, 5981–5986.
23. Mesters, J. R.; Henning, K.; Hilgenfeld, R. Human glutamate carboxypeptidase II inhibition: structures of GCPII in complex with two potent inhibitors, quisqualate and 2-PMPA. *Acta Crystallogr. Sect. D: Biol. Crystallogr.* **2007**, *D63*, 508–513.
24. Lawrence, C. M.; Ray, S.; Babyonyshev, M.; Galluser, R.; Borhani, D. W.; Harrison, S. C. Crystal structure of the ectodomain of human transferrin receptor. *Science* **1999**, *286*, 779–782.
25. Rong, S.-B.; Zhang, J.; Neale Joseph, H.; Wroblewski Jarda, T.; Wang, S.; Kozikowski Alan, P. Molecular modeling of the interactions of glutamate carboxypeptidase II with its potent NAAG-based inhibitors. *J. Med. Chem.* **2002**, *45*, 4140–4152.
26. Mlcochova, P.; Plechanovova, A.; Barinka, C.; Mahadevan, D.; Saldanha, J. W.; Rulisek, L.; Konvalinka, J. Mapping of the active site of glutamate carboxypeptidase II by site-directed mutagenesis. *FEBS J.* **2007**, *274*, 4731–4741.
27. Majer, P.; Hin, B.; Stoermer, D.; Adams, J.; Xu, W.; Duvall, B. R.; Delahanty, G.; Liu, Q.; Stathis, M. J.; Wozniak, K. M.; Slusher, B. S.; Tsukamoto, T. Structural optimization of thiol-based inhibitors of glutamate carboxypeptidase II by modification of the P1′ side chain. *J. Med. Chem.* **2006**, *49*, 2876–2885.
28. Lambert, L. A.; Mitchell, S. L. Molecular evolution of the transferrin receptor/glutamate carboxypeptidase II family. *J. Mol. Evol.* **2007**, *64*, 113–128.
29. Rovenska, M.; Hlouchova, K.; Sacha, P.; Mlcochova, P.; Horak, V.; Zamecnik, J.; Barinka, C.; Konvalinka, J. Tissue expression and enzymologic characterization of human prostate specific membrane antigen and its rat and pig orthologs. *Prostate* **2008**, *68*, 171–182.
30. Barinka, C.; Byun, Y.; Dusich, C. L.; Banerjee, S. R.; Chen, Y.; Castanares, M.; Kozikowski, A. P.; Mease, R. C.; Pomper, M. G.; Lubkowski, J. Interactions between human glutamate carboxypeptidase II and urea-based inhibitors: structural characterization. *J. Med. Chem.* **2008**, *51*, 7737–7743.
31. Kozikowski, A. P.; Zhang, J.; Nan, F.; Petukhov, P. A.; Grajkowska, E.; Wroblewski, J. T.; Yamamoto, T.; Bzdega, T.; Wroblewska, B.; Neale, J. H. Synthesis of urea-based inhibitors as active site probes of glutamate carboxypeptidase II: efficacy as analgesic agents. *J. Med. Chem.* **2004**, *47*, 1729–1738.
32. Chen,Y.; Foss, C. A.; Byun, Y.; Nimmagadda, S.; Pullambhatla, M.; Fox, J. J.; Castanares, M.; Lupold, S. E.; Babich, J. W.; Mease, R. C.; Pomper, M. G. Radiohalogenated

prostate-specific membrane antigen (PSMA)-based ureas as imaging agents for prostate cancer. *J. Med. Chem.* **2008**, *51*, 7933–7943.

33. Cheng,Y.; Prusoff, W. H. Relationship between the inhibition constant (K_1) and the concentration of inhibitor which causes 50 percent inhibition (I50) of an enzymatic reaction. *Biochem. Pharmacol.* **1973**, *22*, 3099–3108.
34. Horoszewicz, J. S.; Kawinski, E.; Murphy, G. P. Monoclonal antibodies to a new antigenic marker in epithelial prostatic cells and serum of prostatic cancer patients. *Anticancer Res.* **1987**, *7*, 927–935.
35. Carter, R. E.; Feldman, A. R.; Coyle, J. T. Prostate-specific membrane antigen is a hydrolase with substrate and pharmacologic characteristics of a neuropeptidase. *Proc. Natl. Acad. Sci. USA* **1996**, *93*, 749–753.
36. O'Keefe, D. S.; Su, S. L.; Bacich, D. J.; Horiguchi, Y.; Luo, Y.; Powell, C. T.; Zandvliet, D.; Russell, P. J.; Molloy, P. L.; Nowak, N. J.; Shows, T. B.; Mullins, C.; Vonder Haar, R. A.; Fair, W. R.; Heston, W. D. Mapping, genomic organization and promoter analysis of the human prostate-specific membrane antigen gene. *Biochim. Biophys. Acta* **1998**, *1443*, 113–127.
37. O'Keefe, D. S.; Bacich, D. J.; Heston,W. D. Comparative analysis of prostate-specific membrane antigen (PSMA) versus a prostate-specific membrane antigen-like gene. *Prostate* **2004**, *58*, 200–210.
38. Israeli, R. S.; Powell, C. T.; Corr, J. G.; Fair, W. R.; Heston, W. D. Expression of the prostate-specific membrane antigen. *Cancer Res.* **1994**, *54*, 1807–1811.
39. Su, S. L.; Huang, I. P.; Fair, W. R.; Powell, C. T.; Heston, W. D. Alternatively spliced variants of prostate-specific membrane antigen RNA: ratio of expression as a potential measurement of progression. *Cancer Res.* **1995**, *55*, 1441–1443.
40. Troyer, J. K.; Beckett, M. L.; Wright, G. L., Jr. Location of prostate-specific membrane antigen in the LNCaP prostate carcinoma cell line. *Prostate* **1997**, *30*, 232–242.
41. Watt, F.; Martorana, A.; Brookes, D. E.; Ho, T.; Kingsley, E.; O'Keefe, D. S.; Russell, P. J.; Heston, W. D.; Molloy, P. L. A tissue-specific enhancer of the prostate-specific membrane antigen gene, FOLH1. *Genomics* **2001**, *73*, 243–254.
42. Jackson, P. F.; Cole, D. C.; Slusher, B. S.; Stetz, S. L.; Ross, L. E.; Donzanti, B. A.; Trainor, D. A. Design, synthesis, and biological activity of a potent inhibitor of the neuropeptidase *N*-acetylated alpha-linked acidic dipeptidase. *J. Med. Chem.* **1996**, *39*, 619–622.
43. Zhou, J.; Neale, J. H.; Pomper, M. G.; Kozikowski, A. P. NAAG peptidase inhibitors and their potential for diagnosis and therapy. *Nat. Rev. Drug Discov.* **2005**, *4*, 1015–1026.
44. Tsukamoto, T.; Wozniak, K. M.; Slusher, B. S. Progress in the discovery and development of glutamate carboxypeptidase II inhibitors. *Drug Discov. Today* **2007**, *12*, 767–776.
45. Baslow, M. H. The astrocyte surface NAAG receptor and NAAG peptidase signaling complex as a therapeutic target. *Drug News Perspect.* **2008**, *21*, 251–257.
46. Tortella, F. C.; Lin, Y.; Ved, H.; Slusher, B. S.; Dave, J. R. Neuroprotection produced by the NAALADase inhibitor 2-PMPA in rat cerebellar neurons. *Eur. J. Pharmacol.* **2000**, *402*, 31–37.
47. Nagel, J.; Belozertseva, I.; Greco, S.; Kashkin, V.; Malyshkin, A.; Jirgensons, A.; Shekunova, E.; Eilbacher, B.; Bespalov, A.; Danysz,W. Effects of NAAG peptidase inhibitor 2-PMPA in model chronic pain: relation to brain concentration. *Neuropharmacology* **2006**, *51*, 1163–1171.

48. Majer, P.; Jackson, P. F.; Delahanty, G.; Grella, B. S.; Ko, Y.-S.; Li, W.; Liu, Q.; Maclin, K. M.; Polakova, J.; Shaffer, K. A.; Stoermer, D.; Vitharana, D.; Wang, E. Y.; Zakrzewski, A.; Rojas, C.; Slusher, B. S.; Wozniak, K. M.; Burak, E.; Limsakun, T.; Tsukamoto, T. Synthesis and biological evaluation of thiol-based inhibitors of glutamate carboxypeptidase II: discovery of an orally active GCP II inhibitor. *J. Med. Chem.* **2003**, *46*, 1989–1996.

49. Tang, H.; Brown, M.; Ye, Y.; Huang, G.; Zhang, Y.; Wang, Y.; Zhai, H.; Chen, X.; Shen, T. Y.; Tenniswood, M. Prostate targeting ligands based on *N*-acetylated alpha-linked acidic dipeptidase. *Biochem. Biophys. Res. Commun.* **2003**, *307*, 8–14.

50. Ding, P.; Helquist, P.; Miller, M. J. Design, synthesis and pharmacological activity of novel enantiomerically pure phosphonic acid-based NAALADase inhibitors. *Org. Biomol. Chem.* **2007**, *5*, 826–831.

51. Williams, A. J.; Lu, X. M.; Slusher, B.; Tortella, F. C. Electroencephalogram analysis and neuroprotective profile of the *N*-acetylated-alpha-linked acidic dipeptidase inhibitor, GPI5232, in normal and brain-injured rats. *J. Pharmacol. Exp. Ther.* **2001**, *299*, 48–57.

52. Oliver, A. J.; Wiest, O.; Helquist, P.; Miller, M. J.; Tenniswood, M. Conformational and SAR analysis of NAALADase and PSMA inhibitors. *Bioorg. Med. Chem.* **2003**, *11*, 4455–4461.

53. Maung, J.; Mallari, J. P.; Girtsman, T. A.; Wu, L. Y.; Rowley, J. A.; Santiago, N. M.; Brunelle, A. N.; Berkman, C. E. Probing for a hydrophobic a binding register in prostate-specific membrane antigen with phenylalkylphosphonamidates. *Bioorg. Med. Chem.* **2004**, *12*, 4969–4979.

54. Liu, T.; Toriyabe, Y.; Kazak, M.; Berkman, C. E. Pseudoirreversible inhibition of prostate-specific membrane antigen by phosphoramidate peptidomimetics. *Biochemistry* **2008**, *47*, 12658–12660.

55. Humblet, V.; Lapidus, R.; Williams Larry, R.; Tsukamoto, T.; Rojas, C.; Majer, P.; Hin, B.; Ohnishi, S.; De Grand Alec, M.; Zaheer, A.; Renze Jurgen, T.; Nakayama, A.; Slusher Barbara, S.; Frangioni John, V. High-affinity near-infrared fluorescent small-molecule contrast agents for *in vivo* imaging of prostate-specific membrane antigen. *Mol. Imaging* **2005**, *4*, 448–462.

56. Liu, T.; Wu, L. Y.; Kazak, M.; Berkman, C. E. Cell-surface labeling and internalization by a fluorescent inhibitor of prostate-specific membrane antigen. *Prostate* **2008**, *68*, 955–964.

57. Misra, P.; Humblet, V.; Pannier, N.; Maison, W.; Frangioni, J. V. Production of multimeric prostate-specific membrane antigen small-molecule radiotracers using a solid-phase 99mTc preloading strategy. *J. Nucl. Med.* **2007**, *48*, 1379–1389.

58. Humblet, V.; Misra, P.; Frangioni, J. V. An HPLC/mass spectrometry platform for the development of multimodality contrast agents and targeted therapeutics: prostate-specific membrane antigen small molecule derivatives. *Contrast Media Mol. Imaging* **2006**, *1*, 196–211.

59. Ding, P.; Helquist, P.; Miller, M. J. Design and synthesis of a siderophore conjugate as a potent PSMA inhibitor and potential diagnostic agent for prostate cancer. *Bioorg. Med. Chem.* **2008**, *16*, 1648–1657.

60. Cushman, D. W.; Ondetti, M. A. Design of angiotensin converting enzyme inhibitors. *Nat. Med.* **1999**, *5*, 1110–1112.

61. Zhang, W.; Murakawa, Y.; Wozniak, K. M.; Slusher, B.; Sima, A. A. F. The preventive and therapeutic effects of GCPII (NAALADase) inhibition on painful and sensory diabetic neuropathy. *J. Neurol. Sci.* **2006**, *247*, 217–223.

62. Tsukamoto, T.; Grella, B.; Majer, P. Preparation of mercaptoethyl indolecarboxylic acids as NAALAdase inhibitors for treating and diagnosing glutamate abnormalities, neurological and other disorders WO Patent 2003/057670, **2003**.

63. Tsukamoto, T.; Majer, P.; Vitharana, D.; Ni, C.; Hin, B.; Lu, X.-C. M.; Thomas, A. G.; Wozniak, K. M.; Calvin, D. C.; Wu, Y.; Slusher, B. S.; Scarpetti, D.; Bonneville, G. W. Enantiospecificity of glutamate carboxypeptidase II inhibition. *J. Med. Chem.* **2005**, *48*, 2319–2324.

64. Kozikowski, A. P.; Nan, F.; Conti, P.; Zhang, J.; Ramadan, E.; Bzdega, T.; Wroblewska, B.; Neale, J. H.; Pshenichkin, S.; Wroblewski, J. T. Design of remarkably simple, yet potent urea-based inhibitors of glutamate carboxypeptidase II (NAALADase). *J. Med. Chem.* **2001**, *44*, 298–301.

65. Foss, C. A.; Mease, R. C.; Fan, H.; Wang, Y.; Ravert, H. T.; Dannals, R. F.; Olszewski, R. T.; Heston, W. D.; Kozikowski, A. P.; Pomper, M. G. Radiolabeled small-molecule ligands for prostate-specific membrane antigen: *in vivo* imaging in experimental models of prostate cancer. *Clin. Cancer Res.* **2005**, *11*, 4022–4028.

66. Mease, R. C.; Dusich, C. L.; Foss, C. A.; Ravert, H. T.; Dannals, R. F.; Seidel, J.; Prideaux, A.; Fox, J. J.; Sgouros, G.; Kozikowski, A. P.; Pomper, M. G. *N*-[*N*-[(*S*)-1,3-dicarboxypropyl]carbamoyl]-4-[18F]fluorobenzyl-L-cysteine, [18F]DCFBC: a new imaging probe for prostate cancer. *Clin. Cancer Res.* **2008**, *14*, 3036–3043.

67. Jayaprakash, S.; Wang, X.; Heston, W. D.; Kozikowski, A. P. Design and synthesis of a PSMA inhibitor-doxorubicin conjugate for targeted prostate cancer therapy. *ChemMedChem* **2006**, *1*, 299–302.

68. Banerjee, S. R.; Foss, C. A.; Castanares, M.; Mease, R. C.; Byun, Y.; Fox, J. J.; Hilton, J.; Lupold, S. E.; Kozikowski, A. P.; Pomper, M. G. Synthesis and evaluation of technetium-99m- and rhenium-labeled inhibitors of the prostate-specific membrane antigen (PSMA). *J. Med. Chem.* **2008**, *51*, 4504–4517.

69. Ding, P.; Miller, M. J.; Chen, Y.; Helquist, P.; Oliver, A. J.; Wiest, O. Syntheses of conformationally constricted molecules as potential NAALADase/PSMA inhibitors. *Org. Lett.* **2004**, *6*, 1805–1808.

70. Anderson, M. O.; Wu, L. Y.; Santiago, N. M.; Moser, J. M.; Rowley, J. A.; Bolstad, E. S. D.; Berkman, C. E. Substrate specificity of prostate-specific membrane antigen. *Bioorg. Med. Chem.* **2007**, *15*, 6678–6686.

71. Mhaka, A.; Gady, A. M.; Rosen, D. M.; Lo, K.-M.; Gillies, S. D.; Denmeade, S. R. Use of methotrexate-based peptide substrates to characterize the substrate specificity of Prostate-Specific Membrane Antigen (PSMA). *Cancer Biol. Ther.* **2004**, *3*, 551–558.

72. Mhaka, A. M. *Prostate Specific Membrane Antigen (PSMA): A Novel Target for Prostate Cancer Prodrug Therapy;* Johns Hopkins University: Baltimore, MD, **2005**.

73. Wynant, G. E.; Murphy, G. P.; Horoszewicz, J. S.; Neal, C. E.; Collier, B. D.; Mitchell, E.; Purnell, G.; Tyson, I.; Heal, A.; Abdel-Nabi, H.; et al. Immunoscintigraphy of prostatic cancer: preliminary results with 111In-labeled monoclonal antibody 7E11-C5.3 (CYT-356). *Prostate* **1991**, *18*, 229–241.

74. Bander, N. H.; Nanus, D. M.; Milowsky, M. I.; Kostakoglu, L.; Vallabahajosula, S.; Goldsmith, S. J. Targeted systemic therapy of prostate cancer with a monoclonal antibody to prostate-specific membrane antigen. *Semin. Oncol.* **2003**, *30*, 667–676.

75. Bander, N. H.; Milowsky, M. I.; Nanus, D. M.; Kostakoglu, L.; Vallabhajosula, S.; Goldsmith, S. J. Phase I trial of 177 lutetium-labeled J591, a monoclonal antibody to

prostate-specific membrane antigen, in patients with androgen-independent prostate cancer. *J. Clin. Oncol.* **2005**, *23*, 4591–4601.

76. Aggarwal, S.; Singh, P.; Topaloglu, O.; Isaacs, J. T.; Denmeade, S. R. A dimeric peptide that binds selectively to prostate-specific membrane antigen and inhibits its enzymatic activity. *Cancer Res.* **2006**, *66*, 9171–9177.
77. Lupold, S. E.; Rodriguez, R. Disulfide-constrained peptides that bind to the extracellular portion of the prostate-specific membrane antigen. *Mol. Cancer Ther.* **2004**, *3*, 597–603.
78. Rege, K.; Patel, S. J.; Megeed, Z.; Yarmush, M. L. Amphipathic peptide-based fusion peptides and immunoconjugates for the targeted ablation of prostate cancer cells. *Cancer Res.* **2007**, *67*, 6368–6375.
79. Lupold, S. E.; Hicke, B. J.; Lin, Y.; Coffey, D. S. Identification and characterization of nuclease-stabilized RNA molecules that bind human prostate cancer cells via the prostate-specific membrane antigen. *Cancer Res.* **2002**, *62*, 4029–4033.
80. Yan, A. C.; Bell, K. M.; Breeden, M. M.; Ellington, A. D. Aptamers: prospects in therapeutics and biomedicine. *Front Biosci.* **2005**, *10*, 1802–1827.
81. Aurup, H.; Williams, D. M.; Eckstein, F. 2′-Fluoro- and 2′-amino-2′-deoxynucleoside 5′-triphosphates as substrates for T7 RNA polymerase. *Biochemistry* **1992**, *31*, 9636–9641.
82. Bagalkot, V.; Farokhzad, O. C.; Langer, R.; Jon, S. An aptamer-doxorubicin physical conjugate as a novel targeted drug-delivery platform. *Angew. Chem. Int. Ed.* **2006**, *45*, 8149–8152.
83. Bagalkot, V.; Zhang, L.; Levy-Nissenbaum, E.; Jon, S.; Kantoff, P. W.; Langer, R.; Farokhzad, O. C. Quantum dot-aptamer conjugates for synchronous cancer imaging, therapy, and sensing of drug delivery based on bi-fluorescence resonance energy transfer. *Nano Lett.* **2007**, *7*, 3065–3070.
84. Cheng, J.; Teply, B. A.; Sherifi, I.; Sung, J.; Luther, G.; Gu, F. X.; Levy-Nissenbaum, E.; Radovic-Moreno, A. F.; Langer, R.; Farokhzad, O. C. Formulation of functionalized PLGA-PEG nanoparticles for *in vivo* targeted drug delivery. *Biomaterials* **2007**, *28*, 869–876.
85. Chu, T. C.; Shieh, F.; Lavery, L. A.; Levy, M.; Richards-Kortum, R.; Korgel, B. A.; Ellington, A. D. Labeling tumor cells with fluorescent nanocrystal-aptamer bioconjugates. *Biosens. Bioelectron.* **2006**, *21*, 1859–1866.
86. Dhar, S.; Gu, F. X.; Langer, R.; Farokhzad, O. C.; Lippard, S. J. Targeted delivery of cisplatin to prostate cancer cells by aptamer functionalized Pt(IV) prodrug-PLGA-PEG nanoparticles. *Proc. Natl. Acad. Sci. USA* **2008**, *105*, 17356–17361.
87. Farokhzad, O. C.; Jon, S.; Khademhosseini, A.; Tran, T. N.; Lavan, D. A.; Langer, R. Nanoparticle-aptamer bioconjugates: a new approach for targeting prostate cancer cells. *Cancer Res.* **2004**, *64*, 7668–7672.
88. Farokhzad, O. C.; Khademhosseini, A.; Jon, S.; Hermmann, A.; Cheng, J.; Chin, C.; Kiselyuk, A.; Teply, B.; Eng, G.; Langer, R. Microfluidic system for studying the interaction of nanoparticles and microparticles with cells. *Anal. Chem.* **2005**, *77*, 5453–5439.
89. Gu, F.; Zhang, L.; Teply, B. A.; Mann, N.; Wang, A.; Radovic-Moreno, A. F.; Langer, R.; Farokhzad, O. C. Precise engineering of targeted nanoparticles by using self-assembled biointegrated block copolymers. *Proc. Natl. Acad. Sci. USA* **2008**, *105*, 2586–2591.
90. Javier, D. J.; Nitin, N.; Levy, M.; Ellington, A.; Richards-Kortum, R. Aptamer-targeted gold nanoparticles as molecular-specific contrast agents for reflectance imaging. *Bioconjug. Chem.* **2008**, *19*, 1309–1312.

91. Murthy, S. C.; Dandin, S. M.; Dandin, A. S.; Patwardan, M. Y. Kikuchi's disease associated with systemic lupus erythematosus. *Indian J. Dermatol. Venereol. Leprol.* **2005**, *71*, 338–341.

92. Wang, A. Z.; Bagalkot, V.; Vasilliou, C. C.; Gu, F.; Alexis, F.; Zhang, L.; Shaikh, M.; Yuet, K.; Cima, M. J.; Langer, R.; Kantoff, P. W.; Bander, N. H.; Jon, S.; Farokhzad, O. C. Superparamagnetic iron oxide nanoparticle-aptamer bioconjugates for combined prostate cancer imaging and therapy. *ChemMedChem* **2008**, *3*, 1311–1315.

93. Chu, T. C.; Twu, K. Y.; Ellington, A. D.; Levy, M. Aptamer mediated siRNA delivery. *Nucleic Acids Res.* **2006**, *34*, e73.

94. McNamara, J. O., 2nd; Andrechek, E. R.; Wang, Y.; Viles, K. D.; Rempel, R. E.; Gilboa, E.; Sullenger, B. A.; Giangrande, P. H. Cell type-specific delivery of siRNAs with aptamer–siRNA chimeras. *Nat. Biotechnol.* **2006**, *24*, 1005–1015.

95. Wullner, U.; Neef, I.; Eller, A.; Kleines, M.; Tur, M. K.; Barth, S. Cell-specific induction of apoptosis by rationally designed bivalent aptamer-siRNA transcripts silencing eukaryotic elongation factor 2. *Curr. Cancer Drug Targets* **2008**, *8*, 554–565.

CHAPTER 37

Targeting HIV-1 Integrase Zinc Binding Motif

MARIO SECHI[1], MAURO CARCELLI[2], DOMINGA ROGOLINO[2], and NOURI NEAMATI[3]

[1]Dipartimento Farmaco Chimico Tossicologico, Università di Sassari, Via Muroni 23/A, 07100 Sassari, Italy

[2]Dipartimento di Chimica Generale ed Inorganica, Chimica Analitica, Chimica Fisica, Università di Parma, V.le Usberti 17/A, Campus Universitario, 43100 Parma, Italy

[3]Department of Pharmacology and Pharmaceutical Sciences, School of Pharmacy, University of Southern California, 1985 Zonal Avenue, PSC 304, Los Angeles, CA 90089, USA

37.1 INTRODUCTION

The acquired immunodeficiency syndrome (AIDS), caused by human immunodeficiency virus type 1 (HIV-1), has become a major epidemic.[1] After 1996, the mortality rate of HIV-infected patients has dramatically decreased in western countries, when highly active antiretroviral combination therapy (HAART) became available.[2–4] HAART has changed the clinical profile of HIV infection from a subacute lethal disease to a chronic ambulatory disease. In fact, chemotherapeutic approaches had a significant impact on the course of this pandemic. Currently, over 25 approved drugs targeting all three virally encoded enzymes, reverse transcriptase, protease, and integrase, are available. Moreover, at least two entry inhibitors have been approved, and many others are under clinical investigation.[5] Because current therapy cannot eradicate HIV infection, the goal of antiretroviral therapy is to achieve and maintain in the plasma an HIV-RNA copy number (viral load) at an undetectable level, using sensitive assays. When an effective therapy is withdrawn, a rapid increase in plasma HIV-RNA level and a decrease in CD4+ blood cell counts to pretherapy values are observed. Although these therapeutic strategies efficiently suppress viral replication for long periods of time, issues of patient adherence, drug toxicity, and multidrug resistance phenotypes[6–8] have highlighted the need to develop new drugs with novel mechanisms of action, targeting critical steps in the retroviral replication processes.[5,9,10]

In this context, HIV-1 integrase (IN), the viral enzyme that catalyzes the integration of proviral cDNA into the host cell genome, has emerged as an attractive target for novel antiretroviral agents.[11–17] Moreover, first IN inhibitor (e.g., raltegravir) was recently approved by the U.S. FDA (Food and Drug Administration),[18] and other IN inhibitors are under clinical trials.[19] The IN inhibitor drug discovery programs have concentrated mainly on high-throughput screening technologies by major pharmaceutical companies and on structure-based drug design, including both ligand-based (e.g., pharmacophore) and target-based (e.g., docking) methods, by nonprofit organizations.

IN is considered a metalloenzyme because it requires divalent metal ions (Mg^{2+} for *in vivo* and Mn^{2+} only for *in vitro* studies) for catalytic activities.[20–22] Because these ions are used as cofactors in biological assays, they represent the main target in the drug design efforts.[23] For these reasons, the structural features and the biology of the catalytic core domain and of its ions have been extensively investigated and reviewed.[24–27] However, there is another ion, Zn^{2+}, that is important for IN function. Zinc binds to the N-terminal domain and plays a crucial role in IN multimerization, and its potential as a new target for IN inhibition should be investigated.

This chapter focuses on the biology of IN, with a particular emphasis on the N-terminal domain, and on the design of prototypes of IN inhibitors targeting this site.

37.2 STRUCTURAL AND FUNCTIONAL DOMAINS: CATALYTIC ROLE OF METAL COFACTORS

IN is a 37-kDa protein encoded by the 3′-end of the *pol* gene. It is first translated as a large component of polyprotein *gag-pol*, from which it is released by the action of protease during maturation. IN belongs to a large family of polynucleotidyl transferases[28,29] that includes transposases and polymerases, and it is composed of a single polypeptide chain of 288 aa that folds into three functional domains[30] (Fig. 37.1):

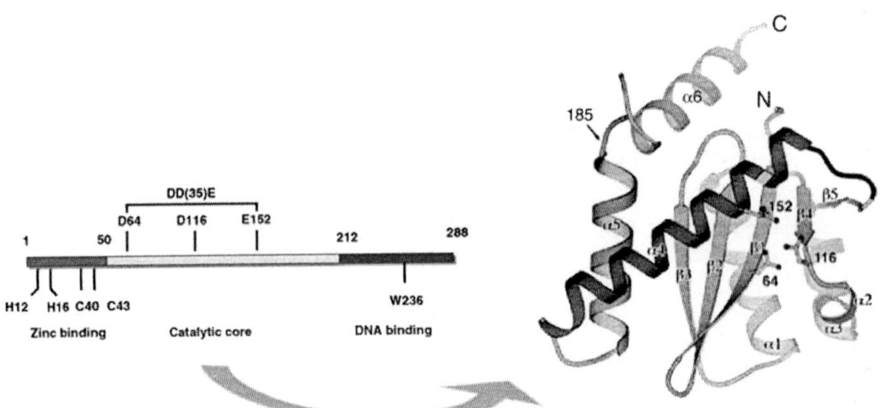

FIGURE 37.1 A schematic diagram of all the three domains spanning residues 1–288 and of the catalytic core of HIV-1 IN containing the DDE motif.

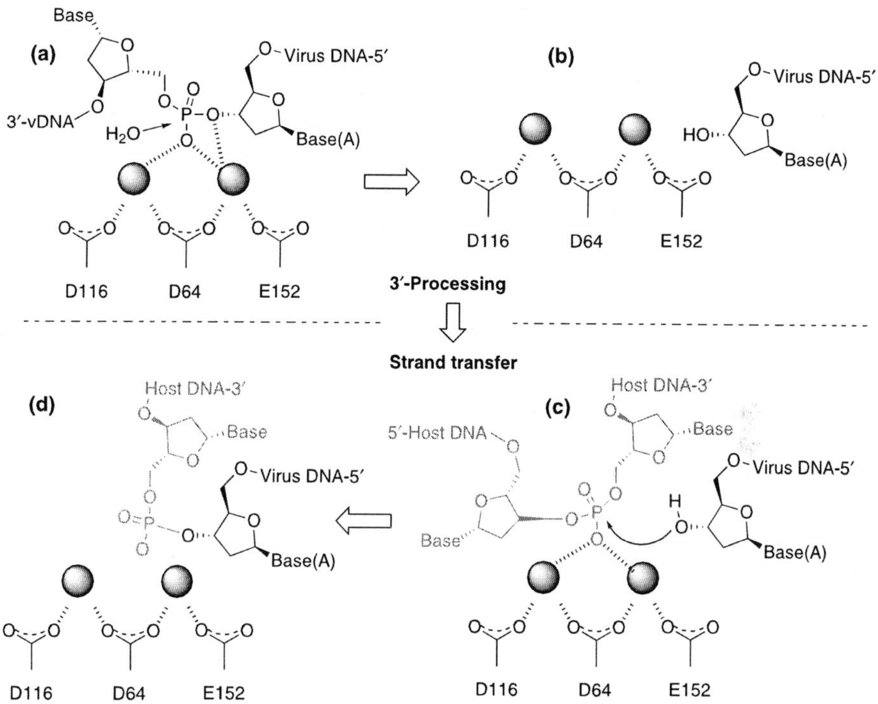

FIGURE 37.2 The two-metal-ion catalysis mechanism. Steps (a) and (b) indicate the 3′-processing reaction, and steps (c) and (d) the strand transfer reaction. (a) The enzyme recognizes the adenine base conserved in the third position from 3′-end of viral DNA, and then activates the next phosphoric ester with the two metals. (b) The activated phosphoric ester is hydrolyzed to excise the terminal dinucleotide and the recognized adenosine is exposed as the new 3′-end, giving a pre-integration complex. (c) The pre-integration complex nonspecifically binds to host DNA to activate a phosphoric ester by the two metals. (d) The activated phosphoric ester is attacked by the recessed 3′-end with an S_N2-like nucleophilic mechanism, and then the viral DNA and the host DNA are joined together.

the N-terminal domain (residues 1–49), the catalytic core domain (residues 50–212), and the C-terminal domain (residues 213–288). To date, the insolubility of HIV-1 IN has been the main barrier to obtaining a crystal structure of the full-length enzyme.

IN functions as a multimer to catalyze the insertion of viral cDNA into the cell genome to form a stable provirus. Integration is catalyzed by IN through two different main reactions, 3′-processing and strand transfer,[21,22,24,27] which are temporally and spatially separated (Fig. 37.2). In particular, a dimeric IN species is required for 3′-processing, whereas a tetrameric arrangement was observed to be the predominant species for the strand transfer process. First, in the cytoplasm, after reverse transcription, IN assembles onto the double-stranded proviral cDNA to form an IN–DNA complex that, together with other viral and cellular proteins, constitutes the pre-integration complex. Next, the enzyme modifies the 3′-terminal ends of the DNA by selectively removing the last two nucleotides (GT) to generate two CA-3′-hydroxyl recessed ends, which are the reactive intermediates required for the next step. Following 3′-processing, IN undergoes a structural change in preparation for the binding of the chromosomal DNA. IN, still bound to the 3′-processed viral DNA, translocates to the nucleus of the infected cell as a part of the pre-integration complex, wherein the terminal 3′-OH of the viral DNA attacks the host DNA. The strand transfer reaction is coordinated in such a way that each of the 3′-hydroxyl ends attacks a DNA phosphodiester bond on each strand of the chromosomal DNA with a five-base-pair stagger across the DNA major groove. At this stage, the proviral DNA is connected to the host DNA, presumably by host cell enzymes, by the removal of the two unpaired nucleotides on the 5′ viral DNA ends, gap filling, and ligation (Fig. 37.2). After transcription of the viral genome, translation of the viral DNA, protease action, and packaging, the new viral progeny is produced and the infection cycle is completed.

The catalytic core domain contains the enzyme active site that is responsible for catalysis. It includes the three amino acids, two aspartates (D64 and D116) and one glutamate (E152), constituting the "D,D(35)E" motif, which is highly conserved among polynucleotidyl transferases[31] (Fig. 37.1). The amino acid residues of the catalytic triad in the catalytic domain coordinate divalent metal ions such as Mg^{2+} or Mn^{2+}. The IN-mediated reactions are thought to proceed by a "two-metal-ion" mechanism similar to the 3′–5′ exonuclease reaction of the Klenow fragment of *Escherichia coli* DNA polymerase I as proposed by Beese and Steitz.[32] This mechanism appears fully capable of providing the necessary rate enhancement, and it can be applied to other polynucleotide polymerases.[33,34] Even if it is clear that IN needs divalent cations to perform its catalytic activity, to date no definitive conclusions have been reached regarding the type, Mg^{2+} or Mn^{2+}, or the number of the required metal ions. However, it is generally accepted that Mg^{2+} is a more likely cofactor for integration in cells ($\sim 10^{-3}$ M),[26,27] given its one million-fold abundance over Mn^{2+} ($< 10^{-7}$ M), which appears to be preferred *in vitro*.

The C-terminal domain is primarily composed of β-strands and is the least conserved domain among the retroviral integrases. This domain has been proposed to be involved in the multimerization of IN.[35, 36] With the exception of feline immunodeficiency virus, deletion of this domain abolishes 3′-processing and strand transfer activities. The crystal structure of a multiple mutant of the two-domain IN, consisting of residues 52–288, has been determined.[37] In addition, this domain has been implicated in protein–protein interactions as it has been shown to interact with reverse transcriptase, an interaction that appears to be crucial for its catalytic activity. Mutational analysis identified some residues in the C-terminal domain critical for oligomerization and DNA binding.[38] A similar function for site-specific DNA binding is exhibited by transposases. In the past, the interactions with the target DNA were attributed to the C-terminal domain; however, recent studies indicate that all the three domains are involved.[39]

37.3 THE N-TERMINAL DOMAIN: Zn^{2+} IS THE OTHER METAL ION IN THE IN ENZYME

The N-terminus contains a pair of histidine (His12, His16) and cysteine (Cys40, Cys43) residues (the HHCC motif) that are highly conserved among retroviral and retrotransposon integrases[40,41] (Fig. 37.1 and Fig. 37.3). It has been proposed that these amino acids might represent a metal binding domain analogous to the zinc finger motifs.[42] Zinc ion plays a key structural role in a large number of metal proteins, and up to now, several classes of zinc binding domains have been identified in many enzyme proteins, such as carbonic anhydrase, histone acetyltransferases, zinc metalloproteases, and so on.[43–46] Their structure and function can be different: some of them are involved in interactions with double-stranded DNA or with single-stranded nucleic acids, while others mediate protein–protein interactions. However, in each case, the zinc ion is coordinated by histidine and cysteine residues.

Interestingly, the HIV-1 nucleocapsid protein NCp7 (nucleocapsid p7),[47] that plays a critical role at different steps of the retrovirus life cycle,[48] contains a similar amino acid motif (two successive CCHC domains are separated by a short basic linker sequence RAPRKKG), and it is characterized by the presence of two aromatic residues crucial for NCp7 activities. The $^1h^1H$-NMR structure[49,50] of NCp7 showed that the N- and C-terminal parts of the protein are relatively flexible, whereas the central domain, encompassing the two zinc fingers and the linker, adopts a preferential globular conformation.

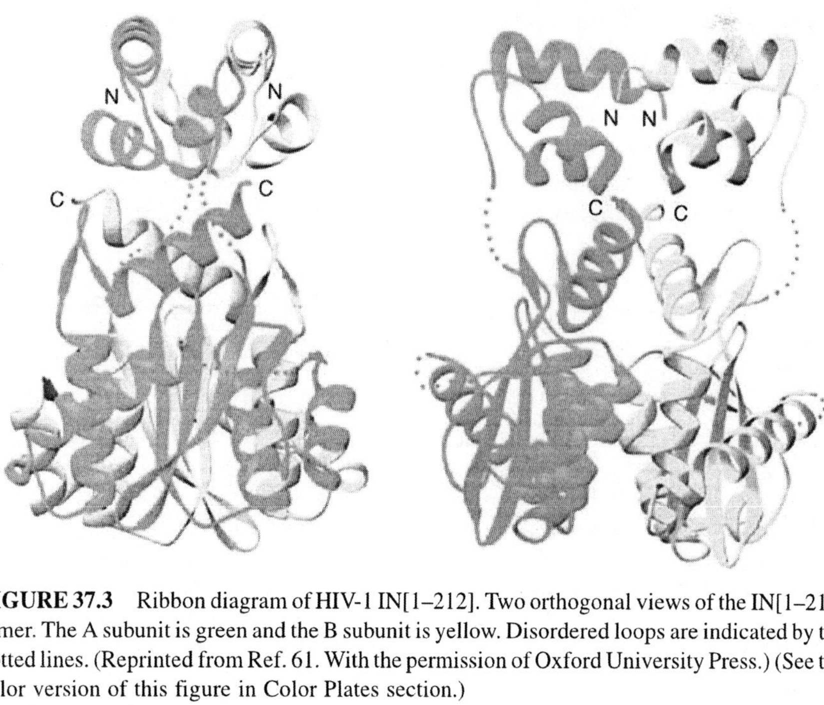

FIGURE 37.3 Ribbon diagram of HIV-1 IN[1–212]. Two orthogonal views of the IN[1–212] dimer. The A subunit is green and the B subunit is yellow. Disordered loops are indicated by the dotted lines. (Reprinted from Ref. 61. With the permission of Oxford University Press.) (See the color version of this figure in Color Plates section.)

The HHCC region of retroviral integrases differs from that of the prototypic zinc finger domain originally reported for the *Xenopus* transcription factor IIIA (TFIIIA).[51] In the latter case, the sequence pattern is Cys–X_{aa4}–Cys–X_{aa12}–His–X_{aa3}–His.[52] The cysteines located near a turn in the β-sheet region and the histidines in the α-helix coordinate a zinc ion. This causes a unique fold in secondary structures of proteins to form a small structural domain. In the corresponding region of IN, the order of the conserved pair of residues is HHCC instead of CCHH, and the number of amino acids between the pairs of conserved HH and CC residues is about twice that of the prototypic zinc finger. Since the formation of a zinc finger depends on the three-dimensional arrangement of the four coordinating ligands rather than on their order in the primary sequence, it has been proposed that the HHCC region of IN can coordinate the zinc metal ion similarly to the CCHH sequence. The HHCC motif binds zinc in a 1:1 ratio and enhances IN oligomerization.

Structural information on IN and its domains provides insights into the functions and mechanism of action of this enzyme. The structures of the catalytic core of the HIV-1, as well as of the simian immunodeficiency and Rous sarcoma virus integrase, have been determined by X-ray diffraction analysis,[28,53–55] and recently the crystal structure of the core domain together with the C-terminal domain has been solved.[37] The solution structures of the N-domain of HIV-1[56–58] and HIV-2[59,60] integrases have been determined by high-resolution two- and three-dimensional NMR techniques. Multidimensional heteronuclear NMR spectroscopy[56] provided evidence that IN [1–55] is dimeric and exists in two interconverting forms, termed D and E, which differ in zinc coordination to histidine residues. Each monomer is composed of four helices, and the zinc metal ion is tetrahedrally bound to His12, His16, Cys40, and Cys43. The relative positions of the two histidines in the E and D forms are reversed, since in the E form, which is predominant below 300 K, His12 coordinates zinc through Nε2 and is buried within the protein interior, while in the D form His12 coordinates through Nδ1 and is located in a loop exposed to the solvent. In both the E and D forms, His16 coordinates the metal ion through its Nδ1 nitrogen atom, but while in the E form it is solvent exposed, in the D form it is buried within the protein interior, such that the relative positions of His12 and His16 are reversed in the two forms. The different arrangement of these two amino acids implies great differences in the conformation of the polypeptide backbone (residues 9–18) around the coordination sphere. The dimeric interface is identical in the two forms and shows interactions with hydrophobic nature. Also the solution structure of a His12 → Cys mutant of the N-terminal domain of HIV-1 bound to cadmium has been solved by NMR spectroscopy.[57] It is very similar to that of the wild type complexed to zinc, except that the mutant exists in a single conformation at low pH. In the same way, another mutant, Y15A, folds correctly in solution, but takes only the E form.[58] In the solution structure of the dimeric HIV-2 IN, it is possible to define four helices, where α1, α2, and α3 form a three-helix bundle stabilized by zinc binding to the HHCC motif.[59,60] The dimer interactions are mainly hydrophobic, and the interface is formed by the residues of the N-terminal tail and the first half of helix α3. In the case of HIV-1 IN, the second halves of α3 and α4 are part of the dimer interface, stabilized by hydrophobic interactions and by a potential salt bridge. These interactions cannot be formed by HIV-2 IN, where the corresponding residues are the positively charged or neutral glutamine, aspargine, and arginine.

Previously, Craigie and coworkers[61] have reported the crystal structure of a fragment of HIV-1 IN comprising the catalytic core and the N-terminal domain (IN[1–212]) that are packed against each other to form almost identical AB or CD dimers (Fig. 37.3). The monomeric zinc finger domain has a structure very similar to that determined by NMR techniques, while strong differences can be seen at the dimer interface, which is smaller and more hydrophobic in the two-domain crystal structure than in the dimer of the isolated domain (Fig. 37.4). It is believed that IN is a tetramer in its active form, and in the crystal structure of IN[1–212] the two dimers, AB and CD, are in fact related by a noncrystallographic twofold axis to form a tetramer: the dimers of dimers interface is considerable. Extensive interactions involve the Zn^{2+} coordinating region of the N-terminal domain of the B subunit and the catalytic core of the D subunit, suggesting a fundamental role of the metal ion in the tetramerization process, responsible for activation of the enzyme.

Several biochemical and physical measurements have been performed with the aim of clarifying the type and nature of the interaction of HIV-1 IN with Zn^{2+}. Gel filtration,[40] atomic absorption, spectroscopic,[62] and zinc blotting assays[41] clearly indicate that an IN monomer binds one equivalent of metal ion. The fact that the domain responsible for zinc binding is indeed the N-terminal domain of IN has been demonstrated by studying deletion mutants, which fail to bind the metal when lacking this region.[40,62] Moreover, the crucial role of the HHCC motif in binding zinc has been further highlighted by spectroscopic investigations[63] and by analyzing the binding capability of mutants bearing substitution in the histidine and cysteine residues.[41]

What is the function of this highly conserved N-terminal domain in the catalytic process, and what is, in particular, the role of the divalent ion Zn^{2+} in the enzymatic activity? On the basis of *in vitro* cell-free assays of IN activity (evaluated in terms of 3′-processing, strand transfer, and disintegration),[64] several functions can be proposed for the HHCC motif and its coordinated ion. IN also binds to the viral DNA, but it is currently unclear which amino acids are involved in this binding.[65,66] It is thought that the HHCC motif is involved in this recognition indirectly, by stabilizing the folding of IN due to the structural function of the metal. Circular dichroism studies[40,63] suggest that isolated N-terminal domain is clearly unfolded in the absence of Zn^{2+}, while the zinc-bound peptide shows an increase in the α-helix content, and neither Mg^{2+} nor Mn^{2+} is able to substitute Zn^{2+} in this structural action. It is therefore possible to say that the Zn^{2+} is able to promote the folding of the isolated domain into secondary elements that are mostly helical. However, the same behavior cannot be clearly inferred for the full-length IN,[63] suggesting that other parts of the protein may help in stabilizing the secondary structure of this domain. Important roles of the HHCC motif and of the zinc ion in IN multimerization have been defined. Through gel filtration and size exclusion chromatography,[40] it has been observed that the apoprotein is predominantly in equilibrium between dimers and tetramers, and zinc has the effect to promote tetramerization or even multimerization of the

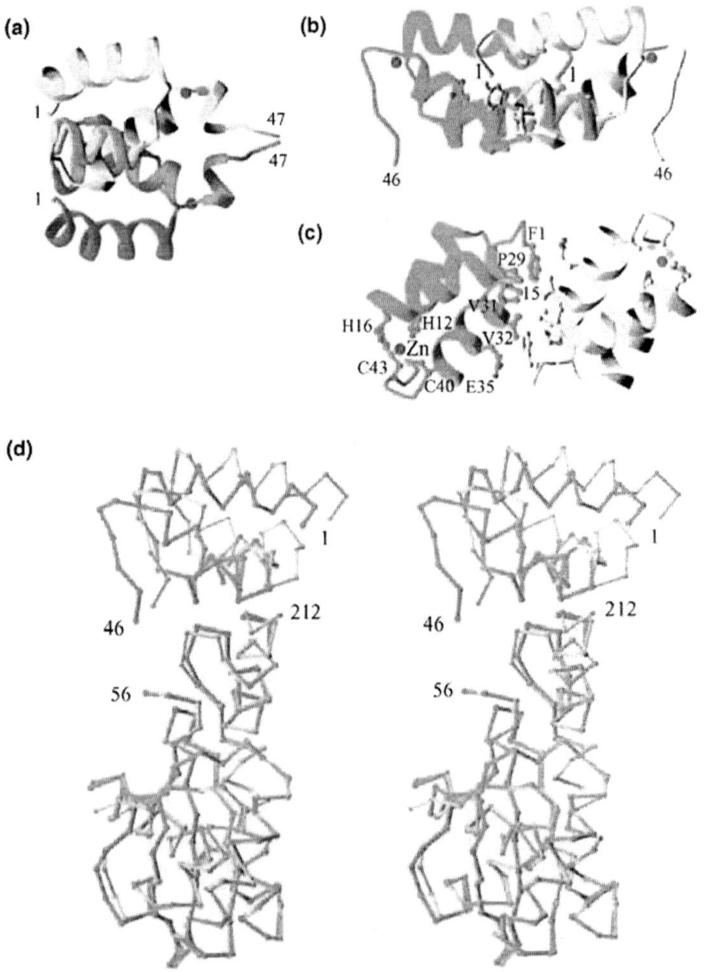

FIGURE 37.4 Analysis of domain interactions. (a) Ribbon diagram of the N-terminal domain structure as an isolated entity determined by NMR (PDB code 1WJA) and (b) in the context of IN[1–212] by X-ray crystallography after superimposing the yellow subunits of the two structures. Residue numbers at the N- and C-termini are labeled. (c) A view near orthogonal to (b) looking down the twofold axis. Residues involved in the dimerization, which are rather hydrophobic, are shown in ball and stick. Zn^{2+} ions are shown as red spheres. (d) The N-terminal domain is oriented differently relative to the core domain in the A and B subunits of IN[1–212] as revealed by superposition of the Cα traces of the core domains. (Reprinted from Ref. 61. With the permission of Oxford University Press.) (See the color version of this figure in Color Plates section.)

protein.[40,62,67] The fact that interaction between domains is essential for the formation of a functional multimeric complex has been proven also by complementation assays with mutant proteins.[68] An N-terminal domain can restore IN activity of a mutant lacking this domain, but only when the zinc ion is coordinated to the HHCC motif. Other studies[69,70] with proteins mutated in different domains indicate that both the N- and C-termini are required for site-specific cleavage and strand transfer reaction, but they are not essential for disintegration, suggesting that the HHCC region is required in an early step of the integration pathway in which an interaction with DNA is established.

It is well known that both Mn^{2+} and Mg^{2+} can support IN 3′-processing and strand transfer *in vitro*. It has been demonstrated[67] that the addition of zinc in *in vitro* assays resulted in activation of cleavage reactions predominantly with Mg^{2+}, while this activation is not observed with Mn^{2+}. Probably Zn^{2+} coordination to the zinc finger domain promotes conformational changes that lead to the protein–protein interactions required for Mg^{2+}-dependent activity. On the other hand, most of the biological processes require interaction between macromolecules due to structural domains that in many cases can fold autonomously, but several of them fold stably only when binding zinc ions.

In summary, although in HIV-1 IN zinc is not directly responsible for the catalytic activity (i.e., 3′-processing and strand transfer) performed by Mg^{2+} of the catalytic core, it is essential for the overall integration process, and it can be regarded as a potential target for the inhibition of this enzyme.

37.4 POTENTIAL TARGETS FOR THE IN INHIBITOR DRUG DISCOVERY

There are several reasons for the slow progress in designing IN inhibitors. The problems can be attributed in part to (a) the complex nature of the integration process, (b) the lack of the complete structure of this enzyme in complex with DNA and with inhibitors, and (c) the lack of knowledge of the exact mechanism of action of the potent IN inhibitors. Despite these obstacles, the discovery and clinical successes of the selective strand transfer inhibitors, represented by the β-diketo acids[11,55,71–73] and their derivatives (see representative compounds in Fig. 37.5), have provided the "driving force" for the drug development process targeting the viral enzyme. In particular, the pyrimidinone carboxamide MK-0518[71–78] (Fig. 37.5) (raltegravir, Isentress™) is the first U.S. FDA approved drug targeting HIV-1 IN and represents a novel chemical class of selective strand transfer inhibitors. The drug received a priority review by the U.S. FDA and was approved in October 2007 for the treatment of HIV infection in combination with existing antiretroviral agents in HAART-experienced adult patients displaying evidence of increased viral production due to HAART-associated drug resistance.

Several small molecules from different chemical classes that display strand transfer specific inhibition toward IN have undergone clinical evaluation.[19,79–86] However, each of these inhibitors displays a similar mechanism of action and likely binds the IN active site region in an overlapping manner.[11] Prolonged clinical use of the strand transfer IN inhibitors will lead to HIV viral strains exhibiting varying degrees of

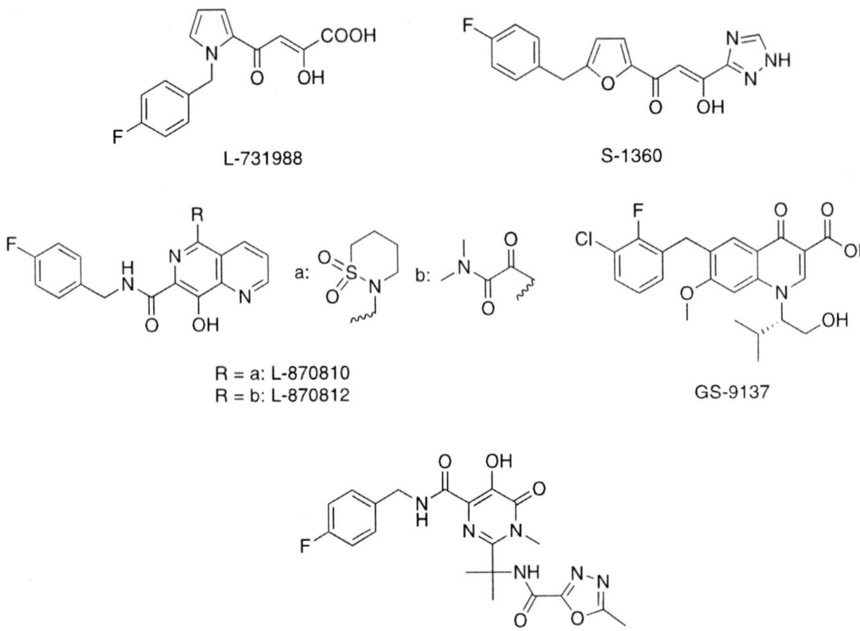

L-731988 S-1360

R = a: L-870810
R = b: L-870812

GS-9137

MK-0518 (raltegravir)

FIGURE 37.5 Structure of representative selective strand transfer inhibitors and raltegravir.

cross-resistance to other strand transfer specific IN compounds. Because of this clinical concern, there is a need to develop inhibitors that bind to IN in a region topographically different from the enzyme active site, resulting in an alternative inhibitory mechanism, commonly referred to as allosteric mode of inhibition.

For specific enzymatic targets, the development of drugs that act by an allosteric binding mechanism may also lead to greater specificity, decreasing the potential toxicity for patient. Furthermore, the design of drugs suitable to bind in an allosteric site is highly significant when the active site of a therapeutic target enzyme is not readily amenable for drug design.[87] Actually, the catalytically essential DD(35)E motif on the catalytic core domain has been the focus for many drug design programs targeting IN. The clinically relevant selective strand transfer inhibitors bind at the active site and are thought to disrupt Mg^{2+} coordination as a possible inhibitory mechanism of action.

To ensure the availability of effective treatment options for selective strand transfer inhibitor-experienced patients, research targeting alternative aspects of IN function *in vivo* is needed. Although the availability of IN structural data is hampered by solubility issues, alternative methods such as affinity labeling coupled with biochemical validation techniques have provided useful information on allosteric inhibitor binding sites that map to each the N-terminal, catalytic core, and C-terminal domains of IN. These allosteric sites can now be exploited by designing novel inhibitors that can act synergistically with current clinically studied IN inhibitors that bind the active site, similarly to what was observed for nucleoside reverse transcriptase inhibitor and nonnucleoside reverse transcriptase inhibitor development.[88]

The N-terminal HHCC motif is a potentially suitable allosteric site to target. As discussed, NCp7, a protein involved in both the early infectious phase and the late synthetic phase of the retroviral cycle, contains a similar structural sequence.[48,89] This protein possesses a three-dimensional structure centered around two highly conserved zinc fingers. Nuclear magnetic resonance studies associated with site-directed mutagenesis have shown that any modification of this structure leads to a complete loss of HIV-1 infectivity. Because the zinc finger motifs are highly conserved among retroviruses, NCp7 appears to be an interesting target for designing new antiviral agents.[90]

As far as IN is concerned, disruption of the IN HHCC motif has also been shown to block HIV replication at two different stages of the viral life cycle.[91] Being an integral component of the reverse transcription complex, IN makes direct physical contacts with reverse transcriptase.[92] Disruption of the HHCC motif may abolish the interaction between IN and reverse transcriptase and/or the role that IN has in the reverse transcription complex during the initiation of viral DNA synthesis.

Allosteric inhibitors and binding sites that map to each IN functional domain have been identified. Hyrtiosal (Fig. 37.6), a natural product derived from a marine sponge, recently reported as protein tyrosine phosphatase 1B inhibitor,[93] competitively

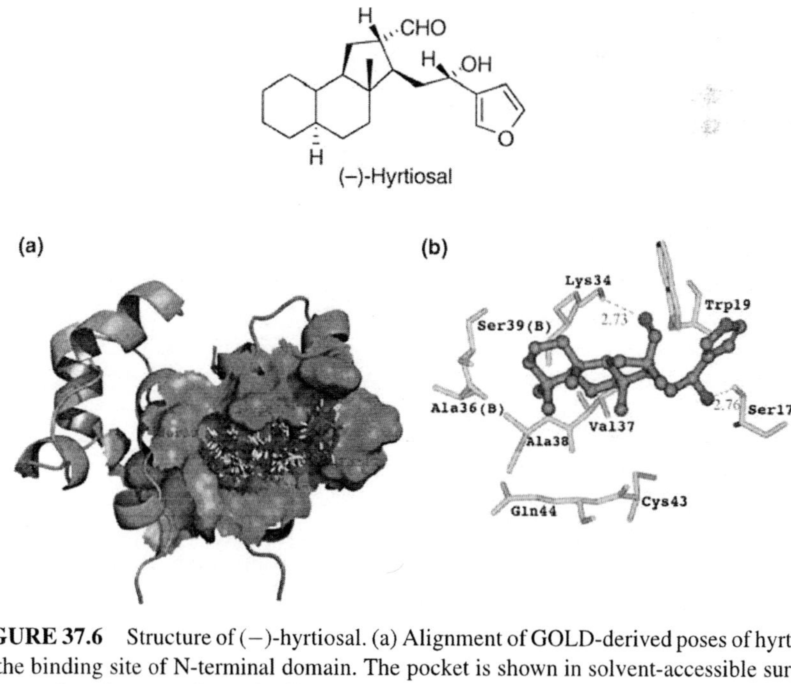

FIGURE 37.6 Structure of (−)-hyrtiosal. (a) Alignment of GOLD-derived poses of hyrtiosal on the binding site of N-terminal domain. The pocket is shown in solvent-accessible surface, whereas the others are shown in ribbon. (b) The detailed interactions of N-terminal domain residues with hyrtiosal (shown in ball and stick and residues in stick only). Dashed lines represent hydrogen bonds. (From Ref. 94. With the permission of Wiley InterScience.)

disrupts IN–viral DNA binding and represents a small molecule that potentially binds the IN N-terminal domain.[94]

There is another potential allosteric-based inhibitory approach to disrupt the IN oligomerization process. In fact, IN functions in a higher ordered multimeric state to catalyze both 3′-processing (as a IN dimeric specie) and strand transfer (in tetrameric stoichiometry).[95] This approach is supported by the promising results in the discovery and design of oligomerization disrupting agents targeting reverse transcriptase and protease, which also require an obligatory multimerization step.[96] Numerous studies have identified small molecules and peptides that interfere with the IN multimerization process.[97] In this context, peptides derived from the IN dimeric interface have proven to be effective IN multimer disrupting agents.[98]

Recently, it was suggested[99] that a different peptide IN inhibitory mechanism through multimer modification, shifts the IN oligomerization equilibrium from the active dimer to an inactive tetramer, that is unable to catalyze 3′-processing, thus blocking the subsequent viral DNA integration or strand transfer. Peptides observed to preferentially bind the IN tetramer, thereby shifting the oligomerization equilibrium, have been termed "shiftides". This IN inhibitory mechanism was first observed with peptides derived from the IN cellular cofactor lens epithelium-derived growth factor (LEDGF)/p75,[100–102] an essential cellular protein for viral infectivity, which mediates IN–chromatin tethering *in vivo*. Solution structure[103] of the IN binding domain in LEDGF/p75 and the crystal structure[104] of the dimeric catalytic core domain of IN complexed to the IN binding domain of LEDGF/p75 (Fig. 37.7a) were recently determined. The catalytic core domain of IN possesses the main determinants for interacting with LEDGF, although the N-terminal domain increases the affinity of the interaction. Three LEDGF/p75 amino acid residues (I365, D366, and F406) appeared to be the most critical residues on IN binding domain (Fig. 37.7b and c). These peptide residues, falling within the regions 353–378 and 402–411, were also effective at disrupting initial IN–DNA binding. This effect was attributed to the shiftide inhibitory mechanism, as the induced IN tetramers are proposed to be 3′-processing catalytically deficient. Moreover, two regions in IN are crucial for the interaction with LEDGF/p75.[105] The first region centers around residues W131 and W137, and the second extends from I161 to E170. In particular, the amino acid residues W131, I161, R166, Q168, and E170 are important for interaction with LEDGF/p75.

In this context, LEDGF and other transcriptional proteins seem to be interesting targets for investigation.[5,106] The benzoic acid derivative D77 (4-[(5-bromo-4-{[2,4-dioxo-3-(2-oxo-2-phenylethyl)-1,3-thiazolidin-5-ylidene]methyl}-2-ethoxyphenoxy)methyl]benzoic acid) (Fig. 37.8) has recently been proposed to target the interaction between the enzyme and cellular LEDGF/p75, and it also showed antiviral activity.[107] Molecular docking with site-directed mutagenesis analysis and surface plasmon resonance binding assays supported the proposed binding mode for D77 (Fig. 37.8).

As detailed,[106,108–110] the integration process requires the recruitment of the pre-integration complex, which contains viral cDNA, viral and cellular proteins such as the

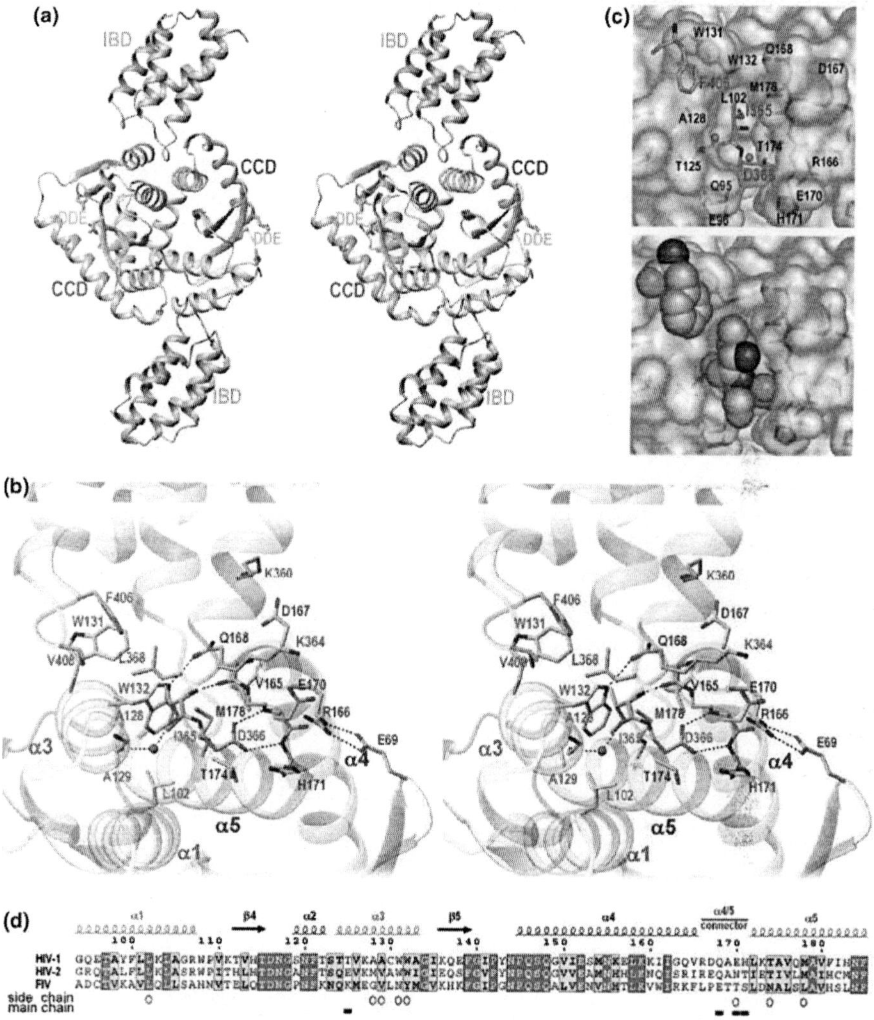

FIGURE 37.7 Molecular mechanism of the IN–LEDGF interaction. (a) The overall structure of the catalytic core domain–IN binding domain complex. IN chains A and B are colored blue and green, respectively; the IN binding domain subunits are violet. The side chains of the DDE catalytic triad are shown as yellow sticks. (b) Key intermolecular contacts at the catalytic core domain–IN binding domain interface. Selected residues are shown as sticks. (c) The pocket at the catalytic core domain dimer interface. LEDGF hotspot residues Ile365, Asp366, and Phe406 are shown as sticks (upper) or in space-fill mode (lower). The IN subunits are shown as semitransparent surfaces. Selected IN residues are indicated. (d) Sequence alignment of HIV-1, HIV-2, and feline immunodeficiency virus INs. Identical residues are white on red background; residues with conserved properties are bold on yellow background. Residue numbering, secondary structure elements, and the position of the α4/5 connector in HIV-1 IN are shown above the alignment; structural elements are colorized as in (a)–(c). Open circles and filled boxes under the alignment indicate residues that make contacts to the LEDGF–IN binding domain through side chain and main chain atoms, respectively. (From Ref. 104. With the permission of the National Academy of Sciences of the United States of America.) (See the color version of this figure in Color Plates section.)

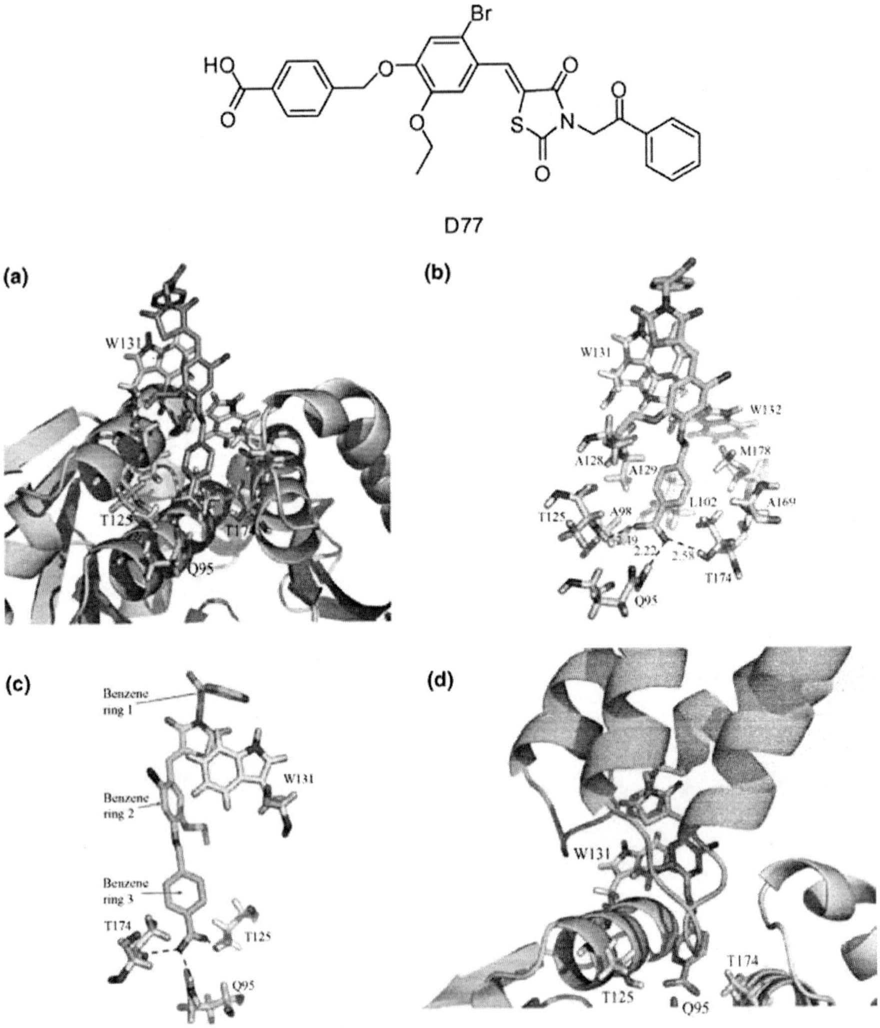

FIGURE 37.8 Structure of D77 and molecular docking into HIV-1 IN catalytic core domain. (a) The binding site of D77 on HIV-1 IN catalytic core domain dimer interface. Yellow and cyan: IN chains; salmon: D77. (b, c) The interaction between D77 and HIV-1 IN catalytic core domain in detail. (b: front view; c: side view) D77 is shown in green and IN residues are colored in yellow. Hydrogen bond is shown in red. (d) Molecular docking of D77 with catalytic core domain–IN binding domain complex. Yellow and cyan: IN chains; green: IN binding domain; salmon: D77. (From Ref. 107. With the permission of Elsevier.) (See the color version of this figure in Color Plates section.)

matrix protein p17, the capsid protein p24, the nucleocapsid protein p7/p9, reverse transcriptase, IN, and the accessory protein Vpr. IN, Vpr, and the matrix protein p17 have been implicated in nuclear import and the nucleocapsid protein in enhancing catalytic activity.

Another approach could involve IN cellular cofactor interaction regions as allosteric inhibitor binding sites. Effective IN function *in vivo* requires numerous direct and indirect interactions with both viral and cellular proteins. Unfortunately, another impediment toward progress in the field is determining whether the identified IN cellular cofactors are required for viral infectivity.

Therefore, according to the paradigm of interfacial inhibitors, protein–protein interactions such as IN monomer–IN monomer, IN–LEDGF/p75, IN–matrix, IN–INI1 (integrase interactor 1), and matrix–BAF (barrier to autointegration factor) and protein–DNA junctions such as IN–viral DNA, BAF–viral DNA, and so on within pre-integration complex are equally important for integration. Since alteration of any of these interfaces may inhibit integration, the prevention of interference of macromolecular contacts among the PIC components is a rational and promising approach for the inhibition of the HIV integration.

37.5 TARGETING Zn^{2+} ION: A SUITABLE WAY TO INHIBIT IN?

Insights derived from studies in this field indicate that several IN-specific inhibitors interfere with metal ions in the active site, and therefore metal chelation at the catalytic core could be exploited in the design of new inhibitors with better selectivity toward IN.[23] Theoretically, such a platform of knowledge could also be used to evaluate the potential of Zn^{2+} as a new biological target.

Many factors should be considered in designing a drug targeting metal ions, taking into account structural, steric, and electronic considerations. As far as the zinc ion is concerned, the chemical features that make it so suitable for structural roles in many important proteins have to be considered. Zn^{2+} is in fact a d^{10} ion, with d orbitals completely occupied. This implies that it has no other oxidation states readily available, especially under physiological conditions, and it cannot undergo undesirable redox reactions. The lack of redox activity makes zinc particularly suitable for structural roles in interactions involving DNA, since it avoids radical processes that result in nucleic acid damage. Moreover, because of its completely occupied d orbitals, Zn^{2+} has no ligand field stabilization energy, independently of the coordination geometry.[111] It can form four-, five-, and six-coordinated complexes and it can undergo changes in the coordination number, without relevant energetic penalty. The preference for certain ligands is qualitatively explained by the hard/soft theory of acids and bases[112]: hard metal ions are small and not easily polarizable Lewis acids, while soft ones are large and fairly polarizable. Ligands with high electronegative donor atoms are hard bases, while polarizable ligands are soft. As a general rule, stable complexes are formed between Lewis acids and bases that are both hard or both soft. Zn^{2+} can be considered a borderline acid, so that it can interact with a variety of donor atoms, including sulfur, nitrogen, and, to a lesser extent, oxygen: it can bind strongly to both histidine and cysteine, or even glutamate, aspartate, and water. Finally, zinc is kinetically labile and undergoes ligand exchange in a relatively rapid manner.

All these considerations have to be kept in mind in designing inhibitors that could selectively chelate the zinc ion in the N-terminal domain of IN. The effectiveness and selectivity of such a drug are in fact partly related to its thermodynamic ability to coordinate the metal, competing with the biological ligands that bind the cofactor in its enzymatic environment. The tetrahedral coordination, with nitrogen or sulfur ligands, is the most common for Zn^{2+}, followed by octahedral and five coordination. Rising from four-coordinated complexes to six or more, the affinity of zinc toward hard bases increases; that is, with a coordination number of 4 it binds preferentially to sulfur, while oxygen is preferred for higher coordination numbers. The disposition of the donor atoms in the ligand can give rise to chelation ring instability. Bite angles at the metal have to be considered as well: matching the bite angles with the size and/or stereospecificity of the metal ion allows selective binding. As previously discussed, drug design targeting the IN HHCC motif would provide therapeutically beneficial results through a range of various inhibitory mechanisms. These interferences could be inducted through competition with DNA or by using the metal chelating approach to Zn^{2+}.

In addition to the structural peptidomimetic inhibitors,[113] the "zinc ejectors" have emerged as a promising class of inhibitors.[90] These include 3-nitrosobenzamide (NOBA),[114] 2,2′-dithiobisbenzamides (DIBAs),[115,116] and their benzisothiazolone derivatives (BITAs),[116] cyclic 2,2′-dithiobisbenzamides (SRR-SB3),[117] *cis*-1,2-dithiane-4,5-diol-1,1-dioxide (dithiane),[118] azodicabonamide (ADA),[119] pyridinioalkanoyl thioesters (PATEs),[120] 2,2′-dithiopyridine (Aldrithiol),[121] and so on (Fig. 37.9a). They were able to remove the zinc ion of NCp7 through several mechanisms. For example, one of them seems to involve a nucleophilic attack at the electron-poor sulfur of the reagent by one zinc-coordinated cysteine sulfur atom.[116] Zinc ejection results in a loss of structural integrity of the NCp7, and it is accompanied by the formation of covalent bond between the reagent and the cysteine of the zinc-depleted NCp7 and by internal or external cross-linking of the free cysteines.[122] In particular, the mechanism of action of two *N*-substituted-*S*-acyl-2-mercaptobenzamide compounds[123] (Fig. 37.9b) that target NCp7 was investigated.[124] It seems that both thioesters (**37.1** and **37.2**) were able to specifically eject the metal from the carboxyl-terminal zinc binding domain of NCp7 by covalent modification of Cys39 (Fig. 37.9c). Exposure of NCp7 to both compounds destroyed its ability to specifically bind RNA, whereas NCp7 already bound to RNA was protected from zinc ejection by the thioesters.

Furthermore, a method for inhibiting infectivity of a lentivirus, comprising contacting a cell that is producing the virus with an antiviral-effective amount of a membrane-permeable Zn chelator, was recently reported.[125] The antiviral-effective amount of the Zn chelator does not substantially inhibit proteins in the cells that contain Zn binding motifs other than lentivirus viral infectivity factor.

Therefore, disrupting the zinc coordination of the HHCC motif would destabilize IN oligomeric structure, resulting in further interruption of reverse transcription complex-mediated viral DNA synthesis, the integration process, and the INI1–IN interaction. Thus, the investigation of this ion in a drug design program is expected.

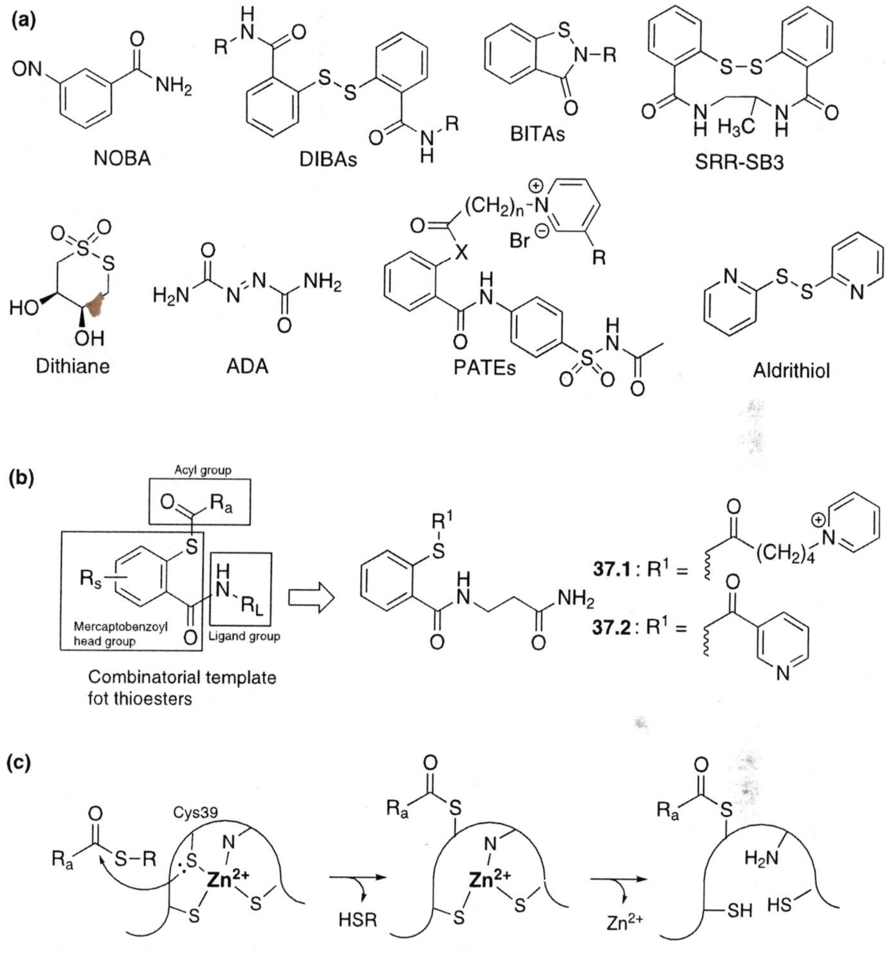

FIGURE 37.9 (a) Examples of zinc ejecting compounds. (b) Chemotype of *S*-acyl-2-mercaptobenzamide thioesters. (c) Scheme of the proposed thioesters mechanism of action.

37.6 CONCLUSIONS AND PERSPECTIVES

The U.S. FDA approval of raltegravir and the clinical success of GS-9137 have validated the targeting of IN as a safe and efficacious approach for the treatment of HIV/AIDS. The addition of IN inhibitors to HAART will have a tremendous impact on treating antiretroviral-experienced patients, thus improving the quality of life of patients not responsive to existing therapeutic protocols.

This milestone for HIV/AIDS treatment is the result of many efforts and strategies undertaken in the drug discovery process, particularly focused on the development of numerous strand transfer specific IN inhibitor chemical classes. Furthermore, both structural studies on HIV-1 IN inhibitors and the biology of metal cofactors have confirmed the important role of metals in the inhibition process, and that the metal binding function is a critically factor in the development of IN inhibitors. In addition to the above-mentioned potent inhibitors, there are a number of validated lead compounds in advanced drug development stages. However, considering this inhibitory similarity, the emergence of cross-resistant HIV strains to multiple clinically viable IN inhibitors is inevitable. As the discovery of the first strand transfer-specific IN inhibitor class (e.g., diketo acids) was a key advance in the field of HIV IN drug design and development, this prompts the need to design second-generation IN inhibitors.

The discovery of IN inhibitors to target other critical aspects of IN activity *in vivo*, such as multimerization disruption and the obstruction of IN–cellular cofactor interactions, will provide other options for potential clinical development. For example, the poorly investigated N-terminal domain of IN would constitute an interesting new target for IN inhibition. In particular, a crystal structure of full-length IN or full-length IN–LEDGF/p75 structures will provide the required structural information to embark on drug design. Analogue- and structure-based drug design strategies targeting this inhibitor binding site and the HHCC motif, as well as the investigation of small molecules that selectively interfere with the Zn^{2+} ion could represent a suitable strategy for the design of novel IN inhibitors.

ACKNOWLEDGMENTS

M.S. is grateful to Fondazione Banco di Sardegna for its partial financial support. We also thank all collaborators, students, and friends who contributed with their great enthusiasm to our research in this field.

REFERENCES

1. Joint United Nations Programme on HIV/AIDS and World Health Organization (UNAIDS). AIDS Epidemic Update, December **2007**.
2. Richman, D. D. HIV chemotherapy. *Nature* **2001**, *410*, 995–1001.
3. Turpin, J. A. The next generation of HIV/AIDS drugs: novel and developmental anti-HIV drugs and targets. *Expert Rev. Anti-Infect. Ther.* **2003**, *1*, 97–128.
4. Barbaro, G.; Scozzafava, A.; Mastrolorenzo, A.; Supuran, C. T. Highly active antiretroviral therapy: current state of the art, new agents and their pharmacological interactions useful for improving therapeutic outcome. *Curr. Pharm. Des.* **2005**, *11*, 1805–1843.
5. Greene, W. C.; Debyser, Z.; Ikeda, Y.; Freed, E. O.; Stephens, E.; Yonemoto, W.; Buckheit, R. W.; Esté, J. A.; Cihlar, T. Novel targets for HIV therapy. *Antiv. Res.* **2008**, *80*, 251–265.
6. Shet, A.; Berry, L.; Mohri, H.; Mehandru, S.; Chung, C.; Kim, A.; Jean-Pierre, P.; Hogan, C.; Simon, V.; Boden, D.; Markowitz, M. Tracking the prevalence of transmitted antiretroviral drug-resistant HIV-1: a decade of experience. *J. Acquir. Immune Defic. Syndr.* **2006**, *41*, 439–446.
7. Shet, A.; Markowitz, M. Transmitted multidrug resistant HIV-1: new and investigational therapeutic approaches. *Curr. Opin. Investig. Drugs* **2006**, *7*, 709–720.

8. Esser, S.; Helbig, D.; Hillen, U.; Dissemond, J.; Grabbe, S. Side effects of HIV therapy. *J. Dtsch. Dermatol. Ges.* **2007**, *5*, 745–754.
9. De Clercq, E. Strategies in the design of antiviral drugs. *Nat. Rev. Drug Discov.* **2002**, *1*, 13–25.
10. De Clercq, E. New approaches toward anti-HIV chemotherapy. *J. Med. Chem.* **2005**, *48*, 1297–1313.
11. Pommier, Y.; Johnson, A. A.; Marchand, C. Integrase inhibitors to treat HIV/AIDS. *Nat. Rev. Drug Discov.* **2005**, *4*, 236–248.
12. Neamati, N.; Marchand, C.; Pommier, Y. HIV-1 integrase inhibitors: past, present, and future. *Adv. Pharmacol.* **2000**, *49*, 147–165.
13. d'Angelo, J.; Mouscadet, J. F.; Desmaele, D.; Zouhiri, F.; Leh, H. HIV-1 integrase: the next target for AIDS therapy? *Pathol. Biol.* **2001**, *49*, 237–246.
14. Neamati, N. Structure-based HIV-1 integrase inhibitor design: a future perspective. *Expert Opin. Investig. Drugs* **2001**, *10*, 281–296.
15. Anthony, N. J. HIV-1 integrase: a target for new AIDS chemotherapeutics. *Curr. Top. Med. Chem.* **2004**, *4*, 979–990.
16. Nair, V.; Chi, G. HIV integrase inhibitors as therapeutic agents in AIDS. *Rev. Med. Virol.* **2007**, *17*, 277–295.
17. Dayam, R.; Al-Mawsawi, L. Q.; Neamati, N. HIV-1 integrase inhibitors: an emerging clinical reality. *Drugs R&D* **2007**, *8*, 155–168.
18. MK-0518 Meeting Transcript. Department of Health and Human Services, Food and Drug Administration, Center for Drug Evaluation and Research, Antiviral Drugs Advisory Committee, **2007**, September 5. Available from www.fda.gov/ohrms/dockets/ac/cder07.htm#AntiviralDrugs.
19. Al-Mawsawi, L. Q.; Al-Safi, R. I.; Neamati, N. Anti-infectives: clinical progress of HIV-1 integrase inhibitors. *Expert Opin. Emerg. Drugs* **2008**, *13*, 213–225.
20. Brown, P. O.; Integration. In *Retroviruses;* Coffin J. C.; Hughes S. H.; Varmus, H. E.,Eds.; Cold Spring Harbor Press: Plainview, NY, **1999**.
21. Bushman, F. D.; Craigie, R. Activities of human immunodeficiency virus (HIV) integration protein *in vitro*: specific cleavage and integration of HIV DNA. *Proc. Natl. Acad. Sci. USA* **1991**, *88*, 1339–1343.
22. Engelman, A.; Mizuuchi, K.; Craigie, R. HIV-1 DNA integration: mechanism of viral DNA cleavage and DNA strand transfer. *Cell* **1991**, *67*, 1211–1221.
23. Sechi, M.; Carcelli, M.; Rogolino, D.; Neamati, N. Role of metals in HIV-1 integrase inhibitor design. In *HIV-1 Integrase: Mechanism of Action and Inhibitor Design;* John Wiley & Sons, in press.
24. Engelman, A.; Craigie, R. Identification of conserved amino acid residues critical for human immunodeficiency virus type 1 integrase function *in vitro*. *J. Virol.* **1992**, *66*, 6361–6369.
25. Esposito, D.; Craigie, R. Integrase structure and function. *Adv. Virus Res.* **1999**, *52*, 319–333.
26. Asante-Appiah, E.; Skalka, A. M. A molecular mechanisms in retrovirus DNA integration. *Antiv. Res.* **1997**, *36*, 139.
27. Asante-Appiah, E.; Skalka, A. M. HIV-1 integrase: structural organization, conformational changes, and catalysis. *Adv. Virus Res.* **1999**, *52*, 351–369.

28. Dyda, F.; Hickman, A. B.; Jenkins, T. M.; Engelman, A.; Craigie, R.; Davies, D. R. Crystal structure of the catalytic domain of HIV-1 integrase: similarity to other polynucleotidyl transferases. *Science* **1994**, *266*, 1981–1986.
29. Rice, P. A.; Baker, T. A. Comparative architecture of transposase and integrase complexes. *Nat. Struct. Biol.* **2001**, *8*, 302–307.
30. Chiu, T. K.; Davies, D. R. Structure and function of HIV-1 integrase. *Curr. Top. Med. Chem.* **2004**, *4*, 965–977.
31. Kulkosky, J.; Jones, K. S.; Katz, R. A.; Mack, J. P.; Skalka, A. M. Residues critical for retroviral integrative recombination in a region that is highly conserved among retroviral/retrotransposon integrases and bacterial insertion sequence transposases. *Mol. Cell. Biol.* **1992**, *12*, 2331–2338.
32. Beese, L. S.; Steitz, T. A. Structural basis for the 3′–5′ exonuclease activity of *Escherichia coli* DNA polymerase I: a two metal ion mechanism. *EMBO J.* **1991**, *10*, 25–33.
33. Steitz, T. A.; Steitz, J. A. A general two-metal-ion mechanism for catalytic RNA. *Proc. Natl. Acad. Sci. USA* **1993**, *90*, 6498–6502.
34. Steitz, T. A. A mechanism for all polymerases. *Nature* **1998**, *391*, 231–232.
35. Engelman, A.; Bushman, F. D.; Craigie, R. Identification of discrete functional domains of HIV-1 integrase and their organization within an active multimeric complex. *EMBO J.* **1993**, *12*, 3769–3775.
36. Jenkins, T. M.; Engelman, A.; Ghirlando, R.; Craigie, R. A soluble active mutant of HIV-1 integrase: involvement of both the core and carboxyl-terminal domains in multimerization. *J. Biol. Chem.* **1996**, *271*, 7712–7718.
37. Chen, J. C.; Krucinski, J.; Miercke, L. J.; Finer-Moore, J. S.; Tang, A. H.; Leavitt, A. D.; Stroud, R. M. Crystal structure of the HIV-1 integrase catalytic core and C-terminal domains: a model for viral DNA binding. *Proc. Natl. Acad. Sci. USA* **2000**, *97*, 8233–8238.
38. Lutzke, R. A.; Plasterk, R. H. Structure-based mutational analysis of the C-terminal DNA-binding domain of human immunodeficiency virus type 1 integrase: critical residues for protein oligomerization and DNA binding. *J. Virol.* **1998**, *72*, 4841–4848.
39. Chen, I. J.; Neamati, N.; MacKerell, A. D., Jr. Structure-based inhibitor design targeting HIV-1 integrase. *Curr. Drug Targets Infect. Disord.* **2002**, *2*, 217–234.
40. Zheng, R.; Jenkins, T. M.; Craigie, R. Zinc folds the N-terminal domain of HIV-1 integrase, promotes multimerization, and enhances catalytic activity. *Proc. Natl. Acad. Sci. USA* **1996**, *93*, 13659–13664.
41. Bushman, F. D.; Engelman, A.; Palmer, I.; Wingfield, P.; Craigie, R. Domains of the integrase protein of human immunodeficiency virus type 1 responsible for polynucleotidyl transfer and zinc binding. *Proc. Natl. Acad. Sci. USA* **1993**, *90*, 3428–3437.
42. Johnson, M. S.; McClure, M. A.; Feng, D. F.; Gray, J.; Doolittle, R. F. Computer analysis of retroviral *pol* genes: assignment of enzymatic functions to specific sequences and homologies with nonviral enzymes. *Proc. Natl. Acad. Sci. USA* **1986**, *83*, 7648–7652.
43. Keilin, D.; Mann, T. Carbonic anhydrase. Purification and nature of the enzyme. *Biochem. J.* **1940**, *34*, 1163–1176.
44. Supuran, C. T.; Scozzafava, A.; Conway, J. *Carbonic Anhydrase: Its Inhibitors and Activators;* CRC Press LLC: Boca Raton, FL, **2004**.
45. Coleman, J. E. Zinc proteins: enzymes, storage proteins, transcription factors, and replication proteins. *Annu. Rev. Biochem.* **1992**, *61*, 897–946.

46. Vallee, B. L.; Auld, D. S. Zinc coordination, function, and structure of zinc enzymes and other proteins. *Biochemistry* **1990**, *29*, 5647–5659.

47. Darlix, J. L.; Lapadat-Tapolsky, M.; de Rocquigny, H.; Roques, B. P. First glimpses at structure–function relationships of the nucleocapsid protein of retroviruses. *Mol. Biol.* **1995**, *254*, 523–537.

48. Druillennec, S.; Roques, B. P. HIV-1 NCp7 as a target for the design of novel antiviral agents. *Drug News Perspect.* **2000**, *13*, 337–349.

49. Summers, M. F.; Henderson, L. E.; Chance, M. R.; Bess, J. W., Jr.; South, T. L.; Blake, P. R.; Sagi, I.; Perez-Alvarado, G.; Sowder, R. C., 3rd; Hare, D. R.; Arthur, L. O. Nucleocapsid zinc fingers detected in retroviruses: EXAFS studies of intact viruses and the solution-state structure of the nucleocapsid protein from HIV-1. *Protein Sci.* **1992**, *1*, 563–574.

50. Morellet, N.; Jullian, N.; De Rocquigny, H.; Maigret, B.; Darlix, J. L.; Roques, B. P. Determination of the structure of the nucleocapsid protein NCp7 from the human immunodeficiency virus type 1 by ^{1}H NMR. *EMBO J.* **1992**, *11*, 3059–3065.

51. Hanas, J. S.; Hazuda, D. J.; Bogenhagen, D. F.; Wu, F. H.-Y.; Wu, C.-W. *Xenopus* transcription factor A requires zinc for binding to the 5S RNA gene. *J. Biol. Chem.* **1983**, *258*, 14120–14125.

52. Miller, J.; McLachlan, A. D.; Klug, A. Repetitive zinc-binding domains in the protein transcription factor IIIA from *Xenopus* oocytes. *EMBO J.* **1985**, *4*, 1609–1614.

53. Goldgur, Y.; Dyda, F.; Hickman, A. B.; Jenkins, T. M.; Craigie, R.; Davies, D. R. Three new structures of the core domain of HIV-1 integrase: an active site that binds magnesium. *Proc. Natl. Acad. Sci. USA* **1998**, *95*, 9150–9154.

54. Maignan, S.; Guilloteau, J. P.; Zhou-Liu, Q.; Clement-Mella, C.; Mikol, V. Crystal structures of the catalytic domain of HIV-1 integrase free and complexed with its metal cofactor: high level of similarity of the active site with other viral integrases. *J. Mol. Biol.* **1998**, *282*, 359–368.

55. Greenwald, J.; Le, V.; Butler, S. L.; Bushman, F. D.; Choe, S. The mobility of an HIV-1 integrase active site loop is correlated with catalytic activity. *Biochemistry* **1999**, *38*, 8892–8898.

56. Cai, M.; Zheng, R.; Caffrey, M.; Craigie, R.; Clore, M.; Gronenborn, A. M. Solution structure of the N-terminal zinc binding domain of HIV-1 integrase. *Nat. Struct. Biol.* **1997**, *4*, 567–577.

57. Cai, M.; Huang, Y.; Caffrey, M.; Zheng, R.; Craigie, R.; Clore, M.; Gronenborn, A. M. Solution structure of the His12→Cys mutant of the N-terminal zinc binding domain of HIV-1 integrase complexed with cadmium. *Protein Sci.* **1998**, *7*, 2669–2674.

58. Nomura, Y.; Masuda, T.; Kawai, G. Structural analysis of a mutant of the HIV-1 integrase zinc finger domain that forms a single conformation. *J. Biochem.* **2006**, *139*, 753–759.

59. Eijkelenboom, A. P. A. M.; van den Ent, F. M. I.; Vos, A.; Doreleijers, J. F.; Hard, K.; Tullius, T. D.; Plasterk, R. H. A.; Kaptein, R.; Boelens, R. The solution structure of the amino-terminal HHCC domain of HIV-2 integrase: a three-helix bundle stabilized by zinc. *Curr. Biol.* **1997**, *7*, 739–746.

60. Eijkelenboom, A. P. A. M.; van den Ent, F. M. I.; Wechselberger, R.; Plasterk, R. H. A.; Kaptein, R.; Boelens, R. Refined solution structure of the dimeric N-terminal HHCC domain of HIV-2 integrase. *J. Biomol. NMR* **2000**, *18*, 119–128.

61. Wang, J.-Y.; Ling, H.; Craigie, R. Structure of a two-domain fragment of HIV-1 integrase: implications for domain organization in the intact protein. *EMBO J.* **2001**, *20*, 7333–7343.
62. Lee, S. P.; Xiao, J.; Knutson, J. R.; Lewis, M. S.; Han, M. K. Zn^{2+} promotes the self-association of human immunodeficiency virus type-1 integrase *in vitro*. *Biochemistry* **1997**, *36*, 173–180.
63. Burke, C. J.; Sanyal, G.; Bruner, M. W.; Ryan, J. A.; LaFemina, R. L.; Robbins, H. L.; Zeft, A. S.; Middaugh, R. C.; Cordingley, M. G. Structural implications of spectroscopic characterization of a putative zinc finger peptide from HIV-1 integrase. *J. Biol. Chem.* **1992**, *14*, 9639–9644.
64. Nakamura, T.; Masuda, T.; Goto, T.; Sano, K.; Nakai, M.; Harada, S. Lack of infectivity of HIV-1 integrase zinc finger-like domain mutant with morphologically normal maturation. *Biochem. Biophys. Res. Commun.* **1997**, *239*, 715–722.
65. Khan, E.; Mack, J. P.; Katz, R. A.; Kulkosky, J.; Skalka, A. M. Retroviral integrase domain: DNA binding and the recognition of LTR sequences. *Nucleic Acids Res.* **1991**, *19*, 851–860.
66. Schauer, M.; Billich, A. The N-terminal region of HIV-1 integrase is required for integration activity, but not for DNA binding. *Biochem. Biophys. Res. Commun.* **1992**, *185*, 874–880.
67. Lee, S. P.; Han, M. K. Zinc stimulates Mg^{2+}-dependent 3′-processing activity of human immunodeficiency virus type 1 integrase *in vitro*. *Biochemistry* **1996**, *35*, 3837–3844.
68. van den Ent, F. M. I.; Vos, A.; Plasterk, R. H. A. Dissecting the role of the N-terminal domain of human immunodeficiency virus integrase by trans-complementation analysis. *J. Virol.* **1999**, *73*, 3176–3183.
69. Engelman, A.; Bushman, F. D.; Craigie, R. Identification of discrete functional domains of HIV-1 integrase and their organization within an active multimeric complex. *EMBO J.* **1993**, *12*, 3769–3775.
70. Vink, C.; Oude Groeneger, A. M.; Plasterk, R. H. Identification of the catalytic and DNA-binding region of the human immunodeficiency virus type I integrase protein. *Nucleic Acids Res.* **1993**, *21*, 1419–1425.
71. Hazuda, D. J.; Felock, P.; Witmer, M.; Wolfe, A.; Stillmock, K.; Grobler, J. A.; Espeseth, A.; Gabryelski, L.; Schleif, W.; Blau, C.; Miller, M. D. Inhibitors of strand transfer that prevent integration and inhibit HIV-1 replication in cells. *Science* **2000**, *287*, 646–650.
72. Goldgur, Y.; Craigie, R.; Cohen, G. H.; Fujiwara, T.; Yoshinaga, T.; Fujishita, T.; Sugimoto, H.; Endo, T.; Murai, H.; Davies, D. R. Structure of the HIV-1 integrase catalytic domain complexed with an inhibitor: a platform for antiviral drug design. *Proc. Natl. Acad. Sci. USA* **1999**, *96*, 13040–13043.
73. Pais, G. C. G.; Burke, T. R. Novel aryl diketo-containing inhibitors of HIV-1 integrase. *Drugs Future* **2002**, *27*, 1101–1111.
74. Egbertson, M. S. HIV integrase inhibitors: from diketoacids to heterocyclic templates: a history of HIV integrase medicinal chemistry at Merck West Point and Merck Rome (IRBM). *Curr. Top. Med. Chem.* **2007**, *7*, 1251–1272.
75. Wang, Y.; Serradell, N.; Bolos, J.; Rosa, E.; MK-0518, HIV integrase inhibitor. *Drugs Future* **2007**, *37*, 118–122.
76. Rowley, M. The discovery of raltegravir, an integrase inhibitor for the treatment of HIV infection. *Prog. Med. Chem.* **2008**, *46*, 1–28.

77. Summa, V.; Petrocchi, A.; Bonelli, F.; Crescenzi, B.; Donghi, M.; Ferrara, M.; Fiore, F.; Gardelli, C.; Gonzalez, Paz O.; Hazuda, D. J.; Jones, P.; Kinzel, O.; Laufer, R.; Monteagudo, E.; Muraglia, E.; Nizi, E.; Orvieto, F.; Pace, P.; Pescatore, G.; Scarpelli, R.; Stillmock, K.; Witmer, M. V.; Rowley, M. Discovery of raltegravir, a potent, selective orally bioavailable HIV-integrase inhibitor for the treatment of HIV-AIDS infection. *J. Med. Chem.* **2008**, *51*, 5843–5855.

78. Serrao, E.; Odde, S.; Ramkumar, K.; Neamati, N., Raltegravir, elvitegravir, and metoogravir: the birth of "me-too" HIV-1 integrase inhibitors. *Retrovirology* **2009**, *6*, 25.

79. Neamati, N. Patented small molecule inhibitors of HIV-1 integrase: a ten-year saga. *Expert Opin. Ther. Patents* **2002**, *12*, 709–724.

80. Gupta, S. P.; Nagappa, A. N. Design and development of integrase inhibitors as anti-HIV agents. *Curr. Med. Chem.* **2003**, *10*, 1779–1794.

81. Dayam, R.; Neamati, N. Small-molecule HIV-1 integrase inhibitors: the 2001–2002 update. *Curr. Pharm. Des.* **2003**, *9*, 1789–1802.

82. Johnson, A. A.; Marchand, C.; Pommier, Y. HIV-1 integrase inhibitors: a decade of research and two drugs in clinical trial. *Curr. Top. Med. Chem.* **2004**, *4*, 671–686.

83. Dayam, R.; Deng, J.; Neamati, N. HIV-1 integrase inhibitors: 2003–2004 update. *Med. Res. Rev.* **2006**, *26*, 271–309.

84. Cotelle, P. Patented HIV-1 integrase inhibitors (1998–2005). *Recent Patents Anti-Infect. Drug Discov.* **2006**, *1*, 1–15.

85. Zeinalipour-Loizidou, E.; Nicolaou, C.; Nicolaides, A.; Kostrikis, L. G. HIV-1 integrase: from biology to chemotherapeutics. *Curr. HIV Res.* **2007**, *5*, 365–388.

86. Dayam, R.; Gundla, R.; Al-Mawsawi, L. Q.; Neamati, N. HIV-1 integrase inhibitors: 2005–2006 update. *Med. Res. Rev.* **2008**, *28*, 118–154.

87. Christopoulos, A. Allosteric binding sites on cell-surface receptors: novel targets for drug discovery. *Nat. Rev. Drug Discov.* **2002**, *1*, 198–210.

88. Debyser, Z.; Pauwels, R.; Andries, K.; Desmyter, J.; Kukla, M.; Janssen, P. A.; De Clercq, E. An antiviral target on reverse transcriptase of human immunodeficiency virus type 1 revealed by tetrahydroimidazo-[4,5,1-*jk*][1,4]benzodiazepin-2(1*H*)-one and -thione derivatives. *Proc. Natl. Acad. Sci. USA* **1991**, *88*, 1451–1455.

89. Thomas, J. A.; Gorelick, R. J. Nucleocapsid protein function in early infection processes. *Virus Res.* **2008**, *134*, 39–63.

90. Rice, W. G.; Supko, J. G.; Malspeis, L.; Buckheit, R. W., Jr.; Clanton, D.; Bu, M.; Graham, L.; Schaeffer, C. A.; Turpin, J. A.; Domagala, J.; Gogliotti, R.; Bader, J. P.; Halliday, S. M.; Coren, L.; Sowder, R. C., 2nd; Arthur, L. O.; Henderson, L. E. Inhibitors of HIV nucleocapsid protein zinc fingers as candidates for the treatment of AIDS. *Science* **1995**, *270*, 1194–1197.

91. Rice, W. G.; Turpin, J. A. Virus-encoded zinc fingers as targets for antiviral chemotherapy. *Rev. Med. Virol.* **1996**, *6*, 187–199.

92. Wu, X.; Liu, H.; Xiao, H.; Conway, J. A.; Hehl, E.; Kalpana, G. V.; Prasad, V.; Kappes, J. C. Human immunodeficiency virus type 1 integrase protein promotes reverse transcription through specific interactions with the nucleoprotein reverse transcription complex. *J. Virol.* **1999**, *73*, 2126–2135.

93. Sun, T.; Wang, Q.; Yu, Z.; Zhang, Y.; Guo, Y.; Chen, K.; Shen, X.; Jiang, H. Hyrtiosal, a PTP1B inhibitor from the marine sponge *Hyrtios erectus*, shows extensive cellular effects on PI3K/AKT activation, glucose transport, and TGFβ/Smad2 signaling. *ChemBioChem* **2007**, *8*, 187–193.

94. Du, L.; Shen, L.; Yu, Z.; Chen, J.; Guo, Y.; Tang, Y.; Shen, X.; Jiang, H. Hyrtiosal, from the marine sponge *Hyrtios erectus*, inhibits HIV-1 integrase binding to viral DNA by a new inhibitor binding site. *ChemMedChem* **2008**, *3*, 173–180.

95. Jones, K. S.; Coleman, J.; Merkel, G. W.; Laue, T. M.; Skalka, A. M. Retroviral integrase functions as a multimer and can turn over catalytically. *J. Biol. Chem.* **1992**, *267*, 16037–16040.

96. Sluis-Cremer, N.; Tachedjian, G. Modulation of the oligomeric structures of HIV-1 retroviral enzymes by synthetic peptides and small molecules. *Eur. J. Biochem.* **2002**, *269*, 5103–5111.

97. Maroun, R. G.; Gayet, S.; Benleulmi, M. S.; Porumb, H.; Zargarian, L.; Merad, H.; Leh, H.; Mouscadet, J. F.; Troalen, F.; Fermandjian, S. Peptide inhibitors of HIV-1 integrase dissociate the enzyme oligomers. *Biochemistry* **2001**, *40*, 13840–13848.

98. Li, H. Y.; Zawahir, Z.; Song, L. D.; Long, Y. Q.; Neamati, N. Sequence-based design and discovery of peptide inhibitors of HIV-1 integrase: insight into the binding mode of the enzyme. *J. Med. Chem.* **2006**, *49*, 4477–4486.

99. Hayouka, Z.; Rosenbluh, J.; Levin, A.; Loya, S.; Lebendiker, M.; Veprintsev, D.; Kotler, M.; Hizi, A.; Loyter, A.; Friedler, A. Inhibiting HIV-1 integrase by shifting its oligomerization equilibrium. *Proc. Natl. Acad. Sci. USA* **2007**, *104*, 8316–8321.

100. Cherepanov, P.; Maertens, G.; Proost, P.; Devreese, B.; Van Beeumen, J.; Engelborghs, Y.; De Clercq, E.; Debyser, Z. HIV-1 integrase forms stable tetramers and associates with LEDGF/p75 protein in human cells. *J. Biol. Chem.* **2003**, *278*, 372–381.

101. Ciuffi, A.; Llano, M.; Poeschla, E.; Hoffmann, C.; Leipzig, J.; Shinn, P.; Ecker, J. R.; Bushman, F. A role for LEDGF/p75 in targeting HIV DNA integration. *Nat. Med.* **2005**, *11*, 1287–1289.

102. Llano, M.; Saenz, D. T.; Meehan, A.; Wongthida, P.; Peretz, M.; Walker, W. H.; Teo, W.; Poeschla, E. M. An essential role for LEDGF/p75 in HIV integration. *Science* **2006**, *314*, 461–464.

103. Cherepanov, P.; Sun, Z. Y.; Rahman, S.; Maertens, G.; Wagner, G.; Engelman, A. Solution structure of the HIV-1 integrase-binding domain in LEDGF/p75. *Nat. Struct. Mol. Biol.* **2005**, *12*, 526–537.

104. Cherepanov, P.; Ambrosio, A. L.; Rahman, S.; Ellenberger, T.; Engelman, A. Structural basis for the recognition between HIV-1 integrase and transcriptional coactivator p75. *Proc. Natl. Acad. Sci. USA* **2005**, *102*, 17308–17313.

105. Busschots, K.; Voet, A.; De Maeyer, M.; Rain, J. C.; Emiliani, S.; Benarous, R.; Desender, L.; Debyser, Z.; Christ, F. Identification of the LEDGF/p75 binding site in HIV-1 integrase. *J. Mol. Biol.* **2007**, *365*, 1480–1492.

106. Al-Mawsawi, L. Q.; Neamati, N. Blocking interactions between HIV-1 integrase and cellular cofactors: an emerging anti-retroviral strategy. *Trends Pharmacol. Sci.* **2007**, *28*, 526–535.

107. Du, L.; Zhao, Y.; Chen, J.; Yang, L.; Zheng, Y.; Tang, Y.; Shen, X.; Jiang, H. D77, one benzoic acid derivative, functions as a novel anti-HIV-1 inhibitor targeting the interaction between integrase and cellular LEDGF/p75. *Biochem. Biophys. Res. Commun.* **2008**, *375*, 139–144.

108. Semenova, E. A.; Marchand, C.; Pommier, Y. HIV-1 integrase inhibitors: update and perspectives. *Adv. Pharmacol.* **2008**, *56*, 199–228.

109. Witvrouw, M.; Van Maele, B.; Vercammen, J.; Hantson, A.; Engelborghs, Y.; De Clercq, E.; Pannecouque, C.; Debyser, Z. Novel inhibitors of HIV-integration. *Curr. Drug Metab.* **2004**, *5*, 291–304.

110. Van Maele, B.; Busschots, K.; Vandekerckhove, L.; Christ, F.; Debyser, Z. Cellular cofactors of HIV-1 integration. *Trends Biochem. Sci.* **2006**, *31*, 98–105.

111. Figgis, B. N. *Introduction to Ligand Fields;* John Wiley & Sons/Interscience: New York, **1966**.

112. Pearson, R. G. Hard and soft acids and bases. *J. Am. Chem. Soc.* **1963**, *85*, 3533–3539.

113. Druillennec, S.; Dong, C. Z.; Escaich, S.; Gresh, N.; Bousseau, A.; Roques, B. P.; Fournié-Zaluski, M. C. A mimic of HIV-1 nucleocapsid protein impairs reverse transcription and displays antiviral activity. *Proc. Natl. Acad. Sci. USA* **1999**, *96*, 4886–4891.

114. Rice, W. G.; Schaeffer, C. A.; Harten, B.; Villinger, F.; South, T. L.; Summers, M. F.; Henderson, L. E.; Bess, J. W., Jr.; Arthur, L. O.; Mcdougal, J. S.; Orloff, S. L.; Mendeleyev, J.; Kun, E. Inhibition of HIV-1 infectivity by zinc-ejecting aromatic C-nitroso compounds. *Nature* **1993**, *361*, 473–475.

115. Rice, W. G.; Turpin, J. A.; Schaeffer, C. A.; Graham, L.; Clanton, D.; Buckheit, R. W., Jr.; Zaharevitz, D.; Summers, M. F.; Wallqvist, A.; Covell, D. G. Evaluation of selected chemotypes in coupled cellular and molecular target-based screens identifies novel HIV-1 zinc finger inhibitors. *J. Med. Chem.* **1996**, *39*, 3606–3616.

116. Loo, J. A.; Holler, T. P.; Sanchez, J.; Gogliotti, R.; Maloney, L.; Reily, M. D. Biophysical characterization of zinc ejection from HIV nucleocapsid protein by anti-HIV 2,2′-dithiobis[benzamides] and benzisothiazolones. *J. Med. Chem.* **1996**, *39*, 4313–4320.

117. Witvrouw, M.; Balzarini, J.; Pannecouque, C.; Jhaumeer-Laulloo, S.; Este, J. A.; Schols, D.; Cherepanov, P.; Schmit, J. C.; Debyser, Z.; Vandamme, A. M.; Desmyter, J.; Ramadas, S. R.; Clercq, E. SRR-SB3, a disulfide-containing macrolide that inhibits a late stage of the replicative cycle of human immunodeficiency virus. *Antimicrob. Agents Chemother.* **1997**, *41*, 262–268.

118. Rice, W. G.; Baker, D. C.; Schaeffer, C. A.; Graham, L.; Bu, M.; Terpening, S.; Clanton, D.; Schultz, R.; Bader, J. P.; Buckheit, R. W., Jr.; Field, L.; Singh, P. K.; Turpin, J. A. Inhibition of multiple phases of human immunodeficiency virus type 1 replication by a dithiane compound that attacks the conserved zinc fingers of retroviral nucleocapsid proteins. *Antimicrob. Agents Chemother.* **1997**, *41*, 419–426.

119. Rice, W. G.; Turpin, J. A.; Huang, M.; Clanton, D.; Buckheit, R. W., Jr.; Covell, D. G.; Wallqvist, A.; McDonnell, N. B.; DeGuzman, R. N.; Summers, M. F.; Zalkow, L.; Bader, J. P.; Haugwitz, R. D.; Sausville, E. A. Azodicarbonamide inhibits HIV-1 replication by targeting the nucleocapsid protein. *Nat. Med.* **1997**, *3*, 341–345.

120. Turpin, J. A.; Song, Y.; Inman, J. K.; Huang, M.; Wallqvist, A.; Maynard, A.; Covell, D. G.; Rice, W. G.; Appella, E. Synthesis and biological properties of novel pyridinioalkanoyl thiolesters (PATE) as anti-HIV-1 agents that target the viral nucleocapsid protein zinc fingers. *J. Med. Chem.* **1999**, *42*, 67–86.

121. Rossio, J. L.; Esser, M. T.; Suryanarayana, K.; Schneider, D. K.; Bess, J. W., Jr.; Vasquez, G. M.; Wiltrout, T. A.; Chertova, E.; Grimes, M. K.; Sattentau, Q.; Arthur, L. O.; Henderson, L. E.; Lifson, J. D. Inactivation of human immunodeficiency virus type 1 infectivity with preservation of conformational and functional integrity of virion surface proteins. *J. Virol.* **1998**, *72*, 7992–8001.

122. Tummino, P. J.; Scholten, J. D.; Harvey, P. J.; Holler, T. P.; Maloney, L.; Gogliotti, R.; Domagala, J.; Hupe, D. The *in vitro* ejection of zinc from human immunodeficiency virus (HIV) type 1 nucleocapsid protein by disulfide benzamides with cellular anti-HIV activity. *Proc. Natl. Acad. Sci. USA* **1996**, *93*, 969–973.

123. Srivastava, P.; Schito, M.; Fattah, R. J.; Hara, T.; Hartman, T.; Buckheit, R. W., Jr.; Turpin, J. A.; Inman, J. K.; Appella, E. Optimization of unique, uncharged, thioesters as inhibitors of HIV replication. *Bioorg. Med. Chem.* **2004**, *12*, 6437–6450.

124. Miller Jenkins, M. J.; Byrd, J. C.; Hara, T.; Srivastava, P.; Mazur, S. J.; Stahl, S. J.; Inman, J. K.; Appella, E.; Omichinski, J. G.; Legault, P. Studies on the mechanism of inactivation of the HIV-1 nucleocapsid protein NCp7 with 2-mercaptobenzamide thioesters. *J. Med. Chem.* **2005**, *48*, 2847–2858.

125. Yu, X.-F.; Erhlich, E. S.; Zuoxiang, X. VIF as a target for HIV inhibition. *U.S. Patent Application* 20080206357, **2008**.

CHAPTER 38

Inhibitors of Histidinol Dehydrogenases as Antibacterial Agents

PASCALE JOSEPH[1], FRANÇOIS TURTAUT[2], STEPHAN KÖHLER[1], and JEAN-YVES WINUM[2]

[1]Centre d'Etudes d'Agents Pathogènes et Biotechnologies pour la Santé (CPBS) UMR 5236 CNRS-UM1-UM2, Université Montpellier II, cc100, Place E. Bataillon, 34095 Montpellier, France

[2]Institut des Biomolécules Max Mousseron (IBMM) UMR 5247 CNRS-UM1-UM2 Bâtiment de Recherche Max Mousseron, Ecole Nationale Supérieure de Chimie de Montpellier, 8 rue de l'Ecole Normale, 34296 Montpellier Cedex, France

38.1 INTRODUCTION

Bacterial infections constitute an alarming health problem worldwide. The widespread emergence of resistance and multiresistance to antibiotics among bacterial pathogens represents a major threat and necessitates a permanent race for new molecules. However, during the past 40 years, only two new classes of antibiotics have been discovered[1] and strains resistant to the molecules used appear in general within a few years following the marketing of the molecule. There are three possibilities to reduce this trend: (1) to improve the prescription method, (2) to find new drug targets, and (3) to design new types of drugs that will reduce the appearance of resistance. The challenge at present is to identify and validate novel pharmaceutical targets in bacteria, which would be the starting point for the discovery of new classes of antibacterial agents that could circumvent the established resistance mechanism. With the accumulation of huge and still growing amounts of data from bacterial genomics, potential new drug targets have been identified. In the last few years, the growing information on the virulence mechanisms of various pathogens in the relationship with their hosts has led to increased interest among scientists from public research institutions and from private companies in the potential definition of virulence factors as novel targets for anti-infectious agents.

As pathogenicity of intracellular bacteria is linked to their capacity to multiply within the host cell, the development of new antibacterials specifically active at the intracellular state would block multiplication of the pathogen without affecting extracellular bacteria. This approach presents the advantage to limit pressure for selection of resistant mutants to intracellular bacteria located inside the host cell and to reduce the secondary effects observed on the commensal flora.

Among the wealth of antibacterial drug discovery targets available from bacterial genomics investigations, metalloenzymes constitute an attractive target for antibacterial chemotherapy and provide an excellent opportunity for mechanism-based drug discovery of novel classes of antibiotics.[2,3] This chapter will focus on a metalloenzyme of the facultative intracellular pathogen *Brucella*: the histidinol dehydrogenase (HDH) as a novel and defined biological target for the development of anti-infectious agents acting specifically inside the host cells.

This metalloprotein has been previously identified as essential for *Brucella* virulence inside the macrophage host cell[4] and is present only in bacteria and plants, hence strongly reducing the risk of secondary effects of potential inhibitors on the host himself. In contrast to classical antibiotic screening programs where large collections of chemical products are systematically tested for an eventual effect on bacterial viability, we describe here an original, focused approach with a defined biological target known to be involved in the virulence of the pathogen of interest. Recently, this strategy has been validated by targeting the acetohydroxyacid synthase (AHAS) of *B. suis* involved in biosynthesis of branched amino acids and crucial for the pathogen's survival within the host cell.[5]

We describe in this chapter the design, synthesis, and evaluation of novel, highly potent, and selective HDH inhibitors and the assessment of their efficacy (1) on the purified target enzyme, (2) on the intact pathogen *in vitro*, and (3) on intracellular bacteria in macrophage infection experiments.

38.2 *Brucella*–HOST INTERACTIONS

38.2.1 Brucellosis

Brucella spp. is the causative agent of brucellosis, the most important anthropozoonosis worldwide[6–8] causing enormous losses in agriculture. In endemic areas (Mediterranean Europe, Middle East, Latin America), incidence of human brucellosis may be as high as 200 per 100,000 habitants. Human brucellosis (or Malta fever) is not considered as contagious disease but it is a highly disabling disease that may last for weeks and may become chronic, eventually causing death. The source of infection always resides in domestic or wild animals, which constitute a huge reservoir of *Brucella* spp. Human infection may result from transmission from infected animals through the consumption of food (unpasteurized milk products), skin lesions, or from inhalation of contaminated dust or aerosols. In human, treatment against brucellosis consists in taking simultaneously two antibiotics (doxycycline and rifampicin or streptomycin) for 6 weeks. *Brucella* is extremely infectious by aerosol and it has been classified as a potential biological weapon,[9] enhancing a growing interest in this pathogen's biology, in particular as a model of complex intracellular pathogen. Furthermore, the appearance of strains resistant to antibiotics has been reported recently in Turkey and in Balkan countries[10] and to date no effective vaccine has been developed for humans. Therefore, it becomes important to identify new potential targets for the development of novel antibrucellosis agents.

38.2.2 Intracellular Trafficking of *Brucella*

Brucella spp. is an intracellular bacterium whose pathogenicity is strictly linked to its capability to enter, survive, and replicate within phagocytic and nonphagocytic cells, among which macrophages are the major target in infected mammals.[11] Schematic presentation of the multiplication kinetics of a *B. suis* strain in a macrophage model of infection is shown in Fig. 38.1.

A decrease in viability of up to 90% is observed 7–10 h after infection and prior to the beginning of multiplication. It has been suggested that this lapse of time is required by the bacteria to build their replicative niche inside the macrophage.[12] During this period of the first few hours of infection, *Brucella* spp. are still sensitive to the bactericidal activity of the macrophage.

Following a LPS-dependent, lipid raft-mediated entry, brucellae are found in vacuoles that interact transiently with early endosomes (Fig. 38.2).

The early brucellae-containing vacuoles (BCVs) mature into acidic intermediate vacuoles avoiding interactions with late endosomes and fusion with lysomes via a LPS- and cyclic β-glucan-dependent mechanism (Fig. 38.2). It has been shown that this transient acidification of the vacuole is required for intracellular multiplication of the pathogen.[13,14] The acidification allows the expression of a set of genes such as the

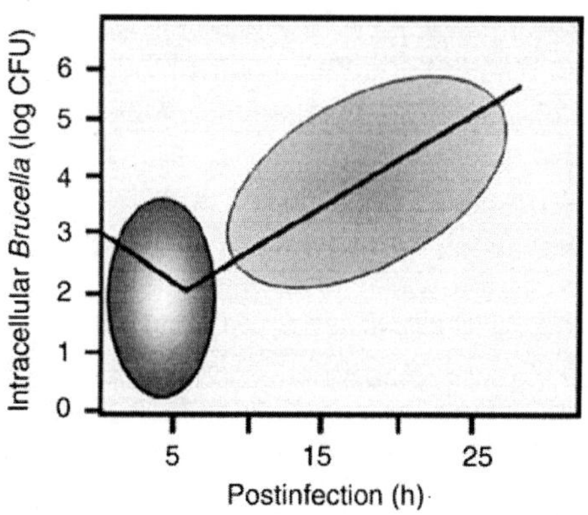

FIGURE 38.1 Typical intracellular growth curves of *B. suis* in macrophages according to Köhler et al. The ellipse at the left represents the phase of the pathogen's adaptation to the host cell environment; the ellipse at the right represents the phase of intramacrophagic replication following localization of the bacteria in their final replicative niche.

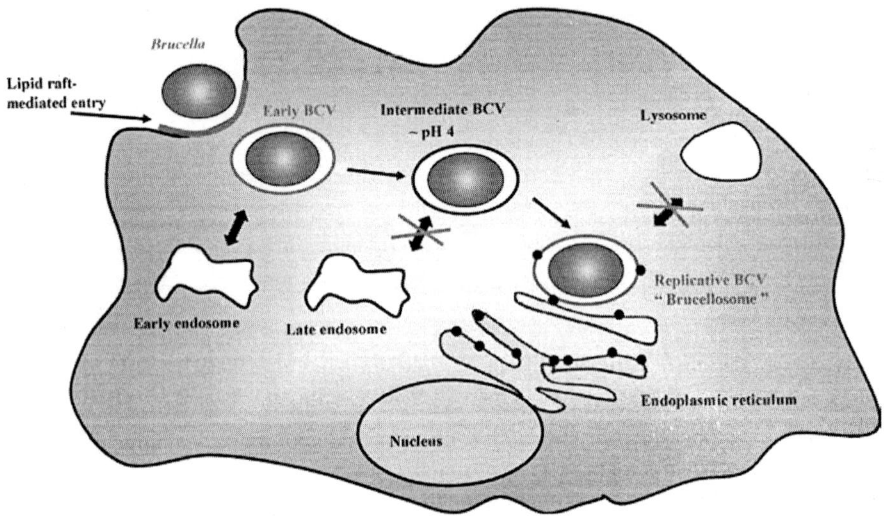

FIGURE 38.2 Intracellular trafficking of *Brucella* in the macrophage host cell.

major virulence factor, the VirB type IV secretion system required in the building of the replicative niche.[11] BCVs subsequently interact with the endoplasmic reticulum (ER) in a manner that leads to fusion between these organelles to generate an ER-derived vacuole permissive for bacterial replication referenced as "brucellosome"[12,15] (Fig. 38.2). As a result of membrane exchanges, BCVs acquire various ER markers. These replicative BCVs do not fuse with lysosomes, and experimental evidence suggests that the intracellular pH also rises within this compartment allowing favorable conditions for the multiplication of the pathogen.[12]

38.2.3 The Virulome Mediates Adaptation of *Brucella* to the Environment of the Brucellosome

The environmental conditions encountered within the "brucellosome," such as the acidification described previously, necessitate a specific adaptation of the pathogen. A genetic screening using transposon mutagenesis identified genes in *Brucella* spp. that are required for adaptation to the environmental conditions encountered in this specific niche, resulting in survival and multiplication. These genes, termed as the intramacrophagic virulome,[4] are a subset of virulence genes.[15] As discussed by Köhler et al.,[12] the analysis of these genes yields information about the nature of the environment encountered by the pathogen within the "brucellosome." This replicative niche is defined as being poor in nutrients, as numerous genes, involved in biosynthetic pathway of nucleotides, lipids, and amino acids, such as leucine, isoleucine, valine, threonine, glycine, and histidine, are required for intramacrophagic multiplication. It is furthermore characterized by low oxygen tension, and nitrate may be used as terminal electron acceptor.

38.3 POTENTIAL TARGETS FOR THE DEVELOPMENT OF NEW ANTIBACTERIAL AGENTS

Several of the enzymes involved in amino acid biosynthesis and encoded by genes of the virulome provide attractive targets for the development of potent antibrucellosis agents. Until now, two such targets have been defined: the AHAS involved in biosynthesis of branched chain amino acids and the HDH catalyzing the last step in the histidine biosynthesis. AHAS can be targeted by sulfonylureas that block the access of the substrate to the active center of the enzyme. The action of these inhibitors has been demonstrated on cell lysates and on live bacteria in minimal medium and at the intramacrophagic state.[5] In contrast, *Brucella* growth is not inhibited in rich medium, as this environment does not require amino acid biosynthesis.

The second target, HDH, which is also essential for intramacrophagic growth of brucellae, is a metalloenzyme. Its properties and the development of anti-infectious agents against HDH will therefore be described in the next section.

38.3.1 HDH as Potential Therapeutic Target and Mechanism of Enzyme Activity

The essential character of histidine biosynthesis for virulence has been demonstrated in other pathogenic bacteria such as *Salmonella typhimurium*[16] and *Burkholderia pseudomallei*.[17] Moreover, the HDH is absent in mammals. Therefore, it has been proposed that HDH represents a selective and promising target for the development of new antibacterial agents.[2] The use of HDH inhibitors as potential antibrucellosis agents would specifically inhibit the growth of the pathogen only in a histidine-poor environment such as the intramacrophagic niche. A major advantage of using these drugs would be the expected absence of side effects on the commensal floral of the host and the reduced rate of resistant bacteria due to the reduced selective pressure limited to intracellular pathogens.

Histidine biosynthesis is an ancient pathway found in bacteria, archaebacteria, fungi, and plants, and it is the most extensively studied biochemically and genetically.[18] This biosynthetic pathway converts 5-phosphorybosyl-1-pyrophosphate to L-histidine in 10 enzymatic reactions. In *E. coli* and *S. typhimurium*, eight genes organized in a single operon encode enzymes involved in the 10 steps of histidine biosynthesis.[18] Three enzymes of this biosynthetic pathway including the histidinol dehydrogenase are bifunctional.

L-Histidinol dehydrogenase catalyses the last two steps in the biosynthesis of L-histidine: sequential NAD-dependent oxidations of L-histidinol to L-histidinaldehyde and then to L-histidine. On the basis of the available data, a catalytic mechanism has been proposed for the conversion of L-histidinol to L-histidine that involves two consecutive oxidoreduction reactions accompanied by a reduction of two NAD + molecules according to a Bi-Uni-Uni-Bi kinetic mechanism,[19] as shown in Fig. 38.3.

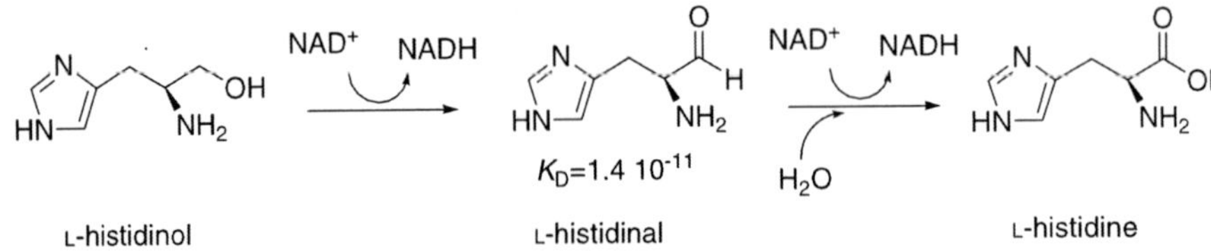

FIGURE 38.3 Enzymatic reaction catalyzed by the histidinol dehydrogenase.

38.3.2 HDH is a Well-Conserved Metalloenzyme

HDH is a metalloenzyme composed of two identical subunits whose molecular weight is approximately 50,000. Each subunit binds one Zn^{2+} cation as cofactor, and removal of Zn^{2+} by metal ion chelator abolishes the enzymatic activity. Although the Zn^{2+} ion is essential for enzyme activity, other metals such as Mn^{2+} or Cd^{2+} can replace Zn^{2+} in forming an active enzyme, indicating that the presence of a divalent cation is essential for enzyme activity.[20] It has been proposed recently that Zn^{2+} plays a crucial role in substrate binding of the *E. coli* HDH, whereas it is not directly involved in catalysis.[19] The three-dimensional structure of *E. coli* HDH in the apo state as well as complexes with Zn^{2+}, L-histidinol, and NAD^+ has been determined.[19] This structure provides a detailed view of the active site, allows the assignment of catalytic roles to specific residues, and shows a mode of NAD^+ binding. As other enzymes of the L-histidine synthesis pathway, the sequence of HDH has been well conserved during evolution from bacteria to fungi to plants. The homologues of HDH from cabbage, *Sacchromyces cerevisiae*, and *E. coli* show approximately 50% amino acids identities, and the homologous enzyme from cabbage is functional in *E. coli*.[21] Although histidine biosynthetic pathway has been elucidated in various organisms, HDH has been cloned and characterized to date only in three species of bacteria, *S. typhimurium*, *E. coli*, and *Brucella suis*.[19,22–24]

38.4 SUBSTITUTED BENZYLIC KETONES AS POTENT INHIBITORS OF *B. suis* HDH

Ten years ago, it was reported that HDH is a suitable target for the development of potential herbicides. The approach developed was to prepare cabbage HDH inhibitors that target the lipophilic binding pocket adjoining the active site of the enzyme.[25] As HDH is absent in mammals, it also constitutes an attractive target for the design of new antibrucellosis agents.

Recently, a series of 15 substituted benzylic ketones derived from histidine has been designed and synthesized as new potential HDH inhibitors (Fig. 38.4).

These drugs, which are substrate analogues, belong to the same class as the previously described cabbage HDH inhibitors[25] and their chemical synthesis has been described lately.[24] The potential inhibitory activity of the benzylic ketones was then evaluated on the purified *B. suis* HDH as well as on the *in vitro* growth and intramacrophagic multiplication of brucellae.

	R
38.1a	-H
38.1b	4-F
38.1c	4-Cl
38.1d	4-Br
38.1e	4-Ph
38.1f	4-CH_3
38.1g	4-OCH_3
38.1h	4-OPh
38.1i	4-OCH_2Ph
38.1j	2,4-F
38.1k	3,4-Cl
38.1l	Perfluoro

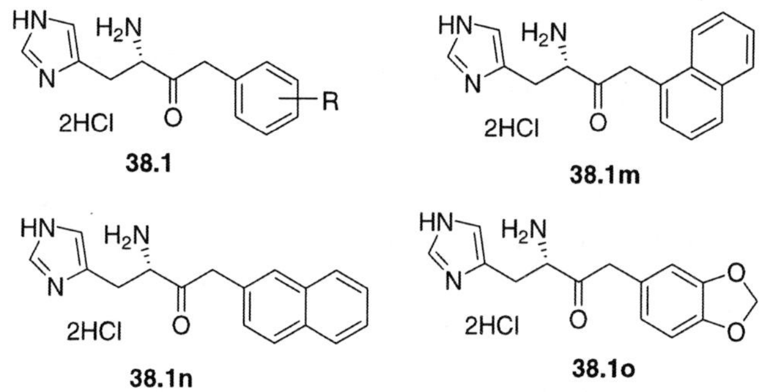

FIGURE 38.4 Structure of the substituted benzylic ketone (HDH inhibitors) as previously described by Abdo et al.[24] The aromatic group of the drugs might interact with the lipophilic binding pocket adjoining the active site of the enzyme.[21]

38.4.1 Inhibition of *B. suis* HDH Activity by Substituted Benzylic Ketones

The *B. suis* HDH encoded by the gene *hisD* (BR0252) has been purified to further characterize its biological activity and to evaluate the inhibitory effect of the newly synthesized drugs. The results show that most of the compounds strongly inhibited *B. suis* HDH, the 50% inhibitory concentration (IC_{50}) being in the nanomolar range (Table 38.1).

TABLE 38.1 Inhibition of *B.suis* Histidinol Dehydrogenase with Substituted Benzylic Ketones 38.1a–38.1o as Previously Described by Abdo et al. [24]

Inhibitors	HDH (*Brucella suis*) IC_{50} (nM)[a]	HDH (Cabbage) IC_{50} (nM)[b]
38.1a	40	100
38.1b	15	–
38.1c	12	300
38.1d	6	40
38.1e	12.5	40
38.1f	15	–
38.1g	150	1000
38.1h	16	–
38.1i	3	–
38.1j	10	–
38.1k	20	–
38.1l	70	–
38.1m	200	–
38.1n	14.5	–
38.1o	25	–

[a] The values are means of three independent assays. Variations were in the range of 5–10% of the shown data.

[b] Reported value as described in Ref. 25.

The five previously described HDH inhibitors (**38.1a**, **38.1c**, **38.1d**, **38.1e**, and **38.1g**, Fig. 38.4) were between 2.5 and 25 times more efficient on *B. suis* HDH than on cabbage HDH.[24,26] Among the 15 drugs, the most effective compound was the newly synthesized **38.1i** structure (4-OCH_2Ph) with an IC_{50} of 3 nM[24] (Table 35.1). Overall, several major observations could be made: (i) the augmentation of the volume of substituent R on position 4 of the phenyl ring increases the inhibitory activity of the drug (**38.1d** > **38.1c** > **38.1b**), (ii) the increase of the lipophilic nature of R allows a better inhibitory activity (**38.1e** > **38.1a** > **38.1g**), (iii) the effect of compound **38.1i**, which was a stronger inhibitor than **38.1h** and characterized by a longer R-chain, may give an indication of the distance between the active site and the lipophilic pocket, and (iv) the flexibility of the structure may play a key role in the adaptation to the catalytic site, as compound **38.1i** showed a better activity compared to compounds **38.1e** and **38.1m** (Fig. 38.4).

38.4.2 Biological Activity of Substituted Benzylic Ketones on Live Brucellae

The biological activity of the drugs has then been investigated on the *in vitro* growth of *B. suis* in a minimal medium culture model that mimicked the presumably nutrient-poor *Brucella*-containing vacuole in the macrophage.[27] The inhibition of HDH by the drugs is therefore expected to specifically abolish the capacity of this pathogen to grow in minimal medium, as under these conditions bacteria require a functional histidine biosynthesis pathway to grow. The results showed that the inhibitors most active on the purified HDH (**38.1b**, **38.1c**, **38.1d**, **38.1e**, **38.1n**, and **38.1i**, Fig. 38.4) were also most effective in blocking *Brucella* growth.[27] The inhibition of the growth of the pathogen was strongly correlated with the inhibition of HDH activity.[27] In a control experiment, the growth of *Brucella* in rich medium (tryptic soy broth) that contains all amino acids was not affected by the drugs.[27] This result was expected, as the bacteria do not need an active histidine biosynthesis pathway under such conditions. The addition of histidine to minimal medium containing HDH inhibitors relieved growth inhibition.[27] Therefore, the inhibitory effect of the drugs on *B. suis* growth is most likely due to the inhibitors' effects on *Brucella* HDH.

38.4.3 Biological Activity of Substituted Benzylic Ketones on Intracellular *Brucella*

The effects of the drugs most active under the conditions described previously were then measured on the intramacrophagic replication of *B. suis* using an infection model with human macrophage-like THP-1 cells.[27] Overall, the inhibition of intramacrophagic growth was observed for low inhibitor concentrations (10–25 μM) and *Brucella* growth was reduced from 50- to 2500-fold compared to that observed in untreated cells.[27] The inhibition of intracellular replication is most likely due to the inhibition of HDH activity, as the authors showed previously a specific biological effect of the drugs on the extracellular growth in minimal medium devoid of histidine and because histidine biosynthesis is essential for the intramacrophagic replication of

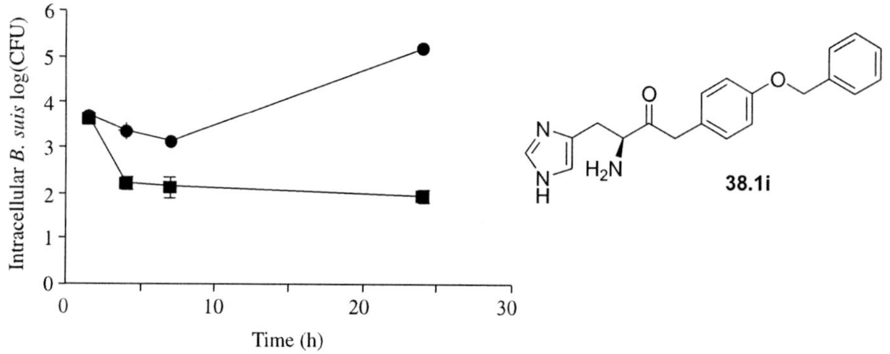

FIGURE 38.5 Effects of lead drug **38.1i** on the intracellular replication of *B. suis* in human macrophage-like THP1 cells according to Joseph et al.[27] *Brucella* growth within untreated cells (circle) or in the presence of 25 μM of the drug (square). The structure of the drug is shown on the right.

B. suis.[4] Addition of compound **38.1i** resulted in drastic reduction in the number of intracellular bacteria at the early stage of macrophage infection that then remained constant throughout the experiment (24 h postinfection) (Fig. 38.5).

This result suggests that the active drug is physically stable and resistant to bacterial degradation under these experimental conditions. Moreover, it can be concluded that the drug efficiently crossed the macrophage membrane, the membrane of the *Brucella*-containing vacuole, and the bacterial membranes to finally reach the cytoplasmic HDH target. Therefore, the molecule **38.1i**, which is also the most effective inhibitor of isolated HDH, constitutes a promising anti-infectious leader molecule against brucellae.

38.4.4 L-Histidinyl Phenylsulfonylhydrazide Derivatives Reveal Important Feature of the HDH Inhibitors

Following this work, the development of new potential HDH inhibitors has been undertaken starting from the compounds described previously as lead molecule. In the newly designed molecules, the methyl group of substituted benzylic ketones has been replaced by a sulfonylhydrazide moiety (SO_2NHNH), leading to the L-histidinyl phenylsulfonylhydrazide derivates[26] (Fig. 38.6).

The presence of SO_2NHNH moiety in other compounds of biological therapeutic interests has been described recently in the literature.[28,29] All the newly synthesized molecules were assayed for their inhibitory activities against purified HDH from *B. suis*. Overall, these compounds were less active than the previously described compounds devoid of sulfonylhydrazide moiety,[26] the 50% inhibitory concentration (IC_{50}) being in the micromolar range (Table 38.2).

Molecule **38.2d**, for example, was totally incapable to inhibit the *in vitro* growth of *B. suis* in minimal medium, even at the high drug concentration of 100 μM (data not shown). The nature of the sulfonylhydrazide moiety might confer to the drug a polarity that decreases its capability to cross the bacterial membranes. Therefore, the linker between

R

38.2a 4-Me
38.2b 4-tBu
38.2c 4-OMe
38.2d 4-Ph
38.2e Perfluoro
38.2f 4-CN
38.2g 4-CF_3
38.2h 4-NO_2
38.2i 4-F
38.2j 4-Cl
38.2k 4-Br
38.2l 4-I

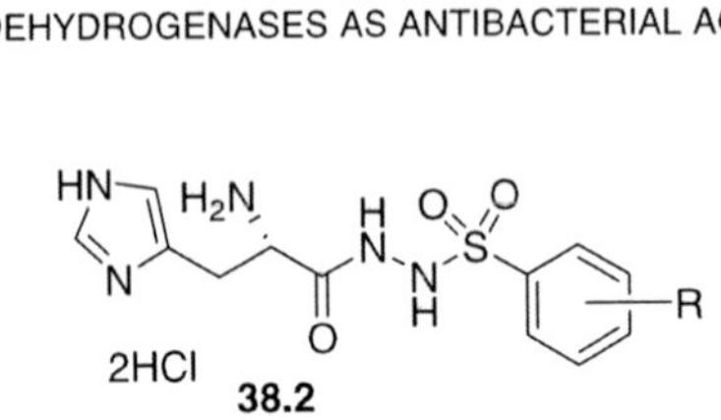

38.2

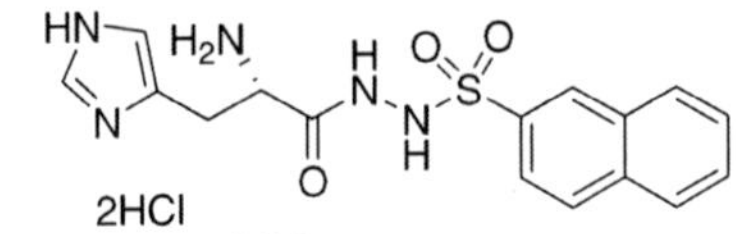

38.2m

FIGURE 38.6 Structure of the different substituted *N*-L-histidinylphenylsulfonyl hydrazide histidinol dehydrogenase inhibitors as described by Abdo et al.[26]

TABLE 38.2 Inhibition of *B. suis* Histidinol Dehydrogenase with L-Histidinyl Phenylsulfonylhydrazide 38.2a–38.2m as Previously Described by Abdo et al. [24]

Inhibitors	HDH (*Brucella suis*) IC_{50} (mM)[a]
38.2a	160
38.2b	375
38.2c	135
38.2d	25
38.2e	>400
38.2f	190
38.2g	200
38.2h	>400
38.2i	>400
38.2j	>400
38.2k	70
38.2l	200
38.2m	140

[a] The values are means of three independent assays. Variations were in the range of 5–10% of the shown data.

the histidinyl moiety and the phenyl ring constitutes an important structural feature that could be the focus of modifications aiming at the design of more potent inhibitors.

38.5 CONCLUSIONS

Metalloenzymes such as HDH may become novel targets of anti-infectious agents. In the intracellular pathogen *Brucella*, HDH is essential for intramacrophagic replication and for growth in nutrient-poor medium only. Therefore, one potential advantage of using HDH inhibitors is that they may limit the selective pressure, that is, the appearance of spontaneously resistant mutants, to the intracellular niche, as they act specifically on *Brucella* inside the host cell. In this context, it is interesting to mention that the appearance of spontaneously resistant mutants in minimal medium in the presence of high concentrations of various drugs (100 μM) and during long-term incubation has been investigated. No spontaneous mutant could be isolated, indicating that enzymatic activity was incompatible with resistance to HDH inhibitors or that the spontaneous mutation rate was low.[26] Another advantage is that HDH inhibitors may cause little or no damage to the bacterial flora in comparison to the classical antimicrobials, which are characterized by permanent, nonselective action on bacteria.

In the future, the use of antibacterial agents targeting factors essential for intracellular replication of certain pathogens combined to classical antibiotics may become conceivable. Among other pathogens of interest, *Mycobacterium tuberculosis*, the causative agent of tuberculosis, must be cited. Tuberculosis has become one of the major health threats worldwide and a growing number of clinical isolates and strains resistant and multiresistant to the currently used antibiotics have been appearing. It is therefore necessary to identify new targets combined with the synthesis of novel inhibitors. Histidinol dehydrogenase and its identified inhibitors are promising candidates and their potential application in the treatment of human infections needs to be further investigated.

REFERENCES

1. Clatworthy, A. E.; Pierson, E.; Hung, D. T. Targeting virulence: a new paradigm for antimicrobial therapy. *Nat. Chem. Biol.* **2007**, *3*, 541–548.
2. Winum, J.-Y.; Köhler, S.; Scozzafava, A.; Montero, J.-L.; Supuran, C. T. Targeting bacterial metalloenzymes: a new strategy for the development of anti-infective agents. *Anti-infect. Agents Med. Chem.* **2008**, *7*, 169–179.
3. White, R. J.; Margolis, P. S.; Trias, J.; Yuan, Z. Targeting metalloenzymes: a strategy that works. *Curr. Opin. Pharmacol.* **2003**, *3*, 502–507.
4. Köhler, S.; Foulongne, V.; Ouahrani-Bettache, S.; Bourg, G.; Teyssier, J.; Ramuz, M.; Liautard, J.-P. The analysis of the intramacrophagic virulome of *Brucella suis* deciphers the environment encountered by the pathogen inside the macrophage host cell. *Proc. Natl. Acad. Sci. USA* **2002**, *99*, 15711–15716.
5. Boigegrain, R.-A.; Liautard, J.-P.; Köhler, S. Targeting of the virulence factor acetohydroxyacid synthase by sulfonylureas results in inhibition of intramacrophagic multiplication of *Brucella suis*. *Antimicrob. Agents Chemother.* **2005**, *49*, 3922–3925.
6. Corbel, M. J. Brucellosis: an overview. *Emerg. Infect. Dis.* **1997**, *3*, 213–221.
7. Pappas, G.; Akritidis, N.; Bosilkovski, M.; Tsianos, E. Brucellosis. *N. Engl. J. Med.* **2005**, *352*, 2325–2336.
8. Godfroid, J.; Cloeckaert, A.; Liautard, J. P.; Köhler, S.; Fretin, D.; Walravens, K.; Garin-Bastuji, B.; Letesson, J. J. From the discovery of the Malta fever's agent to the discovery of a marine mammal reservoir, brucellosis has continuously been a re-emerging zoonosis. *Vet. Res.* **2005**, *36*, 313–326.

9. Pappas, G.; Panagopoulou, P.; Christou, L.; Akritidis, N. *Brucella* as a biological weapon. *Cell Mol. Life Sci.* **2006**, *63*, 2229–2236.
10. Baykam, N.; Esener, H.; Ergonul, O.; Eren, S.; Celikbas, A. K.; Dokuzoguz, B. *In vitro* antimicrobial susceptibility of *Brucella* species. *Int. J. Antimicrob. Agents* **2004**, *23*, 405–407.
11. Celli, J. Surviving inside a macrophage: the many ways of *Brucella*. *Res. Microbiol.* **2006**, *157*, 93–98.
12. Köhler, S.; Michaux-Charachon, S.; Porte, F.; Ramuz, M.; Liautard, J. P. What is the nature of the replicative niche of a stealthy bug named *Brucella*? *Trends Microbiol.* **2003**, *11*, 215–219.
13. Köhler, S.; Porte, F.; Jubier-Maurin, V.; Ouahrani-Bettache, S.; Teyssier, J.; Liautard, J. P. The intramacrophagic environment of *Brucella suis* and bacterial response. *Vet. Microbiol.* **2002**, *90*, 299–309.
14. Porte, F.; Liautard, J. P.; Köhler, S. Early acidification of phagosomes containing *Brucella suis* is essential for intracellular survival in murine macrophages. *Infect. Immun.* **1999**, *67*, 4041–4047.
15. Delrue, R. M.; Lestrate, P.; Tibor, A.; Letesson, J. J.; De Bolle, X. *Brucella* pathogenesis, genes identified from random large-scale screens. *FEMS Microbiol. Lett.* **2004**, *231*, 1–12.
16. Fields, P. I.; Swanson, R. V.; Haidaris, C. G.; Heffron, F. Mutants of *Salmonella typhimurium* that cannot survive within the macrophage are avirulent. *Proc. Natl. Acad. Sci. USA* **1986**, *83*, 5189–5193.
17. Pilatz, S.; Breitbach, K.; Hein, N.; Fehlhaber, B.; Schulze, J.; Brenneke, B.; Eberl, L.; Steinmetz, I. Identification of *Burkholderia pseudomallei* genes required for the intracellular life cycle and *in vivo* virulence. *Infect. Immun.* **2006**, *74*, 3576–3586.
18. Alifano, P.; Fani, R.; Lio, P.; Lazcano, A.; Bazzicalupo, M.; Carlomagno, M. S.; Bruni, C. B. Histidine biosynthetic pathway and genes: structure, regulation, and evolution. *Microbiol. Rev.* **1996**, *60*, 44–69.
19. Barbosa, J. A.; Sivaraman, J.; Li, Y.; Larocque, R.; Matte, A.; Schrag, J. D.; Cygler, M. Mechanism of action and NAD^+-binding mode revealed by the crystal structure of L-histidinol dehydrogenase. *Proc. Natl. Acad. Sci. USA* **2002**, *99*, 1859–1864.
20. Kanaori, K.; Ohta, D.; Nosaka, A. Y. Effect of excess cadmium ion on the metal binding site of cabbage histidinol dehydrogenase studied by 113Cd-NMR spectroscopy. *FEBS Lett.* **1997**, *412*, 301–304.
21. Nagai, A.; Ward, E.; Beck, J.; Tada, S.; Chang, J. Y.; Scheidegger, A.; Ryals, J. Structural and functional conservation of histidinol dehydrogenase between plants and microbes. *Proc. Natl. Acad. Sci. USA* **1991**, *88*, 4133–4137.
22. Teng, H.; Grubmeyer, C. Mutagenesis of histidinol dehydrogenase reveals roles for conserved histidine residues. *Biochemistry* **1999**, *38*, 7363–7371.
23. Teng, H.; Segura, E.; Grubmeyer, C. Conserved cysteine residues of histidinol dehydrogenase are not involved in catalysis. Novel chemistry required for enzymatic aldehyde oxidation. *J. Biol. Chem.* **1993**, *268*, 14182–14188.
24. Abdo, M.-R.; Joseph, P.; Boigegrain, R. A.; Liautard, J. P.; Montero, J.-L.; Köhler, S.; Winum, J. -Y. *Brucella suis* histidinol dehydrogenase: synthesis and inhibition studies of a series of substituted benzylic ketones derived from histidine. *Bioorg. Med. Chem.* **2007**, *15*, 4427–4433.

25. Dancer, J. E.; Ford, J. M.; Hamilton, K.; Kilkelly, M.; Lindell, S. D.; O'Mahony, M. J.; Saville-Stones, E. A. Synthesis of potent inhibitors of histidinol dehydrogenase. *Bioorg. Med. Chem. Lett.* **1996**, *6*, 2131–2136.

26. Abdo, M.-R.; Joseph, P.; Boigegrain, R. A.; Montero, J.-L.; Köhler, S.; Winum, J.-Y. *Brucella suis* histidinol dehydrogenase: synthesis and inhibition studies of substituted *N*-L-histidinylphenylsulfonyl hydrazide. *J. Enzyme Inhib. Med. Chem.* **2008**, *23*, 357–361.

27. Joseph, P.; Abdo, M.-R.; Boigegrain, R. A.; Montero, J.-L.; Winum, J.-Y.; Köhler, S. Targeting of the *Brucella suis* virulence factor histidinol dehydrogenase by histidinol analogues results in inhibition of intramacrophagic multiplication of the pathogen. *Antimicrob. Agents Chemother.* **2007**, *51*, 3752–3755.

28. Siemann, S.; Evanoff, D. P.; Marrone, L.; Clarke, A. J.; Viswanatha, T.; Dmitrienko, G. I. *N*-arylsulfonyl hydrazones as inhibitors of IMP-1 metallo-beta-lactamase. *Antimicrob. Agents Chemother.* **2002**, *46*, 2450–2457.

29. Nofal, Z. M.; Fahmy, H. H.; Mohamed, H. S. Synthesis and antimicrobial activity of new substituted anilinobenzimidazoles. *Arch. Pharm. Res.* **2002**, *25*, 250–257.

CHAPTER 39

Dihydroorotase Inhibitors

MIHWA LEE[1], MEGAN J. MAHER[2], RICHARD I. CHRISTOPHERSON[1], and J. MITCHELL GUSS[1]

[1]School of Molecular and Microbial Biosciences, University of Sydney, Sydney, NSW 2006, Australia
[2]Centenary Institute of Cancer Medicine and Cell Biology, Sydney, NSW 2042, Australia

39.1 DIHYDROOROTASE

Dihydroorotase (DHOase; EC 3.5.2.3) is a Zn metalloenzyme, which catalyses the third step of the pathway for *de novo* biosynthesis of pyrimidine nucleotides, the reversible cyclization of *N*-carbamyl-L-aspartate (CA-asp) to L-dihydroorotate (DHO) (Fig. 39.1, Step 3). Due to the importance of nucleotide biosynthesis, DHOases are found in all three domains of life (eukarya, eubacteria, and archaea). However, the genetic and biophysical organization varies considerably among species. In higher eukaryotes, this enzyme is found as part of a large trifunctional protein, CAD, which also contains the enzymes catalyzing the first two steps of the pathway, carbamyl phosphate synthetase (CPSase) and aspartate transcarbamylase (ATCase).[1] In bacteria and yeast, however, DHOase exists as a monofunctional monomer or homodimer in most cases.[2]

As a part of the pyrimidine biosynthetic pathway, DHOase activity is important in providing pyrimidine nucleotides required for biosynthesis of nucleic acids. Nucleotides can be obtained via two distinct routes: the *de novo* pathways, where small metabolites are assembled into nucleotides, or the salvage pathways, where small metabolites can be recycled from breakdown products of nucleic acids. In resting or fully differentiated human cells, the activity of the *de novo* pyrimidine pathway is, in general, low, where the requirement for pyrimidines is largely met by the salvage pathways. The demand for nucleotides and other cellular components associated with cell proliferation increases in rapidly proliferating cells. Consequently, the activity of the *de novo* pyrimidine pathway is upregulated in rapidly dividing normal cells, cancer cells, or pathogens such as malarial parasites. This suggests that enzymes in the pathway could be attractive targets for anticancer and antiparasitic drug discovery.[3,4]

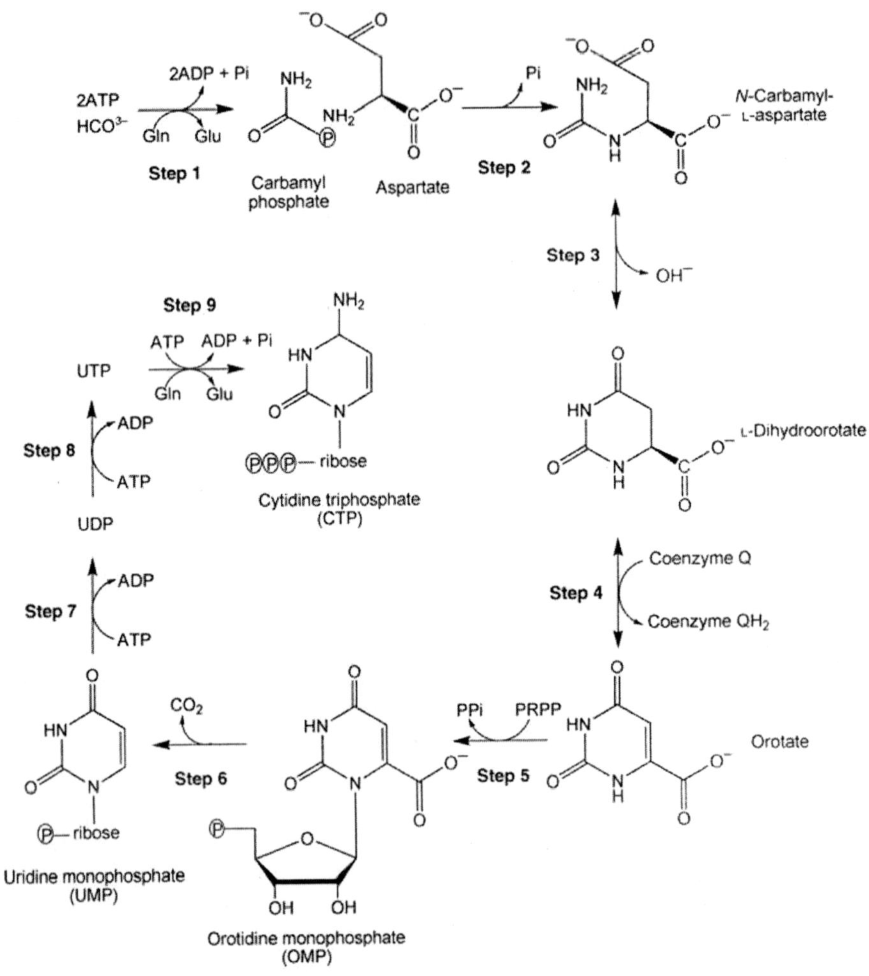

FIGURE 39.1 The *de novo* pathway for pyrimidine biosynthesis (modified from Ref. 8). The enzymes involved in each step are as follows: Step 1, carbamyl phosphate synthetase (CPSase); Step 2, aspartate transcarbamylase (ATCase); Step 3, dihydroorotase (DHOase); Step 4, dihydroorotate dehydrogenase; Step 5, orotate phosphoribosyltransferase (OPRTase); Step 6, OMP decarboxylase (ODCase); Step 7, nucleoside monophosphate kinase; Step 8, nucleoside diphosphate kinase; and Step 9, CTP synthetase.

Unlike the human host in which both the *de novo* and salvage pathways may provide nucleotides, malarial parasites synthesize purine nucleotides via a salvage pathway and pyrimidine nucleotides via the *de novo* pathway.[5,6] The malarial genome does not include genes involved in *de novo* purine synthesis or pyrimidine salvage.[7] The differences in nucleotide synthesis between the host and parasite,[8] the lack of pyrimidine salvage in malarial parasites,[9–11] and the low activity of DHOase relative to other enzymes in the *de novo* pathway[12] make DHOase a suitable target for the design of potent inhibitors as antimalarial drugs.

39.1.1 Occurrence and Evolution of DHOases

The amino acid sequences of DHOases from many species are now available from genome sequencing projects. Genes encoding DHOases have been identified from most genomes, except from fastidious pathogens that depend solely on pyrimidine salvage pathways for their nucleotide requirement, for example, *Mycoplasma mycoides*,[13] *Trichomonas vaginalis*,[14] *Chlamydia psittaci*,[15] and *Cryptosporidium hominis*.[16] Since Lieberman and Kornberg first demonstrated DHOase activity from an anaerobic bacterium, *Clostridium oroticum* (formerly, *Zymobacterium oroticum*) in 1954,[17] DHOases have been purified and characterized from a number of organisms, including *C. oroticum*,[18,19] *Escherichia coli*,[20,21] the parasitic protozoan *Crithidia fasciculate*, and the malarial parasites, *Plasmodium berghei*[22] and *Plasmodium putida*,[23] in addition to the mammalian systems.[24,25] Recently, recombinant DHOases from *E. coli*,[26,27] *Aquifex aeolicus*,[28,29] *Bacillus caldolyticus*,[30] *Toxoplasma gondii*,[31] and *Plasmodium falciparum*,[32] and the recombinant DHOase domain of hamster CAD[33,34] have been characterized. The biochemical properties of purified DHOases are summarized in Table 39.1.

The reversible cyclization of L-CA-asp to L-DHO catalyzed by DHOase is pH dependent. The biosynthetic direction (CA-asp → DHO) is favored at lower pH, whereas the degradative reaction (DHO → CA-asp) is dominant at higher pH. The reported pH optima for *C. oroticum* DHOase are pH 6.0 and 8.2 in the biosynthetic and degradative directions, respectively.[18] Christopherson and Jones reported for mouse DHOase that the maximal rate of cyclization (CA-asp → DHO) occurred on a plateau down to pH 4.2, while the rate for ring-opening approached a maximum at pH 9.2. At pH 6.2, the equilibrium between CA-asp and DHO is unity.[35] A similar pH dependence of the reactions was observed with *E. coli* DHOase.[36]

39.1.2 Type I Versus Type II DHOases

Extensive phylogenetic analysis has classified DHOases into two major subtypes, thought to have arisen from duplication of an ancestral gene.[2] Type I DHOases are found in all three kingdoms of life and may be more ancient. They are also more divergent, being found as monofunctional enzymes, as domains of multifunctional enzymes and in multifunctional enzyme complexes. Type II DHOases have significantly different amino acid sequences but have similar secondary structures to the Type I DHOases. Members of the same subtype of DHOase share more than 40% identity in amino acid sequence, while less than 20% sequence identity is found between Types I and II.

In higher eukaryotes, including *Dictyostelium*, *Drosophila*, and mammals, a Type I DHOase exists as a domain of the trifunctional protein, CAD.[37] DHOases from the archaea and the Gram-positive bacteria with low G + C contents are Type I enzymes and usually exist as monomers or homodimers distinct from other enzymes (e.g., DHOase from *B. caldolyticus*).[30] Some Type I DHOases are fully functional only when they interact with other enzymes in the pyrimidine pathway. For example, the DHOase from *A. aeolicus* is activated several 1000-fold by the presence of aspartate transcarbamylase

TABLE 39.1 Properties of Purified DHOases from Various Organisms

Organism	Molecular Weight (kDa)	Oligomeric State	Recombinant	K_m (mM)		k_{cat} (s^{-1})	
				CA-asp	DHO	CA-asp → DHO	DHO → CA-asp
Aquiflex aeolicus[29]	49	Monomer	Yes		3.03		13.8
Crithidia fasciculata[22]	44	Monomer	No	0.846	0.0258	0.65	4.31
Bacillus caldolyticus[30]	47	Monomer	Yes		0.195		6.3
Clostridium oroticum[18]	110	Homodimer	No	0.13	0.07		
Escherichia coli[21]	76	Homodimer	Yes	1.07	0.0756	195	127
Plasmodium berghei[22]	38	Monomer	No	0.293	0.014	4.06	4.76
Hamster (DHOase domain)[30]	76	Homodimer	Yes		0.022		0.40
Pseudomonas putida[23]	82	Homodimer	No	2.2	0.081	92.9	24.6
Toxoplasma gondii[31]	45.6	Monomer	Yes	0.323	0.0643		

(ATCase).[29] Active DHOases from *Thermus aquaticus*, *Streptomyces griseus*, and *Deinococcus radiophilus* are also found complexed with ATCase.[38–40]

The evolutionarily younger Type II DHOases are found mainly in Gram-negative bacteria, cyanobacteria, fungi (*Saccharomyces* and *Ustilago maydis*), plants (*Arabidopsis thaliana*), and the apicomplexan parasites (*Plasmodium* and *Toxoplasma*). Type II enzymes are smaller than the Type I, being 50 and 10 residues shorter at the amino and carboxy termini, respectively, with several internal insertions and deletions compared with their Type I counterparts. Although there is one report that the Type II DHOase from *Serratia marcescens* interacts with orotate phosphoribosyltransferase and OMP decarboxylase,[41] most known Type II DHOases are monofunctional and independent. Most Type II DHOases are homodimeric with a few monomeric exceptions reported from the parasites, *P. berghei* and *T. gondii*.[22,31] The DHOase from *E. coli* is the most extensively studied Type II DHOase.

Interestingly, a number of organisms have two or more DHOase genes encoding active or inactive DHOases (Table 39.2). In most cases, an active DHOase is found with an inactive DHOase-like protein (designated as PyrC′ protein in Table 39.2). This phenomenon was initially discovered in yeast.[42,43] In these cases, the active DHOase is

TABLE 39.2 Distribution of the PyrC, PyrC2, and PyrC′ Polypeptides[a]

Organism	PyrC (~38 kDa)	PyrC2 (~48 kDa)	PyrC′ (~45 kDa)
Agrobacterium tumefaciens	AAL41420		AAK87098
Azotobacter vinelandii	EAM06897		EAM08491
Bordetella bronchiseptica	CAE34572		CAE34765
Bordetella pertussis	CAE43304		CAE40699
Bradyrhizobium japonicum USDA110		BAC47820	BAC50365
Brucella melitensis 16M		AAL52462	AAL53911
Brucella suis 1330		AAN29597	AAN33789
Burkholderia pseudomallei	CAH36924		CAH36697
Candidatus Pelagibacter ubique		EAS84695	EAS84259
Caulobacter crescentus		AAK22375	AAK24415
Gleobacter violaceus		BAC91610	BAC91763
Mesorhizobium loti		BAB54178	BAB48228
Nitrosomonas europaea ATCC19718	CAD84638		CAD85575
Nostoc sp. PCC7120		BAB74002	BAB75038
Rhodobacter sphaeroides	ABP69231		ABP69132
Rhodpseudomonas palustris CGA009		CAE26512	CAE28559
Ralstonia solanacearum	CAD14015		CAD14209
Pesudomonas aeruginosa	AAG06915	AAG08926	AAG03790
Pseudomonas fluorescens	ABA76256	ABA77399	ABA77052
Pseudomonas putida	AAN66711		AAN70565
Pseudomonas syringae	AAY38924		AAY35551
Synechocystis sp. PCC6803	BAA18539		BAA16951

Modified from Ref. 46.

[a] The NCBI accession number of each homologue identified by BLAST analysis is given.

encoded by the *URA4* gene and is a Type II DHOase, a monofunctional ~80 kDa dimer. The inactive DHOase is encoded by the *URA2* gene as part of a multifunctional protein, homologous with mammalian CAD. This DHOase-like domain may be important in substrate channeling.[44] In general, the inactive DHOase-like proteins lack the key histidine and aspartate residues required for DHOase catalytic activity, but show high sequence identity with active Type I DHOases. These proteins no longer function as enzymes, but the domain may be used for another purpose. In many Gram-negative bacteria, a defective Type I DHOase forms a noncovalently bonded complex with ATCase while the DHOase function is taken over by a monofunctional Type II DHOase. For example, in *Pseudomonas aeruginosa* and *P. putida*, a defective Type I DHOase is required for the correct assembly of active ATCase subunits into the dodecameric holoenzyme.[45]

Recently, some organisms have been reported to have more than one active DHOase. *P. aeruginosa* possesses three *pyrC* genes located in separate regions of the genome.[46] A Type II DHOase (38 kDa, NCBI accession No. PA3527) encoded by the *pyrC* gene was identified by screening a *P. aeruginosa* cosmid library using the *E. coli pyrC* mutant strain X7014a. The *pyrC′* gene encodes a DHOase-like polypeptide (45 kDa, NCBI accession No. PA0401) that is catalytically inactive, but required for the structure and function of *P. aeruginosa* ATCase. The completion of the *P. aeruginosa* genome sequence revealed the presence of a third sequence, designated as *pyrC2* (48 kDa, NCBI accession No. PA5541). The protein encoded by *pyrC2* has significant homology (31% identity and 48% homology) to the DHOase from *Methanobacterium thermoautotrophicum*, a Type I DHOase. An unusual characteristic of this sequence is that the residue corresponding to the first Zn-binding histidine in other active DHOases is replaced by a glutamine. Both DHOases encoded by the *pyrC* and *pyrC2* genes were found to be functional in complementation experiments. Similar findings were reported for *Pseudomonas fluorescens* and are summarized in Table 39.2. Biochemical analyses of the *pyrC2*-encoded DHOases are not reported and the selective advantage for multiple *pyrC* genes in a single organism remains unclear.

39.2 STRUCTURES OF DHOases

Crystal structures of DHOases from four species are available. The first structure, for the Type II DHOase from *E. coli* was reported in 2001 (PDB entry 1J79).[47] The first structure for a Type I DHOase was from *A. aeolicus* published in 2005 (PDB entry 1XRT).[48] Recently, structures of the DHOases from *Porphyromonas gingivalis* (PDB entry 2GWN, June 2006) and *Agrobacterium tumefaciens* (PDB entry 2OGJ, February 2007) have been deposited in the PDB without further publication. The protein for which a structure was deposited as a DHOase from *A. tumefaciens* is annotated as a hypothetical protein (Q8UAV1, Atu3266). The protein has the overall $(\beta/\alpha)_8$-barrel fold and contains a binuclear metal center with a carboxylated lysine residue. However, this protein lacks the signature substrate-binding arginine residue for DHOase and the architecture of the active site is significantly different from the DHOases of *E. coli* and *P. gingivalis*. This protein is unlikely to be a true DHOase and is not described further. The latest addition to the structures of DHOases (*Thermus thermophilus* DHOase, PDB entry 2Z00) was made by another structure genomics group (RIKEN structural genomics/proteomics initiative, RSGI) in November 2007.

39.2.1 Amidohydrolase Superfamily

DHOase belongs to the amidohydrolase superfamily that contains a variety of hydrolytic enzymes of the $(\beta/\alpha)_8$-barrel (or TIM-barrel) fold.[49] The $(\beta/\alpha)_8$-barrel comprises an eightfold repeat of βα subunits, with eight parallel β-strands in the interior of the protein forming a cylindrical core and eight α-helices on the exterior. The members of the amidohydrolase superfamily of enzymes contain a mononuclear or a binuclear metal center, whose main role is to activate the scissile bond (−C(O)−N (H)−) of the substrate for cleavage and to deprotonate a water molecule for nucleophilic attack.[50]

The $(\beta/\alpha)_8$-barrel is the most common fold reported for all enzymes, accounting for ~10% of the structures of enzymes deposited in the PDB. In the most recent release of the Structural Classification of Proteins (SCOP) database (v. 1.73, November 2007), $(\beta/\alpha)_8$-barrel proteins were classified into 33 superfamilies, most of which are enzymes, implying a versatility of function arising from divergence of amino acid sequences. Holm and Sander designated the "metallo-dependent hydrolase" superfamily in SCOP as the "amidohydrolase superfamily."[49]

The enzymes in the amidohydrolase superfamily catalyze hydrolysis of a wide range of substrates with amide or ester functional groups at carbon or phosphorus centers. Residues that determine the substrate specificity and catalytic activity in the $(\beta/\alpha)_8$-barrel enzymes are often contributed either from the C-terminal ends of the β-strands of the barrel or from the loops connecting the β-strands with the α-helices that follow them.[51] In particular, loops 7 and 8 are often involved in contacting the ligands in the active site and determining substrate specificity. Several of these substrates, including DHO, are intermediates in nucleotide metabolism. Holm and Sander suggested that the extended family of urease-related amidohydrolases began to diverge from a common ancestor at a very early evolutionary stage, before the divergence of archaea, prokaryota, and eukaryota.[49] Consequently, the sequence signature, which is required for the basic catalytic mechanism, has remained invariant despite considerable functional specialization.

The amidohydrolase superfamily of enzymes can be divided into two subtypes on the basis of the metal content of the active site. The first subtype with two metal ions per active site includes phosphotriesterase (PTE) and urease (URE). The two metal ions of PTE are bridged by a solvent-derived hydroxide and a post-translationally modified carboxylated lysine residue. Other metal ligands include an aspartate and a cluster of four histidines. The same set of metal ligands as in PTE occurs in the metal center of URE, except this site binds nickel instead of Zn. Other examples in this subtype are hydantoinase (HYD), dihydropyrimidinase (DHPase), and iso-aspartyl dipeptidase (IAD). The inclusion of a carboxylated lysine at a specific location (at the end of strand 4) is unique to enzymes in this subtype although one variation is found in the PTE homology protein (PHP) from *E. coli* with a glutamate residue that bridges the two Zn ions.[52] In contrast, adenosine deaminase (ADA) and cytosine deaminase (CDA) are in the other subtype of the amidohydrolase superfamily with a single divalent cation (Zn or Fe) at the active site.

DHOase was originally proposed as a member of this second subtype, due to reports that DHOases bound a single Zn ion.[21,27] Thoden and colleagues reported the first crystal structure of a DHOase from *E. coli* and showed that the active site contains two Zn ions bridged by a carboxylated lysine.[47] However, the subsequently reported structure of *A. aeolicus* DHOase has only one Zn ion per active site (Section 39.2.4).[48] The recently deposited structures of DHOases from *P. gingivalis* and *T. thermophilus* revealed more variations in the active sites of DHOases (Sections 39.2.5 and 39.2.6). The active site of *P. gingivalis* DHOase is indistinguishable from that of *E. coli* except that one of the Zn-binding histidine residues is substituted with glutamine. The enzyme from *T. thermophilus* contains two Zn ions at the active site, bridged by an aspartate residue instead of a carboxylated lysine as in *E. coli* DHOase. Although the reaction catalyzed by DHOase and the principal features of the enzyme have remained constant, DHOases have diverged, resulting in heterogeneity in the number of Zn ions and the identity of the Zn-bridging ligands.

The identity of the metals in the active sites of enzymes within the amidohydrolase superfamily is not conserved. URE requires nickel for activity while with PTE, HYD, and DHOase, the required metal ions appear to be Zn although PTE and DHOase have been reported to function with Co, Cd, or Mn.[34,36,53,54] CDA requires Fe for activity, whereas the structurally equivalent ADA requires Zn.[55]

Interestingly, some of the enzymes in this superfamily catalyze the reversible hydrolysis of cyclic amide bonds in five- or six-membered rings in reactions of nucleotide metabolism.[56] Allantoinase (ALN) hydrolyzes allantoin to allantoic acid in the purine degradation pathway of plants and microorganisms. DHOase is biosynthetic whereas DHPase catalyzes the hydrolysis of dihydrouracil in the degradation of pyrimidine nucleotides. Microbial hydantoinase (HYD) catalyzes hydrolysis of a variety of hydantoins and is a microbial counterpart of mammalian DHPase. Along with DHOases, these cyclic amidohydrolases show high sequence homology. Microbial HYDs share about 40% identity with mammalian DHPases. DHPases show 19–26% and 7–16% amino acid identities with ALNs and DHOases, respectively.[57] The recently reported 3-dimensional structures of these enzymes have indistinguishable metal centers and overall folds, suggesting that these enzymes are evolutionarily closely related to each other.[47,58,59] Because DHOase is found in all three kingdoms of life, it is considered to be the progenitor for these cyclic amidohydrolase enzymes.[49] Unlike DHOases that show heterogeneity in the number of Zn ions per active site and the identity of the Zn-bridging ligands, structures of other cyclic amidohydrolase enzymes reported so far have two metal ions in each active site and the Zn-binding ligands are composed of four histidine residues, an aspartate and a carboxylated lysine.

39.2.2 Metal Content in DHOases

Sander et al.[60] found Zn to be essential for the activity of DHOase from *C. oroticum*, and it has since been found in the active sites of all DHOases.

However, the number of Zn ions per active site is still controversial. Metal analyses of DHOases conducted by atomic absorption spectroscopy, before the crystal structures were available, reported one or two Zn ions per active site. The DHOase from *C. oroticum* was reported to contain four Zn ions in a dimer, thus two ions per active site,[18,19] whereas the dimeric enzyme from *E. coli* was believed to have one Zn per monomer.[21,27] However, the crystal structure of *E. coli* DHOase revealed the presence of two Zn ions per active site.[47,61] The isolated DHOase domain from hamster CAD was determined to contain one Zn ion per active site.[34] A recent report for the DHOase from *A. aeolicus* indicated one Zn ion per active site,[29] confirmed by the crystal structure of the enzyme.[48]

It appears that the numbers of Zn ions per active site in DHOases do not strictly correlate with the phylogenetic classification given in Section 39.1.2. Although only one structure of a Type II DHOase (*E. coli* DHOase) has been determined to date, the high sequence homology, including the conserved Zn ligand residues and the lysine residue (presumably carboxylated to bridge two Zn ions) found in Type II DHOases, suggests that all Type II DHOases contain two Zn ions per active site. With the exception of *E. coli*, all the DHOases whose metal contents were described earlier are Type I DHOases. At present, there are three Type I DHOases, from *A. aeolicus*, *P. gingivalis*, and *T. thermophilus*, for which structures have been solved. Interestingly, *A. aeolicus* DHOase contains one Zn ion in the active site whereas two Zn ions are found in the structures of DHOases from *P. gingivalis* and *T. thermophilus*. As described previously, Type I DHOases are more widely divergent, being found as monofunctional enzymes and in multifunctional enzymes. It is plausible that Type I DHOases found in CAD or complexed with other enzymes (e.g., *A. aeolicus* DHOase) contain one Zn ion, while independent, monofunctional Type I DHOases possess two Zn ions. The definitive answer for the metal content of DHOases will emerge when more structures of DHOases are solved.

39.2.3 Structure of *E. coli* DHOase

The crystal structure of *E. coli* DHOase was refined to 1.7 Å resolution by Thoden and coworkers.[47] As predicted by Holm and Sander,[49] the enzyme has an overall $(\beta/\alpha)_8$-barrel fold. *E. coli* DHOase is a monofunctional, homodimeric Type II enzyme (Fig. 39.2). The active site contains a binuclear center with a carboxylated lysine as one of the bridging ligands (Fig. 39.3). An interesting asymmetry was found in the active sites of the two monomers in the asymmetric unit of the crystal. The enzyme was crystallized in the presence of a racemic substrate, D,L-CA-asp. One active site contains the substrate, CA-asp while the product, DHO, is bound at the other active site of the dimer (Fig. 39.4). In the DHO-bound subunit, the more buried Zn ion (Zn_α) is ligated by His16, His18, Lys102 (carboxylated lysine), Asp250, and a solvent molecule (a hydroxide ion) in a trigonal bipyramidal arrangement (Fig. 39.3). The more exposed Zn ion (Zn_β) is tetrahedrally ligated by Lys102, His139, His177, and the bridging water molecule. The two Zn ions are separated by 3.5 Å. For the CA-asp-bound subunit, one of the carboxylate groups of CA-asp replaces the bridging solvent and ligates the two Zn ions in a bidentate manner.

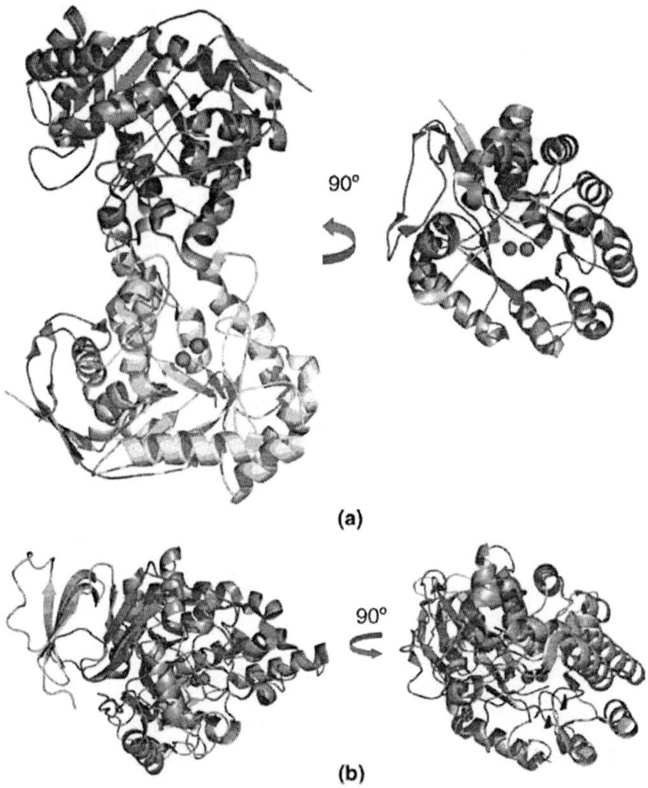

FIGURE 39.2 Overall folds of DHOases. (a) *E. coli* DHOase (Type II) (PDB entry 1XGE). The N-terminal domain (residues 4–10), the main $(\beta/\alpha)_8$-barrel domain (residues 11–307) and the C-terminal domain (residues 308–356) are colored in cyan, gray (or yellow in chain B) and red, respectively. Zinc atoms in the active site are shown as magenta spheres. For clarity, only chain A is shown in the rotated view. (b) *P. gingivalis* DHOase (Type I) (PDB entry 2GWN). The N-terminal domain (residues 1–57), the main $(\beta/\alpha)_8$-barrel domain (residues 58–375) and the C-terminal domain (residues 376–448) are colored in cyan, gray, and red, respectively. Zinc atoms in the active site are shown as magenta spheres. (See the color version of this figure in the Color Plates section.)

Subsequently, our group crystallized *E. coli* DHOase in the presence of L-DHO (PDB entry 1XGE).[61] Analysis of this structure confirmed the many features reported previously (PDB entry 1J79). In particular, we observed asymmetry between active sites in the dimer, with the substrate, L-CA-asp, bound to one active site, and the product, L-DHO bound to the other (Fig. 39.4). More importantly, residues forming part of a surface loop near the active site missing from the structure reported by Thoden et al., were resolved in this structure, revealing an unreported conformational change in *E. coli* DHOase. This surface loop, formed by residues 105–115, displays two

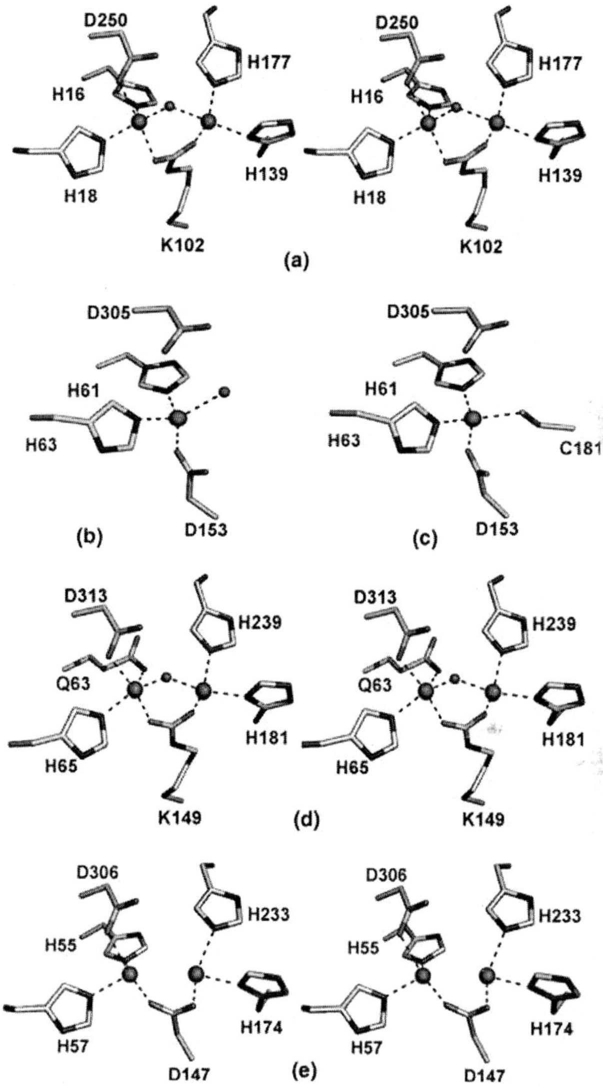

FIGURE 39.3 Stereo views of structures of the active sites of DHOases. (a) *E. coli* DHOase (Type II) (PDB entry 1J79), (b and c) *A. aeolicus* DHOase (Type I) from Crystal form I (PDB entry 1XRF) and Crystal form II (PDB entry 1XRT), (d) *P. gingivalis* DHOase (Type I) (PDB entry 2GWN), and (e) *T. thermophilus* DHOase (Type I) (PDB entry 2Z00).

different conformations depending on the identity of the bound ligand at the active site. With DHO at the active site, this loop is open, forming part of the surface of the protein ("loop-out"). With CA-asp bound at the other subunit, this loop reaches in toward the active site and makes hydrogen-bonding interactions with the bound CA-asp via two threonine residues (Thr109 and Thr110) at the tip of the loop ("loop-in") (Fig. 39.4).

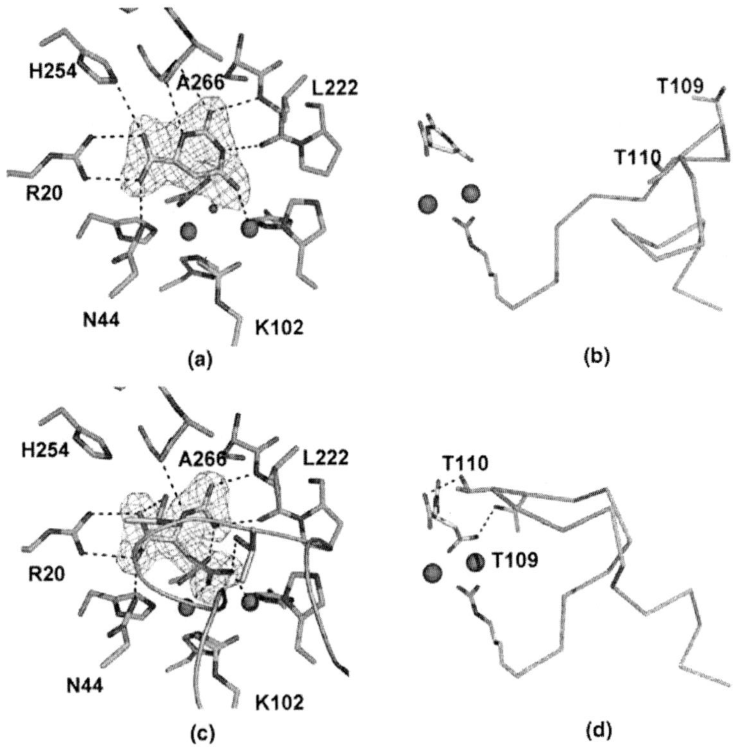

FIGURE 39.4 Interactions of DHO and CA-asp with *E. coli* DHOase in the active site and the conformations of the corresponding flexible loop (residues 105–115) (PDB entry 1XGE). The active sites in the DHO-bound subunit (a and b) and in the CA-asp-bound subunit (c and d). Hydrogen bonds to the bound DHO and CA-asp are depicted as dashed lines. Zinc atoms in each active site are shown as magenta spheres and a solvent molecule in (a) as a red sphere and (b) and (d) show the conformations of the surface loops (residues 105–115) in each active site. (See the color version of this figure in the Color Plates section.)

This unusual asymmetry between two subunits led to the discovery of cooperativity between the two active sites showing how the enzyme functions at the atomic level.

Crystals of a low-activity mutant *E. coli* DHOase (T109S) in the presence of L-DHO, grew first in a tetragonal lattice, dissolved, and then re-grew in an orthorhombic lattice. This unusual behavior can be explained by the active involvement of loop movement in catalysis.[62] The structure of the inherently unstable tetragonal crystal was rendered amenable to structure analysis by replacing L-DHO with a product-like inhibitor, 5-fluoorotate (FOA; **39.11**) in the crystallization medium. In this structure both flexible loops are "loop-out," consistent with FOA, binding in a manner similar to L-DHO at both active sites. The transition of the T109S mutant crystals from one crystal form to the other, in the presence of L-DHO was explained by initial binding of L-DHO to both subunits, isomorphous to the FOA complex, followed by slow conversion to L-CA-asp with consequent movement of the flexible loop, and dissolution of the crystals. Wild-type orthorhombic crystals then grow in the presence of L-DHO and L-CA-asp with one loop out and the other in.

The importance of the conformational change in the catalysis of DHOase was then further established by a combination of X-ray crystallography, site-directed mutagenesis, and enzyme kinetics.[63] A series of mutant DHOases, including deletion of the entire flexible loop, was generated and characterized kinetically and structurally. Mutations of the two threonine residues (Thr109 and Thr110) that make direct hydrogen bonding interactions with the bound substrate, CA-asp, resulted in a significant loss of catalytic activity. The structure of the mutant (Δ107–116), in which the flexible loop is deleted, showed only small differences in positions of other substrate-binding residues and in the binuclear Zn center compared with the native structure, yet the enzyme has negligible activity. Furthermore, there is no significant electron density for DHO in the active site despite the presence of excess DHO in the crystallization medium. Overall, the kinetic and structural analyses of mutants suggested that the two threonine residues in the flexible loop enable productive binding of substrate and stabilize the transition-state intermediate, consequently increasing catalytic activity. Thr109, in particular, is proposed to be involved in the stabilization of the transition state. These analyses allowed a more complete catalytic mechanism for *E. coli* DHOase to be proposed (described in detail in Section 39.3.2).

39.2.4 Structure of *A. aeolicus* DHOase

Martin and coworkers reported the first structure of a Type I DHOase from *A. aeolicus* refined to 1.7 Å in 2005.[48] *A. aeolicus* is an ancient hyperthermophilic bacterium, regarded as the closest to the last common ancestor of bacteria, eukarya, and archaea.[64] The Type I *A. aeolicus* DHOase is larger (49 kDa) than the Type II *E. coli* DHOase (38 kDa). Unlike *E. coli* DHOase, which is homodimeric and fully active alone, *A. aeolicus* DHOase is monomeric and active only when complexed with aspartate transcarbamylase (ATCase). Full activity can be measured when the DHOase forms dodecamers with *A. aeolicus* ATCase.[29]

Although an equal molar ratio of DHOase to ATCase was used for crystallization, the two different crystal forms obtained contained only DHOase.[48] These reported crystal structures of DHOase represent inactive form(s) in the absence of ATCase. Nevertheless, these structures provide insights to the latency of *A. aeolicus* DHOase. Being a member of amidohydrolase superfamily, the overall fold of the enzyme is a $(\beta/\alpha)_8$-barrel. The core barrel composed of eight β-sheets superposes well with that of *E. coli* DHOase. Structures of *A. aeolicus* DHOase have only one Zn ion in their active sites, consistent with atomic absorption spectroscopy.[29]

The two structures of different crystal forms of *A. aeolicus* DHOase have three disordered segments on the C-terminal end of the $(\beta/\alpha)_8$-barrel near the active site, designated flaps A–C. When compared with *E. coli* DHOase, flaps A–C contain the substrate-binding residues. In crystal form I, a Zn ion is ligated by two histidine residues (His63 and His65), one aspartate (Asp153), and a solvent molecule (hydroxide ion) in an approximately tetrahedral arrangement (Fig. 39.3). In crystal form II, a cysteine residue (Cys181) from the flap A is ordered and displaces the bound solvent molecule (hydroxide ion) (Fig. 39.3). On the basis of these observations, Martin et al. proposed a "cysteine switch" that blocks the active site when the DHOase is not associated with ATCase, resulting in the inactive form. They also proposed that association of DHOase with ATCase would order the three flaps containing the substrate-binding residues, displace the cysteine residue, making the active site accessible and active.

39.2.5 Structure of *P. gingivalis* DHOase

The DHOase from *P. gingivalis* is classified as Type I on the basis of size and sequence.[2] From the genome project for *P. gingivalis*,[65] the sequence of DHOase is known, however, biochemical characterization of the enzyme is not published. Analysis of the amino acid sequence of *P. gingivalis* DHOase reveals the presence of a unique substitution of one of the Zn ligand residues. The first Zn-binding histidine of other active DHOases (His16 and His61 for *E. coli* and *A. aeolicus* DHOases, respectively) is replaced by a glutamine. A BLAST search[66] of the primary sequence of the *P. gingivalis* DHOase identified some other Type I DHOases with the same substitution. One of the two active DHOases of *P. aeruginosa* (Section 39.1.2), PyrC2, has this substitution. The effect of the histidine-to-glutamine substitution on the catalytic activity of DHOase is yet to be characterized.

In June 2006, the Midwest Center of Structural Genomics deposited the 3-dimensional structure of *P. gingivalis* DHOase in the PDB without an accompanying publication. As a Type I DHOase, the overall fold is very similar to the extended N-terminal and C-terminal domains and the core $(\beta/\alpha)_8$-barrel domain of *A. aeolicus* DHOase (Fig. 39.2). However, unlike *A. aeolicus* DHOases that have three disordered segments on the C-terminal end of the $(\beta/\alpha)_8$-barrel near the active site, the *P. gingivalis* DHOase has a well-ordered C-terminal end of the barrel, suggesting that this DHOase is fully active alone. The two active sites of these DHOases are quite different, despite having the same overall fold and high sequence similarity (30% sequence identity; 50% sequence similarity). In contrast to the single Zn found in the *A. aeolicus* DHOase, the active site of *P. gingivalis* contains two Zn ions bridged by a carboxylated lysine, very similar to *E. coli* DHOase with the exception of the histidine-to-glutamine substitution described above (Fig. 39.3).

39.2.6 Structure of *T. thermophilus* DHOase

The crystal structure of DHOase from *T. thermophilus* was deposited in the PDB in September 2007 by the RIKEN Structural Genomics/Proteomics Initiative (PDB code 2Z00). The enzyme is composed of 426 amino acids and is Type I. The overall fold of *T. thermophilus* DHOase is similar to the two Type I DHOases (from *A. aeolicus and P. gingivalis*) solved previously. As in *P. gingivalis* DHOase, the C-terminal end of the barrel is well ordered. The main barrel of the two structures superpose well (RMSD = 1.3 Å). The active site of *T. thermophilus* DHOase also has two Zn ions as for *P. gingivalis*. The most important difference between the two Type I DHOases is an aspartate residue used in place of a carboxylated lysine to bridge the two Zn ions in the active site (Fig. 39.3). An association of DHOase with ATCase from *Thermus* strain ZO5 has been reported.[38,67] The two enzymes are active independently at 37°C, but at high temperatures ATCase is inactive when not associated with DHOase. The authors suggested that the association of DHOase with ATCase stabilizes ATCase at high temperatures.

39.3 PROPOSED CATALYTIC MECHANISMS

Before 3-dimensional structures were available, sequence alignments and mutagenesis were used to identify the residues potentially involved in Zn binding, substrate binding, and catalysis.[68,69] Other methods used to elucidate the catalytic mechanism of DHOase include pH-rate profiles,[35] preparation of metal-substituted variants,[27,34] and measurement of isotope effects.[36,70] Most recently, crystal structures of *E. coli* DHOase have provided critical information on how the enzyme functions at the atomic level.[47,61,63] The unusual asymmetry found in the active sites of the homodimeric enzyme, when crystallized in the presence of D,L-CA-asp (PDB code 1J79)[47] and L-DHO (PDB code 1XGE),[61] respectively, provided further insight on the interactions of the enzyme with substrate and product. A reaction mechanism for DHOase with a binuclear metal center (Type II) was proposed. Although the fundamental chemistry of catalysis is essentially the same for Types I and II, since one Zn participates directly in catalysis, the proposed catalytic mechanism of Type I DHOases with a single Zn is described separately.

39.3.1 Mononuclear Metal Center

On the basis of site-directed mutagenesis, Williams et al. proposed a catalytic mechanism for the reverse reaction (DHO → CA-asp) catalyzed by hamster DHOase.[68] The reaction is initiated by an hydroxide ion formed by ionization of a Zn-bound water molecule at the active site. The 4-carbonyl of DHO*, polarized by interaction with the Zn, undergoes nucleophilic attack by the Zn-bound hydroxide, resulting in a tetrahedral oxyanion transition state. This transition state is stabilized by both the Zn ion and a positively charged electrophilic histidine, possibly His234. Arg19 forms an electrostatic bond with the 6-carboxylate of DHO. Asp230 acts as a general acid/base, donating a proton to the ureido N3 atom and abstracting a proton from the –OH group at C4, inducing collapse of the tetrahedral oxyanion transition state to form CA-asp (Fig. 39.5).[68]

Recently, Anderson et al. used primary and secondary ^{15}N and ^{13}C isotope effects on the ring opening of DHO to probe the mechanism of DHOases from hamster and the thermophilic bacterium *B. caldolyticus*.[70] The approximately 1% primary isotope effect suggested that the step in which C–N cleavage or bond formation occurs is the major rate-limiting step. A similar catalytic mechanism for the reverse reaction

*Two different numbering systems for DHO and CA-asp have been employed in the literature. The numbering system used in Ref. 61 was chosen for this review.

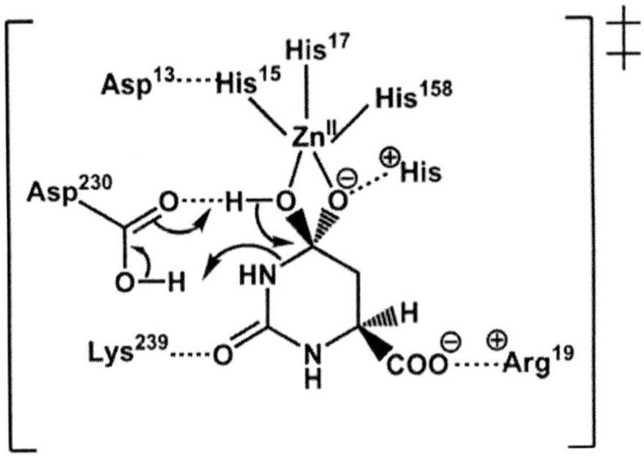

FIGURE 39.5 The proposed enzyme-transition state complex for hamster DHOase domain, (Type I DHOase with a mononuclear center), indicating possible functions for conserved amino acid residues (—Modified from Ref. 68).

(DHO → CA-asp) catalyzed by DHOases from hamster and *B. caldolyticus* was proposed on the basis of these results and those of Williams et al.[68] In this mechanism, the 4-carbonyl group is polarized by interaction with the Zn ion increasing the electrophilicity of C4, which undergoes nucleophilic attack by the coordinated hydroxide. The proton on the hydroxide is simultaneously transferred to an aspartate residue and the tetrahedral transition state is coordinated to the Zn. The proton is then abstracted from the aspartate residue by N3 and the tetrahedral intermediate collapses with C–N bond scission.

39.3.2 Binuclear Metal Center

For DHOases with a binuclear Zn center, the structures of *E. coli* DHOase with D,L-CA-asp (PDB entry 1J79) and L-DHO (PDB entry 1XGE) provide a basis for proposing the catalytic mechanism.[47,61] The Zn ligating residues (His16, His18, Lys102, and Asp 250 for Zn_α; His 139, His177, and Lys102 for Zn_β) and residues involved in substrate binding have also been identified (side chain atoms of Arg20, Asn44, and His254; backbone atoms of Leu222 and Ala266) (Fig. 39.4 in Section 39.2.3).[47,61] Asp250 whose carboxylate group is located 2.5 and 3.0 Å from the bridging solvent molecule and N3 of CA-asp, respectively, plays a dual role in proton abstraction and donation during the interconversion of CA-asp and DHO.

Reaction mechanisms for *E. coli* DHOase have been proposed previously based on X-ray crystallography, pH-rate profiles, site-directed mutagenesis, and solvent isotope effects.[36,47] The loop movement we reported for *E. coli* DHOase interacting with L-CA-asp or L-DHO is an important induced conformational change, enabling a more complete catalytic mechanism for *E. coli* DHOase to be proposed (Fig. 39.6).[61,63] When the active site is empty, the flexible loops of each subunit move in and out dynamically. When CA-asp is bound, it induces closure of the loop over the active site. Movement of the loop results in displacement of His254 $N^{\varepsilon 2}$ by Thr110 $O^{\gamma 1}$ as the

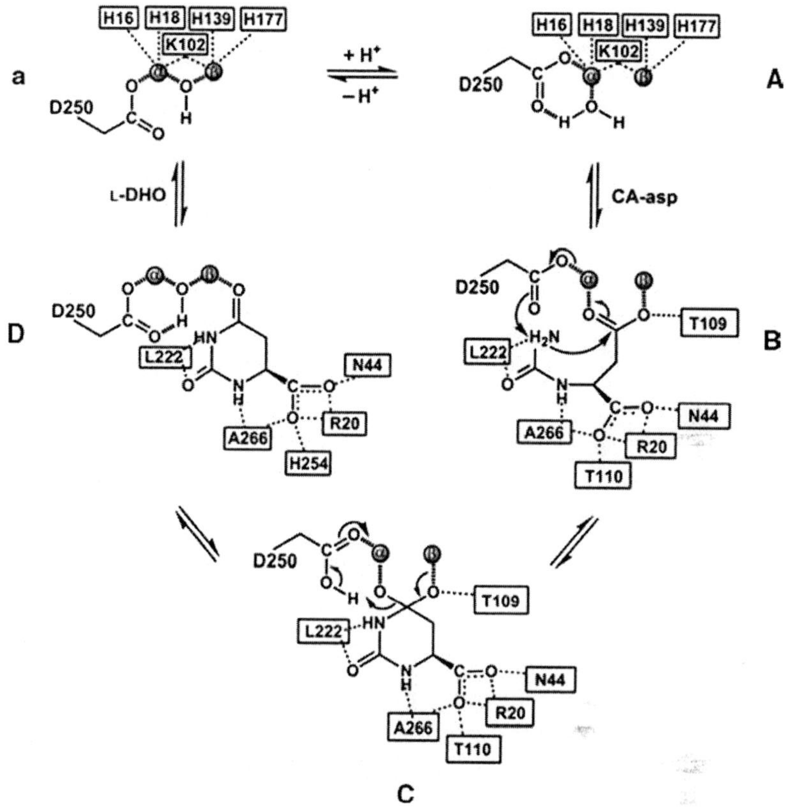

FIGURE 39.6 Proposed catalytic mechanism of Type II DHOases with a binuclear metal center.

hydrogen bonding partner for O62 of CA-asp and Thr109 interacts with O5 of CA-asp to lock the conformation of the substrate (Fig. 39.6, state B). The carboxylate of Asp250 positioned 2.7 Å from the amide nitrogen (N3) of CA-asp at the active site, abstracts a proton from N3 of CA-asp to initiate the nucleophilic attack by the amide nitrogen on the carboxylate group of CA-asp (C4/O4/O5). A tetrahedral transition state is stabilized by interaction with Thr109 and the two Zn ions (Fig. 39.6, state C). Collapse of the tetrahedral intermediate commences with proton donation from the carboxylate of Asp250 to one of the oxygen atoms of the tetrahedral intermediate with cleavage of a C4–O bond forming H_2O. The DHO molecule dissociates from one of the Zn ions, resulting in steric pressure on residue Thr109, and relaxing the loop to the "out" conformation, with release of the product, DHO (Fig. 39.6, state D). States B and D (Fig. 39.6B and D) correspond with the L-CA-asp- and L-DHO-bound subunits in the structure of DHOase crystallized in the presence of L-DHO.[61,71]

For the reverse reaction of DHO hydrolysis, the active enzyme is proposed to have a hydroxide ion bridging the two Zn ions (Fig. 39.6, state a). This conformation corresponds to the structure of ligand-free DHOase crystallized at pH 7.[72] Since this structure was determined at relatively high pH, it corresponds with "state a" rather than with "state A." The water molecule proposed to be coordinated to the β-Zn ion was not observed in this structure and it has been omitted from the proposed mechanism. The nucleophilic attack at C4 of bound DHO by the bridging hydroxide ion is facilitated by transfer of a proton from the hydroxide ion to the carboxylate group of Asp250. Stabilization of the tetrahedral transition state would be achieved via bidentate ligation to both Zn ions and formation of a hydrogen bond between one of the oxygen atoms of the tetrahedral intermediate and Thr109 (Fig. 39.6, state C). Concomitant protonation of the amide nitrogen (N3) by Asp250 assists the cleavage of the N3–C4 bond, leading to collapse of the transition state and formation of CA-asp (Fig. 39.6, state B).

39.4 DHOase-INHIBITOR COMPLEXES

39.4.1 Inhibitors of DHOase

Inhibitors of DHOase have potential use as anticancer or antimalarial drugs. To be of clinical interest, such analogues would need an inhibition constant (K_I value) in the nanomolar range. Most identified inhibitors have been tested against the hamster DHOase domain (Fig. 39.7 and Table 39.3). The most potent inhibitor reported is *cis*-4-carboxy-6-(mercaptomethyl)-3,4,5,6-tetrahydropyridin-2(1*H*)-one (MMDHO; Fig. 39.7, (**39.1**)) with a K_I of 0.14 μM, assayed against hamster CAD.[73] The strong binding of this analogue to DHOase has been attributed to a strong interaction between the thiol substituent of MMDHO and Zn at the active site. A number of transition-state or substrate analogues have been synthesized and assayed for inhibition against hamster DHOase in the Christopherson laboratory.[74–76] L-6-Thiodihydroorotate (TDHO; Fig. 39.7, (**39.2**)), designed as a thiol Zn chelator, is an effective inhibitor ($K_I = 0.85$ μM) and 2-oxo-1,2,3,6-tetra-hydropyrimidine-4,6-dicarboxylate (HDDP; Fig. 39.7, (**39.3**)), synthesized as a transition-state analogue, binds 30 times more strongly than DHO ($K_I = 0.74$ μM).[74] Levenson and Meyer designed and synthesized tetrahedral intermediate analogues based on sulfone and sulfoxide functional groups.[77] 3-Oxo-3,4,5,6-tetrahydro-2*H*-1,2,4-thiadiazine-5-carboxylate 1,1-dioxide (Fig. 39.7, (**39.4**)) and 3-oxo-3,4,5,6-tetrahydro-2*H*-1,4-thiazine-5-carboxylate 1-oxide (Fig. 39.7, (**39.5**)) were the most effective inhibitors. However, these compounds showed moderate inhibition of the hamster DHOase domain with IC_{50} values of 520 and 180 μM, respectively. A phosphadiazine (Fig. 39.7, (**39.6**)) was suggested as a promising transition-state analogue inhibitor by Levenson and Meyer,[77] but has not been synthesized.

Cao et al. synthesized a phosphinic acid transition-state analogue inhibitor, 4-hydroxy-6-oxo-1,4-azaphosphinane-2-carboxylate 4-oxide (Fig. 39.7, (**39.7**))[76] and Manthey et al. subsequently reported mercapto-derivatives of this compound, *cis*- and *trans*-4-mercapto-6-oxo-1,4-azaphosphinane-2-carboxylate-4-oxide (Fig. 39.7, (**39.8**) and (**39.9**)).[75] The K_I values of these compounds were 4.0, 2.9, and 3.1 μM, respectively, indicating that replacement of the phosphinic acid with a phosphinothioic acid provided marginal enhancement of binding.

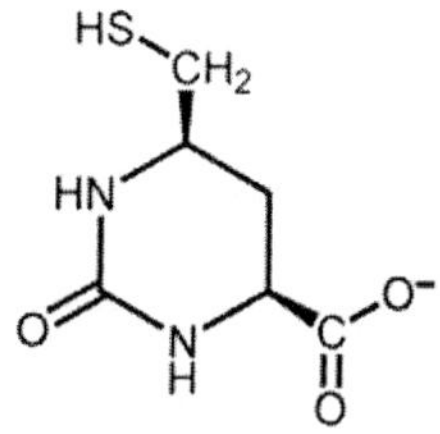

(39.1)

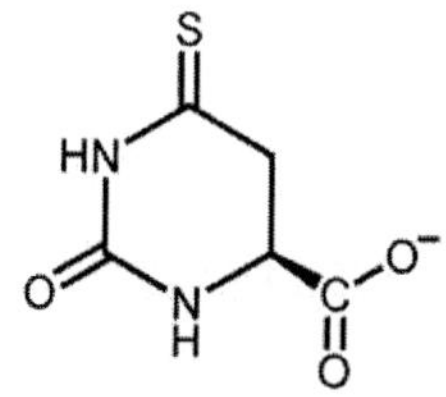

(39.2)

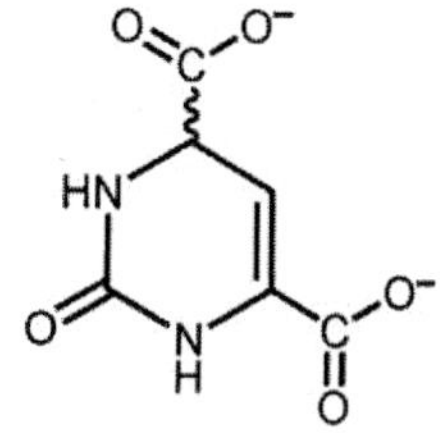

(39.3)

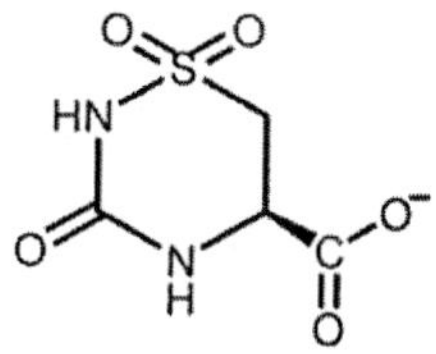

(39.4)

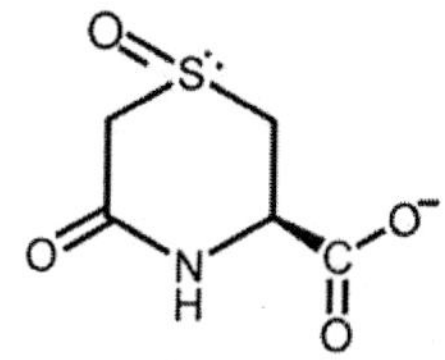

(39.5)

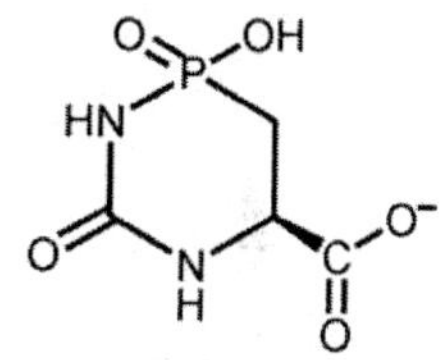

(39.6)

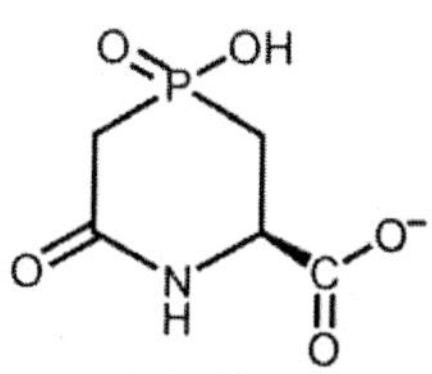

(39.7)

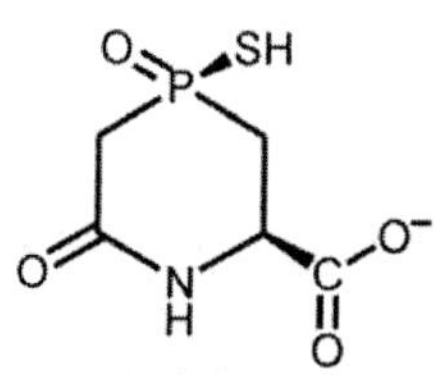

(39.8)

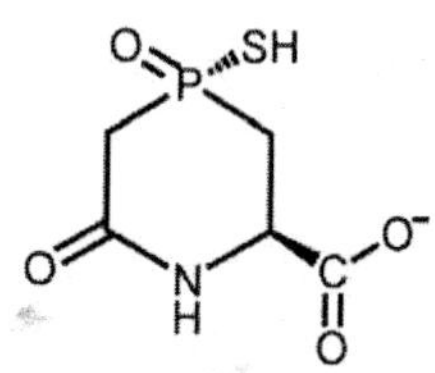

(39.9)

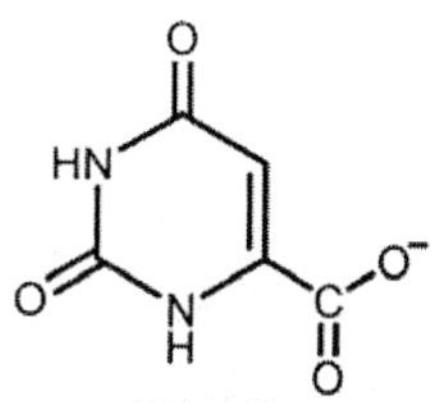

(39.10)

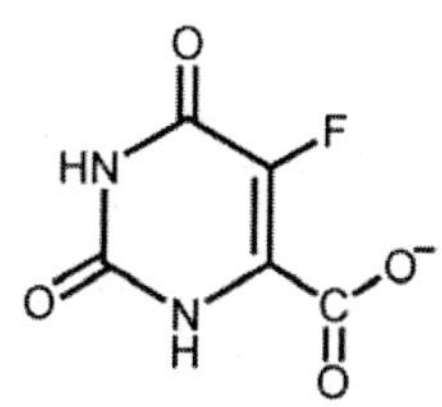

(39.11)

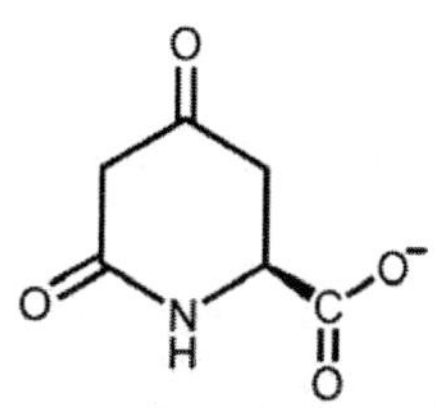

(39.12)

FIGURE 39.7 Inhibitors of DHOase: (**39.1**) *cis*-4-carboxy-6-(mercaptomethyl)-3,4,5,6-tetrahydro-pyrimidine-2(1*H*)-one, (**39.2**) 6-L-thiodihydroorotate, (**39.3**) 2-oxo-1,2,3,6-tetrahydropyrimidine-4,6-dicarboxylate (HDDP), (**39.4**) 3-oxo-3,4,5,6-tetrahydro-2*H*-1,2,4-thiadiazine-5-carboxylate 1,1-dioxide, (**39.5**) 3-oxo-3,4,5,6-tetrahydro-2*H*-1,4-thiazine-5-carboxylate 1-oxide, (**39.6**) 4-hydroxy-6-oxo-1,4-aza-phosphadiazine-2-carboxylate-4-oxide (not synthesized), (**39.7**) 4-hydroxy-6-oxo-1,4-aza-phosphinane-2-carboxylate-4-oxide, (**39.8**) *cis*-4-mercapto-6-oxo-1,4-aza-phosphinane-2-carboxylate-4-oxide, (**39.9**) *trans*-4-mercapto-6-oxo-1,4-azaphosphinane-2-carboxylate-4-oxide, (**39.10**) orotate, (**39.11**) 5-fluoroorotate, (**39.12**) 4,6-dioxo-piperidine-2(*S*)-carboxylate.

TABLE 39.3 Inhibitors of DHOase

Compounds	Tested DHOase	K_I (μM)
1	Hamster CAD[73]	0.14
2	Hamster DHOase domain[74]	0.85
3	Hamster DHOase domain[74]	0.74
3	*E. coli* DHOase[72]	1.3
4	Hamster CAD[77]	520[a]
5	Hamster CAD[77]	180[a]
7	Hamster DHOase domain[76]	4
8	Hamster DHOase domain[75]	2.9
9	Hamster DHOase domain[75]	3.1
10	Hamster DHOase domain[74]	81
11	Hamster DHOase domain[74]	15
11	*E. coli* DHOase[72]	31.8
12	*E. coli* DHOase[83]	76

[a] IC_{50} values.

The substrate analogues, orotate, and 5-fluoroorotate (Fig. 39.7, (**39.10**) and (**39.11**)) also show moderate inhibition against the hamster DHOase domain with reported K_I values of 81 and 15 μM, respectively.[74] FOA is also known as a potent inhibitor of DHOase from the malarial parasite *P. falciparum*, resulting in 50% inhibition of growth of the parasites at concentrations of 6–42 nM *in vitro*.[78–80] Mice infected with the malarial parasites *P. berghei* and *P. yoelii*, have been cleared of parasites with administration of FOA.[80–82] The reported K_I value for FOA, measured for the degradative reaction (DHO → CA-asp) catalyzed by purified DHOase from *P. berghei* is 65 μM.[22] Recently, Li and Raushel reported synthesis of a series of compounds, designed to inhibit *E. coli* DHOase.[83] The most potent inhibitor was 4,6-dioxo-piperidine-2-(*S*)-carboxylate (Fig. 39.7, (**39.12**)) with a K_I value of 76 μM.

39.4.2 Structures of DHOase-Inhibitor Complexes

The only crystal structures of DHOase-inhibitor complexes were published in 2007.[72] Two inhibitors, HDDP (Fig. 39.7, (**39.3**)), designed as a "transition-state analogue," and FOA (Fig. 39.7 (**39.11**)), a product mimic, were cocrystallized with *E. coli* DHOase. Analysis of the structures of these DHOase–inhibitor complexes showed differential effects of loop movement and revealed the details of inhibitor binding.

39.4.2.1 The Complex of E. coli DHOase with HDDP The structure of *E. coli* DHOase complexed with HDDP, refined to 2.0 Å resolution.[72] has a homodimer in the asymmetric unit and each active site contains two Zn ions and one HDDP molecule. The flexible loops (residues 105–115) are ordered in both subunits, but have different conformations for each subunit although HDDP is bound at each active site. In the case of subunit A, the flexible loop (residues 105–115) is in the "out" conformation. Subunit B, has the "loop-in" conformation with Thr109 and Thr110

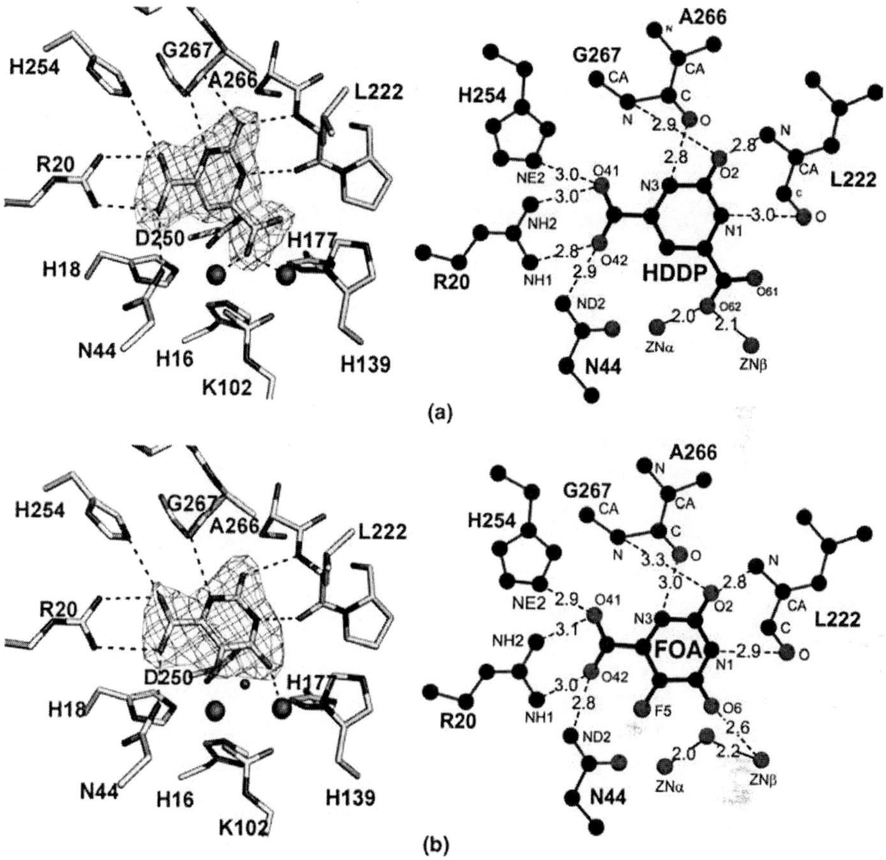

FIGURE 39.8 Interaction of inhibitors with *E. coli* DHOase. (a) HDDP and (b) FOA.

making hydrogen bonds with the HDDP molecule. The asymmetry of the loops conformation corresponds with our previous structure of DHOase crystallized with L-DHO.

HDDP, designed as a "transition-state analogue," has two carboxylate groups. One is joined at C4, coplanar with the ring due to an adjacent double bond (C4–C5), and the other is at C6 (Fig. 39.7, (**39.3**)). Hambley et al.[84] proposed that the coplanar carboxylate group (C41, O41, and O42) would bind to Zn, but the structure analysis shows that the coplanar carboxylate interacts with side chains of the enzyme (Arg20, Asn44, and His254) and it is the other carboxylate group (C61, O61, and O62) that bridges the Zn ions (Fig. 39.8). O62 of the carboxylate (C61, O61, and O62) of HDDP provides a monodentate bridge to the Zn ions at the binuclear center. This is similar to the structure of the product (DHO)-bound subunit where a water molecule (or hydroxide ion) bridges the two Zn ions, but differs from the (CA-asp)-bound structure, where one carboxylate group (C4, O4, and O5) bridges the Zn ions in a bidentate manner.

The interactions made by HDDP in subunits A and B are similar except for the presence of hydrogen bonds from Thr109 and Thr110 of the flexible loop to HDDP.

The carboxylate group (C41, O41, and O42) of HDDP interacts with side chains around the ligand-binding site, and the second carboxylate (C61, O61, and O62) bridges the two Zn ions in a monodentate manner. $O^{\gamma 1}$ of Thr109 hydrogen bonds to O61 of HDDP, and $O^{\gamma 1}$ of Thr110 hydrogen bonds to O41 of HDDP, stabilizing the "loop-in" conformation.

39.4.2.2 The Complex of E. coli DHOase with FOA The structure of *E. coli* DHOase complexed with FOA, refined to 2.2 Å resolution.[72] is a homodimer with a FOA molecule bound to each active site. Unlike the DHOase–HDDP complex, the electron density corresponding to residues 108–114 in subunit B is disordered in the DHOase–FOA complex.

FOA is an analogue of the product, DHO, and contains one carboxylate (Fig. 39.7, (**39.11**)). FOA was bound in both active sites of *E. coli* DHOase with full occupancy in subunit A and partial occupancy in subunit B. FOA binds similarly to DHO at the active site (Fig. 39.8). Although the ring of FOA is approximately planar, the C2, O2, C5, C6, and O6 atoms of FOA lie in positions equivalent those of bound DHO. FOA makes contacts with the same active-site residues as does DHO in the DHO-bound subunit of the DHO/CA-asp complex (Arg20, Asn44, and His254, and the backbone of Leu222, Ala266, and Gly267) (Fig. 39.8). The C5 fluoride atom of FOA is in close contact with Asn44. In both subunits, a water molecule (or hydroxide ion) bridges the two Zn ions. The FOA molecule at the active site of subunit B was refined with a partial occupancy of 0.75. In subunit A the flexible loop (residues 105–115) faces away from the active site ("loop-out"), whereas in subunit B parts of the flexible loop (residues 108–114) are disordered when the active site is partially occupied by FOA. Incompatibility of thc "loop-in" conformation with a product analogue in the active site results in a disordered conformation for this loop and partial occupancy of FOA in subunit B.

39.4.2.3 Implications for the Design of More Effective Inhibitors HDDP binds 25 times more tightly than FOA to *E. coli* DHOase. These kinetic data can be reconciled with the structures of the inhibitor complexes. The divalent Zn center activates the substrate and water molecule for nucleophilic attack. Although FOA is a DHO analogue and interacts with substrate-binding residues, it does not ligate the Zn ions. As is seen in the DHO complex, a water molecule bridges the two Zn ions. In contrast, HDDP uses the additional carboxylate to ligate the Zn ions directly. The carboxylate groups of HDDP stabilize the "loop-in" conformation, increasing the affinity for HDDP. The "loop-in" conformation observed in the CA-asp- and HDDP-bound subunits is presumably energetically favored by formation of hydrogen bonds with residues of the flexible loop, with reduction in the number of ordered water molecules. Effective inhibitors should participate in interactions with residues involved in substrate or product binding, possess a chemical moiety capable of directly ligating the active-site Zn ion(s), and they should stabilize the "loop-in" conformation of the enzyme that closes the active site during catalysis. We can conclude that conformational changes in the flexible loop of DHOase are important for catalytic activity,[63] and for the binding of potent inhibitors of DHOase.

Although HDDP, designed as a transition-state analogue ($K_I = 1.3$ μM),[72] binds more tightly to DHOase than CA-asp or DHO (K_m values 470 μM[85] and 20.8 μM,[61] respectively), the structure of the DHOase-HDDP complex indicates that HDDP is not a true transition-state analogue. The proposed tetrahedral transition-state intermediate (Fig. 39.6) is expected to ligate two Zn ions in a bidentate manner[36,47] but HDDP actually provides a monodentate bridge to the Zn ions (Fig. 39.8). Further, HDDP has an additional carbon atom in comparison to the proposed tetrahedral intermediate. The C6 atom of HDDP makes a very close contact with $O^{\delta 2}$ of Asp250, proposed to act as a general acid/base for proton abstraction or donation during catalysis.[36,47] An inhibitor without the steric hindrance of HDDP with Asp250 should bind even more tightly. Recently, Li and Raushel reported the synthesis of a series of compounds, designed to inhibit *E. coli* DHOase.[83] 4,6-Dioxo-piperidine-2-(*S*)-carboxylic acid (Fig. 39.7, (**39.3**)) was the most potent with $K_I = 76$ μM at pH 7.0. The 25% of the ketone group at C6 that is hydrated at pH 7 is capable of forming a tetrahedral analogue of the proposed transition state, however, the majority exists in DHO-like forms. Increasing the pH resulted in higher K_I values (210 and 440 μM at pH 8 and 9, respectively) due to a decreased proportion of the hydrated form. These K_I values are not sufficiently low to make this analogue of clinical interest.

39.5 CONCLUDING REMARKS

As mentioned above, differences in pyrimidine biosynthesis between human and malarial parasites make DHOase an attractive target for antimalarial chemotherapy. Recently, a recombinant DHOase from *P. falciparum*, that causes most human malaria, has been reported[32] but there is not yet a 3-dimensional structure. The DHOases from *E. coli* and *P. falciparum* are both Type II and have a high sequence similarity (69%) with all substrate- and Zn-binding residues conserved.[72] Therefore, the principles described earlier for design of inhibitors are applicable to the *Plasmodium* DHOase and ultimately for antimalarial chemotherapy.

Most of the inhibitors of DHOase reported have been tested against the DHOase domain from hamster CAD. The 243-kDa mammalian CAD catalyzes the first three steps of the *de novo* pyrimidine biosynthetic pathway, containing carbamyl phosphate synthetase (CPSase), aspartate transcarbamylase (ATCase) and DHOase. The boundary of the DHOase domain has been established by controlled proteolysis[25] and several constructs of the DHOase domain of hamster CAD have been expressed recombinantly and characterized.[33,86] Although the recombinant DHOase domain from hamster CAD has been crystallized, the best crystals diffracted poorly and the structure is yet to be solved.[87] The importance of CAD for pyrimidine biosynthesis makes DHOase a target for anticancer and anti-inflammatory drugs. The structure of human CAD in atomic detail would greatly facilitate drug development. Comparison of the DHOase active sites of human CAD and the *P. falciparum* enzyme would provide a basis for future design of "second-generation" inhibitors of DHOase with potential use as antimalarial drugs.

REFERENCES

1. Kim, H.; Kelly, R. E.; Evans, D. R. The structural organization of the hamster multifunctional protein CAD. Controlled proteolysis, domains, and linkers. *J. Biol. Chem.* **1992**, *267*, 7177–7184.
2. Fields, C.; Brichta, D.; Shepherdson, M.; Farinha, M.; O'Donovan, G. Phylogenetic analysis and classification of dihydroorotases: a complex history for a complex enzyme. *Pathways Pyrimidines: Int. Newslett.* **1999**, *7*, 49–63.
3. Heidelberger, C.; Danenberg, P. V.; Moran, R. G. Fluorinated pyrimidines and their nucleosides. *Adv. Enzymol. Relat. Areas Mol. Biol.* **1983**, *54*, 58–119.
4. Weber, G. Biochemical strategy of cancer cells and the design of chemotherapy: G. H. A. Clowes Memorial Lecture. *Cancer Res.* **1983**, *43*, 3466–3492.
5. Sherman, I. W. Biochemistry of *Plasmodium* (malarial parasites). *Microbiol. Rev.* **1979**, *43*, 453–495.
6. Reyes, P.; Rathod, P. K.; Sanchez, D. J.; Mrema, J. E.; Rieckmann, K. H.; Heidrich, H. G. Enzymes of purine and pyrimidine metabolism from the human malaria parasite, *Plasmodium falciparum*. *Mol. Biochem. Parasitol.* **1982**, *5*, 275–290.
7. Gardner, M. J.; Hall, N.; Fung, E.; White, O.; Berriman, M.; Hyman, R. W.; Carlton, J. M.; Pain, A.; Nelson, K. E.; Bowman, S.; Paulsen, I. T.; James. K.; Eisen, J. A.; Rutherford, K.; Salzberg, S. L.; Craig, A.; Kyes, S.; Chan, M. S.; Nene, V.; Shallom, S. J.; Suh, B.; Peterson, J.; Angiuoli, S.; Pertea, M.; Allen, J.; Selengut, J.; Haft, D.; Mather, M. W.; Vaidya, A. B.; Martin, D. M.; Fairlamb, A. H.; Fraunholz, M. J.; Roos, D. S.; Ralph, S. A.; McFadden, G. I.; Cummings, L. M.; Subramanian, G. M.; Mungall, C.; Venter, J. C.; Carucci, D. J.; Hoffman, S. L.; Newbold, C.; Davis, R. W.; Fraser, C. M.; Barrell, B. Genome sequence of the human malaria parasite *Plasmodium falciparum*. *Nature* **2002**, *419*, 498–511.
8. Christopherson, R. I.; Lyons, S. D.; Wilson, P. K. Inhibitors of *de novo* nucleotide biosynthesis as drugs. *Acc. Chem. Res.* **2002**, *35*, 961–971.
9. Sherman, I. W.; Jones, L. A. *Plasmodium lophurae*: membrane proteins of erythrocyte-free plasmodia and malaria-infected red cells. *J. Protozool.* **1979**, *26*, 489–501.
10. Gero, A. M.; Brown, G. V.; O'Sullivan, W. J. Pyrimidine *de novo* synthesis during the life cycle of the intraerythrocytic stage of *Plasmodium falciparum*. *J. Parasitol.* **1984**, *70*, 536–541.
11. Hyde, J. E. Targeting purine and pyrimidine metabolism in human apicomplexan parasites. *Curr. Drug Targets* **2007**, *8*, 31–47.
12. Kensler, T. W.; Cooney, D. A.; *Advances in Pharmacology and Chemotherapy;* Academic Press: New York, **1981**; Vol. *18*, p 273.
13. Mitchell, A.; Finch, L. R. Pathways of nucleotide biosynthesis in *Mycoplasma mycoides* subsp. *mycoides*. *J. Bacteriol.* **1977**, *130*, 1047–1054.
14. Heyworth, P. G.; Gutteridge, W. E.; Ginger, C. D. Pyrimidine metabolism in *Trichomonas vaginalis*. *FEBS Lett.* **1984**, *176*, 55–60.
15. McClarty, G.; Qin, B. Pyrimidine metabolism by intracellular *Chlamydia psittaci*. *J. Bacteriol.* **1993**, *175*, 4652–4661.
16. Xu, P.; Widmer, G.; Wang, Y.; Ozaki, L. S.; Alves, J. M.; Serrano, M. G.; Puiu, D.; Manque, P.; Akiyoshi, D.; Mackey, A. J.; Pearson, W. R.; Dear, P. H.; Bankier, A. T.; Peterson, D. L.; Abrahamsen, M. S.; Kapur, V.; Tzipori, S.; Buck, G. A. The genome of *Cryptosporidium hominis*. *Nature* **2004**, *431*, 1107–1112.

17. Lieberman, I.; Kornberg, A. Enzymatic synthesis and breakdown of a pyrimidine, orotic acid. I. Dihydroortic acid, ureidosuccinic acid, and 5-carboxymethylhydantoin. *J. Biol. Chem.* **1954**, *207*, 911–924.

18. Taylor, W. H.; Taylor, M. L.; Balch, W. E.; Gilchrist, P. S. Purification of properties of dihydroorotase, a zinc-containing metalloenzyme in *Clostridium oroticum*. *J. Bacteriol.* **1976**, *127*, 863–873.

19. Pettigrew, D. W.; Bidigare, R. R.; Mehta, B. J.; Williams, M. I.; Sander, E. G. Dihydroorotase from *Clostridium oroticum*. Purification and reversible removal of essential zinc. *Biochem. J.* **1985**, *230*, 101–108.

20. Sander, E. G.; Heeb, M. J. Purification and properties of dihydroorotase from *Escherichia coli B*. *Biochim. Biophys. Acta* **1971**, *227*, 442–452.

21. Washabaugh, M. W.; Collins, K. D. Dihydroorotase from *Escherichia coli*. Purification and characterization. *J. Biol. Chem.* **1984**, *259*, 3293–3298.

22. Krungkrai, J.; Cerami, A.; Henderson, G. B. Pyrimidine biosynthesis in parasitic protozoa: purification of a monofunctional dihydroorotase from *Plasmodium berghei* and *Crithidia fasciculata*. *Biochemistry* **1990**, *29*, 6270–6275.

23. Ogawa, J.; Shimizu, S. Purification and characterization of dihydroorotase from *Pseudomonas putida*. *Arch. Microbiol.* **1995**, *164*, 353–357.

24. Christopherson, R. I.; Jones, M. E. The overall synthesis of L-5,6-dihydroorotate by multienzymatic protein pyr1-3 from hamster cells. Kinetic studies, substrate channeling, and the effects of inhibitors. *J. Biol. Chem.* **1980**, *255*, 11381–11395.

25. Kelly, R. E.; Mally, M. I.; Evans, D. R. The dihydroorotase domain of the multifunctional protein CAD. Subunit structure, zinc content, and kinetics. *J. Biol. Chem.* **1986**, *261*, 6073–6083.

26. Brown, D. C.; Collins, K. D. Dihydroorotase from *Escherichia coli*. Cloning the pyrC gene and production of tryptic peptide maps. *J. Biol. Chem.* **1986**, *261*, 5917–5919.

27. Brown, D. C.; Collins, K. D. Dihydroorotase from *Escherichia coli*. Substitution of Co(II) for the active site Zn(II). *J. Biol. Chem.* **1991**, *266*, 1597–1604.

28. Purcarea, C.; Martin, P.; Vickrey, J. F.; Guy, H. I.; Edwards, B. F.; Evans, D. R. Cloning, expression and preliminary X-ray analysis of the dihydroorotase from the hyperthermophilic eubacterium *Aquifex aeolicus*. *Acta Crystallogr., D Biol. Crystallogr.* **2002**, *58*, 154–156.

29. Ahuja, A.; Purcarea, C.; Ebert, R.; Sadecki, S.; Guy, H. I.; Evans, D. R. *Aquifex aeolicus* dihydroorotase: association with aspartate transcarbamoylase switches on catalytic activity. *J. Biol. Chem.* **2004**, *279*, 53136–53144.

30. Huang, D. T.; Kaplan, J.; Menz, R. I.; Katis, V. L.; Wake, R. G.; Zhao, F.; Wolfenden, R.; Christopherson, R. I. Thermodynamic analysis of catalysis by the dihydroorotases from hamster and *Bacillus caldolyticus*, as compared with the uncatalyzed reaction. *Biochemistry* **2006**, *45*, 8275–8283.

31. Robles Lopez, S. M.; Hortua Triana, M. A.; Zimmermann, B. H. Cloning and preliminary characterization of the dihydroorotase from *Toxoplasma gondii*. *Mol. Biochem. Parasitol.* **2006**, *148*, 93–98.

32. Krungkrai, S. R.; Wutipraditkul, N.; Krungkrai, J. Dihydroorotase of human malarial parasite *Plasmodium falciparum* differs from host enzyme. *Biochem. Biophys. Res. Commun.* **2008**, *366*, 821–826.

33. Williams, N. K.; Peide, Y.; Seymour, K. K.; Ralston, G. B.; Christopherson, R. I. Expression of catalytically active hamster dihydroorotase domain in *Escherichia coli*: purification and characterization. *Protein Eng.* **1993**, *6*, 333–340.

34. Huang, D. T.; Thomas, M. A.; Christopherson, R. I. Divalent metal derivatives of the hamster dihydroorotase domain. *Biochemistry* **1999**, *38*, 9964–9970.

35. Christopherson, R. I.; Jones, M. E. Interconversion of carbamayl-L-aspartate and L-dihydroorotate by dihydroorotase from mouse Ehrlich ascites carcinoma. *J. Biol. Chem.* **1979**, *254*, 12506–12512.

36. Porter, T. N.; Li, Y.; Raushel, F. M. Mechanism of the dihydroorotase reaction. *Biochemistry* **2004**, *43*, 16285–16292.

37. Coleman, P. F.; Suttle, D. P.; Stark, G. R. Purification from hamster cells of the multifunctional protein that initiates *de novo* synthesis of pyrimidine nucleotides. *J. Biol. Chem.* **1977**, *252*, 6379–6385.

38. Van de Casteele, M.; Chen, P.; Roovers, M.; Legrain, C.; Glansdorff, N. Structure and expression of a pyrimidine gene cluster from the extreme thermophile *Thermus* strain ZO5. *J. Bacteriol.* **1997**, *179*, 3470–3481.

39. McPhail, D.; Shepherdson, M. The aspartate transcarbamoylase-dihydroorotase complex in *Deinococcus radiophilus* has an active dihydroorotase. *Arch. Microbiol.* **2006**, *185*, 78–81.

40. Hughes, L. E.; Hooshdaran, M. Z.; O'Donovan, G. A. *Streptomyces* aspartate transcarbamoylase is a dodecamer with dihydroorotase activity. *Curr. Microbiol.* **1999**, *39*, 175–179.

41. Wild, J. R.; Belser, W. L. Pyrimidine biosynthesis in *Serratia marcescens*: polypeptide interactions of three nonsequential enzymes. *Biochem. Genet.* **1977**, *15*, 173–193.

42. Guyonvarch, A.; Nguyen-Juilleret, M.; Hubert, J. C.; Lacroute, F. Structure of the *Saccharomyces cerevisiae* URA4 gene encoding dihydroorotase. *Mol. Gen. Genet.* **1988**, *212*, 134–141.

43. Souciet, J. L.; Nagy, M.; Le Gouar, M.; Lacroute, F.; Potier, S. Organization of the yeast *URA2* gene: identification of a defective dihydroorotase-like domain in the multifunctional carbamoylphosphate synthetase-aspartate transcarbamylase complex. *Gene* **1989**, *79*, 59–70.

44. Penverne, B.; Belkaid, M.; Herve, G. *In situ* behavior of the pyrimidine pathway enzymes in *Saccharomyces cerevisiae*. 4. The channeling of carbamylphosphate to aspartate transcarbamylase and its partition in the pyrimidine and arginine pathways. *Arch. Biochem. Biophys.* **1994**, *309*, 85–93.

45. Schurr, M. J.; Vickrey, J. F.; Kumar, A. P.; Campbell, A. L.; Cunin, R.; Benjamin, R. C.; Shanley, M. S.; O'Donovan, G. A. Aspartate transcarbamoylase genes of *Pseudomonas putida*: requirement for an inactive dihydroorotase for assembly into the dodecameric holoenzyme. *J. Bacteriol.* **1995**, *177*, 1751–1759.

46. Brichta, D. M.; Azad, K. N.; Ralli, P.; O'Donovan, G. A. *Pseudomonas aeruginosa* dihydroorotases: a tale of three *pyrCs*. *Arch. Microbiol.* **2004**, *182*, 7–17.

47. Thoden, J. B.; Phillips, G. N., Jr.; Neal, T. M.; Raushel, F. M.; Holden, H. M. Molecular structure of dihydroorotase: a paradigm for catalysis through the use of a binuclear metal center. *Biochemistry* **2001**, *40*, 6989–6997.

48. Martin, P. D.; Purcarea, C.; Zhang, P.; Vaishnav, A.; Sadecki, S.; Guy-Evans, H. I.; Evans, D. R.; Edwards, B. F. The crystal structure of a novel, latent dihydroorotase from *Aquifex aeolicus* at 1.7A resolution. *J. Mol. Biol.* **2005**, *348*, 535–547.

49. Holm, L.; Sander, C. An evolutionary treasure: unification of a broad set of amidohydrolases related to urease. *Proteins* **1997**, *28*, 72–82.

50. Seibert, C. M.; Raushel, F. M. Structural and catalytic diversity within the amidohydrolase superfamily. *Biochemistry* **2005**, *44*, 6383–6391.

51. Reardon, D.; Farber, G. K. The structure and evolution of alpha/beta barrel proteins. *FASEB J.* **1995**, *9*, 497–503.

52. Buchbinder, J. L.; Stephenson, R. C.; Dresser, M. J.; Pitera, J. W.; Scanlan, T. S.; Fletterick, R. J. Biochemical characterization and crystallographic structure of an *Escherichia coli* protein from the phosphotriesterase gene family. *Biochemistry* **1998**, *37*, 5096–5106.

53. Karplus, P. A.; Pearson, M. A.; Hausinger, R. P. 70 years of crystalline urease: what have we learned? *Acc. Chem. Res.* **1997**, *30*, 330–337.

54. Benning, M. M.; Shim, H.; Raushel, F. M.; Holden, H. M. High resolution X-ray structures of different metal-substituted forms of phosphotriesterase from *Pseudomonas diminuta*. *Biochemistry* **2001**, *40*, 2712–2722.

55. Ireton, G. C.; McDermott, G.; Black, M. E.; Stoddard, B. L. The structure of *Escherichia coli* cytosine deaminase. *J. Mol. Biol.* **2002**, *315*, 687–697.

56. Nam, S. H.; Park, H. S.; Kim, H. S. Evolutionary relationship and application of a superfamily of cyclic amidohydrolase enzymes. *Chem. Rec.* **2005**, *5*, 298–307.

57. Kim, G. J.; Kim, H. S. Identification of the structural similarity in the functionally related amidohydrolases acting on the cyclic amide ring. *Biochem. J. 330*,Pt 1 **1998**, 295–302.

58. Abendroth, J.; Niefind, K.; May, O.; Siemann, M.; Syldatk, C.; Schomburg, D. The structure of L-hydantoinase from *Arthobacter aurescens* leads to an understanding of dihydropyrimidinase substrate and enantio specificity. *Biochemistry* **2002**, *41*, 8589–8597.

59. Lohkamp, B.; Andersen, B.; Piskur, J.; Dobritzsch, D. The crystal structures of dihydropyrimidinases reaffirm the close relationship between cyclic amidohydrolases and explain their substrate specificity. *J. Biol. Chem.* **2006**, *281*, 13762–13776.

60. Sander, E. G.; Wright, L. D.; McCormick, D. B. Evidence for function of a metal ion in the activity of dihydroorotase from *Zymobacterium oroticum*. *J. Biol. Chem.* **1965**, *240*, 3628–3630.

61. Lee, M.; Chan, C. W.; Guss, J. M.; Christopherson, R. I.; Maher, M. J. Dihydroorotase from *Escherichia coli*: loop movement and cooperativity between subunits. *J. Mol. Biol.* **2005**, *348*, 523–533.

62. Lee, M.; Maher, M. J.; Guss, J. M. Structure of the T109S mutant of *Escherichia coli* dihydroorotase complexed with the inhibitor 5-fluoroorotate: catalytic activity is reflected by the crystal form. *Acta Crystallogr. Sect. F Struct. Biol. Cryst. Commun.* **2007**, *63*, 154–161.

63. Lee, M.; Maher, M. J.; Christopherson, R. I.; Guss, J. M. Kinetic and structural analysis of mutant *Escherichia coli* dihydroorotases: a flexible loop stabilizes the transition state. *Biochemistry* **2007**, *46*, 10538–10550.

64. Deckert, G.; Warren, P. V.; Gaasterland, T.; Young, W. G.; Lenox, A. L.; Graham, D. E.; Overbeek, R.; Snead, M. A.; Keller, M.; Aujay, M.; Huber, R.; Feldman, R. A.; Short, J. M.; Olsen, G. J.; Swanson, R. V. The complete genome of the hyperthermophilic bacterium *Aquifex aeolicus*. *Nature* **1998**, *392*, 353–358.

65. Nelson, K. E.; Fleischmann, R. D.; DeBoy, R. T.; Paulsen, I. T.; Fouts, D. E.; Eisen, J. A.; Daugherty, S. C.; Dodson, R. J.; Durkin, A. S.; Gwinn, M.; Haft, D. H.; Kolonay, J. F.; Nelson, W. C.; Mason, T.; Tallon, L.; Gray, J.; Granger, D.; Tettelin, H.; Dong, H.; Galvin,

J. L.; Duncan, M. J.; Dewhirst, F. E.; Fraser, C. M. Complete genome sequence of the oral pathogenic bacterium *Porphyromonas gingivalis* strain W83. *J. Bacteriol.* **2003**, *185*, 5591–5601.

66. Altschul, S. F.; Madden, T. L.; Schaffer, A. A.; Zhang, J.; Zhang, Z.; Miller, W.; Lipman, D. J. Gapped BLAST and PSI-BLAST: a new generation of protein database search programs. *Nucleic Acids Res.* **1997**, *25*, 3389–3402.
67. Van de Casteele,M.; Legrain, C.; Desmarez, L.; Chen, P. G.; Pierard, A.; Glansdorff,N. Molecular physiology of carbamoylation under extreme conditions: what can we learn from extreme thermophilic microorganisms? *Comp. Biochem. Physiol. A Physiol.* **1997**, *118*, 463–473.
68. Williams, N. K.; Manthey, M. K.; Hambley, T. W.; O'Donoghue, S. I.; Keegan, M.; Chapman, B. E.; Christopherson, R. I. Catalysis by hamster dihydroorotase: zinc binding, site-directed mutagenesis, and interaction with inhibitors. *Biochemistry* **1995**, *34*, 11344–11352.
69. Zimmermann, B. H.; Kemling, N. M.; Evans, D. R. Function of conserved histidine residues in mammalian dihydroorotase. *Biochemistry* **1995**, *34*, 7038–7046.
70. Anderson, M. A.; Cleland, W. W.; Huang, D. T.; Chan, C.; Shojaei, M.; Christopherson, R. I. ^{13}C and ^{15}N isotope effects for conversion of L-dihydroorotate to *N*-carbamyl-L-aspartate using dihydroorotase from hamster and *Bacillus caldolyticus*. *Biochemistry* **2006**, *45*, 7132–7139.
71. Uhlemann, A. C.; Cameron, A.; Eckstein-Ludwig, U.; Fischbarg, J.; Iserovich, P.; Zuniga, F. A.; East, M.; Lee, A.; Brady, L.; Haynes, R. K.; Krishna, S. A single amino acid residue can determine the sensitivity of SERCAs to artemisinins. *Nat. Struct. Mol. Biol.* **2005**, *12*, 628–629.
72. Lee, M.; Chan, C. W.; Graham, S. C.; Christopherson, R. I.; Guss, J. M.; Maher, M. J. Structures of ligand-free and inhibitor complexes of dihydroorotase from *Escherichia coli*: implications for loop movement in inhibitor design. *J Mol. Biol.* **2007**, *370*, 812–825.
73. Adams, J. L.; Meek, T. D.; Mong, S. M.; Johnson, R. K.; Metcalf, B. W. *cis*-4-Carboxy-6-(mercaptomethyl)-3,4,5,6-tetrahydropyrimidin-2(1 *H*)-one, a potent inhibitor of mammalian dihydroorotase. *J. Med. Chem.* **1988**, *31*, 1355–1359.
74. Christopherson, R. I.; Schmalzl, K. J.; Szabados, E.; Goodridge, R. J.; Harsanyi, M. C.; Sant, M. E.; Algar, E. M.; Anderson, J. E.; Armstrong, A.; Sharma, S. C. et al. Mercaptan and dicarboxylate inhibitors of hamster dihydroorotase. *Biochemistry* **1989**, *28*, 463–470.
75. Manthey, M. K.; Huang, D. T.; Bubb, W. A.; Christopherson, R. I. Synthesis and enzymic evaluation of 4-mercapto-6-oxo-1, 4-azaphosphinane-2-carboxylic acid 4-oxide as an inhibitor of mammalian dihydroorotase. *J. Med. Chem.* **1998**, *41*, 4550–4555.
76. Cao, Y.; Christopherson, R. I.; Elix, J. A.; Gaul, K. L. Synthesis of a phosphinic acid transition state analogue inhibitor of dihydroorotase. *Aust. J. Chem.* **1994**, *47*, 903–911.
77. Levenson, C. H.; Meyer, R. B., Jr. Design and synthesis of tetrahedral intermediate analogues as potential dihydroorotase inhibitors. *J. Med. Chem.* **1984**, *27*, 228–232.
78. Rathod, P. K.; Khatri, A.; Hubbert, T.; Milhous, W. K. Selective activity of 5-fluoroorotic acid against *Plasmodium falciparum in vitro*. *Antimicrob. Agents Chemother.* **1989**, *33*, 1090–1094.

79. Queen, S. A.; Jagt, D. L.; Reyes, P. *In vitro* susceptibilities of *Plasmodium falciparum* to compounds which inhibit nucleotide metabolism. *Antimicrob. Agents Chemother.* **1990**, *34*, 1393–1398.

80. Krungkrai, J.; Krungkrai, S. R.; Phakanont, K. Antimalarial activity of orotate analogs that inhibit dihydroorotase and dihydroorotate dehydrogenase. *Biochem. Pharmacol.* **1992**, *43*, 1295–1301.

81. Gomez, Z. M.; Rathod, P. K. Antimalarial activity of a combination of 5-fluoroorotate and uridine in mice. *Antimicrob. Agents Chemother.* **1990**, *34*, 1371–1375.

82. Rathod, P. K.; Gomez, Z. M. *Plasmodium yoelii*: oral delivery of 5-fluoroorotate to treat malaria in mice. *Exp. Parasitol.* **1991**, *73*, 512–514.

83. Li,Y.; Raushel, F. M. Inhibitors designed for the active site of dihydroorotase. *Bioorg. Chem.* **2005**, *33*, 470–483.

84. Hambley, T. W.; Phillips, L.; Poiner, A. C.; Christopherson, R. I. A crystallographic and molecular mechanics study of inhibitors of dihydroorotase. *Acta Crystallogr. Sect. B* **1993**, *49*, 130–136.

85. Daniel, R.; Kokel, B.; Caminade, E.; Martel, A.; Le Goffic, F. Assay of *Escherichia coli* dihydroorotase with enantiomeric substrate: practical preparation of carbamyl L-aspartate and high-performance liquid chromatography analysis of catalysis product. *Anal. Biochem.* **1996**, *239*, 130–135.

86. Zimmermann, B. H.; Evans, D. R. Cloning, overexpression, and characterization of the functional dihydroorotase domain of the mammalian multifunctional protein CAD. *Biochemistry* **1993**, *32*, 1519–1527.

87. Maher, M. J.; Huang, D. T.; Guss, J. M.; Collyer, C. A.; Christopherson, R. I. Crystallization of hamster dihydroorotase: involvement of a disulfide-linked tetrameric form. *Acta Crystallogr. D Biol. Crystallogr.* **2003**, *59*, 381–384.

CHAPTER 40

APOBEC3G: A Promising Antiviral Target

CLAUDIU T. SUPURAN[1] and JEAN-YVES WINUM[2]

[1]Laboratorio di Chimica Bioinorganica, Università degli Studi di Firenze, Polo Scientifico, Room 188, Via della Lastruccia 3, 50019 Sesto Fiorentino (Florence), Italy
[2]Institut des Biomolécules Max Mousseron (IBMM) UMR 5247 CNRS-UM1-UM2 Bâtiment de Recherche Max Mousseron, Ecole Nationale Supérieure de Chimie de Montpellier, 8 rue de l'Ecole Normale, 34296 Montpellier Cedex, France

40.1 INTRODUCTION

Viral infections represent a major medical problem worldwide. Among them, human immunodeficiency virus (HIV) infection affects more than 42 million people, and although much progress has been registered in the treatment of this infection by the introduction of highly active antiretroviral therapy, HAART (a combination of nucleoside/nucleotide reverse transcriptase inhibitors, nonnucleoside reverse transcriptase inhibitors, aspartic protease inhibitors, and/or integrase inhibitors, eventually combined with fusion inhibitors),[1–6] the massive viral replication (with more than 10^9 virions produced daily) and the high error rate of the reverse transcriptase lead to the emergence of drug-resistant strains and the stringent need of new therapeutic approaches.[1–6] Although viral infections are so widespread, till recently very few antiviral drugs were available.[5] The emergence of the worldwide AIDS epidemic fostered much research and great progress in this area, and currently more than 30 antiviral drugs are available, most of them for the management of HIV infection and AIDS, but also for the treatment of other viral diseases such as hepatitis B, influenza, herpes simplex, varicella-zoster, and cytomegalovirus (HCMV) infections.[5]

In addition to the viral enzymes (exemplified above by the HIV reverse transcriptase, aspartic protease, and/or integrase) that constitute well-known antiviral targets, in the past several years several host proteins have been identified that play a critical role in the viral infection. As retroviruses and their hosts interacted for millions of years, the host organisms have been evolutionarily obliged to develop restriction factors that would help them counteract the viral attack.[7] Such proteins can robustly block infection of sensitive viruses after the pathogen enters the host cell. A high mutation frequency during reverse transcription plays a fundamental role in the genetic variation of primate lentiviral populations, being the main driving force for the generation of drug resistance and the escape from immune surveillance.[8–10] The G-to-A hypermutation is one of the characteristics of primate lentiviruses, as well as other retroviruses, during replication *in vivo* and in cell culture. The molecular mechanisms of this process only recently started to be clarified, with the main player being the DNA deaminase APOBEC3G (apolipoprotein B mRNA editing enzyme, catalytic polypeptide-like 3G), which is a zinc enzyme.[8–12] Here we shall review the literature of this important intrinsic antiviral protein, discovered only in 2002.[8]

40.2 STRUCTURE AND CATALYTIC MECHANISM OF APOBEC3G

Zhang et al. demonstrated that CEM15 (also known as APOBEC3G, abbreviated as A3G) is an endogenous inhibitor of HIV-1 replication, being a cytidine deaminase able to induce G-to-A hypermutation in newly synthesized viral DNA.[11] This effect can be counteracted by the HIV-1 Vif protein. It thus appears that this viral DNA mutator plays a role in viral defense mechanism in host cells that may induce either lethal hypermutation or instability of the incoming nascent viral reverse transcripts, which could account for the Vif-defective phenotype. Importantly, the accumulation of CEM15-mediated nonlethal hypermutation in the replicating viral genome could potently contribute to the genetic variation of primate lentiviral populations.[11] Indeed, when produced in the presence of APOBEC3G, Vif-defective HIV-1 is non-infectious.[13] APOBEC3G is closely related to APOBEC1, the central component of an RNA editing complex that deaminates a cytosine residue in apoB messenger RNA. APOBEC family members also have potent DNA mutator activity through dC deamination.[13] It was demonstrated that APOBEC3G exerts its antiviral effect during reverse transcription to trigger G-to-A hypermutation in the nascent retroviral DNA and that APOBEC3G can act on a broad range of retroviruses in addition to HIV, suggesting that hypermutation by editing is a general innate defense mechanism against this important group of pathogens.[13,14] Thus, Vif facilitates particle infectivity by blocking the inhibitory activity of APOBEC3G. It has been reported that HIV-1 Vif forms a complex with human APOBEC3G that prevents its virion encapsidation.[15] HIV-1 Vif did not efficiently form a complex with mouse APOBEC3G. Vif dramatically reduced the amount of human APOBEC3G encapsidated in HIV-1 virions but did not prevent encapsidation of mouse or AGM APOBEC3G. As a result, these enzymes are potent inhibitors of wild-type HIV-1 replication. The species specificity of this interaction may play a role in restricting HIV-1 infection to humans. Together, these findings suggest that therapeutic intervention that either induced APOBEC3G or blocked its interaction with Vif could be clinically beneficial, but no compounds effective in any of these steps were detected as far as we know up to now.[13] The molecular events regarding the interaction of Vif with APOBEC3G started to be uncovered in more detail: it has been reported very recently that Vif interacts with cellular proteins Cul5, Elongins B and C, and Rbx1 to form an Skp1-Cullin-F-box (SCF)-like complex.[14,15] The ability of Vif to suppress antiviral activity of APOBEC3G specifically depended on Cul5-SCF function, allowing Vif to interact with APOBEC3G and induce its ubiquitination and degradation. A Vif mutant that interacted with APOBEC3G but not with Cul5-SCF was functionally inactive. The Cul5-SCF was also required for Vif function in distantly related SIVmac/HIV-2. These results indicate that the conserved Cul5-SCF pathway used by Vif is a potential new target for antiviral development.[15]

In different cell types, A3G is expressed in high molecular mass (HMM) RNA–protein complexes or low molecular mass (LMM) forms displaying different biological activities. In resting CD4 T cells, a LMM form of A3G potently restricts HIV-1 infection soon after virion entry. However, when T cells are activated, LMM A3G is recruited into HMM complexes that include Staufen-containing RNA granules.[16] These complexes are probably nucleated by the induced expression of Alu/hY retroelement RNAs that accompany T-cell activation. HMM A3G sequesters these retroelement RNAs away from the nuclear long interspersed nuclear element-derived enzymes required for Alu/hY retrotransposition. Human immunodeficiency virus exploits this "window of opportunity" provided by the loss of LMM A3G in activated CD4 T cells to productively infect these cells. During HIV virion formation, newly synthesized LMM A3G is preferentially encapsidated but only under conditions where Vif is absent and thus not able to target A3G for proteasome-mediated degradation. Together, such findings highlight the discrete functions of the different forms of A3G. LMM A3G opposes the external threat posed by exogenous retroviruses, while HMM A3G complexes oppose the internal threat posed by the retrotransposition of select types of retroelements.[16]

A3G is a member of the APOBEC family of cytidine deaminases consisting of APOBEC1, APOBEC2, APOBEC3 (A to H), and AID (activation induced deaminase).[17] During reverse transcription, A3G deaminates dC to dU in nascent minus-strand viral DNA (Scheme 40.1), resulting in G-to-A hypermutation in the plus strand DNA to inhibit the replication of HIV-1. On the contrary, HIV-1 Vif protein counteracts this enzyme by the ubiquitin-proteasome pathway to enable HIV-1 to replicate in target cells. Vif forms an E3 ligase complex with cellular proteins including Cul5,

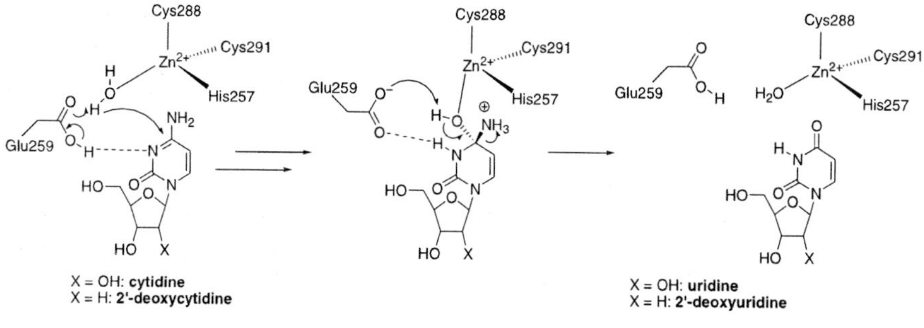

SCHEME 40.1 The cytidine deamination catalyzed by A3G.

Elongin B, and Elongin C (Vif-BC-Cul5) and functions as a substrate recognition subunit of the complex to target APOBEC3G for ubiquitin-proteasome-dependent degradation in virus-producing cells. APOBEC3G has also been shown to have a broad antiviral activity on a wide variety of viruses that include not only retroviruses such as other lentiviruses, murine leukemia virus (MLV), and human T-cell leukemia virus type 1 (HTLV-1) but also other viruses such as hepatitis B virus (HBV) and adeno-associated virus. Furthermore, other members of the APOBEC family also show a broad antiviral activity, but target virus specificities vary among APOBEC members. On the other hand, viruses have their own mechanisms to escape from APOBEC. This expanding evidence suggests that the APOBEC family of cytidine deaminases plays an important role in antiviral innate immunity and might be a novel target for an antiviral therapy. Here we review the present understanding of APOBEC3 proteins as an antiviral innate immunity and battles between APOBEC3 and viruses.[17]

A solution structure (determined by NMR) of A3G,[18] followed shortly thereafter by the X-ray crystal structure of the catalytic domain of this zinc enzyme,[19] allowed us to better understand its catalytic mechanism. The high-resolution crystal structure of the carboxy-terminal deaminase domain of A3G (denominated APOBEC3G-CD2) was purified in recombinant form from *Escherichia coli*. The APOBEC3G-CD2 structure has a five-stranded β-sheet core that is common to all known deaminase structures and closely resembles the structure of another APOBEC protein, APOBEC2.[19] A comparison of APOBEC3G-CD2 with other deaminase structures showed a structural conservation of the active-site loops that are directly involved in substrate binding. In the X-ray structure, these APOBEC3G active-site loops form a continuous "substrate groove" around the active catalytic center. The orientation of this putative substrate groove differs markedly (by 90°) from the groove predicted by the NMR structure.[18]

This X-ray crystal structure[19] revealed that A3G adopts a fold that is very similar to other cytidine deaminases and is characterized by a five-stranded mixed β-sheet and five α helices of which predominantly α1, α3, and α2 define the zinc coordination motif that constitutes the catalytic center. The core structure of the catalytic center is conserved in cytidine deaminases ranging from bacteria to vertebrates.[7] There are, however, major differences with regard to the substrate specificity because other cytidine deaminases act on either free nucleotides or RNA, whereas A3G acts exclusively on single-stranded DNA. Chen et al.[18] have started to address the substrate specificity of A3G by monitoring changes in chemical shift upon titration of single-stranded DNA by means of NMR spectroscopy. Subsequently, residues that showed differential shift were mutated and mutant A3G editing properties assayed. In this manner, it was possible to construct a model for DNA binding, which indicates that contacts with the DNA substrate are predominantly made by arginines in proximity to the catalytic center.[18] Arg residues involved in the binding of substrates are Arg215 and Arg313, which are conserved for all members of the APOBEC family.

The Zn(II) ion in the A3G active site (Scheme 40.1) is coordinated by two Cys (Cys288 and Cys291) and one His residue (His257), the fourth ligand in this tetrahedral geometry of the metal ion being a water molecule that can generate the nucleophilic zinc hydroxide species by proton transfer reactions. A glutamic acid residue (Glu259 in A3G), also conserved for all members of this protein family, is involved in the catalytic mechanism, as it is considered to promote the initial attack at C4 of the cytosine ring by protonating the adjacent N-3 position and deprotonating the zinc-bound water molecule, and then, by using again general acid–base chemistry, to facilitate the breakdown of the tetrahedral intermediate (Scheme 40.1).[20] The overall effect is that the amino group of cytidine is replaced by an OH group (stabilized as the keto tautomer shown in Scheme 40.1).

40.3 INHIBITION OF A3G: TOWARD A NEW CLASS OF ANTI-HIV AGENTS

The HIV-1 protein Vif is essential for *in vivo* viral replication of the virus and, as shown above, targets the human DNA editing enzyme, A3G, which inhibits replication of retroviruses and hepatitis B virus.[12,21] As Vif has no known cellular homologues in the human host, it is an attractive, yet unrealized, target for antiviral intervention. Recently, Nathans et al.[21] screened a large series of derivatives for identifying small molecules able to antagonize Vif function and inhibit HIV-1 replication in the presence of A3G. Several such compounds with low micromolar affinity (1–10 μM) for A3G have been identified, such as RN-3, RN-5, RN-12, and RN-18 among others, belonging to various classes.

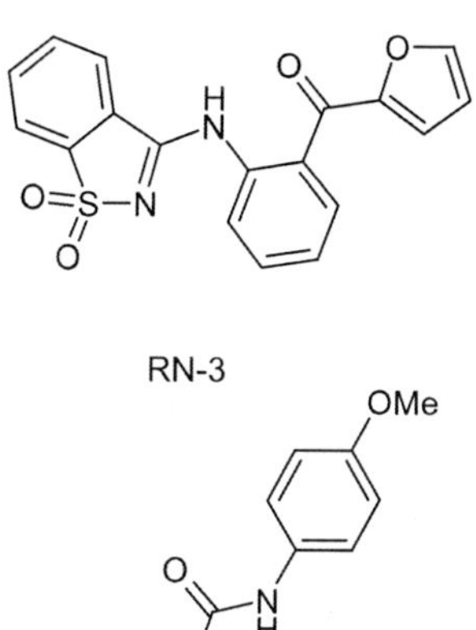

RN-3

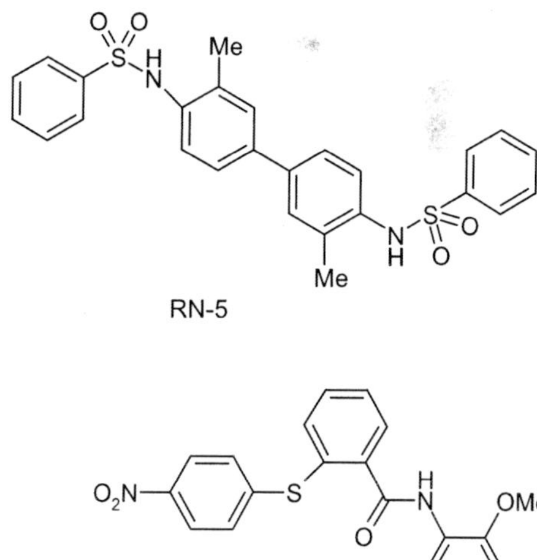

RN-5

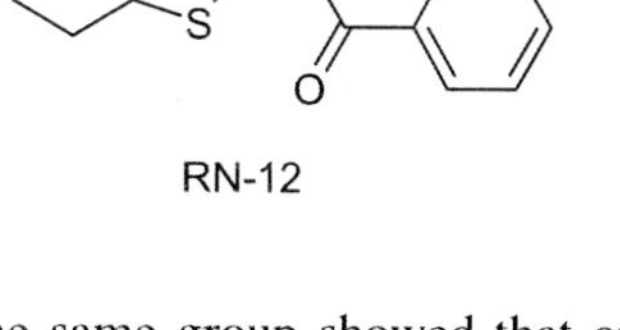

RN-12

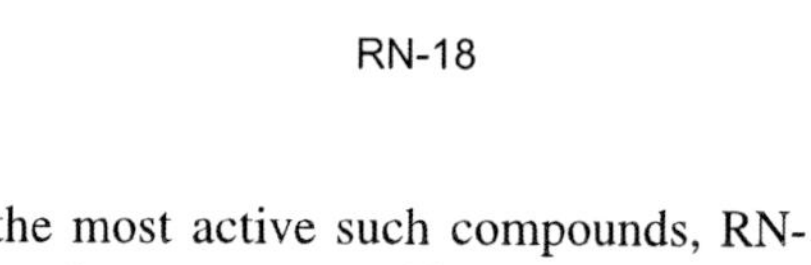

RN-18

The same group showed that one of the most active such compounds, RN-18, increased cellular A3G levels in a Vif-dependent manner and increased A3G incorporation into virions without inhibiting general proteasome-mediated protein degradation.[21] RN-18 enhanced Vif degradation only in the presence of A3G, reduced viral infectivity by increasing A3G incorporation into virions, and enhanced cytidine deamination of the viral genome. Such results demonstrate that the HIV-1 Vif-A3G axis is a valid target for developing small molecule-based new therapies for HIV infection or for enhancing innate immunity against viruses. However, this field is still in its infancy but considering the huge progress made since 2002 when this target was discovered, we are optimistic regarding this very innovative approach for discovering mechanistically new antiviral drugs.

REFERENCES

1. (a) Barbaro, G.; Scozzafava, A.; Mastrolorenzo, A.; Supuran, C. T. Highly active antiretroviral therapy: current state of the art, new agents and their pharmacological interactions useful for improving therapeutic outcome. *Curr. Pharm. Des.* **2005**, *11*, 1805–1843; (b) Rusconi, S.; Scozzafava, A.; Mastrolorenzo, A.; Supuran, C. T. An update in the development of HIV entry inhibitors. *Curr. Top. Med. Chem.* **2007**, *7*, 1273–1289; (c) Mastrolorenzo, A.; Rusconi, S.; Scozzafava, A.; Barbaro, G.; Supuran, C. T. Inhibitors of HIV-1 protease: current state of the art 10 years after their introduction. From antiretroviral drugs to antifungal, antibacterial and antitumor agents based on aspartic protease inhibitors. *Curr. Med. Chem.* **2007**, *14*, 2734–2748.
2. (a) De Clercq, E. Antivirals and antiviral strategies. *Nat. Rev. Microbiol.* **2004**, *2*, 704–720; (b) Richman, D. D. HIV chemotherapy. *Nature* **2001**, *410*, 995–1001.
3. Nabel, G. J. Challenges and opportunities for development of an AIDS vaccine. *Nature* **2001**, *410*, 1002–1007.
4. Scozzafava, A.; Mastrolorenzo, A.; Supuran, C. T. Non-peptidic chemokine receptors antagonists as emerging anti-HIV agents. *J. Enzyme Inhib. Med. Chem.* **2002**, *17*, 69–76.
5. De Clercq, E. New developments in anti-HIV chemotherapy. *Curr. Med. Chem.* **2001**, *8*, 1543–1572.
6. Scozzafava, A.; Owa, T.; Mastrolorenzo, A.; Supuran, C. T. Anticancer and antiviral sulfonamides. *Curr. Med. Chem.* **2003**, *10*, 925–953.
7. Huthoff, H.; Towers, G. J. Restriction of retroviral replication by APOBEC3G/F and TRIM5alpha. *Trends Microbiol.* **2008**, *16*, 612–619.
8. Sheehy, A. M.; Gaddis, N. C.; Choi, J. D.; Malim, M. H. Isolation of a human gene that inhibits HIV-1 infection and is suppressed by the viral Vif protein. *Nature* **2002**, *418*, 646–650.
9. Sheehy, A. M.; Gaddis, N. C.; Malim, M. H. The antiretroviral enzyme APOBEC3G is degraded by the proteasome in response to HIV-1 Vif. *Nat. Med.* **2003**, *9*, 1404–1407.
10. Marin, M.; Rose, K. M.; Kozak, S. L.; Kabat, D. HIV-1 Vif protein binds the editing enzyme APOBEC3G and induces its degradation. *Nat. Med.* **2003**, *9*, 1398–1403.
11. Zhang, H.; Yang, B.; Pomerantz, R. J.; Zhang, C.; Arunachalam, S. C.; Gao, L. The cytidine deaminase CEM15 induces hypermutation in newly synthesized HIV-1 DNA. *Nature* **2003**, *424*, 94–98.
12. Mangeat, B.; Turelli, P.; Caron, G.; Friedli, M.; Perrin, L.; Trono, D. Broad antiretroviral defence by human APOBEC3G through lethal editing of nascent reverse transcripts. *Nature* **2003**, *424*, 99–103.

13. Harris, R. S.; Bishop, K. N.; Sheehy, A. M.; Craig, H. M.; Petersen-Mahrt, S. K.; Watt, I. N. DNA deamination mediates innate immunity to retroviral infection. *Cell* **2003**, *113*, 803–809.

14. Mariani, R.; Chen, D.; Schrofelbauer, B.; Navarro, F.; Konig, R.; Bollman, B. DNA deamination mediates innate immunity to retroviral infection. *Cell* **2003**, *114*, 21–31.

15. Yu, X.; Yu, Y.; Liu, B.; Luo, K.; Kong, W.; Mao, P.; Yu, X. F. Induction of APOBEC3G ubiquitination and degradation by an HIV-1 Vif-Cul5-SCF complex. *Science* **2003**, *302*, 1056–1060.

16. Chiu, Y. L.; Greene, W. C. Review. APOBEC3G: an intracellular centurion. *Philos. Trans. R. Soc. Lond. B Biol. Sci.* **2009**, *364*, 689–703.

17. Izumi, T.; Shirakawa, K.; Takaori-Kondo, A. Cytidine deaminases as a weapon against retroviruses and a new target for antiviral therapy. *Mini Rev. Med. Chem.* **2008**, *8*, 231–238.

18. Chen, K. M.; Harjes, E.; Gross, P. J.; Fahmy, A.; Lu, Y.; Shindo, K.; Harris, R. S.; Matsuo, H. Structure of the DNA deaminase domain of the HIV-1 restriction factor APOBEC3G. *Nature* **2008**, *452*, 116–119.

19. Holden, L. G.; Prochnow, C.; Chang, Y. P.; Bransteitter, R.; Chelico, L.; Sen, U.; Stevens, R. C.; Goodman, M. F.; Chen, X. S. Crystal structure of the anti-viral APOBEC3G catalytic domain and functional implications. *Nature* **2008**, *456*, 121–124.

20. Chung, S. J.; Fromme, J. C.; Verdine, G. L. Structure of human cytidine deaminase bound to a potent inhibitor. *J. Med. Chem.* **2005**, *48*, 658–660.

21. Nathans, R.; Cao, H.; Sharova, N.; Ali, A.; Sharkey, M.; Stranska, R.; Stevenson, M.; Rana, T. M. Small-molecule inhibition of HIV-1 Vif. *Nat. Biotechnol.* **2008**, *26*, 1187–1192.

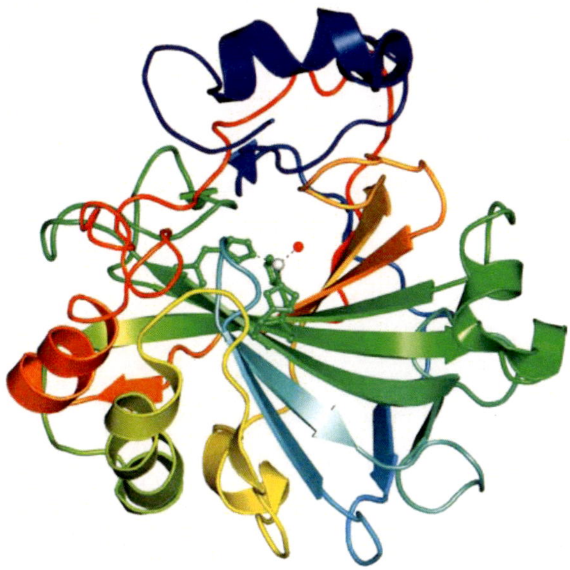

FIGURE 4.1 Ribbon diagram of hCA II, which has been chosen as representative CA isozyme. The active site Zn^{2+} ion coordination is also reported.

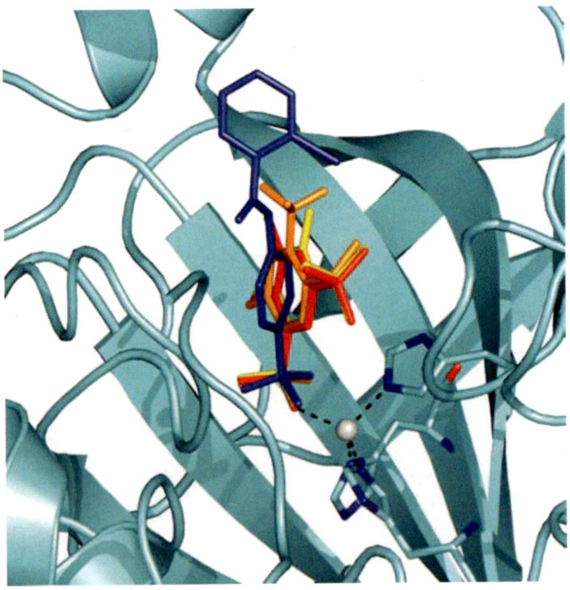

FIGURE 4.14 Superposition of hCA II–inhibitor adducts that highlights the different orientations of the benzene-1,3-disulfonamide inhibitors such as **4.32** (red), **4.33** (yellow), and **4.34** (orange) with respect to a classical monosulfonamide derivative such as **4.38** (blue).

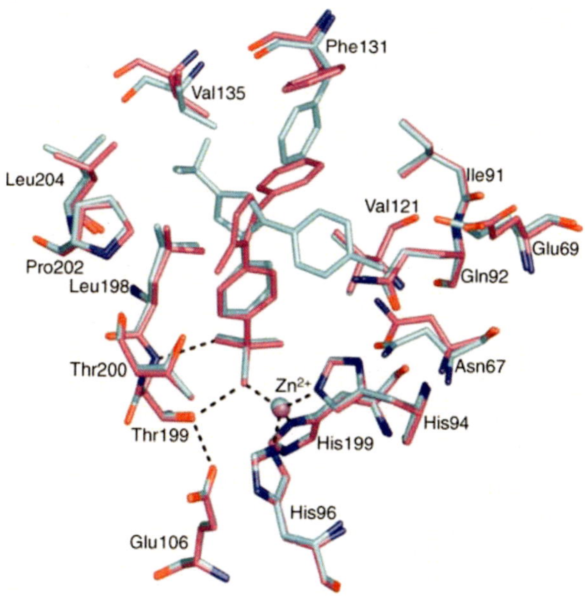

FIGURE 4.18 View of the hCA II active site complexed with valdecoxib **4.42** (magenta) and celecoxib **4.43** (cyan) brought to optimal structural overlay.

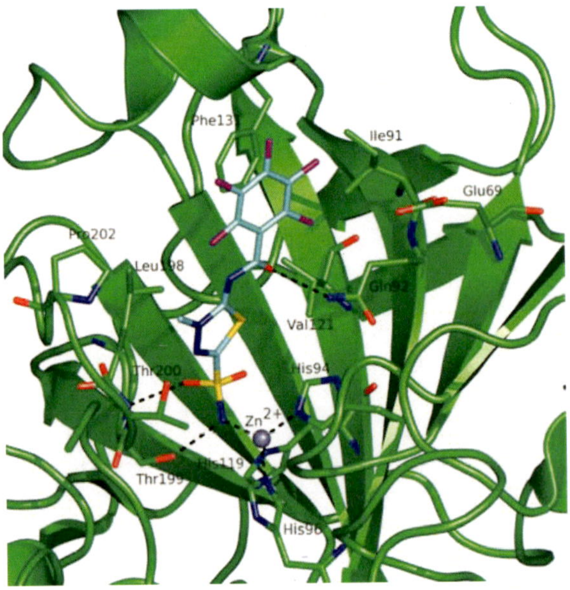

FIGURE 4.21 View of the active site region in the hCA II–**4.52** complex showing the residues participating in recognition of the inhibitor molecule. Hydrogen bonds and the active site Zn^{2+} ion coordination are also shown (dotted lines).

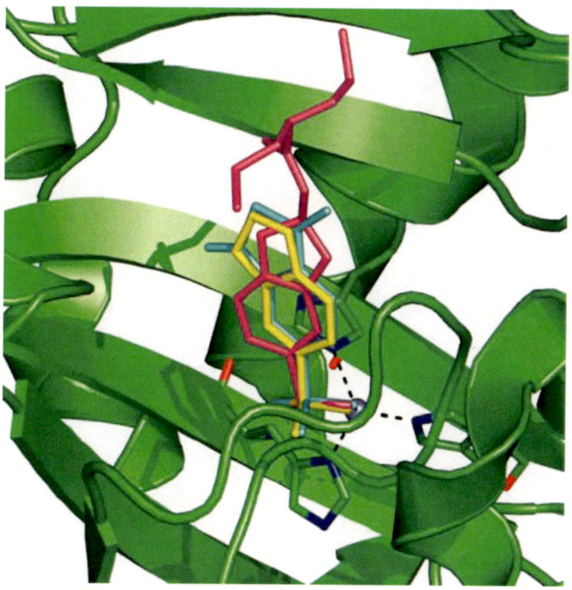

FIGURE 4.26 Superposition of hCA II–inhibitor adducts: **4.68** is reported in yellow, **4.69** in magenta, and **4.70** in cyan.

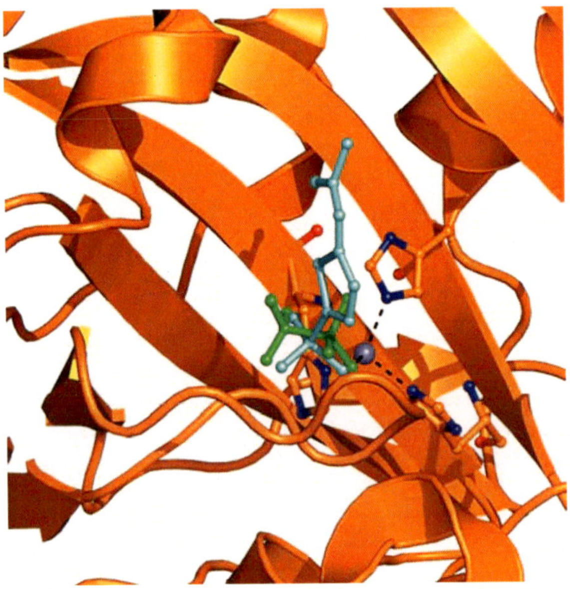

FIGURE 4.27 Superposition of hCA II–inhibitor adducts that highlights the different orientations of trifluoromethane sulfonamide **4.72** (green) with respect to a classical arylsulfonamide derivative such as acetazolamide **4.49** (cyan).

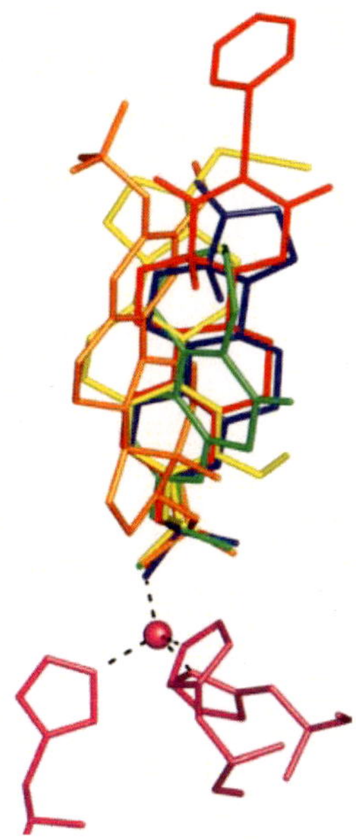

FIGURE 4.31 Superposition of hCA II–inhibitor adducts: **4.78** is reported in blue, **4.79** in green, **4.80** in orange, **4.81** in yellow, and **4.82** in red.

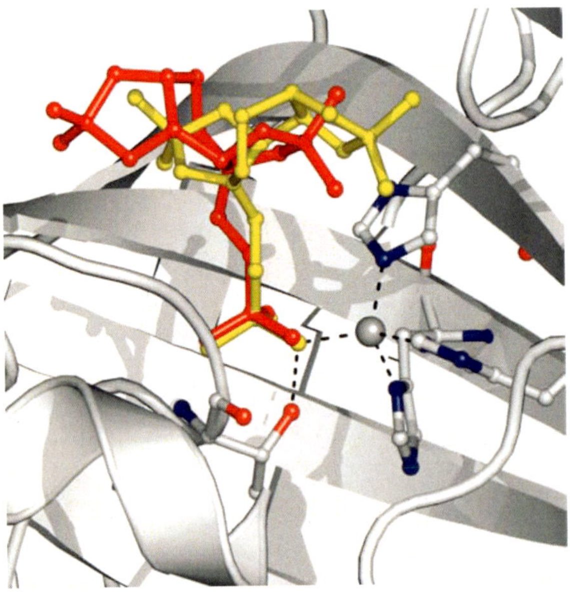

FIGURE 4.35 Superposition of hCA II–inhibitor adducts: **4.94** is reported in yellow and **4.96** in red.

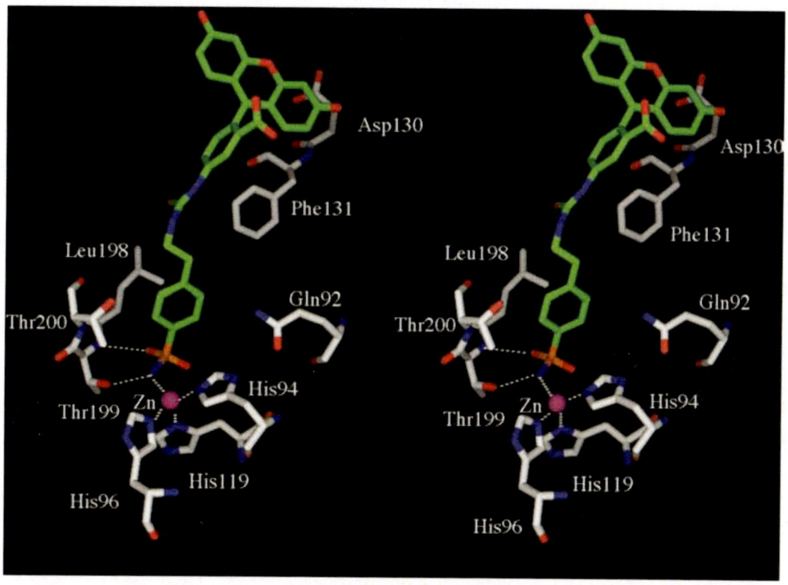

FIGURE 9.2 Stereoview of the Zn(II) coordination sphere and neighboring amino acid residues involved in the binding of compound **9.3** to the hCA II active site.

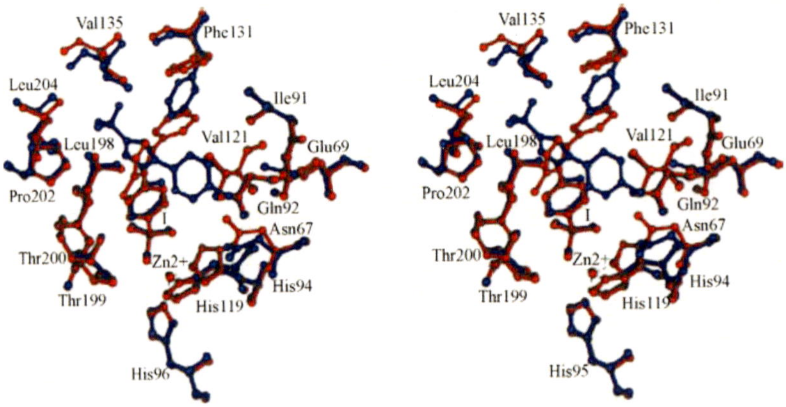

FIGURE 11.6 Stereoview of the hCA II active site complexed with valdecoxib 1 (red) and celecoxib-2 (blue) brought to optimal structural overlay.[33]

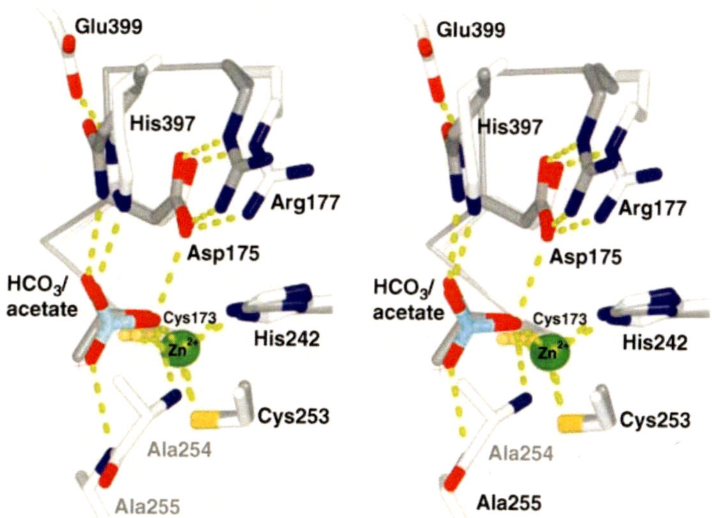

FIGURE 13.6 Stereoview of the active site of CsoSCA from *Halothiobacillus neapolitanus* (shown in light gray) superimposed on the active site of *P. sativum* (shown in dark gray). Bicarbonate was modeled into the active site, based on the position of acetate in the structure from *P. sativum*. Reproduced by permission.[21]

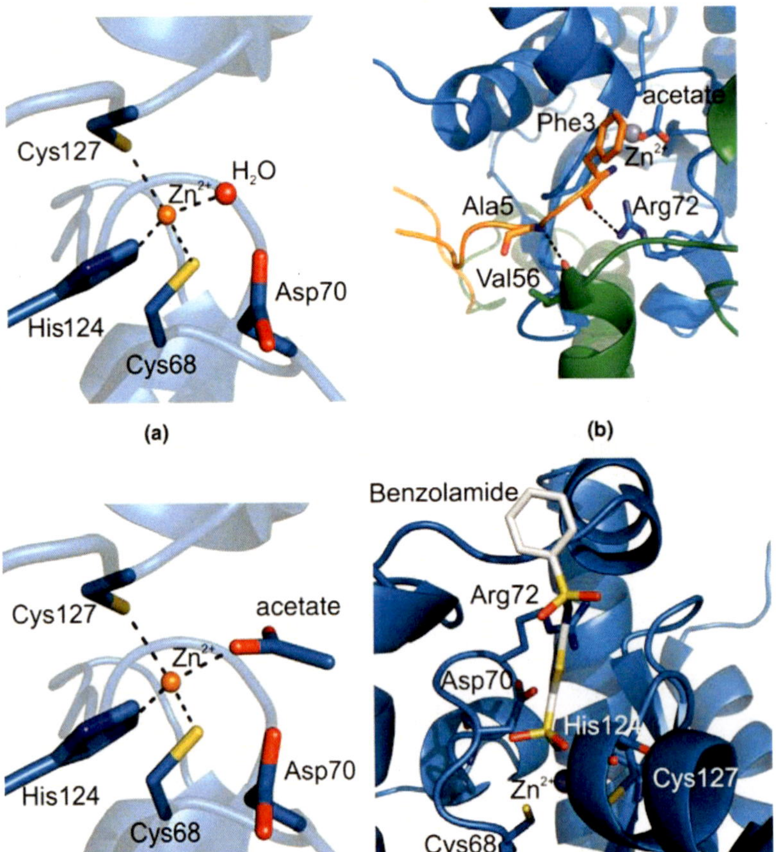

FIGURE 15.3 Active site features of Can2 from *C. neoformans*. (a) Active site of Can2, showing the three Zn(II) coordinating residues and a water molecule as fourth metal ligand. (b) Interaction of the N-terminal extension of Can2, shown in orange, with the active site entrance. The two monomers of the Can2 dimer are indicated in blue and green, respectively. (c) Active site of Can2 in complex with acetate. The inhibitor/product analogue serves as fourth ligand of the active site Zn(II) ion. (d) Modeled complex of Can2 with the inhibitor benzolamide (**15.4**) showing the postulated interactions with the enzyme. The hydrophobic extension of the inhibitor points out of the active site. The N-terminal extension had to be removed to avoid a clash.

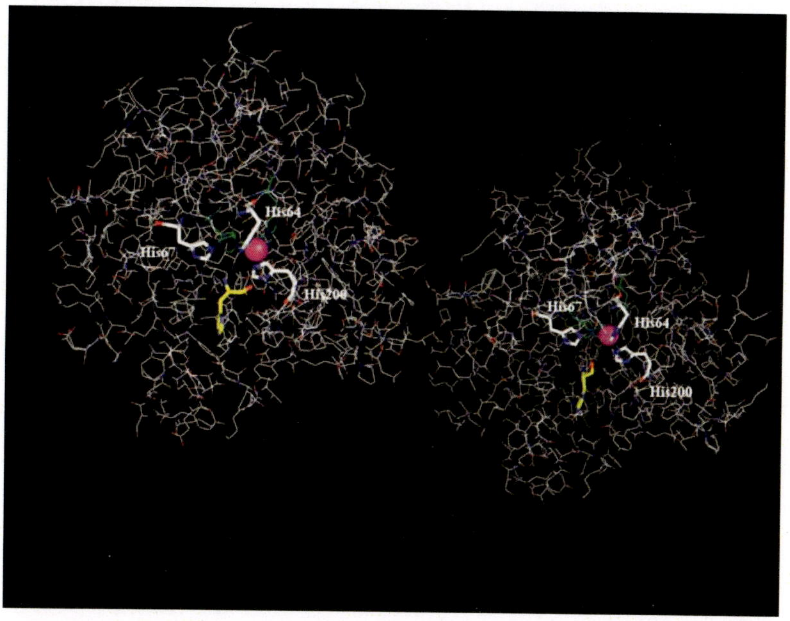

FIGURE 22.3 Overall view of the hCA I/L-His complex. The Zn(II) ion (pink sphere) and the activator molecule (yellow) are shown.

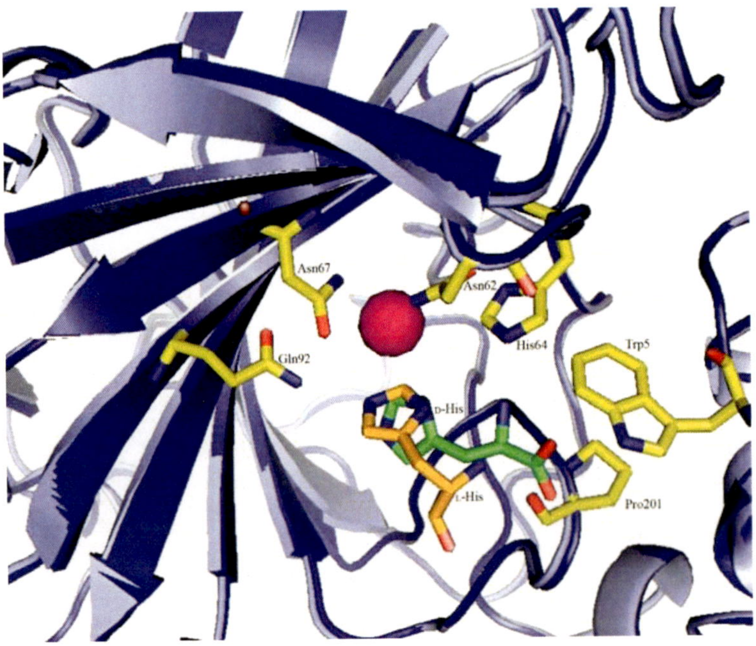

FIGURE 22.4 Superposition of the two hCA II adducts with L/D-His, with the zinc ion and amino acids residues present in the activator binding site.

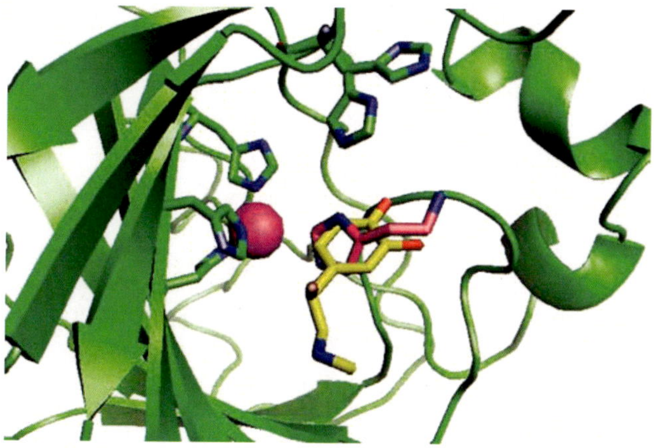

FIGURE 22.6 Superposition of the hCA II–histamine and the hCA II–L-adrenaline complexes. The proton shuttle, His64, is present in both the "in" and the "out" conformations.

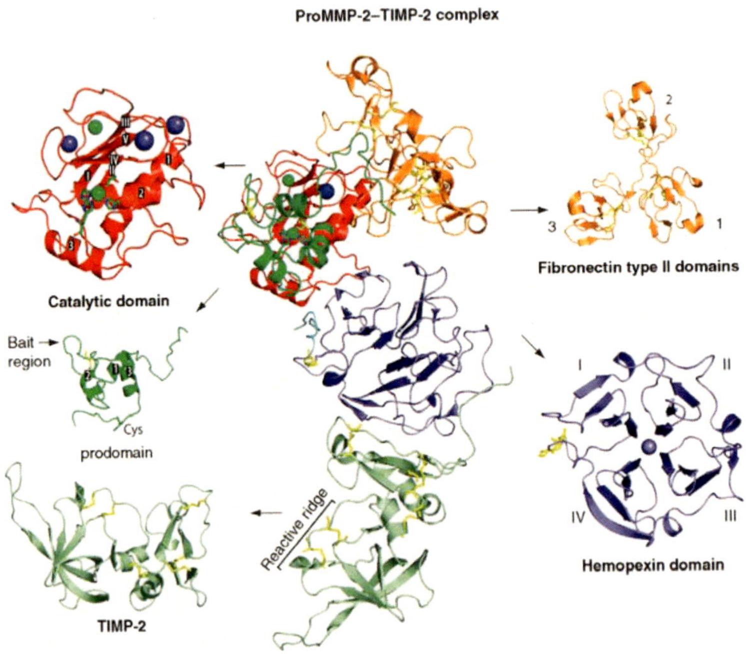

FIGURE 23.2 Three-dimensional structure of the human proMMP-2–TIMP-2 complex. The center shows a ribbon presentation of the complex of proMMP-2 with TIMP-2 (pale green) binding with its C-terminus to the Hpx domain (blue). The prodomain is shown in green, the catalytic domain in red, fibronectin type II domains in orange, the linker region in cyan, and the Hpx domain in blue. The structure of individual domains is also represented as a ribbon structure using the same color scheme. The figure was prepared with Pymol using the PDB ID: 1GXD and 1FBL (for ribbon structure of separate cat and Hpx domains).

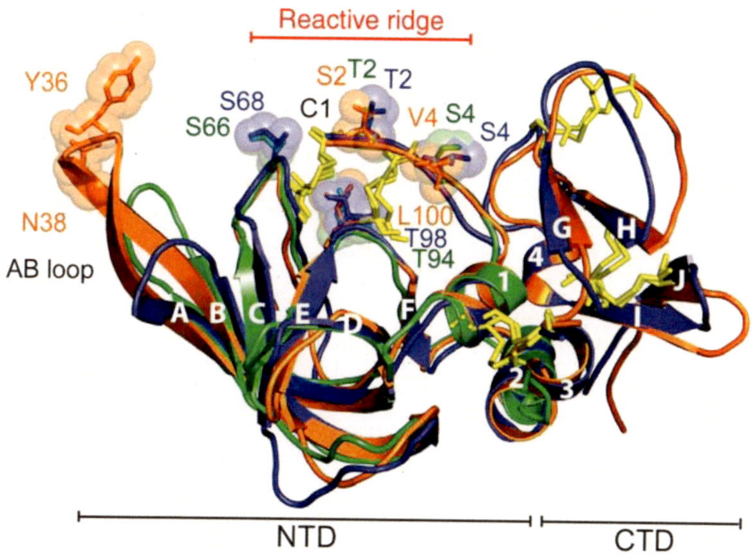

FIGURE 23.4 Superimposition of TIMP-1, TIMP-2, and the N-terminal domain of TIMP-3. Ribbon structure of TIMP-1 is shown in blue, TIMP-2 in orange, and N-TIMP-3[154] in green. Beta-strands are labeled A to J and alpha helices 1–4. Individual residues discussed in the text are shown in corresponding colors as sticks with transparent van der Waals spheres. Disulfide bonds are shown in yellow. Based on PDB IDs: 1UEA for TIMP-1, 1BQQ for TIMP-2, and 3CKI for TIMP-3, superimpositions and figure prepared with Pymol.

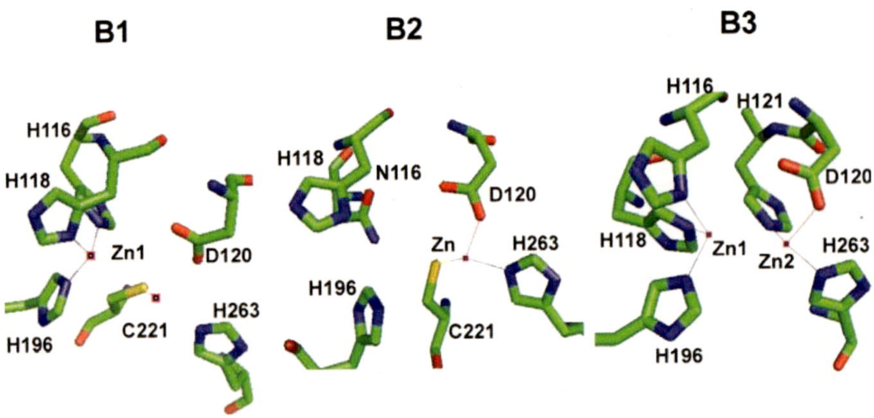

FIGURE 31.1 Metallo-β-lactamase zinc binding sites. Three classes (B1, B2, and B3) of metallo-β-lactamases are depicted with zinc atoms and coordinating residues. The structure of B1 is derived from *B. cereus* BcI1 (protein data bank code, 1BC2), that of B2 is from *A. hydrophila* CphA (1X8G), and that of B3 is from *S. maltophilia* L1 (1SML).

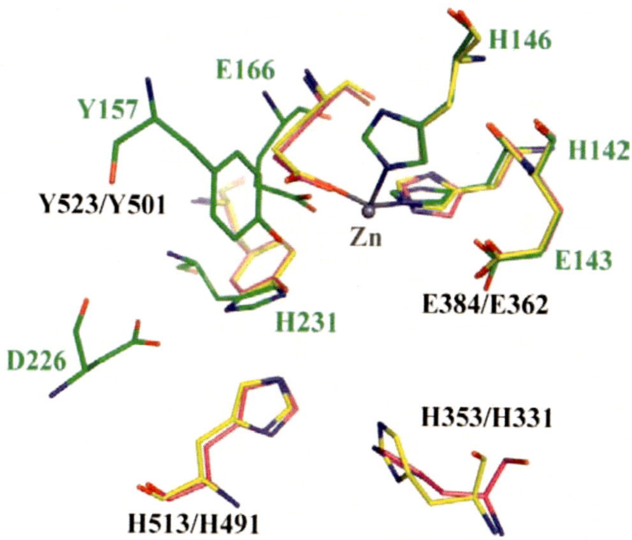

FIGURE 32.2 Active site comparison of C-domain (yellow) and N-domain (cyan) ACE structures with the active site of the thermolysin (green).

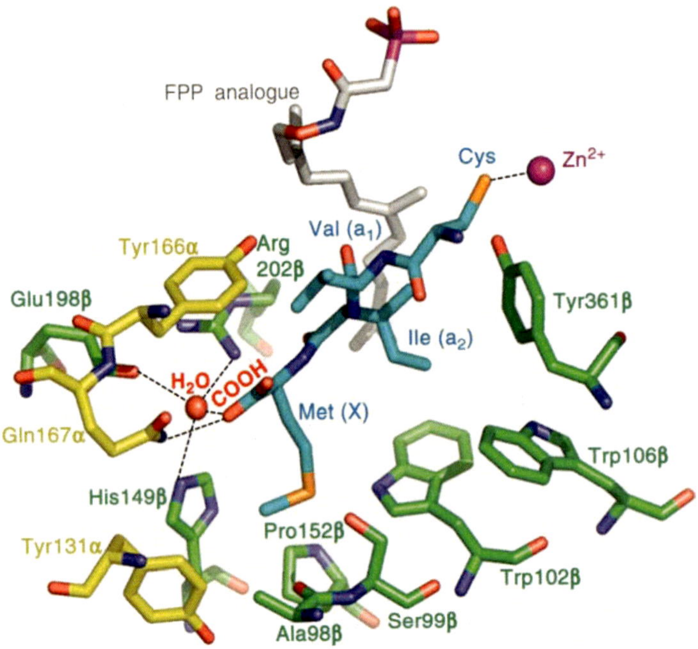

FIGURE 34.6 Binding of a CVIM substrate peptide. Polar interactions are indicated as dotted lines. (Picture based on pdb entry: 1TN6.)

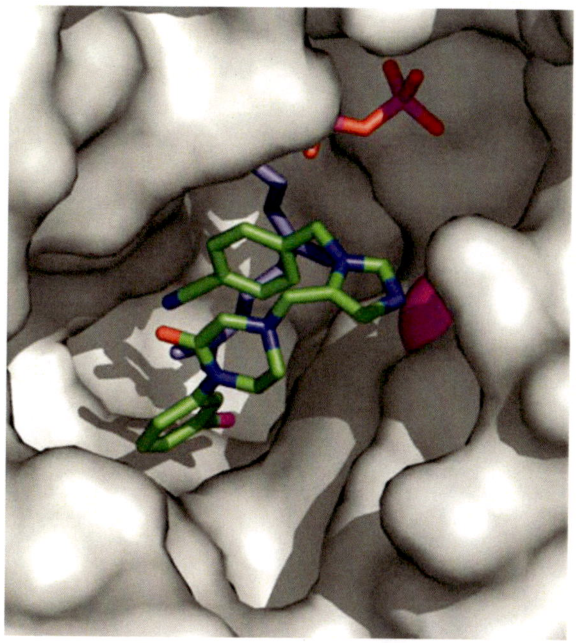

FIGURE 34.14 Crystal structure of L-778,123 (**34.24**) bound to FTase (pdb code: 1S63).[28]

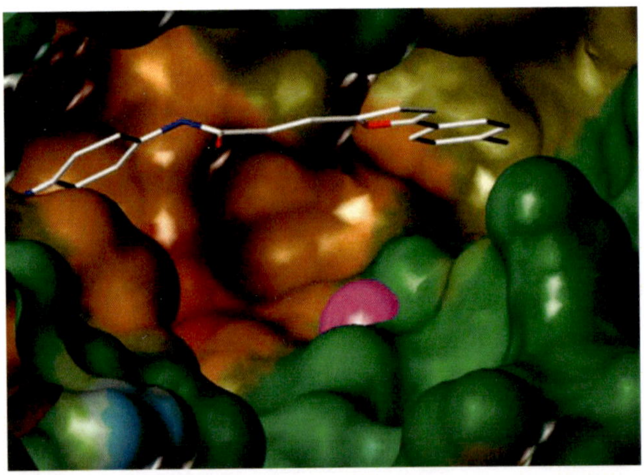

FIGURE 34.22 Benzophenone inhibitor Schl-5199 (**34.48**) that uses the so-called far aryl-binding site adjacent to the active site. Details of farnesyltransferase's active site and the far aryl-binding site. Lipophilic (brown) and hydrophilic (green to blue) properties are displayed on the Connolly surface. The structural zinc is shown as a magenta sphere. Biarylacryloyl-substituted benzophenone inhibitor deeply buries into a lipophilic area in the far aryl-binding site.

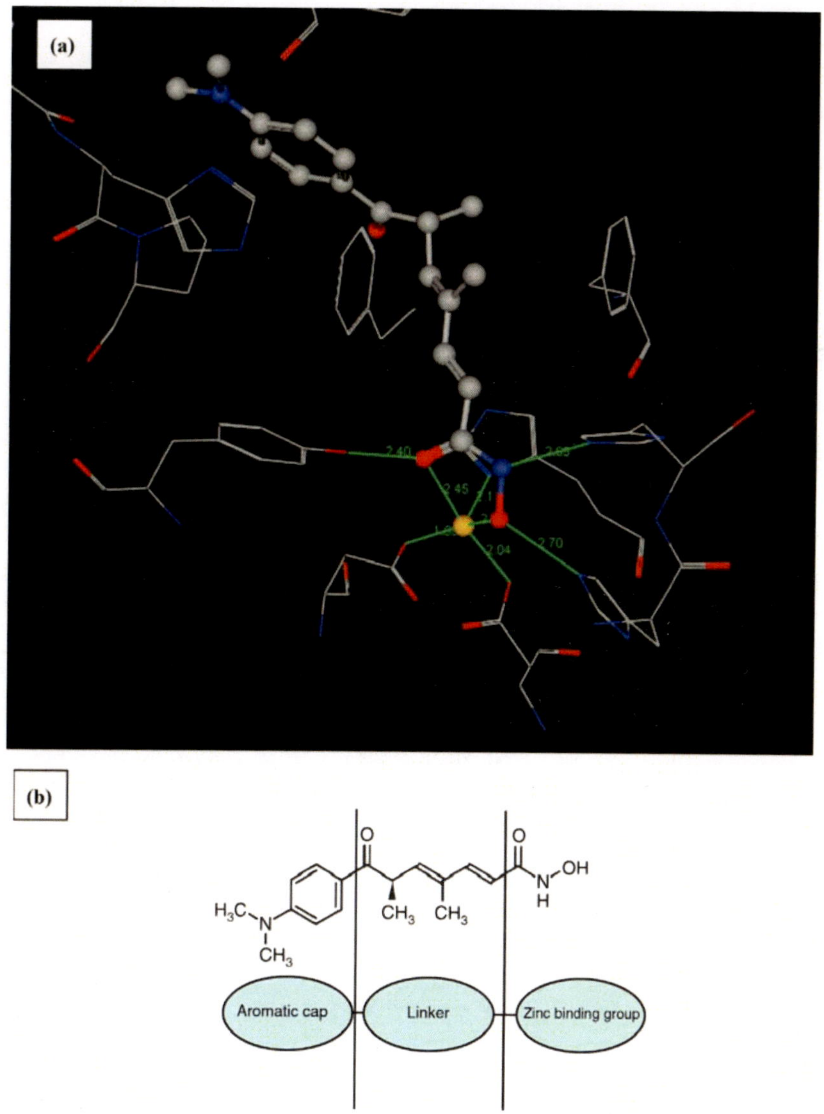

FIGURE 35.2 (a) TSA in the binding site of HDLP. The pentacoordination of the zinc and hydrogen bonding interactions to the hydroxamic acid moiety are shown with green lines; the interactions between the central portion and phenylalanine residues can also be seen. (b) A simple three-component pharmacophore, illustrated for TSA, consisting of a zinc binding group, a linker, and a capping group.

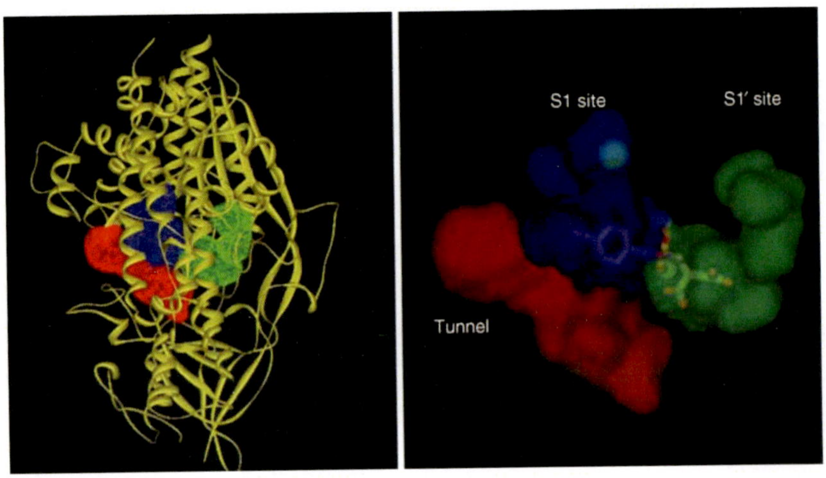

FIGURE 36.2 (a) 3D ectodomain GCPII structure (b) The S1 and S1′ binding pockets and the tunnel region in the active site of GCPII.

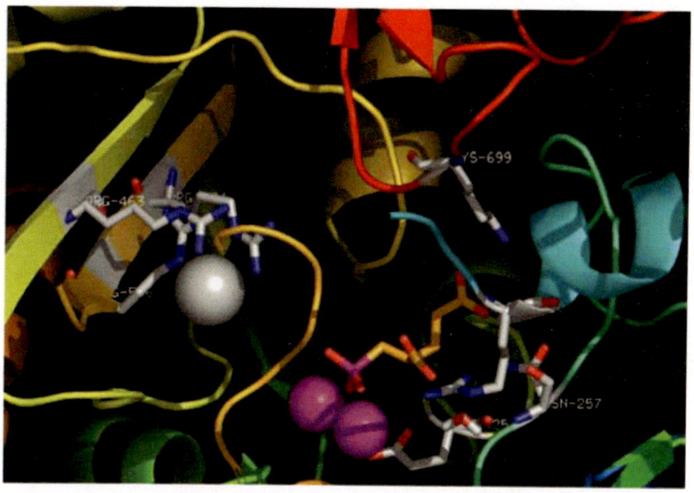

FIGURE 36.3 Active site of GCPII with 2-PMPA (PDB ID: 2PVW). Zinc ions (pink spheres) and chloride ion (white sphere). The carbon atoms backbone of 2-PMPA is shown in yellow. The picture was generated using Pymol program.

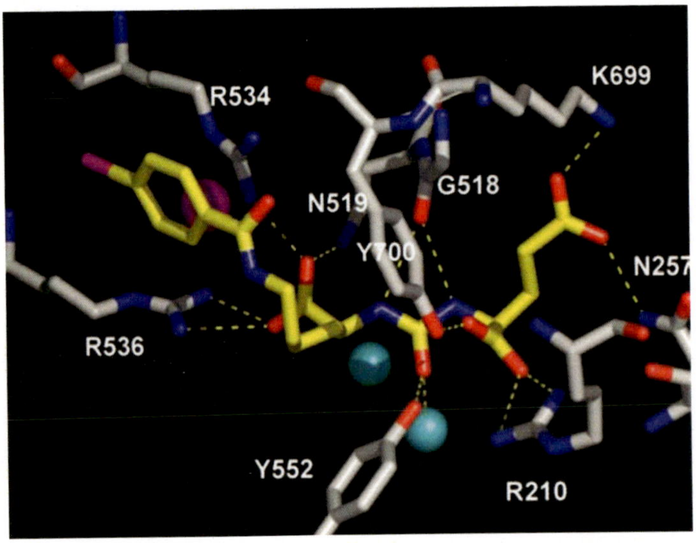

FIGURE 36.12 Crystal structure of GCPII with **36.102**.

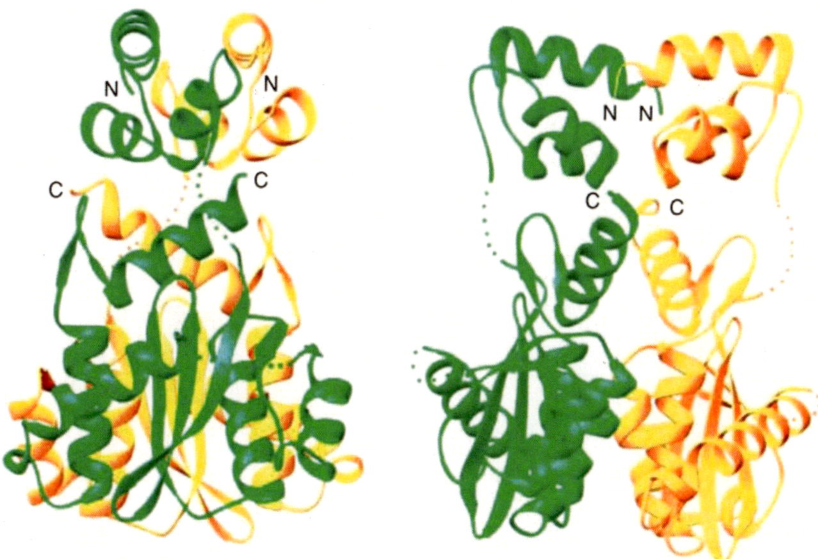

FIGURE 37.3 Ribbon diagram of HIV-1 IN[1–212]. Two orthogonal views of the IN[1–212] dimer. The A subunit is green and the B subunit is yellow. Disordered loops are indicated by the dotted lines. (Reprinted from Ref. 61. With the permission of Oxford University Press.)

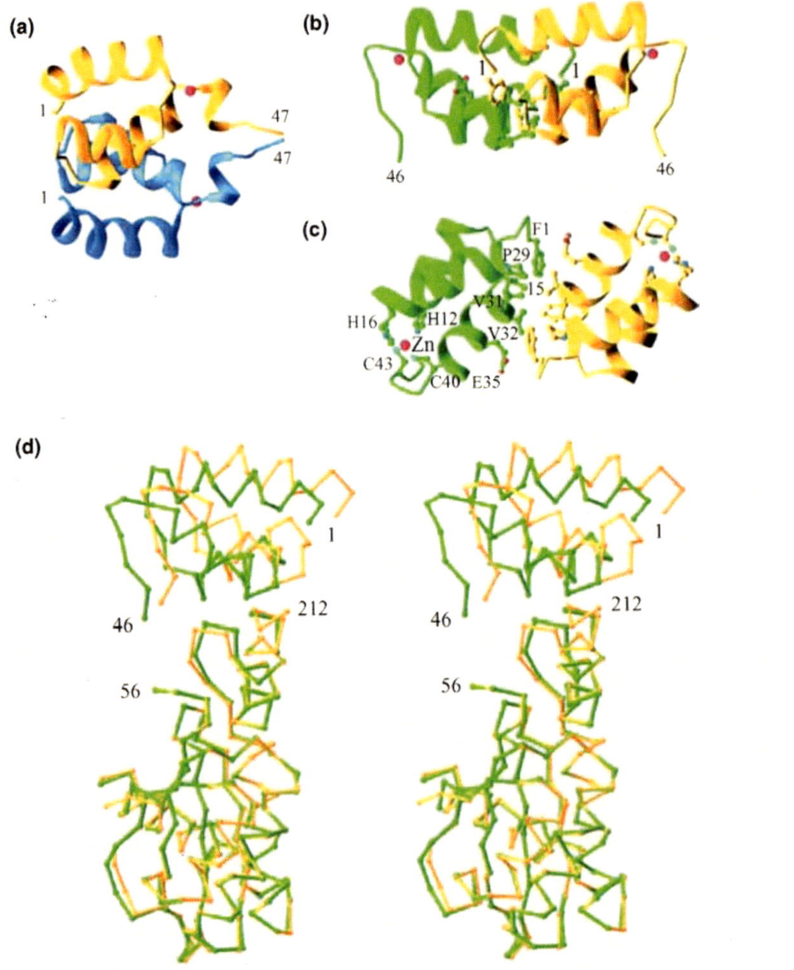

FIGURE 37.4 Analysis of domain interactions. (a) Ribbon diagram of the N-terminal domain structure as an isolated entity determined by NMR (PDB code 1WJA) and (b) in the context of IN[1–212] by X-ray crystallography after superimposing the yellow subunits of the two structures. Residue numbers at the N- and C-termini are labeled. (c) A view near orthogonal to (b) looking down the twofold axis. Residues involved in the dimerization, which are rather hydrophobic, are shown in ball and stick. Zn^{2+} ions are shown as red spheres. (d) The N-terminal domain is oriented differently relative to the core domain in the A and B subunits of IN[1–212] as revealed by superposition of the Cα traces of the core domains. (Reprinted from Ref. 61. With the permission of Oxford University Press.)

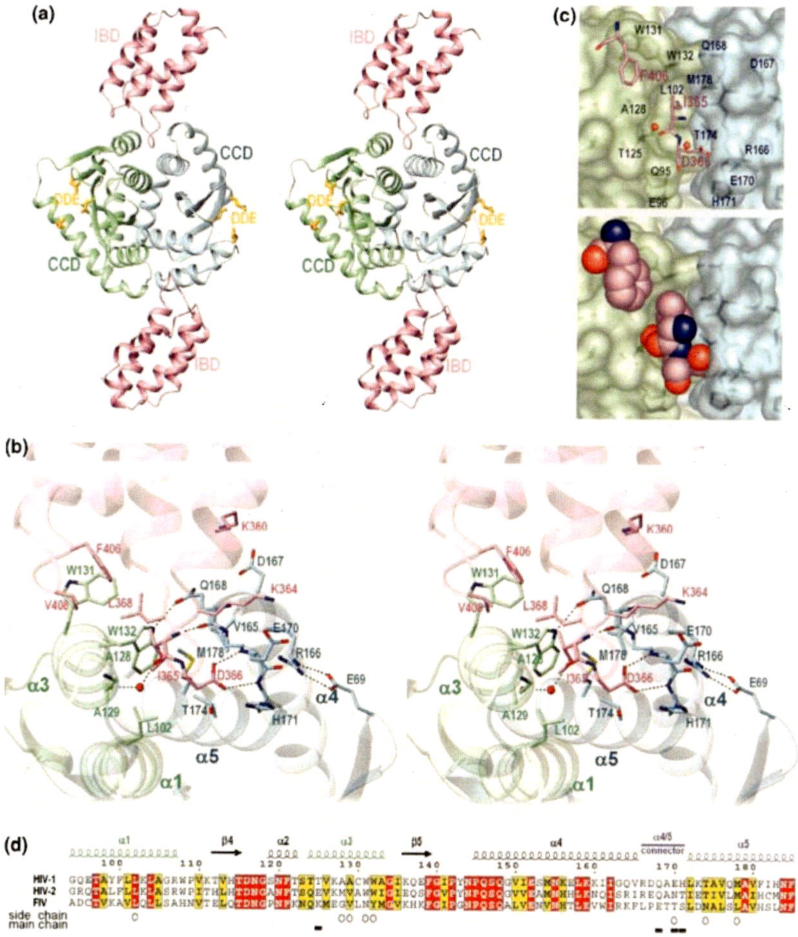

FIGURE 37.7 Molecular mechanism of the IN–LEDGF interaction. (a) The overall structure of the catalytic core domain–IN binding domain complex. IN chains A and B are colored blue and green, respectively; the IN binding domain subunits are violet. The side chains of the DDE catalytic triad are shown as yellow sticks. (b) Key intermolecular contacts at the catalytic core domain–IN binding domain interface. Selected residues are shown as sticks. (c) The pocket at the catalytic core domain dimer interface. LEDGF hotspot residues Ile365, Asp366, and Phe406 are shown as sticks (upper) or in space-fill mode (lower). The IN subunits are shown as semitransparent surfaces. Selected IN residues are indicated. (d) Sequence alignment of HIV-1, HIV-2, and feline immunodeficiency virus INs. Identical residues are white on red background; residues with conserved properties are bold on yellow background. Residue numbering, secondary structure elements, and the position of the α4/5 connector in HIV-1 IN are shown above the alignment; structural elements are colorized as in (a)–(c). Open circles and filled boxes under the alignment indicate residues that make contacts to the LEDGF–IN binding domain through side chain and main chain atoms, respectively. (From Ref. 104. With the permission of the National Academy of Sciences of the United States of America.)

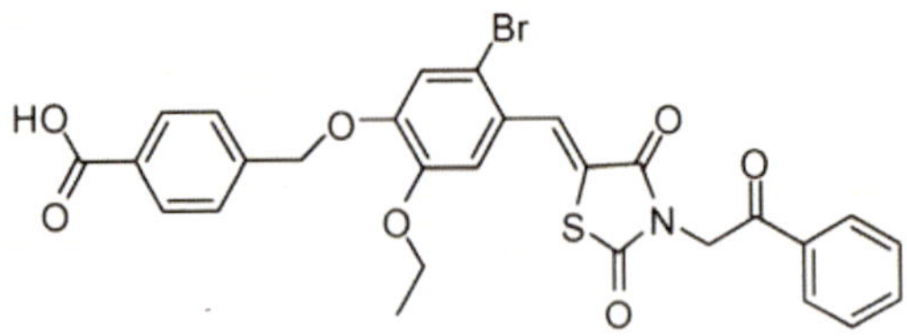

D77

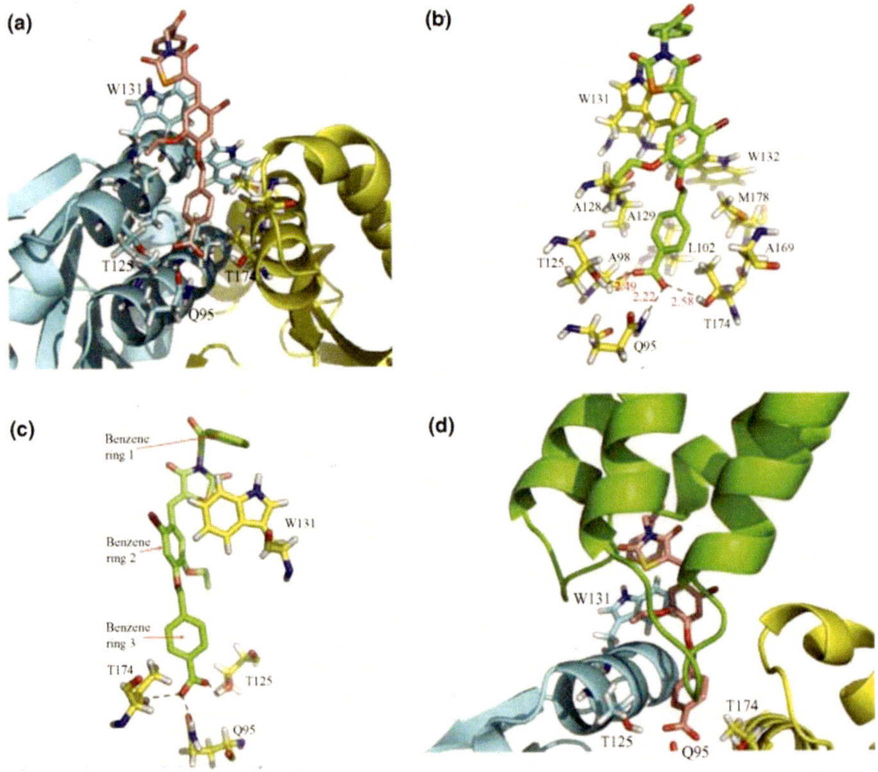

FIGURE 37.8 Structure of D77 and molecular docking into HIV-1 IN catalytic core domain. (a) The binding site of D77 on HIV-1 IN catalytic core domain dimer interface. Yellow and cyan: IN chains; salmon: D77. (b, c) The interaction between D77 and HIV-1 IN catalytic core domain in detail. (b: front view; c: side view) D77 is shown in green and IN residues are colored in yellow. Hydrogen bond is shown in red. (d) Molecular docking of D77 with catalytic core domain–IN binding domain complex. Yellow and cyan: IN chains; green: IN binding domain; salmon: D77. (From Ref. 107. With the permission of Elsevier.)

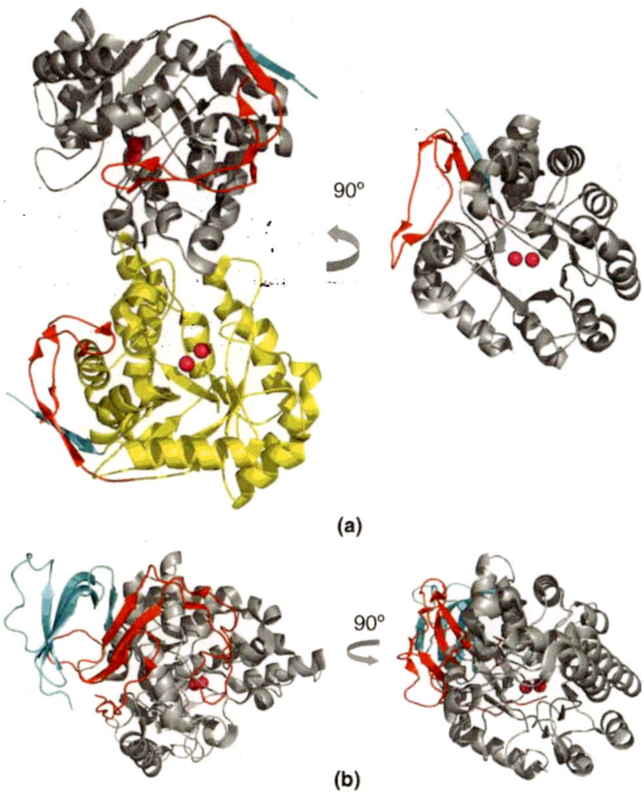

FIGURE 39.2 Overall folds of DHOases. (a) *E. coli* DHOase (Type II) (PDB entry 1XGE). The N-terminal domain (residues 4–10), the main $(\beta/\alpha)_8$-barrel domain (residues 11–307) and the C-terminal domain (residues 308–356) are colored in cyan, gray (or yellow in chain B) and red, respectively. Zinc atoms in the active site are shown as magenta spheres. For clarity, only chain A is shown in the rotated view. (b) *P. gingivalis* DHOase (Type I) (PDB entry 2GWN). The N-terminal domain (residues 1–57), the main $(\beta/\alpha)_8$-barrel domain (residues 58–375) and the C-terminal domain (residues 376–448) are colored in cyan, gray, and red, respectively. Zinc atoms in the active site are shown as magenta spheres.

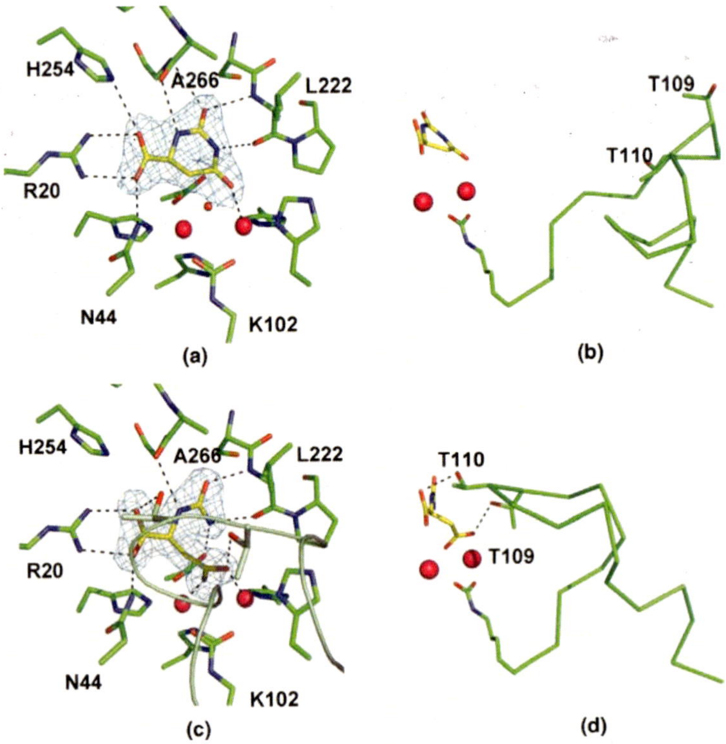

FIGURE 39.4 Interactions of DHO and CA-asp with *E. coli* DHOase in the active site and the conformations of the corresponding flexible loop (residues 105–115) (PDB entry 1XGE). The active sites in the DHO-bound subunit (a and b) and in the CA-asp-bound subunit (c and d). Hydrogen bonds to the bound DHO and CA-asp are depicted as dashed lines. Zinc atoms in each active site are shown as magenta spheres and a solvent molecule in (a) as a red sphere and (b) and (d) show the conformations of the surface loops (residues 105–115) in each active site.

INDEX